PEARSON ALWAYS LEARNING

Jane B. Reece • Lisa A. Urry • Michael L. Cain • Steven A. Wasserman • Peter V. Minorsky • Robert B. Jackson

Campbell Biology

Second Custom Edition for the University of Central Florida
Introduction to Biology 1
BSC 2010

Taken from:
Campbell Biology, Ninth Edition
by Jane B. Reece, Lisa A. Urry, Michael L. Cain, Steven A. Wasserman, Peter V. Minorsky, and Robert B. Jackson

Cover Art: Courtesy of Photodisc/Getty Images.

Taken from:

Campbell Biology, Ninth Edition
by Jane B. Reece, Lisa A. Urry, Michael L. Cain, Steven A. Wasserman,
Peter V. Minorsky, and Robert B. Jackson

Published by Benjamin Cummings
San Francisco, California 94111

This special edition published in cooperation with Pearson Learning Solutions.

Pearson Learning Solutions, 501 Boylston Street, Suite 900,
Boston, MA 02116
A Pearson Education Company
www.pearsoned.com

Printed in the United States of America

1 2 3 4 5 6 7 8 9 10 V363 16 15 14 13 12 11

000200010270771883

KB

ISBN 10: 1-256-28852-7
ISBN 13: 978-1-256-28852-7

Brief Contents

About the Authors

The Ninth Edition author team's contributions reflect their biological expertise as researchers and teaching sensibilities gained from years of experience as instructors at diverse institutions. The team's highly collaborative style continues to be evident in the cohesiveness and consistency of the Ninth Edition.

Jane B. Reece

The head of the Ninth Edition author team, Jane Reece was Neil Campbell's longtime collaborator. She has participated in every edition of *BIOLOGY*. Earlier, Jane taught biology at Middlesex County College and Queensborough Community College. She holds an A.B. in Biology from Harvard University, an M.S. in Microbiology from Rutgers University, and a Ph.D. in Bacteriology from the University of California, Berkeley. Jane's research as a doctoral student and postdoctoral fellow focused on genetic recombination in bacteria. Besides her work on *CAMPBELL BIOLOGY*, she has been a coauthor of *Biology: Concepts & Connections*, *Essential Biology*, and *The World of the Cell*.

Lisa A. Urry

Lisa Urry (Chapter 1 and Units 1–3) is a professor and developmental biologist and recent Chair of the Biology Department at Mills College. After graduating from Tufts University with a double major in Biology and French, Lisa completed her Ph.D. in molecular and developmental biology at Massachusetts Institute of Technology (MIT). She has published a number of research papers, most of them focused on gene expression during embryonic and larval development in sea urchins. Lisa is also deeply committed to promoting opportunities for women in science education and research.

Michael L. Cain

Michael Cain (Units 4 and 5) is an ecologist and evolutionary biologist who is now writing full-time. Michael earned a joint degree in Biology and Math at Bowdoin College, an M.Sc. from Brown University, and a Ph.D. in Ecology and Evolutionary Biology from Cornell University. As a faculty member at New Mexico State University and Rose-Hulman Institute of Technology, he taught a wide range of courses, including introductory biology, ecology, evolution, botany, and conservation biology. Michael is the author of dozens of scientific papers on topics that include foraging behavior in insects and plants, long-distance seed dispersal, and speciation in crickets. In addition to his work on *CAMPBELL BIOLOGY*, Michael is also the lead author of an ecology textbook.

Steven A. Wasserman

Steve Wasserman (Unit 7) is a professor at the University of California, San Diego (UCSD). He earned his A.B. in Biology from Harvard University and his Ph.D. in Biological Sciences from MIT. Through his research on regulatory pathway mechanisms in the fruit fly *Drosophila*, Steve has contributed to the fields of developmental biology, reproduction, and immunity. As a faculty member at the University of Texas Southwestern Medical Center and UCSD, he has taught genetics, development, and physiology to undergraduate, graduate, and medical students. He has also served as the research mentor for more than a dozen doctoral students and more than 50 aspiring scientists at the undergraduate and high school levels. Steve has been the recipient of distinguished scholar awards from both the Markey Charitable Trust and the David and Lucille Packard Foundation. In 2007, he received UCSD's Distinguished Teaching Award for undergraduate teaching.

Peter V. Minorsky

Peter Minorsky (Unit 6) is a professor at Mercy College in New York, where he teaches evolution, ecology, botany, and introductory biology. He received his B.A. in Biology from Vassar College and his Ph.D. in Plant Physiology from Cornell University. He is also the science writer for the journal *Plant Physiology*. After a postdoctoral fellowship at the University of Wisconsin at Madison, Peter taught at Kenyon College, Union College, Western Connecticut State University, and Vassar College. He is an electrophysiologist who studies responses of plants to stress. Peter received the 2008 Award for Teaching Excellence at Mercy College.

Robert B. Jackson

Rob Jackson (Unit 8) is a professor of biology and Nicholas Chair of Environmental Sciences at Duke University. Rob holds a B.S. in Chemical Engineering from Rice University, as well as M.S. degrees in Ecology and Statistics and a Ph.D. in Ecology from Utah State University. Rob directed Duke's Program in Ecology for many years and just finished a term as the Vice President of Science for the Ecological Society of America. Rob has received numerous awards, including a Presidential Early Career Award in Science and Engineering from the National Science Foundation. He also enjoys popular writing, having published a trade book about the environment, *The Earth Remains Forever*, and two books of poetry for children, *Animal Mischief* and *Weekend Mischief*.

Neil A. Campbell

Neil Campbell combined the investigative nature of a research scientist with the soul of an experienced and caring teacher. He earned his M.A. in Zoology from UCLA and his Ph.D. in Plant Biology from the University of California, Riverside, where he received the Distinguished Alumnus Award in 2001. Neil published numerous research articles on desert and coastal plants and how the sensitive plant (*Mimosa*) and other legumes move their leaves. His 30 years of teaching in diverse environments included general biology courses at Cornell University, Pomona College, and San Bernardino Valley College, where he received the college's first Outstanding Professor Award in 1986. Neil was a visiting scholar in the Department of Botany and Plant Sciences at the University of California, Riverside. In addition to his authorship of this book, he coauthored *Biology: Concepts & Connections* and *Essential Biology* with Jane Reece. For the Ninth Edition of this book, we honor Neil's contributions to biology education by adopting the title *CAMPBELL BIOLOGY*.

Preface

Biology is an enormous subject, one that can seem overwhelming to students and scientists alike. Moreover, discoveries are being made at an unprecedented pace—from new kinds of small RNA molecules to the Neanderthal genome, from new biofuels to communities of organisms thriving beneath vast glaciers, from emerging infectious diseases to cancer vaccines. As a result, a general biology course faces a daunting challenge: to keep students from suffocating under an avalanche of information. *CAMPBELL BIOLOGY* addresses this challenge by providing a strong foundation for understanding both current knowledge and new developments in the context of underlying biological concepts.

Key Concepts and Unifying Themes

In each chapter of this textbook, a framework of three to six carefully chosen **Key Concepts** provide context for supporting details, helping students distinguish the "forest" from the "trees." The numbered Key Concepts are presented at the beginning of the chapter and then serve as headings for each chapter section. *Concept Check Questions* at the end of each section provide a hierarchical framework for self-assessment that builds students' confidence and then challenges them to push the limits of their understanding with several types of critical thinking questions. The *Summary of Key Concepts* at the end of the chapter refocuses students on the main points. *CAMPBELL BIOLOGY* also helps students organize and make sense of what they learn on a grander scale by emphasizing **evolution and other unifying themes** that pervade biology. These themes are introduced in Chapter 1 and integrated throughout the book.

New to This Edition: An Emphasis on Making Connections

In addition to Key Concepts and themes, we've created new features for the Ninth Edition that help students see the big picture by *making connections*. These include the following:

New Make Connections Questions: Making connections across chapters

New **Make Connections Questions** help students see how different areas of biology tie together, helping them overcome the tendency to compartmentalize information. Each question challenges students to move beyond memorization and gain a deeper understanding of biological principles by asking them to relate chapter content to material they learned earlier in the course. For example, we ask students to connect

- DNA replication (Chapter 16, see p. 319) to the cell cycle (Chapter 12);
- Soil formation (Chapter 37, see p. 789) to the properties of water (Chapter 3); and
- Aquatic biomes (Chapter 52, see p. 1163) to osmoregulation (Chapter 44).

At least three Make Connections Questions appear in each chapter. In addition, online *Make Connections Tutorials* in MasteringBiology® (see p. xi) connect content from two different chapters using figures from the book.

Expanded Evolution Coverage: Making connections to evolution in every chapter

Evolution is the core theme of biology, and in this edition it is more evident than ever. At least one **Evolution section in every chapter** focuses on evolutionary aspects of the chapter material, highlighted by a new Evolution banner. See, for example, the new discussions of enzyme evolution (p. 157), coevolution of flowers and pollinators (p. 806), and evolution of hormone function in animals (pp. 988–989).

New Impact Figures: Making connections between scientific advances and the real world

Our new **Impact Figures** motivate students by highlighting the dramatic impact of recent discoveries in biology. These figures feature high-interest topics such as induced pluripotent stem cells and regenerative medicine (Chapter 20, p. 417), the discovery of *Tiktaalik* (Chapter 34, p. 710), and the use of forensic ecology to track elephant poaching (Chapter 56, p. 1243). The *Why It Matters* section of each figure explains the relevance of the research to students' lives, global problems, or the field of biology itself. Each Impact Figure ends with a suggestion for *Further Reading* and a *What If?* or *Make Connections Question* to develop critical thinking skills.

New Visual Organizers and 3-D Art: Making connections visually

The new **Visual Organizer** format highlights the main parts of a figure, helping students see the key categories at a glance. See, for instance, Figure 17.24 on types of small-scale mutations (p. 345) or Figure 27.3, Gram staining (p. 557). Throughout the book, selected figures have been rendered in a more **3-D art style** while keeping an appropriate balance between realism and teaching effectiveness. Figure 52.3, Exploring Global Climate Patterns (p. 1146), is one example.

Restructured Chapter Reviews: Making connections at a higher level

In the chapter summaries, each concept section now concludes with a new **Summary of Key Concepts Question** that is tied to a major learning goal. Also, this edition increases student awareness of different levels of thinking by organizing the end-of-chapter questions into three levels based on **Bloom's taxonomy**, which classifies types of thinking that are important in learning. Our levels are (1) Knowledge/Comprehension, (2) Application/Analysis, and (3) Synthesis/Evaluation. (These same levels are used in the Campbell Test Bank.) The range of question types helps students develop critical thinking skills and prepare for the kinds of questions they'll encounter on exams. New **Write About a Theme Questions** give students practice writing short, coherent essays that connect the chapter's content to one of the book's themes. (A suggested grading rubric can be found on p. xv, and sample answers are provided for instructors in the MasteringBiology Instructor Resources area.) A new **MasteringBiology preview section** at the end of each chapter lists *Assignments*—tutorials, activities, and questions that instructors can assign. This section also directs students to the *eText* and *Study Area* for online self study.

New Content: Making connections to advances in science

As in each new edition, the Ninth Edition incorporates **new scientific content** and **organizational improvements**. These are summarized on pp. viii–ix, following this Preface.

MasteringBiology®: Making connections outside of class

MasteringBiology, the most widely used online assessment and tutorial program for biology, provides an extensive library of homework assignments that are graded automatically. In addition to BioFlix® Tutorials, other Tutorials, Activities, Reading Quiz Questions in every chapter, and 4,500 Test Bank Questions, MasteringBiology for the Ninth Edition features an **improved user interface** and the following **new Tutorials and Questions**: *Make Connections Tutorials, Student Misconceptions Questions* for every chapter, *Data Analysis Tutorials, Experimental Inquiry Tutorials, Video Tutor Sessions,* and *MasteringBiology: Virtual Labs*. For more information, see pp. xvi–xix and www.masteringbiology.com.

Our Hallmark Features

Besides our Key Concepts and unifying themes, several other features have contributed to the success of *CAMPBELL BIOLOGY*. Because text and illustrations are equally important for learning biology, **integration of figures and text** has been a hallmark of this book since its inception. Our popular *Exploring Figures* on selected topics epitomize this approach: Each is a learning unit of core content that brings together related illustrations and text. Another example is our *Guided Tour Figures*, which use descriptions in blue type to walk students through complex figures like an instructor would, pointing out key structures, functions, and steps of processes.

To encourage **active learning**, recent editions have incorporated new types of questions: *What If? Questions, Figure Legend Questions*, and *Draw It Questions* that ask students to sketch a structure, annotate a figure, or graph data. In the Ninth Edition, these questions are augmented by the new *Make Connections Questions*. Online, the highly interactive *MasteringBiology tutorials* are sophisticated active-learning tools.

Finally, *CAMPBELL BIOLOGY* features **scientific inquiry**, an essential component of any biology course. Complementing stories of scientific discovery in the text narrative and the unit-opening interviews, *Inquiry Figures* help students understand "how we know what we know" and provide a model of how to think like a scientist. Each one begins with a research question and then describes how researchers designed an experiment, interpreted their results, and drew conclusions. The source article is referenced, and a What If? Question asks students to consider an alternative scenario. Selected Inquiry Figures invite students to read and analyze the original research article in the supplement *Inquiry in Action: Interpreting Scientific Papers* (see p. xxi). At the end of each chapter, *Scientific Inquiry Questions* give students additional opportunities to practice critical thinking by developing hypotheses, designing experiments, and analyzing real research data. Beyond the book, activities involving scientific inquiry are featured in MasteringBiology and other supplements, both print and electronic (see pp. xviii–xxi).

Our Partnership with Instructors

A core value underlying all our work as authors is our belief in the importance of our partnership with instructors. Our primary way of serving instructors, of course, is providing a textbook, supplements, and media resources that serve their students well. In addition, Benjamin Cummings makes available a rich variety of instructor resources, in both print and electronic form (see p. xx). In our continuing efforts to improve the book and its supplements, we benefit tremendously from instructor feedback, not only in formal reviews from hundreds of scientists, but also via informal communication, often by e-mail.

The real test of any textbook is how well it helps instructors teach and students learn. We welcome comments from the students and instructors who use *CAMPBELL BIOLOGY*. Please address your suggestions to any of us:

Jane Reece at janereece@cal.berkeley.edu
Lisa Urry (Chapter 1 and Units 1–3) at lurry@mills.edu
Michael Cain (Units 4 and 5) at mcain@bowdoin.edu
Peter Minorsky (Unit 6) at pminorsky@mercy.edu
Steve Wasserman (Unit 7) at stevenw@ucsd.edu
Rob Jackson (Unit 8) at jackson@duke.edu

New Content

This section provides just a few highlights of new content and organizational improvements in *CAMPBELL BIOLOGY*, Ninth Edition.

CHAPTER 1 Introduction: Themes in the Study of Life

We have added a separate new theme on energy flow while retaining a theme on environmental interactions. Concept 1.3, on the scientific method, has been reframed to more accurately reflect the scientific process, with a focus on observations and hypotheses. A new Concept 1.4 discusses the value of technology to society while emphasizing the cooperative nature of science and the value of diversity among scientists.

UNIT ONE The Chemistry of Life

For this edition, the basic chemistry is enlivened by new content connecting it to evolution, ecology, and other areas of biology. Examples of new material include omega-3 fatty acids, the isomeric forms of methamphetamine, arsenic contamination of groundwater, and the basis of mad cow disease. The burgeoning importance of nucleic acids throughout biology has prompted us to expand our coverage of DNA and RNA structures in this first unit. In fact, a general aim for the first two units is to infuse the chapters with more detail about nucleic acids, genes, and related topics. Another enhancement, in this and the next two units, is the inclusion of more computer models of important proteins in contexts where they support students' understanding of molecular function.

UNIT TWO The Cell

For Chapter 6, we developed an Exploring Figure on microscopy, which includes new types of microscopy, and we added micrographs of various cell types to the Exploring Figure on eukaryotic cells. We also expanded our description of chromosome composition, with the goal of preempting some common student misconceptions about chromosomes and DNA. New connections to evolution include an introduction to the endosymbiont theory in Chapter 6 and some interesting evolutionary adaptations of cell membranes in Chapter 7. We've added a new section to Chapter 8 on the evolution of enzymes with new functions, which not only strengthens enzyme coverage but also provides an early introduction to the concept that mutations contribute to molecular evolution. In Chapter 9, we simplified the glycolysis figure and emphasized pyruvate oxidation as a separate step to help students focus on the main ideas. In keeping with our increased focus on global issues in the Ninth Edition, Chapter 10 has an Impact Figure on biofuels and a discussion of the possible effect of climate change on the distribution of C_3 and C_4 plants. In Chapter 11, we have added an Impact Figure to highlight the importance and medical relevance of G protein-coupled receptors.

UNIT THREE Genetics

In Chapters 13–17, we have added material to stimulate student interest—for example, a new Impact Figure on genetic testing for disease-associated mutations. As done throughout the Ninth Edition, we ask students to make connections between chapters so that they avoid the trap of compartmentalizing the information in each chapter. For instance, Chapter 15 discusses the Philadelphia chromosome associated with chronic myelogenous leukemia and asks students to connect this information to what they learned about signaling in the cell cycle in Chapter 12. Also, we encourage students to connect what they learn about DNA replication and chromosome structure in Chapter 16 to the material on chromosome behavior during the cell cycle in Chapter 12. Chapter 16 has a new figure showing a current 3-D model of the DNA replication complex, with the lagging strand looping back through it.

Chapters 18–21 are extensively updated, with the changes dominated by new genomic sequence data and discoveries about the regulation of gene expression. (The introduction to genes, genomes, and gene expression in Units One and Two should help prepare students for these revisions.) Chapter 18 includes a new section on nuclear architecture, which describes the organization of chromatin in the nucleus in relation to gene expression. The roles of various types of RNA molecules in regulation also receive special attention. In the section on cancer, we describe how technical advances can contribute to personalized cancer treatments based on the molecular characteristics of an individual's tumor. Chapter 19 discusses the 2009 H1N1 flu pandemic. Chapter 20 includes advances in techniques for DNA sequencing and for obtaining induced pluripotent stem (iPS) cells. Finally, the heavily revised Chapter 21 describes what has been learned from the sequencing of many genomes, including those of a number of human individuals.

UNIT FOUR Mechanisms of Evolution

For this edition, we have continued to bolster our presentation of the vast evidence for evolution by adding new examples and figures that illustrate key conceptual points throughout the unit. For example, Chapter 22 now presents research data on adaptive evolution in soapberry bugs, fossil findings that shed light on the origins of cetaceans, and an Impact Figure on the rise of methicillin-resistant *Staphylococcus aureus*. Chapter 23 examines gene flow and adaptation in songbird populations. Chapter 24 incorporates several new examples of speciation research, including reproductive isolation in mosquitofish, speciation in shrimp, and hybridization of bear species. Other changes strengthen the storyline of the unit, ensuring that the chapters flow smoothly and build to a clear overall picture of what evolution is and how it works. For instance, new connections between Chapters 24 and 25 illustrate how differences in speciation and extinction rates shape the broad patterns in the history of life. We've also added earlier and more discussion of "tree thinking," the interpretation and application of phylogenetic trees, beginning in Chapter 22.

Unit Five The Evolutionary History of Biological Diversity

One of our goals for the diversity unit was to expand the coverage of the scientific evidence underlying the evolutionary story told in the chapters. So, for example, Chapter 27 now presents new findings on the evolutionary origin of bacterial flagella. In keeping with our increased emphasis on big-picture "tree thinking," we've added an "evogram" on tetrapod evolution in Chapter 34. (An evogram is a diagram illustrating the multiple lines of evidence that support the hypothesis shown in an evolutionary tree.) In addition, to help engage students, we've included new applications and woven more ecological information into our discussions of groups of organisms. Examples include new material on global growth of photosynthetic protists (Chapter 28), endangered molluscs (Chapter 33), and the impact of a pathogenic chytrid fungus on amphibian population declines (Chapters 31 and 34).

Unit Six Plant Form and Function

Plant biology is in a transitional phase; some professors prefer strong coverage of classical botany while others seek more in-depth coverage of the molecular biology of plants. In developing the Ninth Edition, we have continued to balance the old and the new to provide students with a basic understanding of plant anatomy and function while highlighting dynamic areas of plant research and the many important connections between plants and other organisms. One major revision goal was to provide more explicit discussion of the evolutionary aspects of plant biology, such as the coevolution of insects and animal pollinators (Chapter 38). Updates include new findings in plant development in Concept 35.5 and new material on the dynamism of plant architecture as it relates to resource acquisition in Chapter 36.

Unit Seven Animal Form and Function

In revising this unit, we strove to introduce physiological systems through a comparative approach that underscores how adaptations are linked to shared physiological challenges. In particular, we have highlighted the interrelationship of the endocrine and nervous systems at multiple points in the unit, helping students appreciate how these two forms of communication link tissues, organs, and individuals. Other revisions aim to keep students focused on fundamental concepts amid the details of complex systems. For example, many figures have been reconceived to emphasize key information, including new figures comparing single and double circulation (Chapter 42) and examining the function of antigen receptors (Chapter 43), as well as new Exploring Figures on the vertebrate kidney (Chapter 44) and the structure and function of the eye (Chapter 50). Chapter 43 has been significantly revised to support students' conceptual understanding of basic immunological responses and the key cellular players. Throughout the unit, new state-of-the-art images and material on current and compelling topics—such as circadian rhythms (Chapter 40), novel strains of influenza (Chapter 43), the effects of climate change on animal reproductive cycles (Chapter 46), and advances in understanding brain plasticity and function (Chapter 49)—will help engage students and encourage them to make connections beyond the text.

Unit Eight Ecology

Our revision was informed by the fact that biologists are increasingly asked to apply their knowledge to help solve global problems, such as climate change, that already are profoundly affecting life on Earth. As part of our increased emphasis on global ecology in this edition, we have made significant changes to Unit Eight's organization and content. The organizational changes begin with the introductory chapter of the unit (Chapter 52), which includes a new Key Concept 52.1: "Earth's climate varies by latitude and season and is changing rapidly." Introducing the global nature of climate and its effects on life at the beginning of the chapter provides a logical foundation for the rest of the material. New content in Chapters 53 and 54 highlights factors that limit population growth, the ecological importance of disease, positive interactions among organisms, and biodiversity. Chapter 55 now explores restoration ecology together with ecosystem ecology because successful restoration efforts depend on understanding ecosystem structure and function. Finally, the new title of the unit's capstone, Chapter 56, reflects its emphasis on the combined importance of conservation and our changing Earth: "Conservation Biology and Global Change." Several new Impact Figures in the unit show students how ecologists apply biological knowledge and ecological theory at all scales to understand and solve problems in the world around them.

To the Student: How to Use This Book

Focus on the Key Concepts.

Each chapter is organized around a framework of 3 to 6 **Key Concepts** that will help you stay focused on the big picture and give you a context for the supporting details.

Before you begin reading the chapter, get oriented by reading the **list of Key Concepts,** which introduces the big ideas covered in the chapter.

Each **Key Concept** serves as the heading for a major section of the chapter.

52

An Introduction to Ecology and the Biosphere

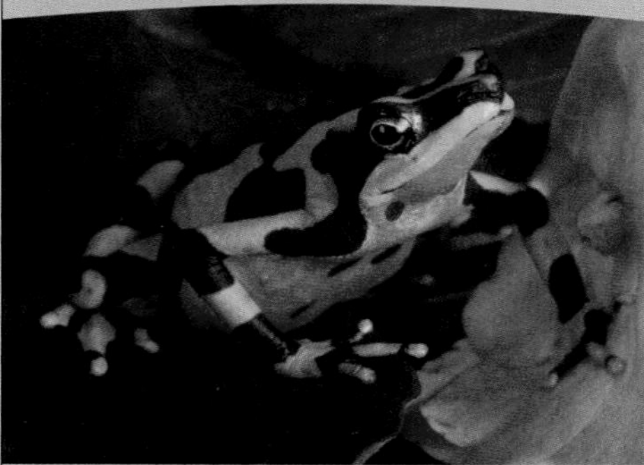

▲ **Figure 52.1 What threatens this amphibian's survival?**

KEY CONCEPTS

52.1 Earth's climate varies by latitude and season and is changing rapidly
52.2 The structure and distribution of terrestrial biomes are controlled by climate and disturbance
52.3 Aquatic biomes are diverse and dynamic systems that cover most of Earth
52.4 Interactions between organisms and the environment limit the distribution of species

OVERVIEW

Discovering Ecology

When University of Delaware undergraduate Justin Yeager spent his summer abroad in Costa Rica, all he wanted was to see the tropical rain forest and to practice his Spanish. Instead, he rediscovered the variable harlequin toad (*Atelopus varius*), a species thought to be extinct in the mountain slopes of Costa Rica and Panama where it once lived (**Figure 52.1**). During the 1980s and 1990s, roughly two-thirds of the 82 known species of harlequin toads vanished. Scientists think that a disease-causing chytrid fungus, *Batrachochytrium dendrobatidis* (see Figure 31.26), contributed to many of these extinctions. Why was the fungus suddenly thriving in the rain forest? Cloudier days and warmer nights associated with global warming appear to have created an environment ideal for its success. As of 2009, the species that Yeager found was surviving as a single known population of fewer than 100 individuals.

What environmental factors limit the geographic distribution of harlequin toads? How do variations in their food supply or interactions with other species, such as pathogens, affect the size of their population? Questions like these are the subject of **ecology** (from the Greek *oikos*, home, and *logos*, study), the scientific study of the interactions between organisms and the environment. Ecological interactions occur at a hierarchy of scales that ecologists study, from single organisms to the globe (**Figure 52.2**).

Ecology's roots are in our basic human interest in observing other organisms. Naturalists, including Aristotle and Darwin, have long studied the living world and systematically recorded their observations. However, modern ecology involves more than observation. It is a rigorous experimental science that requires a breadth of biological knowledge. Ecologists generate hypotheses, manipulate environmental variables, and observe the outcome. In this unit, you will encounter many examples of ecological experiments, whose complex challenges have made ecologists innovators in experimental design and statistical inference.

In addition to providing a conceptual framework for understanding the field of ecology, Figure 52.2 provides the organizational framework for our final unit. In this chapter, we first describe Earth's climate and the importance of climate and other physical factors in determining the location of major life zones on land and in the oceans. We then examine how ecologists determine what controls the distribution and abundance of individual species. The next three chapters investigate population, community, and ecosystem ecology in detail, including approaches for restoring degraded ecosystems. The final chapter explores conservation biology and global ecology as we consider how ecologists apply biological knowledge to predict the global consequences of human activities and to conserve Earth's biodiversity.

CONCEPT **52.1**

Earth's climate varies by latitude and season and is changing rapidly

The most significant influence on the distribution of organisms on land and in the oceans is **climate**, the long-term, prevailing weather conditions in a given area. Four physical

After reading a concept section, check your understanding using the **Concept Check Questions** at the end of the section. Work through these questions on your own or in a study group—they're good practice for the kinds of questions you might be asked on an exam. ▶

What If? Questions ask you to apply what you've learned. New **Make Connections Questions** ask you to relate content in the chapter to a concept you learned earlier in the course.

If you can answer these questions (see Appendix A to check your work), you're ready to move on. ▶

CONCEPT CHECK **52.1**

1. Explain how the sun's unequal heating of Earth's surface leads to the development of deserts around 30° north and south of the equator.
2. What are some of the differences in microclimate between an unplanted agricultural field and a nearby stream corridor with trees?
3. WHAT IF? Changes in Earth's climate at the end of the last ice age happened gradually, taking centuries to thousands of years. If the current global warming happens very quickly, as predicted, how may this rapid climate change affect the ability of long-lived trees to evolve, compared with annual plants, which have much shorter generation times?
4. MAKE CONNECTIONS In Concept 10.4 (pp. 199–201), you learned about the important differences between C_3 and C_4 plants. Focusing just on the effects of temperature, would you expect the global distribution of C_4 plants to expand or contract as Earth becomes warmer? Why?

For suggested answers, see Appendix A.

Make connections across biology.

By relating the content of a chapter to material you learned earlier in the course, new **Make Connections Questions** help you develop a deeper understanding of biological principles.

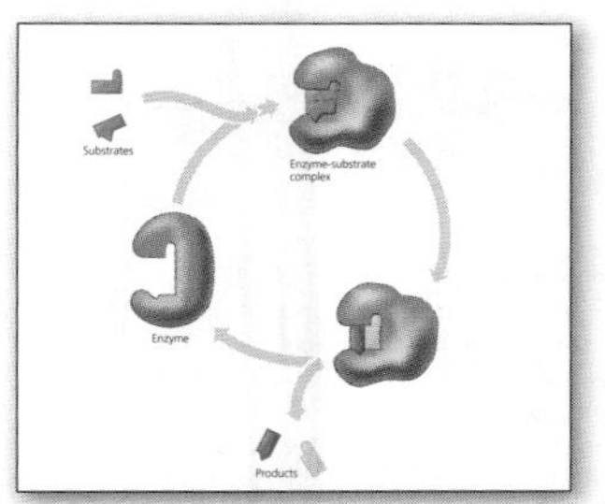

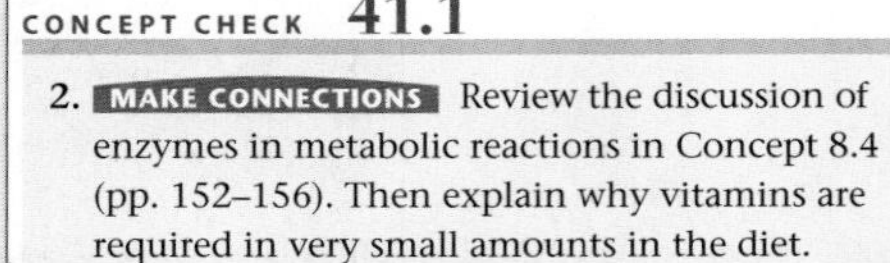

CONCEPT CHECK 41.1

2. **MAKE CONNECTIONS** Review the discussion of enzymes in metabolic reactions in Concept 8.4 (pp. 152–156). Then explain why vitamins are required in very small amounts in the diet.

Enzymes (Chapter 8) ⟷ **Animal nutrition** (Chapter 41)

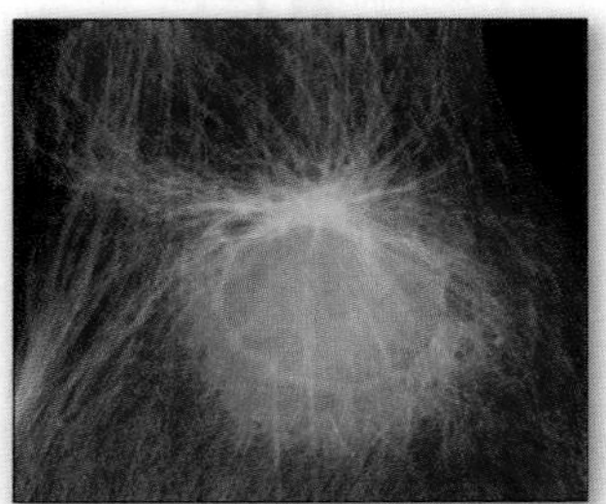

CONCEPT CHECK 16.2

3. **MAKE CONNECTIONS** What is the relationship between DNA replication and the S phase of the cell cycle? See Figure 12.6, page 231.

Cell cycle (Chapter 12) ⟷ **DNA replication** (Chapter 16)

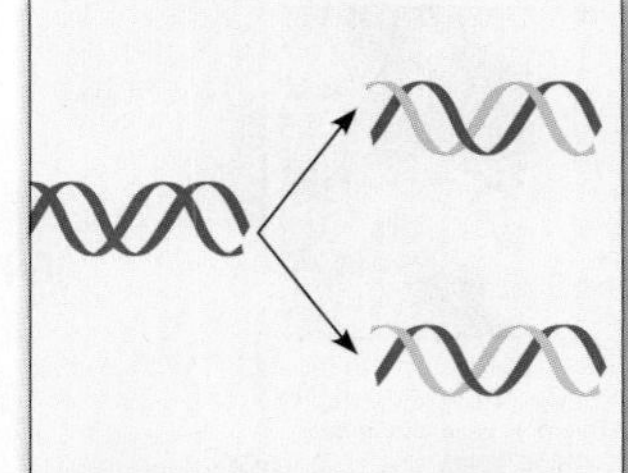

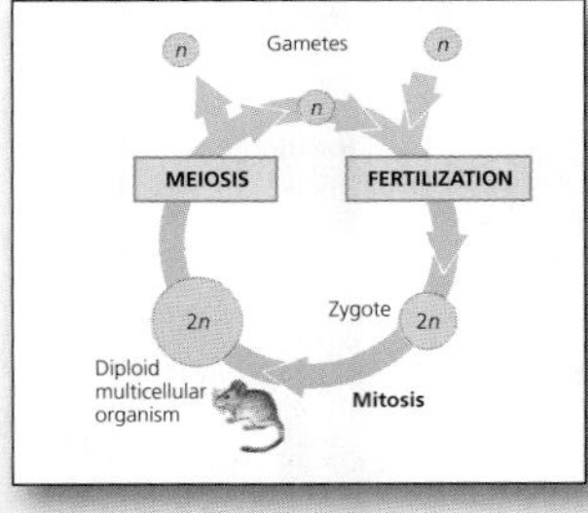

CONCEPT CHECK 31.2

1. **MAKE CONNECTIONS** Compare Figure 31.5 with Figure 13.6 (p. 252). In terms of haploidy versus diploidy, how do the life cycles of fungi and animals differ?

Meiosis (Chapter 13) ⟷ **Fungi** (Chapter 31)

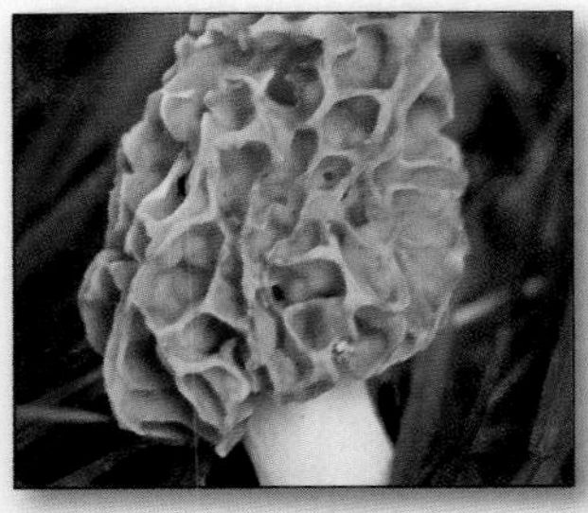

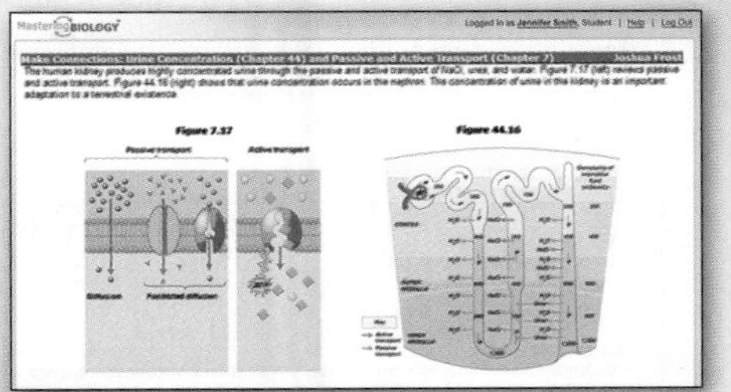

New **Make Connections Tutorials** help you connect biological concepts across chapters in an interactive way.

Make connections to **evolution,** the fundamental theme of biology.

Look for new **Evolution banners** highlighting sections in each chapter that focus on evolutionary aspects of the topic.

The Evolutionary Origins of Mitochondria and Chloroplasts

EVOLUTION Mitochondria and chloroplasts display similarities with bacteria that led to the **endosymbiont theory**, illustrated in Figure 6.16. This theory states that an early ancestor of eukaryotic cells engulfed an oxygen-using nonphotosynthetic prokaryotic cell. Eventually, the engulfed cell formed a relationship with the host cell in which it was enclosed, becoming an *endosymbiont* (a cell living within another cell). Indeed, over the course of evolution, the host cell and its endosymbiont merged into a single organism, a eukaryotic cell with a mitochondrion. At least one of these cells may have then taken up a photosynthetic prokaryote, becoming the ancestor of eukaryotic cells that contain chloroplasts.

This is a widely accepted theory, which we will discuss in more detail in Chapter 25. The model it proposes is consistent ...y structural features of mitochondria and chloro... ...t, rather than being bounded by a single membrane ...elles of the endomembrane system, mitochondria ...al chloroplasts have two membranes surrounding ...loroplasts also have an internal system of membra... ...) There is evidence that the ancestral engulfed ...es had two outer membranes, which became the ...embranes of mitochondria and chloroplasts. Second, ...ryotes, mitochondria and chloroplasts contain ribo... ...well as circular DNA molecules attached to their ...mbranes. The DNA in these organelles programs the ...of some of their own proteins, which are made on ...omes inside the organelles. Third, also consistent ...probable evolutionary origins as cells, mitochondria ...oplasts are autonomous (somewhat independent) or... ...hat grow and reproduce within the cell.

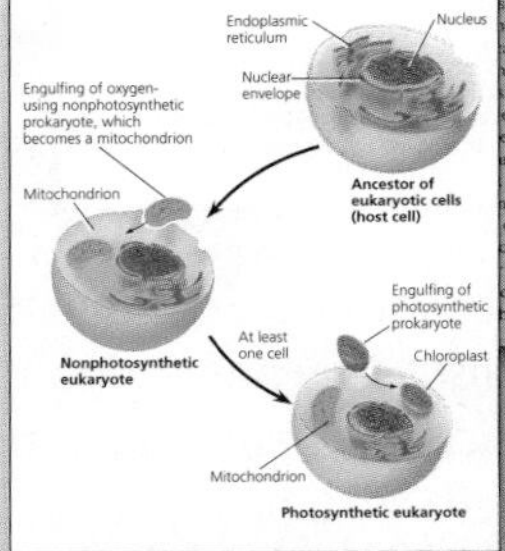

Practice thinking like a scientist.

New **Impact Figures** demonstrate the dramatic impact of recent discoveries in biology and show that biology is constantly changing as new discoveries add to our understanding. ▶

▼ **Figure 56.9**

IMPACT

Forensic Ecology and Elephant Poaching

This array of severed tusks is part of an illegal shipment of 6,000 kg of ivory intercepted on its way from Africa to Singapore in 2002. Investigators wondered whether the elephants slaughtered for the ivory—perhaps as many as 6,500—were killed in the area where the shipment originated, primarily Zambia, or instead were killed across Africa, indicating a broader smuggling ring. Samuel Wasser, of the University of Washington, and colleagues amplified specific segments of DNA from the tusks using the polymerase chain reaction (PCR). These segments included stretches of DNA containing short tandem repeats (STRs; see Concept 20.4, pp. 420–421), the number of which varies among different elephant populations. The researchers then compared alleles at seven or more loci with a reference DNA database they had generated for elephants of known geographic origin. Their results showed conclusively that the elephants came from a narrow east-west band centered on Zambia rather than from across Africa.

Why It Matters explains the relevance of the research. ▶

WHY IT MATTERS The DNA analyses suggested that poaching rates were 30 times higher in Zambia than previously estimated. This news led to improved antipoaching efforts by the Zambian government. Techniques like those used in this study are being employed by conservation biologists to track the harvesting of many endangered species, including whales, sharks, and orchids.

Further Reading directs you to articles to explore. ▶

FURTHER READING S. K. Wasser et al., Forensic tools battle ivory poachers, *Scientific American* 399:68–76 (2009); S. K. Wasser et al., Using DNA to track the origin of the largest ivory seizure since the 1989 trade ban, *Proceedings of the National Academy of Sciences USA* 104:4228–4233 (2007).

A **Make Connections** or **What If? Question** encourages critical thinking. ▶

MAKE CONNECTIONS Figure 26.6 (p. 539) describes another example in which conservation biologists used DNA analyses to compare harvested samples with a reference DNA database. How are these examples similar, and how are they different? What limitations might there be to using such forensic methods in other suspected cases of poaching?

Inquiry Figures reveal "how we know what we know" by highlighting how researchers designed an experiment, interpreted their results, and drew conclusions.
▼

▼ **Figure 37.14** **INQUIRY**

Does the invasive weed garlic mustard disrupt mutualistic associations between native tree seedlings and arbuscular mycorrhizal fungi?

EXPERIMENT Kristina Stinson, of Harvard University, and colleagues investigated the effect of invasive garlic mustard on the growth of native tree seedlings and associated mycorrhizal fungi. In one experiment, they grew seedlings of three North American trees—sugar maple, red maple, and white ash—in four different soils. Two of the soil samples were collected from a location where garlic mustard was growing, and one of these samples was sterilized. The other two soil samples were collected from a location devoid of garlic mustard, and one was then sterilized. After four months of growth, the researchers harvested the shoots and roots and determined the dried biomass. The roots were also analyzed for percent colonization by arbuscular mycorrhizal fungi.

RESULTS Native tree seedlings grew more slowly and were less able to form mycorrhizal associations when grown either in sterilized soil or in unsterilized soil collected from a location that had been invaded by garlic mustard.

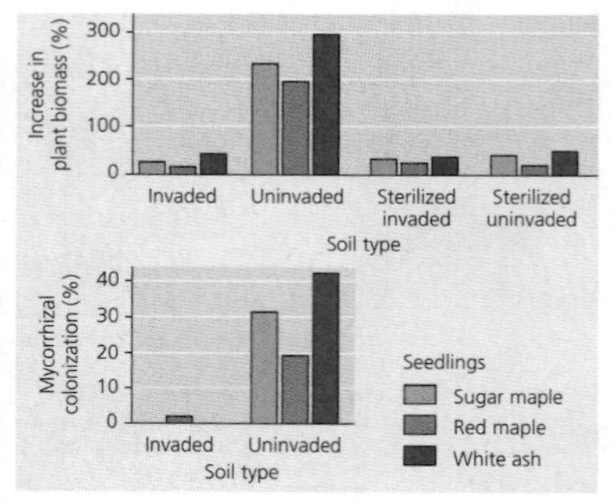

CONCLUSION The data support the hypothesis that garlic mustard suppresses growth of native trees by affecting the soil in a way that disrupts mutualistic associations between the trees and arbuscular mycorrhizal fungi.

SOURCE K. A. Stinson et al., Invasive plant suppresses the growth of native tree seedlings by disrupting belowground mutualisms, *PLoS Biol* (*Public Library of Science: Biology*) 4(5): e140 (2006).

INQUIRY IN ACTION Read and analyze the original paper in *Inquiry in Action: Interpreting Scientific Papers*.

WHAT IF? What effect would applying inorganic phosphate to soil invaded by garlic mustard have on the plant's ability to outcompete native species?

▲
After exploring the featured experiment, test your analytical skills by answering the **What If? Question**. Suggested answers are provided in Appendix A to help you gauge your understanding.

◀ Some **Inquiry Figures** invite you to read and analyze the original research paper in its complete form. You can find the journal article, along with a worksheet guiding you through it, in the separate book ***Inquiry in Action: Interpreting Scientific Papers.***

MasteringBIOLOGY®
www.masteringbiology.com

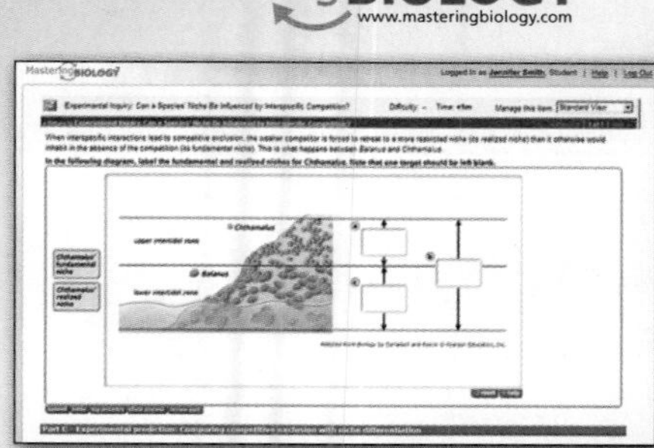

New **Experimental Inquiry Tutorials** give you practice analyzing experimental design and data and drawing conclusions.

Study the figures as you read the text.

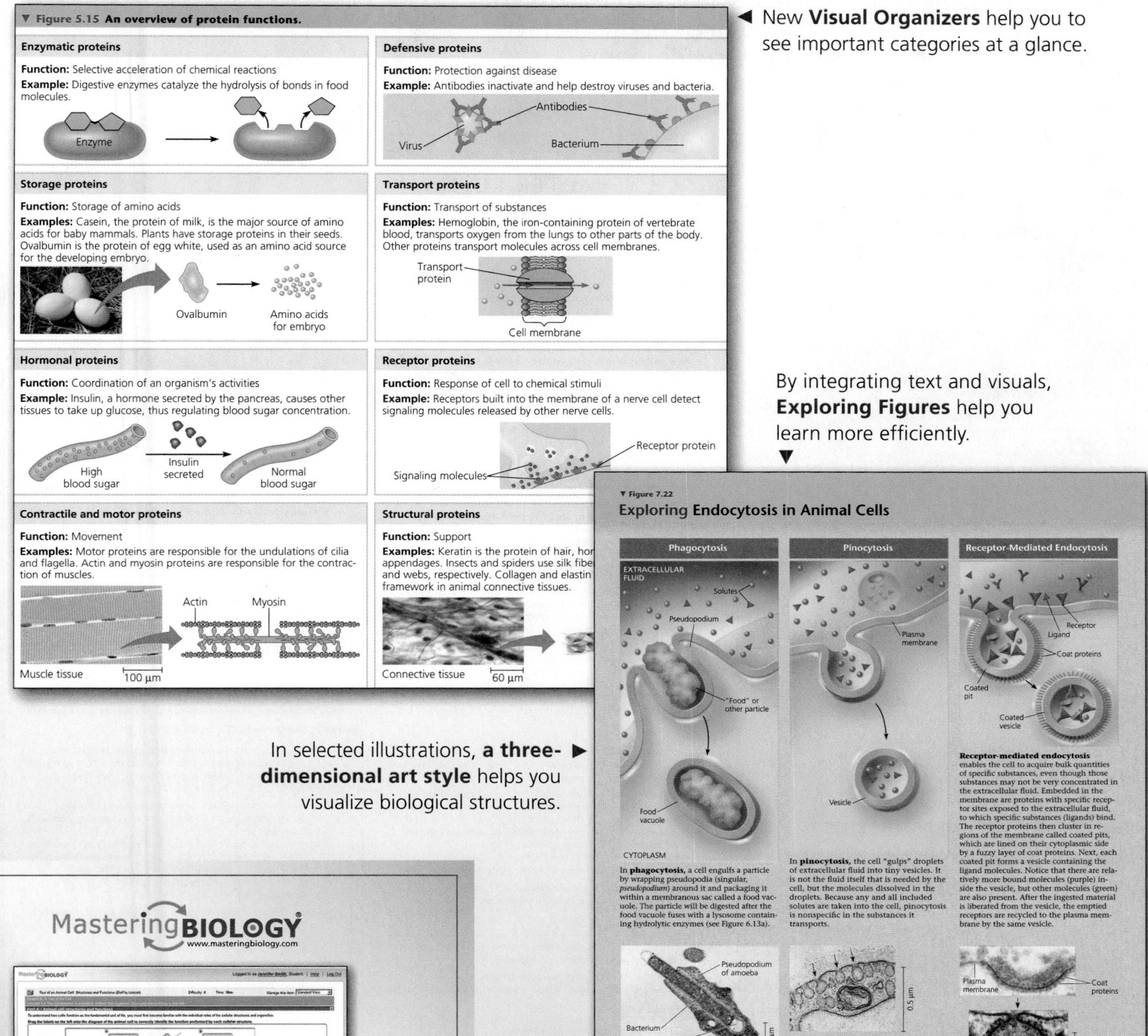

◀ New **Visual Organizers** help you to see important categories at a glance.

By integrating text and visuals, **Exploring Figures** help you learn more efficiently. ▼

In selected illustrations, **a three-dimensional art style** helps you visualize biological structures. ▶

MasteringBIOLOGY®
www.masteringbiology.com

Many **Tutorials** and **Activities** integrate art from the textbook, providing a unified learning experience.

▲ **BioFlix®** icons direct you to high-impact 3-D animations in the Study Area at www.masteringbiology.com.

Review what you've learned.

Chapter Reviews help you efficiently master the chapter content by focusing on the main points of the chapter and offering opportunities to practice for exams.

Summary Figures present key information in a visual way.

23

The Evolution of Populations

▲ **Figure 23.1 Is this finch evolving?**

EVOLUTION

KEY CONCEPTS

23.1 Genetic variation makes evolution possible
23.2 The Hardy-Weinberg equation can be used to test whether a population is evolving
23.3 Natural selection, genetic drift, and gene flow can alter allele frequencies in a population
23.4 Natural selection is the only mechanism that consistently causes adaptive evolution

OVERVIEW

The Smallest Unit of Evolution

One common misconception about evolution is that individual organisms evolve. It is true that natural selection acts on individuals: Each organism's traits affect its survival and reproductive success compared with other individuals. But the evolutionary impact of natural selection is only apparent in the changes in a *population* of organisms over time.

Key Concepts, which were introduced in the beginning of the chapter and developed in the text, are summarized in the Chapter Review.

New **Summary of Key Concepts Questions** appear at the end of each concept summary. Check your answers using Appendix A.

23 CHAPTER REVIEW

SUMMARY OF KEY CONCEPTS

CONCEPT 23.1

Genetic variation makes evolution possible (pp. 469–473)

- **Genetic variation** refers to genetic differences among individuals within a population.
- The nucleotide differences that provide the basis of genetic variation arise by mutation and other processes that produce new alleles and new genes.
- New genetic variants are produced rapidly in organisms with short generation times. In sexually reproducing organisms, most of the genetic differences among individuals result from crossing over, the independent assortment of chromosomes, and fertilization.

? *Why do biologists estimate gene variability and nucleotide variability, and what do these estimates represent?*

CONCEPT 23.2

The Hardy-Weinberg equation can be used to test whether a population is evolving (pp. 473–476)

- A **population**, a localized group of organisms belonging to one species, is united by its **gene pool**, the aggregate of all the alleles in the population.
- The **Hardy-Weinberg principle** states that the allele and genotype frequencies of a population will remain constant if the population is large, mating is random, mutation is negligible, there is no gene flow, and there is no natural selection. For such a population, if p and q represent the frequencies of the only two possible alleles at a particular locus, then p^2 is the frequency of one kind of homozygote, q^2 is the frequency of the other kind of homozygote, and $2pq$ is the frequency of the heterozygous genotype.

? *Is it circular reasoning to calculate* p *and* q *from observed genotype frequencies and then use those values of* p *and* q *to test if the population is in Hardy-Weinberg equilibrium? Explain your answer. (Hint: Consider a specific case, such as a population with 195 individuals of genotype* AA*, 10 of genotype* Aa*, and 195 of genotype* aa.*)*

CONCEPT 23.3

Natural selection, genetic drift, and gene flow can alter allele frequencies in a population (pp. 476–480)

- In natural selection, individuals that have certain inherited traits tend to survive and reproduce at higher rates than other individuals *because of* those traits.
- In **genetic drift**, chance fluctuations in allele frequencies over generations tend to reduce genetic variation.
- **Gene flow**, the transfer of alleles between populations, tends to reduce genetic differences between populations over time.

? *Would two small, geographically isolated populations in very different environments be likely to evolve in similar ways? Explain.*

CONCEPT 23.4

Natural selection is the only mechanism that consistently causes adaptive evolution (pp. 480–485)

- One organism has greater **relative fitness** than a second organism if it leaves more fertile descendants than the second organism. The modes of natural selection differ in how selection acts on phenotype (the white arrows in the summary diagram below represent selective pressure on a population).

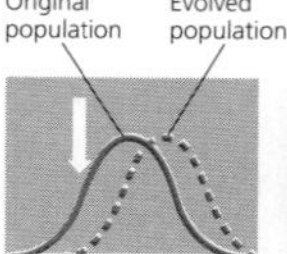

Directional selection

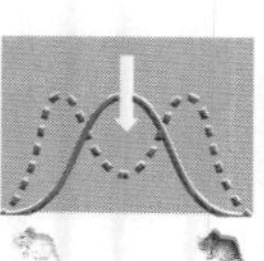

Disruptive selection

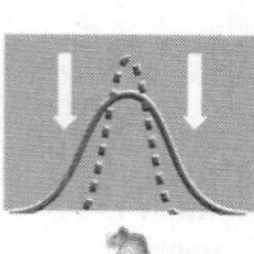

Stabilizing selection

- Unlike genetic drift and gene flow, natural selection consistently increases the frequencies of alleles that enhance survival and reproduction, thus improving the match between organisms and their environment.
- **Sexual selection** influences evolutionary change in secondary sex characteristics that can give individuals advantages in mating.
- Despite the winnowing effects of selection, populations have considerable genetic variation. Some of this variation represents **neutral variation**; additional variation can be maintained by diploidy and balancing selection.
- There are constraints to evolution: Natural selection can act only on available variation; structures result from modified ancestral anatomy; adaptations are often compromises; and chance, natural selection, and the environment interact.

? *How might secondary sex characteristics differ between males and females in a species in which females compete for mates?*

TEST YOUR UNDERSTANDING

LEVEL 1: KNOWLEDGE/COMPREHENSION

1. Natural selection changes allele frequencies because some ______ survive and reproduce more successfully than others.
   a. alleles
   b. loci
   c. gene pools
   d. species
   e. individuals
2. No two people are genetically identical, except for identical twins. The main source of genetic variation among human individuals is
   a. new mutations that occurred in the preceding generation.
   b. genetic drift due to the small size of the population.
   c. the reshuffling of alleles in sexual reproduction.
   d. geographic variation within the population.
   e. environmental effects.
3. Sparrows with average-sized wings survive severe storms better than those with longer or shorter wings, illustrating
   a. the bottleneck effect.
   b. disruptive selection.
   c. frequency-dependent selection.
   d. neutral variation.
   e. stabilizing selection.

486 UNIT FOUR Mechanisms of Evolution

To help you prepare for the various kinds of questions that may appear on a test, the end-of-chapter questions are now organized into three levels based on Bloom's Taxonomy:

Level 1: Knowledge/Comprehension
Level 2: Application/Analysis
Level 3: Synthesis/Evaluation

Evolution Connection Questions in the Chapter Review ask you to think critically about how an aspect of the chapter relates to evolution. ▶

Scientific Inquiry ▶ **Questions** at the end of each chapter give you opportunities to practice scientific thinking by developing hypotheses, designing experiments, and analyzing real research data.

Draw It Exercises in each chapter ask you to put pencil to paper and draw a structure, annotate a figure, or graph experimental data.

LEVEL 2: APPLICATION/ANALYSIS

4. If the nucleotide variability of a locus equals 0%, what is the gene variability and number of alleles at that locus?
   a. gene variability = 0%; number of alleles = 0
   b. gene variability = 0%; number of alleles = 1
   c. gene variability = 0%; number of alleles = 2
   d. gene variability > 0%; number of alleles = 2
   e. Without more information, gene variability and number of alleles cannot be determined.
5. There are 40 individuals in population 1, all with genotype *A1A1*, and there are 25 individuals in population 2, all with genotype *A2A2*. Assume that these populations are located far from each other and that their environmental conditions are very similar. Based on the information given here, the observed genetic variation is most likely an example of
   a. genetic drift.
   b. gene flow.
   c. disruptive selection.
   d. discrete variation.
   e. directional selection.
6. A fruit fly population has a gene with two alleles, *A1* and *A2*. Tests show that 70% of the gametes produced in the population contain the *A1* allele. If the population is in Hardy-Weinberg equilibrium, what proportion of the flies carry both *A1* and *A2*?
   a. 0.7 b. 0.49 c. 0.21 d. 0.42 e. 0.09

LEVEL 3: SYNTHESIS/EVALUATION

7. **EVOLUTION CONNECTION**
   How is the process of evolution revealed by the imperfections of living organisms?
8. **SCIENTIFIC INQUIRY**
   DRAW IT Richard Koehn, of the State University of New York, Stony Brook, and Thomas Hilbish, of the University of South Carolina, studied genetic variation in the marine mussel *Mytilus edulis* around Long Island, New York. They measured the frequency of a particular allele (*lap*^{94}) for an enzyme involved in regulating the mussel's internal saltwater balance. The researchers presented their data as a series of pie charts linked to sampling sites within Long Island Sound, where the salinity is highly variable, and along the coast of the open ocean, where salinity is constant:

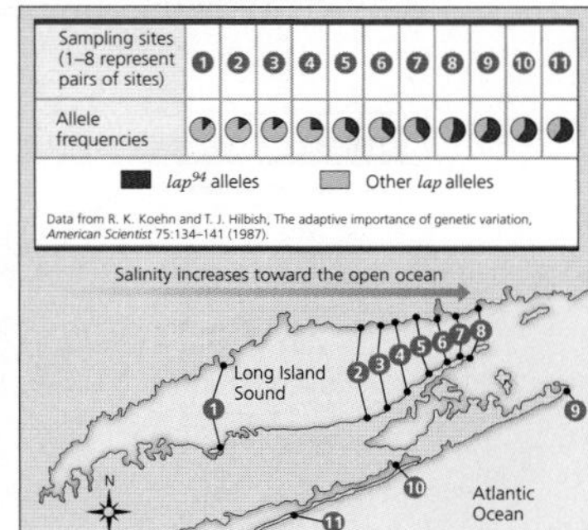

(Question 8, continued)

Create a data table for the 11 sampling sites by estimating the frequency of *lap*^{94} from the pie charts. (*Hint*: Think of each pie chart as a clock face to help you estimate the proportion of the shaded area.) Then graph the frequencies for sites 1–8 to show how the frequency of this allele changes with increasing salinity in Long Island Sound (from southwest to northeast). How do the data from sites 9–11 compare with the data from the sites within the Sound?

Construct a hypothesis that explains the patterns you observe in the data and that accounts for the following observations: (1) the *lap*^{94} allele helps mussels maintain osmotic balance in water with a high salt concentration but is costly to use in less salty water; and (2) mussels produce larvae that can disperse long distances before they settle on rocks and grow into adults.

9. **WRITE ABOUT A THEME**
   Emergent Properties Heterozygotes at the sickle-cell locus produce both normal and abnormal (sickle-cell) hemoglobin (see Concept 14.4). When hemoglobin molecules are packed into a heterozygote's red blood cells, some cells receive relatively large quantities of abnormal hemoglobin, making these cells prone to sickling. In a short essay (approximately 100–150 words), explain how these molecular and cellular events lead to emergent properties at the individual and population levels of biological organization.

For selected answers, see Appendix A.

MasteringBIOLOGY www.masteringbiology.com

1. MasteringBiology® Assignments

Make Connections Tutorial Hardy-Weinberg Principle (Chapter 23) and Inheritance of Alleles (Chapter 14)
Experimental Inquiry Tutorial Did Natural Selection of Ground Finches Occur When the Environment Changed?
BioFlix **Tutorial** Mechanisms of Evolution
Tutorial Hardy-Weinberg Principle
Activities Genetic Variation from Sexual Recombination • The Hardy-Weinberg Principle • Causes of Evolutionary Change • Three Modes of Natural Selection
Questions Student Misconceptions • Reading Quiz • Multiple Choice • End-of-Chapter

2. eText

Read your book online, search, take notes, highlight text, and more.

3. The Study Area

Practice Tests • Cumulative Test • *BioFlix* 3-D Animations • MP3 Tutor Sessions • Videos • Activities • Investigations • Lab Media • Audio Glossary • Word Study Tools • Art

CHAPTER 23 The Evolution of Populations 487

◀ New **Write About a Theme Questions** give you practice writing a short essay that connects the chapter's content to one of the bookwide themes introduced in Chapter 1.

◀ Each chapter now ends with a preview of the **MasteringBiology®** resources that can help you succeed in the course.

Suggested Grading Rubric for "Write About a Theme" Short-Answer Essays

	Understanding of Theme and Relationship to Topic	Use of Supporting Examples or Details	Appropriate Use of Terminology	Quality of Writing
4	Evidence of full and complete understanding	Examples well chosen, details accurate and applied to theme	Accurate scientific terminology enhances the essay	Excellent organization, sentence structure, and grammar
3	Evidence of good understanding	Examples or details are generally well applied to theme	Terminology is correctly used	Good sentence flow, sentence structure, and grammar
2	Evidence of a basic understanding	Supporting examples and details are adequate	Terminology used is not totally accurate or appropriate	Some organizational and grammatical problems
1	Evidence of limited understanding	Examples and details are minimal	Appropriate terminology is not present	Poorly organized. Grammatical and spelling errors detract from essay
0	Essay shows no understanding of theme	Examples lacking or incorrect	Terminology lacking or incorrect	Essay is very poorly written

◀ This **Writing Rubric** explains criteria on which your writing may be graded. The rubric and tips for writing good short-answer essays can be found in the Study Area at www.masteringbiology.com.

To the Student: How to Effectively

The Mastering system empowers you to take charge of your learning—at your convenience, 24/7.

Use the Study Area on your own or in a study group.

BioFlix

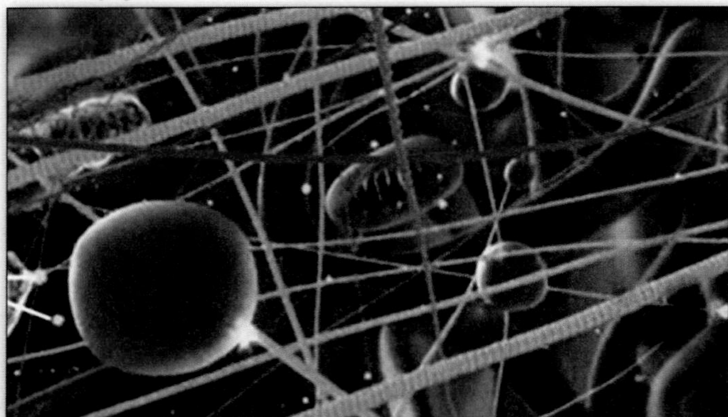

BioFlix® 3-D animations explore the most difficult biology topics, reinforced with tutorials, quizzes, and more.

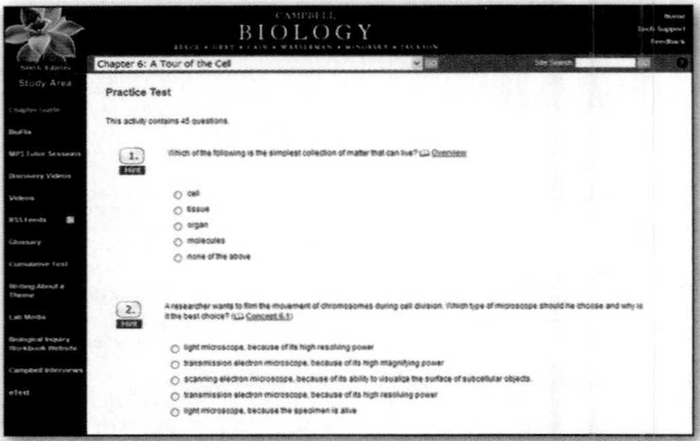

Practice Tests help you assess your understanding of each chapter, providing feedback for right and wrong answers.

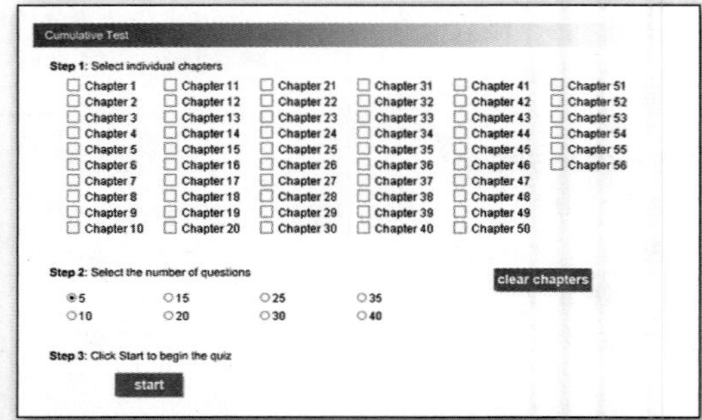

The **Cumulative Test** allows you to build a practice test with questions from multiple chapters.

Access your book online.

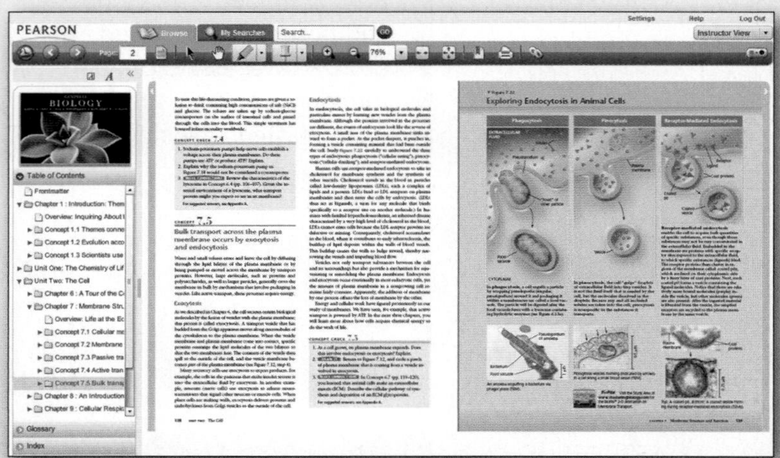

The **Pearson eText** gives you access to the text whenever and wherever you can access the Internet. The eText includes powerful interactive and customization functions:

- write notes
- highlight text
- bookmark pages
- zoom
- click hyperlinked words to view definitions
- search
- link to media activities and quizzes

Your professor can also write notes for the class and highlight important material using a new tool that works like an electronic pen on a whiteboard.

Use MasteringBiology®

Get personalized coaching & feedback.

Your instructor may assign self-paced **MasteringBiology® tutorials** that provide individualized coaching with specific hints and feedback on the toughest topics in the course.

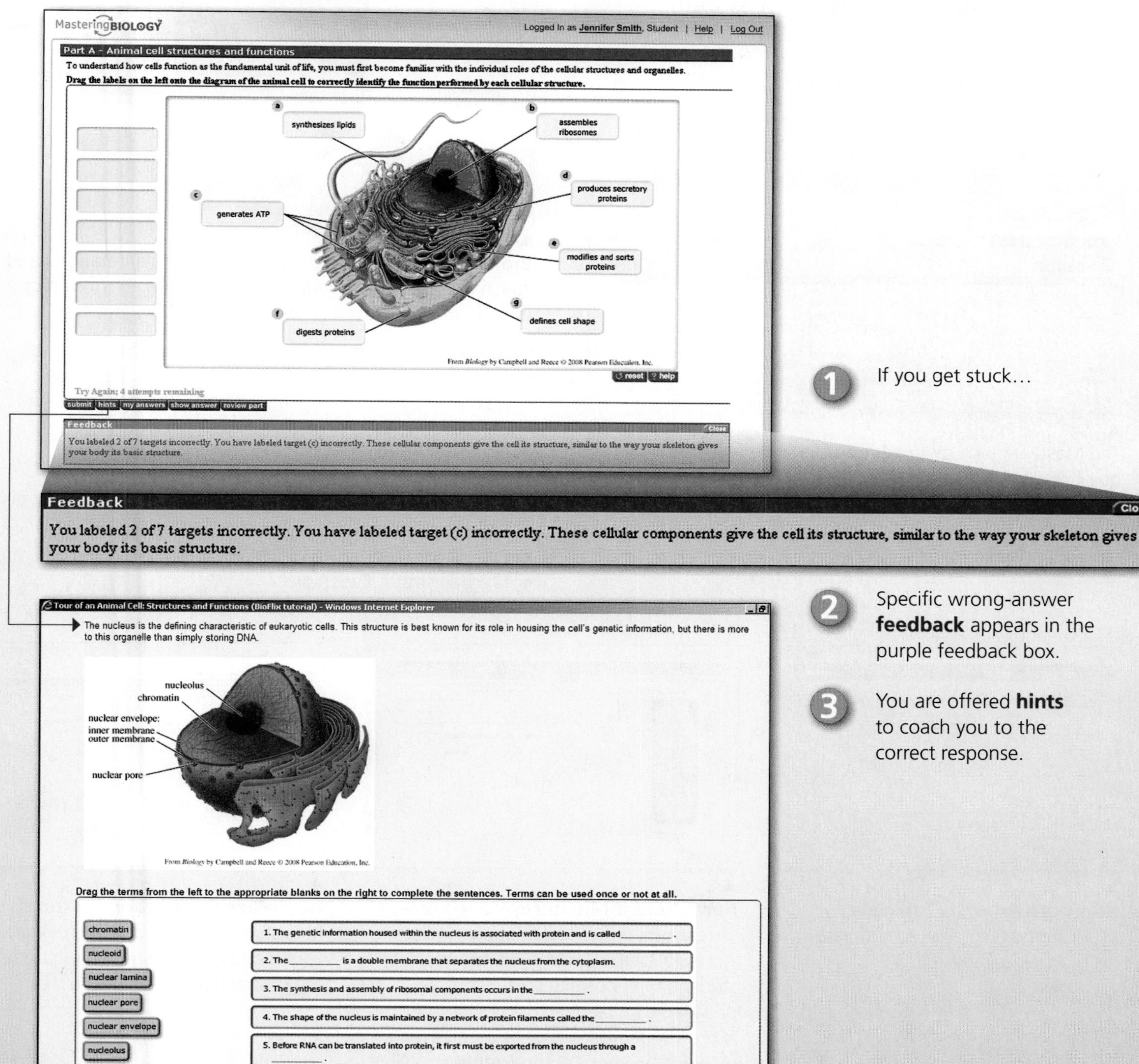

1 If you get stuck…

2 Specific wrong-answer **feedback** appears in the purple feedback box.

3 You are offered **hints** to coach you to the correct response.

To the Instructor: New Content in

MasteringBiology is the most effective and widely used online science tutorial, homework, and assessment system available.

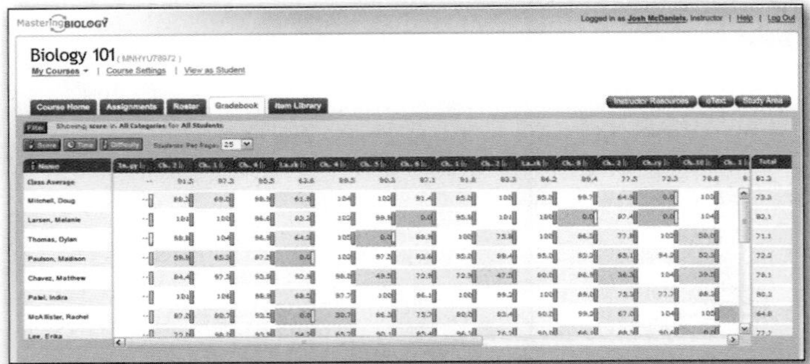

The MasteringBiology **gradebook** provides you with quick results and easy-to-interpret insights into student performance. Every assignment is **automatically graded** and shades of red highlight vulnerable students and challenging assignments.

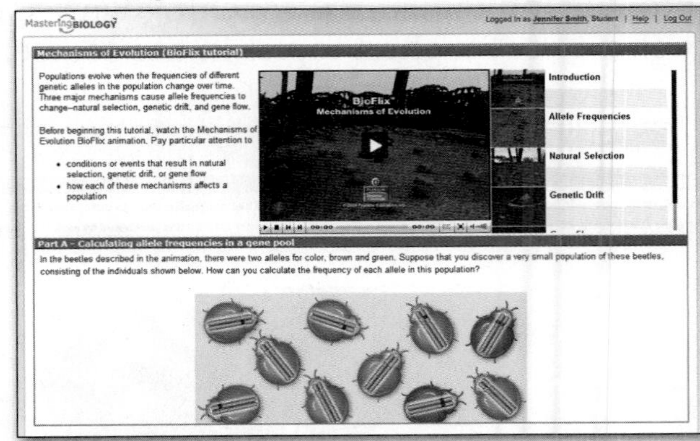

BioFlix® Tutorials use 3-D, movie-quality animations and coaching exercises to help students master tough topics outside of class. Tutorials and animations include:

- A Tour of the Animal Cell
- A Tour of the Plant Cell
- Membrane Transport
- Cellular Respiration
- Photosynthesis
- Mitosis
- Meiosis
- DNA Replication
- Protein Synthesis
- Mechanisms of Evolution
- Water Transport in Plants
- Homeostasis: Regulating Blood Sugar
- Gas Exchange
- How Neurons Work
- How Synapses Work
- Muscle Contraction
- Population Ecology
- The Carbon Cycle

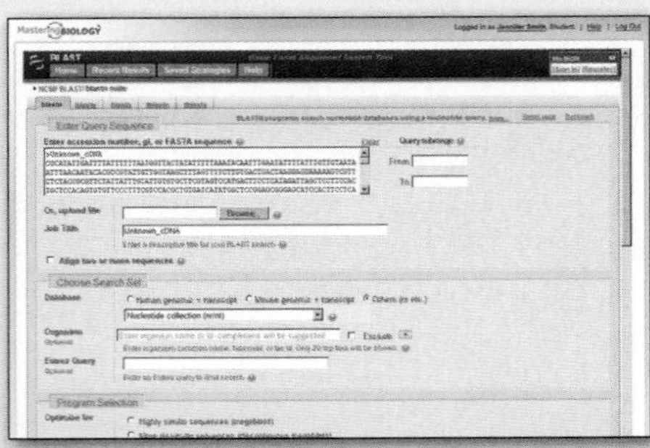

New **Data Analysis Tutorials** allow students to analyze real data from online databases.

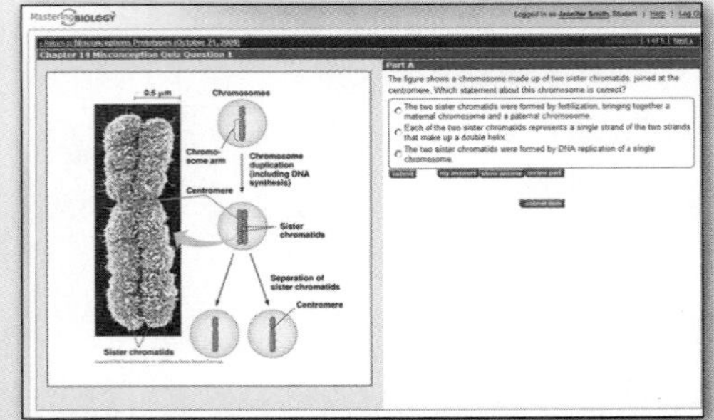

New **Student Misconceptions Questions** provide assignable quizzes for each chapter based on common student misconceptions. Students are provided with feedback, and the instructor is provided with in-class strategies for overcoming these misconceptions.

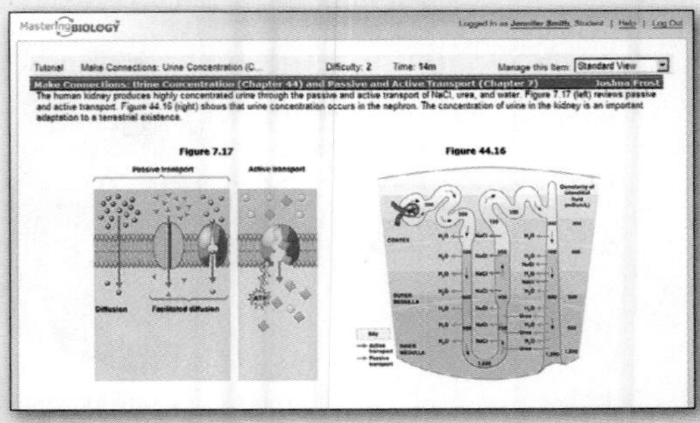

New **Make Connections Tutorials** help students connect what they are learning in one chapter with material they learned in an earlier chapter.

MasteringBiology®

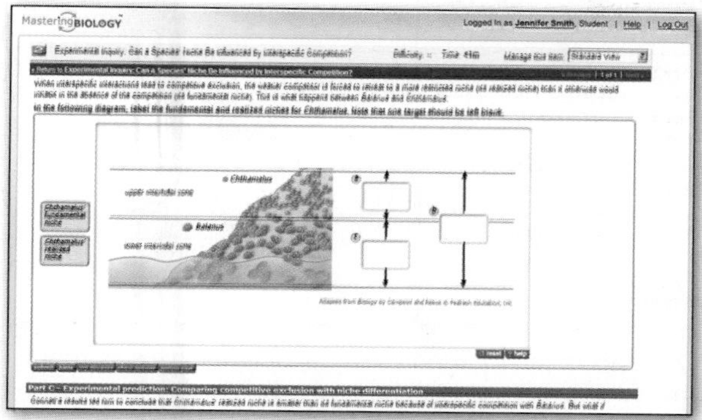

◀ New **Experimental Inquiry Tutorials,** based on some of biology's most influential experiments, give students practice analyzing experimental design and data, and help students understand how to reach conclusions based on collected data. Topics include:

- What Can You Learn About the Process of Science from Investigating a Cricket's Chirp?
- Which Wavelengths of Light Drive Photosynthesis?
- What Is the Inheritance Pattern of Sex-Linked Traits?
- Does DNA Replication Follow the Conservative, Semiconservative, or Dispersive Model?
- Did Natural Selection of Ground Finches Occur When the Environment Changed?
- What Effect Does Auxin Have on Coleoptile Growth?
- What Role Do Genes Play in Appetite Regulation?
- How Do Calcium Ions Help to Prevent Polyspermy During Egg Fertilization?
- Can a Species' Niche Be Influenced by Interspecific Competition?
- What Factors Influence the Loss of Nutrients from a Forest Ecosystem?

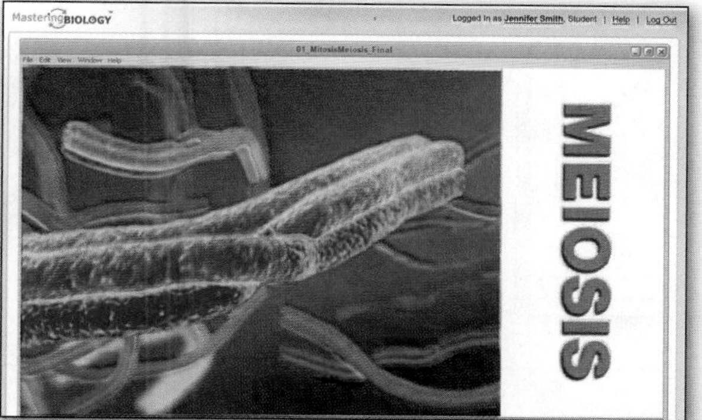

◀ The New **Video Tutor Sessions** walk students through tough topics with clearly explained visuals and demonstrations. Topics include:

- Mitosis and Meiosis
- Sex-Linked Pedigrees
- DNA Structure
- DNA Profiling Techniques
- Biodiversity
- Phylogenetic Trees

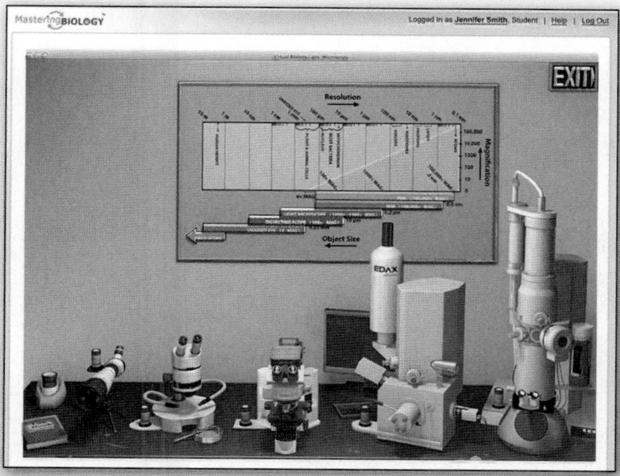

◀ The new **MasteringBiology: Virtual Biology Labs** online environment promotes critical thinking skills using virtual experiments and explorations that may be difficult to perform in a wet lab environment due to time, cost, or safety concerns.

- MasteringBiology: Virtual Biology Labs offer unique learning experiences in microscopy, molecular biology, genetics, ecology, and systematics.
- Choose from 20–30 automatically graded, "pre-set" lab activities that are ready to assign to students, or create your own from scratch.
- Each "pre-set" lab provides an assignable lab report with questions that are automatically graded and recorded in the MasteringBiology gradebook.
- Student subscriptions are available standalone or packaged with the *Campbell Biology* textbook.

Supplements

For Instructors

Instructor Resource DVD

978-0-321-67786-0 • 0-321-67786-2

Assets for each chapter include:

- All figures, photos, and tables in JPEG and PowerPoint®
- Prepared PowerPoint Presentations for each chapter, with lecture notes, editable figures from the text, and links to animations and videos
- Clicker Questions in PowerPoint
- 250+ Instructor Animations, including 3-D BioFlix®
- Discovery Channel™ Videos
- Test Bank questions in TestGen® software and Microsoft® Word

BioFlix® animations invigorate classroom lectures with 3-minute "movie quality" 3-D graphics (see list on p. xviii).

Customizable PowerPoints provide a jumpstart for each lecture.

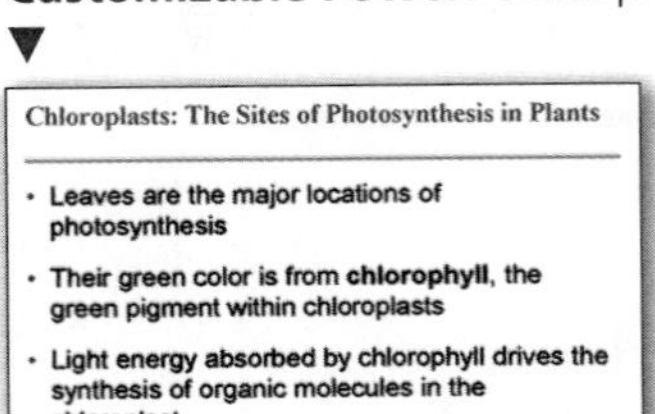

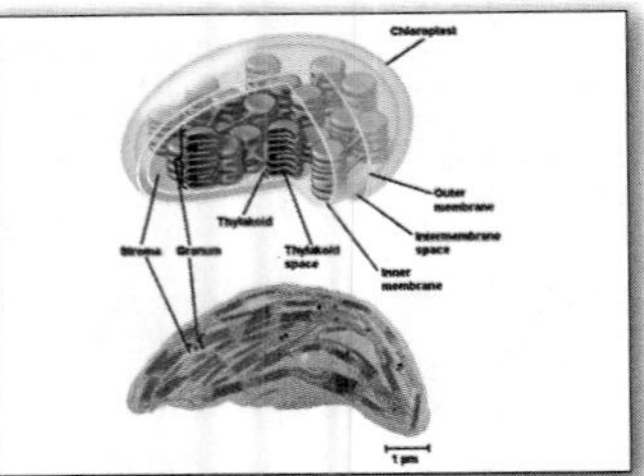

Clicker Questions can be used to stimulate effective classroom discussions (for use with or without clickers).

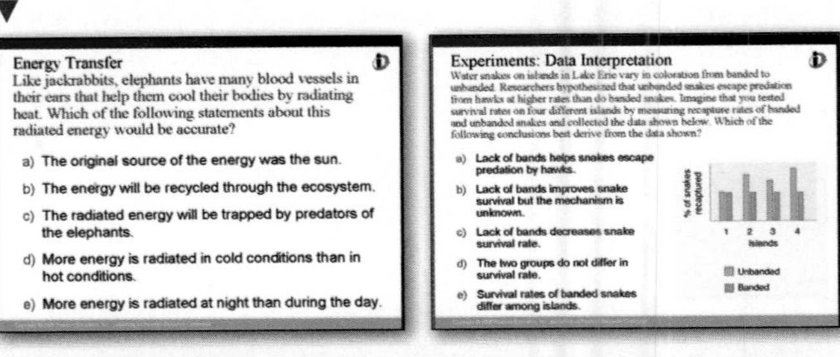

All of the art and photos from the book are provided with customizable labels. More than 1,600 photos from the text and other sources are included.

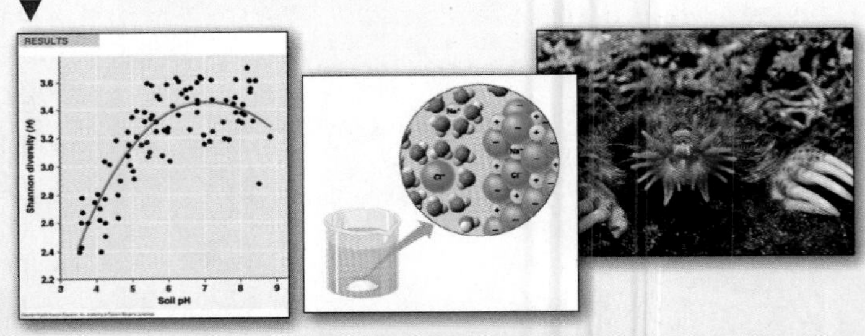

Printed Test Bank

by Louise Paquin, *McDaniel College*, Michael Dini, *Texas Tech University*, John Lepri, *University of North Carolina, Greensboro*, Jung Choi, *Georgia Institute of Technology*, John Zarnetske, *Hoosick Falls Central School*, and Ronald Balsamo, *Villanova University*

978-0-321-69729-5 • 0-321-69729-4

For the Ninth Edition, the Test Bank authors have increased the number of high-level application and synthesis questions.

Transparency Acetates

978-0-321-69708-0 • 0-321-69708-1

Course Management Systems

Test Bank questions, quizzes, and selected content from the Study Area of MasteringBiology are available in these popular course management systems:

CourseCompass™ (www.pearsonhighered.com)

Blackboard (www.pearsonhighered.com)

Instructor Resources Area in MasteringBiology

This area includes:

- PowerPoint® Lectures
- Clicker Questions
- JPEG Images
- Animations
- Videos
- Lecture Outlines
- Learning Objectives
- Strategies for Overcoming Common Student Misconceptions
- Instructor Guides for Supplements
- Write About a Theme: Tips for Grading Short-Answer Essays
- Suggested Answers for Essay Questions
- Test Bank Files
- Lab Media

For Students

***Student Study Guide,* Ninth Edition**
by Martha R. Taylor
978-0-321-62992-0 • 0-321-62992-2

This popular study guide helps students extract key ideas from the textbook and organize their knowledge of biology. Exercises include concept maps for each chapter, chapter summaries, word roots, chapter tests, and a variety of interactive questions in various formats.

Inquiry in Action: Interpreting Scientific Papers,* Second Edition
by Ruth Buskirk, *University of Texas at Austin,* and Christopher M. Gillen, *Kenyon College*
978-0-321-68336-6 • 0-321-68336-6

Selected Inquiry Figures in the Ninth Edition direct students to read and analyze the complete original research paper. In this supplement, those articles are reprinted and accompanied by questions that help students analyze the article. The Inquiry Figures from the book are reprinted in the supplement.

Practicing Biology: A Student Workbook,* Fourth Edition
by Jean Heitz and Cynthia Giffen, *University of Wisconsin, Madison*
978-0-321-68328-1 • 0-321-68328-5

This workbook offers a variety of activities to suit different learning styles. Activities such as modeling and mapping allow students to visualize and understand biological processes. Other activities focus on basic skills, such as reading and drawing graphs.

Biological Inquiry: A Workbook of Investigative Cases,* Third Edition
by Margaret Waterman, *Southeast Missouri State University,* and Ethel Stanley, *BioQUEST Curriculum Consortium and Beloit College*
978-0-321-68320-5 • 0-321-68320-X

This workbook offers ten investigative cases. A student website is in the Study Area at **www.masteringbiology.com.**

***Study Card,* Ninth Edition**
978-0-321-68322-9 • 0-321-68322-6

This quick-reference card provides an overview of the entire field of biology and helps students see the connections between topics.

***Spanish Glossary,* Ninth Edition**
by Laura P. Zanello, *University of California, Riverside*
978-0-321-68321-2 • 0-321-68321-8

This resource provides definitions in Spanish for all the glossary terms.

Into the Jungle: Great Adventures in the Search for Evolution
by Sean B. Carroll, *University of Wisconsin, Madison*
978-0-321-55671-4 • 0-321-55671-2

These nine short tales vividly depict key discoveries in evolutionary biology and the excitement of the scientific process. Online resources available at www.aw-bc.com/carroll.

Get Ready for Biology
978-0-321-50057-1 • 0-321-50057-1

This engaging workbook helps students brush up on important math and study skills and get up to speed on biological terminology and the basics of chemistry and cell biology.

***A Short Guide to Writing About Biology,* Seventh Edition**
by Jan A. Pechenik, *Tufts University*
978-0-321-66838-7 • 0-321-66838-3

This best-selling writing guide teaches students to think as biologists and to express ideas clearly and concisely through their writing.

***An Introduction to Chemistry for Biology Students,* Ninth Edition**
by George I. Sackheim, *University of Illinois, Chicago*
978-0- 8053-9571-6 • 0-8053-9571-7

This text/workbook helps students review and master all the basic facts, concepts, and terminology of chemistry that they need for their life science course.

***The Chemistry of Life CD-ROM, Biology Version,* Second Edition**
by Robert M. Thornton, *University of California, Davis*
978-0-8053-3063-2 • 0-8053-3063-1

This CD-ROM uses animations, interactive simulations, and quizzes with feedback to help students learn or review the chemistry needed to succeed in introductory biology.

For Lab

***Investigating Biology Laboratory Manual,* Seventh Edition**
by Judith G. Morgan, *Emory University,* and M. Eloise Brown Carter, *Oxford College of Emory University*
978-0-321-66821-9 • 0-321-66821-9

The Seventh Edition emphasizes connections to recurring themes in biology, including structure and function, unity and diversity, and the overarching theme of evolution.

***Annotated Instructor Edition for Investigating Biology Laboratory Manual,* Seventh Edition**
by Judith G. Morgan, *Emory University,* and M. Eloise Brown Carter, *Oxford College of Emory University*
978-0-321-67668-9 • 0-321-67668-8

***Preparation Guide for Investigating Biology Laboratory Manual,* Seventh Edition**
by Judith G. Morgan, *Emory University*, and M. Eloise Brown Carter, *Oxford College of Emory University*
978-0-321-67669-6 • 0-321-67669-6

Symbiosis: The Benjamin Cummings Custom Laboratory Program for the Biological Sciences

www.pearsoncustom.com/database/symbiosis/bc.html

MasteringBiology®: Virtual Biology Labs

www.masteringbiology.com
This online environment promotes critical thinking skills using virtual experiments and explorations that may be difficult to perform in a wet lab environment due to time, cost, or safety concerns. Designed to supplement or substitute for existing wet labs, this product offers students unique learning experiences and critical thinking exercises in the areas of microscopy, molecular biology, genetics, ecology, and systematics.

*** An Instructor Guide is available for download in the Instructor Resources Area at www.masteringbiology.com.**

Featured Figures

Impact Figures

Exploring Figures

Inquiry Figures

* The Inquiry Figure, original research paper, and a worksheet to guide you through the paper are provided in *Inquiry in Action: Interpreting Scientific Papers*, Second Edition.
† See the related Experimental Inquiry Tutorial in MasteringBiology® (www.masteringbiology.com).

Research Method Figures

Interviews

Acknowledgments

The authors wish to express their gratitude to the global community of instructors, researchers, students, and publishing professionals who have contributed to this edition.

As authors of this text, we are mindful of the daunting challenge of keeping up to date in all areas of our rapidly expanding subject. We are grateful to the many scientists who helped shape this edition by discussing their research fields with us, answering specific questions in their areas of expertise, and sharing their ideas about biology education. We are especially grateful to the following, listed alphabetically: John Archibald, John Armour, Kristian Axelsen, Scott Bowling, Barbara Bowman, Andy Cameron, Scott Carroll, Amy Cheng-Vollmer, Michele Clamp, David DeRosier, Doug DeSimone, Binh An Diep, David Ehrhardt, Robert Fowler, Peter Fraser, Matt Friedman, Tom Gingeras, Anita Gondor, Ken Halanych, Jeff Hardin, Catherine Hurlbut, Adam Johnson, Dale Kaiser, Patrick Keeling, Emir Khatipov, Chris Killian, Andrew Knoll, Nikos Kyrpides, Teri Liegler, Zhe-Xi Luo, Kent MacDonald, Nick Matzke, Melissa Michael, Nadia Naffakh, Rolf Ohlsson, Aharon Oren, Tom Owens, Kevin Padian, Nathalie Pardigon, Bruce Pavlik, Kevin J. Peterson, Michael Pollock, Rebekah Rasooly, Andrew Roger, Ole Seehausen, Alastair Simpson, Betty Smocovitis, Frank Solomon, Pam Soltis, Hans Thewissen, Mark Uhen, Vance Vredenburg, Elisabeth Wade, Phil Zamore, and Christine Zardecki. In addition, a total of 168 biologists, listed on pages xxvi–xxvii, provided detailed reviews of chapters for this edition, helping us ensure the book's scientific accuracy and improve its pedagogical effectiveness. And finally, we thank Marty Taylor, author of all nine editions of the *Student Study Guide*, for her many contributions to the accuracy, clarity, and consistency of the book.

Thanks also to the other professors and students, from all over the world, who offered suggestions directly to the authors. We alone bear the responsibility for any errors that remain in the text, but the dedication of our consultants, reviewers, and other correspondents makes us confident in the accuracy and effectiveness of this edition.

Interviews with prominent scientists have been a hallmark of CAMPBELL *BIOLOGY* since its inception, and conducting these interviews was again one of the great pleasures of revising the book. To open the eight units of this edition, we are proud to include interviews with Susan Solomon, Bonnie Bassler, Joan Steitz, Geerat Vermeij, Ford Doolittle, Luis Herrera-Estrella, Toto Olivera, and Camille Parmesan.

The value of CAMPBELL *BIOLOGY* as a learning tool is greatly enhanced by the supplementary materials that have been created for instructors and students. We recognize that the dedicated authors of these materials are essentially writing mini (and not so mini) books. We much appreciate the hard work and creativity of all the authors listed, with their creations, on pages xx–xxi. In addition, we are grateful to Joan Sharp (*Lecture Outlines*, *Learning Objectives*, and *Student Misconceptions*) and Erin Barley and Kathleen Fitzpatrick (*PowerPoint® Lectures*).

The electronic media for this text are invaluable teaching and learning aids. We thank the hardworking and creative authors of the new material for this edition: Tom Owens, Joan Sharp, and Jennifer Yeh (*MasteringBiology®*); Richard Cowlishaw, Tod Duncan, and Stephanie Pandolfi (*Study Area Practice Tests*); and Eric Simon (*VideoTutors*). And we thank Brian White for his work. Also, we are grateful to the many other people—biology instructors, editors, artists, production experts, and narrators—who are listed in the credits for these and other elements of the electronic media that accompany the book. And we thank the reviewers and class testers of BioFlix® and MasteringBiology.

CAMPBELL *BIOLOGY*, Ninth Edition, results from an unusually strong synergy between a team of scientists and a team of publishing professionals. Our editorial team at Benjamin Cummings again demonstrated unmatched talents, commitment, and pedagogical insights. Our new Acquisitions Editor, Josh Frost, brought publishing savvy, intelligence, and a much appreciated level head to leading the whole team. The clarity and effectiveness of every page owe much to our extraordinary Supervising Editors Pat Burner and Beth Winickoff, who headed the top-notch development team: Developmental Editors John Burner, Matt Lee, and Mary Catherine Hager; and Developmental Artists Hilair Chism, Carla Simmons, Andrew Recher, and Jay McElroy. Our unsurpassed Senior Editorial Manager Ginnie Simione Jutson, Executive Director of Development Deborah Gale, Assistant Editor Logan Triglia, and Editor-in-Chief Beth Wilbur were indispensable in moving the project in the right direction. We also want to thank Robin Heyden for organizing the annual Biology Leadership Conferences and keeping us in touch with the world of AP Biology.

You would not have this beautiful book in your hands today if not for the work of the book production team: Executive Managing Editor Erin Gregg; Managing Editor Michael Early; Senior Production Project Manager Shannon Tozier; Senior Photo Editor Donna Kalal; Photo Researcher Maureen Spuhler; Copy Editor Janet Greenblatt; Art Editor Laura Murray; Proofreaders Joanna Dinsmore and Pete Shanks; Permissions Editors Sue Ewing and Beth Keister; Senior Project Editor Emily Bush, S4Carlisle; Composition Manager Holly Paige, S4Carlisle; Art Production Manager Kristina Seymour, Precision Graphics; Design Director Mark Ong; Designer Gary Hespenheide; and Manufacturing Buyer Michael Penne. We also thank those who worked on the book's supplements: Susan Berge, Nina Lewallen Hufford, Brady Golden, Jane Brundage, James Bruce, and John Hammett. And for creating the wonderful package of electronic media that accompanies the book, we are grateful to Tania Mlawer, Director of Editorial Content for MasteringBiology, and Jonathan Ballard, Deb Greco, Sarah Jensen, Mary Catherine Hager, Alice Fugate, Juliana Tringali, Josh Gentry, Steve Wright, Kristen Sutton, Katie Foley, Karen Sheh, and David Kokorowski, as well as Director of Media Development Lauren Fogel and Director of Media Strategy Stacy Treco.

For their important roles in marketing the book, we thank Christy Lawrence, Lauren Harp, Scott Dustan, Lillian Carr, Jane Campbell, Jessica Perry, Nora Massuda, and Jessica Tree.

We are grateful to Linda Davis, President of Pearson Math and Science, who has shared our commitment to excellence and provided strong support for five editions now. Moreover, we thank Paul Corey, President of Pearson Benjamin Cummings, and Editorial Director Frank Ruggirello for their enthusiasm, encouragement, and support.

The Pearson sales team, which represents CAMPBELL *BIOLOGY* on campus, is an essential link to the users of the text. They tell us what you like and don't like about the book, communicate the features of the book, and provide prompt service. We thank them for their hard work and professionalism. For representing our book to our international audience, we thank our sales and marketing partners throughout the world. They are all strong allies in biology education.

Finally, we wish to thank our families and friends for their encouragement and patience throughout this long project. Our special thanks to Paul, Dan, Maria, Armelle, and Sean (J.B.R.); Lily, Ross, Lily-too, and Alex (L.A.U.); Debra and Hannah (M.L.C.); Harry, Elga, Aaron, Sophie, Noah, and Gabriele (S.A.W.); Natalie (P.V.M.); and Sally, Robert, David, and Will (R.B.J.). And, as always, thanks to Rochelle, Allison, Jason, and McKay.

Jane Reece, Lisa Urry, Michael Cain,
Steve Wasserman, Peter Minorsky, and Rob Jackson

Reviewers

Ninth Edition Reviewers

Ann Aguanno, *Marymount Manhattan College*
Marc Albrecht, *University of Nebraska*
John Alcock, *Arizona State University*
Eric Alcorn, *Acadia University*
Terry Austin, *Temple College*
Brian Bagatto, *University of Akron*
Virginia Baker, *Chipola College*
Bonnie Baxter, *Westminster College*
Marilee Benore, *University of Michigan, Dearborn*
Catherine Black, *Idaho State University*
William Blaker, *Furman University*
Edward Blumenthal, *Marquette University*
David Bos, *Purdue University*
Scott Bowling, *Auburn University*
Beth Burch, *Huntington University*
Ragan Callaway, *The University of Montana*
Kenneth M. Cameron, *University of Wisconsin, Madison*
Patrick Canary, *Northland Pioneer College*
Cheryl Keller Capone, *Pennsylvania State University*
Karen I. Champ, *Central Florida Community College*
David Champlin, *University of Southern Maine*
Brad Chandler, *Palo Alto College*
Wei-Jen Chang, *Hamilton College*
Jung Choi, *Georgia Institute of Technology*
Steve Christensen, *Brigham Young University, Idaho*
James T. Colbert, *Iowa State University*
William Cushwa, *Clark College*
Shannon Datwyler, *California State University, Sacramento*
Eugene Delay, *University of Vermont*
Daniel DerVartanian, *University of Georgia*
Janet De Souza-Hart, *Massachusetts College of Pharmacy & Health Sciences*
Kathryn A. Durham, *Lorain Community College*
Curt Elderkin, *College of New Jersey*
Mary Ellard-Ivey, *Pacific Lutheran University*
George Ellmore, *Tufts University*
Robert C. Evans, *Rutgers University, Camden*
Sam Fan, *Bradley University*
Paul Farnsworth, *University of New Mexico*
Myriam Alhadeff Feldman, *Cascadia Community College*
Teresa Fischer, *Indian River Community College*
David Fitch, *New York University*
T. Fleming, *Bradley University*
Robert Fowler, *San Jose State University*
Robert Franklin, *College of Charleston*
Art Fredeen, *University of Northern British Columbia*
Matt Friedman, *University of Chicago*
Cynthia M. Galloway, *Texas A&M University, Kingsville*
Simon Gilroy, *University of Wisconsin, Madison*
Jim Goetze, *Laredo Community College*
Lynda Goff, *University of California, Santa Cruz*
Roy Golsteyn, *University of Lethbridge*
Barbara E. Goodman, *University of South Dakota*
David Grise, *Texas A&M University, Corpus Christi*
Devney Hamilton, *Stanford University (student)*
Matthew B. Hamilton, *Georgetown University*
Jeanne M. Harris, *University of Vermont*
Stephanie Harvey, *Georgia Southwestern State University*
Bernard Hauser, *University of Florida*
Andreas Hejnol, *Sars International Centre for Marine Molecular Biology*
Jason Hodin, *Stanford University*
Sara Huang, *Los Angeles Valley College*
Catherine Hurlbut, *Florida State College, Jacksonville*
Diane Husic, *Moravian College*
Thomas Jacobs, *University of Illinois*
Mark Jaffe, *Nova Southeastern University*
Douglas Jensen, *Converse College*
Lance Johnson, *Midland Lutheran College*
Cheryl Jorcyk, *Boise State University*
Caroline Kane, *University of California, Berkeley*
Jennifer Katcher, *Pima Community College*
Eric G. Keeling, *Cary Institute of Ecosystem Studies*
Chris Kennedy, *Simon Fraser University*
Hillar Klandorf, *West Virginia University*
Mark Knauss, *Georgia Highlands College*
Roger Koeppe, *University of Arkansas*
Peter Kourtev, *Central Michigan University*
Eliot Krause, *Seton Hall University*
Steven Kristoff, *Ivy Tech Community College*
William Kroll, *Loyola University*
Rukmani Kuppuswami, *Laredo Community College*
Lee Kurtz, *Georgia Gwinnett College*
Michael P. Labare, *United States Military Academy, West Point*
Ellen Lamb, *University of North Carolina, Greensboro*
William Lamberts, *College of St Benedict and St John's University*
Tali D. Lee, *University of Wisconsin, Eau Claire*
Hugh Lefcort, *Gonzaga University*
Alcinda Lewis, *University of Colorado, Boulder*
Graeme Lindbeck, *Valenica Community College*
Hannah Lui, *University of California, Irvine*
Cindy Malone, *California State University, Northridge*
Julia Marrs, *Barnard College (student)*
Kathleen Marrs, *Indiana University-Purdue University, Indianapolis*
Mike Mayfield, *Ball State University*
Kamau Mbuthia, *Bowling Green State University*
Tanya McGhee, *Craven Community College*
Darcy Medica, *Pennsylvania State University*
Susan Meiers, *Western Illinois University*
Alex Mills, *University of Windsor*
Eli Minkoff, *Bates College*
Subhash Minocha, *University of New Hampshire*
Ivona Mladenovic, *Simon Fraser University*
Courtney Murren, *College of Charleston*
Kimberlyn Nelson, *Pennsylvania State University*
Jacalyn Newman, *University of Pittsburgh*
Kathleen Nolta, *University of Michigan*
Aharon Oren, *The Hebrew University*
Henry R. Owen, *Eastern Illinois University*
Stephanie Pandolfi, *Michigan State University*
Nathalie Pardigon, *Institut Pasteur*
Cindy Paszkowski, *University of Alberta*
Andrew Pease, *Stevenson University*
Nancy Pelaez, *Purdue University*
Irene Perry, *University of Texas of the Permian Basin*
Roger Persell, *Hunter College*
Mark Pilgrim, *College of Coastal Georgia*
Vera M. Piper, *Shenadoah University*
Crima Pogge, *City College of San Francisco*
Michael Pollock, *Mount Royal University*
Roberta Pollock, *Occidental College*
Therese M. Poole, *Georgia State University*
Angela R. Porta, *Kean University*
Robert Powell, *Avila University*
Elena Pravosudova, *University of Nevada, Reno*
Terrell Pritts, *University of Arkansas, Little Rock*
Monica Ranes-Goldberg, *University of California, Berkeley*
Robert S. Rawding, *Gannon University*
Sarah Richart, *Azusa Pacific University*
Kenneth Robinson, *Purdue University*
Heather Roffey, *Marianopolis College*
Patricia Rugaber, *College of Coastal Georgia*
Scott Russell, *University of Oklahoma*
Louis Santiago, *University of California, Riverside*
Tom Sawicki, *Spartanburg Community College*
Thomas W. Schoener, *University of California, Davis*
Patricia Schulte, *University of British Colombia*
Brenda Schumpert, *Valencia Community College*
David Schwartz, *Houston Community College*
Brent Selinger, *University of Lethbridge*
Alison M. Shakarian, *Salve Regina University*
Robin L. Sherman, *Nova Southeastern University*
Sedonia Sipes, *Southern Illinois University, Carbondale*
Joel Stafstrom, *Northern Illinois University*
Alam Stam, *Capital University*
Judy Stone, *Colby College*
Cynthia Surmacz, *Bloomsburg University*
David Tam, *University of North Texas*
Yves Tan, *Cabrillo College*
Emily Taylor, *California Polytechnic State University*
Franklyn Tan Te, *Miami Dade College*
Kent Thomas, *Wichita State University*

Saba Valadkhan, *Center for RNA Molecular Biology*
Sarah VanVickle-Chavez, *Washington University, St. Louis*
William Velhagen, *New York University*
Janice Voltzow, *University of Scranton*
Margaret Voss, *Penn State Erie*
Charles Wade, *C.S. Mott Community College*
Claire Walczak, *Indiana University*
Jerry Waldvogel, *Clemson University*
Robert Lee Wallace, *Ripon College*
Fred Wasserman, *Boston University*
John Weishampel, *University of Central Florida*
Susan Whittemore, *Keene State College*
Janet Wolkenstein, *Hudson Valley Community College*
Grace Wyngaard, *James Madison University*
Paul Yancey, *Whitman College*
Anne D. Yoder, *Duke University*
Nina Zanetti, *Siena College*
Sam Zeveloff, *Weber State University*
Theresa Zucchero, *Methodist University*

Reviewers of Previous Editions

Kenneth Able *(State University of New York, Albany)*
Thomas Adams *(Michigan State University)*
Martin Adamson *(University of British Columbia)*
Dominique Adriaens *(Ghent University)*
Shylaja Akkaraju *(Bronx Community College of CUNY)*
John Alcock *(Arizona State University)*
George R. Aliaga *(Tarrant County College)*
Richard Almon *(State University of New York, Buffalo)*
Bonnie Amos *(Angelo State University)*
Katherine Anderson *(University of California, Berkeley)*
Richard J. Andren *(Montgomery County Community College)*
Estry Ang *(University of Pittsburgh, Greensburg)*
Jeff Appling *(Clemson University)*
J. David Archibald *(San Diego State University)*
David Armstrong *(University of Colorado, Boulder)*
Howard J. Arnott *(University of Texas, Arlington)*
Mary Ashley *(University of Illinois, Chicago)*
Angela S. Aspbury *(Texas State University)*
Robert Atherton *(University of Wyoming)*
Karl Aufderheide *(Texas A&M University)*
Leigh Auleb *(San Francisco State University)*
P. Stephen Baenziger *(University of Nebraska)*
Ellen Baker *(Santa Monica College)*
Katherine Baker *(Millersville University)*
William Barklow *(Framingham State College)*
Susan Barman *(Michigan State University)*
Steven Barnhart *(Santa Rosa Junior College)*
Andrew Barton *(University of Maine Farmington)*
Rebecca A. Bartow *(Western Kentucky University)*
Ron Basmajian *(Merced College)*
David Bass *(University of Central Oklahoma)*
Bonnie Baxter *(Hobart & William Smith)*
Tim Beagley *(Salt Lake Community College)*
Margaret E. Beard *(College of the Holy Cross)*
Tom Beatty *(University of British Columbia)*
Chris Beck *(Emory University)*
Wayne Becker *(University of Wisconsin, Madison)*
Patricia Bedinger *(Colorado State University)*
Jane Beiswenger *(University of Wyoming)*
Anne Bekoff *(University of Colorado, Boulder)*
Marc Bekoff *(University of Colorado, Boulder)*
Tania Beliz *(College of San Mateo)*
Adrianne Bendich *(Hoffman-La Roche, Inc.)*
Barbara Bentley *(State University of New York, Stony Brook)*
Darwin Berg *(University of California, San Diego)*
Werner Bergen *(Michigan State University)*
Gerald Bergstrom *(University of Wisconsin, Milwaukee)*
Anna W. Berkovitz *(Purdue University)*
Dorothy Berner *(Temple University)*
Annalisa Berta *(San Diego State University)*
Paulette Bierzychudek *(Pomona College)*
Charles Biggers *(Memphis State University)*
Kenneth Birnbaum *(New York University)*
Michael W. Black *(California Polytechnic State University, San Luis Obispo)*
Robert Blanchard *(University of New Hampshire)*
Andrew R. Blaustein *(Oregon State University)*
Judy Bluemer *(Morton College)*
Edward Blumenthal *(Marquette University)*
Robert Blystone *(Trinity University)*
Robert Boley *(University of Texas, Arlington)*
Jason E. Bond *(East Carolina University)*
Eric Bonde *(University of Colorado, Boulder)*
Cornelius Bondzi *(Hampton University)*
Richard Boohar *(University of Nebraska, Omaha)*
Carey L. Booth *(Reed College)*
Allan Bornstein *(Southeast Missouri State University)*
Oliver Bossdorf *(State University of New York, Stony Book)*
James L. Botsford *(New Mexico State University)*
Lisa Boucher *(University of Nebraska, Omaha)*
J. Michael Bowes *(Humboldt State University)*
Richard Bowker *(Alma College)*
Robert Bowker *(Glendale Community College, Arizona)*
Barbara Bowman *(Mills College)*
Barry Bowman *(University of California, Santa Cruz)*
Deric Bownds *(University of Wisconsin, Madison)*
Robert Boyd *(Auburn University)*
Sunny Boyd *(University of Notre Dame)*
Jerry Brand *(University of Texas, Austin)*
Edward Braun *(Iowa State University)*
Theodore A. Bremner *(Howard University)*
James Brenneman *(University of Evansville)*
Charles H. Brenner *(Berkeley, California)*
Lawrence Brewer *(University of Kentucky)*
Donald P. Briskin *(University of Illinois, Urbana)*
Paul Broady *(University of Canterbury)*
Chad Brommer *(Emory University)*
Judith L. Bronstein *(University of Arizona)*
Danny Brower *(University of Arizona)*
Carole Browne *(Wake Forest University)*
Mark Browning *(Purdue University)*
David Bruck *(San Jose State University)*
Robb T. Brumfield *(Louisiana State University)*
Herbert Bruneau *(Oklahoma State University)*
Gary Brusca *(Humboldt State University)*
Richard C. Brusca *(University of Arizona, Arizona-Sonora Desert Museum)*
Alan H. Brush *(University of Connecticut, Storrs)*
Howard Buhse *(University of Illinois, Chicago)*
Arthur Buikema *(Virginia Tech)*
Al Burchsted *(College of Staten Island)*
Meg Burke *(University of North Dakota)*
Edwin Burling *(De Anza College)*
William Busa *(Johns Hopkins University)*
Jorge Busciglio *(University of California, Irvine)*
John Bushnell *(University of Colorado)*
Linda Butler *(University of Texas, Austin)*
David Byres *(Florida Community College, Jacksonville)*
Guy A. Caldwell *(University of Alabama)*
Jane Caldwell *(West Virginia University)*
Kim A. Caldwell *(University of Alabama)*
R. Andrew Cameron *(California Institute of Technology)*
Alison Campbell *(University of Waikato)*
Iain Campbell *(University of Pittsburgh)*
W. Zacheus Cande *(University of California, Berkeley)*
Robert E. Cannon *(University of North Carolina, Greensboro)*
Deborah Canington *(University of California, Davis)*
Frank Cantelmo *(St. John's University)*
John Capeheart *(University of Houston, Downtown)*
Gregory Capelli *(College of William and Mary)*
Richard Cardullo *(University of California, Riverside)*
Nina Caris *(Texas A&M University)*
Jeffrey Carmichael *(University of North Dakota)*
Robert Carroll *(East Carolina University)*
Laura L. Carruth *(Georgia State University)*
J. Aaron Cassill *(University of Texas, San Antonio)*
David Champlin *(University of Southern Maine)*
Bruce Chase *(University of Nebraska, Omaha)*
P. Bryant Chase *(Florida State University)*
Doug Cheeseman *(De Anza College)*
Shepley Chen *(University of Illinois, Chicago)*
Giovina Chinchar *(Tougaloo College)*
Joseph P. Chinnici *(Virginia Commonwealth University)*
Jung H. Choi *(Georgia Institute of Technology)*
Geoffrey Church *(Fairfield University)*
Henry Claman *(University of Colorado Health Science Center)*
Anne Clark *(Binghamton University)*
Greg Clark *(University of Texas)*
Patricia J. Clark *(Indiana University-Purdue University, Indianapolis)*
Ross C. Clark *(Eastern Kentucky University)*
Lynwood Clemens *(Michigan State University)*
Janice J. Clymer *(San Diego Mesa College)*
William P. Coffman *(University of Pittsburgh)*
Austin Randy Cohen *(California State University, Northridge)*
J. John Cohen *(University of Colorado Health Science Center)*
Jim Colbert *(Iowa State University)*
Jan Colpaert *(Hasselt University)*

Robert Colvin *(Ohio University)*
Jay Comeaux *(McNeese State University)*
David Cone *(Saint Mary's University)*
Elizabeth Connor *(University of Massachusetts)*
Joanne Conover *(University of Connecticut)*
Gregory Copenhaver *(University of North Carolina, Chapel Hill)*
John Corliss *(University of Maryland)*
James T. Costa *(Western Carolina University)*
Stuart J. Coward *(University of Georgia)*
Charles Creutz *(University of Toledo)*
Bruce Criley *(Illinois Wesleyan University)*
Norma Criley *(Illinois Wesleyan University)*
Joe W. Crim *(University of Georgia)*
Greg Crowther *(University of Washington)*
Karen Curto *(University of Pittsburgh)*
Anne Cusic *(University of Alabama, Birmingham)*
Richard Cyr *(Pennsylvania State University)*
Marymegan Daly *(The Ohio State University)*
W. Marshall Darley *(University of Georgia)*
Cynthia Dassler *(The Ohio State University)*
Marianne Dauwalder *(University of Texas, Austin)*
Larry Davenport *(Samford University)*
Bonnie J. Davis *(San Francisco State University)*
Jerry Davis *(University of Wisconsin, La Crosse)*
Michael A. Davis *(Central Connecticut State University)*
Thomas Davis *(University of New Hampshire)*
John Dearn *(University of Canberra)*
Maria E. de Bellard *(California State University, Northridge)*
Teresa DeGolier *(Bethel College)*
James Dekloe *(University of California, Santa Cruz)*
Patricia A. DeLeon *(University of Delaware)*
Veronique Delesalle *(Gettysburg College)*
T. Delevoryas *(University of Texas, Austin)*
Roger Del Moral *(University of Washington)*
Charles F. Delwiche *(University of Maryland)*
Diane C. DeNagel *(Northwestern University)*
William L. Dentler *(University of Kansas)*
Daniel Dervartanian *(University of Georgia)*
Jean DeSaix *(University of North Carolina, Chapel Hill)*
Biao Ding *(Ohio State University)*
Michael Dini *(Texas Tech University)*
Andrew Dobson *(Princeton University)*
Stanley Dodson *(University of Wisconsin, Madison)*
Mark Drapeau *(University of California, Irvine)*
John Drees *(Temple University School of Medicine)*
Charles Drewes *(Iowa State University)*
Marvin Druger *(Syracuse University)*
Gary Dudley *(University of Georgia)*
Susan Dunford *(University of Cincinnati)*
Betsey Dyer *(Wheaton College)*
Robert Eaton *(University of Colorado)*
Robert S. Edgar *(University of California, Santa Cruz)*
Douglas Eernisse *(California State University, Fullerton)*
Douglas J. Eernisse *(California State University, Fullerton)*
Betty J. Eidemiller *(Lamar University)*
Brad Elder *(Doane College)*
William D. Eldred *(Boston University)*
Michelle Elekonich *(University of Nevada, Las Vegas)*
Mary Ellard-Ivey *(Pacific Lutheran University)*
Norman Ellstrand *(University of California, Riverside)*
Johnny El-Rady *(University of South Florida)*
Dennis Emery *(Iowa State University)*
John Endler *(University of California, Santa Barbara)*
Margaret T. Erskine *(Lansing Community College)*
Gerald Esch *(Wake Forest University)*
Frederick B. Essig *(University of South Florida)*
Mary Eubanks *(Duke University)*
David Evans *(University of Florida)*
Robert C. Evans *(Rutgers University, Camden)*
Sharon Eversman *(Montana State University)*
Olukemi Fadayomi *(Ferris State University)*
Lincoln Fairchild *(Ohio State University)*
Peter Fajer *(Florida State University)*
Bruce Fall *(University of Minnesota)*
Lynn Fancher *(College of DuPage)*
Ellen H. Fanning *(Vanderbilt University)*
Paul Farnsworth *(University of Texas, San Antonio)*
Larry Farrell *(Idaho State University)*
Jerry F. Feldman *(University of California, Santa Cruz)*
Lewis Feldman *(University of California, Berkeley)*
Eugene Fenster *(Longview Community College)*
Russell Fernald *(University of Oregon)*
Rebecca Ferrell *(Metropolitan State College of Denver)*
Kim Finer *(Kent State University)*
Milton Fingerman *(Tulane University)*
Barbara Finney *(Regis College)*
Frank Fish *(West Chester University)*
David Fisher *(University of Hawaii, Manoa)*
Jonathan S. Fisher *(St. Louis University)*
Steven Fisher *(University of California, Santa Barbara)*
Kirk Fitzhugh *(Natural History Museum of Los Angeles County)*
Lloyd Fitzpatrick *(University of North Texas)*
William Fixsen *(Harvard University)*
Abraham Flexer *(Manuscript Consultant, Boulder, Colorado)*
Kerry Foresman *(University of Montana)*
Norma Fowler *(University of Texas, Austin)*
Robert G. Fowler *(San Jose State University)*
David Fox *(University of Tennessee, Knoxville)*
Carl Frankel *(Pennsylvania State University, Hazleton)*
James Franzen *(University of Pittsburgh)*
Bill Freedman *(Dalhousie University)*
Otto Friesen *(University of Virginia)*
Frank Frisch *(Chapman University)*
Virginia Fry *(Monterey Peninsula College)*
Bernard Frye *(University of Texas, Arlington)*
Jed Fuhrman *(University of Southern California)*
Alice Fulton *(University of Iowa)*
Chandler Fulton *(Brandeis University)*
Sara Fultz *(Stanford University)*
Berdell Funke *(North Dakota State University)*
Anne Funkhouser *(University of the Pacific)*
Zofia E. Gagnon *(Marist College)*
Michael Gaines *(University of Miami)*
Arthur W. Galston *(Yale University)*
Stephen Gammie *(University of Wisconsin, Madison)*
Carl Gans *(University of Michigan)*
John Gapter *(University of Northern Colorado)*
Andrea Gargas *(University of Wisconsin, Madison)*
Lauren Garner *(California Polytechnic State University, San Luis Obispo)*
Reginald Garrett *(University of Virginia)*
Patricia Gensel *(University of North Carolina)*
Chris George *(California Polytechnic State University, San Luis Obispo)*
Robert George *(University of Wyoming)*
J. Whitfield Gibbons *(University of Georgia)*
J. Phil Gibson *(Agnes Scott College)*
Frank Gilliam *(Marshall University)*
Simon Gilroy *(Pennsylvania State University)*
Alan D. Gishlick *(Gustavus Adolphus College)*
Todd Gleeson *(University of Colorado)*
Jessica Gleffe *(University of California, Irvine)*
John Glendinning *(Barnard College)*
David Glenn-Lewin *(Wichita State University)*
William Glider *(University of Nebraska)*
Tricia Glidewell *(Marist School)*
Elizabeth A. Godrick *(Boston University)*
Lynda Goff *(University of California, Santa Cruz)*
Elliott Goldstein *(Arizona State University)*
Paul Goldstein *(University of Texas, El Paso)*
Sandra Gollnick *(State University of New York, Buffalo)*
Anne Good *(University of California, Berkeley)*
Judith Goodenough *(University of Massachusetts, Amherst)*
Wayne Goodey *(University of British Columbia)*
Robert Goodman *(University of Wisconsin, Madison)*
Ester Goudsmit *(Oakland University)*
Linda Graham *(University of Wisconsin, Madison)*
Robert Grammer *(Belmont University)*
Joseph Graves *(Arizona State University)*
Phyllis Griffard *(University of Houston, Downtown)*
A. J. F. Griffiths *(University of British Columbia)*
William Grimes *(University of Arizona)*
Mark Gromko *(Bowling Green State University)*
Serine Gropper *(Auburn University)*
Katherine L. Gross *(Ohio State University)*
Gary Gussin *(University of Iowa)*
Mark Guyer *(National Human Genome Research Institute)*
Ruth Levy Guyer *(Bethesda, Maryland)*
R. Wayne Habermehl *(Montgomery County Community College)*
Mac Hadley *(University of Arizona)*
Joel Hagen *(Radford University)*
Jack P. Hailman *(University of Wisconsin)*
Leah Haimo *(University of California, Riverside)*
Ken Halanych *(Auburn University)*
Jody Hall *(Brown University)*
Douglas Hallett, *(Northern Arizona University)*
Rebecca Halyard *(Clayton State College)*
E. William Hamilton *(Washington and Lee University)*
Sam Hammer *(Boston University)*
Penny Hanchey-Bauer *(Colorado State University)*

William F. Hanna *(Massasoit Community College)*
Laszlo Hanzely *(Northern Illinois University)*
Jeff Hardin *(University of Wisconsin, Madison)*
Lisa Harper *(University of California, Berkeley)*
Richard Harrison *(Cornell University)*
Carla Hass *(Pennsylvania State University)*
Chris Haufler *(University of Kansas)*
Bernard A. Hauser *(University of Florida)*
Chris Haynes *(Shelton State Community College)*
Evan B. Hazard *(Bemidji State University, Emeritus)*
H. D. Heath *(California State University, Hayward)*
George Hechtel *(State University of New York, Stony Brook)*
S. Blair Hedges *(Pennsylvania State University)*
Brian Hedlund *(University of Nevada, Las Vegas)*
David Heins *(Tulane University)*
Jean Heitz *(University of Wisconsin, Madison)*
John D. Helmann *(Cornell University)*
Colin Henderson *(University of Montana)*
Susan Hengeveld *(Indiana University)*
Michelle Henricks *(University of California, Los Angeles)*
Caroll Henry *(Chicago State University)*
Frank Heppner *(University of Rhode Island)*
Albert Herrera *(University of Southern California)*
Scott Herrick *(Missouri Western State College)*
Ira Herskowitz *(University of California, San Francisco)*
Paul E. Hertz *(Barnard College)*
David Hibbett *(Clark University)*
R. James Hickey *(Miami University)*
William Hillenius *(College of Charleston)*
Kenneth Hillers *(California Polytechnic State University, San Luis Obispo)*
Ralph Hinegardner *(University of California, Santa Cruz)*
William Hines *(Foothill College)*
Robert Hinrichsen *(Indiana University of Pennsylvania)*
Helmut Hirsch *(State University of New York, Albany)*
Tuan-hua David Ho *(Washington University)*
Carl Hoagstrom *(Ohio Northern University)*
James Hoffman *(University of Vermont)*
A. Scott Holaday *(Texas Tech University)*
N. Michele Holbrook *(Harvard University)*
James Holland *(Indiana State University, Bloomington)*
Charles Holliday *(Lafayette College)*
Lubbock Karl Holte *(Idaho State University)*
Alan R. Holyoak *(Brigham Young University, Idaho)*
Laura Hoopes *(Occidental College)*
Nancy Hopkins *(Massachusetts Institute of Technology)*
Sandra Horikami *(Daytona Beach Community College)*
Kathy Hornberger *(Widener University)*
Pius F. Horner *(San Bernardino Valley College)*
Becky Houck *(University of Portland)*
Margaret Houk *(Ripon College)*
Daniel J. Howard *(New Mexico State University)*
Ronald R. Hoy *(Cornell University)*
Cristin Hulslander *(University of Oregon)*
Donald Humphrey *(Emory University School of Medicine)*
Robert J. Huskey *(University of Virginia)*
Steven Hutcheson *(University of Maryland, College Park)*
Sandra Hsu *(Skyline College)*
Linda L. Hyde *(Gordon College)*
Bradley Hyman *(University of California, Riverside)*
Mark Iked *(San Bernardino Valley College)*
Jeffrey Ihara *(Mira Costa College)*
Cheryl Ingram-Smith *(Clemson University)*
Alice Jacklet *(State University of New York, Albany)*
John Jackson *(North Hennepin Community College)*
John C. Jahoda *(Bridgewater State College)*
Dan Johnson *(East Tennessee State University)*
Lee Johnson *(The Ohio State University)*
Randall Johnson *(University of California, San Diego)*
Stephen Johnson *(William Penn University)*
Wayne Johnson *(Ohio State University)*
Kenneth C. Jones *(California State University, Northridge)*
Russell Jones *(University of California, Berkeley)*
Chad Jordan *(North Carolina State University)*
Alan Journet *(Southeast Missouri State University)*
Walter Judd *(University of Florida)*
Thomas W. Jurik *(Iowa State University)*
Caroline M. Kane *(University of California, Berkeley)*
Thomas C. Kane *(University of Cincinnati)*
Tamos Kapros *(University of Missouri)*
E. L. Karlstrom *(University of Puget Sound)*
Jennifer Katcher *(Pima Community College)*
Laura A. Katz *(Smith College)*
Maureen Kearney *(Field Museum of Natural History)*
Patrick Keeling *(University of British Columbia)*
Elizabeth A. Kellogg *(University of Missouri, St. Louis)*
Norm Kenkel *(University of Manitoba)*
Chris Kennedy *(Simon Fraser University)*
George Khoury *(National Cancer Institute)*
Rebecca T. Kimball *(University of Florida)*
Mark Kirk *(University of Missouri, Columbia)*
Robert Kitchin *(University of Wyoming)*
Attila O. Klein *(Brandeis University)*
Daniel Klionsky *(University of Michigan)*
Jennifer Knight *(University of Colorado)*
Ned Knight *(Linfield College)*
David Kohl *(University of California, Santa Barbara)*
Greg Kopf *(University of Pennsylvania School of Medicine)*
Thomas Koppenheffer *(Trinity University)*
Margareta Krabbe *(Uppsala University)*
Anselm Kratochwil *(Universität Osnabrück)*
Deborah M. Kristan *(California State University, San Marcos)*
William Kroll *(Loyola University, Chicago)*
Janis Kuby *(San Francisco State University)*
Justin P. Kumar *(Indiana University)*
David Kurijaka *(Ohio University)*
Marc-André Lachance *(University of Western Ontario)*
J. A. Lackey *(State University of New York, Oswego)*
Elaine Lai *(Brandeis University)*
Mohamed Lakrim *(Kingsborough Community College)*
Lynn Lamoreux *(Texas A&M University)*
William L'Amoreaux *(College of Staten Island)*
Carmine A. Lanciani *(University of Florida)*
Kenneth Lang *(Humboldt State University)*
Dominic Lannutti *(El Paso Community College)*
Allan Larson *(Washington University)*
John Latto *(University of California, Santa Barbara)*
Diane K. Lavett *(State University of New York, Cortland, and Emory University)*
Charles Leavell *(Fullerton College)*
C. S. Lee *(University of Texas)*
Daewoo Lee *(Ohio University)*
Robert Leonard *(University of California, Riverside)*
Michael R. Leonardo *(Coe College)*
John Lepri *(University of North Carolina, Greensboro)*
Donald Levin *(University of Texas, Austin)*
Mike Levine *(University of California, Berkeley)*
Joseph Levine *(Boston College)*
Bill Lewis *(Shoreline Community College)*
John Lewis *(Loma Linda University)*
Lorraine Lica *(California State University, Hayward)*
Harvey Liftin *(Broward Community College)*
Harvey Lillywhite *(University of Florida, Gainesville)*
Graeme Lindbeck *(Valencia Community College)*
Clark Lindgren *(Grinnell College)*
Diana Lipscomb *(George Washington University)*
Christopher Little *(The University of Texas, Pan American)*
Kevin D. Livingstone *(Trinity University)*
Andrea Lloyd *(Middlebury College)*
Sam Loker *(University of New Mexico)*
Christopher A. Loretz *(State University of New York, Buffalo)*
Jane Lubchenco *(Oregon State University)*
Douglas B. Luckie *(Michigan State University)*
Margaret A. Lynch *(Tufts University)*
Steven Lynch *(Louisiana State University, Shreveport)*
Richard Machemer Jr. *(St. John Fisher College)*
Elizabeth Machunis-Masuoka *(University of Virginia)*
James MacMahon *(Utah State University)*
Christine R. Maher *(University of Southern Maine)*
Linda Maier *(University of Alabama, Huntsville)*
Jose Maldonado *(El Paso Community College)*
Richard Malkin *(University of California, Berkeley)*
Charles Mallery *(University of Miami)*
Keith Malmos *(Valencia Community College, East Campus)*
Cindy Malone *(California State University, Northridge)*
Carol Mapes *(Kutztown University of Pennsylvania)*
William Margolin *(University of Texas Medical School)*
Lynn Margulis *(Boston University)*
Kathleen A. Marrs *(Indiana University-Purdue University, Indianapolis)*
Edith Marsh *(Angelo State University)*
Diane L. Marshall *(University of New Mexico)*
Linda Martin Morris *(University of Washington)*
Karl Mattox *(Miami University of Ohio)*
Joyce Maxwell *(California State University, Northridge)*
Jeffrey D. May *(Marshall University)*
Lee McClenaghan *(San Diego State University)*
Richard McCracken *(Purdue University)*
Andrew McCubbin *(Washington State University)*
Kerry McDonald *(University of Missouri, Columbia)*
Jacqueline McLaughlin *(Pennsylvania State University, Lehigh Valley)*

Neal McReynolds *(Texas A&M International)*
Lisa Marie Meffert *(Rice University)*
Michael Meighan *(University of California, Berkeley)*
Scott Meissner *(Cornell University)*
Paul Melchior *(North Hennepin Community College)*
Phillip Meneely *(Haverford College)*
John Merrill *(Michigan State University)*
Brian Metscher *(University of California, Irvine)*
Ralph Meyer *(University of Cincinnati)*
James Mickle *(North Carolina State University)*
Roger Milkman *(University of Iowa)*
Helen Miller *(Oklahoma State University)*
John Miller *(University of California, Berkeley)*
Kenneth R. Miller *(Brown University)*
John E. Minnich *(University of Wisconsin, Milwaukee)*
Michael J. Misamore *(Texas Christian University)*
Kenneth Mitchell *(Tulane University School of Medicine)*
Alan Molumby *(University of Illinois, Chicago)*
Nicholas Money *(Miami University)*
Russell Monson *(University of Colorado, Boulder)*
Joseph P. Montoya *(Georgia Institute of Technology)*
Frank Moore *(Oregon State University)*
Janice Moore *(Colorado State University)*
Randy Moore *(Wright State University)*
William Moore *(Wayne State University)*
Carl Moos *(Veterans Administration Hospital, Albany, New York)*
Michael Mote *(Temple University)*
Alex Motten *(Duke University)*
Jeanette Mowery *(Madison Area Technical College)*
Deborah Mowshowitz *(Columbia University)*
Rita Moyes *(Texas A&M College Station)*
Darrel L. Murray *(University of Illinois, Chicago)*
John Mutchmor *(Iowa State University)*
Elliot Myerowitz *(California Institute of Technology)*
Gavin Naylor *(Iowa State University)*
John Neess *(University of Wisconsin, Madison)*
Tom Neils *(Grand Rapids Community College)*
Raymond Neubauer *(University of Texas, Austin)*
Todd Newbury *(University of California, Santa Cruz)*
James Newcomb *(New England College)*
Harvey Nichols *(University of Colorado, Boulder)*
Deborah Nickerson *(University of South Florida)*
Bette Nicotri *(University of Washington)*
Caroline Niederman *(Tomball College)*
Maria Nieto *(California State University, Hayward)*
Anders Nilsson *(University of Umeå)*
Greg Nishiyama *(College of the Canyons)*
Charles R. Noback *(College of Physicians and Surgeons, Columbia University)*
Jane Noble-Harvey *(Delaware University)*
Mary C. Nolan *(Irvine Valley College)*
Peter Nonacs *(University of California, Los Angeles)*
Mohamed A. F. Noor *(Duke University)*
Shawn Nordell *(St. Louis University)*
Richard S. Norman *(University of Michigan, Dearborn, Emeritus)*
David O. Norris *(University of Colorado, Boulder)*
Steven Norris *(California State, Channel Islands)*
Gretchen North *(Occidental College)*
Cynthia Norton *(University of Maine, Augusta)*
Steve Norton *(East Carolina University)*
Steve Nowicki *(Duke University)*
Bette H. Nybakken *(Hartnell College)*
Brian O'Conner *(University of Massachusetts, Amherst)*
Gerard O'Donovan *(University of North Texas)*
Eugene Odum *(University of Georgia)*
Mark P. Oemke *(Alma College)*
Linda Ogren *(University of California, Santa Cruz)*
Patricia O'Hern *(Emory University)*
Nathan O. Okia *(Auburn University, Montgomery)*
Jeanette Oliver *(St. Louis Community College, Florissant Valley)*
Gary P. Olivetti *(University of Vermont)*
John Olsen *(Rhodes College)*
Laura J. Olsen *(University of Michigan)*
Sharman O'Neill *(University of California, Davis)*
Wan Ooi *(Houston Community College)*
John Oross *(University of California (Riverside)*
Gay Ostarello *(Diablo Valley College)*
Catherine Ortega *(Fort Lewis College)*
Charissa Osborne *(Butler University)*
Thomas G. Owens *(Cornell University)*
Penny Padgett *(University of North Carolina, Chapel Hill)*
Kevin Padian *(University of California, Berkeley)*
Dianna Padilla *(State University of New York, Stony Brook)*
Anthony T. Paganini *(Michigan State University)*
Barry Palevitz *(University of Georgia)*
Michael A. Palladino *(Monmouth University)*
Daniel Papaj *(University of Arizona)*
Peter Pappas *(County College of Morris)*
Bulah Parker *(North Carolina State University)*
Stanton Parmeter *(Chemeketa Community College)*
Robert Patterson *(San Francisco State University)*
Ronald Patterson *(Michigan State University)*
Crellin Pauling *(San Francisco State University)*
Kay Pauling *(Foothill Community College)*
Daniel Pavuk *(Bowling Green State University)*
Debra Pearce *(Northern Kentucky University)*
Patricia Pearson *(Western Kentucky University)*
Shelley Penrod *(North Harris College)*
Imara Y. Perera *(North Carolina State University)*
Beverly Perry *(Houston Community College)*
David Pfennig *(University of North Carolina, Chapel Hill)*
David S. Pilliod *(California Polytechnic State University, San Luis Obispo)*
J. Chris Pires *(University of Missouri, Columbia)*
Bob Pittman *(Michigan State University)*
James Platt *(University of Denver)*
Martin Poenie *(University of Texas, Austin)*
Scott Poethig *(University of Pennsylvania)*
Jeffrey Pommerville *(Texas A&M University)*
Angela R. Porta *(Kean University)*
Warren Porter *(University of Wisconsin)*
Daniel Potter *(University of California, Davis)*
Donald Potts *(University of California, Santa Cruz)*
Andy Pratt *(University of Canterbury)*
David Pratt *(University of California, Davis)*
Halina Presley *(University of Illinois, Chicago)*
Mary V. Price *(University of California, Riverside)*
Mitch Price *(Pennsylvania State University)*
Rong Sun Pu *(Kean University)*
Rebecca Pyles *(East Tennessee State University)*
Scott Quackenbush *(Florida International University)*
Ralph Quatrano *(Oregon State University)*
Peter Quinby *(University of Pittsburgh)*
Val Raghavan *(Ohio State University)*
Deanna Raineri *(University of Illinois, Champaign-Urbana)*
Talitha Rajah *(Indiana University Southeast)*
Charles Ralph *(Colorado State University)*
Thomas Rand *(Saint Mary's University)*
Robert H. Reaves *(Glendale Community College)*
Kurt Redborg *(Coe College)*
Ahnya Redman *(Pennsylvania State)*
Brian Reeder *(Morehead State University)*
Bruce Reid *(Kean University)*
David Reid *(Blackburn College)*
C. Gary Reiness *(Lewis & Clark College)*
Charles Remington *(Yale University)*
Erin Rempala *(San Diego Mesa College)*
David Reznick *(University of California, Riverside)*
Douglas Rhoads *(University of Arkansas)*
Fred Rhoades *(Western Washington State University)*
Eric Ribbens *(Western Illinois University)*
Christina Richards *(New York University)*
Christopher Riegle *(Irvine Valley College)*
Loren Rieseberg *(University of British Columbia)*
Bruce B. Riley *(Texas A&M University)*
Donna Ritch *(Pennsylvania State University)*
Carol Rivin *(Oregon State University East)*
Laurel Roberts *(University of Pittsburgh)*
Thomas Rodella *(Merced College)*
Rodney Rogers *(Drake University)*
William Roosenburg *(Ohio University)*
Mike Rosenzweig *(Virginia Polytechnic Institute and State University)*
Wayne Rosing *(Middle Tennessee State University)*
Thomas Rost *(University of California, Davis)*
Stephen I. Rothstein *(University of California, Santa Barbara)*
John Ruben *(Oregon State University)*
Albert Ruesink *(Indiana University)*
Neil Sabine *(Indiana University)*
Tyson Sacco *(Cornell University)*
Rowan F. Sage *(University of Toronto)*
Tammy Lynn Sage *(University of Toronto)*
Don Sakaguchi *(Iowa State University)*
Walter Sakai *(Santa Monica College)*
Mark F. Sanders *(University of California, Davis)*
Ted Sargent *(University of Massachusetts, Amherst)*
K. Sathasivan *(University of Texas, Austin)*
Gary Saunders *(University of New Brunswick)*
Thomas R. Sawicki *(Spartanburg Community College)*
Inder Saxena *(University of Texas, Austin)*
Carl Schaefer *(University of Connecticut)*

Maynard H. Schaus *(Virginia Wesleyan College)*
Renate Scheibe *(University of Osnabrück)*
David Schimpf *(University of Minnesota, Duluth)*
William H. Schlesinger *(Duke University)*
Mark Schlissel *(University of California, Berkeley)*
Christopher J. Schneider *(Boston University)*
Thomas W. Schoener *(University of California, Davis)*
Robert Schorr *(Colorado State University)*
Patricia M. Schulte *(University of British Columbia)*
Karen S. Schumaker *(University of Arizona)*
David J. Schwartz *(Houston Community College)*
Christa Schwintzer *(University of Maine)*
Erik P. Scully *(Towson State University)*
Robert W. Seagull *(Hofstra University)*
Edna Seaman *(Northeastern University)*
Duane Sears *(University of California, Santa Barbara)*
Orono Shukdeb Sen *(Bethune-Cookman College)*
Wendy Sera *(Seton Hill University)*
Timothy E. Shannon *(Francis Marion University)*
Joan Sharp *(Simon Fraser University)*
Victoria C. Sharpe *(Blinn College)*
Elaine Shea *(Loyola College, Maryland)*
Stephen Sheckler *(Virginia Polytechnic Institute and State University)*
Richard Sherwin *(University of Pittsburgh)*
Lisa Shimeld *(Crafton Hills College)*
James Shinkle *(Trinity University)*
Barbara Shipes *(Hampton University)*
Richard M. Showman *(University of South Carolina)*
Peter Shugarman *(University of Southern California)*
Alice Shuttey *(DeKalb Community College)*
James Sidie *(Ursinus College)*
Daniel Simberloff *(Florida State University)*
Rebecca Simmons *(University of North Dakota)*
Anne Simon *(University of Maryland, College Park)*
Robert Simons *(University of California (Los Angeles)*
Alastair Simpson *(Dalhousie University)*
Susan Singer *(Carleton College)*
Roger Sloboda *(Dartmouth University)*
John Smarrelli *(Le Moyne College)*
Andrew T. Smith *(Arizona State University)*
Kelly Smith *(University of North Florida)*
Nancy Smith-Huerta *(Miami Ohio University)*
John Smol *(Queen's University)*
Andrew J. Snope *(Essex Community College)*
Julio G. Soto *(San Jose State University)*
Mitchell Sogin *(Woods Hole Marine Biological Laboratory)*
Susan Sovonick-Dunford *(University of Cincinnati)*
Frederick W. Spiegel *(University of Arkansas)*
John Stachowicz *(University of California, Davis)*
Amanda Starnes *(Emory University)*
Karen Steudel *(University of Wisconsin)*
Barbara Stewart *(Swarthmore College)*
Gail A. Stewart *(Camden County College)*
Cecil Still *(Rutgers University, New Brunswick)*
Margery Stinson *(Southwestern College)*
James Stockand *(University of Texas Health Science Center, San Antonio)*
John Stolz *(California Institute of Technology)*
Richard D. Storey *(Colorado College)*
Stephen Strand *(University of California, Los Angeles)*
Eric Strauss *(University of Massachusetts, Boston)*
Antony Stretton *(University of Wisconsin, Madison)*
Russell Stullken *(Augusta College)*
Mark Sturtevant *(University of Michigan, Flint)*
John Sullivan *(Southern Oregon State University)*
Gerald Summers *(University of Missouri)*
Judith Sumner *(Assumption College)*
Marshall D. Sundberg *(Emporia State University)*
Lucinda Swatzell *(Southeast Missouri State University)*
Daryl Sweeney *(University of Illinois, Champaign-Urbana)*
Samuel S. Sweet *(University of California, Santa Barbara)*
Janice Swenson *(University of North Florida)*
Michael A. Sypes *(Pennsylvania State University)*
Lincoln Taiz *(University of California, Santa Cruz)*
Samuel Tarsitano *(Southwest Texas State University)*
David Tauck *(Santa Clara University)*
Emily Taylor *(California Polytechnic State University, San Luis Obispo)*
James Taylor *(University of New Hampshire)*
John W. Taylor *(University of California, Berkeley)*
Martha R. Taylor *(Cornell University)*
Thomas Terry *(University of Connecticut)*
Roger Thibault *(Bowling Green State University)*
William Thomas *(Colby-Sawyer College)*
Cyril Thong *(Simon Fraser University)*
John Thornton *(Oklahoma State University)*
Robert Thornton *(University of California, Davis)*
William Thwaites *(Tillamook Bay Community College)*
Stephen Timme *(Pittsburg State University)*
Eric Toolson *(University of New Mexico)*
Leslie Towill *(Arizona State University)*
James Traniello *(Boston University)*
Paul Q. Trombley *(Florida State University)*
Nancy J. Trun *(Duquesne University)*
Constantine Tsoukas *(San Diego State University)*
Marsha Turell *(Houston Community College)*
Robert Tuveson *(University of Illinois, Urbana)*
Maura G. Tyrrell *(Stonehill College)*
Catherine Uekert *(Northern Arizona University)*
Claudia Uhde-Stone *(California State University, East Bay)*
Gordon Uno *(University of Oklahoma)*
Lisa A. Urry *(Mills College)*
Saba Valadkhan *(Case Western Reserve University School of Medicine)*
James W. Valentine *(University of California, Santa Barbara)*
Joseph Vanable *(Purdue University)*
Theodore Van Bruggen *(University of South Dakota)*
Kathryn VandenBosch *(Texas A&M University)*
Gerald Van Dyke *(North Carolina State University)*
Brandi Van Roo *(Framingham State College)*
Moira Van Staaden *(Bowling Green State)*
Steven D. Verhey *(Central Washington University)*
Kathleen Verville *(Washington College)*
Sara Via *(University of Maryland)*
Frank Visco *(Orange Coast College)*
Laurie Vitt *(University of California, Los Angeles)*
Neal Voelz *(St. Cloud State University)*
Thomas J. Volk *(University of Wisconsin, La Crosse)*
Leif Asbjørn Vøllestad *(University of Oslo)*
Susan D. Waaland *(University of Washington)*
William Wade *(Dartmouth Medical College)*
D. Alexander Wait *(Southwest Missouri State University)*
John Waggoner *(Loyola Marymount University)*
Jyoti Wagle *(Houston Community College)*
Edward Wagner *(University of California, Irvine)*
Dan Walker *(San Jose State University)*
Robert L. Wallace *(Ripon College)*
Jeffrey Walters *(North Carolina State University)*
Linda Walters *(University of Central Florida)*
Nickolas M. Waser *(University of California, Riverside)*
Margaret Waterman *(University of Pittsburgh)*
Charles Webber *(Loyola University of Chicago)*
Peter Webster *(University of Massachusetts, Amherst)*
Terry Webster *(University of Connecticut, Storrs)*
Beth Wee *(Tulane University)*
Andrea Weeks *(George Mason University)*
Peter Wejksnora *(University of Wisconsin, Milwaukee)*
Kentwood Wells *(University of Connecticut)*
David J. Westenberg, *(University of Missouri, Rolla)*
Richard Wetts *(University of California, Irvine)*
Matt White *(Ohio University)*
Ernest H. Williams *(Hamilton College)*
Kathy Williams *(San Diego State University)*
Stephen Williams *(Glendale Community College)*
Elizabeth Willott *(University of Arizona)*
Christopher Wills *(University of California, San Diego)*
Paul Wilson *(California State University, Northridge)*
Fred Wilt *(University of California, Berkeley)*
Peter Wimberger *(University of Puget Sound)*
Robert Winning *(Eastern Michigan University)*
E. William Wischusen *(Louisiana State University)*
Susan Whittemore *(Keene State College)*
Clarence Wolfe *(Northern Virginia Community College)*
Vickie L. Wolfe *(Marshall University)*
Robert T. Woodland *(University of Massachusetts Medical School)*
Joseph Woodring *(Louisiana State University)*
Denise Woodward *(Pennsylvania State University)*
Patrick Woolley *(East Central College)*
Sarah E. Wyatt *(Ohio University)*
Ramin Yadegari *(University of Arizona)*
Paul Yancey *(Whitman College)*
Philip Yant *(University of Michigan)*
Linda Yasui *(Northern Illinois University)*
Hideo Yonenaka *(San Francisco State University)*
Gina M. Zainelli *(Loyola University, Chicago)*
Edward Zalisko *(Blackburn College)*
Zai Ming Zhao *(University of Texas, Austin)*
John Zimmerman *(Kansas State University)*
Miriam Zolan *(Indiana University)*
Uko Zylstra *(Calvin College)*

Detailed Contents

5 The Structure and Function of Large Biological Molecules 68

UNIT 2

The Cell 92

6 A Tour of the Cell 94

7 Membrane Structure and Function 125

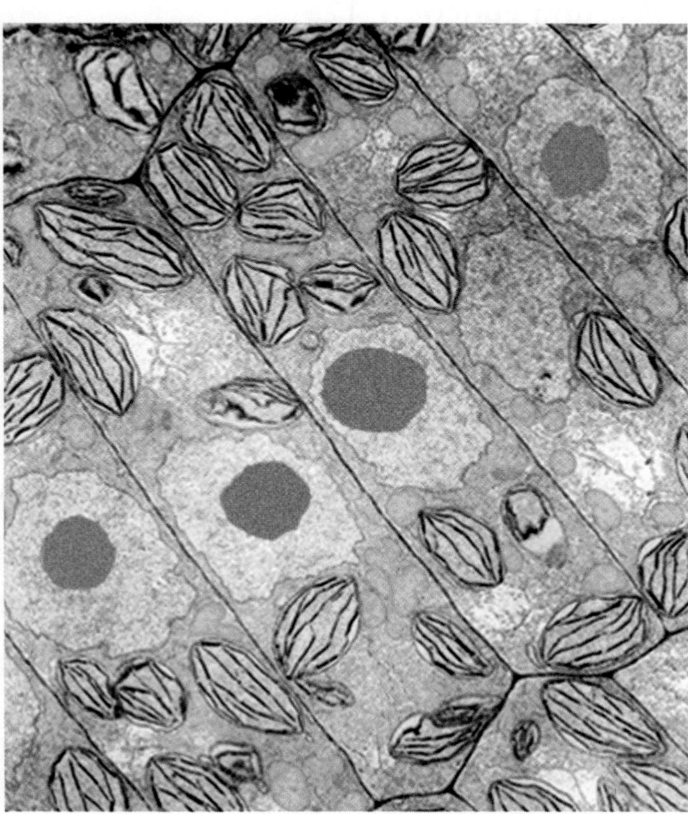

1

Introduction: Themes in the Study of Life

▲ **Figure 1.1 How is the mother-of-pearl plant adapted to its environment?**

KEY CONCEPTS

1.1 The themes of this book make connections across different areas of biology

1.2 The Core Theme: Evolution accounts for the unity and diversity of life

1.3 In studying nature, scientists make observations and then form and test hypotheses

1.4 Science benefits from a cooperative approach and diverse viewpoints

OVERVIEW

Inquiring About Life

The mother-of-pearl plant, or ghost plant (**Figure 1.1** and cover), is native to a single mountain in northeastern Mexico. Its fleshy, succulent leaves and other features allow this plant to store and conserve water. Even when rain falls, the plant's access to water is limited because it grows in crevices of vertical rock walls, where little soil is present to hold rainwater (**Figure 1.2**). The plant's water-conserving characteristics help it survive and thrive in these nooks and crannies. Similar features are found in many plants that live in dry environments, allowing them to eke out a living where rain is unpredictable.

An organism's adaptations to its environment, such as adaptations for conserving water, are the result of **evolution**, the process of change that has transformed life on Earth from its earliest beginnings to the diversity of organisms living today. Evolution is the fundamental organizing principle of biology and the core theme of this book.

Although biologists know a great deal about life on Earth, many mysteries remain. For instance, what exactly led to the origin of flowering among plants such as the one pictured here? Posing questions about the living world and seeking science-based answers—scientific inquiry—are the central activities of **biology**, the scientific study of life. Biologists' questions can be ambitious. They may ask how a single tiny cell becomes a tree or a dog, how the human mind works, or how the different forms of life in a forest interact. Most people wonder about the organisms living around them, and many interesting questions probably occur to you when you are out-of-doors, surrounded by the natural world. When they do, you are already thinking like a biologist. More than anything else, biology is a quest, an ongoing inquiry about the nature of life.

What is life? Even a small child realizes that a dog or a plant is alive, while a rock or a lawn mower is not. Yet the phenomenon we call life defies a simple, one-sentence definition. We recognize life by what living things do. **Figure 1.3**, on the next page, highlights some of the properties and processes we associate with life.

While limited to a handful of images, Figure 1.3 reminds us that the living world is wondrously varied. How do biologists

▲ **Figure 1.2 The mother-of-pearl plant (*Graptopetalum paraguayense*).** This plant's thick leaves hold water, enabling it to live where soil is scarce. The leaves vary in color, as seen here.

▼ **Order.** This close-up of a sunflower illustrates the highly ordered structure that characterizes life.

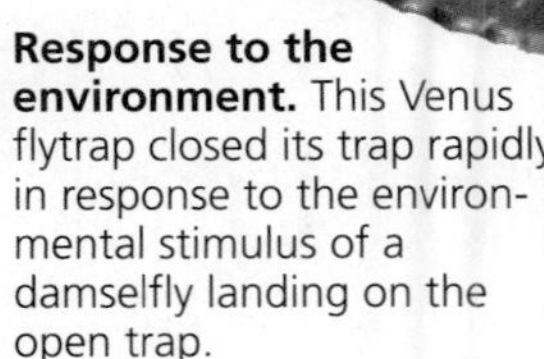

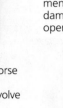

▲ **Evolutionary adaptation.** The appearance of this pygmy sea horse camouflages the animal in its environment. Such adaptations evolve over many generations by the reproductive success of those individuals with heritable traits that are best suited to their environments.

▲ **Response to the environment.** This Venus flytrap closed its trap rapidly in response to the environmental stimulus of a damselfly landing on the open trap.

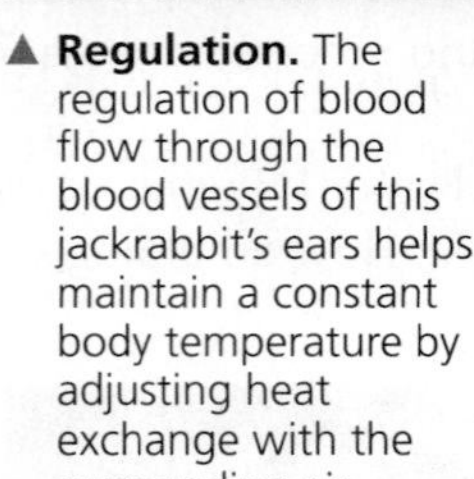

▲ **Regulation.** The regulation of blood flow through the blood vessels of this jackrabbit's ears helps maintain a constant body temperature by adjusting heat exchange with the surrounding air.

▲ **Energy processing.** This hummingbird obtains fuel in the form of nectar from flowers. The hummingbird will use chemical energy stored in its food to power flight and other work.

► **Reproduction.** Organisms (living things) reproduce their own kind. Here, a baby giraffe stands close to its mother.

◄ **Growth and development.** Inherited information carried by genes controls the pattern of growth and development of organisms, such as this Nile crocodile.

▲ **Figure 1.3 Some properties of life.**

make sense of this diversity and complexity? This opening chapter sets up a framework for answering this question. The first part of the chapter provides a panoramic view of the biological "landscape," organized around some unifying themes. We then focus on biology's core theme, evolution, with an introduction to the reasoning that led Charles Darwin to his explanatory theory. Next, we look at scientific inquiry—how scientists raise and attempt to answer questions about the natural world. Finally, we address the culture of science and its effects on society.

CONCEPT 1.1

The themes of this book make connections across different areas of biology

Biology is a subject of enormous scope, and news reports reveal exciting new biological discoveries being made every day. Simply memorizing the factual details of this huge subject is most likely not the best way to develop a coherent view of life. A better approach is to take a more active role by connecting the many things you learn to a set of themes that pervade all of biology. Focusing on a few big ideas—ways of thinking about life that will still hold true decades from now—will help you organize and make sense of all the information you'll encounter as you study biology. To help you, we have selected eight unifying themes to serve as touchstones as you proceed through this book.

Theme: New Properties Emerge at Each Level in the Biological Hierarchy

The study of life extends from the microscopic scale of the molecules and cells that make up organisms to the global scale of the entire living planet. We can divide this enormous range into different levels of biological organization.

Imagine zooming in from space to take a closer and closer look at life on Earth. It is spring in Ontario, Canada, and our destination is a local forest, where we will eventually explore a maple leaf right down to the molecular level. **Figure 1.4**, on the next two pages, narrates this journey into life, with the numbers leading you through the levels of biological organization illustrated by the photographs.

Emergent Properties

If we now zoom back out from the molecular level in Figure 1.4, we can see that novel properties emerge at each step, properties that are not present at the preceding level. These **emergent properties** are due to the arrangement and interactions of parts as complexity increases. For example, although photosynthesis occurs in an intact chloroplast, it will not take place in a disorganized test-tube mixture of chlorophyll and other chloroplast molecules. Photosynthesis requires a specific organization of these molecules in the chloroplast. To take another example, if a blow to the head disrupts the intricate architecture of a human brain, the mind may cease to function properly even though all of the brain tissues are still present. Our thoughts and memories are emergent properties of a complex network of nerve cells. At a much higher level of biological organization—at the ecosystem level—the recycling of chemical elements essential to life, such as carbon, depends on a network of diverse organisms interacting with each other and with the soil, water, and air.

Emergent properties are not unique to life. A box of bicycle parts won't take you anywhere, but if they are arranged in a certain way, you can pedal to your chosen destination. And while the graphite in a pencil "lead" and the diamond in a wedding ring are both pure carbon, they have very different appearances and properties due to the different arrangements of their carbon atoms. Both of these examples point out the importance of arrangement. Compared to such nonliving examples, however, the unrivaled complexity of biological systems makes the emergent properties of life especially challenging to study.

The Power and Limitations of Reductionism

Because the properties of life emerge from complex organization, scientists seeking to understand biological systems confront a dilemma. On the one hand, we cannot fully explain a higher level of order by breaking it down into its parts. A dissected animal no longer functions; a cell reduced to its chemical ingredients is no longer a cell. Disrupting a living system interferes with its functioning. On the other hand, something as complex as an organism or a cell cannot be analyzed without taking it apart.

Reductionism—the approach of reducing complex systems to simpler components that are more manageable to study—is a powerful strategy in biology. For example, by studying the molecular structure of DNA that had been extracted from cells, James Watson and Francis Crick inferred, in 1953, how this molecule could serve as the chemical basis of inheritance. The central role of DNA in cells and organisms became better understood, however, when scientists were able to study the interactions of DNA with other molecules. Biologists must balance the reductionist strategy with the larger-scale, holistic objective of understanding emergent properties—how the parts of cells, organisms, and higher levels of order, such as ecosystems, work together. This is the goal of an approach developed over the last 50 years called systems biology.

Systems Biology

A system is simply a combination of components that function together. A biologist can study a system at any level of organization. A single leaf cell can be considered a system, as can a frog, an ant colony, or a desert ecosytem. To understand how such systems work, it is not enough to have a "parts list," even a complete one. Realizing this, many researchers are now complementing the reductionist approach with new strategies for studying whole systems. This change in perspective is analogous to moving from ground level on a street corner, where you can observe local traffic, to a helicopter high above a city, from which you can see how variables such as time of day, construction projects, accidents, and traffic-signal malfunctions affect traffic throughout the city.

Systems biology is an approach that attempts to model the dynamic behavior of whole biological systems based on a study of the interactions among the system's parts. Successful models enable biologists to predict how a change in one or more variables will affect other components and the whole system. Thus, the systems approach enables us to pose new kinds of questions. How might a drug that lowers blood pressure affect the functioning of organs throughout the human body? How might increasing a crop's water supply affect processes in the plants, such as the storage of molecules essential for human nutrition? How might a gradual increase in atmospheric carbon dioxide alter ecosystems and the entire biosphere? The ultimate aim of systems biology is to answer large-scale questions like the last one.

▼ **Figure 1.4**

Exploring Levels of Biological Organization

◄1 The Biosphere

As soon as we are near enough to Earth to make out its continents and oceans, we begin to see signs of life—in the green mosaic of the planet's forests, for example. This is our first view of the biosphere, which consists of all life on Earth and all the places where life exists—most regions of land, most bodies of water, the atmosphere to an altitude of several kilometers, and even sediments far below the ocean floor and rocks many kilometers below Earth's surface.

◄2 Ecosystems

As we approach Earth's surface for an imaginary landing in Ontario, we can begin to make out a forest with an abundance of trees that lose their leaves in one season and grow new ones in another (deciduous trees). Such a deciduous forest is an example of an ecosystem. Grasslands, deserts, and the ocean's coral reefs are other types of ecosystems. An ecosystem consists of all the living things in a particular area, along with all the nonliving components of the environment with which life interacts, such as soil, water, atmospheric gases, and light. All of Earth's ecosystems combined make up the biosphere.

►3 Communities

The entire array of organisms inhabiting a particular ecosystem is called a biological community. The community in our forest ecosystem includes many kinds of trees and other plants, a diversity of animals, various mushrooms and other fungi, and enormous numbers of diverse microorganisms, which are living forms, such as bacteria, that are too small to see without a microscope. Each of these forms of life is called a *species*.

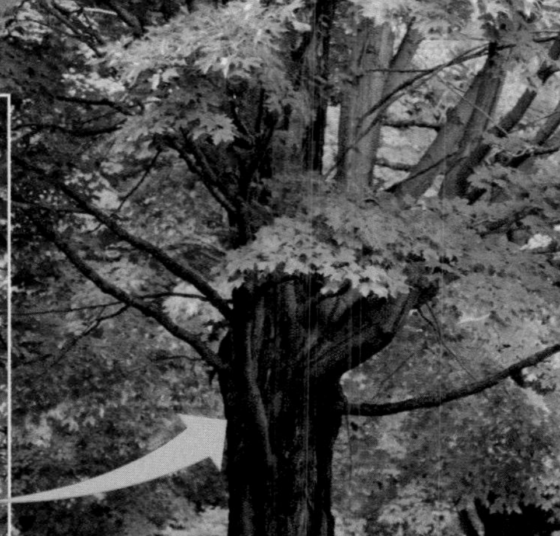

►4 Populations

A population consists of all the individuals of a species living within the bounds of a specified area. For example, our Ontario forest includes a population of sugar maple trees and a population of white-tailed deer. We can now refine our definition of a community as the set of populations that inhabit a particular area.

▲5 Organisms

Individual living things are called organisms. Each of the maple trees and other plants in the forest is an organism, and so is each forest animal—whether deer, squirrel, frog, or beetle. The soil teems with microorganisms such as bacteria.

▼6 Organs and Organ Systems

The structural hierarchy of life continues to unfold as we explore the architecture of the more complex organisms. A maple leaf is an example of an organ, a body part that carries out a particular function in the body. Stems and roots are the other major organs of plants. Examples of human organs are the brain, heart, and kidney. The organs of humans, other complex animals, and plants are organized into organ systems, each a team of organs that cooperate in a larger function. For example, the human digestive system includes such organs as the tongue, stomach, and intestines. Organs consist of multiple tissues.

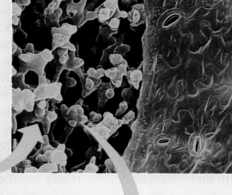

◀7 Tissues

Our next scale change—to see the tissues of a leaf—requires a microscope. Each tissue is made up of a group of cells that work together, performing a specialized function. The leaf shown here has been cut on an angle. The honeycombed tissue in the interior of the leaf (left portion of photo) is the main location of photosynthesis, the process that converts light energy to the chemical energy of sugar and other food. We are viewing the sliced leaf from a perspective that also enables us to see the jigsaw puzzle–like "skin" on the surface of the leaf, a tissue called epidermis (right part of photo). The pores through the epidermis allow the gas carbon dioxide, a raw material for sugar production, to reach the photosynthetic tissue inside the leaf. At this scale, we can also see that each tissue has a distinct cellular structure.

◀8 Cells

The cell is life's fundamental unit of structure and function. Some organisms, such as amoebas and most bacteria, are single cells. Other organisms, including plants and animals, are multicellular. Instead of a single cell performing all the functions of life, a multicellular organism has a division of labor among specialized cells. A human body consists of trillions of microscopic cells of many different kinds, such as muscle cells and nerve cells, which are organized into the various specialized tissues. For example, muscle tissue consists of bundles of muscle cells. In the photo at the upper left, we see a more highly magnified view of some cells in a leaf tissue. One cell is only about 40 micrometers (µm) across. It would take about 500 of these cells to reach across a small coin. As tiny as these cells are, you can see that each contains numerous green structures called chloroplasts, which are responsible for photosynthesis.

▶9 Organelles

Chloroplasts are examples of organelles, the various functional components present in cells. In this image, a very powerful tool called an electron microscope brings a single chloroplast into sharp focus.

▶10 Molecules

Our last scale change drops us into a chloroplast for a view of life at the molecular level. A molecule is a chemical structure consisting of two or more small chemical units called atoms, which are represented as balls in this computer graphic of a chlorophyll molecule. Chlorophyll is the pigment molecule that makes a maple leaf green. One of the most important molecules on Earth, chlorophyll absorbs sunlight during the first step of photosynthesis. Within each chloroplast, millions of chlorophyll molecules, together with accessory molecules, are organized into the equipment that converts light energy to the chemical energy of food.

Systems biology is relevant to the study of life at all levels. During the early years of the 20th century, biologists studying how animal bodies function (animal physiology) began integrating data on how multiple organs coordinate processes such as the regulation of sugar concentration in the blood. And in the 1960s, scientists investigating ecosystems pioneered a more mathematically sophisticated systems approach with elaborate models diagramming the network of interactions between organisms and nonliving components of ecosystems, such as salt marshes. More recently, with the sequencing of DNA from many species, systems biology has taken hold at the cellular and molecular levels, as we'll describe later when we discuss DNA.

Theme: Organisms Interact with Other Organisms and the Physical Environment

Turn back again to Figure 1.4, this time focusing on the forest. In an ecosystem, each organism interacts continuously with its environment, which includes both other organisms and physical factors. The leaves of a tree, for example, absorb light from the sun, take in carbon dioxide from the air, and release oxygen to the air (**Figure 1.5**). Both the organism and the environment are affected by the interactions between them. For example, a plant takes up water and minerals from the soil through its roots, and its roots help form soil by breaking up rocks. On a global scale, plants and other photosynthetic organisms have generated all the oxygen in the air.

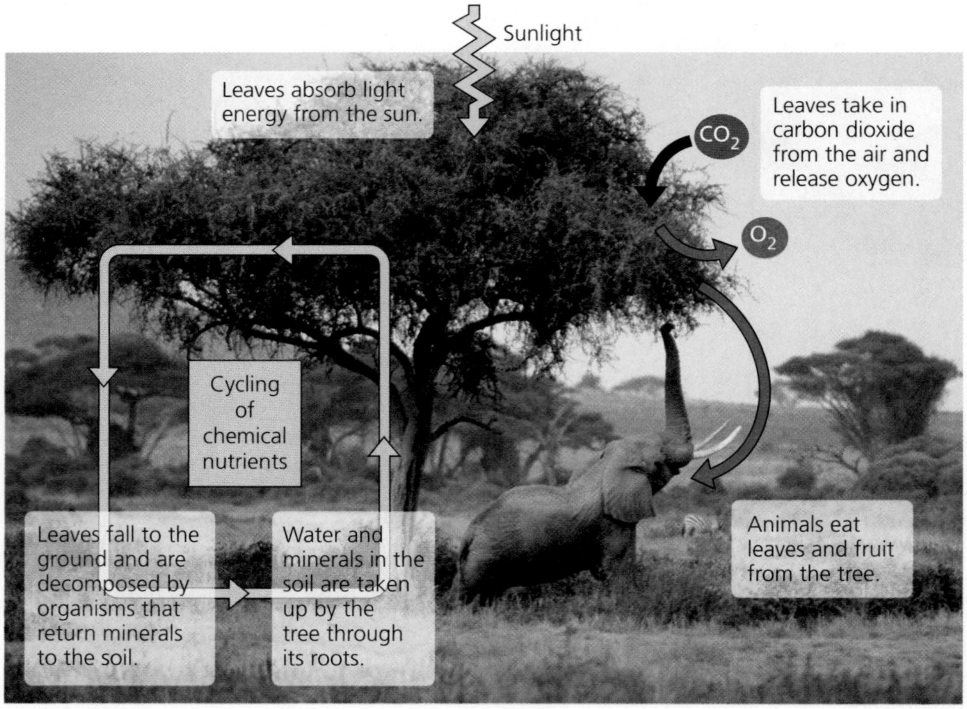

▲ **Figure 1.5 Interactions of an African acacia tree with other organisms and the physical environment.**

A tree also interacts with other organisms, such as soil microorganisms associated with its roots, insects that live in the tree, and animals that eat its leaves and fruit. Interactions between organisms ultimately result in the cycling of nutrients in ecosystems. For example, minerals acquired by a tree will eventually be returned to the soil by other organisms that decompose leaf litter, dead roots, and other organic debris. The minerals are then available to be taken up by plants again.

Like all organisms, we humans interact with our environment. Unfortunately, our interactions sometimes have drastic consequences. For example, since the Industrial Revolution in the 1800s, the burning of fossil fuels (coal, oil, and gas) has been increasing at an ever-accelerating pace. This practice releases gaseous compounds into the atmosphere, including prodigious amounts of carbon dioxide (CO_2). About half the human-generated CO_2 stays in the atmosphere, acting like a layer of glass around the planet that admits radiation that warms the Earth but prevents heat from radiating into outer space. Scientists estimate that the average temperature of the planet has risen 1°C since 1900 due to this "greenhouse effect," and they project an additional rise in average global temperature of at least 3°C over the course of the 21st century.

This global warming, a major aspect of **global climate change**, has already had dire effects on life-forms and their habitats all over planet Earth. Polar bears have lost a significant portion of the ice platform from which they hunt, and there are examples of small rodents and plant species that have shifted their ranges to higher altitudes, as well as bird populations that have altered their migration schedules. Only time will reveal the consequences of these changes. Scientists predict that even if we stopped burning fossil fuels today, it would take several centuries to return to preindustrial CO_2 levels. That scenario is highly improbable, so it is imperative that we learn all we can about the effects of global climate change on Earth and its populations. Acting as the stewards of our planet, we must strive to find ways to address this problem.

Theme: Life Requires Energy Transfer and Transformation

As you saw in Figure 1.5, a tree's leaves absorb sunlight. The input of energy from the sun makes life possible: A fundamental characteristic of living organisms is their use of energy to carry out life's activities. Moving, growing, reproducing, and the other activities of life are work, and work requires energy. In the business of living, organisms often

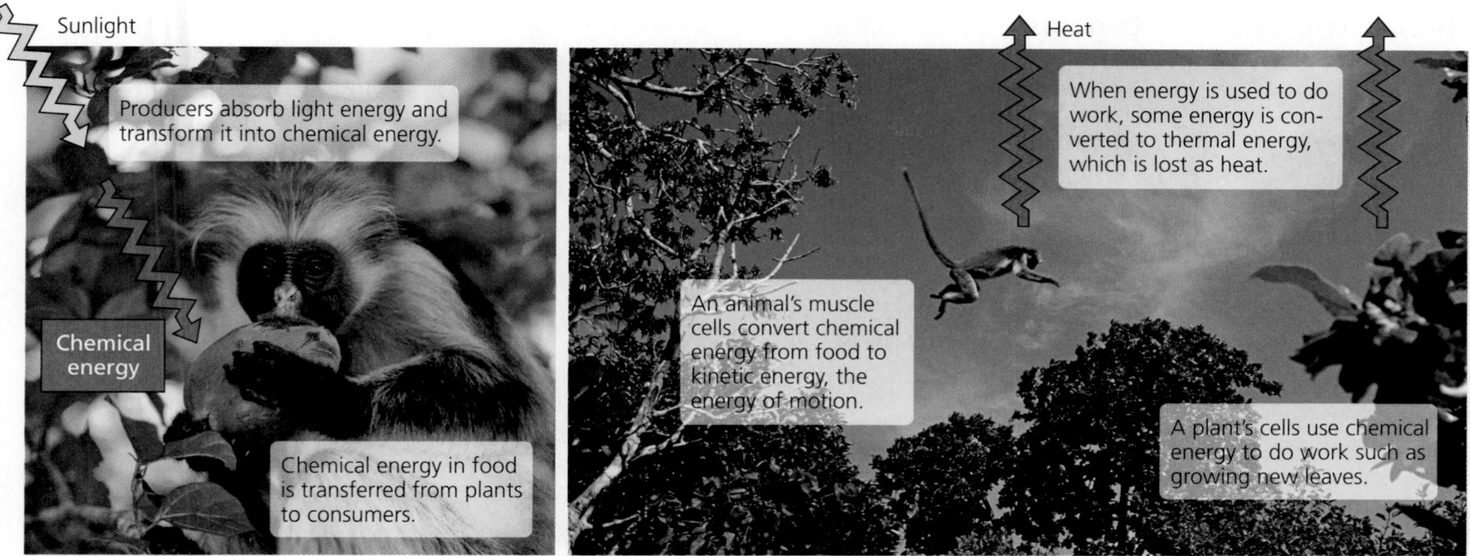

Sunlight
Producers absorb light energy and transform it into chemical energy.
Chemical energy
Chemical energy in food is transferred from plants to consumers.
Heat
When energy is used to do work, some energy is converted to thermal energy, which is lost as heat.
An animal's muscle cells convert chemical energy from food to kinetic energy, the energy of motion.
A plant's cells use chemical energy to do work such as growing new leaves.

(a) Energy flow from sunlight to producers to consumers

(b) Using energy to do work

▲ **Figure 1.6 Energy flow in an ecosystem.** This endangered Red Colobus monkey lives in Tanzania.

transform one form of energy to another. Chlorophyll molecules within the tree's leaves harness the energy of sunlight and use it to drive photosynthesis, converting carbon dioxide and water to sugar and oxygen. The chemical energy in sugar is then passed along by plants and other photosynthetic organisms (producers) to consumers. Consumers are organisms, such as animals, that feed on producers and other consumers (**Figure 1.6a**).

An animal's muscle cells use sugar as fuel to power movements, converting chemical energy to kinetic energy, the energy of motion (**Figure 1.6b**). The cells in a leaf use sugar to drive the process of cell proliferation during leaf growth, transforming stored chemical energy into cellular work. In both cases, some of the energy is converted to thermal energy, which dissipates to the surroundings as heat. In contrast to chemical nutrients, which recycle within an ecosystem, energy flows through an ecosystem, usually entering as light and exiting as heat.

Theme: Structure and Function Are Correlated at All Levels of Biological Organization

Another theme evident in Figure 1.4 is the idea that form fits function, which you'll recognize from everyday life. For example, a screwdriver is suited to tighten or loosen screws, a hammer to pound nails. How a device works is correlated with its structure. Applied to biology, this theme is a guide to the anatomy of life at all its structural levels. An example from Figure 1.4 is seen in the leaf: Its thin, flat shape maximizes the amount of sunlight that can be captured by its chloroplasts. Analyzing a biological structure gives us clues about what it does and how it works. Conversely, knowing the function of something provides insight into its construction. An example from the animal kingdom, the wing of a bird, provides additional instances of the structure-function theme (**Figure 1.7**). In exploring life on its different structural levels, we discover functional beauty at every turn.

(a) A bird's wings have an aerodynamically efficient shape.

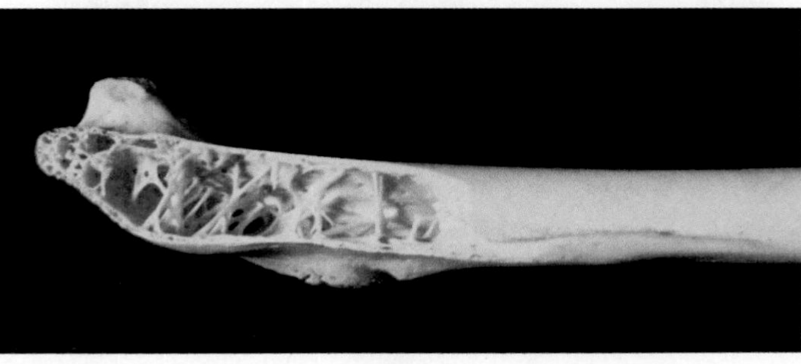

(b) Wing bones have a honeycombed internal structure that is strong but lightweight.

▲ **Figure 1.7 Form fits function in a gull's wing. (a)** The shape of a bird's wings and **(b)** the structure of its bones make flight possible.

? *How does form fit function in a human hand?*

Theme: The Cell Is an Organism's Basic Unit of Structure and Function

In life's structural hierarchy, the cell has a special place as the lowest level of organization that can perform all activities required for life. Moreover, the activities of organisms are all based on the activities of cells. For instance, the movement of your eyes as you read this line is based on activities of muscle and nerve cells. Even a global process such as the recycling of carbon is the cumulative product of cellular activities, including the photosynthesis that occurs in the chloroplasts of leaf cells. Understanding how cells work is a major focus of biological research.

All cells share certain characteristics. For example, every cell is enclosed by a membrane that regulates the passage of materials between the cell and its surroundings. And every cell uses DNA as its genetic information. However, we can distinguish between two main forms of cells: prokaryotic cells and eukaryotic cells. The cells of two groups of microorganisms, called bacteria (singular, *bacterium*) and archaea (singular, *archaean*), are prokaryotic. All other forms of life, including plants and animals, are composed of eukaryotic cells.

A **eukaryotic cell** is subdivided by internal membranes into various membrane-enclosed organelles **(Figure 1.8)**. In most eukaryotic cells, the largest organelle is the nucleus, which contains the cell's DNA. The other organelles are located in the cytoplasm, the entire region between the nucleus and outer membrane of the cell. The chloroplast you saw in Figure 1.4 is an organelle found in eukaryotic cells that carry out photosynthesis. Prokaryotic cells are much simpler and generally smaller than eukaryotic cells, as seen clearly in Figure 1.8. In a **prokaryotic cell**, the DNA is not separated from the rest of the cell by enclosure in a membrane-bounded nucleus. Prokaryotic cells also lack the other kinds of membrane-enclosed organelles that characterize eukaryotic cells. The properties of all organisms, whether prokaryotic or eukaryotic, are based in the structure and function of cells.

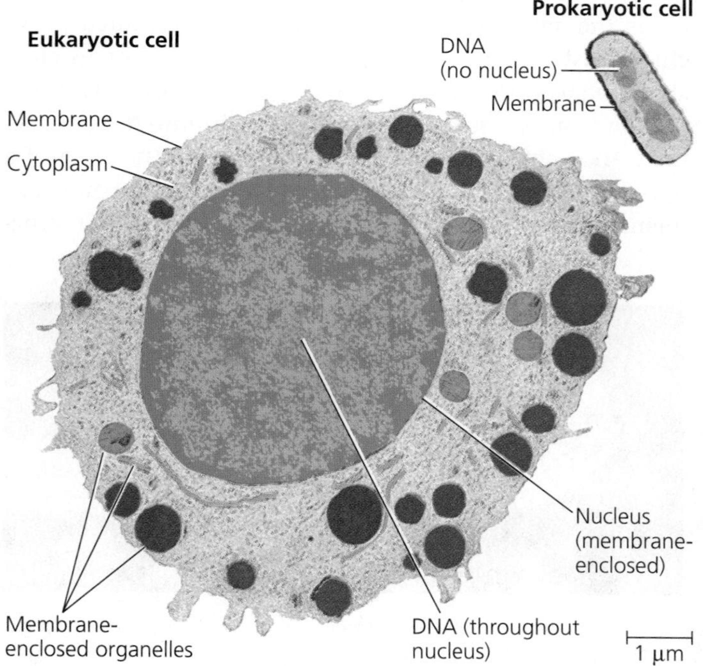

▲ **Figure 1.8 Contrasting eukaryotic and prokaryotic cells in size and complexity.**

Theme: The Continuity of Life Is Based on Heritable Information in the Form of DNA

The division of cells to form new cells is the foundation for all reproduction and for the growth and repair of multicellular organisms. Inside the dividing cell in **Figure 1.9**, you can see structures called chromosomes, which are stained with a blue-glowing dye. The chromosomes have almost all of the cell's genetic material, its **DNA** (short for deoxyribonucleic acid). DNA is the substance of **genes**, the units of inheritance that transmit information from parents to offspring. Your blood group (A, B, AB, or O), for example, is the result of certain genes that you inherited from your parents.

DNA Structure and Function

Each chromosome contains one very long DNA molecule, with hundreds or thousands of genes arranged along its length. The genes encode the information necessary to build other molecules in the cell, most notably proteins. Proteins play structural roles and are also responsible for carrying out cellular work. They thus establish a cell's identity.

The DNA of chromosomes replicates as a cell prepares to divide, and each of the two cellular offspring inherits a complete set of genes, identical to that of the parent cell. Each of us began life as a single cell stocked with DNA inherited from our parents. Replication of that DNA with each round of cell division transmitted copies of the DNA to our trillions of cells. The DNA controls the development and maintenance of the entire organism and, indirectly, everything the organism does **(Figure 1.10)**. The DNA serves as a central database.

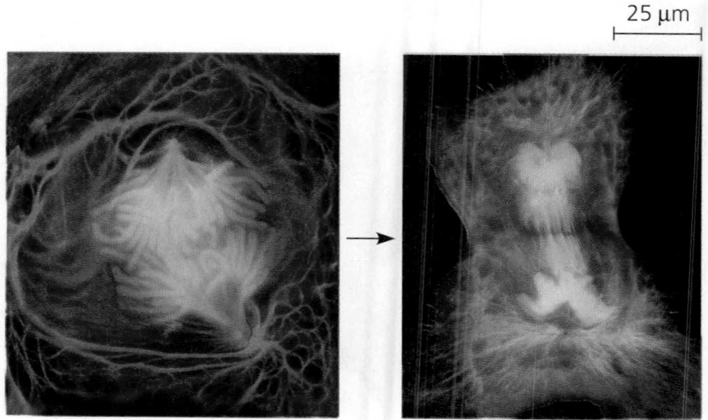

▲ **Figure 1.9 A lung cell from a newt divides into two smaller cells that will grow and divide again.**

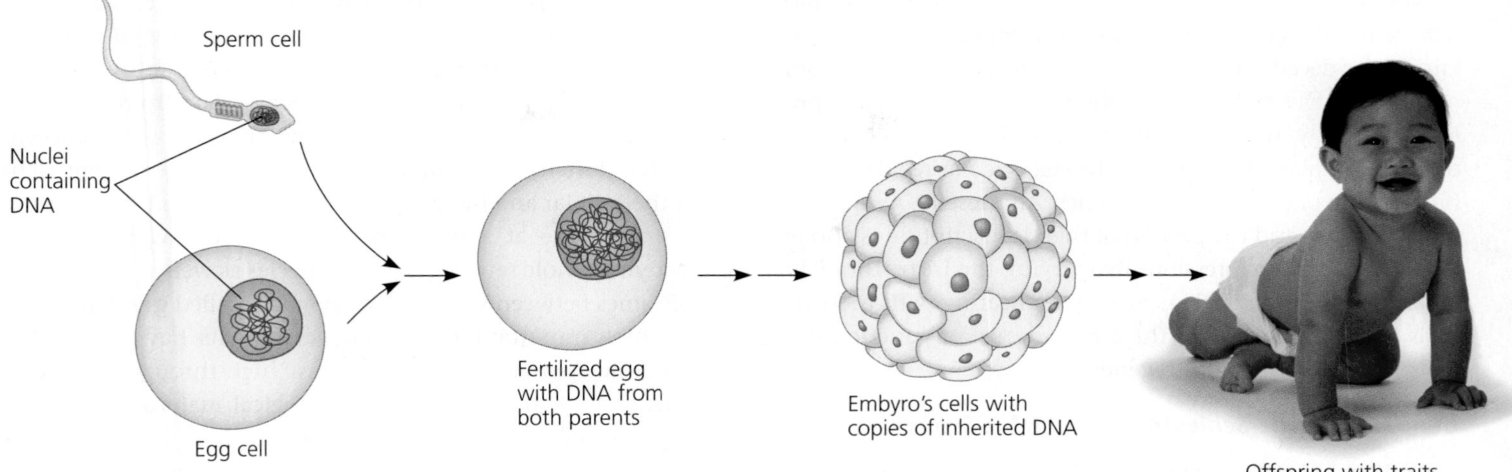

▲ **Figure 1.10 Inherited DNA directs development of an organism.**

The molecular structure of DNA accounts for its ability to store information. Each DNA molecule is made up of two long chains, called strands, arranged in a double helix. Each chain is made up of four kinds of chemical building blocks called nucleotides, abbreviated A, T, C, and G **(Figure 1.11)**. The way DNA encodes information is analogous to how we arrange the letters of the alphabet into precise sequences with specific meanings. The word *rat*, for example, evokes a rodent; the words *tar* and *art*, which contain the same letters, mean very different things. We can think of nucleotides as a four-letter alphabet of inheritance. Specific sequential arrangements of these four nucleotide letters encode the information in genes, which are typically hundreds or thousands of nucleotides long.

DNA provides the blueprints for making proteins, and proteins are the main players in building and maintaining the cell and carrying out its activities. For instance, the information carried in a bacterial gene may specify a certain protein in a bacterial cell membrane, while the information in a human gene may denote a protein hormone that stimulates growth. Other human proteins include proteins in a muscle cell that drive contraction and the defensive proteins called antibodies. Enzymes, which catalyze (speed up) specific chemical reactions, are mostly proteins and are crucial to all cells.

The DNA of genes controls protein production indirectly, using a related kind of molecule called RNA as an intermediary. The sequence of nucleotides along a gene is transcribed into RNA, which is then translated into a specific protein with a unique shape and function. This entire process, by which the information in a gene directs the production of a cellular product, is called **gene expression**. In translating genes into proteins, all forms of life employ essentially the same genetic code. A particular sequence of nucleotides says the same thing in one organism as it does in another. Differences between organisms reflect differences between their nucleotide sequences rather than between their genetic codes.

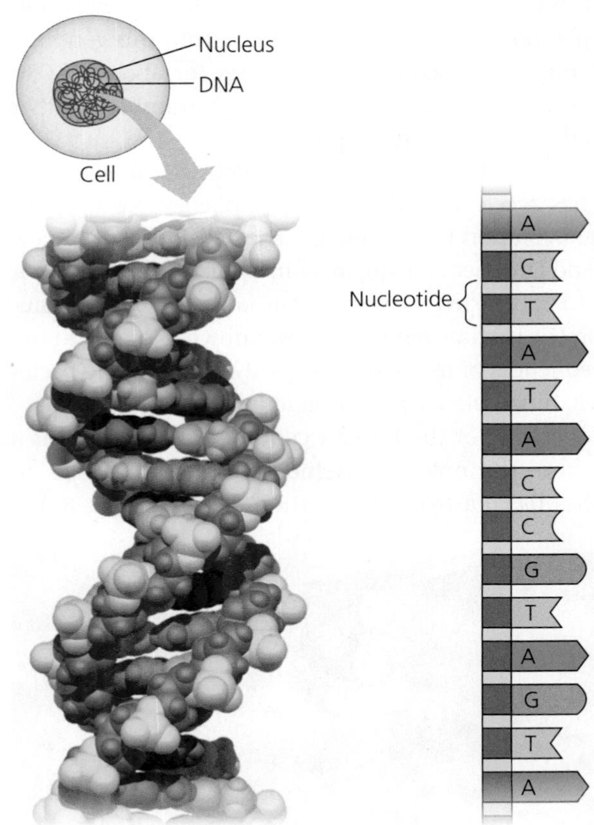

(a) DNA double helix. This model shows each atom in a segment of DNA. Made up of two long chains of building blocks called nucleotides, a DNA molecule takes the three-dimensional form of a double helix.

(b) Single strand of DNA. These geometric shapes and letters are simple symbols for the nucleotides in a small section of one chain of a DNA molecule. Genetic information is encoded in specific sequences of the four types of nucleotides. (Their names are abbreviated A, T, C, and G.)

▲ **Figure 1.11 DNA: The genetic material.**

Not all RNA molecules in the cell are translated into protein; some RNAs carry out other important tasks. We have known for decades that some types of RNA are actually components of the cellular machinery that manufactures proteins. Recently, scientists have discovered whole new classes of RNA that play other roles in the cell, such as regulating the functioning of protein-coding genes. All these RNAs are specified by genes, and the process of their transcription is also referred to as gene expression. By carrying the instructions for making proteins and RNAs and by replicating with each cell division, DNA ensures faithful inheritance of genetic information from generation to generation.

Genomics: Large-Scale Analysis of DNA Sequences

The entire "library" of genetic instructions that an organism inherits is called its **genome**. A typical human cell has two similar sets of chromosomes, and each set has DNA totaling about 3 billion nucleotide pairs. If the one-letter abbreviations for the nucleotides of one strand were written in letters the size of those you are now reading, the genetic text would fill about 600 books the size of this one. Within this genomic library of nucleotide sequences are genes for about 75,000 kinds of proteins and an as yet unknown number of RNA molecules that do not code for proteins.

Since the early 1990s, the pace at which we can sequence genomes has accelerated at an almost unbelievable rate, enabled by a revolution in technology. The development of new methods and DNA-sequencing machines, such as those shown in **Figure 1.12**, have led the charge. The entire sequence of nucleotides in the human genome is now known, along with the genome sequences of many other organisms, including bacteria, archaea, fungi, plants, and other animals.

The sequencing of the human genome was heralded as a scientific and technological achievement comparable to landing the *Apollo* astronauts on the moon in 1969. But it was only the beginning of an even bigger research endeavor, an effort to learn how the activities of the myriad proteins encoded by the DNA are coordinated in cells and whole organisms. To make sense of the deluge of data from genome-sequencing projects and the growing catalog of known protein functions, scientists are applying a systems approach at the cellular and molecular levels. Rather than investigating a single gene at a time, these researchers have shifted to studying whole sets of genes of a species as well as comparing genomes between species—an approach called **genomics**.

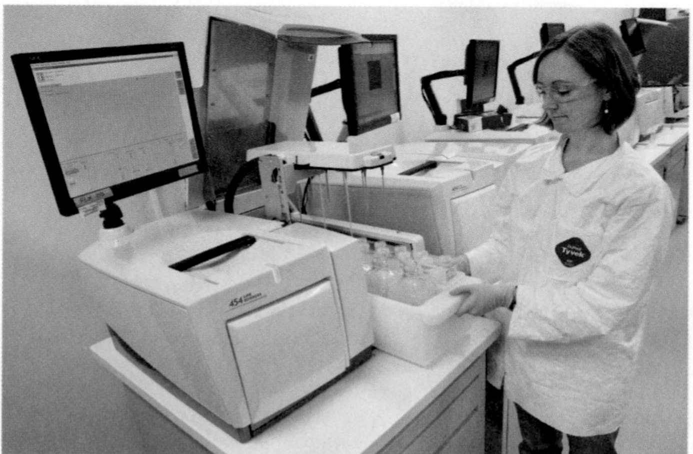

▲ **Figure 1.12 Biology as an information science.** Automatic DNA-sequencing machines and abundant computing power make the sequencing of genomes possible. This facility in Walnut Creek, California, is part of the Joint Genome Institute.

Three important research developments have made the genomic approach possible. One is "high-throughput" technology, tools that can analyze biological materials very rapidly and produce enormous amounts of data. The automatic DNA-sequencing machines that made the sequencing of the human genome possible are examples of high-throughput devices (see Figure 1.12). The second major development is **bioinformatics**, the use of computational tools to store, organize, and analyze the huge volume of data that result from high-throughput methods. The third key development is the formation of interdisciplinary research teams—melting pots of diverse specialists that may include computer scientists, mathematicians, engineers, chemists, physicists, and, of course, biologists from a variety of fields.

Theme: Feedback Mechanisms Regulate Biological Systems

Just as a coordinated control of traffic flow is necessary for a city to function smoothly, regulation of biological processes is crucial to the operation of living systems. Consider your muscles, for instance. When your muscle cells require more energy during exercise, they increase their consumption of the sugar molecules that serve as fuel. In contrast, when you rest, a different set of chemical reactions converts surplus sugar to storage molecules.

Like most of the cell's chemical processes, those that either decompose or store sugar are accelerated, or catalyzed, by proteins called enzymes. Each type of enzyme catalyzes a specific chemical reaction. In many cases, these reactions are linked into chemical pathways, each step with its own enzyme. How does the cell coordinate its various chemical pathways? In our example of sugar management, how does the cell match fuel supply to demand, regulating its opposing pathways of sugar consumption and storage? The key is the ability of many biological processes to self-regulate by a mechanism called feedback.

In feedback regulation, the output, or product, of a process regulates that very process. The most common form of regulation in living systems is **negative feedback**, in which accumulation of an end product of a process slows that process. For example, the cell's breakdown of sugar generates chemical energy in the form of a substance called ATP. When a cell makes more ATP than it can use, the excess ATP "feeds back"

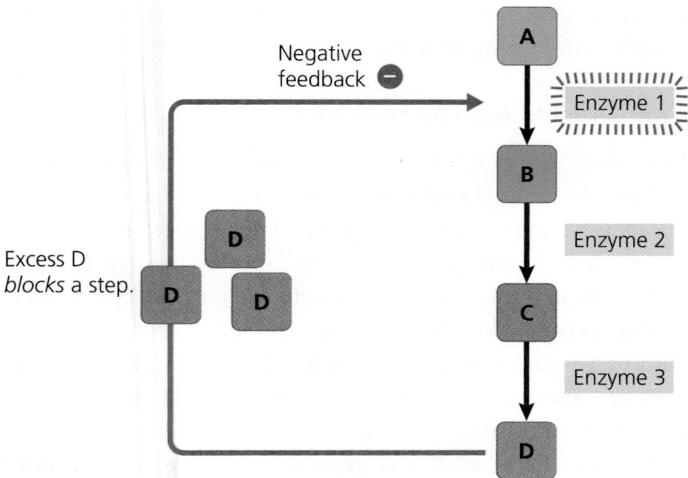

(a) Negative feedback. This three-step chemical pathway converts substance A to substance D. A specific enzyme catalyzes each chemical reaction. Accumulation of the final product (D) inhibits the first enzyme in the sequence, thus slowing down production of more D.

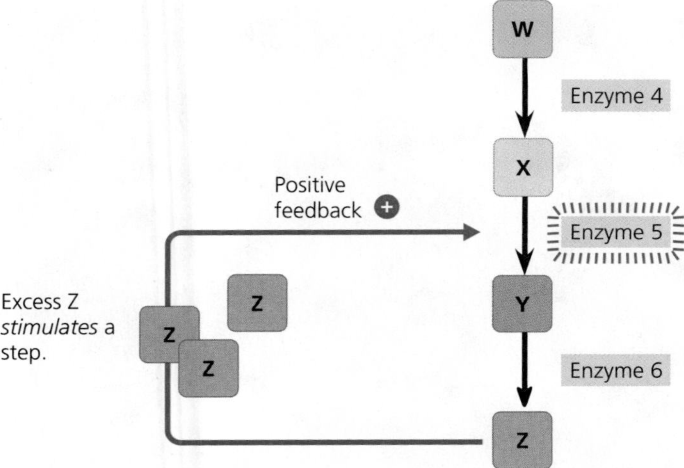

(b) Positive feedback. In a biochemical pathway regulated by positive feedback, a product stimulates an enzyme in the reaction sequence, increasing the rate of production of the product.

▲ **Figure 1.13 Regulation by feedback mechanisms.**

? *What would happen to the feedback system if enzyme 2 were missing?*

and inhibits an enzyme near the beginning of the pathway **(Figure 1.13a)**.

Though less common than processes regulated by negative feedback, there are also many biological processes regulated by **positive feedback**, in which an end product *speeds up* its own production **(Figure 1.13b)**. The clotting of your blood in response to injury is an example. When a blood vessel is damaged, structures in the blood called platelets begin to aggregate at the site. Positive feedback occurs as chemicals released by the platelets attract *more* platelets. The platelet pileup then initiates a complex process that seals the wound with a clot.

Feedback is a regulatory motif common to life at all levels, from the molecular level to ecosystems and the biosphere. Such regulation is an example of the integration that makes living systems much greater than the sum of their parts.

Evolution, the Overarching Theme of Biology

Having considered all the other themes that run through this book, let's now turn to biology's core theme—evolution. Evolution is the one idea that makes sense of everything we know about living organisms. Life has been evolving on Earth for billions of years, resulting in a vast diversity of past and present organisms. But along with the diversity we find many shared features. For example, while the sea horse, jackrabbit, hummingbird, crocodile, and giraffes in Figure 1.3 look very different, their skeletons are basically similar. The scientific explanation for this unity and diversity—and for the suitability of organisms for their environments—is evolution: the idea that the organisms living on Earth today are the modified descendants of common ancestors. In other words, we can explain traits shared by two organisms with the idea that they have descended from a common ancestor, and we can account for differences with the idea that heritable changes have occurred along the way. Many kinds of evidence support the occurrence of evolution and the theory that describes how it takes place. In the next section, we'll consider the fundamental concept of evolution in greater detail.

CONCEPT CHECK 1.1

1. For each biological level in Figure 1.4, write a sentence that includes the next "lower" level. Example: "A community consists of *populations* of the various species inhabiting a specific area."
2. What theme or themes are exemplified by (a) the sharp spines of a porcupine, (b) the cloning of a plant from a single cell, and (c) a hummingbird using sugar to power its flight?
3. **WHAT IF?** For each theme discussed in this section, give an example not mentioned in the book.

For suggested answers, see Appendix A.

CONCEPT 1.2

The Core Theme: Evolution accounts for the unity and diversity of life

EVOLUTION The list of biological themes discussed in Concept 1.1 is not absolute; some people might find a shorter or longer list more useful. There is consensus among biologists, however, as to the core theme of biology: It is evolution. To quote one of the founders of modern evolutionary theory, Theodosius Dobzhansky, "Nothing in biology makes sense except in the light of evolution."

In addition to encompassing a hierarchy of size scales from molecules to the biosphere, biology extends across the great diversity of species that have ever lived on Earth. To understand Dobzhansky's statement, we need to discuss how biologists think about this vast diversity.

Classifying the Diversity of Life

Diversity is a hallmark of life. Biologists have so far identified and named about 1.8 million species. To date, this diversity of life is known to include at least 100,000 species of fungi, 290,000 plant species, 52,000 vertebrate species (animals with backbones), and 1 million insect species (more than half of all known forms of life)—not to mention the myriad types of single-celled organisms. Researchers identify thousands of additional species each year. Estimates of the total number of species range from about 10 million to over 100 million. Whatever the actual number, the enormous variety of life gives biology a very broad scope. Biologists face a major challenge in attempting to make sense of this variety.

Grouping Species: The Basic Idea

There is a human tendency to group diverse items according to their similarities and their relationships to each other. For instance, we may speak of "squirrels" and "butterflies," though we recognize that many different species belong to each group. We may even sort groups into broader categories, such as rodents (which include squirrels) and insects (which include butterflies). Taxonomy, the branch of biology that names and classifies species, formalizes this ordering of species into groups of increasing breadth, based on the degree to which they share characteristics **(Figure 1.14)**. You will learn more about the details of this taxonomic scheme in Chapter 26. For

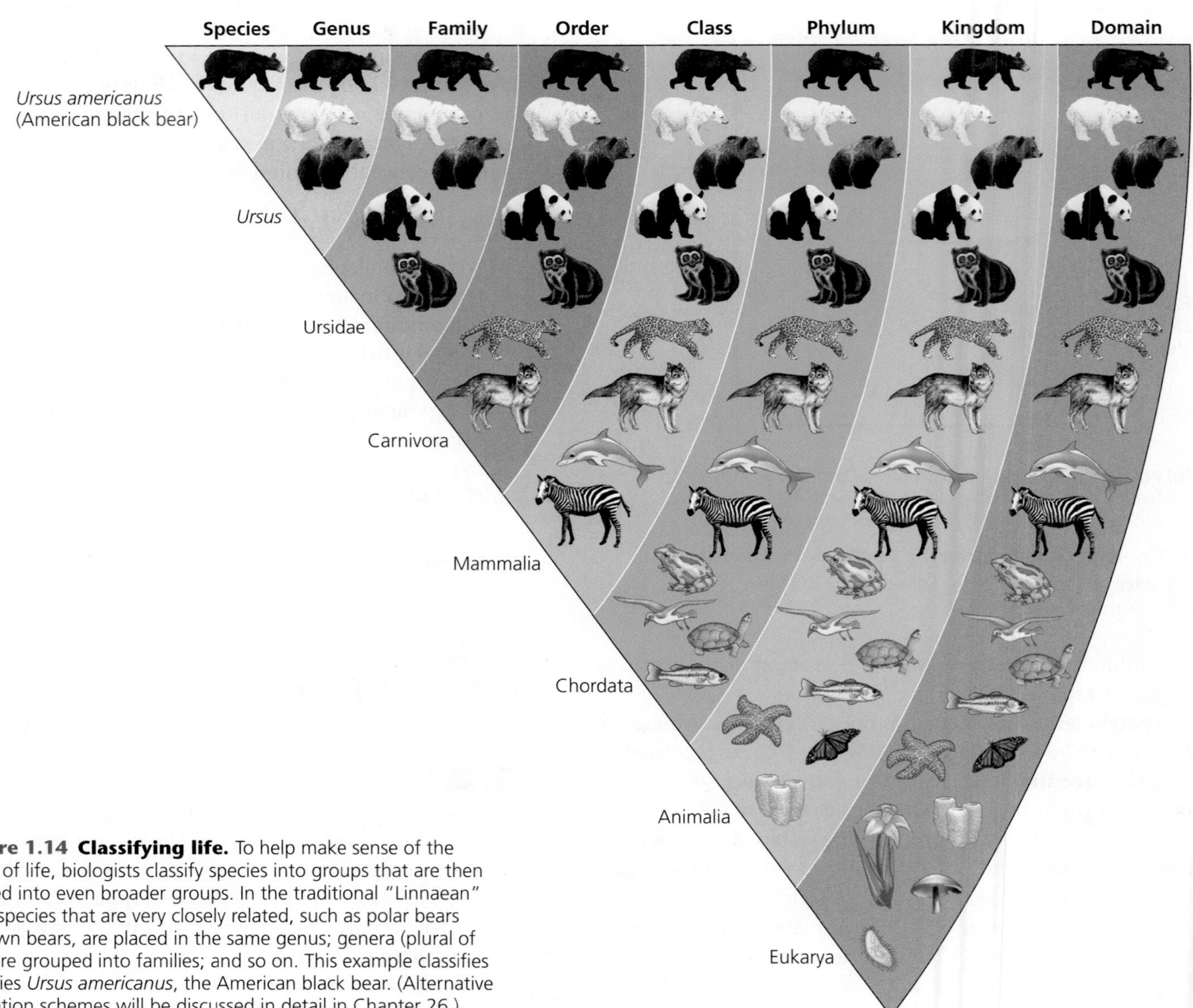

▲ **Figure 1.14 Classifying life.** To help make sense of the diversity of life, biologists classify species into groups that are then combined into even broader groups. In the traditional "Linnaean" system, species that are very closely related, such as polar bears and brown bears, are placed in the same genus; genera (plural of genus) are grouped into families; and so on. This example classifies the species *Ursus americanus*, the American black bear. (Alternative classification schemes will be discussed in detail in Chapter 26.)

now, we will focus on the big picture by considering the broadest units of classification, kingdoms and domains.

The Three Domains of Life

Historically, scientists have classified the diversity of lifeforms into kingdoms and finer groupings by careful comparisons of structure, function, and other obvious features. In the last few decades, new methods of assessing species relationships, such as comparisons of DNA sequences, have led to an ongoing reevaluation of the number and boundaries of kingdoms. Researchers have proposed anywhere from six kingdoms to dozens of kingdoms. While debate continues at the kingdom level, there is consensus among biologists that the kingdoms of life can be grouped into three even higher levels of classification called domains. The three domains are named Bacteria, Archaea, and Eukarya (**Figure 1.15**).

The organisms making up two of the three domains—domain **Bacteria** and domain **Archaea**—are all prokaryotic. Most prokaryotes are single-celled and microscopic. Previously, bacteria and archaea were combined in a single kingdom because they shared the prokaryotic form of cell structure. But much evidence now supports the view that bacteria and archaea represent two very distinct branches of prokaryotic life, different in key ways that you'll learn about in Chapter 27. There is also evidence that archaea are at least as closely related to eukaryotic organisms as they are to bacteria.

All the eukaryotes (organisms with eukaryotic cells) are now grouped in domain **Eukarya**. This domain includes three kingdoms of multicellular eukaryotes: kingdoms Plantae,

▼ **Figure 1.15 The three domains of life.**

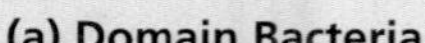

(a) Domain Bacteria

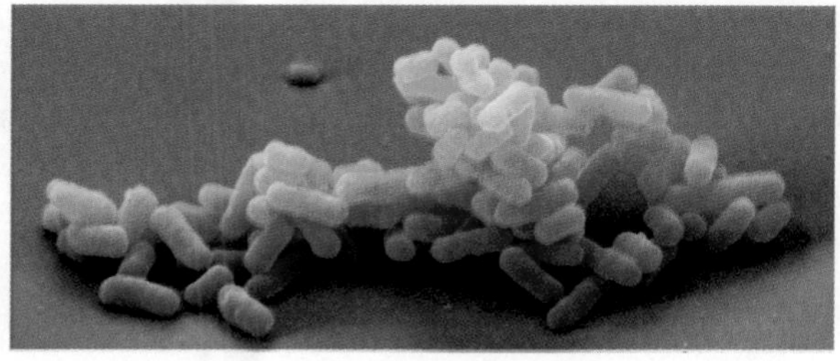

Bacteria are the most diverse and widespread prokaryotes and are now classified into multiple kingdoms. Each rod-shaped structure in this photo is a bacterial cell.

(b) Domain Archaea

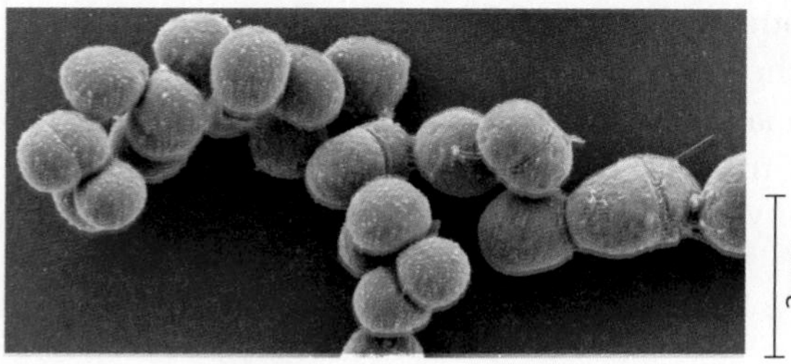

Many of the prokaryotes known as **archaea** live in Earth's extreme environments, such as salty lakes and boiling hot springs. Domain Archaea includes multiple kingdoms. Each round structure in this photo is an archaeal cell.

(c) Domain Eukarya

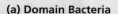

▲ **Kingdom Plantae** consists of terrestrial multicellular eukaryotes (land plants) that carry out photosynthesis, the conversion of light energy to the chemical energy in food.

► **Kingdom Fungi** is defined in part by the nutritional mode of its members (such as this mushroom), which absorb nutrients from outside their bodies.

◄ **Kingdom Animalia** consists of multicellular eukaryotes that ingest other organisms.

100 µm

► **Protists** are mostly unicellular eukaryotes and some relatively simple multicellular relatives. Pictured here is an assortment of protists inhabiting pond water. Scientists are currently debating how to classify protists in a way that accurately reflects their evolutionary relationships.

Fungi, and Animalia. These three kingdoms are distinguished partly by their modes of nutrition. Plants produce their own sugars and other food molecules by photosynthesis. Fungi absorb dissolved nutrients from their surroundings; many decompose dead organisms and organic wastes (such as leaf litter and animal feces) and absorb nutrients from these sources. Animals obtain food by ingestion, which is the eating and digesting of other organisms. Animalia is, of course, the kingdom to which we belong. But neither animals, plants, nor fungi are as numerous or diverse as the single-celled eukaryotes we call protists. Although protists were once placed in a single kingdom, biologists now realize that they do not form a single natural group of species. And recent evidence shows that some protist groups are more closely related to multicellular eukaryotes such as animals and fungi than they are to each other. Thus, the recent taxonomic trend has been to split the protists into several groups.

Unity in the Diversity of Life

As diverse as life is, it also displays remarkable unity. Earlier we mentioned both the similar skeletons of different vertebrate animals and the universal genetic language of DNA (the genetic code). In fact, similarities between organisms are evident at all levels of the biological hierarchy. For example, unity is obvious in many features of cell structure (**Figure 1.16**).

How can we account for life's dual nature of unity and diversity? The process of evolution, explained next, illuminates both the similarities and differences in the world of life and introduces another dimension of biology: historical time.

Charles Darwin and the Theory of Natural Selection

The history of life, as documented by fossils and other evidence, is the saga of a changing Earth billions of years old, inhabited by an evolving cast of living forms (**Figure 1.17**). This evolutionary view of life came into sharp focus in November 1859, when Charles Robert Darwin published one of the most important and influential books ever written. Entitled *On the Origin of Species by Means of Natural Selection*, Darwin's book was an immediate bestseller and soon made "Darwinism," as it was dubbed at the time, almost synonymous with the concept of evolution (**Figure 1.18**).

The Origin of Species articulated two main points. The first point was that contemporary species arose from a succession of ancestors, an idea that Darwin supported with a large amount of evidence. (We will discuss the evidence for evolution in detail in Chapter 22.) Darwin called this evolutionary history of species "descent with modification." It was an insightful phrase, as it captured the duality of life's unity and diversity—unity in the kinship among species that descended

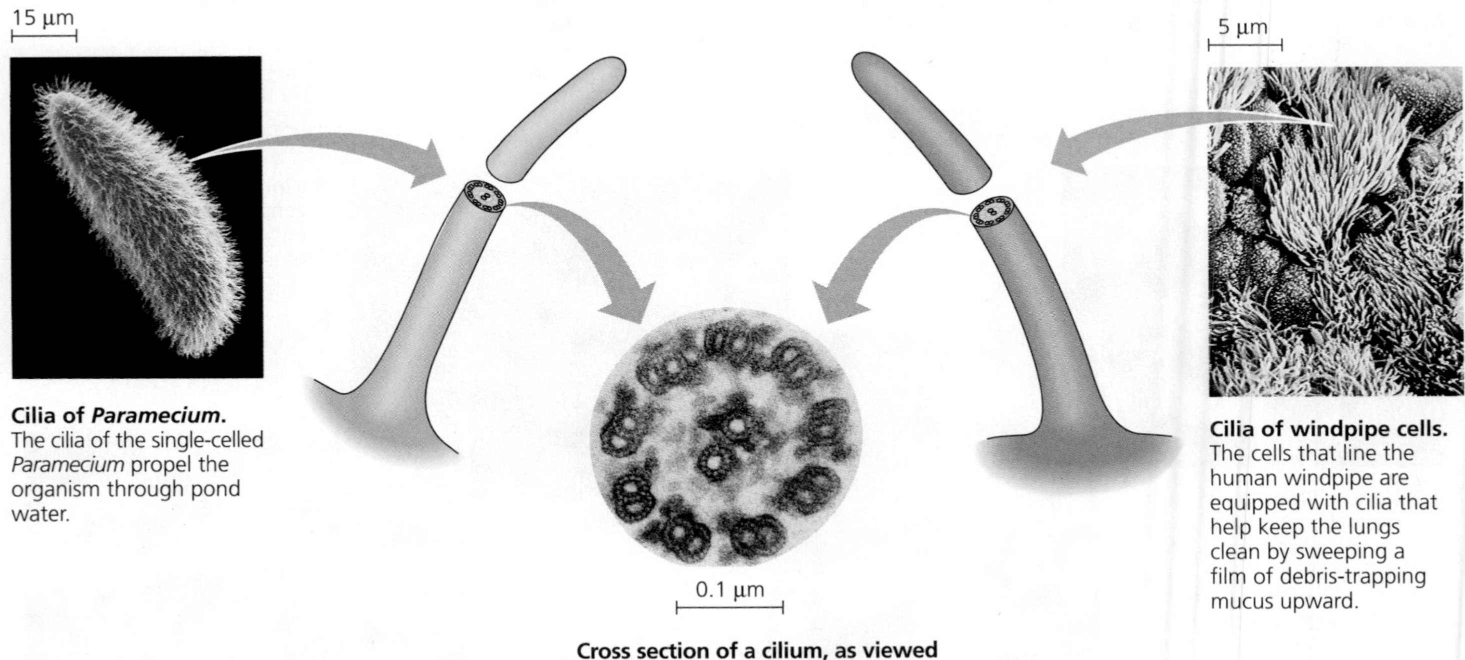

Cilia of *Paramecium*. The cilia of the single-celled *Paramecium* propel the organism through pond water.

Cross section of a cilium, as viewed with an electron microscope

Cilia of windpipe cells. The cells that line the human windpipe are equipped with cilia that help keep the lungs clean by sweeping a film of debris-trapping mucus upward.

▲ **Figure 1.16 An example of unity underlying the diversity of life: the architecture of cilia in eukaryotes.** Cilia (singular, *cilium*) are extensions of cells that function in locomotion. They occur in eukaryotes as diverse as *Paramecium* and humans. Even organisms so different share a common architecture for their cilia, which have an elaborate system of tubules that is striking in cross-sectional views.

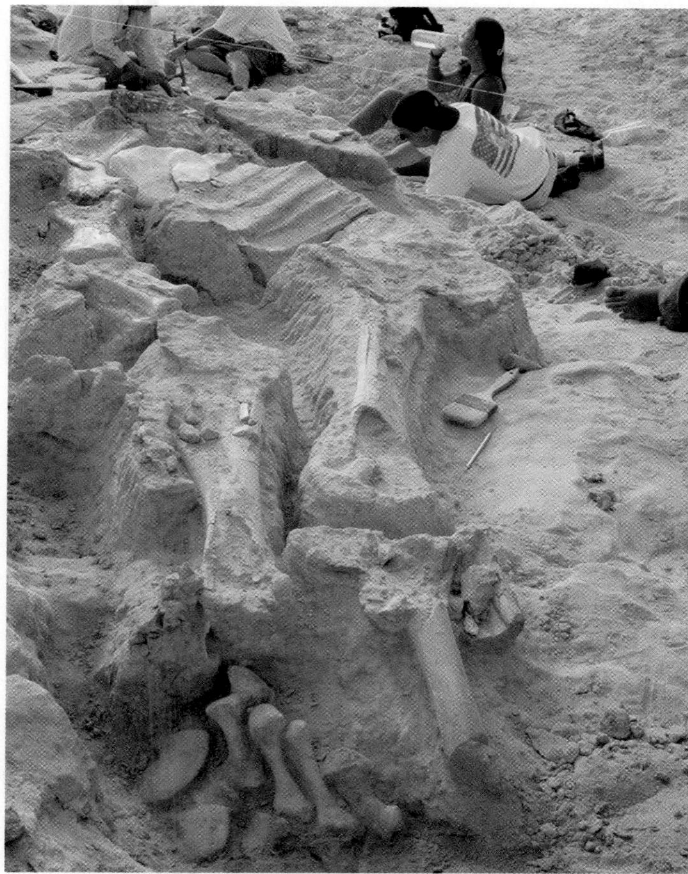

▲ **Figure 1.17 Digging into the past.** Paleontologists carefully excavate the hind leg of a long-necked dinosaur (*Rapetosaurus krausei*) from rocks in Madagascar.

from common ancestors, diversity in the modifications that evolved as species branched from their common ancestors (**Figure 1.19**). Darwin's second main point was a proposed mechanism for descent with modification. He called this evolutionary mechanism "natural selection."

Darwin synthesized his theory of natural selection from observations that by themselves were neither new nor profound. Others had the pieces of the puzzle, but Darwin saw how they fit together. He started with the following three observations from nature: First, individuals in a population vary in their traits, many of which seem to be heritable (passed on from parents to offspring). Second, a population can produce far more offspring than can survive to produce offspring of their own. With more individuals than the environment is able to support, competition is inevitable. Third, species generally suit their environments—in other words, they are adapted to their environments. For instance, a common adaptation among birds with tough seeds as their major food source is that they have especially strong beaks.

▲ **Figure 1.18 Charles Darwin as a young man.**

Darwin made inferences from these observations to arrive at his theory of evolution. He reasoned that individuals with inherited traits that are best suited to the local environment are more likely to survive and reproduce than less suited individuals. Over many generations, a higher and higher proportion of individuals in a population will have the advantageous traits. Evolution occurs as the unequal reproductive success of individuals ultimately leads to adaptation to their environment, as long as the environment remains the same.

Darwin called this mechanism of evolutionary adaptation **natural selection** because the natural environment "selects" for the propagation of certain traits among naturally occurring variant traits in the population. The example

◀ **Figure 1.19 Unity and diversity in the orchid family.** These three orchids are variations on a common floral theme. For example, each of these flowers has a liplike petal that helps attract pollinating insects and provides a landing platform for the pollinators.

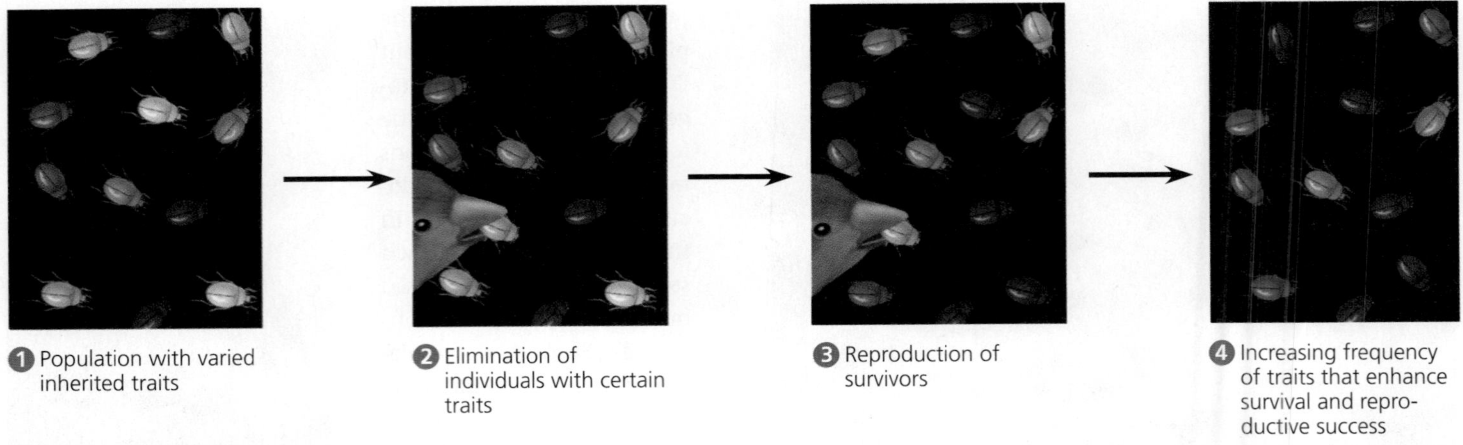

▲ **Figure 1.20 Natural selection.** This imaginary beetle population has colonized a locale where the soil has been blackened by a recent brush fire. Initially, the population varies extensively in the inherited coloration of the individuals, from very light gray to charcoal. For hungry birds that prey on the beetles, it is easiest to spot the beetles that are lightest in color.

in **Figure 1.20** illustrates the ability of natural selection to "edit" a population's heritable variations in color. We see the products of natural selection in the exquisite adaptations of various organisms to the special circumstances of their way of life and their environment. The wings of the bat shown in **Figure 1.21** are an excellent example of adaptation.

The Tree of Life

Take another look at the skeletal architecture of the bat's wings in Figure 1.21. These forelimbs, though adapted for flight, actually have all the same bones, joints, nerves, and blood vessels found in other limbs as diverse as the human arm, the horse's foreleg, and the whale's flipper. Indeed, all mammalian forelimbs are anatomical variations of a common architecture, much as the flowers in Figure 1.19 are variations on an underlying "orchid" theme. Such examples of kinship connect life's unity in diversity to the Darwinian concept of descent with modification. In this view, the unity of mammalian limb anatomy reflects inheritance of that structure from a common ancestor—the "prototype" mammal from which all other mammals descended. The diversity of mammalian forelimbs results from modification by natural selection operating over millions of generations in different environmental contexts. Fossils and other evidence corroborate anatomical unity in supporting this view of mammalian descent from a common ancestor.

▲ **Figure 1.21 Evolutionary adaptation.** Bats, the only mammals capable of active flight, have wings with webbing between extended "fingers." In the Darwinian view of life, such adaptations are refined over time by natural selection.

Darwin proposed that natural selection, by its cumulative effects over long periods of time, could cause an ancestral species to give rise to two or more descendant species. This could occur, for example, if one population fragmented into several subpopulations isolated in different environments. In these separate arenas of natural selection, one species could gradually radiate into multiple species as the geographically isolated populations adapted over many generations to different sets of environmental factors.

The "family tree" of 14 finches in **Figure 1.22** illustrates a famous example of adaptive radiation of new species from a common ancestor. Darwin collected specimens of these birds during his 1835 visit to the remote Galápagos Islands, 900 kilometers (km) off the Pacific coast of South America. These relatively young, volcanic islands are home to many species of plants and animals found nowhere else in the world, though most Galápagos organisms are clearly related to species on the South American mainland. After volcanism built the Galápagos several million years ago, finches probably diversified on the various islands from an ancestral finch species that by chance reached the archipelago from elsewhere. (Once thought to have originated on the mainland of South America like many Galápagos organisms, the ancestral finches are now thought to have come from the West Indies—islands of the Caribbean that were once much closer to the Galápagos than they are now.)

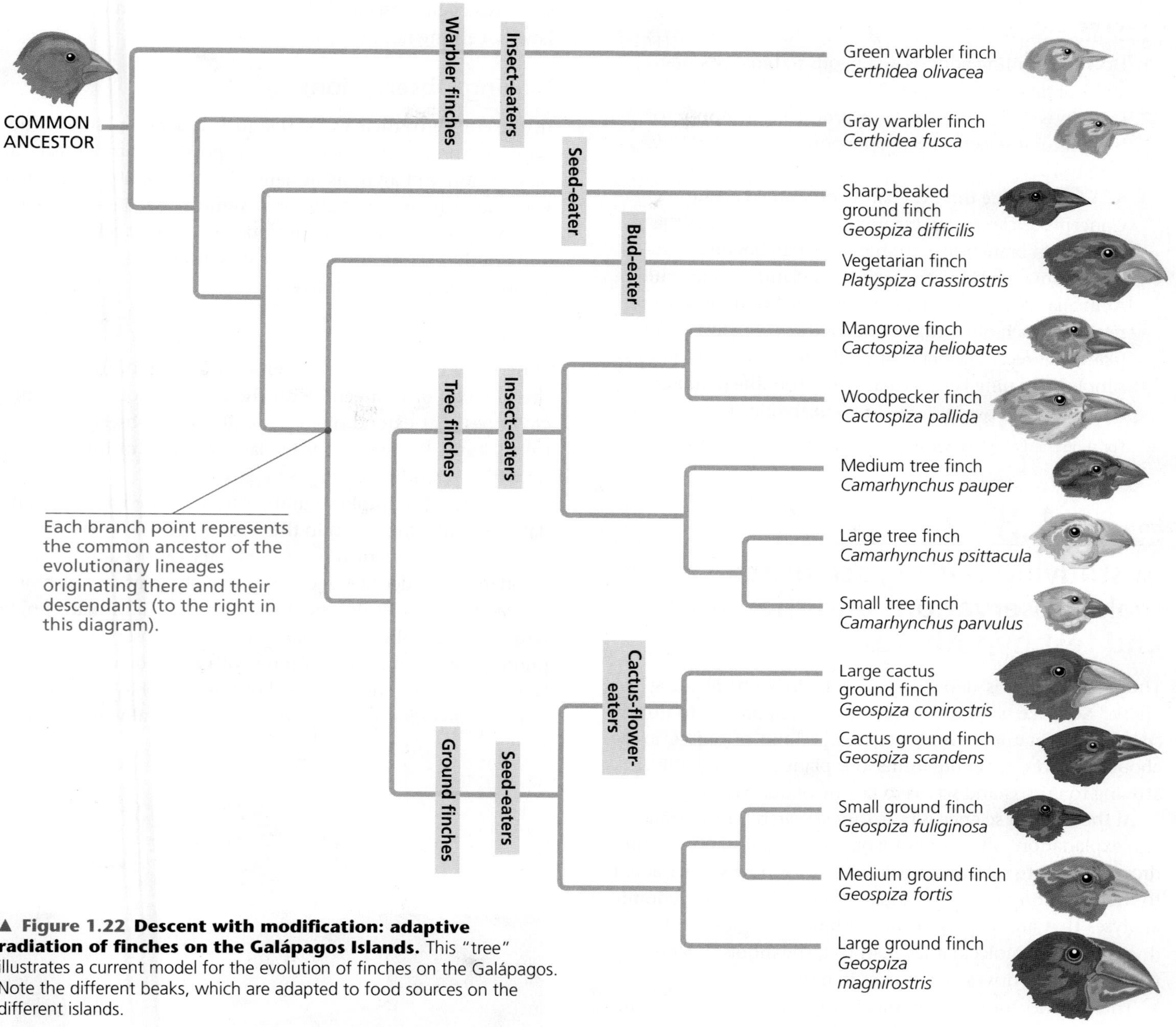

▲ **Figure 1.22 Descent with modification: adaptive radiation of finches on the Galápagos Islands.** This "tree" illustrates a current model for the evolution of finches on the Galápagos. Note the different beaks, which are adapted to food sources on the different islands.

Years after Darwin's collection of Galápagos finches, researchers began to sort out the relationships among the finch species, first from anatomical and geographic data and more recently with the help of DNA sequence comparisons.

Biologists' diagrams of evolutionary relationships generally take treelike forms, though today biologists usually turn the trees sideways as in Figure 1.22. Tree diagrams make sense: Just as an individual has a genealogy that can be diagrammed as a family tree, each species is one twig of a branching tree of life extending back in time through ancestral species more and more remote. Species that are very similar, such as the Galápagos finches, share a common ancestor at a relatively recent branch point on the tree of life. But through an ancestor that lived much farther back in time, finches are related to sparrows, hawks, penguins, and all other birds. And birds, mammals, and all other vertebrates share a common ancestor even more ancient. We find evidence of still broader relationships in such similarities as the identical construction of all eukaryotic cilia (see Figure 1.16). Trace life back far enough, and there are only fossils of the primeval prokaryotes that inhabited Earth over 3.5 billion years ago. We can recognize their vestiges in our own cells—in the universal genetic code, for example. All of life is connected through its long evolutionary history.

CONCEPT CHECK 1.2

1. How is a mailing address analogous to biology's hierarchical taxonomic system?
2. Explain why "editing" is an appropriate metaphor for how natural selection acts on a population's heritable variation.
3. **WHAT IF?** The three domains you learned about in Concept 1.2 can be represented in the tree of life as the three main branches, with three subbranches on the eukaryotic branch being the kingdoms Plantae, Fungi, and Animalia. What if fungi and animals are more closely related to each other than either of these kingdoms is to plants—as recent evidence strongly suggests? Draw a simple branching pattern that symbolizes the proposed relationship between these three eukaryotic kingdoms.

For suggested answers, see Appendix A.

CONCEPT 1.3

In studying nature, scientists make observations and then form and test hypotheses

The word *science* is derived from a Latin verb meaning "to know." **Science** is a way of knowing—an approach to understanding the natural world. It developed out of our curiosity about ourselves, other life-forms, our planet, and the universe. Striving to understand seems to be one of our basic urges.

At the heart of science is **inquiry**, a search for information and explanation, often focusing on specific questions. Inquiry drove Darwin to seek answers in nature for how species adapt to their environments. And today inquiry drives the genomic analyses that are helping us understand biological unity and diversity at the molecular level. In fact, the inquisitive mind is the engine that drives all progress in biology.

There is no formula for successful scientific inquiry, no single scientific method with a rule book that researchers must rigidly follow. As in all quests, science includes elements of challenge, adventure, and luck, along with careful planning, reasoning, creativity, cooperation, competition, patience, and the persistence to overcome setbacks. Such diverse elements of inquiry make science far less structured than most people realize. That said, it is possible to distill certain characteristics that help to distinguish science from other ways of describing and explaining nature.

Scientists attempt to understand how natural phenomena work using a process of inquiry that includes making observations, forming logical hypotheses, and testing them. The process is necessarily repetitive: In testing a hypothesis, more observations may force formation of a new hypothesis or revision of the original one, and further testing. In this way, scientists circle closer and closer to their best estimation of the laws governing nature.

Making Observations

In the course of their work, scientists describe natural structures and processes as accurately as possible through careful observation and analysis of data. The observations are often valuable in their own right. For example, a series of detailed observations have shaped our understanding of cell structure, and another set of observations are currently expanding our databases of genomes of diverse species.

Types of Data

Observation is the use of the senses to gather information, either directly or indirectly with the help of tools such as microscopes that extend our senses. Recorded observations are called **data**. Put another way, data are items of information on which scientific inquiry is based.

The term *data* implies numbers to many people. But some data are *qualitative*, often in the form of recorded descriptions rather than numerical measurements. For example, Jane Goodall spent decades recording her observations of chimpanzee behavior during field research in a Tanzanian jungle (**Figure 1.23**). She also documented her observations with photographs and movies. Along with these qualitative data, Goodall also enriched the field of animal behavior with volumes of *quantitative* data, which are generally recorded as

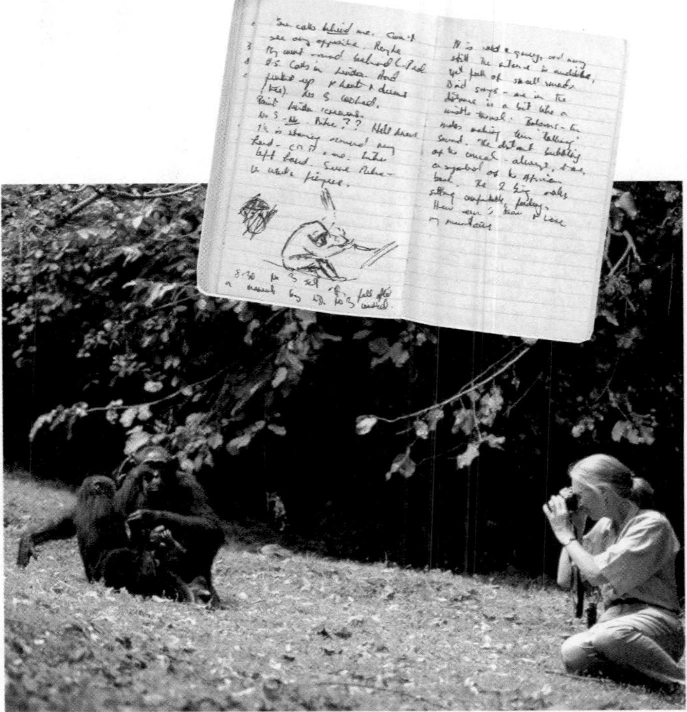

▲ **Figure 1.23 Jane Goodall collecting qualitative data on chimpanzee behavior.** Goodall recorded her observations in field notebooks, often with sketches of the animals' behavior.

measurements. Skim through any of the scientific journals in your college library, and you'll see many examples of quantitative data organized into tables and graphs.

Inductive Reasoning

Collecting and analyzing observations can lead to important conclusions based on a type of logic called **inductive reasoning**. Through induction, we derive generalizations from a large number of specific observations. "The sun always rises in the east" is an example. And so is "All organisms are made of cells." The latter generalization, part of the so-called cell theory, was based on two centuries of microscopic observations by biologists of cells in diverse biological specimens. Careful observations and data analyses, along with the generalizations reached by induction, are fundamental to our understanding of nature.

Forming and Testing Hypotheses

Observations and inductive reasoning stimulate us to seek natural causes and explanations for those observations. What *caused* the diversification of finches on the Galápagos Islands? What *causes* the roots of a plant seedling to grow downward and the leaf-bearing shoot to grow upward? What *explains* the generalization that the sun always rises in the east? In science, such inquiry usually involves the proposing and testing of hypothetical explanations—that is, hypotheses.

The Role of Hypotheses in Inquiry

In science, a **hypothesis** is a tentative answer to a well-framed question—an explanation on trial. It is usually a rational accounting for a set of observations, based on the available data and guided by inductive reasoning. A scientific hypothesis leads to predictions that can be tested by making additional observations or by performing experiments.

We all use hypotheses in solving everyday problems. Let's say, for example, that your flashlight fails during a camp-out. That's an observation. The question is obvious: Why doesn't the flashlight work? Two reasonable hypotheses based on your experience are that (1) the batteries in the flashlight are dead or (2) the bulb is burnt out. Each of these alternative hypotheses leads to predictions you can test with experiments. For example, the dead-battery hypothesis predicts that replacing the batteries will fix the problem. **Figure 1.24** diagrams this campground inquiry. Of course, we rarely dissect our thought processes this way when we are solving a problem using hypotheses, predictions, and experiments. But the hypothesis-based nature of science clearly has its origins in the human tendency to figure things out by trial and error.

Deductive Reasoning and Hypothesis Testing

A type of logic called deduction is built into the use of hypotheses in science. Deduction contrasts with induction,

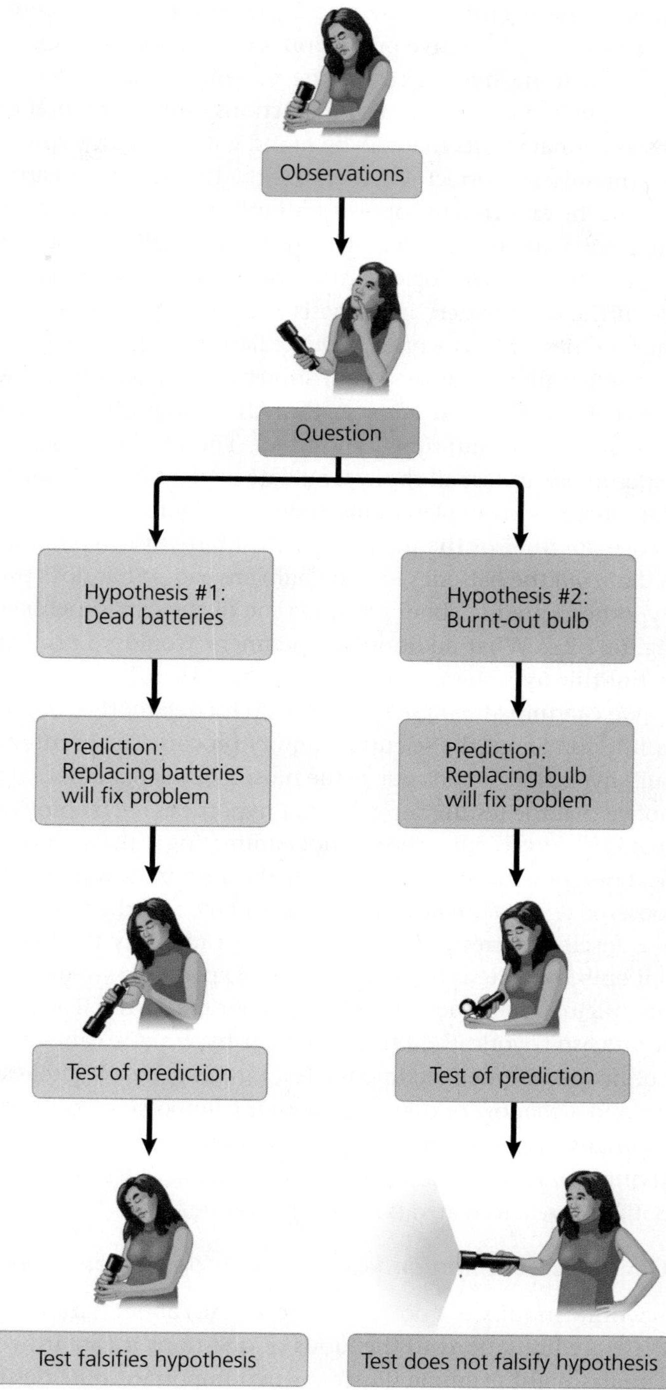

▲ **Figure 1.24 A campground example of hypothesis-based inquiry.**

which, remember, is reasoning from a set of specific observations to reach a general conclusion—a process that feeds into hypothesis formation. **Deductive reasoning** is generally used after the hypothesis has been developed and involves logic that flows in the opposite direction, from the general to the specific. From general premises, we extrapolate to the specific results we should expect if the premises are true. If all organisms are made of cells (premise 1), and humans are organisms (premise 2), then humans are composed of cells (deductive prediction about a specific case).

When using hypotheses in the scientific process, deductions usually take the form of predictions of experimental or observational results that will be found if a particular hypothesis (premise) is correct. We then test the hypothesis by carrying out the experiments or observations to see whether or not the results are as predicted. This deductive testing takes the form of "*If . . . then*" logic. In the case of the flashlight example: *If* the dead-battery hypothesis is correct and you replace the batteries with new ones, *then* the flashlight should work.

The flashlight inquiry demonstrates a key point about the use of hypotheses in science: that the initial observations may give rise to multiple hypotheses. The ideal is to design experiments to test all these candidate explanations. In addition to the two explanations tested in Figure 1.24, for instance, another of the many possible alternative hypotheses is that *both* the batteries *and* the bulb are bad. What does this hypothesis predict about the outcome of the experiments in Figure 1.24? What additional experiment would you design to test this hypothesis of multiple malfunctions?

We can mine the flashlight scenario for yet another important lesson about the scientific inquiry process. The burnt-out bulb hypothesis stands out as the most likely explanation, but notice that the testing supports that hypothesis *not* by proving that it is correct, but rather by not eliminating it through falsification (proving it false). Perhaps the first bulb was simply loose, so it wasn't making electrical contact, and the new bulb was inserted correctly. We could attempt to falsify the burnt-out bulb hypothesis by trying another experiment—removing the original bulb and carefully reinstalling it. If the flashlight still doesn't work, the burnt-out bulb hypothesis can stand. But no amount of experimental testing can *prove* a hypothesis beyond a shadow of doubt, because it is impossible to test *all* alternative hypotheses. A hypothesis gains credibility by surviving multiple attempts to falsify it while alternative hypotheses are eliminated (falsified) by testing.

Questions That Can and Cannot Be Addressed by Science

Scientific inquiry is a powerful way to learn about nature, but there are limitations to the kinds of questions it can answer. The flashlight example illustrates two important qualities of scientific hypotheses. First, a hypothesis must be *testable*; there must be some way to check the validity of the idea. Second, a hypothesis must be *falsifiable*; there must be some observation or experiment that could reveal if such an idea is actually *not* true. The hypothesis that dead batteries are the sole cause of the broken flashlight could be falsified by replacing the old batteries with new ones and finding that the flashlight still doesn't work.

Not all hypotheses meet the criteria of science: You wouldn't be able to devise a test to falsify the hypothesis that invisible campground ghosts are fooling with your flashlight! Because science requires natural explanations for natural phenomena, it can neither support nor falsify hypotheses that angels, ghosts, or spirits, whether benevolent or evil, cause storms, rainbows, illnesses, and cures. Such supernatural explanations are simply outside the bounds of science, as are religious matters, which are issues of personal faith.

The Flexibility of the Scientific Method

The flashlight example of Figure 1.24 traces an idealized process of inquiry called *the scientific method*. We can recognize the elements of this process in most of the research articles published by scientists, but rarely in such structured form. Very few scientific inquiries adhere rigidly to the sequence of steps prescribed by the "textbook" scientific method. For example, a scientist may start to design an experiment, but then backtrack upon realizing that more preliminary observations are necessary. In other cases, puzzling observations simply don't prompt well-defined questions until other research places those observations in a new context. For example, Darwin collected specimens of the Galápagos finches, but it wasn't until years later, as the idea of natural selection began to gel, that biologists began asking key questions about the history of those birds.

Moreover, scientists sometimes redirect their research when they realize they have been asking the wrong question. For example, in the early 20th century, much research on schizophrenia and manic-depressive disorder (now called bipolar disorder) got sidetracked by focusing too much on the question of how life experiences might cause these serious maladies. Research on the causes and potential treatments became more productive when it was refocused on questions of how certain chemical imbalances in the brain contribute to mental illness. To be fair, we acknowledge that such twists and turns in scientific inquiry become more evident with the advantage of historical perspective.

It is important for you to get some experience with the power of the scientific method—by using it for some of the laboratory inquiries in your biology course, for example. But it is also important to avoid stereotyping science as a lock-step adherence to this method.

A Case Study in Scientific Inquiry: Investigating Mimicry in Snake Populations

Now that we have highlighted the key features of scientific inquiry—making observations and forming and testing hypotheses—you should be able to recognize these features in a case study of actual scientific research.

The story begins with a set of observations and inductive generalizations. Many poisonous animals are brightly colored, often with distinctive patterns that stand out against the background. This is called *warning coloration* because it apparently signals "dangerous species" to potential predators. But there are also mimics. These imposters look like poisonous species but are actually harmless. A question that follows from these observations is: What is the function of such mimicry? A reasonable hypothesis is that the "deception" is an evolutionary adaptation that reduces the harmless animal's risk of being eaten because predators mistake it for the poisonous species. This hypothesis was first formulated by British scientist Henry Bates in 1862.

As obvious as this hypothesis may seem, it has been relatively difficult to test, especially with field experiments. But in 2001, biologists David and Karin Pfennig, of the University of North Carolina, along with William Harcombe, an undergraduate, designed a simple but elegant set of field experiments to test Bates's mimicry hypothesis.

The team investigated a case of mimicry among snakes that live in North and South Carolina (**Figure 1.25**). A venomous snake called the eastern coral snake has warning coloration: bold, alternating rings of red, yellow (or white), and black. (The word *venomous* is used when a poisonous species delivers their poison actively, by stinging, stabbing, or biting.) Predators rarely attack these coral snakes. It is unlikely that the predators learn this avoidance behavior by trial and error, as a first encounter with a coral snake is usually deadly. In areas where coral snakes live, natural selection has apparently increased the frequency of predators that have inherited an instinctive avoidance of the coral snake's coloration. A nonvenomous snake named the scarlet kingsnake mimics the ringed coloration of the coral snake.

Both types of snakes live in the Carolinas, but the kingsnakes' geographic range also extends into regions where no coral snakes are found (see Figure 1.25). The geographic distribution of the snakes made it possible to test the key prediction of the mimicry hypothesis. Avoiding snakes with warning coloration is an adaptation we expect to be present only in predator populations that evolved in areas where the venomous coral snakes are present. Therefore, mimicry should help protect kingsnakes from predators *only in regions where coral snakes also live*. The mimicry hypothesis predicts that predators adapted to the warning coloration of coral snakes will attack kingsnakes less frequently than will predators in areas where coral snakes are absent.

Scarlet kingsnake (nonvenomous)

Key

Range of scarlet kingsnake only

Overlapping ranges of scarlet kingsnake and eastern coral snake

North Carolina

South Carolina

Eastern coral snake (venomous)

Scarlet kingsnake (nonvenomous)

▲ **Figure 1.25 The geographic ranges of a venomous snake and its mimic.** The scarlet kingsnake (*Lampropeltis triangulum*) mimics the warning coloration of the venomous eastern coral snake (*Micrurus fulvius*).

Field Experiments with Artificial Snakes

To test the prediction, Harcombe made hundreds of artificial snakes out of wire covered with plasticine. He fashioned two versions of fake snakes: an *experimental group* with the red, black, and white ring pattern of kingsnakes and a *control group* of plain brown artificial snakes as a basis of comparison (**Figure 1.26**).

The researchers placed equal numbers of the two types of artificial snakes in field sites throughout North and South

(a) Artificial kingsnake

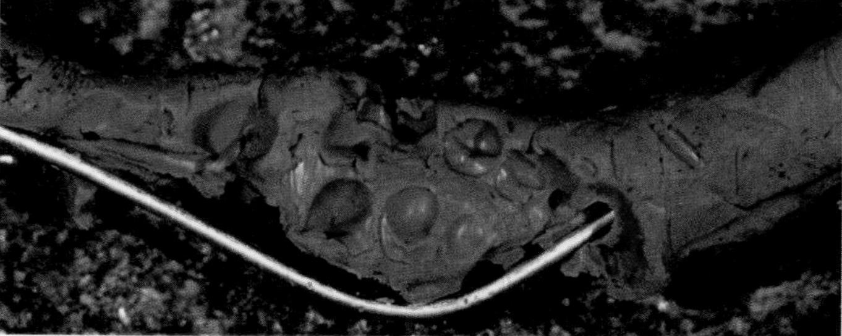

(b) Brown artificial snake that has been attacked

▲ **Figure 1.26 Artificial snakes used in field experiments to test the mimicry hypothesis.** A bear has chewed on the brown artificial snake in **(b)**.

Carolina, including the region where coral snakes are absent. After four weeks, the scientists retrieved the fake snakes and recorded how many had been attacked by looking for bite or claw marks. The most common predators were foxes, coyotes, and raccoons, but black bears also attacked some of the artificial snakes (see Figure 1.26b).

The data fit the key prediction of the mimicry hypothesis. Compared to the brown artificial snakes, the ringed artificial snakes were attacked by predators less frequently only in field sites within the geographic range of the venomous coral snakes. **Figure 1.27** summarizes the field experiments that the researchers carried out. This figure also introduces a format we will use throughout the book for other examples of biological inquiry.

Experimental Controls and Repeatability

The snake mimicry experiment is an example of a **controlled experiment**, one that is designed to compare an experimental group (the artificial kingsnakes, in this case) with a control group (the brown artificial snakes). Ideally, the experimental and control groups differ only in the one factor the experiment is designed to test—in our example, the effect of the snakes' coloration on the behavior of predators. Without the control group, the researchers would not have been able to rule out other factors as causes of the more frequent attacks on the artificial kingsnakes—such as different numbers of predators or different temperatures in the different test areas. The clever experimental design left coloration as the only factor that could account for the low predation rate on the artificial kingsnakes placed within the range of coral snakes. It was not the absolute number of attacks on the artificial kingsnakes that counted, but the difference between that number and the number of attacks on the brown snakes.

A common misconception is that the term *controlled experiment* means that scientists control the experimental environment to keep everything constant except the one variable being tested. But that's impossible in field research and not realistic even in highly regulated laboratory environments. Researchers usually "control" unwanted variables not by *eliminating* them through environmental regulation, but by *canceling out* their effects by using control groups.

Another hallmark of science is that the observations and experimental results must be repeatable. Observations that can't be verified may be interesting or even entertaining, but they cannot count as evidence in scientific inquiry. The headlines of supermarket tabloids would have you believe that humans are occasionally born with the head of a dog and that some of your classmates are extraterrestrials. The unconfirmed eyewitness accounts and the computer-rigged photos are amusing but unconvincing. In science, evidence from observations and experiments is only convincing if it stands up to the criterion of repeatability. The scientists who investigated snake mimicry in the Carolinas obtained similar data when they

▼ **Figure 1.27** INQUIRY

Does the presence of venomous coral snakes affect predation rates on their mimics, kingsnakes?

EXPERIMENT David Pfennig and his colleagues made artificial snakes to test a prediction of the mimicry hypothesis: that kingsnakes benefit from mimicking the warning coloration of venomous coral snakes only in regions where coral snakes are present. The researchers placed equal numbers of artificial kingsnakes (experimental group) and brown artificial snakes (control group) at 14 field sites, half in the area the two snakes cohabit and half in the area where coral snakes are absent. The researchers recovered the artificial snakes after four weeks and tabulated predation data based on teeth and claw marks on the snakes.

RESULTS In field sites where coral snakes are absent, most attacks were on artificial kingsnakes. Where coral snakes were present, most attacks were on brown artificial snakes.

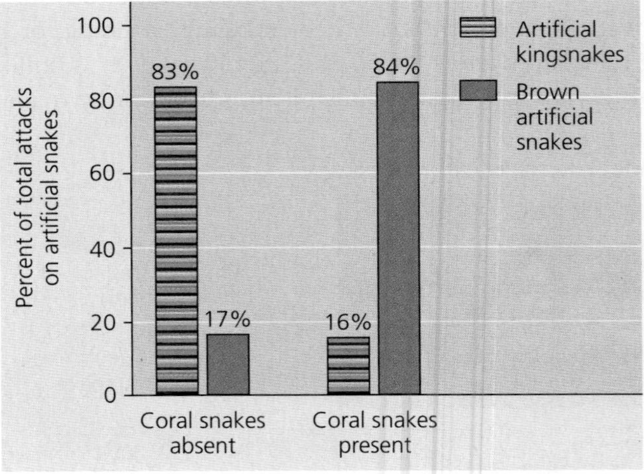

CONCLUSION The field experiments support the mimicry hypothesis by not falsifying the prediction, which was that mimicking coral snakes is effective only in areas where coral snakes are present. The experiments also tested an alternative hypothesis: that predators generally avoid all snakes with brightly colored rings. That hypothesis was falsified by the data showing that in areas without coral snakes, the ringed coloration failed to repel predators. (The fake kingsnakes may have been attacked more often in those areas because their bright pattern made them easier to spot than the brown fakes.)

SOURCE D. W. Pfennig, W. R. Harcombe, and K. S. Pfennig, Frequency-dependent Batesian mimicry, *Nature* 410:323 (2001).

INQUIRY IN ACTION Read and analyze the original paper in *Inquiry in Action: Interpreting Scientific Papers*.

MB See the related Experimental Inquiry Tutorial in MasteringBiology.

WHAT IF? What experimental results would you predict if predators throughout the Carolinas avoided all snakes with brightly colored ring patterns?

repeated their experiments with different species of coral snakes and kingsnakes in Arizona. And *you* should be able to obtain similar results if you were to repeat the snake experiments.

Theories in Science

"It's just a theory!" Our everyday use of the term *theory* often implies an untested speculation. But the term *theory* has a different meaning in science. What is a scientific theory, and how is it different from a hypothesis or from mere speculation?

First, a scientific **theory** is much broader in scope than a hypothesis. *This* is a hypothesis: "Mimicking the coloration of venomous snakes is an adaptation that protects nonvenomous snakes from predators." But *this* is a theory: "Evolutionary adaptations arise by natural selection." Darwin's theory of natural selection accounts for an enormous diversity of adaptations, including mimicry.

Second, a theory is general enough to spin off many new, specific hypotheses that can be tested. For example, two researchers at Princeton University, Peter and Rosemary Grant, were motivated by the theory of natural selection to test the specific hypothesis that the beaks of Galápagos finches evolve in response to changes in the types of available food. (Their results supported their hypothesis; see p. 469.)

And third, compared to any one hypothesis, a theory is generally supported by a much greater body of evidence. Those theories that become widely adopted in science (such as the theory of natural selection) explain a great diversity of observations and are supported by a vast accumulation of evidence. In fact, scrutiny of theories continues through testing of the specific, falsifiable hypotheses they spawn.

In spite of the body of evidence supporting a widely accepted theory, scientists must sometimes modify or even reject theories when new research methods produce results that don't fit. For example, the theory of biological diversity that lumped bacteria and archaea together as a kingdom of prokaryotes began to erode when new methods for comparing cells and molecules made it possible to test some of the hypothetical relationships between organisms that were based on the theory. If there is "truth" in science, it is conditional, based on the preponderance of available evidence.

CONCEPT CHECK 1.3

1. Contrast inductive reasoning with deductive reasoning.
2. In the snake mimicry experiment, what is the variable?
3. Why is natural selection called a theory?
4. **WHAT IF?** Suppose you extended the snake mimicry experiment to an area of Virginia where neither type of snake is known to live. What results would you predict at your field site?

For suggested answers, see Appendix A.

CONCEPT 1.4

Science benefits from a cooperative approach and diverse viewpoints

Movies and cartoons sometimes portray scientists as loners working in isolated labs. In reality, science is an intensely social activity. Most scientists work in teams, which often include both graduate and undergraduate students (**Figure 1.28**). And to succeed in science, it helps to be a good communicator. Research results have no impact until shared with a community of peers through seminars, publications, and websites.

Building on the Work of Others

The great scientist Sir Isaac Newton once said: "To explain all nature is too difficult a task for any one man or even for any one age. 'Tis much better to do a little with certainty, and leave the rest for others that come after you. . . ." Anyone who becomes a scientist, driven by curiosity about how nature works, is sure to benefit greatly from the rich storehouse of discoveries by others who have come before.

Scientists working in the same research field often check one another's claims by attempting to confirm observations or repeat experiments. If experimental results cannot be repeated by scientific colleagues, this failure may reflect some underlying weakness in the original claim, which will then have to be revised. In this sense, science polices itself. Integrity and adherence to high professional standards in reporting results are central to the scientific endeavor. After all, the validity of experimental data is key to designing further lines of inquiry.

It is not unusual for several scientists to converge on the same research question. Some scientists enjoy the challenge of being first with an important discovery or key experiment, while others derive more satisfaction from cooperating with fellow scientists working on the same problem.

▲ **Figure 1.28 Science as a social process.** In laboratory meetings, lab members help each other interpret data, troubleshoot experiments, and plan future lines of inquiry.

Cooperation is facilitated when scientists use the same organism. Often it is a widely used **model organism**—a species that is easy to grow in the lab and lends itself particularly well to the questions being investigated. Because all organisms are evolutionarily related, lessons learned from a model organism are often widely applicable. For example, genetic studies of the fruit fly *Drosophila melanogaster* have taught us a lot about how genes work in other species, including humans. Some other popular model organisms are the mustard plant *Arabidopsis thaliana*, the soil worm *Caenorhabditis elegans*, the zebrafish *Danio rerio*, the mouse *Mus musculus*, and the bacterium *Escherichia coli*. As you read through this book, note the many contributions that these and other model organisms have made to the study of life.

Biologists may come at interesting questions from different angles. Some biologists focus on ecosystems, while others study natural phenomena at the level of organisms or cells. This book is divided into units that look at biology from different levels. Yet any given problem can be addressed from many perspectives, which in fact complement each other.

As a beginning biology student, you can benefit from making connections between the different levels of biology. You can begin to develop this skill by noticing when certain topics crop up again and again in different units. One such topic is sickle-cell disease, a well-understood genetic condition that is prevalent among native inhabitants of Africa and other warm regions and their descendants. Another topic viewed at different levels in this book is global climate change, mentioned earlier in this chapter. Sickle-cell disease and global climate change will appear in several units of the book, each time addressed at a new level. We hope these recurring topics will help you integrate the material you're learning and enhance your enjoyment of biology by helping you keep the "big picture" in mind.

Science, Technology, and Society

The biology community is part of society at large, embedded in the cultural milieu of the times. Some philosophers of science argue that scientists are so influenced by cultural and political values that science is no more objective than other ways of understanding nature. At the other extreme are people who speak of scientific theories as though they were natural laws instead of human interpretations of nature. The reality of science is probably somewhere in between—rarely perfectly objective, but continuously vetted through the expectation that observations and experiments be repeatable and hypotheses be testable and falsifiable.

The relationship of science to society becomes clearer when we add technology to the picture. Though science and technology sometimes employ similar inquiry patterns, their basic goals differ. The goal of science is to understand natural phenomena. In contrast, **technology** generally *applies* scientific knowledge for some specific purpose. Biologists and other scientists usually speak of "discoveries," while engineers and other technologists more usually speak of "inventions." And the beneficiaries of those inventions include scientists, who put new technology to work in their research. Thus, science and technology are interdependent.

▲ **Figure 1.29 DNA technology and crime scene investigation.** In 2008, forensic analysis of DNA samples from a crime scene led to the release of Charles Chatman from prison after he had served nearly 27 years for a rape he didn't commit. The photo shows Judge John Creuzot hugging Mr. Chatman after his conviction was overturned. The details of forensic analysis of DNA will be described in Chapter 20.

The potent combination of science and technology can have dramatic effects on society. Sometimes, the applications of basic research that turn out to be the most beneficial come out of the blue, from completely unanticipated observations in the course of scientific exploration. For example, discovery of the structure of DNA by Watson and Crick 60 years ago and subsequent achievements in DNA science led to the technologies of DNA manipulation that are transforming applied fields such as medicine, agriculture, and forensics (**Figure 1.29**). Perhaps Watson and Crick envisioned that their discovery would someday lead to important applications, but it is unlikely that they could have predicted exactly what all those applications would be.

The directions that technology takes depend less on the curiosity that drives basic science than on the current needs and wants of people and on the social environment of the times. Debates about technology center more on "*should* we do it" than "*can* we do it." With advances in technology come difficult choices. For example, under what circumstances is it acceptable to use DNA technology to find out if particular people have genes for hereditary diseases? Should such tests always be voluntary, or are there circumstances when genetic testing should be mandatory? Should insurance companies or employers have access to the information, as they do for many other types of personal health data? These questions are becoming much more urgent as the sequencing of individual genomes becomes quicker and cheaper.

Such ethical issues have as much to do with politics, economics, and cultural values as with science and technology. All citizens—not only professional scientists—have a responsibility to be informed about how science works and about the potential benefits and risks of technology. The relationship between science, technology, and society increases the significance and value of any biology course.

The Value of Diverse Viewpoints in Science

Many of the technological innovations with the most profound impact on human society originated in settlements along trade routes, where a rich mix of different cultures ignited new ideas. For example, the printing press, which helped spread knowledge to all social classes and ultimately led to the book in your hands, was invented by the German Johannes Gutenberg around 1440. This invention relied on several innovations from China, including paper and ink. Paper traveled along trade routes from China to Baghdad, where technology was developed for its mass production. This technology then migrated to Europe, as did water-based ink from China, which was modified by Gutenberg to become oil-based ink. We have the cross-fertilization of diverse cultures to thank for the printing press, and the same can be said for other important inventions.

Along similar lines, science stands to gain much from embracing a diversity of backgrounds and viewpoints among its practitioners. But just how diverse a population are scientists in relation to gender, race, ethnicity, and other attributes?

The scientific community reflects the cultural standards and behaviors of society at large. It is therefore not surprising that until recently, women and certain minorities have faced huge obstacles in their pursuit to become professional scientists in many countries around the world. Over the past 50 years, changing attitudes about career choices have increased the proportion of women in biology and some other sciences, so that now women constitute roughly half of undergraduate biology majors and biology Ph.D. students. The pace has been slow at higher levels in the profession, however, and women and many racial and ethnic groups are still significantly underrepresented in many branches of science. This lack of diversity hampers the progress of science. The more voices that are heard at the table, the more robust, valuable, and productive the scientific interchange will be. The authors of this textbook welcome all students to the community of biologists, wishing you the joys and satisfactions of this very exciting and satisfying field of science—biology.

CONCEPT CHECK 1.4

1. How does science differ from technology?
2. **WHAT IF?** The gene that causes sickle-cell disease is present in a higher percentage of residents of sub-Saharan Africa than it is among those of African descent living in the United States. The presence of this gene provides some protection from malaria, a serious disease that is widespread in sub-Saharan Africa. Discuss an evolutionary process that could account for the different percentages among residents of the two regions.

For suggested answers, see Appendix A.

1 CHAPTER REVIEW

SUMMARY OF KEY CONCEPTS

CONCEPT 1.1

The themes of this book make connections across different areas of biology (pp. 2–11)

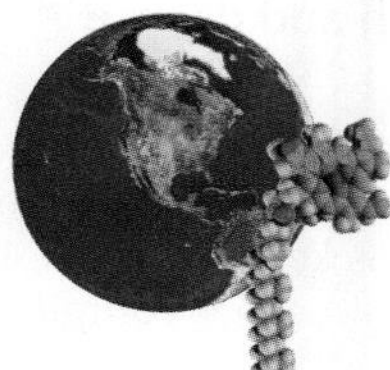

- **Theme: New properties emerge at each level in the biological hierarchy** The hierarchy of life unfolds as follows: biosphere > ecosystem > community > population > organism > organ system > organ > tissue > cell > organelle > molecule > atom. With each step upward from atoms, new properties emerge as a result of interactions among components at the lower levels. In an approach called reductionism, complex systems are broken down to simpler components that are more manageable to study. In **systems biology**, scientists attempt to model the dynamic behavior of whole biological systems based on a study of the interactions among the system's parts.

- **Theme: Organisms interact with other organisms and the physical environment** Plants take up nutrients from the soil and chemicals from the air and use energy from the sun. Interactions between plants and other organisms result in cycling of chemical nutrients within an ecosystem. One harmful outcome of human interactions with the environment has been global climate change, caused by burning of fossil fuels and increasing atmospheric CO_2.

- **Theme: Life requires energy transfer and transformation** Energy flows through an ecosystem. All organisms must perform work, which requires energy. Energy from sunlight is converted to chemical energy by producers, which is then passed on to consumers.

- **Theme: Structure and function are correlated at all levels of biological organization**
  The form of a biological structure suits its function and vice versa.

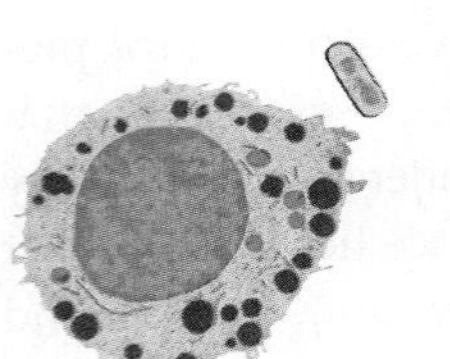

- **Theme: The cell is an organism's basic unit of structure and function**
  The cell is the lowest level of organization that can perform all activities required for life. Cells are either prokaryotic or eukaryotic. **Eukaryotic cells** contain membrane-enclosed organelles, including a DNA-containing nucleus. **Prokaryotic cells** lack such organelles.

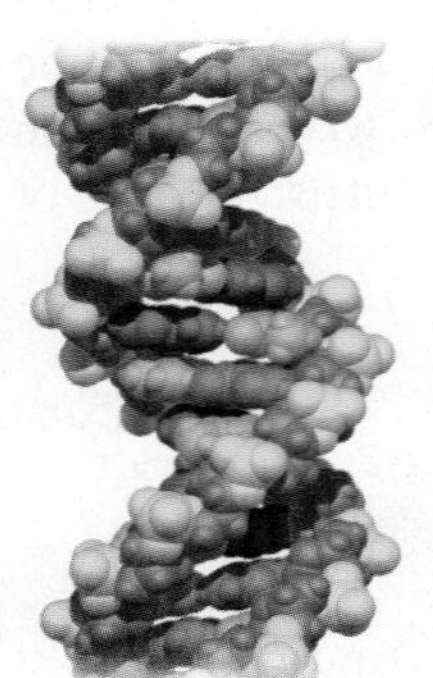

- **Theme: The continuity of life is based on heritable information in the form of DNA**
  Genetic information is encoded in the nucleotide sequences of **DNA**. It is DNA that transmits heritable information from parents to offspring. DNA sequences program a cell's protein production by being transcribed into RNA and then translated into specific proteins, a process called **gene expression**. Gene expression also results in RNAs that are not translated into protein but serve other important functions. **Genomics** is the large-scale analysis of the DNA sequences within a species as well as the comparison of sequences between species.

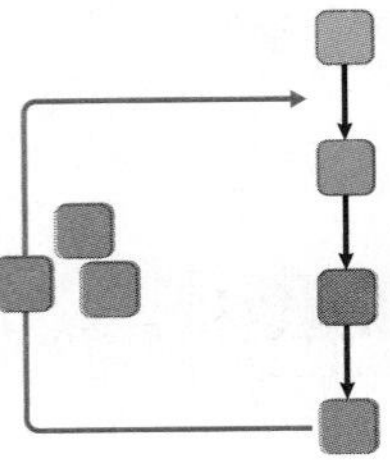

- **Theme: Feedback mechanisms regulate biological systems**
  In **negative feedback**, accumulation of an end product slows the process that makes that product. In **positive feedback**, the end product stimulates the production of more product. Feedback is a type of regulation common to life at all levels, from molecules to ecosystems.

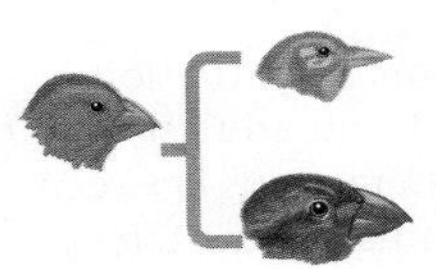

- **Evolution, the Overarching Theme of Biology**
  Evolution accounts for the unity and diversity of life and also for the match of organisms to their environments.

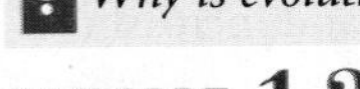

? *Why is evolution considered the core theme of biology?*

CONCEPT 1.2

The Core Theme: Evolution accounts for the unity and diversity of life (pp. 11–18)

- Biologists classify species according to a system of broader and broader groups. Domain **Bacteria** and domain **Archaea** consist of prokaryotes. Domain **Eukarya**, the eukaryotes, includes various groups of protists and the kingdoms Plantae, Fungi, and Animalia. As diverse as life is, there is also evidence of remarkable unity, which is revealed in the similarities between different kinds of organisms.
- Darwin proposed **natural selection** as the mechanism for evolutionary adaptation of populations to their environments.

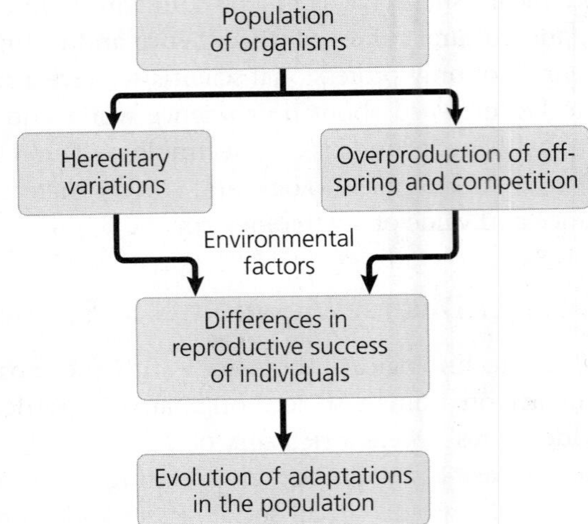

- Each species is one twig of a branching tree of life extending back in time through ancestral species more and more remote. All of life is connected through its long evolutionary history.

? *How could natural selection have led to the evolution of adaptations such as the thick, water-conserving leaves of the mother-of-pearl plant on the cover of this book?*

CONCEPT 1.3

In studying nature, scientists make observations and then form and test hypotheses (pp. 18–23)

- In scientific **inquiry**, scientists make observations (collect **data**) and use **inductive reasoning** to draw a general conclusion, which can be developed into a testable **hypothesis**. **Deductive reasoning** makes predictions that can be used to test hypotheses: If a hypothesis is correct, and we test it, then we can expect the predictions to come true. Hypotheses must be testable and falsifiable; science can address neither the possibility of supernatural phenomena nor the validity of religious beliefs.
- **Controlled experiments**, such as the study investigating mimicry in snake populations, are designed to demonstrate the effect of one variable by testing control groups and experimental groups that differ in only that one variable.
- A scientific **theory** is broad in scope, generates new hypotheses, and is supported by a large body of evidence.

? *What are the roles of inductive and deductive reasoning in the process of scientific inquiry?*

CONCEPT 1.4

Science benefits from a cooperative approach and diverse viewpoints (pp. 23–25)

- Science is a social activity. The work of each scientist builds on the work of others that have come before. Scientists must be able to repeat each other's results, so integrity is key. Biologists approach questions at different levels; their approaches complement each other.
- **Technology** is a method or device that applies scientific knowledge for some specific purpose that affects society. The ultimate impact of basic research is not always immediately obvious.
- Diversity among scientists promotes progress in science.

? *Explain why different approaches and diverse backgrounds among scientists are important.*

TEST YOUR UNDERSTANDING

LEVEL 1: KNOWLEDGE/COMPREHENSION

1. All the organisms on your campus make up
   a. an ecosystem.
   b. a community.
   c. a population.
   d. an experimental group.
   e. a taxonomic domain.

2. Which of the following is a correct sequence of levels in life's hierarchy, proceeding downward from an individual animal?
   a. brain, organ system, nerve cell, nervous tissue
   b. organ system, nervous tissue, brain
   c. organism, organ system, tissue, cell, organ
   d. nervous system, brain, nervous tissue, nerve cell
   e. organ system, tissue, molecule, cell

3. Which of the following is *not* an observation or inference on which Darwin's theory of natural selection is based?
   a. Poorly adapted individuals never produce offspring.
   b. There is heritable variation among individuals.
   c. Because of overproduction of offspring, there is competition for limited resources.
   d. Individuals whose inherited characteristics best fit them to the environment will generally produce more offspring.
   e. A population can become adapted to its environment over time.

4. Systems biology is mainly an attempt to
   a. analyze genomes from different species.
   b. simplify complex problems by reducing the system into smaller, less complex units.
   c. understand the behavior of entire biological systems.
   d. build high-throughput machines for the rapid acquisition of biological data.
   e. speed up the technological application of scientific knowledge.

5. Protists and bacteria are grouped into different domains because
   a. protists eat bacteria.
   b. bacteria are not made of cells.
   c. protists have a membrane-bounded nucleus, which bacterial cells lack.
   d. bacteria decompose protists.
   e. protists are photosynthetic.

6. Which of the following best demonstrates the unity among all organisms?
   a. matching DNA nucleotide sequences
   b. descent with modification
   c. the structure and function of DNA
   d. natural selection
   e. emergent properties

7. A controlled experiment is one that
   a. proceeds slowly enough that a scientist can make careful records of the results.
   b. tests experimental and control groups in parallel.
   c. is repeated many times to make sure the results are accurate.
   d. keeps all variables constant.
   e. is supervised by an experienced scientist.

8. Which of the following statements best distinguishes hypotheses from theories in science?
   a. Theories are hypotheses that have been proved.
   b. Hypotheses are guesses; theories are correct answers.
   c. Hypotheses usually are relatively narrow in scope; theories have broad explanatory power.
   d. Hypotheses and theories are essentially the same thing.
   e. Theories are proved true; hypotheses are often falsified.

LEVEL 2: APPLICATION/ANALYSIS

9. Which of the following is an example of qualitative data?
   a. The temperature decreased from 20°C to 15°C.
   b. The plant's height is 25 centimeters (cm).
   c. The fish swam in a zigzag motion.
   d. The six pairs of robins hatched an average of three chicks.
   e. The contents of the stomach are mixed every 20 seconds.

10. Which of the following best describes the logic of scientific inquiry?
    a. If I generate a testable hypothesis, tests and observations will support it.
    b. If my prediction is correct, it will lead to a testable hypothesis.
    c. If my observations are accurate, they will support my hypothesis.
    d. If my hypothesis is correct, I can expect certain test results.
    e. If my experiments are set up right, they will lead to a testable hypothesis.

11. **DRAW IT** With rough sketches, draw a biological hierarchy similar to the one in Figure 1.4 but using a coral reef as the ecosystem, a fish as the organism, its stomach as the organ, and DNA as the molecule. Include all levels in the hierarchy.

LEVEL 3: SYNTHESIS/EVALUATION

12. **EVOLUTION CONNECTION**
    A typical prokaryotic cell has about 3,000 genes in its DNA, while a human cell has about 20,500 genes. About 1,000 of these genes are present in both types of cells. Based on your understanding of evolution, explain how such different organisms could have this same subset of genes. What sorts of functions might these shared genes have?

13. **SCIENTIFIC INQUIRY**
    Based on the results of the snake mimicry case study, suggest another hypothesis researchers might use to extend the investigation.

14. **WRITE ABOUT A THEME**
    Evolution In a short essay (100–150 words), discuss Darwin's view of how natural selection resulted in both unity and diversity of life on Earth. Include in your discussion some of his evidence. (See p. xv for a suggested grading rubric. The rubric and tips for writing good essays can also be found in the Study Area of MasteringBiology.)

For selected answers, see Appendix A.

MasteringBIOLOGY www.masteringbiology.com

1. MasteringBiology® Assignments

Experimental Inquiry Tutorial What Can You Learn About the Process of Science from Investigating a Cricket's Chirp?
Tutorial The Scientific Method
Activities The Levels of Life Card Game • Form Fits Function: Cells • Heritable Information: DNA • Introduction to Experimental Design • GraphIt!: An Introduction to Graphing
Questions Student Misconceptions • Reading Quiz • Multiple Choice • End-of-Chapter

2. eText

Read your book online, search, take notes, highlight text, and more.

3. The Study Area

Practice Tests • Cumulative Test •  *BioFlix* 3-D Animations • MP3 Tutor Sessions • Videos • Activities • Investigations • Lab Media • Audio Glossary • Word Study Tools • Art

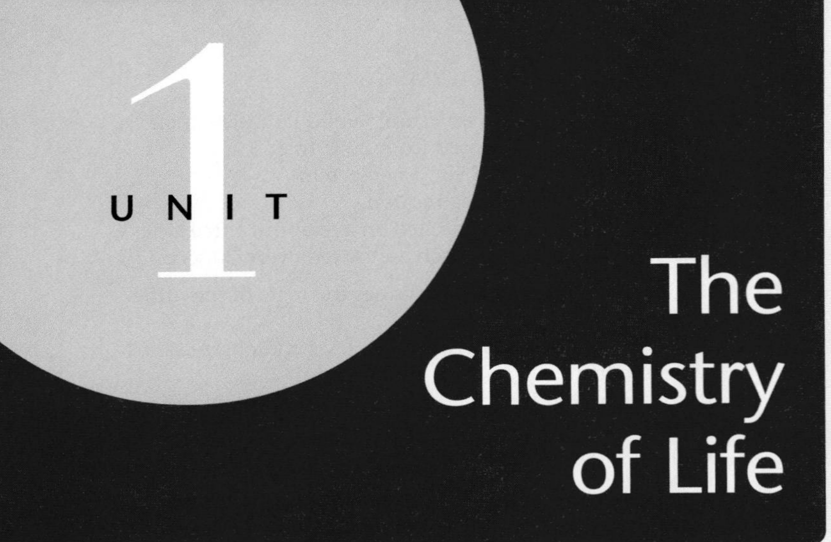

The Chemistry of Life

An Interview with Susan Solomon

Although Susan Solomon is not a biologist, her research as an atmospheric chemist has profound implications for life on Earth. Since earning degrees from the Illinois Institute of Technology and the University of California, Berkeley, Dr. Solomon has been a leader in determining the cause of the Antarctic ozone hole and in producing the 2007 report of the United Nations Intergovernmental Panel on Climate Change (IPCC), which concluded that warming of Earth's climate is unequivocal. These activities have given her a public role in communicating science to policymakers and society at large. In recognition of her scientific accomplishments, she has been awarded the U.S. National Medal of Science, the Blue Planet Prize, and, with Al Gore and the other IPCC members, the Nobel Peace Prize. A member of the U.S. National Academy of Sciences, the European Academy of Sciences, the Academy of Sciences of France, and the Royal Society of the United Kingdom, she works for the National Oceanic and Atmospheric Administration in Boulder, Colorado.

How is Earth's atmosphere important to life?

Life on Earth today could not have evolved without an atmosphere. We all know that we and many other organisms require oxygen (O_2) from the atmosphere, and plants use carbon dioxide (CO_2) to grow. The atmosphere also contains a form of oxygen called ozone that has three oxygen atoms (O_3) instead of two. Organisms would never have been able to leave the ocean and survive on land without the development of an ozone layer in the upper atmosphere. Ozone has the important property of absorbing ultraviolet (UV) light, which would otherwise cause DNA damage. Damage from UV can lead to skin cancer and cataracts; it can also harm many crops and even phytoplankton [small photosynthetic aquatic organisms].

Early in your career, you led an expedition to make measurements of the atmosphere in Antarctica. Tell us about that.

In 1985, the British Antarctic Survey reported a surprising discovery: that the springtime ozone at their station in Antarctica had fallen by 30–50% since the late 1970s, resulting in an "ozone hole"! People had begun to be worried about whether the ozone layer might be vulnerable to changes caused by human activity, but only very minor changes had been expected. In 1986, I had the chance to lead a new Antarctic expedition to help confirm the British data and to study the problem further. We didn't just measure ozone; we measured about a dozen other atmospheric molecules that allowed us to tell *why* the ozone was being destroyed.

What did you find out?

It turns out that the ozone chemistry in Antarctica is extremely different from what it is anywhere else. That's because Antarctica is very cold—it really is the coldest place on Earth. It's so cold that clouds form in its upper stratosphere, about 10–30 kilometers above sea level, and those clouds help convert chemicals from chlorofluorocarbons (CFCs) to ozone-damaging substances.

CFCs are synthetic compounds, made only by humans. They were used back then for a variety of purposes—for example, in refrigeration, as solvents, and as propellants for sprays. Many tons of CFCs were emitted every year. I came up with the idea that the reason an ozone hole developed in Antarctica had to do with chemical reactions that happen between a gas and a surface and that the surface in this case was the small particles that make up those stratospheric clouds. Our data supported this hypothesis. The reactions on those particles make the CFCs hundreds of times more damaging than they would be otherwise. The absence of such clouds in most other parts of the world is why we don't have ozone holes elsewhere, although stratospheric clouds form occasionally in the Arctic and there is significant ozone loss there. Scientists had been concerned since the mid-1970s that human use of CFCs might cause some ozone depletion, but they had expected a loss of only about 3–5% in 100 years.

How do CFCs destroy ozone?

When CFCs arrive at the stratosphere, which typically takes a few years, high-energy radiation up there can break them down, releasing chlorine atoms. The chlorine atoms destroy the ozone catalytically, which means that the atoms don't get used up in the process. So even if only a small amount of CFC is broken down, the tiny bit of chlorine produced can destroy an enormous amount of ozone.

In the U.N. Montreal Protocol of 1987, the nations of the world agreed to stop producing CFCs. However, the CFCs in the atmosphere disappear only very slowly; typically they hang around for 50 to 100 years. What that means is that the CFCs we've already put in the atmosphere will continue to produce an ozone hole for many decades to come, even though we're not using these substances any more. Global emissions are very near zero now, and we're beginning to see the ozone hole slowly diminish. But it will probably not go away completely until around 2060.

While the ozone hole remains, it continues to cause damage. For example, there is evidence that the phytoplankton in the Antarctic Ocean are being affected by increased UV, and the phytoplankton are the base of the main Antarctic food chain: They feed the krill, which feed the fish, which feed the penguins, seals, and whales.

Let's talk about an effect that other atmospheric changes are having—climate change.

There's no question that the planet is getting warmer. We know that, on average, our planet is about 1.4°F (0.8°C) warmer than it was 100 years ago, and this past decade has been the warmest decade in at least the last 100 years. We also know that glaciers worldwide are retreating and that sea level is rising. There's a breadth of scientific data, acquired by different techniques, that tells us that global warming really is unequivocal.

The warming has to do with the greenhouse effect, right?

We're lucky that this planet has a greenhouse effect, because if it didn't, we would be very cold indeed! Our planet is heated by the sun, and much of the infrared radiation (heat) that would otherwise be released back into space is trapped by "greenhouse gases" in the atmosphere. This makes the planet about 30°C hotter than it would be otherwise. But of course anything can be bad if you have too much of it, and what we're doing now is increasing the greenhouse effect of our atmosphere beyond its natural state. If we keep emitting the greenhouse gases that are causing the warming, then we will see some very significant warming in the coming century.

The main greenhouse gas we're adding to the atmosphere is CO_2, from burning fossil fuel and to a lesser extent deforestation. We have increased the atmospheric concentration of CO_2 by about 30% compared to any value that has been found for the last 800,000 years. This has been determined by digging up ice cores in Antarctica and measuring the gases in the air bubbles trapped in the ice. So we know that we have perturbed the atmosphere in a way that the planet hasn't seen in at least 800,000 years.

The CFCs we discussed earlier are actually the third most important greenhouse gas at present, after CO_2 and methane. Pound for pound, CFCs are much more potent as greenhouse gases than CO_2. The phase-out of CFCs since the signing of the Montreal Protocol has not only avoided a lot of ozone destruction that would otherwise have happened, but has also reduced our input of gases that cause climate change.

How is life on Earth being affected by climate change?
There are some things that we can already begin to see and talk about, but there's an enormous amount that we still don't know. We do know that the oceans are getting more acidic because CO_2 is taken up by the ocean and converted to carbonic acid, which can affect the ability of shellfish to make their shells. Other ocean life is also likely to be harmed by the increased acidity, such as the organisms of coral reefs. But there's also emerging evidence that some other marine organisms may do better—lobsters, perhaps.

As a westerner I'm extremely concerned about the greatly increased population of pine beetles in the western United States. These beetles are killing pine trees in unprecedented numbers. There's good evidence that a contributing factor to this explosion of pine beetles is global warming. I think we're going to see more of this kind of thing. Also, it is clear that bird migration is already being affected by global warming. Whether global warming will lead to extinction of some animals is an important question. The signature extinction issue is the polar bear; as the sea ice of the Arctic decreases, the polar bear could become extinct. We don't really know yet how much biological adaptation is possible in the time available. We'll probably find out that there are some winner species out there and some loser species. In agriculture, many crops are sensitive to increasing temperatures. One of the relevant findings about corn is that for every degree of warming, about 10% of crop production is lost—a big change.

Does less precipitation always go along with higher temperatures?
In some places there will probably be less precipitation and in other places more. There's a band of subtropical and tropical regions where we are pretty confident that it will get drier—for example, Mexico, the Mediterranean region, parts of Australia. In the higher latitudes, places like Canada and Norway will likely get wetter. In between, it's harder to predict.

Tell us about the IPCC and your work on it.
The IPCC is fundamentally a mechanism for the communication of information about climate change from the science community to the policy community. It was set up in 1988 when people were beginning to recognize that climate change was a real possibility. Policymakers decided that they needed to get reliable scientific information so they could begin to talk about what to do, if anything. Every six or seven years, scientists are asked by their governments to get together and assess what we know and don't know on the basis of the published scientific literature.

I have been involved in the IPCC since 1992, and in 2001 I was elected by the panel, representing over 100 governments, to co-chair the scientific assessment team. In a process lasting several years, we generated a detailed report summarizing the state of climate science. Our report was then reviewed by dozens of governments and more than 600 scientists. The report itself and every one of their 30,000 comments are available on the Internet. We refined and refined the draft in consideration of those comments and finalized the document in 2007.

What were the main conclusions of your 2007 report?
The first conclusion, based on many independent lines of evidence, was that the Earth is warming. There's no doubt we are now living on a planet that is warmer than a century ago. The second main conclusion was that most—more than half—of the warming is very likely due to increases in greenhouse gases, primarily CO_2. We did a careful analysis of the uncertainties: When we say "very likely," we mean that there's a 90% chance or greater that most of the warming is due to emissions of greenhouse gases by human activity.

What have you learned about working at the interface of science and policy?
It's one of the most difficult things a scientist can do. Science normally takes us into a laboratory or out into the field or into scholarly discussions with colleagues. Getting involved with policy is quite different: It takes us out of the lab and makes us much more aware of the strong emotions around many issues. In that sense, it's a bit daunting. But it's uplifting to see how valuable science can be in helping society make more informed choices. Scientists can help make sure that whatever it is we choose to do as a society we're doing knowingly, not in ignorance. I appreciate all the reasons why people ask tough questions about the science. How much do we know? What really are the uncertainties? Yes, there's a lot at stake here, all the more reason why there has to be really good science going into it.

"There's a breadth of scientific data, acquired by different techniques, that tells us that global warming really is unequivocal."

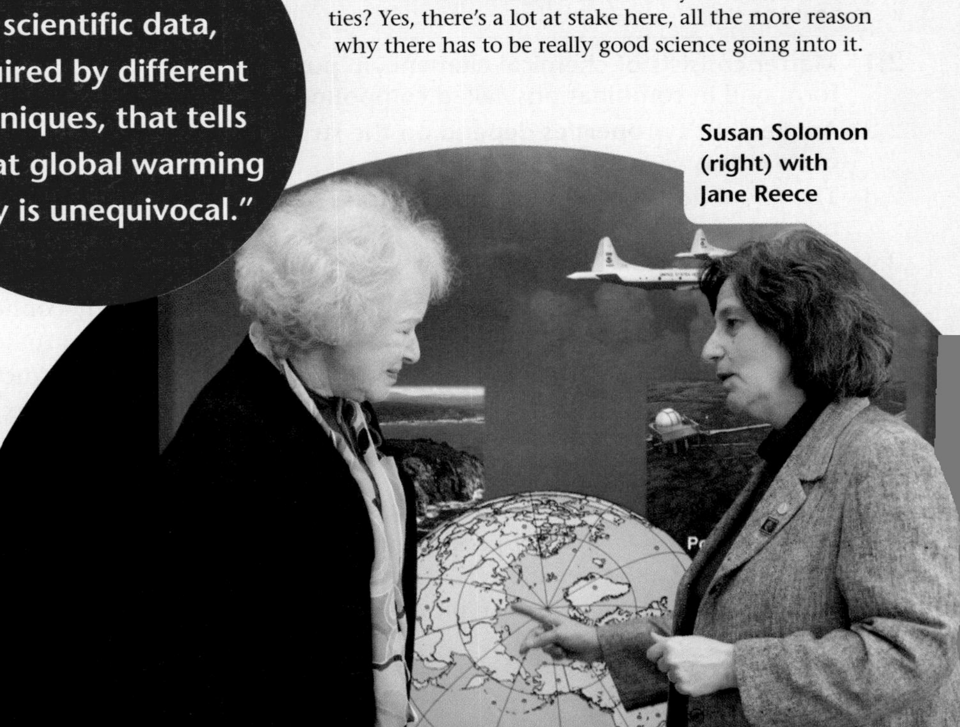

Susan Solomon (right) with Jane Reece

2

The Chemical Context of Life

▲ **Figure 2.1 Who tends this "garden"?**

KEY CONCEPTS

2.1 **Matter consists of chemical elements in pure form and in combinations called compounds**

2.2 **An element's properties depend on the structure of its atoms**

2.3 **The formation and function of molecules depend on chemical bonding between atoms**

2.4 **Chemical reactions make and break chemical bonds**

OVERVIEW

A Chemical Connection to Biology

The Amazon rain forest in South America is a showcase for the diversity of life on Earth. Colorful birds, insects, and other animals live in a densely-packed environment of trees, shrubs, vines, and wildflowers, and an excursion along a waterway or a forest path typically reveals a lush variety of plant life. Visitors traveling near the Amazon's headwaters in Peru are therefore surprised to come across tracts of forest like that seen in the foreground of the photo in **Figure 2.1**. This patch is almost completely dominated by a single plant species—a small flowering tree called *Duroia hirsuta*. Travelers may wonder if the plot of land is planted and maintained by local people, but the indigenous people are as mystified as the visitors. They call these stands of *Duroia* trees "devil's gardens," from a legend attributing them to an evil forest spirit.

Seeking a scientific explanation, a research team at Stanford University recently solved the "devil's garden" mystery. **Figure 2.2** describes their main experiment. The researchers showed that the "farmers" who create and maintain these gardens are actually ants that live in the hollow stems of the *Duroia* trees. The ants do not plant the *Duroia* trees, but they prevent other plant species from growing in the garden by injecting intruders with a poisonous chemical. In this way, the ants create space for the growth of the *Duroia* trees that serve as their home. With the ability to maintain and expand its habitat, a single colony of devil's garden ants can live for hundreds of years.

The chemical used by the ants to weed their garden turns out to be formic acid. This substance is produced by many species of ants and in fact got its name from the Latin word for ant, *formica*. For many ant species, the formic acid probably serves as a disinfectant that protects the ants against microbial parasites. The devil's garden ant is the first ant species found to use formic acid as an herbicide, an important addition to the list of functions mediated by chemicals in the insect world. Scientists have long known that chemicals play a major role in insect communication, attraction of mates, and defense against predators.

Research on devil's gardens is only one example of the relevance of chemistry to the study of life. Unlike a list of college courses, nature is not neatly packaged into the individual natural sciences—biology, chemistry, physics, and so forth. Biologists specialize in the study of life, but organisms and their environments are natural systems to which the concepts of chemistry and physics apply. Biology is a multidisciplinary science.

This unit of chapters introduces some basic concepts of chemistry that apply to the study of life. We will make many connections to the themes introduced in Chapter 1. One of these themes is the organization of life into a hierarchy of structural levels, with additional properties emerging at each successive level. In this unit, we will see how emergent properties are apparent at the lowest levels of biological organization—such as the ordering of atoms into molecules and the interactions of those molecules within cells. Somewhere in the transition from molecules to cells, we will cross the blurry boundary between nonlife and life. This chapter focuses on the chemical components that make up all matter.

▼ **Figure 2.2** **INQUIRY**

What creates "devil's gardens" in the rain forest?

EXPERIMENT Working under Deborah Gordon and with Michael Greene, graduate student Megan Frederickson sought the cause of "devil's gardens," stands of a single species of tree, *Duroia hirsuta*. One hypothesis was that ants living in these trees, *Myrmelachista schumanni*, produce a poisonous chemical that kills trees of other species; another was that the *Duroia* trees themselves kill competing trees, perhaps by means of a chemical.

To test these hypotheses, Frederickson did field experiments in Peru. Two saplings of a local nonhost tree species, *Cedrela odorata*, were planted inside each of ten devil's gardens. At the base of one sapling, a sticky insect barrier was applied; the other was unprotected. Two more *Cedrela* saplings, with and without barriers, were planted about 50 meters outside each garden.

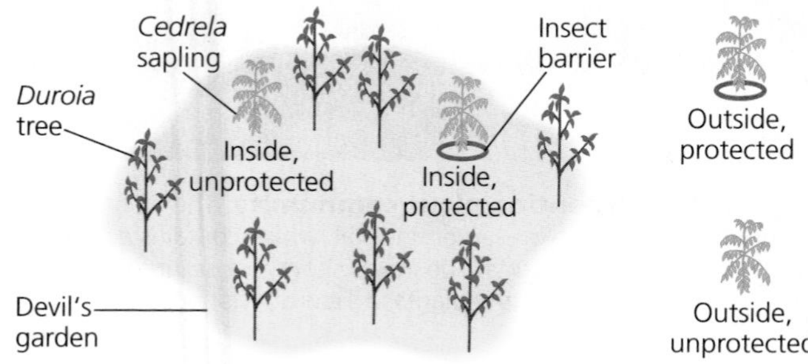

The researchers observed ant activity on the *Cedrela* leaves and measured areas of dead leaf tissue after one day. They also chemically analyzed contents of the ants' poison glands.

RESULTS The ants made injections from the tips of their abdomens into leaves of unprotected saplings in their gardens (see photo). Within one day, these leaves developed dead areas (see graph). The protected saplings were uninjured, as were the saplings planted outside the gardens. Formic acid was the only chemical detected in the poison glands of the ants.

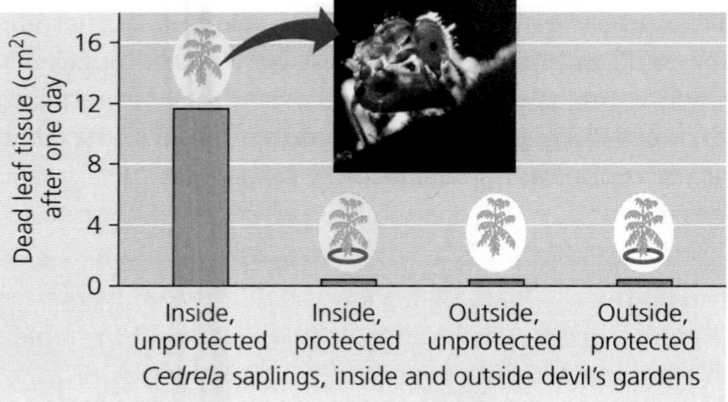

CONCLUSION Ants of the species *Myrmelachista schumanni* kill nonhost trees by injecting the leaves with formic acid, thus creating hospitable habitats (devil's gardens) for the ant colony.

SOURCE M. E. Frederickson, M. J. Greene, and D. M. Gordon, "Devil's gardens" bedevilled by ants, *Nature* 437:495–496 (2005).

INQUIRY IN ACTION Read and analyze the original paper in *Inquiry in Action: Interpreting Scientific Papers*.

WHAT IF? What would be the results if the unprotected saplings' inability to grow in the devil's gardens was caused by a chemical released by the *Duroia* trees rather than by the ants?

CONCEPT **2.1**

Matter consists of chemical elements in pure form and in combinations called compounds

Organisms are composed of **matter**, which is defined as anything that takes up space and has mass.* Matter exists in many diverse forms. Rocks, metals, oils, gases, and humans are just a few examples of what seems an endless assortment of matter.

Elements and Compounds

Matter is made up of elements. An **element** is a substance that cannot be broken down to other substances by chemical reactions. Today, chemists recognize 92 elements occurring in nature; gold, copper, carbon, and oxygen are examples. Each element has a symbol, usually the first letter or two of its name. Some symbols are derived from Latin or German; for instance, the symbol for sodium is Na, from the Latin word *natrium*.

A **compound** is a substance consisting of two or more different elements combined in a fixed ratio. Table salt, for example, is sodium chloride (NaCl), a compound composed of the elements sodium (Na) and chlorine (Cl) in a 1:1 ratio. Pure sodium is a metal, and pure chlorine is a poisonous gas. When chemically combined, however, sodium and chlorine form an edible compound. Water (H_2O), another compound, consists of the elements hydrogen (H) and oxygen (O) in a 2:1 ratio. These are simple examples of organized matter having emergent properties: A compound has characteristics different from those of its elements **(Figure 2.3)**.

▲ **Figure 2.3 The emergent properties of a compound.** The metal sodium combines with the poisonous gas chlorine, forming the edible compound sodium chloride, or table salt.

*Sometimes we substitute the term weight for mass, although the two are not identical. Mass is the amount of matter in an object, whereas the weight of an object is how strongly that mass is pulled by gravity. The weight of an astronaut walking on the moon is approximately 1/6 the astronaut's weight on Earth, but his or her mass is the same. However, as long as we are earthbound, the weight of an object is a measure of its mass; in everyday language, therefore, we tend to use the terms interchangeably.

The Elements of Life

Of the 92 natural elements, about 20–25% are **essential elements** that an organism needs to live a healthy life and reproduce. The essential elements are similar among organisms, but there is some variation—for example, humans need 25 elements, but plants need only 17.

Just four elements—oxygen (O), carbon (C), hydrogen (H), and nitrogen (N)—make up 96% of living matter. Calcium (Ca), phosphorus (P), potassium (K), sulfur (S), and a few other elements account for most of the remaining 4% of an organism's mass. **Trace elements** are required by an organism in only minute quantities. Some trace elements, such as iron (Fe), are needed by all forms of life; others are required only by certain species. For example, in vertebrates (animals with backbones), the element iodine (I) is an essential ingredient of a hormone produced by the thyroid gland. A daily intake of only 0.15 milligram (mg) of iodine is adequate for normal activity of the human thyroid. An iodine deficiency in the diet causes the thyroid gland to grow to abnormal size, a condition called goiter. Where it is available, eating seafood or iodized salt reduces the incidence of goiter. All the elements needed by the human body are listed in **Table 2.1**.

Some naturally occurring elements are toxic to organisms. In humans, for instance, the element arsenic has been linked to numerous diseases and can be lethal. In some areas of the world, arsenic occurs naturally and can make its way into the groundwater. As a result of using water from drilled wells in southern Asia, millions of people have been inadvertently exposed to arsenic-laden water. Efforts are under way to reduce arsenic levels in their water supply.

Table 2.1 Elements in the Human Body

Element	Symbol	Percentage of Body Mass (including water)	
Oxygen	O	65.0%	96.3%
Carbon	C	18.5%	
Hydrogen	H	9.5%	
Nitrogen	N	3.3%	
Calcium	Ca	1.5%	3.7%
Phosphorus	P	1.0%	
Potassium	K	0.4%	
Sulfur	S	0.3%	
Sodium	Na	0.2%	
Chlorine	Cl	0.2%	
Magnesium	Mg	0.1%	

Trace elements (less than 0.01% of mass): Boron (B), chromium (Cr), cobalt (Co), copper (Cu), fluorine (F), iodine (I), iron (Fe), manganese (Mn), molybdenum (Mo), selenium (Se), silicon (Si), tin (Sn), vanadium (V), zinc (Zn)

▲ **Figure 2.4 Serpentine plant community.** The plants in the large photo are growing on serpentine soil, which contains elements that are usually toxic to plants. The insets show a close-up of serpentine rock and one of the plants, a Tiburon Mariposa lily.

Case Study: Evolution of Tolerance to Toxic Elements

EVOLUTION Some species have become adapted to environments containing elements that are usually toxic. A compelling example is found in serpentine plant communities. Serpentine is a jade-like mineral that contains toxic elements such as chromium, nickel, and cobalt. Although most plants cannot survive in soil that forms from serpentine rock, a small number of plant species have adaptations that allow them to do so (**Figure 2.4**). Presumably, variants of ancestral, nonserpentine species arose that could survive in serpentine soils, and subsequent natural selection resulted in the distinctive array of species we see in these areas today.

CONCEPT CHECK 2.1

1. **MAKE CONNECTIONS** Review the discussion of emergent properties in Chapter 1 (p. 3). Explain how table salt has emergent properties.
2. Is a trace element an essential element? Explain.
3. In humans, iron is a trace element required for the proper functioning of hemoglobin, the molecule that carries oxygen in red blood cells. What might be the effects of an iron deficiency?
4. **MAKE CONNECTIONS** Review the discussion of natural selection in Chapter 1 (pp. 14–16) and explain how natural selection might have played a role in the evolution of species that are tolerant of serpentine soils.

For suggested answers, see Appendix A.

CONCEPT 2.2

An element's properties depend on the structure of its atoms

Each element consists of a certain type of atom that is different from the atoms of any other element. An **atom** is the smallest unit of matter that still retains the properties of an element. Atoms are so small that it would take about a million of them to stretch across the period printed at the end of this sentence. We symbolize atoms with the same abbreviation used for the element that is made up of those atoms. For example, the symbol C stands for both the element carbon and a single carbon atom.

Subatomic Particles

Although the atom is the smallest unit having the properties of an element, these tiny bits of matter are composed of even smaller parts, called *subatomic particles*. Physicists have split the atom into more than a hundred types of particles, but only three kinds of particles are relevant here: **neutrons**, **protons**, and **electrons**. Protons and electrons are electrically charged. Each proton has one unit of positive charge, and each electron has one unit of negative charge. A neutron, as its name implies, is electrically neutral.

Protons and neutrons are packed together tightly in a dense core, or **atomic nucleus**, at the center of an atom; protons give the nucleus a positive charge. The electrons form a sort of cloud of negative charge around the nucleus, and it is the attraction between opposite charges that keeps the electrons in the vicinity of the nucleus. **Figure 2.5** shows two commonly used models of the structure of the helium atom as an example.

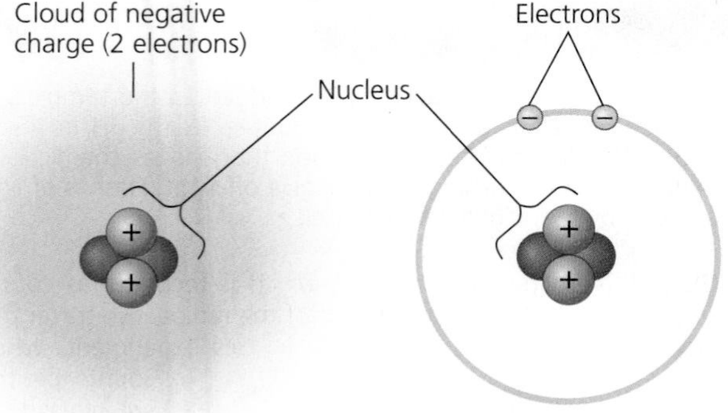

(a) This model represents the two electrons as a cloud of negative charge.

(b) In this more simplified model, the electrons are shown as two small yellow spheres on a circle around the nucleus.

▲ **Figure 2.5 Simplified models of a helium (He) atom.** The helium nucleus consists of 2 neutrons (brown) and 2 protons (pink). Two electrons (yellow) exist outside the nucleus. These models are not to scale; they greatly overestimate the size of the nucleus in relation to the electron cloud.

The neutron and proton are almost identical in mass, each about 1.7×10^{-24} gram (g). Grams and other conventional units are not very useful for describing the mass of objects so minuscule. Thus, for atoms and subatomic particles (and for molecules, too), we use a unit of measurement called the **dalton**, in honor of John Dalton, the British scientist who helped develop atomic theory around 1800. (The dalton is the same as the *atomic mass unit*, or *amu*, a unit you may have encountered elsewhere.) Neutrons and protons have masses close to 1 dalton. Because the mass of an electron is only about 1/2,000 that of a neutron or proton, we can ignore electrons when computing the total mass of an atom.

Atomic Number and Atomic Mass

Atoms of the various elements differ in their number of subatomic particles. All atoms of a particular element have the same number of protons in their nuclei. This number of protons, which is unique to that element, is called the **atomic number** and is written as a subscript to the left of the symbol for the element. The abbreviation $_2He$, for example, tells us that an atom of the element helium has 2 protons in its nucleus. Unless otherwise indicated, an atom is neutral in electrical charge, which means that its protons must be balanced by an equal number of electrons. Therefore, the atomic number tells us the number of protons and also the number of electrons in an electrically neutral atom.

We can deduce the number of neutrons from a second quantity, the **mass number**, which is the sum of protons plus neutrons in the nucleus of an atom. The mass number is written as a superscript to the left of an element's symbol. For example, we can use this shorthand to write an atom of helium as ^4_2He. Because the atomic number indicates how many protons there are, we can determine the number of neutrons by subtracting the atomic number from the mass number: The helium atom, ^4_2He, has 2 neutrons. For sodium (Na):

Mass number = number of protons + neutrons
= 23 for sodium

$^{23}_{11}Na$

Atomic number = number of protons
= number of electrons in a neutral atom
= 11 for sodium

Number of neutrons = mass number − atomic number
= 23 − 11 = 12 for sodium

The simplest atom is hydrogen, ^1_1H, which has no neutrons; it consists of a single proton with a single electron.

As mentioned earlier, the contribution of electrons to mass is negligible. Therefore, almost all of an atom's mass is concentrated in its nucleus. Because neutrons and protons each have a mass very close to 1 dalton, the mass number is an approximation of the total mass of an atom, called its **atomic mass**. So we might say that the atomic mass of sodium ($^{23}_{11}Na$) is 23 daltons, although more precisely it is 22.9898 daltons.

Isotopes

All atoms of a given element have the same number of protons, but some atoms have more neutrons than other atoms of the same element and therefore have greater mass. These different atomic forms of the same element are called **isotopes** of the element. In nature, an element occurs as a mixture of its isotopes. For example, consider the three isotopes of the element carbon, which has the atomic number 6. The most common isotope is carbon-12, $^{12}_{6}C$, which accounts for about 99% of the carbon in nature. The isotope $^{12}_{6}C$ has 6 neutrons. Most of the remaining 1% of carbon consists of atoms of the isotope $^{13}_{6}C$, with 7 neutrons. A third, even rarer isotope, $^{14}_{6}C$, has 8 neutrons. Notice that all three isotopes of carbon have 6 protons; otherwise, they would not be carbon. Although the isotopes of an element have slightly different masses, they behave identically in chemical reactions. (The number usually given as the atomic mass of an element, such as 22.9898 daltons for sodium, is actually an average of the atomic masses of all the element's naturally occurring isotopes.)

Both ^{12}C and ^{13}C are stable isotopes, meaning that their nuclei do not have a tendency to lose particles. The isotope ^{14}C, however, is unstable, or radioactive. A **radioactive isotope** is one in which the nucleus decays spontaneously, giving off particles and energy. When the decay leads to a change in the number of protons, it transforms the atom to an atom of a different element. For example, when a radioactive carbon atom decays, it becomes an atom of nitrogen.

Radioactive isotopes have many useful applications in biology. In Chapter 25, you will learn how researchers use measurements of radioactivity in fossils to date these relics of past life. As shown in **Figure 2.6**, radioactive isotopes are also useful as tracers to follow atoms through metabolism, the chemical processes of an organism. Cells use the radioactive atoms as they would use nonradioactive isotopes of the same element, but the radioactive tracers can be readily detected.

Radioactive tracers are important diagnostic tools in medicine. For example, certain kidney disorders can be diagnosed by injecting small doses of substances containing radioactive isotopes into the blood and then measuring the amount of tracer excreted in the urine. Radioactive tracers are also used in combination with sophisticated imaging instruments. PET scanners, for instance, can monitor chemical processes, such as those involved in cancerous growth, as they actually occur in the body (**Figure 2.7**).

Although radioactive isotopes are very useful in biological research and medicine, radiation from decaying isotopes also poses a hazard to life by damaging cellular molecules. The severity of this damage depends on the type and amount of radiation an organism absorbs. One of the most serious environmental threats is radioactive fallout from nuclear accidents. The doses of most isotopes used in medical diagnosis, however, are relatively safe.

▼ **Figure 2.6** **RESEARCH METHOD**

Radioactive Tracers

APPLICATION Scientists use radioactive isotopes to label certain chemical compounds, creating tracers that allow them to follow a metabolic process or locate the compound within an organism. In this example, radioactive tracers are utilized to determine the effect of temperature on the rate at which cells make copies of their DNA.

TECHNIQUE

Compounds including radioactive tracer (bright blue)

Human cells

Incubators

10°C 15°C 20°C 25°C 30°C 35°C 40°C 45°C 50°C

1 Compounds used by cells to make DNA are added to human cells. One ingredient is labeled with ^{3}H, a radioactive isotope of hydrogen. Nine dishes of cells are incubated at different temperatures. The cells make new DNA, incorporating the radioactive tracer.

2 Cells from each incubator are placed in tubes; their DNA is isolated; and unused labeled compounds are removed.

10° 15° 20° 25° 30° 35° 40° 45° 50°

DNA (old and new)

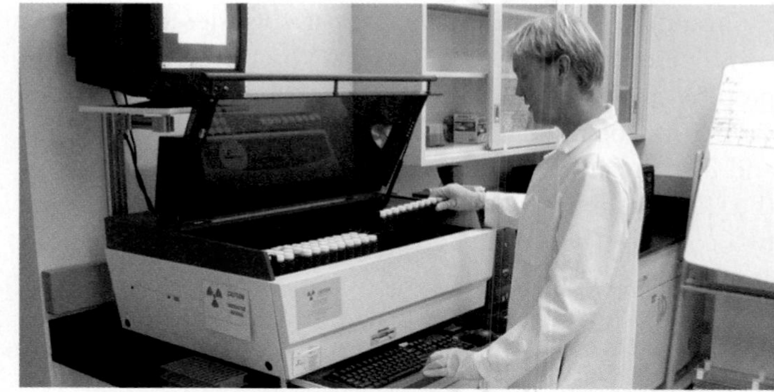

3 A solution called scintillation fluid is added to the samples, which are then placed in a scintillation counter. As the ^{3}H in the newly made DNA decays, it emits radiation that excites chemicals in the scintillation fluid, causing them to give off light. Flashes of light are recorded by the scintillation counter.

RESULTS The frequency of flashes, which is recorded as counts per minute, is proportional to the amount of the radioactive tracer present, indicating the amount of new DNA. In this experiment, when the counts per minute are plotted against temperature, it is clear that temperature affects the rate of DNA synthesis; the most DNA was made at 35°C.

Optimum temperature for DNA synthesis

Counts per minute (× 1,000)

0 10 20 30

10 20 30 40 50

Temperature (°C)

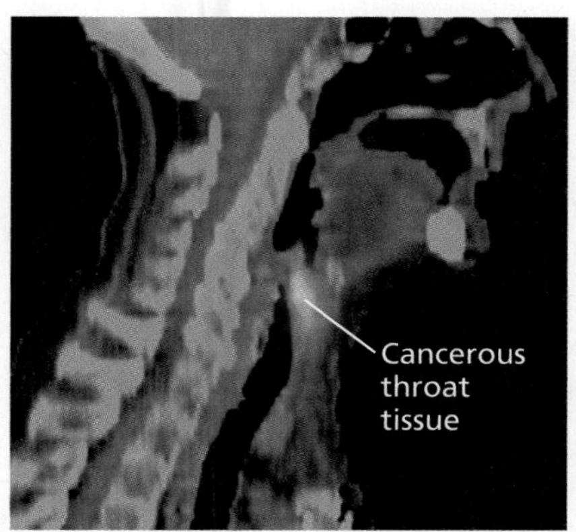

◀ **Figure 2.7 A PET scan, a medical use for radioactive isotopes.** PET, an acronym for positron-emission tomography, detects locations of intense chemical activity in the body. The bright yellow spot marks an area with an elevated level of radioactively labeled glucose, which in turn indicates high metabolic activity, a hallmark of cancerous tissue.

(a) A ball bouncing down a flight of stairs provides an analogy for energy levels of electrons, because the ball can come to rest only on each step, not between steps.

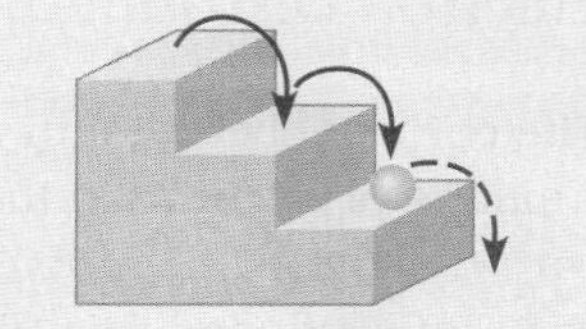

Third shell (highest energy level in this model)

Second shell (higher energy level)

First shell (lowest energy level)

Energy absorbed

Energy lost

Atomic nucleus

(b) An electron can move from one shell to another only if the energy it gains or loses is exactly equal to the difference in energy between the energy levels of the two shells. Arrows in this model indicate some of the stepwise changes in potential energy that are possible.

▲ **Figure 2.8 Energy levels of an atom's electrons.** Electrons exist only at fixed levels of potential energy called electron shells.

The Energy Levels of Electrons

The simplified models of the atom in Figure 2.5 greatly exaggerate the size of the nucleus relative to the volume of the whole atom. If an atom of helium were the size of a typical football stadium, the nucleus would be the size of a pencil eraser in the center of the field. Moreover, the electrons would be like two tiny gnats buzzing around the stadium. Atoms are mostly empty space.

When two atoms approach each other during a chemical reaction, their nuclei do not come close enough to interact. Of the three kinds of subatomic particles we have discussed, only electrons are directly involved in the chemical reactions between atoms.

An atom's electrons vary in the amount of energy they possess. **Energy** is defined as the capacity to cause change—for instance, by doing work. **Potential energy** is the energy that matter possesses because of its location or structure. For example, water in a reservoir on a hill has potential energy because of its altitude. When the gates of the reservoir's dam are opened and the water runs downhill, the energy can be used to do work, such as turning generators. Because energy has been expended, the water has less energy at the bottom of the hill than it did in the reservoir. Matter has a natural tendency to move to the lowest possible state of potential energy; in this example, the water runs downhill. To restore the potential energy of a reservoir, work must be done to elevate the water against gravity.

The electrons of an atom have potential energy because of how they are arranged in relation to the nucleus. The negatively charged electrons are attracted to the positively charged nucleus. It takes work to move a given electron farther away from the nucleus, so the more distant an electron is from the nucleus, the greater its potential energy. Unlike the continuous flow of water downhill, changes in the potential energy of electrons can occur only in steps of fixed amounts. An electron having a certain amount of energy is something like a ball on a staircase **(Figure 2.8a)**. The ball can have different amounts of potential energy, depending on which step it is on, but it cannot spend much time between the steps. Similarly, an electron's potential energy is determined by its energy level. An electron cannot exist between energy levels.

An electron's energy level is correlated with its average distance from the nucleus. Electrons are found in different **electron shells**, each with a characteristic average distance and energy level. In diagrams, shells can be represented by concentric circles **(Figure 2.8b)**. The first shell is closest to the nucleus, and electrons in this shell have the lowest potential energy. Electrons in the second shell have more energy, and electrons in the third shell even more energy. An electron can change the shell it occupies, but only by absorbing or losing an amount of energy equal to the difference in potential energy between its position in the old shell and that in the new shell. When an electron absorbs energy, it moves to a shell farther out from the nucleus. For example, light energy can excite an electron to a higher energy level. (Indeed, this is the first step taken when plants harness the energy of sunlight for photosynthesis, the process that produces food from carbon dioxide and water.) When an electron loses energy, it "falls back" to a shell closer to the nucleus, and the lost energy is usually released to the environment as heat. For example, sunlight excites electrons in the surface of a car to higher energy levels. When the electrons fall back to their original levels, the car's surface heats up. This thermal energy can be transferred to the air or to your hand if you touch the car.

Electron Distribution and Chemical Properties

The chemical behavior of an atom is determined by the distribution of electrons in the atom's electron shells. Beginning with hydrogen, the simplest atom, we can imagine building the atoms of the other elements by adding 1 proton and 1 electron at a time (along with an appropriate number of neutrons). **Figure 2.9**, an abbreviated version of what is called the *periodic table of the elements*, shows this distribution of electrons for the first 18 elements, from hydrogen ($_1$H) to argon ($_{18}$Ar). The elements are arranged in three rows, or periods, corresponding to the number of electron shells in their atoms. The left-to-right sequence of elements in each row corresponds to the sequential addition of electrons and protons. (See Appendix B for the complete periodic table.)

Hydrogen's 1 electron and helium's 2 electrons are located in the first shell. Electrons, like all matter, tend to exist in the lowest available state of potential energy. In an atom, this state is in the first shell. However, the first shell can hold no more than 2 electrons; thus, hydrogen and helium are the only elements in the first row of the table. An atom with more than 2 electrons must use higher shells because the first shell is full. The next element, lithium, has 3 electrons. Two of these electrons fill the first shell, while the third electron occupies the second shell. The second shell holds a maximum of 8 electrons. Neon, at the end of the second row, has 8 electrons in the second shell, giving it a total of 10 electrons.

The chemical behavior of an atom depends mostly on the number of electrons in its *outermost* shell. We call those outer electrons **valence electrons** and the outermost electron shell the **valence shell**. In the case of lithium, there is only 1 valence electron, and the second shell is the valence shell. Atoms with the same number of electrons in their valence shells exhibit similar chemical behavior. For example, fluorine (F) and chlorine (Cl) both have 7 valence electrons, and both form compounds when combined with the element sodium (see Figure 2.3). An atom with a completed valence shell is unreactive; that is, it will not interact readily with other atoms. At the far right of the periodic table are helium, neon, and argon, the only three elements shown in Figure 2.9 that have full valence shells. These elements are said to be *inert*, meaning chemically unreactive. All the other atoms in Figure 2.9 are chemically reactive because they have incomplete valence shells.

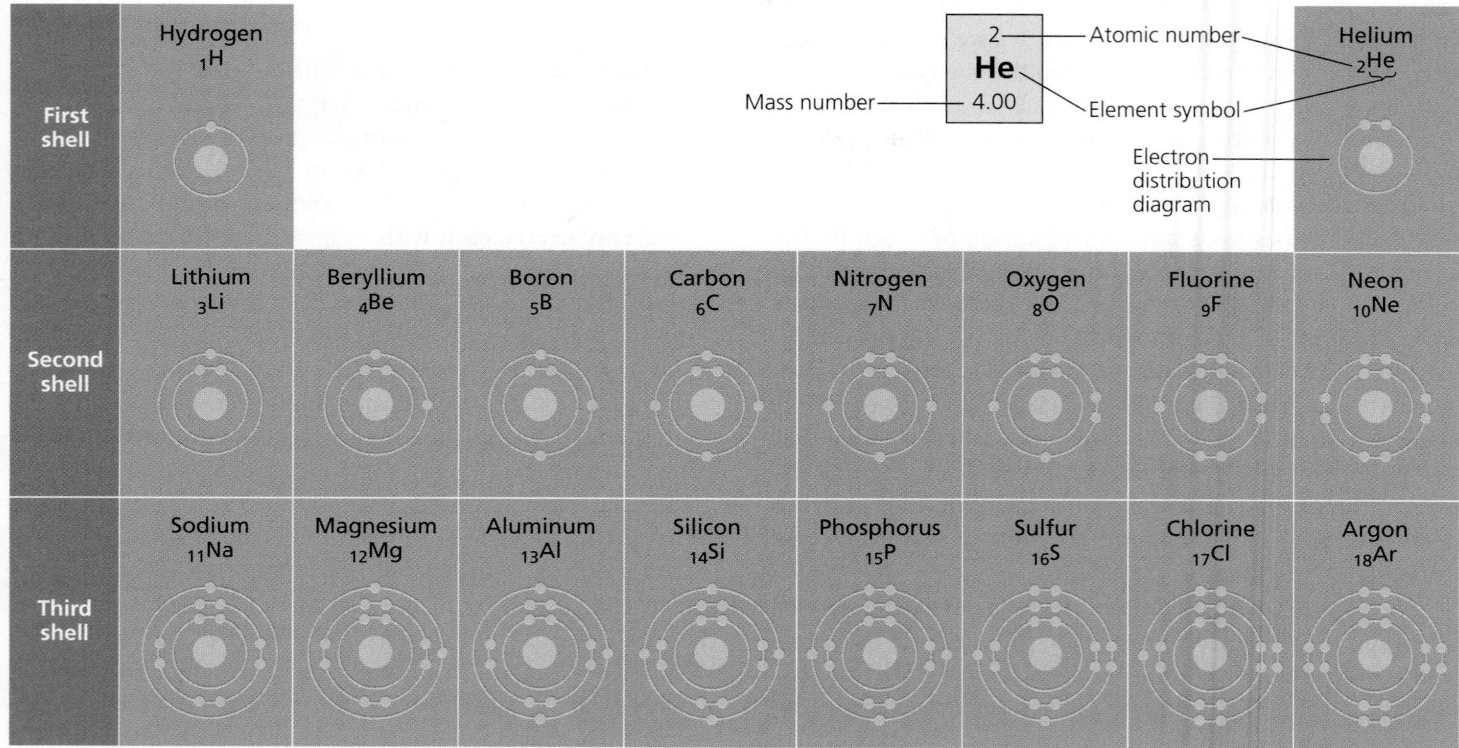

▲ **Figure 2.9 Electron distribution diagrams for the first 18 elements in the periodic table.** In a standard periodic table (see Appendix B), information for each element is presented as shown for helium in the inset. In the diagrams in this table, electrons are represented as yellow dots and electron shells as concentric circles. These diagrams are a convenient way to picture the distribution of an atom's electrons among its electron shells, but these simplified models do not accurately represent the shape of the atom or the location of its electrons. The elements are arranged in rows, each representing the filling of an electron shell. As electrons are added, they occupy the lowest available shell.

? *What is the atomic number of magnesium? How many protons and electrons does it have? How many electron shells? How many valence electrons?*

Electron Orbitals

In the early 1900s, the electron shells of an atom were visualized as concentric paths of electrons orbiting the nucleus, somewhat like planets orbiting the sun. It is still convenient to use two-dimensional concentric-circle diagrams, as in Figure 2.9, to symbolize three-dimensional electron shells. However, you need to remember that each concentric circle represents only the *average* distance between an electron in that shell and the nucleus. Accordingly, the concentric-circle diagrams do not give a real picture of an atom. In reality, we can never know the exact location of an electron. What we can do instead is describe the space in which an electron spends most of its time. The three-dimensional space where an electron is found 90% of the time is called an **orbital**.

Each electron shell contains electrons at a particular energy level, distributed among a specific number of orbitals of distinctive shapes and orientations. **Figure 2.10** shows the orbitals of neon as an example, with its electron distribution diagram for reference. You can think of an orbital as a component of an electron shell. The first electron shell has only one spherical *s* orbital (called 1*s*), but the second shell has four orbitals: one large spherical *s* orbital (called 2*s*) and three dumbbell-shaped *p* orbitals (called 2*p* orbitals). (The third shell and other higher electron shells also have *s* and *p* orbitals, as well as orbitals of more complex shapes.)

No more than 2 electrons can occupy a single orbital. The first electron shell can therefore accommodate up to 2 electrons in its *s* orbital. The lone electron of a hydrogen atom occupies the 1*s* orbital, as do the 2 electrons of a helium atom. The four orbitals of the second electron shell can hold up to 8 electrons, 2 in each orbital. Electrons in each of the four orbitals have nearly the same energy, but they move in different volumes of space.

The reactivity of atoms arises from the presence of unpaired electrons in one or more orbitals of their valence shells. As you will see in the next section, atoms interact in a way that completes their valence shells. When they do so, it is the *unpaired* electrons that are involved.

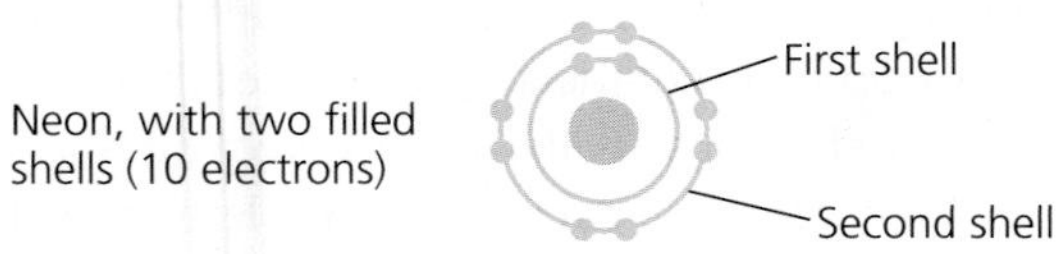

(a) Electron distribution diagram. An electron distribution diagram is shown here for a neon atom, which has a total of 10 electrons. Each concentric circle represents an electron shell, which can be subdivided into electron orbitals.

First shell

Second shell

x

y

z

1*s* orbital

2*s* orbital

Three 2*p* orbitals

(b) Separate electron orbitals. The three-dimensional shapes represent electron orbitals—the volumes of space where the electrons of an atom are most likely to be found. Each orbital holds a maximum of 2 electrons. The first electron shell, on the left, has one spherical (*s*) orbital, designated 1*s*. The second shell, on the right, has one larger *s* orbital (designated 2*s* for the second shell) plus three dumbbell-shaped orbitals called *p* orbitals (2*p* for the second shell). The three 2*p* orbitals lie at right angles to one another along imaginary *x*-, *y*-, and *z*-axes of the atom. Each 2*p* orbital is outlined here in a different color.

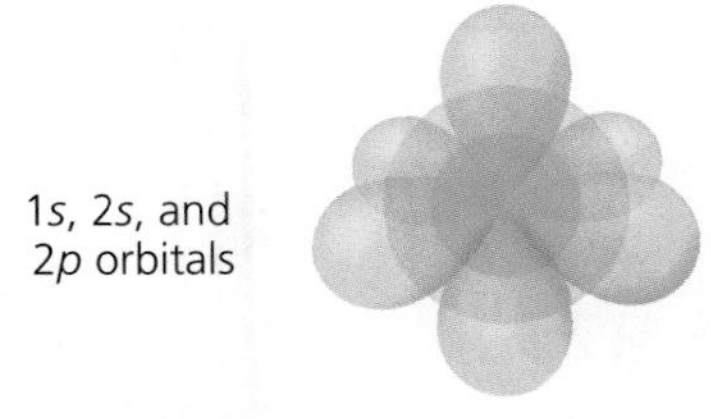

(c) Superimposed electron orbitals. To reveal the complete picture of the electron orbitals of neon, we superimpose the 1*s* orbital of the first shell and the 2*s* and three 2*p* orbitals of the second shell.

▲ **Figure 2.10 Electron orbitals.**

CONCEPT CHECK 2.2

1. A lithium atom has 3 protons and 4 neutrons. What is its atomic mass in daltons?
2. A nitrogen atom has 7 protons, and the most common isotope of nitrogen has 7 neutrons. A radioactive isotope of nitrogen has 8 neutrons. Write the atomic number and mass number of this radioactive nitrogen as a chemical symbol with a subscript and superscript.
3. How many electrons does fluorine have? How many electron shells? Name the orbitals that are occupied. How many electrons are needed to fill the valence shell?
4. **WHAT IF?** In Figure 2.9, if two or more elements are in the same row, what do they have in common? If two or more elements are in the same column, what do they have in common?

For suggested answers, see Appendix A.

CONCEPT 2.3

The formation and function of molecules depend on chemical bonding between atoms

Now that we have looked at the structure of atoms, we can move up the hierarchy of organization and see how atoms combine to form molecules and ionic compounds. Atoms with incomplete valence shells can interact with certain other atoms in such a way that each partner completes its valence shell: The atoms either share or transfer valence electrons. These interactions usually result in atoms staying close together, held by attractions called **chemical bonds**. The strongest kinds of chemical bonds are covalent bonds and ionic bonds.

Covalent Bonds

A **covalent bond** is the sharing of a pair of valence electrons by two atoms. For example, let's consider what happens when two hydrogen atoms approach each other. Recall that hydrogen has 1 valence electron in the first shell, but the shell's capacity is 2 electrons. When the two hydrogen atoms come close enough for their 1*s* orbitals to overlap, they can share their electrons (**Figure 2.11**). Each hydrogen atom is now associated with 2 electrons in what amounts to a completed valence shell. Two or more atoms held together by covalent bonds constitute a **molecule**, in this case a hydrogen molecule.

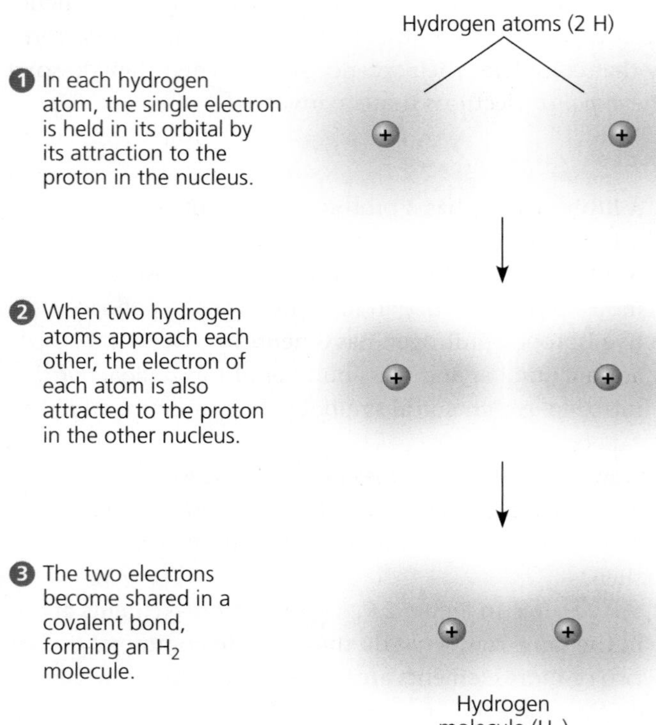

Hydrogen atoms (2 H)
1 In each hydrogen atom, the single electron is held in its orbital by its attraction to the proton in the nucleus.
2 When two hydrogen atoms approach each other, the electron of each atom is also attracted to the proton in the other nucleus.
3 The two electrons become shared in a covalent bond, forming an H_2 molecule.
Hydrogen molecule (H_2)

▲ **Figure 2.11 Formation of a covalent bond.**

Figure 2.12a shows several ways of representing a hydrogen molecule. Its *molecular formula*, H_2, simply indicates that the molecule consists of two atoms of hydrogen. Electron sharing can be depicted by an electron distribution diagram or by a *Lewis dot structure*, in which element symbols are surrounded by dots that represent the valence electrons (H:H). We can also use a *structural formula*, H—H, where the line represents a **single bond**, a pair of shared electrons. A space-filling model comes closest to representing the actual shape of the molecule.

Oxygen has 6 electrons in its second electron shell and therefore needs 2 more electrons to complete its valence shell. Two oxygen atoms form a molecule by sharing *two* pairs of valence electrons (**Figure 2.12b**). The atoms are thus joined by a **double bond** (O=O).

Name and Molecular Formula	Electron Distribution Diagram	Lewis Dot Structure and Structural Formula	Space-Filling Model
(a) Hydrogen (H_2). Two hydrogen atoms share one pair of electrons, forming a single bond.	H H	H:H H—H	
(b) Oxygen (O_2). Two oxygen atoms share two pairs of electrons, forming a double bond.	O O	Ö::Ö O=O	
(c) Water (H_2O). Two hydrogen atoms and one oxygen atom are joined by single bonds, forming a molecule of water.	O H H	:Ö:H H O—H | H	
(d) Methane (CH_4). Four hydrogen atoms can satisfy the valence of one carbon atom, forming methane.	H H C H H	H H:C:H H H | H—C—H | H	

▲ **Figure 2.12 Covalent bonding in four molecules.** The number of electrons required to complete an atom's valence shell generally determines how many covalent bonds that atom will form. This figure shows several ways of indicating covalent bonds.

Each atom that can share valence electrons has a bonding capacity corresponding to the number of covalent bonds the atom can form. When the bonds form, they give the atom a full complement of electrons in the valence shell. The bonding capacity of oxygen, for example, is 2. This bonding capacity is called the atom's **valence** and usually equals the number of unpaired electrons required to complete the atom's outermost (valence) shell. See if you can determine the valences of hydrogen, oxygen, nitrogen, and carbon by studying the electron distribution diagrams in Figure 2.9. You can see that the valence of hydrogen is 1; oxygen, 2; nitrogen, 3; and carbon, 4. However, the situation is more complicated for elements in the third row of the periodic table. Phosphorus, for example, can have a valence of 3, as we would predict from the presence of 3 unpaired electrons in its valence shell. In some molecules that are biologically important, however, phosphorus can form three single bonds and one double bond. Therefore, it can also have a valence of 5.

The molecules H_2 and O_2 are pure elements rather than compounds because a compound is a combination of two or more *different* elements. Water, with the molecular formula H_2O, is a compound. Two atoms of hydrogen are needed to satisfy the valence of one oxygen atom. **Figure 2.12c** shows the structure of a water molecule. Water is so important to life that Chapter 3 is devoted entirely to its structure and behavior.

Methane, the main component of natural gas, is a compound with the molecular formula CH_4. It takes four hydrogen atoms, each with a valence of 1, to complement one atom of carbon, with its valence of 4 **(Figure 2.12d)**. We will look at many other compounds of carbon in Chapter 4.

Atoms in a molecule attract shared electrons to varying degrees, depending on the element. The attraction of a particular atom for the electrons of a covalent bond is called its **electronegativity**. The more electronegative an atom is, the more strongly it pulls shared electrons toward itself. In a covalent bond between two atoms of the same element, the electrons are shared equally because the two atoms have the same electronegativity—the tug-of-war is at a standoff. Such a bond is called a **nonpolar covalent bond**. For example, the single bond of H_2 is nonpolar, as is the double bond of O_2. However, when one atom is bonded to a more electronegative atom, the electrons of the bond are not shared equally. This type of bond is called a **polar covalent bond**. Such bonds vary in their polarity, depending on the relative electronegativity of the two atoms. For example, the bonds between the oxygen and hydrogen atoms of a water molecule are quite polar **(Figure 2.13)**.

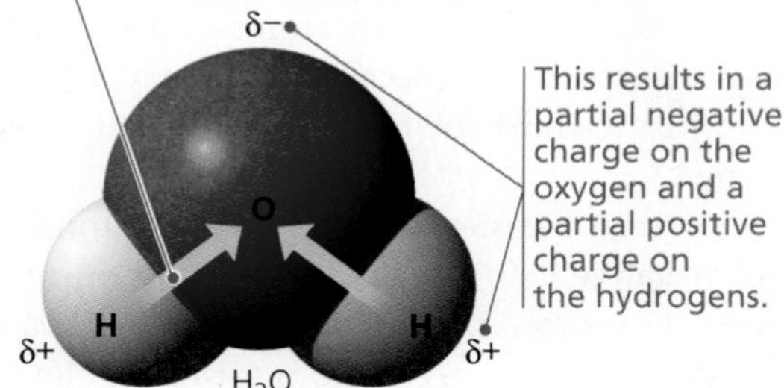
Because oxygen (O) is more electronegative than hydrogen (H), shared electrons are pulled more toward oxygen.
δ–
This results in a partial negative charge on the oxygen and a partial positive charge on the hydrogens.
O
H
H
δ+
δ+
H_2O

▲ **Figure 2.13 Polar covalent bonds in a water molecule.**

Oxygen is one of the most electronegative of all the elements, attracting shared electrons much more strongly than hydrogen does. In a covalent bond between oxygen and hydrogen, the electrons spend more time near the oxygen nucleus than they do near the hydrogen nucleus. Because electrons have a negative charge and are pulled toward oxygen in a water molecule, the oxygen atom has a partial negative charge (indicated by the Greek letter δ with a minus sign, δ–, or "delta minus"), and each hydrogen atom has a partial positive charge (δ+, or "delta plus"). In contrast, the individual bonds of methane (CH_4) are much less polar because the electronegativities of carbon and hydrogen are similar.

Ionic Bonds

In some cases, two atoms are so unequal in their attraction for valence electrons that the more electronegative atom strips an electron completely away from its partner. This is what happens when an atom of sodium ($_{11}Na$) encounters an atom of chlorine ($_{17}Cl$) **(Figure 2.14)**. A sodium atom has a total of 11 electrons, with its single valence electron in the third electron shell. A chlorine atom has a total of 17 electrons,

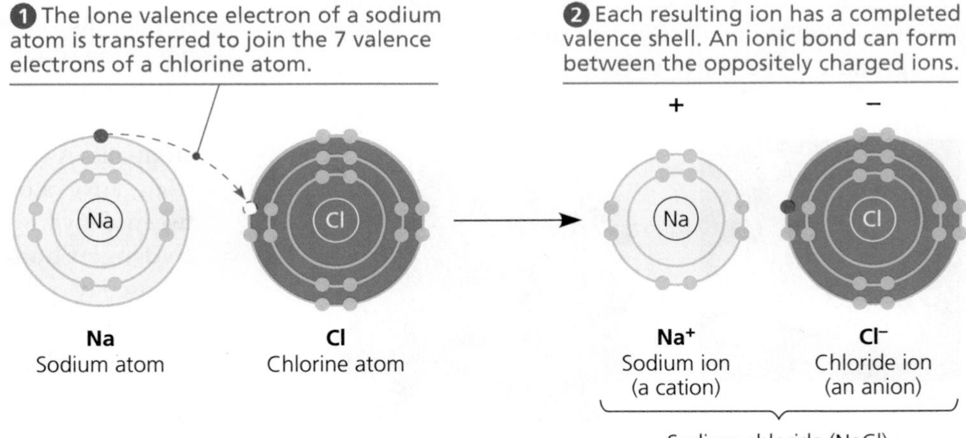
1 The lone valence electron of a sodium atom is transferred to join the 7 valence electrons of a chlorine atom.
2 Each resulting ion has a completed valence shell. An ionic bond can form between the oppositely charged ions.
+
–
Na
Cl
Na
Cl
Na
Sodium atom
Cl
Chlorine atom
Na^+
Sodium ion
(a cation)
Cl^-
Chloride ion
(an anion)
Sodium chloride (NaCl)

▲ **Figure 2.14 Electron transfer and ionic bonding.** The attraction between oppositely charged atoms, or ions, is an ionic bond. An ionic bond can form between any two oppositely charged ions, even if they have not been formed by transfer of an electron from one to the other.

with 7 electrons in its valence shell. When these two atoms meet, the lone valence electron of sodium is transferred to the chlorine atom, and both atoms end up with their valence shells complete. (Because sodium no longer has an electron in the third shell, the second shell is now the valence shell.)

The electron transfer between the two atoms moves one unit of negative charge from sodium to chlorine. Sodium, now with 11 protons but only 10 electrons, has a net electrical charge of 1+. A charged atom (or molecule) is called an **ion**. When the charge is positive, the ion is specifically called a **cation**; the sodium atom has become a cation. Conversely, the chlorine atom, having gained an extra electron, now has 17 protons and 18 electrons, giving it a net electrical charge of 1–. It has become a chloride ion—an **anion**, or negatively charged ion. Because of their opposite charges, cations and anions attract each other; this attraction is called an **ionic bond**. The transfer of an electron is not the formation of a bond; rather, it allows a bond to form because it results in two ions of opposite charge. Any two ions of opposite charge can form an ionic bond. The ions do not need to have acquired their charge by an electron transfer with each other.

Compounds formed by ionic bonds are called **ionic compounds**, or **salts**. We know the ionic compound sodium chloride (NaCl) as table salt **(Figure 2.15)**. Salts are often found in nature as crystals of various sizes and shapes. Each salt crystal is an aggregate of vast numbers of cations and anions bonded by their electrical attraction and arranged in a three-dimensional lattice. Unlike a covalent compound, which consists of molecules having a definite size and number of atoms, an ionic compound does not consist of molecules. The formula for an ionic compound, such as NaCl, indicates only the ratio of elements in a crystal of the salt. "NaCl" by itself is not a molecule.

Not all salts have equal numbers of cations and anions. For example, the ionic compound magnesium chloride ($MgCl_2$) has two chloride ions for each magnesium ion. Magnesium ($_{12}Mg$) must lose 2 outer electrons if the atom is to have a complete valence shell, so it tends to become a cation with a net charge of 2+ (Mg^{2+}). One magnesium cation can therefore form ionic bonds with two chloride anions.

▲ **Figure 2.15 A sodium chloride (NaCl) crystal.** The sodium ions (Na^+) and chloride ions (Cl^-) are held together by ionic bonds. The formula NaCl tells us that the ratio of Na^+ to Cl^- is 1:1.

The term *ion* also applies to entire molecules that are electrically charged. In the salt ammonium chloride (NH_4Cl), for instance, the anion is a single chloride ion (Cl^-), but the cation is ammonium (NH_4^+), a nitrogen atom with four covalently bonded hydrogen atoms. The whole ammonium ion has an electrical charge of 1+ because it is 1 electron short.

Environment affects the strength of ionic bonds. In a dry salt crystal, the bonds are so strong that it takes a hammer and chisel to break enough of them to crack the crystal in two. If the same salt crystal is dissolved in water, however, the ionic bonds are much weaker because each ion is partially shielded by its interactions with water molecules. Most drugs are manufactured as salts because they are quite stable when dry but can dissociate (come apart) easily in water. In the next chapter, you will learn how water dissolves salts.

Weak Chemical Bonds

In organisms, most of the strongest chemical bonds are covalent bonds, which link atoms to form a cell's molecules. But weaker bonding within and between molecules is also indispensable in the cell, contributing greatly to the emergent properties of life. Many large biological molecules are held in their functional form by weak bonds. In addition, when two molecules in the cell make contact, they may adhere temporarily by weak bonds. The reversibility of weak bonding can be an advantage: Two molecules can come together, respond to one another in some way, and then separate.

Several types of weak chemical bonds are important in organisms. One is the ionic bond as it exists between ions dissociated in water, which we just discussed. Hydrogen bonds and van der Waals interactions are also crucial to life.

Hydrogen Bonds

Among the various kinds of weak chemical bonds, hydrogen bonds are so important in the chemistry of life that they deserve special attention. The partial positive charge on a hydrogen atom that is covalently bonded to an electronegative atom allows the hydrogen to be attracted to a different electronegative atom nearby. This noncovalent attraction between a hydrogen and an electronegative atom is called a **hydrogen bond**. In living cells, the electronegative partners are usually oxygen or nitrogen atoms. Refer to **Figure 2.16** to examine the simple case of hydrogen bonding between water (H_2O) and ammonia (NH_3).

Van der Waals Interactions

Even a molecule with nonpolar covalent bonds may have positively and negatively charged regions. Electrons are not always symmetrically distributed in such a molecule; at any

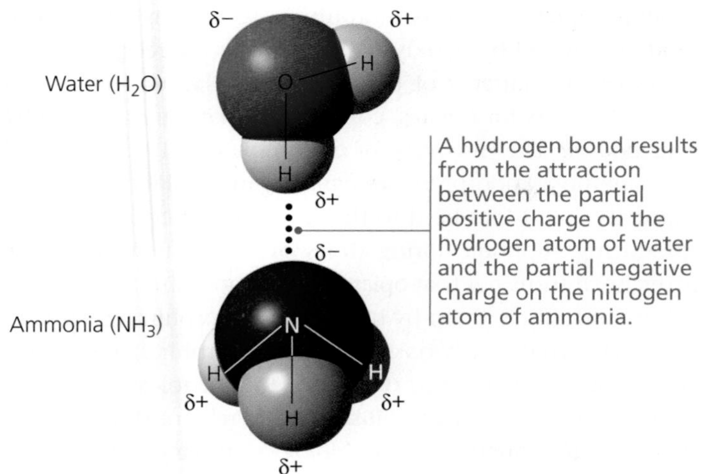

▲ **Figure 2.16 A hydrogen bond.**

DRAW IT *Draw five water molecules using structural formulas and indicating partial charges, and show how they can make hydrogen bonds with each other.*

instant, they may accumulate by chance in one part of the molecule or another. The results are ever-changing regions of positive and negative charge that enable all atoms and molecules to stick to one another. These **van der Waals interactions** are individually weak and occur only when atoms and molecules are very close together. When many such interactions occur simultaneously, however, they can be powerful: Van der Waals interactions are the reason a gecko lizard (right) can walk straight up a wall! Each gecko toe has hundreds of thousands of tiny hairs, with multiple projections at each hair's tip that increase surface area. Apparently, the van der Waals interactions between the hair tip molecules and the molecules of the wall's surface are so numerous that despite their individual weakness, together they can support the gecko's body weight.

Van der Waals interactions, hydrogen bonds, ionic bonds in water, and other weak bonds may form not only between molecules but also between parts of a large molecule, such as a protein. The cumulative effect of weak bonds is to reinforce the three-dimensional shape of the molecule. You will learn more about the very important biological roles of weak bonds in Chapter 5.

Molecular Shape and Function

A molecule has a characteristic size and shape. The precise shape of a molecule is usually very important to its function in the living cell.

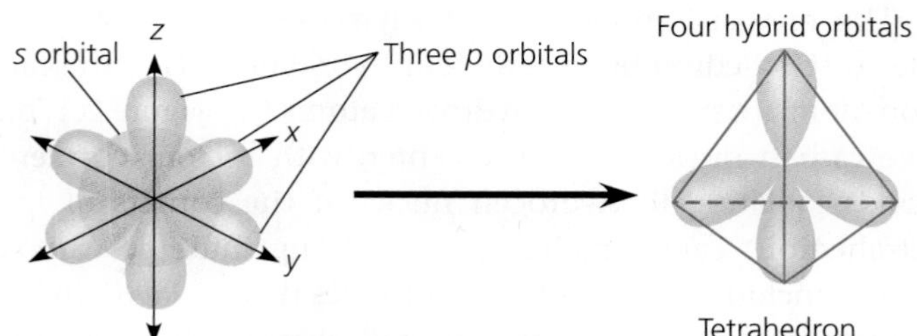

(a) Hybridization of orbitals. The single *s* and three *p* orbitals of a valence shell involved in covalent bonding combine to form four teardrop-shaped hybrid orbitals. These orbitals extend to the four corners of an imaginary tetrahedron (outlined in pink).

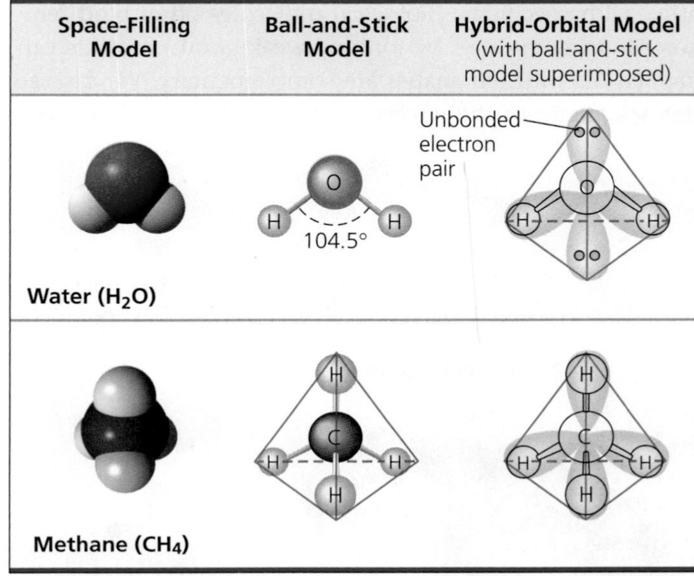

(b) Molecular-shape models. Three models representing molecular shape are shown for water and methane. The positions of the hybrid orbitals determine the shapes of the molecules.

▲ **Figure 2.17 Molecular shapes due to hybrid orbitals.**

A molecule consisting of two atoms, such as H_2 or O_2, is always linear, but most molecules with more than two atoms have more complicated shapes. These shapes are determined by the positions of the atoms' orbitals. When an atom forms covalent bonds, the orbitals in its valence shell undergo rearrangement. For atoms with valence electrons in both *s* and *p* orbitals (review Figure 2.10), the single *s* and three *p* orbitals form four new hybrid orbitals shaped like identical teardrops extending from the region of the atomic nucleus (**Figure 2.17a**). If we connect the larger ends of the teardrops with lines, we have the outline of a geometric shape called a tetrahedron, a pyramid with a triangular base.

For the water molecule (H_2O), two of the hybrid orbitals in the oxygen atom's valence shell are shared with hydrogen atoms (**Figure 2.17b**). The result is a molecule shaped roughly like a V, with its two covalent bonds spread apart at an angle of 104.5°.

The methane molecule (CH_4) has the shape of a completed tetrahedron because all four hybrid orbitals of the carbon atom are shared with hydrogen atoms (see Figure 2.17b). The carbon nucleus is at the center, with its four covalent bonds radiating to hydrogen nuclei at the corners of the tetrahedron. Larger molecules containing multiple carbon atoms, including many of the molecules that make up living matter, have more complex overall shapes. However, the tetrahedral shape of a carbon atom bonded to four other atoms is often a repeating motif within such molecules.

Molecular shape is crucial in biology because it determines how biological molecules recognize and respond to one another with specificity. Biological molecules often bind temporarily to each other by forming weak bonds, but this can happen only if their shapes are complementary. We can see this specificity in the effects of opiates, drugs derived from opium. Opiates, such as morphine and heroin, relieve pain and alter mood by weakly binding to specific receptor molecules on the surfaces of brain cells. Why would brain cells carry receptors for opiates, compounds that are not made by our bodies? The discovery of endorphins in 1975 answered this question. Endorphins are signaling molecules made by the pituitary gland that bind to the receptors, relieving pain and producing euphoria during times of stress, such as intense exercise. It turns out that opiates have shapes similar to endorphins and mimic them by binding to endorphin receptors in the brain. That is why opiates (such as morphine) and endorphins have similar effects (**Figure 2.18**). The role of molecular shape in brain chemistry illustrates the relationship between structure and function, one of biology's unifying themes.

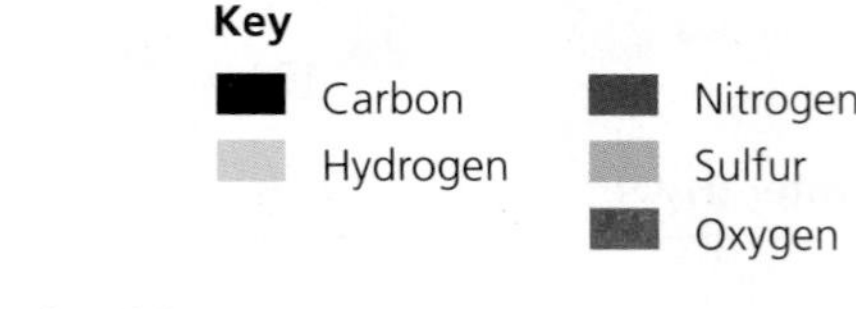

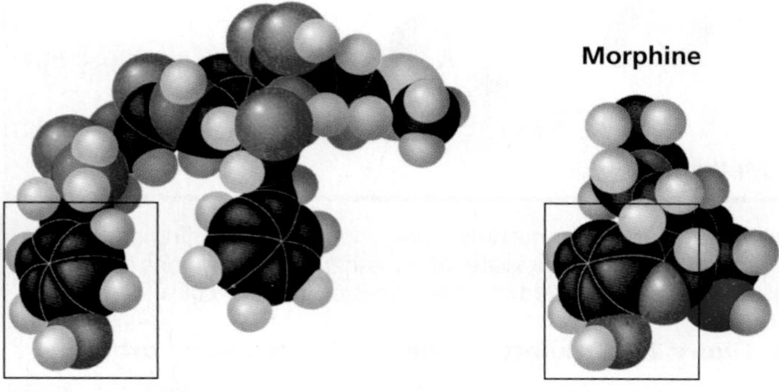

(a) Structures of endorphin and morphine. The boxed portion of the endorphin molecule (left) binds to receptor molecules on target cells in the brain. The boxed portion of the morphine molecule (right) is a close match.

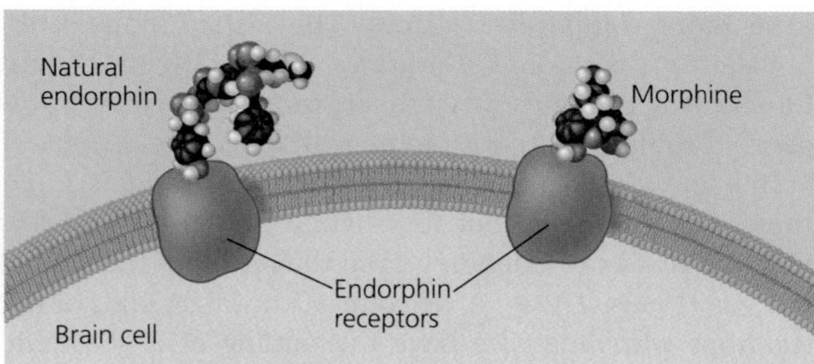

(b) Binding to endorphin receptors. Both endorphin and morphine can bind to endorphin receptors on the surface of a brain cell.

▲ **Figure 2.18 A molecular mimic.** Morphine affects pain perception and emotional state by mimicking the brain's natural endorphins.

CONCEPT CHECK 2.3

1. Why does the structure H—C=C—H fail to make sense chemically?
2. What holds the atoms together in a crystal of magnesium chloride ($MgCl_2$)?
3. **WHAT IF?** If you were a pharmaceutical researcher, why would you want to learn the three-dimensional shapes of naturally occurring signaling molecules?

For suggested answers, see Appendix A.

CONCEPT 2.4

Chemical reactions make and break chemical bonds

The making and breaking of chemical bonds, leading to changes in the composition of matter, are called **chemical reactions**. An example is the reaction between hydrogen and oxygen molecules that forms water:

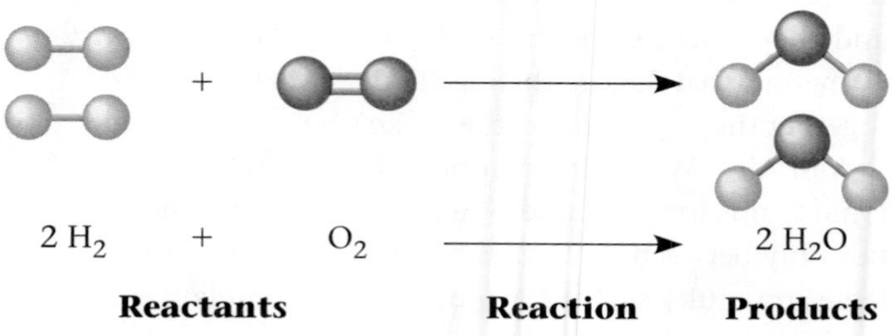

$$2\,H_2 + O_2 \longrightarrow 2\,H_2O$$

Reactants **Reaction** **Products**

This reaction breaks the covalent bonds of H_2 and O_2 and forms the new bonds of H_2O. When we write a chemical reaction, we use an arrow to indicate the conversion of the starting materials, called the **reactants**, to the **products**. The coefficients indicate the number of molecules involved; for example, the coefficient 2 in front of the H_2 means that

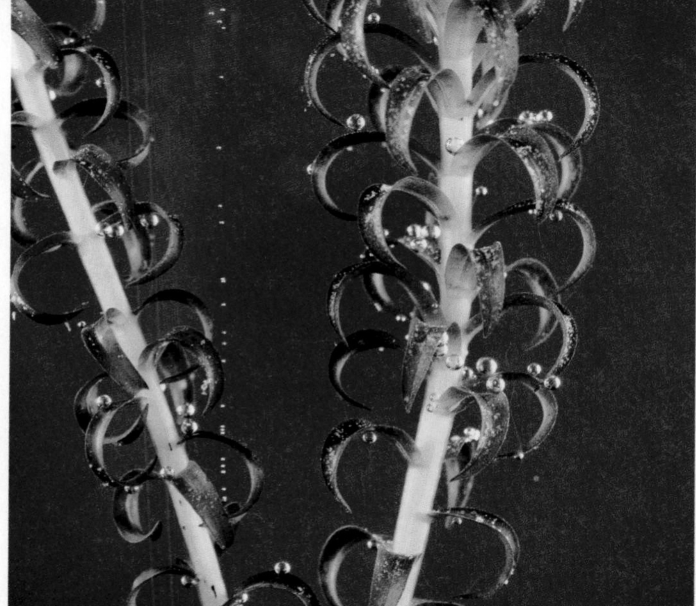

▲ **Figure 2.19 Photosynthesis: a solar-powered rearrangement of matter.** *Elodea*, a freshwater plant, produces sugar by rearranging the atoms of carbon dioxide and water in the chemical process known as photosynthesis, which is powered by sunlight. Much of the sugar is then converted to other food molecules. Oxygen gas (O_2) is a by-product of photosynthesis; notice the bubbles of oxygen escaping from the leaves in the photo.

? *Explain how this photo relates to the reactants and products in the equation for photosynthesis given in the text. (You will learn more about photosynthesis in Chapter 10.)*

the reaction starts with two molecules of hydrogen. Notice that all atoms of the reactants must be accounted for in the products. Matter is conserved in a chemical reaction: Reactions cannot create or destroy matter but can only rearrange it.

Photosynthesis, which takes place within the cells of green plant tissues, is a particularly important example of how chemical reactions rearrange matter. Humans and other animals ultimately depend on photosynthesis for food and oxygen, and this process is at the foundation of almost all ecosystems. The following chemical shorthand summarizes the process of photosynthesis:

$$6\ CO_2 + 6\ H_2O \rightarrow C_6H_{12}O_6 + 6\ O_2$$

The raw materials of photosynthesis are carbon dioxide (CO_2), which is taken from the air, and water (H_2O), which is absorbed from the soil. Within the plant cells, sunlight powers the conversion of these ingredients to a sugar called glucose ($C_6H_{12}O_6$) and oxygen molecules (O_2), a by-product that the plant releases into the surroundings (**Figure 2.19**). Although photosynthesis is actually a sequence of many chemical reactions, we still end up with the same number and types of atoms that we had when we started. Matter has simply been rearranged, with an input of energy provided by sunlight.

All chemical reactions are reversible, with the products of the forward reaction becoming the reactants for the reverse reaction. For example, hydrogen and nitrogen molecules can combine to form ammonia, but ammonia can also decompose to regenerate hydrogen and nitrogen:

$$3\ H_2 + N_2 \rightleftharpoons 2\ NH_3$$

The two opposite-headed arrows indicate that the reaction is reversible.

One of the factors affecting the rate of a reaction is the concentration of reactants. The greater the concentration of reactant molecules, the more frequently they collide with one another and have an opportunity to react and form products. The same holds true for products. As products accumulate, collisions resulting in the reverse reaction become more frequent. Eventually, the forward and reverse reactions occur at the same rate, and the relative concentrations of products and reactants stop changing. The point at which the reactions offset one another exactly is called **chemical equilibrium**. This is a dynamic equilibrium; reactions are still going on, but with no net effect on the concentrations of reactants and products. Equilibrium does *not* mean that the reactants and products are equal in concentration, but only that their concentrations have stabilized at a particular ratio. The reaction involving ammonia reaches equilibrium when ammonia decomposes as rapidly as it forms. In some chemical reactions, the equilibrium point may lie so far to the right that these reactions go essentially to completion; that is, virtually all the reactants are converted to products.

We will return to the subject of chemical reactions after more detailed study of the various types of molecules that are important to life. In the next chapter, we focus on water, the substance in which all the chemical processes of organisms occur.

CONCEPT CHECK 2.4

1. **MAKE CONNECTIONS** Consider the reaction between hydrogen and oxygen that forms water, shown with ball-and-stick models on page 42. Study Figure 2.12 and draw the Lewis dot structures representing this reaction.
2. Which type of chemical reaction occurs faster at equilibrium, the formation of products from reactants or reactants from products?
3. **WHAT IF?** Write an equation that uses the products of photosynthesis as reactants and the reactants of photosynthesis as products. Add energy as another product. This new equation describes a process that occurs in your cells. Describe this equation in words. How does this equation relate to breathing?

For suggested answers, see Appendix A.

2 CHAPTER REVIEW

SUMMARY OF KEY CONCEPTS

CONCEPT 2.1

Matter consists of chemical elements in pure form and in combinations called compounds (pp. 31–32)

- **Elements** cannot be broken down chemically to other substances. A **compound** contains two or more different elements in a fixed ratio. Oxygen, carbon, hydrogen, and nitrogen make up approximately 96% of living matter.

? *In what way does the need for iodine or iron in your diet differ from your need for calcium or phosphorus?*

CONCEPT 2.2

An element's properties depend on the structure of its atoms (pp. 33–37)

- An **atom**, the smallest unit of an element, has the following components:

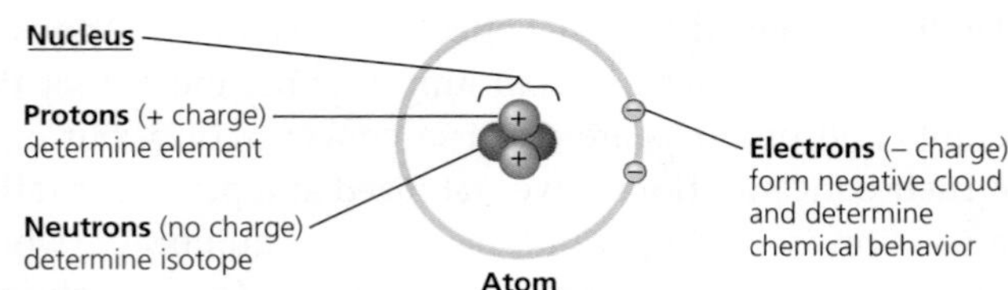

- An electrically neutral atom has equal numbers of electrons and protons; the number of protons determines the **atomic number**. The **atomic mass** is measured in **daltons** and is roughly equal to the sum of protons plus neutrons. **Isotopes** of an element differ from each other in neutron number and therefore mass. Unstable isotopes give off particles and energy as radioactivity.
- In an atom, electrons occupy specific **electron shells**; the electrons in a shell have a characteristic energy level. Electron distribution in shells determines the chemical behavior of an atom. An atom that has an incomplete outer shell, the **valence shell**, is reactive.
- Electrons exist in **orbitals**, three-dimensional spaces with specific shapes that are components of electron shells.

Electron orbitals

DRAW IT *Draw the electron distribution diagrams for neon ($_{10}$Ne) and argon ($_{18}$Ar). Use these diagrams to explain why these elements are chemically unreactive.*

CONCEPT 2.3

The formation and function of molecules depend on chemical bonding between atoms (pp. 38–42)

- **Chemical bonds** form when atoms interact and complete their valence shells. **Covalent bonds** form when pairs of electrons are shared.

H· + H· ⟶ H:H
Single covalent bond

:Ö· + ·Ö: ⟶ Ö::Ö
Double covalent bond

- **Molecules** consist of two or more covalently bonded atoms. The attraction of an atom for the electrons of a covalent bond is its **electronegativity**. If both atoms are the same, they have the same electronegativity and share a **nonpolar covalent bond**. Electrons of a **polar covalent bond** are pulled closer to the more electronegative atom.
- An **ion** forms when an atom or molecule gains or loses an electron and becomes charged. An **ionic bond** is the attraction between two oppositely charged ions.

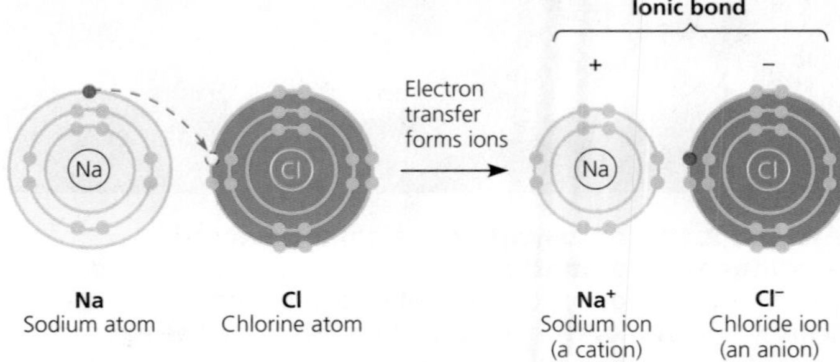

- Weak bonds reinforce the shapes of large molecules and help molecules adhere to each other. A **hydrogen bond** is an attraction between a hydrogen atom carrying a partial positive charge ($\delta+$) and an electronegative atom ($\delta-$). **Van der Waals interactions** occur between transiently positive and negative regions of molecules.
- A molecule's shape is determined by the positions of its atoms' valence orbitals. Covalent bonds result in hybrid orbitals, which are responsible for the shapes of H_2O, CH_4, and many more complex biological molecules. Shape is usually the basis for the recognition of one biological molecule by another.

? *In terms of electron sharing between atoms, compare nonpolar covalent bonds, polar covalent bonds, and the formation of ions.*

CONCEPT 2.4

Chemical reactions make and break chemical bonds (pp. 42–43)

- **Chemical reactions** change **reactants** into **products** while conserving matter. All chemical reactions are theoretically reversible. **Chemical equilibrium** is reached when the forward and reverse reaction rates are equal.

? *What would happen to the concentration of products if more reactants were added to a reaction that was in chemical equilibrium? How would this addition affect the equilibrium?*

TEST YOUR UNDERSTANDING

LEVEL 1: KNOWLEDGE/COMPREHENSION

1. In the term *trace element*, the adjective *trace* means that
   a. the element is required in very small amounts.
   b. the element can be used as a label to trace atoms through an organism's metabolism.
   c. the element is very rare on Earth.
   d. the element enhances health but is not essential for the organism's long-term survival.
   e. the element passes rapidly through the organism.

2. Compared with ^{31}P, the radioactive isotope ^{32}P has
   a. a different atomic number.
   b. a different charge.
   c. one more proton.
   d. one more electron.
   e. one more neutron.

3. The reactivity of an atom arises from
   a. the average distance of the outermost electron shell from the nucleus.
   b. the existence of unpaired electrons in the valence shell.
   c. the sum of the potential energies of all the electron shells.
   d. the potential energy of the valence shell.
   e. the energy difference between the *s* and *p* orbitals.

4. Which statement is true of all atoms that are anions?
   a. The atom has more electrons than protons.
   b. The atom has more protons than electrons.
   c. The atom has fewer protons than does a neutral atom of the same element.
   d. The atom has more neutrons than protons.
   e. The net charge is 1−.

5. Which of the following statements correctly describes any chemical reaction that has reached equilibrium?
   a. The concentrations of products and reactants are equal.
   b. The reaction is now irreversible.
   c. Both forward and reverse reactions have halted.
   d. The rates of the forward and reverse reactions are equal.
   e. No reactants remain.

LEVEL 2: APPLICATION/ANALYSIS

6. We can represent atoms by listing the number of protons, neutrons, and electrons—for example, $2p^+$, $2n^0$, $2e^-$ for helium. Which of the following represents the ^{18}O isotope of oxygen?
   a. $6p^+$, $8n^0$, $6e^-$
   b. $8p^+$, $10n^0$, $8e^-$
   c. $9p^+$, $9n^0$, $9e^-$
   d. $7p^+$, $2n^0$, $9e^-$
   e. $10p^+$, $8n^0$, $9e^-$

7. The atomic number of sulfur is 16. Sulfur combines with hydrogen by covalent bonding to form a compound, hydrogen sulfide. Based on the number of valence electrons in a sulfur atom, predict the molecular formula of the compound.
   a. HS b. HS_2 c. H_2S d. H_3S_2 e. H_4S

8. What coefficients must be placed in the following blanks so that all atoms are accounted for in the products?

   $C_6H_{12}O_6 \rightarrow$ ____ C_2H_6O + ____ CO_2

   a. 1; 2 b. 3; 1 c. 1; 3 d. 1; 1 e. 2; 2

9. **DRAW IT** Draw Lewis dot structures for each hypothetical molecule shown below, using the correct number of valence electrons for each atom. Determine which molecule makes sense because each atom has a complete valence shell and each bond has the correct number of electrons. Explain what makes the other molecules nonsensical, considering the number of bonds each type of atom can make.

   (a) O═C—H

   (b)
   ```
            H   H
            |   |
      H—O—C—C═O
            |
            H
   ```

   (c)
   ```
         H       H
         |       |
      H—C—H—C═O
         |
         H
   ```

   (d)
   ```
           O
           |
      H—N═H
   ```

LEVEL 3: SYNTHESIS/EVALUATION

10. **EVOLUTION CONNECTION**

    The percentages of naturally occurring elements making up the human body (see Table 2.1) are similar to the percentages of these elements found in other organisms. How could you account for this similarity among organisms?

11. **SCIENTIFIC INQUIRY**

    Female silkworm moths (*Bombyx mori*) attract males by emitting chemical signals that spread through the air. A male hundreds of meters away can detect these molecules and fly toward their source. The sensory organs responsible for this behavior are the comblike antennae visible in the photograph shown here. Each filament of an antenna is equipped with thousands of receptor cells that detect the sex attractant. Based on what you learned in this chapter, propose a hypothesis to account for the ability of the male moth to detect a specific molecule in the presence of many other molecules in the air. What predictions does your hypothesis make? Design an experiment to test one of these predictions.

12. **WRITE ABOUT A THEME**

    Emergent Properties While waiting at an airport, Neil Campbell once overheard this claim: "It's paranoid and ignorant to worry about industry or agriculture contaminating the environment with their chemical wastes. After all, this stuff is just made of the same atoms that were already present in our environment." Drawing on your knowledge of electron distribution, bonding, and the theme of emergent properties (pp. 3–5), write a short essay (100–150 words) countering this argument.

For selected answers, see Appendix A.

3

Water and Life

▲ **Figure 3.1 How does the habitat of a polar bear depend on the chemistry of water?**

KEY CONCEPTS

3.1 Polar covalent bonds in water molecules result in hydrogen bonding

3.2 Four emergent properties of water contribute to Earth's suitability for life

3.3 Acidic and basic conditions affect living organisms

OVERVIEW

The Molecule That Supports All of Life

As astronomers study newly discovered planets orbiting distant stars, they hope to find evidence of water on these far-off celestial bodies, for water is the substance that makes possible life as we know it here on Earth. All organisms familiar to us are made mostly of water and live in an environment dominated by water. Water is the biological medium here on Earth, and possibly on other planets as well.

Three-quarters of Earth's surface is covered by water. Although most of this water is in liquid form, water is also present on Earth as a solid (ice) and a gas (water vapor). Water is the only common substance to exist in the natural environment in all three physical states of matter. Furthermore, the solid state of water floats on the liquid, a rare property emerging from the chemistry of the water molecule. Ice can thus provide a hunting platform for the polar bear in **Figure 3.1**.

The abundance of water is a major reason Earth is habitable. In a classic book called *The Fitness of the Environment*, ecologist Lawrence Henderson highlighted the importance of water to life. While acknowledging that life adapts to its environment through natural selection, Henderson emphasized that for life to exist at all, the environment must first be suitable.

Life on Earth began in water and evolved there for 3 billion years before spreading onto land. Modern life, even terrestrial (land-dwelling) life, remains tied to water. All living organisms require water more than any other substance. Human beings, for example, can survive for quite a few weeks without food, but only a week or so without water. Molecules of water participate in many chemical reactions necessary to sustain life. Most cells are surrounded by water, and cells themselves are about 70–95% water.

What properties of the simple water molecule make it so indispensable to life on Earth? In this chapter, you will learn how the structure of a water molecule allows it to interact with other molecules, including other water molecules. This ability leads to water's unique emergent properties that help make Earth suitable for life.

CONCEPT **3.1**

Polar covalent bonds in water molecules result in hydrogen bonding

Water is so common that it is easy to overlook the fact that it is an exceptional substance with many extraordinary qualities. Following the theme of emergent properties, we can trace water's unique behavior to the structure and interactions of its molecules.

Studied on its own, the water molecule is deceptively simple. It is shaped like a wide V, with its two hydrogen atoms joined to the oxygen atom by single covalent bonds. Oxygen is more electronegative than hydrogen, so the electrons of the covalent bonds spend more time closer to oxygen than to hydrogen; these are **polar covalent bonds** (see Figure 2.13). This unequal sharing of electrons and water's V-like shape make it a **polar molecule**, meaning that its overall charge is unevenly distributed: The oxygen region of the molecule has a partial negative charge ($\delta-$), and each hydrogen has a partial positive charge ($\delta+$).

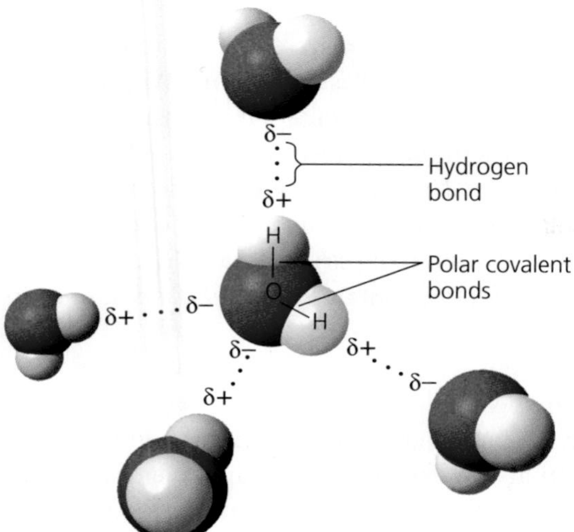

▲ **Figure 3.2 Hydrogen bonds between water molecules.** The charged regions in a water molecule are due to its polar covalent bonds. Oppositely charged regions of neighboring water molecules are attracted to each other, forming hydrogen bonds. Each molecule can hydrogen-bond to multiple partners, and these associations are constantly changing.

DRAW IT *Draw partial charges on all the atoms of the water molecule on the far left above, and draw two more water molecules hydrogen-bonded to it.*

The properties of water arise from attractions between oppositely charged atoms of different water molecules: The slightly positive hydrogen of one molecule is attracted to the slightly negative oxygen of a nearby molecule. The two molecules are thus held together by a hydrogen bond (**Figure 3.2**). When water is in its liquid form, its hydrogen bonds are very fragile, each about 1/20 as strong as a covalent bond. The hydrogen bonds form, break, and re-form with great frequency. Each lasts only a few trillionths of a second, but the molecules are constantly forming new hydrogen bonds with a succession of partners. Therefore, at any instant, a substantial percentage of all the water molecules are hydrogen-bonded to their neighbors. The extraordinary qualities of water are emergent properties resulting in large part from the hydrogen bonding that organizes water molecules into a higher level of structural order.

CONCEPT CHECK 3.1

1. **MAKE CONNECTIONS** What is electronegativity, and how does it affect interactions between water molecules? Review p. 39 and Figure 2.13.
2. Why is it unlikely that two neighboring water molecules would be arranged like this?

   H H
   O   O
   H H

3. **WHAT IF?** What would be the effect on the properties of the water molecule if oxygen and hydrogen had equal electronegativity?

For suggested answers, see Appendix A.

CONCEPT 3.2

Four emergent properties of water contribute to Earth's suitability for life

We will examine four emergent properties of water that contribute to Earth's suitability as an environment for life: cohesive behavior, ability to moderate temperature, expansion upon freezing, and versatility as a solvent.

Cohesion of Water Molecules

Water molecules stay close to each other as a result of hydrogen bonding. Although the arrangement of molecules in a sample of liquid water is constantly changing, at any given moment many of the molecules are linked by multiple hydrogen bonds. These linkages make water more structured than most other liquids. Collectively, the hydrogen bonds hold the substance together, a phenomenon called **cohesion**.

Cohesion due to hydrogen bonding contributes to the transport of water and dissolved nutrients against gravity in plants (**Figure 3.3**). Water from the roots reaches the leaves through a network of water-conducting cells. As water evaporates from a

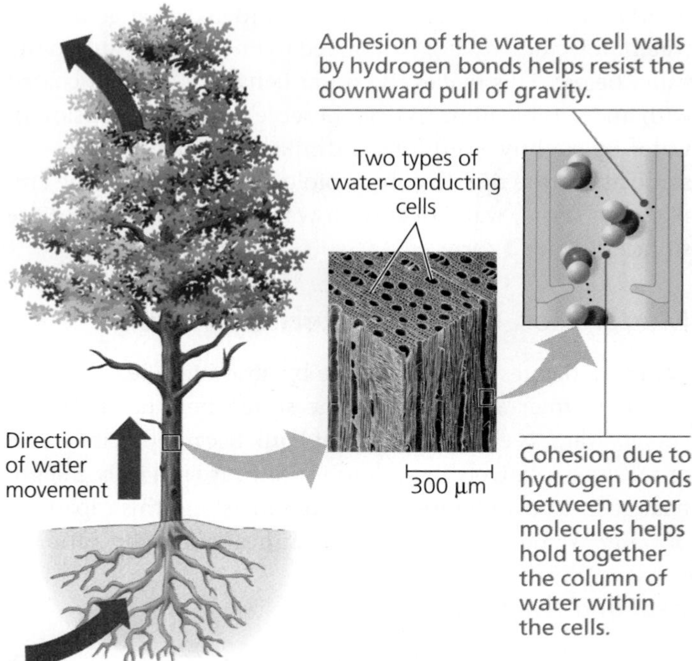

▲ **Figure 3.3 Water transport in plants.** Evaporation from leaves pulls water upward from the roots through water-conducting cells. Because of the properties of cohesion and adhesion, the tallest trees can transport water more than 100 m upward—approximately one-quarter the height of the Empire State Building in New York City.

BioFlix Visit the Study Area at **www.masteringbiology.com** for the BioFlix® 3-D Animation on Water Transport in Plants.

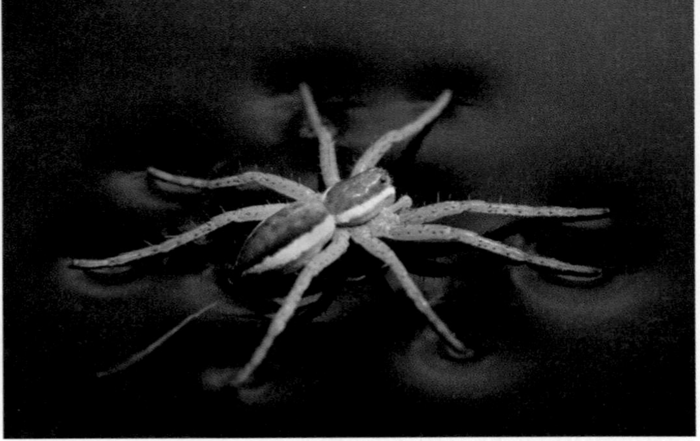

▲ **Figure 3.4 Walking on water.** The high surface tension of water, resulting from the collective strength of its hydrogen bonds, allows this raft spider to walk on the surface of a pond.

leaf, hydrogen bonds cause water molecules leaving the veins to tug on molecules farther down, and the upward pull is transmitted through the water-conducting cells all the way to the roots. **Adhesion**, the clinging of one substance to another, also plays a role. Adhesion of water to cell walls by hydrogen bonds helps counter the downward pull of gravity (see Figure 3.3).

Related to cohesion is **surface tension**, a measure of how difficult it is to stretch or break the surface of a liquid. Water has a greater surface tension than most other liquids. At the interface between water and air is an ordered arrangement of water molecules, hydrogen-bonded to one another and to the water below. This makes the water behave as though coated with an invisible film. You can observe the surface tension of water by slightly overfilling a drinking glass; the water will stand above the rim. In a more biological example, some animals can stand, walk, or run on water without breaking the surface **(Figure 3.4)**.

Moderation of Temperature by Water

Water moderates air temperature by absorbing heat from air that is warmer and releasing the stored heat to air that is cooler. Water is effective as a heat bank because it can absorb or release a relatively large amount of heat with only a slight change in its own temperature. To understand this capability of water, we must first look briefly at heat and temperature.

Heat and Temperature

Anything that moves has **kinetic energy**, the energy of motion. Atoms and molecules have kinetic energy because they are always moving, although not necessarily in any particular direction. The faster a molecule moves, the greater its kinetic energy. **Heat** is a form of energy. For a given body of matter, the amount of heat is a measure of the matter's *total* kinetic energy due to motion of its molecules; thus, heat depends in part on the matter's volume. Although heat is related to temperature, they are not the same thing. **Temperature** is a measure of heat intensity that represents the *average* kinetic energy of the molecules, regardless of volume. When water is heated in a coffeemaker, the average speed of the molecules increases, and the thermometer records this as a rise in temperature of the liquid. The amount of heat also increases in this case. Note, however, that although the pot of coffee has a much higher temperature than, say, the water in a swimming pool, the swimming pool contains more heat because of its much greater volume.

Whenever two objects of different temperature are brought together, heat passes from the warmer to the cooler object until the two are the same temperature. Molecules in the cooler object speed up at the expense of the kinetic energy of the warmer object. An ice cube cools a drink not by adding coldness to the liquid, but by absorbing heat from the liquid as the ice itself melts.

In general, we will use the **Celsius scale** to indicate temperature. (Celsius degrees are abbreviated °C; Appendix C shows how to convert between Celsius and Fahrenheit.) At sea level, water freezes at 0°C and boils at 100°C. The temperature of the human body averages 37°C, and comfortable room temperature is about 20–25°C.

One convenient unit of heat used in this book is the **calorie (cal)**. A calorie is the amount of heat it takes to raise the temperature of 1 g of water by 1°C. Conversely, a calorie is also the amount of heat that 1 g of water releases when it cools by 1°C. A **kilocalorie (kcal)**, 1,000 cal, is the quantity of heat required to raise the temperature of 1 kilogram (kg) of water by 1°C. (The "calories" on food packages are actually kilocalories.) Another energy unit used in this book is the **joule (J)**. One joule equals 0.239 cal; one calorie equals 4.184 J.

Water's High Specific Heat

The ability of water to stabilize temperature stems from its relatively high specific heat. The **specific heat** of a substance is defined as the amount of heat that must be absorbed or lost for 1 g of that substance to change its temperature by 1°C. We already know water's specific heat because we have defined a calorie as the amount of heat that causes 1 g of water to change its temperature by 1°C. Therefore, the specific heat of water is 1 calorie per gram and per degree Celsius, abbreviated as 1 cal/g·°C. Compared with most other substances, water has an unusually high specific heat. For example, ethyl alcohol, the type of alcohol in alcoholic beverages, has a specific heat of 0.6 cal/g·°C; that is, only 0.6 cal is required to raise the temperature of 1 g of ethyl alcohol by 1°C.

Because of the high specific heat of water relative to other materials, water will change its temperature less when it absorbs or loses a given amount of heat. The reason you can burn your fingers by touching the side of an iron pot on the stove when the water in the pot is still lukewarm is that the specific heat of water is ten times greater than that of iron.

▲ **Figure 3.5 Effect of a large body of water on climate.** By absorbing or releasing heat, oceans moderate coastal climates. In this example from an August day in Southern California, the relatively cool ocean reduces coastal air temperatures by absorbing heat.

In other words, the same amount of heat will raise the temperature of 1 g of the iron much faster than it will raise the temperature of 1 g of the water. Specific heat can be thought of as a measure of how well a substance resists changing its temperature when it absorbs or releases heat. Water resists changing its temperature; when it does change its temperature, it absorbs or loses a relatively large quantity of heat for each degree of change.

We can trace water's high specific heat, like many of its other properties, to hydrogen bonding. Heat must be absorbed in order to break hydrogen bonds; by the same token, heat is released when hydrogen bonds form. A calorie of heat causes a relatively small change in the temperature of water because much of the heat is used to disrupt hydrogen bonds before the water molecules can begin moving faster. And when the temperature of water drops slightly, many additional hydrogen bonds form, releasing a considerable amount of energy in the form of heat.

What is the relevance of water's high specific heat to life on Earth? A large body of water can absorb and store a huge amount of heat from the sun in the daytime and during summer while warming up only a few degrees. At night and during winter, the gradually cooling water can warm the air. This is the reason coastal areas generally have milder climates than inland regions (**Figure 3.5**). The high specific heat of water also tends to stabilize ocean temperatures, creating a favorable environment for marine life. Thus, because of its high specific heat, the water that covers most of Earth keeps temperature fluctuations on land and in water within limits that permit life. Also, because organisms are made primarily of water, they are better able to resist changes in their own temperature than if they were made of a liquid with a lower specific heat.

Evaporative Cooling

Molecules of any liquid stay close together because they are attracted to one another. Molecules moving fast enough to overcome these attractions can depart the liquid and enter the air as a gas. This transformation from a liquid to a gas is called vaporization, or *evaporation*. Recall that the speed of molecular movement varies and that temperature is the *average* kinetic energy of molecules. Even at low temperatures, the speediest molecules can escape into the air. Some evaporation occurs at any temperature; a glass of water at room temperature, for example, will eventually evaporate completely. If a liquid is heated, the average kinetic energy of molecules increases and the liquid evaporates more rapidly.

Heat of vaporization is the quantity of heat a liquid must absorb for 1 g of it to be converted from the liquid to the gaseous state. For the same reason that water has a high specific heat, it also has a high heat of vaporization relative to most other liquids. To evaporate 1 g of water at 25°C, about 580 cal of heat is needed—nearly double the amount needed to vaporize a gram of alcohol or ammonia. Water's high heat of vaporization is another emergent property resulting from the strength of its hydrogen bonds, which must be broken before the molecules can make their exodus from the liquid.

The high amount of energy required to vaporize water has a wide range of effects. On a global scale, for example, it helps moderate Earth's climate. A considerable amount of solar heat absorbed by tropical seas is consumed during the evaporation of surface water. Then, as moist tropical air circulates poleward, it releases heat as it condenses and forms rain. On an organismal level, water's high heat of vaporization accounts for the severity of steam burns. These burns are caused by the heat energy released when steam condenses into liquid on the skin.

As a liquid evaporates, the surface of the liquid that remains behind cools down. This **evaporative cooling** occurs because the "hottest" molecules, those with the greatest kinetic energy, are the most likely to leave as gas. It is as if the hundred fastest runners at a college transferred to another school; the average speed of the remaining students would decline.

Evaporative cooling of water contributes to the stability of temperature in lakes and ponds and also provides a mechanism that prevents terrestrial organisms from overheating. For example, evaporation of water from the leaves of a plant helps keep the tissues in the leaves from becoming too warm in the sunlight. Evaporation of sweat from human skin dissipates body heat and helps prevent overheating on a hot day or when excess heat is generated by strenuous activity. High humidity on a hot day increases discomfort because the high concentration of water vapor in the air inhibits the evaporation of sweat from the body.

Floating of Ice on Liquid Water

Water is one of the few substances that are less dense as a solid than as a liquid. In other words, ice floats on liquid water. While other materials contract and become denser when they solidify, water expands. The cause of this exotic behavior is, once again, hydrogen bonding. At temperatures above

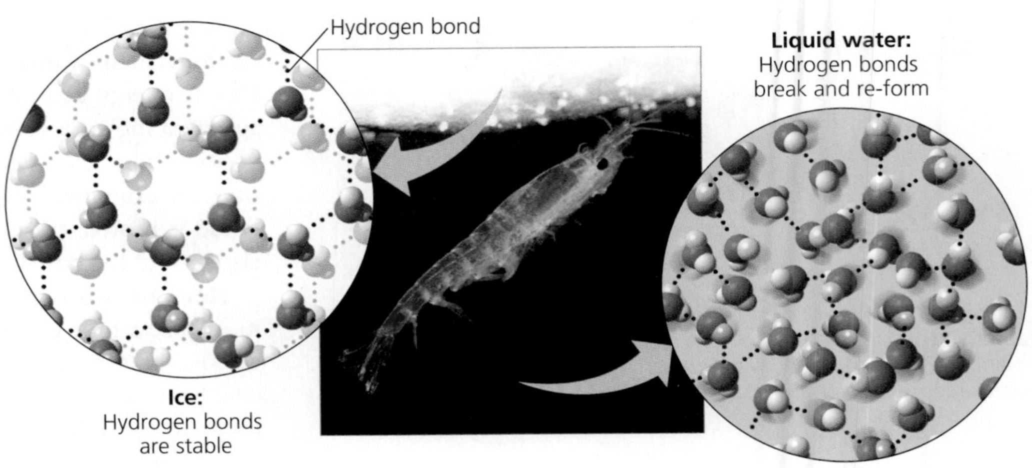

▶ **Figure 3.6 Ice: crystalline structure and floating barrier.** In ice, each molecule is hydrogen-bonded to four neighbors in a three-dimensional crystal. Because the crystal is spacious, ice has fewer molecules than an equal volume of liquid water. In other words, ice is less dense than liquid water. Floating ice becomes a barrier that protects the liquid water below from the colder air. The marine organism shown here is a type of shrimp called krill; it was photographed beneath floating ice in the Southern Ocean near Antarctica.

WHAT IF? *If water did not form hydrogen bonds, what would happen to the shrimp's environment?*

4°C, water behaves like other liquids, expanding as it warms and contracting as it cools. As the temperature falls from 4°C to 0°C, water begins to freeze because more and more of its molecules are moving too slowly to break hydrogen bonds. At 0°C, the molecules become locked into a crystalline lattice, each water molecule hydrogen-bonded to four partners **(Figure 3.6)**. The hydrogen bonds keep the molecules at "arm's length," far enough apart to make ice about 10% less dense (10% fewer molecules for the same volume) than liquid water at 4°C. When ice absorbs enough heat for its temperature to rise above 0°C, hydrogen bonds between molecules are disrupted. As the crystal collapses, the ice melts, and molecules are free to slip closer together. Water reaches its greatest density at 4°C and then begins to expand as the molecules move faster. Even in liquid water, many of the molecules are connected by hydrogen bonds, though only transiently: The hydrogen bonds are constantly breaking and re-forming.

The ability of ice to float due to its lower density is an important factor in the suitability of the environment for life. If ice sank, then eventually all ponds, lakes, and even oceans would freeze solid, making life as we know it impossible on Earth. During summer, only the upper few inches of the ocean would thaw. Instead, when a deep body of water cools, the floating ice insulates the liquid water below, preventing it from freezing and allowing life to exist under the frozen surface, as shown in the photo in Figure 3.6. Besides insulating the water below, ice also provides solid habitat for some animals, such as polar bears and seals (see Figure 3.1).

Along with many other scientists, Susan Solomon, the interviewee for this unit (see pp. 28-29), is worried that these bodies of ice are at risk of disappearing. Global warming, which is caused by carbon dioxide and other "greenhouse" gases in the atmosphere, is having a profound effect on icy environments around the globe. In the Arctic, the average air temperature has risen 1.4°C just since 1961. This temperature increase has affected the seasonal balance between Arctic sea ice and liquid water, causing ice to form later in the year, to melt earlier, and to cover a smaller area. The alarming rate at which glaciers and Arctic sea ice are disappearing is posing an extreme challenge to animals that depend on ice for their survival.

Water: The Solvent of Life

A sugar cube placed in a glass of water will dissolve. The glass will then contain a uniform mixture of sugar and water; the concentration of dissolved sugar will be the same everywhere in the mixture. A liquid that is a completely homogeneous mixture of two or more substances is called a **solution**. The dissolving agent of a solution is the **solvent**, and the substance that is dissolved is the **solute**. In this case, water is the solvent and sugar is the solute. An **aqueous solution** is one in which water is the solvent.

The medieval alchemists tried to find a universal solvent, one that would dissolve anything. They learned that nothing works better than water. Yet, water is not a universal solvent; if it were, it would dissolve any container in which it was stored, including our cells. Water is a very versatile solvent, however, a quality we can trace to the polarity of the water molecule.

Suppose, for example, that a spoonful of table salt, the ionic compound sodium chloride (NaCl), is placed in water **(Figure 3.7)**. At the surface of each grain, or crystal, of salt, the sodium and chloride ions are exposed to the solvent. These ions and the water molecules have a mutual affinity owing to the attraction between opposite charges. The oxygen regions of the water molecules are negatively charged and are attracted to sodium cations. The hydrogen regions are positively charged and are attracted to chloride anions. As a result, water molecules surround the individual sodium and chloride ions, separating and shielding them from one another. The sphere of water molecules around each dissolved ion is called a **hydration shell**. Working inward from the surface of each salt crystal, water eventually dissolves all the ions. The result is a solution of two solutes, sodium cations and chloride anions, homogeneously mixed with water, the solvent. Other ionic compounds also dissolve in water. Seawater, for instance, contains a great variety of dissolved ions, as do living cells.

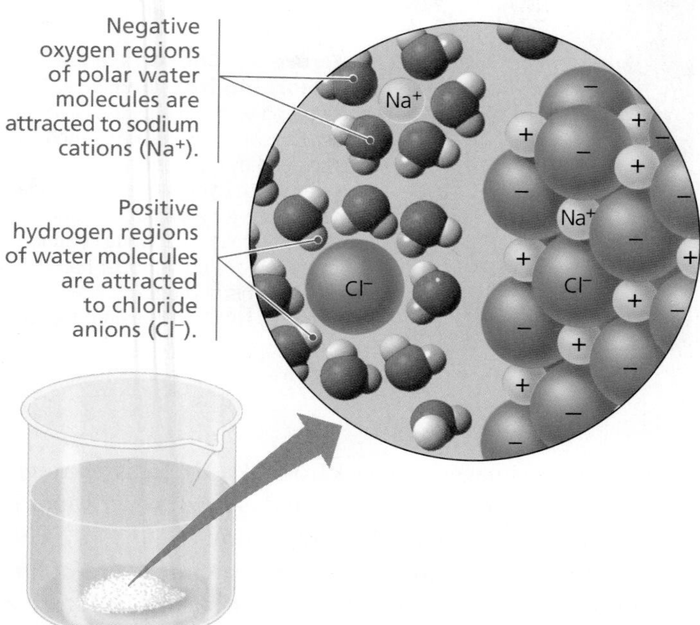

▲ **Figure 3.7 Table salt dissolving in water.** A sphere of water molecules, called a hydration shell, surrounds each solute ion.

WHAT IF? *What would happen if you heated this solution for a long time?*

A compound does not need to be ionic to dissolve in water; many compounds made up of nonionic polar molecules, such as sugars, are also water-soluble. Such compounds dissolve when water molecules surround each of the solute molecules, forming hydrogen bonds with them. Even molecules as large as proteins can dissolve in water if they have ionic and polar regions on their surface **(Figure 3.8)**. Many different kinds of polar compounds are dissolved (along with ions) in the water of such biological fluids as blood, the sap of plants, and the liquid within all cells. Water is the solvent of life.

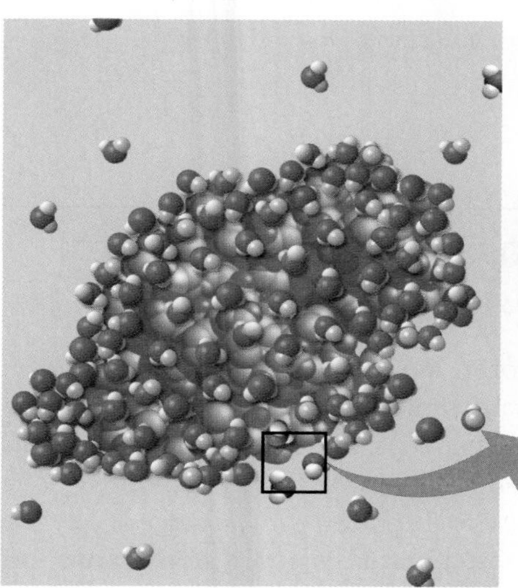

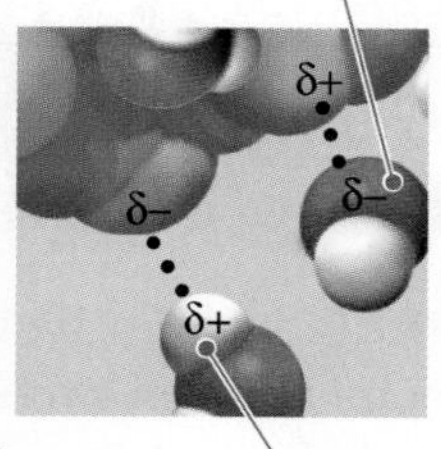

This oxygen is attracted to a slight positive charge on the lysozyme molecule.
δ+
δ–
δ–
δ+
This hydrogen is attracted to a slight negative charge on the lysozyme molecule.

▲ **Figure 3.8 A water-soluble protein.** Human lysozyme is a protein found in tears and saliva that has antibacterial action. This model shows the lysozyme molecule (purple) in an aqueous environment. Ionic and polar regions on the protein's surface attract water molecules.

Hydrophilic and Hydrophobic Substances

Any substance that has an affinity for water is said to be **hydrophilic** (from the Greek *hydro*, water, and *philios*, loving). In some cases, substances can be hydrophilic without actually dissolving. For example, some molecules in cells are so large that they do not dissolve. Instead, they remain suspended in the aqueous liquid of the cell. Such a mixture is an example of a **colloid**, a stable suspension of fine particles in a liquid. Another example of a hydrophilic substance that does not dissolve is cotton, a plant product. Cotton consists of giant molecules of cellulose, a compound with numerous regions of partial positive and partial negative charges that can form hydrogen bonds with water. Water adheres to the cellulose fibers. Thus, a cotton towel does a great job of drying the body, yet it does not dissolve in the washing machine. Cellulose is also present in the walls of water-conducting cells in a plant; you read earlier how the adhesion of water to these hydrophilic walls allows water transport to occur.

There are, of course, substances that do not have an affinity for water. Substances that are nonionic and nonpolar (or otherwise cannot form hydrogen bonds) actually seem to repel water; these substances are said to be **hydrophobic** (from the Greek *phobos*, fearing). An example from the kitchen is vegetable oil, which, as you know, does not mix stably with water-based substances such as vinegar. The hydrophobic behavior of the oil molecules results from a prevalence of relatively nonpolar covalent bonds, in this case bonds between carbon and hydrogen, which share electrons almost equally. Hydrophobic molecules related to oils are major ingredients of cell membranes. (Imagine what would happen to a cell if its membrane dissolved!)

Solute Concentration in Aqueous Solutions

Biological chemistry is "wet" chemistry. Most of the chemical reactions in organisms involve solutes dissolved in water. To understand such reactions, we must know how many atoms and molecules are involved and be able to calculate the concentration of solutes in an aqueous solution (the number of solute molecules in a volume of solution).

When carrying out experiments, we use mass to calculate the number of molecules. We know the mass of each atom in a given molecule, so we can calculate the **molecular mass**, which is simply the sum of the masses of all the atoms in a molecule. As an example, let's calculate the molecular mass of table sugar (sucrose), which has the molecular formula $C_{12}H_{22}O_{11}$. In round numbers of daltons, the mass of a carbon atom is 12, the mass of a hydrogen atom is 1, and the mass of an oxygen atom is 16. Thus, sucrose has a molecular mass of $(12 \times 12) + (22 \times 1) + (11 \times 16) = 342$ daltons. Of course, weighing out small numbers of molecules is not practical. For this reason, we usually measure substances in units called moles. Just as a dozen always means 12 objects, a **mole (mol)** represents an exact number of objects: 6.02×10^{23}, which is called Avogadro's number. Because of the way in which Avogadro's number and the unit *dalton* were originally defined, there are 6.02×10^{23} daltons in 1 g. This is significant because once we determine the molecular mass of a molecule such as sucrose, we can use the same number (342), but with the unit *gram*, to represent the mass of 6.02×10^{23} molecules of sucrose, or 1 mol of sucrose (this is sometimes called the *molar mass*). To obtain 1 mol of sucrose in the lab, therefore, we weigh out 342 g.

The practical advantage of measuring a quantity of chemicals in moles is that a mole of one substance has exactly the same number of molecules as a mole of any other substance. If the molecular mass of substance A is 342 daltons and that of substance B is 10 daltons, then 342 g of A will have the same number of molecules as 10 g of B. A mole of ethyl alcohol (C_2H_6O) also contains 6.02×10^{23} molecules, but its mass is only 46 g because the mass of a molecule of ethyl alcohol is less than that of a molecule of sucrose. Measuring in moles makes it convenient for scientists working in the laboratory to combine substances in fixed ratios of molecules.

How would we make a liter (L) of solution consisting of 1 mol of sucrose dissolved in water? We would measure out 342 g of sucrose and then gradually add water, while stirring, until the sugar was completely dissolved. We would then add enough water to bring the total volume of the solution up to 1 L. At that point, we would have a 1-molar (1 *M*) solution of sucrose. **Molarity**—the number of moles of solute per liter of solution—is the unit of concentration most often used by biologists for aqueous solutions.

Water's capacity as a versatile solvent complements the other properties discussed in this chapter. Since these remarkable properties allow water to support life on Earth so well, scientists who seek life elsewhere in the universe look for water as a sign that a planet might sustain life.

Possible Evolution of Life on Other Planets with Water

EVOLUTION Humans have probably always gazed skyward, wondering whether other living beings exist beyond Earth. And if life has arisen on other planets, into what form or forms has it evolved? Biologists who look for life elsewhere in the universe (known as *astrobiologists*) have concentrated their search on planets that might have water. To date, more than 200 planets have been found outside our solar system, and there is evidence for the presence of water vapor on one or two of them. In our own solar system, Mars has been most compelling to astrobiologists as a focus of study.

Like Earth, Mars has an ice cap at both poles. And in the decades since the age of space exploration began, scientists have found intriguing signs that water may exist elsewhere on Mars. Finally, in 2008, the robotic spacecraft *Phoenix* landed on Mars and began to sample its surface. Years of debate were resolved by the images sent back from *Phoenix*: Ice is definitely present just under Mars's surface, and enough water vapor is in the Martian atmosphere for frost to form (**Figure 3.9**). This exciting finding has reinvigorated the search for signs of life, past or present, on Mars and other planets. If any life-forms or fossils are found, their study will shed light on the process of evolution from an entirely new perspective.

◀ **Figure 3.9 Subsurface ice and morning frost on Mars.** This photograph was taken by the Mars lander *Phoenix* in 2008. The trench was scraped by a robotic arm, uncovering ice (white in rectangle near bottom) below the surface material. Frost also appears as a white coating in several places in the upper half of the image. This photograph was colorized by NASA to highlight the ice.

CONCEPT CHECK 3.2

1. Describe how properties of water contribute to the upward movement of water in a tree.
2. Explain the saying "It's not the heat; it's the humidity."
3. How can the freezing of water crack boulders?
4. The concentration of the appetite-regulating hormone ghrelin is about $1.3 \times 10^{-10} M$ in a fasting person. How many molecules of ghrelin are in 1 L of blood?
5. **WHAT IF?** A water strider (which can walk on water) has legs that are coated with a hydrophobic substance. What might be the benefit? What would happen if the substance were hydrophilic?

For suggested answers, see Appendix A.

CONCEPT 3.3

Acidic and basic conditions affect living organisms

Occasionally, a hydrogen atom participating in a hydrogen bond between two water molecules shifts from one molecule to the other. When this happens, the hydrogen atom leaves its electron behind, and what is actually transferred is a **hydrogen ion** (H^+), a single proton with a charge of 1+. The water molecule that lost a proton is now a **hydroxide ion** (OH^-), which has a charge of 1−. The proton binds to the other water molecule, making that molecule a **hydronium ion** (H_3O^+). We can picture the chemical reaction as shown at the top of the next page.

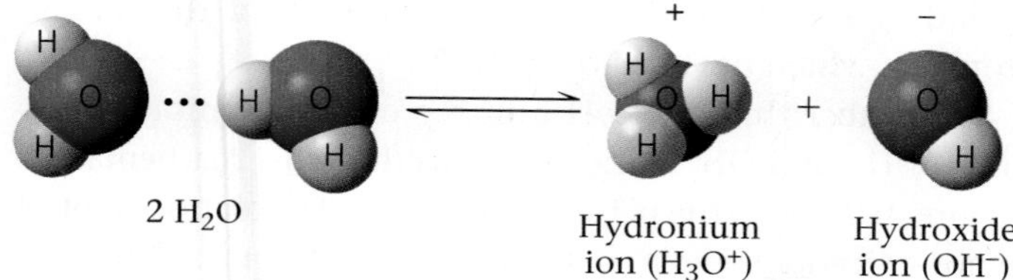

By convention, H^+ (the hydrogen ion) is used to represent H_3O^+ (the hydronium ion), and we follow that practice here. Keep in mind, though, that H^+ does not exist on its own in an aqueous solution. It is always associated with another water molecule in the form of H_3O^+.

As indicated by the double arrows, this is a reversible reaction that reaches a state of dynamic equilibrium when water molecules dissociate at the same rate that they are being re-formed from H^+ and OH^-. At this equilibrium point, the concentration of water molecules greatly exceeds the concentrations of H^+ and OH^-. In pure water, only one water molecule in every 554 million is dissociated; the concentration of each ion in pure water is 10^{-7} *M* (at 25°C). This means there is only one ten-millionth of a mole of hydrogen ions per liter of pure water and an equal number of hydroxide ions.

Although the dissociation of water is reversible and statistically rare, it is exceedingly important in the chemistry of life. H^+ and OH^- are very reactive. Changes in their concentrations can drastically affect a cell's proteins and other complex molecules. As we have seen, the concentrations of H^+ and OH^- are equal in pure water, but adding certain kinds of solutes, called acids and bases, disrupts this balance. Biologists use something called the pH scale to describe how acidic or basic (the opposite of acidic) a solution is. In the remainder of this chapter, you will learn about acids, bases, and pH and why changes in pH can adversely affect organisms.

Acids and Bases

What would cause an aqueous solution to have an imbalance in H^+ and OH^- concentrations? When acids dissolve in water, they donate additional H^+ to the solution. An **acid** is a substance that increases the hydrogen ion concentration of a solution. For example, when hydrochloric acid (HCl) is added to water, hydrogen ions dissociate from chloride ions:

$$HCl \rightarrow H^+ + Cl^-$$

This source of H^+ (dissociation of water is the other source) results in an acidic solution—one having more H^+ than OH^-.

A substance that reduces the hydrogen ion concentration of a solution is called a **base**. Some bases reduce the H^+ concentration directly by accepting hydrogen ions. Ammonia (NH_3), for instance, acts as a base when the unshared electron pair in nitrogen's valence shell attracts a hydrogen ion from the solution, resulting in an ammonium ion (NH_4^+):

$$NH_3 + H^+ \rightleftharpoons NH_4^+$$

Other bases reduce the H^+ concentration indirectly by dissociating to form hydroxide ions, which combine with hydrogen ions and form water. One such base is sodium hydroxide (NaOH), which in water dissociates into its ions:

$$NaOH \rightarrow Na^+ + OH^-$$

In either case, the base reduces the H^+ concentration. Solutions with a higher concentration of OH^- than H^+ are known as basic solutions. A solution in which the H^+ and OH^- concentrations are equal is said to be neutral.

Notice that single arrows were used in the reactions for HCl and NaOH. These compounds dissociate completely when mixed with water, so hydrochloric acid is called a strong acid and sodium hydroxide a strong base. In contrast, ammonia is a relatively weak base. The double arrows in the reaction for ammonia indicate that the binding and release of hydrogen ions are reversible reactions, although at equilibrium there will be a fixed ratio of NH_4^+ to NH_3.

There are also weak acids, which reversibly release and accept back hydrogen ions. An example is carbonic acid:

$$\underset{\text{Carbonic acid}}{H_2CO_3} \rightleftharpoons \underset{\text{Bicarbonate ion}}{HCO_3^-} + \underset{\text{Hydrogen ion}}{H^+}$$

Here the equilibrium so favors the reaction in the left direction that when carbonic acid is added to pure water, only 1% of the molecules are dissociated at any particular time. Still, that is enough to shift the balance of H^+ and OH^- from neutrality.

The pH Scale

In any aqueous solution at 25°C, the *product* of the H^+ and OH^- concentrations is constant at 10^{-14}. This can be written

$$[H^+][OH^-] = 10^{-14}$$

In such an equation, brackets indicate molar concentration. In a neutral solution at room temperature (25°C), $[H^+] = 10^{-7}$ and $[OH^-] = 10^{-7}$, so in this case, 10^{-14} is the product of $10^{-7} \times 10^{-7}$. If enough acid is added to a solution to increase $[H^+]$ to 10^{-5} *M*, then $[OH^-]$ will decline by an equivalent amount to 10^{-9} *M* (note that $10^{-5} \times 10^{-9} = 10^{-14}$). This constant relationship expresses the behavior of acids and bases in an aqueous solution. An acid not only adds hydrogen ions to a solution, but also removes hydroxide ions because of the tendency for H^+ to combine with OH^-, forming water. A base has the opposite effect, increasing OH^- concentration but also reducing H^+ concentration by the formation of water. If enough of a base is added to raise the OH^- concentration to 10^{-4} *M*, it will cause the H^+ concentration to drop to 10^{-10} *M*. Whenever we know the concentration of either H^+ or OH^- in an aqueous solution, we can deduce the concentration of the other ion.

Because the H^+ and OH^- concentrations of solutions can vary by a factor of 100 trillion or more, scientists have

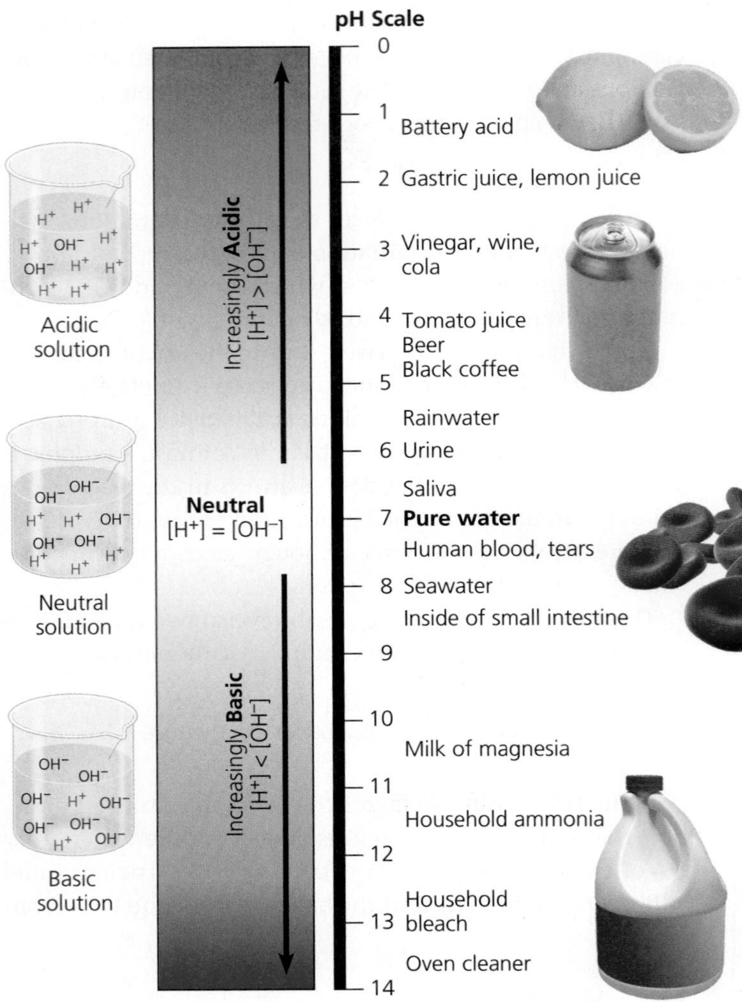

▲ **Figure 3.10 The pH scale and pH values of some aqueous solutions.**

developed a way to express this variation more conveniently than in moles per liter. The pH scale **(Figure 3.10)** compresses the range of H^+ and OH^- concentrations by employing logarithms. The **pH** of a solution is defined as the negative logarithm (base 10) of the hydrogen ion concentration:

$$\text{pH} = -\log [H^+]$$

For a neutral aqueous solution, $[H^+]$ is 10^{-7} *M*, giving us

$$-\log 10^{-7} = -(-7) = 7$$

Notice that pH *declines* as H^+ concentration *increases*. Notice, too, that although the pH scale is based on H^+ concentration, it also implies OH^- concentration. A solution of pH 10 has a hydrogen ion concentration of 10^{-10} *M* and a hydroxide ion concentration of 10^{-4} *M*.

The pH of a neutral aqueous solution at 25°C is 7, the midpoint of the pH scale. A pH value less than 7 denotes an acidic solution; the lower the number, the more acidic the solution. The pH for basic solutions is above 7. Most biological fluids are within the range pH 6–8. There are a few exceptions, however, including the strongly acidic digestive juice of the human stomach, which has a pH of about 2.

Remember that each pH unit represents a tenfold difference in H^+ and OH^- concentrations. It is this mathematical feature that makes the pH scale so compact. A solution of pH 3 is not twice as acidic as a solution of pH 6, but a thousand times ($10 \times 10 \times 10$) more acidic. When the pH of a solution changes slightly, the actual concentrations of H^+ and OH^- in the solution change substantially.

Buffers

The internal pH of most living cells is close to 7. Even a slight change in pH can be harmful, because the chemical processes of the cell are very sensitive to the concentrations of hydrogen and hydroxide ions. The pH of human blood is very close to 7.4, or slightly basic. A person cannot survive for more than a few minutes if the blood pH drops to 7 or rises to 7.8, and a chemical system exists in the blood that maintains a stable pH. If you add 0.01 mol of a strong acid to a liter of pure water, the pH drops from 7.0 to 2.0. If the same amount of acid is added to a liter of blood, however, the pH decrease is only from 7.4 to 7.3. Why does the addition of acid have so much less of an effect on the pH of blood than it does on the pH of water?

The presence of substances called buffers allows biological fluids to maintain a relatively constant pH despite the addition of acids or bases. A **buffer** is a substance that minimizes changes in the concentrations of H^+ and OH^- in a solution. It does so by accepting hydrogen ions from the solution when they are in excess and donating hydrogen ions to the solution when they have been depleted. Most buffer solutions contain a weak acid and its corresponding base, which combine reversibly with hydrogen ions.

There are several buffers that contribute to pH stability in human blood and many other biological solutions. One of these is carbonic acid (H_2CO_3), formed when CO_2 reacts with water in blood plasma. As mentioned earlier, carbonic acid dissociates to yield a bicarbonate ion (HCO_3^-) and a hydrogen ion (H^+):

$$\underset{\substack{H^+ \text{ donor} \\ \text{(acid)}}}{H_2CO_3} \underset{\substack{\text{Response to} \\ \text{a drop in pH}}}{\overset{\substack{\text{Response to} \\ \text{a rise in pH}}}{\rightleftharpoons}} \underset{\substack{H^+ \text{ acceptor} \\ \text{(base)}}}{HCO_3^-} + \underset{\substack{\text{Hydrogen} \\ \text{ion}}}{H^+}$$

The chemical equilibrium between carbonic acid and bicarbonate acts as a pH regulator, the reaction shifting left or right as other processes in the solution add or remove hydrogen ions. If the H^+ concentration in blood begins to fall (that is, if pH rises), the reaction proceeds to the right and more carbonic acid dissociates, replenishing hydrogen ions. But when H^+ concentration in blood begins to rise (when pH drops), the reaction proceeds to the left, with HCO_3^- (the base) removing the hydrogen ions from the solution and forming H_2CO_3. Thus, the carbonic acid–bicarbonate buffering system consists of an acid and a base in equilibrium with each other. Most other buffers are also acid-base pairs.

Acidification: A Threat to Water Quality

Among the many threats to water quality posed by human activities is the burning of fossil fuels, which releases gaseous compounds into the atmosphere. When certain of these compounds react with water, the water becomes more acidic, altering the delicate balance of conditions for life on Earth.

Carbon dioxide is the main product of fossil fuel combustion. About 25% of human-generated CO_2 is absorbed by the oceans. In spite of the huge volume of water in the oceans, scientists worry that the absorption of so much CO_2 will harm marine ecosystems.

Recent data have shown that such fears are well founded. When CO_2 dissolves in seawater, it reacts with water to form carbonic acid, which lowers ocean pH, a process known as **ocean acidification**. Based on measurements of CO_2 levels in air bubbles trapped in ice over thousands of years, scientists calculate that the pH of the oceans is 0.1 pH unit lower now than at any time in the past 420,000 years. Recent studies predict that it will drop another 0.3–0.5 pH unit by the end of this century.

As seawater acidifies, the extra hydrogen ions combine with carbonate ions (CO_3^{2-}) to form bicarbonate ions (HCO_3^-), thereby reducing the carbonate concentration (**Figure 3.11**).

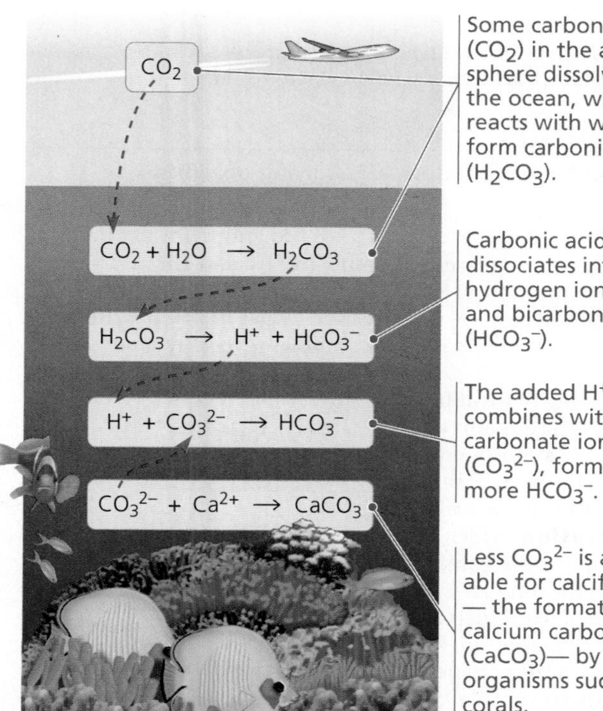

CO_2
$CO_2 + H_2O \rightarrow H_2CO_3$
$H_2CO_3 \rightarrow H^+ + HCO_3^-$
$H^+ + CO_3^{2-} \rightarrow HCO_3^-$
$CO_3^{2-} + Ca^{2+} \rightarrow CaCO_3$
Some carbon dioxide (CO_2) in the atmosphere dissolves in the ocean, where it reacts with water to form carbonic acid (H_2CO_3).
Carbonic acid dissociates into hydrogen ions (H^+) and bicarbonate ions (HCO_3^-).
The added H^+ combines with carbonate ions (CO_3^{2-}), forming more HCO_3^-.
Less CO_3^{2-} is available for calcification — the formation of calcium carbonate ($CaCO_3$)— by marine organisms such as corals.

▲ **Figure 3.11 Atmospheric CO_2 from human activities and its fate in the ocean.**

▼ **Figure 3.12**

IMPACT

The Threat of Ocean Acidification to Coral Reef Ecosystems

Recently, scientists have sounded the alarm about the effects of ocean acidification, the process in which oceans become more acidic due to increased atmospheric carbon dioxide levels (see Figure 3.11). They predict that the resulting decrease in the concentration of carbonate ion (CO_3^{2-}) will take a serious toll on coral reef calcification. Taking many studies into account, and including the effects of ocean warming as well, one group of scientists defined three scenarios for coral reefs during this century, depending on whether the concentration of atmospheric CO_2 (a) stays at today's level, (b) increases at the current rate, or (c) increases more rapidly. The photographs below show coral reefs resembling those predicted under each scenario.

(a)

(b)

(c)

The healthy coral reef in (a) supports a highly diverse group of species and bears little resemblance to the damaged coral reef in (c).

WHY IT MATTERS The disappearance of coral reef ecosystems would be a tragic loss of biological diversity. In addition, coral reefs provide shoreline protection, a feeding ground for many commercial fishery species, and a popular tourist draw, so coastal human communities would suffer from greater wave damage, collapsed fisheries, and reduced tourism.

FURTHER READING O. Hoegh-Guldberg et al., Coral reefs under rapid climate change and ocean acidification, *Science* 318:1737–1742 (2007). S. C. Doney, The dangers of ocean acidification, *Scientific American*, March 2006, 58–65.

WHAT IF? Would lowering the ocean's carbonate concentration have any effect, even indirectly, on organisms that don't form $CaCO_3$? Explain.

Scientists predict that ocean acidification will cause the carbonate concentration to decrease by 40% by the year 2100. This is of great concern because carbonate is required for calcification, the production of calcium carbonate ($CaCO_3$) by many marine organisms, including reef-building corals and animals that build shells. Coral reefs are sensitive ecosystems that act as havens for a great diversity of marine life (**Figure 3.12**).

The burning of fossil fuels is also a major source of sulfur oxides and nitrogen oxides. These compounds react with water in the air to form strong acids, which fall to Earth with rain or snow. **Acid precipitation** refers to rain, snow, or fog with a pH lower (more acidic) than 5.2. (Uncontaminated rain has a pH of about 5.6, which is slightly acidic due to the formation of carbonic acid from CO_2 and water.) Acid precipitation can damage life in lakes and streams, and it adversely affects plants on land by changing soil chemistry. To address this problem, the U.S. Congress amended the Clean Air Act in 1990, and the mandated improvements in industrial technologies have been largely responsible for improving the health of most North American lakes and forests.

If there is any reason for optimism about the future quality of water resources on our planet, it is that we have made progress in learning about the delicate chemical balances in oceans, lakes, and rivers. Continued progress can come only from the actions of informed individuals, like yourselves, who are concerned about environmental quality. This requires understanding the crucial role that water plays in the suitability of the environment for continued life on Earth.

CONCEPT CHECK 3.3

1. Compared with a basic solution at pH 9, the same volume of an acidic solution at pH 4 has ____ times as many hydrogen ions (H^+).
2. HCl is a strong acid that dissociates in water: $HCl \rightarrow H^+ + Cl^-$. What is the pH of 0.01 *M* HCl?
3. Acetic acid (CH_3COOH) can be a buffer, similar to carbonic acid. Write the dissociation reaction, identifying the acid, base, H^+ acceptor, and H^+ donor.
4. **WHAT IF?** Given a liter of pure water and a liter solution of acetic acid, what would happen to the pH if you added 0.01 mol of a strong acid to each? Use the reaction equation from question 3 to explain the result.

For suggested answers, see Appendix A.

3 CHAPTER REVIEW

SUMMARY OF KEY CONCEPTS

CONCEPT 3.1

Polar covalent bonds in water molecules result in hydrogen bonding (pp. 46–47)

- A hydrogen bond forms when the slightly negatively charged oxygen of one water molecule is attracted to the slightly positively charged hydrogen of a nearby water molecule. Hydrogen bonding between water molecules is the basis for water's properties.

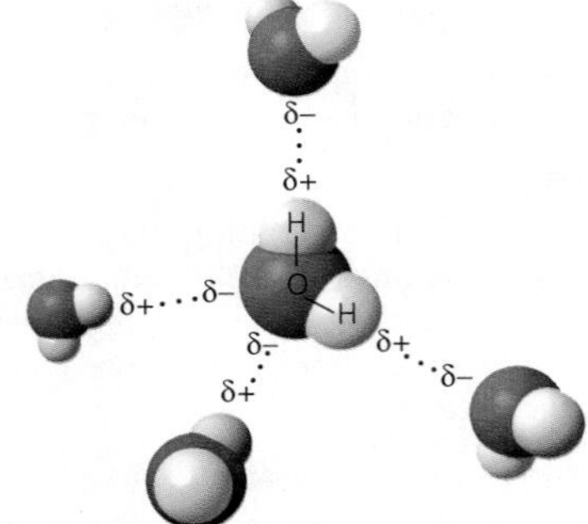

DRAW IT *Label a hydrogen bond and a polar covalent bond in this figure. How many hydrogen bonds can each water molecule make?*

CONCEPT 3.2

Four emergent properties of water contribute to Earth's suitability for life (pp. 47–52)

- Hydrogen bonding keeps water molecules close to each other, and this **cohesion** helps pull water upward in the microscopic water-conducting cells of plants. Hydrogen bonding is also responsible for water's **surface tension**.
- Water has a high **specific heat**: Heat is absorbed when hydrogen bonds break and is released when hydrogen bonds form. This helps keep temperatures relatively steady, within limits that permit life. **Evaporative cooling** is based on water's high **heat of vaporization**. The evaporative loss of the most energetic water molecules cools a surface.
- Ice floats because it is less dense than liquid water. This allows life to exist under the frozen surfaces of lakes and polar seas.

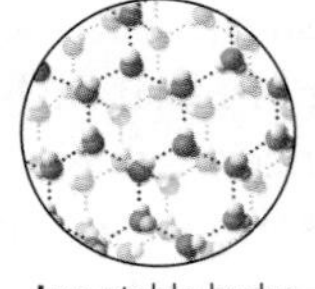

Ice: stable hydrogen bonds

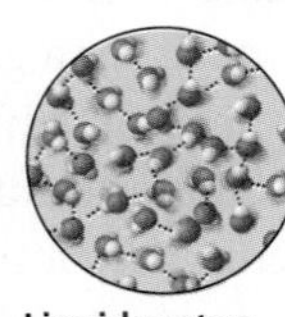

Liquid water: transient hydrogen bonds

- Water is an unusually versatile **solvent** because its polar molecules are attracted to charged and polar substances capable of forming hydrogen bonds. **Hydrophilic** substances have an affinity for water; **hydrophobic** substances do not. **Molarity**, the number of moles of **solute** per liter of **solution**, is used as a measure of solute concentration in solutions. A **mole** is a certain number of molecules of a substance. The mass of a mole of a substance in grams is the same as the **molecular mass** in daltons.
- The emergent properties of water support life on Earth and may contribute to the potential for life to have evolved on other planets.

? *Describe how different types of solutes dissolve in water. Explain the difference between a solution and a colloid.*

CONCEPT 3.3

Acidic and basic conditions affect living organisms (pp. 52–56)

- A water molecule can transfer an H^+ to another water molecule to form H_3O^+ (represented simply by H^+) and OH^-.
- The concentration of H^+ is expressed as **pH**; pH = $-\log [H^+]$. **Buffers** in biological fluids resist changes in pH. A buffer consists of an acid-base pair that combines reversibly with hydrogen ions.
- The burning of fossil fuels increases the amount of CO_2 in the atmosphere. Some CO_2 dissolves in the oceans, causing **ocean acidification**, which has potentially grave consequences for coral reefs. The burning of fossil fuels also releases oxides of sulfur and nitrogen, leading to **acid precipitation**.

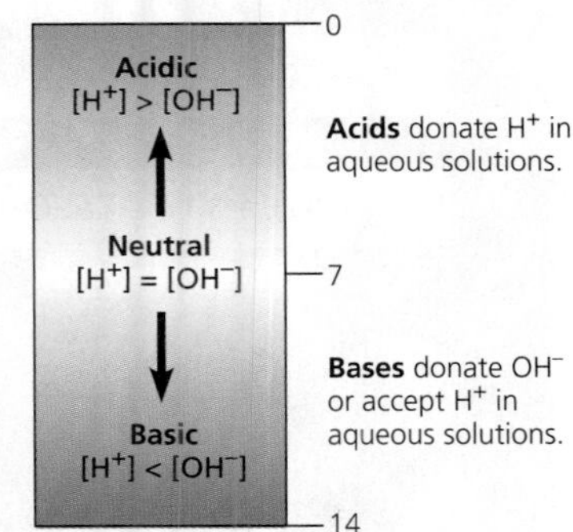

? *Explain how increasing amounts of CO_2 dissolving in the ocean leads to ocean acidification. How does this change in pH affect carbonate ion concentration and the rate of calcification?*

TEST YOUR UNDERSTANDING

LEVEL 1: KNOWLEDGE/COMPREHENSION

1. Many mammals control their body temperature by sweating. Which property of water is most directly responsible for the ability of sweat to lower body temperature?
   a. water's change in density when it condenses
   b. water's ability to dissolve molecules in the air
   c. the release of heat by the formation of hydrogen bonds
   d. the absorption of heat by the breaking of hydrogen bonds
   e. water's high surface tension

2. The bonds that are broken when water vaporizes are
   a. ionic bonds.
   b. hydrogen bonds between water molecules.
   c. covalent bonds between atoms within water molecules.
   d. polar covalent bonds.
   e. nonpolar covalent bonds.

3. Which of the following is a hydrophobic material?
   a. paper d. sugar
   b. table salt e. pasta
   c. wax

4. We can be sure that a mole of table sugar and a mole of vitamin C are equal in their
   a. mass in daltons. d. number of atoms.
   b. mass in grams. e. number of molecules.
   c. volume.

5. Measurements show that the pH of a particular lake is 4.0. What is the hydrogen ion concentration of the lake?
   a. 4.0 *M* b. 10^{-10} *M* c. 10^{-4} *M* d. 10^{4} *M* e. 4%

6. What is the *hydroxide* ion concentration of the lake described in question 5?
   a. 10^{-10} *M* b. 10^{-4} *M* c. 10^{-7} *M* d. 10^{-14}*M* e. 10 *M*

LEVEL 2: APPLICATION/ANALYSIS

7. A slice of pizza has 500 kcal. If we could burn the pizza and use all the heat to warm a 50-L container of cold water, what would be the approximate increase in the temperature of the water? (*Note*: A liter of cold water weighs about 1 kg.)
   a. 50°C b. 5°C c. 1°C d. 100°C e. 10°C

8. How many grams of acetic acid ($C_2H_4O_2$) would you use to make 10 L of a 0.1 *M* aqueous solution of acetic acid? (*Note*: The atomic masses, in daltons, are approximately 12 for carbon, 1 for hydrogen, and 16 for oxygen.)
   a. 10 g b. 0.1 g c. 6.0 g d. 60 g e. 0.6 g

9. **DRAW IT** Draw the hydration shells that form around a potassium ion and a chloride ion when potassium chloride (KCl) dissolves in water. Label the positive, negative, and partial charges on the atoms.

10. **MAKE CONNECTIONS** What do global warming (see Chapter 1, p. 6) and ocean acidification have in common?

LEVEL 3: SYNTHESIS/EVALUATION

11. In agricultural areas, farmers pay close attention to the weather forecast. Right before a predicted overnight freeze, farmers spray water on crops to protect the plants. Use the properties of water to explain how this method works. Be sure to mention why hydrogen bonds are responsible for this phenomenon.

12. **EVOLUTION CONNECTION**
    This chapter explains how the emergent properties of water contribute to the suitability of the environment for life. Until fairly recently, scientists assumed that other physical requirements for life included a moderate range of temperature, pH, atmospheric pressure, and salinity, as well as low levels of toxic chemicals. That view has changed with the discovery of organisms known as extremophiles, which have been found flourishing in hot, acidic sulfur springs, around hydrothermal vents deep in the ocean, and in soils with high levels of toxic metals. Why would astrobiologists be interested in studying extremophiles? What does the existence of life in such extreme environments say about the possibility of life on other planets?

13. **SCIENTIFIC INQUIRY**
    Design a controlled experiment to test the hypothesis that acid precipitation inhibits the growth of *Elodea,* a common freshwater plant (see Figure 2.19, p. 43).

14. **SCIENTIFIC INQUIRY**
    In a study reported in 2000, C. Langdon and colleagues used an artificial coral reef system to test the effect of carbonate concentration on the rate of calcification by reef organisms. The graph on the right presents one set of their results. Describe what these data show. How do these results relate to the ocean acidification that is associated with increasing atmospheric CO_2 levels?

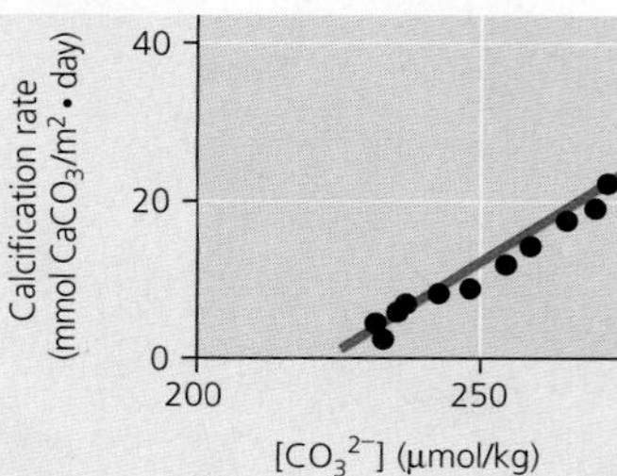

15. **SCIENCE, TECHNOLOGY, AND SOCIETY**
    Agriculture, industry, and the growing populations of cities all compete, through political influence, for water. If you were in charge of water resources in an arid region, what would your priorities be for allocating the limited water supply for various uses? How would you try to build consensus among the different special-interest groups?

16. **WRITE ABOUT A THEME**
    Emergent Properties Several emergent properties of water contribute to the suitability of the environment for life. In a short essay (100–150 words), describe how the ability of water to function as a versatile solvent arises from the structure of water molecules.

For selected answers, see Appendix A.

4

Carbon and the Molecular Diversity of Life

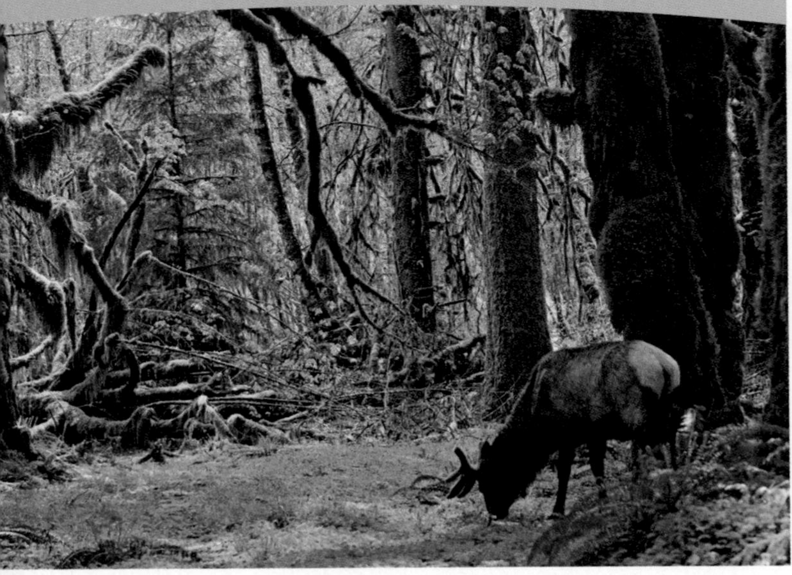

▲ **Figure 4.1 What properties make carbon the basis of all life?**

KEY CONCEPTS

4.1 **Organic chemistry is the study of carbon compounds**

4.2 **Carbon atoms can form diverse molecules by bonding to four other atoms**

4.3 **A few chemical groups are key to the functioning of biological molecules**

OVERVIEW

Carbon: The Backbone of Life

Water is the universal medium for life on Earth, but living organisms, such as the plants and Roosevelt elk in **Figure 4.1**, are made up of chemicals based mostly on the element carbon. Carbon enters the biosphere through the action of plants. Plants use solar energy to transform atmospheric CO_2 into the molecules of life, which are then taken in by plant-eating animals.

Of all chemical elements, carbon is unparalleled in its ability to form molecules that are large, complex, and varied, making possible the diversity of organisms that have evolved on Earth. Proteins, DNA, carbohydrates, and other molecules that distinguish living matter from inanimate material are all composed of carbon atoms bonded to one another and to atoms of other elements. Hydrogen (H), oxygen (O), nitrogen (N), sulfur (S), and phosphorus (P) are other common ingredients of these compounds, but it is the element carbon (C) that accounts for the enormous variety of biological molecules.

Large biological molecules, such as proteins, are the main focus of Chapter 5. In this chapter, we investigate the properties of smaller molecules. We will use these small molecules to illustrate concepts of molecular architecture that will help explain why carbon is so important to life, at the same time highlighting the theme that emergent properties arise from the organization of matter in living organisms.

CONCEPT 4.1

Organic chemistry is the study of carbon compounds

For historical reasons, compounds containing carbon are said to be organic, and the branch of chemistry that specializes in the study of carbon compounds is called **organic chemistry**. Organic compounds range from simple molecules, such as methane (CH_4), to colossal ones, such as proteins, with thousands of atoms. Most organic compounds contain hydrogen atoms in addition to carbon atoms.

The overall percentages of the major elements of life—C, H, O, N, S, and P—are quite uniform from one organism to another. Because of carbon's versatility, however, this limited assortment of atomic building blocks can be used to build an inexhaustible variety of organic molecules. Different species of organisms, and different individuals within a species, are distinguished by variations in their organic molecules.

Since the dawn of human history, people have used other organisms as sources of valued substances—from foods and medicines to fabrics. The science of organic chemistry originated in attempts to purify and improve the yield of such products. By the early 1800s, chemists had learned to make many simple compounds in the laboratory by combining elements under the right conditions. Artificial synthesis of the complex molecules extracted from living matter seemed impossible, however. At that time, the Swedish chemist Jöns Jakob Berzelius made the distinction between organic compounds, those thought to arise only in living organisms, and inorganic compounds, those found only in the nonliving world. *Vitalism*, the belief in a life force outside the jurisdiction of physical and chemical laws, provided the foundation for the new discipline of organic chemistry.

Chemists began to chip away at the support for vitalism when they finally learned to synthesize organic compounds in the laboratory. In 1828, Friedrich Wöhler, a German chemist who had studied with Berzelius, tried to make an "inorganic" salt, ammonium cyanate, by mixing solutions of ammonium ions (NH_4^+) and cyanate ions (CNO^-). Wöhler was astonished to find that instead he had made urea, an organic compound present in the urine of animals. Wöhler challenged the vitalists when he wrote, "I must tell you that I can prepare urea without requiring a kidney or an animal, either man or dog." However, one of the ingredients used in the synthesis, the cyanate, had been extracted from animal blood, and the vitalists were not swayed by Wöhler's discovery. A few years later, however, Hermann Kolbe, a student of Wöhler's, made the organic compound acetic acid from inorganic substances that could be prepared directly from pure elements. Vitalism crumbled completely after several decades of laboratory synthesis of increasingly complex organic compounds.

Organic Molecules and the Origin of Life on Earth

EVOLUTION In 1953, Stanley Miller, a graduate student of Harold Urey's at the University of Chicago, helped bring the abiotic (nonliving) synthesis of organic compounds into the context of evolution. Study **Figure 4.2** to learn about his classic experiment. From his results, Miller concluded that complex organic molecules could arise spontaneously under conditions thought to have existed on the early Earth. Miller also performed experiments designed to mimic volcanic conditions, with roughly similar results. In 2008, a former graduate student of Miller's discovered some samples from these experiments. Reanalyzing them using modern equipment, he identified additional organic compounds that had not been found by Miller. Although the jury is still out, these experiments support the idea that abiotic synthesis of organic compounds, perhaps near volcanoes, could have been an early stage in the origin of life (see Chapter 25).

The pioneers of organic chemistry helped shift the mainstream of biological thought from vitalism to *mechanism*, the view that physical and chemical laws govern all natural phenomena, including the processes of life. Organic chemistry was redefined as the study of carbon compounds, regardless of origin. Organisms produce most of the naturally occurring organic compounds, and these molecules represent a diversity and range of complexity unrivaled by inorganic compounds. However, the rules of chemistry apply to all molecules. The foundation of organic chemistry is not some intangible life force, but the unique chemical versatility of the element carbon.

CONCEPT CHECK 4.1

1. Why was Wöhler astonished to find he had made urea?
2. **WHAT IF?** When Miller tried his experiment without the electrical discharge, no organic compounds were found. What might explain this result?

For suggested answers, see Appendix A.

▼ **Figure 4.2** **INQUIRY**

Can organic molecules form under conditions believed to simulate those on the early Earth?

EXPERIMENT In 1953, Stanley Miller set up a closed system to mimic conditions thought to have existed on the early Earth. A flask of water simulated the primeval sea. The water was heated so that some vaporized and moved into a second, higher flask containing the "atmosphere"—a mixture of gases. Sparks were discharged in the synthetic atmosphere to mimic lightning.

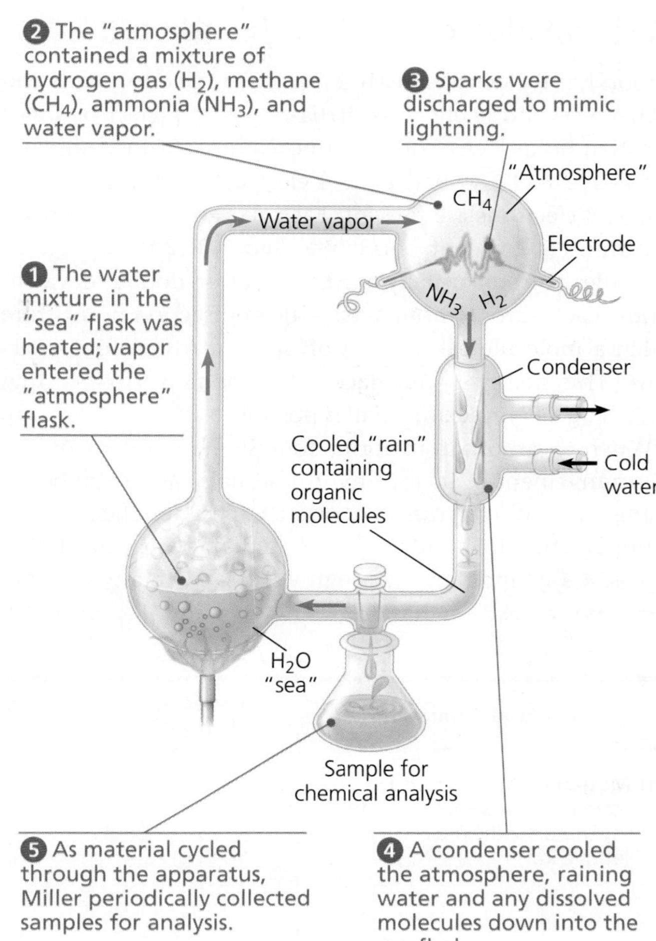

2 The "atmosphere" contained a mixture of hydrogen gas (H2), methane (CH4), ammonia (NH3), and water vapor.
3 Sparks were discharged to mimic lightning.
"Atmosphere"
CH4
Water vapor
Electrode
NH3
H2
1 The water mixture in the "sea" flask was heated; vapor entered the "atmosphere" flask.
Condenser
Cooled "rain" containing organic molecules
Cold water
H2O "sea"
Sample for chemical analysis
5 As material cycled through the apparatus, Miller periodically collected samples for analysis.
4 A condenser cooled the atmosphere, raining water and any dissolved molecules down into the sea flask.

RESULTS Miller identified a variety of organic molecules that are common in organisms. These included simple compounds, such as formaldehyde (CH_2O) and hydrogen cyanide (HCN), and more complex molecules, such as amino acids and long chains of carbon and hydrogen known as hydrocarbons.

CONCLUSION Organic molecules, a first step in the origin of life, may have been synthesized abiotically on the early Earth. (We will explore this hypothesis in more detail in Chapter 25.)

SOURCE S. L. Miller, A production of amino acids under possible primitive Earth conditions, *Science* 117:528–529 (1953).

WHAT IF? If Miller had increased the concentration of NH_3 in his experiment, how might the relative amounts of the products HCN and CH_2O have differed?

CONCEPT **4.2**

Carbon atoms can form diverse molecules by bonding to four other atoms

The key to an atom's chemical characteristics is its electron configuration. This configuration determines the kinds and number of bonds an atom will form with other atoms.

The Formation of Bonds with Carbon

Carbon has 6 electrons, with 2 in the first electron shell and 4 in the second shell; thus, it has 4 valence electrons in a shell that holds 8 electrons. A carbon atom usually completes its valence shell by sharing its 4 electrons with other atoms so that 8 electrons are present. Each pair of shared electrons constitutes a covalent bond (see Figure 2.12d). In organic molecules, carbon usually forms single or double covalent bonds. Each carbon atom acts as an intersection point from which a molecule can branch off in as many as four directions. This ability is one facet of carbon's versatility that makes large, complex molecules possible.

When a carbon atom forms four single covalent bonds, the arrangement of its four hybrid orbitals causes the bonds to angle toward the corners of an imaginary tetrahedron (see Figure 2.17b). The bond angles in methane (CH_4) are 109.5° (**Figure 4.3a**), and they are roughly the same in any group of atoms where carbon has four single bonds. For example, ethane (C_2H_6) is shaped like two overlapping tetrahedrons (**Figure 4.3b**). In molecules with more carbons, every grouping of a carbon bonded to four other atoms has a tetrahedral shape. But when two carbon atoms are joined by a double bond, as in ethene (C_2H_4), the atoms joined to those carbons are in the same plane as the carbons (**Figure 4.3c**). We find it convenient to write molecules as structural formulas, as if the molecules being represented are two-dimensional, but keep in mind that molecules are three-dimensional and that the shape of a molecule often determines its function.

The electron configuration of carbon gives it covalent compatibility with many different elements. **Figure 4.4** shows the valences of carbon and its most frequent partners—hydrogen, oxygen, and nitrogen. These are the four major atomic components of organic molecules. These valences are the basis for the rules of covalent bonding in organic chemistry—the building code for the architecture of organic molecules.

Let's consider how the rules of covalent bonding apply to carbon atoms with partners other than hydrogen. We'll look at two examples, the simple molecules carbon dioxide and urea.

In the carbon dioxide molecule (CO_2), a single carbon atom is joined to two atoms of oxygen by double covalent bonds. The structural formula for CO_2 is shown here:

$$O{=}C{=}O$$

Each line in a structural formula represents a pair of shared electrons. Thus, the two double bonds in CO_2 have the same number of shared electrons as four single bonds. The arrangement completes the valence shells of all atoms in the molecule.

Name and Comment	Molecular Formula	Structural Formula	Ball-and-Stick Model (molecular shape in pink)	Space-Filling Model
(a) Methane. When a carbon atom has four single bonds to other atoms, the molecule is tetrahedral.	CH_4	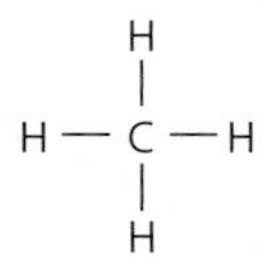 H—C(H)(H)—H	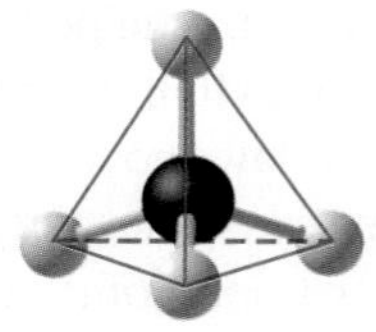	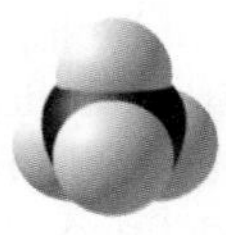
(b) Ethane. A molecule may have more than one tetrahedral group of single-bonded atoms. (Ethane consists of two such groups.)	C_2H_6	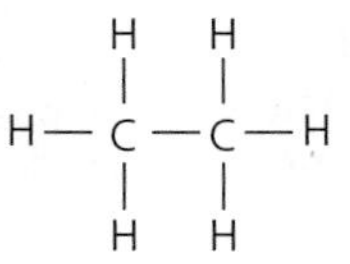 H—C(H)(H)—C(H)(H)—H	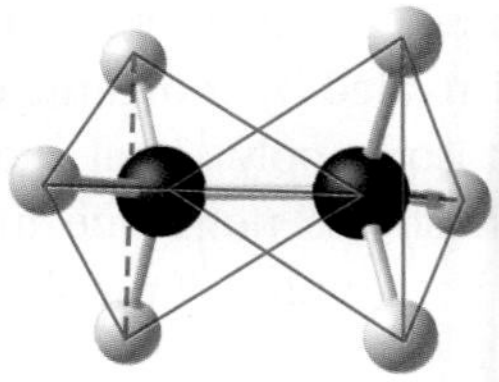	
(c) Ethene (ethylene). When two carbon atoms are joined by a double bond, all atoms attached to those carbons are in the same plane; the molecule is flat.	C_2H_4	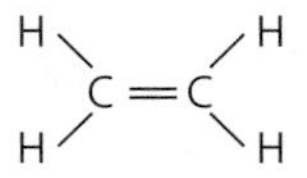 $H_2C{=}CH_2$	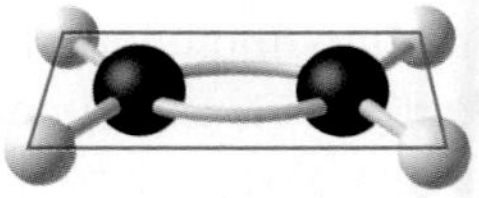	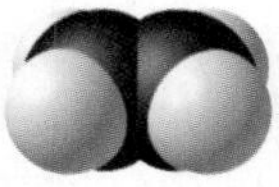

▲ **Figure 4.3 The shapes of three simple organic molecules.**

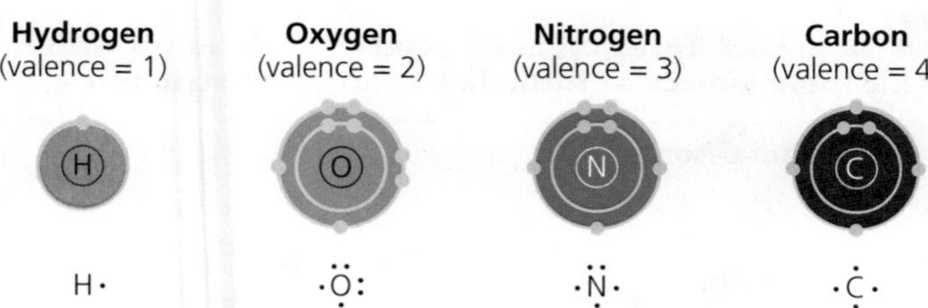

▲ **Figure 4.4 Valences of the major elements of organic molecules.** Valence is the number of covalent bonds an atom can form. It is generally equal to the number of electrons required to complete the valence (outermost) shell (see Figure 2.9). All the electrons are shown for each atom in the electron distribution diagrams (top). Only the valence shell electrons are shown in the Lewis dot structures (bottom). Note that carbon can form four bonds.

MAKE CONNECTIONS *Refer to Figure 2.9 (p. 36) and draw the Lewis dot structures for sodium, phosphorus, sulfur, and chlorine.*

Because CO_2 is a very simple molecule and lacks hydrogen, it is often considered inorganic, even though it contains carbon. Whether we call CO_2 organic or inorganic, however, it is clearly important to the living world as the source of carbon for all organic molecules in organisms.

Urea, $CO(NH_2)_2$, is the organic compound found in urine that Wöhler synthesized in the early 1800s. Again, each atom has the required number of covalent bonds. In this case, one carbon atom participates in both single and double bonds.

Urea

Urea and carbon dioxide are molecules with only one carbon atom. But as Figure 4.3 shows, a carbon atom can also use one or more valence electrons to form covalent bonds to other carbon atoms, linking the atoms into chains of seemingly infinite variety.

Molecular Diversity Arising from Carbon Skeleton Variation

Carbon chains form the skeletons of most organic molecules. The skeletons vary in length and may be straight, branched, or arranged in closed rings (**Figure 4.5**). Some carbon skeletons have double bonds, which vary in number and location. Such variation in carbon skeletons is one important source of the molecular complexity and diversity that characterize living matter. In addition, atoms of other elements can be bonded to the skeletons at available sites.

Hydrocarbons

All of the molecules shown in Figures 4.3 and 4.5 are **hydrocarbons**, organic molecules consisting of only carbon and hydrogen. Atoms of hydrogen are attached to the carbon skeleton wherever electrons are available for covalent bonding. Hydrocarbons are the major components of petroleum, which is called a fossil fuel because it consists of the partially decomposed remains of organisms that lived millions of years ago.

▼ **Figure 4.5 Four ways that carbon skeletons can vary.**

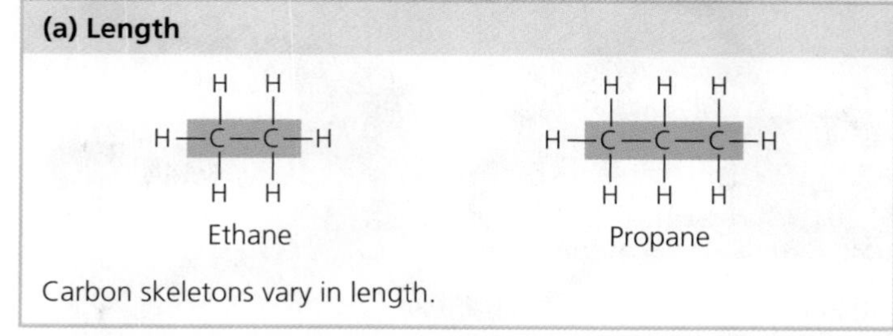

Carbon skeletons vary in length.

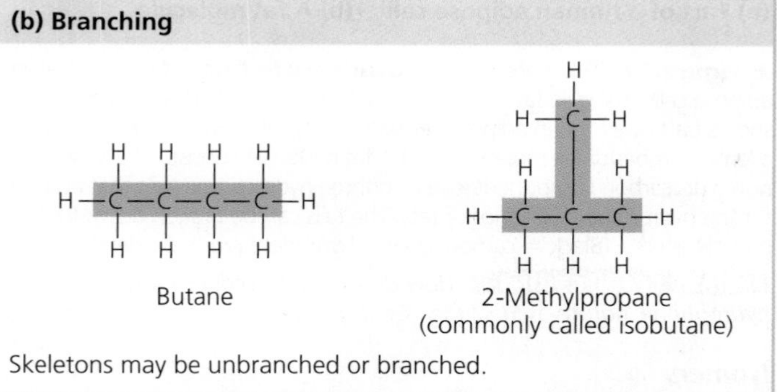

Skeletons may be unbranched or branched.

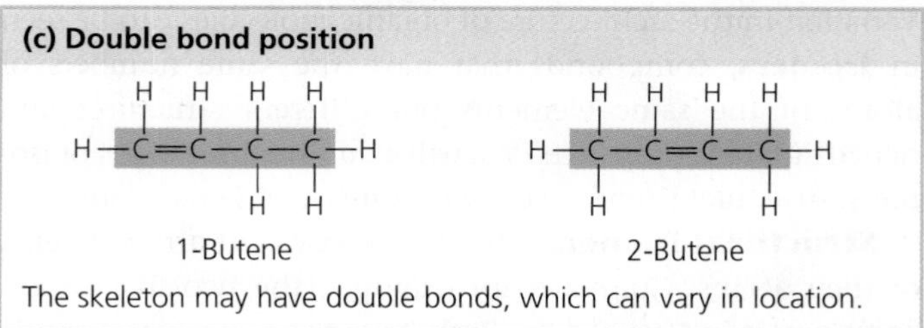

The skeleton may have double bonds, which can vary in location.

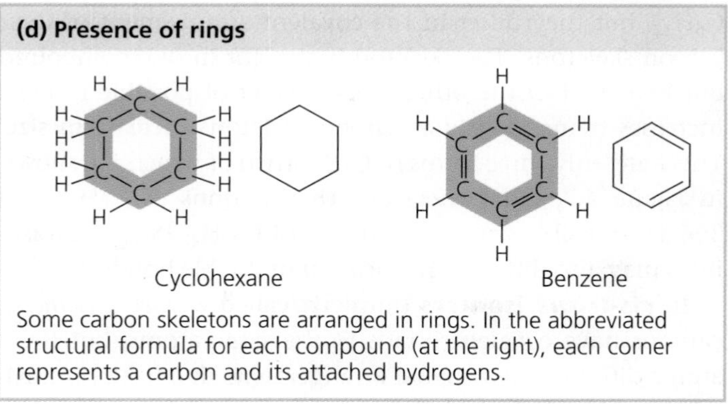

Some carbon skeletons are arranged in rings. In the abbreviated structural formula for each compound (at the right), each corner represents a carbon and its attached hydrogens.

Although hydrocarbons are not prevalent in most living organisms, many of a cell's organic molecules have regions consisting of only carbon and hydrogen. For example, the molecules known as fats have long hydrocarbon tails attached to a nonhydrocarbon component (**Figure 4.6**, on the next page). Neither petroleum nor fat dissolves in water; both are hydrophobic compounds because the great majority of their bonds are relatively nonpolar carbon-to-hydrogen linkages. Another characteristic of hydrocarbons is that they can undergo reactions that release a relatively large amount of energy. The gasoline that fuels a car consists of hydrocarbons, and the hydrocarbon tails of fats serve as stored fuel for animals.

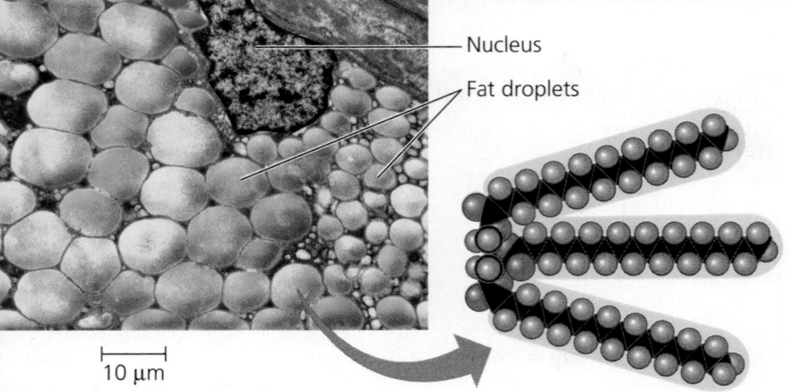

(a) Part of a human adipose cell **(b) A fat molecule**

▲ **Figure 4.6 The role of hydrocarbons in fats.** **(a)** Mammalian adipose cells stockpile fat molecules as a fuel reserve. This colorized TEM shows part of a human adipose cell with many fat droplets, each containing a large number of fat molecules. **(b)** A fat molecule consists of a small, nonhydrocarbon component joined to three hydrocarbon tails that account for the hydrophobic behavior of fats. The tails can be broken down to provide energy. (Black = carbon; gray = hydrogen; red = oxygen.)

MAKE CONNECTIONS *How do the tails account for the hydrophobic nature of fats? (See Concept 3.2, p. 51.)*

Isomers

Variation in the architecture of organic molecules can be seen in **isomers**, compounds that have the same numbers of atoms of the same elements but different structures and hence different properties. We will examine three types of isomers: structural isomers, *cis-trans* isomers, and enantiomers.

Structural isomers differ in the covalent arrangements of their atoms. Compare, for example, the two five-carbon compounds in **Figure 4.7a**. Both have the molecular formula C_5H_{12}, but they differ in the covalent arrangement of their carbon skeletons. The skeleton is straight in one compound but branched in the other. The number of possible isomers increases tremendously as carbon skeletons increase in size. There are only three forms of C_5H_{12} (two of which are shown in Figure 4.7a), but there are 18 variations of C_8H_{18} and 366,319 possible structural isomers of $C_{20}H_{42}$. Structural isomers may also differ in the location of double bonds.

In ***cis-trans* isomers** (formerly called *geometric isomers*), carbons have covalent bonds to the same atoms, but these atoms differ in their spatial arrangements due to the inflexibility of double bonds. Single bonds allow the atoms they join to rotate freely about the bond axis without changing the compound. In contrast, double bonds do not permit such rotation. If a double bond joins two carbon atoms, and each C also has two different atoms (or groups of atoms) attached to it, then two distinct *cis-trans* isomers are possible. Consider a simple molecule with two double-bonded carbons, each of which has an H and an X attached to it **(Figure 4.7b)**. The arrangement with both Xs on the same side of the double bond is called a *cis isomer*, and that with the Xs on opposite sides is called a *trans* isomer. The subtle difference in shape between such isomers can dramatically affect the biological activities of organic molecules. For example, the biochemistry of vision involves a light-induced change of rhodopsin, a chemical compound in the eye, from the *cis* isomer to the *trans* isomer (see Figure 50.17). Another example involves *trans* fats, which are discussed in Chapter 5.

Enantiomers are isomers that are mirror images of each other and that differ in shape due to the presence of an *asymmetric carbon*, one that is attached to four different atoms or groups of atoms. (See the middle carbon in the ball-and-stick models shown in **Figure 4.7c**.) The four groups can be arranged in space around the asymmetric carbon in two different ways that are mirror images. Enantiomers are, in a way, left-handed and right-handed versions of the molecule. Just as your right hand won't fit into a left-handed glove, a "right-handed" molecule won't fit into the same space as the "left-handed" version.

▼ **Figure 4.7 Three types of isomers, compounds with the same molecular formula but different structures.**

(a) Structural isomers

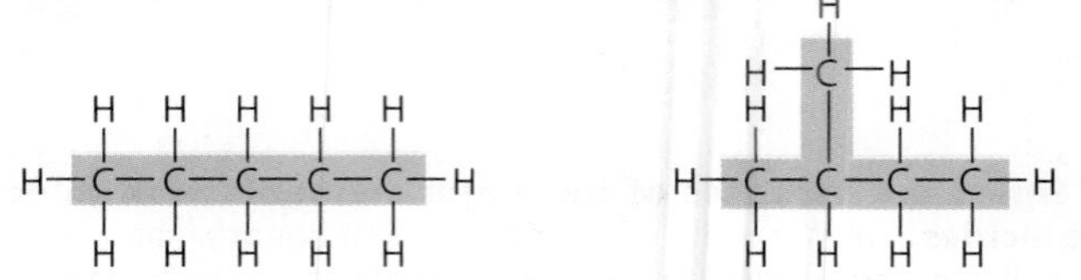

Structural isomers differ in covalent partners, as shown in this example of two isomers of C_5H_{12}: pentane (left) and 2-methyl butane (right).

(b) *Cis-trans* isomers

cis isomer: The two Xs are on the same side.

trans isomer: The two Xs are on opposite sides.

Cis-trans isomers differ in arrangement about a double bond. In these diagrams, X represents an atom or group of atoms attached to a double-bonded carbon.

(c) Enantiomers

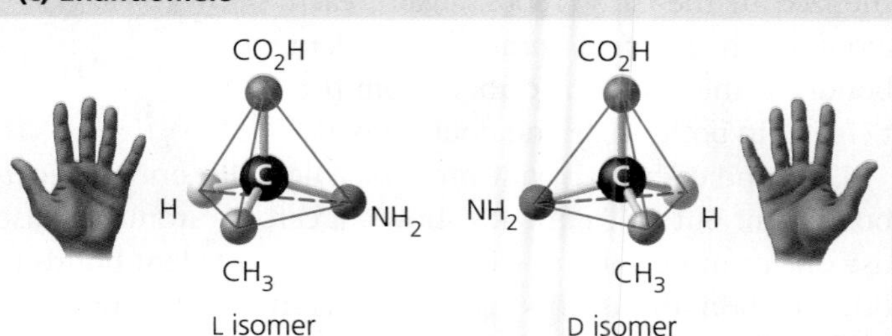

Enantiomers differ in spatial arrangement around an asymmetric carbon, resulting in molecules that are mirror images, like left and right hands. The two isomers are designated the L and D isomers from the Latin for "left" and "right" (*levo* and *dextro*). Enantiomers cannot be superimposed on each other.

DRAW IT *There are three structural isomers of C_5H_{12}; draw the one not shown in (a).*

Drug	Condition	Effective Enantiomer	Ineffective Enantiomer
Ibuprofen	Pain; inflammation	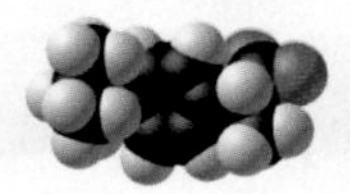 *S*-Ibuprofen	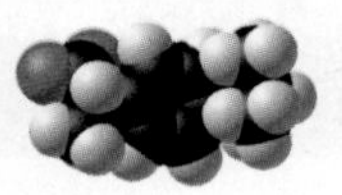 *R*-Ibuprofen
Albuterol	Asthma	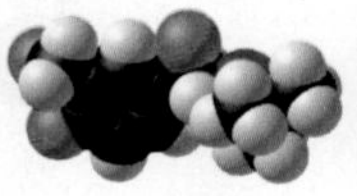 *R*-Albuterol	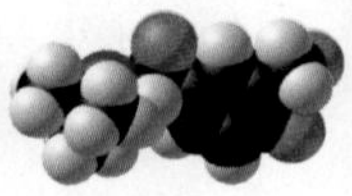 *S*-Albuterol

▲ **Figure 4.8 The pharmacological importance of enantiomers.** Ibuprofen and albuterol are examples of drugs whose enantiomers have different effects. (*S* and *R* are letters used in one system to distinguish between enantiomers.) Ibuprofen reduces inflammation and pain. It is commonly sold as a mixture of the two enantiomers. The *S* enantiomer is 100 times more effective than the other. Albuterol is used to relax bronchial muscles, improving airflow in asthma patients. Only *R*-albuterol is synthesized and sold as a drug; the *S* form counteracts the active *R* form.

Usually, only one isomer is biologically active because only that form can bind to specific molecules in an organism.

The concept of enantiomers is important in the pharmaceutical industry because the two enantiomers of a drug may not be equally effective, as is the case for both ibuprofen and the asthma medication albuterol (**Figure 4.8**). Methamphetamine also occurs in two enantiomers that have very different effects. One enantiomer is the highly addictive stimulant drug known as "crank," sold illegally in the street drug trade. The other has a much weaker effect and is even found as an ingredient in an over-the-counter vapor inhaler for treatment of nasal congestion! The differing effects of enantiomers in the body demonstrate that organisms are sensitive to even the most subtle variations in molecular architecture. Once again, we see that molecules have emergent properties that depend on the specific arrangement of their atoms.

CONCEPT CHECK 4.2

1. **DRAW IT** Draw a structural formula for C_2H_4.
2. Which molecules in Figure 4.5 are isomers? For each pair, identify the type of isomer.
3. How are gasoline and fat chemically similar?
4. **WHAT IF?** Can propane (C_3H_8) form isomers?

For suggested answers, see Appendix A.

CONCEPT 4.3

A few chemical groups are key to the functioning of biological molecules

The distinctive properties of an organic molecule depend not only on the arrangement of its carbon skeleton but also on the chemical groups attached to that skeleton. We can think of hydrocarbons, the simplest organic molecules, as the underlying framework for more complex organic molecules. A number of chemical groups can replace one or more of the hydrogens bonded to the carbon skeleton of the hydrocarbon. (Some groups include atoms of the carbon skeleton, as we will see.) These groups may participate in chemical reactions or may contribute to function indirectly by their effects on molecular shape. The number and arrangement of the groups help give each molecule its unique properties.

The Chemical Groups Most Important in the Processes of Life

Consider the differences between estradiol (a type of estrogen) and testosterone. These compounds are female and male sex hormones, respectively, in humans and other vertebrates. Both are steroids, organic molecules with a common carbon skeleton in the form of four fused rings. These sex hormones differ only in the chemical groups attached to the rings (shown here in abbreviated form); the distinctions in molecular architecture are shaded in blue:

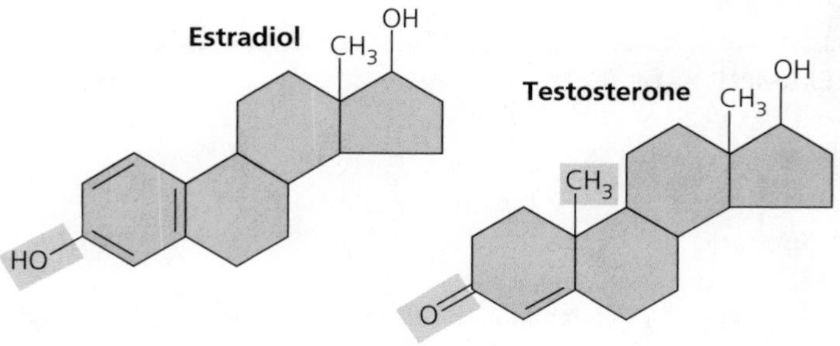

The different actions of these two molecules on many targets throughout the body help produce the contrasting anatomical and physiological features of male and female vertebrates. Thus, even our sexuality has its biological basis in variations of molecular architecture.

In the example of sex hormones, different chemical groups contribute to function by affecting the molecule's shape. In other cases, the chemical groups affect molecular function by being directly involved in chemical reactions; these important chemical groups are known as **functional groups**. Each functional group participates in chemical reactions in a characteristic way from one organic molecule to another.

The seven chemical groups most important in biological processes are the hydroxyl, carbonyl, carboxyl, amino, sulfhydryl, phosphate, and methyl groups. The first six groups can act as functional groups; they are also hydrophilic and thus increase the solubility of organic compounds in water. The methyl group is not reactive, but instead often serves as a recognizable tag on biological molecules. Before reading further, study **Figure 4.9** on the next two pages to familiarize yourself with these biologically important chemical groups.

▼ **Figure 4.9**

Exploring Some Biologically Important Chemical Groups

CHEMICAL GROUP	Hydroxyl	Carbonyl	Carboxyl
STRUCTURE	—OH (may be written HO—)	—C(=O)—	—C(=O)OH
	In a **hydroxyl group** (—OH), a hydrogen atom is bonded to an oxygen atom, which in turn is bonded to the carbon skeleton of the organic molecule. (Do not confuse this functional group with the hydroxide ion, OH^-.)	The **carbonyl group** (>CO) consists of a carbon atom joined to an oxygen atom by a double bond.	When an oxygen atom is double-bonded to a carbon atom that is also bonded to an —OH group, the entire assembly of atoms is called a **carboxyl group** (—COOH).
NAME OF COMPOUND	**Alcohols** (Their specific names usually end in *-ol*.)	**Ketones** if the carbonyl group is within a carbon skeleton **Aldehydes** if the carbonyl group is at the end of the carbon skeleton	**Carboxylic acids**, or organic acids
EXAMPLE	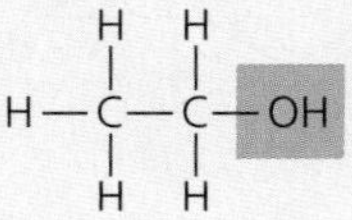 CH_3CH_2OH **Ethanol**, the alcohol present in alcoholic beverages	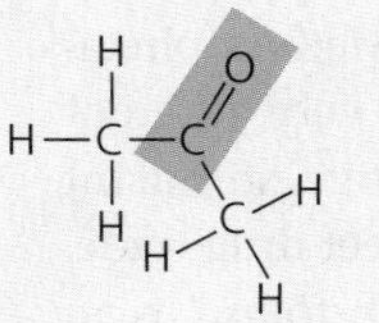 CH_3COCH_3 **Acetone**, the simplest ketone 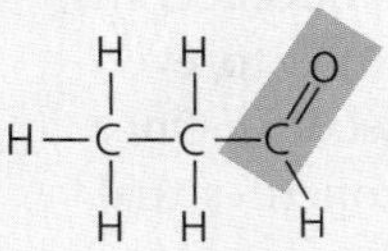 CH_3CH_2CHO **Propanal**, an aldehyde	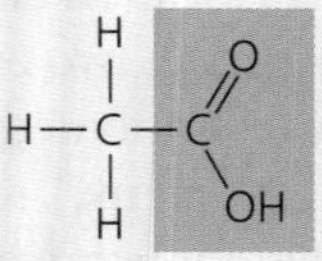 CH_3COOH **Acetic acid**, which gives vinegar its sour taste
FUNCTIONAL PROPERTIES	• Is polar as a result of the electrons spending more time near the electronegative oxygen atom. • Can form hydrogen bonds with water molecules, helping dissolve organic compounds such as sugars. (Sugars are shown in Figure 5.3.)	• A ketone and an aldehyde may be structural isomers with different properties, as is the case for acetone and propanal. • Ketone and aldehyde groups are also found in sugars, giving rise to two major groups of sugars: ketoses (containing ketone groups) and aldoses (containing aldehyde groups).	• Acts as an acid; can donate an H^+ because the covalent bond between oxygen and hydrogen is so polar: 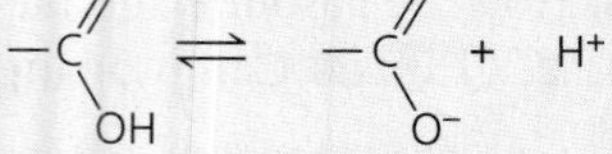 $-COOH \rightleftharpoons -COO^- + H^+$ Nonionized / Ionized • Found in cells in the ionized form with a charge of 1− and called a carboxylate ion.

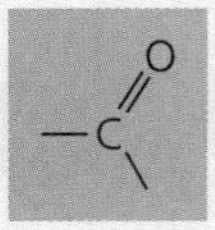

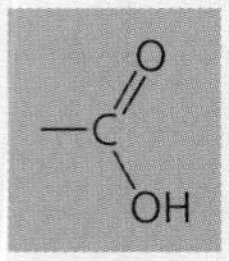

Amino

—N(H)(H)

The **amino group** (—NH_2) consists of a nitrogen atom bonded to two hydrogen atoms and to the carbon skeleton.

Amines

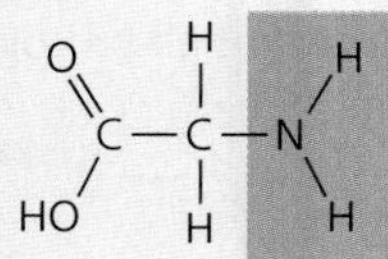

Glycine, a compound that is both an amine and a carboxylic acid because it has both an amino group and a carboxyl group; compounds with both groups are called **amino acids**

- Acts as a base; can pick up an H^+ from the surrounding solution (water, in living organisms):

$$H^+ + —N(H)(H) \rightleftharpoons —{}^+N(H)(H)—H$$

Nonionized   Ionized

- Found in cells in the ionized form with a charge of 1+.

Sulfhydryl

The **sulfhydryl group** (—SH) consists of a sulfur atom bonded to an atom of hydrogen; it resembles a hydroxyl group in shape.

Thiols

O OH
C
H—C—CH_2—SH
N
H H

Cysteine, an important sulfur-containing amino acid

- Two sulfhydryl groups can react, forming a covalent bond. This "cross-linking" helps stabilize protein structure (see Figure 5.20, Tertiary Structure).
- Cross-linking of cysteines in hair proteins maintains the curliness or straightness of hair. Straight hair can be "permanently" curled by shaping it around curlers and then breaking and re-forming the cross-linking bonds.

Phosphate

In the **phosphate group** shown here, a phosphorus atom is bonded to four oxygen atoms; one oxygen is bonded to the carbon skeleton; two oxygens carry negative charges (—OPO_3^{2-}). In this text, Ⓟ represents an attached phosphate group.

Organic phosphates

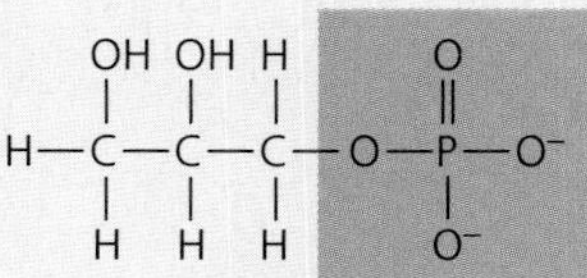

Glycerol phosphate, which takes part in many important chemical reactions in cells; glycerol phosphate also provides the backbone for phospholipids, the most prevalent molecules in cell membranes

- Contributes negative charge to the molecule of which it is a part (2− when at the end of a molecule, as above; 1− when located internally in a chain of phosphates).
- Molecules containing phosphate groups have the potential to react with water, releasing energy.

Methyl

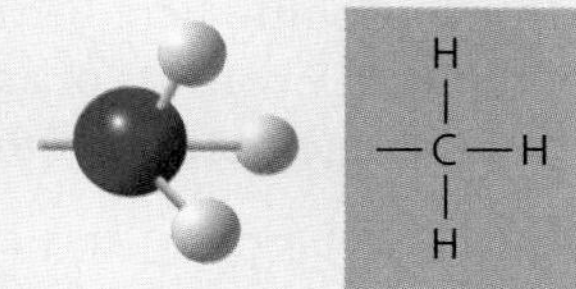

A **methyl group** (—CH_3) consists of a carbon bonded to three hydrogen atoms. The carbon of a methyl group may be attached to a carbon or to a different atom.

Methylated compounds

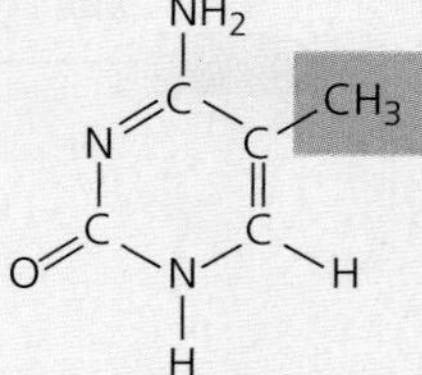

5-Methyl cytidine, a component of DNA that has been modified by addition of a methyl group

- Addition of a methyl group to DNA, or to molecules bound to DNA, affects the expression of genes.
- Arrangement of methyl groups in male and female sex hormones affects their shape and function (see p. 63).

MAKE CONNECTIONS *Given the information in this figure and what you know about the electronegativity of oxygen (see Concept 2.3, p. 39), predict which of the following molecules would be the stronger acid (see Concept 3.3, p. 53). Explain your answer.*

a. H—C(H)(H)—C(H)(H)—C(=O)OH

b. H—C(H)(H)—C(=O)—C(=O)OH

ATP: An Important Source of Energy for Cellular Processes

The "Phosphate" column in Figure 4.9 shows a simple example of an organic phosphate molecule. A more complicated organic phosphate, **adenosine triphosphate**, or **ATP**, is worth mentioning here because its function in the cell is so important. ATP consists of an organic molecule called adenosine attached to a string of three phosphate groups:

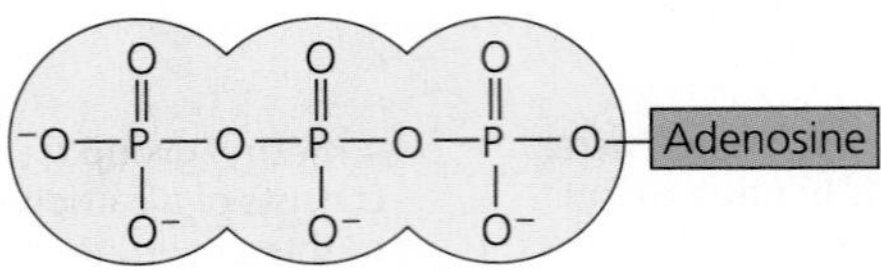

Where three phosphates are present in series, as in ATP, one phosphate may be split off as a result of a reaction with water. This inorganic phosphate ion, $HOPO_3^{2-}$, is often abbreviated $Ⓟ_i$ in this book. Having lost one phosphate, ATP becomes adenosine *di*phosphate, or ADP. Although ATP is sometimes said to store energy, it is more accurate to think of it as storing the potential to react with water. This reaction releases energy that can be used by the cell. You will learn about this in more detail in Chapter 8.

Ⓟ–Ⓟ–Ⓟ–Adenosine (ATP) —Reacts with H_2O→ $Ⓟ_i$ (Inorganic phosphate) + Ⓟ–Ⓟ–Adenosine (ADP) + Energy

CONCEPT CHECK 4.3

1. What does the term *amino acid* signify about the structure of such a molecule?
2. What chemical change occurs to ATP when it reacts with water and releases energy?
3. **WHAT IF?** Suppose you had an organic molecule such as cysteine (see Figure 4.9, sulfhydryl group example), and you chemically removed the $—NH_2$ group and replaced it with —COOH. Draw the structural formula for this molecule and speculate about its chemical properties. Is the central carbon asymmetric before the change? After?

For suggested answers, see Appendix A.

The Chemical Elements of Life: *A Review*

Living matter, as you have learned, consists mainly of carbon, oxygen, hydrogen, and nitrogen, with smaller amounts of sulfur and phosphorus. These elements all form strong covalent bonds, an essential characteristic in the architecture of complex organic molecules. Of all these elements, carbon is the virtuoso of the covalent bond. The versatility of carbon makes possible the great diversity of organic molecules, each with particular properties that emerge from the unique arrangement of its carbon skeleton and the chemical groups appended to that skeleton. At the foundation of all biological diversity lies this variation at the molecular level.

4 CHAPTER REVIEW

SUMMARY OF KEY CONCEPTS

CONCEPT 4.1

Organic chemistry is the study of carbon compounds (pp. 58–59)

- Living matter is made mostly of carbon, oxygen, hydrogen, and nitrogen, with some sulfur and phosphorus. Biological diversity has its molecular basis in carbon's ability to form a huge number of molecules with particular shapes and chemical properties.
- Organic compounds were once thought to arise only within living organisms, but this idea (vitalism) was disproved when chemists were able to synthesize organic compounds in the laboratory.

? *How did Stanley Miller's experiments extend the idea of mechanism to the origin of life?*

CONCEPT 4.2

Carbon atoms can form diverse molecules by bonding to four other atoms (pp. 60–63)

- Carbon, with a valence of 4, can bond to various other atoms, including O, H, and N. Carbon can also bond to other carbon atoms, forming the carbon skeletons of organic compounds. These skeletons vary in length and shape and have bonding sites for atoms of other elements. **Hydrocarbons** consist only of carbon and hydrogen.
- **Isomers** are compounds with the same molecular formula but different structures and properties. Three types of isomers are **structural isomers**, ***cis-trans*** **isomers**, and **enantiomers**.

? *Refer back to Figure 4.9. What type of isomers are acetone and propanal? How many asymmetric carbons are present in acetic acid, glycine, and glycerol phosphate? Can these three molecules exist as forms that are enantiomers?*

CONCEPT 4.3

A few chemical groups are key to the functioning of biological molecules (pp. 63–66)

- Chemical groups attached to the carbon skeletons of organic molecules participate in chemical reactions (**functional groups**) or contribute to function by affecting molecular shape (see Figure 4.9).
- **ATP (adenosine triphosphate)** consists of adenosine attached to three phosphate groups. ATP can react with water, forming inorganic phosphate and ADP (adenosine diphosphate). This reaction releases energy that can be used by the cell (see the equation at the top of the next page).

Reacts with H_2O

Ⓟ-Ⓟ-Ⓟ-Adenosine → $Ⓟ_i$ + Ⓟ-Ⓟ-Adenosine + Energy

ATP — Inorganic phosphate — ADP

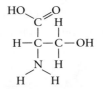

? *In what ways does a methyl group differ chemically from the other six important chemical groups shown in Figure 4.9?*

TEST YOUR UNDERSTANDING

LEVEL 1: KNOWLEDGE/COMPREHENSION

1. Organic chemistry is currently defined as
   a. the study of compounds made only by living cells.
   b. the study of carbon compounds.
   c. the study of vital forces.
   d. the study of natural (as opposed to synthetic) compounds.
   e. the study of hydrocarbons.

2. Which functional group is *not* present in this molecule?
   a. carboxyl
   b. sulfhydryl
   c. hydroxyl
   d. amino

   (Structure: HO–C(=O)–C(H)(NH₂)–C(H)(H)–OH)

3. **MAKE CONNECTIONS** Which chemical group is most likely to be responsible for an organic molecule behaving as a base (see Concept 3.3, p. 53)?
   a. hydroxyl
   b. carbonyl
   c. carboxyl
   d. amino
   e. phosphate

LEVEL 2: APPLICATION/ANALYSIS

4. Which of the following hydrocarbons has a double bond in its carbon skeleton?
   a. C_3H_8 b. C_2H_6 c. CH_4 d. C_2H_4 e. C_2H_2

5. Choose the term that correctly describes the relationship between these two sugar molecules:
   a. structural isomers
   b. *cis-trans* isomers
   c. enantiomers
   d. isotopes

   (Structures: H–C(H)(OH)–C(=O)–C(H)(OH)–H and H–C(=O)–C(H)(OH)–C(H)(OH)–H)

6. Identify the asymmetric carbon in this molecule:

   (Structure: O=C(H)–C(a)(OH)(H)–C(b)(H)(H)–C(c)(H)(H)–C(d)(H)(H)–C(e)(H)(H)–H, carbons labeled a–e)

7. Which action could produce a carbonyl group?
   a. the replacement of the —OH of a carboxyl group with hydrogen
   b. the addition of a thiol to a hydroxyl
   c. the addition of a hydroxyl to a phosphate
   d. the replacement of the nitrogen of an amine with oxygen
   e. the addition of a sulfhydryl to a carboxyl

8. Which of the molecules shown in question 5 has an asymmetric carbon? Which carbon is asymmetric?

LEVEL 3: SYNTHESIS/EVALUATION

9. **EVOLUTION CONNECTION**

   DRAW IT Some scientists think that life elsewhere in the universe might be based on the element silicon, rather than on carbon, as on Earth. Look at the electron distribution diagram for silicon in Figure 2.9 and draw the Lewis dot structure for silicon. What properties does silicon share with carbon that would make silicon-based life more likely than, say, neon-based life or aluminum-based life?

10. **SCIENTIFIC INQUIRY**

    Thalidomide achieved notoriety 50 years ago because of a wave of birth defects among children born to women who took this drug during pregnancy as a treatment for morning sickness. Thalidomide is a mixture of two enantiomers; one reduces morning sickness, but the other causes severe birth defects. (Although the beneficial enantiomer can be synthesized and given to patients, it is converted in the body to the harmful enantiomer.) The U.S. Food and Drug Administration (FDA) withheld approval of thalidomide in 1960. Since then, however, the FDA has approved this drug for the treatment of conditions associated with Hansen's disease (leprosy) and newly diagnosed multiple myeloma, a blood and bone marrow cancer. In clinical trials, thalidomide also shows promise as a treatment for AIDS, tuberculosis, inflammatory diseases, and some other types of cancer. Assuming that molecules related to thalidomide could be synthesized in the laboratory, describe in a broad way the type of experiments you would do to improve the benefits of this drug and minimize its harmful effects.

11. **WRITE ABOUT A THEME**

    Structure and Function In 1918, an epidemic of sleeping sickness caused an unusual rigid paralysis in some survivors, similar to symptoms of advanced Parkinson's disease. Years later, L-dopa (below, left), a chemical used to treat Parkinson's disease, was given to some of these patients, as dramatized in the movie *Awakenings*, starring Robin Williams. L-dopa was remarkably effective at eliminating the paralysis, at least temporarily. However, its enantiomer, D-dopa (right), was subsequently shown to have no effect at all, as is the case for Parkinson's disease. In a short essay (100–150 words), discuss how the effectiveness of one enantiomer and not the other illustrates the theme of structure and function.

L-dopa

D-dopa

For selected answers, see Appendix A.

MasteringBIOLOGY www.masteringbiology.com

1. MasteringBiology® Assignments

Tutorial Carbon Bonding and Functional Groups
Activities Diversity of Carbon-Based Molecules • Isomers • Functional Groups
Questions Student Misconceptions • Reading Quiz • Multiple Choice • End-of-Chapter

2. eText

Read your book online, search, take notes, highlight text, and more.

3. The Study Area

Practice Tests • Cumulative Test • ***BioFlix*** 3-D Animations • MP3 Tutor Sessions • Videos • Activities • Investigations • Lab Media • Audio Glossary • Word Study Tools • Art

5

The Structure and Function of Large Biological Molecules

▲ **Figure 5.1 Why do scientists study the structures of macromolecules?**

KEY CONCEPTS

5.1 **Macromolecules are polymers, built from monomers**

5.2 **Carbohydrates serve as fuel and building material**

5.3 **Lipids are a diverse group of hydrophobic molecules**

5.4 **Proteins include a diversity of structures, resulting in a wide range of functions**

5.5 **Nucleic acids store, transmit, and help express hereditary information**

OVERVIEW

The Molecules of Life

Given the rich complexity of life on Earth, we might expect organisms to have an enormous diversity of molecules. Remarkably, however, the critically important large molecules of all living things—from bacteria to elephants—fall into just four main classes: carbohydrates, lipids, proteins, and nucleic acids. On the molecular scale, members of three of these classes—carbohydrates, proteins, and nucleic acids—are huge and are therefore called **macromolecules**. For example, a protein may consist of thousands of atoms that form a molecular colossus with a mass well over 100,000 daltons. Considering the size and complexity of macromolecules, it is noteworthy that biochemists have determined the detailed structure of so many of them. The scientist in the foreground of **Figure 5.1** is using 3-D glasses to help her visualize the structure of the protein displayed on her screen.

The architecture of a large biological molecule helps explain how that molecule works. Like water and simple organic molecules, large biological molecules exhibit unique emergent properties arising from the orderly arrangement of their atoms. In this chapter, we'll first consider how macromolecules are built. Then we'll examine the structure and function of all four classes of large biological molecules: carbohydrates, lipids, proteins, and nucleic acids.

CONCEPT **5.1**

Macromolecules are polymers, built from monomers

The macromolecules in three of the four classes of life's organic compounds—carbohydrates, proteins, and nucleic acids—are chain-like molecules called polymers (from the Greek *polys*, many, and *meros*, part). A **polymer** is a long molecule consisting of many similar or identical building blocks linked by covalent bonds, much as a train consists of a chain of cars. The repeating units that serve as the building blocks of a polymer are smaller molecules called **monomers** (from the Greek *monos*, single). Some of the molecules that serve as monomers also have other functions of their own.

The Synthesis and Breakdown of Polymers

Although each class of polymer is made up of a different type of monomer, the chemical mechanisms by which cells make and break down polymers are basically the same in all cases. In cells, these processes are facilitated by **enzymes**, specialized macromolecules that speed up chemical reactions. Monomers are connected by a reaction in which two molecules are covalently bonded to each other, with the loss of a water molecule; this is known as a **dehydration reaction** (**Figure 5.2a**). When a bond forms between two monomers, each monomer contributes part of the water molecule that is released during the reaction: One monomer provides a hydroxyl group (—OH), while the other provides a hydrogen (—H). This reaction is repeated as monomers are added to the chain one by one, making a polymer.

Polymers are disassembled to monomers by **hydrolysis**, a process that is essentially the reverse of the dehydration reac-

▼ **Figure 5.2 The synthesis and breakdown of polymers.**

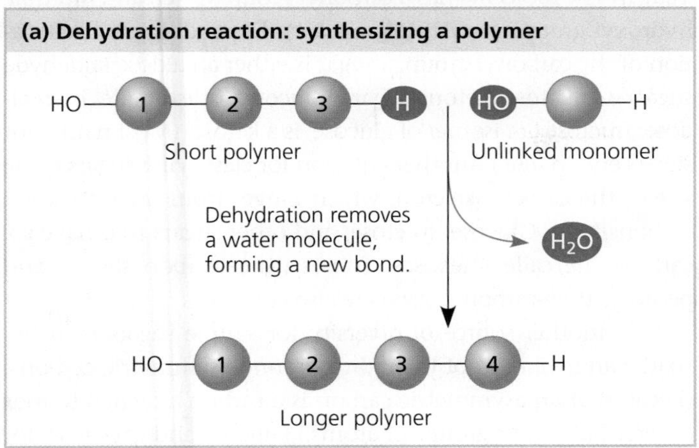

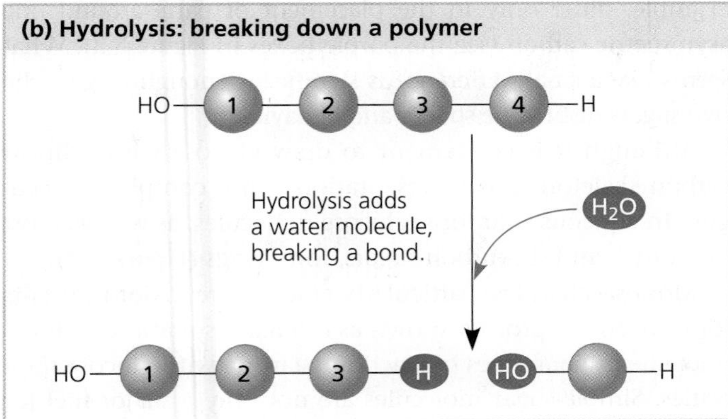

tion (**Figure 5.2b**). Hydrolysis means to break using water (from the Greek *hydro*, water, and *lysis*, break). The bond between the monomers is broken by the addition of a water molecule, with the hydrogen from the water attaching to one monomer and the hydroxyl group attaching to the adjacent monomer. An example of hydrolysis working within our bodies is the process of digestion. The bulk of the organic material in our food is in the form of polymers that are much too large to enter our cells. Within the digestive tract, various enzymes attack the polymers, speeding up hydrolysis. The released monomers are then absorbed into the bloodstream for distribution to all body cells. Those cells can then use dehydration reactions to assemble the monomers into new, different polymers that can perform specific functions required by the cell.

The Diversity of Polymers

Each cell has thousands of different macromolecules; the collection varies from one type of cell to another even in the same organism. The inherent differences between human siblings reflect small variations in polymers, particularly DNA and proteins. Molecular differences between unrelated individuals are more extensive and those between species greater still. The diversity of macromolecules in the living world is vast, and the possible variety is effectively limitless.

What is the basis for such diversity in life's polymers? These molecules are constructed from only 40 to 50 common monomers and some others that occur rarely. Building a huge variety of polymers from such a limited number of monomers is analogous to constructing hundreds of thousands of words from only 26 letters of the alphabet. The key is arrangement—the particular linear sequence that the units follow. However, this analogy falls far short of describing the great diversity of macromolecules because most biological polymers have many more monomers than the number of letters in the longest word. Proteins, for example, are built from 20 kinds of amino acids arranged in chains that are typically hundreds of amino acids long. The molecular logic of life is simple but elegant: Small molecules common to all organisms are ordered into unique macromolecules.

Despite this immense diversity, molecular structure and function can still be grouped roughly by class. Let's examine each of the four major classes of large biological molecules. For each class, the large molecules have emergent properties not found in their individual building blocks.

CONCEPT CHECK 5.1

1. What are the four main classes of large biological molecules? Which class does not consist of polymers?
2. How many molecules of water are needed to completely hydrolyze a polymer that is ten monomers long?
3. **WHAT IF?** Suppose you eat a serving of fish. What reactions must occur for the amino acid monomers in the protein of the fish to be converted to new proteins in your body?

For suggested answers, see Appendix A.

CONCEPT 5.2

Carbohydrates serve as fuel and building material

Carbohydrates include both sugars and polymers of sugars. The simplest carbohydrates are the monosaccharides, or simple sugars; these are the monomers from which more complex carbohydrates are constructed. Disaccharides are double sugars, consisting of two monosaccharides joined by a covalent bond. Carbohydrates also include macromolecules called polysaccharides, polymers composed of many sugar building blocks.

Sugars

Monosaccharides (from the Greek *monos*, single, and *sacchar*, sugar) generally have molecular formulas that are some multiple of the unit CH_2O. Glucose ($C_6H_{12}O_6$), the most common monosaccharide, is of central importance in the chemistry

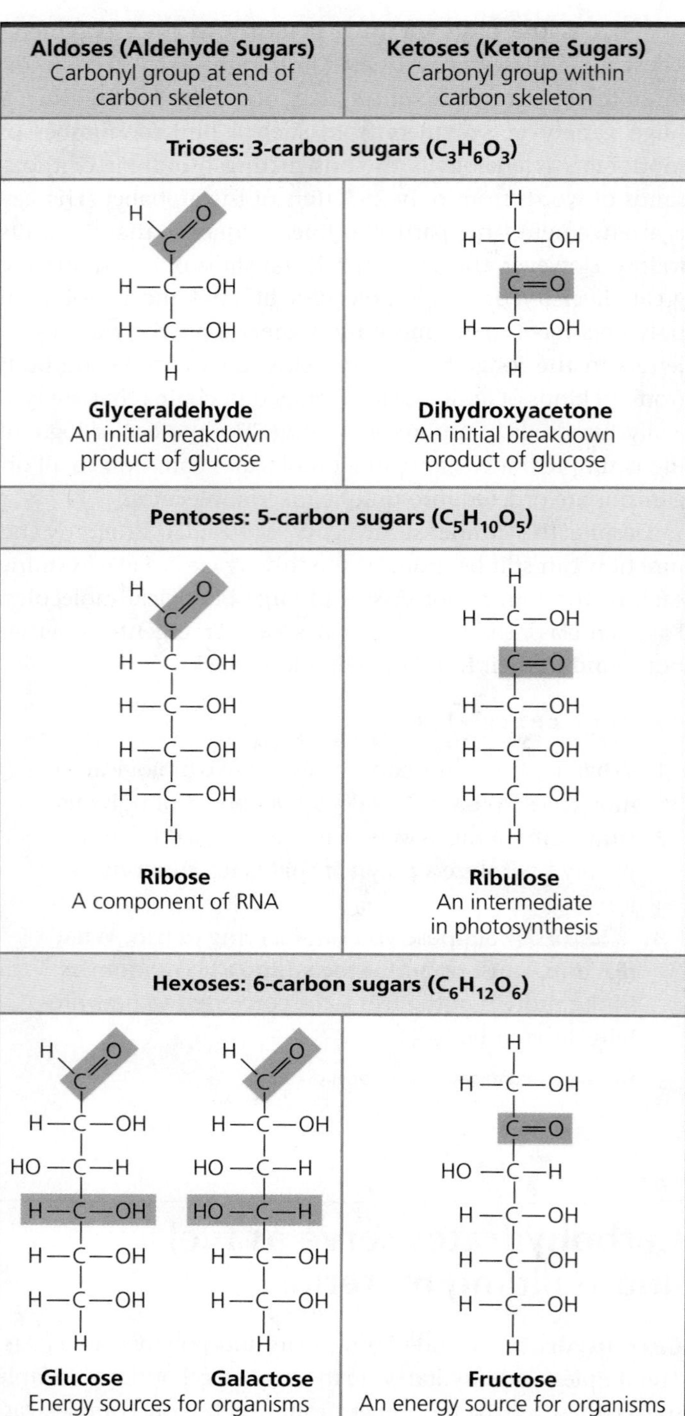

Aldoses (Aldehyde Sugars)
Carbonyl group at end of carbon skeleton
Ketoses (Ketone Sugars)
Carbonyl group within carbon skeleton
Trioses: 3-carbon sugars ($C_3H_6O_3$)
Glyceraldehyde
An initial breakdown product of glucose
Dihydroxyacetone
An initial breakdown product of glucose
Pentoses: 5-carbon sugars ($C_5H_{10}O_5$)
Ribose
A component of RNA
Ribulose
An intermediate in photosynthesis
Hexoses: 6-carbon sugars ($C_6H_{12}O_6$)
Glucose
Galactose
Energy sources for organisms
Fructose
An energy source for organisms

▲ **Figure 5.3 The structure and classification of some monosaccharides.** Sugars vary in the location of their carbonyl groups (orange), the length of their carbon skeletons, and the spatial arrangement around asymmetric carbons (compare, for example, the purple portions of glucose and galactose).

MAKE CONNECTIONS *In the 1970s, a process was developed that converts the glucose in corn syrup to its sweeter isomer, fructose. High-fructose corn syrup, a common ingredient in soft drinks and processed food, is a mixture of glucose and fructose. What type of isomers are glucose and fructose? See Figure 4.7, p. 62.*

of life. In the structure of glucose, we can see the trademarks of a sugar: The molecule has a carbonyl group (C=O) and multiple hydroxyl groups (—OH) **(Figure 5.3)**. Depending on the location of the carbonyl group, a sugar is either an aldose (aldehyde sugar) or a ketose (ketone sugar). Glucose, for example, is an aldose; fructose, an isomer of glucose, is a ketose. (Most names for sugars end in *-ose*.) Another criterion for classifying sugars is the size of the carbon skeleton, which ranges from three to seven carbons long. Glucose, fructose, and other sugars that have six carbons are called hexoses. Trioses (three-carbon sugars) and pentoses (five-carbon sugars) are also common.

Still another source of diversity for simple sugars is in the spatial arrangement of their parts around asymmetric carbons. (Recall that an asymmetric carbon is a carbon attached to four different atoms or groups of atoms.) Glucose and galactose, for example, differ only in the placement of parts around one asymmetric carbon (see the purple boxes in Figure 5.3). What seems like a small difference is significant enough to give the two sugars distinctive shapes and behaviors.

Although it is convenient to draw glucose with a linear carbon skeleton, this representation is not completely accurate. In aqueous solutions, glucose molecules, as well as most other five- and six-carbon sugars, form rings **(Figure 5.4)**.

Monosaccharides, particularly glucose, are major nutrients for cells. In the process known as cellular respiration, cells extract energy in a series of reactions starting with glucose molecules. Simple-sugar molecules are not only a major fuel for cellular work, but their carbon skeletons also serve as raw material for the synthesis of other types of small organic molecules, such as amino acids and fatty acids. Sugar molecules that are not immediately used in these ways are generally incorporated as monomers into disaccharides or polysaccharides.

A **disaccharide** consists of two monosaccharides joined by a **glycosidic linkage**, a covalent bond formed between two monosaccharides by a dehydration reaction. For example, maltose is a disaccharide formed by the linking of two molecules of glucose **(Figure 5.5a)**. Also known as malt sugar, maltose is an ingredient used in brewing beer. The most prevalent disaccharide is sucrose, which is table sugar. Its two monomers are glucose and fructose **(Figure 5.5b)**. Plants generally transport carbohydrates from leaves to roots and other nonphotosynthetic organs in the form of sucrose. Lactose, the sugar present in milk, is another disaccharide, in this case a glucose molecule joined to a galactose molecule.

Polysaccharides

Polysaccharides are macromolecules, polymers with a few hundred to a few thousand monosaccharides joined by glycosidic linkages. Some polysaccharides serve as storage material, hydrolyzed as needed to provide sugar for cells. Other polysaccharides serve as building material for structures that

(a) Linear and ring forms. Chemical equilibrium between the linear and ring structures greatly favors the formation of rings. The carbons of the sugar are numbered 1 to 6, as shown. To form the glucose ring, carbon 1 bonds to the oxygen attached to carbon 5.

(b) Abbreviated ring structure. Each corner represents a carbon. The ring's thicker edge indicates that you are looking at the ring edge-on; the components attached to the ring lie above or below the plane of the ring.

▲ **Figure 5.4 Linear and ring forms of glucose.**

DRAW IT *Start with the linear form of fructose (see Figure 5.3) and draw the formation of the fructose ring in two steps. First, number the carbons starting at the top of the linear structure. Then attach carbon 5 via its oxygen to carbon 2. Compare the number of carbons in the fructose and glucose rings.*

(a) Dehydration reaction in the synthesis of maltose. The bonding of two glucose units forms maltose. The glycosidic linkage joins the number 1 carbon of one glucose to the number 4 carbon of the second glucose. Joining the glucose monomers in a different way would result in a different disaccharide.

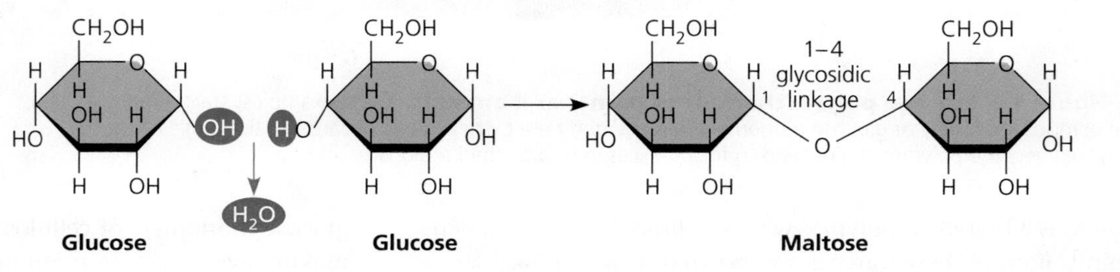

(b) Dehydration reaction in the synthesis of sucrose. Sucrose is a disaccharide formed from glucose and fructose. Notice that fructose, though a hexose like glucose, forms a five-sided ring.

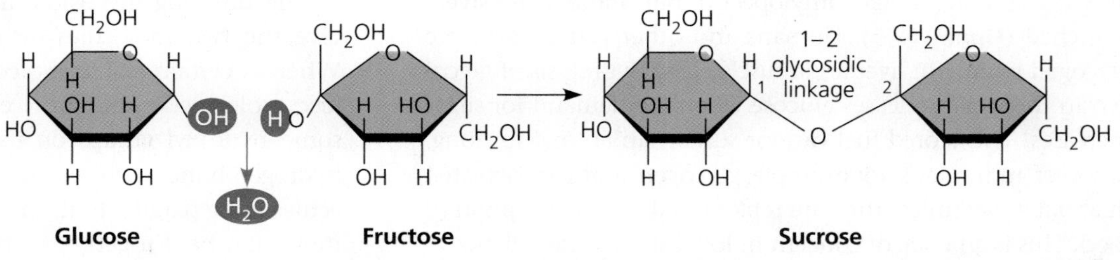

▲ **Figure 5.5 Examples of disaccharide synthesis.**

DRAW IT *Referring to Figure 5.4, number the carbons in each sugar in this figure. Show how the numbering is consistent with the name of the glycosidic linkage in each disaccharide.*

protect the cell or the whole organism. The architecture and function of a polysaccharide are determined by its sugar monomers and by the positions of its glycosidic linkages.

Storage Polysaccharides

Both plants and animals store sugars for later use in the form of storage polysaccharides. Plants store **starch**, a polymer of glucose monomers, as granules within cellular structures known as plastids, which include chloroplasts. Synthesizing starch enables the plant to stockpile surplus glucose. Because glucose is a major cellular fuel, starch represents stored energy. The sugar can later be withdrawn from this carbohydrate "bank" by hydrolysis, which breaks the bonds between the glucose monomers. Most animals, including humans, also have enzymes that can hydrolyze plant starch, making glucose available as a nutrient for cells. Potato tubers and grains—the fruits of wheat, maize (corn), rice, and other grasses—are the major sources of starch in the human diet.

Most of the glucose monomers in starch are joined by 1–4 linkages (number 1 carbon to number 4 carbon), like the glucose units in maltose (see Figure 5.5a). The simplest form of starch, amylose, is unbranched. Amylopectin, a more complex

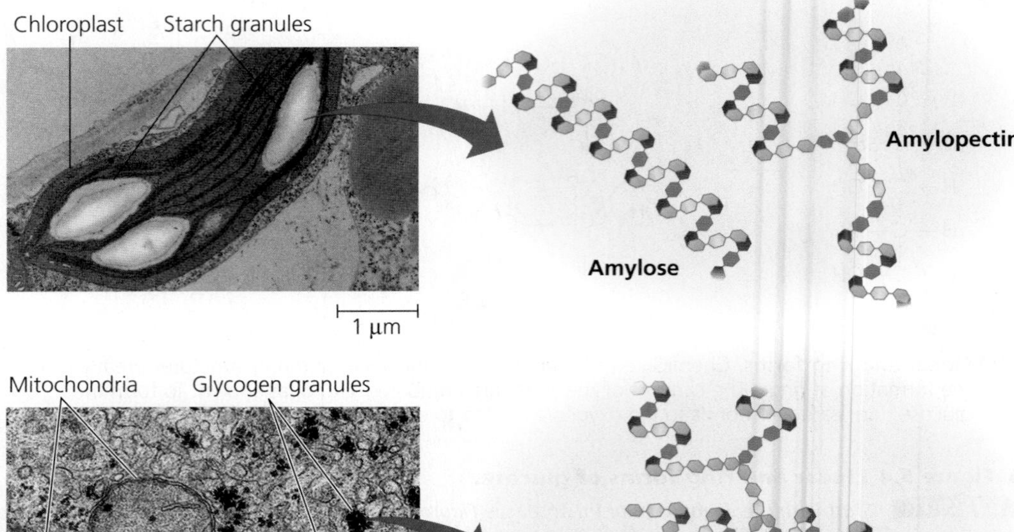

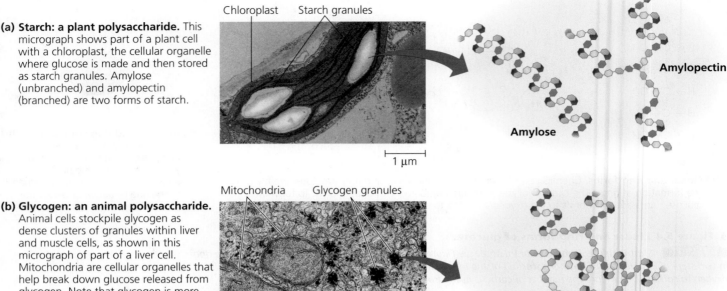

(a) Starch: a plant polysaccharide. This micrograph shows part of a plant cell with a chloroplast, the cellular organelle where glucose is made and then stored as starch granules. Amylose (unbranched) and amylopectin (branched) are two forms of starch.

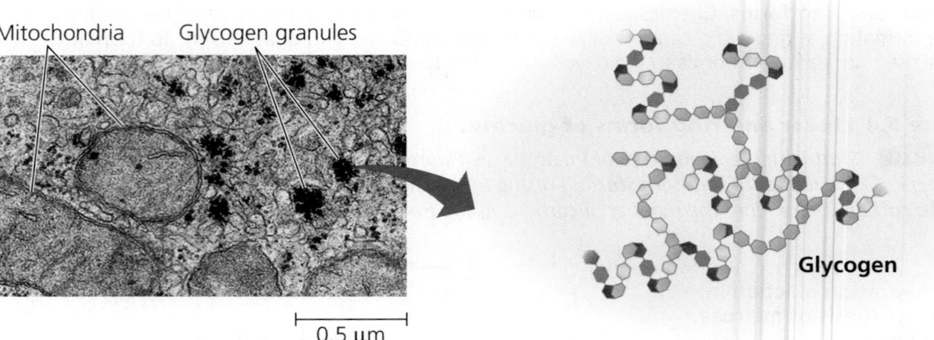

(b) Glycogen: an animal polysaccharide. Animal cells stockpile glycogen as dense clusters of granules within liver and muscle cells, as shown in this micrograph of part of a liver cell. Mitochondria are cellular organelles that help break down glucose released from glycogen. Note that glycogen is more branched than amylopectin starch.

▲ **Figure 5.6 Storage polysaccharides of plants and animals.** These examples, starch and glycogen, are composed entirely of glucose monomers, represented here by hexagons. Because of the angle of the 1–4 linkages, the polymer chains tend to form helices in unbranched regions.

starch, is a branched polymer with 1–6 linkages at the branch points. Both of these starches are shown in **Figure 5.6a**.

Animals store a polysaccharide called **glycogen**, a polymer of glucose that is like amylopectin but more extensively branched **(Figure 5.6b)**. Humans and other vertebrates store glycogen mainly in liver and muscle cells. Hydrolysis of glycogen in these cells releases glucose when the demand for sugar increases. This stored fuel cannot sustain an animal for long, however. In humans, for example, glycogen stores are depleted in about a day unless they are replenished by consumption of food. This is an issue of concern in low-carbohydrate diets.

Structural Polysaccharides

Organisms build strong materials from structural polysaccharides. For example, the polysaccharide called **cellulose** is a major component of the tough walls that enclose plant cells. On a global scale, plants produce almost 10^{14} kg (100 billion tons) of cellulose per year; it is the most abundant organic compound on Earth. Like starch, cellulose is a polymer of glucose, but the glycosidic linkages in these two polymers differ. The difference is based on the fact that there are actually two slightly different ring structures for glucose **(Figure 5.7a)**. When glucose forms a ring, the hydroxyl group attached to the number 1 carbon is positioned either below or above the plane of the ring. These two ring forms for glucose are called alpha (α) and beta (β), respectively. In starch, all the glucose monomers are in the α configuration **(Figure 5.7b)**, the arrangement we saw in Figures 5.4 and 5.5. In contrast, the glucose monomers of cellulose are all in the β configuration, making every glucose monomer "upside down" with respect to its neighbors **(Figure 5.7c)**.

The differing glycosidic linkages in starch and cellulose give the two molecules distinct three-dimensional shapes. Whereas certain starch molecules are largely helical, a cellulose molecule is straight. Cellulose is never branched, and some hydroxyl groups on its glucose monomers are free to hydrogen-bond with the hydroxyls of other cellulose molecules lying parallel to it. In plant cell walls, parallel cellulose molecules held together in this way are grouped into units called microfibrils **(Figure 5.8)**. These cable-like microfibrils are a strong building material for plants and an important substance for humans because cellulose is the major constituent of paper and the only component of cotton.

Enzymes that digest starch by hydrolyzing its α linkages are unable to hydrolyze the β linkages of cellulose because of the distinctly different shapes of these two molecules. In fact, few organisms possess enzymes that can digest cellulose. Animals, including humans, do not; the cellulose in our food passes through the digestive tract and is eliminated with the feces. Along the way, the cellulose abrades the wall of the digestive tract and stimulates the lining to secrete mucus, which aids in the smooth passage of food through the tract. Thus, although cellulose is not a nutrient for humans, it is an important part of a healthful diet. Most fresh fruits, vegetables, and whole grains are rich in cellulose. On food packages, "insoluble fiber" refers mainly to cellulose.

(a) α and β glucose ring structures. These two interconvertible forms of glucose differ in the placement of the hydroxyl group (highlighted in blue) attached to the number 1 carbon.

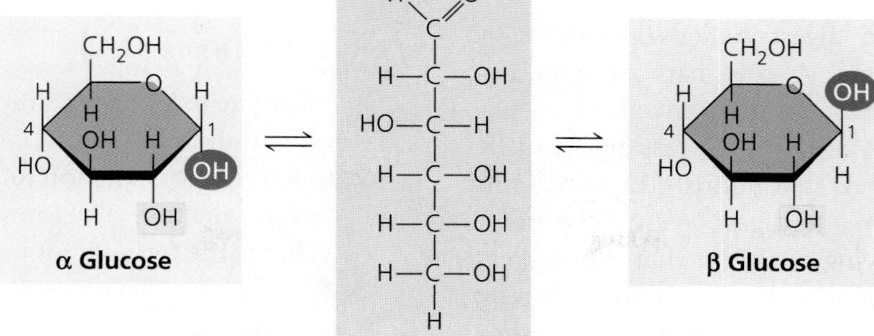

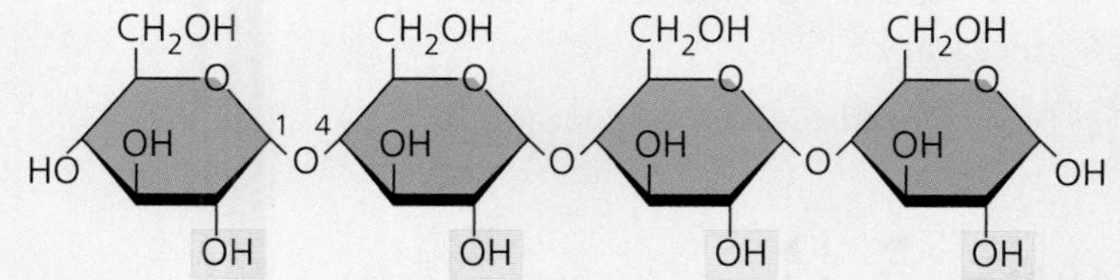

(b) Starch: 1–4 linkage of α glucose monomers. All monomers are in the same orientation. Compare the positions of the —OH groups highlighted in yellow with those in cellulose (c).

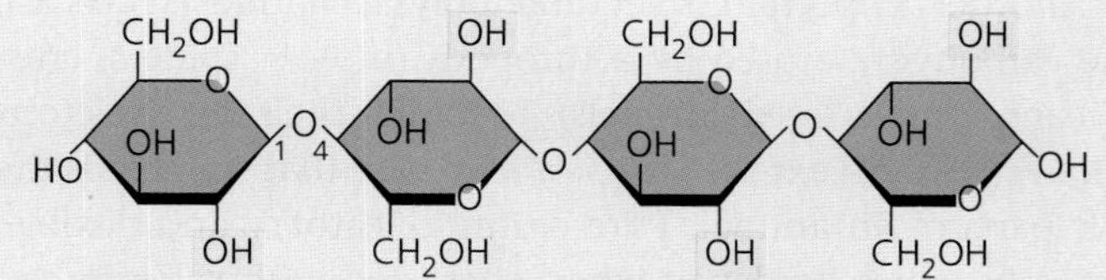

(c) Cellulose: 1–4 linkage of β glucose monomers. In cellulose, every β glucose monomer is upside down with respect to its neighbors.

▲ **Figure 5.7 Starch and cellulose structures.**

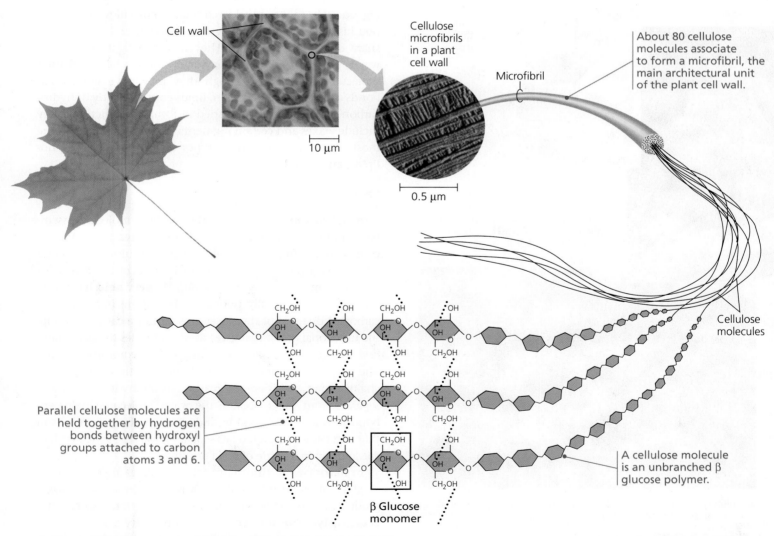

▲ **Figure 5.8 The arrangement of cellulose in plant cell walls.**

Some microorganisms can digest cellulose, breaking it down into glucose monomers. A cow harbors cellulose-digesting prokaryotes and protists in its stomach. These microbes hydrolyze the cellulose of hay and grass and convert the glucose to other compounds that nourish the cow. Similarly, a termite, which is unable to digest cellulose by itself, has prokaryotes or protists living in its gut that can make a meal of wood. Some fungi can also digest cellulose, thereby helping recycle chemical elements within Earth's ecosystems.

Another important structural polysaccharide is **chitin**, the carbohydrate used by arthropods (insects, spiders, crustaceans, and related animals) to build their exoskeletons **(Figure 5.9)**. An exoskeleton is a hard case that surrounds the soft parts of an animal. Pure chitin is leathery and flexible, but it becomes hardened when encrusted with calcium carbonate, a salt. Chitin is also found in many fungi, which use this polysaccharide rather than cellulose as the building material for their cell walls. Chitin is similar to cellulose, with β linkages, except that the glucose monomer of chitin has a nitrogen-containing appendage (see Figure 5.9, top right).

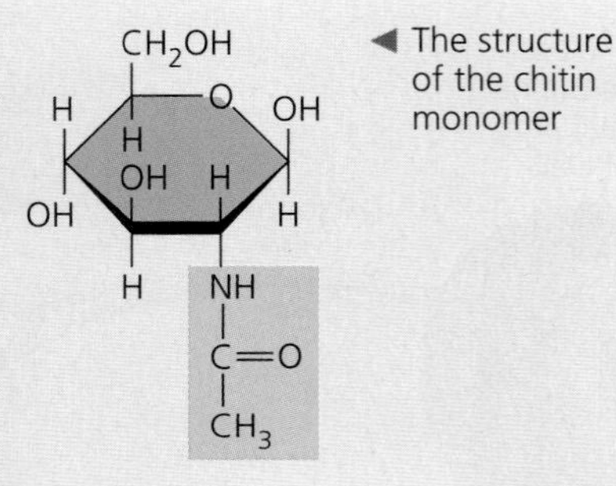

◀ The structure of the chitin monomer

◀ Chitin forms the exoskeleton of arthropods. This cicada is molting, shedding its old exoskeleton and emerging in adult form.

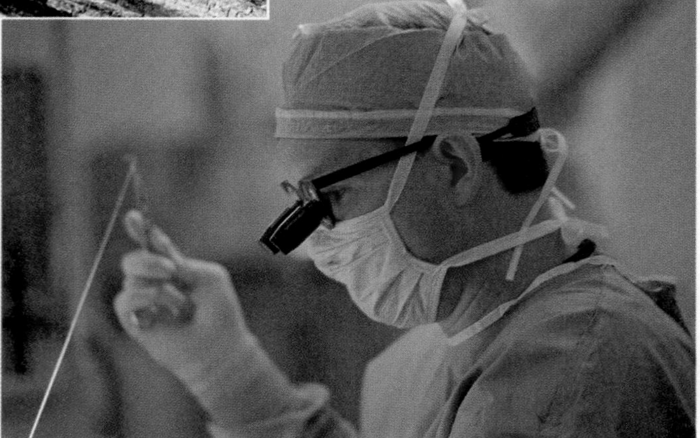

▲ Chitin is used to make a strong and flexible surgical thread that decomposes after the wound or incision heals.

▲ **Figure 5.9 Chitin, a structural polysaccharide.**

CONCEPT CHECK 5.2

1. Write the formula for a monosaccharide that has three carbons.
2. A dehydration reaction joins two glucose molecules to form maltose. The formula for glucose is $C_6H_{12}O_6$. What is the formula for maltose?
3. **WHAT IF?** After a cow is given antibiotics to treat an infection, a vet gives the animal a drink of "gut culture" containing various prokaryotes. Why is this necessary?

For suggested answers, see Appendix A.

CONCEPT 5.3

Lipids are a diverse group of hydrophobic molecules

Lipids are the one class of large biological molecules that does not include true polymers, and they are generally not big enough to be considered macromolecules. The compounds called **lipids** are grouped together because they share one important trait: They mix poorly, if at all, with water. The hydrophobic behavior of lipids is based on their molecular structure. Although they may have some polar bonds associated with oxygen, lipids consist mostly of hydrocarbon regions. Lipids are varied in form and function. They include waxes and certain pigments, but we will focus on the most biologically important types of lipids: fats, phospholipids, and steroids.

Fats

Although fats are not polymers, they are large molecules assembled from smaller molecules by dehydration reactions. A **fat** is constructed from two kinds of smaller molecules: glycerol and fatty acids **(Figure 5.10a)**. Glycerol is an alcohol; each of its three carbons bears a hydroxyl group. A **fatty acid** has a long carbon skeleton, usually 16 or 18 carbon atoms in length. The carbon at one end of the skeleton is part of a carboxyl group, the functional group that gives these molecules the name fatty *acid*. The rest of the skeleton consists of a hydrocarbon chain. The relatively nonpolar C—H bonds in the hydrocarbon chains of fatty acids are the reason fats are hydrophobic. Fats separate from water because the water molecules hydrogen-bond to one another and exclude the fats. This is the reason that vegetable oil (a liquid fat) separates from the aqueous vinegar solution in a bottle of salad dressing.

In making a fat, three fatty acid molecules are each joined to glycerol by an ester linkage, a bond between a hydroxyl group and a carboxyl group. The resulting fat, also called a **triacylglycerol**, thus consists of three fatty acids linked to one glycerol molecule. (Still another name for a fat is

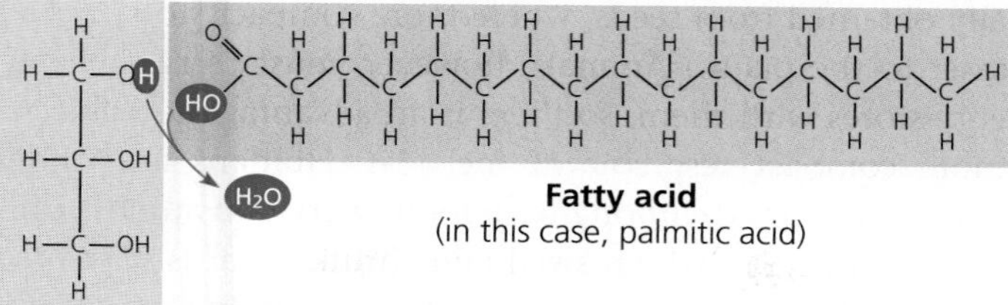

(a) One of three dehydration reactions in the synthesis of a fat

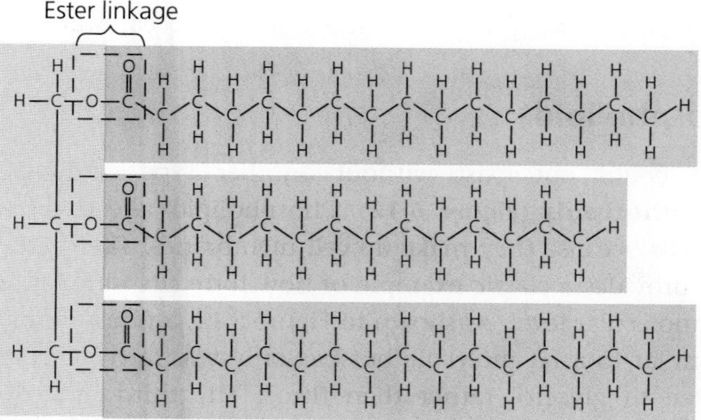

(b) Fat molecule (triacylglycerol)

▲ **Figure 5.10 The synthesis and structure of a fat, or triacylglycerol.** The molecular building blocks of a fat are one molecule of glycerol and three molecules of fatty acids. **(a)** One water molecule is removed for each fatty acid joined to the glycerol. **(b)** A fat molecule with three fatty acid units, two of them identical. The carbons of the fatty acids are arranged zigzag to suggest the actual orientations of the four single bonds extending from each carbon (see Figure 4.3a).

triglyceride, a word often found in the list of ingredients on packaged foods.) The fatty acids in a fat can be the same, or they can be of two or three different kinds, as in **Figure 5.10b**.

The terms *saturated fats* and *unsaturated fats* are commonly used in the context of nutrition **(Figure 5.11)**. These terms refer to the structure of the hydrocarbon chains of the fatty acids. If there are no double bonds between carbon atoms composing a chain, then as many hydrogen atoms as possible are bonded to the carbon skeleton. Such a structure is said to be *saturated* with hydrogen, and the resulting fatty acid therefore called a **saturated fatty acid (Figure 5.11a)**. An **unsaturated fatty acid** has one or more double bonds, with one fewer hydrogen atom on each double-bonded carbon. Nearly all double bonds in naturally occurring fatty acids are *cis* double bonds, which cause a kink in the hydrocarbon chain wherever they occur **(Figure 5.11b)**. (See Figure 4.7 to remind yourself about *cis* and *trans* double bonds.)

A fat made from saturated fatty acids is called a saturated fat. Most animal fats are saturated: The hydrocarbon chains of their fatty acids—the "tails" of the fat molecules—lack double bonds, and their flexibility allows the fat molecules to pack together tightly. Saturated animal fats—such as lard and butter—are solid at room temperature. In contrast, the fats of plants

▼ **Figure 5.11 Saturated and unsaturated fats and fatty acids.**

(a) Saturated fat

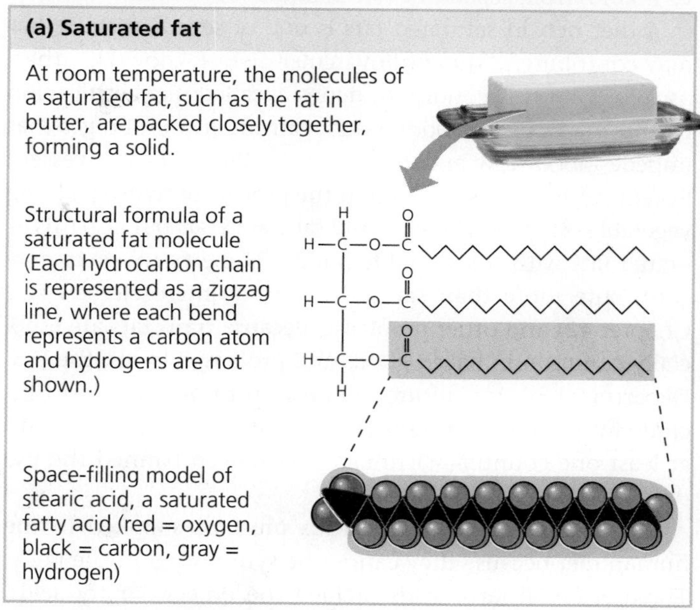

At room temperature, the molecules of a saturated fat, such as the fat in butter, are packed closely together, forming a solid.
Structural formula of a saturated fat molecule (Each hydrocarbon chain is represented as a zigzag line, where each bend represents a carbon atom and hydrogens are not shown.)
Space-filling model of stearic acid, a saturated fatty acid (red = oxygen, black = carbon, gray = hydrogen)

(b) Unsaturated fat

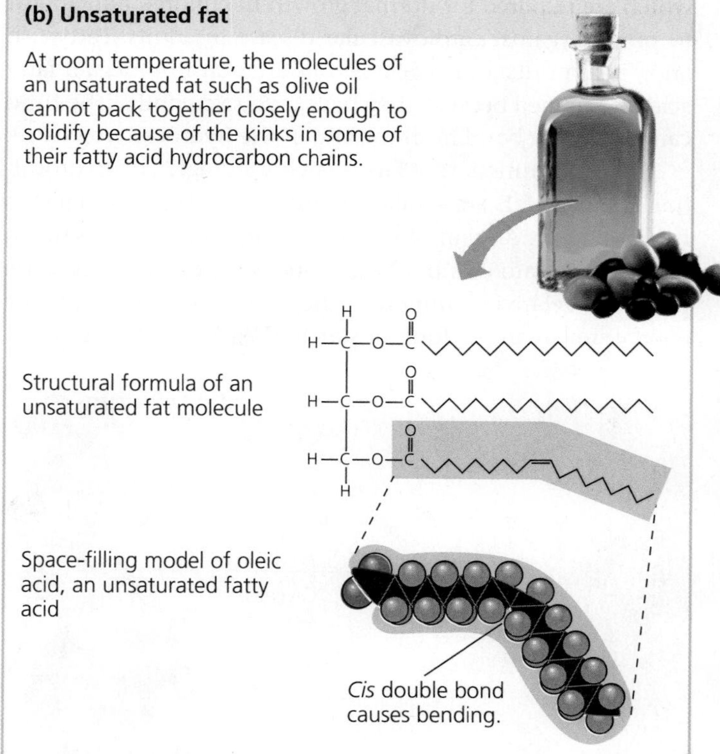

and fishes are generally unsaturated, meaning that they are built of one or more types of unsaturated fatty acids. Usually liquid at room temperature, plant and fish fats are referred to as oils—olive oil and cod liver oil are examples. The kinks where the *cis* double bonds are located prevent the molecules from packing together closely enough to solidify at room temperature. The phrase "hydrogenated vegetable oils" on food labels means that unsaturated fats have been synthetically converted to saturated fats by adding hydrogen. Peanut butter, margarine, and many other products are hydrogenated to prevent lipids from separating out in liquid (oil) form.

A diet rich in saturated fats is one of several factors that may contribute to the cardiovascular disease known as atherosclerosis. In this condition, deposits called plaques develop within the walls of blood vessels, causing inward bulges that impede blood flow and reduce the resilience of the vessels. Recent studies have shown that the process of hydrogenating vegetable oils produces not only saturated fats but also unsaturated fats with *trans* double bonds. These **trans fats** may contribute more than saturated fats to atherosclerosis (see Chapter 42) and other problems. Because trans fats are especially common in baked goods and processed foods, the U.S. Department of Agriculture requires nutritional labels to include information on trans fat content. Some U.S. cities and at least one country—Denmark—have even banned the use of trans fats in restaurants.

Certain unsaturated fatty acids must be supplied in the human diet because they cannot be synthesized in the body. These essential fatty acids include the omega-3 fatty acids, which are required for normal growth in children and appear to protect against cardiovascular disease in adults. Fatty fish and certain nuts and vegetable oils are rich in omega-3 fatty acids (so named because they have a double bond at the third carbon-carbon bond from the end of the hydrocarbon chain).

The major function of fats is energy storage. The hydrocarbon chains of fats are similar to gasoline molecules and just as rich in energy. A gram of fat stores more than twice as much energy as a gram of a polysaccharide, such as starch. Because plants are relatively immobile, they can function with bulky energy storage in the form of starch. (Vegetable oils are generally obtained from seeds, where more compact storage is an asset to the plant.) Animals, however, must carry their energy stores with them, so there is an advantage to having a more compact reservoir of fuel—fat. Humans and other mammals stock their long-term food reserves in adipose cells (see Figure 4.6a), which swell and shrink as fat is deposited and withdrawn from storage. In addition to storing energy, adipose tissue also cushions such vital organs as the kidneys, and a layer of fat beneath the skin insulates the body. This subcutaneous layer is especially thick in whales, seals, and most other marine mammals, protecting them from cold ocean water.

Phospholipids

Cells could not exist without another type of lipid—**phospholipids** (**Figure 5.12**). Phospholipids are essential for cells because they make up cell membranes. Their structure provides a classic example of how form fits function at the molecular level. As shown in Figure 5.12, a phospholipid is similar to a fat molecule but has only two fatty acids attached to glycerol rather than three. The third hydroxyl group of glycerol is joined to a phosphate group, which has a negative electrical charge in the cell. Additional small molecules, which are usually charged or polar, can be linked to the phosphate group to form a variety of phospholipids.

The two ends of phospholipids show different behavior toward water. The hydrocarbon tails are hydrophobic and are excluded from water. However, the phosphate group and its attachments form a hydrophilic head that has an affinity for water. When phospholipids are added to water, they self-assemble into double-layered structures called "bilayers," shielding their hydrophobic portions from water (**Figure 5.13**).

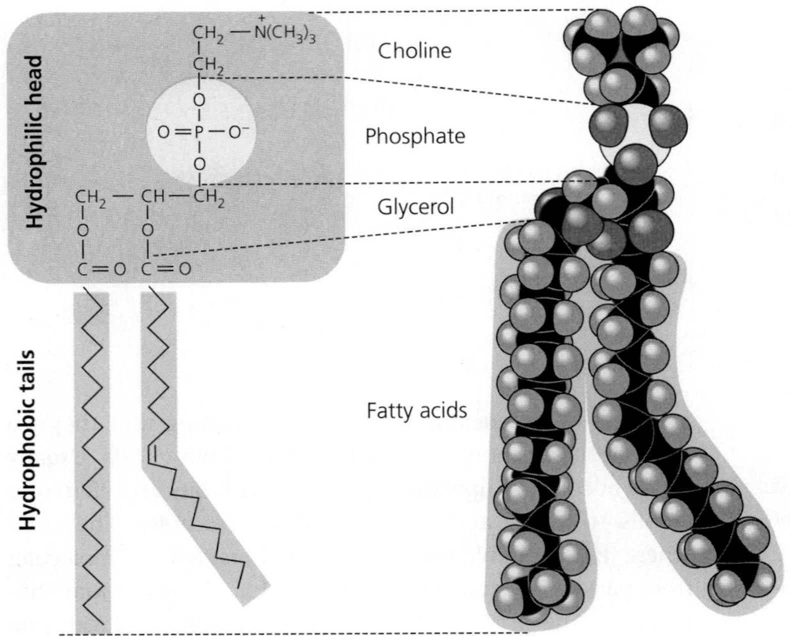

(a) Structural formula

(b) Space-filling model

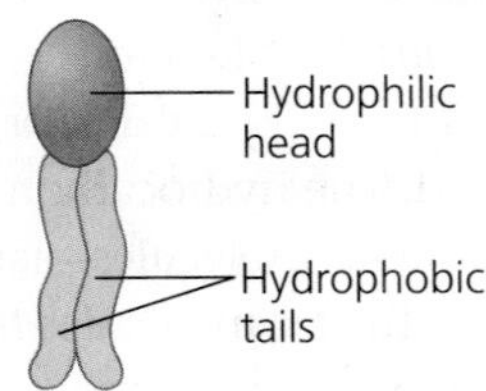

(c) Phospholipid symbol

◄ **Figure 5.12 The structure of a phospholipid.** A phospholipid has a hydrophilic (polar) head and two hydrophobic (nonpolar) tails. Phospholipid diversity is based on differences in the two fatty acids and in the groups attached to the phosphate group of the head. This particular phospholipid, called a phosphatidylcholine, has an attached choline group. The kink in one of its tails is due to a *cis* double bond. Shown here are **(a)** the structural formula, **(b)** the space-filling model (yellow = phosphorus, blue = nitrogen), and **(c)** the symbol for a phospholipid that will appear throughout this book. (In most figures, this symbol will be used to represent a phospholipid with either saturated or unsaturated tails.)

DRAW IT *Draw an oval around the hydrophilic head of the space-filling model.*

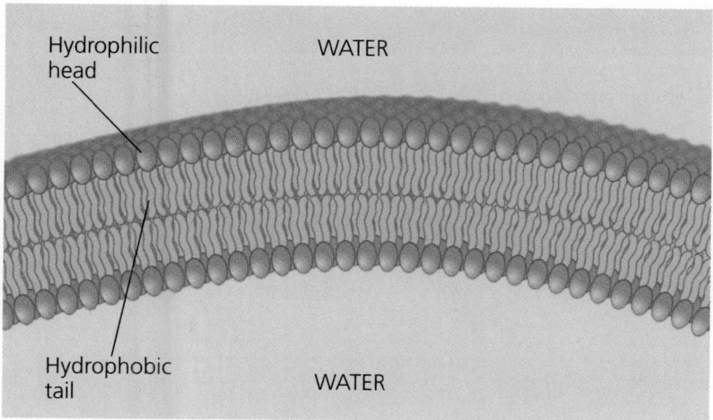

▲ **Figure 5.13 Bilayer structure formed by self-assembly of phospholipids in an aqueous environment.** The phospholipid bilayer shown here is the main fabric of biological membranes. Note that the hydrophilic heads of the phospholipids are in contact with water in this structure, whereas the hydrophobic tails are in contact with each other and remote from water.

At the surface of a cell, phospholipids are arranged in a similar bilayer. The hydrophilic heads of the molecules are on the outside of the bilayer, in contact with the aqueous solutions inside and outside of the cell. The hydrophobic tails point toward the interior of the bilayer, away from the water. The phospholipid bilayer forms a boundary between the cell and its external environment; in fact, cells could not exist without phospholipids.

Steroids

Steroids are lipids characterized by a carbon skeleton consisting of four fused rings. Different steroids, such as cholesterol and the vertebrate sex hormones, are distinguished by the particular chemical groups attached to this ensemble of rings (**Figure 5.14**). **Cholesterol** is a crucial molecule in animals. It is a common component of animal cell membranes and is also the precursor from which other steroids are synthesized. In vertebrates, cholesterol is synthesized in the liver and obtained from the diet. A high level of cholesterol in the blood may contribute to atherosclerosis. In fact, both saturated fats and trans fats exert their negative impact on health by affecting cholesterol levels.

H_3C CH_3 CH_3 CH_3 CH_3 HO

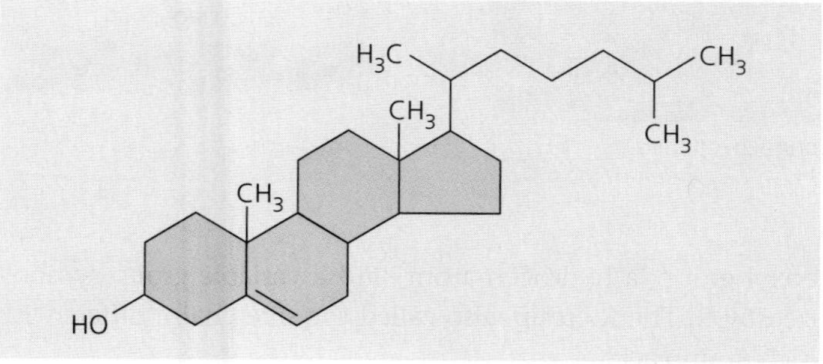

▲ **Figure 5.14 Cholesterol, a steroid.** Cholesterol is the molecule from which other steroids, including the sex hormones, are synthesized. Steroids vary in the chemical groups attached to their four interconnected rings (shown in gold).

MAKE CONNECTIONS *Compare cholesterol with the sex hormones shown in Concept 4.3 on p. 63. Circle the chemical groups that cholesterol has in common with estradiol; put a square around the chemical groups that cholesterol has in common with testosterone.*

CONCEPT CHECK 5.3

1. Compare the structure of a fat (triglyceride) with that of a phospholipid.
2. Why are human sex hormones considered lipids?
3. **WHAT IF?** Suppose a membrane surrounded an oil droplet, as it does in the cells of plant seeds. Describe and explain the form it might take.

For suggested answers, see Appendix A.

CONCEPT 5.4

Proteins include a diversity of structures, resulting in a wide range of functions

Nearly every dynamic function of a living being depends on proteins. In fact, the importance of proteins is underscored by their name, which comes from the Greek word *proteios*, meaning "first," or "primary." Proteins account for more than 50% of the dry mass of most cells, and they are instrumental in almost everything organisms do. Some proteins speed up chemical reactions, while others play a role in defense, storage, transport, cellular communication, movement, or structural support. **Figure 5.15**, on the next page, shows examples of proteins with these functions, which you'll learn more about in later chapters.

Life would not be possible without enzymes, most of which are proteins. Enzymatic proteins regulate metabolism by acting as **catalysts**, chemical agents that selectively speed up chemical reactions without being consumed by the reaction. Because an enzyme can perform its function over and over again, these molecules can be thought of as workhorses that keep cells running by carrying out the processes of life.

A human has tens of thousands of different proteins, each with a specific structure and function; proteins, in fact, are the most structurally sophisticated molecules known. Consistent with their diverse functions, they vary extensively in structure, each type of protein having a unique three-dimensional shape.

Polypeptides

Diverse as proteins are, they are all unbranched polymers constructed from the same set of 20 amino acids. Polymers of amino acids are called **polypeptides**. A **protein** is a biologically functional molecule that consists of one or more polypeptides, each folded and coiled into a specific three-dimensional structure.

▼ Figure 5.15 An overview of protein functions.

Enzymatic proteins

Function: Selective acceleration of chemical reactions

Example: Digestive enzymes catalyze the hydrolysis of bonds in food molecules.

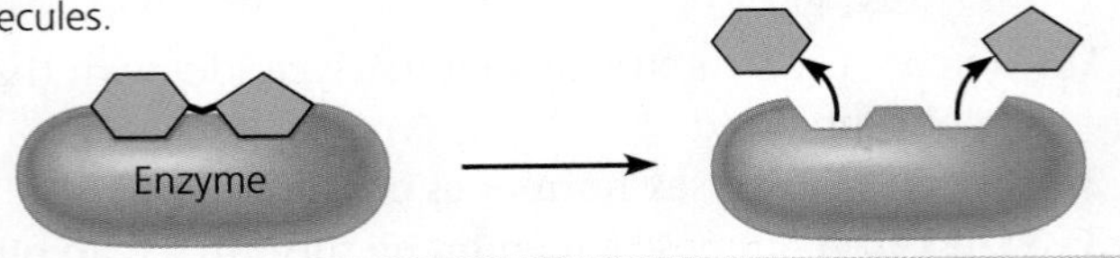

Defensive proteins

Function: Protection against disease

Example: Antibodies inactivate and help destroy viruses and bacteria.

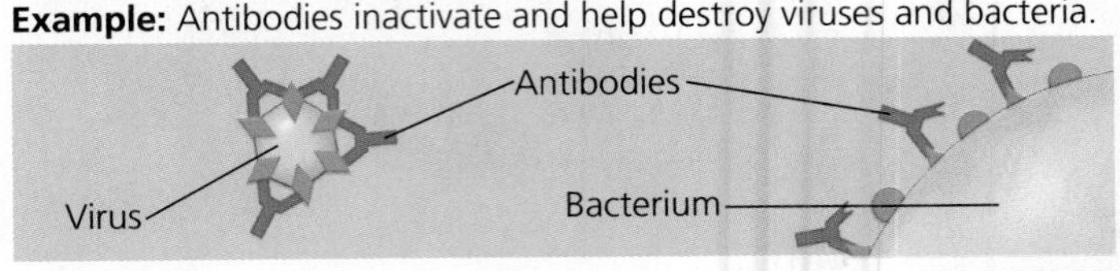

Storage proteins

Function: Storage of amino acids

Examples: Casein, the protein of milk, is the major source of amino acids for baby mammals. Plants have storage proteins in their seeds. Ovalbumin is the protein of egg white, used as an amino acid source for the developing embryo.

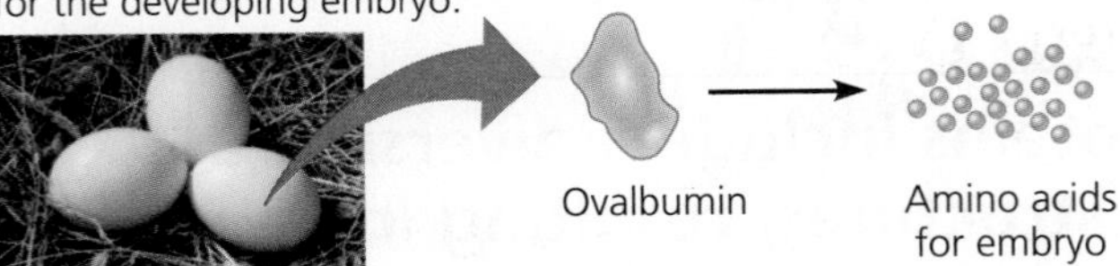

Transport proteins

Function: Transport of substances

Examples: Hemoglobin, the iron-containing protein of vertebrate blood, transports oxygen from the lungs to other parts of the body. Other proteins transport molecules across cell membranes.

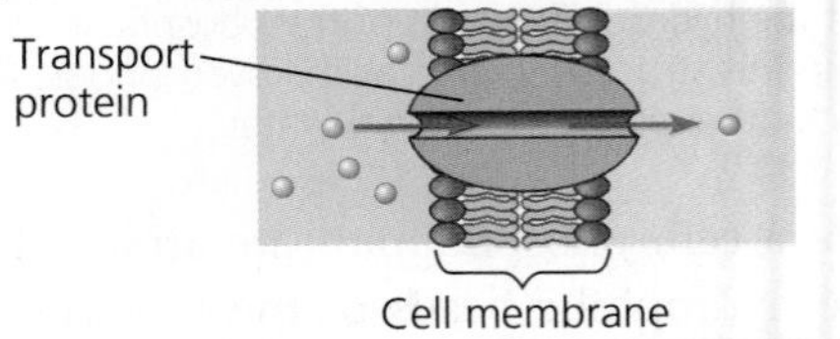

Hormonal proteins

Function: Coordination of an organism's activities

Example: Insulin, a hormone secreted by the pancreas, causes other tissues to take up glucose, thus regulating blood sugar concentration.

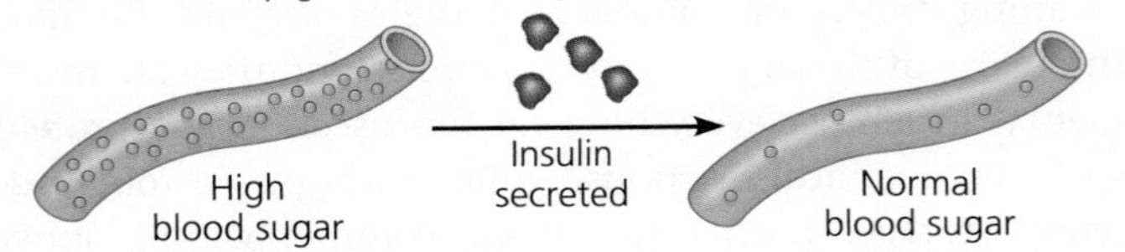

Receptor proteins

Function: Response of cell to chemical stimuli

Example: Receptors built into the membrane of a nerve cell detect signaling molecules released by other nerve cells.

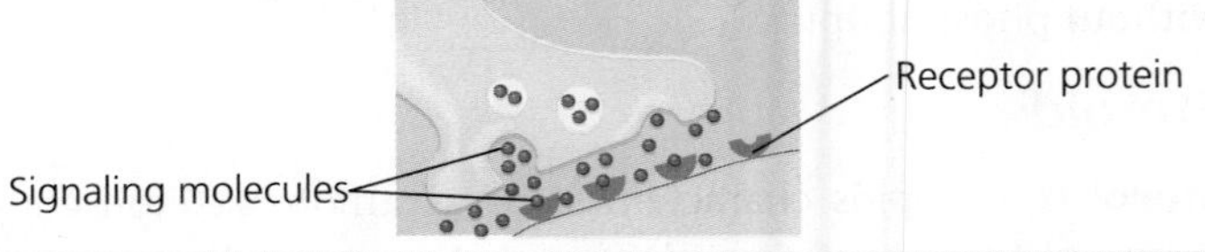

Contractile and motor proteins

Function: Movement

Examples: Motor proteins are responsible for the undulations of cilia and flagella. Actin and myosin proteins are responsible for the contraction of muscles.

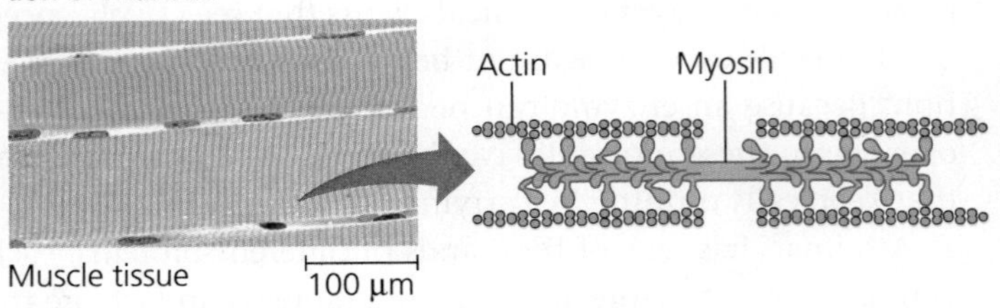

Structural proteins

Function: Support

Examples: Keratin is the protein of hair, horns, feathers, and other skin appendages. Insects and spiders use silk fibers to make their cocoons and webs, respectively. Collagen and elastin proteins provide a fibrous framework in animal connective tissues.

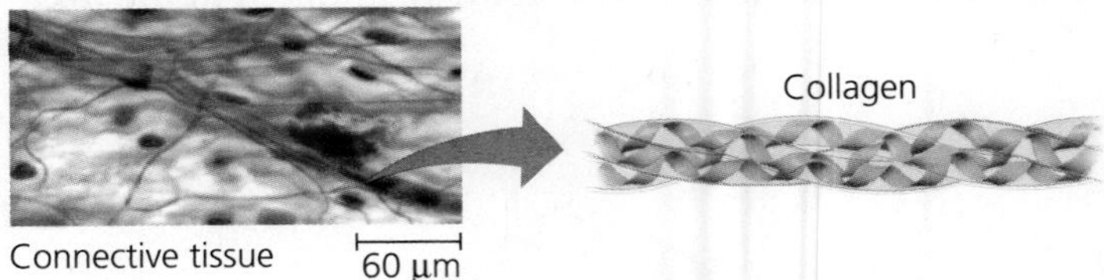

Amino Acid Monomers

All amino acids share a common structure. An **amino acid** is an organic molecule possessing both an amino group and a carboxyl group (see Figure 4.9). The illustration at the right shows the general formula for an amino acid. At the center of the amino acid is an asymmetric carbon atom called the *alpha (α) carbon*. Its four different partners are an amino group, a carboxyl group, a hydrogen atom, and a variable group symbolized by R. The R group, also called the side chain, differs with each amino acid.

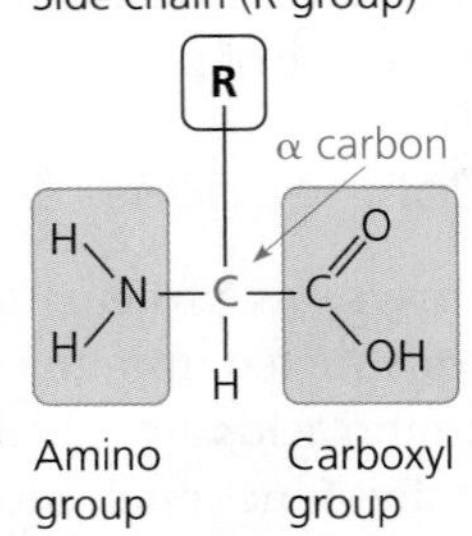

Figure 5.16 shows the 20 amino acids that cells use to build their thousands of proteins. Here the amino groups and carboxyl groups are all depicted in ionized form, the way they usually exist at the pH found in a cell. The side chain (R group) may be as simple as a hydrogen atom, as in the amino acid glycine, or it may be a carbon skeleton with various functional groups attached, as in glutamine.

Figure 5.16 The 20 amino acids of proteins. The amino acids are grouped here according to the properties of their side chains (R groups) and shown in their prevailing ionic forms at pH 7.2, the pH within a cell. The three-letter and one-letter abbreviations for the amino acids are in parentheses. All amino acids used in proteins are L enantiomers, the form shown here (see Figure 4.7).

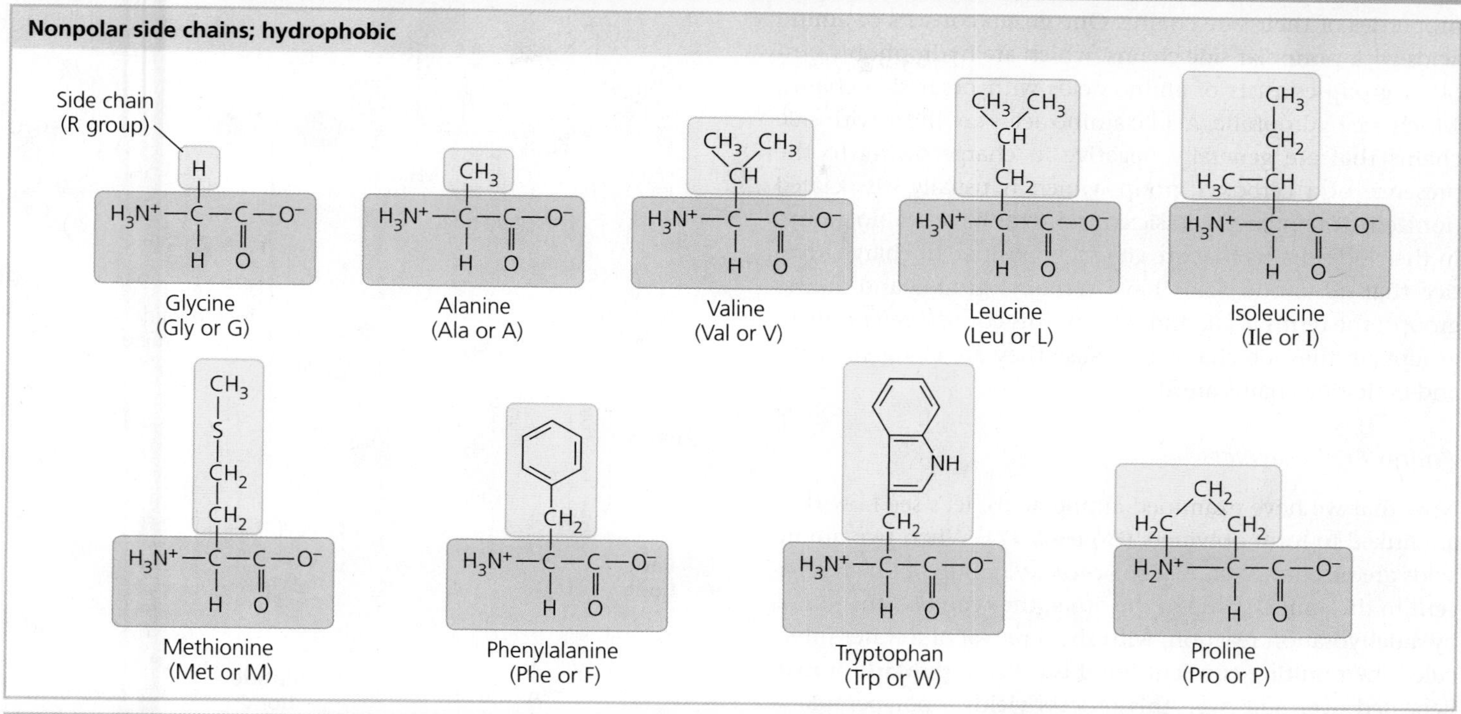

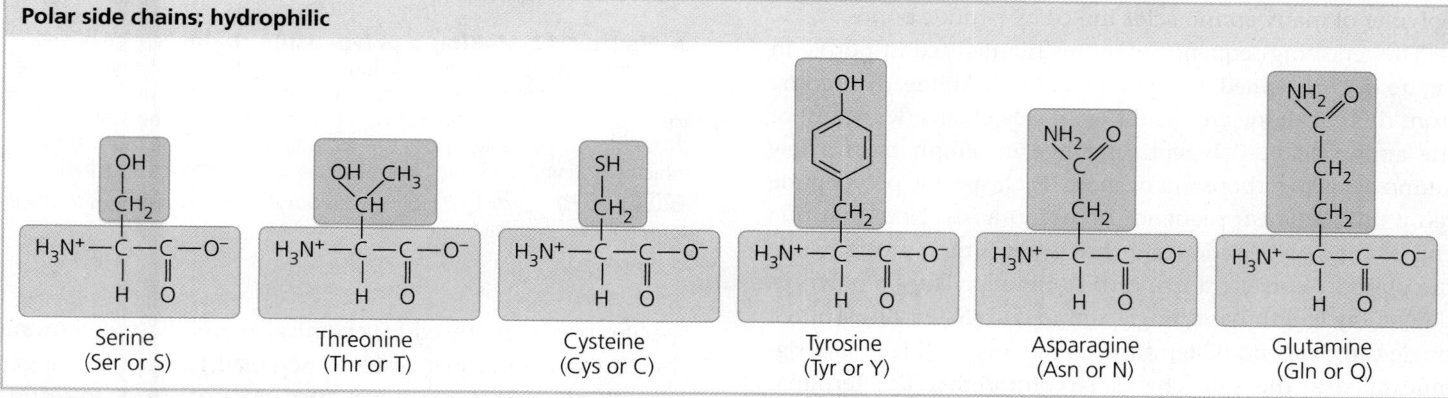

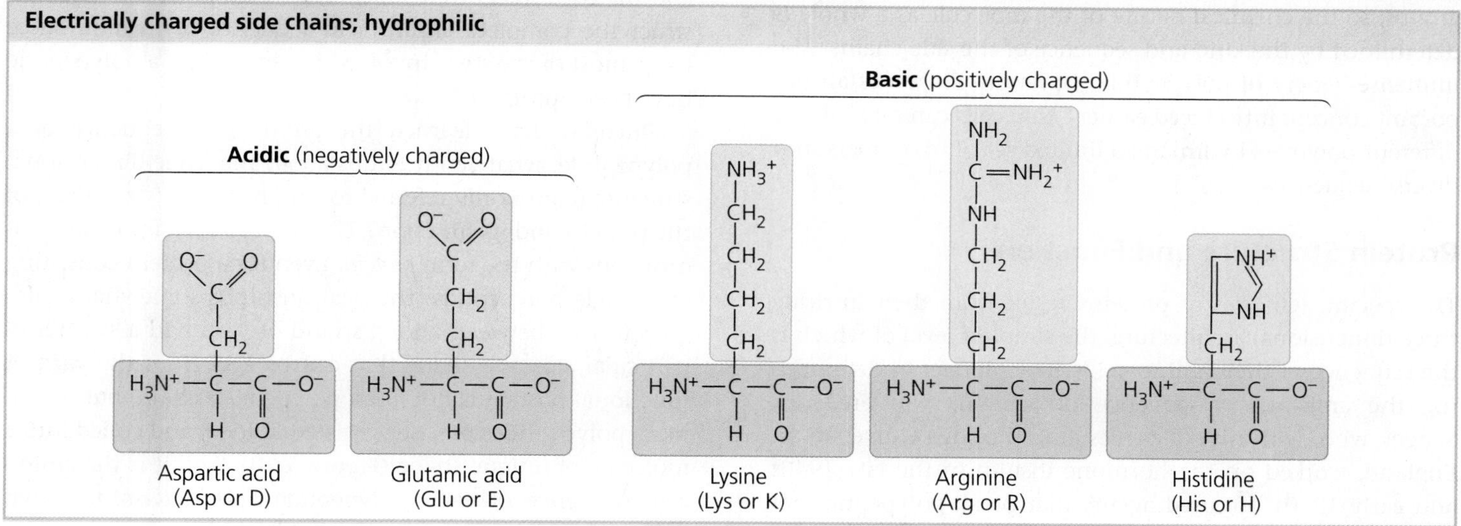

The physical and chemical properties of the side chain determine the unique characteristics of a particular amino acid, thus affecting its functional role in a polypeptide. In Figure 5.16, the amino acids are grouped according to the properties of their side chains. One group consists of amino acids with nonpolar side chains, which are hydrophobic. Another group consists of amino acids with polar side chains, which are hydrophilic. Acidic amino acids are those with side chains that are generally negative in charge owing to the presence of a carboxyl group, which is usually dissociated (ionized) at cellular pH. Basic amino acids have amino groups in their side chains that are generally positive in charge. (Notice that *all* amino acids have carboxyl groups and amino groups; the terms *acidic* and *basic* in this context refer only to groups on the side chains.) Because they are charged, acidic and basic side chains are also hydrophilic.

Amino Acid Polymers

Now that we have examined amino acids, let's see how they are linked to form polymers **(Figure 5.17)**. When two amino acids are positioned so that the carboxyl group of one is adjacent to the amino group of the other, they can become joined by a dehydration reaction, with the removal of a water molecule. The resulting covalent bond is called a **peptide bond**. Repeated over and over, this process yields a polypeptide, a polymer of many amino acids linked by peptide bonds.

The repeating sequence of atoms highlighted in purple in Figure 5.17 is called the polypeptide backbone. Extending from this backbone are the different side chains (R groups) of the amino acids. Polypeptides range in length from a few amino acids to a thousand or more. Each specific polypeptide has a unique linear sequence of amino acids. Note that one end of the polypeptide chain has a free amino group, while the opposite end has a free carboxyl group. Thus, a polypeptide of any length has a single amino end (N-terminus) and a single carboxyl end (C-terminus). In a polypeptide of any significant size, the side chains far outnumber the terminal groups, so the chemical nature of the molecule as a whole is determined by the kind and sequence of the side chains. The immense variety of polypeptides in nature illustrates an important concept introduced earlier—that cells can make many different polymers by linking a limited set of monomers into diverse sequences.

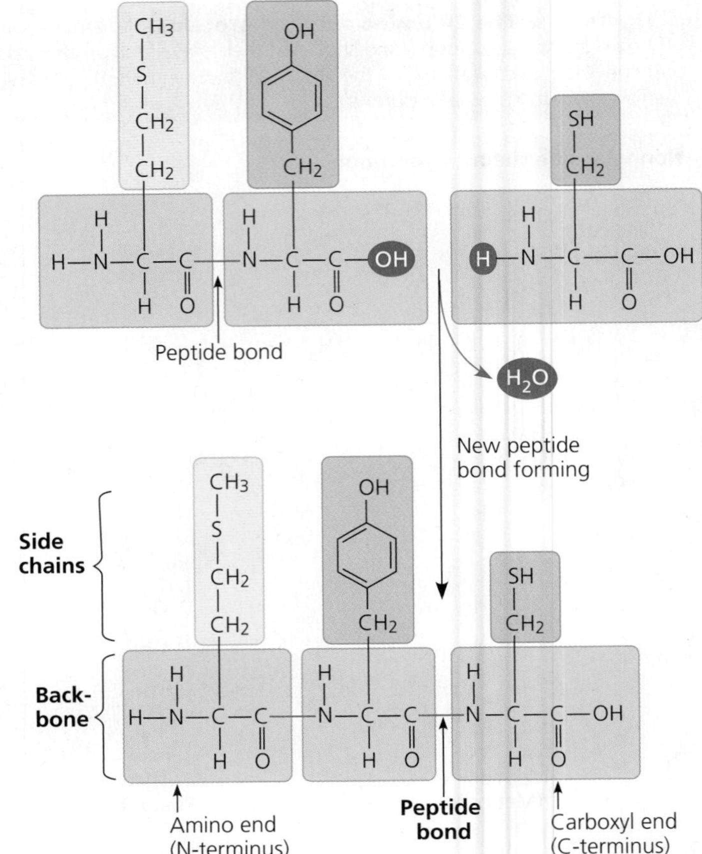

▲ **Figure 5.17 Making a polypeptide chain.** Peptide bonds are formed by dehydration reactions, which link the carboxyl group of one amino acid to the amino group of the next. The peptide bonds are formed one at a time, starting with the amino acid at the amino end (N-terminus). The polypeptide has a repetitive backbone (purple) to which the amino acid side chains (yellow and green) are attached.

DRAW IT *Circle and label the carboxyl and amino groups that will form the new peptide bond.*

Protein Structure and Function

The specific activities of proteins result from their intricate three-dimensional architecture, the simplest level of which is the sequence of their amino acids. The pioneer in determining the amino acid sequence of proteins was Frederick Sanger, who, with his colleagues at Cambridge University in England, worked on the hormone insulin in the late 1940s and early 1950s. He used agents that break polypeptides at specific places, followed by chemical methods to determine the amino acid sequence in these small fragments. Sanger and his co-workers were able, after years of effort, to reconstruct the complete amino acid sequence of insulin. Since then, most of the steps involved in sequencing a polypeptide have been automated.

Once we have learned the amino acid sequence of a polypeptide, what can it tell us about the three-dimensional structure (commonly referred to simply as the "structure") of the protein and its function? The term *polypeptide* is not synonymous with the term *protein*. Even for a protein consisting of a single polypeptide, the relationship is somewhat analogous to that between a long strand of yarn and a sweater of particular size and shape that can be knit from the yarn. A functional protein is not *just* a polypeptide chain, but one or more polypeptides precisely twisted, folded, and coiled into a molecule of unique shape **(Figure 5.18)**. And it is the amino acid sequence of each polypeptide that determines what

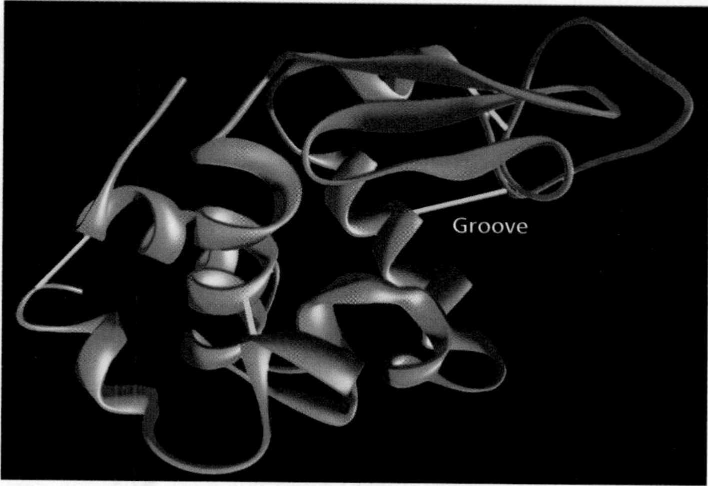

(a) A **ribbon model** shows how the single polypeptide chain folds and coils to form the functional protein. (The yellow lines represent disulfide bridges that stabilize the protein's shape; see Figure 5.20.)

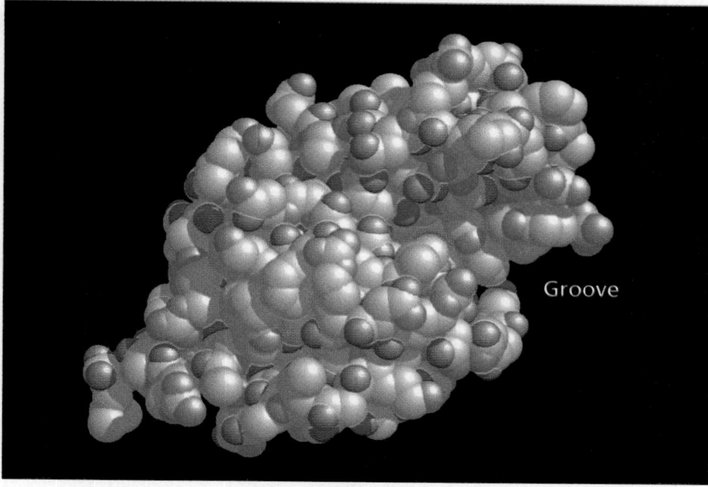

(b) A **space-filling model** shows more clearly the globular shape seen in many proteins, as well as the specific three-dimensional structure unique to lysozyme.

▲ **Figure 5.18 Structure of a protein, the enzyme lysozyme.** Present in our sweat, tears, and saliva, lysozyme is an enzyme that helps prevent infection by binding to and destroying specific molecules on the surface of many kinds of bacteria. The groove is the part of the protein that recognizes and binds to the target molecules on bacterial walls.

three-dimensional structure the protein will have under normal cellular conditions.

When a cell synthesizes a polypeptide, the chain generally folds spontaneously, assuming the functional structure for that protein. This folding is driven and reinforced by the formation of a variety of bonds between parts of the chain, which in turn depends on the sequence of amino acids. Many proteins are roughly spherical (*globular proteins*), while others are shaped like long fibers (*fibrous proteins*). Even within these broad categories, countless variations exist.

A protein's specific structure determines how it works. In almost every case, the function of a protein depends on its ability to recognize and bind to some other molecule. In an especially striking example of the marriage of form and function, **Figure 5.19** shows the exact match of shape between an antibody (a protein in the body) and the particular foreign substance on a flu virus that the antibody binds to and marks for destruction. In Chapter 43, you'll learn more about how the immune system generates antibodies that match the shapes of specific foreign molecules so well. Also, you may recall from Chapter 2 that natural signaling molecules called endorphins bind to specific receptor proteins on the surface of brain cells in humans, producing euphoria and relieving pain. Morphine, heroin, and other opiate drugs are able to mimic endorphins because they all share a similar shape with endorphins and can thus fit into and bind to endorphin receptors in the brain. This fit is very specific, something like a lock and key (see Figure 2.18). Thus, the function of a protein—for instance, the ability of a receptor protein to bind to a particular pain-relieving signaling molecule—is an emergent property resulting from exquisite molecular order.

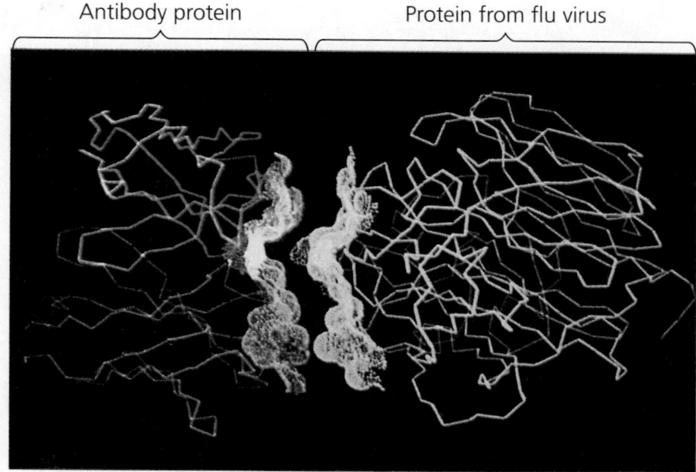

▲ **Figure 5.19 An antibody binding to a protein from a flu virus.** A technique called X-ray crystallography was used to generate a computer model of an antibody protein (blue and orange, left) bound to a flu virus protein (green and yellow, right). Computer software was then used to back the images away from each other, revealing the exact complementarity of shape between the two protein surfaces.

Four Levels of Protein Structure

With the goal of understanding the function of a protein, learning about its structure is often productive. In spite of their great diversity, all proteins share three superimposed levels of structure, known as primary, secondary, and tertiary structure. A fourth level, quaternary structure, arises when a protein consists of two or more polypeptide chains. **Figure 5.20**, on the following two pages, describes these four levels of protein structure. Be sure to study this figure thoroughly before going on to the next section.

▼ **Figure 5.20**

Exploring Levels of Protein Structure

Primary Structure

Linear chain of amino acids

Amino acids

$^{+}H_3N$— Gly Pro Thr Gly Thr Gly Glu Ser Lys Cys Pro Leu Met Val
Amino end

His Val Ala Val Asn Ile Ala Pro Ser Gly Arg Val Ala Asp Leu Val Lys

Val Phe Arg Lys

Ala Ala Asp Asp Thr Trp Glu Pro Phe Ala Ser Gly Lys Thr Ser Glu Ser Gly Glu Leu His

Primary structure of transthyretin

Ile Glu Val Lys Tyr Ile Gly Glu Val Phe Glu Glu Glu Thr Thr Leu Gly

Asp Thr Lys Ser

Tyr Trp Lys Ala Leu Gly Ile Ser Pro Phe His Glu His Ala Glu Val Val Phe Thr Ala Asn

Ser Tyr Pro Ser Leu Leu Ala Ala Ile Thr Tyr Arg Arg Pro Gly Ser Asp

Tyr Ser Thr Thr

Ala Val Val Thr Asn Pro Lys Glu—$C(=O)O^{-}$ Carboxyl end

The **primary structure** of a protein is a linked series of amino acids with a unique sequence. As an example, let's consider transthyretin, a globular blood protein that transports vitamin A and one of the thyroid hormones throughout the body. Transthyretin is made up of four identical polypeptide chains, each composed of 127 amino acids. Shown here is one of these chains unraveled for a closer look at its primary structure. Each of the 127 positions along the chain is occupied by one of the 20 amino acids, indicated here by its three-letter abbreviation.

The primary structure is like the order of letters in a very long word. If left to chance, there would be 20^{127} different ways of making a polypeptide chain 127 amino acids long. However, the precise primary structure of a protein is determined not by the random linking of amino acids, but by inherited genetic information. The primary structure in turn dictates secondary and tertiary structure, due to the chemical nature of the backbone and the side chains (R groups) of the amino acids positioned along the chain.

Secondary Structure

Regions stabilized by hydrogen bonds between atoms of the polypeptide backbone

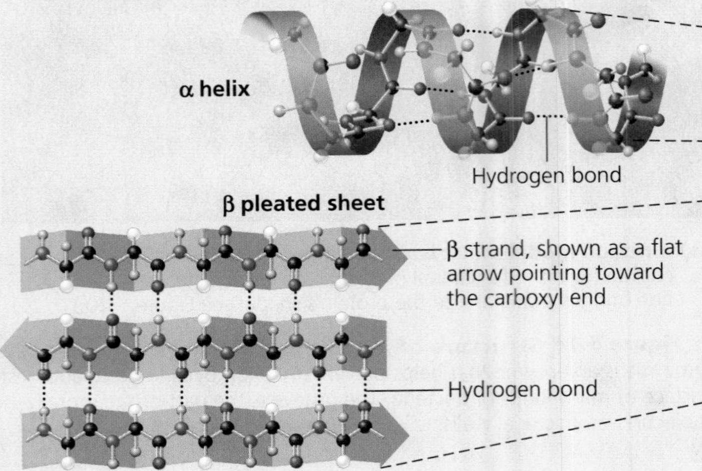

Most proteins have segments of their polypeptide chains repeatedly coiled or folded in patterns that contribute to the protein's overall shape. These coils and folds, collectively referred to as **secondary structure**, are the result of hydrogen bonds between the repeating constituents of the polypeptide backbone (not the amino acid side chains). Within the backbone, the oxygen atoms have a partial negative charge, and the hydrogen atoms attached to the nitrogens have a partial positive charge (see Figure 2.16); therefore, hydrogen bonds can form between these atoms. Individually, these hydrogen bonds are weak, but because they are repeated many times over a relatively long region of the polypeptide chain, they can support a particular shape for that part of the protein.

One such secondary structure is the **α helix**, a delicate coil held together by hydrogen bonding between every fourth amino acid, shown above. Although each transthyretin polypeptide has only one α helix region (see tertiary structure on the next page), other globular proteins have multiple stretches of α helix separated by nonhelical regions (see hemoglobin on the next page). Some fibrous proteins, such as α-keratin, the structural protein of hair, have the α helix formation over most of their length.

The other main type of secondary structure is the **β pleated sheet**. As shown above, in this structure two or more strands of the polypeptide chain lying side by side (called β strands) are connected by hydrogen bonds between parts of the two parallel polypeptide backbones. β pleated sheets make up the core of many globular proteins, as is the case for transthyretin (see tertiary structure on the next page), and dominate some fibrous proteins, including the silk protein of a spider's web. The teamwork of so many hydrogen bonds makes each spider silk fiber stronger than a steel strand of the same weight.

▼ Spiders secrete silk fibers made of a structural protein containing β pleated sheets, which allow the spider web to stretch and recoil.

Tertiary Structure

Three-dimensional shape stabilized by interactions between side chains

Transthyretin polypeptide

Superimposed on the patterns of secondary structure is a protein's tertiary structure, shown above in a ribbon model of the transthyretin polypeptide. While secondary structure involves interactions between backbone constituents, **tertiary structure** is the overall shape of a polypeptide resulting from interactions between the side chains (R groups) of the various amino acids. One type of interaction that contributes to tertiary structure is—somewhat misleadingly—called a **hydrophobic interaction**. As a polypeptide folds into its functional shape, amino acids with hydrophobic (nonpolar) side chains usually end up in clusters at the core of the protein, out of contact with water. Thus, a "hydrophobic interaction" is actually caused by the exclusion of nonpolar substances by water molecules. Once nonpolar amino acid side chains are close together, van der Waals interactions help hold them together. Meanwhile, hydrogen bonds between polar side chains and ionic bonds between positively and negatively charged side chains also help stabilize tertiary structure. These are all weak interactions in the aqueous cellular environment, but their cumulative effect helps give the protein a unique shape.

Covalent bonds called **disulfide bridges** may further reinforce the shape of a protein. Disulfide bridges form where two cysteine monomers, which have sulfhydryl groups (—SH) on their side chains (see Figure 4.9), are brought close together by the folding of the protein. The sulfur of one cysteine bonds to the sulfur of the second, and the disulfide bridge (—S—S—) rivets parts of the protein together (see yellow lines in Figure 5.18a). All of these different kinds of interactions can contribute to the tertiary structure of a protein, as shown here in a small part of a hypothetical protein:

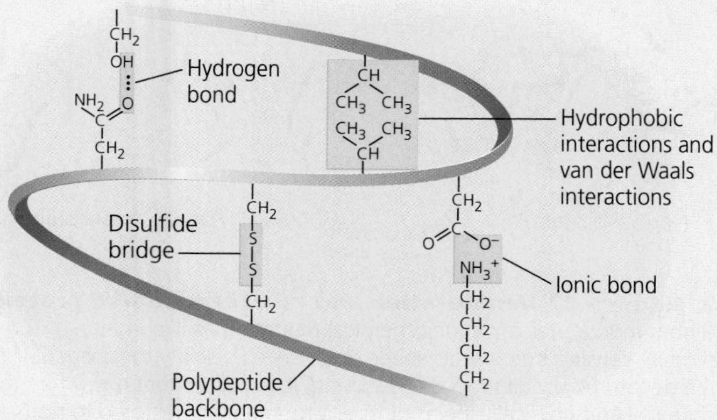

Quaternary Structure

Association of multiple polypeptides, forming a functional protein

Transthyretin protein (four identical polypeptides)

Some proteins consist of two or more polypeptide chains aggregated into one functional macromolecule. **Quaternary structure** is the overall protein structure that results from the aggregation of these polypeptide subunits. For example, shown above is the complete globular transthyretin protein, made up of its four polypeptides.

Another example is collagen, shown below, which is a fibrous protein that has three identical helical polypeptides intertwined into a larger triple helix, giving the long fibers great strength. This suits collagen fibers to their function as the girders of connective tissue in skin, bone, tendons, ligaments, and other body parts. (Collagen accounts for 40% of the protein in a human body.)

Collagen

Hemoglobin, the oxygen-binding protein of red blood cells shown below, is another example of a globular protein with quaternary structure. It consists of four polypeptide subunits, two of one kind (α) and two of another kind (β). Both α and β subunits consist primarily of α-helical secondary structure. Each subunit has a nonpolypeptide component, called heme, with an iron atom that binds oxygen.

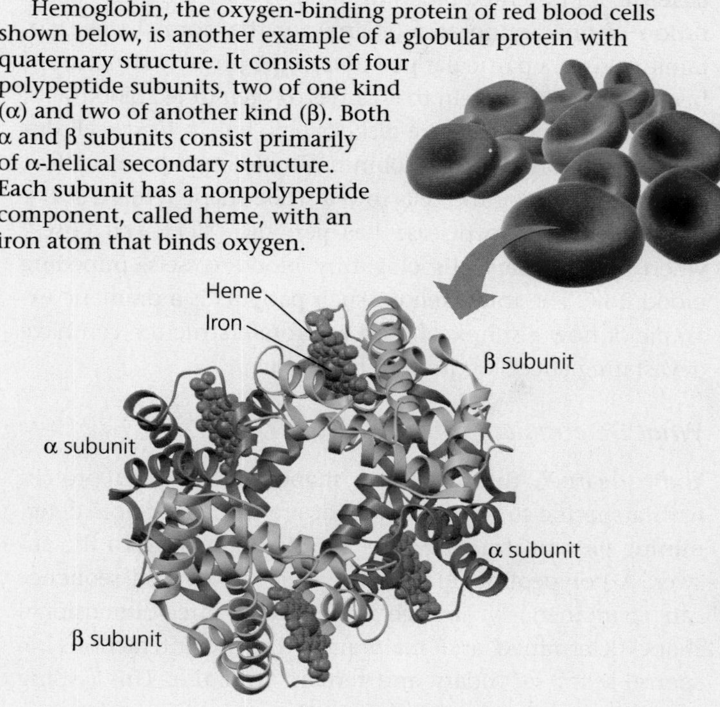

Hemoglobin

	Primary Structure	Secondary and Tertiary Structures	Quaternary Structure	Function	Red Blood Cell Shape
Normal hemoglobin	1 Val 2 His 3 Leu 4 Thr 5 Pro 6 Glu 7 Glu	β subunit	Normal hemoglobin α β α β	Molecules do not associate with one another; each carries oxygen.	Normal red blood cells are full of individual hemoglobin molecules, each carrying oxygen. 10 μm
Sickle-cell hemoglobin	1 Val 2 His 3 Leu 4 Thr 5 Pro 6 Val 7 Glu	Exposed hydrophobic region β subunit	Sickle-cell hemoglobin α β α β	Molecules interact with one another and crystallize into a fiber; capacity to carry oxygen is greatly reduced.	Fibers of abnormal hemoglobin deform red blood cell into sickle shape. 10 μm

▲ **Figure 5.21 A single amino acid substitution in a protein causes sickle-cell disease.**

MAKE CONNECTIONS *Considering the chemical characteristics of the amino acids valine and glutamic acid (see Figure 5.16), propose a possible explanation for the dramatic effect on protein function that occurs when valine is substituted for glutamic acid.*

Sickle-Cell Disease: A Change in Primary Structure

Even a slight change in primary structure can affect a protein's shape and ability to function. For instance, **sickle-cell disease**, an inherited blood disorder, is caused by the substitution of one amino acid (valine) for the normal one (glutamic acid) at a particular position in the primary structure of hemoglobin, the protein that carries oxygen in red blood cells. Normal red blood cells are disk-shaped, but in sickle-cell disease, the abnormal hemoglobin molecules tend to crystallize, deforming some of the cells into a sickle shape (**Figure 5.21**). A person with the disease has periodic "sickle-cell crises" when the angular cells clog tiny blood vessels, impeding blood flow. The toll taken on such patients is a dramatic example of how a simple change in protein structure can have devastating effects on protein function.

What Determines Protein Structure?

You've learned that a unique shape endows each protein with a specific function. But what are the key factors determining protein structure? You already know most of the answer: A polypeptide chain of a given amino acid sequence can spontaneously arrange itself into a three-dimensional shape determined and maintained by the interactions responsible for secondary and tertiary structure. This folding normally occurs as the protein is being synthesized in the crowded environment within a cell, aided by other proteins. However, protein structure also depends on the physical and chemical conditions of the protein's environment. If the pH, salt concentration, temperature, or other aspects of its environment are altered, the weak chemical bonds and interactions within a protein may be destroyed, causing the protein to unravel and lose its native shape, a change called **denaturation** (**Figure 5.22**). Because it is misshapen, the denatured protein is biologically inactive.

Most proteins become denatured if they are transferred from an aqueous environment to a nonpolar solvent, such as

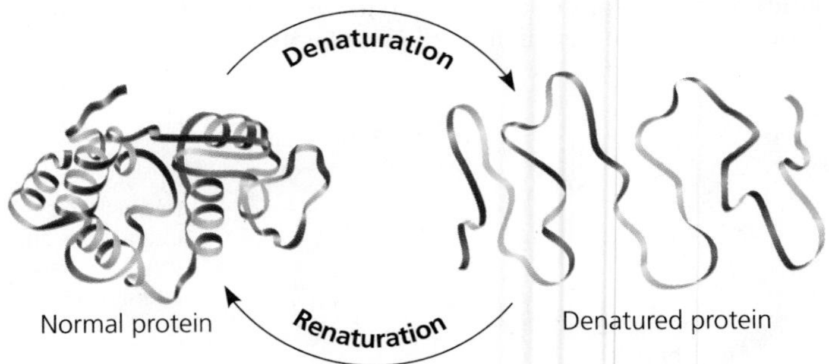

▲ **Figure 5.22 Denaturation and renaturation of a protein.** High temperatures or various chemical treatments will denature a protein, causing it to lose its shape and hence its ability to function. If the denatured protein remains dissolved, it can often renature when the chemical and physical aspects of its environment are restored to normal.

ether or chloroform; the polypeptide chain refolds so that its hydrophobic regions face outward toward the solvent. Other denaturation agents include chemicals that disrupt the hydrogen bonds, ionic bonds, and disulfide bridges that maintain a protein's shape. Denaturation can also result from excessive heat, which agitates the polypeptide chain enough to overpower the weak interactions that stabilize the structure. The white of an egg becomes opaque during cooking because the denatured proteins are insoluble and solidify. This also explains why excessively high fevers can be fatal: Proteins in the blood can denature at very high body temperatures.

When a protein in a test-tube solution has been denatured by heat or chemicals, it can sometimes return to its functional shape when the denaturing agent is removed. We can conclude that the information for building specific shape is intrinsic to the protein's primary structure. The sequence of amino acids determines the protein's shape—where an α helix can form, where β pleated sheets can exist, where disulfide bridges are located, where ionic bonds can form, and so on. But how does protein folding occur in the cell?

Protein Folding in the Cell

Biochemists now know the amino acid sequence for more than 10 million proteins and the three-dimensional shape for more than 20,000. Researchers have tried to correlate the primary structure of many proteins with their three-dimensional structure to discover the rules of protein folding. Unfortunately, however, the protein-folding process is not that simple. Most proteins probably go through several intermediate structures on their way to a stable shape, and looking at the mature structure does not reveal the stages of folding required to achieve that form. However, biochemists have developed methods for tracking a protein through such stages.

Crucial to the folding process are **chaperonins** (also called chaperone proteins), protein molecules that assist in the proper folding of other proteins (**Figure 5.23**). Chaperonins do not specify the final structure of a polypeptide. Instead, they keep the new polypeptide segregated from "bad influences" in the cytoplasmic environment while it folds spontaneously. The chaperonin shown in Figure 5.23, from the bacterium *E. coli*, is a giant multiprotein complex shaped like a hollow cylinder. The cavity provides a shelter for folding polypeptides. In the past decade, researchers have discovered molecular systems that interact with chaperonins and check whether proper folding has occurred. Such systems either refold the misfolded proteins correctly or mark them for destruction.

Misfolding of polypeptides is a serious problem in cells. Many diseases, such as Alzheimer's, Parkinson's, and mad cow disease, are associated with an accumulation of misfolded proteins. In fact, misfolded versions of the transthyretin protein featured in Figure 5.20 have been implicated in several diseases, including one form of senile dementia.

Even when scientists have a correctly folded protein in hand, determining its exact three-dimensional structure is not simple, for a single protein molecule has thousands of atoms. The first 3-D structures were worked out in 1959 for hemoglobin and a related protein. The method that made these feats possible was **X-ray crystallography**, which has since been used to determine the 3-D structure of many other proteins. In a recent example, Roger Kornberg and his colleagues at Stanford University used this method to elucidate the structure of RNA polymerase, an enzyme that plays a crucial role in the expression of genes (**Figure 5.24**, on the next page). Another method for analyzing protein structure is nuclear magnetic resonance (NMR) spectroscopy, which does not require protein crystallization. A still newer approach employs bioinformatics (see Chapter 1) to predict the 3-D structure of polypeptides from their amino acid sequence. X-ray crystallography, NMR spectroscopy, and bioinformatics are complementary approaches to understanding protein structure and function.

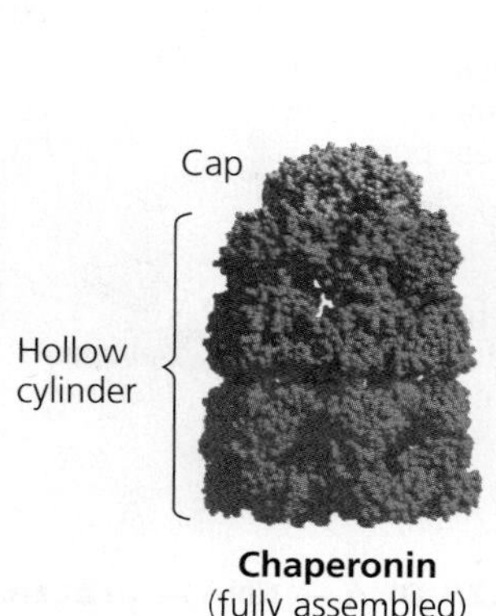

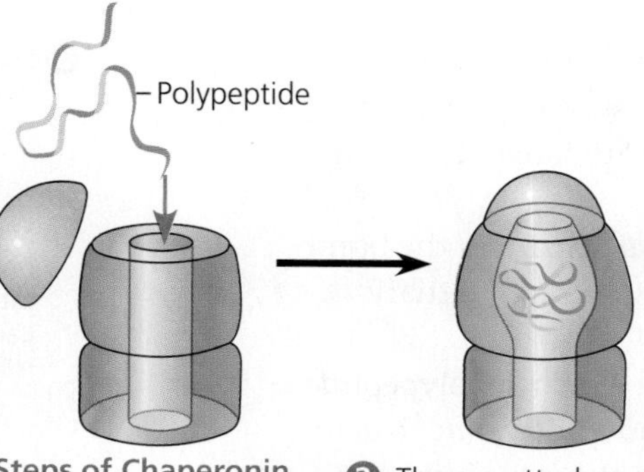

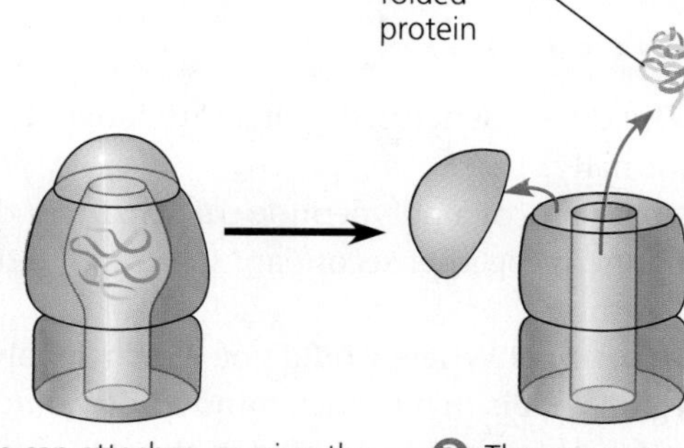

Steps of Chaperonin Action:

1. An unfolded polypeptide enters the cylinder from one end.
2. The cap attaches, causing the cylinder to change shape in such a way that it creates a hydrophilic environment for the folding of the polypeptide.
3. The cap comes off, and the properly folded protein is released.

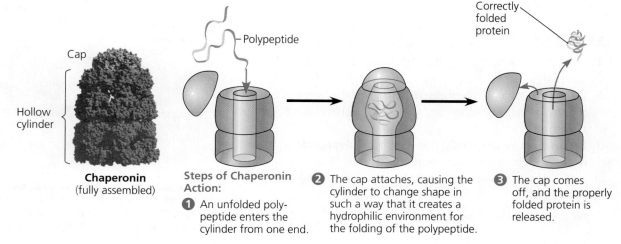

▶ **Figure 5.23 A chaperonin in action.** The computer graphic (left) shows a large chaperonin protein complex from the bacterium *E. coli*. It has an interior space that provides a shelter for the proper folding of newly made polypeptides. The complex consists of two proteins: One protein is a hollow cylinder; the other is a cap that can fit on either end.

▼ **Figure 5.24** INQUIRY

What can the 3-D shape of the enzyme RNA polymerase II tell us about its function?

EXPERIMENT In 2006, Roger Kornberg was awarded the Nobel Prize in Chemistry for using X-ray crystallography to determine the 3-D shape of RNA polymerase II, which binds to the DNA double helix and synthesizes RNA. After crystallizing a complex of all three components, Kornberg and his colleagues aimed an X-ray beam through the crystal. The atoms of the crystal diffracted (bent) the X-rays into an orderly array that a digital detector recorded as a pattern of spots called an X-ray diffraction pattern.

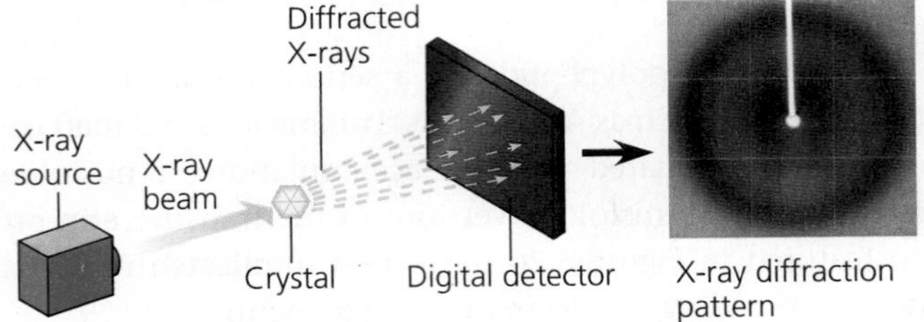

RESULTS Using data from X-ray diffraction patterns, as well as the amino acid sequence determined by chemical methods, Kornberg and colleagues built a 3-D model of the complex with the help of computer software.

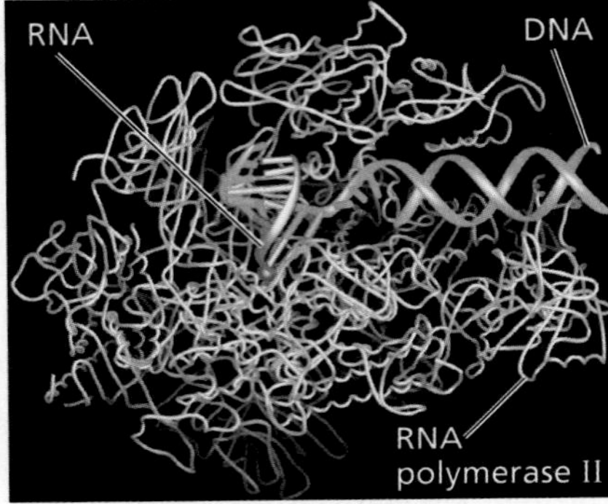

CONCLUSION By analyzing their model, the researchers developed a hypothesis about the functions of different regions of RNA polymerase II. For example, the region above the DNA may act as a clamp that holds the nucleic acids in place. (You'll learn more about this enzyme in Chapter 17.)

SOURCE A. L. Gnatt et al., Structural basis of transcription: an RNA polymerase II elongation complex at 3.3Å, *Science* 292:1876–1882 (2001).

WHAT IF? If you were an author of the paper and were describing the model, what type of protein structure would you call the small polypeptide spirals in RNA polymerase II?

CONCEPT CHECK 5.4

1. Why does a denatured protein no longer function normally?
2. What parts of a polypeptide participate in the bonds that hold together secondary structure? Tertiary structure?
3. **WHAT IF?** Where would you expect a polypeptide region that is rich in the amino acids valine, leucine, and isoleucine to be located in the folded polypeptide? Explain.

For suggested answers, see Appendix A.

CONCEPT 5.5

Nucleic acids store, transmit, and help express hereditary information

If the primary structure of polypeptides determines a protein's shape, what determines primary structure? The amino acid sequence of a polypeptide is programmed by a discrete unit of inheritance known as a **gene**. Genes consist of DNA, which belongs to the class of compounds called nucleic acids. **Nucleic acids** are polymers made of monomers called *nucleotides*.

The Roles of Nucleic Acids

The two types of nucleic acids, **deoxyribonucleic acid (DNA)** and **ribonucleic acid (RNA)**, enable living organisms to reproduce their complex components from one generation to the next. Unique among molecules, DNA provides directions for its own replication. DNA also directs RNA synthesis and, through RNA, controls protein synthesis (**Figure 5.25**).

DNA is the genetic material that organisms inherit from their parents. Each chromosome contains one long DNA molecule, usually carrying several hundred or more genes. When a cell reproduces itself by dividing, its DNA molecules are copied and passed along from one generation of cells to the next. Encoded in the structure of DNA is the information that

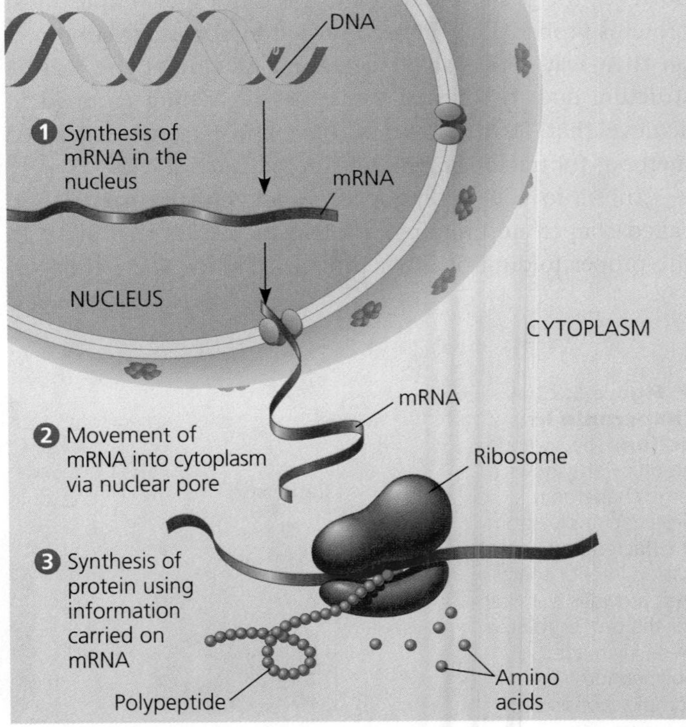

▲ **Figure 5.25 DNA → RNA → protein.** In a eukaryotic cell, DNA in the nucleus programs protein production in the cytoplasm by dictating synthesis of messenger RNA (mRNA). (The cell nucleus is actually much larger relative to the other elements of this figure.)

programs all the cell's activities. The DNA, however, is not directly involved in running the operations of the cell, any more than computer software by itself can print a bank statement or read the bar code on a box of cereal. Just as a printer is needed to print out a statement and a scanner is needed to read a bar code, proteins are required to implement genetic programs. The molecular hardware of the cell—the tools for biological functions—consists mostly of proteins. For example, the oxygen carrier in red blood cells is the protein hemoglobin, not the DNA that specifies its structure.

How does RNA, the other type of nucleic acid, fit into gene expression, the flow of genetic information from DNA to proteins? Each gene along a DNA molecule directs synthesis of a type of RNA called *messenger RNA* (*mRNA*). The mRNA molecule interacts with the cell's protein-synthesizing machinery to direct production of a polypeptide, which folds into all or part of a protein. We can summarize the flow of genetic information as DNA → RNA → protein (see Figure 5.25). The sites of protein synthesis are tiny structures called ribosomes. In a eukaryotic cell, ribosomes are in the cytoplasm, but DNA resides in the nucleus. Messenger RNA conveys genetic instructions for building proteins from the nucleus to the cytoplasm. Prokaryotic cells lack nuclei but still use mRNA to convey a message from the DNA to ribosomes and other cellular equipment that translate the coded information into amino acid sequences. In recent years, the spotlight has been turned on other, previously unknown types of RNA that play many other roles in the cell. As is so often true in biology, the story is still being written! You'll hear more about the newly discovered functions of RNA molecules in Chapter 18.

The Components of Nucleic Acids

Nucleic acids are macromolecules that exist as polymers called **polynucleotides** (**Figure 5.26a**). As indicated by the name, each polynucleotide consists of monomers called **nucleotides**. A nucleotide, in general, is composed of three parts: a nitrogen-containing (nitrogenous) base, a five-carbon sugar (a pentose), and one or more phosphate groups (**Figure 5.26b**). In a polynucleotide, each monomer has only one phosphate group. The portion of a nucleotide without any phosphate groups is called a *nucleoside*.

To build a nucleotide, let's first consider the nitrogenous bases (**Figure 5.26c**). Each nitrogenous base has one or two rings that include nitrogen atoms. (They are called nitrogenous *bases* because the nitrogen atoms tend to take up H^+

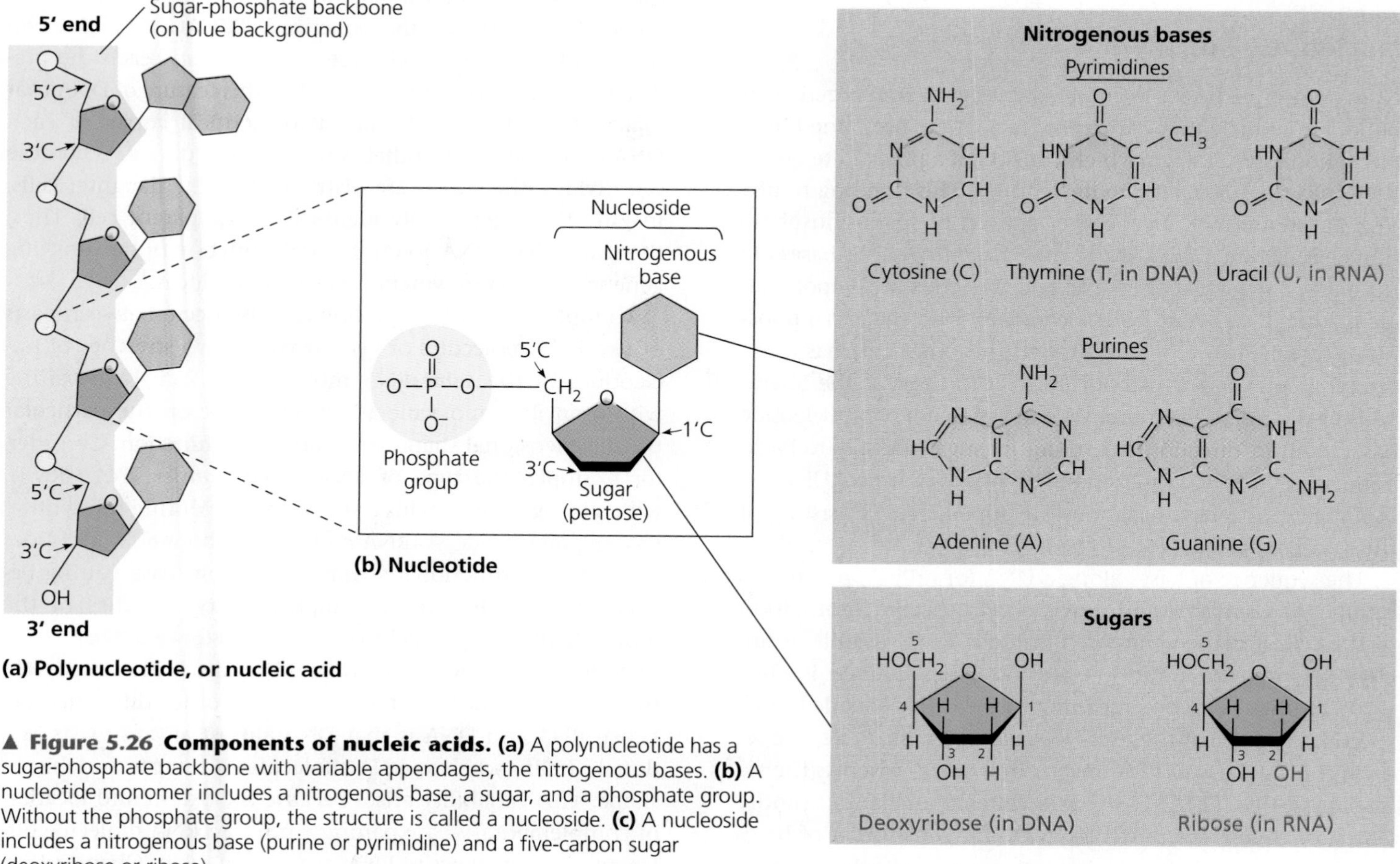

▲ **Figure 5.26 Components of nucleic acids. (a)** A polynucleotide has a sugar-phosphate backbone with variable appendages, the nitrogenous bases. **(b)** A nucleotide monomer includes a nitrogenous base, a sugar, and a phosphate group. Without the phosphate group, the structure is called a nucleoside. **(c)** A nucleoside includes a nitrogenous base (purine or pyrimidine) and a five-carbon sugar (deoxyribose or ribose).

from solution, thus acting as bases.) There are two families of nitrogenous bases: pyrimidines and purines. A **pyrimidine** has one six-membered ring of carbon and nitrogen atoms. The members of the pyrimidine family are cytosine (C), thymine (T), and uracil (U). **Purines** are larger, with a six-membered ring fused to a five-membered ring. The purines are adenine (A) and guanine (G). The specific pyrimidines and purines differ in the chemical groups attached to the rings. Adenine, guanine, and cytosine are found in both DNA and RNA; thymine is found only in DNA and uracil only in RNA.

Now let's add a sugar to the nitrogenous base. In DNA the sugar is **deoxyribose**; in RNA it is **ribose** (see Figure 5.26c). The only difference between these two sugars is that deoxyribose lacks an oxygen atom on the second carbon in the ring; hence the name *deoxy*ribose. To distinguish the numbers of the sugar carbons from those used for the ring atoms of the attached nitrogenous base, the sugar carbon numbers of a nucleoside or nucleotide have a prime (′) after them. Thus, the second carbon in the sugar ring is the 2′ ("2 prime") carbon, and the carbon that sticks up from the ring is called the 5′ carbon.

So far, we have built a nucleoside (nitrogenous base plus sugar). To complete the construction of a nucleotide, we attach a phosphate group to the 5′ carbon of the sugar (see Figure 5.26b). The molecule is now a nucleoside monophosphate, better known as a nucleotide.

Nucleotide Polymers

Now we can see how these nucleotides are linked together to build a polynucleotide. Adjacent nucleotides are joined by a phosphodiester linkage, which consists of a phosphate group that links the sugars of two nucleotides. This bonding results in a backbone with a repeating pattern of sugar-phosphate units (see Figure 5.26a). (Note that the nitrogenous bases are not part of the backbone.) The two free ends of the polymer are distinctly different from each other. One end has a phosphate attached to a 5′ carbon, and the other end has a hydroxyl group on a 3′ carbon; we refer to these as the 5′ end and the 3′ end, respectively. We can say that a polynucleotide has a built-in directionality along its sugar-phosphate backbone, from 5′ to 3′, somewhat like a one-way street. All along this sugar-phosphate backbone are appendages consisting of the nitrogenous bases.

The sequence of bases along a DNA (or mRNA) polymer is unique for each gene and provides very specific information to the cell. Because genes are hundreds to thousands of nucleotides long, the number of possible base sequences is effectively limitless. A gene's meaning to the cell is encoded in its specific sequence of the four DNA bases. For example, the sequence 5′-AGGTAACTT-3′ means one thing, whereas the sequence 5′-CGCTTTAAC-3′ has a different meaning. (Entire genes, of course, are much longer.) The linear order of bases in a gene specifies the amino acid sequence—the primary structure—of a protein, which in turn specifies that protein's three-dimensional structure and its function in the cell.

The Structures of DNA and RNA Molecules

RNA molecules usually exist as single polynucleotide chains like the one shown in Figure 5.26a. In contrast, DNA molecules have two polynucleotides, or "strands," that spiral around an imaginary axis, forming a **double helix** (**Figure 5.27a**). The two sugar-phosphate backbones run in opposite 5′ → 3′ directions from each other; this arrangement is referred to as **antiparallel**, somewhat like a divided highway. The sugar-phosphate backbones are on the outside of the helix, and the nitrogenous bases are paired in the interior of the helix. The two strands are held together by hydrogen bonds between the paired bases (see Figure 5.27a). Most DNA molecules are very long, with thousands or even millions of base pairs. One long DNA double helix includes many genes, each one a particular segment of the molecule.

Only certain bases in the double helix are compatible with each other. Adenine (A) always pairs with thymine (T), and guanine (G) always pairs with cytosine (C). If we were to read the sequence of bases along one strand of the double helix, we would know the sequence of bases along the other strand. If a stretch of one strand has the base sequence 5′-AGGTCCG-3′, then the base-pairing rules tell us that the same stretch of the other strand must have the sequence 3′-TCCAGGC-5′. The two strands of the double helix are *complementary*, each the predictable counterpart of the other. It is this feature of DNA that makes it possible to generate two identical copies of each DNA molecule in a cell that is preparing to divide. When the cell divides, the copies are distributed to the daughter cells, making them genetically identical to the parent cell. Thus, the structure of DNA accounts for its function of transmitting genetic information whenever a cell reproduces.

Complementary base pairing can also occur between parts of two RNA molecules or even between two stretches of nucleotides in the *same* RNA molecule. In fact, base pairing within an RNA molecule allows it to take on the particular three-dimensional shape necessary for its function. Consider, for example, the type of RNA called *transfer RNA* (*tRNA*), which brings amino acids to the ribosome during the synthesis of a polypeptide. A tRNA molecule is about 80 nucleotides in length. Its functional shape results from base pairing between nucleotides where complementary stretches of the molecule run antiparallel to each other (**Figure 5.27b**).

Note that in RNA, adenine (A) pairs with uracil (U); thymine (T) is not present in RNA. Another difference between RNA and DNA is that DNA almost always exists as a double helix, whereas RNA molecules are more variable in shape. This variability arises because the extent and location of complementary base pairing within an RNA molecule differs in different types of RNA, as you will see in Chapter 17.

▶ **Figure 5.27 The structures of DNA and tRNA molecules. (a)** The DNA molecule is usually a double helix, with the sugar-phosphate backbones of the antiparallel polynucleotide strands (symbolized here by blue ribbons) on the outside of the helix. Holding the two strands together are pairs of nitrogenous bases attached to each other by hydrogen bonds. As illustrated here with symbolic shapes for the bases, adenine (A) can pair only with thymine (T), and guanine (G) can pair only with cytosine (C). Each DNA strand in this figure is the structural equivalent of the polynucleotide diagrammed in Figure 5.26a. **(b)** A tRNA molecule has a roughly L-shaped structure, with complementary base pairing of antiparallel stretches of RNA. In RNA, A pairs with U.

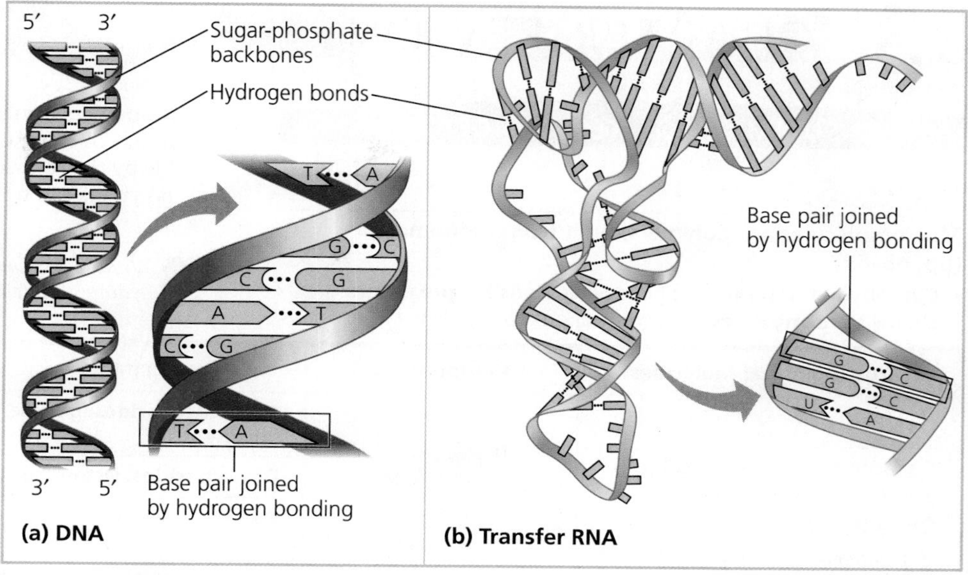

DNA and Proteins as Tape Measures of Evolution

EVOLUTION We are accustomed to thinking of shared traits, such as hair and milk production in mammals, as evidence of shared ancestors. Because we now understand that DNA carries heritable information in the form of genes, we can see that genes and their products (proteins) document the hereditary background of an organism. The linear sequences of nucleotides in DNA molecules are passed from parents to offspring; these sequences determine the amino acid sequences of proteins. Siblings have greater similarity in their DNA and proteins than do unrelated individuals of the same species. If the evolutionary view of life is valid, we should be able to extend this concept of "molecular genealogy" to relationships between species: We should expect two species that appear to be closely related based on fossil and anatomical evidence to also share a greater proportion of their DNA and protein sequences than do more distantly related species. In fact, that is the case. An example is the comparison of the β polypeptide chain of human hemoglobin with the corresponding hemoglobin polypeptide in other vertebrates. In this chain of 146 amino acids, humans and gorillas differ in just 1 amino acid, while humans and frogs differ in 67 amino acids. Molecular biology has added a new tape measure to the toolkit biologists use to assess evolutionary kinship.

The Theme of Emergent Properties in the Chemistry of Life: *A Review*

Recall that life is organized along a hierarchy of structural levels (see Figure 1.4). With each increasing level of order, new properties emerge. In Chapters 2–5, we have dissected the chemistry of life. But we have also begun to develop a more integrated view of life, exploring how properties emerge with increasing order.

We have seen that water's behavior results from the interactions of its molecules, each an ordered arrangement of hydrogen and oxygen atoms. We reduced the complexity and diversity of organic compounds to carbon skeletons and appended chemical groups. We saw that macromolecules are assembled from small organic molecules, taking on new properties. By completing our overview with an introduction to macromolecules and lipids, we have built a bridge to Unit Two, where we will study cell structure and function. We will keep a balance between the need to reduce life to simpler processes and the ultimate satisfaction of viewing those processes in their integrated context.

CONCEPT CHECK 5.5

1. **DRAW IT** Go to Figure 5.26a and, for the top three nucleotides, number all the carbons in the sugars, circle the nitrogenous bases, and star the phosphates.
2. **DRAW IT** In a DNA double helix, a region along one DNA strand has this sequence of nitrogenous bases: 5′-TAGGCCT-3′. Copy this sequence, and write down its complementary strand, clearly indicating the 5′ and 3′ ends of the complementary strand.
3. **WHAT IF?** (a) Suppose a substitution occurred in one DNA strand of the double helix in question 2, resulting in

   5′-TAAGCCT-3′
   3′-ATCCGGA-5′

   Copy these two strands, and circle and label the mismatched bases. (b) If the modified top strand is used by the cell to construct a complementary strand, what would that matching strand be?

For suggested answers, see Appendix A.

5 CHAPTER REVIEW

SUMMARY OF KEY CONCEPTS

CONCEPT 5.1

Macromolecules are polymers, built from monomers (pp. 68–69)

- Carbohydrates, proteins, and nucleic acids are **polymers**, chains of **monomers**. The components of lipids vary.

Monomers form larger molecules by **dehydration reactions**, in which water molecules are released. Polymers can disassemble by the reverse process, **hydrolysis**. An immense variety of polymers can be built from a small set of monomers.

? *What is the fundamental basis for the differences between carbohydrates, proteins, and nucleic acids?*

Large Biological Molecules	Components	Examples	Functions
CONCEPT 5.2 **Carbohydrates serve as fuel and building material (pp. 69–74)** **?** *Compare the composition, structure, and function of starch and cellulose. What role do starch and cellulose play in the human body?*	CH_2OH, H, O, H, H, OH, H, HO, OH, H, OH Monosaccharide monomer	**Monosaccharides:** glucose, fructose	Fuel; carbon sources that can be converted to other molecules or combined into polymers
		Disaccharides: lactose, sucrose	
		Polysaccharides: • Cellulose (plants) • Starch (plants) • Glycogen (animals) • Chitin (animals and fungi)	• Strengthens plant cell walls • Stores glucose for energy • Stores glucose for energy • Strengthens exoskeletons and fungal cell walls
CONCEPT 5.3 **Lipids are a diverse group of hydrophobic molecules (pp. 74–77)** **?** *Why are lipids not considered to be macromolecules or polymers?*	Glycerol 3 fatty acids	**Triacylglycerols** (fats or oils): glycerol + 3 fatty acids	Important energy source
	Head with Ⓟ 2 fatty acids	**Phospholipids:** phosphate group + 2 fatty acids	Lipid bilayers of membranes Hydrophobic tails Hydrophilic heads
	Steroid backbone	**Steroids:** four fused rings with attached chemical groups	• Component of cell membranes (cholesterol) • Signaling molecules that travel through the body (hormones)
CONCEPT 5.4 **Proteins include a diversity of structures, resulting in a wide range of functions (pp. 77–86)** **?** *Proteins are the most structurally and functionally diverse class of biological molecules. Explain the basis for this diversity.*	R, H, N, C, C, O, H, H, OH Amino acid monomer (20 types)	• Enzymes • Structural proteins • Storage proteins • Transport proteins • Hormones • Receptor proteins • Motor proteins • Defensive proteins	• Catalyze chemical reactions • Provide structural support • Store amino acids • Transport substances • Coordinate organismal responses • Receive signals from outside cell • Function in cell movement • Protect against disease
CONCEPT 5.5 **Nucleic acids store, transmit, and help express hereditary information (pp. 86–89)** **?** *What role does complementary base pairing play in the functions of nucleic acids?*	Nitrogenous base Phosphate group Ⓟ—CH_2, O Sugar	**DNA:** • Sugar = deoxyribose • Nitrogenous bases = C, G, A, T • Usually double-stranded	Stores hereditary information
		RNA: • Sugar = ribose • Nitrogenous bases = C, G, A, U • Usually single-stranded	Various functions during gene expression, including carrying instructions from DNA to ribosomes

TEST YOUR UNDERSTANDING

LEVEL 1: KNOWLEDGE/COMPREHENSION

1. Which of the following categories includes all others in the list?
   a. monosaccharide
   b. disaccharide
   c. starch
   d. carbohydrate
   e. polysaccharide

2. The enzyme amylase can break glycosidic linkages between glucose monomers only if the monomers are in the α form. Which of the following could amylase break down?
   a. glycogen, starch, and amylopectin
   b. glycogen and cellulose
   c. cellulose and chitin
   d. starch and chitin
   e. starch, amylopectin, and cellulose

3. Which of the following statements concerning *unsaturated* fats is true?
   a. They are more common in animals than in plants.
   b. They have double bonds in the carbon chains of their fatty acids.
   c. They generally solidify at room temperature.
   d. They contain more hydrogen than do saturated fats having the same number of carbon atoms.
   e. They have fewer fatty acid molecules per fat molecule.

4. The structural level of a protein *least* affected by a disruption in hydrogen bonding is the
   a. primary level.
   b. secondary level.
   c. tertiary level
   d. quaternary level.
   e. All structural levels are equally affected.

5. Enzymes that break down DNA catalyze the hydrolysis of the covalent bonds that join nucleotides together. What would happen to DNA molecules treated with these enzymes?
   a. The two strands of the double helix would separate.
   b. The phosphodiester linkages of the polynucleotide backbone would be broken.
   c. The purines would be separated from the deoxyribose sugars.
   d. The pyrimidines would be separated from the deoxyribose sugars.
   e. All bases would be separated from the deoxyribose sugars.

LEVEL 2: APPLICATION/ANALYSIS

6. The molecular formula for glucose is $C_6H_{12}O_6$. What would be the molecular formula for a polymer made by linking ten glucose molecules together by dehydration reactions?
   a. $C_{60}H_{120}O_{60}$
   b. $C_6H_{12}O_6$
   c. $C_{60}H_{102}O_{51}$
   d. $C_{60}H_{100}O_{50}$
   e. $C_{60}H_{111}O_{51}$

7. Which of the following pairs of base sequences could form a short stretch of a normal double helix of DNA?
   a. 5′-purine-pyrimidine-purine-pyrimidine-3′ with 3′-purine-pyrimidine-purine-pyrimidine-5′
   b. 5′-AGCT-3′ with 5′-TCGA-3′
   c. 5′-GCGC-3′ with 5′-TATA-3′
   d. 5′-ATGC-3′ with 5′-GCAT-3′
   e. All of these pairs are correct.

8. Construct a table that organizes the following terms, and label the columns and rows.

phosphodiester linkages	polypeptides	monosaccharides
peptide bonds	triacylglycerols	nucleotides
glycosidic linkages	polynucleotides	amino acids
ester linkages	polysaccharides	fatty acids

9. **DRAW IT** Copy the polynucleotide strand in Figure 5.26a and label the bases G, T, C, and T, starting from the 5′ end. Assuming this is a DNA polynucleotide, now draw the complementary strand, using the same symbols for phosphates (circles), sugars (pentagons), and bases. Label the bases. Draw arrows showing the 5′ → 3′ direction of each strand. Use the arrows to make sure the second strand is antiparallel to the first. *Hint:* After you draw the first strand vertically, turn the paper upside down; it is easier to draw the second strand from the 5′ toward the 3′ direction as you go from top to bottom.

LEVEL 3: SYNTHESIS/EVALUATION

10. **EVOLUTION CONNECTION**
    Comparisons of amino acid sequences can shed light on the evolutionary divergence of related species. If you were comparing two living species, would you expect all proteins to show the same degree of divergence? Why or why not?

11. **SCIENTIFIC INQUIRY**
    Suppose you are a research assistant in a lab studying DNA-binding proteins. You have been given the amino acid sequences of all the proteins encoded by the genome of a certain species and have been asked to find candidate proteins that could bind DNA. What type of amino acids would you expect to see in such proteins? Why?

12. **SCIENCE, TECHNOLOGY, AND SOCIETY**
    Some amateur and professional athletes take anabolic steroids to help them "bulk up" or build strength. The health risks of this practice are extensively documented. Apart from health considerations, how do you feel about the use of chemicals to enhance athletic performance? Is an athlete who takes anabolic steroids cheating, or is such use part of the preparation required to succeed in competition? Explain.

13. **WRITE ABOUT A THEME**
    Structure and Function Proteins, which have diverse functions in a cell, are all polymers of the same subunits—amino acids. Write a short essay (100–150 words) that discusses how the structure of amino acids allows this one type of polymer to perform so many functions.

For selected answers, see Appendix A.

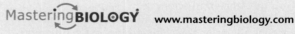

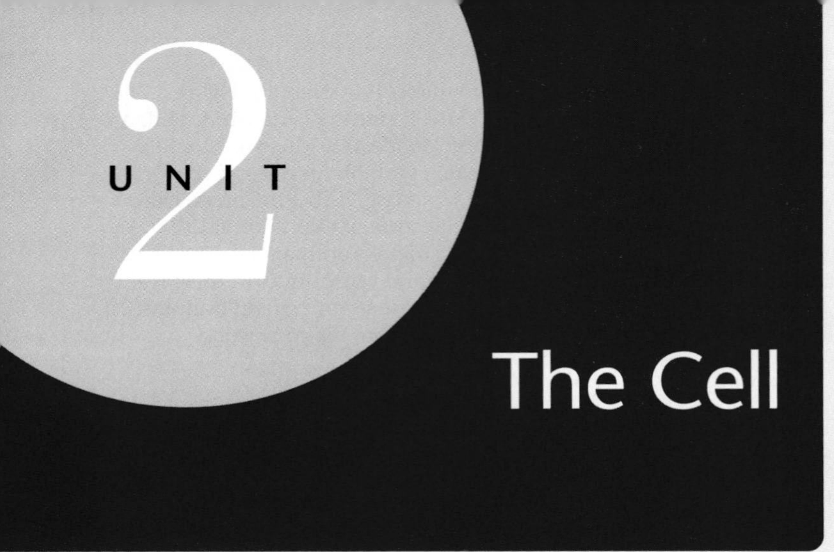

An Interview with

Bonnie L. Bassler

Bonnie Bassler loves her life as a biologist scrutinizing the secret lives of bacteria. For the past 20 years or so, Bonnie and her lab (her "gang," as she calls them) have made momentous discoveries about how bacterial cells use chemicals to communicate with each other in a process called quorum sensing. Dr. Bassler has a B.S. in Biochemistry from the University of California at Davis and a Ph.D. in Biochemistry from The Johns Hopkins University. Among her many awards and honors, she has received a MacArthur Foundation Fellowship and is a member of the National Academy of Sciences. She is the 2010–2011 President of the American Society for Microbiology, the largest specialized life science organization in the world. At Princeton University since 1994, she is currently the Squibb Professor in Molecular Biology and an Investigator of the Howard Hughes Medical Institute.

How did you get started in science?

I've always been interested in nature and animals, and in puzzles and mystery books—I really like figuring things out. As an undergraduate at UC Davis, I worked in a lab on a bacterial project while taking courses in both biochemistry and genetics. Then, as a graduate student at Johns Hopkins, I learned a lot of biochemistry while studying marine bacteria. The bacteria belong to the genus *Vibrio*, and I was working on chemotaxis, movement by cells toward food or away from noxious chemicals in the environment.

What are the advantages of using bacteria for research in cell biology?

Bacteria have been the foundation of molecular biology for the last 100 years because they're accessible. They grow fast, they form clones of identical cells, and they're amenable to biochemical and genetic analyses. Most of what we initially learned about molecular biology—about genes and proteins and other biomolecules—came from work done on bacteria. Because of evolutionary history, the most basic and ancient life processes that happen in bacteria also happen in humans and other higher organisms. Humans have more bells and whistles, more proteins, more sophistication, more complexity. But if you want to understand the basic components of a process, very often you can use bacteria to do that. Also, working with bacteria fits my personality. I prefer having 10 billion bacterial offspring the day after an experiment to having to wait weeks or months for a small number of baby mice. Every morning there's a surprise waiting for me in the incubator!

What is quorum sensing, and how did you first hear about it?

When I was finishing my graduate work, I heard a talk by Mike Silverman, a scientist with the Agouron Institute in San Diego, about how bacteria "talk" to each other, "count" their own numbers, and coordinate their behavior. Mike had been working on a light-producing (bioluminescent) marine bacterium called *Vibrio fischeri* that lives symbiotically inside a variety of marine animals. The animal provides nutrients for the bacteria, which live in an enclosed space within the animal's body. In return, the bacteria provide light that benefits the animal—by scaring away predators or attracting prey or a mate. But if only a small number of bacteria are present, they do not make light—producing light would waste energy because the light wouldn't be visible. The word "quorum" means "the number needed to do something," and bacteria can sense whether there is a quorum or not and act accordingly.

The way quorum sensing works is that bacteria release certain signaling chemicals into the environment. As the bacterial cells increase in number, the molecules reach a concentration at which many of them bind to receptor proteins on the surface of or inside the bacteria. The signaling molecule fits together with the receptor like a key in a lock. In the case of the surface receptors, each receptor molecule has a part on the outside of the cell and a part on the inside. The signaling chemical binds to the outer part of the receptor, "tickling" the protein so that it makes something happen inside the cell. For instance, in *Vibrio fischeri*, binding of signaling molecules ultimately turns on genes that code for enzymes that make light. Mike had worked out this mechanism of how cells of *Vibrio fischeri* turn on light in synchrony.

It's important to understand that back then, we just didn't think about bacteria like that—we thought bacteria ignored each other and did their own thing as solitary cells. I was totally fascinated. I thought, "He's either crazy or he's brilliant—but I just have to work on that." I went up to the podium after his talk and begged him to let me be his postdoc. Finally he said yes, even though he was a geneticist and I was a biochemist! He took a chance on me.

How does a genetic approach differ from a biochemical one?

Geneticists make lots of mutant organisms, then think up clever strategies to find the ones with mutations in the genes they're interested in. In the case of quorum sensing in bioluminescent bacteria, you look for cells that remain dark. If you have mutated genes involved in quorum sensing, you would expect the bacteria not to make light because light emission depends on the cells communicating with each other. Eventually, you would hope to identify the components that function in normal light-emitting bacteria but not in the mutants. Biochemists, on the other hand, start by isolating molecules and studying their properties directly. Genetics and biochemistry are complementary approaches. I'm glad I know both because the combination is more effective than either approach by itself.

What did you learn about quorum sensing as a postdoc?

In Mike's lab, I worked on another species of bioluminescent *Vibrio* called *Vibrio harveyi*. Because these bacteria are free-living in the ocean, we thought their quorum-sensing molecular circuitry might be more complicated than that of *Vibrio fischeri*. What I found was that *Vibrio harveyi* has two parallel systems for quorum sensing, one that senses cells of the same species and one that counts bacteria of other species. Fast-forwarding a decade or so, this second system seems to be present in *many* bacteria, and the second signaling molecule appears to be universal. So, apparently bacteria can measure the ratios of these two signals, and they're saying, "How many of us and how many of you are there?" Then they do different things, depending on who is in the majority. And this isn't just restricted to bioluminescence. Other bacterial behaviors are also controlled by quorum sensing, such as forming an organized thin layer (called a biofilm) on your teeth or coordinating a virulent infection.

Tell us more about biofilms.

We used to think that most bacteria lived as individual cells suspended in liquid environments. But we now understand that in the wild, they live attached to surfaces in biofilms, and they secrete carbohydrates and other molecules that form a protective slime on the biofilm surface. Most of us have noticed the biofilm coating our teeth every morning. Believe it or not, there are about 600 bacterial species in that biofilm just trying to make a living, getting nutrients from us, but the side effect is that we get cavities. And when someone has a lung infection or an implant or heart valve that harbors an infection, the bacteria are growing as a biofilm in the lungs or on the introduced device. So we now understand why these infections are so hard to treat: It's because the slime on the biofilm is providing a protective shield that antibiotics can't penetrate.

What questions are you and your lab asking now?

My group is interested in how information outside an organism gets inside so that the organism does the right thing at the right time. We work on bacteria because they're simple, but we hope that we will have insights for people working on higher organisms. And we're curious about how collective behaviors first evolved on Earth. How did multicellularity come about? We know that the first organisms were bacteria, but how did they begin to do things together? How did groups of cells in your body come to act like a liver or a heart? We're very interested in how the flow of information through networks facilitates multicellularity.

Are there applications for the basic research you do?

When you're asking fundamental questions, you hope that surprises, things you never thought of, will come out of it. Now that we know that bacteria talk to each other and perform group activities, the question is whether we could interfere with these conversations for therapeutic purposes. Could we make molecules that keep bacteria from "talking" or "hearing"? Maybe these would be new antibiotics. Biofilms are a terrible problem in medicine and dental health, and now that we are starting to know the molecular basis for their formation, maybe we can learn how to prevent them from forming.

Bacteria get a lot of bad press for the negative things they do. On the other hand, bacteria also do many miraculous things that keep us alive; they are working for us every instant of our lives. You are covered with a bacterial biofilm that acts as invisible body armor—these good bacteria occupy all the spaces on your skin, preventing invading bacteria from attaching. Throughout your gut you have a huge mass of bacteria, and they're making vitamins for you and helping you digest your food. So all biofilms aren't bad—and for the good biofilms, what if we could find a molecule to make quorum sensing better? Rather than an antibiotic, this would be a probiotic.

Where do you think this field is going?

I think we'll be turning our attention to the possibility of communication between organisms from different kingdoms and different domains. Bacteria have been around for over 4 billion years and have probably been living with multicellular eukaryotic hosts for hundreds of millions of years. So why wouldn't these hosts have evolved strategies to listen in, say, to the conversation being carried out by a group of pathogenic bacteria? Does our immune system "hear" bacterial signaling molecules? Do hosts actively prevent quorum sensing among pathogenic bacteria? Do they tune in and help the good bacteria? I think this is going to be a dialogue, not a monologue.

What do you enjoy most about your life as a scientist?

I love what I work on. I figured out as a postdoc how much fun this life in science is—that it is not about me against other scientists or who is going to discover something first. Instead, it's me against this bacterium, and we are in it head-to-head for the rest of our lives, in a contest of wills between bacteria trying to keep their secrets and me trying to discover them. Also, the basic question of how groups work together fascinates me. I work with a fantastic group of students and we share everything—everybody gives everybody everything, and then we all get more. That's quorum sensing! Both my molecular and nonmolecular lives involve getting the group to do more than the individual. I love that parallelism! My gang of students show me their data, and it's my job to help them figure out the science and get on with their careers. I'm so lucky—having 24 hands and 12 brains is so much better than two hands and one brain. The science is always changing, and trying to keep up with these young and tireless people is hugely challenging and rewarding.

What is your advice to an undergraduate who is considering a career in biology?

For undergraduates who are considering a life in science, my advice is to work on something that you are passionate about. Don't be limited by thinking that bench science is the only thing a scientist can do. There are so many potential careers for a biologist. You could work on Capitol Hill as a scientific advisor or policymaker. You could teach. You could be a lawyer. You could be a writer who helps the public understand science. You could work on science education at the kindergarten level. Figure out your particular combination of personality traits and what you really love doing as a scientist; then make that niche for yourself and bring science to that career. The sky's the limit for biologists because biology is the science of the 21st century and it touches every part of our lives.

"The sky's the limit for biologists because biology is the science of the 21st century and it touches every part of our lives."

Bonnie Bassler (left) with Lisa Urry (center) and Jane Reece

6

A Tour of the Cell

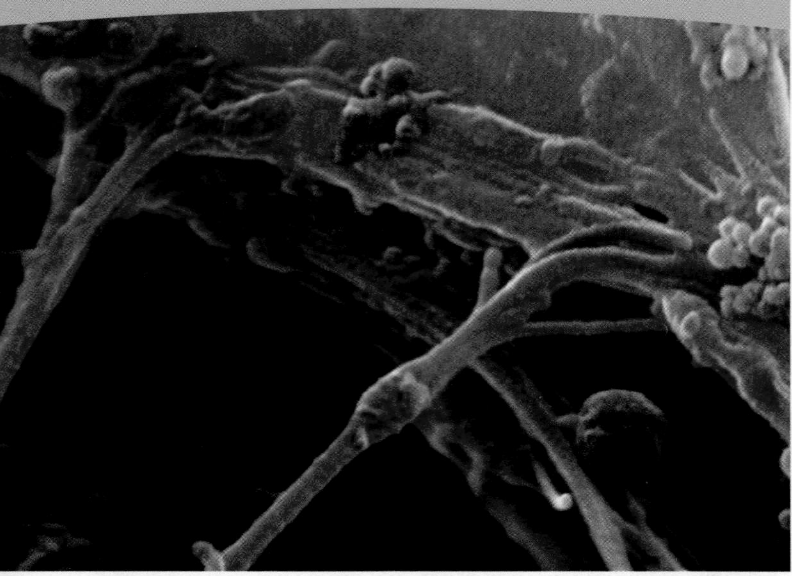

▲ **Figure 6.1 How do your brain cells help you learn about biology?**

KEY CONCEPTS

- **6.1** Biologists use microscopes and the tools of biochemistry to study cells
- **6.2** Eukaryotic cells have internal membranes that compartmentalize their functions
- **6.3** The eukaryotic cell's genetic instructions are housed in the nucleus and carried out by the ribosomes
- **6.4** The endomembrane system regulates protein traffic and performs metabolic functions in the cell
- **6.5** Mitochondria and chloroplasts change energy from one form to another
- **6.6** The cytoskeleton is a network of fibers that organizes structures and activities in the cell
- **6.7** Extracellular components and connections between cells help coordinate cellular activities

OVERVIEW

The Fundamental Units of Life

Given the scope of biology, you may wonder sometimes how you will ever learn all the material in this course! The answer involves cells, which are as fundamental to the living systems of biology as the atom is to chemistry. The contraction of muscle cells moves your eyes as you read this sentence. The words on the page are translated into signals that nerve cells carry to your brain. **Figure 6.1** shows extensions from one nerve cell (purple) making contact with another nerve cell (orange) in the brain. As you study, your goal is to make connections like these that solidify memories and permit learning to occur.

All organisms are made of cells. In the hierarchy of biological organization, the cell is the simplest collection of matter that can be alive. Indeed, many forms of life exist as single-celled organisms. More complex organisms, including plants and animals, are multicellular; their bodies are cooperatives of many kinds of specialized cells that could not survive for long on their own. Even when cells are arranged into higher levels of organization, such as tissues and organs, the cell remains the organism's basic unit of structure and function.

All cells are related by their descent from earlier cells. However, they have been modified in many different ways during the long evolutionary history of life on Earth. But although cells can differ substantially from one another, they share common features. In this chapter, we'll first examine the tools and techniques that allow us to understand cells, then tour the cell and become acquainted with its components.

CONCEPT 6.1

Biologists use microscopes and the tools of biochemistry to study cells

How can cell biologists investigate the inner workings of a cell, usually too small to be seen by the unaided eye? Before we tour the cell, it will be helpful to learn how cells are studied.

Microscopy

The development of instruments that extend the human senses has gone hand in hand with the advance of science. The discovery and early study of cells progressed with the invention of microscopes in 1590 and their refinement during the 1600s. Cell walls were first seen by Robert Hooke in 1665 as he looked through a microscope at dead cells from the bark of an oak tree. But it took the wonderfully crafted lenses of Antoni van Leeuwenhoek to visualize living cells. Imagine Hooke's awe when he visited van Leeuwenhoek in 1674 and the world of microorganisms—what his host called "very little animalcules"—was revealed to him.

The microscopes first used by Renaissance scientists, as well as the microscopes you are likely to use in the laboratory, are all light microscopes. In a **light microscope (LM)**, visible light is passed through the specimen and then through glass lenses. The lenses refract (bend) the light in such a way that the image of the specimen is magnified as it is projected into the eye or into a camera (see Appendix D).

Three important parameters in microscopy are magnification, resolution, and contrast. *Magnification* is the ratio of an object's image size to its real size. Light microscopes can magnify effectively to about 1,000 times the actual size of the specimen; at greater magnifications, additional details cannot be seen clearly. *Resolution* is a measure of the clarity of the image; it is the minimum distance two points can be separated and still be distinguished as two points. For example, what appears to the unaided eye as one star in the sky may be resolved as twin stars with a telescope, which has a higher resolving ability than the eye. Similarly, using standard techniques, the light microscope cannot resolve detail finer than about 0.2 micrometer (μm), or 200 nanometers (nm), regardless of the magnification (**Figure 6.2**). The third parameter, *contrast*, accentuates differences in parts of the sample. Improvements in light microscopy have included new methods for enhancing contrast, such as staining or labeling cell components to stand out visually. **Figure 6.3**, on the next page, shows different types of microscopy; study this figure as you read the rest of this section.

Until recently, the resolution barrier prevented cell biologists from using standard light microscopy to study **organelles**, the membrane-enclosed structures within eukaryotic cells. To see these structures in any detail required the development of a new instrument. In the 1950s, the electron microscope was introduced to biology. Rather than light, the **electron microscope (EM)** focuses a beam of electrons through the specimen or onto its surface (see Appendix D). Resolution is inversely related to the wavelength of the radiation a microscope uses for imaging, and electron beams have much shorter wavelengths than visible light. Modern electron microscopes can theoretically achieve a resolution of about 0.002 nm, though in practice they usually cannot resolve structures smaller than about 2 nm across. Still, this is a hundredfold improvement over the standard light microscope.

The **scanning electron microscope (SEM)** is especially useful for detailed study of the topography of a specimen (see Figure 6.3). The electron beam scans the surface of the sample, usually coated with a thin film of gold. The beam excites electrons on the surface, and these secondary electrons are detected by a device that translates the pattern of electrons into an electronic signal to a video screen. The result is an image of the specimen's surface that appears three-dimensional.

The **transmission electron microscope (TEM)** is used to study the internal structure of cells (see Figure 6.3). The TEM aims an electron beam through a very thin section of the specimen, similar to the way a light microscope transmits light through a slide. The specimen has been stained with atoms of heavy metals, which attach to certain cellular structures, thus enhancing the electron density of some parts of the cell more than others. The electrons passing through the specimen are scattered more in the denser regions, so fewer are transmitted. The image displays the pattern of transmitted electrons. Instead of using glass lenses, the TEM uses electromagnets as lenses to bend the paths of the electrons, ultimately focusing the image onto a monitor for viewing.

Electron microscopes have revealed many organelles and other subcellular structures that were impossible to resolve with the light microscope. But the light microscope offers advantages, especially in studying living cells. A disadvantage of

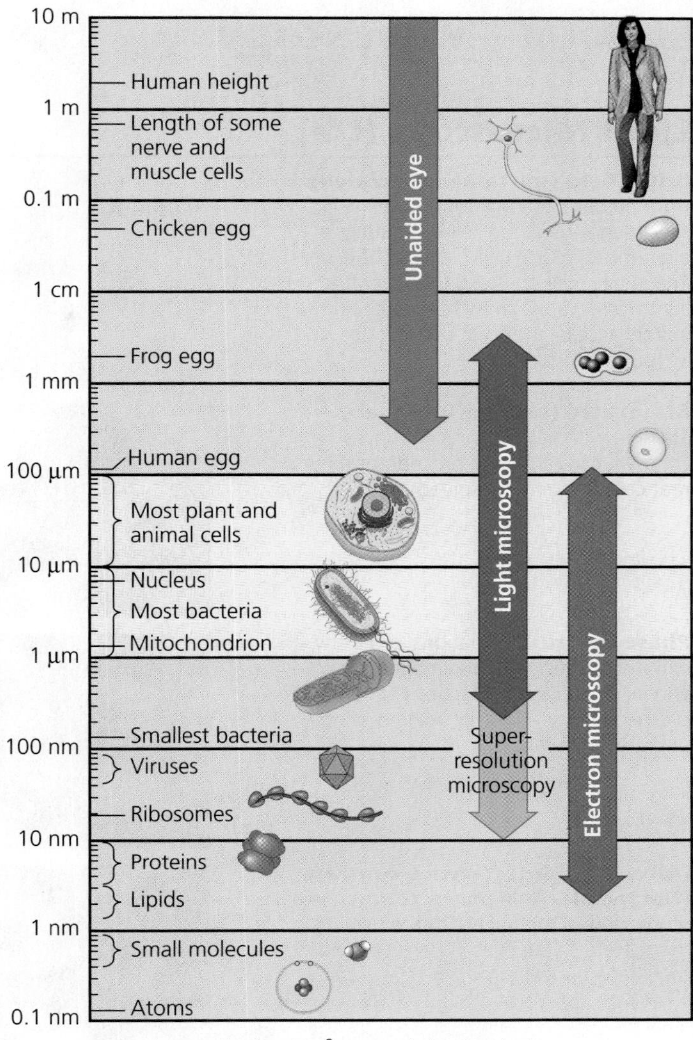

▲ **Figure 6.2 The size range of cells.** Most cells are between 1 and 100 μm in diameter (yellow region of chart) and are therefore visible only under a microscope. Notice that the scale along the left side is logarithmic to accommodate the range of sizes shown. Starting at the top of the scale with 10 m and going down, each reference measurement marks a tenfold decrease in diameter or length. For a complete table of the metric system, see Appendix C.

▼ **Figure 6.3**

Exploring **Microscopy**

Light Microscopy (LM)

Brightfield (unstained specimen). Light passes directly through the specimen. Unless the cell is naturally pigmented or artificially stained, the image has little contrast. (The first four light micrographs show human cheek epithelial cells; the scale bar pertains to all four micrographs.)

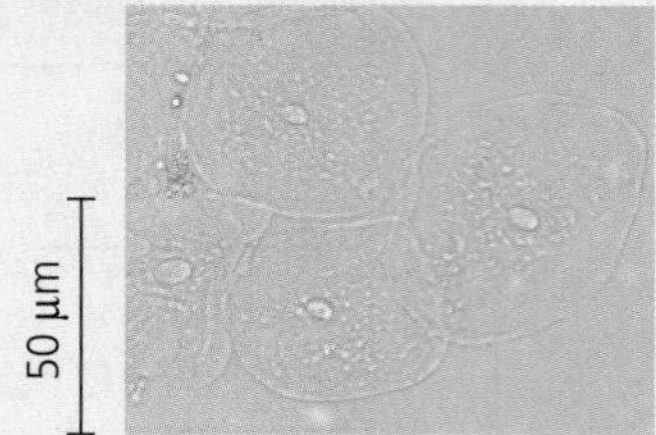

Brightfield (stained specimen). Staining with various dyes enhances contrast. Most staining procedures require that cells be fixed (preserved).

Phase-contrast. Variations in density within the specimen are amplified to enhance contrast in unstained cells, which is especially useful for examining living, unpigmented cells.

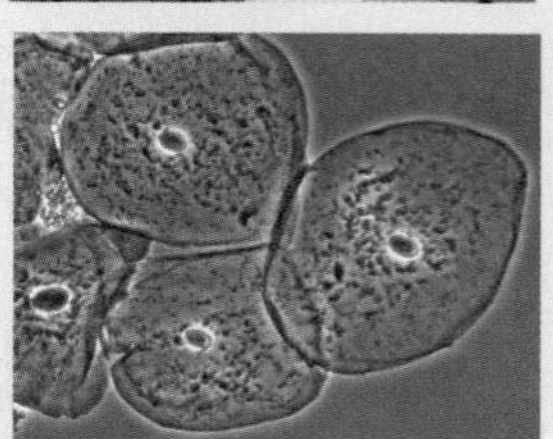

Differential-interference-contrast (Nomarski). As in phase-contrast microscopy, optical modifications are used to exaggerate differences in density, making the image appear almost 3-D.

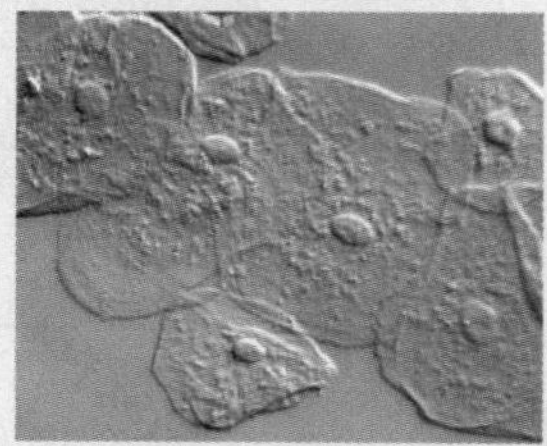

Fluorescence. The locations of specific molecules in the cell can be revealed by labeling the molecules with fluorescent dyes or antibodies; some cells have molecules that fluoresce on their own. Fluorescent substances absorb ultraviolet radiation and emit visible light. In this fluorescently labeled uterine cell, nuclear material is blue, organelles called mitochondria are orange, and the cell's "skeleton" is green.

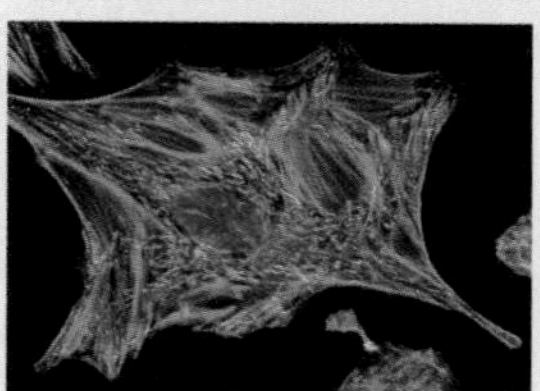

Confocal. The top image is a standard fluorescence micrograph of fluorescently labeled nervous tissue (nerve cells are green, support cells are orange, and regions of overlap are yellow); below it is a confocal image of the same tissue. Using a laser, this "optical sectioning" technique eliminates out-of-focus light from a thick sample, creating a single plane of fluorescence in the image. By capturing sharp images at many different planes, a 3-D reconstruction can be created. The standard image is blurry because out-of-focus light is not excluded.

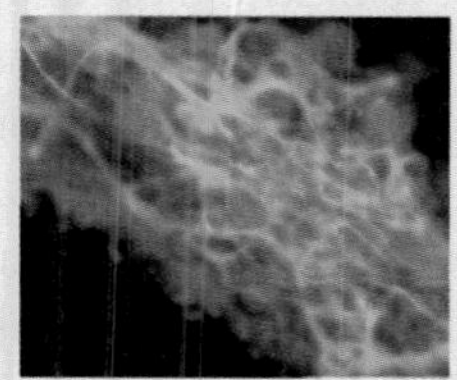

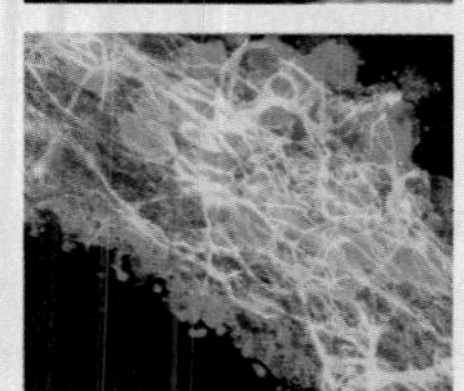

Deconvolution. The top of this split image is a compilation of standard fluorescence micrographs through the depth of a white blood cell. Below is an image of the same cell reconstructed from many blurry images at different planes, each of which was processed using deconvolution software. This process digitally removes out-of-focus light and reassigns it to its source, creating a much sharper 3-D image.

Super-resolution. On the top is a confocal image of part of a nerve cell, using a fluorescent label that binds to a molecule clustered in small sacs in the cell (vesicles) that are 40 nm in diameter. The greenish-yellow spots are blurry because 40 nm is below the 200-nm limit of resolution for standard light microscopy. Below is an image of the same part of the cell, seen using a new "super-resolution" technique. Sophisticated equipment is used to light up individual fluorescent molecules and record their position. Combining information from many molecules in different places "breaks" the limit of resolution, resulting in the sharp greenish-yellow dots seen here. (Each dot is a 40-nm vesicle.)

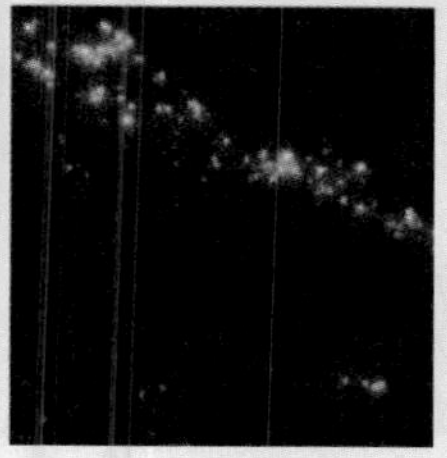

Electron Microscopy (EM)

Scanning electron microscopy (SEM). Micrographs taken with a scanning electron microscope show a 3-D image of the surface of a specimen. This SEM shows the surface of a cell from a trachea (windpipe) covered with cilia. Beating of the cilia helps move inhaled debris upward toward the throat. The SEM and TEM shown here have been artificially colorized. (Electron micrographs are black and white, but are often artificially colorized to highlight particular structures.)

Abbreviations used in this book:
LM = Light Micrograph
SEM = Scanning Electron Micrograph
TEM = Transmission Electron Micrograph

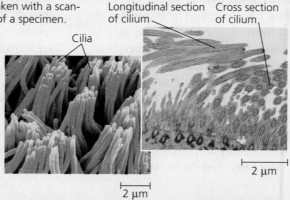

Cilia

2 µm

Longitudinal section of cilium

Cross section of cilium

2 µm

Transmission electron microscopy (TEM). A transmission electron microscope profiles a thin section of a specimen. Here we see a section through a tracheal cell, revealing its internal structure. In preparing the TEM, some cilia were cut along their lengths, creating longitudinal sections, while other cilia were cut straight across, creating cross sections.

electron microscopy is that the methods used to prepare the specimen kill the cells. For all microscopy techniques, in fact, specimen preparation can introduce artifacts, structural features seen in micrographs that do not exist in the living cell.

In the past several decades, light microscopy has been revitalized by major technical advances (see Figure 6.3). Labeling individual cellular molecules or structures with fluorescent markers has made it possible to see such structures with increasing detail. In addition, both confocal and deconvolution microscopy have sharpened images of three-dimensional tissues and cells. Finally, over the past ten years, a group of new techniques and labeling molecules have allowed researchers to "break" the resolution barrier and distinguish subcellular structures as small as 10–20 nm across. As this "super-resolution microscopy" becomes more widespread, the images we'll see of living cells may well be as awe-inspiring to us as van Leeuwenhoek's were to Robert Hooke 350 years ago.

Microscopes are the most important tools of *cytology*, the study of cell structure. To understand the function of each structure, however, required the integration of cytology and *biochemistry*, the study of the chemical processes (metabolism) of cells.

Cell Fractionation

A useful technique for studying cell structure and function is **cell fractionation**, which takes cells apart and separates major organelles and other subcellular structures from one another (**Figure 6.4**). The instrument used is the centrifuge, which spins test tubes holding mixtures of disrupted cells at a series of increasing speeds. At each speed, the resulting force causes a fraction of the cell components to settle to the bottom of the tube, forming a pellet. At lower speeds, the pellet consists of larger components, and higher speeds yield a pellet with smaller components.

Cell fractionation enables researchers to prepare specific cell components in bulk and identify their functions, a task not usually possible with intact cells. For example, on one of the cell fractions, biochemical tests showed the presence of enzymes involved in cellular respiration, while electron microscopy revealed large numbers of the organelles called mitochondria. Together, these data helped biologists determine that mitochondria are the sites of cellular respiration. Biochemistry and cytology thus complement each other in correlating cell function with structure.

CONCEPT CHECK 6.1

1. How do stains used for light microscopy compare with those used for electron microscopy?
2. **WHAT IF?** Which type of microscope would you use to study (a) the changes in shape of a living white blood cell and (b) the details of surface texture of a hair?

For suggested answers, see Appendix A.

▼ **Figure 6.4** RESEARCH METHOD

Cell Fractionation

APPLICATION Cell fractionation is used to isolate (fractionate) cell components based on size and density.

TECHNIQUE Cells are homogenized in a blender to break them up. The resulting mixture (homogenate) is centrifuged. The supernatant (liquid) is poured into another tube and centrifuged at a higher speed for a longer time. This process is repeated several times. This "differential centrifugation" results in a series of pellets, each containing different cell components.

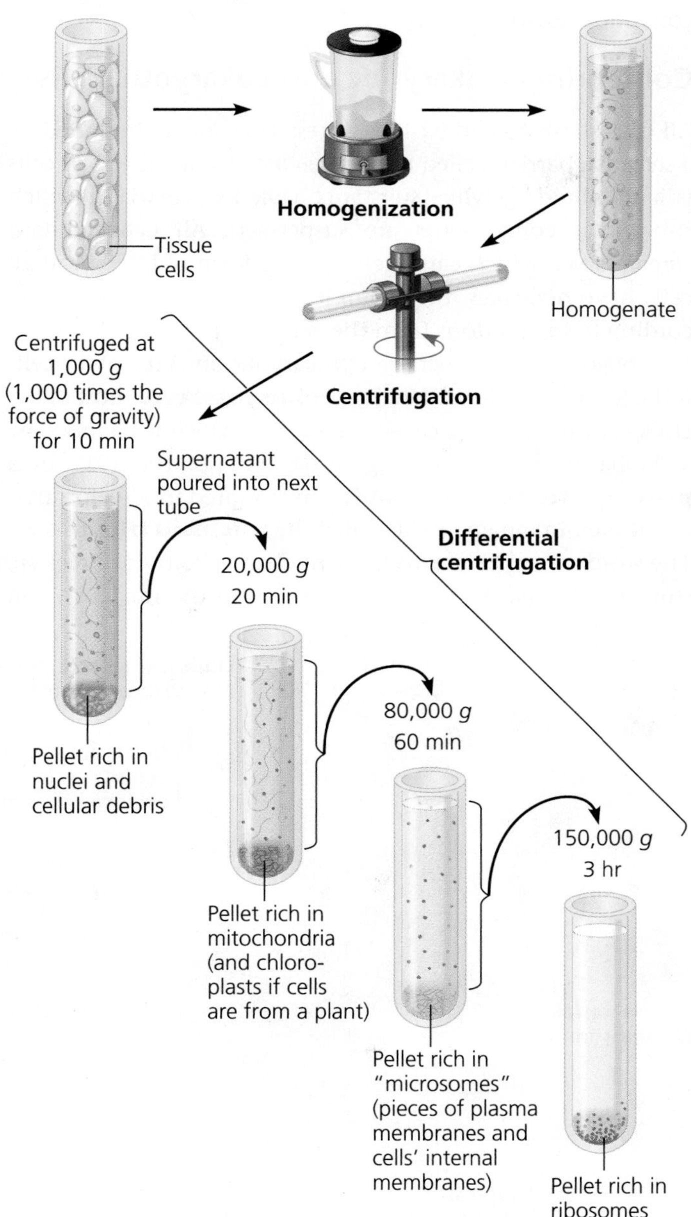

RESULTS In early experiments, researchers used microscopy to identify the organelles in each pellet and biochemical methods to determine their metabolic functions. These identifications established a baseline for this method, enabling today's researchers to know which cell fraction they should collect in order to isolate and study particular organelles.

CONCEPT 6.2

Eukaryotic cells have internal membranes that compartmentalize their functions

Cells—the basic structural and functional units of every organism—are of two distinct types: prokaryotic and eukaryotic. Organisms of the domains Bacteria and Archaea consist of prokaryotic cells. Protists, fungi, animals, and plants all consist of eukaryotic cells.

Comparing Prokaryotic and Eukaryotic Cells

All cells share certain basic features: They are all bounded by a selective barrier, called the *plasma membrane*. Inside all cells is a semifluid, jellylike substance called **cytosol**, in which subcellular components are suspended. All cells contain *chromosomes*, which carry genes in the form of DNA. And all cells have *ribosomes*, tiny complexes that make proteins according to instructions from the genes.

A major difference between prokaryotic and eukaryotic cells is the location of their DNA. In a **eukaryotic cell**, most of the DNA is in an organelle called the *nucleus*, which is bounded by a double membrane (see Figure 6.8, on pp. 100–101). In a **prokaryotic cell**, the DNA is concentrated in a region that is not membrane-enclosed, called the **nucleoid** (**Figure 6.5**). The word *eukaryotic* means "true nucleus" (from the Greek *eu*, true, and *karyon*, kernel, here referring to the nucleus), and the word *prokaryotic* means "before nucleus" (from the Greek *pro*, before), reflecting the fact that prokaryotic cells evolved before eukaryotic cells.

The interior of either type of cell is called the **cytoplasm**; in eukaryotic cells, this term refers only to the region between the nucleus and the plasma membrane. Within the cytoplasm of a eukaryotic cell, suspended in cytosol, are a variety of organelles of specialized form and function. These membrane-bounded structures are absent in prokaryotic cells. Thus, the presence or absence of a true nucleus is just one aspect of the disparity in structural complexity between the two types of cells.

Eukaryotic cells are generally much larger than prokaryotic cells (see Figure 6.2). Size is a general feature of cell structure that relates to function. The logistics of carrying out cellular metabolism sets limits on cell size. At the lower limit, the smallest cells known are bacteria called mycoplasmas, which have diameters between 0.1 and 1.0 μm. These are perhaps the smallest packages with enough DNA to program metabolism and enough enzymes and other cellular equipment to carry out the activities necessary for a cell to sustain itself and reproduce. Typical bacteria are 1–5 μm in diameter, about ten times the size of mycoplasmas. Eukaryotic cells are typically 10–100 μm in diameter.

Metabolic requirements also impose theoretical upper limits on the size that is practical for a single cell. At the boundary of every cell, the **plasma membrane** functions as a selective barrier that allows passage of enough oxygen, nutrients, and wastes to service the entire cell (**Figure 6.6**). For each square micrometer of membrane, only a limited amount of a particular

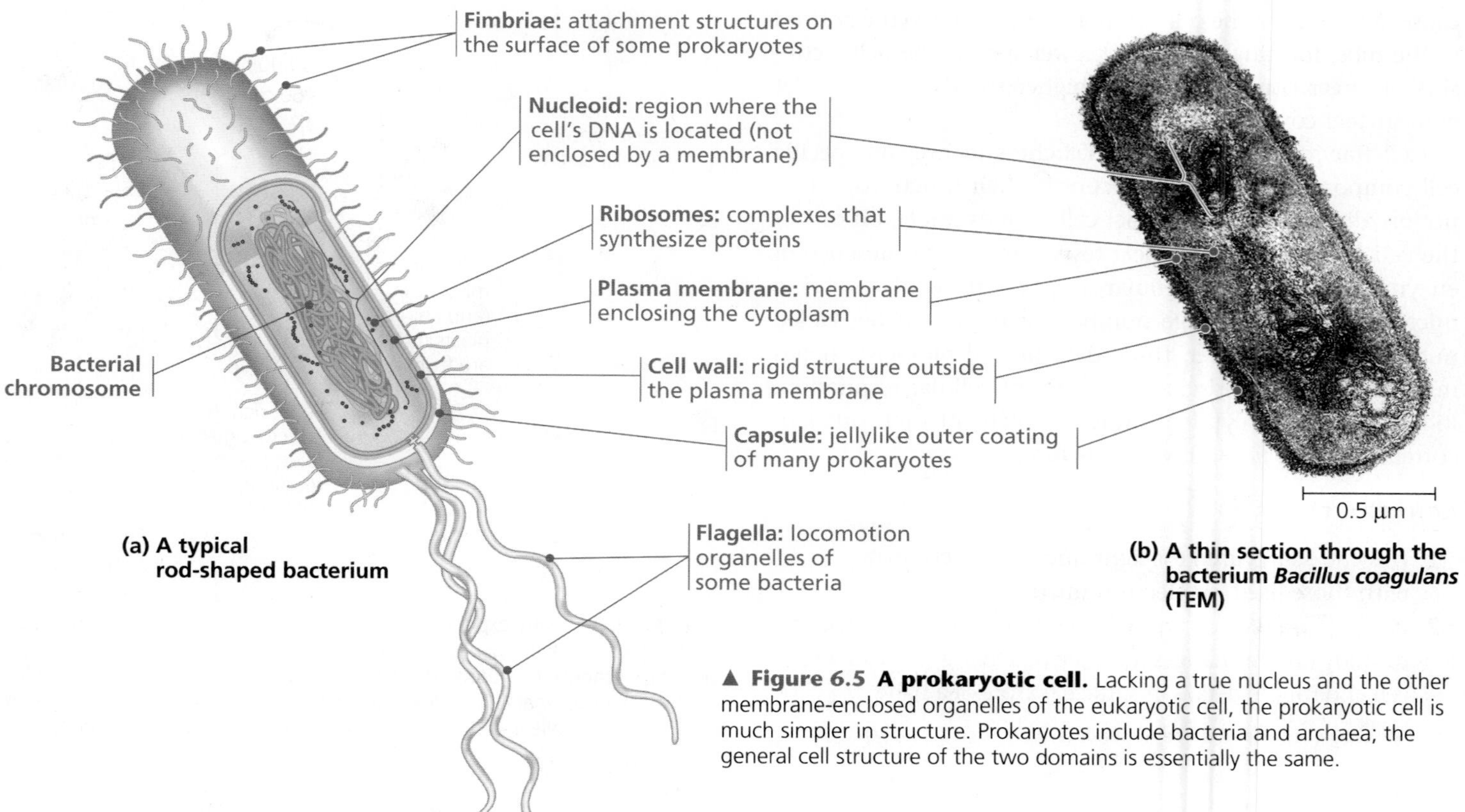

▲ **Figure 6.5 A prokaryotic cell.** Lacking a true nucleus and the other membrane-enclosed organelles of the eukaryotic cell, the prokaryotic cell is much simpler in structure. Prokaryotes include bacteria and archaea; the general cell structure of the two domains is essentially the same.

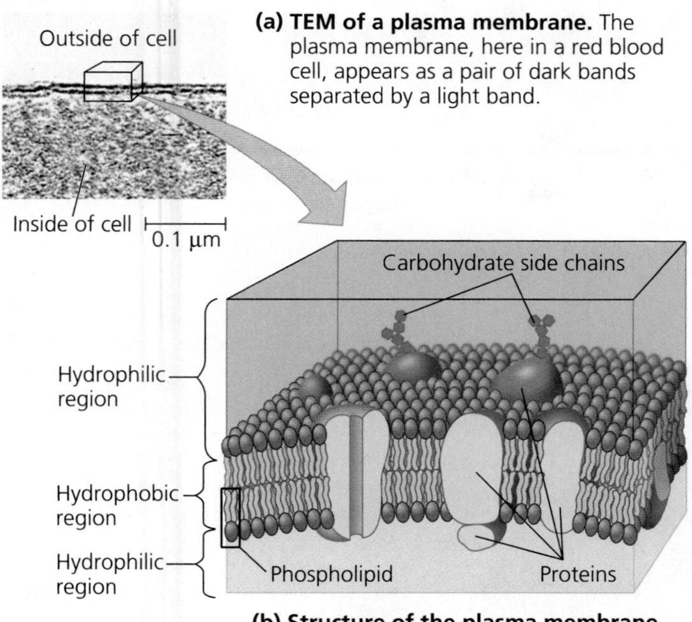

▲ **Figure 6.6 The plasma membrane.** The plasma membrane and the membranes of organelles consist of a double layer (bilayer) of phospholipids with various proteins attached to or embedded in it. The hydrophobic parts, including phospholipid tails and interior portions of membrane proteins, are found in the interior of the membrane. The hydrophilic parts, including phospholipid heads, exterior portions of proteins, and channels of proteins, are in contact with the aqueous solution. Carbohydrate side chains may be attached to proteins or lipids on the outer surface of the plasma membrane.

MAKE CONNECTIONS *Review Figure 5.12 (p. 76) and describe the characteristics of a phospholipid that allow it to function as the major component in the plasma membrane.*

Surface area increases while total volume remains constant

	1	5	1
Total surface area [sum of the surface areas (height × width) of all box sides × number of boxes]	6	150	750
Total volume [height × width × length × number of boxes]	1	125	125
Surface-to-volume (S-to-V) ratio [surface area ÷ volume]	6	1.2	6

▲ **Figure 6.7 Geometric relationships between surface area and volume.** In this diagram, cells are represented as boxes. Using arbitrary units of length, we can calculate the cell's surface area (in square units, or units^2), volume (in cubic units, or units^3), and ratio of surface area to volume. A high surface-to-volume ratio facilitates the exchange of materials between a cell and its environment.

substance can cross per second, so the ratio of surface area to volume is critical. As a cell (or any other object) increases in size, its volume grows proportionately more than its surface area. (Area is proportional to a linear dimension squared, whereas volume is proportional to the linear dimension cubed.) Thus, a smaller object has a greater ratio of surface area to volume (**Figure 6.7**).

The need for a surface area sufficiently large to accommodate the volume helps explain the microscopic size of most cells and the narrow, elongated shapes of others, such as nerve cells. Larger organisms do not generally have *larger* cells than smaller organisms—they simply have *more* cells (see Figure 6.7). A sufficiently high ratio of surface area to volume is especially important in cells that exchange a lot of material with their surroundings, such as intestinal cells. Such cells may have many long, thin projections from their surface called *microvilli*, which increase surface area without an appreciable increase in volume.

The evolutionary relationships between prokaryotic and eukaryotic cells will be discussed later in this chapter, and prokaryotic cells will be described in detail in Chapter 27. Most of the discussion of cell structure that follows in this chapter applies to eukaryotic cells.

A Panoramic View of the Eukaryotic Cell

In addition to the plasma membrane at its outer surface, a eukaryotic cell has extensive and elaborately arranged internal membranes that divide the cell into compartments—the organelles mentioned earlier. The cell's compartments provide different local environments that facilitate specific metabolic functions, so incompatible processes can go on simultaneously inside a single cell. The plasma membrane and organelle membranes also participate directly in the cell's metabolism, because many enzymes are built right into the membranes.

Because membranes are so fundamental to the organization of the cell, Chapter 7 will discuss them in detail. The basic fabric of most biological membranes is a double layer of phospholipids and other lipids. Embedded in this lipid bilayer or attached to its surfaces are diverse proteins (see Figure 6.6). However, each type of membrane has a unique composition of lipids and proteins suited to that membrane's specific functions. For example, enzymes embedded in the membranes of the organelles called mitochondria function in cellular respiration.

Before continuing with this chapter, examine the eukaryotic cells in **Figure 6.8**, on the next two pages. The generalized diagrams of an animal cell and a plant cell introduce the various organelles and highlight the key differences between animal and plant cells. The micrographs at the bottom of the figure give you a glimpse of cells from different types of eukaryotic organisms.

▼ **Figure 6.8**

Exploring Eukaryotic Cells

Animal Cell (cutaway view of generalized cell)

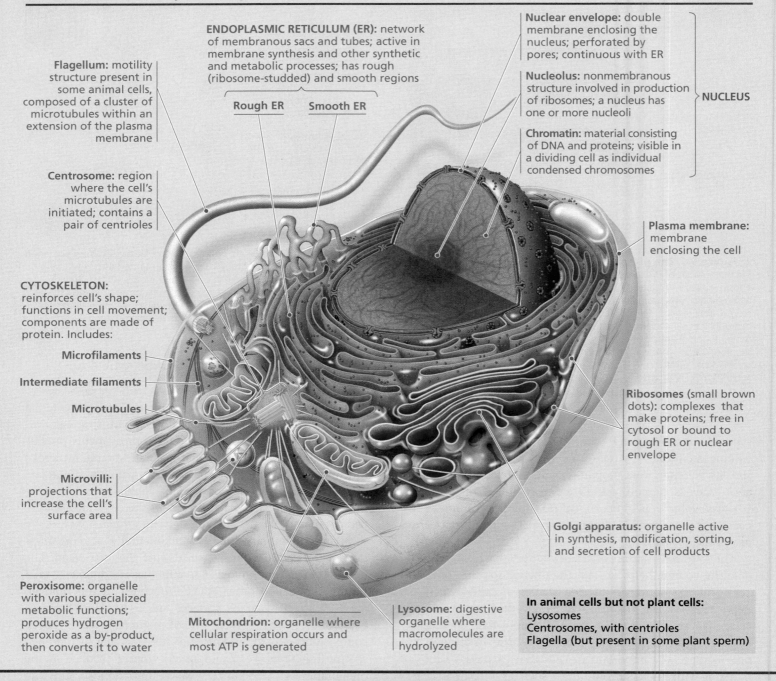

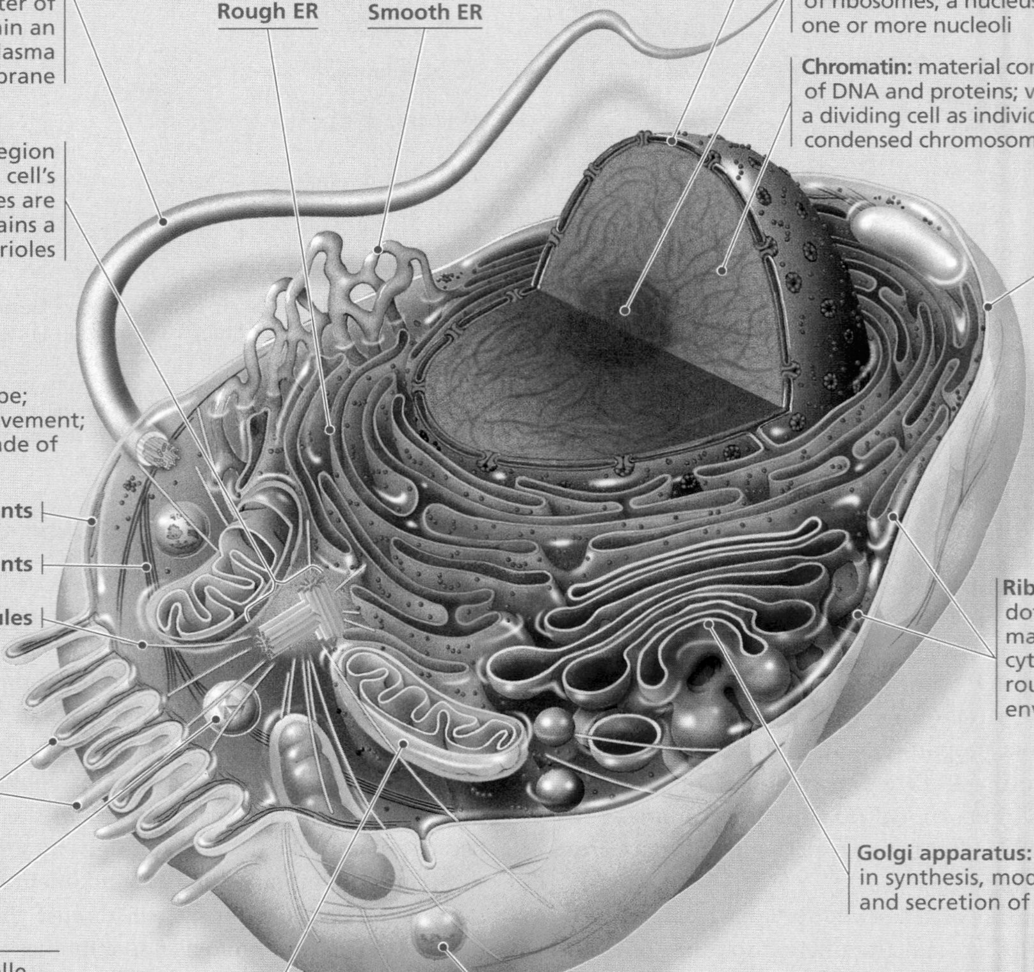

ENDOPLASMIC RETICULUM (ER): network of membranous sacs and tubes; active in membrane synthesis and other synthetic and metabolic processes; has rough (ribosome-studded) and smooth regions
Rough ER
Smooth ER
Flagellum: motility structure present in some animal cells, composed of a cluster of microtubules within an extension of the plasma membrane
Centrosome: region where the cell's microtubules are initiated; contains a pair of centrioles
CYTOSKELETON: reinforces cell's shape; functions in cell movement; components are made of protein. Includes:
Microfilaments
Intermediate filaments
Microtubules
Microvilli: projections that increase the cell's surface area
Peroxisome: organelle with various specialized metabolic functions; produces hydrogen peroxide as a by-product, then converts it to water
Mitochondrion: organelle where cellular respiration occurs and most ATP is generated
Lysosome: digestive organelle where macromolecules are hydrolyzed
Nuclear envelope: double membrane enclosing the nucleus; perforated by pores; continuous with ER
Nucleolus: nonmembranous structure involved in production of ribosomes; a nucleus has one or more nucleoli
Chromatin: material consisting of DNA and proteins; visible in a dividing cell as individual condensed chromosomes
NUCLEUS
Plasma membrane: membrane enclosing the cell
Ribosomes (small brown dots): complexes that make proteins; free in cytosol or bound to rough ER or nuclear envelope
Golgi apparatus: organelle active in synthesis, modification, sorting, and secretion of cell products

In animal cells but not plant cells:
Lysosomes
Centrosomes, with centrioles
Flagella (but present in some plant sperm)

Animal Cells

10 µm
Cell
Nucleus
Nucleolus

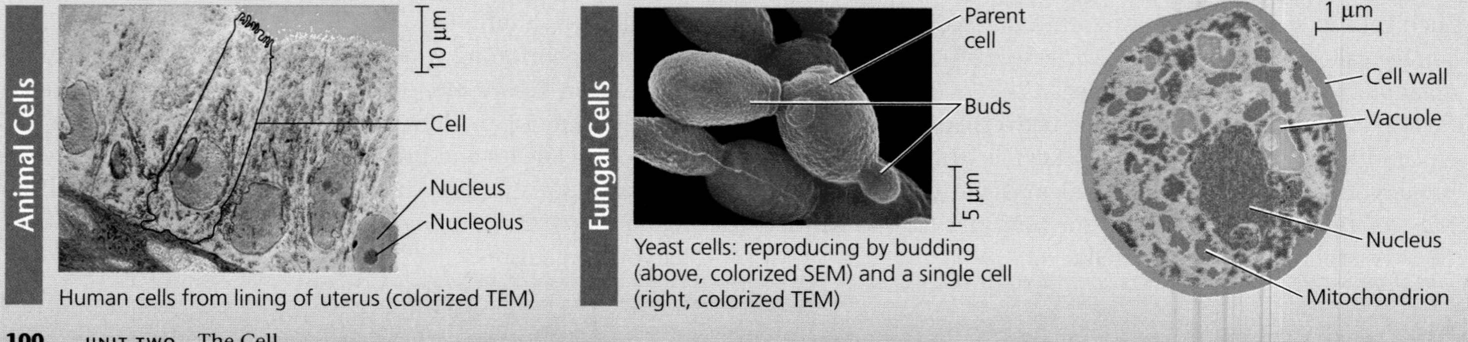

Human cells from lining of uterus (colorized TEM)

Fungal Cells

Parent cell
Buds
5 µm

1 µm
Cell wall
Vacuole
Nucleus
Mitochondrion

Yeast cells: reproducing by budding (above, colorized SEM) and a single cell (right, colorized TEM)

Plant Cell (cutaway view of generalized cell)

ANIMATION **BioFlix** Visit the Study Area at **www.masteringbiology.com** for the BioFlix® 3-D Animations Tour of an Animal Cell and Tour of a Plant Cell.

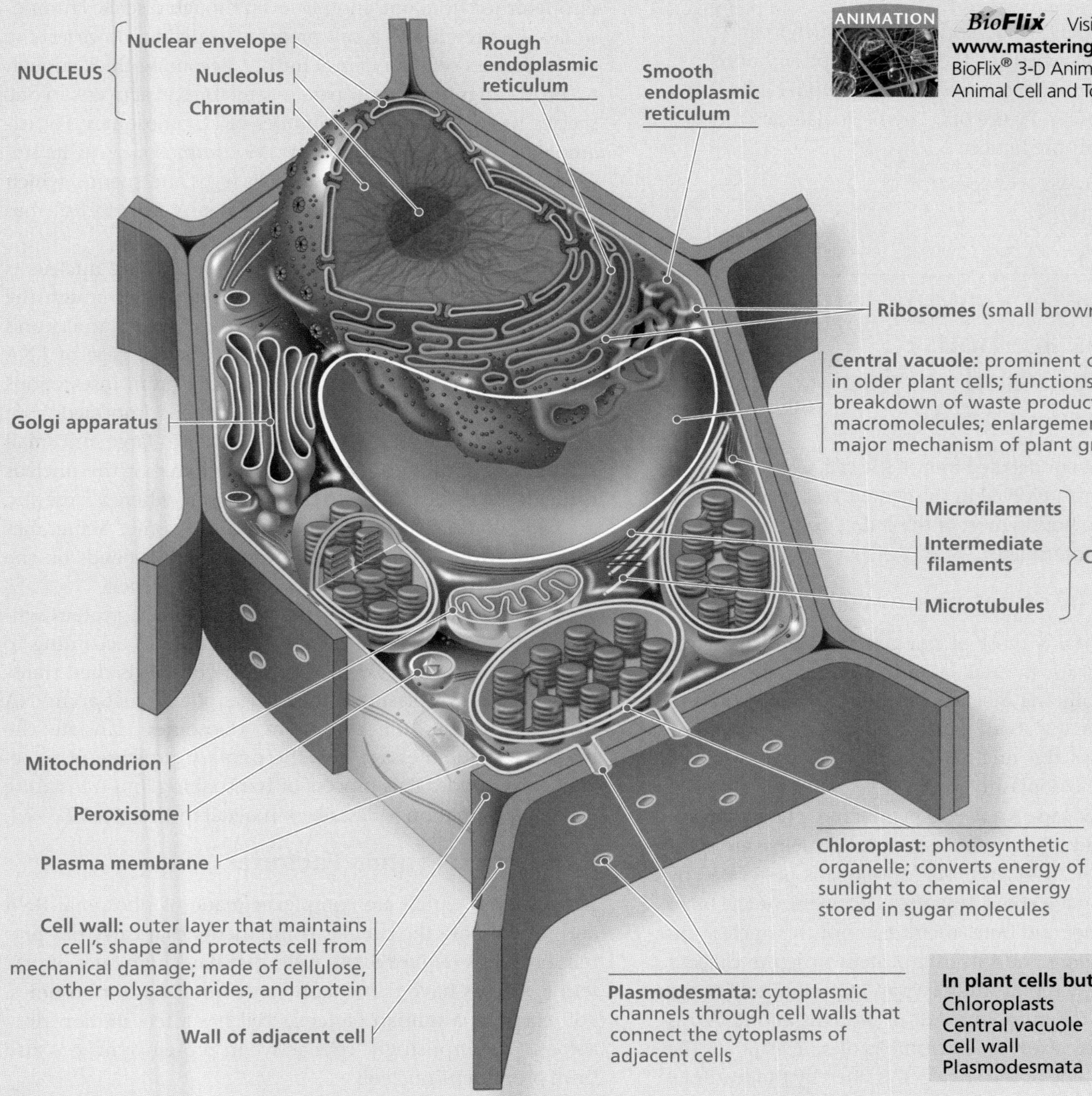

NUCLEUS
Nuclear envelope
Nucleolus
Chromatin
Rough endoplasmic reticulum
Smooth endoplasmic reticulum
Ribosomes (small brown dots)
Central vacuole: prominent organelle in older plant cells; functions include storage, breakdown of waste products, hydrolysis of macromolecules; enlargement of vacuole is a major mechanism of plant growth
Golgi apparatus
Microfilaments
Intermediate filaments
Microtubules
CYTOSKELETON
Mitochondrion
Peroxisome
Plasma membrane
Chloroplast: photosynthetic organelle; converts energy of sunlight to chemical energy stored in sugar molecules
Cell wall: outer layer that maintains cell's shape and protects cell from mechanical damage; made of cellulose, other polysaccharides, and protein
Wall of adjacent cell
Plasmodesmata: cytoplasmic channels through cell walls that connect the cytoplasms of adjacent cells

In plant cells but not animal cells:
Chloroplasts
Central vacuole
Cell wall
Plasmodesmata

Plant Cells

5 µm
Cell
Cell wall
Chloroplast
Mitochondrion
Nucleus
Nucleolus

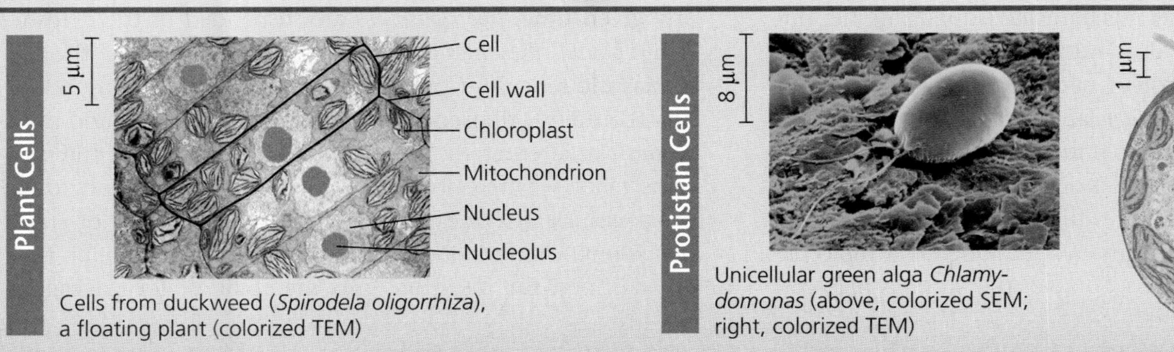

Cells from duckweed (*Spirodela oligorrhiza*), a floating plant (colorized TEM)

Protistan Cells

8 µm

1 µm
Flagella
Nucleus
Nucleolus
Vacuole
Chloroplast
Cell wall

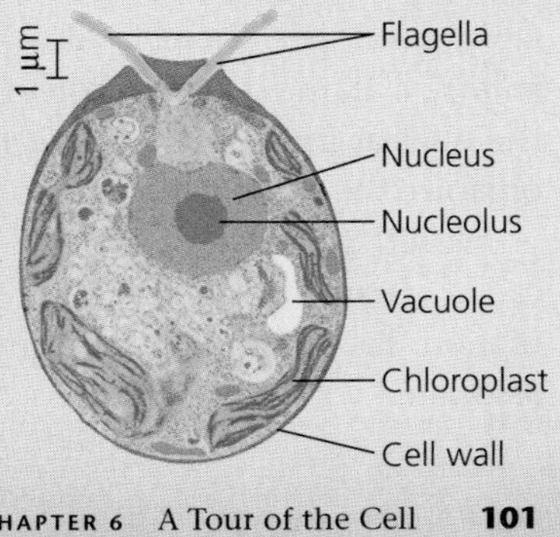

Unicellular green alga *Chlamydomonas* (above, colorized SEM; right, colorized TEM)

CONCEPT CHECK 6.2

1. After carefully reviewing Figure 6.8, briefly describe the structure and function of the nucleus, the mitochondrion, the chloroplast, and the endoplasmic reticulum.
2. **WHAT IF?** Imagine an elongated cell (such as a nerve cell) that measures $125 \times 1 \times 1$ arbitrary units. Predict how its surface-to-volume ratio would compare with those in Figure 6.7. Then calculate the ratio and check your prediction.

For suggested answers, see Appendix A.

CONCEPT **6.3**

The eukaryotic cell's genetic instructions are housed in the nucleus and carried out by the ribosomes

On the first stop of our detailed tour of the cell, let's look at two cellular components involved in the genetic control of the cell: the nucleus, which houses most of the cell's DNA, and the ribosomes, which use information from the DNA to make proteins.

The Nucleus: Information Central

The **nucleus** contains most of the genes in the eukaryotic cell. (Some genes are located in mitochondria and chloroplasts.) It is generally the most conspicuous organelle in a eukaryotic cell, averaging about 5 µm in diameter. The **nuclear envelope** encloses the nucleus (**Figure 6.9**), separating its contents from the cytoplasm.

The nuclear envelope is a *double* membrane. The two membranes, each a lipid bilayer with associated proteins, are separated by a space of 20–40 nm. The envelope is perforated by pore structures that are about 100 nm in diameter. At the lip of each pore, the inner and outer membranes of the nuclear envelope are continuous. An intricate protein structure called a *pore complex* lines each pore and plays an important role in the cell by regulating the entry and exit of proteins and RNAs, as well as large complexes of macromolecules. Except at the pores, the nuclear side of the envelope is lined by the **nuclear lamina**, a netlike array of protein filaments that maintains the shape of the nucleus by mechanically supporting the nuclear envelope. There is also much evidence for a *nuclear matrix*, a framework of protein fibers extending throughout the nuclear interior. The nuclear lamina and matrix may help organize the genetic material so it functions efficiently.

Within the nucleus, the DNA is organized into discrete units called **chromosomes**, structures that carry the genetic information. Each chromosome contains one long DNA molecule associated with many proteins. Some of the proteins help coil the DNA molecule of each chromosome, reducing its length and allowing it to fit into the nucleus. The complex of DNA and proteins making up chromosomes is called **chromatin**. When a cell is not dividing, stained chromatin appears as a diffuse mass in micrographs, and the chromosomes cannot be distinguished from one another, even though discrete chromosomes are present. As a cell prepares to divide, however, the chromosomes coil (condense) further, becoming thick enough to be distinguished as separate structures. Each eukaryotic species has a characteristic number of chromosomes. For example, a typical human cell has 46 chromosomes in its nucleus; the exceptions are the sex cells (eggs and sperm), which have only 23 chromosomes in humans. A fruit fly cell has 8 chromosomes in most cells and 4 in the sex cells.

A prominent structure within the nondividing nucleus is the **nucleolus** (plural, *nucleoli*), which appears through the electron microscope as a mass of densely stained granules and fibers adjoining part of the chromatin. Here a type of RNA called *ribosomal RNA* (rRNA) is synthesized from instructions in the DNA. Also in the nucleolus, proteins imported from the cytoplasm are assembled with rRNA into large and small subunits of ribosomes. These subunits then exit the nucleus through the nuclear pores to the cytoplasm, where a large and a small subunit can assemble into a ribosome. Sometimes there are two or more nucleoli; the number depends on the species and the stage in the cell's reproductive cycle.

As we saw in Figure 5.25, the nucleus directs protein synthesis by synthesizing messenger RNA (mRNA) according to instructions provided by the DNA. The mRNA is then transported to the cytoplasm via the nuclear pores. Once an mRNA molecule reaches the cytoplasm, ribosomes translate the mRNA's genetic message into the primary structure of a specific polypeptide. This process of transcribing and translating genetic information is described in detail in Chapter 17.

Ribosomes: Protein Factories

Ribosomes, which are complexes made of ribosomal RNA and protein, are the cellular components that carry out protein synthesis (**Figure 6.10**). Cells that have high rates of protein synthesis have particularly large numbers of ribosomes. For example, a human pancreas cell has a few million ribosomes. Not surprisingly, cells active in protein synthesis also have prominent nucleoli.

Ribosomes build proteins in two cytoplasmic locales. At any given time, *free ribosomes* are suspended in the cytosol, while *bound ribosomes* are attached to the outside of the endoplasmic reticulum or nuclear envelope (see Figure 6.10). Bound and free ribosomes are structurally identical, and ribosomes can alternate between the two roles. Most of the proteins made on free ribosomes function within the cytosol; examples are enzymes that catalyze the first steps of sugar breakdown. Bound ribosomes generally make proteins that are destined for insertion into membranes, for packaging

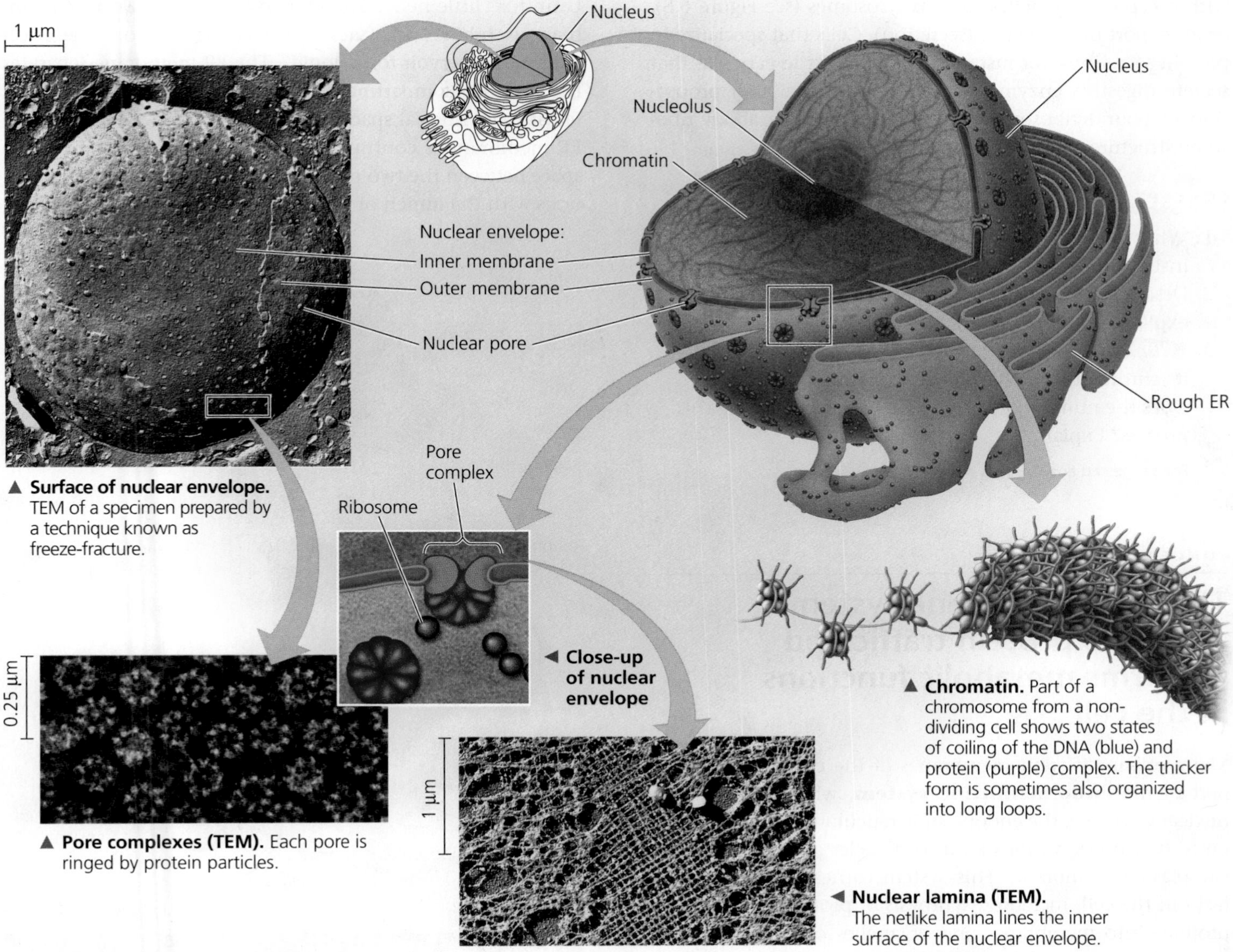

1 µm

Nucleus

Nucleus

Nucleolus

Chromatin

Nuclear envelope:

Inner membrane

Outer membrane

Nuclear pore

Rough ER

▲ **Surface of nuclear envelope.** TEM of a specimen prepared by a technique known as freeze-fracture.

Pore complex

Ribosome

◀ **Close-up of nuclear envelope**

0.25 µm

▲ **Pore complexes (TEM).** Each pore is ringed by protein particles.

1 µm

◀ **Nuclear lamina (TEM).** The netlike lamina lines the inner surface of the nuclear envelope.

▲ **Chromatin.** Part of a chromosome from a non-dividing cell shows two states of coiling of the DNA (blue) and protein (purple) complex. The thicker form is sometimes also organized into long loops.

▲ **Figure 6.9 The nucleus and its envelope.** Within the nucleus are the chromosomes, which appear as a mass of chromatin (DNA and associated proteins), and one or more nucleoli (singular, *nucleolus*), which function in ribosome synthesis. The nuclear envelope, which consists of two membranes separated by a narrow space, is perforated with pores and lined by the nuclear lamina.

MAKE CONNECTIONS *Since the chromosomes contain the genetic material and reside in the nucleus, how does the rest of the cell get access to the information they carry? See Figure 5.25, page 86.*

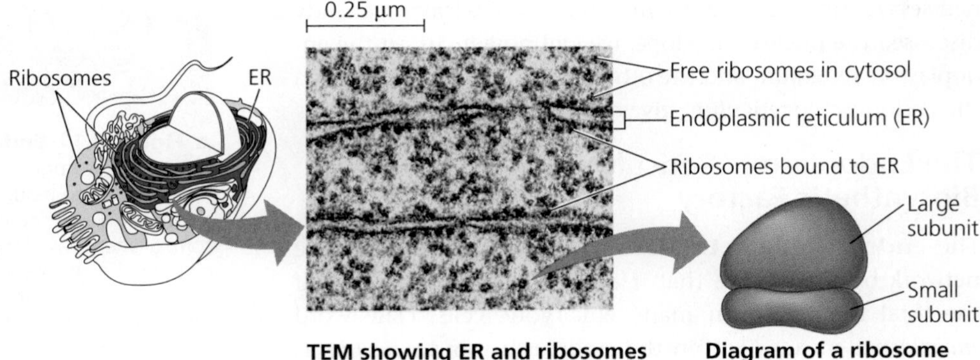

▶ **Figure 6.10 Ribosomes.** This electron micrograph of part of a pancreas cell shows many ribosomes, both free (in the cytosol) and bound (to the endoplasmic reticulum). The simplified diagram of a ribosome shows its two subunits.

DRAW IT *After you have read the section on ribosomes, circle a ribosome in the micrograph that might be making a protein that will be secreted.*

within certain organelles such as lysosomes (see Figure 6.8), or for export from the cell (secretion). Cells that specialize in protein secretion—for instance, the cells of the pancreas that secrete digestive enzymes—frequently have a high proportion of bound ribosomes. You will learn more about ribosome structure and function in Chapter 17.

CONCEPT CHECK 6.3

1. What role do ribosomes play in carrying out genetic instructions?
2. Describe the molecular composition of nucleoli and explain their function.
3. **WHAT IF?** As a cell begins the process of dividing, its chromatin becomes more and more condensed. Does the number of chromosomes change during this process? Explain.

For suggested answers, see Appendix A.

CONCEPT 6.4

The endomembrane system regulates protein traffic and performs metabolic functions in the cell

Many of the different membranes of the eukaryotic cell are part of the **endomembrane system**, which includes the nuclear envelope, the endoplasmic reticulum, the Golgi apparatus, lysosomes, various kinds of vesicles and vacuoles, and the plasma membrane. This system carries out a variety of tasks in the cell, including synthesis of proteins, transport of proteins into membranes and organelles or out of the cell, metabolism and movement of lipids, and detoxification of poisons. The membranes of this system are related either through direct physical continuity or by the transfer of membrane segments as tiny **vesicles** (sacs made of membrane). Despite these relationships, the various membranes are not identical in structure and function. Moreover, the thickness, molecular composition, and types of chemical reactions carried out in a given membrane are not fixed, but may be modified several times during the membrane's life. Having already discussed the nuclear envelope, we will now focus on the endoplasmic reticulum and the other endomembranes to which the endoplasmic reticulum gives rise.

The Endoplasmic Reticulum: Biosynthetic Factory

The **endoplasmic reticulum (ER)** is such an extensive network of membranes that it accounts for more than half the total membrane in many eukaryotic cells. (The word *endoplasmic* means "within the cytoplasm," and *reticulum* is Latin for "little net.") The ER consists of a network of membranous tubules and sacs called cisternae (from the Latin *cisterna*, a reservoir for a liquid). The ER membrane separates the internal compartment of the ER, called the ER lumen (cavity) or cisternal space, from the cytosol. And because the ER membrane is continuous with the nuclear envelope, the space between the two membranes of the envelope is continuous with the lumen of the ER (**Figure 6.11**).

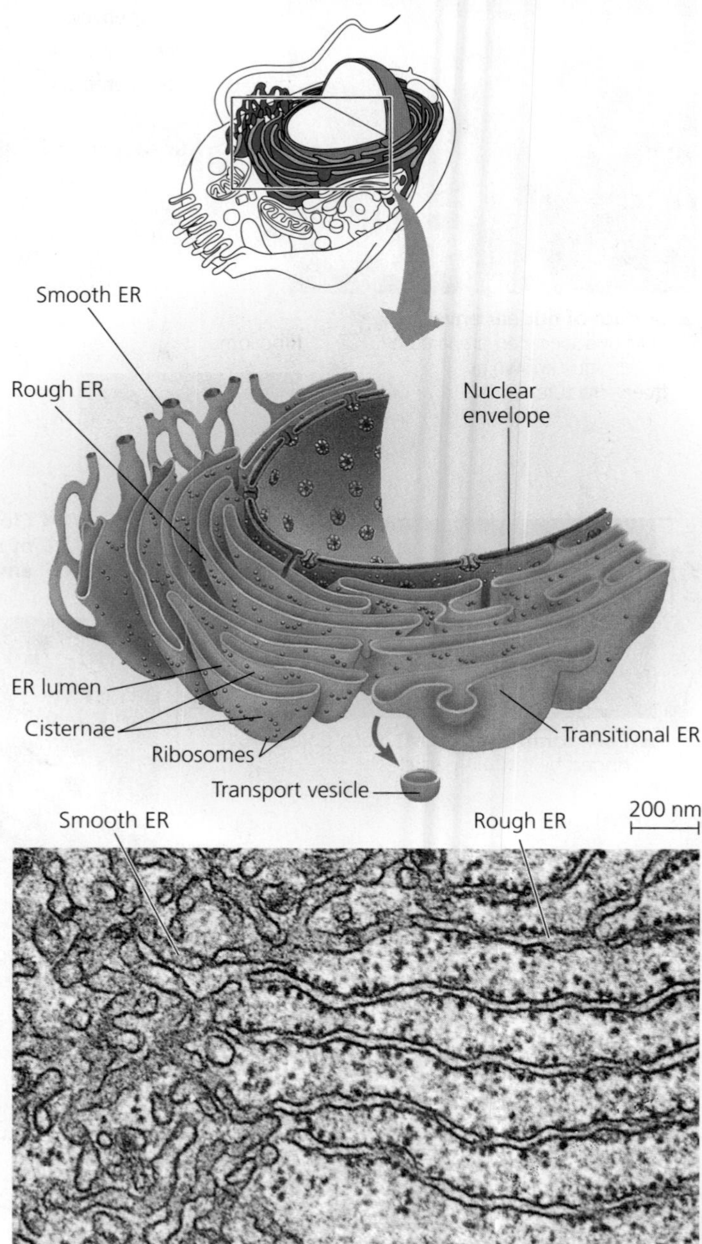

▲ **Figure 6.11 Endoplasmic reticulum (ER).** A membranous system of interconnected tubules and flattened sacs called cisternae, the ER is also continuous with the nuclear envelope. (The drawing is a cutaway view.) The membrane of the ER encloses a continuous compartment called the ER lumen (or cisternal space). Rough ER, which is studded on its outer surface with ribosomes, can be distinguished from smooth ER in the electron micrograph (TEM). Transport vesicles bud off from a region of the rough ER called transitional ER and travel to the Golgi apparatus and other destinations.

There are two distinct, though connected, regions of the ER that differ in structure and function: smooth ER and rough ER. **Smooth ER** is so named because its outer surface lacks ribosomes. **Rough ER** is studded with ribosomes on the outer surface of the membrane and thus appears rough through the electron microscope. As already mentioned, ribosomes are also attached to the cytoplasmic side of the nuclear envelope's outer membrane, which is continuous with rough ER.

Functions of Smooth ER

The smooth ER functions in diverse metabolic processes, which vary with cell type. These processes include synthesis of lipids, metabolism of carbohydrates, detoxification of drugs and poisons, and storage of calcium ions.

Enzymes of the smooth ER are important in the synthesis of lipids, including oils, phospholipids, and steroids. Among the steroids produced by the smooth ER in animal cells are the sex hormones of vertebrates and the various steroid hormones secreted by the adrenal glands. The cells that synthesize and secrete these hormones—in the testes and ovaries, for example—are rich in smooth ER, a structural feature that fits the function of these cells.

Other enzymes of the smooth ER help detoxify drugs and poisons, especially in liver cells. Detoxification usually involves adding hydroxyl groups to drug molecules, making them more soluble and easier to flush from the body. The sedative phenobarbital and other barbiturates are examples of drugs metabolized in this manner by smooth ER in liver cells. In fact, barbiturates, alcohol, and many other drugs induce the proliferation of smooth ER and its associated detoxification enzymes, thus increasing the rate of detoxification. This, in turn, increases tolerance to the drugs, meaning that higher doses are required to achieve a particular effect, such as sedation. Also, because some of the detoxification enzymes have relatively broad action, the proliferation of smooth ER in response to one drug can increase tolerance to other drugs as well. Barbiturate abuse, for example, can decrease the effectiveness of certain antibiotics and other useful drugs.

The smooth ER also stores calcium ions. In muscle cells, for example, the smooth ER membrane pumps calcium ions from the cytosol into the ER lumen. When a muscle cell is stimulated by a nerve impulse, calcium ions rush back across the ER membrane into the cytosol and trigger contraction of the muscle cell. In other cell types, calcium ion release from the smooth ER triggers different responses, such as secretion of vesicles carrying newly synthesized proteins.

Functions of Rough ER

Many types of cells secrete proteins produced by ribosomes attached to rough ER. For example, certain pancreatic cells synthesize the protein insulin in the ER and secrete this hormone into the bloodstream. As a polypeptide chain grows from a bound ribosome, the chain is threaded into the ER lumen through a pore formed by a protein complex in the ER membrane. As the new polypeptide enters the ER lumen, it folds into its native shape. Most secretory proteins are **glycoproteins**, proteins that have carbohydrates covalently bonded to them. The carbohydrates are attached to the proteins in the ER by enzymes built into the ER membrane.

After secretory proteins are formed, the ER membrane keeps them separate from proteins that are produced by free ribosomes and that will remain in the cytosol. Secretory proteins depart from the ER wrapped in the membranes of vesicles that bud like bubbles from a specialized region called transitional ER (see Figure 6.11). Vesicles in transit from one part of the cell to another are called **transport vesicles**; we will discuss their fate shortly.

In addition to making secretory proteins, rough ER is a membrane factory for the cell; it grows in place by adding membrane proteins and phospholipids to its own membrane. As polypeptides destined to be membrane proteins grow from the ribosomes, they are inserted into the ER membrane itself and anchored there by their hydrophobic portions. Like the smooth ER, the rough ER also makes membrane phospholipids; enzymes built into the ER membrane assemble phospholipids from precursors in the cytosol. The ER membrane expands and portions of it are transferred in the form of transport vesicles to other components of the endomembrane system.

The Golgi Apparatus: Shipping and Receiving Center

After leaving the ER, many transport vesicles travel to the **Golgi apparatus**. We can think of the Golgi as a warehouse for receiving, sorting, shipping, and even some manufacturing. Here, products of the ER, such as proteins, are modified and stored and then sent to other destinations. Not surprisingly, the Golgi apparatus is especially extensive in cells specialized for secretion.

The Golgi apparatus consists of flattened membranous sacs—cisternae—looking like a stack of pita bread (**Figure 6.12**, on the next page). A cell may have many, even hundreds, of these stacks. The membrane of each cisterna in a stack separates its internal space from the cytosol. Vesicles concentrated in the vicinity of the Golgi apparatus are engaged in the transfer of material between parts of the Golgi and other structures.

A Golgi stack has a distinct structural directionality, with the membranes of cisternae on opposite sides of the stack differing in thickness and molecular composition. The two sides of a Golgi stack are referred to as the *cis* face and the *trans* face; these act, respectively, as the receiving and shipping departments of the Golgi apparatus. The *cis* face is usually located near the ER. Transport vesicles move material from the ER to the Golgi apparatus. A vesicle that buds from the ER can add its membrane and the contents of its lumen to the *cis* face by fusing with a Golgi membrane. The *trans* face gives rise to vesicles that pinch off and travel to other sites.

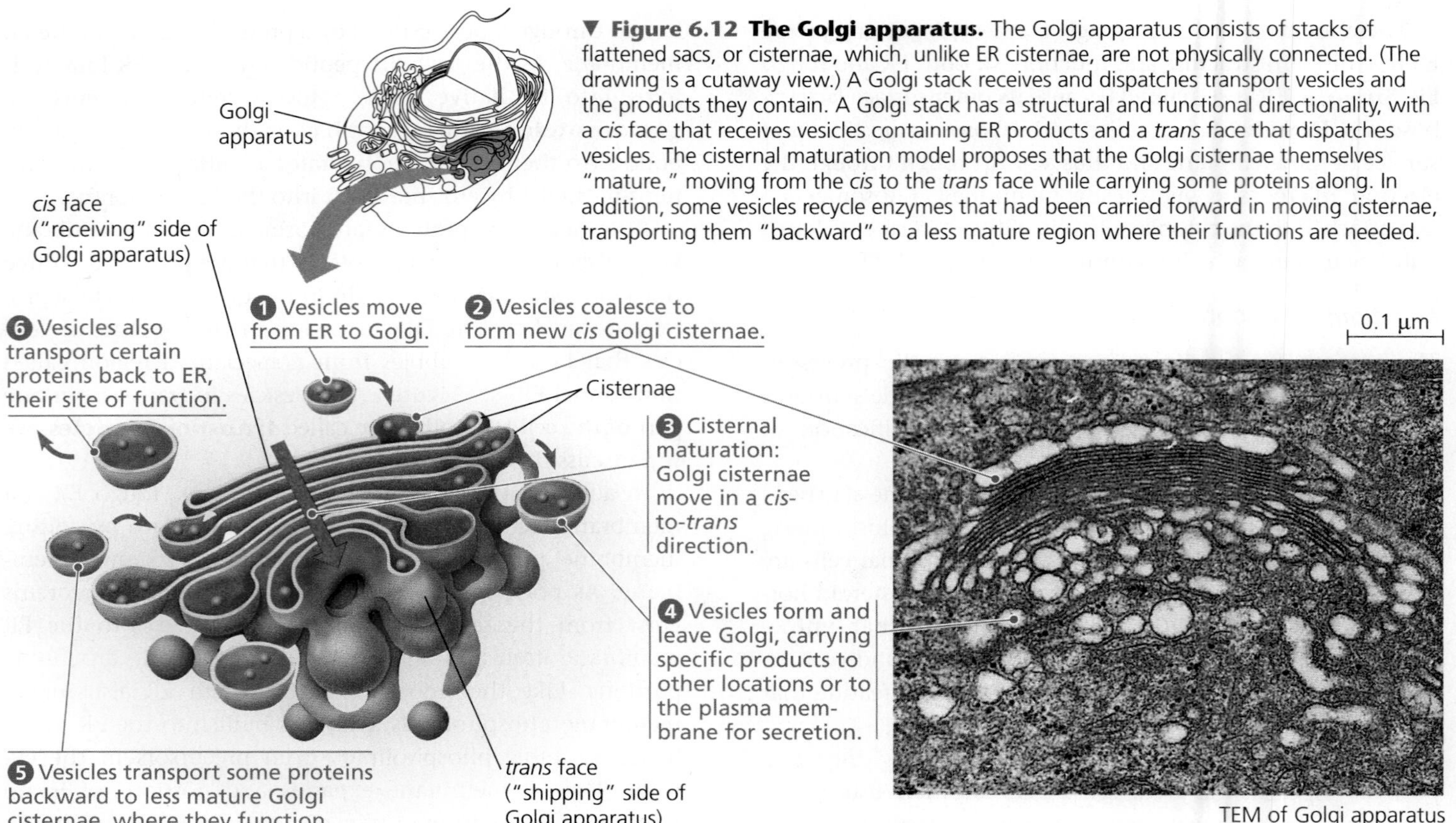

Golgi apparatus
cis face ("receiving" side of Golgi apparatus)
1 Vesicles move from ER to Golgi.
2 Vesicles coalesce to form new *cis* Golgi cisternae.
6 Vesicles also transport certain proteins back to ER, their site of function.
Cisternae
3 Cisternal maturation: Golgi cisternae move in a *cis*-to-*trans* direction.
4 Vesicles form and leave Golgi, carrying specific products to other locations or to the plasma membrane for secretion.
5 Vesicles transport some proteins backward to less mature Golgi cisternae, where they function.
trans face ("shipping" side of Golgi apparatus)
0.1 µm
TEM of Golgi apparatus

▼ **Figure 6.12 The Golgi apparatus.** The Golgi apparatus consists of stacks of flattened sacs, or cisternae, which, unlike ER cisternae, are not physically connected. (The drawing is a cutaway view.) A Golgi stack receives and dispatches transport vesicles and the products they contain. A Golgi stack has a structural and functional directionality, with a *cis* face that receives vesicles containing ER products and a *trans* face that dispatches vesicles. The cisternal maturation model proposes that the Golgi cisternae themselves "mature," moving from the *cis* to the *trans* face while carrying some proteins along. In addition, some vesicles recycle enzymes that had been carried forward in moving cisternae, transporting them "backward" to a less mature region where their functions are needed.

Products of the endoplasmic reticulum are usually modified during their transit from the *cis* region to the *trans* region of the Golgi apparatus. For example, glycoproteins formed in the ER have their carbohydrates modified, first in the ER itself, then as they pass through the Golgi. The Golgi removes some sugar monomers and substitutes others, producing a large variety of carbohydrates. Membrane phospholipids may also be altered in the Golgi.

In addition to its finishing work, the Golgi apparatus also manufactures some macromolecules. Many polysaccharides secreted by cells are Golgi products. For example, pectins and certain other noncellulose polysaccharides are made in the Golgi of plant cells and then incorporated along with cellulose into their cell walls. Like secretory proteins, nonprotein Golgi products that will be secreted depart from the *trans* face of the Golgi inside transport vesicles that eventually fuse with the plasma membrane.

The Golgi manufactures and refines its products in stages, with different cisternae containing unique teams of enzymes. Until recently, biologists viewed the Golgi as a static structure, with products in various stages of processing transferred from one cisterna to the next by vesicles. While this may occur, recent research has given rise to a new model of the Golgi as a more dynamic structure. According to the *cisternal maturation model*, the cisternae of the Golgi actually progress forward from the *cis* to the *trans* face, carrying and modifying their cargo as they move. Figure 6.12 shows the details of this model.

Before a Golgi stack dispatches its products by budding vesicles from the *trans* face, it sorts these products and targets them for various parts of the cell. Molecular identification tags, such as phosphate groups added to the Golgi products, aid in sorting by acting like ZIP codes on mailing labels. Finally, transport vesicles budded from the Golgi may have external molecules on their membranes that recognize "docking sites" on the surface of specific organelles or on the plasma membrane, thus targeting the vesicles appropriately.

Lysosomes: Digestive Compartments

A **lysosome** is a membranous sac of hydrolytic enzymes that an animal cell uses to digest (hydrolyze) macromolecules. Lysosomal enzymes work best in the acidic environment found in lysosomes. If a lysosome breaks open or leaks its contents, the released enzymes are not very active because the cytosol has a neutral pH. However, excessive leakage from a large number of lysosomes can destroy a cell by self-digestion.

Hydrolytic enzymes and lysosomal membrane are made by rough ER and then transferred to the Golgi apparatus for further processing. At least some lysosomes probably arise by budding from the *trans* face of the Golgi apparatus (see Figure 6.12). How are the proteins of the inner surface of the lysosomal membrane and the digestive enzymes themselves spared from destruction? Apparently, the three-dimensional shapes of these proteins protect vulnerable bonds from enzymatic attack.

Lysosomes carry out intracellular digestion in a variety of circumstances. Amoebas and many other protists eat by engulfing smaller organisms or food particles, a process called **phagocytosis** (from the Greek *phagein*, to eat, and *kytos*, vessel, referring here to the cell). The *food vacuole* formed in this way then fuses with a lysosome, whose enzymes digest the food (**Figure 6.13a**, bottom). Digestion products, including simple sugars, amino acids, and other monomers, pass into the cytosol and become nutrients for the cell. Some human cells also carry out phagocytosis. Among them are macrophages, a type of white blood cell that helps defend the body by engulfing and destroying bacteria and other invaders (see Figure 6.13a, top, and Figure 6.33).

Lysosomes also use their hydrolytic enzymes to recycle the cell's own organic material, a process called *autophagy*. During autophagy, a damaged organelle or small amount of cytosol becomes surrounded by a double membrane (of unknown origin), and a lysosome fuses with the outer membrane of this vesicle (**Figure 6.13b**). The lysosomal enzymes dismantle the enclosed material, and the organic monomers are returned to the cytosol for reuse. With the help of lysosomes, the cell continually renews itself. A human liver cell, for example, recycles half of its macromolecules each week.

The cells of people with inherited lysosomal storage diseases lack a functioning hydrolytic enzyme normally present in lysosomes. The lysosomes become engorged with indigestible substrates, which begin to interfere with other cellular activities. In Tay-Sachs disease, for example, a lipid-digesting enzyme is missing or inactive, and the brain becomes impaired by an accumulation of lipids in the cells. Fortunately, lysosomal storage diseases are rare in the general population.

Vacuoles: Diverse Maintenance Compartments

Vacuoles are large vesicles derived from the endoplasmic reticulum and Golgi apparatus. Thus, vacuoles are an integral part of a cell's endomembrane system. Like all cellular membranes, the vacuolar membrane is selective in transporting solutes; as a result, the solution inside a vacuole differs in composition from the cytosol.

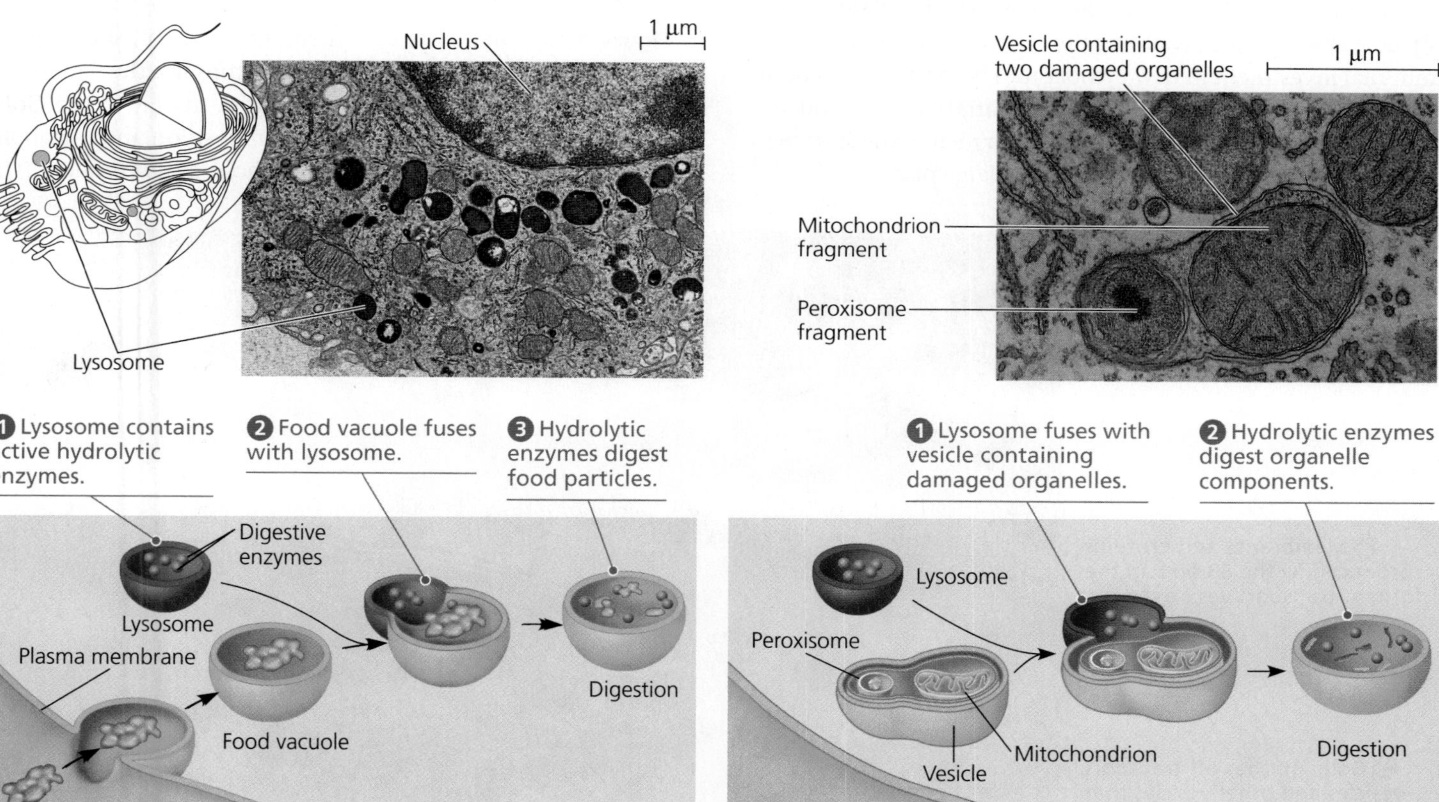

(a) Phagocytosis: lysosome digesting food

(b) Autophagy: lysosome breaking down damaged organelles

▲ **Figure 6.13 Lysosomes.** Lysosomes digest (hydrolyze) materials taken into the cell and recycle intracellular materials. **(a)** *Top*: In this macrophage (a type of white blood cell) from a rat, the lysosomes are very dark because of a stain that reacts with one of the products of digestion within the lysosome (TEM). Macrophages ingest bacteria and viruses and destroy them using lysosomes. *Bottom*: This diagram shows one lysosome fusing with a food vacuole during the process of phagocytosis by a protist. **(b)** *Top*: In the cytoplasm of this rat liver cell is a vesicle containing two disabled organelles; the vesicle will fuse with a lysosome in the process of autophagy (TEM). *Bottom*: This diagram shows fusion of such a vesicle with a lysosome. This type of vesicle has a double membrane of unknown origin. The outer membrane fuses with the lysosome, and the inner membrane is degraded along with the damaged organelles.

Vacuoles perform a variety of functions in different kinds of cells. **Food vacuoles**, formed by phagocytosis, have already been mentioned (see Figure 6.13a). Many freshwater protists have **contractile vacuoles** that pump excess water out of the cell, thereby maintaining a suitable concentration of ions and molecules inside the cell (see Figure 7.16). In plants and fungi, certain vacuoles carry out enzymatic hydrolysis, a function shared by lysosomes in animal cells. (In fact, some biologists consider these hydrolytic vacuoles to be a type of lysosome.) In plants, smaller vacuoles can hold reserves of important organic compounds, such as the proteins stockpiled in the storage cells in seeds. Vacuoles may also help protect the plant against herbivores by storing compounds that are poisonous or unpalatable to animals. Some plant vacuoles contain pigments, such as the red and blue pigments of petals that help attract pollinating insects to flowers.

Mature plant cells generally contain a large **central vacuole** (**Figure 6.14**), which develops by the coalescence of smaller vacuoles. The solution inside the central vacuole, called cell sap, is the plant cell's main repository of inorganic ions, including potassium and chloride. The central vacuole plays a major role in the growth of plant cells, which enlarge as the vacuole absorbs water, enabling the cell to become larger with a minimal investment in new cytoplasm. The cytosol often occupies only a thin layer between the central vacuole and the plasma membrane, so the ratio of plasma membrane surface to cytosolic volume is sufficient, even for a large plant cell.

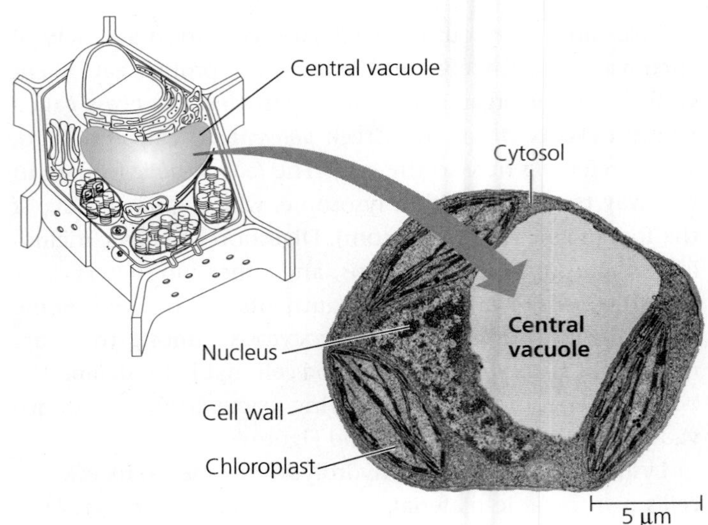

▲ **Figure 6.14 The plant cell vacuole.** The central vacuole is usually the largest compartment in a plant cell; the rest of the cytoplasm is often confined to a narrow zone between the vacuolar membrane and the plasma membrane (TEM).

The Endomembrane System: *A Review*

Figure 6.15 reviews the endomembrane system, showing the flow of membrane lipids and proteins through the various organelles. As the membrane moves from the ER to the Golgi and then elsewhere, its molecular composition and metabolic functions are modified, along with those of its contents. The

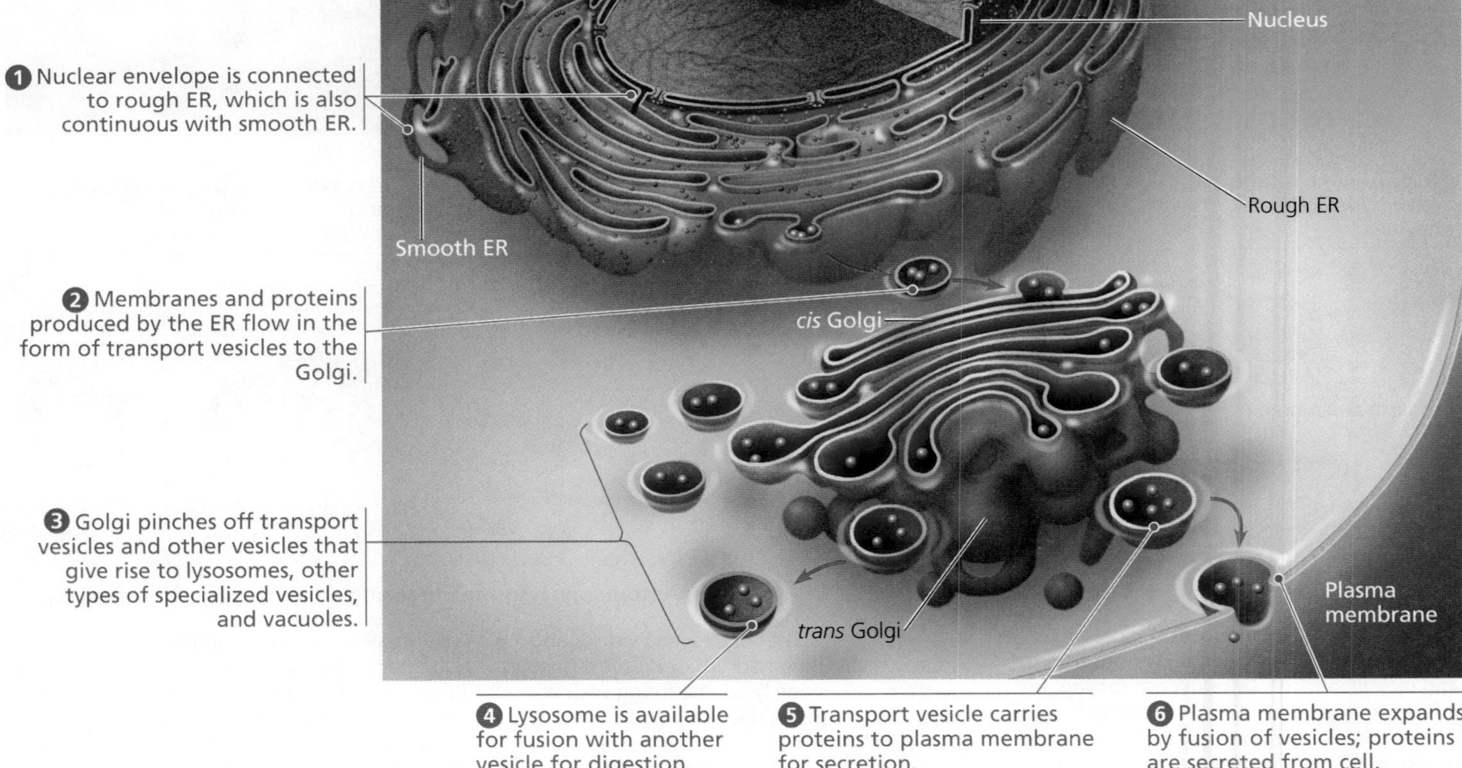

Nucleus
1 Nuclear envelope is connected to rough ER, which is also continuous with smooth ER.
Rough ER
Smooth ER
2 Membranes and proteins produced by the ER flow in the form of transport vesicles to the Golgi.
cis Golgi
3 Golgi pinches off transport vesicles and other vesicles that give rise to lysosomes, other types of specialized vesicles, and vacuoles.
Plasma membrane
trans Golgi
4 Lysosome is available for fusion with another vesicle for digestion.
5 Transport vesicle carries proteins to plasma membrane for secretion.
6 Plasma membrane expands by fusion of vesicles; proteins are secreted from cell.

▲ **Figure 6.15 Review: relationships among organelles of the endomembrane system.** The red arrows show some of the migration pathways for membranes and the materials they enclose.

endomembrane system is a complex and dynamic player in the cell's compartmental organization.

We'll continue our tour of the cell with some organelles that are not closely related to the endomembrane system but play crucial roles in the energy transformations carried out by cells.

CONCEPT CHECK 6.4

1. Describe the structural and functional distinctions between rough and smooth ER.
2. Describe how transport vesicles integrate the endomembrane system.
3. **WHAT IF?** Imagine a protein that functions in the ER but requires modification in the Golgi apparatus before it can achieve that function. Describe the protein's path through the cell, starting with the mRNA molecule that specifies the protein.

For suggested answers, see Appendix A.

CONCEPT 6.5

Mitochondria and chloroplasts change energy from one form to another

Organisms transform the energy they acquire from their surroundings. In eukaryotic cells, mitochondria and chloroplasts are the organelles that convert energy to forms that cells can use for work. **Mitochondria** (singular, *mitochondrion*) are the sites of cellular respiration, the metabolic process that uses oxygen to generate ATP by extracting energy from sugars, fats, and other fuels. **Chloroplasts**, found in plants and algae, are the sites of photosynthesis. These organelles convert solar energy to chemical energy by absorbing sunlight and using it to drive the synthesis of organic compounds such as sugars from carbon dioxide and water.

In addition to having related functions, mitochondria and chloroplasts share a similar evolutionary origin, something we'll discuss briefly before describing their structure. In this section, we will also consider the peroxisome, an oxidative organelle. The evolutionary origin of the peroxisome, as well as its relation to other organelles, is still under debate.

The Evolutionary Origins of Mitochondria and Chloroplasts

EVOLUTION Mitochondria and chloroplasts display similarities with bacteria that led to the **endosymbiont theory**, illustrated in **Figure 6.16**. This theory states that an early ancestor of eukaryotic cells engulfed an oxygen-using nonphotosynthetic prokaryotic cell. Eventually, the engulfed cell formed a relationship with the host cell in which it was enclosed, becoming an *endosymbiont* (a cell living within another cell). Indeed, over the course of evolution, the host cell and its endosymbiont merged into a single organism, a eukaryotic cell with a mitochondrion. At least one of these cells may have then taken up a photosynthetic prokaryote, becoming the ancestor of eukaryotic cells that contain chloroplasts.

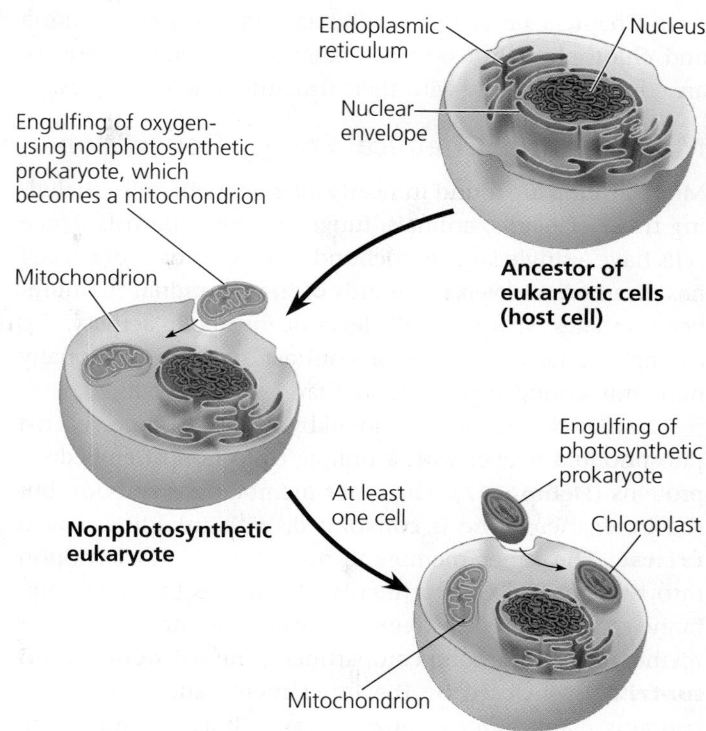

▲ **Figure 6.16 The endosymbiont theory of the origin of mitochondria and chloroplasts in eukaryotic cells.** According to this theory, the proposed ancestors of mitochondria were oxygen-using nonphotosynthetic prokaryotes, while the proposed ancestors of chloroplasts were photosynthetic prokaryotes. The large arrows represent change over evolutionary time; the small arrows inside the cells show the process of the endosymbiont becoming an organelle.

This is a widely accepted theory, which we will discuss in more detail in Chapter 25. The model it proposes is consistent with many structural features of mitochondria and chloroplasts. First, rather than being bounded by a single membrane like organelles of the endomembrane system, mitochondria and typical chloroplasts have two membranes surrounding them. (Chloroplasts also have an internal system of membranous sacs.) There is evidence that the ancestral engulfed prokaryotes had two outer membranes, which became the double membranes of mitochondria and chloroplasts. Second, like prokaryotes, mitochondria and chloroplasts contain ribosomes, as well as circular DNA molecules attached to their inner membranes. The DNA in these organelles programs the synthesis of some of their own proteins, which are made on the ribosomes inside the organelles. Third, also consistent with their probable evolutionary origins as cells, mitochondria and chloroplasts are autonomous (somewhat independent) organelles that grow and reproduce within the cell.

In Chapters 9 and 10, we will focus on how mitochondria and chloroplasts function as energy transformers. Here we are concerned mainly with their structures and their roles.

Mitochondria: Chemical Energy Conversion

Mitochondria are found in nearly all eukaryotic cells, including those of plants, animals, fungi, and most protists. Some cells have a single large mitochondrion, but more often a cell has hundreds or even thousands of mitochondria; the number correlates with the cell's level of metabolic activity. For example, cells that move or contract have proportionally more mitochondria per volume than less active cells.

The mitochondrion is enclosed by two membranes, each a phospholipid bilayer with a unique collection of embedded proteins (**Figure 6.17**). The outer membrane is smooth, but the inner membrane is convoluted, with infoldings called **cristae**. The inner membrane divides the mitochondrion into two internal compartments. The first is the intermembrane space, the narrow region between the inner and outer membranes. The second compartment, the **mitochondrial matrix**, is enclosed by the inner membrane. The matrix contains many different enzymes as well as the mitochondrial DNA and ribosomes. Enzymes in the matrix catalyze some of the steps of cellular respiration. Other proteins that function in respiration, including the enzyme that makes ATP, are built into the inner membrane. As highly folded surfaces, the cristae give the inner mitochondrial membrane a large surface area, thus enhancing the productivity of cellular respiration. This is another example of structure fitting function.

Mitochondria are generally in the range of 1–10 μm long. Time-lapse films of living cells reveal mitochondria moving around, changing their shapes, and fusing or dividing in two, unlike the static structures seen in electron micrographs of dead cells. These observations helped cell biologists understand that mitochondria in a living cell form a branched tubular network, seen in a whole cell in Figure 6.17.

Chloroplasts: Capture of Light Energy

Chloroplasts contain the green pigment chlorophyll, along with enzymes and other molecules that function in the photosynthetic production of sugar. These lens-shaped organelles, about 3–6 μm in length, are found in leaves and other green organs of plants and in algae (**Figure 6.18** and Figure 6.27c).

The contents of a chloroplast are partitioned from the cytosol by an envelope consisting of two membranes separated by a very narrow intermembrane space. Inside the chloroplast is another membranous system in the form of flattened, interconnected sacs called **thylakoids**. In some regions, thylakoids are stacked like poker chips; each stack is called a **granum** (plural, *grana*). The fluid outside the thylakoids is the **stroma**, which contains the chloroplast DNA and ribosomes as well as many enzymes. The membranes of the chloroplast divide the chloroplast space into three compartments: the intermembrane space, the stroma, and the thylakoid space. In Chapter 10, you will learn how this compartmental organization enables the chloroplast to convert light energy to chemical energy during photosynthesis.

As with mitochondria, the static and rigid appearance of chloroplasts in micrographs or schematic diagrams is not true

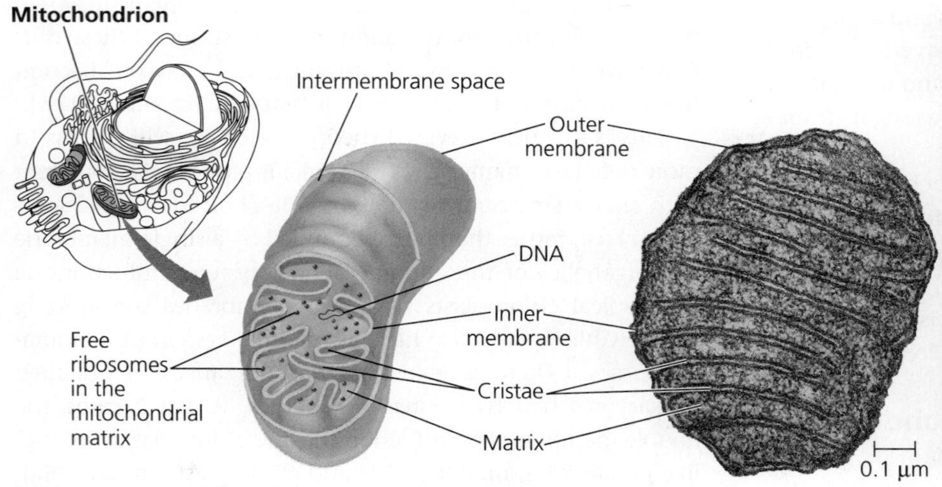

(a) Diagram and TEM of mitochondrion

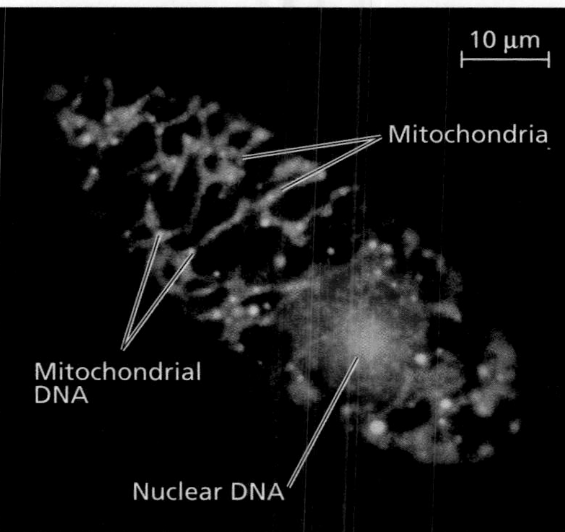

(b) Network of mitochondria in a protist cell (LM)

▲ **Figure 6.17 The mitochondrion, site of cellular respiration. (a)** The inner and outer membranes of the mitochondrion are evident in the drawing and electron micrograph (TEM). The cristae are infoldings of the inner membrane, which increase its surface area. The cutaway drawing shows the two compartments bounded by the membranes: the intermembrane space and the mitochondrial matrix. Many respiratory enzymes are found in the inner membrane and the matrix. Free ribosomes are also present in the matrix. The DNA molecules are usually circular and are attached to the inner mitochondrial membrane. **(b)** The light micrograph shows an entire unicellular protist (*Euglena gracilis*) at a much lower magnification than the TEM. The mitochondrial matrix has been stained green. The mitochondria form a branched tubular network. The nuclear DNA is stained red, and the molecules of mitochondrial DNA appear as bright yellow spots.

▼ **Figure 6.18 The chloroplast, site of photosynthesis. (a)** Many plants have disk-shaped choloroplasts, as shown here. A typical chloroplast has three compartments: the intermembrane space, the stroma, and the thylakoid space. Free ribosomes are present in the stroma, as are copies of chloroplast DNA molecules. **(b)** This fluorescence micrograph shows a cell of the green alga *Spirogyra crassa*, which is named for its spiral chloroplasts. Under natural light the chloroplasts appear green, but under ultraviolet light they naturally fluoresce red, as shown here.

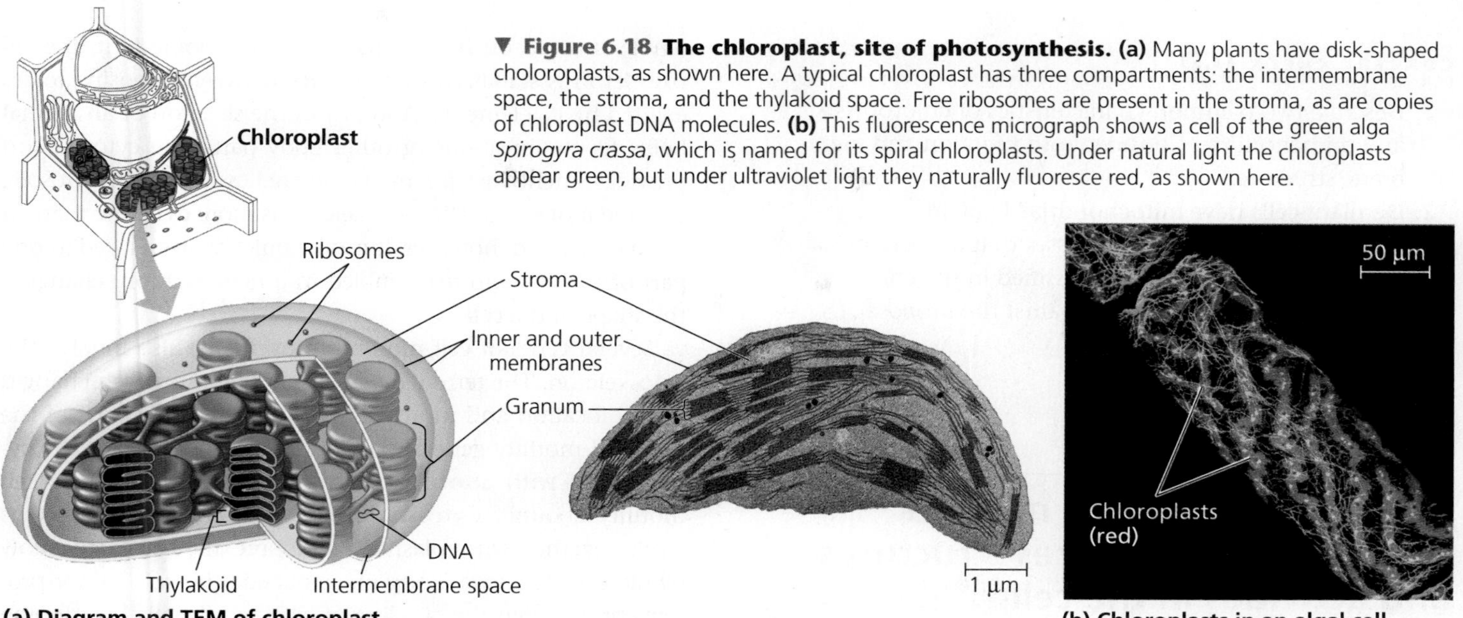

(a) Diagram and TEM of chloroplast

(b) Chloroplasts in an algal cell

to their dynamic behavior in the living cell. Their shape is changeable, and they grow and occasionally pinch in two, reproducing themselves. They are mobile and, with mitochondria and other organelles, move around the cell along tracks of the cytoskeleton, a structural network we will consider later in this chapter.

The chloroplast is a specialized member of a family of closely related plant organelles called **plastids**. One type of plastid, the *amyloplast*, is a colorless organelle that stores starch (amylose), particularly in roots and tubers. Another is the *chromoplast*, which has pigments that give fruits and flowers their orange and yellow hues.

Peroxisomes: Oxidation

The **peroxisome** is a specialized metabolic compartment bounded by a single membrane **(Figure 6.19)**. Peroxisomes contain enzymes that remove hydrogen atoms from various substrates and transfer them to oxygen (O_2), thus producing hydrogen peroxide (H_2O_2) as a by-product (from which the organelle derives its name). These reactions have many different functions. Some peroxisomes use oxygen to break fatty acids down into smaller molecules that are transported to mitochondria and used as fuel for cellular respiration. Peroxisomes in the liver detoxify alcohol and other harmful compounds by transferring hydrogen from the poisons to oxygen. The H_2O_2 formed by peroxisomes is itself toxic, but the organelle also contains an enzyme that converts H_2O_2 to water. This is an excellent example of how the cell's compartmental structure is crucial to its functions: The enzymes that produce hydrogen peroxide and those that dispose of this toxic compound are sequestered away from other cellular components that could be damaged.

Specialized peroxisomes called *glyoxysomes* are found in the fat-storing tissues of plant seeds. These organelles contain enzymes that initiate the conversion of fatty acids to sugar, which the emerging seedling uses as a source of energy and carbon until it can produce its own sugar by photosynthesis.

How peroxisomes are related to other organelles is still an open question. They grow larger by incorporating proteins made in the cytosol and ER, as well as lipids made in the ER and within the peroxisome itself. Peroxisomes may increase in number by splitting in two when they reach a certain size, sparking the suggestion of an endosymbiotic evolutionary origin, but others argue against this scenario. The debate continues.

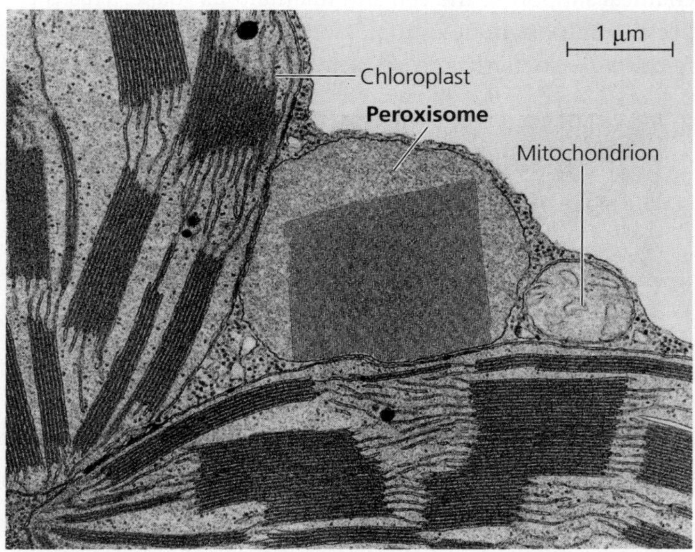

▲ **Figure 6.19 A peroxisome.** Peroxisomes are roughly spherical and often have a granular or crystalline core that is thought to be a dense collection of enzyme molecules. This peroxisome is in a leaf cell (TEM). Notice its proximity to two chloroplasts and a mitochondrion. These organelles cooperate with peroxisomes in certain metabolic functions.

CONCEPT CHECK 6.5

1. Describe two common characteristics of chloroplasts and mitochondria. Consider both function and membrane structure.
2. Do plant cells have mitochondria? Explain.
3. **WHAT IF?** A classmate proposes that mitochondria and chloroplasts should be classified in the endomembrane system. Argue against the proposal.

For suggested answers, see Appendix A.

CONCEPT **6.6**

The cytoskeleton is a network of fibers that organizes structures and activities in the cell

In the early days of electron microscopy, biologists thought that the organelles of a eukaryotic cell floated freely in the cytosol. But improvements in both light microscopy and electron microscopy have revealed the **cytoskeleton**, a network of fibers extending throughout the cytoplasm (**Figure 6.20**). The cytoskeleton, which plays a major role in organizing the structures and activities of the cell, is composed of three types of molecular structures: microtubules, microfilaments, and intermediate filaments.

Roles of the Cytoskeleton: Support and Motility

The most obvious function of the cytoskeleton is to give mechanical support to the cell and maintain its shape. This is especially important for animal cells, which lack walls. The remarkable strength and resilience of the cytoskeleton as a whole is based on its architecture. Like a dome tent, the cytoskeleton is stabilized by a balance between opposing forces exerted by its elements. And just as the skeleton of an animal helps fix the positions of other body parts, the cytoskeleton provides anchorage for many organelles and even cytosolic enzyme molecules. The cytoskeleton is more dynamic than an animal skeleton, however. It can be quickly dismantled in one part of the cell and reassembled in a new location, changing the shape of the cell.

Several types of cell motility (movement) also involve the cytoskeleton. The term *cell motility* encompasses both changes in cell location and more limited movements of parts of the cell. Cell motility generally requires the interaction of the cytoskeleton with **motor proteins**. Examples of such cell motility abound. Cytoskeletal elements and motor proteins work together with plasma membrane molecules to allow whole cells to move along fibers outside the cell. Motor proteins bring about the bending of cilia and flagella by gripping microtubules within those organelles and sliding them against each other. A similar mechanism involving microfilaments causes muscle cells to contract. Inside the cell, vesicles and other organelles often use motor protein "feet" to "walk" to their destinations along a track provided by the cytoskeleton. For example, this is how vesicles containing neurotransmitter molecules migrate to the tips of axons, the long extensions of nerve cells that release these molecules as chemical signals to adjacent nerve cells (**Figure 6.21**). The vesicles that bud off

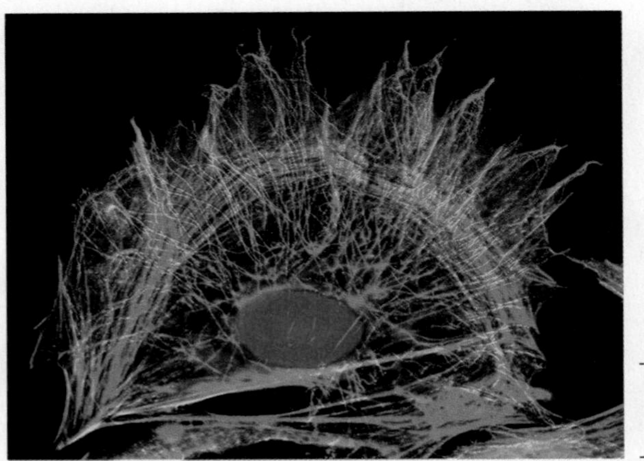

▲ **Figure 6.20 The cytoskeleton.** As shown in this fluorescence micrograph, the cytoskeleton extends throughout the cell. The cytoskeletal elements have been tagged with different fluorescent molecules: green for microtubules and red for microfilaments. A third component of the cytoskeleton, intermediate filaments, is not evident here. (The DNA in the nucleus is blue.)

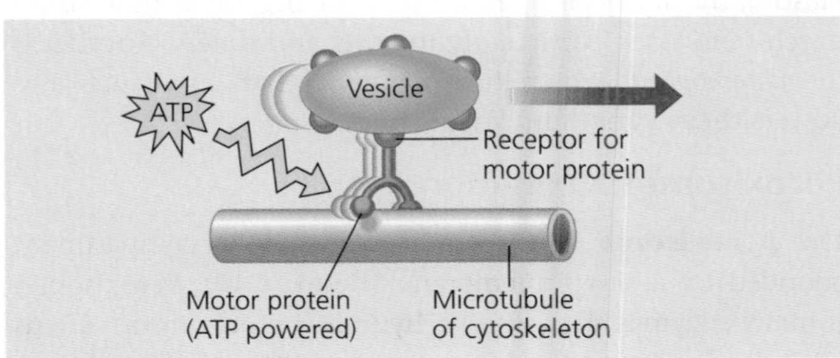

(a) Motor proteins that attach to receptors on vesicles can "walk" the vesicles along microtubules or, in some cases, microfilaments.

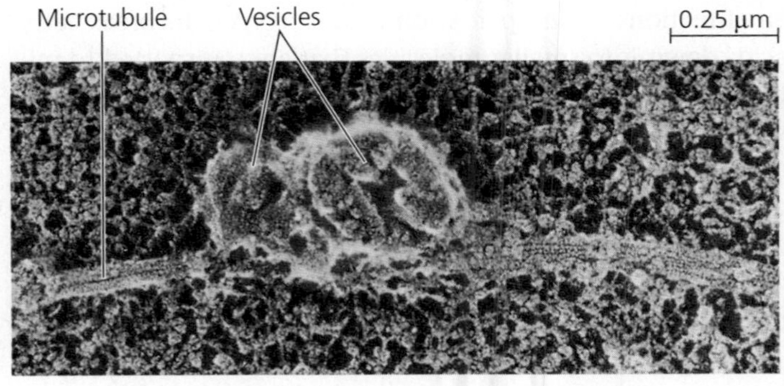

(b) In this SEM of a squid giant axon (a nerve cell extension), two vesicles containing neurotransmitters migrate toward the tip of the axon via the mechanism shown in (a).

▲ **Figure 6.21 Motor proteins and the cytoskeleton.**

Table 6.1 The Structure and Function of the Cytoskeleton

Property	Microtubules (Tubulin Polymers)	Microfilaments (Actin Filaments)	Intermediate Filaments
Structure	Hollow tubes; wall consists of 13 columns of tubulin molecules	Two intertwined strands of actin, each a polymer of actin subunits	Fibrous proteins supercoiled into thicker cables
Diameter	25 nm with 15-nm lumen	7 nm	8–12 nm
Protein subunits	Tubulin, a dimer consisting of α-tubulin and β-tubulin	Actin	One of several different proteins (such as keratins), depending on cell type
Main functions	Maintenance of cell shape (compression-resisting "girders") Cell motility (as in cilia or flagella) Chromosome movements in cell division Organelle movements	Maintenance of cell shape (tension-bearing elements) Changes in cell shape Muscle contraction Cytoplasmic streaming Cell motility (as in pseudopodia) Cell division (cleavage furrow formation)	Maintenance of cell shape (tension-bearing elements) Anchorage of nucleus and certain other organelles Formation of nuclear lamina
Fluorescence micrographs of fibroblasts, a favorite cell type for cell biology studies. In each, the structure of interest has been tagged with fluorescent molecules. In the first and third micrographs, the DNA in the nucleus has also been tagged (blue or orange).	10 μm Column of tubulin dimers 25 nm α β Tubulin dimer	10 μm Actin subunit 7 nm	5 μm Keratin proteins Fibrous subunit (keratins coiled together) 8–12 nm

from the ER travel to the Golgi along cytoskeletal tracks. The cytoskeleton also manipulates the plasma membrane, making it bend inward to form food vacuoles or other phagocytic vesicles. And the streaming of cytoplasm that circulates materials within many large plant cells is yet another kind of cellular movement brought about by the cytoskeleton.

Components of the Cytoskeleton

Now let's look more closely at the three main types of fibers that make up the cytoskeleton: *Microtubules* are the thickest of the three types; *microfilaments* (also called actin filaments) are the thinnest; and *intermediate filaments* are fibers with diameters in a middle range **(Table 6.1)**.

Microtubules

All eukaryotic cells have **microtubules**, hollow rods measuring about 25 nm in diameter and from 200 nm to 25 μm in length. The wall of the hollow tube is constructed from a globular protein called tubulin. Each tubulin protein is a *dimer*, a molecule made up of two subunits. A tubulin dimer consists of two slightly different polypeptides, α-tubulin and β-tubulin. Microtubules grow in length by adding tubulin dimers; they can also be disassembled and their tubulin used to build microtubules elsewhere in the cell. Because of the orientation of tubulin dimers, the two ends of a microtubule are slightly different. One end can accumulate or release tubulin dimers at a much higher rate than the other, thus growing and shrinking significantly during cellular activities. (This is called the "plus end," not because it can only add tubulin proteins but because it's the end where both "on" and "off" rates are much higher.)

Microtubules shape and support the cell and also serve as tracks along which organelles equipped with motor proteins can move. In addition to the example in Figure 6.21, microtubules guide secretory vesicles from the Golgi apparatus to the plasma membrane. Microtubules are also involved in the separation of chromosomes during cell division, which will be discussed in Chapter 12.

Centrosomes and Centrioles In animal cells, microtubules grow out from a **centrosome**, a region that is often located near the nucleus and is considered a "microtubule-organizing center." These microtubules function as compression-resisting girders of the cytoskeleton. Within the centrosome is a pair of **centrioles**, each composed of nine sets of triplet microtubules arranged in a ring **(Figure 6.22)**. Before an animal cell divides, the centrioles replicate. Although centrosomes with centrioles may help organize microtubule assembly in animal cells, they are not essential for this function in all eukaryotes; fungi and almost all plant cells lack centrosomes with centrioles but have well-organized microtubules. Apparently, other microtubule-organizing centers play the role of centrosomes in these cells.

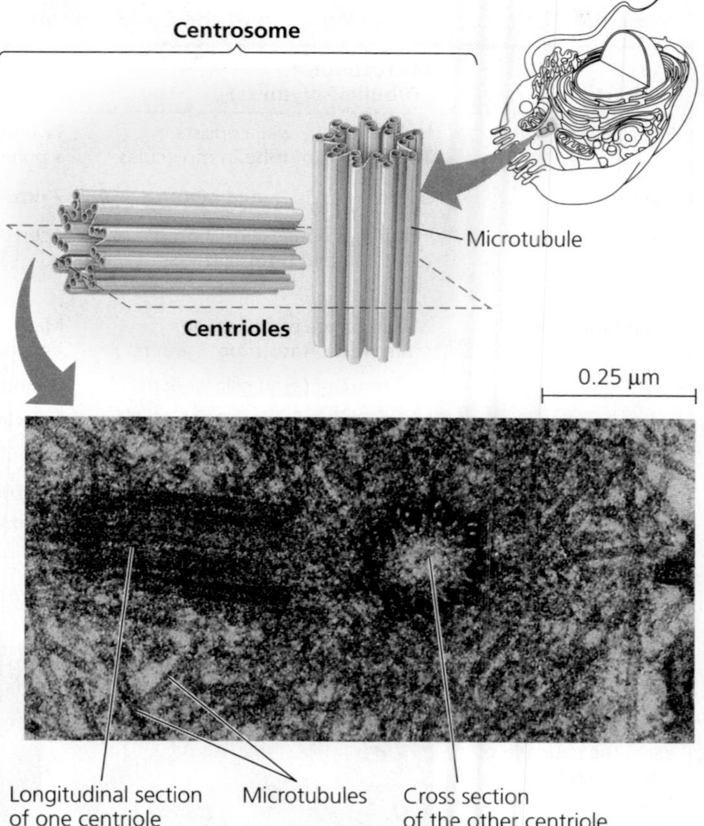

▲ **Figure 6.22 Centrosome containing a pair of centrioles.** Most animal cells have a centrosome, a region near the nucleus where the cell's microtubules are initiated. Within the centrosome is a pair of centrioles, each about 250 nm (0.25 μm) in diameter. The two centrioles are at right angles to each other, and each is made up of nine sets of three microtubules. The blue portions of the drawing represent nontubulin proteins that connect the microtubule triplets (TEM).

? *How many microtubules are in a centrosome? In the drawing, circle and label one microtubule and describe its structure. Circle and label a triplet.*

Cilia and Flagella In eukaryotes, a specialized arrangement of microtubules is responsible for the beating of **flagella** (singular, *flagellum*) and **cilia** (singular, *cilium*), microtubule-containing extensions that project from some cells. (The bacterial flagellum, shown in Figure 6.5, has a completely different structure.) Many unicellular eukaryotes are propelled through water by cilia or flagella that act as locomotor appendages, and the sperm of animals, algae, and some plants have flagella. When cilia or flagella extend from cells that are held in place as part of a tissue layer, they can move fluid over the surface of the tissue. For example, the ciliated lining of the trachea (windpipe) sweeps mucus containing trapped debris out of the lungs (see the EMs in Figure 6.3). In a woman's reproductive tract, the cilia lining the oviducts help move an egg toward the uterus.

Motile cilia usually occur in large numbers on the cell surface. They are about 0.25 μm in diameter and about 2–20 μm long. Flagella are the same diameter but longer, 10–200 μm. Also, flagella are usually limited to just one or a few per cell.

Flagella and cilia differ in their beating patterns **(Figure 6.23)**. A flagellum has an undulating motion that generates force in the same direction as the flagellum's axis, like the tail of a fish. In contrast, cilia work more like oars, with alternating power and recovery strokes generating force in a direction perpendicular to the cilium's axis, much as the oars of a racing crew boat extend outward at a right angle to the boat's forward movement.

A cilium may also act as a signal-receiving "antenna" for the cell. Cilia that have this function are generally nonmotile, and there is only one per cell. (In fact, in vertebrate animals, it appears that almost all cells have such a cilium, which is called a *primary cilium*.) Membrane proteins on this kind of cilium transmit molecular signals from the cell's environment to its interior, triggering signaling pathways that may lead to changes in the cell's activities. Cilium-based signaling appears to be crucial to brain function and to embryonic development.

Though different in length, number per cell, and beating pattern, motile cilia and flagella share a common structure. Each motile cilium and flagellum has a group of microtubules sheathed in an extension of the plasma membrane **(Figure 6.24)**. Nine doublets of microtubules are arranged in a ring; in the center of the ring are two single microtubules.

▶ **Figure 6.23 A comparison of the beating of flagella and motile cilia.**

(a) Motion of flagella. A flagellum usually undulates, its snakelike motion driving a cell in the same direction as the axis of the flagellum. Propulsion of a human sperm cell is an example of flagellate locomotion (LM).

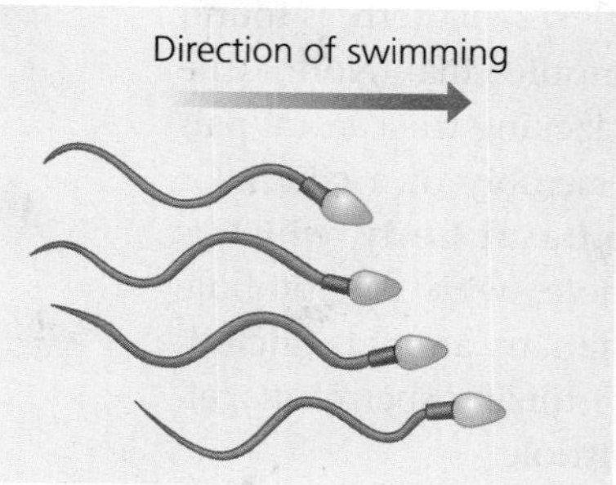

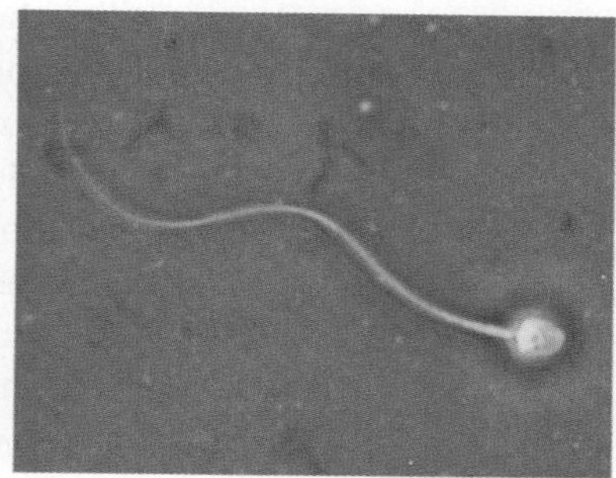

(b) Motion of cilia. Cilia have a back-and-forth motion. The rapid power stroke moves the cell in a direction perpendicular to the axis of the cilium. Then, during the slower recovery stroke, the cilium bends and sweeps sideways, closer to the cell surface. A dense nap of cilia, beating at a rate of about 40 to 60 strokes a second, covers this *Colpidium*, a freshwater protist (colorized SEM).

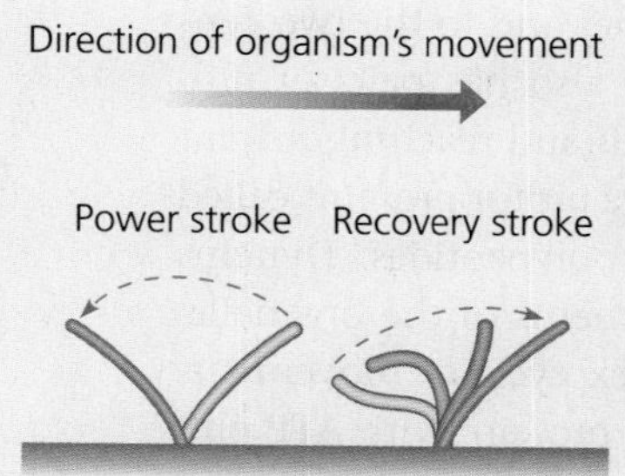

0.5 µm

Microtubules

Plasma membrane

Basal body

(a) A longitudinal section of a motile cilium shows microtubules running the length of the structure (TEM).

0.1 µm

Outer microtubule doublet

Dynein proteins

Central microtubule

Radial spoke

Cross-linking proteins between outer doublets

Plasma membrane

(b) A cross section through a motile cilium shows the "9 + 2" arrangement of microtubules (TEM). The outer microtubule doublets and the two central microtubules are held together by flexible cross-linking proteins (blue in art), including the radial spokes. The doublets also have attached motor proteins called dyneins (red in art).

0.1 µm

Triplet

Cross section of basal body

(c) Basal body: The nine outer doublets of a cilium or flagellum extend into the basal body, where each doublet joins another microtubule to form a ring of nine triplets. Each triplet is connected to the next by nontubulin proteins (thinner blue lines in diagram). This is a "9 + 0" arrangement: the two central microtubules are not present because they terminate above the basal body (TEM).

▲ **Figure 6.24 Structure of a flagellum or motile cilium.**

DRAW IT *In (a), circle the central pair of microtubules. Show where they terminate, and explain why they aren't seen in the cross section of the basal body in (c).*

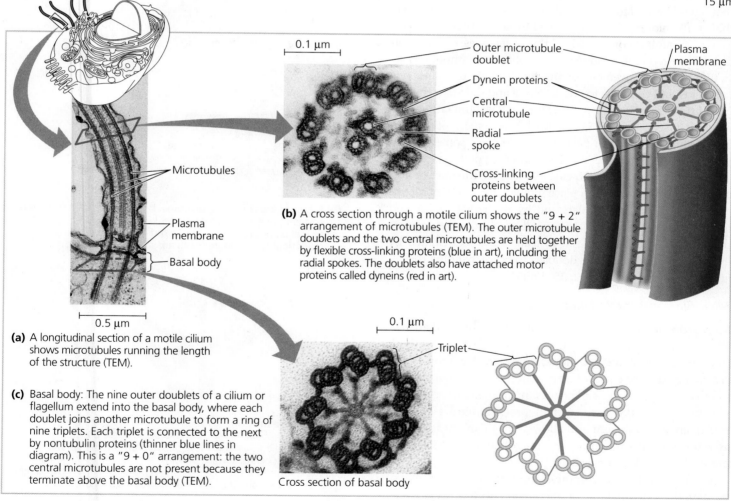

This arrangement, referred to as the "9 + 2" pattern, is found in nearly all eukaryotic flagella and motile cilia. (Nonmotile primary cilia have a "9 + 0" pattern, lacking the central pair of microtubules.) The microtubule assembly of a cilium or flagellum is anchored in the cell by a **basal body**, which is structurally very similar to a centriole, with microtubule triplets in a "9 + 0" pattern. In fact, in many animals (including humans), the basal body of the fertilizing sperm's flagellum enters the egg and becomes a centriole.

In flagella and motile cilia, flexible cross-linking proteins, evenly spaced along the length of the cilium or flagellum, connect the outer doublets to each other and to the two central microtubules. Each outer doublet also has pairs of protruding proteins spaced along its length and reaching toward the neighboring doublet; these are large motor proteins called **dyneins**, each composed of several polypeptides. Dyneins are responsible for the bending movements of the organelle. A dynein molecule performs a complex cycle of movements caused by changes in the shape of the protein, with ATP providing the energy for these changes (**Figure 6.25**).

The mechanics of dynein-based bending involve a process that resembles walking. A typical dynein protein has two "feet" that "walk" along the microtubule of the adjacent doublet, one foot maintaining contact while the other releases and reattaches one step farther along the microtubule. Without any restraints on the movement of the microtubule doublets, one doublet would continue to "walk" along and slide past the surface of the other, elongating the cilium or flagellum rather than bending it (see Figure 6.25a). For lateral movement of a cilium or flagellum, the dynein "walking" must have something to pull against, as when the muscles in your leg pull against your bones to move your knee. In cilia and flagella, the microtubule doublets seem to be held in place by the cross-linking proteins just inside the outer doublets and by the radial spokes and other structural elements. Thus, neighboring doublets cannot slide past each other very far. Instead, the forces exerted by dynein "walking" cause the doublets to curve, bending the cilium or flagellum (see Figure 6.25b and c).

Microfilaments (Actin Filaments)

Microfilaments are solid rods about 7 nm in diameter. They are also called actin filaments because they are built from molecules of **actin**, a globular protein. A microfilament is a twisted double chain of actin subunits (see Table 6.1). Besides occurring as linear filaments, microfilaments can form structural networks when certain proteins bind along the side of an actin filament and allow a new filament to extend as a branch. Like microtubules, microfilaments seem to be present in all eukaryotic cells.

In contrast to the compression-resisting role of microtubules, the structural role of microfilaments in the cytoskeleton is to bear tension (pulling forces). A three-dimensional network formed by microfilaments just inside the plasma

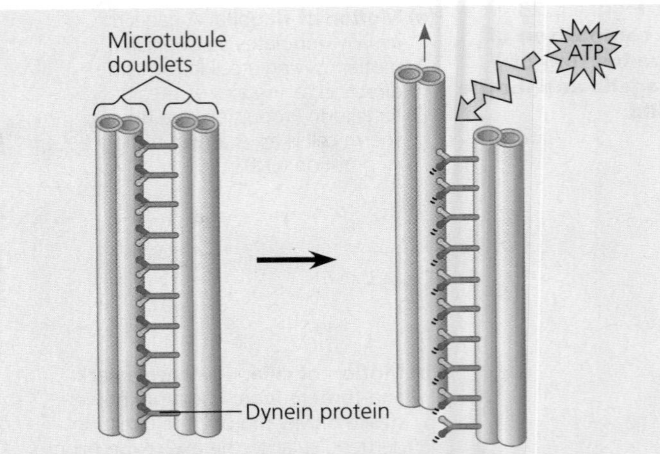

(a) Effect of unrestrained dynein movement. If a cilium or flagellum had no cross-linking proteins, the two feet of each dynein along one doublet (powered by ATP) would alternately grip and release the adjacent doublet. This "walking" motion would push the adjacent doublet up. Instead of bending, the doublets would slide past each other.

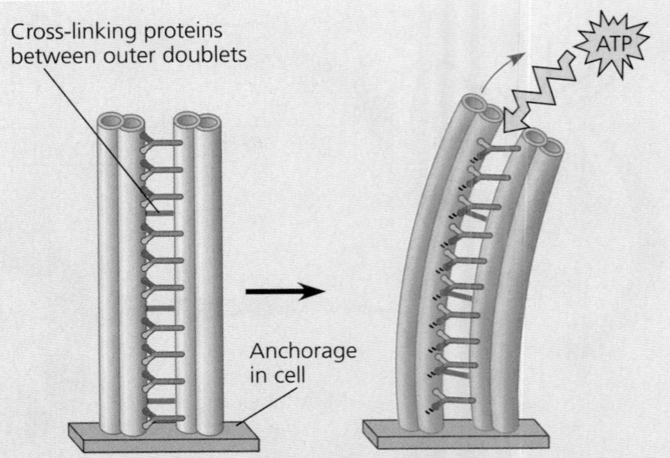

(b) Effect of cross-linking proteins. In a cilium or flagellum, two adjacent doublets cannot slide far because they are physically restrained by proteins, so they bend. (Only two of the nine outer doublets in Figure 6.24b are shown here.)

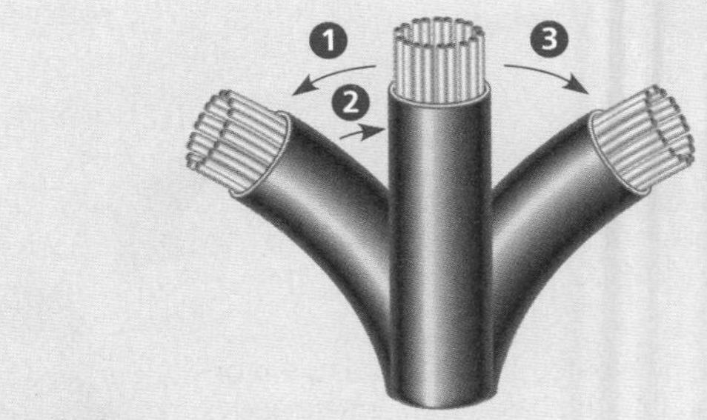

(c) Wavelike motion. Synchronized cycles of movement of many dyneins probably cause a bend to begin at the base of the cilium or flagellum and move outward toward the tip. Many successive bends, such as the ones shown here to the left and right, result in a wavelike motion. In this diagram, the two central microtubules and the cross-linking proteins are not shown.

▲ **Figure 6.25 How dynein "walking" moves flagella and cilia.**

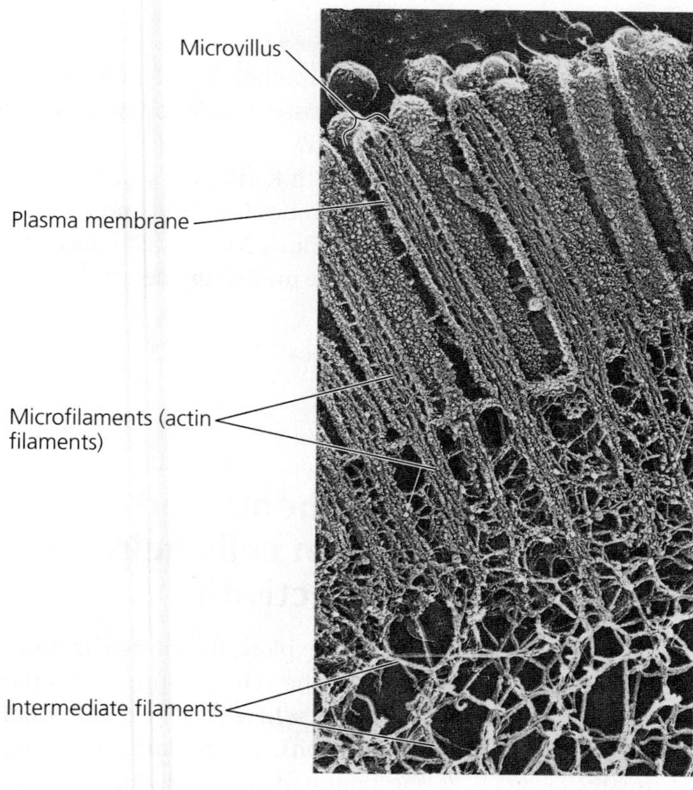

▲ **Figure 6.26 A structural role of microfilaments.** The surface area of this nutrient-absorbing intestinal cell is increased by its many microvilli (singular, *microvillus*), cellular extensions reinforced by bundles of microfilaments. These actin filaments are anchored to a network of intermediate filaments (TEM).

membrane (*cortical microfilaments*) helps support the cell's shape (see Figure 6.8). This network gives the outer cytoplasmic layer of a cell, called the **cortex**, the semisolid consistency of a gel, in contrast with the more fluid (*sol*) state of the interior cytoplasm. In animal cells specialized for transporting materials across the plasma membrane, such as intestinal cells, bundles of microfilaments make up the core of microvilli, delicate projections that increase the cell's surface area **(Figure 6.26)**.

Microfilaments are well known for their role in cell motility, particularly as part of the contractile apparatus of muscle cells. Thousands of actin filaments are arranged parallel to one another along the length of a muscle cell, interdigitated with thicker filaments made of a protein called **myosin** **(Figure 6.27a)**. Like dynein when it interacts with microtubules, myosin acts as a motor protein by means of projections that "walk" along the actin filaments. Contraction of the muscle cell results from the actin and myosin filaments sliding past one another in this way, shortening the cell. In other kinds of cells, actin filaments are associated with myosin in miniature and less elaborate versions of the arrangement in muscle cells. These actin-myosin aggregates are responsible for

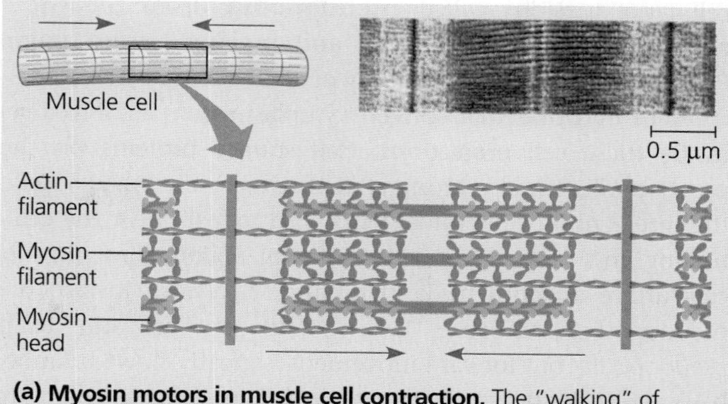

(a) Myosin motors in muscle cell contraction. The "walking" of myosin projections (the so-called heads) drives the parallel myosin and actin filaments past each other so that the actin filaments approach each other in the middle (red arrows). This shortens the muscle cell. Muscle contraction involves the shortening of many muscle cells at the same time (TEM).

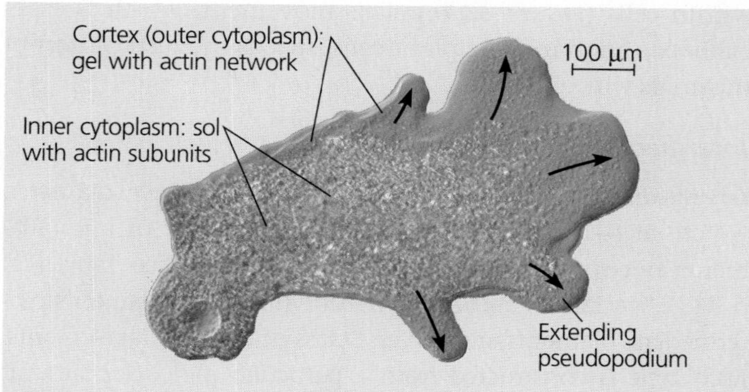

(b) Amoeboid movement. Interaction of actin filaments with myosin causes contraction of the cell, pulling the cell's trailing end (at left) forward (to the right) (LM).

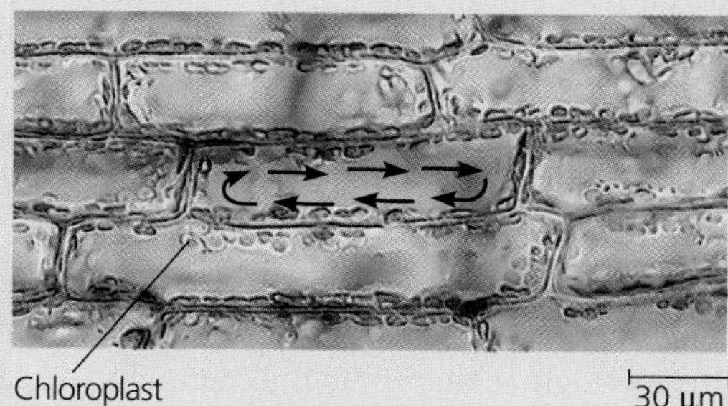

(c) Cytoplasmic streaming in plant cells. A layer of cytoplasm cycles around the cell, moving over a carpet of parallel actin filaments. Myosin motors attached to organelles in the fluid cytosol may drive the streaming by interacting with the actin (LM).

▲ **Figure 6.27 Microfilaments and motility.** In these three examples, interactions between actin filaments and motor proteins bring about cell movement.

localized contractions of cells. For example, a contracting belt of microfilaments forms a cleavage furrow that pinches a dividing animal cell into two daughter cells.

Localized contraction brought about by actin and myosin also plays a role in amoeboid movement **(Figure 6.27b)**. A cell such as an amoeba crawls along a surface by extending cellular extensions called **pseudopodia** (from the Greek *pseudes*, false, and *pod*, foot), and moving toward them. Pseudopodia extend by assembly of actin subunits into microfilament networks that convert cytoplasm from a sol to a gel inside these cell projections. Cell surface proteins on the pseudopodium make strong attachments to the "road." Next, the interaction of microfilaments with myosin near the cell's trailing end causes contraction of that region, loosening its cell-surface attachments and pulling it forward toward the pseudopodia. Amoebae lacking myosin can still form pseudopodia, but forward movement is greatly slowed. Amoebas are not the only cells that move by crawling; so do many cells in the animal body, including some white blood cells.

In plant cells, both actin-myosin interactions and sol-gel transformations brought about by actin may be involved in **cytoplasmic streaming**, a circular flow of cytoplasm within cells (**Figure 6.27c**). This movement, which is especially common in large plant cells, speeds the distribution of materials within the cell.

Intermediate Filaments

Intermediate filaments are named for their diameter, which, at 8–12 nm, is larger than the diameter of microfilaments but smaller than that of microtubules (see Table 6.1, p. 113). Specialized for bearing tension (like microfilaments), intermediate filaments are a diverse class of cytoskeletal elements. Each type is constructed from a particular molecular subunit belonging to a family of proteins whose members include the keratins. Microtubules and microfilaments, in contrast, are consistent in diameter and composition in all eukaryotic cells.

Intermediate filaments are more permanent fixtures of cells than are microfilaments and microtubules, which are often disassembled and reassembled in various parts of a cell. Even after cells die, intermediate filament networks often persist; for example, the outer layer of our skin consists of dead skin cells full of keratin proteins. Chemical treatments that remove microfilaments and microtubules from the cytoplasm of living cells leave a web of intermediate filaments that retains its original shape. Such experiments suggest that intermediate filaments are especially sturdy and that they play an important role in reinforcing the shape of a cell and fixing the position of certain organelles. For instance, the nucleus typically sits within a cage made of intermediate filaments, fixed in location by branches of the filaments that extend into the cytoplasm. Other intermediate filaments make up the nuclear lamina, which lines the interior of the nuclear envelope (see Figure 6.9). By supporting a cell's shape, intermediate filaments help the cell carry out its specific function. For example, the long extensions (axons) of nerve cells that transmit impulses are strengthened by intermediate filaments. Thus, the various kinds of intermediate filaments may function together as the permanent framework of the entire cell.

CONCEPT CHECK 6.6

1. Describe shared features of microtubule-based motion of flagella and microfilament-based muscle contraction.
2. How do cilia and flagella bend?
3. **WHAT IF?** Males afflicted with Kartagener's syndrome are sterile because of immotile sperm, and they tend to suffer from lung infections. This disorder has a genetic basis. Suggest what the underlying defect might be.

For suggested answers, see Appendix A.

CONCEPT 6.7

Extracellular components and connections between cells help coordinate cellular activities

Having crisscrossed the cell to explore its interior components, we complete our tour of the cell by returning to the surface of this microscopic world, where there are additional structures with important functions. The plasma membrane is usually regarded as the boundary of the living cell, but most cells synthesize and secrete materials that are extracellular, or external to the plasma membrane. Although these materials and the structures they form are outside the cell, their study is important to cell biology because they are involved in a great many cellular functions.

Cell Walls of Plants

The **cell wall** is an extracellular structure of plant cells that distinguishes them from animal cells (see Figure 6.8). The wall protects the plant cell, maintains its shape, and prevents excessive uptake of water. On the level of the whole plant, the strong walls of specialized cells hold the plant up against the force of gravity. Prokaryotes, fungi, and some protists also have cell walls, as you saw in Figures 6.5 and 6.8, but we will postpone discussion of them until Unit Five.

Plant cell walls are much thicker than the plasma membrane, ranging from 0.1 μm to several micrometers. The exact chemical composition of the wall varies from species to species and even from one cell type to another in the same plant, but the basic design of the wall is consistent. Microfibrils made of the polysaccharide cellulose (see Figure 5.8) are synthesized by an enzyme called cellulose synthase and secreted to the extracellular space, where they become embedded in a matrix of other polysaccharides and proteins. This combination of materials, strong fibers in a "ground substance" (matrix), is the same basic architectural design found in steel-reinforced concrete and in fiberglass.

A young plant cell first secretes a relatively thin and flexible wall called the **primary cell wall** (**Figure 6.28**). In actively

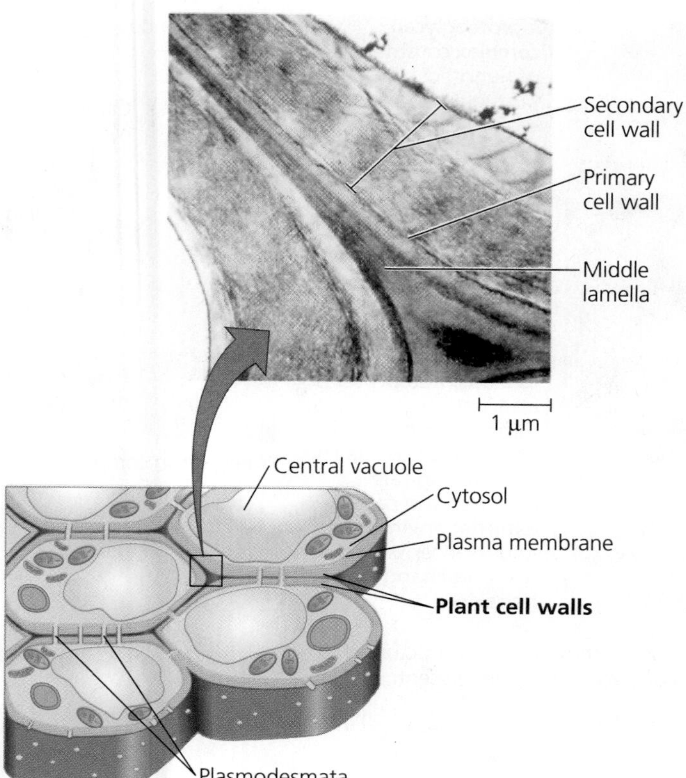

▲ **Figure 6.28 Plant cell walls.** The drawing shows several cells, each with a large vacuole, a nucleus, and several chloroplasts and mitochondria. The transmission electron micrograph shows the cell walls where two cells come together. The multilayered partition between plant cells consists of adjoining walls individually secreted by the cells.

growing cells, the cellulose fibrils are oriented at right angles to the direction of cell expansion. Researchers investigated the role of microtubules in orienting these cellulose fibrils (**Figure 6.29**). Their observations strongly support the idea that microtubules in the cell cortex guide cellulose synthase as it synthesizes and deposits cellulose fibrils. By orienting cellulose deposition, microtubules thus affect the growth pattern of the cells.

Between primary walls of adjacent cells is the **middle lamella**, a thin layer rich in sticky polysaccharides called pectins. The middle lamella glues adjacent cells together (see Figure 6.28). (Pectin is used as a thickening agent in jams and jellies.) When the cell matures and stops growing, it strengthens its wall. Some plant cells do this simply by secreting hardening substances into the primary wall. Other cells add a **secondary cell wall** between the plasma membrane and the primary wall. The secondary wall, often deposited in several laminated layers, has a strong and durable matrix that affords the cell protection and support. Wood, for example, consists mainly of secondary walls. Plant cell walls are usually perforated by channels between adjacent cells called plasmodesmata (see Figure 6.28), which will be discussed shortly.

▼ **Figure 6.29** INQUIRY

What role do microtubules play in orienting deposition of cellulose in cell walls?

EXPERIMENT Previous experiments on preserved plant tissues had shown alignment of microtubules in the cell cortex with cellulose fibrils in the cell wall. Also, drugs that disrupted microtubules were observed to cause disoriented cellulose fibrils. To further investigate the possible role of cortical microtubules in guiding cellulose fibril deposition, David Ehrhardt and colleagues at Stanford University used a type of confocal microscopy to study cell wall deposition in living cells. In these cells, they labeled both cellulose synthase and microtubules with fluorescent markers and observed them over time.

RESULTS Each fluorescence image below represents a combination of 30 images taken over a 5-minute period to detect the movement of cellulose synthase and microtubules. These two coincided highly over time. The labeling molecules caused cellulose synthase to fluoresce green and the microtubules to fluoresce red. The arrowheads indicate prominent areas where the two are seen to align.

10 µm

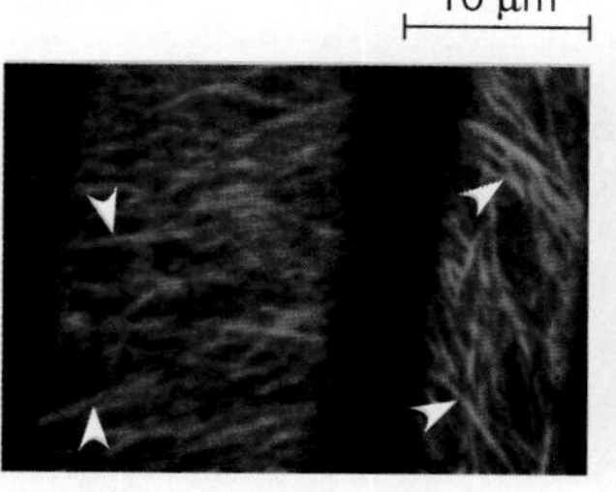

Distribution of cellulose synthase over time

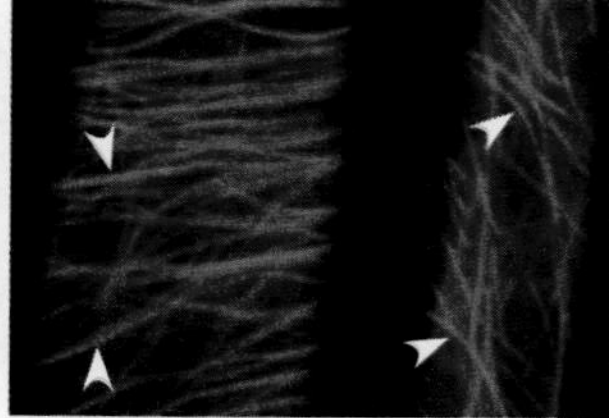

Distribution of microtubules over time

CONCLUSION The organization of microtubules appears to directly guide the path of cellulose synthase as it lays down cellulose, thus determining the orientation of cellulose fibrils.

SOURCE A. R. Paradez et al., Visualization of cellulose synthase demonstrates functional association with microtubules, *Science* 312:1491–1495 (2006).

WHAT IF? In a second experiment, the researchers exposed the plant cells to blue light, previously shown to cause reorientation of microtubules. What events would you predict would follow blue light exposure?

The Extracellular Matrix (ECM) of Animal Cells

Although animal cells lack walls akin to those of plant cells, they do have an elaborate **extracellular matrix (ECM)**. The main ingredients of the ECM are glycoproteins and other carbohydrate-containing molecules secreted by the cells. (Recall that glycoproteins are proteins with covalently bonded carbohydrate, usually short chains of sugars.) The most abundant glycoprotein in the ECM of most animal cells is **collagen**, which forms strong fibers outside the cells (see Figure 5.20). In fact, collagen accounts for about 40% of the total protein in the human body. The collagen fibers are embedded in a network woven out of **proteoglycans** secreted

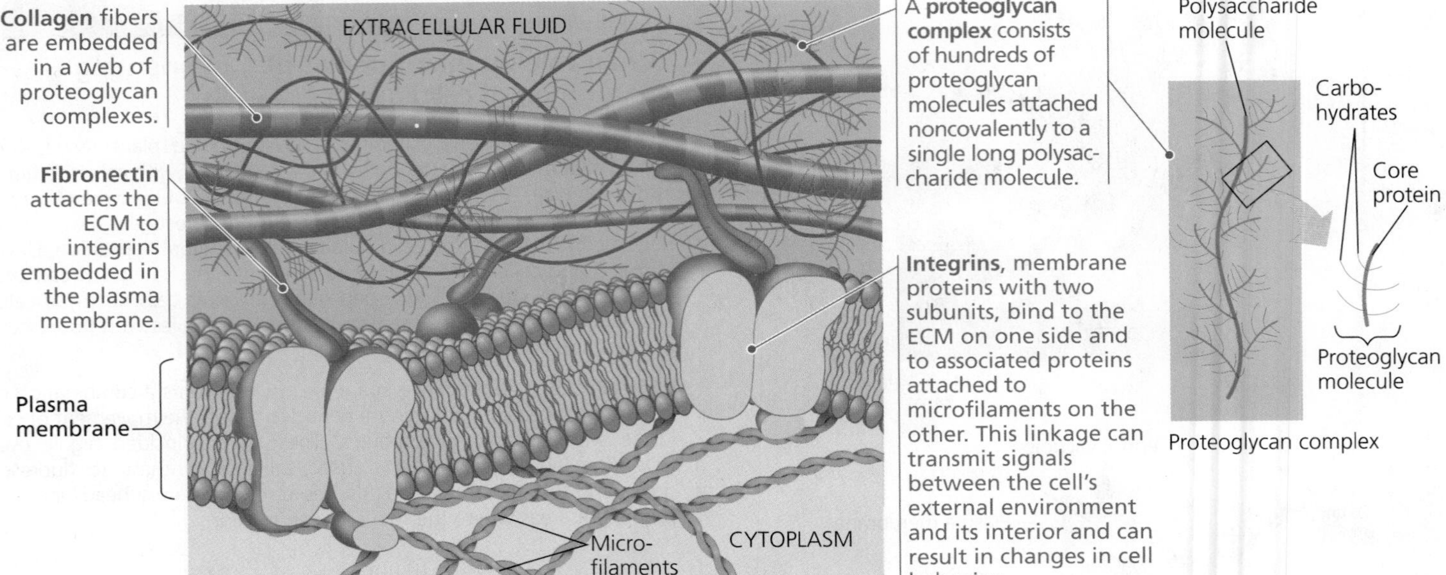
Collagen fibers are embedded in a web of proteoglycan complexes.
EXTRACELLULAR FLUID
A proteoglycan complex consists of hundreds of proteoglycan molecules attached noncovalently to a single long polysaccharide molecule.
Polysaccharide molecule
Carbohydrates
Core protein
Proteoglycan molecule
Proteoglycan complex
Fibronectin attaches the ECM to integrins embedded in the plasma membrane.
Integrins, membrane proteins with two subunits, bind to the ECM on one side and to associated proteins attached to microfilaments on the other. This linkage can transmit signals between the cell's external environment and its interior and can result in changes in cell behavior.
Plasma membrane
Microfilaments
CYTOPLASM

▲ **Figure 6.30 Extracellular matrix (ECM) of an animal cell.** The molecular composition and structure of the ECM vary from one cell type to another. In this example, three different types of glycoproteins are present: proteoglycans, collagen, and fibronectin.

by cells (**Figure 6.30**). A proteoglycan molecule consists of a small core protein with many carbohydrate chains covalently attached, so that it may be up to 95% carbohydrate. Large proteoglycan complexes can form when hundreds of proteoglycan molecules become noncovalently attached to a single long polysaccharide molecule, as shown in Figure 6.30. Some cells are attached to the ECM by still other ECM glycoproteins, such as **fibronectin**. Fibronectin and other ECM proteins bind to cell-surface receptor proteins called **integrins** that are built into the plasma membrane. Integrins span the membrane and bind on their cytoplasmic side to associated proteins attached to microfilaments of the cytoskeleton. The name *integrin* is based on the word *integrate*: Integrins are in a position to transmit signals between the ECM and the cytoskeleton and thus to integrate changes occurring outside and inside the cell.

Current research on fibronectin, other ECM molecules, and integrins is revealing the influential role of the extracellular matrix in the lives of cells. By communicating with a cell through integrins, the ECM can regulate a cell's behavior. For example, some cells in a developing embryo migrate along specific pathways by matching the orientation of their microfilaments to the "grain" of fibers in the extracellular matrix. Researchers have also learned that the extracellular matrix around a cell can influence the activity of genes in the nucleus. Information about the ECM probably reaches the nucleus by a combination of mechanical and chemical signaling pathways. Mechanical signaling involves fibronectin, integrins, and microfilaments of the cytoskeleton. Changes in the cytoskeleton may in turn trigger chemical signaling pathways inside the cell, leading to changes in the set of proteins being made by the cell and therefore changes in the cell's function. In this way, the extracellular matrix of a particular tissue may help coordinate the behavior of all the cells of that tissue. Direct connections between cells also function in this coordination, as we discuss next.

Cell Junctions

Cells in an animal or plant are organized into tissues, organs, and organ systems. Neighboring cells often adhere, interact, and communicate via sites of direct physical contact.

Plasmodesmata in Plant Cells

It might seem that the nonliving cell walls of plants would isolate plant cells from one another. But in fact, as shown in **Figure 6.31**, cell walls are perforated with **plasmodesmata** (singular, *plasmodesma*; from the Greek *desmos*, to bind),

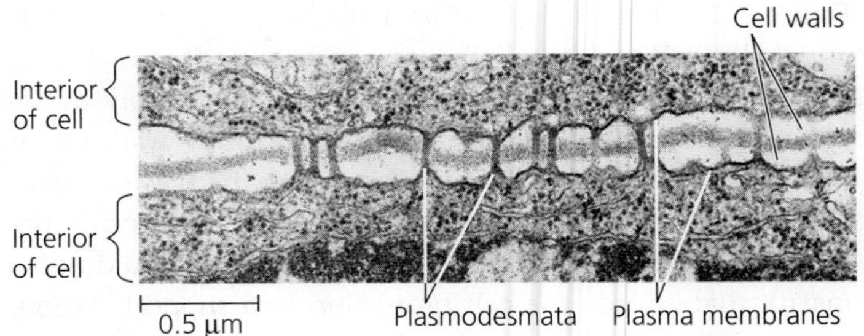
Cell walls
Interior of cell
Interior of cell
0.5 µm
Plasmodesmata
Plasma membranes

▲ **Figure 6.31 Plasmodesmata between plant cells.** The cytoplasm of one plant cell is continuous with the cytoplasm of its neighbors via plasmodesmata, cytoplasmic channels through the cell walls (TEM).

membrane-lined channels filled with cytoplasm. Cytosol passes through the plasmodesmata and joins the internal chemical environments of adjacent cells. These connections unify most of the plant into one living continuum. The plasma membranes of adjacent cells line the channel of each plasmodesma and thus are continuous. Water and small solutes can pass freely from cell to cell, and recent experiments have shown that in some circumstances, certain proteins and RNA molecules can also do this (see Concept 36.6). The macromolecules transported to neighboring cells appear to reach the plasmodesmata by moving along fibers of the cytoskeleton.

Tight Junctions, Desmosomes, and Gap Junctions in Animal Cells

In animals, there are three main types of cell junctions: *tight junctions*, *desmosomes*, and *gap junctions*. (Gap junctions are most like the plasmodesmata of plants, although gap junction pores are not lined with membrane.) All three types of cell junctions are especially common in epithelial tissue, which lines the external and internal surfaces of the body. **Figure 6.32** uses epithelial cells of the intestinal lining to illustrate these junctions.

▼ Figure 6.32

Exploring Cell Junctions in Animal Tissues

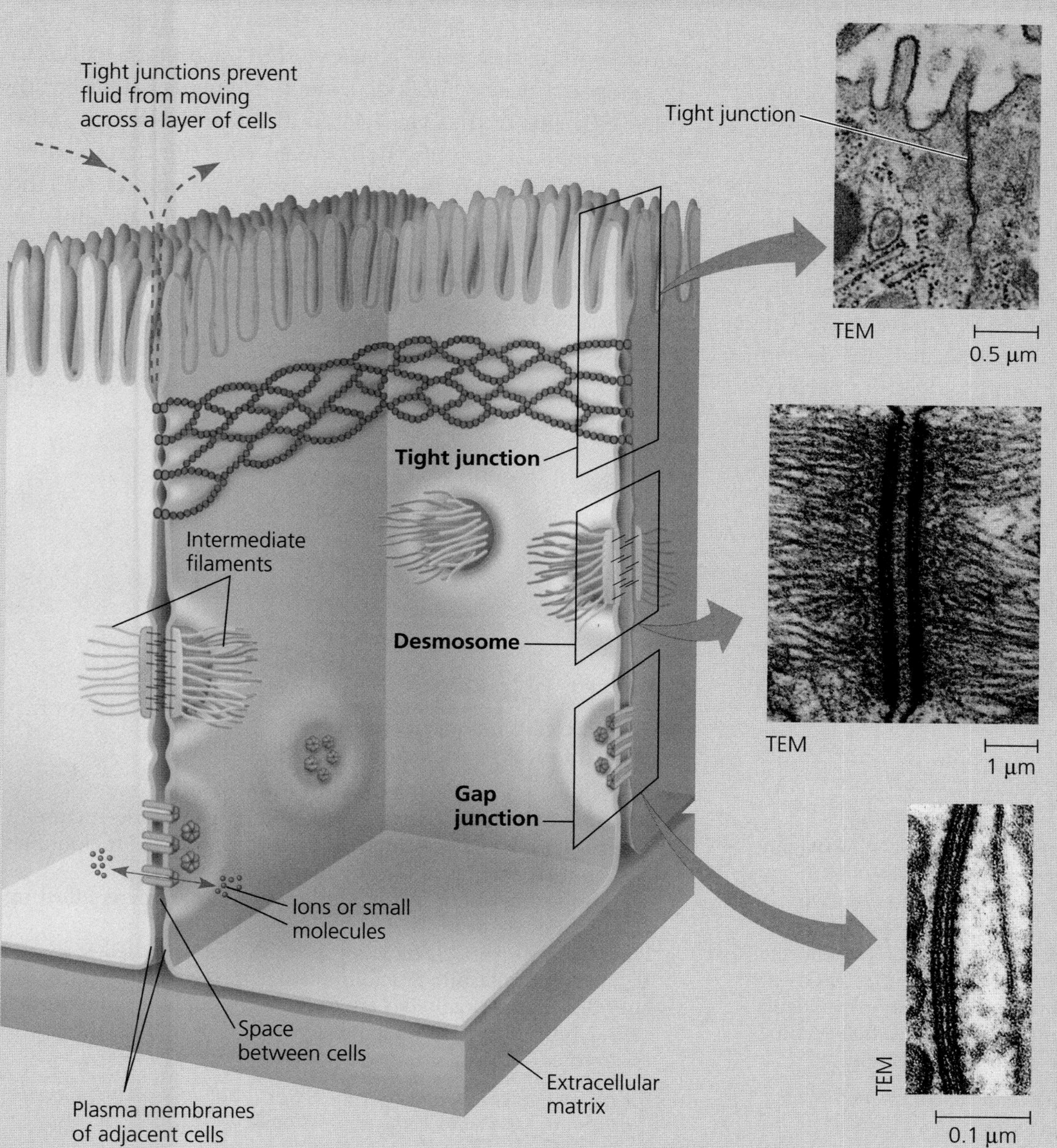

Tight Junctions

At **tight junctions**, the plasma membranes of neighboring cells are very tightly pressed against each other, bound together by specific proteins (purple). Forming continuous seals around the cells, tight junctions prevent leakage of extracellular fluid across a layer of epithelial cells. For example, tight junctions between skin cells make us watertight by preventing leakage between cells in our sweat glands.

Desmosomes

Desmosomes (also called *anchoring junctions*) function like rivets, fastening cells together into strong sheets. Intermediate filaments made of sturdy keratin proteins anchor desmosomes in the cytoplasm. Desmosomes attach muscle cells to each other in a muscle. Some "muscle tears" involve the rupture of desmosomes.

Gap Junctions

Gap junctions (also called *communicating junctions*) provide cytoplasmic channels from one cell to an adjacent cell and in this way are similar in their function to the plasmodesmata in plants. Gap junctions consist of membrane proteins that surround a pore through which ions, sugars, amino acids, and other small molecules may pass. Gap junctions are necessary for communication between cells in many types of tissues, such as heart muscle, and in animal embryos.

CONCEPT CHECK 6.7

1. In what way are the cells of plants and animals structurally different from single-celled eukaryotes?
2. **WHAT IF?** If the plant cell wall or the animal extracellular matrix were impermeable, what effect would this have on cell function?
3. **MAKE CONNECTIONS** The polypeptide chain that makes up a tight junction weaves back and forth through the membrane four times, with two extracellular loops, and one loop plus short C-terminal and N-terminal tails in the cytoplasm. Looking at Figure 5.16 (p. 79), what would you predict about the amino acid sequence of the tight-junction protein?

For suggested answers, see Appendix A.

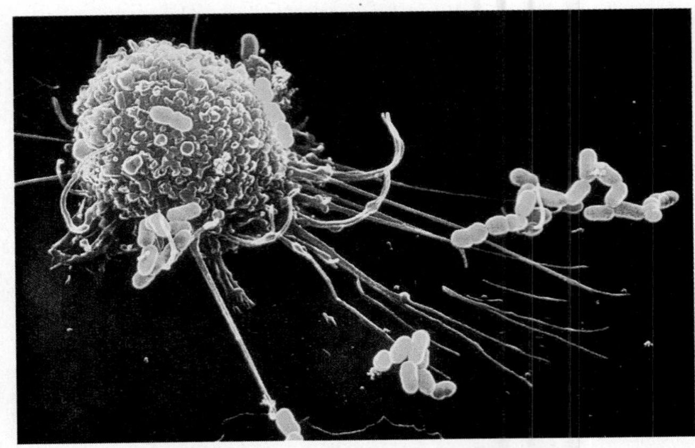

▲ **Figure 6.33 The emergence of cellular functions.** The ability of this macrophage (brown) to recognize, apprehend, and destroy bacteria (yellow) is a coordinated activity of the whole cell. Its cytoskeleton, lysosomes, and plasma membrane are among the components that function in phagocytosis (colorized SEM).

The Cell: A Living Unit Greater Than the Sum of Its Parts

From our panoramic view of the cell's compartmental organization to our close-up inspection of each organelle's architecture, this tour of the cell has provided many opportunities to correlate structure with function. (This would be a good time to review cell structure by returning to Figure 6.8, on pp. 100 and 101.) But even as we dissect the cell, remember that none of its components works alone. As an example of cellular integration, consider the microscopic scene in **Figure 6.33**. The large cell is a macrophage (see Figure 6.13a). It helps defend the mammalian body against infections by ingesting bacteria (the smaller cells) into phagocytic vesicles. The macrophage crawls along a surface and reaches out to the bacteria with thin pseudopodia (called filopodia). Actin filaments interact with other elements of the cytoskeleton in these movements. After the macrophage engulfs the bacteria, they are destroyed by lysosomes. The elaborate endomembrane system produces the lysosomes. The digestive enzymes of the lysosomes and the proteins of the cytoskeleton are all made on ribosomes. And the synthesis of these proteins is programmed by genetic messages dispatched from the DNA in the nucleus. All these processes require energy, which mitochondria supply in the form of ATP. Cellular functions arise from cellular order: The cell is a living unit greater than the sum of its parts.

6 CHAPTER REVIEW

SUMMARY OF KEY CONCEPTS

CONCEPT 6.1

Biologists use microscopes and the tools of biochemistry to study cells (pp. 94–97)

- Improvements in microscopy that affect the parameters of magnification, resolution, and contrast have catalyzed progress in the study of cell structure. **Light microscopy** (LM) and **electron microscopy** (EM), as well as other types, remain important tools.
- Cell biologists can obtain pellets enriched in particular cellular components by centrifuging disrupted cells at sequential speeds, a process known as **cell fractionation**. Larger cellular components are in the pellet after lower-speed centrifugation, and smaller components are in the pellet after higher-speed centrifugation.

? *How do microscopy and biochemistry complement each other to reveal cell structure and function?*

CONCEPT 6.2

Eukaryotic cells have internal membranes that compartmentalize their functions (pp. 98–102)

- All cells are bounded by a **plasma membrane**.
- **Prokaryotic cells** lack nuclei and other membrane-enclosed **organelles**, while **eukaryotic cells** have internal membranes that compartmentalize cellular functions.
- The surface-to-volume ratio is an important parameter affecting cell size and shape.
- Plant and animal cells have most of the same organelles: a nucleus, endoplasmic reticulum, Golgi apparatus, and mitochondria. Some organelles are found only in plant or in animal cells. Chloroplasts are present only in cells of photosynthetic eukaryotes.

? *Explain how the compartmental organization of a eukaryotic cell contributes to its biochemical functioning.*

	Cell Component	Structure	Function
CONCEPT 6.3 **The eukaryotic cell's genetic instructions are housed in the nucleus and carried out by the ribosomes (pp. 102–104)** ? *Describe the relationship between the nucleus and ribosomes.*	Nucleus (ER)	Surrounded by nuclear envelope (double membrane) perforated by nuclear pores; nuclear envelope continuous with endoplasmic reticulum (ER)	Houses chromosomes, which are made of chromatin (DNA and proteins); contains nucleoli, where ribosomal subunits are made; pores regulate entry and exit of materials
	Ribosome	Two subunits made of ribosomal RNA and proteins; can be free in cytosol or bound to ER	Protein synthesis
CONCEPT 6.4 **The endomembrane system regulates protein traffic and performs metabolic functions in the cell (pp. 104–109)** ? *Describe the key role played by transport vesicles in the endomembrane system.*	Endoplasmic reticulum (Nuclear envelope)	Extensive network of membrane-bounded tubules and sacs; membrane separates lumen from cytosol; continuous with nuclear envelope	Smooth ER: synthesis of lipids, metabolism of carbohydrates, Ca^{2+} storage, detoxification of drugs and poisons Rough ER: aids in synthesis of secretory and other proteins from bound ribosomes; adds carbohydrates to proteins to make glycoproteins; produces new membrane
	Golgi apparatus	Stacks of flattened membranous sacs; has polarity (*cis* and *trans* faces)	Modification of proteins, carbohydrates on proteins, and phospholipids; synthesis of many polysaccharides; sorting of Golgi products, which are then released in vesicles
	Lysosome	Membranous sac of hydrolytic enzymes (in animal cells)	Breakdown of ingested substances, cell macromolecules, and damaged organelles for recycling
	Vacuole	Large membrane-bounded vesicle	Digestion, storage, waste disposal, water balance, cell growth, and protection
CONCEPT 6.5 **Mitochondria and chloroplasts change energy from one form to another (pp. 109–112)** ? *What is the endosymbiont theory?*	Mitochondrion	Bounded by double membrane; inner membrane has infoldings (cristae)	Cellular respiration
	Chloroplast	Typically two membranes around fluid stroma, which contains thylakoids stacked into grana (in cells of photosynthetic eukaryotes, including plants)	Photosynthesis
	Peroxisome	Specialized metabolic compartment bounded by a single membrane	Contains enzymes that transfer hydrogen atoms from substrates to oxygen, producing hydrogen peroxide (H_2O_2) as a by-product; H_2O_2 is converted to water by another enzyme

CONCEPT 6.6

The cytoskeleton is a network of fibers that organizes structures and activities in the cell (pp. 112–118)

- The **cytoskeleton** functions in structural support for the cell and in motility and signal transmission.
- **Microtubules** shape the cell, guide organelle movement, and separate chromosomes in dividing cells. **Cilia** and **flagella** are motile appendages containing microtubules. Primary cilia also play sensory and signaling roles. **Microfilaments** are thin rods functioning in muscle contraction, amoeboid movement, **cytoplasmic streaming**, and microvillus support. **Intermediate filaments** support cell shape and fix organelles in place.

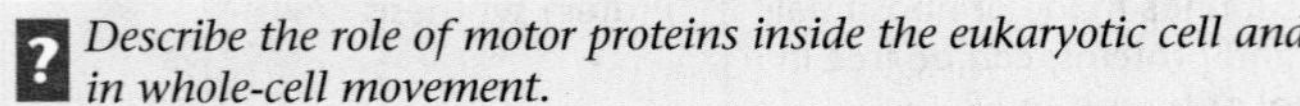

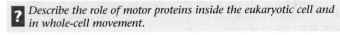

? *Describe the role of motor proteins inside the eukaryotic cell and in whole-cell movement.*

CONCEPT 6.7

Extracellular components and connections between cells help coordinate cellular activities (pp. 118–122)

- Plant **cell walls** are made of cellulose fibers embedded in other polysaccharides and proteins. Cellulose deposition is oriented along microtubules.
- Animal cells secrete glycoproteins that form the **extracellular matrix (ECM)**, which functions in support, adhesion, movement, and regulation.
- Cell junctions connect neighboring cells in plants and animals. Plants have **plasmodesmata** that pass through adjoining cell walls. Animal cells have **tight junctions**, **desmosomes**, and **gap junctions**.

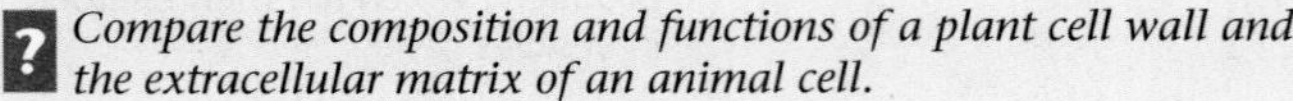

? *Compare the composition and functions of a plant cell wall and the extracellular matrix of an animal cell.*

TEST YOUR UNDERSTANDING

LEVEL 1: KNOWLEDGE/COMPREHENSION

1. Which structure is *not* part of the endomembrane system?
   a. nuclear envelope
   b. chloroplast
   c. Golgi apparatus
   d. plasma membrane
   e. ER
2. Which structure is common to plant *and* animal cells?
   a. chloroplast
   b. wall made of cellulose
   c. central vacuole
   d. mitochondrion
   e. centriole
3. Which of the following is present in a prokaryotic cell?
   a. mitochondrion
   b. ribosome
   c. nuclear envelope
   d. chloroplast
   e. ER
4. Which structure-function pair is *mismatched*?
   a. nucleolus; production of ribosomal subunits
   b. lysosome; intracellular digestion
   c. ribosome; protein synthesis
   d. Golgi; protein trafficking
   e. microtubule; muscle contraction

LEVEL 2: APPLICATION/ANALYSIS

5. Cyanide binds to at least one molecule involved in producing ATP. If a cell is exposed to cyanide, most of the cyanide will be found within the
   a. mitochondria.
   b. ribosomes.
   c. peroxisomes.
   d. lysosomes.
   e. endoplasmic reticulum.
6. What is the most likely pathway taken by a newly synthesized protein that will be secreted by a cell?
   a. ER → Golgi → nucleus
   b. Golgi → ER → lysosome
   c. nucleus → ER → Golgi
   d. ER → Golgi → vesicles that fuse with plasma membrane
   e. ER → lysosomes → vesicles that fuse with plasma membrane
7. Which cell would be best for studying lysosomes?
   a. muscle cell
   b. nerve cell
   c. phagocytic white blood cell
   d. leaf cell of a plant
   e. bacterial cell
8. **DRAW IT** From memory, draw two eukaryotic cells, labeling the structures listed here and showing any physical connections between the internal structures of each cell: nucleus, rough ER, smooth ER, mitochondrion, centrosome, chloroplast, vacuole, lysosome, microtubule, cell wall, ECM, microfilament, Golgi apparatus, intermediate filament, plasma membrane, peroxisome, ribosome, nucleolus, nuclear pore, vesicle, flagellum, microvilli, plasmodesma.

LEVEL 3: SYNTHESIS/EVALUATION

9. **EVOLUTION CONNECTION**
   Which aspects of cell structure best reveal evolutionary unity? What are some examples of specialized modifications?
10. **SCIENTIFIC INQUIRY**
    Imagine protein X, destined to span the plasma membrane. Assume that the mRNA carrying the genetic message for protein X has already been translated by ribosomes in a cell culture. If you fractionate the cells (see Figure 6.4), in which fraction would you find protein X? Explain by describing its transit through the cell.
11. **WRITE ABOUT A THEME**
    Emergent Properties Considering some of the characteristics that define life and drawing on your new knowledge of cellular structures and functions, write a short essay (100–150 words) that discusses this statement: Life is an emergent property that appears at the level of the cell. (Review pp. 3–5 in Chapter 1.)

For selected answers, see Appendix A.

MasteringBIOLOGY www.masteringbiology.com

1. MasteringBiology® Assignments

BioFlix **Tutorials** Tour of an Animal Cell: The Endomembrane System • Structures and Functions; Tour of a Plant Cell: Structures and Functions
Tutorials Connections Between Cells
Activities Metric System Review • Prokaryotic Cell Structure and Function • Discovery Channel Video: Cells • Role of the Nucleus and Ribosomes in Protein Synthesis • Transport into the Nucleus • A Pulse Chase Experiment • The Endomembrane System • Cilia and Flagella • Cell Junctions • Review: Animal Cell Structure and Function
Questions Student Misconceptions • Reading Quiz • Multiple Choice • End-of-Chapter

2. eText

Read your book online, search, take notes, highlight text, and more.

3. The Study Area

Practice Tests • Cumulative Test • ***BioFlix*** 3-D Animations • MP3 Tutor Sessions • Videos • Activities • Investigations • Lab Media • Audio Glossary • Word Study Tools • Art

7

Membrane Structure and Function

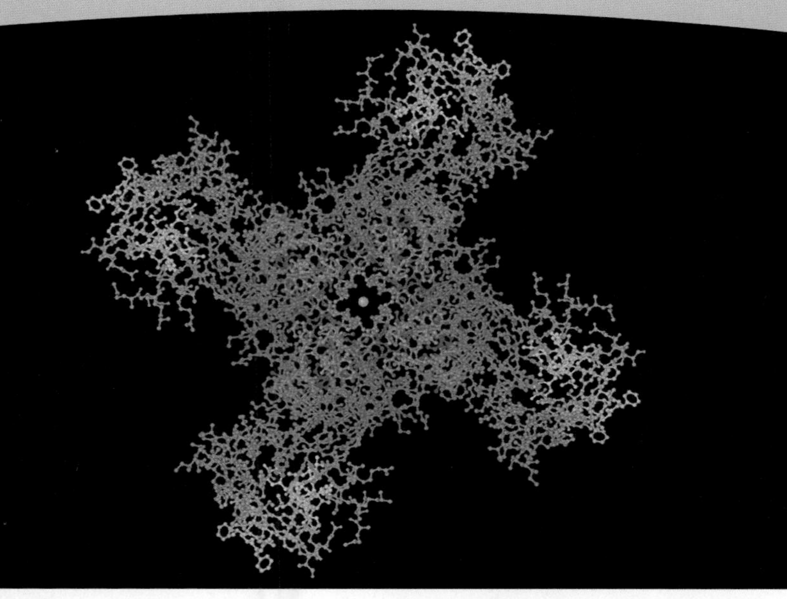

▲ **Figure 7.1 How do cell membrane proteins help regulate chemical traffic?**

KEY CONCEPTS

7.1 Cellular membranes are fluid mosaics of lipids and proteins
7.2 Membrane structure results in selective permeability
7.3 Passive transport is diffusion of a substance across a membrane with no energy investment
7.4 Active transport uses energy to move solutes against their gradients
7.5 Bulk transport across the plasma membrane occurs by exocytosis and endocytosis

OVERVIEW

Life at the Edge

The plasma membrane is the edge of life, the boundary that separates the living cell from its surroundings. A remarkable film only about 8 nm thick—it would take over 8,000 plasma membranes to equal the thickness of this page—the plasma membrane controls traffic into and out of the cell it surrounds. Like all biological membranes, the plasma membrane exhibits **selective permeability**; that is, it allows some substances to cross it more easily than others. One of the earliest episodes in the evolution of life may have been the formation of a membrane that enclosed a solution different from the surrounding solution while still permitting the uptake of nutrients and elimination of waste products. The ability of the cell to discriminate in its chemical exchanges with its environment is fundamental to life, and it is the plasma membrane and its component molecules that make this selectivity possible.

In this chapter, you will learn how cellular membranes control the passage of substances. The image in **Figure 7.1** shows the elegant structure of a eukaryotic plasma membrane protein that plays a crucial role in nerve cell signaling. This protein provides a channel for a stream of potassium ions (K^+) to exit a nerve cell at a precise moment after nerve stimulation, restoring the cell's ability to fire again. (The orange ball in the center represents one potassium ion moving through the channel.) In this way, the plasma membrane and its proteins not only act as an outer boundary but also enable the cell to carry out its functions. The same applies to the many varieties of internal membranes that partition the eukaryotic cell: The molecular makeup of each membrane allows compartmentalized specialization in cells. To understand how membranes work, we'll begin by examining their architecture.

CONCEPT **7.1**

Cellular membranes are fluid mosaics of lipids and proteins

Lipids and proteins are the staple ingredients of membranes, although carbohydrates are also important. The most abundant lipids in most membranes are phospholipids. The ability of phospholipids to form membranes is inherent in their molecular structure. A phospholipid is an **amphipathic** molecule, meaning it has both a hydrophilic region and a hydrophobic region (see Figure 5.12). Other types of membrane lipids are also amphipathic. Furthermore, most of the proteins within membranes have both hydrophobic and hydrophilic regions.

How are phospholipids and proteins arranged in the membranes of cells? In the **fluid mosaic model**, the membrane is a fluid structure with a "mosaic" of various proteins embedded in or attached to a double layer (bilayer) of phospholipids. Scientists propose models as hypotheses, ways of organizing and explaining existing information. Let's explore how the fluid mosaic model was developed.

Membrane Models: *Scientific Inquiry*

Scientists began building molecular models of the membrane decades before membranes were first seen with the electron microscope (in the 1950s). In 1915, membranes isolated from red blood cells were chemically analyzed and found to be composed of lipids and proteins. Ten years later, two Dutch scientists reasoned that cell membranes must be phospholipid bilayers. Such a double layer of molecules could exist as a stable boundary between two aqueous compartments because the molecular arrangement shelters the hydrophobic tails of the phospholipids from water while exposing the hydrophilic heads to water **(Figure 7.2)**.

If a phospholipid bilayer was the main fabric of a membrane, where were the proteins located? Although the heads of phospholipids are hydrophilic, the surface of a pure phospholipid bilayer adheres less strongly to water than does the surface of a biological membrane. Given this difference, Hugh Davson and James Danielli suggested in 1935 that the membrane might be coated on both sides with hydrophilic proteins. They proposed a sandwich model: a phospholipid bilayer between two layers of proteins.

When researchers first used electron microscopes to study cells in the 1950s, the pictures seemed to support the Davson-Danielli model. By the late 1960s, however, many cell biologists recognized two problems with the model. First, inspection of a variety of membranes revealed that membranes with different functions differ in structure and chemical composition. A second, more serious problem became apparent once membrane proteins were better characterized. Unlike proteins dissolved in the cytosol, membrane proteins are not very soluble in water because they are amphipathic. If such proteins were layered on the surface of the membrane, their hydrophobic parts would be in aqueous surroundings.

Taking these observations into account, S. J. Singer and G. Nicolson proposed in 1972 that membrane proteins reside in the phospholipid bilayer with their hydrophilic regions protruding **(Figure 7.3)**. This molecular arrangement would maximize contact of hydrophilic regions of proteins and phospholipids with water in the cytosol and extracellular fluid, while providing their hydrophobic parts with a nonaqueous environment. In this fluid mosaic model, the membrane is a mosaic of protein molecules bobbing in a fluid bilayer of phospholipids.

A method of preparing cells for electron microscopy called freeze-fracture has demonstrated visually that proteins are indeed embedded in the phospholipid bilayer of the membrane **(Figure 7.4)**. Freeze-fracture splits a membrane along the middle of the bilayer, somewhat like pulling apart a chunky peanut butter sandwich. When the membrane layers are viewed in the electron microscope, the interior of the

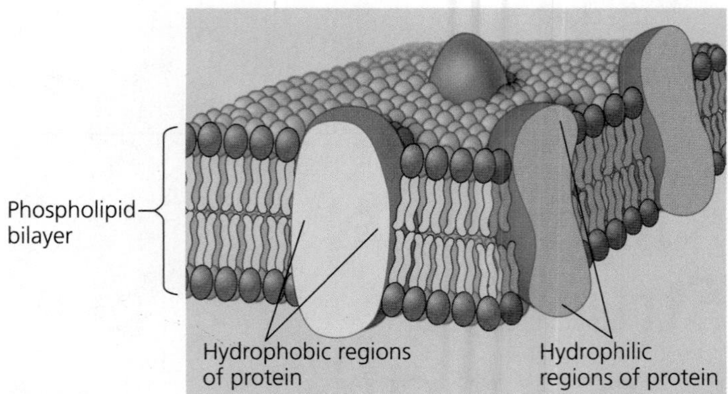

▲ **Figure 7.3 The original fluid mosaic model for membranes.**

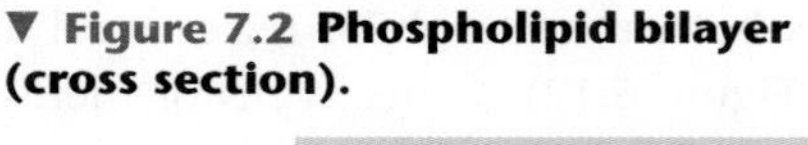

▼ **Figure 7.2 Phospholipid bilayer (cross section).**

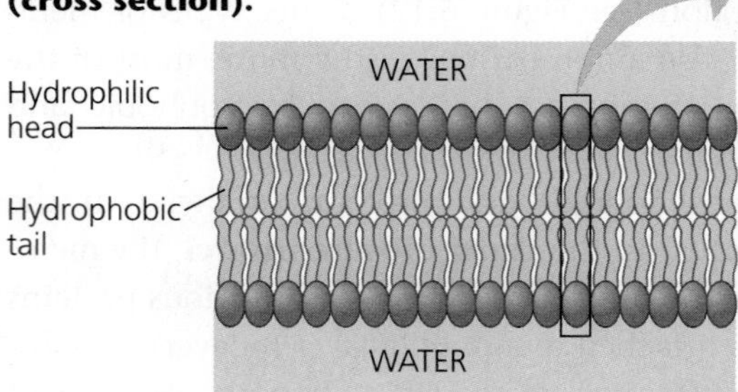

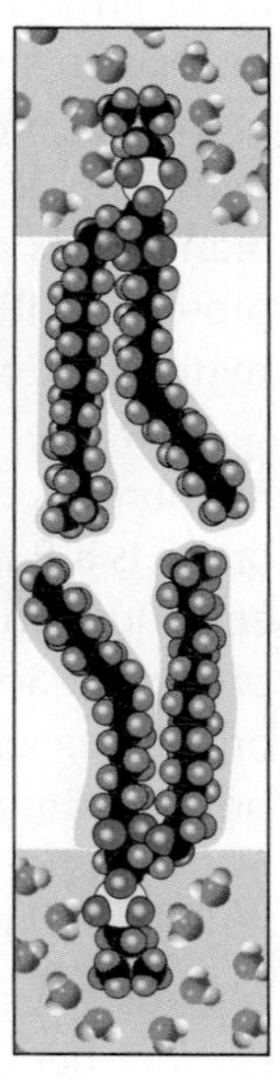

MAKE CONNECTIONS *Consulting Figure 5.12 (p. 76), circle the hydrophilic and hydrophobic portions of the enlarged phospholipids on the right. Explain what each portion contacts when the phospholipids are in the plasma membrane.*

▼ **Figure 7.4 RESEARCH METHOD**

Freeze-fracture

APPLICATION A cell membrane can be split into its two layers, revealing the structure of the membrane's interior.

TECHNIQUE A cell is frozen and fractured with a knife. The fracture plane often follows the hydrophobic interior of a membrane, splitting the phospholipid bilayer into two separated layers. Each membrane protein goes wholly with one of the layers.

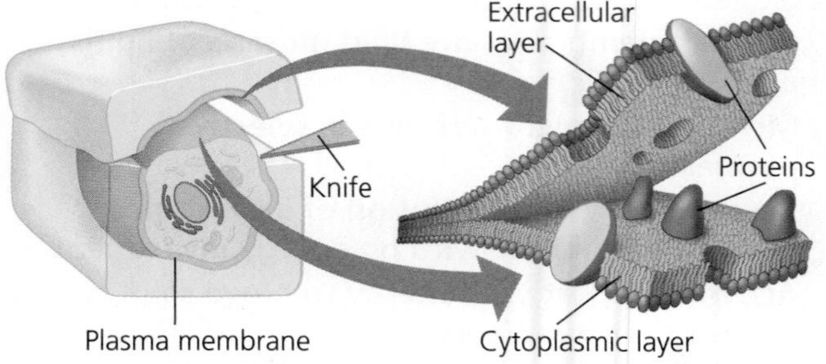

RESULTS These SEMs show membrane proteins (the "bumps") in the two layers, demonstrating that proteins are embedded in the phospholipid bilayer.

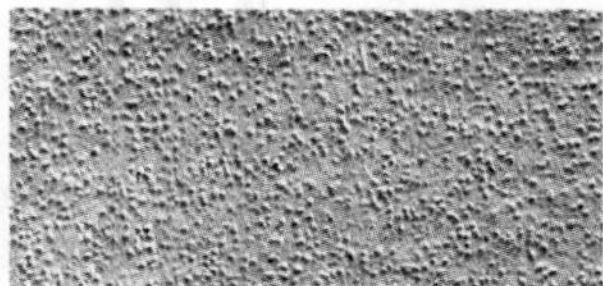

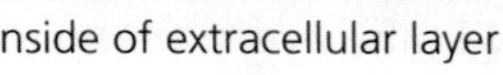

Inside of extracellular layer

Inside of cytoplasmic layer

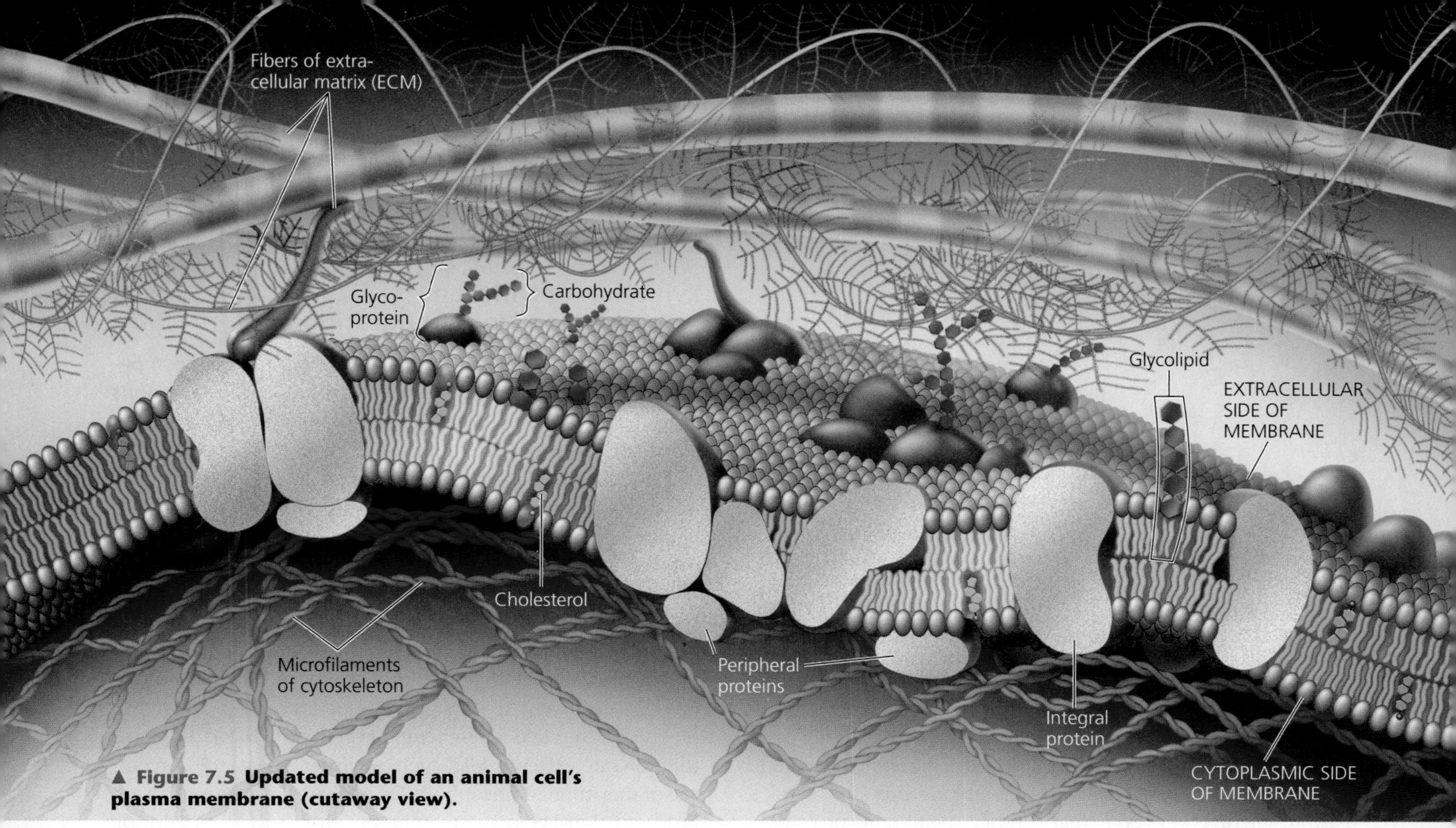

▲ **Figure 7.5 Updated model of an animal cell's plasma membrane (cutaway view).**

bilayer appears cobblestoned, with protein particles interspersed in a smooth matrix, in agreement with the fluid mosaic model. Some proteins remain attached to one layer or the other, like the peanut chunks in the sandwich.

Because models are hypotheses, replacing one model of membrane structure with another does not imply that the original model was worthless. The acceptance or rejection of a model depends on how well it fits observations and explains experimental results. New findings may make a model obsolete; even then, it may not be totally scrapped, but revised to incorporate the new observations. The fluid mosaic model is continually being refined. For example, groups of proteins are often found associated in long-lasting, specialized patches, where they carry out common functions. The lipids themselves appear to form defined regions as well. Also, the membrane may be much more packed with proteins than imagined in the classic fluid mosaic model—compare the updated model in **Figure 7.5** with the original model in Figure 7.3. Let's now take a closer look at membrane structure.

The Fluidity of Membranes

Membranes are not static sheets of molecules locked rigidly in place. A membrane is held together primarily by hydrophobic interactions, which are much weaker than covalent bonds (see Figure 5.20). Most of the lipids and some of the proteins can shift about laterally—that is, in the plane of the membrane, like partygoers elbowing their way through a crowded room (**Figure 7.6**). It is quite rare, however, for a molecule to flip-flop transversely across the membrane, switching from one phospholipid layer to the other; to do so, the hydrophilic part of the molecule must cross the hydrophobic interior of the membrane.

The lateral movement of phospholipids within the membrane is rapid. Adjacent phospholipids switch positions about 10^7 times per second, which means that a phospholipid can travel about 2 μm—the length of many bacterial cells—in 1 second. Proteins are much larger than lipids and move more slowly, but some membrane proteins do drift, as shown in a classic experiment described in **Figure 7.7**, on the next page.

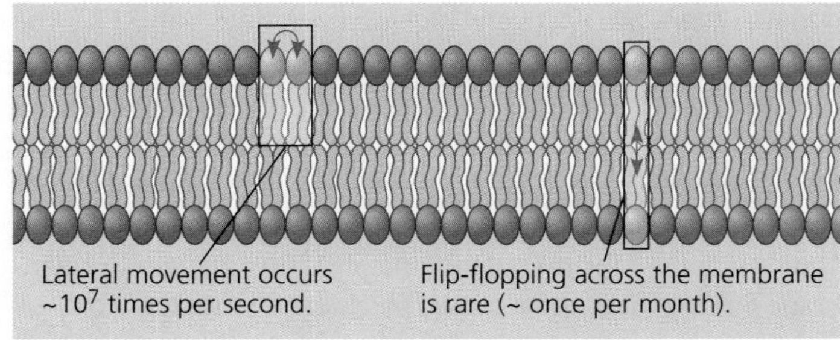
Lateral movement occurs
~10^7 times per second.
Flip-flopping across the membrane
is rare (~ once per month).

▲ **Figure 7.6 The movement of phospholipids.**

▼ Figure 7.7 INQUIRY

Do membrane proteins move?

EXPERIMENT Larry Frye and Michael Edidin, at Johns Hopkins University, labeled the plasma membrane proteins of a mouse cell and a human cell with two different markers and fused the cells. Using a microscope, they observed the markers on the hybrid cell.

RESULTS

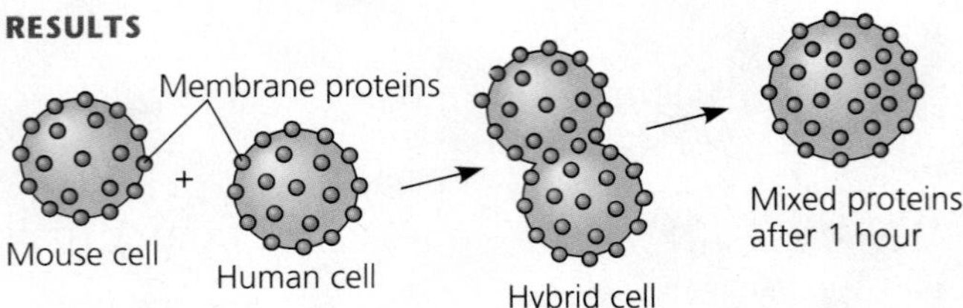

CONCLUSION The mixing of the mouse and human membrane proteins indicates that at least some membrane proteins move sideways within the plane of the plasma membrane.

SOURCE L. D. Frye and M. Edidin, The rapid intermixing of cell surface antigens after formation of mouse-human heterokaryons, *Journal of Cell Science* 7:319 (1970).

WHAT IF? Suppose the proteins did not mix in the hybrid cell, even many hours after fusion. Would you be able to conclude that proteins don't move within the membrane? What other explanation could there be?

And some membrane proteins seem to move in a highly directed manner, perhaps driven along cytoskeletal fibers by motor proteins connected to the membrane proteins' cytoplasmic regions. However, many other membrane proteins seem to be held immobile by their attachment to the cytoskeleton or to the extracellular matrix (see Figure 7.5).

A membrane remains fluid as temperature decreases until finally the phospholipids settle into a closely packed arrangement and the membrane solidifies, much as bacon grease forms lard when it cools. The temperature at which a membrane solidifies depends on the types of lipids it is made of. The membrane remains fluid to a lower temperature if it is rich in phospholipids with unsaturated hydrocarbon tails (see Figures 5.11 and 5.12). Because of kinks in the tails where double bonds are located, unsaturated hydrocarbon tails cannot pack together as closely as saturated hydrocarbon tails, and this makes the membrane more fluid **(Figure 7.8a)**.

The steroid cholesterol, which is wedged between phospholipid molecules in the plasma membranes of animal cells, has different effects on membrane fluidity at different temperatures **(Figure 7.8b)**. At relatively high temperatures—at 37°C, the body temperature of humans, for example—cholesterol makes the membrane less fluid by restraining phospholipid movement. However, because cholesterol also hinders the close packing of phospholipids, it lowers the temperature required for the membrane to solidify. Thus, cholesterol can be thought of as a "fluidity buffer" for the membrane, resisting changes in membrane fluidity that can be caused by changes in temperature.

Membranes must be fluid to work properly; they are usually about as fluid as salad oil. When a membrane solidifies, its permeability changes, and enzymatic proteins in the membrane may become inactive if their activity requires them to be able to move within the membrane. However, membranes that are too fluid cannot support protein function either. Therefore, extreme environments pose a challenge for life, resulting in evolutionary adaptations that include differences in membrane lipid composition.

Evolution of Differences in Membrane Lipid Composition

EVOLUTION Variations in the cell membrane lipid compositions of many species appear to be evolutionary adaptations that maintain the appropriate membrane fluidity under specific environmental conditions. For instance, fishes that live in extreme cold have membranes with a high proportion of unsaturated hydrocarbon tails, enabling their membranes to remain fluid (see Figure 7.8a). At the other extreme, some bacteria and archaea thrive at temperatures greater than 90°C (194°F) in thermal hot springs and geysers. Their membranes include unusual lipids that may prevent excessive fluidity at such high temperatures.

The ability to change the lipid composition of cell membranes in response to changing temperatures has evolved in organisms that live where temperatures vary. In many plants that tolerate extreme cold, such as winter wheat, the percentage of unsaturated phospholipids increases in autumn, an adjustment that keeps the membranes from solidifying during winter. Certain bacteria and archaea can also change the proportion of unsaturated phospholipids in their cell membranes, depending on the temperature at which they are growing. Overall, natural selection has apparently favored organisms whose mix of membrane lipids ensures an appropriate level of membrane fluidity for their environment.

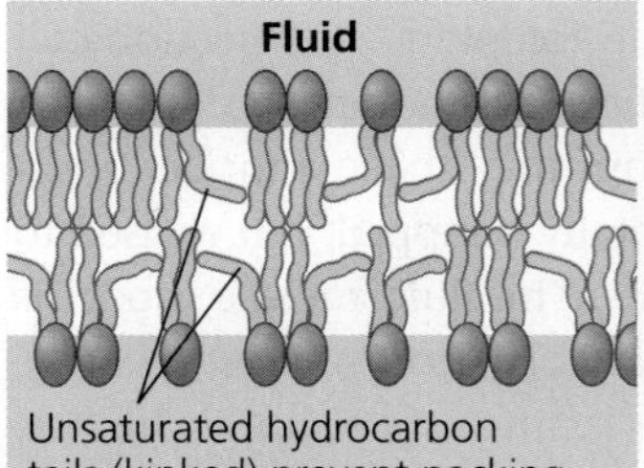

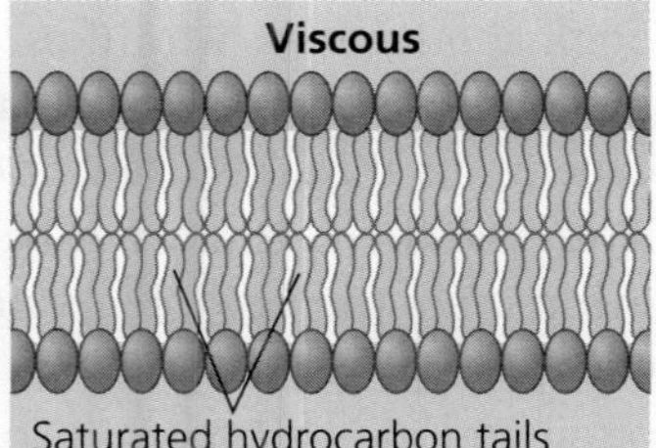

(a) Unsaturated versus saturated hydrocarbon tails.

(b) Cholesterol within the animal cell membrane. Cholesterol reduces membrane fluidity at moderate temperatures by reducing phospholipid movement, but at low temperatures it hinders solidification by disrupting the regular packing of phospholipids.

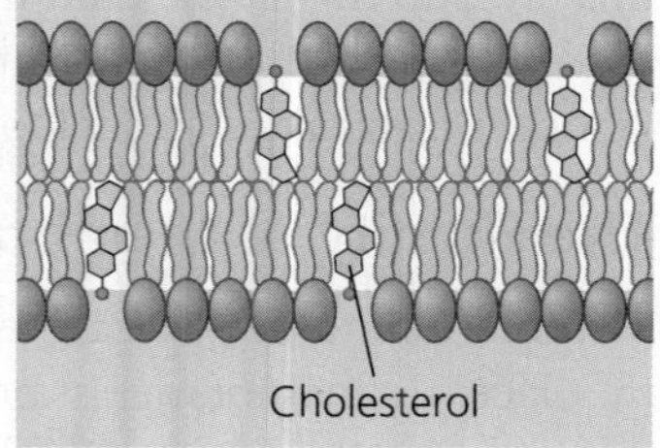

▲ Figure 7.8 Factors that affect membrane fluidity.

Membrane Proteins and Their Functions

Now we come to the *mosaic* aspect of the fluid mosaic model. Somewhat like a tile mosaic, a membrane is a collage of different proteins, often clustered together in groups, embedded in the fluid matrix of the lipid bilayer (see Figure 7.5). More than 50 kinds of proteins have been found so far in the plasma membrane of red blood cells, for example. Phospholipids form the main fabric of the membrane, but proteins determine most of the membrane's functions. Different types of cells contain different sets of membrane proteins, and the various membranes within a cell each have a unique collection of proteins.

Notice in Figure 7.5 that there are two major populations of membrane proteins: integral proteins and peripheral proteins. **Integral proteins** penetrate the hydrophobic interior of the lipid bilayer. The majority are *transmembrane proteins*, which span the membrane; other integral proteins extend only partway into the hydrophobic interior. The hydrophobic regions of an integral protein consist of one or more stretches of nonpolar amino acids (see Figure 5.16), usually coiled into α helices (**Figure 7.9**). The hydrophilic parts of the molecule are exposed to the aqueous solutions on either side of the membrane. Some proteins also have a hydrophilic channel through their center that allows passage of hydrophilic substances (see Figure 7.1). **Peripheral proteins** are not embedded in the lipid bilayer at all; they are appendages loosely bound to the surface of the membrane, often to exposed parts of integral proteins (see Figure 7.5).

On the cytoplasmic side of the plasma membrane, some membrane proteins are held in place by attachment to the cytoskeleton. And on the extracellular side, certain membrane proteins are attached to fibers of the extracellular matrix (see Figure 6.30; *integrins* are one type of integral protein). These attachments combine to give animal cells a stronger framework than the plasma membrane alone could provide.

Figure 7.10 gives an overview of six major functions performed by proteins of the plasma membrane. A single cell

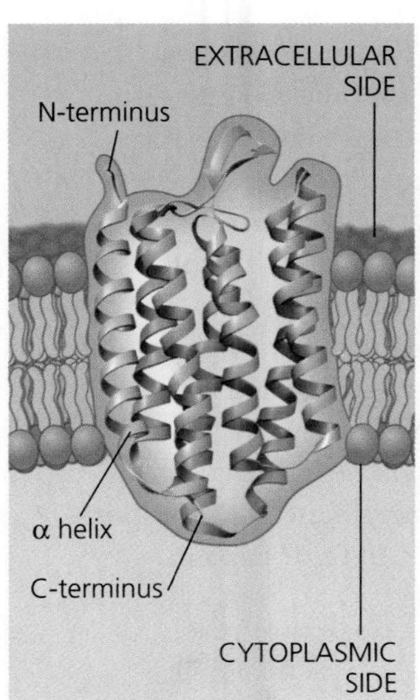

◀ **Figure 7.9 The structure of a transmembrane protein.** Bacteriorhodopsin (a bacterial transport protein) has a distinct orientation in the membrane, with its N-terminus outside the cell and its C-terminus inside. This ribbon model highlights the α-helical secondary structure of the hydrophobic parts, which lie mostly within the hydrophobic interior of the membrane. The protein includes seven transmembrane helices. The nonhelical hydrophilic segments are in contact with the aqueous solutions on the extracellular and cytoplasmic sides of the membrane.

(a) Transport. *Left:* A protein that spans the membrane may provide a hydrophilic channel across the membrane that is selective for a particular solute. *Right:* Other transport proteins shuttle a substance from one side to the other by changing shape (see Figure 7.17). Some of these proteins hydrolyze ATP as an energy source to actively pump substances across the membrane.

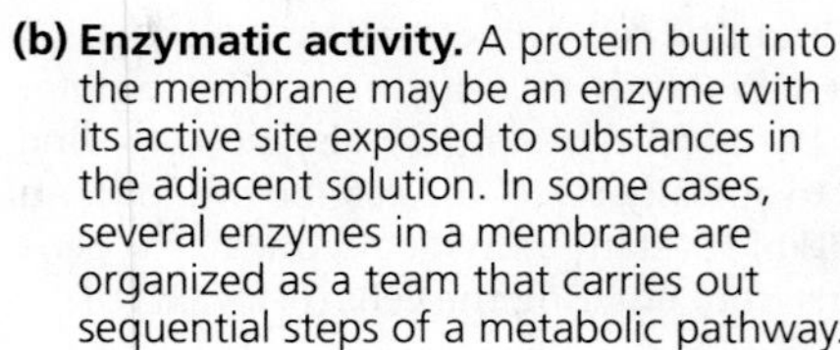

(b) Enzymatic activity. A protein built into the membrane may be an enzyme with its active site exposed to substances in the adjacent solution. In some cases, several enzymes in a membrane are organized as a team that carries out sequential steps of a metabolic pathway.

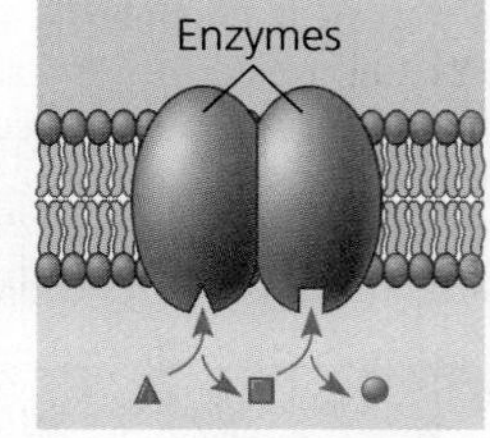

(c) Signal transduction. A membrane protein (receptor) may have a binding site with a specific shape that fits the shape of a chemical messenger, such as a hormone. The external messenger (signaling molecule) may cause the protein to change shape, allowing it to relay the message to the inside of the cell, usually by binding to a cytoplasmic protein (see Figure 11.6).

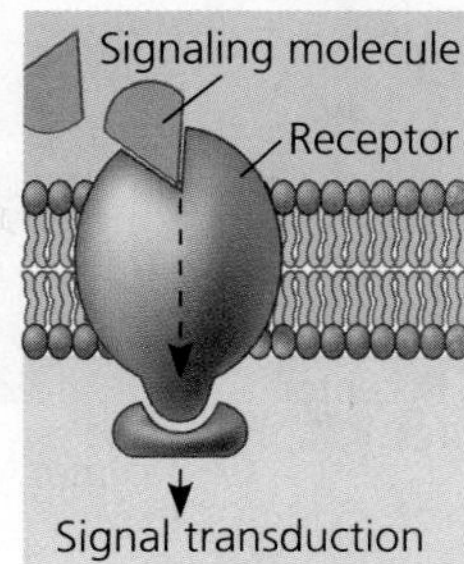

(d) Cell-cell recognition. Some glycoproteins serve as identification tags that are specifically recognized by membrane proteins of other cells. This type of cell-cell binding is usually short-lived compared to that shown in (e).

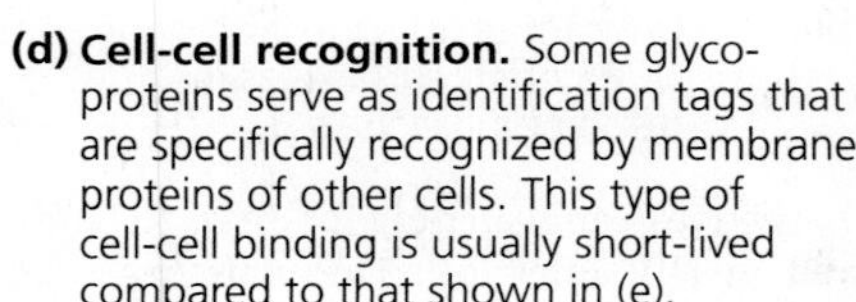

(e) Intercellular joining. Membrane proteins of adjacent cells may hook together in various kinds of junctions, such as gap junctions or tight junctions (see Figure 6.32). This type of binding is more long-lasting than that shown in (d).

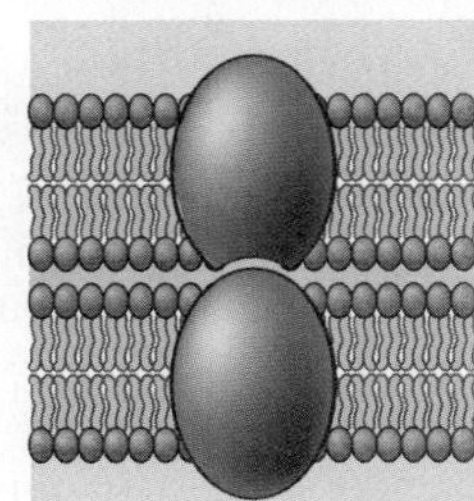

(f) Attachment to the cytoskeleton and extracellular matrix (ECM). Microfilaments or other elements of the cytoskeleton may be noncovalently bound to membrane proteins, a function that helps maintain cell shape and stabilizes the location of certain membrane proteins. Proteins that can bind to ECM molecules can coordinate extracellular and intracellular changes (see Figure 6.30).

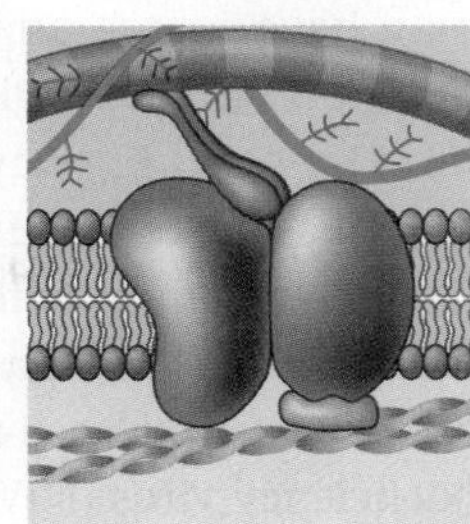

▲ **Figure 7.10 Some functions of membrane proteins.** In many cases, a single protein performs multiple tasks.

? *Some transmembrane proteins can bind to a particular ECM molecule and, when bound, transmit a signal into the cell. Use the proteins shown here to explain how this might occur.*

▼ Figure 7.11
IMPACT

Blocking HIV Entry into Cells as a Treatment for HIV Infections

Despite multiple exposures to HIV, a small number of people do not develop AIDS and show no evidence of HIV-infected cells. Comparing their genes with the genes of infected individuals, researchers discovered that resistant individuals have an unusual form of a gene that codes for an immune cell-surface protein called CCR5. Further work showed that HIV binds to a main protein receptor (CD4) on an immune cell, but most types of HIV also need to bind to CCR5 as a "co-receptor" to actually infect the cell (below, left). An absence of CCR5 on the cells of resistant individuals, due to the gene alteration, prevents the virus from entering the cells (below, right).

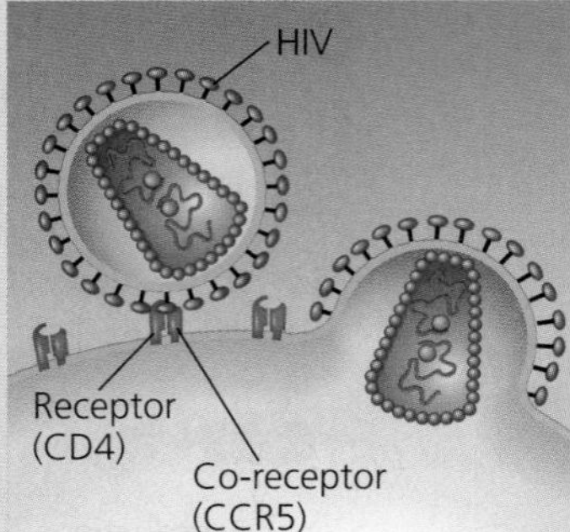

HIV can infect a cell that has CCR5 on its surface, as in most people.

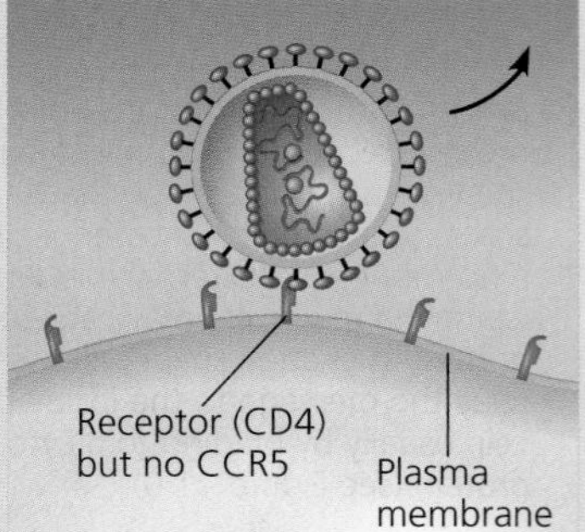

HIV cannot infect a cell lacking CCR5 on its surface, as in resistant individuals.

WHY IT MATTERS Researchers have been searching for drugs to block cell-surface receptors involved in HIV infection. The main receptor protein, CD4, performs many important functions for cells, so interfering with it could cause dangerous side effects. Discovery of the CCR5 co-receptor provided a safer target for development of drugs that mask CCR5 and block HIV entry. One such drug, maraviroc (brand name Selzentry), was approved for treatment of HIV infection in 2007.

FURTHER READING T. Kenakin, New bull's-eyes for drugs, *Scientific American* 293(4):50–57 (2005).

MAKE CONNECTIONS Study Figures 2.18 (p. 42) and 5.19 (p. 81), both of which show pairs of molecules binding to each other. What would you predict about CCR5 that would allow HIV to bind to it? How could a drug molecule interfere with this binding?

may have membrane proteins carrying out several of these functions, and a single membrane protein may have multiple functions. In this way, the membrane is a functional mosaic as well as a structural one.

Proteins on the surface of a cell are important in the medical field because some proteins can help outside agents invade the cell. For example, cell-surface proteins help the human immunodeficiency virus (HIV) infect immune system cells, leading to acquired immune deficiency syndrome (AIDS). (You'll read more about HIV in Chapter 19.) Learning about the proteins that HIV binds to on immune cells has been central to developing a treatment for HIV infection (**Figure 7.11**).

The Role of Membrane Carbohydrates in Cell-Cell Recognition

Cell-cell recognition, a cell's ability to distinguish one type of neighboring cell from another, is crucial to the functioning of an organism. It is important, for example, in the sorting of cells into tissues and organs in an animal embryo. It is also the basis for the rejection of foreign cells by the immune system, an important line of defense in vertebrate animals (see Chapter 43). Cells recognize other cells by binding to molecules, often containing carbohydrates, on the extracellular surface of the plasma membrane (see Figure 7.10d).

Membrane carbohydrates are usually short, branched chains of fewer than 15 sugar units. Some are covalently bonded to lipids, forming molecules called **glycolipids**. (Recall that *glyco* refers to the presence of carbohydrate.) However, most are covalently bonded to proteins, which are thereby **glycoproteins** (see Figure 7.5).

The carbohydrates on the extracellular side of the plasma membrane vary from species to species, among individuals of the same species, and even from one cell type to another in a single individual. The diversity of the molecules and their location on the cell's surface enable membrane carbohydrates to function as markers that distinguish one cell from another. For example, the four human blood types designated A, B, AB, and O reflect variation in the carbohydrate part of glycoproteins on the surface of red blood cells.

Synthesis and Sidedness of Membranes

Membranes have distinct inside and outside faces. The two lipid layers may differ in specific lipid composition, and each protein has directional orientation in the membrane (see Figure 7.9). **Figure 7.12** shows how membrane sidedness arises: The asymmetrical arrangement of proteins, lipids, and their associated carbohydrates in the plasma membrane is determined as the membrane is being built by the endoplasmic reticulum (ER) and Golgi apparatus.

CONCEPT CHECK 7.1

1. The carbohydrates attached to some proteins and lipids of the plasma membrane are added as the membrane is made and refined in the ER and Golgi apparatus. The new membrane then forms transport vesicles that travel to the cell surface. On which side of the vesicle membrane are the carbohydrates?
2. **WHAT IF?** The soil immediately around hot springs is much warmer than that in neighboring regions. Two closely related species of native grasses are found, one in the warmer region and one in the cooler region. If you analyzed their membrane lipid compositions, what would you expect to find? Explain.

For suggested answers, see Appendix A.

▼ **Figure 7.12 Synthesis of membrane components and their orientation in the membrane.** The cytoplasmic (orange) face of the plasma membrane differs from the extracellular (aqua) face. The latter arises from the inside face of ER, Golgi, and vesicle membranes.

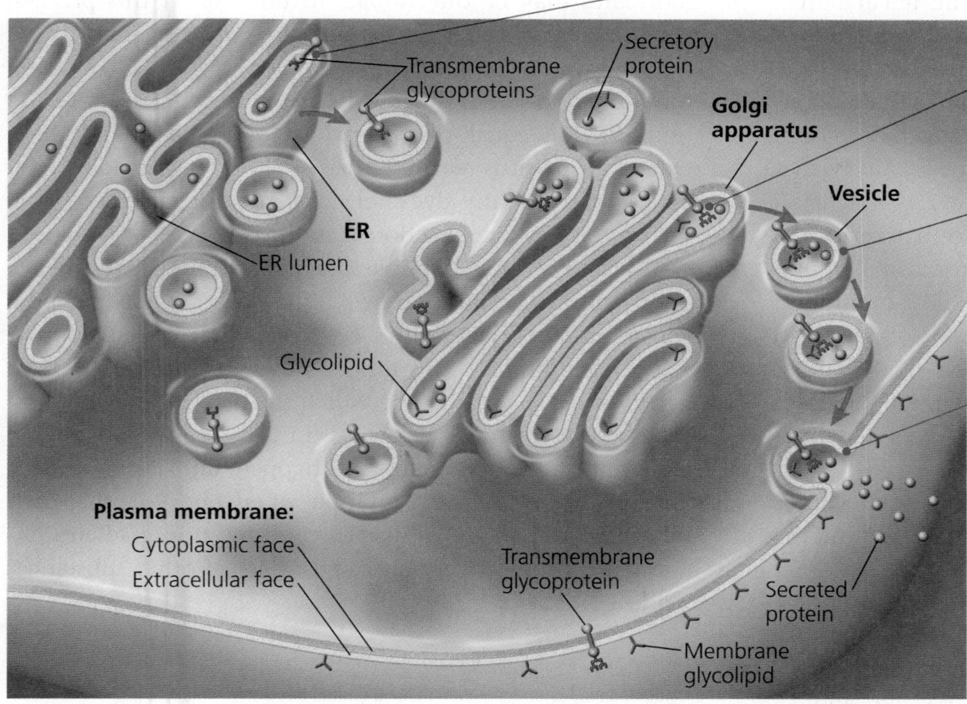

❶ Membrane proteins and lipids are synthesized in the endoplasmic reticulum (ER). Carbohydrates (green) are added to the transmembrane proteins (purple dumbbells), making them glycoproteins. The carbohydrate portions may then be modified.

❷ Inside the Golgi apparatus, the glycoproteins undergo further carbohydrate modification, and lipids acquire carbohydrates, becoming glycolipids.

❸ The glycoproteins, glycolipids, and secretory proteins (purple spheres) are transported in vesicles to the plasma membrane.

❹ As vesicles fuse with the plasma membrane, the outside face of the vesicle becomes continuous with the inside (cytoplasmic) face of the plasma membrane. This releases the secretory proteins from the cell, a process called *exocytosis*, and positions the carbohydrates of membrane glycoproteins and glycolipids on the outside (extracellular) face of the plasma membrane.

DRAW IT *Draw an integral membrane protein extending from partway through the ER membrane into the ER lumen. Next, draw the protein where it would be located in a series of numbered steps ending at the plasma membrane. Would the protein contact the cytoplasm or the extracellular fluid?*

CONCEPT 7.2

Membrane structure results in selective permeability

The biological membrane is an exquisite example of a supramolecular structure—many molecules ordered into a higher level of organization—with emergent properties beyond those of the individual molecules. The remainder of this chapter focuses on one of the most important of those properties: the ability to regulate transport across cellular boundaries, a function essential to the cell's existence. We will see once again that form fits function: The fluid mosaic model helps explain how membranes regulate the cell's molecular traffic.

A steady traffic of small molecules and ions moves across the plasma membrane in both directions. Consider the chemical exchanges between a muscle cell and the extracellular fluid that bathes it. Sugars, amino acids, and other nutrients enter the cell, and metabolic waste products leave it. The cell takes in O_2 for use in cellular respiration and expels CO_2. Also, the cell regulates its concentrations of inorganic ions, such as Na^+, K^+, Ca^{2+}, and Cl^-, by shuttling them one way or the other across the plasma membrane. In spite of heavy traffic through them, cell membranes are selectively permeable, and substances do not cross the barrier indiscriminately. The cell is able to take up some small molecules and ions and exclude others. Also, substances that move through the membrane do so at different rates.

The Permeability of the Lipid Bilayer

Nonpolar molecules, such as hydrocarbons, carbon dioxide, and oxygen, are hydrophobic and can therefore dissolve in the lipid bilayer of the membrane and cross it easily, without the aid of membrane proteins. However, the hydrophobic interior of the membrane impedes the direct passage of ions and polar molecules, which are hydrophilic, through the membrane. Polar molecules such as glucose and other sugars pass only slowly through a lipid bilayer, and even water, an extremely small polar molecule, does not cross very rapidly. A charged atom or molecule and its surrounding shell of water (see Figure 3.7) find the hydrophobic interior of the membrane even more difficult to penetrate. Furthermore, the lipid bilayer is only one aspect of the gatekeeper system responsible for the selective permeability of a cell. Proteins built into the membrane play key roles in regulating transport.

Transport Proteins

Cell membranes *are* permeable to specific ions and a variety of polar molecules. These hydrophilic substances can avoid contact with the lipid bilayer by passing through **transport proteins** that span the membrane.

Some transport proteins, called *channel proteins*, function by having a hydrophilic channel that certain molecules or atomic ions use as a tunnel through the membrane (see Figure 7.10a, left). For example, the passage of water molecules through the membrane in certain cells is greatly facilitated by channel proteins known as **aquaporins**. Each aquaporin allows entry of up to *3 billion* (3×10^9) water molecules per second, passing single file through its central channel, which fits ten at a time. Without aquaporins, only a tiny fraction of these water molecules would pass through the same area of the cell membrane in a second, so the channel protein brings about a tremendous increase in rate. Other transport proteins, called *carrier proteins*, hold onto their passengers and change shape in a way that shuttles them across the membrane (see Figure 7.10a, right). A transport protein is specific for the substance it translocates (moves), allowing only a certain substance (or a small group of related substances) to cross the membrane. For example, a specific carrier protein in the plasma membrane of red blood cells transports glucose across the membrane 50,000 times faster than glucose can pass through on its own. This "glucose transporter" is so selective that it even rejects fructose, a structural isomer of glucose.

Thus, the selective permeability of a membrane depends on both the discriminating barrier of the lipid bilayer and the specific transport proteins built into the membrane. But what establishes the *direction* of traffic across a membrane? At a given time, what determines whether a particular substance will enter the cell or leave the cell? And what mechanisms actually drive molecules across membranes? We will address these questions next as we explore two modes of membrane traffic: passive transport and active transport.

CONCEPT CHECK 7.2

1. Two molecules that can cross a lipid bilayer without help from membrane proteins are O_2 and CO_2. What property allows this to occur?
2. Why is a transport protein needed to move water molecules rapidly and in large quantities across a membrane?
3. **MAKE CONNECTIONS** Aquaporins exclude passage of hydronium ions (H_3O^+; see pp. 52–53). Recent research on fat metabolism has shown that some aquaporins allow passage of glycerol, a three-carbon alcohol (see Figure 5.10, p. 75), as well as H_2O. Since H_3O^+ is much closer in size to water than is glycerol, what do you suppose is the basis of this selectivity?

For suggested answers, see Appendix A.

CONCEPT 7.3

Passive transport is diffusion of a substance across a membrane with no energy investment

Molecules have a type of energy called thermal energy (heat), due to their constant motion. One result of this motion is **diffusion**, the movement of molecules of any substance so that they spread out evenly into the available space. Each molecule moves randomly, yet diffusion of a *population* of molecules may be directional. To understand this process, let's imagine a synthetic membrane separating pure water from a solution of a dye in water. Study **Figure 7.13a** carefully to appreciate how diffusion would result in both solutions having equal concentrations of the dye molecules. Once that point is reached, there will be a dynamic equilibrium, with as many dye molecules crossing the membrane each second in one direction as in the other.

We can now state a simple rule of diffusion: In the absence of other forces, a substance will diffuse from where it is more concentrated to where it is less concentrated. Put another way, any substance will diffuse down its **concentration gradient**, the region along which the density of a chemical substance increases or decreases (in this case, decreases). No work must be done to make this happen; diffusion is a spontaneous process, needing no input of energy. Note that each substance diffuses down its *own* concentration gradient, unaffected by the concentration gradients of other substances (**Figure 7.13b**).

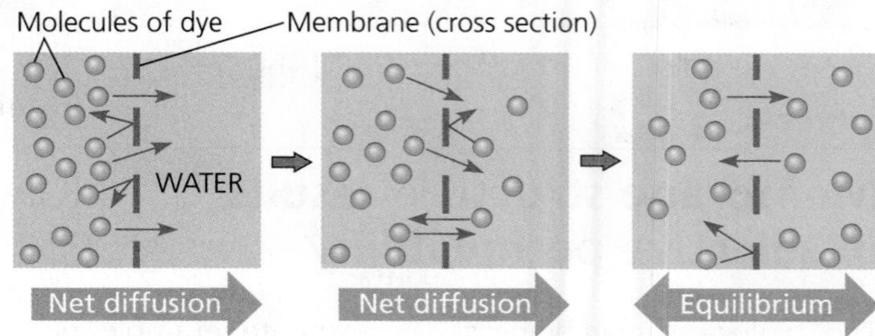

(a) Diffusion of one solute. The membrane has pores large enough for molecules of dye to pass through. Random movement of dye molecules will cause some to pass through the pores; this will happen more often on the side with more dye molecules. The dye diffuses from where it is more concentrated to where it is less concentrated (called diffusing down a concentration gradient). This leads to a dynamic equilibrium: The solute molecules continue to cross the membrane, but at equal rates in both directions.

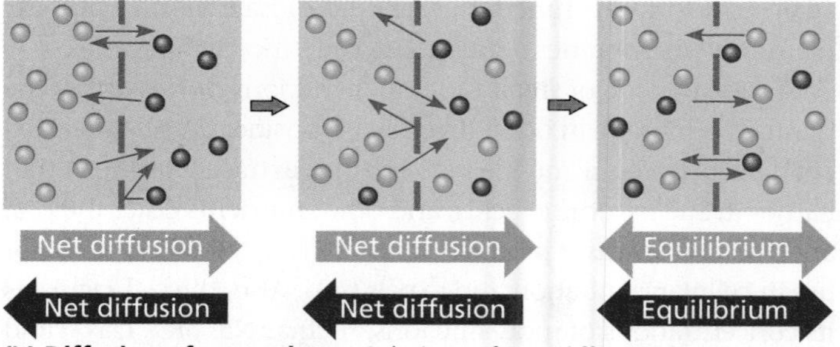

(b) Diffusion of two solutes. Solutions of two different dyes are separated by a membrane that is permeable to both. Each dye diffuses down its own concentration gradient. There will be a net diffusion of the purple dye toward the left, even though the *total* solute concentration was initially greater on the left side.

▲ **Figure 7.13 The diffusion of solutes across a synthetic membrane.** Each of the large arrows under the diagrams shows the net diffusion of the dye molecules of that color.

Much of the traffic across cell membranes occurs by diffusion. When a substance is more concentrated on one side of a membrane than on the other, there is a tendency for the substance to diffuse across the membrane down its concentration gradient (assuming that the membrane is permeable to that substance). One important example is the uptake of oxygen by a cell performing cellular respiration. Dissolved oxygen diffuses into the cell across the plasma membrane. As long as cellular respiration consumes the O_2 as it enters, diffusion into the cell will continue because the concentration gradient favors movement in that direction.

The diffusion of a substance across a biological membrane is called **passive transport** because the cell does not have to expend energy to make it happen. The concentration gradient itself represents potential energy (see Chapter 2, p. 35) and drives diffusion. Remember, however, that membranes are selectively permeable and therefore have different effects on the rates of diffusion of various molecules. In the case of water, aquaporins allow water to diffuse very rapidly across the membranes of certain cells. As we'll see next, the movement of water across the plasma membrane has important consequences for cells.

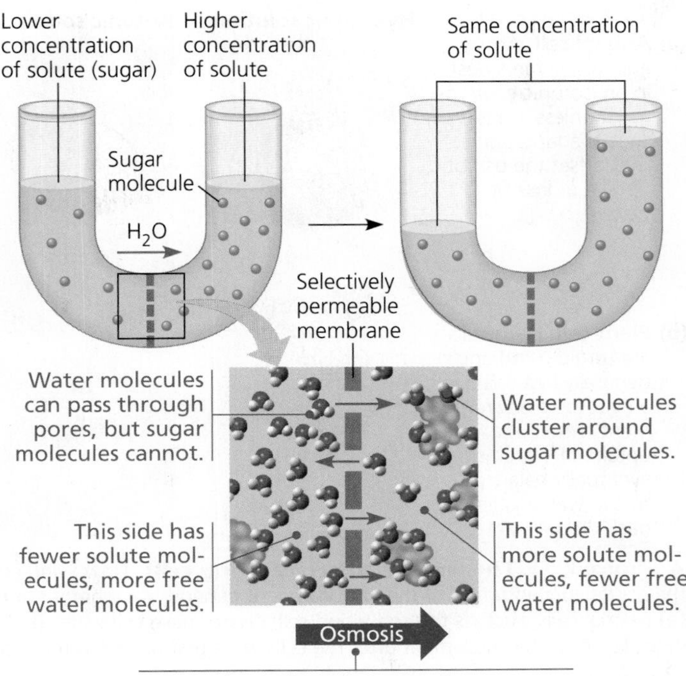

▲ **Figure 7.14 Osmosis.** Two sugar solutions of different concentrations are separated by a membrane that the solvent (water) can pass through but the solute (sugar) cannot. Water molecules move randomly and may cross in either direction, but overall, water diffuses from the solution with less concentrated solute to that with more concentrated solute. This diffusion of water, or osmosis, equalizes the sugar concentrations on both sides.

WHAT IF? *If an orange dye capable of passing through the membrane was added to the left side of the tube above, how would it be distributed at the end of the experiment? (See Figure 7.13.) Would the final solution levels in the tube be affected?*

Effects of Osmosis on Water Balance

To see how two solutions with different solute concentrations interact, picture a U-shaped glass tube with a selectively permeable artificial membrane separating two sugar solutions (**Figure 7.14**). Pores in this synthetic membrane are too small for sugar molecules to pass through but large enough for water molecules. How does this affect the *water* concentration? It seems logical that the solution with the higher concentration of solute would have the lower concentration of water and that water would diffuse into it from the other side for that reason. However, for a dilute solution like most biological fluids, solutes do not affect the water concentration significantly. Instead, tight clustering of water molecules around the hydrophilic solute molecules makes some of the water unavailable to cross the membrane. It is the difference in *free* water concentration that is important. In the end, the effect is the same: Water diffuses across the membrane from the region of lower solute concentration (higher free water concentration) to that of higher solute concentration (lower free water concentration) until the solute concentrations on both sides of the membrane are equal. The diffusion of free water across a selectively permeable membrane, whether artificial or cellular, is called **osmosis**. The movement of water across cell membranes and the balance of water between the cell and its environment are crucial to organisms. Let's now apply to living cells what we have learned about osmosis in artificial systems.

Water Balance of Cells Without Walls

To explain the behavior of a cell in a solution, we must consider both solute concentration and membrane permeability. Both factors are taken into account in the concept of **tonicity**, the ability of a surrounding solution to cause a cell to gain or lose water. The tonicity of a solution depends in part on its concentration of solutes that cannot cross the membrane (nonpenetrating solutes) relative to that inside the cell. If there is a higher concentration of nonpenetrating solutes in the surrounding solution, water will tend to leave the cell, and vice versa.

If a cell without a wall, such as an animal cell, is immersed in an environment that is **isotonic** to the cell (*iso* means "same"), there will be no *net* movement of water across the plasma membrane. Water diffuses across the membrane, but at the same rate in both directions. In an isotonic environment, the volume of an animal cell is stable (**Figure 7.15a**, on the next page).

Now let's transfer the cell to a solution that is **hypertonic** to the cell (*hyper* means "more," in this case referring to nonpenetrating solutes). The cell will lose water, shrivel, and probably die. This is one way an increase in the salinity (saltiness) of a lake can kill animals there; if the lake water becomes hypertonic to the animals' cells, the cells might shrivel and

(a) Animal cell. An animal cell fares best in an isotonic environment unless it has special adaptations that offset the osmotic uptake or loss of water.

(b) Plant cell. Plant cells are turgid (firm) and generally healthiest in a hypotonic environment, where the uptake of water is eventually balanced by the wall pushing back on the cell.

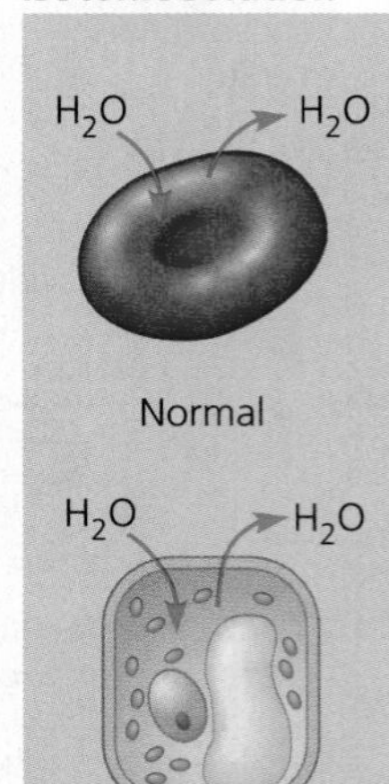

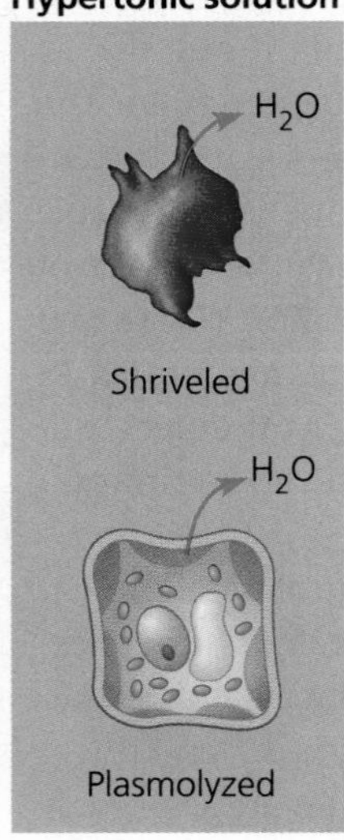

▲ **Figure 7.15 The water balance of living cells.** How living cells react to changes in the solute concentration of their environment depends on whether or not they have cell walls. **(a)** Animal cells, such as this red blood cell, do not have cell walls. **(b)** Plant cells do. (Arrows indicate net water movement after the cells were first placed in these solutions.)

die. However, taking up too much water can be just as hazardous to an animal cell as losing water. If we place the cell in a solution that is **hypotonic** to the cell (*hypo* means "less"), water will enter the cell faster than it leaves, and the cell will swell and lyse (burst) like an overfilled water balloon.

A cell without rigid walls can tolerate neither excessive uptake nor excessive loss of water. This problem of water balance is automatically solved if such a cell lives in isotonic surroundings. Seawater is isotonic to many marine invertebrates. The cells of most terrestrial (land-dwelling) animals are bathed in an extracellular fluid that is isotonic to the cells. In hypertonic or hypotonic environments, however, organisms that lack rigid cell walls must have other adaptations for **osmoregulation**, the control of solute concentrations and water balance. For example, the unicellular protist *Paramecium caudatum* lives in pond water, which is hypotonic to the cell. *P. caudatum* has a plasma membrane that is much less permeable to water than the membranes of most other cells, but this only slows the uptake of water, which continually enters the cell. The *P. caudatum* cell doesn't burst because it is also equipped with a contractile vacuole, an organelle that functions as a bilge pump to force water out of the cell as fast as it enters by osmosis **(Figure 7.16)**. We will examine other evolutionary adaptations for osmoregulation in Chapter 44.

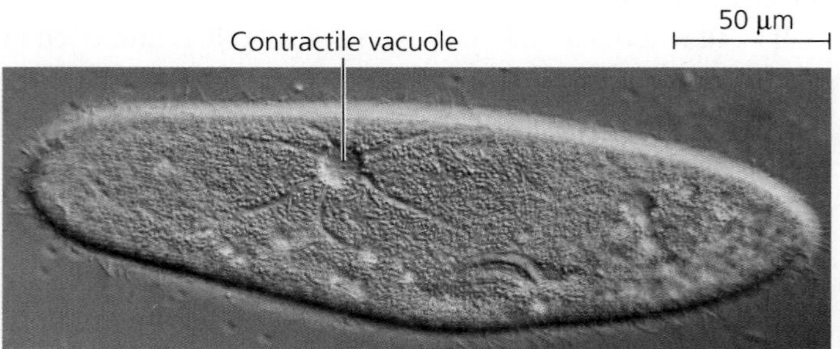

▲ **Figure 7.16 The contractile vacuole of *Paramecium caudatum*.** The vacuole collects fluid from a system of canals in the cytoplasm. When full, the vacuole and canals contract, expelling fluid from the cell (LM).

Water Balance of Cells with Walls

The cells of plants, prokaryotes, fungi, and some protists are surrounded by walls (see Figure 6.28). When such a cell is immersed in a hypotonic solution—bathed in rainwater, for example—the wall helps maintain the cell's water balance. Consider a plant cell. Like an animal cell, the plant cell swells as water enters by osmosis **(Figure 7.15b)**. However, the relatively inelastic wall will expand only so much before it exerts a back pressure on the cell, called *turgor pressure*, that opposes further water uptake. At this point, the cell is **turgid** (very firm), which is the healthy state for most plant cells. Plants that are not woody, such as most houseplants, depend for mechanical support on cells kept turgid by a surrounding hypotonic solution. If a plant's cells and their surroundings are isotonic, there is no net tendency for water to enter, and the cells become **flaccid** (limp).

However, a wall is of no advantage if the cell is immersed in a hypertonic environment. In this case, a plant cell, like an animal cell, will lose water to its surroundings and shrink. As the plant cell shrivels, its plasma membrane pulls away from the wall. This phenomenon, called **plasmolysis**, causes the plant to wilt and can lead to plant death. The walled cells of bacteria and fungi also plasmolyze in hypertonic environments.

Facilitated Diffusion: Passive Transport Aided by Proteins

Let's look more closely at how water and certain hydrophilic solutes cross a membrane. As mentioned earlier, many polar molecules and ions impeded by the lipid bilayer of the membrane diffuse passively with the help of transport proteins that span the membrane. This phenomenon is called **facilitated diffusion**. Cell biologists are still trying to learn exactly how various transport proteins facilitate diffusion. Most transport proteins are very specific: They transport some substances but not others.

As described earlier, the two types of transport proteins are channel proteins and carrier proteins. Channel proteins simply provide corridors that allow specific molecules or ions to cross the membrane **(Figure 7.17a)**. The hydrophilic passageways

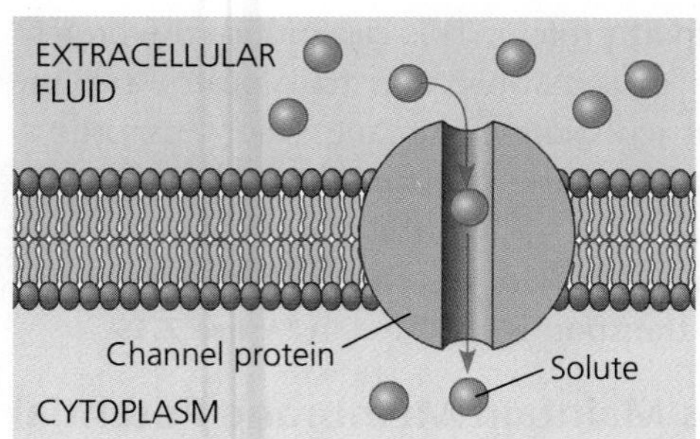

(a) A channel protein (purple) has a channel through which water molecules or a specific solute can pass.

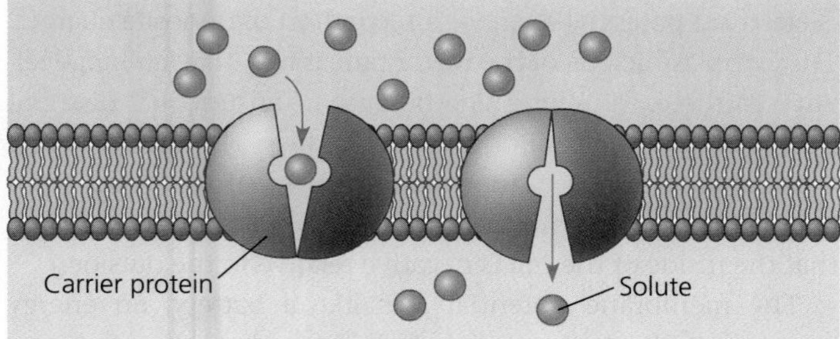

(b) A carrier protein alternates between two shapes, moving a solute across the membrane during the shape change.

▲ **Figure 7.17 Two types of transport proteins that carry out facilitated diffusion.** In both cases, the protein can transport the solute in either direction, but the net movement is down the concentration gradient of the solute.

provided by these proteins can allow water molecules or small ions to diffuse very quickly from one side of the membrane to the other. Aquaporins, the water channel proteins, facilitate the massive amounts of diffusion that occur in plant cells and in animal cells such as red blood cells (see Figure 7.15). Certain kidney cells also have a high number of aquaporins, allowing them to reclaim water from urine before it is excreted. If the kidneys did not perform this function, you would excrete about 180 L of urine per day—and have to drink an equal volume of water!

Channel proteins that transport ions are called **ion channels**. Many ion channels function as **gated channels**, which open or close in response to a stimulus. For some gated channels, the stimulus is electrical. The ion channel shown in Figure 7.1, for example, opens in response to an electrical stimulus, allowing potassium ions to leave the cell. Other gated channels open or close when a specific substance other than the one to be transported binds to the channel. Both types of gated channels are important in the functioning of the nervous system, as you'll learn in Chapter 48.

Carrier proteins, such as the glucose transporter mentioned earlier, seem to undergo a subtle change in shape that somehow translocates the solute-binding site across the membrane (**Figure 7.17b**). Such a change in shape may be triggered by the binding and release of the transported molecule. Like ion channels, carrier proteins involved in facilitated diffusion result in the net movement of a substance down its concentration gradient. No energy input is thus required: This is passive transport.

In certain inherited diseases, specific transport systems are either defective or missing altogether. An example is cystinuria, a human disease characterized by the absence of a carrier protein that transports cysteine and some other amino acids across the membranes of kidney cells. Kidney cells normally reabsorb these amino acids from the urine and return them to the blood, but an individual afflicted with cystinuria develops painful stones from amino acids that accumulate and crystallize in the kidneys.

CONCEPT CHECK 7.3

1. How do you think a cell performing cellular respiration rids itself of the resulting CO_2?
2. In the supermarket, produce is often sprayed with water. Explain why this makes vegetables look crisp.
3. **WHAT IF?** If a *Paramecium caudatum* swims from a hypotonic to an isotonic environment, will its contractile vacuole become more active or less? Why?

For suggested answers, see Appendix A.

CONCEPT 7.4

Active transport uses energy to move solutes against their gradients

Despite the help of transport proteins, facilitated diffusion is considered passive transport because the solute is moving down its concentration gradient, a process that requires no energy. Facilitated diffusion speeds transport of a solute by providing efficient passage through the membrane, but it does not alter the direction of transport. Some transport proteins, however, can move solutes against their concentration gradients, across the plasma membrane from the side where they are less concentrated (whether inside or outside) to the side where they are more concentrated.

The Need for Energy in Active Transport

To pump a solute across a membrane against its gradient requires work; the cell must expend energy. Therefore, this type of membrane traffic is called **active transport**. The transport proteins that move solutes against their concentration gradients are all carrier proteins rather than channel proteins. This makes sense because when channel proteins are open, they merely allow solutes to diffuse down their concentration gradients rather than picking them up and transporting them against their gradients.

Active transport enables a cell to maintain internal concentrations of small solutes that differ from concentrations in its environment. For example, compared with its surroundings,

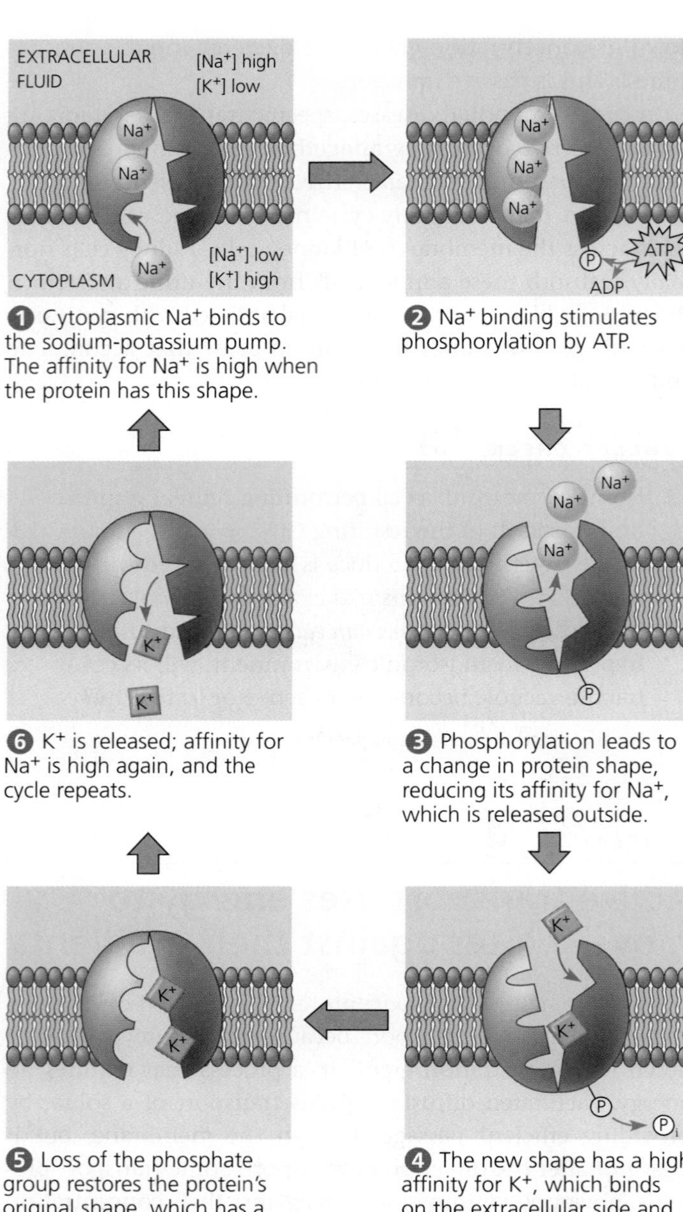

1 Cytoplasmic Na^+ binds to the sodium-potassium pump. The affinity for Na^+ is high when the protein has this shape.

2 Na^+ binding stimulates phosphorylation by ATP.

3 Phosphorylation leads to a change in protein shape, reducing its affinity for Na^+, which is released outside.

4 The new shape has a high affinity for K^+, which binds on the extracellular side and triggers release of the phosphate group.

5 Loss of the phosphate group restores the protein's original shape, which has a lower affinity for K^+.

6 K^+ is released; affinity for Na^+ is high again, and the cycle repeats.

▲ **Figure 7.18 The sodium-potassium pump: a specific case of active transport.** This transport system pumps ions against steep concentration gradients: Sodium ion concentration ($[Na^+]$) is high outside the cell and low inside, while potassium ion concentration ($[K^+]$) is low outside the cell and high inside. The pump oscillates between two shapes in a cycle that moves 3 Na^+ out of the cell for every 2 K^+ pumped into the cell. The two shapes have different affinities for Na^+ and K^+. ATP powers the shape change by transferring a phosphate group to the transport protein (phosphorylating the protein).

an animal cell has a much higher concentration of potassium ions (K^+) and a much lower concentration of sodium ions (Na^+). The plasma membrane helps maintain these steep gradients by pumping Na^+ out of the cell and K^+ into the cell.

As in other types of cellular work, ATP supplies the energy for most active transport. One way ATP can power active transport is by transferring its terminal phosphate group directly to the transport protein. This can induce the protein to change its shape in a manner that translocates a solute bound to the protein across the membrane. One transport system that works this way is the **sodium-potassium pump**, which exchanges Na^+ for K^+ across the plasma membrane of animal cells (**Figure 7.18**). The distinction between passive transport and active transport is reviewed in **Figure 7.19**.

How Ion Pumps Maintain Membrane Potential

All cells have voltages across their plasma membranes. Voltage is electrical potential energy—a separation of opposite charges. The cytoplasmic side of the membrane is negative in charge relative to the extracellular side because of an unequal distribution of anions and cations on the two sides. The voltage across a membrane, called a **membrane potential**, ranges from about -50 to -200 millivolts (mV). (The minus sign indicates that the inside of the cell is negative relative to the outside.)

The membrane potential acts like a battery, an energy source that affects the traffic of all charged substances across the membrane. Because the inside of the cell is negative compared with the outside, the membrane potential favors the passive transport of cations into the cell and anions out of the cell. Thus, *two* forces drive the diffusion of ions across a membrane: a chemical force (the ion's concentration gradient) and an electrical force (the effect of the membrane potential on

▼ **Figure 7.19 Review: passive and active transport.**

Passive transport. Substances diffuse spontaneously down their concentration gradients, crossing a membrane with no expenditure of energy by the cell. The rate of diffusion can be greatly increased by transport proteins in the membrane.

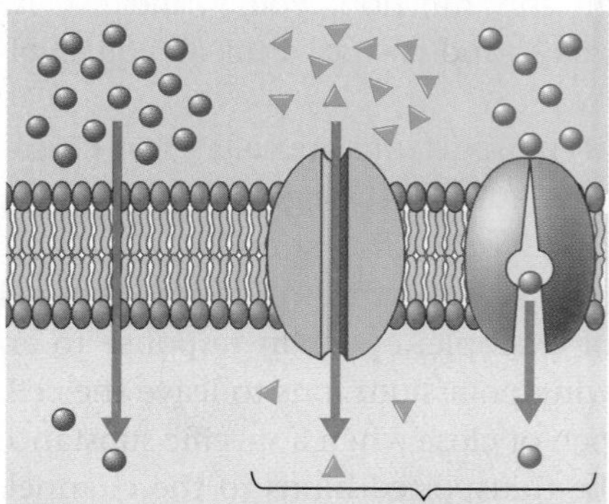

Diffusion. Hydrophobic molecules and (at a slow rate) very small uncharged polar molecules can diffuse through the lipid bilayer.

Facilitated diffusion. Many hydrophilic substances diffuse through membranes with the assistance of transport proteins, either channel proteins (left) or carrier proteins (right).

Active transport. Some transport proteins act as pumps, moving substances across a membrane against their concentration (or electrochemical) gradients. Energy for this work is usually supplied by ATP.

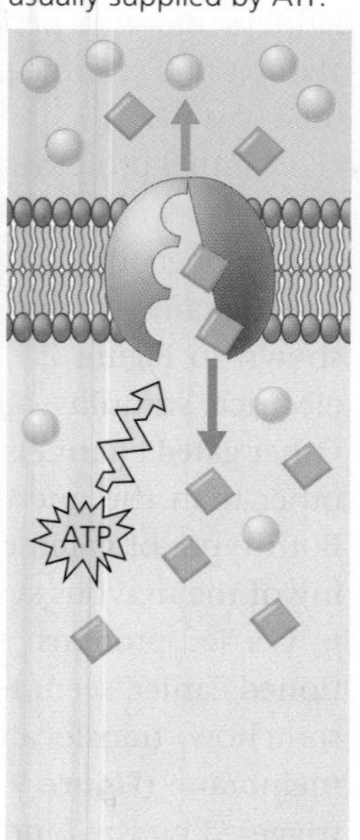

? *For each solute in the right panel, describe its direction of movement, and state whether it is going with or against its concentration gradient.*

the ion's movement). This combination of forces acting on an ion is called the **electrochemical gradient.**

In the case of ions, then, we must refine our concept of passive transport: An ion diffuses not simply down its *concentration* gradient but, more exactly, down its *electrochemical* gradient. For example, the concentration of Na^+ inside a resting nerve cell is much lower than outside it. When the cell is stimulated, gated channels open that facilitate Na^+ diffusion. Sodium ions then "fall" down their electrochemical gradient, driven by the concentration gradient of Na^+ and by the attraction of these cations to the negative side (inside) of the membrane. In this example, both electrical and chemical contributions to the electrochemical gradient act in the same direction across the membrane, but this is not always so. In cases where electrical forces due to the membrane potential oppose the simple diffusion of an ion down its concentration gradient, active transport may be necessary. In Chapter 48, you'll learn about the importance of electrochemical gradients and membrane potentials in the transmission of nerve impulses.

Some membrane proteins that actively transport ions contribute to the membrane potential. An example is the sodium-potassium pump. Notice in Figure 7.18 that the pump does not translocate Na^+ and K^+ one for one, but pumps three sodium ions out of the cell for every two potassium ions it pumps into the cell. With each "crank" of the pump, there is a net transfer of one positive charge from the cytoplasm to the extracellular fluid, a process that stores energy as voltage. A transport protein that generates voltage across a membrane is called an **electrogenic pump**. The sodium-potassium pump appears to be the major electrogenic pump of animal cells. The main electrogenic pump of plants, fungi, and bacteria is a **proton pump**, which actively transports protons (hydrogen ions, H^+) out of the cell. The pumping of H^+ transfers positive charge from the cytoplasm to the extracellular solution **(Figure 7.20)**. By generating voltage across membranes, electrogenic pumps help store energy that can be tapped for cellular work. One important use of proton gradients in the cell is for ATP synthesis during cellular respiration, as you will see in Chapter 9. Another is a type of membrane traffic called cotransport.

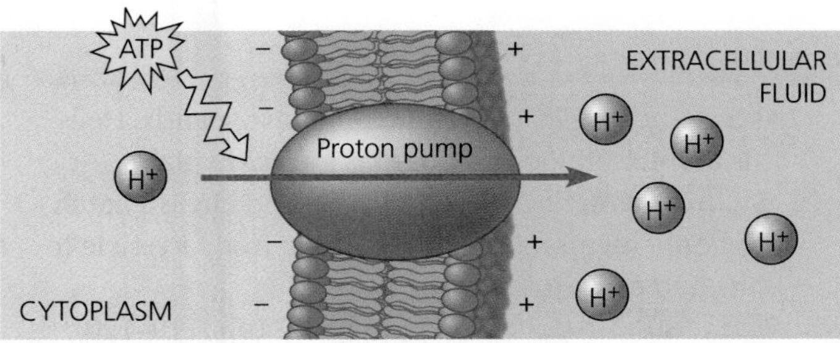

▲ **Figure 7.20 A proton pump.** Proton pumps are electrogenic pumps that store energy by generating voltage (charge separation) across membranes. A proton pump translocates positive charge in the form of hydrogen ions. The voltage and H^+ concentration gradient represent a dual energy source that can drive other processes, such as the uptake of nutrients. Most proton pumps are powered by ATP.

Cotransport: Coupled Transport by a Membrane Protein

A single ATP-powered pump that transports a specific solute can indirectly drive the active transport of several other solutes in a mechanism called **cotransport**. A substance that has been pumped across a membrane can do work as it moves back across the membrane by diffusion, analogous to water that has been pumped uphill and performs work as it flows back down. Another transport protein, a cotransporter separate from the pump, can couple the "downhill" diffusion of this substance to the "uphill" transport of a second substance against its own concentration (or electrochemical) gradient. For example, a plant cell uses the gradient of H^+ generated by its proton pumps to drive the active transport of amino acids, sugars, and several other nutrients into the cell. One transport protein couples the return of H^+ to the transport of sucrose into the cell **(Figure 7.21)**. This protein can translocate sucrose into the cell against a concentration gradient, but only if the sucrose molecule travels in the company of a hydrogen ion. The hydrogen ion uses the transport protein as an avenue to diffuse down the electrochemical gradient maintained by the proton pump. Plants use sucrose-H^+ cotransport to load sucrose produced by photosynthesis into cells in the veins of leaves. The vascular tissue of the plant can then distribute the sugar to nonphotosynthetic organs, such as roots.

What we know about cotransport proteins in animal cells has helped us find more effective treatments for diarrhea, a serious problem in developing countries. Normally, sodium in waste is reabsorbed in the colon, maintaining constant levels in the body, but diarrhea expels waste so rapidly that reabsorption is not possible, and sodium levels fall precipitously.

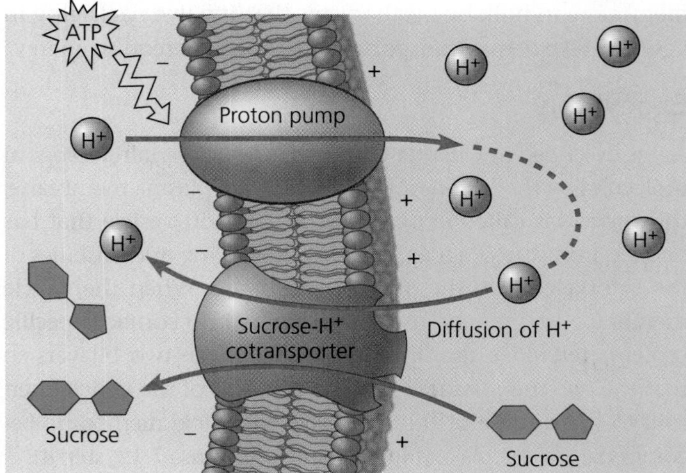

▲ **Figure 7.21 Cotransport: active transport driven by a concentration gradient.** A carrier protein, such as this sucrose-H^+ cotransporter in a plant cell, is able to use the diffusion of H^+ down its electrochemical gradient into the cell to drive the uptake of sucrose. The H^+ gradient is maintained by an ATP-driven proton pump that concentrates H^+ outside the cell, thus storing potential energy that can be used for active transport, in this case of sucrose. Thus, ATP indirectly provides the energy necessary for cotransport. (The cell wall is not shown.)

To treat this life-threatening condition, patients are given a solution to drink containing high concentrations of salt (NaCl) and glucose. The solutes are taken up by sodium-glucose cotransporters on the surface of intestinal cells and passed through the cells into the blood. This simple treatment has lowered infant mortality worldwide.

CONCEPT CHECK 7.4

1. Sodium-potassium pumps help nerve cells establish a voltage across their plasma membranes. Do these pumps use ATP or produce ATP? Explain.
2. Explain why the sodium-potassium pump in Figure 7.18 would not be considered a cotransporter.
3. **MAKE CONNECTIONS** Review the characteristics of the lysosome in Concept 6.4 (pp. 106–107). Given the internal environment of a lysosome, what transport protein might you expect to see in its membrane?

For suggested answers, see Appendix A.

CONCEPT 7.5

Bulk transport across the plasma membrane occurs by exocytosis and endocytosis

Water and small solutes enter and leave the cell by diffusing through the lipid bilayer of the plasma membrane or by being pumped or moved across the membrane by transport proteins. However, large molecules, such as proteins and polysaccharides, as well as larger particles, generally cross the membrane in bulk by mechanisms that involve packaging in vesicles. Like active transport, these processes require energy.

Exocytosis

As we described in Chapter 6, the cell secretes certain biological molecules by the fusion of vesicles with the plasma membrane; this process is called **exocytosis**. A transport vesicle that has budded from the Golgi apparatus moves along microtubules of the cytoskeleton to the plasma membrane. When the vesicle membrane and plasma membrane come into contact, specific proteins rearrange the lipid molecules of the two bilayers so that the two membranes fuse. The contents of the vesicle then spill to the outside of the cell, and the vesicle membrane becomes part of the plasma membrane (see Figure 7.12, step 4).

Many secretory cells use exocytosis to export products. For example, the cells in the pancreas that make insulin secrete it into the extracellular fluid by exocytosis. In another example, neurons (nerve cells) use exocytosis to release neurotransmitters that signal other neurons or muscle cells. When plant cells are making walls, exocytosis delivers proteins and carbohydrates from Golgi vesicles to the outside of the cell.

Endocytosis

In **endocytosis**, the cell takes in biological molecules and particulate matter by forming new vesicles from the plasma membrane. Although the proteins involved in the processes are different, the events of endocytosis look like the reverse of exocytosis. A small area of the plasma membrane sinks inward to form a pocket. As the pocket deepens, it pinches in, forming a vesicle containing material that had been outside the cell. Study **Figure 7.22** carefully to understand the three types of endocytosis: phagocytosis ("cellular eating"), pinocytosis ("cellular drinking"), and receptor-mediated endocytosis.

Human cells use receptor-mediated endocytosis to take in cholesterol for membrane synthesis and the synthesis of other steroids. Cholesterol travels in the blood in particles called low-density lipoproteins (LDLs), each a complex of lipids and a protein. LDLs bind to LDL receptors on plasma membranes and then enter the cells by endocytosis. (LDLs thus act as **ligands**, a term for any molecule that binds specifically to a receptor site on another molecule.) In humans with familial hypercholesterolemia, an inherited disease characterized by a very high level of cholesterol in the blood, LDLs cannot enter cells because the LDL receptor proteins are defective or missing. Consequently, cholesterol accumulates in the blood, where it contributes to early atherosclerosis, the buildup of lipid deposits within the walls of blood vessels. This buildup causes the walls to bulge inward, thereby narrowing the vessels and impeding blood flow.

Vesicles not only transport substances between the cell and its surroundings but also provide a mechanism for rejuvenating or remodeling the plasma membrane. Endocytosis and exocytosis occur continually in most eukaryotic cells, yet the amount of plasma membrane in a nongrowing cell remains fairly constant. Apparently, the addition of membrane by one process offsets the loss of membrane by the other.

Energy and cellular work have figured prominently in our study of membranes. We have seen, for example, that active transport is powered by ATP. In the next three chapters, you will learn more about how cells acquire chemical energy to do the work of life.

CONCEPT CHECK 7.5

1. As a cell grows, its plasma membrane expands. Does this involve endocytosis or exocytosis? Explain.
2. **DRAW IT** Return to Figure 7.12, and circle a patch of plasma membrane that is coming from a vesicle involved in exocytosis.
3. **MAKE CONNECTIONS** In Concept 6.7 (pp. 119–120), you learned that animal cells make an extracellular matrix (ECM). Describe the cellular pathway of synthesis and deposition of an ECM glycoprotein.

For suggested answers, see Appendix A.

▼ **Figure 7.22**

Exploring Endocytosis in Animal Cells

Phagocytosis

EXTRACELLULAR FLUID
Solutes
Pseudopodium
"Food" or other particle
Food vacuole
CYTOPLASM

In **phagocytosis,** a cell engulfs a particle by wrapping pseudopodia (singular, *pseudopodium*) around it and packaging it within a membranous sac called a food vacuole. The particle will be digested after the food vacuole fuses with a lysosome containing hydrolytic enzymes (see Figure 6.13a).

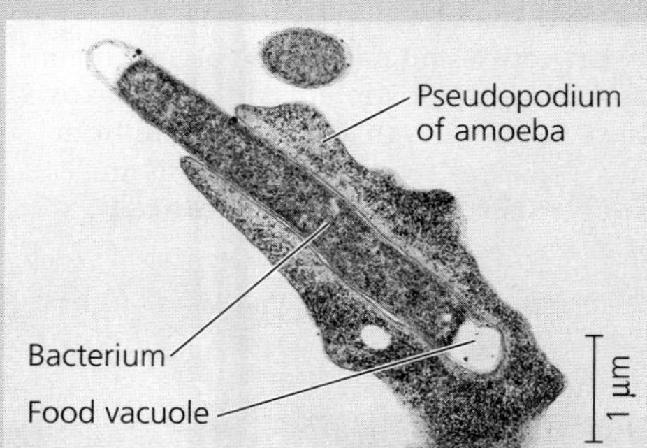

An amoeba engulfing a bacterium via phagocytosis (TEM).

Pinocytosis

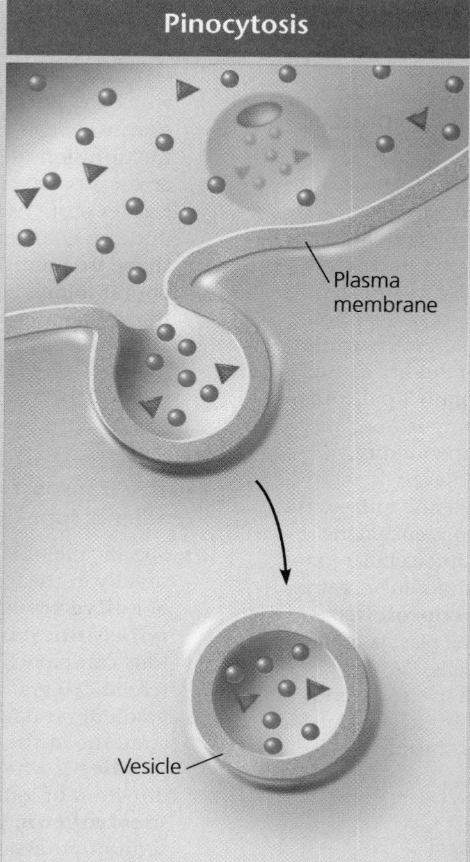

In **pinocytosis,** the cell "gulps" droplets of extracellular fluid into tiny vesicles. It is not the fluid itself that is needed by the cell, but the molecules dissolved in the droplets. Because any and all included solutes are taken into the cell, pinocytosis is nonspecific in the substances it transports.

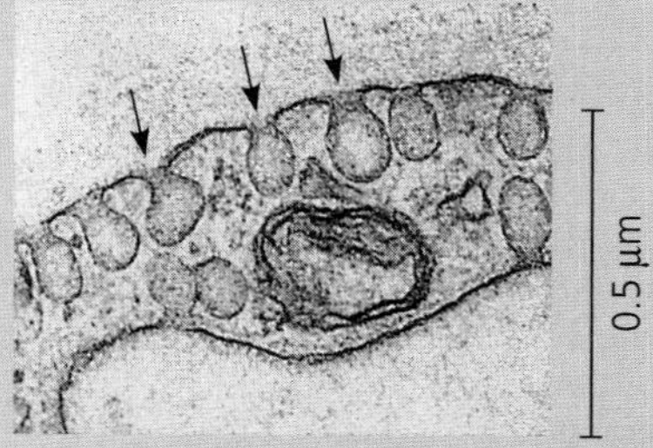

Pinocytosis vesicles forming (indicated by arrows) in a cell lining a small blood vessel (TEM).

BioFlix Visit the Study Area at **www.masteringbiology.com** for the BioFlix® 3-D Animation on Membrane Transport.

Receptor-Mediated Endocytosis

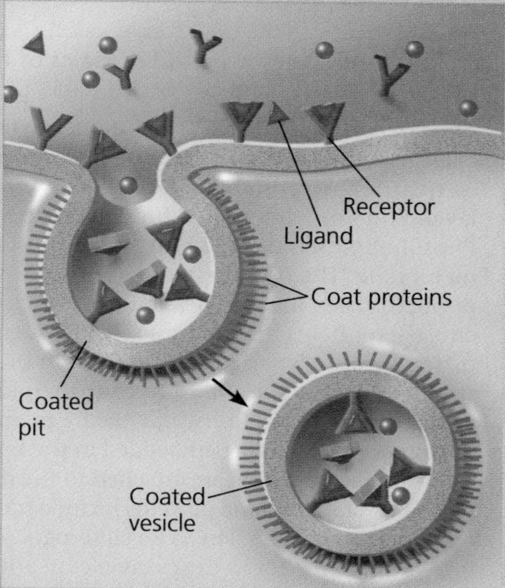

Receptor-mediated endocytosis enables the cell to acquire bulk quantities of specific substances, even though those substances may not be very concentrated in the extracellular fluid. Embedded in the membrane are proteins with specific receptor sites exposed to the extracellular fluid, to which specific substances (ligands) bind. The receptor proteins then cluster in regions of the membrane called coated pits, which are lined on their cytoplasmic side by a fuzzy layer of coat proteins. Next, each coated pit forms a vesicle containing the ligand molecules. Notice that there are relatively more bound molecules (purple) inside the vesicle, but other molecules (green) are also present. After the ingested material is liberated from the vesicle, the emptied receptors are recycled to the plasma membrane by the same vesicle.

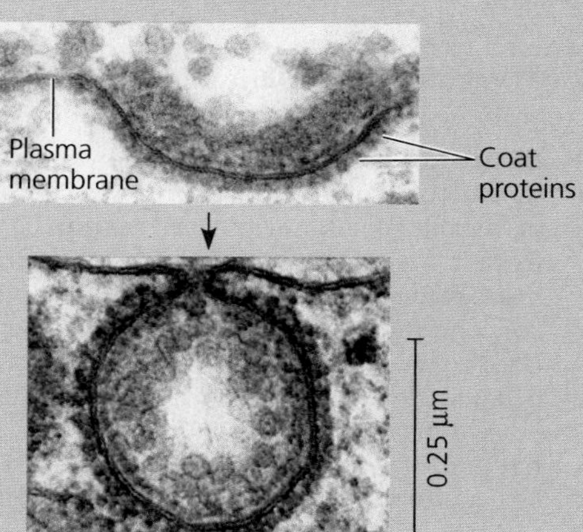

Top: A coated pit. *Bottom:* A coated vesicle forming during receptor-mediated endocytosis (TEMs).

7 CHAPTER REVIEW

SUMMARY OF KEY CONCEPTS

CONCEPT 7.1

Cellular membranes are fluid mosaics of lipids and proteins (pp. 125–131)

- The Davson-Danielli sandwich model of the membrane has been replaced by the **fluid mosaic model**, in which **amphipathic** proteins are embedded in the phospholipid bilayer. Proteins with related functions often cluster in patches.
- Phospholipids and some proteins move laterally within the membrane. The unsaturated hydrocarbon tails of some phospholipids keep membranes fluid at lower temperatures, while cholesterol helps membranes resist changes in fluidity caused by temperature changes. Differences in membrane lipid composition, as well as the ability to change lipid composition, are evolutionary adaptations that ensure membrane fluidity.
- **Integral proteins** are embedded in the lipid bilayer; **peripheral proteins** are attached to the membrane surface. The functions of membrane proteins include transport, enzymatic activity, signal transduction, cell-cell recognition, intercellular joining, and attachment to the cytoskeleton and extracellular matrix. Short chains of sugars linked to proteins (in **glycoproteins**) and lipids (in **glycolipids**) on the exterior side of the plasma membrane interact with surface molecules of other cells.
- Membrane proteins and lipids are synthesized in the ER and modified in the ER and Golgi apparatus. The inside and outside faces of membranes differ in molecular composition.

? *In what ways are membranes crucial to life?*

CONCEPT 7.2

Membrane structure results in selective permeability (pp. 131–132)

- A cell must exchange molecules and ions with its surroundings, a process controlled by the **selective permeability** of the plasma membrane. Hydrophobic substances are soluble in lipid and pass through membranes rapidly, whereas polar molecules and ions generally require specific **transport proteins** to cross the membrane.

? *How do **aquaporins** affect the permeability of a membrane?*

CONCEPT 7.3

Passive transport is diffusion of a substance across a membrane with no energy investment (pp. 132–135)

- **Diffusion** is the spontaneous movement of a substance down its **concentration gradient**. Water diffuses out through the permeable membrane of a cell (**osmosis**) if the solution outside has a higher solute concentration (**hypertonic**) than the cytosol; water enters the cell if the solution has a lower solute concentration (**hypotonic**). If the concentrations are equal (**isotonic**), no net osmosis occurs. Cell survival depends on balancing water uptake and loss. Cells lacking walls (as in animals and some protists) are isotonic with their environments or have adaptations for **osmoregulation**. Plants, prokaryotes, fungi, and some protists have relatively inelastic cell walls, so the cells don't burst in a hypotonic environment.
- In a type of **passive transport** called **facilitated diffusion**, a transport protein speeds the movement of water or a solute across a membrane down its concentration gradient. **Ion channels**, some of which are **gated channels**, facilitate the diffusion of ions across a membrane. Carrier proteins can undergo changes in shape that translocate bound solutes across the membrane.

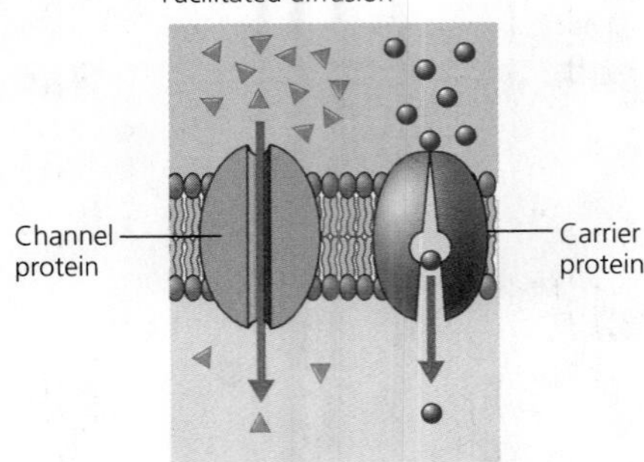

? *What happens to a cell placed in a hypertonic solution? Describe the free water concentration inside and out.*

CONCEPT 7.4

Active transport uses energy to move solutes against their gradients (pp. 135–138)

Active transport

- Specific membrane proteins use energy, usually in the form of ATP, to do the work of **active transport**. The **sodium-potassium pump** is an example.
- Ions can have both a concentration (chemical) gradient and an electrical gradient (voltage). These gradients combine in the **electrochemical gradient**, which determines the net direction of ionic diffusion. **Electrogenic pumps**, such as the sodium-potassium pump and **proton pumps**, are transport proteins that contribute to electrochemical gradients.
- **Cotransport** of two solutes occurs when a membrane protein enables the "downhill" diffusion of one solute to drive the "uphill" transport of the other.

? *ATP is not directly involved in the functioning of a cotransporter. Why, then, is cotransport considered active transport?*

CONCEPT 7.5

Bulk transport across the plasma membrane occurs by exocytosis and endocytosis (p. 138)

- In **exocytosis**, transport vesicles migrate to the plasma membrane, fuse with it, and release their contents. In **endocytosis**, molecules enter cells within vesicles that pinch inward from the plasma membrane. The three types of endocytosis are **phagocytosis**, **pinocytosis**, and **receptor-mediated endocytosis**.

? *Which type of endocytosis involves ligands? What does this type of transport enable a cell to do?*

TEST YOUR UNDERSTANDING

LEVEL 1: KNOWLEDGE/COMPREHENSION

1. In what way do the membranes of a eukaryotic cell vary?
   a. Phospholipids are found only in certain membranes.
   b. Certain proteins are unique to each membrane.

c. Only certain membranes of the cell are selectively permeable.
d. Only certain membranes are constructed from amphipathic molecules.
e. Some membranes have hydrophobic surfaces exposed to the cytoplasm, while others have hydrophilic surfaces facing the cytoplasm.

2. According to the fluid mosaic model of membrane structure, proteins of the membrane are mostly
   a. spread in a continuous layer over the inner and outer surfaces of the membrane.
   b. confined to the hydrophobic interior of the membrane.
   c. embedded in a lipid bilayer.
   d. randomly oriented in the membrane, with no fixed inside-outside polarity.
   e. free to depart from the fluid membrane and dissolve in the surrounding solution.

3. Which of the following factors would tend to increase membrane fluidity?
   a. a greater proportion of unsaturated phospholipids
   b. a greater proportion of saturated phospholipids
   c. a lower temperature
   d. a relatively high protein content in the membrane
   e. a greater proportion of relatively large glycolipids compared with lipids having smaller molecular masses

LEVEL 2: APPLICATION/ANALYSIS

4. Which of the following processes includes all others?
   a. osmosis
   b. diffusion of a solute across a membrane
   c. facilitated diffusion
   d. passive transport
   e. transport of an ion down its electrochemical gradient

5. Based on Figure 7.21, which of these experimental treatments would increase the rate of sucrose transport into the cell?
   a. decreasing extracellular sucrose concentration
   b. decreasing extracellular pH
   c. decreasing cytoplasmic pH
   d. adding an inhibitor that blocks the regeneration of ATP
   e. adding a substance that makes the membrane more permeable to hydrogen ions

6. **DRAW IT** An artificial "cell" consisting of an aqueous solution enclosed in a selectively permeable membrane is immersed in a beaker containing a different solution, the "environment," as shown below. The membrane is permeable to water and to the simple sugars glucose and fructose but impermeable to the disaccharide sucrose.
   a. Draw solid arrows to indicate the net movement of solutes into and/or out of the cell.
   b. Is the solution outside the cell isotonic, hypotonic, or hypertonic?

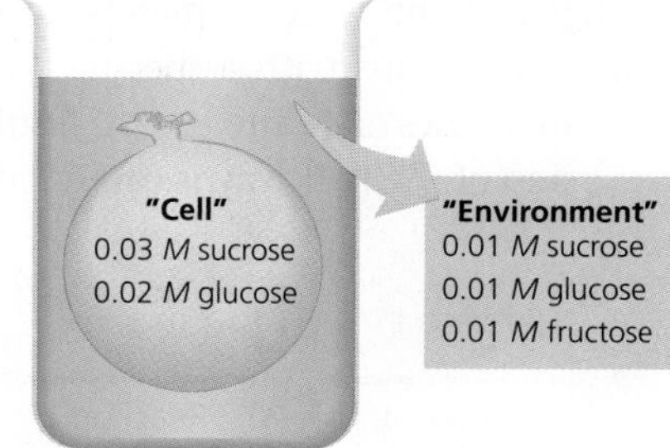

   c. Draw a dashed arrow to show the net osmosis, if any.
   d. Will the artificial cell become more flaccid, more turgid, or stay the same?
   e. Eventually, will the two solutions have the same or different solute concentrations?

LEVEL 3: SYNTHESIS/EVALUATION

7. **EVOLUTION CONNECTION**
   Paramecium and other protists that live in hypotonic environments have cell membranes that limit water uptake, while those living in isotonic environments have membranes that are more permeable to water. What water regulation adaptations might have evolved in protists in hypertonic habitats such as Great Salt Lake? In habitats with changing salt concentration?

8. **SCIENTIFIC INQUIRY**
   An experiment is designed to study the mechanism of sucrose uptake by plant cells. Cells are immersed in a sucrose solution, and the pH of the solution is monitored. Samples of the cells are taken at intervals, and their sucrose concentration is measured. After a decrease in the pH of the solution to a steady, slightly acidic level, sucrose uptake begins. Propose a hypothesis for these results. What do you think would happen if an inhibitor of ATP regeneration by the cell were added to the beaker once the pH is at a steady level? Explain.

9. **SCIENCE, TECHNOLOGY, AND SOCIETY**
   Extensive irrigation in arid regions causes salts to accumulate in the soil. (When water evaporates, salts that were dissolved in the water are left behind in the soil.) Based on what you learned about water balance in plant cells, explain why increased soil salinity (saltiness) might be harmful to crops. Suggest ways to minimize damage. What costs are attached to your solutions?

10. **WRITE ABOUT A THEME**
   Environmental Interactions A human pancreatic cell obtains O_2, fuel molecules such as glucose, and building materials such as amino acids and cholesterol from its environment, and it releases CO_2 as a waste product of cellular respiration. In response to hormonal signals, the cell secretes digestive enzymes. It also regulates its ion concentrations by exchange with its environment. Based on what you have just learned about the structure and function of cellular membranes, write a short essay (100–150 words) that describes how such a cell accomplishes these interactions with its environment.

For selected answers, see Appendix A.

MasteringBIOLOGY www.masteringbiology.com

1. MasteringBiology® Assignments

Make Connections Tutorial Plasma Membranes (Chapter 7) and Phospholipid Structure (Chapter 5)

BioFlix **Tutorials** Membrane Transport: Diffusion and Passive Transport • The Sodium-Potassium Pump • Cotransport • Bulk Transport

Tutorial Osmosis

Activities Membrane Structure • Selective Permeability of Membranes • Diffusion • Diffusion and Osmosis • Facilitated Diffusion • Membrane Transport Proteins • Osmosis and Water Balance in Cells • Active Transport • Exocytosis and Endocytosis

Questions Student Misconceptions • Reading Quiz • Multiple Choice • End-of-Chapter

2. eText

Read your book online, search, take notes, highlight text, and more.

3. The Study Area

Practice Tests • Cumulative Test • ***BioFlix*** 3-D Animations • MP3 Tutor Sessions • Videos • Activities • Investigations • Lab Media • Audio Glossary • Word Study Tools • Art

8

An Introduction to Metabolism

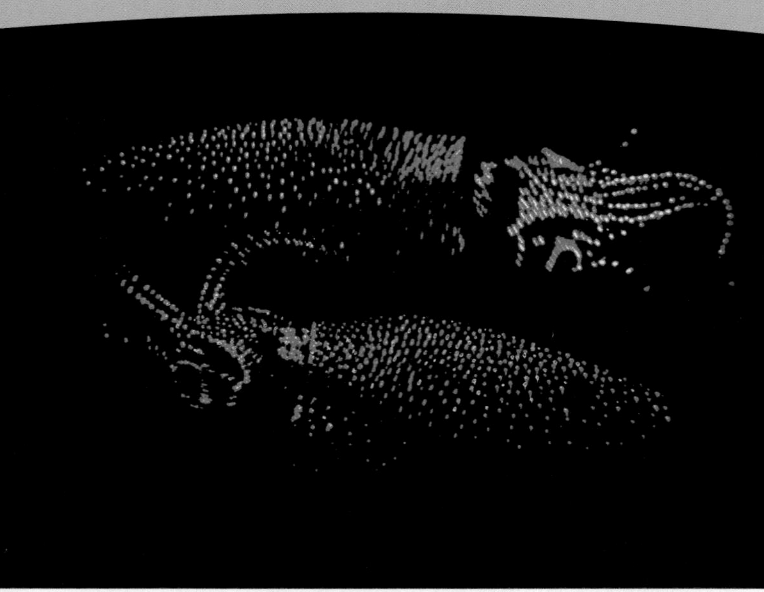

▲ **Figure 8.1 What causes these two squid to glow?**

KEY CONCEPTS

8.1 An organism's metabolism transforms matter and energy, subject to the laws of thermodynamics
8.2 The free-energy change of a reaction tells us whether or not the reaction occurs spontaneously
8.3 ATP powers cellular work by coupling exergonic reactions to endergonic reactions
8.4 Enzymes speed up metabolic reactions by lowering energy barriers
8.5 Regulation of enzyme activity helps control metabolism

OVERVIEW

The Energy of Life

The living cell is a chemical factory in miniature, where thousands of reactions occur within a microscopic space. Sugars can be converted to amino acids that are linked together into proteins when needed, and when food is digested, proteins are dismantled into amino acids that can be converted to sugars. Small molecules are assembled into polymers, which may be hydrolyzed later as the needs of the cell change. In multicellular organisms, many cells export chemical products that are used in other parts of the organism. The process called cellular respiration drives the cellular economy by extracting the energy stored in sugars and other fuels. Cells apply this energy to perform various types of work, such as the transport of solutes across the plasma membrane, which we discussed in Chapter 7. In a more exotic example, cells of the two firefly squid (*Watasenia scintillans*) shown mating in **Figure 8.1** convert the energy stored in certain organic molecules to light, a process called bioluminescence. (The light pattern aids in mate recognition and protection from predators lurking below.) Bioluminescence and other metabolic activities carried out by a cell are precisely coordinated and controlled. In its complexity, its efficiency, and its responsiveness to subtle changes, the cell is peerless as a chemical factory. The concepts of metabolism that you learn in this chapter will help you understand how matter and energy flow during life's processes and how that flow is regulated.

CONCEPT **8.1**

An organism's metabolism transforms matter and energy, subject to the laws of thermodynamics

The totality of an organism's chemical reactions is called **metabolism** (from the Greek *metabole*, change). Metabolism is an emergent property of life that arises from orderly interactions between molecules.

Organization of the Chemistry of Life into Metabolic Pathways

We can picture a cell's metabolism as an elaborate road map of the thousands of chemical reactions that occur in a cell, arranged as intersecting metabolic pathways. A **metabolic pathway** begins with a specific molecule, which is then altered in a series of defined steps, resulting in a certain product. Each step of the pathway is catalyzed by a specific enzyme:

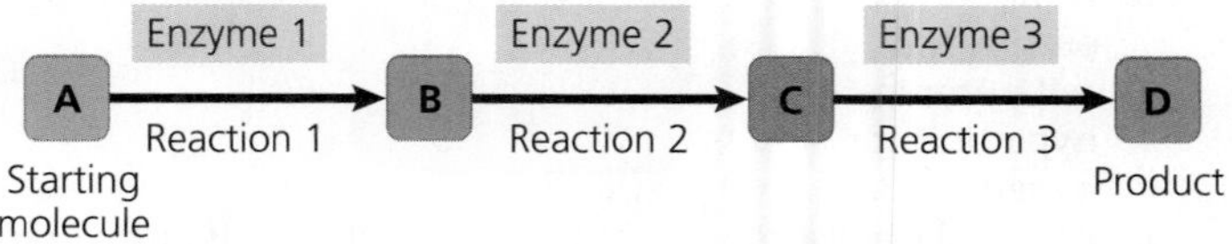

Analogous to the red, yellow, and green stoplights that control the flow of automobile traffic, mechanisms that regulate enzymes balance metabolic supply and demand.

Metabolism as a whole manages the material and energy resources of the cell. Some metabolic pathways release energy by breaking down complex molecules to simpler compounds. These degradative processes are called **catabolic pathways**, or breakdown pathways. A major pathway of catabolism is cellular respiration, in which the sugar glucose and other organic fuels are broken down in the presence of oxygen to carbon dioxide and water. (Pathways can have more than one starting molecule and/or product.) Energy that was stored in the organic molecules becomes available to do the work of the cell, such as ciliary beating or membrane transport. **Anabolic pathways**, in contrast, consume energy to build complicated molecules from simpler ones; they are sometimes called biosynthetic pathways. Examples of anabolism are the synthesis of an amino acid from simpler molecules and the synthesis of a protein from amino acids. Catabolic and anabolic pathways are the "downhill" and "uphill" avenues of the metabolic landscape. Energy released from the downhill reactions of catabolic pathways can be stored and then used to drive the uphill reactions of anabolic pathways.

In this chapter, we will focus on mechanisms common to metabolic pathways. Because energy is fundamental to all metabolic processes, a basic knowledge of energy is necessary to understand how the living cell works. Although we will use some nonliving examples to study energy, the concepts demonstrated by these examples also apply to **bioenergetics**, the study of how energy flows through living organisms.

Forms of Energy

Energy is the capacity to cause change. In everyday life, energy is important because some forms of energy can be used to do work—that is, to move matter against opposing forces, such as gravity and friction. Put another way, energy is the ability to rearrange a collection of matter. For example, you expend energy to turn the pages of this book, and your cells expend energy in transporting certain substances across membranes. Energy exists in various forms, and the work of life depends on the ability of cells to transform energy from one form to another.

Energy can be associated with the relative motion of objects; this energy is called **kinetic energy**. Moving objects can perform work by imparting motion to other matter: A pool player uses the motion of the cue stick to push the cue ball, which in turn moves the other balls; water gushing through a dam turns turbines; and the contraction of leg muscles pushes bicycle pedals. **Heat**, or **thermal energy**, is kinetic energy associated with the random movement of atoms or molecules. Light is also a type of energy that can be harnessed to perform work, such as powering photosynthesis in green plants.

An object not presently moving may still possess energy. Energy that is not kinetic is called **potential energy**; it is energy that matter possesses because of its location or structure. Water behind a dam, for instance, possesses energy because of its altitude above sea level. Molecules possess energy because of the arrangement of electrons in the bonds between their atoms. **Chemical energy** is a term used by biologists to refer to the potential energy available for release in a chemical reaction. Recall that catabolic pathways release energy by breaking down complex molecules. Biologists say that these complex molecules, such as glucose, are high in chemical energy. During a catabolic reaction, some bonds are broken and others formed, releasing energy and resulting in lower-energy breakdown products. This transformation also occurs, for example, in the engine of a car when the hydrocarbons of gasoline react explosively with oxygen, releasing the energy that pushes the pistons and producing exhaust. Although less explosive, a similar reaction of food molecules with oxygen provides chemical energy in biological systems, producing carbon dioxide and water as waste products. Biochemical pathways, carried out in the context of cellular structures, enable cells to release chemical energy from food molecules and use the energy to power life processes.

How is energy converted from one form to another? Consider the divers in **Figure 8.2**. The young woman climbing the ladder to the diving platform is releasing chemical energy from the food she ate for lunch and using some of that energy to perform the work of climbing. The kinetic energy of muscle movement is thus being transformed into potential energy due to her increasing height above the water. The young man diving is converting his potential energy to kinetic energy, which is then transferred to the water as he enters it. A small amount of energy is lost as heat due to friction.

A diver has more potential energy on the platform than in the water.
Diving converts potential energy to kinetic energy.
Climbing up converts the kinetic energy of muscle movement to potential energy.
A diver has less potential energy in the water than on the platform.

▲ **Figure 8.2 Transformations between potential and kinetic energy.**

Now let's go back one step and consider the original source of the organic food molecules that provided the necessary chemical energy for the diver to climb the steps. This chemical energy was itself derived from light energy by plants during photosynthesis. Organisms are energy transformers.

The Laws of Energy Transformation

The study of the energy transformations that occur in a collection of matter is called **thermodynamics**. Scientists use the word *system* to denote the matter under study; they refer to the rest of the universe—everything outside the system—as the *surroundings*. An *isolated system*, such as that approximated by liquid in a thermos bottle, is unable to exchange either energy or matter with its surroundings. In an *open system*, energy and matter can be transferred between the system and its surroundings. Organisms are open systems. They absorb energy—for instance, light energy or chemical energy in the form of organic molecules—and release heat and metabolic waste products, such as carbon dioxide, to the surroundings. Two laws of thermodynamics govern energy transformations in organisms and all other collections of matter.

The First Law of Thermodynamics

According to the **first law of thermodynamics**, the energy of the universe is constant: *Energy can be transferred and transformed, but it cannot be created or destroyed.* The first law is also known as the *principle of conservation of energy*. The electric company does not make energy, but merely converts it to a form that is convenient for us to use. By converting sunlight to chemical energy, a plant acts as an energy transformer, not an energy producer.

The brown bear in **Figure 8.3a** will convert the chemical energy of the organic molecules in its food to kinetic and other forms of energy as it carries out biological processes. What happens to this energy after it has performed work? The second law of thermodynamics helps to answer this question.

The Second Law of Thermodynamics

If energy cannot be destroyed, why can't organisms simply recycle their energy over and over again? It turns out that during every energy transfer or transformation, some energy becomes unavailable to do work. In most energy transformations, more usable forms of energy are at least partly converted to heat, which is the energy associated with the random motion of atoms or molecules. Only a small fraction of the chemical energy from the food in Figure 8.3a is transformed into the motion of the brown bear shown in **Figure 8.3b**; most is lost as heat, which dissipates rapidly through the surroundings.

In the process of carrying out chemical reactions that perform various kinds of work, living cells unavoidably convert other forms of energy to heat. A system can put heat to work only when there is a temperature difference that results in the heat flowing from a warmer location to a cooler one. If temperature is uniform, as it is in a living cell, then the only use for heat energy generated during a chemical reaction is to warm a body of matter, such as the organism. (This can make a room crowded with people uncomfortably warm, as each person is carrying out a multitude of chemical reactions!)

A logical consequence of the loss of usable energy during energy transfer or transformation is that each such event makes the universe more disordered. Scientists use a quantity called **entropy** as a measure of disorder, or randomness.

(a) First law of thermodynamics: Energy can be transferred or transformed but neither created nor destroyed. For example, chemical reactions in this brown bear (*Ursus arctos*) will convert the chemical (potential) energy in the fish into the kinetic energy of running, shown in (b).

(b) Second law of thermodynamics: Every energy transfer or transformation increases the disorder (entropy) of the universe. For example, as it runs, disorder is increased around the bear by the release of heat and small molecules that are the by-products of metabolism. A brown bear can run at speeds up to 35 miles per hour (56 km/hr)—as fast as a racehorse.

▲ **Figure 8.3 The two laws of thermodynamics.**

The more randomly arranged a collection of matter is, the greater its entropy. We can now state the **second law of thermodynamics**: *Every energy transfer or transformation increases the entropy of the universe.* Although order can increase locally, there is an unstoppable trend toward randomization of the universe as a whole.

In many cases, increased entropy is evident in the physical disintegration of a system's organized structure. For example, you can observe increasing entropy in the gradual decay of an unmaintained building. Much of the increasing entropy of the universe is less apparent, however, because it appears as increasing amounts of heat and less ordered forms of matter. As the bear in Figure 8.3b converts chemical energy to kinetic energy, it is also increasing the disorder of its surroundings by producing heat and small molecules, such as the CO_2 it exhales, that are the breakdown products of food.

The concept of entropy helps us understand why certain processes occur without any input of energy. It turns out that for a process to occur on its own, without outside help, it must increase the entropy of the universe. A process that can occur without an input of energy is called a **spontaneous process**. Note that as we're using it here, the word *spontaneous* does not imply that such a process would occur quickly; rather, the word signifies that the process is energetically favorable. (In fact, it may be helpful for you to think of the phrase "energetically favorable" when you read the formal term "spontaneous.") Some spontaneous processes, such as an explosion, may be virtually instantaneous, while others, such as the rusting of an old car over time, are much slower. A process that cannot occur on its own is said to be nonspontaneous; it will happen only if energy is added to the system. We know from experience that certain events occur spontaneously and others do not. For instance, we know that water flows downhill spontaneously but moves uphill only with an input of energy, such as when a machine pumps the water against gravity. This understanding gives us another way to state the second law: *For a process to occur spontaneously, it must increase the entropy of the universe.*

Biological Order and Disorder

Living systems increase the entropy of their surroundings, as predicted by thermodynamic law. It is true that cells create ordered structures from less organized starting materials. For example, simpler molecules are ordered into the more complex structure of an amino acid, and amino acids are ordered into polypeptide chains. At the organismal level as well, complex and beautifully ordered structures result from biological processes that use simpler starting materials (**Figure 8.4**). However, an organism also takes in organized forms of matter and energy from the surroundings and replaces them with less ordered forms. For example, an animal obtains starch, proteins, and other complex molecules from the food it eats. As catabolic pathways break these molecules down, the animal releases carbon dioxide and water—small molecules that possess less chemical energy than the food did. The depletion of chemical energy is accounted for by heat generated during metabolism. On a larger scale, energy flows into most ecosystems in the form of light and exits in the form of heat (see Figure 1.6).

▲ **Figure 8.4 Order as a characteristic of life.** Order is evident in the detailed structures of the sea urchin skeleton and the succulent plant shown here. As open systems, organisms can increase their order as long as the order of their surroundings decreases.

During the early history of life, complex organisms evolved from simpler ancestors. For example, we can trace the ancestry of the plant kingdom from much simpler organisms called green algae to more complex flowering plants. However, this increase in organization over time in no way violates the second law. The entropy of a particular system, such as an organism, may actually decrease as long as the total entropy of the *universe*—the system plus its surroundings—increases. Thus, organisms are islands of low entropy in an increasingly random universe. The evolution of biological order is perfectly consistent with the laws of thermodynamics.

CONCEPT CHECK 8.1

1. **MAKE CONNECTIONS** How does the second law of thermodynamics help explain the diffusion of a substance across a membrane? See Figure 7.13 on page 132.
2. Describe the forms of energy found in an apple as it grows on a tree, then falls, then is digested by someone who eats it.
3. **WHAT IF?** If you place a teaspoon of sugar in the bottom of a glass of water, it will dissolve completely over time. Left longer, eventually the water will disappear and the sugar crystals will reappear. Explain these observations in terms of entropy.

For suggested answers, see Appendix A.

CONCEPT 8.2

The free-energy change of a reaction tells us whether or not the reaction occurs spontaneously

The laws of thermodynamics that we've just discussed apply to the universe as a whole. As biologists, we want to understand the chemical reactions of life—for example, which reactions occur spontaneously and which ones require some input of energy from outside. But how can we know this without assessing the energy and entropy changes in the entire universe for each separate reaction?

Free-Energy Change, ΔG

Recall that the universe is really equivalent to "the system" plus "the surroundings." In 1878, J. Willard Gibbs, a professor at Yale, defined a very useful function called the Gibbs free energy of a system (without considering its surroundings), symbolized by the letter G. We'll refer to the Gibbs free energy simply as free energy. **Free energy** is the portion of a system's energy that can perform work when temperature and pressure are uniform throughout the system, as in a living cell. Let's consider how we determine the free-energy change that occurs when a system changes—for example, during a chemical reaction.

The change in free energy, ΔG, can be calculated for a chemical reaction by applying the following equation:

$$\Delta G = \Delta H - T\Delta S$$

This equation uses only properties of the system (the reaction) itself: ΔH symbolizes the change in the system's *enthalpy* (in biological systems, equivalent to total energy); ΔS is the change in the system's entropy; and T is the absolute temperature in Kelvin (K) units (K = °C + 273; see Appendix C).

Once we know the value of ΔG for a process, we can use it to predict whether the process will be spontaneous (that is, whether it is energetically favorable and will occur without an input of energy). More than a century of experiments has shown that only processes with a negative ΔG are spontaneous. For ΔG to be negative, either ΔH must be negative (the system gives up enthalpy and H decreases) or $T\Delta S$ must be positive (the system gives up order and S increases), or both: When ΔH and $T\Delta S$ are tallied, ΔG has a negative value ($\Delta G < 0$) for all spontaneous processes. In other words, every spontaneous process decreases the system's free energy, and processes that have a positive or zero ΔG are never spontaneous.

This information is immensely interesting to biologists, for it gives us the power to predict which kinds of change can happen without help. Such spontaneous changes can be harnessed to perform work. This principle is very important in the study of metabolism, where a major goal is to determine which reactions can supply energy for cellular work.

Free Energy, Stability, and Equilibrium

As we saw in the previous section, when a process occurs spontaneously in a system, we can be sure that ΔG is negative. Another way to think of ΔG is to realize that it represents the difference between the free energy of the final state and the free energy of the initial state:

$$\Delta G = G_{\text{final state}} - G_{\text{initial state}}$$

Thus, ΔG can be negative only when the process involves a loss of free energy during the change from initial state to final state. Because it has less free energy, the system in its final state is less likely to change and is therefore more stable than it was previously.

We can think of free energy as a measure of a system's instability—its tendency to change to a more stable state. Unstable systems (higher G) tend to change in such a way that they become more stable (lower G). For example, a diver on top of a platform is less stable (more likely to fall) than when floating in the water; a drop of concentrated dye is less stable (more likely to disperse) than when the dye is spread randomly through the liquid; and a glucose molecule is less stable (more likely to break down) than the simpler molecules into which it can be split (**Figure 8.5**). Unless something prevents it, each of these systems will move toward greater stability: The diver falls, the solution becomes uniformly colored, and the glucose molecule is broken down.

Another term that describes a state of maximum stability is *equilibrium*, which you learned about in Chapter 2 in connection with chemical reactions. There is an important relationship between free energy and equilibrium, including chemical equilibrium. Recall that most chemical reactions are reversible and proceed to a point at which the forward and backward reactions occur at the same rate. The reaction is then said to be at chemical equilibrium, and there is no further net change in the relative concentration of products and reactants.

As a reaction proceeds toward equilibrium, the free energy of the mixture of reactants and products decreases. Free energy increases when a reaction is somehow pushed away from equilibrium, perhaps by removing some of the products (and thus changing their concentration relative to that of the reactants). For a system at equilibrium, G is at its lowest possible value in that system. We can think of the equilibrium state as a free-energy valley. Any change from the equilibrium position will have a positive ΔG and will not be spontaneous. For this reason, systems never spontaneously move away from equilibrium. Because a system at equilibrium cannot spontaneously change, it can do no work. *A process is spontaneous and can perform work only when it is moving toward equilibrium.*

- More free energy (higher G)
- Less stable
- Greater work capacity

In a **spontaneous change**
- The free energy of the system decreases ($\Delta G < 0$)
- The system becomes more stable
- The released free energy can be harnessed to do work

- Less free energy (lower G)
- More stable
- Less work capacity

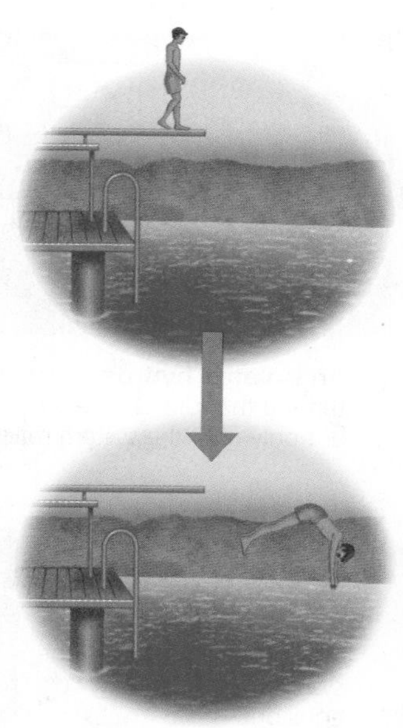

(a) Gravitational motion. Objects move spontaneously from a higher altitude to a lower one.

(b) Diffusion. Molecules in a drop of dye diffuse until they are randomly dispersed.

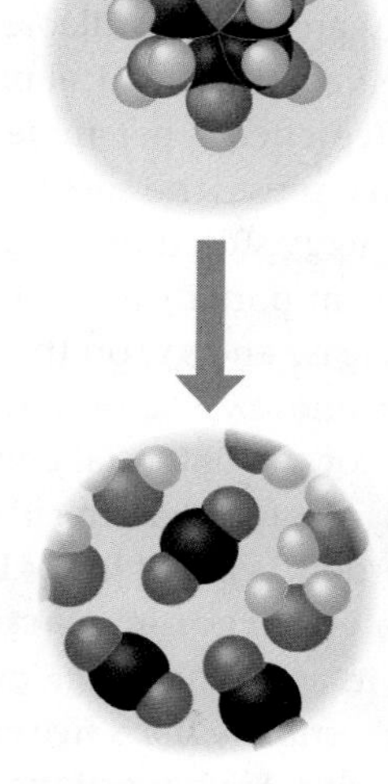

(c) Chemical reaction. In a cell, a glucose molecule is broken down into simpler molecules.

▲ **Figure 8.5 The relationship of free energy to stability, work capacity, and spontaneous change.** Unstable systems (top) are rich in free energy, G. They have a tendency to change spontaneously to a more stable state (bottom), and it is possible to harness this "downhill" change to perform work.

Free Energy and Metabolism

We can now apply the free-energy concept more specifically to the chemistry of life's processes.

Exergonic and Endergonic Reactions in Metabolism

Based on their free-energy changes, chemical reactions can be classified as either exergonic ("energy outward") or endergonic ("energy inward"). An **exergonic reaction** proceeds with a net release of free energy (**Figure 8.6a**). Because the chemical mixture loses free energy (G decreases), ΔG is negative for an exergonic reaction. Using ΔG as a standard for spontaneity, exergonic reactions are those that occur spontaneously. (Remember, the word *spontaneous* implies that it is energetically favorable, not that it will occur rapidly.) The magnitude of ΔG for an exergonic reaction represents the maximum amount of work the reaction can perform.* The greater the decrease in free energy, the greater the amount of work that can be done.

We can use the overall reaction for cellular respiration as an example:

$$C_6H_{12}O_6 + 6\ O_2 \rightarrow 6\ CO_2 + 6\ H_2O$$
$$\Delta G = -686\ \text{kcal/mol}\ (-2{,}870\ \text{kJ/mol})$$

*The word *maximum* qualifies this statement, because some of the free energy is released as heat and cannot do work. Therefore, ΔG represents a theoretical upper limit of available energy.

▼ **Figure 8.6 Free energy changes (ΔG) in exergonic and endergonic reactions.**

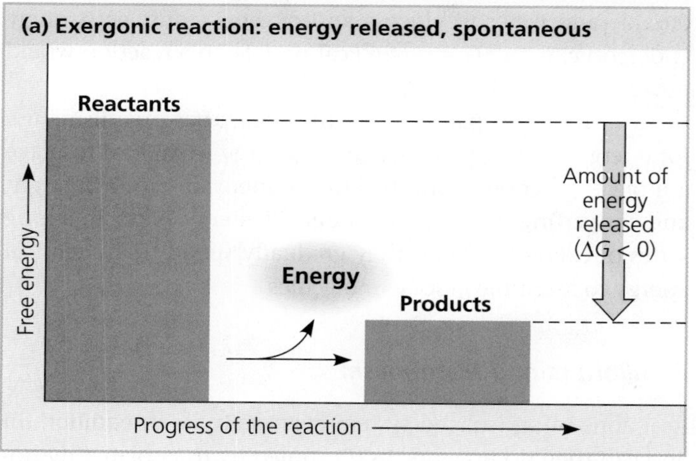

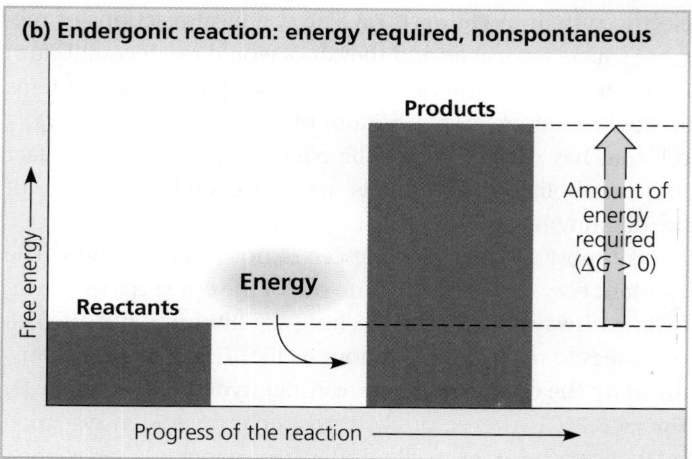

For each mole (180 g) of glucose broken down by respiration under what are called "standard conditions" (1 *M* of each reactant and product, 25°C, pH 7), 686 kcal (2,870 kJ) of energy are made available for work. Because energy must be conserved, the chemical products of respiration store 686 kcal less free energy per mole than the reactants. The products are, in a sense, the spent exhaust of a process that tapped the free energy stored in the bonds of the sugar molecules.

It is important to realize that the breaking of bonds does not release energy; on the contrary, as you will soon see, it requires energy. The phrase "energy stored in bonds" is shorthand for the potential energy that can be released when new bonds are formed after the original bonds break, as long as the products are of lower free energy than the reactants.

An **endergonic reaction** is one that absorbs free energy from its surroundings (**Figure 8.6b**). Because this kind of reaction essentially *stores* free energy in molecules (G increases), ΔG is positive. Such reactions are nonspontaneous, and the magnitude of ΔG is the quantity of energy required to drive the reaction. If a chemical process is exergonic (downhill), releasing energy in one direction, then the reverse process must be endergonic (uphill), using energy. A reversible process cannot be downhill in both directions. If $\Delta G = -686$ kcal/mol for respiration, which converts glucose and oxygen to carbon dioxide and water, then the reverse process—the conversion of carbon dioxide and water to glucose and oxygen—must be strongly endergonic, with $\Delta G = +686$ kcal/mol. Such a reaction would never happen by itself.

How, then, do plants make the sugar that organisms use for energy? Plants get the required energy—686 kcal to make a mole of glucose—from the environment by capturing light and converting its energy to chemical energy. Next, in a long series of exergonic steps, they gradually spend that chemical energy to assemble glucose molecules.

Equilibrium and Metabolism

Reactions in an isolated system eventually reach equilibrium and can then do no work, as illustrated by the isolated hydroelectric system in **Figure 8.7a**. The chemical reactions of metabolism are reversible, and they, too, would reach equilibrium if they occurred in the isolation of a test tube. Because systems at equilibrium are at a minimum of G and can do no work, a cell that has reached metabolic equilibrium is dead! The fact that metabolism as a whole is never at equilibrium is one of the defining features of life.

Like most systems, a living cell is not in equilibrium. The constant flow of materials in and out of the cell keeps the metabolic pathways from ever reaching equilibrium, and the cell continues to do work throughout its life. This principle is illustrated by the open (and more realistic) hydroelectric system in **Figure 8.7b**. However, unlike this simple single-step system, a catabolic pathway in a cell releases free energy in a series of reactions. An example is cellular respiration, illustrated by analogy in **Figure 8.7c**. Some of the reversible reactions of respiration are constantly "pulled" in one direction—that is, they are kept out of equilibrium. The key to maintaining this lack of equilibrium is that the product of a reaction does not accumulate but instead becomes a reactant in the next step; finally, waste products are expelled from the cell. The overall sequence of reactions is kept going by the huge free-energy difference between glucose and oxygen at the top of the energy "hill" and carbon dioxide and water at the "downhill" end. As long as our cells have a steady supply of glucose or other fuels and oxygen and are able to expel waste products to the surroundings, their metabolic pathways never reach equilibrium and can continue to do the work of life.

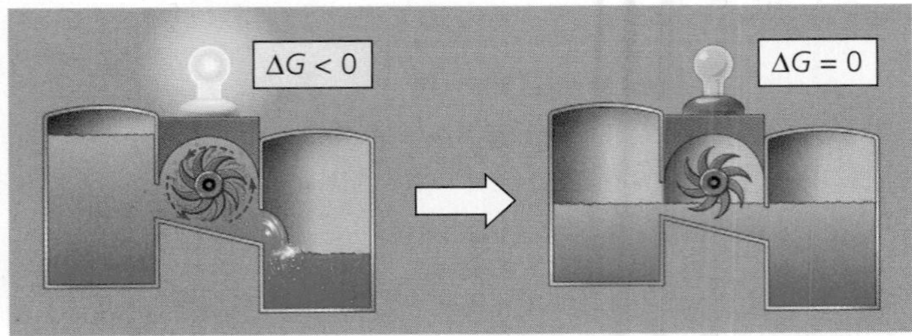

(a) An isolated hydroelectric system. Water flowing downhill turns a turbine that drives a generator providing electricity to a lightbulb, but only until the system reaches equilibrium.

(b) An open hydroelectric system. Flowing water keeps driving the generator because intake and outflow of water keep the system from reaching equilibrium.

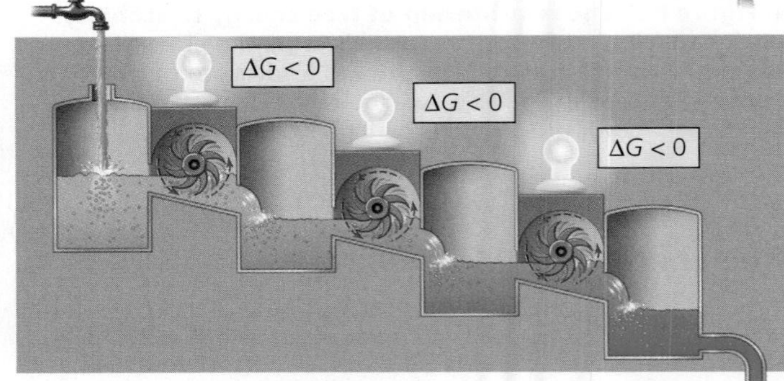

(c) A multistep open hydroelectric system. Cellular respiration is analogous to this system: Glucose is broken down in a series of exergonic reactions that power the work of the cell. The product of each reaction becomes the reactant for the next, so no reaction reaches equilibrium.

▲ **Figure 8.7 Equilibrium and work in isolated and open systems.**

We see once again how important it is to think of organisms as open systems. Sunlight provides a daily source of free energy for an ecosystem's plants and other photosynthetic organisms. Animals and other nonphotosynthetic organisms in an ecosystem must have a source of free energy in the form of the organic products of photosynthesis. Now that we have applied the free-energy concept to metabolism, we are ready to see how a cell actually performs the work of life.

CONCEPT CHECK 8.2

1. Cellular respiration uses glucose and oxygen, which have high levels of free energy, and releases CO_2 and water, which have low levels of free energy. Is cellular respiration spontaneous or not? Is it exergonic or endergonic? What happens to the energy released from glucose?
2. **MAKE CONNECTIONS** As you saw in Figure 7.20 on page 137, a key process in metabolism is the transport of hydrogen ions (H^+) across a membrane to create a concentration gradient. Other processes can result in an equal concentration of H^+ on each side. Which situation allows the H^+ to perform work in this system? How is the answer consistent with what is shown in regard to energy in Figure 7.20?
3. **WHAT IF?** Some night-time partygoers wear glow-in-the-dark necklaces. The necklaces start glowing once they are "activated," which usually involves snapping the necklace in a way that allows two chemicals to react and emit light in the form of chemiluminescence. Is the chemical reaction exergonic or endergonic? Explain your answer.

For suggested answers, see Appendix A.

CONCEPT 8.3

ATP powers cellular work by coupling exergonic reactions to endergonic reactions

A cell does three main kinds of work:

- *Chemical work*, the pushing of endergonic reactions that would not occur spontaneously, such as the synthesis of polymers from monomers (chemical work will be discussed further here and in Chapters 9 and 10)
- *Transport work*, the pumping of substances across membranes against the direction of spontaneous movement (see Chapter 7)
- *Mechanical work*, such as the beating of cilia (see Chapter 6), the contraction of muscle cells, and the movement of chromosomes during cellular reproduction

A key feature in the way cells manage their energy resources to do this work is **energy coupling**, the use of an exergonic process to drive an endergonic one. ATP is responsible for mediating most energy coupling in cells, and in most cases it acts as the immediate source of energy that powers cellular work.

The Structure and Hydrolysis of ATP

ATP (adenosine triphosphate) was introduced in Chapter 4 when we discussed the phosphate group as a functional group. ATP contains the sugar ribose, with the nitrogenous base adenine and a chain of three phosphate groups bonded to it (**Figure 8.8a**). In addition to its role in energy coupling, ATP is also one of the nucleoside triphosphates used to make RNA (see Figure 5.26).

The bonds between the phosphate groups of ATP can be broken by hydrolysis. When the terminal phosphate bond is broken by addition of a water molecule, a molecule of inorganic phosphate ($HOPO_3^{2-}$, abbreviated Ⓟ$_i$ throughout this book) leaves the ATP, which becomes adenosine diphosphate,

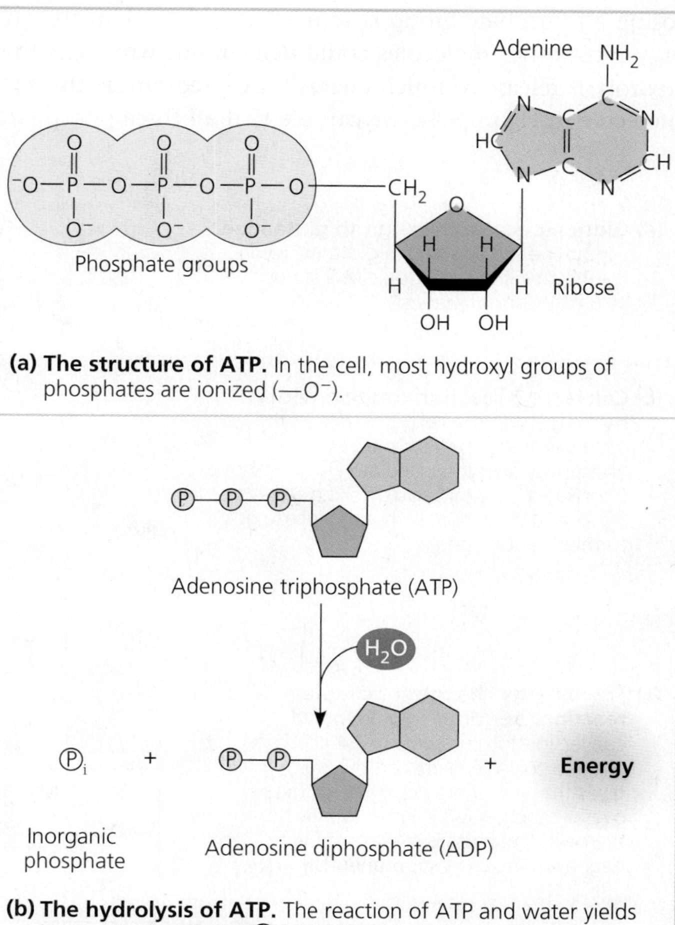

(a) The structure of ATP. In the cell, most hydroxyl groups of phosphates are ionized (—O⁻).

(b) The hydrolysis of ATP. The reaction of ATP and water yields inorganic phosphate (Ⓟ$_i$) and ADP and releases energy.

▲ **Figure 8.8 The structure and hydrolysis of adenosine triphosphate (ATP).**

or ADP **(Figure 8.8b)**. The reaction is exergonic and releases 7.3 kcal of energy per mole of ATP hydrolyzed:

$$ATP + H_2O \rightarrow ADP + Ⓟ_i$$
$$\Delta G = -7.3 \text{ kcal/mol } (-30.5 \text{ kJ/mol})$$

This is the free-energy change measured under standard conditions. In the cell, conditions do not conform to standard conditions, primarily because reactant and product concentrations differ from 1 *M*. For example, when ATP hydrolysis occurs under cellular conditions, the actual ΔG is about -13 kcal/mol, 78% greater than the energy released by ATP hydrolysis under standard conditions.

Because their hydrolysis releases energy, the phosphate bonds of ATP are sometimes referred to as high-energy phosphate bonds, but the term is misleading. The phosphate bonds of ATP are not unusually strong bonds, as "high-energy" may imply; rather, the reactants (ATP and water) themselves have high energy relative to the energy of the products (ADP and Ⓟ$_i$). The release of energy during the hydrolysis of ATP comes from the chemical change to a state of lower free energy, not from the phosphate bonds themselves.

ATP is useful to the cell because the energy it releases on losing a phosphate group is somewhat greater than the energy most other molecules could deliver. But why does this hydrolysis release so much energy? If we reexamine the ATP molecule in Figure 8.8a, we can see that all three phosphate groups are negatively charged. These like charges are crowded together, and their mutual repulsion contributes to the instability of this region of the ATP molecule. The triphosphate tail of ATP is the chemical equivalent of a compressed spring.

How the Hydrolysis of ATP Performs Work

When ATP is hydrolyzed in a test tube, the release of free energy merely heats the surrounding water. In an organism, this same generation of heat can sometimes be beneficial. For instance, the process of shivering uses ATP hydrolysis during muscle contraction to generate heat and warm the body. In most cases in the cell, however, the generation of heat alone would be an inefficient (and potentially dangerous) use of a valuable energy resource. Instead, the cell's proteins harness the energy released during ATP hydrolysis in several ways to perform the three types of cellular work—chemical, transport, and mechanical.

For example, with the help of specific enzymes, the cell is able to use the energy released by ATP hydrolysis directly to drive chemical reactions that, by themselves, are endergonic. If the ΔG of an endergonic reaction is less than the amount of energy released by ATP hydrolysis, then the two reactions can be coupled so that, overall, the coupled reactions are exergonic **(Figure 8.9)**. This usually involves the transfer of a phosphate

(a) Glutamic acid conversion to glutamine. Glutamine synthesis from glutamic acid (Glu) by itself is endergonic (ΔG is positive), so it is not spontaneous.

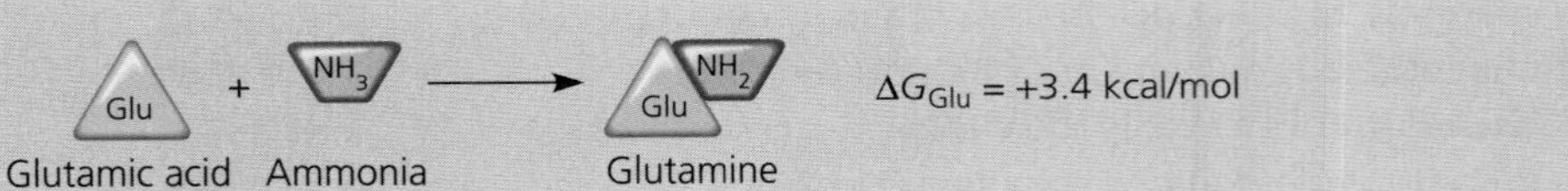

(b) Conversion reaction coupled with ATP hydrolysis. In the cell, glutamine synthesis occurs in two steps, coupled by a phosphorylated intermediate. ❶ ATP phosphorylates glutamic acid, making it less stable. ❷ Ammonia displaces the phosphate group, forming glutamine.

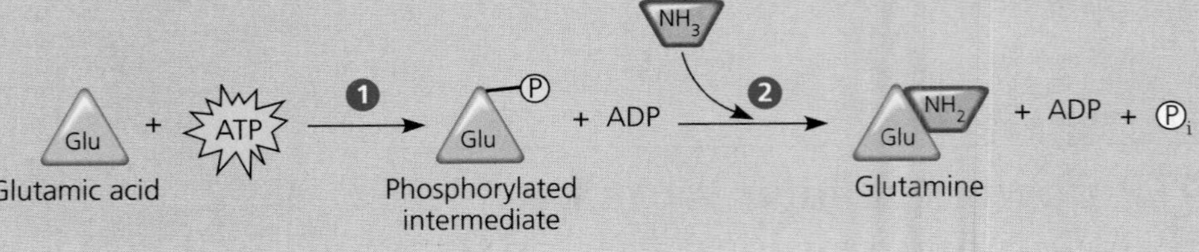

(c) Free-energy change for coupled reaction. ΔG for the glutamic acid conversion to glutamine (+3.4 kcal/mol) plus ΔG for ATP hydrolysis (–7.3 kcal/mol) gives the free-energy change for the overall reaction (–3.9 kcal/mol). Because the overall process is exergonic (net ΔG is negative), it occurs spontaneously.

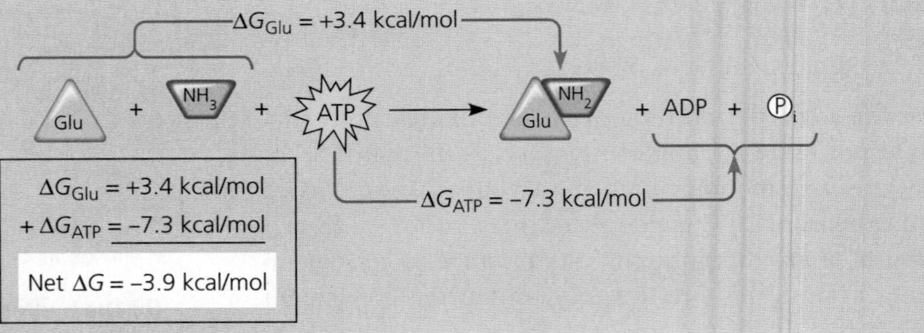

▲ **Figure 8.9 How ATP drives chemical work: Energy coupling using ATP hydrolysis.** In this example, the exergonic process of ATP hydrolysis is used to drive an endergonic process—the cellular synthesis of the amino acid glutamine from glutamic acid and ammonia.

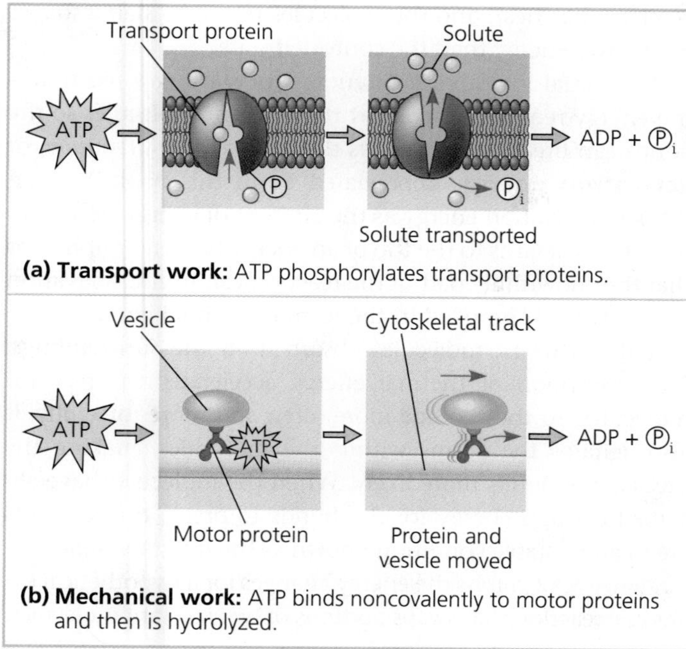

(a) Transport work: ATP phosphorylates transport proteins.

(b) Mechanical work: ATP binds noncovalently to motor proteins and then is hydrolyzed.

▲ **Figure 8.10 How ATP drives transport and mechanical work.** ATP hydrolysis causes changes in the shapes and binding affinities of proteins. This can occur either **(a)** directly, by phosphorylation, as shown for a membrane protein carrying out active transport of a solute (see also Figure 7.18), or **(b)** indirectly, via noncovalent binding of ATP and its hydrolytic products, as is the case for motor proteins that move vesicles (and other organelles) along cytoskeletal "tracks" in the cell (see also Figure 6.21).

group from ATP to some other molecule, such as the reactant. The recipient with the phosphate group covalently bonded to it is then called a **phosphorylated intermediate**. The key to coupling exergonic and endergonic reactions is the formation of this phosphorylated intermediate, which is more reactive (less stable) than the original unphosphorylated molecule.

Transport and mechanical work in the cell are also nearly always powered by the hydrolysis of ATP. In these cases, ATP hydrolysis leads to a change in a protein's shape and often its ability to bind another molecule. Sometimes this occurs via a phosphorylated intermediate, as seen for the transport protein in **Figure 8.10a**. In most instances of mechanical work involving motor proteins "walking" along cytoskeletal elements **(Figure 8.10b)**, a cycle occurs in which ATP is first bound noncovalently to the motor protein. Next, ATP is hydrolyzed, releasing ADP and $Ⓟ_i$. Another ATP molecule can then bind. At each stage, the motor protein changes its shape and ability to bind the cytoskeleton, resulting in movement of the protein along the cytoskeletal track.

The Regeneration of ATP

An organism at work uses ATP continuously, but ATP is a renewable resource that can be regenerated by the addition of phosphate to ADP **(Figure 8.11)**. The free energy required to

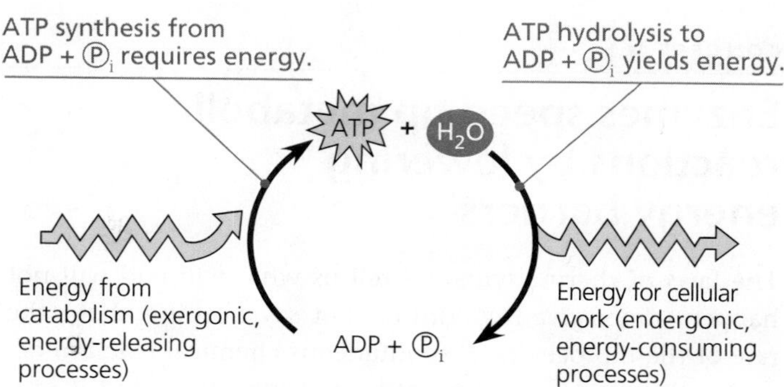

▲ **Figure 8.11 The ATP cycle.** Energy released by breakdown reactions (catabolism) in the cell is used to phosphorylate ADP, regenerating ATP. Chemical potential energy stored in ATP drives most cellular work.

phosphorylate ADP comes from exergonic breakdown reactions (catabolism) in the cell. This shuttling of inorganic phosphate and energy is called the ATP cycle, and it couples the cell's energy-yielding (exergonic) processes to the energy-consuming (endergonic) ones. The ATP cycle proceeds at an astonishing pace. For example, a working muscle cell recycles its entire pool of ATP in less than a minute. That turnover represents 10 million molecules of ATP consumed and regenerated per second per cell. If ATP could not be regenerated by the phosphorylation of ADP, humans would use up nearly their body weight in ATP each day.

Because both directions of a reversible process cannot be downhill, the regeneration of ATP from ADP and $Ⓟ_i$ is necessarily endergonic:

$$\text{ADP} + Ⓟ_i \rightarrow \text{ATP} + H_2O$$
$$\Delta G = +7.3 \text{ kcal/mol } (+30.5 \text{ kJ/mol}) \text{ (standard conditions)}$$

Since ATP formation from ADP and $Ⓟ_i$ is not spontaneous, free energy must be spent to make it occur. Catabolic (exergonic) pathways, especially cellular respiration, provide the energy for the endergonic process of making ATP. Plants also use light energy to produce ATP. Thus, the ATP cycle is a revolving door through which energy passes during its transfer from catabolic to anabolic pathways.

CONCEPT CHECK 8.3

1. How does ATP typically transfer energy from exergonic to endergonic reactions in the cell?
2. Which of the following combinations has more free energy: glutamic acid + ammonia + ATP, or glutamine + ADP + $Ⓟ_i$? Explain your answer.
3. **MAKE CONNECTIONS** Considering what you learned in Concepts 7.3 and 7.4 (pp. 134–136), does Figure 8.10a show passive or active transport? Explain.

For suggested answers, see Appendix A.

CONCEPT 8.4

Enzymes speed up metabolic reactions by lowering energy barriers

The laws of thermodynamics tell us what will and will not happen under given conditions but say nothing about the rate of these processes. A spontaneous chemical reaction occurs without any requirement for outside energy, but it may occur so slowly that it is imperceptible. For example, even though the hydrolysis of sucrose (table sugar) to glucose and fructose is exergonic, occurring spontaneously with a release of free energy ($\Delta G = -7$ kcal/mol), a solution of sucrose dissolved in sterile water will sit for years at room temperature with no appreciable hydrolysis. However, if we add a small amount of the enzyme sucrase to the solution, then all the sucrose may be hydrolyzed within seconds, as shown below:

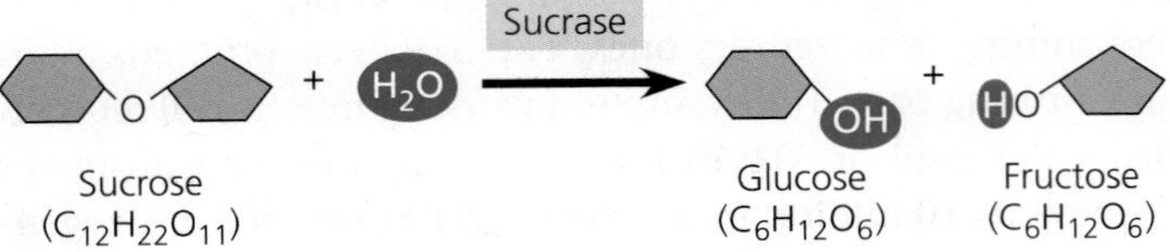

How does the enzyme do this?

An **enzyme** is a macromolecule that acts as a **catalyst**, a chemical agent that speeds up a reaction without being consumed by the reaction. (In this chapter, we are focusing on enzymes that are proteins. RNA enzymes, also called ribozymes, are discussed in Chapters 17 and 25.) Without regulation by enzymes, chemical traffic through the pathways of metabolism would become terribly congested because many chemical reactions would take such a long time. In the next two sections, we will see what prevents a spontaneous reaction from occurring faster and how an enzyme changes the situation.

The Activation Energy Barrier

Every chemical reaction between molecules involves both bond breaking and bond forming. For example, the hydrolysis of sucrose involves breaking the bond between glucose and fructose and one of the bonds of a water molecule and then forming two new bonds, as shown above. Changing one molecule into another generally involves contorting the starting molecule into a highly unstable state before the reaction can proceed. This contortion can be compared to the bending of a metal key ring when you pry it open to add a new key. The key ring is highly unstable in its opened form but returns to a stable state once the key is threaded all the way onto the ring. To reach the contorted state where bonds can change, reactant molecules must absorb energy from their surroundings. When the new bonds of the product molecules form, energy is released as heat, and the molecules return to stable shapes with lower energy than the contorted state.

The initial investment of energy for starting a reaction—the energy required to contort the reactant molecules so the bonds can break—is known as the *free energy of activation*, or **activation energy**, abbreviated E_A in this book. We can think of activation energy as the amount of energy needed to push the reactants to the top of an energy barrier, or uphill, so that the "downhill" part of the reaction can begin. Activation energy is often supplied in the form of thermal energy (heat) that the reactant molecules absorb from the surroundings. The absorption of thermal energy accelerates the reactant molecules, so they collide more often and more forcefully. It also agitates the atoms within the molecules, making the breakage of bonds more likely. When the molecules have absorbed enough energy for the bonds to break, the reactants are in an unstable condition known as the *transition state*.

Figure 8.12 graphs the energy changes for a hypothetical exergonic reaction that swaps portions of two reactant molecules:

$$AB + CD \rightarrow AC + BD$$
Reactants Products

The activation of the reactants is represented by the uphill portion of the graph, in which the free-energy content of the

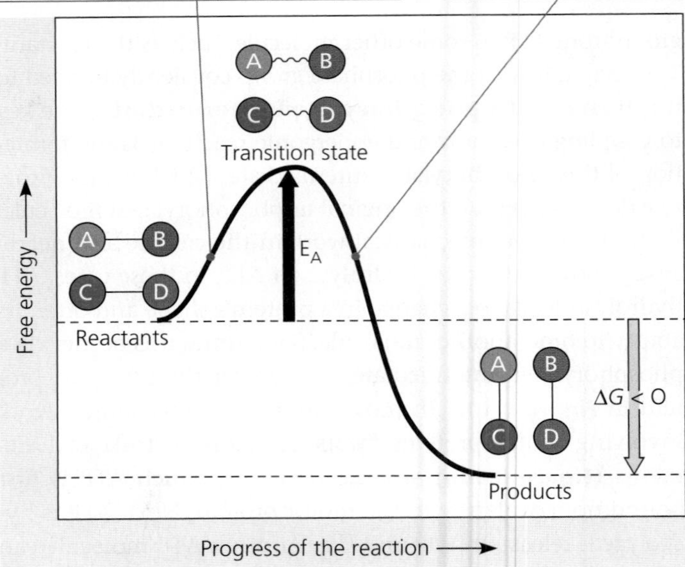

▲ **Figure 8.12 Energy profile of an exergonic reaction.** The "molecules" are hypothetical, with A, B, C, and D representing portions of the molecules. Thermodynamically, this is an exergonic reaction, with a negative ΔG, and the reaction occurs spontaneously. However, the activation energy (E_A) provides a barrier that determines the rate of the reaction.

DRAW IT *Graph the progress of an endergonic reaction in which EF and GH form products EG and FH, assuming that the reactants must pass through a transition state.*

reactant molecules is increasing. At the summit, when energy equivalent to E_A has been absorbed, the reactants are in the transition state: They are activated, and their bonds can be broken. As the atoms then settle into their new, more stable bonding arrangements, energy is released to the surroundings. This corresponds to the downhill part of the curve, which shows the loss of free energy by the molecules. The overall decrease in free energy means that E_A is repaid with interest, as the formation of new bonds releases more energy than was invested in the breaking of old bonds..

The reaction shown in Figure 8.12 is exergonic and occurs spontaneously ($\Delta G < 0$). However, the activation energy provides a barrier that determines the rate of the reaction. The reactants must absorb enough energy to reach the top of the activation energy barrier before the reaction can occur. For some reactions, E_A is modest enough that even at room temperature there is sufficient thermal energy for many of the reactant molecules to reach the transition state in a short time. In most cases, however, E_A is so high and the transition state is reached so rarely that the reaction will hardly proceed at all. In these cases, the reaction will occur at a noticeable rate only if the reactants are heated. For example, the reaction of gasoline and oxygen is exergonic and will occur spontaneously, but energy is required for the molecules to reach the transition state and react. Only when the spark plugs fire in an automobile engine can there be the explosive release of energy that pushes the pistons. Without a spark, a mixture of gasoline hydrocarbons and oxygen will not react because the E_A barrier is too high.

How Enzymes Lower the E_A Barrier

Proteins, DNA, and other complex molecules of the cell are rich in free energy and have the potential to decompose spontaneously; that is, the laws of thermodynamics favor their breakdown. These molecules persist only because at temperatures typical for cells, few molecules can make it over the hump of activation energy. However, the barriers for selected reactions must occasionally be surmounted for cells to carry out the processes needed for life. Heat speeds a reaction by allowing reactants to attain the transition state more often, but this solution would be inappropriate for biological systems. First, high temperature denatures proteins and kills cells. Second, heat would speed up *all* reactions, not just those that are needed. Instead of heat, organisms use catalysis to speed up reactions.

An enzyme catalyzes a reaction by lowering the E_A barrier **(Figure 8.13)**, enabling the reactant molecules to absorb enough energy to reach the transition state even at moderate temperatures. An enzyme cannot change the ΔG for a reaction; it cannot make an endergonic reaction exergonic. Enzymes can only hasten reactions that would eventually occur anyway, but this function makes it possible for the cell to have a dynamic

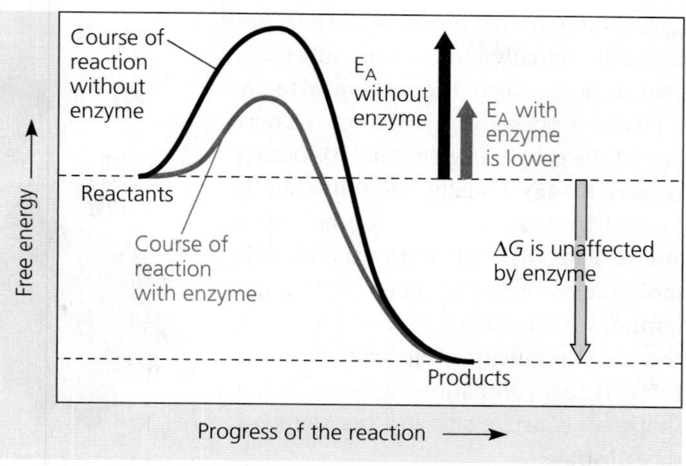

▲ **Figure 8.13 The effect of an enzyme on activation energy.** Without affecting the free-energy change (ΔG) for a reaction, an enzyme speeds the reaction by reducing its activation energy (E_A).

metabolism, routing chemicals smoothly through the cell's metabolic pathways. And because enzymes are very specific for the reactions they catalyze, they determine which chemical processes will be going on in the cell at any particular time.

Substrate Specificity of Enzymes

The reactant an enzyme acts on is referred to as the enzyme's **substrate**. The enzyme binds to its substrate (or substrates, when there are two or more reactants), forming an **enzyme-substrate complex**. While enzyme and substrate are joined, the catalytic action of the enzyme converts the substrate to the product (or products) of the reaction. The overall process can be summarized as follows:

Enzyme + Substrate(s)	⇌	Enzyme-substrate complex	⇌	Enzyme + Product(s)

For example, the enzyme sucrase (most enzyme names end in *-ase*) catalyzes the hydrolysis of the disaccharide sucrose into its two monosaccharides, glucose and fructose (see p. 152):

Sucrase + Sucrose + H_2O	⇌	Sucrase-sucrose-H_2O complex	⇌	Sucrase + Glucose + Fructose

The reaction catalyzed by each enzyme is very specific; an enzyme can recognize its specific substrate even among closely related compounds. For instance, sucrase will act only on sucrose and will not bind to other disaccharides, such as maltose. What accounts for this molecular recognition? Recall that most enzymes are proteins, and proteins are macromolecules with unique three-dimensional configurations. The specificity of an enzyme results from its shape, which is a consequence of its amino acid sequence.

Only a restricted region of the enzyme molecule actually binds to the substrate. This region, called the **active site**, is typically a pocket or groove on the surface of the enzyme where catalysis occurs **(Figure 8.14a)**. Usually, the active site is formed by only a few of the enzyme's amino acids, with the rest of the protein molecule providing a framework that determines the configuration of the active site. The specificity of an enzyme is attributed to a compatible fit between the shape of its active site and the shape of the substrate.

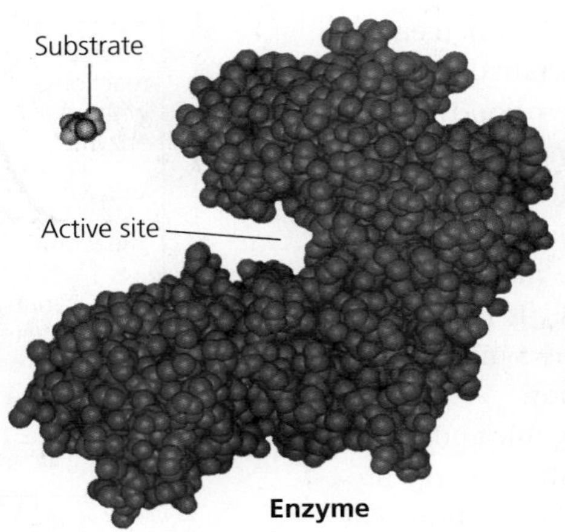

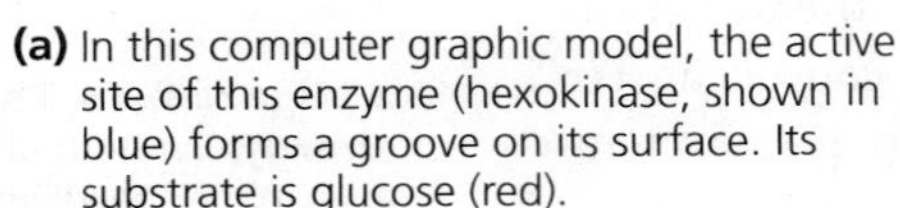

(a) In this computer graphic model, the active site of this enzyme (hexokinase, shown in blue) forms a groove on its surface. Its substrate is glucose (red).

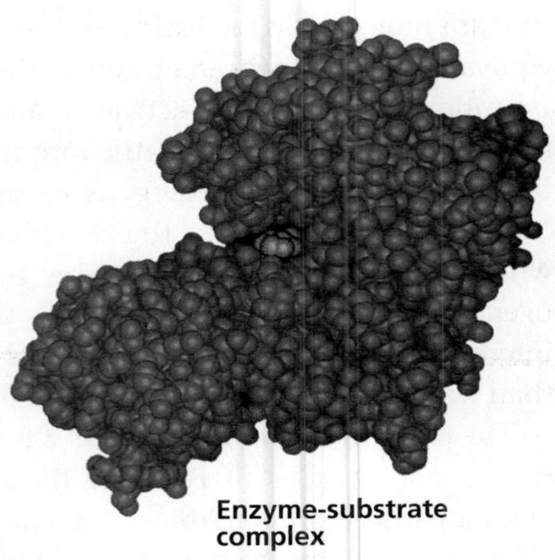

(b) When the substrate enters the active site, it forms weak bonds with the enzyme, inducing a change in the shape of the protein. This change allows additional weak bonds to form, causing the active site to enfold the substrate and hold it in place.

▲ **Figure 8.14 Induced fit between an enzyme and its substrate.**

An enzyme is not a stiff structure locked into a given shape. In fact, recent work by biochemists has shown clearly that enzymes (and other proteins as well) seem to "dance" between subtly different shapes in a dynamic equilibrium, with slight differences in free energy for each "pose." The shape that best fits the substrate isn't necessarily the one with the lowest energy, but during the very short time the enzyme takes on this shape, its active site can bind to the substrate. It has been known for more than 50 years that the active site itself is also not a rigid receptacle for the substrate. As the substrate enters the active site, the enzyme changes shape slightly due to interactions between the substrate's chemical groups and chemical groups on the side chains of the amino acids that form the active site. This shape change makes the active site fit even more snugly around the substrate **(Figure 8.14b)**. This **induced fit** is like a clasping handshake. Induced fit brings chemical groups of the active site into positions that enhance their ability to catalyze the chemical reaction.

Catalysis in the Enzyme's Active Site

In most enzymatic reactions, the substrate is held in the active site by so-called weak interactions, such as hydrogen bonds and ionic bonds. R groups of a few of the amino acids that make up the active site catalyze the conversion of substrate to product, and the product departs from the active site. The enzyme is then free to take another substrate molecule into its active site. The entire cycle happens so fast that a single enzyme molecule typically acts on about a thousand substrate molecules per second. Some enzymes are much faster. Enzymes, like other catalysts, emerge from the reaction in their original form. Therefore, very small amounts of enzyme can have a huge metabolic impact by functioning over and over again in catalytic cycles. **Figure 8.15** shows a catalytic cycle involving two substrates and two products.

Most metabolic reactions are reversible, and an enzyme can catalyze either the forward or the reverse reaction, depending on which direction has a negative ΔG. This in turn depends mainly on the relative concentrations of reactants and products. The net effect is always in the direction of equilibrium.

Enzymes use a variety of mechanisms that lower activation energy and speed up a reaction (see Figure 8.15, step ❸). First, in reactions involving two or more reactants, the active site provides a template on which the substrates can come together in the proper orientation for a reaction to occur between them. Second, as the active site of an enzyme clutches the bound substrates, the enzyme may stretch the substrate molecules toward their transition-state form, stressing and bending critical chemical bonds that must be broken during the reaction. Because E_A is proportional to the difficulty of breaking the bonds, distorting the substrate helps it approach the transition state and thus reduces the amount of free energy that must be absorbed to achieve that state.

Third, the active site may also provide a microenvironment that is more conducive to a particular type of reaction than the solution itself would be without the enzyme. For example, if the active site has amino acids with acidic R groups, the active site may be a pocket of low pH in an otherwise neutral cell. In such cases, an acidic amino acid may facilitate H^+ transfer to the substrate as a key step in catalyzing the reaction.

A fourth mechanism of catalysis is the direct participation of the active site in the chemical reaction. Sometimes this process even involves brief covalent bonding between the substrate and the side chain of an amino acid of the enzyme. Subsequent steps of the reaction restore the side chains to their original states, so that the active site is the same after the reaction as it was before.

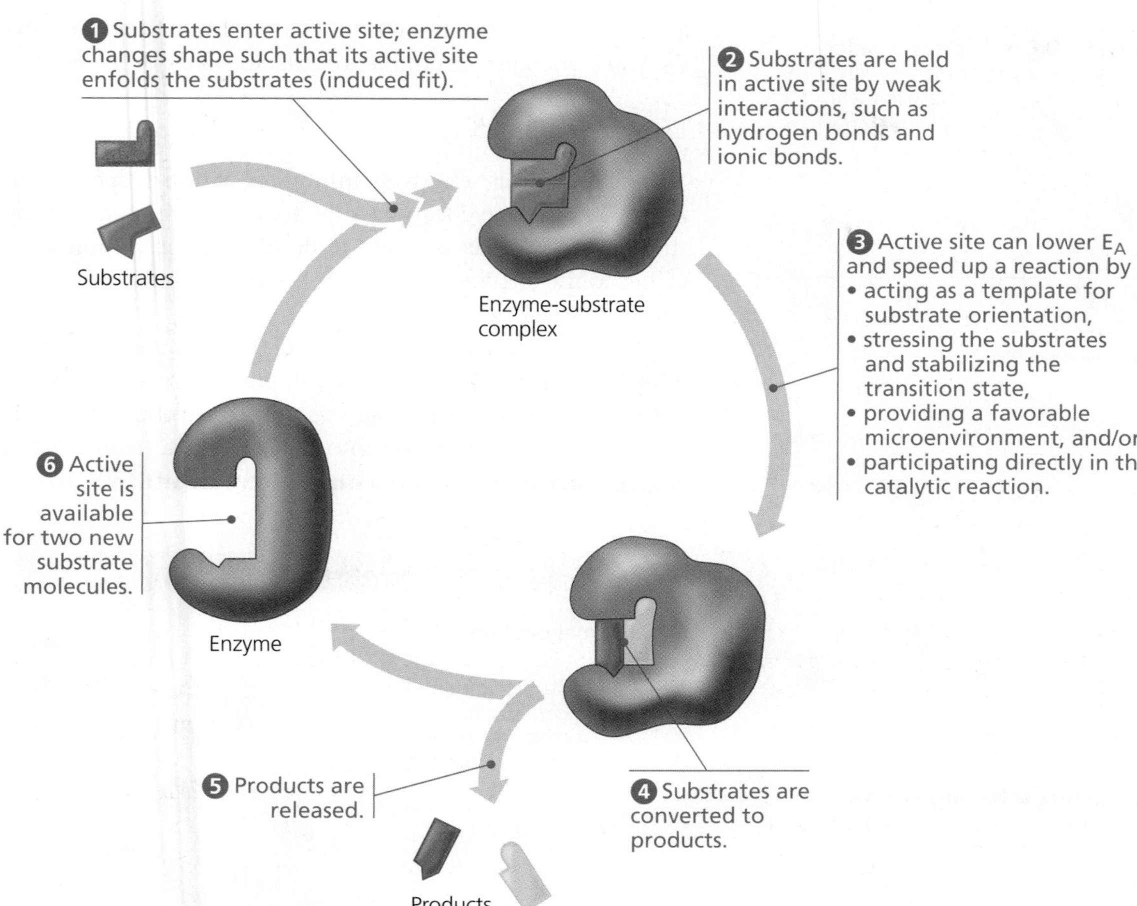
1 Substrates enter active site; enzyme changes shape such that its active site enfolds the substrates (induced fit).
2 Substrates are held in active site by weak interactions, such as hydrogen bonds and ionic bonds.
Substrates
Enzyme-substrate complex
3 Active site can lower E_A and speed up a reaction by
• acting as a template for substrate orientation,
• stressing the substrates and stabilizing the transition state,
• providing a favorable microenvironment, and/or
• participating directly in the catalytic reaction.
6 Active site is available for two new substrate molecules.
Enzyme
5 Products are released.
4 Substrates are converted to products.
Products

◄ **Figure 8.15 The active site and catalytic cycle of an enzyme.** An enzyme can convert one or more reactant molecules to one or more product molecules. The enzyme shown here converts two substrate molecules to two product molecules.

The rate at which a particular amount of enzyme converts substrate to product is partly a function of the initial concentration of the substrate: The more substrate molecules that are available, the more frequently they access the active sites of the enzyme molecules. However, there is a limit to how fast the reaction can be pushed by adding more substrate to a fixed concentration of enzyme. At some point, the concentration of substrate will be high enough that all enzyme molecules have their active sites engaged. As soon as the product exits an active site, another substrate molecule enters. At this substrate concentration, the enzyme is said to be *saturated*, and the rate of the reaction is determined by the speed at which the active site converts substrate to product. When an enzyme population is saturated, the only way to increase the rate of product formation is to add more enzyme. Cells often increase the rate of a reaction by producing more enzyme molecules.

Effects of Local Conditions on Enzyme Activity

The activity of an enzyme—how efficiently the enzyme functions—is affected by general environmental factors, such as temperature and pH. It can also be affected by chemicals that specifically influence that enzyme. In fact, researchers have learned much about enzyme function by employing such chemicals.

Effects of Temperature and pH

Recall from Chapter 5 that the three-dimensional structures of proteins are sensitive to their environment. As a consequence, each enzyme works better under some conditions than under other conditions, because these *optimal conditions* favor the most active shape for the enzyme molecule.

Temperature and pH are environmental factors important in the activity of an enzyme. Up to a point, the rate of an enzymatic reaction increases with increasing temperature, partly because substrates collide with active sites more frequently when the molecules move rapidly. Above that temperature, however, the speed of the enzymatic reaction drops sharply. The thermal agitation of the enzyme molecule disrupts the hydrogen bonds, ionic bonds, and other weak interactions that stabilize the active shape of the enzyme, and the protein molecule eventually denatures. Each enzyme has an optimal temperature at which its reaction rate is greatest. Without denaturing the enzyme, this temperature allows the greatest number of molecular collisions and the fastest conversion of the reactants to product molecules. Most human enzymes have optimal temperatures of about 35–40°C (close to human body temperature). The thermophilic bacteria that live in hot springs contain enzymes with optimal temperatures of 70°C or higher (**Figure 8.16a** on the next page).

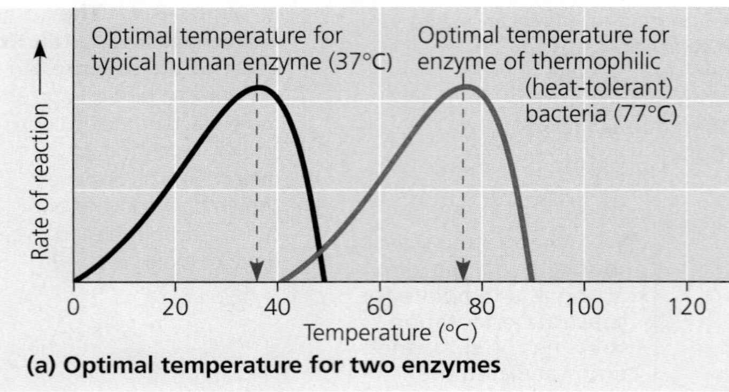
Optimal temperature for typical human enzyme (37°C)
Optimal temperature for enzyme of thermophilic (heat-tolerant) bacteria (77°C)
Rate of reaction
0 20 40 60 80 100 120
Temperature (°C)
(a) Optimal temperature for two enzymes

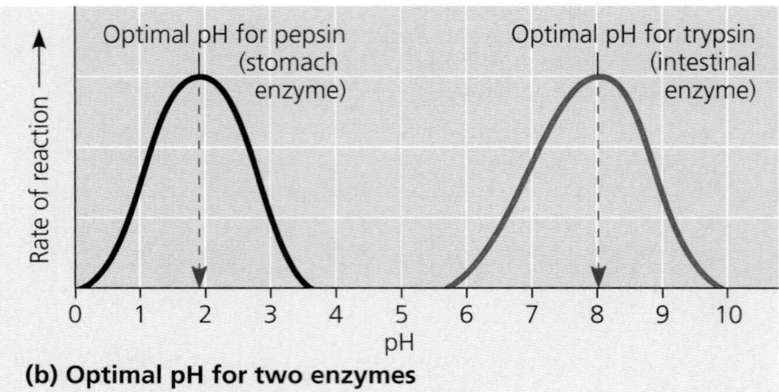
Optimal pH for pepsin (stomach enzyme)
Optimal pH for trypsin (intestinal enzyme)
Rate of reaction
0 1 2 3 4 5 6 7 8 9 10
pH
(b) Optimal pH for two enzymes

▲ **Figure 8.16 Environmental factors affecting enzyme activity.** Each enzyme has an optimal **(a)** temperature and **(b)** pH that favor the most active shape of the protein molecule.

DRAW IT *Given that a mature lysosome has an internal pH of around 4.5, draw a curve in (b) showing what you would predict for a lysosomal enzyme, labeling its optimal pH.*

Just as each enzyme has an optimal temperature, it also has a pH at which it is most active. The optimal pH values for most enzymes fall in the range of pH 6–8, but there are exceptions. For example, pepsin, a digestive enzyme in the human stomach, works best at pH 2. Such an acidic environment denatures most enzymes, but pepsin is adapted to maintain its functional three-dimensional structure in the acidic environment of the stomach. In contrast, trypsin, a digestive enzyme residing in the alkaline environment of the human intestine, has an optimal pH of 8 and would be denatured in the stomach (**Figure 8.16b**).

Cofactors

Many enzymes require nonprotein helpers for catalytic activity. These adjuncts, called **cofactors**, may be bound tightly to the enzyme as permanent residents, or they may bind loosely and reversibly along with the substrate. The cofactors of some enzymes are inorganic, such as the metal atoms zinc, iron, and copper in ionic form. If the cofactor is an organic molecule, it is more specifically called a **coenzyme**. Most vitamins are important in nutrition because they act as coenzymes or raw materials from which coenzymes are made. Cofactors function in various ways, but in all cases where they are used, they perform a crucial chemical function in catalysis. You'll encounter examples of cofactors later in the book.

Enzyme Inhibitors

Certain chemicals selectively inhibit the action of specific enzymes, and we have learned a lot about enzyme function by studying the effects of these molecules. If the inhibitor attaches to the enzyme by covalent bonds, inhibition is usually irreversible.

Many enzyme inhibitors, however, bind to the enzyme by weak interactions, in which case inhibition is reversible. Some reversible inhibitors resemble the normal substrate molecule and compete for admission into the active site (**Figure 8.17a** and **b**). These mimics, called **competitive inhibitors**, reduce

▼ **Figure 8.17 Inhibition of enzyme activity.**

(a) Normal binding

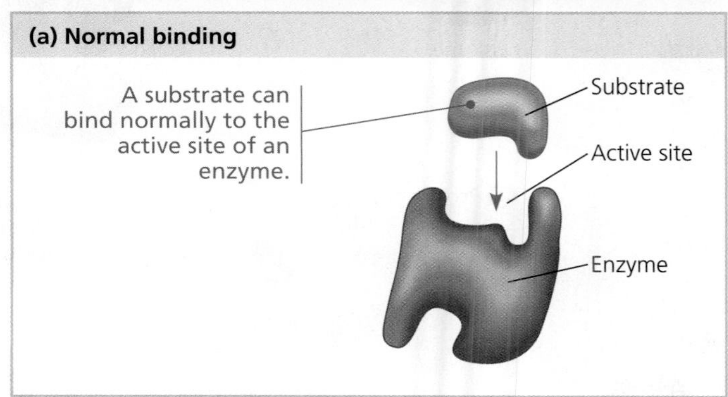
A substrate can bind normally to the active site of an enzyme.
Substrate
Active site
Enzyme

(b) Competitive inhibition

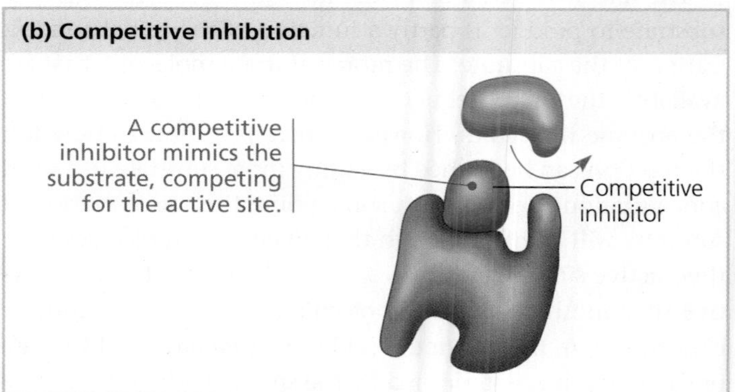
A competitive inhibitor mimics the substrate, competing for the active site.
Competitive inhibitor

(c) Noncompetitive inhibition

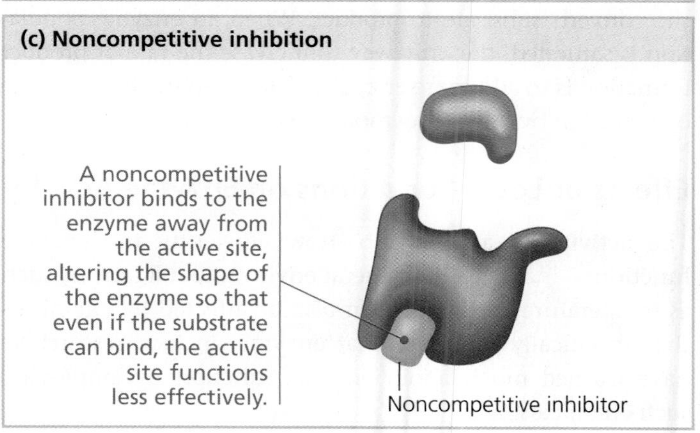
A noncompetitive inhibitor binds to the enzyme away from the active site, altering the shape of the enzyme so that even if the substrate can bind, the active site functions less effectively.
Noncompetitive inhibitor

the productivity of enzymes by blocking substrates from entering active sites. This kind of inhibition can be overcome by increasing the concentration of substrate so that as active sites become available, more substrate molecules than inhibitor molecules are around to gain entry to the sites.

In contrast, **noncompetitive inhibitors** do not directly compete with the substrate to bind to the enzyme at the active site (**Figure 8.17c**). Instead, they impede enzymatic reactions by binding to another part of the enzyme. This interaction causes the enzyme molecule to change its shape in such a way that the active site becomes less effective at catalyzing the conversion of substrate to product.

Toxins and poisons are often irreversible enzyme inhibitors. An example is sarin, a nerve gas that caused the death of several people and injury to many others when it was released by terrorists in the Tokyo subway in 1995. This small molecule binds covalently to the R group on the amino acid serine, which is found in the active site of acetylcholinesterase, an enzyme important in the nervous system. Other examples include the pesticides DDT and parathion, inhibitors of key enzymes in the nervous system. Finally, many antibiotics are inhibitors of specific enzymes in bacteria. For instance, penicillin blocks the active site of an enzyme that many bacteria use to make their cell walls.

Citing enzyme inhibitors that are metabolic poisons may give the impression that enzyme inhibition is generally abnormal and harmful. In fact, molecules naturally present in the cell often regulate enzyme activity by acting as inhibitors. Such regulation—selective inhibition—is essential to the control of cellular metabolism, as we will discuss in Concept 8.5.

The Evolution of Enzymes

EVOLUTION Thus far, biochemists have discovered and named more than 4,000 different enzymes in various species, and this list probably represents the tip of the proverbial iceberg. How did this grand profusion of enzymes arise? Recall that most enzymes are proteins, and proteins are encoded by genes. A permanent change in a gene, known as a *mutation*, can result in a protein with one or more changed amino acids. In the case of an enzyme, if the changed amino acids are in the active site or some other crucial region, the altered enzyme might have a novel activity or might bind to a different substrate. Under environmental conditions where the new function benefits the organism, natural selection would tend to favor the mutated form of the gene, causing it to persist in the population. This simplified model is generally accepted as the main way in which the multitude of different enzymes arose over the past few billion years of life's history.

Data supporting this model have been collected by researchers using a lab procedure that mimics evolution in natural populations. One group tested whether the function of an enzyme called β-galactosidase could change over time in populations of the bacterium *Escherichia coli* (*E. coli*). β-galactosidase breaks down the disaccharide lactose into the simple sugars glucose and galactose. Using molecular techniques, the researchers introduced random mutations into *E. coli* genes and then tested the bacteria for their ability to break down a slightly different disaccharide (one that has the sugar fucose in place of galactose). They selected the mutant bacteria that could do this best and exposed them to another round of mutation and selection. After seven rounds, the "evolved" enzyme bound the new substrate several hundred times more strongly, and broke it down 10 to 20 times more quickly, than did the original enzyme.

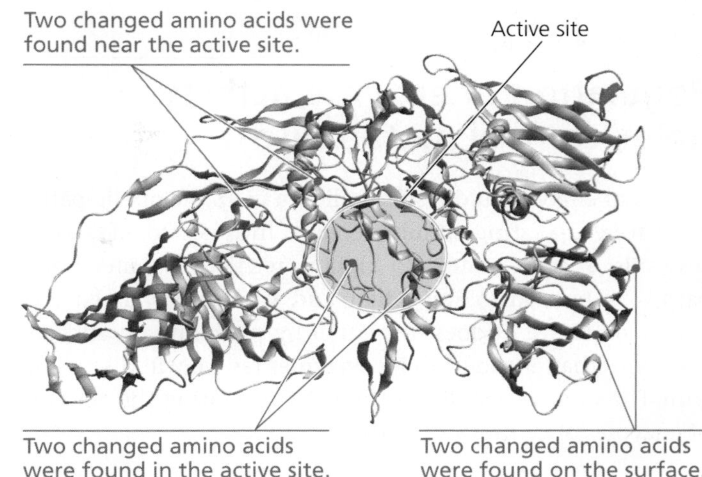

Two changed amino acids were found near the active site.
Active site
Two changed amino acids were found in the active site.
Two changed amino acids were found on the surface.

▲ **Figure 8.18 Mimicking evolution of an enzyme with a new function.** After seven rounds of mutation and selection in a lab, the enzyme β-galactosidase evolved into an enzyme specialized for breaking down a sugar different from lactose. This ribbon model shows one subunit of the altered enzyme; six amino acids were different.

The researchers found that six amino acids had changed in the enzyme altered in this experiment. Two of these changed amino acids were in the active site, two were nearby, and two were on the surface of the protein (**Figure 8.18**). This experiment and others like it strengthen the notion that a few changes can indeed alter enzyme function.

CONCEPT CHECK 8.4

1. Many spontaneous reactions occur very slowly. Why don't all spontaneous reactions occur instantly?
2. Why do enzymes act only on very specific substrates?
3. **WHAT IF?** Malonate is an inhibitor of the enzyme succinate dehydrogenase. How would you determine whether malonate is a competitive or noncompetitive inhibitor?
4. **MAKE CONNECTIONS** In nature, what conditions could lead to natural selection favoring bacteria with enzymes that could break down the fucose-containing disaccharide discussed above? See the discussion of natural selection in Concept 1.2, pages 14–16.

For suggested answers, see Appendix A.

CONCEPT 8.5

Regulation of enzyme activity helps control metabolism

▼ **Figure 8.19 Allosteric regulation of enzyme activity.**

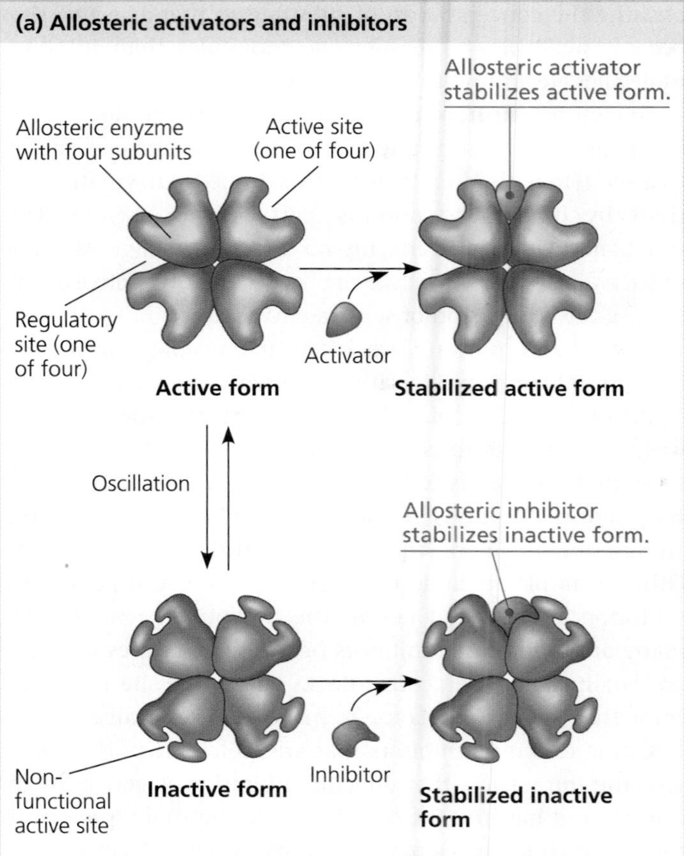

At low concentrations, activators and inhibitors dissociate from the enzyme. The enzyme can then oscillate again.

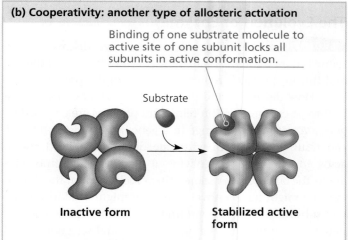

(b) Cooperativity: another type of allosteric activation

Binding of one substrate molecule to active site of one subunit locks all subunits in active conformation.

Substrate

Inactive form

Stabilized active form

The inactive form shown on the left oscillates with the active form when the active form is not stabilized by substrate.

Chemical chaos would result if all of a cell's metabolic pathways were operating simultaneously. Intrinsic to life's processes is a cell's ability to tightly regulate its metabolic pathways by controlling when and where its various enzymes are active. It does this either by switching on and off the genes that encode specific enzymes (as we will discuss in Unit Three) or, as we discuss here, by regulating the activity of enzymes once they are made.

Allosteric Regulation of Enzymes

In many cases, the molecules that naturally regulate enzyme activity in a cell behave something like reversible noncompetitive inhibitors (see Figure 8.17c): These regulatory molecules change an enzyme's shape and the functioning of its active site by binding to a site elsewhere on the molecule, via noncovalent interactions. **Allosteric regulation** is the term used to describe any case in which a protein's function at one site is affected by the binding of a regulatory molecule to a separate site. It may result in either inhibition or stimulation of an enzyme's activity.

Allosteric Activation and Inhibition

Most enzymes known to be allosterically regulated are constructed from two or more subunits, each composed of a polypeptide chain with its own active site. The entire complex oscillates between two different shapes, one catalytically active and the other inactive (**Figure 8.19a**). In the simplest kind of allosteric regulation, an activating or inhibiting regulatory molecule binds to a regulatory site (sometimes called an allosteric site), often located where subunits join. The binding of an *activator* to a regulatory site stabilizes the shape that has functional active sites, whereas the binding of an *inhibitor* stabilizes the inactive form of the enzyme. The subunits of an allosteric enzyme fit together in such a way that a shape change in one subunit is transmitted to all others. Through this interaction of subunits, a single activator or inhibitor molecule that binds to one regulatory site will affect the active sites of all subunits.

Fluctuating concentrations of regulators can cause a sophisticated pattern of response in the activity of cellular enzymes. The products of ATP hydrolysis (ADP and Ⓟ$_i$), for example, play a complex role in balancing the flow of traffic between anabolic and catabolic pathways by their effects on key enzymes. ATP binds to several catabolic enzymes allosterically, lowering their affinity for substrate and thus inhibiting their activity. ADP, however, functions as an activator of the same enzymes. This is logical because catabolism functions in regenerating ATP. If ATP production lags behind its use, ADP accumulates and activates the enzymes that speed up catabolism, producing more ATP. If the supply of ATP exceeds demand, then catabolism slows down as ATP molecules accumulate and bind to the same enzymes, inhibiting them. (You'll see specific examples of this type of regulation when you learn about cellular respiration in the next chapter.) ATP, ADP, and other related molecules also affect key enzymes in anabolic pathways. In this way, allosteric enzymes control the rates of important reactions in both sorts of metabolic pathways.

In another kind of allosteric activation, a *substrate* molecule binding to one active site in a multisubunit enzyme triggers a shape change in all the subunits, thereby increasing catalytic activity at the other active sites **(Figure 8.19b)**. Called **cooperativity**, this mechanism amplifies the response of enzymes to substrates: One substrate molecule primes an enzyme to act on additional substrate molecules more readily. Cooperativity is considered "allosteric" regulation because binding of the substrate to one active site affects catalysis in another active site.

Although the vertebrate oxygen transport protein hemoglobin is not an enzyme, classic studies of cooperative binding in this protein have elucidated the principle of cooperativity. Hemoglobin is made up of four subunits, each of which has an oxygen-binding site (see Figure 5.20). The binding of an oxygen molecule to one binding site increases the affinity for oxygen of the remaining binding sites. Thus, where oxygen is at high levels, such as in the lungs or gills, hemoglobin's affinity for oxygen increases as more binding sites are filled. In oxygen-deprived tissues, however, the release of each oxygen molecule decreases the oxygen affinity of the other binding sites, resulting in the release of oxygen where it is most needed. Cooperativity works similarly in multisubunit enzymes that have been studied.

Identification of Allosteric Regulators

Although allosteric regulation is probably quite widespread, relatively few of the many known metabolic enzymes have been shown to be regulated in this way. Allosteric regulatory molecules are hard to characterize, in part because they tend to bind the enzyme at low affinity and are therefore hard to isolate. Recently, however, pharmaceutical companies have turned their attention to allosteric regulators. These molecules are attractive drug candidates for enzyme regulation because they exhibit higher specificity for particular enzymes than do inhibitors that bind to the active site. (An active site may be similar to the active site in another, related enzyme, whereas allosteric regulatory sites appear to be quite distinct between enzymes.)

Figure 8.20 describes a search for allosteric regulators, carried out as a collaboration between researchers at the University of California at San Francisco and a company called Sunesis Pharmaceuticals. The study was designed to find allosteric inhibitors of *caspases*, protein-digesting enzymes that play an active role in inflammation and cell death. (You'll learn more about caspases and cell death in Chapter 11.) By specifically regulating these enzymes, we may be able to better manage inappropriate inflammatory responses, such as those commonly seen in vascular and neurodegenerative diseases.

▼ Figure 8.20 INQUIRY

Are there allosteric inhibitors of caspase enzymes?

EXPERIMENT In an effort to identify allosteric inhibitors of caspases, Justin Scheer and co-workers screened close to 8,000 compounds for their ability to bind to a possible allosteric binding site in caspase 1 and inhibit the enzyme's activity. Each compound was designed to form a disulfide bond with a cysteine near the site in order to stabilize the low-affinity interaction that is expected of an allosteric inhibitor. As the caspases are known to exist in both active and inactive forms, the researchers hypothesized that this linkage might lock the enzyme in the inactive form.

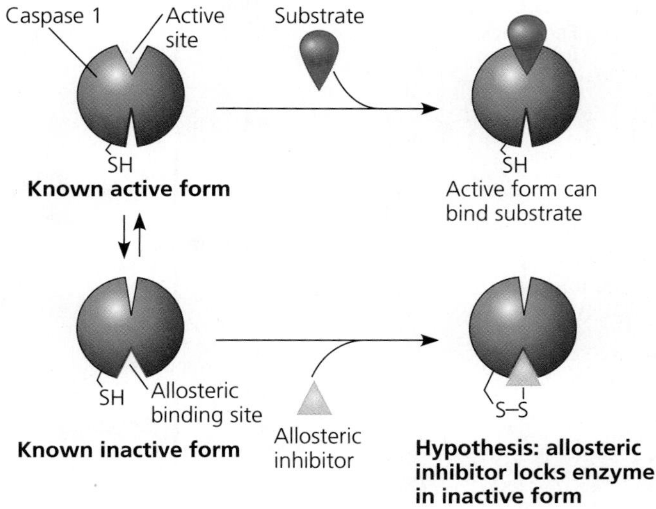

To test this model, X-ray diffraction analysis was used to determine the structure of caspase 1 when bound to one of the inhibitors and to compare it with the active and inactive structures.

RESULTS Fourteen compounds were identified that could bind to the proposed allosteric site (red) of caspase 1 and block enzymatic activity. The enzyme's shape when one such inhibitor was bound resembled the inactive caspase 1 more than the active form.

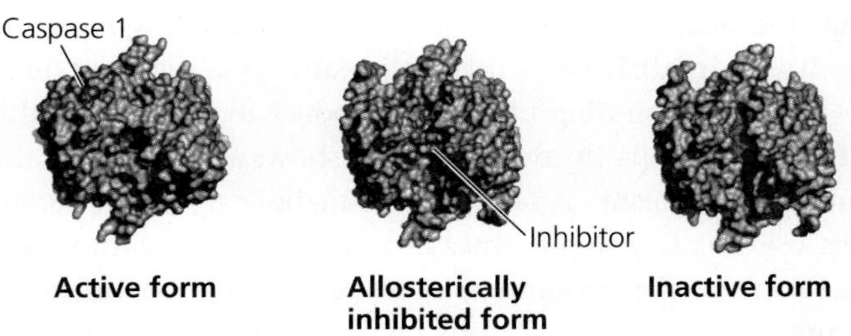

CONCLUSION That particular inhibitory compound apparently locks the enzyme in its inactive form, as expected for a true allosteric regulator. The data therefore support the existence of an allosteric inhibitory site on caspase 1 that can be used to control enzymatic activity.

SOURCE J. M. Scheer et al., A common allosteric site and mechanism in caspases, *Proceedings of the National Academy of Sciences* 103: 7595–7600 (2006).

WHAT IF? As a control, the researchers broke the disulfide linkage between one of the inhibitors and the caspase. Assuming that the experimental solution contains no other inhibitors, how would you expect the caspase 1 activity to be affected?

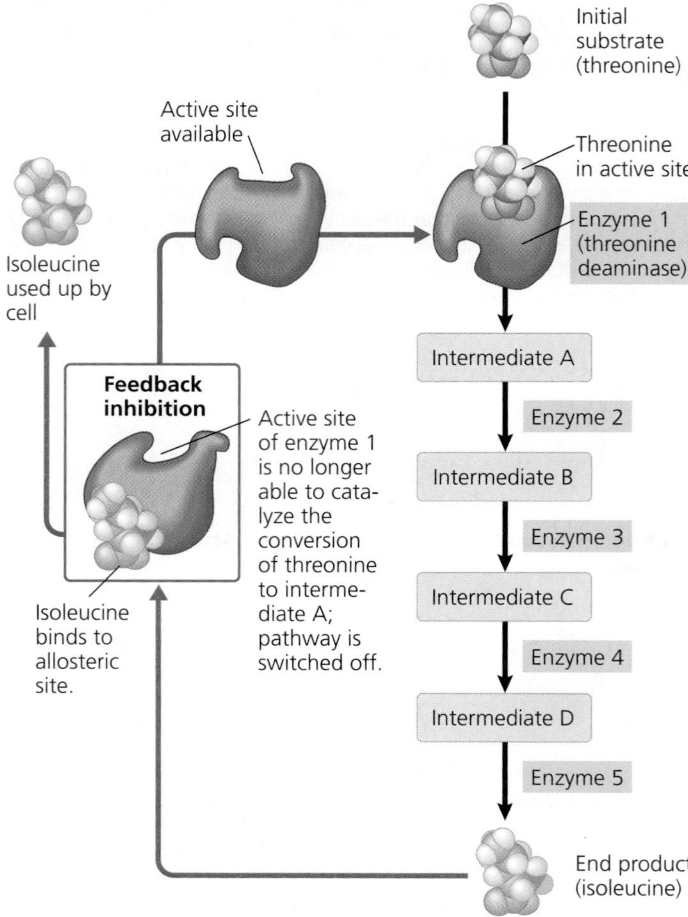

▲ **Figure 8.21 Feedback inhibition in isoleucine synthesis.**

Feedback Inhibition

When ATP allosterically inhibits an enzyme in an ATP-generating pathway, as we discussed earlier, the result is feedback inhibition, a common mode of metabolic control. In **feedback inhibition**, a metabolic pathway is switched off by the inhibitory binding of its end product to an enzyme that acts early in the pathway. **Figure 8.21** shows an example of this control mechanism operating on an anabolic pathway. Certain cells use this five-step pathway to synthesize the amino acid isoleucine from threonine, another amino acid. As isoleucine accumulates, it slows down its own synthesis by allosterically inhibiting the enzyme for the first step of the pathway. Feedback inhibition thereby prevents the cell from wasting chemical resources by making more isoleucine than is necessary.

Specific Localization of Enzymes Within the Cell

The cell is not just a bag of chemicals with thousands of different kinds of enzymes and substrates in a random mix. The cell is compartmentalized, and cellular structures help bring order to metabolic pathways. In some cases, a team of enzymes for several steps of a metabolic pathway are assembled into a multienzyme complex. The arrangement facilitates the sequence of reactions, with the product from the first enzyme becoming the substrate for an adjacent enzyme in the complex, and so on, until the end product is released. Some enzymes and enzyme complexes have fixed locations within the cell and act as structural components of particular membranes. Others are in solution within particular membrane-enclosed eukaryotic organelles, each with its own internal chemical environment. For example, in eukaryotic cells, the enzymes for cellular respiration reside in specific locations within mitochondria (**Figure 8.22**).

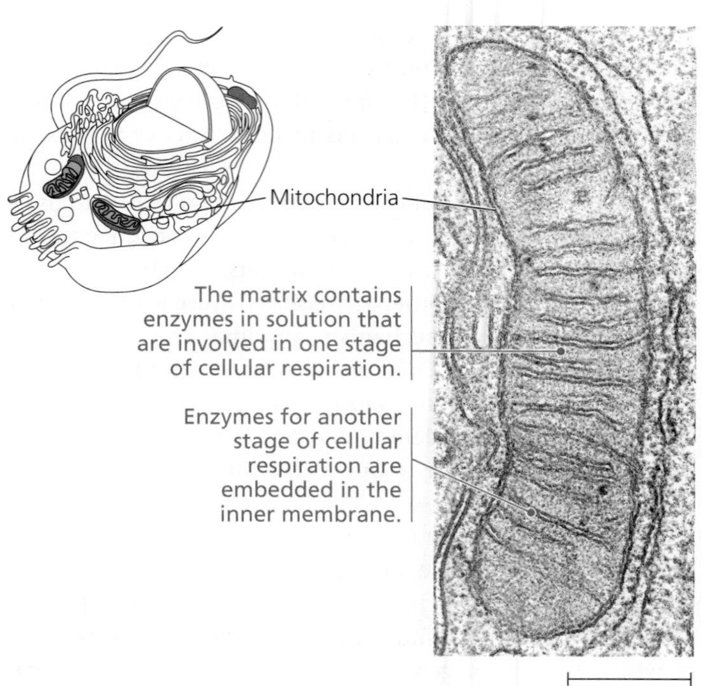

▲ **Figure 8.22 Organelles and structural order in metabolism.** Organelles such as the mitochondrion (TEM) contain enzymes that carry out specific functions, in this case cellular respiration.

In this chapter, you have learned that metabolism, the intersecting set of chemical pathways characteristic of life, is a choreographed interplay of thousands of different kinds of cellular molecules. In the next chapter, we explore cellular respiration, the major catabolic pathway that breaks down organic molecules, releasing energy for the crucial processes of life.

CONCEPT CHECK 8.5

1. How do an activator and an inhibitor have different effects on an allosterically regulated enzyme?
2. **WHAT IF?** Imagine you are a pharmacological researcher who wants to design a drug that inhibits a particular enzyme. Upon reading the scientific literature, you find that the enzyme's active site is similar to that of several other enzymes. What might be a good approach to developing your inhibitor drug?

For suggested answers, see Appendix A.

8 CHAPTER REVIEW

SUMMARY OF KEY CONCEPTS

CONCEPT 8.1

An organism's metabolism transforms matter and energy, subject to the laws of thermodynamics (pp. 142–145)

- **Metabolism** is the collection of chemical reactions that occur in an organism. **Enzymes** catalyze reactions in intersecting **metabolic pathways**, which may be **catabolic** (breaking down molecules, releasing energy) or **anabolic** (building molecules, consuming energy).
- **Energy** is the capacity to cause change; some forms of energy do work by moving matter. **Kinetic energy** is associated with motion and includes **thermal energy (heat)** associated with random motion of atoms or molecules. **Potential energy** is related to the location or structure of matter and includes **chemical energy** possessed by a molecule due to its structure.
- **The first law of thermodynamics**, conservation of energy, states that energy cannot be created or destroyed, only transferred or transformed. The **second law of thermodynamics** states that **spontaneous processes**, those requiring no outside input of energy, increase the **entropy** (disorder) of the universe.

? *Explain how the highly ordered structure of a cell does not conflict with the second law of thermodynamics.*

CONCEPT 8.2

The free-energy change of a reaction tells us whether or not the reaction occurs spontaneously (pp. 146–149)

- A living system's **free energy** is energy that can do work under cellular conditions. The change in free energy (ΔG) during a biological process is related directly to enthalpy change (ΔH) and to the change in entropy (ΔS): $\Delta G = \Delta H - T\Delta S$. Organisms live at the expense of free energy. During a spontaneous change, free energy decreases and the stability of a system increases. At maximum stability, the system is at equilibrium and can do no work.
- In an **exergonic** (spontaneous) chemical reaction, the products have less free energy than the reactants ($-\Delta G$). **Endergonic** (nonspontaneous) reactions require an input of energy ($+\Delta G$). The addition of starting materials and the removal of end products prevent metabolism from reaching equilibrium.

? *Explain the meaning of each component in the equation for the change in free energy of a spontaneous chemical reaction. Why are spontaneous reactions important in the metabolism of a cell?*

CONCEPT 8.3

ATP powers cellular work by coupling exergonic reactions to endergonic reactions (pp. 149–151)

- **ATP** is the cell's energy shuttle. Hydrolysis of its terminal phosphate yields ADP and phosphate and releases free energy.
- Through **energy coupling**, the exergonic process of ATP hydrolysis drives endergonic reactions by transfer of a phosphate group to specific reactants, forming a **phosphorylated intermediate** that is more reactive. ATP hydrolysis (sometimes with protein phosphorylation) also causes changes in the shape and binding affinities of transport and motor proteins.
- Catabolic pathways drive regeneration of ATP from ADP + Ⓟ$_i$.

? *Describe the ATP cycle: How is ATP used and regenerated in a cell?*

CONCEPT 8.4

Enzymes speed up metabolic reactions by lowering energy barriers (pp. 152–157)

- In a chemical reaction, the energy necessary to break the bonds of the reactants is the **activation energy**, E_A.
- **Enzymes** lower the E_A barrier:

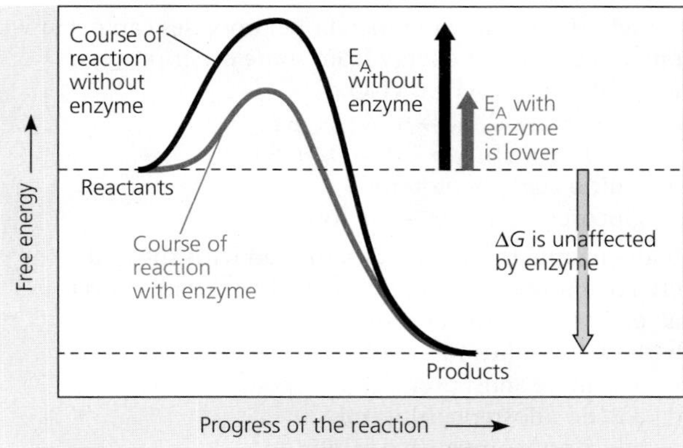

- Each type of enzyme has a unique **active site** that combines specifically with its **substrate(s)**, the reactant molecule(s) on which it acts. The enzyme changes shape slightly when it binds the substrate(s) (**induced fit**).
- The active site can lower an E_A barrier by orienting substrates correctly, straining their bonds, providing a favorable microenvironment, or even covalently bonding with the substrate.
- Each enzyme has an optimal temperature and pH. Inhibitors reduce enzyme function. A **competitive inhibitor** binds to the active site, whereas a **noncompetitive inhibitor** binds to a different site on the enzyme.
- Natural selection, acting on organisms with mutant genes encoding altered enzymes, is a major evolutionary force responsible for the diverse array of enzymes found in organisms.

? *How do both activation energy barriers and enzymes help maintain the structural and metabolic order of life?*

CONCEPT 8.5

Regulation of enzyme activity helps control metabolism (pp. 158–160)

- Many enzymes are subject to **allosteric regulation**: Regulatory molecules, either activators or inhibitors, bind to specific regulatory sites, affecting the shape and function of the enzyme. In **cooperativity**, binding of one substrate molecule can stimulate binding or activity at other active sites. In **feedback inhibition**, the end product of a metabolic pathway allosterically inhibits the enzyme for a previous step in the pathway.
- Some enzymes are grouped into complexes, some are incorporated into membranes, and some are contained inside organelles, increasing the efficiency of metabolic processes.

? *What roles do allosteric regulation and feedback inhibition play in the metabolism of a cell?*

TEST YOUR UNDERSTANDING

LEVEL 1: KNOWLEDGE/COMPREHENSION

1. Choose the pair of terms that correctly completes this sentence: Catabolism is to anabolism as _______ is to _______.
   a. exergonic; spontaneous
   b. exergonic; endergonic
   c. free energy; entropy
   d. work; energy
   e. entropy; enthalpy

2. Most cells cannot harness heat to perform work because
   a. heat is not a form of energy.
   b. cells do not have much heat; they are relatively cool.
   c. temperature is usually uniform throughout a cell.
   d. heat can never be used to do work.
   e. heat must remain constant during work.

3. Which of the following metabolic processes can occur without a net influx of energy from some other process?
   a. ADP + Ⓟ$_i$ $\rightarrow$ ATP + H_2O
   b. $C_6H_{12}O_6 + 6\ O_2 \rightarrow 6\ CO_2 + 6\ H_2O$
   c. $6\ CO_2 + 6\ H_2O \rightarrow C_6H_{12}O_6 + 6\ O_2$
   d. amino acids $\rightarrow$ protein
   e. glucose + fructose $\rightarrow$ sucrose

4. If an enzyme in solution is saturated with substrate, the most effective way to obtain a faster yield of products is to
   a. add more of the enzyme.
   b. heat the solution to 90°C.
   c. add more substrate.
   d. add an allosteric inhibitor.
   e. add a noncompetitive inhibitor.

5. Some bacteria are metabolically active in hot springs because
   a. they are able to maintain a lower internal temperature.
   b. high temperatures make catalysis unnecessary.
   c. their enzymes have high optimal temperatures.
   d. their enzymes are completely insensitive to temperature.
   e. they use molecules other than proteins or RNAs as their main catalysts.

LEVEL 2: APPLICATION/ANALYSIS

6. If an enzyme is added to a solution where its substrate and product are in equilibrium, what will occur?
   a. Additional product will be formed.
   b. Additional substrate will be formed.
   c. The reaction will change from endergonic to exergonic.
   d. The free energy of the system will change.
   e. Nothing; the reaction will stay at equilibrium.

LEVEL 3: SYNTHESIS/EVALUATION

7. **DRAW IT** Using a series of arrows, draw the branched metabolic reaction pathway described by the following statements, and then answer the question at the end. Use red arrows and minus signs to indicate inhibition.

   L can form either M or N.
   M can form O.
   O can form either P or R.
   P can form Q.
   R can form S.
   O inhibits the reaction of L to form M.
   Q inhibits the reaction of O to form P.
   S inhibits the reaction of O to form R.

   Which reaction would prevail if both Q and S were present in the cell in high concentrations?
   a. L $\rightarrow$ M
   b. M $\rightarrow$ O
   c. L $\rightarrow$ N
   d. O $\rightarrow$ P
   e. R $\rightarrow$ S

8. **EVOLUTION CONNECTION**
   A recent revival of the antievolutionary "intelligent design" argument holds that biochemical pathways are too complex to have evolved, because all intermediate steps in a given pathway must be present to produce the final product. Critique this argument. How could you use the diversity of metabolic pathways that produce the same or similar products to support your case?

9. **SCIENTIFIC INQUIRY**
   DRAW IT A researcher has developed an assay to measure the activity of an important enzyme present in liver cells growing in culture. She adds the enzyme's substrate to a dish of cells and then measures the appearance of reaction products. The results are graphed as the amount of product on the *y*-axis versus time on the *x*-axis. The researcher notes four sections of the graph. For a short period of time, no products appear (section A). Then (section B) the reaction rate is quite high (the slope of the line is steep). Next, the reaction gradually slows down (section C). Finally, the graph line becomes flat (section D). Draw and label the graph, and propose a model to explain the molecular events occurring at each stage of this reaction profile.

10. **SCIENCE, TECHNOLOGY, AND SOCIETY**
    Organophosphates (organic compounds containing phosphate groups) are commonly used as insecticides to improve crop yield. Organophosphates typically interfere with nerve signal transmission by inhibiting the enzymes that degrade transmitter molecules. They affect humans and other vertebrates as well as insects. Thus, the use of organophosphate pesticides poses some health risks. On the other hand, these molecules break down rapidly upon exposure to air and sunlight. As a consumer, what level of risk are you willing to accept in exchange for an abundant and affordable food supply?

11. **WRITE ABOUT A THEME**
    Energy Transfer Life requires energy. In a short essay (100–150 words), describe the basic principles of bioenergetics in an animal cell. How is the flow and transformation of energy different in a photosynthesizing cell? Include the role of ATP and enzymes in your discussion.

For selected answers, see Appendix A.

9

Cellular Respiration and Fermentation

▲ **Figure 9.1 How do these leaves power the work of life for this chimpanzee?**

KEY CONCEPTS

9.1 Catabolic pathways yield energy by oxidizing organic fuels

9.2 Glycolysis harvests chemical energy by oxidizing glucose to pyruvate

9.3 After pyruvate is oxidized, the citric acid cycle completes the energy-yielding oxidation of organic molecules

9.4 During oxidative phosphorylation, chemiosmosis couples electron transport to ATP synthesis

9.5 Fermentation and anaerobic respiration enable cells to produce ATP without the use of oxygen

9.6 Glycolysis and the citric acid cycle connect to many other metabolic pathways

OVERVIEW

Life Is Work

Living cells require transfusions of energy from outside sources to perform their many tasks—for example, assembling polymers, pumping substances across membranes, moving, and reproducing. The chimpanzee in **Figure 9.1** obtains energy for its cells by eating plants; some animals feed on other organisms that eat plants. The energy stored in the organic molecules of food ultimately comes from the sun. Energy flows into an ecosystem as sunlight and exits as heat; in contrast, the chemical elements essential to life are recycled (**Figure 9.2**). Photosynthesis generates oxygen and organic molecules used by the mitochondria of eukaryotes (including plants and algae) as fuel for cellular respiration. Respiration breaks this fuel down, generating ATP. The waste products of this type of respiration, carbon dioxide and water, are the raw materials for photosynthesis. In this chapter, we consider how cells harvest the chemical energy stored in organic molecules and use it to generate ATP, the molecule that drives most cellular work. After presenting some basics about respiration, we will focus on three key pathways of respiration: glycolysis, the citric acid cycle, and oxidative phosphorylation. We'll also consider fermentation, a somewhat simpler pathway coupled to glycolysis that has deep evolutionary roots.

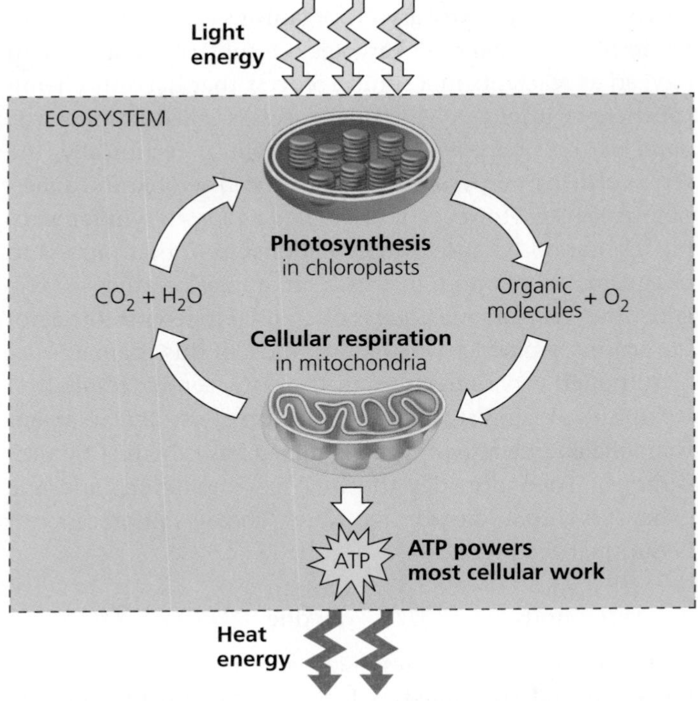

▲ **Figure 9.2 Energy flow and chemical recycling in ecosystems.** Energy flows into an ecosystem as sunlight and ultimately leaves as heat, while the chemical elements essential to life are recycled.

BioFlix Visit the Study Area at **www.masteringbiology.com** for the BioFlix® 3-D Animation on The Carbon Cycle.

CONCEPT 9.1

Catabolic pathways yield energy by oxidizing organic fuels

As you learned in Chapter 8, metabolic pathways that release stored energy by breaking down complex molecules are called catabolic pathways. Electron transfer plays a major role in these pathways. In this section, we consider these processes, which are central to cellular respiration.

Catabolic Pathways and Production of ATP

Organic compounds possess potential energy as a result of the arrangement of electrons in the bonds between their atoms. Compounds that can participate in exergonic reactions can act as fuels. With the help of enzymes, a cell systematically degrades complex organic molecules that are rich in potential energy to simpler waste products that have less energy. Some of the energy taken out of chemical storage can be used to do work; the rest is dissipated as heat.

One catabolic process, **fermentation**, is a partial degradation of sugars or other organic fuel that occurs without the use of oxygen. However, the most prevalent and efficient catabolic pathway is **aerobic respiration**, in which oxygen is consumed as a reactant along with the organic fuel (*aerobic* is from the Greek *aer*, air, and *bios*, life). The cells of most eukaryotic and many prokaryotic organisms can carry out aerobic respiration. Some prokaryotes use substances other than oxygen as reactants in a similar process that harvests chemical energy without oxygen; this process is called *anaerobic respiration* (the prefix *an-* means "without"). Technically, the term **cellular respiration** includes both aerobic and anaerobic processes. However, it originated as a synonym for aerobic respiration because of the relationship of that process to organismal respiration, in which an animal breathes in oxygen. Thus, *cellular respiration* is often used to refer to the aerobic process, a practice we follow in most of this chapter.

Although very different in mechanism, aerobic respiration is in principle similar to the combustion of gasoline in an automobile engine after oxygen is mixed with the fuel (hydrocarbons). Food provides the fuel for respiration, and the exhaust is carbon dioxide and water. The overall process can be summarized as follows:

$$\text{Organic compounds} + \text{Oxygen} \rightarrow \text{Carbon dioxide} + \text{Water} + \text{Energy}$$

Although carbohydrates, fats, and proteins can all be processed and consumed as fuel, it is helpful to learn the steps of cellular respiration by tracking the degradation of the sugar glucose ($C_6H_{12}O_6$):

$$C_6H_{12}O_6 + 6\ O_2 \rightarrow 6\ CO_2 + 6\ H_2O + \text{Energy (ATP + heat)}$$

Glucose is the fuel that cells most often use; we will discuss other organic molecules contained in foods later in the chapter.

This breakdown of glucose is exergonic, having a free-energy change of −686 kcal (2,870 kJ) per mole of glucose decomposed ($\Delta G = -686$ kcal/mol). Recall that a negative ΔG indicates that the products of the chemical process store less energy than the reactants and that the reaction can happen spontaneously—in other words, without an input of energy.

Catabolic pathways do not directly move flagella, pump solutes across membranes, polymerize monomers, or perform other cellular work. Catabolism is linked to work by a chemical drive shaft—ATP, which you learned about in Chapter 8. To keep working, the cell must regenerate its supply of ATP from ADP and Ⓟ$_i$ (see Figure 8.11). To understand how cellular respiration accomplishes this, let's examine the fundamental chemical processes known as oxidation and reduction.

Redox Reactions: Oxidation and Reduction

How do the catabolic pathways that decompose glucose and other organic fuels yield energy? The answer is based on the transfer of electrons during the chemical reactions. The relocation of electrons releases energy stored in organic molecules, and this energy ultimately is used to synthesize ATP.

The Principle of Redox

In many chemical reactions, there is a transfer of one or more electrons (e^-) from one reactant to another. These electron transfers are called oxidation-reduction reactions, or **redox reactions** for short. In a redox reaction, the loss of electrons from one substance is called **oxidation**, and the addition of electrons to another substance is known as **reduction**. (Note that *adding* electrons is called *reduction*; negatively charged electrons added to an atom *reduce* the amount of positive charge of that atom.) To take a simple, nonbiological example, consider the reaction between the elements sodium (Na) and chlorine (Cl) that forms table salt:

becomes oxidized (loses electron)

$$\text{Na} + \text{Cl} \longrightarrow \text{Na}^+ + \text{Cl}^-$$

becomes reduced (gains electron)

We could generalize a redox reaction this way:

becomes oxidized

$$Xe^- + Y \longrightarrow X + Ye^-$$

becomes reduced

In the generalized reaction, substance Xe^-, the electron donor, is called the **reducing agent**; it reduces Y, which accepts the donated electron. Substance Y, the electron acceptor, is the **oxidizing agent**; it oxidizes Xe^- by removing its electron. Because an electron transfer requires both a donor and an acceptor, oxidation and reduction always go together.

Not all redox reactions involve the complete transfer of electrons from one substance to another; some change the degree of electron sharing in covalent bonds. The reaction

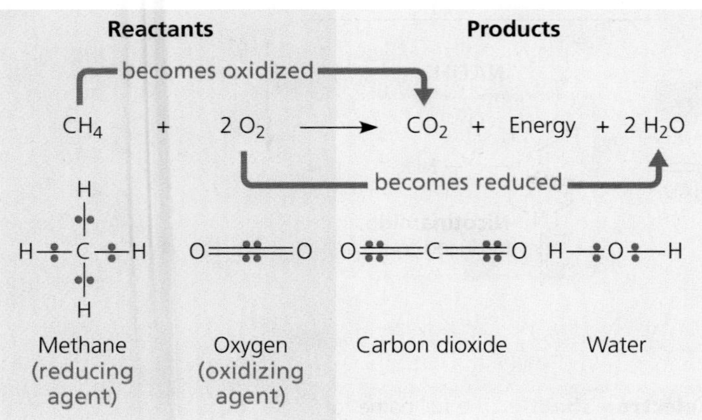

▲ **Figure 9.3 Methane combustion as an energy-yielding redox reaction.** The reaction releases energy to the surroundings because the electrons lose potential energy when they end up being shared unequally, spending more time near electronegative atoms such as oxygen.

between methane and oxygen, shown in **Figure 9.3**, is an example. As explained in Chapter 2, the covalent electrons in methane are shared nearly equally between the bonded atoms because carbon and hydrogen have about the same affinity for valence electrons; they are about equally electronegative. But when methane reacts with oxygen, forming carbon dioxide, electrons end up shared less equally between the carbon atom and its new covalent partners, the oxygen atoms, which are very electronegative. In effect, the carbon atom has partially "lost" its shared electrons; thus, methane has been oxidized.

Now let's examine the fate of the reactant O_2. The two atoms of the oxygen molecule (O_2) share their electrons equally. But when oxygen reacts with the hydrogen from methane, forming water, the electrons of the covalent bonds spend more time near the oxygen (see Figure 9.3). In effect, each oxygen atom has partially "gained" electrons, so the oxygen molecule has been reduced. Because oxygen is so electronegative, it is one of the most potent of all oxidizing agents.

Energy must be added to pull an electron away from an atom, just as energy is required to push a ball uphill. The more electronegative the atom (the stronger its pull on electrons), the more energy is required to take an electron away from it. An electron loses potential energy when it shifts from a less electronegative atom toward a more electronegative one, just as a ball loses potential energy when it rolls downhill. A redox reaction that moves electrons closer to oxygen, such as the burning (oxidation) of methane, therefore releases chemical energy that can be put to work.

Oxidation of Organic Fuel Molecules During Cellular Respiration

The oxidation of methane by oxygen is the main combustion reaction that occurs at the burner of a gas stove. The combustion of gasoline in an automobile engine is also a redox reaction; the energy released pushes the pistons. But the energy-yielding redox process of greatest interest to biologists is respiration: the oxidation of glucose and other molecules in food. Examine again the summary equation for cellular respiration, but this time think of it as a redox process:

$$\underbrace{C_6H_{12}O_6 + 6\,O_2 \longrightarrow 6\,CO_2 + 6\,H_2O + \text{Energy}}$$

(becomes oxidized: $C_6H_{12}O_6 \rightarrow 6\,CO_2$; becomes reduced: $6\,O_2 \rightarrow 6\,H_2O$)

As in the combustion of methane or gasoline, the fuel (glucose) is oxidized and oxygen is reduced. The electrons lose potential energy along the way, and energy is released.

In general, organic molecules that have an abundance of hydrogen are excellent fuels because their bonds are a source of "hilltop" electrons, whose energy may be released as these electrons "fall" down an energy gradient when they are transferred to oxygen. The summary equation for respiration indicates that hydrogen is transferred from glucose to oxygen. But the important point, not visible in the summary equation, is that the energy state of the electron changes as hydrogen (with its electron) is transferred to oxygen. In respiration, the oxidation of glucose transfers electrons to a lower energy state, liberating energy that becomes available for ATP synthesis.

The main energy-yielding foods, carbohydrates and fats, are reservoirs of electrons associated with hydrogen. Only the barrier of activation energy holds back the flood of electrons to a lower energy state (see Figure 8.12). Without this barrier, a food substance like glucose would combine almost instantaneously with O_2. If we supply the activation energy by igniting glucose, it burns in air, releasing 686 kcal (2,870 kJ) of heat per mole of glucose (about 180 g). Body temperature is not high enough to initiate burning, of course. Instead, if you swallow some glucose, enzymes in your cells will lower the barrier of activation energy, allowing the sugar to be oxidized in a series of steps.

Stepwise Energy Harvest via NAD^+ and the Electron Transport Chain

If energy is released from a fuel all at once, it cannot be harnessed efficiently for constructive work. For example, if a gasoline tank explodes, it cannot drive a car very far. Cellular respiration does not oxidize glucose in a single explosive step either. Rather, glucose and other organic fuels are broken down in a series of steps, each one catalyzed by an enzyme. At key steps, electrons are stripped from the glucose. As is often the case in oxidation reactions, each electron travels with a proton—thus, as a hydrogen atom. The hydrogen atoms are not transferred directly to oxygen, but instead are usually passed first to an electron carrier, a coenzyme called **NAD^+** (nicotinamide adenine dinucleotide, a derivative of the vitamin niacin). NAD^+ is well suited as an electron carrier because it can cycle easily between oxidized (NAD^+) and reduced (NADH) states. As an electron acceptor, NAD^+ functions as an oxidizing agent during respiration.

How does NAD^+ trap electrons from glucose and other organic molecules? Enzymes called dehydrogenases remove a

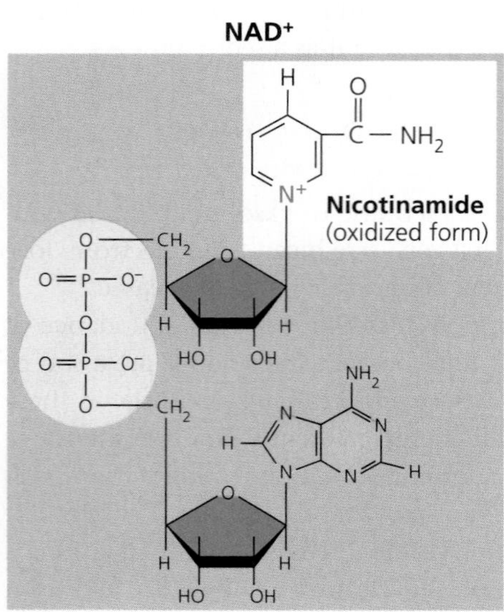

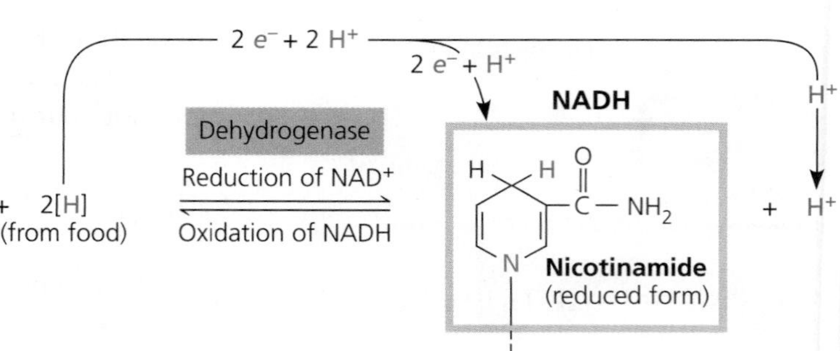

◀ **Figure 9.4 NAD$^+$ as an electron shuttle.** The full name for NAD$^+$, nicotinamide adenine dinucleotide, describes its structure: The molecule consists of two nucleotides joined together at their phosphate groups (shown in yellow). (Nicotinamide is a nitrogenous base, although not one that is present in DNA or RNA; see Figure 5.26.) The enzymatic transfer of 2 electrons and 1 proton (H$^+$) from an organic molecule in food to NAD$^+$ reduces the NAD$^+$ to NADH; the second proton (H$^+$) is released. Most of the electrons removed from food are transferred initially to NAD$^+$.

pair of hydrogen atoms (2 electrons and 2 protons) from the substrate (glucose, in this example), thereby oxidizing it. The enzyme delivers the 2 electrons along with 1 proton to its coenzyme, NAD$^+$ (**Figure 9.4**). The other proton is released as a hydrogen ion (H$^+$) into the surrounding solution:

$$\text{H}-\overset{|}{\underset{|}{\text{C}}}-\text{OH} + \text{NAD}^+ \xrightarrow{\text{Dehydrogenase}} \overset{|}{\underset{|}{\text{C}}}=\text{O} + \text{NADH} + \text{H}^+$$

By receiving 2 negatively charged electrons but only 1 positively charged proton, NAD$^+$ has its charge neutralized when it is reduced to NADH. The name NADH shows the hydrogen that has been received in the reaction. NAD$^+$ is the most versatile electron acceptor in cellular respiration and functions in several of the redox steps during the breakdown of glucose.

Electrons lose very little of their potential energy when they are transferred from glucose to NAD$^+$. Each NADH molecule formed during respiration represents stored energy that can be tapped to make ATP when the electrons complete their "fall" down an energy gradient from NADH to oxygen.

How do electrons that are extracted from glucose and stored as potential energy in NADH finally reach oxygen? It will help to compare the redox chemistry of cellular respiration to a much simpler reaction: the reaction between hydrogen and oxygen to form water (**Figure 9.5a**). Mix H_2 and O_2, provide a spark for activation energy, and the gases combine explosively. In fact, combustion of liquid H_2 and O_2 is harnessed to power the main engines of the space shuttle after it is launched, boosting it into orbit. The explosion represents a release of energy as the electrons of hydrogen "fall" closer to the electronegative oxygen atoms. Cellular respiration also brings hydrogen and oxygen together to form water, but there are two important differences. First, in cellular respiration, the hydrogen that reacts with oxygen is derived from organic molecules rather than H_2. Second, instead of occurring in one explosive reaction, respiration uses an

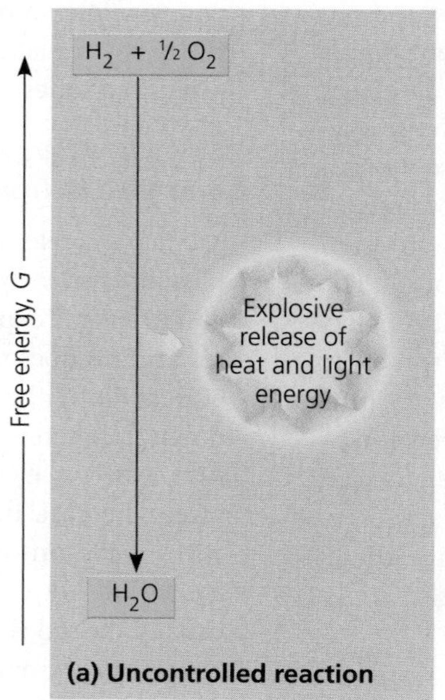

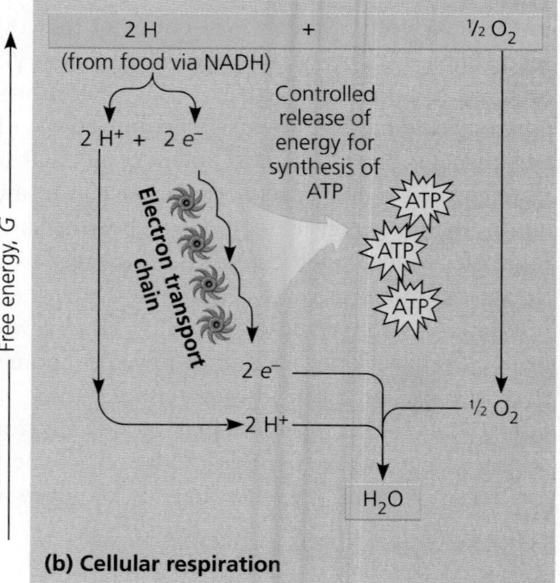

▲ **Figure 9.5 An introduction to electron transport chains.** **(a)** The one-step exergonic reaction of hydrogen with oxygen to form water releases a large amount of energy in the form of heat and light: an explosion. **(b)** In cellular respiration, the same reaction occurs in stages: An electron transport chain breaks the "fall" of electrons in this reaction into a series of smaller steps and stores some of the released energy in a form that can be used to make ATP. (The rest of the energy is released as heat.)

electron transport chain to break the fall of electrons to oxygen into several energy-releasing steps (**Figure 9.5b**). An **electron transport chain** consists of a number of molecules, mostly proteins, built into the inner membrane of the mitochondria of eukaryotic cells and the plasma membrane of aerobically respiring prokaryotes. Electrons removed from glucose are shuttled by NADH to the "top," higher-energy end of the chain. At the "bottom," lower-energy end, O_2 captures these electrons along with hydrogen nuclei (H^+), forming water.

Electron transfer from NADH to oxygen is an exergonic reaction with a free-energy change of −53 kcal/mol (−222 kJ/mol). Instead of this energy being released and wasted in a single explosive step, electrons cascade down the chain from one carrier molecule to the next in a series of redox reactions, losing a small amount of energy with each step until they finally reach oxygen, the terminal electron acceptor, which has a very great affinity for electrons. Each "downhill" carrier is more electronegative than, and thus capable of oxidizing, its "uphill" neighbor, with oxygen at the bottom of the chain. Therefore, the electrons removed from glucose by NAD^+ fall down an energy gradient in the electron transport chain to a far more stable location in the electronegative oxygen atom. Put another way, oxygen pulls electrons down the chain in an energy-yielding tumble analogous to gravity pulling objects downhill.

In summary, during cellular respiration, most electrons travel the following "downhill" route: glucose → NADH → electron transport chain → oxygen. Later in this chapter, you will learn more about how the cell uses the energy released from this exergonic electron fall to regenerate its supply of ATP. For now, having covered the basic redox mechanisms of cellular respiration, let's look at the entire process by which energy is harvested from organic fuels.

The Stages of Cellular Respiration: *A Preview*

The harvesting of energy from glucose by cellular respiration is a cumulative function of three metabolic stages:

1. Glycolysis (color-coded teal throughout the chapter)
2. Pyruvate oxidation and the citric acid cycle (color-coded salmon)
3. Oxidative phosphorylation: electron transport and chemiosmosis (color-coded violet)

Biochemists usually reserve the term *cellular respiration* for stages 2 and 3. We include glycolysis, however, because most respiring cells deriving energy from glucose use glycolysis to produce the starting material for the citric acid cycle.

As diagrammed in **Figure 9.6**, glycolysis and pyruvate oxidation followed by the citric acid cycle are the catabolic pathways that break down glucose and other organic fuels. **Glycolysis**, which occurs in the cytosol, begins the degradation process by breaking glucose into two molecules of a compound called pyruvate. In eukaryotes, pyruvate enters the mitochondrion and is oxidized to a compound called acetyl CoA, which enters the **citric acid cycle**. There, the breakdown of glucose to carbon dioxide is completed. (In prokaryotes, these processes take place in the cytosol.) Thus, the carbon dioxide produced by respiration represents fragments of oxidized organic molecules.

Some of the steps of glycolysis and the citric acid cycle are redox reactions in which dehydrogenases transfer electrons from substrates to NAD^+, forming NADH. In the third stage of respiration, the electron transport chain accepts electrons from the breakdown products of the first two stages (most often via NADH) and passes these electrons from one molecule to another. At the end of the chain, the electrons are combined with molecular oxygen and hydrogen ions (H^+), forming water (see

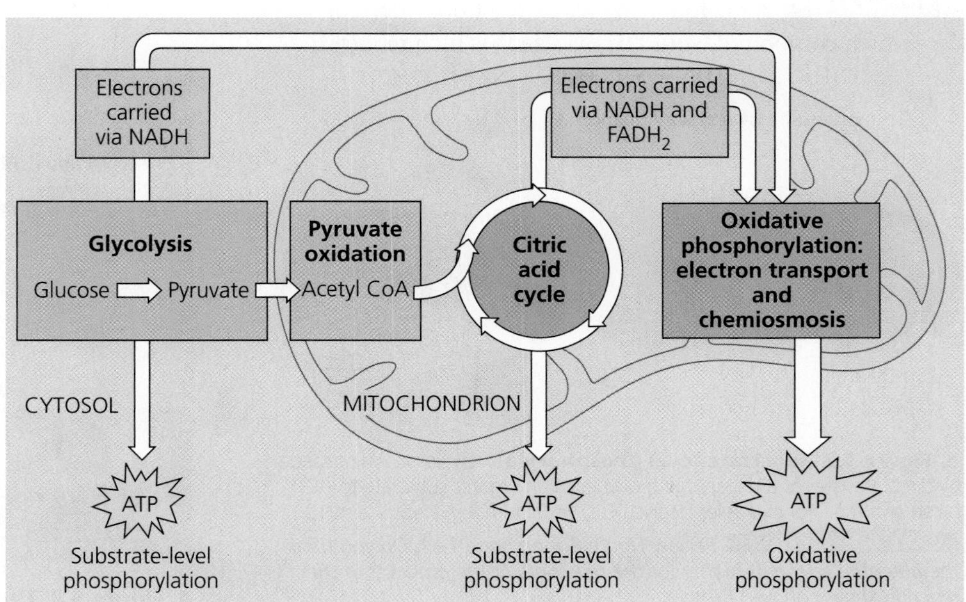

▶ **Figure 9.6 An overview of cellular respiration.** During glycolysis, each glucose molecule is broken down into two molecules of the compound pyruvate. In eukaryotic cells, as shown here, the pyruvate enters the mitochondrion. There it is oxidized to acetyl CoA, which is further oxidized to CO_2 in the citric acid cycle. NADH and a similar electron carrier, a coenzyme called $FADH_2$, transfer electrons derived from glucose to electron transport chains, which are built into the inner mitochondrial membrane. (In prokaryotes, the electron transport chains are located in the plasma membrane.) During oxidative phosphorylation, electron transport chains convert the chemical energy to a form used for ATP synthesis in the process called chemiosmosis.

ANIMATION **BioFlix** Visit the Study Area at **www.masteringbiology.com** for the BioFlix® 3-D Animation on Cellular Respiration.

Figure 9.5b). The energy released at each step of the chain is stored in a form the mitochondrion (or prokaryotic cell) can use to make ATP from ADP. This mode of ATP synthesis is called **oxidative phosphorylation** because it is powered by the redox reactions of the electron transport chain.

In eukaryotic cells, the inner membrane of the mitochondrion is the site of electron transport and chemiosmosis, the processes that together constitute oxidative phosphorylation. (In prokaryotes, these processes take place in the plasma membrane.) Oxidative phosphorylation accounts for almost 90% of the ATP generated by respiration. A smaller amount of ATP is formed directly in a few reactions of glycolysis and the citric acid cycle by a mechanism called **substrate-level phosphorylation** (**Figure 9.7**). This mode of ATP synthesis occurs when an enzyme transfers a phosphate group from a substrate molecule to ADP, rather than adding an inorganic phosphate to ADP as in oxidative phosphorylation. "Substrate molecule" here refers to an organic molecule generated as an intermediate during the catabolism of glucose.

For each molecule of glucose degraded to carbon dioxide and water by respiration, the cell makes up to about 32 molecules of ATP, each with 7.3 kcal/mol of free energy. Respiration cashes in the large denomination of energy banked in a single molecule of glucose (686 kcal/mol) for the small change of many molecules of ATP, which is more practical for the cell to spend on its work.

This preview has introduced you to how glycolysis, the citric acid cycle, and oxidative phosphorylation fit into the process of cellular respiration. We are now ready to take a closer look at each of these three stages of respiration.

CONCEPT CHECK 9.1

1. Compare and contrast aerobic and anaerobic respiration.
2. **WHAT IF?** If the following redox reaction occurred, which compound would be oxidized? Which reduced?

$$C_4H_6O_5 + NAD^+ \longrightarrow C_4H_4O_5 + NADH + H^+$$

For suggested answers, see Appendix A.

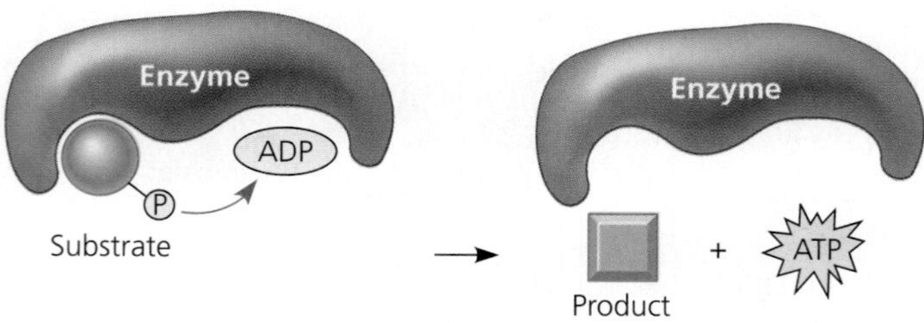

▲ **Figure 9.7 Substrate-level phosphorylation.** Some ATP is made by direct transfer of a phosphate group from an organic substrate to ADP by an enzyme. (For examples in glycolysis, see Figure 9.9, steps 7 and 10.)

MAKE CONNECTIONS *Review Figure 8.8 on page 149. Do you think the potential energy is higher for the reactants or the products in the reaction shown above? Explain.*

CONCEPT 9.2

Glycolysis harvests chemical energy by oxidizing glucose to pyruvate

The word *glycolysis* means "sugar splitting," and that is exactly what happens during this pathway. Glucose, a six-carbon sugar, is split into two three-carbon sugars. These smaller sugars are then oxidized and their remaining atoms rearranged to form two molecules of pyruvate. (Pyruvate is the ionized form of pyruvic acid.)

As summarized in **Figure 9.8**, glycolysis can be divided into two phases: energy investment and energy payoff. During the energy investment phase, the cell actually spends ATP. This investment is repaid with interest during the energy payoff phase, when ATP is produced by substrate-level phosphorylation and NAD^+ is reduced to NADH by electrons released from the oxidation of glucose. The net energy yield from glycolysis, per glucose molecule, is 2 ATP plus 2 NADH. The ten steps of the glycolytic pathway are shown in **Figure 9.9**.

All of the carbon originally present in glucose is accounted for in the two molecules of pyruvate; no carbon is released as CO_2 during glycolysis. Glycolysis occurs whether or not O_2 is present. However, if O_2 *is* present, the chemical energy stored in pyruvate and NADH can be extracted by pyruvate oxidation, the citric acid cycle, and oxidative phosphorylation.

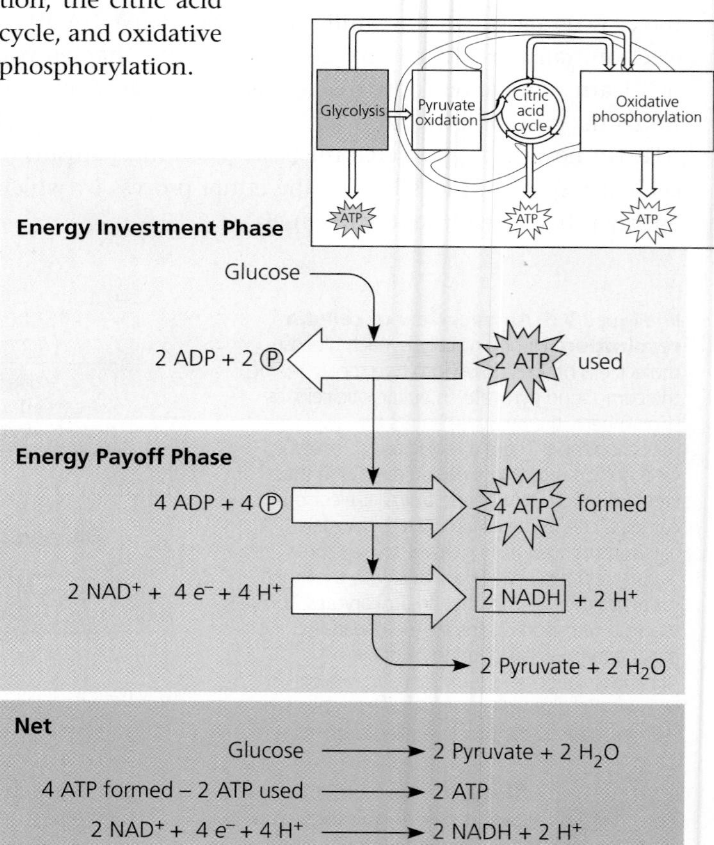

▲ **Figure 9.8 The energy input and output of glycolysis.**

▼ **Figure 9.9 A closer look at glycolysis.** The orientation diagram on the left relates glycolysis to the entire process of respiration. Note that glycolysis is a source of ATP and NADH.

WHAT IF? *What would happen if you removed the dihydroxyacetone phosphate generated in step 4 as fast as it was produced?*

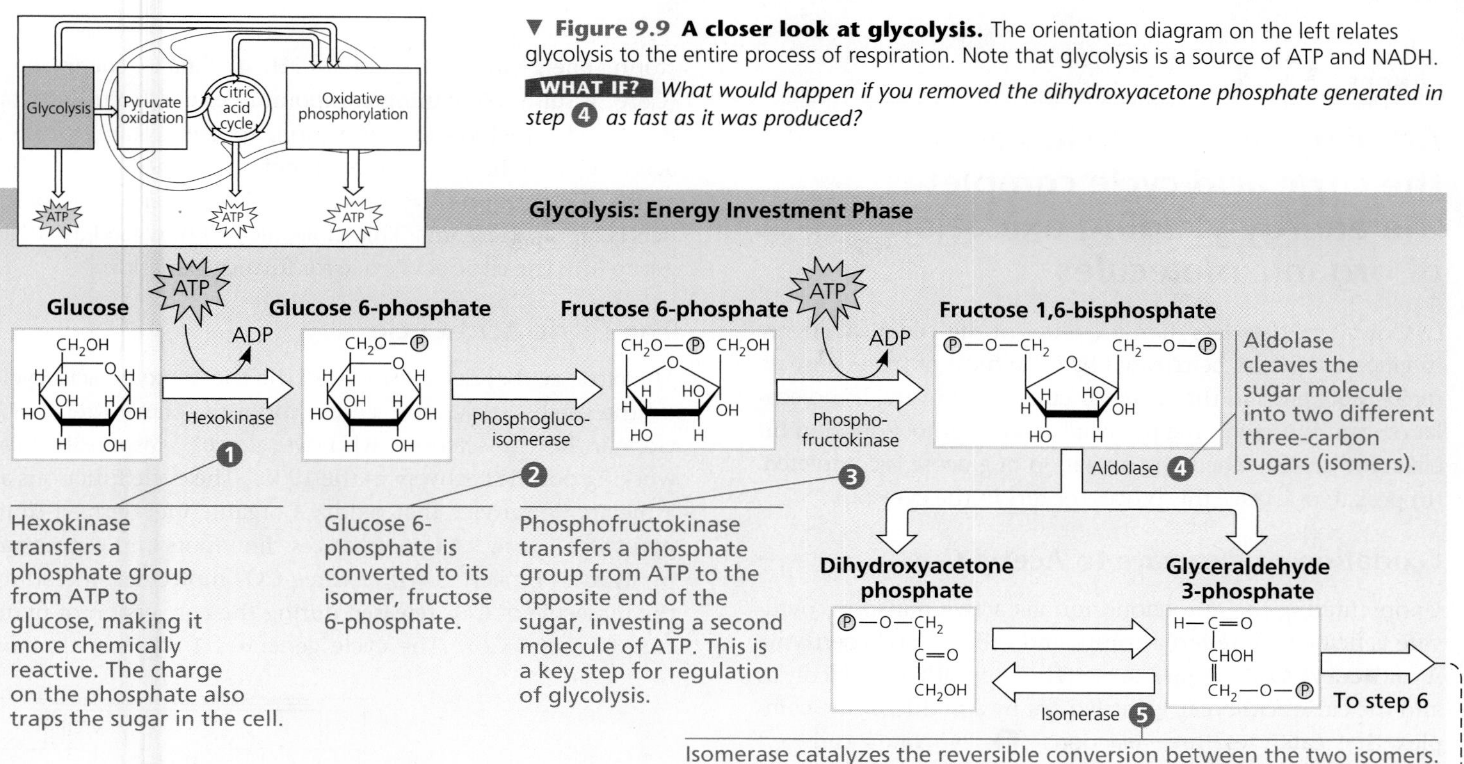

Glycolysis
Pyruvate oxidation
Citric acid cycle
Oxidative phosphorylation
ATP
ATP
ATP
Glycolysis: Energy Investment Phase
Glucose
ATP
ADP
Hexokinase
Glucose 6-phosphate
Phosphogluco-isomerase
Fructose 6-phosphate
ATP
ADP
Phospho-fructokinase
Fructose 1,6-bisphosphate
Aldolase
Aldolase cleaves the sugar molecule into two different three-carbon sugars (isomers).
Hexokinase transfers a phosphate group from ATP to glucose, making it more chemically reactive. The charge on the phosphate also traps the sugar in the cell.
Glucose 6-phosphate is converted to its isomer, fructose 6-phosphate.
Phosphofructokinase transfers a phosphate group from ATP to the opposite end of the sugar, investing a second molecule of ATP. This is a key step for regulation of glycolysis.
Dihydroxyacetone phosphate
Glyceraldehyde 3-phosphate
Isomerase
To step 6
Isomerase catalyzes the reversible conversion between the two isomers. This reaction never reaches equilibrium: Glyceraldehyde 3-phosphate is used as the substrate of the next reaction (step 6) as fast as it forms.

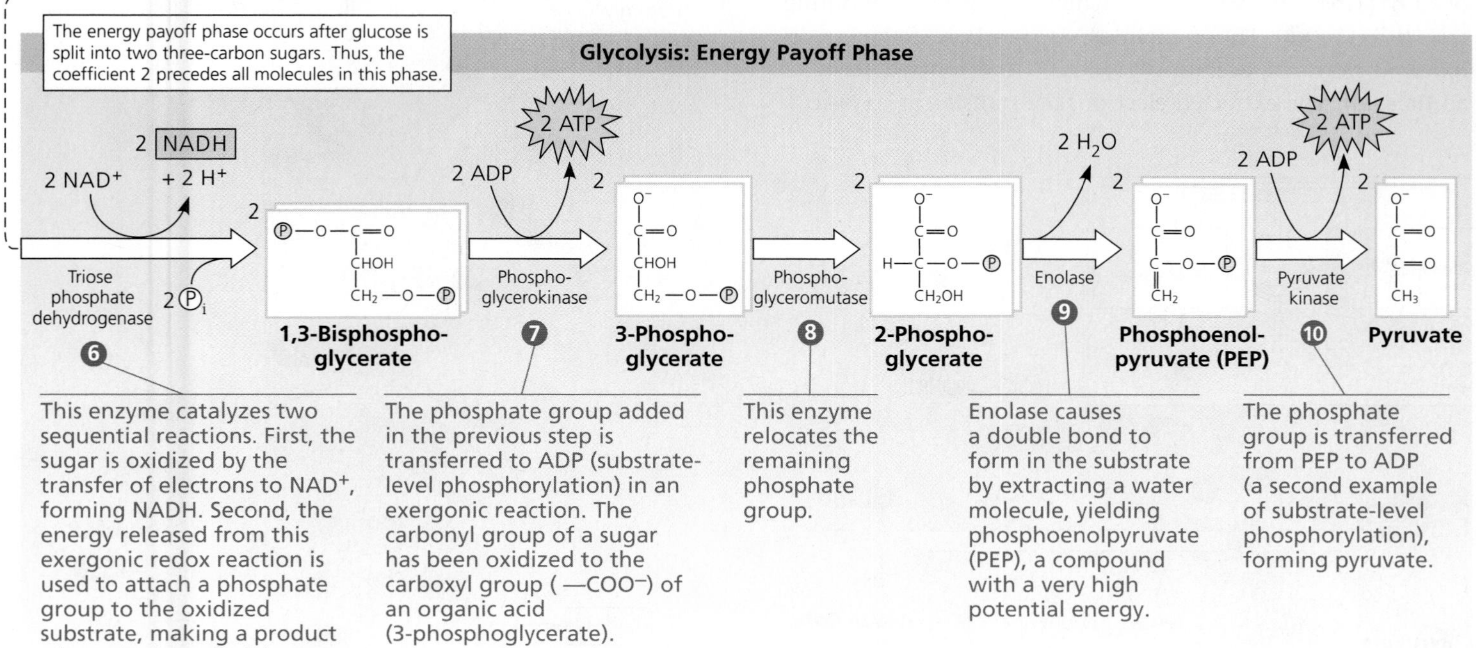

The energy payoff phase occurs after glucose is split into two three-carbon sugars. Thus, the coefficient 2 precedes all molecules in this phase.
Glycolysis: Energy Payoff Phase
2 NADH
2 NAD⁺
+ 2 H⁺
Triose phosphate dehydrogenase
2 Pᵢ
1,3-Bisphospho-glycerate
2 ADP
2 ATP
Phospho-glycerokinase
3-Phospho-glycerate
Phospho-glyceromutase
2-Phospho-glycerate
2 H₂O
Enolase
Phosphoenol-pyruvate (PEP)
2 ADP
2 ATP
Pyruvate kinase
Pyruvate
This enzyme catalyzes two sequential reactions. First, the sugar is oxidized by the transfer of electrons to NAD⁺, forming NADH. Second, the energy released from this exergonic redox reaction is used to attach a phosphate group to the oxidized substrate, making a product of very high potential energy.
The phosphate group added in the previous step is transferred to ADP (substrate-level phosphorylation) in an exergonic reaction. The carbonyl group of a sugar has been oxidized to the carboxyl group (—COO⁻) of an organic acid (3-phosphoglycerate).
This enzyme relocates the remaining phosphate group.
Enolase causes a double bond to form in the substrate by extracting a water molecule, yielding phosphoenolpyruvate (PEP), a compound with a very high potential energy.
The phosphate group is transferred from PEP to ADP (a second example of substrate-level phosphorylation), forming pyruvate.

CONCEPT CHECK 9.2

1. During the redox reaction in glycolysis (step 6 in Figure 9.9), which molecule acts as the oxidizing agent? The reducing agent?
2. **MAKE CONNECTIONS** Step 3 in Figure 9.9 is a major point of regulation of glycolysis. The enzyme phosphofructokinase is allosterically regulated by ATP and related molecules (see Concept 8.5, p. 158). Considering the overall result of glycolysis, would you expect ATP to inhibit or stimulate activity of this enzyme? (*Hint*: Make sure you consider the role of ATP as an allosteric regulator, not as a substrate of the enzyme.)

For suggested answers, see Appendix A.

CONCEPT **9.3**

After pyruvate is oxidized, the citric acid cycle completes the energy-yielding oxidation of organic molecules

Glycolysis releases less than a quarter of the chemical energy in glucose that can be released by cells; most of the energy remains stockpiled in the two molecules of pyruvate. If molecular oxygen is present, the pyruvate enters a mitochondrion (in eukaryotic cells), where the oxidation of glucose is completed. (In prokaryotic cells, this process occurs in the cytosol.)

Oxidation of Pyruvate to Acetyl CoA

Upon entering the mitochondrion via active transport, pyruvate is first converted to a compound called acetyl coenzyme A, or **acetyl CoA** (**Figure 9.10**). This step, linking glycolysis and the citric acid cycle, is carried out by a multienzyme complex that catalyzes three reactions: ❶ Pyruvate's carboxyl group (—COO^-), which is already fully oxidized and thus has little chemical energy, is removed and given off as a molecule of CO_2. (This is the first step in which CO_2 is released during respiration.) ❷ The remaining two-carbon fragment is oxidized, forming acetate (CH_3COO^-, the ionized form of acetic acid). The extracted electrons are transferred to NAD^+, storing energy in the form of NADH. ❸ Finally, coenzyme A (CoA), a sulfur-containing compound derived from a B vitamin, is attached via its sulfur atom to the acetate, forming acetyl CoA, which has a high potential energy; in other words, the reaction of acetyl CoA to yield lower-energy products is highly exergonic. This molecule will now feed its acetyl group into the citric acid cycle for further oxidation.

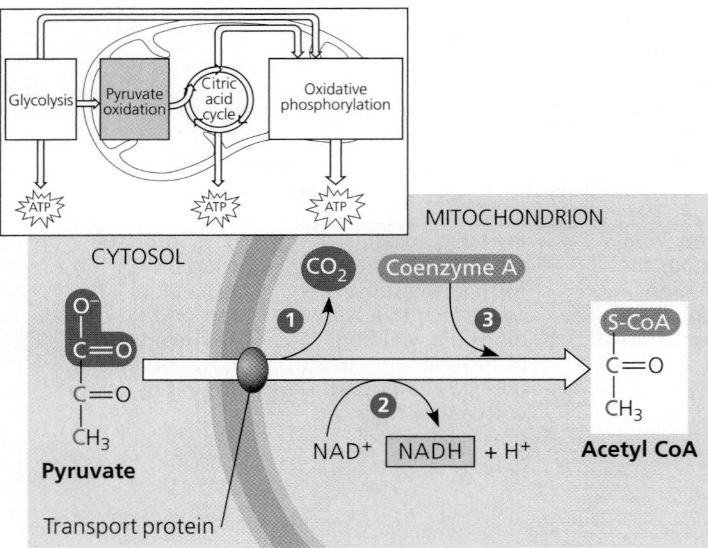

▲ **Figure 9.10 Oxidation of pyruvate to acetyl CoA, the step before the citric acid cycle.** Pyruvate is a charged molecule, so in eukaryotic cells it must enter the mitochondrion via active transport, with the help of a transport protein. Next, a complex of several enzymes (the pyruvate dehydrogenase complex) catalyzes the three numbered steps, which are described in the text. The acetyl group of acetyl CoA will enter the citric acid cycle. The CO_2 molecule will diffuse out of the cell. By convention, coenzyme A is abbreviated S-CoA when it is attached to a molecule, emphasizing the sulfur atom (S).

The Citric Acid Cycle

The citric acid cycle is also called the tricarboxylic acid cycle or the Krebs cycle, the latter honoring Hans Krebs, the German-British scientist who was largely responsible for working out the pathway in the 1930s. The cycle functions as a metabolic furnace that oxidizes organic fuel derived from pyruvate. **Figure 9.11** summarizes the inputs and outputs as pyruvate is broken down to three CO_2 molecules, including the molecule of CO_2 released during the conversion of pyruvate to acetyl CoA. The cycle generates 1 ATP per turn by

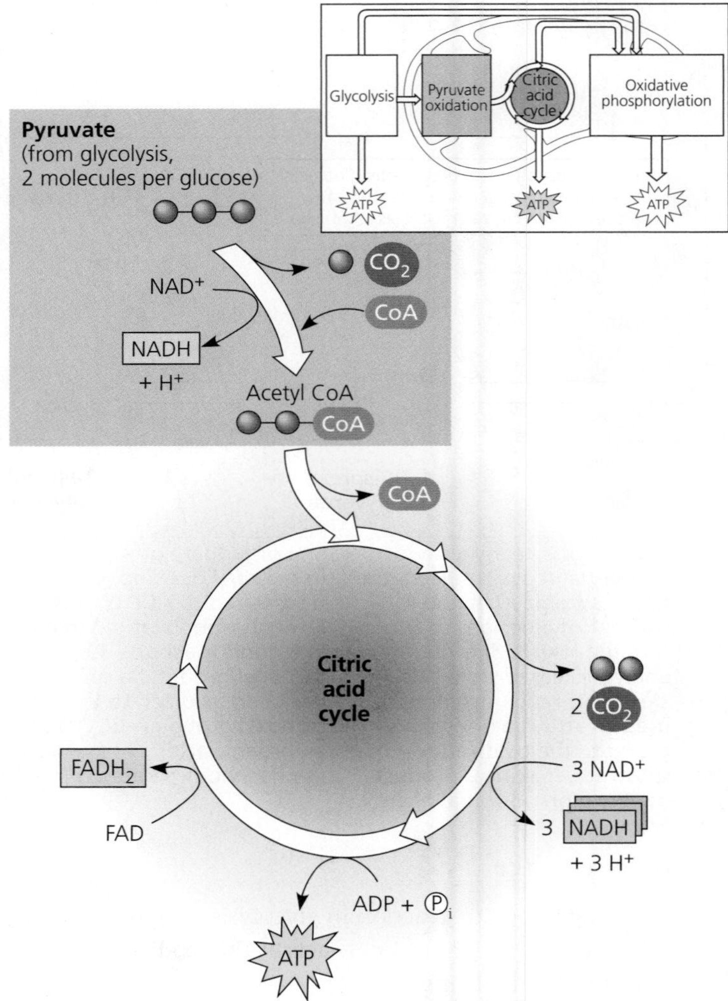

▲ **Figure 9.11 An overview of pyruvate oxidation and the citric acid cycle.** The inputs and outputs per pyruvate molecule are shown. To calculate on a per-glucose basis, multiply by 2, because each glucose molecule is split during glycolysis into two pyruvate molecules.

substrate-level phosphorylation, but most of the chemical energy is transferred to NAD^+ and a related electron carrier, the coenzyme FAD (flavin adenine dinucleotide, derived from riboflavin, a B vitamin), during the redox reactions. The reduced coenzymes, NADH and $FADH_2$, shuttle their cargo of high-energy electrons into the electron transport chain.

Now let's look at the citric acid cycle in more detail. The cycle has eight steps, each catalyzed by a specific enzyme. You can see in **Figure 9.12** that for each turn of the citric acid cycle, two carbons (red) enter in the relatively reduced form of an acetyl group (step 1), and two different carbons (blue) leave in the completely oxidized form of CO_2 molecules

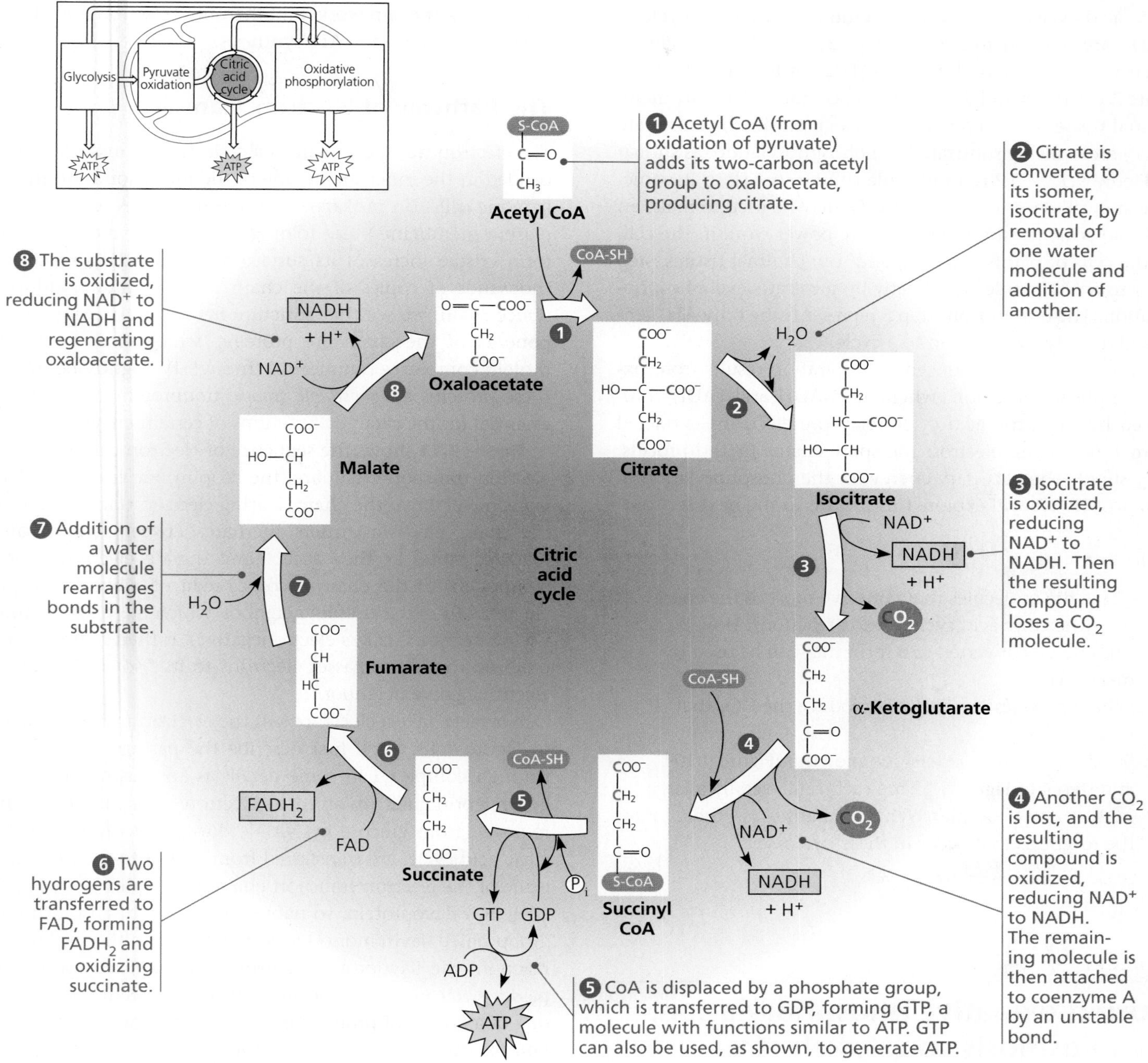

▲ **Figure 9.12 A closer look at the citric acid cycle.** In the chemical structures, red type traces the fate of the two carbon atoms that enter the cycle via acetyl CoA (step 1), and blue type indicates the two carbons that exit the cycle as CO_2 in steps 3 and 4. (The red labeling goes only through step 5 because the succinate molecule is symmetrical; the two ends cannot be distinguished from each other.) Notice that the carbon atoms that enter the cycle from acetyl CoA do not leave the cycle in the same turn. They remain in the cycle, occupying a different location in the molecules on their next turn, after another acetyl group is added. As a consequence, the oxaloacetate that is regenerated at step 8 is composed of different carbon atoms each time around. In eukaryotic cells, all the citric acid cycle enzymes are located in the mitochondrial matrix except for the enzyme that catalyzes step 6, which resides in the inner mitochondrial membrane. Carboxylic acids are represented in their ionized forms, as $—COO^-$, because the ionized forms prevail at the pH within the mitochondrion. For example, citrate is the ionized form of citric acid.

(steps 3 and 4). The acetyl group of acetyl CoA joins the cycle by combining with the compound oxaloacetate, forming citrate (step 1). (Citrate is the ionized form of citric acid, for which the cycle is named.) The next seven steps decompose the citrate back to oxaloacetate. It is this regeneration of oxaloacetate that makes this process a *cycle*.

Now let's tally the energy-rich molecules produced by the citric acid cycle. For each acetyl group entering the cycle, 3 NAD^+ are reduced to NADH (steps 3, 4, and 8). In step 6, electrons are transferred not to NAD^+, but to FAD, which accepts 2 electrons and 2 protons to become $FADH_2$. In many animal tissue cells, step 5 produces a guanosine triphosphate (GTP) molecule by substrate-level phosphorylation, as shown in Figure 9.12. GTP is a molecule similar to ATP in its structure and cellular function. This GTP may be used to make an ATP molecule (as shown) or directly power work in the cell. In the cells of plants, bacteria, and some animal tissues, step 5 forms an ATP molecule directly by substrate-level phosphorylation. The output from step 5 represents the only ATP generated directly by the citric acid cycle.

Most of the ATP produced by respiration results from oxidative phosphorylation, when the NADH and $FADH_2$ produced by the citric acid cycle relay the electrons extracted from food to the electron transport chain. In the process, they supply the necessary energy for the phosphorylation of ADP to ATP. We will explore this process in the next section.

CONCEPT CHECK 9.3

1. Name the molecules that conserve most of the energy from the citric acid cycle's redox reactions. How is this energy converted to a form that can be used to make ATP?
2. What processes in your cells produce the CO_2 that you exhale?
3. **WHAT IF?** The conversions shown in Figure 9.10 and step 4 of Figure 9.12 are each catalyzed by a large multienzyme complex. What similarities are there in the reactions that occur in these two cases?

For suggested answers, see Appendix A.

CONCEPT 9.4

During oxidative phosphorylation, chemiosmosis couples electron transport to ATP synthesis

Our main objective in this chapter is to learn how cells harvest the energy of glucose and other nutrients in food to make ATP. But the metabolic components of respiration we have dissected so far, glycolysis and the citric acid cycle, produce only 4 ATP molecules per glucose molecule, all by substrate-level phosphorylation: 2 net ATP from glycolysis and 2 ATP from the citric acid cycle. At this point, molecules of NADH (and $FADH_2$) account for most of the energy extracted from the glucose. These electron escorts link glycolysis and the citric acid cycle to the machinery of oxidative phosphorylation, which uses energy released by the electron transport chain to power ATP synthesis. In this section, you will learn first how the electron transport chain works and then how electron flow down the chain is coupled to ATP synthesis.

The Pathway of Electron Transport

The electron transport chain is a collection of molecules embedded in the inner membrane of the mitochondrion in eukaryotic cells. (In prokaryotes, these molecules reside in the plasma membrane.) The folding of the inner membrane to form cristae increases its surface area, providing space for thousands of copies of the chain in each mitochondrion. (Once again, we see that structure fits function.) Most components of the chain are proteins, which exist in multiprotein complexes numbered I through IV. Tightly bound to these proteins are *prosthetic groups*, nonprotein components essential for the catalytic functions of certain enzymes.

Figure 9.13 shows the sequence of electron carriers in the electron transport chain and the drop in free energy as electrons travel down the chain. During electron transport along the chain, electron carriers alternate between reduced and oxidized states as they accept and donate electrons. Each component of the chain becomes reduced when it accepts electrons from its "uphill" neighbor, which has a lower affinity for electrons (is less electronegative). It then returns to its oxidized form as it passes electrons to its "downhill," more electronegative neighbor.

Now let's take a closer look at the electron transport chain in Figure 9.13. We'll first describe the passage of electrons through complex I in some detail, as an illustration of the general principles involved in electron transport. Electrons removed from glucose by NAD^+, during glycolysis and the citric acid cycle, are transferred from NADH to the first molecule of the electron transport chain in complex I. This molecule is a flavoprotein, so named because it has a prosthetic group called flavin mononucleotide (FMN). In the next redox reaction, the flavoprotein returns to its oxidized form as it passes electrons to an iron-sulfur protein (Fe·S in complex I), one of a family of proteins with both iron and sulfur tightly bound. The iron-sulfur protein then passes the electrons to a compound called ubiquinone (Q in Figure 9.13). This electron carrier is a small hydrophobic molecule, the only member of the electron transport chain that is not a protein. Ubiquinone is individually mobile within the membrane rather than residing in a particular complex. (Another name for ubiquinone is coenzyme Q, or CoQ; you may have seen it sold as a nutritional supplement.)

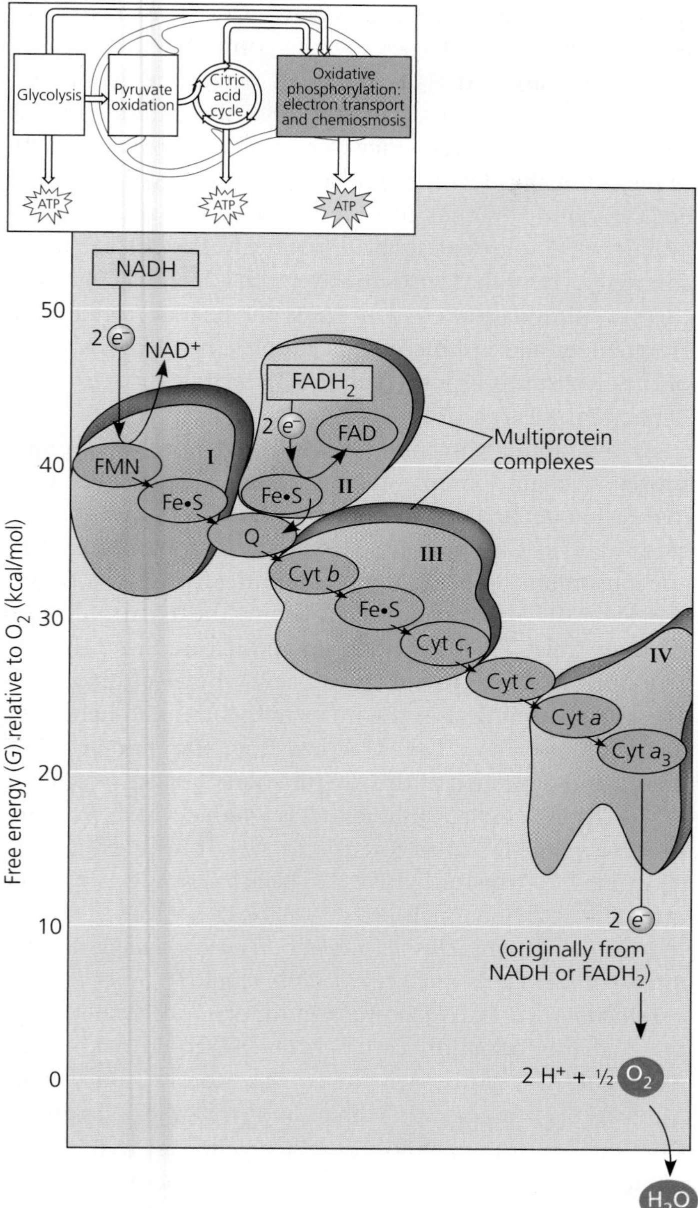

▲ **Figure 9.13 Free-energy change during electron transport.** The overall energy drop (ΔG) for electrons traveling from NADH to oxygen is 53 kcal/mol, but this "fall" is broken up into a series of smaller steps by the electron transport chain. (An oxygen atom is represented here as ½ O_2 to emphasize that the electron transport chain reduces molecular oxygen, O_2, not individual oxygen atoms.)

Most of the remaining electron carriers between ubiquinone and oxygen are proteins called **cytochromes**. Their prosthetic group, called a heme group, has an iron atom that accepts and donates electrons. (It is similar to the heme group in hemoglobin, the protein of red blood cells, except that the iron in hemoglobin carries oxygen, not electrons.) The electron transport chain has several types of cytochromes, each a different protein with a slightly different electron-carrying heme group. The last cytochrome of the chain, cyt a_3, passes its electrons to oxygen, which is *very* electronegative. Each oxygen atom also picks up a pair of hydrogen ions from the aqueous solution, forming water.

Another source of electrons for the transport chain is $FADH_2$, the other reduced product of the citric acid cycle. Notice in Figure 9.13 that $FADH_2$ adds its electrons to the electron transport chain from within complex II, at a lower energy level than NADH does. Consequently, although NADH and $FADH_2$ each donate an equivalent number of electrons (2) for oxygen reduction, the electron transport chain provides about one-third less energy for ATP synthesis when the electron donor is $FADH_2$ rather than NADH. We'll see why in the next section.

The electron transport chain makes no ATP directly. Instead, it eases the fall of electrons from food to oxygen, breaking a large free-energy drop into a series of smaller steps that release energy in manageable amounts. How does the mitochondrion (or the prokaryotic plasma membrane) couple this electron transport and energy release to ATP synthesis? The answer is a mechanism called chemiosmosis.

Chemiosmosis: The Energy-Coupling Mechanism

Populating the inner membrane of the mitochondrion or the prokaryotic plasma membrane are many copies of a protein complex called **ATP synthase**, the enzyme that actually makes ATP from ADP and inorganic phosphate. ATP synthase works like an ion pump running in reverse. Recall from Chapter 7 that ion pumps usually use ATP as an energy source to transport ions against their gradients. In fact, the proton pump shown in Figure 7.20 is an ATP synthase. As we mentioned in Chapter 8, enzymes can catalyze a reaction in either direction, depending on the ΔG for the reaction, which is affected by the local concentrations of reactants and products. Rather than hydrolyzing ATP to pump protons against their concentration gradient, under the conditions of cellular respiration ATP synthase uses the energy of an existing ion gradient to power ATP synthesis. The power source for the ATP synthase is a difference in the concentration of H^+ on opposite sides of the inner mitochondrial membrane. (We can also think of this gradient as a difference in pH, since pH is a measure of H^+ concentration.) This process, in which energy stored in the form of a hydrogen ion gradient across a membrane is used to drive cellular work such as the synthesis of ATP, is called **chemiosmosis** (from the Greek *osmos*, push). We have previously used the word *osmosis* in discussing water transport, but here it refers to the flow of H^+ across a membrane.

From studying the structure of ATP synthase, scientists have learned how the flow of H^+ through this large enzyme powers ATP generation. ATP synthase is a multisubunit complex with four main parts, each made up of multiple polypeptides. Protons move one by one into binding sites on one of the parts (the rotor), causing it to spin in a way that catalyzes ATP production from ADP and inorganic phosphate (**Figure 9.14**). The flow of protons thus behaves somewhat like a rushing stream that turns a waterwheel. ATP synthase is the smallest molecular rotary motor known in nature.

How does the inner mitochondrial membrane or the prokaryotic plasma membrane generate and maintain the H^+ gradient that drives ATP synthesis by the ATP synthase protein complex? Establishing the H^+ gradient is a major function of the electron transport chain, which is shown in its mitochondrial location in **Figure 9.15**. The chain is an energy converter that uses the exergonic flow of electrons from NADH and $FADH_2$ to pump H^+ across the membrane, from the mitochondrial matrix into the intermembrane space. The H^+ has a tendency to move back across the membrane, diffusing down its gradient. And the ATP synthases are the only sites that provide a route through the membrane for H^+. As we described previously, the passage of H^+ through ATP synthase uses the exergonic flow of H^+ to drive the phosphorylation of ADP. Thus, the energy stored in an H^+ gradient across a membrane couples the redox reactions of the electron transport chain to ATP synthesis, an example of chemiosmosis.

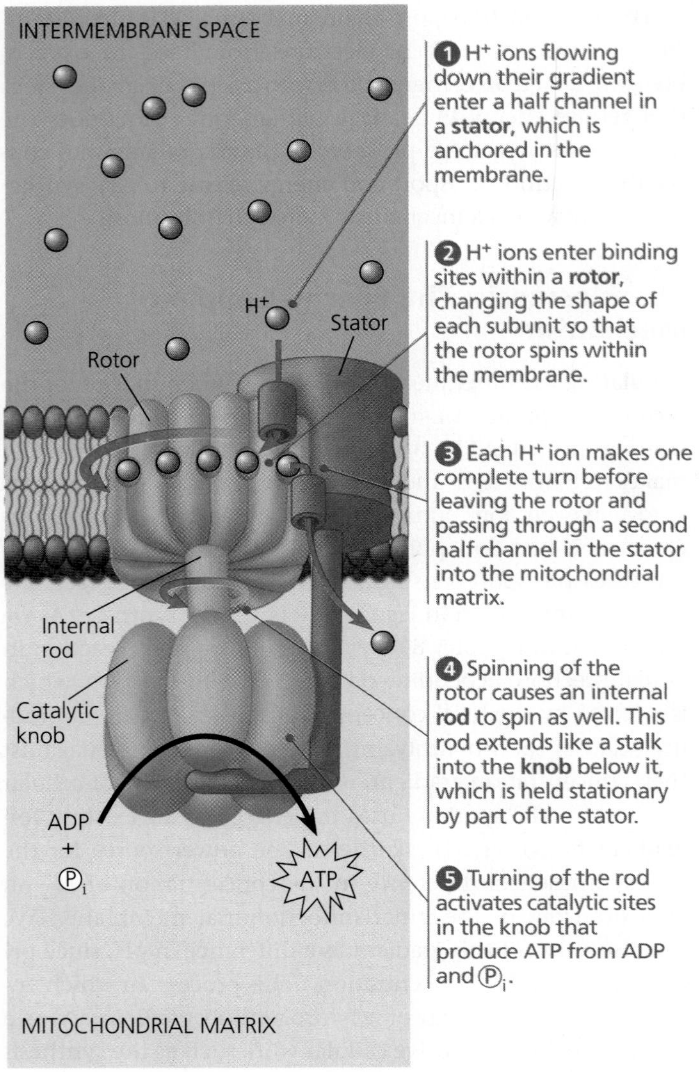

INTERMEMBRANE SPACE

Rotor

H^+

Stator

Internal rod

Catalytic knob

ADP + Ⓟ$_i$

ATP

MITOCHONDRIAL MATRIX

1 H^+ ions flowing down their gradient enter a half channel in a **stator**, which is anchored in the membrane.

2 H^+ ions enter binding sites within a **rotor**, changing the shape of each subunit so that the rotor spins within the membrane.

3 Each H^+ ion makes one complete turn before leaving the rotor and passing through a second half channel in the stator into the mitochondrial matrix.

4 Spinning of the rotor causes an internal **rod** to spin as well. This rod extends like a stalk into the **knob** below it, which is held stationary by part of the stator.

5 Turning of the rod activates catalytic sites in the knob that produce ATP from ADP and Ⓟ$_i$.

▲ **Figure 9.14 ATP synthase, a molecular mill.** The ATP synthase protein complex functions as a mill, powered by the flow of hydrogen ions. Multiple copies of this complex reside in mitochondrial and chloroplast membranes of eukaryotes and in the plasma membranes of prokaryotes. Each of the four parts of ATP synthase consists of a number of polypeptide subunits.

At this point, you may be wondering how the electron transport chain pumps hydrogen ions. Researchers have found that certain members of the electron transport chain accept and release protons (H^+) along with electrons. (The aqueous solutions inside and surrounding the cell are a ready source of H^+.) At certain steps along the chain, electron transfers cause H^+ to be taken up and released into the surrounding solution. In eukaryotic cells, the electron carriers are spatially arranged in the inner mitochondrial membrane in such a way that H^+ is accepted from the mitochondrial matrix and deposited in the intermembrane space (see Figure 9.15). The H^+ gradient that results is referred to as a **proton-motive force**, emphasizing the capacity of the gradient to perform work. The force drives H^+ back across the membrane through the H^+ channels provided by ATP synthases.

In general terms, *chemiosmosis is an energy-coupling mechanism that uses energy stored in the form of an* H^+ *gradient across a membrane to drive cellular work*. In mitochondria, the energy for gradient formation comes from exergonic redox reactions, and ATP synthesis is the work performed. But chemiosmosis also occurs elsewhere and in other variations. Chloroplasts use chemiosmosis to generate ATP during photosynthesis; in these organelles, light (rather than chemical energy) drives both electron flow down an electron transport chain and the resulting H^+ gradient formation. Prokaryotes, as already mentioned, generate H^+ gradients across their plasma membranes. They then tap the proton-motive force not only to make ATP inside the cell but also to rotate their flagella and to pump nutrients and waste products across the membrane. Because of its central importance to energy conversions in prokaryotes and eukaryotes, chemiosmosis has helped unify the study of bioenergetics. Peter Mitchell was awarded the Nobel Prize in 1978 for originally proposing the chemiosmotic model.

An Accounting of ATP Production by Cellular Respiration

In the last few sections, we have looked rather closely at the key processes of cellular respiration. Now let's take a step

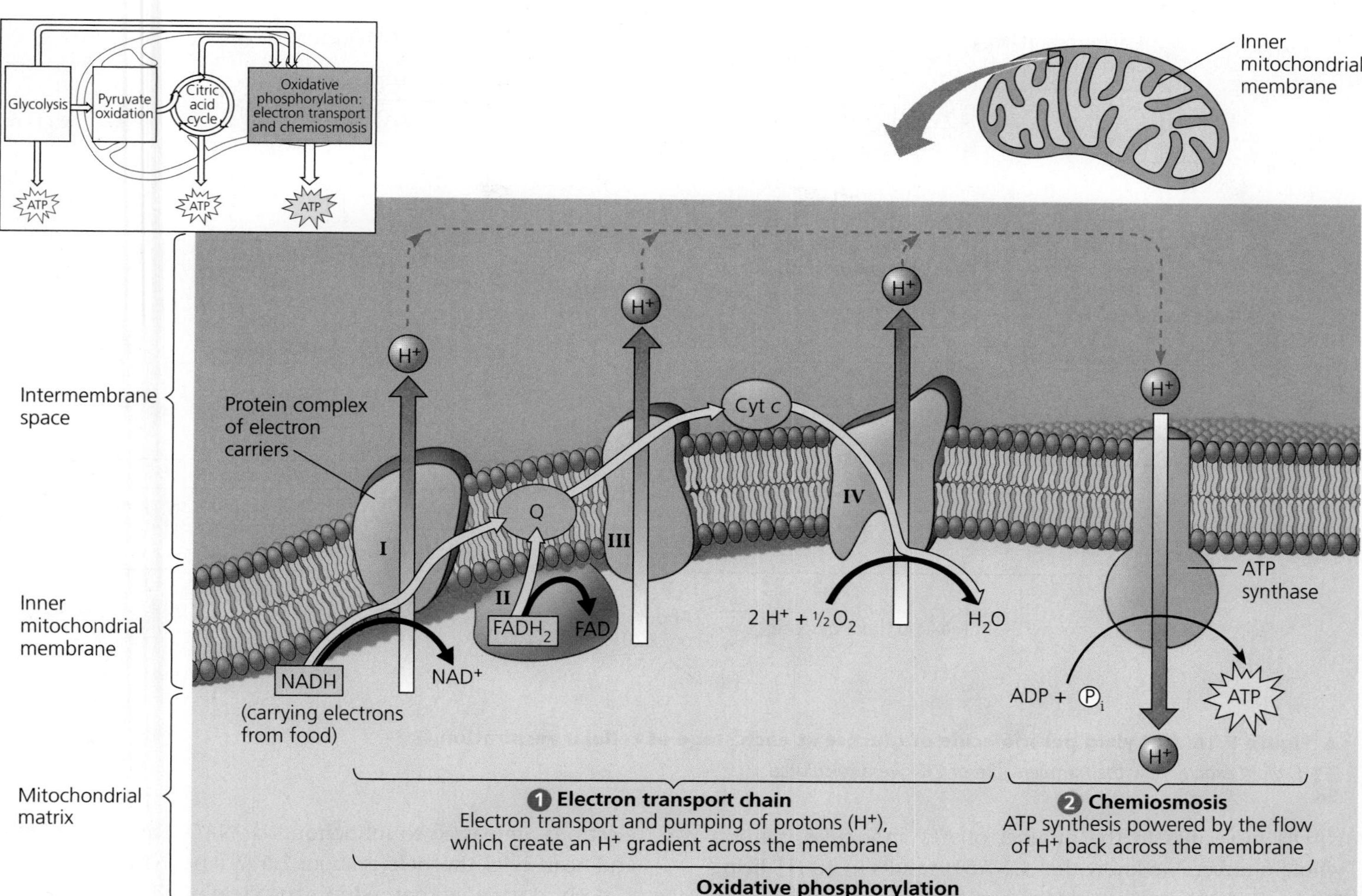

▲ **Figure 9.15 Chemiosmosis couples the electron transport chain to ATP synthesis.** ❶ NADH and $FADH_2$ shuttle high-energy electrons extracted from food during glycolysis and the citric acid cycle into an electron transport chain built into the inner mitochondrial membrane. The gold arrows trace the transport of electrons, which finally pass to oxygen at the "downhill" end of the chain, forming water. As Figure 9.13 showed, most of the electron carriers of the chain are grouped into four complexes. Two mobile carriers, ubiquinone (Q) and cytochrome *c* (Cyt *c*), move rapidly, ferrying electrons between the large complexes. As complexes I, III, and IV accept and then donate electrons, they pump protons from the mitochondrial matrix into the intermembrane space. (In prokaryotes, protons are pumped outside the plasma membrane.) Note that $FADH_2$ deposits its electrons via complex II and so results in fewer protons being pumped into the intermembrane space than occurs with NADH. Chemical energy originally harvested from food is transformed into a proton-motive force, a gradient of H^+ across the membrane. ❷ During chemiosmosis, the protons flow back down their gradient via ATP synthase, which is built into the membrane nearby. The ATP synthase harnesses the proton-motive force to phosphorylate ADP, forming ATP. Together, electron transport and chemiosmosis make up oxidative phosphorylation.

WHAT IF? *If complex IV were nonfunctional, could chemiosmosis produce any ATP, and if so, how would the rate of synthesis differ?*

back and remind ourselves of its overall function: harvesting the energy of glucose for ATP synthesis.

During respiration, most energy flows in this sequence: glucose → NADH → electron transport chain → proton-motive force → ATP. We can do some bookkeeping to calculate the ATP profit when cellular respiration oxidizes a molecule of glucose to six molecules of carbon dioxide. The three main departments of this metabolic enterprise are glycolysis, the citric acid cycle, and the electron transport chain, which drives oxidative phosphorylation. **Figure 9.16**, on the next page, gives a detailed accounting of the ATP yield per glucose molecule oxidized. The tally adds the 4 ATP produced directly by substrate-level phosphorylation during glycolysis and the citric acid cycle to the many more molecules of ATP generated by oxidative phosphorylation. Each NADH that transfers a pair of electrons from glucose to the electron transport chain contributes enough to the proton-motive force to generate a maximum of about 3 ATP.

Why are the numbers in Figure 9.16 inexact? There are three reasons we cannot state an exact number of ATP molecules generated by the breakdown of one molecule of glucose. First, phosphorylation and the redox reactions are not directly coupled to each other, so the ratio of the number of

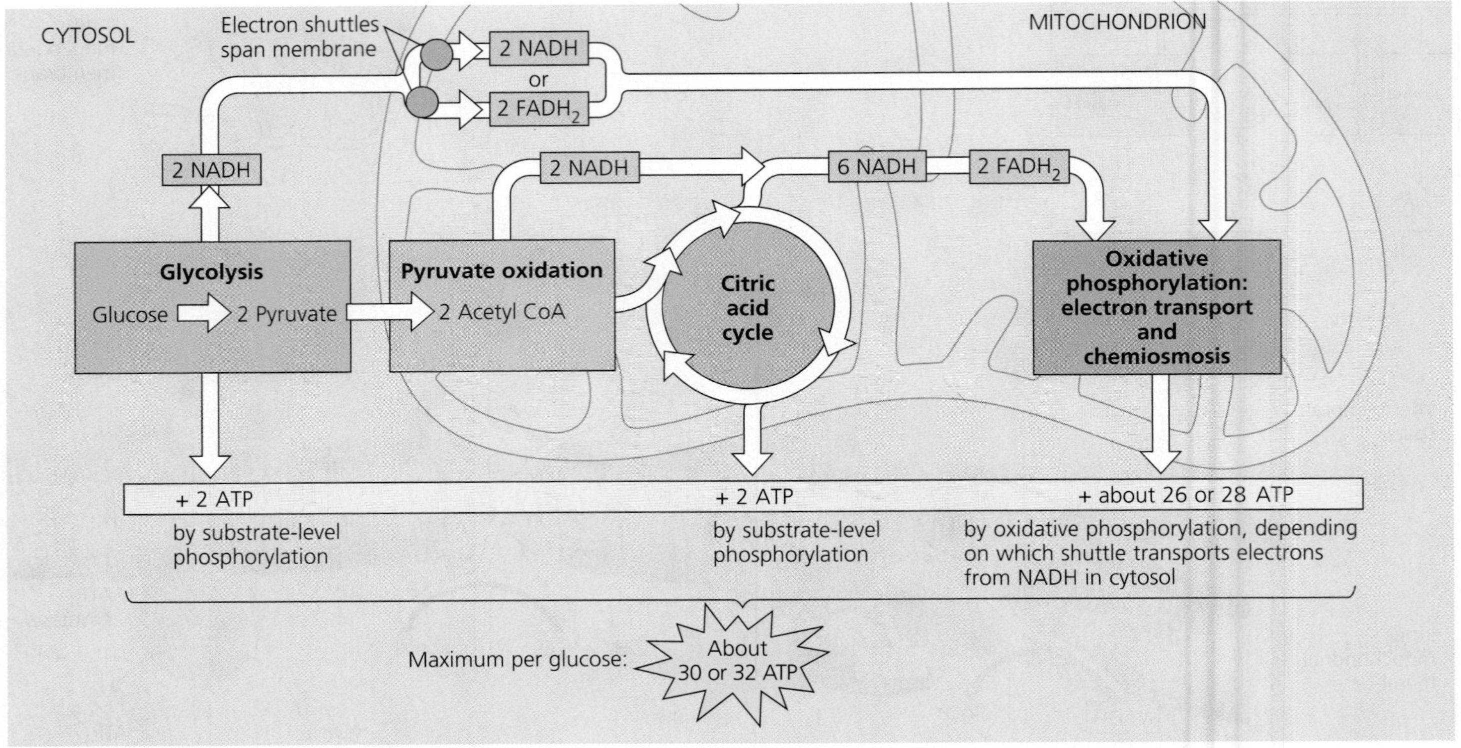

▲ **Figure 9.16 ATP yield per molecule of glucose at each stage of cellular respiration.**

? *Explain exactly how the numbers "26 or 28" were calculated.*

NADH molecules to the number of ATP molecules is not a whole number. We know that 1 NADH results in 10 H^+ being transported out across the inner mitochondrial membrane, but the exact number of H^+ that must reenter the mitochondrial matrix via ATP synthase to generate 1 ATP has long been debated. Based on experimental data, however, most biochemists now agree that the most accurate number is 4 H^+. Therefore, a single molecule of NADH generates enough proton-motive force for the synthesis of 2.5 ATP. The citric acid cycle also supplies electrons to the electron transport chain via $FADH_2$, but since its electrons enter later in the chain, each molecule of this electron carrier is responsible for transport of only enough H^+ for the synthesis of 1.5 ATP. These numbers also take into account the slight energetic cost of moving the ATP formed in the mitochondrion out into the cytosol, where it will be used.

Second, the ATP yield varies slightly depending on the type of shuttle used to transport electrons from the cytosol into the mitochondrion. The mitochondrial inner membrane is impermeable to NADH, so NADH in the cytosol is segregated from the machinery of oxidative phosphorylation. The 2 electrons of NADH captured in glycolysis must be conveyed into the mitochondrion by one of several electron shuttle systems. Depending on the kind of shuttle in a particular cell type, the electrons are passed either to NAD^+ or to FAD in the mitochondrial matrix (see Figure 9.16). If the electrons are passed to FAD, as in brain cells, only about 1.5 ATP can result from each NADH that was originally generated in the cytosol. If the electrons are passed to mitochondrial NAD^+, as in liver cells and heart cells, the yield is about 2.5 ATP per NADH.

A third variable that reduces the yield of ATP is the use of the proton-motive force generated by the redox reactions of respiration to drive other kinds of work. For example, the proton-motive force powers the mitochondrion's uptake of pyruvate from the cytosol. However, if *all* the proton-motive force generated by the electron transport chain were used to drive ATP synthesis, one glucose molecule could generate a maximum of 28 ATP produced by oxidative phosphorylation plus 4 ATP (net) from substrate-level phosphorylation to give a total yield of about 32 ATP (or only about 30 ATP if the less efficient shuttle were functioning).

We can now roughly estimate the efficiency of respiration—that is, the percentage of chemical energy in glucose that has been transferred to ATP. Recall that the complete oxidation of a mole of glucose releases 686 kcal of energy under standard conditions ($\Delta G = -686$ kcal/mol). Phosphorylation of ADP to form ATP stores at least 7.3 kcal per mole of ATP. Therefore, the efficiency of respiration is 7.3 kcal per mole of ATP times 32 moles of ATP per mole of glucose divided by 686 kcal per mole of glucose, which equals 0.34. Thus, about 34% of the potential chemical energy in glucose has been transferred to ATP; the actual percentage is probably higher because ΔG is lower under cellular conditions. Cellular respiration is remarkably efficient in its energy conversion. By comparison, the most efficient automobile converts only about 25% of the energy stored in gasoline to energy that moves the car.

The rest of the energy stored in glucose is lost as heat. We humans use some of this heat to maintain our relatively high body temperature (37°C), and we dissipate the rest through sweating and other cooling mechanisms.

Under certain conditions, it may be beneficial to reduce the efficiency of cellular respiration. A remarkable adaptation is shown by hibernating mammals, which overwinter in a state of inactivity and lowered metabolism. Although their internal body temperature is lower than normal, it still must be kept significantly higher than the external air temperature. One type of tissue, called brown fat, is made up of cells packed full of mitochondria. The inner mitochondrial membrane contains a channel protein called the uncoupling protein, which allows protons to flow back down their concentration gradient without generating ATP. Activation of these proteins in hibernating mammals results in ongoing oxidation of stored fuel stores (fats), generating heat without any ATP production. In the absence of such an adaptation, the ATP level would build up to a point that cellular respiration would be shut down due to regulatory mechanisms to be discussed later.

CONCEPT CHECK 9.4

1. What effect would an absence of O_2 have on the process shown in Figure 9.15?
2. **WHAT IF?** In the absence of O_2, as in question 1, what do you think would happen if you decreased the pH of the intermembrane space of the mitochondrion? Explain your answer.
3. **MAKE CONNECTIONS** In Concept 7.1 (pp. 127–128), you learned that membranes must be fluid to function properly. How does the operation of the electron transport chain support that assertion?

For suggested answers, see Appendix A.

CONCEPT 9.5

Fermentation and anaerobic respiration enable cells to produce ATP without the use of oxygen

Because most of the ATP generated by cellular respiration is due to the work of oxidative phosphorylation, our estimate of ATP yield from aerobic respiration is contingent on an adequate supply of oxygen to the cell. Without the electronegative oxygen to pull electrons down the transport chain, oxidative phosphorylation eventually ceases. However, there are two general mechanisms by which certain cells can oxidize organic fuel and generate ATP *without* the use of oxygen: anaerobic respiration and fermentation. The distinction between these two is that an electron transport chain is used in anaerobic respiration but not in fermentation. (The electron transport chain is also called the respiratory chain because of its role in both types of cellular respiration.)

We have already mentioned anaerobic respiration, which takes place in certain prokaryotic organisms that live in environments without oxygen. These organisms have an electron transport chain but do not use oxygen as a final electron acceptor at the end of the chain. Oxygen performs this function very well because it is extremely electronegative, but other, less electronegative substances can also serve as final electron acceptors. Some "sulfate-reducing" marine bacteria, for instance, use the sulfate ion (SO_4^{2-}) at the end of their respiratory chain. Operation of the chain builds up a proton-motive force used to produce ATP, but H_2S (hydrogen sulfide) is produced as a by-product rather than water. The rotten-egg odor you may have smelled while walking through a salt marsh or a mudflat signals the presence of sulfate-reducing bacteria.

Fermentation is a way of harvesting chemical energy without using either oxygen or any electron transport chain—in other words, without cellular respiration. How can food be oxidized without cellular respiration? Remember, oxidation simply refers to the loss of electrons to an electron acceptor, so it does not need to involve oxygen. Glycolysis oxidizes glucose to two molecules of pyruvate. The oxidizing agent of glycolysis is NAD^+, and neither oxygen nor any electron transfer chain is involved. Overall, glycolysis is exergonic, and some of the energy made available is used to produce 2 ATP (net) by substrate-level phosphorylation. If oxygen *is* present, then additional ATP is made by oxidative phosphorylation when NADH passes electrons removed from glucose to the electron transport chain. But glycolysis generates 2 ATP whether oxygen is present or not—that is, whether conditions are aerobic or anaerobic.

As an alternative to respiratory oxidation of organic nutrients, fermentation is an extension of glycolysis that allows continuous generation of ATP by the substrate-level phosphorylation of glycolysis. For this to occur, there must be a sufficient supply of NAD^+ to accept electrons during the oxidation step of glycolysis. Without some mechanism to recycle NAD^+ from NADH, glycolysis would soon deplete the cell's pool of NAD^+ by reducing it all to NADH and would shut itself down for lack of an oxidizing agent. Under aerobic conditions, NAD^+ is recycled from NADH by the transfer of electrons to the electron transport chain. An anaerobic alternative is to transfer electrons from NADH to pyruvate, the end product of glycolysis.

Types of Fermentation

Fermentation consists of glycolysis plus reactions that regenerate NAD^+ by transferring electrons from NADH to pyruvate or derivatives of pyruvate. The NAD^+ can then be reused to oxidize sugar by glycolysis, which nets two molecules of ATP by substrate-level phosphorylation. There are many types of fermentation, differing in the end products formed from pyruvate. Two common types are alcohol fermentation and lactic acid fermentation.

In **alcohol fermentation** **(Figure 9.17a)**, pyruvate is converted to ethanol (ethyl alcohol) in two steps. The first step releases carbon dioxide from the pyruvate, which is converted to the two-carbon compound acetaldehyde. In the second step, acetaldehyde is reduced by NADH to ethanol. This regenerates the supply of NAD^+ needed for the continuation of glycolysis. Many bacteria carry out alcohol fermentation under anaerobic conditions. Yeast (a fungus) also carries out alcohol fermentation. For thousands of years, humans have used yeast in brewing, winemaking, and baking. The CO_2 bubbles generated by baker's yeast during alcohol fermentation allow bread to rise.

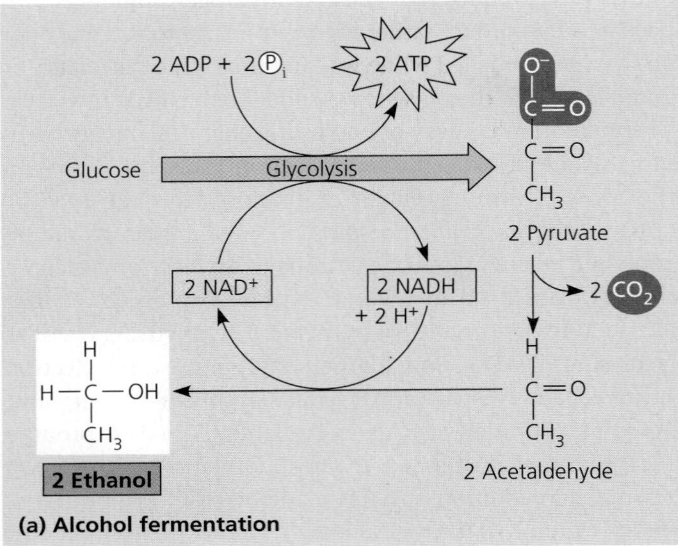

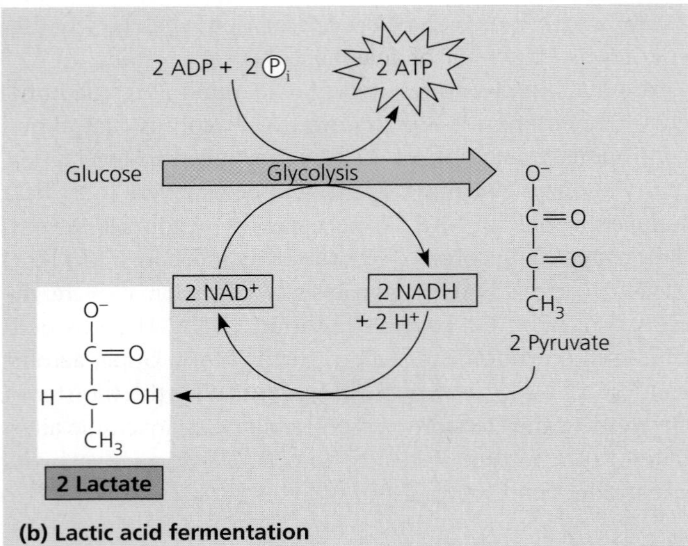

▲ **Figure 9.17 Fermentation.** In the absence of oxygen, many cells use fermentation to produce ATP by substrate-level phosphorylation. Pyruvate, the end product of glycolysis, serves as an electron acceptor for oxidizing NADH back to NAD^+, which can then be reused in glycolysis. Two of the common end products formed from fermentation are **(a)** ethanol and **(b)** lactate, the ionized form of lactic acid.

During **lactic acid fermentation** **(Figure 9.17b)**, pyruvate is reduced directly by NADH to form lactate as an end product, with no release of CO_2. (Lactate is the ionized form of lactic acid.) Lactic acid fermentation by certain fungi and bacteria is used in the dairy industry to make cheese and yogurt.

Human muscle cells make ATP by lactic acid fermentation when oxygen is scarce. This occurs during the early stages of strenuous exercise, when sugar catabolism for ATP production outpaces the muscle's supply of oxygen from the blood. Under these conditions, the cells switch from aerobic respiration to fermentation. The lactate that accumulates was previously thought to cause muscle fatigue and pain, but recent research suggests instead that increased levels of potassium ions (K^+) may be to blame, while lactate appears to enhance muscle performance. In any case, the excess lactate is gradually carried away by the blood to the liver, where it is converted back to pyruvate by liver cells. Because oxygen is available, this pyruvate can then enter the mitochondria in liver cells and complete cellular respiration.

Comparing Fermentation with Anaerobic and Aerobic Respiration

Fermentation, anaerobic respiration, and aerobic respiration are three alternative cellular pathways for producing ATP by harvesting the chemical energy of food. All three use glycolysis to oxidize glucose and other organic fuels to pyruvate, with a net production of 2 ATP by substrate-level phosphorylation. And in all three pathways, NAD^+ is the oxidizing agent that accepts electrons from food during glycolysis.

A key difference among the three pathways is the contrasting mechanisms for oxidizing NADH back to NAD^+, which is required to sustain glycolysis. In fermentation, the final electron acceptor is an organic molecule such as pyruvate (lactic acid fermentation) or acetaldehyde (alcohol fermentation). In cellular respiration, by contrast, electrons carried by NADH are transferred to an electron transport chain, where they move stepwise down a series of redox reactions to a final electron acceptor. In aerobic respiration, the final electron acceptor is oxygen; in anaerobic respiration, the final acceptor is another molecule that is electronegative (although invariably less so than oxygen). Passage of electrons from NADH to the electron transport chain not only regenerates the NAD^+ required for glycolysis but pays an ATP bonus when the stepwise electron transport from this NADH to oxygen drives oxidative phosphorylation. An even bigger ATP payoff comes from the oxidation of pyruvate in the mitochondrion, which is unique to respiration. Without an electron transport chain, the energy still stored in pyruvate is unavailable to most cells. Thus, cellular respiration harvests much more energy from each sugar molecule than fermentation can. In fact, aerobic respiration yields up to 16 times as much ATP per glucose molecule as does fermentation—up to 32 molecules of ATP for respiration, compared with 2 molecules of ATP produced by substrate-level phosphorylation in fermentation.

Some organisms, called **obligate anaerobes**, carry out only fermentation or anaerobic respiration. In fact, these organisms cannot survive in the presence of oxygen, some forms of which can actually be toxic if protective systems are not present in the cell. A few cell types, such as cells of the vertebrate brain, can carry out only aerobic oxidation of pyruvate, not fermentation. Other organisms, including yeasts and many bacteria, can make enough ATP to survive using either fermentation or respiration. Such species are called **facultative anaerobes**. On the cellular level, our muscle cells behave as facultative anaerobes. In such cells, pyruvate is a fork in the metabolic road that leads to two alternative catabolic routes (**Figure 9.18**). Under aerobic conditions, pyruvate can be converted to acetyl CoA, and oxidation continues in the citric acid cycle via aerobic respiration. Under anaerobic conditions, lactic acid fermentation occurs: Pyruvate is diverted from the citric acid cycle, serving instead as an electron acceptor to recycle NAD^+. To make the same amount of ATP, a facultative anaerobe has to consume sugar at a much faster rate when fermenting than when respiring.

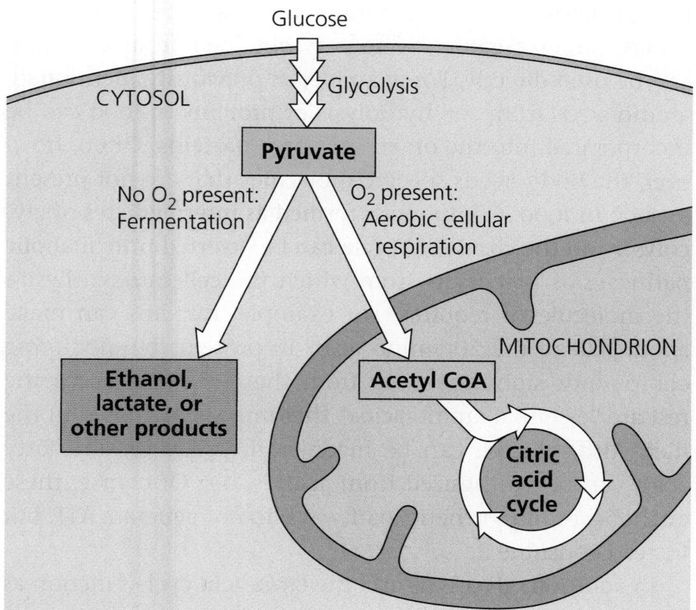

▲ **Figure 9.18 Pyruvate as a key juncture in catabolism.** Glycolysis is common to fermentation and cellular respiration. The end product of glycolysis, pyruvate, represents a fork in the catabolic pathways of glucose oxidation. In a facultative anaerobe or a muscle cell, which are capable of both aerobic cellular respiration and fermentation, pyruvate is committed to one of those two pathways, usually depending on whether or not oxygen is present.

The Evolutionary Significance of Glycolysis

EVOLUTION The role of glycolysis in both fermentation and respiration has an evolutionary basis. Ancient prokaryotes are thought to have used glycolysis to make ATP long before oxygen was present in Earth's atmosphere. The oldest known fossils of bacteria date back 3.5 billion years, but appreciable quantities of oxygen probably did not begin to accumulate in the atmosphere until about 2.7 billion years ago. Cyanobacteria produced this O_2 as a by-product of photosynthesis. Therefore, early prokaryotes may have generated ATP exclusively from glycolysis. The fact that glycolysis is today the most widespread metabolic pathway among Earth's organisms suggests that it evolved very early in the history of life. The cytosolic location of glycolysis also implies great antiquity; the pathway does not require any of the membrane-bounded organelles of the eukaryotic cell, which evolved approximately 1 billion years after the prokaryotic cell. Glycolysis is a metabolic heirloom from early cells that continues to function in fermentation and as the first stage in the breakdown of organic molecules by respiration.

CONCEPT CHECK 9.5

1. Consider the NADH formed during glycolysis. What is the final acceptor for its electrons during fermentation? What is the final acceptor for its electrons during aerobic respiration?
2. **WHAT IF?** A glucose-fed yeast cell is moved from an aerobic environment to an anaerobic one. How would its rate of glucose consumption change if ATP were to be generated at the same rate?

For suggested answers, see Appendix A.

CONCEPT 9.6

Glycolysis and the citric acid cycle connect to many other metabolic pathways

So far, we have treated the oxidative breakdown of glucose in isolation from the cell's overall metabolic economy. In this section, you will learn that glycolysis and the citric acid cycle are major intersections of the cell's catabolic and anabolic (biosynthetic) pathways.

The Versatility of Catabolism

Throughout this chapter, we have used glucose as the fuel for cellular respiration. But free glucose molecules are not common in the diets of humans and other animals. We obtain most of our calories in the form of fats, proteins, sucrose and other disaccharides, and starch, a polysaccharide. All these

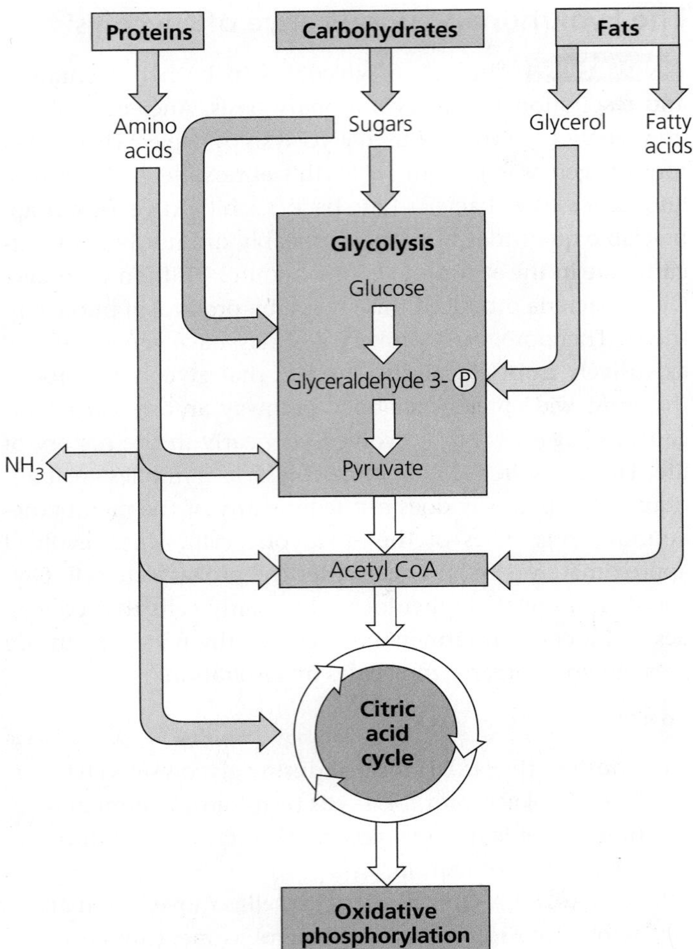

▲ **Figure 9.19 The catabolism of various molecules from food.** Carbohydrates, fats, and proteins can all be used as fuel for cellular respiration. Monomers of these molecules enter glycolysis or the citric acid cycle at various points. Glycolysis and the citric acid cycle are catabolic funnels through which electrons from all kinds of organic molecules flow on their exergonic fall to oxygen.

organic molecules in food can be used by cellular respiration to make ATP (**Figure 9.19**).

Glycolysis can accept a wide range of carbohydrates for catabolism. In the digestive tract, starch is hydrolyzed to glucose, which can then be broken down in the cells by glycolysis and the citric acid cycle. Similarly, glycogen, the polysaccharide that humans and many other animals store in their liver and muscle cells, can be hydrolyzed to glucose between meals as fuel for respiration. The digestion of disaccharides, including sucrose, provides glucose and other monosaccharides as fuel for respiration.

Proteins can also be used for fuel, but first they must be digested to their constituent amino acids. Many of the amino acids are used by the organism to build new proteins. Amino acids present in excess are converted by enzymes to intermediates of glycolysis and the citric acid cycle. Before amino acids can feed into glycolysis or the citric acid cycle, their amino groups must be removed, a process called *deamination*. The nitrogenous refuse is excreted from the animal in the form of ammonia (NH_3), urea, or other waste products.

Catabolism can also harvest energy stored in fats obtained either from food or from storage cells in the body. After fats are digested to glycerol and fatty acids, the glycerol is converted to glyceraldehyde 3-phosphate, an intermediate of glycolysis. Most of the energy of a fat is stored in the fatty acids. A metabolic sequence called **beta oxidation** breaks the fatty acids down to two-carbon fragments, which enter the citric acid cycle as acetyl CoA. NADH and $FADH_2$ are also generated during beta oxidation; they can enter the electron transport chain, leading to further ATP production. Fats make excellent fuel, in large part due to their chemical structure and the high energy level of their electrons (equally shared between carbon and hydrogen) compared to those of carbohydrates. A gram of fat oxidized by respiration produces more than twice as much ATP as a gram of carbohydrate. Unfortunately, this also means that a person trying to lose weight must work hard to use up fat stored in the body because so many calories are stockpiled in each gram of fat.

Biosynthesis (Anabolic Pathways)

Cells need substance as well as energy. Not all the organic molecules of food are destined to be oxidized as fuel to make ATP. In addition to calories, food must also provide the carbon skeletons that cells require to make their own molecules. Some organic monomers obtained from digestion can be used directly. For example, as previously mentioned, amino acids from the hydrolysis of proteins in food can be incorporated into the organism's own proteins. Often, however, the body needs specific molecules that are not present as such in food. Compounds formed as intermediates of glycolysis and the citric acid cycle can be diverted into anabolic pathways as precursors from which the cell can synthesize the molecules it requires. For example, humans can make about half of the 20 amino acids in proteins by modifying compounds siphoned away from the citric acid cycle; the rest are "essential amino acids" that must be obtained in the diet. Also, glucose can be made from pyruvate, and fatty acids can be synthesized from acetyl CoA. Of course, these anabolic, or biosynthetic, pathways do not generate ATP, but instead consume it.

In addition, glycolysis and the citric acid cycle function as metabolic interchanges that enable our cells to convert some kinds of molecules to others as we need them. For example, an intermediate compound generated during glycolysis, dihydroxyacetone phosphate (see Figure 9.9, step 5), can be converted to one of the major precursors of fats. If we eat more food than we need, we store fat even if our diet is fat-free. Metabolism is remarkably versatile and adaptable.

Regulation of Cellular Respiration via Feedback Mechanisms

Basic principles of supply and demand regulate the metabolic economy. The cell does not waste energy making more of a particular substance than it needs. If there is a glut of a certain amino acid, for example, the anabolic pathway that synthesizes that amino acid from an intermediate of the citric acid cycle is switched off. The most common mechanism for this control is feedback inhibition: The end product of the anabolic pathway inhibits the enzyme that catalyzes an early step of the pathway (see Figure 8.21). This prevents the needless diversion of key metabolic intermediates from uses that are more urgent.

The cell also controls its catabolism. If the cell is working hard and its ATP concentration begins to drop, respiration speeds up. When there is plenty of ATP to meet demand, respiration slows down, sparing valuable organic molecules for other functions. Again, control is based mainly on regulating the activity of enzymes at strategic points in the catabolic pathway. As shown in **Figure 9.20**, one important switch is phosphofructokinase, the enzyme that catalyzes step 3 of glycolysis (see Figure 9.9). That is the first step that commits the substrate irreversibly to the glycolytic pathway. By controlling the rate of this step, the cell can speed up or slow down the entire catabolic process. Phosphofructokinase can thus be considered the pacemaker of respiration.

Phosphofructokinase is an allosteric enzyme with receptor sites for specific inhibitors and activators. It is inhibited by ATP and stimulated by AMP (adenosine monophosphate), which the cell derives from ADP. As ATP accumulates, inhibition of the enzyme slows down glycolysis. The enzyme becomes active again as cellular work converts ATP to ADP (and AMP) faster than ATP is being regenerated. Phosphofructokinase is also sensitive to citrate, the first product of the citric acid cycle. If citrate accumulates in mitochondria, some of it passes into the cytosol and inhibits phosphofructokinase. This mechanism helps synchronize the rates of glycolysis and the citric acid cycle. As citrate accumulates, glycolysis slows down, and the supply of acetyl groups to the citric acid cycle decreases. If citrate consumption increases, either because of a demand for more ATP or because anabolic pathways are draining off intermediates of the citric acid cycle, glycolysis accelerates and meets the demand. Metabolic balance is augmented by the control of enzymes that catalyze other key steps of glycolysis and the citric acid cycle. Cells are thrifty, expedient, and responsive in their metabolism.

Cellular respiration and metabolic pathways play a role of central importance in organisms. Examine Figure 9.2 again to put cellular respiration into the broader context of energy flow and chemical cycling in ecosystems. The energy that keeps us alive is *released*, not *produced*, by cellular respiration. We are tapping energy that was stored in food by photosynthesis. In the next chapter, you will learn how photosynthesis captures light and converts it to chemical energy.

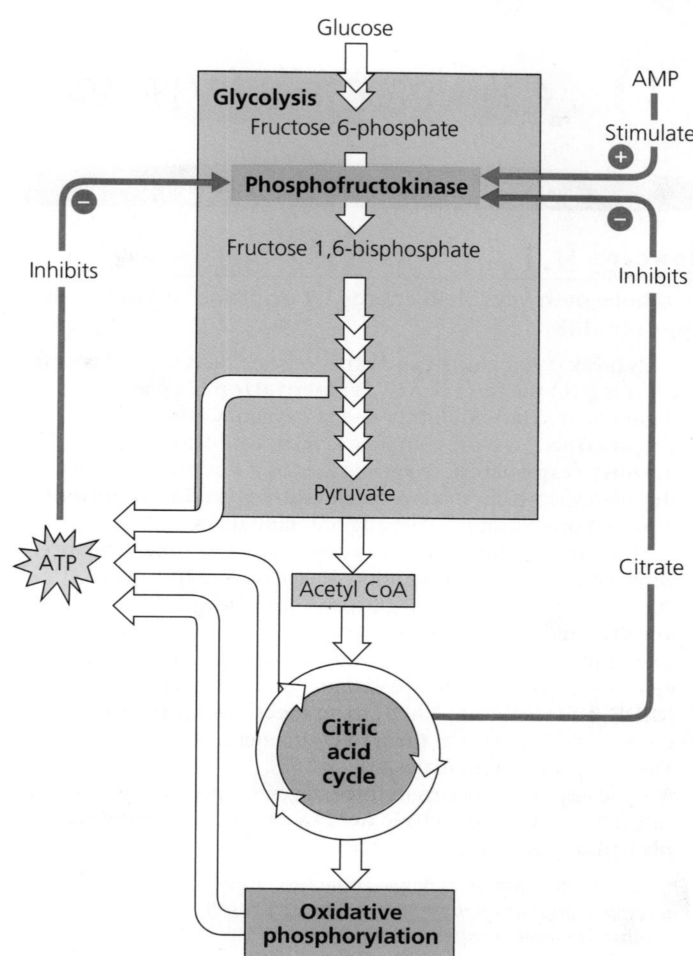

▲ **Figure 9.20 The control of cellular respiration.** Allosteric enzymes at certain points in the respiratory pathway respond to inhibitors and activators that help set the pace of glycolysis and the citric acid cycle. Phosphofructokinase, which catalyzes an early step in glycolysis (see Figure 9.9), is one such enzyme. It is stimulated by AMP (derived from ADP) but is inhibited by ATP and by citrate. This feedback regulation adjusts the rate of respiration as the cell's catabolic and anabolic demands change.

CONCEPT CHECK 9.6

1. **MAKE CONNECTIONS** Compare the structure of a fat (see Figure 5.10, p. 75) with that of a carbohydrate (see Figure 5.3, p. 70). What features of their structures make fat a much better fuel?
2. Under what circumstances might your body synthesize fat molecules?
3. **MAKE CONNECTIONS** Return to Figure 5.6b on page 72 and look at the arrangement of glycogen and mitochondria in the micrograph. What is the connection between glycogen and mitochondria?
4. **WHAT IF?** What will happen in a muscle cell that has used up its supply of oxygen and ATP? (Review Figures 9.18 and 9.20.)
5. **WHAT IF?** During intense exercise, can a muscle cell use fat as a concentrated source of chemical energy? Explain. (Review Figures 9.18 and 9.19.)

For suggested answers, see Appendix A.

9 CHAPTER REVIEW

SUMMARY OF KEY CONCEPTS

CONCEPT 9.1

Catabolic pathways yield energy by oxidizing organic fuels (pp. 164–168)

- Cells break down glucose and other organic fuels to yield chemical energy in the form of ATP. **Fermentation** is a partial degradation of glucose without the use of oxygen. **Cellular respiration** is a more complete breakdown of glucose; in **aerobic respiration**, oxygen is used as a reactant. The cell taps the energy stored in food molecules through **redox reactions**, in which one substance partially or totally shifts electrons to another. **Oxidation** is the loss of electrons from one substance, while **reduction** is the addition of electrons to the other.
- During aerobic respiration, glucose ($C_6H_{12}O_6$) is oxidized to CO_2, and O_2 is reduced to H_2O. Electrons lose potential energy during their transfer from glucose or other organic compounds to oxygen. Electrons are usually passed first to **NAD^+**, reducing it to NADH, and then from NADH to an **electron transport chain**, which conducts them to O_2 in energy-releasing steps. The energy is used to make ATP.
- Aerobic respiration occurs in three stages: (1) **glycolysis**, (2) pyruvate oxidation and the **citric acid cycle**, and (3) **oxidative phosphorylation** (electron transport and chemiosmosis).

? *Describe the difference between the two processes in cellular respiration that produce ATP: oxidative phosphorylation and substrate-level phosphorylation.*

CONCEPT 9.2

Glycolysis harvests chemical energy by oxidizing glucose to pyruvate (pp. 168–169)

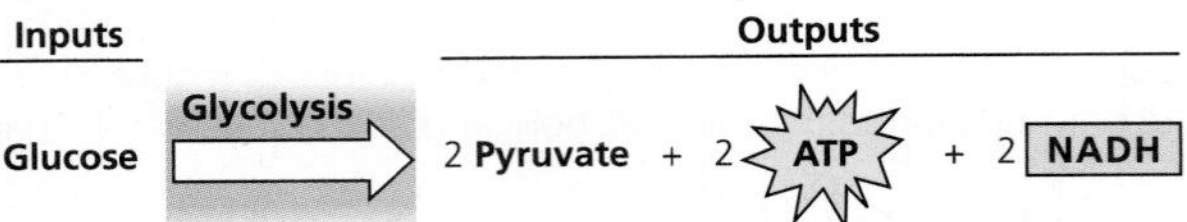

? *What is the source of energy for the formation of ATP and NADH in glycolysis?*

CONCEPT 9.3

After pyruvate is oxidized, the citric acid cycle completes the energy-yielding oxidation of organic molecules (pp. 170–172)

- In eukaryotic cells, pyruvate enters the mitochondrion and is oxidized to **acetyl CoA**, which is further oxidized in the citric acid cycle.

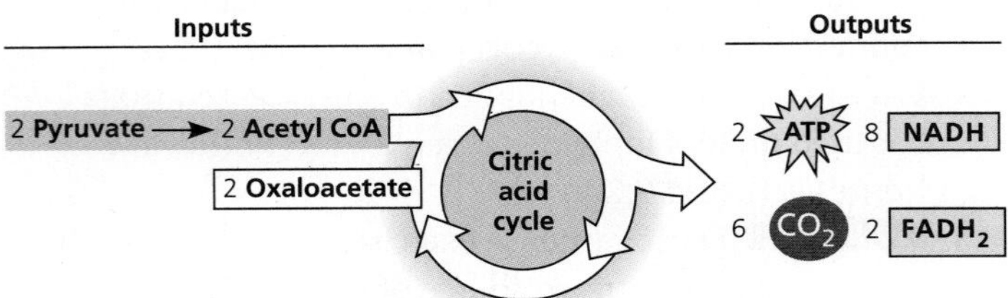

? *What molecular products indicate the complete oxidation of glucose during cellular respiration?*

CONCEPT 9.4

During oxidative phosphorylation, chemiosmosis couples electron transport to ATP synthesis (pp. 172–177)

- NADH and $FADH_2$ transfer electrons to the electron transport chain. Electrons move down the chain, losing energy in several energy-releasing steps. Finally, electrons are passed to O_2, reducing it to H_2O.

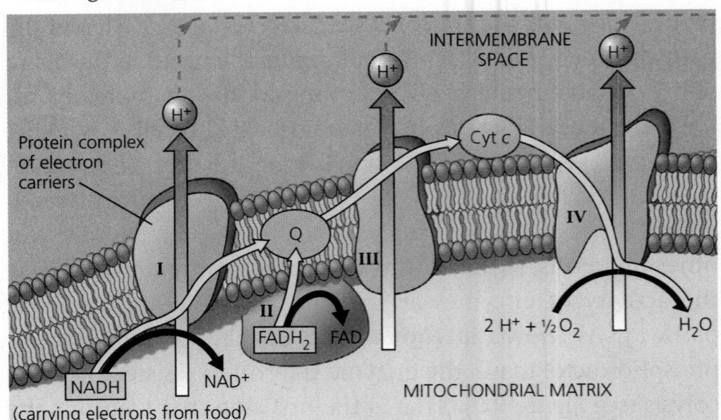

- At certain steps along the electron transport chain, electron transfer causes protein complexes to move H^+ from the mitochondrial matrix (in eukaryotes) to the intermembrane space, storing energy as a **proton-motive force** (H^+ gradient). As H^+ diffuses back into the matrix through **ATP synthase**, its passage drives the phosphorylation of ADP, a process called **chemiosmosis**.

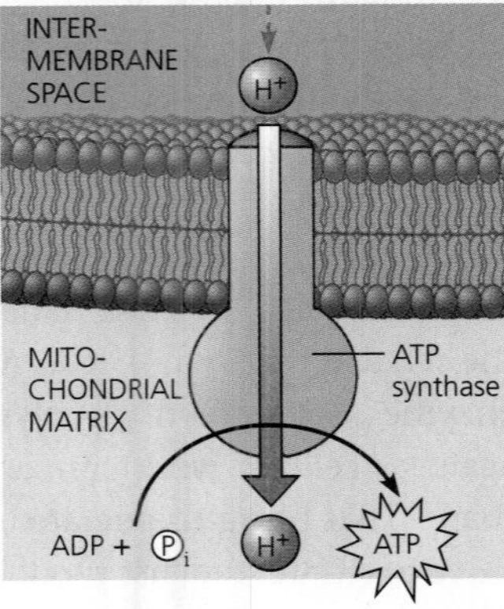

- About 34% of the energy stored in a glucose molecule is transferred to ATP during cellular respiration, producing a maximum of about 32 ATP.

? *Briefly explain the mechanism by which ATP synthase produces ATP. List three locations in which ATP synthases are found.*

CONCEPT 9.5

Fermentation and anaerobic respiration enable cells to produce ATP without the use of oxygen (pp. 177–179)

- Glycolysis nets 2 ATP by substrate-level phosphorylation, whether oxygen is present or not. Under anaerobic conditions, either anaerobic respiration or fermentation can take place. In anaerobic respiration, an electron transport chain is present with a final electron acceptor other than oxygen. In fermentation, the electrons from NADH are passed to pyruvate or a derivative of pyruvate, regenerating the NAD^+ required to oxidize more glucose. Two common types of fermentation are **alcohol fermentation** and **lactic acid fermentation**.
- Fermentation and anaerobic or aerobic respiration all use glycolysis to oxidize glucose, but they differ in their final electron acceptor and whether an electron transport chain is used (respiration) or not (fermentation). Respiration yields more ATP;

aerobic respiration, with O_2 as the final electron acceptor, yields about 16 times as much ATP as does fermentation.

- Glycolysis occurs in nearly all organisms and is thought to have evolved in ancient prokaryotes before there was O_2 in the atmosphere.

? *Which process yields more ATP, fermentation or anaerobic respiration? Explain.*

CONCEPT 9.6

Glycolysis and the citric acid cycle connect to many other metabolic pathways (pp. 179–181)

- Catabolic pathways funnel electrons from many kinds of organic molecules into cellular respiration. Many carbohydrates can enter glycolysis, most often after conversion to glucose. Amino acids of proteins must be deaminated before being oxidized. The fatty acids of fats undergo **beta oxidation** to two-carbon fragments and then enter the citric acid cycle as acetyl CoA. Anabolic pathways can use small molecules from food directly or build other substances using intermediates of glycolysis or the citric acid cycle.
- Cellular respiration is controlled by allosteric enzymes at key points in glycolysis and the citric acid cycle.

? *Describe how the catabolic pathways of glycolysis and the citric acid cycle intersect with anabolic pathways in the metabolism of a cell.*

TEST YOUR UNDERSTANDING

LEVEL 1: KNOWLEDGE/COMPREHENSION

1. The *immediate* energy source that drives ATP synthesis by ATP synthase during oxidative phosphorylation is the
   a. oxidation of glucose and other organic compounds.
   b. flow of electrons down the electron transport chain.
   c. affinity of oxygen for electrons.
   d. H^+ concentration across the membrane holding ATP synthase.
   e. transfer of phosphate to ADP.

2. Which metabolic pathway is common to both fermentation and cellular respiration of a glucose molecule?
   a. the citric acid cycle
   b. the electron transport chain
   c. glycolysis
   d. synthesis of acetyl CoA from pyruvate
   e. reduction of pyruvate to lactate

3. In mitochondria, exergonic redox reactions
   a. are the source of energy driving prokaryotic ATP synthesis.
   b. are directly coupled to substrate-level phosphorylation.
   c. provide the energy that establishes the proton gradient.
   d. reduce carbon atoms to carbon dioxide.
   e. are coupled via phosphorylated intermediates to endergonic processes.

4. The final electron acceptor of the electron transport chain that functions in aerobic oxidative phosphorylation is
   a. oxygen. b. water. c. NAD^+. d. pyruvate. e. ADP.

LEVEL 2: APPLICATION/ANALYSIS

5. What is the oxidizing agent in the following reaction?
   Pyruvate + NADH + H^+ $\rightarrow$ Lactate + NAD^+
   a. oxygen b. NADH c. NAD^+ d. lactate e. pyruvate

6. When electrons flow along the electron transport chains of mitochondria, which of the following changes occurs?
   a. The pH of the matrix increases.
   b. ATP synthase pumps protons by active transport.
   c. The electrons gain free energy.
   d. The cytochromes phosphorylate ADP to form ATP.
   e. NAD^+ is oxidized.

7. Most CO_2 from catabolism is released during
   a. glycolysis.
   b. the citric acid cycle.
   c. lactate fermentation.
   d. electron transport.
   e. oxidative phosphorylation.

LEVEL 3: SYNTHESIS/EVALUATION

8. **DRAW IT** The graph here shows the pH difference across the inner mitochondrial membrane over time in an actively respiring cell. At the time indicated by the vertical arrow, a metabolic poison is added that specifically and completely inhibits all function of mitochondrial ATP synthase. Draw what you would expect to see for the rest of the graphed line.

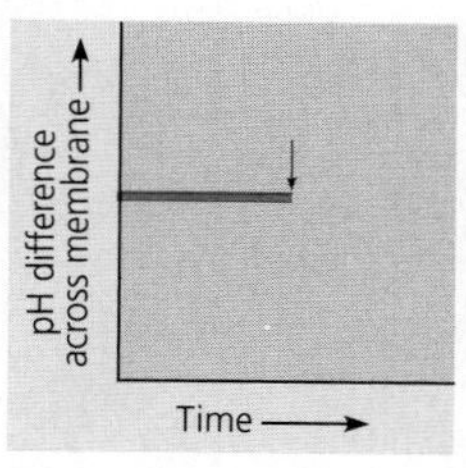

9. **EVOLUTION CONNECTION**
   ATP synthases are found in the prokaryotic plasma membrane and in mitochondria and chloroplasts. What does this suggest about the evolutionary relationship of these eukaryotic organelles to prokaryotes? How might the amino acid sequences of the ATP synthases from the different sources support or refute your hypothesis?

10. **SCIENTIFIC INQUIRY**
    In the 1930s, some physicians prescribed low doses of a compound called dinitrophenol (DNP) to help patients lose weight. This unsafe method was abandoned after some patients died. DNP uncouples the chemiosmotic machinery by making the lipid bilayer of the inner mitochondrial membrane leaky to H^+. Explain how this could cause weight loss and death.

11. **WRITE ABOUT A THEME**
    Emergent Properties In a short essay (100–150 words), explain how oxidative phosphorylation—the production of ATP using energy derived from the redox reactions of a spatially organized electron transport chain followed by chemiosmosis—is an example of how new properties emerge at each level of the biological hierarchy.

For selected answers, see Appendix A.

10

Photosynthesis

▲ **Figure 10.1 How can sunlight, seen here as a spectrum of colors in a rainbow, power the synthesis of organic substances?**

KEY CONCEPTS

10.1 Photosynthesis converts light energy to the chemical energy of food

10.2 The light reactions convert solar energy to the chemical energy of ATP and NADPH

10.3 The Calvin cycle uses the chemical energy of ATP and NADPH to reduce CO_2 to sugar

10.4 Alternative mechanisms of carbon fixation have evolved in hot, arid climates

OVERVIEW

The Process That Feeds the Biosphere

Life on Earth is solar powered. The chloroplasts of plants capture light energy that has traveled 150 million kilometers from the sun and convert it to chemical energy that is stored in sugar and other organic molecules. This conversion process is called **photosynthesis**. Let's begin by placing photosynthesis in its ecological context.

Photosynthesis nourishes almost the entire living world directly or indirectly. An organism acquires the organic compounds it uses for energy and carbon skeletons by one of two major modes: autotrophic nutrition or heterotrophic nutrition. **Autotrophs** are "self-feeders" (*auto-* means "self," and *trophos* means "feeder"); they sustain themselves without eating anything derived from other living beings. Autotrophs produce their organic molecules from CO_2 and other inorganic raw materials obtained from the environment. They are the ultimate sources of organic compounds for all nonautotrophic organisms, and for this reason, biologists refer to autotrophs as the *producers* of the biosphere.

Almost all plants are autotrophs; the only nutrients they require are water and minerals from the soil and carbon dioxide from the air. Specifically, plants are *photo*autotrophs, organisms that use light as a source of energy to synthesize organic substances (**Figure 10.1**). Photosynthesis also occurs in algae, certain other protists, and some prokaryotes (**Figure 10.2**). In this chapter, we will touch on these other groups in passing, but our emphasis will be on plants. Variations in autotrophic nutrition that occur in prokaryotes and algae will be described in Chapters 27 and 28.

Heterotrophs obtain their organic material by the second major mode of nutrition. Unable to make their own food, they live on compounds produced by other organisms (*hetero-* means "other"). Heterotrophs are the biosphere's *consumers*. The most obvious form of this "other-feeding" occurs when an animal eats plants or other animals. But heterotrophic nutrition may be more subtle. Some heterotrophs consume the remains of dead organisms by decomposing and feeding on organic litter such as carcasses, feces, and fallen leaves; they are known as decomposers. Most fungi and many types of prokaryotes get their nourishment this way. Almost all heterotrophs, including humans, are completely dependent, either directly or indirectly, on photoautotrophs for food—and also for oxygen, a by-product of photosynthesis.

The Earth's supply of fossil fuels was formed from remains of organisms that died hundreds of millions of years ago. In a sense, then, fossil fuels represent stores of the sun's energy from the distant past. Because these resources are being used at a much higher rate than they are replenished, researchers

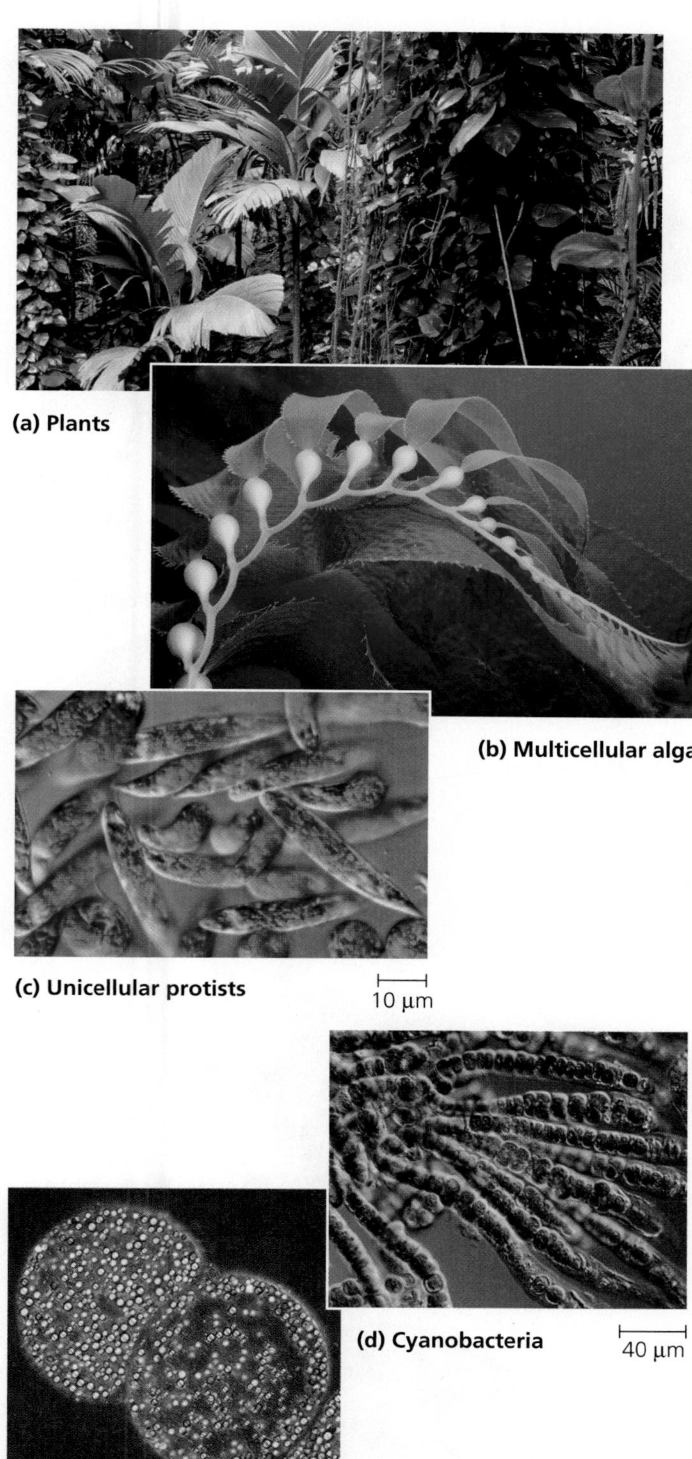

▲ **Figure 10.2 Photoautotrophs.** These organisms use light energy to drive the synthesis of organic molecules from carbon dioxide and (in most cases) water. They feed themselves and the entire living world. **(a)** On land, plants are the predominant producers of food. In aquatic environments, photoautotrophs include unicellular and **(b)** multicellular algae, such as this kelp; **(c)** some non-algal unicellular protists, such as *Euglena*; **(d)** the prokaryotes called cyanobacteria; and **(e)** other photosynthetic prokaryotes, such as these purple sulfur bacteria, which produce sulfur (the yellow globules within the cells) (c–e, LMs).

▼ **Figure 10.3**

IMPACT

Alternative Fuels from Plants and Algae

Biofuels from crops such as corn, soybeans, and cassava have been proposed as a supplement or even replacement for fossil fuels. To produce "bioethanol," the starch made naturally by the plants is simply converted to glucose and then fermented to ethanol by microorganisms. Alternatively, a simple chemical process can yield "biodiesel" from plant oils. Either product can be mixed with gasoline or used alone to power vehicles. Some species of unicellular algae are especially prolific oil producers, and they can be easily cultured in containers such as the tubular plastic bags shown below.

WHY IT MATTERS The rate of fossil fuel use by humans far outpaces its formation in the earth: Fossil fuels are a nonrenewable source of energy. Tapping the power of sunlight by using products of photosynthesis to generate energy is a sustainable alternative if cost-effective techniques can be developed. It is generally agreed that using algae is preferable to growing crops for this purpose because this use of cropland diminishes the food supply and drives up food prices.

FURTHER READING A. L. Haag, Algae bloom again, *Nature* 447:520–521 (2007).

WHAT IF? The main product of fossil fuel combustion is CO_2, and this combustion is the source of the increase in atmospheric CO_2 concentration. Scientists have proposed strategically situating containers of these algae near industrial plants, as shown above, or near highly congested city streets. Why does this arrangement make sense?

are exploring methods of capitalizing on the photosynthetic process to provide alternative fuels (**Figure 10.3**).

In this chapter, you will learn how photosynthesis works. After a discussion of the general principles of photosynthesis, we will consider the two stages of photosynthesis: the light reactions, in which solar energy is captured and transformed into chemical energy; and the Calvin cycle, in which the chemical energy is used to make organic molecules of food. Finally, we will consider a few aspects of photosynthesis from an evolutionary perspective.

CONCEPT 10.1

Photosynthesis converts light energy to the chemical energy of food

The remarkable ability of an organism to harness light energy and use it to drive the synthesis of organic compounds emerges from structural organization in the cell: Photosynthetic enzymes and other molecules are grouped together in a biological membrane, enabling the necessary series of chemical reactions to be carried out efficiently. The process of photosynthesis most likely originated in a group of bacteria that had infolded regions of the plasma membrane containing clusters of such molecules. In existing photosynthetic bacteria, infolded photosynthetic membranes function similarly to the internal membranes of the chloroplast, a eukaryotic organelle. According to the endosymbiont theory, the original chloroplast was a photosynthetic prokaryote that lived inside an ancestor of eukaryotic cells. (You learned about this theory in Chapter 6 and it will be described more fully in Chapter 25.) Chloroplasts are present in a variety of photosynthesizing organisms (see Figure 10.2), but here we will focus on plants.

Chloroplasts: The Sites of Photosynthesis in Plants

All green parts of a plant, including green stems and unripened fruit, have chloroplasts, but the leaves are the major sites of photosynthesis in most plants (**Figure 10.4**). There are about half a million chloroplasts in a chunk of leaf with a top surface area of 1 mm^2. Chloroplasts are found mainly in the cells of the **mesophyll**, the tissue in the interior of the leaf. Carbon dioxide enters the leaf, and oxygen exits, by way of microscopic pores called **stomata** (singular, *stoma*; from the Greek, meaning "mouth"). Water absorbed by the roots is delivered to the leaves in veins. Leaves also use veins to export sugar to roots and other nonphotosynthetic parts of the plant.

A typical mesophyll cell has about 30–40 chloroplasts, each organelle measuring about 2–4 μm by 4–7 μm. A chloroplast has an envelope of two membranes surrounding a dense fluid called the **stroma**. Suspended within the stroma is a third membrane system, made up of sacs called **thylakoids**, which segregates the stroma from the *thylakoid space* inside these sacs. In some places, thylakoid sacs are stacked in columns called *grana* (singular, *granum*). **Chlorophyll**, the green pigment that gives leaves their color, resides in the thylakoid membranes of the chloroplast. (The internal photosynthetic membranes of some prokaryotes are also called thylakoid membranes; see Figure 27.7b.) It is the light energy absorbed

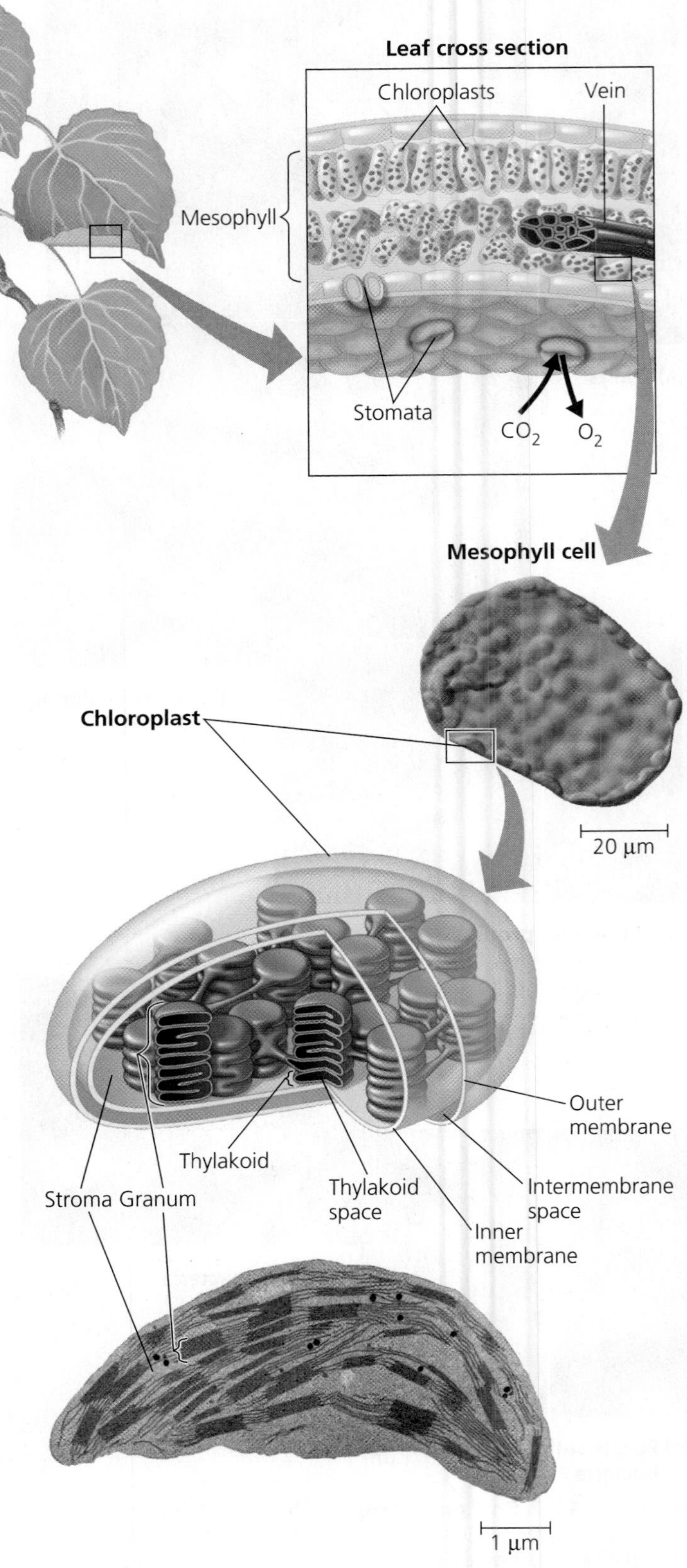

▲ **Figure 10.4 Zooming in on the location of photosynthesis in a plant.** Leaves are the major organs of photosynthesis in plants. These pictures take you into a leaf, then into a cell, and finally into a chloroplast, the organelle where photosynthesis occurs (middle, LM; bottom, TEM).

by chlorophyll that drives the synthesis of organic molecules in the chloroplast. Now that we have looked at the sites of photosynthesis in plants, we are ready to look more closely at the process of photosynthesis.

Tracking Atoms Through Photosynthesis: *Scientific Inquiry*

Scientists have tried for centuries to piece together the process by which plants make food. Although some of the steps are still not completely understood, the overall photosynthetic equation has been known since the 1800s: In the presence of light, the green parts of plants produce organic compounds and oxygen from carbon dioxide and water. Using molecular formulas, we can summarize the complex series of chemical reactions in photosynthesis with this chemical equation:

$$6\ CO_2 + 12\ H_2O + \text{Light energy} \rightarrow C_6H_{12}O_6 + 6\ O_2 + 6\ H_2O$$

We use glucose ($C_6H_{12}O_6$) here to simplify the relationship between photosynthesis and respiration, but the direct product of photosynthesis is actually a three-carbon sugar that can be used to make glucose. Water appears on both sides of the equation because 12 molecules are consumed and 6 molecules are newly formed during photosynthesis. We can simplify the equation by indicating only the net consumption of water:

$$6\ CO_2 + 6\ H_2O + \text{Light energy} \rightarrow C_6H_{12}O_6 + 6\ O_2$$

Writing the equation in this form, we can see that the overall chemical change during photosynthesis is the reverse of the one that occurs during cellular respiration. Both of these metabolic processes occur in plant cells. However, as you will soon learn, chloroplasts do not synthesize sugars by simply reversing the steps of respiration.

Now let's divide the photosynthetic equation by 6 to put it in its simplest possible form:

$$CO_2 + H_2O \rightarrow [CH_2O] + O_2$$

Here, the brackets indicate that CH_2O is not an actual sugar but represents the general formula for a carbohydrate. In other words, we are imagining the synthesis of a sugar molecule one carbon at a time. Six repetitions would theoretically produce a glucose molecule. Let's now use this simplified formula to see how researchers tracked the elements C, H, and O from the reactants of photosynthesis to the products.

The Splitting of Water

One of the first clues to the mechanism of photosynthesis came from the discovery that the O_2 given off by plants is derived from H_2O and not from CO_2. The chloroplast splits water into hydrogen and oxygen. Before this discovery, the prevailing hypothesis was that photosynthesis split carbon dioxide ($CO_2 \rightarrow C + O_2$) and then added water to the carbon ($C + H_2O \rightarrow [CH_2O]$). This hypothesis predicted that the O_2 released during photosynthesis came from CO_2. This idea was challenged in the 1930s by C. B. van Niel, of Stanford University. Van Niel was investigating photosynthesis in bacteria that make their carbohydrate from CO_2 but do not release O_2. He concluded that, at least in these bacteria, CO_2 is not split into carbon and oxygen. One group of bacteria used hydrogen sulfide (H_2S) rather than water for photosynthesis, forming yellow globules of sulfur as a waste product (these globules are visible in Figure 10.2e). Here is the chemical equation for photosynthesis in these sulfur bacteria:

$$CO_2 + 2\ H_2S \rightarrow [CH_2O] + H_2O + 2\ S$$

Van Niel reasoned that the bacteria split H_2S and used the hydrogen atoms to make sugar. He then generalized that idea, proposing that all photosynthetic organisms require a hydrogen source but that the source varies:

$$\begin{aligned} \text{Sulfur bacteria:}\quad & CO_2 + 2\ H_2S \rightarrow [CH_2O] + H_2O + 2\ S \\ \text{Plants:}\quad & CO_2 + 2\ H_2O \rightarrow [CH_2O] + H_2O + O_2 \\ \text{General:}\quad & CO_2 + 2\ H_2X \rightarrow [CH_2O] + H_2O + 2\ X \end{aligned}$$

Thus, van Niel hypothesized that plants split H_2O as a source of electrons from hydrogen atoms, releasing O_2 as a by-product.

Nearly 20 years later, scientists confirmed van Niel's hypothesis by using oxygen-18 (^{18}O), a heavy isotope, as a tracer to follow the fate of oxygen atoms during photosynthesis. The experiments showed that the O_2 from plants was labeled with ^{18}O *only* if water was the source of the tracer (experiment 1). If the ^{18}O was introduced to the plant in the form of CO_2, the label did not turn up in the released O_2 (experiment 2). In the following summary, red denotes labeled atoms of oxygen (^{18}O):

$$\begin{aligned} \text{Experiment 1: } & CO_2 + 2\ H_2O \rightarrow [CH_2O] + H_2O + O_2 \\ \text{Experiment 2: } & CO_2 + 2\ H_2O \rightarrow [CH_2O] + H_2O + O_2 \end{aligned}$$

A significant result of the shuffling of atoms during photosynthesis is the extraction of hydrogen from water and its incorporation into sugar. The waste product of photosynthesis, O_2, is released to the atmosphere. **Figure 10.5** shows the fates of all atoms in photosynthesis.

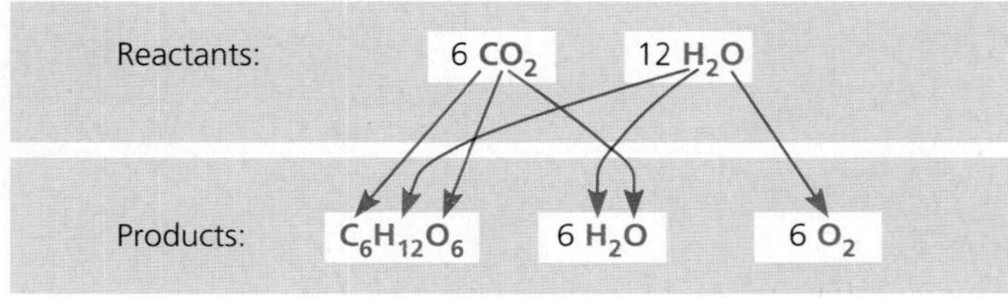

▲ **Figure 10.5 Tracking atoms through photosynthesis.** The atoms from CO_2 are shown in magenta, and the atoms from H_2O are shown in blue.

Photosynthesis as a Redox Process

Let's briefly compare photosynthesis with cellular respiration. Both processes involve redox reactions. During cellular respiration, energy is released from sugar when electrons associated with hydrogen are transported by carriers to oxygen, forming water as a by-product (see p. 164). The electrons lose potential energy as they "fall" down the electron transport chain toward electronegative oxygen, and the mitochondrion harnesses that energy to synthesize ATP (see Figure 9.15). Photosynthesis reverses the direction of electron flow. Water is split, and electrons are transferred along with hydrogen ions from the water to carbon dioxide, reducing it to sugar.

$$\text{Energy} + 6\,CO_2 + 6\,H_2O \longrightarrow C_6H_{12}O_6 + 6\,O_2$$

(CO_2 becomes reduced; H_2O becomes oxidized)

Because the electrons increase in potential energy as they move from water to sugar, this process requires energy—in other words is endergonic. This energy boost is provided by light.

The Two Stages of Photosynthesis: *A Preview*

The equation for photosynthesis is a deceptively simple summary of a very complex process. Actually, photosynthesis is not a single process, but two processes, each with multiple steps. These two stages of photosynthesis are known as the **light reactions** (the *photo* part of photosynthesis) and the **Calvin cycle** (the *synthesis* part) **(Figure 10.6)**.

The light reactions are the steps of photosynthesis that convert solar energy to chemical energy. Water is split, providing a source of electrons and protons (hydrogen ions, H^+) and giving off O_2 as a by-product. Light absorbed by chlorophyll drives a transfer of the electrons and hydrogen ions from water to an acceptor called **$NADP^+$** (nicotinamide adenine dinucleotide phosphate), where they are temporarily stored. The electron acceptor $NADP^+$ is first cousin to NAD^+, which functions as an electron carrier in cellular respiration; the two molecules differ only by the presence of an extra phosphate group in the $NADP^+$ molecule. The light reactions use solar power to reduce $NADP^+$ to NADPH by adding a pair of electrons along with an H^+. The light reactions also generate ATP, using chemiosmosis to power the addition of a phosphate group to ADP, a process called **photophosphorylation**. Thus, light energy is initially converted to chemical energy in the form of two compounds: NADPH, a source of electrons as "reducing power" that can be passed along to an electron acceptor, reducing it, and ATP, the versatile energy currency of cells. Notice that the light reactions produce no sugar; that happens in the second stage of photosynthesis, the Calvin cycle.

The Calvin cycle is named for Melvin Calvin, who, along with his colleagues, began to elucidate its steps in the late 1940s. The cycle begins by incorporating CO_2 from the air

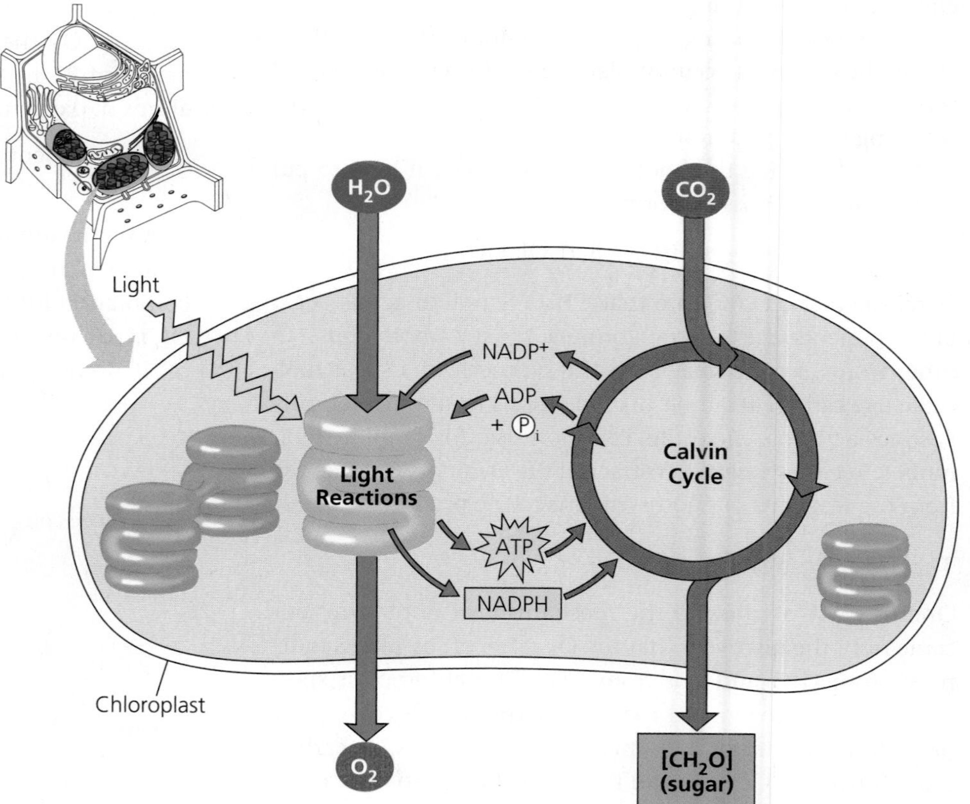

▶ **Figure 10.6 An overview of photosynthesis: cooperation of the light reactions and the Calvin cycle.** In the chloroplast, the thylakoid membranes are the sites of the light reactions, whereas the Calvin cycle occurs in the stroma. The light reactions use solar energy to make ATP and NADPH, which supply chemical energy and reducing power, respectively, to the Calvin cycle. The Calvin cycle incorporates CO_2 into organic molecules, which are converted to sugar. (Recall that most simple sugars have formulas that are some multiple of CH_2O.)

BioFlix Visit the Study Area at **www.masteringbiology.com** for the BioFlix® 3-D Animation on Photosynthesis.

into organic molecules already present in the chloroplast. This initial incorporation of carbon into organic compounds is known as **carbon fixation**. The Calvin cycle then reduces the fixed carbon to carbohydrate by the addition of electrons. The reducing power is provided by NADPH, which acquired its cargo of electrons in the light reactions. To convert CO_2 to carbohydrate, the Calvin cycle also requires chemical energy in the form of ATP, which is also generated by the light reactions. Thus, it is the Calvin cycle that makes sugar, but it can do so only with the help of the NADPH and ATP produced by the light reactions. The metabolic steps of the Calvin cycle are sometimes referred to as the dark reactions, or light-independent reactions, because none of the steps requires light *directly*. Nevertheless, the Calvin cycle in most plants occurs during daylight, for only then can the light reactions provide the NADPH and ATP that the Calvin cycle requires. In essence, the chloroplast uses light energy to make sugar by coordinating the two stages of photosynthesis.

As Figure 10.6 indicates, the thylakoids of the chloroplast are the sites of the light reactions, while the Calvin cycle occurs in the stroma. On the outside of the thylakoids, molecules of $NADP^+$ and ADP pick up electrons and phosphate, respectively, and NADPH and ATP are then released to the stroma, where they play crucial roles in the Calvin cycle. The two stages of photosynthesis are treated in this figure as metabolic modules that take in ingredients and crank out products. In the next two sections, we'll look more closely at how the two stages work, beginning with the light reactions.

CONCEPT CHECK 10.1

1. How do the reactant molecules of photosynthesis reach the chloroplasts in leaves?
2. How did the use of an oxygen isotope help elucidate the chemistry of photosynthesis?
3. **WHAT IF?** The Calvin cycle requires ATP and NADPH, products of the light reactions. If a classmate asserted that the light reactions don't depend on the Calvin cycle and, with continual light, could just keep on producing ATP and NADPH, how would you respond?

For suggested answers, see Appendix A.

CONCEPT 10.2

The light reactions convert solar energy to the chemical energy of ATP and NADPH

Chloroplasts are chemical factories powered by the sun. Their thylakoids transform light energy into the chemical energy of ATP and NADPH. To understand this conversion better, we need to know about some important properties of light.

The Nature of Sunlight

Light is a form of energy known as electromagnetic energy, also called electromagnetic radiation. Electromagnetic energy travels in rhythmic waves analogous to those created by dropping a pebble into a pond. Electromagnetic waves, however, are disturbances of electric and magnetic fields rather than disturbances of a material medium such as water.

The distance between the crests of electromagnetic waves is called the **wavelength**. Wavelengths range from less than a nanometer (for gamma rays) to more than a kilometer (for radio waves). This entire range of radiation is known as the **electromagnetic spectrum** (**Figure 10.7**). The segment most important to life is the narrow band from about 380 nm to 750 nm in wavelength. This radiation is known as **visible light** because it can be detected as various colors by the human eye.

The model of light as waves explains many of light's properties, but in certain respects light behaves as though it consists of discrete particles, called **photons**. Photons are not tangible objects, but they act like objects in that each of them has a fixed quantity of energy. The amount of energy is inversely related to the wavelength of the light: the shorter the wavelength, the greater the energy of each photon of that light. Thus, a photon of violet light packs nearly twice as much energy as a photon of red light.

Although the sun radiates the full spectrum of electromagnetic energy, the atmosphere acts like a selective window, allowing visible light to pass through while screening out a substantial fraction of other radiation. The part of the spectrum we can see—visible light—is also the radiation that drives photosynthesis.

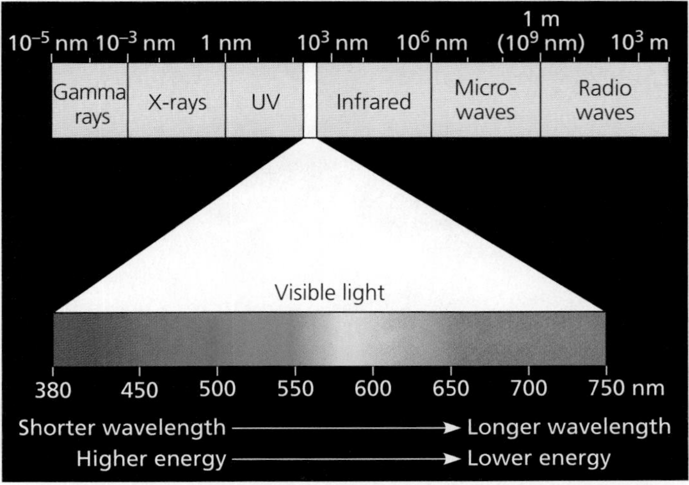

▲ **Figure 10.7 The electromagnetic spectrum.** White light is a mixture of all wavelengths of visible light. A prism can sort white light into its component colors by bending light of different wavelengths at different angles. (Droplets of water in the atmosphere can act as prisms, forming a rainbow; see Figure 10.1.) Visible light drives photosynthesis.

Photosynthetic Pigments: The Light Receptors

When light meets matter, it may be reflected, transmitted, or absorbed. Substances that absorb visible light are known as *pigments*. Different pigments absorb light of different wavelengths, and the wavelengths that are absorbed disappear. If a pigment is illuminated with white light, the color we see is the color most reflected or transmitted by the pigment. (If a pigment absorbs all wavelengths, it appears black.) We see green when we look at a leaf because chlorophyll absorbs violet-blue and red light while transmitting and reflecting green light (**Figure 10.8**). The ability of a pigment to absorb various wavelengths of light can be measured with an instrument called a **spectrophotometer**. This machine directs beams of light of different wavelengths through a solution of the pigment and measures the fraction of the light transmitted at each wavelength. A graph plotting a pigment's light absorption versus wavelength is called an **absorption spectrum** (**Figure 10.9**).

The absorption spectra of chloroplast pigments provide clues to the relative effectiveness of different wavelengths for driving photosynthesis, since light can perform work in chloroplasts only if it is absorbed. **Figure 10.10a** shows the absorption spectra of three types of pigments in chloroplasts: **chlorophyll *a***, which participates directly in the light reactions; the accessory pigment *chlorophyll b*; and a group of accessory pigments called carotenoids. The spectrum of chlorophyll *a* suggests that violet-blue and red light work best for photosynthesis, since they are absorbed, while green is the least effective color. This is confirmed by an **action spectrum** for photosynthesis (**Figure 10.10b**), which profiles the relative effectiveness of different wavelengths of radiation in driving the process. An action spectrum is prepared by illuminating chloroplasts with light of different colors and then plotting wavelength against some measure of photosynthetic rate, such as CO_2 consumption or O_2 release. The action spectrum for photosynthesis was first demonstrated by Theodor W. Engelmann, a German botanist, in 1883. Before equipment for measuring O_2 levels had even been invented, Engelmann performed a

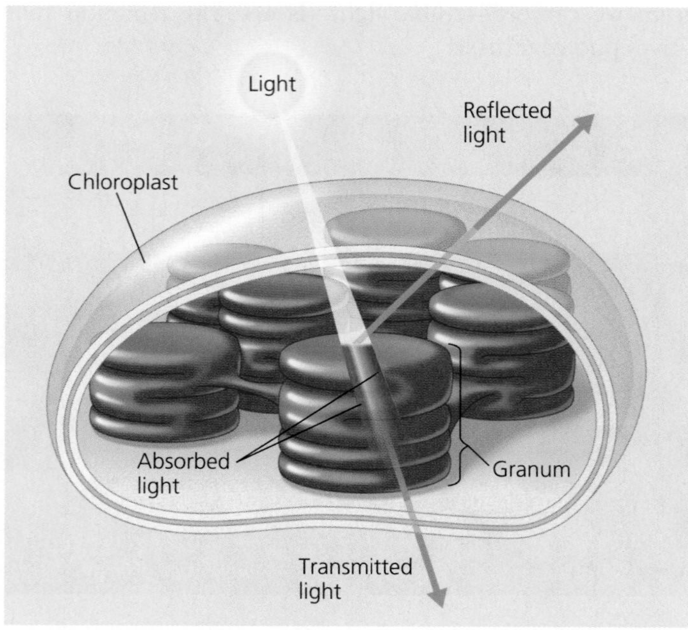

▲ **Figure 10.8 Why leaves are green: interaction of light with chloroplasts.** The chlorophyll molecules of chloroplasts absorb violet-blue and red light (the colors most effective in driving photosynthesis) and reflect or transmit green light. This is why leaves appear green.

▼ **Figure 10.9** RESEARCH METHOD

Determining an Absorption Spectrum

APPLICATION An absorption spectrum is a visual representation of how well a particular pigment absorbs different wavelengths of visible light. Absorption spectra of various chloroplast pigments help scientists decipher each pigment's role in a plant.

TECHNIQUE A spectrophotometer measures the relative amounts of light of different wavelengths absorbed and transmitted by a pigment solution.

1. White light is separated into colors (wavelengths) by a prism.
2. One by one, the different colors of light are passed through the sample (chlorophyll in this example). Green light and blue light are shown here.
3. The transmitted light strikes a photoelectric tube, which converts the light energy to electricity.
4. The electric current is measured by a galvanometer. The meter indicates the fraction of light transmitted through the sample, from which we can determine the amount of light absorbed.

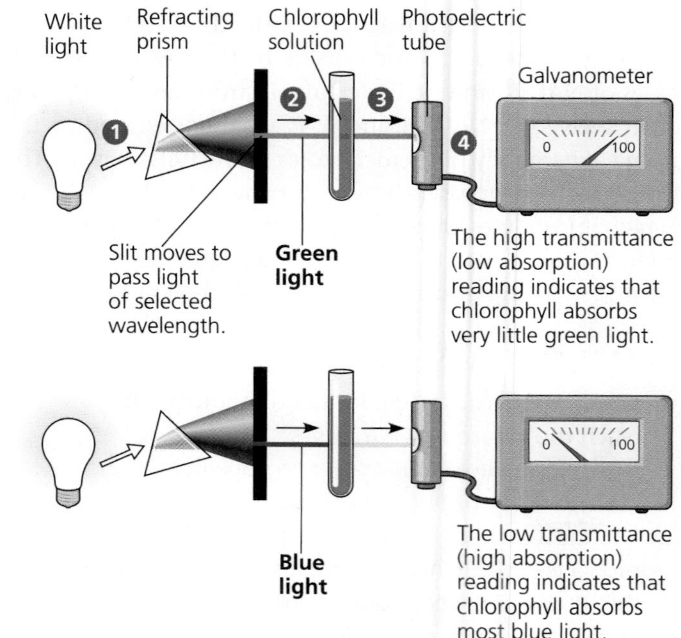

RESULTS See Figure 10.10a for absorption spectra of three types of chloroplast pigments.

▼ Figure 10.10 INQUIRY

Which wavelengths of light are most effective in driving photosynthesis?

EXPERIMENT Absorption and action spectra, along with a classic experiment by Theodor W. Engelmann, reveal which wavelengths of light are photosynthetically important.

RESULTS

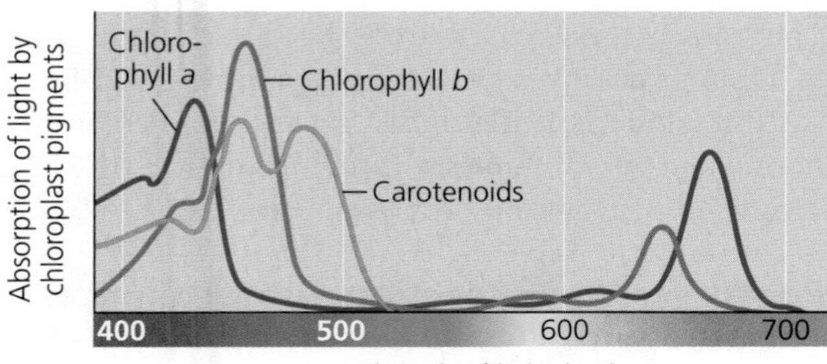

(a) Absorption spectra. The three curves show the wavelengths of light best absorbed by three types of chloroplast pigments.

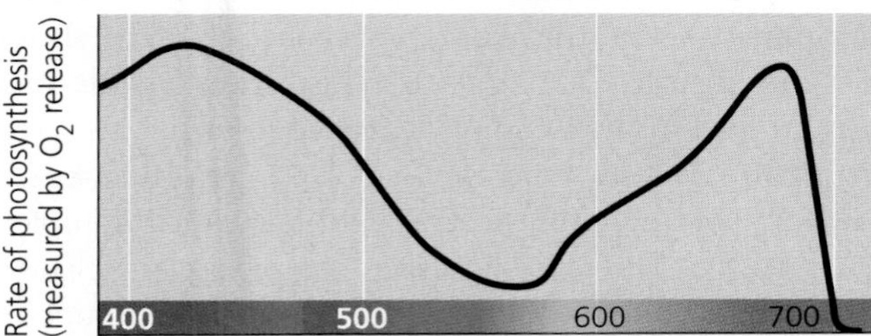

(b) Action spectrum. This graph plots the rate of photosynthesis versus wavelength. The resulting action spectrum resembles the absorption spectrum for chlorophyll *a* but does not match exactly (see part a). This is partly due to the absorption of light by accessory pigments such as chlorophyll *b* and carotenoids.

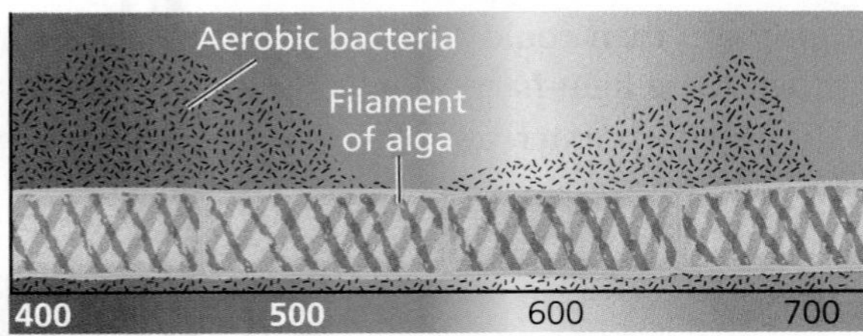

(c) Engelmann's experiment. In 1883, Theodor W. Engelmann illuminated a filamentous alga with light that had been passed through a prism, exposing different segments of the alga to different wavelengths. He used aerobic bacteria, which concentrate near an oxygen source, to determine which segments of the alga were releasing the most O_2 and thus photosynthesizing most. Bacteria congregated in greatest numbers around the parts of the alga illuminated with violet-blue or red light.

CONCLUSION Light in the violet-blue and red portions of the spectrum is most effective in driving photosynthesis.

SOURCE T. W. Engelmann, *Bacterium photometricum.* Ein Betrag zur vergleichenden Physiologie des Licht- und farbensinnes, *Archiv. für Physiologie* 30:95–124 (1883).

See the related Experimental Inquiry Tutorial in MasteringBiology.

WHAT IF? If Engelmann had used a filter that allowed only red light to pass through, how would the results have differed?

clever experiment in which he used bacteria to measure rates of photosynthesis in filamentous algae (**Figure 10.10c**). His results are a striking match to the modern action spectrum shown in Figure 10.10b.

Notice by comparing Figures 10.10a and 10.10b that the action spectrum for photosynthesis does not exactly match the absorption spectrum of chlorophyll *a*. The absorption spectrum of chlorophyll *a* alone underestimates the effectiveness of certain wavelengths in driving photosynthesis. This is partly because accessory pigments with different absorption spectra are also photosynthetically important in chloroplasts and broaden the spectrum of colors that can be used for photosynthesis. **Figure 10.11** shows the structure of chlorophyll *a* compared with that of **chlorophyll *b***. A slight structural difference between them is enough to cause the two pigments to absorb at slightly different wavelengths in the red and blue parts of the spectrum (see Figure 10.10a). As a result, chlorophyll *a* is blue green and chlorophyll *b* is olive green.

Other accessory pigments include **carotenoids**, hydrocarbons that are various shades of yellow and orange because they absorb violet and blue-green light (see Figure 10.10a). Carotenoids may broaden the spectrum of colors that can drive photosynthesis. However, a more important function of at least some carotenoids seems to be *photoprotection*: These

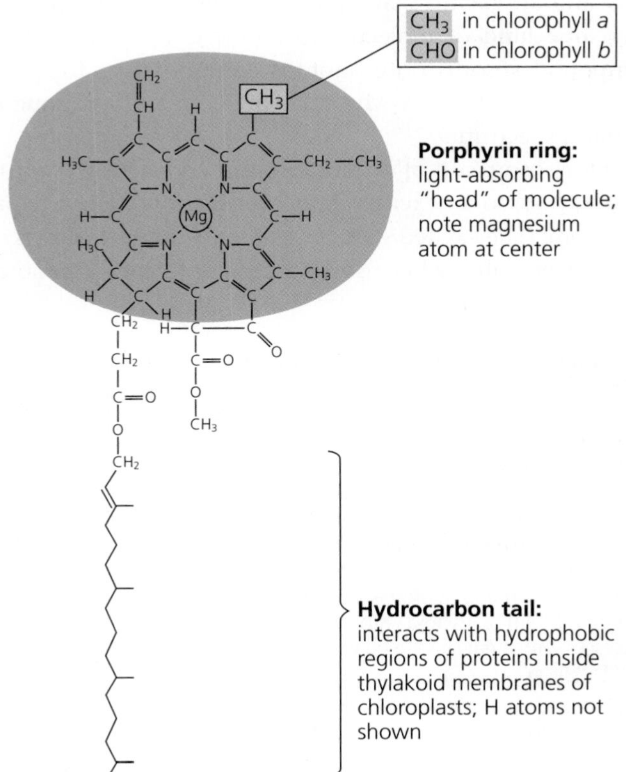

▲ Figure 10.11 Structure of chlorophyll molecules in chloroplasts of plants. Chlorophyll *a* and chlorophyll *b* differ only in one of the functional groups bonded to the porphyrin ring. (Also see the space-filling model of chlorophyll in Figure 1.4, p. 5.)

compounds absorb and dissipate excessive light energy that would otherwise damage chlorophyll or interact with oxygen, forming reactive oxidative molecules that are dangerous to the cell. Interestingly, carotenoids similar to the photoprotective ones in chloroplasts have a photoprotective role in the human eye. These and related molecules, often found in health food products, are valued as "phytochemicals" (from the Greek *phyton*, plant), compounds with antioxidant properties. Plants can synthesize all the antioxidants they require, but humans and other animals must obtain some of them from their diets.

Excitation of Chlorophyll by Light

What exactly happens when chlorophyll and other pigments absorb light? The colors corresponding to the absorbed wavelengths disappear from the spectrum of the transmitted and reflected light, but energy cannot disappear. When a molecule absorbs a photon of light, one of the molecule's electrons is elevated to an orbital where it has more potential energy. When the electron is in its normal orbital, the pigment molecule is said to be in its ground state. Absorption of a photon boosts an electron to an orbital of higher energy, and the pigment molecule is then said to be in an excited state. The only photons absorbed are those whose energy is exactly equal to the energy difference between the ground state and an excited state, and this energy difference varies from one kind of molecule to another. Thus, a particular compound absorbs only photons corresponding to specific wavelengths, which is why each pigment has a unique absorption spectrum.

Once absorption of a photon raises an electron from the ground state to an excited state, the electron cannot remain there long. The excited state, like all high-energy states, is unstable. Generally, when isolated pigment molecules absorb light, their excited electrons drop back down to the ground-state orbital in a billionth of a second, releasing their excess energy as heat. This conversion of light energy to heat is what makes the top of an automobile so hot on a sunny day. (White cars are coolest because their paint reflects all wavelengths of visible light, although it may absorb ultraviolet and other invisible radiation.) In isolation, some pigments, including chlorophyll, emit light as well as heat after absorbing photons. As excited electrons fall back to the ground state, photons are given off. This afterglow is called fluorescence. If a solution of chlorophyll isolated from chloroplasts is illuminated, it will fluoresce in the red-orange part of the spectrum and also give off heat (**Figure 10.12**).

A Photosystem: A Reaction-Center Complex Associated with Light-Harvesting Complexes

Chlorophyll molecules excited by the absorption of light energy produce very different results in an intact chloroplast than they do in isolation (see Figure 10.12). In their native environment of the thylakoid membrane, chlorophyll molecules are organized along with other small organic molecules and proteins into complexes called photosystems.

A **photosystem** is composed of a **reaction-center complex** surrounded by several light-harvesting complexes (**Figure 10.13**). The reaction-center complex is an organized association of proteins holding a special pair of chlorophyll *a* molecules. Each **light-harvesting complex** consists of various pigment molecules (which may include chlorophyll *a*, chlorophyll *b*, and carotenoids) bound to proteins. The number and variety of pigment molecules enable a photosystem to harvest light over a larger surface area and a larger portion of the spectrum than could any single pigment molecule alone. Together, these light-harvesting complexes act as an antenna for the reaction-center complex. When a pigment molecule

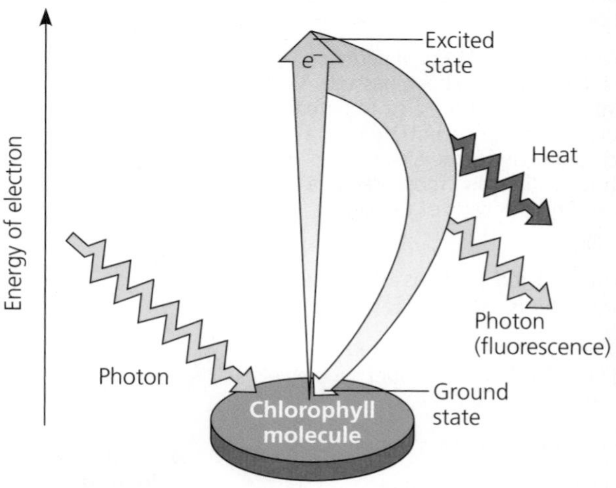

(a) Excitation of isolated chlorophyll molecule

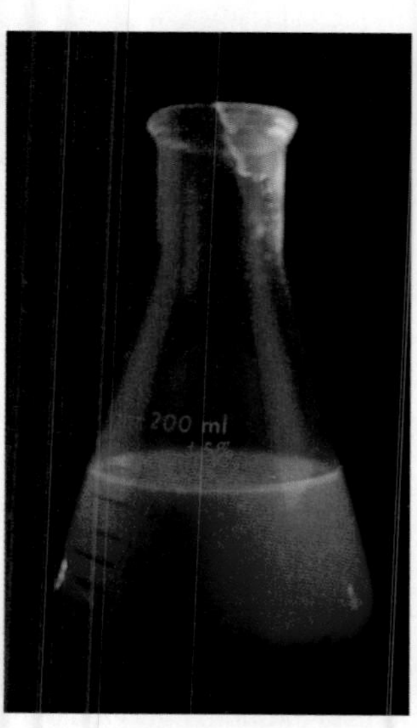

(b) Fluorescence

▶ **Figure 10.12 Excitation of isolated chlorophyll by light. (a)** Absorption of a photon causes a transition of the chlorophyll molecule from its ground state to its excited state. The photon boosts an electron to an orbital where it has more potential energy. If the illuminated molecule exists in isolation, its excited electron immediately drops back down to the ground-state orbital, and its excess energy is given off as heat and fluorescence (light). **(b)** A chlorophyll solution excited with ultraviolet light fluoresces with a red-orange glow.

WHAT IF? *If a leaf containing a similar concentration of chlorophyll as the solution was exposed to the same ultraviolet light, no fluorescence would be seen. Explain the difference in fluorescence emission between the solution and the leaf.*

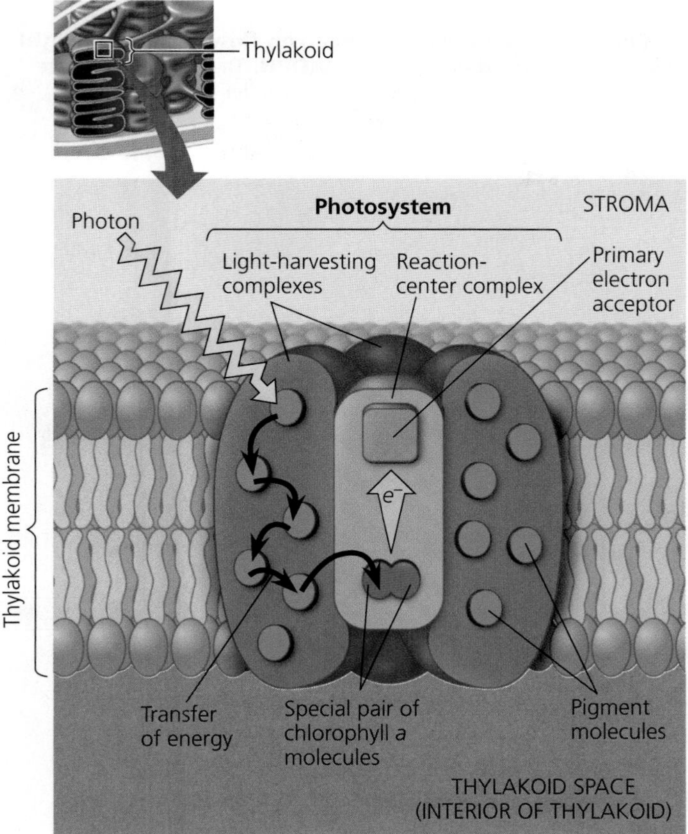

(a) How a photosystem harvests light. When a photon strikes a pigment molecule in a light-harvesting complex, the energy is passed from molecule to molecule until it reaches the reaction-center complex. Here, an excited electron from the special pair of chlorophyll *a* molecules is transferred to the primary electron acceptor.

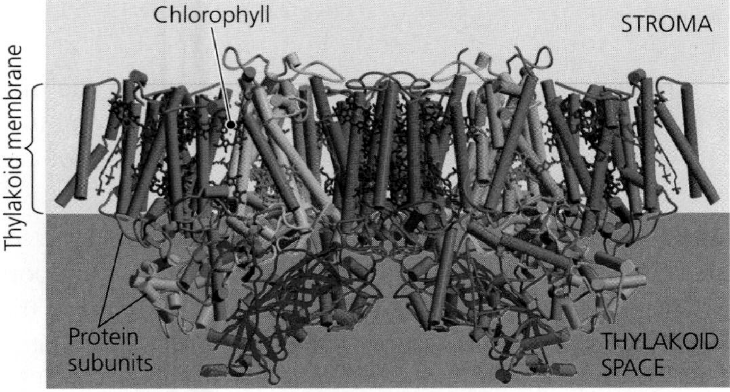

(b) Structure of photosystem II. This computer model of photosystem II, based on X-ray crystallography, shows two photosystem complexes side by side. Chlorophyll molecules (small green ball-and-stick models) are interspersed with protein subunits (cylinders and ribbons). For simplicity, photosystem II will be shown as a single complex in the rest of the chapter.

▲ **Figure 10.13 The structure and function of a photosystem.**

absorbs a photon, the energy is transferred from pigment molecule to pigment molecule within a light-harvesting complex, somewhat like a human "wave" at a sports arena, until it is passed into the reaction-center complex. The reaction-center complex also contains a molecule capable of accepting electrons and becoming reduced; this is called the **primary electron acceptor**. The pair of chlorophyll *a* molecules in the reaction-center complex are special because their molecular environment—their location and the other molecules with which they are associated—enables them to use the energy from light not only to boost one of their electrons to a higher energy level, but also to transfer it to a different molecule—the primary electron acceptor.

The solar-powered transfer of an electron from the reaction-center chlorophyll *a* pair to the primary electron acceptor is the first step of the light reactions. As soon as the chlorophyll electron is excited to a higher energy level, the primary electron acceptor captures it; this is a redox reaction. In the flask shown in Figure 10.12, isolated chlorophyll fluoresces because there is no electron acceptor, so electrons of photoexcited chlorophyll drop right back to the ground state. In the structured environment of a chloroplast, however, an electron acceptor is readily available, and the potential energy represented by the excited electron is not dissipated as light and heat. Thus, each photosystem—a reaction-center complex surrounded by light-harvesting complexes—functions in the chloroplast as a unit. It converts light energy to chemical energy, which will ultimately be used for the synthesis of sugar.

The thylakoid membrane is populated by two types of photosystems that cooperate in the light reactions of photosynthesis. They are called **photosystem II (PS II)** and **photosystem I (PS I)**. (They were named in order of their discovery, but photosystem II functions first in the light reactions.) Each has a characteristic reaction-center complex—a particular kind of primary electron acceptor next to a special pair of chlorophyll *a* molecules associated with specific proteins. The reaction-center chlorophyll *a* of photosystem II is known as P680 because this pigment is best at absorbing light having a wavelength of 680 nm (in the red part of the spectrum). The chlorophyll *a* at the reaction-center complex of photosystem I is called P700 because it most effectively absorbs light of wavelength 700 nm (in the far-red part of the spectrum). These two pigments, P680 and P700, are nearly identical chlorophyll *a* molecules. However, their association with different proteins in the thylakoid membrane affects the electron distribution in the two pigments and accounts for the slight differences in their light-absorbing properties. Now let's see how the two photosystems work together in using light energy to generate ATP and NADPH, the two main products of the light reactions.

Linear Electron Flow

Light drives the synthesis of ATP and NADPH by energizing the two photosystems embedded in the thylakoid membranes of chloroplasts. The key to this energy transformation is a flow of electrons through the photosystems and other molecular components built into the thylakoid membrane. This is called

▼ **Figure 10.14 How linear electron flow during the light reactions generates ATP and NADPH.** The gold arrows trace the current of light-driven electrons from water to NADPH.

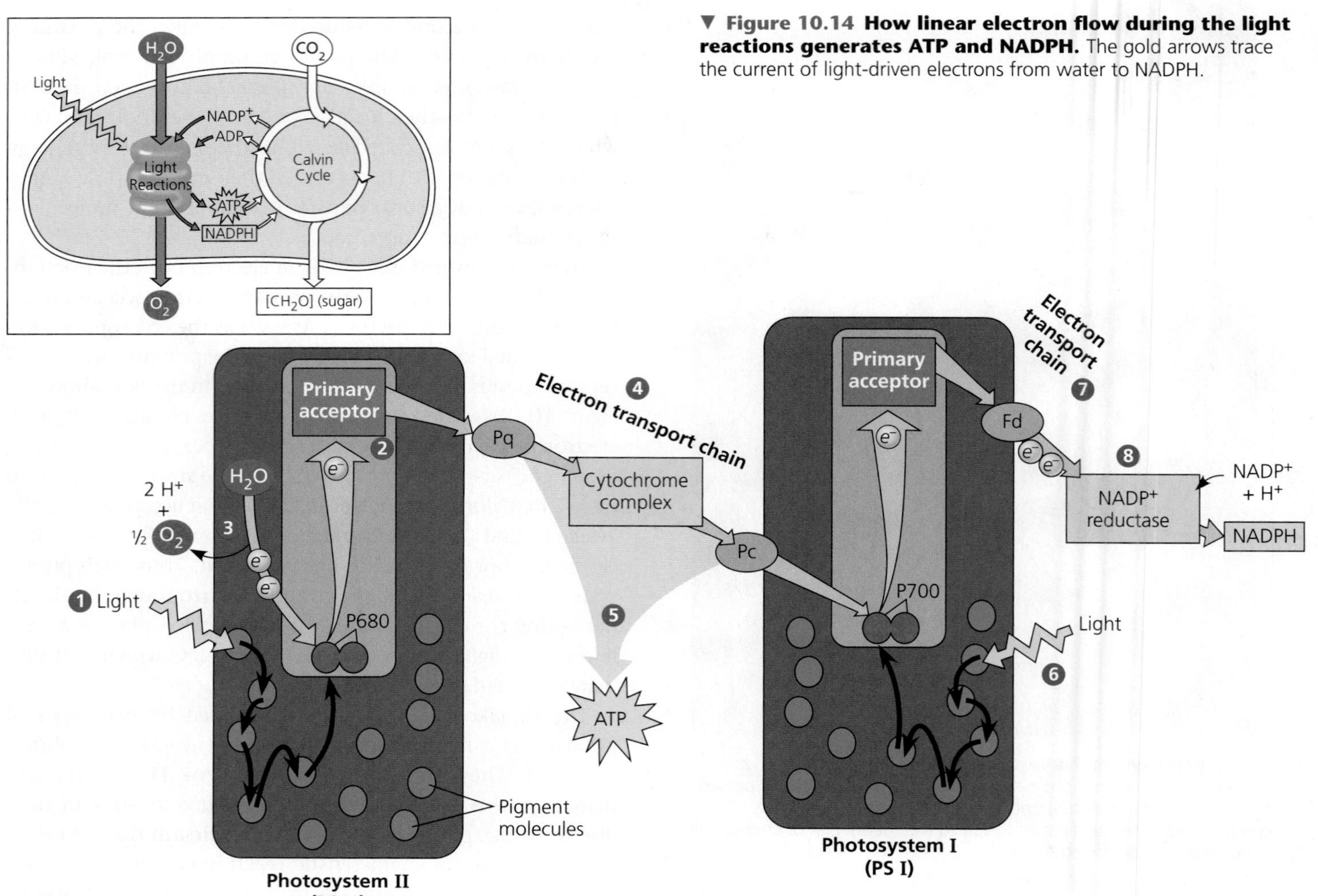

linear electron flow, and it occurs during the light reactions of photosynthesis, as shown in **Figure 10.14**. The following steps correspond to the numbered steps in the figure.

1. A photon of light strikes a pigment molecule in a light-harvesting complex of PS II, boosting one of its electrons to a higher energy level. As this electron falls back to its ground state, an electron in a nearby pigment molecule is simultaneously raised to an excited state. The process continues, with the energy being relayed to other pigment molecules until it reaches the P680 pair of chlorophyll *a* molecules in the PS II reaction-center complex. It excites an electron in this pair of chlorophylls to a higher energy state.
2. This electron is transferred from the excited P680 to the primary electron acceptor. We can refer to the resulting form of P680, missing an electron, as $P680^+$.
3. An enzyme catalyzes the splitting of a water molecule into two electrons, two hydrogen ions (H^+), and an oxygen atom. The electrons are supplied one by one to the $P680^+$ pair, each electron replacing one transferred to the primary electron acceptor. ($P680^+$ is the strongest biological oxidizing agent known; its electron "hole" must be filled. This greatly facilitates the transfer of electrons from the split water molecule.) The H^+ are released into the thylakoid lumen. The oxygen atom immediately combines with an oxygen atom generated by the splitting of another water molecule, forming O_2.
4. Each photoexcited electron passes from the primary electron acceptor of PS II to PS I via an electron transport chain, the components of which are similar to those of the electron transport chain that functions in cellular respiration. The electron transport chain between PS II and PS I is made up of the electron carrier plastoquinone (Pq), a cytochrome complex, and a protein called plastocyanin (Pc).
5. The exergonic "fall" of electrons to a lower energy level provides energy for the synthesis of ATP. As electrons pass through the cytochrome complex, H^+ are pumped into the thylakoid lumen, contributing to the proton gradient that is subsequently used in chemiosmosis.
6. Meanwhile, light energy has been transferred via light-harvesting complex pigments to the PS I reaction-center complex, exciting an electron of the P700 pair of chlorophyll *a* molecules located there. The photoexcited electron was then transferred to PS I's primary electron acceptor, creating an electron "hole" in the P700—which

we now can call $P700^+$. In other words, $P700^+$ can now act as an electron acceptor, accepting an electron that reaches the bottom of the electron transport chain from PS II.

7. Photoexcited electrons are passed in a series of redox reactions from the primary electron acceptor of PS I down a second electron transport chain through the protein ferredoxin (Fd). (This chain does not create a proton gradient and thus does not produce ATP.)

8. The enzyme $NADP^+$ reductase catalyzes the transfer of electrons from Fd to $NADP^+$. Two electrons are required for its reduction to NADPH. This molecule is at a higher energy level than water, and its electrons are more readily available for the reactions of the Calvin cycle than were those of water. This process also removes an H^+ from the stroma.

As complicated as the scheme shown in Figure 10.14 is, do not lose track of its functions. The light reactions use solar power to generate ATP and NADPH, which provide chemical energy and reducing power, respectively, to the carbohydrate-synthesizing reactions of the Calvin cycle. The energy changes of electrons during their linear flow through the light reactions are shown in a mechanical analogy in **Figure 10.15**.

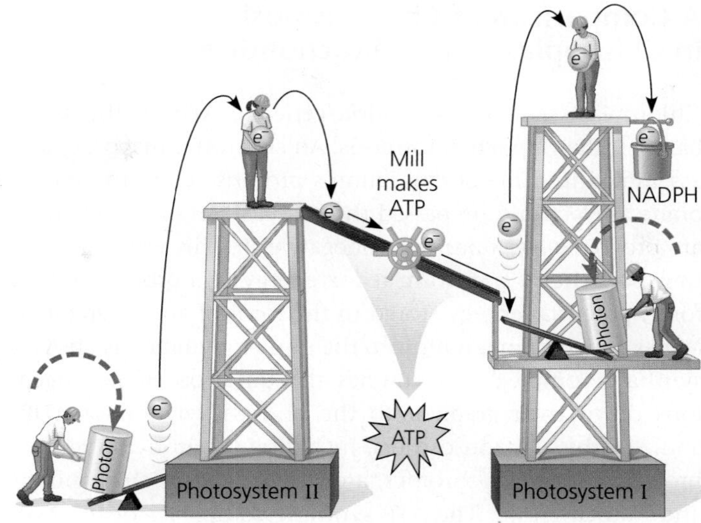

▲ **Figure 10.15 A mechanical analogy for linear electron flow during the light reactions.**

Cyclic Electron Flow

In certain cases, photoexcited electrons can take an alternative path called **cyclic electron flow**, which uses photosystem I but not photosystem II. You can see in **Figure 10.16** that cyclic flow is a short circuit: The electrons cycle back from ferredoxin (Fd) to the cytochrome complex and from there continue on to a P700 chlorophyll in the PS I reaction-center complex. There is no production of NADPH and no release of oxygen. Cyclic flow does, however, generate ATP.

Several of the currently existing groups of photosynthetic bacteria are known to have photosystem I but not photosystem II; for these species, which include the purple sulfur bacteria (see Figure 10.2e), cyclic electron flow is the sole means of generating ATP in photosynthesis. Evolutionary biologists hypothesize that these bacterial groups are descendants of the bacteria in which photosynthesis first evolved, in a form similar to cyclic electron flow.

Cyclic electron flow can also occur in photosynthetic species that possess both photosystems; this includes some prokaryotes, such as the cyanobacteria shown in Figure 10.2d, as well as the eukaryotic photosynthetic species that have been tested so far. Although the process is probably in part an "evolutionary leftover," it clearly plays at least one beneficial role for these organisms. Mutant plants that are not able to carry out cyclic electron flow are capable of growing well in low light, but do not grow well where light is intense. This is evidence for the idea that cyclic electron flow may be photoprotective. Later you'll learn more about cyclic electron flow as it relates to a particular adaptation of photosynthesis (C_4 plants; see Concept 10.4).

Whether ATP synthesis is driven by linear or cyclic electron flow, the actual mechanism is the same. Before we move on to consider the Calvin cycle, let's review chemiosmosis, the process that uses membranes to couple redox reactions to ATP production.

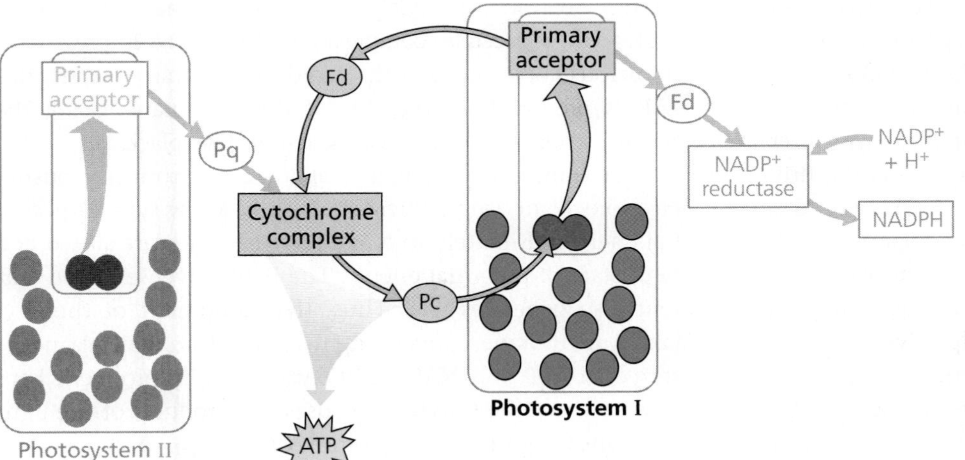

◀ **Figure 10.16 Cyclic electron flow.** Photoexcited electrons from PS I are occasionally shunted back from ferredoxin (Fd) to chlorophyll via the cytochrome complex and plastocyanin (Pc). This electron shunt supplements the supply of ATP (via chemiosmosis) but produces no NADPH. The "shadow" of linear electron flow is included in the diagram for comparison with the cyclic route. The two ferredoxin molecules shown in this diagram are actually one and the same—the final electron carrier in the electron transport chain of PS I.

? *Look at Figure 10.15, and explain how you would alter it to show a mechanical analogy for cyclic electron flow.*

A Comparison of Chemiosmosis in Chloroplasts and Mitochondria

Chloroplasts and mitochondria generate ATP by the same basic mechanism: chemiosmosis. An electron transport chain assembled in a membrane pumps protons across the membrane as electrons are passed through a series of carriers that are progressively more electronegative. In this way, electron transport chains transform redox energy to a proton-motive force, potential energy stored in the form of an H^+ gradient across a membrane. Built into the same membrane is an ATP synthase complex that couples the diffusion of hydrogen ions down their gradient to the phosphorylation of ADP. Some of the electron carriers, including the iron-containing proteins called cytochromes, are very similar in chloroplasts and mitochondria. The ATP synthase complexes of the two organelles are also very much alike. But there are noteworthy differences between oxidative phosphorylation in mitochondria and photophosphorylation in chloroplasts. In mitochondria, the high-energy electrons dropped down the transport chain are extracted from organic molecules (which are thus oxidized), while in chloroplasts, the source of electrons is water. Chloroplasts do not need molecules from food to make ATP; their photosystems capture light energy and use it to drive the electrons from water to the top of the transport chain. In other words, mitochondria use chemiosmosis to transfer chemical energy from food molecules to ATP, whereas chloroplasts transform light energy into chemical energy in ATP.

Although the spatial organization of chemiosmosis differs slightly between chloroplasts and mitochondria, it is easy to see similarities in the two (**Figure 10.17**). The inner membrane of the mitochondrion pumps protons from the mitochondrial matrix out to the intermembrane space, which then serves as a reservoir of hydrogen ions. The thylakoid membrane of the chloroplast pumps protons from the stroma into the thylakoid space (interior of the thylakoid), which functions as the H^+ reservoir. If you imagine the cristae of mitochondria pinching off from the inner membrane, this may help you see how the thylakoid space and the intermembrane space are comparable spaces in the two organelles, while the mitochondrial matrix is analogous to the stroma of the chloroplast. In the mitochondrion, protons diffuse down their concentration gradient from the intermembrane space through ATP synthase to the matrix, driving ATP synthesis. In the chloroplast, ATP is synthesized as the hydrogen ions diffuse from the thylakoid space back to the stroma through ATP synthase complexes, whose catalytic knobs are on the stroma side of the membrane. Thus, ATP forms in the stroma, where it is used to help drive sugar synthesis during the Calvin cycle (**Figure 10.18**).

The proton (H^+) gradient, or pH gradient, across the thylakoid membrane is substantial. When chloroplasts in an experimental setting are illuminated, the pH in the thylakoid space drops to about 5 (the H^+ concentration increases), and the pH in the stroma increases to about 8 (the H^+ concentration decreases). This gradient of three pH units corresponds to a thousandfold difference in H^+ concentration. If in the laboratory the lights are turned off, the pH gradient is abolished, but it can quickly be restored by turning the lights back on. Experiments such as this provided strong evidence in support of the chemiosmotic model.

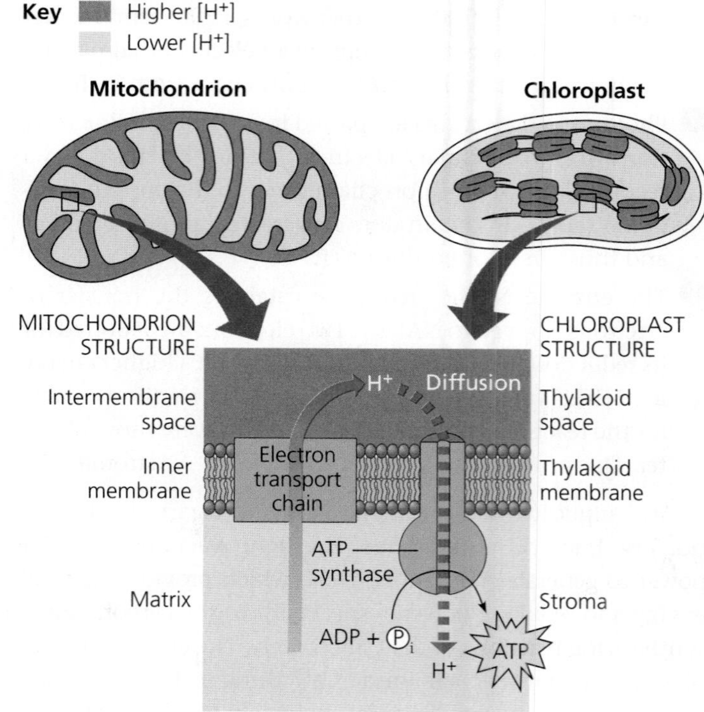

▲ **Figure 10.17 Comparison of chemiosmosis in mitochondria and chloroplasts.** In both kinds of organelles, electron transport chains pump protons (H^+) across a membrane from a region of low H^+ concentration (light gray in this diagram) to one of high H^+ concentration (dark gray). The protons then diffuse back across the membrane through ATP synthase, driving the synthesis of ATP.

Based on studies in several laboratories, Figure 10.18 shows a current model for the organization of the light-reaction "machinery" within the thylakoid membrane. Each of the molecules and molecular complexes in the figure is present in numerous copies in each thylakoid. Notice that NADPH, like ATP, is produced on the side of the membrane facing the stroma, where the Calvin cycle reactions take place.

Let's summarize the light reactions. Electron flow pushes electrons from water, where they are at a low state of potential energy, ultimately to NADPH, where they are stored at a high state of potential energy. The light-driven electron current also generates ATP. Thus, the equipment of the thylakoid membrane converts light energy to chemical energy stored in ATP and NADPH. (Oxygen is a by-product.) Let's now see how the Calvin cycle uses the products of the light reactions to synthesize sugar from CO_2.

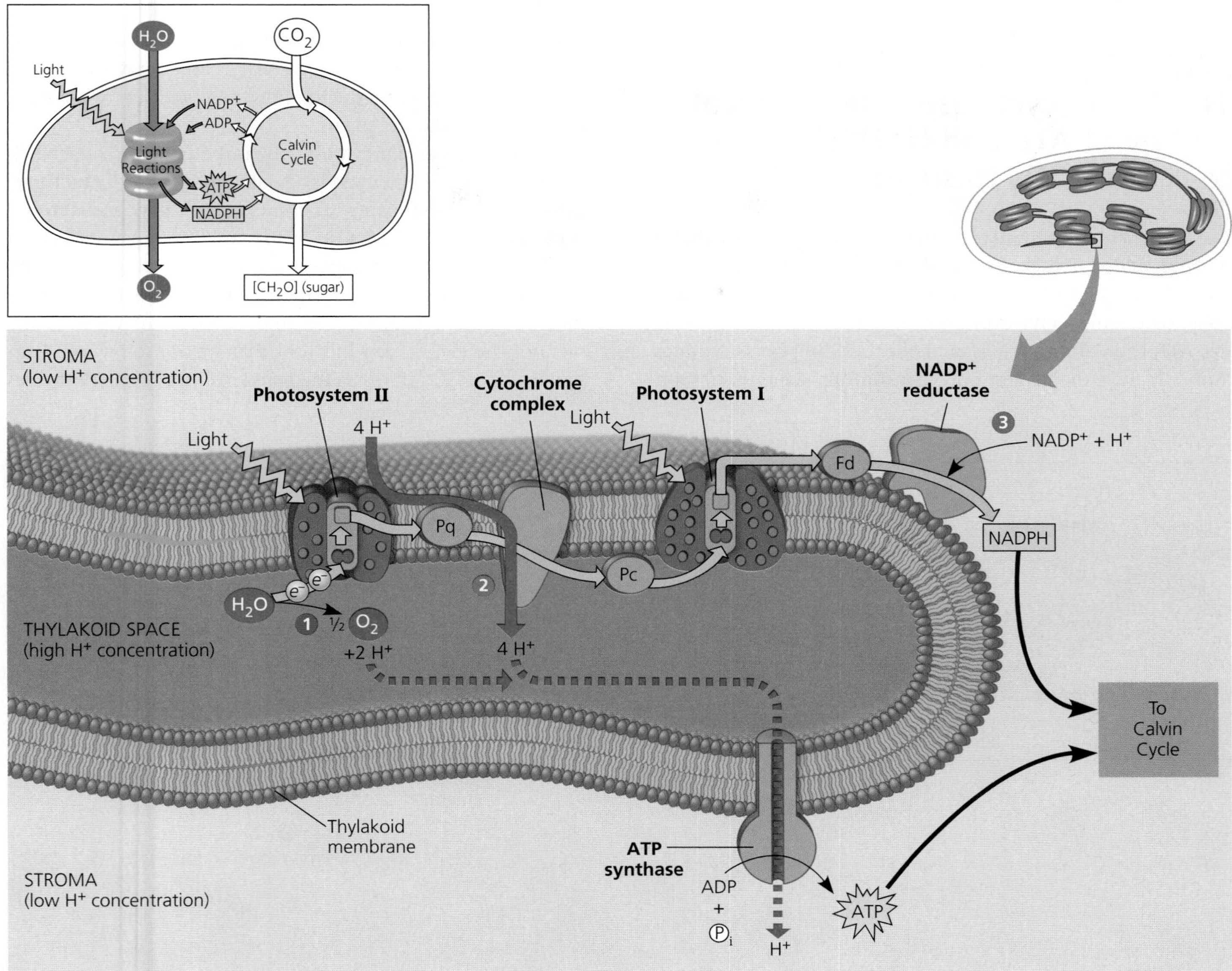

▲ **Figure 10.18 The light reactions and chemiosmosis: the organization of the thylakoid membrane.** This diagram shows a current model for the organization of the thylakoid membrane. The gold arrows track the linear electron flow outlined in Figure 10.14. As electrons pass from carrier to carrier in redox reactions, hydrogen ions removed from the stroma are deposited in the thylakoid space, storing energy as a proton-motive force (H^+ gradient). At least three steps in the light reactions contribute to the proton gradient: ❶ Water is split by photosystem II on the side of the membrane facing the thylakoid space; ❷ as plastoquinone (Pq), a mobile carrier, transfers electrons to the cytochrome complex, four protons are translocated across the membrane into the thylakoid space; and ❸ a hydrogen ion is removed from the stroma when it is taken up by $NADP^+$. Notice that in step 2, hydrogen ions are being pumped from the stroma into the thylakoid space, as in Figure 10.17. The diffusion of H^+ from the thylakoid space back to the stroma (along the H^+ concentration gradient) powers the ATP synthase. These light-driven reactions store chemical energy in NADPH and ATP, which shuttle the energy to the carbohydrate-producing Calvin cycle.

CONCEPT CHECK 10.2

1. What color of light is *least* effective in driving photosynthesis? Explain.
2. Compared to a solution of isolated chlorophyll, why do intact chloroplasts release less heat and fluorescence when illuminated?
3. In the light reactions, what is the initial electron donor? Where do the electrons finally end up?
4. **WHAT IF?** In an experiment, isolated chloroplasts placed in an illuminated solution with the appropriate chemicals can carry out ATP synthesis. Predict what would happen to the rate of synthesis if a compound is added to the solution that makes membranes freely permeable to hydrogen ions.

For suggested answers, see Appendix A.

CONCEPT 10.3

The Calvin cycle uses the chemical energy of ATP and NADPH to reduce CO_2 to sugar

The Calvin cycle is similar to the citric acid cycle in that a starting material is regenerated after molecules enter and leave the cycle. However, while the citric acid cycle is catabolic, oxidizing acetyl CoA and using the energy to synthesize ATP, the Calvin cycle is anabolic, building carbohydrates from smaller molecules and consuming energy. Carbon enters the Calvin cycle in the form of CO_2 and leaves in the form of sugar. The cycle spends ATP as an energy source and consumes NADPH as reducing power for adding high-energy electrons to make the sugar.

As we mentioned previously, the carbohydrate produced directly from the Calvin cycle is actually not glucose, but a three-carbon sugar; the name of this sugar is **glyceraldehyde 3-phosphate (G3P)**. For the net synthesis of one molecule of G3P, the cycle must take place three times, fixing three molecules of CO_2. (Recall that carbon fixation refers to the initial incorporation of CO_2 into organic material.) As we trace the steps of the cycle, keep in mind that we are following three molecules of CO_2 through the reactions. **Figure 10.19** divides

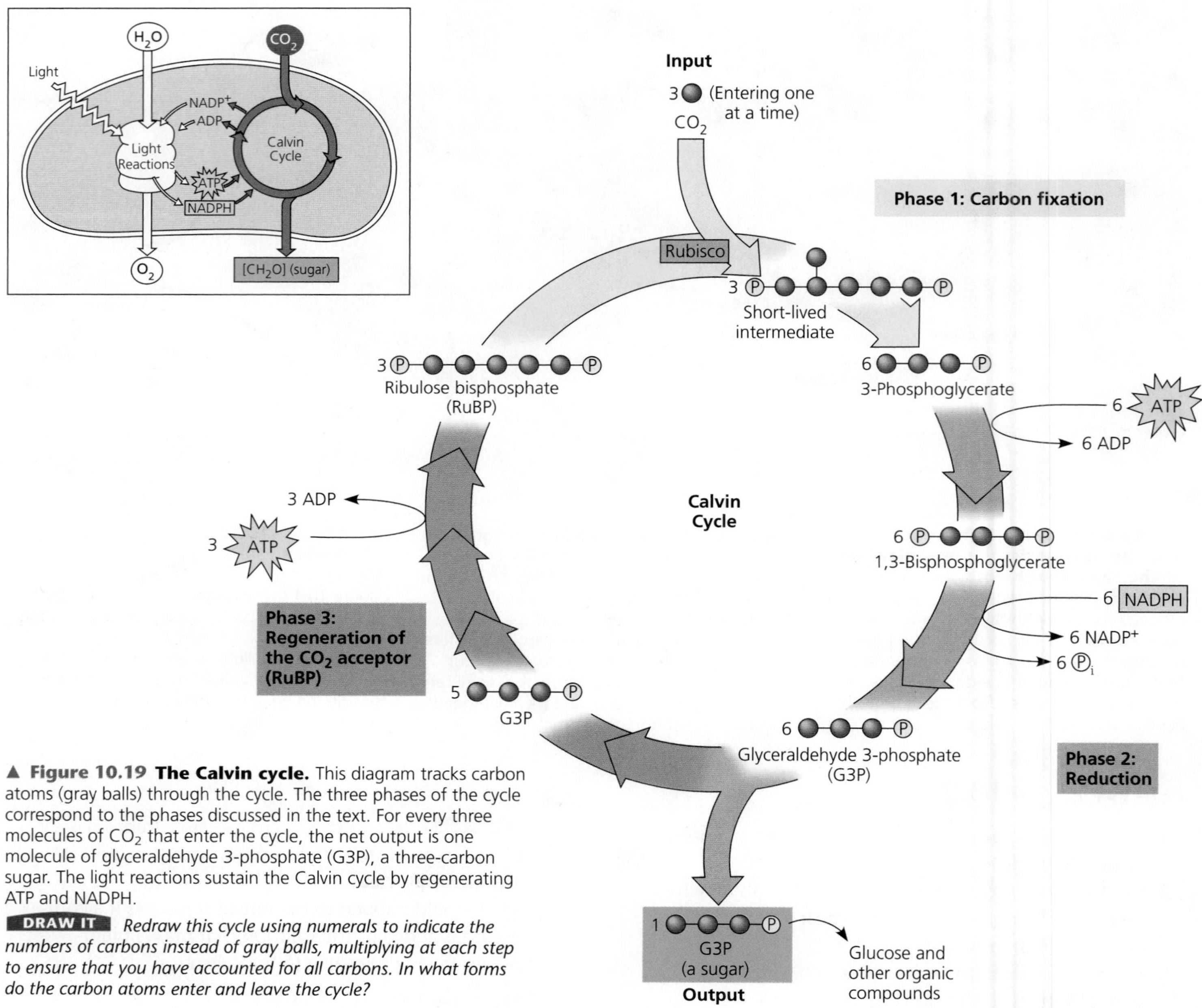

▲ **Figure 10.19 The Calvin cycle.** This diagram tracks carbon atoms (gray balls) through the cycle. The three phases of the cycle correspond to the phases discussed in the text. For every three molecules of CO_2 that enter the cycle, the net output is one molecule of glyceraldehyde 3-phosphate (G3P), a three-carbon sugar. The light reactions sustain the Calvin cycle by regenerating ATP and NADPH.

DRAW IT *Redraw this cycle using numerals to indicate the numbers of carbons instead of gray balls, multiplying at each step to ensure that you have accounted for all carbons. In what forms do the carbon atoms enter and leave the cycle?*

the Calvin cycle into three phases: carbon fixation, reduction, and regeneration of the CO_2 acceptor.

Phase 1: Carbon fixation. The Calvin cycle incorporates each CO_2 molecule, one at a time, by attaching it to a five-carbon sugar named ribulose bisphosphate (abbreviated RuBP). The enzyme that catalyzes this first step is RuBP carboxylase, or **rubisco**. (This is the most abundant protein in chloroplasts and is also thought to be the most abundant protein on Earth.) The product of the reaction is a six-carbon intermediate so unstable that it immediately splits in half, forming two molecules of 3-phosphoglycerate (for each CO_2 fixed).

Phase 2: Reduction. Each molecule of 3-phosphoglycerate receives an additional phosphate group from ATP, becoming 1,3-bisphosphoglycerate. Next, a pair of electrons donated from NADPH reduces 1,3-bisphosphoglycerate, which also loses a phosphate group, becoming G3P. Specifically, the electrons from NADPH reduce a carboyxl group on 1,3-bisphosphoglycerate to the aldehyde group of G3P, which stores more potential energy. G3P is a sugar—the same three-carbon sugar formed in glycolysis by the splitting of glucose (see Figure 9.9). Notice in Figure 10.19 that for every *three* molecules of CO_2 that enter the cycle, there are *six* molecules of G3P formed. But only one molecule of this three-carbon sugar can be counted as a net gain of carbohydrate. The cycle began with 15 carbons' worth of carbohydrate in the form of three molecules of the five-carbon sugar RuBP. Now there are 18 carbons' worth of carbohydrate in the form of six molecules of G3P. One molecule exits the cycle to be used by the plant cell, but the other five molecules must be recycled to regenerate the three molecules of RuBP.

Phase 3: Regeneration of the CO_2 acceptor (RuBP). In a complex series of reactions, the carbon skeletons of five molecules of G3P are rearranged by the last steps of the Calvin cycle into three molecules of RuBP. To accomplish this, the cycle spends three more molecules of ATP. The RuBP is now prepared to receive CO_2 again, and the cycle continues.

For the net synthesis of one G3P molecule, the Calvin cycle consumes a total of nine molecules of ATP and six molecules of NADPH. The light reactions regenerate the ATP and NADPH. The G3P spun off from the Calvin cycle becomes the starting material for metabolic pathways that synthesize other organic compounds, including glucose and other carbohydrates. Neither the light reactions nor the Calvin cycle alone can make sugar from CO_2. Photosynthesis is an emergent property of the intact chloroplast, which integrates the two stages of photosynthesis.

CONCEPT CHECK 10.3

1. To synthesize one glucose molecule, the Calvin cycle uses ________ molecules of CO_2, ________ molecules of ATP, and ________ molecules of NADPH.
2. Explain why the large numbers of ATP and NADPH molecules used during the Calvin cycle are consistent with the high value of glucose as an energy source.
3. **WHAT IF?** Explain why a poison that inhibits an enzyme of the Calvin cycle will also inhibit the light reactions.
4. **MAKE CONNECTIONS** Review Figures 9.9 (p. 169) and 10.19. Discuss the roles of intermediate and product played by glyceraldehyde 3-phosphate (G3P) in the two processes shown in these figures.

For suggested answers, see Appendix A.

CONCEPT 10.4

Alternative mechanisms of carbon fixation have evolved in hot, arid climates

EVOLUTION Ever since plants first moved onto land about 475 million years ago, they have been adapting to the problems of terrestrial life, particularly the problem of dehydration. In Chapters 29 and 36, we will consider anatomical adaptations that help plants conserve water, while in this chapter we are concerned with metabolic adaptations. The solutions often involve trade-offs. An important example is the compromise between photosynthesis and the prevention of excessive water loss from the plant. The CO_2 required for photosynthesis enters a leaf via stomata, the pores on the leaf surface (see Figure 10.4). However, stomata are also the main avenues of transpiration, the evaporative loss of water from leaves. On a hot, dry day, most plants close their stomata, a response that conserves water. This response also reduces photosynthetic yield by limiting access to CO_2. With stomata even partially closed, CO_2 concentrations begin to decrease in the air spaces within the leaf, and the concentration of O_2 released from the light reactions begins to increase. These conditions within the leaf favor an apparently wasteful process called photorespiration.

Photorespiration: An Evolutionary Relic?

In most plants, initial fixation of carbon occurs via rubisco, the Calvin cycle enzyme that adds CO_2 to ribulose bisphosphate. Such plants are called **C_3 plants** because the first organic product of carbon fixation is a three-carbon compound,

3-phosphoglycerate (see Figure 10.19). Rice, wheat, and soybeans are C_3 plants that are important in agriculture. When their stomata partially close on hot, dry days, C_3 plants produce less sugar because the declining level of CO_2 in the leaf starves the Calvin cycle. In addition, rubisco can bind O_2 in place of CO_2. As CO_2 becomes scarce within the air spaces of the leaf, rubisco adds O_2 to the Calvin cycle instead of CO_2. The product splits, and a two-carbon compound leaves the chloroplast. Peroxisomes and mitochondria rearrange and split this compound, releasing CO_2. The process is called **photorespiration** because it occurs in the light (*photo*) and consumes O_2 while producing CO_2 (*respiration*). However, unlike normal cellular respiration, photorespiration generates no ATP; in fact, photorespiration consumes ATP. And unlike photosynthesis, photorespiration produces no sugar. In fact, photorespiration *decreases* photosynthetic output by siphoning organic material from the Calvin cycle and releasing CO_2 that would otherwise be fixed.

How can we explain the existence of a metabolic process that seems to be counterproductive for the plant? According to one hypothesis, photorespiration is evolutionary baggage—a metabolic relic from a much earlier time when the atmosphere had less O_2 and more CO_2 than it does today. In the ancient atmosphere that prevailed when rubisco first evolved, the inability of the enzyme's active site to exclude O_2 would have made little difference. The hypothesis suggests that modern rubisco retains some of its chance affinity for O_2, which is now so concentrated in the atmosphere that a certain amount of photorespiration is inevitable.

We now know that, at least in some cases, photorespiration plays a protective role in plants. Plants that are impaired in their ability to carry out photorespiration (due to defective genes) are more susceptible to damage induced by excess light. Researchers consider this clear evidence that photorespiration acts to neutralize the otherwise damaging products of the light reactions, which build up when a low CO_2 concentration limits the progress of the Calvin cycle. Whether there are other benefits of photorespiration is still unknown. In many types of plants—including a significant number of crop plants—photorespiration drains away as much as 50% of the carbon fixed by the Calvin cycle. As heterotrophs that depend on carbon fixation in chloroplasts for our food, we naturally view photorespiration as wasteful. Indeed, if photorespiration could be reduced in certain plant species without otherwise affecting photosynthetic productivity, crop yields and food supplies might increase.

In some plant species, alternate modes of carbon fixation have evolved that minimize photorespiration and optimize the Calvin cycle—even in hot, arid climates. The two most important of these photosynthetic adaptations are C_4 photosynthesis and crassulacean acid metabolism (CAM).

C_4 Plants

The **C_4 plants** are so named because they preface the Calvin cycle with an alternate mode of carbon fixation that forms a four-carbon compound as its first product. Several thousand species in at least 19 plant families use the C_4 pathway. Among the C_4 plants important to agriculture are sugarcane and corn, members of the grass family.

A unique leaf anatomy is correlated with the mechanism of C_4 photosynthesis (**Figure 10.20**; compare with Figure 10.4). In C_4 plants, there are two distinct types of photosynthetic cells: bundle-sheath cells and mesophyll cells. **Bundle-sheath cells** are arranged into tightly packed sheaths around the veins of the leaf. Between the bundle sheath and the leaf surface are the more loosely arranged mesophyll cells. The Calvin cycle is confined to the chloroplasts of the bundle-sheath cells. However, the Calvin cycle is preceded by incorporation of CO_2 into organic compounds in the mesophyll cells. See the numbered steps in Figure 10.20, which are also described here:

1. The first step is carried out by an enzyme present only in mesophyll cells called **PEP carboxylase**. This enzyme adds CO_2 to phosphoenolpyruvate (PEP), forming the four-carbon product oxaloacetate. PEP carboxylase has a much higher affinity for CO_2 than does rubisco and no affinity for O_2. Therefore, PEP carboxylase can fix carbon efficiently when rubisco cannot—that is, when it is hot and dry and stomata are partially closed, causing CO_2 concentration in the leaf to fall and O_2 concentration to rise.
2. After the C_4 plant fixes carbon from CO_2, the mesophyll cells export their four-carbon products (malate in the example shown in Figure 10.20) to bundle-sheath cells through plasmodesmata (see Figure 6.31).
3. Within the bundle-sheath cells, the four-carbon compounds release CO_2, which is reassimilated into organic material by rubisco and the Calvin cycle. The same reaction regenerates pyruvate, which is transported to mesophyll cells. There, ATP is used to convert pyruvate to PEP, allowing the reaction cycle to continue; this ATP can be thought of as the "price" of concentrating CO_2 in the bundle-sheath cells. To generate this extra ATP, bundle-sheath cells carry out cyclic electron flow, the process described earlier in this chapter (see Figure 10.16). In fact, these cells contain PS I but no PS II, so cyclic electron flow is their only photosynthetic mode of generating ATP.

In effect, the mesophyll cells of a C_4 plant pump CO_2 into the bundle sheath, keeping the CO_2 concentration in the bundle-sheath cells high enough for rubisco to bind carbon

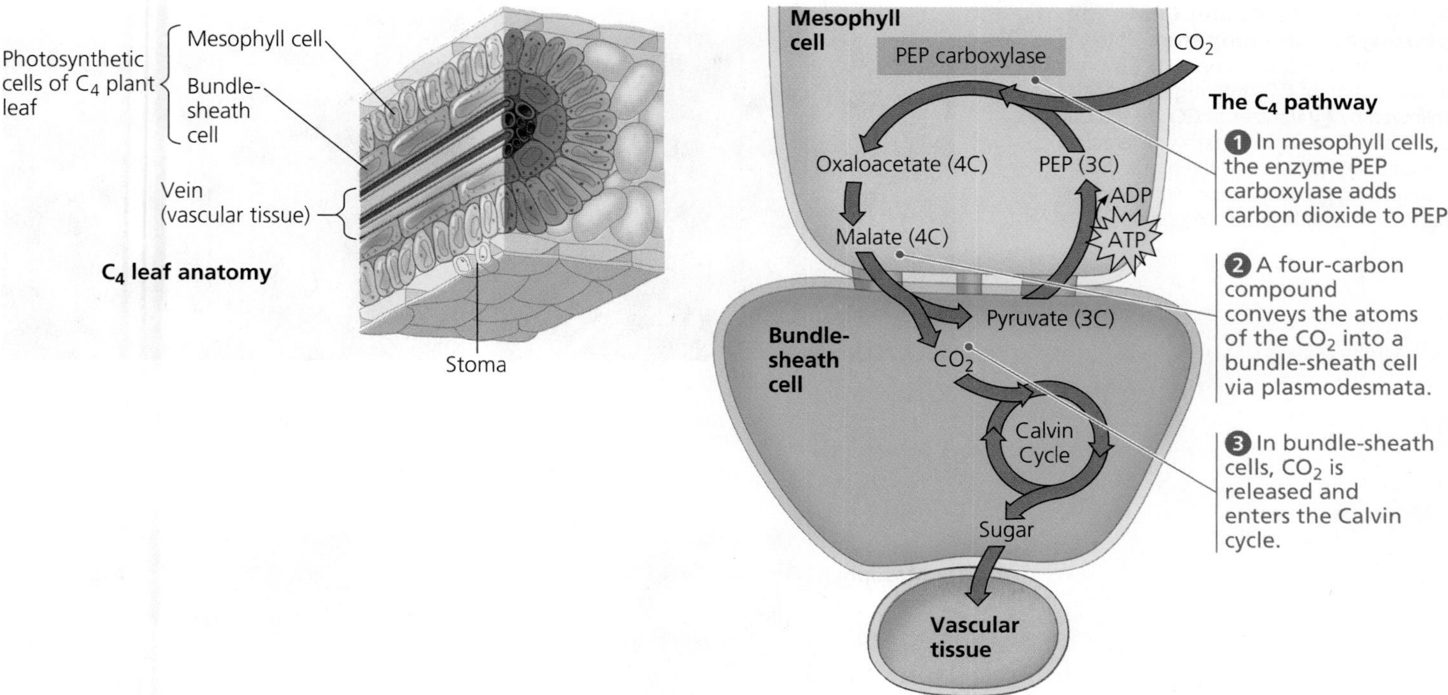

▲ **Figure 10.20 C_4 leaf anatomy and the C_4 pathway.** The structure and biochemical functions of the leaves of C_4 plants are an evolutionary adaptation to hot, dry climates. This adaptation maintains a CO_2 concentration in the bundle sheath that favors photosynthesis over photorespiration.

dioxide rather than oxygen. The cyclic series of reactions involving PEP carboxylase and the regeneration of PEP can be thought of as a CO_2-concentrating pump that is powered by ATP. In this way, C_4 photosynthesis minimizes photorespiration and enhances sugar production. This adaptation is especially advantageous in hot regions with intense sunlight, where stomata partially close during the day, and it is in such environments that C_4 plants evolved and thrive today.

Since the Industrial Revolution began in the 1800s, human activities such as the burning of fossil fuels have drastically increased the concentration of CO_2 in the atmosphere. The resulting global climate change, including an increase in average temperatures around the planet, may have far-reaching effects on plant species. Scientists are concerned that increasing CO_2 concentration and temperature may affect C_3 and C_4 plants differently, thus changing the relative abundance of these species in a given plant community.

Which type of plant would stand to gain more from increasing CO_2 levels? Recall that in C_3 plants, the binding of O_2 rather than CO_2 by rubisco leads to photorespiration, lowering the efficiency of photosynthesis. C_4 plants overcome this problem by concentrating CO_2 in the bundle-sheath cells at the cost of ATP. Rising CO_2 levels should benefit C_3 plants by lowering the amount of photorespiration that occurs. At the same time, rising temperatures have the opposite effect, increasing photorespiration. (Other factors such as water availability may also come into play.) In contrast, many C_4 plants could be largely unaffected by increasing CO_2 levels or temperature. In different regions, the particular combination of these two factors is likely to alter the balance of C_3 and C_4 plants in varying ways. The effects of such a widespread and variable change in community structure are unpredictable and thus a cause of legitimate concern.

CAM Plants

A second photosynthetic adaptation to arid conditions has evolved in many succulent (water-storing) plants, numerous cacti, pineapples, and representatives of several other plant families. These plants open their stomata during the night and close them during the day, just the reverse of how other plants behave. Closing stomata during the day helps desert plants conserve water, but it also prevents CO_2 from entering the leaves. During the night, when their stomata are open, these plants take up CO_2 and incorporate it into a variety

▶ **Figure 10.21 C_4 and CAM photosynthesis compared.** Both adaptations are characterized by ❶ preliminary incorporation of CO_2 into organic acids, followed by ❷ transfer of CO_2 to the Calvin cycle. The C_4 and CAM pathways are two evolutionary solutions to the problem of maintaining photosynthesis with stomata partially or completely closed on hot, dry days.

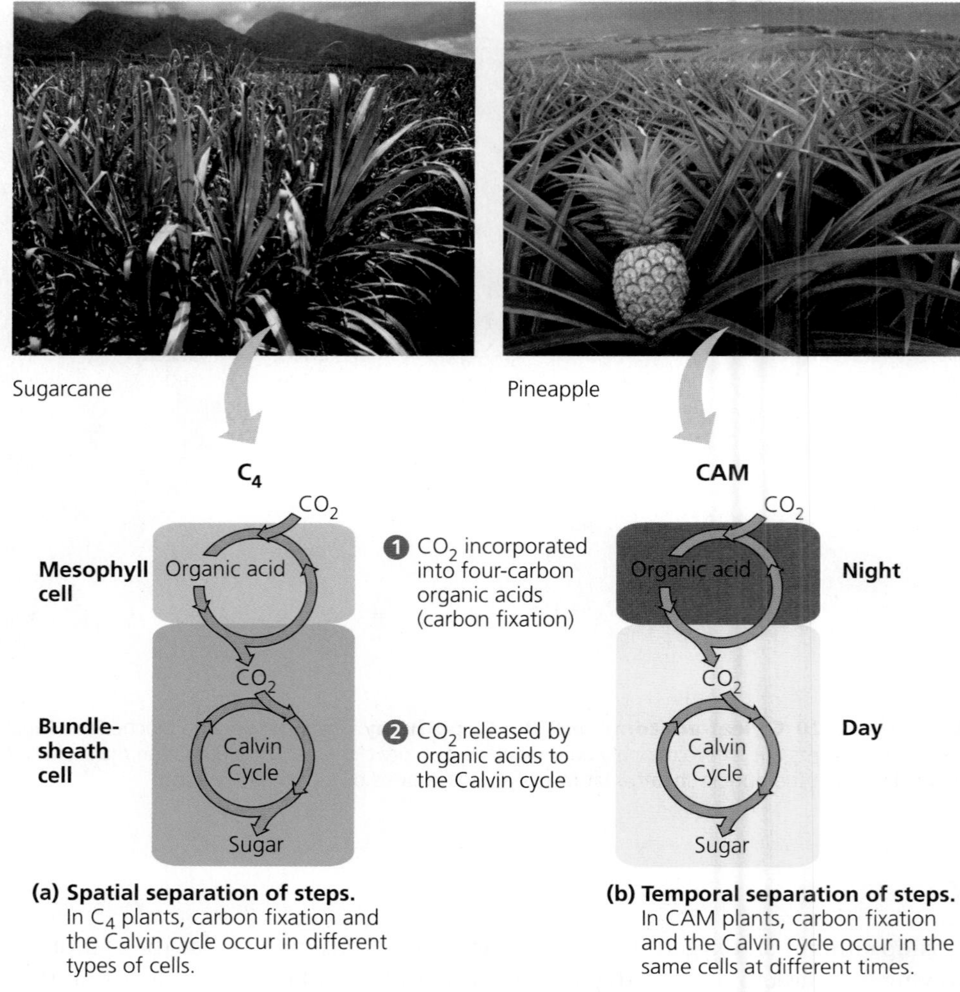
Sugarcane
Pineapple
C4
CAM
CO2
Mesophyll cell
Organic acid
CO2
Bundle-sheath cell
Calvin Cycle
Sugar
❶ CO2 incorporated into four-carbon organic acids (carbon fixation)
❷ CO2 released by organic acids to the Calvin cycle
Night
Day
(a) Spatial separation of steps. In C4 plants, carbon fixation and the Calvin cycle occur in different types of cells.
(b) Temporal separation of steps. In CAM plants, carbon fixation and the Calvin cycle occur in the same cells at different times.

of organic acids. This mode of carbon fixation is called **crassulacean acid metabolism**, or **CAM**, after the plant family Crassulaceae, the succulents in which the process was first discovered. The mesophyll cells of **CAM plants** store the organic acids they make during the night in their vacuoles until morning, when the stomata close. During the day, when the light reactions can supply ATP and NADPH for the Calvin cycle, CO_2 is released from the organic acids made the night before to become incorporated into sugar in the chloroplasts.

Notice in **Figure 10.21** that the CAM pathway is similar to the C_4 pathway in that carbon dioxide is first incorporated into organic intermediates before it enters the Calvin cycle. The difference is that in C_4 plants, the initial steps of carbon fixation are separated structurally from the Calvin cycle, whereas in CAM plants, the two steps occur at separate times but within the same cell. (Keep in mind that CAM, C_4, and C_3 plants all eventually use the Calvin cycle to make sugar from carbon dioxide.)

CONCEPT CHECK 10.4

1. Explain why photorespiration lowers photosynthetic output for plants.
2. The presence of only PS I, not PS II, in the bundle-sheath cells of C_4 plants has an effect on O_2 concentration. What is that effect, and how might that benefit the plant?
3. **MAKE CONNECTIONS** Refer to the discussion of ocean acidification in Concept 3.3 (p. 55). Ocean acidification and changes in the distribution of C_3 and C_4 plants may seem to be two very different problems, but what do they have in common? Explain.
4. **WHAT IF?** How would you expect the relative abundance of C_3 versus C_4 and CAM species to change in a geographic region whose climate becomes much hotter and drier, with no change in CO_2 concentration?

For suggested answers, see Appendix A.

The Importance of Photosynthesis: *A Review*

In this chapter, we have followed photosynthesis from photons to food. The light reactions capture solar energy and use it to make ATP and transfer electrons from water to $NADP^+$, forming NADPH. The Calvin cycle uses the ATP and NADPH to produce sugar from carbon dioxide. The energy that enters the chloroplasts as sunlight becomes stored as chemical energy in organic compounds. See **Figure 10.22** for a review of the entire process.

What are the fates of photosynthetic products? The sugar made in the chloroplasts supplies the entire plant with chemical energy and carbon skeletons for the synthesis of all the major organic molecules of plant cells. About 50% of the organic material made by photosynthesis is consumed as fuel for cellular respiration in the mitochondria of the plant cells. Sometimes there is a loss of photosynthetic products to photorespiration.

Technically, green cells are the only autotrophic parts of the plant. The rest of the plant depends on organic molecules exported from leaves via veins. In most plants, carbohydrate is transported out of the leaves in the form of sucrose, a disaccharide. After arriving at nonphotosynthetic cells, the sucrose provides raw material for cellular respiration and a multitude of anabolic pathways that synthesize proteins, lipids, and other products. A considerable amount of sugar in the form of glucose is linked together to make the polysaccharide cellulose, especially in plant cells that are still growing and maturing. Cellulose, the main ingredient of cell walls, is the most abundant organic molecule in the plant—and probably on the surface of the planet.

Most plants manage to make more organic material each day than they need to use as respiratory fuel and precursors for biosynthesis. They stockpile the extra sugar by synthesizing starch, storing some in the chloroplasts themselves and some in storage cells of roots, tubers, seeds, and fruits. In accounting for the consumption of the food molecules produced by photosynthesis, let's not forget that most plants lose leaves, roots, stems, fruits, and sometimes their entire bodies to heterotrophs, including humans.

On a global scale, photosynthesis is the process responsible for the presence of oxygen in our atmosphere. Furthermore, in terms of food production, the collective productivity of the minuscule chloroplasts is prodigious: Photosynthesis makes an estimated 160 billion metric tons of carbohydrate per year (a metric ton is 1,000 kg, about 1.1 tons). That's organic matter equivalent in mass to a stack of about 60 trillion copies of this textbook—17 stacks of books reaching from Earth to the sun! No other chemical process on the planet can match the output of photosynthesis. And as we mentioned earlier, researchers are seeking ways to capitalize on photosynthetic production to produce alternative fuels. No process is more important than photosynthesis to the welfare of life on Earth.

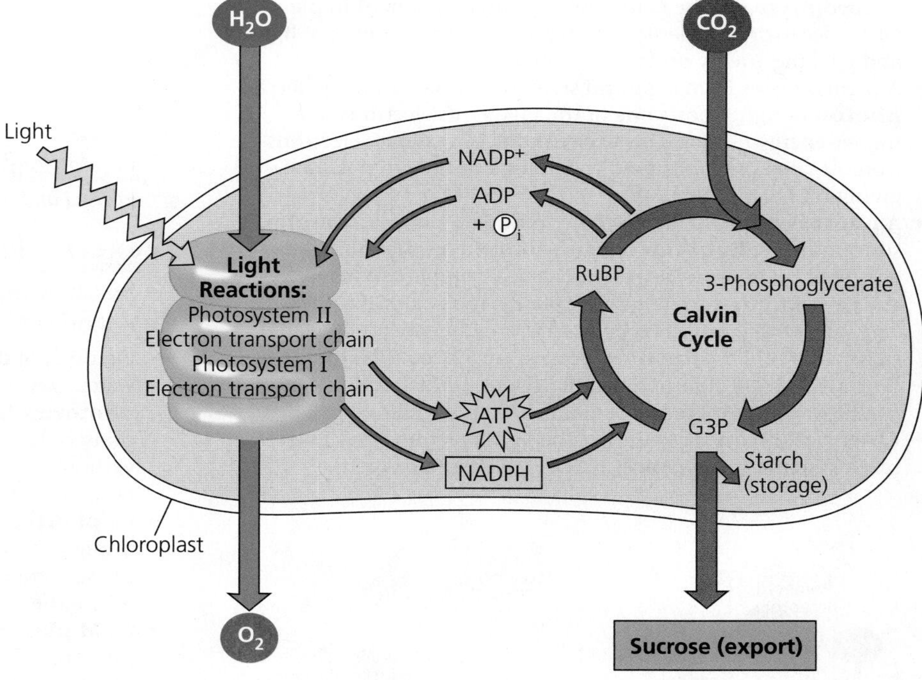

Light Reactions:

- Are carried out by molecules in the thylakoid membranes
- Convert light energy to the chemical energy of ATP and NADPH
- Split H_2O and release O_2 to the atmosphere

Calvin Cycle Reactions:

- Take place in the stroma
- Use ATP and NADPH to convert CO_2 to the sugar G3P
- Return ADP, inorganic phosphate, and $NADP^+$ to the light reactions

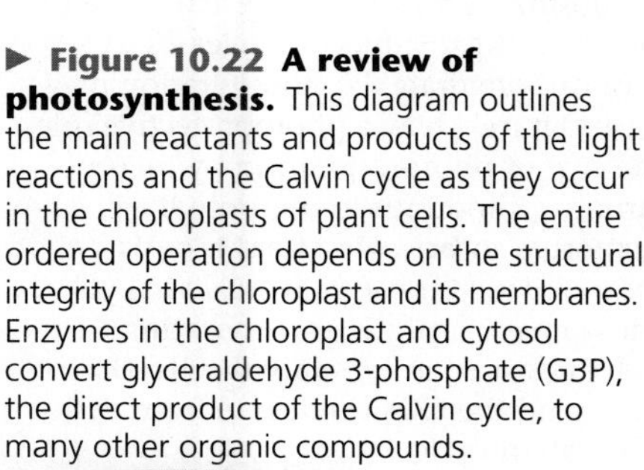

▶ **Figure 10.22 A review of photosynthesis.** This diagram outlines the main reactants and products of the light reactions and the Calvin cycle as they occur in the chloroplasts of plant cells. The entire ordered operation depends on the structural integrity of the chloroplast and its membranes. Enzymes in the chloroplast and cytosol convert glyceraldehyde 3-phosphate (G3P), the direct product of the Calvin cycle, to many other organic compounds.

MAKE CONNECTIONS *Return to the micrograph in Figure 5.6a, on page 72. Label and describe where the light reactions and the Calvin cycle take place. Also explain where the starch granules in the micrograph came from.*

10 CHAPTER REVIEW

SUMMARY OF KEY CONCEPTS

CONCEPT 10.1

Photosynthesis converts light energy to the chemical energy of food (pp. 186–189)

- In **autotrophic** eukaryotes, photosynthesis occurs in **chloroplasts**, organelles containing **thylakoids**. Stacks of thylakoids form grana. **Photosynthesis** is summarized as

  $6\ CO_2 + 12\ H_2O + \text{Light energy} \rightarrow C_6H_{12}O_6 + 6\ O_2 + 6\ H_2O.$

  Chloroplasts split water into hydrogen and oxygen, incorporating the electrons of hydrogen into sugar molecules. Photosynthesis is a redox process: H_2O is oxidized, and CO_2 is reduced. The **light reactions** in the thylakoid membranes split water, releasing O_2, producing ATP, and forming **NADPH**. The **Calvin cycle** in the **stroma** forms sugar from CO_2, using ATP for energy and NADPH for reducing power.

? *Compare and describe the roles of CO_2 and H_2O in respiration and photosynthesis.*

CONCEPT 10.2

The light reactions convert solar energy to the chemical energy of ATP and NADPH (pp. 189–197)

- Light is a form of electromagnetic energy. The colors we see as **visible light** include those **wavelengths** that drive photosynthesis. A pigment absorbs light of specific wavelengths; **chlorophyll *a*** is the main photosynthetic pigment in plants. Other accessory pigments absorb different wavelengths of light and pass the energy on to chlorophyll *a*.
- A pigment goes from a ground state to an excited state when a **photon** of light boosts one of the pigment's electrons to a higher-energy orbital. This excited state is unstable. Electrons from isolated pigments tend to fall back to the ground state, giving off heat and/or light.
- A **photosystem** is composed of a **reaction-center complex** surrounded by **light-harvesting complexes** that funnel the energy of photons to the reaction-center complex. When a special pair of reaction-center chlorophyll *a* molecules absorbs energy, one of its electrons is boosted to a higher energy level and transferred to the **primary electron acceptor**. **Photosystem II** contains P680 chlorophyll *a* molecules in the reaction-center complex; **photosystem I** contains P700 molecules.
- **Linear electron flow** during the light reactions uses both photosystems and produces NADPH, ATP, and oxygen:

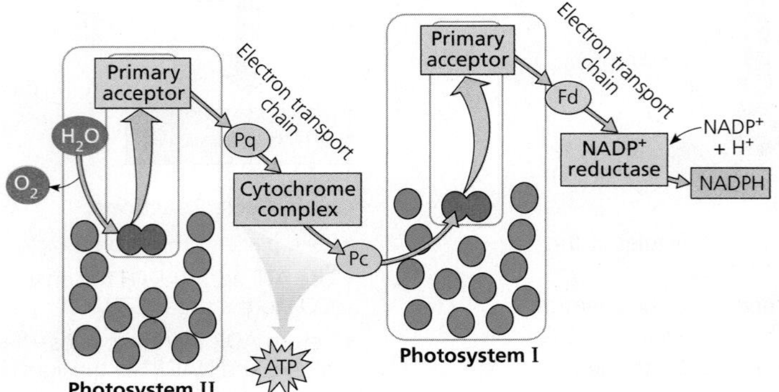

- **Cyclic electron flow** employs only photosystem I, producing ATP but no NADPH or O_2.
- During chemiosmosis in both mitochondria and chloroplasts, electron transport chains generate an H^+ gradient across a membrane. ATP synthase uses this proton-motive force to make ATP.

? *The absorption spectrum of chlorophyll* a *differs from the action spectrum of photosynthesis. Explain this observation.*

CONCEPT 10.3

The Calvin cycle uses the chemical energy of ATP and NADPH to reduce CO_2 to sugar (pp. 198–199)

- The Calvin cycle occurs in the stroma, using electrons from NADPH and energy from ATP. One molecule of **G3P** exits the cycle per three CO_2 molecules fixed and is converted to glucose and other organic molecules.

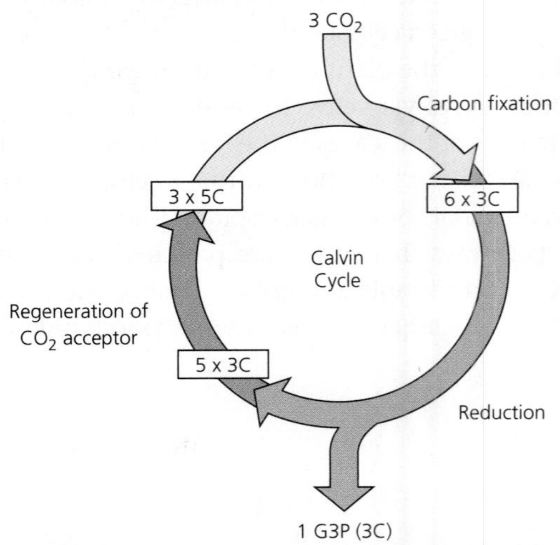

DRAW IT *On the diagram above, draw where ATP and NADPH are used and where rubisco functions. Describe these steps.*

CONCEPT 10.4

Alternative mechanisms of carbon fixation have evolved in hot, arid climates (pp. 199–202)

- On dry, hot days, **C_3 plants** close their stomata, conserving water. Oxygen from the light reactions builds up. In **photorespiration**, O_2 substitutes for CO_2 in the active site of rubisco. This process consumes organic fuel and releases CO_2 without producing ATP or carbohydrate. Photorespiration may be an evolutionary relic, and it may play a photoprotective role.
- **C_4 plants** minimize the cost of photorespiration by incorporating CO_2 into four-carbon compounds in mesophyll cells. These compounds are exported to **bundle-sheath cells**, where they release carbon dioxide for use in the Calvin cycle.
- **CAM plants** open their stomata at night, incorporating CO_2 into organic acids, which are stored in mesophyll cells. During the day, the stomata close, and the CO_2 is released from the organic acids for use in the Calvin cycle.
- Organic compounds produced by photosynthesis provide the energy and building material for ecosystems.

? *Why are C_4 and CAM photosynthesis more energetically expensive than C_3 photosynthesis? What climate conditions would favor C_4 and CAM plants?*

TEST YOUR UNDERSTANDING

LEVEL 1: KNOWLEDGE/COMPREHENSION

1. The light reactions of photosynthesis supply the Calvin cycle with
   a. light energy.
   b. CO_2 and ATP.
   c. H_2O and NADPH.
   d. ATP and NADPH.
   e. sugar and O_2.

2. Which of the following sequences correctly represents the flow of electrons during photosynthesis?
   a. NADPH → O_2 → CO_2
   b. H_2O → NADPH → Calvin cycle
   c. NADPH → chlorophyll → Calvin cycle
   d. H_2O → photosystem I → photosystem II
   e. NADPH → electron transport chain → O_2

3. How is photosynthesis similar in C_4 plants and CAM plants?
   a. In both cases, only photosystem I is used.
   b. Both types of plants make sugar without the Calvin cycle.
   c. In both cases, rubisco is not used to fix carbon initially.
   d. Both types of plants make most of their sugar in the dark.
   e. In both cases, thylakoids are not involved in photosynthesis.

4. Which of the following statements is a correct distinction between autotrophs and heterotrophs?
   a. Only heterotrophs require chemical compounds from the environment.
   b. Cellular respiration is unique to heterotrophs.
   c. Only heterotrophs have mitochondria.
   d. Autotrophs, but not heterotrophs, can nourish themselves beginning with CO_2 and other nutrients that are inorganic.
   e. Only heterotrophs require oxygen.

5. Which of the following does *not* occur during the Calvin cycle?
   a. carbon fixation
   b. oxidation of NADPH
   c. release of oxygen
   d. regeneration of the CO_2 acceptor
   e. consumption of ATP

LEVEL 2: APPLICATION/ANALYSIS

6. In mechanism, photophosphorylation is most similar to
   a. substrate-level phosphorylation in glycolysis.
   b. oxidative phosphorylation in cellular respiration.
   c. the Calvin cycle.
   d. carbon fixation.
   e. reduction of $NADP^+$.

7. Which process is most directly driven by light energy?
   a. creation of a pH gradient by pumping protons across the thylakoid membrane
   b. carbon fixation in the stroma
   c. reduction of $NADP^+$ molecules
   d. removal of electrons from chlorophyll molecules
   e. ATP synthesis

LEVEL 3: SYNTHESIS/EVALUATION

8. **EVOLUTION CONNECTION**
   Photorespiration can decrease soybeans' photosynthetic output by about 50%. Would you expect this figure to be higher or lower in wild relatives of soybeans? Why?

9. **SCIENTIFIC INQUIRY**
   MAKE CONNECTIONS **DRAW IT** The following diagram represents an experiment with isolated thylakoids. The thylakoids were first made acidic by soaking them in a solution at pH 4. After the thylakoid space reached pH 4, the thylakoids were transferred to a basic solution at pH 8. The thylakoids then made ATP in the dark. (See Concept 3.3, pp. 53–54, to review pH).

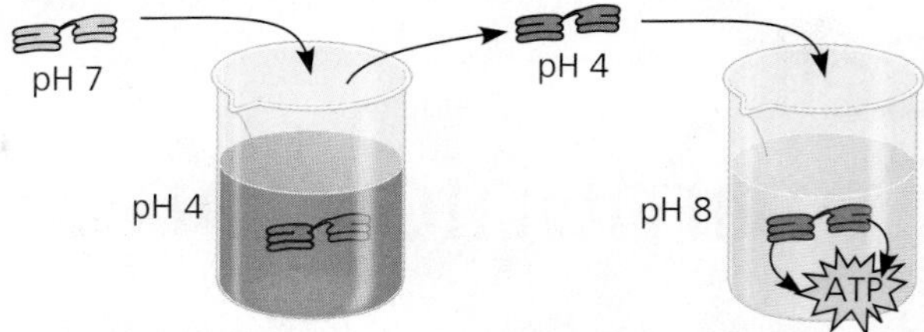

   Draw an enlargement of part of the thylakoid membrane in the beaker with the solution at pH 8. Draw ATP synthase. Label the areas of high H^+ concentration and low H^+ concentration. Show the direction protons flow through the enzyme, and show the reaction where ATP is synthesized. Would ATP end up in the thylakoid or outside of it? Explain why the thylakoids in the experiment were able to make ATP in the dark.

10. **SCIENCE, TECHNOLOGY, AND SOCIETY**
    Scientific evidence indicates that the CO_2 added to the air by the burning of wood and fossil fuels is contributing to global warming, a rise in global temperature. Tropical rain forests are estimated to be responsible for approximately 20% of global photosynthesis, yet the consumption of large amounts of CO_2 by living trees is thought to make little or no *net* contribution to reduction of global warming. Why might this be? (*Hint*: What processes in both living and dead trees produce CO_2?)

11. **WRITE ABOUT A THEME**
    Energy Transfer Life is solar powered. Almost all the producers of the biosphere depend on energy from the sun to produce the organic molecules that supply the energy and carbon skeletons needed for life. In a short essay (100–150 words), describe how the process of photosynthesis in the chloroplasts of plants transforms the energy of sunlight into the chemical energy of sugar molecules.

For selected answers, see Appendix A.

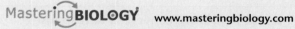

11

Cell Communication

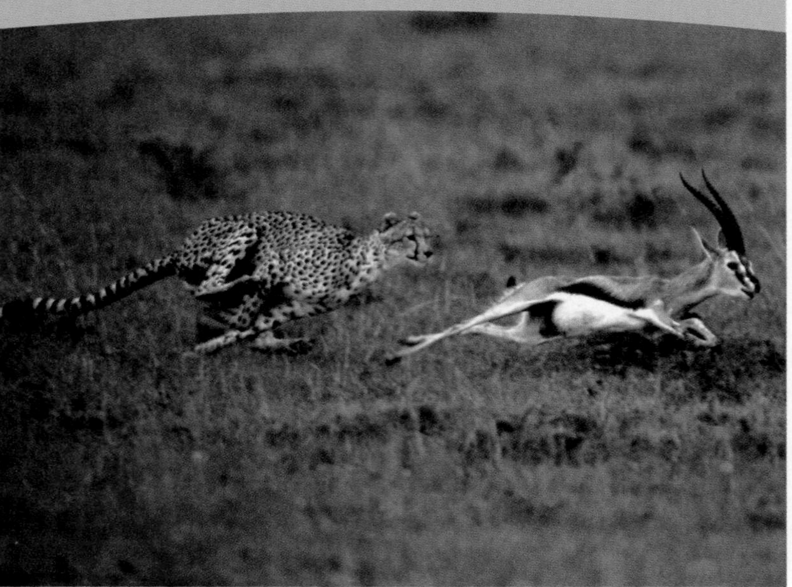

▲ **Figure 11.1 How does cell signaling trigger the desperate flight of this gazelle?**

KEY CONCEPTS

11.1 External signals are converted to responses within the cell

11.2 Reception: A signaling molecule binds to a receptor protein, causing it to change shape

11.3 Transduction: Cascades of molecular interactions relay signals from receptors to target molecules in the cell

11.4 Response: Cell signaling leads to regulation of transcription or cytoplasmic activities

11.5 Apoptosis integrates multiple cell-signaling pathways

OVERVIEW

Cellular Messaging

The Thomson's gazelle in **Figure 11.1** is fleeing for its life, seeking to escape the predatory cheetah nipping at its heels. The gazelle's heart is racing, its breathing accelerated and its muscles performing at their highest level. These physiological functions are all part of the "fight-or-flight" response, driven by hormones released from the adrenal glands at times of stress—in this case, when the gazelle first sensed the cheetah. Hormonal signaling and the subsequent response by cells and tissues throughout the gazelle's body illustrate how cell-to-cell communication allows the trillions of cells in a multicellular organism to "talk" to each other, coordinating their activities. Communication between cells is essential not only for multicellular organisms such as gazelles and oak trees but for many unicellular organisms as well.

In studying how cells signal to each other and how they interpret the signals they receive, biologists have discovered some universal mechanisms of cellular regulation, additional evidence for the evolutionary relatedness of all life. The same small set of cell-signaling mechanisms shows up again and again in diverse species, in biological processes ranging from hormone action to embryonic development to cancer. The signals received by cells, whether originating from other cells or from changes in the physical environment, take various forms, including light and touch. However, cells most often communicate with each other by chemical signals. For instance, the fight-or-flight response is triggered by a signaling molecule called epinephrine. In this chapter, we focus on the main mechanisms by which cells receive, process, and respond to chemical signals sent from other cells. We will also take a look at *apoptosis*, a type of programmed cell death that integrates input from multiple signaling pathways.

CONCEPT **11.1**

External signals are converted to responses within the cell

What does a "talking" cell say to a "listening" cell, and how does the latter cell respond to the message? Let's approach these questions by first looking at communication among microorganisms, for microbes living today provide a glimpse into the role of cell signaling in the evolution of life on Earth.

Evolution of Cell Signaling

EVOLUTION One topic of cell "conversation" is sex—at least for the yeast *Saccharomyces cerevisiae*, which people have used for millennia to make bread, wine, and beer. Researchers have learned that cells of this yeast identify their mates by chemical signaling. There are two sexes, or mating types, called **a** and α (**Figure 11.2**). Cells of mating type **a** secrete a signaling

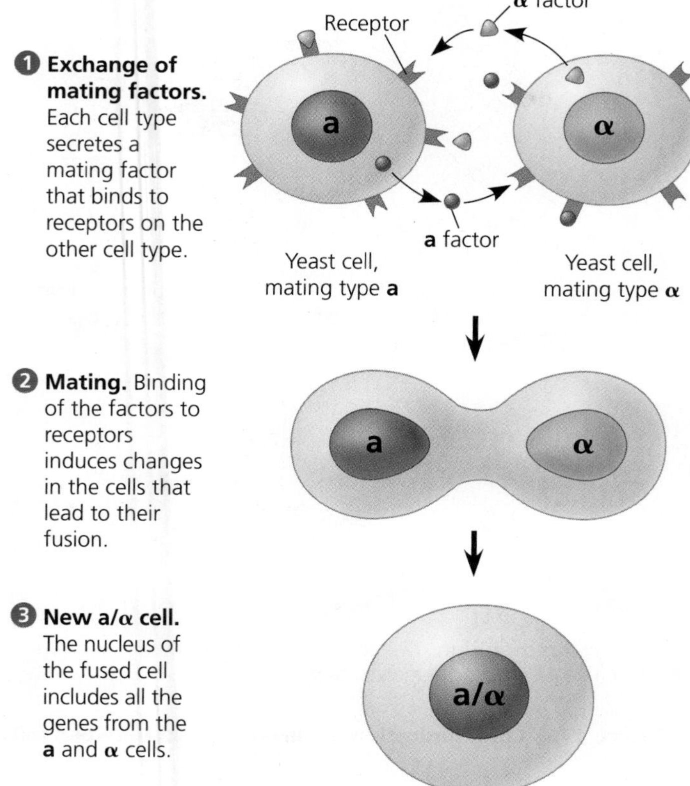

α factor
Receptor
1 Exchange of mating factors. Each cell type secretes a mating factor that binds to receptors on the other cell type.
a
α
a factor
Yeast cell, mating type a
Yeast cell, mating type α
2 Mating. Binding of the factors to receptors induces changes in the cells that lead to their fusion.
3 New a/α cell. The nucleus of the fused cell includes all the genes from the a and α cells.
a/α

▲ **Figure 11.2 Communication between mating yeast cells.** *Saccharomyces cerevisiae* cells use chemical signaling to identify cells of opposite mating type and initiate the mating process. The two mating types and their corresponding chemical signaling molecules, or mating factors, are called **a** and **α**.

molecule called **a** factor, which can bind to specific receptor proteins on nearby **α** cells. At the same time, **α** cells secrete **α** factor, which binds to receptors on **a** cells. Without actually entering the cells, the two mating factors cause the cells to grow toward each other and also bring about other cellular changes. The result is the fusion, or mating, of two cells of opposite type. The new **a**/**α** cell contains all the genes of both original cells, a combination of genetic resources that provides advantages to the cell's descendants, which arise by subsequent cell divisions.

Once received at the yeast cell surface, how is the mating signal changed, or *transduced*, into a form that brings about the cellular response of mating? The received signal is converted to a specific cellular response in a series of steps called a **signal transduction pathway**. Many such pathways have been extensively studied in both yeast and animal cells. Amazingly, the molecular details of signal transduction in yeast and mammals are strikingly similar, even though the last common ancestor of these two groups of organisms lived over a billion years ago. These similarities—and others more recently uncovered between signaling systems in bacteria and plants—suggest that early versions of today's cell-signaling mechanisms evolved well before the first multicellular creatures appeared on Earth.

Scientists such as Bonnie Bassler, the interviewee for Unit 2 (see pp. 92–93), think that signaling mechanisms first evolved in ancient prokaryotes and single-celled eukaryotes, then were adopted for new uses by their multicellular descendants. Cell signaling is critical in the microbial world; a classic example in one bacterial species is shown in **Figure 11.3**. Bacterial cells secrete small molecules that can be detected by other bacterial cells. The concentration of such signaling molecules, sensed by the bacteria, allows them to monitor the local density of cells, a phenomenon called *quorum sensing*. Quorum sensing allows bacterial populations to coordinate their behaviors so they can carry out activities that are only productive when performed by a given number of cells in synchrony. One example is formation of a *biofilm*, an aggregation of bacterial cells adhered to a surface; the cells in the biofilm generally derive nutrition from the surface they are on. You have probably encountered biofilms many times, perhaps without realizing it. The slimy coating on a fallen log or on leaves lying on a forest path, or on your teeth each morning, are examples of bacterial biofilms. Biofilms are responsible for cavities—a good argument for daily tooth brushing and flossing to disrupt them!

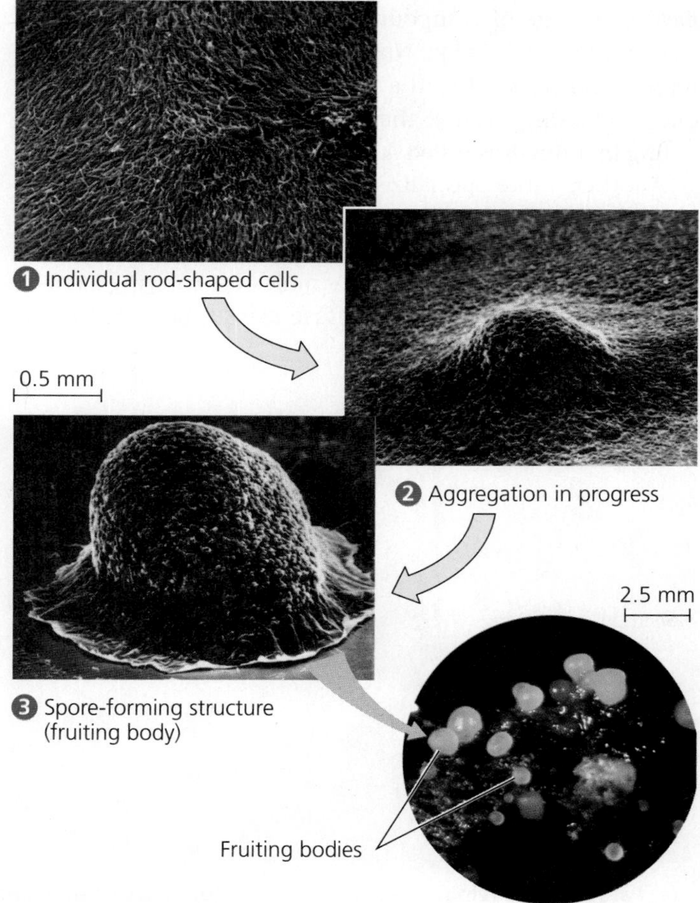

▲ **Figure 11.3 Communication among bacteria.** Soil-dwelling bacteria called myxobacteria ("slime bacteria") use chemical signals to share information about nutrient availability. When food is scarce, starving cells secrete a molecule that stimulates neighboring cells to aggregate. The cells form a structure, called a fruiting body, that produces thick-walled spores capable of surviving until the environment improves. The bacteria shown here are *Myxococcus xanthus* (steps 1–3, SEMs; lower photo, LM).

Local and Long-Distance Signaling

Like bacteria or yeast cells, cells in a multicellular organism usually communicate via chemical messengers targeted for cells that may or may not be immediately adjacent. As we saw in Chapters 6 and 7, eukaryotic cells may communicate by direct contact (**Figure 11.4**), one type of local signaling. Both animals and plants have cell junctions that, where present, directly connect the cytoplasms of adjacent cells (**Figure 11.4a**). In these cases, signaling substances dissolved in the cytosol can pass freely between adjacent cells. Moreover, animal cells may communicate via direct contact between membrane-bound cell-surface molecules in a process called cell-cell recognition (**Figure 11.4b**). This sort of local signaling is important in embryonic development and the immune response.

In many other cases of local signaling, messenger molecules are secreted by the signaling cell. Some of these travel only short distances; such **local regulators** influence cells in the vicinity. One class of local regulators in animals, *growth factors*, consists of compounds that stimulate nearby target cells to grow and divide. Numerous cells can simultaneously receive and respond to the molecules of growth factor produced by a single cell in their vicinity. This type of local signaling in animals is called *paracrine signaling* (**Figure 11.5a**).

Another, more specialized type of local signaling called *synaptic signaling* occurs in the animal nervous system (**Figure 11.5b**). An electrical signal along a nerve cell triggers the secretion of neurotransmitter molecules carrying a chemical signal. These molecules diffuse across the synapse, the narrow space between the nerve cell and its target cell (often another nerve cell), triggering a response in the target cell.

Beyond communication through plasmodesmata (plant cell junctions), local signaling in plants is not as well understood. Because of their cell walls, plants use mechanisms somewhat different from those operating locally in animals.

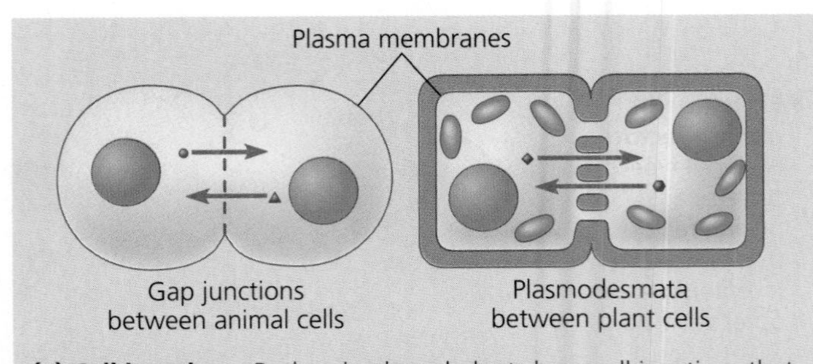

(a) Cell junctions. Both animals and plants have cell junctions that allow molecules to pass readily between adjacent cells without crossing plasma membranes.

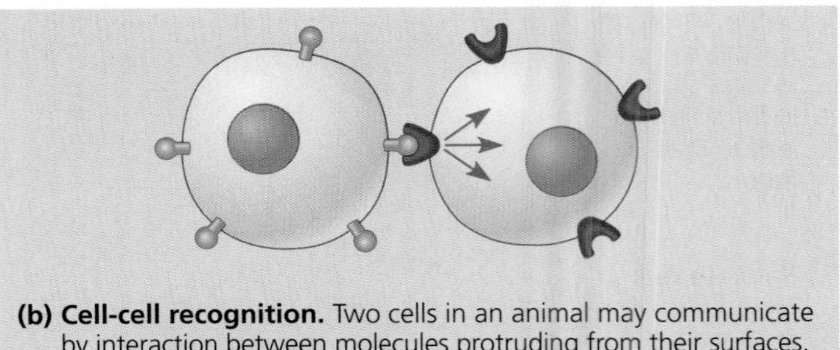

(b) Cell-cell recognition. Two cells in an animal may communicate by interaction between molecules protruding from their surfaces.

▲ **Figure 11.4 Communication by direct contact between cells.**

Local signaling

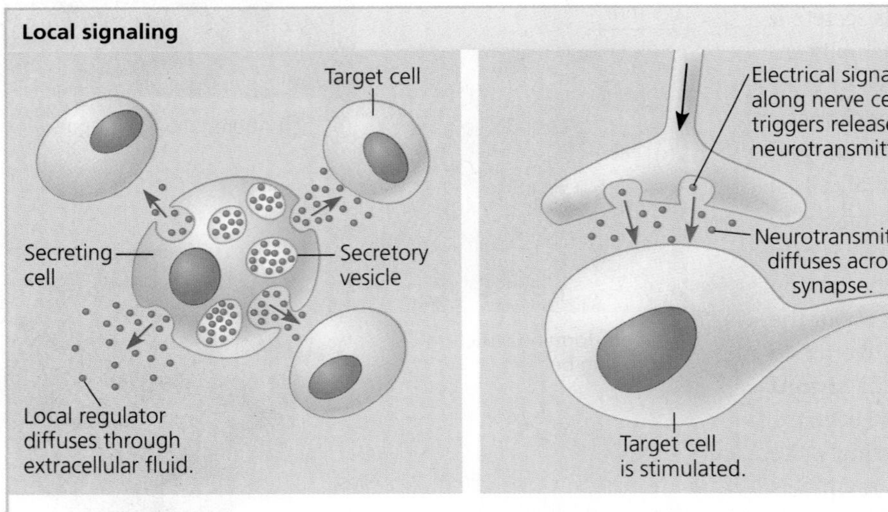

(a) Paracrine signaling. A secreting cell acts on nearby target cells by discharging molecules of a local regulator (a growth factor, for example) into the extracellular fluid.

(b) Synaptic signaling. A nerve cell releases neurotransmitter molecules into a synapse, stimulating the target cell.

Long-distance signaling

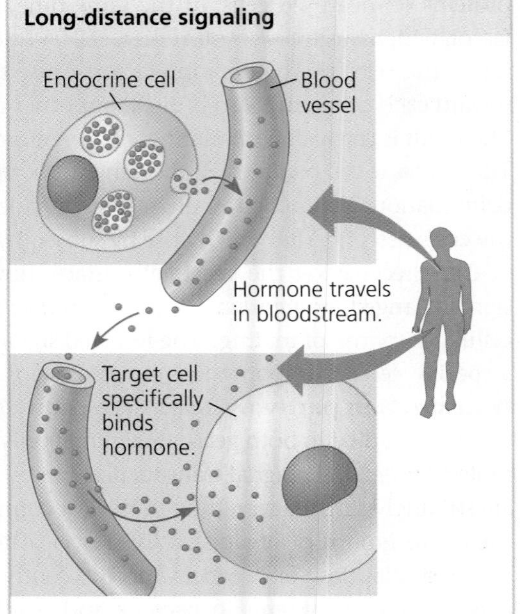

(c) Endocrine (hormonal) signaling. Specialized endocrine cells secrete hormones into body fluids, often blood. Hormones reach virtually all body cells, but are bound only by some cells.

▲ **Figure 11.5 Local and long-distance cell signaling by secreted molecules in animals.** In both local and long-distance signaling, only specific target cells that can recognize a given signaling molecule will respond to it.

Both animals and plants use chemicals called **hormones** for long-distance signaling. In hormonal signaling in animals, also known as *endocrine signaling*, specialized cells release hormone molecules, which travel via the circulatory system to other parts of the body, where they reach target cells that can recognize and respond to the hormones (**Figure 11.5c**). Plant hormones (often called *plant growth regulators*) sometimes travel in vessels but more often reach their targets by moving through cells or by diffusing through the air as a gas (see Chapter 39). Hormones vary widely in molecular size and type, as do local regulators. For instance, the plant hormone ethylene, a gas that promotes fruit ripening and helps regulate growth, is a hydrocarbon of only six atoms (C_2H_4), small enough to pass through cell walls. In contrast, the mammalian hormone insulin, which regulates sugar levels in the blood, is a protein with thousands of atoms.

The transmission of a signal through the nervous system can also be considered an example of long-distance signaling. An electrical signal travels the length of a nerve cell and is then converted back to a chemical signal when a signaling molecule is released and crosses the synapse to another nerve cell. Here it is converted back to an electrical signal. In this way, a nerve signal can travel along a series of nerve cells. Because some nerve cells are quite long, the nerve signal can quickly travel great distances—from your brain to your big toe, for example. This type of long-distance signaling will be covered in detail in Chapter 48.

What happens when a cell encounters a secreted signaling molecule? The ability of a cell to respond is determined by whether it has a specific receptor molecule that can bind to the signaling molecule. The information conveyed by this binding, the signal, must then be changed into another form—transduced—inside the cell before the cell can respond. The remainder of the chapter discusses this process, primarily as it occurs in animal cells.

The Three Stages of Cell Signaling: *A Preview*

Our current understanding of how chemical messengers act via signal transduction pathways had its origins in the pioneering work of Earl W. Sutherland, whose research led to a Nobel Prize in 1971. Sutherland and his colleagues at Vanderbilt University were investigating how the animal hormone epinephrine (also called adrenaline) stimulates the breakdown of the storage polysaccharide glycogen within liver cells and skeletal muscle cells. Glycogen breakdown releases the sugar glucose 1-phosphate, which the cell converts to glucose 6-phosphate. The cell (a liver cell, for example) can then use this compound, an early intermediate in glycolysis, for energy production. Alternatively, the compound can be stripped of phosphate and released from the liver cell into the blood as glucose, which can fuel cells throughout the body. Thus, one effect of epinephrine is the mobilization of fuel reserves, which can be used by the animal to either defend itself (fight) or escape whatever elicited a scare (flight). (The gazelle in Figure 11.1 is clearly engaged in the latter.)

Sutherland's research team discovered that epinephrine stimulates glycogen breakdown by somehow activating a cytosolic enzyme, glycogen phosphorylase. However, when epinephrine was added to a test-tube mixture containing the enzyme and its substrate, glycogen, no breakdown occurred. Epinephrine could activate glycogen phosphorylase only when the hormone was added to a solution containing *intact* cells. This result told Sutherland two things. First, epinephrine does not interact directly with the enzyme responsible for glycogen breakdown; an intermediate step or series of steps must be occurring inside the cell. Second, the plasma membrane is somehow involved in transmitting the signal.

Sutherland's early work suggested that the process going on at the receiving end of a cellular conversation can be dissected into three stages: reception, transduction, and response (**Figure 11.6**):

❶ **Reception.** Reception is the target cell's detection of a signaling molecule coming from outside the cell. A chemical signal is "detected" when the signaling molecule binds to a receptor protein located at the cell's surface or inside the cell.

▶ **Figure 11.6 Overview of cell signaling.** From the perspective of the cell receiving the message, cell signaling can be divided into three stages: signal reception, signal transduction, and cellular response. When reception occurs at the plasma membrane, as shown here, the transduction stage is usually a pathway of several steps, with each relay molecule in the pathway bringing about a change in the next molecule. The final molecule in the pathway triggers the cell's response. The three stages are explained in more detail in the text.

? *How does the epinephrine in Sutherland's experiment fit into this diagram of cell signaling?*

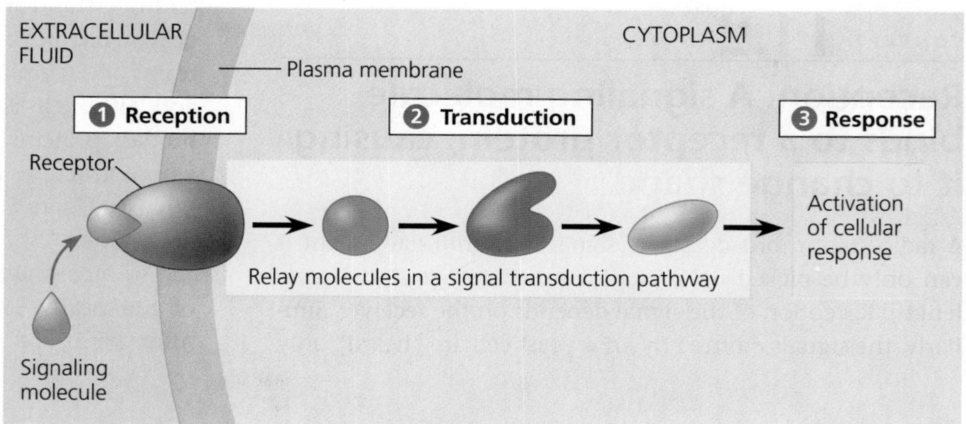

2. **Transduction.** The binding of the signaling molecule changes the receptor protein in some way, initiating the process of transduction. The transduction stage converts the signal to a form that can bring about a specific cellular response. In Sutherland's system, the binding of epinephrine to a receptor protein in a liver cell's plasma membrane leads to activation of glycogen phosphorylase. Transduction sometimes occurs in a single step but more often requires a sequence of changes in a series of different molecules—a *signal transduction pathway*. The molecules in the pathway are often called relay molecules.

3. **Response.** In the third stage of cell signaling, the transduced signal finally triggers a specific cellular response. The response may be almost any imaginable cellular activity—such as catalysis by an enzyme (for example, glycogen phosphorylase), rearrangement of the cytoskeleton, or activation of specific genes in the nucleus. The cell-signaling process helps ensure that crucial activities like these occur in the right cells, at the right time, and in proper coordination with the activities of other cells of the organism. We'll now explore the mechanisms of cell signaling in more detail, including a discussion of fine-tuning and termination of the process.

CONCEPT CHECK 11.1

1. Explain how signaling is involved in ensuring that yeast cells fuse only with cells of the opposite mating type.
2. Explain how nerve cells provide examples of both local and long-distance signaling.
3. **WHAT IF?** When epinephrine is mixed with glycogen phosphorylase and glycogen in a test tube, is glucose 1-phosphate generated? Why or why not?
4. In liver cells, glycogen phosphorylase acts in which of the three stages of the signaling pathway associated with an epinephrine-initiated signal?

For suggested answers, see Appendix A.

CONCEPT 11.2

Reception: A signaling molecule binds to a receptor protein, causing it to change shape

A radio station broadcasts its signal indiscriminately, but it can only be picked up by radios tuned to the right wavelength: Reception of the signal depends on the receiver. Similarly, the signals emitted by an **a** yeast cell are "heard" only by its prospective mates, **α** cells. In the case of epinephrine, the hormone encounters many types of cells as it circulates in the blood, but only certain target cells detect and react to the hormone molecule. A receptor protein on or in the target cell allows the cell to "hear" the signal and respond to it. The signaling molecule is complementary in shape to a specific site on the receptor and attaches there, like a key in a lock or a substrate in the catalytic site of an enzyme. The signaling molecule behaves as a **ligand**, the term for a molecule that specifically binds to another molecule, often a larger one. Ligand binding generally causes a receptor protein to undergo a change in shape. For many receptors, this shape change directly activates the receptor, enabling it to interact with other cellular molecules. For other kinds of receptors, the immediate effect of ligand binding is to cause the aggregation of two or more receptor molecules, which leads to further molecular events inside the cell.

Most signal receptors are plasma membrane proteins. Their ligands are water-soluble and generally too large to pass freely through the plasma membrane. Other signal receptors, however, are located inside the cell. We discuss both of these types next.

Receptors in the Plasma Membrane

Most water-soluble signaling molecules bind to specific sites on receptor proteins that span the cell's plasma membrane. Such a transmembrane receptor transmits information from the extracellular environment to the inside of the cell by changing shape or aggregating when a specific ligand binds to it. We can see how cell-surface transmembrane receptors work by looking at three major types: G protein-coupled receptors, receptor tyrosine kinases, and ion channel receptors. These receptors are discussed and illustrated in **Figure 11.7**, on the next three pages; study this figure before going on.

Cell-surface receptor molecules play crucial roles in the biological systems of animals, and not surprisingly, their malfunctions are associated with many human diseases, including cancer, heart disease, and asthma. Working out the structure and function of these receptors will allow us to better understand and treat these conditions. Therefore, this endeavor has been a major focus of both university research teams and the pharmaceutical industry. In spite of this effort, and although cell-surface receptors make up 30% of all human proteins, they make up only 1% of the proteins whose structures have been determined by X-ray crystallography (see Figure 5.24): Their structures are very challenging to determine.

The largest family of human cell-surface receptors consists of the nearly 1,000 G protein-coupled receptors (GPCRs). After persistent efforts, researchers have made significant

▼ **Figure 11.7**

Exploring Cell-Surface Transmembrane Receptors

G Protein-Coupled Receptors

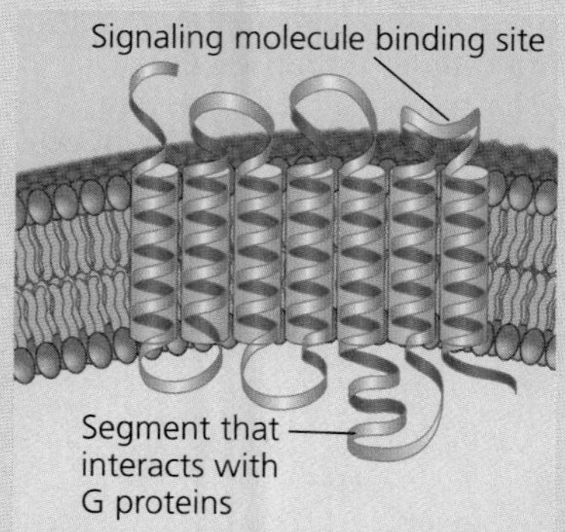

G protein-coupled receptor

A **G protein-coupled receptor** (GPCR) is a cell-surface transmembrane receptor that works with the help of a **G protein**, a protein that binds the energy-rich molecule GTP. Many different signaling molecules, including yeast mating factors, epinephrine and many other hormones, and neurotransmitters, use G protein-coupled receptors. These receptors vary in the binding sites for their signaling molecules (often referred to as their ligands) and also for different types of G proteins inside the cell. Nevertheless, G protein-coupled receptor proteins are all remarkably similar in structure.

In fact, they make up a large family of eukaryotic receptor proteins with a secondary structure in which the single polypeptide, represented here as a ribbon, has seven transmembrane α helices, outlined with cylinders and depicted in a row for clarity. Specific loops between the helices form binding sites for signaling and G protein molecules.

G protein-coupled receptor systems are extremely widespread and diverse in their functions, including roles in embryonic development and sensory reception. In humans, for example, vision, smell, and taste depend on such systems. Similarities in structure in G proteins and G protein-coupled receptors in diverse organisms suggest that G proteins and associated receptors evolved very early.

G protein systems are involved in many human diseases, including bacterial infections. The bacteria that cause cholera, pertussis (whooping cough), and botulism, among others, make their victims ill by producing toxins that interfere with G protein function. Pharmacologists now realize that up to 60% of all medicines used today exert their effects by influencing G protein pathways.

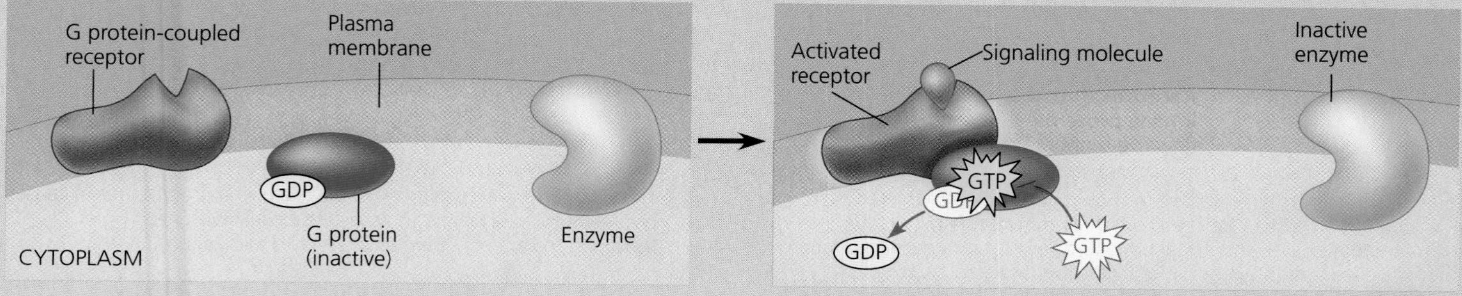

1 Loosely attached to the cytoplasmic side of the membrane, the G protein functions as a molecular switch that is either on or off, depending on which of two guanine nucleotides is attached, GDP or GTP—hence the term *G protein*. (GTP, or guanosine triphosphate, is similar to ATP.) When GDP is bound to the G protein, as shown above, the G protein is inactive. The receptor and G protein work together with another protein, usually an enzyme.

2 When the appropriate signaling molecule binds to the extracellular side of the receptor, the receptor is activated and changes shape. Its cytoplasmic side then binds an inactive G protein, causing a GTP to displace the GDP. This activates the G protein.

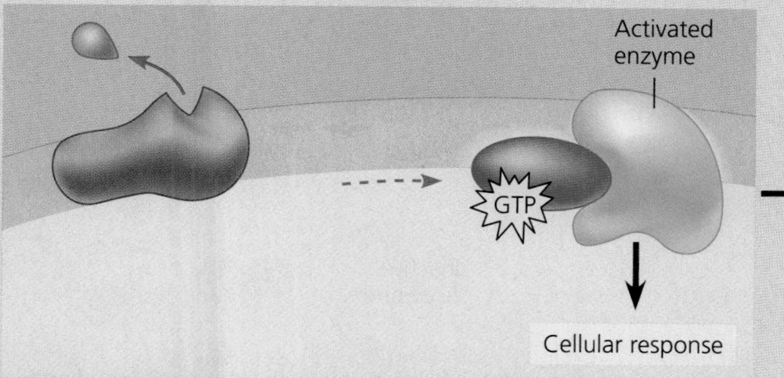

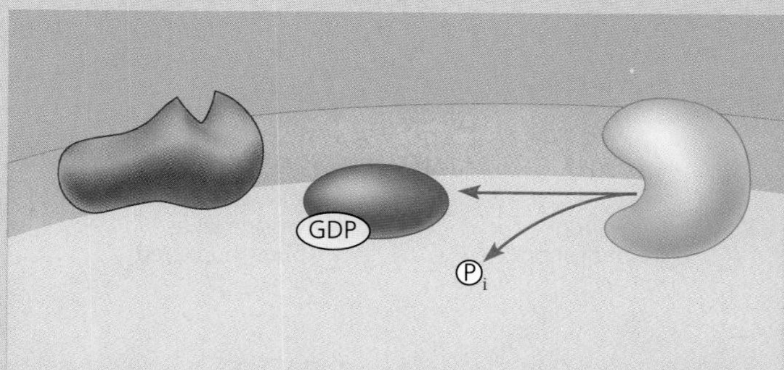

3 The activated G protein dissociates from the receptor, diffuses along the membrane, and then binds to an enzyme, altering the enzyme's shape and activity. Once activated, the enzyme can trigger the next step leading to a cellular response. (Binding of signaling molecules is reversible: Like other ligands, they bind and dissociate many times. The ligand concentration outside the cell determines how often a ligand is bound and causes signaling.)

4 The changes in the enzyme and G protein are only temporary because the G protein also functions as a GTPase enzyme—in other words, it then hydrolyzes its bound GTP to GDP. Now inactive again, the G protein leaves the enzyme, which returns to its original state. The G protein is now available for reuse. The GTPase function of the G protein allows the pathway to shut down rapidly when the signaling molecule is no longer present.

Continued on next page

▼ **Figure 11.7 (continued)**

Exploring Cell-Surface Transmembrane Receptors

Receptor Tyrosine Kinases

Receptor tyrosine kinases (RTKs) belong to a major class of plasma membrane receptors characterized by having enzymatic activity. A *kinase* is an enzyme that catalyzes the transfer of phosphate groups. The part of the receptor protein extending into the cytoplasm functions as a tyrosine kinase, an enzyme that catalyzes the transfer of a phosphate group from ATP to the amino acid tyrosine on a substrate protein. Thus, receptor tyrosine kinases are membrane receptors that attach phosphates to tyrosines.

One receptor tyrosine kinase complex may activate ten or more different transduction pathways and cellular responses. Often, more than one signal transduction pathway can be triggered at once, helping the cell regulate and coordinate many aspects of cell growth and cell reproduction. The ability of a single ligand-binding event to trigger so many pathways is a key difference between receptor tyrosine kinases and G protein-coupled receptors. Abnormal receptor tyrosine kinases that function even in the absence of signaling molecules are associated with many kinds of cancer.

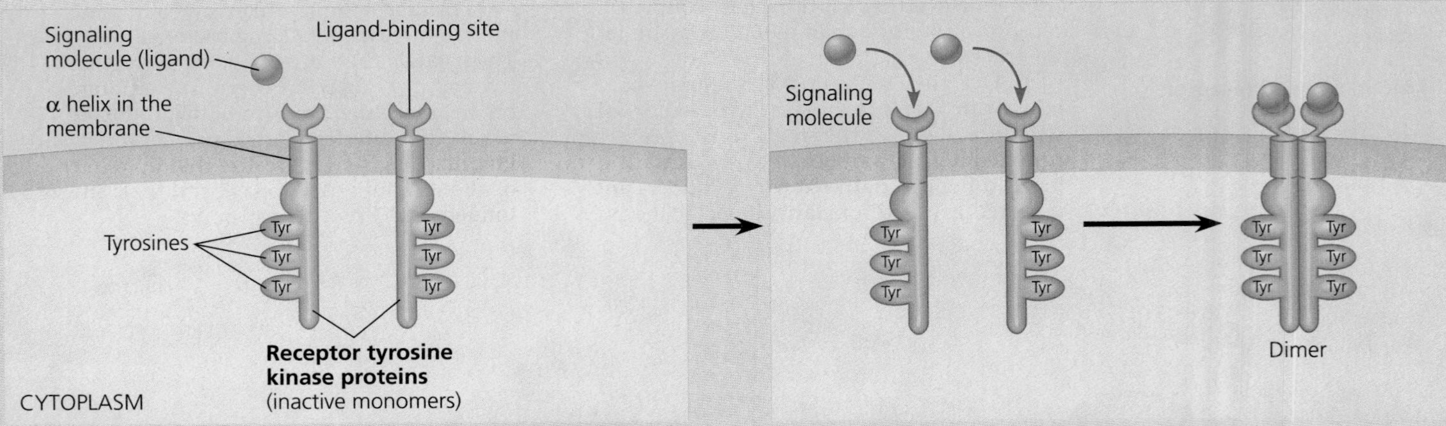

❶ Many receptor tyrosine kinases have the structure depicted schematically here. Before the signaling molecule binds, the receptors exist as individual units referred to as monomers. Notice that each has an extracellular ligand-binding site, an α helix spanning the membrane, and an intracellular tail containing multiple tyrosines.

❷ The binding of a signaling molecule (such as a growth factor) causes two receptor monomers to associate closely with each other, forming a complex known as a dimer (dimerization).

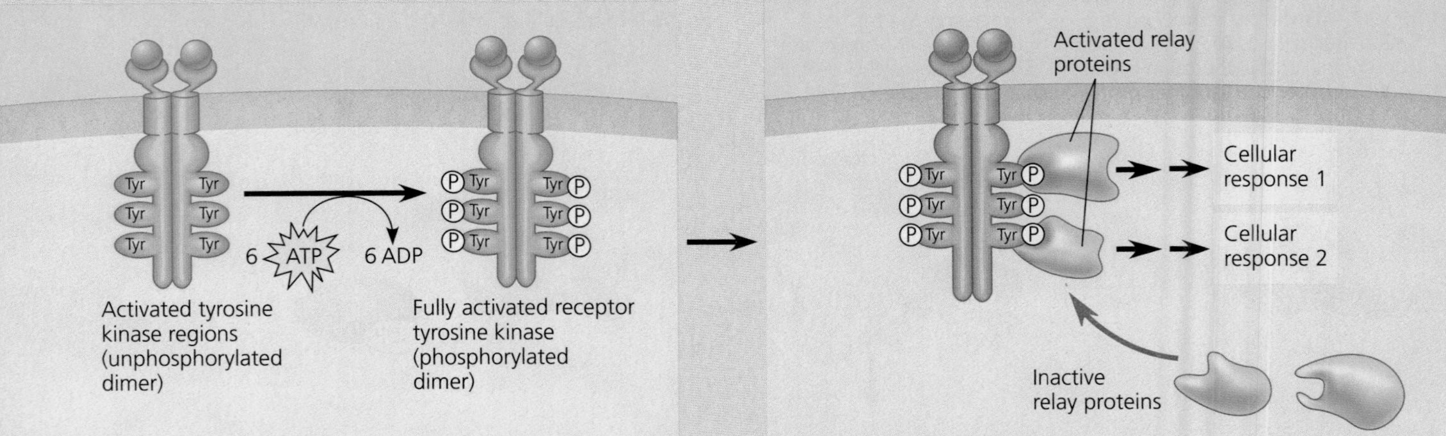

❸ Dimerization activates the tyrosine kinase region of each monomer; each tyrosine kinase adds a phosphate from an ATP molecule to a tyrosine on the tail of the other monomer.

❹ Now that the receptor is fully activated, it is recognized by specific relay proteins inside the cell. Each such protein binds to a specific phosphorylated tyrosine, undergoing a resulting structural change that activates the bound protein. Each activated protein triggers a transduction pathway, leading to a cellular response.

Ion Channel Receptors

A **ligand-gated ion channel** is a type of membrane receptor containing a region that can act as a "gate" when the receptor changes shape. When a signaling molecule binds as a ligand to the receptor protein, the gate opens or closes, allowing or blocking the flow of specific ions, such as Na^+ or Ca^{2+}, through a channel in the receptor. Like the other receptors we have discussed, these proteins bind the ligand at a specific site on their extracellular sides.

1 Here we show a ligand-gated ion channel receptor in which the gate remains closed until a ligand binds to the receptor.

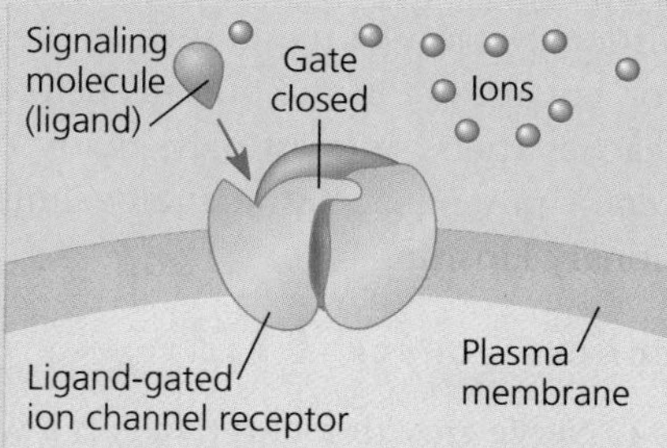

2 When the ligand binds to the receptor and the gate opens, specific ions can flow through the channel and rapidly change the concentration of that particular ion inside the cell. This change may directly affect the activity of the cell in some way.

Gate open
Cellular response

3 When the ligand dissociates from this receptor, the gate closes and ions no longer enter the cell.

Gate closed

Ligand-gated ion channels are very important in the nervous system. For example, the neurotransmitter molecules released at a synapse between two nerve cells (see Figure 11.5b) bind as ligands to ion channels on the receiving cell, causing the channels to open. Ions flow in (or, in some cases, out), triggering an electrical signal that propagates down the length of the receiving cell. Some gated ion channels are controlled by electrical signals instead of ligands; these *voltage-gated ion channels* are also crucial to the functioning of the nervous system, as we will discuss in Chapter 48.

MAKE CONNECTIONS *Examine the ion channel protein in Figure 7.1 (p. 125) and the discussion of it on page 135. What type of stimulus opens that ion channel? According to the information given above, what type of ion channel is described?*

▼ **Figure 11.8**

IMPACT

Determining the Structure of a G Protein-Coupled Receptor (GPCR)

GPCRs are flexible and inherently unstable, so they have been difficult to crystallize, a required step in determining their structure by X-ray crystallography. Recently, however, researchers have crystallized the human β_2-adrenergic receptor in the presence of a ligand similar to the natural one (green in the model below) and cholesterol (orange), which stabilized the receptor enough for its structure to be determined. Two receptor molecules (blue) are shown here as ribbon models in a side view within the plasma membrane.

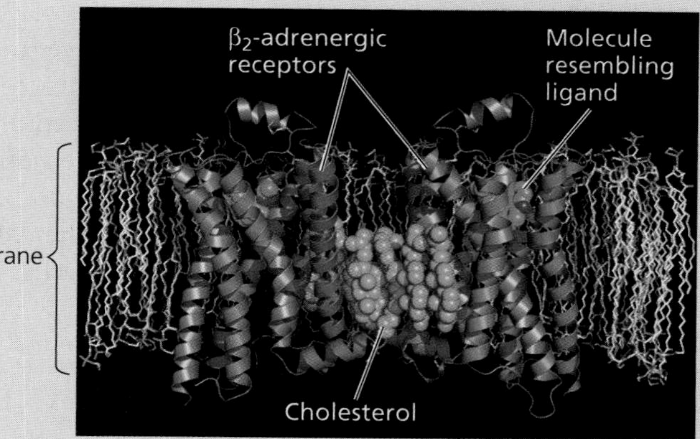

WHY IT MATTERS The β_2-adrenergic receptor is found on smooth muscle cells throughout the body, and abnormal forms of it are associated with diseases such as asthma, hypertension, and heart failure. Current drugs used for these conditions produce unwanted side effects, and further research may yield better drugs. Also, since GPCRs share structural similarities, this work on the β_2-adrenergic receptor will aid development of treatments for diseases associated with other GPCRs.

FURTHER READING R. Ranganathan, Signaling across the cell membrane, *Science* 318:1253–1254 (2007).

WHAT IF? The model shown above represents the receptor in an inactive state, not bound to a G protein. Can you suggest conditions for crystallizing the protein that would reveal the structure of the receptor while it is actively signaling to the inside of the cell?

breakthroughs in elucidating the structure of several G protein-coupled receptors over the past few years (**Figure 11.8**).

Abnormal functioning of receptor tyrosine kinases (RTKs) is associated with many types of cancers. For example, patients with breast cancer cells that have excessive levels of a receptor tyrosine kinase called HER2 have a poor prognosis. Using molecular biological techniques, researchers have developed a protein called Herceptin that binds to HER2 on cells and inhibits their growth, thus thwarting further tumor development. In some clinical studies, treatment with Herceptin improved patient survival rates by more than one-third. One goal of ongoing research into these cell-surface receptors and other cell-signaling proteins is development of additional successful treatments.

Intracellular Receptors

Intracellular receptor proteins are found in either the cytoplasm or nucleus of target cells. To reach such a receptor, a chemical messenger passes through the target cell's plasma membrane. A number of important signaling molecules can do this because they are either hydrophobic enough or small enough to cross the hydrophobic interior of the membrane. Such hydrophobic chemical messengers include the steroid hormones and thyroid hormones of animals. Another chemical signaling molecule with an intracellular receptor is nitric oxide (NO), a gas; its very small molecules readily pass between the membrane phospholipids.

The behavior of testosterone is representative of steroid hormones. In males, the hormone is secreted by cells of the testes. It then travels through the blood and enters cells all over the body. However, only cells that contain receptor molecules for testosterone respond. In these cells, the hormone binds to the receptor protein, activating it (**Figure 11.9**). With the hormone attached, the active form of the receptor protein then enters the nucleus and turns on specific genes that control male sex characteristics.

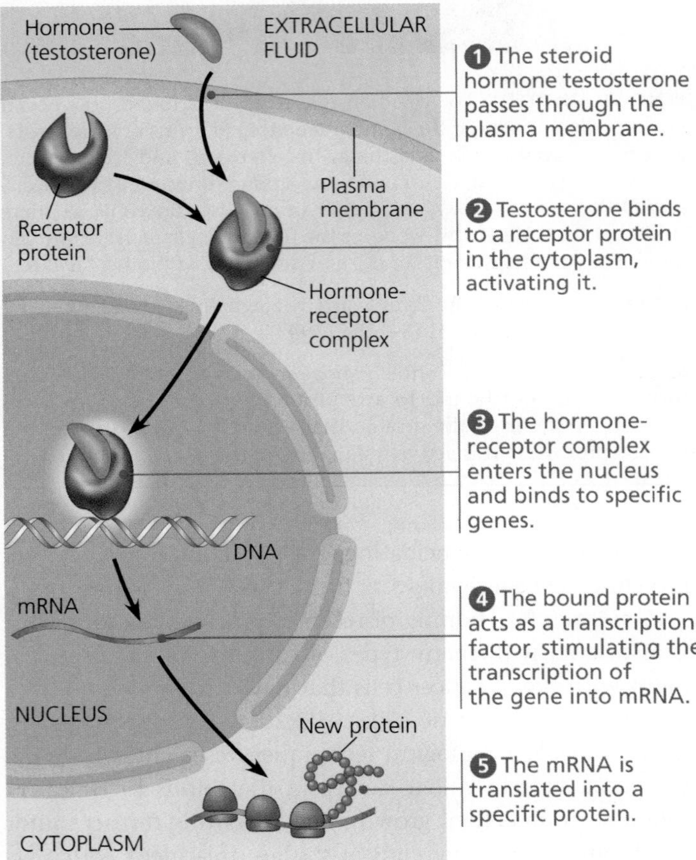

Hormone (testosterone)
EXTRACELLULAR FLUID
1 The steroid hormone testosterone passes through the plasma membrane.
Plasma membrane
Receptor protein
2 Testosterone binds to a receptor protein in the cytoplasm, activating it.
Hormone-receptor complex
3 The hormone-receptor complex enters the nucleus and binds to specific genes.
DNA
mRNA
4 The bound protein acts as a transcription factor, stimulating the transcription of the gene into mRNA.
NUCLEUS
New protein
5 The mRNA is translated into a specific protein.
CYTOPLASM

▲ **Figure 11.9 Steroid hormone interacting with an intracellular receptor.**

? *Why is a cell-surface receptor protein not required for this steroid hormone to enter the cell?*

How does the activated hormone-receptor complex turn on genes? Recall that the genes in a cell's DNA function by being transcribed and processed into messenger RNA (mRNA), which leaves the nucleus and is translated into a specific protein by ribosomes in the cytoplasm (see Figure 5.25). Special proteins called *transcription factors* control which genes are turned on—that is, which genes are transcribed into mRNA—in a particular cell at a particular time. The testosterone receptor, when activated, acts as a transcription factor that turns on specific genes.

By acting as a transcription factor, the testosterone receptor itself carries out the complete transduction of the signal. Most other intracellular receptors function in the same way, although many of them, such as the thyroid hormone receptor, are already in the nucleus before the signaling molecule reaches them. Interestingly, many of these intracellular receptor proteins are structurally similar, suggesting an evolutionary kinship.

CONCEPT CHECK 11.2

1. Nerve growth factor (NGF) is a water-soluble signaling molecule. Would you expect the receptor for NGF to be intracellular or in the plasma membrane? Why?
2. **WHAT IF?** What would the effect be if a cell made defective receptor tyrosine kinase proteins that were unable to dimerize?
3. **MAKE CONNECTIONS** How is ligand binding similar to the process of allosteric regulation of enzymes? See Figure 8.19 on page 158.

For suggested answers, see Appendix A.

CONCEPT 11.3

Transduction: Cascades of molecular interactions relay signals from receptors to target molecules in the cell

When receptors for signaling molecules are plasma membrane proteins, like most of those we have discussed, the transduction stage of cell signaling is usually a multistep pathway. Steps often include activation of proteins by addition or removal of phosphate groups or release of other small molecules or ions that act as messengers. One benefit of multiple steps is the possibility of greatly amplifying a signal. If some of the molecules in a pathway transmit the signal to numerous molecules at the next step in the series, the result can be a large number of activated molecules at the end of the pathway. Moreover, multistep pathways provide more opportunities for coordination and regulation than simpler systems do. This allows fine-tuning of the response, in both unicellular and multicellular organisms, as we'll discuss later in the chapter.

Signal Transduction Pathways

The binding of a specific signaling molecule to a receptor in the plasma membrane triggers the first step in the chain of molecular interactions—the signal transduction pathway—that leads to a particular response within the cell. Like falling dominoes, the signal-activated receptor activates another molecule, which activates yet another molecule, and so on, until the protein that produces the final cellular response is activated. The molecules that relay a signal from receptor to response, which we call relay molecules in this book, are often proteins. The interaction of proteins is a major theme of cell signaling. Indeed, protein interaction is a unifying theme of all regulation at the cellular level.

Keep in mind that the original signaling molecule is not physically passed along a signaling pathway; in most cases, it never even enters the cell. When we say that the signal is relayed along a pathway, we mean that certain information is passed on. At each step, the signal is transduced into a different form, commonly a shape change in a protein. Very often, the shape change is brought about by phosphorylation.

Protein Phosphorylation and Dephosphorylation

Previous chapters introduced the concept of activating a protein by adding one or more phosphate groups to it (see Figure 8.10a). In Figure 11.7, we have already seen how phosphorylation is involved in the activation of receptor tyrosine kinases. In fact, the phosphorylation and dephosphorylation of proteins is a widespread cellular mechanism for regulating protein activity. An enzyme that transfers phosphate groups from ATP to a protein is generally known as a **protein kinase**. Recall that a receptor tyrosine kinase phosphorylates tyrosines on the other receptor tyrosine kinase in a dimer. Most cytoplasmic protein kinases, however, act on proteins different from themselves. Another distinction is that most cytoplasmic protein kinases phosphorylate either of two other amino acids, serine or threonine, rather than tyrosine. Such serine/threonine kinases are widely involved in signaling pathways in animals, plants, and fungi.

Many of the relay molecules in signal transduction pathways are protein kinases, and they often act on other protein kinases in the pathway. **Figure 11.10** depicts a hypothetical

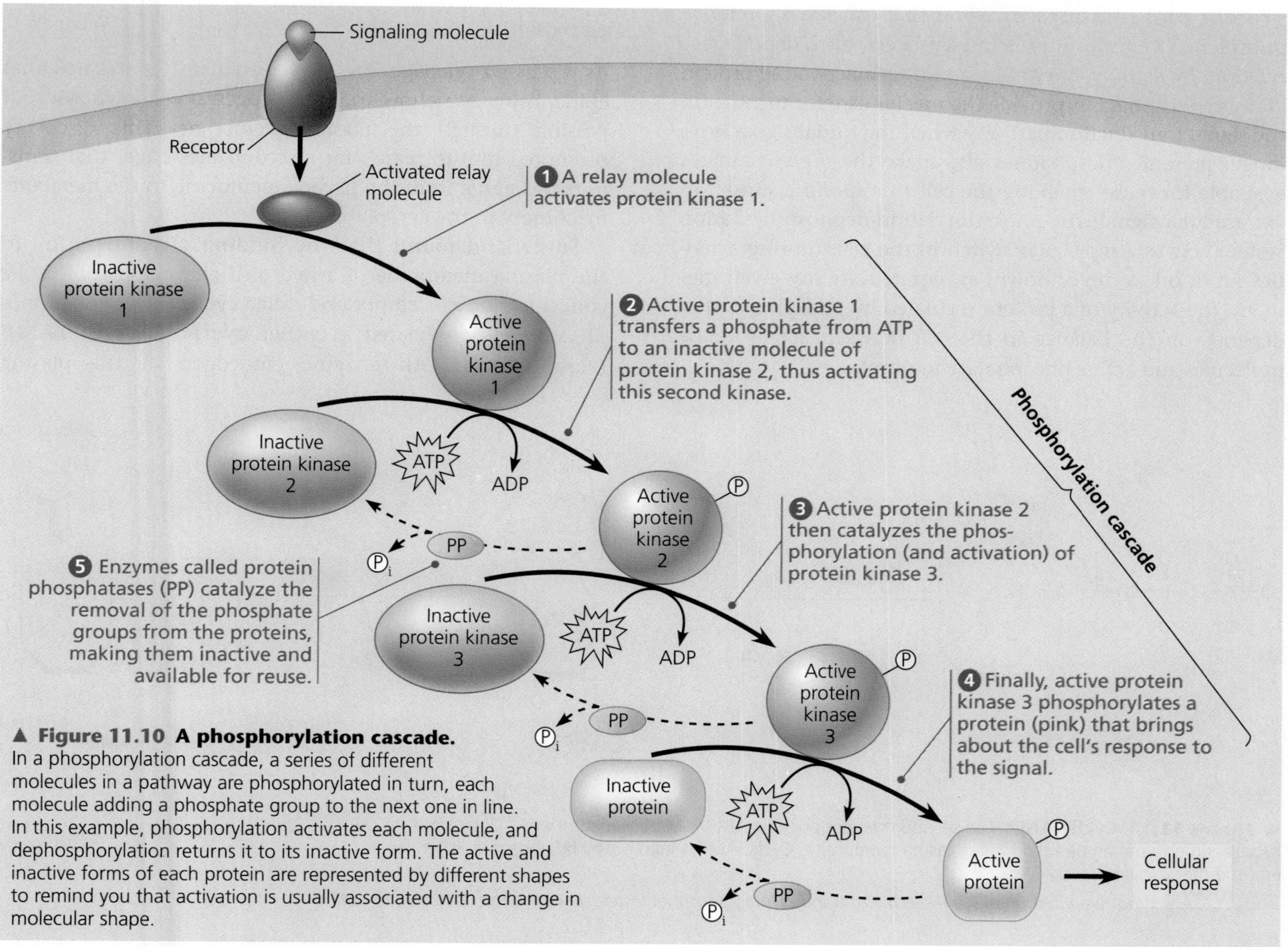

▲ **Figure 11.10 A phosphorylation cascade.** In a phosphorylation cascade, a series of different molecules in a pathway are phosphorylated in turn, each molecule adding a phosphate group to the next one in line. In this example, phosphorylation activates each molecule, and dephosphorylation returns it to its inactive form. The active and inactive forms of each protein are represented by different shapes to remind you that activation is usually associated with a change in molecular shape.

? *Which protein is responsible for activation of protein kinase 3?*

pathway containing three different protein kinases that create a "phosphorylation cascade." The sequence shown is similar to many known pathways, including those triggered in yeast by mating factors and in animal cells by many growth factors. The signal is transmitted by a cascade of protein phosphorylations, each bringing with it a shape change. Each such shape change results from the interaction of the newly added phosphate groups with charged or polar amino acids (see Figure 5.16). The addition of phosphate groups often changes a protein from an inactive form to an active form. In other cases, though, phosphorylation *decreases* the activity of the protein.

The importance of protein kinases can hardly be overstated. About 2% of our own genes are thought to code for protein kinases. A single cell may have hundreds of different kinds, each specific for a different substrate protein. Together, they probably regulate a large proportion of the thousands of proteins in a cell. Among these are most of the proteins that, in turn, regulate cell reproduction. Abnormal activity of such a kinase can cause abnormal cell growth and contribute to the development of cancer.

Equally important in the phosphorylation cascade are the **protein phosphatases**, enzymes that can rapidly remove phosphate groups from proteins, a process called dephosphorylation. By dephosphorylating and thus inactivating protein kinases, phosphatases provide the mechanism for turning off the signal transduction pathway when the initial signal is no longer present. Phosphatases also make the protein kinases available for reuse, enabling the cell to respond again to an extracellular signal. The phosphorylation-dephosphorylation system acts as a molecular switch in the cell, turning activities on or off, or up or down, as required. At any given moment, the activity of a protein regulated by phosphorylation depends on the balance in the cell between active kinase molecules and active phosphatase molecules.

Small Molecules and Ions as Second Messengers

Not all components of signal transduction pathways are proteins. Many signaling pathways also involve small, nonprotein, water-soluble molecules or ions called **second messengers**. (This term is used because the pathway's "first messenger" is considered to be the extracellular signaling molecule—the ligand—that binds to the membrane receptor.) Because second messengers are small and water-soluble, they can readily spread throughout the cell by diffusion. For example, as we'll see shortly, a second messenger called cyclic AMP carries the signal initiated by epinephrine from the plasma membrane of a liver or muscle cell into the cell's interior, where the signal eventually brings about glycogen breakdown. Second messengers participate in pathways that are initiated by both G protein-coupled receptors and receptor tyrosine kinases. The two most widely used second messengers are cyclic AMP and calcium ions, Ca^{2+}. A large variety of relay proteins are sensitive to the cytosolic concentration of one or the other of these second messengers.

Cyclic AMP

As discussed on page 209, Earl Sutherland established that epinephrine somehow causes glycogen breakdown without passing through the plasma membrane. This discovery prompted him to search for a second messenger that transmits the signal from the plasma membrane to the metabolic machinery in the cytoplasm.

Sutherland found that the binding of epinephrine to the plasma membrane of a liver cell elevates the cytosolic concentration of a compound called cyclic adenosine monophosphate, abbreviated as either **cyclic AMP** or **cAMP** (**Figure 11.11**). An enzyme embedded in the plasma

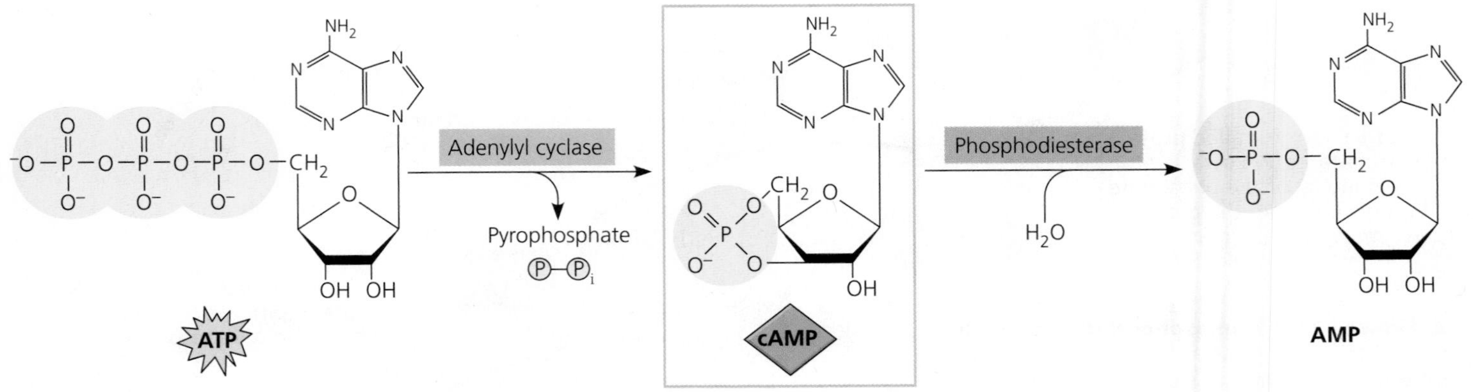

▲ **Figure 11.11 Cyclic AMP.** The second messenger cyclic AMP (cAMP) is made from ATP by adenylyl cyclase, an enzyme embedded in the plasma membrane. Cyclic AMP is inactivated by phosphodiesterase, an enzyme that converts it to AMP.

WHAT IF? *What would happen if a molecule that inactivated phosphodiesterase were introduced into the cell?*

membrane, **adenylyl cyclase**, converts ATP to cAMP in response to an extracellular signal—in this case, provided by epinephrine. But epinephrine doesn't stimulate adenylyl cyclase directly. When epinephrine outside the cell binds to a specific receptor protein, the protein activates adenylyl cyclase, which in turn can catalyze the synthesis of many molecules of cAMP. In this way, the normal cellular concentration of cAMP can be boosted 20-fold in a matter of seconds. The cAMP broadcasts the signal to the cytoplasm. It does not persist for long in the absence of the hormone because another enzyme, called phosphodiesterase, converts cAMP to AMP. Another surge of epinephrine is needed to boost the cytosolic concentration of cAMP again.

Subsequent research has revealed that epinephrine is only one of many hormones and other signaling molecules that trigger the formation of cAMP. It has also brought to light the other components of cAMP pathways, including G proteins, G protein-coupled receptors, and protein kinases **(Figure 11.12)**. The immediate effect of cAMP is usually the activation of a serine/threonine kinase called *protein kinase A*. The activated protein kinase A then phosphorylates various other proteins, depending on the cell type. (The complete pathway for epinephrine's stimulation of glycogen breakdown is shown later, in Figure 11.16.)

Further regulation of cell metabolism is provided by other G protein systems that *inhibit* adenylyl cyclase. In these systems, a different signaling molecule activates a different receptor, which in turn activates an *inhibitory* G protein.

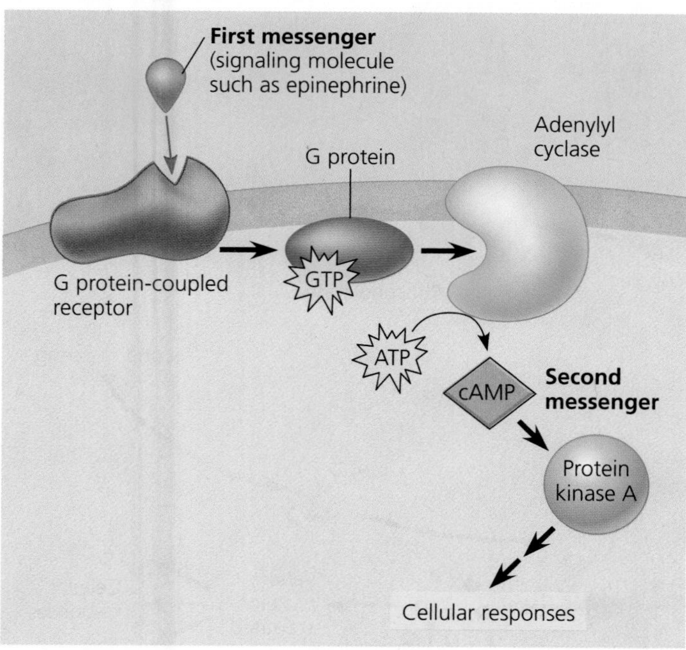

▲ **Figure 11.12 cAMP as a second messenger in a G protein signaling pathway.** The first messenger activates a G protein-coupled receptor, which activates a specific G protein. In turn, the G protein activates adenylyl cyclase, which catalyzes the conversion of ATP to cAMP. The cAMP then acts as a second messenger and activates another protein, usually protein kinase A, leading to cellular responses.

Now that we know about the role of cAMP in G protein signaling pathways, we can explain in molecular detail how certain microbes cause disease. Consider cholera, a disease that is frequently epidemic in places where the water supply is contaminated with human feces. People acquire the cholera bacterium, *Vibrio cholerae*, by drinking contaminated water. The bacteria form a biofilm on the lining of the small intestine and produce a toxin. The cholera toxin is an enzyme that chemically modifies a G protein involved in regulating salt and water secretion. Because the modified G protein is unable to hydrolyze GTP to GDP, it remains stuck in its active form, continuously stimulating adenylyl cyclase to make cAMP. The resulting high concentration of cAMP causes the intestinal cells to secrete large amounts of salts into the intestines, with water following by osmosis. An infected person quickly develops profuse diarrhea and if left untreated can soon die from the loss of water and salts.

Our understanding of signaling pathways involving cyclic AMP or related messengers has allowed us to develop treatments for certain conditions in humans. In one pathway, *cyclic GMP*, or *cGMP*, acts as a signaling molecule whose effects include relaxation of smooth muscle cells in artery walls. A compound that inhibits the hydrolysis of cGMP to GMP, thus prolonging the signal, was originally prescribed for chest pains because it increased blood flow to the heart muscle. Under the trade name Viagra, this compound is now widely used as a treatment for erectile dysfunction in human males. Because Viagra leads to dilation of blood vessels, it also allows increased blood flow to the penis, optimizing physiological conditions for penile erections.

Calcium Ions and Inositol Trisphosphate (IP_3)

Many signaling molecules in animals, including neurotransmitters, growth factors, and some hormones, induce responses in their target cells via signal transduction pathways that increase the cytosolic concentration of calcium ions (Ca^{2+}). Calcium is even more widely used than cAMP as a second messenger. Increasing the cytosolic concentration of Ca^{2+} causes many responses in animal cells, including muscle cell contraction, secretion of certain substances, and cell division. In plant cells, a wide range of hormonal and environmental stimuli can cause brief increases in cytosolic Ca^{2+} concentration, triggering various signaling pathways, such as the pathway for greening in response to light (see Figure 39.4). Cells use Ca^{2+} as a second messenger in both G protein and receptor tyrosine kinase pathways.

Although cells always contain some Ca^{2+}, this ion can function as a second messenger because its concentration in the cytosol is normally much lower than the concentration

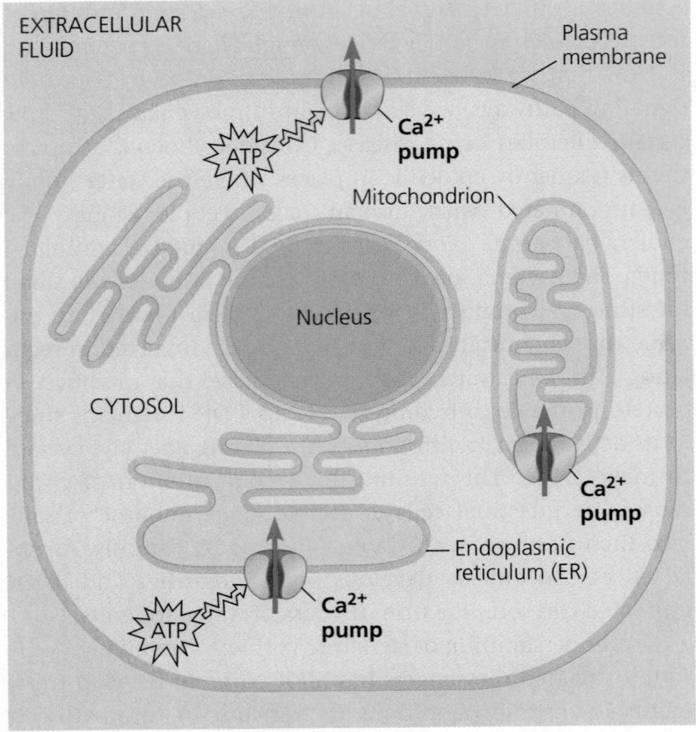
EXTRACELLULAR FLUID
Plasma membrane
ATP
Ca^{2+} pump
Mitochondrion
Nucleus
CYTOSOL
Ca^{2+} pump
Endoplasmic reticulum (ER)
ATP
Ca^{2+} pump
Key
High [Ca^{2+}]
Low [Ca^{2+}]

▲ **Figure 11.13 The maintenance of calcium ion concentrations in an animal cell.** The Ca^{2+} concentration in the cytosol is usually much lower (beige) than in the extracellular fluid and ER (blue). Protein pumps in the plasma membrane and the ER membrane, driven by ATP, move Ca^{2+} from the cytosol into the extracellular fluid and into the lumen of the ER. Mitochondrial pumps, driven by chemiosmosis (see Chapter 9), move Ca^{2+} into mitochondria when the calcium level in the cytosol rises significantly.

outside the cell **(Figure 11.13)**. In fact, the level of Ca^{2+} in the blood and extracellular fluid of an animal is often more than 10,000 times higher than that in the cytosol. Calcium ions are actively transported out of the cell and are actively imported from the cytosol into the endoplasmic reticulum (and, under some conditions, into mitochondria and chloroplasts) by various protein pumps. As a result, the calcium concentration in the ER is usually much higher than that in the cytosol. Because the cytosolic calcium level is low, a small change in absolute numbers of ions represents a relatively large percentage change in calcium concentration.

In response to a signal relayed by a signal transduction pathway, the cytosolic calcium level may rise, usually by a mechanism that releases Ca^{2+} from the cell's ER. The pathways leading to calcium release involve still other second messengers, **inositol trisphosphate (IP_3)** and **diacylglycerol (DAG)**. These two messengers are produced by cleavage of a certain kind of phospholipid in the plasma membrane. **Figure 11.14** shows how this occurs and how IP_3 stimulates the release of calcium from the ER. Because IP_3 acts before calcium in these pathways, calcium could be considered a *"third* messenger." However, scientists use the term *second messenger* for all small, nonprotein components of signal transduction pathways.

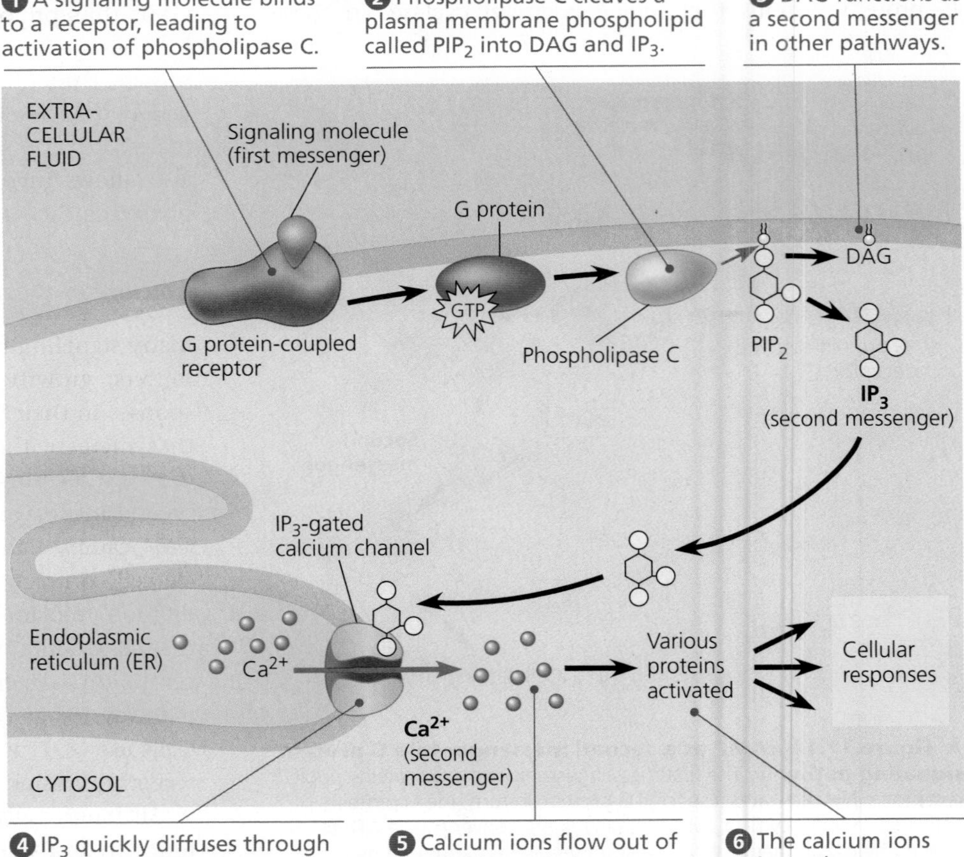
1 A signaling molecule binds to a receptor, leading to activation of phospholipase C.
2 Phospholipase C cleaves a plasma membrane phospholipid called PIP_2 into DAG and IP_3.
3 DAG functions as a second messenger in other pathways.
EXTRA-CELLULAR FLUID
Signaling molecule (first messenger)
G protein
DAG
GTP
G protein-coupled receptor
Phospholipase C
PIP_2
IP_3 (second messenger)
IP_3-gated calcium channel
Endoplasmic reticulum (ER)
Ca^{2+}
Various proteins activated
Cellular responses
Ca^{2+} (second messenger)
CYTOSOL
4 IP_3 quickly diffuses through the cytosol and binds to an IP_3-gated calcium channel in the ER membrane, causing it to open.
5 Calcium ions flow out of the ER (down their concentration gradient), raising the Ca^{2+} level in the cytosol.
6 The calcium ions activate the next protein in one or more signaling pathways.

▶ **Figure 11.14 Calcium and IP_3 in signaling pathways.** Calcium ions (Ca^{2+}) and inositol trisphosphate (IP_3) function as second messengers in many signal transduction pathways. In this figure, the process is initiated by the binding of a signaling molecule to a G protein-coupled receptor. A receptor tyrosine kinase could also initiate this pathway by activating phospholipase C.

CONCEPT CHECK 11.3

1. What is a protein kinase, and what is its role in a signal transduction pathway?
2. When a signal transduction pathway involves a phosphorylation cascade, how does the cell's response get turned off?
3. What is the actual "signal" that is being transduced in any signal transduction pathway, such as those shown in Figures 11.6 and 11.10? In what way is this information being passed from the exterior to the interior of the cell?
4. **WHAT IF?** Upon activation of phospholipase C by the binding of a ligand to a receptor, what effect does the IP_3-gated calcium channel have on Ca^{2+} concentration in the cytosol?

For suggested answers, see Appendix A.

CONCEPT 11.4

Response: Cell signaling leads to regulation of transcription or cytoplasmic activities

We now take a closer look at the cell's subsequent response to an extracellular signal—what some researchers call the "output response." What is the nature of the final step in a signaling pathway?

Nuclear and Cytoplasmic Responses

Ultimately, a signal transduction pathway leads to the regulation of one or more cellular activities. The response at the end of the pathway may occur in the nucleus of the cell or in the cytoplasm.

Many signaling pathways ultimately regulate protein synthesis, usually by turning specific genes on or off in the nucleus. Like an activated steroid receptor (see Figure 11.9), the final activated molecule in a signaling pathway may function as a transcription factor. **Figure 11.15** shows an example in which a signaling pathway activates a transcription factor that turns a gene on: The response to the growth factor signal is transcription, the synthesis of mRNA, which will be translated in the cytoplasm into a specific protein. In other cases, the transcription factor might regulate a gene by turning it off. Often a transcription factor regulates several different genes.

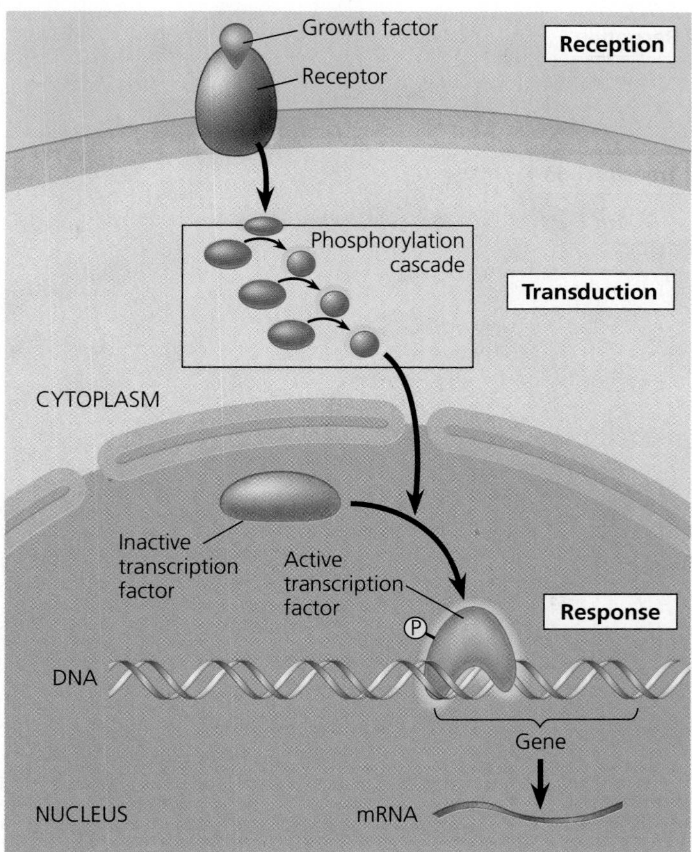

▲ **Figure 11.15 Nuclear responses to a signal: the activation of a specific gene by a growth factor.** This diagram is a simplified representation of a typical signaling pathway that leads to the regulation of gene activity in the cell nucleus. The initial signaling molecule, a local regulator called a growth factor, triggers a phosphorylation cascade, as in Figure 11.10. (The ATP molecules and phosphate groups are not shown.) Once phosphorylated, the last kinase in the sequence enters the nucleus and there activates a gene-regulating protein, a transcription factor. This protein stimulates transcription of a specific gene (or genes). The resulting mRNA then directs the synthesis of a particular protein in the cytoplasm.

Sometimes a signaling pathway may regulate the *activity* of proteins rather than their *synthesis*, directly affecting proteins that function outside the nucleus. For example, a signal may cause the opening or closing of an ion channel in the plasma membrane or a change in cell metabolism. As we have seen, the response of liver cells to the hormone epinephrine helps regulate cellular energy metabolism by affecting the activity of an enzyme. The final step in the signaling pathway that begins with epinephrine binding activates the enzyme that catalyzes the breakdown of glycogen. **Figure 11.16**, on the next page, shows the complete pathway leading to the release of glucose 1-phosphate molecules from glycogen. Notice that as each molecule is activated, the response is amplified, a subject we'll return to shortly.

In addition to controlling enzymes, signaling events may regulate other cellular attributes, even activities of the cell as a whole. An example of the latter can be found in the processes leading to the mating of yeast cells (see Figure 11.2). Yeast cells are not motile; their mating process depends on the growth of localized projections of one cell toward a cell of the

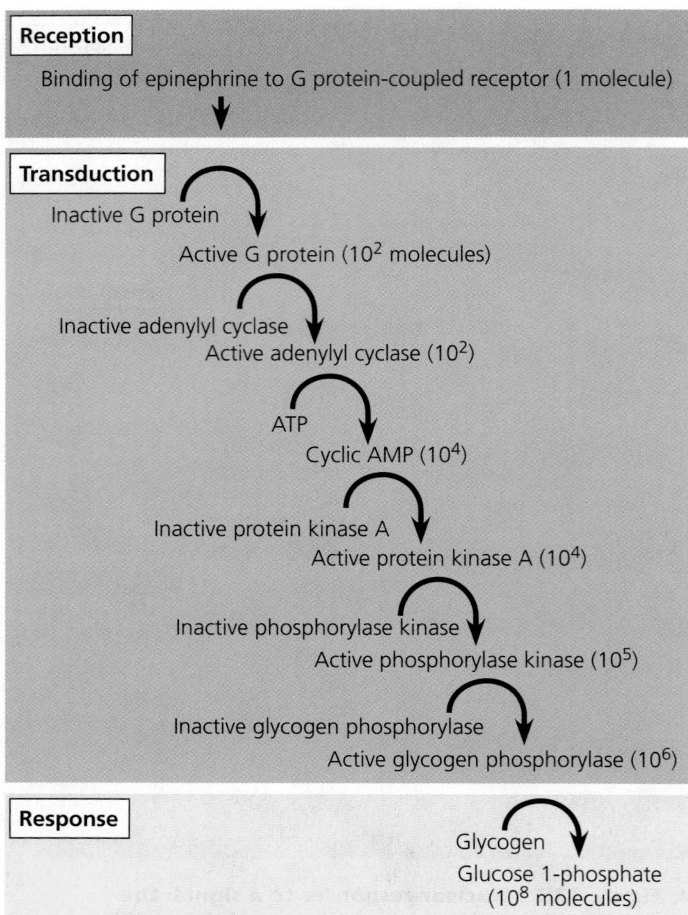

▲ **Figure 11.16 Cytoplasmic response to a signal: the stimulation of glycogen breakdown by epinephrine.** In this signaling system, the hormone epinephrine acts through a G protein-coupled receptor to activate a succession of relay molecules, including cAMP and two protein kinases (see also Figure 11.12). The final protein activated is the enzyme glycogen phosphorylase, which uses inorganic phosphate to release glucose monomers from glycogen in the form of glucose 1-phosphate molecules. This pathway amplifies the hormonal signal: One receptor protein can activate about 100 molecules of G protein, and each enzyme in the pathway, once activated, can act on many molecules of its substrate, the next molecule in the cascade. The number of activated molecules given for each step is approximate.

opposite mating type. As shown in **Figure 11.17**, binding of the mating factor causes this directional growth. When the mating factor binds, it activates signaling pathway kinases that affect the growth and orientation of cytoskeletal microfilaments. Because activation of signaling kinases is coupled in this way to cytoskeletal dynamics, cell projections emerge from regions of the plasma membrane exposed to the highest concentration of the mating factor. As a result, these projections are oriented toward the cell of the opposite mating type, which is the source of the signaling molecule.

The signal receptors, relay molecules, and second messengers introduced so far in this chapter participate in a variety of pathways, leading to both nuclear and cytoplasmic responses. Some of these pathways lead to cell division. The molecular messengers that initiate cell division pathways include growth factors and certain plant and animal hormones. Malfunctioning of growth factor pathways like the one in Figure 11.15 can contribute to the development of cancer, as we will see in Chapter 18.

Fine-Tuning of the Response

Regardless of whether the response occurs in the nucleus or in the cytoplasm, it is fine-tuned at multiple points rather than simply being turned "on" or "off." Here we'll consider four aspects of fine-tuning. First, as mentioned earlier, a signaling pathway with numerous steps between the initial signaling event at the cell surface and the cell's response results in amplification of the signal and thus the response. Second, such a multistep pathway has many different points at which the cell's response can be regulated, contributing to the specificity of the response and allowing coordination with other signaling pathways. Third, the overall efficiency of the response is enhanced by the presence of proteins known as scaffolding proteins. Finally, a crucial point in fine-tuning the response is the termination of the signal.

Signal Amplification

Elaborate enzyme cascades amplify the cell's response to a signal. At each catalytic step in the cascade, the number of activated products is much greater than in the preceding step. For example, in the epinephrine-triggered pathway in Figure 11.16, each adenylyl cyclase molecule catalyzes the formation of many cAMP molecules, each molecule of protein kinase A phosphorylates many molecules of the next kinase in the pathway, and so on. The amplification effect stems from the fact that these proteins persist in the active form long enough to process numerous molecules of substrate before they become inactive again. As a result of the signal's amplification, a small number of epinephrine molecules binding to receptors on the surface of a liver cell or muscle cell can lead to the release of hundreds of millions of glucose molecules from glycogen.

The Specificity of Cell Signaling and Coordination of the Response

Consider two different cells in your body—a liver cell and a heart muscle cell, for example. Both are in contact with your bloodstream and are therefore constantly exposed to many different hormone molecules, as well as to local regulators secreted by nearby cells. Yet the liver cell responds to some signals but ignores others, and the same is true for the heart cell. And some kinds of signals trigger responses in both cells—but different responses. For instance, epinephrine stimulates the liver cell to break down glycogen, but the main response of the heart cell to epinephrine is contraction, leading to a more rapid heartbeat. How do we account for this difference?

▼ **Figure 11.17** INQUIRY

How do signals induce directional cell growth during mating in yeast?

EXPERIMENT When a yeast cell binds mating factor molecules from a cell of the opposite mating type, a signaling pathway causes it to grow a projection toward the potential mate. The cell with the projection is called a "shmoo" because it resembles a 1950s cartoon character by that name. Dina Matheos and colleagues in Mark Rose's lab at Princeton University sought to determine how mating factor signaling is linked to this asymmetrical growth. Previous work had shown that activation of Fus3, one of the kinases in the signaling cascade, caused it to move to the membrane near where the factor bound. Preliminary experiments by these researchers identified formin, a protein that directs the construction of microfilaments, as a phosphorylation target of Fus3 kinase. To examine the role of Fus3 and formin in shmoo formation, the researchers generated two mutant yeast strains: one that no longer had the kinase (this strain is called ΔFus3) and one that lacked the formin (Δformin). To observe the effects of these mutations on cell growth induced by the mating factor, the cell walls of each strain were first stained with a green fluorescent dye. These green-stained cells were then exposed to mating factor and stained with a red fluorescent dye that labeled new cell wall growth. Images taken of the cells after the staining procedure were then compared with a similarly treated strain that expressed Fus3 and formin (the wild type).

RESULTS The cells of the wild-type strain showed shmoo projections, whose walls were stained red, while the rest of their cell walls were green, indicating asymmetrical growth. Cells of both the ΔFus3 and Δformin strains showed no shmoo formation, and their cell walls were stained almost uniformly yellow. This color resulted from merged green and red stains, indicating symmetrical growth, characteristic of cells not exposed to mating factor.

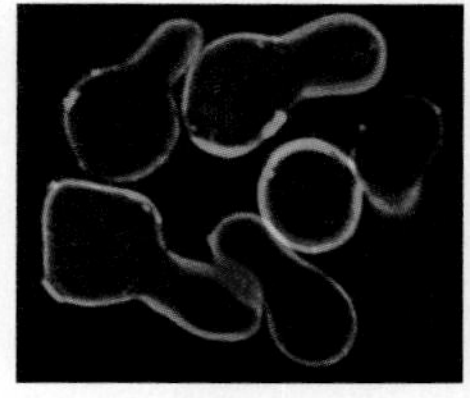

Wild type (with shmoos)

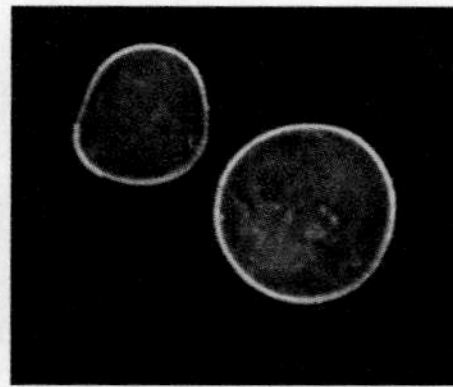

ΔFus3

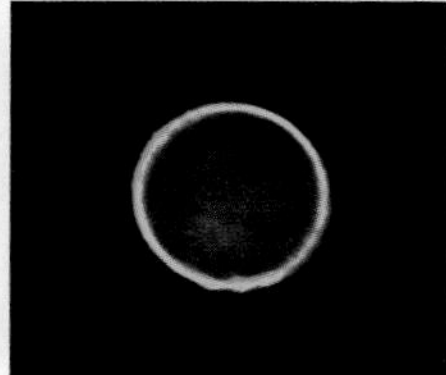

Δformin

CONCLUSION The similar defect (lack of ability to form shmoos) in strains lacking either Fus3 or formin suggests that both proteins are required for shmoo formation. These results led the investigators to propose the model shown here for the induction of asymmetrical growth in the receiving cell directed toward the cell of the opposite mating type.

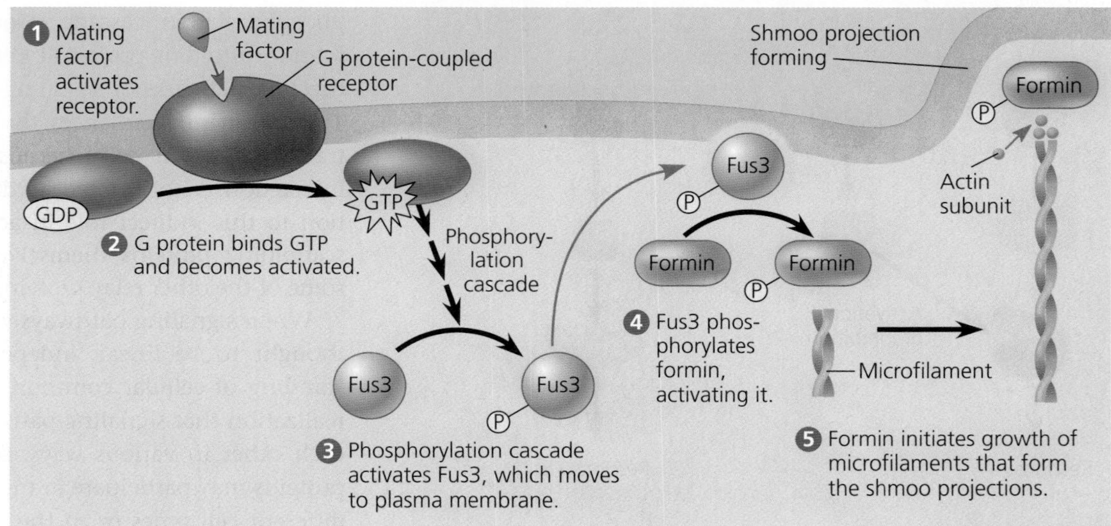

1 Mating factor activates receptor.
Mating factor
G protein-coupled receptor
GDP
GTP
2 G protein binds GTP and becomes activated.
Phosphorylation cascade
Fus3
Fus3
P
3 Phosphorylation cascade activates Fus3, which moves to plasma membrane.
Shmoo projection forming
Fus3
P
Formin
Formin
P
4 Fus3 phosphorylates formin, activating it.
Formin
P
Actin subunit
Microfilament
5 Formin initiates growth of microfilaments that form the shmoo projections.

SOURCE D. Matheos et al., Pheromone-induced polarization is dependent on the Fus3p MAPK acting through the formin Bni1p, *Journal of Cell Biology* 165:99–109 (2004).

WHAT IF? Based on these results and the proposed model from this work, what would happen to a cell if its Fus3 kinase were not able to associate with the membrane upon activation?

The explanation for the specificity exhibited in cellular responses to signals is the same as the basic explanation for virtually all differences between cells: Because different kinds of cells turn on different sets of genes, *different kinds of cells have different collections of proteins* (**Figure 11.18**, on the next page). The response of a particular cell to a signal depends on its particular collection of signal receptor proteins, relay proteins, and proteins needed to carry out the response. A liver cell, for example, is poised to respond appropriately to epinephrine by having the proteins listed in Figure 11.16 as well as those needed to manufacture glycogen.

Thus, two cells that respond differently to the same signal differ in one or more of the proteins that handle and respond to the signal. Notice in Figure 11.18 that different pathways may have some molecules in common. For example, cells A, B, and C all use the same receptor protein for the red signaling molecule; differences in other proteins account for their differing responses. In cell D, a different receptor protein is used for the same signaling molecule, leading to yet another response. In cell B, a pathway that is triggered by a single kind of signal diverges to produce two responses; such branched pathways often involve receptor tyrosine kinases (which can activate multiple relay proteins) or second messengers (which can regulate numerous proteins). In cell C, two pathways triggered by separate signals converge to modulate a single response. Branching of pathways and "cross-talk" (interaction)

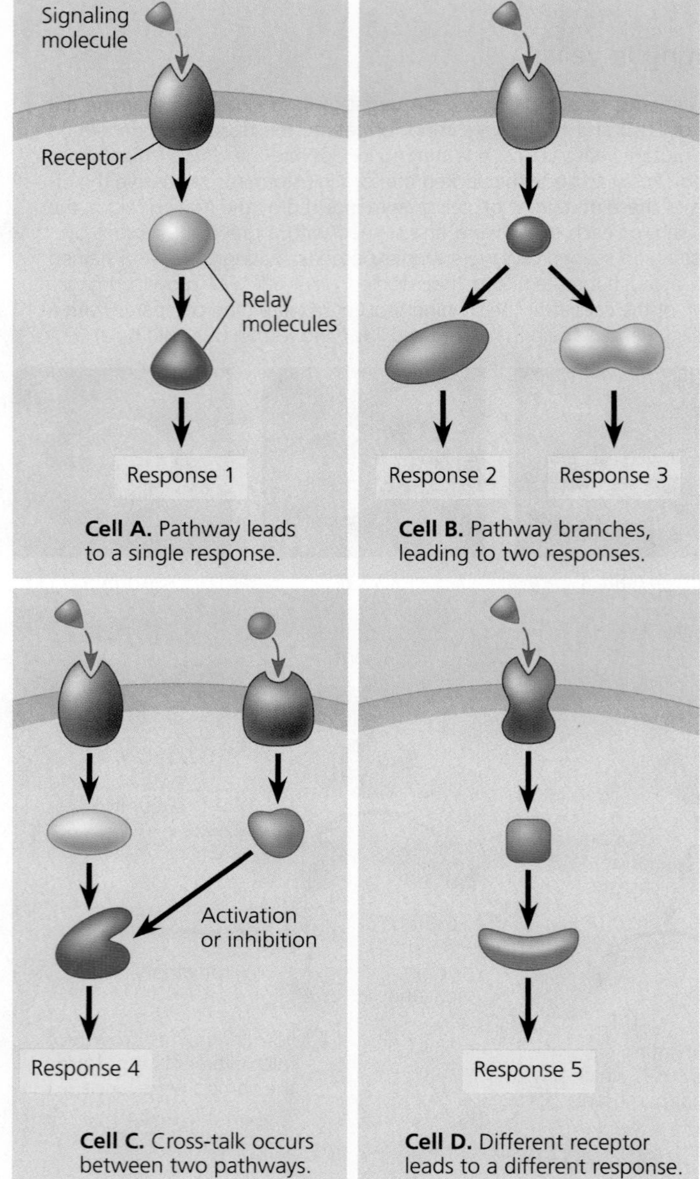

▲ **Figure 11.18 The specificity of cell signaling.** The particular proteins a cell possesses determine what signaling molecules it responds to and the nature of the response. The four cells in these diagrams respond to the same signaling molecule (red) in different ways because each has a different set of proteins (purple and teal). Note, however, that the same kinds of molecules can participate in more than one pathway.

MAKE CONNECTIONS *Study the signaling pathway shown in Figure 11.14 (p. 218), and explain how the situation pictured for cell B above could apply to that pathway.*

Signaling Efficiency: Scaffolding Proteins and Signaling Complexes

The illustrations of signaling pathways in Figure 11.18 (as well as diagrams of other pathways in this chapter) are greatly simplified. The diagrams show only a few relay molecules and, for clarity's sake, display these molecules spread out in the cytosol. If this were true in the cell, signaling pathways would operate very inefficiently because most relay molecules are proteins, and proteins are too large to diffuse quickly through the viscous cytosol. How does a particular protein kinase, for instance, find its substrate?

In many cases, the efficiency of signal transduction is apparently increased by the presence of **scaffolding proteins**, large relay proteins to which several other relay proteins are simultaneously attached. For example, one scaffolding protein isolated from mouse brain cells holds three protein kinases and carries these kinases with it when it binds to an appropriately activated membrane receptor; it thus facilitates a specific phosphorylation cascade (**Figure 11.19**). Researchers have found scaffolding proteins in brain cells that *permanently* hold together networks of signaling pathway proteins at synapses. This hardwiring enhances the speed and accuracy of signal transfer between cells, because the rate of protein-protein interaction is not limited by diffusion. Furthermore, in addition to this indirect role in activation of relay proteins, the scaffolding proteins themselves may more directly activate some of the other relay proteins.

When signaling pathways were first discovered, they were thought to be linear, independent pathways. Our understanding of cellular communication has benefited from the realization that signaling-pathway components interact with each other in various ways. As seen in Figure 11.18, some proteins may participate in more than one pathway, either in different cell types or in the same cell at different times or under different conditions. These observations underscore the importance of permanent or transient protein complexes in the functioning of a cell.

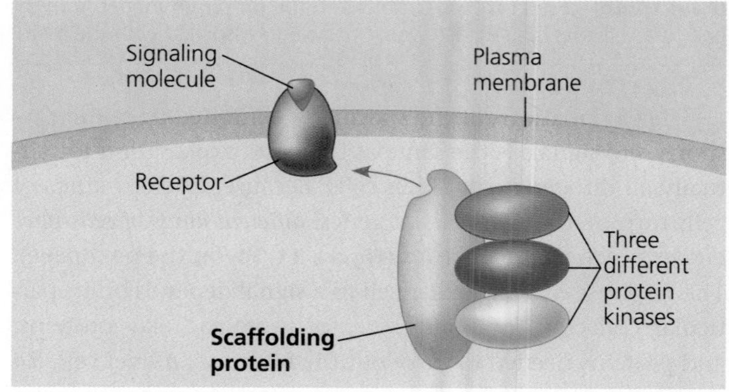

▲ **Figure 11.19 A scaffolding protein.** The scaffolding protein shown here (pink) simultaneously binds to a specific activated membrane receptor and three different protein kinases. This physical arrangement facilitates signal transduction by these molecules and may directly activate relay molecules in some cases.

between pathways are important in regulating and coordinating a cell's responses to information coming in from different sources in the body. (You'll learn more about this coordination in Concept 11.5.) Moreover, the use of some of the same proteins in more than one pathway allows the cell to economize on the number of different proteins it must make.

the importance of transient—or, in some cases, permanent—protein complexes in the process of cell signaling.

The importance of the relay proteins that serve as points of branching or intersection in signaling pathways is highlighted by the problems arising when these proteins are defective or missing. For instance, in an inherited disorder called Wiskott-Aldrich syndrome (WAS), the absence of a single relay protein leads to such diverse effects as abnormal bleeding, eczema, and a predisposition to infections and leukemia. These symptoms are thought to arise primarily from the absence of the protein in cells of the immune system. By studying normal cells, scientists found that the WAS protein is located just beneath the cell surface. The protein interacts both with microfilaments of the cytoskeleton and with several different components of signaling pathways that relay information from the cell surface, including pathways regulating immune cell proliferation. This multifunctional relay protein is thus both a branch point and an important intersection point in a complex signal transduction network that controls immune cell behavior. When the WAS protein is absent, the cytoskeleton is not properly organized and signaling pathways are disrupted, leading to the WAS symptoms.

Termination of the Signal

To keep Figure 11.18 simple, we did not indicate the *inactivation* mechanisms that are an essential aspect of cell signaling. For a cell of a multicellular organism to remain capable of responding to incoming signals, each molecular change in its signaling pathways must last only a short time. As we saw in the cholera example, if a signaling pathway component becomes locked into one state, whether active or inactive, consequences for the organism can be dire.

The ability of a cell to receive new signals depends on reversibility of the changes produced by prior signals. The binding of signaling molecules to receptors is reversible. As the external concentration of signaling molecules falls, fewer receptors are bound at any given moment, and the unbound receptors revert to their inactive form. The cellular response occurs only when the concentration of receptors with bound signaling molecules is above a certain threshold. When the number of active receptors falls below that threshold, the cellular response ceases. Then, by a variety of means, the relay molecules return to their inactive forms: The GTPase activity intrinsic to a G protein hydrolyzes its bound GTP; the enzyme phosphodiesterase converts cAMP to AMP; protein phosphatases inactivate phosphorylated kinases and other proteins; and so forth. As a result, the cell is soon ready to respond to a fresh signal.

In this section, we explored the complexity of signaling initiation and termination in a single pathway, and we saw the potential for pathways to intersect with each other. In the next section, we'll consider one especially important network of interacting pathways in the cell.

CONCEPT CHECK 11.4

1. How can a target cell's response to a single hormone molecule result in a response that affects a million other molecules?
2. **WHAT IF?** If two cells have different scaffolding proteins, explain how they might behave differently in response to the same signaling molecule.
3. **MAKE CONNECTIONS** Review the discussion of protein phosphatases on page 216, and see Figure 11.10 on page 215. Some human diseases are associated with malfunctioning protein phosphatases. How would such proteins affect signaling pathways?

For suggested answers, see Appendix A.

CONCEPT 11.5

Apoptosis integrates multiple cell-signaling pathways

To be or not to be? One of the most elaborate networks of signaling pathways in the cell seems to ask and answer this question posed by Hamlet. Cells that are infected, damaged, or have reached the end of their functional life span often undergo "programmed cell death." The best-understood type of this controlled cell suicide is **apoptosis** (from the Greek, meaning "falling off," and used in a classic Greek poem to refer to leaves falling from a tree). During this process, cellular agents chop up the DNA and fragment the organelles and other cytoplasmic components. The cell shrinks and becomes lobed (a change called "blebbing"; **Figure 11.20**), and the cell's parts are packaged up in vesicles that are engulfed and digested by specialized scavenger cells, leaving no trace. Apoptosis protects neighboring cells from damage that they would otherwise suffer if a dying cell merely leaked out all its contents, including its many digestive enzymes.

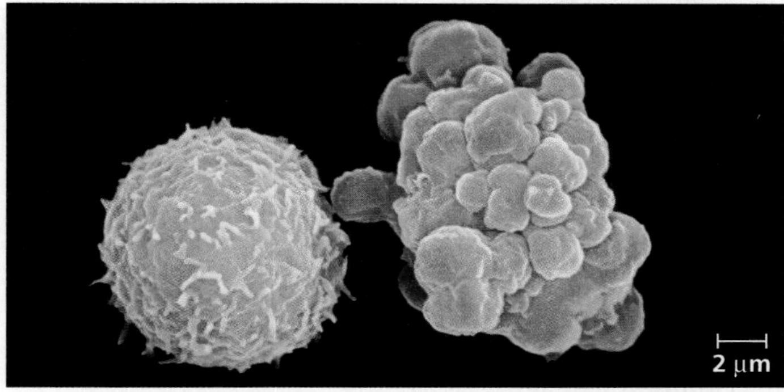

▲ **Figure 11.20 Apoptosis of a human white blood cell.** We can compare a normal white blood cell (left) with a white blood cell undergoing apoptosis (right). The apoptotic cell is shrinking and forming lobes ("blebs"), which eventually are shed as membrane-bounded cell fragments (colorized SEMs).

Apoptosis in the Soil Worm *Caenorhabditis elegans*

Embryonic development is a period during which apoptosis is widespread and plays a crucial role. The molecular mechanisms underlying apoptosis were worked out in detail by researchers studying embryonic development of a small soil worm, a nematode called *Caenorhabditis elegans*. Because the adult worm has only about a thousand cells, the researchers were able to work out the entire ancestry of each cell. The timely suicide of cells occurs exactly 131 times during normal development of *C. elegans*, at precisely the same points in the cell lineage of each worm. In worms and other species, apoptosis is triggered by signals that activate a cascade of "suicide" proteins in the cells destined to die.

Genetic research on *C. elegans* has revealed two key apoptosis genes, called *ced-3* and *ced-4* (*ced* stands for "cell death"), which encode proteins essential for apoptosis. The proteins are called Ced-3 and Ced-4, respectively. These and most other proteins involved in apoptosis are continually present in cells, but in inactive form; thus, regulation occurs at the level of protein activity rather than through gene activity and protein synthesis. In *C. elegans*, a protein in the outer mitochondrial membrane, called Ced-9 (the product of the *ced-9* gene), serves as a master regulator of apoptosis, acting as a brake in the absence of a signal promoting apoptosis (**Figure 11.21**). When a death signal is received by the cell, it overrides the brake, and the apoptotic pathway activates proteases and nucleases, enzymes that cut up the proteins and DNA of the cell. The main proteases of apoptosis are called *caspases*; in the nematode, the chief caspase is Ced-3.

Apoptotic Pathways and the Signals That Trigger Them

In humans and other mammals, several different pathways, involving about 15 different caspases, can carry out apoptosis. The pathway that is used depends on the type of cell and on the particular signal that initiates apoptosis. One major pathway involves certain mitochondrial proteins that are triggered to form molecular pores in the mitochondrial outer membrane, causing it to leak and release other proteins that promote apoptosis. Surprisingly, these latter include cytochrome *c*, which functions in mitochondrial electron transport in healthy cells (see Figure 9.15) but acts as a cell death factor when released from mitochondria. The process of mitochondrial apoptosis in mammals uses proteins similar to the nematode proteins Ced-3, Ced-4, and Ced-9. These can be thought of as relay proteins capable of transducing the apoptotic signal.

At key gateways into the apoptotic program, relay proteins integrate signals from several different sources and can send a cell down an apoptotic pathway. Often, the signal originates outside the cell, like the death-signaling molecule depicted in Figure 11.21b, which presumably was released by a neighboring cell. When a death-signaling ligand occupies a cell-surface receptor, this binding leads to activation of caspases and other enzymes that carry out apoptosis, without involving the mitochondrial pathway. This process of signal reception, transduction, and response is similar to what we discussed earlier in this chapter. In a twist on the classic scenario, two other types of alarm signals that can lead to apoptosis originate from *inside* the cell rather than from a cell-surface receptor.

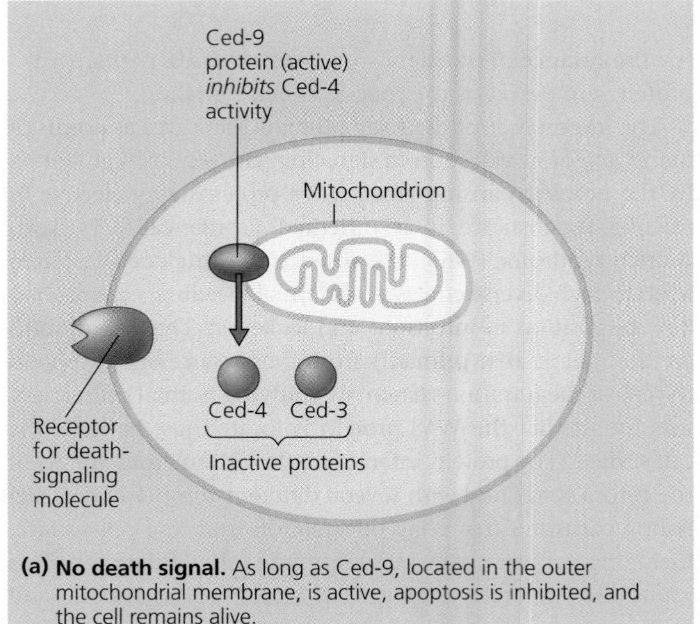

(a) No death signal. As long as Ced-9, located in the outer mitochondrial membrane, is active, apoptosis is inhibited, and the cell remains alive.

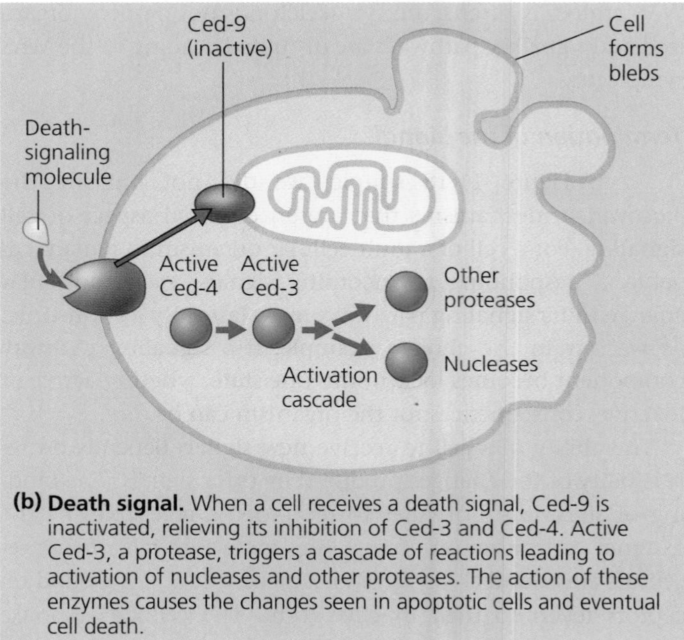

(b) Death signal. When a cell receives a death signal, Ced-9 is inactivated, relieving its inhibition of Ced-3 and Ced-4. Active Ced-3, a protease, triggers a cascade of reactions leading to activation of nucleases and other proteases. The action of these enzymes causes the changes seen in apoptotic cells and eventual cell death.

▲ **Figure 11.21 Molecular basis of apoptosis in *C. elegans*.** Three proteins, Ced-3, Ced-4, and Ced-9, are critical to apoptosis and its regulation in the nematode. Apoptosis is more complicated in mammals but involves proteins similar to those in the nematode.

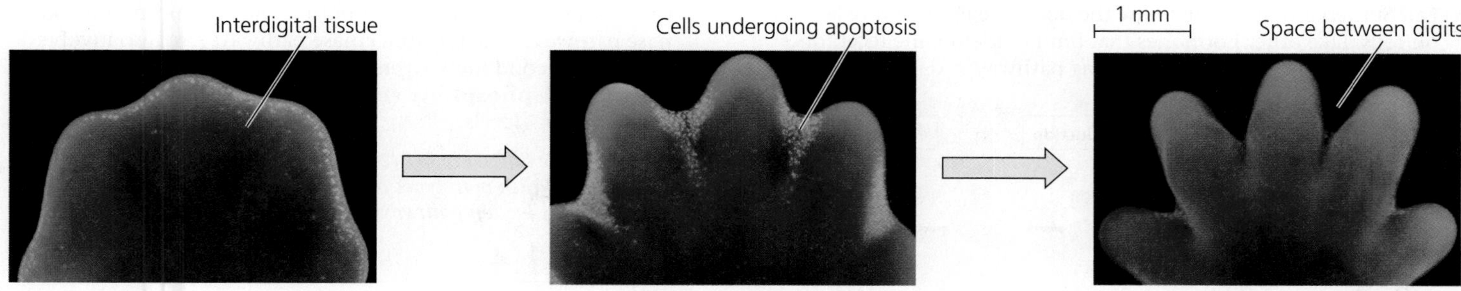

▲ **Figure 11.22 Effect of apoptosis during paw development in the mouse.** In mice, humans, other mammals, and land birds, the embryonic region that develops into feet or hands initially has a solid, platelike structure. Apoptosis eliminates the cells in the interdigital regions, thus forming the digits. The embryonic mouse paws shown in these fluorescence light micrographs are stained so that cells undergoing apoptosis appear a bright yellowish green. Apoptosis of cells begins at the margin of each interdigital region (left), peaks as the tissue in these regions is reduced (middle), and is no longer visible when the interdigital tissue has been eliminated (right).

One signal comes from the nucleus, generated when the DNA has suffered irreparable damage, and a second comes from the endoplasmic reticulum when excessive protein misfolding occurs. Mammalian cells make life-or-death "decisions" by somehow integrating the death signals and life signals they receive from these external and internal sources.

A built-in cell suicide mechanism is essential to development and maintenance in all animals. The similarities between apoptosis genes in nematodes and mammals, as well as the observation that apoptosis occurs in multicellular fungi and even in single-celled yeasts, indicate that the basic mechanism evolved early in the evolution of eukaryotes. In vertebrates, apoptosis is essential for normal development of the nervous system, for normal operation of the immune system, and for normal morphogenesis of hands and feet in humans and paws in other mammals (**Figure 11.22**). The level of apoptosis between the developing digits is lower in the webbed feet of ducks and other water birds than in the nonwebbed feet of land birds, such as chickens. In the case of humans, the failure of appropriate apoptosis can result in webbed fingers and toes.

Significant evidence points to the involvement of apoptosis in certain degenerative diseases of the nervous system, such as Parkinson's disease and Alzheimer's disease. Also, cancer can result from a failure of cell suicide; some cases of human melanoma, for example, have been linked to faulty forms of the human version of the *C. elegans* Ced-4 protein. It is not surprising, therefore, that the signaling pathways feeding into apoptosis are quite elaborate. After all, the life-or-death question is the most fundamental one imaginable for a cell.

This chapter has introduced you to many of the general mechanisms of cell communication, such as ligand binding, protein-protein interactions and shape changes, cascades of interactions, and protein phosphorylation. As you continue through the text, you will encounter numerous examples of cell signaling.

CONCEPT CHECK 11.5

1. Give an example of apoptosis during embryonic development, and explain its function in the developing embryo.
2. **WHAT IF?** What types of protein defects could result in apoptosis occurring when it should not? What types could result in apoptosis not occurring when it should?

For suggested answers, see Appendix A.

11 CHAPTER REVIEW

SUMMARY OF KEY CONCEPTS

CONCEPT 11.1

External signals are converted to responses within the cell (pp. 206–210)

- **Signal transduction pathways** are crucial for many processes, including the mating of yeast cells. In fact, signaling in microbes has much in common with processes in multicellular organisms, suggesting an early evolutionary origin of signaling mechanisms. Bacterial cells can sense the local density of bacterial cells (quorum sensing) by binding molecules secreted by other cells. In some cases, such signals lead to aggregation of these cells into biofilms.
- In local signaling, animal cells may communicate by direct contact or by secreting **local regulators**, such as growth factors or neurotransmitters. For long-distance signaling, both animals and plants use **hormones**; animals also pass signals electrically.

- Earl Sutherland discovered how the hormone epinephrine acts on cells. Like other hormones that bind to membrane receptors, it triggers a three-stage cell-signaling pathway:

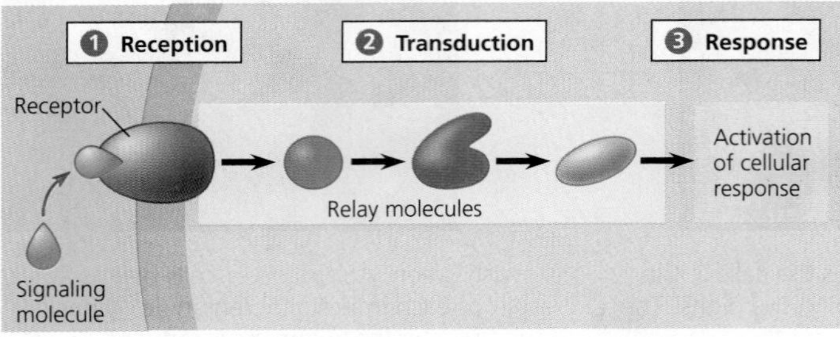

? *What determines* whether *a cell responds to a hormone such as epinephrine? What determines* how *a cell responds to such a hormone?*

CONCEPT 11.2

Reception: A signaling molecule binds to a receptor protein, causing it to change shape (pp. 210–214)

- The binding between signaling molecule (**ligand**) and receptor is highly specific. A specific shape change in a receptor is often the initial transduction of the signal.
- There are three major types of cell-surface transmembrane receptors: (1) **G protein-coupled receptors (GPCRs)** work with the help of cytoplasmic **G proteins**. Ligand binding activates the receptor, which then activates a specific G protein, which activates yet another protein, thus propagating the signal along a signal transduction pathway. (2) **Receptor tyrosine kinases (RTKs)** react to the binding of signaling molecules by forming dimers and then adding phosphate groups to tyrosines on the cytoplasmic part of the other monomer making up the dimer. Relay proteins in the cell can then be activated by binding to different phosphorylated tyrosines, allowing this receptor to trigger several pathways at once. (3) **Ligand-gated ion channels** open or close in response to binding by specific signaling molecules, regulating the flow of specific ions across the membrane.
- The activity of all three types of receptors is crucial to proper cell functioning, and abnormal GPCRs and RTKs are associated with many human diseases.
- Intracellular receptors are cytoplasmic or nuclear proteins. Signaling molecules that are hydrophobic or small enough to cross the plasma membrane bind to these receptors inside the cell.

? *How are the structures of a G protein-coupled receptor and a receptor tyrosine kinase similar? In what key way does the triggering of signal transduction pathways differ for these two types of receptors?*

CONCEPT 11.3

Transduction: Cascades of molecular interactions relay signals from receptors to target molecules in the cell (pp. 214–219)

- At each step in a signal transduction pathway, the signal is transduced into a different form, which commonly involves a shape change in a protein. Many signal transduction pathways include phosphorylation cascades, in which a series of **protein kinases** each add a phosphate group to the next one in line, activating it. Enzymes called **protein phosphatases** remove the phosphate groups. The balance between phosphorylation and dephosphorylation regulates the activity of proteins involved in the sequential steps of a signal transduction pathway.
- **Second messengers**, such as the small molecule **cyclic AMP (cAMP)** and the ion Ca^{2+}, diffuse readily through the cytosol and thus help broadcast signals quickly. Many G proteins activate **adenylyl cyclase**, which makes cAMP from ATP. Cells use Ca^{2+} as a second messenger in both G protein and tyrosine kinase pathways. The tyrosine kinase pathways can also involve two other second messengers, **diacylglycerol (DAG)** and **inositol trisphosphate (IP_3)**. IP_3 can trigger a subsequent increase in Ca^{2+} levels.

? *What is the difference between a protein kinase and a second messenger? Can both types of molecules operate in the same signal transduction pathway?*

CONCEPT 11.4

Response: Cell signaling leads to regulation of transcription or cytoplasmic activities (pp. 219–223)

- Some pathways lead to a nuclear response: Specific genes are turned on or off by activation of proteins called transcription factors. In other pathways, the response involves cytoplasmic regulation, including cytoskeletal rearrangement (which can lead to cell shape changes) or changes in enzyme activity.
- Cellular responses are not simply on or off; they are fine-tuned at many steps in the process. Each catalytic protein in a signaling pathway amplifies the signal by activating multiple copies of the next component of the pathway; for long pathways, the total amplification may be a millionfold or more. The particular combination of proteins in a cell gives the cell great specificity in both the signals it detects and the responses it carries out. **Scaffolding proteins** can increase signal transduction efficiency. Pathway branching and cross-talk further help the cell coordinate incoming signals and responses. Signal response is terminated quickly by the reversal of ligand binding.

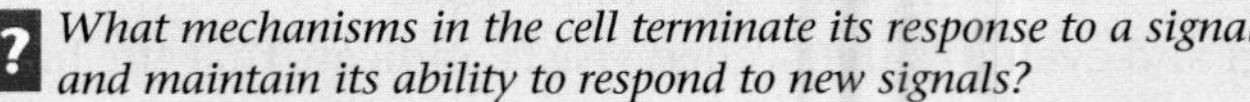

? *What mechanisms in the cell terminate its response to a signal and maintain its ability to respond to new signals?*

CONCEPT 11.5

Apoptosis integrates multiple cell-signaling pathways (pp. 223–225)

- **Apoptosis** is a type of programmed cell death in which cell components are disposed of in an orderly fashion, without damage to neighboring cells. Studies of the soil worm *Caenorhabditis elegans* showed that apoptosis occurs at defined times during embryonic development and clarified molecular details of the signaling pathway involved in the process. A protein (Ced-9) in the mitochondrial membrane acts as a brake; when released by a death signal, it allows activation of caspases, the main proteases that carry out apoptosis, and nucleases.
- Several apoptotic signaling pathways exist in the cells of humans and other mammals, and these pathways may be triggered in several ways. A major pathway involves pore formation in the outer mitochondrial membrane, which leads to release of factors that activate caspases. Signals eliciting this response can originate from outside or inside the cell.

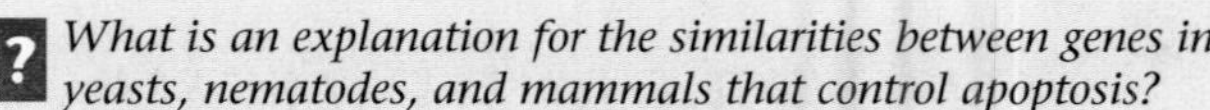

? *What is an explanation for the similarities between genes in yeasts, nematodes, and mammals that control apoptosis?*

TEST YOUR UNDERSTANDING

LEVEL 1: KNOWLEDGE/COMPREHENSION

1. Phosphorylation cascades involving a series of protein kinases are useful for cellular signal transduction because
   a. they are species specific.
   b. they always lead to the same cellular response.
   c. they amplify the original signal manyfold.
   d. they counter the harmful effects of phosphatases.
   e. the number of molecules used is small and fixed.

2. Binding of a signaling molecule to which type of receptor leads directly to a change in the distribution of ions on opposite sides of the membrane?
   a. receptor tyrosine kinase
   b. G protein-coupled receptor
   c. phosphorylated receptor tyrosine kinase dimer
   d. ligand-gated ion channel
   e. intracellular receptor

3. The activation of receptor tyrosine kinases is characterized by
   a. dimerization and phosphorylation.
   b. dimerization and IP_3 binding.
   c. a phosphorylation cascade.
   d. GTP hydrolysis.
   e. channel protein shape change.

4. Lipid-soluble signaling molecules, such as testosterone, cross the membranes of all cells but affect only target cells because
   a. only target cells retain the appropriate DNA segments.
   b. intracellular receptors are present only in target cells.
   c. most cells lack the Y chromosome required.
   d. only target cells possess the cytosolic enzymes that transduce the testosterone.
   e. only in target cells is testosterone able to initiate the phosphorylation cascade leading to activated transcription factor.

5. Consider this pathway: epinephrine → G protein-coupled receptor → G protein → adenylyl cyclase → cAMP. Identify the second messenger.
   a. cAMP
   b. G protein
   c. GTP
   d. adenylyl cyclase
   e. G protein-coupled receptor

6. Apoptosis involves all but which of the following?
   a. fragmentation of the DNA
   b. cell-signaling pathways
   c. activation of cellular enzymes
   d. lysis of the cell
   e. digestion of cellular contents by scavenger cells

LEVEL 2: APPLICATION/ANALYSIS

7. Which observation suggested to Sutherland the involvement of a second messenger in epinephrine's effect on liver cells?
   a. Enzymatic activity was proportional to the amount of calcium added to a cell-free extract.
   b. Receptor studies indicated that epinephrine was a ligand.
   c. Glycogen breakdown was observed only when epinephrine was administered to intact cells.
   d. Glycogen breakdown was observed when epinephrine and glycogen phosphorylase were combined.
   e. Epinephrine was known to have different effects on different types of cells.

8. Protein phosphorylation is commonly involved with all of the following *except*
   a. regulation of transcription by extracellular signaling molecules.
   b. enzyme activation.
   c. activation of G protein-coupled receptors.
   d. activation of receptor tyrosine kinases.
   e. activation of protein kinase molecules.

LEVEL 3: SYNTHESIS/EVALUATION

9. **DRAW IT** Draw the following apoptotic pathway, which operates in human immune cells. A death signal is received when a molecule called Fas binds its cell-surface receptor. The binding of many Fas molecules to receptors causes receptor clustering. The intracellular regions of the receptors, when together, bind proteins called adaptor proteins. These in turn bind to inactive molecules of caspase-8, which become activated and then activate caspase-3. Once activated, caspase-3 initiates apoptosis.

10. **EVOLUTION CONNECTION**
    What evolutionary mechanisms might account for the origin and persistence of cell-to-cell signaling systems in unicellular prokaryotes?

11. **SCIENTIFIC INQUIRY**
    Epinephrine initiates a signal transduction pathway that involves production of cyclic AMP (cAMP) and leads to the breakdown of glycogen to glucose, a major energy source for cells. But glycogen breakdown is actually only part of the fight-or-flight response that epinephrine brings about; the overall effect on the body includes increased heart rate and alertness, as well as a burst of energy. Given that caffeine blocks the activity of cAMP phosphodiesterase, propose a mechanism by which caffeine ingestion leads to heightened alertness and sleeplessness.

12. **SCIENCE, TECHNOLOGY, AND SOCIETY**
    The aging process is thought to be initiated at the cellular level. Among the changes that can occur after a certain number of cell divisions is the loss of a cell's ability to respond to growth factors and other chemical signals. Much research into aging is aimed at understanding such losses, with the ultimate goal of significantly extending the human life span. Not everyone, however, agrees that this is a desirable goal. If life expectancy were greatly increased, what might be the social and ecological consequences?

13. **WRITE ABOUT A THEME**
    Emergent Properties The property of life emerges at the biological level of the cell. The highly regulated process of apoptosis is not simply the destruction of a cell; it is also an emergent property. Write a short essay (about 100–150 words) that briefly explains the role of apoptosis in the development and proper functioning of an animal and then describes how this form of programmed cell death is a process that emerges from the orderly integration of signaling pathways.

For selected answers, see Appendix A.

12

The Cell Cycle

▲ **Figure 12.1 How do a cell's chromosomes change during cell division?**

KEY CONCEPTS

12.1 Most cell division results in genetically identical daughter cells

12.2 The mitotic phase alternates with interphase in the cell cycle

12.3 The eukaryotic cell cycle is regulated by a molecular control system

OVERVIEW

The Key Roles of Cell Division

The ability of organisms to produce more of their own kind is the one characteristic that best distinguishes living things from nonliving matter. This unique capacity to procreate, like all biological functions, has a cellular basis. Rudolf Virchow, a German physician, put it this way in 1855: "Where a cell exists, there must have been a preexisting cell, just as the animal arises only from an animal and the plant only from a plant." He summarized this concept with the Latin axiom "*Omnis cellula e cellula*," meaning "Every cell from a cell." The continuity of life is based on the reproduction of cells, or **cell division**. The series of fluorescence micrographs in **Figure 12.1** follows an animal cell's chromosomes, from lower left to lower right, as one cell divides into two.

Cell division plays several important roles in life. The division of one prokaryotic cell reproduces an entire organism. The same is true of a unicellular eukaryote **(Figure 12.2a)**. Cell division also enables multicellular eukaryotes to develop from a single cell, like the fertilized egg that gave rise to the two-celled embryo in **Figure 12.2b**. And after such an organism is fully grown, cell division continues to function in renewal and repair, replacing cells that die from normal wear and tear or accidents. For example, dividing cells in your bone marrow continuously make new blood cells **(Figure 12.2c)**.

The cell division process is an integral part of the **cell cycle**, the life of a cell from the time it is first formed from a dividing parent cell until its own division into two daughter cells. (Our use of the words *daughter* or *sister* in relation to cells is not meant to imply gender.) Passing identical genetic material to cellular offspring is a crucial function of cell division. In this chapter, you will learn how this process occurs. After studying the cellular mechanics of cell division in eukaryotes and bacteria, you will learn about the molecular control system that regulates progress through the eukaryotic cell cycle and what happens when the control system malfunctions. Because a breakdown in cell cycle control plays a major role in cancer development, this aspect of cell biology is an active area of research.

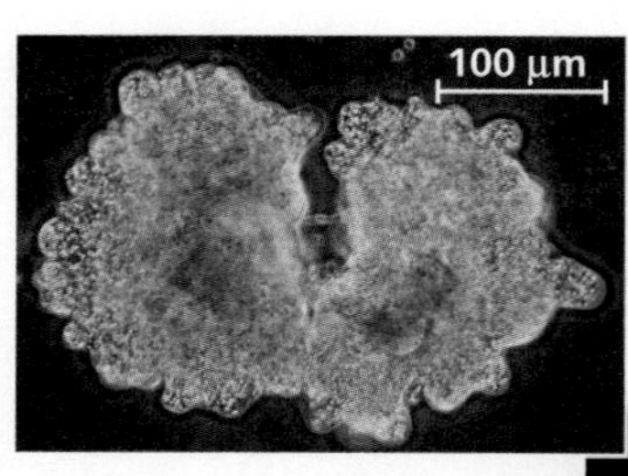

◀ **(a) Reproduction.** An amoeba, a single-celled eukaryote, is dividing into two cells. Each new cell will be an individual organism (LM).

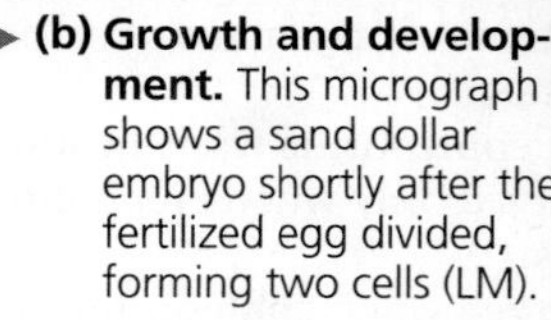

▶ **(b) Growth and development.** This micrograph shows a sand dollar embryo shortly after the fertilized egg divided, forming two cells (LM).

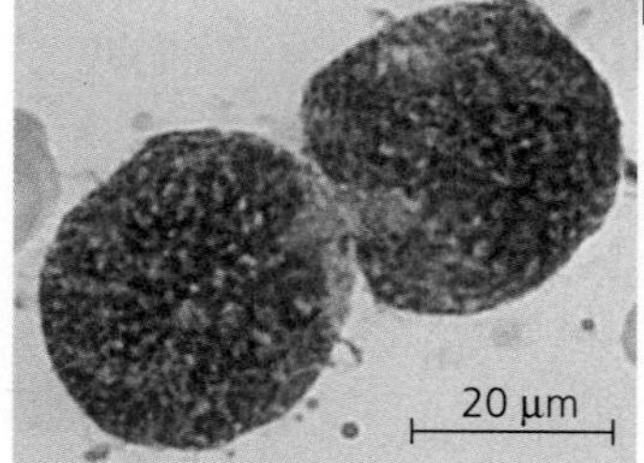

◀ **(c) Tissue renewal.** These dividing bone marrow cells will give rise to new blood cells (LM).

▲ **Figure 12.2 The functions of cell division.**

CONCEPT **12.1**

Most cell division results in genetically identical daughter cells

The reproduction of an assembly as complex as a cell cannot occur by a mere pinching in half; a cell is not like a soap bubble that simply enlarges and splits in two. In both prokaryotes and eukaryotes, most cell division involves the distribution of identical genetic material—DNA—to two daughter cells. (The exception is meiosis, the special type of eukaryotic cell division that can produce sperm and eggs.) What is most remarkable about cell division is the fidelity with which the DNA is passed along from one generation of cells to the next. A dividing cell duplicates its DNA, allocates the two copies to opposite ends of the cell, and only then splits into daughter cells. After we describe the distribution of DNA during cell division in animal and plant cells, we'll consider the process in other eukaryotes as well as in bacteria.

Cellular Organization of the Genetic Material

A cell's endowment of DNA, its genetic information, is called its **genome**. Although a prokaryotic genome is often a single DNA molecule, eukaryotic genomes usually consist of a number of DNA molecules. The overall length of DNA in a eukaryotic cell is enormous. A typical human cell, for example, has about 2 m of DNA—a length about 250,000 times greater than the cell's diameter. Yet before the cell can divide to form genetically identical daughter cells, all of this DNA must be copied, or replicated, and then the two copies must be separated so that each daughter cell ends up with a complete genome.

The replication and distribution of so much DNA is manageable because the DNA molecules are packaged into structures called **chromosomes**, so named because they take up certain dyes used in microscopy (from the Greek *chroma*, color, and *soma*, body) (**Figure 12.3**). Each eukaryotic chromosome consists of one very long, linear DNA molecule associated with many proteins (see Figure 6.9). The DNA molecule carries several hundred to a few thousand genes, the units of information that specify an organism's inherited traits. The associated proteins maintain the structure of the chromosome and help control the activity of the genes. Together, the entire complex of DNA and proteins that is the building material of chromosomes is referred to as **chromatin**. As you will soon see, the chromatin of a chromosome varies in its degree of condensation during the process of cell division.

Every eukaryotic species has a characteristic number of chromosomes in each cell nucleus. For example, the nuclei of human **somatic cells** (all body cells except the reproductive cells) each contain 46 chromosomes, made up of two sets of 23, one set inherited from each parent. Reproductive cells, or **gametes**—sperm and eggs—have half as many chromosomes as somatic cells, or one set of 23 chromosomes in humans. The number of chromosomes in somatic cells varies widely among species: 18 in cabbage plants, 48 in chimpanzees, 56 in elephants, 90 in hedgehogs, and 148 in one species of alga. We'll now consider how these chromosomes behave during cell division.

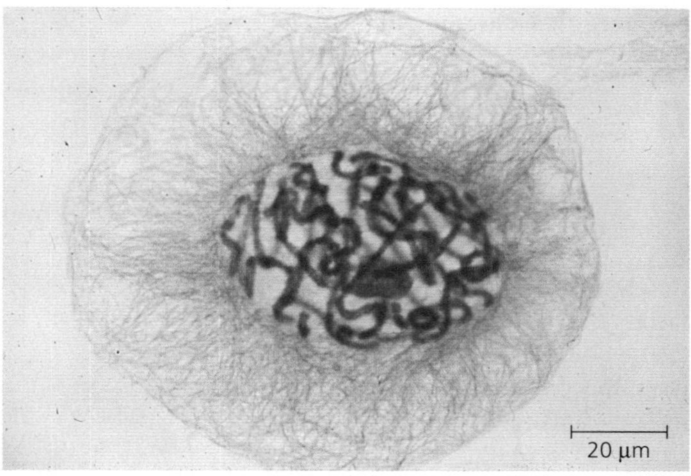

▲ **Figure 12.3 Eukaryotic chromosomes.** Chromosomes (stained purple) are visible within the nucleus of this cell from an African blood lily. The thinner red threads in the surrounding cytoplasm are the cytoskeleton. The cell is preparing to divide (LM).

Distribution of Chromosomes During Eukaryotic Cell Division

When a cell is not dividing, and even as it replicates its DNA in preparation for cell division, each chromosome is in the form of a long, thin chromatin fiber. After DNA replication, however, the chromosomes condense as a part of cell division: Each chromatin fiber becomes densely coiled and folded, making the chromosomes much shorter and so thick that we can see them with a light microscope.

Each duplicated chromosome has two **sister chromatids**, which are joined copies of the original chromosome (**Figure 12.4**). The two chromatids, each containing an identical DNA molecule, are initially attached all along their lengths by protein complexes called *cohesins*; this attachment is known as *sister chromatid cohesion*. Each sister chromatid has a **centromere**, a region containing specific DNA sequences

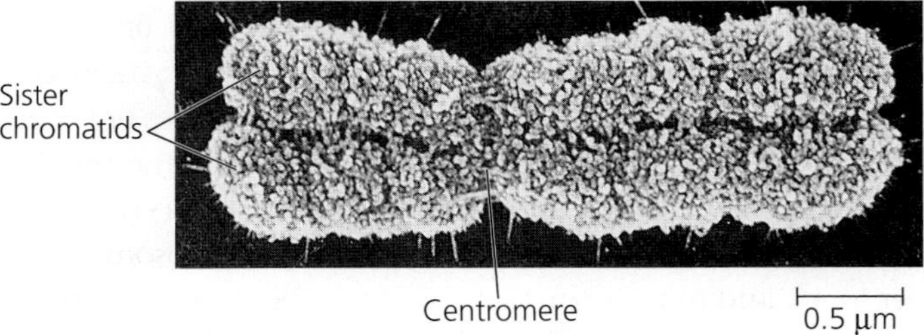

▲ **Figure 12.4 A highly condensed, duplicated human chromosome (SEM).**

DRAW IT *Circle one sister chromatid of the chromosome in this micrograph.*

▶ **Figure 12.5 Chromosome duplication and distribution during cell division.**

? *How many chromatid arms does the chromosome in ② have?*

① One of the multiple chromosomes in a eukaryotic cell is represented here, not yet duplicated. Normally it would be a long, thin chromatin fiber containing one DNA molecule and associated proteins; here its condensed form is shown for illustration purposes only.

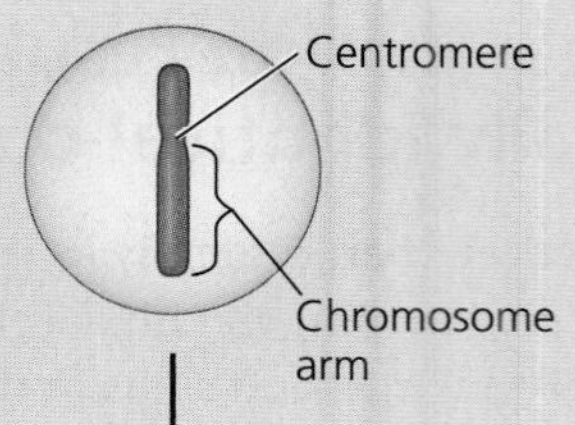

Chromosome duplication (including DNA replication) and condensation

② Once duplicated, a chromosome consists of two sister chromatids connected along their entire lengths by sister chromatid cohesion. Each chromatid contains a copy of the DNA molecule.

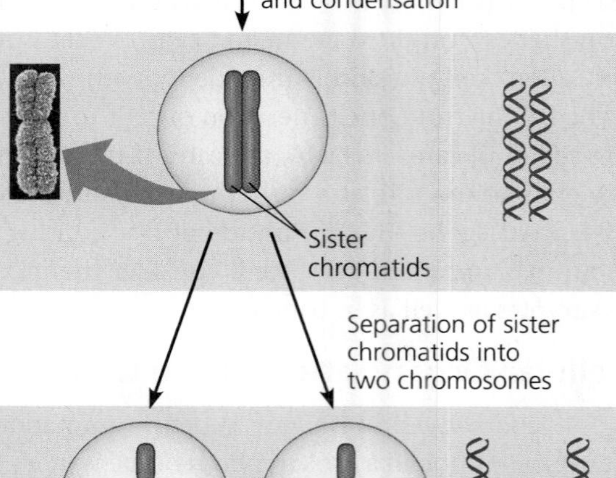

③ Molecular and mechanical processes separate the sister chromatids into two chromosomes and distribute them to two daughter cells.

where the chromatid is attached most closely to its sister chromatid. This attachment is mediated by proteins bound to the centromeric DNA sequences and gives the condensed, duplicated chromosome a narrow "waist." The part of a chromatid on either side of the centromere is referred to as an *arm* of the chromatid. (An uncondensed, unduplicated chromosome has a single centromere and two arms.)

Later in the cell division process, the two sister chromatids of each duplicated chromosome separate and move into two new nuclei, one forming at each end of the cell. Once the sister chromatids separate, they are no longer called sister chromatids but are considered individual chromosomes. Thus, each new nucleus receives a collection of chromosomes identical to that of the parent cell (**Figure 12.5**). **Mitosis**, the division of the genetic material in the nucleus, is usually followed immediately by **cytokinesis**, the division of the cytoplasm. One cell has become two, each the genetic equivalent of the parent cell.

What happens to the chromosome number as we follow the human life cycle through the generations? You inherited 46 chromosomes, one set of 23 from each parent. They were combined in the nucleus of a single cell when a sperm from your father united with an egg from your mother, forming a fertilized egg, or zygote. Mitosis and cytokinesis produced the 200 trillion somatic cells that now make up your body, and the same processes continue to generate new cells to replace dead and damaged ones. In contrast, you produce gametes—eggs or sperm—by a variation of cell division called *meiosis*, which yields nonidentical daughter cells that have only one set of chromosomes, half as many chromosomes as the parent cell. Meiosis in humans occurs only in the gonads (ovaries or testes). In each generation, meiosis reduces the chromosome number from 46 (two sets of chromosomes) to 23 (one set). Fertilization fuses two gametes together and returns the chromosome number to 46, and mitosis conserves that number in every somatic cell nucleus of the new individual. In Chapter 13, we will examine the role of meiosis in reproduction and inheritance in more detail. In the remainder of this chapter, we focus on mitosis and the rest of the cell cycle in eukaryotes.

CONCEPT CHECK 12.1

1. How many chromatids are in a duplicated chromosome?
2. **WHAT IF?** A chicken has 78 chromosomes in its somatic cells. How many chromosomes did the chicken inherit from each parent? How many chromosomes are in each of the chicken's gametes? How many chromosomes will be in each somatic cell of the chicken's offspring?

For suggested answers, see Appendix A.

CONCEPT 12.2

The mitotic phase alternates with interphase in the cell cycle

In 1882, a German anatomist named Walther Flemming developed dyes that allowed him to observe, for the first time, the behavior of chromosomes during mitosis and cytokinesis. (In fact, Flemming coined the terms *mitosis* and *chromatin*.) During the period between one cell division and the next, it appeared to Flemming that the cell was simply growing larger. But we now know that many critical events occur during this stage in the life of a cell.

Phases of the Cell Cycle

Mitosis is just one part of the cell cycle **(Figure 12.6)**. In fact, the **mitotic (M) phase**, which includes both mitosis and cytokinesis, is usually the shortest part of the cell cycle. Mitotic cell division alternates with a much longer stage called **interphase**, which often accounts for about 90% of the cycle. During interphase, a cell that is about to divide grows and copies its chromosomes in preparation for cell division. Interphase can be divided into subphases: the **G_1 phase** ("first gap"), the **S phase** ("synthesis"), and the **G_2 phase** ("second gap"). During all three subphases, a cell that will eventually divide grows by producing proteins and cytoplasmic organelles such as mitochondria and endoplasmic reticulum. However, chromosomes are duplicated only during the S phase. (We will discuss synthesis of DNA in Chapter 16.) Thus, a cell grows (G_1), continues to grow as it copies its chromosomes (S), grows more as it completes preparations for cell division (G_2), and divides (M). The daughter cells may then repeat the cycle.

A particular human cell might undergo one division in 24 hours. Of this time, the M phase would occupy less than 1 hour, while the S phase might occupy about 10–12 hours, or about half the cycle. The rest of the time would be apportioned between the G_1 and G_2 phases. The G_2 phase usually takes 4–6 hours; in our example, G_1 would occupy about 5–6 hours. G_1 is the most variable in length in different types of cells. Some cells in a multicellular organism divide very infrequently or not at all. These cells spend their time in G_1 (or a related phase called G_0) doing their job in the organism—a nerve cell carries impulses, for example.

Mitosis is conventionally broken down into five stages: **prophase**, **prometaphase**, **metaphase**, **anaphase**, and **telophase**. Overlapping with the latter stages of mitosis, cytokinesis completes the mitotic phase. **Figure 12.7**, on the next two pages, describes these stages in an animal cell. Study this figure thoroughly before progressing to the next two sections, which examine mitosis and cytokinesis more closely.

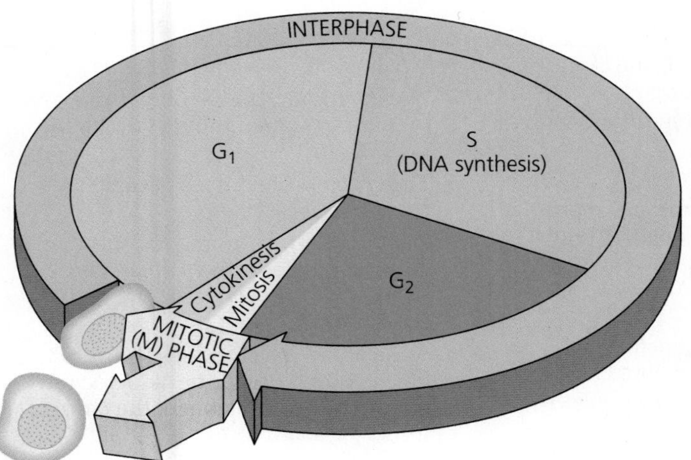

▲ **Figure 12.6 The cell cycle.** In a dividing cell, the mitotic (M) phase alternates with interphase, a growth period. The first part of interphase (G_1) is followed by the S phase, when the chromosomes duplicate; G_2 is the last part of interphase. In the M phase, mitosis distributes the daughter chromosomes to daughter nuclei, and cytokinesis divides the cytoplasm, producing two daughter cells. The relative durations of G_1, S, and G_2 may vary.

The Mitotic Spindle: *A Closer Look*

Many of the events of mitosis depend on the **mitotic spindle**, which begins to form in the cytoplasm during prophase. This structure consists of fibers made of microtubules and associated proteins. While the mitotic spindle assembles, the other microtubules of the cytoskeleton partially disassemble, providing the material used to construct the spindle. The spindle microtubules elongate (polymerize) by incorporating more subunits of the protein tubulin (see Table 6.1) and shorten (depolymerize) by losing subunits.

In animal cells, the assembly of spindle microtubules starts at the **centrosome**, a subcellular region containing material that functions throughout the cell cycle to organize the cell's microtubules. (It is also called the *microtubule-organizing center.*) A pair of centrioles is located at the center of the centrosome, but they are not essential for cell division: If the centrioles are destroyed with a laser microbeam, a spindle nevertheless forms during mitosis. In fact, centrioles are not even present in plant cells, which do form mitotic spindles.

During interphase in animal cells, the single centrosome duplicates, forming two centrosomes, which remain together near the nucleus. The two centrosomes move apart during prophase and prometaphase of mitosis as spindle microtubules grow out from them. By the end of prometaphase, the two centrosomes, one at each pole of the spindle, are at opposite ends of the cell. An **aster**, a radial array of short microtubules, extends from each centrosome. The spindle includes the centrosomes, the spindle microtubules, and the asters.

Each of the two sister chromatids of a duplicated chromosome has a **kinetochore**, a structure of proteins associated with specific sections of chromosomal DNA at each centromere. The chromosome's two kinetochores face in opposite directions. During prometaphase, some of the spindle microtubules attach to the kinetochores; these are called kinetochore microtubules. (The number of microtubules attached to a kinetochore varies among species, from one microtubule in yeast cells to 40 or so in some mammalian cells.) When one of a chromosome's kinetochores is "captured" by microtubules, the chromosome begins to move toward the pole from which those microtubules extend. However, this movement is checked as soon as microtubules from the opposite pole attach to the other kinetochore. What happens next is like a tug-of-war that ends in a draw. The chromosome moves first in one direction, then the other, back and forth, finally settling midway between the two ends of the cell. At metaphase, the centromeres of all the duplicated chromosomes are on a plane midway between the spindle's

▼ **Figure 12.7**

Exploring Mitosis in an Animal Cell

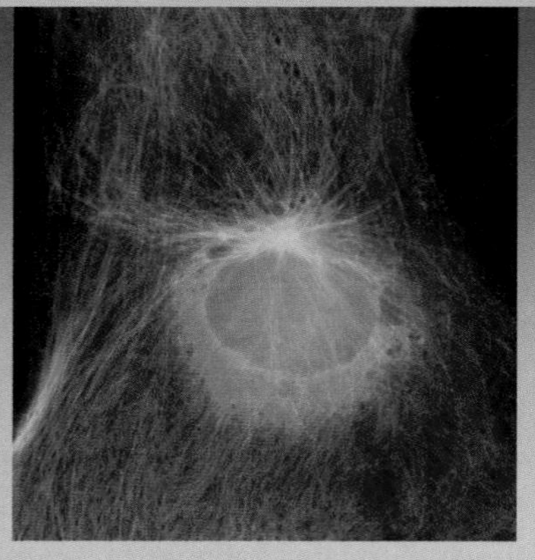

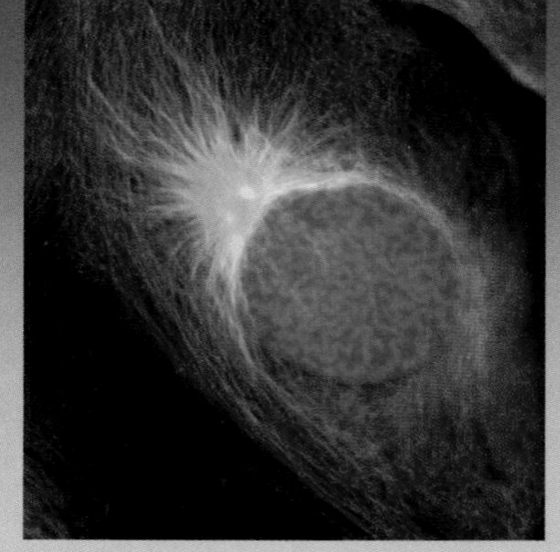

G_2 of Interphase

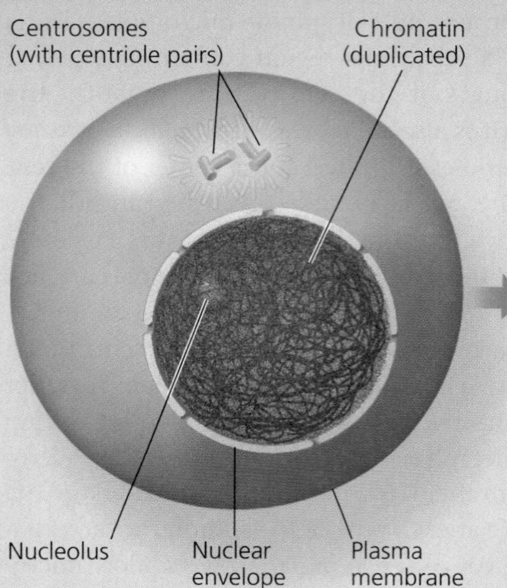

Prophase

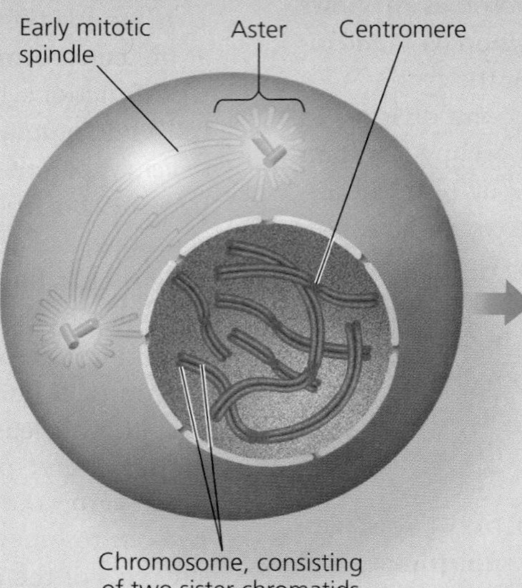

Prometaphase

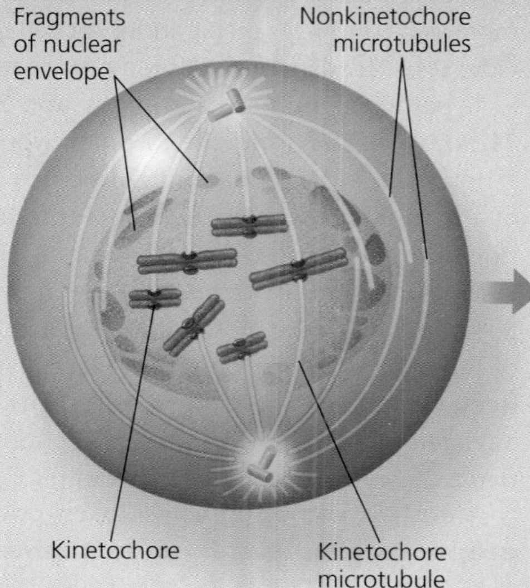

G_2 of Interphase

- A nuclear envelope encloses the nucleus.
- The nucleus contains one or more nucleoli (singular, *nucleolus*).
- Two centrosomes have formed by duplication of a single centrosome. Centrosomes are regions in animal cells that organize the microtubules of the spindle. Each centrosome contains two centrioles.
- Chromosomes, duplicated during S phase, cannot be seen individually because they have not yet condensed.

The light micrographs show dividing lung cells from a newt, which has 22 chromosomes in its somatic cells. Chromosomes appear blue, microtubules green, and intermediate filaments red. For simplicity, the drawings show only 6 chromosomes.

Prophase

- The chromatin fibers become more tightly coiled, condensing into discrete chromosomes observable with a light microscope.
- The nucleoli disappear.
- Each duplicated chromosome appears as two identical sister chromatids joined at their centromeres and, in some species, all along their arms by cohesins (sister chromatid cohesion).
- The mitotic spindle (named for its shape) begins to form. It is composed of the centrosomes and the microtubules that extend from them. The radial arrays of shorter microtubules that extend from the centrosomes are called asters ("stars").
- The centrosomes move away from each other, propelled partly by the lengthening microtubules between them.

Prometaphase

- The nuclear envelope fragments.
- The microtubules extending from each centrosome can now invade the nuclear area.
- The chromosomes have become even more condensed.
- Each of the two chromatids of each chromosome now has a kinetochore, a specialized protein structure at the centromere.
- Some of the microtubules attach to the kinetochores, becoming "kinetochore microtubules," which jerk the chromosomes back and forth.
- Nonkinetochore microtubules interact with those from the opposite pole of the spindle.

? *How many molecules of DNA are in the prometaphase drawing? How many molecules per chromosome? How many double helices are there per chromosome? Per chromatid?*

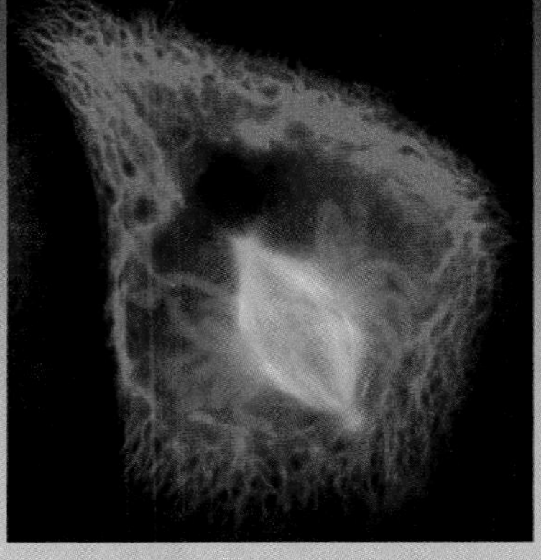

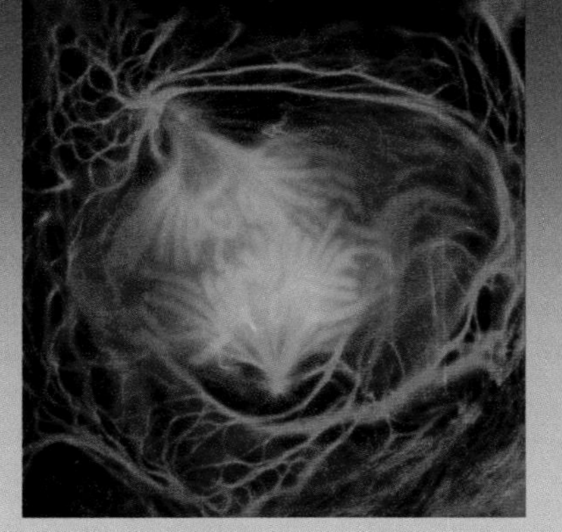

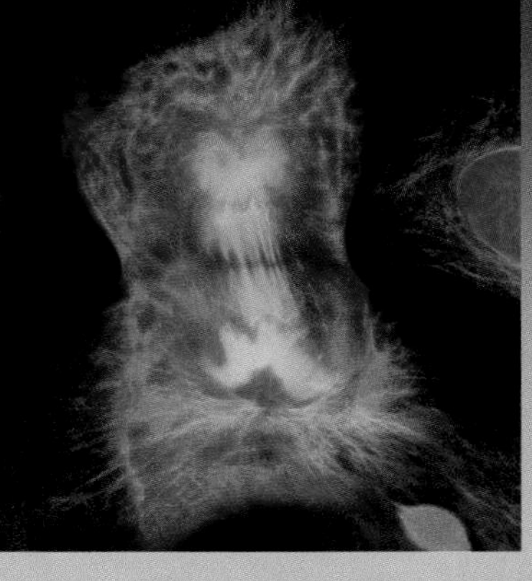

Metaphase

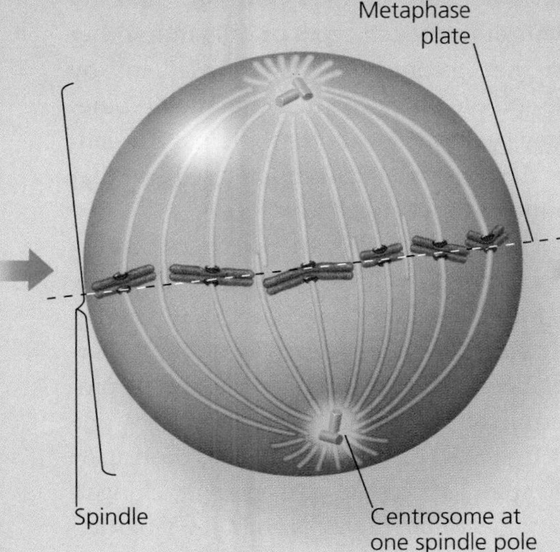

Anaphase

Telophase and Cytokinesis

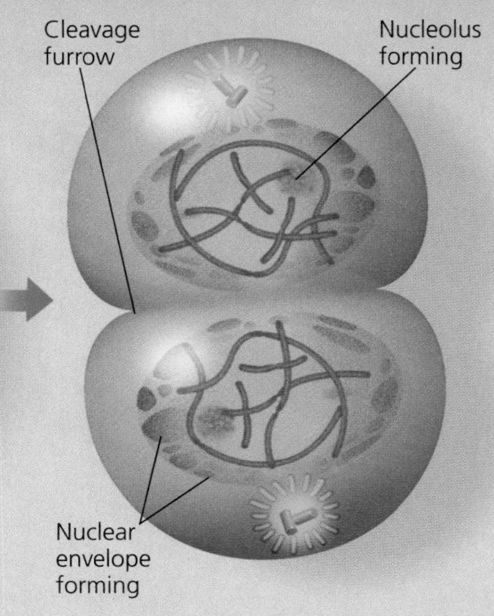

Metaphase

- The centrosomes are now at opposite poles of the cell.
- The chromosomes convene at the *metaphase plate*, a plane that is equidistant between the spindle's two poles. The chromosomes' centromeres lie at the metaphase plate.
- For each chromosome, the kinetochores of the sister chromatids are attached to kinetochore microtubules coming from opposite poles.

Anaphase

- Anaphase is the shortest stage of mitosis, often lasting only a few minutes.
- Anaphase begins when the cohesin proteins are cleaved. This allows the two sister chromatids of each pair to part suddenly. Each chromatid thus becomes a full-fledged chromosome.
- The two liberated daughter chromosomes begin moving toward opposite ends of the cell as their kinetochore microtubules shorten. Because these microtubules are attached at the centromere region, the chromosomes move centromere first (at about 1 μm/min).
- The cell elongates as the nonkinetochore microtubules lengthen.
- By the end of anaphase, the two ends of the cell have equivalent—and complete—collections of chromosomes.

Telophase

- Two daughter nuclei form in the cell. Nuclear envelopes arise from the fragments of the parent cell's nuclear envelope and other portions of the endomembrane system.
- Nucleoli reappear.
- The chromosomes become less condensed.
- Any remaining spindle microtubules are depolymerized.
- Mitosis, the division of one nucleus into two genetically identical nuclei, is now complete.

Cytokinesis

- The division of the cytoplasm is usually well under way by late telophase, so the two daughter cells appear shortly after the end of mitosis.
- In animal cells, cytokinesis involves the formation of a cleavage furrow, which pinches the cell in two.

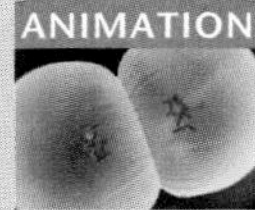

BioFlix Visit the Study Area at **www.masteringbiology.com** for the BioFlix® 3-D Animation on Mitosis.

two poles. This plane is called the **metaphase plate**, which is an imaginary rather than an actual cellular structure **(Figure 12.8)**. Meanwhile, microtubules that do not attach to kinetochores have been elongating, and by metaphase they overlap and interact with other nonkinetochore microtubules from the opposite pole of the spindle. (These are sometimes called "polar" microtubules.) By metaphase, the microtubules of the asters have also grown and are in contact with the plasma membrane. The spindle is now complete.

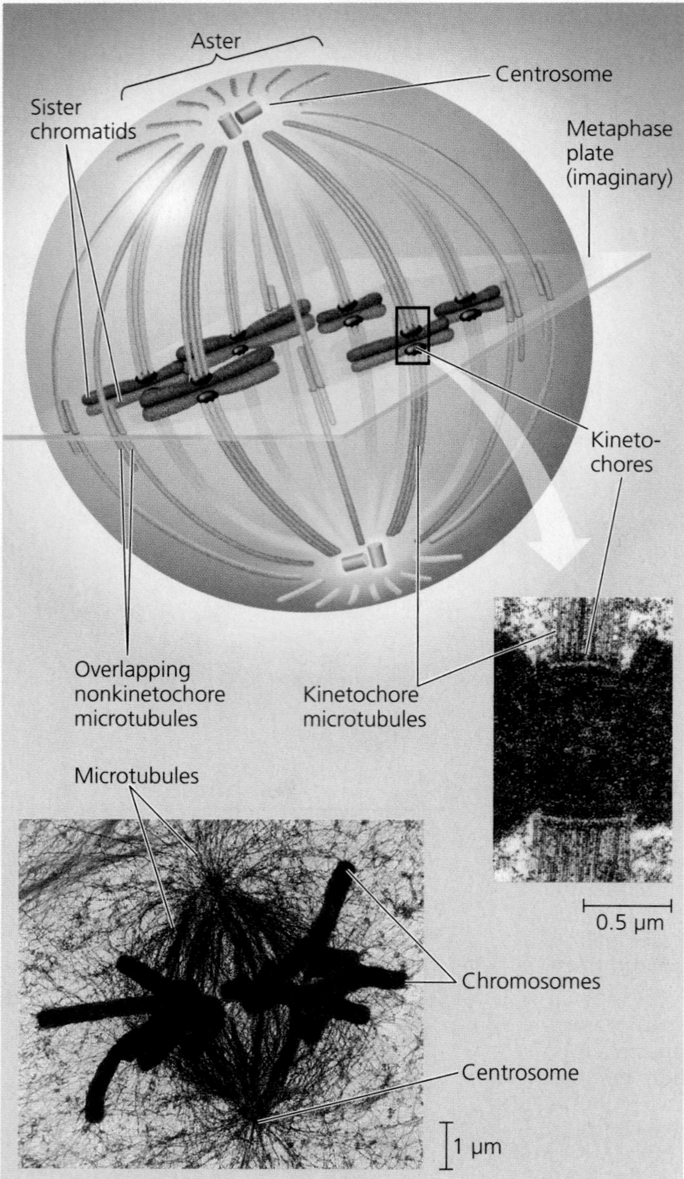

▲ **Figure 12.8 The mitotic spindle at metaphase.** The kinetochores of each chromosome's two sister chromatids face in opposite directions. Here, each kinetochore is attached to a *cluster* of kinetochore microtubules extending from the nearest centrosome. Nonkinetochore microtubules overlap at the metaphase plate (TEMs).

DRAW IT *On the lower micrograph, draw a line indicating the position of the metaphase plate. Circle an aster. Draw arrows indicating the directions of chromosome movement once anaphase begins.*

The structure of the completed spindle correlates well with its function during anaphase. Anaphase commences suddenly when the cohesins holding together the sister chromatids of each chromosome are cleaved by an enzyme called *separase*. Once the chromatids become separate, full-fledged chromosomes, they move toward opposite ends of the cell.

How do the kinetochore microtubules function in this poleward movement of chromosomes? Apparently, two mechanisms are in play, both involving motor proteins. (To review how motor proteins move an object along a microtubule, see Figure 6.21.) A clever experiment carried out in 1987 suggested that motor proteins on the kinetochores "walk" the chromosomes along the microtubules, which depolymerize at their kinetochore ends after the motor proteins have passed **(Figure 12.9)**. (This is referred to as the "Pacman" mechanism because of its resemblance to the arcade game character that moves by eating all the dots in its path.) However, other researchers, working with different cell types or cells from other species, have shown that chromosomes are "reeled in" by motor proteins at the spindle poles and that the microtubules depolymerize after they pass by these motor proteins. The general consensus now is that both mechanisms are used and that their relative contributions vary among cell types.

In a dividing animal cell, the nonkinetochore microtubules are responsible for elongating the whole cell during anaphase. Nonkinetochore microtubules from opposite poles overlap each other extensively during metaphase (see Figure 12.8). During anaphase, the region of overlap is reduced as motor proteins attached to the microtubules walk them away from one another, using energy from ATP. As the microtubules push apart from each other, their spindle poles are pushed apart, elongating the cell. At the same time, the microtubules lengthen somewhat by the addition of tubulin subunits to their overlapping ends. As a result, the microtubules continue to overlap.

At the end of anaphase, duplicate groups of chromosomes have arrived at opposite ends of the elongated parent cell. Nuclei re-form during telophase. Cytokinesis generally begins during anaphase or telophase, and the spindle eventually disassembles by depolymerization of microtubules.

Cytokinesis: *A Closer Look*

In animal cells, cytokinesis occurs by a process known as **cleavage**. The first sign of cleavage is the appearance of a **cleavage furrow**, a shallow groove in the cell surface near the old metaphase plate **(Figure 12.10a)**. On the cytoplasmic side of the furrow is a contractile ring of actin microfilaments associated with molecules of the protein myosin. The actin microfilaments interact with the myosin molecules, causing the ring to contract. The contraction of the dividing cell's ring of microfilaments is like the pulling of a drawstring. The cleavage furrow deepens until the parent cell is pinched in two, producing two completely separated cells, each with its own nucleus and share of cytosol, organelles, and other subcellular structures.

▼ **Figure 12.9** INQUIRY

At which end do kinetochore microtubules shorten during anaphase?

EXPERIMENT Gary Borisy and colleagues at the University of Wisconsin wanted to determine whether kinetochore microtubules depolymerize at the kinetochore end or the pole end as chromosomes move toward the poles during mitosis. First they labeled the microtubules of a pig kidney cell in early anaphase with a yellow fluorescent dye.

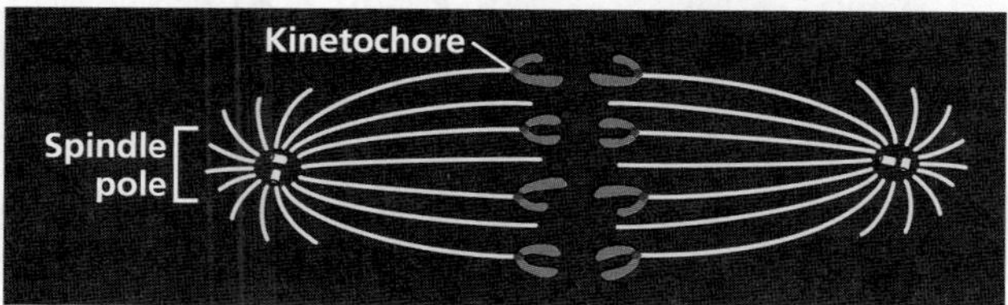

Then they marked a region of the kinetochore microtubules between one spindle pole and the chromosomes by using a laser to eliminate the fluorescence from that region, while leaving the microtubules intact (see below). As anaphase proceeded, they monitored the changes in microtubule length on either side of the mark.

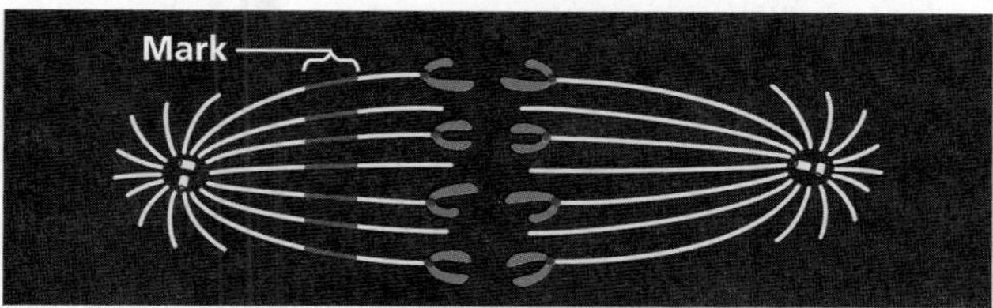

RESULTS As the chromosomes moved poleward, the microtubule segments on the kinetochore side of the mark shortened, while those on the spindle pole side stayed the same length.

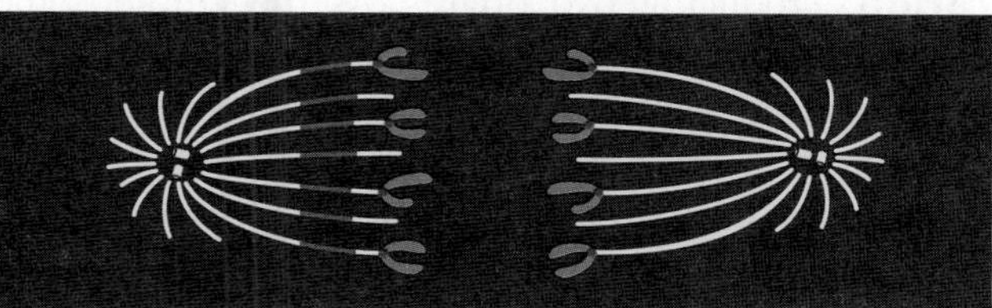

CONCLUSION During anaphase in this cell type, chromosome movement is correlated with kinetochore microtubules shortening at their kinetochore ends and not at their spindle pole ends. This experiment supports the hypothesis that during anaphase, a chromosome is walked along a microtubule as the microtubule depolymerizes at its kinetochore end, releasing tubulin subunits.

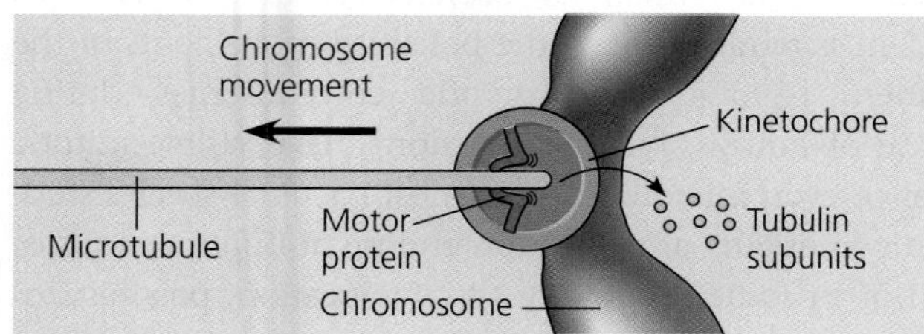

SOURCE G. J. Gorbsky, P. J. Sammak, and G. G. Borisy, Chromosomes move poleward in anaphase along stationary microtubules that coordinately disassemble from their kinetochore ends, *Journal of Cell Biology* 104:9–18 (1987).

WHAT IF? If this experiment had been done on a cell type in which "reeling in" at the poles was the main cause of chromosome movement, how would the mark have moved relative to the poles? How would the microtubule lengths have changed?

▼ **Figure 12.10 Cytokinesis in animal and plant cells.**

(a) Cleavage of an animal cell (SEM)

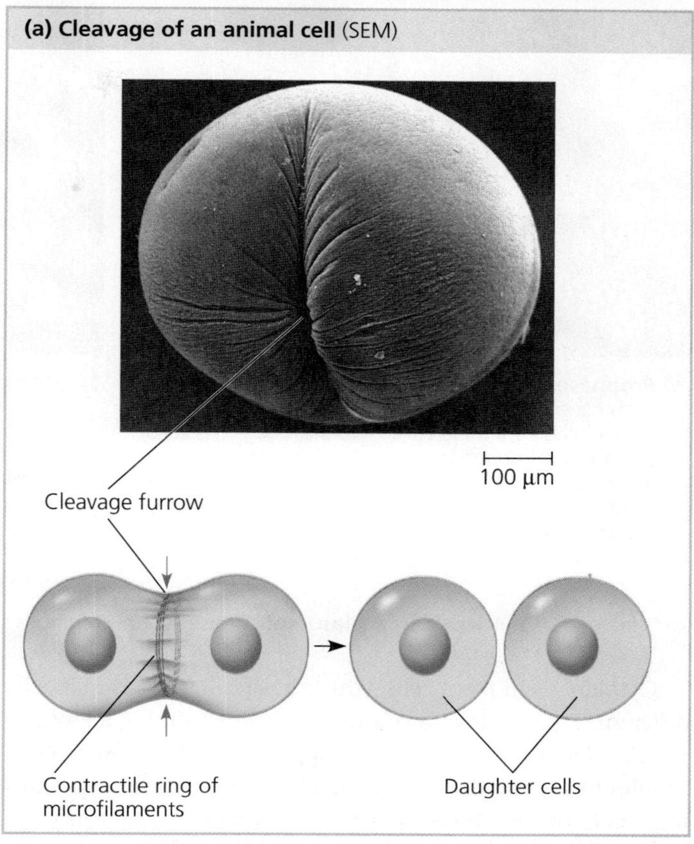

(b) Cell plate formation in a plant cell (TEM)

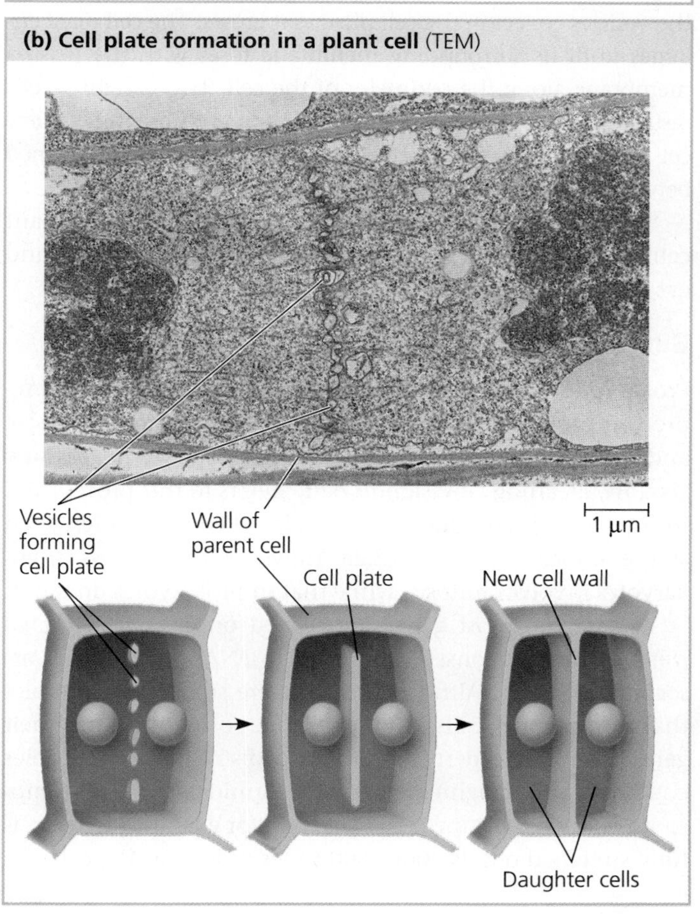

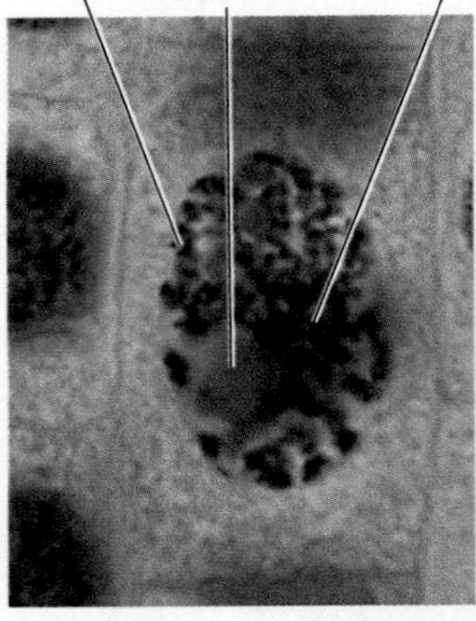

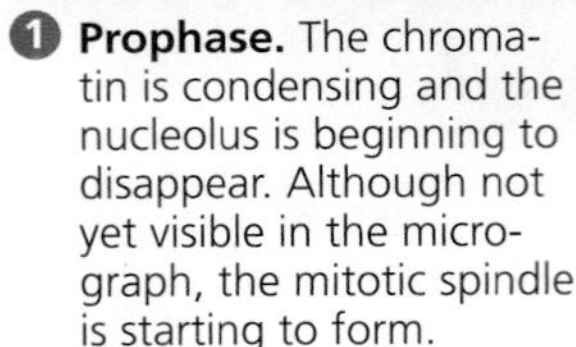

1 **Prophase.** The chromatin is condensing and the nucleolus is beginning to disappear. Although not yet visible in the micrograph, the mitotic spindle is starting to form.

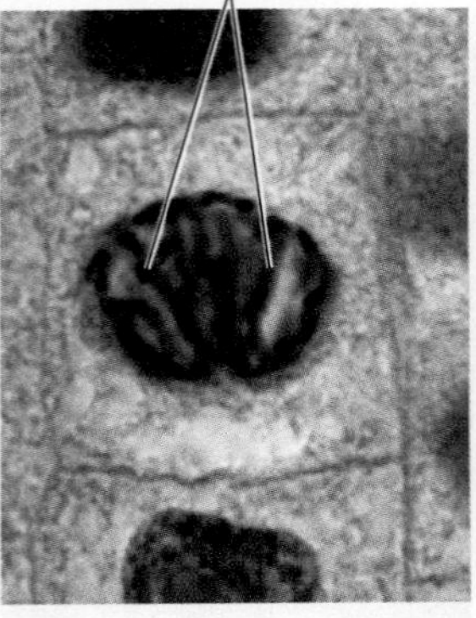

2 **Prometaphase.** Discrete chromosomes are now visible; each consists of two aligned, identical sister chromatids. Later in prometaphase, the nuclear envelope will fragment.

3 **Metaphase.** The spindle is complete, and the chromosomes, attached to microtubules at their kinetochores, are all at the metaphase plate.

4 **Anaphase.** The chromatids of each chromosome have separated, and the daughter chromosomes are moving to the ends of the cell as their kinetochore microtubules shorten.

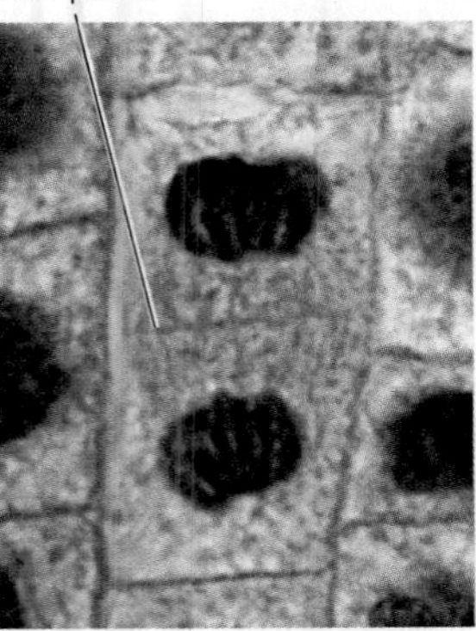

5 **Telophase.** Daughter nuclei are forming. Meanwhile, cytokinesis has started: The cell plate, which will divide the cytoplasm in two, is growing toward the perimeter of the parent cell.

▲ **Figure 12.11 Mitosis in a plant cell.** These light micrographs show mitosis in cells of an onion root.

Cytokinesis in plant cells, which have cell walls, is markedly different. There is no cleavage furrow. Instead, during telophase, vesicles derived from the Golgi apparatus move along microtubules to the middle of the cell, where they coalesce, producing a **cell plate** (**Figure 12.10b**). Cell wall materials carried in the vesicles collect in the cell plate as it grows. The cell plate enlarges until its surrounding membrane fuses with the plasma membrane along the perimeter of the cell. Two daughter cells result, each with its own plasma membrane. Meanwhile, a new cell wall arising from the contents of the cell plate has formed between the daughter cells.

Figure 12.11 is a series of micrographs of a dividing plant cell. Examining this figure will help you review mitosis and cytokinesis.

Binary Fission in Bacteria

Prokaryotes (bacteria and archaea) can undergo a type of reproduction in which the cell grows to roughly double its size and then divides to form two cells. The term **binary fission**, meaning "division in half," refers to this process and to the asexual reproduction of single-celled eukaryotes, such as the amoeba in Figure 12.2a. However, the process in eukaryotes involves mitosis, while that in prokaryotes does not.

In bacteria, most genes are carried on a single *bacterial chromosome* that consists of a circular DNA molecule and associated proteins. Although bacteria are smaller and simpler than eukaryotic cells, the challenge of replicating their genomes in an orderly fashion and distributing the copies equally to two daughter cells is still formidable. The chromosome of the bacterium *Escherichia coli*, for example, when it is fully stretched out, is about 500 times as long as the cell. For such a long chromosome to fit within the cell requires that it be highly coiled and folded.

In *E. coli*, the process of cell division is initiated when the DNA of the bacterial chromosome begins to replicate at a specific place on the chromosome called the **origin of replication**, producing two origins. As the chromosome continues to replicate, one origin moves rapidly toward the opposite end of the cell (**Figure 12.12**). While the chromosome is replicating, the cell elongates. When replication is complete and the bacterium has reached about twice its initial size, its plasma membrane pinches inward, dividing the parent *E. coli* cell into two daughter cells. In this way, each cell inherits a complete genome.

Using the techniques of modern DNA technology to tag the origins of replication with molecules that glow green in fluorescence microscopy (see Figure 6.3), researchers have directly observed the movement of bacterial chromosomes. This movement is reminiscent of the poleward movements of the centromere regions of eukaryotic chromosomes during anaphase of mitosis, but bacteria don't have visible mitotic spindles or even microtubules. In most bacterial species studied, the two origins of replication end up at opposite ends of the cell or in some other very specific location, possibly anchored there by one or more proteins. How bacterial chromosomes move and how their specific location is established and maintained are still not fully understood. However, several proteins have been identified that play important roles: One resembling eukaryotic actin apparently functions in bacterial chromosome movement during cell division, and another that is related to tubulin seems to help pinch the plasma membrane inward, separating the two bacterial daughter cells.

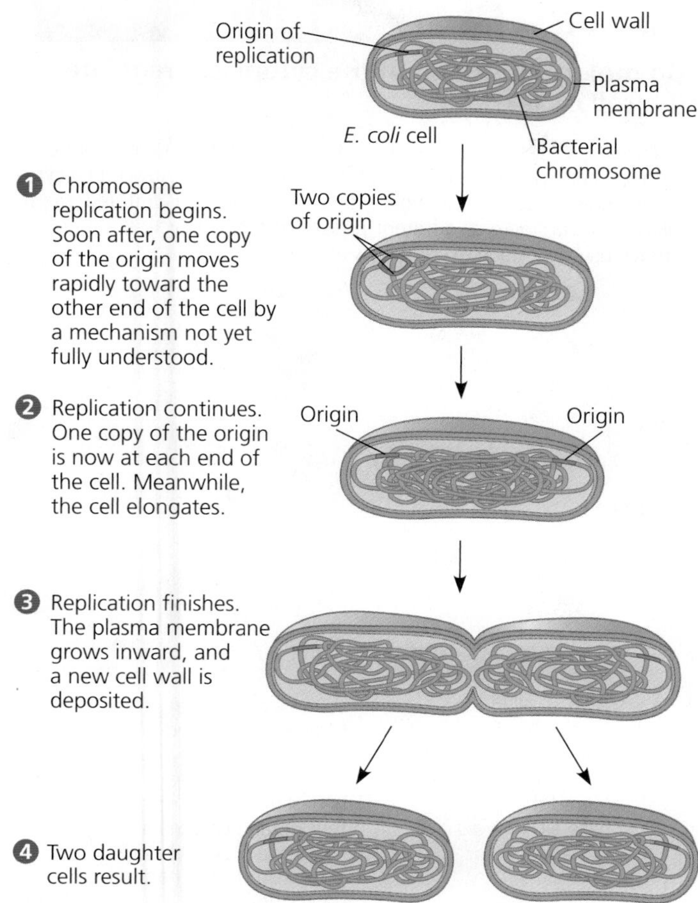

Origin of replication
Cell wall
Plasma membrane
E. coli cell
Bacterial chromosome
1 Chromosome replication begins. Soon after, one copy of the origin moves rapidly toward the other end of the cell by a mechanism not yet fully understood.
Two copies of origin
2 Replication continues. One copy of the origin is now at each end of the cell. Meanwhile, the cell elongates.
Origin
Origin
3 Replication finishes. The plasma membrane grows inward, and a new cell wall is deposited.
4 Two daughter cells result.

▲ **Figure 12.12 Bacterial cell division by binary fission.** The bacterium *E. coli*, shown here, has a single, circular chromosome.

The Evolution of Mitosis

EVOLUTION Given that prokaryotes preceded eukaryotes on Earth by more than a billion years, we might hypothesize that mitosis evolved from simpler prokaryotic mechanisms of cell reproduction. The fact that some of the proteins involved in bacterial binary fission are related to eukaryotic proteins that function in mitosis supports that hypothesis.

As eukaryotes evolved, along with their larger genomes and nuclear envelopes, the ancestral process of binary fission, seen today in bacteria, somehow gave rise to mitosis. **Figure 12.13** shows some variations on cell division in different groups of organisms. These processes may be similar to mechanisms used by ancestral species and thus may resemble steps in the evolution of mitosis from a binary fission-like process presumably carried out by very early bacteria. Possible intermediate stages are suggested by two unusual types of nuclear division found today in certain unicellular eukaryotes—dinoflagellates, diatoms, and some yeasts. These two modes of nuclear division are thought to be cases where ancestral mechanisms have remained relatively unchanged over evolutionary time. In both types, the nuclear envelope remains intact, in contrast to what happens in most eukaryotic cells.

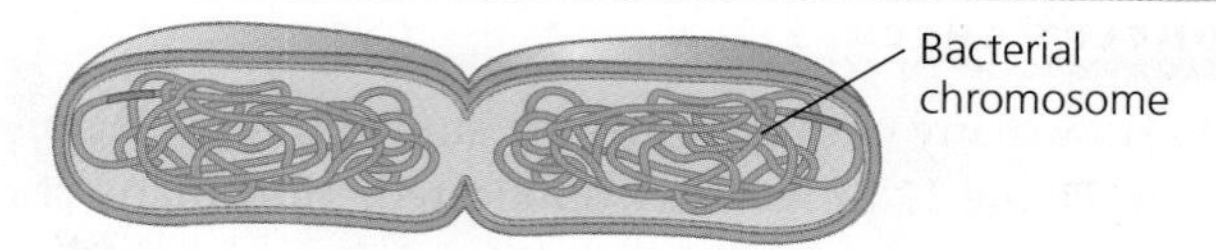

(a) Bacteria. During binary fission in bacteria, the origins of the daughter chromosomes move to opposite ends of the cell. The mechanism is not fully understood, but proteins may anchor the daughter chromosomes to specific sites on the plasma membrane.

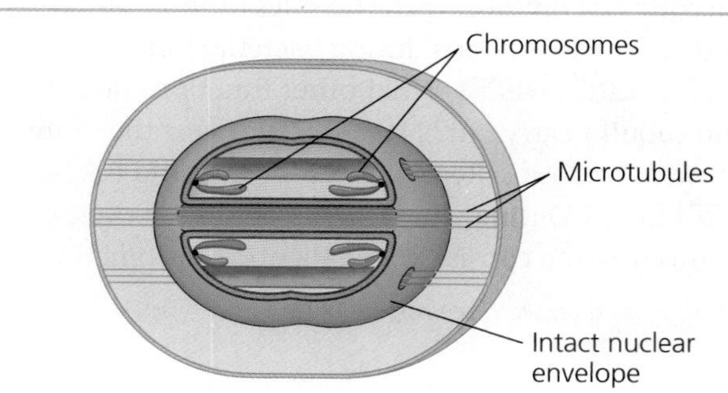

(b) Dinoflagellates. In unicellular protists called dinoflagellates, the chromosomes attach to the nuclear envelope, which remains intact during cell division. Microtubules pass through the nucleus inside cytoplasmic tunnels, reinforcing the spatial orientation of the nucleus, which then divides in a process reminiscent of bacterial binary fission.

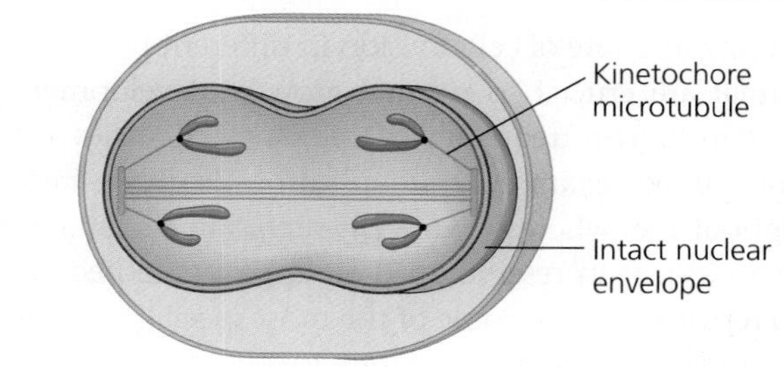

(c) Diatoms and some yeasts. In two other groups of unicellular protists, diatoms and some yeasts, the nuclear envelope also remains intact during cell division. In these organisms, the microtubules form a spindle *within* the nucleus. Microtubules separate the chromosomes, and the nucleus splits into two daughter nuclei.

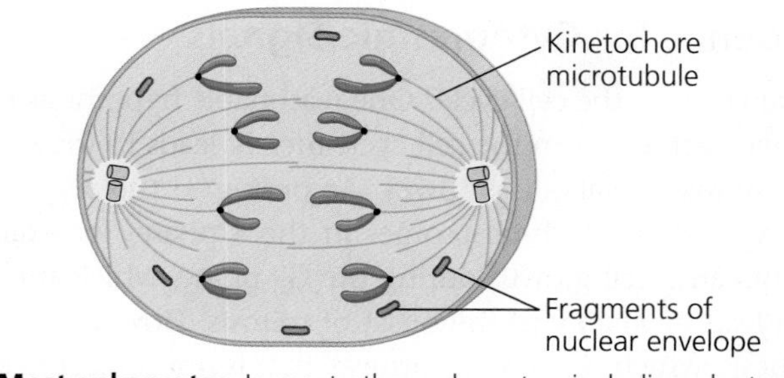

(d) Most eukaryotes. In most other eukaryotes, including plants and animals, the spindle forms outside the nucleus, and the nuclear envelope breaks down during mitosis. Microtubules separate the chromosomes, and two nuclear envelopes then form.

▲ **Figure 12.13 Mechanisms of cell division in several groups of organisms.** Some unicellular eukaryotes existing today have mechanisms of cell division that may resemble intermediate steps in the evolution of mitosis. Except for (a), these schematic diagrams do not show cell walls.

CONCEPT CHECK 12.2

1. How many chromosomes are shown in the diagram in Figure 12.8? Are they duplicated? How many chromatids are shown?
2. Compare cytokinesis in animal cells and plant cells.
3. What is the function of nonkinetochore microtubules?
4. Compare the roles of tubulin and actin during eukaryotic cell division with the roles of tubulin-like and actin-like proteins during bacterial binary fission.
5. **MAKE CONNECTIONS** What other functions do actin and tubulin carry out? Name the proteins they interact with to do so. (Review Figures 6.21a and 6.27a.)
6. **WHAT IF?** During which stages of the cell cycle does a chromosome consist of two identical chromatids?

For suggested answers, see Appendix A.

CONCEPT 12.3

The eukaryotic cell cycle is regulated by a molecular control system

The timing and rate of cell division in different parts of a plant or animal are crucial to normal growth, development, and maintenance. The frequency of cell division varies with the type of cell. For example, human skin cells divide frequently throughout life, whereas liver cells maintain the ability to divide but keep it in reserve until an appropriate need arises—say, to repair a wound. Some of the most specialized cells, such as fully formed nerve cells and muscle cells, do not divide at all in a mature human. These cell cycle differences result from regulation at the molecular level. The mechanisms of this regulation are of intense interest, not only for understanding the life cycles of normal cells but also for understanding how cancer cells manage to escape the usual controls.

Evidence for Cytoplasmic Signals

What controls the cell cycle? One reasonable hypothesis might be that each event in the cell cycle merely leads to the next, as in a simple metabolic pathway. According to this hypothesis, the replication of chromosomes in the S phase, for example, might cause cell growth during the G_2 phase, which might in turn lead inevitably to the onset of mitosis. However, this hypothesis, which proposes a pathway that is not subject to either internal or external regulation, turns out to be incorrect.

In the early 1970s, a variety of experiments led to an alternative hypothesis: that the cell cycle is driven by specific signaling molecules present in the cytoplasm. Some of the first strong evidence for this hypothesis came from experiments with mammalian cells grown in culture. In these experiments, two cells in different phases of the cell cycle were fused to form a single cell with two nuclei. If one of the original cells was in the S phase and the other was in G_1, the G_1 nucleus immediately entered the S phase, as though stimulated by signaling molecules present in the cytoplasm of the first cell. Similarly, if a cell undergoing mitosis (M phase) was fused with another cell in any stage of its cell cycle, even G_1, the second nucleus immediately entered mitosis, with condensation of the chromatin and formation of a mitotic spindle **(Figure 12.14)**.

▼ **Figure 12.14** INQUIRY

Do molecular signals in the cytoplasm regulate the cell cycle?

EXPERIMENT Researchers at the University of Colorado wondered whether a cell's progression through the cell cycle is controlled by cytoplasmic molecules. To investigate this, they selected cultured mammalian cells that were at different phases of the cell cycle and induced them to fuse. Two such experiments are shown here.

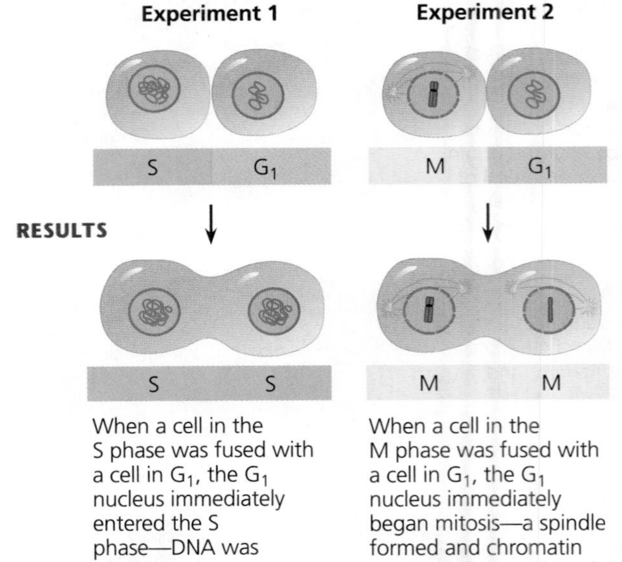

When a cell in the S phase was fused with a cell in G_1, the G_1 nucleus immediately entered the S phase—DNA was synthesized.

When a cell in the M phase was fused with a cell in G_1, the G_1 nucleus immediately began mitosis—a spindle formed and chromatin condensed, even though the chromosome had not been duplicated.

CONCLUSION The results of fusing a G_1 cell with a cell in the S or M phase of the cell cycle suggest that molecules present in the cytoplasm during the S or M phase control the progression to those phases.

SOURCE R. T. Johnson and P. N. Rao, Mammalian cell fusion: Induction of premature chromosome condensation in interphase nuclei, *Nature* 226:717–722 (1970).

WHAT IF? If the progression of phases did not depend on cytoplasmic molecules and each phase began when the previous one was complete, how would the results have differed?

The Cell Cycle Control System

The experiment shown in Figure 12.14 and other experiments on animal cells and yeasts demonstrated that the sequential events of the cell cycle are directed by a distinct **cell cycle control system**, a cyclically operating set of molecules in the cell that both triggers and coordinates key events

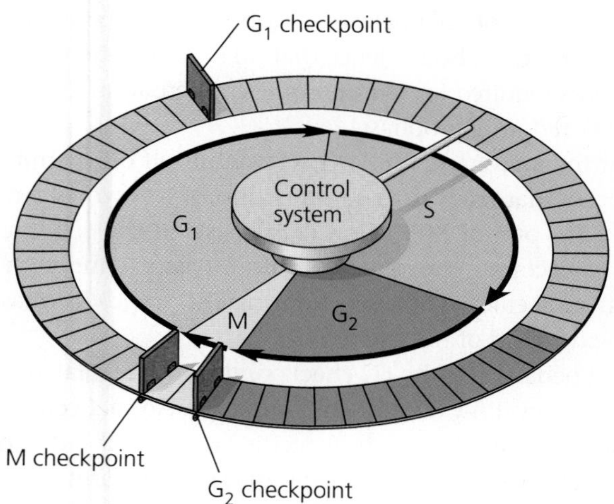

▲ **Figure 12.15 Mechanical analogy for the cell cycle control system.** In this diagram of the cell cycle, the flat "stepping stones" around the perimeter represent sequential events. Like the control device of an automatic washer, the cell cycle control system proceeds on its own, driven by a built-in clock. However, the system is subject to internal and external regulation at various checkpoints, of which three are shown (red).

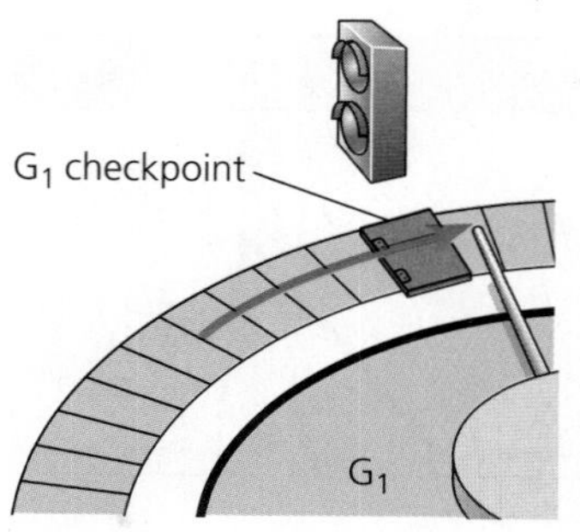

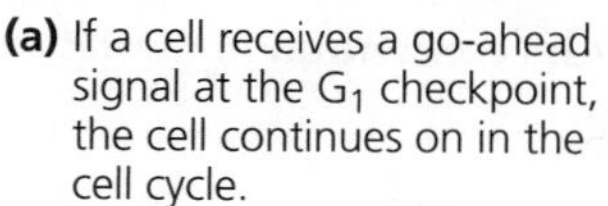

(a) If a cell receives a go-ahead signal at the G_1 checkpoint, the cell continues on in the cell cycle.

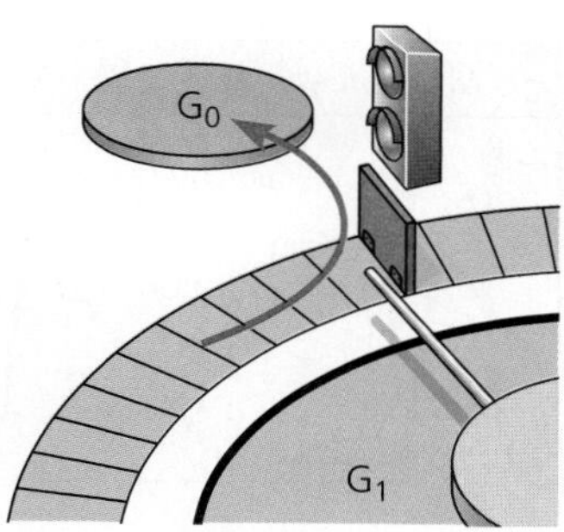

(b) If a cell does not receive a go-ahead signal at the G_1 checkpoint, the cell exits the cell cycle and goes into G_0, a nondividing state.

▲ **Figure 12.16 The G_1 checkpoint.**

WHAT IF? *What might be the result if the cell ignored the checkpoint and progressed through the cell cycle?*

in the cell cycle. The cell cycle control system has been compared to the control device of an automatic washing machine (**Figure 12.15**). Like the washer's timing device, the cell cycle control system proceeds on its own, according to a built-in clock. However, just as a washer's cycle is subject to both internal control (such as the sensor that detects when the tub is filled with water) and external adjustment (such as activation of the start mechanism), the cell cycle is regulated at certain checkpoints by both internal and external signals.

A **checkpoint** in the cell cycle is a control point where stop and go-ahead signals can regulate the cycle. (The signals are transmitted within the cell by the kinds of signal transduction pathways discussed in Chapter 11.) Animal cells generally have built-in stop signals that halt the cell cycle at checkpoints until overridden by go-ahead signals. Many signals registered at checkpoints come from cellular surveillance mechanisms inside the cell. These signals report whether crucial cellular processes that should have occurred by that point have in fact been completed correctly and thus whether or not the cell cycle should proceed. Checkpoints also register signals from outside the cell, as we will discuss later. Three major checkpoints are found in the G_1, G_2, and M phases (see Figure 12.15).

For many cells, the G_1 checkpoint—dubbed the "restriction point" in mammalian cells—seems to be the most important. If a cell receives a go-ahead signal at the G_1 checkpoint, it will usually complete the G_1, S, G_2, and M phases and divide. If it does not receive a go-ahead signal at that point, it will exit the cycle, switching into a nondividing state called the **G_0 phase** (**Figure 12.16**). Most cells of the human body are actually in the G_0 phase. As mentioned earlier, mature nerve cells and muscle cells never divide. Other cells, such as liver cells, can be "called back" from the G_0 phase to the cell cycle by external cues, such as growth factors released during injury.

To understand how cell cycle checkpoints work, we first need to see what kinds of molecules make up the cell cycle control system (the molecular basis for the cell cycle clock) and how a cell progresses through the cycle. Then we will consider the internal and external checkpoint signals that can make the clock pause or continue.

The Cell Cycle Clock: Cyclins and Cyclin-Dependent Kinases

Rhythmic fluctuations in the abundance and activity of cell cycle control molecules pace the sequential events of the cell cycle. These regulatory molecules are mainly proteins of two types: protein kinases and cyclins. Protein kinases are enzymes that activate or inactivate other proteins by phosphorylating them (see Chapter 11). Particular protein kinases give the go-ahead signals at the G_1 and G_2 checkpoints.

Many of the kinases that drive the cell cycle are actually present at a constant concentration in the growing cell, but much of the time they are in an inactive form. To be active, such a kinase must be attached to a **cyclin**, a protein that gets its name from its cyclically fluctuating concentration in the cell. Because of this requirement, these kinases are called **cyclin-dependent kinases**, or **Cdks**. The activity of a Cdk rises and falls with changes in the concentration of its cyclin partner. **Figure 12.17a**, on the next page, shows the fluctuating activity of **MPF**, the cyclin-Cdk complex that was discovered first (in frog eggs). Note that the peaks of MPF activity correspond to the peaks of cyclin concentration. The cyclin level rises during the S and G_2 phases and then falls abruptly during M phase.

The initials MPF stand for "maturation-promoting factor," but we can think of MPF as "M-phase-promoting factor" because it triggers the cell's passage past the G_2 checkpoint into

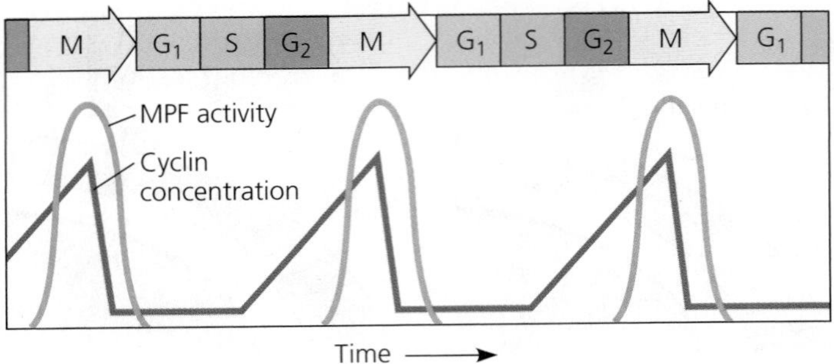

(a) Fluctuation of MPF activity and cyclin concentration during the cell cycle

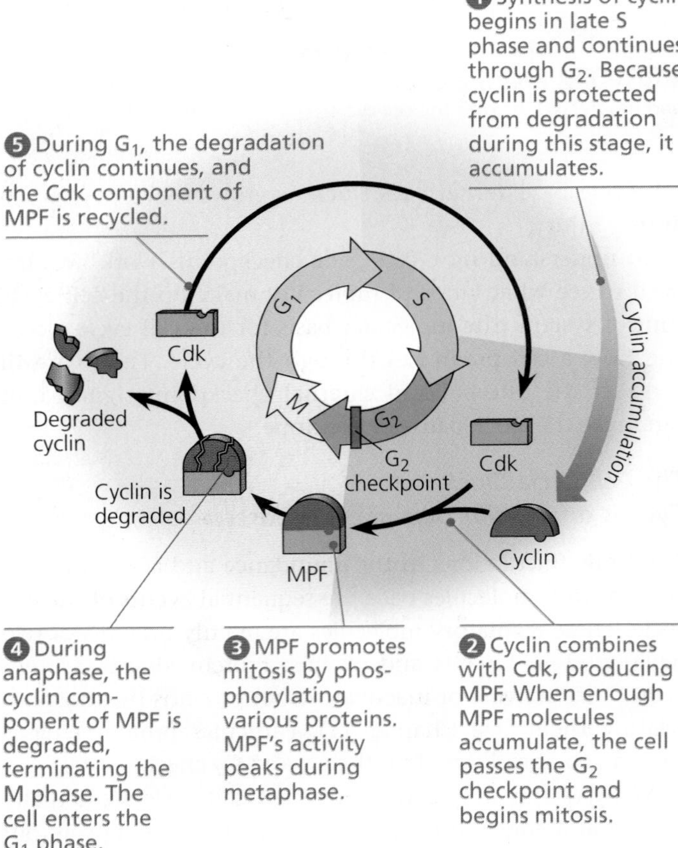

1 Synthesis of cyclin begins in late S phase and continues through G2. Because cyclin is protected from degradation during this stage, it accumulates.
5 During G1, the degradation of cyclin continues, and the Cdk component of MPF is recycled.
G1
S
M
G2
G2 checkpoint
Cdk
Degraded cyclin
Cyclin is degraded
MPF
Cdk
Cyclin
Cyclin accumulation
4 During anaphase, the cyclin component of MPF is degraded, terminating the M phase. The cell enters the G1 phase.
3 MPF promotes mitosis by phosphorylating various proteins. MPF's activity peaks during metaphase.
2 Cyclin combines with Cdk, producing MPF. When enough MPF molecules accumulate, the cell passes the G2 checkpoint and begins mitosis.

(b) Molecular mechanisms that help regulate the cell cycle

▲ **Figure 12.17 Molecular control of the cell cycle at the G_2 checkpoint.** The steps of the cell cycle are timed by rhythmic fluctuations in the activity of cyclin-dependent kinases (Cdks). Here we focus on a cyclin-Cdk complex in animal cells called MPF, which acts at the G_2 checkpoint as a go-ahead signal, triggering the events of mitosis.

? *Explain how the events in the diagram in (b) are related to the "Time" axis of the graph in (a).*

M phase (**Figure 12.17b**). When cyclins that accumulate during G_2 associate with Cdk molecules, the resulting MPF complex phosphorylates a variety of proteins, initiating mitosis. MPF acts both directly as a kinase and indirectly by activating other kinases. For example, MPF causes phosphorylation of various proteins of the nuclear lamina (see Figure 6.9), which promotes fragmentation of the nuclear envelope during prometaphase of mitosis. There is also evidence that MPF contributes to molecular events required for chromosome condensation and spindle formation during prophase.

During anaphase, MPF helps switch itself off by initiating a process that leads to the destruction of its own cyclin. The noncyclin part of MPF, the Cdk, persists in the cell, inactive until it becomes part of MPF again by associating with new cyclin molecules synthesized during the S and G_2 phases of the next round of the cycle.

Cell behavior at the G_1 checkpoint is also regulated by the activity of cyclin-Cdk protein complexes. Animal cells appear to have at least three Cdk proteins and several different cyclins that operate at this checkpoint. The fluctuating activities of different cyclin-Cdk complexes are of major importance in controlling all the stages of the cell cycle.

Stop and Go Signs: Internal and External Signals at the Checkpoints

Research scientists are currently working out the pathways that link signals originating inside and outside the cell with the responses by cyclin-dependent kinases and other proteins. An example of an internal signal occurs at the third important checkpoint, the M phase checkpoint. Anaphase, the separation of sister chromatids, does not begin until all the chromosomes are properly attached to the spindle at the metaphase plate. Researchers have learned that as long as some kinetochores are unattached to spindle microtubules, the sister chromatids remain together, delaying anaphase. Only when the kinetochores of all the chromosomes are properly attached to the spindle does the appropriate regulatory protein complex become activated. (In this case, the regulatory molecule is not a cyclin-Cdk complex but, instead, a different complex made up of several proteins.) Once activated, the complex sets off a chain of molecular events that activates the enzyme separase, which cleaves the cohesins, allowing the sister chromatids to separate. This mechanism ensures that daughter cells do not end up with missing or extra chromosomes.

Studies using animal cells in culture have led to the identification of many external factors, both chemical and physical, that can influence cell division. For example, cells fail to divide if an essential nutrient is lacking in the culture medium. (This is analogous to trying to run an automatic washing machine without the water supply hooked up; an internal sensor won't allow the machine to continue past the point where water is needed.) And even if all other conditions are favorable, most types of mammalian cells divide in culture only if the growth medium includes specific growth factors. As mentioned in Chapter 11, a **growth factor** is a protein released by certain cells that stimulates other cells to divide. Researchers have discovered more than 50 growth factors. Different cell types respond specifically to different growth factors or combinations of growth factors.

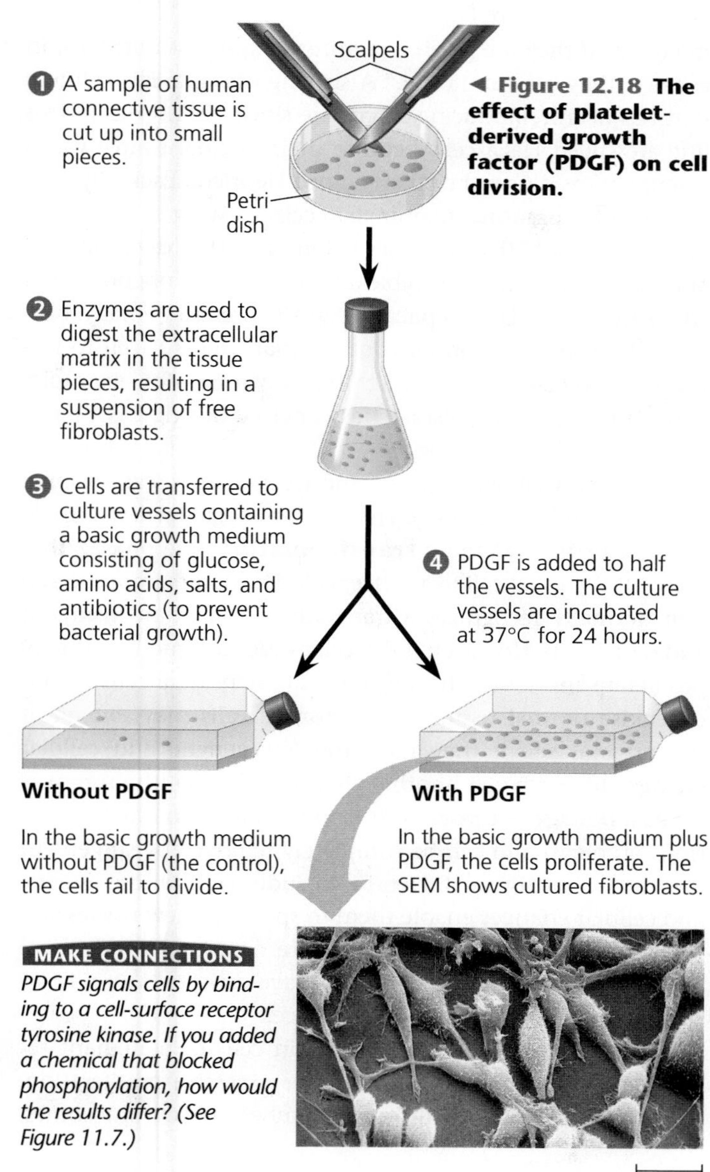

Scalpels
1 A sample of human connective tissue is cut up into small pieces.
Petri dish
2 Enzymes are used to digest the extracellular matrix in the tissue pieces, resulting in a suspension of free fibroblasts.
3 Cells are transferred to culture vessels containing a basic growth medium consisting of glucose, amino acids, salts, and antibiotics (to prevent bacterial growth).
4 PDGF is added to half the vessels. The culture vessels are incubated at 37°C for 24 hours.
Without PDGF
In the basic growth medium without PDGF (the control), the cells fail to divide.
With PDGF
In the basic growth medium plus PDGF, the cells proliferate. The SEM shows cultured fibroblasts.
MAKE CONNECTIONS
PDGF signals cells by binding to a cell-surface receptor tyrosine kinase. If you added a chemical that blocked phosphorylation, how would the results differ? (See Figure 11.7.)
10 μm

◄ **Figure 12.18 The effect of platelet-derived growth factor (PDGF) on cell division.**

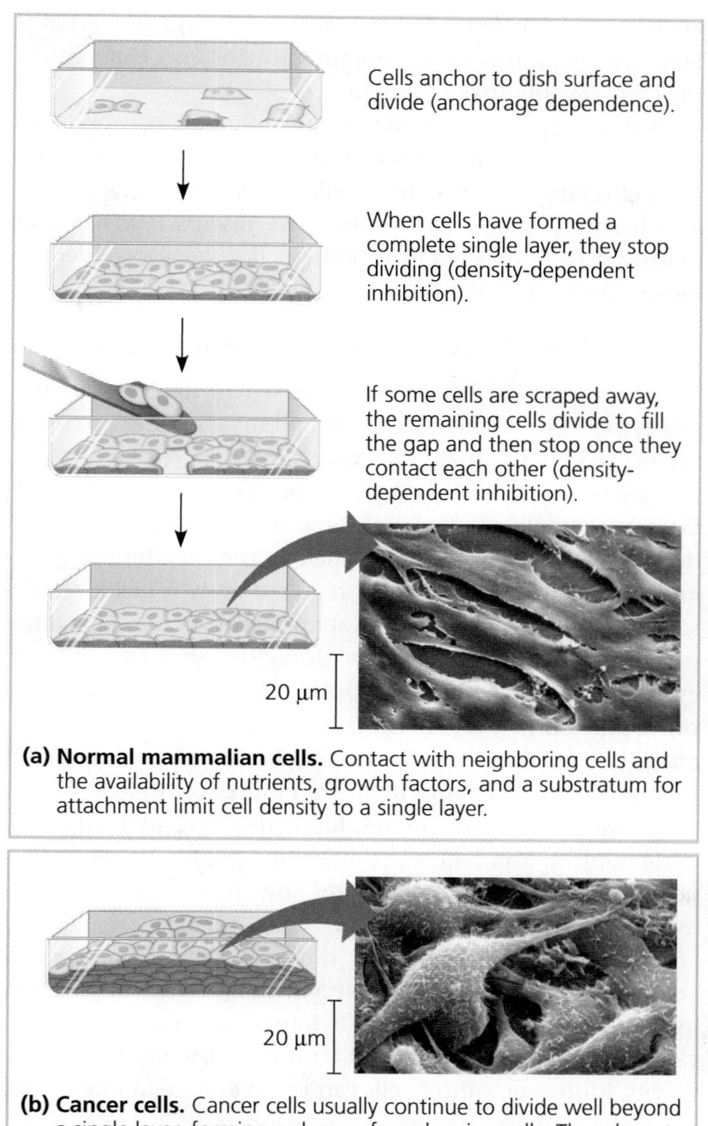

▲ **Figure 12.19 Density-dependent inhibition and anchorage dependence of cell division.** Individual cells are shown disproportionately large in the drawings.

Consider, for example, *platelet-derived growth factor (PDGF)*, which is made by blood cell fragments called platelets. The experiment illustrated in **Figure 12.18** demonstrates that PDGF is required for the division of cultured fibroblasts, a type of connective tissue cell. Fibroblasts have PDGF receptors on their plasma membranes. The binding of PDGF molecules to these receptors (which are receptor tyrosine kinases; see Chapter 11) triggers a signal transduction pathway that allows the cells to pass the G_1 checkpoint and divide. PDGF stimulates fibroblast division not only in the artificial conditions of cell culture, but also in an animal's body. When an injury occurs, platelets release PDGF in the vicinity. The resulting proliferation of fibroblasts helps heal the wound.

The effect of an external physical factor on cell division is clearly seen in **density-dependent inhibition**, a phenomenon in which crowded cells stop dividing (**Figure 12.19a**). As first observed many years ago, cultured cells normally divide until they form a single layer of cells on the inner surface of the culture container, at which point the cells stop dividing. If some cells are removed, those bordering the open space begin dividing again and continue until the vacancy is filled. Follow-up studies revealed that the binding of a cell-surface protein to its counterpart on an adjoining cell sends a growth-inhibiting signal to both cells, preventing them from moving forward in the cell cycle, even in the presence of growth factors.

Most animal cells also exhibit **anchorage dependence** (see Figure 12.19a). To divide, they must be attached to a substratum, such as the inside of a culture jar or the extracellular matrix of a tissue. Experiments suggest that like cell density, anchorage is signaled to the cell cycle control system via pathways involving plasma membrane proteins and elements of the cytoskeleton linked to them.

Density-dependent inhibition and anchorage dependence appear to function in the body's tissues as well as in cell culture, checking the growth of cells at some optimal density and location. Cancer cells, which we discuss next, exhibit neither density-dependent inhibition nor anchorage dependence **(Figure 12.19b)**.

Loss of Cell Cycle Controls in Cancer Cells

Cancer cells do not heed the normal signals that regulate the cell cycle. They divide excessively and invade other tissues. If unchecked, they can kill the organism.

Cancer cells in culture do not stop dividing when growth factors are depleted. A logical hypothesis is that cancer cells do not need growth factors in their culture medium to grow and divide. They may make a required growth factor themselves, or they may have an abnormality in the signaling pathway that conveys the growth factor's signal to the cell cycle control system even in the absence of that factor. Another possibility is an abnormal cell cycle control system. In all of these scenarios, the underlying basis of the abnormality is almost always a change in one or more genes that alters the function of their protein products, resulting in faulty cell cycle control. You will learn more in Chapter 18 about the genetic bases of these changes and how these conditions may lead to cancer.

There are other important differences between normal cells and cancer cells that reflect derangements of the cell cycle. If and when they stop dividing, cancer cells do so at random points in the cycle, rather than at the normal checkpoints. Moreover, cancer cells can go on dividing indefinitely in culture if they are given a continual supply of nutrients; in essence, they are "immortal." A striking example is a cell line that has been reproducing in culture since 1951. Cells of this line are called HeLa cells because their original source was a tumor removed from a woman named Henrietta Lacks. By contrast, nearly all normal mammalian cells growing in culture divide only about 20 to 50 times before they stop dividing, age, and die. (We'll see a possible reason for this phenomenon when we discuss DNA replication in Chapter 16.) Finally, cancer cells evade the normal controls that trigger a cell to undergo apoptosis when something is wrong—for example, when an irreparable mistake has occurred during DNA replication preceding mitosis.

The abnormal behavior of cancer cells can be catastrophic when it occurs in the body. The problem begins when a single cell in a tissue undergoes **transformation**, the process that converts a normal cell to a cancer cell. The body's immune system normally recognizes a transformed cell as an insurgent and destroys it. However, if the cell evades destruction, it may proliferate and form a tumor, a mass of abnormal cells within otherwise normal tissue. The abnormal cells may remain at the original site if they have too few genetic and cellular changes to survive at another site. In that case, the tumor is called a **benign tumor**. Most benign tumors do not cause serious problems and can be completely removed by surgery. In contrast, a **malignant tumor** includes cells whose genetic and cellular changes enable them to spread to new tissues and impair the functions of one or more organs. An individual with a malignant tumor is said to have cancer; **Figure 12.20** shows the development of breast cancer.

The changes that have occurred in cells of malignant tumors show up in many ways besides excessive proliferation. These cells may have unusual numbers of chromosomes,

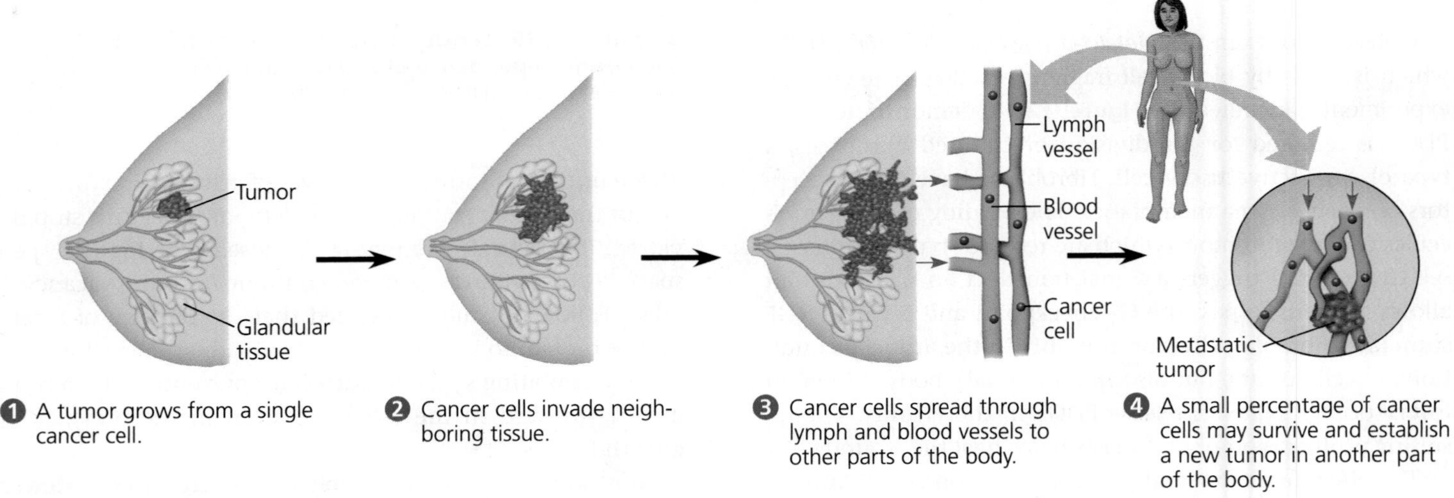

Tumor
Glandular tissue
Lymph vessel
Blood vessel
Cancer cell
Metastatic tumor
1 A tumor grows from a single cancer cell.
2 Cancer cells invade neighboring tissue.
3 Cancer cells spread through lymph and blood vessels to other parts of the body.
4 A small percentage of cancer cells may survive and establish a new tumor in another part of the body.

▲ **Figure 12.20 The growth and metastasis of a malignant breast tumor.** The cells of malignant (cancerous) tumors grow in an uncontrolled way and can spread to neighboring tissues and, via lymph and blood vessels, to other parts of the body. The spread of cancer cells beyond their original site is called metastasis.

though whether this is a cause or an effect of transformation is a current topic of debate. Their metabolism may be disabled, and they may cease to function in any constructive way. Abnormal changes on the cell surface cause cancer cells to lose attachments to neighboring cells and the extracellular matrix, allowing them to spread into nearby tissues. Cancer cells may also secrete signaling molecules that cause blood vessels to grow toward the tumor. A few tumor cells may separate from the original tumor, enter blood vessels and lymph vessels, and travel to other parts of the body. There, they may proliferate and form a new tumor. This spread of cancer cells to locations distant from their original site is called **metastasis** (see Figure 12.20).

A tumor that appears to be localized may be treated with high-energy radiation, which damages DNA in cancer cells much more than it does in normal cells, apparently because the majority of cancer cells have lost the ability to repair such damage. To treat known or suspected metastatic tumors, chemotherapy is used, in which drugs that are toxic to actively dividing cells are administered through the circulatory system. As you might expect, chemotherapeutic drugs interfere with specific steps in the cell cycle. For example, the drug Taxol freezes the mitotic spindle by preventing microtubule depolymerization, which stops actively dividing cells from proceeding past metaphase. The side effects of chemotherapy are due to the drugs' effects on normal cells that divide often. For example, nausea results from chemotherapy's effects on intestinal cells, hair loss from effects on hair follicle cells, and susceptibility to infection from effects on immune system cells.

Over the past several decades, researchers have produced a flood of valuable information about cell-signaling pathways and how their malfunction contributes to the development of cancer through effects on the cell cycle. Coupled with new molecular techniques, such as the ability to rapidly sequence the DNA of cells in a particular tumor, medical treatments for cancer are beginning to become more "personalized" to a particular patient's tumor. Breast cancer provides a good example. Basic research on the processes described in Chapters 11 and 12 has augmented our understanding of the molecular events underlying development of breast cancer. Proteins functioning in cell signaling pathways that affect the cell cycle are often found to be altered in breast cancer cells. Analyzing the level and sequences of such proteins has allowed physicians to better tailor the treatment to the cancers of some individuals, as shown in **Figure 12.21**.

One of the big lessons we've learned about the development of cancer, though, is how very complex the process is. There are many areas that remain to be explored. Perhaps the reason we have so many unanswered questions about cancer cells is that there is still so much to learn about how normal cells function. The cell, life's basic unit of structure and function, holds enough secrets to engage researchers well into the future.

▼ **Figure 12.21**

IMPACT

Advances in Treatment of Breast Cancer

Cancer cells, such as the breast cancer cell shown below, are analyzed by DNA sequencing and other molecular techniques to look for alterations in the level or sequence of specific proteins associated with cancer. For example, the cells of roughly 20–25% of breast cancer tumors show abnormally high amounts of a cell-surface receptor tyrosine kinase called HER2, and many show an increase in the number of estrogen receptor (ER) molecules, intracellular receptors that can trigger cell division. Based on lab findings, a physician can prescribe chemotherapy with a molecule that blocks the function of the specific protein (Herceptin for HER2 and tamoxifen for ERs). Treatment using these agents, when appropriate, has led to increased survival rates and fewer cancer recurrences.

WHY IT MATTERS Approximately one out of every eight women will develop breast cancer, the most common cancer among women. Worldwide, the incidence of breast cancer has been increasing annually. However, the mortality rate from this disease is falling in the United States and elsewhere, probably a result of earlier detection and improved treatment. Furthermore, what we are learning from the study of breast cancer also enhances our understanding of the development and treatment of other types of cancer.

FURTHER READING F. J. Esteva and G. N. Hortobagyi, Gaining ground on breast cancer, *Scientific American* 298:58–65 (2008).

MAKE CONNECTIONS Review the material in Chapter 11 on receptor tyrosine kinases and intracellular receptors (Figures 11.7 and 11.9 on pp. 212–214). Explain in general how these receptors might function in triggering cell division.

CONCEPT CHECK 12.3

1. In Figure 12.14, why do the nuclei resulting from experiment 2 contain different amounts of DNA?
2. How does MPF allow a cell to pass the G_2 phase checkpoint and enter mitosis? (See Figure 12.17.)
3. What phase are most of your body cells in?
4. Compare and contrast a benign tumor and a malignant tumor.
5. **WHAT IF?** What would happen if you performed the experiment in Figure 12.18 with cancer cells?

For suggested answers, see Appendix A.

12 CHAPTER REVIEW

SUMMARY OF KEY CONCEPTS

- Unicellular organisms reproduce by **cell division**; multicellular organisms depend on cell division for their development from a fertilized egg and for growth and repair. Cell division is part of the **cell cycle**, an ordered sequence of events in the life of a cell from its origin until it divides into daughter cells.

CONCEPT 12.1

Most cell division results in genetically identical daughter cells (pp. 229–230)

- The genetic material (DNA) of a cell—its **genome**—is partitioned among **chromosomes**. Each eukaryotic chromosome consists of one DNA molecule associated with many proteins that maintain chromosome structure and help control the activity of genes. Together, the complex of DNA and associated proteins is called **chromatin**. The chromatin of a chromosome exists in different states of condensation at different times. In animals, **gametes** have one set of chromosomes and **somatic cells** have two sets.
- Cells replicate their genetic material before they divide, ensuring that each daughter cell can receive a copy of the DNA. In preparation for cell division, chromosomes are duplicated, each one then consisting of two identical **sister chromatids** joined along their lengths by sister chromatid cohesion and held most tightly together at a constricted region at the **centromeres** of the chromatids. When this cohesion is broken, the chromatids separate during cell division, becoming the chromosomes of the new daughter cells. Eukaryotic cell division consists of **mitosis** (division of the nucleus) and **cytokinesis** (division of the cytoplasm).

? *Differentiate between these terms: chromosome, chromatin, and chromatid.*

CONCEPT 12.2

The mitotic phase alternates with interphase in the cell cycle (pp. 230–238)

- Between divisions, a cell is in **interphase**: the $\mathbf{G_1}$, **S**, and $\mathbf{G_2}$ phases. The cell grows throughout interphase, but DNA is replicated only during the synthesis (S) phase. Mitosis and cytokinesis make up the **mitotic (M) phase** of the cell cycle.

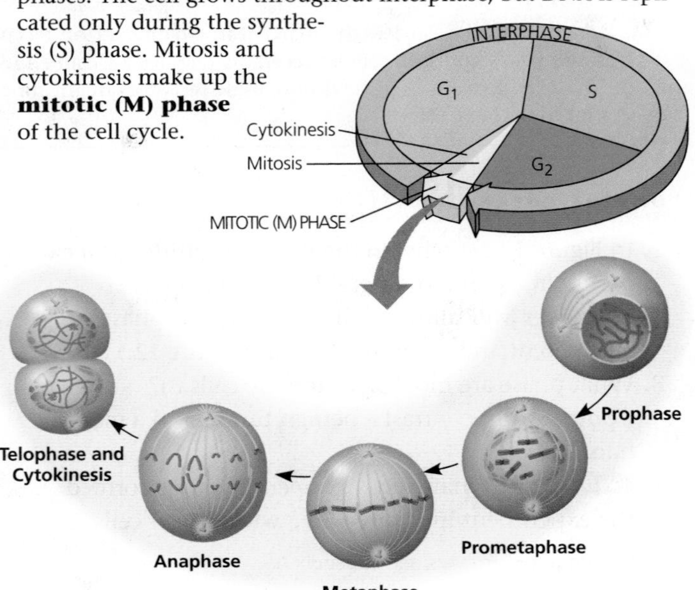

- The **mitotic spindle** is an apparatus of microtubules that controls chromosome movement during mitosis. In animal cells, the spindle arises from the **centrosomes** and includes spindle microtubules and **asters**. Some spindle microtubules attach to the **kinetochores** of chromosomes and move the chromosomes to the **metaphase plate**. In anaphase, sister chromatids separate, and motor proteins move them along the kinetochore microtubules toward opposite ends of the cell. Meanwhile, motor proteins push nonkinetochore microtubules from opposite poles away from each other, elongating the cell. In telophase, genetically identical daughter nuclei form at opposite ends of the cell.
- Mitosis is usually followed by cytokinesis. Animal cells carry out cytokinesis by **cleavage**, and plant cells form a **cell plate**.
- During **binary fission** in bacteria, the chromosome replicates and the two daughter chromosomes actively move apart. Some of the proteins involved in bacterial binary fission are related to eukaryotic actin and tubulin.
- Since prokaryotes preceded eukaryotes by more than a billion years, it is likely that mitosis evolved from prokaryotic cell division. Certain unicellular eukaryotes exhibit mechanisms of cell division that may be similar to those of ancestors of existing eukaryotes. Such mechanisms might have been intermediate steps in the evolution of mitosis from bacterial binary fission.

? *In which of the three subphases of interphase and the stages of mitosis do chromosomes exist as single DNA molecules?*

CONCEPT 12.3

The eukaryotic cell cycle is regulated by a molecular control system (pp. 238–243)

- Signaling molecules present in the cytoplasm regulate progress through the cell cycle.
- The **cell cycle control system** is molecularly based. Cyclic changes in regulatory proteins work as a cell cycle clock. The key molecules are **cyclins** and **cyclin-dependent kinases (Cdks)**. The clock has specific **checkpoints** where the cell cycle stops until a go-ahead signal is received. Cell culture has enabled researchers to study the molecular details of cell division. Both internal signals and external signals control the cell cycle checkpoints via signal transduction pathways. Most cells exhibit **density-dependent inhibition** of cell division as well as **anchorage dependence**.
- Cancer cells elude normal cell cycle regulation and divide out of control, forming tumors. **Malignant tumors** invade surrounding tissues and can undergo **metastasis**, exporting cancer cells to other parts of the body, where they may form secondary tumors. Recent advances in understanding the cell cycle and cell signaling, as well as techniques for sequencing DNA, have allowed improvements in cancer treatment.

? *Explain the significance of the G_1, G_2, and M checkpoints and the go-ahead signals involved in the cell cycle control system.*

TEST YOUR UNDERSTANDING

LEVEL 1: KNOWLEDGE/COMPREHENSION

1. Through a microscope, you can see a cell plate beginning to develop across the middle of a cell and nuclei forming on either side of the cell plate. This cell is most likely
   a. an animal cell in the process of cytokinesis.
   b. a plant cell in the process of cytokinesis.

c. an animal cell in the S phase of the cell cycle.
d. a bacterial cell dividing.
e. a plant cell in metaphase.

2. Vinblastine is a standard chemotherapeutic drug used to treat cancer. Because it interferes with the assembly of microtubules, its effectiveness must be related to
   a. disruption of mitotic spindle formation.
   b. inhibition of regulatory protein phosphorylation.
   c. suppression of cyclin production.
   d. myosin denaturation and inhibition of cleavage furrow formation.
   e. inhibition of DNA synthesis.

3. One difference between cancer cells and normal cells is that cancer cells
   a. are unable to synthesize DNA.
   b. are arrested at the S phase of the cell cycle.
   c. continue to divide even when they are tightly packed together.
   d. cannot function properly because they are affected by density-dependent inhibition.
   e. are always in the M phase of the cell cycle.

4. The decline of MPF activity at the end of mitosis is due to
   a. the destruction of the protein kinase Cdk.
   b. decreased synthesis of Cdk.
   c. the degradation of cyclin.
   d. the accumulation of cyclin.
   e. synthesis of DNA.

5. In the cells of some organisms, mitosis occurs without cytokinesis. This will result in
   a. cells with more than one nucleus.
   b. cells that are unusually small.
   c. cells lacking nuclei.
   d. destruction of chromosomes.
   e. cell cycles lacking an S phase.

6. Which of the following does *not* occur during mitosis?
   a. condensation of the chromosomes
   b. replication of the DNA
   c. separation of sister chromatids
   d. spindle formation
   e. separation of the spindle poles

LEVEL 2: APPLICATION/ANALYSIS

7. In the light micrograph below of dividing cells near the tip of an onion root, identify a cell in each of the following stages: prophase, prometaphase, metaphase, anaphase, and telophase. Describe the major events occurring at each stage.

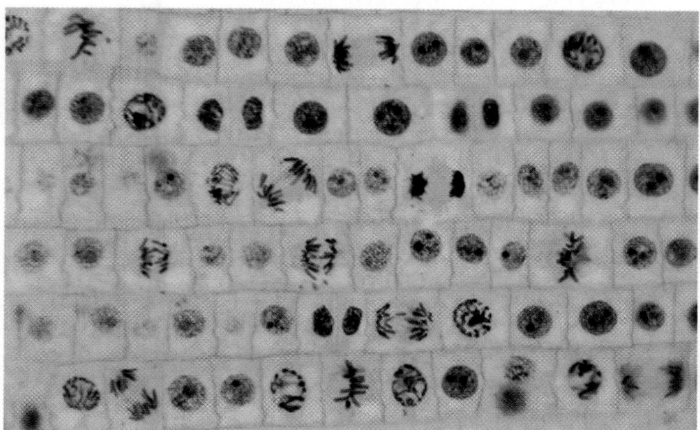

8. A particular cell has half as much DNA as some other cells in a mitotically active tissue. The cell in question is most likely in
   a. G_1.
   b. G_2.
   c. prophase.
   d. metaphase.
   e. anaphase.

9. The drug cytochalasin B blocks the function of actin. Which of the following aspects of the animal cell cycle would be most disrupted by cytochalasin B?
   a. spindle formation
   b. spindle attachment to kinetochores
   c. DNA synthesis
   d. cell elongation during anaphase
   e. cleavage furrow formation and cytokinesis

10. **DRAW IT** Draw one eukaryotic chromosome as it would appear during interphase, during each of the stages of mitosis, and during cytokinesis. Also draw and label the nuclear envelope and any microtubules attached to the chromosome(s).

LEVEL 3: SYNTHESIS/EVALUATION

11. **EVOLUTION CONNECTION**
    The result of mitosis is that the daughter cells end up with the same number of chromosomes that the parent cell had. Another way to maintain the number of chromosomes would be to carry out cell division first and then duplicate the chromosomes in each daughter cell. Do you think this would be an equally good way of organizing the cell cycle? Why do you suppose that evolution has not led to this alternative?

12. **SCIENTIFIC INQUIRY**
    Although both ends of a microtubule can gain or lose subunits, one end (called the plus end) polymerizes and depolymerizes at a higher rate than the other end (the minus end). For spindle microtubules, the plus ends are in the center of the spindle, and the minus ends are at the poles. Motor proteins that move along microtubules specialize in walking either toward the plus end or toward the minus end; the two types are called plus end–directed and minus end–directed motor proteins, respectively. Given what you know about chromosome movement and spindle changes during anaphase, predict which type of motor proteins would be present on (a) kinetochore microtubules and (b) nonkinetochore microtubules.

13. **WRITE ABOUT A THEME**
    The Genetic Basis of Life The continuity of life is based on heritable information in the form of DNA. In a short essay (100–150 words), explain how the process of mitosis faithfully parcels out exact copies of this heritable information in the production of genetically identical daughter cells.

For selected answers, see Appendix A.

MasteringBIOLOGY www.masteringbiology.com

1. MasteringBiology® Assignments
   BioFlix **Tutorials** Mitosis: Mitosis and the Cell Cycle • Mechanism of Mitosis • Comparing Cell Division in Animals, Plants, and Bacteria
   Activities The Cell Cycle • Mitosis and Cytokinesis Animation • The Phases of Mitosis • Four Phases of the Cell Cycle • Causes of Cancer
   Questions Student Misconceptions • Reading Quiz • Multiple Choice • End-of-Chapter
2. eText
   Read your book online, search, take notes, highlight text, and more.
3. The Study Area
   Practice Tests • Cumulative Test • ***BioFlix*** 3-D Animations • MP3 Tutor Sessions • Videos · Activities • Investigations • Lab Media • Audio Glossary • Word Study Tools • Art

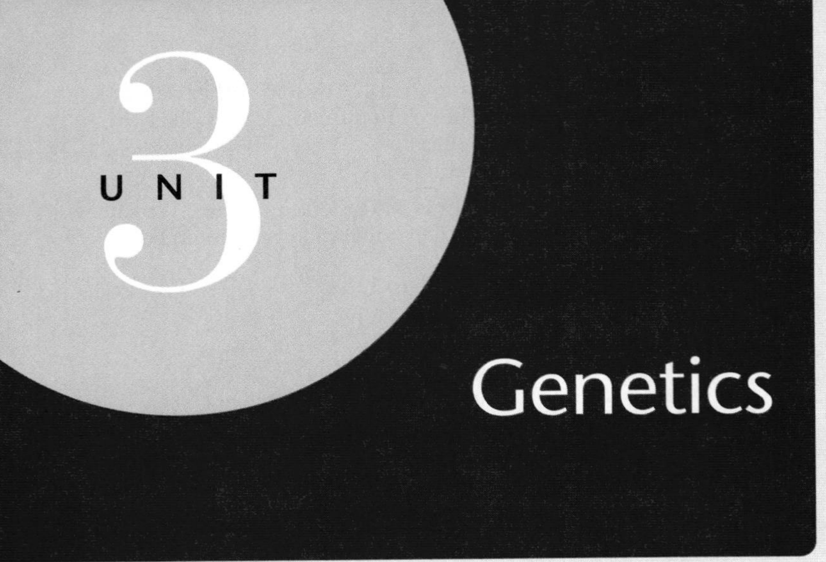

An Interview with

Joan A. Steitz

RNA is Joan Steitz's favorite molecule, and her research into its structures and functions has made contributions of enormous importance to our understanding of genetics at the molecular level. Raised in Minnesota, Dr. Steitz has a B.S. in Chemistry from Antioch College and a Ph.D. in Biochemistry and Molecular Biology from Harvard, where she worked in the laboratory of James D. Watson. Among her many awards and honors are the National Medal of Science, the Gairdner International Award, and 12 honorary doctorates. She is a member of the National Academy of Sciences and the Institute of Medicine. A teacher and researcher at Yale University since 1970, she is now Sterling Professor of Molecular Biophysics and Biochemistry and an Investigator of the Howard Hughes Medical Institute.

How did you get started in molecular genetics?

I first learned about the structure of DNA in my third year of college, during a co-op job at MIT. I was enthralled with the idea that DNA might be the molecular basis for all of the genetics—red hair, wrinkled peas, and so forth—that I had learned about in high school. After that, I worked in a molecular biology lab in Germany as a student abroad. Nevertheless, I decided to go to medical school.

I didn't apply to a Ph.D. program because I'd never seen a woman heading up a research lab, and it didn't enter my mind that I could do that. But I did know some women physicians, so I applied to medical school and was admitted to Harvard. However, the summer before I was supposed to enter, I ended up working in the lab of cell biologist Joe Gall, then at the University of Minnesota. For the first time, I had my own project, and I loved it. By August 1st, I decided that I didn't care if I would never be the head of a lab; I just wanted to do research. Luckily, I was able to switch from the medical school to a graduate program at Harvard.

How did you end up as a graduate student of Jim Watson?

I was interested in the question of whether all cellular organelles have DNA, like mitochondria do. So I first approached a cell biologist, a famous microscopist who nevertheless reserved a bench in the corner of his lab for biochemistry. He conceded that his lab might be suitable, then gave me an unencouraging look and said, "But you're a woman. What are you going to do when you get married and have kids?" I barely made it out of his office before bursting into tears. Then I went to my second-choice thesis advisor, Jim Watson. I had done very well in his course, and he accepted me into his lab. So I became his first female graduate student, something I didn't discover until months later.

What was it like being in Watson's lab?

The Watson lab was a very exciting place at that time, in 1964. We knew that genes in DNA were transcribed into complementary RNA (a process called transcription) and that RNA called messenger RNA (mRNA) was translated into protein by ribosomes (translation). Besides mRNA, the only kinds of cellular RNA that were known were transfer RNA (tRNA) and ribosomal RNA (rRNA), although it was also known that some viruses had RNA instead of DNA as their genomes. But when I started grad school, we didn't yet know the genetic code—how the nucleotide sequence in mRNA corresponds to the amino acid sequence in protein—or much of anything about how transcription or translation occurred.

Jim would go off to meetings, and when he came back, everybody would crowd around him in the hall to find out what was new. Imagine the excitement when we heard, at an international biochemistry congress I actually attended, that the genetic code had been figured out! Or when someone in our lab discovered that a special kind of tRNA initiated protein synthesis. Things were happening very, very rapidly! The atmosphere was fiercely competitive but paradoxically collegial—the three or four labs that were working on the mechanisms of transcription and translation were all in contact with each other.

What was your research as a graduate student?

I worked on a newly discovered virus, R17, that infects the bacterium *E. coli*. Like other simple viruses, R17 is just a small amount of nucleic acid inside a protein coat. Throughout that era, molecular biologists fervently believed that unless you worked on something really simple, you would never figure out the molecular basis of life. So a virus that had only three genes (later found to be four) was the perfect thing to study.

The nucleic acid of R17, its genome, is RNA. This RNA gets into bacterial cells, and about an hour later out come 10,000 copies of the virus. So lots of things are happening in those cells. I studied a viral protein called the A-protein. For my thesis, I characterized the A-protein and what happened if there were mutations in its gene: You got virus particles that looked normal in the electron microscope but couldn't infect a bacterium. It turned out that the A-protein was needed for the virus to attach to the cell.

What did you do after graduate school?

I was married by then, and my husband had arranged to do a postdoc at the Medical Research Council (MRC) at Cambridge University, a mecca for structural and molecular biology. Jim Watson had written to Francis Crick asking him to find a place for me, but when I arrived at Cambridge, Francis suggested I do library research. Eventually, however, I found a bit of bench space for a lab project.

Fred Sanger's lab was nearby, and he was just working out his method for sequencing RNA. There was a lot of interchange with the people in Fred's lab, and they were very interested in the sequence of the R17 genome. Since it was very small, it was a really good molecule to work on. Previously, a paper had been published describing a method for isolating the particular stretches of mRNA bound to a functioning ribosome: You treated the mRNA-ribosome complex with ribonuclease, an enzyme that breaks down unprotected RNA, and you ended up with the part of the mRNA that had been bound and therefore protected by the ribosome, about 30 nucleotides long. The project I took on was to make ribosomes bind to R17 RNA (which functions as mRNA in normal virus infection) under conditions where they start but do not elongate proteins, and then isolate the ribosome-bound RNA segments. I would then determine the sequence of the parts of this RNA where translation started. Other people had considered and rejected this project. They were all male postdocs with wives and children who knew that in two years they would have to interview for tenure-track jobs, and this project had little chance of quick success. But since I thought I couldn't aim higher than a research position in somebody else's lab, I felt free to take on a risky project. (So, being a woman determined the two most important decisions of my early scientific career: ending up in Watson's lab and choosing my project at Cambridge.)

I determined the RNA nucleotide sequences at the beginning of the three R17 genes known at the time. These sequences included AUG, already known to be the "start codon" in mRNA (the first nucleotide triplet translated). And the sequences that followed AUG fit what was already known about the protein sequences, according to the genetic code. We also established that there were spaces between genes in the viral genome. And we figured out that sometimes the virus RNA folded into secondary structures that were important in regulating how many ribosomes would get on at a particular start site. This work at Cambridge—and better academic opportunities for women in the United States—led to my faculty position at Yale.

When you arrived at Yale, what was your first big discovery?
I found out how ribosomes locate the regions on mRNA where they attach and start translation. At Cambridge I had worked out the three 30-nucleotide sequences where ribosomes bind to R17 RNA, but it still wasn't clear how ribosomes honed in on these sequences out of the virus's 3,500 nucleotides. One idea was that a stretch of mRNA rich in purines, just upstream of where translation actually starts, would base-pair with the 3′ end of the rRNA molecule in the small ribosomal subunit of bacteria. So I went to work testing that hypothesis. I soon had direct evidence that there actually is a physical interaction between the end of the "16S" rRNA molecule and the regions of mRNA that are bound by ribosomes. So this RNA-RNA base pairing, along with RNA-RNA base pairing between tRNA and mRNA, is the basis of polypeptide initiation.

You then turned to eukaryotic mRNA. What is different about mRNA production in eukaryotic cells, compared with bacteria?
The main difference comes from the fact that the genes of humans and other eukaryotes have interruptions in them, stretches of nucleic acid that are not translated. These interruptions, called *introns*, have to be removed from the RNA transcript before it is translated. But we didn't know this when I got interested in the subject. At that time, all we knew was that only 5–10% of the RNA transcribed from eukaryotic genes got out of the nucleus as mRNA. I was intrigued by this mystery and decided to switch from prokaryotes to eukaryotes to try to study it. Then, when introns were discovered, the reason for the loss of RNA became clear—though not how the extra RNA was removed. To make mRNA, somehow the introns have to be precisely removed and the coding bits have to be glued back together—a process called RNA splicing.

"I started working on RNA while I was a student, and it has continued to be my favorite molecule!"

What have you learned since then about RNA splicing?
The most important molecular players are small RNA molecules that base-pair with sequences at the ends of RNA introns. This base pairing initiates the assembly of a ribosome-sized machine called a *spliceosome* made of RNA-protein subunits called snRNPs (pronounced "snurps") and other proteins. A spliceosome removes introns and joins together the protein-coding pieces. So RNA-RNA base pairing is the basis of the whole splicing process, just like it's the basis of the initiation of translation. Now there is more and more evidence that the RNAs are the catalytic components of the spliceosome, with the proteins playing supporting roles.

Does your research have any medical relevance?
We learned early on that people with lupus, an autoimmune disease, make antibodies to snRNPs, the RNA-protein subunits of spliceosomes. This discovery has been useful for the diagnosis of a number of autoimmune diseases and even for the prognosis of individual patients—although it hasn't led to cures. What we do in my lab, however, is very basic research. Somebody's got to figure out the basics in order for somebody else to figure out how to apply it.

What's going on now in the RNA field?
Lots of new classes of small RNA molecules have been discovered that, like rRNA, tRNA, and the RNAs in snRNPs, do not themselves code for protein. All these RNAs are important in getting information out of the DNA and into the functioning proteins of the cell. For instance, tiny RNAs called microRNAs, which associate with particular proteins, are involved in regulating translation. Again, it's RNA-RNA base pairing that determines the specificity. The theme of my research over my entire career has been finding out how RNAs interact with other RNAs to provide specificity along the pathway of gene expression. Proteins play important auxiliary roles, but it's basically been one RNA interacting with another RNA. I started working on RNA while I was a student, and it has continued to be my favorite molecule! There's enough to learn to last for many more lifetimes.

What do the discoveries about RNA suggest about the early stages of life on Earth?
Most biologists think that RNA was the first and most important genetic material, probably serving the first cells as both genome and the means by which the information in the genome directed cellular functions. Over time, cells have replaced the RNA genome with DNA, and many of the other RNA molecules with proteins. But the crucial processes of gene expression and its regulation are still dependent on various RNAs—4 billion years after life first arose!

Joan Steitz (center) with Lisa Urry (right) and Jane Reece

13

Meiosis and Sexual Life Cycles

▲ **Figure 13.1 What accounts for family resemblance?**

KEY CONCEPTS

- **13.1** Offspring acquire genes from parents by inheriting chromosomes
- **13.2** Fertilization and meiosis alternate in sexual life cycles
- **13.3** Meiosis reduces the number of chromosome sets from diploid to haploid
- **13.4** Genetic variation produced in sexual life cycles contributes to evolution

OVERVIEW

Variations on a Theme

Most people who send out birth announcements mention the sex of the baby, but they don't feel the need to specify that their offspring is a human being! One of the characteristics of life is the ability of organisms to reproduce their own kind—elephants produce little elephants, and oak trees generate oak saplings. Exceptions to this rule show up only as sensational but highly suspect stories in tabloid newspapers.

Another rule often taken for granted is that offspring resemble their parents more than they do unrelated individuals. If you examine the family members shown in **Figure 13.1**—actress Sissy Spacek and her husband Jack Fisk with daughters Madison and Schuyler Fisk—you can pick out some similar features among them. The transmission of traits from one generation to the next is called inheritance, or **heredity** (from the Latin *heres*, heir). However, sons and daughters are not identical copies of either parent or of their siblings. Along with inherited similarity, there is also **variation**. Farmers have exploited the principles of heredity and variation for thousands of years, breeding plants and animals for desired traits. But what are the biological mechanisms leading to the hereditary similarity and variation that we call a "family resemblance"? The answer to this question eluded biologists until the advance of genetics in the 20th century.

Genetics is the scientific study of heredity and hereditary variation. In this unit, you will learn about genetics at multiple levels, from organisms to cells to molecules. On the practical side, you will see how genetics continues to revolutionize medicine and agriculture, and you will be asked to consider some social and ethical questions raised by our ability to manipulate DNA, the genetic material. At the end of the unit, you will be able to stand back and consider the whole genome, an organism's entire complement of DNA. Rapid acquisition and analysis of the genome sequences of many species, including our own, have taught us a great deal about evolution on the molecular level—in other words, evolution of the genome itself. In fact, genetic methods and discoveries are catalyzing progress in all areas of biology, from cell biology to physiology, developmental biology, behavior, and even ecology.

We begin our study of genetics in this chapter by examining how chromosomes pass from parents to offspring in sexually reproducing organisms. The processes of meiosis (a special type of cell division) and fertilization (the fusion of sperm and egg) maintain a species' chromosome count during the sexual life cycle. We will describe the cellular mechanics of meiosis and explain how this process differs from mitosis. Finally, we will consider how both meiosis and fertilization contribute to genetic variation, such as the variation obvious in the family shown in Figure 13.1.

CONCEPT 13.1

Offspring acquire genes from parents by inheriting chromosomes

Family friends may tell you that you have your mother's freckles or your father's eyes. Of course, parents do not, in any literal sense, give their children freckles, eyes, hair, or any other traits. What, then, *is* actually inherited?

Inheritance of Genes

Parents endow their offspring with coded information in the form of hereditary units called **genes**. The genes we inherit from our mothers and fathers are our genetic link to our parents, and they account for family resemblances such as shared eye color or freckles. Our genes program the specific traits that emerge as we develop from fertilized eggs into adults.

The genetic program is written in the language of DNA, the polymer of four different nucleotides you learned about in Chapters 1 and 5. Inherited information is passed on in the form of each gene's specific sequence of DNA nucleotides, much as printed information is communicated in the form of meaningful sequences of letters. In both cases, the language is symbolic. Just as your brain translates the word *apple* into a mental image of the fruit, cells translate genes into freckles and other features. Most genes program cells to synthesize specific enzymes and other proteins, whose cumulative action produces an organism's inherited traits. The programming of these traits in the form of DNA is one of the unifying themes of biology.

The transmission of hereditary traits has its molecular basis in the precise replication of DNA, which produces copies of genes that can be passed from parents to offspring. In animals and plants, reproductive cells called **gametes** are the vehicles that transmit genes from one generation to the next. During fertilization, male and female gametes (sperm and eggs) unite, thereby passing on genes of both parents to their offspring.

Except for small amounts of DNA in mitochondria and chloroplasts, the DNA of a eukaryotic cell is packaged into chromosomes within the nucleus. Every species has a characteristic number of chromosomes. For example, humans have 46 chromosomes in their **somatic cells**—all cells of the body except the gametes and their precursors. Each chromosome consists of a single long DNA molecule elaborately coiled in association with various proteins. One chromosome includes several hundred to a few thousand genes, each of which is a specific sequence of nucleotides within the DNA molecule. A gene's specific location along the length of a chromosome is called the gene's **locus** (plural, *loci*; from the Latin, meaning "place"). Our genetic endowment consists of the genes that are part of the chromosomes we inherited from our parents.

Comparison of Asexual and Sexual Reproduction

Only organisms that reproduce asexually have offspring that are exact genetic copies of themselves. In **asexual reproduction**, a single individual is the sole parent and passes copies of all its genes to its offspring without the fusion of gametes. For example, single-celled eukaryotic organisms can reproduce asexually by mitotic cell division, in which DNA is copied and allocated equally to two daughter cells. The genomes of the offspring are virtually exact copies of the parent's genome. Some multicellular organisms are also capable of reproducing

(a) Hydra (b) Redwoods

▲ **Figure 13.2 Asexual reproduction in two multicellular organisms. (a)** This relatively simple animal, a hydra, reproduces by budding. The bud, a localized mass of mitotically dividing cells, develops into a small hydra, which detaches from the parent (LM). **(b)** All the trees in this circle of redwoods arose asexually from a single parent tree, whose stump is in the center of the circle.

asexually (**Figure 13.2**). Because the cells of the offspring are derived by mitosis in the parent, the "chip off the old block" is usually genetically identical to its parent. An individual that reproduces asexually gives rise to a **clone**, a group of genetically identical individuals. Genetic differences occasionally arise in asexually reproducing organisms as a result of changes in the DNA called mutations, which we will discuss in Chapter 17.

In **sexual reproduction**, two parents give rise to offspring that have unique combinations of genes inherited from the two parents. In contrast to a clone, offspring of sexual reproduction vary genetically from their siblings and both parents: They are variations on a common theme of family resemblance, not exact replicas. Genetic variation like that shown in Figure 13.1 is an important consequence of sexual reproduction. What mechanisms generate this genetic variation? The key is the behavior of chromosomes during the sexual life cycle.

CONCEPT CHECK 13.1

1. Explain what causes the traits of parents (such as hair color) to show up in their offspring.
2. How do asexually reproducing organisms produce offspring that are genetically identical to each other and to their parents?
3. **WHAT IF?** A horticulturalist breeds orchids, trying to obtain a plant with a unique combination of desirable traits. After many years, she finally succeeds. To produce more plants like this one, should she crossbreed it with another plant or clone it? Why?

For suggested answers, see Appendix A.

CONCEPT 13.2

Fertilization and meiosis alternate in sexual life cycles

A **life cycle** is the generation-to-generation sequence of stages in the reproductive history of an organism, from conception to production of its own offspring. In this section, we use humans as an example to track the behavior of chromosomes through the sexual life cycle. We begin by considering the chromosome count in human somatic cells and gametes. We will then explore how the behavior of chromosomes relates to the human life cycle and other types of sexual life cycles.

Sets of Chromosomes in Human Cells

In humans, each somatic cell has 46 chromosomes. During mitosis, the chromosomes become condensed enough to be visible under a light microscope. At this point, they can be distinguished from one another by their size, the positions of their centromeres, and the pattern of colored bands produced by certain stains.

Careful examination of a micrograph of the 46 human chromosomes from a single cell in mitosis reveals that there are two chromosomes of each of 23 types. This becomes clear when images of the chromosomes are arranged in pairs, starting with the longest chromosomes. The resulting ordered display is called a **karyotype** (**Figure 13.3**). The two chromosomes composing a pair have the same length, centromere position, and staining pattern: These are called **homologous chromosomes**, or homologs. Both chromosomes of each pair carry genes controlling the same inherited characters. For example, if a gene for eye color is situated at a particular locus on a certain chromosome, then the homolog of that chromosome will also have a version of the same gene specifying eye color at the equivalent locus.

The two distinct chromosomes referred to as X and Y are an important exception to the general pattern of homologous chromosomes in human somatic cells. Human females have a homologous pair of X chromosomes (XX), but males have one X and one Y chromosome (XY). Only small parts of the X and Y are homologous. Most of the genes carried on the X chromosome do not have counterparts on the tiny Y, and the Y chromosome has genes lacking on the X. Because they determine an individual's sex, the X and Y chromosomes are called **sex chromosomes**. The other chromosomes are called **autosomes**.

The occurrence of pairs of homologous chromosomes in each human somatic cell is a consequence of our sexual origins. We inherit one chromosome of each pair from each parent. Thus, the 46 chromosomes in our somatic cells are actually two sets of 23 chromosomes—a maternal set (from our mother) and a paternal set (from our father). The number of chromosomes in

▼ **Figure 13.3** RESEARCH METHOD

Preparing a Karyotype

APPLICATION A karyotype is a display of condensed chromosomes arranged in pairs. Karyotyping can be used to screen for defective chromosomes or abnormal numbers of chromosomes associated with certain congenital disorders, such as Down syndrome.

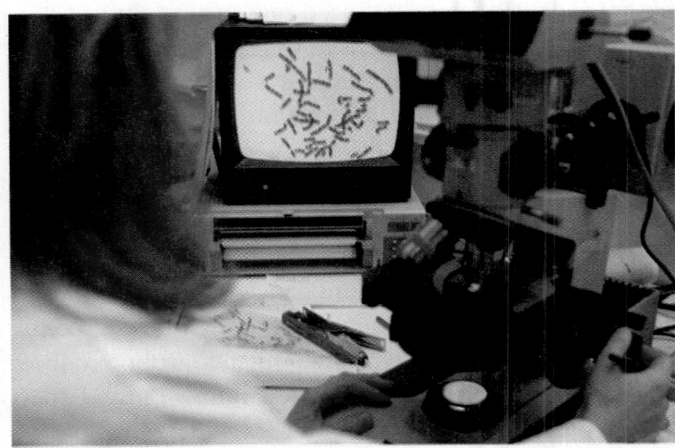

TECHNIQUE Karyotypes are prepared from isolated somatic cells, which are treated with a drug to stimulate mitosis and then grown in culture for several days. Cells arrested in metaphase, when chromosomes are most highly condensed, are stained and then viewed with a microscope equipped with a digital camera. A photograph of the chromosomes is displayed on a computer monitor, and the images of the chromosomes are arranged into pairs according to their appearance.

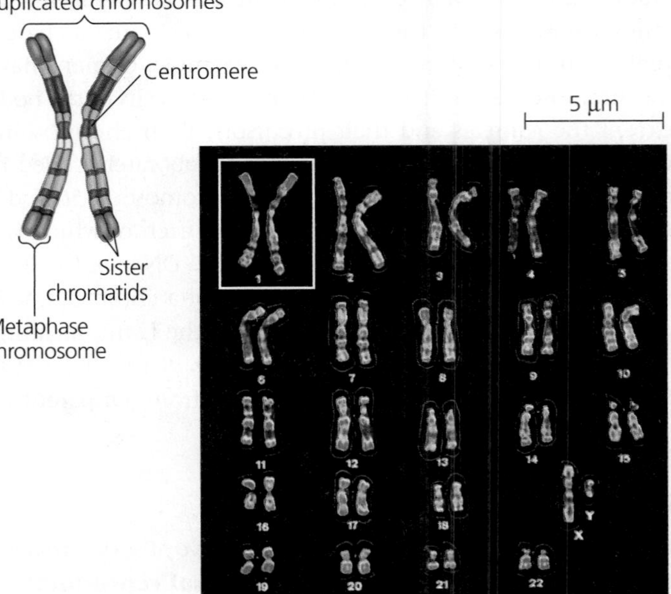

RESULTS This karyotype shows the chromosomes from a normal human male. The size of the chromosome, position of the centromere, and pattern of stained bands help identify specific chromosomes. Although difficult to discern in the karyotype, each metaphase chromosome consists of two closely attached sister chromatids (see the diagram of a pair of homologous duplicated chromosomes).

a single set is represented by *n*. Any cell with two chromosome sets is called a **diploid cell** and has a diploid number of chromosomes, abbreviated $2n$. For humans, the diploid number is 46 ($2n = 46$), the number of chromosomes in our somatic cells. In a cell in which DNA synthesis has occurred, all the chromosomes are duplicated, and therefore each consists of two identical sister chromatids, associated closely at the centromere and along the arms. **Figure 13.4** helps clarify the various terms that we use to describe duplicated chromosomes in a diploid cell. Study this figure so that you understand the differences between homologous chromosomes, sister chromatids, nonsister chromatids, and chromosome sets.

Unlike somatic cells, gametes contain a single set of chromosomes. Such cells are called **haploid cells**, and each has a haploid number of chromosomes (n). For humans, the haploid number is 23 ($n = 23$). The set of 23 consists of the 22 autosomes plus a single sex chromosome. An unfertilized egg contains an X chromosome, but a sperm may contain an X or a Y chromosome.

Note that each sexually reproducing species has a characteristic diploid number and haploid number. For example, the fruit fly, *Drosophila melanogaster*, has a diploid number ($2n$) of 8 and a haploid number (n) of 4, while dogs have a diploid number of 78 and a haploid number of 39.

Now that you have learned the concepts of diploid and haploid numbers of chromosomes, let's consider chromosome behavior during sexual life cycles. We'll use the human life cycle as an example.

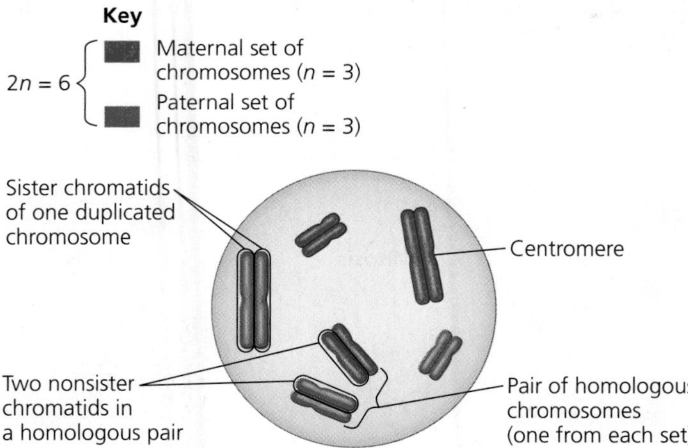

▲ **Figure 13.4 Describing chromosomes.** A cell from an organism with a diploid number of 6 ($2n = 6$) is depicted here following chromosome duplication and condensation. Each of the six duplicated chromosomes consists of two sister chromatids associated closely along their lengths. Each homologous pair is composed of one chromosome from the maternal set (red) and one from the paternal set (blue). Each set is made up of three chromosomes in this example. Nonsister chromatids are any two chromatids in a pair of homologous chromosomes that are not sister chromatids—in other words, one maternal and one paternal chromatid.

? *What is the haploid number of this cell? Is a "set" of chromosomes haploid or diploid?*

Behavior of Chromosome Sets in the Human Life Cycle

The human life cycle begins when a haploid sperm from the father fuses with a haploid egg from the mother. This union of gametes, culminating in fusion of their nuclei, is called **fertilization**. The resulting fertilized egg, or **zygote**, is diploid because it contains two haploid sets of chromosomes bearing genes representing the maternal and paternal family lines. As a human develops into a sexually mature adult, mitosis of the zygote and its descendant cells generates all the somatic cells of the body. Both chromosome sets in the zygote and all the genes they carry are passed with precision to the somatic cells.

The only cells of the human body not produced by mitosis are the gametes, which develop from specialized cells called *germ cells* in the gonads—ovaries in females and testes in males **(Figure 13.5)**. Imagine what would happen if human gametes were made by mitosis: They would be diploid like the somatic cells. At the next round of fertilization, when two gametes fused, the normal chromosome number of 46 would

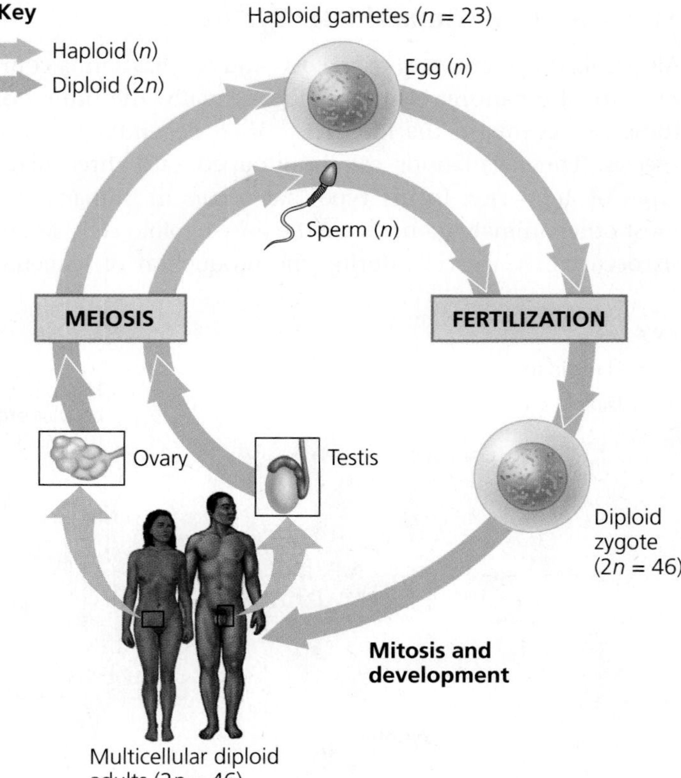

▲ **Figure 13.5 The human life cycle.** In each generation, the number of chromosome sets doubles at fertilization but is halved during meiosis. For humans, the number of chromosomes in a haploid cell is 23, consisting of one set ($n = 23$); the number of chromosomes in the diploid zygote and all somatic cells arising from it is 46, consisting of two sets ($2n = 46$).

This figure introduces a color code that will be used for other life cycles later in this book. The aqua arrows identify haploid stages of a life cycle, and the tan arrows identify diploid stages.

double to 92, and each subsequent generation would double the number of chromosomes yet again. This does not happen, however, because in sexually reproducing organisms, gamete formation involves a type of cell division called **meiosis**. This type of cell division reduces the number of sets of chromosomes from two to one in the gametes, counterbalancing the doubling that occurs at fertilization. In animals, meiosis occurs only in germ cells, which are in the ovaries or testes. As a result of meiosis, each human sperm and egg is haploid ($n = 23$). Fertilization restores the diploid condition by combining two haploid sets of chromosomes, and the human life cycle is repeated, generation after generation (see Figure 13.5). You will learn more about the production of sperm and eggs in Chapter 46.

In general, the steps of the human life cycle are typical of many sexually reproducing animals. Indeed, the processes of fertilization and meiosis are the hallmarks of sexual reproduction in plants, fungi, and protists as well as in animals. Fertilization and meiosis alternate in sexual life cycles, maintaining a constant number of chromosomes in each species from one generation to the next.

The Variety of Sexual Life Cycles

Although the alternation of meiosis and fertilization is common to all organisms that reproduce sexually, the timing of these two events in the life cycle varies, depending on the species. These variations can be grouped into three main types of life cycles. In the type that occurs in humans and most other animals, gametes are the only haploid cells. Meiosis occurs in germ cells during the production of gametes, which undergo no further cell division prior to fertilization. After fertilization, the diploid zygote divides by mitosis, producing a multicellular organism that is diploid **(Figure 13.6a)**.

Plants and some species of algae exhibit a second type of life cycle called **alternation of generations**. This type includes both diploid and haploid stages that are multicellular. The multicellular diploid stage is called the *sporophyte*. Meiosis in the sporophyte produces haploid cells called *spores*. Unlike a gamete, a haploid spore doesn't fuse with another cell but divides mitotically, generating a multicellular haploid stage called the *gametophyte*. Cells of the gametophyte give rise to gametes by mitosis. Fusion of two haploid gametes at fertilization results in a diploid zygote, which develops into the next sporophyte generation. Therefore, in this type of life cycle, the sporophyte generation produces a gametophyte as its offspring, and the gametophyte generation produces the next sporophyte generation **(Figure 13.6b)**. Clearly, the term *alternation of generations* is a fitting name for this type of life cycle.

A third type of life cycle occurs in most fungi and some protists, including some algae. After gametes fuse and form a diploid zygote, meiosis occurs without a multicellular diploid offspring developing. Meiosis produces not gametes but haploid cells that then divide by mitosis and give rise to either unicellular descendants or a haploid multicellular adult organism. Subsequently, the haploid organism carries out further mitoses, producing the cells that develop into gametes. The only diploid stage found in these species is the single-celled zygote **(Figure 13.6c)**.

Note that *either* haploid or diploid cells can divide by mitosis, depending on the type of life cycle. Only diploid cells,

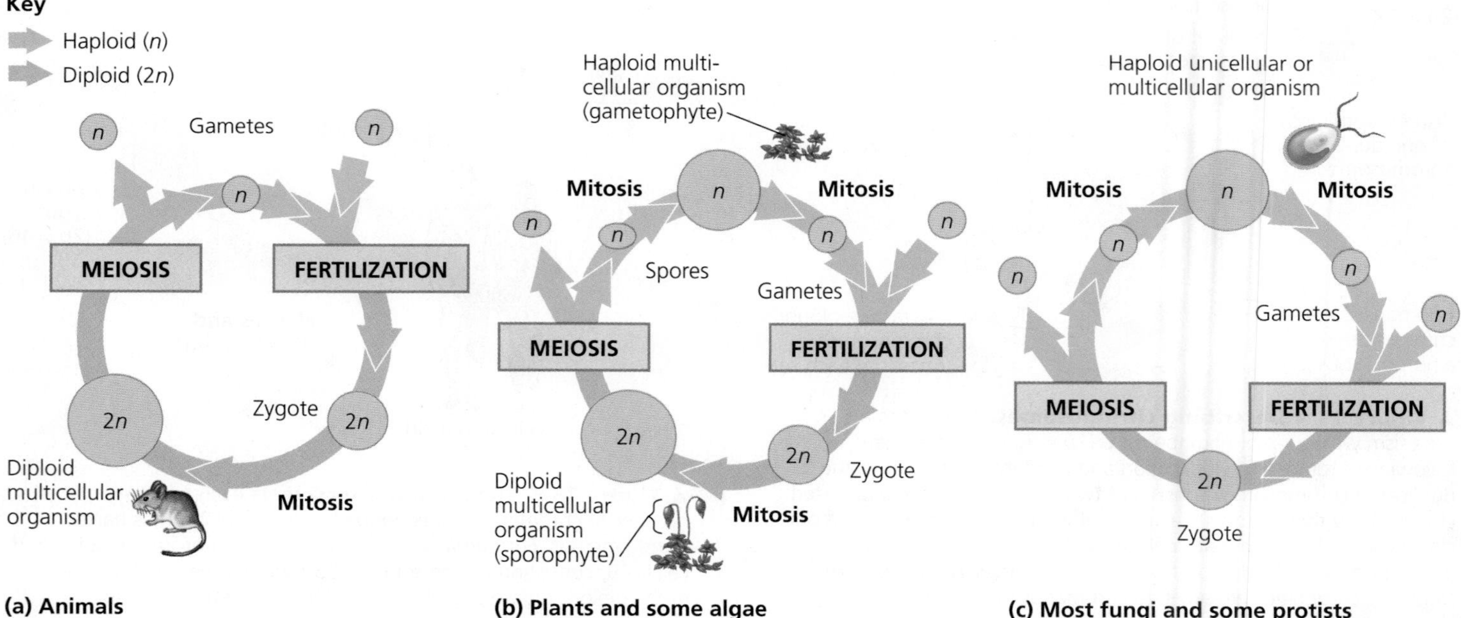

▲ **Figure 13.6 Three types of sexual life cycles.** The common feature of all three cycles is the alternation of meiosis and fertilization, key events that contribute to genetic variation among offspring. The cycles differ in the timing of these two key events.

however, can undergo meiosis because haploid cells have a single set of chromosomes that cannot be further reduced. Though the three types of sexual life cycles differ in the timing of meiosis and fertilization, they share a fundamental result: genetic variation among offspring. A closer look at meiosis will reveal the sources of this variation.

CONCEPT CHECK 13.2

1. **MAKE CONNECTIONS** In Figure 13.4, how many DNA molecules (double helices) are present (see Figure 12.5)?
2. How does the alternation of meiosis and fertilization in the life cycles of sexually reproducing organisms maintain the normal chromosome count for each species?
3. Each sperm of a pea plant contains seven chromosomes. What are the haploid and diploid numbers for this species?
4. **WHAT IF?** A certain eukaryote lives as a unicellular organism, but during environmental stress, it produces gametes. The gametes fuse, and the resulting zygote undergoes meiosis, generating new single cells. What type of organism could this be?

For suggested answers, see Appendix A.

CONCEPT 13.3

Meiosis reduces the number of chromosome sets from diploid to haploid

Many of the steps of meiosis closely resemble corresponding steps in mitosis. Meiosis, like mitosis, is preceded by the duplication of chromosomes. However, this single duplication is followed by not one but two consecutive cell divisions, called **meiosis I** and **meiosis II**. These two divisions result in four daughter cells (rather than the two daughter cells of mitosis), each with only half as many chromosomes as the parent cell.

The Stages of Meiosis

The overview of meiosis in **Figure 13.7** shows, for a single pair of homologous chromosomes in a diploid cell, that both members of the pair are duplicated and the copies sorted into four haploid daughter cells. Recall that sister chromatids are two copies of *one* chromosome, closely associated all along their lengths; this association is called *sister chromatid cohesion*. Together, the sister chromatids make up one duplicated chromosome (see Figure 13.4). In contrast, the two chromosomes of a homologous pair are individual chromosomes that were inherited from different parents. Homologs appear alike in the microscope, but they may have different versions of genes, each called an *allele*, at corresponding loci (for example, an allele for freckles on one chromosome and an allele for the absence of freckles at the same locus on the homolog). Homologs are not associated with each other in any obvious way except during meiosis, as you will soon see.

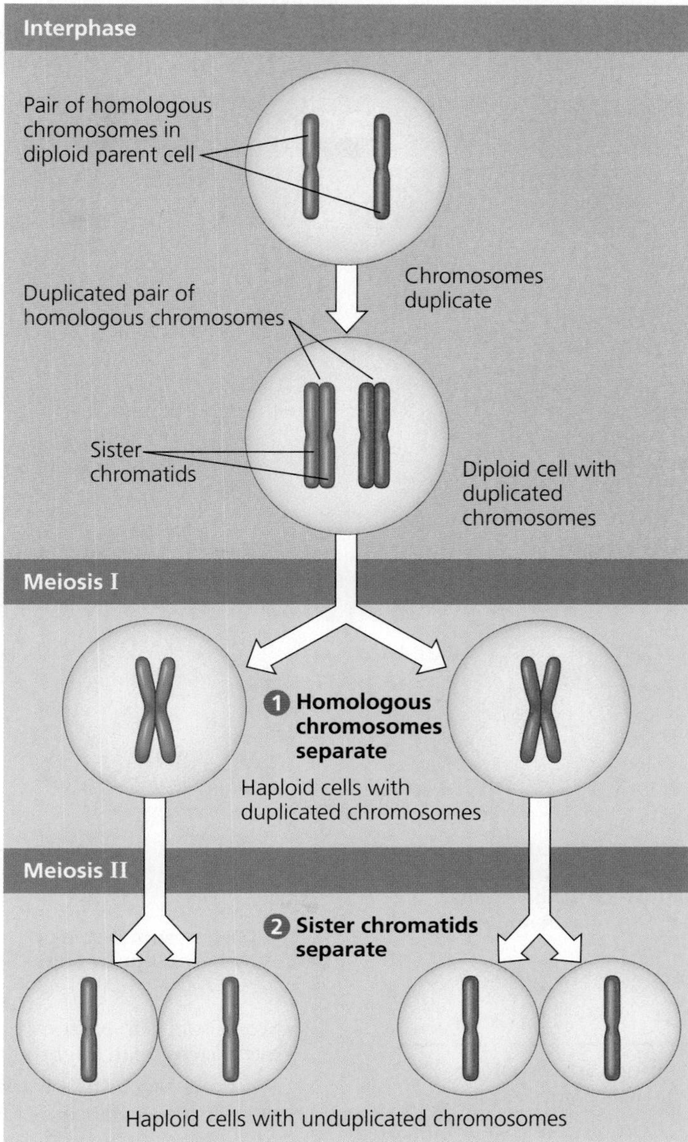

▲ **Figure 13.7 Overview of meiosis: how meiosis reduces chromosome number.** After the chromosomes duplicate in interphase, the diploid cell divides *twice*, yielding four haploid daughter cells. This overview tracks just one pair of homologous chromosomes, which for the sake of simplicity are drawn in the condensed state throughout. (They would not normally be condensed during interphase.) The red chromosome was inherited from the female parent, the blue chromosome from the male parent.

DRAW IT *Redraw the cells in this figure using a simple double helix to represent each DNA molecule.*

Figure 13.8, on the next two pages, describes in detail the stages of the two divisions of meiosis for an animal cell whose diploid number is 6. Meiosis halves the total number of chromosomes in a very specific way, reducing the number of sets from two to one, with each daughter cell receiving one set of chromosomes. Study Figure 13.8 thoroughly before going on.

▼ **Figure 13.8**

Exploring Meiosis in an Animal Cell

MEIOSIS I: Separates homologous chromosomes

Prophase I	Metaphase I	Anaphase I	Telophase I and Cytokinesis

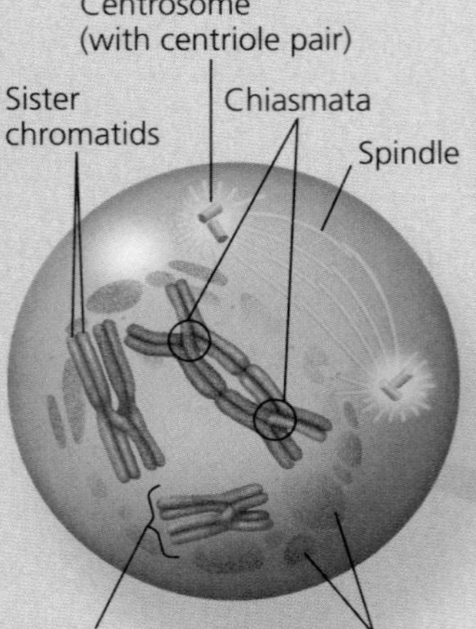

Duplicated homologous chromosomes (red and blue) pair and exchange segments; $2n = 6$ in this example.

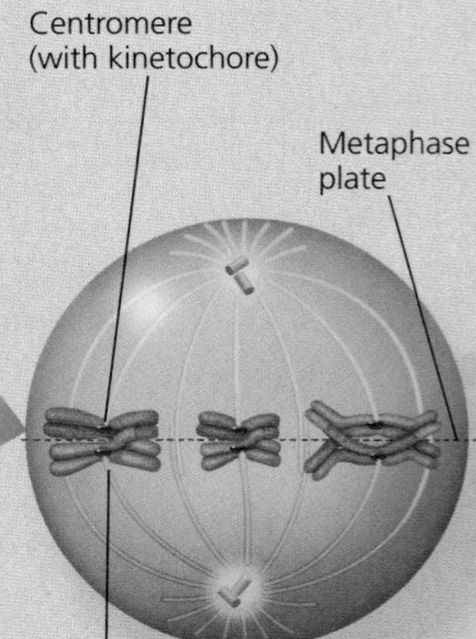

Chromosomes line up by homologous pairs.

Each pair of homologous chromosomes separates.

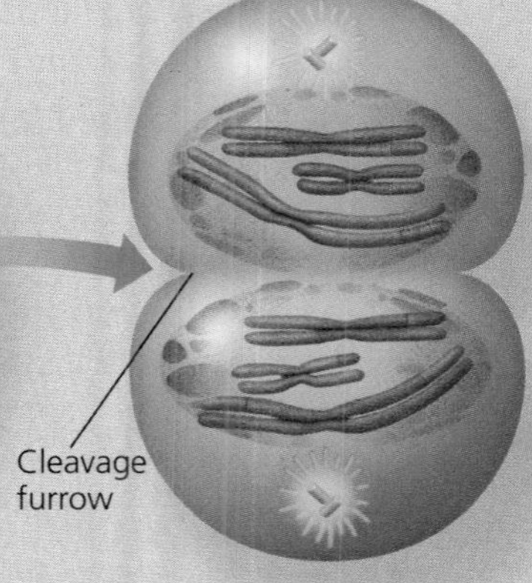

Two haploid cells form; each chromosome still consists of two sister chromatids.

Prophase I

During early prophase I, before the stage shown above:

- Chromosomes begin to condense, and homologs loosely pair along their lengths, aligned gene by gene.
- Paired homologs become physically connected to each other along their lengths by a zipper-like protein structure, the *synaptonemal complex;* this state is called **synapsis.**
- **Crossing over**, a genetic rearrangement between nonsister chromatids involving the exchange of corresponding segments of DNA molecules, begins during pairing and synaptonemal complex formation, and is completed while homologs are in synapsis.

At the stage shown above:

- Synapsis has ended with the disassembly of the synaptonemal complex in mid-prophase, and the chromosomes in each pair have moved apart slightly.
- Each homologous pair has one or more X-shaped regions called **chiasmata** (singular, *chiasma*). A chiasma exists at the point where a crossover has occurred. It appears as a cross because sister chromatid cohesion still holds the two original sister chromatids together, even in regions beyond the crossover point, where one chromatid is now part of the other homolog.
- Centrosome movement, spindle formation, and nuclear envelope breakdown occur as in mitosis.

Later in prophase I, after the stage shown above:

- Microtubules from one pole or the other attach to the two kinetochores, protein structures at the centromeres of the two homologs. The homologous pairs then move toward the metaphase plate.

Metaphase I

- Pairs of homologous chromosomes are now arranged at the metaphase plate, with one chromosome in each pair facing each pole.
- Both chromatids of one homolog are attached to kinetochore microtubules from one pole; those of the other homolog are attached to microtubules from the opposite pole.

Anaphase I

- Breakdown of proteins responsible for sister chromatid cohesion along chromatid arms allows homologs to separate.
- The homologs move toward opposite poles, guided by the spindle apparatus.
- Sister chromatid cohesion persists at the centromere, causing chromatids to move as a unit toward the same pole.

Telophase I and Cytokinesis

- At the beginning of telophase I, each half of the cell has a complete haploid set of duplicated chromosomes. Each chromosome is composed of two sister chromatids; one or both chromatids include regions of nonsister chromatid DNA.
- Cytokinesis (division of the cytoplasm) usually occurs simultaneously with telophase I, forming two haploid daughter cells.
- In animal cells like these, a cleavage furrow forms. (In plant cells, a cell plate forms.)
- In some species, chromosomes decondense and nuclear envelopes form.
- No chromosome duplication occurs between meiosis I and meiosis II.

MEIOSIS II: Separates sister chromatids

Prophase II | Metaphase II | Anaphase II | Telophase II and Cytokinesis

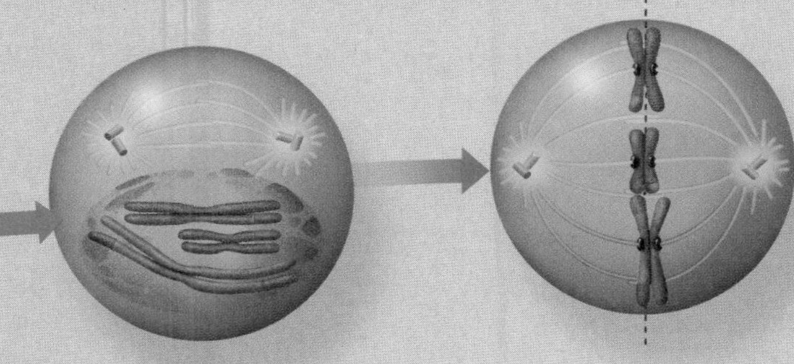

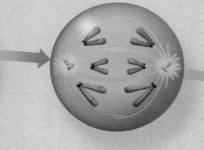

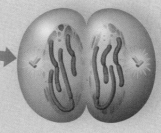

During another round of cell division, the sister chromatids finally separate; four haploid daughter cells result, containing unduplicated chromosomes.

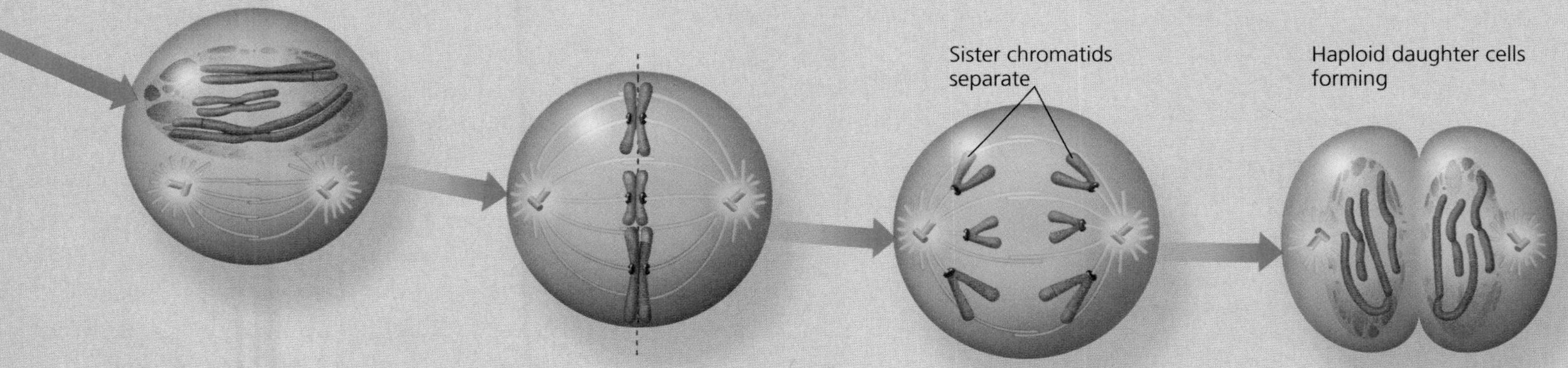

Prophase II

- A spindle apparatus forms.
- In late prophase II (not shown here), chromosomes, each still composed of two chromatids associated at the centromere, move toward the metaphase II plate.

Metaphase II

- The chromosomes are positioned at the metaphase plate as in mitosis.
- Because of crossing over in meiosis I, the two sister chromatids of each chromosome are *not* genetically identical.
- The kinetochores of sister chromatids are attached to microtubules extending from opposite poles.

Anaphase II

- Breakdown of proteins holding the sister chromatids together at the centromere allows the chromatids to separate. The chromatids move toward opposite poles as individual chromosomes.

Telophase II and Cytokinesis

- Nuclei form, the chromosomes begin decondensing, and cytokinesis occurs.
- The meiotic division of one parent cell produces four daughter cells, each with a haploid set of (unduplicated) chromosomes.
- The four daughter cells are genetically distinct from one another and from the parent cell.

MAKE CONNECTIONS *Look at Figure 12.7 and imagine the two daughter cells undergoing another round of mitosis, yielding four cells. Compare the number of chromosomes in each of those four cells, after mitosis, with the number in each cell in Figure 13.8, after meiosis. What is it about the process of meiosis that accounts for this difference, even though meiosis also includes two cell divisions?*

BioFlix Visit the Study Area at **www.masteringbiology.com** for the BioFlix® 3-D Animation on Meiosis.

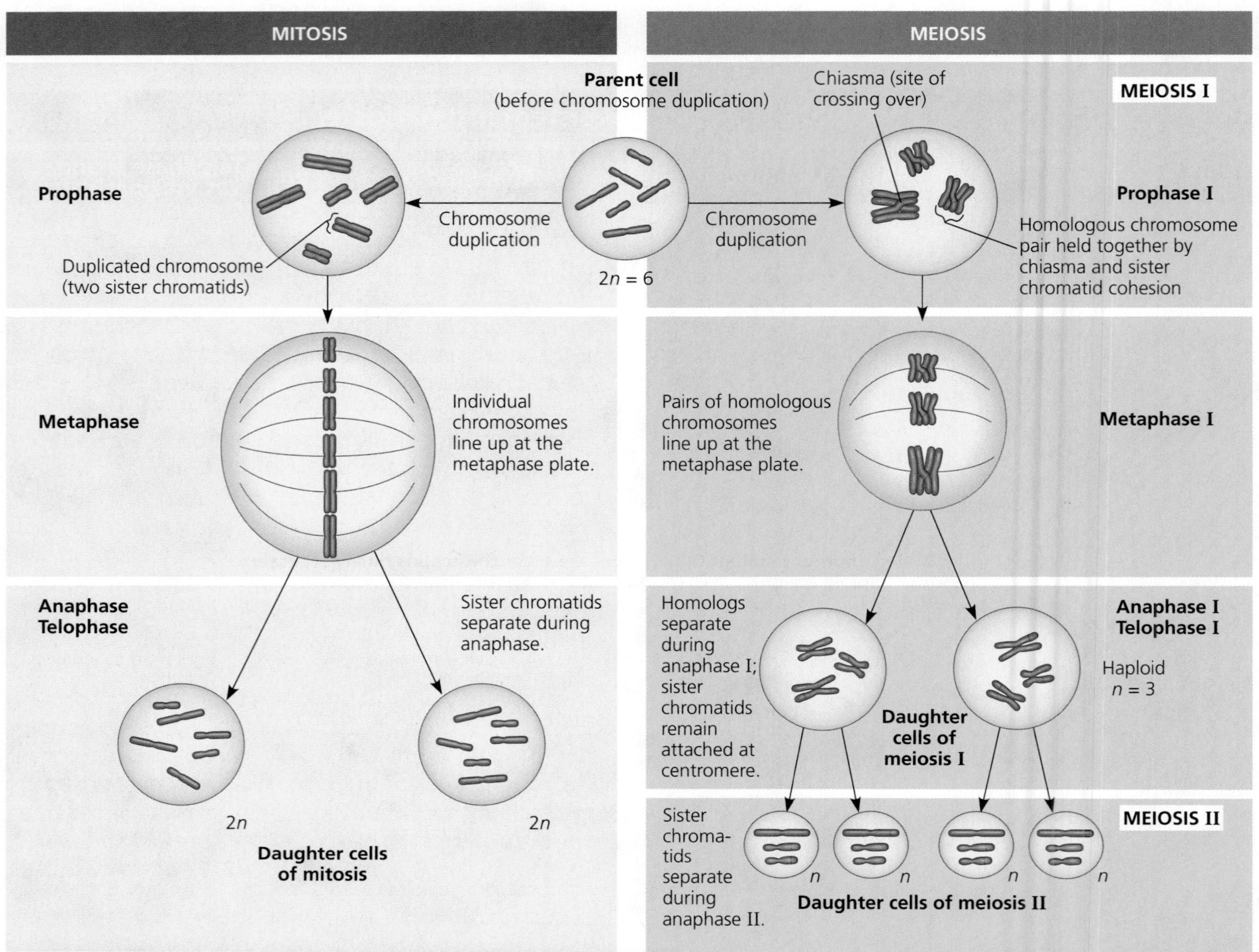

SUMMARY

Property	Mitosis	Meiosis
DNA replication	Occurs during interphase before mitosis begins	Occurs during interphase before meiosis I begins
Number of divisions	One, including prophase, metaphase, anaphase, and telophase	Two, each including prophase, metaphase, anaphase, and telophase
Synapsis of homologous chromosomes	Does not occur	Occurs during prophase I along with crossing over between nonsister chromatids; resulting chiasmata hold pairs together due to sister chromatid cohesion
Number of daughter cells and genetic composition	Two, each diploid (2*n*) and genetically identical to the parent cell	Four, each haploid (*n*), containing half as many chromosomes as the parent cell; genetically different from the parent cell and from each other
Role in the animal body	Enables multicellular adult to arise from zygote; produces cells for growth, repair, and, in some species, asexual reproduction	Produces gametes; reduces number of chromosomes by half and introduces genetic variability among the gametes

▲ **Figure 13.9 A comparison of mitosis and meiosis in diploid cells.**

DRAW IT *Could any other combinations of chromosomes be generated during meiosis II from the specific cells shown in telophase I? Explain. (Hint: Draw the cells as they would appear in metaphase II.)*

A Comparison of Mitosis and Meiosis

Figure 13.9 summarizes the key differences between meiosis and mitosis in diploid cells. Basically, meiosis reduces the number of chromosome sets from two (diploid) to one (haploid), whereas mitosis conserves the number of chromosome sets. Therefore, meiosis produces cells that differ genetically from their parent cell and from each other, whereas mitosis produces daughter cells that are genetically identical to their parent cell and to each other.

Three events unique to meiosis occur during meiosis I:

1. **Synapsis and crossing over.** During prophase I, duplicated homologs pair up, and the formation of the synaptonemal complex between them holds them in synapsis. Crossing over also occurs during prophase I. Synapsis and crossing over normally do not occur during prophase of mitosis.
2. **Homologous pairs at the metaphase plate.** At metaphase I of meiosis, chromosomes are positioned at the metaphase plate as pairs of homologs, rather than individual chromosomes, as in metaphase of mitosis.
3. **Separation of homologs.** At anaphase I of meiosis, the duplicated chromosomes of each homologous pair move toward opposite poles, but the sister chromatids of each duplicated chromosome remain attached. In anaphase of mitosis, by contrast, sister chromatids separate.

How do sister chromatids stay together through meiosis I but separate from each other in meiosis II and mitosis? Sister chromatids are attached along their lengths by protein complexes called *cohesins*. In mitosis, this attachment lasts until the end of metaphase, when enzymes cleave the cohesins, freeing the sister chromatids to move to opposite poles of the cell. In meiosis, sister chromatid cohesion is released in two steps, one at the start of anaphase I and one at anaphase II. In metaphase I, homologs are held together by cohesion between sister chromatid arms in regions beyond points of crossing over, where stretches of sister chromatids now belong to different chromosomes. As shown in Figure 13.8, the combination of crossing over and sister chromatid cohesion along the arms results in the formation of a chiasma. Chiasmata hold homologs together as the spindle forms for the first meiotic division. At the onset of anaphase I, the release of cohesion along sister chromatid arms allows homologs to separate. At anaphase II, the release of sister chromatid cohesion at the centromeres allows the sister chromatids to separate. Thus, sister chromatid cohesion and crossing over, acting together, play an essential role in the lining up of chromosomes by homologous pairs at metaphase I.

Meiosis I is called the *reductional division* because it halves the number of chromosome sets per cell—a reduction from two sets (the diploid state) to one set (the haploid state). During the second meiotic division, meiosis II (sometimes called the *equational division*), the sister chromatids separate, producing haploid daughter cells. The mechanism for separating sister chromatids is virtually identical in meiosis II and mitosis. The molecular basis of chromosome behavior during meiosis continues to be a focus of intense research.

CONCEPT CHECK 13.3

1. **MAKE CONNECTIONS** How are the chromosomes in a cell at metaphase of mitosis similar to and different from the chromosomes in a cell at metaphase of meiosis II? (Compare Figures 12.7 and 13.8.)
2. **WHAT IF?** Given that the synaptonemal complex disappears by the end of prophase, how would the two homologs be associated if crossing over did not occur? What effect might this ultimately have on gamete formation?

For suggested answers, see Appendix A.

CONCEPT 13.4

Genetic variation produced in sexual life cycles contributes to evolution

How do we account for the genetic variation illustrated in Figure 13.1? As you will learn in more detail in later chapters, mutations are the original source of genetic diversity. These changes in an organism's DNA create the different versions of genes known as *alleles*. Once these differences arise, reshuffling of the alleles during sexual reproduction produces the variation that results in each member of a sexually reproducing population having a unique combination of traits.

Origins of Genetic Variation Among Offspring

In species that reproduce sexually, the behavior of chromosomes during meiosis and fertilization is responsible for most of the variation that arises in each generation. Let's examine three mechanisms that contribute to the genetic variation arising from sexual reproduction: independent assortment of chromosomes, crossing over, and random fertilization.

Independent Assortment of Chromosomes

One aspect of sexual reproduction that generates genetic variation is the random orientation of pairs of homologous chromosomes at metaphase of meiosis I. At metaphase I, the homologous pairs, each consisting of one maternal and one paternal chromosome, are situated at the metaphase plate. (Note that the terms *maternal* and *paternal* refer, respectively, to the mother and father of the individual whose cells are undergoing meiosis.) Each pair may orient with either its maternal or paternal homolog closer to a given pole—its orientation is as random as the flip of a coin. Thus, there is a 50% chance that a particular daughter cell of meiosis I will get the maternal chromosome of a certain homologous pair and a 50% chance that it will get the paternal chromosome.

Because each pair of homologous chromosomes is positioned independently of the other pairs at metaphase I, the first meiotic division results in each pair sorting its maternal and paternal homologs into daughter cells independently of every other pair. This is called *independent assortment*. Each daughter cell represents one outcome of all possible combinations of maternal and paternal chromosomes. As shown in **Figure 13.10**, the number of combinations possible for daughter cells formed by meiosis of a diploid cell with $n = 2$ (two pairs of homologous chromosomes) is four: two possible arrangements for the first pair times two possible arrangements for the second pair. Note that only two of the four combinations of daughter cells shown in the figure would result from meiosis of a *single* diploid cell, because a single parent cell would have one or the other possible chromosomal arrangement at metaphase I, but not both. However, the population of daughter cells resulting from meiosis of a large number of diploid cells contains all four types in approximately equal numbers. In the case of $n = 3$, eight combinations of chromosomes are possible for daughter cells. More generally, the number of possible combinations when chromosomes sort independently during meiosis is 2^n, where n is the haploid number of the organism.

In the case of humans ($n = 23$), the number of possible combinations of maternal and paternal chromosomes in the resulting gametes is 2^{23}, or about 8.4 million. Each gamete that you produce in your lifetime contains one of roughly 8.4 million possible combinations of chromosomes.

Crossing Over

As a consequence of the independent assortment of chromosomes during meiosis, each of us produces a collection of gametes differing greatly in their combinations of the chromosomes we inherited from our two parents. Figure 13.10 suggests that each chromosome in a gamete is exclusively maternal or paternal in origin. In fact, this is *not* the case, because crossing over produces **recombinant chromosomes**, individual chromosomes that carry genes (DNA) derived from two different parents (**Figure 13.11**). In meiosis in humans, an average of one to three crossover events occur per chromosome pair, depending on the size of the chromosomes and the position of their centromeres.

Crossing over begins very early in prophase I as homologous chromosomes pair loosely along their lengths. Each gene on one homolog is aligned precisely with the corresponding gene on the other homolog. In a single crossover event, the DNA of two *nonsister* chromatids—one maternal and one paternal chromatid of a homologous pair—is broken by specific proteins at precisely corresponding points, and the two segments beyond the crossover point are each joined to the other chromatid. Thus, a paternal chromatid is joined to a piece of maternal chromatid beyond the crossover point, and vice versa. In this way, crossing over produces chromosomes with new combinations of maternal and paternal alleles (see Figure 13.11).

At metaphase II, chromosomes that contain one or more recombinant chromatids can be oriented in two alternative, nonequivalent ways with respect to other chromosomes, because their sister chromatids are no longer identical. The different possible arrangements of nonidentical sister chromatids during meiosis II further increase the number of genetic types of daughter cells that can result from meiosis.

You will learn more about crossing over in Chapter 15. The important point for now is that crossing over, by combining DNA inherited from two parents into a single chromosome, is an important source of genetic variation in sexual life cycles.

Random Fertilization

The random nature of fertilization adds to the genetic variation arising from meiosis. In humans, each male and female gamete represents one of about 8.4 million (2^{23}) possible chromosome combinations due to independent assortment. The fusion of a male gamete with a female gamete during

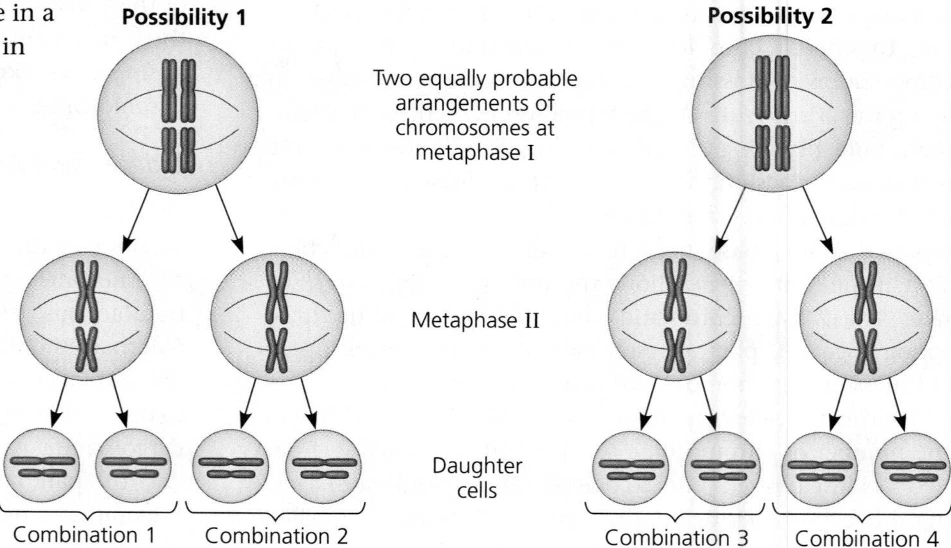

▶ **Figure 13.10 The independent assortment of homologous chromosomes in meiosis.**

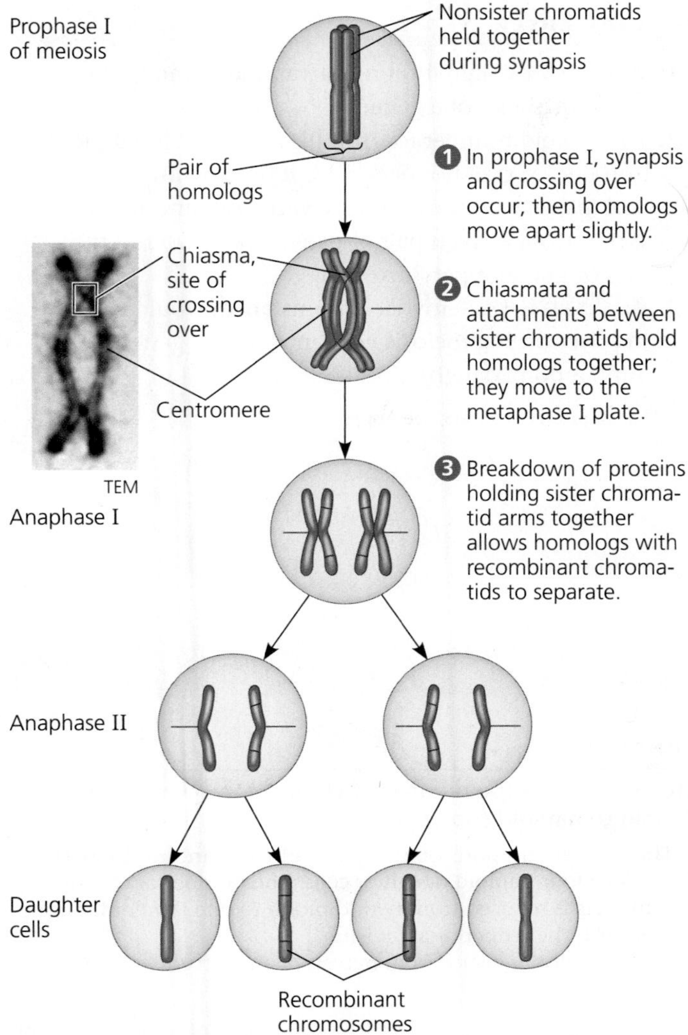
Prophase I
of meiosis
Nonsister chromatids
held together
during synapsis
Pair of
homologs
1 In prophase I, synapsis
and crossing over
occur; then homologs
move apart slightly.
Chiasma,
site of
crossing
over
Centromere
TEM
2 Chiasmata and
attachments between
sister chromatids hold
homologs together;
they move to the
metaphase I plate.
Anaphase I
3 Breakdown of proteins
holding sister chroma-
tid arms together
allows homologs with
recombinant chroma-
tids to separate.
Anaphase II
Daughter
cells
Recombinant
chromosomes

▲ **Figure 13.11 The results of crossing over during meiosis.**

fertilization will produce a zygote with any of about 70 trillion ($2^{23} \times 2^{23}$) diploid combinations. If we factor in the variation brought about by crossing over, the number of possibilities is truly astronomical. It may sound trite, but you really *are* unique.

The Evolutionary Significance of Genetic Variation Within Populations

EVOLUTION Now that you've learned how new combinations of genes arise among offspring in a sexually reproducing population, let's see how the genetic variation in a population relates to evolution. Darwin recognized that a population evolves through the differential reproductive success of its variant members. On average, those individuals best suited to the local environment leave the most offspring, thereby transmitting their genes. Thus, natural selection results in the accumulation of genetic variations favored by the environment. As the environment changes, the population may survive if, in each generation, at least some of its members can cope effectively with the new conditions. Mutations are the original source of different alleles, which are then mixed and matched during meiosis. New and different combinations of alleles may work better than those that previously prevailed. The ability of sexual reproduction to generate genetic diversity is one of the most commonly proposed explanations for the evolutionary persistence of this process.

On the other hand, in a stable environment, asexual reproduction would seem to be more advantageous, because it ensures perpetuation of successful combinations of alleles. Furthermore, asexual reproduction is less expensive; its energy costs to the organism are lower than those of sexual reproduction, for reasons that will be discussed in Chapter 46.

In spite of these apparent disadvantages, sexual reproduction is almost universal among animals as far as we know. While a few species are capable of reproducing asexually under unusual circumstances, animals that always reproduce asexually are quite rare. The best-established example, to date, is a group of microscopic animals called bdelloid rotifers (the "b" in "bdelloid" is silent), shown in **Figure 13.12**. This group includes about 400 species that live in a great variety of environments around the world. They inhabit streams, lake bottoms, puddles, lichens, tree bark, and masses of decaying vegetation. Recent studies have provided convincing evidence that these animals reproduce only asexually and probably haven't engaged in sex in the 40 million years since their evolutionary origins!

Does the discovery of the evolutionarily successful, asexually reproducing bdelloid rotifer cast doubt on the advantage of genetic variation arising from sexual reproduction? On the contrary, this group may be considered an exception that proves the rule. In studies of bdelloid rotifers, biologists have found mechanisms other than sexual reproduction that increase genetic diversity in these organisms. For example, they live in environments that can dry up for long periods of time, during which they can enter a state of suspended animation. In this state, their cell membranes may crack in places, allowing entry of DNA from other rotifers and even other species. Evidence suggests that this DNA can become incorporated into the genome of the rotifer, leading to increased genetic diversity. (You'll learn more about this process, called *horizontal gene transfer*, in Chapter 26.) Taken as a whole, these studies support the idea that genetic variation is evolutionarily advantageous and that a

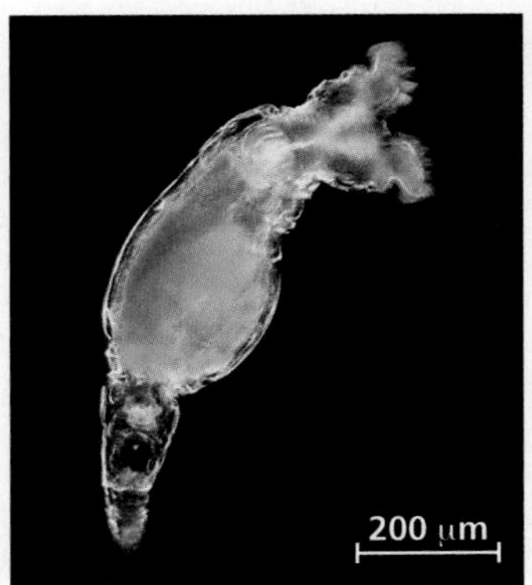
200 μm

▲ **Figure 13.12 A bdelloid rotifer, an animal that reproduces only asexually.**

different mechanism to generate genetic variation has evolved in bdelloid rotifers.

In this chapter, we have seen how sexual reproduction greatly increases the genetic variation present in a population. Although Darwin realized that heritable variation is what makes evolution possible, he could not explain why offspring resemble—but are not identical to—their parents. Ironically, Gregor Mendel, a contemporary of Darwin, published a theory of inheritance that helps explain genetic variation, but his discoveries had no impact on biologists until 1900, more than 15 years after Darwin (1809–1882) and Mendel (1822–1884) had died. In the next chapter, you will learn how Mendel discovered the basic rules governing the inheritance of specific traits.

CONCEPT CHECK 13.4

1. What is the original source of variation among the different alleles of a gene?
2. The diploid number for fruit flies is 8, and the diploid number for grasshoppers is 46. If no crossing over took place, would the genetic variation among offspring from a given pair of parents be greater in fruit flies or grasshoppers? Explain.
3. **WHAT IF?** Under what circumstances would crossing over during meiosis *not* contribute to genetic variation among daughter cells?

For suggested answers, see Appendix A.

13 CHAPTER REVIEW

SUMMARY OF KEY CONCEPTS

CONCEPT 13.1

Offspring acquire genes from parents by inheriting chromosomes (pp. 248–249)

- Each **gene** in an organism's DNA exists at a specific **locus** on a certain chromosome. We inherit one set of chromosomes from our mother and one set from our father.
- In **asexual reproduction**, a single parent produces genetically identical offspring by mitosis. **Sexual reproduction** combines sets of genes from two different parents, leading to genetically diverse offspring.

? *Explain why human offspring resemble their parents but are not identical to them.*

CONCEPT 13.2

Fertilization and meiosis alternate in sexual life cycles (pp. 250–253)

- As seen in a **karyotype**, normal human **somatic cells** are **diploid**. They have 46 chromosomes made up of two sets of 23—one set from each parent. In human diploid cells, there are 22 **homologous** pairs of **autosomes**, each with a maternal and a paternal homolog. The 23rd pair, the **sex chromosomes**, determines whether the person is female (XX) or male (XY).
- At sexual maturity in the human **life cycle**, ovaries and testes (the gonads) produce **haploid gametes** by **meiosis**, each gamete containing a single set of 23 chromosomes ($n = 23$). During **fertilization**, an egg and sperm unite, forming a diploid ($2n = 46$) single-celled **zygote**, which develops into a multicellular organism by mitosis.
- Sexual life cycles differ in the timing of meiosis relative to fertilization and in the point(s) of the cycle at which a multicellular organism is produced by mitosis.

? *Compare the life cycles of animals and plants, mentioning their similarities and differences.*

CONCEPT 13.3

Meiosis reduces the number of chromosome sets from diploid to haploid (pp. 253–257)

- The two cell divisions of meiosis, **meiosis I** and **meiosis II**, produce four haploid daughter cells. The number of chromosome sets is reduced from two (diploid) to one (haploid) during meiosis I, the reductional division.
- Meiosis is distinguished from mitosis by three events of meiosis I:

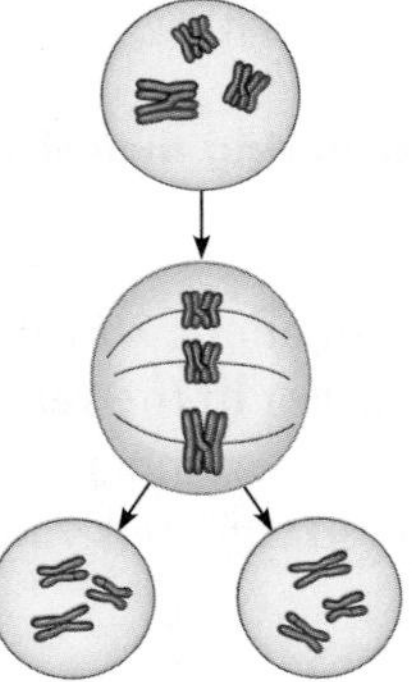

Prophase I: Each homologous pair undergoes **synapsis** and **crossing over** between nonsister chromatids with the subsequent appearance of **chiasmata**.

Metaphase I: Chromosomes line up as homologous pairs on the metaphase plate.

Anaphase I: Homologs separate from each other; sister chromatids remain joined at the centromere.

Meiosis II separates the sister chromatids.
- The combination of sister chromatid cohesion and crossing over leads to chiasmata, which hold homologs together until anaphase I. Cohesins are cleaved along the chromatid arms at anaphase I, allowing the homologs to separate, and at the centromeres in anaphase II, allowing sister chromatids to separate.

? *During prophase I, homologous chromosomes pair up and undergo synapsis and crossing over. Explain why this cannot also occur during prophase II.*

CONCEPT 13.4

Genetic variation produced in sexual life cycles contributes to evolution (pp. 257–260)

- Three events in sexual reproduction contribute to genetic variation in a population: independent assortment of chromosomes during meiosis, crossing over during meiosis I, and random fertilization of egg cells by sperm. Crossing over involves breakage and rejoining of the DNA of nonsister chromatids in a homologous pair, resulting in recombinant chromatids that will become **recombinant chromosomes**.

- Genetic variation is the raw material for evolution by natural selection. Mutations are the original source of this variation; the production of new combinations of variant genes in sexual reproduction generates additional genetic diversity. Animals that reproduce only asexually are quite rare, underscoring the apparently great advantage of genetic diversity.

? *Explain how three processes unique to meiosis generate a great deal of genetic variation.*

TEST YOUR UNDERSTANDING

LEVEL 1: KNOWLEDGE/COMPREHENSION

1. A human cell containing 22 autosomes and a Y chromosome is
   a. a sperm.
   b. an egg.
   c. a zygote.
   d. a somatic cell of a male.
   e. a somatic cell of a female.

2. Which life cycle stage is found in plants but not animals?
   a. gamete
   b. zygote
   c. multicellular diploid
   d. multicellular haploid
   e. unicellular diploid

3. Homologous chromosomes move toward opposite poles of a dividing cell during
   a. mitosis.
   b. meiosis I.
   c. meiosis II.
   d. fertilization.
   e. binary fission.

LEVEL 2: APPLICATION/ANALYSIS

4. Meiosis II is similar to mitosis in that
   a. sister chromatids separate during anaphase.
   b. DNA replicates before the division.
   c. the daughter cells are diploid.
   d. homologous chromosomes synapse.
   e. the chromosome number is reduced.

5. If the DNA content of a diploid cell in the G_1 phase of the cell cycle is x, then the DNA content of the same cell at metaphase of meiosis I would be
   a. $0.25x$.
   b. $0.5x$.
   c. x.
   d. $2x$.
   e. $4x$.

6. If we continued to follow the cell lineage from question 5, then the DNA content of a single cell at metaphase of meiosis II would be
   a. $0.25x$.
   b. $0.5x$.
   c. x.
   d. $2x$.
   e. $4x$.

7. How many different combinations of maternal and paternal chromosomes can be packaged in gametes made by an organism with a diploid number of 8 ($2n = 8$)?
   a. 2
   b. 4
   c. 8
   d. 16
   e. 32

8. **DRAW IT** The diagram at right shows a cell in meiosis.

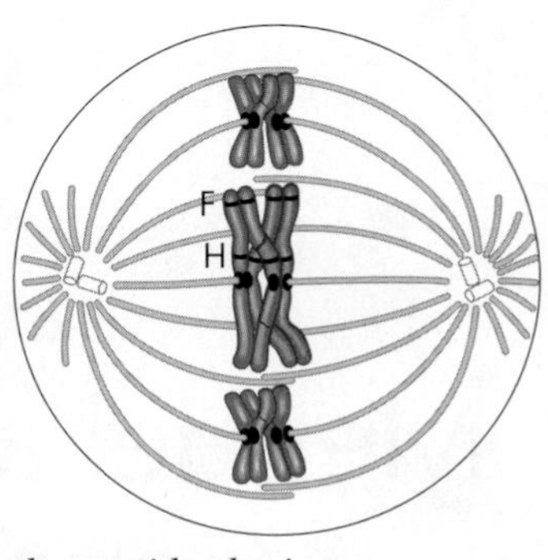

   (a) Copy the drawing to a separate sheet of paper and label appropriate structures with these terms, drawing lines or brackets as needed: chromosome (label as duplicated or unduplicated), centromere, kinetochore, sister chromatids, nonsister chromatids, homologous pair, homologs, chiasma, sister chromatid cohesion.
   (b) Describe the makeup of a haploid set and a diploid set.
   (c) Identify the stage of meiosis shown.

LEVEL 3: SYNTHESIS/EVALUATION

9. How can you tell the cell in question 8 is undergoing meiosis, not mitosis?

10. **EVOLUTION CONNECTION**
    Many species can reproduce either asexually or sexually. What might be the evolutionary significance of the switch from asexual to sexual reproduction that occurs in some organisms when the environment becomes unfavorable?

11. **SCIENTIFIC INQUIRY**
    The diagram above represents a meiotic cell in a certain individual. A previous study has shown that the freckles gene is located at the locus marked F, and the hair-color gene is located at the locus marked H, both on the long chromosome. The individual from whom this cell was taken has inherited different alleles for each gene ("freckles" and "black hair" from one parent, and "no freckles" and "blond hair" from the other). Predict allele combinations in the gametes resulting from this meiotic event. (It will help if you draw out the rest of meiosis, labeling alleles by name.) List other possible combinations of these alleles in this individual's gametes.

12. **WRITE ABOUT A THEME**
    The Genetic Basis of Life The continuity of life is based on heritable information in the form of DNA. In a short essay (100–150 words), explain how chromosome behavior during sexual reproduction in animals ensures perpetuation of parental traits in offspring and, at the same time, genetic variation among offspring.

For selected answers, see Appendix A.

www.masteringbiology.com

1. MasteringBiology® Assignments

Video Tutor Session Mitosis and Meiosis

BioFlix **Tutorials** Meiosis: Genes, Chromosomes, and Sexual Reproduction • The Mechanism • Determinants of Heredity and Genetic Variation

Activities Asexual and Sexual Life Cycles • Meiosis • Meiosis Animation • Origins of Genetic Variation

Questions Student Misconceptions • Reading Quiz • Multiple Choice • End-of-Chapter

2. eText

Read your book online, search, take notes, highlight text, and more.

3. The Study Area

Practice Tests • Cumulative Test • ***BioFlix*** 3-D Animations • MP3 Tutor Sessions • Videos • Activities • Investigations • Lab Media • Audio Glossary • Word Study Tools • Art

14

Mendel and the Gene Idea

▲ **Figure 14.1 What principles of inheritance did Gregor Mendel discover by breeding garden pea plants?**

KEY CONCEPTS

14.1 Mendel used the scientific approach to identify two laws of inheritance
14.2 The laws of probability govern Mendelian inheritance
14.3 Inheritance patterns are often more complex than predicted by simple Mendelian genetics
14.4 Many human traits follow Mendelian patterns of inheritance

OVERVIEW

Drawing from the Deck of Genes

If you spotted a woman with bright purple hair walking down the street, you would probably deduce that she hadn't inherited her striking hair color from either parent. Consciously or not, you have transformed a lifetime of observations of hair color and other features into a list of possible variations that occur naturally among people. Brown, blue, green, or gray eyes; black, brown, blond, or red hair—these are just a few examples of heritable variations that we may observe among individuals in a population. What are the genetic principles that account for the transmission of such traits from parents to offspring in humans and other organisms?

The explanation of heredity most widely in favor during the 1800s was the "blending" hypothesis, the idea that genetic material contributed by the two parents mixes in a manner analogous to the way blue and yellow paints blend to make green. This hypothesis predicts that over many generations, a freely mating population will give rise to a uniform population of individuals. However, our everyday observations and the results of breeding experiments with animals and plants contradict that prediction. The blending hypothesis also fails to explain other phenomena of inheritance, such as traits reappearing after skipping a generation.

An alternative to the blending model is a "particulate" hypothesis of inheritance: the gene idea. According to this model, parents pass on discrete heritable units—genes—that retain their separate identities in offspring. An organism's collection of genes is more like a deck of cards than a pail of paint. Like playing cards, genes can be shuffled and passed along, generation after generation, in undiluted form.

Modern genetics had its genesis in an abbey garden, where a monk named Gregor Mendel documented a particulate mechanism for inheritance. **Figure 14.1** shows Mendel (back row, holding a sprig of fuchsia) with his fellow monks. Mendel developed his theory of inheritance several decades before chromosomes were observed under the microscope and the significance of their behavior was understood. In this chapter, we will step into Mendel's garden to re-create his experiments and explain how he arrived at his theory of inheritance. We will also explore inheritance patterns more complex than those observed by Mendel in garden peas. Finally, we will see how the Mendelian model applies to the inheritance of human variations, including hereditary disorders such as sickle-cell disease.

CONCEPT **14.1**

Mendel used the scientific approach to identify two laws of inheritance

Mendel discovered the basic principles of heredity by breeding garden peas in carefully planned experiments. As we retrace his work, you will recognize the key elements of the scientific process that were introduced in Chapter 1.

Mendel's Experimental, Quantitative Approach

Mendel grew up on his parents' small farm in a region of Austria that is now part of the Czech Republic. In this agricultural area, Mendel and the other children received agricultural training in school along with their basic education. As an adolescent, Mendel overcame financial hardship and illness to excel in high school and, later, at the Olmutz Philosophical Institute.

In 1843, at the age of 21, Mendel entered an Augustinian monastery, a reasonable choice at that time for someone who valued the life of the mind. He considered becoming a teacher but failed the necessary examination. In 1851, he left the monastery to pursue two years of study in physics and chemistry at the University of Vienna. These were very important years for Mendel's development as a scientist, in large part due to the strong influence of two professors. One was the physicist Christian Doppler, who encouraged his students to learn science through experimentation and trained Mendel to use mathematics to help explain natural phenomena. The other was a botanist named Franz Unger, who aroused Mendel's interest in the causes of variation in plants. The instruction Mendel received from these two mentors later played a critical role in his experiments with garden peas.

After attending the university, Mendel returned to the monastery and was assigned to teach at a local school, where several other instructors were enthusiastic about scientific research. In addition, his fellow monks shared a long-standing fascination with the breeding of plants. The monastery therefore provided fertile soil in more ways than one for Mendel's scientific endeavors. Around 1857, Mendel began breeding garden peas in the abbey garden to study inheritance. Although the question of heredity had long been a focus of curiosity at the monastery, Mendel's fresh approach allowed him to deduce principles that had remained elusive to others.

One reason Mendel probably chose to work with peas is that they are available in many varieties. For example, one variety has purple flowers, while another variety has white flowers. A heritable feature that varies among individuals, such as flower color, is called a **character**. Each variant for a character, such as purple or white color for flowers, is called a **trait**.

Other advantages of using peas are their short generation time and the large number of offspring from each mating. Furthermore, Mendel could strictly control mating between plants. The reproductive organs of a pea plant are in its flowers, and each pea flower has both pollen-producing organs (stamens) and an egg-bearing organ (carpel).* In nature, pea plants usually self-fertilize: Pollen grains from the stamens land on the carpel of the same flower, and sperm released from the pollen grains fertilize eggs present in the carpel. To achieve cross-pollination (fertilization between different plants), Mendel removed the immature stamens of a plant before they produced pollen and then dusted pollen from another plant onto the altered flowers (**Figure 14.2**). Each resulting zygote then developed into a plant embryo encased in a seed (pea). Mendel could thus always be sure of the parentage of new seeds.

*As you learned in Figure 13.6b, meiosis in plants produces spores, not gametes. In flowering plants like the pea, each spore develops into a microscopic haploid gametophyte that contains only a few cells and is located on the parent plant. The gametophyte produces sperm, in pollen grains, and eggs, in the carpel. For simplicity, we will not include the gametophyte stage in our discussion of fertilization in plants.

▼ **Figure 14.2** **RESEARCH METHOD**

Crossing Pea Plants

APPLICATION By crossing (mating) two true-breeding varieties of an organism, scientists can study patterns of inheritance. In this example, Mendel crossed pea plants that varied in flower color.

TECHNIQUE

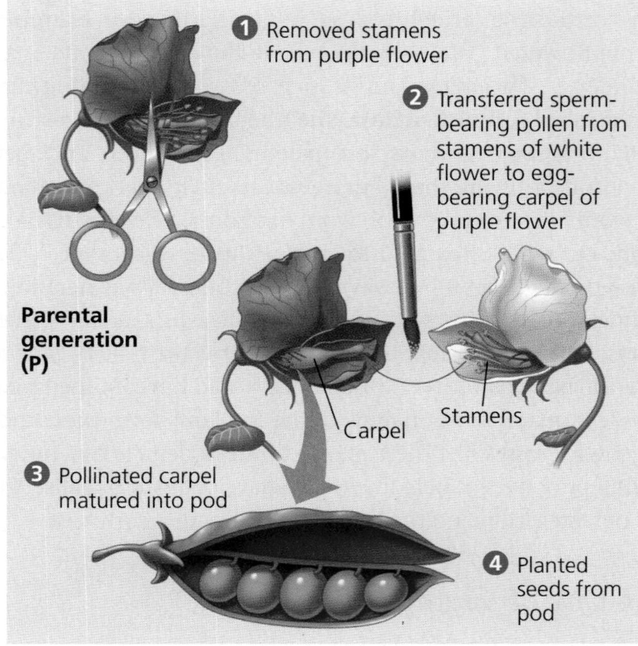

1 Removed stamens from purple flower
2 Transferred sperm-bearing pollen from stamens of white flower to egg-bearing carpel of purple flower
Parental generation (P)
Carpel
Stamens
3 Pollinated carpel matured into pod
4 Planted seeds from pod

RESULTS When pollen from a white flower was transferred to a purple flower, the first-generation hybrids all had purple flowers. The result was the same for the reciprocal cross, which involved the transfer of pollen from purple flowers to white flowers.

5 Examined offspring: all purple flowers
First filial generation offspring (F_1)

Mendel chose to track only those characters that occurred in two distinct, alternative forms. For example, his plants had either purple flowers or white flowers; there were no colors intermediate between these two varieties. Had Mendel focused instead on characters that varied in a continuum among individuals—seed weight, for example—he would not have discovered the particulate nature of inheritance. (You'll learn why later.)

Mendel also made sure that he started his experiments with varieties that, over many generations of self-pollination, had produced only the same variety as the parent plant. Such plants are said to be **true-breeding**. For example, a plant with purple flowers is true-breeding if the seeds produced by self-pollination in successive generations all give rise to plants that also have purple flowers.

In a typical breeding experiment, Mendel cross-pollinated two contrasting, true-breeding pea varieties—for example, purple-flowered plants and white-flowered plants (see Figure 14.2). This mating, or *crossing*, of two true-breeding varieties is called **hybridization**. The true-breeding parents are referred to as the **P generation** (parental generation), and their hybrid offspring are the **F_1 generation** (first filial generation, the word *filial* from the Latin word for "son"). Allowing these F_1 hybrids to self-pollinate (or to cross-pollinate with other F_1 hybrids) produces an **F_2 generation** (second filial generation). Mendel usually followed traits for at least the P, F_1, and F_2 generations. Had Mendel stopped his experiments with the F_1 generation, the basic patterns of inheritance would have escaped him. Mendel's quantitative analysis of the F_2 plants from thousands of genetic crosses like these allowed him to deduce two fundamental principles of heredity, which have come to be called the law of segregation and the law of independent assortment.

The Law of Segregation

If the blending model of inheritance were correct, the F_1 hybrids from a cross between purple-flowered and white-flowered pea plants would have pale purple flowers, a trait intermediate between those of the P generation. Notice in Figure 14.2 that the experiment produced a very different result: All the F_1 offspring had flowers just as purple as the purple-flowered parents. What happened to the white-flowered plants' genetic contribution to the hybrids? If it were lost, then the F_1 plants could produce only purple-flowered offspring in the F_2 generation. But when Mendel allowed the F_1 plants to self-pollinate and planted their seeds, the white-flower trait reappeared in the F_2 generation.

Mendel used very large sample sizes and kept accurate records of his results: 705 of the F_2 plants had purple flowers, and 224 had white flowers. These data fit a ratio of approximately three purple to one white (**Figure 14.3**). Mendel reasoned that the heritable factor for white flowers did not disappear in the F_1 plants, but was somehow hidden, or masked, when the purple-flower factor was present. In Mendel's terminology, purple flower color is a *dominant* trait, and white flower color is a *recessive* trait. The reappearance of white-flowered plants in the F_2 generation was evidence that the heritable factor causing white flowers had not been diluted or destroyed by coexisting with the purple-flower factor in the F_1 hybrids.

Mendel observed the same pattern of inheritance in six other characters, each represented by two distinctly different traits (**Table 14.1**). For example, when Mendel crossed a true-breeding variety that produced smooth, round pea seeds with one that produced wrinkled seeds, all the F_1 hybrids produced round seeds; this is the dominant trait for seed shape. In the F_2 generation, approximately 75% of the seeds were round and 25% were wrinkled—a 3:1 ratio, as in Figure 14.3. Now let's see how Mendel deduced the law of

▼ Figure 14.3 INQUIRY

When F_1 hybrid pea plants self- or cross-pollinate, which traits appear in the F_2 generation?

EXPERIMENT Around 1860, in a monastery garden in Brünn, Austria, Gregor Mendel used the character of flower color in pea plants to follow traits through two generations. He crossed true-breeding purple-flowered plants and white-flowered plants (crosses are symbolized by ×). The resulting F_1 hybrids were allowed to self-pollinate or were cross-pollinated with other F_1 hybrids. The F_2 generation plants were then observed for flower color.

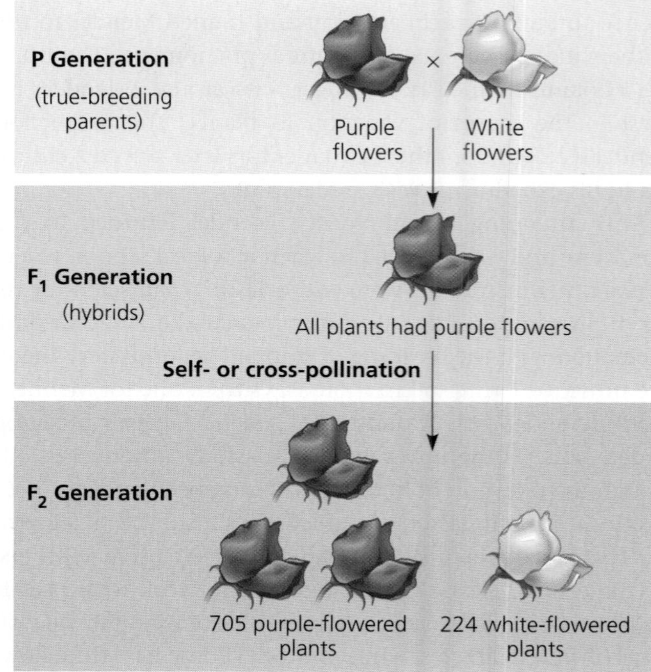

RESULTS Both purple-flowered and white-flowered plants appeared in the F_2 generation, in a ratio of approximately 3:1.

CONCLUSION The "heritable factor" for the recessive trait (white flowers) had not been destroyed, deleted, or "blended" in the F_1 generation but was merely masked by the presence of the factor for purple flowers, which is the dominant trait.

SOURCE G. Mendel, Experiments in plant hybridization, *Proceedings of the Natural History Society of Brünn* 4:3–47 (1866).

WHAT IF? If you mated two purple-flowered plants from the P generation, what ratio of traits would you expect to observe in the offspring? Explain.

Table 14.1 The Results of Mendel's F_1 Crosses for Seven Characters in Pea Plants

Character	Dominant Trait	×	Recessive Trait	F_2 Generation Dominant: Recessive	Ratio
Flower color	Purple	×	White	705:224	3.15:1
Flower position	Axial	×	Terminal	651:207	3.14:1
Seed color	Yellow	×	Green	6,022:2,001	3.01:1
Seed shape	Round	×	Wrinkled	5,474:1,850	2.96:1
Pod shape	Inflated	×	Constricted	882:299	2.95:1
Pod color	Green	×	Yellow	428:152	2.82:1
Stem length	Tall	×	Dwarf	787:277	2.84:1

segregation from his experimental results. In the discussion that follows, we will use modern terms instead of some of the terms used by Mendel. (For example, we'll use "gene" instead of Mendel's "heritable factor.")

Mendel's Model

Mendel developed a model to explain the 3:1 inheritance pattern that he consistently observed among the F_2 offspring in his pea experiments. We describe four related concepts making up this model, the fourth of which is the law of segregation.

First, *alternative versions of genes account for variations in inherited characters.* The gene for flower color in pea plants, for example, exists in two versions, one for purple flowers and the other for white flowers. These alternative versions of a gene are

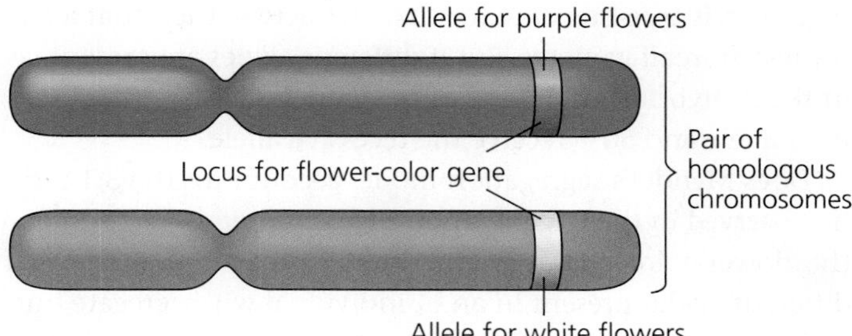

▲ **Figure 14.4 Alleles, alternative versions of a gene.** A somatic cell has two copies of each chromosome (forming a homologous pair) and thus two copies of each gene; the alleles may be identical or different. This figure depicts a pair of homologous chromosomes in an F_1 hybrid pea plant. The paternally inherited chromosome (blue), which was present in the sperm within a pollen grain, has an allele for purple flowers, and the maternally inherited chromosome (red), which was present in an egg within a carpel, has an allele for white flowers.

called **alleles** (**Figure 14.4**). Today, we can relate this concept to chromosomes and DNA. As noted in Chapter 13, each gene is a sequence of nucleotides at a specific place, or locus, along a particular chromosome. The DNA at that locus, however, can vary slightly in its nucleotide sequence and hence in its information content. The purple-flower allele and the white-flower allele are two DNA sequence variations possible at the flower-color locus on one of a pea plant's chromosomes.

Second, *for each character, an organism inherits two copies of a gene, one from each parent.* (These are also called alleles of that gene.) Remarkably, Mendel made this deduction without knowing about the role, or even the existence, of chromosomes. Recall from Chapter 13 that each somatic cell in a diploid organism has two sets of chromosomes, one set inherited from each parent. Thus, a genetic locus is actually represented twice in a diploid cell, once on each homolog of a specific pair of chromosomes. The two alleles at a particular locus may be identical, as in the true-breeding plants of Mendel's P generation. Or the alleles may differ, as in the F_1 hybrids (see Figure 14.4).

Third, *if the two alleles at a locus differ, then one, the* ***dominant allele****, determines the organism's appearance; the other, the* ***recessive allele****, has no noticeable effect on the organism's appearance.* Accordingly, Mendel's F_1 plants had purple flowers because the allele for that trait is dominant and the allele for white flowers is recessive.

The fourth and final part of Mendel's model, the **law of segregation**, states that *the two alleles for a heritable character segregate (separate from each other) during gamete formation and end up in different gametes.* Thus, an egg or a sperm gets only one of the two alleles that are present in the somatic cells of the organism making the gamete. In terms of chromosomes, this segregation corresponds to the distribution of the two members of a pair of homologous chromosomes to different gametes in meiosis (see Figure 13.7). Note that if an organism has identical alleles for a particular character—that is, the organism is true-breeding for that character—then that allele is present in all gametes. But if different alleles are present, as in the F_1 hybrids, then 50% of the gametes receive the dominant allele and 50% receive the recessive allele.

Does Mendel's segregation model account for the 3:1 ratio he observed in the F_2 generation of his numerous crosses? For the flower-color character, the model predicts that the two different alleles present in an F_1 individual will segregate into gametes such that half the gametes will have the purple-flower allele and half will have the white-flower allele. During self-pollination, gametes of each class unite randomly. An egg with a purple-flower allele has an equal chance of being fertilized by a sperm with a purple-flower allele or one with a white-flower allele. Since the same is true for an egg with a white-flower allele, there are four equally likely combinations of sperm and egg. **Figure 14.5** illustrates these combinations using a **Punnett square**, a handy diagrammatic device for predicting the allele composition of offspring from a cross between individuals of known genetic makeup. Notice that we use a capital letter to symbolize a dominant allele and a lowercase letter for a recessive allele. In our example, *P* is the purple-flower allele, and *p* is the white-flower allele; the gene itself is sometimes referred to as the *P*/*p* gene.

In the F_2 offspring, what color will the flowers be? One-fourth of the plants have inherited two purple-flower alleles; clearly, these plants will have purple flowers. One-half of the F_2 offspring have inherited one purple-flower allele and one white-flower allele; these plants will also have purple flowers, the dominant trait. Finally, one-fourth of the F_2 plants have inherited two white-flower alleles and will express the recessive trait. Thus, Mendel's model accounts for the 3:1 ratio of traits that he observed in the F_2 generation.

Useful Genetic Vocabulary

An organism that has a pair of identical alleles for a character is said to be **homozygous** for the gene controlling that character. In the parental generation in Figure 14.5, the purple pea plant is homozygous for the dominant allele (*PP*), while the white plant is homozygous for the recessive allele (*pp*). Homozygous plants "breed true" because all of their gametes contain the same allele—either *P* or *p* in this example. If we cross dominant homozygotes with recessive homozygotes, every offspring will have two different alleles—*Pp* in the case of the F_1 hybrids of our flower-color experiment (see Figure 14.5). An organism that has two different alleles for a gene is said to be **heterozygous** for that gene. Unlike homozygotes, heterozygotes produce gametes with different alleles, so they are not true-breeding. For example, *P*- and *p*-containing gametes are both produced by our F_1 hybrids. Self-pollination of the F_1 hybrids thus produces both purple-flowered and white-flowered offspring.

Because of the different effects of dominant and recessive alleles, an organism's traits do not always reveal its genetic composition. Therefore, we distinguish between an organism's appearance or observable traits, called its **phenotype**, and its genetic makeup, its **genotype**. In the case of flower color in pea plants, *PP* and *Pp* plants have the same phenotype (purple) but different genotypes. **Figure 14.6** reviews these terms. Note that "phenotype" refers to physiological traits as well as traits that relate directly to appearance.

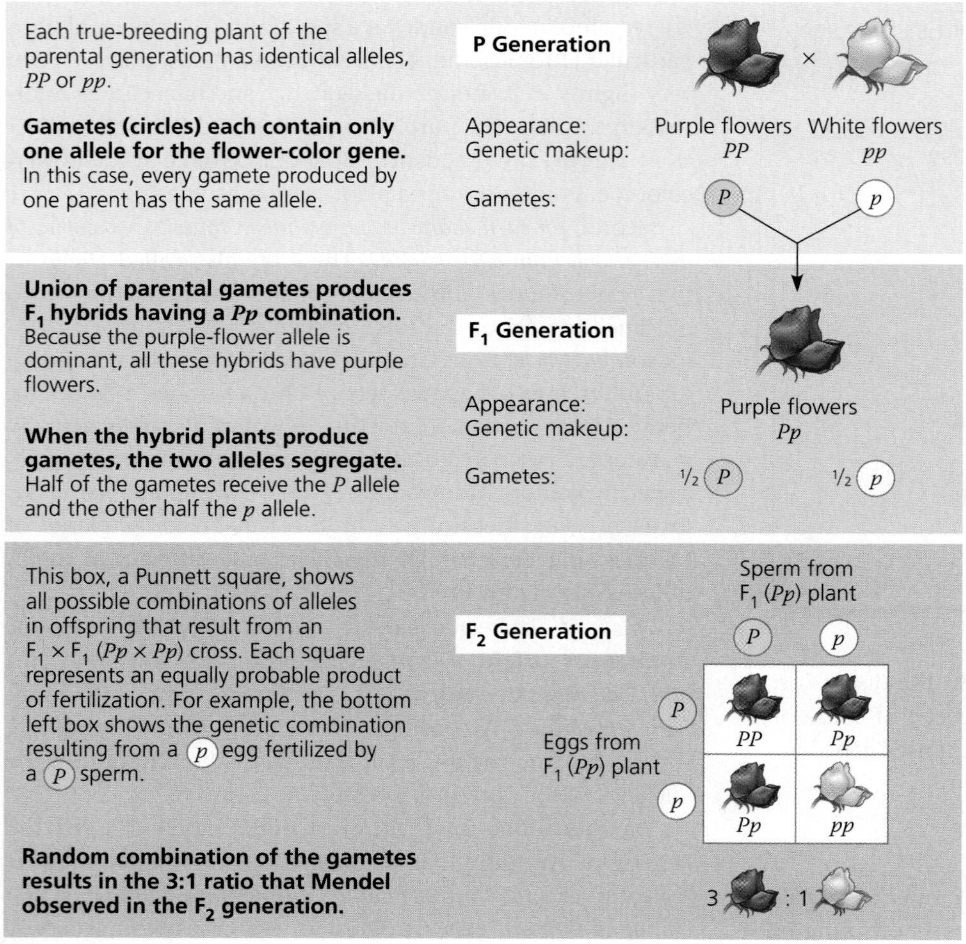

Each true-breeding plant of the parental generation has identical alleles, *PP* or *pp*.

Gametes (circles) each contain only one allele for the flower-color gene. In this case, every gamete produced by one parent has the same allele.

P Generation

Appearance: Purple flowers × White flowers
Genetic makeup: *PP* *pp*
Gametes: *P* *p*

Union of parental gametes produces F_1 hybrids having a *Pp* combination. Because the purple-flower allele is dominant, all these hybrids have purple flowers.

When the hybrid plants produce gametes, the two alleles segregate. Half of the gametes receive the *P* allele and the other half the *p* allele.

F_1 Generation

Appearance: Purple flowers
Genetic makeup: *Pp*
Gametes: 1/2 *P* 1/2 *p*

This box, a Punnett square, shows all possible combinations of alleles in offspring that result from an $F_1 \times F_1$ (*Pp* × *Pp*) cross. Each square represents an equally probable product of fertilization. For example, the bottom left box shows the genetic combination resulting from a *p* egg fertilized by a *P* sperm.

Random combination of the gametes results in the 3:1 ratio that Mendel observed in the F_2 generation.

F_2 Generation

Sperm from F_1 (*Pp*) plant

Eggs from F_1 (*Pp*) plant

	P	*p*
P	*PP*	*Pp*
p	*Pp*	*pp*

3 : 1

▲ **Figure 14.5 Mendel's law of segregation.** This diagram shows the genetic makeup of the generations in Figure 14.3. It illustrates Mendel's model for inheritance of the alleles of a single gene. Each plant has two alleles for the gene controlling flower color, one allele inherited from each of the plant's parents. To construct a Punnett square that predicts the F_2 generation offspring, we list all the possible gametes from one parent (here, the F_1 female) along the left side of the square and all the possible gametes from the other parent (here, the F_1 male) along the top. The boxes represent the offspring resulting from all the possible unions of male and female gametes.

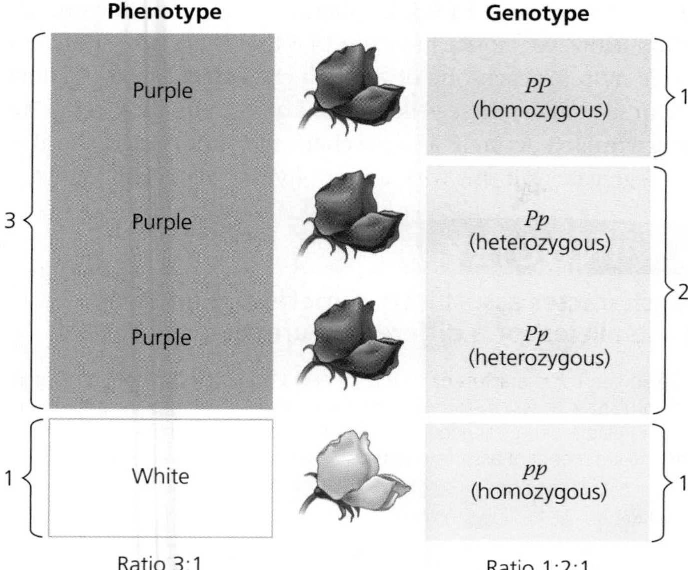

▲ **Figure 14.6 Phenotype versus genotype.** Grouping F_2 offspring from a cross for flower color according to phenotype results in the typical 3:1 phenotypic ratio. In terms of genotype, however, there are actually two categories of purple-flowered plants, *PP* (homozygous) and *Pp* (heterozygous), giving a 1:2:1 genotypic ratio.

For example, there is a pea variety that lacks the normal ability to self-pollinate. This physiological variation (non-self-pollination) is a phenotypic trait.

The Testcross

Suppose we have a "mystery" pea plant that has purple flowers. We cannot tell from its flower color if this plant is homozygous (*PP*) or heterozygous (*Pp*) because both genotypes result in the same purple phenotype. To determine the genotype, we can cross this plant with a white-flowered plant (*pp*), which will make only gametes with the recessive allele (*p*). The allele in the gamete contributed by the mystery plant will therefore determine the appearance of the offspring (**Figure 14.7**). If all the offspring of the cross have purple flowers, then the purple-flowered mystery plant must be homozygous for the dominant allele, because a *PP* × *pp* cross produces all *Pp* offspring. But if both the purple and the white phenotypes appear among the offspring, then the purple-flowered parent must be heterozygous. The offspring of a *Pp* × *pp* cross will be expected to have a 1:1 phenotypic ratio. Breeding an organism of unknown genotype with a recessive homozygote is called a **testcross** because it can reveal the genotype of that organism. The testcross was devised by Mendel and continues to be an important tool of geneticists.

▼ **Figure 14.7** **RESEARCH METHOD**

The Testcross

APPLICATION An organism that exhibits a dominant trait, such as purple flowers in pea plants, can be either homozygous for the dominant allele or heterozygous. To determine the organism's genotype, geneticists can perform a testcross.

TECHNIQUE In a testcross, the individual with the unknown genotype is crossed with a homozygous individual expressing the recessive trait (white flowers in this example), and Punnett squares are used to predict the possible outcomes.

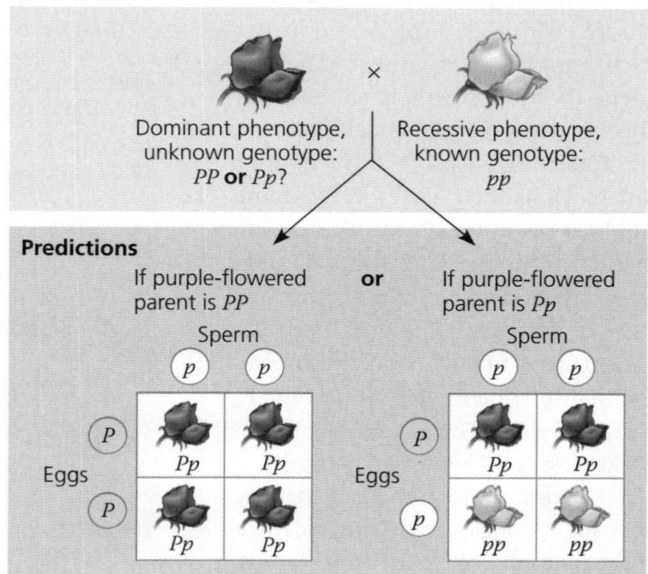

RESULTS Matching the results to either prediction identifies the unknown parental genotype (either *PP* or *Pp* in this example). In this testcross, we transferred pollen from a white-flowered plant to the carpels of a purple-flowered plant; the opposite (reciprocal) cross would have led to the same results.

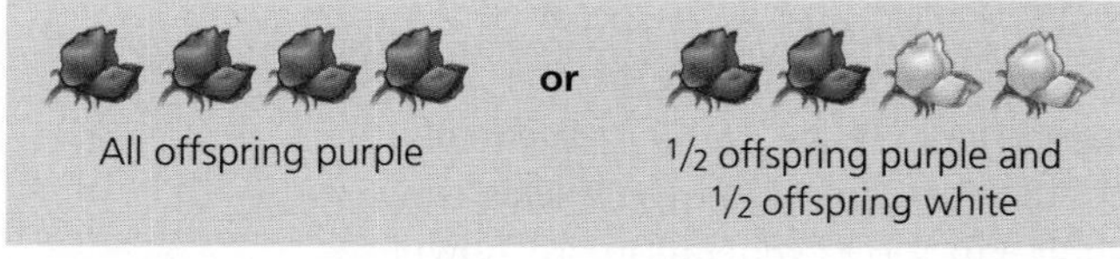

The Law of Independent Assortment

Mendel derived the law of segregation from experiments in which he followed only a *single* character, such as flower color. All the F_1 progeny produced in his crosses of true-breeding parents were **monohybrids**, meaning that they were heterozygous for the one particular character being followed in the cross. We refer to a cross between such heterozygotes as a **monohybrid cross**.

Mendel identified his second law of inheritance by following *two* characters at the same time, such as seed color and seed shape. Seeds (peas) may be either yellow or green. They also may be either round (smooth) or wrinkled. From single-character crosses, Mendel knew that the allele for yellow seeds is dominant (*Y*), and the allele for green seeds is recessive (*y*). For the seed-shape character, the allele for round is dominant (*R*), and the allele for wrinkled is recessive (*r*).

Imagine crossing two true-breeding pea varieties that differ in *both* of these characters—a cross between a plant with yellow-round seeds (*YYRR*) and a plant with green-wrinkled seeds (*yyrr*).

The F_1 plants will be **dihybrids**, individuals heterozygous for the two characters being followed in the cross (*YyRr*). But are these two characters transmitted from parents to offspring as a package? That is, will the *Y* and *R* alleles always stay together, generation after generation? Or are seed color and seed shape inherited independently? **Figure 14.8** shows how a **dihybrid cross**, a cross between F_1 dihybrids, can determine which of these two hypotheses is correct.

The F_1 plants, of genotype *YyRr*, exhibit both dominant phenotypes, yellow seeds with round shapes, no matter which hypothesis is correct. The key step in the experiment is to see what happens when F_1 plants self-pollinate and produce F_2 offspring. If the hybrids must transmit their alleles in the same combinations in which the alleles were inherited from the P generation, then the F_1 hybrids will produce only two classes of gametes: *YR* and *yr*. This "dependent assortment" hypothesis predicts that the phenotypic ratio of the F_2 generation will be 3:1, just as in a monohybrid cross (see Figure 14.8, left side).

The alternative hypothesis is that the two pairs of alleles segregate independently of each other. In other words, genes are packaged into gametes in all possible allelic combinations, as long as each gamete has one allele for each gene. In our example, an F_1 plant will produce four classes of gametes in equal quantities: *YR*, *Yr*, *yR*, and *yr*. If sperm of the four classes fertilize eggs of the four classes, there will be 16 (4 × 4) equally probable ways in which the alleles can combine in the F_2 generation, as shown in Figure 14.8, right side. These combinations result in four phenotypic categories with a ratio of 9:3:3:1 (nine yellow-round to three green-round to three yellow-wrinkled to one green-wrinkled). When Mendel did the experiment and classified the F_2 offspring, his results were close to the predicted 9:3:3:1 phenotypic ratio, supporting the hypothesis that the alleles for one gene—controlling seed color or seed shape, in this example—are sorted into gametes independently of the alleles of other genes.

Mendel tested his seven pea characters in various dihybrid combinations and always observed a 9:3:3:1 phenotypic ratio in the F_2 generation. However, notice in Figure 14.8 that there is a 3:1 phenotypic ratio for each one of the two characters if you consider them separately: three yellow to one green, and three round to one wrinkled. As far as a single character is concerned, the alleles segregate as if this were a monohybrid cross. The results of

▼ **Figure 14.8** **INQUIRY**

Do the alleles for one character assort into gametes dependently or independently of the alleles for a different character?

EXPERIMENT Gregor Mendel followed the characters of seed color and seed shape through the F_2 generation. He crossed a true-breeding plant with yellow-round seeds with a true-breeding plant with green-wrinkled seeds, producing dihybrid F_1 plants. Self-pollination of the F_1 dihybrids produced the F_2 generation. The two hypotheses (dependent and independent assortment) predict different phenotypic ratios.

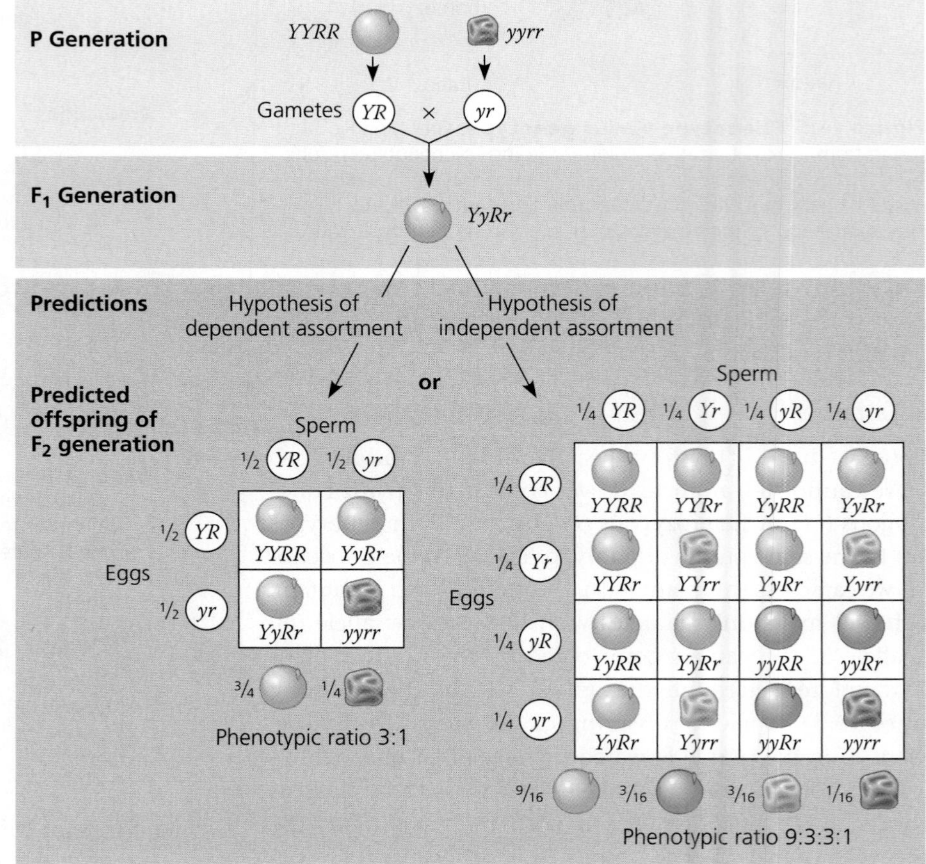

RESULTS

CONCLUSION Only the hypothesis of independent assortment predicts the appearance of two of the observed phenotypes: green-round seeds and yellow-wrinkled seeds (see the right-hand Punnett square). The alleles for seed color and seed shape sort into gametes independently of each other.

SOURCE G. Mendel, Experiments in plant hybridization, *Proceedings of the Natural History Society of Brünn* 4:3–47 (1866).

WHAT IF? Suppose Mendel had transferred pollen from an F_1 plant to the carpel of a plant that was homozygous recessive for both genes. Set up the cross and draw Punnett squares that predict the offspring for both hypotheses. Would this cross have supported the hypothesis of independent assortment equally well?

Mendel's dihybrid experiments are the basis for what we now call the **law of independent assortment**, which states that *each pair of alleles segregates independently of each other pair of alleles during gamete formation.*

This law applies only to genes (allele pairs) located on different chromosomes—that is, on chromosomes that are not homologous—or very far apart on the same chromosome. (The latter case will be explained in Chapter 15, along with the more complex inheritance patterns of genes located near each other, which tend to be inherited together.) All the pea characters Mendel chose for analysis were controlled by genes on different chromosomes (or far apart on one chromosome); this situation greatly simplified interpretation of his multicharacter pea crosses. All the examples we consider in the rest of this chapter involve genes located on different chromosomes.

CONCEPT CHECK 14.1

1. **DRAW IT** Pea plants heterozygous for flower position and stem length (*AaTt*) are allowed to self-pollinate, and 400 of the resulting seeds are planted. Draw a Punnett square for this cross. How many offspring would be predicted to have terminal flowers and be dwarf? (See Table 14.1.)
2. **WHAT IF?** List all gametes that could be made by a pea plant heterozygous for seed color, seed shape, and pod shape (*YyRrIi*; see Table 14.1). How large a Punnett square would you need to draw to predict the offspring of a self-pollination of this "trihybrid"?
3. **MAKE CONNECTIONS** In some pea plant crosses, the plants are self-pollinated. Refer back to Concept 13.1 (pp. 248–249) and explain whether self-pollination is considered asexual or sexual reproduction.

For suggested answers, see Appendix A.

CONCEPT 14.2

The laws of probability govern Mendelian inheritance

Mendel's laws of segregation and independent assortment reflect the same rules of probability that apply to tossing coins, rolling dice, and drawing cards from a deck. The probability scale ranges from 0 to 1. An event that is certain to occur has a probability of 1, while an event that is certain *not* to occur has a probability of 0. With a coin that has heads on both sides, the probability of tossing heads is 1, and the probability of tossing tails is 0. With a normal coin, the chance of tossing heads is ½, and the chance of tossing tails is ½. The probability of drawing the ace of spades from a 52-card deck is 1/52. The probabilities of all possible outcomes for an event must add up to 1. With a deck of cards, the chance of picking a card other than the ace of spades is 51/52.

Tossing a coin illustrates an important lesson about probability. For every toss, the probability of heads is ½. The outcome of any particular toss is unaffected by what has happened on previous trials. We refer to phenomena such as coin tosses as independent events. Each toss of a coin, whether done sequentially with one coin or simultaneously with many, is independent of every other toss. And like two separate coin tosses, the alleles of one gene segregate into gametes independently of another gene's alleles (the law of independent assortment). Two basic rules of probability can help us predict the outcome of the fusion of such gametes in simple monohybrid crosses and more complicated crosses.

The Multiplication and Addition Rules Applied to Monohybrid Crosses

How do we determine the probability that two or more independent events will occur together in some specific combination? For example, what is the chance that two coins tossed simultaneously will both land heads up? The **multiplication rule** states that to determine this probability, we multiply the probability of one event (one coin coming up heads) by the probability of the other event (the other coin coming up heads). By the multiplication rule, then, the probability that both coins will land heads up is ½ × ½ = ¼.

We can apply the same reasoning to an F_1 monohybrid cross. With seed shape in pea plants as the heritable character, the genotype of F_1 plants is *Rr*. Segregation in a heterozygous plant is like flipping a coin in terms of calculating the probability of each outcome: Each egg produced has a ½ chance of carrying the dominant allele (*R*) and a ½ chance of carrying the recessive allele (*r*). The same odds apply to each sperm cell produced. For a particular F_2 plant to have wrinkled seeds, the recessive trait, both the egg and the sperm that come together must carry the *r* allele. The probability that an *r* allele will be present in both gametes at fertilization is found by multiplying ½ (the probability that the egg will have an *r*) × ½ (the probability that the sperm will have an *r*). Thus, the multiplication rule tells us that the probability of an F_2 plant having wrinkled seeds (*rr*) is ¼ (**Figure 14.9**, on the next page). Likewise, the probability of an F_2 plant carrying both dominant alleles for seed shape (*RR*) is ¼.

To figure out the probability that an F_2 plant from a monohybrid cross will be heterozygous rather than homozygous, we need to invoke a second rule. Notice in Figure 14.9 that the dominant allele can come from the egg and the recessive allele from the sperm, or vice versa. That is, F_1 gametes can combine to produce *Rr* offspring in two *mutually exclusive* ways: For any particular heterozygous F_2 plant, the dominant allele can come from the egg *or* the sperm, but not from both. According to the **addition rule**, the probability that any one of two or more mutually exclusive events will occur is calculated by adding their individual probabilities. As we have just seen, the multiplication rule gives us the individual

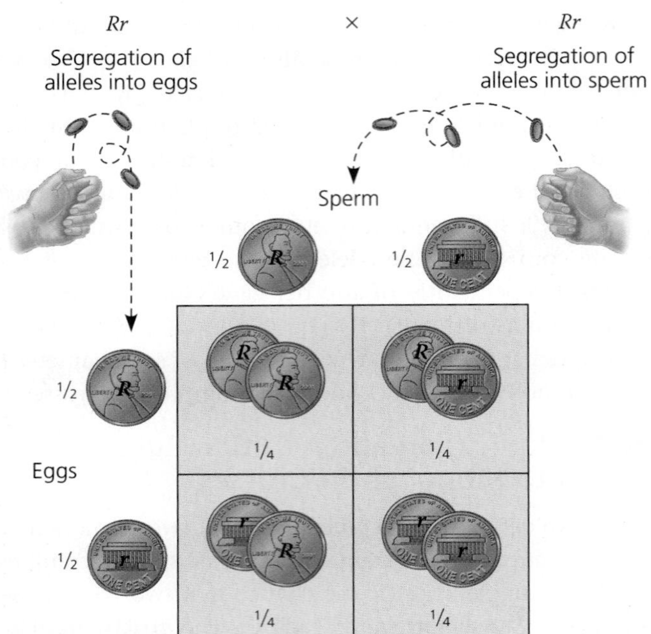

▲ **Figure 14.9 Segregation of alleles and fertilization as chance events.** When a heterozygote (*Rr*) forms gametes, whether a particular gamete ends up with an *R* or an *r* is like the toss of a coin. We can determine the probability for any genotype among the offspring of two heterozygotes by multiplying together the individual probabilities of an egg and sperm having a particular allele (*R* or *r* in this example).

probabilities that we will now add together. The probability for one possible way of obtaining an F_2 heterozygote—the dominant allele from the egg and the recessive allele from the sperm—is ¼. The probability for the other possible way—the recessive allele from the egg and the dominant allele from the sperm—is also ¼ (see Figure 14.9). Using the rule of addition, then, we can calculate the probability of an F_2 heterozygote as ¼ + ¼ = ½.

Solving Complex Genetics Problems with the Rules of Probability

We can also apply the rules of probability to predict the outcome of crosses involving multiple characters. Recall that each allelic pair segregates independently during gamete formation (the law of independent assortment). Thus, a dihybrid or other multicharacter cross is equivalent to two or more independent monohybrid crosses occurring simultaneously. By applying what we have learned about monohybrid crosses, we can determine the probability of specific genotypes occurring in the F_2 generation without having to construct unwieldy Punnett squares.

Consider the dihybrid cross between *YyRr* heterozygotes shown in Figure 14.8. We will focus first on the seed-color character. For a monohybrid cross of *Yy* plants, we can use a simple Punnett square to determine that the probabilities of the offspring genotypes are ¼ for *YY*, ½ for *Yy*, and ¼ for *yy*. We can draw a second Punnett square to determine that the same probabilities apply to the offspring genotypes for seed shape: ¼ *RR*, ½ *Rr*, and ¼ *rr*. Knowing these probabilities, we can simply use the multiplication rule to determine the probability of each of the genotypes in the F_2 generation. To give two examples, the calculations for finding the probabilities of two of the possible F_2 genotypes (*YYRR* and *YyRR*) are shown below:

Probability of *YYRR* = 1/4 (probability of *YY*) × 1/4 (*RR*) = 1/16

Probability of *YyRR* = 1/2 (*Yy*) × 1/4 (*RR*) = 1/8

The *YYRR* genotype corresponds to the upper left box in the larger Punnett square in Figure 14.8 (one box = 1/16). Looking closely at the larger Punnett square in Figure 14.8, you will see that 2 of the 16 boxes (⅛) correspond to the *YyRR* genotype.

Now let's see how we can combine the multiplication and addition rules to solve even more complex problems in Mendelian genetics. Imagine a cross of two pea varieties in which we track the inheritance of three characters. Let's cross a trihybrid with purple flowers and yellow, round seeds (heterozygous for all three genes) with a plant with purple flowers and green, wrinkled seeds (heterozygous for flower color but homozygous recessive for the other two characters). Using Mendelian symbols, our cross is *PpYyRr* × *Ppyyrr*. What fraction of offspring from this cross are predicted to exhibit the recessive phenotypes for *at least two* of the three characters?

To answer this question, we can start by listing all genotypes that fulfill this condition: *ppyyRr*, *ppYyrr*, *Ppyyrr*, *PPyyrr*, and *ppyyrr*. (Because the condition is *at least two* recessive traits, it includes the last genotype, which shows all three recessive traits.) Next, we calculate the probability for each of these genotypes resulting from our *PpYyRr* × *Ppyyrr* cross by multiplying together the individual probabilities for the allele pairs, just as we did in our dihybrid example. Note that in a cross involving heterozygous and homozygous allele pairs (for example, *Yy* × *yy*), the probability of heterozygous offspring is ½ and the probability of homozygous offspring is ½. Finally, we use the addition rule to add the probabilities for all the different genotypes that fulfill the condition of at least two recessive traits, as shown below:

ppyyRr	1/4 (probability of *pp*) × 1/2 (*yy*) × 1/2 (*Rr*)	= 1/16
ppYyrr	1/4 × 1/2 × 1/2	= 1/16
Ppyyrr	1/2 × 1/2 × 1/2	= 2/16
PPyyrr	1/4 × 1/2 × 1/2	= 1/16
ppyyrr	1/4 × 1/2 × 1/2	= 1/16
Chance of *at least two* recessive traits		= 6/16 or 3/8

In time, you'll be able to solve genetics problems faster by using the rules of probability than by filling in Punnett squares.

We cannot predict with certainty the exact numbers of progeny of different genotypes resulting from a genetic cross. But the rules of probability give us the *chance* of various outcomes. Usually, the larger the sample size, the closer the results will conform to our predictions. The reason Mendel counted so many offspring from his crosses is that he understood this statistical feature of inheritance and had a keen sense of the rules of chance.

CONCEPT CHECK 14.2

1. For any gene with a dominant allele *A* and recessive allele *a*, what proportions of the offspring from an $AA \times Aa$ cross are expected to be homozygous dominant, homozygous recessive, and heterozygous?
2. Two organisms, with genotypes *BbDD* and *BBDd*, are mated. Assuming independent assortment of the *B/b* and *D/d* genes, write the genotypes of all possible offspring from this cross and use the rules of probability to calculate the chance of each genotype occurring.
3. **WHAT IF?** Three characters (flower color, seed color, and pod shape) are considered in a cross between two pea plants ($PpYyIi \times ppYyii$). What fraction of offspring are predicted to be homozygous recessive for at least two of the three characters?

For suggested answers, see Appendix A.

CONCEPT **14.3**

Inheritance patterns are often more complex than predicted by simple Mendelian genetics

In the 20th century, geneticists extended Mendelian principles not only to diverse organisms, but also to patterns of inheritance more complex than those described by Mendel. For the work that led to his two laws of inheritance, Mendel chose pea plant characters that turn out to have a relatively simple genetic basis: Each character is determined by one gene, for which there are only two alleles, one completely dominant and the other completely recessive. (There is one exception: Mendel's pod-shape character is actually determined by two genes.) Not all heritable characters are determined so simply, and the relationship between genotype and phenotype is rarely so straightforward. Mendel himself realized that he could not explain the more complicated patterns he observed in crosses involving other pea characters or other plant species. This does not diminish the utility of Mendelian genetics (also called Mendelism), however, because the basic principles of segregation and independent assortment apply even to more complex patterns of inheritance. In this section, we will extend Mendelian genetics to hereditary patterns that were not reported by Mendel.

Extending Mendelian Genetics for a Single Gene

The inheritance of characters determined by a single gene deviates from simple Mendelian patterns when alleles are not completely dominant or recessive, when a particular gene has more than two alleles, or when a single gene produces multiple phenotypes. We will describe examples of each of these situations in this section.

Degrees of Dominance

Alleles can show different degrees of dominance and recessiveness in relation to each other. In Mendel's classic pea crosses, the F_1 offspring always looked like one of the two parental varieties because one allele in a pair showed **complete dominance** over the other. In such situations, the phenotypes of the heterozygote and the dominant homozygote are indistinguishable.

For some genes, however, neither allele is completely dominant, and the F_1 hybrids have a phenotype somewhere between those of the two parental varieties. This phenomenon, called **incomplete dominance**, is seen when red snapdragons are crossed with white snapdragons: All the F_1 hybrids have pink flowers (**Figure 14.10**). This third, intermediate phenotype results from flowers of the heterozygotes having

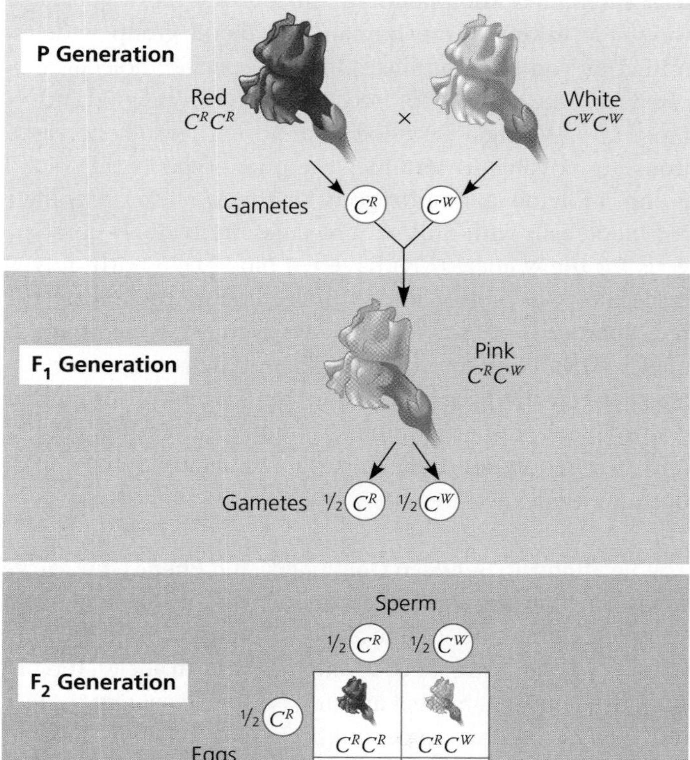

▲ **Figure 14.10 Incomplete dominance in snapdragon color.** When red snapdragons are crossed with white ones, the F_1 hybrids have pink flowers. Segregation of alleles into gametes of the F_1 plants results in an F_2 generation with a 1:2:1 ratio for both genotype and phenotype. Neither allele is dominant, so rather than using upper- and lowercase letters, we use the letter *C* with a superscript to indicate an allele for flower color: C^R for red and C^W for white.

? *Suppose a classmate argues that this figure supports the blending hypothesis for inheritance. What might your classmate say, and how would you respond?*

less red pigment than the red homozygotes. (This is unlike the case of Mendel's pea plants, where the *Pp* heterozygotes make enough pigment for the flowers to be purple, indistinguishable from those of *PP* plants.)

At first glance, incomplete dominance of either allele seems to provide evidence for the blending hypothesis of inheritance, which would predict that the red or white trait could never be retrieved from the pink hybrids. In fact, interbreeding F_1 hybrids produces F_2 offspring with a phenotypic ratio of one red to two pink to one white. (Because heterozygotes have a separate phenotype, the genotypic and phenotypic ratios for the F_2 generation are the same, 1:2:1.) The segregation of the red-flower and white-flower alleles in the gametes produced by the pink-flowered plants confirms that the alleles for flower color are heritable factors that maintain their identity in the hybrids; that is, inheritance is particulate.

Another variation on dominance relationships between alleles is called **codominance**; in this variation, the two alleles each affect the phenotype in separate, distinguishable ways. For example, the human MN blood group is determined by codominant alleles for two specific molecules located on the surface of red blood cells, the M and N molecules. A single gene locus, at which two allelic variations are possible, determines the phenotype of this blood group. Individuals homozygous for the *M* allele (*MM*) have red blood cells with only M molecules; individuals homozygous for the *N* allele (*NN*) have red blood cells with only N molecules. But *both* M and N molecules are present on the red blood cells of individuals heterozygous for the *M* and *N* alleles (*MN*). Note that the MN phenotype is *not* intermediate between the M and N phenotypes, which distinguishes codominance from incomplete dominance. Rather, *both* M and N phenotypes are exhibited by heterozygotes, since both molecules are present.

The Relationship Between Dominance and Phenotype We've now seen that the relative effects of two alleles range from complete dominance of one allele, through incomplete dominance of either allele, to codominance of both alleles. It is important to understand that an allele is called *dominant* because it is seen in the phenotype, not because it somehow subdues a recessive allele. Alleles are simply variations in a gene's nucleotide sequence. When a dominant allele coexists with a recessive allele in a heterozygote, they do not actually interact at all. It is in the pathway from genotype to phenotype that dominance and recessiveness come into play.

To illustrate the relationship between dominance and phenotype, we can use one of the characters Mendel studied—round versus wrinkled pea seed shape. The dominant allele (round) codes for an enzyme that helps convert an unbranched form of starch to a branched form in the seed. The recessive allele (wrinkled) codes for a defective form of this enzyme, leading to an accumulation of unbranched starch, which causes excess water to enter the seed by osmosis. Later, when the seed dries, it wrinkles. If a dominant allele is present, no excess water enters the seed and it does not wrinkle when it dries. One dominant allele results in enough of the enzyme to synthesize adequate amounts of branched starch, which means that dominant homozygotes and heterozygotes have the same phenotype: round seeds.

A closer look at the relationship between dominance and phenotype reveals an intriguing fact: For any character, the observed dominant/recessive relationship of alleles depends on the level at which we examine phenotype. **Tay-Sachs disease**, an inherited disorder in humans, provides an example. The brain cells of a child with Tay-Sachs disease cannot metabolize certain lipids because a crucial enzyme does not work properly. As these lipids accumulate in brain cells, the child begins to suffer seizures, blindness, and degeneration of motor and mental performance and dies within a few years.

Only children who inherit two copies of the Tay-Sachs allele (homozygotes) have the disease. Thus, at the *organismal* level, the Tay-Sachs allele qualifies as recessive. However, the activity level of the lipid-metabolizing enzyme in heterozygotes is intermediate between that in individuals homozygous for the normal allele and that in individuals with Tay-Sachs disease. The intermediate phenotype observed at the *biochemical* level is characteristic of incomplete dominance of either allele. Fortunately, the heterozygote condition does not lead to disease symptoms, apparently because half the normal enzyme activity is sufficient to prevent lipid accumulation in the brain. Extending our analysis to yet another level, we find that heterozygous individuals produce equal numbers of normal and dysfunctional enzyme molecules. Thus, at the *molecular* level, the normal allele and the Tay-Sachs allele are codominant. As you can see, whether alleles appear to be completely dominant, incompletely dominant, or codominant depends on the level at which the phenotype is analyzed.

Frequency of Dominant Alleles Although you might assume that the dominant allele for a particular character would be more common in a population than the recessive allele, this is not a given. For example, about one baby out of 400 in the United States is born with extra fingers or toes, a condition known as polydactyly. Some cases are caused by the presence of a dominant allele. The low frequency of polydactyly indicates that the recessive allele, which results in five digits per appendage, is far more prevalent than the dominant allele in the population. In Chapter 23, you will learn how relative frequencies of alleles in a population are affected by natural selection.

Multiple Alleles

Only two alleles exist for the pea characters that Mendel studied, but most genes exist in more than two allelic forms. The ABO blood groups in humans, for instance, are determined by three alleles of a single gene: I^A, I^B, and *i*. A person's blood

(a) The three alleles for the ABO blood groups and their carbohydrates. Each allele codes for an enzyme that may add a specific carbohydrate (designated by the superscript on the allele and shown as a triangle or circle) to red blood cells.

Allele	I^A	I^B	i
Carbohydrate	A △	B ○	none

(b) Blood group genotypes and phenotypes. There are six possible genotypes, resulting in four different phenotypes.

Genotype	I^AI^A or I^Ai	I^BI^B or I^Bi	I^AI^B	ii
Red blood cell appearance				
Phenotype (blood group)	A	B	AB	O

▲ **Figure 14.11 Multiple alleles for the ABO blood groups.** The four blood groups result from different combinations of three alleles.

? *Based on the surface carbohydrate phenotype in (b), what are the dominance relationships among the alleles?*

group (phenotype) may be one of four types: A, B, AB, or O. These letters refer to two carbohydrates—A and B—that may be found on the surface of red blood cells. A person's blood cells may have carbohydrate A (type A blood), carbohydrate B (type B), both (type AB), or neither (type O), as shown schematically in **Figure 14.11**. Matching compatible blood groups is critical for safe blood transfusions (see Chapter 43).

Pleiotropy

So far, we have treated Mendelian inheritance as though each gene affects only one phenotypic character. Most genes, however, have multiple phenotypic effects, a property called **pleiotropy** (from the Greek *pleion*, more). In humans, for example, pleiotropic alleles are responsible for the multiple symptoms associated with certain hereditary diseases, such as cystic fibrosis and sickle-cell disease, discussed later in this chapter. In the garden pea, the gene that determines flower color also affects the color of the coating on the outer surface of the seed, which can be gray or white. Given the intricate molecular and cellular interactions responsible for an organism's development and physiology, it isn't surprising that a single gene can affect a number of characteristics in an organism.

Extending Mendelian Genetics for Two or More Genes

Dominance relationships, multiple alleles, and pleiotropy all have to do with the effects of the alleles of a single gene. We now consider two situations in which two or more genes are involved in determining a particular phenotype.

Epistasis

In **epistasis** (from the Greek for "standing upon"), the phenotypic expression of a gene at one locus alters that of a gene at a second locus. An example will help clarify this concept. In Labrador retrievers (commonly called "Labs"), black coat color is dominant to brown. Let's designate B and b as the two alleles for this character. For a Lab to have brown fur, its genotype must be bb; these dogs are called chocolate Labs. But there is more to the story. A second gene determines whether or not pigment will be deposited in the hair. The dominant allele, symbolized by E, results in the deposition of either black or brown pigment, depending on the genotype at the first locus. But if the Lab is homozygous recessive for the second locus (ee), then the coat is yellow, regardless of the genotype at the black/brown locus. In this case, the gene for pigment deposition (E/e) is said to be epistatic to the gene that codes for black or brown pigment (B/b).

What happens if we mate black Labs that are heterozygous for both genes ($BbEe$)? Although the two genes affect the same phenotypic character (coat color), they follow the law of independent assortment. Thus, our breeding experiment represents an F_1 dihybrid cross, like those that produced a 9:3:3:1 ratio in Mendel's experiments. We can use a Punnett square to represent the genotypes of the F_2 offspring (**Figure 14.12**). As a result

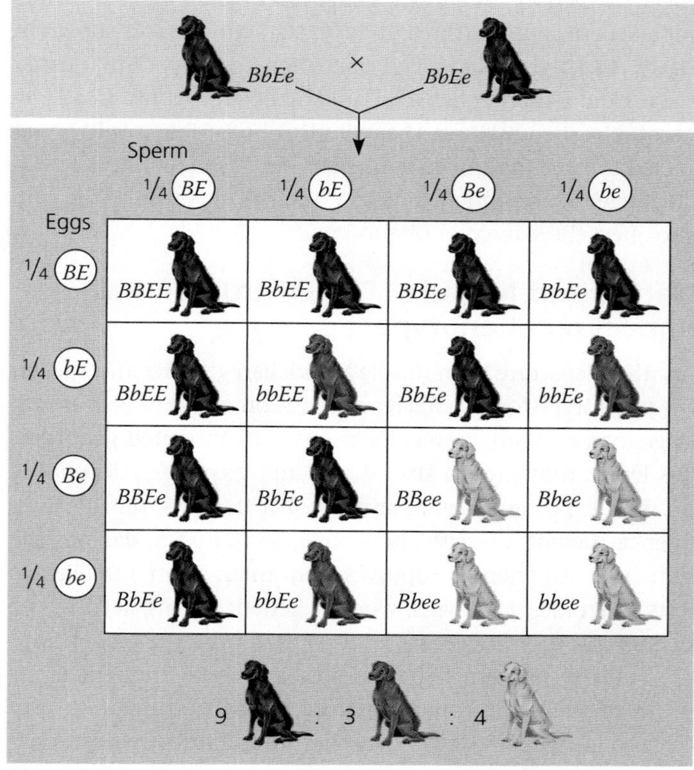

▲ **Figure 14.12 An example of epistasis.** This Punnett square illustrates the genotypes and phenotypes predicted for offspring of matings between two black Labrador retrievers of genotype *BbEe*. The *E/e* gene, which is epistatic to the *B/b* gene coding for hair pigment, controls whether or not pigment of any color will be deposited in the hair.

of epistasis, the phenotypic ratio among the F_2 offspring is nine black to three chocolate (brown) to four yellow. Other types of epistatic interactions produce different ratios, but all are modified versions of 9:3:3:1.

Polygenic Inheritance

Mendel studied characters that could be classified on an either-or basis, such as purple versus white flower color. But for many characters, such as human skin color and height, an either-or classification is impossible because the characters vary in the population in gradations along a continuum. These are called **quantitative characters**. Quantitative variation usually indicates **polygenic inheritance**, an additive effect of two or more genes on a single phenotypic character (the converse of pleiotropy, where a single gene affects several phenotypic characters).

There is evidence, for instance, that skin pigmentation in humans is controlled by at least three separately inherited genes (probably more, but we will simplify). Let's consider three genes, with a dark-skin allele for each gene (*A*, *B*, or *C*) contributing one "unit" of darkness (also a simplification) to the phenotype and being incompletely dominant to the other allele (*a*, *b*, or *c*). An *AABBCC* person would be very dark, while an *aabbcc* individual would be very light. An *AaBbCc* person would have skin of an intermediate shade. Because the alleles have a cumulative effect, the genotypes *AaBbCc* and *AABbcc* would make the same genetic contribution (three units) to skin darkness. As shown in **Figure 14.13**, there are seven skin-color phenotypes that could result from a mating between *AaBbCc* heterozygotes. In a large number of such matings, the majority of offspring would be expected to have intermediate phenotypes (skin color in the middle range). Environmental factors, such as exposure to the sun, also affect the skin-color phenotype.

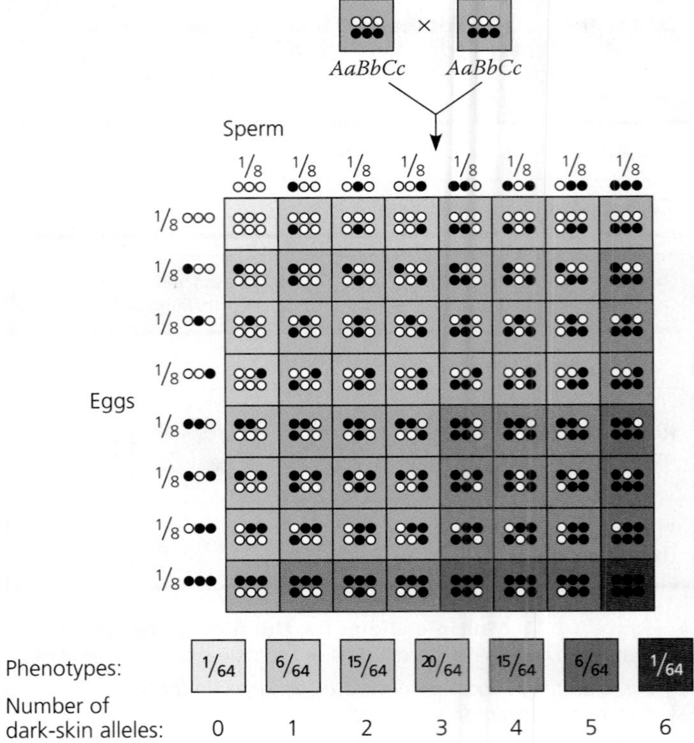

▲ **Figure 14.13 A simplified model for polygenic inheritance of skin color.** According to this model, three separately inherited genes affect the darkness of skin. The heterozygous individuals (*AaBbCc*) represented by the two rectangles at the top of this figure each carry three dark-skin alleles (black circles, which represent *A*, *B*, or *C*) and three light-skin alleles (white circles, which represent *a*, *b*, or *c*). The Punnett square shows all the possible genetic combinations in gametes and in offspring of a large number of hypothetical matings between these heterozygotes. The results are summarized by the phenotypic ratios under the Punnett square.

DRAW IT *Make a bar graph of the results, with skin color (number of dark-skin alleles) along the x-axis and fraction of offspring along the y-axis. Draw a rough curve corresponding to the results and discuss what it shows about the relative proportions of different phenotypes among the offspring.*

Nature and Nurture: The Environmental Impact on Phenotype

Another departure from simple Mendelian genetics arises when the phenotype for a character depends on environment as well as genotype. A single tree, locked into its inherited genotype, has leaves that vary in size, shape, and greenness, depending on their exposure to wind and sun. For humans, nutrition influences height, exercise alters build, sun-tanning darkens the skin, and experience improves performance on intelligence tests. Even identical twins, who are genetic equals, accumulate phenotypic differences as a result of their unique experiences.

Whether human characteristics are more influenced by genes or the environment—nature versus nurture—is a very old and hotly contested debate that we will not attempt to settle here. We can say, however, that a genotype generally is not associated with a rigidly defined phenotype, but rather with a range of phenotypic possibilities due to environmental influences. This phenotypic range is called the **norm of reaction** for a genotype (**Figure 14.14**). For some characters, such as the

▲ **Figure 14.14 The effect of environment on phenotype.** The outcome of a genotype lies within its norm of reaction, a phenotypic range that depends on the environment in which the genotype is expressed. For example, hydrangea flowers of the same genetic variety range in color from blue-violet to pink, with the shade and intensity of color depending on the acidity and aluminum content of the soil.

ABO blood group system, the norm of reaction has no breadth whatsoever; that is, a given genotype mandates a very specific phenotype. Other characteristics, such as a person's blood count of red and white cells, vary quite a bit, depending on such factors as the altitude, the customary level of physical activity, and the presence of infectious agents.

Generally, norms of reaction are broadest for polygenic characters. Environment contributes to the quantitative nature of these characters, as we have seen in the continuous variation of skin color. Geneticists refer to such characters as **multifactorial**, meaning that many factors, both genetic and environmental, collectively influence phenotype.

Integrating a Mendelian View of Heredity and Variation

We have now broadened our view of Mendelian inheritance by exploring degrees of dominance as well as multiple alleles, pleiotropy, epistasis, polygenic inheritance, and the phenotypic impact of the environment. How can we integrate these refinements into a comprehensive theory of Mendelian genetics? The key is to make the transition from the reductionist emphasis on single genes and phenotypic characters to the emergent properties of the organism as a whole, one of the themes of this book.

The term *phenotype* can refer not only to specific characters, such as flower color and blood group, but also to an organism in its entirety—*all* aspects of its physical appearance, internal anatomy, physiology, and behavior. Similarly, the term *genotype* can refer to an organism's entire genetic makeup, not just its alleles for a single genetic locus. In most cases, a gene's impact on phenotype is affected by other genes and by the environment. In this integrated view of heredity and variation, an organism's phenotype reflects its overall genotype and unique environmental history.

Considering all that can occur in the pathway from genotype to phenotype, it is indeed impressive that Mendel could uncover the fundamental principles governing the transmission of individual genes from parents to offspring. Mendel's two laws, segregation and independent assortment, explain heritable variations in terms of alternative forms of genes (hereditary "particles," now known as the alleles of genes) that are passed along, generation after generation, according to simple rules of probability. This theory of inheritance is equally valid for peas, flies, fishes, birds, and human beings—indeed, for any organism with a sexual life cycle. Furthermore, by extending the principles of segregation and independent assortment to help explain such hereditary patterns as epistasis and quantitative characters, we begin to see how broadly Mendelism applies. From Mendel's abbey garden came a theory of particulate inheritance that anchors modern genetics. In the last section of this chapter, we will apply Mendelian genetics to human inheritance, with emphasis on the transmission of hereditary diseases.

CONCEPT CHECK 14.3

1. *Incomplete dominance* and *epistasis* are both terms that define genetic relationships. What is the most basic distinction between these terms?
2. If a man with type AB blood marries a woman with type O, what blood types would you expect in their children? What fraction would you expect of each type?
3. **WHAT IF?** A rooster with gray feathers and a hen of the same phenotype produce 15 gray, 6 black, and 8 white chicks. What is the simplest explanation for the inheritance of these colors in chickens? What phenotypes would you expect in the offspring of a cross between a gray rooster and a black hen?

For suggested answers, see Appendix A.

CONCEPT **14.4**

Many human traits follow Mendelian patterns of inheritance

Peas are convenient subjects for genetic research, but humans are not. The human generation span is long—about 20 years—and human parents produce many fewer offspring than peas and most other species. Even more important, it wouldn't be ethical to ask pairs of humans to breed so that the phenotypes of their offspring could be analyzed! In spite of these constraints, the study of human genetics continues, spurred on by our desire to understand our own inheritance. New molecular biological techniques have led to many breakthrough discoveries, as we will see in Chapter 20, but basic Mendelism endures as the foundation of human genetics.

Pedigree Analysis

Unable to manipulate the mating patterns of people, geneticists must analyze the results of matings that have already occurred. They do so by collecting information about a family's history for a particular trait and assembling this information into a family tree describing the traits of parents and children across the generations—the family **pedigree**.

Figure 14.15a, on the next page, shows a three-generation pedigree that traces the occurrence of a pointed contour of the hairline on the forehead. This trait, called a widow's peak, is due to a dominant allele, *W*. Because the widow's-peak allele is dominant, all individuals who lack a widow's peak must be homozygous recessive (*ww*). The two grandparents with widow's peaks must have the *Ww* genotype, since some of their offspring are homozygous recessive. The offspring in the second generation who *do* have widow's peaks must also be heterozygous, because they are the products of *Ww* $\times$ *ww* matings. The third generation in this pedigree consists of two sisters. The one

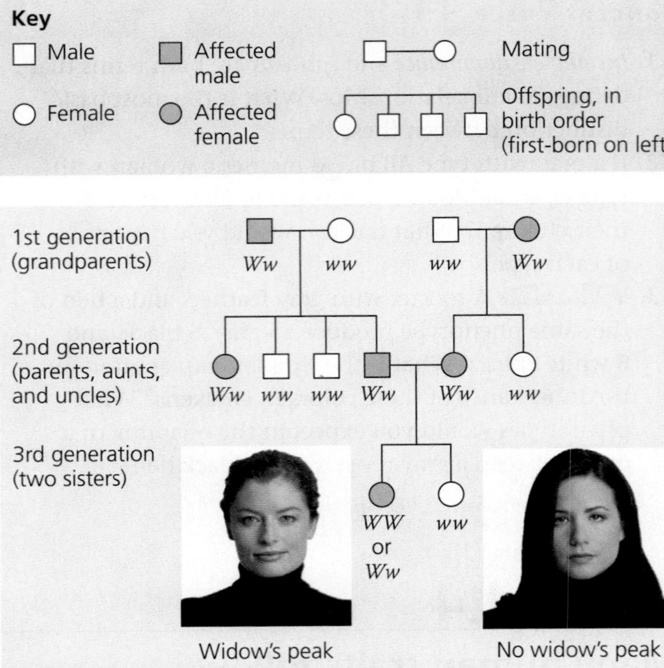

(a) Is a widow's peak a dominant or recessive trait?
Tips for pedigree analysis: Notice in the third generation that the second-born daughter lacks a widow's peak, although both of her parents had the trait. Such a pattern of inheritance supports the hypothesis that the trait is due to a dominant allele. If the trait were due to a *recessive* allele, and both parents had the recessive phenotype, then *all* of their offspring would also have the recessive phenotype.

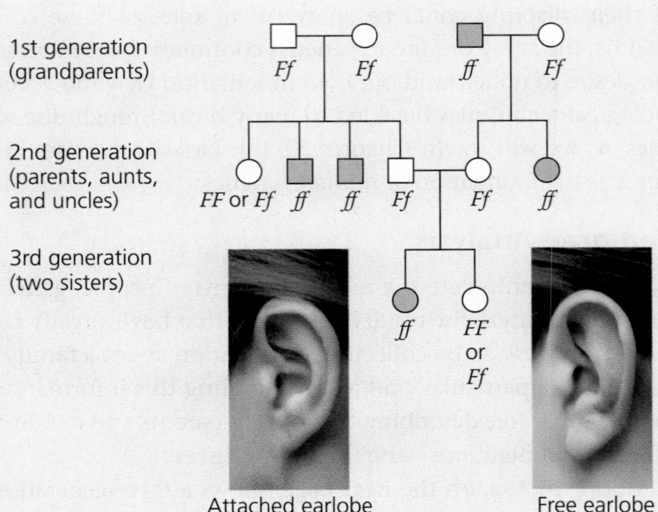

(b) Is an attached earlobe a dominant or recessive trait?
Tips for pedigree analysis: Notice that the first-born daughter in the third generation has attached earlobes, although both of her parents lack that trait (they have free earlobes). Such a pattern is easily explained if the attached-lobe phenotype is due to a recessive allele. If it were due to a *dominant* allele, then at least one parent would also have had the trait.

▲ **Figure 14.15 Pedigree analysis.** Each of these pedigrees traces a trait through three generations of the same family. The two traits have different inheritance patterns, as seen by analysis of the pedigrees.

who has a widow's peak could be either homozygous (*WW*) or heterozygous (*Ww*), given what we know about the genotypes of her parents (both *Ww*).

Figure 14.15b is a pedigree of the same family, but this time we focus on a recessive trait, attached earlobes. We'll use *f* for the recessive allele and *F* for the dominant allele, which results in free earlobes. As you work your way through the pedigree, notice once again that you can apply what you have learned about Mendelian inheritance to understand the genotypes shown for the family members.

An important application of a pedigree is to help us calculate the probability that a future child will have a particular genotype and phenotype. Suppose that the couple represented in the second generation of Figure 14.15 decides to have one more child. What is the probability that the child will have a widow's peak? This is equivalent to a Mendelian F_1 monohybrid cross (*Ww* × *Ww*), and thus the probability that a child will inherit a dominant allele and have a widow's peak is $\frac{3}{4}$ ($\frac{1}{4}$ *WW* + $\frac{1}{2}$ *Ww*). What is the probability that the child will have attached earlobes? Again, we can treat this as a monohybrid cross (*Ff* × *Ff*), but this time we want to know the chance that the offspring will be homozygous recessive (*ff*). That probability is $\frac{1}{4}$. Finally, what is the chance that the child will have a widow's peak *and* attached earlobes? Assuming that the genes for these two characters are on different chromosomes, the two pairs of alleles will assort independently in this dihybrid cross (*WwFf* × *WwFf*). Thus, we can use the multiplication rule: $\frac{3}{4}$ (chance of widow's peak) × $\frac{1}{4}$ (chance of attached earlobes) = $\frac{3}{16}$ (chance of widow's peak and attached earlobes).

Pedigrees are a more serious matter when the alleles in question cause disabling or deadly diseases instead of innocuous human variations such as hairline or earlobe configuration. However, for disorders inherited as simple Mendelian traits, the same techniques of pedigree analysis apply.

Recessively Inherited Disorders

Thousands of genetic disorders are known to be inherited as simple recessive traits. These disorders range in severity from relatively mild, such as albinism (lack of pigmentation, which results in susceptibility to skin cancers and vision problems), to life-threatening, such as cystic fibrosis.

The Behavior of Recessive Alleles

How can we account for the behavior of alleles that cause recessively inherited disorders? Recall that genes code for proteins of specific function. An allele that causes a genetic disorder (let's call it allele *a*) codes for either a malfunctioning protein or no protein at all. In the case of disorders classified as recessive, heterozygotes (*Aa*) are typically normal in phenotype because one copy of the normal allele (*A*) produces a sufficient amount of the specific protein. Thus, a recessively inherited disorder shows up only in the homozygous individuals (*aa*)

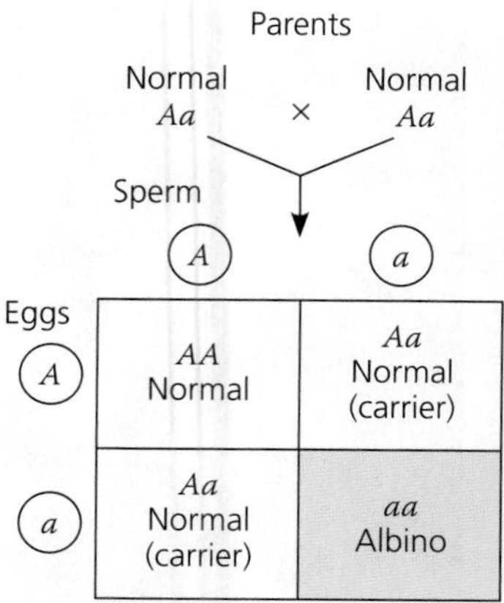

▲ **Figure 14.16 Albinism: a recessive trait.** One of the two sisters shown here has normal coloration; the other is albino. Most recessive homozygotes are born to parents who are carriers of the disorder but themselves have a normal phenotype, the case shown in the Punnett square.

? *What is the probability that the sister with normal coloration is a carrier of the albinism allele?*

who inherit one recessive allele from each parent. Although phenotypically normal with regard to the disorder, heterozygotes may transmit the recessive allele to their offspring and thus are called **carriers**. **Figure 14.16** illustrates these ideas using albinism as an example.

Most people who have recessive disorders are born to parents who are carriers of the disorder but have a normal phenotype, as is the case shown in the Punnett square in Figure 14.16. A mating between two carriers corresponds to a Mendelian F_1 monohybrid cross, so the predicted genotypic ratio for the offspring is 1 *AA* : 2 *Aa* : 1 *aa*. Thus, each child has a ¼ chance of inheriting a double dose of the recessive allele; in the case of albinism, such a child will be albino. From the genotypic ratio, we also can see that out of three offspring with the *normal* phenotype (one *AA* plus two *Aa*), two are predicted to be heterozygous carriers, a ⅔ chance. Recessive homozygotes could also result from *Aa* × *aa* and *aa* × *aa* matings, but if the disorder is lethal before reproductive age or results in sterility (neither of which is true for albinism), no *aa* individuals will reproduce. Even if recessive homozygotes are able to reproduce, such individuals will still account for a much smaller percentage of the population than heterozygous carriers (for reasons we will examine in Chapter 23).

In general, genetic disorders are not evenly distributed among all groups of people. For example, the incidence of Tay-Sachs disease, which we described earlier in this chapter, is disproportionately high among Ashkenazic Jews, Jewish people whose ancestors lived in central Europe. In that population, Tay-Sachs disease occurs in one out of 3,600 births, an incidence about 100 times greater than that among non-Jews or Mediterranean (Sephardic) Jews. This uneven distribution results from the different genetic histories of the world's peoples during less technological times, when populations were more geographically (and hence genetically) isolated.

When a disease-causing recessive allele is rare, it is relatively unlikely that two carriers of the same harmful allele will meet and mate. However, if the man and woman are close relatives (for example, siblings or first cousins), the probability of passing on recessive traits increases greatly. These are called consanguineous ("same blood") matings, and they are indicated in pedigrees by double lines. Because people with recent common ancestors are more likely to carry the same recessive alleles than are unrelated people, it is more likely that a mating of close relatives will produce offspring homozygous for recessive traits—including harmful ones. Such effects can be observed in many types of domesticated and zoo animals that have become inbred.

There is debate among geneticists about the extent to which human consanguinity increases the risk of inherited diseases. Many deleterious alleles have such severe effects that a homozygous embryo spontaneously aborts long before birth. Still, most societies and cultures have laws or taboos forbidding marriages between close relatives. These rules may have evolved out of empirical observation that in most populations, stillbirths and birth defects are more common when parents are closely related. Social and economic factors have also influenced the development of customs and laws against consanguineous marriages.

Cystic Fibrosis

The most common lethal genetic disease in the United States is **cystic fibrosis**, which strikes one out of every 2,500 people of European descent but is much rarer in other groups. Among people of European descent, one out of 25 (4%) are carriers of the cystic fibrosis allele. The normal allele for this gene codes for a membrane protein that functions in the transport of chloride ions between certain cells and the extracellular fluid. These chloride transport channels are defective or absent in the plasma membranes of children who inherit two recessive alleles for cystic fibrosis. The result is an abnormally high concentration of extracellular chloride, which causes the mucus that coats certain cells to become thicker and stickier than normal. The mucus builds up in the pancreas, lungs, digestive tract, and other organs, leading to multiple (pleiotropic) effects, including poor absorption of nutrients from the intestines, chronic bronchitis, and recurrent bacterial infections.

If untreated, most children with cystic fibrosis die before their 5th birthday. But daily doses of antibiotics to prevent infection, gentle pounding on the chest to clear mucus from clogged airways, and other preventive treatments can prolong life. In the United States, more than half of those with cystic fibrosis now survive into their late 20s or even 30s and beyond.

Sickle-Cell Disease: A Genetic Disorder with Evolutionary Implications

EVOLUTION The most common inherited disorder among people of African descent is **sickle-cell disease**, which affects one out of 400 African-Americans. Sickle-cell disease is caused by the substitution of a single amino acid in the hemoglobin protein of red blood cells; in homozygous individuals, all hemoglobin is of the sickle-cell (abnormal) variety. When the oxygen content of an affected individual's blood is low (at high altitudes or under physical stress, for instance), the sickle-cell hemoglobin molecules aggregate into long rods that deform the red cells into a sickle shape (see Figure 5.21). Sickled cells may clump and clog small blood vessels, often leading to other symptoms throughout the body, including physical weakness, pain, organ damage, and even paralysis. Regular blood transfusions can ward off brain damage in children with sickle-cell disease, and new drugs can help prevent or treat other problems, but there is no cure.

Although two sickle-cell alleles are necessary for an individual to manifest full-blown sickle-cell disease, the presence of one sickle-cell allele can affect the phenotype. Thus, at the organismal level, the normal allele is incompletely dominant to the sickle-cell allele. Heterozygotes (carriers), said to have *sickle-cell trait*, are usually healthy, but they may suffer some sickle-cell symptoms during prolonged periods of reduced blood oxygen. At the molecular level, the two alleles are codominant; both normal and abnormal (sickle-cell) hemoglobins are made in heterozygotes.

About one out of ten African-Americans have sickle-cell trait, an unusually high frequency of heterozygotes for an allele with severe detrimental effects in homozygotes. Why haven't evolutionary processes resulted in the disappearance of this allele among this population? One explanation is that having a single copy of the sickle-cell allele reduces the frequency and severity of malaria attacks, especially among young children. The malaria parasite spends part of its life cycle in red blood cells (see Figure 28.10), and the presence of even heterozygous amounts of sickle-cell hemoglobin results in lower parasite densities and hence reduced malaria symptoms. Thus, in tropical Africa, where infection with the malaria parasite is common, the sickle-cell allele confers an advantage to heterozygotes even though it is harmful in the homozygous state. (The balance between these two effects will be discussed in Chapter 23, p. 484.) The relatively high frequency of African-Americans with sickle-cell trait is a vestige of their African roots.

Dominantly Inherited Disorders

Although many harmful alleles are recessive, a number of human disorders are due to dominant alleles. One example is *achondroplasia*, a form of dwarfism that occurs in one of every 25,000 people. Heterozygous individuals have the dwarf phenotype **(Figure 14.17)**. Therefore, all people who are not achondroplastic dwarfs—99.99% of the population—are homozygous for the recessive allele. Like the presence of extra fingers or toes mentioned earlier, achondroplasia is a trait for which the recessive allele is much more prevalent than the corresponding dominant allele.

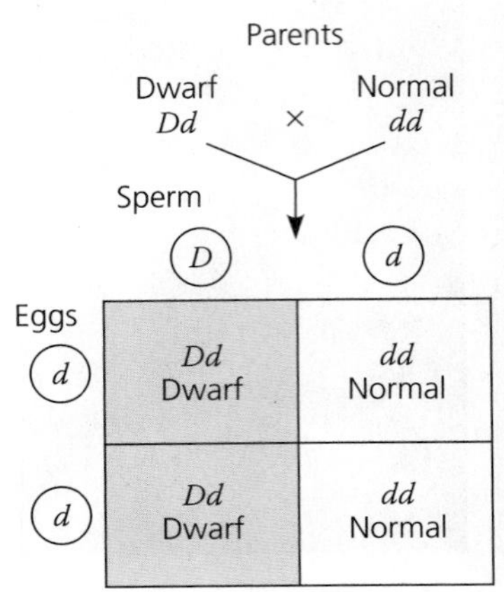

▲ **Figure 14.17 Achondroplasia: a dominant trait.** Dr. Michael C. Ain has achondroplasia, a form of dwarfism caused by a dominant allele. This has inspired his work: He is a specialist in the repair of bone defects caused by achondroplasia and other disorders. The dominant allele (*D*) might have arisen as a mutation in the egg or sperm of a parent or could have been inherited from an affected parent, as shown for an affected father in the Punnett square.

Dominant alleles that cause a lethal disease are much less common than recessive alleles that have lethal effects. All lethal alleles arise by mutations (changes to the DNA) in cells that produce sperm or eggs; presumably, such mutations are equally likely to be recessive or dominant. A lethal recessive allele can be passed from one generation to the next by heterozygous carriers because the carriers themselves have normal phenotypes. A lethal dominant allele, however, often causes the death of afflicted individuals before they can mature and reproduce, so the allele is not passed on to future generations.

Huntington's Disease: A Late-Onset Lethal Disease

The timing of onset of a disease significantly affects its inheritance. A lethal dominant allele is able to be passed on if it causes death at a relatively advanced age. By the time symptoms are evident, the individual with the allele may have already transmitted it to his or her children. For example, **Huntington's disease**, a degenerative disease of the nervous system, is caused by a lethal dominant allele that has no obvious phenotypic effect until the individual is about 35 to 45 years old. Once the deterioration of the nervous system begins, it is irreversible and inevitably fatal. As with other dominant traits, a child born to a parent with the Huntington's disease allele has a 50% chance of inheriting the allele and the disorder (see the Punnett square in Figure 14.17). In the United States, this devastating disease afflicts about one in 10,000 people.

At one time, the onset of symptoms was the only way to know if a person had inherited the Huntington's allele, but this is no longer the case. By analyzing DNA samples from a large family with a high incidence of the disorder, geneticists tracked the Huntington's allele to a locus near the tip of chromosome 4, and the gene was sequenced in 1993. This information led to the development of a test that could detect the presence of the Huntington's allele in an individual's genome. (The methods that make such tests possible are discussed in Chapter 20.) The availability of this test poses an agonizing dilemma for those with a family history of Huntington's disease. Some individuals may want to be tested for this disease before planning a family, whereas others may decide it would be too stressful to find out. Clearly, this is a highly personal decision.

Multifactorial Disorders

The hereditary diseases we have discussed so far are sometimes described as simple Mendelian disorders because they result from abnormality of one or both alleles at a single genetic locus. Many more people are susceptible to diseases that have a multifactorial basis—a genetic component plus a significant environmental influence. Heart disease, diabetes, cancer, alcoholism, certain mental illnesses such as schizophrenia and bipolar disorder, and many other diseases are multifactorial. In many cases, the hereditary component is polygenic. For example, many genes affect cardiovascular health, making some of us more prone than others to heart attacks and strokes. No matter what our genotype, however, our lifestyle has a tremendous effect on phenotype for cardiovascular health and other multifactorial characters. Exercise, a healthful diet, abstinence from smoking, and an ability to handle stressful situations all reduce our risk of heart disease and some types of cancer.

At present, so little is understood about the genetic contributions to most multifactorial diseases that the best public health strategy is to educate people about the importance of environmental factors and to promote healthful behavior.

Genetic Testing and Counseling

Avoiding simple Mendelian disorders is possible when the risk of a particular genetic disorder can be assessed before a child is conceived or during the early stages of the pregnancy. Many hospitals have genetic counselors who can provide information to prospective parents concerned about a family history for a specific disease.

Counseling Based on Mendelian Genetics and Probability Rules

Consider the case of a hypothetical couple, John and Carol. Each had a brother who died from the same recessively inherited lethal disease. Before conceiving their first child, John and Carol seek genetic counseling to determine the risk of having a child with the disease. From the information about their brothers, we know that both parents of John and both parents of Carol must have been carriers of the recessive allele. Thus, John and Carol are both products of $Aa \times Aa$ crosses, where a symbolizes the allele that causes this particular disease. We also know that John and Carol are not homozygous recessive (aa), because they do not have the disease. Therefore, their genotypes are either AA or Aa.

Given a genotypic ratio of 1 AA : 2 Aa : 1 aa for offspring of an $Aa \times Aa$ cross, John and Carol each have a $\frac{2}{3}$ chance of being carriers (Aa). According to the rule of multiplication, the overall probability of their firstborn having the disorder is $\frac{2}{3}$ (the chance that John is a carrier) times $\frac{2}{3}$ (the chance that Carol is a carrier) times $\frac{1}{4}$ (the chance of two carriers having a child with the disease), which equals $\frac{1}{9}$. Suppose that Carol and John decide to have a child—after all, there is an $\frac{8}{9}$ chance that their baby will not have the disorder. If, despite these odds, their child is born with the disease, then we would know that *both* John and Carol are, in fact, carriers (Aa genotype). If both John and Carol are carriers, there is a $\frac{1}{4}$ chance that any subsequent child this couple has will have the disease. The probability is higher for subsequent children because the diagnosis of the disease in the first child established that both parents are carriers, not because the genotype of the first child affects in any way that of future children.

When we use Mendel's laws to predict possible outcomes of matings, it is important to remember that each child represents an independent event in the sense that its genotype is unaffected by the genotypes of older siblings. Suppose that John and Carol have three more children, and *all three* have the hypothetical hereditary disease. There is only one chance in 64 ($\frac{1}{4} \times \frac{1}{4} \times \frac{1}{4}$) that such an outcome will occur. Despite this run of misfortune, the chance that still another child of this couple will have the disease remains $\frac{1}{4}$.

Tests for Identifying Carriers

Most children with recessive disorders are born to parents with normal phenotypes. The key to accurately assessing the genetic risk for a particular disease is therefore to find out whether the prospective parents are heterozygous carriers of the recessive allele. For an increasing number of heritable disorders, tests are available that can distinguish individuals of normal phenotype who are dominant homozygotes from those who are heterozygous carriers (**Figure 14.18**, on the next page). There are now tests that can identify carriers of the alleles for Tay-Sachs disease, sickle-cell disease, and the most common form of cystic fibrosis.

These tests for identifying carriers enable people with family histories of genetic disorders to make informed decisions about having children, but raise other issues. Could carriers be denied health or life insurance or lose the jobs providing those benefits, even though they themselves are healthy? The Genetic Information Nondiscrimination Act, signed into law in the United States in 2008, allays these concerns by prohibiting discrimination in employment or insurance coverage based on genetic test results. A question that remains is whether sufficient genetic counseling is available to help large numbers of individuals understand their genetic test results. Even when test results are clearly understood, affected individuals may still face difficult decisions. Advances in biotechnology offer

▼ **Figure 14.18**

IMPACT

Genetic Testing

Since the sequencing of the human genome was completed in 2003, there has been a virtual explosion in the number and kinds of DNA-based genetic tests. As of 2010, genetic testing for over 2,000 different disease-causing alleles is available.

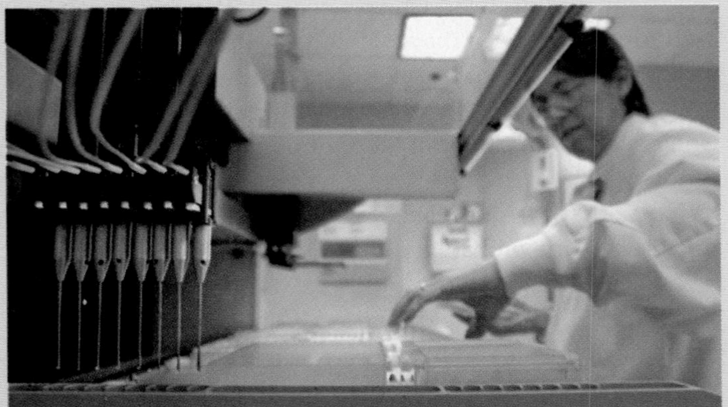

WHY IT MATTERS For prospective parents with a family history of a recessive or late-onset dominant disorder, deciding whether to have children can be a difficult decision. Genetic testing can eliminate some of the uncertainty and allow better predictions of the probabilities and risks involved.

FURTHER READING Designing rules for designer babies, *Scientific American* 300:29 (2009).

WHAT IF? If one parent tests positive and the other tests negative for a recessive allele associated with a disorder, what is the probability that their first child will have the disorder? That their first child will be a carrier? That, if their first child is a carrier, the second will also be a carrier?

the potential to reduce human suffering, but along with them come ethical issues that require conscientious deliberation.

Fetal Testing

Suppose a couple expecting a child learns that they are both carriers of the Tay-Sachs allele. In the 14th–16th week of pregnancy, tests performed along with a technique called **amniocentesis** can determine whether the developing fetus has Tay-Sachs disease **(Figure 14.19a)**. In this procedure, a physician inserts a needle into the uterus and extracts about 10 mL of amniotic fluid, the liquid that bathes the fetus. Some genetic disorders can be detected from the presence of certain molecules in the amniotic fluid itself. Tests for other disorders, including Tay-Sachs disease, are performed on the DNA of cells cultured in the laboratory, descendants of fetal cells sloughed off into the amniotic fluid. A karyotype of these cultured cells can also identify certain chromosomal defects (see Figure 13.3).

In an alternative technique called **chorionic villus sampling (CVS)**, a physician inserts a narrow tube through the cervix into the uterus and suctions out a tiny sample of tissue from the placenta, the organ that transmits nutrients and fetal wastes between the fetus and the mother **(Figure 14.19b)**. The cells of the chorionic villi of the placenta, the portion sampled, are derived from the fetus and have the same genotype and DNA sequence as the new individual. These cells are proliferating rapidly enough to allow karyotyping to be carried out immediately. This rapid analysis represents an advantage over amniocentesis, in which the cells must be cultured for several weeks before karyotyping. Another advantage of CVS is that it can be performed as early as the 8th–10th week of pregnancy.

Recently, medical scientists have developed methods for isolating fetal cells, or even fetal DNA, that have escaped into the mother's blood. Although very few are present, the cells can be cultured and tested, and the fetal DNA can also be analyzed.

Imaging techniques allow a physician to examine a fetus directly for major anatomical abnormalities that might not show up in genetic tests. In the *ultrasound* technique, reflected sound waves are used to produce an image of the fetus by a simple noninvasive procedure. In *fetoscopy*, a needle-thin tube containing a viewing scope and fiber optics (to transmit light) is inserted into the uterus.

Ultrasound and isolation of fetal cells or DNA from maternal blood pose no known risk to either mother or fetus, while the other procedures can cause complications in a small percentage of cases. Amniocentesis or CVS for diagnostic testing is generally offered to women over age 35, due to their increased risk of bearing a child with Down syndrome, and may also be offered to younger women if there are known concerns. If the fetal tests reveal a serious disorder, the parents face the difficult choice of either terminating the pregnancy or preparing to care for a child with a genetic disorder.

Newborn Screening

Some genetic disorders can be detected at birth by simple biochemical tests that are now routinely performed in most hospitals in the United States. One common screening program is for phenylketonuria (PKU), a recessively inherited disorder that occurs in about one out of every 10,000–15,000 births in the United States. Children with this disease cannot properly metabolize the amino acid phenylalanine. This compound and its by-product, phenylpyruvate, can accumulate to toxic levels in the blood, causing severe intellectual disability (mental retardation). However, if PKU is detected in the newborn, a special diet low in phenylalanine will usually allow normal development. (Among many other substances, this diet excludes the artificial sweetener aspartame, which contains phenylalanine.) Unfortunately, few other genetic disorders are treatable at present.

Fetal and newborn screening for serious inherited diseases, tests for identifying carriers, and genetic counseling all rely on the Mendelian model of inheritance. We owe the "gene idea"—the concept of heritable factors transmitted according

▼ **Figure 14.19 Testing a fetus for genetic disorders.** Biochemical tests may detect substances associated with particular disorders, and genetic testing can detect many genetic abnormalities. Karyotyping shows whether the chromosomes of the fetus are normal in number and appearance.

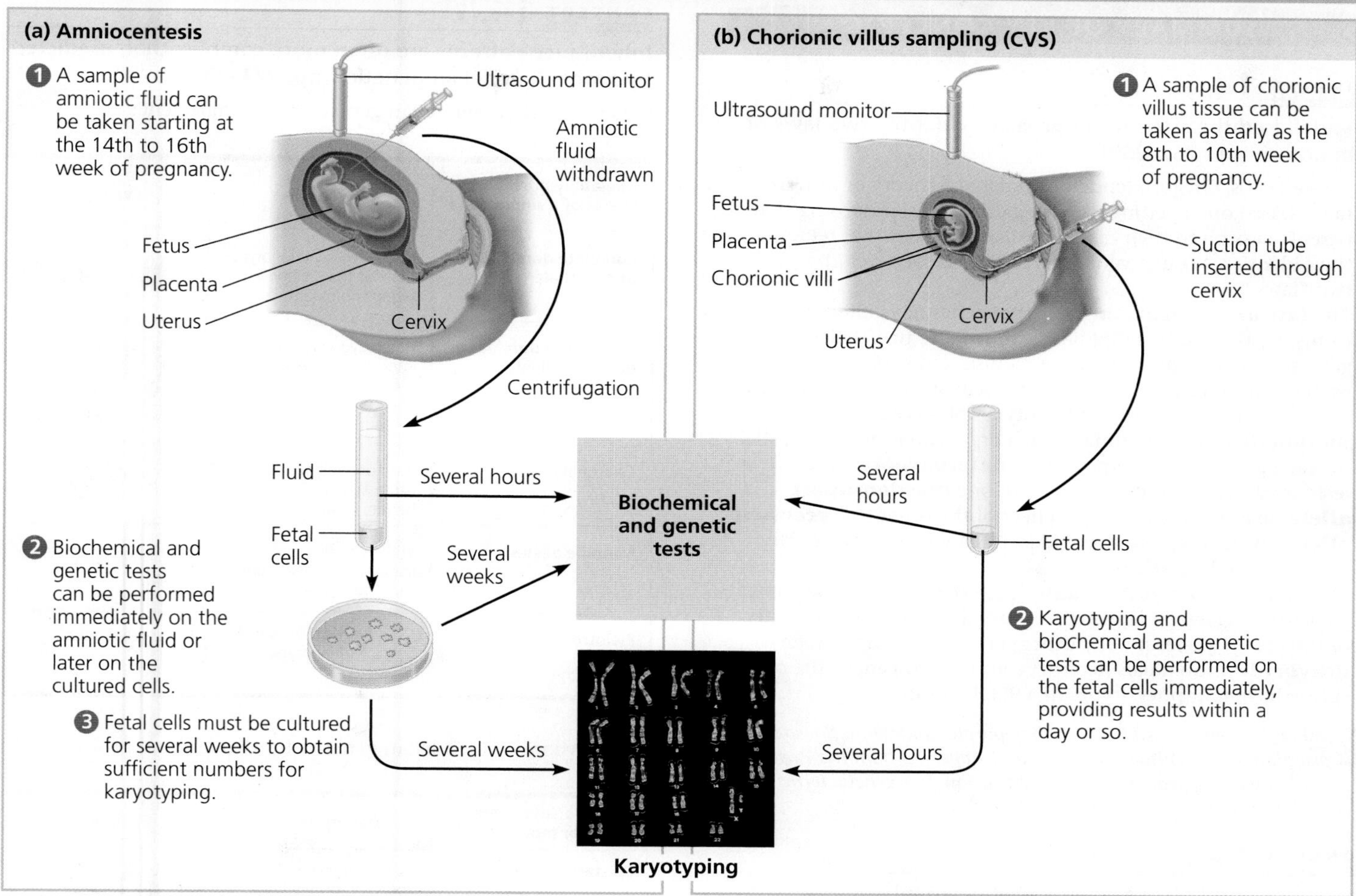

(a) Amniocentesis
1 A sample of amniotic fluid can be taken starting at the 14th to 16th week of pregnancy.
Ultrasound monitor
Amniotic fluid withdrawn
Fetus
Placenta
Uterus
Cervix
Centrifugation
Fluid
Fetal cells
Several hours
Several weeks
2 Biochemical and genetic tests can be performed immediately on the amniotic fluid or later on the cultured cells.
3 Fetal cells must be cultured for several weeks to obtain sufficient numbers for karyotyping.
Several weeks
Biochemical and genetic tests
Karyotyping
(b) Chorionic villus sampling (CVS)
Ultrasound monitor
1 A sample of chorionic villus tissue can be taken as early as the 8th to 10th week of pregnancy.
Fetus
Placenta
Chorionic villi
Uterus
Cervix
Suction tube inserted through cervix
Several hours
Fetal cells
2 Karyotyping and biochemical and genetic tests can be performed on the fetal cells immediately, providing results within a day or so.
Several hours

to simple rules of chance—to the elegant quantitative experiments of Gregor Mendel. The importance of his discoveries was overlooked by most biologists until early in the 20th century, decades after he reported his findings. In the next chapter, you will learn how Mendel's laws have their physical basis in the behavior of chromosomes during sexual life cycles and how the synthesis of Mendelism and a chromosome theory of inheritance catalyzed progress in genetics.

CONCEPT CHECK 14.4

1. Beth and Tom each have a sibling with cystic fibrosis, but neither Beth nor Tom nor any of their parents have the disease. Calculate the probability that if this couple has a child, the child will have cystic fibrosis. What would be the probability if a test revealed that Tom is a carrier but Beth is not? Explain your answers.
2. **MAKE CONNECTIONS** Review Figures 5.16, 5.20, and 5.21 (pp. 79 and 82–84). Explain how the change of a single amino acid in hemoglobin leads to the aggregation of hemoglobin into long rods.
3. Joan was born with six toes on each foot, a dominant trait called polydactyly. Two of her five siblings and her mother, but not her father, also have extra digits. What is Joan's genotype for the number-of-digits character? Explain your answer. Use *D* and *d* to symbolize the alleles for this character.
4. **MAKE CONNECTIONS** In Table 14.1 (p. 265), note the phenotypic ratio of the dominant to recessive trait in the F_2 generation for the monohybrid cross involving flower color. Then determine the phenotypic ratio for the offspring of the second-generation couple in Figure 14.15b. What accounts for the difference in the two ratios?

For suggested answers, see Appendix A.

14 CHAPTER REVIEW

SUMMARY OF KEY CONCEPTS

CONCEPT 14.1

Mendel used the scientific approach to identify two laws of inheritance (pp. 262–269)

- In the 1860s, Gregor Mendel formulated a theory of inheritance based on experiments with garden peas, proposing that parents pass on to their offspring discrete genes that retain their identity through generations. This theory includes two "laws."
- The **law of segregation** states that genes have alternative forms, or **alleles**. In a diploid organism, the two alleles of a gene segregate (separate) during meiosis and gamete formation; each sperm or egg carries only one allele of each pair. This law explains the 3:1 ratio of F_2 phenotypes observed when **monohybrids** self-pollinate. Each organism inherits one allele for each gene from each parent. In **heterozygotes**, the two alleles are different, and expression of one (the **dominant allele**) masks the phenotypic effect of the other (the **recessive allele**). **Homozygotes** have identical alleles of a given gene and are **true-breeding**.
- The **law of independent assortment** states that the pair of alleles for a given gene segregates into gametes independently of the pair of alleles for any other gene. In a cross between **dihybrids** (individuals heterozygous for two genes), the offspring have four phenotypes in a 9:3:3:1 ratio.

? *When Mendel crossed true-breeding purple- and white-flowered pea plants, the white-flowered trait disappeared from the F_1 generation but reappeared in the F_2 generation. Use genetic terms to explain why that happened.*

CONCEPT 14.2

The laws of probability govern Mendelian inheritance (pp. 269–271)

- The **multiplication rule** states that the probability of two or more events occurring together is equal to the product of the individual probabilities of the independent single events. The **addition rule** states that the probability of an event that can occur in two or more independent, mutually exclusive ways is the sum of the individual probabilities.
- The rules of probability can be used to solve complex genetics problems. A dihybrid or other multicharacter cross is equivalent to two or more independent monohybrid crosses occurring simultaneously. In calculating the chances of the various offspring genotypes from such crosses, each character is first considered separately and then the individual probabilities are multiplied.

DRAW IT *Redraw the Punnett square on the right side of Figure 14.8 as two smaller monohybrid Punnett squares, one for each gene. Below each square, list the fractions of each phenotype produced. Use the rule of multiplication to compute the overall fraction of each possible dihybrid phenotype. What is the phenotypic ratio?*

CONCEPT 14.3

Inheritance patterns are often more complex than predicted by simple Mendelian genetics (pp. 271–275)

- Extensions of Mendelian genetics for a single gene:

Relationship among alleles of a single gene	Description	Example
Complete dominance of one allele	Heterozygous phenotype same as that of homozygous dominant	*PP* *Pp*
Incomplete dominance of either allele	Heterozygous phenotype intermediate between the two homozygous phenotypes	C^RC^R C^RC^W C^WC^W
Codominance	Both phenotypes expressed in heterozygotes	I^AI^B
Multiple alleles	In the whole population, some genes have more than two alleles	ABO blood group alleles I^A, I^B, i
Pleiotropy	One gene is able to affect multiple phenotypic characters	Sickle-cell disease

- Extensions of Mendelian genetics for two or more genes:

Relationship among two or more genes	Description	Example
Epistasis	The phenotypic expression of one gene affects that of another	*BbEe* × *BbEe*; *BE*, *bE*, *Be*, *be*; 9 : 3 : 4
Polygenic inheritance	A single phenotypic character is affected by two or more genes	*AaBbCc* × *AaBbCc*

- The expression of a genotype can be affected by environmental influences, the "nurture" in nature versus nurture. The phenotypic range of a particular genotype is called its **norm of reaction**.

Polygenic characters that are also influenced by the environment are called **multifactorial** characters.

- An organism's overall phenotype, including its physical appearance, internal anatomy, physiology, and behavior, reflects its overall genotype and unique environmental history. Even in more complex inheritance patterns, Mendel's fundamental laws of segregation and independent assortment still apply.

? *Which of the following are demonstrated by the inheritance patterns of the ABO blood group alleles: complete dominance, incomplete dominance, codominance, multiple alleles, pleiotropy, epistasis, and/or polygenic inheritance? Explain how, for each of your answers.*

CONCEPT 14.4

Many human traits follow Mendelian patterns of inheritance (pp. 275–281)

- Analysis of family **pedigrees** can be used to deduce the possible genotypes of individuals and make predictions about future offspring. Predictions are statistical probabilities rather than certainties.
- Many genetic disorders are inherited as simple recessive traits. Most affected (homozygous recessive) individuals are children of phenotypically normal, heterozygous **carriers**.
- Lethal dominant alleles are eliminated from the population if affected people die before reproducing. Nonlethal dominant alleles and lethal ones that strike relatively late in life can be inherited in a Mendelian way.
- Many human diseases are multifactorial—that is, they have both genetic and environmental components and do not follow simple Mendelian inheritance patterns.
- Using family histories, genetic counselors help couples determine the probability that their children will have genetic disorders. Genetic testing of prospective parents to reveal whether they are carriers of recessive alleles associated with specific disorders has become widely available. **Amniocentesis** and **chorionic villus sampling** can indicate whether a suspected genetic disorder is present in a fetus. Other genetic tests can be performed after birth.

? *Both members of a couple know that they are carriers of the cystic fibrosis allele. None of their three children has cystic fibrosis, but any one of them might be a carrier. They would like to have a fourth child but are worried that it would very likely have the disease, since the first three do not. What would you tell the couple? Would it remove some more uncertainty in their prediction if they could find out from genetic tests whether the three children are carriers?*

TIPS FOR GENETICS PROBLEMS

1. Write down symbols for the alleles. (These may be given in the problem.) When represented by single letters, the dominant allele is uppercase and the recessive is lowercase.
2. Write down the possible genotypes, as determined by the phenotype.
   a. If the phenotype is that of the dominant trait (for example, purple flowers), then the genotype is either homozygous dominant or heterozygous (*PP* or *Pp*, in this example).
   b. If the phenotype is that of the recessive trait, the genotype must be homozygous recessive (for example, *pp*).
   c. If the problem says "true-breeding," the genotype is homozygous.
3. Determine what the problem is asking for. If asked to do a cross, write it out in the form [Genotype] × [Genotype], using the alleles you've decided on.
4. To figure out the outcome of a cross, set up a Punnett square.
   a. Put the gametes of one parent at the top and those of the other on the left. To determine the allele(s) in each gamete for a given genotype, set up a systematic way to list all the possibilities. (Remember, each gamete has one allele of each gene.) Note that there are 2^n possible types of gametes, where n is the number of gene loci that are heterozygous. For example, an individual with genotype *AaBbCc* would produce $2^3 = 8$ types of gametes. Write the genotypes of the gametes in circles above the columns and to the left of the rows.
   b. Fill in the Punnett square as if each possible sperm were fertilizing each possible egg, making all of the possible offspring. In a cross of *AaBbCc* × *AaBbCc*, for example, the Punnett square would have 8 columns and 8 rows, so there are 64 different offspring; you would know the genotype of each and thus the phenotype. Count genotypes and phenotypes to obtain the genotypic and phenotypic ratios. Because the Punnett square is so large, this method is not the most efficient. Instead, see tip 5.
5. You can use the rules of probability if the Punnett square would be too big. (For example, see the question at the end of the summary for Concept 14.2 and question 7 on the next page.) You can consider each gene separately (see pp. 270–271).
6. If, instead, the problem gives you the phenotypic ratios of offspring, but not the genotypes of the parents in a given cross, the phenotypes can help you deduce the parents' unknown genotypes.
   a. For example, if ½ of the offspring have the recessive phenotype and ½ the dominant, you know that the cross was between a heterozygote and a homozygous recessive.
   b. If the ratio is 3:1, the cross was between two heterozygotes.
   c. If two genes are involved and you see a 9:3:3:1 ratio in the offspring, you know that each parent is heterozygous for both genes. Caution: Don't assume that the reported numbers will exactly equal the predicted ratios. For example, if there are 13 offspring with the dominant trait and 11 with the recessive, assume that the ratio is one dominant to one recessive.
7. For pedigree problems, use the tips in Figure 14.15 and below to determine what kind of trait is involved.
   a. If parents without the trait have offspring with the trait, the trait must be recessive and the parents both carriers.
   b. If the trait is seen in every generation, it is most likely dominant (see the next possibility, though).
   c. If both parents have the trait, then in order for it to be recessive, all offspring must show the trait.
   d. To determine the likely genotype of a certain individual in a pedigree, first label the genotypes of all the family members you can. Even if some of the genotypes are incomplete, label what you do know. For example, if an individual has the dominant phenotype, the genotype must be *AA* or *Aa*; you can write this as *A_*. Try different possibilities to see which fits the results. Use the rules of probability to calculate the probability of each possible genotype being the correct one.

TEST YOUR UNDERSTANDING

LEVEL 1: KNOWLEDGE/COMPREHENSION

1. Match each term on the left with a statement on the right.

Term	Statement
__ Gene	a. Has no effect on phenotype in a heterozygote
__ Allele	b. A variant for a character
__ Character	c. Having two identical alleles for a gene
__ Trait	d. A cross between individuals heterozygous for a single character
__ Dominant allele	e. An alternative version of a gene
__ Recessive allele	f. Having two different alleles for a gene
__ Genotype	g. A heritable feature that varies among individuals
__ Phenotype	h. An organism's appearance or observable traits
__ Homozygous	i. A cross between an individual with an unknown genotype and a homozygous recessive individual
__ Heterozygous	j. Determines phenotype in a heterozygote
__ Testcross	k. The genetic makeup of an individual
__ Monohybrid cross	l. A heritable unit that determines a character; can exist in different forms

2. **DRAW IT** Two pea plants heterozygous for the characters of pod color and pod shape are crossed. Draw a Punnett square to determine the phenotypic ratios of the offspring.

3. In some plants, a true-breeding, red-flowered strain gives all pink flowers when crossed with a white-flowered strain: C^RC^R (red) $\times$ C^WC^W (white) $\rightarrow$ C^RC^W (pink). If flower position (axial or terminal) is inherited as it is in peas (see Table 14.1), what will be the ratios of genotypes and phenotypes of the F_1 generation resulting from the following cross: axial-red (true-breeding) $\times$ terminal-white? What will be the ratios in the F_2 generation?

4. A man with type A blood marries a woman with type B blood. Their child has type O blood. What are the genotypes of these three individuals? What genotypes, and in what frequencies, would you expect in future offspring from this marriage?

5. A man has six fingers on each hand and six toes on each foot. His wife and their daughter have the normal number of digits. Remember that extra digits is a dominant trait. What fraction of this couple's children would be expected to have extra digits?

6. **DRAW IT** A pea plant heterozygous for inflated pods (*Ii*) is crossed with a plant homozygous for constricted pods (*ii*). Draw a Punnett square for this cross. Assume that pollen comes from the *ii* plant.

LEVEL 2: APPLICATION/ANALYSIS

7. Flower position, stem length, and seed shape are three characters that Mendel studied. Each is controlled by an independently assorting gene and has dominant and recessive expression as follows:

Character	Dominant	Recessive
Flower position	Axial (*A*)	Terminal (*a*)
Stem length	Tall (*T*)	Dwarf (*t*)
Seed shape	Round (*R*)	Wrinkled (*r*)

If a plant that is heterozygous for all three characters is allowed to self-fertilize, what proportion of the offspring would you expect to be as follows? (*Note*: Use the rules of probability instead of a huge Punnett square.)

(a) homozygous for the three dominant traits
(b) homozygous for the three recessive traits
(c) heterozygous for all three characters
(d) homozygous for axial and tall, heterozygous for seed shape

8. A black guinea pig crossed with an albino guinea pig produces 12 black offspring. When the albino is crossed with a second black one, 7 blacks and 5 albinos are obtained. What is the best explanation for this genetic outcome? Write genotypes for the parents, gametes, and offspring.

9. In sesame plants, the one-pod condition (*P*) is dominant to the three-pod condition (*p*), and normal leaf (*L*) is dominant to wrinkled leaf (*l*). Pod type and leaf type are inherited independently. Determine the genotypes for the two parents for all possible matings producing the following offspring:

(a) 318 one-pod, normal leaf and 98 one-pod, wrinkled leaf
(b) 323 three-pod, normal leaf and 106 three-pod, wrinkled leaf
(c) 401 one-pod, normal leaf
(d) 150 one-pod, normal leaf, 147 one-pod, wrinkled leaf, 51 three-pod, normal leaf, and 48 three-pod, wrinkled leaf
(e) 223 one-pod, normal leaf, 72 one-pod, wrinkled leaf, 76 three-pod, normal leaf, and 27 three-pod, wrinkled leaf

10. Phenylketonuria (PKU) is an inherited disease caused by a recessive allele. If a woman and her husband, who are both carriers, have three children, what is the probability of each of the following?

(a) All three children are of normal phenotype.
(b) One or more of the three children have the disease.
(c) All three children have the disease.
(d) At least one child is phenotypically normal.
(*Note*: It will help to remember that the probabilities of all possible outcomes always add up to 1.)

11. The genotype of F_1 individuals in a tetrahybrid cross is *AaBbCcDd*. Assuming independent assortment of these four genes, what are the probabilities that F_2 offspring will have the following genotypes?

(a) *aabbccdd*
(b) *AaBbCcDd*
(c) *AABBCCDD*
(d) *AaBBccDd*
(e) *AaBBCCdd*

12. What is the probability that each of the following pairs of parents will produce the indicated offspring? (Assume independent assortment of all gene pairs.)

(a) *AABBCC* $\times$ *aabbcc* $\rightarrow$ *AaBbCc*
(b) *AABbCc* $\times$ *AaBbCc* $\rightarrow$ *AAbbCC*
(c) *AaBbCc* $\times$ *AaBbCc* $\rightarrow$ *AaBbCc*
(d) *aaBbCC* $\times$ *AABbcc* $\rightarrow$ *AaBbCc*

13. Karen and Steve each have a sibling with sickle-cell disease. Neither Karen nor Steve nor any of their parents have the disease, and none of them have been tested to see if they have the sickle-cell trait. Based on this incomplete information, calculate the probability that if this couple has a child, the child will have sickle-cell disease.

14. In 1981, a stray black cat with unusual rounded, curled-back ears was adopted by a family in California. Hundreds of descendants of the cat have since been born, and cat fanciers hope to develop the curl cat into a show breed. Suppose you

owned the first curl cat and wanted to develop a true-breeding variety. How would you determine whether the curl allele is dominant or recessive? How would you obtain true-breeding curl cats? How could you be sure they are true-breeding?

15. Imagine that a newly discovered, recessively inherited disease is expressed only in individuals with type O blood, although the disease and blood group are independently inherited. A normal man with type A blood and a normal woman with type B blood have already had one child with the disease. The woman is now pregnant for a second time. What is the probability that the second child will also have the disease? Assume that both parents are heterozygous for the gene that causes the disease.

16. In tigers, a recessive allele causes an absence of fur pigmentation (a white tiger) and a cross-eyed condition. If two phenotypically normal tigers that are heterozygous at this locus are mated, what percentage of their offspring will be cross-eyed? What percentage of cross-eyed tigers will be white?

17. In maize (corn) plants, a dominant allele *I* inhibits kernel color, while the recessive allele *i* permits color when homozygous. At a different locus, the dominant allele *P* causes purple kernel color, while the homozygous recessive genotype *pp* causes red kernels. If plants heterozygous at both loci are crossed, what will be the phenotypic ratio of the offspring?

18. The pedigree below traces the inheritance of alkaptonuria, a biochemical disorder. Affected individuals, indicated here by the colored circles and squares, are unable to metabolize a substance called alkapton, which colors the urine and stains body tissues. Does alkaptonuria appear to be caused by a dominant allele or by a recessive allele? Fill in the genotypes of the individuals whose genotypes can be deduced. What genotypes are possible for each of the other individuals?

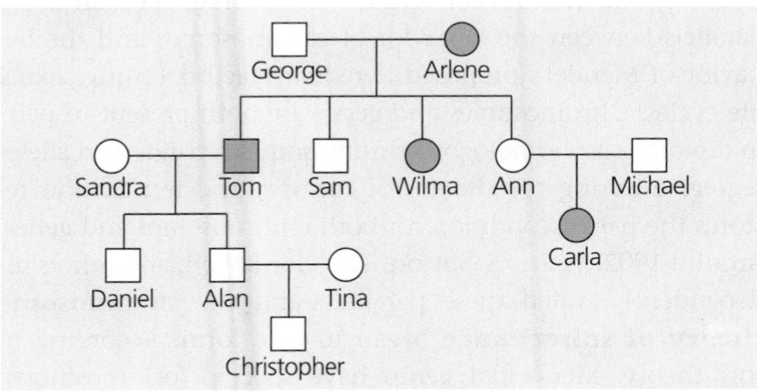

19. Imagine that you are a genetic counselor, and a couple planning to start a family comes to you for information. Charles was married once before, and he and his first wife had a child with cystic fibrosis. The brother of his current wife, Elaine, died of cystic fibrosis. What is the probability that Charles and Elaine will have a baby with cystic fibrosis? (Neither Charles, Elaine, nor their parents have cystic fibrosis.)

20. In mice, black fur (*B*) is dominant to white (*b*). At a different locus, a dominant allele (*A*) produces a band of yellow just below the tip of each hair in mice with black fur. This gives a frosted appearance known as agouti. Expression of the recessive allele (*a*) results in a solid coat color. If mice that are heterozygous at both loci are crossed, what is the expected phenotypic ratio of their offspring?

LEVEL 3: SYNTHESIS/EVALUATION

21. **EVOLUTION CONNECTION**
Over the past half century, there has been a trend in the United States and other developed countries for people to marry and start families later in life than did their parents and grandparents. What effects might this trend have on the incidence (frequency) of late-acting dominant lethal alleles in the population?

22. **SCIENTIFIC INQUIRY**
You are handed a mystery pea plant with tall stems and axial flowers and asked to determine its genotype as quickly as possible. You know that the allele for tall stems (*T*) is dominant to that for dwarf stems (*t*) and that the allele for axial flowers (*A*) is dominant to that for terminal flowers (*a*).
    (a) What are *all* the possible genotypes for your mystery plant?
    (b) Describe the *one* cross you would do, out in your garden, to determine the exact genotype of your mystery plant.
    (c) While waiting for the results of your cross, you predict the results for each possible genotype listed in part a. How do you do this? Why is this not called "performing a cross"?
    (d) Explain how the results of your cross and your predictions will help you learn the genotype of your mystery plant.

23. **SCIENCE, TECHNOLOGY, AND SOCIETY**
Imagine that one of your parents has Huntington's disease. What is the probability that you, too, will someday manifest the disease? There is no cure for Huntington's. Would you want to be tested for the Huntington's allele? Why or why not?

24. **WRITE ABOUT A THEME**
The Genetic Basis of Life The continuity of life is based on heritable information in the form of DNA. In a short essay (100–150 words), explain how the passage of genes from parents to offspring, in the form of particular alleles, ensures perpetuation of parental traits in offspring and, at the same time, genetic variation among offspring. Use genetic terms in your explanation.

For selected answers, see Appendix A.

MasteringBIOLOGY www.masteringbiology.com

1. MasteringBiology® Assignments

Tutorials Determining Genotype: Pea Pod Color • Inheritance of Fur Color in Mice • Pedigree Analysis: Galactosemia
Activities Monohybrid Cross • Dihybrid Cross • Mendel's Experiments • The Principle of Independent Assortment • Gregor's Garden • Incomplete Dominance
Questions Student Misconceptions • Reading Quiz • Multiple Choice • End-of-Chapter

2. eText

Read your book online, search, take notes, highlight text, and more.

3. The Study Area

Practice Tests • Cumulative Test • *BioFlix* 3-D Animations • MP3 Tutor Sessions • Videos • Activities • Investigations • Lab Media • Audio Glossary • Word Study Tools • Art

15

The Chromosomal Basis of Inheritance

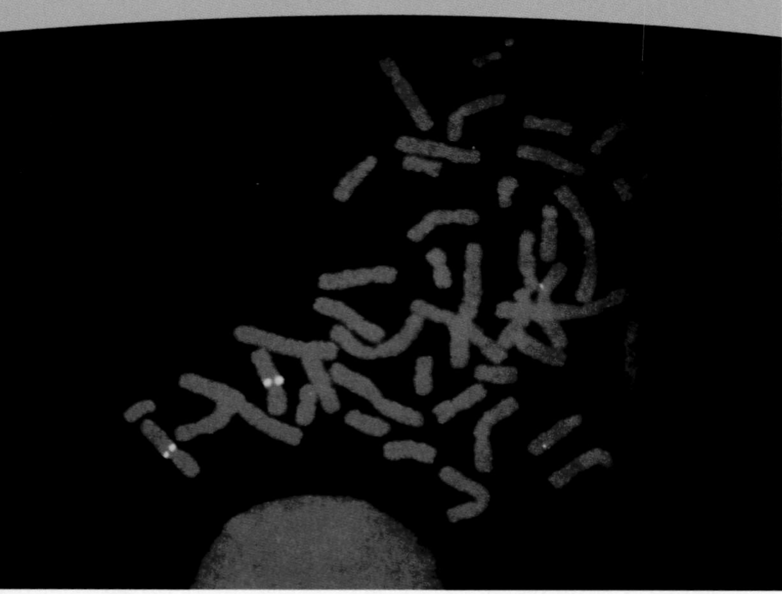

▲ **Figure 15.1 Where are Mendel's hereditary factors located in the cell?**

KEY CONCEPTS

- 15.1 Mendelian inheritance has its physical basis in the behavior of chromosomes
- 15.2 Sex-linked genes exhibit unique patterns of inheritance
- 15.3 Linked genes tend to be inherited together because they are located near each other on the same chromosome
- 15.4 Alterations of chromosome number or structure cause some genetic disorders
- 15.5 Some inheritance patterns are exceptions to standard Mendelian inheritance

OVERVIEW

Locating Genes Along Chromosomes

Gregor Mendel's "hereditary factors" were purely an abstract concept when he proposed their existence in 1860. At that time, no cellular structures were known that could house these imaginary units. Even after chromosomes were first observed, many biologists remained skeptical about Mendel's laws of segregation and independent assortment until there was sufficient evidence that these principles of heredity had a physical basis in chromosomal behavior.

Today, we know that genes—Mendel's "factors"—are located along chromosomes. We can see the location of a particular gene by tagging chromosomes with a fluorescent dye that highlights that gene. For example, the four yellow dots in **Figure 15.1** mark the locus of a specific gene on the sister chromatids of a homologous pair of replicated human chromosomes. This chapter extends what you learned in the past two chapters: We describe the chromosomal basis for the transmission of genes from parents to offspring, along with some important exceptions to the standard mode of inheritance.

CONCEPT 15.1

Mendelian inheritance has its physical basis in the behavior of chromosomes

Using improved techniques of microscopy, cytologists worked out the process of mitosis in 1875 and meiosis in the 1890s. Cytology and genetics converged when biologists began to see parallels between the behavior of chromosomes and the behavior of Mendel's proposed hereditary factors during sexual life cycles: Chromosomes and genes are both present in pairs in diploid cells; homologous chromosomes separate and alleles segregate during the process of meiosis; and fertilization restores the paired condition for both chromosomes and genes. Around 1902, Walter S. Sutton, Theodor Boveri, and others independently noted these parallels, and the **chromosome theory of inheritance** began to take form. According to this theory, Mendelian genes have specific loci (positions) along chromosomes, and it is the chromosomes that undergo segregation and independent assortment.

Figure 15.2 shows that the behavior of homologous chromosomes during meiosis can account for the segregation of the alleles at each genetic locus to different gametes. The figure also shows that the behavior of nonhomologous chromosomes can account for the independent assortment of the alleles for two or more genes located on different chromosomes. By carefully studying this figure, which traces the same dihybrid pea cross you learned about in Figure 14.8, you can see how the behavior of chromosomes during meiosis in the F_1 generation and subsequent random fertilization give rise to the F_2 phenotypic ratio observed by Mendel.

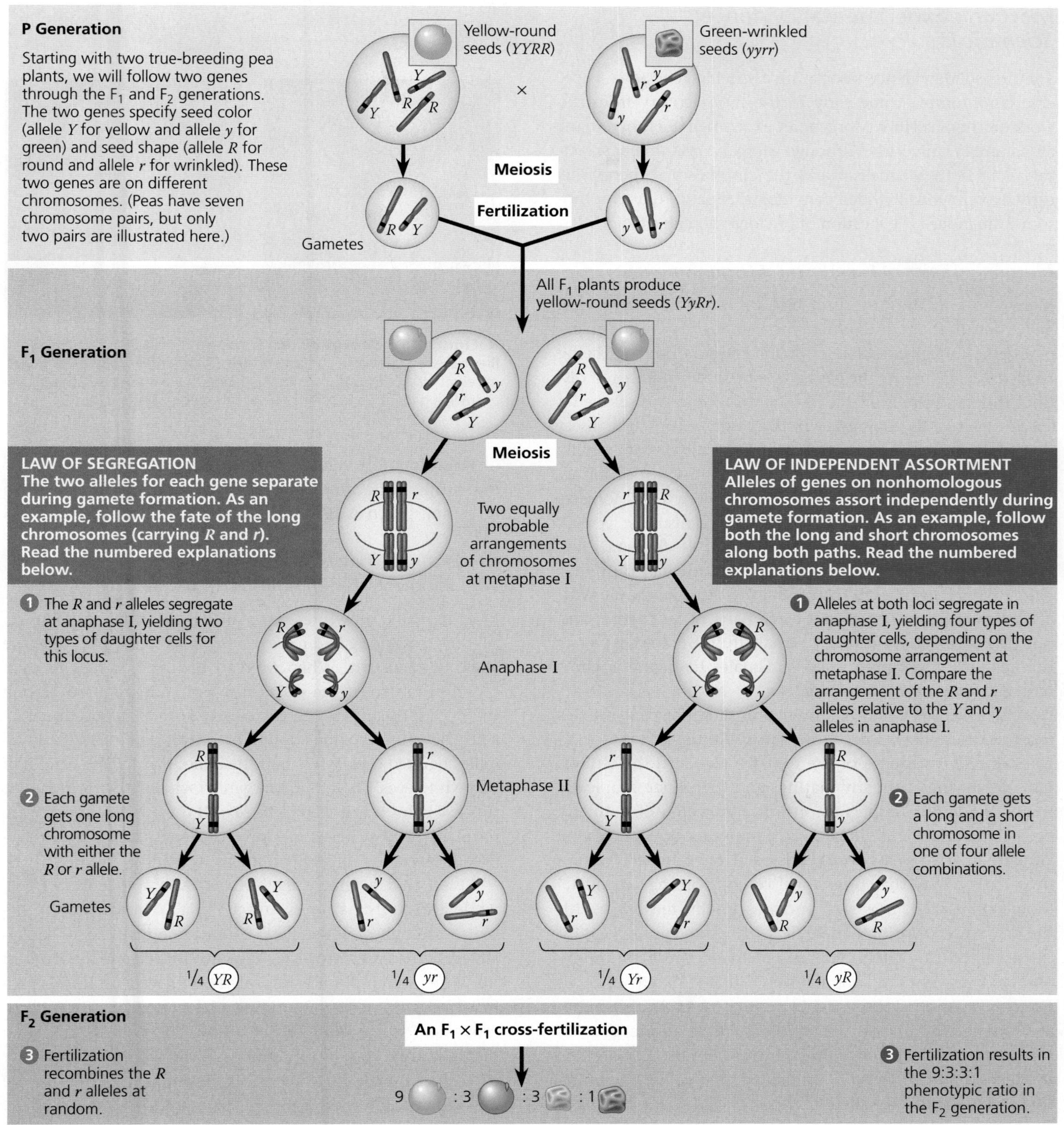

▲ **Figure 15.2 The chromosomal basis of Mendel's laws.** Here we correlate the results of one of Mendel's dihybrid crosses (see Figure 14.8) with the behavior of chromosomes during meiosis (see Figure 13.8). The arrangement of chromosomes at metaphase I of meiosis and their movement during anaphase I account for the segregation and independent assortment of the alleles for seed color and shape. Each cell that undergoes meiosis in an F_1 plant produces two kinds of gametes. If we count the results for all cells, however, each F_1 plant produces equal numbers of all four kinds of gametes because the alternative chromosome arrangements at metaphase I are equally likely.

? *If you crossed an F_1 plant with a plant that was homozygous recessive for both genes (*yyrr*), how would the phenotypic ratio of the offspring compare with the 9:3:3:1 ratio seen here?*

Morgan's Experimental Evidence: *Scientific Inquiry*

The first solid evidence associating a specific gene with a specific chromosome came early in the 20th century from the work of Thomas Hunt Morgan, an experimental embryologist at Columbia University. Although Morgan was initially skeptical about both Mendelism and the chromosome theory, his early experiments provided convincing evidence that chromosomes are indeed the location of Mendel's heritable factors.

Morgan's Choice of Experimental Organism

Many times in the history of biology, important discoveries have come to those insightful or lucky enough to choose an experimental organism suitable for the research problem being tackled. Mendel chose the garden pea because a number of distinct varieties were available. For his work, Morgan selected a species of fruit fly, *Drosophila melanogaster*, a common insect that feeds on the fungi growing on fruit. Fruit flies are prolific breeders; a single mating will produce hundreds of offspring, and a new generation can be bred every two weeks. Morgan's laboratory began using this convenient organism for genetic studies in 1907 and soon became known as "the fly room."

Another advantage of the fruit fly is that it has only four pairs of chromosomes, which are easily distinguishable with a light microscope. There are three pairs of autosomes and one pair of sex chromosomes. Female fruit flies have a pair of homologous X chromosomes, and males have one X chromosome and one Y chromosome.

While Mendel could readily obtain different pea varieties from seed suppliers, Morgan was probably the first person to want different varieties of the fruit fly. He faced the tedious task of carrying out many matings and then microscopically inspecting large numbers of offspring in search of naturally occurring variant individuals. After many months of this, he lamented, "Two years' work wasted. I have been breeding those flies for all that time and I've got nothing out of it." Morgan persisted, however, and was finally rewarded with the discovery of a single male fly with white eyes instead of the usual red. The phenotype for a character most commonly observed in natural populations, such as red eyes in *Drosophila*, is called the **wild type** (**Figure 15.3**). Traits that are alternatives to the wild type, such as white eyes in *Drosophila*, are called *mutant phenotypes* because they are due to alleles assumed to have originated as changes, or mutations, in the wild-type allele.

Morgan and his students invented a notation for symbolizing alleles in *Drosophila* that is still widely used for fruit flies. For a given character in flies, the gene takes its symbol from the first mutant (non–wild type) discovered. Thus, the allele for white eyes in *Drosophila* is symbolized by *w*. A superscript + identifies the allele for the wild-type trait—w^+ for the allele for red eyes, for example. Over the years, a variety of gene notation systems have been developed for different organisms. For example, human genes are usually written in all capitals, such as *HD* for the allele for Huntington's disease.

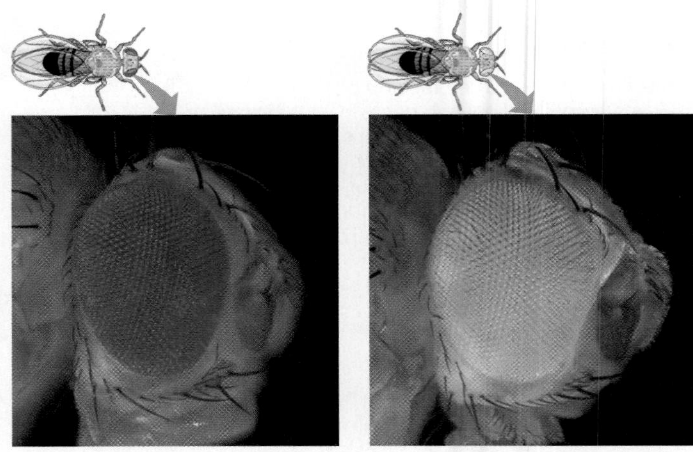

▲ **Figure 15.3 Morgan's first mutant.** Wild-type *Drosophila* flies have red eyes (left). Among his flies, Morgan discovered a mutant male with white eyes (right). This variation made it possible for Morgan to trace a gene for eye color to a specific chromosome (LMs).

Correlating Behavior of a Gene's Alleles with Behavior of a Chromosome Pair

Morgan mated his white-eyed male fly with a red-eyed female. All the F_1 offspring had red eyes, suggesting that the wild-type allele is dominant. When Morgan bred the F_1 flies to each other, he observed the classical 3:1 phenotypic ratio among the F_2 offspring. However, there was a surprising additional result: The white-eye trait showed up only in males. All the F_2 females had red eyes, while half the males had red eyes and half had white eyes. Therefore, Morgan concluded that somehow a fly's eye color was linked to its sex. (If the eye-color gene were unrelated to sex, he would have expected half of the white-eyed flies to be male and half female.)

Recall that a female fly has two X chromosomes (XX), while a male fly has an X and a Y (XY). The correlation between the trait of white eye color and the male sex of the affected F_2 flies suggested to Morgan that the gene involved in his white-eyed mutant was located exclusively on the X chromosome, with no corresponding allele present on the Y chromosome. His reasoning can be followed in **Figure 15.4**. For a male, a single copy of the mutant allele would confer white eyes; since a male has only one X chromosome, there can be no wild-type allele (w^+) present to mask the recessive allele. On the other hand, a female could have white eyes only if both her X chromosomes carried the recessive mutant allele (*w*). This was impossible for the F_2 females in Morgan's experiment because all the F_1 fathers had red eyes.

Morgan's finding of the correlation between a particular trait and an individual's sex provided support for the chromosome theory of inheritance: namely, that a specific gene is

▼ Figure 15.4 **INQUIRY**

In a cross between a wild-type female fruit fly and a mutant white-eyed male, what color eyes will the F_1 and F_2 offspring have?

EXPERIMENT Thomas Hunt Morgan wanted to analyze the behavior of two alleles of a fruit fly eye-color gene. In crosses similar to those done by Mendel with pea plants, Morgan and his colleagues mated a wild-type (red-eyed) female with a mutant white-eyed male.

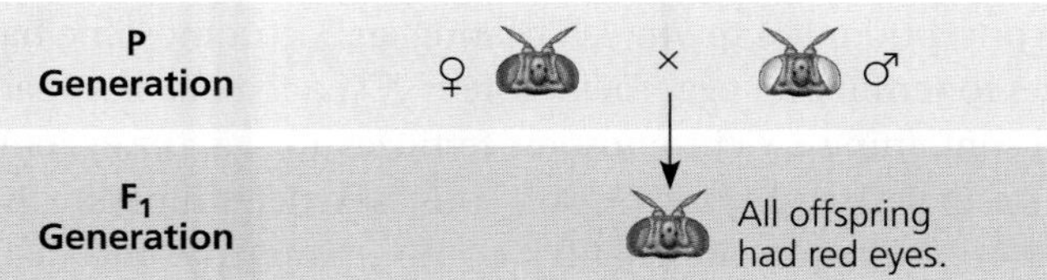

Morgan then bred an F_1 red-eyed female to an F_1 red-eyed male to produce the F_2 generation.

RESULTS The F_2 generation showed a typical Mendelian ratio of 3 red-eyed flies : 1 white-eyed fly. However, no females displayed the white-eye trait; all white-eyed flies were males.

CONCLUSION All F_1 offspring had red eyes, so the mutant white-eye trait (w) must be recessive to the wild-type red-eye trait (w^+). Since the recessive trait—white eyes—was expressed only in males in the F_2 generation, Morgan deduced that this eye-color gene is located on the X chromosome and that there is no corresponding locus on the Y chromosome.

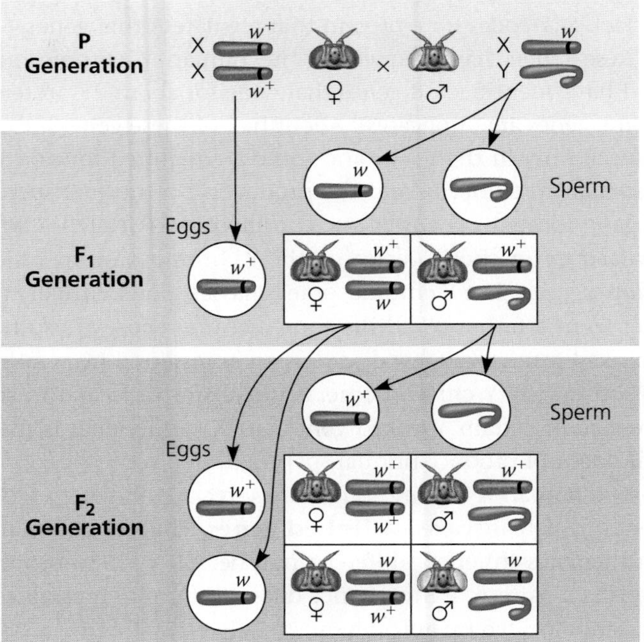

SOURCE: T. H. Morgan, Sex-limited inheritance in *Drosophila, Science* 32:120–122 (1910).

(MB) See the related Experimental Inquiry Tutorial in MasteringBiology.

WHAT IF? Suppose this eye-color gene were located on an autosome. Predict the phenotypes (including gender) of the F_2 flies in this hypothetical cross. (*Hint*: Draw a Punnett square.)

carried on a specific chromosome (in this case, an eye-color gene on the X chromosome). In addition, Morgan's work indicated that genes located on a sex chromosome exhibit unique inheritance patterns, which we will discuss in the next section. Recognizing the importance of Morgan's early work, many bright students were attracted to his fly room.

CONCEPT CHECK 15.1

1. Which one of Mendel's laws relates to the inheritance of alleles for a single character? Which law relates to the inheritance of alleles for two characters in a dihybrid cross?
2. **MAKE CONNECTIONS** Review the description of meiosis in Figure 13.8 (pp. 254–255) and Mendel's two laws in Concept 14.1 (pp. 264–269). What is the physical basis for each of Mendel's laws?
3. **WHAT IF?** Propose a possible reason that the first naturally occurring mutant fruit fly Morgan saw involved a gene on a sex chromosome.

For suggested answers, see Appendix A.

CONCEPT **15.2**

Sex-linked genes exhibit unique patterns of inheritance

As you just learned, Morgan's discovery of a trait (white eyes) that correlated with the sex of flies was a key episode in the development of the chromosome theory of inheritance. Because the identity of the sex chromosomes in an individual could be inferred by observing the sex of the fly, the behavior of the two members of the pair of sex chromosomes could be correlated with the behavior of the two alleles of the eye-color gene. In this section, we consider the role of sex chromosomes in inheritance in more detail. We begin by reviewing the chromosomal basis of sex determination in humans and some other animals.

The Chromosomal Basis of Sex

Whether we are male or female is one of our more obvious phenotypic characters. Although the anatomical and physiological differences between women and men are numerous, the chromosomal basis for determining sex is rather simple. In humans and other mammals, there are two varieties of sex chromosomes, designated X and Y. The Y chromosome is much smaller than the X chromosome (**Figure 15.5**). A person who inherits two

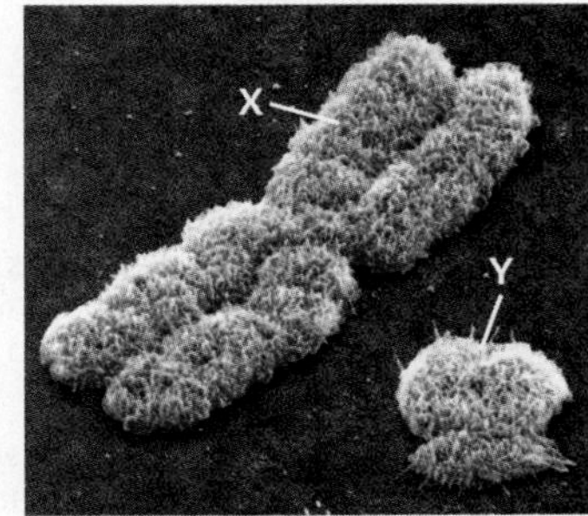

▲ **Figure 15.5 Human sex chromosomes.**

X chromosomes, one from each parent, usually develops as a female. A male develops from a zygote containing one X chromosome and one Y chromosome (**Figure 15.6a**). Short segments at either end of the Y chromosome are the only regions that are homologous with corresponding regions of the X.

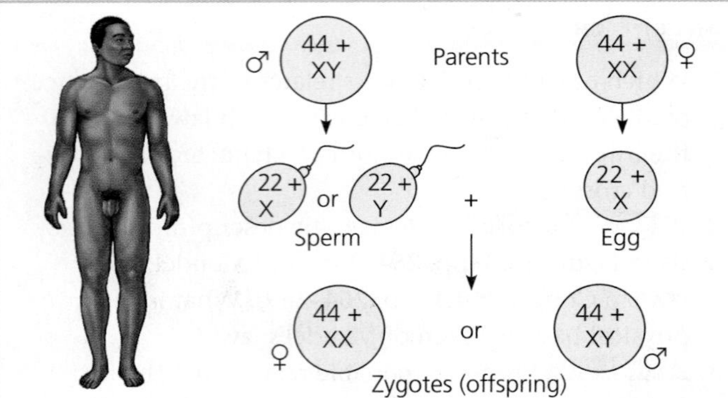

(a) The X-Y system. In mammals, the sex of an offspring depends on whether the sperm cell contains an X chromosome or a Y.

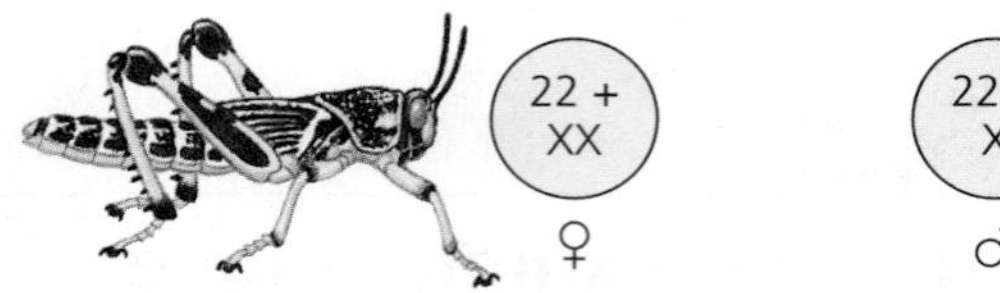

(b) The X-0 system. In grasshoppers, cockroaches, and some other insects, there is only one type of sex chromosome, the X. Females are XX; males have only one sex chromosome (X0). Sex of the offspring is determined by whether the sperm cell contains an X chromosome or no sex chromosome.

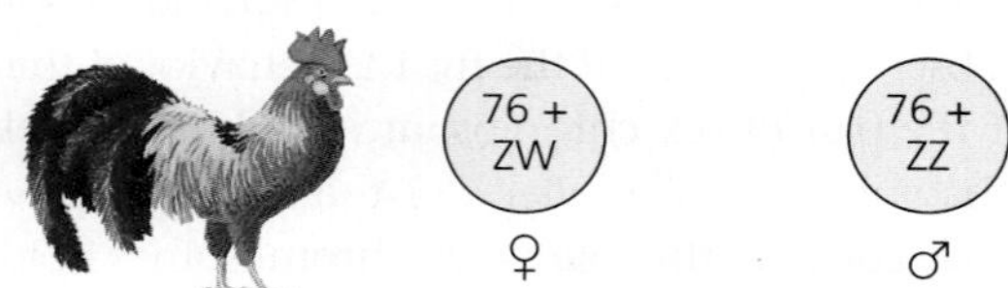

(c) The Z-W system. In birds, some fishes, and some insects, the sex chromosomes present in the egg (not the sperm) determine the sex of offspring. The sex chromosomes are designated Z and W. Females are ZW and males are ZZ.

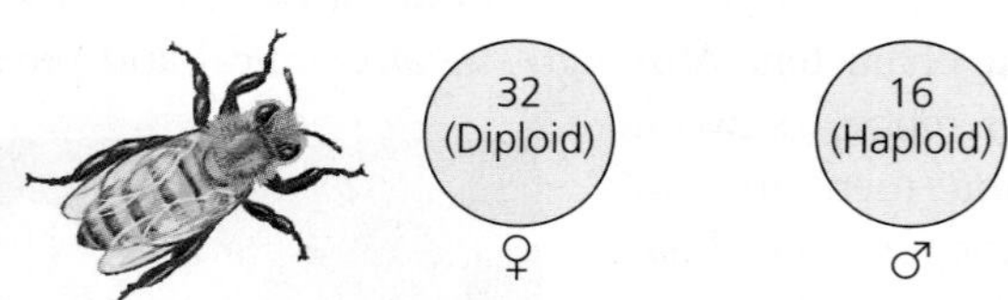

(d) The haplo-diploid system. There are no sex chromosomes in most species of bees and ants. Females develop from fertilized eggs and are thus diploid. Males develop from unfertilized eggs and are haploid; they have no fathers.

▲ **Figure 15.6 Some chromosomal systems of sex determination.** Numerals indicate the number of autosomes in the species pictured. In *Drosophila*, males are XY, but sex depends on the ratio between the number of X chromosomes and the number of autosome sets, not simply on the presence of a Y chromosome.

These homologous regions allow the X and Y chromosomes in males to pair and behave like homologous chromosomes during meiosis in the testes.

In mammalian testes and ovaries, the two sex chromosomes segregate during meiosis, and each gamete receives one. Each egg contains one X chromosome. In contrast, sperm fall into two categories: Half the sperm cells a male produces contain an X chromosome, and half contain a Y chromosome. We can trace the sex of each offspring to the events of conception: If a sperm cell bearing an X chromosome happens to fertilize an egg, the zygote is XX, a female; if a sperm cell containing a Y chromosome fertilizes an egg, the zygote is XY, a male (see Figure 15.6a). Thus, sex determination is a matter of chance—a fifty-fifty chance. Note that the mammalian X-Y system isn't the only chromosomal system for determining sex. **Figure 15.6b–d** illustrates three other systems.

In humans, the anatomical signs of sex begin to emerge when the embryo is about 2 months old. Before then, the rudiments of the gonads are generic—they can develop into either testes or ovaries, depending on whether or not a Y chromosome is present. In 1990, a British research team identified a gene on the Y chromosome required for the development of testes. They named the gene *SRY*, for sex-determining region of Y. In the absence of *SRY*, the gonads develop into ovaries. The biochemical, physiological, and anatomical features that distinguish males and females are complex, and many genes are involved in their development. In fact, *SRY* codes for a protein that regulates other genes.

Researchers have sequenced the human Y chromosome and have identified 78 genes that code for about 25 proteins (some genes are duplicates). About half of these genes are expressed only in the testis, and some are required for normal testicular functioning and the production of normal sperm. A gene located on either sex chromosome is called a **sex-linked gene**; those located on the Y chromosome are called *Y-linked genes*. The Y chromosome is passed along virtually intact from a father to all his sons. Because there are so few Y-linked genes, very few disorders are transferred from father to son on the Y chromosome. A rare example is that in the absence of certain Y-linked genes, an XY individual is male but does not produce normal sperm.

The human X chromosome contains approximately 1,100 genes, which are called **X-linked genes**. The fact that males and females inherit a different number of X chromosomes leads to a pattern of inheritance different from that produced by genes located on autosomes.

Inheritance of X-Linked Genes

While most Y-linked genes help determine sex, the X chromosomes have genes for many characters unrelated to sex. X-linked genes in humans follow the same pattern of inheritance that Morgan observed for the eye-color locus he studied in *Drosophila* (see Figure 15.4). Fathers pass X-linked

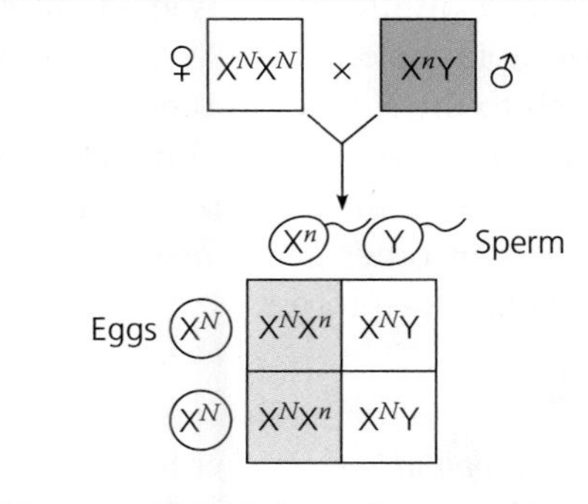

(a) A color-blind father will transmit the mutant allele to all daughters but to no sons. When the mother is a dominant homozygote, the daughters will have the normal phenotype but will be carriers of the mutation.

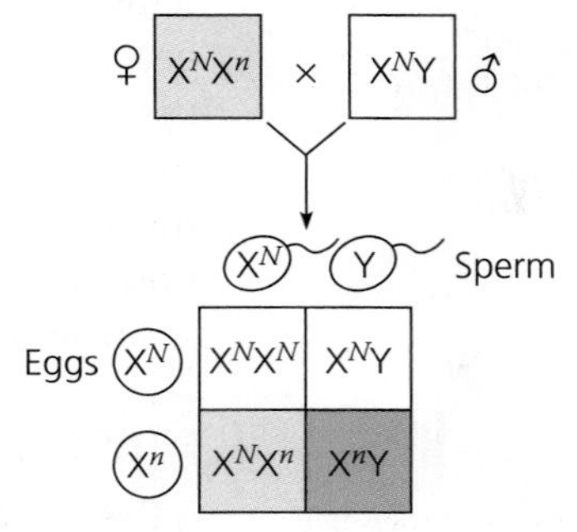

(b) If a carrier mates with a male who has normal color vision, there is a 50% chance that each daughter will be a carrier like her mother and a 50% chance that each son will have the disorder.

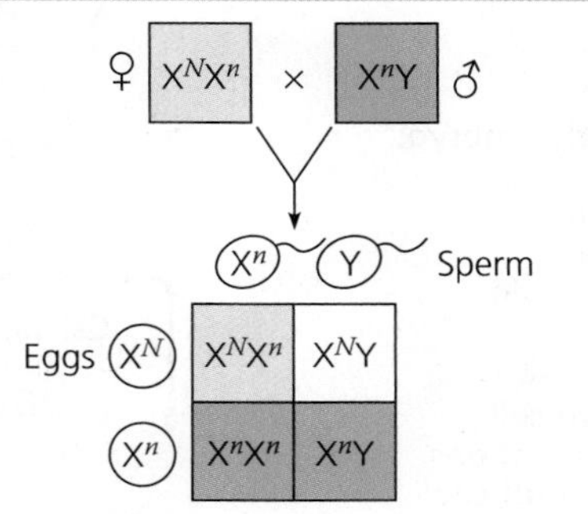

(c) If a carrier mates with a color-blind male, there is a 50% chance that each child born to them will have the disorder, regardless of sex. Daughters who have normal color vision will be carriers, whereas males who have normal color vision will be free of the recessive allele.

▲ **Figure 15.7 The transmission of X-linked recessive traits.** In this diagram, color blindness is used as an example. The superscript *N* represents the dominant allele for normal color vision carried on the X chromosome, and the superscript *n* represents the recessive allele, which has a mutation causing color blindness. White boxes indicate unaffected individuals, light orange boxes indicate carriers, and dark orange boxes indicate color-blind individuals.

? *If a color-blind woman married a man who had normal color vision, what would be the probable phenotypes of their children?*

alleles to all of their daughters but to none of their sons. In contrast, mothers can pass X-linked alleles to both sons and daughters, as shown in **Figure 15.7**.

If an X-linked trait is due to a recessive allele, a female will express the phenotype only if she is homozygous for that allele. Because males have only one locus, the terms *homozygous* and *heterozygous* lack meaning for describing their X-linked genes; the term *hemizygous* is used in such cases. Any male receiving the recessive allele from his mother will express the trait. For this reason, far more males than females have X-linked recessive disorders. However, even though the chance of a female inheriting a double dose of the mutant allele is much less than the probability of a male inheriting a single dose, there *are* females with X-linked disorders. For instance, color blindness is a mild disorder almost always inherited as an X-linked trait. A color-blind daughter may be born to a color-blind father whose mate is a carrier (see Figure 15.7c). Because the X-linked allele for color blindness is relatively rare, though, the probability that such a man and woman will mate is low.

A number of human X-linked disorders are much more serious than color blindness. An example is **Duchenne muscular dystrophy**, which affects about one out of every 3,500 males born in the United States. The disease is characterized by a progressive weakening of the muscles and loss of coordination. Affected individuals rarely live past their early 20s. Researchers have traced the disorder to the absence of a key muscle protein called dystrophin and have mapped the gene for this protein to a specific locus on the X chromosome.

Hemophilia is an X-linked recessive disorder defined by the absence of one or more of the proteins required for blood clotting. When a person with hemophilia is injured, bleeding is prolonged because a firm clot is slow to form. Small cuts in the skin are usually not a problem, but bleeding in the muscles or joints can be painful and can lead to serious damage. In the 1800s, hemophilia was widespread among the royal families of Europe. Queen Victoria of England is known to have passed the allele to several of her descendants. Subsequent intermarriage with royal family members of other nations, such as Spain and Russia, further spread this X-linked trait, and its incidence is well documented in royal pedigrees. Today, people with hemophilia are treated as needed with intravenous injections of the protein that is missing.

X Inactivation in Female Mammals

Female mammals, including humans, inherit two X chromosomes—twice the number inherited by males—so you may wonder whether females make twice as much as males of the proteins encoded by X-linked genes. In fact, most of one X chromosome in each cell in female mammals becomes inactivated during early embryonic development. As a result, the cells of females and males have the same effective dose (one copy) of most X-linked genes. The inactive X in each cell of a female condenses into a compact object called a **Barr body** (discovered by Canadian anatomist Murray Barr), which lies along the inside of the nuclear envelope. Most of the genes of the X chromosome that forms the Barr body are not expressed. In the ovaries, Barr-body chromosomes are reactivated in the cells that give rise to eggs, so every female gamete has an active X.

British geneticist Mary Lyon demonstrated that the selection of which X chromosome will form the Barr body occurs randomly and independently in each embryonic cell present at the time of X inactivation. As a consequence, females consist

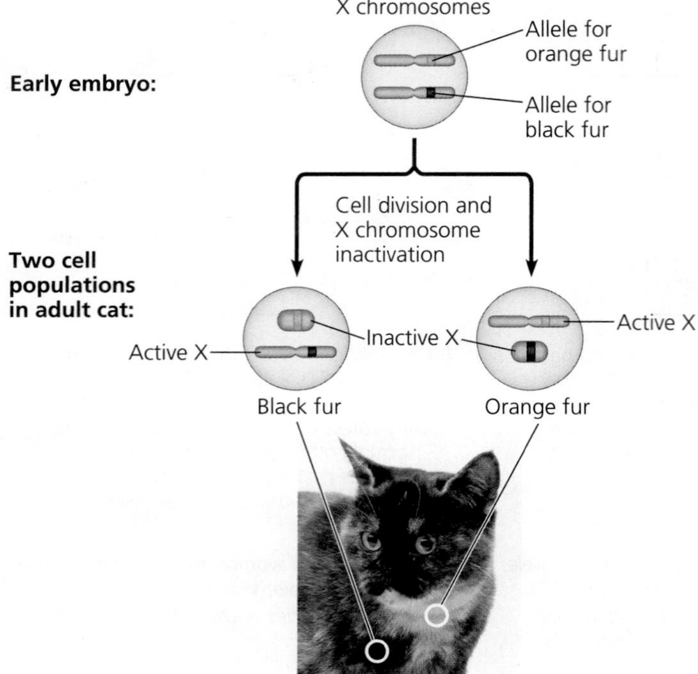

▲ **Figure 15.8 X inactivation and the tortoiseshell cat.** The tortoiseshell gene is on the X chromosome, and the tortoiseshell phenotype requires the presence of two different alleles, one for orange fur and one for black fur. Normally, only females can have both alleles, because only they have two X chromosomes. If a female cat is heterozygous for the tortoiseshell gene, she is tortoiseshell. Orange patches are formed by populations of cells in which the X chromosome with the orange allele is active; black patches have cells in which the X chromosome with the black allele is active. ("Calico" cats also have white areas, which are determined by yet another gene.)

of a *mosaic* of two types of cells: those with the active X derived from the father and those with the active X derived from the mother. After an X chromosome is inactivated in a particular cell, all mitotic descendants of that cell have the same inactive X. Thus, if a female is heterozygous for a sex-linked trait, about half her cells will express one allele, while the others will express the alternate allele. **Figure 15.8** shows how this mosaicism results in the mottled coloration of a tortoiseshell cat. In humans, mosaicism can be observed in a recessive X-linked mutation that prevents the development of sweat glands. A woman who is heterozygous for this trait has patches of normal skin and patches of skin lacking sweat glands.

Inactivation of an X chromosome involves modification of the DNA and the histone proteins bound to it, including attachment of methyl groups (—CH_3) to one of the nitrogenous bases of DNA nucleotides. (The regulatory role of DNA methylation is discussed further in Chapter 18.) A particular region of each X chromosome contains several genes involved in the inactivation process. The two regions, one on each X chromosome, associate briefly with each other in each cell at an early stage of embryonic development. Then one of the genes, called *XIST* (for X-inactive specific transcript) becomes active *only* on the chromosome that will become the Barr body. Multiple copies of the RNA product of this gene apparently attach to the X chromosome on which they are made, eventually almost covering it. Interaction of this RNA with the chromosome seems to initiate X inactivation, and the RNA products of other genes nearby on the X chromosome help to regulate the process.

CONCEPT CHECK 15.2

1. A white-eyed female *Drosophila* is mated with a red-eyed (wild-type) male, the reciprocal cross of the one shown in Figure 15.4. What phenotypes and genotypes do you predict for the offspring?
2. Neither Tim nor Rhoda has Duchenne muscular dystrophy, but their firstborn son does have it. What is the probability that a second child of this couple will have the disease? What is the probability if the second child is a boy? A girl?
3. **MAKE CONNECTIONS** Consider what you learned about dominant and recessive alleles in Concept 14.1 (p. 265). If a disorder were caused by a dominant X-linked allele, how would the inheritance pattern differ from what we see for recessive X-linked disorders?

For suggested answers, see Appendix A.

CONCEPT 15.3

Linked genes tend to be inherited together because they are located near each other on the same chromosome

The number of genes in a cell is far greater than the number of chromosomes; in fact, each chromosome has hundreds or thousands of genes. (The Y chromosome is an exception.) Genes located near each other on the same chromosome tend to be inherited together in genetic crosses; such genes are said to be genetically linked and are called **linked genes**. (Note the distinction between the terms *sex-linked gene*, referring to a single gene on a sex chromosome, and *linked genes*, referring to two or more genes on the same chromosome that tend to be inherited together.) When geneticists follow linked genes in breeding experiments, the results deviate from those expected from Mendel's law of independent assortment.

How Linkage Affects Inheritance

To see how linkage between genes affects the inheritance of two different characters, let's examine another of Morgan's *Drosophila* experiments. In this case, the characters are body color and wing size, each with two different phenotypes.

Wild-type flies have gray bodies and normal-sized wings. In addition to these flies, Morgan had managed to obtain, through breeding, doubly mutant flies with black bodies and wings much smaller than normal, called vestigial wings. The mutant alleles are recessive to the wild-type alleles, and neither gene is on a sex chromosome. In his investigation of these two genes, Morgan carried out the crosses shown in **Figure 15.9**. The first was a P generation cross to generate F_1 dihybrid flies, and the second was a testcross.

The resulting flies had a much higher proportion of the combinations of traits seen in the P generation flies (called parental phenotypes) than would be expected if the two genes assorted independently. Morgan thus concluded that body color and wing size are usually inherited together in

▼ Figure 15.9 INQUIRY

How does linkage between two genes affect inheritance of characters?

EXPERIMENT Morgan wanted to know whether the genes for body color and wing size were genetically linked, and if so, how this affected their inheritance. The alleles for body color are b^+ (gray) and b (black), and those for wing size are vg^+ (normal) and vg (vestigial).

Morgan mated true-breeding P (parental) generation flies—wild-type flies with black, vestigial-winged flies—to produce heterozygous F_1 dihybrids ($b^+ \, b \; vg^+ \, vg$), all of which are wild-type in appearance.

He then mated wild-type F_1 dihybrid females with black, vestigial-winged males. This testcross will reveal the genotype of the eggs made by the dihybrid female.

The male's sperm contributes only recessive alleles, so the phenotype of the offspring reflects the genotype of the female's eggs.

Note: Although only females (with pointed abdomens) are shown, half the offspring in each class would be males (with rounded abdomens).

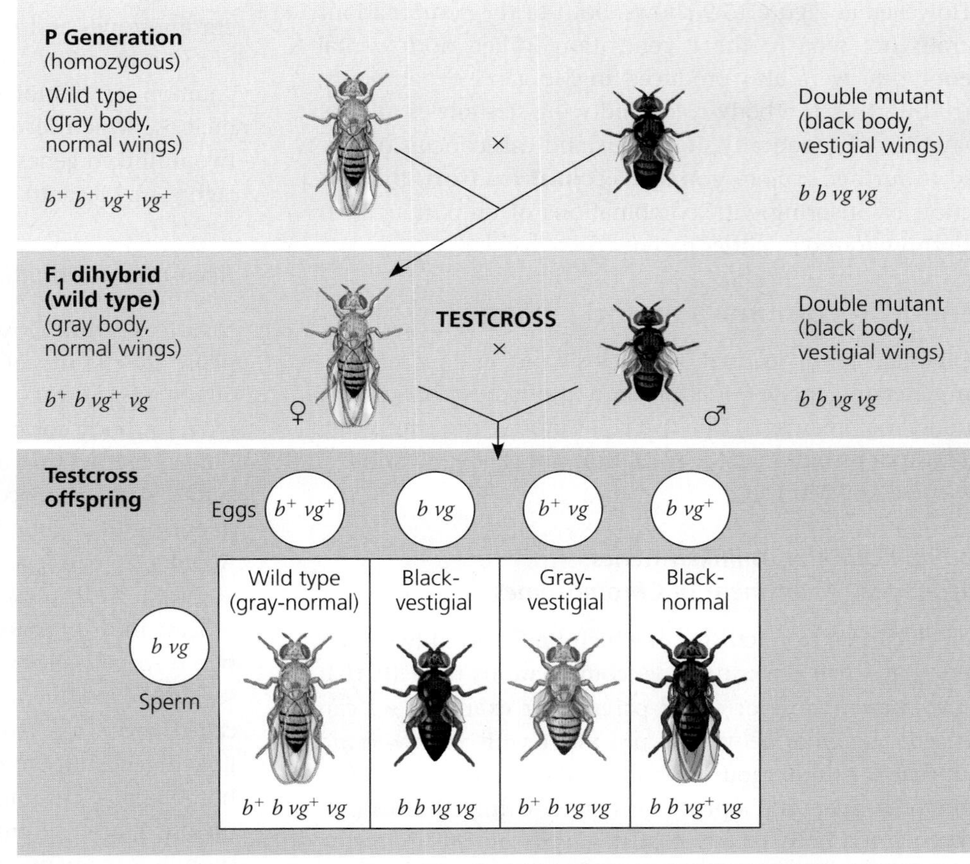

PREDICTED RATIOS

If genes are located on different chromosomes:	1	:	1	:	1	:	1
If genes are located on the same chromosome *and* parental alleles are always inherited together:	1	:	1	:	0	:	0
RESULTS	965	:	944	:	206	:	185

CONCLUSION Since most offspring had a parental (P generation) phenotype, Morgan concluded that the genes for body color and wing size are genetically linked on the same chromosome. However, the production of a relatively small number of offspring with nonparental phenotypes indicated that some mechanism occasionally breaks the linkage between specific alleles of genes on the same chromosome.

SOURCE: T. H. Morgan and C. J. Lynch, The linkage of two factors in *Drosophila* that are not sex-linked, *Biological Bulletin* 23:174–182 (1912).

WHAT IF? If the parental (P generation) flies had been true-breeding for gray body with vestigial wings and black body with normal wings, which phenotypic class(es) would be largest among the testcross offspring?

specific (parental) combinations because the genes for these characters are near each other on the same chromosome:

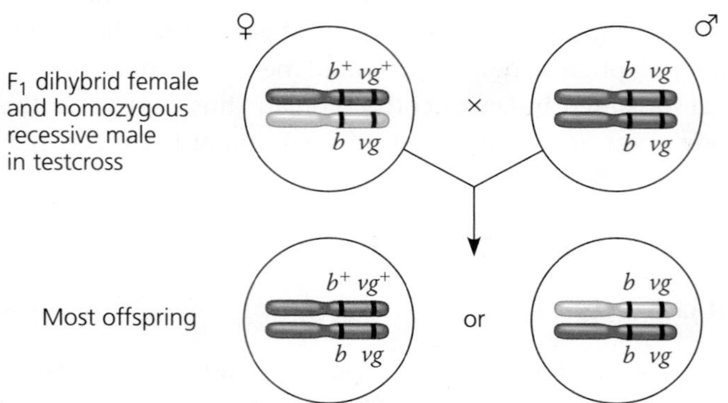

However, as Figure 15.9 shows, both of the combinations of traits not seen in the P generation (called nonparental phenotypes) were also produced in Morgan's experiments, suggesting that the body-color and wing-size alleles are not always linked genetically. To understand this conclusion, we need to further explore **genetic recombination**, the production of offspring with combinations of traits that differ from those found in either parent.

Genetic Recombination and Linkage

In Chapter 13, you learned that meiosis and random fertilization generate genetic variation among offspring of sexually reproducing organisms. Here we will examine the chromosomal basis of recombination in relation to the genetic findings of Mendel and Morgan.

Recombination of Unlinked Genes: Independent Assortment of Chromosomes

Mendel learned from crosses in which he followed two characters that some offspring have combinations of traits that do not match those of either parent. For example, we can represent the cross between a pea plant with yellow-round seeds that is heterozygous for both seed color and seed shape (a dihybrid, *YyRr*) and a plant with green-wrinkled seeds (homozygous for both recessive alleles, *yyrr*) by the following Punnett square:

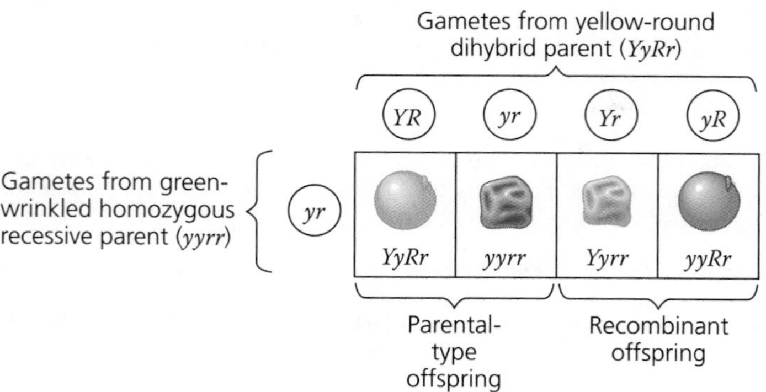

Notice in this Punnett square that one-half of the offspring are expected to inherit a phenotype that matches either of the parental (P generation) phenotypes. These offspring are called **parental types**. But two nonparental phenotypes are also found among the offspring. Because these offspring have new combinations of seed shape and color, they are called **recombinant types**, or **recombinants** for short. When 50% of all offspring are recombinants, as in this example, geneticists say that there is a 50% frequency of recombination. The predicted phenotypic ratios among the offspring are similar to what Mendel actually found in *YyRr* × *yyrr* crosses (a type of testcross because it reveals the genotype of the gametes made by the dihybrid *YyRr* plant).

A 50% frequency of recombination in such testcrosses is observed for any two genes that are located on different chromosomes and thus cannot be linked. The physical basis of recombination between unlinked genes is the random orientation of homologous chromosomes at metaphase I of meiosis, which leads to the independent assortment of the two unlinked genes (see Figure 13.10 and the question in the Figure 15.2 legend).

Recombination of Linked Genes: Crossing Over

Now let's return to Morgan's fly room to see how we can explain the results of the *Drosophila* testcross illustrated in Figure 15.9. Recall that most of the offspring from the testcross for body color and wing size had parental phenotypes. That suggested that the two genes were on the same chromosome, since the occurrence of parental types with a frequency greater than 50% indicates that the genes are linked. About 17% of offspring, however, were recombinants.

Faced with these results, Morgan proposed that some process must occasionally break the physical connection between specific alleles of genes on the same chromosome. Subsequent experiments demonstrated that this process, now called **crossing over**, accounts for the recombination of linked genes. In crossing over, which occurs while replicated homologous chromosomes are paired during prophase of meiosis I, a set of proteins orchestrates an exchange of corresponding segments of one maternal and one paternal chromatid (see Figure 13.11). In effect, end portions of two nonsister chromatids trade places each time a crossover occurs.

Figure 15.10 shows how crossing over in a dihybrid female fly resulted in recombinant eggs and ultimately recombinant offspring in Morgan's testcross. Most of the eggs had a chromosome with either the b^+ vg^+ or *b vg* parental genotype for body color and wing size, but some eggs had a recombinant chromosome (b^+ *vg* or *b* vg^+). Fertilization of these various classes of eggs by homozygous recessive sperm (*b vg*) produced an offspring population in which 17% exhibited a nonparental, recombinant phenotype, reflecting combinations of alleles not seen before in either P generation parent.

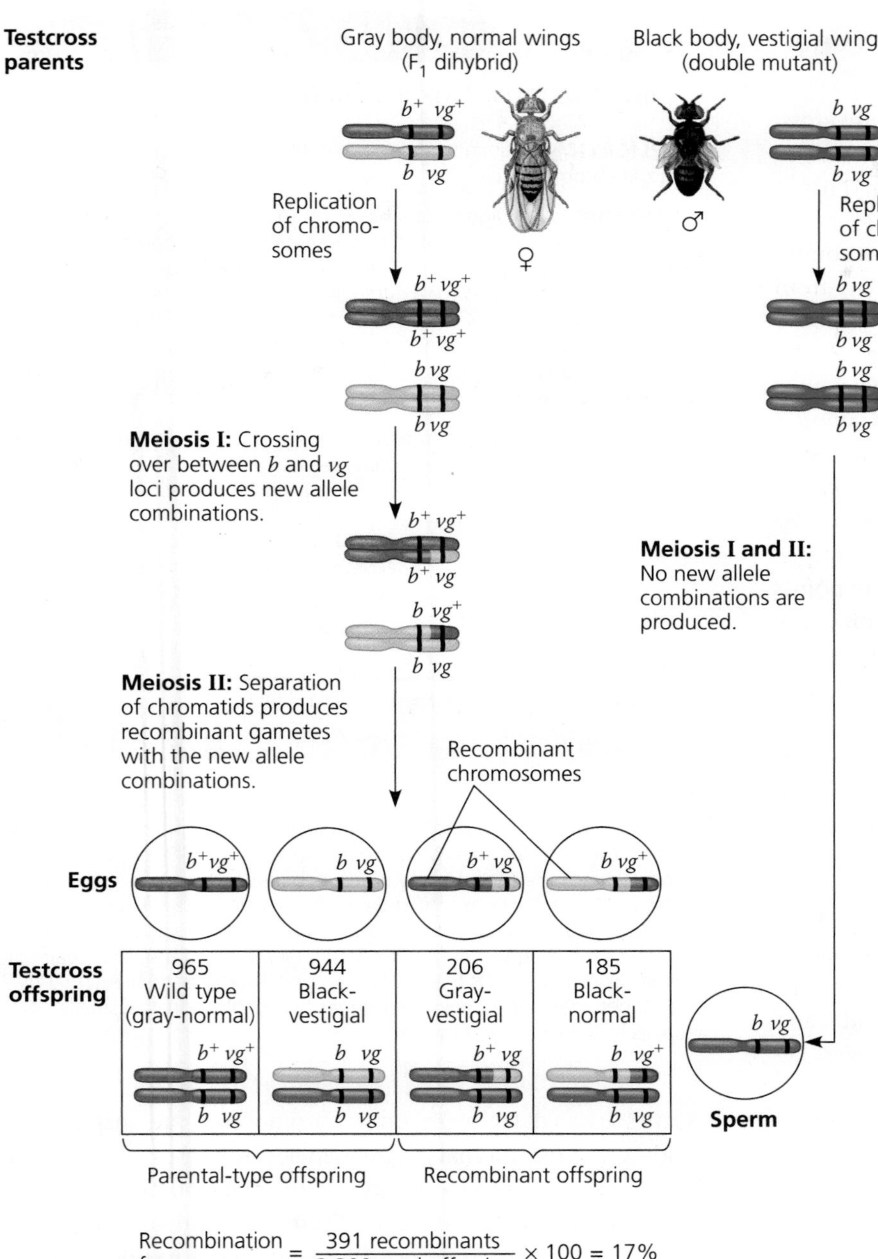

◄ **Figure 15.10 Chromosomal basis for recombination of linked genes.** In these diagrams re-creating the testcross in Figure 15.9, we track chromosomes as well as genes. The maternal chromosomes are color-coded red and pink to distinguish one homolog from the other before any meiotic crossing over has taken place. Because crossing over between the *b* and *vg* loci occurs in some, but not all, egg-producing cells, more eggs with parental-type chromosomes than with recombinant ones are produced in the mating females. Fertilization of the eggs by sperm of genotype *b vg* gives rise to some recombinant offspring. The recombination frequency is the percentage of recombinant flies in the total pool of offspring.

DRAW IT *Suppose, as in the question at the bottom of Figure 15.9, the parental (P generation) flies were true-breeding for gray body with vestigial wings and black body with normal wings. Draw the chromosomes in each of the four possible kinds of eggs from an F_1 female, and label each chromosome as "parental" or "recombinant."*

New Combinations of Alleles: Variation for Natural Selection

EVOLUTION In Chapter 13, you learned how the physical behavior of chromosomes during meiosis contributes to the generation of variation in offspring. Each pair of homologous chromosomes lines up independently of other pairs during metaphase I, and crossing over prior to that, during prophase I, can mix and match parts of maternal and paternal homologs. Chapter 14 described Mendel's elegant experiments showing that the behavior of the abstract entities known as genes (or, more concretely, alleles of genes) also leads to variation in offspring. Now, putting these different ideas together, you can see that the recombinant chromosomes resulting from crossing over may bring alleles together in new combinations, and the subsequent events of meiosis distribute to gametes the recombinant chromosomes in a multitude of combinations, such as the new variants seen in Figures 15.9 and 15.10. Random fertilization then increases even further the number of variant allele combinations that can be created.

This abundance of genetic variation provides the raw material on which natural selection works. If the traits conferred by particular combinations of alleles are better suited for a given environment, organisms possessing those genotypes will be expected to thrive and leave more offspring, ensuring the continuation of their genetic complement. In the next generation, of course, the alleles will be shuffled anew. Ultimately, the interplay between environment and genotype will determine which genetic combinations persist over time.

Mapping the Distance Between Genes Using Recombination Data: *Scientific Inquiry*

The discovery of linked genes and recombination due to crossing over led one of Morgan's students, Alfred H. Sturtevant, to a method for constructing a **genetic map**, an ordered list of the genetic loci along a particular chromosome.

Sturtevant hypothesized that the percentage of recombinant offspring, the *recombination frequency*, calculated from experiments like the one in Figures 15.9 and 15.10, depends on the distance between genes on a chromosome. He assumed that crossing over is a random event, with the chance of crossing over approximately equal at all points along a chromosome. Based on these assumptions, Sturtevant predicted that *the farther apart two genes are, the higher the probability that a crossover will occur between them and therefore the higher the recombination frequency*. His reasoning was simple: The greater the distance between two genes, the more points there are between them where crossing over can occur. Using recombination data from various fruit fly crosses, Sturtevant proceeded to assign relative positions to genes on the same chromosomes—that is, to *map* genes.

A genetic map based on recombination frequencies is called a **linkage map**. **Figure 15.11** shows Sturtevant's linkage map of three genes: the body-color (*b*) and wing-size (*vg*) genes depicted in Figure 15.10 and a third gene, called cinnabar (*cn*). Cinnabar is one of many *Drosophila* genes affecting eye color. Cinnabar eyes, a mutant phenotype, are a brighter red than the wild-type color. The recombination frequency between *cn* and *b* is 9%; that between *cn* and *vg*, 9.5%; and that between *b* and *vg*, 17%. In other words, crossovers between *cn* and *b* and between *cn* and *vg* are about half as frequent as crossovers between *b* and *vg*. Only a map that locates *cn* about midway between *b* and *vg* is consistent with these data, as you can prove to yourself by drawing alternative maps. Sturtevant expressed the distances between genes in **map units**, defining one map unit as equivalent to a 1% recombination frequency.

In practice, the interpretation of recombination data is more complicated than this example suggests. Some genes on a chromosome are so far from each other that a crossover between them is virtually certain. The observed frequency of recombination in crosses involving two such genes can have a maximum value of 50%, a result indistinguishable from that for genes on different chromosomes. In this case, the physical connection between genes on the same chromosome is not reflected in the results of genetic crosses. Despite being on the same chromosome and thus being *physically connected*, the genes are *genetically unlinked*; alleles of such genes assort independently, as if they were on different chromosomes. In fact, at least two of the genes for pea characters that Mendel studied are now known to be on the same chromosome, but the distance between them is so great that linkage is not observed in genetic crosses. Consequently, the two genes behaved as if they were on different chromosomes in Mendel's experiments.

▼ Figure 15.11 RESEARCH METHOD

Constructing a Linkage Map

APPLICATION A linkage map shows the relative locations of genes along a chromosome.

TECHNIQUE A linkage map is based on the assumption that the probability of a crossover between two genetic loci is proportional to the distance separating the loci. The recombination frequencies used to construct a linkage map for a particular chromosome are obtained from experimental crosses, such as the cross depicted in Figures 15.9 and 15.10. The distances between genes are expressed as map units, with one map unit equivalent to a 1% recombination frequency. Genes are arranged on the chromosome in the order that best fits the data.

RESULTS In this example, the observed recombination frequencies between three *Drosophila* gene pairs (*b–cn* 9%, *cn–vg* 9.5%, and *b–vg* 17%) best fit a linear order in which *cn* is positioned about halfway between the other two genes:

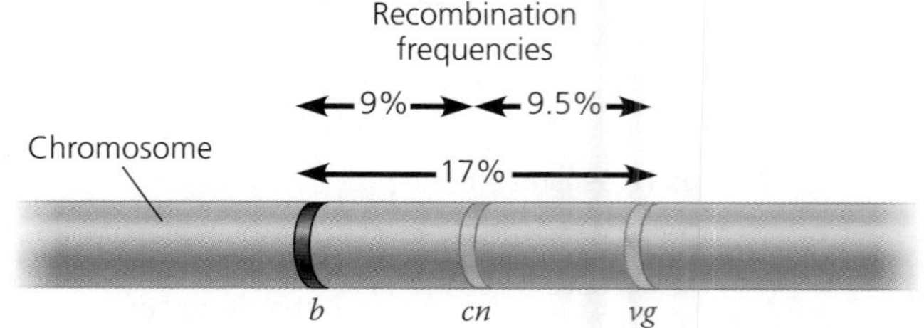

The *b–vg* recombination frequency (17%) is slightly less than the sum of the *b–cn* and *cn–vg* frequencies (9 + 9.5 = 18.5%) because of the few times that one crossover occurs between *b* and *cn* and another crossover occurs between *cn* and *vg*. The second crossover would "cancel out" the first, reducing the observed *b–vg* recombination frequency while contributing to the frequency between each of the closer pairs of genes. The value of 18.5% (18.5 map units) is closer to the actual distance between the genes, so a geneticist would add the smaller distances in constructing a map.

Genes located far apart on a chromosome are mapped by adding the recombination frequencies from crosses involving closer pairs of genes lying between the two distant genes.

Using recombination data, Sturtevant and his colleagues were able to map numerous *Drosophila* genes in linear arrays. They found that the genes clustered into four groups of linked genes (*linkage groups*). Light microscopy had revealed four pairs of chromosomes in *Drosophila*, so the linkage map provided additional evidence that genes are located on chromosomes. Each chromosome has a linear array of specific genes, each gene with its own locus (**Figure 15.12**).

Because a linkage map is based strictly on recombination frequencies, it gives only an approximate picture of a chromosome. The frequency of crossing over is not actually uniform over the length of a chromosome, as Sturtevant assumed, and therefore map units do not correspond to actual physical distances (in nanometers, for instance). A linkage map does portray the order of genes along a chromosome, but it does not accurately portray the precise locations of those genes. Other methods enable geneticists to construct **cytogenetic maps** of chromosomes, which locate genes with respect to chromosomal

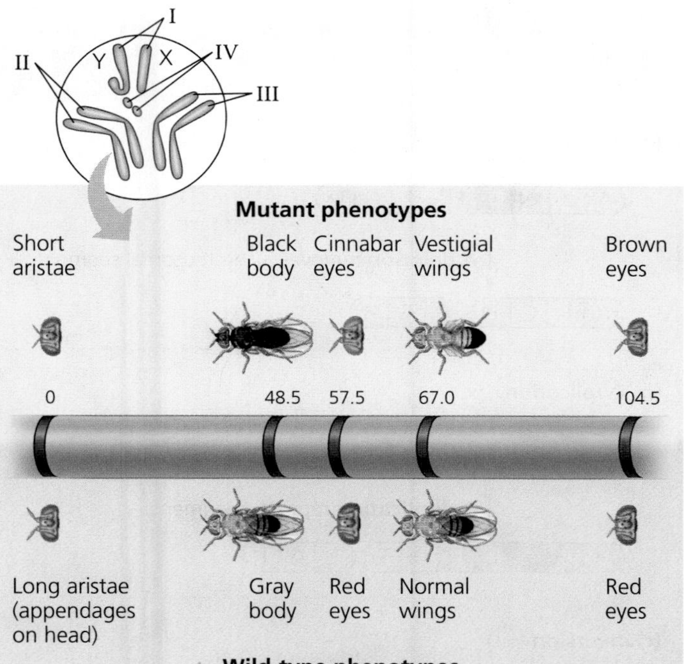

▲ **Figure 15.12 A partial genetic (linkage) map of a *Drosophila* chromosome.** This simplified map shows just a few of the genes that have been mapped on *Drosophila* chromosome II. The number at each gene locus indicates the number of map units between that locus and the locus for arista length (left). Notice that more than one gene can affect a given phenotypic characteristic, such as eye color. Also, note that in contrast to the homologous autosomes (II–IV), the X and Y sex chromosomes (I) have distinct shapes.

features, such as stained bands, that can be seen in the microscope. The ultimate maps, which we will discuss in Chapter 21, display the physical distances between gene loci in DNA nucleotides. Comparing a linkage map with such a physical map or with a cytogenetic map of the same chromosome, we find that the linear order of genes is identical in all the maps, but the spacing between genes is not.

CONCEPT CHECK 15.3

1. When two genes are located on the same chromosome, what is the physical basis for the production of recombinant offspring in a testcross between a dihybrid parent and a double-mutant (recessive) parent?
2. For each type of offspring of the testcross in Figure 15.9, explain the relationship between its phenotype and the alleles contributed by the female parent.
3. **WHAT IF?** Genes *A*, *B*, and *C* are located on the same chromosome. Testcrosses show that the recombination frequency between *A* and *B* is 28% and between *A* and *C* is 12%. Can you determine the linear order of these genes? Explain.

For suggested answers, see Appendix A.

CONCEPT 15.4

Alterations of chromosome number or structure cause some genetic disorders

As you have learned so far in this chapter, the phenotype of an organism can be affected by small-scale changes involving individual genes. Random mutations are the source of all new alleles, which can lead to new phenotypic traits.

Large-scale chromosomal changes can also affect an organism's phenotype. Physical and chemical disturbances, as well as errors during meiosis, can damage chromosomes in major ways or alter their number in a cell. Large-scale chromosomal alterations in humans and other mammals often lead to spontaneous abortion (miscarriage) of a fetus, and individuals born with these types of genetic defects commonly exhibit various developmental disorders. Plants may tolerate such genetic defects better than animals do.

Abnormal Chromosome Number

Ideally, the meiotic spindle distributes chromosomes to daughter cells without error. But there is an occasional mishap, called a **nondisjunction**, in which the members of a pair of homologous chromosomes do not move apart properly during meiosis I or sister chromatids fail to separate during meiosis II (**Figure 15.13**). In these cases, one gamete

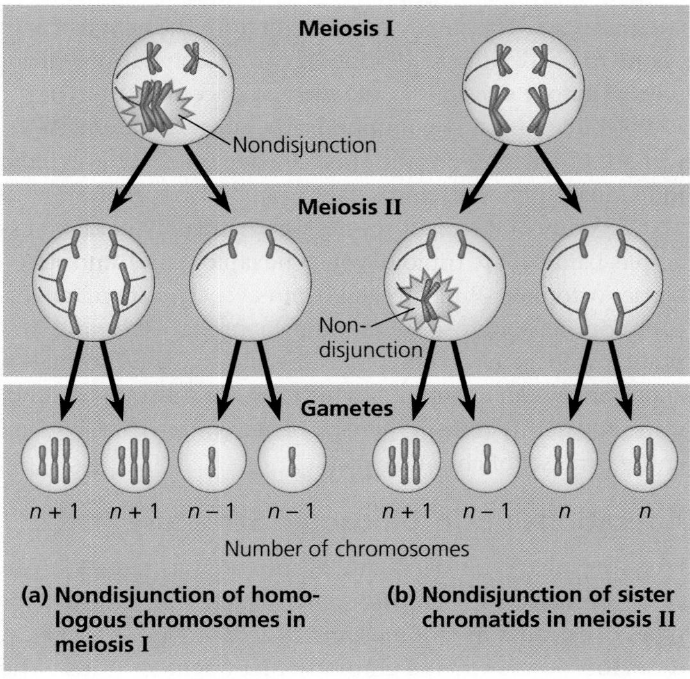

▲ **Figure 15.13 Meiotic nondisjunction.** Gametes with an abnormal chromosome number can arise by nondisjunction in either meiosis I or meiosis II. For simplicity, the figure does not show the spores formed by meiosis in plants. Ultimately, spores form gametes that have the defects shown. (See Figure 13.6.)

receives two of the same type of chromosome and another gamete receives no copy. The other chromosomes are usually distributed normally.

If either of the aberrant gametes unites with a normal one at fertilization, the zygote will also have an abnormal number of a particular chromosome, a condition known as **aneuploidy**. (Aneuploidy may involve more than one chromosome.) Fertilization involving a gamete that has no copy of a particular chromosome will lead to a missing chromosome in the zygote (so that the cell has $2n - 1$ chromosomes); the aneuploid zygote is said to be **monosomic** for that chromosome. If a chromosome is present in triplicate in the zygote (so that the cell has $2n + 1$ chromosomes), the aneuploid cell is **trisomic** for that chromosome. Mitosis will subsequently transmit the anomaly to all embryonic cells. If the organism survives, it usually has a set of traits caused by the abnormal dose of the genes associated with the extra or missing chromosome. Down syndrome is an example of trisomy in humans that will be discussed later. Nondisjunction can also occur during mitosis. If such an error takes place early in embryonic development, then the aneuploid condition is passed along by mitosis to a large number of cells and is likely to have a substantial effect on the organism.

Some organisms have more than two complete chromosome sets in all somatic cells. The general term for this chromosomal alteration is **polyploidy**; the specific terms *triploidy* ($3n$) and *tetraploidy* ($4n$) indicate three or four chromosomal sets, respectively. One way a triploid cell may arise is by the fertilization of an abnormal diploid egg produced by nondisjunction of all its chromosomes. Tetraploidy could result from the failure of a $2n$ zygote to divide after replicating its chromosomes. Subsequent normal mitotic divisions would then produce a $4n$ embryo.

Polyploidy is fairly common in the plant kingdom. As we will see in Chapter 24, the spontaneous origin of polyploid individuals plays an important role in the evolution of plants. Many of the plant species we eat are polyploid; for example, bananas are triploid, wheat hexaploid ($6n$), and strawberries octoploid ($8n$). Polyploid animal species are much less common, although some are found among fishes and amphibians. In general, polyploids are more nearly normal in appearance than aneuploids. One extra (or missing) chromosome apparently disrupts genetic balance more than does an entire extra set of chromosomes.

Alterations of Chromosome Structure

Errors in meiosis or damaging agents such as radiation can cause breakage of a chromosome, which can lead to four types of changes in chromosome structure **(Figure 15.14)**. A **deletion** occurs when a chromosomal fragment is lost. The affected chromosome is then missing certain genes. (If the centromere is deleted, the entire chromosome will be lost.) The "deleted" fragment may become attached as an extra segment to a sister chromatid, producing a **duplication**.

▼ **Figure 15.14 Alterations of chromosome structure.** Red arrows indicate breakage points. Dark purple highlights the chromosomal parts affected by the rearrangements.

(a) Deletion

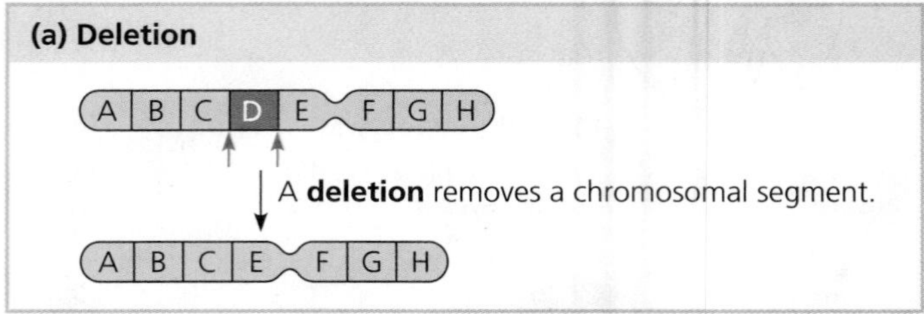
A B C D E F G H
A **deletion** removes a chromosomal segment.
A B C E F G H

(b) Duplication

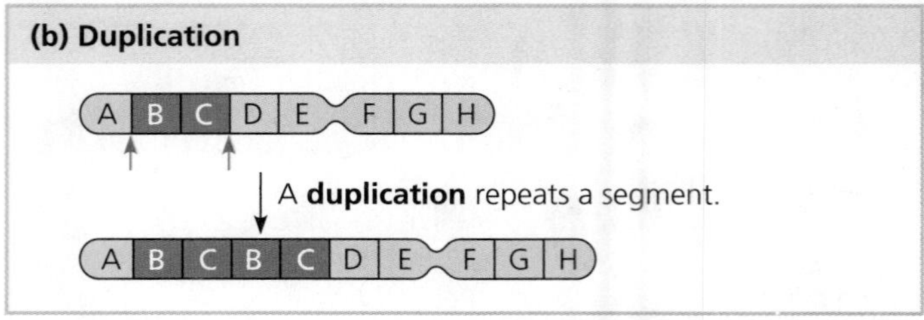
A B C D E F G H
A **duplication** repeats a segment.
A B C B C D E F G H

(c) Inversion

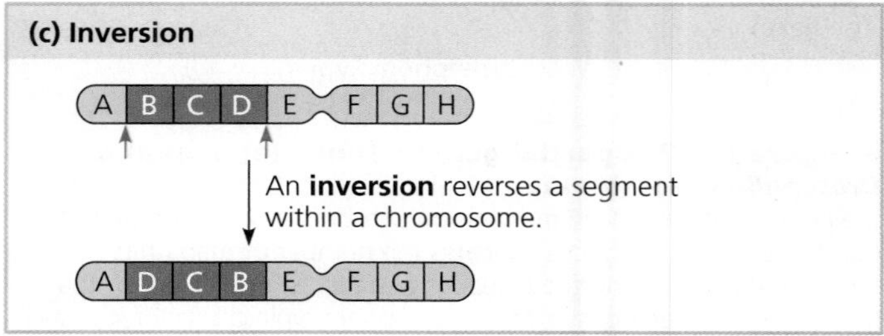
A B C D E F G H
An **inversion** reverses a segment within a chromosome.
A D C B E F G H

(d) Translocation

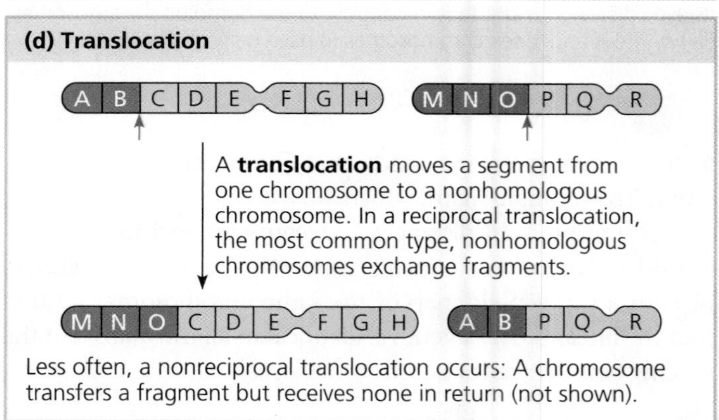
A B C D E F G H
M N O P Q R
A **translocation** moves a segment from one chromosome to a nonhomologous chromosome. In a reciprocal translocation, the most common type, nonhomologous chromosomes exchange fragments.
M N O C D E F G H
A B P Q R

Less often, a nonreciprocal translocation occurs: A chromosome transfers a fragment but receives none in return (not shown).

Alternatively, a detached fragment could attach to a nonsister chromatid of a homologous chromosome. In that case, though, the "duplicated" segments might not be identical because the homologs could carry different alleles of certain genes. A chromosomal fragment may also reattach to the original chromosome but in the reverse orientation, producing an **inversion**. A fourth possible result of chromosomal breakage is for the fragment to join a nonhomologous chromosome, a rearrangement called a **translocation**.

Deletions and duplications are especially likely to occur during meiosis. In crossing over, nonsister chromatids sometimes exchange unequal-sized segments of DNA, so that one partner gives up more genes than it receives. The products of such an unequal crossover are one chromosome with a deletion and one chromosome with a duplication.

A diploid embryo that is homozygous for a large deletion (or has a single X chromosome with a large deletion, in a male) is usually missing a number of essential genes, a condition that is ordinarily lethal. Duplications and translocations also tend to be harmful. In reciprocal translocations, in which segments are exchanged between nonhomologous chromosomes, and in inversions, the balance of genes is not abnormal—all genes are present in their normal doses. Nevertheless, translocations and inversions can alter phenotype because a gene's expression can be influenced by its location among neighboring genes; such events sometimes have devastating effects.

Human Disorders Due to Chromosomal Alterations

Alterations of chromosome number and structure are associated with a number of serious human disorders. As described earlier, nondisjunction in meiosis results in aneuploidy in gametes and any resulting zygotes. Although the frequency of aneuploid zygotes may be quite high in humans, most of these chromosomal alterations are so disastrous to development that the affected embryos are spontaneously aborted long before birth. However, some types of aneuploidy appear to upset the genetic balance less than others, with the result that individuals with certain aneuploid conditions can survive to birth and beyond. These individuals have a set of traits—a *syndrome*—characteristic of the type of aneuploidy. Genetic disorders caused by aneuploidy can be diagnosed before birth by fetal testing (see Figure 14.19).

Down Syndrome (Trisomy 21)

One aneuploid condition, **Down syndrome**, affects approximately one out of every 700 children born in the United States (**Figure 15.15**). Down syndrome is usually the result of an extra chromosome 21, so that each body cell has a total of 47 chromosomes. Because the cells are trisomic for chromosome 21, Down syndrome is often called *trisomy 21*. Down syndrome includes characteristic facial features, short stature, correctable heart defects, and developmental delays. Individuals with Down syndrome have an increased chance of developing leukemia and Alzheimer's disease but have a lower rate of high blood pressure, atherosclerosis (hardening of the arteries), stroke, and many types of solid tumors. Although people with Down syndrome, on average, have a life span shorter than normal, most, with proper medical treatment, live to middle age and beyond. Many live independently or at home with their families, are employed, and are valuable contributors to their communities. Almost all males and about half of females with Down syndrome are sexually underdeveloped and sterile.

The frequency of Down syndrome increases with the age of the mother. While the disorder occurs in just 0.04% of children born to women under age 30, the risk climbs to 0.92% for

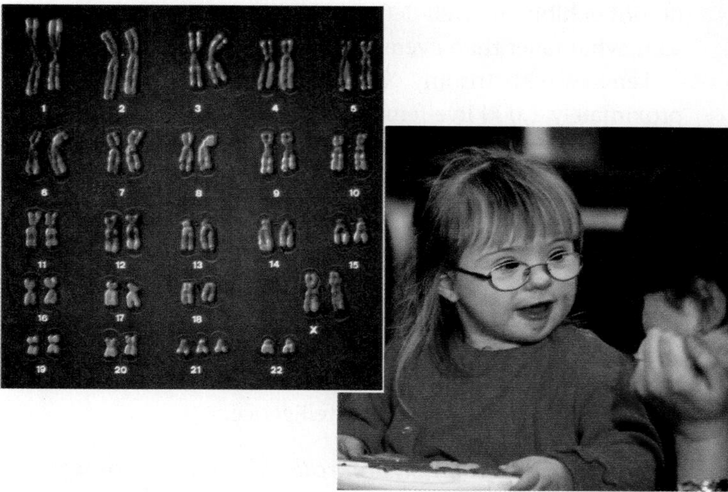

▲ **Figure 15.15 Down syndrome.** The karyotype shows trisomy 21, the most common cause of Down syndrome. The child exhibits the facial features characteristic of this disorder.

mothers at age 40 and is even higher for older mothers. The correlation of Down syndrome with maternal age has not yet been explained. Most cases result from nondisjunction during meiosis I, and some research points to an age-dependent abnormality in a meiosis checkpoint that normally delays anaphase until all the kinetochores are attached to the spindle (like the M phase checkpoint of the mitotic cell cycle; see Chapter 12). Trisomies of some other chromosomes also increase in incidence with maternal age, although infants with other autosomal trisomies rarely survive for long. Due to its low risk and its potential for providing useful information, prenatal screening for trisomies in the embryo is now offered to all pregnant women. In 2008, the Prenatally and Postnatally Diagnosed Conditions Awareness Act was signed into law in the United States. This law stipulates that medical practitioners give accurate, up-to-date information about any prenatal or postnatal diagnosis received by parents and that they connect parents with appropriate support services.

Aneuploidy of Sex Chromosomes

Nondisjunction of sex chromosomes produces a variety of aneuploid conditions. Most of these conditions appear to upset the genetic balance less than aneuploid conditions involving autosomes. This may be because the Y chromosome carries relatively few genes and because extra copies of the X chromosome become inactivated as Barr bodies in somatic cells.

An extra X chromosome in a male, producing XXY, occurs approximately once in every 500 to 1,000 live male births. People with this disorder, called *Klinefelter syndrome*, have male sex organs, but the testes are abnormally small and the man is sterile. Even though the extra X is inactivated, some breast enlargement and other female body characteristics are common. Affected individuals may have subnormal intelligence. About 1 of every 1,000 males is born with an extra Y chromosome (XYY). These males undergo normal sexual development and do not exhibit any well-defined syndrome, but they tend to be somewhat taller than average.

Females with trisomy X (XXX), which occurs once in approximately 1,000 live female births, are healthy and have no unusual physical features other than being slightly taller than average. Triple-X females are at risk for learning disabilities but are fertile. Monosomy X, called *Turner syndrome*, occurs about once in every 2,500 female births and is the only known viable monosomy in humans. Although these X0 individuals are phenotypically female, they are sterile because their sex organs do not mature. When provided with estrogen replacement therapy, girls with Turner syndrome do develop secondary sex characteristics. Most have normal intelligence.

Disorders Caused by Structurally Altered Chromosomes

Many deletions in human chromosomes, even in a heterozygous state, cause severe problems. One such syndrome, known as *cri du chat* ("cry of the cat"), results from a specific deletion in chromosome 5. A child born with this deletion is severely intellectually disabled, has a small head with unusual facial features, and has a cry that sounds like the mewing of a distressed cat. Such individuals usually die in infancy or early childhood.

Chromosomal translocations have been implicated in certain cancers, including *chronic myelogenous leukemia* (CML). This disease occurs when a reciprocal translocation happens during mitosis of cells that will become white blood cells. In these cells, the exchange of a large portion of chromosome 22 with a small fragment from a tip of chromosome 9 produces a much shortened, easily recognized chromosome 22, called the *Philadelphia chromosome* (**Figure 15.16**). Such an exchange causes cancer by activating a gene that leads to uncontrolled cell cycle progression. The mechanism of gene activation will be discussed in Chapter 18.

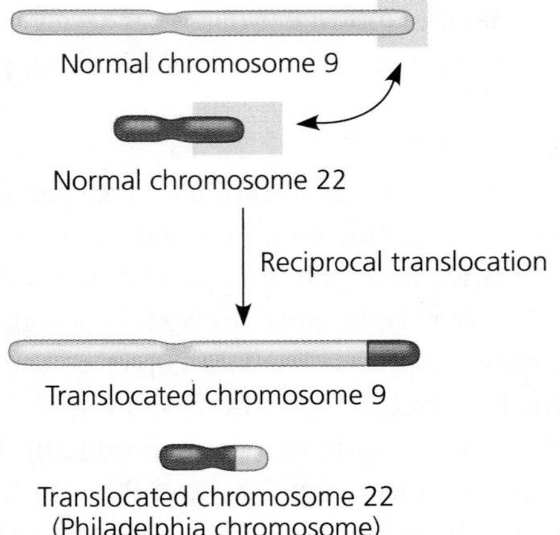

▲ **Figure 15.16 Translocation associated with chronic myelogenous leukemia (CML).** The cancerous cells in nearly all CML patients contain an abnormally short chromosome 22, the so-called Philadelphia chromosome, and an abnormally long chromosome 9. These altered chromosomes result from the reciprocal translocation shown here, which presumably occurred in a single white blood cell precursor undergoing mitosis and was then passed along to all descendant cells.

CONCEPT CHECK 15.4

1. About 5% of individuals with Down syndrome have a chromosomal translocation in which a third copy of chromosome 21 is attached to chromosome 14. If this translocation occurred in a parent's gonad, how could it lead to Down syndrome in a child?
2. **WHAT IF?** The ABO blood type locus has been mapped on chromosome 9. A father who has type AB blood and a mother who has type O blood have a child with trisomy 9 and type A blood. Using this information, can you tell in which parent the nondisjunction occurred? Explain your answer.
3. **MAKE CONNECTIONS** The gene that is activated on the Philadelphia chromosome codes for an intracellular tyrosine kinase. Review the discussion of cell cycle control and cancer in Concept 12.3 (pp. 242–243), and explain how the activation of this gene could contribute to the development of cancer.

For suggested answers, see Appendix A.

CONCEPT 15.5

Some inheritance patterns are exceptions to standard Mendelian inheritance

In the previous section, you learned about deviations from the usual patterns of chromosomal inheritance due to abnormal events in meiosis and mitosis. We conclude this chapter by describing two normally occurring exceptions to Mendelian genetics, one involving genes located in the nucleus and the other involving genes located outside the nucleus. In both cases, the sex of the parent contributing an allele is a factor in the pattern of inheritance.

Genomic Imprinting

Throughout our discussions of Mendelian genetics and the chromosomal basis of inheritance, we have assumed that a given allele will have the same effect whether it was inherited from the mother or the father. This is probably a safe assumption most of the time. For example, when Mendel crossed purple-flowered pea plants with white-flowered pea plants, he observed the same results regardless of whether the purple-flowered parent supplied the eggs or the sperm. In recent years, however, geneticists have identified two to three dozen traits in mammals that depend on which parent passed along the alleles for those traits. Such variation in phenotype depending on whether an allele is inherited from the male or female parent is called **genomic imprinting**. (Note that unlike sex-linked genes, most imprinted genes are on autosomes.)

Genomic imprinting occurs during gamete formation and results in the silencing of a particular allele of certain genes. Because these genes are imprinted differently in sperm and eggs, a zygote expresses only one allele of an imprinted gene, that inherited from either the female or the male parent. The imprints are then transmitted to all body cells during development. In each generation, the old imprints are "erased" in gamete-producing cells, and the chromosomes of the developing gametes are newly imprinted according to the sex of the individual forming the gametes. In a given species, the imprinted genes are always imprinted in the same way. For instance, a gene imprinted for maternal allele expression is always imprinted this way, generation after generation.

Consider, for example, the mouse gene for insulin-like growth factor 2 (*Igf2*), one of the first imprinted genes to be identified. Although this growth factor is required for normal prenatal growth, only the paternal allele is expressed **(Figure 15.17a)**. Evidence that the *Igf2* gene is imprinted came initially from crosses between normal-sized (wild-type) mice and dwarf (mutant) mice homozygous for a recessive mutation in the *Igf2* gene. The phenotypes of heterozygous offspring (with one normal allele and one mutant) differed, depending on whether the mutant allele came from the mother or the father **(Figure 15.17b)**.

What exactly is a genomic imprint? In many cases, it seems to consist of methyl (—CH_3) groups that are added to cytosine nucleotides of one of the alleles. Such methylation may silence the allele, an effect consistent with evidence that heavily methylated genes are usually inactive (see Chapter 18). However, for a few genes, methylation has been shown to *activate* expression of the allele. This is the case for the *Igf2* gene: Methylation of certain cytosines on the paternal chromosome leads to expression of the paternal *Igf2* allele. The apparent inconsistency as to whether methylation activates or silences alleles was resolved in part when researchers found that DNA methylation operates indirectly by recruiting enzymes that modify DNA-associated proteins (histones), leading to condensation of the local DNA. Depending on the original function of the condensed DNA in regulating allele expression, the result is either silencing or activation of a given allele.

Genomic imprinting is thought to affect only a small fraction of the genes in mammalian genomes, but most of the known imprinted genes are critical for embryonic development. In experiments with mice, for example, embryos engineered to inherit both copies of certain chromosomes from the same parent usually die before birth, whether that parent is male or female. A few years ago, however, scientists in Japan combined the genetic material from two eggs in a zygote while allowing expression of the *Igf2* gene from only one of the egg nuclei. The zygote developed into an apparently healthy mouse. Normal development seems to require that embryonic cells have exactly one active copy—not zero, not two—of certain genes. The association of aberrant imprinting with abnormal development and certain cancers has stimulated numerous studies of how different genes are imprinted.

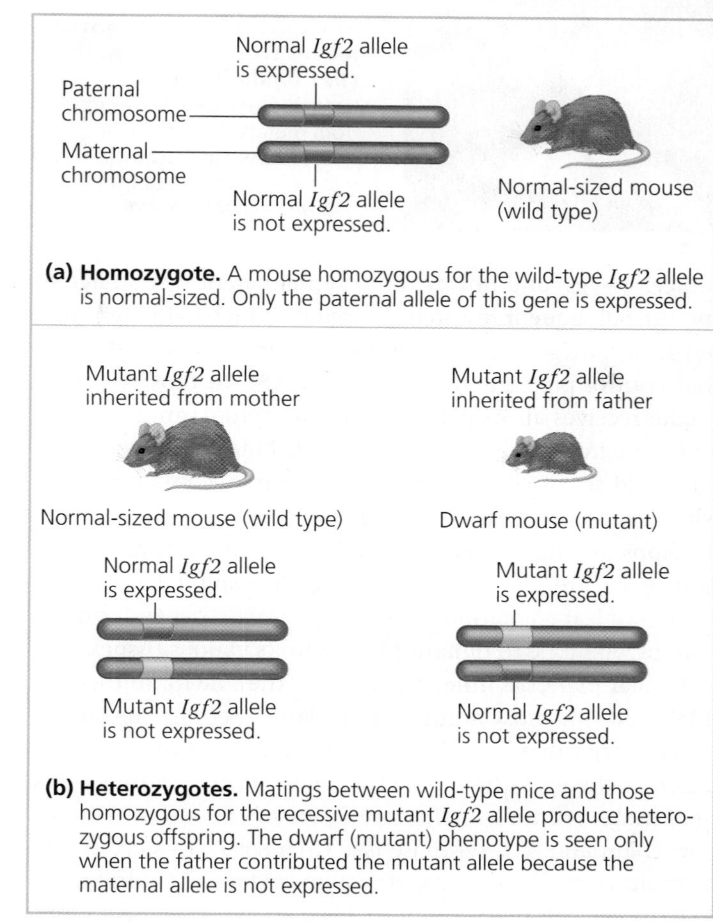

(a) Homozygote. A mouse homozygous for the wild-type *Igf2* allele is normal-sized. Only the paternal allele of this gene is expressed.

(b) Heterozygotes. Matings between wild-type mice and those homozygous for the recessive mutant *Igf2* allele produce heterozygous offspring. The dwarf (mutant) phenotype is seen only when the father contributed the mutant allele because the maternal allele is not expressed.

▲ **Figure 15.17 Genomic imprinting of the mouse *Igf2* gene.**

Inheritance of Organelle Genes

Although our focus in this chapter has been on the chromosomal basis of inheritance, we end with an important amendment: Not all of a eukaryotic cell's genes are located on nuclear chromosomes, or even in the nucleus; some genes are located in organelles in the cytoplasm. Because they are outside the nucleus, these genes are sometimes called *extranuclear genes* or *cytoplasmic genes*. Mitochondria, as well as chloroplasts and other plastids in plants, contain small circular DNA molecules that carry a number of genes. These organelles reproduce themselves and transmit their genes to daughter organelles. Organelle genes are not distributed to offspring according to the same rules that direct the distribution of nuclear chromosomes during meiosis, so they do not display Mendelian inheritance.

The first hint that extranuclear genes exist came from studies by the German scientist Karl Correns on the inheritance of yellow or white patches on the leaves of an otherwise green plant. In 1909, he observed that the coloration of the offspring was determined only by the maternal parent (the

◀ **Figure 15.18 Variegated leaves from English holly (*Ilex aquifolium*).** Variegated (striped or spotted) leaves result from mutations in pigment genes located in plastids, which generally are inherited from the maternal parent.

source of eggs) and not by the paternal parent (the source of sperm). Subsequent research showed that such coloration patterns, or variegation, are due to mutations in plastid genes that control pigmentation **(Figure 15.18)**. In most plants, a zygote receives all its plastids from the cytoplasm of the egg and none from the sperm, which contributes little more than a haploid set of chromosomes. An egg may contain plastids with different alleles for a pigmentation gene. As the zygote develops, plastids containing wild-type or mutant pigmentation genes are distributed randomly to daughter cells. The pattern of leaf coloration exhibited by a plant depends on the ratio of wild-type to mutant plastids in its various tissues.

Similar maternal inheritance is also the rule for mitochondrial genes in most animals and plants, because almost all the mitochondria passed on to a zygote come from the cytoplasm of the egg. The products of most mitochondrial genes help make up the protein complexes of the electron transport chain and ATP synthase (see Chapter 9). Defects in one or more of these proteins, therefore, reduce the amount of ATP the cell can make and have been shown to cause a number of rare human disorders. Because the parts of the body most susceptible to energy deprivation are the nervous system and the muscles, most mitochondrial diseases primarily affect these systems. For example, *mitochondrial myopathy* causes weakness, intolerance of exercise, and muscle deterioration. Another mitochondrial disorder is *Leber's hereditary optic neuropathy*, which can produce sudden blindness in people as young as their 20s or 30s. The four mutations found thus far to cause this disorder affect oxidative phosphorylation during cellular respiration, a crucial function for the cell.

In addition to the rare diseases clearly caused by defects in mitochondrial DNA, mitochondrial mutations inherited from a person's mother may contribute to at least some cases of diabetes and heart disease, as well as to other disorders that commonly debilitate the elderly, such as Alzheimer's disease. In the course of a lifetime, new mutations gradually accumulate in our mitochondrial DNA, and some researchers think that these mutations play a role in the normal aging process.

Wherever genes are located in the cell—in the nucleus or in cytoplasmic organelles—their inheritance depends on the precise replication of DNA, the genetic material. In the next chapter, you will learn how this molecular reproduction occurs.

CONCEPT CHECK 15.5

1. Gene dosage, the number of active copies of a gene, is important to proper development. Identify and describe two processes that establish the proper dosage of certain genes.
2. Reciprocal crosses between two primrose varieties, A and B, produced the following results: A female × B male → offspring with all green (nonvariegated) leaves; B female × A male → offspring with spotted (variegated) leaves. Explain these results.
3. **WHAT IF?** Mitochondrial genes are critical to the energy metabolism of cells, but mitochondrial disorders caused by mutations in these genes are generally not lethal. Why not?

For suggested answers, see Appendix A.

15 CHAPTER REVIEW

SUMMARY OF KEY CONCEPTS

CONCEPT 15.1

Mendelian inheritance has its physical basis in the behavior of chromosomes (pp. 286–289)

- The **chromosome theory of inheritance** states that genes are located on chromosomes and that the behavior of chromosomes during meiosis accounts for Mendel's laws of segregation and independent assortment.
- Morgan's discovery that transmission of the X chromosome in *Drosophila* correlates with inheritance of an eye-color trait was the first solid evidence indicating that a specific gene is associated with a specific chromosome.

? *What characteristic of the sex chromosomes allowed Morgan to correlate their behavior with that of the alleles of the eye-color gene?*

CONCEPT 15.2

Sex-linked genes exhibit unique patterns of inheritance (pp. 289–292)

- Sex is an inherited phenotypic character usually determined by which sex chromosomes are present. Humans and other mammals have an X-Y system in which sex is determined by whether a Y chromosome is present. Other systems are found in birds, fishes, and insects.
- The sex chromosomes carry **sex-linked genes** for some traits that are unrelated to sex characteristics. For instance, recessive alleles causing color blindness are **X-linked** (carried on the X chromosome). Fathers transmit this and other X-linked alleles to all daughters but to no sons. Any male who inherits such an allele from his mother will express the trait.
- In mammalian females, one of the two X chromosomes in each cell is randomly inactivated during early embryonic development,

becoming highly condensed into a **Barr body**. The descendant cells inherit the same inactivated X chromosome. If a female is heterozygous for a particular gene located on the X chromosome, she will be mosaic for that character, with about half her cells expressing the maternal allele and about half expressing the paternal allele.

? *Why are males affected much more often than females by X-linked disorders?*

CONCEPT 15.3

Linked genes tend to be inherited together because they are located near each other on the same chromosome (pp. 292–297)

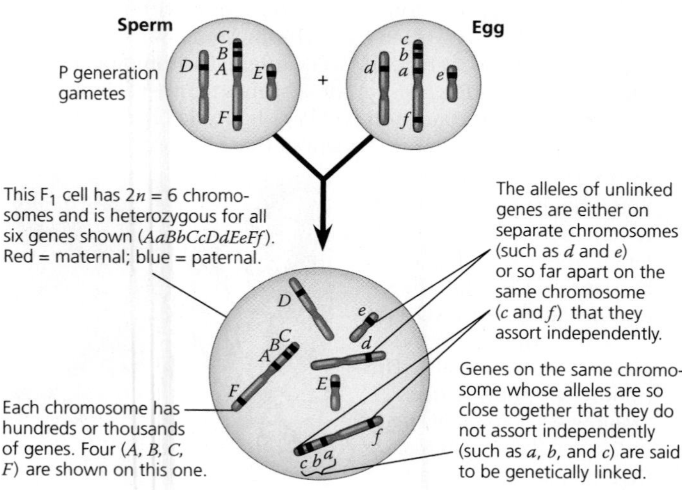

Sperm
Egg
P generation gametes
This F_1 cell has $2n = 6$ chromosomes and is heterozygous for all six genes shown (*AaBbCcDdEeFf*). Red = maternal; blue = paternal.
The alleles of unlinked genes are either on separate chromosomes (such as *d* and *e*) or so far apart on the same chromosome (*c* and *f*) that they assort independently.
Genes on the same chromosome whose alleles are so close together that they do not assort independently (such as *a*, *b*, and *c*) are said to be genetically linked.
Each chromosome has hundreds or thousands of genes. Four (*A, B, C, F*) are shown on this one.

- Among offspring from an F_1 testcross, **parental types** have the same combination of traits as those in the P generation parents. **Recombinant types (recombinants)** exhibit new combinations of traits not seen in either P generation parent. Because of the independent assortment of chromosomes, unlinked genes exhibit a 50% frequency of recombination in the gametes. For genetically **linked genes**, **crossing over** between nonsister chromatids during meiosis I accounts for the observed recombinants, always less than 50% of the total.
- The order of genes on a chromosome and the relative distances between them can be deduced from recombination frequencies observed in genetic crosses. These data allow construction of a **linkage map** (a type of **genetic map**). The farther apart genes are, the more likely their allele combinations will be recombined during crossing over.

? *Why are specific alleles of two genes that are farther apart more likely to show recombination than those of two closer genes?*

CONCEPT 15.4

Alterations of chromosome number or structure cause some genetic disorders (pp. 297–300)

- **Aneuploidy**, an abnormal chromosome number, can result from **nondisjunction** during meiosis. When a normal gamete unites with one containing two copies or no copies of a particular chromosome, the resulting zygote and its descendant cells either have one extra copy of that chromosome (**trisomy**, $2n + 1$) or are missing a copy (**monosomy**, $2n - 1$). **Polyploidy** (more than two complete sets of chromosomes) can result from complete nondisjunction during gamete formation.
- Chromosome breakage can result in alterations of chromosome structure: **deletions**, **duplications**, **inversions**, and **translocations**. Translocations can be reciprocal or nonreciprocal.
- Changes in the number of chromosomes per cell or in the structure of individual chromosomes can affect the phenotype and, in some cases, lead to human disorders. Such alterations cause **Down syndrome** (usually due to trisomy of chromosome 21), certain cancers associated with chromosomal translocations, and various other human disorders.

? *Why are inversions and reciprocal translocations less likely to be lethal than are aneuploidy, duplications, deletions, and nonreciprocal translocations?*

CONCEPT 15.5

Some inheritance patterns are exceptions to standard Mendelian inheritance (pp. 300–302)

- In mammals, the phenotypic effects of a small number of particular genes depend on which allele is inherited from each parent, a phenomenon called **genomic imprinting**. Imprints are formed during gamete production, with the result that one allele (either maternal or paternal) is not expressed in offspring.
- The inheritance of traits controlled by the genes present in mitochondria and plastids depends solely on the maternal parent because the zygote's cytoplasm containing these organelles comes from the egg. Some diseases affecting the nervous and muscular systems are caused by defects in mitochondrial genes that prevent cells from making enough ATP.

? *Explain how genomic imprinting and inheritance of mitochondrial and chloroplast DNA are exceptions to standard Mendelian inheritance.*

TEST YOUR UNDERSTANDING

LEVEL 1: KNOWLEDGE/COMPREHENSION

1. A man with hemophilia (a recessive, sex-linked condition) has a daughter of normal phenotype. She marries a man who is normal for the trait. What is the probability that a daughter of this mating will be a hemophiliac? That a son will be a hemophiliac? If the couple has four sons, what is the probability that all four will be born with hemophilia?
2. Pseudohypertrophic muscular dystrophy is an inherited disorder that causes gradual deterioration of the muscles. It is seen almost exclusively in boys born to apparently normal parents and usually results in death in the early teens. Is this disorder caused by a dominant or a recessive allele? Is its inheritance sex-linked or autosomal? How do you know? Explain why this disorder is almost never seen in girls.
3. A wild-type fruit fly (heterozygous for gray body color and normal wings) is mated with a black fly with vestigial wings. The offspring have the following phenotypic distribution: wild-type, 778; black-vestigial, 785; black-normal, 158; gray-vestigial, 162. What is the recombination frequency between these genes for body color and wing size?
4. What pattern of inheritance would lead a geneticist to suspect that an inherited disorder of cell metabolism is due to a defective mitochondrial gene?
5. A space probe discovers a planet inhabited by creatures that reproduce with the same hereditary patterns seen in humans. Three phenotypic characters are height (T = tall, t = dwarf), head appendages (A = antennae, a = no antennae), and nose morphology (S = upturned snout, s = downturned snout). Since the creatures are not "intelligent," Earth scientists are able to do some controlled breeding experiments using various heterozygotes in testcrosses. For tall heterozygotes with antennae, the offspring are

tall-antennae, 46; dwarf-antennae, 7; dwarf-no antennae, 42; tall-no antennae, 5. For heterozygotes with antennae and an upturned snout, the offspring are antennae-upturned snout, 47; antennae-downturned snout, 2; no antennae-downturned snout, 48; no antennae-upturned snout, 3. Calculate the recombination frequencies for both experiments.

LEVEL 2: APPLICATION/ANALYSIS

6. Using the information from problem 5, scientists do a further testcross using a heterozygote for height and nose morphology. The offspring are: tall-upturned snout, 40; dwarf-upturned snout, 9; dwarf-downturned snout, 42; tall-downturned snout, 9. Calculate the recombination frequency from these data; then use your answer from problem 5 to determine the correct sequence of the three linked genes.
7. Red-green color blindness is caused by a sex-linked recessive allele. A color-blind man marries a woman with normal vision whose father was color-blind. What is the probability that they will have a color-blind daughter? What is the probability that their first son will be color-blind? (Note the different wording in the two questions.)
8. A wild-type fruit fly (heterozygous for gray body color and red eyes) is mated with a black fruit fly with purple eyes. The offspring are wild-type, 721; black-purple, 751; gray-purple, 49; black-red, 45. What is the recombination frequency between these genes for body color and eye color? Using information from problem 3, what fruit flies (genotypes and phenotypes) would you mate to determine the sequence of the body-color, wing-size, and eye-color genes on the chromosome?
9. **DRAW IT** A fruit fly that is true-breeding for gray body with vestigial wings ($b^+ b^+ vg\ vg$) is mated with one that is true-breeding for black body with normal wings ($b\ b\ vg^+ vg^+$).
   (a) Draw the chromosomes for the P generation flies, using red for the gray fly and pink for the black one. Show the position of each allele.
   (b) Draw the chromosomes and label the alleles of an F_1 fly.
   (c) Suppose an F_1 female is testcrossed. Draw the chromosomes of the resulting offspring in a Punnett square.
   (d) Knowing that the distance between these two genes is 17 map units, predict the phenotypic ratios of these offspring.
10. Women born with an extra X chromosome (XXX) are generally healthy and indistinguishable in appearance from normal XX women. What is a likely explanation for this finding? How could you test this explanation?
11. Determine the sequence of genes along a chromosome based on the following recombination frequencies: *A–B*, 8 map units; *A–C*, 28 map units; *A–D*, 25 map units; *B–C*, 20 map units; *B–D*, 33 map units.
12. Assume that genes *A* and *B* are on the same chromosome and are 50 map units apart. An animal heterozygous at both loci is crossed with one that is homozygous recessive at both loci. What percentage of the offspring will show recombinant phenotypes resulting from crossovers? Without knowing these genes are on the same chromosome, how would you interpret the results of this cross?
13. Two genes of a flower, one controlling blue (*B*) versus white (*b*) petals and the other controlling round (*R*) versus oval (*r*) stamens, are linked and are 10 map units apart. You cross a homozygous blue-oval plant with a homozygous white-round plant. The resulting F_1 progeny are crossed with homozygous white-oval plants, and 1,000 F_2 progeny are obtained. How many F_2 plants of each of the four phenotypes do you expect?
14. You design *Drosophila* crosses to provide recombination data for gene *a*, which is located on the chromosome shown in Figure 15.12. Gene *a* has recombination frequencies of 14% with the vestigial-wing locus and 26% with the brown-eye locus. Approximately where is *a* located along the chromosome?

LEVEL 3: SYNTHESIS/EVALUATION

15. Banana plants, which are triploid, are seedless and therefore sterile. Propose a possible explanation.
16. **EVOLUTION CONNECTION**
    You have seen that crossing over, or recombination, is thought to be evolutionarily advantageous because it continually shuffles genetic alleles into novel combinations, allowing evolutionary processes to occur. Until recently, it was thought that the genes on the Y chromosome might degenerate because they lack homologous genes on the X chromosome with which to recombine. However, when the Y chromosome was sequenced, eight large regions were found to be internally homologous to each other, and quite a few of the 78 genes represent duplicates. (Y chromosome researcher David Page has called it a "hall of mirrors.") What might be a benefit of these regions?
17. **SCIENTIFIC INQUIRY**

    Butterflies have an X-Y sex determination system that is different from that of flies or humans. Female butterflies may be either XY or XO, while butterflies with two or more X chromosomes are males. This photograph shows a tiger swallowtail *gynandromorph*, an individual that is half male (left side) and half female (right side). Given that the first division of the zygote divides the embryo into the future right and left halves of the butterfly, propose a hypothesis that explains how nondisjunction during the first mitosis might have produced this unusual-looking butterfly.
18. **WRITE ABOUT A THEME**
    The Genetic Basis of Life The continuity of life is based on heritable information in the form of DNA. In a short essay (100–150 words), relate the structure and behavior of chromosomes to inheritance in both asexually and sexually reproducing species.

For selected answers, see Appendix A.

16

The Molecular Basis of Inheritance

▲ **Figure 16.1 How was the structure of DNA determined?**

KEY CONCEPTS

16.1 DNA is the genetic material

16.2 Many proteins work together in DNA replication and repair

16.3 A chromosome consists of a DNA molecule packed together with proteins

OVERVIEW

Life's Operating Instructions

In April 1953, James Watson and Francis Crick shook the scientific world with an elegant double-helical model for the structure of deoxyribonucleic acid, or DNA. **Figure 16.1** shows Watson (left) and Crick admiring their DNA model, which they built from tin and wire. Over the past 60 years or so, their model has evolved from a novel proposition to an icon of modern biology. Mendel's heritable factors and Morgan's genes on chromosomes are, in fact, composed of DNA. Chemically speaking, your genetic endowment is the DNA you inherited from your parents. DNA, the substance of inheritance, is the most celebrated molecule of our time.

Of all nature's molecules, nucleic acids are unique in their ability to direct their own replication from monomers. Indeed, the resemblance of offspring to their parents has its basis in the precise replication of DNA and its transmission from one generation to the next. Hereditary information is encoded in the chemical language of DNA and reproduced in all the cells of your body. It is this DNA program that directs the development of your biochemical, anatomical, physiological, and, to some extent, behavioral traits. In this chapter, you will discover how biologists deduced that DNA is the genetic material and how Watson and Crick worked out its structure. You will also learn about **DNA replication**, the process by which a DNA molecule is copied, and how cells repair their DNA. Finally, you will explore how a molecule of DNA is packaged together with proteins in a chromosome.

CONCEPT 16.1

DNA is the genetic material

Today, even schoolchildren have heard of DNA, and scientists routinely manipulate DNA in the laboratory, often to change the heritable traits of cells in their experiments. Early in the 20th century, however, identifying the molecules of inheritance loomed as a major challenge to biologists.

The Search for the Genetic Material: *Scientific Inquiry*

Once T. H. Morgan's group showed that genes exist as parts of chromosomes (described in Chapter 15), the two chemical components of chromosomes—DNA and protein—became the candidates for the genetic material. Until the 1940s, the case for proteins seemed stronger, especially since biochemists had identified them as a class of macromolecules with great heterogeneity and specificity of function, essential requirements for the hereditary material. Moreover, little was known about nucleic acids, whose physical and chemical properties seemed far too uniform to account for the multitude of specific inherited traits exhibited by every organism. This view gradually changed as experiments with microorganisms yielded unexpected results. As with the work of Mendel and Morgan, a key factor in determining the identity of the genetic material was the choice of appropriate experimental organisms. The role of DNA in heredity was first worked out while studying bacteria and the viruses that infect them, which are far simpler than pea plants, fruit flies, or humans. In this section, we will trace the search for the genetic material in some detail as a case study in scientific inquiry.

Evidence That DNA Can Transform Bacteria

The discovery of the genetic role of DNA dates back to 1928. While attempting to develop a vaccine against pneumonia, a British medical officer named Frederick Griffith was studying *Streptococcus pneumoniae*, a bacterium that causes pneumonia in mammals. Griffith had two strains (varieties) of the bacterium, one pathogenic (disease-causing) and one nonpathogenic (harmless). He was surprised to find that when he killed the pathogenic bacteria with heat and then mixed the cell remains with living bacteria of the nonpathogenic strain, some of the living cells became pathogenic (**Figure 16.2**). Furthermore, this newly acquired trait of pathogenicity was inherited by all the descendants of the transformed bacteria. Clearly, some chemical component of the dead pathogenic cells caused this heritable change, although the identity of the substance was not known. Griffith called the phenomenon **transformation**, now defined as a change in genotype and phenotype due to the assimilation of external DNA by a cell. (This use of the word *transformation* should not be confused with the conversion of a normal animal cell to a cancerous one, discussed near the end of Concept 12.3)

Griffith's work set the stage for a 14-year effort by American bacteriologist Oswald Avery to identify the transforming substance. Avery focused on three main candidates: DNA, RNA (the other nucleic acid in cells), and protein. Avery broke open the heat-killed pathogenic bacteria and extracted the cellular contents. He treated each of three samples with an agent that inactivated one type of molecule, then tested the sample for its ability to transform live nonpathogenic bacteria. Only when DNA was allowed to remain active did transformation occur. In 1944, Avery and his colleagues Maclyn McCarty and Colin MacLeod announced that the transforming agent was DNA. Their discovery was greeted with interest but considerable skepticism, in part because of the lingering belief that proteins were better candidates for the genetic material. Moreover, many biologists were not convinced that the genes of bacteria would be similar in composition and function to those of more complex organisms. But the major reason for the continued doubt was that so little was known about DNA.

▼ Figure 16.2 INQUIRY

Can a genetic trait be transferred between different bacterial strains?

EXPERIMENT Frederick Griffith studied two strains of the bacterium *Streptococcus pneumoniae*. Bacteria of the S (smooth) strain can cause pneumonia in mice; they are pathogenic because they have an outer capsule that protects them from an animal's defense system. Bacteria of the R (rough) strain lack a capsule and are nonpathogenic. To test for the trait of pathogenicity, Griffith injected mice with the two strains:

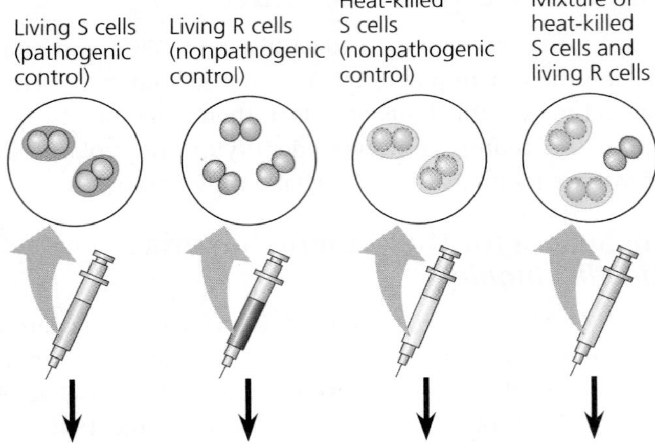

RESULTS

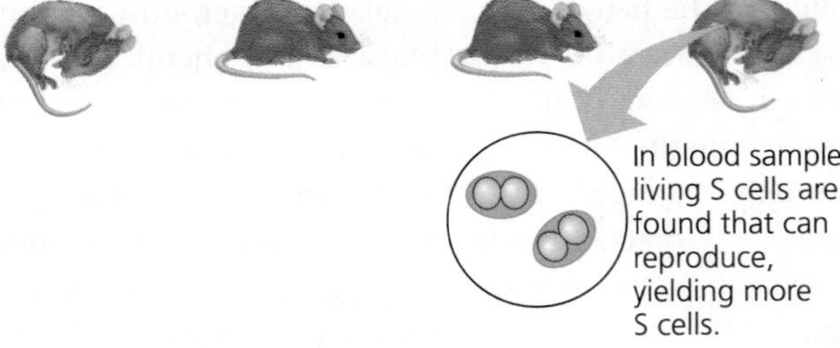

CONCLUSION Griffith concluded that the living R bacteria had been transformed into pathogenic S bacteria by an unknown, heritable substance from the dead S cells that allowed the R cells to make capsules.

SOURCE F. Griffith, The significance of pneumococcal types, *Journal of Hygiene* 27:113–159 (1928).

WHAT IF? How did this experiment rule out the possibility that the R cells could have simply used the capsules of the dead S cells to become pathogenic?

Evidence That Viral DNA Can Program Cells

Additional evidence for DNA as the genetic material came from studies of viruses that infect bacteria (**Figure 16.3**). These viruses are called **bacteriophages** (meaning "bacteria-eaters"), or **phages** for short. Viruses are much simpler than

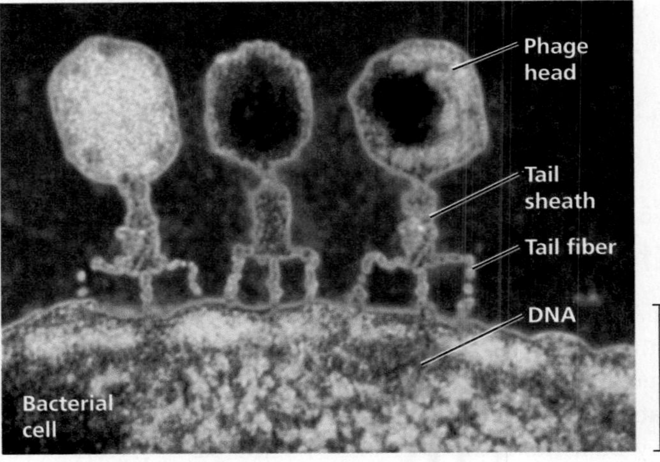

▲ **Figure 16.3 Viruses infecting a bacterial cell.** Phages called T2 attach to the host cell and inject their genetic material through the plasma membrane while the head and tail parts remain on the outer bacterial surface (colorized TEM).

cells. A **virus** is little more than DNA (or sometimes RNA) enclosed by a protective coat, which is often simply protein. To produce more viruses, a virus must infect a cell and take over the cell's metabolic machinery.

Phages have been widely used as tools by researchers in molecular genetics. In 1952, Alfred Hershey and Martha Chase performed experiments showing that DNA is the genetic material of a phage known as T2. This is one of many phages that infect *Escherichia coli* (*E. coli*), a bacterium that normally lives in the intestines of mammals and is a model organism for molecular biologists. At that time, biologists already knew that T2, like many other phages, was composed almost entirely of DNA and protein. They also knew that the T2 phage could quickly turn an *E. coli* cell into a T2-producing factory that released many copies when the cell ruptured. Somehow, T2 could reprogram its host cell to produce viruses. But which viral component—protein or DNA—was responsible?

Hershey and Chase answered this question by devising an experiment showing that only one of the two components of T2 actually enters the *E. coli* cell during infection (**Figure 16.4**). In their experiment, they used a radioactive isotope of sulfur to tag protein in one batch of T2 and a radioactive isotope of

▼ Figure 16.4 **INQUIRY**

Is protein or DNA the genetic material of phage T2?

EXPERIMENT Alfred Hershey and Martha Chase used radioactive sulfur and phosphorus to trace the fates of protein and DNA, respectively, of T2 phages that infected bacterial cells. They wanted to see which of these molecules entered the cells and could reprogram them to make more phages.

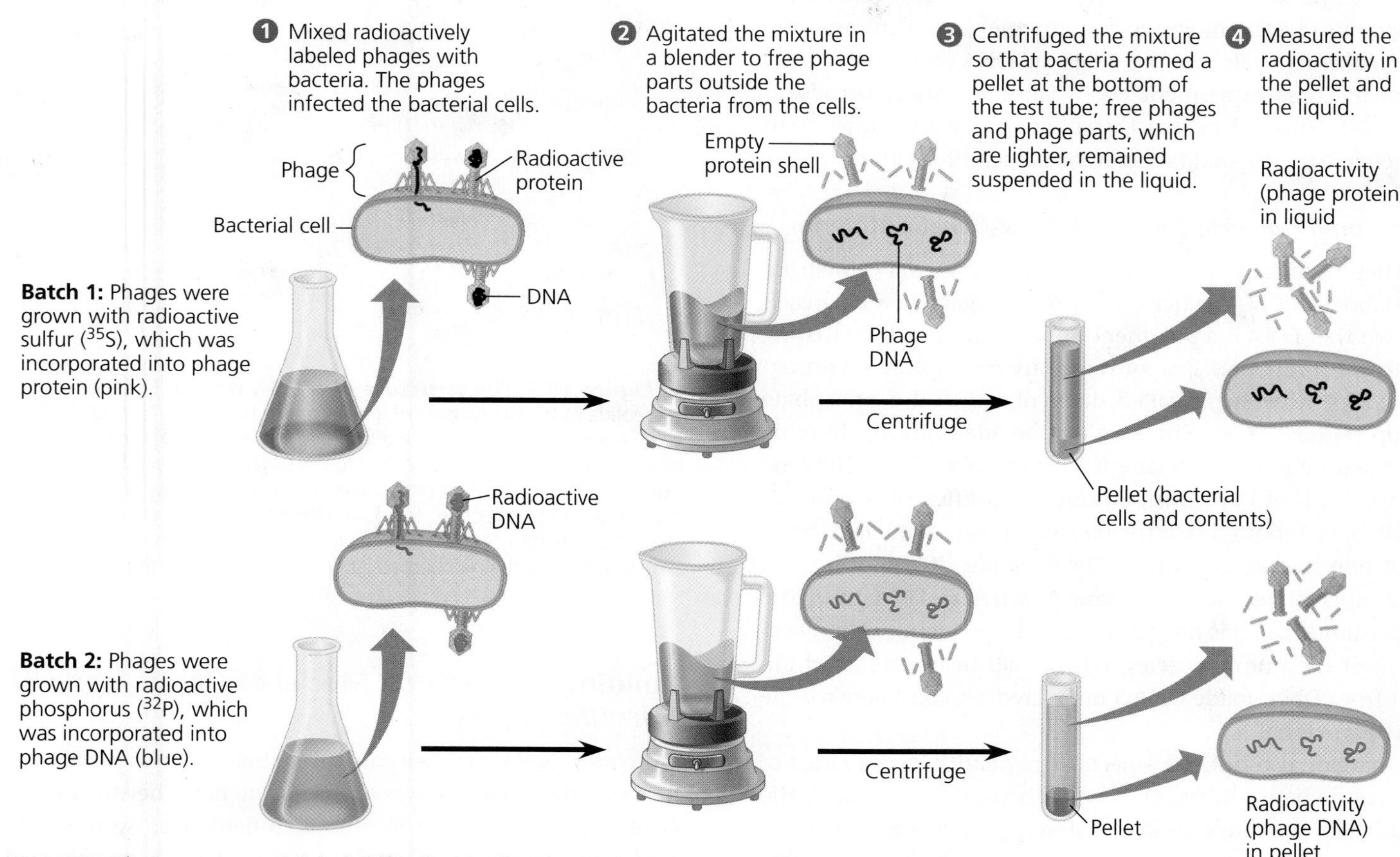

RESULTS When proteins were labeled (batch 1), radioactivity remained outside the cells; but when DNA was labeled (batch 2), radioactivity was found inside the cells. Bacterial cells with radioactive phage DNA released new phages with some radioactive phosphorus.

CONCLUSION Phage DNA entered bacterial cells, but phage proteins did not. Hershey and Chase concluded that DNA, not protein, functions as the genetic material of phage T2.

SOURCE A. D. Hershey and M. Chase, Independent functions of viral protein and nucleic acid in growth of bacteriophage, *Journal of General Physiology* 36:39–56 (1952).

WHAT IF? How would the results have differed if proteins carried the genetic information?

phosphorus to tag DNA in a second batch. Because protein, but not DNA, contains sulfur, radioactive sulfur atoms were incorporated only into the protein of the phage. In a similar way, the atoms of radioactive phosphorus labeled only the DNA, not the protein, because nearly all the phage's phosphorus is in its DNA. In the experiment, separate samples of nonradioactive *E. coli* cells were allowed to be infected by the protein-labeled and DNA-labeled batches of T2. The researchers then tested the two samples shortly after the onset of infection to see which type of molecule—protein or DNA—had entered the bacterial cells and would therefore be capable of reprogramming them.

Hershey and Chase found that the phage DNA entered the host cells but the phage protein did not. Moreover, when these bacteria were returned to a culture medium, the infection ran its course, and the *E. coli* released phages that contained some radioactive phosphorus, further showing that the DNA inside the cell played an ongoing role during the infection process.

Hershey and Chase concluded that the DNA injected by the phage must be the molecule carrying the genetic information that makes the cells produce new viral DNA and proteins. The Hershey-Chase experiment was a landmark study because it provided powerful evidence that nucleic acids, rather than proteins, are the hereditary material, at least for viruses.

Additional Evidence That DNA Is the Genetic Material

Further evidence that DNA is the genetic material came from the laboratory of biochemist Erwin Chargaff. It was already known that DNA is a polymer of nucleotides, each consisting of three components: a nitrogenous (nitrogen-containing) base, a pentose sugar called deoxyribose, and a phosphate group (**Figure 16.5**). The base can be adenine (A), thymine (T), guanine (G), or cytosine (C). Chargaff analyzed the base composition of DNA from a number of different organisms. In 1950, he reported that the base composition of DNA varies from one species to another. For example, 30.3% of human DNA nucleotides have the base A, whereas DNA from the bacterium *E. coli* has only 26.0% A. This evidence of molecular diversity among species, which had been presumed absent from DNA, made DNA a more credible candidate for the genetic material.

Chargaff also noticed a peculiar regularity in the ratios of nucleotide bases. In the DNA of each species he studied, the number of adenines approximately equaled the number of thymines, and the number of guanines approximately equaled the number of cytosines. In human DNA, for example, the four bases are present in these percentages: A = 30.3% and T = 30.3%; G = 19.5% and C = 19.9%.

These two findings became known as *Chargaff's rules*: (1) the base composition varies between species, and (2) within a species, the number of A and T bases are equal and the number of G and C bases are equal. The basis for these rules remained unexplained until the discovery of the double helix.

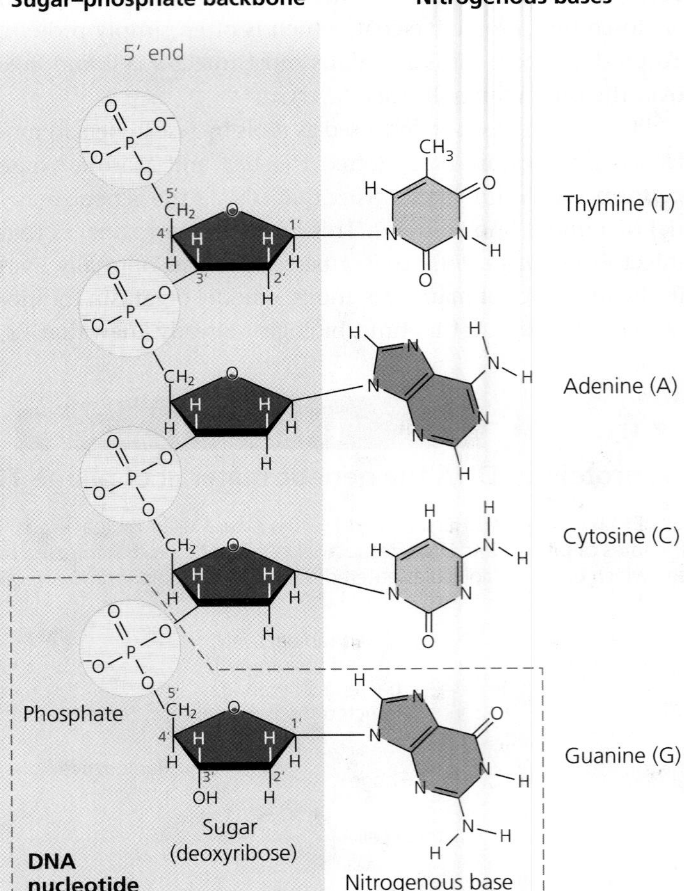

▲ **Figure 16.5 The structure of a DNA strand.** Each DNA nucleotide monomer consists of a nitrogenous base (T, A, C, or G), the sugar deoxyribose (blue), and a phosphate group (yellow). The phosphate group of one nucleotide is attached to the sugar of the next, forming a "backbone" of alternating phosphates and sugars from which the bases project. The polynucleotide strand has directionality, from the 5′ end (with the phosphate group) to the 3′ end (with the —OH group of the sugar). 5′ and 3′ refer to the numbers assigned to the carbons in the sugar ring.

Building a Structural Model of DNA: *Scientific Inquiry*

Once most biologists were convinced that DNA was the genetic material, the challenge was to determine how the structure of DNA could account for its role in inheritance. By the early 1950s, the arrangement of covalent bonds in a nucleic acid polymer was well established (see Figure 16.5), and researchers focused on discovering the three-dimensional structure of DNA. Among the scientists working on the problem were Linus Pauling, at the California Institute of Technology, and Maurice Wilkins and Rosalind Franklin, at King's College in London. First to come up with the correct answer, however, were two scientists who were relatively unknown at the time—the American James Watson and the Englishman Francis Crick.

The brief but celebrated partnership that solved the puzzle of DNA structure began soon after Watson journeyed to Cambridge University, where Crick was studying protein structure with a technique called X-ray crystallography (see Figure 5.24). While visiting the laboratory of Maurice Wilkins, Watson saw an X-ray diffraction image of DNA produced by Wilkins's accomplished colleague Rosalind Franklin (**Figure 16.6a**). Images produced by X-ray crystallography are not actually pictures of molecules. The spots and smudges in **Figure 16.6b** were produced by X-rays that were diffracted (deflected) as they passed through aligned fibers of purified DNA. Watson was familiar with the type of X-ray diffraction pattern that helical molecules produce, and an examination of the photo that Wilkins showed him confirmed that DNA was helical in shape. It also augmented earlier data obtained by Franklin and others suggesting the width of the helix and the spacing of the nitrogenous bases along it. The pattern in this photo implied that the helix was made up of two strands, contrary to a three-stranded model that Linus Pauling had proposed a short time earlier. The presence of two strands accounts for the now-familiar term **double helix** (**Figure 16.7**).

Watson and Crick began building models of a double helix that would conform to the X-ray measurements and what was then known about the chemistry of DNA, including Chargaff's rule of base equivalences. Having also read an unpublished annual report summarizing Franklin's work, they knew she had concluded that the sugar-phosphate backbones were on the outside of the DNA molecule, contrary to their working model. Franklin's arrangement was appealing because it put the

(a) Rosalind Franklin

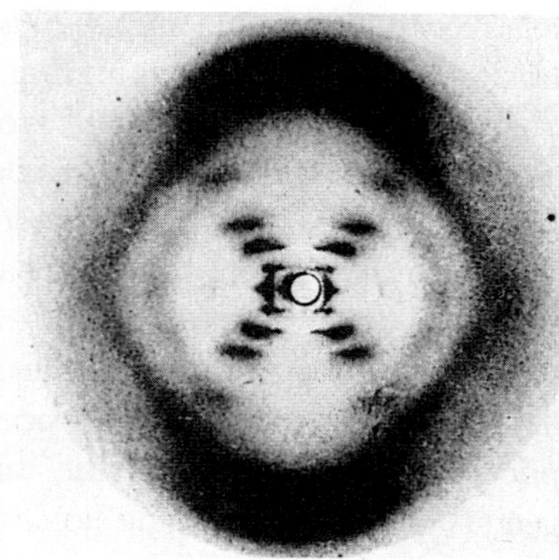

(b) Franklin's X-ray diffraction photograph of DNA

▲ **Figure 16.6 Rosalind Franklin and her X-ray diffraction photo of DNA.** Franklin, a very accomplished X-ray crystallographer, conducted critical experiments resulting in the photograph that allowed Watson and Crick to deduce the double-helical structure of DNA.

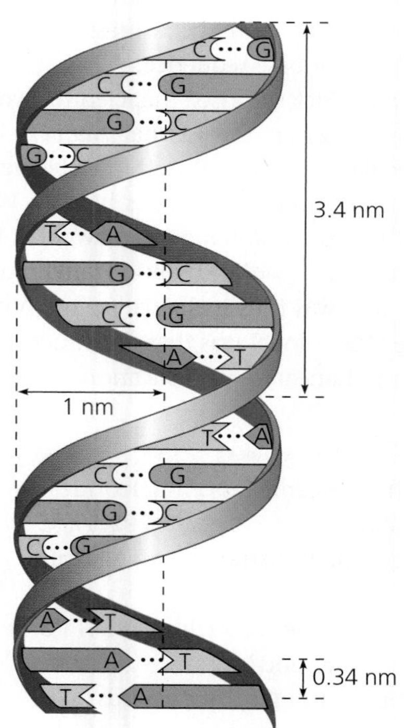

(a) Key features of DNA structure. The "ribbons" in this diagram represent the sugar-phosphate backbones of the two DNA strands. The helix is "right-handed," curving up to the right. The two strands are held together by hydrogen bonds (dotted lines) between the nitrogenous bases, which are paired in the interior of the double helix.

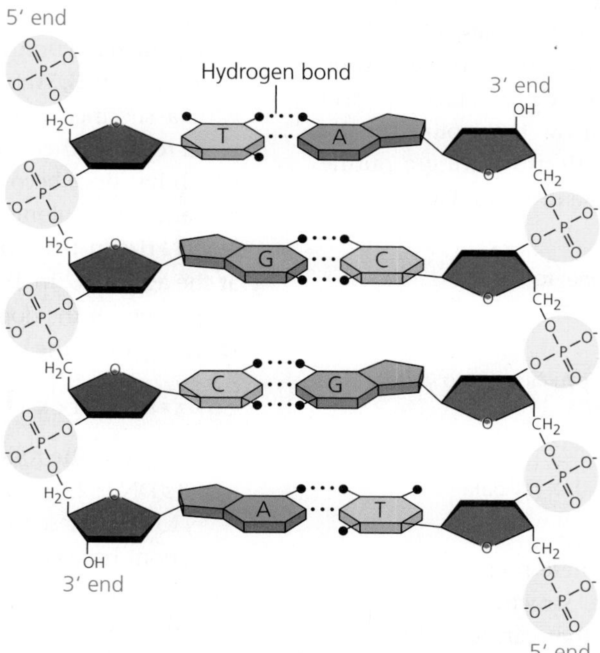

(b) Partial chemical structure. For clarity, the two DNA strands are shown untwisted in this partial chemical structure. Strong covalent bonds link the units of each strand, while weaker hydrogen bonds hold one strand to the other. Notice that the strands are antiparallel, meaning that they are oriented in opposite directions.

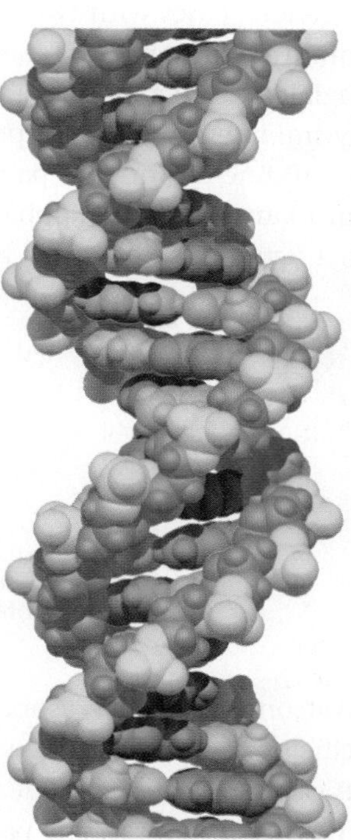

(c) Space-filling model. The tight stacking of the base pairs is clear in this computer model. Van der Waals interactions between the stacked pairs play a major role in holding the molecule together (see Chapter 2).

▲ **Figure 16.7 The double helix.**

relatively hydrophobic nitrogenous bases in the molecule's interior, away from the surrounding aqueous solution, and the negatively charged phosphate groups wouldn't be forced together in the interior. Watson constructed a model with the nitrogenous bases facing the interior of the double helix. In this model, the two sugar-phosphate backbones are **antiparallel**—that is, their subunits run in opposite directions (see Figure 16.7). You can imagine the overall arrangement as a rope ladder with rigid rungs. The side ropes represent the sugar-phosphate backbones, and the rungs represent pairs of nitrogenous bases. Now imagine holding one end of the ladder and twisting the other end, forming a spiral. Franklin's X-ray data indicated that the helix makes one full turn every 3.4 nm along its length. With the bases stacked just 0.34 nm apart, there are ten layers of base pairs, or rungs of the ladder, in each full turn of the helix.

The nitrogenous bases of the double helix are paired in specific combinations: adenine (A) with thymine (T), and guanine (G) with cytosine (C). It was mainly by trial and error that Watson and Crick arrived at this key feature of DNA. At first, Watson imagined that the bases paired like with like—for example, A with A and C with C. But this model did not fit the X-ray data, which suggested that the double helix had a uniform diameter. Why is this requirement inconsistent with like-with-like pairing of bases? Adenine and guanine are purines, nitrogenous bases with two organic rings, while cytosine and thymine are nitrogenous bases called pyrimidines, which have a single ring. Thus, purines (A and G) are about twice as wide as pyrimidines (C and T). A purine-purine pair is too wide and a pyrimidine-pyrimidine pair too narrow to account for the 2-nm diameter of the double helix. Always pairing a purine with a pyrimidine, however, results in a uniform diameter:

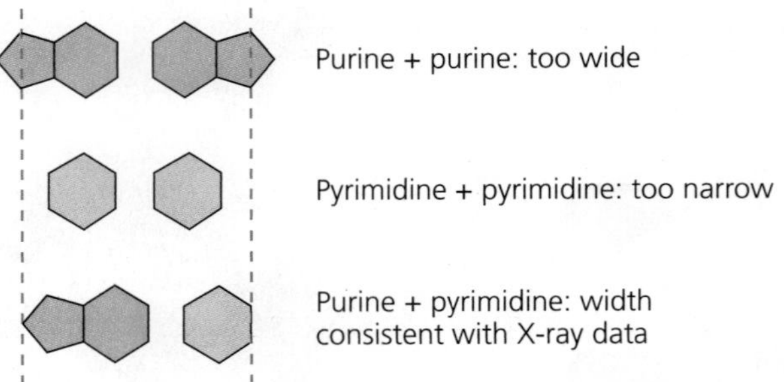

Watson and Crick reasoned that there must be additional specificity of pairing dictated by the structure of the bases. Each base has chemical side groups that can form hydrogen bonds with its appropriate partner: Adenine can form two hydrogen bonds with thymine and only thymine; guanine forms three hydrogen bonds with cytosine and only cytosine. In shorthand, A pairs with T, and G pairs with C (**Figure 16.8**).

The Watson-Crick model took into account Chargaff's ratios and ultimately explained them. Wherever one strand of a DNA molecule has an A, the partner strand has a T. And a G in one strand is always paired with a C in the complementary strand. Therefore, in the DNA of any organism, the amount of

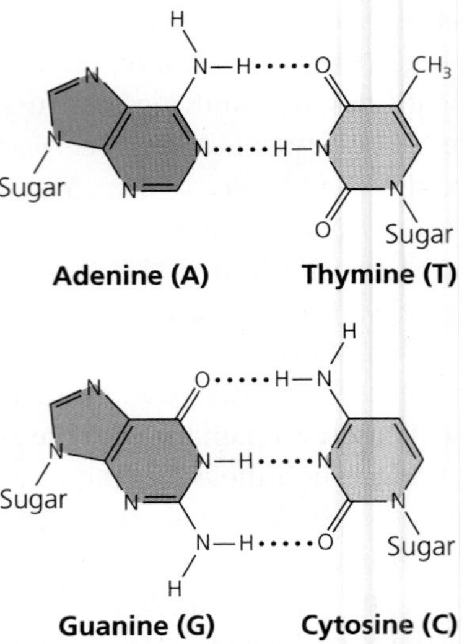

▲ **Figure 16.8 Base pairing in DNA.** The pairs of nitrogenous bases in a DNA double helix are held together by hydrogen bonds, shown here as black dotted lines.

adenine equals the amount of thymine, and the amount of guanine equals the amount of cytosine. Although the base-pairing rules dictate the combinations of nitrogenous bases that form the "rungs" of the double helix, they do not restrict the sequence of nucleotides *along* each DNA strand. The linear sequence of the four bases can be varied in countless ways, and each gene has a unique order, or base sequence.

In April 1953, Watson and Crick surprised the scientific world with a succinct, one-page paper in the journal *Nature*.* The paper reported their molecular model for DNA: the double helix, which has since become the symbol of molecular biology. Watson and Crick, along with Maurice Wilkins, were awarded the Nobel Prize in 1962 for this work. (Sadly, Rosalind Franklin died at the age of 38, in 1958, and was thus ineligible for the prize.) The beauty of the double helix model was that the structure of DNA suggested the basic mechanism of its replication.

CONCEPT CHECK 16.1

1. A fly has the following percentages of nucleotides in its DNA: 27.3% A, 27.6% T, 22.5% G, and 22.5% C. How do these numbers demonstrate Chargaff's rule about base ratios?
2. Given a polynucleotide sequence such as GAATTC, can you tell which is the 5′ end? If not, what further information do you need to identify the ends? (See Figure 16.5.)
3. **WHAT IF?** Griffith did not expect transformation to occur in his experiment. What results was he expecting? Explain.

For suggested answers, see Appendix A.

*J. D. Watson and F. H. C. Crick, Molecular structure of nucleic acids: a structure for deoxyribose nucleic acids, *Nature* 171:737–738 (1953).

CONCEPT 16.2

Many proteins work together in DNA replication and repair

The relationship between structure and function is manifest in the double helix. The idea that there is specific pairing of nitrogenous bases in DNA was the flash of inspiration that led Watson and Crick to the double helix. At the same time, they saw the functional significance of the base-pairing rules. They ended their classic paper with this wry statement: "It has not escaped our notice that the specific pairing we have postulated immediately suggests a possible copying mechanism for the genetic material." In this section, you will learn about the basic principle of DNA replication, as well as some important details of the process.

The Basic Principle: Base Pairing to a Template Strand

In a second paper, Watson and Crick stated their hypothesis for how DNA replicates:

> Now our model for deoxyribonucleic acid is, in effect, a pair of templates, each of which is complementary to the other. We imagine that prior to duplication the hydrogen bonds are broken, and the two chains unwind and separate. Each chain then acts as a template for the formation onto itself of a new companion chain, so that eventually we shall have two pairs of chains, where we only had one before. Moreover, the sequence of the pairs of bases will have been duplicated exactly.*

Figure 16.9 illustrates Watson and Crick's basic idea. To make it easier to follow, we show only a short section of double helix in untwisted form. Notice that if you cover one of the two DNA strands of Figure 16.9a, you can still determine its linear sequence of nucleotides by referring to the uncovered strand and applying the base-pairing rules. The two strands are complementary; each stores the information necessary to reconstruct the other. When a cell copies a DNA molecule, each strand serves as a template for ordering nucleotides into a new, complementary strand. Nucleotides line up along the template strand according to the base-pairing rules and are linked to form the new strands. Where there was one double-stranded DNA molecule at the beginning of the process, there are soon two, each an exact replica of the "parental" molecule. The copying mechanism is analogous to using a photographic negative to make a positive image, which can in turn be used to make another negative, and so on.

This model of DNA replication remained untested for several years following publication of the DNA structure. The requisite experiments were simple in concept but difficult to perform. Watson and Crick's model predicts that when a double helix replicates, each of the two daughter molecules will have one old strand, from the parental molecule, and one newly made strand. This **semiconservative model** can be distinguished from a conservative model of replication, in which the two parental strands somehow come back together after the process (that is, the parental molecule is conserved). In yet a third model, called the dispersive model, all four strands of DNA following replication have a mixture of old and new DNA. These three models are shown in

*F. H. C. Crick and J. D. Watson, The complementary structure of deoxyribonucleic acid, *Proceedings of the Royal Society of London A* 223:80 (1954).

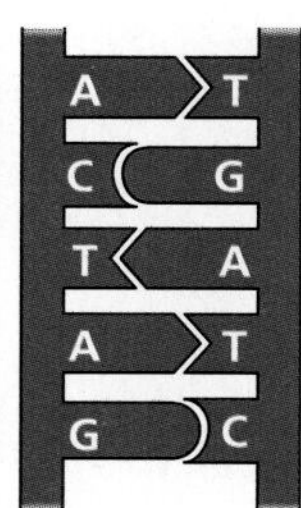

(a) The parental molecule has two complementary strands of DNA. Each base is paired by hydrogen bonding with its specific partner, A with T and G with C.

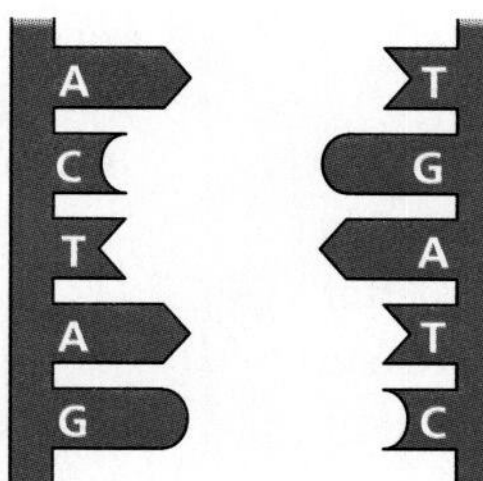

(b) The first step in replication is separation of the two DNA strands. Each parental strand can now serve as a template that determines the order of nucleotides along a new, complementary strand.

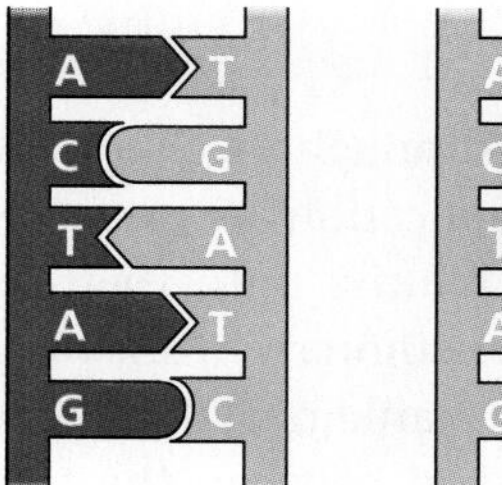

(c) The complementary nucleotides line up and are connected to form the sugar-phosphate backbones of the new strands. Each "daughter" DNA molecule consists of one parental strand (dark blue) and one new strand (light blue).

▲ **Figure 16.9 A model for DNA replication: the basic concept.** In this simplified illustration, a short segment of DNA has been untwisted into a structure that resembles a ladder. The side rails of the ladder are the sugar-phosphate backbones of the two DNA strands; the rungs are the pairs of nitrogenous bases. Simple shapes symbolize the four kinds of bases. Dark blue represents DNA strands present in the parental molecule; light blue represents newly synthesized DNA.

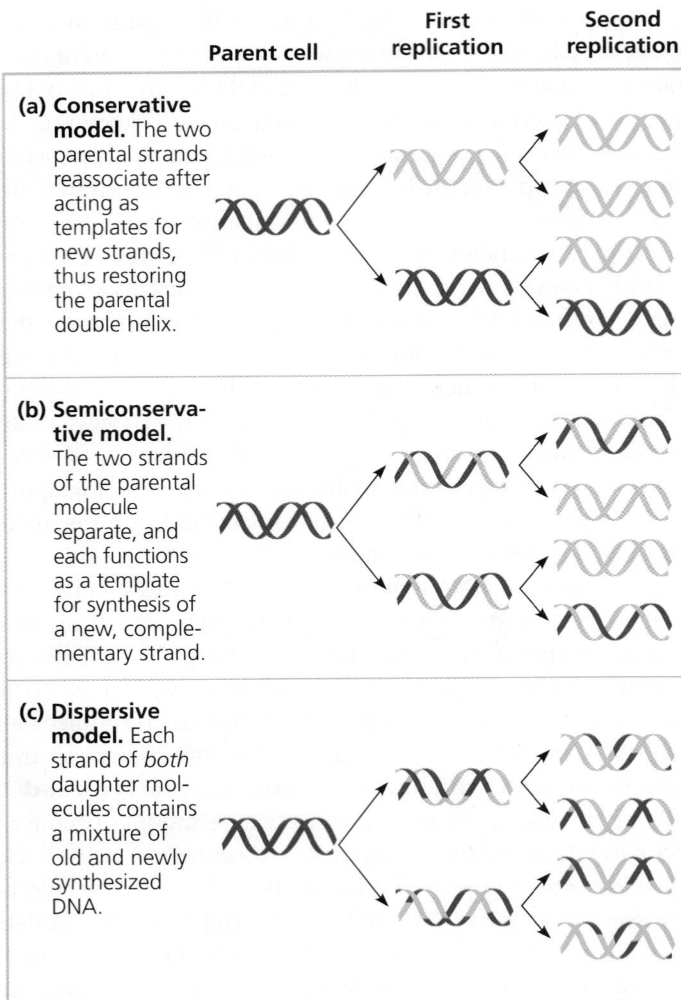

Parent cell
First replication
Second replication
(a) Conservative model. The two parental strands reassociate after acting as templates for new strands, thus restoring the parental double helix.
(b) Semiconservative model. The two strands of the parental molecule separate, and each functions as a template for synthesis of a new, complementary strand.
(c) Dispersive model. Each strand of *both* daughter molecules contains a mixture of old and newly synthesized DNA.

▲ **Figure 16.10 Three alternative models of DNA replication.** Each short segment of double helix symbolizes the DNA within a cell. Beginning with a parent cell, we follow the DNA for two more generations of cells—two rounds of DNA replication. Newly made DNA is light blue.

Figure 16.10. Although mechanisms for conservative or dispersive DNA replication are not easy to devise, these models remained possibilities until they could be ruled out. After two years of preliminary work in the late 1950s, Matthew Meselson and Franklin Stahl devised a clever experiment that distinguished between the three models, described in detail in **Figure 16.11**. Their experiment supported the semiconservative model of DNA replication, as predicted by Watson and Crick, and is widely acknowledged among biologists to be a classic example of elegant experimental design.

The basic principle of DNA replication is conceptually simple. However, the actual process involves some complicated biochemical gymnastics, as we will now see.

DNA Replication: *A Closer Look*

The bacterium *E. coli* has a single chromosome of about 4.6 million nucleotide pairs. In a favorable environment, an *E. coli* cell

---

▼ **Figure 16.11** INQUIRY

Does DNA replication follow the conservative, semiconservative, or dispersive model?

EXPERIMENT At the California Institute of Technology, Matthew Meselson and Franklin Stahl cultured *E. coli* for several generations in a medium containing nucleotide precursors labeled with a heavy isotope of nitrogen, ^{15}N. They then transferred the bacteria to a medium with only ^{14}N, a lighter isotope. A sample was taken after DNA replicated once; another sample was taken after DNA replicated again. They extracted DNA from the bacteria in the samples and then centrifuged each DNA sample to separate DNA of different densities.

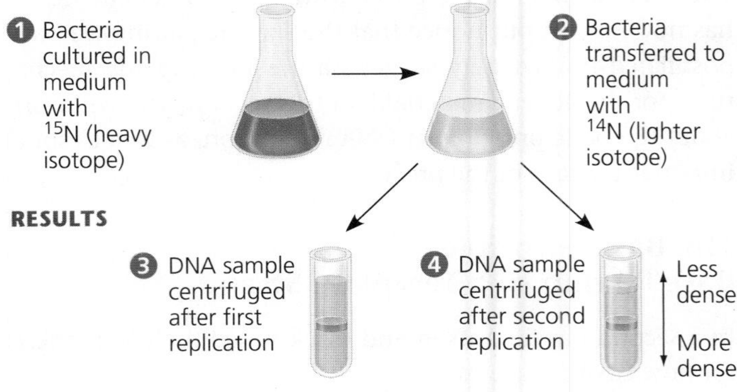

CONCLUSION Meselson and Stahl compared their results to those predicted by each of the three models in Figure 16.10, as shown below. The first replication in the ^{14}N medium produced a band of hybrid (^{15}N-^{14}N) DNA. This result eliminated the conservative model. The second replication produced both light and hybrid DNA, a result that refuted the dispersive model and supported the semiconservative model. They therefore concluded that DNA replication is semiconservative.

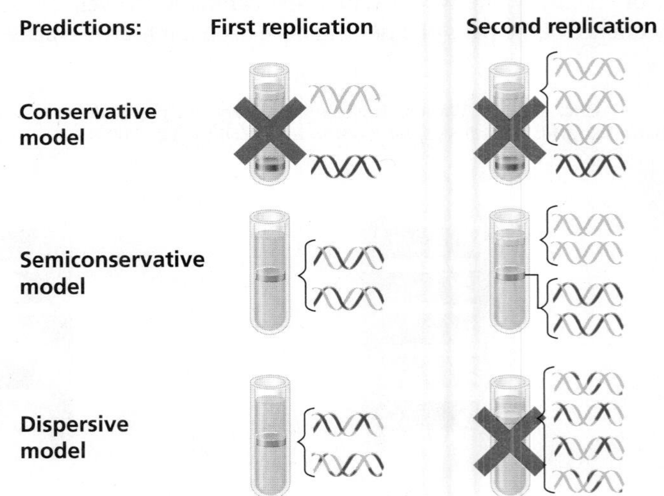

SOURCE M. Meselson and F. W. Stahl, The replication of DNA in *Escherichia coli, Proceedings of the National Academy of Sciences USA* 44:671–682 (1958).

INQUIRY IN ACTION Read and analyze the original paper in *Inquiry in Action: Interpreting Scientific Papers.*

MB See the related Experimental Inquiry Tutorial in MasteringBiology.

WHAT IF? If Meselson and Stahl had first grown the cells in ^{14}N-containing medium and then moved them into ^{15}N-containing medium before taking samples, what would have been the result?

can copy all this DNA and divide to form two genetically identical daughter cells in less than an hour. Each of *your* cells has 46 DNA molecules in its nucleus, one long double-helical molecule per chromosome. In all, that represents about 6 billion nucleotide pairs, or over a thousand times more DNA than is found in a bacterial cell. If we were to print the one-letter symbols for these bases (A, G, C, and T) the size of the type you are now reading, the 6 billion nucleotide pairs of information in a diploid human cell would fill about 1,200 books as thick as this text. Yet it takes one of your cells just a few hours to copy all of this DNA. This replication of an enormous amount of genetic information is achieved with very few errors—only about one per 10 billion nucleotides. The copying of DNA is remarkable in its speed and accuracy.

More than a dozen enzymes and other proteins participate in DNA replication. Much more is known about how this "replication machine" works in bacteria (such as *E. coli*) than in eukaryotes, and we will describe the basic steps of the process for *E. coli*, except where otherwise noted. What scientists have learned about eukaryotic DNA replication suggests, however, that most of the process is fundamentally similar for prokaryotes and eukaryotes.

Getting Started

The replication of a DNA molecule begins at particular sites called **origins of replication**, short stretches of DNA having a specific sequence of nucleotides. The *E. coli* chromosome, like many other bacterial chromosomes, is circular and has a single origin. Proteins that initiate DNA replication recognize this sequence and attach to the DNA, separating the two strands and opening up a replication "bubble." Replication of DNA then proceeds in both directions until the entire molecule is copied (**Figure 16.12a**). In contrast to a bacterial chromosome, a eukaryotic chromosome may have hundreds

▼ **Figure 16.12 Origins of replication in *E. coli* and eukaryotes.** The red arrows indicate the movement of the replication forks and thus the overall directions of DNA replication within each bubble.

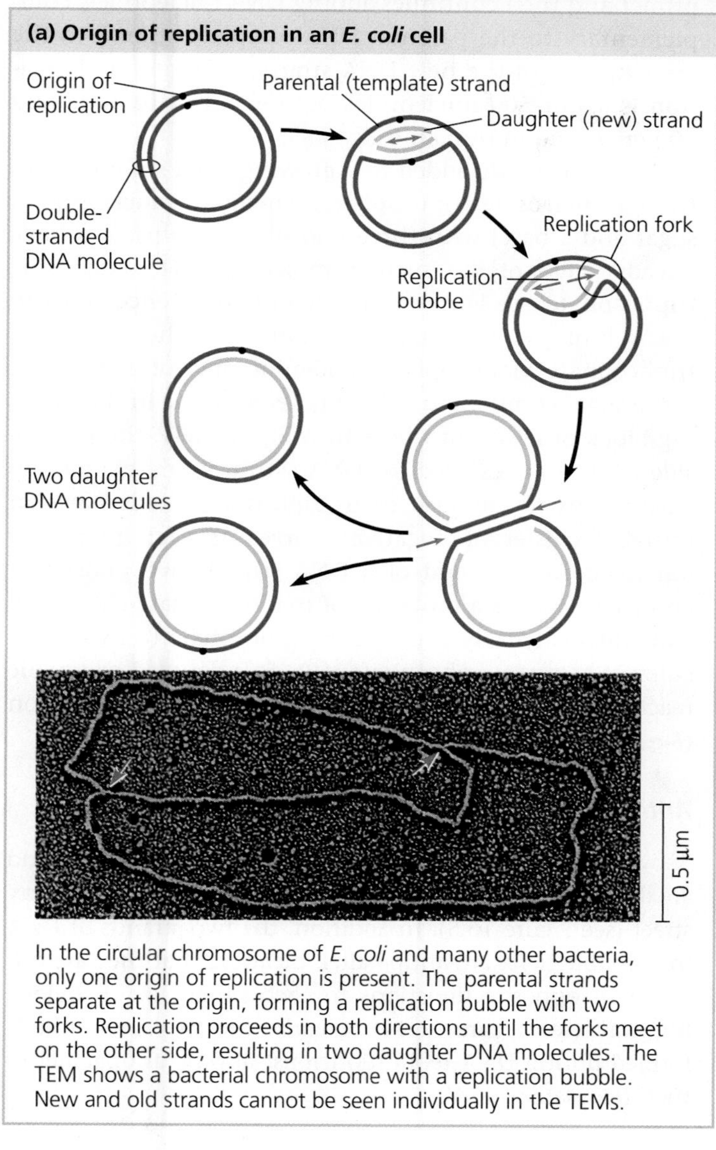

In the circular chromosome of *E. coli* and many other bacteria, only one origin of replication is present. The parental strands separate at the origin, forming a replication bubble with two forks. Replication proceeds in both directions until the forks meet on the other side, resulting in two daughter DNA molecules. The TEM shows a bacterial chromosome with a replication bubble. New and old strands cannot be seen individually in the TEMs.

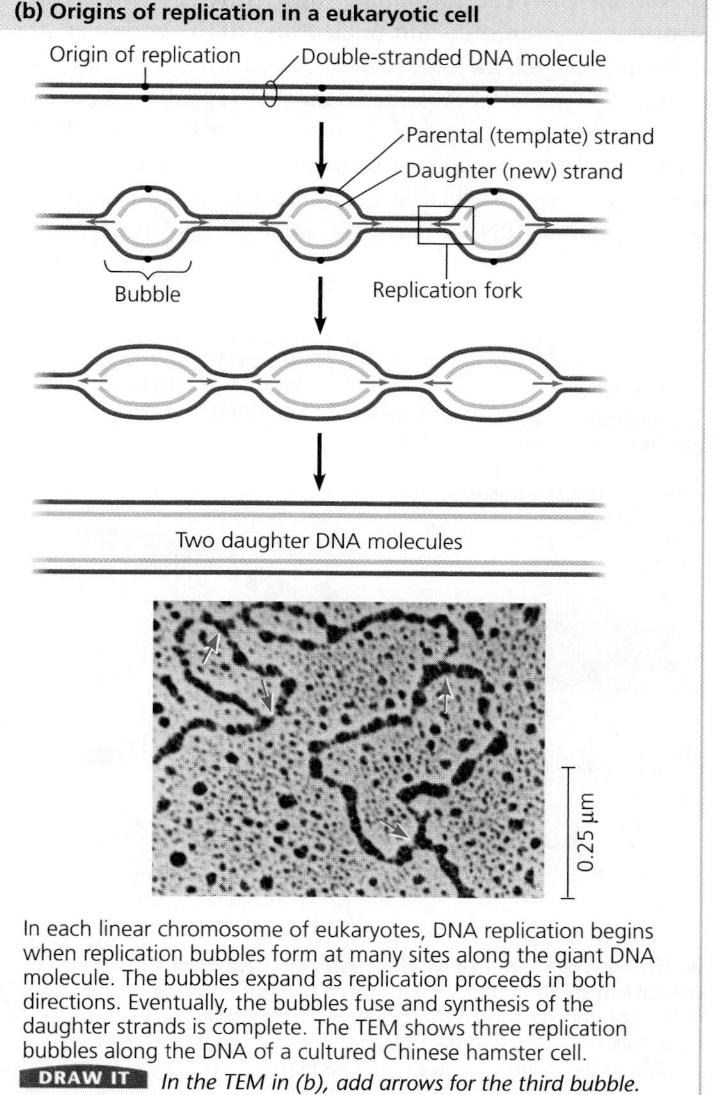

In each linear chromosome of eukaryotes, DNA replication begins when replication bubbles form at many sites along the giant DNA molecule. The bubbles expand as replication proceeds in both directions. Eventually, the bubbles fuse and synthesis of the daughter strands is complete. The TEM shows three replication bubbles along the DNA of a cultured Chinese hamster cell.

DRAW IT *In the TEM in (b), add arrows for the third bubble.*

or even a few thousand replication origins. Multiple replication bubbles form and eventually fuse, thus speeding up the copying of the very long DNA molecules (**Figure 16.12b**). As in bacteria, eukaryotic DNA replication proceeds in both directions from each origin.

At each end of a replication bubble is a **replication fork**, a Y-shaped region where the parental strands of DNA are being unwound. Several kinds of proteins participate in the unwinding (**Figure 16.13**). **Helicases** are enzymes that untwist the double helix at the replication forks, separating the two parental strands and making them available as template strands. After the parental strands separate, **single-strand binding proteins** bind to the unpaired DNA strands, keeping them from re-pairing. The untwisting of the double helix causes tighter twisting and strain ahead of the replication fork. **Topoisomerase** helps relieve this strain by breaking, swiveling, and rejoining DNA strands.

The unwound sections of parental DNA strands are now available to serve as templates for the synthesis of new complementary DNA strands. However, the enzymes that synthesize DNA cannot *initiate* the synthesis of a polynucleotide; they can only add nucleotides to the end of an already existing chain that is base-paired with the template strand. The initial nucleotide chain that is produced during DNA synthesis is actually a short stretch of RNA, not DNA. This RNA chain is called a **primer** and is synthesized by the enzyme **primase** (see Figure 16.13). Primase starts a complementary RNA chain from a single RNA nucleotide, adding RNA nucleotides one at a time, using the parental DNA strand as a template. The completed primer, generally 5–10 nucleotides long, is thus base-paired to the template strand. The new DNA strand will start from the 3′ end of the RNA primer.

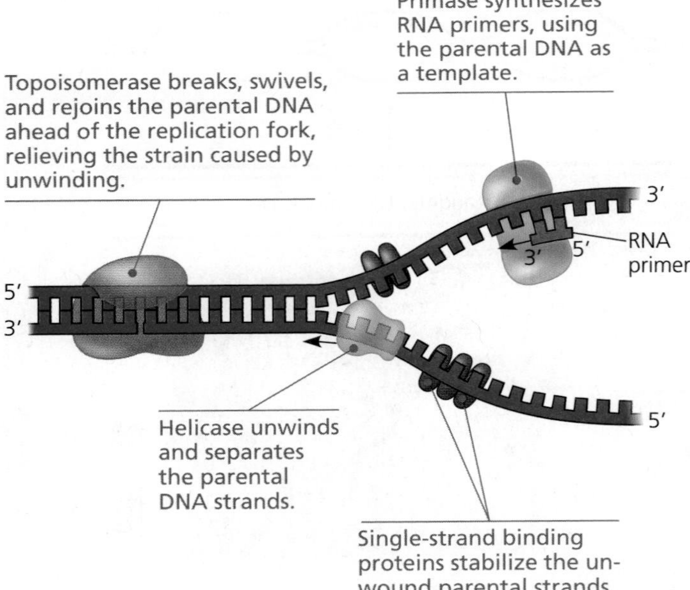
Primase synthesizes RNA primers, using the parental DNA as a template.
Topoisomerase breaks, swivels, and rejoins the parental DNA ahead of the replication fork, relieving the strain caused by unwinding.
3′
RNA primer
3′
5′
5′
3′
5′
Helicase unwinds and separates the parental DNA strands.
Single-strand binding proteins stabilize the unwound parental strands.

▲ **Figure 16.13 Some of the proteins involved in the initiation of DNA replication.** The same proteins function at both replication forks in a replication bubble. For simplicity, only the left-hand fork is shown, and the DNA bases are drawn much larger in relation to the proteins than they are in reality.

Synthesizing a New DNA Strand

Enzymes called **DNA polymerases** catalyze the synthesis of new DNA by adding nucleotides to a preexisting chain. In *E. coli*, there are several different DNA polymerases, but two appear to play the major roles in DNA replication: DNA polymerase III and DNA polymerase I. The situation in eukaryotes is more complicated, with at least 11 different DNA polymerases discovered so far; however, the general principles are the same.

Most DNA polymerases require a primer and a DNA template strand, along which complementary DNA nucleotides line up. In *E. coli*, DNA polymerase III (abbreviated DNA pol III) adds a DNA nucleotide to the RNA primer and then continues adding DNA nucleotides, complementary to the parental DNA template strand, to the growing end of the new DNA strand. The rate of elongation is about 500 nucleotides per second in bacteria and 50 per second in human cells.

Each nucleotide added to a growing DNA strand comes from a nucleoside triphosphate, which is a nucleoside (a sugar and a base) with three phosphate groups. You have already encountered such a molecule—ATP (adenosine triphosphate; see Figure 8.8). The only difference between the ATP of energy metabolism and dATP, the nucleoside triphosphate that supplies an adenine nucleotide to DNA, is the sugar component, which is deoxyribose in the building block of DNA but ribose in ATP. Like ATP, the nucleoside triphosphates used for DNA synthesis are chemically reactive, partly because their triphosphate tails have an unstable cluster of negative charge. As each monomer joins the growing end of a DNA strand, two phosphate groups are lost as a molecule of pyrophosphate (Ⓟ—Ⓟ$_i$). Subsequent hydrolysis of the pyrophosphate to two molecules of inorganic phosphate (Ⓟ$_i$) is a coupled exergonic reaction that helps drive the polymerization reaction (**Figure 16.14**).

Antiparallel Elongation

As we have noted previously, the two ends of a DNA strand are different, giving each strand directionality, like a one-way street (see Figure 16.5). In addition, the two strands of DNA in a double helix are antiparallel, meaning that they are oriented in opposite directions to each other, like a divided highway (see Figure 16.14). Clearly, the two new strands formed during DNA replication must also be antiparallel to their template strands.

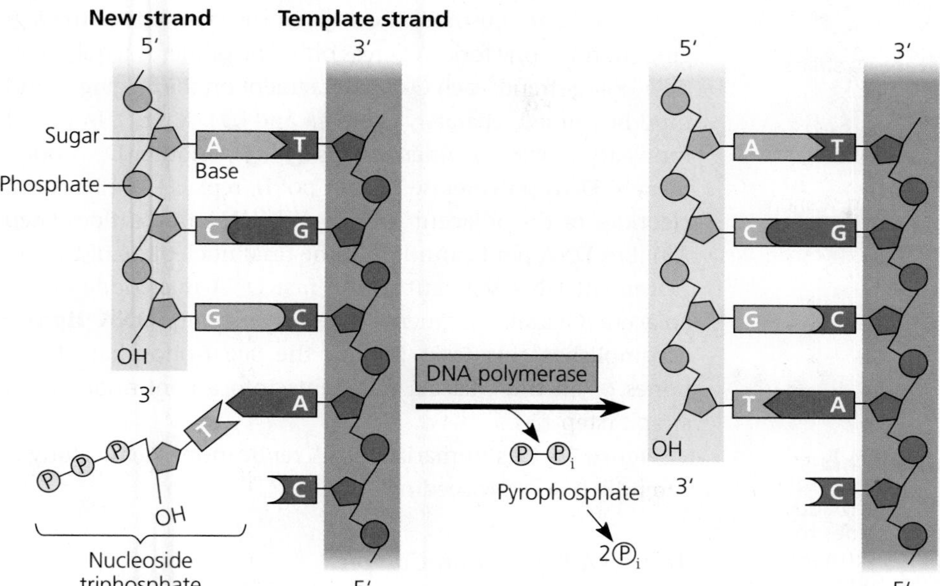

◄ **Figure 16.14 Incorporation of a nucleotide into a DNA strand.** DNA polymerase catalyzes the addition of a nucleoside triphosphate to the 3′ end of a growing DNA strand, with the release of two phosphates.

? *Use this diagram to explain what we mean when we say that each DNA strand has directionality.*

How does the antiparallel arrangement of the double helix affect replication? Because of their structure, DNA polymerases can add nucleotides only to the free 3′ end of a primer or growing DNA strand, never to the 5′ end (see Figure 16.14). Thus, a new DNA strand can elongate only in the 5′ → 3′ direction. With this in mind, let's examine one of the two replication forks in a bubble (**Figure 16.15**). Along one template strand, DNA polymerase III can synthesize a complementary strand continuously by elongating the new DNA in the mandatory 5′ → 3′ direction. DNA pol III remains in the replication fork on that template strand and continuously adds nucleotides to the new complementary strand as the fork progresses. The DNA strand made by this mechanism is called the **leading strand**. Only one primer is required for DNA pol III to synthesize the leading strand (see Figure 16.15).

To elongate the other new strand of DNA in the mandatory 5′ → 3′ direction, DNA pol III must work along the other template strand in the direction *away from* the replication fork. The DNA strand elongating in this direction is called the **lagging strand**.* In contrast to the leading strand, which elongates continuously, the lagging strand is synthesized discontinuously, as a series of segments. These segments of the lagging strand are called **Okazaki fragments**, after the Japanese scientist who discovered them. The fragments are about 1,000–2,000 nucleotides long in *E. coli* and 100–200 nucleotides long in eukaryotes.

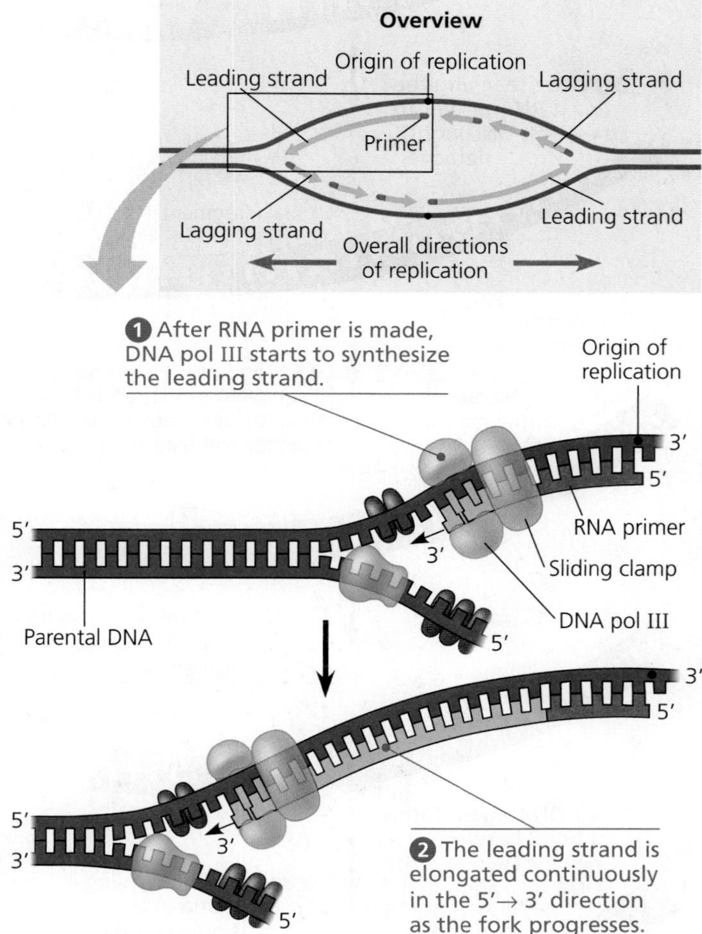

▲ **Figure 16.15 Synthesis of the leading strand during DNA replication.** This diagram focuses on the left replication fork shown in the overview box. DNA polymerase III (DNA pol III), shaped like a cupped hand, is shown closely associated with a protein called the "sliding clamp" that encircles the newly synthesized double helix like a doughnut. The sliding clamp moves DNA pol III along the DNA template strand.

*Synthesis of the leading strand and synthesis of the lagging strand occur concurrently and at the same rate. The lagging strand is so named because its synthesis is delayed slightly relative to synthesis of the leading strand; each new fragment of the lagging strand cannot be started until enough template has been exposed at the replication fork.

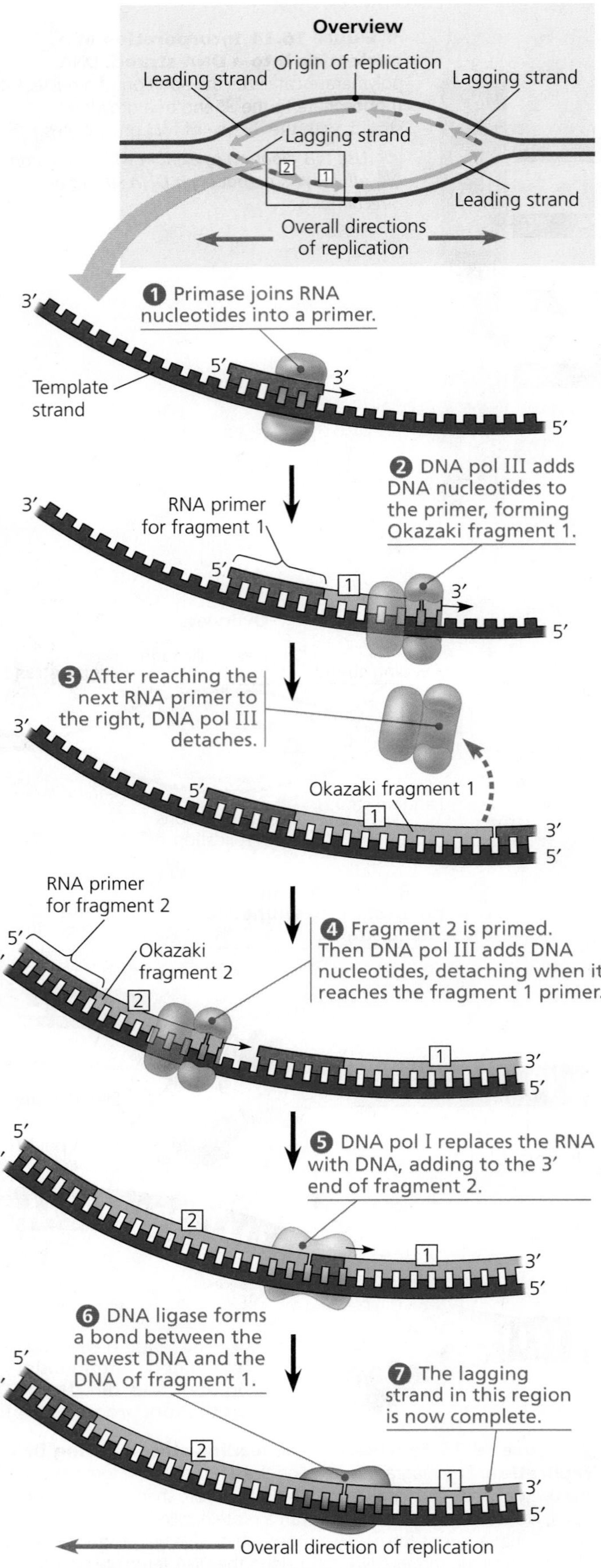
Overview
Origin of replication
Leading strand
Lagging strand
Lagging strand
2
1
Leading strand
Overall directions
of replication
3′
1 Primase joins RNA nucleotides into a primer.
5′
3′
Template strand
5′
2 DNA pol III adds DNA nucleotides to the primer, forming Okazaki fragment 1.
3′
RNA primer for fragment 1
5′
1
3′
5′
3 After reaching the next RNA primer to the right, DNA pol III detaches.
3′
5′
Okazaki fragment 1
1
3′
5′
RNA primer for fragment 2
5′
3′
Okazaki fragment 2
2
4 Fragment 2 is primed. Then DNA pol III adds DNA nucleotides, detaching when it reaches the fragment 1 primer.
1
3′
5′
5′
3′
5 DNA pol I replaces the RNA with DNA, adding to the 3′ end of fragment 2.
2
1
3′
5′
6 DNA ligase forms a bond between the newest DNA and the DNA of fragment 1.
5′
3′
7 The lagging strand in this region is now complete.
2
1
3′
5′
Overall direction of replication

▲ **Figure 16.16 Synthesis of the lagging strand.**

Figure 16.16 illustrates the steps in the synthesis of the lagging strand at one fork. Whereas only one primer is required on the leading strand, each Okazaki fragment on the lagging strand must be primed separately (steps 1 and 4). After DNA pol III forms an Okazaki fragment (steps 2–4), another DNA polymerase, DNA polymerase I (DNA pol I), replaces the RNA nucleotides of the adjacent primer with DNA nucleotides (step 5). But DNA pol I cannot join the final nucleotide of this replacement DNA segment to the first DNA nucleotide of the adjacent Okazaki fragment. Another enzyme, **DNA ligase**, accomplishes this task, joining the sugar-phosphate backbones of all the Okazaki fragments into a continuous DNA strand (step 6).

Figure 16.17 summarizes DNA replication. Please study it carefully before proceeding.

The DNA Replication Complex

It is traditional—and convenient—to represent DNA polymerase molecules as locomotives moving along a DNA "railroad track," but such a model is inaccurate in two important ways. First, the various proteins that participate in DNA replication actually form a single large complex, a "DNA replication machine." Many protein-protein interactions facilitate the efficiency of this complex. For example, by interacting with other proteins at the fork, primase apparently acts as a molecular brake, slowing progress of the replication fork and coordinating the placement of primers and the rates of replication on the leading and lagging strands. Second, the DNA replication complex may not move along the DNA; rather, the DNA may move through the complex during the replication process. In eukaryotic cells, multiple copies of the complex, perhaps grouped into "factories," may be anchored to the nuclear matrix, a framework of fibers extending through the interior of the nucleus. Recent studies support a model in which two DNA polymerase molecules, one on each template strand, "reel in" the parental DNA and extrude newly made daughter DNA molecules. Additional evidence suggests that the lagging strand is looped back through the complex (**Figure 16.18**).

Proofreading and Repairing DNA

We cannot attribute the accuracy of DNA replication solely to the specificity of base pairing. Although errors in the completed DNA molecule amount to only one in 10^{10} (10 billion) nucleotides, initial pairing errors between incoming nucleotides and those in the template strand are 100,000 times more common—an error rate of one in 10^5 nucleotides. During DNA replication, DNA polymerases proofread each nucleotide against its template as soon as it is added to the growing strand. Upon finding an incorrectly paired nucleotide, the polymerase removes the nucleotide and then resumes synthesis. (This action is similar to fixing a typing error by deleting the wrong letter and then entering the correct letter.)

Overview

Origin of replication

Leading strand

Lagging strand

Lagging strand

Leading strand

Overall directions of replication

① Helicase unwinds the parental double helix.

② Molecules of single-strand binding protein stabilize the unwound template strands.

③ The leading strand is synthesized continuously in the 5′ → 3′ direction by DNA pol III.

Leading strand

5′ 3′ Parental DNA

DNA pol III 3′

Primer 5′ Primase 3′

④ Primase begins synthesis of the RNA primer for the fifth Okazaki fragment.

⑤ DNA pol III is completing synthesis of the fourth fragment. When it reaches the RNA primer on the third fragment, it will detach and begin adding DNA nucleotides to the 3′ end of the fifth fragment primer in the replication fork.

DNA pol III 5′ 4 3′ 5′

Lagging strand

DNA pol I

DNA ligase

3 2 1 3′ 5′

⑥ DNA pol I removes the primer from the 5′ end of the second fragment, replacing it with DNA nucleotides that it adds one by one to the 3′ end of the third fragment. The replacement of the last RNA nucleotide with DNA leaves the sugar-phosphate backbone with a free 3′ end.

⑦ DNA ligase joins the 3′ end of the second fragment to the 5′ end of the first fragment.

▲ **Figure 16.17 A summary of bacterial DNA replication.** The detailed diagram shows the left-hand replication fork of the replication bubble in the overview (upper right). Viewing each daughter strand in its entirety in the overview, you can see that half of it is made continuously as the leading strand, while the other half (on the other side of the origin) is synthesized in fragments as the lagging strand.

DRAW IT *Draw a similar diagram showing the right-hand fork of this bubble, numbering the Okazaki fragments appropriately.*

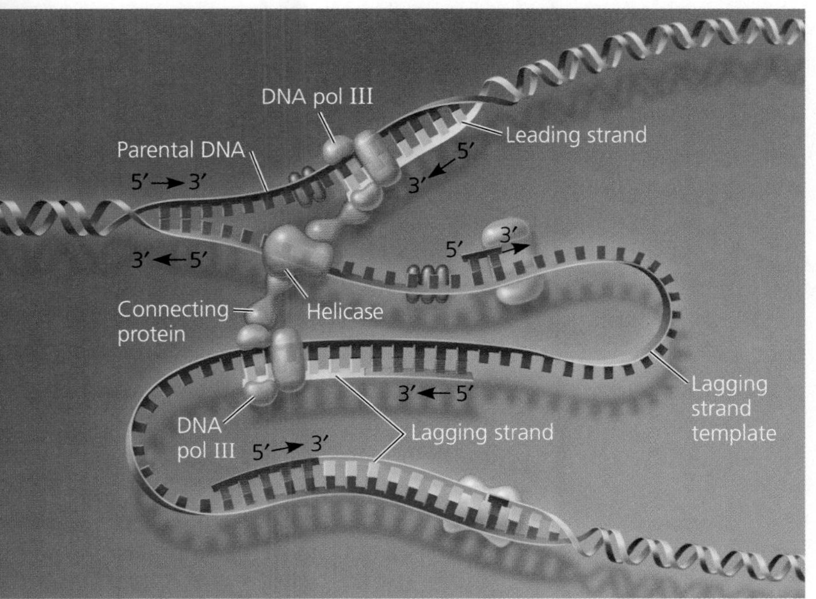

▲ **Figure 16.18 A current model of the DNA replication complex.** Two DNA polymerase III molecules work together in a complex, one on each template strand. The lagging strand template DNA loops through the complex.

BioFlix Visit the Study Area at **www.masteringbiology.com** for the BioFlix® 3-D Animation on DNA Replication.

Mismatched nucleotides sometimes evade proofreading by a DNA polymerase. In **mismatch repair**, other enzymes remove and replace incorrectly paired nucleotides that have resulted from replication errors. Researchers spotlighted the importance of such repair enzymes when they found that a hereditary defect in one of them is associated with a form of colon cancer. Apparently, this defect allows cancer-causing errors to accumulate in the DNA faster than normal.

Incorrectly paired or altered nucleotides can also arise after replication. In fact, maintenance of the genetic information encoded in DNA requires frequent repair of various kinds of damage to existing DNA. DNA molecules are constantly subjected to potentially harmful chemical and physical agents, such as cigarette smoke and X-rays, as we'll discuss in Chapter 17. In addition, DNA bases often undergo spontaneous chemical changes under normal cellular conditions. However, these changes in DNA are usually corrected before they become permanent changes—*mutations*—perpetuated through successive replications. Each cell continuously monitors and repairs its genetic material. Because repair of damaged DNA is so important to the survival of an organism, it is no surprise that many different DNA repair enzymes have evolved. Almost 100 are known in *E. coli*, and about 130 have been identified so far in humans.

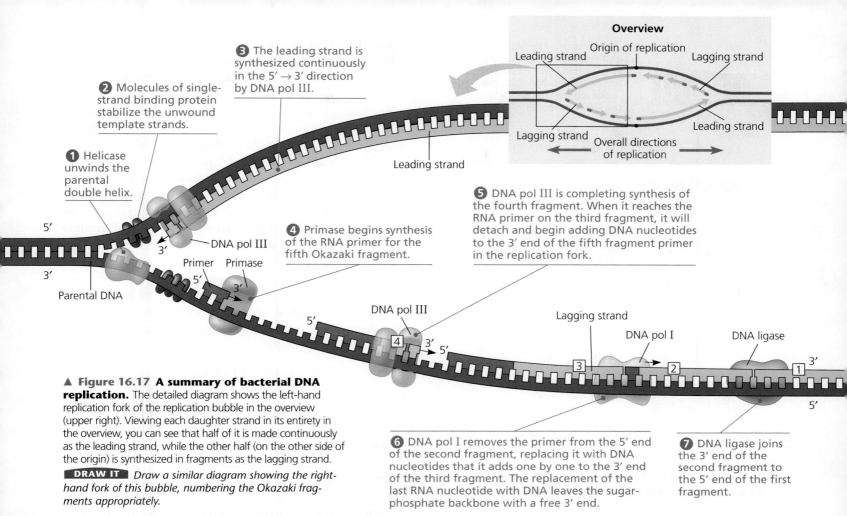

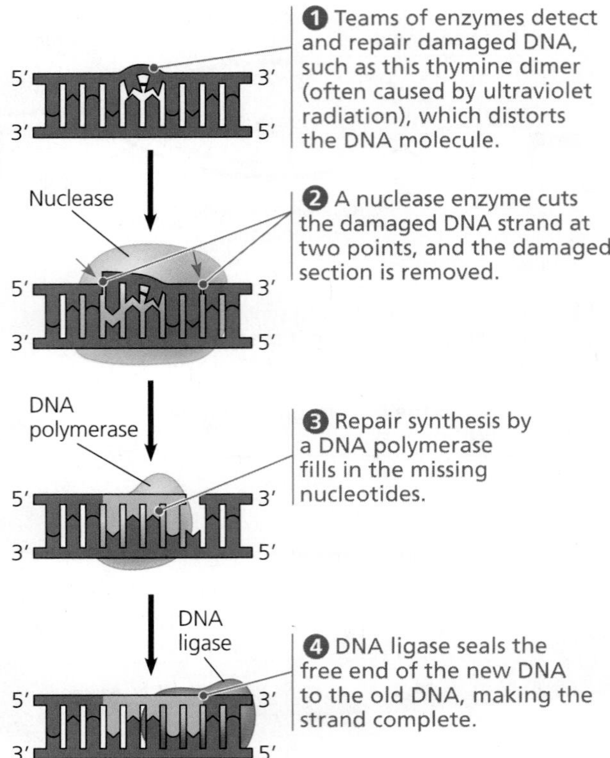

1 Teams of enzymes detect and repair damaged DNA, such as this thymine dimer (often caused by ultraviolet radiation), which distorts the DNA molecule.
5′
3′
3′
5′
Nuclease
2 A nuclease enzyme cuts the damaged DNA strand at two points, and the damaged section is removed.
DNA polymerase
3 Repair synthesis by a DNA polymerase fills in the missing nucleotides.
DNA ligase
4 DNA ligase seals the free end of the new DNA to the old DNA, making the strand complete.

▲ **Figure 16.19 Nucleotide excision repair of DNA damage.**

Most cellular systems for repairing incorrectly paired nucleotides, whether they are due to DNA damage or to replication errors, use a mechanism that takes advantage of the base-paired structure of DNA. In many cases, a segment of the strand containing the damage is cut out (excised) by a DNA-cutting enzyme—a **nuclease**—and the resulting gap is then filled in with nucleotides, using the undamaged strand as a template. The enzymes involved in filling the gap are a DNA polymerase and DNA ligase. One such DNA repair system is called **nucleotide excision repair** (**Figure 16.19**).

An important function of the DNA repair enzymes in our skin cells is to repair genetic damage caused by the ultraviolet rays of sunlight. One type of damage, shown in Figure 16.19, is the covalent linking of thymine bases that are adjacent on a DNA strand. Such *thymine dimers* cause the DNA to buckle and interfere with DNA replication. The importance of repairing this kind of damage is underscored by the disorder xeroderma pigmentosum, which in most cases is caused by an inherited defect in a nucleotide excision repair enzyme. Individuals with this disorder are hypersensitive to sunlight; mutations in their skin cells caused by ultraviolet light are left uncorrected, resulting in skin cancer.

Evolutionary Significance of Altered DNA Nucleotides

EVOLUTION Faithful replication of the genome and repair of DNA damage are important for the functioning of the organism and for passing on a complete, accurate genome to the next generation. The error rate after proofreading and repair is extremely low, but rare mistakes do slip through. Once a mismatched nucleotide pair is replicated, the sequence change is permanent in the daughter molecule that has the incorrect nucleotide as well as in any subsequent copies. As you know, a permanent change in the DNA sequence is called a mutation.

As you'll learn in Chapter 17, mutations can change the phenotype of an organism. And if they occur in germ cells (which give rise to gametes), mutations can be passed on from generation to generation. The vast majority of such changes are harmful, but a very small percentage can be beneficial. In either case, mutations are the source of the variation on which natural selection operates during evolution and are ultimately responsible for the appearance of new species. (You'll learn more about this process in Unit Four.) The balance between complete fidelity of DNA replication or repair and a low mutation rate has, over long periods of time, allowed the evolution of the rich diversity of species we see on Earth today.

Replicating the Ends of DNA Molecules

In spite of the impressive capabilities of DNA polymerases, there is a small portion of the cell's DNA that DNA polymerases can neither replicate nor repair. For linear DNA, such as the DNA of eukaryotic chromosomes, the fact that a DNA polymerase can add nucleotides only to the 3′ end of a preexisting polynucleotide leads to what might appear to be a problem. The usual replication machinery provides no way to complete the 5′ ends of daughter DNA strands. Even if an Okazaki fragment can be started with an RNA primer bound to the very end of the template strand, once that primer is removed, it cannot be replaced with DNA because there is no 3′ end available for nucleotide addition (**Figure 16.20**). As a result, repeated rounds of replication produce shorter and shorter DNA molecules with uneven ("staggered") ends.

Most prokaryotes have a circular chromosome, with no ends, so the shortening of DNA does not occur. But what protects the genes of linear eukaryotic chromosomes from being eroded away during successive rounds of DNA replication? It turns out that eukaryotic chromosomal DNA molecules have special nucleotide sequences called **telomeres** at their ends (**Figure 16.21**). Telomeres do not contain genes; instead, the DNA typically consists of multiple repetitions of one short nucleotide sequence. In each human telomere, for example, the six-nucleotide sequence TTAGGG is repeated between 100 and 1,000 times. Telomeric DNA acts as a kind of buffer zone that protects the organism's genes. In addition, specific proteins associated with telomeric DNA prevent the staggered ends of the daughter molecule from activating the cell's systems for monitoring DNA damage. (Staggered ends of a DNA molecule, which often result from double-strand breaks, can trigger signal transduction pathways leading to cell cycle arrest or cell death.)

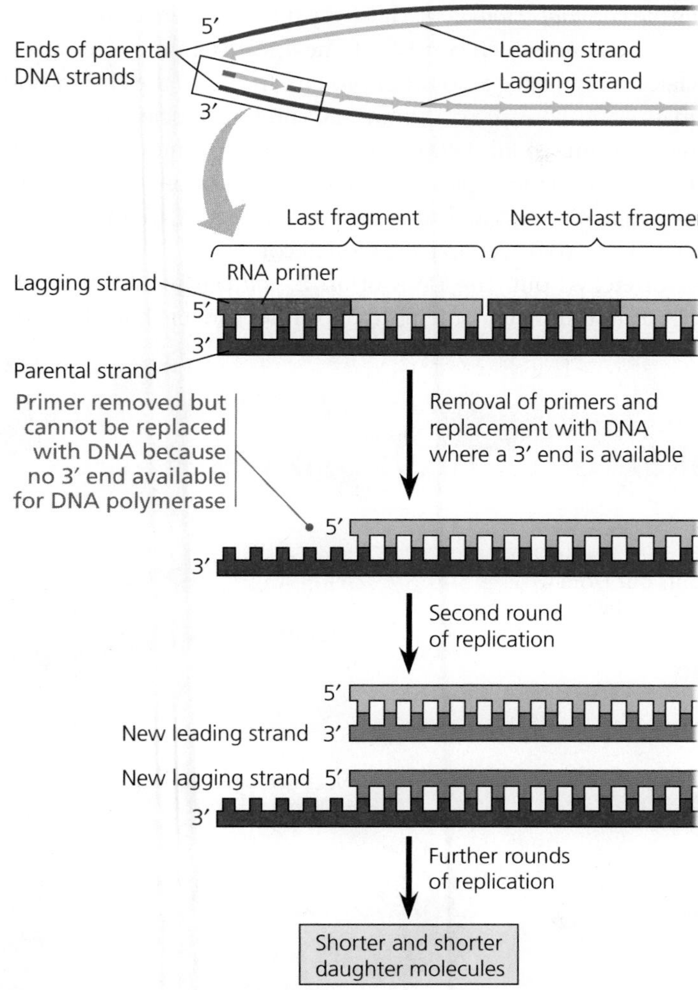

▲ **Figure 16.20 Shortening of the ends of linear DNA molecules.** Here we follow the end of one strand of a DNA molecule through two rounds of replication. After the first round, the new lagging strand is shorter than its template. After a second round, both the leading and lagging strands have become shorter than the original parental DNA. Although not shown here, the other ends of these DNA molecules also become shorter.

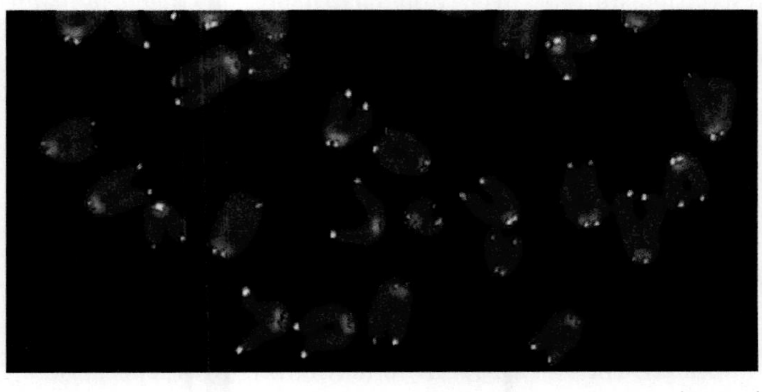

▲ **Figure 16.21 Telomeres.** Eukaryotes have repetitive, noncoding sequences called telomeres at the ends of their DNA. Telomeres are stained orange in these mouse chromosomes (LM).

Telomeres provide their protective function by postponing the erosion of genes located near the ends of DNA molecules. As shown in Figure 16.20, telomeres become shorter during every round of replication. As we would expect, telomeric DNA tends to be shorter in dividing somatic cells of older individuals and in cultured cells that have divided many times. It has been proposed that shortening of telomeres is somehow connected to the aging process of certain tissues and even to aging of the organism as a whole.

But what about cells whose genome must persist virtually unchanged from an organism to its offspring over many generations? If the chromosomes of germ cells became shorter in every cell cycle, essential genes would eventually be missing from the gametes they produce. However, this does not occur: An enzyme called **telomerase** catalyzes the lengthening of telomeres in eukaryotic germ cells, thus restoring their original length and compensating for the shortening that occurs during DNA replication. Telomerase is not active in most human somatic cells, but its activity in germ cells results in telomeres of maximum length in the zygote.

Normal shortening of telomeres may protect organisms from cancer by limiting the number of divisions that somatic cells can undergo. Cells from large tumors often have unusually short telomeres, as we would expect for cells that have undergone many cell divisions. Further shortening would presumably lead to self-destruction of the tumor cells. Intriguingly, researchers have found telomerase activity in cancerous somatic cells, suggesting that its ability to stabilize telomere length may allow these cancer cells to persist. Many cancer cells do seem capable of unlimited cell division, as do immortal strains of cultured cells (see Chapter 12). If telomerase is indeed an important factor in many cancers, it may provide a useful target for both cancer diagnosis and chemotherapy.

Thus far in this chapter, you have learned about the structure and replication of a DNA molecule. In the next section, we'll take a step back and examine how DNA is packaged into chromosomes, the structures that carry the genetic information.

CONCEPT CHECK 16.2

1. What role does complementary base pairing play in the replication of DNA?
2. Make a table listing the functions of seven proteins involved in DNA replication in *E. coli.*
3. **MAKE CONNECTIONS** What is the relationship between DNA replication and the S phase of the cell cycle? See Figure 12.6, page 231.
4. **WHAT IF?** If the DNA pol I in a given cell were nonfunctional, how would that affect the synthesis of a *leading* strand? In the overview box in Figure 16.17, point out where DNA pol I would normally function on the top leading strand.

For suggested answers, see Appendix A.

CONCEPT 16.3

A chromosome consists of a DNA molecule packed together with proteins

The main component of the genome in most bacteria is one double-stranded, circular DNA molecule that is associated with a small amount of protein. Although we refer to this structure as the *bacterial chromosome*, it is very different from a eukaryotic chromosome, which consists of one linear DNA molecule associated with a large amount of protein. In *E. coli*, the chromosomal DNA consists of about 4.6 million nucleotide pairs, representing about 4,400 genes. This is 100 times more DNA than is found in a typical virus, but only about one-thousandth as much DNA as in a human somatic cell. Still, that is a lot of DNA to be packaged in such a small container.

Stretched out, the DNA of an *E. coli* cell would measure about a millimeter in length, 500 times longer than the cell.

▼ **Figure 16.22**

Exploring Chromatin Packing in a Eukaryotic Chromosome

This series of diagrams and transmission electron micrographs depicts a current model for the progressive levels of DNA coiling and folding. The illustration zooms out from a single molecule of DNA to a metaphase chromosome, which is large enough to be seen with a light microscope.

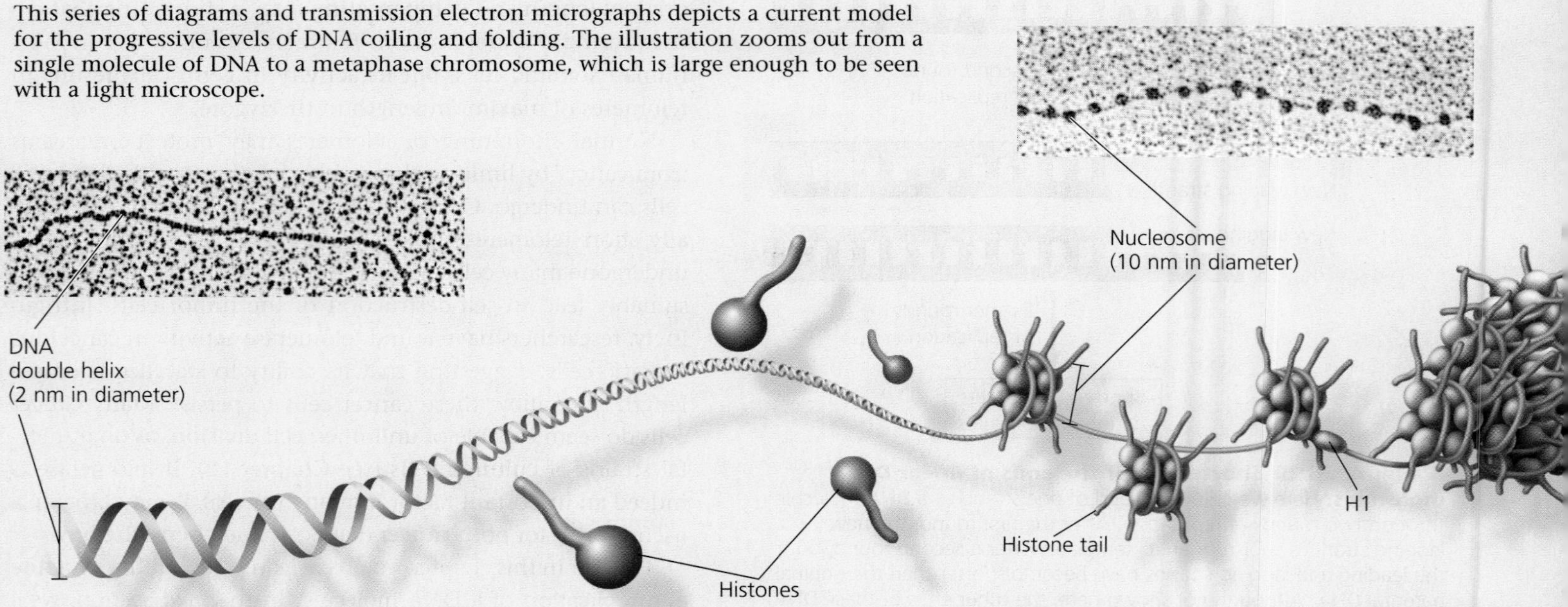

DNA, the double helix

Shown here is a ribbon model of DNA, with each ribbon representing one of the sugar-phosphate backbones. As you will recall from Figure 16.7, the phosphate groups along the backbone contribute a negative charge along the outside of each strand. The TEM shows a molecule of naked DNA; the double helix alone is 2 nm across.

Histones

Proteins called **histones** are responsible for the first level of DNA packing in chromatin. Although each histone is small—containing only about 100 amino acids—the total mass of histone in chromatin approximately equals the mass of DNA. More than a fifth of a histone's amino acids are positively charged (lysine or arginine) and therefore bind tightly to the negatively charged DNA.

Four types of histones are most common in chromatin: H2A, H2B, H3, and H4. The histones are very similar among eukaryotes; for example, all but two of the amino acids in cow H4 are identical to those in pea H4. The apparent conservation of histone genes during evolution probably reflects the important role of histones in organizing DNA within cells.

The four main types of histones are critical to the next level of DNA packing. (A fifth type of histone, called H1, is involved in a further stage of packing.)

Nucleosomes, or "beads on a string" (10-nm fiber)

In electron micrographs, unfolded chromatin is 10 nm in diameter (the *10-nm fiber*). Such chromatin resembles beads on a string (see the TEM). Each "bead" is a **nucleosome**, the basic unit of DNA packing; the "string" between beads is called *linker DNA*.

A nucleosome consists of DNA wound twice around a protein core composed of two molecules each of the four main histone types. The amino end (N-terminus) of each histone (the *histone tail*) extends outward from the nucleosome.

In the cell cycle, the histones leave the DNA only briefly during DNA replication. Generally, they do the same during transcription, another process that requires access to the DNA by the cell's molecular machinery. Chapter 18 will discuss some recent findings about the role of histone tails and nucleosomes in the regulation of gene expression.

Within a bacterium, however, certain proteins cause the chromosome to coil and "supercoil," densely packing it so that it fills only part of the cell. Unlike the nucleus of a eukaryotic cell, this dense region of DNA in a bacterium, called the **nucleoid**, is not bounded by membrane (see Figure 6.5).

Eukaryotic chromosomes each contain a single linear DNA double helix that, in humans, averages about 1.5×10^8 nucleotide pairs. This is an enormous amount of DNA relative to a chromosome's condensed length. If completely stretched out, such a DNA molecule would be about 4 cm long, thousands of times the diameter of a cell nucleus—and that's not even considering the DNA of the other 45 human chromosomes!

In the cell, eukaryotic DNA is precisely combined with a large amount of protein. Together, this complex of DNA and protein, called **chromatin**, fits into the nucleus through an elaborate, multilevel system of packing. Our current view of the successive levels of DNA packing in a chromosome is outlined in **Figure 16.22**. Study this figure carefully before reading further.

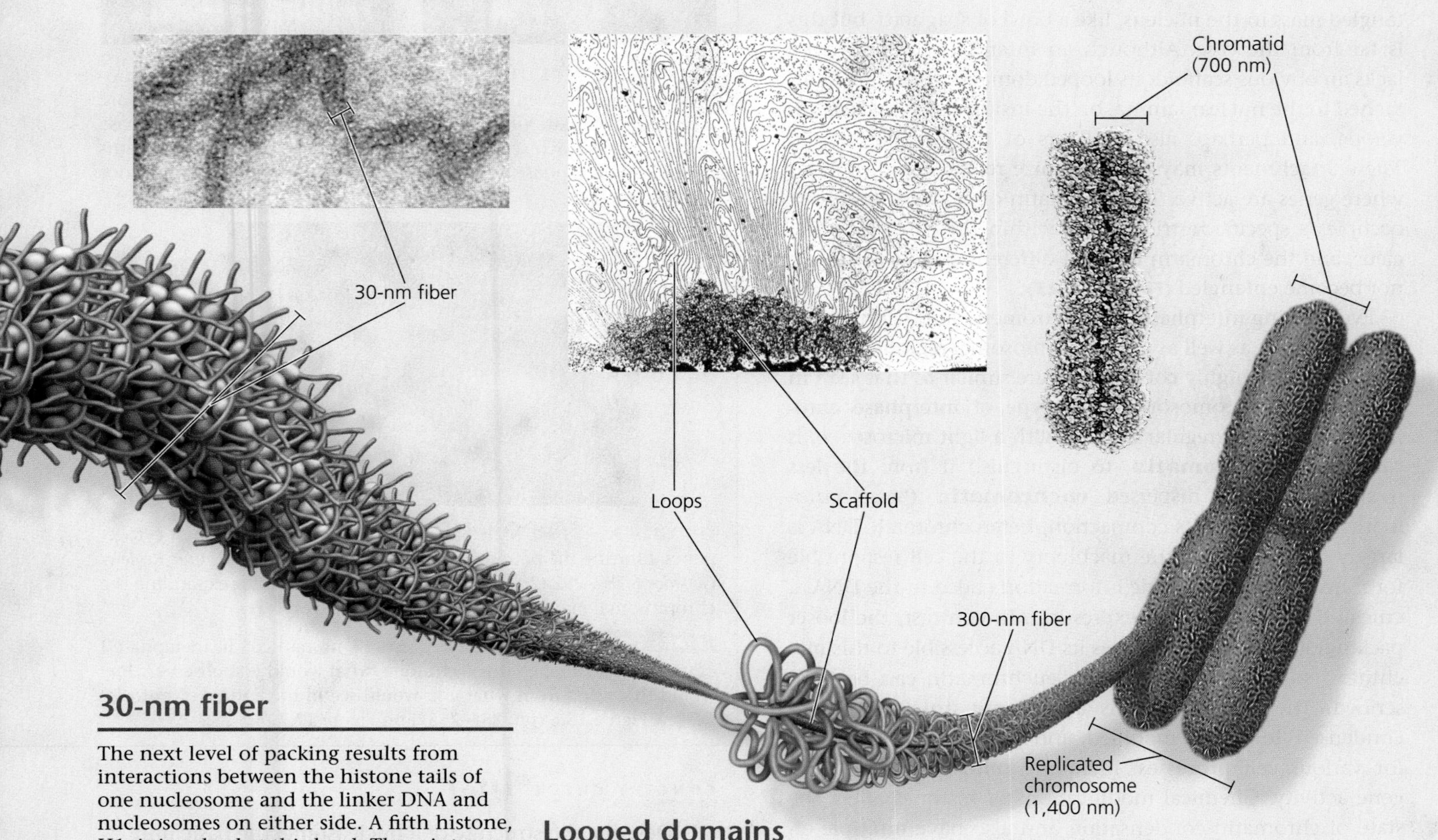

30-nm fiber

The next level of packing results from interactions between the histone tails of one nucleosome and the linker DNA and nucleosomes on either side. A fifth histone, H1, is involved at this level. These interactions cause the extended 10-nm fiber to coil or fold, forming a chromatin fiber roughly 30 nm in thickness, the *30-nm fiber*. Although the 30-nm fiber is quite prevalent in the interphase nucleus, the packing arrangement of nucleosomes in this form of chromatin is still a matter of some debate.

Looped domains (300-nm fiber)

The 30-nm fiber, in turn, forms loops called *looped domains* attached to a chromosome scaffold made of proteins, thus making up a *300-nm fiber*. The scaffold is rich in one type of topoisomerase, and H1 molecules also appear to be present.

Metaphase chromosome

In a mitotic chromosome, the looped domains themselves coil and fold in a manner not yet fully understood, further compacting all the chromatin to produce the characteristic metaphase chromosome shown in the micrograph above. The width of one chromatid is 700 nm. Particular genes always end up located at the same places in metaphase chromosomes, indicating that the packing steps are highly specific and precise.

Chromatin undergoes striking changes in its degree of packing during the course of the cell cycle (see Figure 12.7). In interphase cells stained for light microscopy, the chromatin usually appears as a diffuse mass within the nucleus, suggesting that the chromatin is highly extended. As a cell prepares for mitosis, its chromatin coils and folds up (condenses), eventually forming a characteristic number of short, thick metaphase chromosomes that are distinguishable from each other with the light microscope.

Though interphase chromatin is generally much less condensed than the chromatin of mitotic chromosomes, it shows several of the same levels of higher-order packing. Some of the chromatin comprising a chromosome seems to be present as a 10-nm fiber, but much is compacted into a 30-nm fiber, which in some regions is further folded into looped domains. Early on, biologists assumed that interphase chromatin was a tangled mass in the nucleus, like a bowl of spaghetti, but this is far from the case. Although an interphase chromosome lacks an obvious scaffold, its looped domains appear to be attached to the nuclear lamina, on the inside of the nuclear envelope, and perhaps also to fibers of the nuclear matrix. These attachments may help organize regions of chromatin where genes are active. The chromatin of each chromosome occupies a specific restricted area within the interphase nucleus, and the chromatin fibers of different chromosomes do not become entangled (**Figure 16.23**).

Even during interphase, the centromeres and telomeres of chromosomes, as well as other chromosomal regions in some cells, exist in a highly condensed state similar to that seen in a metaphase chromosome. This type of interphase chromatin, visible as irregular clumps with a light microscope, is called **heterochromatin**, to distinguish it from the less compacted, more dispersed **euchromatin** ("true chromatin"). Because of its compaction, heterochromatic DNA is largely inaccessible to the machinery in the cell responsible for transcribing the genetic information coded in the DNA, a crucial early step in gene expression. In contrast, the looser packing of euchromatin makes its DNA accessible to this machinery, so the genes present in euchromatin can be transcribed. The chromosome is a dynamic structure that is condensed, loosened, modified, and remodeled as necessary for various cell processes, including mitosis, meiosis, and gene activity. Chemical modifications of histones affect the state of chromatin condensation and also have multiple effects on gene activity, as you will see in Chapter 18.

In this chapter, you have learned how DNA molecules are arranged in chromosomes and how DNA replication provides the copies of genes that parents pass to offspring. However, it is not enough that genes be copied and transmitted; the information they carry must be used by the cell. In other words, genes must also be "expressed." In the next chapter, we will examine how the cell expresses the genetic information encoded in DNA.

▼ Figure 16.23

IMPACT

Painting Chromosomes

Using techniques you'll learn about in Chapter 20, researchers have been able to treat human chromosomes with special molecular tags, such that each chromosome pair can be seen as a different color. Below on the left is a spread of chromosomes treated in this way; on the right they are organized into a karyotype.

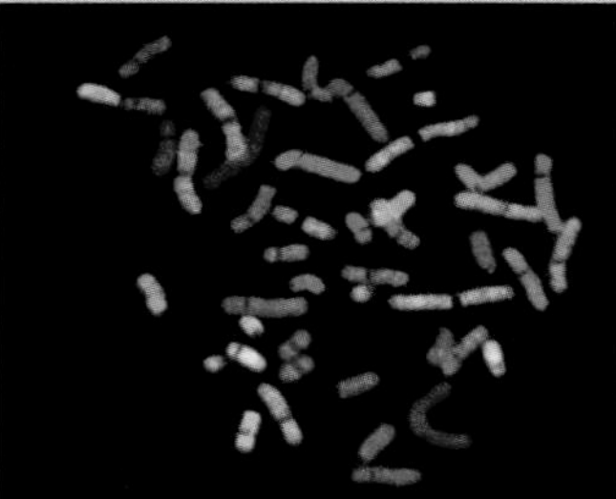

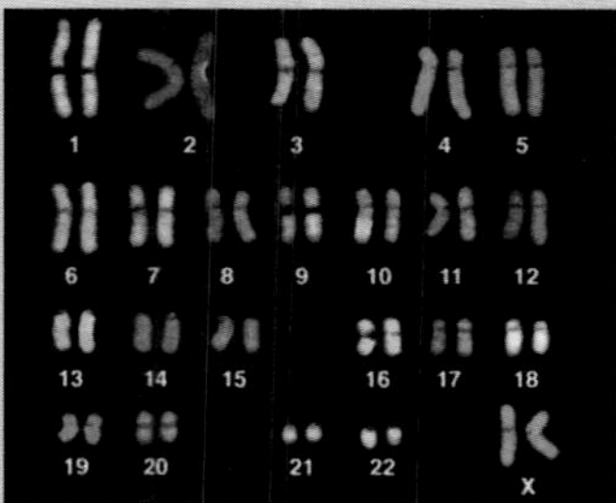

WHY IT MATTERS The ability to visually distinguish among chromosomes has allowed researchers to see how the chromosomes are arranged in the interphase nucleus. As you can see in the interphase nucleus below, each chromosome appears to occupy a specific territory during interphase. In general, the two homologs of a pair are not located together.

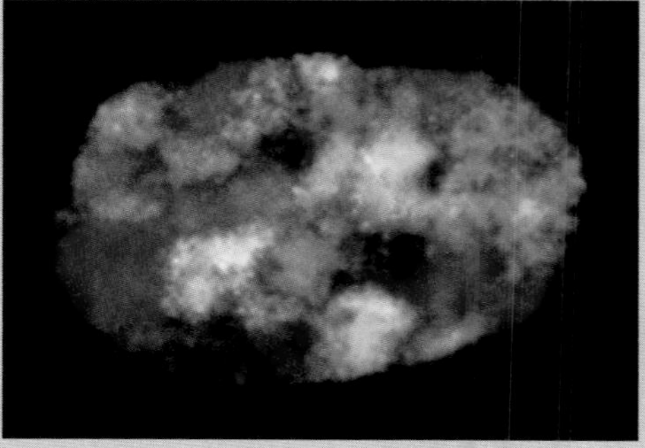

5 μm

FURTHER READING M. R. Speicher and N. P. Carter, The new cytogenetics: blurring the boundaries with molecular biology, *Nature Reviews Genetics* 6:782–792 (2005); J. L. Marx, New methods for expanding the chromosomal paint kit, *Science* 273:430 (1996).

MAKE CONNECTIONS If you arrested a human cell in metaphase I of meiosis and applied this technique, what would you observe? How would this differ from what you would see in metaphase of mitosis? Review Figure 13.8 (pp. 254–255) and Figure 12.7 (pp. 232–233).

CONCEPT CHECK 16.3

1. Describe the structure of a nucleosome, the basic unit of DNA packing in eukaryotic cells.
2. What two properties, one structural and one functional, distinguish heterochromatin from euchromatin?
3. **MAKE CONNECTIONS** Interphase chromosomes appear to be attached to the nuclear lamina and perhaps also the nuclear matrix. Describe these two structures. See page 102 and Figure 6.9 on page 103.

For suggested answers, see Appendix A.

16 CHAPTER REVIEW

SUMMARY OF KEY CONCEPTS

CONCEPT 16.1

DNA is the genetic material (pp. 305–310)

- Experiments with bacteria and with **phages** provided the first strong evidence that the genetic material is DNA.
- Watson and Crick deduced that DNA is a **double helix** and built a structural model. Two **antiparallel** sugar-phosphate chains wind around the outside of the molecule; the nitrogenous bases project into the interior, where they hydrogen-bond in specific pairs, A with T, G with C.

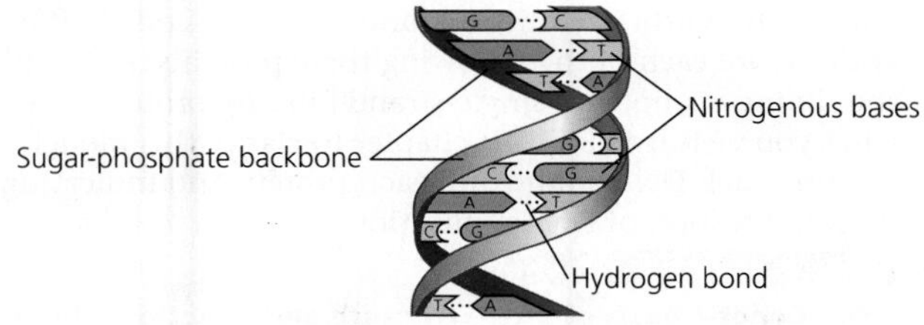
Nitrogenous bases
Sugar-phosphate backbone
Hydrogen bond

? *What does it mean when we say that the two DNA strands in the double helix are antiparallel? What would an end of the double helix look like if the strands were parallel?*

CONCEPT 16.2

Many proteins work together in DNA replication and repair (pp. 311–319)

- The Meselson-Stahl experiment showed that **DNA replication** is **semiconservative**: The parental molecule unwinds, and each strand then serves as a template for the synthesis of a new strand according to base-pairing rules.
- DNA replication at one **replication fork** is summarized here:

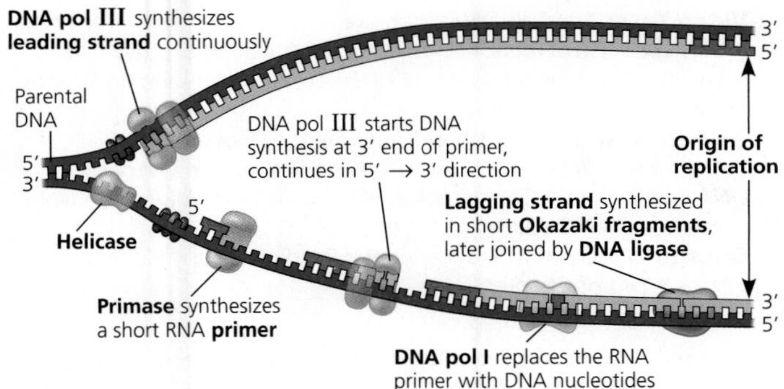
DNA pol III synthesizes leading strand continuously
Parental DNA
5′
3′
DNA pol III starts DNA synthesis at 3′ end of primer, continues in 5′ → 3′ direction
Origin of replication
Lagging strand synthesized in short Okazaki fragments, later joined by DNA ligase
Helicase
5′
Primase synthesizes a short RNA primer
DNA pol I replaces the RNA primer with DNA nucleotides
3′
5′
3′
5′

- DNA polymerases proofread new DNA, replacing incorrect nucleotides. In **mismatch repair**, enzymes correct errors that persist. **Nucleotide excision repair** is a general process by which **nucleases** cut out and replace damaged stretches of DNA.
- The ends of eukaryotic chromosomal DNA get shorter with each round of replication. The presence of **telomeres**, repetitive sequences at the ends of linear DNA molecules, postpones the erosion of genes. **Telomerase** catalyzes the lengthening of telomeres in germ cells.

? *Compare DNA replication on the leading and lagging strands, including both similarities and differences.*

CONCEPT 16.3

A chromosome consists of a DNA molecule packed together with proteins (pp. 320–322)

- The bacterial chromosome is usually a circular DNA molecule with some associated proteins, making up the **nucleoid** of the cell. Eukaryotic **chromatin** making up a chromosome is composed of DNA, **histones**, and other proteins. The histones bind to each other and to the DNA to form **nucleosomes**, the most basic units of DNA packing. Histone tails extend outward from each bead-like nucleosome core. Additional coiling and folding leads ultimately to the highly condensed chromatin of the metaphase chromosome. In interphase cells, most chromatin is less compacted (**euchromatin**), but some remains highly condensed (**heterochromatin**). Euchromatin, but not heterochromatin, is generally accessible for transcription of genes.

? *Describe the levels of chromatin packing you'd expect to see in an interphase nucleus.*

TEST YOUR UNDERSTANDING

LEVEL 1: KNOWLEDGE/COMPREHENSION

1. In his work with pneumonia-causing bacteria and mice, Griffith found that
   a. the protein coat from pathogenic cells was able to transform nonpathogenic cells.
   b. heat-killed pathogenic cells caused pneumonia.
   c. some substance from pathogenic cells was transferred to nonpathogenic cells, making them pathogenic.
   d. the polysaccharide coat of bacteria caused pneumonia.
   e. bacteriophages injected DNA into bacteria.
2. What is the basis for the difference in how the leading and lagging strands of DNA molecules are synthesized?
   a. The origins of replication occur only at the 5′ end.
   b. Helicases and single-strand binding proteins work at the 5′ end.
   c. DNA polymerase can join new nucleotides only to the 3′ end of a growing strand.
   d. DNA ligase works only in the 3′ → 5′ direction.
   e. Polymerase can work on only one strand at a time.
3. In analyzing the number of different bases in a DNA sample, which result would be consistent with the base-pairing rules?
   a. A = G
   b. A + G = C + T
   c. A + T = G + T
   d. A = C
   e. G = T
4. The elongation of the leading strand during DNA synthesis
   a. progresses away from the replication fork.
   b. occurs in the 3′ → 5′ direction.
   c. produces Okazaki fragments.
   d. depends on the action of DNA polymerase.
   e. does not require a template strand.
5. In a nucleosome, the DNA is wrapped around
   a. polymerase molecules.
   b. ribosomes.
   c. histones.
   d. a thymine dimer.
   e. satellite DNA.

LEVEL 2: APPLICATION/ANALYSIS

6. *E. coli* cells grown on ^{15}N medium are transferred to ^{14}N medium and allowed to grow for two more generations (two rounds of DNA replication). DNA extracted from these cells is centrifuged. What density distribution of DNA would you expect in this experiment?
   a. one high-density and one low-density band
   b. one intermediate-density band
   c. one high-density and one intermediate-density band
   d. one low-density and one intermediate-density band
   e. one low-density band
7. A biochemist isolates, purifies, and combines in a test tube a variety of molecules needed for DNA replication. When she adds some DNA to the mixture, replication occurs, but each DNA molecule consists of a normal strand paired with numerous segments of DNA a few hundred nucleotides long. What has she probably left out of the mixture?
   a. DNA polymerase
   b. DNA ligase
   c. nucleotides
   d. Okazaki fragments
   e. primase
8. The spontaneous loss of amino groups from adenine in DNA results in hypoxanthine, an uncommon base, opposite thymine. What combination of proteins could repair such damage?
   a. nuclease, DNA polymerase, DNA ligase
   b. telomerase, primase, DNA polymerase
   c. telomerase, helicase, single-strand binding protein
   d. DNA ligase, replication fork proteins, adenylyl cyclase
   e. nuclease, telomerase, primase
9. **MAKE CONNECTIONS** Although the proteins that cause the *E. coli* chromosome to coil are not histones, what property would you expect them to share with histones, given their ability to bind to DNA (see Figure 5.16, p. 79)?

LEVEL 3: SYNTHESIS/EVALUATION

10. The table below shows the base composition of DNA in several species. Explain how these data demonstrate Chargaff's rules.

Source	Adenine	Guanine	Cytosine	Thymine
E. coli	24.7%	26.0%	25.7%	23.6%
Wheat	28.1	21.8	22.7	27.4
Sea urchin	32.8	17.7	17.3	32.1
Salmon	29.7	20.8	20.4	29.1
Human	30.4	19.6	19.9	30.1
Ox	29.0	21.2	21.2	28.7

11. **EVOLUTION CONNECTION**
    Some bacteria may be able to respond to environmental stress by increasing the rate at which mutations occur during cell division. How might this be accomplished? Might there be an evolutionary advantage of this ability? Explain.

12. **SCIENTIFIC INQUIRY**

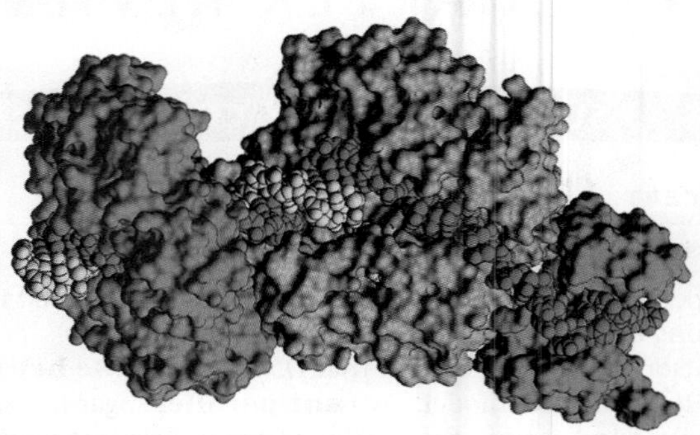

DRAW IT Model building can be an important part of the scientific process. The illustration shown above is a computer-generated model of a DNA replication complex. The parental and newly synthesized DNA strands are color-coded differently, as are each of the following three proteins: DNA pol III, the sliding clamp, and single-strand binding protein. Use what you've learned in this chapter to clarify this model by labeling each DNA strand and each protein and indicating the overall direction of DNA replication.

13. **WRITE ABOUT A THEME**
    The Genetic Basis of Life; Structure and Function The continuity of life is based on heritable information in the form of DNA, and structure and function are correlated at all levels of biological organization. In a short essay (100–150 words), describe how the structure of DNA is correlated with its role as the molecular basis of inheritance.

For selected answers, see Appendix A.

17

From Gene to Protein

▲ **Figure 17.1 How does a single faulty gene result in the dramatic appearance of an albino deer?**

KEY CONCEPTS

17.1 Genes specify proteins via transcription and translation

17.2 Transcription is the DNA-directed synthesis of RNA: *a closer look*

17.3 Eukaryotic cells modify RNA after transcription

17.4 Translation is the RNA-directed synthesis of a polypeptide: *a closer look*

17.5 Mutations of one or a few nucleotides can affect protein structure and function

17.6 While gene expression differs among the domains of life, the concept of a gene is universal

OVERVIEW

The Flow of Genetic Information

In 2006, a young albino deer seen frolicking with several brown deer in the mountains of eastern Germany elicited a public outcry (**Figure 17.1**). A local hunting organization announced that the albino deer suffered from a "genetic disorder" and should be shot. Some argued that the deer should merely be prevented from mating with other deer to safeguard the population's gene pool. Others favored relocating the albino deer to a nature reserve because they worried that it might be more noticeable to predators if left in the wild. A German rock star even held a benefit concert to raise funds for the relocation. What led to the striking phenotype of this deer, the cause of this lively debate?

You learned in Chapter 14 that inherited traits are determined by genes and that the trait of albinism is caused by a recessive allele of a pigmentation gene. The information content of genes is in the form of specific sequences of nucleotides along strands of DNA, the genetic material. But how does this information determine an organism's traits? Put another way, what does a gene actually say? And how is its message translated by cells into a specific trait, such as brown hair, type A blood, or, in the case of an albino deer, a total lack of pigment? The albino deer has a faulty version of a key protein, an enzyme required for pigment synthesis, and this protein is faulty because the gene that codes for it contains incorrect information.

This example illustrates the main point of this chapter: The DNA inherited by an organism leads to specific traits by dictating the synthesis of proteins and of RNA molecules involved in protein synthesis. In other words, proteins are the link between genotype and phenotype. **Gene expression** is the process by which DNA directs the synthesis of proteins (or, in some cases, just RNAs). The expression of genes that code for proteins includes two stages: transcription and translation. This chapter describes the flow of information from gene to protein in detail and explains how genetic mutations affect organisms through their proteins. Understanding the processes of gene expression, which are similar in all three domains of life, will allow us to revisit the concept of the gene in more detail at the end of the chapter.

CONCEPT **17.1**

Genes specify proteins via transcription and translation

Before going into the details of how genes direct protein synthesis, let's step back and examine how the fundamental relationship between genes and proteins was discovered.

Evidence from the Study of Metabolic Defects

In 1902, British physician Archibald Garrod was the first to suggest that genes dictate phenotypes through enzymes that catalyze specific chemical reactions in the cell. Garrod postulated that the symptoms of an inherited disease reflect a person's inability to make a particular enzyme. He later referred to such diseases as "inborn errors of metabolism." Garrod gave as one example the hereditary condition called alkaptonuria. In this disorder, the urine is black because it contains the chemical alkapton, which darkens upon exposure to air. Garrod reasoned that most people have an enzyme that metabolizes alkapton, whereas people with alkaptonuria have inherited an inability to make that enzyme.

Garrod may have been the first to recognize that Mendel's principles of heredity apply to humans as well as peas. Garrod's realization was ahead of its time, but research several decades later supported his hypothesis that a gene dictates the production of a specific enzyme. Biochemists accumulated much evidence that cells synthesize and degrade most organic molecules via metabolic pathways, in which each chemical reaction in a sequence is catalyzed by a specific enzyme (see p. 142). Such metabolic pathways lead, for instance, to the synthesis of the pigments that give the brown deer in Figure 17.1 their fur color or fruit flies (*Drosophila*) their eye color (see Figure 15.3). In the 1930s, the American biochemist and geneticist George Beadle and his French colleague Boris Ephrussi speculated that in *Drosophila*, each of the various mutations affecting eye color blocks pigment synthesis at a specific step by preventing production of the enzyme that catalyzes that step. But neither the chemical reactions nor the enzymes that catalyze them were known at the time.

Nutritional Mutants in Neurospora: Scientific Inquiry

A breakthrough in demonstrating the relationship between genes and enzymes came a few years later at Stanford University, where Beadle and Edward Tatum began working with a bread mold, *Neurospora crassa*. They bombarded *Neurospora* with X-rays, shown in the 1920s to cause genetic changes, and then looked among the survivors for mutants that differed in their nutritional needs from the wild-type bread mold. Wild-type *Neurospora* has modest food requirements. It can grow in the laboratory on a simple solution of inorganic salts, glucose, and the vitamin biotin, incorporated into agar, a support medium. From this *minimal medium*, the mold cells use their metabolic pathways to produce all the other molecules they need. Beadle and Tatum identified mutants that could not survive on minimal medium, apparently because they were unable to synthesize certain essential molecules from the minimal ingredients. To ensure survival of these nutritional mutants, Beadle and Tatum allowed them to grow on a *complete growth medium*, which consisted of minimal medium supplemented with all 20 amino acids and a few other nutrients. The complete growth medium could support any mutant that couldn't synthesize one of the supplements.

To characterize the metabolic defect in each nutritional mutant, Beadle and Tatum took samples from the mutant growing on complete medium and distributed them to a number of different vials. Each vial contained minimal medium plus a single additional nutrient. The particular supplement that allowed growth indicated the metabolic defect. For example, if the only supplemented vial that supported growth of the mutant was the one fortified with the amino acid arginine, the researchers could conclude that the mutant was defective in the biochemical pathway that wild-type cells use to synthesize arginine.

In fact, such arginine-requiring mutants were obtained and studied by two colleagues of Beadle and Tatum, Adrian Srb and Norman Horowitz, who wanted to investigate the biochemical pathway for arginine synthesis in *Neurospora* (**Figure 17.2**). Srb and Horowitz pinned down each mutant's defect more specifically, using additional tests to distinguish among three classes of arginine-requiring mutants. Mutants in each class required a different set of compounds along the arginine-synthesizing pathway, which has three steps. These results, and those of many similar experiments done by Beadle and Tatum, suggested that each class was blocked at a different step in this pathway because mutants in that class lacked the enzyme that catalyzes the blocked step.

Because each mutant was defective in a single gene, Beadle and Tatum saw that, taken together, the collected results provided strong support for a working hypothesis they had proposed earlier. The *one gene–one enzyme hypothesis*, as they dubbed it, states that the function of a gene is to dictate the production of a specific enzyme. Further support for this hypothesis came from experiments that identified the specific enzymes lacking in the mutants. Beadle and Tatum shared a Nobel Prize in 1958 for "their discovery that genes act by regulating definite chemical events" (in the words of the Nobel committee).

The Products of Gene Expression: A Developing Story

As researchers learned more about proteins, they made revisions to the one gene–one enzyme hypothesis. First of all, not all proteins are enzymes. Keratin, the structural protein of animal hair, and the hormone insulin are two examples of nonenzyme proteins. Because proteins that are not enzymes are nevertheless gene products, molecular biologists began to think in terms of one gene–one protein. However, many proteins are constructed from two or more different polypeptide chains, and each polypeptide is specified by its own gene. For example, hemoglobin, the oxygen-transporting protein of vertebrate red blood cells, contains two kinds of polypeptides, and thus two genes code for this protein (see Figure 5.20). Beadle and Tatum's idea was therefore restated as the *one gene–one polypeptide hypothesis*. Even this description is not entirely accurate, though. First, many eukaryotic genes can each code for a set of closely related polypeptides via a process called alternative splicing, which you will learn about later in this chapter. Second, quite a few genes code for RNA molecules that have important functions in cells

▼ Figure 17.2 INQUIRY

Do individual genes specify the enzymes that function in a biochemical pathway?

EXPERIMENT Working with the mold *Neurospora crassa*, Adrian Srb and Norman Horowitz, then at Stanford University, used Beadle and Tatum's experimental approach to isolate mutants that required arginine in their growth medium. The researchers showed that these mutants fell into three classes, each defective in a different gene. From other considerations, they suspected that the metabolic pathway of arginine biosynthesis involved a precursor nutrient and the intermediate molecules ornithine and citrulline. Their most famous experiment, shown here, tested both the one gene–one enzyme hypothesis and their postulated arginine-synthesizing pathway. In this experiment, they grew their three classes of mutants under the four different conditions shown in the Results section below. They included minimal medium (MM) as a control because they knew that wild-type cells could grow on MM but mutant cells could not. (See test tubes on the right.)

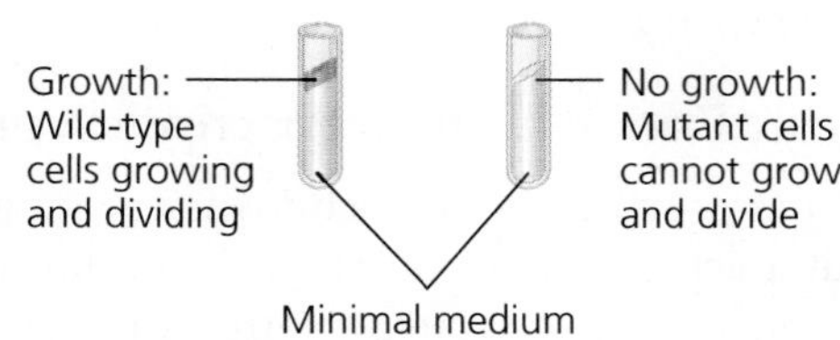

RESULTS The wild-type strain was capable of growth under all experimental conditions, requiring only the minimal medium. The three classes of mutants each had a specific set of growth requirements. For example, class II mutants could not grow when ornithine alone was added but could grow when either citrulline or arginine was added.

Condition	Classes of *Neurospora crassa*: Wild type	Class I mutants	Class II mutants	Class III mutants
Minimal medium (MM) (control)	growth	no growth	no growth	no growth
MM + ornithine	growth	growth	no growth	no growth
MM + citrulline	growth	growth	growth	no growth
MM + arginine (control)	growth	growth	growth	growth
Summary of results	Can grow with or without any supplements	Can grow on ornithine, citrulline, or arginine	Can grow only on citrulline or arginine	Require arginine to grow

CONCLUSION From the growth requirements of the mutants, Srb and Horowitz deduced that each class of mutant was unable to carry out one step in the pathway for synthesizing arginine, presumably because it lacked the necessary enzyme. Because each of their mutants was mutated in a single gene, they concluded that each mutated gene must normally dictate the production of one enzyme. Their results supported the one gene–one enzyme hypothesis proposed by Beadle and Tatum and also confirmed that the arginine pathway described in the mammalian liver also operates in *Neurospora*. (Notice in the Results that a mutant can grow only if supplied with a compound made *after* the defective step because this bypasses the defect.)

Gene (codes for enzyme)	Wild type	Class I mutants (mutation in gene *A*)	Class II mutants (mutation in gene *B*)	Class III mutants (mutation in gene *C*)
	Precursor	Precursor	Precursor	Precursor
Gene *A* →	Enzyme A	Enzyme A ✕	Enzyme A	Enzyme A
	Ornithine	Ornithine	Ornithine	Ornithine
Gene *B* →	Enzyme B	Enzyme B	Enzyme B ✕	Enzyme B
	Citrulline	Citrulline	Citrulline	Citrulline
Gene *C* →	Enzyme C	Enzyme C	Enzyme C	Enzyme C ✕
	Arginine	Arginine	Arginine	Arginine

SOURCE A. M. Srb and N. H. Horowitz, The ornithine cycle in *Neurospora* and its genetic control, *Journal of Biological Chemistry* 154:129–139 (1944).

WHAT IF? Suppose the experiment had shown that class I mutants could grow only in MM supplemented by ornithine or arginine and that class II mutants could grow in MM supplemented by citrulline, ornithine, or arginine. What conclusions would the researchers have drawn from those results regarding the biochemical pathway and the defect in class I and class II mutants?

even though they are never translated into protein. For now, we will focus on genes that do code for polypeptides. (Note that it is common to refer to these gene products as proteins—a practice you will encounter in this book—rather than more precisely as polypeptides.)

Basic Principles of Transcription and Translation

Genes provide the instructions for making specific proteins. But a gene does not build a protein directly. The bridge between DNA and protein synthesis is the nucleic acid RNA. You learned in Chapter 5 that RNA is chemically similar to DNA except that it contains ribose instead of deoxyribose as its sugar and has the nitrogenous base uracil rather than thymine (see Figure 5.26). Thus, each nucleotide along a DNA strand has A, G, C, or T as its base, and each nucleotide along an RNA strand has A, G, C, or U as its base. An RNA molecule usually consists of a single strand.

It is customary to describe the flow of information from gene to protein in linguistic terms because both nucleic acids and proteins are polymers with specific sequences of monomers that convey information, much as specific sequences of letters communicate information in a language like English. In DNA or RNA, the monomers are the four types of nucleotides, which differ in their nitrogenous bases. Genes are typically hundreds or thousands of nucleotides long, each gene having a specific sequence of nucleotides. Each polypeptide of a protein also has monomers arranged in a particular linear order (the protein's primary structure), but its monomers are amino acids. Thus, nucleic acids and proteins contain information written in two different chemical languages. Getting from DNA to protein requires two major stages: transcription and translation.

Transcription is the synthesis of RNA using information in the DNA. The two nucleic acids are written in different forms of the same language, and the information is simply transcribed, or "rewritten," from DNA to RNA. Just as a DNA strand provides a template for making a new complementary strand during DNA replication, it also can serve as a template for assembling a complementary sequence of RNA nucleotides. For a protein-coding gene, the resulting RNA molecule is a faithful transcript of the gene's protein-building instructions. This type of RNA molecule is called **messenger RNA (mRNA)** because it carries a genetic message from the DNA to the protein-synthesizing machinery of the cell. (Transcription is the general term for the synthesis of *any* kind of RNA on a DNA template. Later, you will learn about some other types of RNA produced by transcription.)

Translation is the synthesis of a polypeptide using the information in the mRNA. During this stage, there is a change in language: The cell must translate the nucleotide sequence of an mRNA molecule into the amino acid sequence of a polypeptide. The sites of translation are **ribosomes**, complex particles that facilitate the orderly linking of amino acids into polypeptide chains.

Transcription and translation occur in all organisms, both those that lack a membrane-bounded nucleus (bacteria and archaea) and those that have one (eukaryotes). Because most studies of transcription and translation have used bacteria and eukaryotic cells, these are our main focus in this chapter. Our understanding of transcription and translation in archaea lags behind, but in the last section of the chapter we will discuss a few aspects of archaeal gene expression.

The basic mechanics of transcription and translation are similar for bacteria and eukaryotes, but there is an important difference in the flow of genetic information within the cells. Because bacteria do not have nuclei, their DNA is not separated by nuclear membranes from ribosomes and the other protein-synthesizing equipment **(Figure 17.3a)**. As you will see later, this lack of compartmentalization allows translation of an mRNA to begin while its transcription is still in progress. In a eukaryotic cell, by contrast, the nuclear envelope separates transcription from translation in space and time **(Figure 17.3b)**. Transcription occurs in the nucleus, and mRNA is then transported to the cytoplasm, where translation occurs. But before eukaryotic RNA transcripts from protein-coding genes can leave the nucleus, they are modified in various ways to produce the final, functional mRNA. The transcription of a protein-coding eukaryotic gene results in *pre-mRNA*, and further processing yields the finished mRNA. The initial RNA transcript from any gene, including those specifying RNA that is not translated into protein, is more generally called a **primary transcript**.

To summarize: Genes program protein synthesis via genetic messages in the form of messenger RNA. Put another way, cells are governed by a molecular chain of command with a directional flow of genetic information, shown here by arrows:

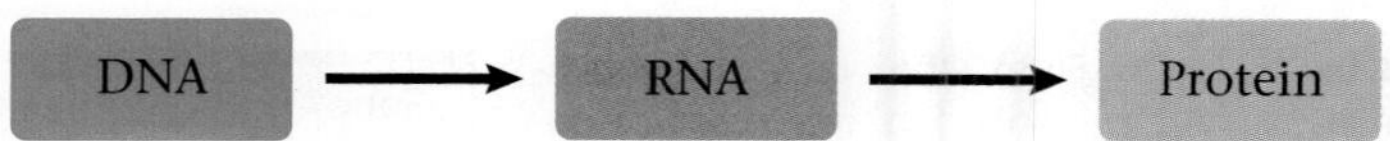

This concept was dubbed the *central dogma* by Francis Crick in 1956. How has the concept held up over time? In the 1970s, scientists were surprised to discover that some RNA molecules can act as templates for DNA synthesis, a process you'll read about in Chapter 19. However, these exceptions do not invalidate the idea that, in general, genetic information flows from DNA to RNA to protein. In the next section, we discuss how the instructions for assembling amino acids into a specific order are encoded in nucleic acids.

The Genetic Code

When biologists began to suspect that the instructions for protein synthesis were encoded in DNA, they recognized a problem: There are only four nucleotide bases to specify 20 amino acids. Thus, the genetic code cannot be a language like Chinese, where each written symbol corresponds to a word. How many nucleotides, then, correspond to an amino acid?

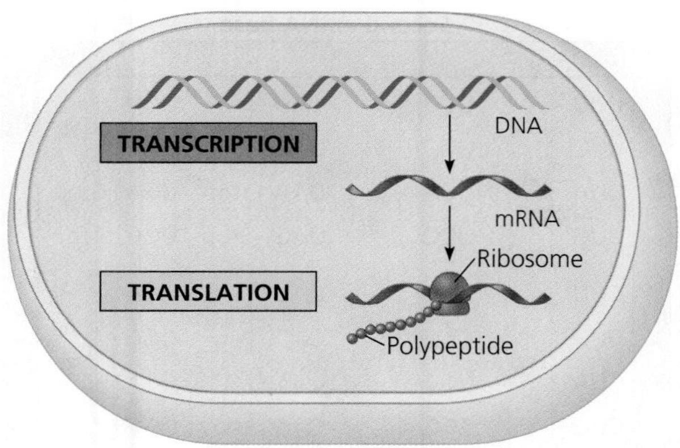

(a) Bacterial cell. In a bacterial cell, which lacks a nucleus, mRNA produced by transcription is immediately translated without additional processing.

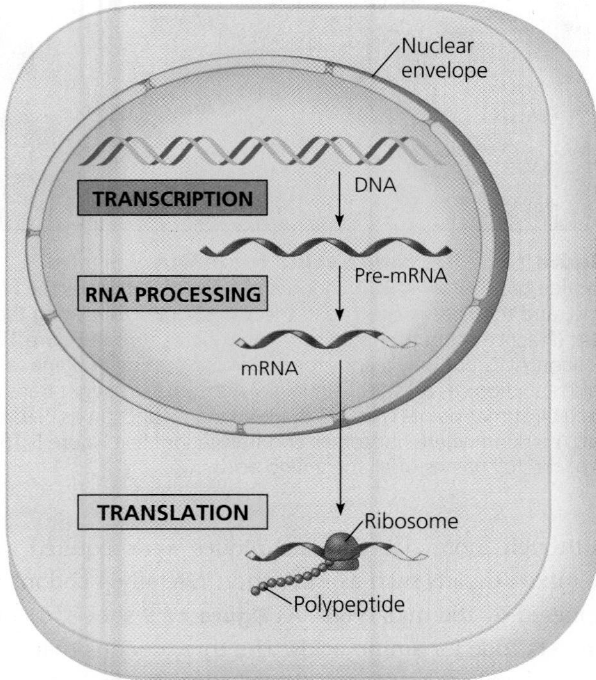

(b) Eukaryotic cell. The nucleus provides a separate compartment for transcription. The original RNA transcript, called pre-mRNA, is processed in various ways before leaving the nucleus as mRNA.

▲ **Figure 17.3 Overview: the roles of transcription and translation in the flow of genetic information.** In a cell, inherited information flows from DNA to RNA to protein. The two main stages of information flow are transcription and translation. A miniature version of part (a) or (b) accompanies several figures later in the chapter as an orientation diagram to help you see where a particular figure fits into the overall scheme.

Codons: Triplets of Nucleotides

If each kind of nucleotide base were translated into an amino acid, only 4 of the 20 amino acids could be specified. Would a language of two-letter code words suffice? The two-nucleotide sequence AG, for example, could specify one amino acid, and GT could specify another. Since there are four possible nucleotide bases in each position, this would give us 16 (that is, 4^2) possible arrangements—still not enough to code for all 20 amino acids.

Triplets of nucleotide bases are the smallest units of uniform length that can code for all the amino acids. If each arrangement of three consecutive nucleotide bases specifies an amino acid, there can be 64 (that is, 4^3) possible code words—more than enough to specify all the amino acids. Experiments have verified that the flow of information from gene to protein is based on a **triplet code**: The genetic instructions for a polypeptide chain are written in the DNA as a series of nonoverlapping, three-nucleotide words. The series of words in a gene is transcribed into a complementary series of nonoverlapping, three-nucleotide words in mRNA, which is then translated into a chain of amino acids **(Figure 17.4)**.

During transcription, the gene determines the sequence of nucleotide bases along the length of the RNA molecule that is being synthesized. For each gene, only one of the two DNA strands is transcribed. This strand is called the **template strand** because it provides the pattern, or template, for the sequence of nucleotides in an RNA transcript. For any given gene, the same strand is used as the template every time the gene is transcribed. For other genes on the same DNA molecule, however, the opposite strand may be the one that always functions as the template.

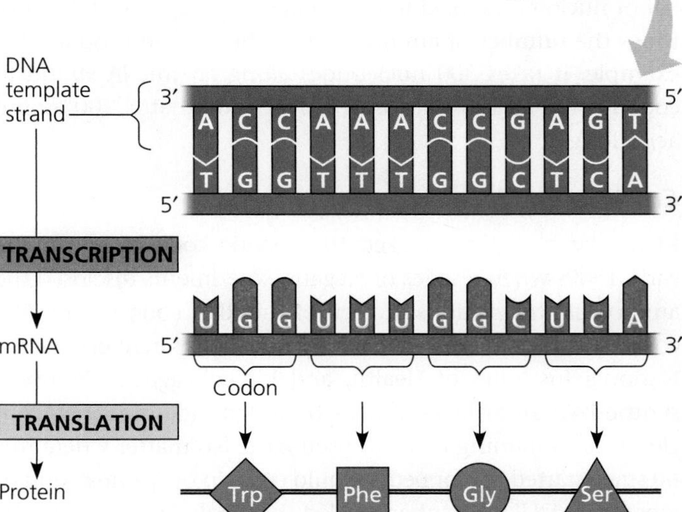

▲ **Figure 17.4 The triplet code.** For each gene, one DNA strand functions as a template for transcription of RNAs, such as mRNA. The base-pairing rules for DNA synthesis also guide transcription, except that uracil (U) takes the place of thymine (T) in RNA. During translation, the mRNA is read as a sequence of nucleotide triplets, called codons. Each codon specifies an amino acid to be added to the growing polypeptide chain. The mRNA is read in the 5′ → 3′ direction.

? *Compare the sequence of the mRNA to that of the nontemplate DNA strand, in both cases reading from 5′ → 3′.*

An mRNA molecule is complementary rather than identical to its DNA template because RNA nucleotides are assembled on the template according to base-pairing rules (see Figure 17.4). The pairs are similar to those that form during DNA replication, except that U, the RNA substitute for T, pairs with A and the mRNA nucleotides contain ribose instead of deoxyribose. Like a new strand of DNA, the RNA molecule is synthesized in an antiparallel direction to the template strand of DNA. (To review what is meant by "antiparallel" and the 5′ and 3′ ends of a nucleic acid chain, see Figure 16.7.) In the example in Figure 17.4, the nucleotide triplet ACC along the DNA (written as 3′-ACC-5′) provides a template for 5′-UGG-3′ in the mRNA molecule. The mRNA nucleotide triplets are called **codons**, and they are customarily written in the 5′ → 3′ direction. In our example, UGG is the codon for the amino acid tryptophan (abbreviated Trp). The term *codon* is also used for the DNA nucleotide triplets along the *nontemplate* strand. These codons are complementary to the template strand and thus identical in sequence to the mRNA, except that they have T instead of U. (For this reason, the nontemplate DNA strand is sometimes called the "coding strand.")

During translation, the sequence of codons along an mRNA molecule is decoded, or translated, into a sequence of amino acids making up a polypeptide chain. The codons are read by the translation machinery in the 5′ → 3′ direction along the mRNA. Each codon specifies which one of the 20 amino acids will be incorporated at the corresponding position along a polypeptide. Because codons are nucleotide triplets, the number of nucleotides making up a genetic message must be three times the number of amino acids in the protein product. For example, it takes 300 nucleotides along an mRNA strand to code for the amino acids in a polypeptide that is 100 amino acids long.

Cracking the Code

Molecular biologists cracked the genetic code of life in the early 1960s when a series of elegant experiments disclosed the amino acid translations of each of the RNA codons. The first codon was deciphered in 1961 by Marshall Nirenberg, of the National Institutes of Health, and his colleagues. Nirenberg synthesized an artificial mRNA by linking identical RNA nucleotides containing uracil as their base. No matter where this message started or stopped, it could contain only one codon in repetition: UUU. Nirenberg added this "poly-U" to a test-tube mixture containing amino acids, ribosomes, and the other components required for protein synthesis. His artificial system translated the poly-U into a polypeptide containing many units of the amino acid phenylalanine (Phe), strung together as a long polyphenylalanine chain. Thus, Nirenberg determined that the mRNA codon UUU specifies the amino acid phenylalanine. Soon, the amino acids specified by the codons AAA, GGG, and CCC were also determined.

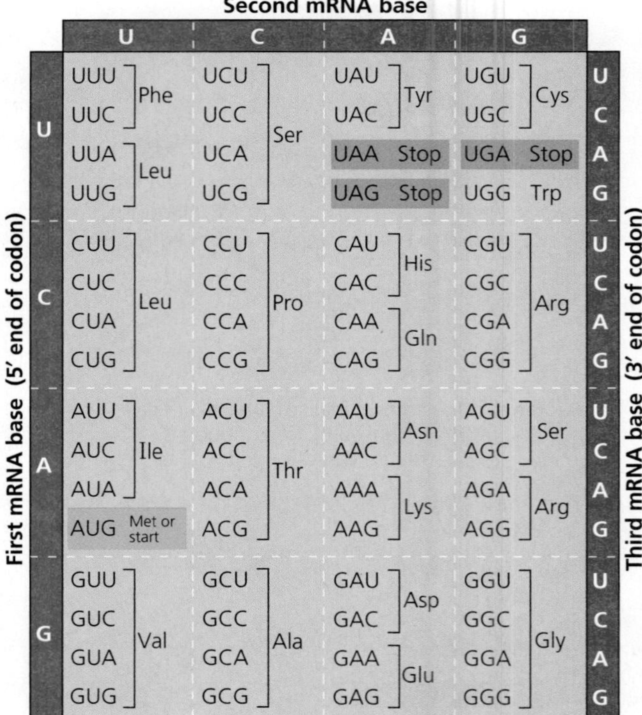

First mRNA base (5′ end of codon)	Second mRNA base: U	C	A	G	Third mRNA base (3′ end of codon)
U	UUU Phe	UCU Ser	UAU Tyr	UGU Cys	U
U	UUC Phe	UCC Ser	UAC Tyr	UGC Cys	C
U	UUA Leu	UCA Ser	UAA Stop	UGA Stop	A
U	UUG Leu	UCG Ser	UAG Stop	UGG Trp	G
C	CUU Leu	CCU Pro	CAU His	CGU Arg	U
C	CUC Leu	CCC Pro	CAC His	CGC Arg	C
C	CUA Leu	CCA Pro	CAA Gln	CGA Arg	A
C	CUG Leu	CCG Pro	CAG Gln	CGG Arg	G
A	AUU Ile	ACU Thr	AAU Asn	AGU Ser	U
A	AUC Ile	ACC Thr	AAC Asn	AGC Ser	C
A	AUA Ile	ACA Thr	AAA Lys	AGA Arg	A
A	AUG Met or start	ACG Thr	AAG Lys	AGG Arg	G
G	GUU Val	GCU Ala	GAU Asp	GGU Gly	U
G	GUC Val	GCC Ala	GAC Asp	GGC Gly	C
G	GUA Val	GCA Ala	GAA Glu	GGA Gly	A
G	GUG Val	GCG Ala	GAG Glu	GGG Gly	G

▲ **Figure 17.5 The codon table for mRNA.** The three nucleotide bases of an mRNA codon are designated here as the first, second, and third bases, reading in the 5′ → 3′ direction along the mRNA. (Practice using this table by finding the codons in Figure 17.4.) The codon AUG not only stands for the amino acid methionine (Met) but also functions as a "start" signal for ribosomes to begin translating the mRNA at that point. Three of the 64 codons function as "stop" signals, marking where ribosomes end translation. See Figure 5.16 for a list of the full names of all the amino acids.

Although more elaborate techniques were required to decode mixed triplets such as AUA and CGA, all 64 codons were deciphered by the mid-1960s. As **Figure 17.5** shows, 61 of the 64 triplets code for amino acids. The three codons that do not designate amino acids are "stop" signals, or termination codons, marking the end of translation. Notice that the codon AUG has a dual function: It codes for the amino acid methionine (Met) and also functions as a "start" signal, or initiation codon. Genetic messages usually begin with the mRNA codon AUG, which signals the protein-synthesizing machinery to begin translating the mRNA at that location. (Because AUG also stands for methionine, polypeptide chains begin with methionine when they are synthesized. However, an enzyme may subsequently remove this starter amino acid from the chain.)

Notice in Figure 17.5 that there is redundancy in the genetic code, but no ambiguity. For example, although codons GAA and GAG both specify glutamic acid (redundancy), neither of them ever specifies any other amino acid (no ambiguity). The redundancy in the code is not altogether random. In many cases, codons that are synonyms for a particular amino acid differ only in the third nucleotide base of the triplet. We will consider a possible benefit of this redundancy later in the chapter.

Our ability to extract the intended message from a written language depends on reading the symbols in the correct groupings—that is, in the correct **reading frame**. Consider this statement: "The red dog ate the bug." Group the letters incorrectly by starting at the wrong point, and the result will probably be gibberish: for example, "her edd oga tet heb ug." The reading frame is also important in the molecular language of cells. The short stretch of polypeptide shown in Figure 17.4, for instance, will be made correctly only if the mRNA nucleotides are read from left to right (5′ → 3′) in the groups of three shown in the figure: UGG UUU GGC UCA. Although a genetic message is written with no spaces between the codons, the cell's protein-synthesizing machinery reads the message as a series of nonoverlapping three-letter words. The message is *not* read as a series of overlapping words—UGGUUU, and so on—which would convey a very different message.

Evolution of the Genetic Code

EVOLUTION The genetic code is nearly universal, shared by organisms from the simplest bacteria to the most complex plants and animals. The RNA codon CCG, for instance, is translated as the amino acid proline in all organisms whose genetic code has been examined. In laboratory experiments, genes can be transcribed and translated after being transplanted from one species to another, sometimes with quite striking results, as shown in **Figure 17.6**! Bacteria can be programmed by the insertion of human genes to synthesize certain human proteins for medical use, such as insulin. Such applications have produced many exciting developments in the area of biotechnology (see Chapter 20).

Exceptions to the universality of the genetic code include translation systems in which a few codons differ from the standard ones. Slight variations in the genetic code exist in certain unicellular eukaryotes and in the organelle genes of some species. Despite these exceptions, the evolutionary significance of the code's *near* universality is clear. A language shared by all living things must have been operating very early in the history of life—early enough to be present in the common ancestor of all present-day organisms. A shared genetic vocabulary is a reminder of the kinship that bonds all life on Earth.

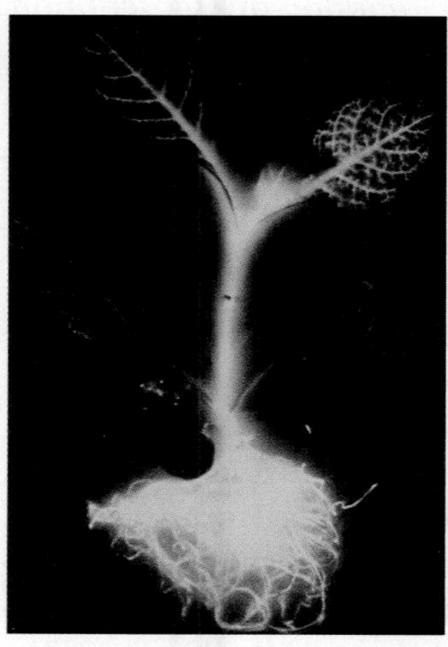

(a) Tobacco plant expressing a firefly gene. The yellow glow is produced by a chemical reaction catalyzed by the protein product of the firefly gene.

(b) Pig expressing a jellyfish gene. Researchers injected the gene for a fluorescent protein into fertilized pig eggs. One of the eggs developed into this fluorescent pig.

▲ **Figure 17.6 Expression of genes from different species.** Because diverse forms of life share a common genetic code, one species can be programmed to produce proteins characteristic of a second species by introducing DNA from the second species into the first.

CONCEPT CHECK 17.1

1. **MAKE CONNECTIONS** In a research article about alkaptonuria published in 1902, Garrod suggested that humans inherit two "characters" (alleles) for a particular enzyme and that both parents must contribute a faulty version for the offspring to have the disorder. Today, would this disorder be called dominant or recessive? See Concept 14.4, pages 276–278.
2. What polypeptide product would you expect from a poly-G mRNA that is 30 nucleotides long?
3. **DRAW IT** The template strand of a gene contains the sequence 3′-TTCAGTCGT-5′. Draw the nontemplate sequence and the mRNA sequence, indicating 5′ and 3′ ends of each. Compare the two sequences.
4. **WHAT IF?** **DRAW IT** Imagine that the nontemplate sequence in question 3 was transcribed instead of the template sequence. Draw the mRNA sequence and translate it using Figure 17.5. (Be sure to pay attention to the 5′ and 3′ ends.) Predict how well the protein synthesized from the nontemplate strand would function, if at all.

For suggested answers, see Appendix A.

CONCEPT 17.2

Transcription is the DNA-directed synthesis of RNA: *a closer look*

Now that we have considered the linguistic logic and evolutionary significance of the genetic code, we are ready to reexamine transcription, the first stage of gene expression, in more detail.

Molecular Components of Transcription

Messenger RNA, the carrier of information from DNA to the cell's protein-synthesizing machinery, is transcribed from the template strand of a gene. An enzyme called an **RNA polymerase** pries the two strands of DNA apart and joins together RNA nucleotides complementary to the DNA template strand, thus elongating the RNA polynucleotide (**Figure 17.7**). Like the DNA polymerases that function in DNA replication, RNA polymerases can assemble a polynucleotide only in its 5′ → 3′ direction. Unlike DNA polymerases, however, RNA polymerases are able to start a chain from scratch; they don't need a primer.

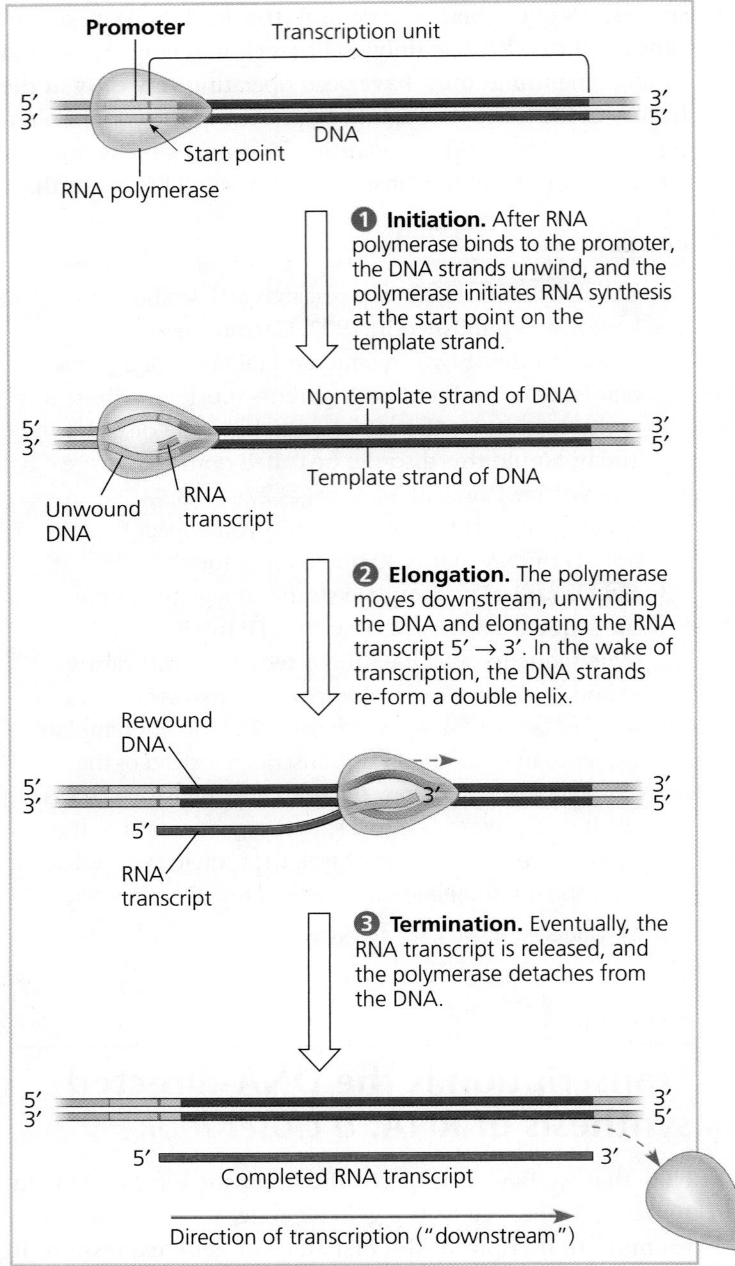

Promoter
Transcription unit
5′
3′
3′
5′
Start point
DNA
RNA polymerase
1 Initiation. After RNA polymerase binds to the promoter, the DNA strands unwind, and the polymerase initiates RNA synthesis at the start point on the template strand.
Nontemplate strand of DNA
Template strand of DNA
Unwound DNA
RNA transcript
2 Elongation. The polymerase moves downstream, unwinding the DNA and elongating the RNA transcript 5′ → 3′. In the wake of transcription, the DNA strands re-form a double helix.
Rewound DNA
RNA transcript
3 Termination. Eventually, the RNA transcript is released, and the polymerase detaches from the DNA.
Completed RNA transcript
Direction of transcription ("downstream")

▲ **Figure 17.7 The stages of transcription: initiation, elongation, and termination.** This general depiction of transcription applies to both bacteria and eukaryotes, but the details of termination differ, as described in the text. Also, in a bacterium, the RNA transcript is immediately usable as mRNA; in a eukaryote, the RNA transcript must first undergo processing.

MAKE CONNECTIONS *Compare the use of a template strand during transcription and replication. See Figure 16.17, page 317.*

Specific sequences of nucleotides along the DNA mark where transcription of a gene begins and ends. The DNA sequence where RNA polymerase attaches and initiates transcription is known as the **promoter**; in bacteria, the sequence that signals the end of transcription is called the **terminator**. (The termination mechanism is different in eukaryotes; we'll describe it later.) Molecular biologists refer to the direction of transcription as "downstream" and the other direction as "upstream." These terms are also used to describe the positions of nucleotide sequences within the DNA or RNA. Thus, the promoter sequence in DNA is said to be upstream from the terminator. The stretch of DNA that is transcribed into an RNA molecule is called a **transcription unit**.

Bacteria have a single type of RNA polymerase that synthesizes not only mRNA but also other types of RNA that function in protein synthesis, such as ribosomal RNA. In contrast, eukaryotes have at least three types of RNA polymerase in their nuclei. The one used for mRNA synthesis is called RNA polymerase II. The other RNA polymerases transcribe RNA molecules that are not translated into protein. In the discussion of transcription that follows, we start with the features of mRNA synthesis common to both bacteria and eukaryotes and then describe some key differences.

Synthesis of an RNA Transcript

The three stages of transcription, as shown in Figure 17.7 and described next, are initiation, elongation, and termination of the RNA chain. Study Figure 17.7 to familiarize yourself with the stages and the terms used to describe them.

RNA Polymerase Binding and Initiation of Transcription

The promoter of a gene includes within it the transcription **start point** (the nucleotide where RNA synthesis actually begins) and typically extends several dozen or more nucleotide pairs upstream from the start point. RNA polymerase binds in a precise location and orientation on the promoter, therefore determining where transcription starts and which of the two strands of the DNA helix is used as the template.

Certain sections of a promoter are especially important for binding RNA polymerase. In bacteria, the RNA polymerase itself specifically recognizes and binds to the promoter. In eukaryotes, a collection of proteins called **transcription factors** mediate the binding of RNA polymerase and the initiation of transcription. Only after transcription factors are attached to the promoter does RNA polymerase II bind to it. The whole complex of transcription factors and RNA polymerase II bound to the promoter is called a **transcription initiation complex**. **Figure 17.8** shows the role of transcription factors and a crucial promoter DNA sequence called a **TATA box** in forming the initiation complex at a eukaryotic promoter.

The interaction between eukaryotic RNA polymerase II and transcription factors is an example of the importance of protein-protein interactions in controlling eukaryotic

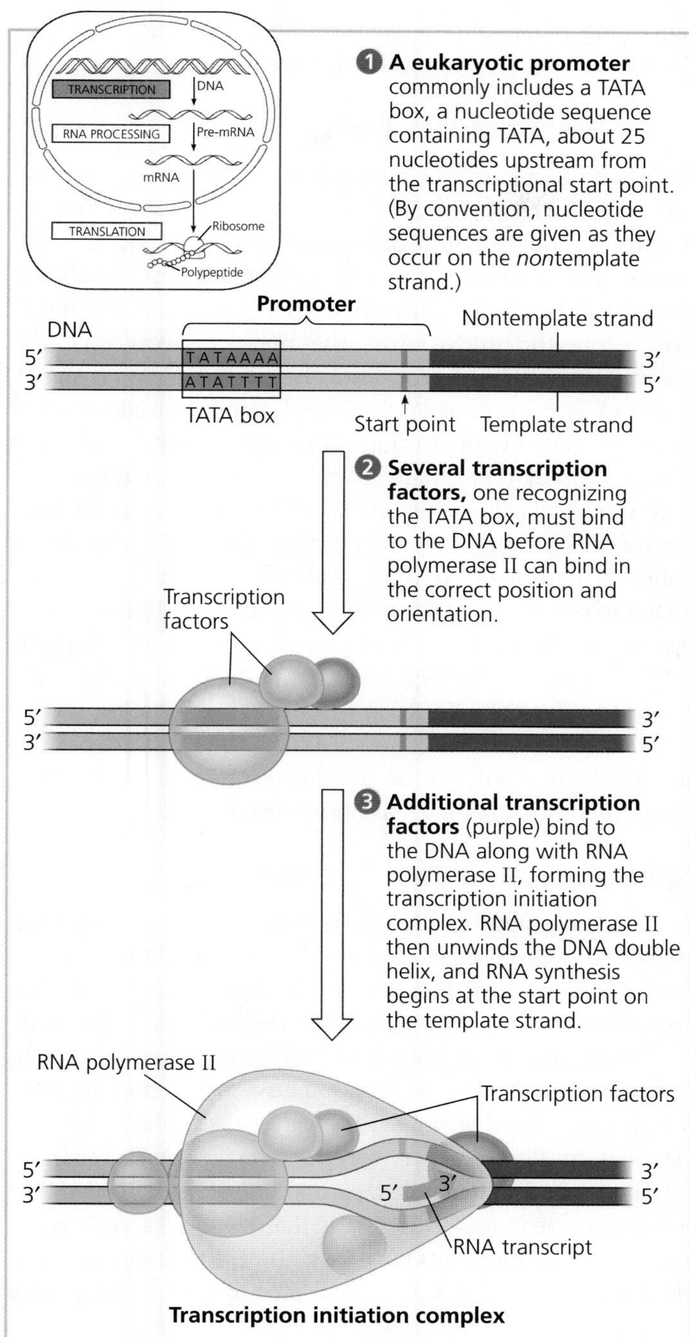
TRANSCRIPTION
DNA
RNA PROCESSING
Pre-mRNA
mRNA
TRANSLATION
Ribosome
Polypeptide
1 A eukaryotic promoter commonly includes a TATA box, a nucleotide sequence containing TATA, about 25 nucleotides upstream from the transcriptional start point. (By convention, nucleotide sequences are given as they occur on the nontemplate strand.)
Promoter
Nontemplate strand
DNA
5′
3′
TATAAAA
ATATTTT
3′
5′
TATA box
Start point
Template strand
2 Several transcription factors, one recognizing the TATA box, must bind to the DNA before RNA polymerase II can bind in the correct position and orientation.
Transcription factors
3 Additional transcription factors (purple) bind to the DNA along with RNA polymerase II, forming the transcription initiation complex. RNA polymerase II then unwinds the DNA double helix, and RNA synthesis begins at the start point on the template strand.
RNA polymerase II
Transcription factors
RNA transcript
Transcription initiation complex

▲ **Figure 17.8 The initiation of transcription at a eukaryotic promoter.** In eukaryotic cells, proteins called transcription factors mediate the initiation of transcription by RNA polymerase II.

? *Explain how the interaction of RNA polymerase with the promoter would differ if the figure showed transcription initiation for bacteria.*

transcription. (And as you learned in Figure 16.22, the DNA of a eukaryotic chromosome is complexed with histones and other proteins in the form of chromatin. The roles of these proteins in making the DNA accessible to transcription factors will be discussed in Chapter 18). Once the appropriate transcription factors are firmly attached to the promoter DNA and the polymerase is bound in the correct orientation, the enzyme unwinds the two DNA strands and starts transcribing the template strand.

Elongation of the RNA Strand

As RNA polymerase moves along the DNA, it continues to untwist the double helix, exposing about 10–20 DNA nucleotides at a time for pairing with RNA nucleotides (**Figure 17.9**). The enzyme adds nucleotides to the 3′ end of the growing RNA molecule as it continues along the double helix. In the wake of this advancing wave of RNA synthesis, the new RNA molecule peels away from its DNA template, and the DNA double helix re-forms. Transcription progresses at a rate of about 40 nucleotides per second in eukaryotes.

A single gene can be transcribed simultaneously by several molecules of RNA polymerase following each other like trucks in a convoy. A growing strand of RNA trails off from each polymerase, with the length of each new strand reflecting how far along the template the enzyme has traveled from the start point (see the mRNA molecules in Figure 17.25). The congregation of many polymerase molecules simultaneously transcribing a single gene increases the amount of mRNA transcribed from it, which helps the cell make the encoded protein in large amounts.

Termination of Transcription

The mechanism of termination differs between bacteria and eukaryotes. In bacteria, transcription proceeds through a terminator sequence in the DNA. The transcribed terminator (an RNA sequence) functions as the termination signal,

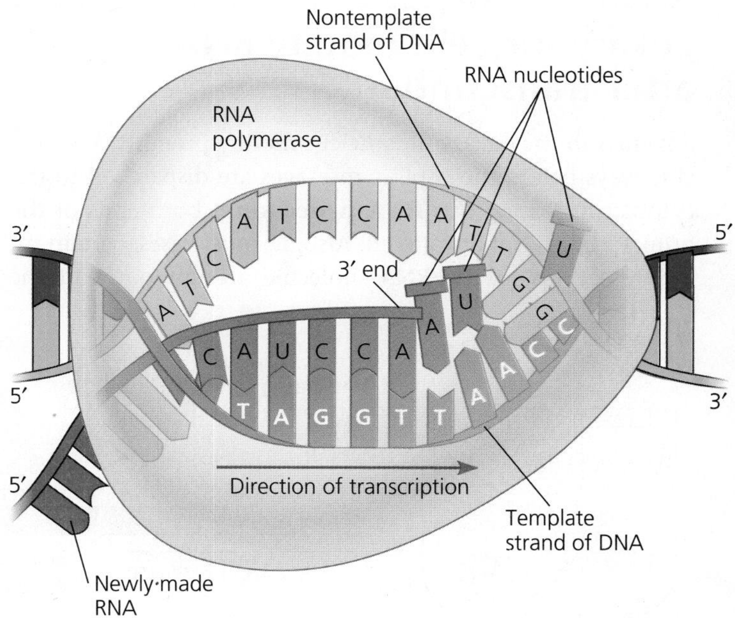
Nontemplate strand of DNA
RNA nucleotides
RNA polymerase
3′ end
Direction of transcription
Template strand of DNA
Newly-made RNA

▲ **Figure 17.9 Transcription elongation.** RNA polymerase moves along the DNA template strand, joining complementary RNA nucleotides to the 3′ end of the growing RNA transcript. Behind the polymerase, the new RNA peels away from the template strand, which re-forms a double helix with the nontemplate strand.

causing the polymerase to detach from the DNA and release the transcript, which requires no further modification before translation. In eukaryotes, RNA polymerase II transcribes a sequence on the DNA called the polyadenylation signal sequence, which codes for a polyadenylation signal (AAUAAA) in the pre-mRNA. Then, at a point about 10–35 nucleotides downstream from the AAUAAA signal, proteins associated with the growing RNA transcript cut it free from the polymerase, releasing the pre-mRNA. The pre-mRNA then undergoes processing, the topic of the next section.

CONCEPT CHECK 17.2

1. **MAKE CONNECTIONS** Compare DNA polymerase and RNA polymerase in terms of how they function, the requirement for a template and primer, the direction of synthesis, and the type of nucleotides used. See Figure 16.17, page 317.
2. What is a promoter, and is it located at the upstream or downstream end of a transcription unit?
3. What enables RNA polymerase to start transcribing a gene at the right place on the DNA in a bacterial cell? In a eukaryotic cell?
4. **WHAT IF?** Suppose X-rays caused a sequence change in the TATA box of a particular gene's promoter. How would that affect transcription of the gene? (See Figure 17.8.)

For suggested answers, see Appendix A.

CONCEPT **17.3**

Eukaryotic cells modify RNA after transcription

Enzymes in the eukaryotic nucleus modify pre-mRNA in specific ways before the genetic messages are dispatched to the cytoplasm. During this **RNA processing**, both ends of the primary transcript are altered. Also, in most cases, certain interior sections of the RNA molecule are cut out and the remaining parts spliced together. These modifications produce an mRNA molecule ready for translation.

Alteration of mRNA Ends

Each end of a pre-mRNA molecule is modified in a particular way (**Figure 17.10**). The 5′ end is synthesized first; it receives a **5′ cap**, a modified form of a guanine (G) nucleotide added onto the 5′ end after transcription of the first 20–40 nucleotides. The 3′ end of the pre-mRNA molecule is also modified before the mRNA exits the nucleus. Recall that the pre-mRNA is released soon after the polyadenylation signal, AAUAAA, is transcribed. At the 3′ end, an enzyme adds 50–250 more adenine (A) nucleotides, forming a **poly-A tail**. The 5′ cap and poly-A tail share several important functions. First, they seem to facilitate the export of the mature mRNA from the nucleus. Second, they help protect the mRNA from degradation by hydrolytic enzymes. And third, they help ribosomes attach to the 5′ end of the mRNA once the mRNA reaches the cytoplasm. Figure 17.10 shows a diagram of a eukaryotic mRNA molecule with cap and tail. The figure also shows the untranslated regions (UTRs) at the 5′ and 3′ ends of the mRNA (referred to as the 5′ UTR and 3′ UTR). The UTRs are parts of the mRNA that will not be translated into protein, but they have other functions, such as ribosome binding.

Split Genes and RNA Splicing

A remarkable stage of RNA processing in the eukaryotic nucleus is the removal of large portions of the RNA molecule that is initially synthesized—a cut-and-paste job called **RNA splicing**, similar to editing a video (**Figure 17.11**). The average length of a transcription unit along a human DNA molecule is about 27,000 nucleotide pairs, so the primary RNA transcript is also that long. However, it takes only 1,200 nucleotides in RNA to code for the average-sized protein of 400 amino acids. (Remember, each amino acid is encoded by a *triplet* of nucleotides.) This means that most eukaryotic genes and their RNA transcripts have long noncoding stretches of nucleotides, regions that are not translated. Even more surprising

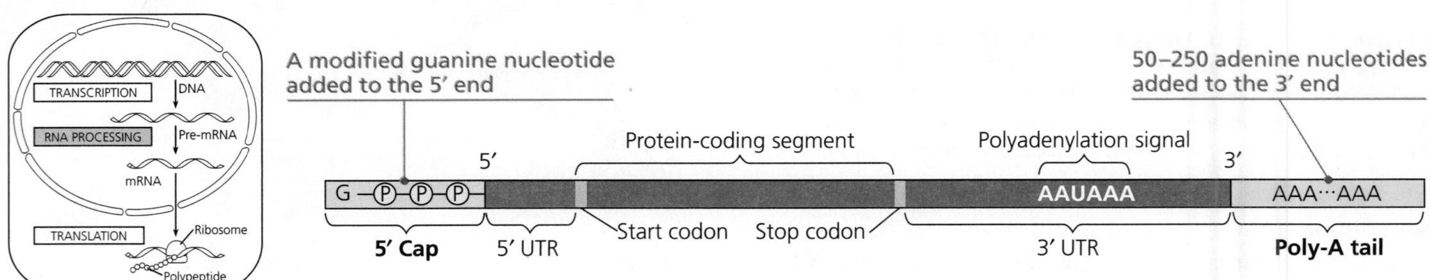

▲ **Figure 17.10 RNA processing: Addition of the 5′ cap and poly-A tail.** Enzymes modify the two ends of a eukaryotic pre-mRNA molecule. The modified ends may promote the export of mRNA from the nucleus, and they help protect the mRNA from degradation. When the mRNA reaches the cytoplasm, the modified ends, in conjunction with certain cytoplasmic proteins, facilitate ribosome attachment. The 5′ cap and poly-A tail are not translated into protein, nor are the regions called the 5′ untranslated region (5′ UTR) and 3′ untranslated region (3′ UTR).

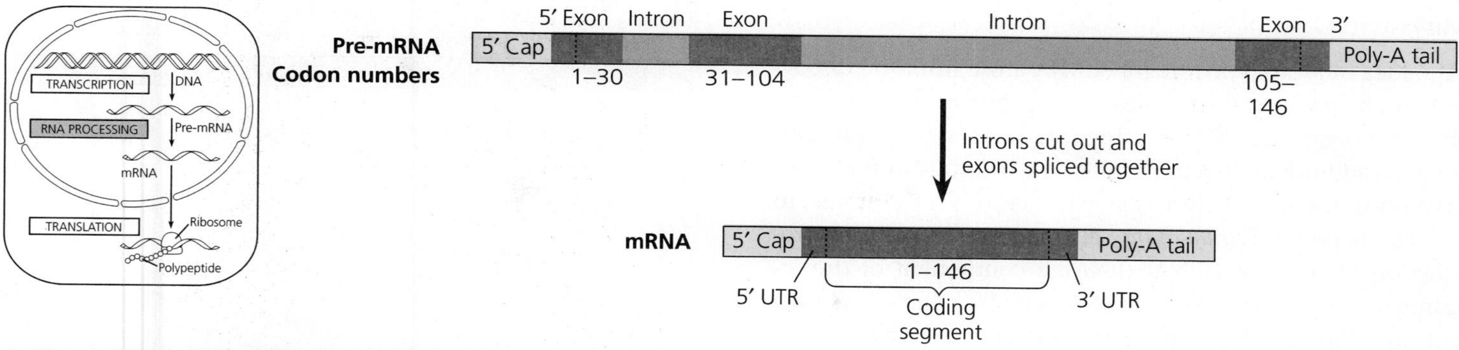

▲ **Figure 17.11 RNA processing: RNA splicing.** The RNA molecule shown here codes for β-globin, one of the polypeptides of hemoglobin. The numbers under the RNA refer to codons; β-globin is 146 amino acids long. The β-globin gene and its pre-mRNA transcript have three exons, corresponding to sequences that will leave the nucleus as mRNA. (The 5′ UTR and 3′ UTR are parts of exons because they are included in the mRNA; however, they do not code for protein.) During RNA processing, the introns are cut out and the exons spliced together. In many genes, the introns are much larger than the exons.

is that most of these noncoding sequences are interspersed between coding segments of the gene and thus between coding segments of the pre-mRNA. In other words, the sequence of DNA nucleotides that codes for a eukaryotic polypeptide is usually not continuous; it is split into segments. The noncoding segments of nucleic acid that lie between coding regions are called *int*ervening sequences, or **introns**. The other regions are called **exons**, because they are eventually *ex*pressed, usually by being translated into amino acid sequences. (Exceptions include the UTRs of the exons at the ends of the RNA, which make up part of the mRNA but are not translated into protein. Because of these exceptions, you may find it helpful to think of exons as sequences of RNA that *exit* the nucleus.) The terms *intron* and *exon* are used for both RNA sequences and the DNA sequences that encode them.

In making a primary transcript from a gene, RNA polymerase II transcribes both introns and exons from the DNA, but the mRNA molecule that enters the cytoplasm is an abridged version. The introns are cut out from the molecule and the exons joined together, forming an mRNA molecule with a continuous coding sequence. This is the process of RNA splicing.

How is pre-mRNA splicing carried out? Researchers have learned that the signal for RNA splicing is a short nucleotide sequence at each end of an intron. Joan Steitz, our interviewee for this unit (see pp. 246–247), discovered in 1979 that particles called *small nuclear ribonucleoproteins*, abbreviated *snRNPs* (pronounced "snurps"), recognize these splice sites. As the full name implies, snRNPs are located in the cell nucleus and are composed of RNA and protein molecules. The RNA in a snRNP particle is called a *small nuclear RNA (snRNA)*; each snRNA molecule is about 150 nucleotides long. Several different snRNPs join with additional proteins to form an even larger assembly called a **spliceosome**, which is almost as big as a ribosome. The spliceosome interacts with certain sites along an intron, releasing the intron, which is rapidly degraded, and joining together the two exons that flanked the intron (**Figure 17.12**). It turns out that snRNAs catalyze these processes, as well as participating in spliceosome assembly and splice site recognition.

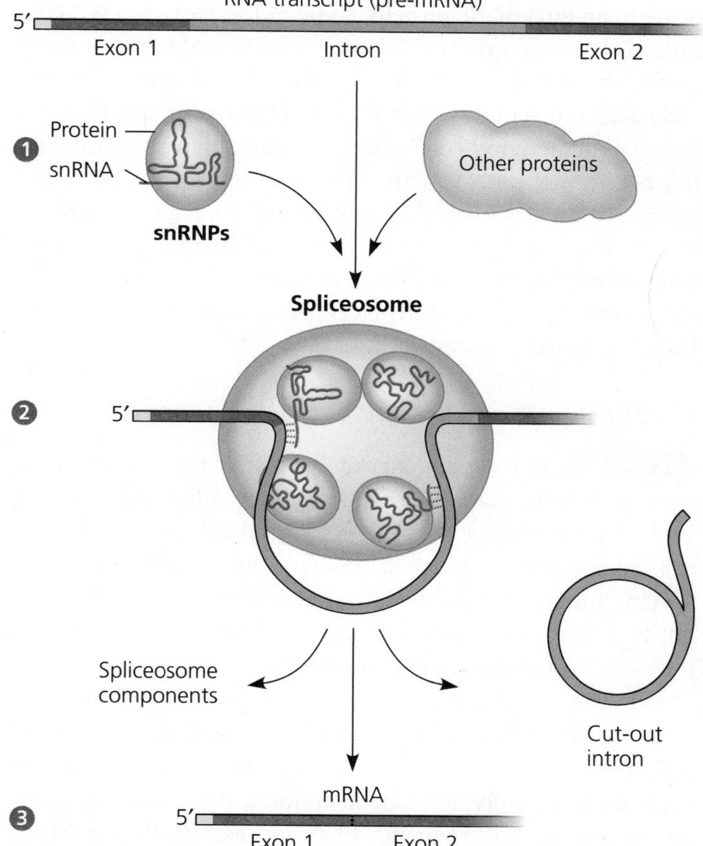

▲ **Figure 17.12 The roles of snRNPs and spliceosomes in pre-mRNA splicing.** The diagram shows only a portion of the pre-mRNA transcript; additional introns and exons lie downstream from the ones pictured here. ❶ Small nuclear ribonucleoproteins (snRNPs) and other proteins form a molecular complex called a spliceosome on a pre-mRNA molecule containing exons and introns. ❷ Within the spliceosome, snRNA base-pairs with nucleotides at specific sites along the intron. ❸ The spliceosome cuts the pre-mRNA, releasing the intron for rapid degradation, and at the same time splices the exons together. The spliceosome then comes apart, releasing mRNA, which now contains only exons.

Ribozymes

The idea of a catalytic role for snRNA arose from the discovery of **ribozymes**, RNA molecules that function as enzymes. In some organisms, RNA splicing can occur without proteins or even additional RNA molecules: The intron RNA functions as a ribozyme and catalyzes its own excision! For example, in the ciliate protist *Tetrahymena*, self-splicing occurs in the production of ribosomal RNA (rRNA), a component of the organism's ribosomes. The pre-rRNA actually removes its own introns. The discovery of ribozymes rendered obsolete the idea that all biological catalysts are proteins.

Three properties of RNA enable some RNA molecules to function as enzymes. First, because RNA is single-stranded, a region of an RNA molecule may base-pair with a complementary region elsewhere in the same molecule, which gives the molecule a particular three-dimensional structure. A specific structure is essential to the catalytic function of ribozymes, just as it is for enzymatic proteins. Second, like certain amino acids in an enzymatic protein, some of the bases in RNA contain functional groups that may participate in catalysis. Third, the ability of RNA to hydrogen-bond with other nucleic acid molecules (either RNA or DNA) adds specificity to its catalytic activity. For example, complementary base pairing between the RNA of the spliceosome and the RNA of a primary RNA transcript precisely locates the region where the ribozyme catalyzes splicing. Later in this chapter, you will see how these properties of RNA also allow it to perform important noncatalytic roles in the cell, such as recognition of the three-nucleotide codons on mRNA.

The Functional and Evolutionary Importance of Introns

EVOLUTION Whether or not RNA splicing and the presence of introns have provided selective advantages during evolutionary history is a matter of some debate. In any case, it is informative to consider their possible adaptive benefits. Specific functions have not been identified for most introns, but at least some contain sequences that regulate gene expression, and many affect gene products.

One important consequence of the presence of introns in genes is that a single gene can encode more than one kind of polypeptide. Many genes are known to give rise to two or more different polypeptides, depending on which segments are treated as exons during RNA processing; this is called **alternative RNA splicing** (see Figure 18.13). For example, sex differences in fruit flies are largely due to differences in how males and females splice the RNA transcribed from certain genes. Results from the Human Genome Project (discussed in Chapter 21) suggest that alternative RNA splicing is one reason humans can get along with about the same number of genes as a nematode (roundworm). Because of alternative splicing, the number of different protein products an organism produces can be much greater than its number of genes.

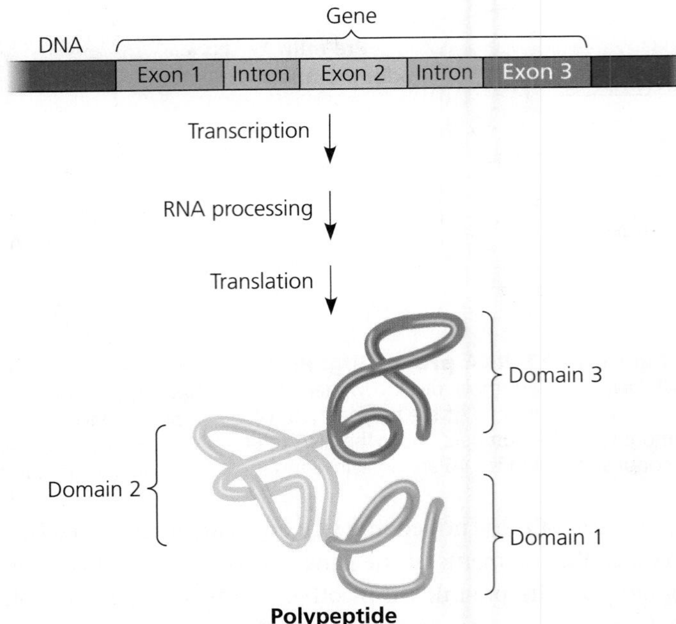

▲ **Figure 17.13 Correspondence between exons and protein domains.**

Proteins often have a modular architecture consisting of discrete structural and functional regions called **domains**. One domain of an enzyme, for example, might include the active site, while another might allow the enzyme to bind to a cellular membrane. In quite a few cases, different exons code for the different domains of a protein (**Figure 17.13**).

The presence of introns in a gene may facilitate the evolution of new and potentially beneficial proteins as a result of a process known as *exon shuffling*. Introns increase the probability of crossing over between the exons of alleles of a gene—simply by providing more terrain for crossovers without interrupting coding sequences. This might result in new combinations of exons and proteins with altered structure and function. We can also imagine the occasional mixing and matching of exons between completely different (nonallelic) genes. Exon shuffling of either sort could lead to new proteins with novel combinations of functions. While most of the shuffling would result in nonbeneficial changes, occasionally a beneficial variant might arise.

CONCEPT CHECK 17.3

1. How can human cells make 75,000–100,000 different proteins, given that there are about 20,000 human genes?
2. How is RNA splicing similar to editing a video? What would introns correspond to in this analogy?
3. **WHAT IF?** What would be the effect of treating cells with an agent that removed the cap from mRNAs?

For suggested answers, see Appendix A.

CONCEPT 17.4

Translation is the RNA-directed synthesis of a polypeptide: *a closer look*

We will now examine in greater detail how genetic information flows from mRNA to protein—the process of translation. As we did for transcription, we'll concentrate on the basic steps of translation that occur in both bacteria and eukaryotes, while pointing out key differences.

Molecular Components of Translation

In the process of translation, a cell "reads" a genetic message and builds a polypeptide accordingly. The message is a series of codons along an mRNA molecule, and the translator is called **transfer RNA (tRNA)**. The function of tRNA is to transfer amino acids from the cytoplasmic pool of amino acids to a growing polypeptide in a ribosome. A cell keeps its cytoplasm stocked with all 20 amino acids, either by synthesizing them from other compounds or by taking them up from the surrounding solution. The ribosome, a structure made of proteins and RNAs, adds each amino acid brought to it by tRNA to the growing end of a polypeptide chain **(Figure 17.14)**.

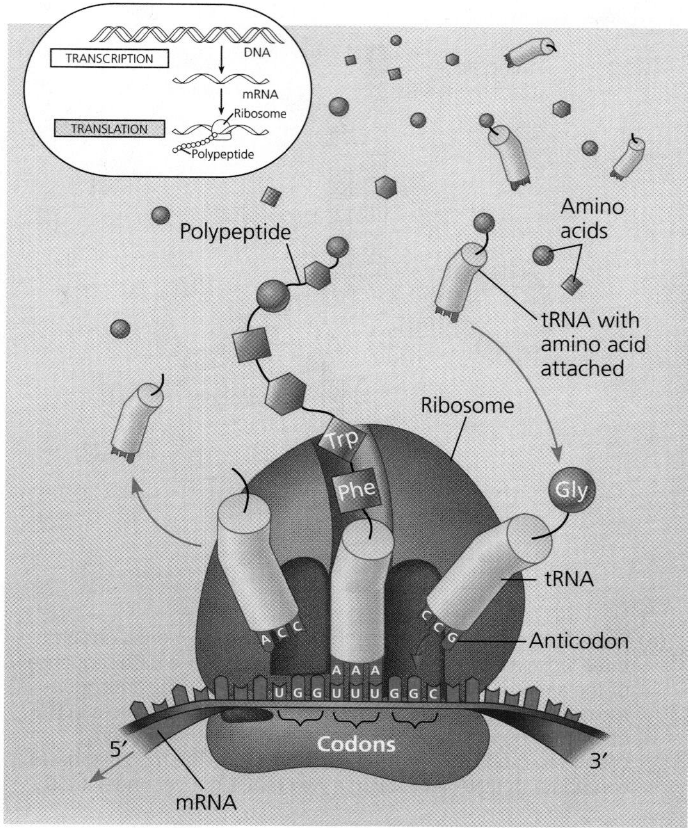

▲ **Figure 17.14 Translation: the basic concept.** As a molecule of mRNA is moved through a ribosome, codons are translated into amino acids, one by one. The interpreters are tRNA molecules, each type with a specific anticodon at one end and a corresponding amino acid at the other end. A tRNA adds its amino acid cargo to a growing polypeptide chain when the anticodon hydrogen-bonds to a complementary codon on the mRNA. The figures that follow show some of the details of translation in a bacterial cell.

BioFlix Visit the Study Area at **www.masteringbiology.com** for the BioFlix® 3-D Animation on Protein Synthesis.

Translation is simple in principle but complex in its biochemistry and mechanics, especially in the eukaryotic cell. In dissecting translation, we'll concentrate on the slightly less complicated version of the process that occurs in bacteria. We'll begin by looking at the major players in this cellular process and then see how they act together in making a polypeptide.

The Structure and Function of Transfer RNA

The key to translating a genetic message into a specific amino acid sequence is the fact that molecules of tRNA are not all identical, and each type of tRNA molecule translates a particular mRNA codon into a particular amino acid. A tRNA molecule arrives at a ribosome bearing a specific amino acid at one end. At the other end of the tRNA is a nucleotide triplet called an **anticodon**, which base-pairs with a complementary codon on mRNA. For example, consider the mRNA codon GGC, which is translated as the amino acid glycine. The tRNA that base-pairs with this codon by hydrogen bonding has CCG as its anticodon and carries glycine at its other end (see the incoming tRNA approaching the ribosome in Figure 17.14). As an mRNA molecule is moved through a ribosome, glycine will be added to the polypeptide chain whenever the codon GGC is presented for translation. Codon by codon, the genetic message is translated as tRNAs deposit amino acids in the order prescribed, and the ribosome joins the amino acids into a chain. The tRNA molecule is a translator in the sense that it can read a nucleic acid word (the mRNA codon) and interpret it as a protein word (the amino acid).

Like mRNA and other types of cellular RNA, transfer RNA molecules are transcribed from DNA templates. In a eukaryotic cell, tRNA, like mRNA, is made in the nucleus and then travels from the nucleus to the cytoplasm, where translation occurs. In both bacterial and eukaryotic cells, each tRNA molecule is used repeatedly, picking up its designated amino acid in the cytosol, depositing this cargo onto a polypeptide chain at the ribosome, and then leaving the ribosome, ready to pick up another amino acid.

A tRNA molecule consists of a single RNA strand that is only about 80 nucleotides long (compared to hundreds of nucleotides for most mRNA molecules). Because of the presence of complementary stretches of nucleotide bases that can hydrogen-bond to each other, this single strand can fold back upon itself

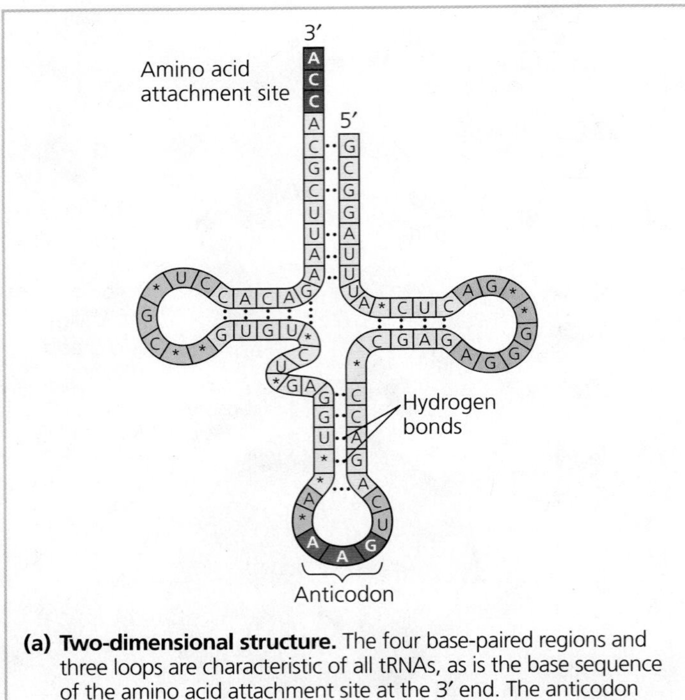

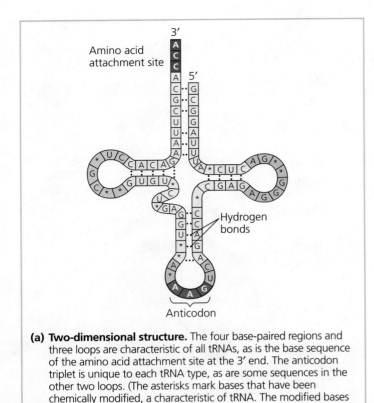

(a) Two-dimensional structure. The four base-paired regions and three loops are characteristic of all tRNAs, as is the base sequence of the amino acid attachment site at the 3′ end. The anticodon triplet is unique to each tRNA type, as are some sequences in the other two loops. (The asterisks mark bases that have been chemically modified, a characteristic of tRNA. The modified bases contribute to tRNA function in a way that is not yet understood.)

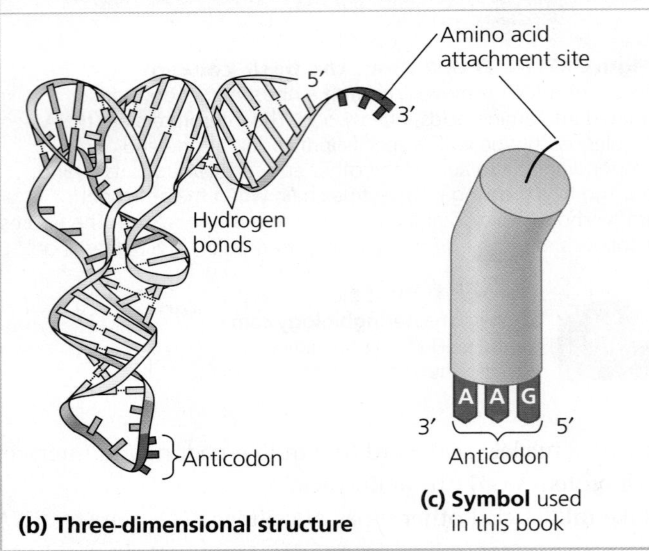

(b) Three-dimensional structure

(c) Symbol used in this book

▲ **Figure 17.15 The structure of transfer RNA (tRNA).** Anticodons are conventionally written 3′ → 5′ to align properly with codons written 5′ → 3′ (see Figure 17.14). For base pairing, RNA strands must be antiparallel, like DNA. For example, anticodon 3′-AAG-5′ pairs with mRNA codon 5′-UUC-3′.

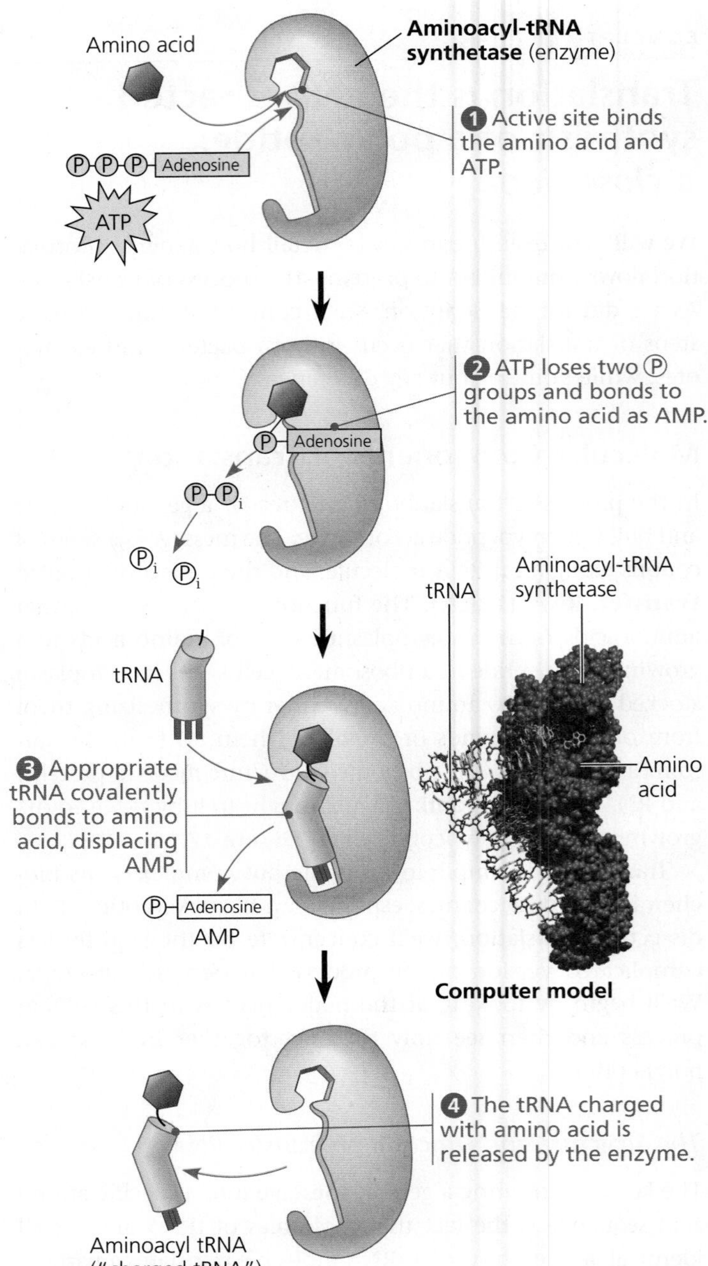

▲ **Figure 17.16 An aminoacyl-tRNA synthetase joining a specific amino acid to a tRNA.** Linkage of the tRNA and amino acid is an endergonic process that occurs at the expense of ATP. The ATP loses two phosphate groups, becoming AMP (adenosine monophosphate).

and form a molecule with a three-dimensional structure. Flattened into one plane to clarify this base pairing, a tRNA molecule looks like a cloverleaf (**Figure 17.15a**). The tRNA actually twists and folds into a compact three-dimensional structure that is roughly L-shaped (**Figure 17.15b**). The loop extending from one end of the L includes the anticodon, the particular nucleotide triplet that base-pairs to a specific mRNA codon. From the other end of the L-shaped tRNA molecule protrudes its 3′ end, which is the attachment site for an amino acid. Thus, the structure of a tRNA molecule fits its function.

The accurate translation of a genetic message requires two instances of molecular recognition. First, a tRNA that binds to an mRNA codon specifying a particular amino acid must carry that amino acid, and no other, to the ribosome. The correct matching up of tRNA and amino acid is carried out by a family of related enzymes called **aminoacyl-tRNA synthetases** (**Figure 17.16**). The active site of each type of aminoacyl-tRNA synthetase fits only a specific combination of amino acid and tRNA. (Regions of both the amino acid attachment end and the anticodon end of the tRNA are instrumental in ensuring the specific fit.) There are 20 different synthetases, one for each amino acid; each synthetase is able to bind all the different tRNAs that code for its particular amino acid. The synthetase catalyzes the covalent attachment of the amino acid to its tRNA in a process driven by the hydrolysis of ATP. The resulting aminoacyl tRNA, also called a charged tRNA, is released from the enzyme and is then available to deliver its amino acid to a growing polypeptide chain on a ribosome.

The second instance of molecular recognition is the pairing of the tRNA anticodon with the appropriate mRNA codon. If one tRNA variety existed for each mRNA codon specifying an amino acid, there would be 61 tRNAs (see Figure 17.5). In fact, there are only about 45, signifying that some tRNAs must be able to bind to more than one codon. Such versatility is possible because the rules for base pairing between the third nucleotide base of a codon and the corresponding base of a tRNA anticodon are relaxed compared to those at other codon positions. For example, the nucleotide base U at the 5′ end of a tRNA anticodon can pair with either A or G in the third position (at the 3′ end) of an mRNA codon. The flexible base pairing at this codon position is called **wobble**. Wobble explains why the synonymous codons for a given amino acid most often differ in their third nucleotide base, but not in the other bases. For example, a tRNA with the anticodon 3′-UCU-5′ can base-pair with either the mRNA codon 5′-AGA-3′ or 5′-AGG-3′, both of which code for arginine (see Figure 17.5).

Ribosomes

Ribosomes facilitate the specific coupling of tRNA anticodons with mRNA codons during protein synthesis. A ribosome consists of a large subunit and a small subunit, each made up of proteins and one or more **ribosomal RNAs (rRNAs)** (**Figure 17.17**). In eukaryotes, the subunits are made in the nucleolus. Ribosomal RNA genes are transcribed, and the RNA is processed and assembled with proteins imported from the cytoplasm. The resulting ribosomal subunits are then exported via nuclear pores to the cytoplasm. In both bacteria and eukaryotes, large and small subunits join to form a functional ribosome only when they attach to an mRNA molecule. About one-third of the mass of a ribosome is made up of proteins; the rest consists of rRNAs, either three molecules (in bacteria) or four (in eukaryotes). Because most cells contain thousands of ribosomes, rRNA is the most abundant type of cellular RNA.

Although the ribosomes of bacteria and eukaryotes are very similar in structure and function, those of eukaryotes are slightly larger and differ somewhat from bacterial ribosomes in their molecular composition. The differences are medically significant. Certain antibiotic drugs can inactivate bacterial ribosomes without inhibiting the ability of eukaryotic ribosomes to make proteins. These drugs, including tetracycline and streptomycin, are used to combat bacterial infections.

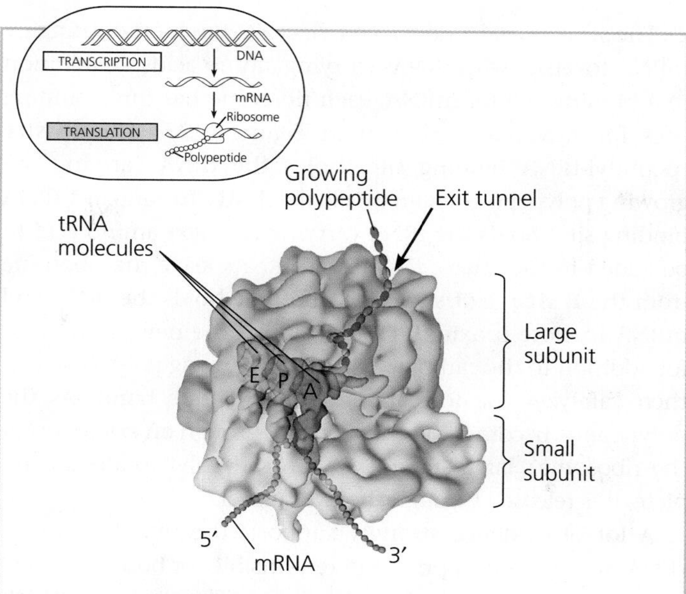

(a) Computer model of functioning ribosome. This is a model of a bacterial ribosome, showing its overall shape. The eukaryotic ribosome is roughly similar. A ribosomal subunit is a complex of ribosomal RNA molecules and proteins.

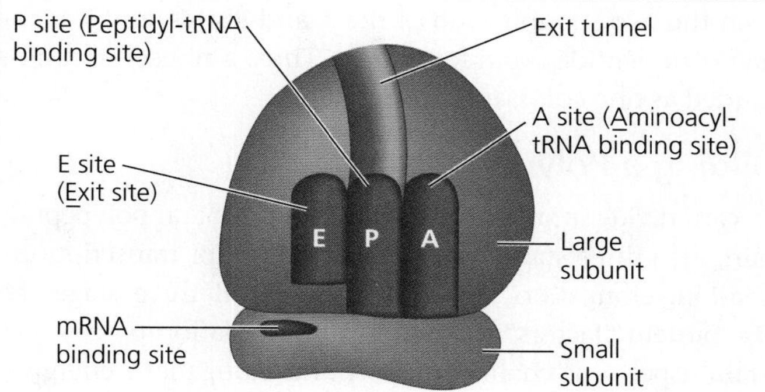

(b) Schematic model showing binding sites. A ribosome has an mRNA binding site and three tRNA binding sites, known as the A, P, and E sites. This schematic ribosome will appear in later diagrams.

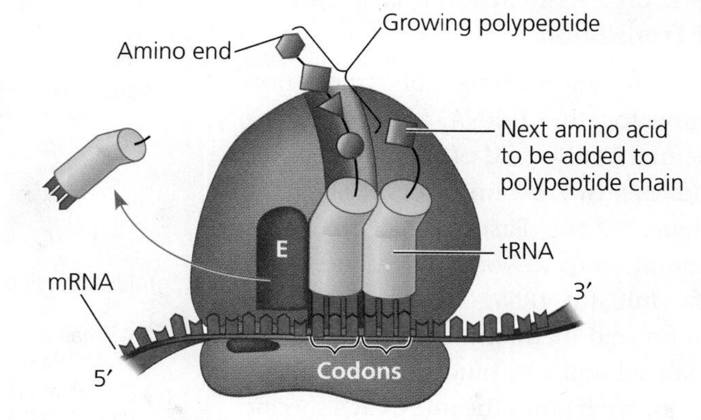

(c) Schematic model with mRNA and tRNA. A tRNA fits into a binding site when its anticodon base-pairs with an mRNA codon. The P site holds the tRNA attached to the growing polypeptide. The A site holds the tRNA carrying the next amino acid to be added to the polypeptide chain. Discharged tRNA leaves from the E site.

▲ **Figure 17.17 The anatomy of a functioning ribosome.**

The structure of a ribosome reflects its function of bringing mRNA together with tRNAs carrying amino acids. In addition to a binding site for mRNA, each ribosome has three binding sites for tRNA, as described in Figure 17.17. The **P site** (**p**eptidyl-tRNA binding site) holds the tRNA carrying the growing polypeptide chain, while the **A site** (**a**minoacyl-tRNA binding site) holds the tRNA carrying the next amino acid to be added to the chain. Discharged tRNAs leave the ribosome from the **E site** (**e**xit site). The ribosome holds the tRNA and mRNA in close proximity and positions the new amino acid for addition to the carboxyl end of the growing polypeptide. It then catalyzes the formation of the peptide bond. As the polypeptide becomes longer, it passes through an *exit tunnel* in the ribosome's large subunit. When the polypeptide is complete, it is released through the exit tunnel.

A lot of evidence strongly supports the hypothesis that rRNA, not protein, is primarily responsible for both the structure and the function of the ribosome. The proteins, which are largely on the exterior, support the shape changes of the rRNA molecules as they carry out catalysis during translation. Ribosomal RNA is the main constituent of the interface between the two subunits and of the A and P sites, and it is the catalyst of peptide bond formation. Thus, a ribosome can be regarded as one colossal ribozyme!

Building a Polypeptide

We can divide translation, the synthesis of a polypeptide chain, into three stages (analogous to those of transcription): initiation, elongation, and termination. All three stages require protein "factors" that aid in the translation process. For certain aspects of chain initiation and elongation, energy is also required. It is provided by the hydrolysis of guanosine triphosphate (GTP), a molecule closely related to ATP.

Ribosome Association and Initiation of Translation

The initiation stage of translation brings together mRNA, a tRNA bearing the first amino acid of the polypeptide, and the two subunits of a ribosome (**Figure 17.18**). First, a small ribosomal subunit binds to both mRNA and a specific initiator tRNA, which carries the amino acid methionine. In bacteria, the small subunit can bind these two in either order; it binds the mRNA at a specific RNA sequence, just upstream of the start codon, AUG. (Joan Steitz, our Unit Three interviewee, discovered the binding site on the mRNA and showed that complementary base pairing between this site and a ribosomal RNA was involved.) In eukaryotes, the small subunit, with the initiator tRNA already bound, binds to the 5′ cap of the mRNA and then moves, or *scans*, downstream along the mRNA until it reaches the start codon; the initiator tRNA then hydrogen-bonds to the AUG start codon. In either case, the start codon signals the start of translation; this is important because it establishes the codon reading frame for the mRNA.

The union of mRNA, initiator tRNA, and a small ribosomal subunit is followed by the attachment of a large ribosomal subunit, completing the *translation initiation complex*. Proteins called *initiation factors* are required to bring all these components together. The cell also expends energy obtained by hydrolysis of a GTP molecule to form the initiation complex. At the completion of the initiation process, the initiator tRNA sits in the P site of the ribosome, and the vacant A site is ready for the next aminoacyl tRNA. Note that a polypeptide is always synthesized in one direction, from the initial methionine at the amino end, also called the N-terminus, toward the final amino acid at the carboxyl end, also called the C-terminus (see Figure 5.17).

Elongation of the Polypeptide Chain

In the elongation stage of translation, amino acids are added one by one to the previous amino acid at the C-terminus of the growing chain. Each addition involves the participation of several proteins called *elongation factors* and occurs in a three-step cycle described in **Figure 17.19**. Energy expenditure occurs in the first and third steps. Codon recognition requires hydrolysis of one molecule of GTP, which increases the accuracy and efficiency of this step. One more GTP is hydrolyzed to provide energy for the translocation step.

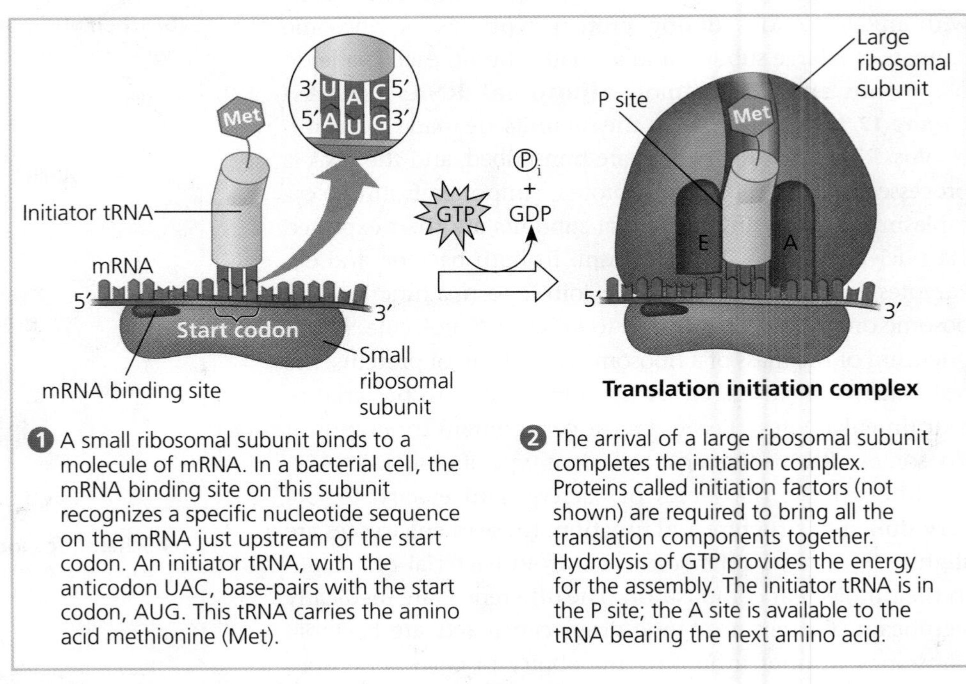

3′ U A C 5′
5′ A U G 3′
Met
Initiator tRNA
mRNA
5′
3′
Start codon
mRNA binding site
Small ribosomal subunit
Ⓟi
+
GDP
GTP
Large ribosomal subunit
P site
Met
E
A
5′
3′
Translation initiation complex
1 A small ribosomal subunit binds to a molecule of mRNA. In a bacterial cell, the mRNA binding site on this subunit recognizes a specific nucleotide sequence on the mRNA just upstream of the start codon. An initiator tRNA, with the anticodon UAC, base-pairs with the start codon, AUG. This tRNA carries the amino acid methionine (Met).
2 The arrival of a large ribosomal subunit completes the initiation complex. Proteins called initiation factors (not shown) are required to bring all the translation components together. Hydrolysis of GTP provides the energy for the assembly. The initiator tRNA is in the P site; the A site is available to the tRNA bearing the next amino acid.

▲ **Figure 17.18 The initiation of translation.**

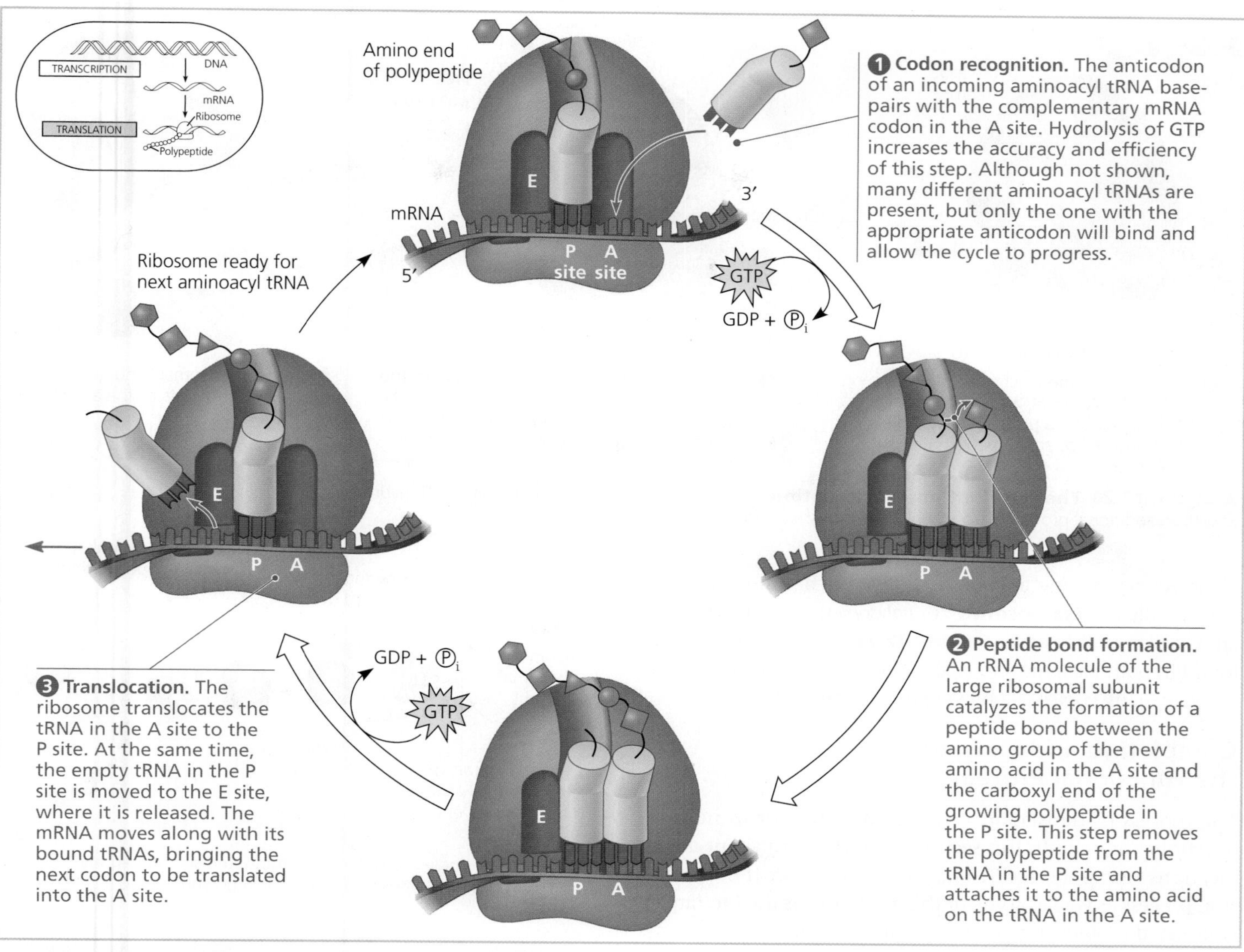

TRANSCRIPTION
DNA
mRNA
Ribosome
TRANSLATION
Polypeptide
Amino end
of polypeptide
mRNA
5′
3′
E
P site
A site
Ribosome ready for
next aminoacyl tRNA
GTP
GDP + Ⓟi
1 Codon recognition. The anticodon of an incoming aminoacyl tRNA base-pairs with the complementary mRNA codon in the A site. Hydrolysis of GTP increases the accuracy and efficiency of this step. Although not shown, many different aminoacyl tRNAs are present, but only the one with the appropriate anticodon will bind and allow the cycle to progress.
2 Peptide bond formation. An rRNA molecule of the large ribosomal subunit catalyzes the formation of a peptide bond between the amino group of the new amino acid in the A site and the carboxyl end of the growing polypeptide in the P site. This step removes the polypeptide from the tRNA in the P site and attaches it to the amino acid on the tRNA in the A site.
3 Translocation. The ribosome translocates the tRNA in the A site to the P site. At the same time, the empty tRNA in the P site is moved to the E site, where it is released. The mRNA moves along with its bound tRNAs, bringing the next codon to be translated into the A site.
P
A

▲ **Figure 17.19 The elongation cycle of translation.** The hydrolysis of GTP plays an important role in the elongation process. Not shown are the proteins called elongation factors.

The mRNA is moved through the ribosome in one direction only, 5′ end first; this is equivalent to the ribosome moving 5′ → 3′ on the mRNA. The important point is that the ribosome and the mRNA move relative to each other, unidirectionally, codon by codon. The elongation cycle takes less than a tenth of a second in bacteria and is repeated as each amino acid is added to the chain until the polypeptide is completed.

Termination of Translation

The final stage of translation is termination (**Figure 17.20**, on the next page). Elongation continues until a stop codon in the mRNA reaches the A site of the ribosome. The nucleotide base triplets UAG, UAA, and UGA do not code for amino acids but instead act as signals to stop translation. A *release factor*, a protein shaped like an aminoacyl tRNA, binds directly to the stop codon in the A site. The release factor causes the addition of a water molecule instead of an amino acid to the polypeptide chain. (There are plenty of water molecules available in the aqueous cellular environment.) This reaction breaks (hydrolyzes) the bond between the completed polypeptide and the tRNA in the P site, releasing the polypeptide through the exit tunnel of the ribosome's large subunit. The remainder of the translation assembly then comes apart in a multistep process, aided by other protein factors. Breakdown of the translation assembly requires the hydrolysis of two more GTP molecules.

Polyribosomes

A single ribosome can make an average-sized polypeptide in less than a minute. Typically, however, multiple ribosomes translate an mRNA at the same time; that is, a single mRNA is used to make many copies of a polypeptide simultaneously. Once a ribosome is far enough past the start codon, a second ribosome can attach to the mRNA, eventually resulting in a number

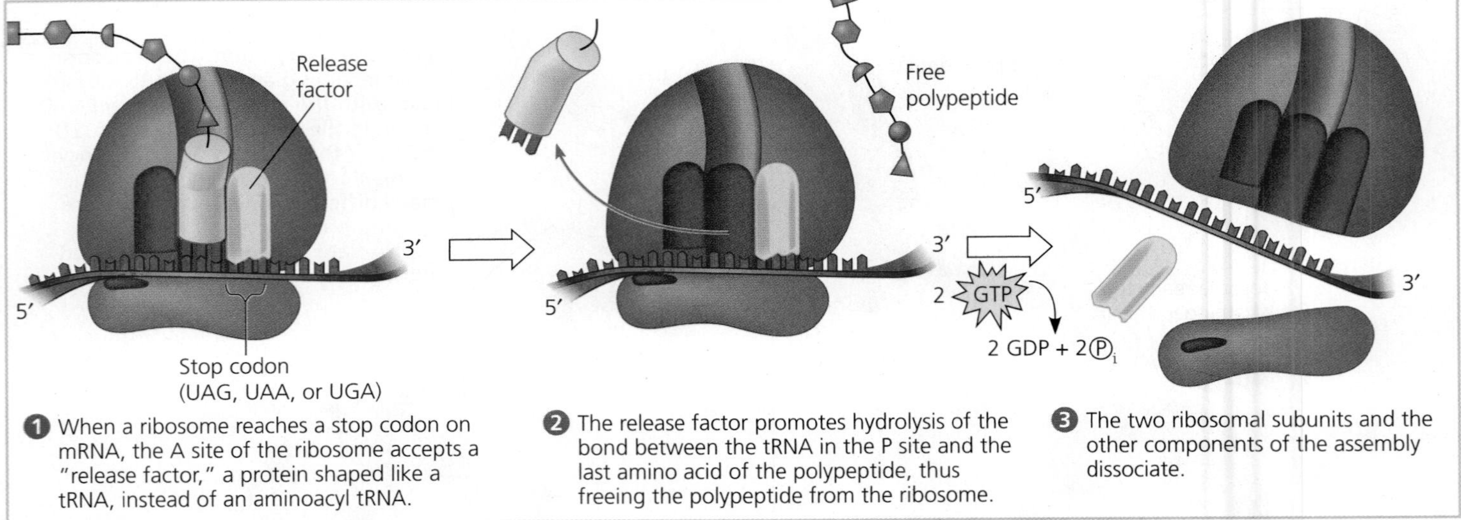
Release
factor
3′
5′
Stop codon
(UAG, UAA, or UGA)
Free
polypeptide
3′
5′
2 GTP
2 GDP + 2 Ⓟi
5′
3′
1 When a ribosome reaches a stop codon on mRNA, the A site of the ribosome accepts a "release factor," a protein shaped like a tRNA, instead of an aminoacyl tRNA.
2 The release factor promotes hydrolysis of the bond between the tRNA in the P site and the last amino acid of the polypeptide, thus freeing the polypeptide from the ribosome.
3 The two ribosomal subunits and the other components of the assembly dissociate.

▲ **Figure 17.20 The termination of translation.** Like elongation, termination requires GTP hydrolysis as well as additional protein factors, which are not shown here.

of ribosomes trailing along the mRNA. Such strings of ribosomes, called **polyribosomes** (or *polysomes*), can be seen with an electron microscope **(Figure 17.21)**. Polyribosomes are found in both bacterial and eukaryotic cells. They enable a cell to make many copies of a polypeptide very quickly.

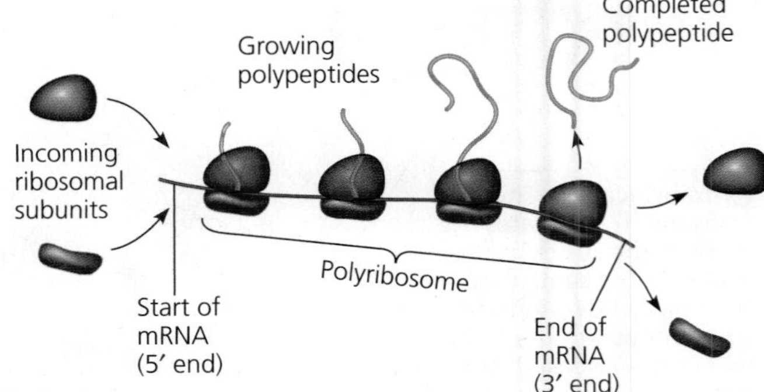
Completed
polypeptide
Growing
polypeptides
Incoming
ribosomal
subunits
Polyribosome
Start of
mRNA
(5′ end)
End of
mRNA
(3′ end)

(a) An mRNA molecule is generally translated simultaneously by several ribosomes in clusters called polyribosomes.

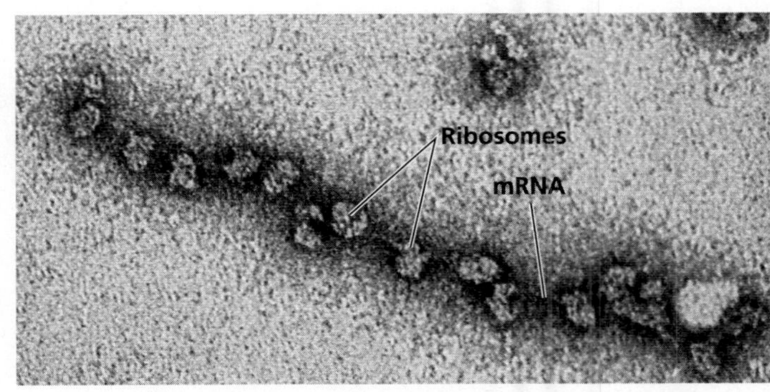
Ribosomes
mRNA
0.1 μm

(b) This micrograph shows a large polyribosome in a bacterial cell. Growing polypeptides are not visible here (TEM).

▲ **Figure 17.21 Polyribosomes.**

Completing and Targeting the Functional Protein

The process of translation is often not sufficient to make a functional protein. In this section, you will learn about modifications that polypeptide chains undergo after the translation process as well as some of the mechanisms used to target completed proteins to specific sites in the cell.

Protein Folding and Post-Translational Modifications

During its synthesis, a polypeptide chain begins to coil and fold spontaneously as a consequence of its amino acid sequence (primary structure), forming a protein with a specific shape: a three-dimensional molecule with secondary and tertiary structure (see Figure 5.20). Thus, a gene determines primary structure, and primary structure in turn determines shape. In many cases, a chaperone protein (chaperonin) helps the polypeptide fold correctly (see Figure 5.23).

Additional steps—*post-translational modifications*—may be required before the protein can begin doing its particular job in the cell. Certain amino acids may be chemically modified by the attachment of sugars, lipids, phosphate groups, or other additions. Enzymes may remove one or more amino acids from the leading (amino) end of the polypeptide chain. In some cases, a polypeptide chain may be enzymatically cleaved into two or more pieces. For example, the protein insulin is first synthesized as a single polypeptide chain but becomes active only after an enzyme cuts out a central part of the chain, leaving a protein made up of two polypeptide chains connected by disulfide bridges. In other cases, two or more polypeptides that are synthesized separately may come together, becoming the subunits of a protein that has quaternary structure. A familiar example is hemoglobin (see Figure 5.20).

Targeting Polypeptides to Specific Locations

In electron micrographs of eukaryotic cells active in protein synthesis, two populations of ribosomes (and polyribosomes) are evident: free and bound (see Figure 6.10). Free ribosomes are suspended in the cytosol and mostly synthesize proteins that stay in the cytosol and function there. In contrast, bound ribosomes are attached to the cytosolic side of the endoplasmic reticulum (ER) or to the nuclear envelope. Bound ribosomes make proteins of the endomembrane system (the nuclear envelope, ER, Golgi apparatus, lysosomes, vacuoles, and plasma membrane) as well as proteins secreted from the cell, such as insulin. It is important to note that the ribosomes themselves are identical and can switch their status from free to bound.

What determines whether a ribosome is free in the cytosol or bound to rough ER? Polypeptide synthesis always begins in the cytosol as a free ribosome starts to translate an mRNA molecule. There the process continues to completion—*unless* the growing polypeptide itself cues the ribosome to attach to the ER. The polypeptides of proteins destined for the endomembrane system or for secretion are marked by a **signal peptide**, which targets the protein to the ER **(Figure 17.22)**. The signal peptide, a sequence of about 20 amino acids at or near the leading end (N-terminus) of the polypeptide, is recognized as it emerges from the ribosome by a protein-RNA complex called a **signal-recognition particle (SRP)**. This particle functions as an escort that brings the ribosome to a receptor protein built into the ER membrane. The receptor is part of a multiprotein translocation complex. Polypeptide synthesis continues there, and the growing polypeptide snakes across the membrane into the ER lumen via a protein pore. The signal peptide is usually removed by an enzyme. The rest of the completed polypeptide, if it is to be secreted from the cell, is released into solution within the ER lumen (as in Figure 17.22). Alternatively, if the polypeptide is to be a membrane protein, it remains partially embedded in the ER membrane.

Other kinds of signal peptides are used to target polypeptides to mitochondria, chloroplasts, the interior of the nucleus, and other organelles that are not part of the endomembrane system. The critical difference in these cases is that translation is completed in the cytosol before the polypeptide is imported into the organelle. The mechanisms of translocation also vary, but in all cases studied to date, the "postal zip codes" that address proteins for secretion or to cellular locations are signal peptides of some sort. Bacteria also employ signal peptides to target proteins to the plasma membrane for secretion.

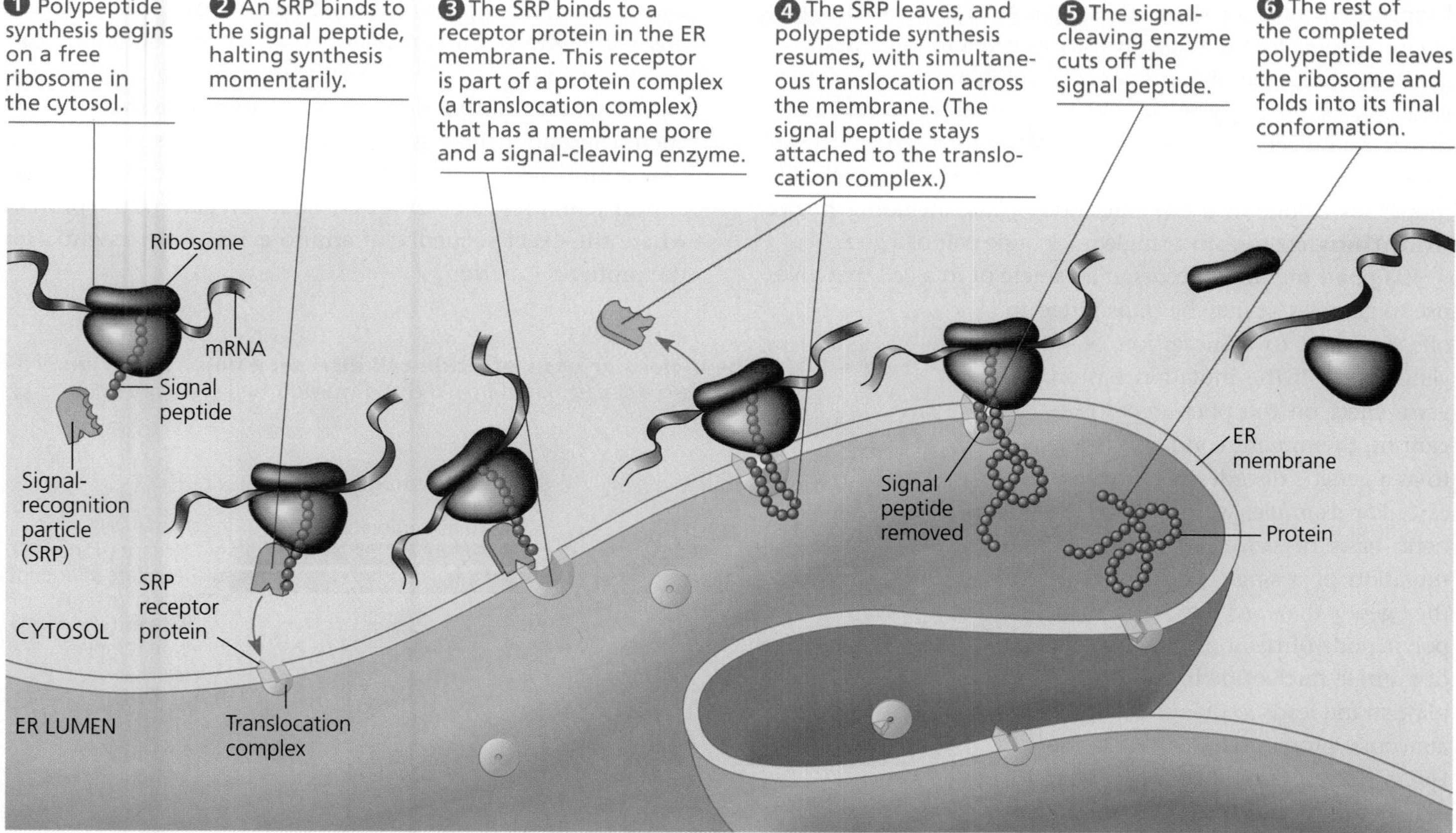

1 Polypeptide synthesis begins on a free ribosome in the cytosol.
2 An SRP binds to the signal peptide, halting synthesis momentarily.
3 The SRP binds to a receptor protein in the ER membrane. This receptor is part of a protein complex (a translocation complex) that has a membrane pore and a signal-cleaving enzyme.
4 The SRP leaves, and polypeptide synthesis resumes, with simultaneous translocation across the membrane. (The signal peptide stays attached to the translocation complex.)
5 The signal-cleaving enzyme cuts off the signal peptide.
6 The rest of the completed polypeptide leaves the ribosome and folds into its final conformation.
Ribosome
mRNA
Signal peptide
Signal-recognition particle (SRP)
SRP receptor protein
CYTOSOL
ER LUMEN
Translocation complex
Signal peptide removed
ER membrane
Protein

▲ **Figure 17.22 The signal mechanism for targeting proteins to the ER.** A polypeptide destined for the endomembrane system or for secretion from the cell begins with a signal peptide, a series of amino acids that targets it for the ER. This figure shows the synthesis of a secretory protein and its simultaneous import into the ER. In the ER and then in the Golgi, the protein will be processed further. Finally, a transport vesicle will convey it to the plasma membrane for release from the cell (see Figure 7.12).

CONCEPT CHECK 17.4

1. What two processes ensure that the correct amino acid is added to a growing polypeptide chain?
2. Discuss the ways in which rRNA structure likely contributes to ribosomal function.
3. Describe how a polypeptide to be secreted is transported to the endomembrane system.
4. **WHAT IF? DRAW IT** If a tRNA has the anticodon 3′-CGU-5′, what two different codons could it bind to? Draw each on an mRNA, labeling all 5′ and 3′ ends. Add the amino acid carried by this tRNA.

For suggested answers, see Appendix A.

CONCEPT 17.5

Mutations of one or a few nucleotides can affect protein structure and function

Now that you have explored the process of gene expression, you are ready to understand the effects of changes to the genetic information of a cell (or virus). These changes, called **mutations**, are responsible for the huge diversity of genes found among organisms because mutations are the ultimate source of new genes. In Figure 15.14, we considered chromosomal rearrangements that affect long segments of DNA, which can be considered large-scale mutations. Here we examine small-scale mutations of one or a few nucleotide pairs, including **point mutations**, changes in a single nucleotide pair of a gene.

If a point mutation occurs in a gamete or in a cell that gives rise to gametes, it may be transmitted to offspring and to a succession of future generations. If the mutation has an adverse effect on the phenotype of an organism, the mutant condition is referred to as a genetic disorder or hereditary disease. For example, we can trace the genetic basis of sickle-cell disease to the mutation of a single nucleotide pair in the gene that encodes the β-globin polypeptide of hemoglobin. The change of a single nucleotide in the DNA's template strand leads to the production of an abnormal protein (**Figure 17.23**; also see Figure 5.21). In individuals who are homozygous for the mutant allele, the sickling of red blood cells caused by the altered hemoglobin produces the multiple symptoms associated with sickle-cell disease (see Chapter 14). Another disorder caused by a point mutation is a heart condition, familial cardiomyopathy, that is responsible for some incidents of sudden death in young athletes. Point mutations in several genes have been identified, any of which can lead to this disorder.

▼ **Figure 17.23 The molecular basis of sickle-cell disease: a point mutation.** The allele that causes sickle-cell disease differs from the wild-type (normal) allele by a single DNA nucleotide pair.

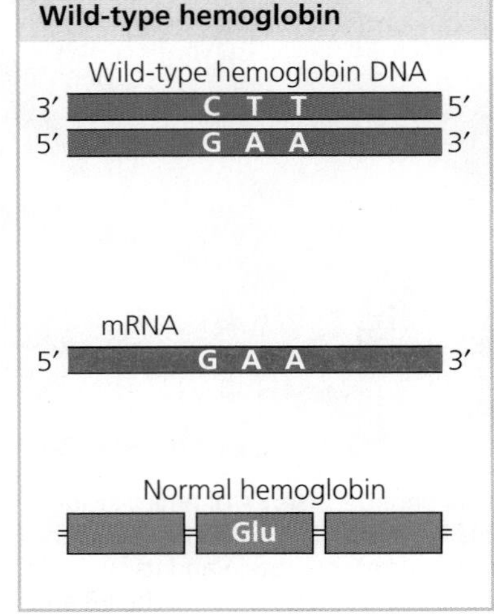

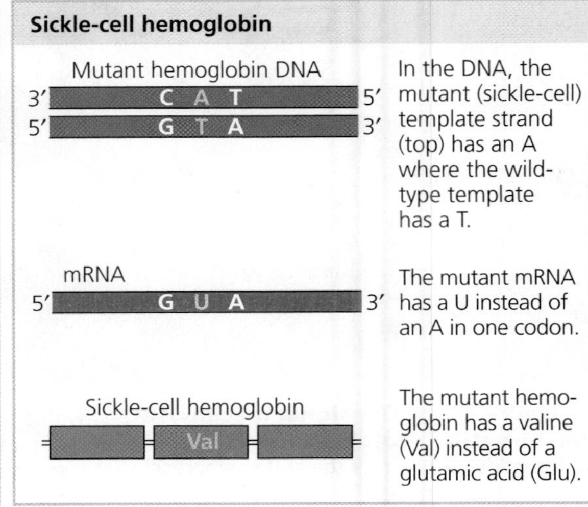

Types of Small-Scale Mutations

Let's now consider how small-scale mutations affect proteins. Point mutations within a gene can be divided into two general categories: (1) single nucleotide-pair substitutions and (2) nucleotide-pair insertions or deletions. Insertions and deletions can involve one or more nucleotide pairs.

Substitutions

A **nucleotide-pair substitution** is the replacement of one nucleotide and its partner with another pair of nucleotides (**Figure 17.24a**). Some substitutions have no effect on the encoded protein, owing to the redundancy of the genetic code. For example, if 3′-CCG-5′ on the template strand mutated to 3′-CCA-5′, the mRNA codon that used to be GGC would become GGU, but a glycine would still be inserted at the proper location in the protein (see Figure 17.5). In other words, a change in a nucleotide pair may transform one codon into another that is translated into the same amino acid. Such a change is an example of a **silent mutation**, which has no observable effect on the phenotype. (Silent mutations can occur outside genes as well.) Substitutions that change one amino acid to another one are called **missense mutations**. Such a mutation may have little effect on the protein: The new amino acid may have properties similar to those of the amino acid it replaces, or it may be in a region of the protein where the exact sequence of amino acids is not essential to the protein's function.

▼ **Figure 17.24 Types of small-scale mutations that affect mRNA sequence.** All but one of the types shown here also affect the amino acid sequence of the encoded polypeptide.

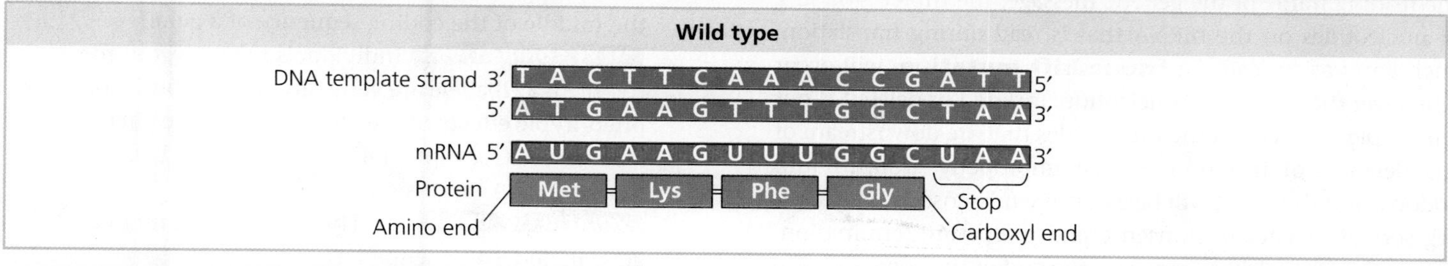

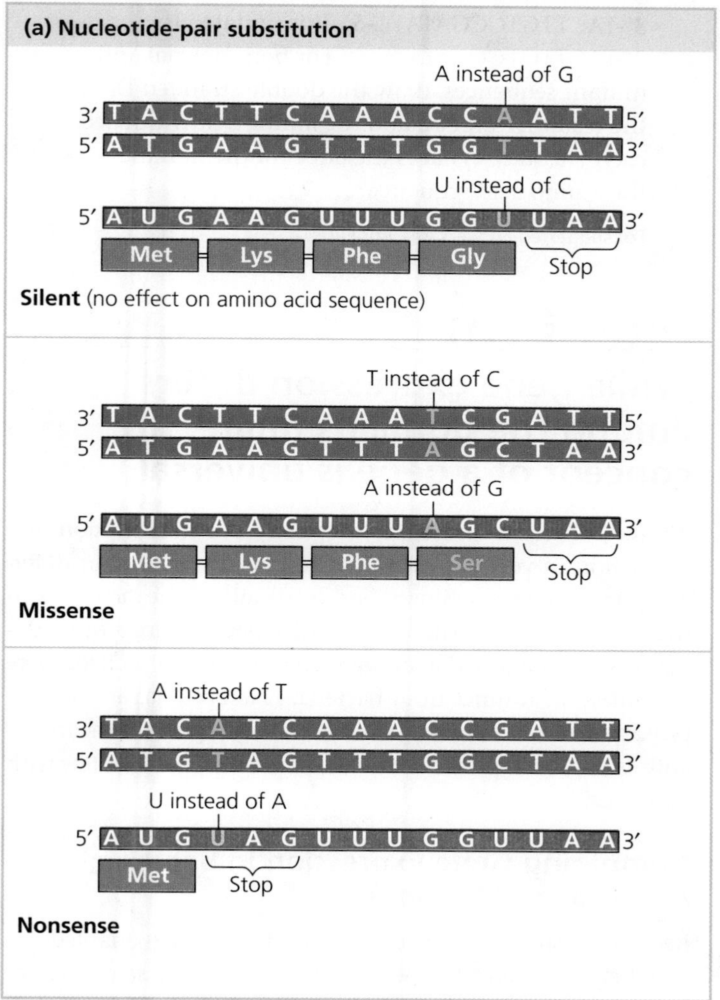

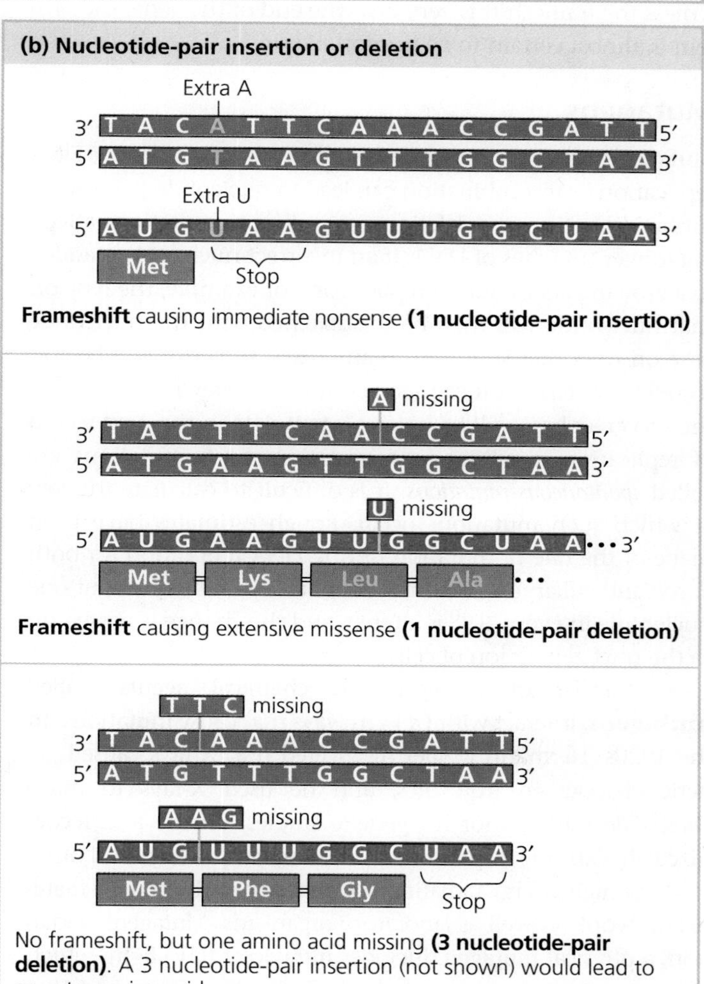

However, the nucleotide-pair substitutions of greatest interest are those that cause a major change in a protein. The alteration of a single amino acid in a crucial area of a protein—such as in the part of hemoglobin shown in Figure 17.23 or in the active site of an enzyme as shown in Figure 8.18—will significantly alter protein activity. Occasionally, such a mutation leads to an improved protein or one with novel capabilities, but much more often such mutations are detrimental, leading to a useless or less active protein that impairs cellular function.

Substitution mutations are usually missense mutations; that is, the altered codon still codes for an amino acid and thus makes sense, although not necessarily the *right* sense. But a point mutation can also change a codon for an amino acid into a stop codon. This is called a **nonsense mutation**, and it causes translation to be terminated prematurely; the resulting polypeptide will be shorter than the polypeptide encoded by the normal gene. Nearly all nonsense mutations lead to nonfunctional proteins.

Insertions and Deletions

Insertions and **deletions** are additions or losses of nucleotide pairs in a gene **(Figure 17.24b)**. These mutations have a disastrous effect on the resulting protein more often than substitutions do. Insertion or deletion of nucleotides may alter the reading frame of the genetic message, the triplet grouping of nucleotides on the mRNA that is read during translation. Such a mutation, called a **frameshift mutation**, will occur whenever the number of nucleotides inserted or deleted is not a multiple of three. All the nucleotides that are downstream of the deletion or insertion will be improperly grouped into codons, and the result will be extensive missense, usually ending sooner or later in nonsense and premature termination. Unless the frameshift is very near the end of the gene, the protein is almost certain to be nonfunctional.

Mutagens

Mutations can arise in a number of ways. Errors during DNA replication or recombination can lead to nucleotide-pair substitutions, insertions, or deletions, as well as to mutations affecting longer stretches of DNA. If an incorrect nucleotide is added to a growing chain during replication, for example, the base on that nucleotide will then be mismatched with the nucleotide base on the other strand. In many cases, the error will be corrected by systems you learned about in Chapter 16. Otherwise, the incorrect base will be used as a template in the next round of replication, resulting in a mutation. Such mutations are called *spontaneous mutations*. It is difficult to calculate the rate at which such mutations occur. Rough estimates have been made of the rate of mutation during DNA replication for both *E. coli* and eukaryotes, and the numbers are similar: About one nucleotide in every 10^{10} is altered, and the change is passed on to the next generation of cells.

A number of physical and chemical agents, called **mutagens**, interact with DNA in ways that cause mutations. In the 1920s, Hermann Muller discovered that X-rays caused genetic changes in fruit flies, and he used X-rays to make *Drosophila* mutants for his genetic studies. But he also recognized an alarming implication of his discovery: X-rays and other forms of high-energy radiation pose hazards to the genetic material of people as well as laboratory organisms. Mutagenic radiation, a physical mutagen, includes ultraviolet (UV) light, which can cause disruptive thymine dimers in DNA (see Figure 16.19).

Chemical mutagens fall into several categories. Nucleotide analogs are chemicals that are similar to normal DNA nucleotides but that pair incorrectly during DNA replication. Some other chemical mutagens interfere with correct DNA replication by inserting themselves into the DNA and distorting the double helix. Still other mutagens cause chemical changes in bases that change their pairing properties.

Researchers have developed a variety of methods to test the mutagenic activity of chemicals. A major application of these tests is the preliminary screening of chemicals to identify those that may cause cancer. This approach makes sense because most carcinogens (cancer-causing chemicals) are mutagenic, and conversely, most mutagens are carcinogenic.

CONCEPT CHECK 17.5

1. What happens when one nucleotide pair is lost from the middle of the coding sequence of a gene?
2. **MAKE CONNECTIONS** Individuals heterozygous for the sickle-cell allele are generally healthy but show phenotypic effects of the allele under some circumstances; see Concept 14.4, pages 277–278. Explain in terms of gene expression.
3. **WHAT IF?** **DRAW IT** The template strand of a gene includes this sequence: 3′-TACTTGTCCGATATC-5′. It is mutated to 3′-TACTTGTCCAATATC-5′. For both normal and mutant sequences, draw the double-stranded DNA, the resulting mRNA, and the amino acid sequence each encodes. What is the effect of the mutation on the amino acid sequence?

For suggested answers, see Appendix A.

CONCEPT 17.6

While gene expression differs among the domains of life, the concept of a gene is universal

Although bacteria and eukaryotes carry out transcription and translation in very similar ways, we have noted certain differences in cellular machinery and in details of the processes in these two domains. The division of organisms into three domains was established about 40 years ago, when archaea were recognized as distinct from bacteria. Like bacteria, archaea are prokaryotes. However, archaea share many aspects of the mechanisms of gene expression with eukaryotes, as well as a few with bacteria.

Comparing Gene Expression in Bacteria, Archaea, and Eukarya

Recent advances in molecular biology have enabled researchers to determine the complete nucleotide sequences of hundreds of genomes, including many genomes from each domain. This wealth of data allows us to compare gene and protein sequences across domains. Foremost among genes of interest are those that encode components of such fundamental biological processes as transcription and translation.

Bacterial and eukaryotic RNA polymerases differ significantly from each other, while the single archaeal RNA polymerase resembles the three eukaryotic ones. Archaea and eukaryotes use a complex set of transcription factors, unlike the smaller set of accessory proteins in bacteria. Transcription is terminated differently in bacteria and eukaryotes. The little that is known about archaeal transcription termination suggests that it is similar to the eukaryotic process.

As far as translation is concerned, archaeal ribosomes are the same size as bacterial ribosomes, but their sensitivities to chemical inhibitors more closely match those of eukaryotic ribosomes. We mentioned earlier that initiation of translation is slightly different in bacteria and eukaryotes. In this respect, the archaeal process is more like that of bacteria.

The most important differences between bacteria and eukaryotes with regard to gene expression arise from the bacterial cell's lack of compartmental organization. Like a one-room workshop, a bacterial cell ensures a streamlined operation. In the absence of a nucleus, it can simultaneously transcribe and translate the same gene **(Figure 17.25)**, and the newly made protein can quickly diffuse to its site of function. Most researchers suspect that transcription and translation are coupled like this in archaeal cells as well, since archaea lack a nuclear envelope. In contrast, the eukaryotic cell's nuclear envelope segregates transcription from translation and provides a compartment for extensive RNA processing. This processing stage includes additional steps whose regulation can help coordinate the eukaryotic cell's elaborate activities (see Chapter 18).

Learning more about the proteins and RNAs involved in archaeal transcription and translation will tell us much about the evolution of these processes in all three domains. In spite of the differences in gene expression cataloged here, however, the idea of the gene itself is a unifying concept among all forms of life.

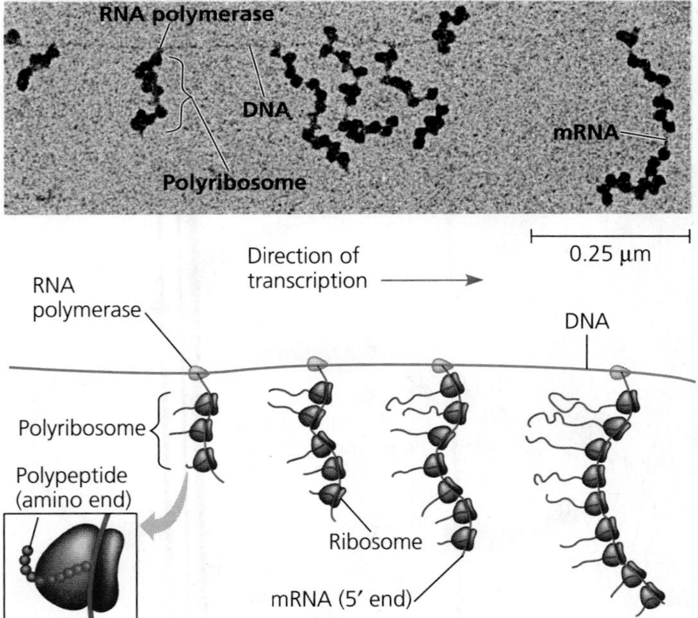

▲ **Figure 17.25 Coupled transcription and translation in bacteria.** In bacterial cells, the translation of mRNA can begin as soon as the leading (5′) end of the mRNA molecule peels away from the DNA template. The micrograph (TEM) shows a strand of *E. coli* DNA being transcribed by RNA polymerase molecules. Attached to each RNA polymerase molecule is a growing strand of mRNA, which is already being translated by ribosomes. The newly synthesized polypeptides are not visible in the micrograph but are shown in the diagram.

? *Which one of the mRNA molecules started transcription first? On that mRNA, which ribosome started translating first?*

What Is a Gene? *Revisiting the Question*

Our definition of a gene has evolved over the past few chapters, as it has through the history of genetics. We began with the Mendelian concept of a gene as a discrete unit of inheritance that affects a phenotypic character (Chapter 14). We saw that Morgan and his colleagues assigned such genes to specific loci on chromosomes (Chapter 15). We went on to view a gene as a region of specific nucleotide sequence along the length of the DNA molecule of a chromosome (Chapter 16). Finally, in this chapter, we have considered a functional definition of a gene as a DNA sequence that codes for a specific polypeptide chain. (**Figure 17.26**, on the next page, summarizes the path from gene to polypeptide in a eukaryotic cell.) All these definitions are useful, depending on the context in which genes are being studied.

Clearly, the statement that a gene codes for a polypeptide is too simple. Most eukaryotic genes contain noncoding segments (such as introns), so large portions of these genes have no corresponding segments in polypeptides. Molecular biologists also often include promoters and certain other regulatory regions of DNA within the boundaries of a gene. These DNA sequences are not transcribed, but they can be considered part of the functional gene because they must be present for transcription to occur. Our definition of a gene must also be broad enough to include the DNA that is transcribed into rRNA, tRNA, and other RNAs that are not translated. These genes have no polypeptide products but play crucial roles in the cell. Thus, we arrive at the following definition: *A gene is a region of DNA that can be expressed to produce a final functional product that is either a polypeptide or an RNA molecule.*

When considering phenotypes, however, it is often useful to start by focusing on genes that code for polypeptides. In this chapter, you have learned in molecular terms how a typical gene is expressed—by transcription into RNA and then translation into a polypeptide that forms a protein of specific structure and function. Proteins, in turn, bring about an organism's observable phenotype.

A given type of cell expresses only a subset of its genes. This is an essential feature in multicellular organisms: You'd be in trouble if the lens cells in your eyes started expressing the genes for hair proteins, which are normally expressed only in hair follicle cells! Gene expression is precisely regulated. We'll explore gene regulation in the next chapter, beginning with the simpler case of bacteria and continuing with eukaryotes.

CONCEPT CHECK 17.6

1. Would the coupling of processes shown in Figure 17.25 be found in a eukaryotic cell? Explain.
2. **WHAT IF?** In eukaryotic cells, mRNAs have been found to have a circular arrangement in which proteins hold the poly-A tail near the 5′ cap. How might this increase translation efficiency?

For suggested answers, see Appendix A.

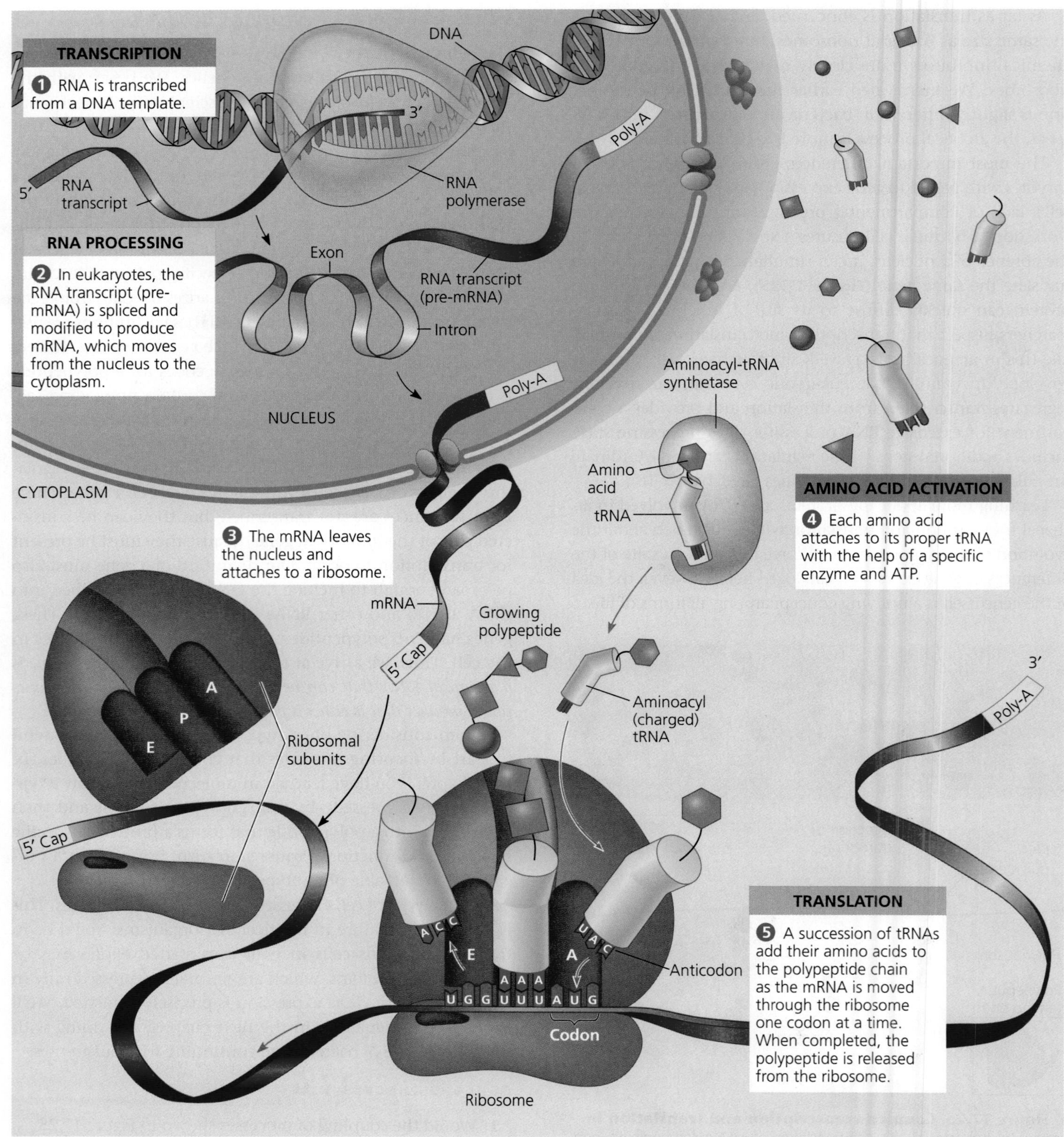

TRANSCRIPTION
1 RNA is transcribed from a DNA template.
DNA
3′
5′
RNA transcript
RNA polymerase
Poly-A
RNA PROCESSING
2 In eukaryotes, the RNA transcript (pre-mRNA) is spliced and modified to produce mRNA, which moves from the nucleus to the cytoplasm.
Exon
RNA transcript (pre-mRNA)
Intron
Poly-A
NUCLEUS
CYTOPLASM
Aminoacyl-tRNA synthetase
Amino acid
tRNA
AMINO ACID ACTIVATION
4 Each amino acid attaches to its proper tRNA with the help of a specific enzyme and ATP.
3 The mRNA leaves the nucleus and attaches to a ribosome.
mRNA
5′ Cap
Growing polypeptide
Aminoacyl (charged) tRNA
3′
Poly-A
A
P
E
Ribosomal subunits
5′ Cap
E
A
ACC
UAC
AAA
UGGUUUAUG
Anticodon
Codon
TRANSLATION
5 A succession of tRNAs add their amino acids to the polypeptide chain as the mRNA is moved through the ribosome one codon at a time. When completed, the polypeptide is released from the ribosome.
Ribosome

▲ **Figure 17.26 A summary of transcription and translation in a eukaryotic cell.** This diagram shows the path from one gene to one polypeptide. Keep in mind that each gene in the DNA can be transcribed repeatedly into many identical RNA molecules and that each mRNA can be translated repeatedly to yield many identical polypeptide molecules. (Also, remember that the final products of some genes are not polypeptides but RNA molecules, including tRNA and rRNA.) In general, the steps of transcription and translation are similar in bacterial, archaeal, and eukaryotic cells. The major difference is the occurrence of RNA processing in the eukaryotic nucleus. Other significant differences are found in the initiation stages of both transcription and translation and in the termination of transcription.

17 CHAPTER REVIEW

SUMMARY OF KEY CONCEPTS

CONCEPT 17.1

Genes specify proteins via transcription and translation (pp. 325–331)

- DNA controls metabolism by directing cells to make specific enzymes and other proteins, via the process of **gene expression**. Beadle and Tatum's studies of mutant strains of *Neurospora* led to the one gene–one polypeptide hypothesis. Genes code for polypeptide chains or specify RNA molecules.
- **Transcription** is the synthesis of RNA complementary to a **template strand** of DNA, providing a nucleotide-to-nucleotide transfer of information. **Translation** is the synthesis of a polypeptide whose amino acid sequence is specified by the nucleotide sequence in **mRNA**; this informational transfer thus involves a change of language, from that of nucleotides to that of amino acids.
- Genetic information is encoded as a sequence of nonoverlapping nucleotide triplets, or **codons**. A codon in messenger RNA (mRNA) either is translated into an amino acid (61 of the 64 codons) or serves as a stop signal (3 codons). Codons must be read in the correct **reading frame**.

? *Describe the process of gene expression, by which a gene affects the phenotype of an organism.*

CONCEPT 17.2

Transcription is the DNA-directed synthesis of RNA: *a closer look* (pp. 331–334)

- RNA synthesis is catalyzed by **RNA polymerase**, which links together RNA nucleotides complementary to a DNA template strand. This process follows the same base-pairing rules as DNA replication, except that in RNA, uracil substitutes for thymine.

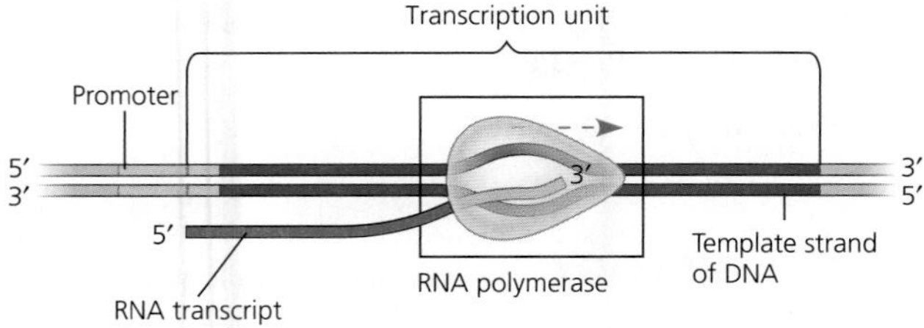

- The three stages of transcription are initiation, elongation, and termination. A **promoter**, often including a **TATA box** in eukaryotes, establishes where RNA synthesis is initiated. **Transcription factors** help eukaryotic RNA polymerase recognize promoter sequences, forming a **transcription initiation complex**. The mechanisms of termination are different in bacteria and eukaryotes.

? *What are the similarities and differences in the initiation of gene transcription in bacteria and eukaryotes?*

CONCEPT 17.3

Eukaryotic cells modify RNA after transcription (pp. 334–336)

- Before leaving the nucleus, eukaryotic mRNA molecules undergo **RNA processing**, which includes RNA splicing, the addition of a modified nucleotide **5′ cap** to the 5′ end, and the addition of a **poly-A tail** to the 3′ end.

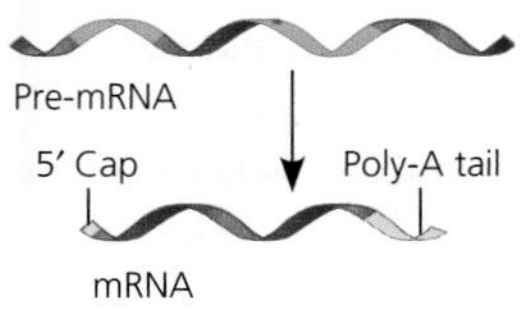

- Most eukaryotic genes are split into segments: They have **introns** interspersed among the **exons** (the regions included in the mRNA). In **RNA splicing**, introns are removed and exons joined. RNA splicing is typically carried out by **spliceosomes**, but in some cases, RNA alone catalyzes its own splicing. The catalytic ability of some RNA molecules, called **ribozymes**, derives from the inherent properties of RNA. The presence of introns allows for **alternative RNA splicing**.

? *What function do the 5′ cap and the poly-A tail serve on a eukaryotic mRNA?*

CONCEPT 17.4

Translation is the RNA-directed synthesis of a polypeptide: *a closer look* (pp. 337–344)

- A cell translates an mRNA message into protein using **transfer RNAs (tRNAs)**. After being bound to a specific amino acid by an **aminoacyl-tRNA synthetase**, a tRNA lines up via its **anticodon** at the complementary codon on mRNA. A **ribosome**, made up of **ribosomal RNAs (rRNAs)** and proteins, facilitates this coupling with binding sites for mRNA and tRNA.
- Ribosomes coordinate the three stages of translation: initiation, elongation, and termination. The formation of peptide bonds between amino acids is catalyzed by rRNA as tRNAs move through the **A** and **P sites** and exit through the **E site**.
- A single mRNA molecule can be translated simultaneously by a number of ribosomes, forming a **polyribosome**.
- After translation, modifications to proteins can affect their three-dimensional shape. Free ribosomes in the cytosol initiate synthesis of all proteins, but proteins destined for the endomembrane system or for secretion are transported into the ER. Such proteins have a **signal peptide** to which a **signal-recognition particle (SRP)** binds, enabling the translating ribosome to bind to the ER.

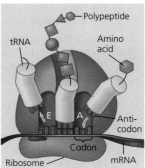

Polypeptide
tRNA
Amino acid
E
A
Anti-codon
Codon
Ribosome
mRNA

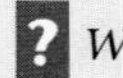 *What function do tRNAs serve in the process of translation?*

CONCEPT 17.5

Mutations of one or a few nucleotides can affect protein structure and function (pp. 344–346)

- Small-scale **mutations** include **point mutations**, changes in one DNA nucleotide pair, which may lead to production of nonfunctional proteins. **Nucleotide-pair substitutions** can cause **missense** or **nonsense mutations**. Nucleotide-pair **insertions** or **deletions** may produce **frameshift mutations**.
- Spontaneous mutations can occur during DNA replication, recombination, or repair. Chemical and physical **mutagens** cause DNA damage that can alter genes.

? *What will be the results of chemically modifying one nucleotide base of a gene? What role is played by DNA repair systems in the cell?*

CONCEPT 17.6

While gene expression differs among the domains of life, the concept of a gene is universal (pp. 346–348)

- There are some differences in gene expression among bacteria, archaea, and eukaryotes. Because bacterial cells lack a nuclear envelope, translation can begin while transcription is still in progress. Archaeal cells show similarities to both eukaryotic and bacterial cells in their processes of gene expression. In a eukaryotic cell, the nuclear envelope separates transcription from translation, and extensive RNA processing occurs in the nucleus.
- A gene is a region of DNA whose final functional product is either a polypeptide or an RNA molecule.

? *How does the presence of a nuclear envelope affect gene expression in eukaryotes?*

TEST YOUR UNDERSTANDING

LEVEL 1: KNOWLEDGE/COMPREHENSION

1. In eukaryotic cells, transcription cannot begin until
   a. the two DNA strands have completely separated and exposed the promoter.
   b. several transcription factors have bound to the promoter.
   c. the 5′ caps are removed from the mRNA.
   d. the DNA introns are removed from the template.
   e. DNA nucleases have isolated the transcription unit.
2. Which of the following is *not* true of a codon?
   a. It consists of three nucleotides.
   b. It may code for the same amino acid as another codon.
   c. It never codes for more than one amino acid.
   d. It extends from one end of a tRNA molecule.
   e. It is the basic unit of the genetic code.
3. The anticodon of a particular tRNA molecule is
   a. complementary to the corresponding mRNA codon.
   b. complementary to the corresponding triplet in rRNA.
   c. the part of tRNA that bonds to a specific amino acid.
   d. changeable, depending on the amino acid that attaches to the tRNA.
   e. catalytic, making the tRNA a ribozyme.
4. Which of the following is *not* true of RNA processing?
   a. Exons are cut out before mRNA leaves the nucleus.
   b. Nucleotides may be added at both ends of the RNA.
   c. Ribozymes may function in RNA splicing.
   d. RNA splicing can be catalyzed by spliceosomes.
   e. A primary transcript is often much longer than the final RNA molecule that leaves the nucleus.
5. Which component is *not* directly involved in translation?
   a. mRNA b. DNA c. tRNA d. ribosomes e. GTP

LEVEL 2: APPLICATION/ANALYSIS

6. Using Figure 17.5, identify a 5′ → 3′ sequence of nucleotides in the DNA template strand for an mRNA coding for the polypeptide sequence Phe-Pro-Lys.
   a. 5′-UUUGGGAAA-3′
   b. 5′-GAACCCCTT-3′
   c. 5′-AAAACCTTT-3′
   d. 5′-CTTCGGGAA-3′
   e. 5′-AAACCCUUU-3′
7. Which of the following mutations would be *most* likely to have a harmful effect on an organism?
   a. a nucleotide-pair substitution
   b. a deletion of three nucleotides near the middle of a gene
   c. a single nucleotide deletion in the middle of an intron
   d. a single nucleotide deletion near the end of the coding sequence
   e. a single nucleotide insertion downstream of, and close to, the start of the coding sequence
8. **DRAW IT** Fill in the following table:

Type of RNA	Functions
Messenger RNA (mRNA)	
Transfer RNA (tRNA)	
	Plays catalytic (ribozyme) roles and structural roles in ribosomes
Primary transcript	
Small nuclear RNA (snRNA)	

LEVEL 3: SYNTHESIS/EVALUATION

9. **EVOLUTION CONNECTION** Most amino acids are coded for by a set of similar codons (see Figure 17.5). What evolutionary explanations can you give for this pattern? (*Hint*: There is one explanation relating to ancestry, and some less obvious ones of a "form-fits-function" type.)
10. **SCIENTIFIC INQUIRY**
    Knowing that the genetic code is almost universal, a scientist uses molecular biological methods to insert the human β-globin gene (shown in Figure 17.11) into bacterial cells, hoping the cells will express it and synthesize functional β-globin protein. Instead, the protein produced is nonfunctional and is found to contain many fewer amino acids than does β-globin made by a eukaryotic cell. Explain why.
11. **WRITE ABOUT A THEME**
    Evolution and The Genetic Basis of Life Evolution accounts for the unity and diversity of life, and the continuity of life is based on heritable information in the form of DNA. In a short essay (100–150 words), discuss how the fidelity with which DNA is inherited is related to the processes of evolution. (Review the discussion of proofreading and DNA repair in Concept 16.2, pp. 316–318.)

For selected answers, see Appendix A.

18

Regulation of Gene Expression

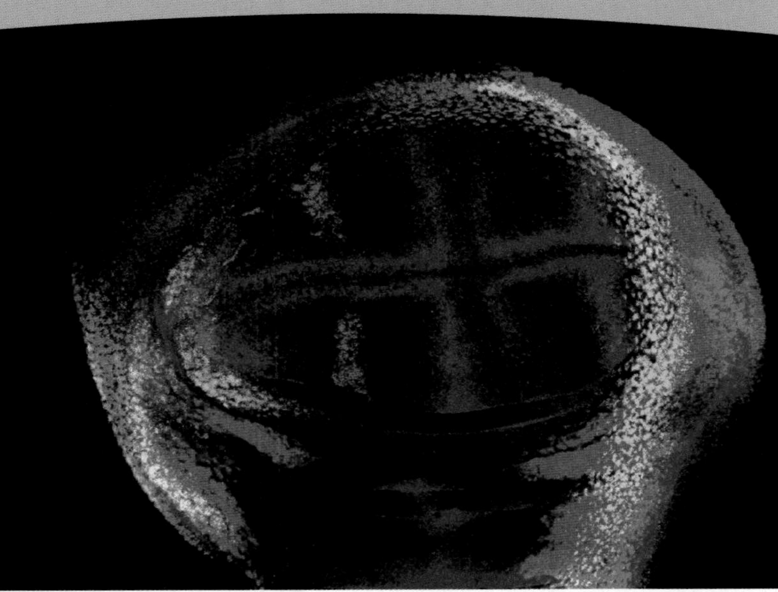

▲ **Figure 18.1 What regulates the precise pattern of gene expression in the developing wing of a fly embryo?**

KEY CONCEPTS

18.1 Bacteria often respond to environmental change by regulating transcription

18.2 Eukaryotic gene expression is regulated at many stages

18.3 Noncoding RNAs play multiple roles in controlling gene expression

18.4 A program of differential gene expression leads to the different cell types in a multicellular organism

18.5 Cancer results from genetic changes that affect cell cycle control

OVERVIEW

Conducting the Genetic Orchestra

It's almost concert time! Dissonance reigns as the orchestra members individually tune their instruments. Then, after a brief hush, the conductor's baton rises, pauses, and begins a series of elaborate movements, directing specific instruments to join in and others to raise or lower their volume at defined moments. Properly balanced and timed, discordant sounds are thus transformed into a beautiful symphony that enraptures the audience.

In a similar way, cells intricately and precisely regulate their gene expression. Both prokaryotes and eukaryotes must alter their patterns of gene expression in response to changes in environmental conditions. Multicellular eukaryotes must also develop and maintain multiple cell types. Each cell type contains the same genome but expresses a different subset of genes, a significant challenge in gene regulation.

An adult fruit fly, for example, develops from a single fertilized egg, passing through a wormlike stage called a larva. At every stage, gene expression is carefully regulated, ensuring that the right genes are expressed only at the correct time and place. In the larva, the adult wing forms in a disk-shaped pocket of several thousand cells, shown in **Figure 18.1**. The tissue in this image has been treated to reveal the mRNA for three genes—labeled red, blue, and green—using techniques covered in Chapter 20. (Red and green together appear yellow.) The intricate pattern of expression for each gene is the same from larva to larva at this stage, and it provides a graphic display of the precision of gene regulation. But what is the molecular basis for this pattern? Why is one particular gene expressed only in the few hundred cells that appear blue in this image and not in the other cells?

In this chapter, we first explore how bacteria regulate expression of their genes in response to different environmental conditions. We then examine how eukaryotes regulate gene expression to maintain different cell types. Gene expression in eukaryotes, as in bacteria, is often regulated at the stage of transcription, but control at other stages is also important. In recent years, researchers have been surprised to discover the many roles played by RNA molecules in regulating eukaryotic gene expression, a topic we cover next. We then consider what happens when a complex program of gene regulation works properly during embryonic development: A single cell—the fertilized egg—becomes a fully functioning organism made up of many different cell types. Finally, we investigate how cancer can result when gene regulation goes awry. Orchestrating proper gene expression by all cells is crucial to the functions of life.

CONCEPT **18.1**

Bacteria often respond to environmental change by regulating transcription

Bacterial cells that can conserve resources and energy have a selective advantage over cells that are unable to do so. Thus, natural selection has favored bacteria that express only the genes whose products are needed by the cell.

Consider, for instance, an individual *E. coli* cell living in the erratic environment of a human colon, dependent for its nutrients on the whimsical eating habits of its host. If the environment is lacking in the amino acid tryptophan, which the bacterium needs to survive, the cell responds by activating a metabolic pathway that makes tryptophan from another compound. Later, if the human host eats a tryptophan-rich meal, the bacterial cell stops producing tryptophan, thus saving itself from squandering its resources to produce a substance that is available from the surrounding solution in prefabricated form. This is just one example of how bacteria tune their metabolism to changing environments.

Metabolic control occurs on two levels, as shown for the synthesis of tryptophan in **Figure 18.2**. First, cells can adjust the activity of enzymes already present. This is a fairly fast response, which relies on the sensitivity of many enzymes to chemical cues that increase or decrease their catalytic activity (see Chapter 8). The activity of the first enzyme in the tryptophan synthesis pathway is inhibited by the pathway's end product **(Figure 18.2a)**. Thus, if tryptophan accumulates in a cell, it shuts down the synthesis of more tryptophan by inhibiting enzyme activity. Such *feedback inhibition*, typical of anabolic (biosynthetic) pathways, allows a cell to adapt to short-term fluctuations in the supply of a substance it needs.

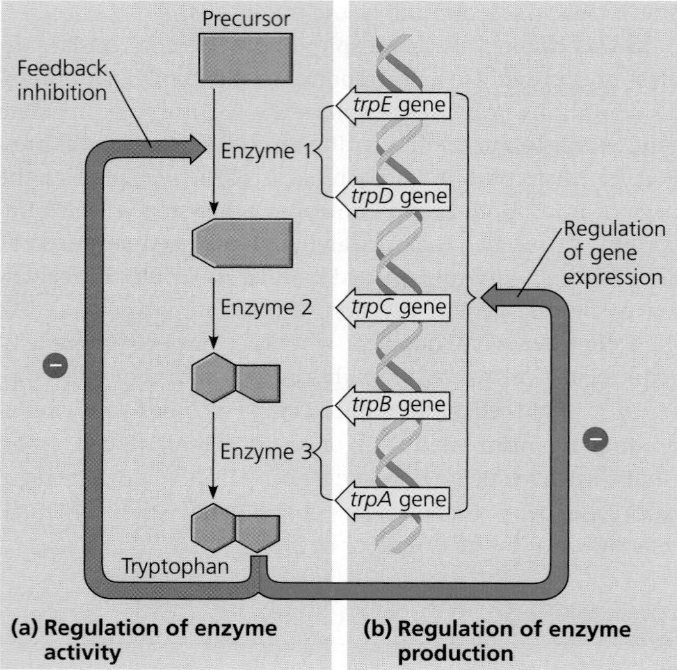

▲ **Figure 18.2 Regulation of a metabolic pathway.** In the pathway for tryptophan synthesis, an abundance of tryptophan can both **(a)** inhibit the activity of the first enzyme in the pathway (feedback inhibition), a rapid response, and **(b)** repress expression of the genes encoding all subunits of the enzymes in the pathway, a longer-term response. Genes *trpE* and *trpD* encode the two subunits of enzyme 1, and genes *trpB* and *trpA* encode the two subunits of enzyme 3. (The genes were named before the order in which they functioned in the pathway was determined.) The ⊖ symbol stands for inhibition.

Second, cells can adjust the production level of certain enzymes; that is, they can regulate the expression of the genes encoding the enzymes. If, in our example, the environment provides all the tryptophan the cell needs, the cell stops making the enzymes that catalyze the synthesis of tryptophan **(Figure 18.2b)**. In this case, the control of enzyme production occurs at the level of transcription, the synthesis of messenger RNA coding for these enzymes. More generally, many genes of the bacterial genome are switched on or off by changes in the metabolic status of the cell. One basic mechanism for this control of gene expression in bacteria, described as the *operon model*, was discovered in 1961 by François Jacob and Jacques Monod at the Pasteur Institute in Paris. Let's see what an operon is and how it works, using the control of tryptophan synthesis as our first example.

Operons: The Basic Concept

E. coli synthesizes the amino acid tryptophan from a precursor molecule in the multistep pathway shown in Figure 18.2. Each reaction in the pathway is catalyzed by a specific enzyme, and the five genes that code for the subunits of these enzymes are clustered together on the bacterial chromosome. A single promoter serves all five genes, which together constitute a transcription unit. (Recall from Chapter 17 that a promoter is a site where RNA polymerase can bind to DNA and begin transcription.) Thus, transcription gives rise to one long mRNA molecule that codes for the five polypeptides making up the enzymes in the tryptophan pathway. The cell can translate this one mRNA into five separate polypeptides because the mRNA is punctuated with start and stop codons that signal where the coding sequence for each polypeptide begins and ends.

A key advantage of grouping genes of related function into one transcription unit is that a single "on-off switch" can control the whole cluster of functionally related genes; in other words, these genes are *coordinately controlled*. When an *E. coli* cell must make tryptophan for itself because the nutrient medium lacks this amino acid, all the enzymes for the metabolic pathway are synthesized at one time. The switch is a segment of DNA called an **operator**. Both its location and name suit its function: Positioned within the promoter or, in some cases, between the promoter and the enzyme-coding genes, the operator controls the access of RNA polymerase to the genes. All together, the operator, the promoter, and the genes they control—the entire stretch of DNA required for enzyme production for the tryptophan pathway—constitute an **operon**. The *trp* operon (*trp* for tryptophan) is one of many operons in the *E. coli* genome **(Figure 18.3)**.

If the operator is the operon's switch for controlling transcription, how does this switch work? By itself, the *trp* operon is turned on; that is, RNA polymerase can bind to the promoter and transcribe the genes of the operon. The operon

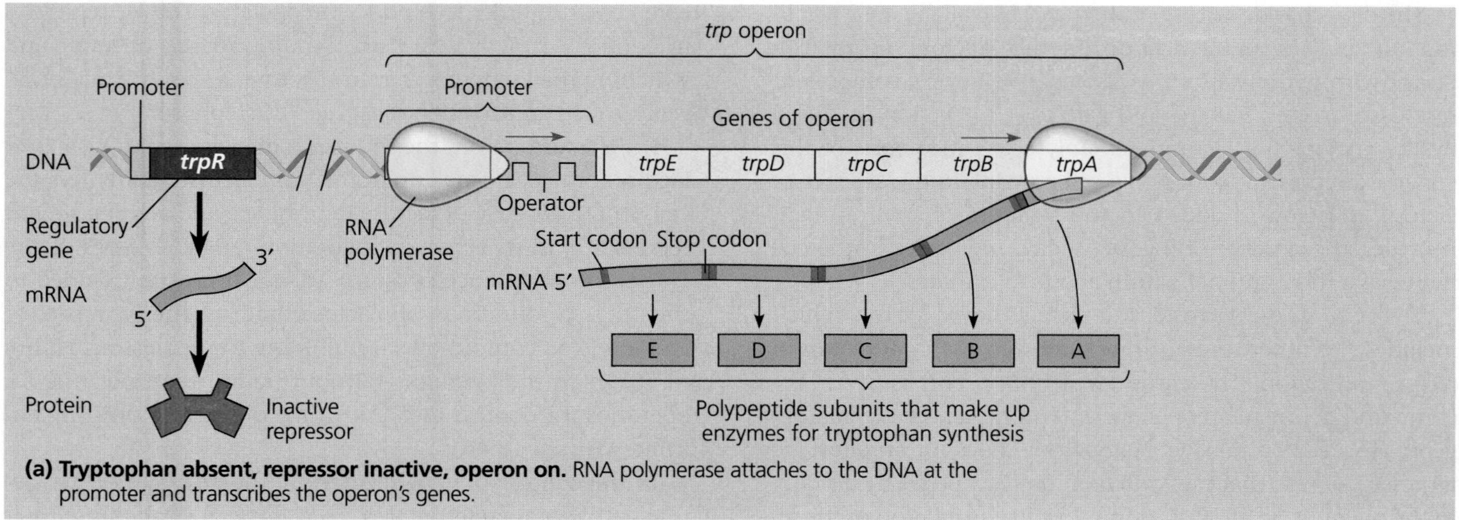

(a) Tryptophan absent, repressor inactive, operon on. RNA polymerase attaches to the DNA at the promoter and transcribes the operon's genes.

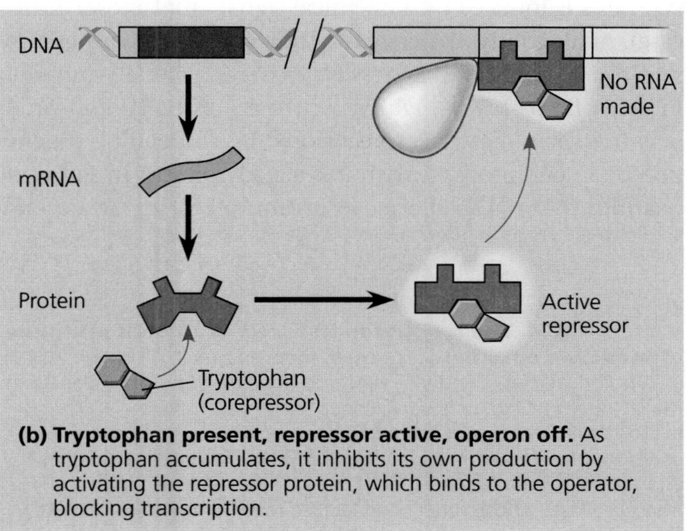

(b) Tryptophan present, repressor active, operon off. As tryptophan accumulates, it inhibits its own production by activating the repressor protein, which binds to the operator, blocking transcription.

▲ **Figure 18.3 The *trp* operon in *E. coli*: regulated synthesis of repressible enzymes.** Tryptophan is an amino acid produced by an anabolic pathway catalyzed by repressible enzymes. **(a)** The five genes encoding the polypeptide subunits of the enzymes in this pathway (see Figure 18.2) are grouped, along with a promoter, into the *trp* operon. The *trp* operator (the repressor binding site) is located within the *trp* promoter (the RNA polymerase binding site). **(b)** Accumulation of tryptophan, the end product of the pathway, represses transcription of the *trp* operon, thus blocking synthesis of all the enzymes in the pathway and shutting down tryptophan production.

? *Describe what happens to the* trp *operon as the cell uses up its store of tryptophan.*

can be switched off by a protein called the *trp* **repressor**. The repressor binds to the operator and blocks attachment of RNA polymerase to the promoter, preventing transcription of the genes. A repressor protein is specific for the operator of a particular operon. For example, the repressor that switches off the *trp* operon by binding to the *trp* operator has no effect on other operons in the *E. coli* genome.

The *trp* repressor is the protein product of a **regulatory gene** called *trpR*, which is located some distance from the *trp* operon and has its own promoter. Regulatory genes are expressed continuously, although at a low rate, and a few *trp* repressor molecules are always present in *E. coli* cells. Why, then, is the *trp* operon not switched off permanently? First, the binding of repressors to operators is reversible. An operator vacillates between two states: one without the repressor bound and one with the repressor bound. The relative duration of each state depends on the number of active repressor molecules around. Second, the *trp* repressor, like most regulatory proteins, is an allosteric protein, with two alternative shapes, active and inactive (see Figure 8.20). The *trp* repressor is synthesized in an inactive form with little affinity for the *trp* operator. Only if tryptophan binds to the *trp* repressor at an allosteric site does the repressor protein change to the active form that can attach to the operator, turning the operon off.

Tryptophan functions in this system as a **corepressor**, a small molecule that cooperates with a repressor protein to switch an operon off. As tryptophan accumulates, more tryptophan molecules associate with *trp* repressor molecules, which can then bind to the *trp* operator and shut down production of the tryptophan pathway enzymes. If the cell's tryptophan level drops, transcription of the operon's genes resumes. The *trp* operon is one example of how gene expression can respond to changes in the cell's internal and external environment.

Repressible and Inducible Operons: Two Types of Negative Gene Regulation

The *trp* operon is said to be a *repressible operon* because its transcription is usually on but can be inhibited (repressed) when a specific small molecule (in this case, tryptophan) binds allosterically to a regulatory protein. In contrast, an *inducible operon* is usually off but can be stimulated (induced) when a specific small molecule interacts with a regulatory protein. The classic example of an inducible operon is the *lac* operon (*lac* for lactose), which was the subject of Jacob and Monod's pioneering research.

The disaccharide lactose (milk sugar) is available to *E. coli* in the human colon if the host drinks milk. Lactose metabolism begins with hydrolysis of the disaccharide into its component monosaccharides, glucose and galactose, a reaction catalyzed by the enzyme β-galactosidase. Only a few molecules of this enzyme are present in an *E. coli* cell growing in the absence of lactose. If lactose is added to the bacterium's environment, however, the number of β-galactosidase molecules in the cell increases a thousandfold within about 15 minutes.

The gene for β-galactosidase is part of the *lac* operon, which includes two other genes coding for enzymes that function in lactose utilization. The entire transcription unit is under the command of one main operator and promoter. The regulatory gene, *lacI*, located outside the operon, codes for an allosteric repressor protein that can switch off the *lac* operon by binding to the operator. So far, this sounds just like regulation of the *trp* operon, but there is one important difference. Recall that the *trp* repressor is inactive by itself and requires tryptophan as a corepressor in order to bind to the operator. The *lac* repressor, in contrast, is active by itself, binding to the operator and switching the *lac* operon off. In this case, a specific small molecule, called an **inducer**, *inactivates* the repressor.

For the *lac* operon, the inducer is allolactose, an isomer of lactose formed in small amounts from lactose that enters the cell. In the absence of lactose (and hence allolactose), the *lac* repressor is in its active configuration, and the genes of the *lac* operon are silenced **(Figure 18.4a)**. If lactose is added to the cell's surroundings, allolactose binds to the *lac* repressor and alters its conformation, nullifying the repressor's ability to attach to the operator. Without bound repressor, the *lac* operon is transcribed into mRNA for the lactose-utilizing enzymes **(Figure 18.4b)**.

In the context of gene regulation, the enzymes of the lactose pathway are referred to as *inducible enzymes* because their synthesis is induced by a chemical signal (allolactose, in this case). Analogously, the enzymes for tryptophan synthesis are said to be repressible. *Repressible enzymes* generally function in anabolic pathways, which synthesize essential end products from raw materials (precursors). By suspending production of an end product when it is already present in sufficient quantity, the cell can allocate its organic precursors and energy

▼ **Figure 18.4 The *lac* operon in *E. coli*: regulated synthesis of inducible enzymes.** *E. coli* uses three enzymes to take up and metabolize lactose. The genes for these three enzymes are clustered in the *lac* operon. One gene, *lacZ*, codes for β-galactosidase, which hydrolyzes lactose to glucose and galactose. The second gene, *lacY*, codes for a permease, the membrane protein that transports lactose into the cell. The third gene, *lacA*, codes for an enzyme called transacetylase, whose function in lactose metabolism is still unclear. The gene for the *lac* repressor, *lacI*, happens to be adjacent to the *lac* operon, an unusual situation. The function of the teal region at the upstream end of the promoter (the left end in these diagrams) will be revealed in Figure 18.5.

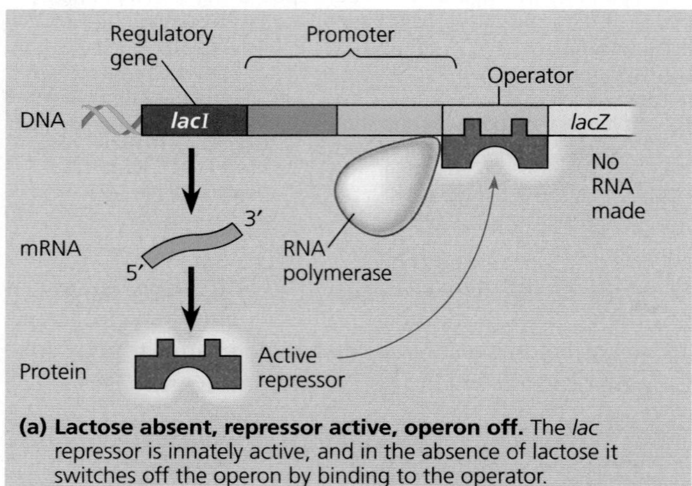

(a) Lactose absent, repressor active, operon off. The *lac* repressor is innately active, and in the absence of lactose it switches off the operon by binding to the operator.

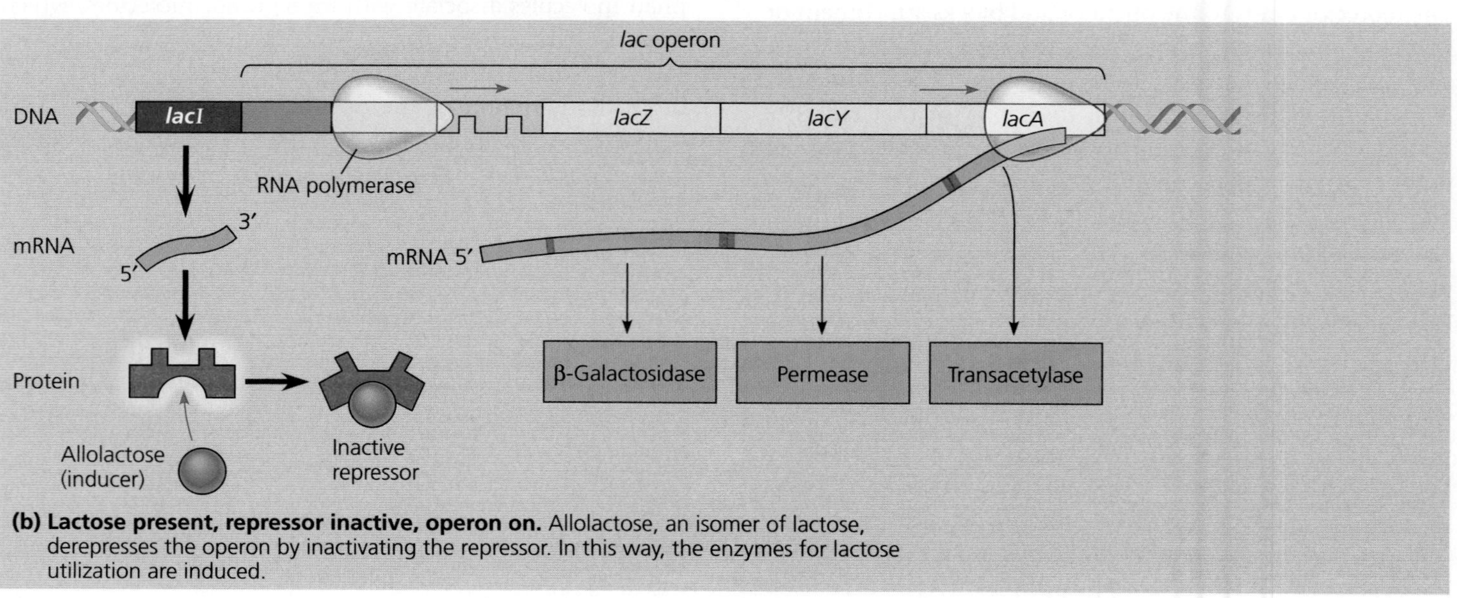

(b) Lactose present, repressor inactive, operon on. Allolactose, an isomer of lactose, derepresses the operon by inactivating the repressor. In this way, the enzymes for lactose utilization are induced.

for other uses. In contrast, inducible enzymes usually function in catabolic pathways, which break down a nutrient to simpler molecules. By producing the appropriate enzymes only when the nutrient is available, the cell avoids wasting energy and precursors making proteins that are not needed.

Regulation of both the *trp* and *lac* operons involves the *negative* control of genes, because the operons are switched off by the active form of the repressor protein. It may be easier to see this for the *trp* operon, but it is also true for the *lac* operon. Allolactose induces enzyme synthesis not by acting directly on the genome, but by freeing the *lac* operon from the negative effect of the repressor. Gene regulation is said to be *positive* only when a regulatory protein interacts directly with the genome to switch transcription on. Let's look at an example of the positive control of genes, again involving the *lac* operon.

Positive Gene Regulation

When glucose and lactose are both present in its environment, *E. coli* preferentially uses glucose. The enzymes for glucose breakdown in glycolysis (see Figure 9.9) are continually present. Only when lactose is present *and* glucose is in short supply does *E. coli* use lactose as an energy source, and only then does it synthesize appreciable quantities of the enzymes for lactose breakdown.

How does the *E. coli* cell sense the glucose concentration and relay this information to the genome? Again, the mechanism depends on the interaction of an allosteric regulatory protein with a small organic molecule, in this case **cyclic AMP (cAMP)**, which accumulates when glucose is scarce (see Figure 11.11 for the structure of cAMP). The regulatory protein, called *catabolite activator protein (CAP)*, is an **activator**, a protein that binds to DNA and stimulates transcription of a gene. When cAMP binds to this regulatory protein, CAP assumes its active shape and can attach to a specific site at the upstream end of the *lac* promoter **(Figure 18.5a)**. This attachment increases the affinity of RNA polymerase for the promoter, which is actually rather low even when no repressor is bound to the operator. By facilitating the binding of RNA polymerase to the promoter and thereby increasing the rate of transcription, the attachment of CAP to the promoter directly stimulates gene expression. Therefore, this mechanism qualifies as positive regulation.

If the amount of glucose in the cell increases, the cAMP concentration falls, and without cAMP, CAP detaches from the operon. Because CAP is inactive, RNA polymerase binds less efficiently to the promoter, and transcription of the *lac* operon proceeds at only a low level, even in the presence of lactose **(Figure 18.5b)**. Thus, the *lac* operon is under dual control: negative control by the *lac* repressor and positive control by CAP. The state of the *lac* repressor (with or without bound allolactose) determines whether or not transcription of the *lac* operon's genes occurs at all; the state of CAP (with or without bound cAMP) controls the *rate* of transcription if the operon is repressor-free. It is as though the operon has both an on-off switch and a volume control.

In addition to regulating the *lac* operon, CAP helps regulate other operons that encode enzymes used in catabolic pathways. All told, it may affect the expression of more than 100 genes in *E. coli*. When glucose is plentiful and CAP is inactive, the synthesis of enzymes that catabolize compounds other than glucose generally slows down. The ability to catabolize other compounds, such as lactose, enables a cell deprived of glucose to survive. The compounds present in the cell at the moment determine which operons are switched on—the result of simple interactions of activator and repressor proteins with the promoters of the genes in question.

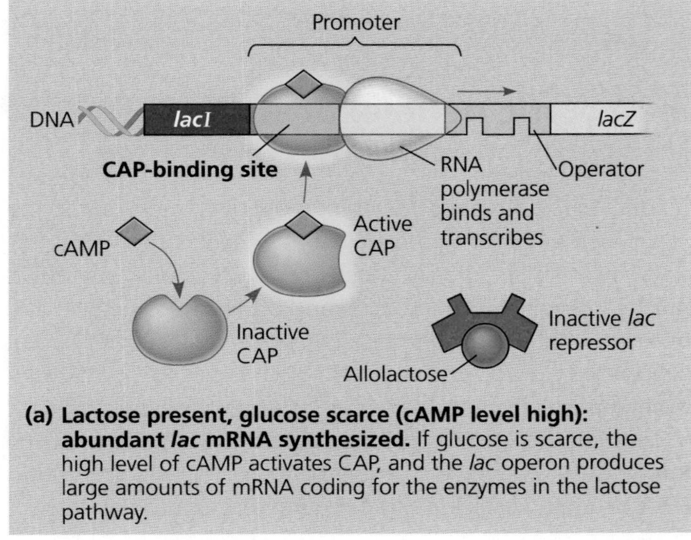

(a) Lactose present, glucose scarce (cAMP level high): abundant *lac* mRNA synthesized. If glucose is scarce, the high level of cAMP activates CAP, and the *lac* operon produces large amounts of mRNA coding for the enzymes in the lactose pathway.

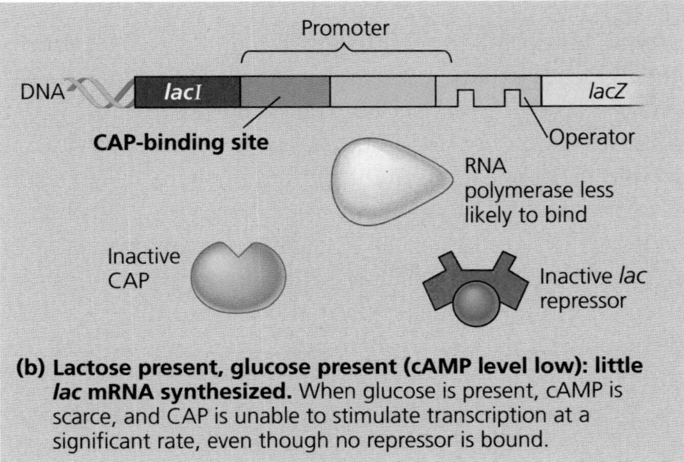

(b) Lactose present, glucose present (cAMP level low): little *lac* mRNA synthesized. When glucose is present, cAMP is scarce, and CAP is unable to stimulate transcription at a significant rate, even though no repressor is bound.

▲ **Figure 18.5 Positive control of the *lac* operon by catabolite activator protein (CAP).** RNA polymerase has high affinity for the *lac* promoter only when catabolite activator protein (CAP) is bound to a DNA site at the upstream end of the promoter. CAP attaches to its DNA site only when associated with cyclic AMP (cAMP), whose concentration in the cell rises when the glucose concentration falls. Thus, when glucose is present, even if lactose also is available, the cell preferentially catabolizes glucose and makes very little of the lactose-utilizing enzymes.

CONCEPT CHECK 18.1

1. How does binding of the *trp* corepressor and the *lac* inducer to their respective repressor proteins alter repressor function and transcription in each case?
2. Describe the binding of RNA polymerase, repressors, and activators to the *lac* operon when both lactose and glucose are scarce. What is the effect of these scarcities on transcription of the *lac* operon?
3. **WHAT IF?** A certain mutation in *E. coli* changes the *lac* operator so that the active repressor cannot bind. How would this affect the cell's production of β-galactosidase?

For suggested answers, see Appendix A.

CONCEPT 18.2

Eukaryotic gene expression is regulated at many stages

All organisms, whether prokaryotes or eukaryotes, must regulate which genes are expressed at any given time. Both unicellular organisms and the cells of multicellular organisms must continually turn genes on and off in response to signals from their external and internal environments. Regulation of gene expression is also essential for cell specialization in multicellular organisms, which are made up of different types of cells, each with a distinct role. To perform its role, each cell type must maintain a specific program of gene expression in which certain genes are expressed and others are not.

Differential Gene Expression

A typical human cell might express about 20% of its protein-coding genes at any given time. Highly differentiated cells, such as muscle or nerve cells, express an even smaller fraction of their genes. Almost all the cells in an organism contain an identical genome. (Cells of the immune system are one exception, as you will see in Chapter 43.) However, the subset of genes expressed in the cells of each type is unique, allowing these cells to carry out their specific function. The differences between cell types, therefore, are due not to different genes being present, but to **differential gene expression**, the expression of different genes by cells with the same genome.

The function of any cell, whether a single-celled eukaryote or a particular cell type in a multicellular organism, depends on the appropriate set of genes being expressed. The transcription factors of a cell must locate the right genes at the right time, a task on a par with finding a needle in a haystack. When gene expression proceeds abnormally, serious imbalances and diseases, including cancer, can arise.

Figure 18.6 summarizes the entire process of gene expression in a eukaryotic cell, highlighting key stages in the expression of a protein-coding gene. Each stage depicted in

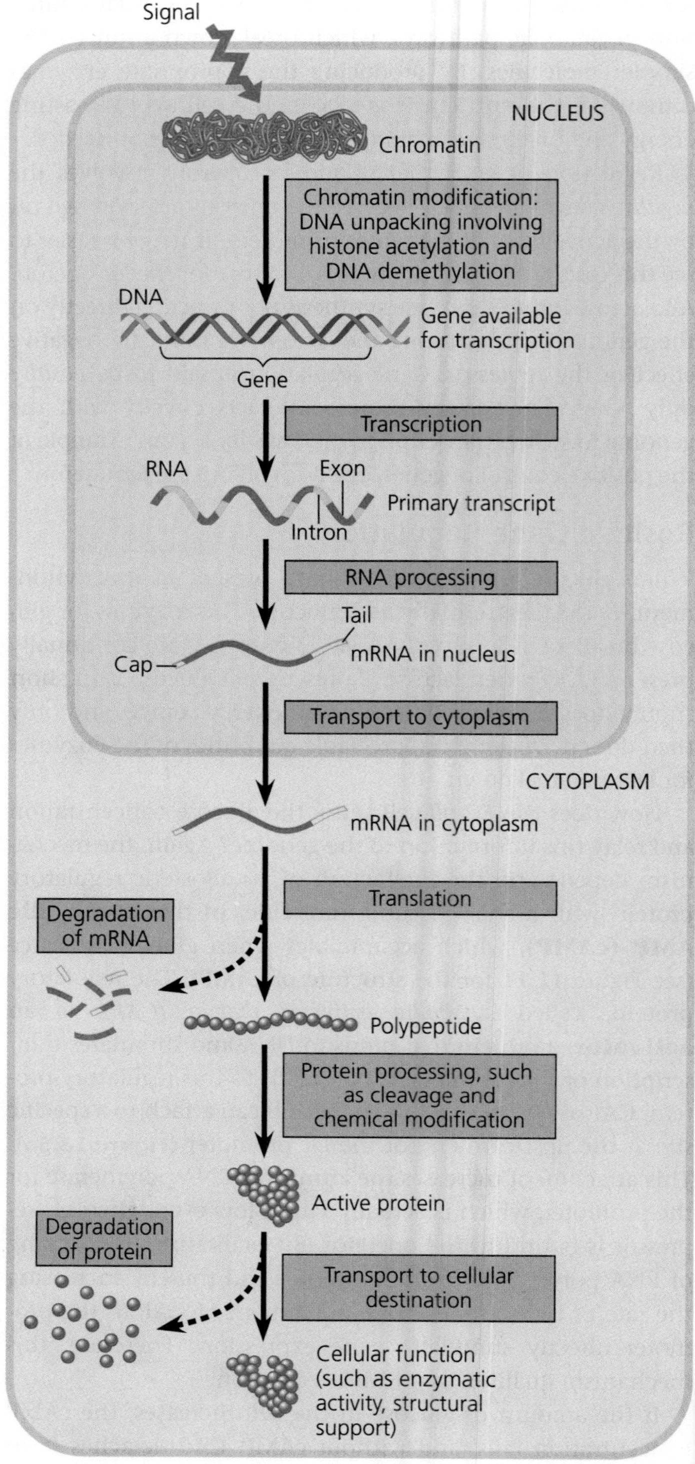

▲ **Figure 18.6 Stages in gene expression that can be regulated in eukaryotic cells.** In this diagram, the colored boxes indicate the processes most often regulated; each color indicates the type of molecule that is affected (blue = DNA, orange = RNA, purple = protein). The nuclear envelope separating transcription from translation in eukaryotic cells offers an opportunity for post-transcriptional control in the form of RNA processing that is absent in prokaryotes. In addition, eukaryotes have a greater variety of control mechanisms operating before transcription and after translation. The expression of any given gene, however, does not necessarily involve every stage shown; for example, not every polypeptide is cleaved.

Figure 18.6 is a potential control point at which gene expression can be turned on or off, accelerated, or slowed down.

Only 50 years ago, an understanding of the mechanisms that control gene expression in eukaryotes seemed almost hopelessly out of reach. Since then, new research methods, notably advances in DNA technology (see Chapter 20), have enabled molecular biologists to uncover many of the details of eukaryotic gene regulation. In all organisms, a common control point for gene expression is at transcription; regulation at this stage often occurs in response to signals coming from outside the cell, such as hormones or other signaling molecules. For this reason, the term *gene expression* is often equated with transcription for both bacteria and eukaryotes. While this is most often the case for bacteria, the greater complexity of eukaryotic cell structure and function provides opportunities for regulating gene expression at many additional stages (see Figure 18.6). In the remainder of this section, we'll examine some of the important control points of eukaryotic gene expression more closely.

Regulation of Chromatin Structure

Recall that the DNA of eukaryotic cells is packaged with proteins in an elaborate complex known as chromatin, the basic unit of which is the nucleosome (see Figure 16.22). The structural organization of chromatin not only packs a cell's DNA into a compact form that fits inside the nucleus, but also helps regulate gene expression in several ways. The location of a gene's promoter relative to nucleosomes and to the sites where the DNA attaches to the chromosome scaffold or nuclear lamina can affect whether the gene is transcribed. In addition, genes within heterochromatin, which is highly condensed, are usually not expressed. Lastly, certain chemical modifications to the histone proteins and to the DNA of chromatin can influence both chromatin structure and gene expression. Here we examine the effects of these modifications, which are catalyzed by specific enzymes.

Histone Modifications

There is abundant evidence that chemical modifications to histones, the proteins around which the DNA is wrapped in nucleosomes, play a direct role in the regulation of gene transcription. The N-terminus of each histone molecule in a nucleosome protrudes outward from the nucleosome (**Figure 18.7a**). These histone tails are accessible to various modifying enzymes that catalyze the addition or removal of specific chemical groups.

In **histone acetylation,** acetyl groups (—$COCH_3$) are attached to lysines in histone tails; deacetylation is the removal of acetyl groups. When the lysines are acetylated, their positive charges are neutralized and the histone tails no longer bind to neighboring nucleosomes (**Figure 18.7b**). Such binding promotes the folding of chromatin into a more compact structure; when this binding does not occur, chromatin has a looser structure. As a result, transcription proteins have easier

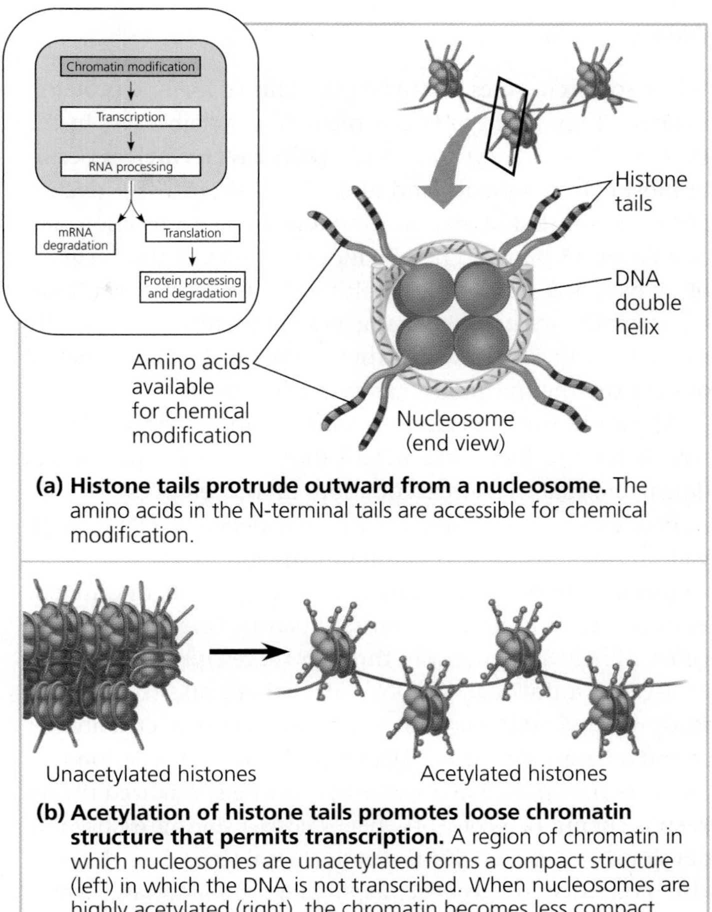

(a) Histone tails protrude outward from a nucleosome. The amino acids in the N-terminal tails are accessible for chemical modification.

(b) Acetylation of histone tails promotes loose chromatin structure that permits transcription. A region of chromatin in which nucleosomes are unacetylated forms a compact structure (left) in which the DNA is not transcribed. When nucleosomes are highly acetylated (right), the chromatin becomes less compact, and the DNA is accessible for transcription.

▲ **Figure 18.7 A simple model of histone tails and the effect of histone acetylation.** In addition to acetylation, histones can undergo several other types of modifications that also help determine the chromatin configuration in a region.

access to genes in an acetylated region. Researchers have shown that some enzymes that acetylate or deacetylate histones are closely associated with or even components of the transcription factors that bind to promoters (see Figure 17.8). These observations suggest that histone acetylation enzymes may promote the initiation of transcription not only by remodeling chromatin structure, but also by binding to and thus "recruiting" components of the transcription machinery.

Other chemical groups, such as methyl and phosphate groups, can be reversibly attached to amino acids in histone tails. Addition of methyl groups (—CH_3) to histone tails (histone methylation) can promote condensation of the chromatin, while addition of a phosphate group (phosphorylation) to an amino acid next to a methylated amino acid can have the opposite effect. The recent discovery that modifications to histone tails can affect chromatin structure and gene expression has led to the *histone code hypothesis*. This hypothesis proposes that specific combinations of modifications, as well as the order in which they have occurred, help determine the chromatin configuration, which in turn influences transcription.

DNA Methylation

While some enzymes methylate the tails of histone proteins, a different set of enzymes can methylate certain bases in the DNA itself, usually cytosine. Such **DNA methylation** occurs in most plants, animals, and fungi. Long stretches of inactive DNA, such as that of inactivated mammalian X chromosomes (see Figure 15.8), are generally more methylated than regions of actively transcribed DNA, although there are exceptions. On a smaller scale, individual genes are usually more heavily methylated in cells in which they are not expressed. Removal of the extra methyl groups can turn on some of these genes.

At least in some species, DNA methylation seems to be essential for the long-term inactivation of genes that occurs during normal cell differentiation in the embryo. For instance, experiments have shown that deficient DNA methylation (due to lack of a methylating enzyme) leads to abnormal embryonic development in organisms as different as mice and *Arabidopsis* (a mustard plant). Once methylated, genes usually stay that way through successive cell divisions in a given individual. At DNA sites where one strand is already methylated, enzymes methylate the correct daughter strand after each round of DNA replication. Methylation patterns are thus passed on, and cells forming specialized tissues keep a chemical record of what occurred during embryonic development. A methylation pattern maintained in this way also accounts for *genomic imprinting* in mammals, where methylation permanently regulates expression of either the maternal or paternal allele of particular genes at the start of development (see Figure 15.17).

Epigenetic Inheritance

The chromatin modifications that we have just discussed do not entail a change in the DNA sequence, yet they may be passed along to future generations of cells. Inheritance of traits transmitted by mechanisms not directly involving the nucleotide sequence is called **epigenetic inheritance**. Whereas mutations in the DNA are permanent changes, modifications to the chromatin can be reversed, by processes that are not yet fully understood. The molecular systems for chromatin modification may well interact with each other in a regulated way. In *Drosophila*, for example, experiments have suggested that a particular histone-modifying enzyme recruits a DNA methylation enzyme to one region and that the two enzymes collaborate to silence a particular set of genes. Working in the opposite order, proteins have also been found that first bind to methylated DNA and then recruit histone deacetylation enzymes. Thus, a dual mechanism, involving both DNA methylation and histone deacetylation, can repress transcription.

Researchers are amassing more and more evidence for the importance of epigenetic information in the regulation of gene expression. Epigenetic variations might help explain why one identical twin acquires a genetically based disease, such as schizophrenia, but the other does not, despite their identical genomes. Alterations in normal patterns of DNA methylation are seen in some cancers, where they are associated with inappropriate gene expression. Evidently, enzymes that modify chromatin structure are integral parts of the eukaryotic cell's machinery for regulating transcription.

Regulation of Transcription Initiation

Chromatin-modifying enzymes provide initial control of gene expression by making a region of DNA either more or less able to bind the transcription machinery. Once the chromatin of a gene is optimally modified for expression, the initiation of transcription is the next major step at which gene expression is regulated. As in bacteria, the regulation of transcription initiation in eukaryotes involves proteins that bind to DNA and either facilitate or inhibit binding of RNA polymerase. The process is more complicated in eukaryotes, however. Before looking at how eukaryotic cells control their transcription, let's review the structure of a typical eukaryotic gene and its transcript.

Organization of a Typical Eukaryotic Gene

A eukaryotic gene and the DNA elements (segments) that control it are typically organized as shown in **Figure 18.8**, which extends what you learned about eukaryotic genes in Chapter 17. Recall that a cluster of proteins called a *transcription initiation complex* assembles on the promoter sequence at the "upstream" end of the gene. One of these proteins, RNA polymerase II, then proceeds to transcribe the gene, synthesizing a primary RNA transcript (pre-mRNA). RNA processing includes enzymatic addition of a 5′ cap and a poly-A tail, as well as splicing out of introns, to yield a mature mRNA. Associated with most eukaryotic genes are multiple **control elements**, segments of noncoding DNA that serve as binding sites for the proteins called transcription factors, which in turn regulate transcription. Control elements and the transcription factors they bind are critical to the precise regulation of gene expression seen in different cell types.

The Roles of Transcription Factors

To initiate transcription, eukaryotic RNA polymerase requires the assistance of transcription factors. Some transcription factors, such as those illustrated in Figure 17.8, are essential for the transcription of *all* protein-coding genes; therefore, they are often called *general transcription factors*. Only a few general transcription factors independently bind a DNA sequence, such as the TATA box within the promoter; the others primarily bind proteins, including each other and RNA polymerase II. Protein-protein interactions are crucial to the initiation of eukaryotic transcription. Only when the complete initiation complex has assembled can the polymerase begin to move along the DNA template strand, producing a complementary strand of RNA.

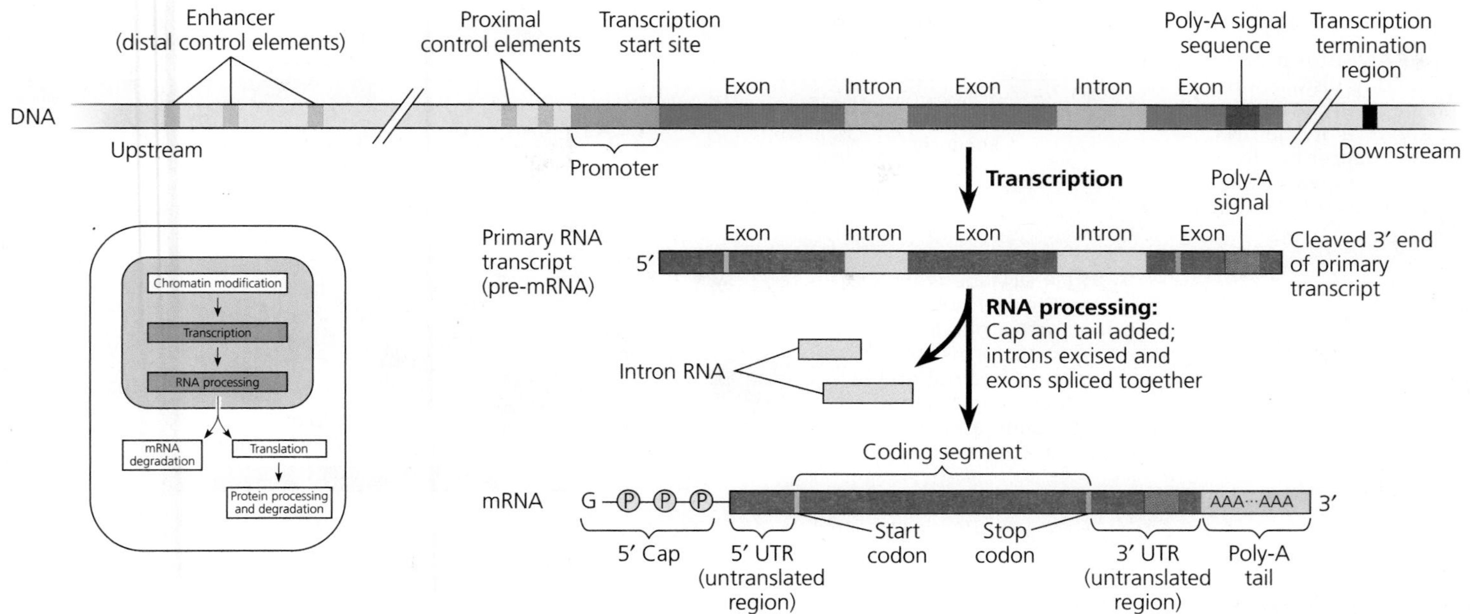

▲ **Figure 18.8 A eukaryotic gene and its transcript.** Each eukaryotic gene has a promoter, a DNA sequence where RNA polymerase binds and starts transcription, proceeding "downstream." A number of control elements (gold) are involved in regulating the initiation of transcription; these are DNA sequences located near (proximal to) or far from (distal to) the promoter. Distal control elements can be grouped together as enhancers, one of which is shown for this gene. A polyadenylation (poly-A) signal sequence in the last exon of the gene is transcribed into an RNA sequence that signals where the transcript is cleaved and the poly-A tail added. Transcription may continue for hundreds of nucleotides beyond the poly-A signal before terminating. RNA processing of the primary transcript into a functional mRNA involves three steps: addition of the 5′ cap, addition of the poly-A tail, and splicing. In the cell, the 5′ cap is added soon after transcription is initiated; splicing and poly-A tail addition may also occur while transcription is still under way (see Figure 17.10).

The interaction of general transcription factors and RNA polymerase II with a promoter usually leads to only a low rate of initiation and production of few RNA transcripts. In eukaryotes, high levels of transcription of particular genes at the appropriate time and place depend on the interaction of control elements with another set of proteins, which can be thought of as *specific transcription factors*.

Enhancers and Specific Transcription Factors As you can see in Figure 18.8, some control elements, named *proximal control elements*, are located close to the promoter. (Although some biologists consider proximal control elements part of the promoter, in this book we do not.) The more distant *distal control elements*, groupings of which are called **enhancers**, may be thousands of nucleotides upstream or downstream of a gene or even within an intron. A given gene may have multiple enhancers, each active at a different time or in a different cell type or location in the organism. Each enhancer, however, is generally associated with only that gene and no other.

In eukaryotes, the rate of gene expression can be strongly increased or decreased by the binding of specific transcription factors, either activators or repressors, to the control elements of enhancers. Hundreds of transcription activators have been discovered in eukaryotes; the structure of one example is shown in **Figure 18.9**. Researchers have identified two common structural elements in a large number of activator proteins: a DNA-binding domain—a part of the protein's three-dimensional structure that binds to DNA—and one or more activation domains. Activation domains bind other regulatory proteins or components of the transcription machinery, facilitating a series of protein-protein interactions that result in transcription of a given gene.

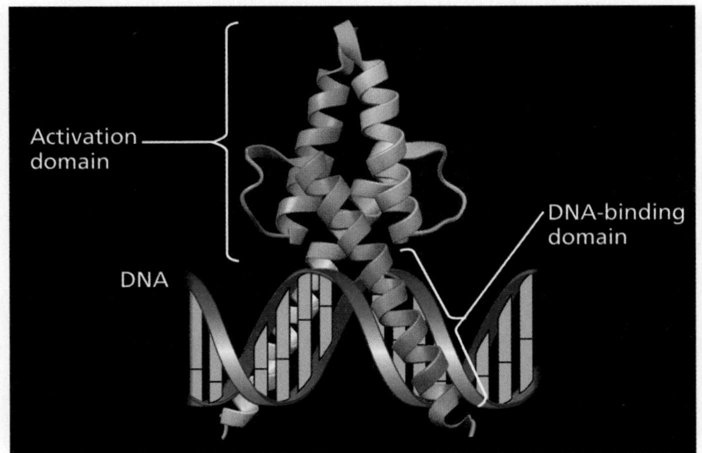

▲ **Figure 18.9 The structure of MyoD, a specific transcription factor that acts as an activator.** The MyoD protein is made up of two subunits (purple and salmon) with extensive regions of α helix. Each subunit has a DNA-binding domain and an activation domain (indicated by brackets for the purple subunit). The activation domain includes binding sites for the other subunit as well as other proteins. MyoD is involved in muscle development in vertebrate embryos and will be discussed further in Concept 18.4.

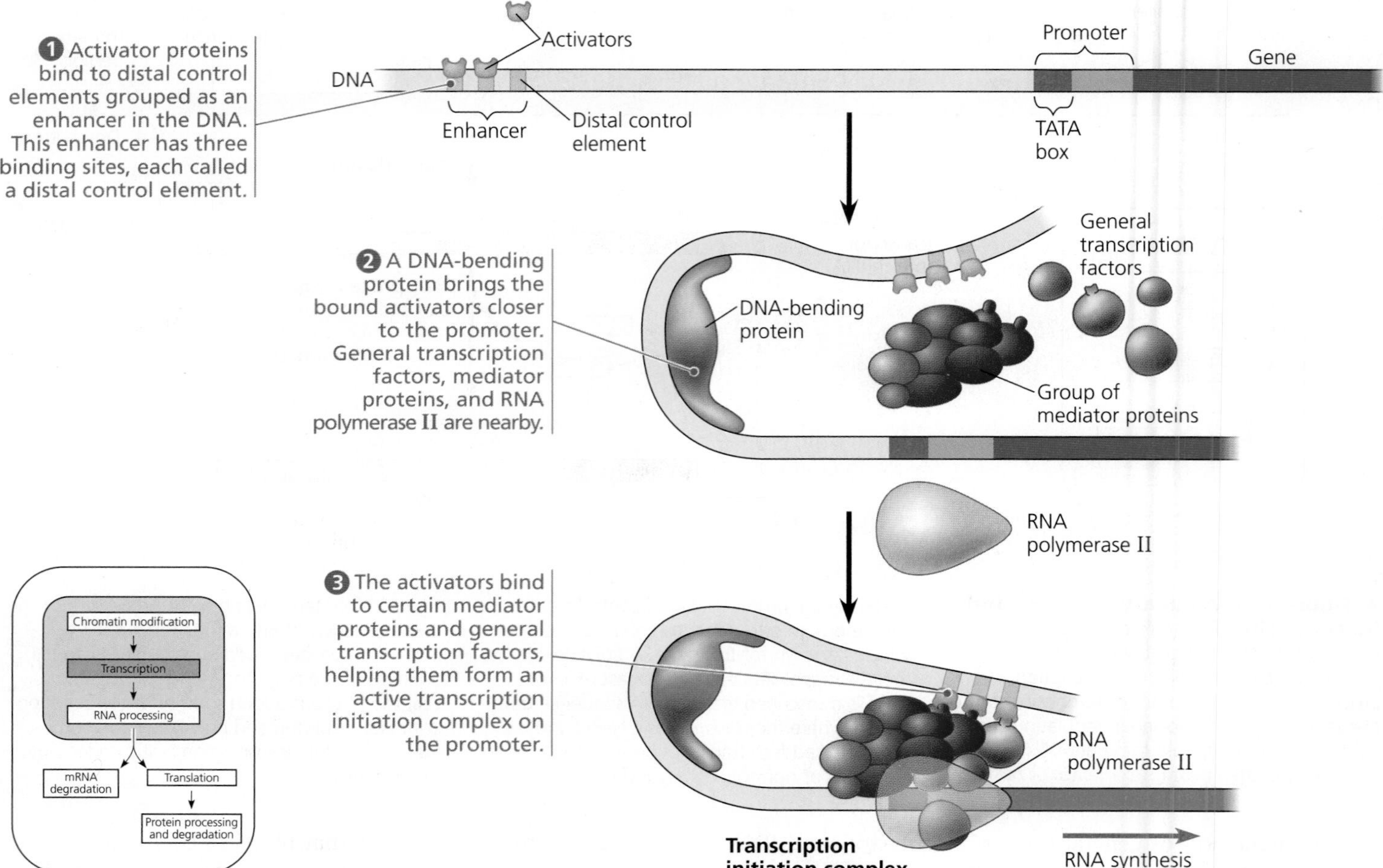

1 Activator proteins bind to distal control elements grouped as an enhancer in the DNA. This enhancer has three binding sites, each called a distal control element.
DNA
Activators
Enhancer
Distal control element
Promoter
Gene
TATA box
2 A DNA-bending protein brings the bound activators closer to the promoter. General transcription factors, mediator proteins, and RNA polymerase II are nearby.
DNA-bending protein
General transcription factors
Group of mediator proteins
RNA polymerase II
3 The activators bind to certain mediator proteins and general transcription factors, helping them form an active transcription initiation complex on the promoter.
RNA polymerase II
Transcription initiation complex
RNA synthesis
Chromatin modification
Transcription
RNA processing
mRNA degradation
Translation
Protein processing and degradation

▲ **Figure 18.10 A model for the action of enhancers and transcription activators.** Bending of the DNA by a protein enables enhancers to influence a promoter hundreds or even thousands of nucleotides away. Specific transcription factors called activators bind to the enhancer DNA sequences and then to a group of mediator proteins, which in turn bind to general transcription factors, assembling the transcription initiation complex. These protein-protein interactions facilitate the correct positioning of the complex on the promoter and the initiation of RNA synthesis. Only one enhancer (with three orange control elements) is shown here, but a gene may have several enhancers that act at different times or in different cell types.

Figure 18.10 shows a current model for how binding of activators to an enhancer located far from the promoter can influence transcription. Protein-mediated bending of the DNA is thought to bring the bound activators into contact with a group of *mediator proteins*, which in turn interact with proteins at the promoter. These multiple protein-protein interactions help assemble and position the initiation complex on the promoter. Support for this model includes a study showing that the proteins regulating a mouse globin gene contact both the gene's promoter and an enhancer located about 50,000 nucleotides upstream. Evidently, these two regions in the DNA must come together in a very specific fashion for this interaction to occur.

Specific transcription factors that function as repressors can inhibit gene expression in several different ways. Some repressors bind directly to control element DNA (in enhancers or elsewhere), blocking activator binding or, in some cases, turning off transcription even when activators are bound. Other repressors block the binding of activators to proteins that allow the activators to bind to DNA.

In addition to influencing transcription directly, some activators and repressors act indirectly by affecting chromatin structure. Studies using yeast and mammalian cells show that some activators recruit proteins that acetylate histones near the promoters of specific genes, thus promoting transcription (see Figure 18.7). Similarly, some repressors recruit proteins that deacetylate histones, leading to reduced transcription, a phenomenon referred to as *silencing*. Indeed, recruitment of chromatin-modifying proteins seems to be the most common mechanism of repression in eukaryotes.

Combinatorial Control of Gene Activation In eukaryotes, the precise control of transcription depends largely on the binding of activators to DNA control elements. Considering the great number of genes that must be regulated in a typical animal or plant cell, the number of completely different nucleotide sequences found in control elements is surprisingly small. A dozen or so short nucleotide sequences appear again and again in the control elements for different genes. On average, each enhancer is composed of about ten control elements, each of which can bind only one or two specific transcription factors. It is the particular *combination* of control elements in an enhancer associated with a gene, rather than the presence of a single unique control element, that is important in regulating transcription of the gene.

Even with only a dozen control element sequences available, a very large number of combinations are possible. A particular combination of control elements will be able to activate transcription only when the appropriate activator proteins are present, which may occur at a precise time during development or in a particular cell type. **Figure 18.11** illustrates how the use of different combinations of just a few control elements can allow differential regulation of transcription in two cell types. This can occur because each cell type contains a different group of activator proteins. How these groups came to differ will be explored in Concept 18.4.

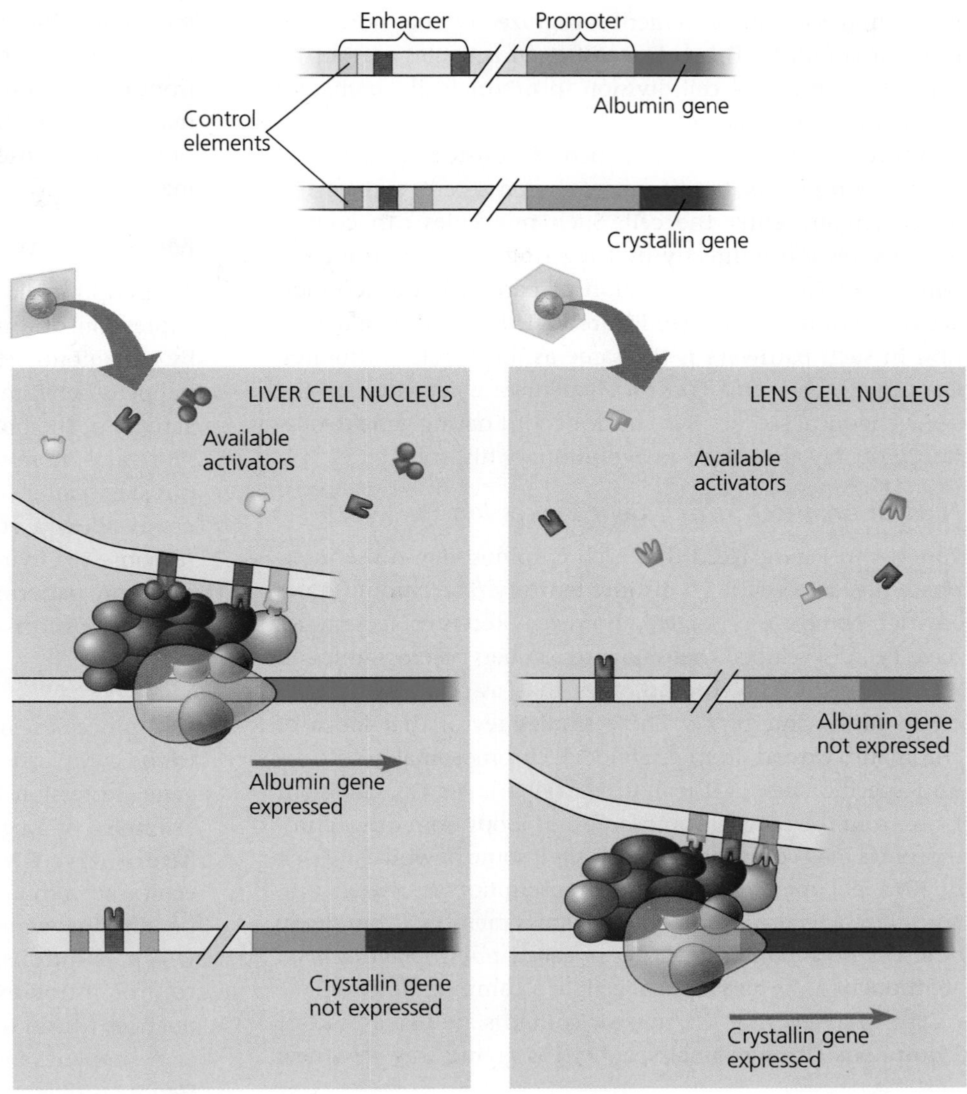

(a) Liver cell. The albumin gene is expressed, and the crystallin gene is not.

(b) Lens cell. The crystallin gene is expressed, and the albumin gene is not.

▲ **Figure 18.11 Cell type–specific transcription.** Both liver cells and lens cells have the genes for making the proteins albumin and crystallin, but only liver cells make albumin (a blood protein) and only lens cells make crystallin (the main protein of the lens of the eye). The specific transcription factors made in a cell determine which genes are expressed. In this example, the genes for albumin and crystallin are shown at the top, each with an enhancer made up of three different control elements. Although the enhancers for the two genes share one control element (gray), each enhancer has a unique combination of elements. All the activators required for high-level expression of the albumin gene are present only in liver cells (a), whereas the activators needed for expression of the crystallin gene are present only in lens cells (b). For simplicity, we consider only the role of activators here, although the presence or absence of repressors may also influence transcription in certain cell types.

? *Describe the enhancer for the albumin gene in each cell. How would the nucleotide sequence of this enhancer in the liver cell compare with that in the lens cell?*

Coordinately Controlled Genes in Eukaryotes

How does the eukaryotic cell deal with genes of related function that need to be turned on or off at the same time? Earlier in this chapter, you learned that in bacteria, such coordinately controlled genes are often clustered into an operon, which is regulated by a single promoter and transcribed into a single mRNA molecule. Thus, the genes are expressed together, and the encoded proteins are produced concurrently. With a few minor exceptions, operons that work in this way have *not* been found in eukaryotic cells.

Co-expressed eukaryotic genes, such as genes coding for the enzymes of a metabolic pathway, are typically scattered over different chromosomes. In these cases, coordinate gene expression depends on the association of a specific combination of control elements with every gene of a dispersed group. The presence of these elements can be compared to the raised flags on a few mailboxes out of many, signaling to the mail carrier to check those boxes. Copies of the activators that recognize the control elements bind to them, promoting simultaneous transcription of the genes, no matter where they are in the genome.

Coordinate control of dispersed genes in a eukaryotic cell often occurs in response to chemical signals from outside the cell. A steroid hormone, for example, enters a cell and binds to a specific intracellular receptor protein, forming a hormone-receptor complex that serves as a transcription activator (see Figure 11.9). Every gene whose transcription is stimulated by a particular steroid hormone, regardless of its chromosomal location, has a control element recognized by that hormone-receptor complex. This is how estrogen activates a group of genes that stimulate cell division in uterine cells, preparing the uterus for pregnancy.

Many signaling molecules, such as nonsteroid hormones and growth factors, bind to receptors on a cell's surface and never actually enter the cell. Such molecules can control gene expression indirectly by triggering signal transduction pathways that lead to activation of particular transcription activators or repressors (see Figure 11.15). Coordinate regulation in such pathways is the same as for steroid hormones: Genes with the same control elements are activated by the same chemical signals. Systems for coordinating gene regulation probably arose early in evolutionary history.

Nuclear Architecture and Gene Expression

You saw in Figure 16.23 that each chromosome in the interphase nucleus occupies a distinct territory. The chromosomes are not completely isolated, however. Recently, techniques have been developed that allow researchers to cross-link and identify regions of chromosomes that associate with each other during interphase. These studies reveal that loops of chromatin extend from individual chromosomal territories into specific sites in the nucleus **(Figure 18.12)**. Different loops from the same chromosome and loops from other chromosomes may congregate in such sites, some of which are rich in RNA polymerases and other transcription-associated proteins. Like a recreation center that draws members from many different neighborhoods, these so-called *transcription factories* are thought to be areas specialized for a common function.

The old view that the nuclear contents are like a bowl of amorphous chromosomal spaghetti is giving way to a new model of a nucleus with a defined architecture and regulated movements of chromatin. Relocation of particular genes from their chromosomal territories to transcription factories may be part of the process of readying genes for transcription. This is an exciting area of current research that raises many fascinating questions for future study.

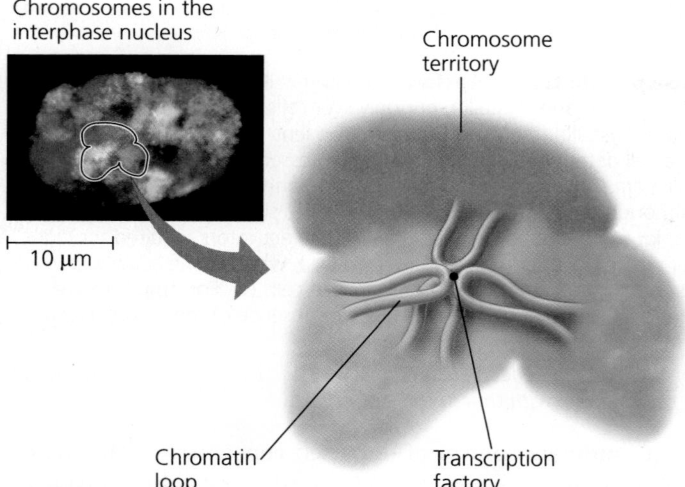

▲ **Figure 18.12 Chromosomal interactions in the interphase nucleus.** Although each chromosome has its own territory (see Figure 16.23), loops of chromatin may extend into other sites in the nucleus. Some of these sites are transcription factories that are occupied by multiple chromatin loops from the same chromosome (blue loops) or other chromosomes (red and green loops).

Mechanisms of Post-Transcriptional Regulation

Transcription alone does not constitute gene expression. The expression of a protein-coding gene is ultimately measured by the amount of functional protein a cell makes, and much happens between the synthesis of the RNA transcript and the activity of the protein in the cell. Researchers are discovering more and more regulatory mechanisms that operate at various stages after transcription (see Figure 18.6). These mechanisms allow a cell to fine-tune gene expression rapidly in response to environmental changes without altering its transcription patterns. Here we discuss how cells can regulate gene expression once a gene has been transcribed.

RNA Processing

RNA processing in the nucleus and the export of mature RNA to the cytoplasm provide several opportunities for regulating gene expression that are not available in prokaryotes. One example of regulation at the RNA-processing level is **alternative RNA splicing**, in which different mRNA molecules are produced from the same primary transcript, depending on which RNA segments are treated as exons and which as introns. Regulatory proteins specific to a cell type control intron-exon choices by binding to regulatory sequences within the primary transcript.

A simple example of alternative RNA splicing is shown in **Figure 18.13** for the troponin T gene, which encodes two different (though related) proteins. Other genes offer possibilities for far greater numbers of products. For instance, researchers have found a *Drosophila* gene with enough alternatively spliced exons to generate about 19,000 membrane proteins that have different extracellular domains. At least 17,500 (94%) of the alternative mRNAs are actually synthesized. Each developing nerve cell in the fly appears to synthesize a unique form of the protein, which acts as an identification badge on the cell surface.

It is clear that alternative RNA splicing can significantly expand the repertoire of a eukaryotic genome. In fact, alternative splicing was proposed as one explanation for the surprisingly low number of human genes counted when the human genome was sequenced about ten years ago. The number of human genes was found to be similar to that of a soil worm (nematode), a mustard plant, or a sea anemone. This discovery prompted questions about what, if not the number of genes, accounts for the more complex morphology (external form) of humans. It turns out that 75–100% of human genes that have multiple exons probably undergo alternative splicing. Thus, the extent of alternative splicing

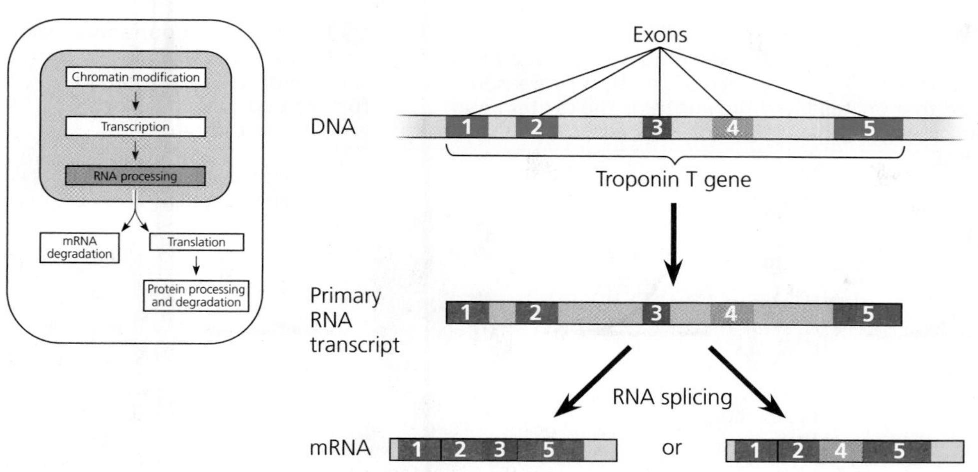

◀ **Figure 18.13 Alternative RNA splicing of the troponin T gene.** The primary transcript of this gene can be spliced in more than one way, generating different mRNA molecules. Notice that one mRNA molecule has ended up with exon 3 (green) and the other with exon 4 (purple). These two mRNAs are translated into different but related muscle proteins.

greatly multiplies the number of possible human proteins, which may be better correlated with complexity of form.

mRNA Degradation

The life span of mRNA molecules in the cytoplasm is important in determining the pattern of protein synthesis in a cell. Bacterial mRNA molecules typically are degraded by enzymes within a few minutes of their synthesis. This short life span of mRNAs is one reason bacteria can change their patterns of protein synthesis so quickly in response to environmental changes. In contrast, mRNAs in multicellular eukaryotes typically survive for hours, days, or even weeks. For instance, the mRNAs for the hemoglobin polypeptides (α-globin and β-globin) in developing red blood cells are unusually stable, and these long-lived mRNAs are translated repeatedly in these cells.

Nucleotide sequences that affect how long an mRNA remains intact are often found in the untranslated region (UTR) at the 3′ end of the molecule (see Figure 18.8). In one experiment, researchers transferred such a sequence from the short-lived mRNA for a growth factor to the 3′ end of a normally stable globin mRNA. The globin mRNA was quickly degraded.

During the past few years, other mechanisms that degrade or block expression of mRNA molecules have come to light. These mechanisms involve an important group of newly discovered RNA molecules that regulate gene expression at several levels, and we will discuss them later in this chapter.

Initiation of Translation

Translation presents another opportunity for regulating gene expression; such regulation occurs most commonly at the initiation stage (see Figure 17.18). For some mRNAs, the initiation of translation can be blocked by regulatory proteins that bind to specific sequences or structures within the untranslated region at the 5′ or 3′ end (5′ or 3′ UTR), preventing the attachment of ribosomes. (Recall from Chapter 17 that both the 5′ cap and the poly-A tail of an mRNA molecule are important for ribosome binding.) A different mechanism for blocking translation is seen in a variety of mRNAs present in the eggs of many organisms: Initially, these stored mRNAs lack poly-A tails of sufficient length to allow translation initiation. At the appropriate time during embryonic development, however, a cytoplasmic enzyme adds more adenine (A) nucleotides, prompting translation to begin.

Alternatively, translation of *all* the mRNAs in a cell may be regulated simultaneously. In a eukaryotic cell, such "global" control usually involves the activation or inactivation of one or more of the protein factors required to initiate translation. This mechanism plays a role in starting translation of mRNAs that are stored in eggs. Just after fertilization, translation is triggered by the sudden activation of translation initiation factors. The response is a burst of synthesis of the proteins encoded by the stored mRNAs. Some plants and algae store mRNAs during periods of darkness; light then triggers the reactivation of the translational apparatus.

Protein Processing and Degradation

The final opportunities for controlling gene expression occur after translation. Often, eukaryotic polypeptides must be processed to yield functional protein molecules. For instance, cleavage of the initial insulin polypeptide (pro-insulin) forms the active hormone. In addition, many proteins undergo chemical modifications that make them functional. Regulatory proteins are commonly activated or inactivated by the reversible addition of phosphate groups, and proteins destined for the surface of animal cells acquire sugars. Cell-surface proteins and many others must also be transported to target destinations in the cell in order to function. Regulation might occur at any of the steps involved in modifying or transporting a protein.

Finally, the length of time each protein functions in the cell is strictly regulated by means of selective degradation. Many proteins, such as the cyclins involved in regulating the cell cycle, must be relatively short-lived if the cell is to function appropriately (see Figure 12.17). To mark a particular protein for destruction, the cell commonly attaches molecules of a small protein called ubiquitin to the protein. Giant protein complexes called **proteasomes** then recognize the

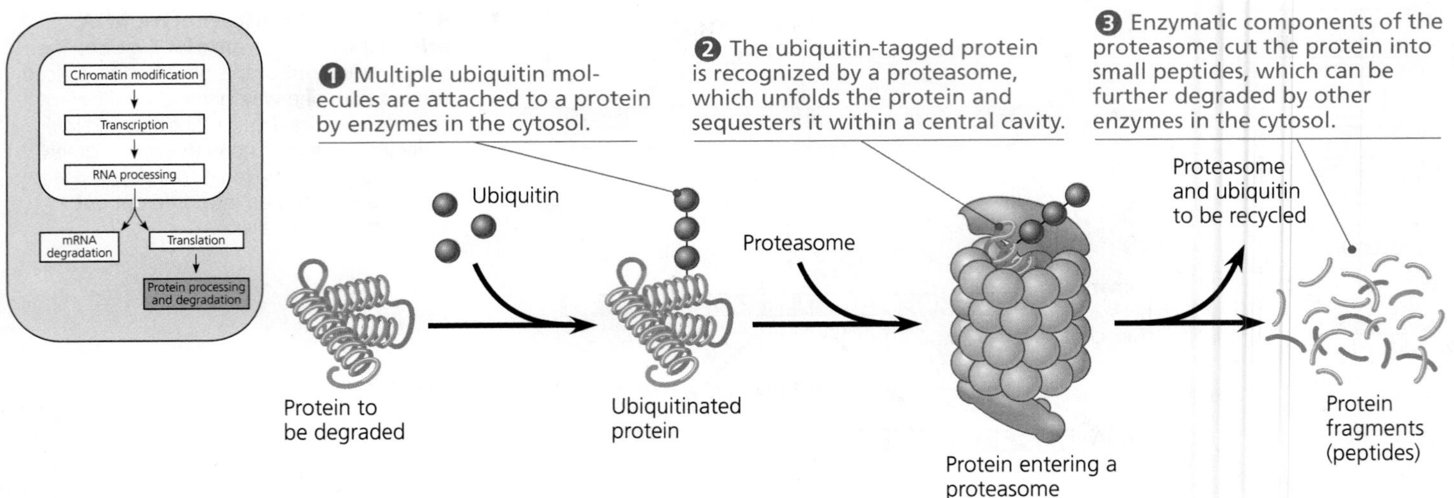
Chromatin modification
Transcription
RNA processing
mRNA degradation
Translation
Protein processing and degradation
1 Multiple ubiquitin molecules are attached to a protein by enzymes in the cytosol.
2 The ubiquitin-tagged protein is recognized by a proteasome, which unfolds the protein and sequesters it within a central cavity.
3 Enzymatic components of the proteasome cut the protein into small peptides, which can be further degraded by other enzymes in the cytosol.
Ubiquitin
Proteasome
Proteasome and ubiquitin to be recycled
Protein to be degraded
Ubiquitinated protein
Protein entering a proteasome
Protein fragments (peptides)

▲ **Figure 18.14 Degradation of a protein by a proteasome.** A proteasome, an enormous protein complex shaped like a trash can, chops up unneeded proteins in the cell. In most cases, the proteins attacked by a proteasome have been tagged with short chains of ubiquitin, a small protein. Steps 1 and 3 require ATP. Eukaryotic proteasomes are as massive as ribosomal subunits and are distributed throughout the cell. Their shape somewhat resembles that of chaperone proteins, which protect protein structure rather than destroy it (see Figure 5.23).

ubiquitin-tagged proteins and degrade them **(Figure 18.14)**. The importance of proteasomes is underscored by the finding that mutations making specific cell cycle proteins impervious to proteasome degradation can lead to cancer. The 2004 Nobel Prize in Chemistry was awarded to three scientists—two from Israel and one from the United States—who worked out the regulated process of protein degradation.

CONCEPT CHECK 18.2

1. In general, what is the effect of histone acetylation and DNA methylation on gene expression?
2. Compare the roles of general and specific transcription factors in regulating gene expression.
3. Suppose you compared the nucleotide sequences of the distal control elements in the enhancers of three genes that are expressed only in muscle tissue. What would you expect to find? Why?
4. Once mRNA encoding a particular protein reaches the cytoplasm, what are four mechanisms that can regulate the amount of the protein that is active in the cell?
5. **WHAT IF?** Examine Figure 18.11 and suggest a mechanism by which the yellow activator protein comes to be present in the liver cell but not in the lens cell.

For suggested answers, see Appendix A.

CONCEPT 18.3

Noncoding RNAs play multiple roles in controlling gene expression

Genome sequencing has revealed that protein-coding DNA accounts for only 1.5% of the human genome and a similarly small percentage of the genomes of many other multicellular eukaryotes. A very small fraction of the non-protein-coding DNA consists of genes for RNAs such as ribosomal RNA and transfer RNA. Until recently, most of the remaining DNA was assumed to be untranscribed. The idea was that since it didn't specify proteins or the few known types of RNA, such DNA didn't contain meaningful genetic information. However, a flood of recent data has contradicted this idea. For example, an in-depth study of a region comprising 1% of the human genome showed that more than 90% of that region was transcribed. Introns accounted for only a fraction of this transcribed, nontranslated RNA. These and other results suggest that a significant amount of the genome may be transcribed into non-protein-coding RNAs (also called *noncoding RNAs*, or *ncRNAs*), including a variety of small RNAs. While many questions about the functions of these RNAs remain unanswered, researchers are uncovering more evidence of their biological roles every day.

Biologists are excited about these recent discoveries, which hint at a large, diverse population of RNA molecules in the cell that play crucial roles in regulating gene expression—and have gone largely unnoticed until now. Clearly, we must revise our long-standing view that because mRNAs code for proteins, they are the most important RNAs functioning in the cell. This represents a major shift in the thinking of biologists, one that you are witnessing as students entering this field of study. It's as if our exclusive focus on a famous rock star has blinded us to the many backup musicians and songwriters working behind the scenes.

Regulation by both small and large ncRNAs is known to occur at several points in the pathway of gene expression, including mRNA translation and chromatin modification. We will focus mainly on two types of small ncRNAs that have been extensively studied in the past few years; the importance of these RNAs was acknowledged when they were the focus of the 2006 Nobel Prize in Physiology or Medicine.

Effects on mRNAs by MicroRNAs and Small Interfering RNAs

Since 1993, a number of research studies have uncovered small single-stranded RNA molecules, called **microRNAs (miRNAs)**, that are capable of binding to complementary sequences in mRNA molecules. The miRNAs are made from longer RNA precursors that fold back on themselves, forming one or more short double-stranded hairpin structures, each held together by hydrogen bonds (**Figure 18.15**). After each hairpin is cut away from the precursor, it is trimmed by an enzyme (fittingly called Dicer) into a short double-stranded fragment of about 22 nucleotide pairs. One of the two strands is degraded, while the other strand, which is the miRNA, forms a complex with one or more proteins; the miRNA allows the complex to bind to any mRNA molecule with 7–8 nucleotides of complementary sequence. The miRNA-protein complex then either degrades the target mRNA or blocks its translation. It has been estimated that expression of at least one-half of all human genes may be regulated by miRNAs, a remarkable figure given that the existence of miRNAs was unknown a mere two decades ago.

A growing understanding of the miRNA pathway provided an explanation for a perplexing observation: Researchers had found that injecting double-stranded RNA molecules into a cell somehow turned off expression of a gene with the same sequence as the RNA. They called this experimental phenomenon **RNA interference (RNAi)**. It was later shown to be due to **small interfering RNAs (siRNAs)**, which are similar in size and function to miRNAs. In fact, subsequent research showed that the same cellular machinery generates miRNAs and siRNAs and that both can associate with the same proteins, producing similar results. The distinction between miRNAs and siRNAs is based on the nature of the precursor molecule for each. While an miRNA is usually formed from a single hairpin in a precursor RNA (see Figure 18.15), multiple siRNAs are formed from a much longer, linear, double-stranded RNA molecule.

We mentioned that laboratory investigators had injected double-stranded RNAs into cells, and you may wonder whether such molecules are ever found naturally. As you will learn in Chapter 19, some viruses have double-stranded RNA genomes. Because the cellular RNAi pathway can lead to the destruction of RNAs with sequences complementary to those found in double-stranded RNAs, this pathway may have evolved as a natural defense against infection by such viruses. However, the fact that RNAi can also affect the expression of nonviral cellular genes may reflect a different evolutionary origin for the RNAi pathway. Moreover, many species, including mammals, apparently produce their own long, double-stranded RNA precursors to small RNAs such as siRNAs. Once produced, these RNAs can interfere with gene expression at stages other than translation, as we'll discuss next.

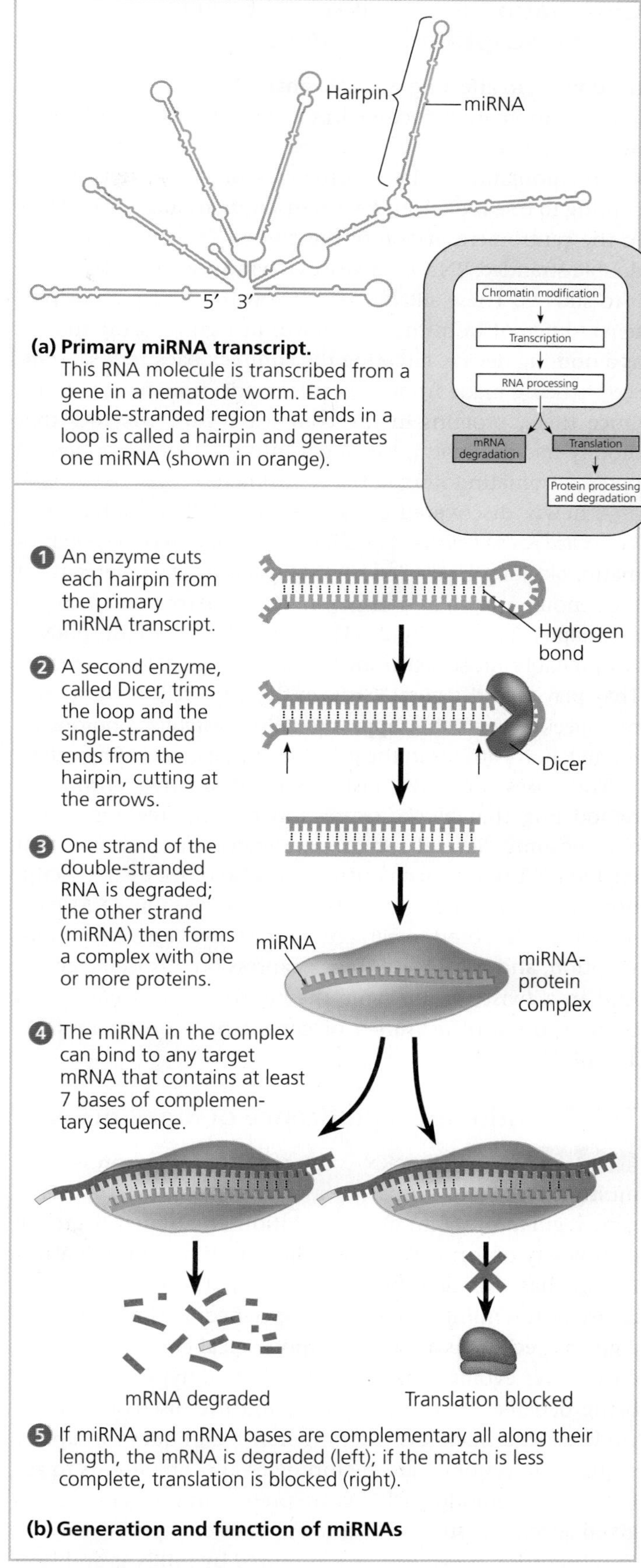

Hairpin
miRNA
5′
3′
(a) Primary miRNA transcript.
This RNA molecule is transcribed from a gene in a nematode worm. Each double-stranded region that ends in a loop is called a hairpin and generates one miRNA (shown in orange).
Chromatin modification
Transcription
RNA processing
mRNA degradation
Translation
Protein processing and degradation
1 An enzyme cuts each hairpin from the primary miRNA transcript.
Hydrogen bond
2 A second enzyme, called Dicer, trims the loop and the single-stranded ends from the hairpin, cutting at the arrows.
Dicer
3 One strand of the double-stranded RNA is degraded; the other strand (miRNA) then forms a complex with one or more proteins.
miRNA
miRNA-protein complex
4 The miRNA in the complex can bind to any target mRNA that contains at least 7 bases of complementary sequence.
mRNA degraded
Translation blocked
5 If miRNA and mRNA bases are complementary all along their length, the mRNA is degraded (left); if the match is less complete, translation is blocked (right).
(b) Generation and function of miRNAs

▲ **Figure 18.15 Regulation of gene expression by miRNAs.**

Chromatin Remodeling and Effects on Transcription by ncRNAs

In addition to affecting mRNAs, small RNAs can cause remodeling of chromatin structure. In some yeasts, siRNAs produced by the yeast cells themselves are required for the formation of heterochromatin at the centromeres of chromosomes. According to one model, an RNA transcript produced from DNA in the centromeric region of the chromosome is copied into double-stranded RNA by a yeast enzyme and then processed into siRNAs. These siRNAs associate with a complex of proteins (different from the one shown in Figure 18.15) and act as a homing device, targeting the complex back to RNA transcripts being made from the centromeric sequences of DNA. Once there, proteins in the complex recruit enzymes that modify the chromatin, turning it into the highly condensed heterochromatin found at the centromere.

A newly discovered class of small ncRNAs called *piwi-associated RNAs* (*piRNAs*) also induce formation of heterochromatin, blocking expression of some parasitic DNA elements in the genome known as transposons. (Transposons are discussed in Chapter 21.) Usually 24–31 nucleotides in length, piRNAs are probably processed from single-stranded RNA precursors. They play an indispensable role in the germ cells of many animal species, where they appear to help re-establish appropriate methylation patterns in the genome during gamete formation.

The cases we have just described involve chromatin remodeling that blocks expression of large regions of the chromosome. Several recent experiments have shown that related RNA-based mechanisms may also block the transcription of specific genes. For instance, some plant miRNAs have sequences that bind to gene promoters and can repress transcription, and piRNAs can block expression of specific genes. And in a twist on the same theme, some cases have even been reported of *activation* of gene expression by miRNAs and piRNAs.

The Evolutionary Significance of Small ncRNAs

EVOLUTION Small ncRNAs can regulate gene expression at multiple steps and in many ways. In general, extra levels of gene regulation might allow evolution of a higher degree of complexity of form. Therefore, the versatility of miRNA regulation has led some biologists to hypothesize that an increase in the number of miRNAs specified by the genome of a given species has allowed morphological complexity to increase over evolutionary time. While this hypothesis is still being debated, it is logical to expand the discussion to include all small ncRNAs. Exciting new techniques for rapidly sequencing genomes are beginning to allow biologists to ask how many genes for ncRNAs are present in the genome of a given species. A survey of different species supports the notion that siRNAs evolved first, followed by miRNAs and later piRNAs, which are found only in animals. And while there are hundreds of types of miRNAs, there appear to be many thousands of types of piRNAs, allowing the potential for very sophisticated gene regulation by piRNAs.

Given the extensive functions of ncRNAs, it is not surprising that many of the ncRNAs characterized thus far play important roles in embryonic development—the topic we turn to in the next section. Embryonic development is perhaps the ultimate example of precisely regulated gene expression.

CONCEPT CHECK 18.3

1. Compare and contrast miRNAs and siRNAs.
2. **WHAT IF?** If the mRNA being degraded in Figure 18.15 coded for a protein that promotes cell division in a multicellular organism, what would happen if a mutation disabled the gene encoding the miRNA that triggers this degradation?
3. **MAKE CONNECTIONS** In Concept 15.2 (pp. 291–292), you learned about inactivation of one of the X chromosomes in female mammals. Reread those pages, and suggest a model for how the *XIST* noncoding RNA functions to cause Barr body formation.

For suggested answers, see Appendix A.

CONCEPT 18.4

A program of differential gene expression leads to the different cell types in a multicellular organism

In the embryonic development of multicellular organisms, a fertilized egg (a zygote) gives rise to cells of many different types, each with a different structure and corresponding function. Typically, cells are organized into tissues, tissues into organs, organs into organ systems, and organ systems into the whole organism. Thus, any developmental program must produce cells of different types that form higher-level structures arranged in a particular way in three dimensions. The processes that occur during development in plants and animals are detailed in Chapters 35 and 47, respectively. In this chapter, we focus instead on the program of regulation of gene expression that orchestrates development, using a few animal species as examples.

A Genetic Program for Embryonic Development

The photos in **Figure 18.16** illustrate the dramatic difference between a zygote and the organism it becomes. This remarkable transformation results from three interrelated processes: cell division, cell differentiation, and morphogenesis. Through a succession of mitotic cell divisions, the zygote gives rise to a

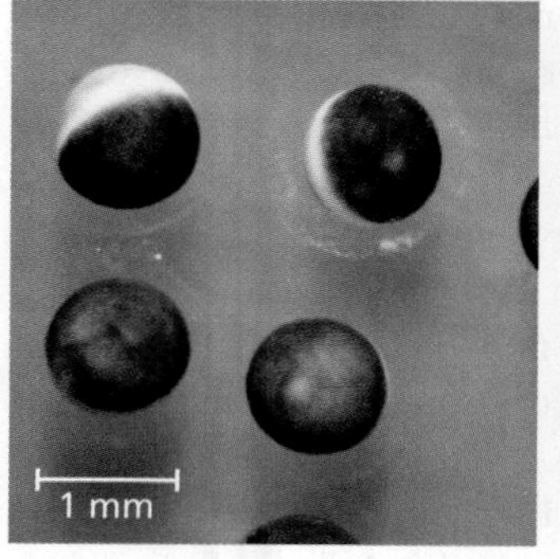

(a) Fertilized eggs of a frog

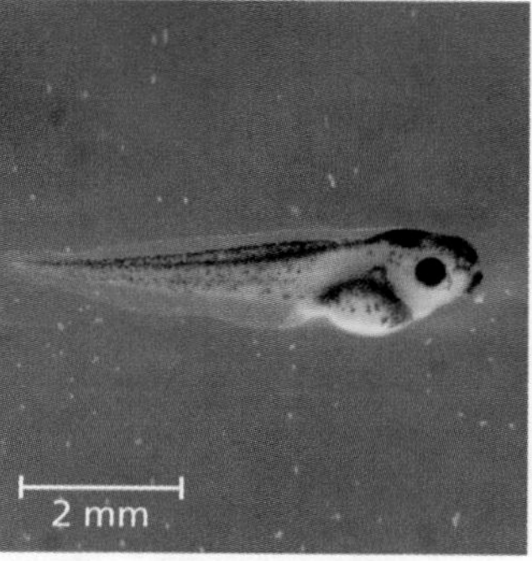

(b) Newly hatched tadpole

▲ **Figure 18.16 From fertilized egg to animal: What a difference four days makes.** It takes just four days for cell division, differentiation, and morphogenesis to transform each of the fertilized frog eggs shown in (a) into a tadpole like the one in (b).

large number of cells. Cell division alone, however, would merely produce a great ball of identical cells, nothing like a tadpole. During embryonic development, cells not only increase in number, but also undergo cell **differentiation**, the process by which cells become specialized in structure and function. Moreover, the different kinds of cells are not randomly distributed but are organized into tissues and organs in a particular three-dimensional arrangement. The physical processes that give an organism its shape constitute **morphogenesis**, meaning "creation of form."

All three processes have their basis in cellular behavior. Even morphogenesis, the shaping of the organism, can be traced back to changes in the shape, motility, and other characteristics of the cells that make up various regions of the embryo. As you have seen, the activities of a cell depend on the genes it expresses and the proteins it produces. Almost all cells in an organism have the same genome; therefore, differential gene expression results from the genes being regulated differently in each cell type.

In Figure 18.11, you saw a simplified view of how differential gene expression occurs in two cell types, a liver cell and a lens cell. Each of these fully differentiated cells has a particular mix of specific activators that turn on the collection of genes whose products are required in the cell. The fact that both cells arose through a series of mitoses from a common fertilized egg inevitably leads to a question: How do different sets of activators come to be present in the two cells?

It turns out that materials placed into the egg by the mother set up a sequential program of gene regulation that is carried out as cells divide, and this program makes the cells become different from each other in a coordinated fashion. To understand how this works, we will consider two basic developmental processes: First, we'll explore how cells that arise from early embryonic mitoses develop the differences that start each cell along its own differentiation pathway. Second, we'll see how cellular differentiation leads to one particular cell type, using muscle development as an example.

Cytoplasmic Determinants and Inductive Signals

What generates the first differences among cells in an early embryo? And what controls the differentiation of all the various cell types as development proceeds? By this point in the chapter, you can probably deduce the answer: The specific genes expressed in any particular cell of a developing organism determine its path. Two sources of information, used to varying extents in different species, "tell" a cell which genes to express at any given time during embryonic development.

One important source of information early in development is the egg's cytoplasm, which contains both RNA and proteins encoded by the mother's DNA. The cytoplasm of an unfertilized egg is not homogeneous. Messenger RNA, proteins, other substances, and organelles are distributed unevenly in the unfertilized egg, and this unevenness has a profound impact on the development of the future embryo in many species. Maternal substances in the egg that influence the course of early development are called **cytoplasmic determinants** (**Figure 18.17a**, on the next page). After fertilization, early mitotic divisions distribute the zygote's cytoplasm into separate cells. The nuclei of these cells may thus be exposed to different cytoplasmic determinants, depending on which portions of the zygotic cytoplasm a cell received. The combination of cytoplasmic determinants in a cell helps determine its developmental fate by regulating expression of the cell's genes during the course of cell differentiation.

The other major source of developmental information, which becomes increasingly important as the number of embryonic cells increases, is the environment around a particular cell. Most influential are the signals impinging on an embryonic cell from other embryonic cells in the vicinity, including contact with cell-surface molecules on neighboring cells and the binding of growth factors secreted by neighboring cells. Such signals cause changes in the target cells, a process called **induction** (**Figure 18.17b**). The molecules conveying these signals within the target cell are cell-surface receptors and other proteins expressed by the embryo's own genes. In general, the signaling molecules send a cell down a specific developmental path by causing changes in its gene expression that eventually result in observable cellular changes. Thus, interactions between embryonic cells help induce differentiation of the many specialized cell types making up a new organism.

Sequential Regulation of Gene Expression During Cellular Differentiation

As the tissues and organs of an embryo develop and their cells differentiate, the cells become noticeably different in structure and function. These observable changes are actually the outcome of a cell's developmental history beginning at the first mitotic division of the zygote, as we have just seen. The

▼ Figure 18.17 Sources of developmental information for the early embryo.

(a) Cytoplasmic determinants in the egg

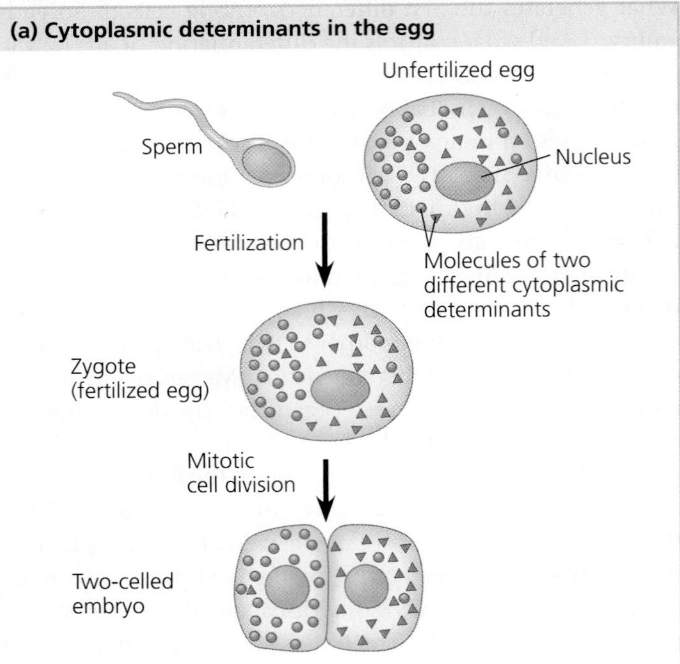

The unfertilized egg has molecules in its cytoplasm, encoded by the mother's genes, that influence development. Many of these cytoplasmic determinants, like the two shown here, are unevenly distributed in the egg. After fertilization and mitotic division, the cell nuclei of the embryo are exposed to different sets of cytoplasmic determinants and, as a result, express different genes.

(b) Induction by nearby cells

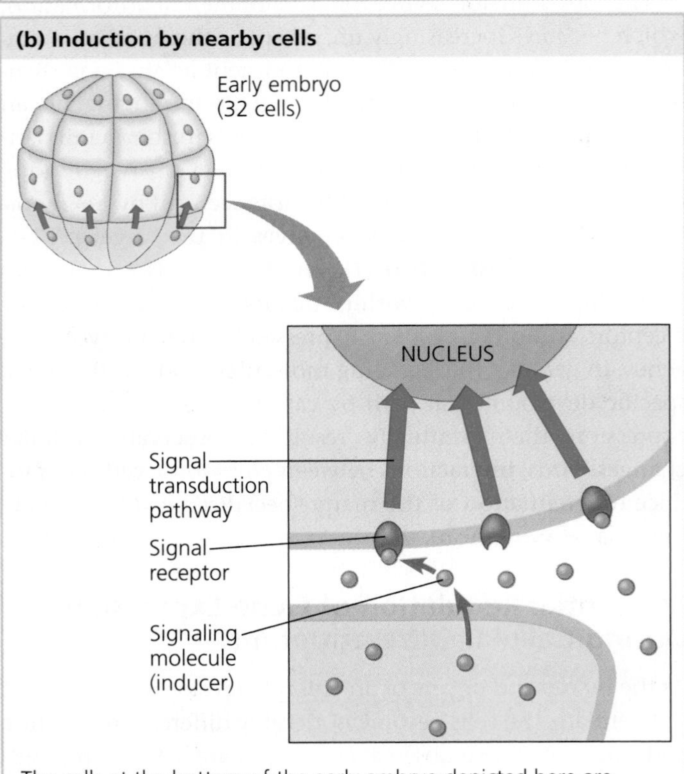

The cells at the bottom of the early embryo depicted here are releasing chemicals that signal nearby cells to change their gene expression.

earliest changes that set a cell on a path to specialization are subtle ones, showing up only at the molecular level. Before biologists knew much about the molecular changes occurring in embryos, they coined the term **determination** to refer to the events that lead to the observable differentiation of a cell. Once it has undergone determination, an embryonic cell is irreversibly committed to its final fate. If a committed cell is experimentally placed in another location in the embryo, it will still differentiate into the cell type that is its normal fate.

Today we understand determination in terms of molecular changes. The outcome of determination, observable cell differentiation, is marked by the expression of genes for *tissue-specific proteins*. These proteins are found only in a specific cell type and give the cell its characteristic structure and function. The first evidence of differentiation is the appearance of mRNAs for these proteins. Eventually, differentiation is observable with a microscope as changes in cellular structure. On the molecular level, different sets of genes are sequentially expressed in a regulated manner as new cells arise from division of their precursors. A number of the steps in gene expression may be regulated during differentiation, with transcription among the most important. In the fully differentiated cell, transcription remains the principal regulatory point for maintaining appropriate gene expression.

Differentiated cells are specialists at making tissue-specific proteins. For example, as a result of transcriptional regulation, liver cells specialize in making albumin, and lens cells specialize in making crystallin (see Figure 18.11). Skeletal muscle cells in vertebrates are another instructive example. Each of these cells is a long fiber containing many nuclei within a single plasma membrane. Skeletal muscle cells have high concentrations of muscle-specific versions of the contractile proteins myosin and actin, as well as membrane receptor proteins that detect signals from nerve cells.

Muscle cells develop from embryonic precursor cells that have the potential to develop into a number of cell types, including cartilage cells and fat cells, but particular conditions commit them to becoming muscle cells. Although the committed cells appear unchanged under the microscope, determination has occurred, and they are now *myoblasts*. Eventually, myoblasts start to churn out large amounts of muscle-specific proteins and fuse to form mature, elongated, multinucleate skeletal muscle cells **(Figure 18.18)**.

Researchers have worked out what happens at the molecular level during muscle cell determination by growing myoblasts in culture and analyzing them using molecular biological techniques you will learn about in Chapter 20. In a series of experiments, they isolated different genes, caused each to be expressed in a separate embryonic precursor cell, and then looked for differentiation into myoblasts and muscle cells. In this way, they identified several so-called "master regulatory genes" whose protein products commit the cells to becoming skeletal muscle. Thus, in the case of muscle cells,

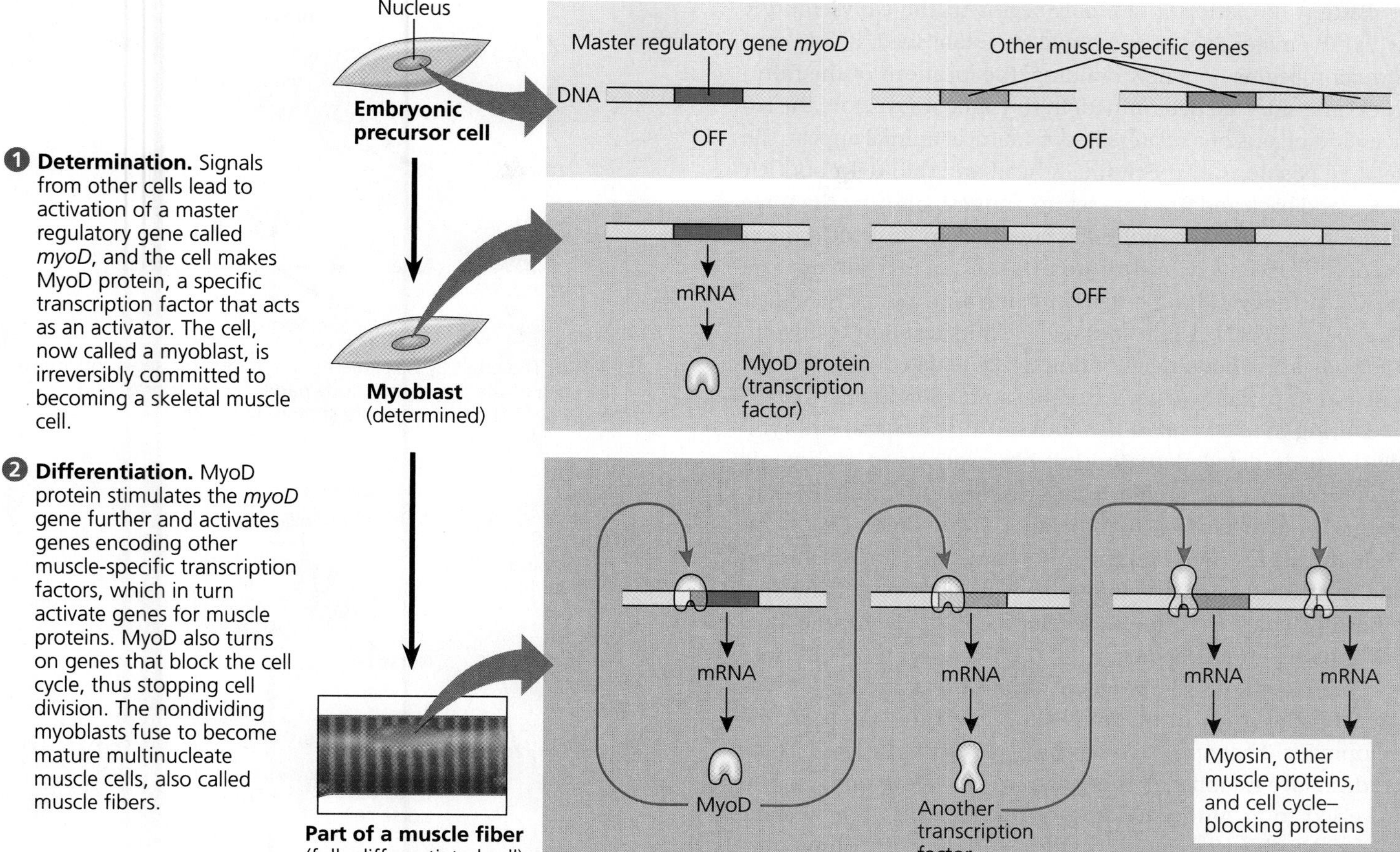

Nucleus
Master regulatory gene myoD
Other muscle-specific genes
DNA
Embryonic precursor cell
OFF
OFF
① Determination. Signals from other cells lead to activation of a master regulatory gene called myoD, and the cell makes MyoD protein, a specific transcription factor that acts as an activator. The cell, now called a myoblast, is irreversibly committed to becoming a skeletal muscle cell.
mRNA
OFF
MyoD protein (transcription factor)
Myoblast (determined)
② Differentiation. MyoD protein stimulates the myoD gene further and activates genes encoding other muscle-specific transcription factors, which in turn activate genes for muscle proteins. MyoD also turns on genes that block the cell cycle, thus stopping cell division. The nondividing myoblasts fuse to become mature multinucleate muscle cells, also called muscle fibers.
mRNA
mRNA
mRNA
mRNA
MyoD
Another transcription factor
Myosin, other muscle proteins, and cell cycle–blocking proteins
Part of a muscle fiber (fully differentiated cell)

▲ **Figure 18.18 Determination and differentiation of muscle cells.** Skeletal muscle cells arise from embryonic cells as a result of changes in gene expression. (In this depiction, the process of gene activation is greatly simplified.)

WHAT IF? *What would happen if a mutation in the* myoD *gene resulted in a MyoD protein that could not activate the* myoD *gene?*

the molecular basis of determination is the expression of one or more of these master regulatory genes.

To understand more about how commitment occurs in muscle cell differentiation, let's focus on the master regulatory gene called *myoD* (see Figure 18.18). This gene encodes MyoD protein, a transcription factor that binds to specific control elements in the enhancers of various target genes and stimulates their expression (see Figure 18.9). Some target genes for MyoD encode still other muscle-specific transcription factors. MyoD also stimulates expression of the *myoD* gene itself, thus perpetuating its effect in maintaining the cell's differentiated state. Presumably, all the genes activated by MyoD have enhancer control elements recognized by MyoD and are thus coordinately controlled. Finally, the secondary transcription factors activate the genes for proteins such as myosin and actin that confer the unique properties of skeletal muscle cells.

The MyoD protein deserves its designation as a master regulatory gene. Researchers have shown that it is even capable of changing some kinds of fully differentiated nonmuscle cells, such as fat cells and liver cells, into muscle cells. Why doesn't it work on *all* kinds of cells? One likely explanation is that activation of the muscle-specific genes is not solely dependent on MyoD but requires a particular *combination* of regulatory proteins, some of which are lacking in cells that do not respond to MyoD. The determination and differentiation of other kinds of tissues may play out in a similar fashion.

We have now seen how different programs of gene expression that are activated in the fertilized egg can result in differentiated cells and tissues. But for the tissues to function effectively in the organism as a whole, the organism's *body plan*—its overall three-dimensional arrangement—must be established and superimposed on the differentiation process. Next we'll investigate the molecular basis for the establishment of the body plan, using the well-studied *Drosophila* as an example.

Pattern Formation: Setting Up the Body Plan

Cytoplasmic determinants and inductive signals both contribute to the development of a spatial organization in which the tissues and organs of an organism are all in their characteristic places. This process is called **pattern formation**.

Pattern formation in animals begins in the early embryo, when the major axes of an animal are established. Before construction begins on a new building, the locations of the front, back, and sides are determined. In the same way, before the tissues and organs of a bilaterally symmetrical animal appear, the relative positions of the animal's head and tail, right and left sides, and back and front are set up, thus establishing the three major body axes. The molecular cues that control pattern formation, collectively called **positional information**, are provided by cytoplasmic determinants and inductive signals (see Figure 18.17). These cues tell a cell its location relative to the body axes and to neighboring cells and determine how the cell and its progeny will respond to future molecular signals.

During the first half of the 20th century, classical embryologists made detailed anatomical observations of embryonic development in a number of species and performed experiments in which they manipulated embryonic tissues. Although this research laid the groundwork for understanding the mechanisms of development, it did not reveal the specific molecules that guide development or determine how patterns are established.

Then, in the 1940s, scientists began using the genetic approach—the study of mutants—to investigate *Drosophila* development. That approach has had spectacular success. These studies have established that genes control development and have led to an understanding of the key roles that specific molecules play in defining position and directing differentiation. By combining anatomical, genetic, and biochemical approaches to the study of *Drosophila* development, researchers have discovered developmental principles common to many other species, including humans.

The Life Cycle of Drosophila

Fruit flies and other arthropods have a modular construction, an ordered series of segments. These segments make up the body's three major parts: the head, the thorax (the midbody, from which the wings and legs extend), and the abdomen (**Figure 18.19a**). Like other bilaterally symmetrical animals, *Drosophila* has an anterior-posterior (head-to-tail) axis, a dorsal-ventral (back-to-belly) axis, and a right-left axis. In *Drosophila*, cytoplasmic determinants that are localized in the unfertilized egg provide positional information for the placement of anterior-posterior and dorsal-ventral axes even before fertilization. We'll focus here on the molecules involved in establishing the anterior-posterior axis.

The *Drosophila* egg develops in the female's ovary, surrounded by ovarian cells called nurse cells and follicle cells (**Figure 18.19b**, top). These support cells supply the egg with nutrients, mRNAs, and other substances needed for development and make the egg shell. After fertilization and laying of the egg, embryonic development results in the formation of a segmented larva, which goes through three larval stages. Then, in a process much like that by which a caterpillar

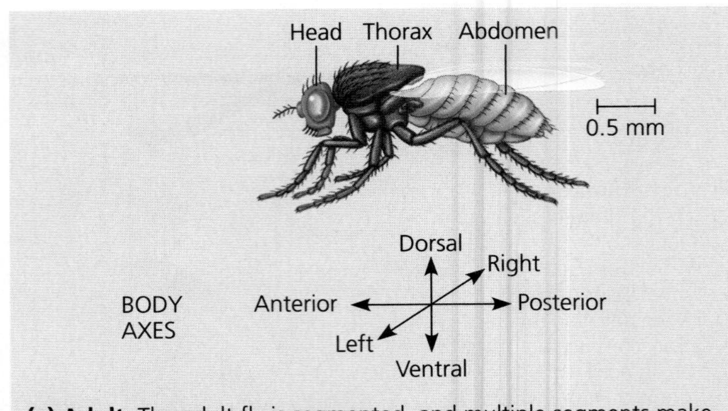

(a) Adult. The adult fly is segmented, and multiple segments make up each of the three main body parts—head, thorax, and abdomen. The body axes are shown by arrows.

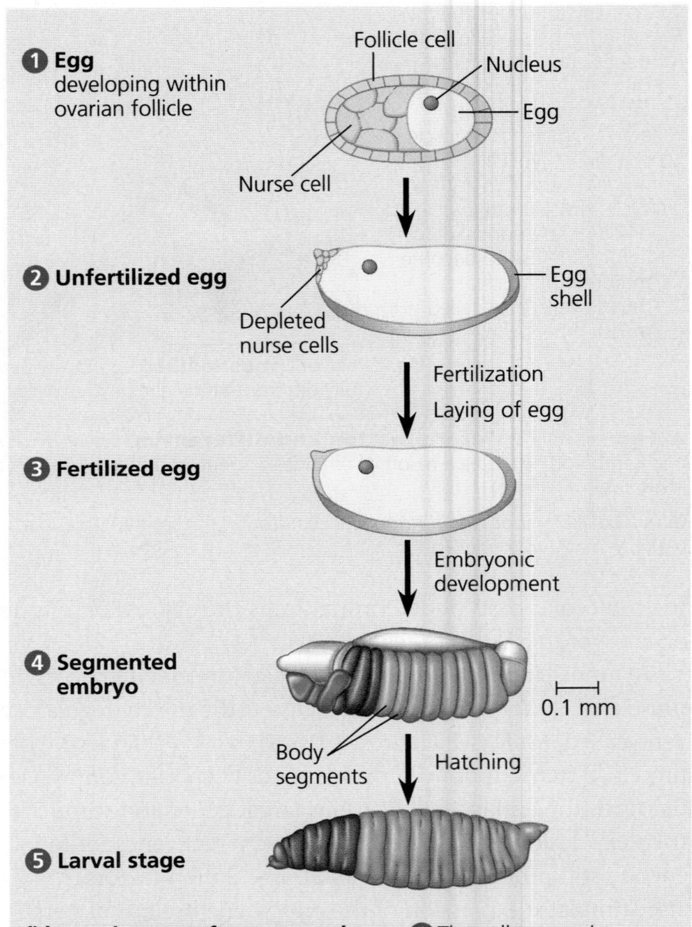

(b) Development from egg to larva. 1 The yellow egg is surrounded by other cells that form a structure called the follicle within one of the mother's ovaries. 2 The nurse cells shrink as they supply nutrients and mRNAs to the developing egg, which grows larger. Eventually, the mature egg fills the egg shell that is secreted by the follicle cells. 3 The egg is fertilized within the mother and then laid. 4 Embryonic development forms 5 a larva, which goes through three stages. The third stage forms a cocoon (not shown), within which the larva metamorphoses into the adult shown in (a).

▲ **Figure 18.19 Key developmental events in the life cycle of *Drosophila*.**

becomes a butterfly, the fly larva forms a cocoon in which it metamorphoses into the adult fly pictured in Figure 18.19a.

Genetic Analysis of Early Development: Scientific Inquiry

Edward B. Lewis was a visionary American biologist who, in the 1940s, first showed the value of the genetic approach to studying embryonic development in *Drosophila*. Lewis studied bizarre mutant flies with developmental defects that led to extra wings or legs in the wrong place (**Figure 18.20**). He located the mutations on the fly's genetic map, thus connecting the developmental abnormalities to specific genes. This research supplied the first concrete evidence that genes somehow direct the developmental processes studied by embryologists. The genes Lewis discovered, called **homeotic genes**, control pattern formation in the late embryo, larva, and adult.

Insight into pattern formation during early embryonic development did not come for another 30 years, when two researchers in Germany, Christiane Nüsslein-Volhard and Eric Wieschaus, set out to identify *all* the genes that affect segment formation in *Drosophila*. The project was daunting for three reasons. The first was the sheer number of *Drosophila* genes, now known to total about 13,700. The genes affecting segmentation might be just a few needles in a haystack or might be so numerous and varied that the scientists would be unable to make sense of them. Second, mutations affecting a process as fundamental as segmentation would surely be **embryonic lethals**, mutations with phenotypes causing death at the embryonic or larval stage. Because organisms with embryonic lethal mutations never reproduce, they cannot be bred for study. The researchers dealt with this problem by looking for recessive mutations, which can be propagated in heterozygous flies that act as genetic carriers. Third, cytoplasmic determinants in the egg were known to play a role in axis formation, so the researchers knew they would have to study the mother's genes as well as those of the embryo. It is the mother's genes that we will discuss further as we focus on how the anterior-posterior body axis is set up in the developing egg.

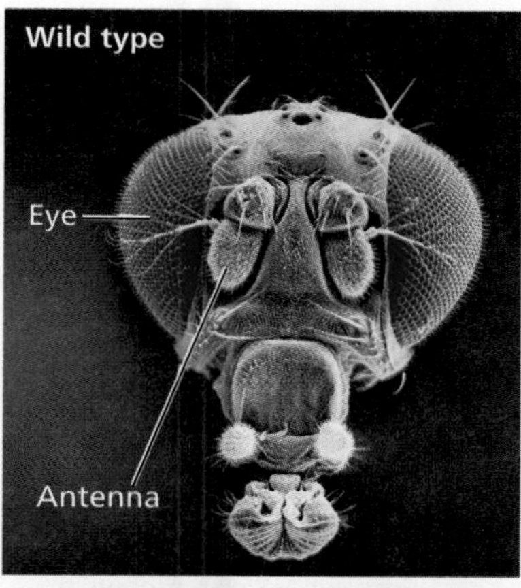

◀ **Figure 18.20 Abnormal pattern formation in *Drosophila*.** Mutations in certain regulatory genes, called homeotic genes, cause a misplacement of structures in an animal. These scanning electron micrographs contrast the head of a wild-type fly, bearing a pair of small antennae, with that of a homeotic mutant (a fly with a mutation in a single gene), bearing a pair of legs in place of antennae.

Nüsslein-Volhard and Wieschaus began their search for segmentation genes by exposing flies to a mutagenic chemical that affected the flies' gametes. They mated the mutagenized flies and then scanned their descendants for dead embryos or larvae with abnormal segmentation or other defects. For example, to find genes that might set up the anterior-posterior axis, they looked for embryos or larvae with abnormal ends, such as two heads or two tails, predicting that such abnormalities would arise from mutations in maternal genes required for correctly setting up the offspring's head or tail end.

Using this approach, Nüsslein-Volhard and Wieschaus eventually identified about 1,200 genes essential for pattern formation during embryonic development. Of these, about 120 were essential for normal segmentation. Over several years, the researchers were able to group these segmentation genes by general function, to map them, and to clone many of them for further study in the lab. The result was a detailed molecular understanding of the early steps in pattern formation in *Drosophila*.

When the results of Nüsslein-Volhard and Wieschaus were combined with Lewis's earlier work, a coherent picture of *Drosophila* development emerged. In recognition of their discoveries, the three researchers were awarded a Nobel Prize in 1995.

Let's consider further the genes that Nüsslein-Volhard, Wieschaus, and co-workers found for cytoplasmic determinants deposited in the egg by the mother. These genes set up the initial pattern of the embryo by regulating gene expression in broad regions of the early embryo.

Axis Establishment

As we mentioned earlier, cytoplasmic determinants in the egg are the substances that initially establish the axes of the *Drosophila* body. These substances are encoded by genes of the mother, fittingly called maternal effect genes. A **maternal effect gene** is a gene that, when mutant in the mother, results in a mutant phenotype in the offspring, regardless of the offspring's own genotype. In fruit fly development, the mRNA or protein products of maternal effect genes are placed in the egg while it is still in the mother's ovary. When the mother has a mutation in such a gene, she makes a defective gene product (or none at all), and her eggs are defective; when these eggs are fertilized, they fail to develop properly.

Because they control the orientation (polarity) of the egg and consequently of the fly, maternal effect genes are also called **egg-polarity genes**. One group of these genes sets up the anterior-posterior axis of the embryo, while a second group establishes the dorsal-ventral axis. Like mutations in segmentation genes, mutations in maternal effect genes are generally embryonic lethals.

Bicoid: A Morphogen Determining Head Structures To see how maternal effect genes determine the body axes of the offspring, we will focus on one such gene, called ***bicoid***, a term meaning "two-tailed." An embryo whose mother has two mutant alleles of the *bicoid* gene lacks the front half of its body and has posterior structures at both ends (**Figure 18.21**). This phenotype suggested to Nüsslein-Volhard and her colleagues that the product of the mother's *bicoid* gene is essential for setting up the anterior end of the fly and might be concentrated at the future anterior end of the embryo. This hypothesis is an example of the *morphogen gradient hypothesis* first proposed by embryologists a century ago; in this hypothesis, gradients of substances called **morphogens** establish an embryo's axes and other features of its form.

DNA technology and other modern biochemical methods enabled the researchers to test whether the *bicoid* product, a protein called Bicoid, is in fact a morphogen that determines the anterior end of the fly. The first question they asked was whether the mRNA and protein products of these genes are located in the egg in a position consistent with the hypothesis.

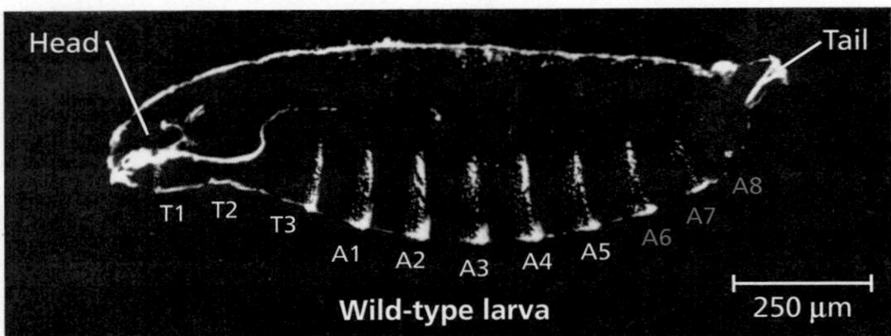

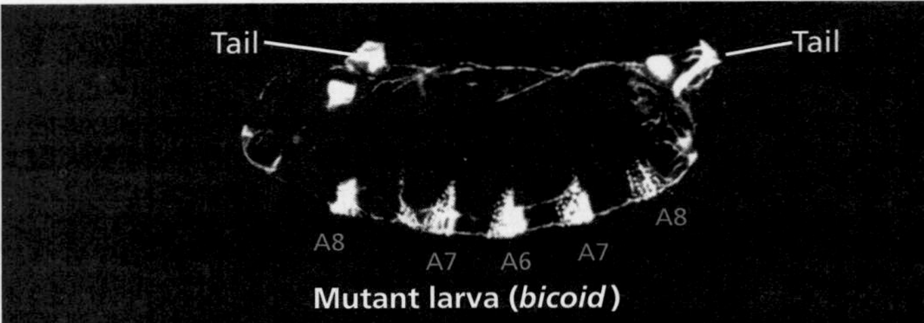

▲ **Figure 18.21 Effect of the *bicoid* gene on *Drosophila* development.** A wild-type fruit fly larva has a head, three thoracic (T) segments, eight abdominal (A) segments, and a tail. A larva whose mother has two mutant alleles of the *bicoid* gene has two tails and lacks all anterior structures (LMs).

They found that *bicoid* mRNA is highly concentrated at the extreme anterior end of the mature egg, as predicted by the hypothesis (**Figure 18.22**). After the egg is fertilized, the mRNA is translated into protein. The Bicoid protein then diffuses from the anterior end toward the posterior, resulting in a gradient of protein within the early embryo, with the highest concentration at the anterior end. These results are consistent with the hypothesis that Bicoid protein specifies the fly's anterior end. To test the hypothesis more specifically, scientists injected pure *bicoid* mRNA into various regions of early embryos. The protein that resulted from its translation caused anterior structures to form at the injection sites.

▼ **Figure 18.22** INQUIRY

Is Bicoid a morphogen that determines the anterior end of a fruit fly?

EXPERIMENT Using a genetic approach to study *Drosophila* development, Christiane Nüsslein-Volhard and colleagues at the European Molecular Biology Laboratory in Heidelberg, Germany, analyzed expression of the *bicoid* gene. The researchers hypothesized that *bicoid* normally codes for a morphogen that specifies the head (anterior) end of the embryo. To test this hypothesis, they used molecular techniques to determine where the mRNA and protein encoded by this gene were found in the fertilized egg and early embryo of wild-type flies.

RESULTS *Bicoid* mRNA (dark blue) was confined to the anterior end of the unfertilized egg. Later in development, Bicoid protein (dark orange) was seen to be concentrated in cells at the anterior end of the embryo.

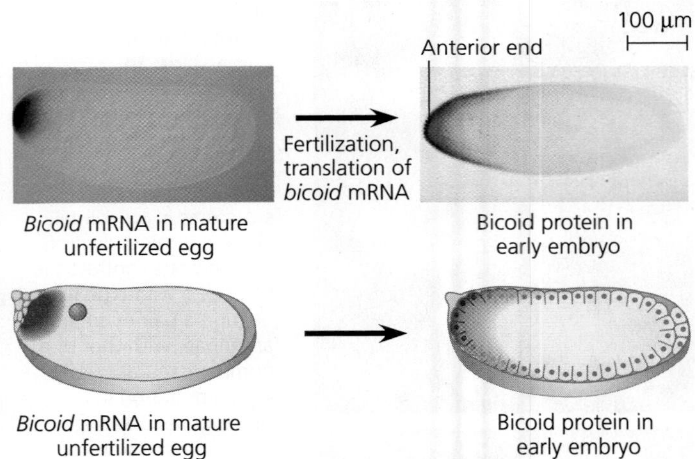

CONCLUSION The location of *bicoid* mRNA and the diffuse gradient of Bicoid protein seen later support the hypothesis that Bicoid protein is a morphogen specifying formation of head-specific structures.

SOURCE: C. Nüsslein-Volhard et al., Determination of anteroposterior polarity in *Drosophila*, *Science* 238:1675–1681 (1987); W. Driever and C. Nüsslein-Volhard, A gradient of *bicoid* protein in *Drosophila* embryos, *Cell* 54:83–93 (1988); T. Berleth et al., The role of localization of *bicoid* RNA in organizing the anterior pattern of the *Drosophila* embryo, *EMBO Journal* 7:1749–1756 (1988).

WHAT IF? If the hypothesis is correct, predict what would happen if you injected *bicoid* mRNA into the anterior end of an egg from a female with a mutation disabling the *bicoid* gene.

The *bicoid* research was groundbreaking for several reasons. First, it led to the identification of a specific protein required for some of the earliest steps in pattern formation. It thus helped us understand how different regions of the egg can give rise to cells that go down different developmental pathways. Second, it increased our understanding of the mother's critical role in the initial phases of embryonic development. Finally, the principle that a gradient of morphogens can determine polarity and position has proved to be a key developmental concept for a number of species, just as early embryologists had thought.

Maternal mRNAs are crucial during development of many species. In *Drosophila*, gradients of specific proteins encoded by maternal mRNAs determine the posterior and anterior ends and establish the dorsal-ventral axis. As the fly embryo grows, it reaches a point when the embryonic program of gene expression takes over, and the maternal mRNAs must be destroyed. (This process involves miRNAs in *Drosophila* and other species.) Later, positional information encoded by the embryo's genes, operating on an ever finer scale, establishes a specific number of correctly oriented segments and triggers the formation of each segment's characteristic structures. When the genes operating in this final step are abnormal, the pattern of the adult is abnormal, as you saw in Figure 18.20.

In this section, we have seen how a carefully orchestrated program of sequential gene regulation controls the transformation of a fertilized egg into a multicellular organism. The program is carefully balanced between turning on the genes for differentiation in the right place and turning off other genes. Even when an organism is fully developed, gene expression is regulated in a similarly fine-tuned manner. In the final section of the chapter, we'll consider how fine this tuning is by looking at how specific changes in expression of one or a few genes can lead to the development of cancer.

CONCEPT CHECK 18.4

1. As you learned in Chapter 12, mitosis gives rise to two daughter cells that are genetically identical to the parent cell. Yet you, the product of many mitotic divisions, are not composed of identical cells. Why?
2. **MAKE CONNECTIONS** Explain how the signaling molecules released by an embryonic cell can induce changes in a neighboring cell without entering the cell. (See Figures 11.15 and 11.16, pp. 219 and 220.)
3. Why are fruit fly maternal effect genes also called egg-polarity genes?
4. **WHAT IF?** In the blowup box in Figure 18.17b, the lower cell is synthesizing signaling molecules, whereas the upper cell is expressing receptors for these molecules. In terms of gene regulation, explain how these cells came to synthesize different molecules.

For suggested answers, see Appendix A.

CONCEPT 18.5

Cancer results from genetic changes that affect cell cycle control

In Chapter 12, we considered cancer as a set of diseases in which cells escape from the control mechanisms that normally limit their growth. Now that we have discussed the molecular basis of gene expression and its regulation, we are ready to look at cancer more closely. The gene regulation systems that go wrong during cancer turn out to be the very same systems that play important roles in embryonic development, the immune response, and many other biological processes. Thus, research into the molecular basis of cancer has both benefited from and informed many other fields of biology.

Types of Genes Associated with Cancer

The genes that normally regulate cell growth and division during the cell cycle include genes for growth factors, their receptors, and the intracellular molecules of signaling pathways. (To review the cell cycle, see Chapter 12.) Mutations that alter any of these genes in somatic cells can lead to cancer. The agent of such change can be random spontaneous mutation. However, it is likely that many cancer-causing mutations result from environmental influences, such as chemical carcinogens, X-rays and other high-energy radiation, and some viruses.

Cancer research led to the discovery of cancer-causing genes called **oncogenes** (from the Greek *onco*, tumor) in certain types of viruses (see Chapter 19). Subsequently, close counterparts of viral oncogenes were found in the genomes of humans and other animals. The normal versions of the cellular genes, called **proto-oncogenes**, code for proteins that stimulate normal cell growth and division.

How might a proto-oncogene—a gene that has an essential function in normal cells—become an oncogene, a cancer-causing gene? In general, an oncogene arises from a genetic change that leads to an increase either in the amount of the proto-oncogene's protein product or in the intrinsic activity of each protein molecule. The genetic changes that convert proto-oncogenes to oncogenes fall into three main categories: movement of DNA within the genome, amplification of a proto-oncogene, and point mutations in a control element or in the proto-oncogene itself (**Figure 18.23**, on the next page).

Cancer cells are frequently found to contain chromosomes that have broken and rejoined incorrectly, translocating fragments from one chromosome to another (see Figure 15.14). Now that you have learned how gene expression is regulated, you can understand the possible consequences of such translocations. If a translocated proto-oncogene ends up near an especially active promoter (or other control element), its transcription may increase, making it an oncogene. The second main type of genetic change, amplification, increases the

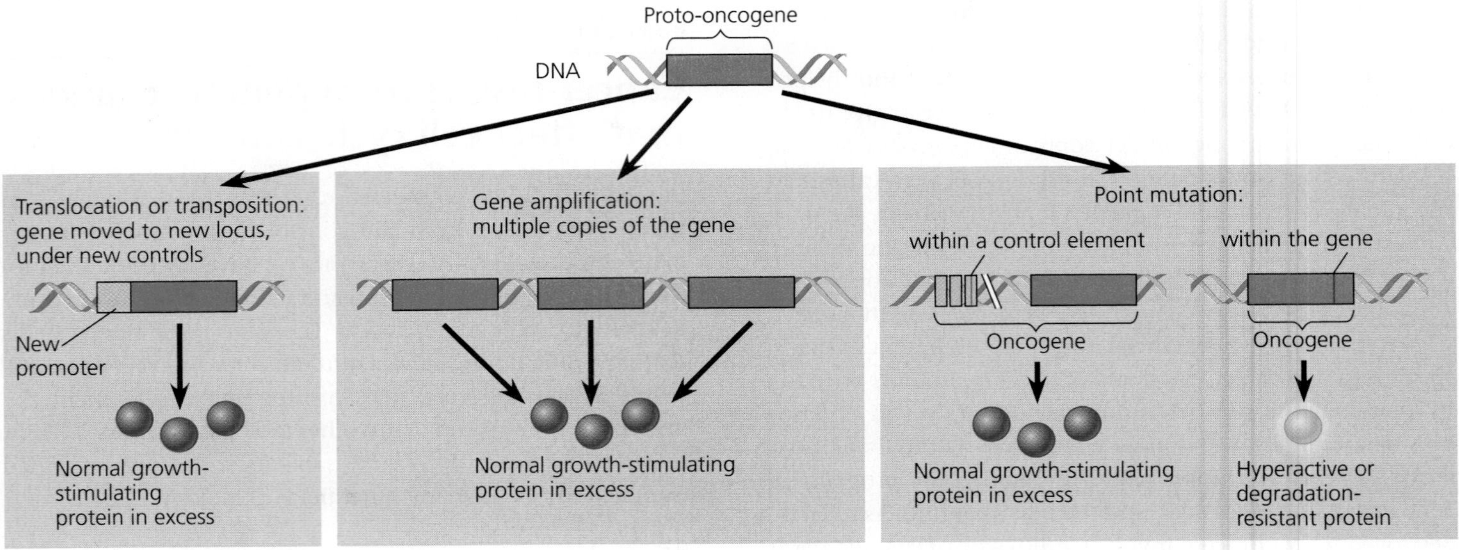

▲ **Figure 18.23 Genetic changes that can turn proto-oncogenes into oncogenes.**

number of copies of the proto-oncogene in the cell through repeated gene duplication (discussed in Chapter 21). The third possibility is a point mutation either (1) in the promoter or an enhancer that controls a proto-oncogene, causing an increase in its expression, or (2) in the coding sequence, changing the gene's product to a protein that is more active or more resistant to degradation than the normal protein. All these mechanisms can lead to abnormal stimulation of the cell cycle and put the cell on the path to malignancy.

Tumor-Suppressor Genes

In addition to genes whose products normally promote cell division, cells contain genes whose normal products *inhibit* cell division. Such genes are called **tumor-suppressor genes** because the proteins they encode help prevent uncontrolled cell growth. Any mutation that decreases the normal activity of a tumor-suppressor protein may contribute to the onset of cancer, in effect stimulating growth through the absence of suppression.

The protein products of tumor-suppressor genes have various functions. Some tumor-suppressor proteins repair damaged DNA, a function that prevents the cell from accumulating cancer-causing mutations. Other tumor-suppressor proteins control the adhesion of cells to each other or to the extracellular matrix; proper cell anchorage is crucial in normal tissues—and is often absent in cancers. Still other tumor-suppressor proteins are components of cell-signaling pathways that inhibit the cell cycle.

Interference with Normal Cell-Signaling Pathways

The proteins encoded by many proto-oncogenes and tumor-suppressor genes are components of cell-signaling pathways. Let's take a closer look at how such proteins function in normal cells and what goes wrong with their function in cancer cells. We will focus on the products of two key genes, the *ras* proto-oncogene and the *p53* tumor-suppressor gene. Mutations in *ras* occur in about 30% of human cancers, and mutations in *p53* in more than 50%.

The Ras protein, encoded by the ***ras* gene** (named for rat sarcoma, a connective tissue cancer), is a G protein that relays a signal from a growth factor receptor on the plasma membrane to a cascade of protein kinases (see Figure 11.7). The cellular response at the end of the pathway is the synthesis of a protein that stimulates the cell cycle **(Figure 18.24a)**. Normally, such a pathway will not operate unless triggered by the appropriate growth factor. But certain mutations in the *ras* gene can lead to production of a hyperactive Ras protein that triggers the kinase cascade even in the absence of growth factor, resulting in increased cell division. In fact, hyperactive versions or excess amounts of any of the pathway's components can have the same outcome: excessive cell division.

Figure 18.24b shows a pathway in which a signal leads to the synthesis of a protein that suppresses the cell cycle. In this case, the signal is damage to the cell's DNA, perhaps as the result of exposure to ultraviolet light. Operation of this signaling pathway blocks the cell cycle until the damage has been repaired. Otherwise, the damage might contribute to tumor formation by causing mutations or chromosomal abnormalities. Thus, the genes for the components of the pathway act as tumor-suppressor genes. The ***p53* gene**, named for the 53,000-dalton molecular weight of its protein product, is a tumor-suppressor gene. The protein it encodes is a specific transcription factor that promotes the synthesis of cell cycle–inhibiting proteins. That is why a mutation that knocks out the *p53* gene, like a mutation that leads to a hyperactive Ras protein, can lead to excessive cell growth and cancer **(Figure 18.24c)**.

(a) Cell cycle–stimulating pathway. This pathway is triggered by ❶ a growth factor that binds to ❷ its receptor in the plasma membrane. The signal is relayed to ❸ a G protein called Ras. Like all G proteins, Ras is active when GTP is bound to it. Ras passes the signal to ❹ a series of protein kinases. The last kinase activates ❺ a transcription activator that turns on one or more genes for proteins that stimulate the cell cycle. If a mutation makes Ras or any other pathway component abnormally active, excessive cell division and cancer may result.

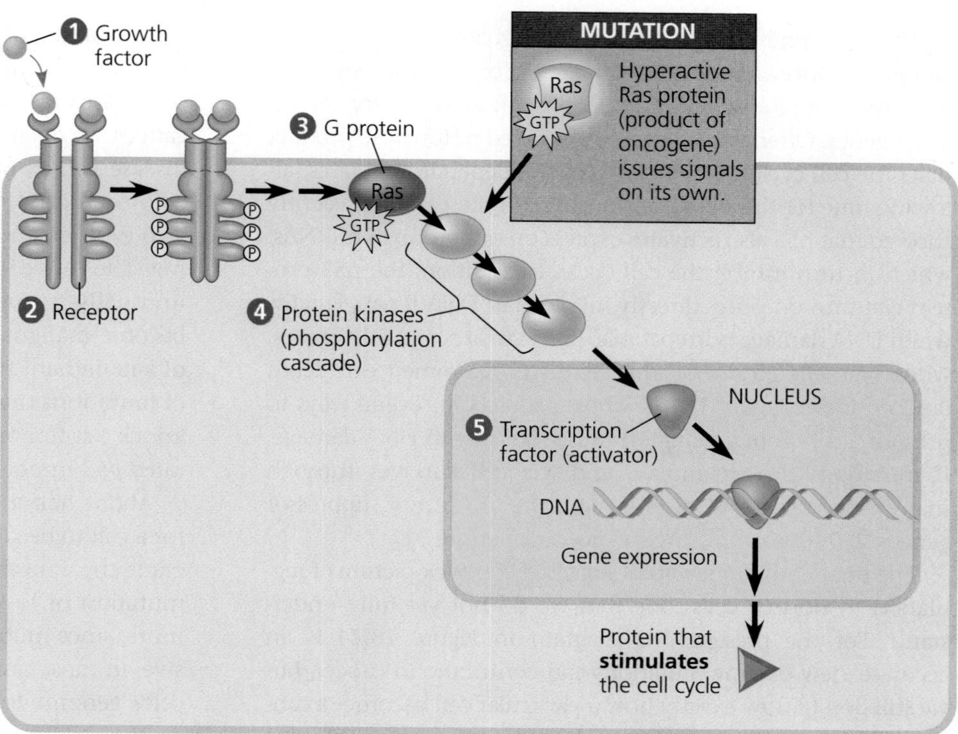

(b) Cell cycle–inhibiting pathway. In this pathway, ❶ DNA damage is an intracellular signal that is passed via ❷ protein kinases and leads to activation of ❸ p53. Activated p53 promotes transcription of the gene for a protein that inhibits the cell cycle. The resulting suppression of cell division ensures that the damaged DNA is not replicated. If the DNA damage is irreparable, the p53 signal leads to programmed cell death (apoptosis). Mutations causing deficiencies in any pathway component can contribute to the development of cancer.

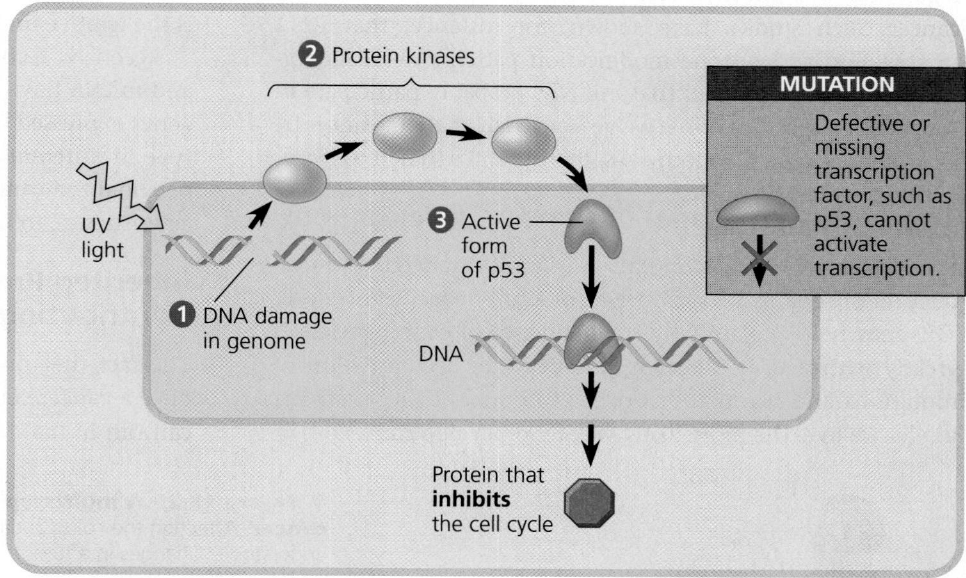

(c) Effects of mutations. Increased cell division, possibly leading to cancer, can result if the cell cycle is overstimulated, as in (a), or not inhibited when it normally would be, as in (b).

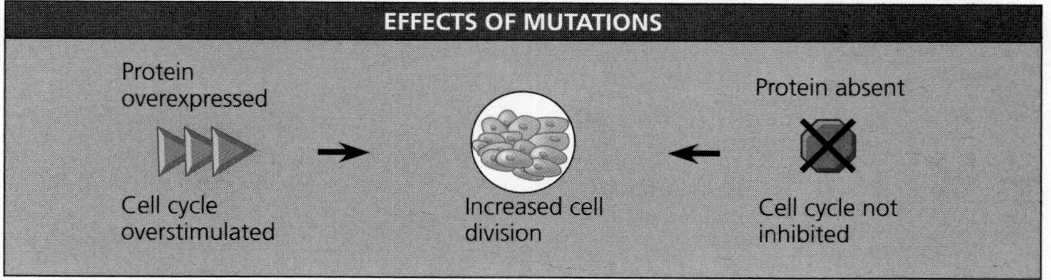

▲ **Figure 18.24 Signaling pathways that regulate cell division.** Both stimulatory and inhibitory pathways regulate the cell cycle, commonly by influencing transcription. Cancer can result from aberrations in such pathways, which may be caused by mutations, either spontaneous or environmentally triggered.

? *Looking at the pathway in (b), explain whether a cancer-causing mutation in a tumor-suppressor gene, such as p53, is more likely to be a recessive or a dominant mutation.*

The *p53* gene has been called the "guardian angel of the genome." Once the gene is activated—for example, by DNA damage—the p53 protein functions as an activator for several other genes. Often it activates a gene called *p21*, whose product halts the cell cycle by binding to cyclin-dependent kinases, allowing time for the cell to repair the DNA. Researchers recently showed that p53 also activates expression of a group of miRNAs, which in turn inhibit the cell cycle. In addition, the p53 protein can turn on genes directly involved in DNA repair. Finally, when DNA damage is irreparable, p53 activates "suicide" genes, whose protein products bring about programmed cell death (apoptosis; see Figure 11.21). Thus, p53 acts in several ways to prevent a cell from passing on mutations due to DNA damage. If mutations do accumulate and the cell survives through many divisions—as is more likely if the *p53* tumor-suppressor gene is defective or missing—cancer may ensue.

The many functions of p53 suggest a complex picture of regulation in normal cells, one that we do not yet fully understand. For the present, the diagram in Figure 18.24 is an accurate view of how mutations can contribute to cancer, but we still don't know exactly how a particular cell becomes a cancer cell. As we discover previously unknown aspects of gene regulation, it is informative to study their role in the onset of cancer. Such studies have shown, for instance, that DNA methylation and histone modification patterns differ in normal and cancer cells and that miRNAs probably participate in cancer development. While we've learned a lot about cancer by studying cell-signaling pathways, there is still a lot left to learn.

The Multistep Model of Cancer Development

More than one somatic mutation is generally needed to produce all the changes characteristic of a full-fledged cancer cell. This may help explain why the incidence of cancer increases greatly with age. If cancer results from an accumulation of mutations and if mutations occur throughout life, then the longer we live, the more likely we are to develop cancer.

The model of a multistep path to cancer is well supported by studies of one of the best-understood types of human cancer, colorectal cancer. About 135,000 new cases of colorectal cancer are diagnosed each year in the United States, and the disease causes 60,000 deaths each year. Like most cancers, colorectal cancer develops gradually (**Figure 18.25**). The first sign is often a polyp, a small, benign growth in the colon lining. The cells of the polyp look normal, although they divide unusually frequently. The tumor grows and may eventually become malignant, invading other tissues. The development of a malignant tumor is paralleled by a gradual accumulation of mutations that convert proto-oncogenes to oncogenes and knock out tumor-suppressor genes. A *ras* oncogene and a mutated *p53* tumor-suppressor gene are often involved.

About half a dozen changes must occur at the DNA level for a cell to become fully cancerous. These changes usually include the appearance of at least one active oncogene and the mutation or loss of several tumor-suppressor genes. Furthermore, since mutant tumor-suppressor alleles are usually recessive, in most cases mutations must knock out *both* alleles in a cell's genome to block tumor suppression. (Most oncogenes, on the other hand, behave as dominant alleles.) The order in which these changes must occur is still under investigation, as is the relative importance of different mutations.

Recently, technical advances in the sequencing of DNA and mRNA have allowed medical researchers to compare the genes expressed by different types of tumors and by the same type in different individuals. These comparisons have led to personalized cancer treatments based on the molecular characteristics of an individual's tumor (see Figure 12.21.)

Inherited Predisposition and Other Factors Contributing to Cancer

The fact that multiple genetic changes are required to produce a cancer cell helps explain the observation that cancers can run in families. An individual inheriting an oncogene or

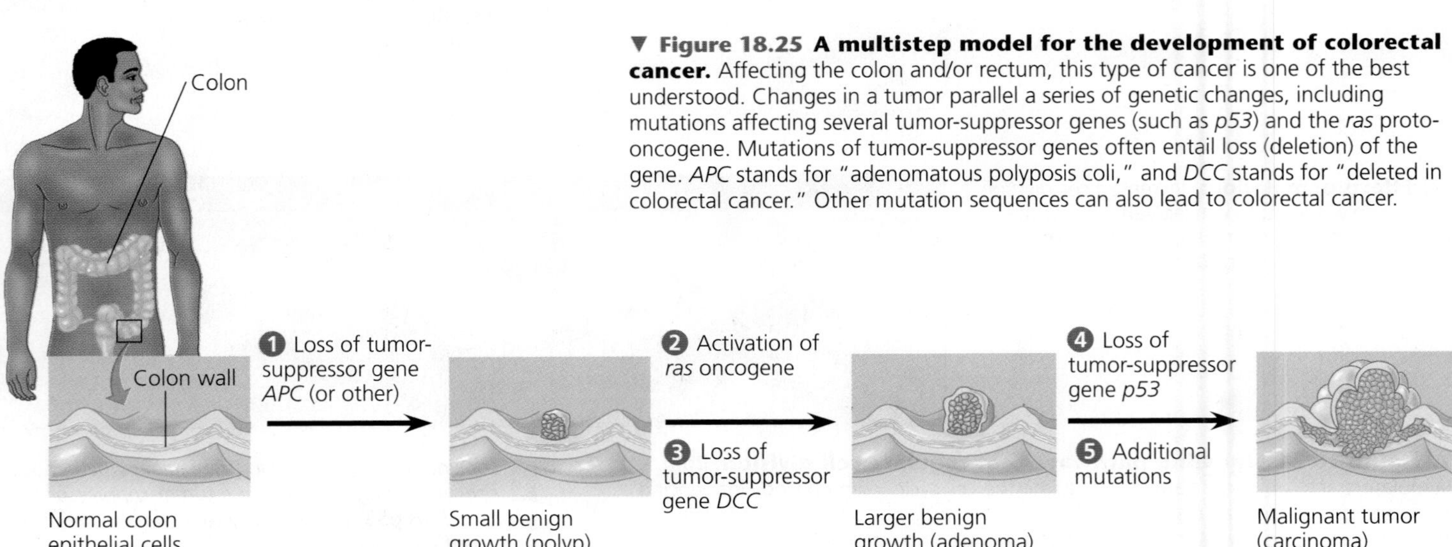

▼ **Figure 18.25 A multistep model for the development of colorectal cancer.** Affecting the colon and/or rectum, this type of cancer is one of the best understood. Changes in a tumor parallel a series of genetic changes, including mutations affecting several tumor-suppressor genes (such as *p53*) and the *ras* proto-oncogene. Mutations of tumor-suppressor genes often entail loss (deletion) of the gene. *APC* stands for "adenomatous polyposis coli," and *DCC* stands for "deleted in colorectal cancer." Other mutation sequences can also lead to colorectal cancer.

a mutant allele of a tumor-suppressor gene is one step closer to accumulating the necessary mutations for cancer to develop than is an individual without any such mutations.

Geneticists are devoting much effort to identifying inherited cancer alleles so that predisposition to certain cancers can be detected early in life. About 15% of colorectal cancers, for example, involve inherited mutations. Many of these affect the tumor-suppressor gene called *adenomatous polyposis coli*, or *APC* (see Figure 18.25). This gene has multiple functions in the cell, including regulation of cell migration and adhesion. Even in patients with no family history of the disease, the *APC* gene is mutated in 60% of colorectal cancers. In these individuals, new mutations must occur in both *APC* alleles before the gene's function is lost. Since only 15% of colorectal cancers are associated with known inherited mutations, researchers continue in their efforts to identify "markers" that could predict the risk of developing this type of cancer.

There is evidence of a strong inherited predisposition in 5–10% of patients with breast cancer. This is the second most common type of cancer in the United States, striking over 180,000 women (and some men) annually and killing 40,000 each year. In 1990, after 16 years of research, geneticist Mary-Claire King convincingly demonstrated that mutations in one gene—*BRCA1*—were associated with increased susceptibility to breast cancer, a finding that flew in the face of medical opinion at the time. (*BRCA* stands for *br*east *ca*ncer.) Mutations in that gene or the related *BRCA2* gene are found in at least half of inherited breast cancers, and tests using DNA sequencing can detect these mutations (**Figure 18.26**). A woman who inherits one mutant *BRCA1* allele has a 60% probability of developing breast cancer before the age of 50, compared with only a 2% probability for an individual homozygous for the normal allele. Both *BRCA1* and *BRCA2* are considered tumor-suppressor genes because their wild-type alleles protect against breast cancer and their mutant alleles are recessive. Apparently, the BRCA1 and BRCA2 proteins both function in the cell's DNA damage repair pathway. More is known about BRCA2, which, in association with another protein, helps repair breaks that occur in both strands of DNA; it is crucial for maintaining undamaged DNA in a cell's nucleus.

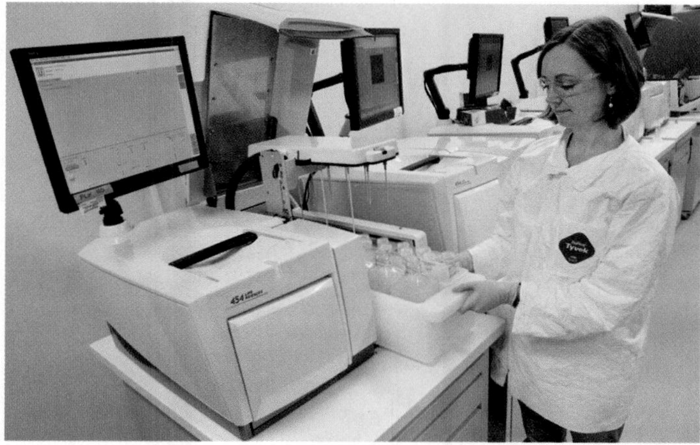

▲ **Figure 18.26 Testing for mutations in *BRCA1* and *BRCA2*.** Genetic testing for mutations that increase the risk of breast cancer is available for individuals with a family history of breast cancer. New "high-throughput" sequencing techniques can sequence many DNA samples at once, as shown here.

Because DNA breakage can contribute to cancer, it makes sense that the risk of cancer can be lowered by minimizing exposure to DNA-damaging agents, such as the ultraviolet radiation in sunlight and chemicals found in cigarette smoke. Novel methods for early diagnosis and treatment of specific cancers are being developed that rely on new techniques for analyzing, and perhaps interfering with, gene expression in tumors. Ultimately, such approaches may lower the death rate from cancer.

The study of genes associated with cancer, inherited or not, increases our basic understanding of how disruption of normal gene regulation can result in this disease. In addition to the mutations and other genetic alterations described in this section, a number of *tumor viruses* can cause cancer in various animals, including humans. In fact, one of the earliest breakthroughs in understanding cancer came in 1911, when Peyton Rous, an American pathologist, discovered a virus that causes cancer in chickens. The Epstein-Barr virus, which causes infectious mononucleosis, has been linked to several types of cancer in humans, notably Burkitt's lymphoma. Papillomaviruses are associated with cancer of the cervix, and a virus called HTLV-1 causes a type of adult leukemia. Worldwide, viruses seem to play a role in about 15% of the cases of human cancer.

Viruses may at first seem very different from mutations as a cause of cancer. However, we now know that viruses can interfere with gene regulation in several ways if they integrate their genetic material into the DNA of a cell. Viral integration may donate an oncogene to the cell, disrupt a tumor-suppressor gene, or convert a proto-oncogene to an oncogene. In addition, some viruses produce proteins that inactivate p53 and other tumor-suppressor proteins, making the cell more prone to becoming cancerous. Viruses are powerful biological agents, and you'll learn more about their function in Chapter 19.

CONCEPT CHECK 18.5

1. **MAKE CONNECTIONS** The p53 protein can activate genes involved in apoptosis, or programmed cell death. Review Concept 11.5 (pp. 223–225) and discuss how mutations in genes coding for proteins that function in apoptosis could contribute to cancer.
2. Under what circumstances is cancer considered to have a hereditary component?
3. **WHAT IF?** Explain how the types of mutations that lead to cancer are different for a proto-oncogene and a tumor-suppressor gene, in terms of the effect of the mutation on the activity of the gene product.

For suggested answers, see Appendix A.

18 CHAPTER REVIEW

SUMMARY OF KEY CONCEPTS

CONCEPT 18.1

Bacteria often respond to environmental change by regulating transcription (pp. 351–356)

- Cells control metabolism by regulating enzyme activity or the expression of genes coding for enzymes. In bacteria, genes are often clustered into **operons,** with one promoter serving several adjacent genes. An **operator** site on the DNA switches the operon on or off, resulting in coordinate regulation of the genes.

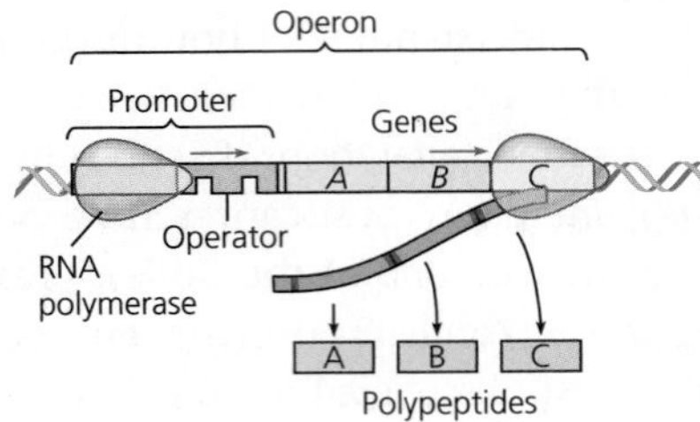

- Both repressible and inducible operons are examples of negative gene regulation. In either type of operon, binding of a specific **repressor** protein to the operator shuts off transcription. (The repressor is encoded by a separate **regulatory gene.**) In a repressible operon, the repressor is active when bound to a **corepressor**, usually the end product of an anabolic pathway.

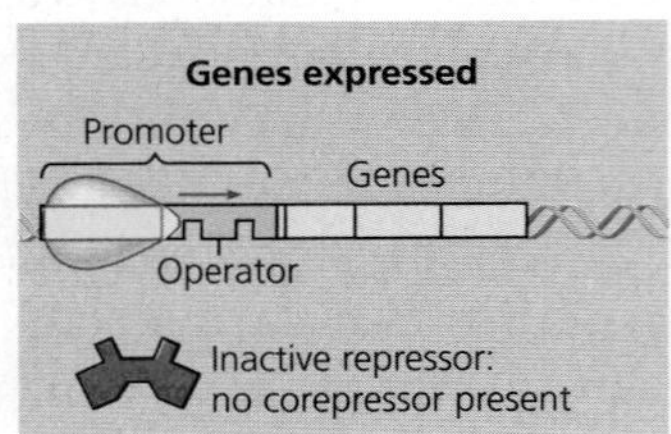

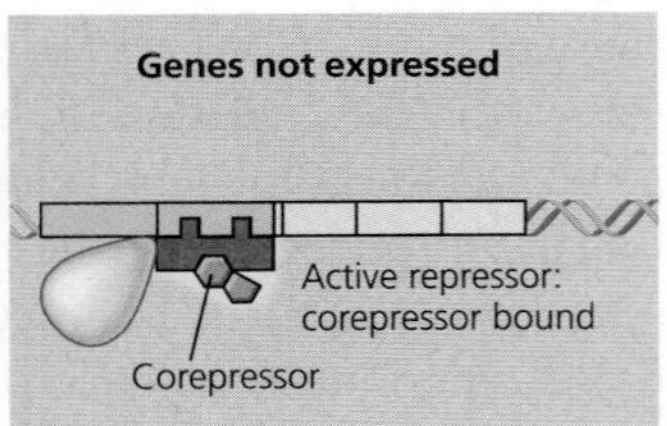

In an inducible operon, binding of an **inducer** to an innately active repressor inactivates the repressor and turns on transcription. Inducible enzymes usually function in catabolic pathways.

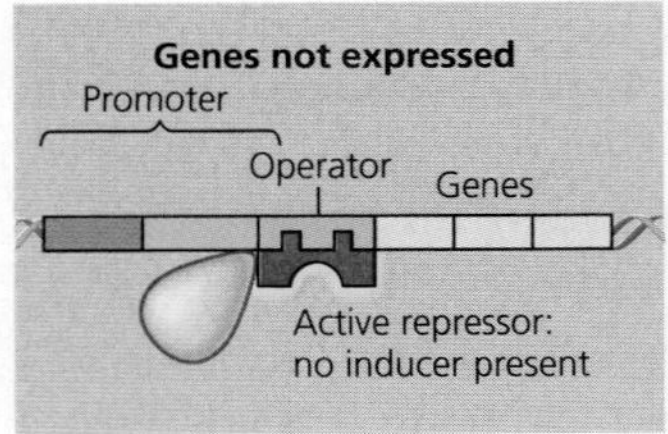

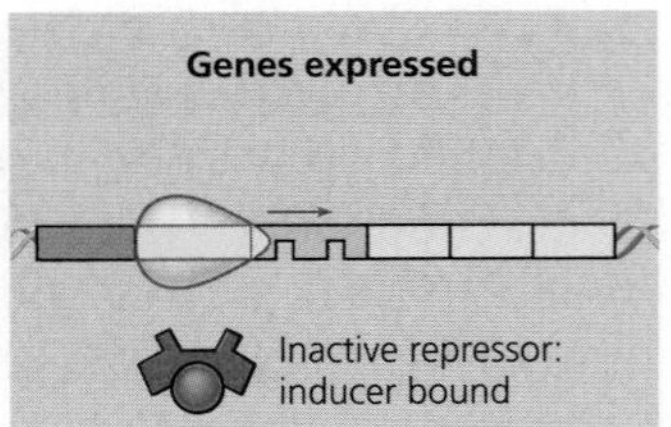

- Some operons are also subject to positive gene regulation via a stimulatory **activator** protein, such as catabolite activator protein (CAP), which, when activated by **cyclic AMP**, binds to a site within the promoter and stimulates transcription.

? *Compare and contrast the roles of the corepressor and the inducer in negative regulation of an operon.*

CONCEPT 18.2

Eukaryotic gene expression is regulated at many stages (pp. 356–364)

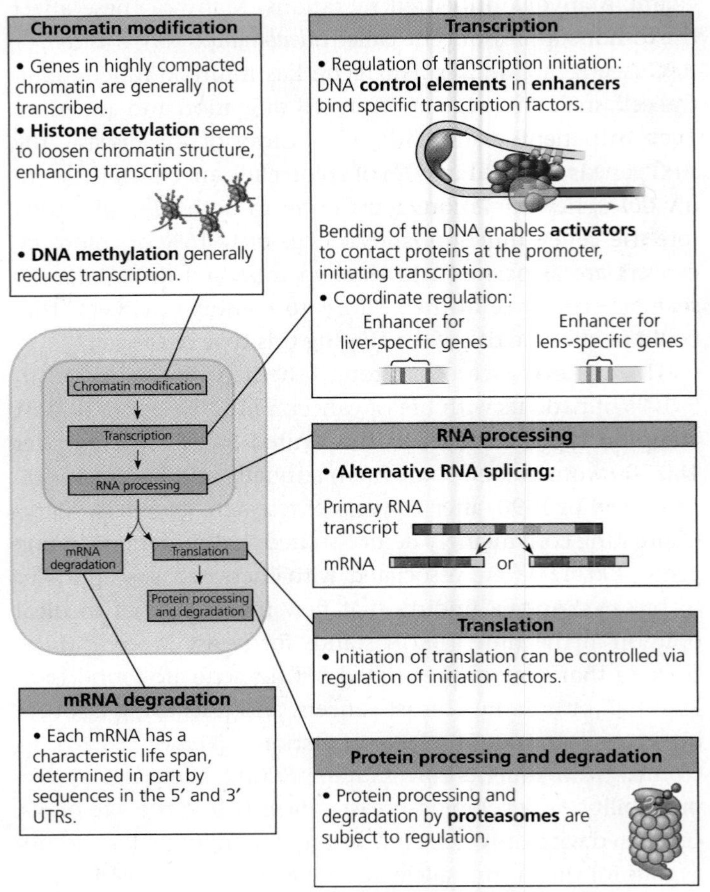

Chromatin modification
• Genes in highly compacted chromatin are generally not transcribed.
• Histone acetylation seems to loosen chromatin structure, enhancing transcription.
• DNA methylation generally reduces transcription.
Transcription
• Regulation of transcription initiation: DNA control elements in enhancers bind specific transcription factors.
Bending of the DNA enables activators to contact proteins at the promoter, initiating transcription.
• Coordinate regulation:
Enhancer for liver-specific genes
Enhancer for lens-specific genes
Chromatin modification
Transcription
RNA processing
mRNA degradation
Translation
Protein processing and degradation
RNA processing
• Alternative RNA splicing:
Primary RNA transcript
mRNA
or
Translation
• Initiation of translation can be controlled via regulation of initiation factors.
mRNA degradation
• Each mRNA has a characteristic life span, determined in part by sequences in the 5′ and 3′ UTRs.
Protein processing and degradation
• Protein processing and degradation by proteasomes are subject to regulation.

? *Describe what must happen for a cell-type-specific gene to be transcribed in a cell of that type.*

CONCEPT 18.3

Noncoding RNAs play multiple roles in controlling gene expression (pp. 364–366)

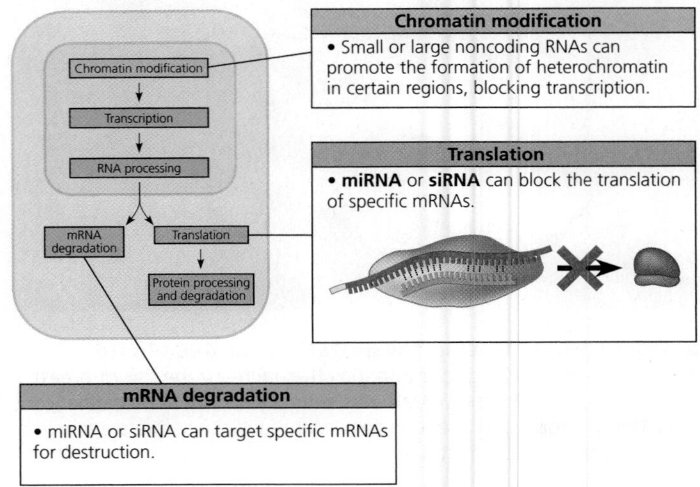

Chromatin modification
• Small or large noncoding RNAs can promote the formation of heterochromatin in certain regions, blocking transcription.
Chromatin modification
Transcription
RNA processing
mRNA degradation
Translation
Protein processing and degradation
Translation
• miRNA or siRNA can block the translation of specific mRNAs.
mRNA degradation
• miRNA or siRNA can target specific mRNAs for destruction.

? *Why are miRNAs called noncoding RNAs? Explain how they participate in gene regulation.*

CONCEPT 18.4

A program of differential gene expression leads to the different cell types in a multicellular organism (pp. 366–373)

- Embryonic cells undergo **differentiation**, becoming specialized in structure and function. **Morphogenesis** encompasses the processes that give shape to the organism and its various parts. Cells differ in structure and function not because they contain different genes but because they express different portions of a common genome.
- **Cytoplasmic determinants** in the unfertilized egg regulate the expression of genes in the zygote that affect the developmental fate of embryonic cells. In the process called **induction**, signaling molecules from embryonic cells cause transcriptional changes in nearby target cells.
- Differentiation is heralded by the appearance of tissue-specific proteins, which enable differentiated cells to carry out their specialized roles.
- In animals, **pattern formation**, the development of a spatial organization of tissues and organs, begins in the early embryo. **Positional information**, the molecular cues that control pattern formation, tells a cell its location relative to the body's axes and to other cells. In *Drosophila*, gradients of **morphogens** encoded by **maternal effect genes** determine the body axes. For example, the gradient of **Bicoid** protein determines the anterior-posterior axis.

? *Describe the two main processes that cause embryonic cells to head down different pathways to their final fates.*

CONCEPT 18.5

Cancer results from genetic changes that affect cell cycle control (pp. 373–377)

- The products of **proto-oncogenes** and **tumor-suppressor genes** control cell division. A DNA change that makes a proto-oncogene excessively active converts it to an **oncogene**, which may promote excessive cell division and cancer. A tumor-suppressor gene encodes a protein that inhibits abnormal cell division. A mutation in such a gene that reduces the activity of its protein product may also lead to excessive cell division and possibly to cancer.
- Many proto-oncogenes and tumor-suppressor genes encode components of growth-stimulating and growth-inhibiting signaling pathways, respectively, and mutations in these genes can interfere with normal cell-signaling pathways. A hyperactive version of a protein in a stimulatory pathway, such as **Ras** (a G protein), functions as an oncogene protein. A defective version of a protein in an inhibitory pathway, such as **p53** (a transcription activator), fails to function as a tumor suppressor.
- In the multistep model of cancer development, normal cells are converted to cancer cells by the accumulation of mutations affecting proto-oncogenes and tumor-suppressor genes. Technical advances in DNA and mRNA sequencing are enabling cancer treatments that are more individually based.
- Individuals who inherit a mutant oncogene or tumor-suppressor allele have a predisposition to develop a particular cancer. Certain viruses promote cancer by integration of viral DNA into a cell's genome.

? *Compare the usual functions of proteins encoded by proto-oncogenes with the functions of proteins encoded by tumor-suppressor genes.*

TEST YOUR UNDERSTANDING

LEVEL 1: KNOWLEDGE/COMPREHENSION

1. If a particular operon encodes enzymes for making an essential amino acid and is regulated like the *trp* operon, then
   a. the amino acid inactivates the repressor.
   b. the enzymes produced are called inducible enzymes.
   c. the repressor is active in the absence of the amino acid.
   d. the amino acid acts as a corepressor.
   e. the amino acid turns on transcription of the operon.
2. Muscle cells differ from nerve cells mainly because they
   a. express different genes.
   b. contain different genes.
   c. use different genetic codes.
   d. have unique ribosomes.
   e. have different chromosomes.
3. The functioning of enhancers is an example of
   a. transcriptional control of gene expression.
   b. a post-transcriptional mechanism to regulate mRNA.
   c. the stimulation of translation by initiation factors.
   d. post-translational control that activates certain proteins.
   e. a eukaryotic equivalent of prokaryotic promoter functioning.
4. Cell differentiation always involves
   a. the production of tissue-specific proteins, such as muscle actin.
   b. the movement of cells.
   c. the transcription of the *myoD* gene.
   d. the selective loss of certain genes from the genome.
   e. the cell's sensitivity to environmental cues, such as light or heat.
5. Which of the following is an example of post-transcriptional control of gene expression?
   a. the addition of methyl groups to cytosine bases of DNA
   b. the binding of transcription factors to a promoter
   c. the removal of introns and alternative splicing of exons
   d. gene amplification contributing to cancer
   e. the folding of DNA to form heterochromatin

LEVEL 2: APPLICATION/ANALYSIS

6. What would occur if the repressor of an inducible operon were mutated so it could not bind the operator?
   a. irreversible binding of the repressor to the promoter
   b. reduced transcription of the operon's genes
   c. buildup of a substrate for the pathway controlled by the operon
   d. continuous transcription of the operon's genes
   e. overproduction of catabolite activator protein (CAP)
7. Absence of *bicoid* mRNA from a *Drosophila* egg leads to the absence of anterior larval body parts and mirror-image duplication of posterior parts. This is evidence that the product of the *bicoid* gene
   a. is transcribed in the early embryo.
   b. normally leads to formation of tail structures.
   c. normally leads to formation of head structures.
   d. is a protein present in all head structures.
   e. leads to programmed cell death.
8. Which of the following statements about the DNA in one of your brain cells is true?
   a. Most of the DNA codes for protein.
   b. The majority of genes are likely to be transcribed.
   c. Each gene lies immediately adjacent to an enhancer.
   d. Many genes are grouped into operon-like clusters.
   e. It is the same as the DNA in one of your heart cells.

9. Within a cell, the amount of protein made using a given mRNA molecule depends partly on
   a. the degree of DNA methylation.
   b. the rate at which the mRNA is degraded.
   c. the presence of certain transcription factors.
   d. the number of introns present in the mRNA.
   e. the types of ribosomes present in the cytoplasm.

10. Proto-oncogenes can change into oncogenes that cause cancer. Which of the following best explains the presence of these potential time bombs in eukaryotic cells?
   a. Proto-oncogenes first arose from viral infections.
   b. Proto-oncogenes normally help regulate cell division.
   c. Proto-oncogenes are genetic "junk."
   d. Proto-oncogenes are mutant versions of normal genes.
   e. Cells produce proto-oncogenes as they age.

LEVEL 3: SYNTHESIS/EVALUATION

11. **DRAW IT** The diagram below shows five genes, including their enhancers, from the genome of a certain species. Imagine that orange, blue, green, black, red, and purple activator proteins exist that can bind to the appropriately color-coded control elements in the enhancers of these genes.

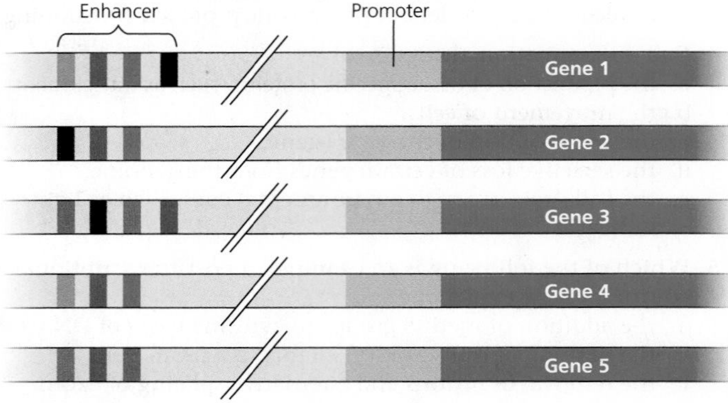

   a. Draw an X above enhancer elements (of all the genes) that would have activators bound in a cell in which only gene 5 is transcribed. Which colored activators would be present?
   b. Draw a dot above all enhancer elements that would have activators bound in a cell in which the green, blue, and orange activators are present. Which gene(s) would be transcribed?
   c. Imagine that genes 1, 2, and 4 code for nerve-specific proteins, and genes 3 and 5 are skin specific. Which activators would have to be present in each cell type to ensure transcription of the appropriate genes?

12. **EVOLUTION CONNECTION**
DNA sequences can act as "tape measures of evolution" (see Chapter 5). Scientists analyzing the human genome sequence were surprised to find that some of the regions of the human genome that are most highly conserved (similar to comparable regions in other species) don't code for proteins. Propose a possible explanation for this observation.

13. **SCIENTIFIC INQUIRY**
Prostate cells usually require testosterone and other androgens to survive. But some prostate cancer cells thrive despite treatments that eliminate androgens. One hypothesis is that estrogen, often considered a female hormone, may be activating genes normally controlled by an androgen in these cancer cells. Describe one or more experiments to test this hypothesis. (See Figure 11.9, p. 214, to review the action of these steroid hormones.)

14. **SCIENCE, TECHNOLOGY, AND SOCIETY**
Trace amounts of dioxin were present in Agent Orange, a defoliant sprayed on vegetation during the Vietnam War. Animal tests suggest that dioxin can cause birth defects, cancer, liver and thymus damage, and immune system suppression, sometimes leading to death. But the animal tests are equivocal; a hamster is not affected by a dose that can kill a guinea pig. Dioxin acts somewhat like a steroid hormone, entering a cell and binding to a receptor protein that then attaches to the cell's DNA. How might this mechanism help explain the variety of dioxin's effects on different body systems and in different animals? How might you determine whether a type of illness is related to dioxin exposure? How might you determine whether a particular individual became ill as a result of exposure to dioxin? Which would be more difficult to demonstrate? Why?

15. **WRITE ABOUT A THEME**
Feedback Regulation In a short essay (100–150 words), discuss how the processes shown in Figure 18.24a and b are examples of feedback mechanisms regulating biological systems.

For selected answers, see Appendix A.

MasteringBIOLOGY www.masteringbiology.com

1. MasteringBiology® Assignments

Tutorials Regulation of Gene Expression in Bacteria • Regulation of Gene Expression in Eukaryotes • Pattern Formation
Activities The *lac* Operon • The *lac* Operon in *E. coli* • Transcription Initiation in Eukaryotes • Overview: Control of Gene Expression • Control of Transcription • Review: Control of Gene Expression • Early Pattern Formation in *Drosophila* • Role of *bicoid* Gene in *Drosophila* Development
Questions Student Misconceptions • Reading Quiz • Multiple Choice • End-of-Chapter

2. eText

Read your book online, search, take notes, highlight text, and more.

3. The Study Area

Practice Tests • Cumulative Test • ***BioFlix*** 3-D Animations • MP3 Tutor Sessions • Videos • Activities • Investigations • Lab Media • Audio Glossary • Word Study Tools • Art

19

Viruses

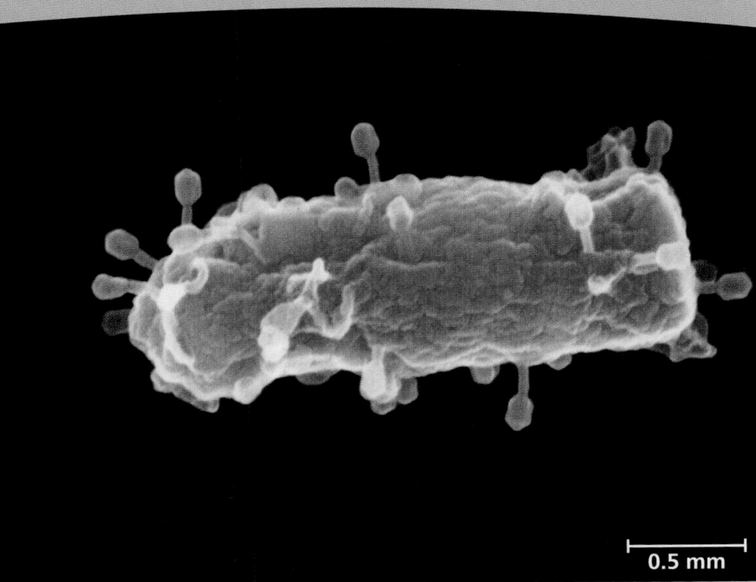

▲ **Figure 19.1 Are the tiny viruses infecting this *E. coli* cell alive?**

KEY CONCEPTS

19.1 A virus consists of a nucleic acid surrounded by a protein coat

19.2 Viruses replicate only in host cells

19.3 Viruses, viroids, and prions are formidable pathogens in animals and plants

OVERVIEW

A Borrowed Life

The photo in **Figure 19.1** shows a remarkable event: the attack of a bacterial cell by numerous structures that resemble miniature lollipops. These structures, a type of virus called T4 bacteriophage, are seen infecting the bacterium *Escherichia coli* in this colorized SEM. By injecting its DNA into the cell, the virus sets in motion a genetic takeover of the bacterium, recruiting cellular machinery to mass-produce many new viruses.

Recall that bacteria and other prokaryotes are cells much smaller and more simply organized than the cells of eukaryotes, such as plants and animals. Viruses are smaller and simpler still. Lacking the structures and metabolic machinery found in a cell, a **virus** is an infectious particle consisting of little more than genes packaged in a protein coat.

Are viruses living or nonliving? Early on, they were considered biological chemicals; in fact, the Latin root for the word *virus* means "poison." Because viruses are capable of causing a wide variety of diseases and can be spread between organisms, researchers in the late 1800s saw a parallel with bacteria and proposed that viruses were the simplest of living forms. However, viruses cannot reproduce or carry out metabolic activities outside of a host cell. Most biologists studying viruses today would probably agree that they are not alive but exist in a shady area between life-forms and chemicals. The simple phrase used recently by two researchers describes them aptly enough: Viruses lead "a kind of borrowed life."

To a large extent, molecular biology was born in the laboratories of biologists studying viruses that infect bacteria. Experiments with viruses provided important evidence that genes are made of nucleic acids, and they were critical in working out the molecular mechanisms of the fundamental processes of DNA replication, transcription, and translation.

Beyond their value as experimental systems, viruses have unique genetic mechanisms that are interesting in their own right and that also help us understand how viruses cause disease. In addition, the study of viruses has led to the development of techniques that enable scientists to manipulate genes and transfer them from one organism to another. These techniques play an important role in basic research, biotechnology, and medical applications. For instance, viruses are used as agents of gene transfer in gene therapy (see Chapter 20).

In this chapter, we will explore the biology of viruses. We will begin with the structure of these simplest of all genetic systems and then describe the cycles by which they replicate. Next, we will discuss the role of viruses as disease-causing agents, or pathogens, and conclude by considering some even simpler infectious agents, viroids and prions.

CONCEPT 19.1

A virus consists of a nucleic acid surrounded by a protein coat

Scientists were able to detect viruses indirectly long before they were actually able to see them. The story of how viruses were discovered begins near the end of the 19th century.

The Discovery of Viruses: *Scientific Inquiry*

Tobacco mosaic disease stunts the growth of tobacco plants and gives their leaves a mottled, or mosaic, coloration. In 1883, Adolf Mayer, a German scientist, discovered that he could transmit the disease from plant to plant by rubbing sap extracted from diseased leaves onto healthy plants. After an unsuccessful search for an infectious microbe in the sap, Mayer suggested that the disease was caused by unusually small bacteria that were invisible under a microscope. This hypothesis was tested a decade later by Dimitri Ivanowsky, a Russian biologist who passed sap from infected tobacco leaves through a filter designed to remove bacteria. After filtration, the sap still produced mosaic disease.

But Ivanowsky clung to the hypothesis that bacteria caused tobacco mosaic disease. Perhaps, he reasoned, the bacteria were small enough to pass through the filter or made a toxin that could do so. The second possibility was ruled out when the Dutch botanist Martinus Beijerinck carried out a classic series of experiments that showed that the infectious agent in the filtered sap could replicate **(Figure 19.2)**.

In fact, the pathogen replicated only within the host it infected. In further experiments, Beijerinck showed that unlike bacteria used in the lab at that time, the mysterious agent of mosaic disease could not be cultivated on nutrient media in test tubes or petri dishes. Beijerinck imagined a replicating particle much smaller and simpler than a bacterium, and he is generally credited with being the first scientist to voice the concept of a virus. His suspicions were confirmed in 1935 when the American scientist Wendell Stanley crystallized the infectious particle, now known as tobacco mosaic virus (TMV). Subsequently, TMV and many other viruses were actually seen with the help of the electron microscope.

▼ **Figure 19.2** **INQUIRY**

What causes tobacco mosaic disease?

EXPERIMENT In the late 1800s, Martinus Beijerinck, of the Technical School in Delft, the Netherlands, investigated the properties of the agent that causes tobacco mosaic disease (then called spot disease).

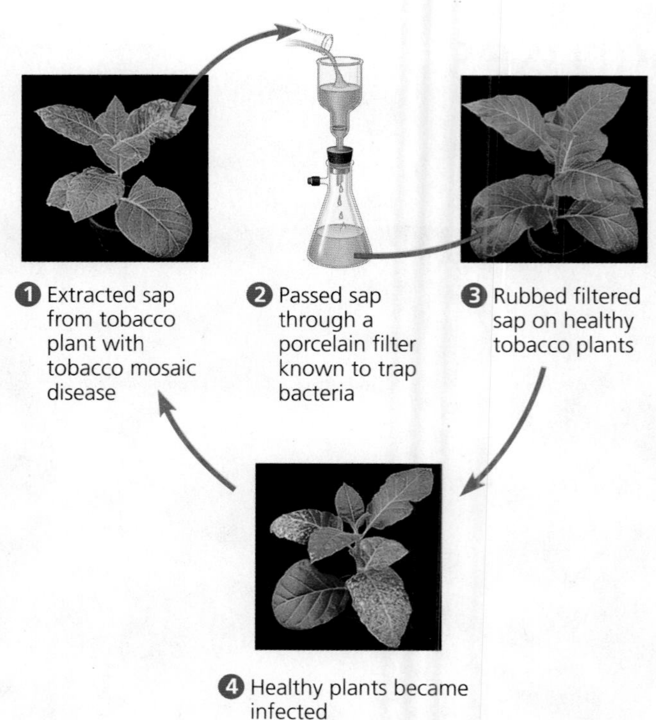

1 Extracted sap from tobacco plant with tobacco mosaic disease
2 Passed sap through a porcelain filter known to trap bacteria
3 Rubbed filtered sap on healthy tobacco plants
4 Healthy plants became infected

RESULTS When the filtered sap was rubbed on healthy plants, they became infected. Their sap, when extracted and filtered, could then act as the source of infection for another group of plants. Each successive group of plants developed the disease to the same extent as earlier groups.

CONCLUSION The infectious agent was apparently not a bacterium because it could pass through a bacterium-trapping filter. The pathogen must have been replicating in the plants because its ability to cause disease was undiluted after several transfers from plant to plant.

SOURCE M. J. Beijerinck, Concerning a *contagium vivum fluidum* as cause of the spot disease of tobacco leaves, *Verhandelingen der Koninkyke akademie Wettenschappen te Amsterdam* 65:3–21 (1898). Translation published in English as Phytopathological Classics Number 7 (1942), American Phytopathological Society Press, St. Paul, MN.

WHAT IF? If Beijerinck had observed that the infection of each group was weaker than that of the previous group and that ultimately the sap could no longer cause disease, what might he have concluded?

Structure of Viruses

The tiniest viruses are only 20 nm in diameter—smaller than a ribosome. Millions could easily fit on a pinhead. Even the largest known virus, which has a diameter of several hundred nanometers, is barely visible under the light microscope. Stanley's discovery that some viruses could be crystallized was exciting and puzzling news. Not even the simplest of cells can aggregate into regular crystals. But if viruses are not cells, then what are they? Examining the structure of a virus more closely reveals that it is an infectious particle consisting of nucleic acid enclosed in a protein coat and, for some viruses, surrounded by a membranous envelope.

Viral Genomes

We usually think of genes as being made of double-stranded DNA—the conventional double helix—but many viruses defy this convention. Their genomes may consist of double-stranded DNA, single-stranded DNA, double-stranded RNA, or single-stranded RNA, depending on the type of virus. A virus is called a DNA virus or an RNA virus, based on the kind of nucleic acid that makes up its genome. In either case, the genome is usually organized as a single linear or circular molecule of nucleic acid, although the genomes of some viruses consist of multiple molecules of nucleic acid. The smallest viruses known have only four genes in their genome, while the largest have several hundred to a thousand. For comparison, bacterial genomes contain about 200 to a few thousand genes.

Capsids and Envelopes

The protein shell enclosing the viral genome is called a **capsid**. Depending on the type of virus, the capsid may be rod-shaped, polyhedral, or more complex in shape (like T4). Capsids are built from a large number of protein subunits called *capsomeres*, but the number of different *kinds* of proteins in a capsid is usually small. Tobacco mosaic virus has a rigid, rod-shaped capsid made from over a thousand molecules of a single type of protein arranged in a helix; rod-shaped viruses are commonly called *helical viruses* for this reason (**Figure 19.3a**). Adenoviruses, which infect the respiratory tracts of animals, have 252 identical protein molecules arranged in a polyhedral capsid with 20 triangular facets—an icosahedron; thus, these and other similarly shaped viruses are referred to as *icosahedral viruses* (**Figure 19.3b**).

Some viruses have accessory structures that help them infect their hosts. For instance, a membranous envelope surrounds the capsids of influenza viruses and many other viruses found in animals (**Figure 19.3c**). These **viral envelopes**, which are derived from the membranes of the host cell, contain host cell phospholipids and membrane proteins. They also contain proteins and glycoproteins of viral origin. (Glycoproteins are proteins with carbohydrates covalently attached.) Some viruses carry a few viral enzyme molecules within their capsids.

Many of the most complex capsids are found among the viruses that infect bacteria, called **bacteriophages**, or simply **phages**. The first phages studied included seven that infect *E. coli*. These seven phages were named type 1 (T1), type 2 (T2), and so forth, in the order of their discovery. The three T-even phages (T2, T4, and T6) turned out to be very similar in structure. Their capsids have elongated icosahedral heads

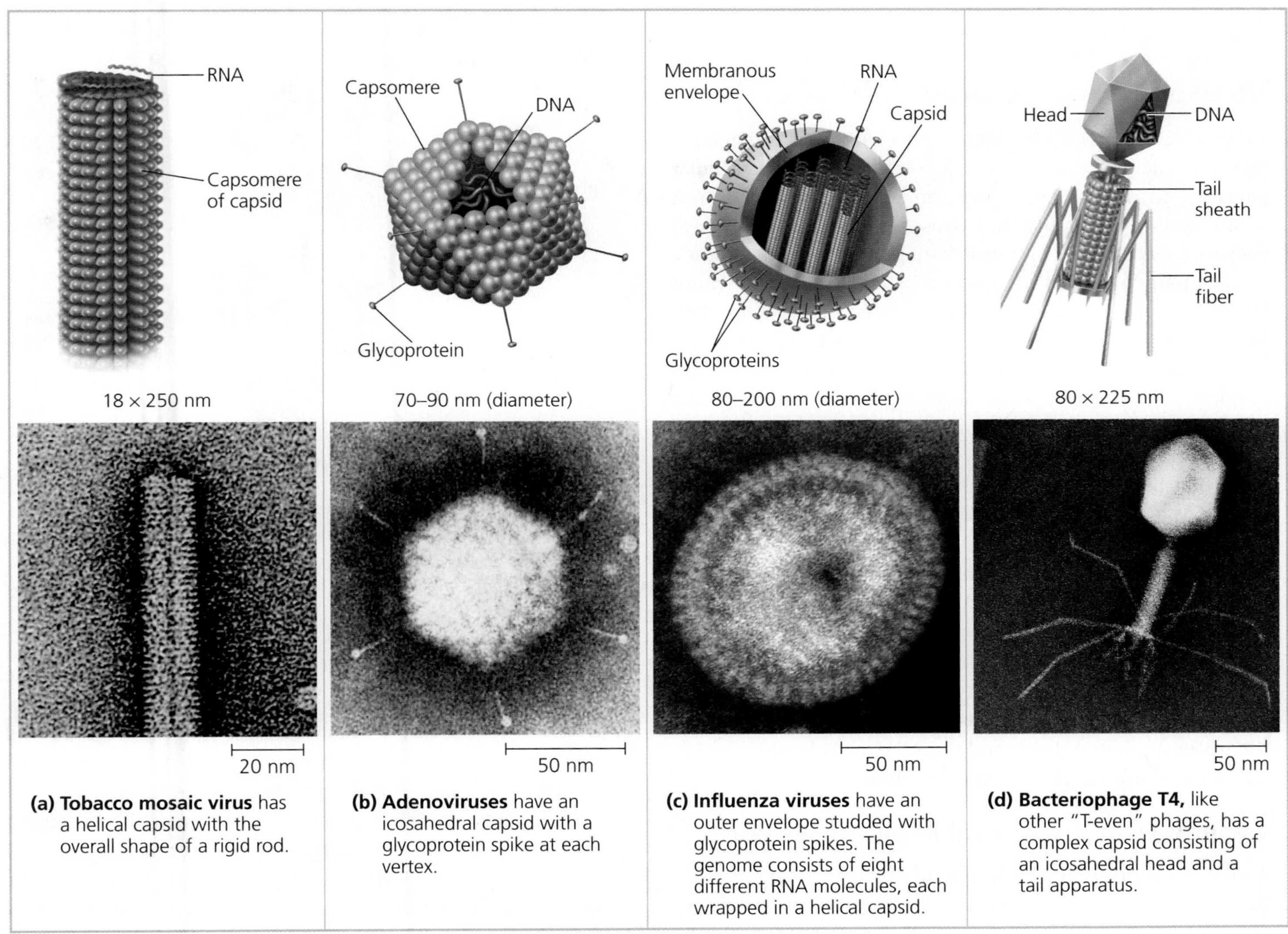

(a) Tobacco mosaic virus has a helical capsid with the overall shape of a rigid rod.

(b) Adenoviruses have an icosahedral capsid with a glycoprotein spike at each vertex.

(c) Influenza viruses have an outer envelope studded with glycoprotein spikes. The genome consists of eight different RNA molecules, each wrapped in a helical capsid.

(d) Bacteriophage T4, like other "T-even" phages, has a complex capsid consisting of an icosahedral head and a tail apparatus.

▲ **Figure 19.3 Viral structure.** Viruses are made up of nucleic acid (DNA or RNA) enclosed in a protein coat (the capsid) and sometimes further wrapped in a membranous envelope. The individual protein subunits making up the capsid are called capsomeres. Although diverse in size and shape, viruses have many common structural features. (All micrographs are colorized TEMs.)

enclosing their DNA. Attached to the head is a protein tail piece with fibers by which the phages attach to a bacterium (**Figure 19.3d**). In the next section, we'll examine how these few viral parts function together with cellular components to produce large numbers of viral progeny.

CONCEPT CHECK 19.1

1. Compare the structures of tobacco mosaic virus (TMV) and influenza virus (see Figure 19.3).
2. **MAKE CONNECTIONS** In Figure 16.4 (p. 307), you learned how bacteriophages were used to provide evidence that DNA carries genetic information. Briefly describe the experiment carried out by Hershey and Chase, including in your description why the researchers chose to use phages.

For suggested answers, see Appendix A.

CONCEPT 19.2

Viruses replicate only in host cells

Viruses lack metabolic enzymes and equipment for making proteins, such as ribosomes. They are obligate intracellular parasites; in other words, they can replicate only within a host cell. It is fair to say that viruses in isolation are merely packaged sets of genes in transit from one host cell to another.

Each particular virus can infect cells of only a limited number of host species, called the **host range** of the virus. This host specificity results from the evolution of recognition systems by the virus. Viruses usually identify host cells by a "lock-and-key" fit between viral surface proteins and specific receptor molecules on the outside of cells. (According to one model, such receptor molecules originally carried out functions that benefited the host cell but were co-opted later by viruses as portals of entry.) Some viruses have broad host ranges. For example, West Nile virus and equine encephalitis virus are distinctly different viruses that can each infect mosquitoes, birds, horses, and humans. Other viruses have host ranges so narrow that they infect only a single species. Measles virus, for instance, can infect only humans. Furthermore, viral infection of multicellular eukaryotes is usually limited to particular tissues. Human cold viruses infect only the cells lining the upper respiratory tract, and the AIDS virus binds to receptors present only on certain types of white blood cells.

General Features of Viral Replicative Cycles

A viral infection begins when a virus binds to a host cell and the viral genome makes its way inside (**Figure 19.4**). The mechanism of genome entry depends on the type of virus and the type of host cell. For example, T-even phages use their elaborate tail apparatus to inject DNA into a bacterium (see Figure 19.3d). Other viruses are taken up by endocytosis or, in the case of

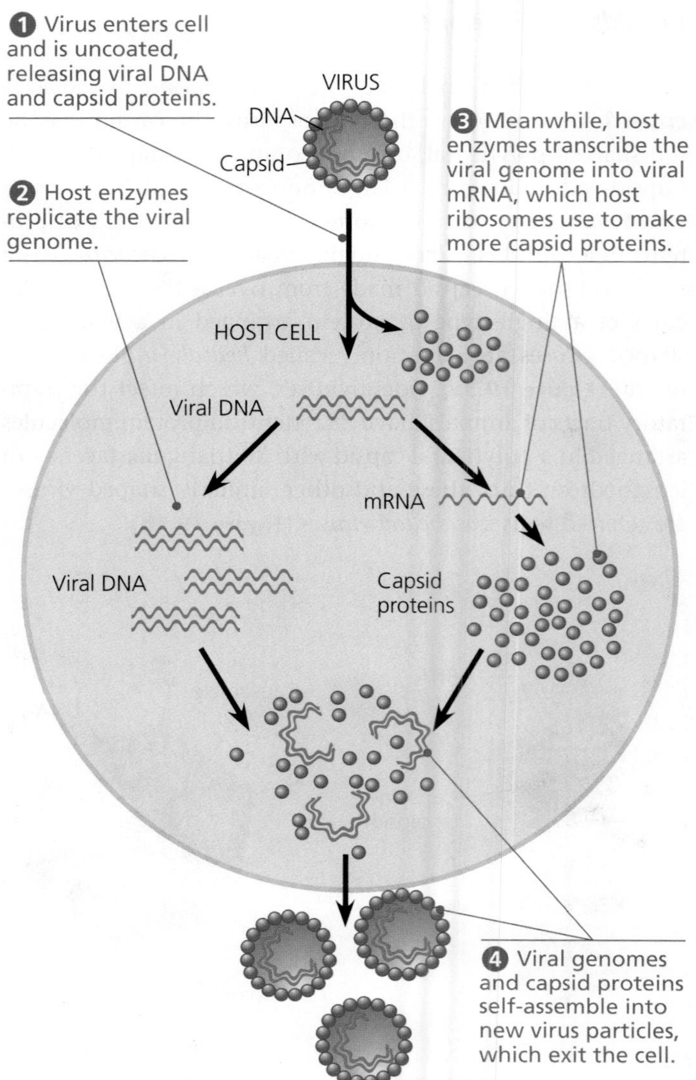

1 Virus enters cell and is uncoated, releasing viral DNA and capsid proteins.
VIRUS
DNA
Capsid
3 Meanwhile, host enzymes transcribe the viral genome into viral mRNA, which host ribosomes use to make more capsid proteins.
2 Host enzymes replicate the viral genome.
HOST CELL
Viral DNA
mRNA
Viral DNA
Capsid proteins
4 Viral genomes and capsid proteins self-assemble into new virus particles, which exit the cell.

▲ **Figure 19.4 A simplified viral replicative cycle.** A virus is an obligate intracellular parasite that uses the equipment and small molecules of its host cell to replicate. In this simplest of viral cycles, the parasite is a DNA virus with a capsid consisting of a single type of protein.

MAKE CONNECTIONS *Label each of the straight black arrows with one word representing the name of the process that is occurring. Review Figure 17.26 on page 348.*

enveloped viruses, by fusion of the viral envelope with the plasma membrane. Once the viral genome is inside, the proteins it encodes can commandeer the host, reprogramming the cell to copy the viral nucleic acid and manufacture viral proteins. The host provides the nucleotides for making viral nucleic acids, as well as enzymes, ribosomes, tRNAs, amino acids, ATP, and other components needed for making the viral proteins. Many DNA viruses use the DNA polymerases of the host cell to synthesize new genomes along the templates provided by the viral DNA. In contrast, to replicate their genomes, RNA viruses use virally encoded RNA polymerases that can use RNA as a template. (Uninfected cells generally make no enzymes for carrying out this process.)

After the viral nucleic acid molecules and capsomeres are produced, they spontaneously self-assemble into new viruses. In fact, researchers can separate the RNA and capsomeres of TMV and then reassemble complete viruses simply by mixing the components together under the right conditions. The simplest type of viral replicative cycle ends with the exit of hundreds or thousands of viruses from the infected host cell, a process that often damages or destroys the cell. Such cellular damage and death, as well as the body's responses to this destruction, cause many of the symptoms associated with viral infections. The viral progeny that exit a cell have the potential to infect additional cells, spreading the viral infection.

There are many variations on the simplified viral replicative cycle we have just described. We will now take a look at some of these variations in bacterial viruses (phages) and animal viruses; later in the chapter, we will consider plant viruses.

Replicative Cycles of Phages

Phages are the best understood of all viruses, although some of them are also among the most complex. Research on phages led to the discovery that some double-stranded DNA viruses can replicate by two alternative mechanisms: the lytic cycle and the lysogenic cycle.

The Lytic Cycle

A phage replicative cycle that culminates in death of the host cell is known as a **lytic cycle**. The term refers to the last stage of infection, during which the bacterium lyses (breaks open) and releases the phages that were produced within the cell. Each of these phages can then infect a healthy cell, and a few successive lytic cycles can destroy an entire bacterial population in just a few hours. A phage that replicates only by a lytic cycle is a **virulent phage**. **Figure 19.5** illustrates the major steps in the lytic cycle of T4, a typical virulent phage. Study this figure before proceeding.

After reading about the lytic cycle, you may wonder why phages haven't exterminated all bacteria. In fact, phage treatments have been used medically in some countries to help control bacterial infections in humans. Bacteria are not defenseless, however. First, natural selection favors bacterial mutants with receptors that are no longer recognized by a particular type of phage. Second, when phage DNA successfully enters a bacterium, the DNA often is identified as foreign and cut up by cellular enzymes called **restriction enzymes**, which are so named because their activity *restricts* the ability of the phage to infect the bacterium. The bacterial cell's own DNA is methylated in a way that prevents attack

▶ **Figure 19.5 The lytic cycle of phage T4, a virulent phage.** Phage T4 has almost 300 genes, which are transcribed and translated using the host cell's machinery. One of the first phage genes translated after the viral DNA enters the host cell codes for an enzyme that degrades the host cell's DNA (step 2); the phage DNA is protected from breakdown because it contains a modified form of cytosine that is not recognized by the enzyme. The entire lytic cycle, from the phage's first contact with the cell surface to cell lysis, takes only 20–30 minutes at 37°C.

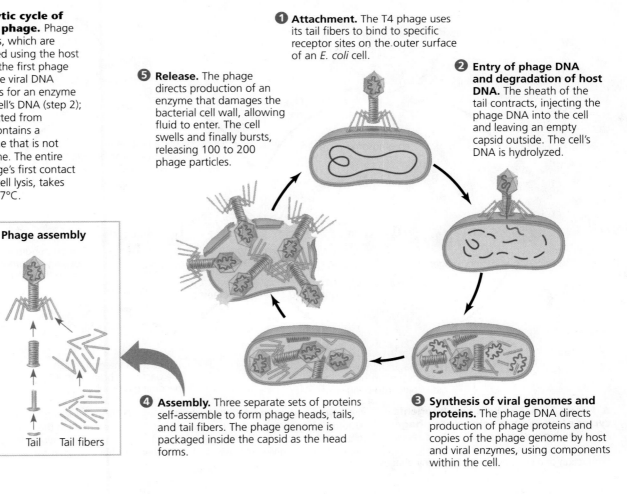

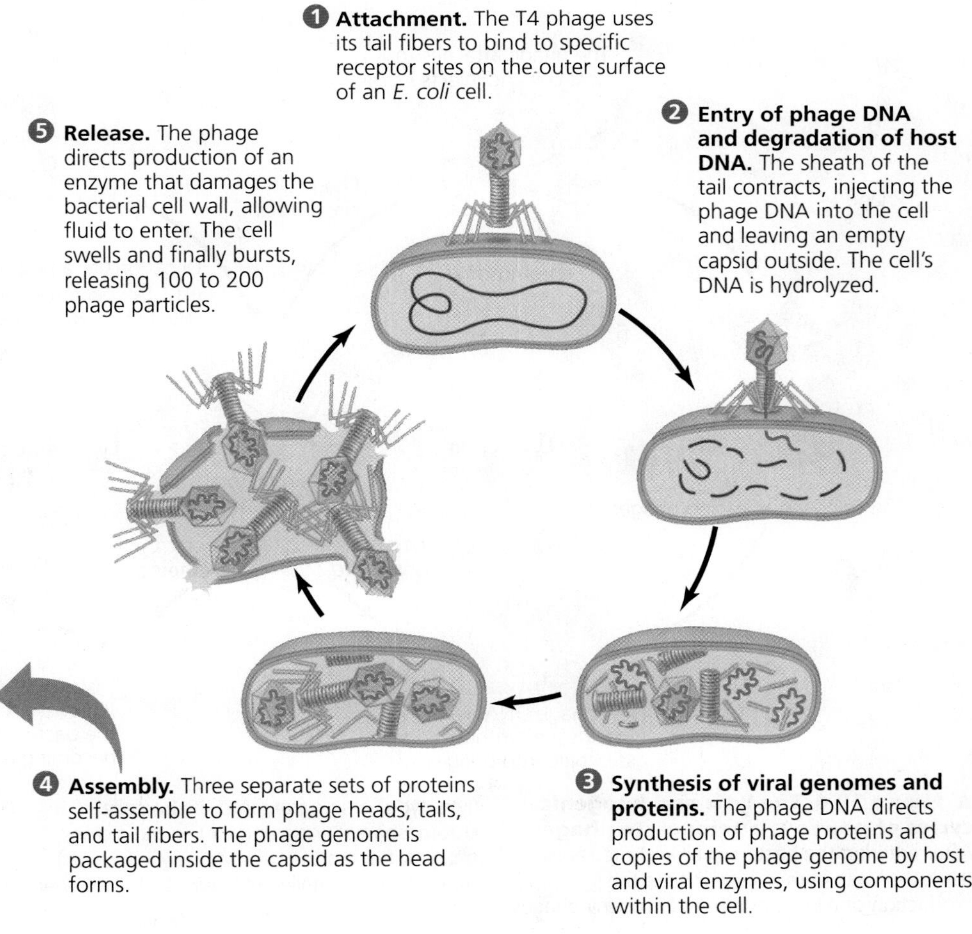

❶ **Attachment.** The T4 phage uses its tail fibers to bind to specific receptor sites on the outer surface of an *E. coli* cell.
❷ **Entry of phage DNA and degradation of host DNA.** The sheath of the tail contracts, injecting the phage DNA into the cell and leaving an empty capsid outside. The cell's DNA is hydrolyzed.
❸ **Synthesis of viral genomes and proteins.** The phage DNA directs production of phage proteins and copies of the phage genome by host and viral enzymes, using components within the cell.
❹ **Assembly.** Three separate sets of proteins self-assemble to form phage heads, tails, and tail fibers. The phage genome is packaged inside the capsid as the head forms.
❺ **Release.** The phage directs production of an enzyme that damages the bacterial cell wall, allowing fluid to enter. The cell swells and finally bursts, releasing 100 to 200 phage particles.
Phage assembly
Head
Tail
Tail fibers

by its own restriction enzymes. But just as natural selection favors bacteria with mutant receptors or effective restriction enzymes, it also favors phage mutants that can bind the altered receptors or are resistant to particular restriction enzymes. Thus, the parasite-host relationship is in constant evolutionary flux.

There is yet a third important reason bacteria have been spared from extinction as a result of phage activity. Instead of lysing their host cells, many phages coexist with them in a state called lysogeny, which we'll now discuss.

The Lysogenic Cycle

In contrast to the lytic cycle, which kills the host cell, the **lysogenic cycle** allows replication of the phage genome without destroying the host. Phages capable of using both modes of replicating within a bacterium are called **temperate phages**. A temperate phage called lambda, written with the Greek letter λ, is widely used in biological research. Phage λ resembles T4, but its tail has only one short tail fiber.

Infection of an *E. coli* cell by phage λ begins when the phage binds to the surface of the cell and injects its linear DNA genome **(Figure 19.6)**. Within the host, the λ DNA molecule forms a circle. What happens next depends on the replicative mode: lytic cycle or lysogenic cycle. During a lytic cycle, the viral genes immediately turn the host cell into a λ-producing factory, and the cell soon lyses and releases its viral products. During a lysogenic cycle, however, the λ DNA molecule is incorporated into a specific site on the *E. coli* chromosome by viral proteins that break both circular DNA molecules and join them to each other. When integrated into the bacterial chromosome in this way, the viral DNA is known as a **prophage**. One prophage gene codes for a protein that prevents transcription of most of the other prophage genes. Thus, the phage genome is mostly silent within the bacterium. Every time the *E. coli* cell prepares to divide, it replicates the phage DNA along with its own and passes the copies on to daughter cells. A single infected cell can quickly give rise to a large population of bacteria carrying the virus in prophage form. This mechanism enables viruses to propagate without killing the host cells on which they depend.

The term *lysogenic* implies that prophages are capable of generating active phages that lyse their host cells. This occurs when the λ genome is induced to exit the bacterial chromosome and initiate a lytic cycle. An environmental signal, such as a certain chemical or high-energy radiation, usually triggers the switchover from the lysogenic to the lytic mode.

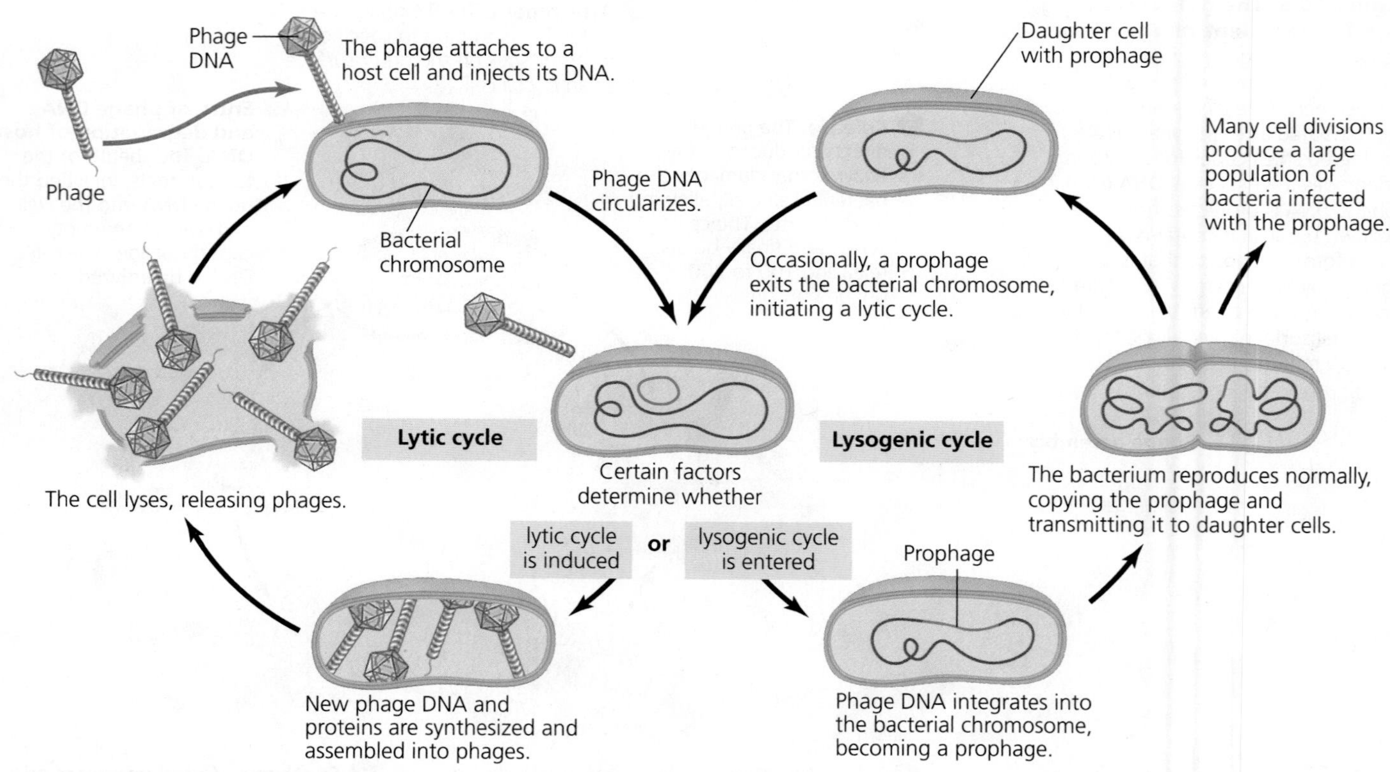

▲ **Figure 19.6 The lytic and lysogenic cycles of phage λ, a temperate phage.** After entering the bacterial cell and circularizing, the λ DNA can immediately initiate the production of a large number of progeny phages (lytic cycle) or integrate into the bacterial chromosome (lysogenic cycle). In most cases, phage λ follows the lytic pathway, which is similar to that detailed in Figure 19.5. However, once a lysogenic cycle begins, the prophage may be carried in the host cell's chromosome for many generations. Phage λ has one main tail fiber, which is short.

Table 19.1 Classes of Animal Viruses

Class/Family	Envelope	Examples That Cause Human Diseases
I. Double-Stranded DNA (dsDNA)		
Adenovirus (see Figure 19.3b)	No	Respiratory viruses; tumor-causing viruses
Papovavirus	No	Papillomavirus (warts, cervical cancer); polyomavirus (tumors)
Herpesvirus	Yes	Herpes simplex I and II (cold sores, genital sores); varicella zoster (shingles, chicken pox); Epstein-Barr virus (mononucleosis, Burkitt's lymphoma)
Poxvirus	Yes	Smallpox virus; cowpox virus
II. Single-Stranded DNA (ssDNA)		
Parvovirus	No	B19 parvovirus (mild rash)
III. Double-Stranded RNA (dsRNA)		
Reovirus	No	Rotavirus (diarrhea); Colorado tick fever virus
IV. Single-Stranded RNA (ssRNA); Serves as mRNA		
Picornavirus	No	Rhinovirus (common cold); poliovirus; hepatitis A virus; other enteric (intestinal) viruses
Coronavirus	Yes	Severe acute respiratory syndrome (SARS)
Flavivirus	Yes	Yellow fever virus; West Nile virus; hepatitis C virus
Togavirus	Yes	Rubella virus; equine encephalitis viruses
V. ssRNA; Template for mRNA Synthesis		
Filovirus	Yes	Ebola virus (hemorrhagic fever)
Orthomyxovirus (see Figures 19.3c and 19.9a)	Yes	Influenza virus
Paramyxovirus	Yes	Measles virus; mumps virus
Rhabdovirus	Yes	Rabies virus
VI. ssRNA; Template for DNA Synthesis		
Retrovirus (see Figure 19.8)	Yes	Human immunodeficiency virus (HIV/AIDS); RNA tumor viruses (leukemia)

In addition to the gene for the transcription-preventing protein, a few other prophage genes may be expressed during lysogeny. Expression of these genes may alter the host's phenotype, a phenomenon that can have important medical significance. For example, the three species of bacteria that cause the human diseases diphtheria, botulism, and scarlet fever would not be so harmful to humans without certain prophage genes that cause the host bacteria to make toxins. And the difference between the *E. coli* strain that resides in our intestines and the O157:H7 strain that has caused several deaths by food poisoning appears to be the presence of prophages in the O157:H7 strain.

Replicative Cycles of Animal Viruses

Everyone has suffered from viral infections, whether cold sores, influenza, or the common cold. Like all viruses, those that cause illness in humans and other animals can replicate only inside host cells. Many variations on the basic scheme of viral infection and replication are represented among the animal viruses. One key variable is the nature of the viral genome: Is it composed of DNA or RNA? Is it double-stranded or single-stranded? The nature of the genome is the basis for the common classification of viruses shown in **Table 19.1**. Single-stranded RNA viruses are further classified into three classes (IV–VI) according to how the RNA genome functions in a host cell.

Whereas few bacteriophages have an envelope or RNA genome, many animal viruses have both. In fact, nearly all animal viruses with RNA genomes have an envelope, as do some with DNA genomes (see Table 19.1). Rather than consider all the mechanisms of viral infection and replication, we will focus on the roles of viral envelopes and on the functioning of RNA as the genetic material of many animal viruses.

Viral Envelopes

An animal virus equipped with an envelope—that is, an outer membrane—uses it to enter the host cell. Protruding from the outer surface of this envelope are viral glycoproteins that bind to specific receptor molecules on the surface of a host cell. **Figure 19.7**, on the next page, outlines the events in the replicative cycle of an enveloped virus with an RNA genome. Ribosomes bound to the endoplasmic reticulum (ER) of the host cell make the protein parts of the envelope glycoproteins; cellular enzymes in the ER and Golgi apparatus then add the sugars. The resulting viral glycoproteins, embedded in host cell–derived membrane, are transported to the cell surface. In a process much like exocytosis, new viral capsids are wrapped in membrane as they bud from the cell. In other words, the viral envelope is derived from the host cell's plasma membrane, although some of the molecules of this membrane are specified by viral genes. The enveloped viruses are now free to infect other cells. This replicative cycle does not necessarily kill the host cell, in contrast to the lytic cycles of phages.

Some viruses have envelopes that are not derived from plasma membrane. Herpesviruses, for example, are temporarily cloaked in membrane derived from the nuclear envelope of the host; they then shed this membrane in the cytoplasm and

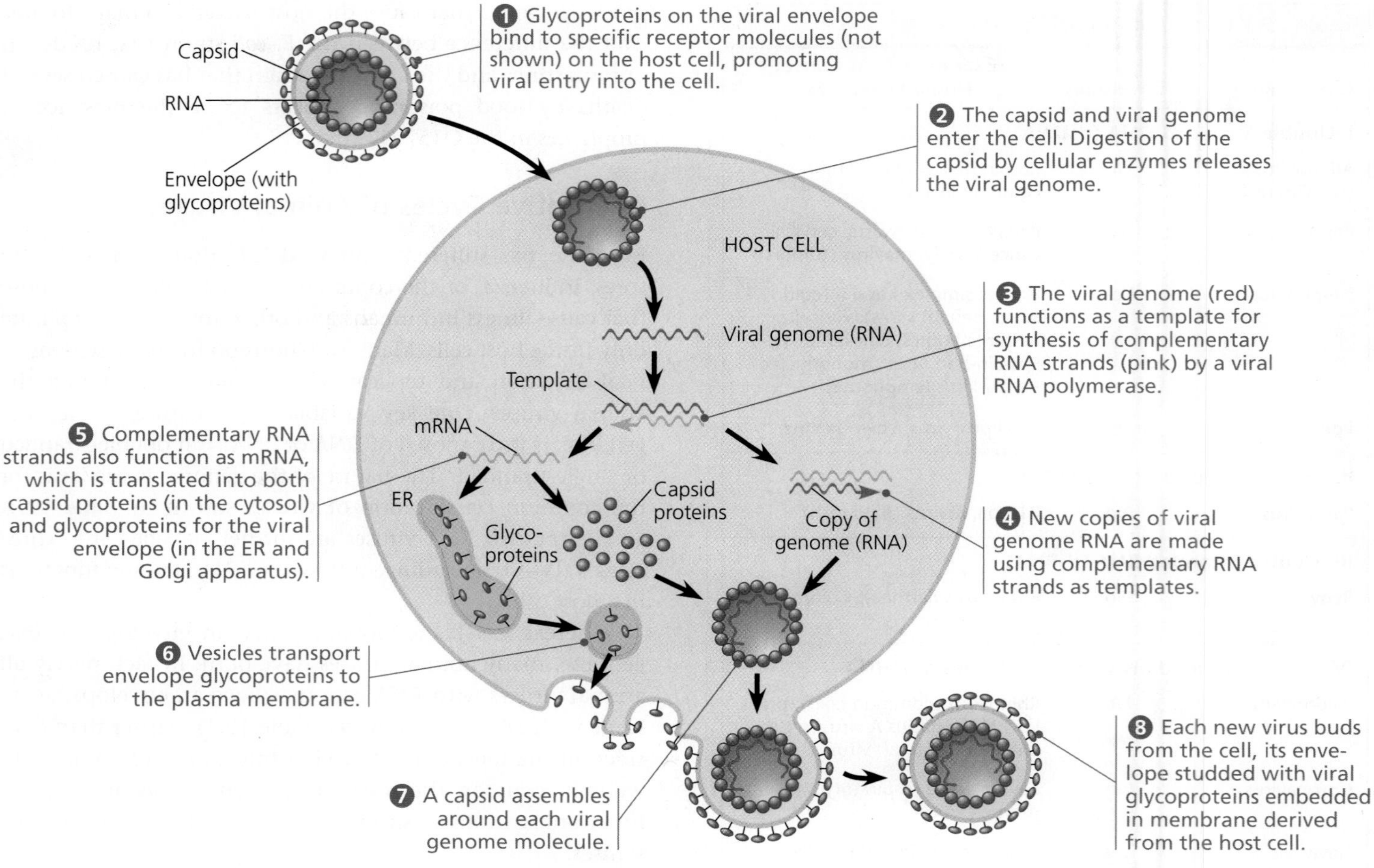

Capsid
RNA
Envelope (with glycoproteins)
1 Glycoproteins on the viral envelope bind to specific receptor molecules (not shown) on the host cell, promoting viral entry into the cell.
2 The capsid and viral genome enter the cell. Digestion of the capsid by cellular enzymes releases the viral genome.
HOST CELL
Viral genome (RNA)
3 The viral genome (red) functions as a template for synthesis of complementary RNA strands (pink) by a viral RNA polymerase.
Template
mRNA
5 Complementary RNA strands also function as mRNA, which is translated into both capsid proteins (in the cytosol) and glycoproteins for the viral envelope (in the ER and Golgi apparatus).
ER
Glyco-proteins
Capsid proteins
Copy of genome (RNA)
4 New copies of viral genome RNA are made using complementary RNA strands as templates.
6 Vesicles transport envelope glycoproteins to the plasma membrane.
7 A capsid assembles around each viral genome molecule.
8 Each new virus buds from the cell, its envelope studded with viral glycoproteins embedded in membrane derived from the host cell.

▲ **Figure 19.7 The replicative cycle of an enveloped RNA virus.** Shown here is a virus with a single-stranded RNA genome that functions as a template for synthesis of mRNA. Some enveloped viruses enter the host cell by fusion of the envelope with the cell's plasma membrane; others enter by endocytosis. For all enveloped RNA viruses, the formation of new envelopes for progeny viruses occurs by the mechanism depicted in this figure.

? *Name a virus that has infected you and has a replicative cycle matching this one. (Hint: See Table 19.1.)*

acquire a new envelope made from membrane of the Golgi apparatus. These viruses have a double-stranded DNA genome and replicate within the host cell nucleus, using a combination of viral and cellular enzymes to replicate and transcribe their DNA. In the case of herpesviruses, copies of the viral DNA can remain behind as mini-chromosomes in the nuclei of certain nerve cells. There they remain latent until some sort of physical or emotional stress triggers a new round of active virus production. The infection of other cells by these new viruses causes the blisters characteristic of herpes, such as cold sores or genital sores. Once someone acquires a herpesvirus infection, flare-ups may recur throughout the person's life.

RNA as Viral Genetic Material

Although some phages and most plant viruses are RNA viruses, the broadest variety of RNA genomes is found among the viruses that infect animals. Among the three types of single-stranded RNA genomes found in animal viruses, the genome of class IV viruses can directly serve as mRNA and thus can be translated into viral protein immediately after infection. Figure 19.7 shows a virus of class V, in which the RNA genome serves as a *template* for mRNA synthesis. The RNA genome is transcribed into complementary RNA strands, which function both as mRNA and as templates for the synthesis of additional copies of genomic RNA. All viruses that require RNA → RNA synthesis to make mRNA use a viral enzyme capable of carrying out this process; there are no such enzymes in most cells. The viral enzyme is packaged with the genome inside the viral capsid.

The RNA animal viruses with the most complicated replicative cycles are the **retroviruses** (class VI). These viruses are equipped with an enzyme called **reverse transcriptase**, which transcribes an RNA template into DNA, providing an RNA → DNA information flow, the opposite of the usual direction. This unusual phenomenon is the source of the name retroviruses (*retro* means "backward"). Of particular medical importance is **HIV (human immunodeficiency virus)**, the retrovirus that causes **AIDS (acquired immunodeficiency syndrome)**. HIV and other retroviruses are enveloped viruses that contain two identical molecules of single-stranded RNA and two molecules of reverse transcriptase.

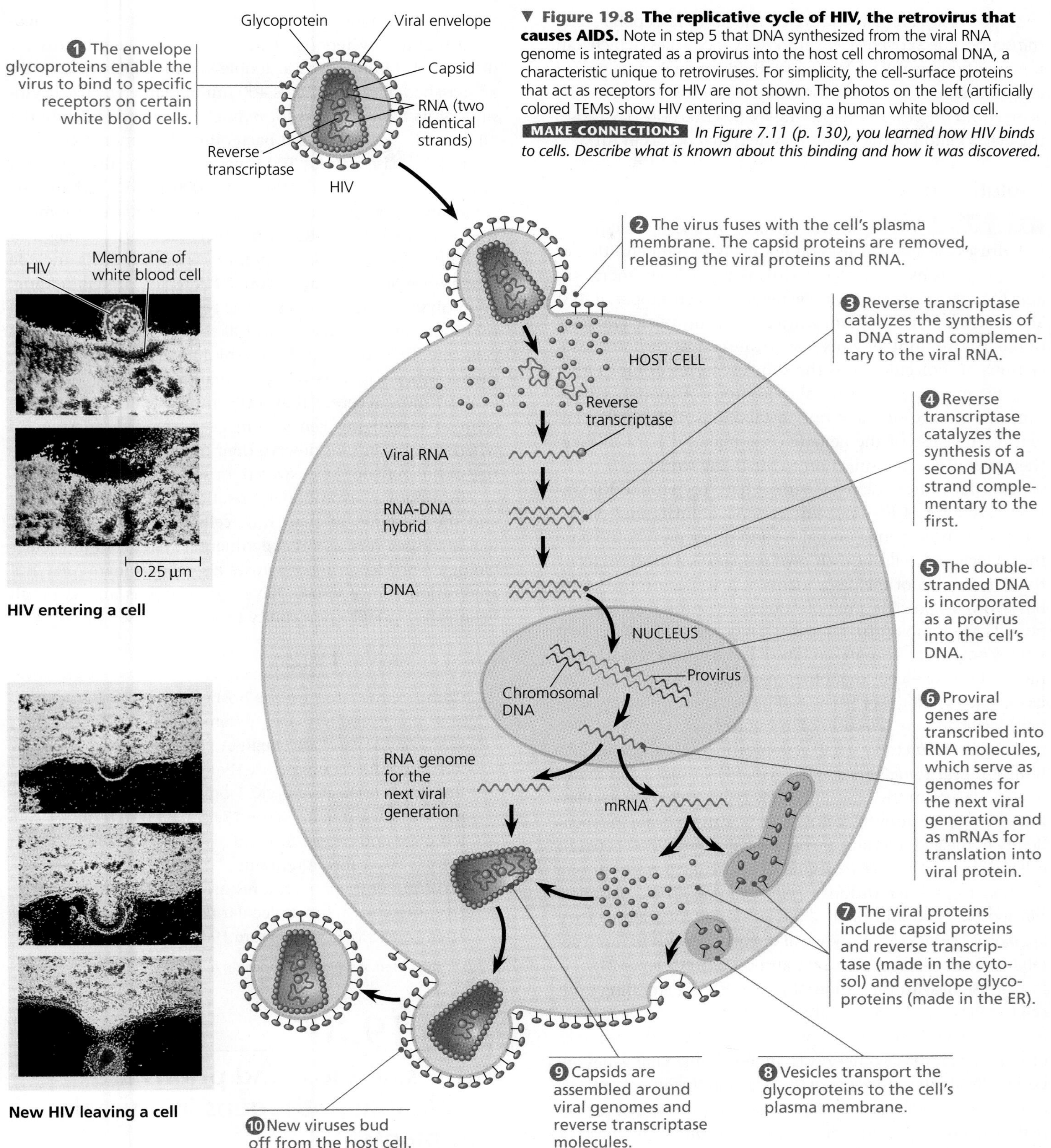
Glycoprotein
Viral envelope
1 The envelope glycoproteins enable the virus to bind to specific receptors on certain white blood cells.
Capsid
RNA (two identical strands)
Reverse transcriptase
HIV
HIV
Membrane of white blood cell
0.25 µm
HIV entering a cell
2 The virus fuses with the cell's plasma membrane. The capsid proteins are removed, releasing the viral proteins and RNA.
3 Reverse transcriptase catalyzes the synthesis of a DNA strand complementary to the viral RNA.
HOST CELL
Reverse transcriptase
Viral RNA
RNA-DNA hybrid
4 Reverse transcriptase catalyzes the synthesis of a second DNA strand complementary to the first.
DNA
NUCLEUS
5 The double-stranded DNA is incorporated as a provirus into the cell's DNA.
Provirus
Chromosomal DNA
6 Proviral genes are transcribed into RNA molecules, which serve as genomes for the next viral generation and as mRNAs for translation into viral protein.
RNA genome for the next viral generation
mRNA
7 The viral proteins include capsid proteins and reverse transcriptase (made in the cytosol) and envelope glycoproteins (made in the ER).
New HIV leaving a cell
10 New viruses bud off from the host cell.
9 Capsids are assembled around viral genomes and reverse transcriptase molecules.
8 Vesicles transport the glycoproteins to the cell's plasma membrane.

▼ **Figure 19.8 The replicative cycle of HIV, the retrovirus that causes AIDS.** Note in step 5 that DNA synthesized from the viral RNA genome is integrated as a provirus into the host cell chromosomal DNA, a characteristic unique to retroviruses. For simplicity, the cell-surface proteins that act as receptors for HIV are not shown. The photos on the left (artificially colored TEMs) show HIV entering and leaving a human white blood cell.

MAKE CONNECTIONS *In Figure 7.11 (p. 130), you learned how HIV binds to cells. Describe what is known about this binding and how it was discovered.*

Figure 19.8 traces the HIV replicative cycle, which is typical of a retrovirus. After HIV enters a host cell, its reverse transcriptase molecules are released into the cytoplasm, where they catalyze synthesis of viral DNA. The newly made viral DNA then enters the cell's nucleus and integrates into the DNA of a chromosome. The integrated viral DNA, called a **provirus**, never leaves the host's genome, remaining a permanent resident of the cell. (Recall that a prophage, in contrast, leaves the host's genome at the start of a lytic cycle.) The host's RNA polymerase transcribes the proviral DNA into RNA molecules, which can function both as mRNA for the synthesis of viral proteins and as genomes for the new viruses that will be assembled and released from the cell. In Chapter 43, we describe how HIV causes the deterioration of the immune system that occurs in AIDS.

Evolution of Viruses

EVOLUTION We began this chapter by asking whether or not viruses are alive. Viruses do not really fit our definition of living organisms. An isolated virus is biologically inert, unable to replicate its genes or regenerate its own supply of ATP. Yet it has a genetic program written in the universal language of life. Do we think of viruses as nature's most complex associations of molecules or as the simplest forms of life? Either way, we must bend our usual definitions. Although viruses cannot replicate or carry out metabolic activities independently, their use of the genetic code makes it hard to deny their evolutionary connection to the living world.

How did viruses originate? Viruses have been found that infect every form of life—not just bacteria, animals, and plants, but also archaea, fungi, and algae and other protists. Because they depend on cells for their own propagation, it seems likely that viruses are not the descendants of precellular forms of life but evolved—possibly multiple times—*after* the first cells appeared. Most molecular biologists favor the hypothesis that viruses originated from naked bits of cellular nucleic acids that moved from one cell to another, perhaps via injured cell surfaces. The evolution of genes coding for capsid proteins may have facilitated the infection of uninjured cells. Candidates for the original sources of viral genomes include plasmids and transposons. *Plasmids* are small, circular DNA molecules found in bacteria and in the unicellular eukaryotes called yeasts. Plasmids exist apart from the cell's genome, can replicate independently of the genome, and are occasionally transferred between cells. *Transposons* are DNA segments that can move from one location to another within a cell's genome. Thus, plasmids, transposons, and viruses all share an important feature: They are *mobile genetic elements*. We will discuss plasmids in more detail in Chapters 20 and 27, and transposons in Chapter 21.

Consistent with this vision of pieces of DNA shuttling from cell to cell is the observation that a viral genome can have more in common with the genome of its host than with the genomes of viruses that infect other hosts. Indeed, some viral genes are essentially identical to genes of the host. On the other hand, recent sequencing of many viral genomes has shown that the genetic sequences of some viruses are quite similar to those of seemingly distantly related viruses; for example, some animal viruses share similar sequences with plant viruses. This genetic similarity may reflect the persistence of groups of viral genes that were favored by natural selection during the early evolution of viruses and the eukaryotic cells that served as their hosts.

The debate about the origin of viruses has been reinvigorated recently by reports of mimivirus, the largest virus yet discovered. Mimivirus is a double-stranded DNA virus with an icosahedral capsid that is 400 nm in diameter. (The beginning of its name is short for *mi*micking *mi*crobe because the virus is the size of a small bacterium.) Its genome contains 1.2 million bases (about 100 times as many as the influenza virus genome) and an estimated 1,000 genes. Perhaps the most surprising aspect of mimivirus, however, is that some of the genes appear to code for products previously thought to be hallmarks of cellular genomes. These products include proteins involved in translation, DNA repair, protein folding, and polysaccharide synthesis. The researchers who described mimivirus propose that it most likely evolved *before* the first cells and then developed an exploitative relationship with them. Other scientists disagree, maintaining that the virus evolved more recently than cells and has simply been efficient at scavenging genes from its hosts. The question of whether some viruses deserve their own early branch on the tree of life may not be answered for some time.

The ongoing evolutionary relationship between viruses and the genomes of their host cells is an association that makes viruses very useful experimental systems in molecular biology. Knowledge about viruses also allows many practical applications, since viruses have a tremendous impact on all organisms through their ability to cause disease.

CONCEPT CHECK 19.2

1. Compare the effect on the host cell of a lytic (virulent) phage and a lysogenic (temperate) phage.
2. **MAKE CONNECTIONS** The RNA virus in Figure 19.7 has a viral RNA polymerase that functions in step 3 of the virus's replicative cycle. Compare this RNA polymerase to the one in Figure 17.9 (p. 333) in terms of template and overall function.
3. Why is HIV called a retrovirus?
4. **WHAT IF?** If you were a researcher trying to combat HIV infection, what molecular processes could you attempt to block? (See Figure 19.8.)

For suggested answers, see Appendix A.

CONCEPT 19.3

Viruses, viroids, and prions are formidable pathogens in animals and plants

Diseases caused by viral infections afflict humans, agricultural crops, and livestock worldwide. Other smaller, less complex entities known as viroids and prions also cause disease in plants and animals, respectively.

Viral Diseases in Animals

A viral infection can produce symptoms by a number of different routes. Viruses may damage or kill cells by causing the release of hydrolytic enzymes from lysosomes. Some viruses cause infected cells to produce toxins that lead to disease symptoms, and some have molecular components that are toxic, such as envelope proteins. How much damage a virus causes depends partly on the ability of the infected tissue to regenerate by cell division. People usually recover completely from colds because the epithelium of the respiratory tract, which the viruses infect, can efficiently repair itself. In contrast, damage inflicted by poliovirus to mature nerve cells is permanent because these cells do not divide and usually cannot be replaced. Many of the temporary symptoms associated with viral infections, such as fever and aches, actually result from the body's own efforts at defending itself against infection rather than from cell death caused by the virus.

The immune system is a complex and critical part of the body's natural defenses (see Chapter 43). It is also the basis for the major medical tool for preventing viral infections—vaccines. A **vaccine** is a harmless variant or derivative of a pathogen that stimulates the immune system to mount defenses against the harmful pathogen. Smallpox, a viral disease that was at one time a devastating scourge in many parts of the world, was eradicated by a vaccination program carried out by the World Health Organization (WHO). The very narrow host range of the smallpox virus—it infects only humans—was a critical factor in the success of this program. Similar worldwide vaccination campaigns are currently under way to eradicate polio and measles. Effective vaccines are also available to protect against rubella, mumps, hepatitis B, and a number of other viral diseases.

Although vaccines can prevent certain viral illnesses, medical technology can do little, at present, to cure most viral infections once they occur. The antibiotics that help us recover from bacterial infections are powerless against viruses. Antibiotics kill bacteria by inhibiting enzymes specific to bacteria but have no effect on eukaryotic or virally encoded enzymes. However, the few enzymes that are encoded by viruses have provided targets for other drugs. Most antiviral drugs resemble nucleosides and as a result interfere with viral nucleic acid synthesis. One such drug is acyclovir, which impedes herpesvirus replication by inhibiting the viral polymerase that synthesizes viral DNA. Similarly, azidothymidine (AZT) curbs HIV replication by interfering with the synthesis of DNA by reverse transcriptase. In the past two decades, much effort has gone into developing drugs against HIV. Currently, multidrug treatments, sometimes called "cocktails," have been found to be most effective. Such treatments commonly include a combination of two nucleoside mimics and a protease inhibitor, which interferes with an enzyme required for assembly of the viruses.

Emerging Viruses

Viruses that suddenly become apparent are often referred to as *emerging viruses*. HIV, the AIDS virus, is a classic example: This virus appeared in San Francisco in the early 1980s, seemingly out of nowhere, although later studies uncovered a case in the Belgian Congo in 1959. The deadly Ebola virus, recognized initially in 1976 in central Africa, is one of several emerging viruses that cause *hemorrhagic fever*, an often fatal syndrome (set of symptoms) characterized by fever, vomiting, massive bleeding, and circulatory system collapse. A number of other dangerous emerging viruses cause encephalitis, inflammation of the brain. One example is the West Nile virus, which appeared in North America for the first time in 1999 and has spread to all 48 contiguous states in the United States.

In April 2009, a general outbreak, or **epidemic**, of a flu-like illness appeared in Mexico and the United States. The infectious agent was quickly identifed as an influenza virus related to viruses that cause the seasonal flu (**Figure 19.9a**). This particular virus was named H1N1 for reasons that will be

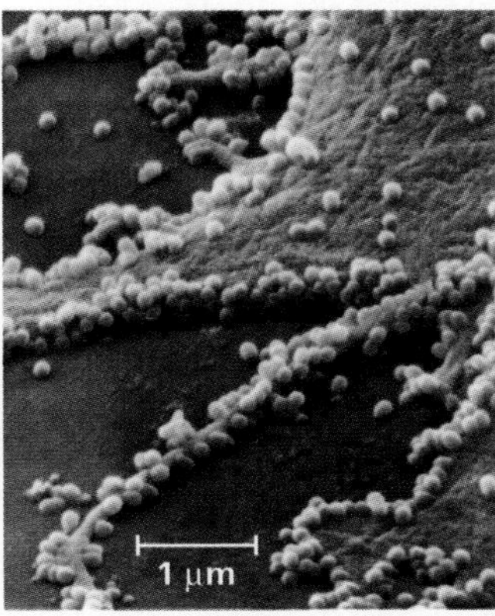

(a) 2009 pandemic H1N1 influenza A virus. Viruses (blue) are seen on an infected cell (green) in this colorized SEM.

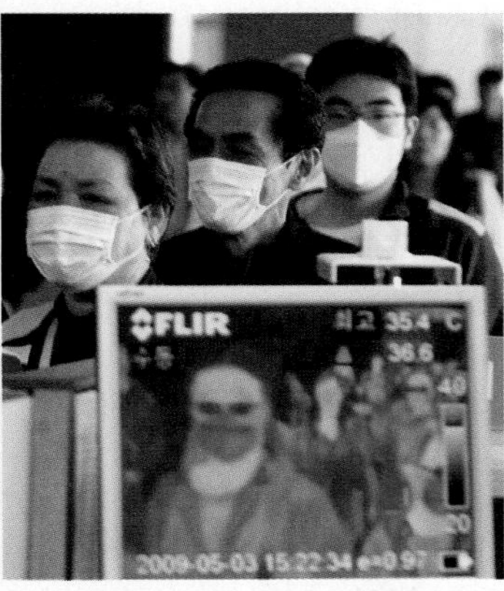

(b) 2009 pandemic screening. At a South Korean airport, thermal scans were used to detect passengers with a fever who might have the H1N1 flu.

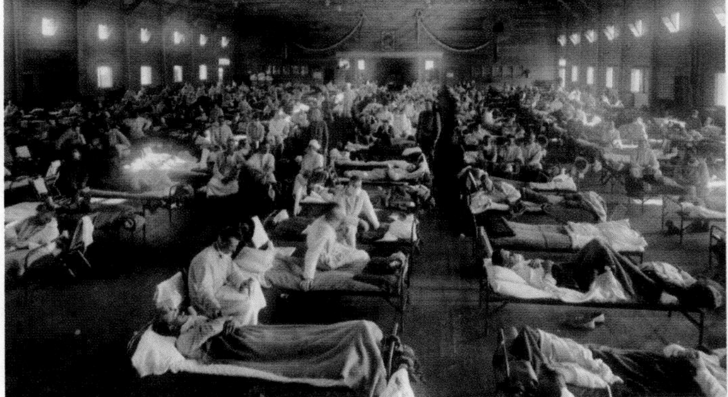

(c) 1918 flu pandemic. Many of those infected during the worst flu epidemic in the last 100 years were treated in large makeshift hospitals, such as this one.

▲ **Figure 19.9 Influenza in humans.**

explained shortly. The viral disease spread rapidly, prompting WHO to declare a global epidemic, or **pandemic**, in June 2009. By November, the disease had reached 207 countries, infecting over 600,000 people and killing almost 8,000. Public health agencies responded rapidly with guidelines for shutting down schools and other public places, and vaccine development and screening efforts were accelerated **(Figure 19.9b)**.

How do such viruses burst on the human scene, giving rise to harmful diseases that were previously rare or even unknown? Three processes contribute to the emergence of viral diseases. The first, and perhaps most important, is the mutation of existing viruses. RNA viruses tend to have an unusually high rate of mutation because errors in replicating their RNA genomes are not corrected by proofreading. Some mutations change existing viruses into new genetic varieties (strains) that can cause disease, even in individuals who are immune to the ancestral virus. For instance, seasonal flu epidemics are caused by new strains of influenza virus genetically different enough from earlier strains that people have little immunity to them.

A second process that can lead to the emergence of viral diseases is the dissemination of a viral disease from a small, isolated human population. For instance, AIDS went unnamed and virtually unnoticed for decades before it began to spread around the world. In this case, technological and social factors, including affordable international travel, blood transfusions, sexual promiscuity, and the abuse of intravenous drugs, allowed a previously rare human disease to become a global scourge.

A third source of new viral diseases in humans is the spread of existing viruses from other animals. Scientists estimate that about three-quarters of new human diseases originate in this way. Animals that harbor and can transmit a particular virus but are generally unaffected by it are said to act as a natural reservoir for that virus. For example, the 2009 flu pandemic mentioned earlier was likely passed to humans from pigs; for this reason, it was originally called "swine flu."

In general, flu epidemics provide an instructive example of the effects of viruses moving between species. There are three types of influenza virus: types B and C, which infect only humans and have never caused an epidemic, and type A, which infects a wide range of animals, including birds, pigs, horses, and humans. Influenza A strains have caused four major flu epidemics among humans in the last 100 years. The worst was the first one, the "Spanish flu" pandemic of 1918–1919, which killed about 40 million people, including many World War I soldiers **(Figure 19.9c)**.

Different strains of influenza A are given standardized names; for example, both the strain that caused the 1918 flu and the one that caused the 2009 pandemic flu are called H1N1. The name identifies which forms of two viral surface proteins are present: hemagglutinin (H) and neuraminidase (N). There are 16 different types of hemagglutinin, a protein that helps the flu virus attach to host cells, and 9 types of neuraminidase, an enzyme that helps release new virus particles from infected cells. Waterbirds have been found that carry viruses with all possible combinations of H and N.

A likely scenario for the 1918 pandemic and others is that the virus mutated as it passed from one host species to another. When an animal like a pig or a bird is infected with more than one strain of flu virus, the different strains can undergo genetic recombination if the RNA molecules making up their genomes mix and match during viral assembly. Pigs are thought to have been the breeding ground for the 2009 flu virus, which contains sequences from bird, pig, and human flu viruses. Coupled with mutation, these reassortments can lead to the emergence of a viral strain that is capable of infecting human cells. Humans who have never been exposed to that particular strain before will lack immunity, and the recombinant virus has the potential to be highly pathogenic. If such a flu virus recombines with viruses that circulate widely among humans, it may acquire the ability to spread easily from person to person, dramatically increasing the potential for a major human outbreak.

Although the 2009 H1N1 flu was declared a pandemic, its toll in lives was significantly lower than that of the 1918 flu. Significantly, however, 79% of the confirmed H1N1 cases in 2009 occurred in people under 30 years of age, and the highest mortality rates occurred in people under 64, opposite to patterns seen for seasonal flu. Some scientists hypothesize that the 1918 flu virus was the ancestor of most subsequent H1N1 epidemic-causing viruses, including that responsible for the 2009 pandemic. Older people are likely to have been exposed to earlier H1N1 viruses and have probably built up immunity to them. This could explain why contracting the 2009 H1N1 virus was more deadly for younger people, who were less likely to have been exposed to H1N1 viruses and to have built up immune defenses.

Perhaps a greater long-term threat is the avian flu caused by an H5N1 virus carried by wild and domestic birds. The first documented transmission to humans was in 1997, when 18 people in Hong Kong were infected and 6 subsequently died. While the 2009 H1N1 flu virus spread easily from human to human, reports of human-to-human transmission of the H5N1 avian flu are quite rare. More alarming, however, is the overall mortality rate of the H5N1 virus, which is greater than 50%. Furthermore, the host range of H5N1 is expanding, which provides increasing opportunities for different strains of the virus to reassort their genetic material and for new strains to emerge. If the H5N1 avian flu virus evolves so that it can spread easily from person to person, it could represent a major global health threat akin to that of the 1918 pandemic.

As we have seen, emerging viruses are generally not new; rather, they are existing viruses that mutate, disseminate more widely in the current host species, or spread to new host species. Changes in host behavior or environmental changes can increase the viral traffic responsible for emerging diseases. For example, new roads built through remote areas can allow viruses to spread between previously isolated human populations. Also, the destruction of forests to expand cropland can bring humans into contact with other animals that may host viruses capable of infecting humans.

Viral Diseases in Plants

More than 2,000 types of viral diseases of plants are known, and together they account for an estimated annual loss of $15 billion worldwide due to their destruction of agricultural and horticultural crops. Common signs of viral infection include bleached or brown spots on leaves and fruits, stunted growth, and damaged flowers or roots, all tending to diminish the yield and quality of crops (**Figure 19.10**).

Plant viruses have the same basic structure and mode of replication as animal viruses. Most plant viruses discovered thus far, including tobacco mosaic virus (TMV), have an RNA genome. Many have a helical capsid, like TMV, while others have an icosahedral capsid (see Figure 19.3).

Viral diseases of plants spread by two major routes. In the first route, called *horizontal transmission*, a plant is infected from an external source of the virus. Because the invading virus must get past the plant's outer protective layer of cells (the epidermis), a plant becomes more susceptible to viral infections if it has been damaged by wind, injury, or herbivores. Herbivores, especially insects, pose a double threat because they can also act as carriers of viruses, transmitting disease from plant to plant. Moreover, farmers and gardeners may transmit plant viruses inadvertently on pruning shears and other tools. The other route of viral infection is *vertical transmission*, in which a plant inherits a viral infection from a parent. Vertical transmission can occur in asexual propagation (for example, through cuttings) or in sexual reproduction via infected seeds.

Once a virus enters a plant cell and begins replicating, viral genomes and associated proteins can spread throughout the plant by means of plasmodesmata, the cytoplasmic connections that penetrate the walls between adjacent plant cells (see Figure 36.20). The passage of viral macromolecules from cell to cell is facilitated by virally encoded proteins that cause enlargement of plasmodesmata. Scientists have not yet devised cures for most viral plant diseases. Consequently, research efforts are focused largely on reducing the transmission of such diseases and on breeding resistant varieties of crop plants.

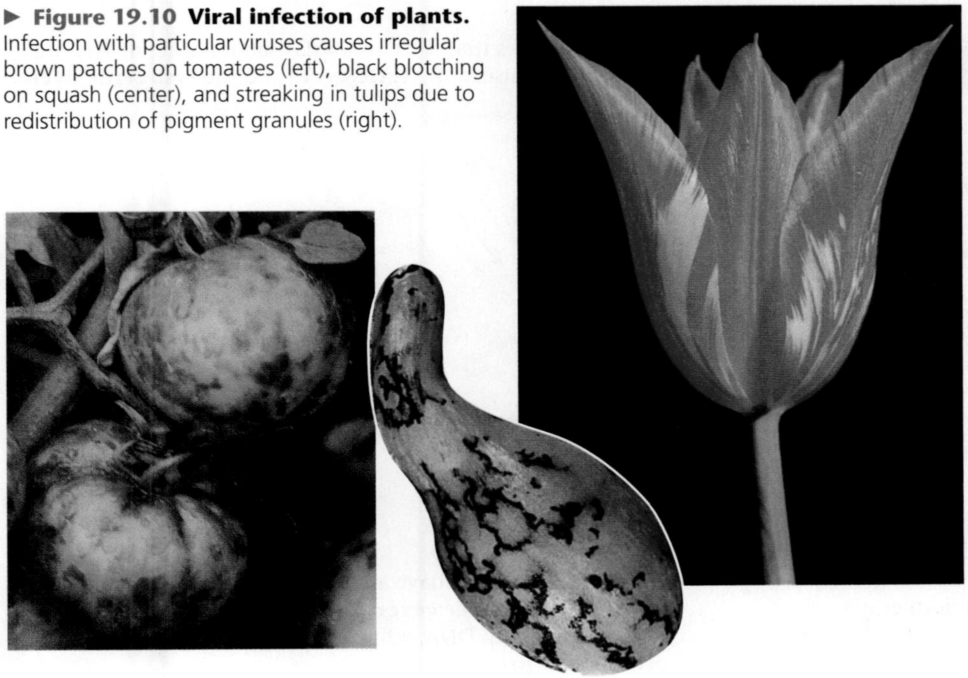

▶ **Figure 19.10 Viral infection of plants.** Infection with particular viruses causes irregular brown patches on tomatoes (left), black blotching on squash (center), and streaking in tulips due to redistribution of pigment granules (right).

Viroids and Prions: The Simplest Infectious Agents

As small and simple as viruses are, they dwarf another class of pathogens: **viroids**. These are circular RNA molecules, only a few hundred nucleotides long, that infect plants. Viroids do not encode proteins but can replicate in host plant cells, apparently using host cell enzymes. These small RNA molecules seem to cause errors in the regulatory systems that control plant growth; the typical signs of viroid diseases are abnormal development and stunted growth. One viroid disease, called cadang-cadang, has killed more than 10 million coconut palms in the Philippines.

An important lesson from viroids is that a single molecule can be an infectious agent that spreads a disease. But viroids are nucleic acids, whose ability to be replicated is well known. Even more surprising is the evidence for infectious *proteins*, called **prions**, which appear to cause a number of degenerative brain diseases in various animal species. These diseases include scrapie in sheep; mad cow disease, which has plagued the European beef industry in recent years; and Creutzfeldt-Jakob disease in humans, which has caused the death of some 150 people in Great Britain over the past decade. Prions are most likely transmitted in food, as may occur when people eat prion-laden beef from cattle with mad cow disease. Kuru, another human disease caused by prions, was identified in the early 1900s among the South Fore natives of New Guinea. A kuru epidemic peaked there in the 1960s, puzzling scientists, who at first thought the disease had a genetic basis. Eventually, however, anthropological investigations ferreted out how the disease was spread: ritual cannibalism, a widespread practice among South Fore natives at that time.

Two characteristics of prions are especially alarming. First, prions act very

▶ **Figure 19.11 Model for how prions propagate.** Prions are misfolded versions of normal brain proteins. When a prion contacts a normally folded version of the same protein, it may induce the normal protein to assume the abnormal shape. The resulting chain reaction may continue until high levels of prion aggregation cause cellular malfunction and eventual degeneration of the brain.

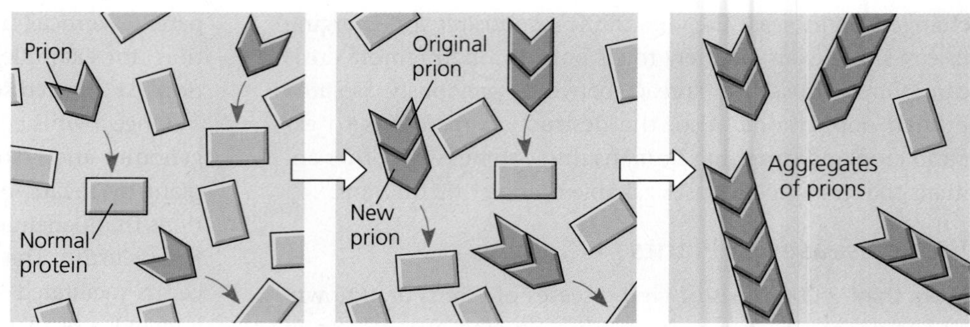

Prion
Normal protein
Original prion
New prion
Aggregates of prions

slowly, with an incubation period of at least ten years before symptoms develop. The lengthy incubation period prevents sources of infection from being identified until long after the first cases appear, allowing many more infections to occur. Second, prions are virtually indestructible; they are not destroyed or deactivated by heating to normal cooking temperatures. To date, there is no known cure for prion diseases, and the only hope for developing effective treatments lies in understanding the process of infection.

How can a protein, which cannot replicate itself, be a transmissible pathogen? According to the leading model, a prion is a misfolded form of a protein normally present in brain cells. When the prion gets into a cell containing the normal form of the protein, the prion somehow converts normal protein molecules to the misfolded prion versions. Several prions then aggregate into a complex that can convert other normal proteins to prions, which join the chain (**Figure 19.11**). Prion aggregation interferes with normal cellular functions and causes disease symptoms. This model was greeted with much skepticism when it was first proposed by Stanley Prusiner in the early 1980s, but it is now widely accepted. Prusiner was awarded the Nobel Prize in 1997 for his work on prions.

CONCEPT CHECK 19.3

1. Describe two ways a preexisting virus can become an emerging virus.
2. Contrast horizontal and vertical transmission of viruses in plants.
3. **WHAT IF?** TMV has been isolated from virtually all commercial tobacco products. Why, then, is TMV infection not an additional hazard for smokers?

For suggested answers, see Appendix A.

19 CHAPTER REVIEW

SUMMARY OF KEY CONCEPTS

CONCEPT 19.1

A virus consists of a nucleic acid surrounded by a protein coat (pp. 381–384)

- Researchers discovered viruses in the late 1800s by studying a plant disease, tobacco mosaic disease.
- A **virus** is a small nucleic acid genome enclosed in a protein **capsid** and sometimes a membranous **viral envelope** containing viral proteins that help viruses enter cells. The genome may be single- or double-stranded DNA or RNA.

? *Are viruses generally considered living or nonliving? Explain.*

CONCEPT 19.2

Viruses replicate only in host cells (pp. 384–390)

- Viruses use enzymes, ribosomes, and small molecules of host cells to synthesize progeny viruses during replication. Each type of virus has a characteristic **host range**.
- **Phages** (viruses that infect bacteria) can replicate by two alternative mechanisms: the **lytic cycle** and the **lysogenic cycle**.

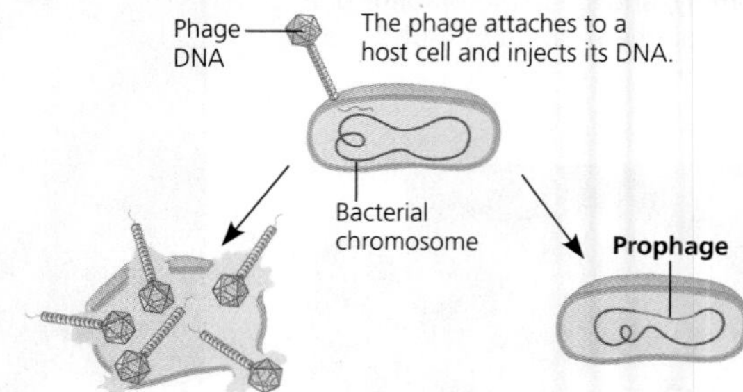

Phage DNA
The phage attaches to a host cell and injects its DNA.
Bacterial chromosome
Prophage
Lytic cycle
• Virulent or temperate phage
• Destruction of host DNA
• Production of new phages
• Lysis of host cell causes release of progeny phages
Lysogenic cycle
• Temperate phage only
• Genome integrates into bacterial chromosome as prophage, which (1) is replicated and passed on to daughter cells and (2) can be induced to leave the chromosome and initiate a lytic cycle

- Many animal viruses have an envelope. **Retroviruses** (such as **HIV**) use the enzyme **reverse transcriptase** to copy their RNA genome into DNA, which can be integrated into the host genome as a **provirus**.

- Since viruses can replicate only within cells, they probably evolved after the first cells appeared, perhaps as packaged fragments of cellular nucleic acid. The origin of viruses is still being debated.

? *Describe enzymes that are not found in most cells but are necessary for the replication of viruses of certain types.*

CONCEPT 19.3

Viruses, viroids, and prions are formidable pathogens in animals and plants (pp. 390–394)

- Symptoms of viral diseases may be caused by direct viral harm to cells or by the body's immune response. **Vaccines** stimulate the immune system to defend the host against specific viruses.
- Outbreaks of "new" viral diseases in humans are usually caused by existing viruses that expand their host territory. The H1N1 2009 flu virus was a new combination of pig, human, and avian viral genes that caused a pandemic. The H5N1 avian flu virus has the potential to cause a high-mortality flu pandemic.
- Viruses enter plant cells through damaged cell walls (horizontal transmission) or are inherited from a parent (vertical transmission).
- **Viroids** are naked RNA molecules that infect plants and disrupt their growth. **Prions** are slow-acting, virtually indestructible infectious proteins that cause brain diseases in mammals.

? *What aspect of an RNA virus makes it more likely than a DNA virus to become an emerging virus?*

TEST YOUR UNDERSTANDING

LEVEL 1: KNOWLEDGE/COMPREHENSION

1. Which of the following characteristics, structures, or processes is common to both bacteria and viruses?
a. metabolism
b. ribosomes
c. genetic material composed of nucleic acid
d. cell division
e. independent existence

2. Emerging viruses arise by
a. mutation of existing viruses.
b. the spread of existing viruses to new host species.
c. the spread of existing viruses more widely within their host species.
d. all of the above
e. none of the above

3. To cause a human pandemic, the H5N1 avian flu virus would have to
a. spread to primates such as chimpanzees.
b. develop into a virus with a different host range.
c. become capable of human-to-human transmission.
d. arise independently in chickens in North and South America.
e. become much more pathogenic.

LEVEL 2: APPLICATION/ANALYSIS

4. A bacterium is infected with an experimentally constructed bacteriophage composed of the T2 phage protein coat and T4 phage DNA. The new phages produced would have
a. T2 protein and T4 DNA.
b. T2 protein and T2 DNA.
c. a mixture of the DNA and proteins of both phages.
d. T4 protein and T4 DNA.
e. T4 protein and T2 DNA.

5. RNA viruses require their own supply of certain enzymes because
a. host cells rapidly destroy the viruses.
b. host cells lack enzymes that can replicate the viral genome.
c. these enzymes translate viral mRNA into proteins.
d. these enzymes penetrate host cell membranes.
e. these enzymes cannot be made in host cells.

6. **DRAW IT** Redraw Figure 19.7 to show the replicative cycle of a virus with a single-stranded genome that can function as mRNA (a class IV virus).

LEVEL 3: SYNTHESIS/EVALUATION

7. EVOLUTION CONNECTION
The success of some viruses lies in their ability to evolve rapidly within the host. Such a virus evades the host's defenses by mutating and producing many altered progeny viruses before the body can mount an attack. Thus, the viruses present late in infection differ from those that initially infected the body. Discuss this as an example of evolution in microcosm. Which viral lineages tend to predominate?

8. SCIENTIFIC INQUIRY
When bacteria infect an animal, the number of bacteria in the body increases in an exponential fashion (graph A). After infection by a virulent animal virus with a lytic replicative cycle, there is no evidence of infection for a while. Then the number of viruses rises suddenly and subsequently increases in a series of steps (graph B). Explain the difference in the curves.

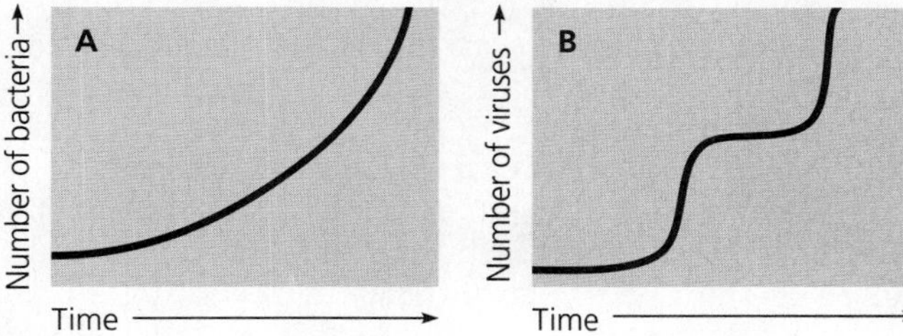

9. 

Structure and Function While viruses are considered by most scientists to be nonliving, they do show some characteristics of life, including the correlation of structure and function. In a short essay (100–150 words), discuss how the structure of a virus correlates with its function.

For selected answers, see Appendix A.

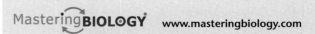

20

Biotechnology

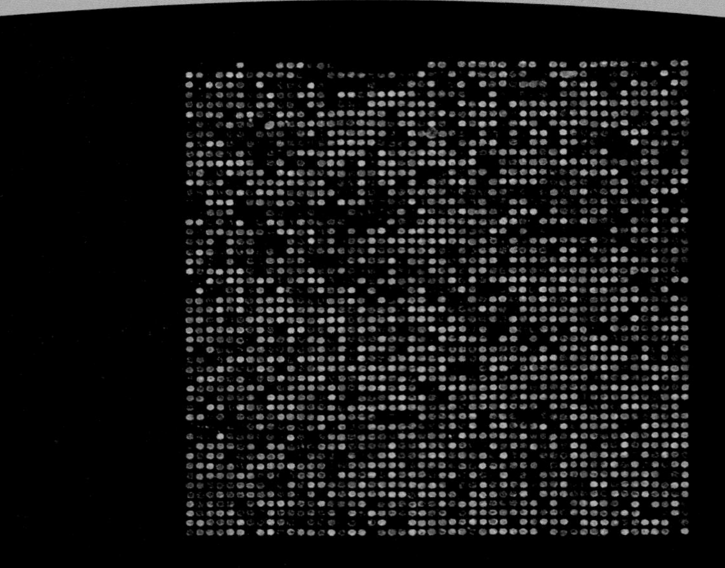

▲ **Figure 20.1 How can this array of spots be used to compare normal and cancerous tissues?**

KEY CONCEPTS

- **20.1** DNA cloning yields multiple copies of a gene or other DNA segment
- **20.2** DNA technology allows us to study the sequence, expression, and function of a gene
- **20.3** Cloning organisms may lead to production of stem cells for research and other applications
- **20.4** The practical applications of DNA technology affect our lives in many ways

OVERVIEW

The DNA Toolbox

In 2001, a major scientific milestone was announced: Researchers had completed a "first draft" sequence of all 3 billion base pairs of the human genome—only the fourth eukaryotic genome to be sequenced. This news electrified the scientific community. Few among them would have dared to dream that a mere nine years later, genome sequencing would be under way for more than 7,000 species. By 2010, researchers had completed sequencing more than 1,000 bacterial, 80 archaeal, and 100 eukaryotic genomes, with many more in progress.

Ultimately, these achievements are attributable to advances in DNA technology—methods of working with and manipulating DNA—that had their roots in the 1970s. A key accomplishment was the invention of techniques for making **recombinant DNA**, DNA molecules formed when segments of DNA from two different sources—often different species—are combined *in vitro* (in a test tube). This advance gave rise to the development of powerful techniques for analyzing genes and gene expression. How scientists prepare recombinant DNA and use DNA technology to answer fundamental biological questions are one focus of this chapter. In the next chapter (Chapter 21), we'll see how these techniques have allowed the sequencing of whole genomes, and we'll consider what we've learned from these sequences about the evolution of species and of the genome itself.

Another focus of this chapter is how our lives are affected by **biotechnology**, the manipulation of organisms or their components to make useful products. Biotechnology includes such early practices as selective breeding of farm animals and using microorganisms to make wine and cheese. Today, biotechnology also encompasses **genetic engineering**, the direct manipulation of genes for practical purposes. Genetic engineering has launched a revolution in biotechnology, greatly expanding the scope of its potential applications. Tools from the DNA toolbox are now applied in ways that affect everything from agriculture to criminal law to medical research. For instance, on the DNA microarray in **Figure 20.1**, the colored spots represent the relative level of expression of 2,400 human genes in normal and cancerous tissue. Using microarray analysis, researchers can quickly compare gene expression in different samples, such as those tested here. The knowledge gained from such gene expression studies is making a significant contribution to the study of cancer and other diseases.

In this chapter, we'll first describe the main techniques for manipulating DNA and analyzing gene expression and function. Next, we'll explore advances in cloning organisms and producing stem cells, techniques that have both expanded our basic understanding of biology and enhanced our ability to apply this understanding to global problems. Finally, we'll survey the practical applications of biotechnology and consider some of the social and ethical issues that arise as biotechnology becomes more pervasive in our lives.

CONCEPT 20.1

DNA cloning yields multiple copies of a gene or other DNA segment

The molecular biologist studying a particular gene faces a challenge. Naturally occurring DNA molecules are very long, and a single molecule usually carries many genes. Moreover, in many eukaryotic genomes, genes occupy only a small proportion of the chromosomal DNA, the rest being noncoding nucleotide sequences. A single human gene, for example, might constitute only 1/100,000 of a chromosomal DNA molecule. As a further complication, the distinctions between a gene and the surrounding DNA are subtle, consisting only of differences in nucleotide sequence. To work directly with specific genes, scientists have developed methods for preparing well-defined segments of DNA in multiple identical copies, a process called *DNA cloning*.

DNA Cloning and Its Applications: *A Preview*

Most methods for cloning pieces of DNA in the laboratory share certain general features. One common approach uses bacteria, most often *Escherichia coli*. Recall from Figure 16.12 that the *E. coli* chromosome is a large circular molecule of DNA. In addition, *E. coli* and many other bacteria have **plasmids**, small circular DNA molecules that replicate separately from the bacterial chromosome. A plasmid has only a small number of genes; these genes may be useful when the bacterium is in a particular environment but may not be required for survival or reproduction under most conditions.

To clone pieces of DNA in the laboratory, researchers first obtain a plasmid (originally isolated from a bacterial cell and genetically engineered for efficient cloning) and insert DNA from another source ("foreign" DNA) into it **(Figure 20.2)**. The resulting plasmid is now a recombinant DNA molecule. The plasmid is then returned to a bacterial cell, producing a *recombinant bacterium*. This single cell reproduces through repeated cell divisions to form a clone of cells, a population of genetically identical cells. Because the dividing bacteria replicate the recombinant plasmid and pass it on to their descendants, the foreign DNA and any genes it carries are cloned at the same time. The production of multiple copies of a single gene is called **gene cloning**.

Gene cloning is useful for two basic purposes: to make many copies of, or *amplify*, a particular gene and to produce a protein product. Researchers can isolate copies of a cloned gene from bacteria for use in basic research or to endow an organism with a new metabolic capability, such as pest resistance. For example, a resistance gene present in one crop species might be cloned and transferred into plants of another species. Alternatively, a protein with medical uses, such as human growth hormone, can be harvested in large quantities from cultures of bacteria carrying the cloned gene for the protein.

A single gene is usually a very small part of the total DNA in a cell. For example, a typical gene makes up only about one-millionth of the DNA in a human cell. The ability to

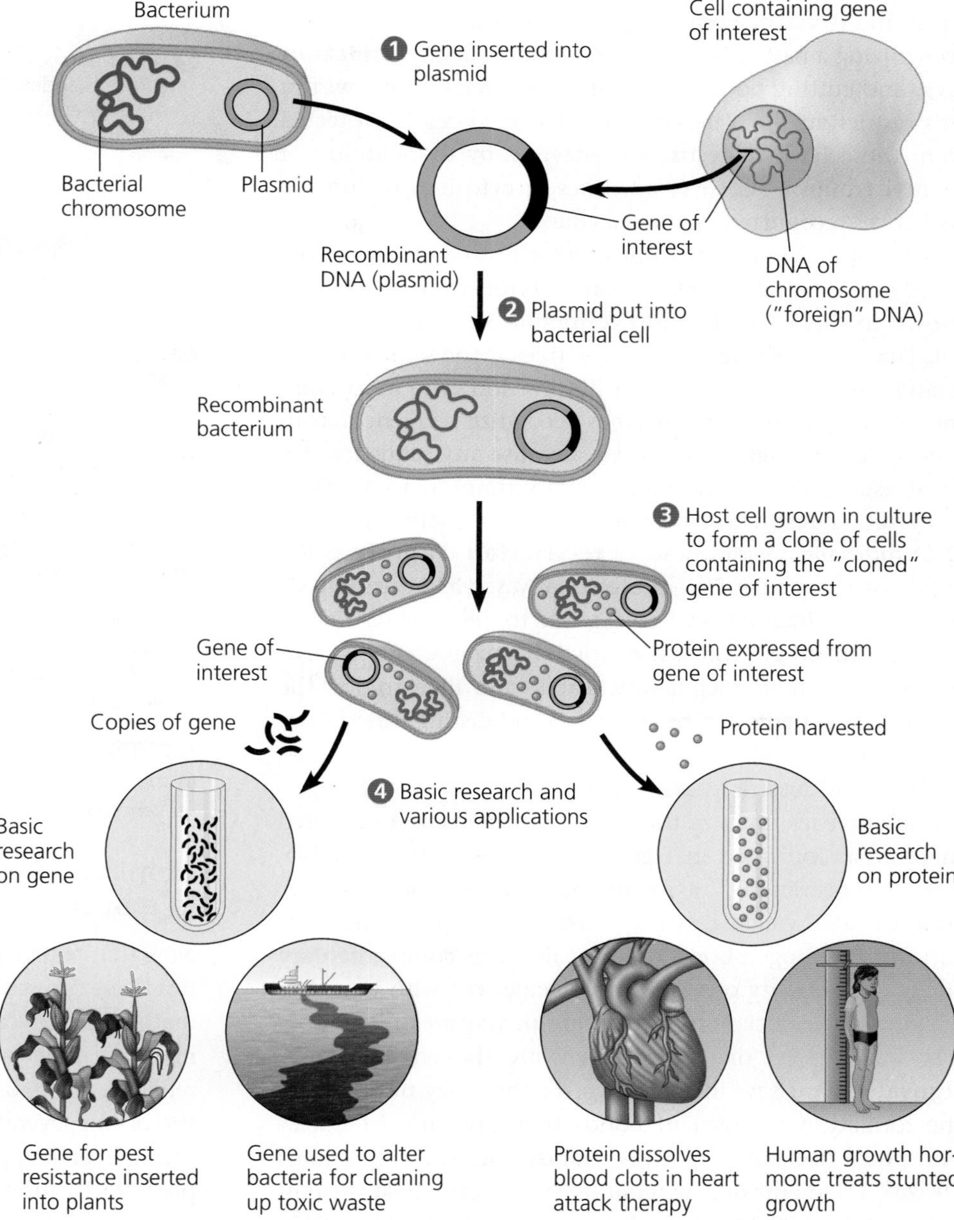

▲ **Figure 20.2 A preview of gene cloning and some uses of cloned genes.** In this simplified diagram of gene cloning, we start with a plasmid (originally isolated from a bacterial cell) and a gene of interest from another organism. Only one plasmid and one copy of the gene of interest are shown at the top of the figure, but the starting materials would include many of each.

amplify such rare DNA fragments is therefore crucial for any application involving a single gene.

Using Restriction Enzymes to Make Recombinant DNA

Gene cloning and genetic engineering rely on the use of enzymes that cut DNA molecules at a limited number of specific locations. These enzymes, called restriction endonucleases, or **restriction enzymes**, were discovered in the late 1960s by biologists doing basic research on bacteria. Restriction enzymes protect the bacterial cell by cutting up foreign DNA from other organisms or phages (see Chapter 19).

Hundreds of different restriction enzymes have been identified and isolated. Each restriction enzyme is very specific, recognizing a particular short DNA sequence, or **restriction site**, and cutting both DNA strands at precise points within this restriction site. The DNA of a bacterial cell is protected from the cell's own restriction enzymes by the addition of methyl groups (—CH_3) to adenines or cytosines within the sequences recognized by the enzymes.

The top of **Figure 20.3** illustrates a restriction site recognized by a particular restriction enzyme from *E. coli.* As shown in this example, most restriction sites are symmetrical. That is, the sequence of nucleotides is the same on both strands when read in the 5′ → 3′ direction. The most commonly used restriction enzymes recognize sequences containing four to eight nucleotides. Because any sequence this short usually occurs (by chance) many times in a long DNA molecule, a restriction enzyme will make many cuts in a DNA molecule, yielding a set of **restriction fragments**. All copies of a particular DNA molecule always yield the same set of restriction fragments when exposed to the same restriction enzyme. In other words, a restriction enzyme cuts a DNA molecule in a reproducible way. (Later you will learn how the different fragments can be separated and distinguished from each other.)

The most useful restriction enzymes cleave the sugar-phosphate backbones in the two DNA strands in a staggered manner, as indicated in Figure 20.3. The resulting double-stranded restriction fragments have at least one single-stranded end, called a **sticky end**. These short extensions can form hydrogen-bonded base pairs with complementary sticky ends on any other DNA molecules cut with the same enzyme. The associations formed in this way are only temporary but can be made permanent by the enzyme **DNA ligase**. As you saw in Figure 16.16, this enzyme catalyzes the formation of covalent bonds that close up the sugar-phosphate backbones of DNA strands; for example, it joins Okazaki fragments during replication. You can see at the bottom of Figure 20.3 that the ligase-catalyzed joining of DNA from two different sources produces a stable recombinant DNA molecule.

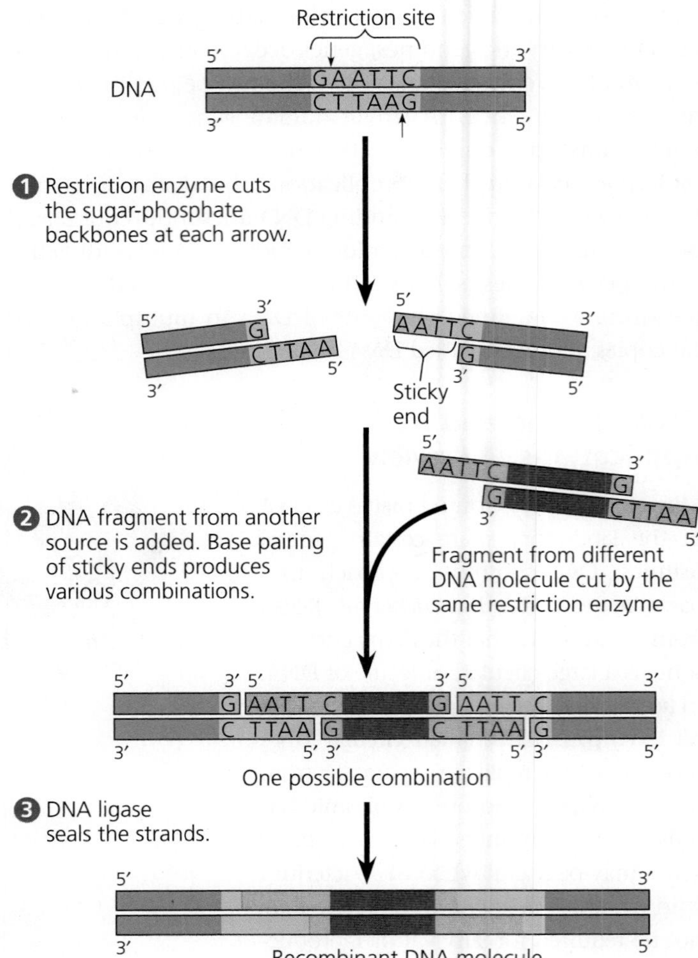

Restriction site
DNA
5′ GAATTC 3′
3′ CTTAAG 5′
1 Restriction enzyme cuts the sugar-phosphate backbones at each arrow.
5′ G 3′
3′ CTTAA 5′
5′ AATTC 3′
3′ G 5′
Sticky end
2 DNA fragment from another source is added. Base pairing of sticky ends produces various combinations.
Fragment from different DNA molecule cut by the same restriction enzyme
G AATT C G AATT C
C TTAA G C TTAA G
One possible combination
3 DNA ligase seals the strands.
Recombinant DNA molecule

▲ **Figure 20.3 Using a restriction enzyme and DNA ligase to make recombinant DNA.** The restriction enzyme in this example (called *Eco*RI) recognizes a specific six-base-pair sequence, the restriction site, and makes staggered cuts in the sugar-phosphate backbones within this sequence, producing fragments with sticky ends. Any fragments with complementary sticky ends can base-pair, including the two original fragments. If the fragments come from different DNA molecules, the ligated product is recombinant DNA.

DRAW IT *The restriction enzyme* Hind*III recognizes the sequence 5′-AAGCTT-3′, cutting between the two 'A's. Draw the double-stranded sequence before and after the enzyme cuts.*

Cloning a Eukaryotic Gene in a Bacterial Plasmid

Now that you've learned about restriction enzymes and DNA ligase, we can see how genes are cloned in plasmids. The original plasmid is called a **cloning vector**, defined as a DNA molecule that can carry foreign DNA into a host cell and replicate there. Bacterial plasmids are widely used as cloning vectors for several reasons: They can be readily obtained from commercial suppliers, manipulated to form recombinant plasmids by insertion of foreign DNA *in vitro*, and then introduced into bacterial cells. Moreover, recombinant bacterial plasmids (and the foreign DNA they carry) multiply rapidly owing to the high reproductive rate of their host cells.

Producing Clones of Cells Carrying Recombinant Plasmids

Let's say we are researchers interested in studying the β-globin gene in a particular species of hummingbird. We start by cloning all the hummingbird genes; later we'll isolate the β-globin gene from all the others, a task very much like finding a needle in a haystack. **Figure 20.4** details one method for cloning hummingbird genes using a bacterial plasmid as the cloning vector.

❶ First, we isolate hummingbird genomic DNA from hummingbird cells. We also obtain our chosen vector, a particular bacterial plasmid from *E. coli* cells. The plasmid has been engineered to carry two genes that will later prove useful: amp^R, which makes *E. coli* cells resistant to the antibiotic ampicillin, and *lacZ*, which encodes the enzyme β-galactosidase, which hydrolyzes lactose (see p. 354). This enzyme can also hydrolyze a similar synthetic molecule called X-gal to form a blue product. The plasmid contains only one copy of the restriction site recognized by the restriction enzyme used in the next step, and that site is within the *lacZ* gene.

❷ Both the plasmid and the hummingbird DNA are cut with the same restriction enzyme, and then ❸ the fragments are mixed together, allowing base pairing between their complementary sticky ends. We then add DNA ligase, which covalently bonds the sugar-phosphate backbones of the fragments whose sticky ends have base-paired. Many of the resulting recombinant plasmids contain single hummingbird DNA fragments (three are shown in Figure 20.4), and at least one of them is expected to carry all or part of the β-globin gene. This step will also generate other products, such as plasmids containing multiple hummingbird DNA fragments, a combination of two plasmids, or a rejoined, nonrecombinant version of the original plasmid.

❹ The DNA mixture is then added to bacteria that have a mutation in the *lacZ* gene on their own chromosome, making them unable to hydrolyze

▼ **Figure 20.4** RESEARCH METHOD

Cloning Genes in Bacterial Plasmids

APPLICATION Gene cloning is a process that produces many copies of a gene of interest. These copies can be used in sequencing the gene, in producing its encoded protein, or in basic research or other applications.

TECHNIQUE In this example, hummingbird genes are inserted into plasmids from *E. coli*. Only three plasmids and three hummingbird DNA fragments are shown, but millions of copies of the plasmid and a mixture of millions of different hummingbird DNA fragments would be present in the samples.

❶ Obtain engineered plasmid DNA and DNA from hummingbird cells. The hummingbird DNA contains the gene of interest.

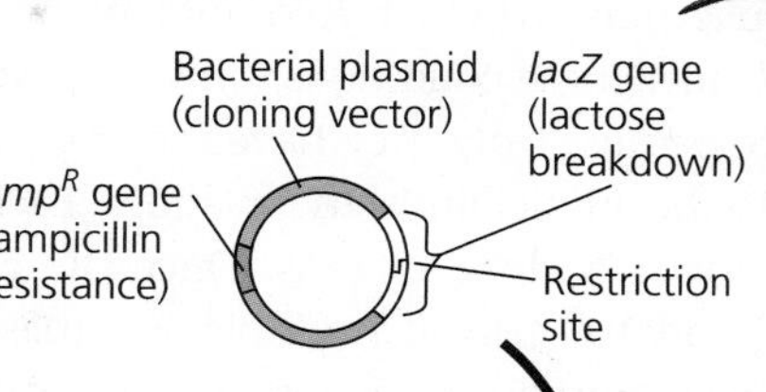

❷ Cut both DNA samples with the same restriction enzyme, one that makes a single cut within the *lacZ* gene and many cuts within the hummingbird DNA.

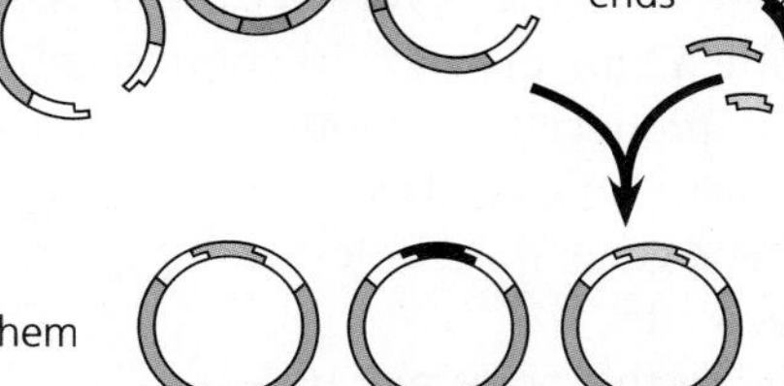

❸ Mix the cut plasmids and DNA fragments. Some join by base pairing; add DNA ligase to seal them together. The products include recombinant plasmids and many nonrecombinant plasmids.

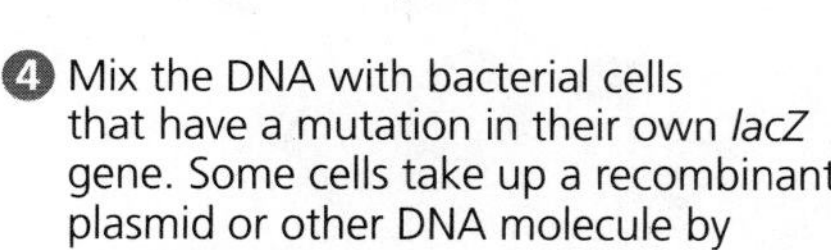

❹ Mix the DNA with bacterial cells that have a mutation in their own *lacZ* gene. Some cells take up a recombinant plasmid or other DNA molecule by transformation.

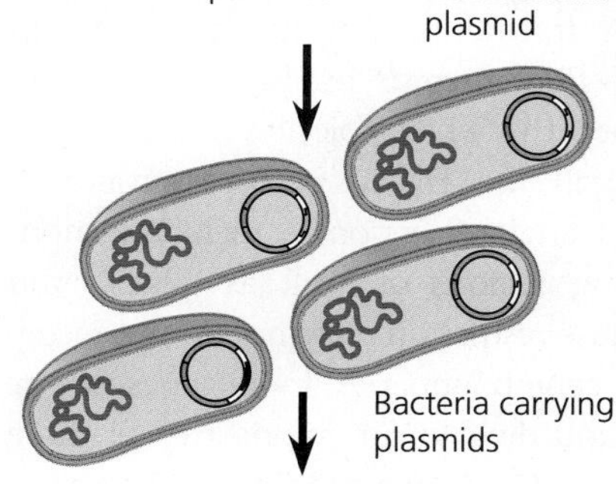

❺ Plate the bacteria on nutrient-containing agar medium supplemented with ampicillin and X-gal, a molecule resembling lactose. Incubate until colonies grow.

RESULTS Only a cell that took up a plasmid, which has the amp^R gene, will reproduce and form a colony. Colonies with nonrecombinant plasmids will be blue because they can hydrolyze X-gal, forming a blue product. Colonies with recombinant plasmids, in which *lacZ* is disrupted, will be white because they cannot hydrolyze X-gal.

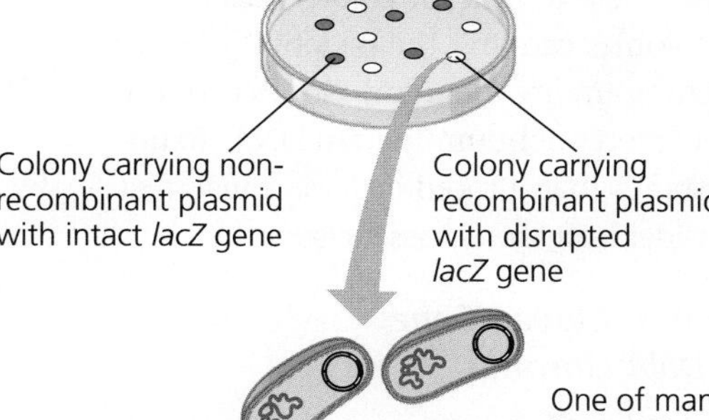

WHAT IF? If the medium used in step 5 did not contain ampicillin, what other colonies would grow? What color would they be?

lactose or X-gal. Under suitable experimental conditions, the cells take up foreign DNA by transformation (see p. 306). Some cells acquire a recombinant plasmid carrying a gene, while others may take up a nonrecombinant plasmid, a fragment of noncoding hummingbird DNA, or nothing at all. The *amp*^{R} and *lacZ* genes on the plasmid can help us sort out these possibilities.

❺ First, plating out all the bacteria on solid nutrient medium containing ampicillin allows us to distinguish the cells that have taken up plasmids, whether recombinant or not, from the other cells. Under these conditions, only cells with a plasmid will reproduce because only they have the *amp*^{R} gene conferring resistance to the ampicillin in the medium. Each reproducing bacterium forms a clone of cells. Once the clone contains between 10^5 and 10^8 cells, it is visible as a mass, or *colony*, on the agar. As cells reproduce, any foreign genes carried by recombinant plasmids are also copied (cloned).

Second, the presence of X-gal in the medium allows us to distinguish colonies with recombinant plasmids from those with nonrecombinant plasmids. Colonies containing nonrecombinant plasmids have the *lacZ* gene intact and will produce functional β-galactosidase. These colonies will be blue because the enzyme hydrolyzes the X-gal in the medium, forming a blue product. In contrast, no functional β-galactosidase is produced in colonies containing recombinant plasmids with foreign DNA inserted into the *lacZ* gene; these colonies will therefore be white.

The procedure to this point will have cloned many different hummingbird DNA fragments, not just the β-globin gene that interests us. In fact, taken together, the white colonies should represent all the DNA sequences from the hummingbird genome, including noncoding regions as well as genes. And because restriction enzymes do not recognize gene boundaries, some genes will be cut and divided up among two or more clones. Shortly, we will discuss the procedure we use to find the colony (cell clone) or colonies carrying the β-globin gene sequences among the many clones carrying other pieces of hummingbird DNA. To understand that procedure, we must first consider how the clones are stored.

Storing Cloned Genes in DNA Libraries

The cloning procedure in Figure 20.4, which starts with a mixture of fragments from the entire genome of an organism, is called a "shotgun" approach; no single gene is targeted for cloning. Numerous different recombinant plasmids are produced in step 3, and a clone of cells carrying each type of plasmid ends up as a white colony in step 5. The complete set of plasmid-containing cell clones, each carrying copies of a particular segment from the initial genome, is referred to as a **genomic library** (**Figure 20.5a**). Each "plasmid clone" in the library is like a book containing specific information. Today, scientists often obtain such libraries (or even particular cloned genes) from another researcher, a commercial source, or a sequencing center.

Historically, certain bacteriophages have also been used as cloning vectors for making genomic libraries. Fragments of foreign DNA can be spliced into a trimmed-down version of a phage genome, as into a plasmid, by using a restriction enzyme and DNA ligase. The normal infection process allows production of many new phage particles, each carrying the foreign DNA insert. Today, phages are generally used for making genomic libraries only in special cases.

Another type of vector widely used in library construction is a **bacterial artificial chromosome (BAC)**. In spite of the name, these are simply large plasmids, trimmed down so they contain just the genes necessary to ensure replication. An advantage of using BACs as vectors is that while a standard plasmid can carry a DNA insert no larger than 10,000 base pairs (10 kb), a BAC can carry an insert of 100–300 kb

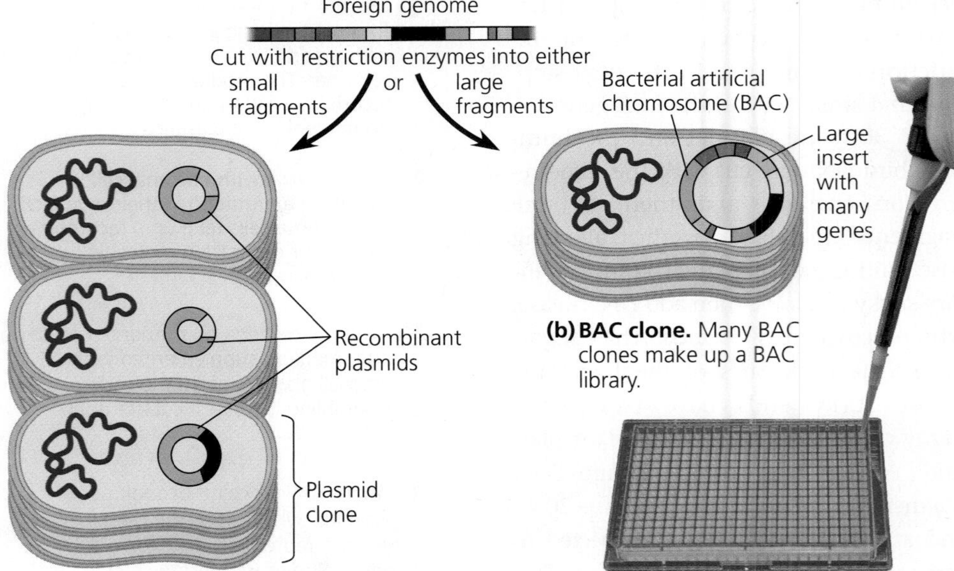

(a) Plasmid library. Shown are three of the thousands of "books" in a plasmid library. Each "book" is a clone of bacterial cells, which contain copies of a particular foreign genome fragment (pink, yellow, black segments) in their recombinant plasmids.

(b) BAC clone. Many BAC clones make up a BAC library.

(c) Storing genome libraries. Both plasmid and BAC genomic libraries are usually stored in a "multiwell" plastic plate; a 384-well plate is shown here. Each clone occupies one well. (The library of an entire genome would require many such plates.)

▲ **Figure 20.5 Genomic libraries.** A genomic library is a collection of many clones. Each clone carries copies of a particular DNA segment from a foreign genome, integrated into an appropriate DNA vector, such as a plasmid or a bacterial artificial chromosome (BAC). In a complete genomic library, the foreign DNA segments cover the entire genome of an organism. Note that the bacterial chromosomes are not drawn to scale; they are actually about 1,000 times larger than the vectors.

(Figure 20.5b). The very large insert size minimizes the number of clones needed to make up the genomic library, but it also makes them more challenging to work with in the lab, so the insert may later be cut up into smaller pieces that are "subcloned" into plasmid vectors.

Clones are usually stored in multiwelled plastic plates, with one clone per well **(Figure 20.5c)**. This orderly storage of clones, identified by their location in the plate, makes screening for the gene of interest very efficient, as you will see.

In a genomic library, the cloned β-globin gene would include not just exons containing the coding sequence, but also the promoter, untranslated regions, and any introns. Some biologists might be interested in the β-globin protein itself—they might wonder, for instance, if this oxygen-carrying protein is different from its counterpart in other, less metabolically active species. Such researchers can make another kind of DNA library by starting with fully processed mRNA extracted from cells where the gene is expressed **(Figure 20.6)**. The enzyme reverse transcriptase (obtained from retroviruses) is used *in vitro* to make a single-stranded DNA *reverse transcript* of each mRNA molecule. Recall that the 3′ end of the mRNA has a stretch of adenine (A)

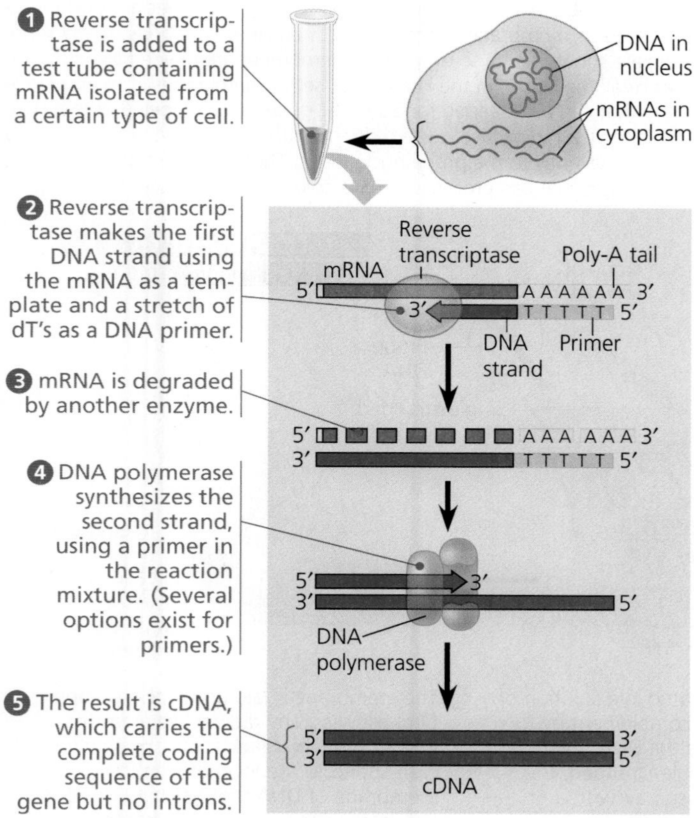

1 Reverse transcriptase is added to a test tube containing mRNA isolated from a certain type of cell.
DNA in nucleus
mRNAs in cytoplasm
2 Reverse transcriptase makes the first DNA strand using the mRNA as a template and a stretch of dT's as a DNA primer.
Reverse transcriptase
Poly-A tail
mRNA
5′ AAAAAA 3′
3′ TTTTT 5′
DNA strand
Primer
3 mRNA is degraded by another enzyme.
5′ AAA AAA 3′
3′ TTTTT 5′
4 DNA polymerase synthesizes the second strand, using a primer in the reaction mixture. (Several options exist for primers.)
5′ 3′
3′ 5′
DNA polymerase
5 The result is cDNA, which carries the complete coding sequence of the gene but no introns.
5′ 3′
3′ 5′
cDNA

▲ **Figure 20.6 Making complementary DNA (cDNA) from eukaryotic genes.** Complementary DNA is DNA made *in vitro* using mRNA as a template for the first strand. Because the mRNA contains only exons, the resulting double-stranded cDNA carries the complete coding sequence of the gene but no introns. Although only one mRNA is shown here, the final collection of cDNAs would reflect all the mRNAs that were present in the cell.

ribonucleotides called a poly-A tail. This feature allows use of a short strand of thymine deoxyribonucleotides (dT's) as a primer for the reverse transcriptase. Following enzymatic degradation of the mRNA, a second DNA strand, complementary to the first, is synthesized by DNA polymerase. The resulting double-stranded DNA is called **complementary DNA (cDNA)**. To create a library, the researchers must now modify the cDNA by adding restriction enzyme recognition sequences at each end. Then the cDNA is inserted into vector DNA in a manner similar to the insertion of genomic DNA fragments. The extracted mRNA is a mixture of all the mRNA molecules in the original cells, transcribed from many different genes. Therefore, the cDNAs that are cloned make up a **cDNA library** containing a collection of genes. However, a cDNA library represents only part of the genome—only the subset of genes that were transcribed in the cells from which the mRNA was isolated.

Genomic and cDNA libraries each have advantages, depending on what is being studied. If you want to clone a gene but don't know what cell type expresses it or cannot obtain enough cells of the appropriate type, a genomic library is almost certain to contain the gene. Also, if you are interested in the regulatory sequences or introns associated with a gene, a genomic library is necessary, since these sequences are absent from mRNAs used in making a cDNA library. On the other hand, to study a specific protein (like β-globin), a cDNA library made from cells expressing the gene (like red blood cells) is ideal. A cDNA library can also be used to study sets of genes expressed in particular cell types, such as brain or liver cells. Finally, by making cDNA from cells of the same type at different times in the life of an organism, researchers can trace changes in patterns of gene expression during development.

Screening a Library for Clones Carrying a Gene of Interest

Now, returning to the results in Figure 20.4, we're ready to screen all the colonies with recombinant plasmids (the white colonies) for a clone of cells containing the hummingbird β-globin gene. We can detect this gene's DNA by its ability to base-pair with a complementary sequence on another nucleic acid molecule, using **nucleic acid hybridization**. The complementary molecule, a short, single-stranded nucleic acid that can be either RNA or DNA, is called a **nucleic acid probe**. If we know at least part of the nucleotide sequence of the gene of interest (perhaps from knowing the amino acid sequence of the protein it encodes or, as in our case, the gene's nucleotide sequence in a closely related species), we can synthesize a probe complementary to it. For example, if part of the sequence on one strand of the desired gene were

5′ ···CTCATCACCGGC··· 3′

then we would synthesize this probe:

3′ GAGTAGTGGCCG 5′

Each probe molecule, which will hydrogen-bond specifically to a complementary sequence in the desired gene, is labeled with a radioactive isotope, a fluorescent tag, or another molecule so we can track it.

Recall that the clones in our hummingbird genomic library have been stored in a multiwell plate (see Figure 20.5c). If we transfer a few cells from each well to a defined location on a membrane made of nylon or nitrocellulose, we can screen a large number of clones simultaneously for the presence of DNA complementary to our DNA probe (**Figure 20.7**).

After we've identified the location of a clone carrying the β-globin gene, we can grow some cells from that colony in liquid culture in a large tank and then easily isolate many copies of the gene for our studies. We can also use the cloned gene as a probe to identify similar or identical genes in DNA from other sources, such as other species of birds.

Expressing Cloned Eukaryotic Genes

Once a particular gene has been cloned in host cells, its protein product can be produced in large amounts for research purposes or valuable practical applications, which we'll explore in Concept 20.4. Cloned genes can be expressed as protein in either bacterial or eukaryotic cells; each option has advantages and disadvantages.

Bacterial Expression Systems

Getting a cloned eukaryotic gene to function in bacterial host cells can be difficult because certain aspects of gene expression are different in eukaryotes and bacteria. To overcome differences in promoters and other DNA control sequences, scientists usually employ an **expression vector**, a cloning vector that contains a highly active bacterial promoter just upstream of a restriction site where the eukaryotic gene can be inserted in the correct reading frame. The bacterial host cell will recognize the promoter and proceed to express the foreign gene now linked to that promoter. Such expression vectors allow the synthesis of many eukaryotic proteins in bacterial cells.

▼ Figure 20.7 RESEARCH METHOD

Detecting a Specific DNA Sequence by Hybridization with a Nucleic Acid Probe

APPLICATION Hybridization with a complementary nucleic acid probe detects a specific DNA sequence within a mixture of DNA molecules. In this example, a collection of bacterial clones from a hummingbird genomic library is screened to identify clones that carry a recombinant plasmid bearing the gene of interest. The library is stored in many multiwell plates, with one clone per well (see Figure 20.5c).

TECHNIQUE Cells from each clone are applied to a special nylon membrane. Each membrane has room for thousands of clones (many more than are shown here), so only a few membranes are needed to hold samples of all the clones in the library. This set of membranes is an *arrayed library* that can be screened for a specific gene using a labeled probe. Here the label is a radioactive nucleotide, but other labels are also commonly linked covalently to the probe nucleotides. These include fluorescent tags or enzymes that can produce either a colored or luminescent product.

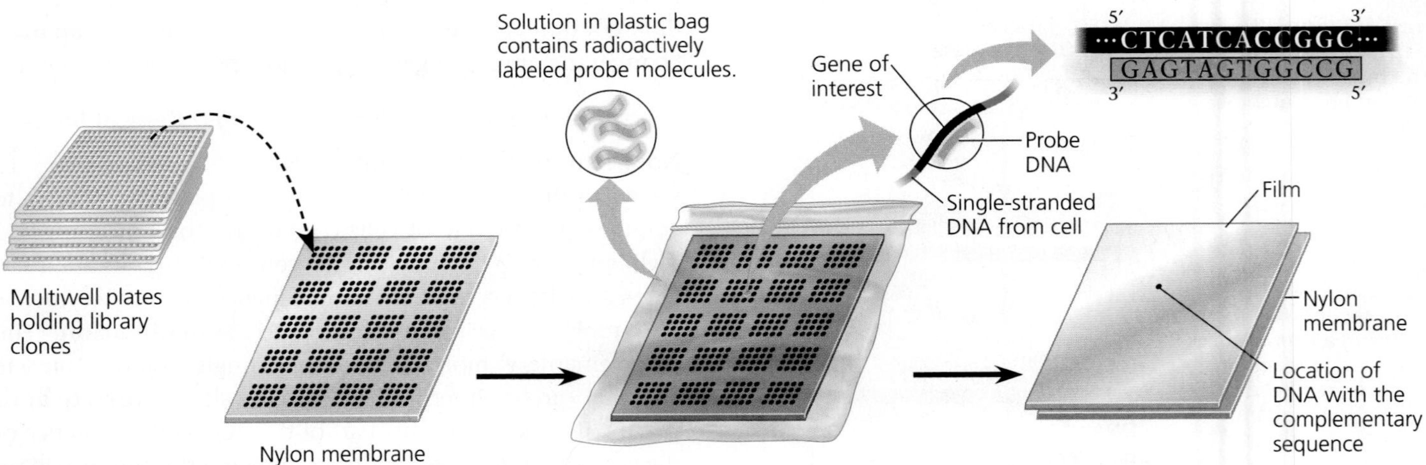

❶ Plate by plate, cells from each well, representing one clone, are transferred to a defined spot on a special nylon membrane. The nylon membrane is treated to break open the cells and denature their DNA; the resulting single-stranded DNA molecules stick to the membrane.

❷ The membrane is then incubated in a solution of radioactive probe molecules complementary to the gene of interest. Because the DNA immobilized on the membrane is single-stranded, the single-stranded probe can base-pair with any complementary DNA on the membrane. Excess DNA is then rinsed off. (One spot with radioactive probe–DNA hybrids is shown here in orange but would not be distinguishable yet.)

❸ The membrane is laid under photographic film, allowing any radioactive areas to expose the film. Black spots on the film correspond to the locations on the membrane of DNA that has hybridized to the probe. Each spot can be traced back to the original well containing the bacterial clone that holds the gene of interest.

RESULTS For a radioactive probe, the location of the black spot on a piece of photographic film identifies the clone containing the gene of interest. (Probes labeled in other ways use other detection systems.) By using probes with different nucleotide sequences in different experiments, researchers can screen the collection of bacterial clones for different genes.

Another problem with expressing cloned eukaryotic genes in bacteria is the presence of noncoding regions (introns) in most eukaryotic genes. Introns can make a eukaryotic gene very long and unwieldy, and they prevent correct expression of the gene by bacterial cells, which do not have RNA-splicing machinery. This problem can be surmounted by using a cDNA form of the gene, which includes only the exons.

Eukaryotic Cloning and Expression Systems

Molecular biologists can avoid eukaryotic-bacterial incompatibility by using eukaryotic cells such as yeasts, rather than bacteria, as hosts for cloning and/or expressing eukaryotic genes of interest. Yeasts, single-celled fungi, offer two advantages: They are as easy to grow as bacteria, and they have plasmids, a rarity among eukaryotes. Scientists have even constructed recombinant plasmids that combine yeast and bacterial DNA and can replicate in either type of cell.

Another reason to use eukaryotic host cells for expressing a cloned eukaryotic gene is that many eukaryotic proteins will not function unless they are modified after translation, for example, by the addition of carbohydrate (glycosylation) or lipid groups. Bacterial cells cannot carry out these modifications, and if the gene product requiring such processing is from a mammal, even yeast cells may not be able to modify the protein correctly. Several cultured cell types have proved successful as host cells for this purpose, including some mammalian cell lines and an insect cell line that can be infected by a virus (called baculovirus) carrying recombinant DNA.

Besides using vectors, scientists have developed a variety of other methods for introducing recombinant DNA into eukaryotic cells. In **electroporation**, a brief electrical pulse applied to a solution containing cells creates temporary holes in their plasma membranes, through which DNA can enter. (This technique is now commonly used for bacteria as well.) Alternatively, scientists can inject DNA directly into single eukaryotic cells using microscopically thin needles. To get DNA into plant cells, the soil bacterium *Agrobacterium* can be used, as well as other methods you will learn about later. If the introduced DNA is incorporated into a cell's genome by genetic recombination, then it may be expressed by the cell.

Cross-Species Gene Expression and Evolutionary Ancestry

EVOLUTION The ability to express eukaryotic proteins in bacteria (even if the proteins aren't glycosylated properly) is quite remarkable when we consider how different eukaryotic and bacterial cells are. Examples abound of genes that are taken from one species and function perfectly well when transferred into another very different species. These observations underscore the shared evolutionary ancestry of species living today.

One example involves a gene called *Pax-6*, which has been found in animals as diverse as vertebrates and fruit flies. The vertebrate *Pax-6* gene product (the PAX-6 protein) triggers a complex program of gene expression resulting in formation of the vertebrate eye, which has a single lens. The fly *Pax-6* gene leads to formation of the compound fly eye, which is quite different from the vertebrate eye. When scientists cloned the mouse *Pax-6* gene and introduced it into a fly embryo, they were surprised to see that it led to formation of a compound fly eye (see Figure 50.16). Conversely, when the fly *Pax-6* gene was transferred into a vertebrate embryo—a frog, in this case—a frog eye formed. Although the genetic programs triggered in vertebrates and flies generate very different eyes, both versions of the *Pax-6* gene can substitute for each other, evidence of their evolution from a gene in a common ancestor.

Simpler examples are seen in Figure 17.6, where a firefly gene is expressed in a tobacco plant, and a jellyfish gene product is seen in a pig. The basic mechanisms of gene expression have ancient evolutionary roots, which is the basis of many recombinant DNA techniques described in this chapter.

Amplifying DNA *in Vitro*: The Polymerase Chain Reaction (PCR)

DNA cloning in cells remains the best method for preparing large quantities of a particular gene or other DNA sequence. However, when the source of DNA is scanty or impure, the **polymerase chain reaction**, or **PCR**, is quicker and more selective. In this technique, any specific target segment within one or many DNA molecules can be quickly amplified in a test tube. With automation, PCR can make billions of copies of a target segment of DNA in a few hours, significantly faster than the days it would take to obtain the same number of copies by screening a DNA library for a clone with the desired gene and letting it replicate within host cells. In fact, PCR is being used increasingly to make enough of a specific DNA fragment to insert it directly into a vector, entirely skipping the steps of making and screening a library. To continue our literary analogy, PCR is like photocopying a few pages rather than checking out a book from the library.

In the PCR procedure, a three-step cycle brings about a chain reaction that produces an exponentially growing population of identical DNA molecules. During each cycle, the reaction mixture is heated to denature (separate) the DNA strands and then cooled to allow annealing (hydrogen bonding) of short, single-stranded DNA primers complementary to sequences on opposite strands at each end of the target sequence; finally, a heat-stable DNA polymerase extends the primers in the $5' \rightarrow 3'$ direction. If a standard DNA polymerase were used, the protein would be denatured along with the DNA during the first heating step and would have to be replaced after each cycle. The key to automating PCR was the discovery of an unusual heat-stable DNA polymerase called Taq polymerase, named after the bacterial species from which it was first isolated. This bacterial species, *Thermus aquaticus*, lives in hot springs, so natural selection has resulted in a heat-stable DNA polymerase that can withstand the heat at the start of each cycle. **Figure 20.8** illustrates the steps in PCR.

Just as impressive as the speed of PCR is its specificity. Only minute amounts of DNA need be present in the starting material, and this DNA can be in a partially degraded state, as long as a few molecules contain the complete target sequence. The key to this high specificity is the primers, which hydrogen-bond *only* to sequences at opposite ends of the target segment. (For high specificity, the primers must be at least 15 or so nucleotides long.) By the end of the third cycle, one-fourth of the molecules are identical to the target segment, with both strands the appropriate length. With each successive cycle, the number of target segment molecules of the correct length doubles, so the number of molecules equals 2^n, where n is the number of cycles. After 30 more cycles, about a billion copies of the target sequence are present!

Despite its speed and specificity, PCR amplification cannot substitute for gene cloning in cells when large amounts of a gene are desired. Occasional errors during PCR replication impose limits on the number of good copies that can be made by this method. When PCR is used to provide the specific DNA fragment for cloning, the resulting clones are sequenced to select clones with error-free inserts. PCR errors also impose limits on the length of DNA fragments that can be copied.

Devised in 1985, PCR has had a major impact on biological research and biotechnology. PCR has been used to amplify DNA from a wide variety of sources: fragments of ancient DNA from a 40,000-year-old frozen woolly mammoth; DNA from fingerprints or from tiny amounts of blood, tissue, or semen found at crime scenes; DNA from single embryonic cells for rapid prenatal diagnosis of genetic disorders; and DNA of viral genes from cells infected with viruses that are difficult to detect, such as HIV. We'll return to applications of PCR later in the chapter.

▼ Figure 20.8 **RESEARCH METHOD**

The Polymerase Chain Reaction (PCR)

APPLICATION With PCR, any specific segment—the target sequence—within a DNA sample can be copied many times (amplified), completely *in vitro*.

TECHNIQUE PCR requires double-stranded DNA containing the target sequence, a heat-resistant DNA polymerase, all four nucleotides, and two 15- to 20-nucleotide DNA strands that serve as primers. One primer is complementary to one end of the target sequence on one strand; the second primer is complementary to the other end of the sequence on the other strand.

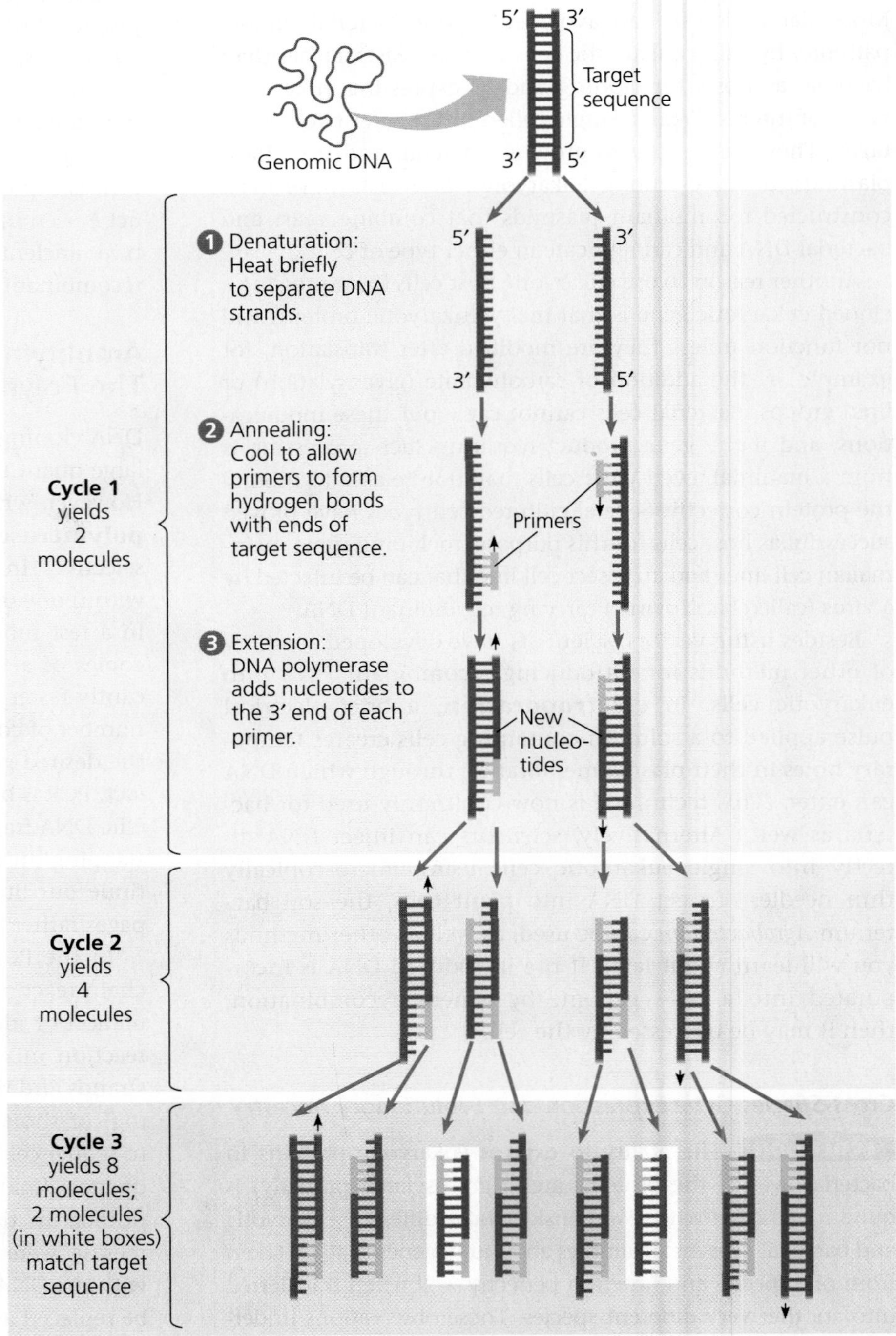

RESULTS After 3 cycles, two molecules match the target sequence exactly. After 30 more cycles, over 1 billion (10^9) molecules match the target sequence.

CONCEPT CHECK 20.1

1. The restriction site for an enzyme called *PvuI* is the following sequence:

   5′-C G A T C G-3′
   3′-G C T A G C-5′

   Staggered cuts are made between the T and C on each strand. What type of bonds are being cleaved?
2. **DRAW IT** One strand of a DNA molecule has the following sequence: 5′-CCTTGACGATCGTTACCG-3′. Draw the other strand. Will *PvuI* cut this molecule? If so, draw the products.
3. What are some potential difficulties in using plasmid vectors and bacterial host cells to produce large quantities of proteins from cloned eukaryotic genes?
4. **MAKE CONNECTIONS** Compare Figure 20.8 with Figure 16.20 (p. 319). How does replication of DNA ends during PCR proceed without shortening the fragments each time?

For suggested answers, see Appendix A.

CONCEPT 20.2

DNA technology allows us to study the sequence, expression, and function of a gene

Once DNA cloning has provided us with large quantities of specific DNA segments, we can tackle some interesting questions about a particular gene and its function. For example, does the sequence of the hummingbird β-globin gene suggest a protein structure that can carry oxygen more efficiently than its counterpart in less metabolically active species? Does a particular human gene differ from person to person, and are certain alleles of that gene associated with a hereditary disorder? Where in the body and when is a given gene expressed? And, ultimately, what role does a certain gene play in an organism?

Before we can begin to address such compelling questions, we must consider a few standard laboratory techniques that are used to analyze the DNA of genes.

Gel Electrophoresis and Southern Blotting

Many approaches for studying DNA molecules involve **gel electrophoresis**. This technique uses a gel made of a polymer, such as the polysaccharide *agarose*. The gel acts as a molecular sieve to separate nucleic acids or proteins on the basis of size, electrical charge, and other physical properties (**Figure 20.9**). Because nucleic acid molecules carry negative charges on their phosphate groups, they all travel toward the positive pole in an

▼ **Figure 20.9** RESEARCH METHOD

Gel Electrophoresis

APPLICATION Gel electrophoresis is used for separating nucleic acids or proteins that differ in size, electrical charge, or other physical properties. DNA molecules are separated by gel electrophoresis in restriction fragment analysis of both cloned genes (see Figure 20.10) and genomic DNA (see Figure 20.11).

TECHNIQUE Gel electrophoresis separates macromolecules on the basis of their rate of movement through an agarose gel in an electric field: The distance a DNA molecule travels is inversely proportional to its length. A mixture of DNA molecules, usually fragments produced by restriction enzyme digestion (cutting) or PCR amplification, is separated into bands. Each band contains thousands of molecules of the same length.

1. Each sample, a mixture of DNA molecules, is placed in a separate well near one end of a thin slab of agarose gel. The gel is set into a small plastic support and immersed in an aqueous, buffered solution in a tray with electrodes at each end.

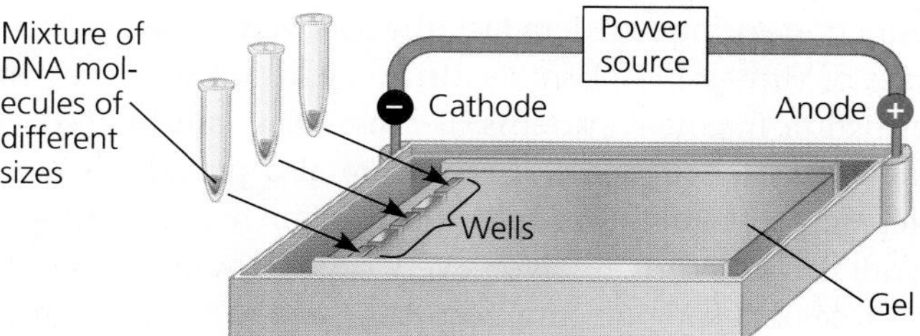

2. When the current is turned on, the negatively charged DNA molecules move toward the positive electrode, with shorter molecules moving faster than longer ones. Bands are shown here in blue, but in an actual gel, the bands would not be visible at this time.

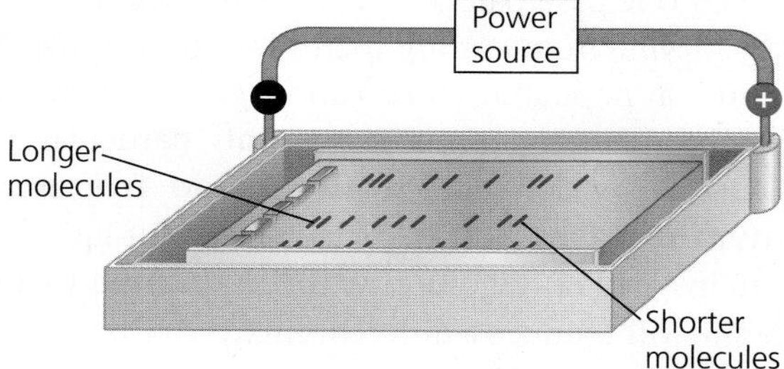

RESULTS After the current is turned off, a DNA-binding dye (ethidium bromide) is added. This dye fluoresces pink in ultraviolet light, revealing the separated bands to which it binds. In the gel below, the pink bands correspond to DNA fragments of different lengths separated by electrophoresis. If all the samples were initially cut with the same restriction enzyme, then the different band patterns indicate that they came from different sources.

electric field. As they move, the thicket of agarose fibers impedes longer molecules more than it does shorter ones, separating them by length. Thus, gel electrophoresis separates a mixture of linear DNA molecules into bands, each band consisting of many thousands of DNA molecules of the same length.

One historically useful application of this technique has been *restriction fragment analysis*, which rapidly provides information about DNA sequences. With advances in sequencing technology, the approach taken in labs today is often simply to sequence the DNA sample in question. However, restriction fragment analysis is still done in some cases, and understanding how it is done will give you a better grasp of recombinant DNA technology. In this type of analysis, the DNA fragments produced by restriction enzyme digestion (cutting) of a DNA molecule are separated by gel electrophoresis. When the mixture of restriction fragments undergoes electrophoresis, it yields a band pattern characteristic of the starting molecule and the restriction enzyme used. In fact, the relatively small DNA molecules of viruses and plasmids can be identified simply by their restriction fragment patterns. Because DNA can be recovered undamaged from gels, the procedure also provides a way to prepare pure samples of individual fragments—assuming the bands can be clearly resolved. (Very large DNA molecules, such as those of eukaryotic chromosomes, yield so many fragments that they appear as a smear instead of distinct bands.)

Restriction fragment analysis can be used to compare two different DNA molecules—for example, two alleles of a gene—if the nucleotide difference affects a restriction site. A change in even one base pair of that sequence will prevent a restriction enzyme from cutting there. Variations in DNA sequence among a population are called *polymorphisms* (from the Greek for "many forms"), and this particular type of sequence change is called a **restriction fragment length polymorphism** (**RFLP**, pronounced "Rif-lip"). If one allele contains a RFLP, digestion with the enzyme that recognizes the site will produce a different mixture of fragments for each of the two alleles. Each mixture will give its own band pattern in gel electrophoresis.

For example, sickle-cell disease is caused by mutation of a single nucleotide located within a restriction sequence (a RFLP) in the human β-globin gene (see pp. 277–278 and Figure 17.23). Consequently, while other assays are preferred today, restriction fragment analysis was used for many years to distinguish the normal and sickle-cell alleles of the β-globin gene, as shown in **Figure 20.10**.

The starting materials in Figure 20.10 are samples of the cloned and purified β-globin alleles. But how could this test be done if we didn't have purified alleles to start with? If we wanted to determine whether a person is a heterozygous carrier of the mutant allele for sickle-cell disease, we would directly compare the genomic DNA from that person with DNA from both a person who has sickle-cell disease (and is homozygous for the mutant allele) and a person who is

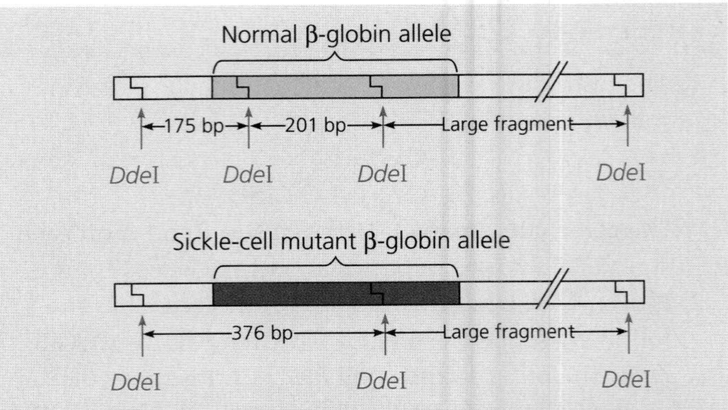

(a) *Dde*I restriction sites in normal and sickle-cell alleles of the β-globin gene. Shown here are the cloned alleles, separated from the vector DNA but including some DNA next to the coding sequence. The normal allele contains two sites within the coding sequence recognized by the *Dde*I restriction enzyme. The sickle-cell allele lacks one of these sites.

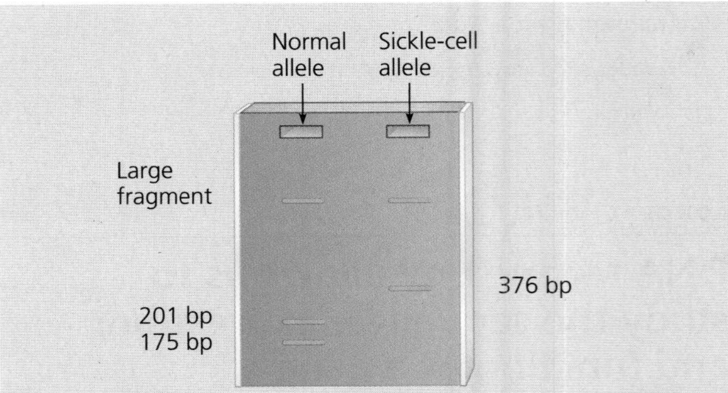

(b) Electrophoresis of restriction fragments from normal and sickle-cell alleles. Samples of each purified allele were cut with the *Dde*I enzyme and then subjected to gel electrophoresis, resulting in three bands for the normal allele and two bands for the sickle-cell allele. (The tiny fragments on the ends of both initial DNA molecules are identical and are not seen here.)

▲ **Figure 20.10 Using restriction fragment analysis to distinguish the normal and sickle-cell alleles of the human β-globin gene. (a)** The sickle-cell mutation destroys one of the *Dde*I restriction sites within the gene. **(b)** As a result, digestion with the *Dde*I enzyme generates different sets of fragments from the normal and sickle-cell alleles.

WHAT IF? *Given bacterial clones with recombinant plasmids carrying each of these alleles, how would you isolate the pure DNA samples run on the gel in (b)? (Hint: Study Figures 20.4 and 20.9.)*

homozygous for the normal allele. As we mentioned already, electrophoresis of genomic DNA digested with a restriction enzyme and stained with a DNA-binding dye yields too many bands to distinguish them individually. However, a classic method called **Southern blotting** (developed by British biochemist Edwin Southern), which combines gel electrophoresis and nucleic acid hybridization, allows us to detect just those bands that include parts of the β-globin gene. The principle is the same as in nucleic acid hybridization for screening bacterial clones (see Figure 20.7). In Southern blotting, the probe is usually a radioactively or otherwise labeled single-stranded DNA molecule that is complementary to the gene of interest. **Figure 20.11** outlines the entire procedure and demonstrates how it can differentiate a heterozygote (in this case, for the sickle-cell allele) from an individual homozygous for the normal allele.

The identification of carriers of mutant alleles associated with genetic diseases is only one of the ways Southern blotting has been used. In fact, this technique was a laboratory workhorse for many years. Recently, however, it has been supplanted by more rapid methods, often involving PCR amplification of the specific parts of genomes that may differ.

DNA Sequencing

Once a gene is cloned, its complete nucleotide sequence can be determined. Today, sequencing is automated, carried out by sequencing machines (see Figure 1.12). The first automated procedure was based on a technique called the *dideoxyribonucleotide*

▼ Figure 20.11 RESEARCH METHOD

Southern Blotting of DNA Fragments

APPLICATION Researchers can detect specific nucleotide sequences within a complex DNA sample with this method. In particular, Southern blotting can be used to compare the restriction fragments produced from different samples of genomic DNA.

TECHNIQUE In this example, we compare genomic DNA samples from three individuals: a homozygote for the normal β-globin allele (I), a homozygote for the mutant sickle-cell allele (II), and a heterozygote (III). As in Figure 20.7, we show a radioactively labeled probe, but other methods of probe labeling and detection are also used.

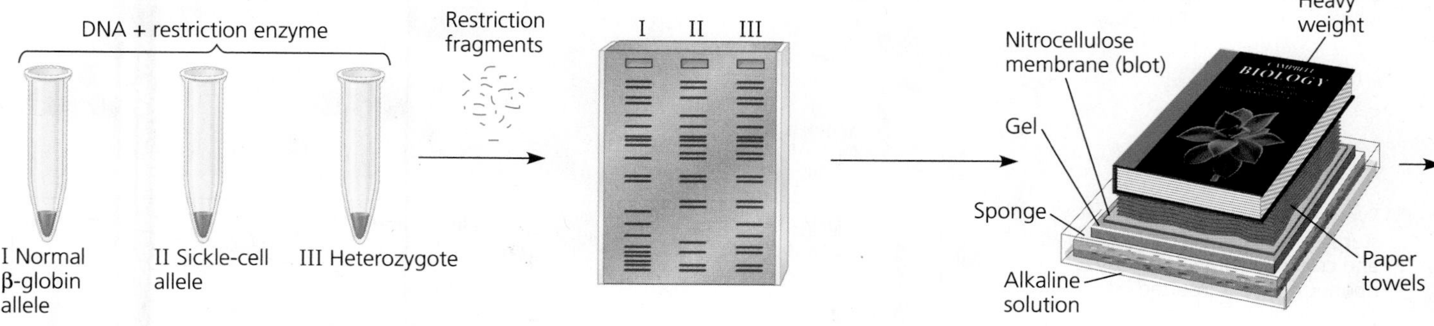

1 Preparation of restriction fragments. Each DNA sample is mixed with the same restriction enzyme, in this case *Dde*I. Digestion of each sample yields a mixture of thousands of restriction fragments.

2 Gel electrophoresis. The restriction fragments in each sample are separated by electrophoresis, forming a characteristic pattern of bands. (In reality, there would be many more bands than shown here, and they would be invisible unless stained.)

3 DNA transfer (blotting). With the gel arranged as shown above, capillary action pulls the alkaline solution upward through the gel, denaturing and transferring the DNA to a nitrocellulose membrane. This produces a blot with a pattern of DNA bands exactly like that of the gel.

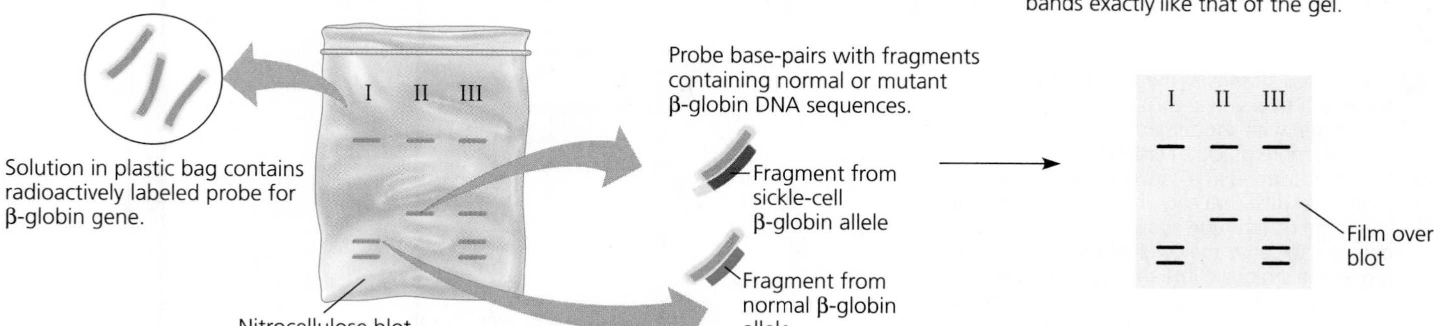

4 Hybridization with labeled probe. The nitrocellulose blot is exposed to a solution containing a probe labeled in some way. In this example, the probe is radioactively labeled, single-stranded DNA complementary to the β-globin gene. Probe molecules attach by base-pairing to any restriction fragments containing a part of the β-globin gene. (The bands would not be visible yet.)

5 Probe detection. A sheet of photographic film is laid over the blot. The radioactivity in the bound probe exposes the film to form an image corresponding to those bands containing DNA that base-paired with the probe.

RESULTS The band patterns for the three samples are clearly different, so this method can be used to identify heterozygous carriers of the sickle-cell allele (III), as well as those with the disease, who have two mutant alleles (II), and unaffected individuals, who have two normal alleles (I). Band patterns for samples I and II resemble those seen for the purified normal and mutant alleles, respectively, seen in Figure 20.10b. The band pattern for the sample from the heterozygote (III) is a combination of the patterns for the two homozygotes (I and II).

▼ **Figure 20.12** RESEARCH METHOD

Dideoxy Chain Termination Method for Sequencing DNA

APPLICATION The sequence of nucleotides in any cloned DNA fragment of up to 800–1,000 base pairs in length can be determined rapidly with machines that carry out sequencing reactions and separate the labeled reaction products by length.

TECHNIQUE This method synthesizes a set of DNA strands complementary to the original DNA fragment. Each strand starts with the same primer and ends with a dideoxyribonucleotide (ddNTP), a modified nucleotide. Incorporation of a ddNTP terminates a growing DNA strand because it lacks a 3′ —OH group, the site for attachment of the next nucleotide (see Figure 16.14). In the set of strands synthesized, each nucleotide position along the original sequence is represented by strands ending at that point with the complementary ddNTP. Because each type of ddNTP is tagged with a distinct fluorescent label, the identity of the ending nucleotides of the new strands, and ultimately the entire original sequence, can be determined.

1. The fragment of DNA to be sequenced is denatured into single strands and incubated in a test tube with the necessary ingredients for DNA synthesis: a primer designed to base-pair with the known 3′ end of the template strand, DNA polymerase, the four deoxyribonucleotides, and the four dideoxyribonucleotides, each tagged with a specific fluorescent molecule.

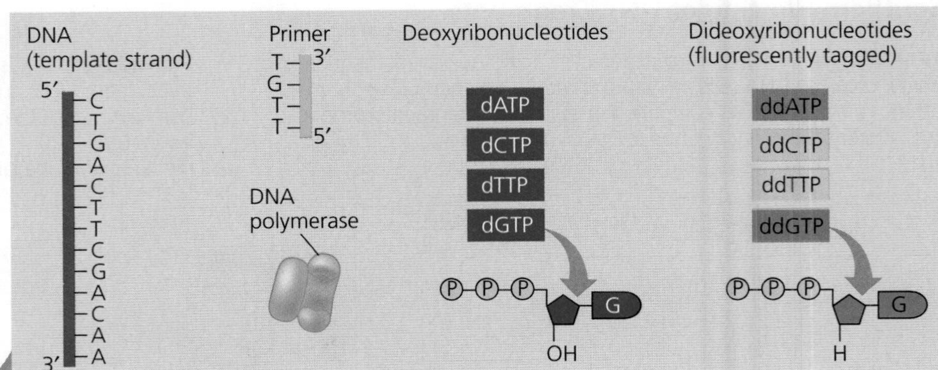

2. Synthesis of each new strand starts at the 3′ end of the primer and continues until a dideoxyribonucleotide is inserted, at random, instead of the normal equivalent deoxyribonucleotide. This prevents further elongation of the strand. Eventually, a set of labeled strands of various lengths is generated, with the color of the tag representing the last nucleotide in the sequence.

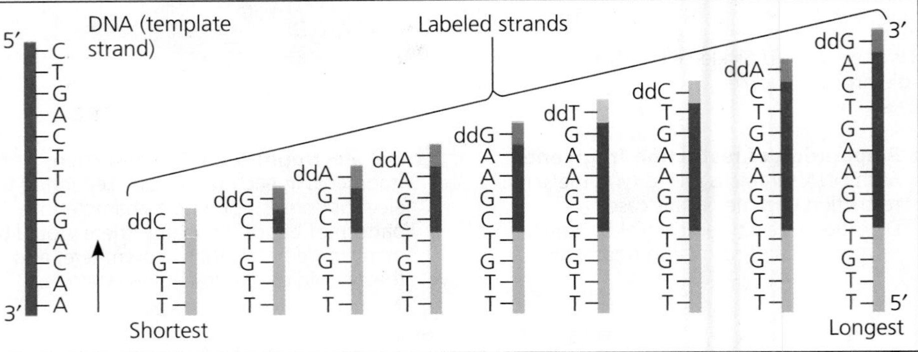

3. The labeled strands in the mixture are separated by passage through a polyacrylamide gel, with shorter strands moving through more quickly. For DNA sequencing, the gel is formed in a capillary tube rather than a slab like that shown in Figure 20.9. The small size of the tube allows a fluorescence detector to sense the color of each fluorescent tag as the strands come through. Strands differing in length by as little as one nucleotide can be distinguished from each other.

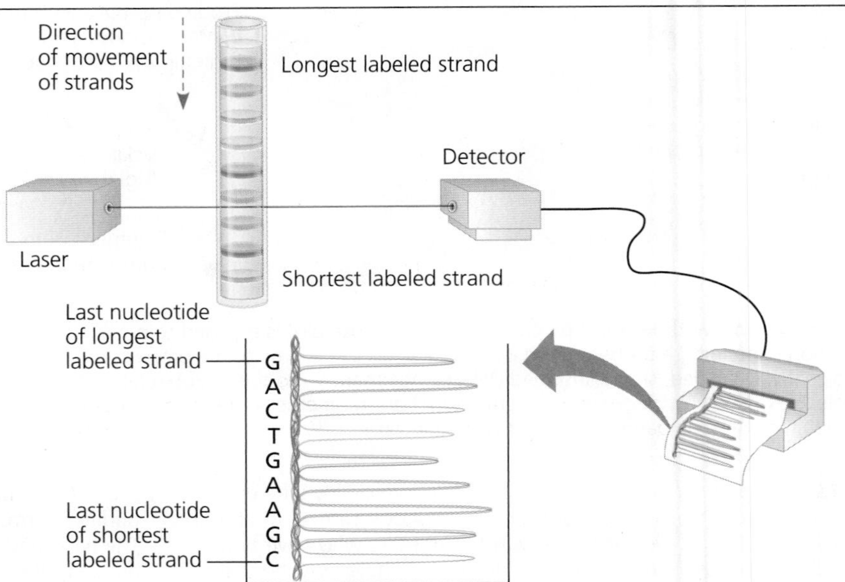

RESULTS The color of the fluorescent tag on each strand indicates the identity of the nucleotide at its end. The results can be printed out as a spectrogram, and the sequence, which is complementary to the template strand, can then be read from bottom (shortest strand) to top (longest strand). (Notice that the sequence here begins after the primer.)

(or *dideoxy*) *chain termination method*, for reasons you can see in **Figure 20.12**. This method was developed by British biochemist Frederick Sanger, who received the Nobel Prize in 1980 for this accomplishment. (One of only four people to win two Nobel Prizes, Sanger also won one in 1975 for determining the amino acid sequence of insulin.)

In the last ten years, "next-generation sequencing" techniques have been developed that do not rely on chain termination. Instead, a single template strand is immobilized, and reagents are added that allow so-called *sequencing by synthesis* of a complementary strand, one nucleotide at a time. A chemical trick enables electronic monitors to identify which of the four nucleotides is added, allowing determination of the sequence. Further technical changes have given rise to "third-generation sequencing," with each new technique being faster and less expensive than the previous. In Chapter 21, you'll learn more about how this rapid acceleration of sequencing technology has enhanced our study of genes and whole genomes.

Knowing the sequence of a gene allows researchers to compare it directly with genes in other species, where the function of the gene product may be known. If two genes from different species are quite similar in sequence, it is reasonable to suppose that their gene products perform similar functions. In this way, sequence comparisons provide clues to a gene's function, a topic we'll return to shortly. Another set of clues is provided by experimental approaches that analyze when and where a gene is expressed.

Analyzing Gene Expression

Having cloned a given gene, researchers can make labeled nucleic acid probes that will hybridize with mRNAs transcribed from the gene. The probes can provide information about when or where in the organism the gene is transcribed. Transcription levels are commonly used as a measure of gene expression.

Studying the Expression of Single Genes

Suppose we want to find out how the expression of the β-globin gene changes during the embryonic development of the hummingbird. There are at least two ways to do this.

The first is called **Northern blotting** (a play on words based on this method's close similarity to Southern blotting). In this method, we carry out gel electrophoresis on samples of mRNA from hummingbird embryos at different stages of development, transfer the samples to a nitrocellulose membrane, and then allow the mRNAs on the membrane to hybridize with a labeled probe recognizing β-globin mRNA. If we expose a film to the membrane, the resulting image will look similar to the Southern blot in Figure 20.11, with one band of a given size showing up in each sample. If the mRNA band is seen at a particular stage, we can hypothesize that the protein functions during events taking place at that stage. Like Southern blotting, Northern blotting has been a mainstay over the years, but it is being supplanted in many labs by other techniques.

A method that is quicker and more sensitive than Northern blotting (because it requires less mRNA) and therefore becoming more widely used is the **reverse transcriptase–polymerase chain reaction**, or **RT-PCR** (**Figure 20.13**). Analysis of hummingbird β-globin gene expression with RT-PCR begins similarly to Northern blotting, with the isolation of mRNAs from different developmental stages of hummingbird embryos. Reverse transcriptase is added next to make cDNA, which then serves as a template for PCR amplification using primers from the β-globin gene. When the products are run on a gel, copies of the amplified region will be observed as bands only in samples that originally contained the β-globin mRNA. In the case of hummingbird β-globin, for instance, we might expect to see a band appear at the stage when red blood cells begin forming, with all subsequent stages

▼ **Figure 20.13** RESEARCH METHOD

RT-PCR Analysis of the Expression of Single Genes

APPLICATION RT-PCR uses the enzyme reverse transcriptase (RT) in combination with PCR and gel electrophoresis. RT-PCR can be used to compare gene expression between samples—for instance, in different embryonic stages, in different tissues, or in the same type of cell under different conditions.

TECHNIQUE In this example, samples containing mRNAs from six embryonic stages of hummingbird were processed as shown below. (The mRNA from only one stage is shown.)

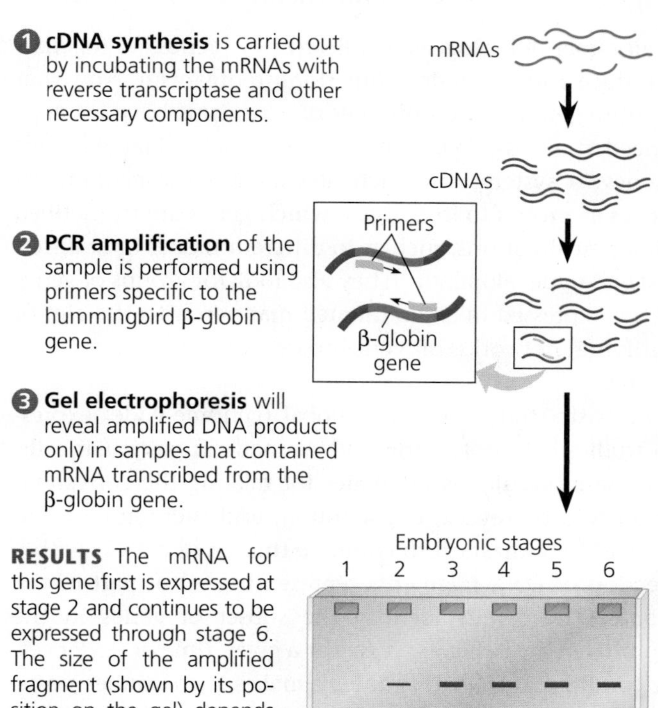

1 **cDNA synthesis** is carried out by incubating the mRNAs with reverse transcriptase and other necessary components.
mRNAs
cDNAs
2 **PCR amplification** of the sample is performed using primers specific to the hummingbird β-globin gene.
Primers
β-globin gene
3 **Gel electrophoresis** will reveal amplified DNA products only in samples that contained mRNA transcribed from the β-globin gene.
Embryonic stages
1 2 3 4 5 6

RESULTS The mRNA for this gene first is expressed at stage 2 and continues to be expressed through stage 6. The size of the amplified fragment (shown by its position on the gel) depends on the distance between the primers that were used.

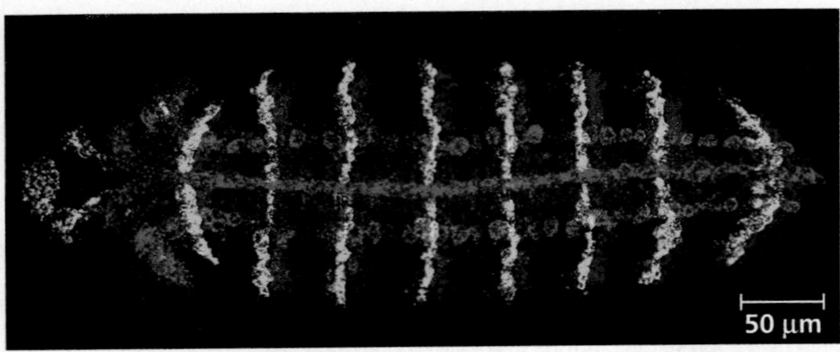

▲ **Figure 20.14 Determining where genes are expressed by *in situ* hybridization analysis.** This *Drosophila* embryo was incubated in a solution containing probes for five different mRNAs, each probe labeled with a different fluorescently colored tag. The embryo was then viewed using fluorescence microscopy. Each color marks where a specific gene is expressed as mRNA.

showing the same band. RT-PCR can also be carried out with mRNAs collected from different tissues at one time to discover which tissue is producing a specific mRNA.

An alternative way to determine which tissues or cells are expressing certain genes is to track down the location of specific mRNAs using labeled probes in place, or *in situ*, in the intact organism. This technique, called ***in situ* hybridization**, is most often carried out with probes labeled by attachment of fluorescent dyes (see Chapter 6). Different probes can be labeled with different dyes, sometimes with strikingly beautiful results (**Figure 20.14**).

Studying the Expression of Interacting Groups of Genes

A major goal of biologists is to learn how genes act together to produce and maintain a functioning organism. Now that the entire genomes of a number of organisms have been sequenced, it is possible to study the expression of large groups of genes—a systems approach. Researchers use genome sequences as probes to investigate which genes are transcribed in different situations, such as in different tissues or at different stages of development. They also look for groups of genes that are expressed in a coordinated manner, with the aim of identifying networks of gene expression across an entire genome.

The basic strategy in such global (genome-wide) expression studies is to isolate the mRNAs made in particular cells, use these molecules as templates for making the corresponding cDNAs by reverse transcription, and then employ nucleic acid hybridization to compare this set of cDNAs with a collection of DNA fragments representing all or part of the genome. The results identify the subset of genes in the genome that are being expressed at a given time or under certain conditions. DNA technology makes such studies possible; with automation, they are easily performed on a large scale. Scientists can now measure the expression of thousands of genes at one time.

Genome-wide expression studies are made possible by **DNA microarray assays**. A DNA microarray consists of tiny amounts of a large number of single-stranded DNA fragments representing different genes fixed to a glass slide in a tightly spaced array, or grid (see Figure 20.1). (The microarray is also called a *DNA chip* by analogy to a computer chip.) Ideally, these fragments represent all the genes of an organism. **Figure 20.15** outlines how the DNA fragments on a microarray are tested for hybridization with cDNA molecules that have been prepared from the mRNAs in particular cells of interest and labeled with fluorescent dyes.

Using this technique, researchers have performed DNA microarray assays on more than 90% of the genes of the nematode *Caenorhabditis elegans* during every stage of its life cycle. The results show that expression of nearly 60% of *C. elegans* genes changes dramatically during development and that many genes are expressed in a sex-specific pattern. This study supports the model held by most developmental biologists that embryonic development involves a complex and elaborate program of gene expression, rather than simply the expression of a small number of important genes. This example illustrates the ability of DNA microarrays to reveal general profiles of gene expression over the lifetime of an organism.

In addition to uncovering gene interactions and providing clues to gene function, DNA microarray assays may contribute to a better understanding of diseases and suggest new diagnostic techniques or therapies. For instance, comparing patterns of gene expression in breast cancer tumors and noncancerous breast tissue has already resulted in more informed and effective treatment protocols. Ultimately, information from DNA microarray assays should provide a grander view of how ensembles of genes interact to form an organism and maintain its vital systems.

Determining Gene Function

How do scientists determine the function of a gene identified by the techniques described thus far in the chapter? Perhaps the most common approach is to disable the gene and then observe the consequences in the cell or organism. In one application of this approach, called ***in vitro* mutagenesis**, specific mutations are introduced into a cloned gene, and then the mutated gene is returned to a cell in such a way that it disables ("knocks out") the normal cellular copies of the same gene. If the introduced mutations alter or destroy the function of the gene product, the phenotype of the mutant cell may help reveal the function of the missing normal protein. Using molecular and genetic techniques worked out in the 1980s, researchers can even generate mice with any given gene disabled, in order to study the role of that gene in development and in the adult. Mario Capecchi, Martin Evans, and Oliver Smithies received the Nobel Prize in 2007 for first accomplishing this feat.

▼ **Figure 20.15** **RESEARCH METHOD**

DNA Microarray Assay of Gene Expression Levels

APPLICATION With this method, researchers can test thousands of genes simultaneously to determine which ones are expressed in a particular tissue, under different environmental conditions, in various disease states, or at different developmental stages. They can also look for coordinated gene expression.

TECHNIQUE

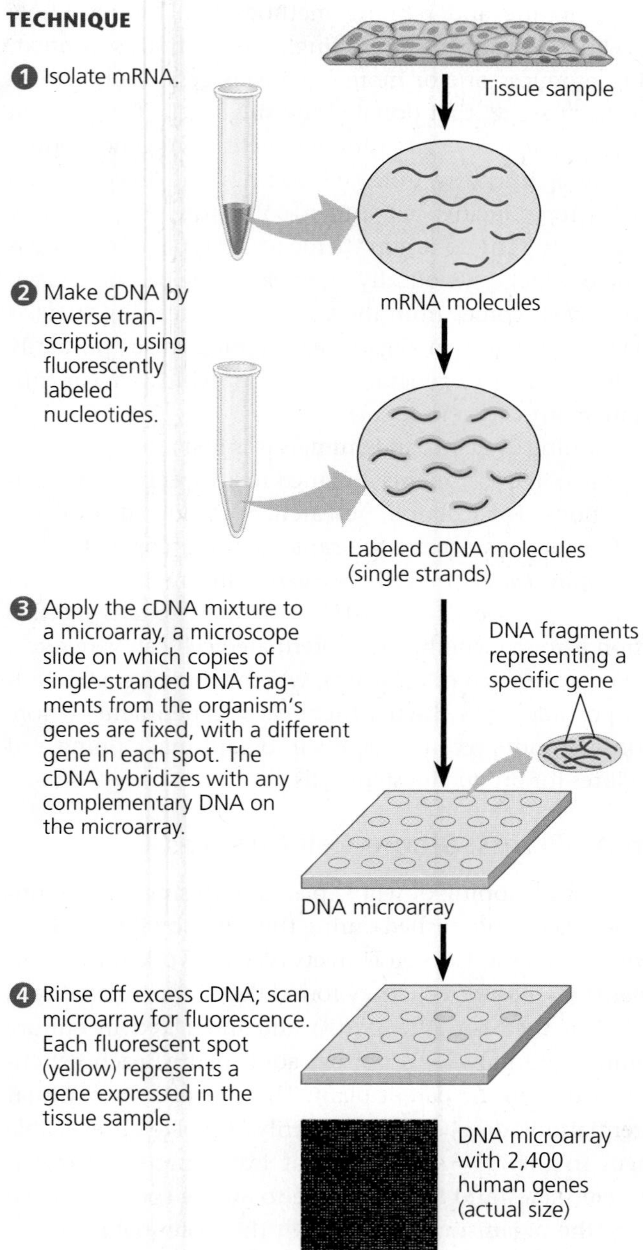

1 Isolate mRNA.
Tissue sample
mRNA molecules
2 Make cDNA by reverse transcription, using fluorescently labeled nucleotides.
Labeled cDNA molecules (single strands)
3 Apply the cDNA mixture to a microarray, a microscope slide on which copies of single-stranded DNA fragments from the organism's genes are fixed, with a different gene in each spot. The cDNA hybridizes with any complementary DNA on the microarray.
DNA fragments representing a specific gene
DNA microarray
4 Rinse off excess cDNA; scan microarray for fluorescence. Each fluorescent spot (yellow) represents a gene expressed in the tissue sample.
DNA microarray with 2,400 human genes (actual size)

RESULTS The intensity of fluorescence at each spot is a measure of the expression in the tissue sample of the gene represented by that spot. Most often, as in the actual microarray above, two different samples are tested together by labeling the cDNAs prepared from each sample with labels of different colors, often green and red. The resulting color at a spot reveals the relative levels of expression of a particular gene in the two samples: Green indicates expression in one sample, red in the other, yellow in both, and black in neither. (See Figure 20.1 for a larger view.)

A newer method for silencing expression of selected genes exploits the phenomenon of **RNA interference (RNAi)**, described in Chapter 18. This experimental approach uses synthetic double-stranded RNA molecules matching the sequence of a particular gene to trigger breakdown of the gene's messenger RNA or to block its translation. In organisms such as the nematode and the fruit fly, RNAi has already proved valuable for analyzing the functions of genes on a large scale. In one study, RNAi was used to prevent expression of 86% of the genes in early nematode embryos, one gene at a time. Analysis of the phenotypes of the worms that developed from these embryos allowed the researchers to classify most of the genes into a small number of groups by function. This type of analysis, in which the functions of multiple genes are considered in a single study, is sure to become more common as research focuses on the importance of interactions between genes in the system as a whole. This is the basis of systems biology (see Chapter 21).

In humans, ethical considerations prohibit knocking out genes to determine their functions. An alternative approach is to analyze the genomes of large numbers of people with a certain phenotypic condition or disease, such as heart disease or diabetes, to try to find differences they all share compared with people without that condition. These large-scale analyses, called **genome-wide association studies**, do not require complete sequencing of all the genomes in the two groups. Instead, researchers test for *genetic markers*, DNA sequences that vary in the population. In a gene, such sequence variation is the basis of different alleles, as we saw earlier for sickle-cell disease. And just like coding sequences, noncoding DNA at a specific locus on a chromosome may exhibit small nucleotide differences (polymorphisms) among individuals.

Among the most useful of these genetic markers are single base-pair variations in the genomes of the human population. A single base-pair site where variation is found in at least 1% of the population is called a **single nucleotide polymorphism** (**SNP**, pronounced "snip"). A few million SNPs occur in the human genome, about once in 100–300 base pairs of both coding and noncoding DNA sequences. (Roughly 98.5% of our genome doesn't code for protein, as you will learn in Chapter 21.) It isn't necessary to sequence the DNA of multiple individuals to find SNPs; today they can be detected by very sensitive microarray analysis or by PCR.

Once a region is found that has a SNP shared by affected but not unaffected people, researchers focus on that region and sequence it. In the vast majority of cases, the SNP itself does not contribute to the disease, and most SNPs are in noncoding regions. Instead, if the SNP and a disease-causing allele are close enough, scientists can take advantage of the fact that crossing over between the marker and the gene is very unlikely during gamete formation. Therefore, the marker and gene will almost always be inherited together, even though the marker is not

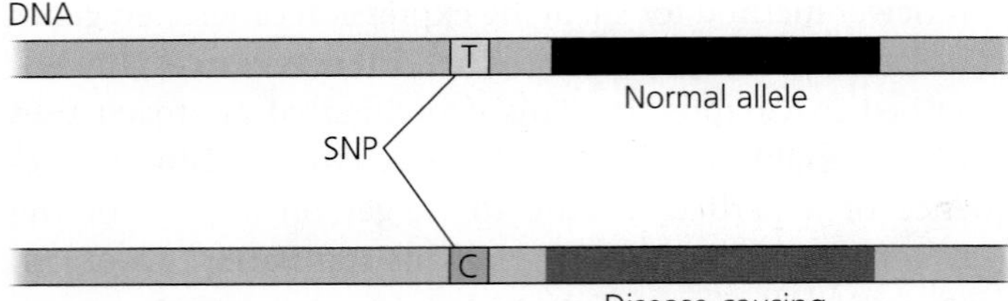

▲ **Figure 20.16 Single nucleotide polymorphisms (SNPs) as genetic markers for disease-causing alleles.** This diagram depicts homologous segments of DNA from two groups of individuals, those in one group having a particular disease or condition with a genetic basis. Unaffected people have a T at a particular SNP locus, while affected people have a C at that locus. A SNP that varies in this way is likely to be closely linked to one or more alleles of genes that contribute to the disease in question. (Here, only a single strand is shown for each DNA molecule.)

MAKE CONNECTIONS *What does it mean for a SNP to be "closely linked" to a disease-causing allele, and how does this allow the SNP to be used as a genetic marker? (See Concept 15.3, p. 296.)*

part of the gene (**Figure 20.16**). SNPs have been found that correlate with diabetes, heart disease, and several types of cancer, and the search is on for genes that might be involved.

The techniques and experimental approaches you have learned about thus far have already taught us a great deal about genes and the functions of their products. This research is now being augmented by the development of powerful techniques for cloning whole multicellular organisms. An aim of this work is to obtain special types of cells, called stem cells, that give rise to all the different kinds of tissues. On a basic level, stem cells would allow scientists to use the DNA-based methods previously discussed to study the process of cell differentiation. On a more applied level, recombinant DNA techniques could be used to alter stem cells for the treatment of disease. Methods involving the cloning of organisms and production of stem cells are the subject of the next section.

CONCEPT CHECK 20.2

1. If you isolated DNA from human cells, treated it with a restriction enzyme, and analyzed the sample by gel electrophoresis, what would you see? Explain.
2. Describe the role of complementary base pairing during Southern blotting, DNA sequencing, Northern blotting, RT-PCR, and microarray analysis.
3. Distinguish between a SNP and a RFLP.
4. **WHAT IF?** Consider the microarray in Figure 20.1, a larger image of the one in Figure 20.15. If a sample from normal tissue is labeled with a green fluorescent dye, and a sample from cancerous tissue is labeled red, what can you conclude about a spot that is green? Red? Yellow? Black? Which genes would you be interested in examining further if you were studying cancer? Explain.

For suggested answers, see Appendix A.

CONCEPT 20.3

Cloning organisms may lead to production of stem cells for research and other applications

In parallel with advances in DNA technology, scientists have been developing and refining methods for cloning whole multicellular organisms from single cells. In this context, cloning produces one or more organisms genetically identical to the "parent" that donated the single cell. This is often called *organismal cloning* to differentiate it from gene cloning and, more significantly, from cell cloning—the division of an asexually reproducing cell into a collection of genetically identical cells. (The common theme for all types of cloning is that the product is genetically identical to the parent. In fact, the word *clone* comes from the Greek *klon*, meaning "twig.") The current interest in organismal cloning arises primarily from its potential to generate stem cells, which can in turn generate many different tissues.

The cloning of plants and animals was first attempted over 50 years ago in experiments designed to answer basic biological questions. For example, researchers wondered if all the cells of an organism have the same genes (a concept called *genomic equivalence*) or if cells lose genes during the process of differentiation (see Chapter 18). One way to answer this question is to see whether a differentiated cell can generate a whole organism—in other words, whether cloning an organism is possible. Let's discuss these early experiments before we consider more recent progress in organismal cloning and procedures for producing stem cells.

Cloning Plants: Single-Cell Cultures

The successful cloning of whole plants from single differentiated cells was accomplished during the 1950s by F. C. Steward and his students at Cornell University, who worked with carrot plants (**Figure 20.17**). They found that differentiated cells taken from the root (the carrot) and incubated in culture medium could grow into normal adult plants, each genetically identical to the parent plant. These results showed that differentiation does not necessarily involve irreversible changes in the DNA. In plants, at least, mature cells can "dedifferentiate" and then give rise to all the specialized cell types of the organism. Any cell with this potential is said to be **totipotent**.

Plant cloning is now used extensively in agriculture. For some plants, such as orchids, cloning is the only commercially practical means of reproducing plants. In other cases, cloning has been used to reproduce a plant with valuable characteristics, such as the ability to resist a plant pathogen. In fact, you yourself may be a plant cloner: If you have ever grown a new plant from a cutting, you have practiced cloning!

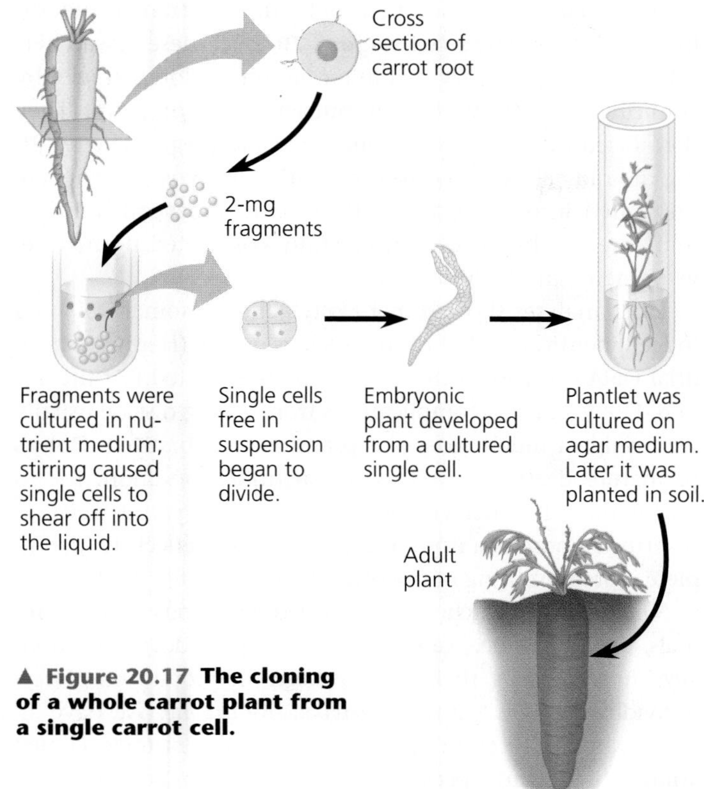

Cross section of carrot root
2-mg fragments
Fragments were cultured in nutrient medium; stirring caused single cells to shear off into the liquid.
Single cells free in suspension began to divide.
Embryonic plant developed from a cultured single cell.
Plantlet was cultured on agar medium. Later it was planted in soil.
Adult plant

▲ **Figure 20.17 The cloning of a whole carrot plant from a single carrot cell.**

Cloning Animals: Nuclear Transplantation

Differentiated cells from animals generally do not divide in culture, much less develop into the multiple cell types of a new organism. Therefore, early researchers had to use a different approach to the question of whether differentiated animal cells can be totipotent. Their approach was to remove the nucleus of an unfertilized or fertilized egg and replace it with the nucleus of a differentiated cell, a procedure called *nuclear transplantation*. If the nucleus from the differentiated donor cell retains its full genetic capability, then it should be able to direct development of the recipient cell into all the tissues and organs of an organism.

Such experiments were conducted on one species of frog (*Rana pipiens*) by Robert Briggs and Thomas King in the 1950s and on another (*Xenopus laevis*) by John Gurdon in the 1970s. These researchers transplanted a nucleus from an embryonic or tadpole cell into an enucleated (nucleus-lacking) egg of the same species. In Gurdon's experiments, the transplanted nucleus was often able to support normal development of the egg into a tadpole (**Figure 20.18**). However, he found that the potential of a transplanted nucleus to direct normal development was inversely related to the age of the donor: the older the donor nucleus, the lower the percentage of normally developing tadpoles.

From these results, Gurdon concluded that something in the nucleus *does* change as animal cells differentiate. In frogs and most other animals, nuclear potential tends to be restricted more and more as embryonic development and cell differentiation progress.

▼ **Figure 20.18** **INQUIRY**

Can the nucleus from a differentiated animal cell direct development of an organism?

EXPERIMENT John Gurdon and colleagues at Oxford University, in England, destroyed the nuclei of frog (*Xenopus laevis*) eggs by exposing the eggs to ultraviolet light. They then transplanted nuclei from cells of frog embryos and tadpoles into the enucleated eggs.

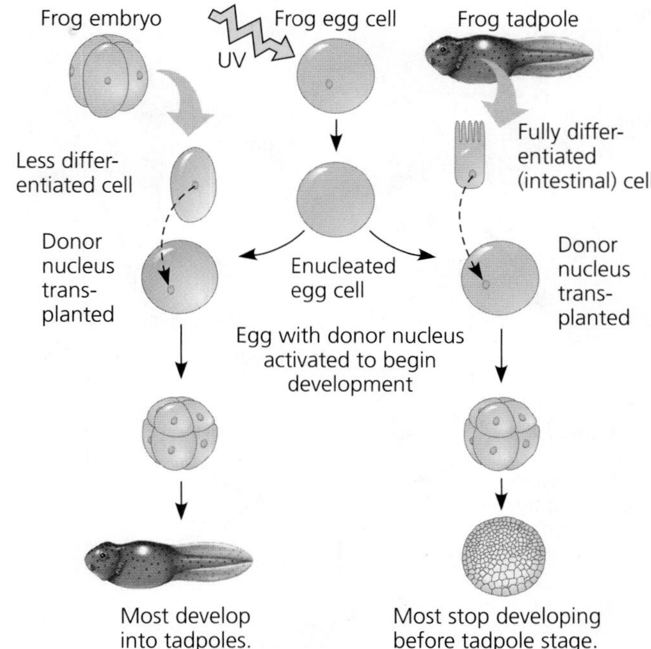

RESULTS When the transplanted nuclei came from an early embryo, whose cells are relatively undifferentiated, most of the recipient eggs developed into tadpoles. But when the nuclei came from the fully differentiated intestinal cells of a tadpole, fewer than 2% of the eggs developed into normal tadpoles, and most of the embryos stopped developing at a much earlier stage.

CONCLUSION The nucleus from a differentiated frog cell can direct development of a tadpole. However, its ability to do so decreases as the donor cell becomes more differentiated, presumably because of changes in the nucleus.

SOURCE J. B. Gurdon et al., The developmental capacity of nuclei transplanted from keratinized cells of adult frogs, *Journal of Embryology and Experimental Morphology* 34:93–112 (1975).

WHAT IF? If each cell in a four-cell embryo was already so specialized that it was not totipotent, what results would you predict for the experiment on the left side of the figure?

Reproductive Cloning of Mammals

In addition to cloning frogs, researchers have long been able to clone mammals by transplanting nuclei or cells from a variety of early embryos. But it was not known whether a nucleus from a fully differentiated cell could be reprogrammed to succeed in acting as a donor nucleus. In 1997, however, researchers at the Roslin Institute in Scotland captured newspaper headlines when they announced the birth of Dolly, a

▼ Figure 20.19 RESEARCH METHOD

Reproductive Cloning of a Mammal by Nuclear Transplantation

APPLICATION This method is used to produce cloned animals whose nuclear genes are identical to those of the animal supplying the nucleus.

TECHNIQUE Shown here is the procedure used to produce Dolly, the first reported case of a mammal cloned using the nucleus of a differentiated cell.

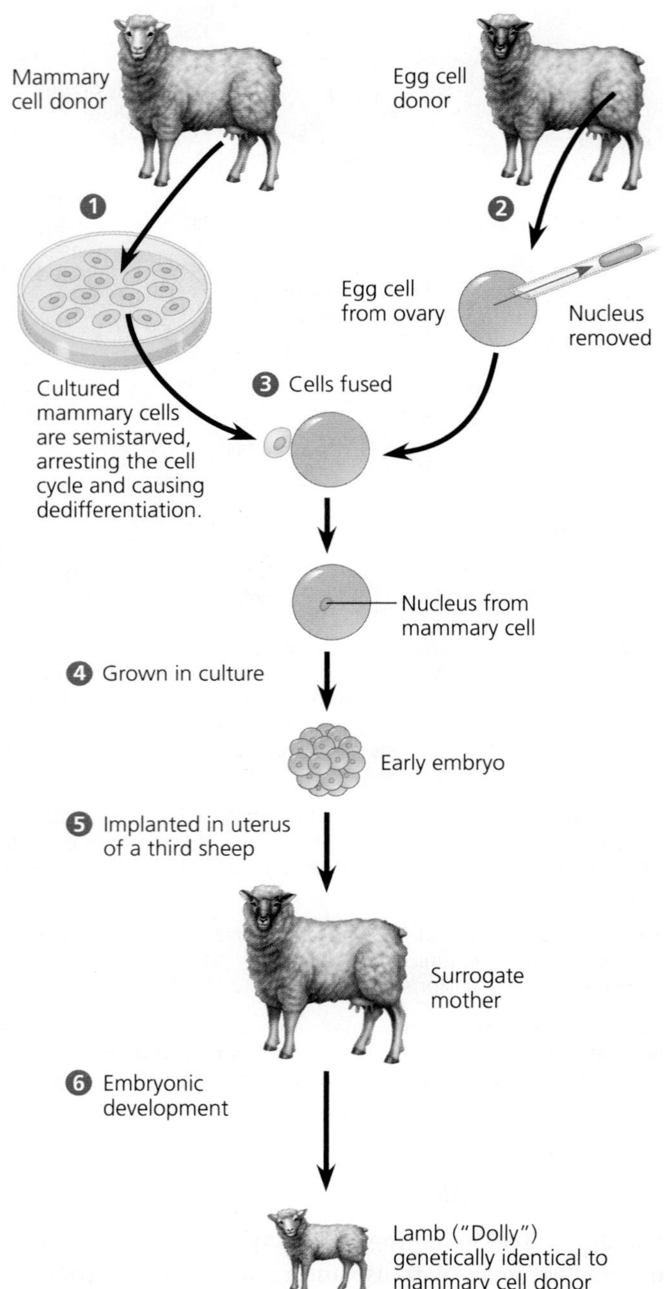

RESULTS The genetic makeup of the cloned animal is identical to that of the animal supplying the nucleus but differs from that of the egg donor and surrogate mother. (The latter two are "Scottish blackface" sheep, with dark faces.)

lamb cloned from an adult sheep by nuclear transplantation from a differentiated cell (**Figure 20.19**). These researchers achieved the necessary dedifferentiation of donor nuclei by culturing mammary cells in nutrient-poor medium. They then fused these cells with enucleated sheep eggs. The resulting diploid cells divided to form early embryos, which were implanted into surrogate mothers. Out of several hundred implanted embryos, one successfully completed normal development, and Dolly was born.

Later analyses showed that Dolly's chromosomal DNA was indeed identical to that of the nucleus donor. (Her mitochondrial DNA came from the egg donor, as expected.) At the age of 6, Dolly suffered complications from a lung disease usually seen only in much older sheep and was euthanized. Dolly's premature death, as well as her arthritic condition, led to speculation that her cells were in some way not quite as healthy as those of a normal sheep, possibly reflecting incomplete reprogramming of the original transplanted nucleus.

Since 1997, researchers have cloned numerous other mammals, including mice, cats, cows, horses, pigs, dogs, and monkeys. In most cases, their goal has been the production of new individuals; this is known as *reproductive cloning*. We have already learned a lot from such experiments. For example, cloned animals of the same species do *not* always look or behave identically. In a herd of cows cloned from the same line of cultured cells, certain cows are dominant in behavior and others are more submissive. Another example of nonidentity in clones is the first cloned cat, named CC for Carbon Copy (**Figure 20.20**). She has a calico coat, like her single female parent, but the color and pattern are different because of random X chromosome inactivation, which is a normal occurrence during embryonic development (see Figure 15.8). And identical human twins, which

▲ Figure 20.20 CC, the first cloned cat, and her single parent. Rainbow (left) donated the nucleus in a cloning procedure that resulted in CC (right). However, the two cats are not identical: Rainbow is a classic calico cat with orange patches on her fur and has a "reserved personality," while CC has a gray and white coat and is more playful.

are naturally occurring "clones," are always slightly different. Clearly, environmental influences and random phenomena can play a significant role during development.

The successful cloning of so many mammals has heightened speculation about the cloning of humans. Scientists in several labs around the world have tackled the first steps of human cloning. In the most common approach, nuclei from differentiated human cells are transplanted into unfertilized enucleated eggs, and the eggs are stimulated to divide. In 2001, a research group at a biotechnology company in Massachusetts observed a few early cell divisions in such an experiment. A few years later, researchers at Seoul National University, in South Korea, reported cloning embryos to an early stage called the blastocyst stage, but the scientists were later found guilty of research misconduct and data fabrication. This episode sent shock waves through the scientific community. In 2007, the first primate (macaque) embryos were cloned by researchers at the Oregon National Primate Research Center; these clones reached the blastocyst stage. This achievement has moved the field one step closer to human cloning, the prospect of which raises unprecedented ethical issues.

Problems Associated with Animal Cloning

In most nuclear transplantation studies thus far, only a small percentage of cloned embryos develop normally to birth. And like Dolly, many cloned animals exhibit defects. Cloned mice, for instance, are prone to obesity, pneumonia, liver failure, and premature death. Scientists assert that even cloned animals that appear normal are likely to have subtle defects.

In recent years, we have begun to uncover some reasons for the low efficiency of cloning and the high incidence of abnormalities. In the nuclei of fully differentiated cells, a small subset of genes is turned on and expression of the rest is repressed. This regulation often is the result of epigenetic changes in chromatin, such as acetylation of histones or methylation of DNA (see Figure 18.7). During the nuclear transfer procedure, many of these changes must be reversed in the later-stage nucleus from a donor animal for genes to be expressed or repressed appropriately in early stages of development. Researchers have found that the DNA in cells from cloned embryos, like that of differentiated cells, often has more methyl groups than does the DNA in equivalent cells from normal embryos of the same species. This finding suggests that the reprogramming of donor nuclei requires chromatin restructuring, which occurs incompletely during cloning procedures. Because DNA methylation helps regulate gene expression, misplaced methyl groups in the DNA of donor nuclei may interfere with the pattern of gene expression necessary for normal embryonic development. In fact, the success of a cloning attempt may depend in large part on whether or not the chromatin in the donor nucleus can be artificially modified to resemble that of a newly fertilized egg.

Stem Cells of Animals

The major goal of cloning human embryos is not reproduction, but the production of stem cells for treating human diseases. A **stem cell** is a relatively unspecialized cell that can both reproduce itself indefinitely and, under appropriate conditions, differentiate into specialized cells of one or more types. Thus, stem cells are able both to replenish their own population and to generate cells that travel down specific differentiation pathways.

Many early animal embryos contain stem cells capable of giving rise to differentiated embryonic cells of any type. Stem cells can be isolated from early embryos at a stage called the blastula stage or its human equivalent, the blastocyst stage (**Figure 20.21**). In culture, these *embryonic stem (ES) cells*

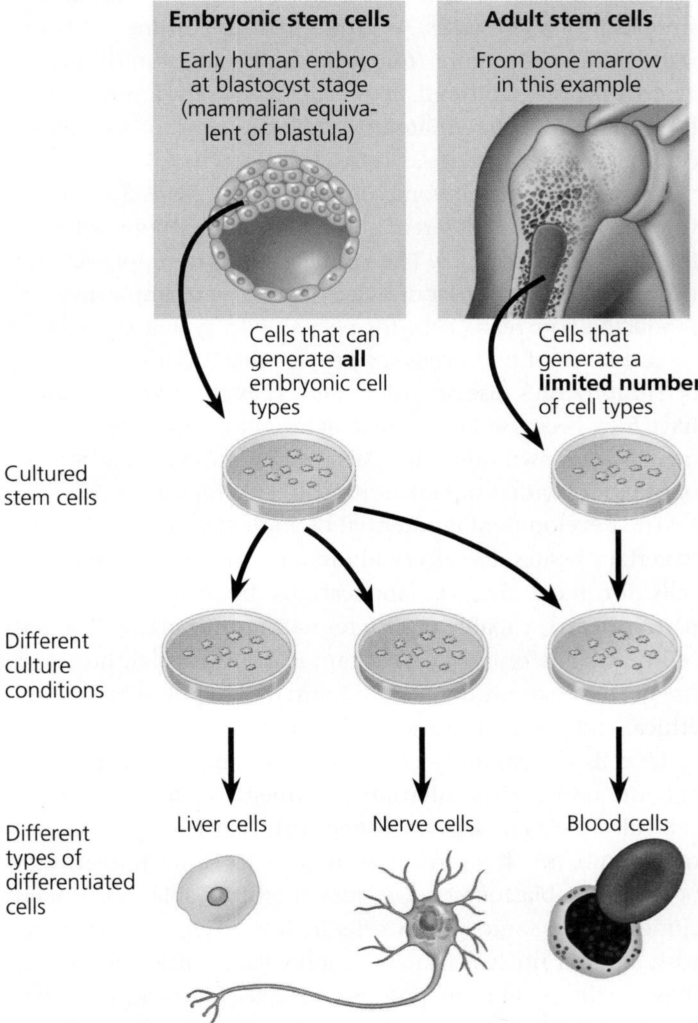

▲ **Figure 20.21 Working with stem cells.** Animal stem cells, which can be isolated from early embryos or adult tissues and grown in culture, are self-perpetuating, relatively undifferentiated cells. Embryonic stem cells are easier to grow than adult stem cells and can theoretically give rise to *all* types of cells in an organism. The range of cell types that can arise from adult stem cells is not yet fully understood.

reproduce indefinitely; and depending on culture conditions, they can be made to differentiate into a wide variety of specialized cells, including even eggs and sperm.

The adult body also has stem cells, which serve to replace nonreproducing specialized cells as needed. In contrast to ES cells, *adult stem cells* are not able to give rise to all cell types in the organism, though they can generate multiple types. For example, one of the several types of stem cells in bone marrow can generate all the different kinds of blood cells (see Figure 20.21), and another can differentiate into bone, cartilage, fat, muscle, and the linings of blood vessels. To the surprise of many, the adult brain has been found to contain stem cells that continue to produce certain kinds of nerve cells there. And recently, researchers have reported finding stem cells in skin, hair, eyes, and dental pulp. Although adult animals have only tiny numbers of stem cells, scientists are learning to identify and isolate these cells from various tissues and, in some cases, to grow them in culture. With the right culture conditions (for instance, the addition of specific growth factors), cultured stem cells from adult animals have been made to differentiate into multiple types of specialized cells, although none are as versatile as ES cells.

Research with embryonic or adult stem cells is a source of valuable data about differentiation and has enormous potential for medical applications. The ultimate aim is to supply cells for the repair of damaged or diseased organs: for example, insulin-producing pancreatic cells for people with type 1 diabetes or certain kinds of brain cells for people with Parkinson's disease or Huntington's disease. Adult stem cells from bone marrow have long been used as a source of immune system cells in patients whose own immune systems are nonfunctional because of genetic disorders or radiation treatments for cancer.

The developmental potential of adult stem cells is limited to certain tissues. ES cells hold more promise than adult stem cells for most medical applications because ES cells are **pluripotent**, capable of differentiating into many different cell types. The only way to obtain ES cells thus far, however, has been to harvest them from human embryos, which raises ethical and political issues.

ES cells are currently obtained from embryos donated by patients undergoing infertility treatment or from long-term cell cultures originally established with cells isolated from donated embryos. If scientists were able to clone human embryos to the blastocyst stage, they might be able to use such clones as the source of ES cells in the future. Furthermore, with a donor nucleus from a person with a particular disease, they might be able to produce ES cells for treatment that match the patient and are thus not rejected by his or her immune system. When the main aim of cloning is to produce ES cells to treat disease, the process is called *therapeutic cloning*. Although most people believe that reproductive cloning of humans is unethical, opinions vary about the morality of therapeutic cloning.

Resolving the debate now seems less imperative because researchers have been able to turn back the clock in fully differentiated cells, reprogramming them to act like ES cells. The accomplishment of this feat, which posed formidable obstacles, was announced in 2007, first by labs using mouse skin cells and then by additional groups using cells from human skin and other organs or tissues. In all these cases, researchers transformed the differentiated cells into ES cells by using retroviruses to introduce extra cloned copies of four "stem cell" master regulatory genes. All the tests that were carried out at the time indicated that the transformed cells, known as *induced pluripotent stem (iPS) cells*, could do everything ES cells can do. More recently, however, several research groups have uncovered differences between iPS and ES cells in gene expression and other cellular functions, such as cell division. At least until these differences are fully understood, the study of ES cells will continue to make important contributions to the development of stem cell therapies. (In fact, ES cells will likely always be a focus of basic research as well.) In the meantime, work is proceeding using the iPS cells in hand.

There are two major potential uses for human iPS cells. First, cells from patients suffering from diseases can be reprogrammed to become iPS cells, which can act as model cells for studying the disease and potential treatments. Human iPS cell lines have already been developed from individuals with type 1 diabetes, Parkinson's disease, and at least a dozen other diseases. Second, in the field of regenerative medicine, a patient's own cells could be reprogrammed into iPS cells and then used to replace nonfunctional tissues (**Figure 20.22**). Developing techniques that direct iPS cells to become specific cell types for this purpose is an area of intense research, one that has already seen some success. The iPS cells created in this way could eventually provide tailor-made "replacement" cells for patients without using any human eggs or embryos, thus circumventing most ethical objections.

CONCEPT CHECK 20.3

1. Based on current knowledge, how would you explain the difference in the percentage of tadpoles that developed from the two kinds of donor nuclei in Figure 20.18?
2. If you were to clone a carrot using the technique shown in Figure 20.17, would all the progeny plants ("clones") look identical? Why or why not?
3. **WHAT IF?** If you were a doctor who wanted to use iPS cells to treat a patient with severe type 1 diabetes, what new technique would have to be developed?
4. **MAKE CONNECTIONS** Compare an individual carrot cell in Figure 20.17 with the fully differentiated muscle cell in Figure 18.18 (p. 369) in terms of their potential to develop into different cell types.

For suggested answers, see Appendix A.

▼ **Figure 20.22**

IMPACT

The Impact of Induced Pluripotent Stem (iPS) Cells on Regenerative Medicine

While embryonic stem (ES) cells can generate every cell in an organism, the use of human embryos as their source is controversial. Several research groups have developed similar procedures for reprogramming fully differentiated cells to become induced pluripotent stem (iPS) cells, which act like ES cells. The technique is based on introducing transcription factors that are characteristic of stem cells into differentiated cells, such as skin cells.

WHY IT MATTERS Patients with diseases such as heart disease, diabetes, or Alzheimer's could have their own skin cells reprogrammed to become iPS cells. Once procedures have been developed for converting iPS cells into heart, pancreatic, or nervous system cells, the patients' own iPS cells might be used to treat their disease. This technique has already been used successfully to treat sickle-cell disease in a mouse that had been genetically engineered to have the disease. Shown below is how this therapy could work in humans, once researchers learn how iPS cells can be triggered to differentiate as desired (step 3).

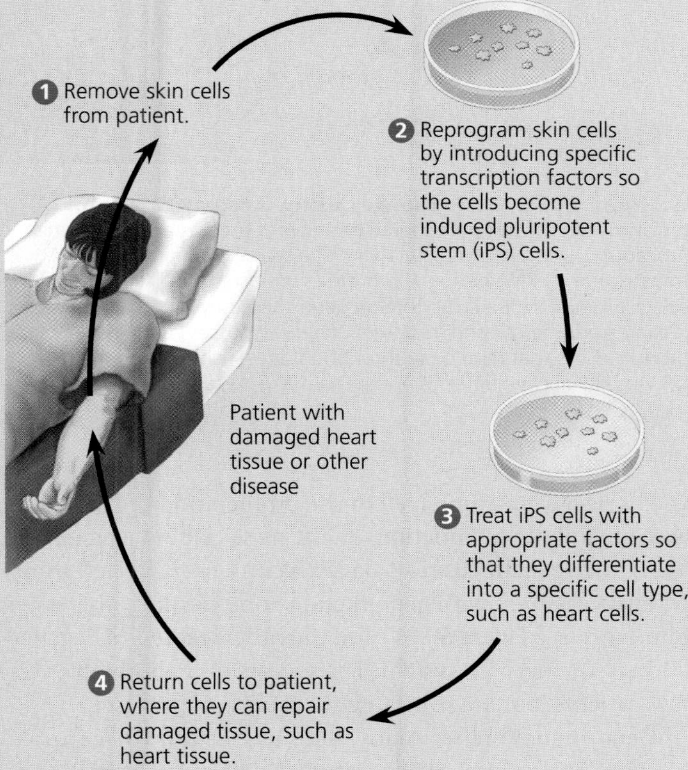

1 Remove skin cells from patient.
2 Reprogram skin cells by introducing specific transcription factors so the cells become induced pluripotent stem (iPS) cells.
Patient with damaged heart tissue or other disease
3 Treat iPS cells with appropriate factors so that they differentiate into a specific cell type, such as heart cells.
4 Return cells to patient, where they can repair damaged tissue, such as heart tissue.

FURTHER READING G. Vogel and C. Holden, Field leaps forward with new stem cell advances, *Science* 318:1224–1225 (2007); K. Hochedlinger, Your inner healers, *Scientific American* 302:46–53 (2010).

WHAT IF? When organs are transplanted from a donor to a diseased recipient, the recipient's immune system may reject the transplant, a condition with serious and often fatal consequences. Would using converted iPS cells be expected to carry the same risk? Why or why not?

CONCEPT **20.4**

The practical applications of DNA technology affect our lives in many ways

DNA technology is in the news almost every day. Most often, the topic is a new and promising application in medicine, but this is just one of numerous fields benefiting from DNA technology and genetic engineering.

Medical Applications

One important use of DNA technology is the identification of human genes whose mutation plays a role in genetic diseases. These discoveries may lead to ways of diagnosing, treating, and even preventing such conditions. DNA technology is also contributing to our understanding of "nongenetic" diseases, from arthritis to AIDS, since a person's genes influence susceptibility to these diseases. Furthermore, diseases of all sorts involve changes in gene expression within the affected cells and often within the patient's immune system. By using DNA microarray assays or other techniques to compare gene expression in healthy and diseased tissues, as seen in Figure 20.1, researchers hope to find many of the genes that are turned on or off in particular diseases. These genes and their products are potential targets for prevention or therapy.

Diagnosis and Treatment of Diseases

A new chapter in the diagnosis of infectious diseases has been opened by DNA technology, in particular the use of PCR and labeled nucleic acid probes to track down pathogens. For example, because the sequence of the RNA genome of HIV is known, RT-PCR can be used to amplify, and thus detect, HIV RNA in blood or tissue samples (see Figure 20.13). RT-PCR is often the best way to detect an otherwise elusive infective agent.

Medical scientists can now diagnose hundreds of human genetic disorders by using PCR with primers that target the genes associated with these disorders. The amplified DNA product is then sequenced to reveal the presence or absence of the disease-causing mutation. Among the genes for human diseases that have been identified are those for sickle-cell disease, hemophilia, cystic fibrosis, Huntington's disease, and Duchenne muscular dystrophy. Individuals afflicted with such diseases can often be identified before the onset of symptoms, even before birth. PCR can also be used to identify symptomless carriers of potentially harmful recessive alleles, essentially replacing Southern blotting for this purpose.

As you learned earlier, genome-wide association studies have pinpointed SNPs (single nucleotide polymorphisms) that are linked to disease-causing alleles. Individuals can be tested for the presence of a SNP that indicates the presence of the abnormal allele. The presence of particular SNPs is correlated with increased risk for conditions such as heart disease, Alzheimer's, and some types of cancer. Companies that offer individual genetic testing for risk factors like these are looking for the presence of previously identified, linked SNPs. It may be helpful for an individual to learn about their health risks, with the understanding that such genetic tests merely reflect correlations and do not make predictions.

The techniques described in this chapter have also prompted improvements in disease treatments. By analyzing the expression of many genes in breast cancer patients, researchers carrying out one genome-wide association study were able to identify 70 genes whose expression pattern could be correlated with the likelihood that the cancer would recur. Given that low-risk patients have a 96% survival rate over a ten-year period with no treatment, gene expression analysis allows doctors and patients access to valuable information when they are considering treatment options.

Many envision a future of "personalized medicine" where each person's genetic health profile can inform them about diseases or conditions for which they are especially at risk and help them make treatment choices. A genetic profile is currently taken to mean a set of genetic markers such as SNPs, but ultimately it could mean the complete DNA sequence of each individual—once sequencing becomes inexpensive enough.

Human Gene Therapy

Gene therapy—introducing genes into an afflicted individual for therapeutic purposes—holds great potential for treating the relatively small number of disorders traceable to a single defective gene. In theory, a normal allele of the defective gene could be inserted into the somatic cells of the tissue affected by the disorder.

For gene therapy of somatic cells to be permanent, the cells that receive the normal allele must be ones that multiply throughout the patient's life. Bone marrow cells, which include the stem cells that give rise to all the cells of the blood and immune system, are prime candidates. **Figure 20.23** outlines one possible procedure for gene therapy of an individual whose bone marrow cells do not produce a vital enzyme because of a single defective gene. One type of severe combined immunodeficiency (SCID) is caused by just this kind of defect. If the treatment is successful, the patient's bone marrow cells will begin producing the missing protein, and the patient may be cured.

The procedure shown in Figure 20.23 has been used in gene therapy trials for SCID. In a trial begun in France in 2000, ten young children with SCID were treated by the same procedure. Nine of these patients showed significant, definitive improvement after two years, the first indisputable success of gene therapy. However, three of the patients subsequently developed leukemia, a type of blood cell cancer, and one of them died. Two factors may have contributed to the development of leukemia: the insertion of the retroviral vector near a gene involved in the proliferation of blood cells and an unknown function of the replacement gene itself. Two other genetic diseases have recently been treated somewhat successfully with gene therapy: one causing progressive blindness (see Figure 50.21) and the other leading to degeneration of the nervous system. The successful trials involve very few patients, but are still cause for cautious optimism.

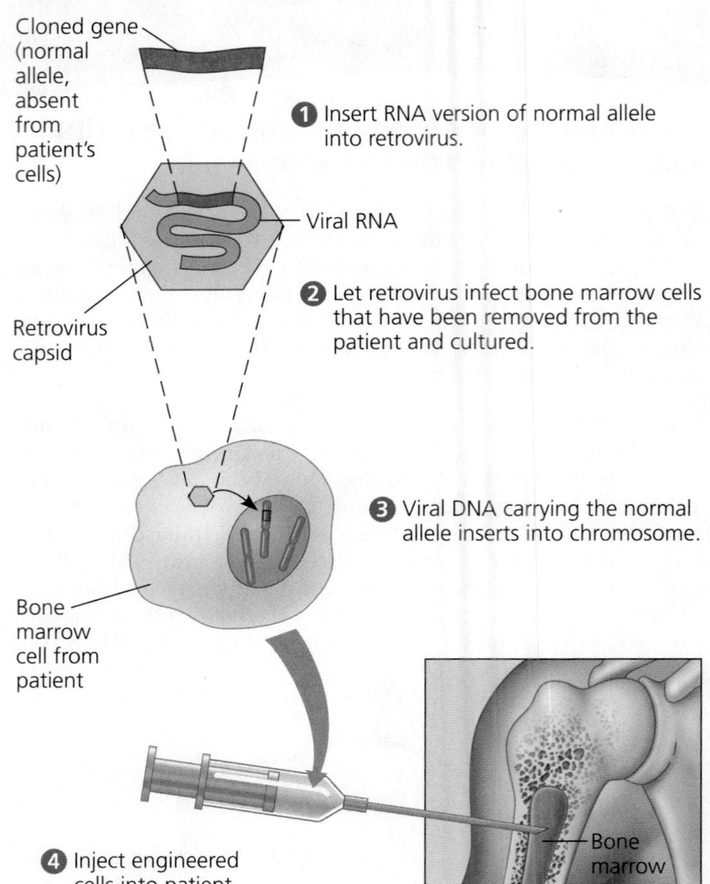
Cloned gene (normal allele, absent from patient's cells)
1 Insert RNA version of normal allele into retrovirus.
Viral RNA
Retrovirus capsid
2 Let retrovirus infect bone marrow cells that have been removed from the patient and cultured.
3 Viral DNA carrying the normal allele inserts into chromosome.
Bone marrow cell from patient
Bone marrow
4 Inject engineered cells into patient.

▲ **Figure 20.23 Gene therapy using a retroviral vector.** A retrovirus that has been rendered harmless is used as a vector in this procedure, which exploits the ability of a retrovirus to insert a DNA transcript of its RNA genome into the chromosomal DNA of its host cell (see Figure 19.8). If the foreign gene carried by the retroviral vector is expressed, the cell and its descendants will possess the gene product. Cells that reproduce throughout life, such as bone marrow cells, are ideal candidates for gene therapy.

Gene therapy raises many technical issues. For example, how can the activity of the transferred gene be controlled so that cells make appropriate amounts of the gene product at the right time and in the right place? How can we be sure that the insertion of the therapeutic gene does not harm some other necessary cell function? As more is learned about DNA control elements and gene interactions, researchers may be able to answer such questions.

In addition to technical challenges, gene therapy provokes ethical questions. Some critics believe that tampering with human genes in any way is immoral. Other observers see no fundamental difference between the transplantation of genes into somatic cells and the transplantation of organs. You might wonder whether scientists are considering engineering human germ-line cells in the hope of correcting a defect in future generations. At present, no one in the mainstream scientific community is pursuing this goal—it is considered much too risky. Such genetic engineering is routinely done in laboratory mice, though, and the technical problems relating to similar genetic engineering in humans will eventually be solved. Under what circumstances, if any, should we alter the genomes of human germ lines? Would this inevitably lead to the practice of eugenics, a deliberate effort to control the genetic makeup of human populations? While we may not have to resolve these questions right now, considering them is worthwhile because they will probably arise at some point in the future.

Pharmaceutical Products

The pharmaceutical industry derives significant benefit from advances in DNA technology and genetic research, applying them to the development of useful drugs to treat diseases. Pharmaceutical products are synthesized using methods of either organic chemistry or biotechnology, depending on the nature of the product.

Synthesis of Small Molecules for Use as Drugs Determining the sequence and structure of proteins crucial for tumor cell survival has led to the identification of small molecules that combat certain cancers by blocking the function of these proteins. One drug, imatinib (trade name Gleevec), is a small molecule that inhibits a specific receptor tyrosine kinase (see Figure 11.7). The overexpression of this receptor, resulting from a chromosomal translocation, is instrumental in causing chronic myelogenous leukemia (CML; see Figure 15.16). Patients in the early stages of CML who are treated with imatinib have exhibited nearly complete, sustained remission from the cancer. Drugs that work like this have also been used with success to treat a few types of lung and breast cancers. This approach is feasible only for cancers for which the molecular basis is fairly well understood.

Pharmaceutical products that are proteins can be synthesized on a large scale, using cells or whole organisms. Cell cultures are more widely used at present.

Protein Production in Cell Cultures You learned earlier in the chapter about DNA cloning and gene expression systems for producing large quantities of proteins that are present naturally in only minute amounts. The host cells used in such expression systems can even be engineered to secrete a protein as it is made, thereby simplifying the task of purifying it by traditional biochemical methods.

Among the first pharmaceutical products "manufactured" in this way were human insulin and human growth hormone (HGH). Some 2 million people with diabetes in the United States depend on insulin treatment to control their disease. Human growth hormone has been a boon to children born with a form of dwarfism caused by inadequate amounts of HGH. Another important pharmaceutical product produced by genetic engineering is tissue plasminogen activator (TPA). If administered shortly after a heart attack, TPA helps dissolve blood clots and reduces the risk of subsequent heart attacks.

Protein Production by "Pharm" Animals In some cases, instead of using cell systems to produce large quantities of protein products, pharmaceutical scientists can use whole animals. They can introduce a gene from an animal of one genotype into the genome of another individual, often of a different species. This individual is then called a **transgenic** animal. To do this, they first remove eggs from a female of the recipient species and fertilize them *in vitro*. Meanwhile, they have cloned the desired gene from the donor organism. They then inject the cloned DNA directly into the nuclei of the fertilized eggs. Some of the cells integrate the foreign DNA, the *transgene*, into their genomes and are able to express the foreign gene. The engineered embryos are then surgically implanted in a surrogate mother. If an embryo develops successfully, the result is a transgenic animal that expresses its new, "foreign" gene.

Assuming that the introduced gene encodes a protein desired in large quantities, these transgenic animals can act as pharmaceutical "factories." For example, a transgene for a human blood protein such as antithrombin can be inserted into the genome of a goat in such a way that the transgene's product is secreted in the animal's milk (**Figure 20.24**). The protein is then purified from the milk (which is easier than purification from a cell culture). Researchers have also engineered transgenic chickens that express large amounts of the transgene's product in eggs. Biotechnology companies consider the characteristics of candidate animals in deciding which to use for engineering. For

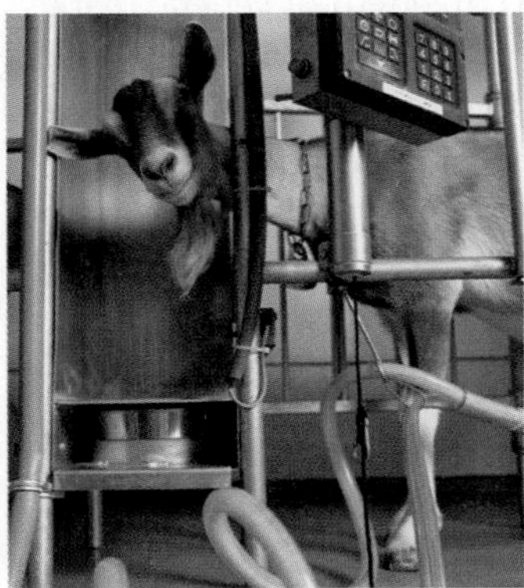

▲ **Figure 20.24 Goats as "pharm" animals.** This transgenic goat carries a gene for a human blood protein, antithrombin, which she secretes in her milk. Patients with a rare hereditary disorder in which this protein is lacking suffer from formation of blood clots in their blood vessels. Easily purified from the goat's milk, the protein has been approved in the United States and Europe for treating these patients.

example, goats reproduce faster than cows, and it is possible to harvest more protein from goat milk than from the milk of other rapidly reproducing mammals, such as rabbits.

Human proteins produced in transgenic farm animals for use in humans may differ in some ways from the naturally produced human proteins, possibly because of subtle differences in protein modification. Therefore, such proteins must be tested very carefully to ensure that they (or contaminants from the farm animals) will not cause allergic reactions or other adverse effects in patients who receive them.

Forensic Evidence and Genetic Profiles

In violent crimes, body fluids or small pieces of tissue may be left at the scene or on the clothes or other possessions of the victim or assailant. If enough blood, semen, or tissue is available, forensic laboratories can determine the blood type or tissue type by using antibodies to detect specific cell-surface proteins. However, such tests require fairly fresh samples in relatively large amounts. Also, because many people have the same blood or tissue type, this approach can only exclude a suspect; it cannot provide strong evidence of guilt.

DNA testing, on the other hand, can identify the guilty individual with a high degree of certainty, because the DNA sequence of every person is unique (except for identical twins). Genetic markers that vary in the population can be analyzed for a given person to determine that individual's unique set of genetic markers, or **genetic profile**. (This term is preferred over "DNA fingerprint" by forensic scientists, who want to emphasize the heritable aspect of these markers rather than the fact that they produce a pattern on a gel that, like a fingerprint, is visually recognizable.) The FBI started applying DNA technology to forensics in 1988, using RFLP analysis by Southern blotting to detect similarities and differences in DNA samples. This method required much smaller samples of blood or tissue than earlier methods—only about 1,000 cells.

Today, forensic scientists use an even more sensitive method that takes advantage of variations in length of genetic markers called **short tandem repeats (STRs)**. These are tandemly repeated units of two- to five-base sequences in specific regions of the genome. The number of repeats present in these regions is highly variable from person to person (polymorphic), and for one individual, the two alleles of an STR may even differ from each other. For example, one individual may have the sequence ACAT repeated 30 times at one genome locus and 15 times at the same locus on the other homolog, whereas another individual may have 18 repeats at this locus on each homolog. (These two genotypes can be expressed by the two repeat numbers: 30,15 and 18,18.) PCR is used to amplify particular STRs, using sets of primers that are labeled with different-colored fluorescent tags; the length of the region, and thus the number of repeats, can then be determined by electrophoresis. Because Southern blotting is not required, this method is quicker than RFLP analysis. And the PCR step allows use of the method even when the DNA is in poor condition or available only in minute quantities. A tissue sample containing as few as 20 cells can be sufficient for PCR amplification.

In a murder case, for example, this method can be used to compare DNA samples from the suspect, the victim, and a small amount of blood found at the crime scene. The forensic scientist tests only a few selected portions of the DNA—usually 13 STR markers. However, even this small set of markers can provide a forensically useful genetic profile because the probability that two people (who are not identical twins) would have exactly the same set of STR markers is vanishingly small. The Innocence Project, a nonprofit organization dedicated to overturning wrongful convictions, uses STR analysis of archived samples from crime scenes to revisit old cases. As of 2010, more than 250 innocent people had been released from prison as a result of forensic and legal work by this group (**Figure 20.25**).

Genetic profiles can also be useful for other purposes. A comparison of the DNA of a mother, her child, and the purported father can conclusively settle a question of paternity. Sometimes paternity is of historical interest: Genetic profiles

(a) In 1984, Earl Washington was convicted and sentenced to death for the 1982 rape and murder of Rebecca Williams. His sentence was commuted to life in prison in 1993 due to new doubts about the evidence. In 2000, STR analysis by forensic scientists associated with The Innocence Project showed conclusively that he was innocent. This photo shows Washington just before his release in 2001, after 17 years in prison.

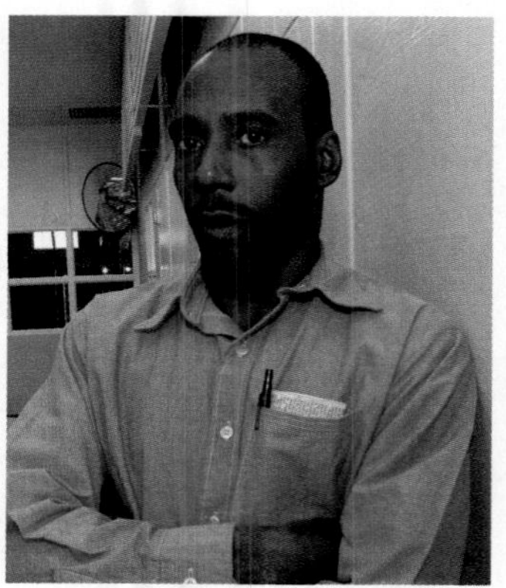

Source of sample	STR marker 1	STR marker 2	STR marker 3
Semen on victim	17,19	13,16	12,12
Earl Washington	16,18	14,15	11,12
Kenneth Tinsley	17,19	13,16	12,12

(b) In STR analysis, selected STR markers in a DNA sample are amplified by PCR, and the PCR products are separated by electrophoresis. The procedure reveals how many repeats are present for each STR locus in the sample. An individual has two alleles per STR locus, each with a certain number of repeats. This table shows the number of repeats for three STR markers in three samples: from semen found on the victim, from Washington, and from another man named Kenneth Tinsley, who was in prison because of an unrelated conviction. These and other STR data (not shown) exonerated Washington and led Tinsley to plead guilty to the murder.

▲ **Figure 20.25 STR analysis used to release an innocent man from prison.**

provided strong evidence that Thomas Jefferson or one of his close male relatives fathered at least one of the children of his slave Sally Hemings. Genetic profiles can also identify victims of mass casualties. The largest such effort occurred after the attack on the World Trade Center in 2001; more than 10,000 samples of victims' remains were compared with DNA samples from personal items, such as toothbrushes, provided by families. Ultimately, forensic scientists succeeded in identifying almost 3,000 victims using these methods.

Just how reliable is a genetic profile? The greater the number of markers examined in a DNA sample, the more likely it is that the profile is unique to one individual. In forensic cases using STR analysis with 13 markers, the probability of two people having identical DNA profiles is somewhere between one chance in 10 billion and one in several trillion. (For comparison, the world's population in 2009 was about 6.8 billion.) The exact probability depends on the frequency of those markers in the general population. Information on how common various markers are in different ethnic groups is critical because these marker frequencies may vary considerably among ethnic groups and between a particular ethnic group and the population as a whole. With the increasing availability of frequency data, forensic scientists can make extremely accurate statistical calculations. Thus, despite problems that can still arise from insufficient data, human error, or flawed evidence, genetic profiles are now accepted as compelling evidence by legal experts and scientists alike.

Environmental Cleanup

Increasingly, the remarkable ability of certain microorganisms to transform chemicals is being exploited for environmental cleanup. If the growth needs of such microbes make them unsuitable for direct use, scientists can now transfer the genes for their valuable metabolic capabilities into other microorganisms, which can then be used to treat environmental problems. For example, many bacteria can extract heavy metals, such as copper, lead, and nickel, from their environments and incorporate the metals into compounds such as copper sulfate or lead sulfate, which are readily recoverable. Genetically engineered microbes may become important in both mining minerals (especially as ore reserves are depleted) and cleaning up highly toxic mining wastes. Biotechnologists are also trying to engineer microbes that can degrade chlorinated hydrocarbons and other harmful compounds. These microbes could be used in wastewater treatment plants or by manufacturers before the compounds are ever released into the environment.

Agricultural Applications

Scientists are working to learn more about the genomes of agriculturally important plants and animals. For a number of years, they have been using DNA technology in an effort to improve agricultural productivity. The selective breeding of both livestock (animal husbandry) and crops has exploited naturally occurring mutations and genetic recombination for thousands of years.

As we described earlier, DNA technology enables scientists to produce transgenic animals, which speeds up the selective breeding process. The goals of creating a transgenic animal are often the same as the goals of traditional breeding—for instance, to make a sheep with better quality wool, a pig with leaner meat, or a cow that will mature in a shorter time. Scientists might, for example, identify and clone a gene that causes the development of larger muscles (muscles make up most of the meat we eat) in one breed of cattle and transfer it to other cattle or even to sheep. However, problems such as low fertility or increased susceptibility to disease are not uncommon among farm animals carrying genes from other species. Animal health and welfare are important issues to consider when developing transgenic animals.

Agricultural scientists have already endowed a number of crop plants with genes for desirable traits, such as delayed ripening and resistance to spoilage and disease. In one striking way, plants are easier to genetically engineer than most animals. For many plant species, a single tissue cell grown in culture can give rise to an adult plant (see Figure 20.17). Thus, genetic manipulations can be performed on an ordinary somatic cell and the cell then used to generate an organism with new traits.

The most commonly used vector for introducing new genes into plant cells is a plasmid, called the **Ti plasmid**, from the soil bacterium *Agrobacterium tumefaciens*. This plasmid integrates a segment of its DNA, known as T DNA, into the chromosomal DNA of its host plant cells. For vector purposes, researchers work with versions of the plasmid that do not cause disease (unlike the wild-type version) and that have been engineered to carry genes of interest within the borders of the T DNA. **Figure 20.26** (on the next page) outlines one method for using the Ti plasmid to produce transgenic plants.

Genetic engineering is rapidly replacing traditional plant-breeding programs, especially for useful traits, such as herbicide or pest resistance, determined by one or a few genes. Crops engineered with a bacterial gene making the plants resistant to herbicides can grow while weeds are destroyed, and genetically engineered crops that can resist destructive insects reduce the need for chemical insecticides. In India, the insertion of a salinity resistance gene from a coastal mangrove plant into the genomes of several rice varieties has resulted in rice plants that can grow in water three times as salty as seawater. The research foundation that carried out this feat of genetic engineering estimates that one-third of all irrigated land has high salinity owing to overirrigation and intensive use of chemical fertilizers, representing a serious threat to the food supply. Thus, salinity-resistant crop plants would be enormously valuable worldwide.

▼ **Figure 20.26** **RESEARCH METHOD**

Using the Ti Plasmid to Produce Transgenic Plants

APPLICATION Genes conferring useful traits, such as pest resistance, herbicide resistance, delayed ripening, and increased nutritional value, can be transferred from one plant variety or species to another using the Ti plasmid as a vector.

TECHNIQUE

1. The Ti plasmid is isolated from the bacterium *Agrobacterium tumefaciens*. The segment of the plasmid that integrates into the genome of host cells is called T DNA.

2. The foreign gene of interest is inserted into the middle of the T DNA using methods shown in Figure 20.4.

3. Recombinant plasmids can be introduced into cultured plant cells by electroporation. Or plasmids can be returned to *Agrobacterium*, which is then applied as a liquid suspension to the leaves of susceptible plants, infecting them. Once a plasmid is taken into a plant cell, its T DNA integrates into the cell's chromosomal DNA.

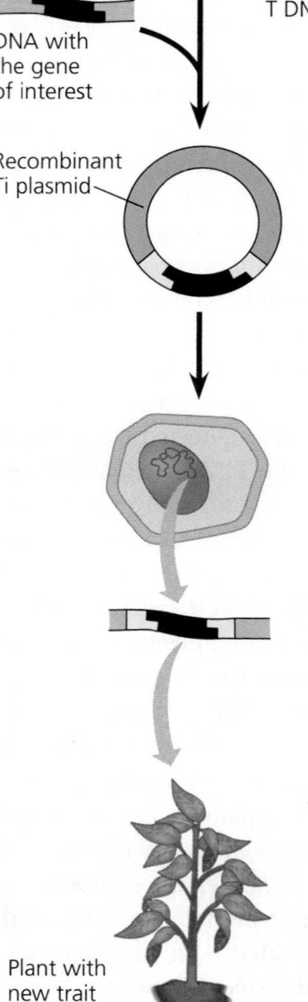

RESULTS Transformed cells carrying the transgene of interest can regenerate complete plants that exhibit the new trait conferred by the transgene.

Safety and Ethical Questions Raised by DNA Technology

Early concerns about potential dangers associated with recombinant DNA technology focused on the possibility that hazardous new pathogens might be created. What might happen, for instance, if cancer cell genes were transferred into bacteria or viruses? To guard against such rogue microbes, scientists developed a set of guidelines that were adopted as formal government regulations in the United States and some other countries. One safety measure is a set of strict laboratory procedures designed to protect researchers from infection by engineered microbes and to prevent the microbes from accidentally leaving the laboratory. In addition, strains of microorganisms to be used in recombinant DNA experiments are genetically crippled to ensure that they cannot survive outside the laboratory. Finally, certain obviously dangerous experiments have been banned.

Today, most public concern about possible hazards centers not on recombinant microbes but on **genetically modified (GM) organisms** used as food. A GM organism is one that has acquired by artificial means one or more genes from another species or even from another variety of the same species. Some salmon, for example, have been genetically modified by addition of a more active salmon growth hormone gene. However, the majority of the GM organisms that contribute to our food supply are not animals, but crop plants.

GM crops are widespread in the United States, Argentina, and Brazil; together these countries account for over 80% of the world's acreage devoted to such crops. In the United States, most corn, soybean, and canola crops are genetically modified, and GM products are not required to be labeled. However, the same foods are an ongoing subject of controversy in Europe, where the GM revolution has met with strong opposition. Many Europeans are concerned about the safety of GM foods and the possible environmental consequences of growing GM plants. Early in 2000, negotiators from 130 countries agreed on a Biosafety Protocol that requires exporters to identify GM organisms present in bulk food shipments and allows importing countries to decide whether the products pose environmental or health risks. (Although the United States declined to sign the agreement, it went into effect anyway because the majority of countries were in favor of it.) Since then, European countries have, on occasion, refused crops from the United States and other countries, leading to trade disputes. Although a small number of GM crops have been grown on European soil, these products have generally failed in local markets, and the future of GM crops in Europe is uncertain.

Advocates of a cautious approach toward GM crops fear that transgenic plants might pass their new genes to close relatives in nearby wild areas. We know that lawn and crop grasses, for example, commonly exchange genes with wild relatives via pollen transfer. If crop plants carrying genes for resistance to herbicides, diseases, or insect pests pollinated wild ones, the offspring might become "super weeds" that are very difficult to control. Another worry concerns possible risks to human health from GM foods. Some people fear that the protein products of transgenes might lead to allergic reactions. Although there is some evidence that this could happen, advocates claim that these proteins could be tested in advance to avoid producing ones that cause allergic reactions.

Today, governments and regulatory agencies throughout the world are grappling with how to facilitate the use of biotechnology in agriculture, industry, and medicine while ensuring that new products and procedures are safe. In the United States, such applications of biotechnology are evaluated for potential risks by various regulatory agencies, including the Food and Drug Administration, the Environmental Protection Agency, the National Institutes of Health, and the Department of Agriculture. Meanwhile, these same agencies and the public must consider the ethical implications of biotechnology.

Advances in biotechnology have allowed us to obtain complete genome sequences for humans and many other species, providing a vast treasure trove of information about genes. We can ask how certain genes differ from species to species, as well as how genes and, ultimately, entire genomes have evolved. (These are the subjects of Chapter 21.) At the same time, the increasing speed and falling cost of sequencing the genomes of individuals are raising significant ethical questions. Who should have the right to examine someone else's genetic information? How should that information be used? Should a person's genome be a factor in determining eligibility for a job or insurance? Ethical considerations, as well as concerns about potential environmental and health hazards, will likely slow some applications of biotechnology. There is always a danger that too much regulation will stifle basic research and its potential benefits. However, the power of DNA technology and genetic engineering—our ability to profoundly and rapidly alter species that have been evolving for millennia—demands that we proceed with humility and caution.

CONCEPT CHECK 20.4

1. What is the advantage of using stem cells for gene therapy?
2. List at least three different properties that have been acquired by crop plants via genetic engineering.
3. **WHAT IF?** As a physician, you have a patient with symptoms that suggest a hepatitis A infection, but you have not been able to detect viral proteins in the blood. Knowing that hepatitis A is an RNA virus, what lab tests could you perform to support your diagnosis? Explain what the results would mean.

For suggested answers, see Appendix A.

20 CHAPTER REVIEW

SUMMARY OF KEY CONCEPTS

CONCEPT 20.1

DNA cloning yields multiple copies of a gene or other DNA segment (pp. 396–405)

- **Gene cloning** and other techniques, collectively termed DNA technology, can be used to manipulate and analyze DNA and to produce useful new products and organisms.
- In **genetic engineering**, bacterial **restriction enzymes** are used to cut DNA molecules within short, specific nucleotide sequences (**restriction sites**), yielding a set of double-stranded DNA fragments with single-stranded **sticky ends**.

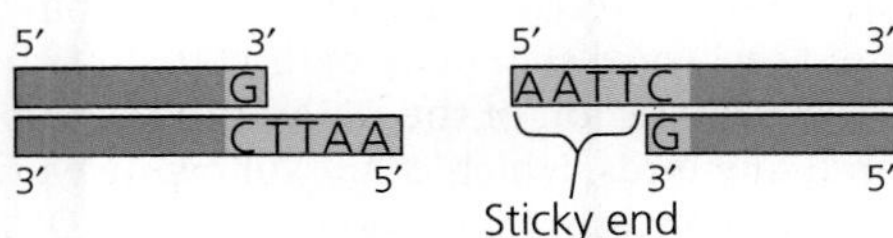

- The sticky ends on **restriction fragments** from one DNA source can base-pair with complementary sticky ends on fragments from other DNA molecules; sealing the base-paired fragments with **DNA ligase** produces **recombinant DNA** molecules.
- Cloning a eukaryotic gene in a bacterial plasmid:

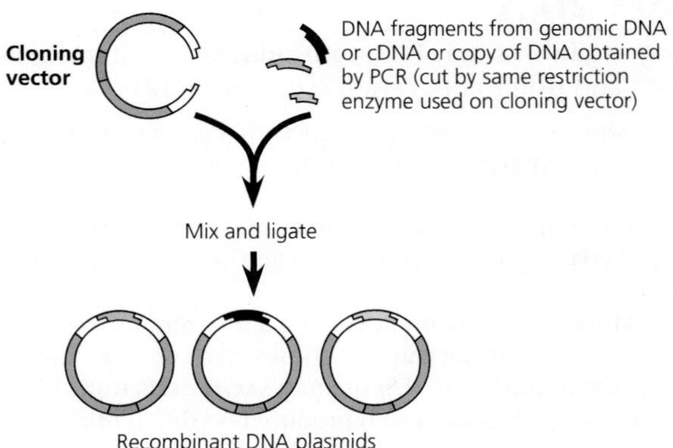

Cloning vectors include **plasmids** and **bacterial artificial chromosomes (BACs)**. Recombinant plasmids are returned to host cells, each of which divides to form a clone of cells. Collections of clones are stored as **genomic** or **complementary DNA (cDNA) libraries**. Libraries can be screened for a gene of interest using **nucleic acid hybridization** with a **nucleic acid probe**.
- Several technical difficulties hinder the expression of cloned eukaryotic genes in bacterial host cells. The use of cultured

eukaryotic cells (such as yeasts, insect cells, or cultured mammalian cells) as host cells, coupled with appropriate **expression vectors**, helps avoid these problems.
- The **polymerase chain reaction (PCR)** can produce many copies of (amplify) a specific target segment of DNA *in vitro*, using primers that bracket the desired sequence and a heat-resistant DNA polymerase.

? *Describe how the process of gene cloning results in a cell clone containing a recombinant plasmid.*

CONCEPT 20.2

DNA technology allows us to study the sequence, expression, and function of a gene (pp. 405–412)

- DNA restriction fragments of different lengths can be separated by **gel electrophoresis**. Specific fragments can be identified by **Southern blotting**, using labeled probes that hybridize to the DNA immobilized on a "blot" of the gel. Historically, **restriction fragment length polymorphisms (RFLPs)** were used to screen for some disease-causing alleles, such as the sickle-cell allele.
- Relatively short DNA fragments can be sequenced by the dideoxy chain termination method, which can be performed in automated sequencing machines. The rapid development of faster and cheaper methods is ongoing.
- Expression of a gene can be investigated using hybridization with labeled probes to look for specific mRNAs, either on a gel (**Northern blotting**) or in a whole organism (***in situ* hybridization**). Also, RNA can be transcribed into cDNA by reverse transcriptase and the cDNA amplified by PCR with specific primers (**RT-PCR**). **DNA microarrays** allow researchers to compare the expression of many genes at once in different tissues, at different times, or under different conditions.
- For a gene of unknown function, experimental inactivation of the gene and observation of the resulting phenotypic effects can provide clues to its function. In humans, **genome-wide association studies** use **single nucleotide polymorphisms (SNPs)** as genetic markers for alleles that are associated with particular conditions.

? *Complementary base pairing is the basis of most procedures used to analyze gene expression. Explain.*

CONCEPT 20.3

Cloning organisms may lead to production of stem cells for research and other applications (pp. 412–417)

- Studies showing genomic equivalence (that an organism's cells have the same genome) provided the first examples of organismal cloning.
- Single differentiated cells from mature plants are often **totipotent**: capable of generating all the tissues of a complete new plant.
- Transplantation of the nucleus from a differentiated animal cell into an enucleated egg can sometimes give rise to a new animal.
- Certain embryonic stem (ES) or adult **stem cells** from animal embryos or adult tissues can reproduce and differentiate *in vitro* as well as *in vivo*, offering the potential for medical use. ES cells are **pluripotent** but difficult to acquire. Induced pluripotent stem (iPS) cells resemble ES cells in their capacity to differentiate; they can be generated by reprogramming differentiated cells. iPS cells hold promise for medical research and regenerative medicine.

? *Describe how a researcher could carry out organismal cloning, production of ES cells, and generation of iPS cells, focusing on how the cells are reprogrammed and using mice as an example. (The procedures are basically the same in humans and mice.)*

CONCEPT 20.4

The practical applications of DNA technology affect our lives in many ways (pp. 417–423)

- DNA technology, including the analysis of genetic markers such as SNPs, is increasingly being used in the diagnosis of genetic and other diseases and offers potential for better treatment of genetic disorders (or even permanent cures through **gene therapy**), as well as more informed cancer therapies. Large-scale production of protein hormones and other proteins with therapeutic uses is possible with DNA technology. Some therapeutic proteins are being produced in **transgenic** "pharm" animals.
- Analysis of genetic markers such as **short tandem repeats (STRs)** in DNA isolated from tissue or body fluids found at crime scenes leads to a **genetic profile** that can provide definitive evidence that a suspect is innocent or strong evidence of guilt. Such analysis is also useful in parenthood disputes and in identifying the remains of crime victims.
- Genetically engineered microorganisms can be used to extract minerals from the environment or degrade various types of toxic waste materials.
- The aims of developing transgenic plants and animals are to improve agricultural productivity and food quality.
- The potential benefits of genetic engineering must be carefully weighed against the potential for harm to humans or the environment.

? *What factors affect whether a given genetic disease would be a good candidate for successful gene therapy?*

TEST YOUR UNDERSTANDING

LEVEL 1: KNOWLEDGE/COMPREHENSION

1. Which of the following tools of recombinant DNA technology is *incorrectly* paired with its use?
   a. restriction enzyme—analysis of RFLPs
   b. DNA ligase—cutting DNA, creating sticky ends of restriction fragments
   c. DNA polymerase—polymerase chain reaction to amplify sections of DNA
   d. reverse transcriptase—production of cDNA from mRNA
   e. electrophoresis—separation of DNA fragments
2. Plants are more readily manipulated by genetic engineering than are animals because
   a. plant genes do not contain introns.
   b. more vectors are available for transferring recombinant DNA into plant cells.
   c. a somatic plant cell can often give rise to a complete plant.
   d. genes can be inserted into plant cells by microinjection.
   e. plant cells have larger nuclei.
3. A paleontologist has recovered a bit of tissue from the 400-year-old preserved skin of an extinct dodo (a bird). To compare a specific region of the DNA from the sample with DNA from living birds, which of the following would be most useful for increasing the amount of dodo DNA available for testing?
   a. RFLP analysis
   b. polymerase chain reaction (PCR)
   c. electroporation
   d. gel electrophoresis
   e. Southern blotting

4. DNA technology has many medical applications. Which of the following is *not* done routinely at present?
   a. production of hormones for treating diabetes and dwarfism
   b. production of microbes that can metabolize toxins
   c. introduction of genetically engineered genes into human gametes
   d. prenatal identification of genetic disease alleles
   e. genetic testing for carriers of harmful alleles

5. In recombinant DNA methods, the term *vector* can refer to
   a. the enzyme that cuts DNA into restriction fragments.
   b. the sticky end of a DNA fragment.
   c. a SNP marker.
   d. a plasmid used to transfer DNA into a living cell.
   e. a DNA probe used to identify a particular gene.

LEVEL 2: APPLICATION/ANALYSIS

6. Which of the following would *not* be true of cDNA produced using human brain tissue as the starting material?
   a. It could be amplified by the polymerase chain reaction.
   b. It could be used to create a complete genomic library.
   c. It was produced from mRNA using reverse transcriptase.
   d. It could be used as a probe to detect genes expressed in the brain.
   e. It lacks the introns of the human genes.

7. Expression of a cloned eukaryotic gene in a bacterial cell involves many challenges. The use of mRNA and reverse transcriptase is part of a strategy to solve the problem of
   a. post-transcriptional processing.
   b. electroporation.
   c. post-translational processing.
   d. nucleic acid hybridization.
   e. restriction fragment ligation.

8. Which of the following sequences in double-stranded DNA is most likely to be recognized as a cutting site for a restriction enzyme?

a. AAGG	b. AGTC	c. GGCC	d. ACCA	e. AAAA
TTCC	TCAG	CCGG	TGGT	TTTT

9. **DRAW IT** You are making a genomic library for the aardvark, using a bacterial plasmid as a vector. The green diagram below shows the plasmid, which contains the restriction site for the enzyme used in Figure 20.3. Above the plasmid is a segment of linear aardvark DNA. Diagram your cloning procedure, showing what would happen to these two molecules during each step. Use one color for the aardvark DNA and its bases and another color for those of the plasmid. Label each step and all 5′ and 3′ ends.

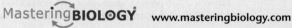

5′ TCCATGAATTCTAAAGCGCTTATGAATTCACGGC 3′
3′ AGGTACTTAAGATTTCGCGAATACTTAAGTGCCG 5′

Aardvark DNA

Plasmid

LEVEL 3: SYNTHESIS/EVALUATION

10. **WHAT IF?** Imagine you want to study one of the human crystallins, proteins present in the lens of the eye. To obtain a sufficient amount of the protein of interest, you decide to clone the gene that codes for it. Would you construct a genomic library or a cDNA library? What material would you use as a source of DNA or RNA?

11. **EVOLUTION CONNECTION**
Ethical considerations aside, if DNA-based technologies became widely used, how might they change the way evolution proceeds, as compared with the natural evolutionary mechanisms of the past 4 billion years?

12. **SCIENTIFIC INQUIRY**
You hope to study a gene that codes for a neurotransmitter protein produced in human brain cells. You know the amino acid sequence of the protein. Explain how you might (a) identify what genes are expressed in a specific type of brain cell, (b) identify (isolate) the neurotransmitter gene, (c) produce multiple copies of the gene for study, and (d) produce large quantities of the neurotransmitter for evaluation as a potential medication.

13. **SCIENCE, TECHNOLOGY, AND SOCIETY**
Is there danger of discrimination based on testing for "harmful" genes? What policies can you suggest that would prevent such abuses?

14. **SCIENCE, TECHNOLOGY, AND SOCIETY**
Government funding of embryonic stem cell research has been a contentious political issue. Why has this debate been so heated? Summarize the arguments for and against embryonic stem cell research, and explain your own position on the issue.

15. **WRITE ABOUT A THEME**
The Genetic Basis of Life In a short essay (100–150 words), discuss how the genetic basis of life plays a central role in biotechnology.

For selected answers, see Appendix A.

www.masteringbiology.com

1. MasteringBiology® Assignments

Video Tutor Session DNA Profiling Techniques
Tutorial Restriction Enzymes, Recombinant DNA, and Gene Cloning
Activities Cloning a Gene in Bacteria • Producing Human Growth Hormone • Restriction Enzymes • The Polymerase Chain Reaction • Gel Electrophoresis of DNA • Analyzing DNA Fragments Using Gel Electrophoresis • Discovery Channel Video: Cloning • DNA Fingerprinting • Making Decisions About DNA Technology: Golden Rice • Discovery Channel Videos: DNA Forensics; Transgenics
Questions Student Misconceptions • Reading Quiz • Multiple Choice • End-of-Chapter

2. eText

Read your book online, search, take notes, highlight text, and more.

3. The Study Area

Practice Tests • Cumulative Test • ***BioFlix*** 3-D Animations • MP3 Tutor Sessions • Videos • Activities • Investigations • Lab Media • Audio Glossary • Word Study Tools • Art

21

Genomes and Their Evolution

▲ **Figure 21.1 What genomic information distinguishes a human from a chimpanzee?**

KEY CONCEPTS

- **21.1** New approaches have accelerated the pace of genome sequencing
- **21.2** Scientists use bioinformatics to analyze genomes and their functions
- **21.3** Genomes vary in size, number of genes, and gene density
- **21.4** Multicellular eukaryotes have much noncoding DNA and many multigene families
- **21.5** Duplication, rearrangement, and mutation of DNA contribute to genome evolution
- **21.6** Comparing genome sequences provides clues to evolution and development

OVERVIEW

Reading the Leaves from the Tree of Life

The chimpanzee (*Pan troglodytes*) is our closest living relative on the evolutionary tree of life. The boy in **Figure 21.1** and his chimpanzee companion are intently studying the same leaf, but only one of them is able to talk about it. What accounts for this difference between two primates that share so much of their evolutionary history? With the advent of recent techniques for rapidly sequencing complete genomes, we can now start to address the genetic basis of intriguing questions like this.

The chimpanzee genome was sequenced in 2005, two years after sequencing of the human genome was largely completed. Now that we can compare our genome with that of the chimpanzee base by base, we can tackle the more general issue of what differences in the genetic information account for the distinct characteristics of these two species of primates.

In addition to determining the sequences of the human and chimpanzee genomes, researchers have obtained complete genome sequences for *E. coli* and numerous other prokaryotes, as well as many eukaryotes, including *Zea mays* (corn), *Drosophila melanogaster* (fruit fly), *Mus musculus* (house mouse), and *Macaca mulatta* (rhesus macaque). In 2010, a draft sequence was announced for the genome of *Homo neanderthalensis,* an extinct species closely related to present-day humans. These whole and partial genomes are of great interest in their own right and are also providing important insights into evolution and other biological processes. Broadening the human-chimpanzee comparison to the genomes of other primates and more distantly related animals should reveal the sets of genes that control group-defining characteristics. Beyond that, comparisons with the genomes of bacteria, archaea, fungi, protists, and plants should enlighten us about the long evolutionary history of shared ancient genes and their products.

With the genomes of many species fully sequenced, scientists can study whole sets of genes and their interactions, an approach called **genomics**. The sequencing efforts that feed this approach have generated, and continue to generate, enormous volumes of data. The need to deal with this ever-increasing flood of information has spawned the field of **bioinformatics**, the application of computational methods to the storage and analysis of biological data.

We will begin this chapter by discussing two approaches to genome sequencing and some of the advances in bioinformatics and its applications. We will then summarize what has been learned from the genomes that have been sequenced thus far. Next, we will describe the composition of the human genome as a representative genome of a complex multicellular eukaryote. Finally, we will explore current ideas about how genomes evolve and about how the evolution of developmental mechanisms could have generated the great diversity of life on Earth today.

CONCEPT 21.1

New approaches have accelerated the pace of genome sequencing

Sequencing of the human genome, an ambitious undertaking, officially began as the **Human Genome Project** in 1990. Organized by an international, publicly funded consortium of scientists at universities and research institutes, the project involved 20 large sequencing centers in six countries plus a host of other labs working on small projects.

After sequencing of the human genome was largely completed in 2003, the sequence of each chromosome was carefully analyzed and described in a series of papers, the last of which covered chromosome 1 and was published in 2006. With this refinement, researchers termed the sequencing "virtually complete." To reach these milestones, the project proceeded through three stages that provided progressively more detailed views of the human genome: linkage mapping, physical mapping, and DNA sequencing.

Three-Stage Approach to Genome Sequencing

Even before the Human Genome Project began, earlier research had sketched a rough picture of the organization of the genomes of many organisms. For instance, the karyotyping of many species had revealed their chromosome numbers and banding patterns (see Figure 13.3). And some human genes had already been located on a particular region of a chromosome by fluorescence *in situ* hybridization (FISH), a method in which fluorescently labeled nucleic acid probes are allowed to hybridize to an immobilized array of whole chromosomes (see Figure 15.1). Cytogenetic maps based on this type of information provided the starting point for more detailed mapping of the human genome.

With these cytogenetic maps of the chromosomes in hand, the initial stage in sequencing the human genome was to construct a **linkage map** (a type of genetic map; see Figure 15.11) of several thousand genetic markers spaced throughout the chromosomes (**Figure 21.2**, stage 1). The order of the markers and the relative distances between them on such a map are based on recombination frequencies. The markers can be genes or any other identifiable sequences in the DNA, such as RFLPs or short tandem repeats (STRs), both discussed in Chapter 20. By 1992, researchers had compiled a human linkage map with some 5,000 markers. Such a map enabled them to locate other markers, including genes, by testing for genetic linkage to the known markers. It was also valuable as a framework for organizing more detailed maps of particular regions. Remember from Chapter 15, however, that absolute distances between genes cannot be determined using this approach.

The next stage was the physical mapping of the human genome. In a **physical map**, the distances between markers are expressed by some physical measure, usually the number of base pairs along the DNA. For whole-genome mapping, a physical map is made by cutting the DNA of each chromosome into a number of restriction fragments and then determining the original order of the fragments in the chromosomal DNA. The key is to make fragments that overlap and then use probes or automated nucleotide sequencing of the ends to find the overlaps (see Figure 21.2, stage 2). In this way, fragments can be assigned to a sequential order that corresponds to their order in a chromosome.

The DNA fragments used for physical mapping were prepared by DNA cloning. With such a large genome, researchers had to carry out several rounds of DNA cutting, cloning, and physical mapping. In this approach, the first cloning vector was often a yeast artificial chromosome (YAC), which can carry inserted fragments a million base pairs long, or a bacterial artificial chromosome (BAC), which typically carries inserts of 100,000–300,000 base pairs. After such long fragments were put in order, each fragment was

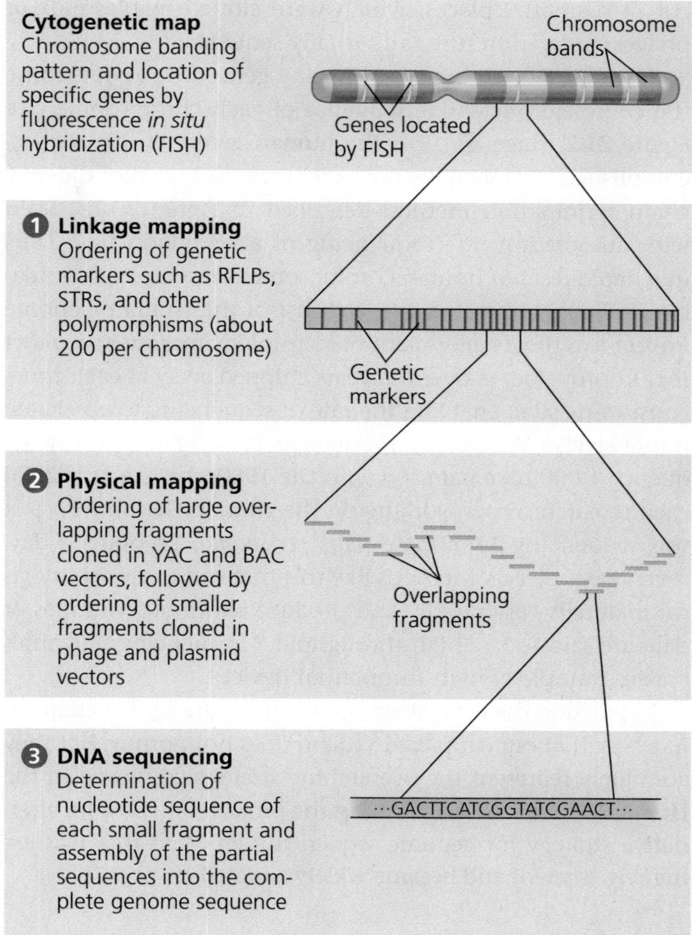

Cytogenetic map
Chromosome banding pattern and location of specific genes by fluorescence *in situ* hybridization (FISH)
Chromosome bands
Genes located by FISH
1 Linkage mapping
Ordering of genetic markers such as RFLPs, STRs, and other polymorphisms (about 200 per chromosome)
Genetic markers
2 Physical mapping
Ordering of large overlapping fragments cloned in YAC and BAC vectors, followed by ordering of smaller fragments cloned in phage and plasmid vectors
Overlapping fragments
3 DNA sequencing
Determination of nucleotide sequence of each small fragment and assembly of the partial sequences into the complete genome sequence
···GACTTCATCGGTATCGAACT···

▲ **Figure 21.2 Three-stage approach to sequencing an entire genome.** Starting with a cytogenetic map of each chromosome, researchers with the Human Genome Project proceeded through three stages to reach the ultimate goal, the virtually complete nucleotide sequence of every chromosome.

cut into smaller pieces, which were cloned in plasmids or phages, ordered in turn, and finally sequenced.

The ultimate goal in mapping any genome is to determine the complete nucleotide sequence of each chromosome (see Figure 21.2, stage 3). For the human genome, this was accomplished by sequencing machines, using the dideoxy chain termination method described in Figure 20.12. Even with automation, the sequencing of all 3 billion base pairs in a haploid set of human chromosomes presented a formidable challenge. In fact, a major thrust of the Human Genome Project was the development of technology for faster sequencing. Improvements over the years chipped away at each time-consuming step, enabling the rate of sequencing to accelerate impressively: Whereas a productive lab could typically sequence 1,000 base pairs a day in the 1980s, by the year 2000 each research center working on the Human Genome Project was sequencing 1,000 base pairs *per second*, 24 hours a day, seven days a week. Methods like this that can analyze biological materials very rapidly and produce enormous volumes of data are said to be "high-throughput." Sequencing machines are an example of high-throughput devices.

In practice, the three stages shown in Figure 21.2 overlapped in a way that our simplified version does not portray, but they accurately represent the overarching strategy employed in the Human Genome Project. During the project, however, an alternative strategy for genome sequencing emerged that was extremely efficient and became widely adopted.

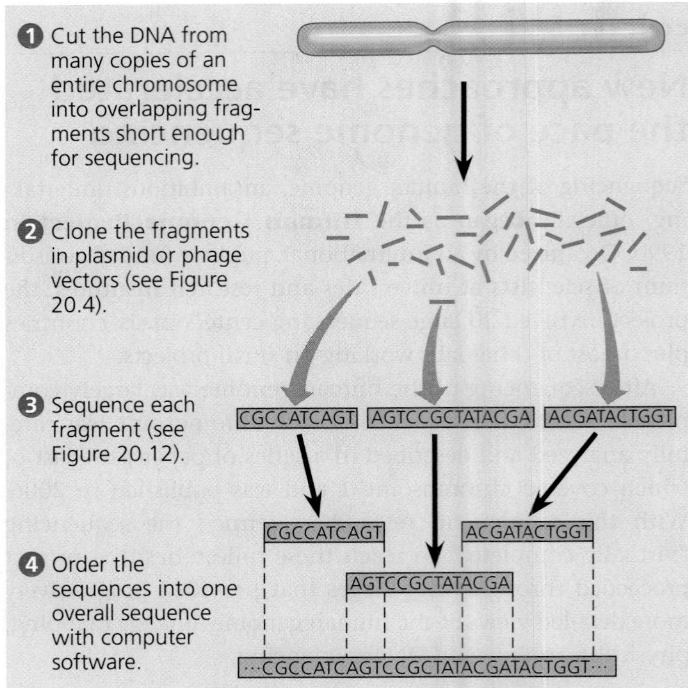

1 Cut the DNA from many copies of an entire chromosome into overlapping fragments short enough for sequencing.
2 Clone the fragments in plasmid or phage vectors (see Figure 20.4).
3 Sequence each fragment (see Figure 20.12).
CGCCATCAGT
AGTCCGCTATACGA
ACGATACTGGT
4 Order the sequences into one overall sequence with computer software.
CGCCATCAGT
ACGATACTGGT
AGTCCGCTATACGA
···CGCCATCAGTCCGCTATACGATACTGGT···

▲ **Figure 21.3 Whole-genome shotgun approach to sequencing.** In this approach, developed by Craig Venter and colleagues at the company he founded, Celera Genomics, random DNA fragments are sequenced and then ordered relative to each other. Compare this approach with the hierarchical, three-stage approach shown in Figure 21.2.

? *The fragments in stage 2 of this figure are depicted as scattered, whereas those in stage 2 of Figure 21.2 are drawn in a much more orderly fashion. How do these depictions reflect the two approaches?*

Whole-Genome Shotgun Approach to Genome Sequencing

In 1992, emboldened by advances in sequencing and computer technology, molecular biologist J. Craig Venter devised an alternative approach to the sequencing of whole genomes. Called the *whole-genome shotgun approach*, it essentially skips the linkage mapping and physical mapping stages and starts directly with the sequencing of DNA fragments from randomly cut DNA. Powerful computer programs then assemble the resulting very large number of overlapping short sequences into a single continuous sequence (**Figure 21.3**). In 1998, despite the skepticism of many scientists, Venter set up a company (Celera Genomics) and declared his intention to sequence the entire human genome. Five years later, and 13 years after the Human Genome Project began, Celera Genomics and the public consortium jointly announced that sequencing of the human genome was largely complete.

Representatives of the public consortium point out that Celera's accomplishment relied heavily on the consortium's maps and sequence data and that the infrastructure established by their approach was a tremendous aid to Celera's efforts. Venter, on the other hand, has argued for the efficiency and economy of Celera's methods, and indeed, the public consortium made some use of them as well. Evidently, both approaches made valuable contributions.

Today, the whole-genome shotgun approach is widely used. Also, the development of newer sequencing techniques, generally called *sequencing by synthesis* (see Chapter 20), has resulted in massive increases in speed and decreases in the cost of sequencing entire genomes. In these new techniques, many very small fragments (fewer than 100 base pairs) are sequenced at the same time, and computer software rapidly assembles the complete sequence. Because of the sensitivity of these techniques, the fragments can be sequenced directly; the cloning step (stage 2 in Figure 21.3) is unnecessary. Whereas sequencing the first human genome took 13 years and cost $100 million, James Watson's genome was sequenced during four months in 2007 for about $1 million, and a group of researchers reported in 2010 that they had rapidly sequenced three human genomes for approximately $4,400 each!

These technological advances have also facilitated an approach called **metagenomics** (from the Greek *meta*, beyond), in which DNA from a group of species (a *metagenome*) is collected from an environmental sample and sequenced. Again, computer software accomplishes the task of sorting out the partial sequences and assembling them into specific genomes. So far, this approach has been applied to microbial communities found in environments as diverse as the Sargasso Sea and the human intestine. The ability to sequence the DNA of mixed populations eliminates the need to culture each species separately in the lab, a difficulty that has limited the study of many microbial species.

At first glance, genome sequences of humans and other organisms are simply dry lists of nucleotide bases—millions of A's, T's, C's, and G's in mind-numbing succession. Crucial to making sense of this massive amount of data have been new analytical approaches, which we discuss next.

CONCEPT CHECK 21.1

1. What is the major difference between a linkage map and a physical map of a chromosome?
2. In general, how does the approach to genome mapping used in the Human Genome Project differ from the whole-genome shotgun approach?

For suggested answers, see Appendix A.

CONCEPT 21.2

Scientists use bioinformatics to analyze genomes and their functions

Each of the 20 or so sequencing centers around the world working on the Human Genome Project churned out voluminous amounts of DNA sequence day after day. As the data began to accumulate, the need to coordinate efforts to keep track of all the sequences became clear. Thanks to the foresight of research scientists and government officials involved in the Human Genome Project, its goals included the establishment of banks of data, or databases, and the refining of analytical software. These databases and software programs would then be centralized and made readily accessible on the Internet. Accomplishing this aim has accelerated progress in DNA sequence analysis by making bioinformatics resources available to researchers worldwide and by speeding up the dissemination of information.

Centralized Resources for Analyzing Genome Sequences

Government-funded agencies carried out their mandate to establish databases and provide software with which scientists could analyze the sequence data. For example, in the United States, a joint endeavor between the National Library of Medicine and the National Institutes of Health (NIH) created the National Center for Biotechnology Information (NCBI), which maintains a website (www.ncbi.nlm.nih.gov) with extensive bioinformatics resources. On this site are links to databases, software, and a wealth of information about genomics and related topics. Similar websites have also been established by the European Molecular Biology Laboratory, the DNA Data Bank of Japan, and BGI (formerly known as the Beijing Genome Institute) in Shenzhen, China, three genome centers with which the NCBI collaborates. These large, comprehensive websites are complemented by others maintained by individual or small groups of laboratories. Smaller websites often provide databases and software designed for a narrower purpose, such as studying genetic and genomic changes in one particular type of cancer.

The NCBI database of sequences is called GenBank. As of May 2010, it included the sequences of 119 million fragments of genomic DNA, totaling 114 billion base pairs! GenBank is constantly updated, and the amount of data it contains is estimated to double approximately every 18 months. Any sequence in the database can be retrieved and analyzed using software from the NCBI website or elsewhere.

One software program available on the NCBI website, called BLAST, allows the visitor to compare a DNA sequence with every sequence in GenBank, base by base, to look for similar regions. Another program allows comparison of predicted protein sequences. Yet a third can search any protein sequence for common stretches of amino acids (domains) for which a function is known or suspected, and it can show a three-dimensional model of the domain alongside other relevant information (**Figure 21.4**, on the next page). There is even a software program that can compare a collection of sequences, either nucleic acids or polypeptides, and diagram them in the form of an evolutionary tree based on the sequence relationships. (One such diagram is shown in Figure 21.16.)

Two research institutions, Rutgers University and the University of California, San Diego, also maintain a worldwide Protein Data Bank, a database of all three-dimensional protein structures that have been determined. (The database is accessible at www.wwpdb.org.) These structures can be rotated by the viewer to show all sides of the protein.

There is a vast array of resources available for researchers anywhere in the world to use. Let us now consider the types of questions scientists can address using these resources.

Identifying Protein-Coding Genes and Understanding Their Functions

Using available DNA sequences, geneticists can study genes directly, without having to infer genotype from phenotype as in classical genetics. But this approach, called *reverse genetics*, poses a new challenge: determining the phenotype from the genotype. Given a long DNA sequence from a database such as GenBank, the aim of scientists is to identify all protein-coding genes in the sequence and ultimately their functions. This process is called **gene annotation**.

In the past, gene annotation was carried out laboriously by individual scientists interested in particular genes, but the process has now been largely automated. The usual approach is to use software to scan the stored sequences for transcriptional and translational start and stop signals, for RNA-splicing sites, and for other telltale signs of protein-coding genes. The software also looks for certain short sequences that specify known mRNAs. Thousands of such sequences, called

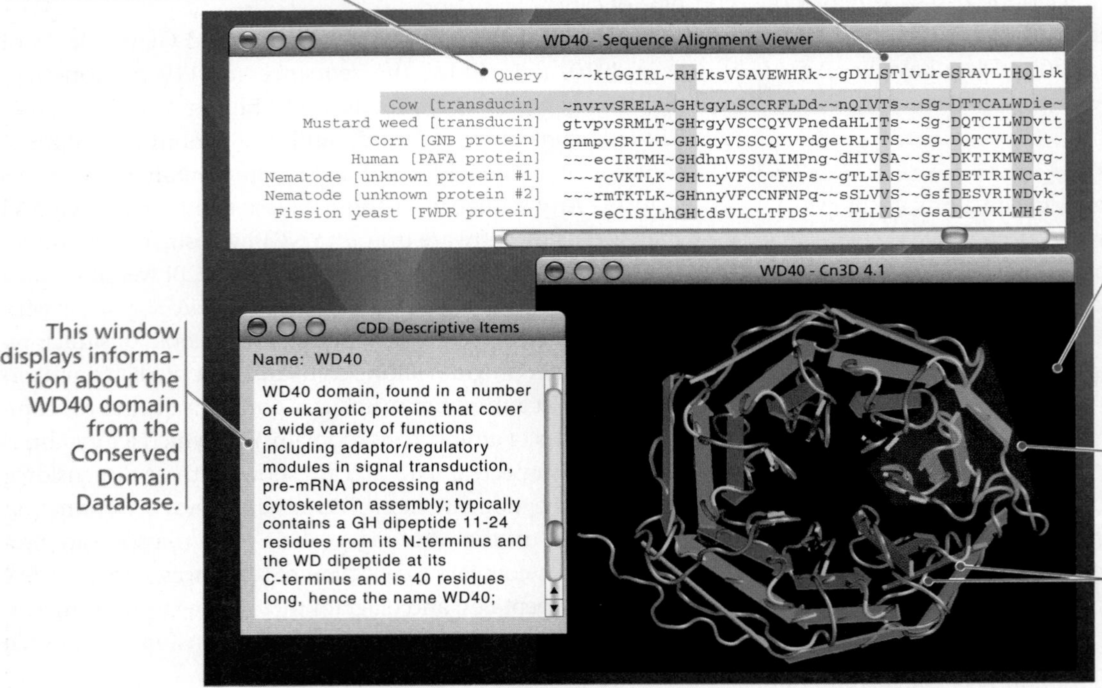
In this window, a partial amino acid sequence from an unknown muskmelon protein ("Query") is aligned with sequences from other proteins that the computer program found to be similar. Each sequence represents a domain called WD40.
Four hallmarks of the WD40 domain are highlighted in yellow. (Sequence similarity is based on chemical aspects of the amino acids, so the amino acids in each hallmark region are not always identical.)
WD40 - Sequence Alignment Viewer
Query ~~~ktGGIRL~RHfksVSAVEWHRk~~gDYLSTlvLreSRAVLIHQlsk
Cow [transducin] ~nvrvSRELA~GHtgyLSCCRFLDd~~nQIVTs~~Sg~DTTCALWDie~
Mustard weed [transducin] gtvpvSRMLT~GHrgyVSCCQYVPnedaHLITs~~Sg~DQTCILWDvtt
Corn [GNB protein] gnmpvSRILT~GHkgyVSSCQYVPdgetRLITS~~Sg~DQTCVLWDvt~
Human [PAFA protein] ~~~ecIRTMH~GHdhnVSSVAIMPng~dHIVSA~~Sr~DKTIKMWEvg~
Nematode [unknown protein #1] ~~~rcVKTLK~GHtnyVFCCCFNPs~~gTLIAS~~GsfDETIRIWCar~
Nematode [unknown protein #2] ~~~rmTKTLK~GHnnyVFCCNFNPq~~sSLVVS~~GsfDESVRIWDvk~
Fission yeast [FWDR protein] ~~~seCISILhGHtdsVLCLTFDS~~~~TLLVS~~GsaDCTVKLWHfs~
The Cn3D program displays a three-dimensional ribbon model of cow transducin (the protein highlighted in purple in the Sequence Alignment Viewer). This protein is the only one of those shown for which a structure has been determined. The sequence similarity of the other proteins to cow transducin suggests that their structures are likely to be similar.
WD40 - Cn3D 4.1
This window displays information about the WD40 domain from the Conserved Domain Database.
CDD Descriptive Items
Name: WD40
WD40 domain, found in a number of eukaryotic proteins that cover a wide variety of functions including adaptor/regulatory modules in signal transduction, pre-mRNA processing and cytoskeleton assembly; typically contains a GH dipeptide 11-24 residues from its N-terminus and the WD dipeptide at its C-terminus and is 40 residues long, hence the name WD40;
Cow transducin contains seven WD40 domains, one of which is highlighted here in gray.
The yellow segments correspond to the WD40 hallmarks highlighted in yellow in the window above.

▲ **Figure 21.4 Bioinformatics tools available on the Internet.** A website maintained by the National Center for Biotechnology Information allows scientists and the public to access DNA and protein sequences and other stored data. The site includes a link to a protein structure database (Conserved Domain Database, CDD) that can find and describe similar domains in related proteins, as well as software (Cn3D, "See in 3D") that displays three-dimensional models of domains for which the structure has been determined. Some results are shown from a search for regions of proteins similar to an amino acid sequence in a muskmelon protein.

expressed sequence tags, or *ESTs*, have been collected from cDNA sequences and are cataloged in computer databases. This type of analysis identifies sequences that may be previously unknown protein-coding genes.

The identities of about half of the human genes were known before the Human Genome Project began. But what about the others, the previously unknown genes revealed by analysis of DNA sequences? Clues about their identities and functions come from comparing sequences that might be genes with known genes from other organisms, using the software described previously. Due to redundancy in the genetic code, the DNA sequence itself may vary more than the protein sequence does. Thus, scientists interested in proteins often compare the predicted amino acid sequence of a protein to that of other proteins.

Sometimes a newly identified sequence will match, at least partially, the sequence of a gene or protein whose function is well known. For example, part of a new gene may match a known gene that encodes an important signaling pathway protein such as a protein kinase (see Chapter 11), suggesting that the new gene does, too. Alternatively, the new gene sequence may be similar to a previously encountered sequence whose function is still unknown. Another possibility is that the sequence is entirely unlike anything ever seen before. This was true for about a third of the genes of *E. coli* when its genome was sequenced. In the last case, protein function is usually deduced through a combination of biochemical and functional studies. The biochemical approach aims to determine the three-dimensional structure of the protein as well as other attributes, such as potential binding sites for other molecules. Functional studies usually involve blocking or disabling the gene to see how the phenotype is affected. RNAi, described in Chapter 20, is an example of an experimental technique used to block gene function.

Understanding Genes and Gene Expression at the Systems Level

The impressive computational power provided by the tools of bioinformatics allows the study of whole sets of genes and their interactions, as well as the comparison of genomes from different species. Genomics is a rich source of new insights into fundamental questions about genome organization, regulation of gene expression, growth and development, and evolution.

One informative approach has been taken by a research project called ENCODE (Encyclopedia of DNA Elements), which began in 2003. First, researchers focused intensively on 1% of the human genome and attempted to learn all they could about the functionally important elements in that sequence. They looked for protein-coding genes and genes for noncoding RNAs as well as sequences that regulate DNA replication, gene expression (such as enhancers and promoters), and chromatin modifications. The pilot project was completed in 2007, yielding a wealth of information. One big surprise, discussed in Concept 18.3, was that over 90% of the region was transcribed into RNA, even though less than 2% codes for proteins. The success of this approach has led to two follow-up studies, one extending the analysis to the entire human genome and the other analyzing in a similar fashion the genomes of two model organisms, the soil nematode *Caenorhabditis elegans* and the fruit fly *Drosophila melanogaster*. Because genetic and molecular biological experiments can be performed on these species, testing the activities of potentially functional DNA elements in their genomes will reveal much about how the human genome works.

The success in sequencing genomes and studying entire sets of genes has encouraged scientists to attempt similar systematic study of the full protein sets (*proteomes*) encoded by genomes, an approach called **proteomics**. Proteins, not the genes that encode them, actually carry out most of the activities of the cell. Therefore, we must study when and where proteins are produced in an organism, as well as how they interact in networks, if we are to understand the functioning of cells and organisms.

How Systems Are Studied: An Example

Genomics and proteomics are enabling molecular biologists to approach the study of life from an increasingly global perspective. Using the tools we have described, biologists have begun to compile catalogs of genes and proteins—listings of all the "parts" that contribute to the operation of cells, tissues, and organisms. With such catalogs in hand, researchers have shifted their attention from the individual parts to their functional integration in biological systems. As you may recall, in Chapter 1 we discussed this systems biology approach, which aims to model the dynamic behavior of whole biological systems.

One important use of the systems biology approach is to define gene circuits and protein interaction networks. To map the protein interaction network in the yeast *Saccharomyces cerevisiae*, for instance, researchers used sophisticated techniques to knock out (disable) pairs of genes, one pair at a time, creating doubly mutant cells. They then compared the fitness of each double mutant (based in part on the size of the cell colony it formed) to that predicted from the fitnesses of the two single mutants. The researchers reasoned that if the observed fitness matched the prediction, then the products of the two genes didn't interact with each other, but if the observed fitness was greater or less than predicted, then the gene products interacted in the cell. Computer software then mapped genes based on the similarity of their interactions; a network-like "functional map" of these genetic interactions is displayed in **Figure 21.5**. To process the vast number of protein-protein interactions generated by this experiment and integrate them

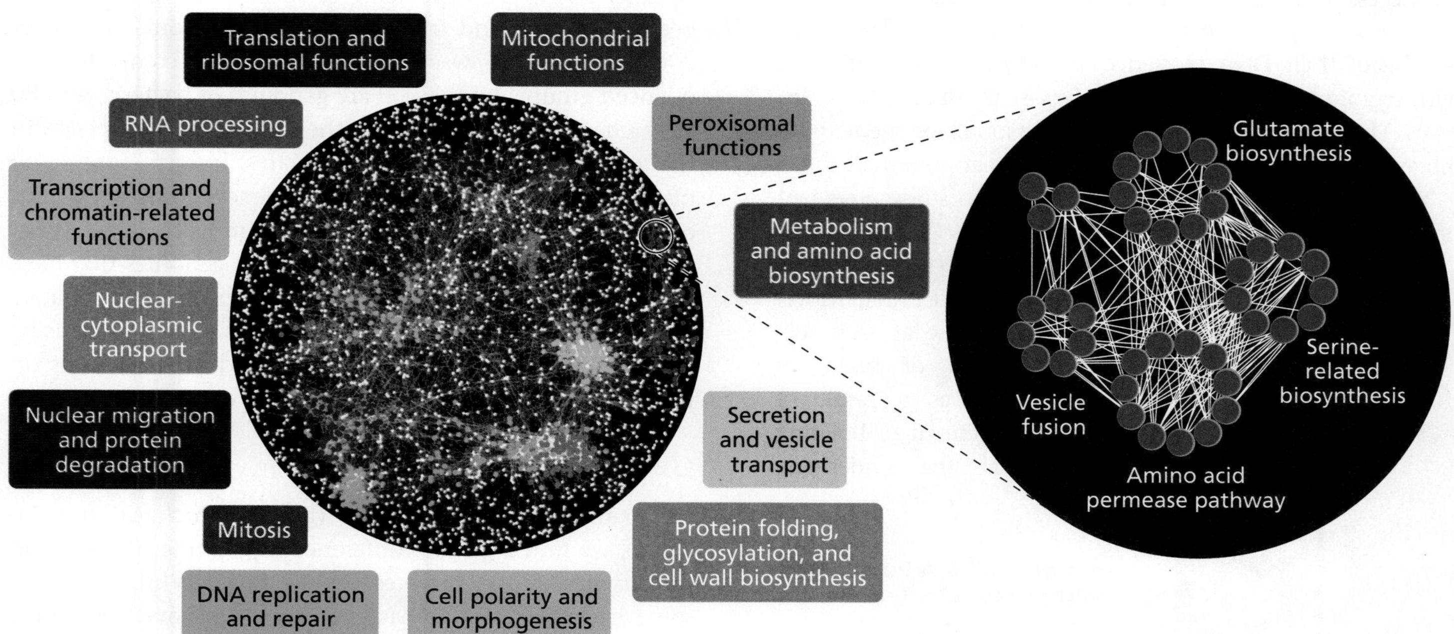

▲ **Figure 21.5 The systems biology approach to protein interactions.** This global protein interaction map shows the likely interactions (lines) among about 4,500 gene products (circles) in the yeast *Saccharomyces cerevisiae*. Circles of the same color represent gene products involved in one of the 13 cellular functions listed around the map. The blowup shows additional details of one map region where the gene products (blue circles) carry out amino acid biosynthesis, uptake, and related functions.

into the completed map required powerful computers, mathematical tools, and newly developed software. Thus, the systems biology approach has really been made possible by advances in computer technology and bioinformatics.

Application of Systems Biology to Medicine

The Cancer Genome Atlas is another example of systems biology in which a large group of interacting genes and gene products are analyzed together. This project, under the joint leadership of the National Cancer Institute and the NIH, aims to determine how changes in biological systems lead to cancer. A three-year pilot project beginning in 2007 set out to find all the common mutations in three types of cancer—lung cancer, ovarian cancer, and glioblastoma of the brain—by comparing gene sequences and patterns of gene expression in cancer cells with those in normal cells. Work on glioblastoma has confirmed the role of several suspected genes and identified a few unknown ones, suggesting possible new targets for therapies. The approach has proved so fruitful for these three types of cancer that it has been extended to ten other types, chosen because they are common and often lethal in humans.

Systems biology has tremendous potential in human medicine that is just starting to be explored. Silicon and glass "chips" have been developed that hold a microarray of most of the known human genes (**Figure 21.6**). Such chips are being used to analyze gene expression patterns in patients suffering from various cancers and other diseases, with the eventual aim of tailoring their treatment to their unique genetic makeup and the specifics of their cancers. This approach has had modest success in characterizing subsets of several cancers.

Ultimately, people may carry with their medical records a catalog of their DNA sequence, a sort of genetic bar code, with regions highlighted that predispose them to specific diseases. The use of such sequences for personalized medicine—disease prevention and treatment—has great potential.

Systems biology is a very efficient way to study emergent properties at the molecular level. Recall from Chapter 1 that according to the theme of emergent properties, novel properties arise at each successive level of biological complexity as a result of the arrangement of building blocks at the underlying level. The more we can learn about the arrangement and interactions of the components of genetic systems, the deeper will be our understanding of whole organisms. The rest of this chapter will survey what we've learned from genomic studies thus far.

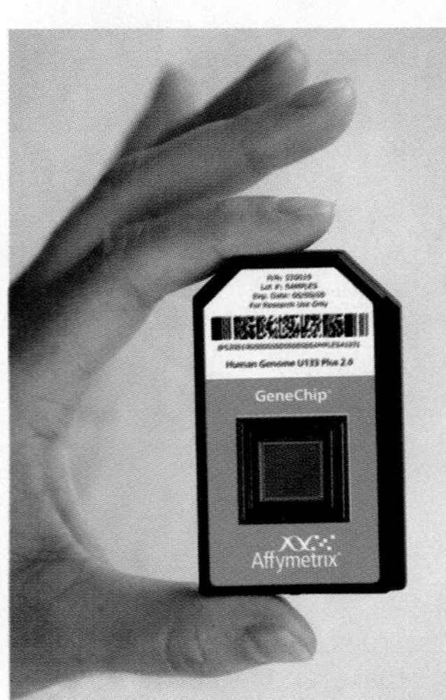

◀ **Figure 21.6 A human gene microarray chip.** Tiny spots of DNA arranged in a grid on this silicon wafer represent almost all of the genes in the human genome. Using this chip, researchers can analyze expression patterns for all these genes at the same time.

CONCEPT CHECK 21.2

1. What role does the Internet play in current genomics and proteomics research?
2. Explain the advantage of the systems biology approach to studying cancer versus the approach of studying a single gene at a time.
3. **MAKE CONNECTIONS** The ENCODE pilot project found that more than 90% of the genomic region being studied was transcribed into RNAs, far more than could be accounted for by protein-coding genes. Review Concept 18.3 (pp. 364–366) and suggest some roles that these RNAs might play.
4. **MAKE CONNECTIONS** In Concept 20.2 (p. 411), you learned about genome-wide association studies. Explain how these studies use the systems biology approach.

For suggested answers, see Appendix A.

CONCEPT 21.3

Genomes vary in size, number of genes, and gene density

By early 2010, the sequencing of about 1,200 genomes had been completed and that of over 5,500 genomes and over 200 metagenomes was in progress. In the completely sequenced group, about 1,000 are genomes of bacteria, and 80 are archaeal genomes. Among the 124 eukaryotic species in the group are vertebrates, invertebrates, protists, fungi, and plants. The accumulated genome sequences contain a wealth of information that we are now beginning to mine. What have we learned so far by comparing the genomes that have been sequenced? In this section, we will examine the characteristics of genome size, number of genes, and gene density. Because these characteristics are so broad, we will focus on general trends, for which there are often exceptions.

Genome Size

Comparing the three domains (Bacteria, Archaea, and Eukarya), we find a general difference in genome size between prokaryotes and eukaryotes (**Table 21.1**). While there are some exceptions, most bacterial genomes have between 1 and 6 million base pairs (Mb); the genome of *E. coli*, for instance, has 4.6 Mb. Genomes of archaea are, for the most part, within the size range of bacterial genomes. (Keep in mind, however, that many fewer archaeal genomes have

Table 21.1 Genome Sizes and Estimated Numbers of Genes*

Organism	Haploid Genome Size (Mb)	Number of Genes	Genes per Mb
Bacteria			
Haemophilus influenzae	1.8	1,700	940
Escherichia coli	4.6	4,400	950
Archaea			
Archaeoglobus fulgidus	2.2	2,500	1,130
Methanosarcina barkeri	4.8	3,600	750
Eukaryotes			
Saccharomyces cerevisiae (yeast, a fungus)	12	6,300	525
Caenorhabditis elegans (nematode)	100	20,100	200
Arabidopsis thaliana (mustard family plant)	120	27,000	225
Drosophila melanogaster (fruit fly)	165	13,700	83
Oryza sativa (rice)	430	42,000	98
Zea mays (corn)	2,300	32,000	14
Mus musculus (house mouse)	2,600	22,000	11
Ailuropoda melanoleuca (giant panda)	2,400	21,000	9
Homo sapiens (human)	3,000	$<$21,000	7
Fritillaria assyriaca (lily family plant)	124,000	ND	ND

*Some values given here are likely to be revised as genome analysis continues. Mb = million base pairs. ND = not determined.

been completely sequenced, so this picture may change.) Eukaryotic genomes tend to be larger: The genome of the single-celled yeast *Saccharomyces cerevisiae* (a fungus) has about 12 Mb, while most animals and plants, which are multicellular, have genomes of at least 100 Mb. There are 165 Mb in the fruit fly genome, while humans have 3,000 Mb, about 500 to 3,000 times as many as a typical bacterium.

Aside from this general difference between prokaryotes and eukaryotes, a comparison of genome sizes among eukaryotes fails to reveal any systematic relationship between genome size and the organism's phenotype. For instance, the genome of *Fritillaria assyriaca*, a flowering plant in the lily family, contains 124 billion base pairs (124,000 Mb), about 40 times the size of the human genome. Even more striking, there is a single-celled amoeba, *Polychaos dubia*, whose genome size has been estimated at 670,000 Mb. (This genome has not yet been sequenced.) On a finer scale, comparing two insect species, the cricket (*Anabrus simplex*) genome turns out to have 11 times as many base pairs as the *Drosophila melanogaster* genome. There is a wide range of genome sizes within the groups of protists, insects, amphibians, and plants and less of a range within mammals and reptiles.

Number of Genes

The number of genes also varies between prokaryotes and eukaryotes: Bacteria and archaea, in general, have fewer genes than eukaryotes. Free-living bacteria and archaea have from 1,500 to 7,500 genes, while the number of genes in eukaryotes ranges from about 5,000 for unicellular fungi to at least 40,000 for some multicellular eukaryotes (see Table 21.1).

Within the eukaryotes, the number of genes in a species is often lower than expected from simply considering the size of its genome. Looking at Table 21.1, you can see that the genome of the nematode *C. elegans* is 100 Mb in size and contains roughly 20,000 genes. The *Drosophila* genome, in comparison, is much bigger (165 Mb) but has about two-thirds the number of genes—only 13,700 genes.

Considering an example closer to home, we noted that the human genome contains 3,000 Mb, well over ten times the size of either the *Drosophila* or *C. elegans* genome. At the outset of the Human Genome Project, biologists expected somewhere between 50,000 and 100,000 genes to be identified in the completed sequence, based on the number of known human proteins. As the project progressed, the estimate was revised downward several times, and in 2010, the most reliable count placed the number at fewer than 21,000. This relatively low number, similar to the number of genes in the nematode *C. elegans*, has surprised biologists, who had clearly expected many more human genes.

What genetic attributes allow humans (and other vertebrates) to get by with no more genes than nematodes? An important factor is that vertebrate genomes "get more bang for the buck" from their coding sequences because of extensive alternative splicing of RNA transcripts. Recall that this process generates more than one functional protein from a single gene (see Figure 18.13). A typical human gene contains about ten exons, and an estimated 93% or so of these multi-exon genes are spliced in at least two different ways. Some genes are expressed in hundreds of alternatively spliced forms, others in just two. It is not yet possible to catalog all of the different forms, but it is clear that the number of different proteins encoded in the human genome far exceeds the proposed number of genes.

Additional polypeptide diversity could result from post-translational modifications such as cleavage or the addition of carbohydrate groups in different cell types or at different developmental stages. Finally, the discovery of miRNAs and other small RNAs that play regulatory roles have added a new variable to the mix (see Concept 18.3). Some scientists think that this added level of regulation, when present, may contribute to greater organismal complexity for a given number of genes.

Gene Density and Noncoding DNA

In addition to genome size and number of genes, we can compare gene density in different species—in other words, how many genes there are in a given length of DNA. When we compare the genomes of bacteria, archaea, and eukaryotes, we see that eukaryotes generally have larger genomes but fewer genes in a given number of base pairs. Humans have hundreds or thousands of times as many base pairs in their genome as most bacteria, as we already noted, but only 5 to 15 times as many genes; thus, gene density is lower in humans (see Table 21.1). Even unicellular eukaryotes, such as yeasts, have fewer genes per million base pairs than bacteria and archaea. Among the genomes that have been sequenced completely thus far, humans and other mammals have the lowest gene density.

In all bacterial genomes studied so far, most of the DNA consists of genes for protein, tRNA, or rRNA; the small amount remaining consists mainly of nontranscribed regulatory sequences, such as promoters. The sequence of nucleotides along a bacterial protein-coding gene proceeds from start to finish without interruption by noncoding sequences (introns). In eukaryotic genomes, by contrast, most of the DNA neither encodes protein nor is transcribed into RNA molecules of known function, and the DNA includes more complex regulatory sequences. In fact, humans have 10,000 times as much noncoding DNA as bacteria. Some of this DNA in multicellular eukaryotes is present as introns within genes. Indeed, introns account for most of the difference in average length between human genes (27,000 base pairs) and bacterial genes (1,000 base pairs).

In addition to introns, multicellular eukaryotes have a vast amount of non-protein-coding DNA between genes. In the next section, we will describe the composition and arrangement of these great stretches of DNA in the human genome.

CONCEPT CHECK 21.3

1. According to the best current estimate, the human genome contains fewer than 21,000 genes. However, there is evidence that human cells produce many more than 21,000 different polypeptides. What processes might account for this discrepancy?
2. The number of sequenced genomes is constantly being updated. Go to www.genomesonline.org to find the current number of completed genomes for each domain as well as the number of genomes whose sequencing is in progress. (*Hint*: Click on "Enter GOLD," and then click on "Published Complete Genomes" for extra information.)
3. **WHAT IF?** What evolutionary processes might account for prokaryotes having smaller genomes than eukaryotes?

For suggested answers, see Appendix A.

CONCEPT 21.4

Multicellular eukaryotes have much noncoding DNA and many multigene families

We have spent most of this chapter, and indeed this unit, focusing on genes that code for proteins. Yet the coding regions of these genes and the genes for RNA products such as rRNA, tRNA, and miRNA make up only a small portion of the genomes of most multicellular eukaryotes. The bulk of many eukaryotic genomes consists of DNA sequences that neither code for proteins nor are transcribed to produce RNAs with known functions; this noncoding DNA was often described in the past as "junk DNA." However, much evidence is accumulating that this DNA plays important roles in the cell, an idea supported by its persistence in diverse genomes over many hundreds of generations. For example, comparison of the genomes of humans, rats, and mice has revealed the presence of almost 500 regions of noncoding DNA that are identical in sequence in all three species. This is a higher level of sequence conservation than is seen for protein-coding regions in these species, strongly suggesting that the noncoding regions have important functions. In this section, we examine how genes and noncoding DNA sequences are organized within genomes of multicellular eukaryotes, using the human genome as our main example. Genome organization tells us much about how genomes have evolved and continue to evolve, the next subject we'll consider.

Once the sequencing of the human genome was completed, it became clear that only a tiny part—1.5%—codes for proteins or is transcribed into rRNAs or tRNAs. **Figure 21.7** shows what is known about the makeup of the remaining 98.5%. Gene-related regulatory sequences and introns account, respectively, for 5% and about 20% of the human genome. The rest, located between functional genes, includes some unique noncoding DNA, such as gene fragments and **pseudogenes**, former genes that have accumulated mutations over a long time and no longer produce functional proteins. (The genes that produce small noncoding RNAs are a tiny percentage of the genome, distributed between the 20% introns and the 15% unique noncoding DNA.) Most intergenic DNA, however, is **repetitive DNA**, which consists of sequences that are present in multiple copies in the genome. Somewhat surprisingly, about 75% of this repetitive DNA (44% of the entire human genome) is made up of units called transposable elements and sequences related to them.

Transposable Elements and Related Sequences

Both prokaryotes and eukaryotes have stretches of DNA that can move from one location to another within the genome. These stretches are known as *transposable genetic elements*, or

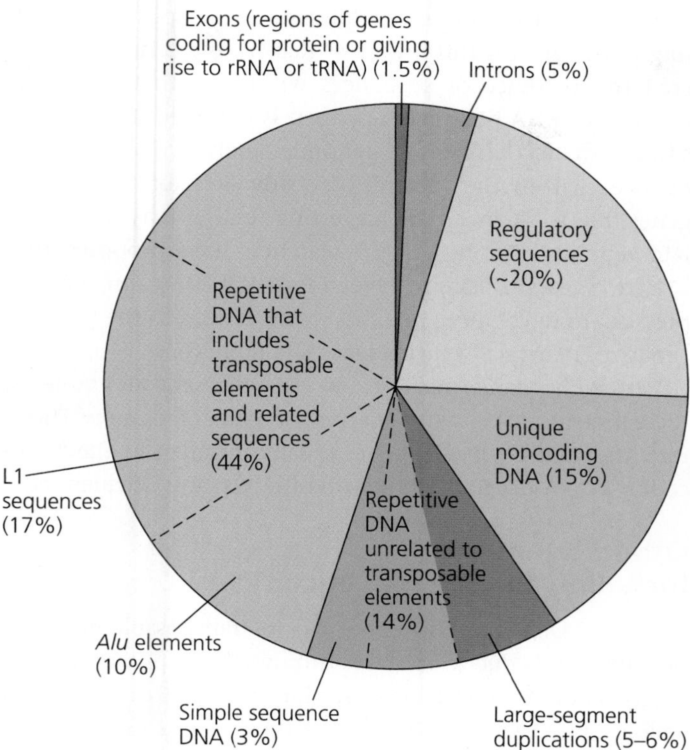

▲ **Figure 21.7 Types of DNA sequences in the human genome.** The gene sequences that code for proteins or are transcribed into rRNA or tRNA molecules make up only about 1.5% of the human genome (dark purple in the pie chart), while introns and regulatory sequences associated with genes (light purple) make up about a quarter. The vast majority of the human genome does not code for proteins or give rise to known RNAs, and much of it is repetitive DNA (dark and light green). Because repetitive DNA is the most difficult to sequence and analyze, classification of some portions is tentative, and the percentages given here may shift slightly as genome analysis proceeds. The genes that are transcribed into small noncoding RNAs such as miRNAs, which were recently discovered, are found among unique noncoding DNA sequences and within introns and thus would be included in two segments of this chart.

simply **transposable elements**. During the process called *transposition*, a transposable element moves from one site in a cell's DNA to a different target site by a type of recombination process. Transposable elements are sometimes called "jumping genes," but it should be kept in mind that they never completely detach from the cell's DNA. Instead, the original and new DNA sites are brought together by enzymes and other proteins that bend the DNA.

The first evidence for wandering DNA segments came from American geneticist Barbara McClintock's breeding experiments with Indian corn (maize) in the 1940s and 1950s **(Figure 21.8)**. As she tracked corn plants through multiple generations, McClintock identified changes in the color of corn kernels that made sense only if she postulated the existence of genetic elements capable of moving from other locations in the genome into the genes for kernel color, disrupting the genes so that the kernel color was changed. McClintock's discovery was met with great skepticism and virtually discounted at the time. Her

▲ **Figure 21.8 The effect of transposable elements on corn kernel color.** Barbara McClintock first proposed the idea of mobile genetic elements after observing variegations in corn kernel color (right).

careful work and insightful ideas were finally validated many years later when transposable elements were found in bacteria. In 1983, at the age of 81, McClintock received the Nobel Prize for her pioneering research.

Movement of Transposons and Retrotransposons

Eukaryotic transposable elements are of two types. The first type are **transposons**, which move within a genome by means of a DNA intermediate. Transposons can move by a "cut-and-paste" mechanism, which removes the element from the original site, or by a "copy-and-paste" mechanism, which leaves a copy behind **(Figure 21.9)**. Both mechanisms require an enzyme called *transposase*, which is generally encoded by the transposon.

Most transposable elements in eukaryotic genomes are of the second type, **retrotransposons**, which move by means of an RNA intermediate that is a transcript of the retrotransposon DNA. Retrotransposons always leave a copy at the original site during transposition, since they are initially transcribed

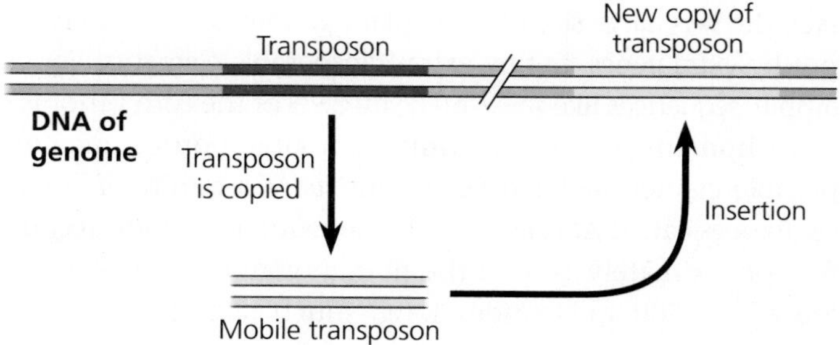

▲ **Figure 21.9 Transposon movement.** Movement of transposons by either the cut-and-paste mechanism or the copy-and-paste mechanism (shown here) involves a double-stranded DNA intermediate that is inserted into the genome.

? *How would this figure differ if it showed the cut-and-paste mechanism?*

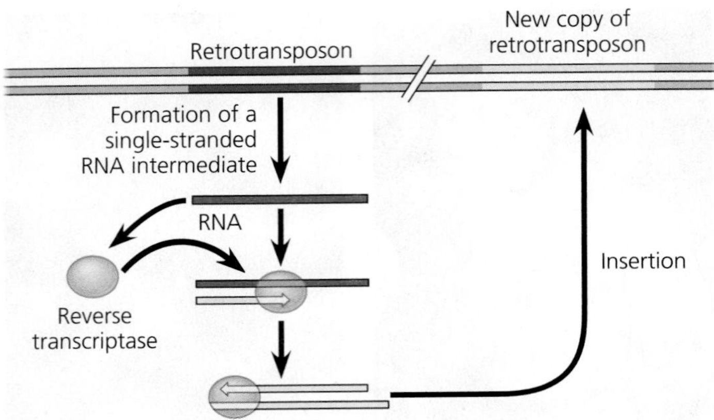

▲ **Figure 21.10 Retrotransposon movement.** Movement begins with formation of a single-stranded RNA intermediate. The remaining steps are essentially identical to part of the retrovirus replicative cycle (see Figure 19.8).

into an RNA intermediate (**Figure 21.10**). To insert at another site, the RNA intermediate is first converted back to DNA by reverse transcriptase, an enzyme encoded by the retrotransposon. (Reverse transcriptase is also encoded by retroviruses, as you learned in Chapter 19. In fact, retroviruses may have evolved from retrotransposons.) Another cellular enzyme catalyzes insertion of the reverse-transcribed DNA at a new site.

Sequences Related to Transposable Elements

Multiple copies of transposable elements and sequences related to them are scattered throughout eukaryotic genomes. A single unit is usually hundreds to thousands of base pairs long, and the dispersed "copies" are similar but usually not identical to each other. Some of these are transposable elements that can move; the enzymes required for this movement may be encoded by any transposable element, including the one that is moving. Others are related sequences that have lost the ability to move altogether. Transposable elements and related sequences make up 25–50% of most mammalian genomes (see Figure 21.7) and even higher percentages in amphibians and many plants. In fact, the very large size of some plant genomes is accounted for not by extra genes, but by extra transposable elements. For example, sequences like these make up 85% of the corn genome!

In humans and other primates, a large portion of transposable element–related DNA consists of a family of similar sequences called *Alu elements*. These sequences alone account for approximately 10% of the human genome. *Alu* elements are about 300 nucleotides long, much shorter than most functional transposable elements, and they do not code for any protein. However, many *Alu* elements are transcribed into RNA; its cellular function, if any, is currently unknown.

An even larger percentage (17%) of the human genome is made up of a type of retrotransposon called *LINE-1*, or *L1*. These sequences are much longer than *Alu* elements—about 6,500 base pairs—and have a low rate of transposition. What might account for this low rate? Recent research has uncovered the presence of sequences within L1 that block the progress of RNA polymerase, which is necessary for transposition. An accompanying genomic analysis found L1 sequences within the introns of nearly 80% of the human genes that were analyzed, suggesting that L1 may help regulate gene expression. Other researchers have proposed that L1 retrotransposons may have differential effects on gene expression in developing neurons, contributing to the great diversity of neuronal cell types (see Chapter 48).

Although many transposable elements encode proteins, these proteins do not carry out normal cellular functions. Therefore, transposable elements are usually included in the "noncoding" DNA category, along with other repetitive sequences.

Other Repetitive DNA, Including Simple Sequence DNA

Repetitive DNA that is not related to transposable elements probably arises due to mistakes during DNA replication or recombination. Such DNA accounts for about 14% of the human genome (see Figure 21.7). About a third of this (5–6% of the human genome) consists of duplications of long stretches of DNA, with each unit ranging from 10,000 to 300,000 base pairs. The large segments seem to have been copied from one chromosomal location to another site on the same or a different chromosome and probably include some functional genes.

In contrast to scattered copies of long sequences, **simple sequence DNA** contains many copies of tandemly repeated short sequences, as in the following example (showing one DNA strand only):

. . . GTTACGTTACGTTACGTTACGTTACGTTAC . . .

In this case, the repeated unit (GTTAC) consists of 5 nucleotides. Repeated units may contain as many as 500 nucleotides, but often contain fewer than 15 nucleotides, as in this example. When the unit contains 2–5 nucleotides, the series of repeats is called a **short tandem repeat**, or **STR**; we discussed the use of STR analysis in preparing genetic profiles in Chapter 20. The number of copies of the repeated unit can vary from site to site within a given genome. There could be as many as several hundred thousand repetitions of the GTTAC unit at one site, but only half that number at another. STR analysis is performed on sites selected because they have relatively few repeats. The repeat number can vary from person to person, and since humans are diploid, each person has two alleles per site, which can differ. This diversity produces the variation represented in the genetic profiles that result from STR analysis. Altogether, simple sequence DNA makes up 3% of the human genome.

Much of a genome's simple sequence DNA is located at chromosomal telomeres and centromeres, suggesting that this DNA plays a structural role for chromosomes. The DNA at centromeres is essential for the separation of chromatids in cell division (see Chapter 12). Centromeric DNA, along with simple sequence DNA located elsewhere, may also help organize the chromatin within the interphase nucleus. The simple sequence DNA located at telomeres, at the tips of chromosomes, prevents genes from being lost as the DNA shortens with each round of replication (see Chapter 16). Telomeric DNA also binds proteins that protect the ends of a chromosome from degradation and from joining to other chromosomes.

Genes and Multigene Families

We finish our discussion of the various types of DNA sequences in eukaryotic genomes with a closer look at genes. Recall that DNA sequences that code for proteins or give rise to tRNA or rRNA compose a mere 1.5% of the human genome (see Figure 21.7). If we include introns and regulatory sequences associated with genes, the total amount of DNA that is gene-related—coding and noncoding—constitutes about 25% of the human genome. Put another way, only about 6% (1.5% out of 25%) of the length of the average gene is represented in the final gene product.

Like the genes of bacteria, many eukaryotic genes are present as unique sequences, with only one copy per haploid set of chromosomes. But in the human genome and the genomes of many other animals and plants, solitary genes make up less than half of the total gene-related DNA. The rest occur in **multigene families**, collections of two or more identical or very similar genes.

In multigene families that consist of *identical* DNA sequences, those sequences are usually clustered tandemly and, with the notable exception of the genes for histone proteins, have RNAs as their final products. An example is the family of identical DNA sequences that are the genes for the three largest rRNA molecules (**Figure 21.11a**). These rRNA molecules are transcribed from a single transcription unit that is repeated tandemly hundreds to thousands of times in one or several clusters in the genome of a multicellular eukaryote. The many copies of this rRNA transcription unit help cells to quickly make the millions of ribosomes needed for active protein synthesis. The primary transcript is cleaved to yield the three rRNA molecules, which combine with proteins and one other kind of rRNA (5S rRNA) to form ribosomal subunits.

The classic examples of multigene families of *nonidentical* genes are two related families of genes that encode globins, a group of proteins that include the α and β polypeptide subunits of hemoglobin. One family, located on chromosome 16 in humans, encodes various forms of α-globin; the other, on chromosome 11, encodes forms of β-globin (**Figure 21.11b**). The different forms of each globin subunit are expressed at different times in development, allowing hemoglobin to function effectively in the changing environment of the developing animal. In humans, for example, the embryonic and fetal forms of hemoglobin have a higher affinity for oxygen

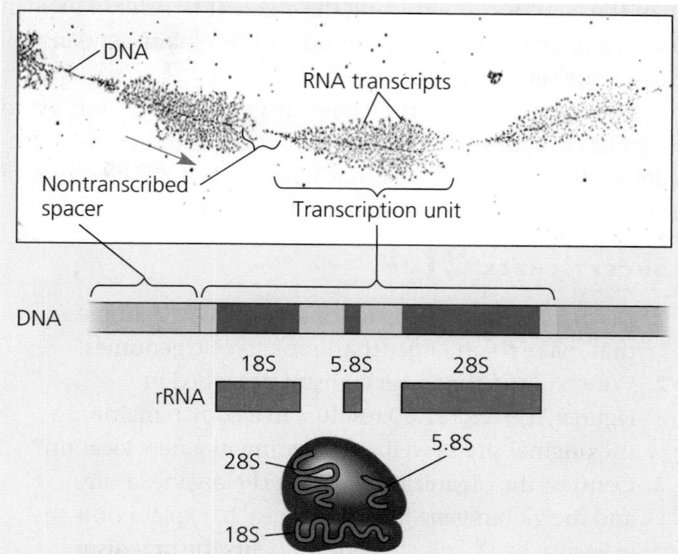

(a) Part of the ribosomal RNA gene family. The TEM at the top shows three of the hundreds of copies of rRNA transcription units in a salamander genome. Each "feather" corresponds to a single unit being transcribed by about 100 molecules of RNA polymerase (dark dots along the DNA), moving left to right (red arrow). The growing RNA transcripts extend from the DNA. In the diagram of a trascription unit below the TEM, the genes for three types of rRNA (blue) are adjacent to regions that are transcribed but later removed (yellow). A single trascript is processed to yield one of each of the three rRNAs (red), key components of the ribosome.

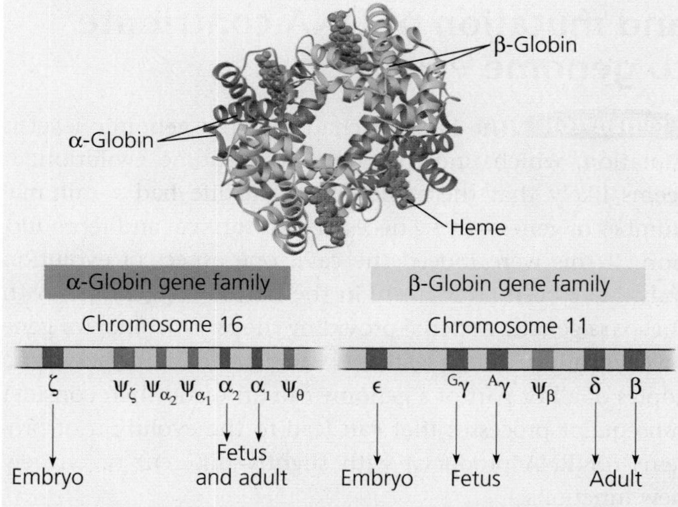

(b) The human α-globin and β-globin gene families. Adult hemoglobin is composed of two α-globin and two β-globin polypeptide subunits, as shown in the molecular model. The genes (dark blue) encoding α- and β-globins are found in two families, organized as shown here. The noncoding DNA separating the functional genes within each family includes pseudogenes (ψ; green), versions of the functional genes that no longer produce functional proteins. Genes and pseudogenes are named with Greek letters. Some genes are expressed only in the embryo or fetus.

▲ **Figure 21.11 Gene families.**

? *In (a), how could you determine the direction of transcription if it wasn't indicated by the red arrow?*

than the adult forms, ensuring the efficient transfer of oxygen from mother to fetus. Also found in the globin gene family clusters are several pseudogenes.

The arrangement of the genes in gene families has given biologists insight into the evolution of genomes. We will consider some of the processes that have shaped the genomes of different species over evolutionary time in the next section.

CONCEPT CHECK 21.4

1. Discuss the characteristics of mammalian genomes that make them larger than prokaryotic genomes.
2. Which of the three mechanisms described in Figures 21.9 and 21.10 result(s) in a copy remaining at the original site as well as appearing in a new location?
3. Contrast the organizations of the rRNA gene family and the globin gene families. For each, explain how the existence of a family of genes benefits the organism.
4. **MAKE CONNECTIONS** Assign each DNA segment at the top of Figure 18.8 (p. 359) to a sector in the pie chart in Figure 21.7.

For suggested answers, see Appendix A.

CONCEPT 21.5

Duplication, rearrangement, and mutation of DNA contribute to genome evolution

EVOLUTION The basis of change at the genomic level is mutation, which underlies much of genome evolution. It seems likely that the earliest forms of life had a minimal number of genes—those necessary for survival and reproduction. If this were indeed the case, one aspect of evolution must have been an increase in the size of the genome, with the extra genetic material providing the raw material for gene diversification. In this section, we will first describe how extra copies of all or part of a genome can arise and then consider subsequent processes that can lead to the evolution of proteins (or RNA products) with slightly different or entirely new functions.

Duplication of Entire Chromosome Sets

An accident in meiosis can result in one or more extra sets of chromosomes, a condition known as polyploidy. Although such accidents would most often be lethal, in rare cases they could facilitate the evolution of genes. In a polyploid organism, one set of genes can provide essential functions for the organism. The genes in the one or more extra sets can diverge by accumulating mutations; these variations may persist if the organism carrying them survives and reproduces. In this way, genes with novel functions can evolve. As long as one copy of an essential gene is expressed, the divergence of another copy can lead to its encoded protein acting in a novel way, thereby changing the organism's phenotype. The outcome of this accumulation of mutations may be the branching off of a new species, as happens often in flowering plants (see Chapter 24). Polyploid animals also exist, but they are much rarer; the tetraploid model organism *Xenopus laevis*, the African clawed frog, is an example.

Alterations of Chromosome Structure

Scientists have long known that sometime in the last 6 million years, when the ancestors of humans and chimpanzees diverged as species, the fusion of two ancestral chromosomes in the human line led to different haploid numbers for humans ($n = 23$) and chimpanzees ($n = 24$). The banding patterns in stained chromosomes suggested that the ancestral versions of current chimp chromosomes 12 and 13 fused end to end, forming chromosome 2 in an ancestor of the human lineage. With the recent explosion in genomic sequence information, we can now compare the chromosomal organizations of many different species on a much finer scale. This information allows us to make inferences about the evolutionary processes that shape chromosomes and may drive speciation. Sequencing and analysis of human chromosome 2 in 2005 provided very strong supporting evidence for the model we have just described (**Figure 21.12a**).

In another study of broader scope, researchers compared the DNA sequence of each human chromosome with the whole-genome sequence of the mouse. **Figure 21.12b** shows the results of this comparison for human chromosome 16: Large blocks of genes on this chromosome are found on four mouse chromosomes, indicating that the genes in each block stayed together during the evolution of the mouse and human lineages.

Performing the same comparative analysis between chromosomes of humans and six other mammalian species allowed the researchers to reconstruct the evolutionary history of chromosomal rearrangements in these eight species. They found many duplications and inversions of large portions of chromosomes, the result of mistakes during meiotic recombination in which the DNA broke and was rejoined incorrectly. The rate of these events seems to have accelerated about 100 million years ago, around the time large dinosaurs became extinct and the number of mammalian species increased rapidly. The apparent coincidence is interesting because chromosomal rearrangements are thought to contribute to the generation of new species. Although two individuals with different arrangements could still mate and produce offspring, the offspring would have two nonequivalent sets of chromosomes, making meiosis inefficient or even impossible. Thus, chromosomal rearrangements would lead to two populations that could not successfully mate with each other, a step on the way to their becoming two separate species. (You'll learn more about this in Chapter 24.)

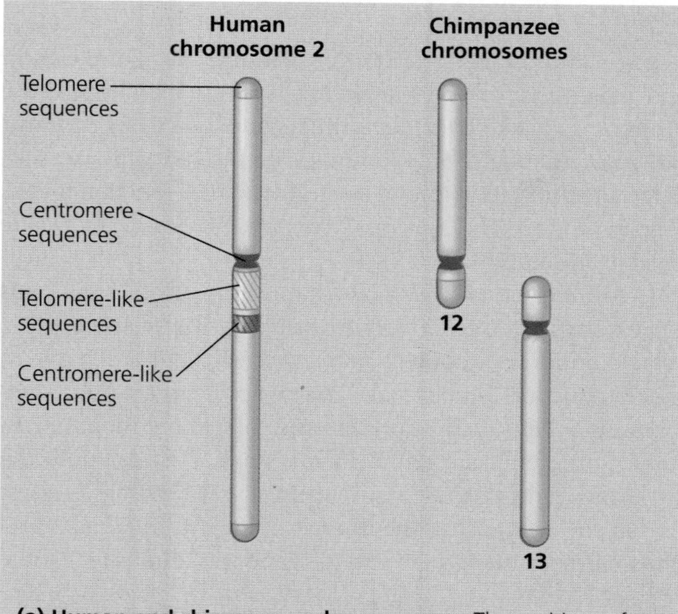

(a) Human and chimpanzee chromosomes. The positions of telomere-like and centromere-like sequences on human chromosome 2 (left) match those of telomeres on chimp chromosomes 12 and 13 and the centromere on chimp chromosome 13 (right). This suggests that chromosomes 12 and 13 in a human ancestor fused end to end to form human chromosome 2. The centromere from ancestral chromosome 12 remained functional on human chromosome 2, while the one from ancestral chromosome 13 did not. (Chimp chromosomes 12 and 13 have been renamed 2a and 2b, respectively.)

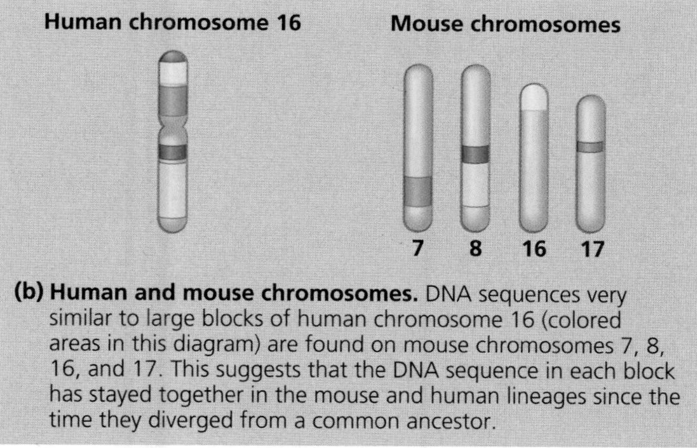

(b) Human and mouse chromosomes. DNA sequences very similar to large blocks of human chromosome 16 (colored areas in this diagram) are found on mouse chromosomes 7, 8, 16, and 17. This suggests that the DNA sequence in each block has stayed together in the mouse and human lineages since the time they diverged from a common ancestor.

▲ **Figure 21.12 Related chromosome sequences among mammals.**

Somewhat unexpectedly, the same study also unearthed a pattern with medical relevance. Analysis of the chromosomal breakage points associated with the rearrangements showed that they were not randomly distributed; specific sites were used over and over again. A number of these recombination "hot spots" correspond to locations of chromosomal rearrangements within the human genome that are associated with congenital diseases. Researchers are, of course, looking at the other sites as well for their possible association with as yet unidentified diseases.

Duplication and Divergence of Gene-Sized Regions of DNA

Errors during meiosis can also lead to the duplication of chromosomal regions that are smaller than the ones we've just discussed, including segments the length of individual genes. Unequal crossing over during prophase I of meiosis, for instance, can result in one chromosome with a deletion and another with a duplication of a particular gene. As illustrated in **Figure 21.13**, transposable elements can provide homologous sites where nonsister chromatids can cross over, even when other chromatid regions are not correctly aligned.

Also, slippage can occur during DNA replication, such that the template shifts with respect to the new complementary strand, and a part of the template strand is either skipped by the replication machinery or used twice as a template. As a result, a segment of DNA is deleted or duplicated. It is easy to imagine how such errors could occur in regions of repeats. The variable number of repeated units of simple sequence DNA at a given site, used for STR analysis, is probably due to errors like these. Evidence that unequal crossing over and template slippage during DNA replication lead to duplication of genes is found in the existence of multigene families, such as the globin family.

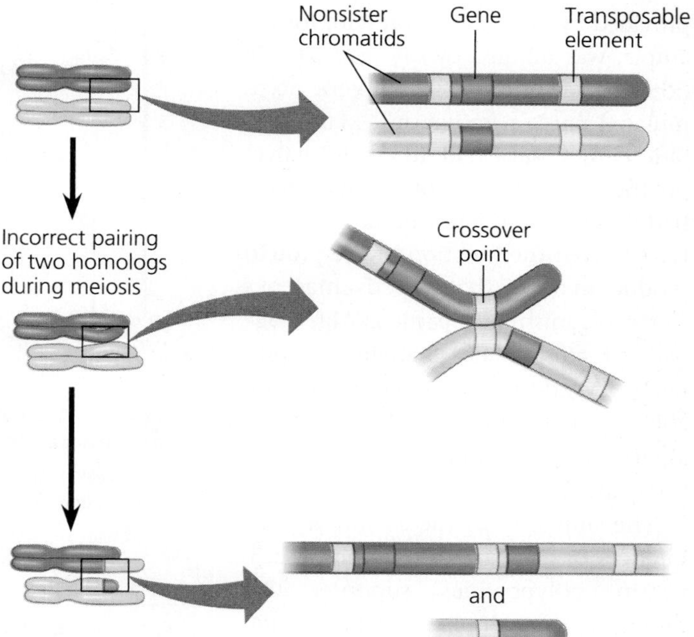

▲ **Figure 21.13 Gene duplication due to unequal crossing over.** One mechanism by which a gene (or other DNA segment) can be duplicated is recombination during meiosis between copies of a transposable element flanking the gene. Such recombination between misaligned nonsister chromatids of homologous chromosomes produces one chromatid with two copies of the gene and one chromatid with no copy.

MAKE CONNECTIONS *Examine how crossing over occurs in Figure 13.11 (p. 259). In the middle panel above, draw a line through the portions that result in the upper chromatid in the bottom panel. Use a different color to do the same for the other chromatid.*

Evolution of Genes with Related Functions: The Human Globin Genes

Duplication events can lead to the evolution of genes with related functions, such as those of the α-globin and β-globin gene families (see Figure 21.11b). A comparison of gene sequences within a multigene family can suggest the order in which the genes arose. This approach to re-creating the evolutionary history of the globin genes indicates that they all evolved from one common ancestral globin gene that underwent duplication and divergence into the α-globin and β-globin ancestral genes about 450–500 million years ago **(Figure 21.14)**. Each of these genes was later duplicated several times, and the copies then diverged from each other in sequence, yielding the current family members. In fact, the common ancestral globin gene also gave rise to the oxygen-binding muscle protein myoglobin and to the plant protein leghemoglobin. The latter two proteins function as monomers, and their genes are included in a "globin superfamily."

After the duplication events, the differences between the genes in the globin families undoubtedly arose from mutations that accumulated in the gene copies over many generations. The current model is that the necessary function provided by an α-globin protein, for example, was fulfilled by one gene, while other copies of the α-globin gene accumulated random mutations. Many mutations may have had an adverse effect on the organism and others may have had no effect, but a few mutations must have altered the function of the protein product in a way that was advantageous to the organism at a particular life stage without substantially changing the protein's oxygen-carrying function. Presumably, natural selection acted on these altered genes, maintaining them in the population.

The similarity in the amino acid sequences of the various α-globin and β-globin polypeptides supports this model of gene duplication and mutation **(Table 21.2)**. The amino acid sequences of the β-globins, for instance, are much more similar to each other than to the α-globin sequences. The existence of several pseudogenes among the functional globin genes provides additional evidence for this model (see Figure 21.11b): Random mutations in these "genes" over evolutionary time have destroyed their function.

Evolution of Genes with Novel Functions

In the evolution of the globin gene families, gene duplication and subsequent divergence produced family members whose protein products performed similar functions (oxygen transport). Alternatively, one copy of a duplicated gene can undergo alterations that lead to a completely new function for the protein product. The genes for lysozyme and α-lactalbumin are good examples.

Lysozyme is an enzyme that helps protect animals against bacterial infection by hydrolyzing bacterial cell walls; α-lactalbumin is a nonenzymatic protein that plays a role in milk production in mammals. The two proteins are quite similar in their amino acid sequences and three-dimensional structures. Both genes are found in mammals, whereas only the lysozyme gene is present in birds. These findings suggest that at some time after the lineages leading to mammals and birds had separated, the lysozyme gene was duplicated in the

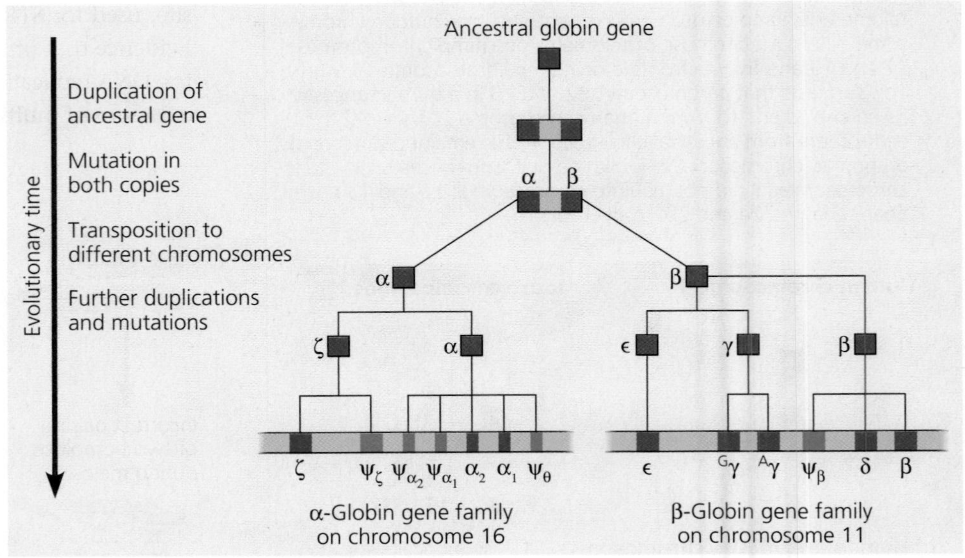

▲ **Figure 21.14 A model for the evolution of the human α-globin and β-globin gene families from a single ancestral globin gene.**

? *The green elements are pseudogenes. Explain how they could have arisen after gene duplication.*

Table 21.2 Percentage of Similarity in Amino Acid Sequence Between Human Globin Proteins

		α-Globins		β-Globins		
		α	ζ	β	γ	ε
α-Globins	α	—	58	42	39	37
	ζ	58	—	34	38	37
β-Globins	β	42	34	—	73	75
	γ	39	38	73	—	80
	ε	37	37	75	80	—

mammalian lineage but not in the avian lineage. Subsequently, one copy of the duplicated lysozyme gene evolved into a gene encoding α-lactalbumin, a protein with a completely different function.

Rearrangements of Parts of Genes: Exon Duplication and Exon Shuffling

Rearrangement of existing DNA sequences within genes has also contributed to genome evolution. The presence of introns in most genes of multicellular eukaryotes may have promoted the evolution of new and potentially useful proteins by facilitating the duplication or repositioning of exons in the genome. Recall from Chapter 17 that an exon often codes for a domain, a distinct structural or functional region of a protein.

We've already seen that unequal crossing over during meiosis can lead to duplication of a gene on one chromosome and its loss from the homologous chromosome (see Figure 21.13). By a similar process, a particular exon within a gene could be duplicated on one chromosome and deleted from the other. The gene with the duplicated exon would code for a protein containing a second copy of the encoded domain. This change in the protein's structure could augment its function by increasing its stability, enhancing its ability to bind a particular ligand, or altering some other property. Quite a few protein-coding genes have multiple copies of related exons, which presumably arose by duplication and then diverged. The gene encoding the extracellular matrix protein collagen is a good example. Collagen is a structural protein with a highly repetitive amino acid sequence, which is reflected in the repetitive pattern of exons in the collagen gene.

Alternatively, we can imagine the occasional mixing and matching of different exons either within a gene or between two different (nonallelic) genes owing to errors in meiotic recombination. This process, termed *exon shuffling*, could lead to new proteins with novel combinations of functions. As an example, let's consider the gene for tissue plasminogen activator (TPA). The TPA protein is an extracellular protein that helps control blood clotting. It has four domains of three types, each encoded by an exon; one exon is present in two copies. Because each type of exon is also found in other proteins, the gene for TPA is thought to have arisen by several instances of exon shuffling and duplication **(Figure 21.15)**.

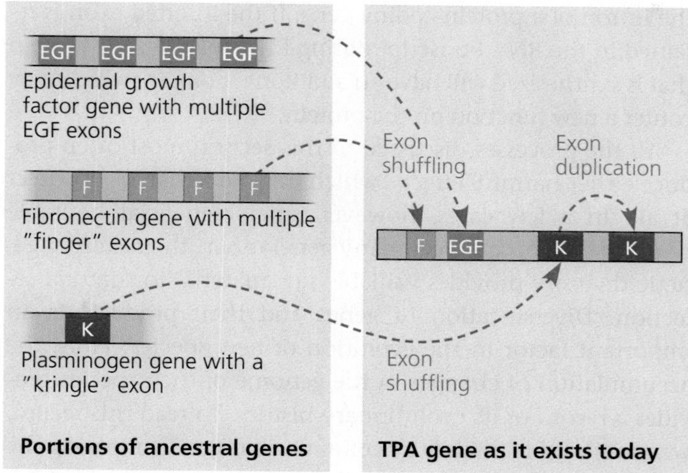

▲ **Figure 21.15 Evolution of a new gene by exon shuffling.** Exon shuffling could have moved exons, each encoding a particular domain, from ancestral forms of the genes for epidermal growth factor, fibronectin, and plasminogen (left) into the evolving gene for tissue plasminogen activator, TPA (right). Duplication of the "kringle" exon from the plasminogen gene after its movement could account for the two copies of this exon in the TPA gene.

? *How could the presence of transposable elements in introns have facilitated the exon shuffling shown here?*

How Transposable Elements Contribute to Genome Evolution

The persistence of transposable elements as a large fraction of some eukaryotic genomes is consistent with the idea that they play an important role in shaping a genome over evolutionary time. These elements can contribute to the evolution of the genome in several ways. They can promote recombination, disrupt cellular genes or control elements, and carry entire genes or individual exons to new locations.

Transposable elements of similar sequence scattered throughout the genome facilitate recombination between different chromosomes by providing homologous regions for crossing over. Most such recombination events are probably detrimental, causing chromosomal translocations and other changes in the genome that may be lethal to the organism. But over the course of evolutionary time, an occasional recombination event of this sort may be advantageous to the organism. (For the change to be heritable, of course, it must happen in a cell that will give rise to a gamete.)

The movement of a transposable element can have a variety of consequences. For instance, if a transposable element "jumps" into the middle of a protein-coding sequence, it will prevent the production of a normal transcript of the gene. If a transposable element inserts within a regulatory sequence, the transposition may lead to increased or decreased production of one or more proteins. Transposition caused both types of effects on the genes coding for pigment-synthesizing enzymes in McClintock's corn kernels. Again, while such changes are usually harmful, in the long run some may prove beneficial by providing a survival advantage.

During transposition, a transposable element may carry along a gene or group of genes to a new position in the genome. This mechanism probably accounts for the location of the α-globin and β-globin gene families on different human chromosomes, as well as the dispersion of the genes of certain other gene families. By a similar tag-along process, an exon from one gene may be inserted into another gene in a mechanism similar to that of exon shuffling during recombination. For example, an exon may be inserted by transposition into the intron of a protein-coding gene. If the inserted exon is retained in the RNA transcript during RNA splicing, the protein that is synthesized will have an additional domain, which may confer a new function on the protein.

All the processes discussed in this section most often produce either harmful effects, which may be lethal, or no effect at all. In a few cases, however, small beneficial heritable changes may occur. Over many generations, the resulting genetic diversity provides valuable raw material for natural selection. Diversification of genes and their products is an important factor in the evolution of new species. Thus, the accumulation of changes in the genome of each species provides a record of its evolutionary history. To read this record, we must be able to identify genomic changes. Comparing the genomes of different species allows us to do that and has increased our understanding of how genomes evolve. You will learn more about these topics in the final section.

CONCEPT CHECK 21.5

1. Describe three examples of errors in cellular processes that lead to DNA duplications.
2. Explain how multiple exons might have arisen in the ancestral EGF and fibronectin genes shown in Figure 21.15 (left).
3. What are three ways that transposable elements are thought to contribute to genome evolution?
4. **WHAT IF?** In 2005, Icelandic scientists reported finding a large chromosomal inversion present in 20% of northern Europeans, and they noted that Icelandic women with this inversion had significantly more children than women without it. What would you expect to happen to the frequency of this inversion in the Icelandic population in future generations?

For suggested answers, see Appendix A.

CONCEPT 21.6

Comparing genome sequences provides clues to evolution and development

EVOLUTION One researcher has likened the current state of biology to the Age of Exploration in the 15th century after major improvements in navigation and the building of faster ships. In the last 25 years, we have seen rapid advances in genome sequencing and data collection, new techniques for assessing gene activity across the whole genome, and refined approaches for understanding how genes and their products work together in complex systems. We are truly poised on the brink of a new world.

Comparisons of genome sequences from different species reveal much about the evolutionary history of life, from very ancient to more recent. Similarly, comparative studies of the genetic programs that direct embryonic development in different species are beginning to clarify the mechanisms that generated the great diversity of life-forms present today. In this final section of the chapter, we will discuss what has been learned from these two approaches.

Comparing Genomes

The more similar in sequence the genes and genomes of two species are, the more closely related those species are in their evolutionary history. Comparing genomes of closely related species sheds light on more recent evolutionary events, whereas comparing genomes of very distantly related species helps us understand ancient evolutionary history. In either case, learning about characteristics that are shared or divergent between groups enhances our picture of the evolution of life-forms and biological processes. As you learned in Chapter 1, the evolutionary relationships between species can be represented by a diagram in the form of a tree (often turned sideways), where each branch point marks the divergence of two lineages. **Figure 21.16** shows the evolutionary relationships of some groups and species we will be discussing. We will consider comparisons between distantly related species first.

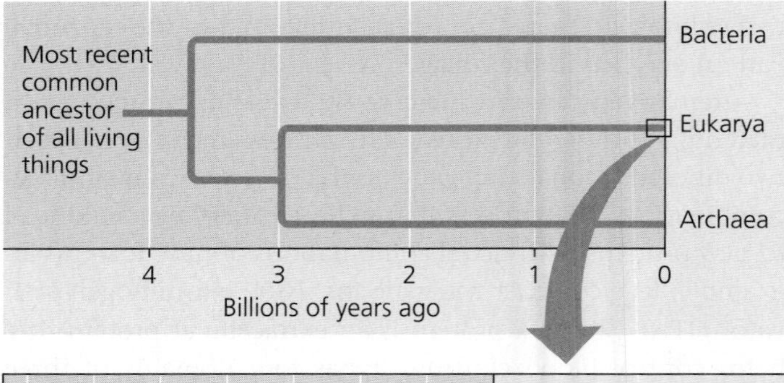

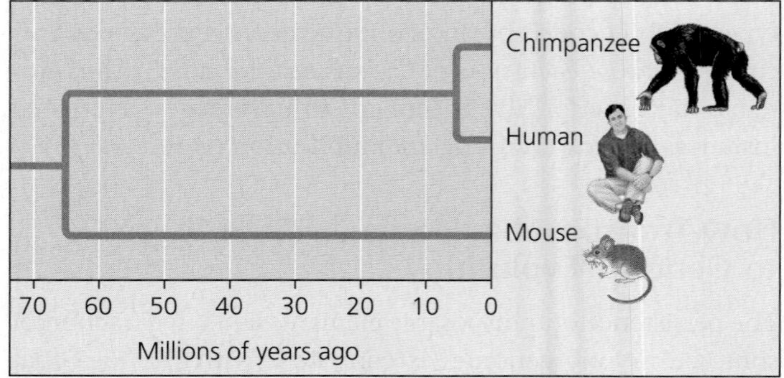

▲ **Figure 21.16 Evolutionary relationships of the three domains of life.** This tree diagram shows the ancient divergence of bacteria, archaea, and eukaryotes. A portion of the eukaryote lineage is expanded in the inset to show the more recent divergence of three mammalian species discussed in this chapter.

Comparing Distantly Related Species

Determining which genes have remained similar—that is, are *highly conserved*—in distantly related species can help clarify evolutionary relationships among species that diverged from each other long ago. Indeed, comparisons of the complete genome sequences of bacteria, archaea, and eukaryotes indicate that these three groups diverged between 2 and 4 billion years ago and strongly support the theory that they are the fundamental domains of life (see Figure 21.16).

In addition to their value in evolutionary biology, comparative genomic studies confirm the relevance of research on model organisms to our understanding of biology in general and human biology in particular. Genes that evolved a very long time ago can still be surprisingly similar in disparate species. As a case in point, several genes in yeast are so similar to certain human disease genes that researchers have deduced the functions of the disease genes by studying their yeast counterparts. This striking similarity underscores the common origin of these two distantly related species.

Comparing Closely Related Species

The genomes of two closely related species are likely to be organized similarly because of their relatively recent divergence. As we mentioned earlier, this allows the fully sequenced genome of one species to be used as a scaffold for assembling the genomic sequences of a closely related species, accelerating mapping of the second genome. For instance, using the human genome sequence as a guide, researchers were able to quickly sequence the chimpanzee genome.

The recent divergence of two closely related species also underlies the small number of gene differences that are found when their genomes are compared. The particular genetic differences can therefore be more easily correlated with phenotypic differences between the two species. An exciting application of this type of analysis is seen as researchers compare the human genome with the genomes of the chimpanzee, mouse, rat, and other mammals. Identifying the genes shared by all of these species but not by nonmammals should give clues about what it takes to make a mammal, while finding the genes shared by chimpanzees and humans but not by rodents should tell us something about primates. And, of course, comparing the human genome with that of the chimpanzee should help us answer the tantalizing question we asked at the beginning of the chapter: What genomic information makes a human or a chimpanzee?

An analysis of the overall composition of the human and chimpanzee genomes, which are thought to have diverged only about 6 million years ago (see Figure 21.16), reveals some general differences. Considering single nucleotide substitutions, the two genomes differ by only 1.2%. When researchers looked at longer stretches of DNA, however, they were surprised to find a further 2.7% difference due to insertions or deletions of larger regions in the genome of one or the other species; many of the insertions were duplications or other repetitive DNA. In fact, a third of the human duplications are not present in the chimpanzee genome, and some of these duplications contain regions associated with human diseases. There are more *Alu* elements in the human genome than in the chimpanzee genome, and the latter contains many copies of a retroviral provirus not present in humans. All of these observations provide clues to the forces that might have swept the two genomes along different paths, but we don't have a complete picture yet. We also don't know how these differences might account for the distinct characteristics of each species.

To discover the basis for the phenotypic differences between the two species, biologists are studying specific genes and types of genes that differ between humans and chimpanzees and comparing them with their counterparts in other mammals. This approach has revealed a number of genes that are apparently changing (evolving) faster in the human than in either the chimpanzee or the mouse. Among them are genes involved in defense against malaria and tuberculosis and at least one gene that regulates brain size. When genes are classified by function, the genes that seem to be evolving the fastest are those that code for transcription factors. This discovery makes sense because transcription factors regulate gene expression and thus play a key role in orchestrating the overall genetic program.

One transcription factor whose gene shows evidence of rapid change in the human lineage is called FOXP2. Several lines of evidence suggest that the *FOXP2* gene functions in vocalization in vertebrates. For one thing, mutations in this gene can produce severe speech and language impairment in humans. Moreover, the *FOXP2* gene is expressed in the brains of zebra finches and canaries at the time when these songbirds are learning their songs. But perhaps the strongest evidence comes from a "knock-out" experiment in which researchers disrupted the *FOXP2* gene in mice and analyzed the resulting phenotype (**Figure 21.17**, on the next page). The homozygous mutant mice had malformed brains and failed to emit normal ultrasonic vocalizations, and mice with one faulty copy of the gene also showed significant problems with vocalization. These results support the idea that the *FOXP2* gene product turns on genes involved in vocalization.

Expanding on this analysis, another research group more recently replaced the *FOXP2* gene in mice with a "humanized" copy coding for the human versions of two amino acids that differ between human and chimp; these are the changes potentially responsible for a human's ability to speak. Although the mice were generally healthy, they had subtly different vocalizations and showed changes in brain cells in circuits associated with speech in human brains.

The *FOXP2* story is an excellent example of how different approaches can complement each other in uncovering biological phenomena of widespread importance. The

▼ **Figure 21.17** INQUIRY

What is the function of a gene (*FOXP2*) that is rapidly evolving in the human lineage?

EXPERIMENT Several lines of evidence support a role for the *FOXP2* gene in the development of speech and language in humans and of vocalization in other vertebrates. In 2005, Joseph Buxbaum and collaborators at the Mount Sinai School of Medicine and several other institutions tested the function of *FOXP2*. They used the mouse, a model organism in which genes can be easily knocked out, as a representative vertebrate that vocalizes: Mice produce ultrasonic squeaks (whistles) to communicate stress. The researchers used genetic engineering to produce mice in which one or both copies of *FOXP2* were disrupted.

Wild type: two normal copies of *FOXP2*

Heterozygote: one copy of *FOXP2* disrupted

Homozygote: both copies of *FOXP2* disrupted

They then compared the phenotypes of these mice. Two of the characters they examined are included here: brain anatomy and vocalization.

Experiment 1: Researchers cut thin sections of brain and stained them with reagents that allow visualization of brain anatomy in a UV fluorescence microscope.

Experiment 2: Researchers separated each newborn pup from its mother and recorded the number of ultrasonic whistles produced by the pup.

RESULTS

Experiment 1: Disruption of both copies of *FOXP2* led to brain abnormalities in which the cells were disorganized. Phenotypic effects on the brain of heterozygotes, with one disrupted copy, were less severe. (Each color reveals a different cell or tissue type.)

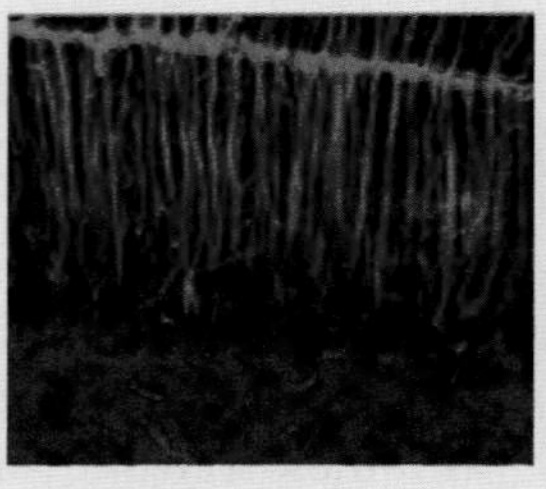

Wild type

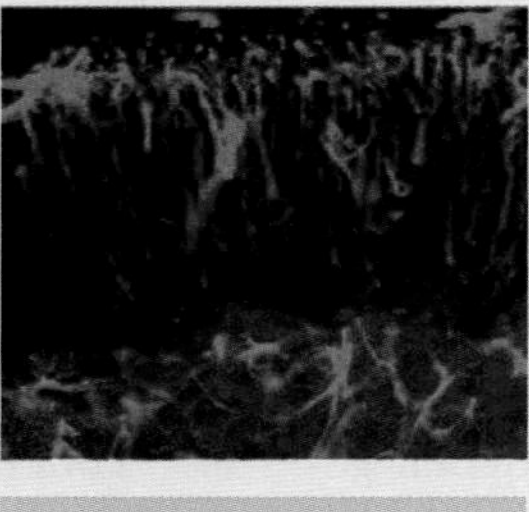

Heterozygote

Homozygote

Experiment 2: Disruption of both copies of *FOXP2* led to an absence of ultrasonic vocalization in response to stress. The effect on vocalization in the heterozygote was also extreme.

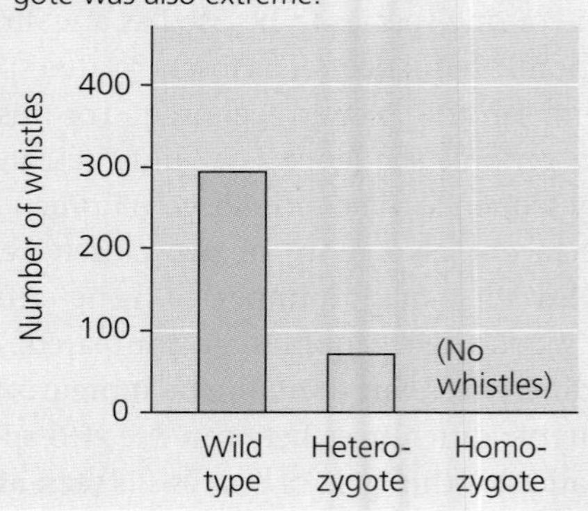

CONCLUSION *FOXP2* plays a significant role in the development of functional communication systems in mice. The results augment evidence from studies of birds and humans, supporting the hypothesis that *FOXP2* may act similarly in diverse organisms.

SOURCE W. Shu et al., Altered ultrasonic vocalization in mice with a disruption in the *Foxp2* gene, *Proceedings of the National Academy of Sciences* 102:9643–9648 (2005).

WHAT IF? Since the results support a role for mouse *FOXP2* in vocalization, you might wonder whether the human FOXP2 protein is a key regulator of speech. If you were given the amino acid sequences of wild-type and mutant human FOXP2 proteins and the wild-type chimpanzee FOXP2 protein, how would you investigate this question? What further clues could you obtain by comparing these sequences to that of the mouse FOXP2 protein?

FOXP2 experiments used mice as a model for humans because it would be unethical (as well as impractical) to carry out such experiments in humans. Mice and humans diverged about 65.5 million years ago (see Figure 21.16) and share about 85% of their genes. This genetic similarity can be exploited in studying human genetic disorders. If researchers know the organ or tissue that is affected by a particular genetic disorder, they can look for genes that are expressed in these locations in mice.

Further research efforts are under way to extend genomic studies to many more microbial species, additional primates, and neglected species from diverse branches of the tree of life.

These studies will advance our understanding of all aspects of biology, including health and ecology as well as evolution.

Comparing Genomes Within a Species

Another exciting consequence of our ability to analyze genomes is our growing understanding of the spectrum of genetic variation in humans. Because the history of the human species is so short—probably about 200,000 years—the amount of DNA variation among humans is small compared to that of many other species. Much of our diversity seems to be in the form of single nucleotide polymorphisms (SNPs, described in Chapter 20), usually detected by DNA sequencing. In the human genome, SNPs occur on average about once in 100–300 base pairs. Scientists have already identified the location of several million SNP sites in the human genome and continue to find more.

In the course of this search, they have also found other variations—including inversions, deletions, and duplications. The most surprising discovery has been the widespread occurrence of *copy-number variants (CNVs)*, loci where some individuals have one or multiple copies of a particular gene or genetic region, rather than the standard two copies (one on each homolog). CNVs result from regions of the genome being duplicated or deleted inconsistently within the population. A 2010 study of 40 people found more than 8,000 CNVs involving 13% of the genes in the genome, and these CNVs probably represent just a small subset of the total. Since these variants encompass much longer stretches of DNA than the single nucleotides of SNPs, CNVs are more likely to have phenotypic consequences and to play a role in complex diseases and disorders. At the very least, the high incidence of copy-number variation casts doubt on the meaning of the phrase "a normal human genome."

Copy-number variants, SNPs, and variations in repetitive DNA such as short tandem repeats (STRs) will be useful genetic markers for studying human evolution. In 2010, the genomes of two Africans from different communities were sequenced: Archbishop Desmond Tutu, the South African civil rights advocate and a member of the Bantu tribe, the majority population in southern Africa; and !Gubi, a hunter-gatherer from the Khoisan community in Namibia, a minority African population that is probably the human group with the oldest known lineage. The comparison revealed many differences, as you might expect. The analysis was then broadened to compare the protein-coding regions of !Gubi's genome with those of three other Khoisan community members (self-identified Bushmen) living nearby. Remarkably, these four genomes differed more from each other than a European would from an Asian. These data highlight the extensive diversity among African genomes. Extending this approach will help us answer important questions about the differences between human populations and the migratory routes of human populations throughout history.

Comparing Developmental Processes

Biologists in the field of evolutionary developmental biology, or **evo-devo** as it is often called, compare developmental processes of different multicellular organisms. Their aim is to understand how these processes have evolved and how changes in them can modify existing organismal features or lead to new ones. With the advent of molecular techniques and the recent flood of genomic information, we are beginning to realize that the genomes of related species with strikingly different forms may have only minor differences in gene sequence or regulation. Discovering the molecular basis of these differences in turn helps us understand the origins of the myriad diverse forms that cohabit this planet, thus informing our study of evolution.

Widespread Conservation of Developmental Genes Among Animals

In Chapter 18, you learned about the homeotic genes in *Drosophila*, which specify the identity of body segments in the fruit fly (see Figure 18.20). Molecular analysis of the homeotic genes in *Drosophila* has shown that they all include a 180-nucleotide sequence called a **homeobox**, which specifies a 60-amino-acid *homeodomain* in the encoded proteins. An identical or very similar nucleotide sequence has been discovered in the homeotic genes of many invertebrates and vertebrates. The sequences are so similar between humans and fruit flies, in fact, that one researcher has whimsically referred to flies as "little people with wings." The resemblance even extends to the organization of these genes: The vertebrate genes homologous to the homeotic genes of fruit flies have kept the same chromosomal arrangement (**Figure 21.18**, on the next page). Homeobox-containing sequences have also been found in regulatory genes of much more distantly related eukaryotes, including plants and yeasts. From these similarities, we can deduce that the homeobox DNA sequence evolved very early in the history of life and was sufficiently valuable to organisms to have been conserved in animals and plants virtually unchanged for hundreds of millions of years.

Homeotic genes in animals were named *Hox* genes, short for *h*omeob*ox*-containing genes, because homeotic genes were the first genes found to have this sequence. Other homeobox-containing genes were later found that do not act as homeotic genes; that is, they do not directly control the identity of body parts. However, most of these genes, in animals at least, are associated with development, suggesting their ancient and fundamental importance in that process. In *Drosophila*, for example, homeoboxes are present not only in the homeotic genes but also in the egg-polarity gene *bicoid* (see Figures 18.21 and 18.22), in several of the segmentation genes, and in a master regulatory gene for eye development.

Researchers have discovered that the homeobox-encoded homeodomain is the part of a protein that binds to DNA

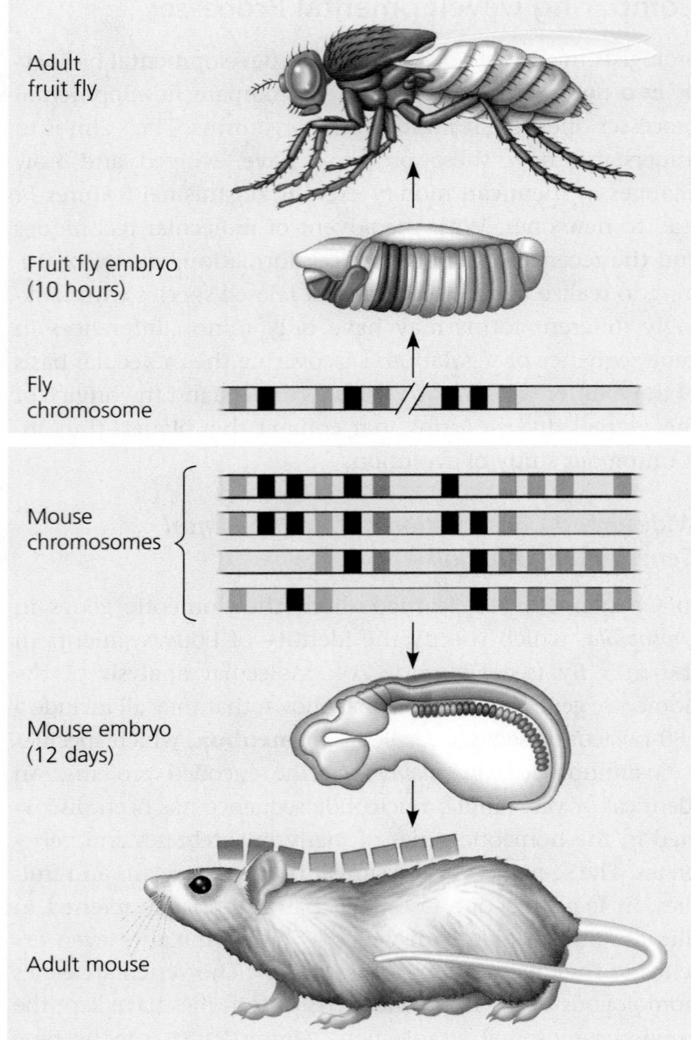

▲ **Figure 21.18 Conservation of homeotic genes in a fruit fly and a mouse.** Homeotic genes that control the form of anterior and posterior structures of the body occur in the same linear sequence on chromosomes in *Drosophila* and mice. Each colored band on the chromosomes shown here represents a homeotic gene. In fruit flies, all homeotic genes are found on one chromosome. The mouse and other mammals have the same or similar sets of genes on four chromosomes. The color code indicates the parts of the embryos in which these genes are expressed and the adult body regions that result. All of these genes are essentially identical in flies and mice, except for those represented by black bands, which are less similar in the two animals.

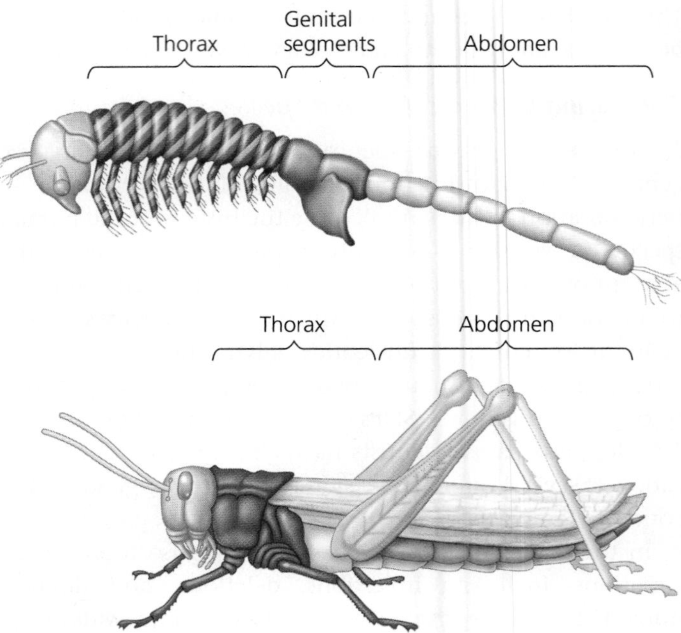

▲ **Figure 21.19 Effect of differences in *Hox* gene expression in crustaceans and insects.** Changes in the expression patterns of *Hox* genes have occurred over evolutionary time. These changes account in part for the different body plans of the brine shrimp *Artemia*, a crustacean (top), and the grasshopper, an insect. Shown here are regions of the adult body color-coded for expression of four *Hox* genes that determine formation of particular body parts during embryonic development. Each color represents a specific *Hox* gene. Colored stripes on the thorax of *Artemia* indicate co-expression of three *Hox* genes.

when the protein functions as a transcriptional regulator. However, the shape of the homeodomain allows it to bind to any DNA segment; its own structure is not specific for a particular sequence. Instead, other, more variable domains in a homeodomain-containing protein determine which genes the protein regulates. Interaction of these variable domains with still other transcription factors helps a homeodomain-containing protein recognize specific enhancers in the DNA. Proteins with homeodomains probably regulate development by coordinating the transcription of batteries of developmental genes, switching them on or off. In embryos of *Drosophila* and other animal species, different combinations of homeobox genes are active in different parts of the embryo. This selective expression of regulatory genes, varying over time and space, is central to pattern formation.

Developmental biologists have found that in addition to homeotic genes, many other genes involved in development are highly conserved from species to species. These include numerous genes encoding components of signaling pathways. The extraordinary similarity among particular developmental genes in different animal species raises a question: How can the same genes be involved in the development of animals whose forms are so very different from each other?

Ongoing studies are suggesting answers to this question. In some cases, small changes in regulatory sequences of particular genes cause changes in gene expression patterns that can lead to major changes in body form. For example, the differing patterns of expression of the *Hox* genes along the body axis in insects and crustaceans can explain the variation in number of leg-bearing segments among these segmented animals (**Figure 21.19**). Also, recent research suggests that the same *Hox* gene product may have subtly dissimilar effects in different species, turning on new genes or turning on the same genes at higher or lower levels. In other cases, similar genes direct differing developmental processes in different organisms, resulting in diverse body shapes. Several *Hox* genes, for instance, are expressed in the embryonic and larval stages of the sea urchin, a nonsegmented animal that has a body plan quite different from those of insects and mice. Sea urchin adults make the pincushion-shaped shells you may have seen on the beach (see Figure 8.4). They are among the organisms long used in classical embryological studies (see Chapter 47).

Comparison of Animal and Plant Development

The last common ancestor of animals and plants was probably a single-celled eukaryote that lived hundreds of millions of years ago, so the processes of development must have evolved independently in the two multicellular lineages of organisms. Plants evolved with rigid cell walls, which rule out the morphogenetic movements of cells and tissues that are so important in animals. Instead, morphogenesis in plants relies primarily on differing planes of cell division and on selective cell enlargement. (You will learn about these processes in Chapter 35.) But despite the differences between animals and plants, there are similarities in the molecular mechanisms of development, which are legacies of their shared unicellular origin.

In both animals and plants, development relies on a cascade of transcriptional regulators turning on or turning off genes in a finely tuned series. For example, work on the small flowering plant *Arabidopsis thaliana* has shown that establishing the radial pattern of flower parts, like setting up the head-to-tail axis in *Drosophila*, involves a cascade of transcription factors (see Chapter 35). The genes that direct these processes, however, differ considerably in animals and plants. While quite a few of the master regulatory switches in *Drosophila* are homeobox-containing *Hox* genes, those in *Arabidopsis* belong to a completely different family of genes, called the *Mads-box* genes. And although homeobox-containing genes can be found in plants and *Mads-box* genes in animals, in neither case do they perform the same major roles in development that they do in the other group. Thus, molecular evidence supports the supposition that developmental programs evolved separately in animals and plants.

In this final chapter of the genetics unit, you have learned how studying genomic composition and comparing the genomes of different species can disclose much about how genomes evolve. Further, comparing developmental programs, we can see that the unity of life is reflected in the similarity of molecular and cellular mechanisms used to establish body pattern, although the genes directing development may differ among organisms. The similarities between genomes reflect the common ancestry of life on Earth. But the differences are also crucial, for they have created the huge diversity of organisms that have evolved. In the remainder of the book, we expand our perspective beyond the level of molecules, cells, and genes to explore this diversity on the organismal level.

CONCEPT CHECK 21.6

1. Would you expect the genome of the macaque (a monkey) to be more similar to the mouse genome or the human genome? Why?
2. The DNA sequences called homeoboxes, which help homeotic genes in animals direct development, are common to flies and mice. Given this similarity, explain why these animals are so different.
3. **WHAT IF?** There are three times as many *Alu* elements in the human genome as in the chimpanzee genome. How do you think these extra *Alu* elements arose in the human genome? Propose a role they might have played in the divergence of these two species.

For suggested answers, see Appendix A.

21 CHAPTER REVIEW

SUMMARY OF KEY CONCEPTS

CONCEPT 21.1

New approaches have accelerated the pace of genome sequencing (pp. 427–429)

- The **Human Genome Project** began in 1990, using a three-stage approach. In **linkage mapping**, the order of genes and other inherited markers in the genome and the relative distances between them can be determined from recombination frequencies. Next, **physical mapping** uses overlaps between DNA fragments to order the fragments and determine the distance in base pairs between markers. Finally, the ordered fragments are sequenced, providing the finished genome sequence.
- In the whole-genome shotgun approach, the whole genome is cut into many small, overlapping fragments that are sequenced; computer software then assembles the complete sequence. Correct assembly is made easier when mapping information is also available.

? *Why has the whole-genome shotgun approach been widely adopted for genome-sequencing projects?*

CONCEPT 21.2

Scientists use bioinformatics to analyze genomes and their functions (pp. 429–432)

- Websites on the Internet provide centralized access to genome sequence databases, analytical tools, and genome-related information.
- Computer analysis of genome sequences aids **gene annotation**, the identification of protein-coding sequences and determination of their function. Methods for determining gene function include comparing the sequences of newly discovered genes with those of known genes in other species and observing the phenotypic effects of experimentally inactivating genes of unknown function.
- In systems biology, scientists use the computer-based tools of **bioinformatics** to compare genomes and study sets of genes and proteins as whole systems (**genomics** and **proteomics**). Studies

include large-scale analyses of protein interactions, functional DNA elements, and genes contributing to medical conditions.

? *What was the most significant finding of the ENCODE pilot project? Why has the project been expanded to include other species?*

CONCEPT 21.3

Genomes vary in size, number of genes, and gene density (pp. 432–434)

	Bacteria	Archaea	Eukarya
Genome size	Most are 1–6 Mb		Most are 10–4,000 Mb, but a few are much larger
Number of genes	1,500–7,500		5,000–40,000
Gene density	Higher than in eukaryotes		Lower than in prokaryotes (Within eukaryotes, lower density is correlated with larger genomes.)
Introns	None in protein-coding genes	Present in some genes	Unicellular eukaryotes: present, but prevalent only in some species Multicellular eukaryotes: present in most genes
Other noncoding DNA	Very little		Can be large amounts; generally more repetitive noncoding DNA in multicellular eukaryotes

? *Compare genome size, gene number, and gene density (a) in the three domains and (b) among eukaryotes.*

CONCEPT 21.4

Multicellular eukaryotes have much noncoding DNA and many multigene families (pp. 434–438)

- Only 1.5% of the human genome codes for proteins or gives rise to rRNAs or tRNAs; the rest is noncoding DNA, including **pseudogenes** and **repetitive DNA** of unknown function.

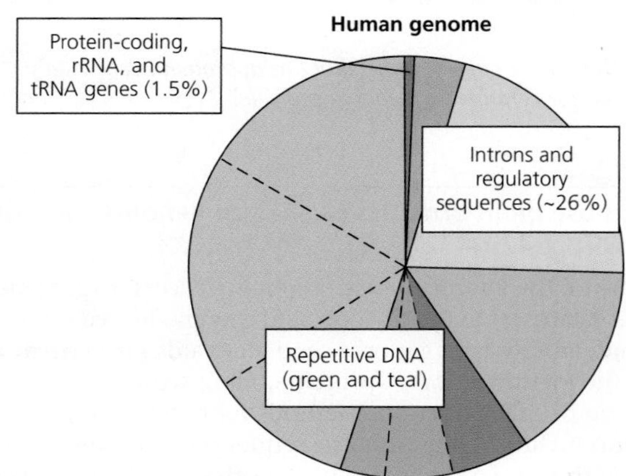

- The most abundant type of repetitive DNA in multicellular eukaryotes consists of **transposable elements** and related sequences. In eukaryotes, there are two types of transposable elements: **transposons**, which move via a DNA intermediate, and **retrotransposons**, which are more prevalent and move via an RNA intermediate.
- Other repetitive DNA includes short noncoding sequences that are tandemly repeated thousands of times (**simple sequence DNA**, which includes **STRs**); these sequences are especially prominent in centromeres and telomeres, where they probably play structural roles in the chromosome.
- Though many eukaryotic genes are present in one copy per haploid chromosome set, others (most, in some species) are members of a family of related genes, such as the human globin gene families:

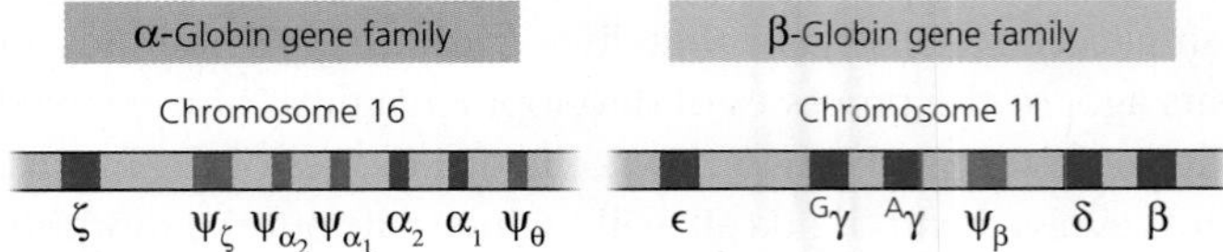

? *Explain how the function of transposable elements might account for their prevalence in human noncoding DNA.*

CONCEPT 21.5

Duplication, rearrangement, and mutation of DNA contribute to genome evolution (pp. 438–442)

- Accidents in cell division can lead to extra copies of all or part of entire chromosome sets, which may then diverge if one set accumulates sequence changes.
- The chromosomal organization of genomes can be compared among species, providing information about evolutionary relationships. Within a given species, rearrangements of chromosomes are thought to contribute to the emergence of new species.
- The genes encoding the various globin proteins evolved from one common ancestral globin gene, which duplicated and diverged into α-globin and β-globin ancestral genes. Subsequent duplication and random mutation gave rise to the present globin genes, all of which code for oxygen-binding proteins. The copies of some duplicated genes have diverged so much that the functions of their encoded proteins (such as lysozyme and α-lactalbumin) are now substantially different.
- Rearrangement of exons within and between genes during evolution has led to genes containing multiple copies of similar exons and/or several different exons derived from other genes.
- Movement of transposable elements or recombination between copies of the same element occasionally generates new sequence combinations that are beneficial to the organism. Such mechanisms can alter the functions of genes or their patterns of expression and regulation.

? *How could chromosomal rearrangements lead to the emergence of new species?*

CONCEPT 21.6

Comparing genome sequences provides clues to evolution and development (pp. 442–447)

- Comparative studies of genomes from widely divergent and closely related species provide valuable information about ancient and more recent evolutionary history, respectively. Human and chimpanzee sequences show about 4% difference, mostly due to insertions, deletions, and duplications in one lineage. Along with nucleotide variations in specific genes (such as *FOXP2*, a gene affecting speech), these differences may account for the distinct characteristics of the two species. Analysis of single nucleotide polymorphisms (SNPs) and copy-number variants (CNVs) among individuals in a species can also yield information about the evolution of that species.

• Evolutionary developmental (**evo-devo**) biologists have shown that homeotic genes and some other genes associated with animal development contain a **homeobox** region whose sequence is highly conserved among diverse species. Related sequences are present in the genes of plants and yeasts. During embryonic development in both plants and animals, a cascade of transcription regulators turns genes on or off in a carefully regulated sequence. However, the genes that direct analogous developmental processes differ in plants and animals as a result of their remote ancestry.

? *What type of information can be obtained by comparing the genomes of closely related species? Of very distantly related species?*

TEST YOUR UNDERSTANDING

LEVEL 1: KNOWLEDGE/COMPREHENSION

1. Bioinformatics includes all of the following *except*
   a. using computer programs to align DNA sequences.
   b. analyzing protein interactions in a species.
   c. using molecular biology to combine DNA from two different sources in a test tube.
   d. developing computer-based tools for genome analysis.
   e. using mathematical tools to make sense of biological systems.
2. One of the characteristics of retrotransposons is that
   a. they code for an enzyme that synthesizes DNA using an RNA template.
   b. they are found only in animal cells.
   c. they generally move by a cut-and-paste mechanism.
   d. they contribute a significant portion of the genetic variability seen within a population of gametes.
   e. their amplification is dependent on a retrovirus.
3. Homeotic genes
   a. encode transcription factors that control the expression of genes responsible for specific anatomical structures.
   b. are found only in *Drosophila* and other arthropods.
   c. are the only genes that contain the homeobox domain.
   d. encode proteins that form anatomical structures in the fly.
   e. are responsible for patterning during plant development.

LEVEL 2: APPLICATION/ANALYSIS

4. Two eukaryotic proteins have one domain in common but are otherwise very different. Which of the following processes is most likely to have contributed to this similarity?
   a. gene duplication
   b. RNA splicing
   c. exon shuffling
   d. histone modification
   e. random point mutations
5. **DRAW IT** Below are the amino acid sequences (using the single-letter code; see Figure 5.16) of four short segments of the FOXP2 protein from six species: chimpanzee, orangutan, gorilla, rhesus macaque, mouse, and human. These segments contain all of the amino acid differences between the FOXP2 proteins of these species.

   1. ATETI...PKSSD...TSSTT...NARRD
   2. ATETI...PKSSE...TSSTT...NARRD
   3. ATETI...PKSSD...TSSTT...NARRD
   4. ATETI...PKSSD...TSSNT...SARRD
   5. ATETI...PKSSD...TSSTT...NARRD
   6. VTETI...PKSSD...TSSTT...NARRD

   Use a highlighter to color any amino acid that varies among the species. (Color that amino acid in all sequences.) Then answer the questions at the top of the next column.

   (a) The chimpanzee, gorilla, and rhesus macaque (C, G, R) sequences are identical. Which lines correspond to those sequences?
   (b) The human sequence differs from that of the C, G, R species at two amino acids. Which line corresponds to the human sequence? Underline the two differences.
   (c) The orangutan sequence differs from the C, G, R sequence at one amino acid (having valine instead of alanine) and from the human sequence at three amino acids. Which line corresponds to the orangutan sequence?
   (d) How many amino acid differences are there between the mouse and the C, G, R species? Circle the amino acid(s) that differ(s) in the mouse. How many amino acid differences are there between the mouse and the human? Draw a square around the amino acid(s) that differ(s) in the mouse.
   (e) Primates and rodents diverged between 60 and 100 million years ago, and chimpanzees and humans diverged about 6 million years ago. Knowing that, what can you conclude by comparing the amino acid differences between the mouse and the C, G, R species with the differences between the human and the C, G, R species?

LEVEL 3: SYNTHESIS/EVALUATION

6. **EVOLUTION CONNECTION**
   Genes important in the embryonic development of animals, such as homeobox-containing genes, have been relatively well conserved during evolution; that is, they are more similar among different species than are many other genes. Why is this?
7. **SCIENTIFIC INQUIRY**
   The scientists mapping the SNPs in the human genome noticed that groups of SNPs tended to be inherited together, in blocks known as haplotypes, ranging in length from 5,000 to 200,000 base pairs. There are as few as four or five commonly occurring combinations of SNPs per haplotype. Propose an explanation for this observation, integrating what you've learned throughout this chapter and this unit.
8. **WRITE ABOUT A THEME**
   The Genetic Basis of Life The continuity of life is based on heritable information in the form of DNA. In a short essay (100–150 words), explain how mutations in protein-coding genes and regulatory DNA contribute to evolution.

For selected answers, see Appendix A.

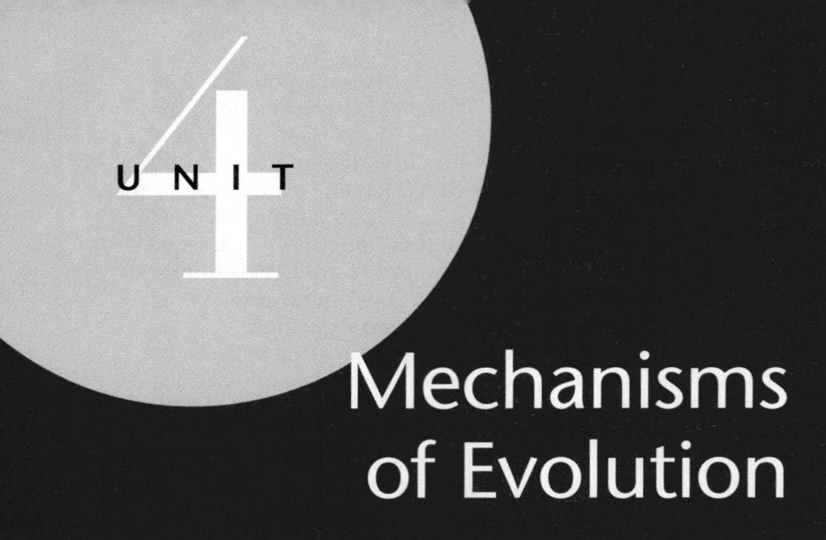

Mechanisms of Evolution

An Interview with Geerat J. Vermeij

Born in the Netherlands, Geerat Vermeij (pronounced "ver-may") lost his sight at the age of 3. Undeterred, he went on to earn degrees from Princeton and Yale and is now a Distinguished Professor at the University of California, Davis. A member of the Department of Geology, he nevertheless focuses on biology—the structure, evolution, and ecology of marine molluscs, both living and extinct. He is particularly well known for his work on the evolutionary "arms race" between long-extinct molluscs and their predators and more generally the roles of organismal interactions in evolution, although his many publications reflect much wider-ranging interests. (One of his books, *Nature: An Economic History*, relates the principles of evolution to the principles of economics; he has also written a memoir, *Privileged Hands: A Scientific Life*.) Dr. Vermeij has received numerous awards, including the MacArthur Award and the Daniel Giraud Elliot Medal from the National Academy of Sciences. His office at UC Davis features a large collection of marine shells and fossils and an extensive library. Jane Reece and Michael Cain spoke with him there.

How did you first become interested in biology?

As far back as I can recall, back to my earliest childhood, I've always liked natural history. When I was a child in the Netherlands, I liked pinecones and seeds and shells on the beach and leaves. I liked the whole ambience of being outside! Also, my parents were very good observers, and they spent a lot of time describing the world to me and letting me touch as much as I possibly could. When I moved to the U.S. at the age of 9, I found myself in a completely different environment. In New Jersey where we lived, there were wild forests full of huge vines, noisy crickets, cicadas, and strange birds, and I found this environment so different from the one I had left behind that I began to ask myself why this was.

When I was in the fourth grade, I had a wonderful teacher who brought shells from Florida to her classroom. And I explored these things and fell in love with them. And again, I wondered why these things were so different from anything I had collected in Holland. They were beautiful, with lovely shapes and wonderful contrast between the outside surface and the inside. I was smitten. And from then on I knew I was going to do something scientific.

Much of your work focuses on marine molluscs. Please tell us about these animals.

Molluscs include snails, clams, squids, octopuses, and many lesser-known groups. There are something like 100,000 species living, and we know of fossils of tens of thousands of extinct ones, dating all the way back to at least 540 million years ago. Molluscs are a major animal group on the tree of life. Found on land and in fresh water and the sea, they do just about everything you can imagine—they range from top predators (such as squids) to suspension feeders, herbivores, detritus feeders, and parasites. Originally, all molluscs had some kind of mantle covering the major organs of the body. They probably started off as pretty simple creatures without shells, but shells soon evolved. Most living molluscs have shells, though some have lost their shell in the course of evolution.

How do you identify the shells you are studying?

Entirely by touch. You know, shells differ in size, shape, and texture, all of which are readily discernable by the fingers, and the same is true of fossils. Shells are ideally suited for a blind person like me.

In your research over the years, what are the main questions you have been trying to answer?

The questions have changed over the course of my career, but the overarching ones have been, What are the pathways of adaptation by which all the different lineages of organisms—not just molluscs, but all of life—have gotten here? How have the conditions to which organisms are exposed changed over time? How have organisms affected those conditions over time? I'm very interested in the history of life and how this history has been shaped.

What makes molluscs a good research focus for answering your questions?

Molluscs have several huge advantages. For me personally, of course, they are accessible. Most of them don't run away—squids and octopuses being exceptions. Their shells are extremely easy to observe with the hand, and importantly from a paleontologist's point of view, these hard mineralized objects have left a very good fossil record. That's a gigantic advantage. Not only can we trace molluscs all the way back through time, but because we know so much about how shells work, we can figure out how the extinct animals lived—even those that lived hundreds of millions of years ago.

What kinds of evolutionary insights can fossils provide that cannot be extracted from DNA evidence?

First of all, I should say that my collaborators and I do use DNA sequences ourselves to reconstruct the order of branching in evolutionary trees. But to estimate *when* these evolutionary lineages arose, we need to calibrate the tree with fossils of known age. Moreover, you can only get DNA from living things and a few rather recent fossils; so if you go back far enough, you find many lineages that no longer exist and for which, therefore, DNA evidence is simply unavailable. And yet these animals often have combinations of traits that we never see in living animals. Fossils give us a very good idea of what the ancestral organisms were like, which you couldn't get solely from DNA sequences of living organisms. So if you're trying to reconstruct early branches in the tree of life, fossils are very helpful.

How does your research relate to the mechanisms of evolution—to the principles as opposed to the pattern?

That's an important question. I do distinguish between describing what actually happened and the mechanisms that account for evolutionary events over time. A lot of our work is descriptive, figuring out what happened and what extinct animals were like. But we also try to determine the mechanisms that account for the phenomena. And given that I work on adaptive characteristics and on the fit between animals and plants and their environments, I am particularly interested in the mechanisms by which organisms become adapted to their environments. That's not simply natural selection; it's also the modification of environments by the animals and plants that reside there.

Tell us how you go about your work.

I have done a lot of fieldwork all over the world. In the field I observe molluscs in nature and occasionally do some experiments with them. I want to understand how the molluscs relate to their environments, including their predators. I am interested in how they live—for example, how quickly they move—and how their performance levels compare with the performance of the agents that are out there making the world tough for them.

Recently I've spent more time in museum collections. I also maintain a very large research collection, most of which I've collected myself over the years. All of these collections are critical for learning what the shapes of organisms are in different evolutionary groups. I also visit and learn from other scientists. And I do an enormous amount of reading, because I like to synthesize information and ideas, to put things together. I read hundreds of papers a year about a very wide variety of subjects, everything from biology to geology to economics and history, so that I can place the particular work that I do into a larger context. As a scientist, you can never read enough.

When you do this kind of work, whether in collections of specimens or in the field or library, you always come across wonderful surprises—perhaps a shell with a feature you've never seen or even a book you didn't know about. Every single day for me is like that.

You have written about the "arms race" of evolution. What do you mean by that, and how has it played out in the creatures you've studied?

All living things are exposed to competition for resources and also to predation, where one animal eats another animal or part of another animal. The animals I am working on mostly don't move very fast, and one of the typical results of predation is that armor in the form of shells evolves in the prey; the mollusc shell probably first evolved as armor. But as predators become more powerful thanks to competition among themselves, the performance criteria for an effective shell also escalate. Nowadays, in order to survive in tropical shallow water environments where there are lots of predators, a mollusc needs a very well-armored shell—one that has thick walls, bumps all over the place, a narrow opening, and many other features. In fact, if you look at shell architecture over geological time, keeping habitat as constant as you can, you find that some of these protective features (the narrow opening, for example) are found only in the more recent evolutionary lineages and don't appear at all in the first couple of hundred million years of mollusc history. Meanwhile, all sorts of ways of overcoming mollusc defenses have evolved in predators. They have developed stronger, more powerful jaws or claws. They have "learned" how to drill a hole through the wall of a shell. They swallow or envelop larger prey. That tells us that there has been an arms race, an escalation of improvements in both shell architecture and methods of attack by predators.

> "When you do this kind of work . . . you always come across wonderful surprises. . . . Every single day for me is like that."

In addition to arms races, what other kinds of ecological interactions have influenced evolutionary history?

Competition and predation are fundamental and inevitable, but the history of life from the very beginning is also a story of cooperation. The reason for that I think is simple: By cooperating, you can do things that neither party can do by itself. So, cooperation or some other kind of mutually beneficial relationship is a wonderful way to compete. Biology is absolutely filled with examples of social animals, mutually beneficial relationships between individuals of different species, and so forth. Cooperation is in some sense an emergent property of life as a whole. The interactions of organisms with one another give rise to properties that the individual components don't have. For example, a lichen, which is an alga and a fungus living together, has properties different from those of either participant.

How do the things you've been saying about the effects of ecological interactions on evolutionary history fit in with Darwin's main ideas?

Darwin was an incredibly smart guy. One of the many, many things that he got right was that natural selection is often brought about by the interactions of organisms with other organisms, as well as interactions of organisms with their physical environment. Natural selection isn't some nebulous agency out there that's choosing survivors over nonsurvivors.

Why is it important for people to understand evolution?

There are many reasons. An understanding of evolution is certainly of practical importance in medicine and agriculture. But an understanding of evolution also gives us a closer connection to the rest of life. Also, it's very important for people to understand that the theory of evolution, like all scientific theories, is a body of explanation and fact that explains natural phenomena and can predict them. A lot of the resistance to evolution comes down to the idea some people have that somehow evolution makes life meaningless or purposeless. To this I reply that meaning and purpose is an emergent property of evolution! It is our own responsibility to make life meaningful.

Geerat Vermeij (right) with Michael Cain (center) and Jane Reece

22

Descent with Modification: A Darwinian View of Life

▲ **Figure 22.1 How can this beetle survive in the desert, and what is it doing?**

EVOLUTION

KEY CONCEPTS

22.1 **The Darwinian revolution challenged traditional views of a young Earth inhabited by unchanging species**

22.2 **Descent with modification by natural selection explains the adaptations of organisms and the unity and diversity of life**

22.3 **Evolution is supported by an overwhelming amount of scientific evidence**

OVERVIEW

Endless Forms Most Beautiful

In the coastal Namib desert of southwestern Africa, a land where fog is common but virtually no rain falls, lives the beetle *Onymacris unguicularis*. To obtain the water it needs to survive, this insect relies on a peculiar "headstanding" behavior (**Figure 22.1**). Tilting head-downward, the beetle faces into the winds that blow fog across the dunes. Droplets of moisture from the fog collect on the beetle's body and run down into its mouth.

Interesting in its own right, this headstander beetle is also a member of an astonishingly diverse group: the more than 350,000 species of beetles. In fact, nearly one of every five known species is a beetle. These beetles all share similar features, such as six pairs of legs, a hard outer surface, and two pairs of wings. But they also differ from one another. How did there come to be so many beetles, and what causes their similarities and differences?

The headstander beetle and its many close relatives illustrate three key observations about life:

- the striking ways in which organisms are suited for life in their environments*
- the many shared characteristics (unity) of life
- the rich diversity of life

A century and a half ago, Charles Darwin was inspired to develop a scientific explanation for these three broad observations. When he published his hypothesis in *The Origin of Species*, Darwin ushered in a scientific revolution—the era of evolutionary biology.

For now, we will define **evolution** as *descent with modification*, a phrase Darwin used in proposing that Earth's many species are descendants of ancestral species that were different from the present-day species. Evolution can also be defined more narrowly as a change in the genetic composition of a population from generation to generation, as discussed further in Chapter 23.

Whether it is defined broadly or narrowly, we can view evolution in two related but different ways: as a pattern and as a process. The *pattern* of evolutionary change is revealed by data from a range of scientific disciplines, including biology, geology, physics, and chemistry. These data are facts—they are observations about the natural world. The *process* of evolution consists of the mechanisms that produce the observed pattern of change. These mechanisms represent natural causes of the natural phenomena we observe. Indeed, the power of evolution as a unifying theory is its ability to explain and connect a vast array of observations about the living world.

As with all general theories in science, we continue to test our understanding of evolution by examining whether it can account for new observations and experimental results. In this and the following chapters, we'll examine how ongoing discoveries shape what we know about the pattern and process of evolution. To set the stage, we'll first retrace Darwin's quest to explain the adaptations, unity, and diversity of what he called life's "endless forms most beautiful."

*Here and throughout this book, the term *environment* refers to other organisms as well as to the physical aspects of an organism's surroundings.

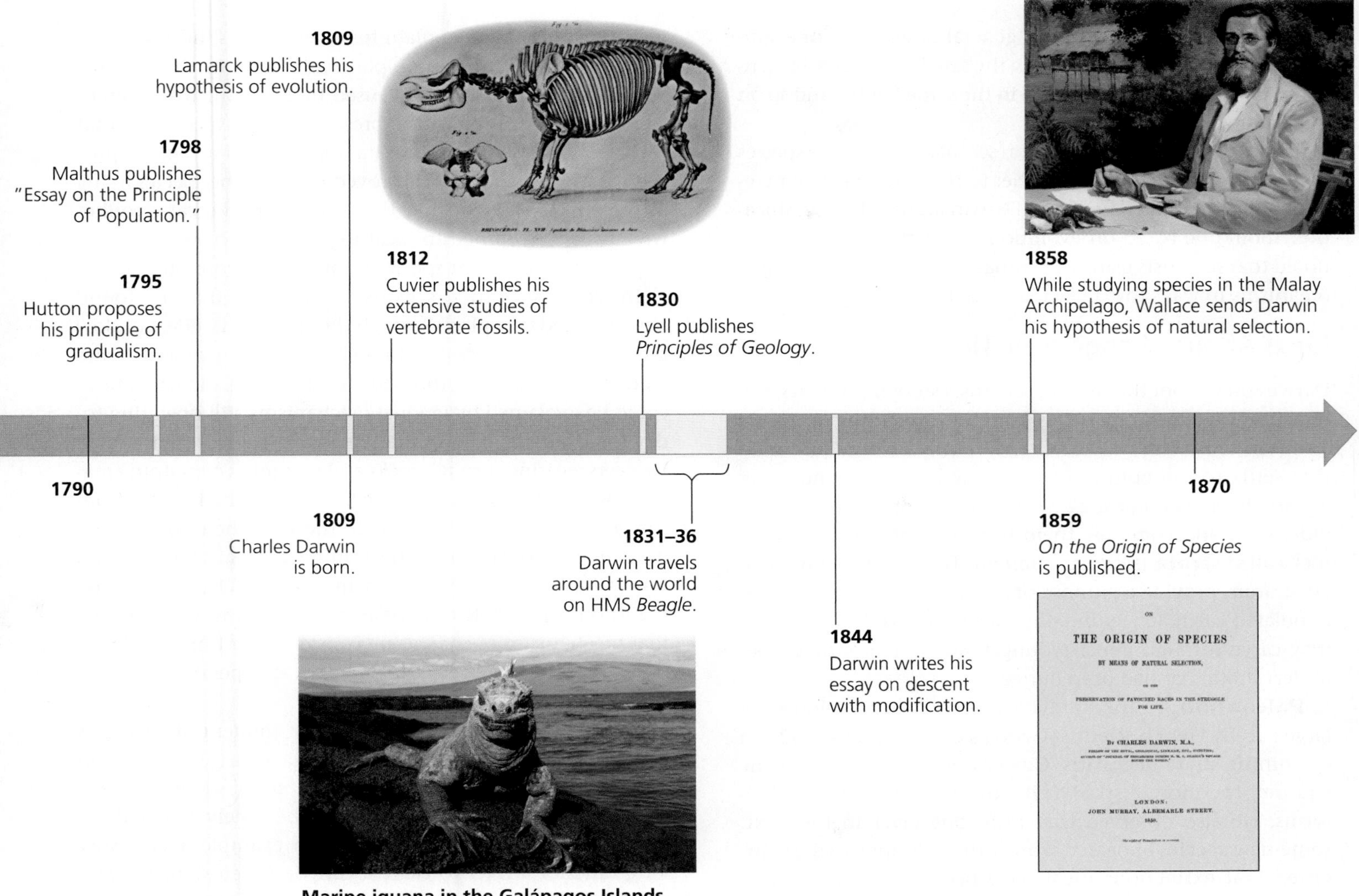

1809
Lamarck publishes his hypothesis of evolution.
1798
Malthus publishes "Essay on the Principle of Population."
1795
Hutton proposes his principle of gradualism.
1812
Cuvier publishes his extensive studies of vertebrate fossils.
1830
Lyell publishes *Principles of Geology*.
1858
While studying species in the Malay Archipelago, Wallace sends Darwin his hypothesis of natural selection.
1790
1870
1809
Charles Darwin is born.
1831–36
Darwin travels around the world on HMS *Beagle*.
1844
Darwin writes his essay on descent with modification.
1859
On the Origin of Species is published.
ON
THE ORIGIN OF SPECIES
BY MEANS OF NATURAL SELECTION,
LONDON:
JOHN MURRAY, ALBEMARLE STREET.
Marine iguana in the Galápagos Islands

▲ **Figure 22.2 The intellectual context of Darwin's ideas.**

CONCEPT **22.1**

The Darwinian revolution challenged traditional views of a young Earth inhabited by unchanging species

What impelled Darwin to challenge the prevailing views about Earth and its life? Darwin's revolutionary proposal developed over time, influenced by the work of others and by his travels **(Figure 22.2)**. As we'll see, his ideas had deep historical roots.

Scala Naturae and Classification of Species

Long before Darwin was born, several Greek philosophers suggested that life might have changed gradually over time. But one philosopher who greatly influenced early Western science, Aristotle (384–322 BCE), viewed species as fixed (unchanging). Through his observations of nature, Aristotle recognized certain "affinities" among organisms. He concluded that life-forms could be arranged on a ladder, or scale, of increasing complexity, later called the *scala naturae* ("scale of nature"). Each form of life, perfect and permanent, had its allotted rung on this ladder.

These ideas were generally consistent with the Old Testament account of creation, which holds that species were individually designed by God and therefore perfect. In the 1700s, many scientists interpreted the often remarkable match of organisms to their environment as evidence that the Creator had designed each species for a particular purpose.

One such scientist was Carolus Linnaeus (1707–1778), a Swedish physician and botanist who sought to classify life's diversity, in his words, "for the greater glory of God." Linnaeus developed the two-part, or *binomial*, format for naming species (such as *Homo sapiens* for humans) that is still used today. In contrast to the linear hierarchy of the *scala naturae*, Linnaeus adopted a nested classification system, grouping similar species into increasingly general categories. For example, similar species are grouped in the same genus, similar genera (plural of genus) are grouped in the same family, and so on (see Figure 1.14).

Linnaeus did not ascribe the resemblances among species to evolutionary kinship, but rather to the pattern of their creation. A century later, however, Darwin argued that classification should be based on evolutionary relationships. He also noted that scientists using the Linnaean system often grouped organisms in ways that reflected those relationships.

Ideas About Change over Time

Darwin drew from the work of scientists studying **fossils**, the remains or traces of organisms from the past. Many fossils are found in sedimentary rocks formed from the sand and mud that settle to the bottom of seas, lakes, swamps, and other aquatic habitats **(Figure 22.3)**. New layers of sediment cover older ones and compress them into superimposed layers of rock called **strata** (singular, *stratum*). The fossils in a particular stratum provide a glimpse of some of the organisms that populated Earth at the time that layer formed. Later, erosion may carve through upper (younger) strata, revealing deeper (older) strata that had been buried.

Paleontology, the study of fossils, was developed in large part by French scientist Georges Cuvier (1769–1832). In examining strata near Paris, Cuvier noted that the older the stratum, the more dissimilar its fossils were to current life-forms. He also observed that from one layer to the next, some new species appeared while others disappeared. He inferred that extinctions must have been a common occurrence in the history of life. Yet Cuvier staunchly opposed the

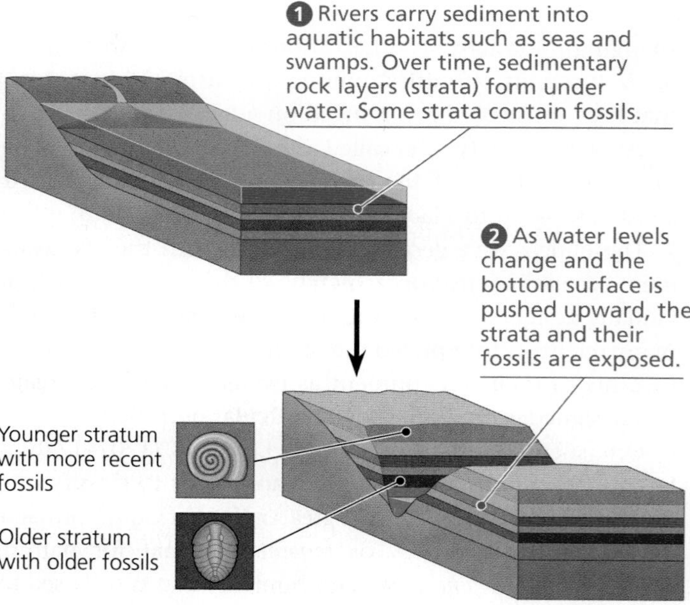

1 Rivers carry sediment into aquatic habitats such as seas and swamps. Over time, sedimentary rock layers (strata) form under water. Some strata contain fossils.
2 As water levels change and the bottom surface is pushed upward, the strata and their fossils are exposed.
Younger stratum with more recent fossils
Older stratum with older fossils

▲ **Figure 22.3 Formation of sedimentary strata with fossils.**

idea of evolution. To explain his observations, he advocated **catastrophism**, the principle that events in the past occurred suddenly and were caused by mechanisms different from those operating in the present. Cuvier speculated that each boundary between strata represented a catastrophe, such as a flood, that had destroyed many of the species living at that time. He proposed that these periodic catastrophes were usually confined to local regions, which were later repopulated by different species immigrating from other areas.

In contrast, other scientists suggested that profound change could take place through the cumulative effect of slow but continuous processes. In 1795, Scottish geologist James Hutton (1726–1797) proposed that Earth's geologic features could be explained by gradual mechanisms still operating today. For example, he suggested that valleys were often formed by rivers wearing through rocks and that rocks containing marine fossils were formed when sediments that had eroded from the land were carried by rivers to the sea, where they buried dead marine organisms. The leading geologist of Darwin's time, Charles Lyell (1797–1875), incorporated Hutton's thinking into his principle of **uniformitarianism**, which stated that mechanisms of change are constant over time. Lyell proposed that the same geologic processes are operating today as in the past, and at the same rate.

Hutton and Lyell's ideas strongly influenced Darwin's thinking. Darwin agreed that if geologic change results from slow, continuous actions rather than from sudden events, then Earth must be much older than the widely accepted age of a few thousand years. It would, for example, take a very long time for a river to carve a canyon by erosion. He later reasoned that perhaps similarly slow and subtle processes could produce substantial biological change. Darwin was not the first to apply the idea of gradual change to biological evolution, however.

Lamarck's Hypothesis of Evolution

During the 18th century, several naturalists (including Darwin's grandfather, Erasmus Darwin) suggested that life evolves as environments change. But only one of Charles Darwin's predecessors proposed a mechanism for *how* life changes over time: French biologist Jean-Baptiste de Lamarck (1744–1829). Alas, Lamarck is primarily remembered today *not* for his visionary recognition that evolutionary change explains patterns in fossils and the match of organisms to their environments, but for the incorrect mechanism he proposed to explain how evolution occurs.

Lamarck published his hypothesis in 1809, the year Darwin was born. By comparing living species with fossil forms, Lamarck had found what appeared to be several lines of descent, each a chronological series of older to younger fossils leading to a living species. He explained his findings using two principles that were widely accepted at the time. The first was *use and disuse*, the idea that parts of the body that are used

▲ **Figure 22.4 Acquired traits cannot be inherited.** This bonsai tree was "trained" to grow as a dwarf by pruning and shaping. However, seeds from this tree would produce offspring of normal size.

extensively become larger and stronger, while those that are not used deteriorate. Among many examples, he cited a giraffe stretching its neck to reach leaves on high branches. The second principle, *inheritance of acquired characteristics*, stated that an organism could pass these modifications to its offspring. Lamarck reasoned that the long, muscular neck of the living giraffe had evolved over many generations as giraffes stretched their necks ever higher.

Lamarck also thought that evolution happens because organisms have an innate drive to become more complex. Darwin rejected this idea, but he, too, thought that variation was introduced into the evolutionary process in part through inheritance of acquired characteristics. Today, however, our understanding of genetics refutes this mechanism: Experiments show that traits acquired by use during an individual's life are not inherited in the way proposed by Lamarck (**Figure 22.4**).

Lamarck was vilified in his own time, especially by Cuvier, who denied that species ever evolve. In retrospect, however, Lamarck did recognize that the match of organisms to their environments can be explained by gradual evolutionary change, and he did propose a testable explanation for how this change occurs.

CONCEPT CHECK 22.1

1. How did Hutton's and Lyell's ideas influence Darwin's thinking about evolution?
2. **MAKE CONNECTIONS** In Concept 1.3 (pp. 19–20), you read that scientific hypotheses must be testable and falsifiable. Applying these criteria, are Cuvier's explanation of the fossil record and Lamarck's hypothesis of evolution scientific? Explain your answer in each case.

For suggested answers, see Appendix A.

CONCEPT 22.2

Descent with modification by natural selection explains the adaptations of organisms and the unity and diversity of life

As the 19th century dawned, it was generally thought that species had remained unchanged since their creation. A few clouds of doubt about the permanence of species were beginning to gather, but no one could have forecast the thundering storm just beyond the horizon. How did Charles Darwin become the lightning rod for a revolutionary view of life?

Darwin's Research

Charles Darwin (1809–1882) was born in Shrewsbury, in western England. Even as a boy, he had a consuming interest in nature. When he was not reading nature books, he was fishing, hunting, and collecting insects. Darwin's father, a physician, could see no future for his son as a naturalist and sent him to medical school in Edinburgh. But Charles found medicine boring and surgery before the days of anesthesia horrifying. He quit medical school and enrolled at Cambridge University, intending to become a clergyman. (At that time in England, many scholars of science belonged to the clergy.)

At Cambridge, Darwin became the protégé of the Reverend John Henslow, a botany professor. Soon after Darwin graduated, Henslow recommended him to Captain Robert FitzRoy, who was preparing the survey ship HMS *Beagle* for a long voyage around the world. Darwin would pay his own way and serve as a conversation partner to the young captain. FitzRoy, who was himself an accomplished scientist, accepted Darwin because he was a skilled naturalist and because they were of the same social class and close in age.

The Voyage of the Beagle

Darwin embarked from England on the *Beagle* in December 1831. The primary mission of the voyage was to chart poorly known stretches of the South American coastline. While the ship's crew surveyed the coast, Darwin spent most of his time on shore, observing and collecting thousands of South American plants and animals. He noted the characteristics of plants and animals that made them well suited to such diverse environments as the humid jungles of Brazil, the expansive grasslands of Argentina, and the towering peaks of the Andes.

Darwin observed that the plants and animals in temperate regions of South America more closely resembled species living in the South American tropics than species living in temperate regions of Europe. Furthermore, the fossils he found, though clearly different from living species, were distinctly South American in their resemblance to the living organisms of that continent.

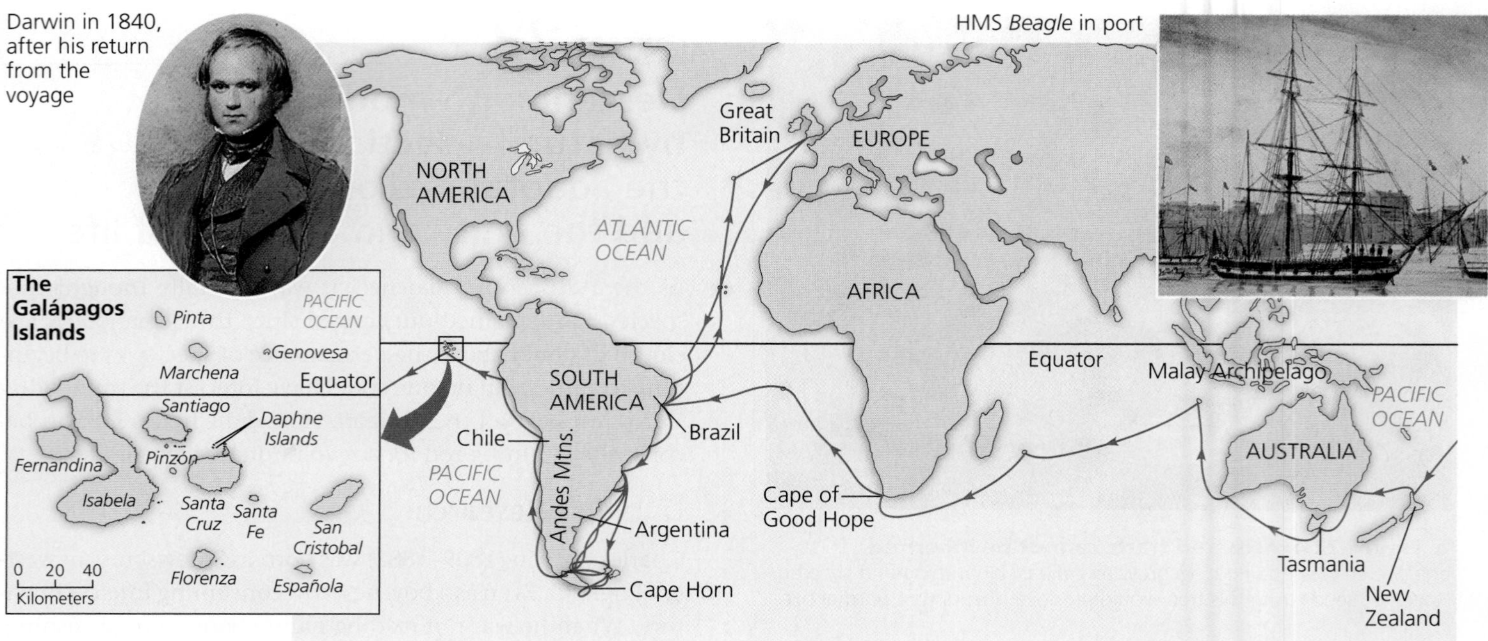

▲ **Figure 22.5 The voyage of HMS *Beagle*.**

Darwin also spent much time thinking about geology. Despite bouts of seasickness, he read Lyell's *Principles of Geology* while aboard the *Beagle*. He experienced geologic change firsthand when a violent earthquake rocked the coast of Chile, and he observed afterward that rocks along the coast had been thrust upward by several feet. Finding fossils of ocean organisms high in the Andes, Darwin inferred that the rocks containing the fossils must have been raised there by many similar earthquakes. These observations reinforced what he had learned from Lyell: The physical evidence did not support the traditional view that Earth was only a few thousand years old.

Darwin's interest in the geographic distribution of species was further stimulated by the *Beagle*'s stop at the Galápagos, a group of volcanic islands located near the equator about 900 km west of South America (**Figure 22.5**). Darwin was fascinated by the unusual organisms there. The birds he collected included the finches mentioned in Chapter 1 and several kinds of mockingbirds. These mockingbirds, though similar to each other, seemed to be different species. Some were unique to individual islands, while others lived on two or more adjacent islands. Furthermore, although the animals on the Galápagos resembled species living on the South American mainland, most of the Galápagos species were not known from anywhere else in the world. Darwin hypothesized that the Galápagos had been colonized by organisms that had strayed from South America and then diversified, giving rise to new species on the various islands.

Darwin's Focus on Adaptation

During the voyage of the *Beagle*, Darwin observed many examples of **adaptations**, inherited characteristics of organisms that enhance their survival and reproduction in specific environments. Later, as he reassessed his observations, he began to perceive adaptation to the environment and the origin of new species as closely related processes. Could a new species arise from an ancestral form by the gradual accumulation of adaptations to a different environment? From studies made years after Darwin's voyage, biologists have concluded that this is indeed what happened to the diverse group of Galápagos finches (see Figure 1.22). The finches' various beaks and behaviors are adapted to the specific foods available on their home islands (**Figure 22.6**). Darwin realized that explaining such adaptations was essential to understanding evolution. As we'll explore further, his explanation of how adaptations arise centered on **natural selection**, a process in which individuals that have certain inherited traits tend to survive and reproduce at higher rates than other individuals *because of* those traits.

By the early 1840s, Darwin had worked out the major features of his hypothesis. He set these ideas on paper in 1844, when he wrote a long essay on descent with modification and its underlying mechanism, natural selection. Yet he was still reluctant to publish his ideas, apparently because he anticipated the uproar they would cause. During this time, Darwin continued to compile evidence in support of his hypothesis. By the mid-1850s, he had described his ideas to Lyell and a few others. Lyell, who was not yet convinced of evolution, nevertheless urged Darwin to publish on the subject before someone else came to the same conclusions and published first.

In June 1858, Lyell's prediction came true. Darwin received a manuscript from Alfred Russel Wallace (1823–1913), a British naturalist working in the South Pacific islands of the Malay

(a) Cactus-eater. The long, sharp beak of the cactus ground finch (*Geospiza scandens*) helps it tear and eat cactus flowers and pulp.

(b) Insect-eater. The green warbler finch (*Certhidea olivacea*) uses its narrow, pointed beak to grasp insects.

(c) Seed-eater. The large ground finch (*Geospiza magnirostris*) has a large beak adapted for cracking seeds that fall from plants to the ground.

▲ **Figure 22.6 Three examples of beak variation in Galápagos finches.** The Galápagos Islands are home to more than a dozen species of closely related finches, some found only on a single island. The most striking differences among them are their beaks, which are adapted for specific diets.

MAKE CONNECTIONS *Review Figure 1.22 (p. 17). To which of the other two species shown above is the cactus-eater more closely related (that is, with which does it share a more recent common ancestor)?*

Archipelago (see Figure 22.2). Wallace had developed a hypothesis of natural selection nearly identical to Darwin's. He asked Darwin to evaluate his paper and forward it to Lyell if it merited publication. Darwin complied, writing to Lyell: "Your words have come true with a vengeance. . . . I never saw a more striking coincidence . . . so all my originality, whatever it may amount to, will be smashed." On July 1, 1858, Lyell and a colleague presented Wallace's paper, along with extracts from Darwin's unpublished 1844 essay, to the Linnean Society of London. Darwin quickly finished his book, titled *On the Origin of Species by Means of Natural Selection* (commonly referred to as *The Origin of Species*), and published it the next year. Although Wallace had submitted his ideas for publication first, he admired Darwin and thought that Darwin had developed the idea of natural selection so extensively that he should be known as its main architect.

Within a decade, Darwin's book and its proponents had convinced most scientists that life's diversity is the product of evolution. Darwin succeeded where previous evolutionists had failed, mainly by presenting a plausible scientific mechanism with immaculate logic and an avalanche of evidence.

The Origin of Species

In his book, Darwin amassed evidence that descent with modification by natural selection explains the three broad observations about nature listed in the Overview: the unity of life, the diversity of life, and the match between organisms and their environments.

Descent with Modification

In the first edition of *The Origin of Species*, Darwin never used the word *evolution* (although the final word of the book is "evolved"). Rather, he discussed *descent with modification*, a phrase that summarized his view of life. Organisms share many characteristics, leading Darwin to perceive unity in life. He attributed the unity of life to the descent of all organisms from an ancestor that lived in the remote past. He also thought that as the descendants of that ancestral organism lived in various habitats over millions of years, they accumulated diverse modifications, or adaptations, that fit them to specific ways of life. Darwin reasoned that over long periods of time, descent with modification eventually led to the rich diversity of life we see today.

Darwin viewed the history of life as a tree, with multiple branchings from a common trunk out to the tips of the youngest twigs (**Figure 22.7**). The tips of the twigs represent the diversity of organisms living in the present. Each fork of the tree represents the most recent common ancestor of all the lines of evolution that subsequently branch from that point. As an example, consider the three living species of elephants: the Asian elephant (*Elephas maximus*) and African elephants

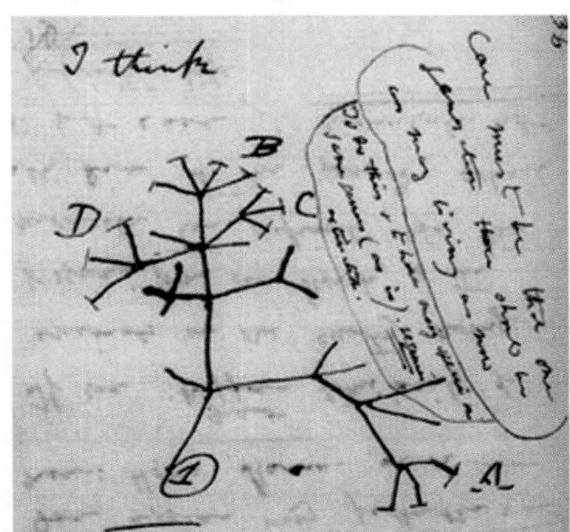

◀ **Figure 22.7 "I think. . ."** In this 1837 sketch, Darwin envisioned the branching pattern of evolution.

(*Loxodonta africana* and *L. cyclotis*). These closely related species are very similar because they shared the same line of descent until a relatively recent split from their common ancestor, as shown in the tree diagram in **Figure 22.8**. Note that seven lineages related to elephants have become extinct over the past 32 million years. As a result, there are no living species that fill the gap between the elephants and their nearest relatives today, the hyraxes and manatees. Such extinctions are not uncommon. In fact, many evolutionary branches, even some major ones, are dead ends: Scientists estimate that over 99% of all species that have ever lived are now extinct. As in Figure 22.8, fossils of extinct species can document the divergence of present-day groups by "filling in" gaps between them.

In his efforts at classification, Linnaeus had realized that some organisms resemble each other more closely than others, but he had not linked these resemblances to evolution. Nonetheless, because he had recognized that the great diversity of organisms could be organized into "groups subordinate to groups" (Darwin's phrase), Linnaeus's system meshed well with Darwin's hypothesis. To Darwin, the Linnaean hierarchy reflected the branching history of life, with organisms at the various levels related through descent from common ancestors.

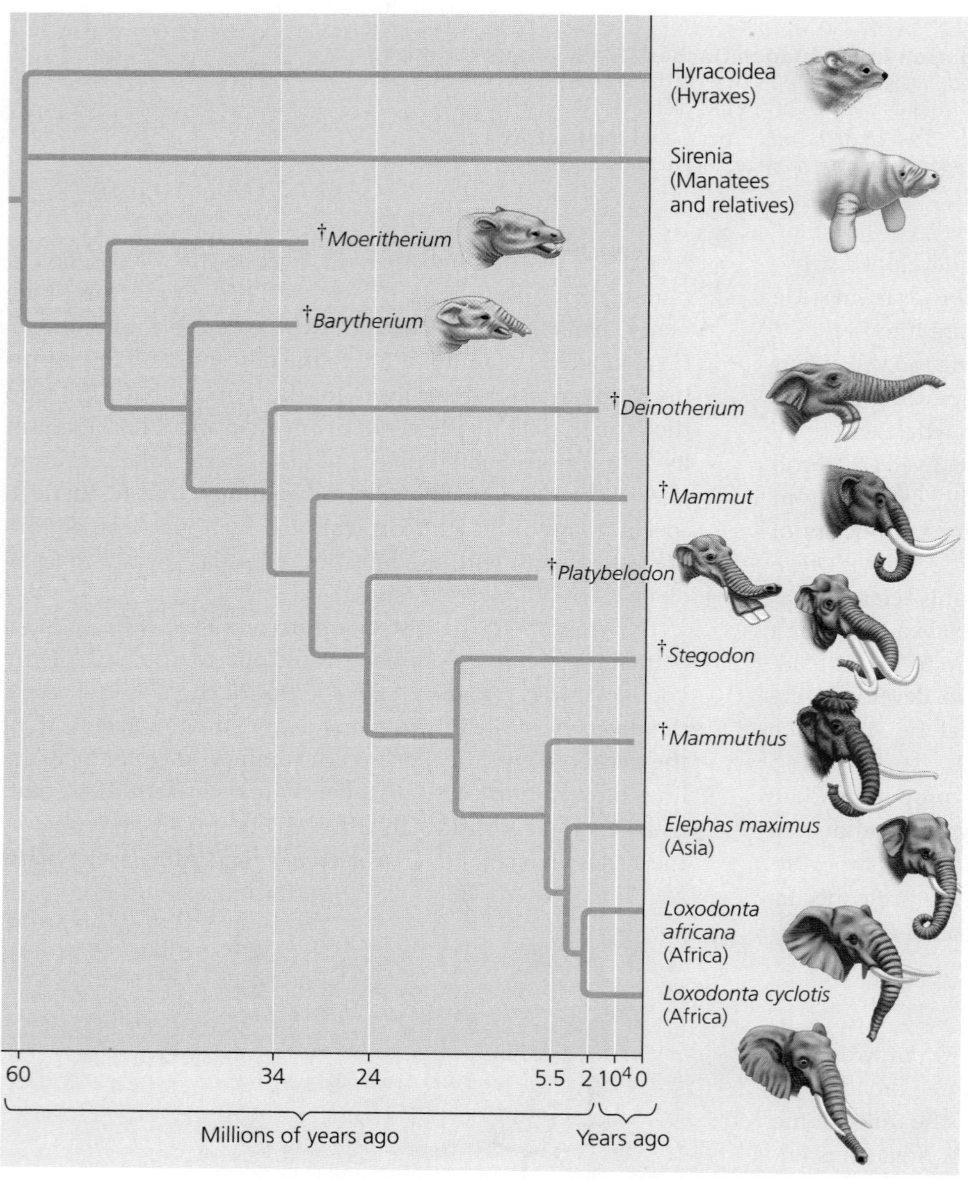

▲ **Figure 22.8 Descent with modification.** This evolutionary tree of elephants and their relatives is based mainly on fossils—their anatomy, order of appearance in strata, and geographic distribution. Note that most branches of descent ended in extinction (denoted by the dagger symbol †). (Time line not to scale.)

? *Based on the tree shown here, approximately when did the most recent ancestor shared by* Mammuthus *(woolly mammoths), Asian elephants, and African elephants live?*

Artificial Selection, Natural Selection, and Adaptation

Darwin proposed the mechanism of natural selection to explain the observable patterns of evolution. He crafted his argument carefully, to persuade even the most skeptical readers. First he discussed familiar examples of selective breeding of domesticated plants and animals. Humans have modified other species over many generations by selecting and breeding individuals that possess desired traits, a process called **artificial selection** (**Figure 22.9**). As a result of artificial selection, crops, livestock animals, and pets often bear little resemblance to their wild ancestors.

Darwin then argued that a similar process occurs in nature. He based his argument on two observations, from which he drew two inferences:

Observation #1: Members of a population often vary in their inherited traits (**Figure 22.10**).

Observation #2: All species can produce more offspring than their environment can support (**Figure 22.11**), and many of these offspring fail to survive and reproduce.

Inference #1: Individuals whose inherited traits give them a higher probability of surviving and reproducing in a given environment tend to leave more offspring than other individuals.

Inference #2: This unequal ability of individuals to survive and reproduce will lead to the accumulation of favorable traits in the population over generations.

Darwin saw an important connection between natural selection and the

▶ **Figure 22.9 Artificial selection.** These different vegetables have all been selected from one species of wild mustard. By selecting variations in different parts of the plant, breeders have obtained these divergent results.

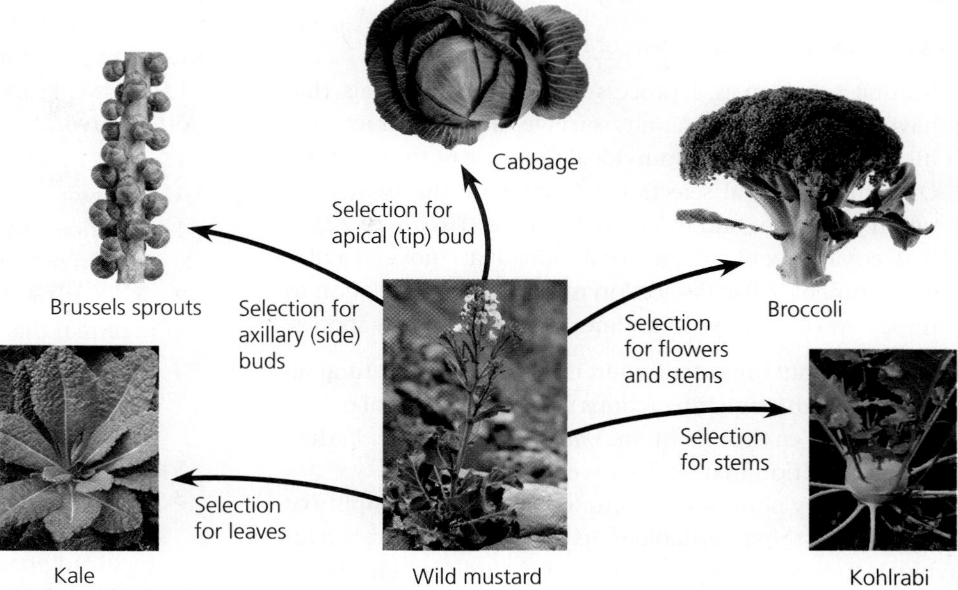

▲ **Figure 22.10 Variation in a population.** Individuals in this population of Asian ladybird beetles vary in color and spot pattern. Natural selection may act on these variations only if (1) they are heritable and (2) they affect the beetles' ability to survive and reproduce.

▶ **Figure 22.11 Overproduction of offspring.** A single puffball fungus can produce billions of offspring. If all of these offspring and their descendants survived to maturity, they would carpet the surrounding land surface.

capacity of organisms to "overreproduce." He began to make this connection after reading an essay by economist Thomas Malthus, who contended that much of human suffering—disease, famine, and war—was the inescapable consequence of the human population's potential to increase faster than food supplies and other resources. Darwin realized that the capacity to overreproduce was characteristic of all species. Of the many eggs laid, young born, and seeds spread, only a tiny fraction complete their development and leave offspring of their own. The rest are eaten, starved, diseased, unmated, or unable to tolerate physical conditions of the environment such as salinity or temperature.

An organism's heritable traits can influence not only its own performance, but also how well its offspring cope with environmental challenges. For example, an organism might have a trait that gives its offspring an advantage in escaping predators, obtaining food, or tolerating physical conditions. When such advantages increase the number of offspring that survive and reproduce, the traits that are favored will likely appear at a greater frequency in the next generation. Thus, over time, natural selection resulting from factors such as predators, lack of food, or adverse physical conditions can lead to an increase in the proportion of favorable traits in a population.

How rapidly do such changes occur? Darwin reasoned that if artificial selection can bring about dramatic change in a relatively short period of time, then natural selection should be capable of substantial modification of species over many hundreds of generations. Even if the advantages of some heritable traits over others are slight, the advantageous variations will gradually accumulate in the population, and less favorable variations will diminish. Over time, this process will increase the frequency of individuals with favorable adaptations and hence refine the match between organisms and their environment (see Figure 1.20).

Natural Selection: A Summary

Let's now recap the main ideas of natural selection:

- Natural selection is a process in which individuals that have certain heritable traits survive and reproduce at a higher rate than other individuals because of those traits.
- Over time, natural selection can increase the match between organisms and their environment (**Figure 22.12**).
- If an environment changes, or if individuals move to a new environment, natural selection may result in adaptation to these new conditions, sometimes giving rise to new species.

One subtle but important point is that although natural selection occurs through interactions between individual organisms and their environment, *individuals do not evolve*. Rather, it is the population that evolves over time.

A second key point is that natural selection can amplify or diminish only those heritable traits that differ among the individuals in a population. Thus, even if a trait is heritable, if all the individuals in a population are genetically identical for that trait, evolution by natural selection cannot occur.

Third, remember that environmental factors vary from place to place and over time. A trait that is favorable in one place or time may be useless—or even detrimental—in other places or times. Natural selection is always operating, but which traits are favored depends on the context in which a species lives and mates.

Next, we'll survey the wide range of observations that support a Darwinian view of evolution by natural selection.

CONCEPT CHECK 22.2

1. How does the concept of descent with modification explain both the unity and diversity of life?
2. **WHAT IF?** If you discovered a fossil of an extinct mammal that lived high in the Andes, would you predict that it would more closely resemble present-day mammals from South American jungles or present-day mammals that live high in African mountains? Explain.
3. **MAKE CONNECTIONS** Review Figures 14.4 and 14.6 (pp. 265 and 267) on the relationship between genotype and phenotype. In a particular pea population, suppose that flowers with the white phenotype are favored by natural selection. Predict what would happen over time to the frequency of the *p* allele in the population, and explain your reasoning.

For suggested answers, see Appendix A.

(a) A flower mantid in Malaysia

(b) A leaf mantid in Borneo

▲ **Figure 22.12 Camouflage as an example of evolutionary adaptation.** Related species of the insects called mantids have diverse shapes and colors that evolved in different environments.

? *Explain how these mantids demonstrate the three key observations about life introduced in this chapter's Overview: the match between organisms and their environments, unity, and diversity.*

CONCEPT 22.3

Evolution is supported by an overwhelming amount of scientific evidence

In *The Origin of Species*, Darwin marshaled a broad range of evidence to support the concept of descent with modification. Still—as he readily acknowledged—there were instances in which key evidence was lacking. For example, Darwin referred to the origin of flowering plants as an "abominable mystery," and he lamented the lack of fossils showing how earlier groups of organisms gave rise to new groups.

In the last 150 years, new discoveries have filled many of the gaps that Darwin identified. The origin of flowering plants, for example, is much better understood (see Chapter 30), and many fossils have been discovered that signify the origin of new groups of organisms (see Chapter 25). In this section, we'll consider four types of data that document the pattern of evolution and illuminate the processes by which it occurs: direct observations of evolution, homology, the fossil record, and biogeography.

Direct Observations of Evolutionary Change

Biologists have documented evolutionary change in thousands of scientific studies. We'll examine many such studies throughout this unit, but let's look at two examples here.

Natural Selection in Response to Introduced Plant Species

Animals that eat plants, called herbivores, often have adaptations that help them feed efficiently on their primary food sources. What happens when herbivores begin to feed on a plant species with different characteristics than their usual food source?

An opportunity to study this question in nature is provided by soapberry bugs, which use their "beak," a hollow, needle-like mouthpart, to feed on seeds located within the fruits of various plants. In southern Florida, the soapberry bug *Jadera haematoloma* feeds on the seeds of a native plant, the balloon vine (*Cardiospermum corindum*). In central Florida, however, balloon vines have become rare. Instead, soapberry bugs in that region now feed on goldenrain tree (*Koelreuteria elegans*), a species recently introduced from Asia.

Soapberry bugs feed most effectively when their beak length closely matches the depth at which the seeds are found within the fruit. Goldenrain tree fruit consists of three flat lobes, and its seeds are much closer to the fruit surface than the seeds of the plump, round native balloon vine fruit. Researchers at the University of Utah predicted that in populations that feed on goldenrain tree, natural selection would result in beaks that are *shorter* than those in populations that feed on balloon vine (**Figure 22.13**). Indeed, beak lengths are shorter in the populations that feed on goldenrain tree.

Researchers have also studied beak length evolution in soapberry bug populations that feed on plants introduced to Louisiana, Oklahoma, and Australia. In each of these locations, the fruit of the introduced plants is larger than the fruit of the native plant. Thus, in populations feeding on introduced species in these regions, the researchers predicted that natural selection would result in the evolution of *longer* beak length. Again, data collected in field studies upheld this prediction.

The adaptation observed in these soapberry bug populations had important consequences: In Australia, for example, the increase in beak length nearly doubled the success with which soapberry bugs could eat the seeds of the introduced species. Furthermore, since historical data show that the goldenrain tree reached central Florida just 35 years before the scientific studies were initiated, the results demonstrate that natural selection can cause rapid evolution in a wild population.

The Evolution of Drug-Resistant Bacteria

An example of ongoing natural selection that dramatically affects humans is the evolution of drug-resistant pathogens (disease-causing organisms and viruses). This is a particular problem with bacteria and viruses because resistant strains of these pathogens can proliferate very quickly.

Consider the evolution of drug resistance in the bacterium *Staphylococcus aureus*. About one in three people harbor this species on their skin or in their nasal passages with no negative effects. However, certain genetic varieties (strains) of this species, known as methicillin-resistant *S. aureus* (MRSA), are

▼ **Figure 22.13** INQUIRY

Can a change in a population's food source result in evolution by natural selection?

FIELD STUDY Soapberry bugs (*Jadera haematoloma*) feed most effectively when the length of their "beak" closely matches the depth within the fruits of the seeds they eat. Scott Carroll and his colleagues measured beak lengths in soapberry bug populations in southern Florida feeding on the native balloon vine. They also measured beak lengths in populations in central Florida feeding on the introduced goldenrain tree, which has a flatter fruit shape than the balloon vine. The researchers then compared the measurements to those of museum specimens collected in the two areas before the goldenrain tree was introduced.

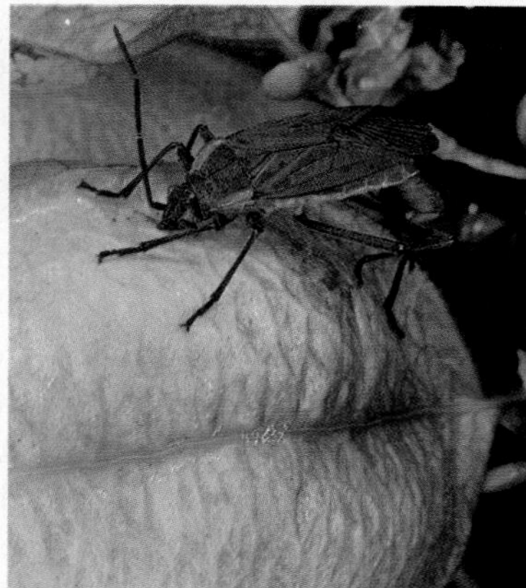

Soapberry bug with beak inserted in balloon vine fruit

RESULTS Beak lengths were shorter in populations feeding on the introduced species than in populations feeding on the native species, in which the seeds are buried more deeply. The average beak length in museum specimens from each population (indicated by red arrows) was similar to beak lengths in populations feeding on native species.

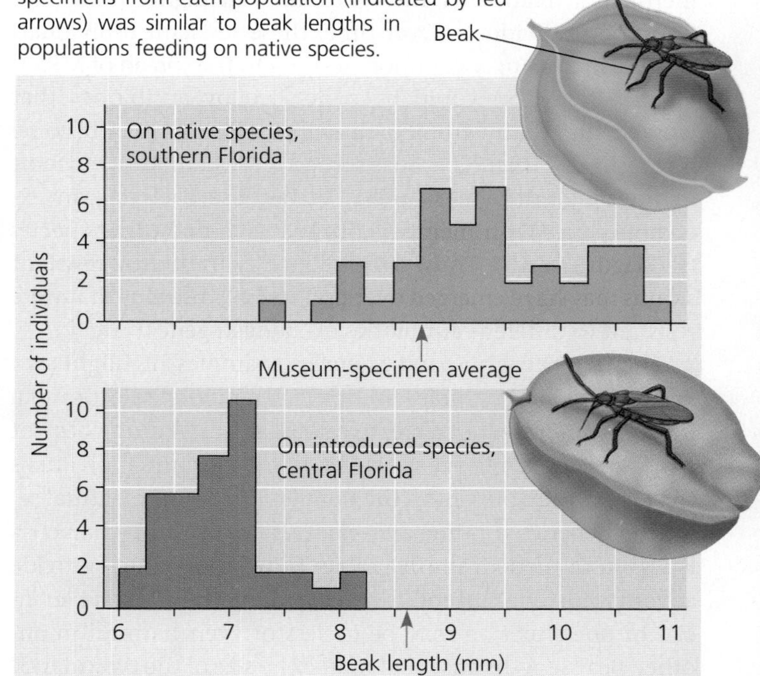

CONCLUSION Museum specimens and contemporary data suggest that a change in the size of the soapberry bug's food source can result in evolution by natural selection for matching beak size.

SOURCE S. P. Carroll and C. Boyd, Host race radiation in the soapberry bug: natural history with the history, *Evolution* 46: 1052–1069 (1992).

WHAT IF? When soapberry bug eggs from a population fed on balloon vine fruits were reared on goldenrain tree fruits (or vice versa), the beak lengths of the adult insects matched those in the population from which the eggs were obtained. Interpret these results.

formidable pathogens. The past decade has seen an alarming increase in virulent forms of MRSA such as clone USA300, a strain that can cause "flesh-eating disease" and potentially fatal infections (**Figure 22.14**). How did clone USA300 and other strains of MRSA become so dangerous?

The story begins in 1943, when penicillin became the first widely used antibiotic. Since then, penicillin and other antibiotics have saved millions of lives. However, by 1945, more than 20% of the *S. aureus* strains seen in hospitals were already resistant to penicillin. These bacteria had an enzyme, penicillinase, that could destroy penicillin. Researchers responded by developing antibiotics that were not destroyed by penicillinase, but some *S. aureus* populations developed resistance to each new drug within a few years.

In 1959, doctors began using the powerful antibiotic methicillin, but within two years, methicillin-resistant strains of *S. aureus* appeared. How did these resistant strains emerge? Methicillin works by deactivating a protein that bacteria use to synthesize their cell walls. However, *S. aureus* populations exhibited variations in how strongly their members were affected by the drug. In particular, some individuals were able to synthesize their cell walls using a different protein that was not affected by methicillin. These individuals survived the methicillin treatments and reproduced at higher rates than did other individuals. Over time, these resistant individuals became increasingly common, leading to the spread of MRSA.

Initially, MRSA could be controlled by antibiotics that worked differently from methicillin. But this has become increasingly difficult because some MRSA strains are resistant to multiple antibiotics—probably because bacteria can exchange genes with members of their own and other species (see Figure 27.13). Thus, the present-day multidrug-resistant strains may have emerged over time as MRSA strains that were resistant to different antibiotics exchanged genes.

The soapberry bug and *S. aureus* examples highlight two key points about natural selection. First, natural selection is a process of editing, not a creative mechanism. A drug does not *create* resistant pathogens; it *selects for* resistant individuals that are already present in the population. Second, natural selection depends on time and place. It favors those characteristics in a genetically variable population that provide advantage in the current, local environment. What is beneficial in one situation may be useless or even harmful in another. Beak lengths arise that match the size of the typical fruit eaten by a particular soapberry bug population. However, a beak length suitable for fruit of one size can be disadvantageous when the bug is feeding on fruit of another size.

▼ **Figure 22.14**

IMPACT

The Rise of MRSA

Most methicillin-resistant *Staphylococcus aureus* (MRSA) infections are caused by recently appearing strains such as clone USA300. Resistant to multiple antibiotics and highly contagious, this strain and its close relatives can cause lethal infections of the skin, lungs, and blood. Researchers have identified key areas of the USA300 genome that code for its particularly virulent properties.

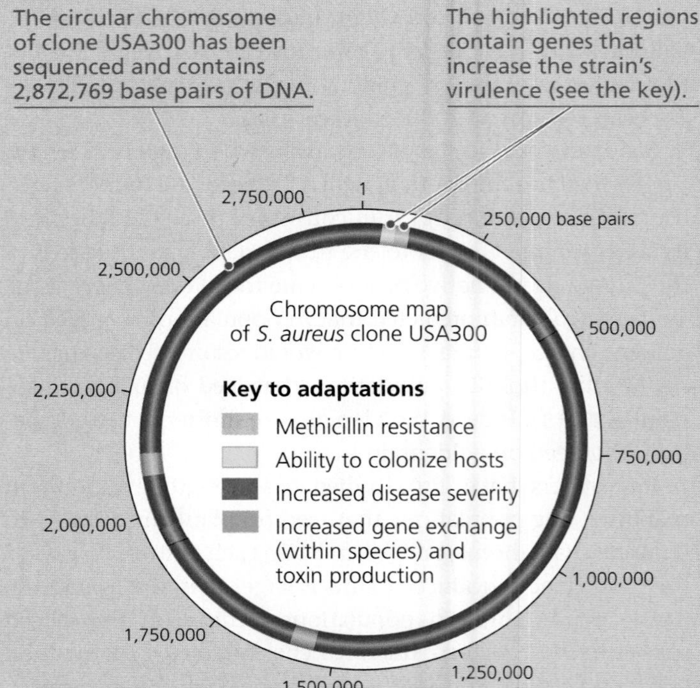

WHY IT MATTERS MRSA infections have proliferated dramatically in the past few decades, and the annual death toll in the United States is in the tens of thousands. There is grave concern about the continuing evolution of drug resistance and the resulting difficulty of treating MRSA infections. Ongoing studies of how MRSA strains colonize their hosts and cause disease may help scientists develop drugs to combat MRSA.

FURTHER READING General information about MRSA can be found on the Centers for Disease Control and Prevention (CDC) website (www.cdc.gov/mrsa) and in G. Taubes, The bacteria fight back, *Science* 321:356–361 (2008).

WHAT IF? Efforts are underway to develop drugs that target *S. aureus* specifically and to develop drugs that slow the growth of MRSA but do not kill it. Based on how natural selection works and on the fact that bacterial species can exchange genes, explain why each of these strategies might be effective.

Homology

A second type of evidence for evolution comes from analyzing similarities among different organisms. As we've discussed, evolution is a process of descent with modification: Characteristics present in an ancestral organism are altered (by natural selection) in its descendants over time as they face different environmental conditions. As a result, related species can have characteristics that have an underlying similarity yet function differently. Similarity resulting from common ancestry is known as **homology**. As this section will explain, an understanding of homology can be used to make testable predictions and explain observations that are otherwise puzzling.

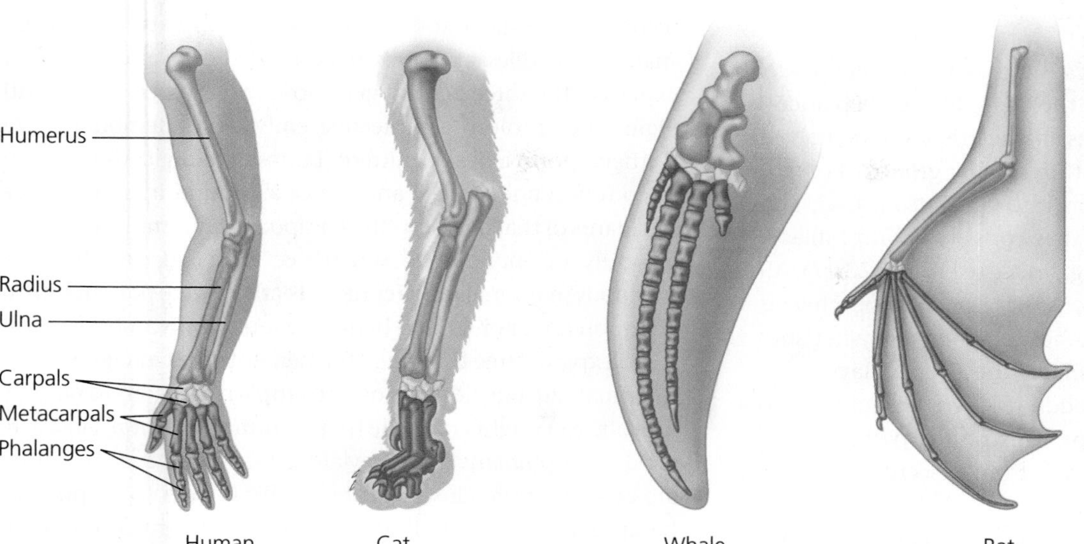

◀ **Figure 22.15 Mammalian forelimbs: homologous structures.** Even though they have become adapted for different functions, the forelimbs of all mammals are constructed from the same basic skeletal elements: one large bone (purple), attached to two smaller bones (orange and tan), attached to several small bones (gold), attached to several metacarpals (green), attached to approximately five digits, or phalanges (blue).

Anatomical and Molecular Homologies

The view of evolution as a remodeling process leads to the prediction that closely related species should share similar features—and they do. Of course, closely related species share the features used to determine their relationship, but they also share many other features. Some of these shared features make little sense except in the context of evolution. For example, the forelimbs of all mammals, including humans, cats, whales, and bats, show the same arrangement of bones from the shoulder to the tips of the digits, even though these appendages have very different functions: lifting, walking, swimming, and flying (**Figure 22.15**). Such striking anatomical resemblances would be highly unlikely if these structures had arisen anew in each species. Rather, the underlying skeletons of the arms, forelegs, flippers, and wings of different mammals are **homologous structures** that represent variations on a structural theme that was present in their common ancestor.

Comparing early stages of development in different animal species reveals additional anatomical homologies not visible in adult organisms. For example, at some point in their development, all vertebrate embryos have a tail located posterior to (behind) the anus, as well as structures called pharyngeal (throat) pouches (**Figure 22.16**). These homologous throat pouches ultimately develop into structures with very different functions, such as gills in fishes and parts of the ears and throat in humans and other mammals.

Some of the most intriguing homologies concern "leftover" structures of marginal, if any, importance to the organism. These **vestigial structures** are remnants of features that served important functions in the organism's ancestors. For instance, the skeletons of some snakes retain vestiges of the pelvis and leg bones of walking ancestors. Another example is provided by eye remnants that are buried under scales in blind species of cave fishes. We would not expect to see these vestigial structures if snakes and blind cave fishes had origins separate from other vertebrate animals.

Biologists also observe similarities among organisms at the molecular level. All forms of life use the same genetic language of DNA and RNA, and the genetic code is essentially universal. Thus, it is likely that all species descended from common ancestors that used this code. But molecular homologies go beyond a shared code. For example, organisms as dissimilar as humans and bacteria share genes inherited from a very distant common ancestor. Some of these homologous genes have acquired new functions, while others, such as those coding for the ribosomal subunits used in protein synthesis (see Figure 17.17), have retained their original functions. It is also common for organisms to have genes that have lost their function, even though the homologous genes in related species may be fully functional. Like vestigial structures, it appears that such inactive "pseudogenes" may be present simply because a common ancestor had them.

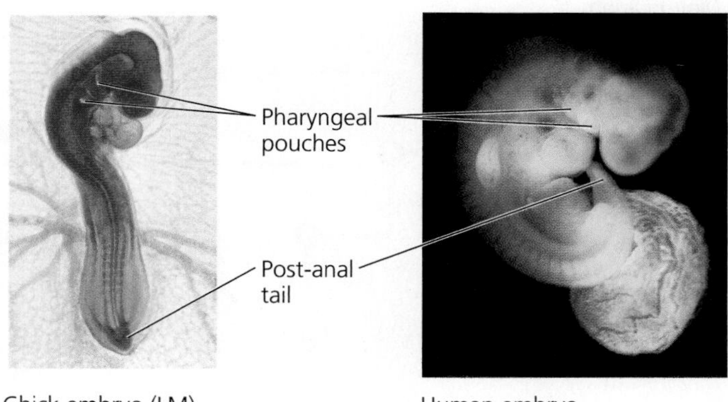

▲ **Figure 22.16 Anatomical similarities in vertebrate embryos.** At some stage in their embryonic development, all vertebrates have a tail located posterior to the anus (referred to as a post-anal tail), as well as pharyngeal (throat) pouches. Descent from a common ancestor can explain such similarities.

Homologies and "Tree Thinking"

Some homologous characteristics, such as the genetic code, are shared by all species because they date to the deep ancestral past. In contrast, homologous characteristics that evolved more recently are shared only within smaller groups of organisms. Consider the *tetrapods* (from the Greek *tetra*, four, and *pod*, foot), the vertebrate group that consists of amphibians, mammals, and reptiles (including birds—see Figure 22.17). All tetrapods have limbs with digits (see Figure 22.15), whereas other vertebrates do not. Thus, homologous characteristics form a nested pattern: All life shares the deepest layer, and each successive smaller group adds its own homologies to those it shares with larger groups. This nested pattern is exactly what we would expect to result from descent with modification from a common ancestor.

Biologists often represent the pattern of descent from common ancestors and the resulting homologies with an **evolutionary tree**, a diagram that reflects evolutionary relationships among groups of organisms. We will explore in detail how evolutionary trees are constructed in Chapter 26, but for now, let's consider how we can interpret and use such trees.

Figure 22.17 is an evolutionary tree of tetrapods and their closest living relatives, the lungfishes. In this diagram, each branch point represents the common ancestor of all species that descended from it. For example, lungfishes and all tetrapods descended from ancestor ❶, whereas mammals, lizards and snakes, crocodiles, and birds all descended from ancestor ❸. As expected, the three homologies shown on the tree—limbs with digits, the amnion (a protective embryonic membrane), and feathers—form a nested pattern. Limbs with digits were present in common ancestor ❷ and hence are found in all of the descendants of that ancestor (the tetrapods). The amnion was present only in ancestor ❸ and hence is shared only by some tetrapods (mammals and reptiles). Feathers were present only in common ancestor ❻ and hence are found only in birds.

To explore "tree thinking" further, note that in Figure 22.17, mammals are positioned closer to amphibians than to birds. As a result, you might conclude that mammals are more closely related to amphibians than they are to birds. However, mammals are actually more closely related to birds than to amphibians because mammals and birds share a more recent common ancestor (ancestor ❸) than do mammals and amphibians (ancestor ❷). Ancestor ❷ is also the most recent common ancestor of birds and amphibians, making mammals and birds equally related to amphibians. Finally, note that the tree in Figure 22.17 shows the relative timing of evolutionary events but not their actual dates. Thus, we can conclude that ancestor ❷ lived before ancestor ❸, but we do not know when that was.

Evolutionary trees are hypotheses that summarize our current understanding of patterns of descent. Our confidence in these relationships, as with any hypothesis, depends on the strength of the supporting data. In the case of Figure 22.17, the tree is supported by a variety of independent data sets, including both anatomical and DNA sequence data. As a result, biologists feel confident that it accurately reflects evolutionary history. As you will read in Chapter 26, scientists can use such well-supported evolutionary trees to make specific and sometimes surprising predictions about organisms.

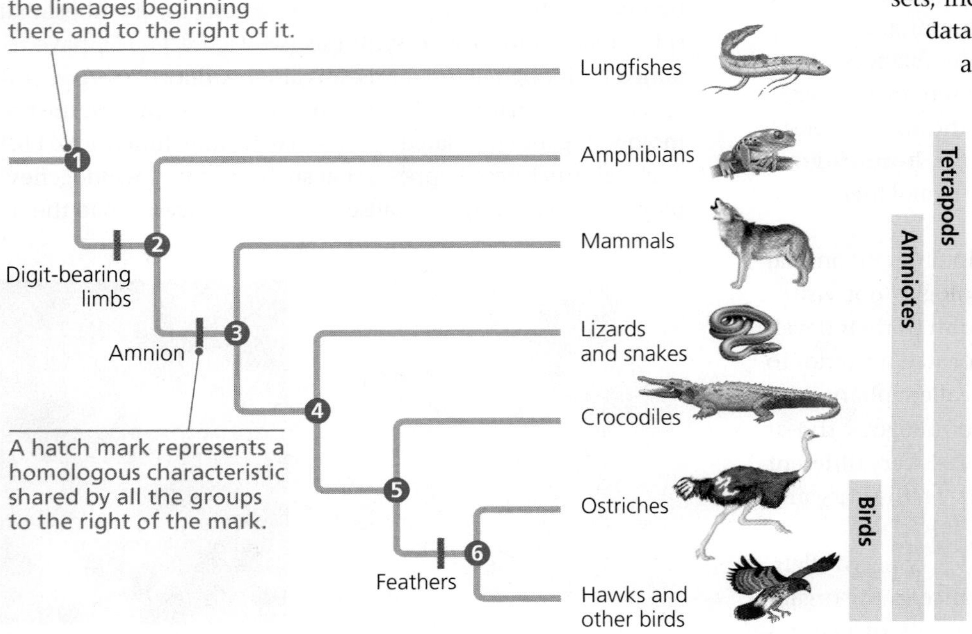

▲ **Figure 22.17 Tree thinking: information provided in an evolutionary tree.** This evolutionary tree for tetrapods and their closest living relatives, the lungfishes, is based on anatomical and DNA sequence data. The purple bars indicate the origin of three important homologies, each of which evolved only once. Birds are nested within and evolved from reptiles; hence, the group of organisms called "reptiles" technically includes birds.

? *Are crocodiles more closely related to lizards or birds? Explain your answer.*

A Different Cause of Resemblance: Convergent Evolution

Although organisms that are closely related share characteristics because of common descent, distantly related organisms can resemble one another for a different reason: **convergent evolution**, the independent evolution of similar features in different lineages. Consider marsupial mammals, many of which live in Australia. Marsupials are distinct from another group of mammals—the eutherians—few of which live in Australia. (Eutherians complete their embryonic development in the uterus, whereas marsupials

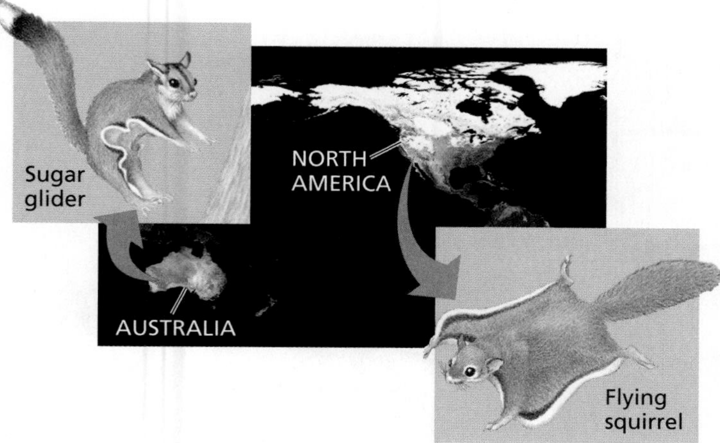

▲ **Figure 22.18 Convergent evolution.** The ability to glide through the air evolved independently in these two distantly related mammals.

are born as embryos and complete their development in an external pouch.) Some Australian marsupials have eutherian look-alikes with superficially similar adaptations. For instance, a forest-dwelling Australian marsupial called the sugar glider is superficially very similar to flying squirrels, gliding eutherians that live in North American forests (**Figure 22.18**). But the sugar glider has many other characteristics that make it a marsupial, much more closely related to kangaroos and other Australian marsupials than to flying squirrels or other eutherians. Once again, our understanding of evolution can explain these observations. Although they evolved independently from different ancestors, these two mammals have adapted to similar environments in similar ways. In such examples in which species share features because of convergent evolution, the resemblance is said to be **analogous**, not homologous. Analogous features share similar function, but not common ancestry, while homologous features share common ancestry, but not necessarily similar function.

The Fossil Record

A third type of evidence for evolution comes from fossils. As Chapter 25 discusses in more detail, the fossil record documents the pattern of evolution, showing that past organisms differed from present-day organisms and that many species have become extinct. Fossils also show the evolutionary changes that have occurred in various groups of organisms. To give one of hundreds of possible examples, researchers found that the pelvic bone in fossil stickleback fish became greatly reduced in size over time in a number of different lakes. The consistent nature of this change suggests that the reduction in the size of the pelvic bone may have been driven by natural selection.

Fossils can also shed light on the origins of new groups of organisms. An example is the fossil record of cetaceans, the mammalian order that includes whales, dolphins, and porpoises. Some of these fossils provided an unexpected line of support for a hypothesis based on DNA data: that cetaceans are closely related to even-toed ungulates, a group that includes deer, pigs, camels, and cows (**Figure 22.19**). What else can fossils tell us about cetacean origins? The earliest cetaceans lived 50–60 million years ago. The fossil record indicates that prior to that time, most mammals were terrestrial. Although scientists had long realized that whales and other cetaceans originated from land mammals, few fossils had been found that revealed how cetacean limb structure had changed over time, leading eventually to the loss of hind limbs and the development of flippers and tail flukes. In the past few decades, however, a series of remarkable fossils have been discovered in Pakistan, Egypt, and North America. These fossils document steps in the transition from life on land to life in the sea, filling in some of the gaps between ancestral and living cetaceans (**Figure 22.20**, on the next page).

Collectively, the recent fossil discoveries document the formation of new species and the origin of a major new group of mammals, the cetaceans. These discoveries also show that cetaceans and their close living relatives (hippopotamuses, pigs, deer, and other even-toed ungulates) are much more different from each other than were *Pakicetus* and early even-toed ungulates, such as *Diacodexis*. Similar patterns are seen in fossils documenting the origins of other major new groups of organisms, including

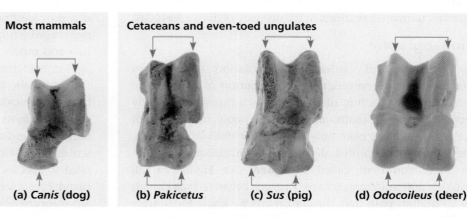

▲ ***Diacodexis*, an early even-toed ungulate**

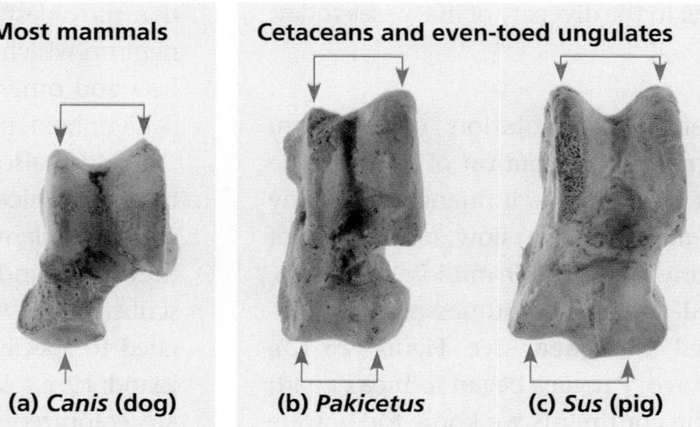

▲ **Figure 22.19 Ankle bones: one piece of the puzzle.** Comparing fossils and present-day examples of the astragalus (a type of ankle bone) provides one line of evidence that cetaceans are closely related to even-toed ungulates. **(a)** In most mammals, the astragalus is shaped like that of a dog, with a double hump on one end (indicated by the red arrows) but not at the opposite end (blue arrow). **(b)** Fossils show that the early cetacean *Pakicetus* had an astragalus with double humps at both ends, a unique shape that is otherwise found only in even-toed ungulates, as shown here for **(c)** a pig and **(d)** a deer.

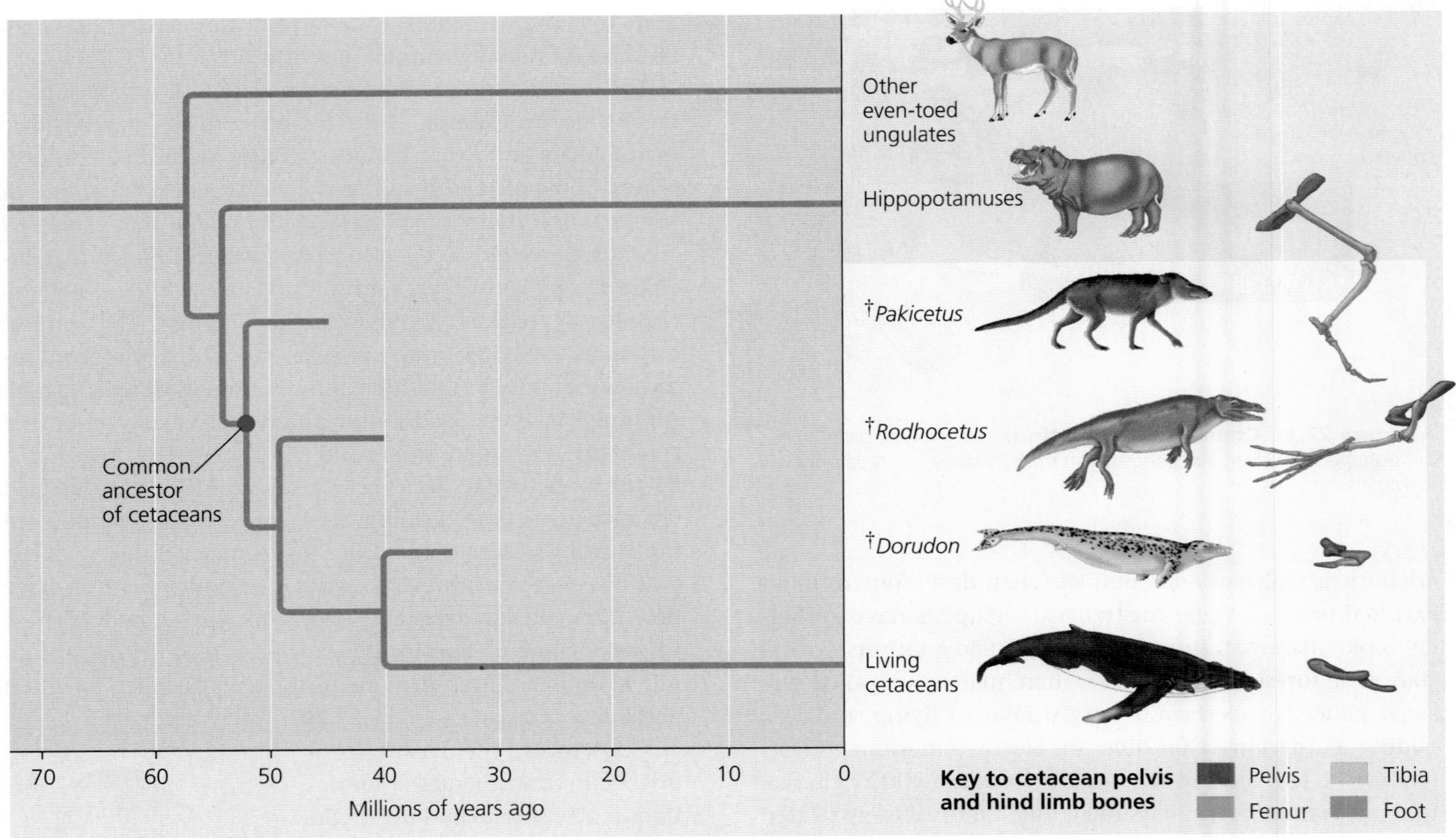

▲ **Figure 22.20 The transition to life in the sea.** Multiple lines of evidence support the hypothesis that cetaceans evolved from terrestrial mammals. Fossils document the reduction over time in the pelvis and hind limb bones of extinct cetacean ancestors, including *Pakicetus*, *Rodhocetus*, and *Dorudon*. DNA sequence data support the hypothesis that cetaceans are most closely related to hippopotamuses, even-toed ungulates.

? *Which happened first during the evolution of cetaceans: changes in hind limb structure or the origin of tail flukes?*

mammals (see Chapter 25), flowering plants (see Chapter 30), and tetrapods (see Chapter 34). In each of these cases, the fossil record shows that over time, descent with modification produced increasingly large differences among related groups of organisms, ultimately resulting in the diversity of life we see today.

Biogeography

A fourth type of evidence for evolution comes from **biogeography**, the geographic distribution of species. The geographic distribution of organisms is influenced by many factors, including *continental drift*, the slow movement of Earth's continents over time. About 250 million years ago, these movements united all of Earth's landmasses into a single large continent called **Pangaea** (see Figure 25.13). Roughly 200 million years ago, Pangaea began to break apart; by 20 million years ago, the continents we know today were within a few hundred kilometers of their present locations.

We can use our understanding of evolution and continental drift to predict where fossils of different groups of organisms might be found. For example, scientists have constructed evolutionary trees for horses based on anatomical data. These trees and the ages of fossils of horse ancestors suggest that present-day horse species originated 5 million years ago in North America. At that time, North and South America were close to their present locations, but they were not yet connected, making it difficult for horses to travel between them. Thus, we would predict that the oldest horse fossils should be found only on the continent on which horses originated—North America. This prediction and others like it for different groups of organisms have been upheld, providing more evidence for evolution.

We can also use our understanding of evolution to explain biogeographic data. For example, islands generally have many species of plants and animals that are **endemic**, which means they are found nowhere else in the world. Yet, as Darwin described in *The Origin of Species*, most island species are closely related to species from the nearest mainland or a neighboring island. He explained this observation by suggesting that islands are colonized by species from the nearest mainland. These colonists eventually give rise to new species as they adapt to their new environments. Such a process also explains why two islands with similar environments in distant parts of the world tend to be populated not by species that are closely related to each other, but rather by species related to those of the nearest mainland, where the environment is often quite different.

What Is Theoretical About Darwin's View of Life?

Some people dismiss Darwin's ideas as "just a theory." However, as we have seen, the *pattern* of evolution—the observation that life has evolved over time—has been documented directly and is supported by a great deal of evidence. In addition, Darwin's explanation of the *process* of evolution—that natural selection is the primary cause of the observed pattern of evolutionary change—makes sense of massive amounts of data. The effects of natural selection also can be observed and tested in nature.

What, then, is theoretical about evolution? Keep in mind that the scientific meaning of the term *theory* is very different from its meaning in everyday use. The colloquial use of the word *theory* comes close to what scientists mean by a hypothesis. In science, a theory is more comprehensive than a hypothesis. A theory, such as the theory of evolution by natural selection, accounts for many observations and explains and integrates a great variety of phenomena. Such a unifying theory does not become widely accepted unless its predictions stand up to thorough and continual testing by experiment and additional observation (see Chapter 1). As the next three chapters demonstrate, this has certainly been the case with the theory of evolution by natural selection.

The skepticism of scientists as they continue to test theories prevents these ideas from becoming dogma. For example, although Darwin thought that evolution was a very slow process, we now know that this isn't always true. New species can form in relatively short periods of time (a few thousand years or less; see Chapter 24). Furthermore, as we'll explore throughout this unit, evolutionary biologists now recognize that natural selection is not the only mechanism responsible for evolution. Indeed, the study of evolution today is livelier than ever as scientists find more ways to test predictions based on natural selection and other evolutionary mechanisms.

Although Darwin's theory attributes the diversity of life to natural processes, the diverse products of evolution nevertheless remain elegant and inspiring. As Darwin wrote in the final sentence of *The Origin of Species*, "There is grandeur in this view of life . . . [in which] endless forms most beautiful and most wonderful have been, and are being, evolved."

CONCEPT CHECK 22.3

1. Explain how the following statement is inaccurate: "Antibiotics have created drug resistance in MRSA."
2. How does evolution account for (a) the similar mammalian forelimbs with different functions shown in Figure 22.15 and (b) the similar lifestyle of the two distantly related mammals shown in Figure 22.18?
3. **WHAT IF?** The fossil record shows that dinosaurs originated 200–250 million years ago. Would you expect the geographic distribution of early dinosaur fossils to be broad (on many continents) or narrow (on one or a few continents only)? Explain.

For suggested answers, see Appendix A.

22 CHAPTER REVIEW

SUMMARY OF KEY CONCEPTS

CONCEPT 22.1

The Darwinian revolution challenged traditional views of a young Earth inhabited by unchanging species (pp. 453–455)

- Darwin proposed that life's diversity arose from ancestral species through natural selection, a departure from prevailing views.
- In contrast to **catastrophism** (the principle that events in the past occurred suddenly by mechanisms not operating today), Hutton and Lyell thought that geologic change results from mechanisms that operated in the past in the same manner as at the present time (**uniformitarianism**).
- Lamarck hypothesized that species evolve, but the underlying mechanisms he proposed are not supported by evidence.

? *Why was the age of Earth important for Darwin's ideas about evolution?*

CONCEPT 22.2

Descent with modification by natural selection explains the adaptations of organisms and the unity and diversity of life (pp. 455–460)

- Darwin's experiences during the voyage of the *Beagle* gave rise to his idea that new species originate from ancestral forms through the accumulation of **adaptations**. He refined his theory for many years and finally published it in 1859 after learning that Wallace had come to the same idea.
- In *The Origin of Species*, Darwin proposed that evolution occurs by **natural selection**.

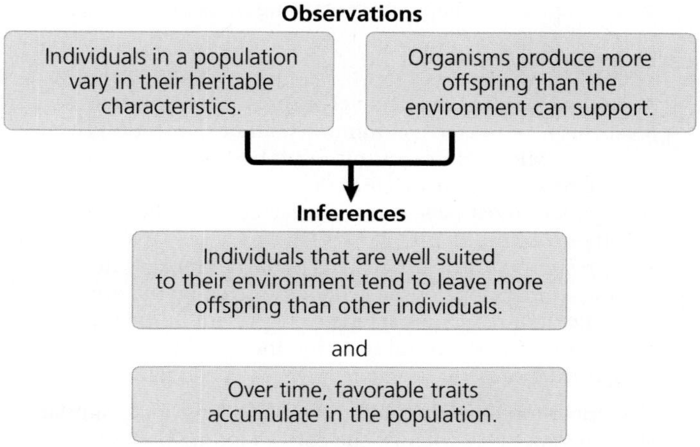

Observations

Individuals in a population vary in their heritable characteristics.

Organisms produce more offspring than the environment can support.

Inferences

Individuals that are well suited to their environment tend to leave more offspring than other individuals.

and

Over time, favorable traits accumulate in the population.

? *Describe how overreproduction and heritable variation relate to evolution by natural selection.*

CONCEPT 22.3

Evolution is supported by an overwhelming amount of scientific evidence (pp. 460–467)

- Researchers have directly observed natural selection leading to adaptive evolution in many studies, including research on soapberry bug populations and on MRSA.
- Organisms share characteristics because of common descent (**homology**) or because natural selection affects independently evolving species in similar environments in similar ways (**convergent evolution**).
- Fossils show that past organisms differed from living organisms, that many species have become extinct, and that species have evolved over long periods of time; fossils also document the origin of major new groups of organisms.
- Evolutionary theory can explain biogeographic patterns.

? *Summarize the different lines of evidence supporting the hypothesis that cetaceans descended from land mammals and are closely related to even-toed ungulates.*

TEST YOUR UNDERSTANDING

Level 1: Knowledge/Comprehension

1. Which of the following is *not* an observation or inference on which natural selection is based?
   a. There is heritable variation among individuals.
   b. Poorly adapted individuals never produce offspring.
   c. Species produce more offspring than the environment can support.
   d. Individuals whose characteristics are best suited to the environment generally leave more offspring than those whose characteristics are less well suited.
   e. Only a fraction of an individual's offspring may survive.

2. Which of the following observations helped Darwin shape his concept of descent with modification?
   a. Species diversity declines farther from the equator.
   b. Fewer species live on islands than on the nearest continents.
   c. Birds can be found on islands located farther from the mainland than the birds' maximum nonstop flight distance.
   d. South American temperate plants are more similar to the tropical plants of South America than to the temperate plants of Europe.
   e. Earthquakes reshape life by causing mass extinctions.

Level 2: Application/Analysis

3. Within six months of effectively using methicillin to treat *S. aureus* infections in a community, all new infections were caused by MRSA. How can this result best be explained?
   a. *S. aureus* can resist vaccines.
   b. A patient must have become infected with MRSA from another community.
   c. In response to the drug, *S. aureus* began making drug-resistant versions of the protein targeted by the drug.
   d. Some drug-resistant bacteria were present at the start of treatment, and natural selection increased their frequency.
   e. The drug caused the *S. aureus* DNA to change.

4. The upper forelimbs of humans and bats have fairly similar skeletal structures, whereas the corresponding bones in whales have very different shapes and proportions. However, genetic data suggest that all three kinds of organisms diverged from a common ancestor at about the same time. Which of the following is the most likely explanation for these data?
   a. Humans and bats evolved by natural selection, and whales evolved by Lamarckian mechanisms.
   b. Forelimb evolution was adaptive in people and bats, but not in whales.
   c. Natural selection in an aquatic environment resulted in significant changes to whale forelimb anatomy.
   d. Genes mutate faster in whales than in humans or bats.
   e. Whales are not properly classified as mammals.

5. DNA sequences in many human genes are very similar to the sequences of corresponding genes in chimpanzees. The most likely explanation for this result is that
   a. humans and chimpanzees share a relatively recent common ancestor.
   b. humans evolved from chimpanzees.
   c. chimpanzees evolved from humans.
   d. convergent evolution led to the DNA similarities.
   e. humans and chimpanzees are not closely related.

Level 3: Synthesis/Evaluation

6. **EVOLUTION CONNECTION**
   Explain why anatomical and molecular features often fit a similar nested pattern. In addition, describe a process that can cause this not to be the case.

7. **SCIENTIFIC INQUIRY**
   DRAW IT Mosquitoes resistant to the pesticide DDT first appeared in India in 1959, but now are found throughout the world. (a) Graph the data in the table below. (b) Examining the graph, hypothesize why the percentage of mosquitoes resistant to DDT rose rapidly. (c) Suggest an explanation for the global spread of DDT resistance.

Month	0	8	12
Mosquitoes Resistant* to DDT	4%	45%	77%

Source: C. F. Curtis et al., Selection for and against insecticide resistance and possible methods of inhibiting the evolution of resistance in mosquitoes, *Ecological Entomology* 3:273–287 (1978).

*Mosquitoes were considered resistant if they were not killed within 1 hour of receiving a dose of 4% DDT.

8. **WRITE ABOUT A THEME**
   Environmental Interactions Write a short essay (about 100–150 words) evaluating whether changes to an organism's physical environment are likely to result in evolutionary change. Use an example to support your reasoning.

For selected answers, see Appendix A.

23

The Evolution of Populations

▲ **Figure 23.1 Is this finch evolving?**

EVOLUTION

KEY CONCEPTS

- **23.1** Genetic variation makes evolution possible
- **23.2** The Hardy-Weinberg equation can be used to test whether a population is evolving
- **23.3** Natural selection, genetic drift, and gene flow can alter allele frequencies in a population
- **23.4** Natural selection is the only mechanism that consistently causes adaptive evolution

OVERVIEW

The Smallest Unit of Evolution

One common misconception about evolution is that individual organisms evolve. It is true that natural selection acts on individuals: Each organism's traits affect its survival and reproductive success compared with other individuals. But the evolutionary impact of natural selection is only apparent in the changes in a *population* of organisms over time.

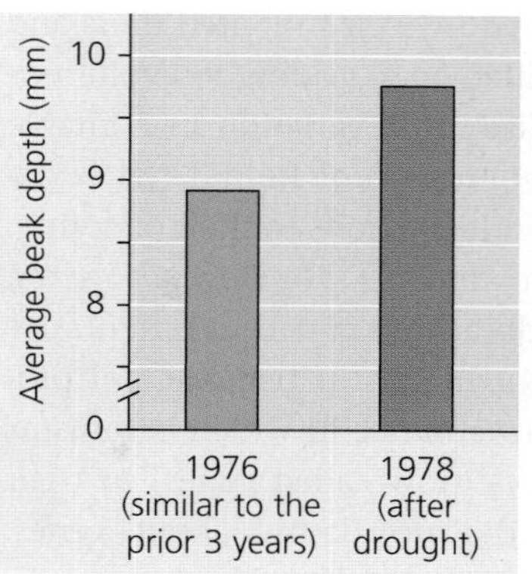

◄ **Figure 23.2 Evidence of selection by food source.** The data represent adult beak depth measurements of medium ground finches hatched in the generations before and after the 1977 drought. Beak sizes remained large until 1983, when changing conditions no longer favored large-beaked birds.

MB See the related Experimental Inquiry Tutorial in MasteringBiology.

Consider the medium ground finch (*Geospiza fortis*), a seed-eating bird that inhabits the Galápagos Islands (**Figure 23.1**). In 1977, the *G. fortis* population on the island of Daphne Major was decimated by a long period of drought: Of some 1,200 birds, only 180 survived. Researchers Peter and Rosemary Grant observed that during the drought, small, soft seeds were in short supply. The finches mostly fed on large, hard seeds that were more plentiful. Birds with larger, deeper beaks were better able to crack and eat these larger seeds, and they survived at a higher rate than finches with smaller beaks. Since beak depth is an inherited trait in these birds, the average beak depth in the next generation of *G. fortis* was greater than it had been in the pre-drought population (**Figure 23.2**). The finch population had evolved by natural selection. However, the *individual* finches did not evolve. Each bird had a beak of a particular size, which did not grow larger during the drought. Rather, the proportion of large beaks in the population increased from generation to generation: The population evolved, not its individual members.

Focusing on evolutionary change in populations, we can define evolution on its smallest scale, called **microevolution**, as change in allele frequencies in a population over generations. As we will see in this chapter, natural selection is not the only cause of microevolution. In fact, there are three main mechanisms that can cause allele frequency change: natural selection, genetic drift (chance events that alter allele frequencies), and gene flow (the transfer of alleles between populations). Each of these mechanisms has distinctive effects on the genetic composition of populations. However, only natural selection consistently improves the match between organisms and their environment (adaptation). Before we examine natural selection and adaptation more closely, let's revisit a prerequisite for these processes in a population: genetic variation.

CONCEPT **23.1**

Genetic variation makes evolution possible

In *The Origin of Species*, Darwin provided abundant evidence that life on Earth has evolved over time, and he proposed natural selection as the primary mechanism for that change. He observed that individuals differed in their inherited traits and that selection acted on such differences, leading to evolutionary change. Thus, Darwin realized that variation in heritable traits was a prerequisite for evolution, but he did not know precisely how organisms pass heritable traits to their offspring.

Just a few years after Darwin published *The Origin of Species*, Gregor Mendel wrote a groundbreaking paper on inheritance in pea plants (see Chapter 14). In that paper, Mendel proposed a particulate model of inheritance in which organisms transmit discrete heritable units (now called genes) to their offspring. Although Darwin did not know about genes, Mendel's paper set the stage for understanding the genetic differences on which evolution is based. Here we'll examine such genetic differences and how they are produced.

Genetic Variation

You probably have no trouble recognizing your friends in a crowd. Each person is unique, exhibiting differences in their facial features, height, and voice. Indeed, individual variation occurs in all species. In addition to the differences that we can see or hear, individuals vary extensively at the molecular level. For example, you cannot identify a person's blood group (A, B, AB, or O) from his or her appearance, but this and many other molecular traits vary among individuals.

Individual variations often reflect **genetic variation**, differences among individuals in the composition of their genes or other DNA segments. As you read in earlier chapters, however, some phenotypic variation is not heritable (see **Figure 23.3** for a striking example in a caterpillar of the southwestern United States). Phenotype is the product of an inherited genotype and many environmental influences. In a human example, bodybuilders alter their phenotypes dramatically but do not pass their huge muscles on to the next generation. In general, only the genetically determined part of phenotypic variation can have evolutionary consequences. As such, genetic variation provides the raw material for evolutionary change: Without genetic variation, evolution cannot occur.

Variation Within a Population

Characters that vary within a population may be discrete or quantitative. *Discrete characters*, such as the purple or white flower colors of Mendel's pea plants (see Figure 14.3), can be classified on an either-or basis (each plant has flowers that are either purple or white). Many discrete characters are determined by a single gene locus with different alleles that produce distinct phenotypes. However, most heritable variation involves *quantitative characters*, which vary along a continuum within a population. Heritable quantitative variation usually results from the influence of two or more genes on a single phenotypic character.

For both discrete and quantitative characters, biologists often need to describe how much genetic variation there is in a particular population. We can measure genetic variation at the whole-gene level (*gene variability*) and at the molecular level of DNA (*nucleotide variability*). Gene variability can be quantified as the **average heterozygosity**, the average percentage of loci that are heterozygous. (Recall that a heterozygous individual has two different alleles for a given locus, whereas a homozygous individual has two identical alleles for that locus.) As an example, on average the fruit fly *Drosophila melanogaster* is heterozygous for about 1,920 of its 13,700 loci (14%) and homozygous for all the rest. We can therefore say that a *D. melanogaster* population has an average heterozygosity of 14%. Analyses of this and many other species show that this level of genetic variation provides ample raw material for natural selection to operate, resulting in evolutionary change.

When determining gene variability, how do scientists identify heterozygous loci? One method is to survey the protein products of genes using gel electrophoresis (see Figure 20.9). However, this approach cannot detect silent mutations that

▲ **Figure 23.3 Nonheritable variation.** These caterpillars of the moth *Nemoria arizonaria* owe their different appearances to chemicals in their diets, not to differences in their genotypes. Caterpillars raised on a diet of oak flowers resembled the flowers **(a)**, whereas their siblings raised on oak leaves resembled oak twigs **(b)**.

alter the DNA sequence of a gene but not the amino acid sequence of the protein (see Figure 17.24). To include such silent mutations in their estimates of average heterozygosity, researchers must use other approaches, such as PCR-based methods and restriction fragment analyses (see Chapter 20).

To measure nucleotide variability, biologists compare the DNA sequences of two individuals in a population and then average the data from many such comparisons. The genome of *D. melanogaster* has about 180 million nucleotides, and the sequences of any two fruit flies differ on average by approximately 1.8 million (1%) of their nucleotides. Thus, the nucleotide variability of *D. melanogaster* populations is about 1%.

As in this example, gene variability tends to exceed nucleotide variability. Why is this true? Remember that a gene can consist of thousands of nucleotides. A difference at only one of these nucleotides can be sufficient to make two alleles of that gene different, increasing gene variability.

Variation Between Populations

In addition to variation observed within a population, species also exhibit **geographic variation**, differences in the genetic composition of separate populations. **Figure 23.4** illustrates geographic variation in populations of house mice (*Mus musculus*) separated by mountains on the Atlantic island of Madeira. Inadvertently introduced by Portuguese settlers in the 15th century, several populations of mice have evolved in isolation from one another. Researchers have observed differences in the karyotypes (chromosome sets) of these isolated populations. In certain populations, some of the chromosomes have become fused. However, the patterns of fused chromosomes differ from one population to another. Because these chromosome-level changes leave genes intact, their phenotypic effects on the mice seem to be neutral. Thus, the variation between these populations appears to have resulted from chance events (drift) rather than natural selection.

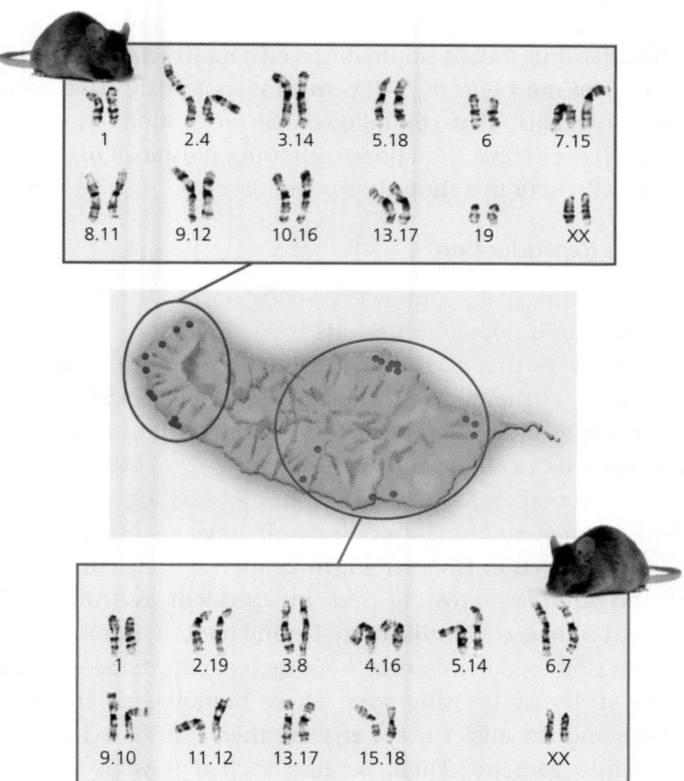

▲ **Figure 23.4 Geographic variation in isolated mouse populations on Madeira.** The number pairs represent fused chromosomes. For example, "2.4" indicates fusion of chromosome 2 and chromosome 4. Mice in the areas indicated by the blue dots have the set of fused chromosomes in the blue box; mice in the red-dot locales have the set of fused chromosomes in the red box.

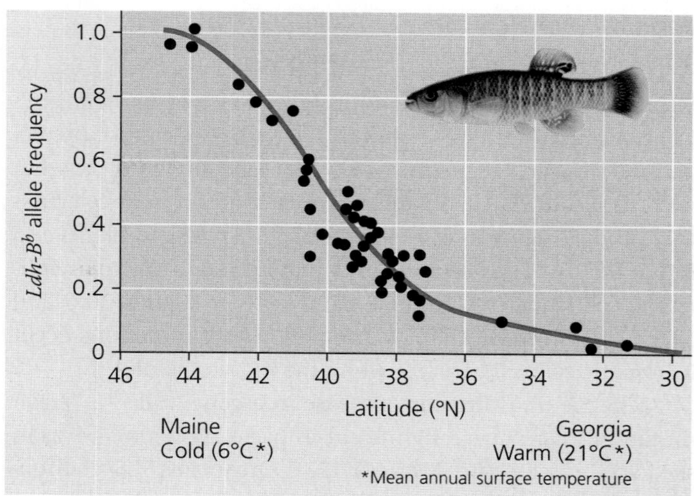

▲ **Figure 23.5 A cline determined by temperature.** In mummichog fish, the frequency of the $Ldh\text{-}B^b$ allele for the enzyme lactate dehydrogenase-B (which functions in metabolism) decreases in fish sampled from Maine to Georgia. The $Ldh\text{-}B^b$ allele codes for a form of the enzyme that is a better catalyst in cold water than are other versions of the enzyme. Individuals with the $Ldh\text{-}B^b$ allele can swim faster in cold water than can individuals with other alleles.

Other examples of geographic variation occur as a **cline**, a graded change in a character along a geographic axis. Some clines are produced by a gradation in an environmental variable, as illustrated by the impact of temperature on the frequency of a cold-adaptive allele in mummichog fish (*Fundulus heteroclitus*). Clines such as the one depicted in **Figure 23.5** probably result from natural selection—otherwise there would be no reason to expect a close association between the environmental variable and the frequency of the allele. But selection can only operate if multiple alleles exist for a given locus. Such variation in alleles can arise in several ways.

Sources of Genetic Variation

The genetic variation on which evolution depends originates when mutation, gene duplication, or other processes produce new alleles and new genes. Many new genetic variants can be produced in short periods of time in organisms that reproduce rapidly. Sexual reproduction can also result in genetic variation as existing genes are arranged in new ways.

Formation of New Alleles

As described in Chapters 17 and 21, new alleles can arise by *mutation*, a change in the nucleotide sequence of an organism's DNA. A mutation is like a shot in the dark—we cannot predict accurately which segments of DNA will be altered or in what way. In multicellular organisms, only mutations in cell lines that produce gametes can be passed to offspring. In plants and fungi, this is not as limiting as it may sound, since many different cell lines can produce gametes (see Figures 30.6 and 31.17). But in most animals, the majority of mutations occur in somatic cells and are lost when the individual dies.

A change of as little as one base in a gene, called a "point mutation," can have a significant impact on phenotype, as in sickle-cell disease (see Figure 17.23). Organisms reflect thousands of generations of past selection, and hence their phenotypes generally provide a close match to their environment. As a result, it's unlikely that a new mutation that alters a phenotype will improve it. In fact, most such mutations are at least slightly harmful. But much of the DNA in eukaryotic genomes does not code for protein products, and point mutations in these noncoding regions are often harmless. Also, because of the redundancy in the genetic code, even a point mutation in a gene that encodes a protein will have no effect on the protein's function if the amino acid composition is not changed. And even where there is a change in the amino acid, it may not affect the protein's shape and function. However, as will be discussed later in this chapter, a mutant allele may on rare occasions actually make its bearer better suited to the environment, enhancing reproductive success.

Altering Gene Number or Position

Chromosomal changes that delete, disrupt, or rearrange many loci at once are usually harmful. However, when such large-scale changes leave genes intact, their effects on organisms may be neutral (as in the case of the Madeira mice described in Figure 23.4). In rare cases, chromosomal rearrangements may even be beneficial. For example, the translocation of part of one chromosome to a different chromosome could link DNA segments in a way that results in a positive effect.

An important source of variation begins when genes are duplicated due to errors in meiosis (such as unequal crossing over), slippage during DNA replication, or the activities of transposable elements (see Chapters 15 and 21). Duplications of large chromosome segments, like other chromosomal aberrations, are often harmful, but the duplication of smaller pieces of DNA may not be. Gene duplications that do not have severe effects can persist over generations, allowing mutations to accumulate. The result is an expanded genome with new genes that may take on new functions.

Such beneficial increases in gene number appear to have played a major role in evolution. For example, the remote ancestors of mammals had a single gene for detecting odors that has since been duplicated many times. As a result, humans today have about 1,000 olfactory receptor genes, and mice have 1,300. This dramatic proliferation of olfactory genes probably helped early mammals, enabling them to detect faint odors and to distinguish among many different smells. More recently, about 60% of human olfactory receptor genes have been inactivated by mutations, whereas mice have lost only 20% of theirs. Since mutation rates in humans and mice are similar, this difference is likely due to strong selection against mice with mutations that inactivate their olfactory genes. A versatile sense of smell appears to be much more important to mice than to humans!

Rapid Reproduction

Mutation rates tend to be low in plants and animals, averaging about one mutation in every 100,000 genes per generation, and they are often even lower in prokaryotes. But prokaryotes typically have short generation spans, so mutations can quickly generate genetic variation in populations of these organisms. The same is true of viruses. For instance, HIV has a generation span of about two days. It also has an RNA genome, which has a much higher mutation rate than a typical DNA genome because of the lack of RNA repair mechanisms in host cells (see Chapter 19). For this reason, it is unlikely that a single-drug treatment would ever be effective against HIV; mutant forms of the virus that are resistant to a particular drug would no doubt proliferate in relatively short order. The most effective AIDS treatments to date have been drug "cocktails" that combine several medications. It is less likely that multiple mutations conferring resistance to *all* the drugs will occur in a short time period.

Sexual Reproduction

In organisms that reproduce sexually, most of the genetic variation in a population results from the unique combination of alleles that each individual receives from its parents. Of course, at the nucleotide level, all the differences among these alleles have originated from past mutations and other processes that can produce new alleles. But it is the mechanism of sexual reproduction that shuffles existing alleles and deals them at random to produce individual genotypes.

As described in Chapter 13, three mechanisms contribute to this shuffling: crossing over, independent assortment of chromosomes, and fertilization. During meiosis, homologous chromosomes, one inherited from each parent, trade some of their alleles by crossing over. These homologous chromosomes and the alleles they carry are then distributed at random into gametes. Then, because myriad possible mating combinations exist in a population, fertilization brings together gametes that are likely to have different genetic backgrounds. The combined effects of these three mechanisms ensure that sexual reproduction rearranges existing alleles into fresh combinations each generation, providing much of the genetic variation that makes evolution possible.

CONCEPT CHECK 23.1

1. (a) Explain why genetic variation within a population is a prerequisite for evolution. (b) What factors can produce genetic differences between populations?
2. Of all the mutations that occur in a population, why do only a small fraction become widespread?
3. **MAKE CONNECTIONS** If a population stopped reproducing sexually (but still reproduced asexually), how would its genetic variation be affected over time? Explain. (See Concept 13.4, pp. 257–259.)

For suggested answers, see Appendix A.

CONCEPT 23.2

The Hardy-Weinberg equation can be used to test whether a population is evolving

Although the individuals in a population must differ genetically for evolution to occur, the presence of genetic variation does not guarantee that a population will evolve. For that to happen, one of the factors that cause evolution must be at work. In this section, we'll explore one way to test whether evolution is occurring in a population. The first step in this process is to clarify what we mean by a population.

Gene Pools and Allele Frequencies

A **population** is a group of individuals of the same species that live in the same area and interbreed, producing fertile offspring. Different populations of a single species may be isolated geographically from one another, thus exchanging genetic material only rarely. Such isolation is common for species that live on widely separated islands or in different lakes. But not all populations are isolated, nor must populations have sharp boundaries (**Figure 23.6**). Still, members of a population typically breed with one another and thus on average are more closely related to each other than to members of other populations.

We can characterize a population's genetic makeup by describing its **gene pool**, which consists of all copies of every type of allele at every locus in all members of the population. If only one allele exists for a particular locus in a population, that allele is said to be *fixed* in the gene pool, and all individuals are homozygous for that allele. But if there are two or more alleles for a particular locus in a population, individuals may be either homozygous or heterozygous.

Each allele has a frequency (proportion) in the population. For example, imagine a population of 500 wildflower plants with two alleles, C^R and C^W, for a locus that codes for flower pigment. These alleles show incomplete dominance (see Figure 14.10); thus, each genotype has a distinct phenotype.

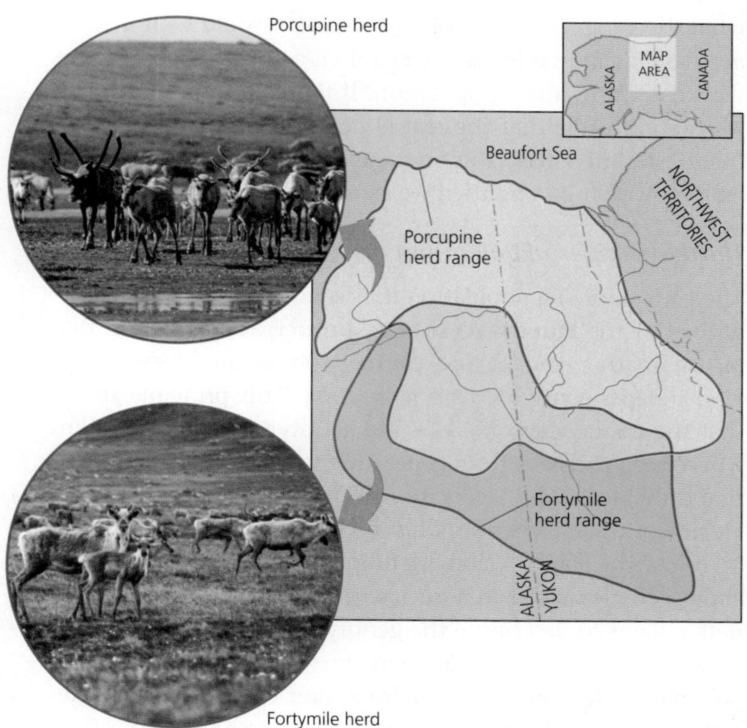

▲ **Figure 23.6 One species, two populations.** These two caribou populations in the Yukon are not totally isolated; they sometimes share the same area. Still, members of either population are most likely to breed within their own population.

Plants homozygous for the C^R allele (C^RC^R) produce red pigment and have red flowers; plants homozygous for the C^W allele (C^WC^W) produce no red pigment and have white flowers; and heterozygotes (C^RC^W) produce some red pigment and have pink flowers.

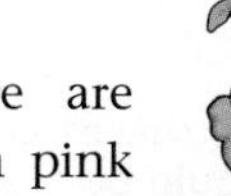

In our population, suppose there are 320 plants with red flowers, 160 with pink flowers, and 20 with white flowers. Because these are diploid organisms, there are a total of 1,000 copies of the gene for flower color in the population of 500 individuals. The C^R allele accounts for 800 of these copies ($320 \times 2 = 640$ for C^RC^R plants, plus $160 \times 1 = 160$ for C^RC^W plants).

When studying a locus with two alleles, the convention is to use p to represent the frequency of one allele and q to represent the frequency of the other allele. Thus, p, the frequency of the C^R allele in the gene pool of this population, is $800/1{,}000 = 0.8 = 80\%$. And because there are only two alleles for this gene, the frequency of the C^W allele, represented by q, must be $200/1{,}000 = 0.2 = 20\%$. For loci that have more than two alleles, the sum of all allele frequencies must still equal 1 (100%).

Next we'll see how allele and genotype frequencies can be used to test whether evolution is occurring in a population.

The Hardy-Weinberg Principle

One way to assess whether natural selection or other factors are causing evolution at a particular locus is to determine what the genetic makeup of a population would be if it were *not* evolving at that locus. We can then compare that scenario with data from a real population. If there are no differences, we can conclude that the real population is not evolving. If there are differences, this suggests that the real population may be evolving—and then we can try to figure out why.

Hardy-Weinberg Equilibrium

The gene pool of a population that is not evolving can be described by the **Hardy-Weinberg principle**, named for the British mathematician and German physician, respectively, who independently derived it in 1908. This principle states that the frequencies of alleles and genotypes in a population will remain constant from generation to generation, provided that only Mendelian segregation and recombination of alleles are at work. Such a gene pool is in *Hardy-Weinberg equilibrium*.

To use the Hardy-Weinberg principle, it is helpful to think about genetic crosses in a new way. Previously, we used Punnett squares to determine the genotypes of offspring in a genetic cross (see Figure 14.5). Here, instead of considering the possible allele combinations from one cross, consider the combination of alleles in *all* of the crosses in a population.

Imagine that all the alleles for a given locus from all the individuals in a population were placed in a large bin **(Figure 23.7)**.

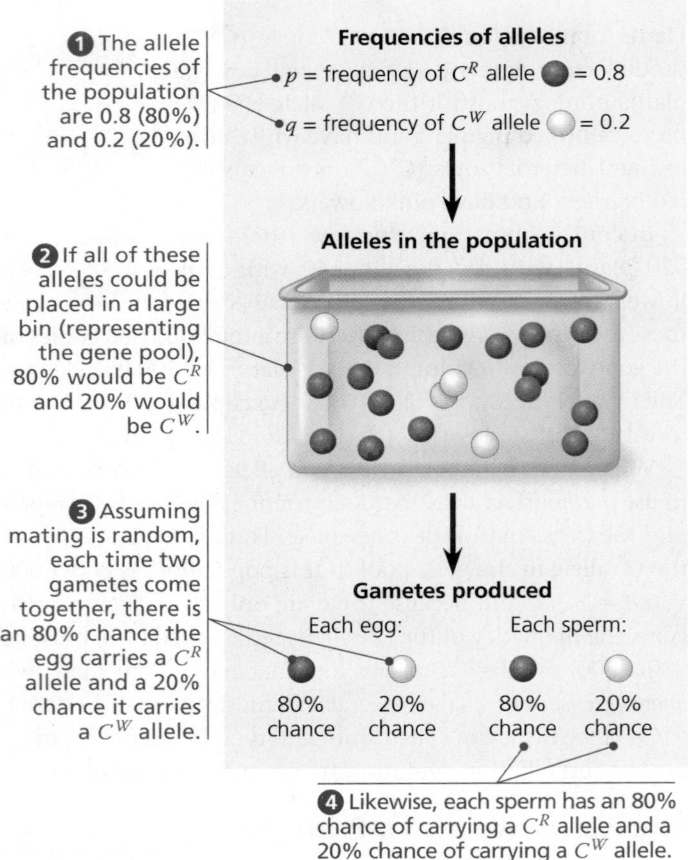

1 The allele frequencies of the population are 0.8 (80%) and 0.2 (20%).
Frequencies of alleles
p = frequency of C^R allele = 0.8
q = frequency of C^W allele = 0.2
Alleles in the population
2 If all of these alleles could be placed in a large bin (representing the gene pool), 80% would be C^R and 20% would be C^W.
3 Assuming mating is random, each time two gametes come together, there is an 80% chance the egg carries a C^R allele and a 20% chance it carries a C^W allele.
Gametes produced
Each egg:
Each sperm:
80% chance
20% chance
80% chance
20% chance
4 Likewise, each sperm has an 80% chance of carrying a C^R allele and a 20% chance of carrying a C^W allele.

▲ **Figure 23.7 Selecting alleles at random from a gene pool.**

We can think of this bin as holding the population's gene pool for that locus. "Reproduction" occurs by selecting alleles at random from the bin; somewhat similar events occur in nature when fish release sperm and eggs into the water or when pollen (containing plant sperm) is blown about by the wind. By viewing reproduction as a process of randomly selecting and combining alleles from the bin (the gene pool), we are in effect assuming that mating occurs at random—that is, that all male-female matings are equally likely.

Let's apply the bin analogy to the hypothetical wildflower population discussed earlier. In that population of 500 flowers, the frequency of the allele for red flowers (C^R) is $p = 0.8$, and the frequency of the allele for white flowers (C^W) is $q = 0.2$. Thus, a bin holding all 1,000 copies of the flower-color gene in the population contains 800 C^R alleles and 200 C^W alleles. Assuming that gametes are formed by selecting alleles at random from the bin, the probability that an egg or sperm contains a C^R or C^W allele is equal to the frequency of these alleles in the bin. Thus, as shown in Figure 23.7, each egg has an 80% chance of containing a C^R allele and a 20% chance of containing a C^W allele; the same is true for each sperm.

Using the rule of multiplication (see Figure 14.9), we can now calculate the frequencies of the three possible genotypes, assuming random unions of sperm and eggs. The probability that two C^R alleles will come together is $p \times p = p^2 = 0.8 \times 0.8 = 0.64$. Thus, about 64% of the plants in the next generation will have the genotype C^RC^R. The frequency of C^WC^W individuals is expected to be about $q \times q = q^2 = 0.2 \times 0.2 = 0.04$, or 4%. C^RC^W heterozygotes can arise in two different ways. If the sperm provides the C^R allele and the egg provides the C^W allele, the resulting heterozygotes will be $p \times q = 0.8 \times 0.2 = 0.16$, or 16% of the total. If the sperm provides the C^W allele and the egg the C^R allele, the heterozygous offspring will make up $q \times p = 0.2 \times 0.8 = 0.16$, or 16%. The frequency of heterozygotes is thus the sum of these possibilities: $pq + qp = 2pq = 0.16 + 0.16 = 0.32$, or 32%.

As shown in **Figure 23.8** on the facing page, the genotype frequencies in the next generation must add up to 1 (100%). Thus, the equation for Hardy-Weinberg equilibrium states that at a locus with two alleles, the three genotypes will appear in the following proportions:

p^2	+	$2pq$	+	q^2	=	1
Expected frequency of genotype C^RC^R		Expected frequency of genotype C^RC^W		Expected frequency of genotype C^WC^W		

Note that for a locus with two alleles, only three genotypes are possible (in this case, C^RC^R, C^RC^W, and C^WC^W). As a result, the sum of the frequencies of the three genotypes must equal 1 (100%) in *any* population—regardless of whether the population is in Hardy-Weinberg equilibrium. A population is in Hardy-Weinberg equilibrium only if the genotype frequencies are such that the actual frequency of

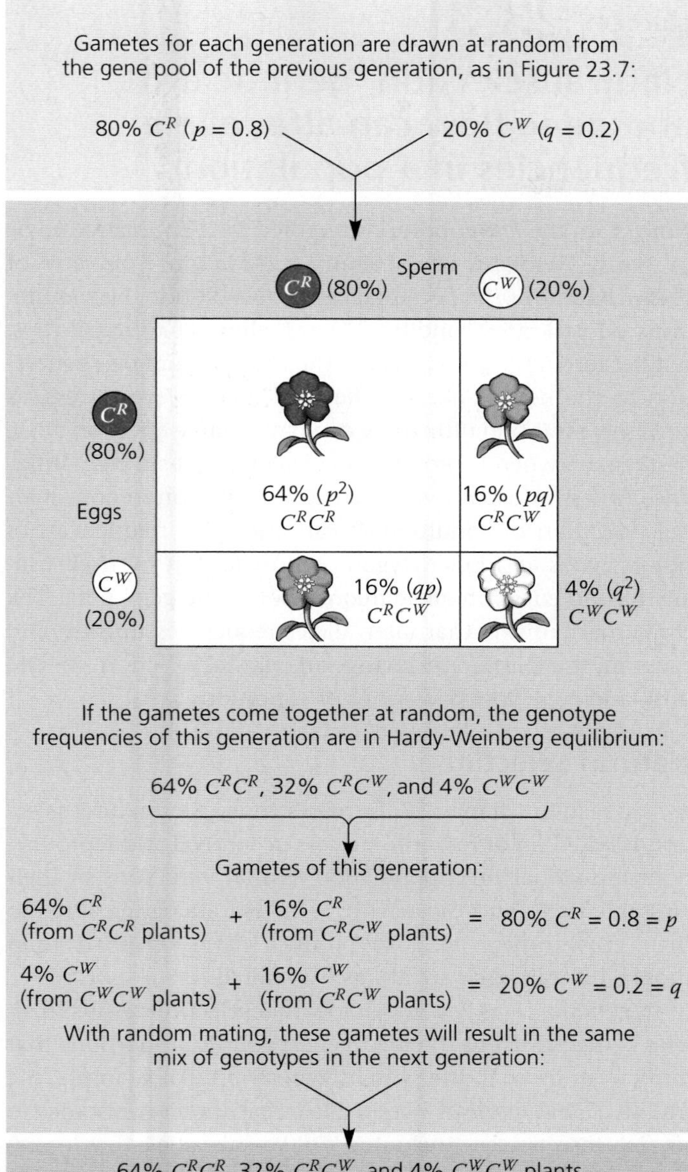
Gametes for each generation are drawn at random from the gene pool of the previous generation, as in Figure 23.7:
80% C^R (p = 0.8)
20% C^W (q = 0.2)
Sperm
C^R (80%)
C^W (20%)
C^R (80%)
Eggs
C^W (20%)
64% (p^2) C^RC^R
16% (pq) C^RC^W
16% (qp) C^RC^W
4% (q^2) C^WC^W
If the gametes come together at random, the genotype frequencies of this generation are in Hardy-Weinberg equilibrium:
64% C^RC^R, 32% C^RC^W, and 4% C^WC^W
Gametes of this generation:
64% C^R (from C^RC^R plants) + 16% C^R (from C^RC^W plants) = 80% C^R = 0.8 = p
4% C^W (from C^WC^W plants) + 16% C^W (from C^RC^W plants) = 20% C^W = 0.2 = q
With random mating, these gametes will result in the same mix of genotypes in the next generation:
64% C^RC^R, 32% C^RC^W, and 4% C^WC^W plants

▲ **Figure 23.8 The Hardy-Weinberg principle.** In our wildflower population, the gene pool remains constant from one generation to the next. Mendelian processes alone do not alter frequencies of alleles or genotypes.

? *If the frequency of the* C^R *allele is 60%, predict the frequencies of the* C^RC^R*,* C^RC^W*, and* C^WC^W *genotypes.*

one homozygote is p^2, the actual frequency of the other homozygote is q^2, and the actual frequency of heterozygotes is $2pq$. Finally, as suggested by Figure 23.8, if a population such as our wildflowers is in Hardy-Weinberg equilibrium and its members continue to mate randomly generation after generation, allele and genotype frequencies will remain constant. The system operates somewhat like a deck of cards: No matter how many times the deck is reshuffled to deal out new hands, the deck itself remains the same. Aces do not grow more numerous than jacks. And the repeated shuffling of a population's gene pool over the generations cannot, in itself, change the frequency of one allele relative to another.

Conditions for Hardy-Weinberg Equilibrium

The Hardy-Weinberg principle describes a hypothetical population that is not evolving. But in real populations, the allele and genotype frequencies often *do* change over time. Such changes can occur when at least one of the following five conditions of Hardy-Weinberg equilibrium is not met:

1. **No mutations.** The gene pool is modified if mutations alter alleles or if entire genes are deleted or duplicated.
2. **Random mating.** If individuals mate preferentially within a subset of the population, such as their close relatives (inbreeding), random mixing of gametes does not occur, and genotype frequencies change.
3. **No natural selection.** Differences in the survival and reproductive success of individuals carrying different genotypes can alter allele frequencies.
4. **Extremely large population size.** The smaller the population, the more likely it is that allele frequencies will fluctuate by chance from one generation to the next (a process called genetic drift).
5. **No gene flow.** By moving alleles into or out of populations, gene flow can alter allele frequencies.

Departure from these conditions usually results in evolutionary change, which, as we've already described, is common in natural populations. But it is also common for natural populations to be in Hardy-Weinberg equilibrium for specific genes. This apparent contradiction occurs because a population can be evolving at some loci, yet simultaneously be in Hardy-Weinberg equilibrium at other loci. In addition, some populations evolve so slowly that the changes in their allele and genotype frequencies are difficult to distinguish from those predicted for a nonevolving population.

Applying the Hardy-Weinberg Principle

The Hardy-Weinberg equation is often used as an initial test of whether evolution is occurring in a population (you'll encounter an example in Concept Check 23.2, question 3). The equation also has medical applications, such as estimating the percentage of a population carrying the allele for an inherited disease. For example, consider phenylketonuria (PKU), a metabolic disorder that results from homozygosity for a recessive allele and occurs in about one out of every 10,000 babies born in the United States. Left untreated, PKU results in mental disability and other problems. (Newborns are now tested for PKU, and symptoms can be largely avoided with a diet very low in phenylalanine. For this reason, products that contain phenylalanine, such as diet colas, carry warning labels.)

To apply the Hardy-Weinberg equation, we must assume that no new PKU mutations are being introduced into the population (condition 1), and that people neither choose their mates on the basis of whether or not they carry this gene nor generally mate with close relatives (condition 2). We must also ignore any effects of differential survival and reproductive success among PKU genotypes (condition 3) and assume that there are no effects of genetic drift (condition 4) or of gene flow from other populations into the United States (condition 5). These assumptions are reasonable: The mutation rate for the PKU gene is low, inbreeding is not common in the United States, selection occurs only against the rare homozygotes (and then only if dietary restrictions are not followed), the U.S. population is very large, and populations outside the country have PKU allele frequencies similar to those seen in the United States. If all these assumptions hold, then the frequency of individuals in the population born with PKU will correspond to q^2 in the Hardy-Weinberg equation (q^2 = frequency of homozygotes). Because the allele is recessive, we must estimate the number of heterozygotes rather than counting them directly as we did with the pink flowers. Since we know there is one PKU occurrence per 10,000 births ($q^2 = 0.0001$), the frequency of the recessive allele for PKU is

$$q = \sqrt{0.0001} = 0.01$$

and the frequency of the dominant allele is

$$p = 1 - q = 1 - 0.01 = 0.99$$

The frequency of carriers, heterozygous people who do not have PKU but may pass the PKU allele to offspring, is

$$2pq = 2 \times 0.99 \times 0.01 = 0.0198$$
(approximately 2% of the U.S. population)

Remember, the assumption of Hardy-Weinberg equilibrium yields an approximation; the real number of carriers may differ. Still, our calculations suggest that harmful recessive alleles at this and other loci can be concealed in a population because they are carried by healthy heterozygotes.

CONCEPT CHECK 23.2

1. Suppose a population of organisms with 20,000 gene loci is fixed at half of these loci and has two alleles at each of the other loci. How many different types of alleles are found in its entire gene pool? Explain.
2. If p is the frequency of allele A, use the Hardy-Weinberg equation to predict the frequency of individuals that have at least one A allele.
3. **WHAT IF?** A locus that affects susceptibility to a degenerative brain disease has two alleles, A and a. In a population, 16 people have genotype AA, 92 have genotype Aa, and 12 have genotype aa. Is this population evolving? Explain.

For suggested answers, see Appendix A.

CONCEPT 23.3

Natural selection, genetic drift, and gene flow can alter allele frequencies in a population

Note again the five conditions required for a population to be in Hardy-Weinberg equilibrium. A deviation from any of these conditions is a potential cause of evolution. New mutations (violation of condition 1) can alter allele frequencies, but because mutations are rare, the change from one generation to the next is likely to be very small. Nevertheless, as we'll see, mutation ultimately can have a large effect on allele frequencies when it produces new alleles that strongly influence fitness in a positive or negative way. Nonrandom mating (violation of condition 2) can affect the frequencies of homozygous and heterozygous genotypes but by itself usually has no effect on allele frequencies in the gene pool. The three mechanisms that alter allele frequencies directly and cause most evolutionary change are natural selection, genetic drift, and gene flow (violations of conditions 3–5).

Natural Selection

As you read in Chapter 22, Darwin's concept of natural selection is based on differential success in survival and reproduction: Individuals in a population exhibit variations in their heritable traits, and those with traits that are better suited to their environment tend to produce more offspring than those with traits that are not as well suited.

In genetic terms, we now know that selection results in alleles being passed to the next generation in proportions that differ from those in the present generation. For example, the fruit fly *D. melanogaster* has an allele that confers resistance to several insecticides, including DDT. This allele has a frequency of 0% in laboratory strains of *D. melanogaster* established from flies collected in the wild in the early 1930s, prior to DDT use. However, in strains established from flies collected after 1960 (following 20 or more years of DDT use), the allele frequency is 37%. We can infer that this allele either arose by mutation between 1930 and 1960 or that it was present in 1930, but very rare. In any case, the rise in frequency of this allele most likely occurred because DDT is a powerful poison that is a strong selective force in exposed fly populations.

As the *D. melanogaster* example shows, an allele that confers insecticide resistance will increase in frequency in a population exposed to that insecticide. Such changes are not coincidental. By consistently favoring some alleles over others, natural selection can cause *adaptive evolution* (evolution that results in a better match between organisms and their environment). We'll explore this process in more detail a little later in this chapter.

Genetic Drift

If you flip a coin 1,000 times, a result of 700 heads and 300 tails might make you suspicious about that coin. But if you flip a coin only 10 times, an outcome of 7 heads and 3 tails would not be surprising. The smaller the number of coin flips, the more likely it is that chance alone will cause a deviation from the predicted result. (In this case, the prediction is an equal number of heads and tails.) Chance events can also cause allele frequencies to fluctuate unpredictably from one generation to the next, especially in small populations—a process called **genetic drift**.

Figure 23.9 models how genetic drift might affect a small population of our wildflowers. In this example, an allele is lost from the gene pool, but it is a matter of chance that the C^W allele is lost and not the C^R allele. Such unpredictable changes in allele frequencies can be caused by chance events associated with survival and reproduction. Perhaps a large animal such as a moose stepped on the three C^WC^W individuals in generation 2, killing them and increasing the chance that only the C^R allele would be passed to the next generation. Allele frequencies can also be affected by chance events that occur during fertilization. For example, suppose two individuals of genotype C^RC^W had a small number of offspring. By chance alone, every egg and sperm pair that generated offspring could happen to have carried the C^R allele and not the C^W allele.

Certain circumstances can result in genetic drift having a significant impact on a population. Two examples are the founder effect and the bottleneck effect.

The Founder Effect

When a few individuals become isolated from a larger population, this smaller group may establish a new population whose gene pool differs from the source population; this is called the **founder effect**. The founder effect might occur, for example, when a few members of a population are blown by a storm to a new island. Genetic drift, in which chance events alter allele frequencies, will occur in such a case if the storm indiscriminately transports some individuals (and their alleles), but not others, from the source population.

The founder effect probably accounts for the relatively high frequency of certain inherited disorders among isolated human populations. For example, in 1814, 15 British colonists founded a settlement on Tristan da Cunha, a group of small islands in the Atlantic Ocean midway between Africa and South America. Apparently, one of the colonists carried a recessive allele for retinitis pigmentosa, a progressive form of blindness that afflicts homozygous individuals. Of the founding colonists' 240 descendants on the island in the late 1960s, 4 had retinitis pigmentosa. The frequency of the allele that causes this disease is ten times

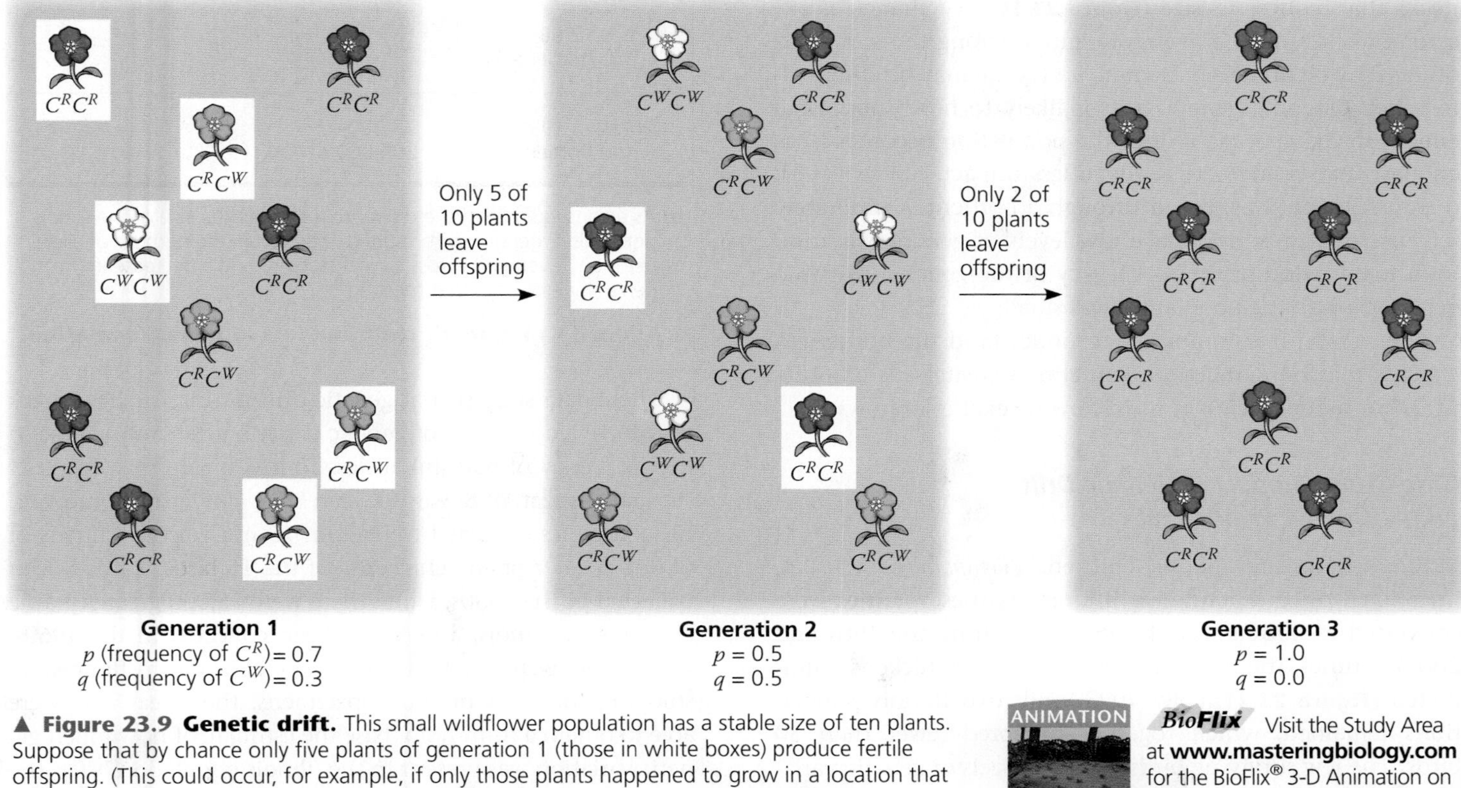

▲ **Figure 23.9 Genetic drift.** This small wildflower population has a stable size of ten plants. Suppose that by chance only five plants of generation 1 (those in white boxes) produce fertile offspring. (This could occur, for example, if only those plants happened to grow in a location that provided enough nutrients to support the production of offspring.) Again by chance, only two plants of generation 2 leave fertile offspring. As a result, by chance the frequency of the C^W allele first increases in generation 2, then falls to zero in generation 3.

ANIMATION **BioFlix** Visit the Study Area at **www.masteringbiology.com** for the BioFlix® 3-D Animation on Mechanisms of Evolution.

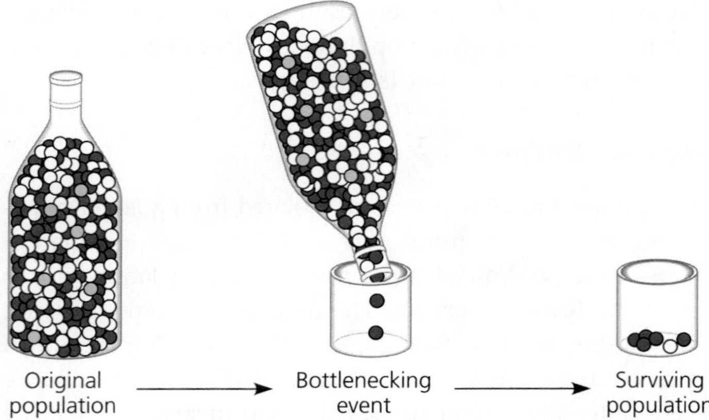

▲ **Figure 23.10 The bottleneck effect.** Shaking just a few marbles through the narrow neck of a bottle is analogous to a drastic reduction in the size of a population. By chance, blue marbles are overrepresented in the surviving population and gold marbles are absent.

higher on Tristan da Cunha than in the populations from which the founders came.

The Bottleneck Effect

A sudden change in the environment, such as a fire or flood, may drastically reduce the size of a population. A severe drop in population size can cause the **bottleneck effect**, so named because the population has passed through a "bottleneck" that reduces its size **(Figure 23.10)**. By chance alone, certain alleles may be overrepresented among the survivors, others may be underrepresented, and some may be absent altogether. Ongoing genetic drift is likely to have substantial effects on the gene pool until the population becomes large enough that chance events have less impact. But even if a population that has passed through a bottleneck ultimately recovers in size, it may have low levels of genetic variation for a long period of time—a legacy of the genetic drift that occurred when the population was small.

One reason it is important to understand the bottleneck effect is that human actions sometimes create severe bottlenecks for other species, as the following example shows.

Case Study: *Impact of Genetic Drift on the Greater Prairie Chicken*

Millions of greater prairie chickens (*Tympanuchus cupido*) once lived on the prairies of Illinois. As these prairies were converted to farmland and other uses during the 19th and 20th centuries, the number of greater prairie chickens plummeted **(Figure 23.11a)**. By 1993, only two Illinois populations remained, which together harbored fewer than 50 birds. The few surviving birds had low levels of genetic variation, and less than 50% of their eggs hatched, compared with much higher hatching rates of the larger populations in Kansas and Nebraska **(Figure 23.11b)**.

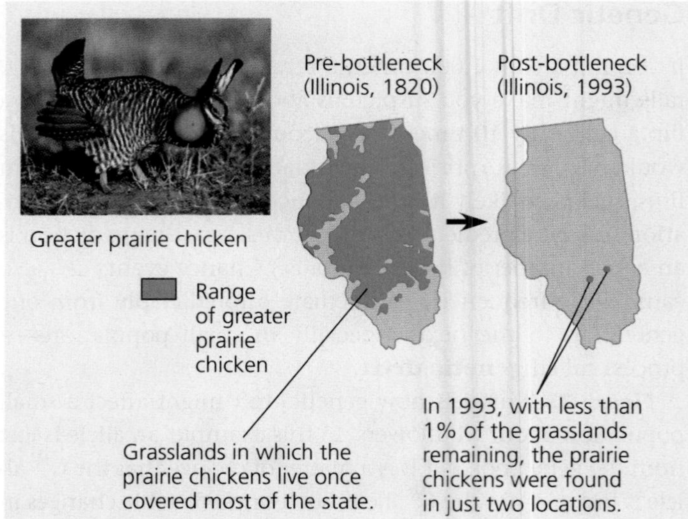

(a) The Illinois population of greater prairie chickens dropped from millions of birds in the 1800s to fewer than 50 birds in 1993.

Location	Population size	Number of alleles per locus	Percentage of eggs hatched
Illinois			
1930–1960s	1,000–25,000	5.2	93
1993	<50	3.7	<50
Kansas, 1998 (no bottleneck)	750,000	5.8	99
Nebraska, 1998 (no bottleneck)	75,000–200,000	5.8	96

(b) As a consequence of the drastic reduction in the size of the Illinois population, genetic drift resulted in a drop in the number of alleles per locus (averaged across six loci studied) and a decrease in the percentage of eggs that hatched.

▲ **Figure 23.11 Genetic drift and loss of genetic variation.**

These data suggest that genetic drift during the bottleneck may have led to a loss of genetic variation and an increase in the frequency of harmful alleles. To investigate this hypothesis, Juan Bouzat, of Bowling Green State University, Ohio, and his colleagues extracted DNA from 15 museum specimens of Illinois greater prairie chickens. Of the 15 birds, 10 had been collected in the 1930s, when there were 25,000 greater prairie chickens in Illinois, and 5 had been collected in the 1960s, when there were 1,000 greater prairie chickens in Illinois. By studying the DNA of these specimens, the researchers were able to obtain a minimum, baseline estimate of how much genetic variation was present in the Illinois population *before* the population shrank to extremely low numbers. This baseline estimate is a key piece of information that is not usually available in cases of population bottlenecks.

The researchers surveyed six loci and found that the 1993 Illinois greater prairie chicken population had lost nine alleles that were present in the museum specimens. The 1993 population also had fewer alleles per locus than the pre-bottleneck Illinois or the current Kansas and Nebraska populations (see Figure 23.11b). Thus, as predicted, drift had reduced the genetic variation of the small 1993 population. Drift may also have increased the frequency of harmful alleles, leading to the low egg-hatching rate. To counteract these negative effects, 271 birds from neighboring states were added to the Illinois population over four years. This strategy succeeded: New alleles entered the population, and the egg-hatching rate improved to over 90%. Overall, studies on the Illinois greater prairie chicken illustrate the powerful effects of genetic drift in small populations and provide hope that in at least some populations, these effects can be reversed.

Effects of Genetic Drift: A Summary

The examples we've described highlight four key points:

1. **Genetic drift is significant in small populations.** Chance events can cause an allele to be disproportionately over- or underrepresented in the next generation. Although chance events occur in populations of all sizes, they tend to alter allele frequencies substantially only in small populations.
2. **Genetic drift can cause allele frequencies to change at random.** Because of genetic drift, an allele may increase in frequency one year, then decrease the next; the change from year to year is not predictable. Thus, unlike natural selection, which in a given environment consistently favors some alleles over others, genetic drift causes allele frequencies to change at random over time.
3. **Genetic drift can lead to a loss of genetic variation within populations.** By causing allele frequencies to fluctuate randomly over time, genetic drift can eliminate alleles from a population. Because evolution depends on genetic variation, such losses can influence how effectively a population can adapt to a change in the environment.
4. **Genetic drift can cause harmful alleles to become fixed.** Alleles that are neither harmful nor beneficial can be lost or become fixed entirely by chance through genetic drift. In very small populations, genetic drift can also cause alleles that are slightly harmful to become fixed. When this occurs, the population's survival can be threatened (as for the greater prairie chicken).

Gene Flow

Natural selection and genetic drift are not the only phenomena affecting allele frequencies. Allele frequencies can also change by **gene flow**, the transfer of alleles into or out of a population due to the movement of fertile individuals or their gametes. For example, suppose that near our original hypothetical wildflower population there is another population consisting primarily of white-flowered individuals (C^WC^W). Insects carrying pollen from these plants may fly to and pollinate plants in our original population. The introduced C^W alleles would modify our original population's allele frequencies in the next generation. Because alleles are exchanged between populations, gene flow tends to reduce the genetic differences between populations. In fact, if it is extensive enough, gene flow can result in two populations combining into a single population with a common gene pool.

Alleles transferred by gene flow can also affect how well populations are adapted to local environmental conditions. Researchers studying the songbird *Parus major* (great tit) on the small Dutch island of Vlieland noted survival differences between two populations on the island. Females born in the eastern population survive twice as well as females born in the central population, regardless of where the females eventually settle and raise offspring **(Figure 23.12)**. This finding suggests that females born in the eastern population are better adapted to life on the island than females born in the

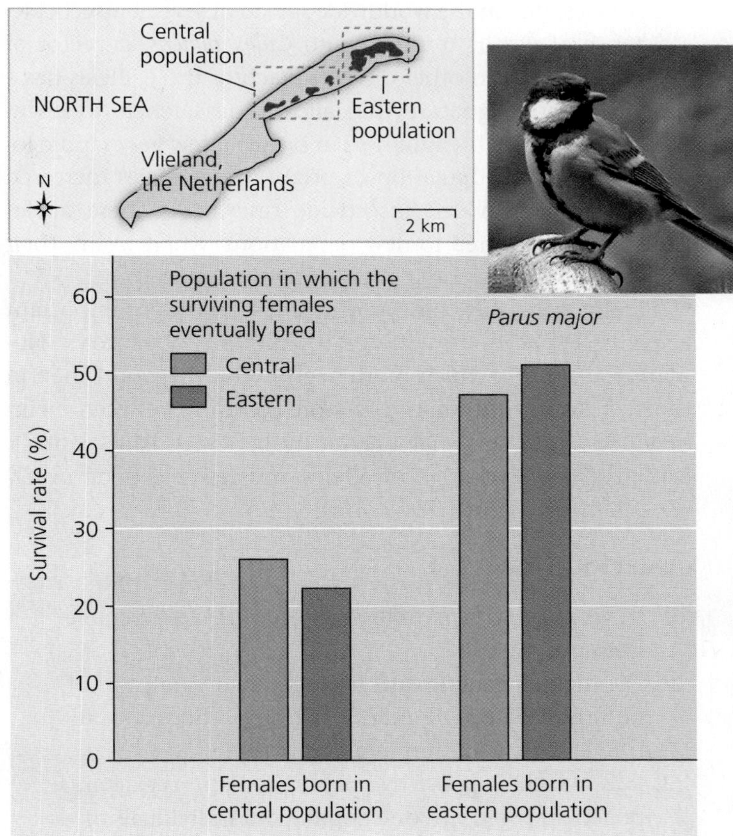

▲ **Figure 23.12 Gene flow and local adaptation.** In *Parus major* populations on Vlieland, the yearly survival rate of females born in the eastern population is higher than that of females born in the central population. Gene flow from the mainland to the central population is 3.3 times higher than it is to the eastern population, and birds from the mainland are selected against in both populations. These data suggest that gene flow from the mainland has prevented the central population from adapting fully to its local conditions.

central population. But extensive field studies also showed that the two populations are connected by high levels of gene flow (mating), which should reduce genetic differences between them. So how can the eastern population be better adapted to life on Vlieland than the central population? The answer lies in the unequal amounts of gene flow from the mainland. In any given year, 43% of the first-time breeders in the central population are immigrants from the mainland, compared with only 13% in the eastern population. Birds with mainland genotypes survive and reproduce poorly on Vlieland, and in the eastern population, selection reduces the frequency of these genotypes. In the central population, however, gene flow from the mainland is so high that it overwhelms the effects of selection. As a result, females born in the central population have many immigrant genes, reducing the degree to which members of that population are adapted to life on the island. Researchers are currently investigating why gene flow is so much higher in the central population and why birds with mainland genotypes have low fitness on Vlieland.

Gene flow can also transfer alleles that improve the ability of populations to adapt to local conditions. For example, gene flow has resulted in the worldwide spread of several insecticide-resistance alleles in the mosquito *Culex pipiens*, a vector of West Nile virus and other diseases. Each of these alleles has a unique genetic signature that allowed researchers to document that it arose by mutation in one or a few geographic locations. In their population of origin, these alleles increased because they provided insecticide resistance. These alleles were then transferred to new populations, where again, their frequencies increased as a result of natural selection.

Finally, gene flow has become an increasingly important agent of evolutionary change in human populations. Humans today move much more freely about the world than in the past. As a result, mating is more common between members of populations that previously had very little contact, leading to an exchange of alleles and fewer genetic differences between those populations.

CONCEPT CHECK 23.3

1. In what sense is natural selection more "predictable" than genetic drift?
2. Distinguish genetic drift from gene flow in terms of (a) how they occur and (b) their implications for future genetic variation in a population.
3. **WHAT IF?** Suppose two plant populations exchange pollen and seeds. In one population, individuals of genotype *AA* are most common (9,000 *AA*, 900 *Aa*, 100 *aa*), while the opposite is true in the other population (100 *AA*, 900 *Aa*, 9,000 *aa*). If neither allele has a selective advantage, what will happen over time to the allele and genotype frequencies of these populations?

For suggested answers, see Appendix A.

CONCEPT 23.4

Natural selection is the only mechanism that consistently causes adaptive evolution

Evolution by natural selection is a blend of chance and "sorting": chance in the creation of new genetic variations (as in mutation) and sorting as natural selection favors some alleles over others. Because of this favoring process, the outcome of natural selection is *not* random. Instead, natural selection consistently increases the frequencies of alleles that provide reproductive advantage and thus leads to adaptive evolution.

A Closer Look at Natural Selection

In examining how natural selection brings about adaptive evolution, we'll begin with the concept of relative fitness and the different ways that an organism's phenotype is subject to natural selection.

Relative Fitness

The phrases "struggle for existence" and "survival of the fittest" are commonly used to describe natural selection, but these expressions are misleading if taken to mean direct competitive contests among individuals. There *are* animal species in which individuals, usually the males, lock horns or otherwise do combat to determine mating privilege. But reproductive success is generally more subtle and depends on many factors besides outright battle. For example, a barnacle that is more efficient at collecting food than its neighbors may have greater stores of energy and hence be able to produce a larger number of eggs. A moth may have more offspring than other moths in the same population because its body colors more effectively conceal it from predators, improving its chance of surviving long enough to produce more offspring. These examples illustrate how in a given environment, certain traits can lead to greater **relative fitness**: the contribution an individual makes to the gene pool of the next generation *relative to* the contributions of other individuals.

Although we often refer to the relative fitness of a genotype, remember that the entity that is subjected to natural selection is the whole organism, not the underlying genotype. Thus, selection acts more directly on the phenotype than on the genotype; it acts on the genotype indirectly, via how the genotype affects the phenotype.

Directional, Disruptive, and Stabilizing Selection

Natural selection can alter the frequency distribution of heritable traits in three ways, depending on which phenotypes in a population are favored. These three modes of selection are called directional selection, disruptive selection, and stabilizing selection.

Directional selection occurs when conditions favor individuals exhibiting one extreme of a phenotypic range, thereby shifting a population's frequency curve for the phenotypic character in one direction or the other (**Figure 23.13a**). Directional selection is common when a population's environment changes or when members of a population migrate to a new (and different) habitat. For instance, an increase in the size of seeds available as food led to an increase in beak depth in a population of Galápagos finches (see Figure 23.1).

Disruptive selection (**Figure 23.13b**) occurs when conditions favor individuals at both extremes of a phenotypic range over individuals with intermediate phenotypes. One example is a population of black-bellied seedcracker finches in Cameroon whose members display two distinctly different beak sizes. Small-billed birds feed mainly on soft seeds, whereas large-billed birds specialize in cracking hard seeds. It appears that birds with intermediate-sized bills are relatively inefficient at cracking both types of seeds and thus have lower relative fitness.

Stabilizing selection (**Figure 23.13c**) acts against both extreme phenotypes and favors intermediate variants. This mode of selection reduces variation and tends to maintain the status quo for a particular phenotypic character. For example, the birth weights of most human babies lie in the range of 3–4 kg (6.6–8.8 pounds); babies who are either much smaller or much larger suffer higher rates of mortality.

Regardless of the mode of selection, however, the basic mechanism remains the same. Selection favors individuals whose heritable phenotypic traits provide higher reproductive success than do the traits of other individuals.

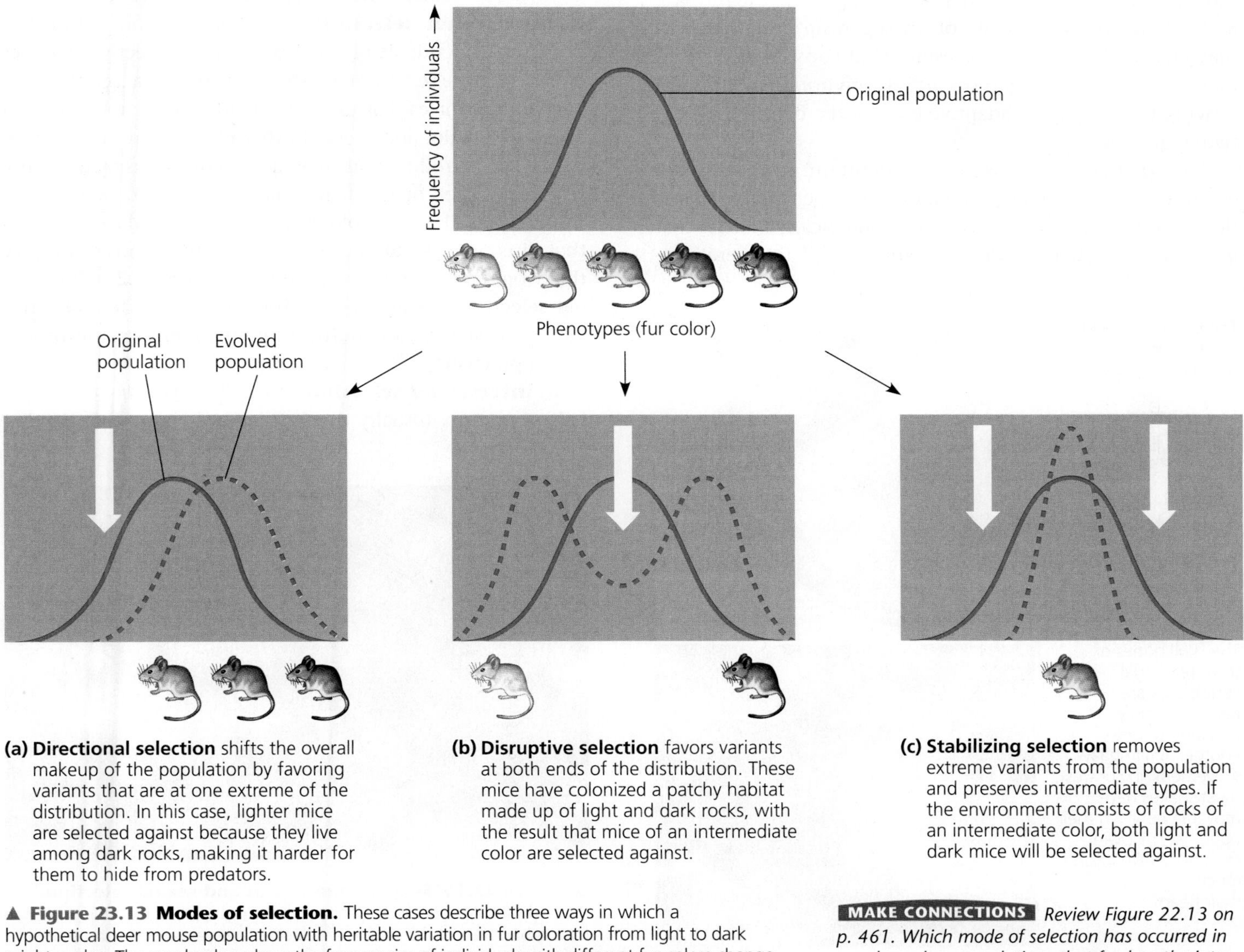

(a) Directional selection shifts the overall makeup of the population by favoring variants that are at one extreme of the distribution. In this case, lighter mice are selected against because they live among dark rocks, making it harder for them to hide from predators.

(b) Disruptive selection favors variants at both ends of the distribution. These mice have colonized a patchy habitat made up of light and dark rocks, with the result that mice of an intermediate color are selected against.

(c) Stabilizing selection removes extreme variants from the population and preserves intermediate types. If the environment consists of rocks of an intermediate color, both light and dark mice will be selected against.

▲ **Figure 23.13 Modes of selection.** These cases describe three ways in which a hypothetical deer mouse population with heritable variation in fur coloration from light to dark might evolve. The graphs show how the frequencies of individuals with different fur colors change over time. The large white arrows symbolize selective pressures against certain phenotypes.

MAKE CONNECTIONS *Review Figure 22.13 on p. 461. Which mode of selection has occurred in soapberry bug populations that feed on the introduced goldenrain tree? Explain.*

The Key Role of Natural Selection in Adaptive Evolution

The adaptations of organisms include many striking examples. Cuttlefish, for example, have the ability to change color rapidly, enabling them to blend into different backgrounds. Another example is the remarkable jaws of snakes (**Figure 23.14**), which allow them to swallow prey much larger than their own head (a feat analogous to a person swallowing a whole watermelon). Other adaptations, such as a version of an enzyme that shows improved function in cold environments (see Figure 23.5), may be less visually dramatic but just as important for survival and reproduction.

Such adaptations can arise gradually over time as natural selection increases the frequencies of alleles that enhance survival and reproduction. As the proportion of individuals that have favorable traits increases, the match between a species and its environment improves; that is, adaptive evolution occurs. However, as we saw in Chapter 22, the physical and biological components of an organism's environment may change over time. As a result, what constitutes a "good match" between an organism and its environment can be a moving target, making adaptive evolution a continuous, dynamic process.

And what about the two other important mechanisms of evolutionary change in populations, genetic drift and gene flow? Both can, in fact, increase the frequencies of alleles that improve the match between organisms and their environment, but neither does so consistently. Genetic drift can cause the frequency of a slightly beneficial allele to increase, but it also can cause the frequency of such an allele to decrease. Similarly, gene flow may introduce alleles that are advantageous or ones that are disadvantageous. Natural selection is the only evolutionary mechanism that consistently leads to adaptive evolution.

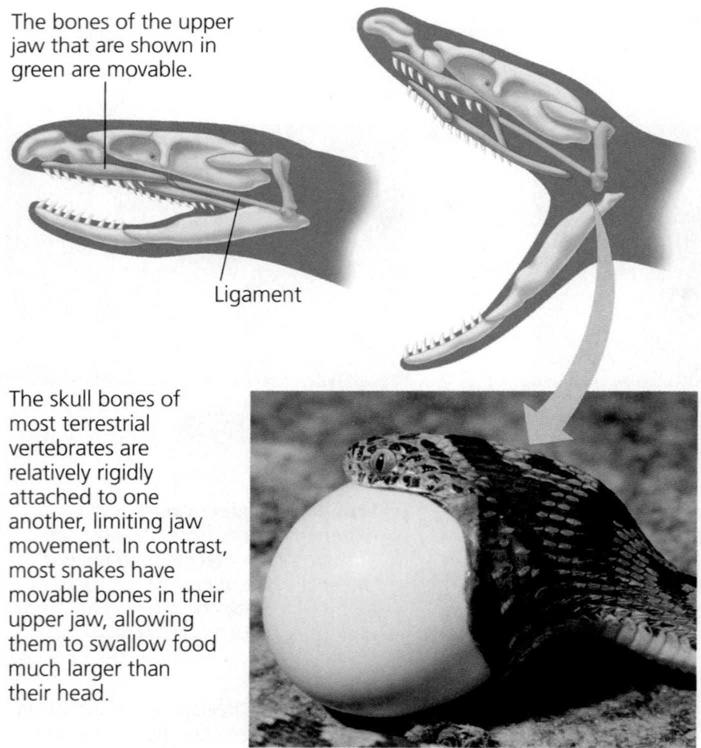

The bones of the upper jaw that are shown in green are movable.
Ligament
The skull bones of most terrestrial vertebrates are relatively rigidly attached to one another, limiting jaw movement. In contrast, most snakes have movable bones in their upper jaw, allowing them to swallow food much larger than their head.

▲ **Figure 23.14 Movable jaw bones in snakes.**

Sexual Selection

Charles Darwin was the first to explore the implications of **sexual selection**, a form of selection in which individuals with certain inherited characteristics are more likely than other individuals to obtain mates. Sexual selection can result in **sexual dimorphism**, a difference between the two sexes in secondary sexual characteristics (**Figure 23.15**). These distinctions include differences in size, color, ornamentation, and behavior.

How does sexual selection operate? There are several ways. In **intrasexual selection**, meaning selection within the same sex, individuals of one sex compete directly for mates of the opposite sex. In many species, intrasexual selection occurs among males. For example, a single male may patrol a group of females and prevent other males from mating with them. The patrolling male may defend his status by defeating smaller, weaker, or less fierce males in combat. More often, this male is the psychological victor in ritualized displays that discourage would-be competitors but do not risk injury that would reduce his own fitness (see Figure 51.22). Intrasexual selection has also been observed among females in a variety of species, including ring-tailed lemurs and broad-nosed pipefish.

In **intersexual selection**, also called *mate choice*, individuals of one sex (usually the females) are choosy in selecting

▲ **Figure 23.15 Sexual dimorphism and sexual selection.** Peacocks (above left) and peahens (above right) show extreme sexual dimorphism. There is intrasexual selection between competing males, followed by intersexual selection when the females choose among the showiest males.

their mates from the other sex. In many cases, the female's choice depends on the showiness of the male's appearance or behavior (see Figure 23.15). What intrigued Darwin about mate choice is that male showiness may not seem adaptive in any other way and may in fact pose some risk. For example, bright plumage may make male birds more visible to predators. But if such characteristics help a male gain a mate, and if this benefit outweighs the risk from predation, then both the bright plumage and the female preference for it will be reinforced because they enhance overall reproductive success.

How do female preferences for certain male characteristics evolve in the first place? One hypothesis is that females prefer male traits that are correlated with "good genes." If the trait preferred by females is indicative of a male's overall genetic quality, both the male trait and female preference for it should increase in frequency. **Figure 23.16** describes one experiment testing this hypothesis in gray tree frogs (*Hyla versicolor*).

Other researchers have shown that in several bird species, the traits preferred by females are related to overall male health. Here, too, female preference appears to be based on traits that reflect "good genes," in this case alleles indicative of a robust immune system.

The Preservation of Genetic Variation

Some of the genetic variation in populations represents **neutral variation**, differences in DNA sequence that do not confer a selective advantage or disadvantage. But variation is also found at loci affected by selection. What prevents natural selection from reducing genetic variation at those loci by culling all unfavorable alleles? The tendency for directional and stabilizing selection to reduce variation is countered by mechanisms that preserve or restore it.

Diploidy

In diploid eukaryotes, a considerable amount of genetic variation is hidden from selection in the form of recessive alleles. Recessive alleles that are less favorable than their dominant counterparts, or even harmful in the current environment, can persist by propagation in heterozygous individuals. This latent variation is exposed to natural selection only when both parents carry the same recessive allele and two copies end up in the same zygote. This happens only rarely if the frequency of the recessive allele is very low. Heterozygote protection maintains a huge pool of alleles that might not be favored under present conditions, but which could bring new benefits if the environment changes.

Balancing Selection

Selection itself may preserve variation at some loci. **Balancing selection** occurs when natural selection maintains two or

▼ **Figure 23.16** INQUIRY

Do females select mates based on traits indicative of "good genes"?

EXPERIMENT Female gray tree frogs (*Hyla versicolor*) prefer to mate with males that give long mating calls. Allison Welch and colleagues, at the University of Missouri, tested whether the genetic makeup of long-calling (LC) males is superior to that of short-calling (SC) males. The researchers fertilized half the eggs of each female with sperm from an LC male and fertilized the remaining eggs with sperm from an SC male. The resulting half-sibling offspring were raised in a common environment, and several measures of their "performance" were tracked for two years.

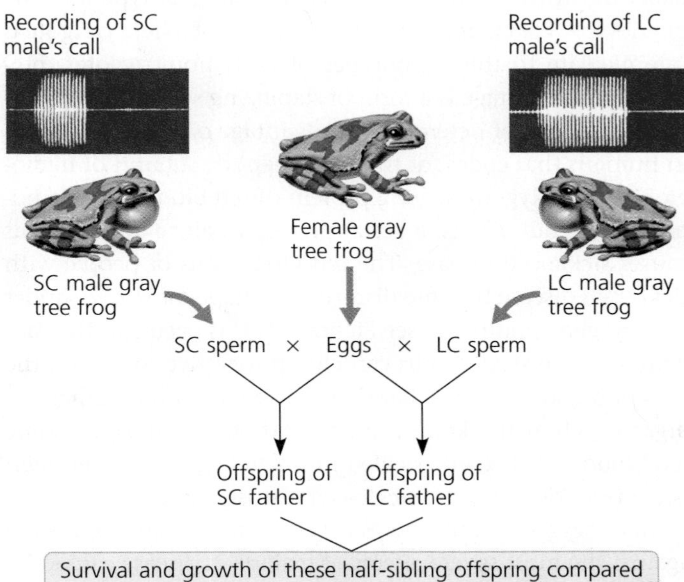

RESULTS

Offspring Performance	1995	1996
Larval survival	LC better	NSD
Larval growth	NSD	LC better
Time to metamorphosis	LC better (shorter)	LC better (shorter)

NSD = no significant difference; LC better = offspring of LC males superior to offspring of SC males.

CONCLUSION Because offspring fathered by an LC male outperformed their half-siblings fathered by an SC male, the team concluded that the duration of a male's mating call is indicative of the male's overall genetic quality. This result supports the hypothesis that female mate choice can be based on a trait that indicates whether the male has "good genes."

SOURCE A. M. Welch et al., Call duration as an indicator of genetic quality in male gray tree frogs, *Science* 280:1928–1930 (1998).

INQUIRY IN ACTION Read and analyze the original paper in *Inquiry in Action: Interpreting Scientific Papers*.

WHAT IF? Why did the researchers split each female frog's eggs into two batches for fertilization by different males? Why didn't they mate each female with a single male frog?

more forms in a population. This type of selection includes heterozygote advantage and frequency-dependent selection.

Heterozygote Advantage If individuals who are heterozygous at a particular locus have greater fitness than do both kinds of homozygotes, they exhibit **heterozygote advantage**. In such a case, natural selection tends to maintain two or more alleles at that locus. Note that heterozygote advantage is defined in terms of *genotype*, not phenotype. Thus, whether heterozygote advantage represents stabilizing or directional selection depends on the relationship between the genotype and the phenotype. For example, if the phenotype of a heterozygote is intermediate to the phenotypes of both homozygotes, heterozygote advantage is a form of stabilizing selection.

An example of heterozygote advantage occurs at the locus in humans that codes for the β polypeptide subunit of hemoglobin, the oxygen-carrying protein of red blood cells. In homozygous individuals, a certain recessive allele at that locus causes sickle-cell disease. The red blood cells of people with sickle-cell disease become distorted in shape, or *sickled*, under low-oxygen conditions (see Figure 5.21), as occurs in the capillaries. These sickled cells can clump together and block the flow of blood in the capillaries, resulting in serious damage to organs such as the kidney, heart, and brain. Although some red blood cells become sickled in heterozygotes, not enough become sickled to cause sickle-cell disease.

Heterozygotes for the sickle-cell allele are protected against the most severe effects of malaria, a disease caused by a parasite that infects red blood cells (see Figure 28.10). This partial protection occurs because the body destroys sickled red blood cells rapidly, killing the parasites they harbor (but not affecting parasites inside normal red blood cells). Protection against malaria is important in tropical regions where the disease is a major killer. In such regions, selection favors heterozygotes over homozygous dominant individuals, who are more vulnerable to the effects of malaria, and also over homozygous recessive individuals, who develop sickle-cell disease. The frequency of the sickle-cell allele in Africa is generally highest in areas where the malaria parasite is most common (**Figure 23.17**). In some populations, it accounts for 20% of the hemoglobin alleles in the gene pool, a very high frequency for such a harmful allele.

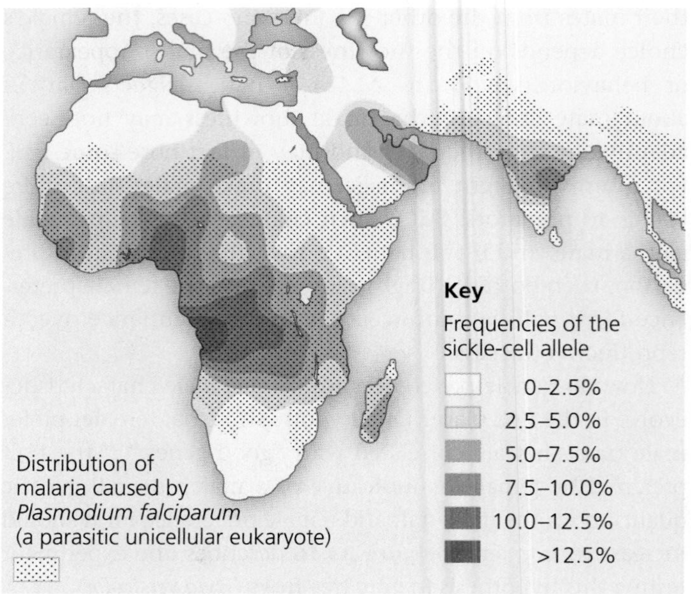

▲ **Figure 23.17 Mapping malaria and the sickle-cell allele.** The sickle-cell allele is most common in Africa, but it is not the only case of heterozygote advantage providing protection against malaria. Alleles at other loci (not shown on this map) are also favored by heterozygote advantage in populations near the Mediterranean Sea and in southeast Asia where malaria is widespread.

Frequency-Dependent Selection In **frequency-dependent selection**, the fitness of a phenotype depends on how common it is in the population. Consider the scale-eating fish (*Perissodus microlepis*) of Lake Tanganyika, in Africa. These fish attack other fish from behind, darting in to remove a few scales from the flank of their prey. Of interest here is a peculiar feature of the scale-eating fish: Some are "left-mouthed" and some are "right-mouthed." Simple Mendelian inheritance determines these phenotypes, with the right-mouthed allele being dominant to the left-mouthed allele. Because their mouth twists to the left, left-mouthed fish always attack their prey's right flank (**Figure 23.18**). (To see why, twist your lower jaw and lips to the left and imagine trying to take a bite from the left side of a fish, approaching it from behind.) Similarly, right-mouthed fish always attack from the left. Prey species guard against attack from whatever phenotype of scale-eating fish is most common in the lake. Thus, from year to year, selection favors whichever mouth phenotype is least common. As a result, the frequency of left- and right-mouthed fish oscillates over time, and balancing selection (due to frequency dependence) keeps the frequency of each phenotype close to 50%.

Why Natural Selection Cannot Fashion Perfect Organisms

Though natural selection leads to adaptation, nature abounds with examples of organisms that are less than ideally "engineered" for their lifestyles. There are several reasons why.

1. **Selection can act only on existing variations.** Natural selection favors only the fittest phenotypes among those currently in the population, which may not be the ideal traits. New advantageous alleles do not arise on demand.
2. **Evolution is limited by historical constraints.** Each species has a legacy of descent with modification from ancestral forms. Evolution does not scrap the

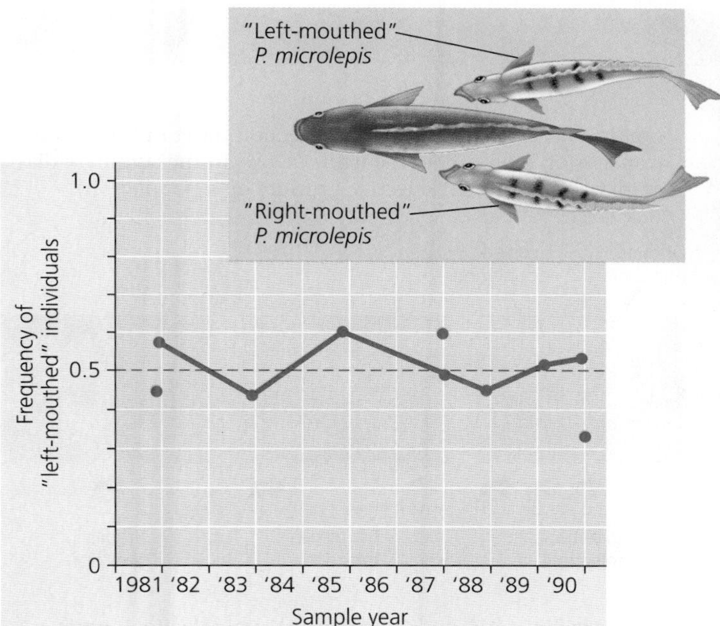

▲ **Figure 23.18 Frequency-dependent selection in scale-eating fish (*Perissodus microlepis*).** Michio Hori, of Kyoto University, Japan, noted that the frequency of left-mouthed individuals rises and falls in a regular manner. At each of three time periods when the phenotypes of breeding adults were assessed, adults that reproduced (represented by green dots) had the opposite phenotype of that which was most common in the population. Thus, it appeared that right-mouthed individuals were favored by selection when left-mouthed individuals were more common, and vice versa.

? *What did the researchers measure to determine which phenotype was favored by selection? Are any assumptions implied by this choice? Explain.*

ancestral anatomy and build each new complex structure from scratch; rather, evolution co-opts existing structures and adapts them to new situations. We could imagine that if a terrestrial animal were to adapt to an environment in which flight would be advantageous, it might be best just to grow an extra pair of limbs that would serve as wings. However, evolution does not work this way; instead, it operates on the traits an organism already has. Thus, in birds and bats, an existing pair of limbs took on new functions for flight as these organisms evolved from nonflying ancestors.

3. **Adaptations are often compromises.** Each organism must do many different things. A seal spends part of its time on rocks; it could probably walk better if it had legs instead of flippers, but then it would not swim nearly as well. We humans owe much of our versatility and athleticism to our prehensile hands and flexible limbs, but these also make us prone to sprains, torn ligaments, and dislocations: Structural reinforcement has been compromised for agility. **Figure 23.19** depicts another example of evolutionary compromise.

▲ **Figure 23.19 Evolutionary compromise.** The loud call that enables a Túngara frog to attract mates also attracts more dangerous characters in the neighborhood—in this case, a bat about to seize a meal.

4. **Chance, natural selection, and the environment interact.** Chance events can affect the subsequent evolutionary history of populations. For instance, when a storm blows insects or birds hundreds of kilometers over an ocean to an island, the wind does not necessarily transport those individuals that are best suited to the new environment. Thus, not all alleles present in the founding population's gene pool are better suited to the new environment than the alleles that are "left behind." In addition, the environment at a particular location may change unpredictably from year to year, again limiting the extent to which adaptive evolution results in a close match between the organism and current environmental conditions.

With these four constraints, evolution does not tend to craft perfect organisms. Natural selection operates on a "better than" basis. We can, in fact, see evidence for evolution in the many imperfections of the organisms it produces.

CONCEPT CHECK 23.4

1. What is the relative fitness of a sterile mule? Explain.
2. Explain why natural selection is the only evolutionary mechanism that consistently leads to adaptive evolution.
3. **WHAT IF?** Consider a population in which heterozygotes at a certain locus have an extreme phenotype (such as being larger than homozygotes) that confers a selective advantage. Does such a situation represent directional, disruptive, or stabilizing selection? Explain your answer.
4. **WHAT IF?** Would individuals who are heterozygous for the sickle-cell allele be selected for or against in a region free from malaria? Explain.

For suggested answers, see Appendix A.

23 CHAPTER REVIEW

SUMMARY OF KEY CONCEPTS

CONCEPT 23.1

Genetic variation makes evolution possible (pp. 469–473)

- **Genetic variation** refers to genetic differences among individuals within a population.
- The nucleotide differences that provide the basis of genetic variation arise by mutation and other processes that produce new alleles and new genes.
- New genetic variants are produced rapidly in organisms with short generation times. In sexually reproducing organisms, most of the genetic differences among individuals result from crossing over, the independent assortment of chromosomes, and fertilization.

? *Why do biologists estimate gene variability and nucleotide variability, and what do these estimates represent?*

CONCEPT 23.2

The Hardy-Weinberg equation can be used to test whether a population is evolving (pp. 473–476)

- A **population**, a localized group of organisms belonging to one species, is united by its **gene pool**, the aggregate of all the alleles in the population.
- The **Hardy-Weinberg principle** states that the allele and genotype frequencies of a population will remain constant if the population is large, mating is random, mutation is negligible, there is no gene flow, and there is no natural selection. For such a population, if p and q represent the frequencies of the only two possible alleles at a particular locus, then p^2 is the frequency of one kind of homozygote, q^2 is the frequency of the other kind of homozygote, and $2pq$ is the frequency of the heterozygous genotype.

? *Is it circular reasoning to calculate* p *and* q *from observed genotype frequencies and then use those values of* p *and* q *to test if the population is in Hardy-Weinberg equilibrium? Explain your answer. (Hint: Consider a specific case, such as a population with 195 individuals of genotype* AA*, 10 of genotype* Aa*, and 195 of genotype* aa*.)*

CONCEPT 23.3

Natural selection, genetic drift, and gene flow can alter allele frequencies in a population (pp. 476–480)

- In natural selection, individuals that have certain inherited traits tend to survive and reproduce at higher rates than other individuals *because of* those traits.
- In **genetic drift**, chance fluctuations in allele frequencies over generations tend to reduce genetic variation.
- **Gene flow**, the transfer of alleles between populations, tends to reduce genetic differences between populations over time.

? *Would two small, geographically isolated populations in very different environments be likely to evolve in similar ways? Explain.*

CONCEPT 23.4

Natural selection is the only mechanism that consistently causes adaptive evolution (pp. 480–485)

- One organism has greater **relative fitness** than a second organism if it leaves more fertile descendants than the second organism. The modes of natural selection differ in how selection acts on phenotype (the white arrows in the summary diagram below represent selective pressure on a population).

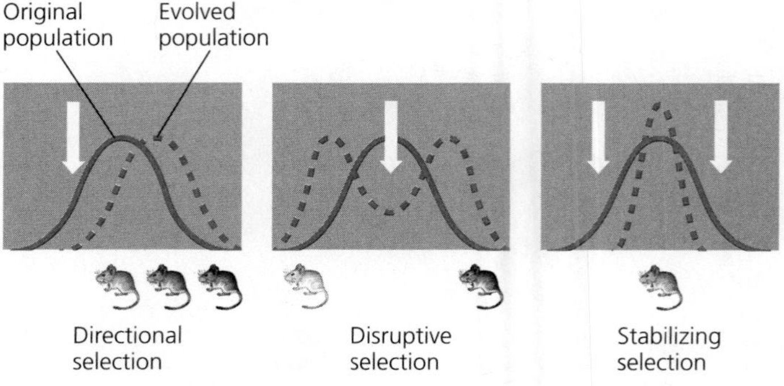

- Unlike genetic drift and gene flow, natural selection consistently increases the frequencies of alleles that enhance survival and reproduction, thus improving the match between organisms and their environment.
- **Sexual selection** influences evolutionary change in secondary sex characteristics that can give individuals advantages in mating.
- Despite the winnowing effects of selection, populations have considerable genetic variation. Some of this variation represents **neutral variation**; additional variation can be maintained by diploidy and balancing selection.
- There are constraints to evolution: Natural selection can act only on available variation; structures result from modified ancestral anatomy; adaptations are often compromises; and chance, natural selection, and the environment interact.

? *How might secondary sex characteristics differ between males and females in a species in which females compete for mates?*

TEST YOUR UNDERSTANDING

LEVEL 1: KNOWLEDGE/COMPREHENSION

1. Natural selection changes allele frequencies because some _______ survive and reproduce more successfully than others.
   a. alleles
   b. loci
   c. gene pools
   d. species
   e. individuals

2. No two people are genetically identical, except for identical twins. The main source of genetic variation among human individuals is
   a. new mutations that occurred in the preceding generation.
   b. genetic drift due to the small size of the population.
   c. the reshuffling of alleles in sexual reproduction.
   d. geographic variation within the population.
   e. environmental effects.

3. Sparrows with average-sized wings survive severe storms better than those with longer or shorter wings, illustrating
   a. the bottleneck effect.
   b. disruptive selection.
   c. frequency-dependent selection.
   d. neutral variation.
   e. stabilizing selection.

LEVEL 2: APPLICATION/ANALYSIS

4. If the nucleotide variability of a locus equals 0%, what is the gene variability and number of alleles at that locus?
   a. gene variability = 0%; number of alleles = 0
   b. gene variability = 0%; number of alleles = 1
   c. gene variability = 0%; number of alleles = 2
   d. gene variability > 0%; number of alleles = 2
   e. Without more information, gene variability and number of alleles cannot be determined.

5. There are 40 individuals in population 1, all with genotype *A1A1*, and there are 25 individuals in population 2, all with genotype *A2A2*. Assume that these populations are located far from each other and that their environmental conditions are very similar. Based on the information given here, the observed genetic variation is most likely an example of
   a. genetic drift.
   b. gene flow.
   c. disruptive selection.
   d. discrete variation.
   e. directional selection.

6. A fruit fly population has a gene with two alleles, *A1* and *A2*. Tests show that 70% of the gametes produced in the population contain the *A1* allele. If the population is in Hardy-Weinberg equilibrium, what proportion of the flies carry both *A1* and *A2*?
   a. 0.7 b. 0.49 c. 0.21 d. 0.42 e. 0.09

LEVEL 3: SYNTHESIS/EVALUATION

7. **EVOLUTION CONNECTION**
   How is the process of evolution revealed by the imperfections of living organisms?

8. **SCIENTIFIC INQUIRY**
   DRAW IT Richard Koehn, of the State University of New York, Stony Brook, and Thomas Hilbish, of the University of South Carolina, studied genetic variation in the marine mussel *Mytilus edulis* around Long Island, New York. They measured the frequency of a particular allele (lap^{94}) for an enzyme involved in regulating the mussel's internal saltwater balance. The researchers presented their data as a series of pie charts linked to sampling sites within Long Island Sound, where the salinity is highly variable, and along the coast of the open ocean, where salinity is constant:

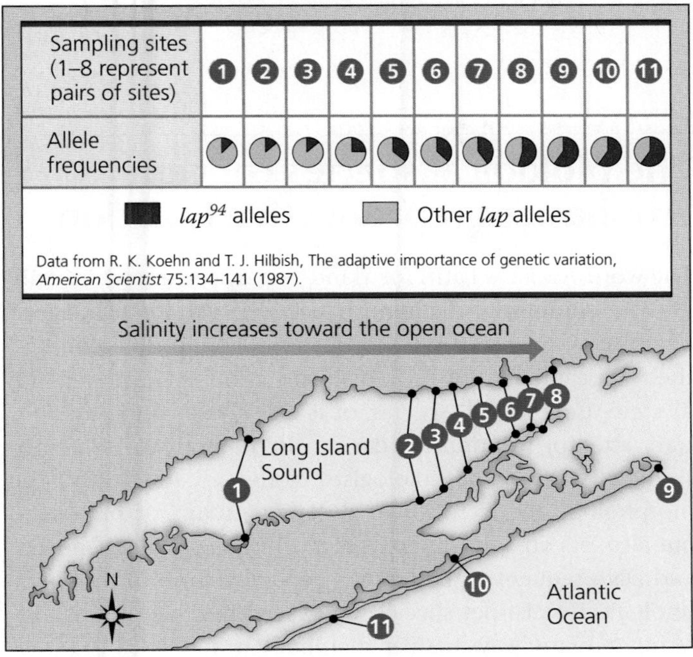

(Question 8, continued)
Create a data table for the 11 sampling sites by estimating the frequency of lap^{94} from the pie charts. (*Hint*: Think of each pie chart as a clock face to help you estimate the proportion of the shaded area.) Then graph the frequencies for sites 1–8 to show how the frequency of this allele changes with increasing salinity in Long Island Sound (from southwest to northeast). How do the data from sites 9–11 compare with the data from the sites within the Sound?

Construct a hypothesis that explains the patterns you observe in the data and that accounts for the following observations: (1) the lap^{94} allele helps mussels maintain osmotic balance in water with a high salt concentration but is costly to use in less salty water; and (2) mussels produce larvae that can disperse long distances before they settle on rocks and grow into adults.

9. **WRITE ABOUT A THEME**
   Emergent Properties Heterozygotes at the sickle-cell locus produce both normal and abnormal (sickle-cell) hemoglobin (see Concept 14.4). When hemoglobin molecules are packed into a heterozygote's red blood cells, some cells receive relatively large quantities of abnormal hemoglobin, making these cells prone to sickling. In a short essay (approximately 100–150 words), explain how these molecular and cellular events lead to emergent properties at the individual and population levels of biological organization.

For selected answers, see Appendix A.

24

The Origin of Species

▲ **Figure 24.1 How did this flightless bird come to live on the isolated Galápagos Islands?**

EVOLUTION

KEY CONCEPTS

24.1 The biological species concept emphasizes reproductive isolation

24.2 Speciation can take place with or without geographic separation

24.3 Hybrid zones reveal factors that cause reproductive isolation

24.4 Speciation can occur rapidly or slowly and can result from changes in few or many genes

OVERVIEW

That "Mystery of Mysteries"

Darwin came to the Galápagos Islands eager to explore landforms newly emerged from the sea. He noted that these volcanic islands, despite their geologic youth, were teeming with plants and animals found nowhere else in the world (**Figure 24.1**). Later he realized that these species, like the islands, were relatively new. He wrote in his diary: "Both in space and time, we seem to be brought somewhat near to that great fact—that mystery of mysteries—the first appearance of new beings on this Earth."

The "mystery of mysteries" that captivated Darwin is **speciation**, the process by which one species splits into two or more species. Speciation fascinated Darwin (and many biologists since) because it is responsible for the tremendous diversity of life, repeatedly yielding new species that differ from existing ones. Speciation explains not only differences between species, but also similarities between them (the unity of life). When one species splits into two, the species that result share many characteristics because they are descended from this common ancestral species. For example, DNA similarities indicate that the flightless cormorant (*Phalacrocorax harrisi*) in Figure 24.1 is closely related to flying cormorants found in the Americas. This suggests that the flightless cormorant may have originated from an ancestral cormorant species that migrated from the mainland to the Galápagos.

Speciation also forms a conceptual bridge between **microevolution**, changes over time in allele frequencies in a population, and **macroevolution**, the broad pattern of evolution above the species level. An example of macroevolutionary change is the origin of new groups of organisms, such as mammals or flowering plants, through a series of speciation events. We examined microevolutionary mechanisms (mutation, natural selection, genetic drift, and gene flow) in Chapter 23, and we'll turn to macroevolution in Chapter 25. In this chapter, we will explore the "bridge"—the mechanisms by which new species originate from existing ones. First, however, we need to establish what we actually mean when we talk about "species."

CONCEPT 24.1

The biological species concept emphasizes reproductive isolation

The word *species* is Latin for "kind" or "appearance." In daily life, we commonly distinguish between various "kinds" of organisms—dogs and cats, for instance—from differences in their appearance. But are organisms truly divided into the discrete units we call species, or is this classification an arbitrary attempt to impose order on the natural world? To answer this question, biologists compare not only the morphology (body form) of different groups of organisms but also less obvious differences in physiology, biochemistry, and DNA sequences. The results generally confirm that morphologically distinct species are indeed discrete groups, differing in many ways besides their body forms.

The Biological Species Concept

The primary definition of species used in this textbook is the **biological species concept**. According to this concept, a **species** is a group of populations whose members have the potential to interbreed in nature and produce viable, fertile offspring—but do not produce viable, fertile offspring with members of other such groups (**Figure 24.2**). Thus, the members of a biological species are united by being reproductively compatible, at least potentially. All human beings, for example, belong to the same species. A businesswoman in Manhattan may be unlikely to meet a dairy farmer in Mongolia, but if the two should happen to meet and mate, they could have viable babies that develop into fertile adults. In contrast, humans and chimpanzees remain distinct biological species even where they share territory, because many factors keep them from interbreeding and producing fertile offspring.

(a) Similarity between different species. The eastern meadowlark (*Sturnella magna*, left) and the western meadowlark (*Sturnella neglecta*, right) have similar body shapes and colorations. Nevertheless, they are distinct biological species because their songs and other behaviors are different enough to prevent interbreeding should they meet in the wild.

(b) Diversity within a species. As diverse as we may be in appearance, all humans belong to a single biological species (*Homo sapiens*), defined by our capacity to interbreed successfully.

▲ **Figure 24.2 The biological species concept is based on the potential to interbreed rather than on physical similarity.**

What holds the gene pool of a species together, causing its members to resemble each other more than they resemble other species? To answer this question, we need to return to the evolutionary mechanism called *gene flow*, the transfer of alleles between populations (see Chapter 23). Typically, gene flow occurs between the different populations of a species. This ongoing exchange of alleles tends to hold the populations together genetically. As we'll explore in the following sections, the absence of gene flow plays a key role in the formation of new species, as well as in keeping them apart once their potential to interbreed has been reduced.

Reproductive Isolation

Because biological species are defined in terms of reproductive compatibility, the formation of a new species hinges on **reproductive isolation**—the existence of biological factors (barriers) that impede members of two species from interbreeding and producing viable, fertile offspring. Such barriers block gene flow between the species and limit the formation of **hybrids**, offspring that result from an interspecific mating. Although a single barrier may not prevent all gene flow, a combination of several barriers can effectively isolate a species' gene pool.

Clearly, a fly cannot mate with a frog or a fern, but the reproductive barriers between more closely related species are not so obvious. These barriers can be classified according to whether they contribute to reproductive isolation before or after fertilization. **Prezygotic barriers** ("before the zygote") block fertilization from occurring. Such barriers typically act in one of three ways: by impeding members of different species from attempting to mate, by preventing an attempted mating from being completed successfully, or by hindering fertilization if mating is completed successfully. If a sperm cell from one species overcomes prezygotic barriers and fertilizes an ovum from another species, a variety of **postzygotic barriers** ("after the zygote") may contribute to reproductive isolation after the hybrid zygote is formed. For example, developmental errors may reduce survival among hybrid embryos. Or problems after birth may cause hybrids to be infertile or may decrease their chance of surviving long enough to reproduce. **Figure 24.3**, on the next two pages, describes prezygotic and postzygotic barriers in more detail.

▼ **Figure 24.3**

Exploring Reproductive Barriers

Prezygotic barriers impede mating or hinder fertilization if mating does occur

Individuals of different species

MATING ATTEMPT

Habitat Isolation

Two species that occupy different habitats within the same area may encounter each other rarely, if at all, even though they are not isolated by obvious physical barriers, such as mountain ranges.

Example: Two species of garter snakes in the genus *Thamnophis* occur in the same geographic areas, but one lives mainly in water (a) while the other is primarily terrestrial (b).

(a)

(b)

Temporal Isolation

Species that breed during different times of the day, different seasons, or different years cannot mix their gametes.

Example: In North America, the geographic ranges of the eastern spotted skunk (*Spilogale putorius*) (c) and the western spotted skunk (*Spilogale gracilis*) (d) overlap, but *S. putorius* mates in late winter and *S. gracilis* mates in late summer.

(c)

(d)

Behavioral Isolation

Courtship rituals that attract mates and other behaviors unique to a species are effective reproductive barriers, even between closely related species. Such behavioral rituals enable *mate recognition*—a way to identify potential mates of the same species.

Example: Blue-footed boobies, inhabitants of the Galápagos, mate only after a courtship display unique to their species. Part of the "script" calls for the male to high-step (e), a behavior that calls the female's attention to his bright blue feet.

(e)

Mechanical Isolation

Mating is attempted, but morphological differences prevent its successful completion.

Example: The shells of two species of snails in the genus *Bradybaena* spiral in different directions: Moving inward to the center, one spirals in a counterclockwise direction (f, left), the other in a clockwise direction (f, right). As a result, the snails' genital openings (indicated by arrows) are not aligned, and mating cannot be completed.

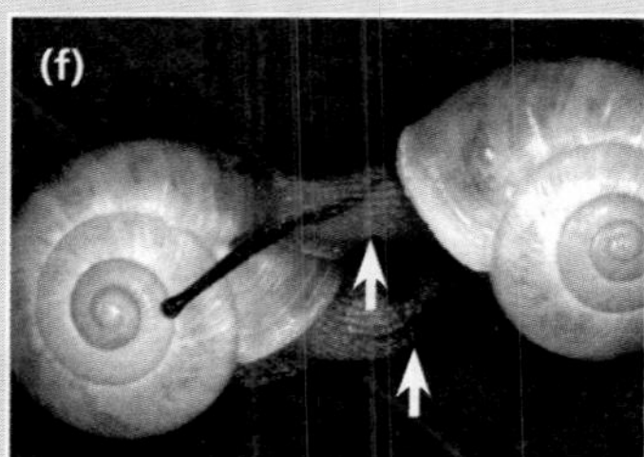
(f)

Postzygotic barriers prevent a hybrid zygote from developing into a viable, fertile adult

Gametic Isolation

Reduced Hybrid Viability

Reduced Hybrid Fertility

Hybrid Breakdown

FERTILIZATION

VIABLE, FERTILE OFFSPRING

Sperm of one species may not be able to fertilize the eggs of another species. For instance, sperm may not be able to survive in the reproductive tract of females of the other species, or biochemical mechanisms may prevent the sperm from penetrating the membrane surrounding the other species' eggs.

Example: Gametic isolation separates certain closely related species of aquatic animals, such as sea urchins (g). Sea urchins release their sperm and eggs into the surrounding water, where they fuse and form zygotes. It is difficult for gametes of different species, such as the red and purple urchins shown here, to fuse because proteins on the surfaces of the eggs and sperm bind very poorly to each other.

(g)

The genes of different parent species may interact in ways that impair the hybrid's development or survival in its environment.

Example: Some salamander subspecies of the genus *Ensatina* live in the same regions and habitats, where they may occasionally hybridize. But most of the hybrids do not complete development, and those that do are frail (h).

(h)

Even if hybrids are vigorous, they may be sterile. If the chromosomes of the two parent species differ in number or structure, meiosis in the hybrids may fail to produce normal gametes. Since the infertile hybrids cannot produce offspring when they mate with either parent species, genes cannot flow freely between the species.

Example: The hybrid offspring of a male donkey (i) and a female horse (j) is a mule (k), which is robust but sterile. A "hinny" (not shown), the offspring of a female donkey and a male horse, is also sterile.

(i)

(j)

(k)

Some first-generation hybrids are viable and fertile, but when they mate with one another or with either parent species, offspring of the next generation are feeble or sterile.

Example: Strains of cultivated rice have accumulated different mutant recessive alleles at two loci in the course of their divergence from a common ancestor. Hybrids between them are vigorous and fertile (l, left and right), but plants in the next generation that carry too many of these recessive alleles are small and sterile (l, center). Although these rice strains are not yet considered different species, they have begun to be separated by postzygotic barriers.

(l)

Limitations of the Biological Species Concept

One strength of the biological species concept is that it directs our attention to how speciation occurs: by the evolution of reproductive isolation. However, the number of species to which this concept can be usefully applied is limited. There is, for example, no way to evaluate the reproductive isolation of fossils. The biological species concept also does not apply to organisms that reproduce asexually all or most of the time, such as prokaryotes. (Many prokaryotes do transfer genes among themselves, as we will discuss in Chapter 27, but this is not part of their reproductive process.) Furthermore, in the biological species concept, species are designated by the *absence* of gene flow. But there are many pairs of species that are morphologically and ecologically distinct, and yet gene flow occurs between them. An example is the grizzly bear (*Ursus arctos*) and polar bear (*Ursus maritimus*), whose hybrid offspring have been dubbed "grolar bears" **(Figure 24.4)**. As we'll discuss, natural selection can cause such species to remain distinct even though some gene flow occurs between them. This observation has led some researchers to argue that the biological species concept overemphasizes gene flow and downplays the role of natural selection. Because of the limitations to the biological species concept, alternative species concepts are useful in certain situations.

◀ Grizzly bear (*U. arctos*)

▼ Polar bear (*U. maritimus*)

▲ Hybrid "grolar bear"

▲ **Figure 24.4 Hybridization between two species of bears in the genus *Ursus*.**

Other Definitions of Species

While the biological species concept emphasizes the *separateness* of species from one another due to reproductive barriers, several other definitions emphasize the *unity within* a species. For example, the **morphological species concept** characterizes a species by body shape and other structural features. The morphological species concept can be applied to asexual and sexual organisms, and it can be useful even without information on the extent of gene flow. In practice, this is how scientists distinguish most species. One disadvantage, however, is that this definition relies on subjective criteria; researchers may disagree on which structural features distinguish a species.

The **ecological species concept** views a species in terms of its ecological niche, the sum of how members of the species interact with the nonliving and living parts of their environment (see Chapter 54). For example, two species of salamanders might be similar in appearance but differ in the foods they eat or in their ability to tolerate dry conditions. Unlike the biological species concept, the ecological species concept can accommodate asexual as well as sexual species. It also emphasizes the role of disruptive natural selection as organisms adapt to different environmental conditions.

The **phylogenetic species concept** defines a species as the smallest group of individuals that share a common ancestor, forming one branch on the tree of life. Biologists trace the phylogenetic history of a species by comparing its characteristics, such as morphology or molecular sequences, with those of other organisms. Such analyses can distinguish groups of individuals that are sufficiently different to be considered separate species. Of course, the difficulty with this species concept is determining the degree of difference required to indicate separate species.

In addition to those discussed here, more than 20 other species definitions have been proposed. The usefulness of each definition depends on the situation and the research questions being asked. For our purposes of studying how species originate, the biological species concept, with its focus on reproductive barriers, is particularly helpful.

CONCEPT CHECK 24.1

1. (a) Which species concept(s) could you apply to both asexual and sexual species? (b) Which would be most useful for identifying species in the field? Explain.
2. **WHAT IF?** Suppose you are studying two bird species that live in a forest and are not known to interbreed. One species feeds and mates in the treetops and the other on the ground. But in captivity, the birds can interbreed and produce viable, fertile offspring. What type of reproductive barrier most likely keeps these species separate in nature? Explain.

For suggested answers, see Appendix A.

CONCEPT 24.2

Speciation can take place with or without geographic separation

Now that we have a clearer sense of what constitutes a unique species, let's return to our discussion of the process by which such species arise from existing species. Speciation can occur in two main ways, depending on how gene flow is interrupted between populations of the existing species (**Figure 24.5**).

Allopatric ("Other Country") Speciation

In **allopatric speciation** (from the Greek *allos*, other, and *patra*, homeland), gene flow is interrupted when a population is divided into geographically isolated subpopulations. For example, the water level in a lake may subside, resulting in two or more smaller lakes that are now home to separated populations (see Figure 24.5a). Or a river may change course and divide a population of animals that cannot cross it. Allopatric speciation can also occur without geologic remodeling, such as when individuals colonize a remote area and their descendants become geographically isolated from the parent population. The flightless cormorant shown in Figure 24.1 most likely originated in this way from an ancestral flying species that reached the Galápagos Islands.

The Process of Allopatric Speciation

How formidable must a geographic barrier be to promote allopatric speciation? The answer depends on the ability of the organisms to move about. Birds, mountain lions, and coyotes can cross rivers and canyons—as can the windblown pollen of pine trees and the seeds of many flowering plants. In contrast, small rodents may find a wide river or deep canyon a formidable barrier (**Figure 24.6**).

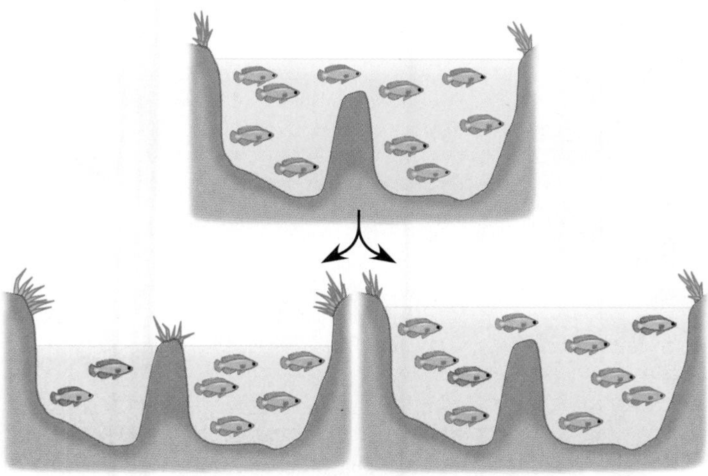

(a) **Allopatric speciation.** A population forms a new species while geographically isolated from its parent population.

(b) **Sympatric speciation.** A subset of a population forms a new species without geographic separation.

▲ **Figure 24.5 Two main modes of speciation.**

▲ **Figure 24.6 Allopatric speciation of antelope squirrels on opposite rims of the Grand Canyon.** Harris's antelope squirrel (*Ammospermophilus harrisii*) inhabits the canyon's south rim (left). Just a few kilometers away on the north rim (right) lives the closely related white-tailed antelope squirrel (*Ammospermophilus leucurus*). Birds and other organisms that can disperse easily across the canyon have not diverged into different species on the two rims.

Once geographic separation has occurred, the separated gene pools may diverge. Different mutations arise, and natural selection and genetic drift may alter allele frequencies in different ways in the separated populations. Reproductive isolation may then arise as a by-product of selection or drift having caused the populations to diverge genetically. For example, on Andros Island, in the Bahamas, populations of the mosquitofish *Gambusia hubbsi* colonized a series of ponds that later became isolated from one another. Genetic analyses indicate that little or no gene flow currently occurs between the ponds. The environments of these ponds are very similar except that some contain many predatory fishes, while others do not. In the "high-predation" ponds, selection has favored the evolution of a mosquitofish body shape that enables rapid bursts of speed (**Figure 24.7**). In ponds lacking

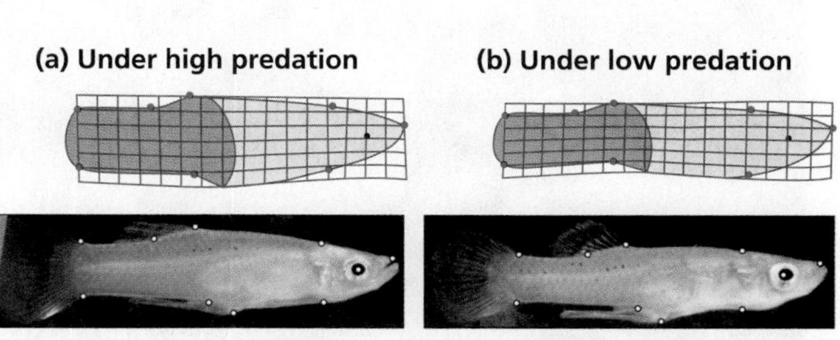

In ponds with predatory fishes, the head region of the mosquitofish is streamlined and the tail region is powerful, enabling rapid bursts of speed.

In ponds without predatory fishes, mosquitofish have a different body shape that favors long, steady swimming.

▲ **Figure 24.7 Reproductive isolation as a by-product of selection.** After bringing together mosquitofish from different ponds, researchers concluded that selection for traits that enable mosquitofish in high-predation ponds to avoid predators has isolated them reproductively from mosquitofish in low-predation ponds.

predatory fishes, selection has favored a different body shape, one that improves the ability to swim for long periods of time. How have these different selective pressures affected the evolution of reproductive barriers? Researchers answered this question by bringing together mosquitofish from the two types of ponds. They found that female mosquitofish prefer to mate with males whose body shape is similar to their own. This preference establishes a barrier to reproduction between mosquitofish from ponds with predators and those from ponds without predators. Thus, as a by-product of selection for avoiding predators, reproductive barriers have started to form in these allopatric populations.

Evidence of Allopatric Speciation

Many studies provide evidence that speciation can occur in allopatric populations. Consider the 30 species of snapping shrimp in the genus *Alpheus* that live off the Isthmus of Panama, the land bridge that connects South and North America (**Figure 24.8**). Fifteen of these species live on the Atlantic side of the isthmus, while the other 15 live on the Pacific side. Before the isthmus formed, gene flow could occur between the Atlantic and Pacific populations of snapping shrimp. Did the species on different sides of the isthmus originate by allopatric speciation? Morphological and genetic data group these shrimp into 15 pairs of *sibling species*, pairs whose member species are each other's closest relative. In each of these 15 pairs, one of the sibling species lives on the Atlantic side of the isthmus, while the other lives on the Pacific side, strongly suggesting that the two species arose as a consequence of geographic separation. Furthermore, genetic analyses indicate that the *Alpheus* species originated from 9 to 3 million years ago, with the sibling species that live in the deepest water diverging first. These divergence times are consistent with geologic evidence that the isthmus formed gradually, starting 10 million years ago, and closing completely about 3 million years ago.

The importance of allopatric speciation is also suggested by the fact that regions that are isolated or highly subdivided by barriers typically have more species than do otherwise similar regions that lack such features. For example, many unique plants and animals are found on the geographically isolated Hawaiian Islands (we'll return to the origin of Hawaiian species in Chapter 25). Similarly, unusually high numbers of butterfly species are found in South American regions that are subdivided by many rivers.

Laboratory and field tests also provide evidence that reproductive isolation between two populations generally increases as the distance between them increases. In a study of dusky salamanders (*Desmognathus ochrophaeus*), biologists brought individuals from different populations into the laboratory and tested their ability to produce viable, fertile offspring (**Figure 24.9**). The researchers observed little reproductive isolation in salamanders from neighboring populations. In contrast, salamanders from widely separated populations often failed to reproduce, a result consistent with allopatric speciation. In other studies, researchers have tested whether intrinsic reproductive barriers develop when populations are isolated

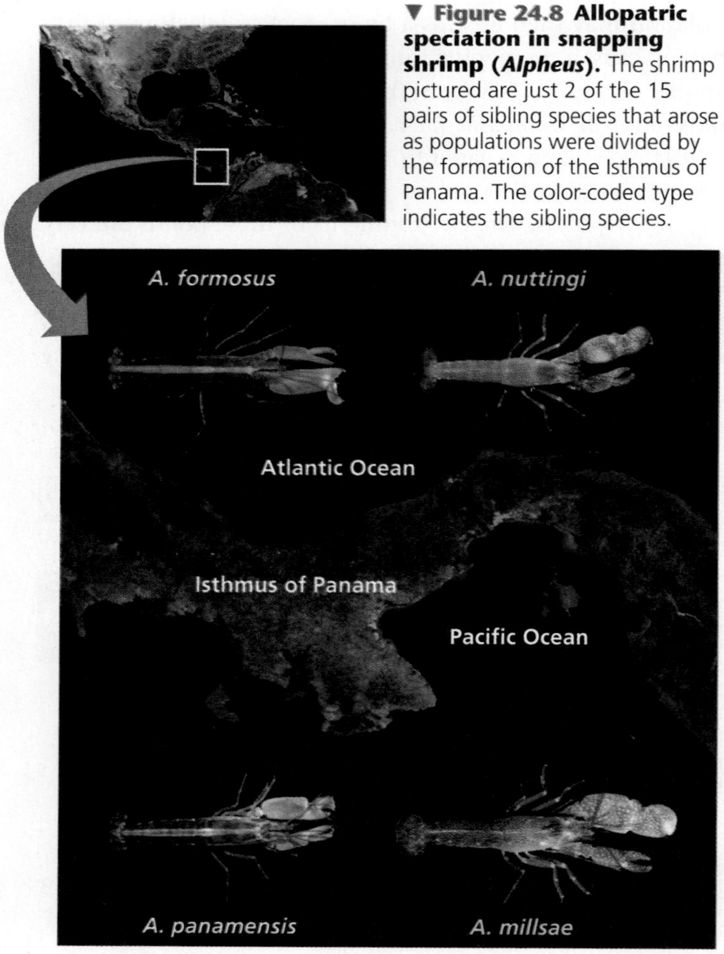

▼ **Figure 24.8 Allopatric speciation in snapping shrimp (*Alpheus*).** The shrimp pictured are just 2 of the 15 pairs of sibling species that arose as populations were divided by the formation of the Isthmus of Panama. The color-coded type indicates the sibling species.

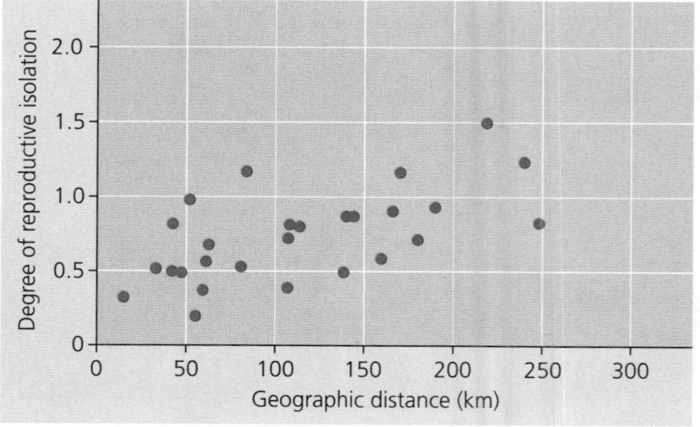

▲ **Figure 24.9 Reproductive isolation increases with distance in populations of dusky salamanders.** The degree of reproductive isolation is represented here by an index ranging from 0 (no isolation) to 2 (complete isolation).

▼ **Figure 24.10** INQUIRY

Can divergence of allopatric populations lead to reproductive isolation?

EXPERIMENT Diane Dodd, then at Yale University, divided a laboratory population of the fruit fly *Drosophila pseudoobscura*, raising some flies on a starch medium and others on a maltose medium. After one year (about 40 generations), natural selection resulted in divergent evolution: Populations raised on starch digested starch more efficiently, while those raised on maltose digested maltose more efficiently. Dodd then put flies from the same or different populations in mating cages and measured mating frequencies. All flies used in the mating preference tests were reared for one generation on a standard cornmeal medium.

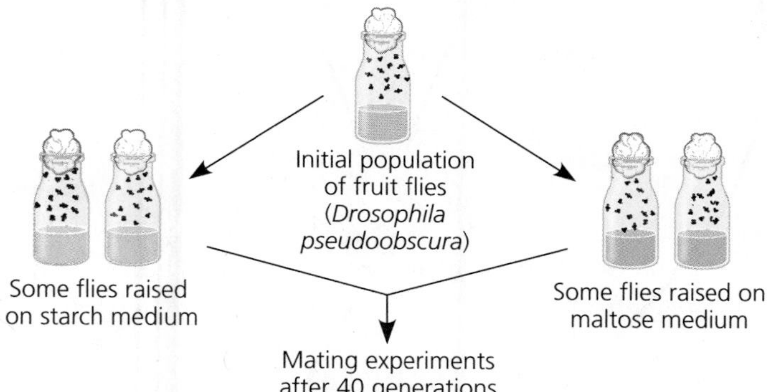

RESULTS Mating patterns among populations of flies raised on different media are shown below. When flies from "starch populations" were mixed with flies from "maltose populations," the flies tended to mate with like partners. But in the control group (shown on the right), flies from different populations adapted to starch were about as likely to mate with each other as with flies from their own population; similar results were obtained for control groups adapted to maltose.

Male \ Female	Starch	Maltose
Starch	22	9
Maltose	8	20

Number of matings in experimental group

Male \ Female	Starch population 1	Starch population 2
Starch population 1	18	15
Starch population 2	12	15

Number of matings in control group

CONCLUSION In the experimental group, the strong preference of "starch flies" and "maltose flies" to mate with like-adapted flies indicates that a reproductive barrier was forming between these fly populations. Although this reproductive barrier was not absolute (some mating between starch flies and maltose flies did occur), after 40 generations it appeared to be under way. This barrier may have been caused by differences in courtship behavior that arose as an incidental by-product of differing selective pressures as these allopatric populations adapted to different sources of food.

SOURCE D. M. B. Dodd, Reproductive isolation as a consequence of adaptive divergence in *Drosophila pseudoobscura*, *Evolution* 43:1308–1311 (1989).

WHAT IF? Why were all flies used in the mating preference tests reared on a standard medium (rather than on a starch or maltose medium)?

experimentally and subjected to different environmental conditions. In such cases, too, the results provide strong support for allopatric speciation (**Figure 24.10**).

We need to emphasize here that although geographic isolation prevents interbreeding between allopatric populations, physical separation is not a biological barrier to reproduction. Biological reproductive barriers such as those described in Figure 24.3 are intrinsic to the organisms themselves. Hence, it is biological barriers that can prevent interbreeding when members of different populations come into contact with one another.

Sympatric ("Same Country") Speciation

In **sympatric speciation** (from the Greek *syn*, together), speciation occurs in populations that live in the same geographic area. How can reproductive barriers form between sympatric populations while their members remain in contact with each other? Although such contact (and the ongoing gene flow that results) makes sympatric speciation less common than allopatric speciation, sympatric speciation can occur if gene flow is reduced by such factors as polyploidy, habitat differentiation, and sexual selection. (Note that these factors can also promote allopatric speciation.)

Polyploidy

A species may originate from an accident during cell division that results in extra sets of chromosomes, a condition called **polyploidy**. Polyploid speciation occasionally occurs in animals; for example, the gray tree frog *Hyla versicolor* (see Figure 23.16) is thought to have originated in this way. However, polyploidy is far more common in plants. Botanists estimate that more than 80% of the plant species alive today are descended from ancestors that formed by polyploid speciation.

Two distinct forms of polyploidy have been observed in plant (and a few animal) populations. An **autopolyploid** (from the Greek *autos*, self) is an individual that has more than two chromosome sets that are all derived from a single species. In plants, for example, a failure of cell division could double a cell's chromosome number from the diploid number ($2n$) to a tetraploid number ($4n$).

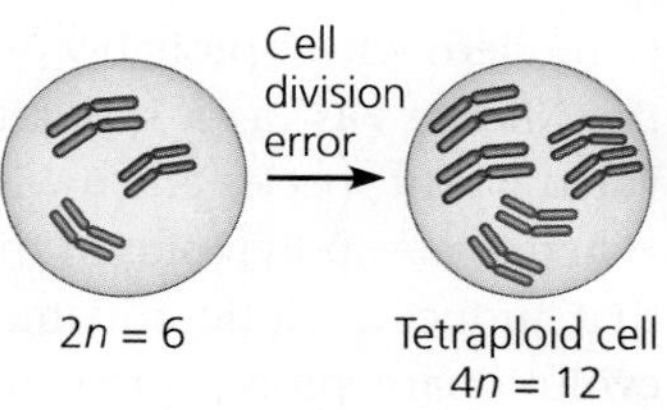

A tetraploid can produce fertile tetraploid offspring by self-pollinating or by mating with other tetraploids. In addition, the tetraploids are reproductively isolated from diploid plants of the original population, because the

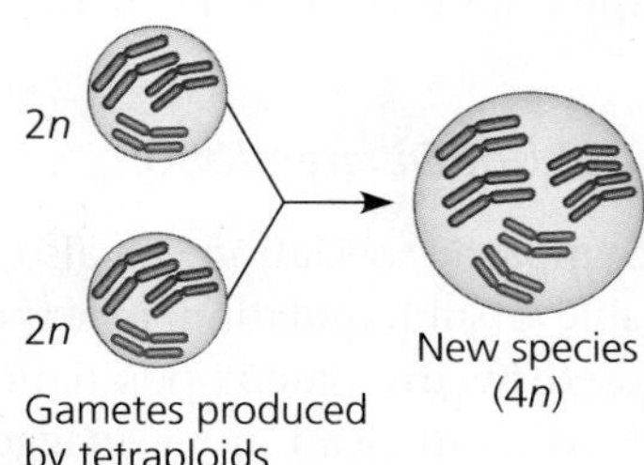

triploid (3*n*) offspring of such unions have reduced fertility. Thus, in just one generation, autopolyploidy can generate reproductive isolation without any geographic separation.

A second form of polyploidy can occur when two different species interbreed and produce hybrid offspring. Most such hybrids are sterile because the set of chromosomes from one species cannot pair during meiosis with the set of chromosomes from the other species. However, an infertile hybrid may be able to propagate itself asexually (as many plants can do). In subsequent generations, various mechanisms can change a sterile hybrid into a fertile polyploid called an **allopolyploid** (**Figure 24.11**). The allopolyploids are fertile when mating with each other but cannot interbreed with either parent species; thus, they represent a new biological species.

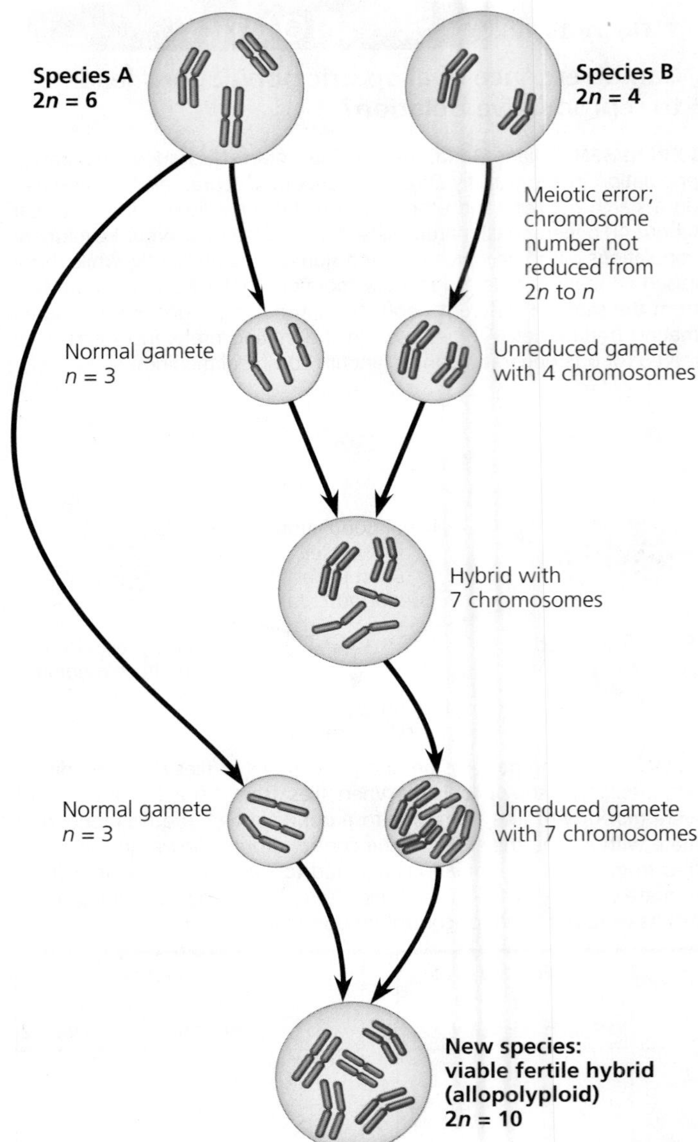

▲ **Figure 24.11 One mechanism for allopolyploid speciation in plants.** Most hybrids are sterile because their chromosomes are not homologous and cannot pair during meiosis. However, such a hybrid may be able to reproduce asexually. This diagram traces one mechanism that can produce fertile hybrids (allopolyploids) as new species. The new species has a diploid chromosome number equal to the sum of the diploid chromosome numbers of the two parent species.

Although polyploid speciation is relatively rare, even in plants, scientists have documented that at least five new plant species have originated in this way since 1850. One of these examples involves the origin of a new species of goatsbeard plant (genus *Tragopogon*) in the Pacific Northwest. *Tragopogon* first arrived in the region when humans introduced three European species in the early 1900s. These three species are now common weeds in abandoned parking lots and other urban sites. In 1950, a new *Tragopogon* species was discovered near the Idaho-Washington border, a region where all three European species also were found. Genetic analyses revealed that this new species, *Tragopogon miscellus*, is a tetraploid hybrid of two of the European species. Although the *T. miscellus* population grows mainly by reproduction of its own members, additional episodes of hybridization between the parent species continue to add new members to the *T. miscellus* population—just one of many examples in which scientists have observed speciation in progress.

Many important agricultural crops—such as oats, cotton, potatoes, tobacco, and wheat—are polyploids. The wheat used for bread, *Triticum aestivum*, is an allohexaploid (six sets of chromosomes, two sets from each of three different species). The first of the polyploidy events that eventually led to modern wheat probably occurred about 8,000 years ago in the Middle East as a spontaneous hybrid of an early cultivated wheat species and a wild grass. Today, plant geneticists generate new polyploids in the laboratory by using chemicals that induce meiotic and mitotic errors. By harnessing the evolutionary process, researchers can produce new hybrid species with desired qualities, such as a hybrid that combines the high yield of wheat with the hardiness of rye.

Habitat Differentiation

Sympatric speciation can also occur when genetic factors enable a subpopulation to exploit a habitat or resource not used by the parent population. Such is the case with the North American apple maggot fly (*Rhagoletis pomonella*), a pest of apples. The fly's original habitat was the native hawthorn tree, but about 200 years ago, some populations colonized apple trees that had been introduced by European settlers. As apples mature more quickly than hawthorn fruit, natural selection has favored apple-feeding flies with rapid development. These apple-feeding populations now show temporal isolation from the hawthorn-feeding *R. pomonella*, providing a prezygotic restriction to gene flow between the two populations. Researchers also have identified alleles that benefit the flies that use one host plant but harm the flies that use the other host plant. As a result, natural selection operating on these alleles provides a postzygotic barrier to reproduction, further limiting gene flow. Altogether, although the two populations are still classified as subspecies rather than separate species, sympatric speciation appears to be well under way.

Sexual Selection

There is evidence that sympatric speciation can also be driven by sexual selection. Clues to how this can occur have been found in cichlid fish from one of Earth's hot spots of animal speciation, East Africa's Lake Victoria. This lake was once home to as many as 600 species of cichlids. Genetic data indicate that these species originated within the last 100,000 years from a small number of colonizing species that arrived from rivers and lakes located elsewhere. How did so many species—more than double the number of freshwater fish species known in all of Europe—originate within a single lake?

One hypothesis is that subgroups of the original cichlid populations adapted to different food sources and that the resulting genetic divergence contributed to speciation in Lake Victoria. But sexual selection, in which (typically) females select males based on their appearance (see Chapter 23), may also have been a factor. Researchers have studied two closely related sympatric species of cichlids that differ mainly in the coloration of breeding males: Breeding *Pundamilia pundamilia* males have a blue-tinged back, whereas breeding *Pundamilia nyererei* males have a red-tinged back (**Figure 24.12**). Their results suggest that mate choice based on male breeding coloration is the main reproductive barrier that normally keeps the gene pools of these two species separate.

▼ **Figure 24.12** **INQUIRY**

Does sexual selection in cichlids result in reproductive isolation?

EXPERIMENT Ole Seehausen and Jacques van Alphen, then at the University of Leiden, placed males and females of *Pundamilia pundamilia* and *P. nyererei* together in two aquarium tanks, one with natural light and one with a monochromatic orange lamp. Under normal light, the two species are noticeably different in male breeding coloration; under monochromatic orange light, the two species are very similar in color. The researchers then observed the mate choices of the females in each tank.

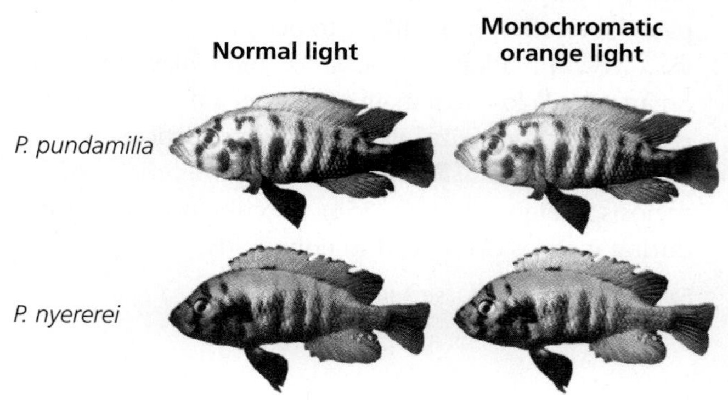

RESULTS Under normal light, females of each species strongly preferred males of their own species. But under orange light, females of each species responded indiscriminately to males of both species. The resulting hybrids were viable and fertile.

CONCLUSION Seehausen and van Alphen concluded that mate choice by females based on male breeding coloration is the main reproductive barrier that normally keeps the gene pools of these two species separate. Since the species can still interbreed when this prezygotic behavioral barrier is breached in the laboratory, the genetic divergence between the species is likely to be small. This suggests that speciation in nature has occurred relatively recently.

SOURCE O. Seehausen and J. J. M. van Alphen, The effect of male coloration on female mate choice in closely related Lake Victoria cichlids (*Haplochromis nyererei* complex), *Behavioral Ecology and Sociobiology* 42:1–8 (1998).

WHAT IF? If changing the light to orange had not affected the mating behavior of the cichlids, how might the researchers' conclusion have been different?

Allopatric and Sympatric Speciation: *A Review*

Now let's recap the two main modes by which new species form. In allopatric speciation, a new species forms in geographic isolation from its parent population. Geographic isolation severely restricts gene flow. As a result, other reproductive barriers from the ancestral species may arise as a byproduct of genetic changes that occur within the isolated population. Many different processes can produce such genetic changes, including natural selection under different environmental conditions, genetic drift, and sexual selection. Once formed, intrinsic reproductive barriers that arise in allopatric populations can prevent interbreeding with the parent population even if the populations come back into contact.

Sympatric speciation, in contrast, requires the emergence of a reproductive barrier that isolates a subset of a population from the remainder of the population in the same area. Though rarer than allopatric speciation, sympatric speciation can occur when gene flow to and from the isolated subpopulation is blocked. This can occur as a result of polyploidy, a condition in which an organism has extra sets of chromosomes. Sympatric speciation also can occur when a subset of a population becomes reproductively isolated because of natural selection that results from a switch to a habitat or food source not used by the parent population. Finally, sympatric speciation can result from sexual selection.

Having reviewed the geographic context in which species originate, we'll next explore in more detail what can happen when new or partially formed species come into contact.

CONCEPT CHECK 24.2

1. Summarize key differences between allopatric and sympatric speciation. Which type of speciation is more common, and why?
2. Describe two mechanisms that can decrease gene flow in sympatric populations, thereby making sympatric speciation more likely to occur.
3. **WHAT IF?** Is allopatric speciation more likely to occur on an island close to a mainland or on a more isolated island of the same size? Explain your prediction.
4. **MAKE CONNECTIONS** After reviewing the process of meiosis in Figure 13.8 (p. 254), describe how an error during meiosis could lead to polyploidy.

For suggested answers, see Appendix A.

CONCEPT 24.3

Hybrid zones reveal factors that cause reproductive isolation

What happens if species with incomplete reproductive barriers come into contact with one another? One possible outcome is the formation of a **hybrid zone**, a region in which members of different species meet and mate, producing at least some offspring of mixed ancestry. In this section, we'll explore hybrid zones and what they reveal about factors that cause the evolution of reproductive isolation.

Patterns Within Hybrid Zones

Some hybrid zones form as narrow bands, such as the one depicted in **Figure 24.13** for two species of toads in the genus *Bombina*, the yellow-bellied toad (*B. variegata*) and the fire-bellied toad (*B. bombina*). This hybrid zone, represented by the red line on the map, extends for 4,000 km but is less than 10 km wide in most places. The hybrid zone occurs where the

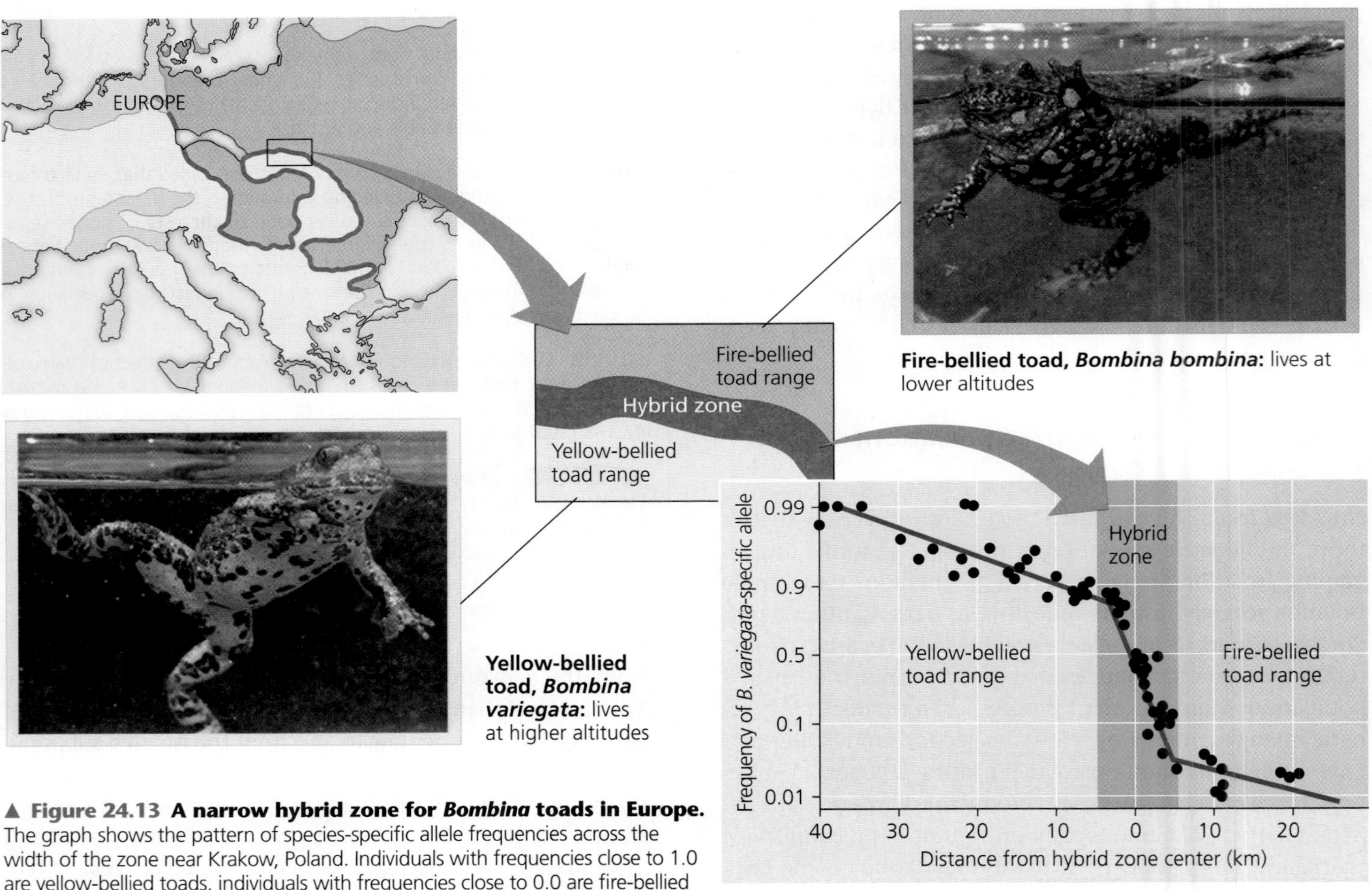

▲ **Figure 24.13 A narrow hybrid zone for *Bombina* toads in Europe.** The graph shows the pattern of species-specific allele frequencies across the width of the zone near Krakow, Poland. Individuals with frequencies close to 1.0 are yellow-bellied toads, individuals with frequencies close to 0.0 are fire-bellied toads, and individuals with intermediate frequencies are considered hybrids.

? *Does the graph indicate that gene flow is spreading fire-bellied toad alleles into the range of the yellow-bellied toad? Explain.*

higher-altitude habitat of the yellow-bellied toad meets the lowland habitat of the fire-bellied toad. Across a given "slice" of the zone, the frequency of alleles specific to yellow-bellied toads typically decreases from close to 100% at the edge where only yellow-bellied toads are found, to 50% in the central portion of the zone, to 0% at the edge where only fire-bellied toads are found.

What causes such a pattern of allele frequencies across a hybrid zone? We can infer that there is an obstacle to gene flow—otherwise, alleles from one parent species would also be common in the gene pool of the other parent species. Are geographic barriers reducing gene flow? Not in this case, since the toads can move throughout the hybrid zone. A more important factor is that hybrid toads have increased rates of embryonic mortality and a variety of morphological abnormalities, including ribs that are fused to the spine and malformed tadpole mouthparts. Because the hybrids have poor survival and reproduction, they produce few viable offspring with members of the parent species. As a result, hybrid individuals rarely serve as a stepping-stone from which alleles are passed from one species to the other. Outside the hybrid zone, additional obstacles to gene flow may be provided by natural selection in the different environments in which the parent species live.

Other hybrid zones have more complicated spatial patterns. For example, many plant species only occur in locations that have a very particular set of environmental conditions. Favorable "patches" that have such conditions are often scattered irregularly across the landscape and are isolated from one another. When two such plant species interbreed, the hybrid zone occurs in a group of disconnected patches, a more complex spatial pattern than the continuous band shown in Figure 24.13. But regardless of whether they have complex or simple spatial patterns, hybrid zones form when two species lacking complete barriers to reproduction come into contact. Once formed, how does a hybrid zone change over time?

Hybrid Zones over Time

Studying a hybrid zone is like observing a natural experiment on speciation. Will the hybrids become reproductively isolated from their parents and form a new species, as occurred by polyploidy in the goatsbeard plant of the Pacific Northwest? If not, there are three possible outcomes for the hybrid zone over time: reinforcement of barriers, fusion of species, or stability **(Figure 24.14)**. Reproductive barriers between species may be reinforced over time (limiting the formation of hybrids) or weakened over time (causing the separating species to fuse into one species). Or hybrids may continue to be produced, creating a long-term and stable hybrid zone. Let's examine what studies in the field suggest about these three possibilities.

Reinforcement: Strengthening Reproductive Barriers

When hybrids are less fit than members of their parent species, as in the *Bombina* example, we might expect natural selection to strengthen prezygotic barriers to reproduction, thus reducing

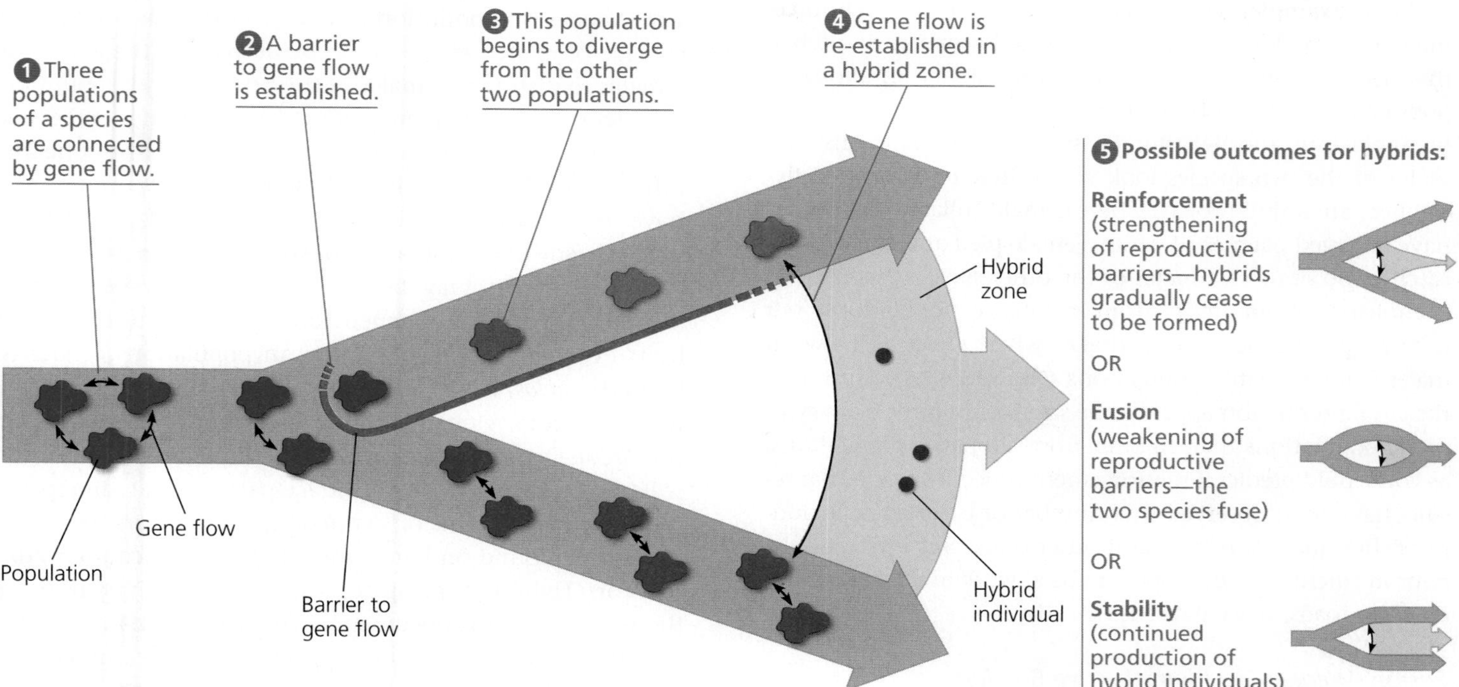

1 Three populations of a species are connected by gene flow.
2 A barrier to gene flow is established.
3 This population begins to diverge from the other two populations.
4 Gene flow is re-established in a hybrid zone.
5 Possible outcomes for hybrids:
Reinforcement (strengthening of reproductive barriers—hybrids gradually cease to be formed)
OR
Fusion (weakening of reproductive barriers—the two species fuse)
OR
Stability (continued production of hybrid individuals)
Hybrid zone
Gene flow
Population
Barrier to gene flow
Hybrid individual

▲ **Figure 24.14 Formation of a hybrid zone and possible outcomes for hybrids over time.** The thick colored arrows represent the passage of time.

WHAT IF? *Predict what might happen if gene flow were re-established at step 3 in this process.*

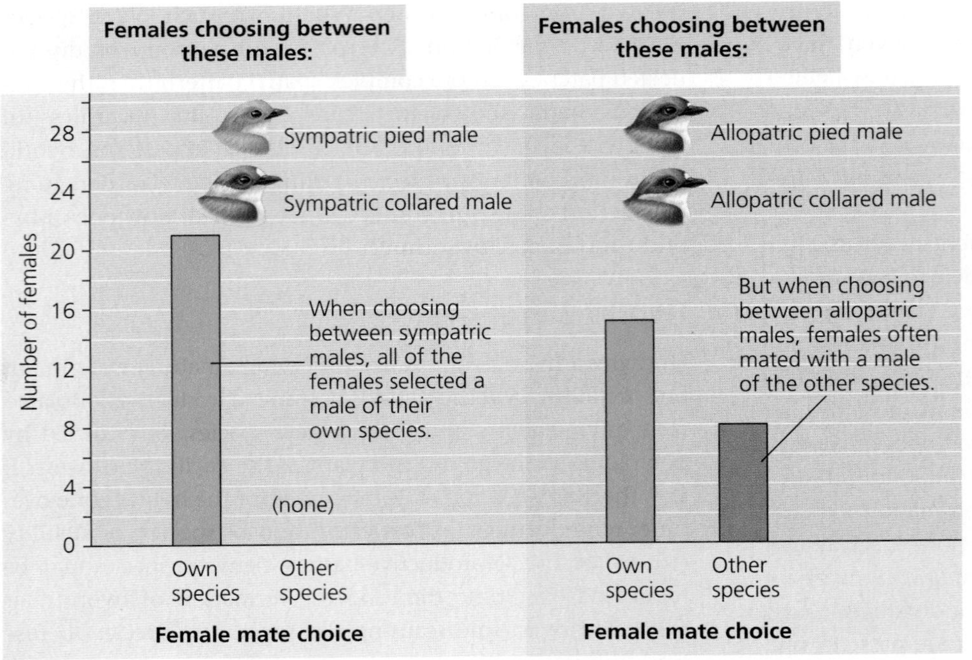

▲ **Figure 24.15 Reinforcement of barriers to reproduction in closely related species of European flycatchers.**

the formation of unfit hybrids. Because this process involves *reinforcing* reproductive barriers, it is called **reinforcement**. If reinforcement is occurring, a logical prediction is that barriers to reproduction between species should be stronger for sympatric populations than for allopatric populations.

As an example, let's consider the evidence for reinforcement in two closely related species of European flycatcher, the pied flycatcher and the collared flycatcher. In allopatric populations of these birds, males of the two species closely resemble one another. But in sympatric populations, the males of the two species look very different: Male pied flycatchers are a dull brown, whereas male collared flycatchers have enlarged patches of white. Female pied and collared flycatchers do not select males of the other species when given a choice between males from sympatric populations, but they frequently do make mistakes when selecting between males from allopatric populations (**Figure 24.15**). Thus, barriers to reproduction appear to be stronger in birds from sympatric populations than in birds from allopatric populations, as you would predict if reinforcement is occurring. Similar results have been observed in a number of organisms, including fishes, insects, plants, and other birds. But interestingly, reinforcement does *not* appear to be at work in the case of the *Bombina* toads, as we'll discuss shortly.

Fusion: Weakening Reproductive Barriers

Next let's consider the case in which two species contact one another in a hybrid zone, but the barriers to reproduction are not strong. So much gene flow may occur that reproductive barriers weaken further and the gene pools of the two species become increasingly alike. In effect, the speciation process reverses, eventually causing the two hybridizing species to fuse into a single species.

Such a situation may be occurring among some of the Lake Victoria cichlids we discussed earlier. In the past 30 years, about 200 of the former 600 species of Lake Victoria cichlids have vanished. Some of these species were driven to extinction by an introduced predator, the Nile perch. But many species not eaten by Nile perch also have disappeared—perhaps in some cases by species fusion. Many pairs of ecologically similar cichlid species are reproductively isolated because the females of one species prefer to mate with males of one color, while females of the other species prefer to mate with males of a different color (see Figure 24.12). Researchers think that murky waters caused by pollution may have reduced the ability of females to use color to distinguish males of their own species from males of closely related species. If further evidence supports this hypothesis, it would seem that pollution in Lake Victoria has produced a cascade of related effects. First, by decreasing the ability of females to distinguish males of their own species, pollution has increased the frequency of mating between members of species that had been isolated reproductively from one another. Second, as a result of these matings, many hybrids have been produced, leading to fusion of the parent species' gene pools and a loss of species (**Figure 24.16**).

Similar events may be affecting the polar bear (*Ursus maritimus*). Fossils and genetic analyses indicate that polar bears evolved from North American populations of grizzly bears (*U. arctos*) between 100,000 and 200,000 years ago. In recent decades, global warming has reduced the extent of the Arctic Ocean ice packs from which polar bears hunt for seals and other prey. As their ice-pack habitat disappears (placing them at risk of extinction), polar bears are more likely to be found on land, where they may encounter grizzly bears. Hybrid offspring of polar bears and grizzly bears in the wild have been documented (see Figure 24.4). As polar bear habitat continues to disappear, increasing numbers of such hybrids may cause the gene pools of these species to begin to fuse, thus contributing to the possible eventual extinction of the polar bear.

Pundamilia nyererei **Pundamilia pundamilia**

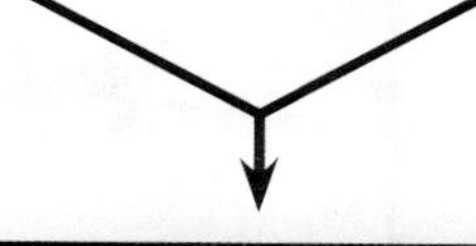

***Pundamilia* "turbid water,"**
hybrid offspring from a location
with turbid water

▲ **Figure 24.16 Fusion: The breakdown of reproductive barriers.** Increasingly cloudy water in Lake Victoria over the past 30 years may have weakened reproductive barriers between *P. nyererei* and *P. pundamilia*. In areas of cloudy water, the two species have hybridized extensively, causing their gene pools to fuse.

Stability: Continued Formation of Hybrid Individuals

Many hybrid zones are stable in the sense that hybrids continue to be produced. In some cases, this occurs because the hybrids survive or reproduce better than members of either parent species, at least in certain habitats or years. But stable hybrid zones have also been observed in cases where the hybrids are selected *against*—an unexpected result.

Recall that hybrids are at a strong disadvantage in the *Bombina* hybrid zone. As a result, the offspring of individuals that prefer to mate with members of their own species should survive or reproduce better than the unfit hybrid offspring of individuals that mate indiscriminately with members of the other species. This suggests that reinforcement should occur, strengthening reproductive barriers and thereby limiting the production of hybrid toads. But in more than 20 years of study, no evidence for reinforcement has been found, and hybrids continue to be produced.

What could explain this finding? One possibility relates to the narrowness of the *Bombina* hybrid zone (see Figure 24.13). Evidence suggests that members of both parent species migrate into the zone from the parent populations located outside the zone. Such movements lead to the continued production of hybrids, potentially overwhelming the selection for increased reproductive isolation inside the zone. If the hybrid zone were wider, this would be less likely to occur, since the center of the zone would receive little gene flow from distant parent populations located outside the hybrid zone.

In short, sometimes the outcomes in hybrid zones match our predictions (European flycatchers and cichlid fishes), and sometimes they don't (*Bombina*). But whether our predictions are upheld or not, events in hybrid zones can shed light on how barriers to reproduction between closely related species change over time. In the next section, we'll examine how interactions between hybridizing species can also provide a glimpse into the speed and genetic control of speciation.

CONCEPT CHECK 24.3

1. What are hybrid zones, and why can they be viewed as "natural laboratories" in which to study speciation?
2. **WHAT IF?** Consider two species that diverged while geographically separated but resumed contact before reproductive isolation was complete. Predict what would happen over time if the two species mated indiscriminately and (a) hybrid offspring survived and reproduced more poorly than offspring from intraspecific matings or (b) hybrid offspring survived and reproduced as well as offspring from intraspecific matings.

For suggested answers, see Appendix A.

CONCEPT **24.4**

Speciation can occur rapidly or slowly and can result from changes in few or many genes

Darwin faced many unanswered questions when he began to ponder that "mystery of mysteries"—speciation. As you read in Chapter 22, he found answers to some of those questions when he realized that evolution by natural selection helped explain both the diversity of life and the adaptations of organisms. But biologists since Darwin have continued to ask fundamental questions about speciation. For example, how long does it take for new species to form? And how many genes change when one species splits into two? Answers to these questions are also beginning to emerge.

The Time Course of Speciation

We can gather information about how long it takes new species to form from broad patterns in the fossil record and from studies that use morphological data (including fossils) or molecular data to assess the time interval between speciation events in particular groups of organisms.

(a) In a punctuated pattern, new species change most as they branch from a parent species and then change little for the rest of their existence.

(b) Other species diverge from one another much more gradually over time.

▲ **Figure 24.17 Two models for the tempo of speciation.**

Patterns in the Fossil Record

The fossil record includes many episodes in which new species appear suddenly in a geologic stratum, persist essentially unchanged through several strata, and then disappear. For example, there are dozens of species of marine invertebrates that make their debut in the fossil record with novel morphologies, but then change little for millions of years before becoming extinct. Paleontologists Niles Eldredge, of the American Museum of Natural History, and Stephen Jay Gould, of Harvard University, coined the term **punctuated equilibria** to describe these periods of apparent stasis punctuated by sudden change **(Figure 24.17a)**. Other species do not show a punctuated pattern; instead, they change more gradually over long periods of time **(Figure 24.17b)**.

What do punctuated and gradual patterns tell us about how long it takes new species to form? Suppose that a species survived for 5 million years, but most of the morphological changes that caused it to be designated a new species occurred during the first 50,000 years of its existence—just 1% of its total lifetime. Time periods this short (in geologic terms) often cannot be distinguished in fossil strata, in part because the rate of sediment accumulation is too slow to separate layers this close in time. Thus, based on its fossils, the species would seem to have appeared suddenly and then lingered with little or no change before becoming extinct. Even though such a species may have originated more slowly than its fossils suggest (in this case taking 50,000 years), a punctuated pattern indicates that speciation occurred relatively rapidly. For species whose fossils changed much more gradually, we also cannot tell exactly when a new biological species formed, since information about reproductive isolation does not fossilize. However, it is likely that speciation in such groups occurred relatively slowly, perhaps taking millions of years.

Speciation Rates

The punctuated pattern suggests that once the process of speciation begins, it can be completed relatively rapidly—a suggestion supported by a growing number of studies.

For example, research conducted at Indiana University suggests that rapid speciation produced the wild sunflower *Helianthus anomalus*. Genetic evidence indicates that this species originated by the hybridization of two other sunflower species, *H. annuus* and *H. petiolaris*. The hybrid species *H. anomalus* is ecologically distinct and reproductively isolated from both parent species **(Figure 24.18)**. Unlike the outcome of allopolyploid speciation, in which there is a change in chromosome number after hybridization, in these sunflowers the two parent species and the hybrid all have the same number of chromosomes ($2n = 34$). How, then, did speciation occur? To answer this question, the researchers performed an

▲ **Figure 24.18 A hybrid sunflower species and its dry sand dune habitat.** The wild sunflower *Helianthus anomalus* originated via the hybridization of two other sunflowers, *H. annuus* and *H. petiolaris*, which live in nearby but moister environments.

▼ Figure 24.19 INQUIRY

How does hybridization lead to speciation in sunflowers?

EXPERIMENT At Indiana University, Loren Rieseberg and his colleagues crossed the two parent sunflower species, *H. annuus* and *H. petiolaris*, to produce experimental hybrids in the laboratory (for each gamete, only two of the $n = 17$ chromosomes are shown).

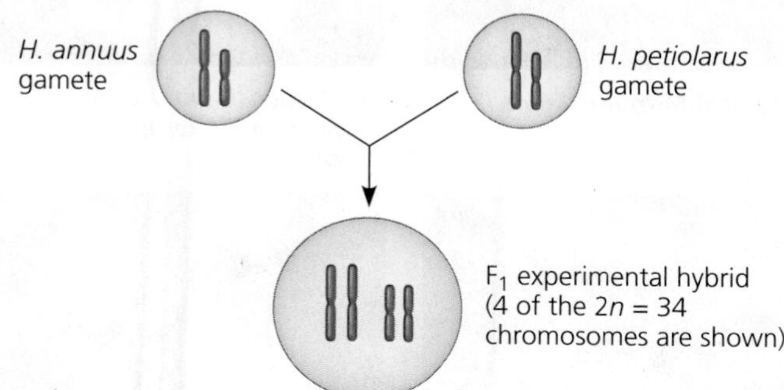

Note that in the first (F_1) generation, each chromosome of the experimental hybrids consisted entirely of DNA from one or the other parent species. The researchers then tested whether the F_1 and subsequent generations of experimental hybrids were fertile. They also used species-specific genetic markers to compare the chromosomes in the experimental hybrids with the chromosomes in the naturally occurring hybrid *H. anomalus*.

RESULTS Although only 5% of the F_1 experimental hybrids were fertile, after just four more generations the hybrid fertility rose to more than 90%. The chromosomes of individuals from this fifth hybrid generation differed from those in the F_1 generation but were similar to those in *H. anomalus* individuals from natural populations:

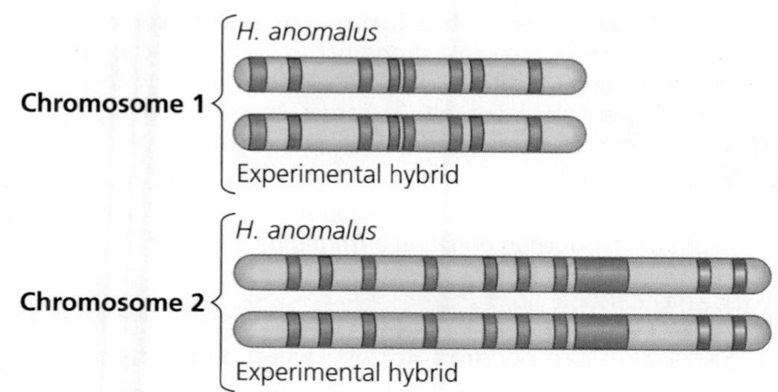

CONCLUSION Over time, the chromosomes in the population of experimental hybrids became similar to the chromosomes of *H. anomalus* individuals from natural populations. This suggests that the observed rise in the fertility of the experimental hybrids occurred as selection eliminated regions of DNA from the parent species that were not compatible with one another. Overall, it appeared that the initial steps of the speciation process occurred rapidly and could be mimicked in a laboratory experiment.

SOURCE L. H. Rieseberg et al., Role of gene interactions in hybrid speciation: evidence from ancient and experimental hybrids, *Science* 272:741–745 (1996).

WHAT IF? *The increased fertility of the experimental hybrids could have resulted from natural selection for thriving under laboratory conditions. Evaluate this alternative explanation for the results.*

experiment designed to mimic events in nature (**Figure 24.19**). Their results indicated that natural selection could produce extensive genetic changes in hybrid populations over short periods of time. These changes appear to have caused the hybrids to diverge reproductively from their parents and form a new species, *H. anomalus*.

The sunflower example, along with the apple maggot fly, Lake Victoria cichlid, and fruit fly examples discussed earlier, suggests that new species can arise rapidly *once divergence begins*. But what is the total length of time between speciation events? This interval consists of the time that elapses before populations of a newly formed species start to diverge from one another plus the time it takes for speciation to be complete once divergence begins. It turns out that the total time between speciation events varies considerably. For example, in a survey of data from 84 groups of plants and animals, the interval between speciation events ranged from 4,000 years (in cichlids of Lake Nabugabo, Uganda) to 40 million years (in some beetles). Overall, the time between speciation events averaged 6.5 million years and rarely took less than 500,000 years.

What can we learn from such data? First, the data suggest that on average, millions of years may pass before a newly formed species will itself give rise to another new species. As we'll see in Chapter 25, this result has implications for how long it takes life on Earth to recover from mass extinction events. Second, the extreme variability in the time it takes new species to form indicates that organisms do not have a "speciation clock" ticking inside them, causing them to produce new species at regular time intervals. Instead, speciation begins only after gene flow between populations is interrupted, perhaps by changing environmental conditions or by unpredictable events, such as a storm that transports a few individuals to an isolated area. Furthermore, once gene flow is interrupted, the populations must diverge genetically to such an extent that they become reproductively isolated—all before other events cause gene flow to resume, possibly reversing the speciation process (see Figure 24.16).

Studying the Genetics of Speciation

Studies of ongoing speciation (as in hybrid zones) can reveal traits that cause reproductive isolation. By identifying the genes that control those traits, scientists can explore a fundamental question of evolutionary biology: How many genes change when a new species forms?

In a few cases, the evolution of reproductive isolation is due to a change in a single gene. For example, in Japanese snails of the genus *Euhadra*, a change in a single gene can result in a mechanical barrier to reproduction. This gene controls the direction in which the shells spiral. When their shells spiral in different directions, the snails' genitalia are oriented in a manner that prevents mating (Figure 24.3f shows a similar example).

A major barrier to reproduction between two closely related species of monkey flower, *Mimulus cardinalis* and *M. lewisii*, also appears to be influenced by a relatively small number of genes. These two species are isolated by several prezygotic and postzygotic barriers. Of these, one prezygotic barrier, pollinator choice, accounts for most of the isolation: In a hybrid zone between *M. cardinalis* and *M. lewisii*, nearly 98% of pollinator visits were restricted to one species or the other.

The two monkey flower species are visited by different pollinators: Hummingbirds prefer the red-flowered *M. cardinalis*, and bumblebees prefer the pink-flowered *M. lewisii*. Douglas Schemske, of Michigan State University, and colleagues have shown that pollinator choice is affected by at least two loci in the monkey flowers, one of which, the "yellow upper," or *yup*, locus, influences flower color (**Figure 24.20**). By crossing the two parent species to produce F_1 hybrids and then performing repeated backcrosses of these F_1 hybrids to each parent species, Schemske and colleagues succeeded in transferring the *M. cardinalis* allele at this locus into *M. lewisii*, and vice versa. In a field experiment, *M. lewisii* plants with the *M. cardinalis yup* allele received 68-fold more visits from hummingbirds than did wild-type *M. lewisii*. Similarly, *M. cardinalis* plants with the *M. lewisii yup* allele received 74-fold more visits from bumblebees than did wild-type *M. cardinalis*. Thus, a mutation at a single locus can influence pollinator preference and hence contribute to reproductive isolation in monkey flowers.

In other organisms, the speciation process is influenced by larger numbers of genes and gene interactions. For example, hybrid sterility between two subspecies of the fruit fly *Drosophila pseudoobscura* results from gene interactions among at least four loci, and postzygotic isolation in the sunflower hybrid zone discussed earlier is influenced by at least 26 chromosome segments (and an unknown number of genes). Overall, studies suggest that few or many genes can influence the evolution of reproductive isolation and hence the emergence of a new species.

From Speciation to Macroevolution

As you've seen, speciation may begin with differences as seemingly small as the color on a cichlid's back. However, as speciation occurs again and again, such differences can accumulate and become more pronounced, eventually leading to the formation of new groups of organisms that differ greatly from their ancestors (as in the origin of whales from land-dwelling mammals; see Figure 22.20). Furthermore, as one group of organisms increases in size by producing many new species, another group of organisms may shrink, losing species to extinction. The cumulative effects of many such speciation and extinction events have helped shape the sweeping evolutionary changes that are documented in the fossil record. In the next chapter, we turn to such large-scale evolutionary changes as we begin our study of macroevolution.

(a) Typical *Mimulus lewisii*

(b) *M. lewisii* with an *M. cardinalis* flower-color allele

(c) Typical *Mimulus cardinalis*

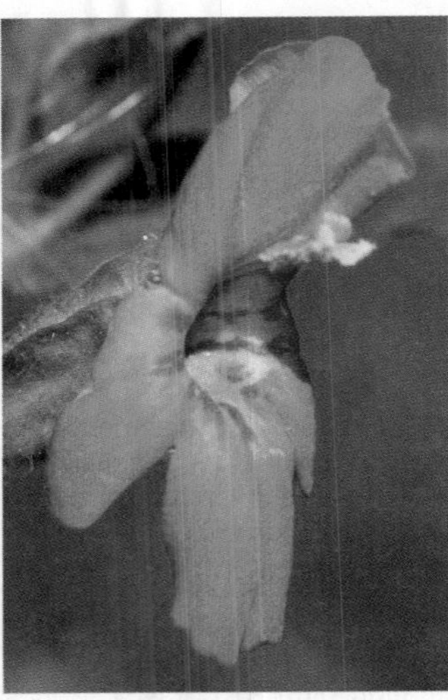

(d) *M. cardinalis* with an *M. lewisii* flower-color allele

▲ **Figure 24.20 A locus that influences pollinator choice.** Pollinator preferences provide a strong barrier to reproduction between *Mimulus lewisii* and *M. cardinalis*. After transferring the *M. lewisii* allele for a flower-color locus into *M. cardinalis* and vice versa, researchers observed a shift in some pollinators' preferences.

WHAT IF? *If* M. cardinalis *individuals that had the* M. lewisii yup *allele were planted in an area that housed both monkey flower species, how might the production of hybrid offspring be affected?*

CONCEPT CHECK 24.4

1. Speciation can occur rapidly between diverging populations, yet the length of time between speciation events is often more than a million years. Explain this apparent contradiction.
2. Summarize evidence that the *yup* locus acts as a prezygotic barrier to reproduction in two species of monkey flowers. Do these results demonstrate that the *yup* locus alone controls barriers to reproduction between these species? Explain.
3. **MAKE CONNECTIONS** Compare Figure 13.11 (p. 259) with Figure 24.19 (p. 503). What cellular process could cause the hybrid chromosomes to contain DNA from both parent species? Explain.

For suggested answers, see Appendix A.

24 CHAPTER REVIEW

SUMMARY OF KEY CONCEPTS

CONCEPT 24.1

The biological species concept emphasizes reproductive isolation (pp. 488–492)

- A biological **species** is a group of populations whose individuals have the potential to interbreed and produce viable, fertile offspring with each other but not with members of other species. The **biological species concept** emphasizes reproductive isolation through prezygotic and postzygotic barriers that separate gene pools.
- Although helpful in thinking about how speciation occurs, the biological species concept has limitations. For instance, it cannot be applied to organisms known only as fossils or to organisms that reproduce only asexually. Thus, scientists use other species concepts, such as the **morphological species concept**, in certain circumstances.

? *Explain the importance of gene flow to the biological species concept.*

CONCEPT 24.2

Speciation can take place with or without geographic separation (pp. 493–498)

- In **allopatric speciation**, gene flow is reduced when two populations of one species become geographically separated from each other. One or both populations may undergo evolutionary change during the period of separation, resulting in the establishment of prezygotic or postzygotic barriers to reproduction.
- In **sympatric speciation**, a new species originates while remaining in the same geographic area as the parent species. Plant species (and, more rarely, animal species) have evolved sympatrically through polyploidy. Sympatric speciation can also result from habitat shifts and sexual selection.

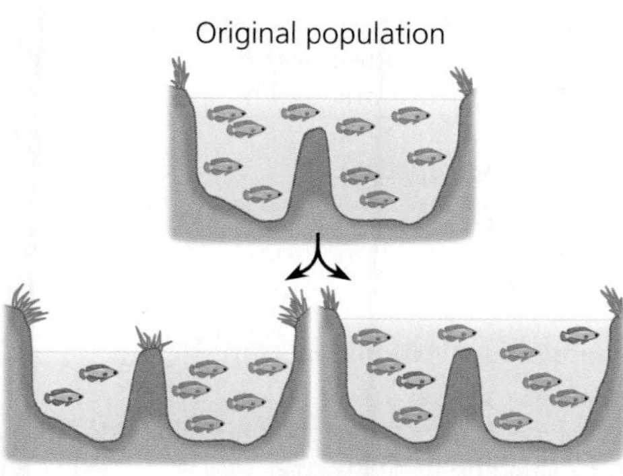

? *Can factors that cause sympatric speciation also cause allopatric speciation? Explain.*

CONCEPT 24.3

Hybrid zones reveal factors that cause reproductive isolation (pp. 498–501)

- Many groups of organisms form **hybrid zones** in which members of different species meet and mate, producing at least some offspring of mixed ancestry.
- Many hybrid zones are *stable* in that hybrid offspring continue to be produced over time. In others, **reinforcement** strengthens prezygotic barriers to reproduction, thus decreasing the formation of unfit hybrids. In still other hybrid zones, barriers to reproduction may weaken over time, resulting in the *fusion* of the species' gene pools (reversing the speciation process).

? *What factors can support the long-term stability of a hybrid zone if the parent species live in different environments?*

CONCEPT 24.4

Speciation can occur rapidly or slowly and can result from changes in few or many genes (pp. 501–504)

- New species can form rapidly once divergence begins—but it can take millions of years for that to happen. The time interval between speciation events varies considerably, from a few thousand years to tens of millions of years.
- New developments in genetics have enabled researchers to identify specific genes involved in some cases of speciation. Results show that speciation can be driven by few or many genes.

? *Is speciation something that happened only in the distant past, or are new species continuing to arise today? Explain.*

TEST YOUR UNDERSTANDING

LEVEL 1: KNOWLEDGE/COMPREHENSION

1. The *largest* unit within which gene flow can readily occur is a
   a. population.
   b. species.
   c. genus.
   d. hybrid.
   e. phylum.
2. Males of different species of the fruit fly *Drosophila* that live in the same parts of the Hawaiian Islands have different elaborate courtship rituals. These rituals involve fighting other males and making stylized movements that attract females. What type of reproductive isolation does this represent?
   a. habitat isolation
   b. temporal isolation
   c. behavioral isolation
   d. gametic isolation
   e. postzygotic barriers
3. According to the punctuated equilibria model,
   a. natural selection is unimportant as a mechanism of evolution.
   b. given enough time, most existing species will branch gradually into new species.
   c. most new species accumulate their unique features relatively rapidly as they come into existence, then change little for the rest of their duration as a species.
   d. most evolution occurs in sympatric populations.
   e. speciation is usually due to a single mutation.

LEVEL 2: APPLICATION/ANALYSIS

4. Bird guides once listed the myrtle warbler and Audubon's warbler as distinct species. Recently, these birds have been classified as eastern and western forms of a single species, the

yellow-rumped warbler. Which of the following pieces of evidence, if true, would be cause for this reclassification?
a. The two forms interbreed often in nature, and their offspring have good survival and reproduction.
b. The two forms live in similar habitats.
c. The two forms have many genes in common.
d. The two forms have similar food requirements.
e. The two forms are very similar in coloration.

5. Which of the following factors would *not* contribute to allopatric speciation?
a. A population becomes geographically isolated from the parent population.
b. The separated population is small, and genetic drift occurs.
c. The isolated population is exposed to different selection pressures than the ancestral population.
d. Different mutations begin to distinguish the gene pools of the separated populations.
e. Gene flow between the two populations is extensive.

6. Plant species A has a diploid number of 12. Plant species B has a diploid number of 16. A new species, C, arises as an allopolyploid from A and B. The diploid number for species C would probably be
a. 12. b. 14. c. 16. d. 28. e. 56.

LEVEL 3: SYNTHESIS/EVALUATION

7. Suppose that a group of male pied flycatchers migrated from a region where there were no collared flycatchers to a region where both species were present (see Figure 24.15). Assuming events like this are very rare, which of the following scenarios is *least* likely?
a. The frequency of hybrid offspring would increase.
b. Migrant pied males would produce fewer offspring than would resident pied males.
c. Pied females would rarely mate with collared males.
d. Migrant males would mate with collared females more often than with pied females.
e. The frequency of hybrid offspring would decrease.

8. **EVOLUTION CONNECTION**
What is the biological basis for assigning all human populations to a single species? Can you think of a scenario by which a second human species could originate in the future?

9. **SCIENCE, TECHNOLOGY, AND SOCIETY**
In the United States, the rare red wolf (*Canis lupus*) has been known to hybridize with coyotes (*Canis latrans*), which are much more numerous. Although red wolves and coyotes differ in terms of morphology, DNA, and behavior, genetic evidence suggests that living red wolf individuals are actually hybrids. Red wolves are designated as an endangered species and hence receive legal protection under the Endangered Species Act. Some people think that their endangered status should be withdrawn because the remaining red wolves are hybrids, not members of a "pure" species. Do you agree? Why or why not?

10. **SCIENTIFIC INQUIRY**
DRAW IT In this chapter, you read that bread wheat (*Triticum aestivum*) is an allohexaploid, containing two sets of chromosomes from each of three different parent species. Genetic analysis suggests that the three species pictured following this question each contributed chromosome sets to *T. aestivum*. (The capital letters here represent sets of chromosomes rather than individual genes.) Evidence also indicates that the first polyploidy event was a spontaneous hybridization of the early cultivated wheat species *T. monococcum* and a wild *Triticum* grass species. Based on this information, draw a diagram of one possible chain of events that could have produced the allohexaploid *T. aestivum*.

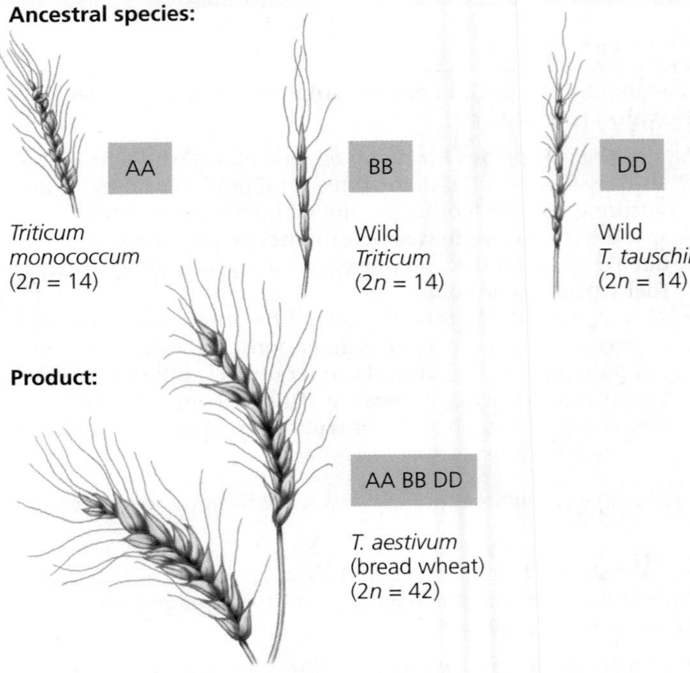

11. 
WRITE ABOUT A THEME
The Genetic Basis of Life In sexually reproducing species, each individual begins life with DNA inherited from both parent organisms. In a short essay (100–150 words), apply this idea to what occurs when organisms of two species that have homologous chromosomes mate and produce (F_1) hybrid offspring. What percentage of the DNA in the F_1 hybrids' chromosomes comes from each parent species? As the hybrids mate and produce F_2 and later-generation hybrid offspring, describe how recombination and natural selection may affect whether the DNA in hybrid chromosomes is derived from one parent species or the other.

For selected answers, see Appendix A.

MasteringBIOLOGY www.masteringbiology.com

1. MasteringBiology® Assignments
Tutorial Defining Species
Activities Overview of Macroevolution • Allopatric Speciation • Speciation by Changes in Ploidy
Questions Student Misconceptions • Reading Quiz • Multiple Choice • End-of-Chapter

2. eText
Read your book online, search, take notes, highlight text, and more.

3. The Study Area
Practice Tests • Cumulative Test • *BioFlix* 3-D Animations • MP3 Tutor Sessions • Videos • Activities • Investigations • Lab Media • Audio Glossary • Word Study Tools • Art

APPENDIX A Answers

Chapter 1

Figure Questions

Figure 1.7 The arrangement of fingers and opposable thumb in the human hand, combined with fingernails and a complex system of nerves and muscles, allows the hand to grasp and manipulate objects with great dexterity. **Figure 1.13** Substance B would be made continuously and would accumulate in large amounts. Neither C nor D would be made, so D would not be able to inhibit Enzyme 1 and regulate the pathway. **Figure 1.27** The percentage of brown artificial snakes attacked would probably be higher than the percentage of artificial kingsnakes attacked in all areas (whether or not inhabited by coral snakes).

Concept Check 1.1

1. Examples: A molecule consists of *atoms* bonded together. Each organelle has an orderly arrangement of *molecules*. Photosynthetic plant cells contain *organelles* called chloroplasts. A tissue consists of a group of similar *cells*. Organs such as the heart are constructed from several *tissues*. A complex multicellular organism, such as a plant, has several types of *organs*, such as leaves and roots. A population is a set of *organisms* of the same species. A community consists of *populations* of the various species inhabiting a specific area. An ecosystem consists of a biological *community* along with the nonliving factors important to life, such as air, soil, and water. The biosphere is made up of all of Earth's *ecosystems*. **2.** (a) Structure and function are correlated. (b) Cells are an organism's basic units, *and* the continuity of life is based on heritable information in the form of DNA. (c) Organisms interact with other organisms and with the physical environment, *and* life requires energy transfer and transformation. **3.** Some possible answers: *Emergent properties*: The ability of a human heart to pump blood requires an intact heart; it is not a capability of any of the heart's tissues or cells working alone. *Environmental interactions*: A mouse eats food, such as nuts or grasses, and deposits some of the food material as feces and urine. Construction of a nest rearranges the physical environment and may hasten degradation of some of its components. The mouse may also act as food for a predator. *Energy transfer*: A plant, such as a grass, absorbs energy from the sun and transforms it into molecules that act as stored fuel. Animals can eat parts of the plant and use the food for energy to carry out their activities. *Structure and function*: The strong, sharp teeth of a wolf are well suited to grasping and dismembering its prey. *The cellular basis of life*: The digestion of food is made possible by chemicals (chiefly enzymes) made by cells of the digestive tract. *The genetic basis of life*: Human eye color is determined by the combination of genes inherited from the two parents. *Feedback regulation*: When your stomach is full, it signals your brain to decrease your appetite. *Evolution*: All plants have chloroplasts, indicating their descent from a common ancestor.

Concept Check 1.2

1. An address pinpoints a location by tracking from broader to narrower categories—a state, city, zip, street, and building number. This is analogous to the groups-subordinate-to-groups structure of biological taxonomy. **2.** The naturally occurring heritable variation in a population is "edited" by natural selection because individuals with heritable traits better suited to the environment survive and reproduce more successfully than others. Over time, better-suited individuals persist and their percentage in the population increases, while less suited individuals become less prevalent—a type of population editing.
3.

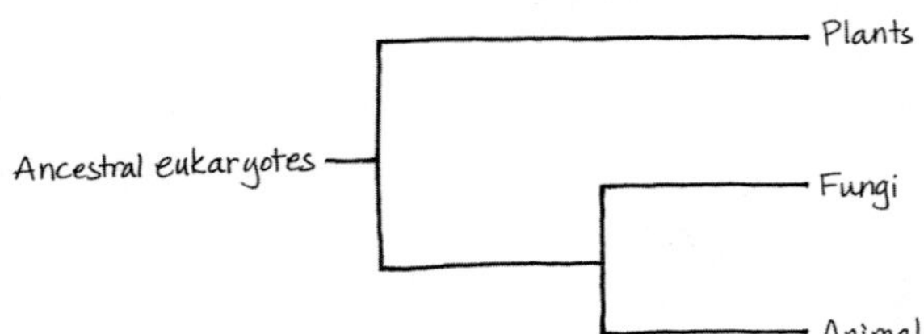

Concept Check 1.3

1. Inductive reasoning derives generalizations from specific cases; deductive reasoning predicts specific outcomes from general premises. **2.** The coloration pattern on the snakes **3.** Compared to a hypothesis, a scientific theory is usually more general and substantiated by a much greater amount of evidence. Natural selection is an explanatory idea that applies to all kinds of organisms and is supported by vast amounts of evidence of various kinds. **4.** Based on the results shown in Figure 1.27, you might predict that the colorful artificial snakes would be attacked more often than the brown ones, simply because they are easier to see. This prediction assumes that the area in Virginia where you are working has predators that attack snakes but no poisonous snakes that resemble the colorful artificial snakes.

Concept Check 1.4

1. Science aims to understand natural phenomena and how they work, while technology involves application of scientific discoveries for a particular purpose or to solve a specific problem. **2.** Natural selection could be operating. Malaria is present in sub-Saharan Africa, so there might be an advantage to people with the sickle-cell disease form of the gene that makes them more able to survive and pass on their genes to offspring. Among those of African descent living in the United States, where malaria is absent, there would be no advantage, so they would be selected against more strongly, resulting in fewer individuals with the sickle-cell disease form of the gene.

Summary of Key Concepts Questions

1.1 Evolution explains the most fundamental aspects of all life on earth. It accounts for the common features shared by all forms of life due to descent from a common ancestor, while also providing an explanation for how the great diversity of living organisms on the planet has arisen. **1.2** Ancestors of this plant may have exhibited variation in how well their leaves conserved water. Because not much soil is present in the crevices where these plants are found, the variant plants that could conserve water may have survived better and been able to produce more offspring. Over time, a higher and higher proportion of individuals in the population would have had the adaptation of thick, water-conserving leaves. **1.3** Inductive reasoning is used in forming hypotheses, while deductive reasoning leads to predictions that are used to test hypotheses. **1.4** Different approaches taken by scientists studying natural phenomena at different levels complement each other, so more is learned about each problem being studied. A diversity of backgrounds among scientists may lead to fruitful ideas in the same way that important innovations have often arisen where a mix of cultures coexist.

Test Your Understanding

1. b **2.** d **3.** a **4.** c **5.** c **6.** c **7.** b **8.** c **9.** c **10.** d
11. Your figure should show: (1) For the biosphere, the Earth with an arrow coming out of a tropical ocean; (2) for the ecosystem, a distant view of a coral reef; (3) for the community, a collection of reef animals and algae, with corals, fishes, some seaweed, and any other organisms you can think of; (4) for the population, a group of fish of the same species; (5) for the organism, one fish from your population; (6) for the organ, the fish's stomach, and for the organ system, the whole digestive tract (see Chapter 41 for help); (7) for a tissue, a group of similar cells from the stomach; (8) for a cell, one cell from the tissue, showing its nucleus and a few other organelles; (9) for an organelle, the nucleus, where most of the cell's DNA is located; and (10) for a molecule, a DNA double helix. Your sketches can be very rough!

Chapter 2

Figure Questions

Figure 2.2 The most significant difference in the results would be that the two *Cedrela* saplings inside each garden would show similar amounts of dying leaf tissue because a poisonous chemical released from the *Duroia* trees would presumably reach the saplings via the air or soil and would not be blocked by the insect barrier. The *Cedrela* saplings planted outside the gardens would not show damage unless *Duroia* trees were nearby. Also, any ants present on the unprotected *Cedrela* saplings inside the gardens would probably not be observed making injections into the leaves. However, formic acid would likely still be found in the ants' glands, as it is for most species of ants. **Figure 2.9** Atomic number = 12; 12 protons, 12 electrons; 3 electron shells; 2 valence electrons **Figure 2.16** One possible answer:

Figure 2.19 The plant is submerged in water (H_2O), in which the CO_2 is dissolved. The sun's energy is used to make sugar, which is found in the plant and can act as food for the plant itself, as well as for animals that eat the plant. The oxygen (O_2) is present in the bubbles.

Concept Check 2.1

1. Table salt (sodium chloride) is made up of sodium and chlorine. We are able to eat the compound, showing that it has different properties from those of a metal (sodium) and a poisonous gas (chlorine). **2.** Yes, because an organism requires trace elements, even though only in small amounts **3.** A person with an iron deficiency will probably show fatigue and other effects of a low oxygen level in the blood. (The condition is called anemia and can also result from too few red blood cells or abnormal hemoglobin.) **4.** Variant ancestral plants that could tolerate the toxic elements could grow and reproduce in serpentine soils. (Plants that were well adapted to nonserpentine soils would not be expected to survive in serpentine areas.) The offspring of the variants would also vary, with those most capable of thriving under serpentine conditions growing best and reproducing most. Over many generations, this probably led to the serpentine-adapted species we see today.

Concept Check 2.2

1. 7 **2.** $^{15}_{7}N$ **3.** 9 electrons; two electron shells; 1*s*, 2*s*, 2*p* (three orbitals); 1 electron is needed to fill the valence shell. **4.** The elements in a row all have the same number of electron shells. In a column, all the elements have the same number of electrons in their valence shells.

Concept Check 2.3

1. Each carbon atom has only three covalent bonds instead of the required four. **2.** The attraction between oppositely charged ions, forming ionic bonds **3.** If you could synthesize molecules that mimic these shapes, you might be able to treat diseases or conditions caused by the inability of affected individuals to synthesize such molecules.

Concept Check 2.4

1.

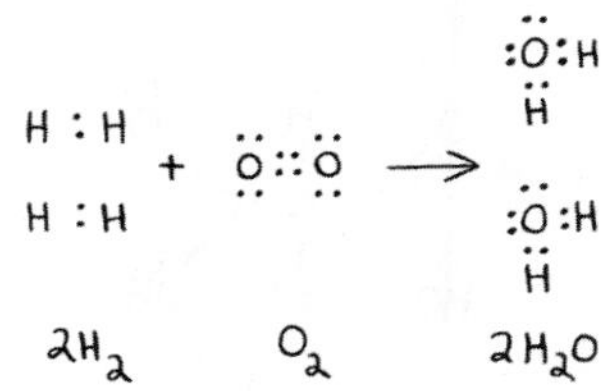

2. At equilibrium, the forward and reverse reactions occur at the same rate. **3.** $C_6H_{12}O_6 + 6\ O_2 \rightarrow 6\ CO_2 + 6\ H_2O$ + Energy. Glucose and oxygen react to form carbon dioxide and water, releasing energy. We breathe in oxygen because we need it for this reaction to occur, and we breathe out carbon dioxide because it is a by-product of this reaction. (This reaction is called cellular respiration, and you will learn more about it in Chapter 9.)

Summary of Key Concepts Questions

2.1 Iodine (part of a thyroid hormone) and iron (part of hemoglobin in blood) are both trace elements, required in minute quantities. Calcium and phosphorus (components of bones and teeth) are needed by the body in much greater quantities.
2.2

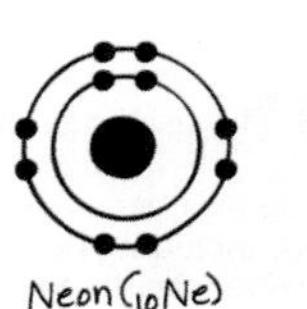

Both neon and argon have completed valence shells, containing 8 electrons. They do not have unpaired electrons that could participate in chemical bonds. **2.3** Electrons are shared equally between the two atoms in a nonpolar covalent bond. In a polar covalent bond, the electrons are drawn closer to the more electronegative atom. In the formation of ions, an electron is completely transferred from one atom to a much more electronegative atom. **2.4** The concentration of products would increase as the added reactants were converted to products. Eventually, an equilibrium would again be reached in which the forward and reverse reactions were proceeding at the same rate and the relative concentrations of reactants and products returned to where they were before the addition of more reactants.

Test Your Understanding

1. a **2.** e **3.** b **4.** a **5.** d **6.** b **7.** c **8.** e

9. a. O::C:H This structure doesn't make sense because the valence shell of carbon is incomplete; carbon can form 4 bonds.

b. H:O:C:C::O (with H, H above and H below) This structure makes sense because all valence shells are complete, and all bonds have the correct number of electrons.

c. H:C:H.C::O (with H, H above and H below) This structure doesn't make sense because H has only 1 electron to share, so it cannot form bonds with 2 atoms.

d. This structure doesn't make sense for several reasons:
:O: The valence shell of oxygen is incomplete; oxygen can form 2 bonds.
H:N..H H has only 1 electron to share, so it cannot form a double bond.
Nitrogen usually makes only 3 bonds. It does not have enough electrons to make 2 single bonds, make a double bond, and complete its valence shell.

Chapter 3

Figure Questions

Figure 3.2 One possible answer:

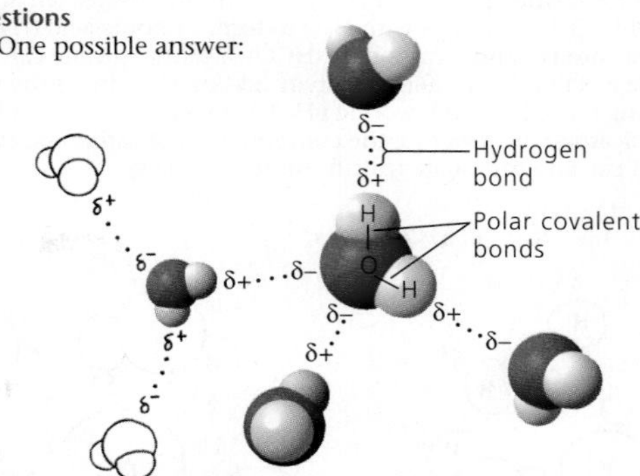

Figure 3.6 Without hydrogen bonds, water would behave like other small molecules, and the solid phase (ice) would be denser than liquid water. The ice would sink to the bottom and would no longer insulate the whole body of water, which would eventually freeze because the average annual temperature at the South Pole is −50°C. The krill could not survive. **Figure 3.7** Heating the solution would cause the water to evaporate faster than it is evaporating at room temperature. At a certain point, there wouldn't be enough water molecules to dissolve the salt ions. The salt would start coming out of solution and re-forming crystals. Eventually, all the water would evaporate, leaving behind a pile of salt like the original pile. **Figure 3.12** By causing the loss of coral reefs, a decrease in the ocean's carbonate concentration would have a ripple effect on noncalcifying organisms. Some of these organisms depend on the reef structure for protection, while others feed on species associated with reefs.

Concept Check 3.1

1. Electronegativity is the attraction of an atom for the electrons of a covalent bond. Because oxygen is more electronegative than hydrogen, the oxygen atom in H_2O pulls electrons toward itself, resulting in a partial negative charge on the oxygen atom and partial positive charges on the hydrogen atoms. Atoms in neighboring water molecules with opposite partial charges are attracted to each other, forming a hydrogen bond. **2.** The hydrogen atoms of one molecule, with their partial positive charges, would repel the hydrogen atoms of the adjacent molecule. **3.** The covalent bonds of water molecules would not be polar, and water molecules would not form hydrogen bonds with each other.

Concept Check 3.2

1. Hydrogen bonds hold neighboring water molecules together. This cohesion helps the chain of water molecules move upward against gravity in water-conducting cells as water evaporates from the leaves. Adhesion between water molecules and the walls of the water-conducting cells also helps counter gravity. **2.** High humidity hampers cooling by suppressing the evaporation of sweat. **3.** As water freezes, it expands because water molecules move farther apart in forming ice crystals. When there is water in a crevice of a boulder, expansion due to freezing may crack the boulder. **4.** A liter of blood would contain 7.8×10^{13} molecules of ghrelin (1.3×10^{-10} moles per liter $\times\ 6.02 \times 10^{23}$ molecules per mole). **5.** The hydrophobic substance repels water, perhaps helping to keep the ends of the legs from becoming coated with water and breaking through the surface. If the legs were coated with a hydrophilic substance, water would be drawn up them, possibly making it more difficult for the water strider to walk on water.

Concept Check 3.3

1. 10^5, or 100,000 **2.** $[H^+] = 0.01\ M = 10^{-2}\ M$, so pH = 2. **3.** $CH_3COOH \rightarrow CH_3COO^- + H^+$. CH_3COOH is the acid (the H^+ donor), and CH_3COO^- is the base (the H^+ acceptor). **4.** The pH of the water should decrease from 7 to about 2; the pH of the acetic acid solution will decrease only a small amount, because the reaction shown for question 3 will shift to the left, with CH_3COO^- accepting the influx of H^+ and becoming CH_3COOH molecules.

Summary of Key Concepts Questions

3.1

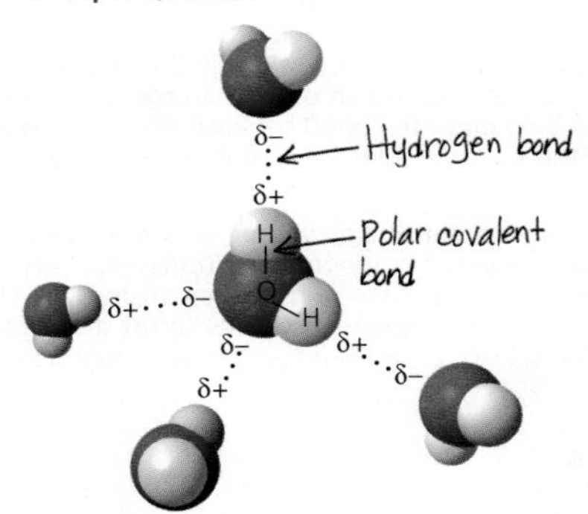

Each water molecule can make four hydrogen bonds with neighboring molecules. **3.2** Ions dissolve in water when polar water molecules form a hydration shell around them. Polar molecules dissolve as water molecules form hydrogen bonds with them and surround them. Solutions are homogeneous mixtures of solute and solvent. Colloids form when particles that are too large to dissolve remain suspended in a liquid. **3.3** CO_2 reacts with H_2O to form carbonic acid (H_2CO_3), which dissociates into H^+ and bicarbonate (HCO_3^-). Although the carbonic acid–bicarbonate reaction is a buffering system, adding CO_2 drives the reaction to the right, releasing more H^+ and lowering pH. The excess protons combine with CO_3^{2-} to form bicarbonate, lowering the concentration of carbonate available for the formation of calcium carbonate (calcification) by corals.

Test Your Understanding

1. d **2.** b **3.** c **4.** e **5.** c **6.** a **7.** e **8.** d

9.

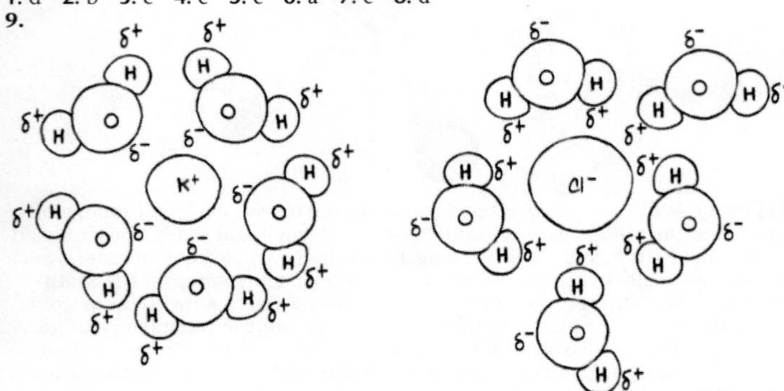

10. Both global warming and ocean acidification are caused by increasing levels of carbon dioxide in the atmosphere, the result of burning fossil fuels. **11.** Due to intermolecular hydrogen bonds, water has a high specific heat (the amount of heat required to increase the temperature of water by 1°C). When water is heated, much of the heat is absorbed in breaking hydrogen bonds before the water molecules increase their motion and the temperature increases. Conversely, when water is cooled, many H bonds are formed, which releases a significant amount of heat. This release of heat can provide some protection against freezing of the plants' leaves, thus protecting the cells from damage.

Chapter 4

Figure Questions

Figure 4.2 Because the concentration of the reactants influences the equilibrium (as discussed in Chapter 2), there might have been more HCN relative to CH_2O, since there would have been a higher concentration of the reactant gas containing nitrogen.

Figure 4.4

Na· ·P· ·S: ·Cl:

Figure 4.6 The tails of fats contain only carbon-hydrogen bonds, which are relatively nonpolar. Because the tails occupy the bulk of a fat molecule, they make the molecule as a whole nonpolar and therefore incapable of forming hydrogen bonds with water.

Figure 4.7

```
        H
        |
      H-C-H
   H    |    H
   |    |    |
 H-C----C----C-H
   |    |    |
   H    |    H
      H-C-H
        |
        H
```

Figure 4.9 Molecule b, because there are not only the two electronegative oxygens of the carboxyl group, but also an oxygen on the next (carbonyl) carbon. All of these oxygens help make the bond between the O and H of the —OH group more polar, thus making the dissociation of H+ more likely.

Concept Check 4.1

1. Prior to Wöhler's experiment, the prevailing view was that only living organisms could synthesize "organic" compounds. Wöhler made urea, an organic compound, without the involvement of living organisms. **2.** The spark provided energy needed for the inorganic molecules in the atmosphere to react with each other. (You'll learn more about energy and chemical reactions in Chapter 8.)

Concept Check 4.2

1.

```
H       H
 \     /
  C = C
 /     \
H       H
```

2. The forms of C_4H_{10} in (b) are structural isomers, as are the butenes in (c). **3.** Both consist largely of hydrocarbon chains. **4.** No. There is not enough diversity in the atoms. It can't form structural isomers because there is only one way for three carbons to attach to each other (in a line). There are no double bonds, so *cis-trans* isomers are not possible. Each carbon has at least two hydrogens attached to it, so the molecule is symmetrical and cannot have enantiomers.

Concept Check 4.3

1. It has both an amino group (—NH_2), which makes it an amine, and a carboxyl group (—COOH), which makes it a carboxylic acid. **2.** The ATP molecule loses a phosphate, becoming ADP. **3.** A chemical group that can act as a base has been replaced with a group that can act as an acid, increasing the acidic properties of the molecule. The shape of the molecule would also change, likely changing the molecules with which it can interact. The original cysteine molecule has an asymmetric carbon in the center. After replacement of the amino group with a carboxyl group, this carbon is no longer asymmetric.

```
 O     H     O
  \\   |   //
    C--C--C
   /   |   \
 HO   CH2   OH
       |
       SH
```

Summary of Key Concepts Questions

4.1 Miller showed that organic molecules could form under the physical and chemical conditions believed to have been present on early Earth. This abiotic synthesis of organic molecules would have been a first step in the origin of life. **4.2** Acetone and propanal are structural isomers. Acetic acid and glycine have no asymmetric carbons, whereas glycerol phosphate has one. Therefore, glycerol phosphate can exist as forms that are enantiomers, but acetic acid and glycine cannot. **4.3** The methyl group is nonpolar and not reactive. The other six groups are called functional groups. They are each hydrophilic, increasing the solubility of organic compounds in water, and can participate in chemical reactions.

Test Your Understanding

1. b **2.** b **3.** d **4.** d **5.** a **6.** b **7.** a **8.** The molecule on the right; the middle carbon is asymmetric.

9. ·Si· (with 4 dots) Si has 4 valence electrons, the same number as carbon. Therefore, silicon would be able to form long chains, including branches, that could act as skeletons for large molecules. It would clearly do this much better than neon (with no valence electrons) or aluminum (with 3 valence electrons).

Chapter 5

Figure Questions

Figure 5.3 Glucose and fructose are structural isomers.

Figure 5.4

Linear form → Ring forming → Ring form

Note that the oxygen on carbon 5 lost its proton and that the oxygen on carbon 2, which used to be the carbonyl oxygen, gained a proton. Four carbons are in the fructose ring, and two are not. (The latter two carbons are attached to carbons 2 and 5, which are in the ring.) The fructose ring differs from the glucose ring, which has five carbons in the ring and one that is not. (Note that the orientation of this fructose molecule is flipped relative to that of the one in Figure 5.5b.)

Figure 5.5

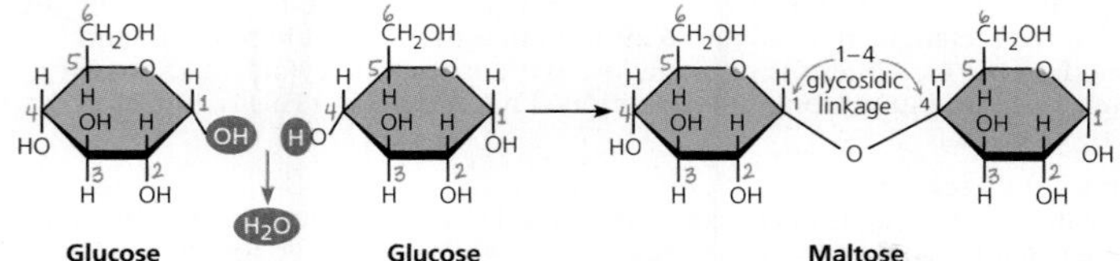

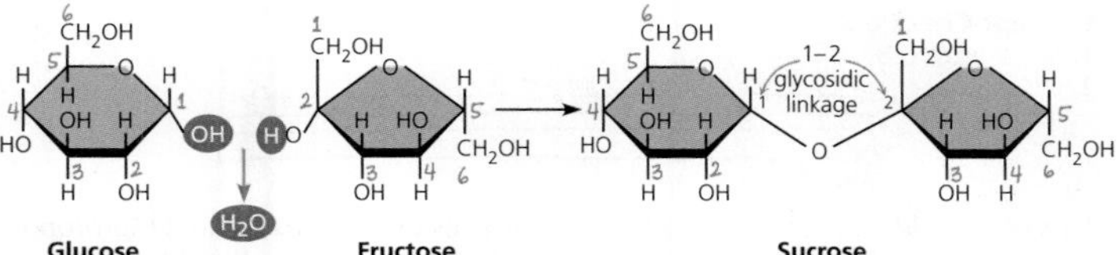

Figure 5.12

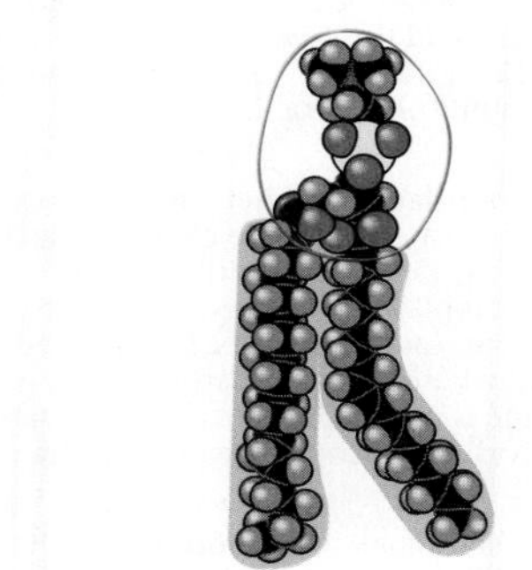

Figure 5.14

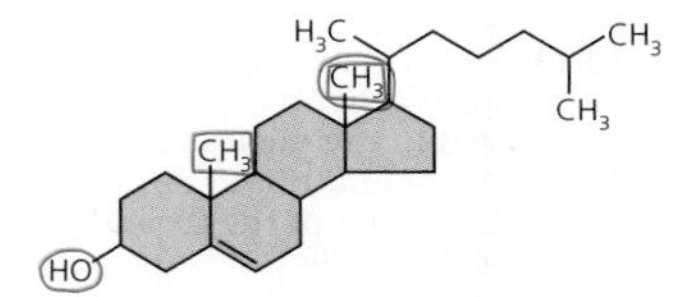

Figure 5.17

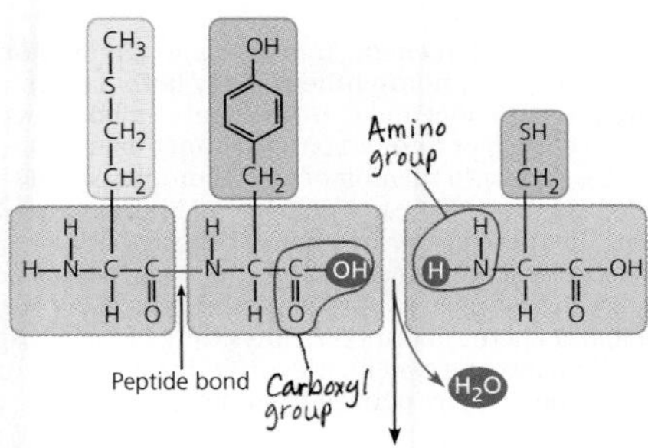

Figure 5.21 The R group on glutamic acid is acidic and hydrophilic, whereas that on valine is nonpolar and hydrophobic. Therefore, it is unlikely that valine can participate in the same intramolecular interactions that glutamic acid can. A change in these interactions causes a disruption of molecular structure.
Figure 5.24 The spirals are α helices.

Concept Check 5.1

1. The four main classes are proteins, carbohydrates, lipids, and nucleic acids. Lipids are not polymers. **2.** Nine, with one water molecule required to hydrolyze each connected pair of monomers **3.** The amino acids in the fish protein must be released in hydrolysis reactions and incorporated into other proteins in dehydration reactions.

Concept Check 5.2

1. $C_3H_6O_3$ **2.** $C_{12}H_{22}O_{11}$ **3.** The antibiotic treatment is likely to have killed the cellulose-digesting prokaryotes in the cow's stomach. The absence of these prokaryotes would hamper the cow's ability to obtain energy from food and could lead to weight loss and possibly death. Thus, prokaryotic species are reintroduced, in appropriate combinations, in the gut culture given to treated cows.

Concept Check 5.3

1. Both have a glycerol molecule attached to fatty acids. The glycerol of a fat has three fatty acids attached, whereas the glycerol of a phospholipid is attached to two fatty acids and one phosphate group. **2.** Human sex hormones are steroids, a type of hydrophobic compound. **3.** The oil droplet membrane could consist of a single layer of phospholipids rather than a bilayer, because an arrangement in which the hydrophobic tails of the membrane phospholipids were in contact with the hydrocarbon regions of the oil molecules would be more stable.

Concept Check 5.4

1. The function of a protein is a consequence of its specific shape, which is lost when a protein becomes denatured. **2.** Secondary structure involves hydrogen bonds between atoms of the polypeptide backbone. Tertiary structure involves interactions between atoms of the side chains of the amino acid subunits. **3.** These are all nonpolar amino acids, so you would expect this region to be located in the interior of the folded polypeptide, where it would not contact the aqueous environment inside the cell.

Concept Check 5.5

1.

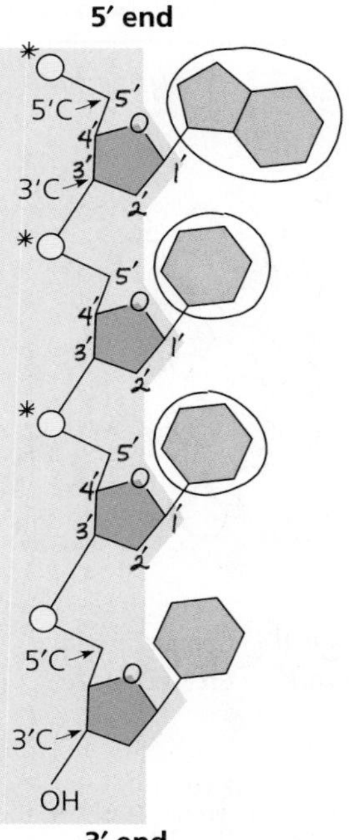

2. 5′–T A G G C C T–3′
3′–A T C C G G A–5′

3. a. Mismatch
5′–T A A G C C T–3′
3′–A T C C G G A–5′

b. 3′–A T T C G G A–5′

Summary of Key Concepts Questions

Concept 5.1 The polymers of carbohydrates, proteins, and nucleic acids are built from three different types of monomers: monosaccharides, amino acids, and nucleotides, respectively. **Concept 5.2** Both starch and cellulose are polymers of glucose, but the glucose monomers are in the α configuration in starch and the β configuration in cellulose. The glycosidic linkages thus have different geometries, giving the polymers different shapes and thus different properties. Starch is an energy-storage compound in plants; cellulose is a structural component of plant cell walls. Humans can hydrolyze starch to provide energy but cannot hydrolyze cellulose. Cellulose aids in the passage of food through the digestive tract. **Concept 5.3** Lipids are not polymers because they do not exist as a chain of linked monomers. They are not considered macromolecules because they do not reach the giant size of many polysaccharides, proteins, and nucleic acids. **Concept 5.4** A polypeptide, which may consist of hundreds of amino acids in a specific sequence (primary structure), has regions of coils and pleats (secondary structure), which are then folded into irregular contortions (tertiary structure) and may be noncovalently associated with other polypeptides (quaternary structure). The linear order of amino acids, with the varying properties of their side chains (R groups), determines what secondary and tertiary structures will form to produce a protein. The resulting unique three-dimensional shapes of proteins are key to their specific and diverse functions. **Concept 5.5** The complementary base pairing of the two strands of DNA makes possible the precise replication of DNA every time a cell divides, ensuring that genetic information is faithfully transmitted. In some types of RNA, complementary base pairing enables

RNA molecules to assume specific three-dimensional shapes that facilitate diverse functions.

Test Your Understanding

1. d **2.** a **3.** b **4.** a **5.** b **6.** c **7.** d

8.

	Monomers or Components	Polymer or larger molecule	Type of linkage
Carbohydrates	Monosaccharides	Polysaccharides	Glycosidic linkages
Lipids	Fatty acids	Triacylglycerols	Ester linkages
Proteins	Amino acids	Polypeptides	Peptide bonds
Nucleic acids	Nucleotides	Polynucleotides	Phosphodiester linkages

9.

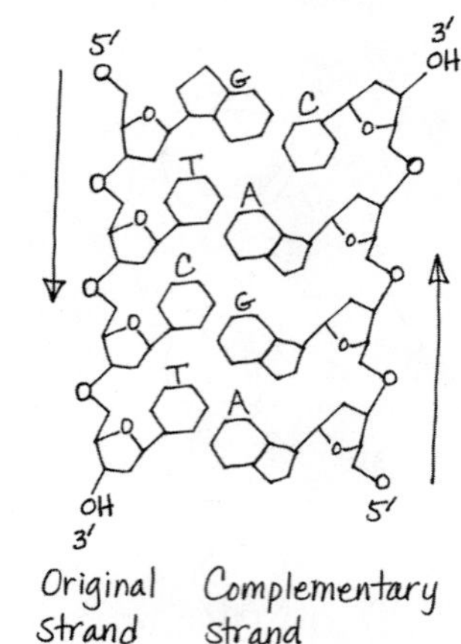

Chapter 6

Figure Questions

Figure 6.6 A phospholipid is a lipid, consisting of a glycerol molecule joined to two fatty acids and one phosphate group. Together, the glycerol and phosphate end of the phospholipid form the "head," which is hydrophilic, while the hydrocarbon chains on the fatty acids form hydrophobic "tails." The presence in a single molecule of both a hydrophilic and a hydrophobic region makes the molecule ideal as the main building block of a membrane. **Figure 6.9** The DNA in a chromosome dictates synthesis of a messenger RNA (mRNA) molecule, which then moves out to the cytoplasm. There, the information is used for the production, on ribosomes, of proteins that carry out cellular functions. **Figure 6.10** Any of the bound ribosomes (attached to the endoplasmic reticulum) could be circled, because any could be making a protein that will be secreted. **Figure 6.22** Each centriole has 9 sets of 3 microtubules, so the entire centrosome (two centrioles) has 54 microtubules. Each microtubule consists of a helical array of tubulin dimers (as shown in Table 6.1).

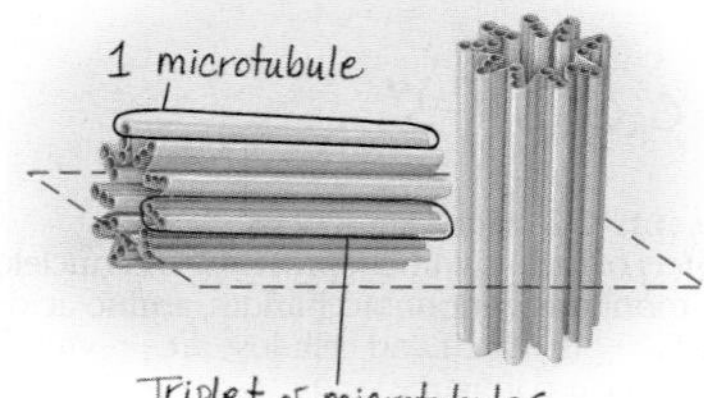

Figure 6.24

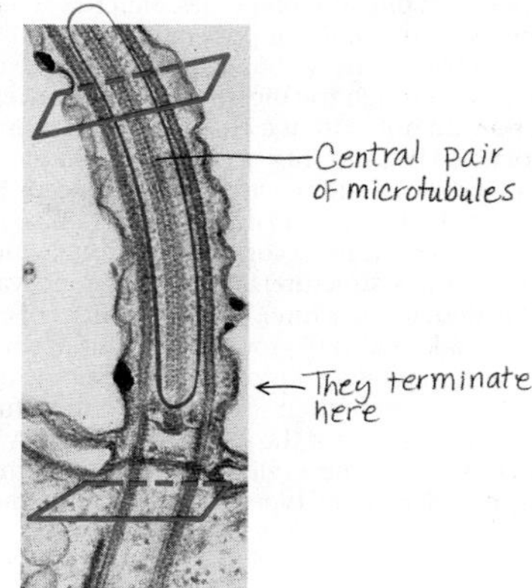

The two central microtubules terminate above the basal body, so they aren't present at the level of the cross section through the basal body, indicated by the lower red rectangle in (a). **Figure 6.29** The microtubules would reorient, and based on the earlier results, the cellulose synthase proteins would also change their path, orienting along the repositioned microtubules. (This is, in fact, what was observed.)

Concept Check 6.1

1. Stains used for light microscopy are colored molecules that bind to cell components, affecting the light passing through, while stains used for electron microscopy involve heavy metals that affect the beams of electrons passing through. **2.** (a) Light microscope, (b) scanning electron microscope

Concept Check 6.2

1. See Figure 6.8.

2.

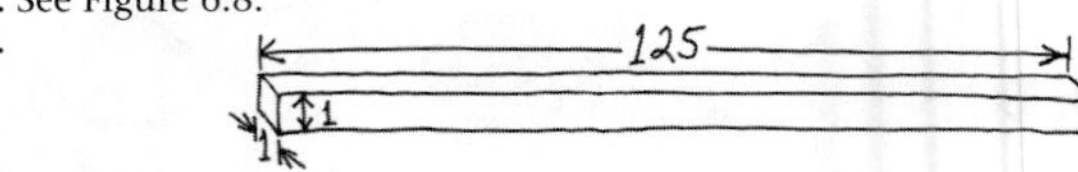

This cell would have the same volume as the cells in columns 2 and 3 but proportionally more surface area than that in column 2 and less than that in column 3. Thus, the surface-to-volume ratio should be greater than 1.2 but less than 6. To obtain the surface area, you would add the area of the six sides (the top, bottom, sides, and ends): 125 + 125 + 125 + 125 + 1 + 1 = 502. The surface-to-volume ratio equals 502 divided by a volume of 125, or 4.0.

Concept Check 6.3

1. Ribosomes in the cytoplasm translate the genetic message, carried from the DNA in the nucleus by mRNA, into a polypeptide chain. **2.** Nucleoli consist of DNA and the ribosomal RNA (rRNA) made according to its instructions, as well as proteins imported from the cytoplasm. Together, the rRNA and proteins are assembled into large and small ribosomal subunits. (These are exported through nuclear pores to the cytoplasm, where they will participate in polypeptide synthesis.) **3.** No. Each chromosome is present whether its chromatin is relatively diffuse (when the cell is not dividing) or condensed (when the cell is dividing).

Concept Check 6.4

1. The primary distinction between rough and smooth ER is the presence of bound ribosomes on the rough ER. Both types of ER make phospholipids, but membrane proteins and secretory proteins are all produced on the ribosomes of the rough ER. The smooth ER also functions in detoxification, carbohydrate metabolism, and storage of calcium ions. **2.** Transport vesicles move membranes and substances they enclose between other components of the endomembrane system. **3.** The mRNA is synthesized in the nucleus and then passes out through a nuclear pore to be translated on a bound ribosome, attached to the rough ER. The protein is synthesized into the lumen of the ER and perhaps modified there. A transport vesicle carries the protein to the Golgi apparatus. After further modification in the Golgi, another transport vesicle carries it back to the ER, where it will perform its cellular function.

Concept Check 6.5

1. Both organelles are involved in energy transformation, mitochondria in cellular respiration and chloroplasts in photosynthesis. They both have multiple membranes that separate their interiors into compartments. In both organelles, the innermost membranes—cristae, or infoldings of the inner membrane, in mitochondria, and the thylakoid membranes in chloroplasts—have large surface areas with embedded enzymes that carry out their main functions. **2.** Yes. Plant cells are able to make their own sugar by photosynthesis, but mitochondria in these eukaryotic cells are the organelles that are able to generate energy from sugars, a function required in all cells. **3.** Mitochondria and chloroplasts are not derived from the ER, nor are they connected physically or via transport vesicles to organelles of the endomembrane system. Mitochondria and chloroplasts are structurally quite different from vesicles derived from the ER, which are bounded by a single membrane.

Concept Check 6.6

1. Both systems of movement involve long filaments that are moved in relation to each other by motor proteins that grip, release, and grip again adjacent polymers. **2.** Dynein arms, powered by ATP, move neighboring doublets of microtubules relative to each other. Because they are anchored within the organelle and with respect to one another, the doublets bend instead of sliding past each other. Synchronized bending of the nine microtubule doublets brings about bending of both structures. **3.** Such individuals have defects in the microtubule-based movement of cilia and flagella. Thus, the sperm can't move because of malfunctioning or nonexistent flagella, and the airways are compromised because cilia that line the trachea malfunction or don't exist, and so mucus cannot be cleared from the lungs.

Concept Check 6.7

1. The most obvious difference is the presence of direct cytoplasmic connections between cells of plants (plasmodesmata) and animals (gap junctions). These connections result in the cytoplasm being continuous between adjacent cells. **2.** The cell would not be able to function properly and would probably soon die, as the cell wall or ECM must be permeable to allow the exchange of matter between the cell and its external environment. Molecules involved in energy production and use must be allowed entry, as well as those that provide information about the cell's environment. Other molecules, such as products synthesized by the cell for export and the by-products of cellular respiration, must be allowed to exit. **3.** The parts of the protein that face aqueous regions would be expected to have polar or charged (hydrophilic) amino acids, while the parts that go through the membrane would be expected to have nonpolar (hydrophobic) amino acids. You would predict polar or charged amino acids at each end (tail), in the region of the cytoplasmic loop, and in the regions of the two extracellular loops. You would predict nonpolar amino acids in the four regions that go through the membrane between the tails and loops.

Summary of Key Concepts Questions

6.1 Both light and electron microscopy allow cells to be studied visually, thus helping us understand internal cellular structure and the arrangement of cell components. Cell fractionation techniques separate out different groups of cell components, which can then be analyzed biochemically to determine their function. Performing microscopy on the same cell fraction helps to correlate the biochemical function of the cell with the cell component responsible. **6.2** The separation of different functions in different organelles has several advantages. Reactants and enzymes can be concentrated in one area instead of spread throughout the cell. Reactions that require specific conditions, such as a lower pH, can be compartmentalized. And enzymes for specific reactions are often embedded in the membranes that enclose or partition an organelle. **6.3** The nucleus contains the genetic material of the cell in the form of DNA, which codes for messenger RNA, which in turn provides instructions for the synthesis of proteins (including the proteins that make up part of the ribosomes). DNA also codes for ribosomal RNA, which is combined with proteins in the nucleolus into the subunits of ribosomes. Within the cytoplasm, ribosomes join with mRNA to build polypeptides, using the genetic information in the mRNA. **6.4** Transport vesicles move proteins and membranes synthesized by the rough ER to the Golgi for further processing and then to the plasma membrane, lysosomes, or other locations in the cell, including back to the ER. **6.5** According to the endosymbiont theory, mitochondria originated from an oxygen-using prokaryotic cell that was engulfed by an ancestral eukaryotic cell. Over time, the host and endosymbiont evolved into a single organism. Chloroplasts originated when at least one of these eukaryotic cells containing mitochondria engulfed and then retained a photosynthetic prokaryote. **6.6** Inside the cell, motor proteins interact with components of the cytoskeleton to move cellular parts. Motor proteins may "walk" vesicles along microtubules. The movement of cytoplasm within a cell involves interactions of the motor protein myosin and microfilaments (actin filaments). Whole cells can be moved by the rapid bending of flagella or cilia, which is caused by the motor-protein-powered sliding of microtubules within these structures. Cell movement can also occur when pseudopodia form at one end of a cell (caused by actin polymerization into a filamentous network), followed by contraction of the cell toward that end; this is powered by interactions of microfilaments with myosin. Interactions of motor proteins and microfilaments in muscle cells can propel whole organisms. **6.7** A plant cell wall is primarily composed of microfibrils of cellulose embedded in other polysaccharides and proteins. The ECM of animal cells is primarily composed of collagen and other glycoprotein fibers, such as fibronectins. These fibers are embedded in a network of carbohydrate-rich proteoglycans. A plant cell wall provides structural support for the cell and, collectively, for the plant body. In addition to giving support, the ECM of an animal cell allows for communication of environmental changes into the cell.

Test Your Understanding

1. b **2.** d **3.** b **4.** e **5.** a **6.** d **7.** c **8.** See Figure 6.8.

Chapter 7

Figure Questions

Figure 7.2

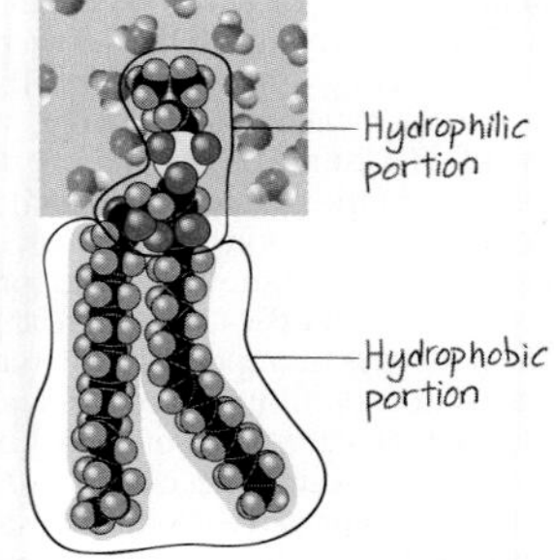

The hydrophilic portion is in contact with an aqueous environment (cytosol or extracellular fluid), and the hydrophobic portion is in contact with the hydrophobic portions of other phospholipids in the interior of the bilayer. **Figure 7.7** You couldn't rule out movement of proteins within membranes of the same species. You might propose that the membrane lipids and proteins from one species weren't able to mingle with those from the other species because of some incompatibility. **Figure 7.10** A transmembrane protein like the dimer in (f) might change its shape upon binding to a particular ECM molecule. The new shape might enable the interior portion of the protein to bind to a second, cytoplasmic protein that would relay the message to the inside of the cell, as shown in (c). **Figure 7.11** The shape of a protein on the HIV surface is likely to be complementary to the shape of the receptor (CD4) and also to that of the co-receptor (CCR5). A molecule that was a similar shape to the HIV surface protein could bind CCR5, blocking HIV binding. (Another alternative would be a molecule that bound to CCR5 and changed the shape of CCR5 so it could no longer bind HIV.)

Figure 7.12

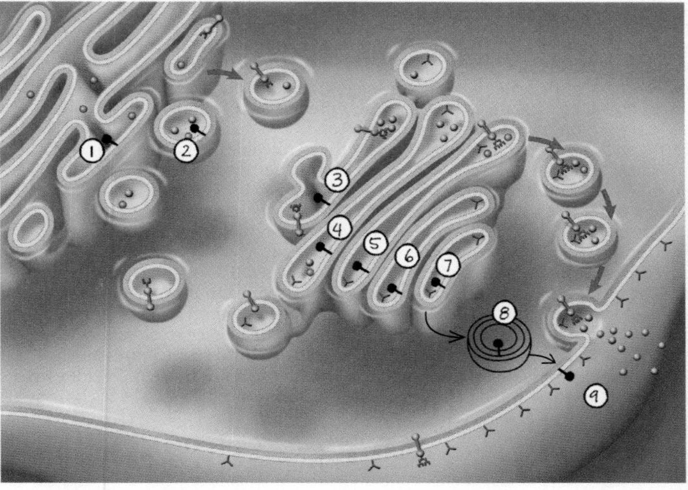

The protein would contact the extracellular fluid. **Figure 7.14** The orange dye would be evenly distributed throughout the solution on both sides of the membrane. The solution levels would not be affected because the orange dye can diffuse through the membrane and equalize its concentration. Thus, no additional osmosis would take place in either direction. **Figure 7.19** The diamond solutes are moving into the cell (down), and the round solutes are moving out of the cell (up); both are moving against their concentration gradient.

Concept Check 7.1

1. They are on the inner side of the transport vesicle membrane. **2.** The grasses living in the cooler region would be expected to have more unsaturated fatty acids in their membranes because those fatty acids remain fluid at lower temperatures. The grasses living immediately adjacent to the hot springs would be expected to have more saturated fatty acids, which would allow the fatty acids to "stack" more closely, making the membranes less fluid and therefore helping them to stay intact at higher temperatures. (Cholesterol could not be used to moderate the effects of temperature on membrane fluidity because it is not found within plant cell membranes.)

Concept Check 7.2

1. O_2 and CO_2 are both nonpolar molecules that can easily pass through the hydrophobic interior of a membrane. **2.** Water is a polar molecule, so it cannot pass very rapidly through the hydrophobic region in the middle of a phospholipid bilayer. **3.** The hydronium ion is charged, while glycerol is not. Charge is probably more significant than size as a basis for exclusion by the aquaporin channel.

Concept Check 7.3

1. CO_2 is a nonpolar molecule that can diffuse through the plasma membrane. As long as it diffuses away so that the concentration remains low outside the cell, it will continue to exit the cell in this way. (This is the opposite of the case for O_2, described in this section.) **2.** The water is hypotonic to the plant cells, so the plant cells take up water. Thus, the cells of the vegetable remain turgid rather than plasmolyzing, and the vegetable (for example, lettuce or spinach) remains crisp and not wilted. **3.** The activity of *Paramecium caudatum*'s contractile vacuole will decrease. The vacuole pumps out excess water that accumulates in the cell; this accumulation occurs only in a hypotonic environment.

Concept Check 7.4

1. The pump uses ATP. To establish a voltage, ions have to be pumped against their gradients, which requires energy. **2.** Each ion is being transported against its electrochemical gradient. If either ion were transported down its electrochemical gradient, this *would* be considered cotransport. **3.** The internal environment of a lysosome is acidic, so it has a higher concentration of H^+ than does the cytoplasm. Therefore, you might expect the membrane of the lysosome to have a proton pump such as that shown in Figure 7.20 to pump H^+ into the lysosome.

Concept Check 7.5

1. Exocytosis. When a transport vesicle fuses with the plasma membrane, the vesicle membrane becomes part of the plasma membrane.

2.

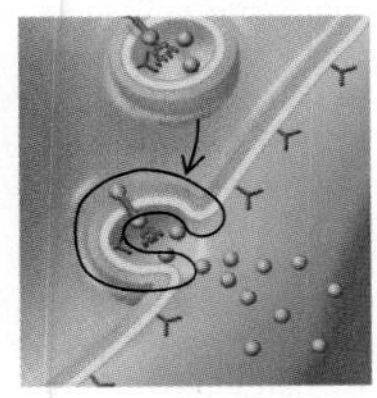

3. The glycoprotein would be synthesized in the ER lumen, move through the Golgi apparatus, and then travel in a vesicle to the plasma membrane, where it would undergo exocytosis and become part of the ECM.

Appendix A Answers

Summary of Key Concepts Questions

7.1 Plasma membranes define the cell by separating the cellular components from the external environment. This allows conditions inside cells to be controlled by membrane proteins, which regulate entry and exit of molecules and even cell function (see Figure 7.10). The processes of life can be carried out inside the controlled environment of the cell, so membranes are crucial. In eukaryotes, membranes also function to subdivide the cytoplasm into different compartments where distinct processes can occur, even under differing conditions such as pH. **7.2** Aquaporins are channel proteins that greatly increase the permeability of a membrane to water molecules, which are polar and therefore do not readily diffuse through the hydrophobic interior of the membrane. **7.3** There will be a net diffusion of water out of a cell into a hypertonic solution. The free water concentration is higher inside the cell than in the solution (where water molecules are not free, but are clustered around the higher concentration of solute particles). **7.4** One of the solutes moved by the cotransporter is actively transported against its concentration gradient. The energy for this transport comes from the concentration gradient of the other solute, which was established by an electrogenic pump that used energy to transport the other solute across the membrane. **7.5** In receptor-mediated endocytosis, specific molecules act as ligands when they bind to receptors on the plasma membrane. The cell can acquire bulk quantities of those molecules when a coated pit forms a vesicle and carries the bound molecules into the cell.

Test Your Understanding

1. b **2.** c **3.** a **4.** d **5.** b

6. (a)

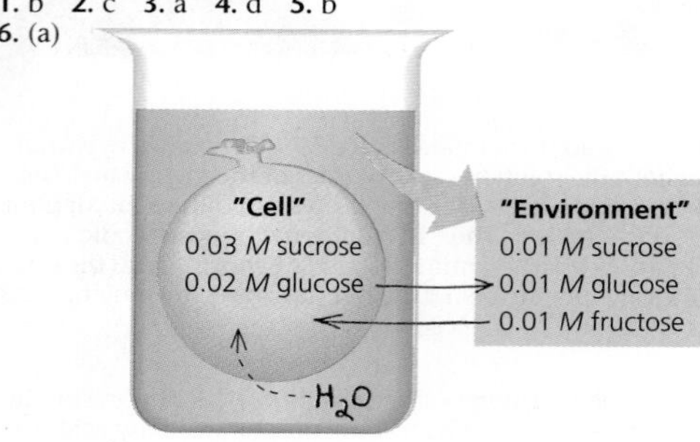

(b) The solution outside is hypotonic. It has less sucrose, which is a nonpenetrating solute. (c) See answer for (a). (d) The artificial cell will become more turgid. (e) Eventually, the two solutions will have the same solute concentrations. Even though sucrose can't move through the membrane, water flow (osmosis) will lead to isotonic conditions.

Chapter 8

Figure Questions

Figure 8.12 **Figure 8.16**

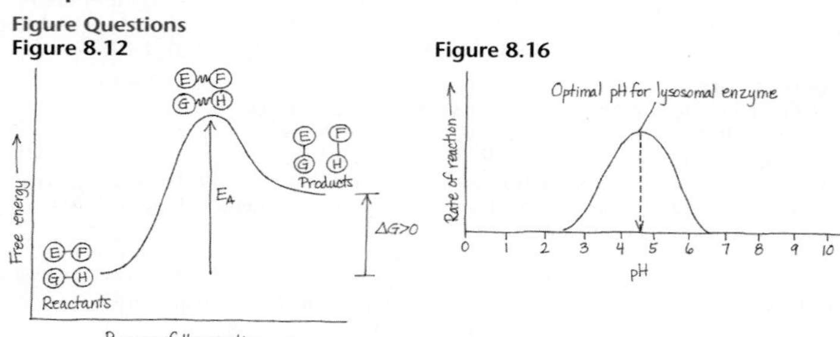

Figure 8.20 Because the affinity of the caspase for the inhibitor is very low (as is expected of an allosterically inhibited enzyme), the inhibitor is likely to diffuse away. Because no additional source of the inhibitory compound is present and the concentration of inhibitor is very low, the inhibitor is unlikely to bind again to the allosteric binding site once the covalent linkage is broken. Therefore, the activity of the enzyme would most likely be normal. (In fact, this is what the researchers observed when they broke the disulfide linkage.)

Concept Check 8.1

1. The second law is the trend toward randomization, or increasing entropy. When the concentrations of a substance on both sides of a membrane are equal, the distribution is more random than when they are unequal. Diffusion of a substance to a region where it is initially less concentrated increases entropy, making it an energetically favorable (spontaneous) process as described by the second law. This explains the process seen in Figure 7.13. **2.** The apple has potential energy in its position hanging on the tree, and the sugars and other nutrients it contains have chemical energy. The apple has kinetic energy as it falls from the tree to the ground. Finally, when the apple is digested and its molecules broken down, some of the chemical energy is used to do work, and the rest is lost as thermal energy. **3.** The sugar crystals become less ordered (entropy increases) as they dissolve and become randomly spread out in the water. Over time, the water evaporates, and the crystals form again because the water volume is insufficient to keep them in solution. While the reappearance of sugar crystals may represent a "spontaneous" increase in order (decrease in entropy), it is balanced by the decrease in order (increase in entropy) of the water molecules, which changed from a relatively compact arrangement as liquid water to a much more dispersed and disordered form as water vapor.

Concept Check 8.2

1. Cellular respiration is a spontaneous and exergonic process. The energy released from glucose is used to do work in the cell or is lost as heat. **2.** When the H^+ concentrations are the same, the system is at equilibrium and can do no work. Hydrogen ions can perform work only if their concentrations on each side of a membrane differ—in other words, when a gradient is present. This is consistent with Figure 7.20, which shows that an energy input (provided by ATP hydrolysis) is required to establish the concentration gradient (the H^+ gradient) that can in turn perform work. **3.** The reaction is exergonic because it releases energy—in this case, in the form of light. (This is a nonbiological version of the bioluminescence seen in Figure 8.1.)

Concept Check 8.3

1. ATP usually transfers energy to endergonic processes by phosphorylating (adding phosphate groups to) other molecules. (Exergonic processes phosphorylate ADP to regenerate ATP.) **2.** A set of coupled reactions can transform the first combination into the second. Since this is an exergonic process overall, ΔG is negative and the first combination must have more free energy (see Figure 8.9). **3.** Active transport: The solute is being transported against its concentration gradient, which requires energy, provided by ATP hydrolysis.

Concept Check 8.4

1. A spontaneous reaction is a reaction that is exergonic. However, if it has a high activation energy that is rarely attained, the rate of the reaction may be low. **2.** Only the specific substrate(s) will fit properly into the active site of an enzyme, the part of the enzyme that carries out catalysis. **3.** In the presence of malonate, increase the concentration of the normal substrate (succinate) and see whether the rate of reaction increases. If it does, malonate is a competitive inhibitor. **4.** If lactose wasn't present in the environment as a source of food and the fucose-containing disaccharide was available, bacteria that could digest the latter would be better able to grow and multiply than those that could not.

Concept Check 8.5

1. The activator binds in such a way that it stabilizes the active form of an enzyme, whereas the inhibitor stabilizes the inactive form. **2.** An inhibitor that binds to the active site of the enzyme you want to inhibit could also bind to and block the enzymes with similar structures, causing significant side effects. For this reason, you would be better off choosing to screen chemical compounds that bind allosterically to the enzyme in question, because allosteric regulatory sites are less likely to share similarity with other enzymes.

Summary of Key Concepts Questions

8.1 The process of "ordering" a cell's structure is accompanied by an increase in the entropy or disorder of the universe. For example, an animal cell takes in highly ordered organic molecules as the source of matter and energy used to build and maintain its structures. In the same process, however, the cell releases heat and the simple molecules of carbon dioxide and water to the surroundings. The increase in entropy of the latter process offsets the entropy decrease in the former. **8.2** A spontaneous reaction has a negative ΔG and is exergonic. For a chemical reaction to proceed with a net release of free energy ($-\Delta G$), the enthalpy or total energy of the system must decrease ($-\Delta H$), and/or the entropy or disorder must increase (yielding a more negative term, $-T\Delta S$). Spontaneous reactions supply the energy to perform cellular work. **8.3** The free energy released from the hydrolysis of ATP may drive endergonic reactions through the transfer of a phosphate group to a reactant molecule, forming a more reactive phosphorylated intermediate. ATP hydrolysis also powers the mechanical and transport work of a cell, often by powering shape changes in the relevant motor proteins. Cellular respiration, the catabolic breakdown of glucose, provides the energy for the endergonic regeneration of ATP from ADP and Ⓟ$_i$. **8.4** Activation energy barriers prevent the complex molecules of the cell, which are rich in free energy, from spontaneously breaking down to less ordered, more stable molecules. Enzymes permit a regulated metabolism by binding to specific substrates and forming enzyme-substrate complexes that selectively lower the E_A for the chemical reactions in a cell. **8.5** A cell tightly regulates its metabolic pathways in response to fluctuating needs for energy and materials. The binding of activators or inhibitors to regulatory sites on allosteric enzymes stabilizes either the active or inactive form of the subunits. For example, the binding of ATP to a catabolic enzyme in a cell with excess ATP would inhibit that pathway. Such types of feedback inhibition preserve chemical resources within a cell. If ATP supplies are depleted, binding of ADP to the regulatory site of catabolic enzymes would activate that pathway.

Test Your Understanding

1. b **2.** c **3.** b **4.** a **5.** c **6.** e **7.** c

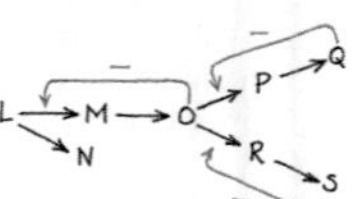

9.

Amount of product
Time
A B C D

A. The substrate molecules are entering the cells, so no product is made yet.

B. There is sufficient substrate, so the reaction is proceeding at a maximum rate.

C. As the substrate is used up, the rate decreases (the slope is less steep).

D. The line is flat because no new substrate remains and thus no new product appears.

Chapter 9

Figure Questions

Figure 9.7 Because there is no external source of energy for the reaction, it must be exergonic, and the reactants must be at a higher energy level than the products. **Figure 9.9** The removal would probably stop glycolysis, or at least slow it down, since it would push the equilibrium for step 5 toward the left. If less (or no) glyceraldehyde 3-phosphate were available, step 6 would slow down (or be unable to occur). **Figure 9.15** At first, some ATP could be made, since electron transport could proceed as far as complex III, and a small H^+ gradient could be built up. Soon, however, no more electrons could be passed to complex III because it could not be reoxidized by passing its electrons to complex IV. **Figure 9.16** First, there are 2 NADH from the oxidation of pyruvate plus 6 NADH from the citric acid cycle (CAC); 8 NADH $\times$ 2.5 ATP/NADH = 20 ATP. Second, there are 2 $FADH_2$ from the CAC; 2 $FADH_2 \times 1.5$ ATP/$FADH_2$ = 3 ATP. Third, the 2 NADH from glycolysis enter the mitochondrion through one of two types of shuttle. They pass their electrons either to 2 FAD, which become $FADH_2$ and result in 3 ATP, or to 2 NAD^+, which become NADH and result in 5 ATP. Thus, 20 + 3 + 3 = 26 ATP, or 20 + 3 + 5 = 28 ATP from all NADH and $FADH_2$.

Concept Check 9.1

1. Both processes include glycolysis, the citric acid cycle, and oxidative phosphorylation. In aerobic respiration, the final electron acceptor is molecular oxygen (O_2); in anaerobic respiration, the final electron acceptor is a different substance. **2.** $C_4H_6O_5$ would be oxidized and NAD^+ would be reduced.

Concept Check 9.2

1. NAD^+ acts as the oxidizing agent in step 6, accepting electrons from glyceraldehyde 3-phosphate, which thus acts as the reducing agent. **2.** Since the overall process of glycolysis results in net production of ATP, it would make sense for the process to slow down when ATP levels have increased substantially. Thus, we would expect ATP to allosterically inhibit phosphofructokinase.

Concept Check 9.3

1. NADH and $FADH_2$; they will donate electrons to the electron transport chain. **2.** CO_2 is released from the pyruvate that is the end product of glycolysis, and CO_2 is also released during the citric acid cycle. **3.** In both cases, the precursor molecule loses a CO_2 molecule and then donates electrons to an electron carrier in an oxidation step. Also, the product has been activated due to the attachment of a CoA group.

Concept Check 9.4

1. Oxidative phosphorylation would eventually stop entirely, resulting in no ATP production by this process. Without oxygen to "pull" electrons down the electron transport chain, H^+ would not be pumped into the mitochondrion's intermembrane space and chemiosmosis would not occur. **2.** Decreasing the pH means addition of H^+. This would establish a proton gradient even without the function of the electron transport chain, and we would expect ATP synthase to function and synthesize ATP. (In fact, it was experiments like this that provided support for chemiosmosis as an energy-coupling mechanism.) **3.** One of the components of the electron transport chain, ubiquinone (Q), must be able to diffuse within the membrane. It could not do so if the membrane were locked rigidly into place.

Concept Check 9.5

1. A derivative of pyruvate, such as acetaldehyde during alcohol fermentation, or pyruvate itself during lactic acid fermentation; oxygen **2.** The cell would need to consume glucose at a rate about 16 times the consumption rate in the aerobic environment (2 ATP are generated by fermentation versus up to 32 ATP by cellular respiration).

Concept Check 9.6

1. The fat is much more reduced; it has many $—CH_2—$ units, and in all these bonds the electrons are equally shared. The electrons present in a carbohydrate molecule are already somewhat oxidized (shared unequally in bonds), as quite a few of them are bound to oxygen. **2.** When we consume more food than necessary for metabolic processes, our body synthesizes fat as a way of storing energy for later use. **3.** Glycogen is a storage polysaccharide in liver and muscle cells. When energy is needed, glucose units are hydrolyzed from glycogen. Glycolysis in the cytosol breaks down glucose to two pyruvate molecules, which are transported into the mitochondrion. Here they are further oxidized, ultimately producing the needed ATP. **4.** AMP will accumulate, stimulating phosphofructokinase, and thus increasing the rate of glycolysis. Since oxygen is not present, the cell will convert pyruvate to lactate in lactic acid fermentation, providing a supply of ATP. **5.** When oxygen is present, the fatty acid chains containing most of the energy of a fat are oxidized and fed into the citric acid cycle and the electron transport chain. During intense exercise, however, oxygen is scarce in muscle cells, so ATP must be generated by glycolysis alone. A very small part of the fat molecule, the glycerol backbone, can be oxidized via glycolysis, but the amount of energy released by this portion is insignificant compared to that released by the fatty acid chains. (This is why moderate exercise, staying below 70% maximum heart rate, is better for burning fat—because enough oxygen remains available to the muscles.)

Summary of Key Concepts Questions

9.1 Most of the ATP produced in cellular respiration comes from oxidative phosphorylation, in which the energy released from redox reactions in an electron transport chain is used to produce ATP. In substrate-level phosphorylation, an enzyme directly transfers a phosphate group to ADP from an intermediate substrate. All ATP production in glycolysis occurs by substrate-level phosphorylation; this form of ATP production also occurs at one step in the citric acid cycle. **9.2** The oxidation of the three-carbon sugar, glyceraldehyde 3-phosphate, yields energy. In this oxidation, electrons and H^+ are transferred to NAD^+, forming NADH, and a phosphate group is attached to the oxidized substrate. ATP is then formed by substrate-level phosphorylation when this phosphate group is transferred to ADP. **9.3** The release of six molecules of CO_2 represents the complete oxidation of glucose. During the processing of two pyruvates to acetyl CoA, the fully oxidized carboxyl group ($—COO^-$) is given off as CO_2. The remaining four carbons are released as CO_2 in the citric acid cycle as citrate is oxidized back to oxaloacetate. **9.4** The flow of H^+ through the ATP synthase complex causes the rotor and attached rod to rotate, exposing catalytic sites in the knob portion that produce ATP from ADP and Ⓟ$_i$. ATP synthases are found in the inner mitochondrial membrane, the plasma membrane of prokaryotes, and membranes within chloroplasts. **9.5** Anaerobic respiration yields more ATP. The 2 ATP produced by substrate-level phosphorylation in glycolysis represent the total energy yield of fermentation. NADH passes its "high-energy" electrons to pyruvate or a derivative of pyruvate, recycling NAD^+ and allowing glycolysis to continue. Anaerobic respiration uses an electron transport chain to capture the energy of the electrons in NADH via a series of redox reactions; ultimately, the electrons are transferred to an electronegative molecule other than oxygen. And additional molecules of NADH are produced in anaerobic respiration as pyruvate is oxidized. **9.6** The ATP produced by catabolic pathways is used to drive anabolic pathways. Also, many of the intermediates of glycolysis and the citric acid cycle are used in the biosynthesis of a cell's molecules.

Test Your Understanding

1. d **2.** c **3.** c **4.** a **5.** e **6.** a **7.** b

8.

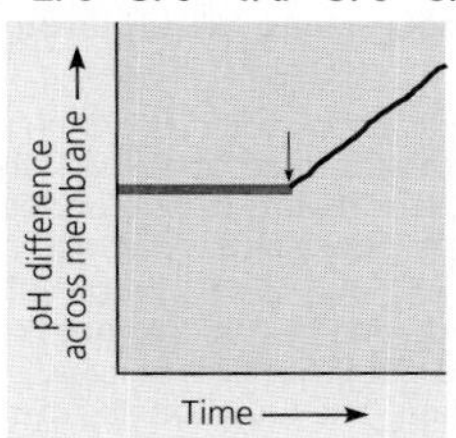

Chapter 10

Figure Questions

Figure 10.3 Situating containers of algae near sources of CO_2 emissions makes sense because algae need CO_2 to carry out photosynthesis. The higher their rate of photosynthesis, the more plant oil they will produce. At the same time, algae would be absorbing the CO_2 emitted from industrial plants or from car engines, reducing the amount of CO_2 entering the atmosphere. **Figure 10.10** Red, but not violet-blue, wavelengths would pass through the filter, so the bacteria would not congregate where the violet-blue light normally comes through. Therefore, the left "peak" of bacteria would not be present, but the right peak would be observed because the red wavelengths passing through the filter would be used for photosynthesis. **Figure 10.12** In the leaf, most of the chlorophyll electrons excited by photon absorption are used to power the reactions of photosynthesis. **Figure 10.16** The person at the top of the photosystem I tower would not turn and throw his electron into the bucket. Instead, he would throw it onto the top of the ramp right next to the photosystem II tower. The electron would then roll down the ramp, get energized by a photon, and return to him. This cycle would continue as long as light was available. (This is why it's called cyclic electron flow.) **Figure 10.19**

3 × 1 (=3)
3 × 6 (=18)
6 × 3 (=18)
6 × 3 (=18)
6 × 3 (=18)
1 × 3 (=3)
5 × 3 (=15)
3 × 5 (=15)

Three carbon atoms enter the cycle, one by one, as individual CO_2 molecules, and leave the cycle in one three-carbon molecule (G3P) per three turns of the cycle. **Figure 10.22**

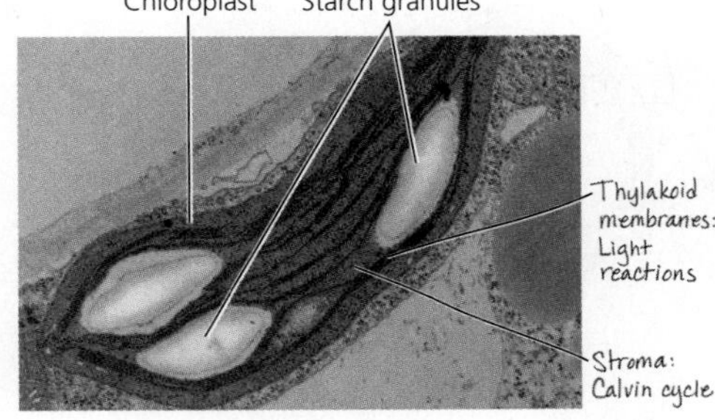

The photosystems that carry out the light reactions are embedded in the thylakoid membranes, and the ATP and NADPH that are formed are released into the stroma. There, they are used for the reactions of the Calvin cycle, which produces G3P. Excess sugar molecules that are not used by the plant can be converted to glucose, then stored in the form of starch.

Concept Check 10.1

1. CO_2 enters leaves via stomata, and water enters via roots and is carried to leaves through veins. **2.** Using ^{18}O, a heavy isotope of oxygen, as a label, researchers were able to confirm van Niel's hypothesis that the oxygen produced during photosynthesis originates in water, not in carbon dioxide. **3.** The light reactions could *not* keep producing NADPH and ATP without the $NADP^+$, ADP, and Ⓟ$_i$ that the Calvin cycle generates. The two cycles are interdependent.

Concept Check 10.2

1. Green, because green light is mostly transmitted and reflected—not absorbed—by photosynthetic pigments **2.** In chloroplasts, light-excited electrons are trapped by a primary electron acceptor, which prevents them from dropping back to the ground state. In isolated chlorophyll, there is no electron acceptor, so the photoexcited electrons immediately drop back down to the ground state, with the emission of light and heat. **3.** Water (H_2O) is the initial electron donor; $NADP^+$ accepts electrons at the end of the electron transport chain, becoming reduced to NADPH. **4.** In this experiment, the rate of ATP synthesis would slow and eventually stop. Because the added compound would not allow a proton gradient to build up across the membrane, ATP synthase could not catalyze ATP production.

Concept Check 10.3

1. 6, 18, 12 **2.** The more potential energy a molecule stores, the more energy and reducing power is required for the formation of that molecule. Glucose is a valuable energy source because it is highly reduced, storing lots of potential energy in its electrons. To reduce CO_2 to glucose, much energy and reducing power are required in the form of large numbers of ATP and NADPH molecules, respectively. **3.** The light reactions require ADP and $NADP^+$, which would not be formed in sufficient quantities from ATP and NADPH if the Calvin cycle stopped. **4.** In glycolysis, G3P acts as an intermediate. The 6-carbon sugar fructose 1,6-bisphosphate is cleaved into two 3-carbon sugars, one of which is G3P. The other is an isomer called dihydroxyacetone phosphate, which can be converted to G3P by an isomerase. Because G3P is the substrate for the next enzyme, it is constantly removed, and the reaction equilibrium is pulled in the direction of conversion of dihydroxyacetone phosphate to more G3P. In the Calvin cycle, G3P acts as both an intermediate and a product. For every three CO_2 molecules that enter the cycle, six G3P molecules are formed, five of which must remain in the cycle and become rearranged to regenerate three 5-carbon RuBP molecules. The one remaining G3P is a product, which can be thought of as the result of "reducing" the three CO_2 molecules that entered the cycle into a 3-carbon sugar that can later be used to generate energy.

Concept Check 10.4

1. Photorespiration decreases photosynthetic output by adding oxygen, instead of carbon dioxide, to the Calvin cycle. As a result, no sugar is generated (no carbon is fixed), and O_2 is used rather than generated. **2.** Without PS II, no O_2 is generated in bundle-sheath cells. This avoids the problem of O_2 competing with CO_2 for binding to rubisco in these cells. **3.** Both problems are caused by a drastic change in Earth's atmosphere due to burning of fossil fuels. The increase in CO_2 concentration affects ocean chemistry by decreasing pH, thus affecting calcification by marine organisms. On land, CO_2 concentration and air temperature are conditions that plants have become adapted to, and changes in these characteristics have a strong effect on photosynthesis by plants. Thus, alteration of these two fundamental factors could have critical effects on organisms all around the planet, in all different habitats. **4.** C_4 and CAM species would replace many of the C_3 species.

Summary of Key Concepts Questions

10.1 CO_2 and H_2O are the products of respiration; they are the reactants in photosynthesis. In respiration, glucose is oxidized to CO_2 as electrons are passed through an electron transfer chain from glucose to O_2, producing H_2O. In photosynthesis, H_2O is the source of electrons, which are energized by light, temporarily stored in NADPH, and used to reduce CO_2 to carbohydrate. **10.2** The action spectrum of photosynthesis shows that some wavelengths of light that are not absorbed by chlorophyll *a* are still effective at promoting photosynthesis. The light-harvesting complexes of photosystems contain accessory pigments such as chlorophyll *b* and carotenoids, which absorb different wavelengths and pass the energy to chlorophyll *a*, broadening the spectrum of light useful for photosynthesis.

10.3

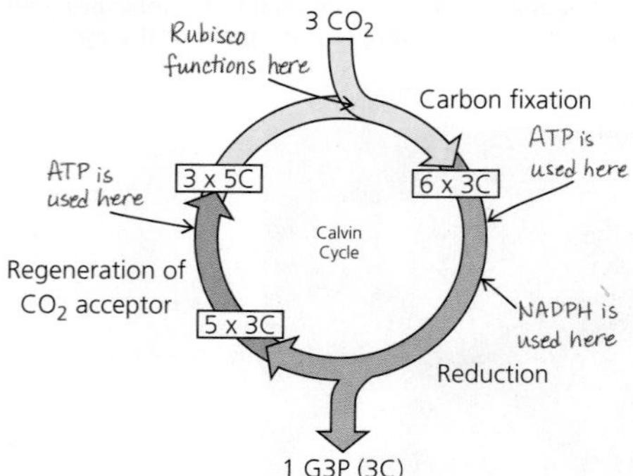

In the reduction phase of the Calvin cycle, ATP phosphorylates a 3-carbon compound, and NADPH then reduces this compound to G3P. ATP is also used in the regeneration phase, when five molecules of G3P are converted to three molecules of the 5-carbon compound RuBP. Rubisco catalyzes the first step of carbon fixation—the addition of CO_2 to RuBP. **10.4** Both C_4 and CAM photosynthesis involve initial fixation of CO_2 to produce a 4-carbon compound (in mesophyll cells in C_4 plants and at night in CAM plants). These compounds are then broken down to release CO_2 (in the bundle-sheath cells in C_4 plants and during the day in CAM plants). ATP is required for recycling the molecule that is used initially to combine with CO_2. These pathways avoid the photorespiration that consumes ATP and reduces the photosynthetic output of C_3 plants when they close stomata on hot, dry, bright days. Thus, hot, arid climates would favor C_4 and CAM plants.

Test Your Understanding

1. d **2.** b **3.** c **4.** d **5.** c **6.** b **7.** d

9.

High H^+ concentration (pH=4)
H^+
Low H^+ concentration (pH=8)
ADP + Ⓟ$_i$
H^+
ATP

The ATP would end up outside the thylakoid. The thylakoids were able to make ATP in the dark because the researchers set up an artificial proton concentration gradient across the thylakoid membrane; thus, the light reactions were not necessary to establish the H^+ gradient required for ATP synthesis by ATP synthase.

Chapter 11

Figure Questions

Figure 11.6 Epinephrine is a signaling molecule; presumably, it binds to a cell-surface receptor protein. **Figure 11.7** Figure 7.1 shows a potassium channel, which, according to the description on p. 135, opens in response to an electrical stimulus, allowing potassium ions to rush out of the cell. Thus, it is a voltage-gated ion channel. **Figure 11.8** When the receptor is actively transmitting a signal to the inside of the cell, it is bound to a G protein. To determine a structure corresponding to that state, it might work to crystallize the receptor in the presence of many copies of the G protein. (In fact, the researchers planned to try this approach next. Another research group also used this approach successfully with a related G protein-coupled receptor the following year.) **Figure 11.9** The testosterone molecule is hydrophobic and can therefore pass directly through the lipid bilayer of the plasma membrane into the cell. (Hydrophilic molecules cannot do this.) **Figure 11.10** The active form of protein kinase 2 **Figure 11.11** The signaling molecule (cAMP) would remain in its active form and would continue to signal. **Figure 11.17** In the model, the directionality of growth is determined by the association of Fus3 with the membrane near the site of receptor activation. Thus, the development of shmoos would be severely compromised, and the affected cell would likely resemble the ΔFus3 and Δformin cells. **Figure 11.18** The signaling pathway shown in Figure 11.14 leads to the splitting of PIP_2 into the second messengers DAG and IP_3, which produce different responses. (The response elicited by DAG is mentioned but not shown.) The pathway shown for cell B is similar in that it branches and leads to two responses.

Concept Check 11.1

1. The two cells of opposite mating type (**a** and **α**) each secrete a certain signaling molecule, which can only be bound by receptors carried on cells of the opposite mating type. Thus, the **a** mating factor cannot bind to another **a** cell and cause it to grow toward the first **a** cell. Only an **α** cell can "receive" the signaling molecule and respond by directed growth (see Figure 11.17 for more information). **2.** The secretion of neurotransmitter molecules at a synapse is an example of local signaling. The electrical signal that travels along a very long nerve cell and is passed to the next nerve cell can be considered an example of long-distance signaling. (Note, however, that local signaling at the synapse between two cells is necessary for the signal to pass from one cell to the next.) **3.** Glucose 1-phosphate is not generated, because the activation of the enzyme requires an intact cell, with an intact receptor in the membrane and an intact signal transduction pathway. The enzyme cannot be activated directly by interaction with the signaling molecule in the test tube. **4.** Glycogen phosphorylase acts in the third stage, the response to epinephrine signaling.

Concept Check 11.2

1. NGF is water-soluble (hydrophilic), so it cannot pass through the lipid membrane to reach intracellular receptors, as steroid hormones can. Therefore, you'd expect the NGF receptor to be in the plasma membrane—which is, in fact, the case. **2.** The cell with the faulty receptor would not be able to respond appropriately to the signaling molecule when it was present. This would most likely have dire consequences for the cell, since regulation of the cell's activities by this receptor would not occur appropriately. **3.** Binding of a ligand to a receptor changes the shape of the receptor, altering the ability of the receptor to transmit a signal. Binding of an allosteric regulator to an enzyme changes the shape of the enzyme, either promoting or inhibiting enzyme activity.

Concept Check 11.3

1. A protein kinase is an enzyme that transfers a phosphate group from ATP to a protein, usually activating that protein (often a second type of protein kinase). Many signal transduction pathways include a series of such interactions, in which each phosphorylated protein kinase in turn phosphorylates the next protein kinase in the series. Such phosphorylation cascades carry a signal from outside the cell to the cellular protein(s) that will carry out the response. **2.** Protein phosphatases reverse the effects of the kinases. **3.** The signal that is being transduced is the information that a signaling molecule is bound to the cell-surface receptor. Information is transduced by way of sequential protein-protein interactions that change protein shapes, causing them to function in a way that passes the signal along. **4.** The IP_3-gated channel opens, allowing calcium ions to flow out of the ER, which raises the cytosolic Ca^{2+} concentration.

Concept Check 11.4

1. At each step in a cascade of sequential activations, one molecule or ion may activate numerous molecules functioning in the next step. **2.** Scaffolding proteins hold molecular components of signaling pathways in a complex with each other. Different scaffolding proteins would assemble different collections of proteins, leading to different cellular responses in the two cells. **3.** A malfunctioning protein phosphatase would not be able to dephosphorylate a particular receptor or relay protein. As a result, the signaling pathway, once activated, would not be able to be terminated. (In fact, one study found altered protein phosphatases in cells from 25% of colorectal tumors.)

Concept Check 11.5

1. In formation of the hand or paw in mammals, cells in the regions between the digits are programmed to undergo apoptosis. This serves to shape the digits of the hand or paw so that they are not webbed. **2.** If a receptor protein for a death-signaling molecule were defective such that it was activated even in the absence of the death signal, this would lead to apoptosis when it wouldn't normally occur. Similar defects in any of the proteins in the signaling pathway, which would activate these relay or response proteins in the absence of interaction with the previous protein or second messenger in the pathway, would have the same effect. Conversely, if any protein in the pathway were defective in its ability to respond to an interaction with an early protein or other molecule or ion, apoptosis would not occur when it normally should. For example, a receptor protein for a death-signaling ligand might not be able to be activated, even when ligand was bound. This would stop the signal from being transduced into the cell.

Summary of Key Concepts Questions

11.1 A cell is able to respond to a hormone only if it has a receptor protein on the cell surface or inside the cell that can bind to the hormone. The response to a hormone depends on the specific cellular activity that a signal transduction pathway triggers within the cell. The response can vary for different types of cells. **11.2** Both GPCRs and RTKs have an extracellular binding site for a signaling molecule (ligand) and an α helix region of the polypeptide that spans the membrane. GPCRs usually trigger a single transduction pathway, whereas the multiple activated tyrosines on an RTK dimer may trigger several different transduction pathways at the same time. **11.3** A protein kinase is an enzyme that adds a phosphate group to another protein. Protein kinases are often part of a phosphorylation cascade that transduces a signal. A second messenger is a small, nonprotein molecule or ion that rapidly diffuses and relays a signal throughout a cell. Both protein kinases and second messengers can operate in the same pathway. For example, the second messenger cAMP often activates protein kinase A, which then phosphorylates other proteins. **11.4** In G protein-coupled pathways, the GTPase portion of a G protein converts GTP to GDP and inactivates the G protein. Protein phosphatases remove phosphate groups from activated proteins, thus stopping a phosphorylation cascade of protein kinases. Phosphodiesterase converts cAMP to AMP, thus reducing the effect of cAMP in a signal transduction pathway. **11.5** The basic mechanism of controlled cell suicide evolved early in eukaryotic evolution, and the genetic basis for these pathways has been conserved during animal evolution. Such a mechanism is essential to the development and maintenance of all animals.

Test Your Understanding

1. c **2.** d **3.** a **4.** b **5.** a **6.** d **7.** c **8.** c **9.** This is one possible drawing of the pathway. (Similar drawings would also be correct.)

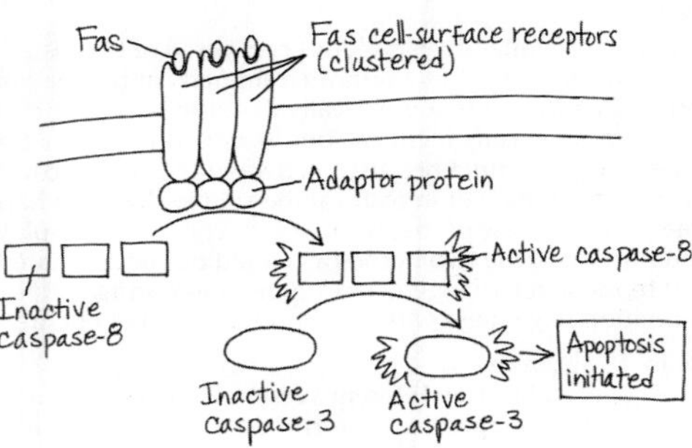

Chapter 12

Figure Questions

Figure 12.4

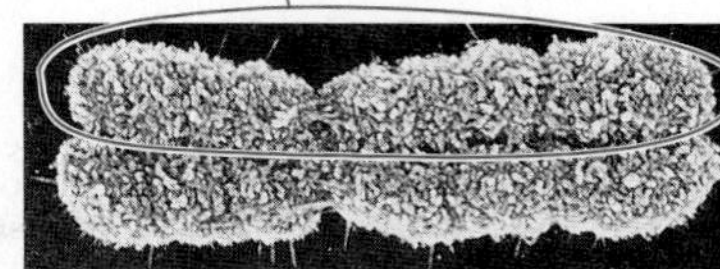

Circling the other chromatid instead would also be correct. **Figure 12.5** The chromosome has four arms. **Figure 12.7** 12; 2; 2; 1

Figure 12.8

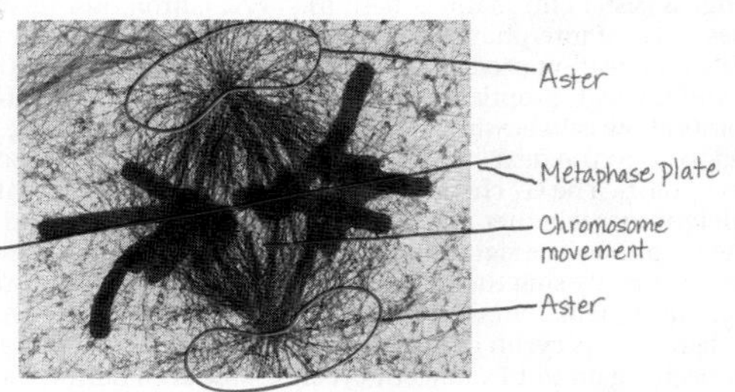

Figure 12.9 The mark would have moved toward the nearer pole. The lengths of fluorescent microtubules between that pole and the mark would have decreased, while the lengths between the chromosomes and the mark would have remained the same. **Figure 12.14** In both cases, the G_1 nucleus would have remained in G_1 until the time it normally would have entered the S phase. Chromosome condensation and spindle formation would not have occurred until the S and G_2 phases had been completed. **Figure 12.16** The cell would divide under conditions where it was inappropriate to do so. If the daughter cells and their descendants also ignored the checkpoint and divided, there would soon be an abnormal mass of cells. (This type of inappropriate cell division can contribute to the development of cancer.) **Figure 12.17** Passing the G_2 checkpoint in the diagram corresponds to the beginning of the "Time" axis of the graph, and entry into the mitotic phase (yellow background on the diagram) corresponds to the peaks of MPF activity and cyclin concentration on the graph (see the yellow M banner over the peaks). During G_1 and S phase in the diagram, Cdk is present without cyclin, so on the graph both cyclin concentration and MPF activity are low. The curved purple arrow in the diagram shows increasing cyclin concentration, seen on the graph during the end of S phase and throughout G_2 phase. Then the cell cycle begins again. **Figure 12.18** The cells in the vessel with PDGF would not be able to respond to the growth factor signal and thus would not divide. The culture would resemble that without the added PDGF. **Figure 12.21** The intracellular estrogen receptor, once activated, would be able to act as a transcription factor in the nucleus, turning on genes that may cause the cell to pass a checkpoint and divide. The HER2 receptor, when activated by a ligand, would form a dimer, and each subunit of the dimer would phosphorylate the other. This would lead to a series of signal transduction steps, ultimately turning on genes in the nucleus. As in the case of the estrogen receptor, the genes would code for proteins necessary to commit the cell to divide.

Concept Check 12.1

1. 2 **2.** 39; 39; 78

Concept Check 12.2

1. 6 chromosomes, duplicated; 12 chromatids **2.** Following mitosis, cytokinesis results in two genetically identical daughter cells in both plant cells and animal cells. However, the mechanism of dividing the cytoplasm is different in animals and plants. In an animal cell, cytokinesis occurs by cleavage, which divides the parent cell in two with a contractile ring of actin filaments. In a plant cell, a cell plate forms in the middle of the cell and grows until its membrane fuses with the plasma membrane of the parent cell. A new cell wall grows inside the cell plate. **3.** They elongate the cell during anaphase. **4.** During eukaryotic cell division, tubulin is involved in spindle formation and chromosome movement, while actin functions during cytokinesis. In bacterial binary fission, it's the opposite: Tubulin-like molecules are thought to act in daughter cell separation, and actin-like molecules are thought to move the daughter bacterial chromosomes to opposite ends of the cell. **5.** Microtubules made up of tubulin in the cell provide "rails" along which vesicles and other organelles can travel, based on interactions of motor proteins with tubulin in the microtubules. In muscle cells, actin in microfilaments interacts with myosin filaments to cause muscle contraction. **6.** From the end of S phase in interphase through the end of metaphase in mitosis

Concept Check 12.3

1. The nucleus on the right was originally in the G_1 phase; therefore, it had not yet duplicated its chromosome. The nucleus on the left was in the M phase, so it had already duplicated its chromosome. **2.** A sufficient amount of MPF has to exist for a cell to pass the G_2 checkpoint; this occurs through the accumulation of cyclin proteins, which combine with Cdk to form MPF. **3.** Most body cells are in a nondividing state called G_0. **4.** Both types of tumors consist of abnormal cells, but their characteristics are different. A benign tumor stays at the original site and can usually be surgically removed; the cells have some genetic and cellular changes from normal, non-tumor cells. Cancer cells from a malignant tumor have more significant genetic and cellular changes, can spread from the original site by metastasis, and may impair the functions of one or more organs. **5.** The cells might divide even in the absence of PDGF. In addition, they would not stop when the surface of the culture vessel was covered; they would continue to divide, piling on top of one another.

Summary of Key Concepts Questions

12.1 The DNA of a eukaryotic cell is packaged into structures called *chromosomes*. Each chromosome is a long molecule of DNA, which carries hundreds to thousands of genes, with associated proteins that maintain chromosome structure and help control gene activity. This DNA-protein complex is called *chromatin*. The chromatin of each chromosome is long and thin when the cell is not dividing. Prior to cell division, each chromosome is duplicated, and the resulting sister *chromatids* are attached to each other by proteins at the centromeres and, for many species, all along their lengths (sister chromatid cohesion). **12.2** Chromosomes exist as single DNA molecules in G_1 of interphase and in anaphase and telophase of mitosis. During S phase, DNA replication produces sister chromatids, which persist during G_2 of interphase and through prophase, prometaphase, and metaphase of mitosis. **12.3** Checkpoints allow cellular surveillance mechanisms to determine whether the cell is prepared to go to the next stage. Internal and external signals move a cell past these checkpoints. The G_1 checkpoint, called the "restriction point" in mammalian cells, determines whether a cell will complete the cell cycle and divide or switch into the G_0 phase. The signals to pass this checkpoint often are external—such as growth factors. Passing the G_2 checkpoint requires sufficient numbers of active MPF complexes, which in turn orchestrate several mitotic events. MPF also initiates degradation of its cyclin component, terminating the M phase. The M phase will not begin again until sufficient cyclin is produced during the next S and G_2 phases. The signal to pass the M phase checkpoint is not activated until all chromosomes are attached to kinetochore fibers and are aligned at the metaphase plate. Only then will sister chromatid separation occur.

Test Your Understanding

1. b **2.** a **3.** c **4.** c **5.** a **6.** b
7. See Figure 12.7 for a description of major events.

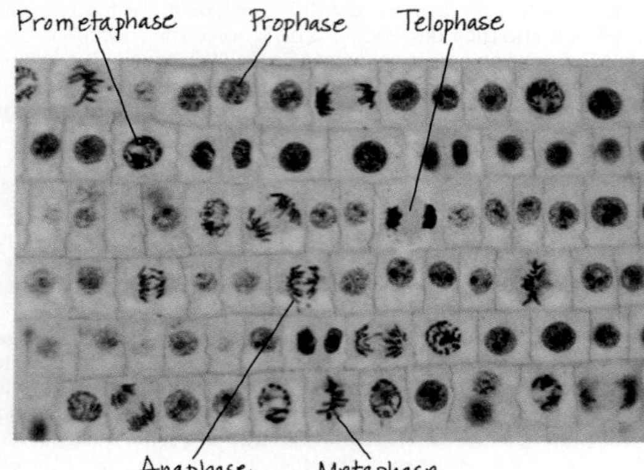

Only one cell is indicated for each stage, but other correct answers are also present in this micrograph.

8. a **9.** e
10.

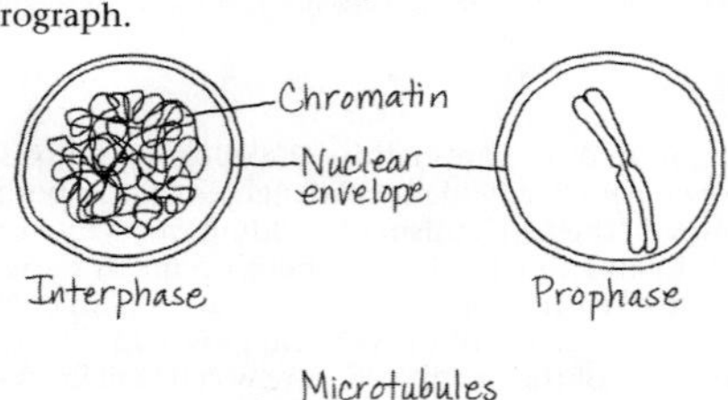

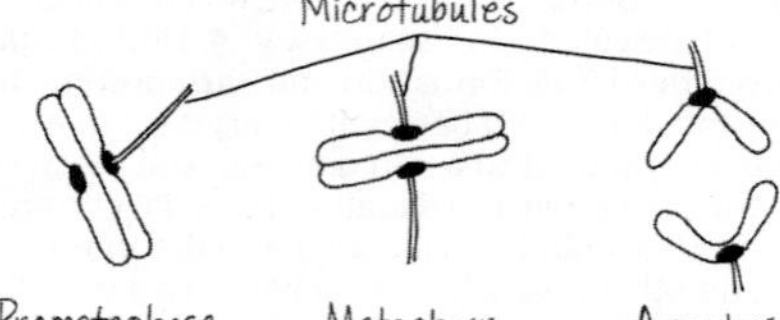

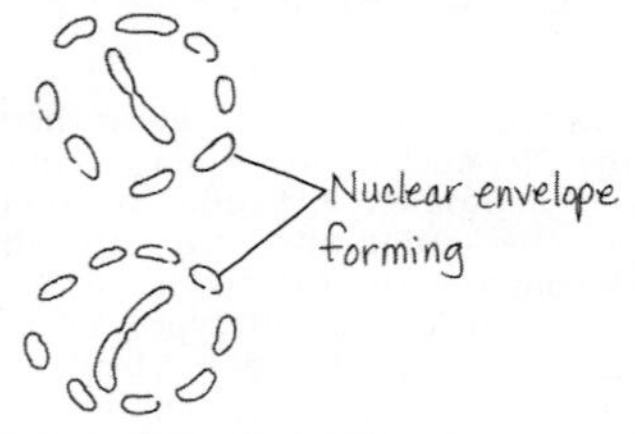

Chapter 13

Figure Questions

Figure 13.4 The haploid number, *n*, is 3. A set is always haploid.
Figure 13.7

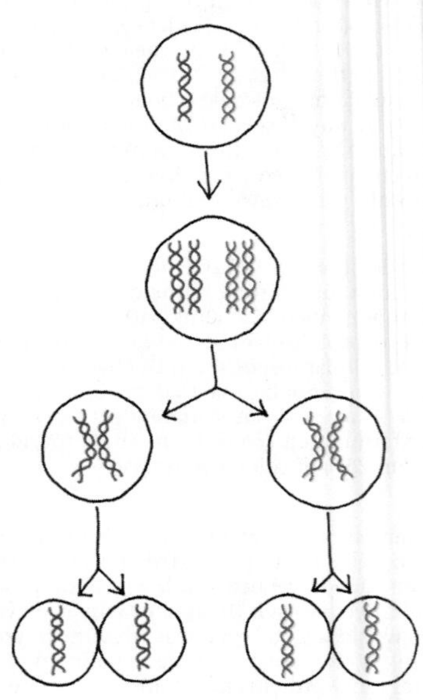

(A short strand of DNA is shown here for simplicity, but each chromosome or chromatid contains a very long coiled and folded DNA molecule.)
Figure 13.8 If the two cells in Figure 12.7 underwent another round of mitosis, each of the four resulting cells would have six chromosomes, while the four cells resulting from meiosis in Figure 13.8 each have three chromosomes. In mitosis, DNA replication (and thus chromosome duplication) precedes each prophase, ensuring that daughter cells have the same number of chromosomes as the parent cell. In meiosis, in contrast, DNA replication occurs only before prophase I (not prophase II). Thus, in two rounds of mitosis, the chromosomes duplicate twice and divide twice, while in meiosis, the chromosomes duplicate once and divide twice. **Figure 13.9** Yes. Each of the six chromosomes (three per cell) shown in telophase I has one nonrecombinant chromatid and one recombinant chromatid. Therefore, eight possible sets of chromosomes can be generated for the cell on the left and eight for the cell on the right.

Concept Check 13.1

1. Parents pass genes to their offspring; the genes program cells to make specific enzymes and other proteins, whose cumulative action produces an individual's inherited traits. **2.** Such organisms reproduce by mitosis, which generates offspring whose genomes are exact copies of the parent's genome (in the absence of mutation). **3.** She should clone it. Cross-breeding it with another plant would generate offspring that have additional variation, which she no longer desires now that she has obtained her ideal orchid.

Concept Check 13.2

1. Each of the six chromosomes is duplicated, so each contains two DNA double helices. Therefore, there are 12 DNA molecules in the cell. **2.** In meiosis, the chromosome count is reduced from diploid to haploid; the union of two haploid gametes in fertilization restores the diploid chromosome count. **3.** The haploid number (n) is 7; the diploid number ($2n$) is 14. **4.** This organism has the life cycle shown in Figure 13.6c. Therefore, it must be a fungus or a protist, perhaps an alga.

Concept Check 13.3

1. The chromosomes are similar in that each is composed of two sister chromatids, and the individual chromosomes are positioned similarly at the metaphase plate. The chromosomes differ in that in a mitotically dividing cell, sister chromatids of each chromosome are genetically identical, but in a meiotically dividing cell, sister chromatids are genetically distinct because of crossing over in meiosis I. Moreover, the chromosomes in metaphase of mitosis can be a diploid set or a haploid set, but the chromosomes in metaphase of meiosis II always consist of a haploid set. **2.** If crossing over did not occur, the two homologs would not be associated in any way. This might result in incorrect arrangement of homologs during metaphase I and ultimately in formation of gametes with an abnormal number of chromosomes.

Concept Check 13.4

1. Mutations in a gene lead to the different versions (alleles) of that gene.
2. Without crossing over, independent assortment of chromosomes during meiosis I theoretically can generate 2^n possible haploid gametes, and random fertilization can produce $2^n \times 2^n$ possible diploid zygotes. Because the haploid number (n) of grasshoppers is 23 and that of fruit flies is 4, two grasshoppers would be expected to produce a greater variety of zygotes than would two fruit flies. **3.** If the segments of the maternal and paternal chromatids that undergo

crossing over are genetically identical and thus have the same two alleles for every gene, then the recombinant chromosomes will be genetically equivalent to the parental chromosomes. Crossing over contributes to genetic variation only when it involves the rearrangement of different alleles.

Summary of Key Concepts Questions

13.1 Genes program specific traits, and offspring inherit their genes from each parent, accounting for similarities in their appearance to one or the other parent. Humans reproduce sexually, which ensures new combinations of genes (and thus traits) in the offspring. Consequently, the offspring are not clones of their parents (which would be the case if humans reproduced asexually). **13.2** Animals and plants both reproduce sexually, alternating meiosis with fertilization. Both have haploid gametes that unite to form a diploid zygote, which then goes on to divide mitotically, forming a diploid multicellular organism. In animals, haploid cells become gametes and don't undergo mitosis, while in plants, the haploid cells resulting from meiosis undergo mitosis to form a haploid multicellular organism, the gametophyte. This organism then goes on to generate haploid gametes. (In plants such as trees, the gametophyte is quite reduced in size and not obvious to the casual observer.) **13.3** At the end of meiosis I, the two members of a homologous pair end up in different cells, so they cannot pair up and undergo crossing over. **13.4** First, during independent assortment in metaphase I, each pair of homologous chromosomes lines up independent of each other pair at the metaphase plate, so a daughter cell of meiosis I randomly inherits either a maternal or paternal chromosome. Second, due to crossing over, each chromosome is not exclusively maternal or paternal, but includes regions at the ends of the chromatid from a nonsister chromatid (a chromatid of the other homolog). (The nonsister segment can also be in an internal region of the chromatid if a second crossover occurs beyond the first one before the end of the chromatid.) This provides much additional diversity in the form of new combinations of alleles. Third, random fertilization ensures even more variation, since any sperm of a large number containing many possible genetic combinations can fertilize any egg of a similarly large number of possible combinations.

Test Your Understanding

1. a **2.** d **3.** b **4.** a **5.** d **6.** c **7.** d
8. (a)

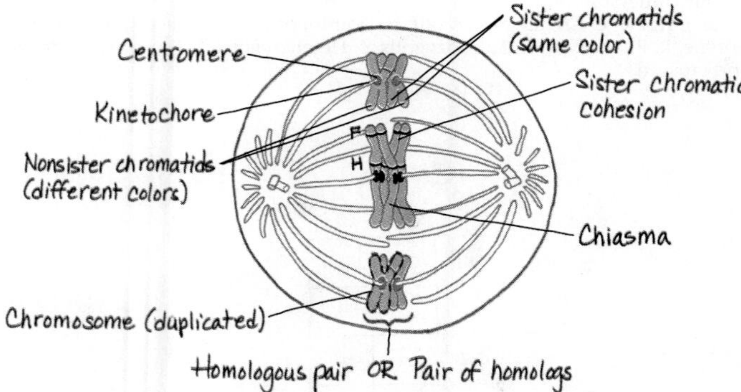

(b) The chromosomes of one color make up a haploid set. (In cases where crossovers have occurred, a haploid set of one color may include segments of chromatids of the other color.) All red and blue chromosomes together make up a diploid set.
(c) Metaphase I **9.** This cell must be undergoing meiosis because homologous chromosomes are associated with each other at the metaphase plate; this does not occur in mitosis.

Chapter 14

Figure Questions

Figure 14.3 All offspring would have purple flowers. (The ratio would be one purple to zero white.) The P generation plants are true-breeding, so mating two purple-flowered plants produces the same result as self-pollination: All the offspring have the same trait.

Figure 14.8

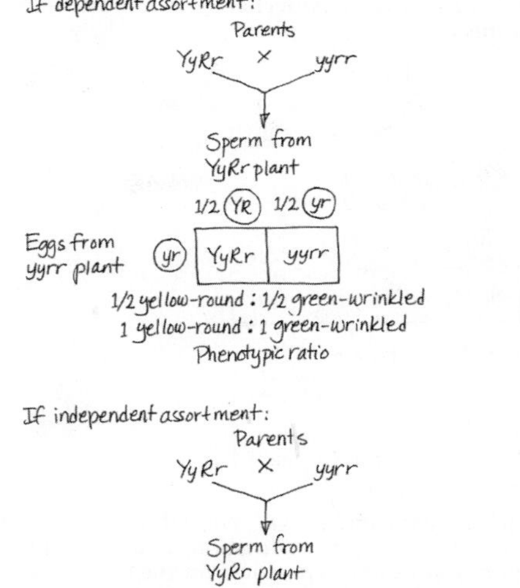

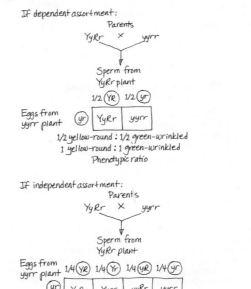

Yes, this cross would also have allowed Mendel to make different predictions for the two hypotheses, thereby allowing him to distinguish the correct one.
Figure 14.10 Your classmate would probably point out that the F_1 generation hybrids show an intermediate phenotype between those of the homozygous parents, which supports the blending hypothesis. You could respond that crossing the F_1 hybrids results in the reappearance of the white phenotype, rather than identical pink offspring, which fails to support the idea of traits blending during inheritance.
Figure 14.11 Both the I^A and I^B alleles are dominant to the i allele, which results in no attached carbohydrate. The I^A and I^B alleles are codominant; both are expressed in the phenotype of I^AI^B heterozygotes, who have type AB blood.
Figure 14.13

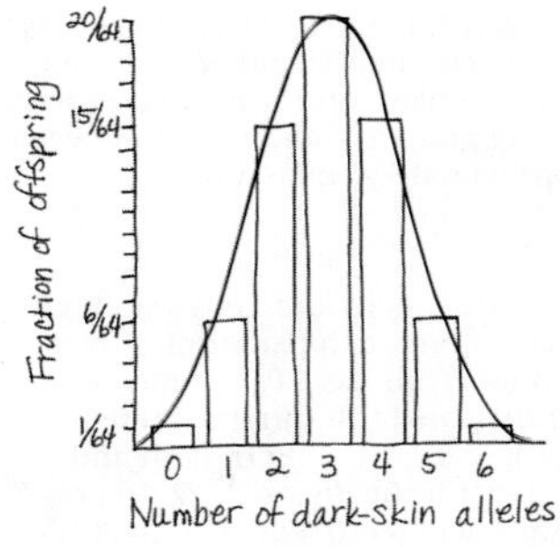

The majority of individuals have intermediate phenotypes (skin color in the middle range), while fewer individuals have phenotypes at either end (very dark or very light skin). (As you may know, this is called a "bell curve" and represents a "normal distribution.") **Figure 14.16** In the Punnett square, two of the three individuals with normal coloration are carriers, so the probability is ⅔. (Note that you must take into account everything you know when you calculate probability: You know she is not *aa*, so there are only three possible genotypes to consider.) **Figure 14.18** If one parent tests negative for the recessive allele, then the probability is zero that the offspring will have the disease and ½ that the offspring will be a carrier. If the first child is a carrier, the probability of the next child being a carrier is still ½ because the two births are independent events.

Concept Check 14.1

1. According to the law of independent assortment, 25 plants (1/16 of the offspring) are predicted to be *aatt*, or recessive for both characters. The actual result is likely to differ slightly from this value.

Parents

AaTt × *AaTt*

Sperm from *AaTt* plant

Eggs from *AaTt* plant	*(AT)*	*(At)*	*(aT)*	*(at)*
(AT)	*AATT*	*AATt*	*AaTT*	*AaTt*
(At)	*AATt*	*AAtt*	*AaTt*	*Aatt*
(aT)	*AaTT*	*AaTt*	*aaTT*	*aaTt*
(at)	*AaTt*	*Aatt*	*aaTt*	*aatt*

2. The plant could make eight different gametes (*YRI, YRi, YrI, Yri, yRI, yRi, yrI,* and *yri*). To fit all the possible gametes in a self-pollination, a Punnett square would need 8 rows and 8 columns. It would have spaces for the 64 possible unions of gametes in the offspring. **3.** Self-pollination is sexual reproduction because meiosis is involved in forming gametes, which unite during fertilization. As a result, the offspring in self-pollination are genetically different from the parent. (As mentioned in the footnote on p. 263, we have simplified the explanation in referring to the single pea plant as a parent. Technically, the gametophytes in the flower are the two "parents.")

Concept Check 14.2

1. ½ homozygous dominant (*AA*), 0 homozygous recessive (*aa*), and ½ heterozygous (*Aa*) **2.** ¼ *BBDD*; ¼ *BbDD*; ¼ *BBDd*; ¼ *BbDd* **3.** The genotypes that fulfill this condition are *ppyyIi, ppYyii, Ppyyii, ppYYii,* and *ppyyii*. Use the multiplication rule to find the probability of getting each genotype, and then use the addition rule to find the overall probability of meeting the conditions of this problem:

ppyyIi	½ (probability of *pp*) × ¼ (*yy*) × ½ (*ii*)	= 1/16
ppYyii	½ (*pp*) × ½ (*Yy*) × ½ (*ii*)	= 2/16
Ppyyii	½ (*Pp*) × ¼ (*yy*) × ½ (*ii*)	= 1/16
ppYYii	½ (*pp*) × ¼ (*YY*) × ½ (*ii*)	= 1/16
ppyyii	½ (*pp*) × ¼ (*yy*) × ½ (*ii*)	= 1/16
Fraction predicted to have at least two recessive traits		= 6/16 or 3/8

Concept Check 14.3

1. Incomplete dominance describes the relationship between two alleles of a single gene, whereas epistasis relates to the genetic relationship between two genes (and the respective alleles of each). **2.** Half of the children would be expected to have type A blood and half type B blood. **3.** The black and white alleles are incompletely dominant, with heterozygotes being gray in color. A cross between a gray rooster and a black hen should yield approximately equal numbers of gray and black offspring.

Concept Check 14.4

1. 1/9 (Since cystic fibrosis is caused by a recessive allele, Beth and Tom's siblings who have CF must be homozygous recessive. Therefore, each parent must be a carrier of the recessive allele. Since neither Beth nor Tom has CF, this means they each have a ⅔ chance of being a carrier. If they are both carriers, there is a ¼ chance that they will have a child with CF. ⅔ × ⅔ × ¼ = 1/9); 0 (Both Beth and Tom would have to be carriers to produce a child with the disease.) **2.** In normal hemoglobin, the sixth amino acid is glutamic acid (Glu), which is acidic (has a negative charge on its side chain). In sickle-cell hemoglobin, Glu is replaced by valine (Val), which is a nonpolar amino acid, very different from Glu. The primary structure of a protein (its amino acid sequence) ultimately determines the shape of the protein and thus its function. The substitution of Val for Glu enables the hemoglobin molecules to interact with each other and form long fibers, leading to the protein's deficient function and the deformation of the red blood cell. **3.** Joan's genotype is *Dd*. Because the allele for polydactyly (*D*) is dominant to the allele for five digits per appendage (*d*), the trait is expressed in people with either the *DD* or *Dd* genotype. But because Joan's father does not have polydactyly, his genotype must be *dd*, which means that Joan inherited a *d* allele from him. Therefore Joan, who does have the trait, must be heterozygous. **4.** In the monohybrid cross involving flower color, the ratio is 3.15 purple : 1 white, while in the human family in the pedigree, the ratio in the third generation is 1 free : 1 attached earlobe. The difference is due to the small sample size (two offspring) in the human family. If the second-generation couple in this pedigree were able to have 929 offspring as in the pea plant cross, the ratio would likely be closer to 3:1. (Note that none of the pea plant crosses in Table 14.1 yielded *exactly* a 3:1 ratio.)

Summary of Key Concepts Questions

14.1 Alternative versions of genes, called alleles, are passed from parent to offspring during sexual reproduction. In a cross between purple- and white-flowered homozygous parents, the F_1 offspring are all heterozygous, each inheriting a purple allele from one parent and a white allele from the other. Because the purple allele is dominant, it determines the phenotype of the F_1 offspring to be purple, and the expression of the white allele is masked. Only in the F_2 generation is it possible for a white allele to exist in a homozygous state, which causes the white trait to be expressed.

14.2

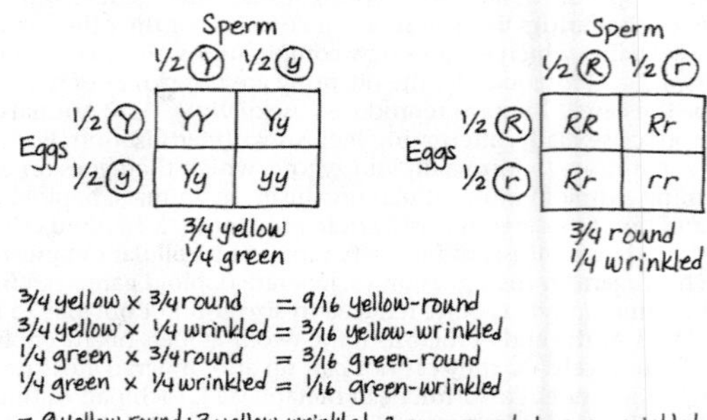

14.3 The ABO blood group is an example of multiple alleles because this single gene has more than two alleles (I^A, I^B, and *i*). Two of the alleles, I^A and I^B, exhibit codominance, since both carbohydrates (A and B) are present when these two alleles exist together in a genotype. I^A and I^B each exhibit complete dominance over the *i* allele. This situation is not an example of incomplete dominance because each allele affects the phenotype in a distinguishable way, so the result is not intermediate between the two phenotypes. Because this situation involves a single gene, it is not an example of epistasis or polygenic inheritance. **14.4** The chance of the fourth child having cystic fibrosis is ¼, as it was for each of the other children, because each birth is an independent event. We already know both parents are carriers, so whether their first three children are carriers or not has no bearing on the probability that their next child will have the disease. The parents' genotypes provide the only relevant information.

Test Your Understanding

1. Gene, l. Allele, e. Character, g. Trait, b. Dominant allele, j. Recessive allele, a. Genotype, k. Phenotype, h. Homozygous, c. Heterozygous, f. Testcross, i. Monohybrid cross, d.

2.

Parents

GgIi × *GgIi*

Sperm

Eggs	*(GI)*	*(Gi)*	*(gI)*	*(gi)*
(GI)	*GGII*	*GGIi*	*GgII*	*GgIi*
(Gi)	*GGIi*	*GGii*	*GgIi*	*Ggii*
(gI)	*GgII*	*GgIi*	*ggII*	*ggIi*
(gi)	*GgIi*	*Ggii*	*ggIi*	*ggii*

9 green-inflated : 3 green-constricted : 3 yellow-inflated : 1 yellow-constricted

3. Parental cross is $AAC^RC^R \times aaC^WC^W$. F_1 genotype is AaC^RC^W, phenotype is all axial-pink. F_2 genotypes are 1 AAC^RC^R : 2 AAC^RC^W : 1 AAC^WC^W : 2 AaC^RC^R : 4 AaC^RC^W : 2 AaC^WC^W : 1 aaC^RC^R : 2 aaC^RC^W : 1 aaC^WC^W. F_2 phenotypes are 3 axial-red : 6 axial-pink : 3 axial-white : 1 terminal-red : 2 terminal-pink : 1 terminal-white. **4.** Man I^Ai; woman I^Bi; child *ii*. Genotypes for future children are predicted to be ¼ I^AI^B, ¼ I^Ai, ¼ I^Bi, ¼ *ii*. **5.** ½ **6.** A cross of *Ii* × *ii* would yield offspring with a genotypic ratio of 1 *Ii* : 1 *ii* (2:2 is an equivalent answer) and a phenotypic ratio of 1 inflated : 1 constricted (2:2 is equivalent).

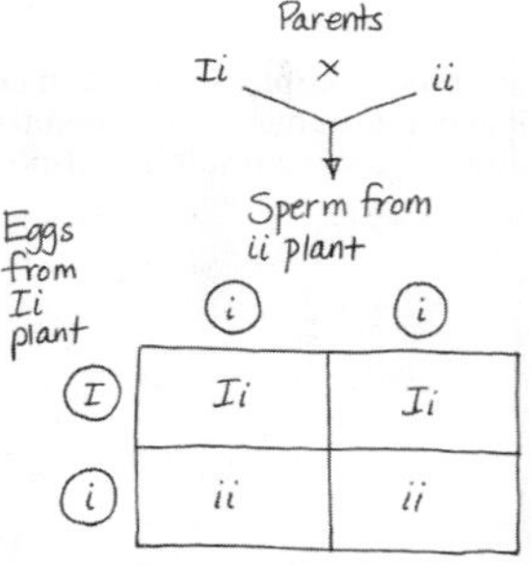

7. (a) 1/64; (b) 1/64; (c) 1/8; (d) 1/32 **8.** Albino (*b*) is a recessive trait; black (*B*) is dominant. First cross: parents *BB* × *bb*; gametes *B* and *b*; offspring all *Bb* (black coat). The black guinea pig in the second cross is a heterozygote. Second cross: parents *Bb* × *bb*; gametes ½ *B* and ½ *b* (heterozygous parent) and *b*; offspring ½ *Bb* and ½ *bb*. **9.** (a) *PPLl* × *PPLl*, *PPLl* × *PpLl*, or *PPLl* × *ppLl*; (b) *ppLl* × *ppLl*; (c) *PPLL* × any of the 9 possible genotypes or *PPll* × *ppLL*; (d) *PpLl* × *Ppll*; (e) *PpLl* × *PpLl* **10.** (a) ¾ × ¾ × ¾ = 27/64; (b) 1 − 27/64 = 37/64; (c) ¼ × ¼ × ¼ = 1/64; (d) 1 − 1/64 = 63/64 **11.** (a) 1/256; (b) 1/16; (c) 1/256; (d) 1/64; (e) 1/128 **12.** (a) 1; (b) 1/32; (c) 1/8; (d) ½ **13.** 1/9 **14.** Matings of the original mutant cat with true-breeding noncurl cats will produce both curl and noncurl F_1 offspring if the curl allele is dominant, but only noncurl offspring if the curl allele is recessive. You would obtain some true-breeding offspring homozygous for the curl allele from matings between the F_1 cats resulting from the original curl × noncurl crosses whether the curl trait is dominant or recessive. You know that cats are true-breeding when curl × curl matings produce only curl offspring. As it turns out, the allele that causes curled ears is dominant. **15.** 1/16 **16.** 25%, or ¼, will be cross-eyed; all (100%) of the cross-eyed offspring will also be white. **17.** The dominant allele *I* is epistatic to the *P/p* locus, and thus the genotypic ratio for the F_1 generation will be 9 *I_P_* (colorless) : 3 *I_pp* (colorless) : 3 *iiP_* (purple) : 1 *iipp* (red). Overall, the phenotypic ratio is 12 colorless : 3 purple : 1 red. **18.** Recessive. All affected individuals (Arlene, Tom, Wilma, and Carla) are homozygous recessive *aa*. George is *Aa*, since some of his children with Arlene are affected. Sam, Ann, Daniel, and Alan are each *Aa*, since they are all unaffected children with one affected parent. Michael also is *Aa*, since he has an affected child (Carla) with his heterozygous wife Ann. Sandra, Tina, and Christopher can each have either the *AA* or *Aa* genotype. **19.** 1/6 **20.** 9 *B_A_* (agouti) : 3 *B_aa* (black) : 3 *bbA_* (white) : 1 *bbaa* (white). Overall, 9 agouti : 3 black : 4 white.

Chapter 15

Figure Questions

Figure 15.2 The ratio would be 1 yellow-round : 1 green-round : 1 yellow-wrinkled : 1 green-wrinkled. **Figure 15.4** About ¾ of the F_2 offspring would have red eyes and about ¼ would have white eyes. About half of the white-eyed flies would be female and half would be male; similarly, about half of the red-eyed flies would be female and half would be male. **Figure 15.7** All the males would be color-blind, and all the females would be carriers. **Figure 15.9** The two largest classes would still be the parental-type offspring (offspring with the phenotypes of the true-breeding P generation flies), but now they would be gray-vestigial and black-normal because those were the specific allele combinations in the P generation. **Figure 15.10** The two chromosomes below, left, are like the two chromosomes inherited by the F_1 female, one from each P generation fly. They are passed by the F_1 female intact to the offspring and thus could be called "parental" chromosomes. The other two chromosomes result from crossing over during meiosis in the F_1 female. Because they have combinations of alleles not seen in either of the F_1 female's chromosomes, they can be called "recombinant" chromosomes. (Note that in this example, the alleles on the recombinant chromosomes, $b^+ vg^+$ and *b vg*, are the allele combinations that were on the parental chromosomes in the cross shown in Figures 15.9 and 15.10. The basis for calling them parental chromosomes is the combination of alleles that was present on the P generation chromosomes.)

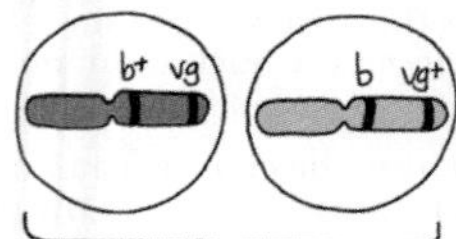

Concept Check 15.1

1. The law of segregation relates to the inheritance of alleles for a single character. The law of independent assortment of alleles relates to the inheritance of alleles for two characters. **2.** The physical basis for the law of segregation is the separation of homologs in anaphase I. The physical basis for the law of independent assortment is the alternative arrangements of homologous chromosome pairs in metaphase I. **3.** To show the mutant phenotype, a male needs to possess only one mutant allele. If this gene had been on a pair of autosomes, *two* mutant alleles would have had to be present for an individual to show the recessive mutant phenotype, a much less probable situation.

Concept Check 15.2

1. Because the gene for this eye-color character is located on the X chromosome, all female offspring will be red-eyed and heterozygous ($X^{w^+} X^w$); all male offspring will inherit a Y chromosome from the father and be white-eyed (X^wY). **2.** ¼ (½ chance that the child will inherit a Y chromosome from the father and be male × ½ chance that he will inherit the X carrying the disease allele from his mother) If the child is a boy, there is a ½ chance he will have the disease; a female would have zero chance (but ½ chance of being a carrier). **3.** With a disorder caused by a dominant allele, there is no such thing as a "carrier," since those with the allele have the disorder. Because the allele is dominant, the females lose any "advantage" in having two X chromosomes, since one disorder-associated allele is sufficient to result in the disorder. All fathers who have the dominant allele will pass it along to *all* their daughters, who will also have the disorder. A mother who has the allele (and thus the disorder) will pass it to half of her sons and half of her daughters.

Concept Check 15.3

1. Crossing over during meiosis I in the heterozygous parent produces some gametes with recombinant genotypes for the two genes. Offspring with a recombinant phenotype arise from fertilization of the recombinant gametes by homozygous recessive gametes from the double-mutant parent. **2.** In each case, the alleles contributed by the female parent determine the phenotype of the offspring because the male in this cross contributes only recessive alleles. **3.** No. The order could be *A-C-B* or *C-A-B*. To determine which possibility is correct, you need to know the recombination frequency between *B* and *C*.

Concept Check 15.4

1. In meiosis, a combined 14-21 chromosome will behave as one chromosome. If a gamete receives the combined 14-21 chromosome and a normal copy of chromosome 21, trisomy 21 will result when this gamete combines with a normal gamete during fertilization. **2.** No. The child can be either I^AI^Ai or I^Aii. A sperm of genotype I^AI^A could result from nondisjunction in the father during meiosis II, while an egg with the genotype *ii* could result from nondisjunction in the mother during either meiosis I or meiosis II. **3.** Activation of this gene could lead to the production of too much of this kinase. If the kinase is involved in a signaling pathway that triggers cell division, too much of it could trigger unrestricted cell division, which in turn could contribute to the development of a cancer (in this case, a cancer of one type of white blood cell).

Concept Check 15.5

1. Inactivation of an X chromosome in females and genomic imprinting. Because of X inactivation, the effective dose of genes on the X chromosome is the same in males and females. As a result of genomic imprinting, only one allele of certain genes is phenotypically expressed. **2.** The genes for leaf coloration are located in plastids within the cytoplasm. Normally, only the maternal parent transmits plastid genes to offspring. Since variegated offspring are produced only when the female parent is of the B variety, we can conclude that variety B contains both the wild-type and mutant alleles of pigment genes, producing variegated leaves. (Variety A contains only the wild-type allele of pigment genes.) **3.** The situation is similar to that for chloroplasts. Each cell contains numerous mitochondria, and in affected individuals, most cells contain a variable mixture of normal and mutant mitochondria. The normal mitochondria carry out enough cellular respiration for survival.

Summary of Key Concepts Questions

15.1 Because the sex chromosomes are different from each other and because they determine the sex of the offspring, Morgan could use the sex of the offspring as a phenotypic characteristic to follow the parental chromosomes. (He could also have followed them under a microscope, as the X and Y chromosomes look different.) At the same time, he could record eye color to follow the eye-color alleles. **15.2** Males have only one X chromosome, along with a Y chromosome, while females have two X chromosomes. The Y chromosome has very few genes on it, while the X has about 1,000. When a recessive X-linked allele that causes a disorder is inherited by a male on the X from his mother, there isn't a second allele present on the Y (males are hemizygous), so the male has the disorder. Because females have two X chromosomes, they must inherit two recessive alleles in order to have the disorder, a rarer occurrence. **15.3** Crossing over results in new combinations of alleles. Crossing over is a random occurrence, and the more distance there is between two genes, the more chances there are for crossing over to occur, leading to a new allele combination. **15.4** In inversions and reciprocal translocations, the same genetic material is present in the same relative amount but just organized differently. In aneuploidy, duplications, deletions, and nonreciprocal translocations, the balance of genetic material is upset, as large segments are either missing or present in more than one copy. Apparently, this type of imbalance is very damaging to the organism. (Although it isn't lethal in the developing embryo, the reciprocal translocation that produces the Philadelphia chromosome can lead to a serious condition, cancer, by altering the expression of important genes.) **15.5** In these cases, the sex of the parent contributing an allele affects the inheritance pattern. For imprinted genes, either the paternal or the maternal allele is expressed, depending on the imprint. For mitochondrial and chloroplast genes, only the maternal contribution will affect offspring phenotype because the offspring inherit these organelles from the mother, via the egg cytoplasm.

Test Your Understanding

1. 0; ½; 1/16 **2.** Recessive; if the disorder were dominant, it would affect at least one parent of a child born with the disorder. The disorder's inheritance is sex-linked because it is seen only in boys. For a girl to have the disorder, she would have to inherit recessive alleles from *both* parents. This would be very rare, since males with the recessive allele on their X chromosome die in their early teens. **3.** 17% **4.** The disorder would always be inherited from the mother. **5.** Between *T* and *A*, 12%; between *A* and *S*, 5% **6.** Between *T* and *S*, 18%; sequence of genes is *T–A–S* **7.** ¼ for each daughter (½ chance that child will be female × ½ chance of a homozygous recessive genotype); ½ for first son. **8.** 6%; wild-type heterozygous for normal wings and red eyes) × recessive homozygous for vestigial wings and purple eyes
9.

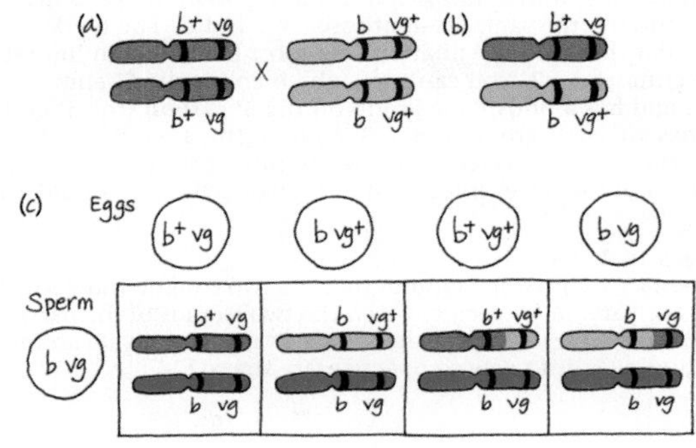

(d) 41.5% gray body, vestigial wings
41.5% black body, normal wings
8.5% gray body, normal wings
8.5% black body, vestigial wings

10. The inactivation of two X chromosomes in XXX women would leave them with one genetically active X, as in women with the normal number of chromosomes. Microscopy should reveal two Barr bodies in XXX women. **11.** *D–A–B–C* **12.** Fifty percent of the offspring will show phenotypes resulting from crossovers. These results would be the same as those from a cross where *A* and *B* were *not* on the same chromosome. Further crosses involving other genes on the same chromosome would reveal the genetic linkage and map distances. **13.** 450 each of blue-oval and white-round (parentals) and 50 each of blue-round and white-oval (recombinants) **14.** About one-third of the distance from the vestigial-wing locus to the brown-eye locus **15.** Because bananas are triploid, homologous pairs cannot line up during meiosis. Therefore, it is not possible to generate gametes that can fuse to produce a zygote with the triploid number of chromosomes.

Chapter 16

Figure Questions

Figure 16.2 The living S cells found in the blood sample were able to reproduce to yield more S cells, indicating that the S trait is a permanent, heritable change, rather than just a one-time use of the dead S cells' capsules. **Figure 16.4** The radioactivity would have been found in the pellet when proteins were labeled (batch 1) because proteins would have had to enter the bacterial cells to program them with genetic instructions. It's hard for us to imagine now, but the DNA might have played a structural role that allowed some of the proteins to be injected while it remained outside the bacterial cell (thus no radioactivity in the pellet in batch 2). **Figure 16.11** The tube from the first replication would look the same, with a middle band of hybrid ^{15}N-^{14}N DNA, but the second tube would not have the upper band of two light blue strands. Instead it would have a bottom band of two dark blue strands, like the bottom band in the result predicted after one replication in the conservative model. **Figure 16.12** In the bubble at the top in (b), arrows should be drawn pointing left and right to indicate the two replication forks. **Figure 16.14** Looking at any of the DNA strands, we see that one end is called the 5′ end and the other the 3′ end. If we proceed from the 5′ end to the 3′ end on the left-most strand, for example, we list the components in this order: phosphate group → 5′ C of the sugar → 3′ C → phosphate → 5′ C → 3′ C. Going in the opposite direction on the same strand, the components proceed in the reverse order: 3′ C → 5′ C → phosphate. Thus, the two directions are distinguishable, which is what we mean when we say that the strands have directionality. (Review Figure 16.5 if necessary.)
Figure 16.17

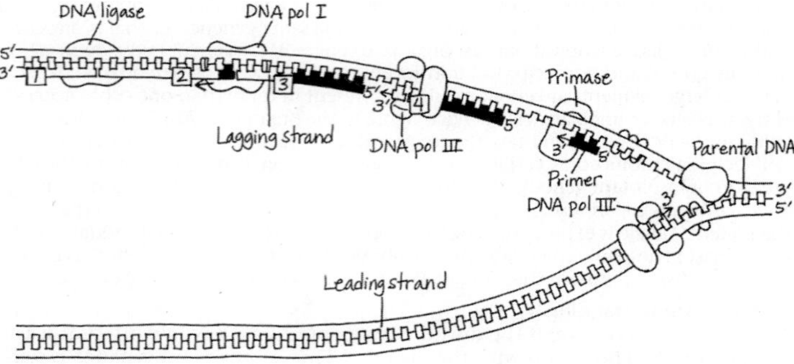

Figure 16.23 The two members of a homologous pair (which would be the same color) would be associated tightly together at the metaphase plate. In metaphase of mitosis, however, each chromosome would be lined up individually, so the two chromosomes of the same color would be in different places at the metaphase plate.

Concept Check 16.1

1. Chargaff's rule about base ratios states that in DNA, the percentages of A and T are essentially the same, as are those of G and C. The fly data are consistent with that rule. (Slight variations are most likely due to limitations of analytical technique.) **2.** You can't tell which end is the 5′ end. You need to know which end has a phosphate group on the 5′ carbon (the 5′ end) or which end has an —OH group on the 3′ carbon (the 3′ end). **3.** He was expecting that the mouse injected with the mixture of heat-killed S cells and living R cells would survive, since neither type of cell alone would kill the mouse.

Concept Check 16.2

1. Complementary base pairing ensures that the two daughter molecules are exact copies of the parental molecule. When the two strands of the parental molecule separate, each serves as a template on which nucleotides are arranged, by the base-pairing rules, into new complementary strands.

2.

Protein	Function
Helicase	Unwinds parental double helix at replication forks
Single-strand binding protein	Binds to and stabilizes single-stranded DNA until it can be used as a template
Topoisomerase	Relieves "overwinding" strain ahead of replication forks by breaking, swiveling, and rejoining DNA strands
Primase	Synthesizes an RNA primer at 5′ end of leading strand and at 5′ end of each Okazaki fragment of lagging strand
DNA pol III	Using parental DNA as a template, synthesizes new DNA strand by covalently adding nucleotides to 3′ end of a pre-existing DNA strand or RNA primer
DNA pol I	Removes RNA nucleotides of primer from 5′ end and replaces them with DNA nucleotides
DNA ligase	Joins 3′ end of DNA that replaces primer to rest of leading strand and joins Okazaki fragments of lagging strand

3. In the cell cycle, DNA synthesis occurs during the S phase, between the G_1 and G_2 phases of interphase. DNA replication is therefore complete before the mitotic phase begins. **4.** Synthesis of the leading strand is initiated by an RNA primer, which must be removed and replaced with DNA, a task that could not be performed if the cell's DNA pol I were nonfunctional. In the overview box in Figure 16.17, just to the left of the top origin of replication, a functional DNA pol I would replace the RNA primer of the leading strand (shown in red) with DNA nucleotides (blue).

Concept Check 16.3

1. A nucleosome is made up of eight histone proteins, two each of four different types, around which DNA is wound. Linker DNA runs from one nucleosome to the next. **2.** Euchromatin is chromatin that becomes less compacted during interphase and is accessible to the cellular machinery responsible for gene activity. Heterochromatin, on the other hand, remains quite condensed during interphase and contains genes that are largely inaccessible to this machinery. **3.** The nuclear lamina is a netlike array of protein filaments that provides mechanical support just inside the nuclear envelope and thus maintains the shape of the nucleus. Considerable evidence also supports the existence of a nuclear matrix, a framework of protein fibers extending throughout the nuclear interior.

Summary of Key Concepts Questions

16.1 Each strand in the double helix has polarity, the end with a phosphate group on the 5′ carbon of the sugar being called the 5′ end, and the end with an —OH group on the 3′ carbon of the sugar being called the 3′ end. The two strands run in opposite directions, so each end of the molecule has both a 5′ and a 3′ end. This arrangement is called "antiparallel." If the strands were parallel, they would both run 5′ → 3′ in the same direction, so an end of the molecule would have either two 5′ ends or two 3′ ends. **16.2** On both the leading and lagging strands, DNA polymerase adds onto the 3′ end of an RNA primer synthesized by primase, synthesizing DNA in the 5′ → 3′ direction. Because the parental strands are antiparallel, however, only on the leading strand does synthesis proceed continuously into the replication fork. The lagging strand is synthesized bit by bit in the direction away from the fork as a series of shorter Okazaki fragments, which are later joined together by DNA ligase. Each fragment is initiated by synthesis of an RNA primer by primase as soon as a given stretch of single-stranded template strand is opened up. Although both strands are synthesized at the same rate, synthesis of the lagging strand is delayed because initiation of each fragment begins only when sufficient template strand is available. **16.3** Most of the chromatin in an interphase nucleus is uncondensed. Much is present as the 30-nm fiber, with some in the form of the 10-nm fiber and some as looped domains of the 30-nm fiber. (These different levels of chromatin packing may reflect differences in gene expression occurring in these regions.) Also, a small percentage of the chromatin, such as that at the centromeres and telomeres, is highly condensed heterochromatin.

Test Your Understanding

1. c **2.** c **3.** b **4.** d **5.** c **6.** d **7.** b **8.** a
9. Like histones, the *E. coli* proteins would be expected to contain many basic (positively charged) amino acids, such as lysine and arginine, which can form weak bonds with the negatively charged phosphate groups on the sugar-phosphate backbone of the DNA molecule. **10.** Each species' DNA has a slightly different percentage of a given base. For example, the percentage of A ranges from 24.7% for *E. coli* to 32.8% for sea urchin. This illustrates Chargaff's rule that the DNA of different species varies in its base composition. Chargaff's other rule states

that in any given species, the percentage of A is roughly equal to that of T, and the percentage of C is roughly equal to that of G. For example, sea urchins have about 32–33% each of A and T and about 17% of G and C. (Your answer may use any similar examples from the table.)

12.

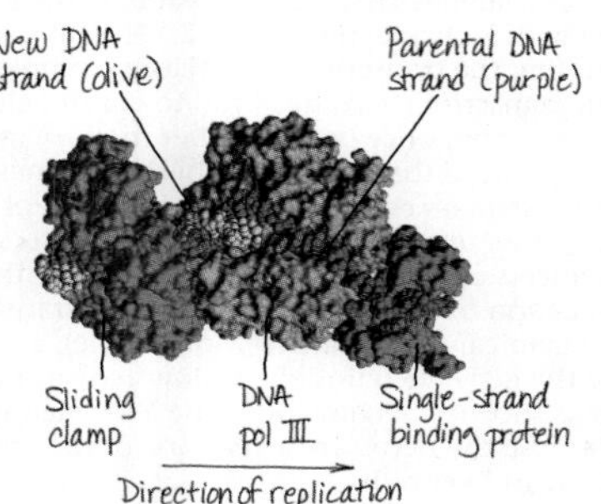

Chapter 17

Figure Questions

Figure 17.2 The previously presumed pathway would have been wrong. The new results would support this pathway: precursor $\rightarrow$ citrulline $\rightarrow$ ornithine $\rightarrow$ arginine. They would also indicate that class I mutants have a defect in the second step and class II mutants have a defect in the first step. **Figure 17.4** The mRNA sequence (5′-UGGUUUGGCUCA-3′) is the same as the nontemplate DNA strand sequence (5′-TGGTTTGGCTCA-3′), except there is U in the mRNA and T in the DNA. **Figure 17.7** The processes are similar in that polymerases form polynucleotides complementary to an antiparallel DNA template strand. In replication, however, both strands act as templates, whereas in transcription, only one DNA strand acts as a template. **Figure 17.8** The RNA polymerase would bind directly to the promoter, rather than depending on the previous binding of other factors. **Figure 17.25** The mRNA on the right (the longest one) started transcription first. The ribosome at the top, closest to the DNA, started translating first and thus has the longest polypeptide.

Concept Check 17.1

1. Recessive **2.** A polypeptide made up of 10 Gly (glycine) amino acids

3. Template sequence (from problem): 3′-TTCAGTCGT-5′

Nontemplate sequence: 5′-AAGTCAGCA-3′

mRNA sequence: 5′-AAGUCAGCA-3′

The nontemplate and mRNA nucleotide sequences are the same except that there is T in the nontemplate strand of DNA wherever there is U in the mRNA.

4. "Template sequence" (from nontemplate sequence in problem, written 3′ $\rightarrow$ 5′): 3′-ACGACTGAA-5′

mRNA sequence: 5′-UGCUGACUU-3′

Translated: Cys-STOP-Leu

(Remember that the mRNA is antiparallel to the DNA strand.) A protein translated from the nontemplate sequence would have a completely different amino acid sequence and would most likely be nonfunctional. (It would also be shorter because of the stop signal shown in the mRNA sequence above—and possibly others earlier in the mRNA sequence.)

Concept Check 17.2

1. Both assemble nucleic acid chains from monomer nucleotides whose order is determined by complementary base pairing to a template strand. Both synthesize in the 5′ $\rightarrow$ 3′ direction, antiparallel to the template. DNA polymerase requires a primer, but RNA polymerase can start a nucleotide chain from scratch. DNA polymerase uses nucleotides with the sugar deoxyribose and the base T, whereas RNA polymerase uses nucleotides with the sugar ribose and the base U. **2.** The promoter is the region of DNA to which RNA polymerase binds to begin transcription, and it is at the upstream end of the gene (transcription unit). **3.** In a bacterial cell, RNA polymerase recognizes the gene's promoter and binds to it. In a eukaryotic cell, transcription factors mediate the binding of RNA polymerase to the promoter. In both cases, sequences in the promoter bind precisely to the RNA polymerase, so the enzyme is in the right location and orientation. **4.** The transcription factor that recognizes the TATA sequence would be unable to bind, so RNA polymerase could not bind and transcription of that gene probably would not occur.

Concept Check 17.3

1. Due to alternative splicing of exons, each gene can result in multiple different mRNAs and can thus direct synthesis of multiple different proteins. **2.** In editing a video, segments are cut out and discarded (like introns), and the remaining segments are joined together (like exons) so that the regions of joining ("splicing") are not noticeable. **3.** Once the mRNA has exited the nucleus, the cap prevents it from being degraded by hydrolytic enzymes and facilitates its attachment to ribosomes. If the cap were removed from all mRNAs, the cell would no longer be able to synthesize any proteins and would probably die.

Concept Check 17.4

1. First, each aminoacyl-tRNA synthetase specifically recognizes a single amino acid and attaches it only to an appropriate tRNA. Second, a tRNA charged with its specific amino acid binds only to an mRNA codon for that amino acid. **2.** The structure and function of the ribosome seem to depend more on the rRNAs than on the ribosomal proteins. Because it is single-stranded, an RNA molecule can hydrogen-bond with itself and with other RNA molecules. RNA molecules make up the interface between the two ribosomal subunits, so presumably RNA-RNA binding helps hold the ribosome together. The binding site for mRNA in the ribosome includes rRNA that can bind the mRNA. Also, complementary bonding within an RNA molecule allows it to assume a particular three-dimensional shape and, along with the RNA's functional groups, presumably enables rRNA to catalyze peptide bond formation during translation. **3.** A signal peptide on the leading end of the polypeptide being synthesized is recognized by a signal-recognition particle that brings the ribosome to the ER membrane. There the ribosome attaches and continues to synthesize the polypeptide, depositing it in the ER lumen. **4.** Because of wobble, the tRNA could bind to either 5′-GCA-3′ or 5′-GCG-3′, both of which code for alanine (Ala). Alanine would be attached to the tRNA.

Ala

3′ CGU 5′

5′-GCA-3′ 5′-GCG-3′

Concept Check 17.5

1. In the mRNA, the reading frame downstream from the deletion is shifted, leading to a long string of incorrect amino acids in the polypeptide, and in most cases, a stop codon will arise, leading to premature termination. The polypeptide will most likely be nonfunctional. **2.** Heterozygous individuals, said to have sickle-cell trait, have a copy each of the wild-type allele and the sickle-cell allele. Both alleles will be expressed, so these individuals will have both normal and sickle-cell hemoglobin molecules. Apparently, having a mix of the two forms of β-globin has no effect under most conditions, but during prolonged periods of low blood oxygen (such as at higher altitudes), these individuals can show some signs of sickle-cell disease.

3.

Normal DNA sequence (template strand is on top): 3′-TACTTGTCCGATATC-5′
5′-ATGAACAGGCTATAG-3′

mRNA sequence: 5′-AUGAACAGGCUAUAG-3′

Amino acid sequence: Met-Asn-Arg-Leu-STOP

Mutated DNA sequence (template strand is on top): 3′-TACTTGTCCAATATC-5′
5′-ATGAACAGGTTATAG-3′

mRNA sequence: 5′-AUGAACAGGUUAUAG-3′

Amino acid sequence: Met-Asn-Arg-Leu-STOP

No effect: The amino acid sequence is Met-Asn-Arg-Leu both before and after the mutation because the mRNA codons 5′-CUA-3′ and 5′-UUA-3′ both code for Leu. (The fifth codon is a stop codon.)

Concept Check 17.6

1. No, transcription and translation are separated in space and time in a eukaryotic cell, a result of the eukaryotic cell's nuclear compartment. **2.** When one ribosome terminates translation and dissociates, the two subunits would be very close to the cap. This could facilitate their rebinding and initiating synthesis of a new polypeptide, thus increasing the efficiency of translation.

Summary of Key Concepts Questions

17.1 A gene contains genetic information in the form of a nucleotide sequence. The gene is first transcribed into an RNA molecule, and a messenger RNA molecule is ultimately translated into a polypeptide. The polypeptide makes up part or all of a protein, which performs a function in the cell and contributes to the phenotype of the organism. **17.2** Both bacterial and eukaryotic genes have promoters, regions where RNA polymerase ultimately binds and begins transcription. In bacteria, RNA polymerase binds directly to the promoter; in eukaryotes, transcription factors bind first to the promoter, and then RNA polymerase binds to the transcription factors and promoter together. **17.3** Both the 5′ cap and the poly-A tail help the mRNA exit from the nucleus and then, in the cytoplasm, help ensure mRNA stability and allow it to bind to ribosomes. **17.4** tRNAs function as translators between the nucleotide-based language of mRNA and the amino-acid-based language of polypeptides. A tRNA carries a specific amino acid, and the anticodon on the tRNA is complementary to the codon on the mRNA that codes for that amino acid. In the ribosome, the tRNA binds to the A site, where the polypeptide being synthesized is joined to the new amino acid, which becomes the new (C-terminal) end of the polypeptide. Next, the tRNA moves to the P site. When the next amino acid is added via transfer of the polypeptide to the new tRNA, the now empty tRNA moves to the E site, where it exits the ribosome. **17.5** When a nucleotide base is altered chemically, its base-pairing characteristics may be changed. When that happens, an incorrect nucleotide is likely to be incorporated into the complementary strand during the next replication of the DNA, and successive rounds of replication will perpetuate the mutation. Once the gene is transcribed, the mutated codon may code for a different amino acid that inhibits or changes the function of a protein. If the chemical change in the base is detected and repaired by the DNA repair system before the next replication, no mutation will result. **17.6** The presence of a nuclear envelope in eukaryotes means that transcription and translation are separated in space and therefore in time. This separation allows other processes (specifically, RNA processing) to occur and provides other steps at which gene expression can be regulated.

Test Your Understanding

1. b **2.** d **3.** a **4.** a **5.** b **6.** d **7.** e

8.

Type of RNA	Functions
Messenger RNA (mRNA)	Carries information specifying amino acid sequences of proteins from DNA to ribosomes
Transfer RNA (tRNA)	Serves as translator molecule in protein synthesis; translates mRNA codons into amino acids
Ribosomal RNA (rRNA)	Plays catalytic (ribozyme) roles and structural roles in ribosomes
Primary transcript	Is a precursor to mRNA, rRNA, or tRNA, before being processed; some intron RNA acts as a ribozyme, catalyzing its own splicing
Small nuclear RNA (snRNA)	Plays structural and catalytic roles in spliceosomes, the complexes of protein and RNA that splice pre-mRNA

Chapter 18

Figure Questions

Figure 18.3 As the concentration of tryptophan in the cell falls, eventually there will be none bound to repressor molecules. These will then take on their inactive shapes and dissociate from the operator, allowing transcription of the operon to resume. The enzymes for tryptophan synthesis will be made, and they will begin to synthesize tryptophan again in the cell. **Figure 18.11** The albumin gene enhancer has the three control elements colored yellow, gray, and red. The sequences in the liver and lens cells would be identical, since the cells are in the same organism. **Figure 18.18** Even if the mutant MyoD protein couldn't activate the *myoD* gene, it could still turn on genes for the other proteins in the pathway (other transcription factors, which would turn on the genes for muscle-specific proteins, for example). Therefore, some differentiation would occur. But unless there were other activators that could compensate for the loss of the MyoD protein's activation of the *myoD* gene, the cell would not be able to maintain its differentiated state. **Figure 18.22** Normal Bicoid protein would be made in the anterior end and compensate for the presence of mutant *bicoid* mRNA put into the egg by the mother. Development should be normal, with a head present. **Figure 18.24** The mutation is likely to be recessive because it is more likely to have an effect if both copies of the gene are mutated and code for nonfunctional proteins. If one normal copy of the gene is present, its product could inhibit the cell cycle. (However, there are also known cases of dominant *p53* mutations.)

Concept Check 18.1

1. Binding by the *trp* corepressor (tryptophan) activates the *trp* repressor, shutting off transcription of the *trp* operon; binding by the *lac* inducer (allolactose) inactivates the *lac* repressor, leading to transcription of the *lac* operon. **2.** When glucose is scarce, cAMP is bound to CAP and CAP is bound to the promoter, favoring the binding of RNA polymerase. However, in the absence of lactose, the repressor is bound to the operator, blocking RNA polymerase binding to the promoter. Therefore, the operon genes are not transcribed. **3.** The cell would continuously produce β-galactosidase and the two other enzymes for lactose utilization, even in the absence of lactose, thus wasting cell resources.

Concept Check 18.2

1. Histone acetylation is generally associated with gene expression, while DNA methylation is generally associated with lack of expression. **2.** General transcription factors function in assembling the transcription initiation complex at the promoters for all genes. Specific transcription factors bind to control elements associated with a particular gene and, once bound, either increase (activators) or decrease (repressors) transcription of that gene. **3.** The three genes should have some similar or identical sequences in the control elements of their enhancers. Because of this similarity, the same specific transcription factors in muscle cells could bind to the enhancers of all three genes and stimulate their expression coordinately. **4.** Degradation of the mRNA, regulation of translation, activation of the protein (by chemical modification, for example), and protein degradation **5.** Expression of the gene encoding the yellow activator (YA) must be regulated at one of the steps shown in Figure 18.6. The YA gene might be transcribed only in liver cells because the necessary activators for the enhancer of the YA gene are found only in liver cells.

Concept Check 18.3

1. Both miRNAs and siRNAs are small, single-stranded RNAs that associate with a complex of proteins and then can base-pair with mRNAs that have a complementary sequence. This base pairing leads to either degradation of the mRNA or blockage of its translation. Some siRNAs, in association with other proteins, can bind back to the chromatin in a certain region, causing chromatin changes that affect transcription. Both miRNAs and siRNAs are processed from double-stranded RNA precursors by the enzyme Dicer. All miRNAs are specified by genes in the cell's genome, and the single transcript folds back on itself to form one or more double-stranded hairpins, each of which is processed into an miRNA. In contrast, siRNAs arise from a longer stretch of linear double-stranded RNA, which may be introduced into the cell by a virus or an experimenter. Alternatively, in some cases, a cellular gene specifies one RNA strand of the precursor molecule, and an enzyme then synthesizes the complementary strand. **2.** The mRNA would persist and be translated into the cell division–promoting protein, and the cell would probably divide. If the intact miRNA is necessary for inhibition of cell division, then division of this cell might be inappropriate. Uncontrolled cell division could lead to formation of a mass of cells (tumor) that prevents proper functioning of the organism and could contribute to the development of cancer. **3.** The *XIST* RNA is transcribed from the *XIST* gene on the X chromosome that will be inactivated. It then binds to that chromosome and induces heterochromatin formation. A likely model is that the *XIST* RNA somehow recruits chromatin modification enzymes that lead to formation of heterochromatin.

Concept Check 18.4

1. Cells undergo differentiation during embryonic development, becoming different from each other; in the adult organism, there are many highly specialized cell types. **2.** By binding to a receptor on the receiving cell's surface and triggering a signal transduction pathway, involving intracellular molecules such as second messengers and transcription factors that affect gene expression **3.** Because their products, made and deposited into the egg by the mother, determine the head and tail ends, as well as the back and belly, of the embryo (and eventually the adult fly) **4.** The lower cell is synthesizing signaling molecules because the gene encoding them is activated, meaning that the appropriate specific transcription factors are binding to the gene's enhancer. The genes encoding these specific transcription factors are also being expressed in this cell because the transcriptional activators that can turn them on were expressed in the precursor to this cell. A similar explanation also applies to the cells expressing the receptor proteins. This scenario began with specific cytoplasmic determinants localized in specific regions of the egg. These cytoplasmic determinants were distributed unevenly to daughter cells, resulting in cells going down different developmental pathways.

Concept Check 18.5

1. Apoptosis is signaled by p53 protein when a cell has extensive DNA damage, so apoptosis plays a protective role in eliminating a cell that might contribute to cancer. If mutations in the genes in the apoptotic pathway blocked apoptosis, a cell with such damage could continue to divide and might lead to tumor formation. **2.** When an individual has inherited an oncogene or a mutant allele of a tumor-suppressor gene **3.** A cancer-causing mutation in a proto-oncogene usually makes the gene product overactive, whereas a cancer-causing mutation in a tumor-suppressor gene usually makes the gene product nonfunctional.

Summary of Key Concepts Questions

18.1 A corepressor and an inducer are both small molecules that bind to the repressor protein in an operon, causing the repressor to change shape. In the case of a corepressor (like tryptophan), this shape change allows the repressor to bind to the operator, blocking transcription. In contrast, an inducer causes the repressor to dissociate from the operator, allowing transcription to begin. **18.2** The chromatin must not be tightly condensed because it must be accessible to transcription factors. The appropriate specific transcription factors (activators) must bind to the control elements in the enhancer of the gene, while repressors must not be bound. The DNA must be bent by a bending protein so the activators can contact the mediator proteins and form a complex with general transcription factors at the promoter. Then RNA polymerase must bind and begin transcription. **18.3** miRNAs do not "code" for the amino acids of a protein—they are never translated. Each miRNA is cleaved from its hairpin RNA structure and then trimmed by Dicer. Next, one strand is degraded while the other associates with a group of proteins to form a complex. Binding of the complex to an mRNA with a complementary sequence causes that mRNA to be degraded or blocks its translation. This is considered gene regulation because it controls the amount of a particular mRNA that can be translated into a functional protein. **18.4** The first process involves cytoplasmic determinants, including mRNAs and proteins, placed into specific locations in the egg by the mother. The cells that are formed from different regions in the egg during early cell divisions will have different proteins in them, which will direct different programs of gene expression. The second process involves the cell in question responding to signaling molecules secreted by neighboring cells. The signaling pathway in the responding cell also leads to a different pattern of gene expression. The coordination of these two processes results in each cell following a unique pathway in the developing embryo. **18.5** The protein product of a proto-oncogene is usually involved in a pathway that stimulates cell division. The protein product of a tumor-suppressor gene is usually involved in a pathway that inhibits cell division.

Test Your Understanding

1. d **2.** a **3.** a **4.** a **5.** c **6.** d **7.** c **8.** e **9.** b **10.** b

11. a.

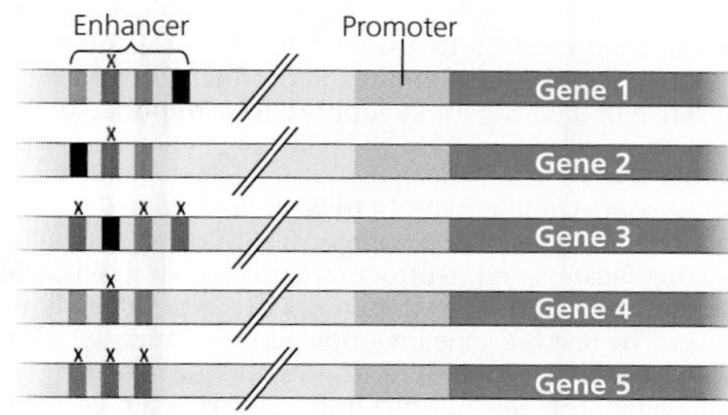

The purple, blue, and red activator proteins would be present.

b.

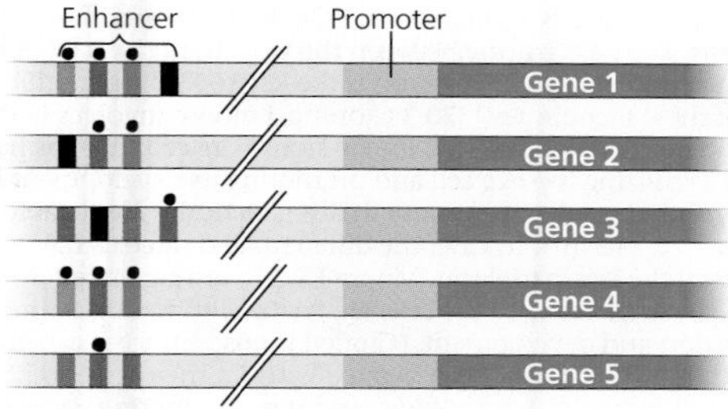

Only gene 4 would be transcribed.

c. In nerve cells, the orange, blue, green, and black activators would have to be present, thus activating transcription of genes 1, 2, and 4. In skin cells, the red, black, purple, and blue activators would have to be present, thus activating genes 3 and 5.

Chapter 19

Figure Questions

Figure 19.2 Beijerinck might have concluded that the agent was a toxin produced by the plant that was able to pass through a filter but that became more and more dilute. In this case, he would have concluded that the infectious agent could not replicate. **Figure 19.4** Top vertical arrow: Infection. Left upper arrow: Replication. Right upper arrow: Transcription. Right middle arrow: Translation. Lower left and right arrows: Self-assembly. Bottom middle arrow: Exit. **Figure 19.7** Any class V virus, including the viruses that cause influenza (flu), measles, and mumps **Figure 19.8** The main protein on the cell surface that HIV binds to is called CD4. However, HIV also requires a "co-receptor," which in many cases is a protein called CCR5. HIV binds to both of these proteins together and then is taken into the cell. Researchers discovered this requirement by studying individuals who seemed to be resistant to HIV infection, despite multiple exposures. These individuals turned out to have mutations in the gene that encodes CCR5 such that the protein apparently cannot act as a co-receptor, and so HIV can't enter and infect cells.

Concept Check 19.1

1. TMV consists of one molecule of RNA surrounded by a helical array of proteins. The influenza virus has eight molecules of RNA, each surrounded by a helical array of proteins, similar to the arrangement of the single RNA molecule in TMV. Another difference between the viruses is that the influenza virus has an outer envelope and TMV does not. **2.** The T2 phages were an excellent choice for use in the Hershey-Chase experiment because they consist of only DNA surrounded by a protein coat, and DNA and protein were the two candidates for macromolecules that carried genetic information. Hershey and Chase were able to radioactively label each type of molecule alone and follow it during separate infections of *E. coli* cells with T2. Only the DNA entered the bacterial cell during infection, and only labeled DNA showed up in some of the progeny phage. Hershey and Chase concluded that the DNA must carry the genetic information necessary for the phage to reprogram the cell and produce progeny phages.

Concept Check 19.2

1. Lytic phages can only carry out lysis of the host cell, whereas lysogenic phages may either lyse the host cell or integrate into the host chromosome. In the latter case, the viral DNA (prophage) is simply replicated along with the host chromosome. Under certain conditions, a prophage may exit the host chromosome and initiate a lytic cycle. **2.** Both the viral RNA polymerase and the RNA polymerase in Figure 17.9 synthesize an RNA molecule complementary to a template strand. However, the RNA polymerase in Figure 17.9 uses one of the strands of the DNA double helix as a template, whereas the viral RNA polymerase uses the RNA of the viral genome as a template. **3.** Because it synthesizes DNA from its RNA genome. This is the reverse ("retro") of the usual DNA → RNA information flow. **4.** There are many steps that could be interfered with: binding of the virus to the cell, reverse transcriptase function, integration into the host cell chromosome, genome synthesis (in this case, transcription of RNA from the integrated provirus), assembly of the virus inside the cell, and budding of the virus. (Many of these, if not all, are targets of actual medical strategies to block progress of the infection in HIV-infected people.)

Concept Check 19.3

1. Mutations can lead to a new strain of a virus that can no longer be effectively fought by the immune system, even if an animal had been exposed to the original strain; a virus can jump from one species to a new host; and a rare virus can spread if a host population becomes less isolated. **2.** In horizontal transmission, a plant is infected from an external source of virus, which could enter through a break in the plant's epidermis due to damage by herbivores. In vertical transmission, a plant inherits viruses from its parent either via infected seeds (sexual reproduction) or via an infected cutting (asexual reproduction). **3.** Humans are not within the host range of TMV, so they can't be infected by the virus.

Summary of Key Concepts Questions

19.1 Viruses are generally considered nonliving, because they are not capable of replicating outside of a host cell. To replicate, they depend completely on host enzymes and resources. **19.2** Single-stranded RNA viruses require an RNA polymerase that can make RNA using an RNA template. (Cellular RNA polymerases make RNA using a DNA template.) Retroviruses require reverse transcriptases to make DNA using an RNA template. (Once the first DNA strand has been made, the same enzyme can promote synthesis of the second DNA strand.) **19.3** The mutation rate of RNA viruses is higher than that of DNA viruses because RNA polymerase has no proofreading function, so errors in replication are not corrected. Their higher mutation rate means that RNA viruses change faster than DNA viruses, leading to their being able to have an altered host range and to evade immune defenses in possible hosts.

Test Your Understanding

1. c **2.** d **3.** c **4.** d **5.** b

6. As shown below, the viral genome would be translated into capsid proteins and envelope glycoproteins directly, rather than after a complementary RNA copy was made. A complementary RNA strand would still be made, however, that could be used as a template for many new copies of the viral genome.

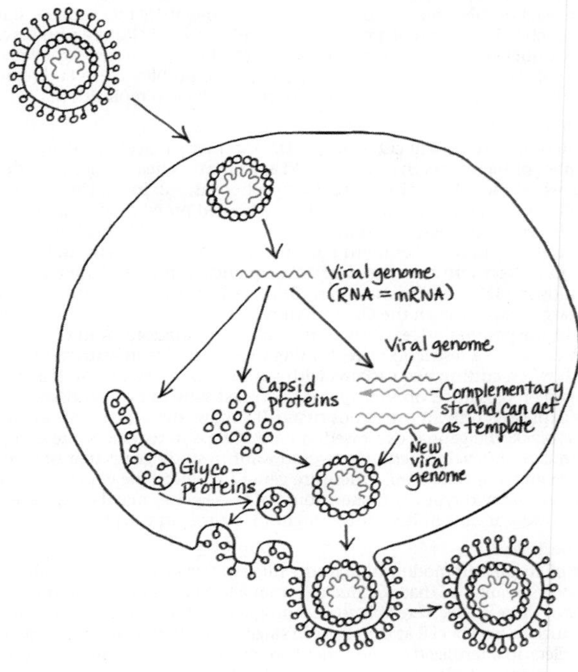

Chapter 20

Figure Questions

Figure 20.3

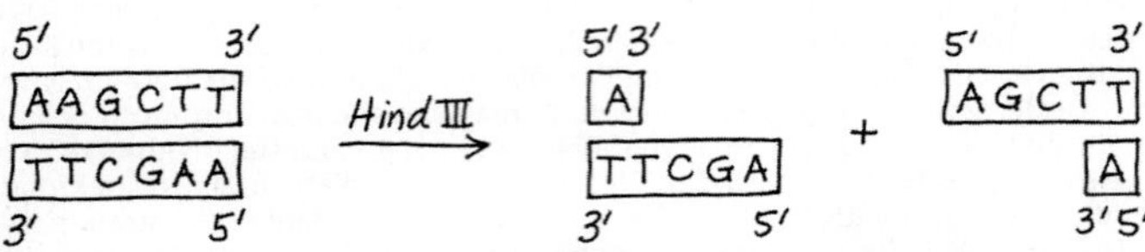

Figure 20.4 Cells containing no plasmid at all would be able to grow; these colonies would be white because they would lack functional *lacZ* genes. **Figure 20.10** Grow each clone of cells in culture. Isolate the plasmids from each and cut them with the restriction enzyme originally used to make the clone (see Figure 20.4). Run each sample on an electrophoretic gel, and recover the DNA of the insert from the gel band. **Figure 20.16** Crossing over, which causes recombination, is a random event. The chance of crossing over occurring between two loci increases as the distance between them increases. The SNP is located very close to an unknown disease-causing allele, and therefore crossing over rarely occurs between the SNP and the allele, so the SNP is a genetic marker indicating the presence of the particular allele. **Figure 20.18** None of the eggs with the transplanted nuclei from the four-cell embryo at the upper left would have developed into a tadpole. Also, the result might include only some of the tissues of a tadpole, which might differ, depending on which nucleus was transplanted. (This assumes that there was some way to tell the four cells apart, as one can in some frog species.) **Figure 20.22** Using converted iPS cells would not carry the same risk, which is its major advantage. Because the donor cells would come from the patient, they would be perfectly matched. The patient's immune system would recognize them as "self" cells and would not mount an attack (which is what leads to rejection).

Concept Check 20.1

1. The covalent sugar-phosphate bonds of the DNA strands **2.** Yes, *PvuI* will cut the molecule.

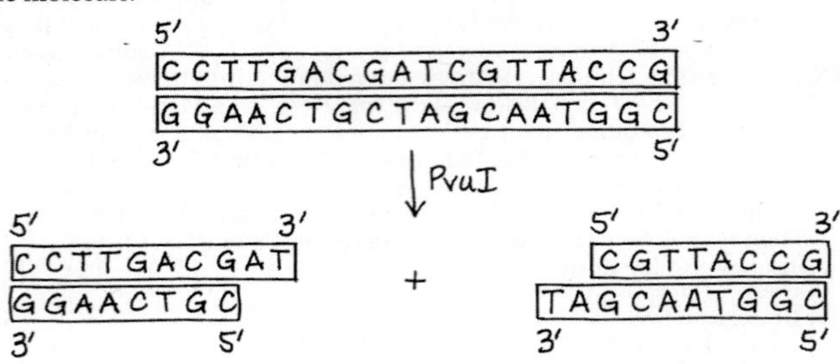

3. Some human genes are too large to be incorporated into bacterial plasmids. Bacterial cells lack the means to process RNA transcripts into mRNA, and even if the need for RNA processing is avoided by using cDNA, bacteria lack enzymes to catalyze the post-translational processing that many human proteins require to function properly. **4.** During the replication of the ends of linear DNA molecules (see Figure 16.20), an RNA primer is used at the 5′ end of each new strand. The RNA must be replaced by DNA nucleotides, but DNA polymerase is incapable of starting from scratch at the 5′ end of a new DNA strand. During PCR, the primers are made of DNA nucleotides already, so they don't need to be replaced—they just remain as part of each new strand. Therefore, there is no problem with end replication during PCR, and the fragments don't shorten with each replication.

Concept Check 20.2

1. Any restriction enzyme will cut genomic DNA in many places, generating such a large number of fragments that they would appear as a smear rather than distinct bands when the gel is stained after electrophoresis. **2.** In Southern blotting, Northern blotting, and microarray analysis, the labeled probe binds only to the specific target sequence owing to complementary nucleic acid hybridization (DNA-DNA hybridization in Southern blotting and microarray analysis, DNA-RNA hybridization in Northern blotting). In DNA sequencing, primers base-pair to the template, allowing DNA synthesis to start. In RT-PCR, the primers must base-pair with their target sequences in the DNA mixture. **3.** A SNP is a single nucleotide that varies in the population, existing in two or more variations. A RFLP is a type of SNP that occurs in a restriction site, leading to a difference in restriction fragment length when cutting two variants with a restriction enzyme. **4.** If a spot is green, the gene represented on that spot is expressed only in normal tissue. If red, the gene is expressed only in cancerous tissue. If yellow, the gene is expressed in both. And if black, the gene is expressed in neither type of tissue. As a researcher interested in cancer development, you would want to study genes represented by spots that are green or red because these are genes for which the expression level differs between the two types of tissues. Some of these genes may be expressed differently as a result of cancer, but others might play a role in causing cancer.

Concept Check 20.3

1. The state of chromatin modification in the nucleus from the intestinal cell was undoubtedly less similar to that of a nucleus from a fertilized egg, explaining why many fewer of these nuclei were able to be reprogrammed. In contrast, the chromatin in a nucleus from a cell at the four-cell stage would have been much more like that of a nucleus in a fertilized egg and therefore much more easily programmed to direct development. **2.** No, primarily because of subtle (and perhaps not so subtle) differences in their environments **3.** A technique would have to be worked out for turning a human iPS cell into a pancreatic cell (probably by inducing expression of pancreas-specific regulatory genes in the cell). **4.** The carrot cell has much more potential. The cloning experiment shows that an individual carrot cell can generate all the tissues of an adult plant. The muscle cell, on the other hand, will always remain a muscle cell because of its genetic program (it expresses the *myoD* gene, which ensures continued differentiation). The muscle cell is like other fully differentiated animal cells: It will remain fully differentiated on its own unless it is reprogrammed into an iPS cell using the new techniques described here. (This would be quite difficult to accomplish because a muscle cell has multiple nuclei.)

Concept Check 20.4

1. Stem cells continue to reproduce themselves. **2.** Herbicide resistance, pest resistance, disease resistance, salinity resistance, and delayed ripening **3.** Because hepatitis A is an RNA virus, you could isolate RNA from the blood and try to detect copies of hepatitis A RNA by one of three methods. First, you could run the RNA on a gel and then do a Northern blot using probes complementary to hepatitis A genome sequences. A second approach would be to use reverse transcriptase to make cDNA from the RNA in the blood, run the cDNA on a gel, and do a Southern blot using the same probe. However, neither of these methods would be as sensitive as RT-PCR, in which you would reverse-transcribe the blood RNA into cDNA and then use PCR to amplify the cDNA, using primers specific to hepatitis A sequences. If you then ran the products on an electrophoretic gel, the presence of a band would support your hypothesis.

Summary of Key Concepts Questions

20.1 A plasmid vector and a source of foreign DNA to be cloned are both cut with the same restriction enzyme, generating restriction fragments with sticky ends. These fragments are mixed together, ligated, and reintroduced into bacterial cells, which are then grown on the antibiotic ampicillin. The plasmid has two genes that allow selection of recombinant clones. The first is a gene for ampicillin resistance, which only allows cells to grow if they have taken up a plasmid. The second is a gene for β-galactosidase, which can generate a blue product if the gene is intact. The cloning site is within this gene, so only nonrecombinant colonies will be blue, and recombinant plasmids will be found in cells that are in white colonies. **20.2** Many techniques used to analyze genes and their expression involve nucleic acid hybridization: Southern and Northern blotting, DNA sequencing, PCR, *in situ* hybridization, and DNA microarray analysis. The base pairing between the two strands of a DNA molecule or between a DNA strand and an RNA strand is the key to finding specific nucleic acid sequences in all of these techniques. **20.3** Cloning a mouse involves transplanting a nucleus from a differentiated mouse cell into a mouse egg cell that has had its own nucleus removed. Fertilizing the egg cell and promoting its development into an embryo in a surrogate mother results in a mouse that is genetically identical to the mouse that donated the nucleus. In this case, the differentiated nucleus has been reprogrammed by factors in the egg cytoplasm. Mouse ES cells are generated from inner cells in mouse blastocysts, so in this case the cells are "naturally" reprogrammed by the process of reproduction and development. (Cloned mouse embryos can also be used as a source of ES cells.) iPS cells can be generated without the use of embryos from a differentiated adult mouse cell, by adding certain transcription factors into the cell. In this case, the transcription factors are reprogramming the cells to become pluripotent. **20.4** First, the disease must be caused by a single gene, and the molecular basis of the problem must be understood. Second, the cells that are going to be introduced into the patient must be cells that will integrate into body tissues and continue to multiply (and provide the needed gene product). Third, the gene must be able to be introduced into the cells in question in a safe way, as there have been instances of cancer resulting from some gene therapy trials. (Note that this will require testing the procedure in mice; moreover, the factors that determine a safe vector are not yet well understood. Maybe one of you will go on to solve this problem!)

Test Your Understanding

1. b **2.** c **3.** b **4.** c **5.** d **6.** b **7.** a **8.** c
9.

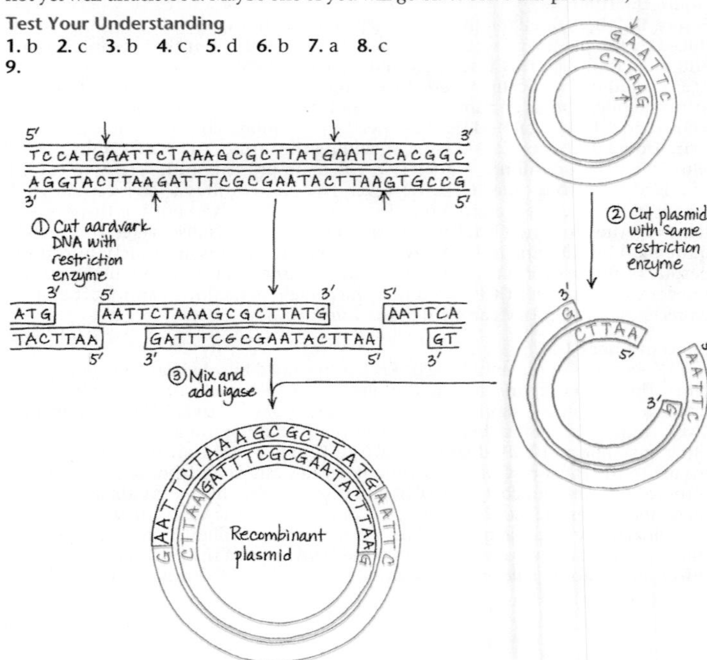

10. A cDNA library, made using mRNA from human lens cells, which would be expected to contain many copies of mRNA for the crystallin of interest

Chapter 21

Figure Questions

Figure 21.3 The fragments in stage 2 of this figure are like those in stage 2 of Figure 21.2, but in this figure their order relative to each other is not known and will be determined later by computer. The order of the fragments in Figure 21.2 is completely known before sequencing begins. (Determining the order takes longer but makes the eventual sequence assembly much easier.) **Figure 21.9** The transposon would be cut out of the DNA at the original site rather than copied, so the figure would show the original stretch of DNA without the transposon after the mobile transposon had been cut out. **Figure 21.11** The RNA transcripts extending from the DNA in each transcription unit are shorter on the left and longer on the right. This means that RNA polymerase must be starting on the left end of the unit and moving toward the right.

Figure 21.13

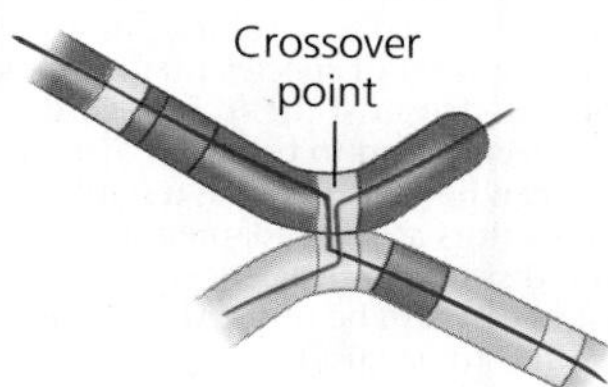

Figure 21.14 Pseudogenes are nonfunctional. They could have arisen by any mutations in the second copy that made the gene product unable to function. Examples would be base changes that introduce stop codons in the sequence, alter amino acids, or change a region of the gene promoter so that the gene can no longer be expressed. **Figure 21.15** Let's say a transposable element (TE) existed in the intron to the left of the indicated EGF exon in the EGF gene, and the same TE was present in the intron to the right of the indicated F exon in the fibronectin gene. During meiotic recombination, these TEs could cause nonsister chromatids on homologous chromosomes to pair up incorrectly, as seen in Figure 21.13. One gene might end up with an F exon next to an EGF exon. Further mistakes in pairing over many generations might result in these two exons being separated from the rest of the gene and placed next to a single or duplicated K exon. In general, the presence of repeated sequences in introns and between genes facilitates these processes because it allows incorrect pairing of nonsister chromatids, leading to novel exon combinations. **Figure 21.17** Since you know that chimpanzees do not speak but humans do, you'd probably want to know how many amino acid differences there are between the human wild-type FOXP2 protein and that of the chimpanzee and whether these changes affect the function of the protein. (As we explain later in the text, there are two amino acid differences.) You know that humans with mutations in this gene have severe language impairment. You would want to learn more about the human mutations by checking whether they affect the same amino acids in the gene product that the chimpanzee sequence differences affect. If so, those amino acids might play an important role in the function of the protein in language. Going further, you could analyze the differences between the chimpanzee and mouse FOXP2 proteins. You might ask: Are they more similar than the chimpanzee and human proteins? (It turns out that the chimpanzee and mouse proteins have only one amino acid difference and thus are more similar than the chimpanzee and human proteins, which have two differences, and more similar than the human and mouse proteins, which have three differences.)

Concept Check 21.1

1. In a linkage map, genes and other markers are ordered with respect to each other, but only the relative distances between them are known. In a physical map, the actual distances between markers, expressed in base pairs, are known. **2.** The three-stage approach employed in the Human Genome Project involves linkage mapping, physical mapping, and then sequencing of short, overlapping fragments that previously have been ordered relative to each other (see Figure 21.2). The whole-genome shotgun approach eliminates the linkage mapping and physical mapping stages; instead, short fragments generated by multiple restriction enzymes are sequenced and then ordered by computer programs that identify overlapping regions (see Figure 21.3).

Concept Check 21.2

1. The Internet allows centralization of databases such as GenBank and software resources such as BLAST, making them freely accessible. Having all the data in a central database, easily accessible on the Internet, minimizes the possibility of errors and of researchers working with different data. It streamlines the process of science, since all researchers are able to use the same software programs, rather than each having to obtain their own, possibly different, software. It speeds up dissemination of data and ensures as much as possible that errors are corrected in a timely fashion. These are just a few answers; you can probably think of more. **2.** Cancer is a disease caused by multiple factors. To focus on a single gene or a single defect would ignore other factors that may influence the cancer and even the behavior of the single gene being studied. The systems approach, because it takes into account many factors at the same time, is more likely to lead to an understanding of the causes and most useful treatments for cancer. **3.** Some of the transcribed region is accounted for by introns. The rest is transcribed into noncoding RNAs, including small RNAs, such as microRNAs (miRNAs). These RNAs help regulate gene expression by blocking translation, causing degradation of mRNA, binding to the promoter and repressing transcription, or causing remodeling of chromatin structure. The functions of the remainder are not yet known. **4.** Genome-wide association studies use the systems biology approach in that they consider the correlation of many single nucleotide polymorphisms (SNPs) with particular diseases, such as heart disease and diabetes, in an attempt to find patterns of SNPs that correlate with each disease.

Concept Check 21.3

1. Alternative splicing of RNA transcripts from a gene and post-translational processing of polypeptides **2.** The total number of completed genomes is found by clicking on "Published Complete Genomes." Add the figures for bacterial, archaeal, and eukaryotic "ongoing genomes" to get the number "in progress." Finally, look at the top of the Published Complete Genomes page to get numbers of completed genomes for each domain. (*Note*: You can click on the "Size" column and the table will be re-sorted by genome size. Scroll down to get an idea of relative sizes of genomes in the three domains. Remember, though, that most of the sequenced genomes are bacterial.) **3.** Prokaryotes are generally smaller cells than eukaryotic cells, and they reproduce by binary fission. The evolutionary process involved is natural selection for more quickly reproducing cells: The faster they can replicate their DNA and divide, the more likely they will be able to dominate a population of prokaryotes. The less DNA they have to replicate, then, the faster they will reproduce.

Concept Check 21.4

1. The number of genes is higher in mammals, and the amount of noncoding DNA is greater. Also, the presence of introns in mammalian genes makes them larger, on average, than prokaryotic genes. **2.** In the copy-and-paste transposon mechanism and in retrotransposition **3.** In the rRNA gene family, identical transcription units for the three different RNA products are present in long, tandemly repeated arrays. The large number of copies of the rRNA genes enable organisms to produce the rRNA for enough ribosomes to carry out active protein synthesis, and the single transcription unit ensures that the relative amounts of the different rRNA molecules produced are correct. Each globin gene family consists of a relatively small number of nonidentical genes. The differences in the globin proteins encoded by these genes result in production of hemoglobin molecules adapted to particular developmental stages of the organism. **4.** The exons would be classified as exons (1.5%); the enhancer region containing the distal control elements, the region closer to the promoter containing the proximal control elements, and the promoter itself would be classified as regulatory sequences (5%); and the introns would be classified as introns (20%).

Concept Check 21.5

1. If meiosis is faulty, two copies of the entire genome can end up in a single cell. Errors in crossing over during meiosis can lead to one segment being duplicated while another is deleted. During DNA replication, slippage backward along the template strand can result in segment duplication. **2.** For either gene, a mistake in crossing over during meiosis could have occurred between the two copies of that gene, such that one ended up with a duplicated exon. This could have happened several times, resulting in the multiple copies of a particular exon in each gene. **3.** Homologous transposable elements scattered throughout the genome provide sites where recombination can occur between different chromosomes. Movement of these elements into coding or regulatory sequences may change expression of genes. Transposable elements also can carry genes with them, leading to dispersion of genes and in some cases different patterns of expression. Transport of an exon during transposition and its insertion into a gene may add a new functional domain to the originally encoded protein, a type of exon shuffling. (For any of these changes to be heritable, they must happen in germ cells, cells that will give rise to gametes.) **4.** Because more offspring are born to women who have this inversion, it must provide some advantage. It would be expected to persist and spread in the population. (In fact, evidence in the study allowed the researchers to conclude that it has been increasing in proportion in the population. You'll learn more about population genetics in the next unit.)

Concept Check 21.6

1. Because both humans and macaques are primates, their genomes are expected to be more similar than the macaque and mouse genomes are. The mouse lineage diverged from the primate lineage before the human and macaque lineages diverged. **2.** Homeotic genes differ in their *non*homeobox sequences, which determine the interactions of homeotic gene products with other transcription factors and hence which genes are regulated by the homeotic genes. These nonhomeobox sequences differ in the two organisms, as do the expression patterns of the homeobox genes. **3.** *Alu* elements must have undergone transposition more actively in the human genome for some reason. Their increased numbers may have then allowed more recombination errors in the human genome, resulting in more or different duplications. The divergence of the organization and content of the two genomes presumably made the chromosomes of each genome less homologous to those of the other, thus accelerating divergence of the two species by making matings less and less likely to result in fertile offspring.

Summary of Key Concepts Questions

21.1 Considering the sequencing of the human genome as an example, less time was required to sequence the first human genome using the whole-genome shotgun approach. Although this approach relied in part on data resulting from the three-stage approach used by the public consortium, the whole-genome shotgun approach was (and still is) faster and more efficient than the more labor-intensive three-stage process. The whole-genome shotgun approach was facilitated in large part by significant advances in computing power. **21.2** The most significant finding was that more than 90% of the human genomic region studied was transcribed, which suggested that the transcribed RNA (and thus the DNA from which it was produced) was performing some unknown functions. The project has been expanded to include other species because to determine the functions of these transcribed DNA elements, it is necessary to carry out this type of analysis on the genomes of species that can be used in laboratory experiments. **21.3** (a) In general, bacteria and archaea have smaller genomes, lower numbers of genes, and higher gene density than eukaryotes. (b) Among eukaryotes, there is no apparent systematic relationship between genome size and phenotype. The number of genes is often lower than would be expected from the size of the genome—in other words, the gene density is often lower in larger genomes. (Humans are an example.) **21.4** Transposable element-related sequences can move from place to place in the genome, and a subset of these sequences make a new copy of themselves when they do so. Thus, it is not surprising that they make up a significant percentage of the genome, and this percentage might be expected to increase over evolutionary time. **21.5** Chromosomal rearrangements within a species lead to some individuals having different chromosomal arrangements. Each of these individuals could still undergo meiosis and produce gametes, and fertilization involving gametes with different chromosomal arrangements could result in viable offspring. However, during meiosis in the offspring, the maternal and paternal chromosomes might not be able to pair up, causing gametes with incomplete sets of chromosomes to form. Most often, when zygotes are produced from such gametes, they do not survive. Ultimately, a new species could form if two different chromosomal arrangements became prevalent within a population and individuals could mate successfully only with other individuals having the same arrangement. **21.6** Comparing the genomes of two closely related species can reveal information about more recent evolutionary events, perhaps events that resulted in the distinguishing characteristics of the two species. Comparing the genomes of very distantly related species can tell us about evolutionary events that occurred a very long time ago. For example, genes that are shared between two distantly related species must have arisen before the two species diverged.

Test Your Understanding

1. c **2.** a **3.** a **4.** c

5.

1. ATETI...PKSSD...TSSTT...NARRD
2. ATETI...PKSSE...TSSTT...NARRD
3. ATETI...PKSSD...TSSTT...NARRD
4. ATETI...PKSSD...TSSNT...SARRD
5. ATETI...PKSSD...TSSTT...NARRD
6. VTETI...PKSSD...TSSTT...NARRD

(a) Lines 1, 3, and 5 are the C, G, R species. (b) Line 4 is the human sequence. (c) Line 6 is the orangutan sequence. (d) There is one amino acid difference between the mouse (the E on line 2) and the C, G, R species (which have a D in that position). There are three amino acid differences between the mouse and the human. (The E, T, and N in the mouse sequence are instead D, N, and S, respectively, in the human sequence.) (e) Because only one amino acid difference arose during the 60–100 million years since the mouse and C, G, R species diverged, it is somewhat surprising that two additional amino acid differences resulted during the 6 million years since chimpanzees and humans diverged. This indicates that the *FOXP2* gene has been evolving faster in the human lineage than in the lineages of other primates.

Chapter 22

Figure Questions

Figure 22.6 The cactus-eater is more closely related to the seed-eater; Figure 1.22 shows that they share a more recent common ancestor (a seed-eater) than the cactus-eater shares with the insect-eater. **Figure 22.8** More than 5.5 million years ago. **Figure 22.12** The colors and body forms of these mantids allow them to blend into their surroundings, providing an example of how organisms are well matched to life in their environments. The mantids also share features with one another (and with all other mantids), such as six legs, grasping forelimbs, and large eyes. These shared features illustrate another key observation about life: the unity of life that results from descent from a common ancestor. Over time, as these mantids diverged from a common ancestor, they accumulated different adaptations that made them well suited for life in their different environments. Eventually, these differences became large enough that new species were formed, thus contributing to the great diversity of life. **Figure 22.13** These results show that being reared from the egg stage on one plant species or the other did not result in the adult having a beak length appropriate for that host; instead, adult beak lengths were determined primarily by the population from which the eggs were obtained. Because an egg from a balloon vine population likely had long-beaked parents, while an egg from a goldenrain tree population likely had short-beaked parents, these results indicate that beak length is an inherited trait. **Figure 22.14** Both strategies should increase the time it takes *S. aureus* to become resistant to a new drug. If a drug that harms *S. aureus* does not harm other bacteria, natural selection will not favor resistance to that drug in the other species. This would decrease the chance that *S. aureus* would acquire resistance genes from other bacteria—thus slowing the evolution of resistance. Similarly, selection for resistance to a drug that slows the growth but does not kill *S. aureus* is much weaker than selection for resistance to a drug the kills *S. aureus*—again slowing the evolution of resistance. **Figure 22.17** Based on this evolutionary tree, crocodiles are more closely related to birds than to lizards because they share a more recent common ancestor with birds (ancestor 5) than with lizards (ancestor 4). **Figure 22.20** Hind limb structure changed first. *Rodhocetus* lacked flukes, but its pelvic bones and hind limbs had changed substantially from how those bones were shaped and arranged in *Pakicetus*. For example, in *Rodhocetus*, the pelvis and hind limbs appear to be oriented for paddling, whereas they were oriented for walking in *Pakicetus*.

Concept Check 22.1

1. Hutton and Lyell proposed that events in the past were caused by the same processes operating today. This principle suggested that Earth must be much older than a few thousand years, the age that was widely accepted at that time. Hutton and Lyell also thought that geologic change occurs gradually, stimulating Darwin to reason that the slow accumulation of small changes could ultimately produce the profound changes documented in the fossil record. In this context, the age of Earth was important to Darwin, because unless Earth was very old, he could not envision how there would have been enough time for evolution to occur. **2.** By these criteria, Cuvier's explanation of the fossil record and Lamarck's hypothesis of evolution are both scientific. Cuvier thought that species did not evolve over time. He also suggested that catastrophes and the resulting extinctions were usually confined to local regions and that such regions were later repopulated by a different set of species that immigrated from other areas. These assertions can be tested against the fossil record, and his assertion that species do not evolve has been found to be false. With respect to Lamarck, his principle of use and disuse can be used to make testable predictions for fossils of groups such as whale ancestors as they adapted to a new habitat. Lamarck's principle of use and disuse and his associated principle of the inheritance of acquired characteristics can also be tested directly in living organisms (these principles have been found to be false).

Concept Check 22.2

1. Organisms share characteristics (the unity of life) because they share common ancestors; the great diversity of life occurs because new species have repeatedly formed when descendant organisms gradually adapted to different environments, becoming different from their ancestors. **2.** The fossil mammal species (or its ancestors) would most likely have colonized the Andes from within South America, whereas ancestors of mammals currently found in African mountains would most likely have colonized those mountains from other parts of Africa. As a result, the Andes fossil species would share a more recent common ancestor with South American mammals than with mammals in Africa. Thus, for many of its traits, the fossil mammal species would probably more closely resemble mammals that live in South American jungles than mammals that live on African mountains. It is also possible, however, that the fossil mammal species could resemble the African mountain mammals by convergent evolution (even though they were only distantly related to one another). **3.** As long as the white phenotype (encoded by the genotype *pp*) continues to be favored by natural selection, the frequency of the *p* allele will likely increase over time in the population. The explanation is that if the proportion of white individuals increases relative to purple individuals, the frequency of the recessive *p* allele will also increase relative to that of the *P* allele, which only appears in purple individuals (some of whom also carry a *p* allele).

Concept Check 22.3

1. An environmental factor such as a drug does not create new traits, such as drug resistance, but rather selects for traits among those that are already present in the population. **2.** (a) Despite their different functions, the forelimbs of different mammals are structurally similar because they all represent modifications of a structure found in the common ancestor. (b) Convergent evolution: The similarities between the sugar glider and flying squirrel indicate that similar environments selected for similar adaptations despite different ancestry. **3.** At the time that dinosaurs originated, Earth's landmasses formed a single large continent, Pangaea. Because many dinosaurs were large and mobile, it is likely that early members of these groups lived on many different parts of Pangaea. When Pangaea broke apart, fossils of these organisms would have moved with the rocks in which they were deposited. As a result, we would predict that fossils of early dinosaurs would have a broad geographic distribution (this prediction has been upheld).

Summary of Key Concepts Questions

Concept 22.1 Darwin thought that descent with modification occurred as a gradual, steplike process. The age of Earth was important to him because if Earth were only a few thousand years old (as conventional wisdom suggested), there wouldn't have been sufficient time for major evolutionary change. **Concept 22.2** All species have the potential to produce more offspring (overreproduce) than can be supported by the environment. This ensures that there will be what Darwin called a "struggle for existence" in which many of the offspring are eaten, starved, diseased, or unable to reproduce for a variety of other reasons. Members of a population exhibit a range of heritable variations, some of which make it likely that their bearers will leave more offspring than other individuals (for example, the bearer may escape predators more effectively or be more tolerant of the physical conditions of the environment). Over time, natural selection resulting from factors such as predators, lack of food, or the physical conditions of the environment can increase the proportion of individuals with favorable traits in a population (evolutionary adaptation). **Concept 22.3** The hypothesis that cetaceans originated from a terrestrial mammal and are closely related to even-toed ungulates is supported by several lines of evidence. For example, fossils document that early cetaceans had hind limbs, as expected for organisms that descended from a land mammal; these fossils also show that cetacean hind limbs became reduced over time. Other fossils show that early cetaceans had a type of ankle bone that is otherwise found only in even-toed ungulates, providing strong evidence that even-toed ungulates are the land mammals to which cetaceans are most closely related. DNA sequence data also indicate that even-toed ungulates are the land mammals to which cetaceans are most closely related.

Test Your Understanding

1. b **2.** d **3.** d **4.** c **5.** a
7. (a)

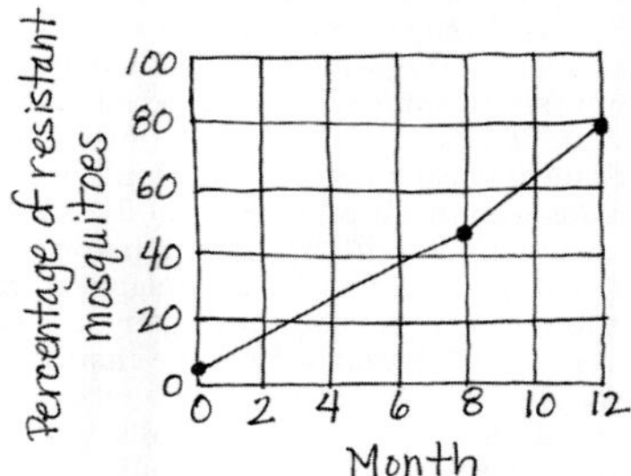

(b) The rapid rise in the percentage of mosquitoes resistant to DDT was most likely caused by natural selection in which mosquitoes resistant to DDT could survive and reproduce while other mosquitoes could not. (c) In India—where DDT resistance first appeared—natural selection would have caused the frequency of resistant mosquitoes to increase over time. If resistant mosquitoes then migrated from India (for example, transported by wind or in planes, trains, or ships) to other parts of the world, the frequency of DDT resistance would increase there as well.

Chapter 23

Figure Questions

Figure 23.8 The predicted frequencies are 36% C^RC^R, 48% C^RC^W, and 16% C^WC^W. **Figure 23.12** Local survival should increase in both populations. The increase would probably occur more rapidly among birds born into the central population because gene flow from the mainland is higher in that population. **Figure 23.13** Directional selection. Goldenrain tree has smaller fruit than does the native host, balloon vine. Thus, in soapberry bug populations feeding on goldenrain tree, bugs with shorter beaks had an advantage, resulting in directional selection for shorter beak length. **Figure 23.16** Crossing a single female's eggs with both an SC and an LC male's sperm allowed the researchers to directly compare the effects of the males' contribution to the next generation, since both batches of offspring had the same maternal contribution. This isolation of the male's impact enabled researchers to draw conclusions about differences in genetic "quality" between the SC and LC males. **Figure 23.18** The researchers measured the percentages of successfully reproducing adults in the breeding adult population that had each phenotype. This approach of determining which phenotype was favored by selection assumes that reproduction was a sufficient indicator of relative fitness (as opposed to counting the number of eggs laid or offspring hatched, for example) and that mouth phenotype was the driving factor determining the fish's ability to reproduce.

Concept Check 23.1

1. (a) Within a population, genetic differences among individuals provide the raw material on which natural selection and other mechanisms can act. Without such differences, allele frequencies could not change over time—and hence the population could not evolve. (b) Genetic differences between separate populations can result from natural selection if different alleles are favored in different populations; this might occur, for example, if the different populations experienced different environmental conditions (as in Figure 23.4). Genetic differences between populations can also result from chance events (genetic drift) if the genetic changes have few or no phenotypic effects (as in Figure 23.3). **2.** Many mutations occur in somatic cells, which do not produce gametes and so are lost when the organism dies. Of mutations that do occur in cell lines that produce gametes, many do not have a phenotypic effect on which natural selection can act. Others have a harmful effect and are thus unlikely to increase in frequency because they decrease the reproductive success of their bearers. **3.** Its genetic variation (whether measured at the level of the gene or at the level of nucleotide sequences) would probably drop over time. During meiosis, crossing over and the independent assortment of chromosomes produce many new combinations of alleles. In addition, a population contains a vast number of possible mating combinations, and fertilization brings together the gametes of individuals with different genetic backgrounds. Thus, via crossing over, independent assortment of chromosomes, and fertilization, sexual reproduction reshuffles alleles into fresh combinations each generation. Without sexual reproduction, the rate of forming new combinations of alleles would be vastly reduced, causing the overall amount of genetic variation to drop.

Concept Check 23.2

1. 30,000. Half the loci (10,000) are fixed, meaning only one type of allele exists for each locus: $10{,}000 \times 1 = 10{,}000$. There are two types of alleles each for the other loci: $10{,}000 \times 2 = 20{,}000$. $10{,}000 + 20{,}000 = 30{,}000$. **2.** $p^2 + 2pq$; p^2 represents homozygotes with two *A* alleles, and $2pq$ represents heterozygotes with one *A* allele. **3.** There are 120 individuals in the population, so there are 240 alleles. Of these, there are 124 *A* alleles—32 from the 16 *AA* individuals and 92 from the 92 *Aa* individuals. Thus, the frequency of the *A* allele is $p = 124/240 = 0.52$; hence, the frequency of the *a* allele is $q = 0.48$. Based on the Hardy-Weinberg equation, if the population were not evolving, the frequency of genotype *AA* should be $p^2 = 0.52 \times 0.52 = 0.27$; the frequency of genotype *Aa* should be $2pq = 2 \times 0.52 \times 0.48 = 0.5$; and the frequency of genotype *aa* should be $q^2 = 0.48 \times 0.48 = 0.23$. In a population of 120 individuals, these expected genotype frequencies lead us to predict that there would be 32 *AA* individuals (0.27×120), 60 *Aa* individuals (0.5×120), and 28 *aa* individuals (0.23×120). The actual numbers for the population (16 *AA*, 92 *Aa*, 12 *aa*) deviate from these expectations (fewer homozygotes and more heterozygotes than expected). This indicates that the population is not in Hardy-Weinberg equilibrium and hence may be evolving at this locus.

Concept Check 23.3

1. Natural selection is more "predictable" in that it alters allele frequencies in a nonrandom way: It tends to increase the frequency of alleles that increase the organism's reproductive success in its environment and decrease the frequency of alleles that decrease the organism's reproductive success. Alleles subject to genetic drift increase or decrease in frequency by chance alone, whether or not they are advantageous. **2.** Genetic drift results from chance events that cause allele frequencies to fluctuate at random from generation to generation; within a population, this process tends to decrease genetic variation over time. Gene flow is the exchange of alleles between populations, a process that can introduce new alleles to a population and hence may increase its genetic variation (albeit slightly, since rates of gene flow are often low). **3.** Selection is not important at this locus; furthermore, the populations are not small, and hence the effects of genetic drift should not be pronounced. Gene flow is occurring via the movement of pollen and seeds. Thus, allele and genotype frequencies in these populations should become more similar over time as a result of gene flow.

Concept Check 23.4

1. Zero, because fitness includes reproductive contribution to the next generation, and a sterile mule cannot produce offspring. **2.** Although both gene flow and genetic drift can increase the frequency of advantageous alleles in a population, they can also decrease the frequency of advantageous alleles or increase the frequency of harmful alleles. Only natural selection *consistently* results in an increase in the frequency of alleles that enhance survival or reproduction. Thus, natural selection is the only mechanism that consistently causes adaptive evolution. **3.** The three modes of natural selection (directional, stabilizing, and disruptive) are defined in terms of the selective advantage of different *phenotypes*, not different genotypes. Thus, the type of selection represented by heterozygote advantage depends on the phenotype of the heterozygotes. In this question, because heterozygous individuals have a more extreme phenotype than either homozygote, heterozygote advantage represents directional selection. **4.** Under prolonged low-oxygen conditions, some of the red blood cells of a heterozygote may sickle, leading to harmful effects (see Chapter 14). This does not occur in individuals with two normal hemoglobin alleles, suggesting that there may be selection against heterozygotes in malaria-free regions (where heterozygote advantage does not occur). However, since heterozygotes are healthy under most conditions, selection against them is unlikely to be strong.

Summary of Key Concepts Questions

23.1 One reason biologists estimate gene variability and nucleotide variability is to assess whether populations have enough genetic variation for evolution to occur. Gene variability indicates the extent to which individuals differ genetically at the whole-gene level. Nucleotide variability provides a measure of genetic variation at the DNA sequence level. **23.2** No, this is not an example of circular reasoning. Calculating *p* and *q* from observed genotype frequencies does not imply that those genotype frequencies must be in Hardy-Weinberg equilibrium. Consider a population that has 195 individuals of genotype *AA*, 10 of genotype *Aa*, and 195 of genotype *aa*. Calculating *p* and *q* from these values yields $p = q = 0.5$. Using the Hardy-Weinberg equation, the predicted equilibrium frequencies are $p^2 = 0.25$ for genotype *AA*, $2pq = 0.5$ for genotype *Aa*, and $q^2 = 0.25$ for genotype *aa*. Since there are 400 individuals in the population, these predicted genotype frequencies indicate there should be 100 *AA* individuals, 200 *Aa* individuals, and 100 *aa* individuals—numbers that differ greatly from the values that we used to calculate *p* and *q*. **23.3** It is unlikely that two such populations would evolve in similar ways. Since their environments are very different, the alleles favored by natural selection would probably differ between the two populations; although genetic drift may have important effects in each of these small populations, drift causes unpredictable changes in allele frequencies, so it is unlikely that drift would cause the populations to evolve in similar ways; both populations are geographically isolated, suggesting that little gene flow will occur between them (again making it less likely that they will evolve in similar ways). **23.4** Compared to males, it is likely that the females of such species would be larger, more colorful, endowed with more elaborate ornamentation (for example, a large morphological feature such as the peacock's tail), and more apt to engage in behaviors intended to attract mates or prevent other members of their sex from obtaining mates.

Test Your Understanding

1. e **2.** c **3.** e **4.** b **5.** a **6.** d

7. Although natural selection can improve the match between organisms and their environments, the evolutionary process can also lead to imperfections in organisms. A central reason for this is that evolution does not design organisms from scratch to match their environments and ways of life but works instead by a process of descent with modification: Organisms inherit a basic form from their ancestors, and that form is modified by natural selection over time. As a result, a flying mammal such as a bat has wings that are not perfectly designed, but rather represent modifications of forelimbs that bat ancestors used for walking. Imperfections in organisms result from a variety of other constraints, such as a lack of genetic variation for the trait in question, and the fact that adaptations often represent compromises (since organisms must do many different things, and a "perfect" design for one activity might impair the performance of another activity). **8.** The frequency of the *lap*^{94} allele forms a cline, decreasing as one moves from southwest to northeast across Long Island Sound.

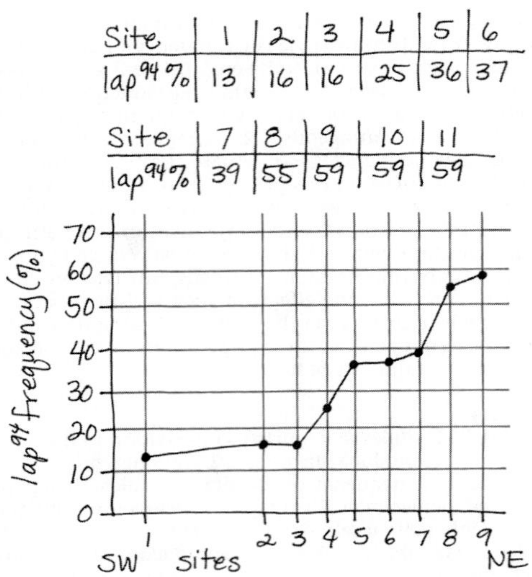

Site	1	2	3	4	5	6
lap^{94} %	13	16	16	25	36	37

Site	7	8	9	10	11
lap^{94} %	39	55	59	59	59

A hypothesis that explains the cline and accounts for the observations stated in the question is that the cline is maintained by an interaction between selection and gene flow. Under this hypothesis, in the southwest portion of the Sound, salinity is relatively low, and selection against the *lap*^{94} allele is strong. Moving toward the northeast and into the open ocean, where salinity is relatively high, selection favors a high frequency of the *lap*^{94} allele. However, because mussel larvae disperse long distances, gene flow prevents the *lap*^{94} allele from becoming fixed in the open ocean or from declining to zero in the southwestern portion of Long Island Sound.

Chapter 24

Figure Questions

Figure 24.10 This was done to remove the possibility that the flies could differentiate among potential mates by detecting what those potential mates had eaten as larvae. If this had not been done, the strong preference of "starch flies" and "maltose flies" to mate with like-adapted flies could have occurred simply because the flies could detect (for example, by sense of smell) what their potential mates had eaten as larvae—and they preferred to mate with flies that had a similar smell to their own. **Figure 24.12** Such results would suggest that mate choice based on coloration does not provide a reproductive barrier between these two cichlid species. **Figure 24.13** The graph suggests there has been gene flow of some fire-bellied toad alleles into the range of the yellow-bellied toad. Otherwise, all individuals located to the left of the hybrid zone portion of the graph would have allele frequencies close to 1.0. **Figure 24.14** Because the populations had only just begun to diverge from one another at this point in the process, it is likely that any existing barriers to reproduction would weaken over time. **Figure 24.19** No. Over time, the chromosomes of the experimental hybrids came to resemble those of *H. anomalus*. This occurred even though conditions in the laboratory differed greatly from conditions in the field, where *H. anomalus* is found, suggesting that selection for laboratory conditions was not strong. Thus, it is unlikely that the observed rise in the fertility of the experimental hybrids was due to selection for life under laboratory conditions. **Figure 24.20** The presence of *M. cardinalis* plants that carry the *M. lewisii yup* allele would make it more likely that bumblebees would transfer pollen between the two monkey flower species. As a result, we would expect the number of hybrid offspring to increase.

Concept Check 24.1

1. (a) All except the biological species concept can be applied to both asexual and sexual species because they define species on the basis of characteristics other than ability to reproduce. In contrast, the biological species concept can be applied only to sexual species. (b) The easiest species concept to apply in the field would be the morphological species concept because it is based only on the appearance of the organism. Additional information about its ecological habits, evolutionary history, and reproduction are not required. **2.** Because these birds live in fairly similar environments and can breed successfully in captivity, the reproductive barrier in nature is probably prezygotic; given the species' differences in habitat preference, this barrier could result from habitat isolation.

Concept Check 24.2

1. In allopatric speciation, a new species forms while in geographic isolation from its parent species; in sympatric speciation, a new species forms in the absence of geographic isolation. Geographic isolation greatly reduces gene flow between populations, whereas ongoing gene flow is more likely in sympatric populations. As a result, sympatric speciation is less common than allopatric speciation. **2.** Gene flow between subsets of a population that live in the same area can be reduced in a variety of ways. In some species—especially plants—changes in chromosome number can block gene flow and establish reproductive isolation in a single generation. Gene flow can also be reduced in sympatric populations by habitat differentiation (as seen in the apple maggot fly, *Rhagoletis*) and sexual selection (as seen in Lake Victoria cichlids). **3.** Allopatric speciation would be less likely to occur on a nearby island than on an isolated island of the same size. The reason we expect this result is that continued gene flow between mainland populations and those on a nearby island reduces the chance that enough genetic divergence will take place for allopatric speciation to occur. **4.** If all of the homologs failed to separate during anaphase I of meiosis, some gametes would end up with an extra set of chromosomes (and others would end up with no chromosomes). If a gamete with an extra set of chromosomes fused with a normal gamete, a triploid would result; if two gametes with an extra set of chromosomes fused with each other, a tetraploid would result.

Concept Check 24.3

1. Hybrid zones are regions in which members of different species meet and mate, producing some offspring of mixed ancestry. Such regions can be viewed as "natural laboratories" in which to study speciation because scientists can directly observe factors that cause (or fail to cause) reproductive isolation. **2.** (a) If hybrids consistently survive and reproduce poorly compared to the offspring of intraspecific matings, reinforcement could occur. If it did, natural selection would cause prezygotic barriers to reproduction between the parent species to strengthen over time, decreasing the production of unfit hybrids and leading to a completion of the speciation process. (b) If hybrid offspring survived and reproduced as well as the offspring of intraspecific matings, indiscriminate mating between the parent species would lead to the production of large numbers of hybrid offspring. As these hybrids mated with each other and with members of both parent species, the gene pools of the parent species could fuse over time, reversing the speciation process.

Concept Check 24.4

1. The time between speciation events includes (1) the length of time that it takes for populations of a newly formed species to begin diverging reproductively from one another and (2) the time it takes for speciation to be complete once this divergence begins. Although speciation can occur rapidly once populations have begun to diverge from one another, it may take millions of years for that divergence to begin. **2.** Investigators transferred alleles at the *yup* locus (which influences flower color) from each parent species to the other. *M. lewisii* plants with an *M. cardinalis yup* allele received many more visits from hummingbirds than usual; hummingbirds usually pollinate *M. cardinalis* but avoid *M. lewisii*. Similarly, *M. cardinalis* plants with an *M. lewisii yup* allele received many more visits from bumblebees than usual; bumblebees usually pollinate *M. lewisii* and avoid *M. cardinalis*. Thus, alleles at the *yup* locus can influence pollinator choice, which in these species provides the primary barrier to interspecific mating. Nevertheless, the experiment does not prove that the *yup* locus alone controls barriers to reproduction between *M. lewisii* and *M. cardinalis*; other genes might enhance the effect of the *yup* locus (by modifying flower color) or cause entirely different barriers to reproduction (for example, gametic isolation or a postzygotic barrier). **3.** Crossing over. If crossing over did not occur, each chromosome in an experimental hybrid would remain as in the F_1 generation: composed entirely of DNA from one parent species or the other.

Summary of Key Concepts Questions

24.1 According to the biological species concept, a species is a group of populations whose members interbreed and produce viable, fertile offspring; thus, gene flow occurs between populations of a species. In contrast, members of different species do not interbreed and hence no gene flow occurs between their populations. Overall, then, in the biological species concept, species can be viewed as designated by the *absence* of gene flow—making gene flow of central importance to the biological species concept. **24.2** Sympatric speciation can be promoted by factors such as polyploidy, habitat shifts, and sexual selection, all of which can reduce gene flow between the subpopulations of a larger population. But such factors can also occur in allopatric populations and hence can also promote allopatric speciation. **24.3** If the hybrids are selected against, the hybrid zone could persist if individuals from the parent species regularly travel into the zone, where they mate to produce hybrid offspring. If hybrids are not selected against, there is no cost to the continued production of hybrids, and large numbers of hybrid offspring may be produced. However, natural selection for life in different environments may keep the gene pools of the two parent species distinct—thus preventing the loss (by fusion) of the parent species and once again causing the hybrid zone to be stable over time. **24.4** As the goatsbeard plant, Bahamas mosquitofish, and apple maggot fly examples illustrate, speciation continues to happen today. A new species can begin to form whenever gene flow is reduced between populations of the parent species. Such reductions in gene flow can occur in many ways: A new, geographically isolated population may be founded by a few colonists; some members of the parent species may begin to utilize new habitat; and sexual selection may isolate formerly connected populations or subpopulations. These and many other such events are happening today.

Test Your Understanding
1. b **2.** c **3.** c **4.** a **5.** e **6.** d **7.** e
10. Here is one possibility:

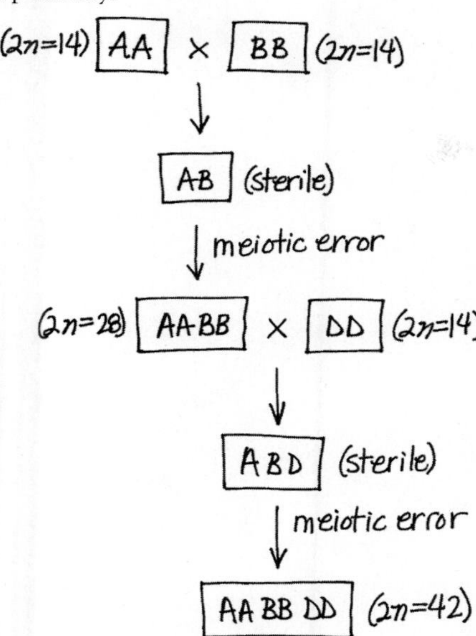

Appendix A Answers

APPENDIX B Periodic Table of the Elements

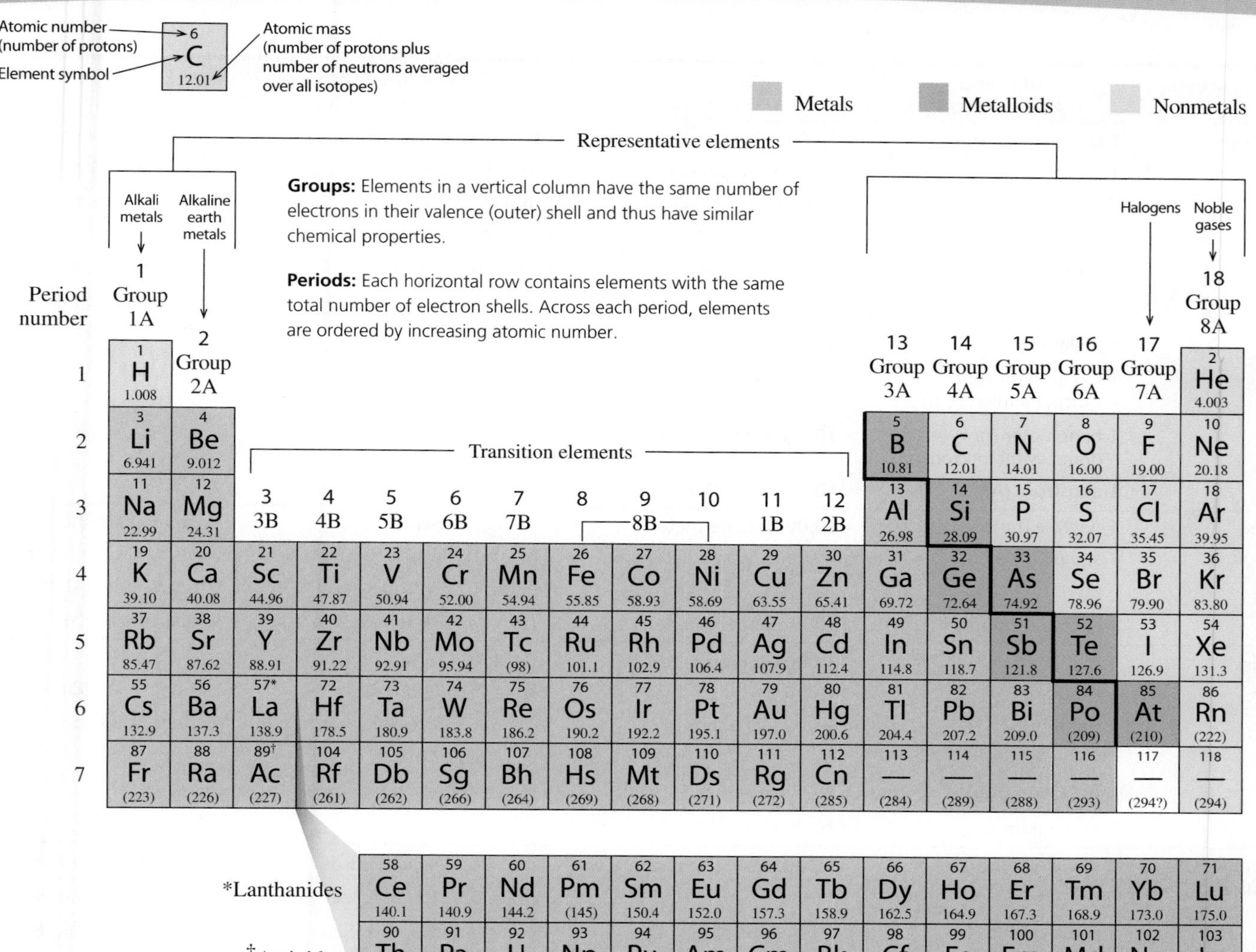

Name (Symbol)	Atomic Number	Name (Symbol)	Atomic Number	Name (Symbol)	Atomic Number	Name (Symbol)	Atomic Number	Name (Symbol)	Atomic Number
Actinium (Ac)	89	Copernicium (Cn)	112	Iridium (Ir)	77	Palladium (Pd)	46	Sodium (Na)	11
Aluminum (Al)	13	Copper (Cu)	29	Iron (Fe)	26	Phosphorus (P)	15	Strontium (Sr)	38
Americium (Am)	95	Curium (Cm)	96	Krypton (Kr)	36	Platinum (Pt)	78	Sulfur (S)	16
Antimony (Sb)	51	Darmstadtium (Ds)	110	Lanthanum (La)	57	Plutonium (Pu)	94	Tantalum (Ta)	73
Argon (Ar)	18	Dubnium (Db)	105	Lawrencium (Lr)	103	Polonium (Po)	84	Technetium (Tc)	43
Arsenic (As)	33	Dysprosium (Dy)	66	Lead (Pb)	82	Potassium (K)	19	Tellurium (Te)	52
Astatine (At)	85	Einsteinium (Es)	99	Lithium (Li)	3	Praseodymium (Pr)	59	Terbium (Tb)	65
Barium (Ba)	56	Erbium (Er)	68	Lutetium (Lu)	71	Promethium (Pm)	61	Thallium (Tl)	81
Berkelium (Bk)	97	Europium (Eu)	63	Magnesium (Mg)	12	Protactinium (Pa)	91	Thorium (Th)	90
Beryllium (Be)	4	Fermium (Fm)	100	Manganese (Mn)	25	Radium (Ra)	88	Thulium (Tm)	69
Bismuth (Bi)	83	Fluorine (F)	9	Meitnerium (Mt)	109	Radon (Rn)	86	Tin (Sn)	50
Bohrium (Bh)	107	Francium (Fr)	87	Mendelevium (Md)	101	Rhenium (Re)	75	Titanium (Ti)	22
Boron (B)	5	Gadolinium (Gd)	64	Mercury (Hg)	80	Rhodium (Rh)	45	Tungsten (W)	74
Bromine (Br)	35	Gallium (Ga)	31	Molybdenum (Mo)	42	Roentgenium (Rg)	111	Uranium (U)	92
Cadmium (Cd)	48	Germanium (Ge)	32	Neodymium (Nd)	60	Rubidium (Rb)	37	Vanadium (V)	23
Calcium (Ca)	20	Gold (Au)	79	Neon (Ne)	10	Ruthenium (Ru)	44	Xenon (Xe)	54
Californium (Cf)	98	Hafnium (Hf)	72	Neptunium (Np)	93	Rutherfordium (Rf)	104	Ytterbium (Yb)	70
Carbon (C)	6	Hassium (Hs)	108	Nickel (Ni)	28	Samarium (Sm)	62	Yttrium (Y)	39
Cerium (Ce)	58	Helium (He)	2	Niobium (Nb)	41	Scandium (Sc)	21	Zinc (Zn)	30
Cesium (Cs)	55	Holmium (Ho)	67	Nitrogen (N)	7	Seaborgium (Sg)	106	Zirconium (Zr)	40
Chlorine (Cl)	17	Hydrogen (H)	1	Nobelium (No)	102	Selenium (Se)	34		
Chromium (Cr)	24	Indium (In)	49	Osmium (Os)	76	Silicon (Si)	14		
Cobalt (Co)	27	Iodine (I)	53	Oxygen (O)	8	Silver (Ag)	47		

APPENDIX C

The Metric System

Measurement	Unit and Abbreviation	Metric Equivalent	Metric-to-English Conversion Factor	English-to-Metric Conversion Factor
Length	1 kilometer (km)	= 1,000 (10^3) meters	1 km = 0.62 mile	1 mile = 1.61 km
	1 meter (m)	= 100 (10^2) centimeters	1 m = 1.09 yards	1 yard = 0.914 m
		= 1,000 millimeters	1 m = 3.28 feet	1 foot = 0.305 m
			1 m = 39.37 inches	
	1 centimeter (cm)	= 0.01 (10^{-2}) meter	1 cm = 0.394 inch	1 foot = 30.5 cm
				1 inch = 2.54 cm
	1 millimeter (mm)	= 0.001 (10^{-3}) meter	1 mm = 0.039 inch	
	1 micrometer (μm) (formerly micron, μ)	= 10^{-6} meter (10^{-3} mm)		
	1 nanometer (nm) (formerly millimicron, mμ)	= 10^{-9} meter (10^{-3} μm)		
	1 angstrom (Å)	= 10^{-10} meter (10^{-4} μm)		
Area	1 hectare (ha)	= 10,000 square meters	1 ha = 2.47 acres	1 acre = 0.405 ha
	1 square meter (m^2)	= 10,000 square centimeters	1 m^2 = 1.196 square yards	1 square yard = 0.8361 m^2
			1 m^2 = 10.764 square feet	1 square foot = 0.0929 m^2
	1 square centimeter (cm^2)	= 100 square millimeters	1 cm^2 = 0.155 square inch	1 square inch = 6.4516 cm^2
Mass	1 metric ton (t)	= 1,000 kilograms	1 t = 1.103 tons	1 ton = 0.907 t
	1 kilogram (kg)	= 1,000 grams	1 kg = 2.205 pounds	1 pound = 0.4536 kg
	1 gram (g)	= 1,000 milligrams	1 g = 0.0353 ounce	1 ounce = 28.35 g
			1 g = 15.432 grains	
	1 milligram (mg)	= 10^{-3} gram	1 mg = approx. 0.015 grain	
	1 microgram (μg)	= 10^{-6} gram		
Volume (solids)	1 cubic meter (m^3)	= 1,000,000 cubic centimeters	1 m^3 = 1.308 cubic yards	1 cubic yard = 0.7646 m^3
			1 m^3 = 35.315 cubic feet	1 cubic foot = 0.0283 m^3
	1 cubic centimeter (cm^3 or cc)	= 10^{-6} cubic meter	1 cm^3 = 0.061 cubic inch	1 cubic inch = 16.387 cm^3
	1 cubic millimeter (mm^3)	= 10^{-9} cubic meter = 10^{-3} cubic centimeter		
Volume (liquids and gases)	1 kiloliter (kL or kl)	= 1,000 liters	1 kL = 264.17 gallons	
	1 liter (L or l)	= 1,000 milliliters	1 L = 0.264 gallons	1 gallon = 3.785 L
			1 L = 1.057 quarts	1 quart = 0.946 L
	1 milliliter (mL or ml)	= 10^{-3} liter	1 mL = 0.034 fluid ounce	1 quart = 946 mL
		= 1 cubic centimeter	1 mL = approx. ¼ teaspoon	1 pint = 473 mL
			1 mL = approx. 15–16 drops (gtt.)	1 fluid ounce = 29.57 mL
				1 teaspoon = approx. 5 mL
	1 microliter (μL or μl)	= 10^{-6} liter (10^{-3} milliliters)		
Time	1 second (s or sec)	= $\frac{1}{60}$ minute		
	1 millisecond (ms or msec)	= 10^{-3} second		
Temperature	Degrees Celsius (°C) (Absolute zero, when all molecular motion ceases, is −273.15°C. The Kelvin [K] scale, which has the same size degrees as Celsius, has its zero point at absolute zero. Thus, 0 K = −273.15°C.)		°F = $\frac{9}{5}$°C + 32	°C = $\frac{5}{9}$ (°F − 32)

APPENDIX D A Comparison of the Light Microscope and the Electron Microscope

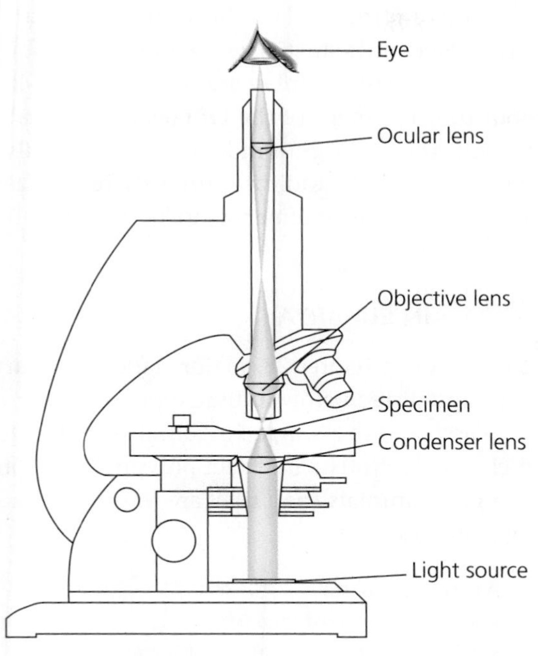

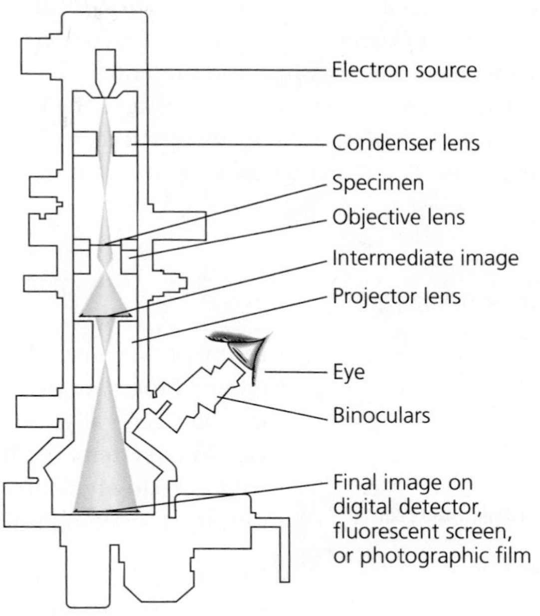

Light Microscope

In light microscopy, light is focused on a specimen by a glass condenser lens; the image is then magnified by an objective lens and an ocular lens, for projection on the eye, digital camera, digital video camera, or photographic film.

Electron Microscope

In electron microscopy, a beam of electrons (top of the microscope) is used instead of light, and electromagnets are used instead of glass lenses. The electron beam is focused on the specimen by a condenser lens; the image is magnified by an objective lens and a projector lens for projection on a digital detector, fluorescent screen, or photographic film.

APPENDIX E Classification of Life

This appendix presents a taxonomic classification for the major extant groups of organisms discussed in this text; not all phyla are included. The classification presented here is based on the three-domain system, which assigns the two major groups of prokaryotes, bacteria and archaea, to separate domains (with eukaryotes making up the third domain).

Various alternative classification schemes are discussed in Unit Five of the text. The taxonomic turmoil includes debates about the number and boundaries of kingdoms and about the alignment of the Linnaean classification hierarchy with the findings of modern cladistic analysis. In this review, asterisks (*) indicate currently recognized phyla thought by some systematists to be paraphyletic.

DOMAIN BACTERIA

- **Proteobacteria**
- **Chlamydia**
- **Spirochetes**
- **Gram-positive Bacteria**
- **Cyanobacteria**

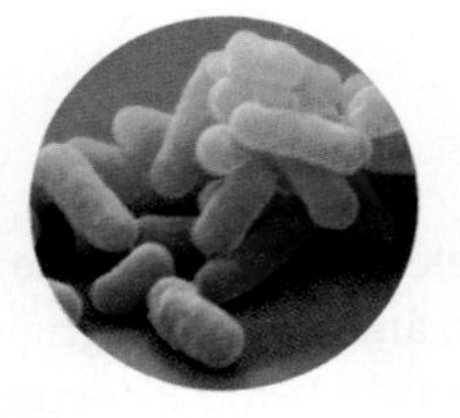

DOMAIN ARCHAEA

- **Korarchaeota**
- **Euryarchaeota**
- **Crenarchaeota**
- **Nanoarchaeota**

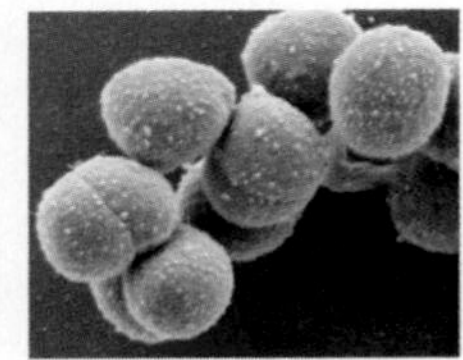

DOMAIN EUKARYA

In the phylogenetic hypothesis we present in Chapter 28, major clades of eukaryotes are grouped together in the five "supergroups" listed below in blue type. Formerly, all the eukaryotes generally called protists were assigned to a single kingdom, Protista. However, advances in systematics have made it clear that Protista is in fact polyphyletic: Some protists are more closely related to plants, fungi, or animals than they are to other protists. As a result, the kingdom Protista has been abandoned.

Excavata

- Diplomonadida (diplomonads)
- Parabasala (parabasalids)
- Euglenozoa (euglenozoans)
  - Kinetoplastida (kinetoplastids)
  - Euglenophyta (euglenids)

Chromalveolata

- Alveolata (alveolates)
  - Dinoflagellata (dinoflagellates)
  - Apicomplexa (apicomplexans)
  - Ciliophora (ciliates)
- Stramenopila (stramenopiles)
  - Bacillariophyta (diatoms)
  - Chrysophyta (golden algae)
  - Phaeophyta (brown algae)
  - Oomycota (water molds)

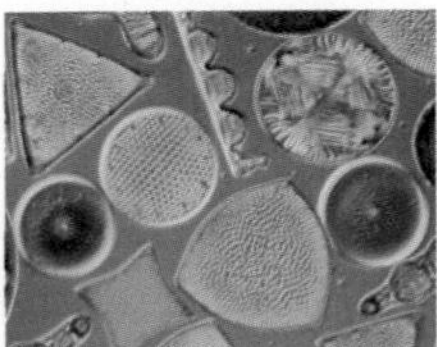

Archaeplastida

- Rhodophyta (red algae)
- Chlorophyta (green algae: chlorophytes)
- Charophyta (green algae: charophytes)
- Plantae
  - Nonvascular plants (bryophytes):
    - Phylum Hepatophyta (liverworts)
    - Phylum Bryophyta (mosses)
    - Phylum Anthocerophyta (hornworts)
  - Seedless vascular plants:
    - Phylum Lycophyta (lycophytes)
    - Phylum Pterophyta (ferns, horsetails, whisk ferns)
  - Seed plants:
    - Gymnosperms:
      - Phylum Ginkgophyta (ginkgo)
      - Phylum Cycadophyta (cycads)
      - Phylum Gnetophyta (gnetophytes)
      - Phylum Coniferophyta (conifers)
    - Angiosperms:
      - Phylum Anthophyta (flowering plants)

Rhizaria

- Radiolaria (radiolarians)
- Foraminifera (forams)
- Cercozoa (cercozoans)

DOMAIN EUKARYA, continued

Unikonta

- Amoebozoa (amoebozoans)
  - Myxogastrida (plasmodial slime molds)
  - Dictyostelida (cellular slime molds)
  - Gymnamoeba (gymnamoebas)
  - Entamoeba (entamoebas)
- Nucleariida (nucleariids)
- Fungi
  - *Phylum Chytridiomycota (chytrids)
  - *Phylum Zygomycota (zygomycetes)
  - Phylum Glomeromycota (glomeromycetes)
  - Phylum Basidiomycota (club fungi)
  - Phylum Ascomycota (sac fungi)

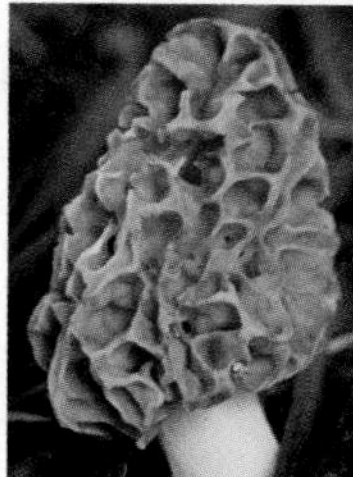

- Choanoflagellata (choanoflagellates)
- Animalia
  - Phylum Porifera (sponges)
  - Phylum Ctenophora (comb jellies)
  - Phylum Cnidaria (cnidarians)
    - Hydrozoa (hydrozoans)
    - Scyphozoa (jellies)
    - Cubozoa (box jellies and sea wasps)
    - Anthozoa (sea anemones and most corals)
  - Phylum Acoela (acoel flatworms)
  - Phylum Placozoa (placozoans)

<u>Lophotrochozoa (lophotrochozoans)</u>

- Phylum Kinorhyncha (kinorhynchs)
- Phylum Platyhelminthes (flatworms)
  - Catenulida (chain worms)
  - Rhabditophora (planarians, flukes, tapeworms)
- Phylum Nemertea (proboscis worms)
- Phylum Ectoprocta (ectoprocts)
- Phylum Brachiopoda (brachiopods)
- Phylum Phoronida (phoronids)
- Phylum Rotifera (rotifers)
- Phylum Cycliophora (cycliophorans)
- Phylum Mollusca (molluscs)
  - Polyplacophora (chitons)
  - Gastropoda (gastropods)
  - Bivalvia (bivalves)
  - Cephalopoda (cephalopods)
- Phylum Annelida (segmented worms)
  - Polychaeta (polychaetes)
  - Oligochaeta (oligochaetes)
- Phylum Acanthocephala (spiny-headed worms)

<u>Ecdysozoa (ecdysozoans)</u>

- Phylum Loricifera (loriciferans)
- Phylum Priapula (priapulans)
- Phylum Nematoda (roundworms)
- Phylum Arthropoda (This survey groups arthropods into a single phylum, but some zoologists now split the arthropods into multiple phyla.)
  - Subphylum Chelicerata (horseshoe crabs, arachnids)
  - Subphylum Myriapoda (millipedes, centipedes)
  - Subphylum Hexapoda (insects, springtails)
  - Subphylum Crustacea (crustaceans)
- Phylum Tardigrada (tardigrades)
- Phylum Onychophora (velvet worms)

<u>Deuterostomia (deuterostomes)</u>

- Phylum Hemichordata (hemichordates)
- Phylum Echinodermata (echinoderms)
  - Asteroidea (sea stars, sea daisies)
  - Ophiuroidea (brittle stars)
  - Echinoidea (sea urchins and sand dollars)
  - Crinoidea (sea lilies)
  - Holothuroidea (sea cucumbers)
- Phylum Chordata (chordates)
  - Subphylum Cephalochordata (cephalochordates: lancelets)
  - Subphylum Urochrodata (urochordates: tunicates)
  - Subphylum Craniata (craniates)
    - Myxini (hagfishes)
    - Petromyzontida (lampreys)
    - Chondrichthyes (sharks, rays, chimaeras)
    - Actinopterygii (ray-finned fishes)
    - Actinistia (coelacanths)
    - Dipnoi (lungfishes)
    - Amphibia (amphibians)
    - Reptilia (tuataras, lizards, snakes, turtles, crocodilians, birds)
    - Mammalia (mammals)

(Petromyzontida through Mammalia: Vertebrates)

Credits

Photo Credits

Cover Image "Succulent I" © 2005 Amy Lamb, www.amylamb.com

Unit Opening Interviews UNIT 1 Justina Thorsen, Showcase Reflections; **UNIT 2** Brian Wilson, Princeton University; **UNIT 3** William K. Sacco, Yale ITS Photo + Design Services; **UNIT 4** Steve Bonnel, Bonnel Photography; **UNIT 5** Jacob Mailman; **UNIT 6** Michael Starghill, Michael Starghill Photography; **UNIT 7** River Healey, Rusty Healey Photography; **UNIT 8** Marsha Miller, Director of Photography, University of Texas at Austin.

About the Author Photos River Healey, Rusty Healey Photography.

Detailed Contents UNIT I Photo by T. Naeser, from Patrick Cramer Laboratory, Gene Center Munich, Ludwig-Maximilians-Universität München, Munich, Germany; **UNIT II** Biophoto Associates/Photo Researchers; **UNIT III** From: Multicolor Spectral Karyotyping of Human Chromosomes. E. Schröck, S. du Manoir, T. Veldman, B. Schoell, J. Wienberg, M. A. Ferguson-Smith, Y. Ning, D. H. Ledbetter, I. Bar-Am, D. Soenksen, Y. Garini, T. Ried. *Science*. 1996 Jul 26;273(5274):494-7.; **UNIT IV** © Ted Daeschler/ Academy of Natural Sciences/VIREO; **UNIT V** Yufeng Zhou/iStockphoto; **UNIT VI** PD photo: John Walker (www.fourmilab.ch); **UNIT VII** David Wall/Alamy; David Doubilet/Getty Images; Kenneth Catania; **UNIT VIII** Daniel Mosquin.

Chapter 1 1.1 Amy Lamb Studio; **1.2** Walter Teague; **1.3.2** R.Dirscherl/FLPA; **1.3.3** Kim Taylor and Jane Burton/Dorling Kindersley; **1.3.4** Malcolm Schuyl/FLPA; **1.3.5** Frans Lanting/Corbis; **1.3.6** Michael & Patricia Fogden/Corbis; **1.3.7** Joe McDonald/ Corbis; **1.3.8** ImageState/International Stock Photography Ltd.; **1.4.2** WorldSat International/Photo Researchers, Inc.; **1.4.3** Bill Brooks/Alamy; **1.4.4** Linda Freshwaters Arndt/Alamy; **1.4.5** Michael Orton/Photographer's Choice/Getty Images; **1.4.6** Ross M. Horowitz/Getty Images; **1.4.7** Photodisc/Getty Images; **1.4.9** Jeremy Burgess/ SPL/Photo Researchers; **1.4.11** John Durham/Photo Researchers; **1.4.13** Electron micrograph by Wm. P. Wergin, courtesy of E. H. Newcomb, University of Wisconsin; **1.5.2** James Balog/Aurora Creative/Getty Images; **1.6.2** Anup Shah/Nature Picture Library; **1.6.3** Anup Shah/Nature Picture Library; **1.7.2** Photodisc/Getty Images; **1.7.3** Janice Sheldon; **1.8.2** S. C. Holt/Biological Photo Service; **1.8.3** Steve Gschmeissner/ Photo Researchers; **1.9.2** Conly L. Rieder; **1.9.3** Conly L. Rieder; **1.10.2** Camille Tokerud/ Stone/Getty Images; **1.11.2** Photodisc/Getty Images; **1.12** Roy Kaltschmidt, Lawrence Berkeley National Laboratory; **1.15.2** Oliver Meckes/Nicole Ottawa/Photo Researchers; **1.15.3** Eye of Science/Photo Researchers; **1.15.4** Kunst & Scheidulin/AGE Fotostock; **1.15.5** Peter Lilja/Taxi/Getty Images; **1.15.6** Anup Shah/Nature Picture Library; **1.15.7** D. P. Wilson/Photo Researchers; **1.16.2** VVG/SPL/Photo Researchers; **1.16.3** W. L. Dentler/Biological Photo Service; **1.16.4** OMIKRON/Photo Researchers; **1.17** Photo by Dede Randrianarisata, from Kristi Curry Rogers, Macalester College, St. Paul, MN; **1.18** ARCHIV/Photo Researchers; **1.19.2** Michael P. Fogden/Bruce Coleman/ Alamy; **1.19.3** Matt T. Lee; **1.19.4** Hal Horwitz/Corbis; **1.21** Frank Greenaway/Dorling Kindersley; **1.23.2** Karl Ammann/Corbis; **1.23.3** Tim Ridley/Dorling Kindersley, Courtesy of the Jane Goodall Institute, Clarendon Park, Hampshire; **1.25.2** Breck P. Kent; **1.25.3** Photo Researchers; **1.26.2** David Pfennig; **1.26.3** David Pfennig; **1.28** Gary L. Firestone's Lab researchers meeting in the Dept. of Molecular and Cell Biology, University of California at Berkeley. Photo: Seelevel.com, Pearson Science; **1.29** Tim Sharp/AP Images; **p. 25, top right** James Balog/Aurora Creative/Getty Images; **p. 25, bottom left** Anup Shah/Nature Picture Library; **p. 25, bottom left** Anup Shah/Nature Picture Library; **p.26, top** PhotoDisc/Getty Images; **p. 26, 2nd from top** S. C. Holt/ Biological Photo Service; **p. 26, 3rd from top** Steve Gschmeissner/Photo Researchers.

Chapter 2 2.1 Martin Dohrn/BBC Natural History Unit; **2.2.3** Martin Dohrn/BBC Natural History Unit; **2.3.2** Chip Clark; **2.4.2** C. Michael Hogan; **2.4.3** Rick York and California Native Plant Society (www.cnps.org); **2.4.4** Andrew Alden; **2.6.2** Clayton T. Hamilton, Stanford University; **2.7.2** National Library of Medicine; **p. 41, left** Jerry Young/Dorling Kindersley; **2.19** Nigel Cattlin/Photo Researchers; **p. 45** Rolf Nussbaumer/Nature Picture Library.

Chapter 3 3.1 Alexander/Fotoloia; **3.3.2** N.C. Brown Center for Ultrastructure Studies, SUNY-Environmental Science & Forestry, Syracuse, NY; **3.4** iStockphoto; **3.6.2** Jan van Franeker, IMARES, Alfred Wegener Institute for Polar and Marine Research; **3.9** NASA/JPL-Caltech/University of Arizona/Texas A&M University; **3.10.2** Jakub Semeniuk/ iStockphoto; **3.10.3** Feng Yu/iStockphoto; **3.10.4** Monika Wisniewska/iStockphoto; **3.10.5** Beth Van Trees/Shutterstock; **3.12.2** From "Coral Reefs Under Rapid Climate Change and Ocean Acidification." O. Hoegh-Guldberg, et al. *Science* 14 December 2007: 318 (5857):1737–1742. Photos by Ove Hoegh-Guldberg, Centre for Marine Studies, The University of Queensland; **3.12.3** From "Coral Reefs Under Rapid Climate Change and Ocean Acidification." O. Hoegh-Guldberg, et al. *Science* 14 December 2007: 318 (5857):1737–1742. Photos by Ove Hoegh-Guldberg, Centre for Marine Studies, The University of Queensland; **3.12.4** From "Coral Reefs Under Rapid Climate Change and Ocean Acidification." O. Hoegh-Guldberg, et al. *Science* 14 December 2007: 318 (5857):1737–1742. Photos by Ove Hoegh-Guldberg, Centre for Marine Studies, The University of Queensland.

Chapter 4 4.1 Shutterstock; **4.6.2** David M. Phillips/Photo Researchers.

Chapter 5 5.1 Photo by T. Naeser, from Patrick Cramer Laboratory, Gene Center Munich, Ludwig-Maximilians-Universität München, Munich, Germany; **5.6.2** John N. A. Lott/Biological Photo Service; **5.6.3** H. Shio and P. B. Lazarow; **5.8.2** Alexey Repka/iStockphoto; **5.8.3** John Durham/Photo Researchers; **5.8.4** Biophoto Associates/Photo Researchers; **5.9.2** F. Collet/Photo Researchers; **5.9.3** Corbis; **5.11.2** Dorling Kindersley; **5.11.3** Dorling Kindersley; **5.15.2** Andrey Stratilatov/ Shutterstock; **5.15.3** Nina Zanetti; **5.15.4** Nina Zanetti; **5.19.2** Reproduced by permission from Tulip WR, Varghese JN, Laver WG, Webster RG, Colman PM. Refined crystal structure of the influenza virus N9 neuraminidase-NC41 Fab complex. *J Mol Biol*. September 5; 227(1):122–48. Copyright © 1992 by Elsevier Science Ltd. **5.20.3** Dieter Hopf/AGE Fotostock; **5.20.8** Monika Wisniewska/iStockphoto; **5.21.2** Eye of Science/Photo Researchers; **5.21.3** Eye of Science/Photo Researchers; **5.23.2** Reprinted by permission from *Nature*. P. B. Sigler from Z. Xu, A. L. Horwich, and P. B. Sigler. 388:741–750 Copyright © 1997 Macmillan Magazines Limited; **5.24.2** Dave Bushnell; **5.24.4** Dave Bushnell; **p. 90** Dorling Kindersley.

Chapter 6 6.1 Eye of Science/Photo Researchers; **6.3.2** Elisabeth Pierson, FNWI-Radboud University Nijmegen, Pearson Science; **6.3.3** Elisabeth Pierson, FNWI-Radboud University Nijmegen, Pearson Science; **6.3.4** Elisabeth Pierson, FNWI-Radboud University Nijmegen, Pearson Science; **6.3.5** Elisabeth Pierson, FNWI-Radboud University Nijmegen, Pearson Science; **6.3.6** Michael W. Davidson/The Florida State University Research Foundation; **6.3.7** Karl Garsha, Beckman Institute for Advanced Science and Technology, University of Illinois; **6.3.9** Macrophage fluorescently stained for tubulin (yellow), actin (red) and the nucleus (DAPI, blue). Top part of the image: data recorded with a widefield microscope and visualized with the Simulated Fluorescence Process (SFP) volume rendering algorithm. Bottom part: the same dataset, deconvolved using Huygens Professional (Scientific Volume Imaging, Hilversum, The Netherlands) and again rendered with the SFP algorithm. Data courtesy Dr. James G. Evans, Whitehead Institute, MIT Boston MA, USA; **6.3.11** From "Motile Cilia of Human Airway Epithelia Are Chemosensory." Alok S. Shah, Yehuda Ben-Shahar, T.s O. Moninger, J. N. Kline, M. J. Welsh. Science Express on 23 July 2009. *Science* 28 August 2009:325 (5944): 1131–1134 (cover). Pseudocolored scanning electron micrograph by Tom Moninger (epithelia generated by Phil Karp); **6.3.12** William Dentler/Biological Photo Service; **6.3.13** From "STED microscopy reveals that synaptotagmin remains clustered after synaptic vesicle exocytosis." Katrin I. Willig, Silvio O. Rizzoli, Volker Westphal, Reinhard Jahn & . . . Stefan W. Hell. *Nature*, 440 (13) April 2006. doi:10.1038/nature04592, Letters; **6.5.2** S. C. Holt/Biological Photo Service; **6.6.2** Daniel Friend; **6.8.4** S. Cinti/Photo Researchers; **6.8.6** SPL/Photo Researchers; **6.8.8** A. Barry Dowsett/Photo Researchers; **6.8.10** Biophoto Associates/ Photo Researchers; **6.8.12** SPL/Photo Researchers; **6.8.14** From "Flagellar Microtubule Dynamics in *Chlamydomonas:* Cytochalasin D Induces Periods of Microtubule Shortening and Elongation; and Colchicine Induces Disassembly of the Distal, but Not Proximal, Half of the Flagellum." W. L. Dentler and C. Adams. *The Journal of Cell Biology*, 117(6): 1289–1298, Copyright © 1992 by The Rockefeller University Press; **6.9.2** Reproduced by permission from L.Orci and A.Perelet, Freeze-Etch Histology. (Heidelberg: Springer-Verlag, 1975). Copyright ©1975 by Springer-Verlag GmbH & Co KG; **6.9.3** Reproduced by permission from A. C. Faberge, *Cell Tiss. Res.* 151 Copyright © 1974 by Springer-Verlag GmbH & Co KG; **6.9.4** Reprinted by permission from *Nature* 323. U. Aebi et al. Copyright © 1996: 560–564, figure 1a. Used with permission. Macmillan Magazines Limited; **6.10.2** D. W. Fawcett/Photo Researchers; **6.11.2** R. Bolender, D. Fawcett/Photo Researchers; **6.12.2** Don W. Fawcett/ Photo Researchers; **6.13.2** Daniel S. Friend; **6.13.3** Daniel S. Friend; **6.14.2** E. H. Newcomb; **6.17.2** Daniel S. Friend; **6.17.3** From "The shape of mitochondria and the number of mitochondrial nucleoids during the cell cycle of *Euglena gracilis*." Y. Hayashi and K. Ueda. *Journal of Cell Science*, 93:565–570, Copyright © 1989 by Company of Biologists; **6.18.2** Courtesy of W.P. Wergin and E.H. Newcomb, University of Wisconsin/Biological Photo Service; **6.18.3** Franz Grolig, Philipps-University Marburg, Germany. Image acquired with the confocal microscope Leica TCS SP2; **6.19.2** From S. E. Fredrick and E. H. Newcomb, *The Journal of Cell Biology* 43 (1969):343. Provided by E. H. Newcomb; **6.20** Albert Tousson, High Resolution Imaging Facility, University of Alabama at Birmingham; **6.21.2** Dr. Bruce J. Schnapp; **p. 113, left** Mary Osborn; **p. 113, middle** Frank Solomon; **p. 113, right** Mark S. Ladinsky and J. Richard McIntosh, University of Colorado; **6.22.2** Kent L. McDonald; **6.23.2** Biophoto Associates/Photo Researchers; **6.23.3** Oliver Meckes & Nicole Ottawa/Eye of Science/Photo Researchers; **6.24.2** OMIKRON/Science Source/Photo Researchers; **6.24.3** W. L. Dentler/Biological Photo Service; **6.24.4** Linck RW, Stephens RE. Functional protofilament numbering of ciliary, flagellar, and centriolar microtubules. *Cell Motil Cytoskeleton*. 2007 Jul;64(7):489–95; cover. Micrograph by D. Woodrum Hensley; **6.26.2** From Hirokawa Nobutaka, *The Journal of Cell Biology* 94 (1982):425 by copyright permission of The Rockefeller University Press; **6.27.2** Clara Franzini-Armstrong, University of Pennsylvania; **6.27.3** M. I. Walker/Photo Researchers; **6.27.4** Michael Clayton, University of Wisconsin-Madison; **6.28.2** G. F. Leedale/Photo Researchers; **6.29.2** David Ehrhardt; **6.29.3** From "Visualization of cellulose synthase demonstrates functional association with microtubules." A. R. Paredez, C. R. Somerville, D. W. Ehrhardt. *Science*. 2006 Jun 9;312(5779):1491–5. Epub 2006 Apr 20; **6.31.2** Micrograph by W. P. Wergin, provided by E. H. Newcomb; **6.32.2** From Douglas J. Kelly, *The Journal of Cell Biology* 28 (1966): 51. Fig.17. Reproduced by copyright permission of The Rockefeller University Press; **6.32.3** From Douglas J. Kelly, *The Journal of Cell Biology* 28 (1966):51 by copyright permission of The Rockefeller University Press; **6.32.5** Reproduced by permission from L. Orci and A. Perrelet, Freeze-Etch Histology. (Heidelberg: Springer-Verlag). Copyright © 1975 by Springer-Verlag GmbH & Co KG; **6.32.7** From C. Peracchia and A. F. Dulhunty, *The Journal of Cell Biology* 70 (1976):419 by copyright permission of The Rockefeller University Press; **6.33.2** Lennart Nilsson/ Scanpix; **p. 123, top** Courtesy E. H. Newcomb; **p. 123, bottom** From S. E. Fredrick and E. H. Newcomb, *The Journal of Cell Biology* 43 (1969): 343.

Chapter 7 7.1 Roderick Mackinnon; **7.4.3** D. W. Fawcett/Photo Researchers; **7.4.4** D. W. Fawcett/Photo Researchers; **7.16.2.** Michael Abbey/Photo Researchers; **7.22.3** H. S. Pankratz, T.C. Beaman & P. Gerhardt/Biological Photo Service; **7.22.6** D. W. Fawcett/Photo Researchers; **7.22.9** M. M. Perry and A. B. Gilbert, *J. Cell Science* 39 (1979) 257. Copyright 1979 by The Company of Biologists Ltd. **7.22.10** M. M. Perry and A. B. Gilbert, *J. Cell Science* 39 (1979) 257. Copyright 1979 by The Company of Biologists Ltd.

Chapter 8 8.1 Photoshot/NHPA; **8.2.2** Jupiter Images; **8.3.2** Robert N. Johnson/RnJ Photography; **8.3.3** Robert N. Johnson/RnJ Photography; **8.4.1** Brandon Blinkenberg/iStockphoto; **8.4.2** Bridget Lazenby/iStockphoto; **8.14.2** Thomas A. Steitz, Yale University; **8.14.3** Thomas A. Steitz, Yale University; **8.20.3** Scheer JM, Romanowski MJ, Wells JA. A common allosteric site and mechanism in caspases. *Proc Natl Acad Sci U S A*. 2006 May 16;103(20):7595–600; Fig. 4a; **8.22.2** Nicolae Simionescu.

Chapter 9 **9.1** Anup Shah/Nature Picture Library.

Chapter 10 **10.1** Bob Rowan, Progressive Image/Corbis; **10.2.2** Jean-Paul Nacivet/AGE Fotostock; **10.2.3** Lawrence Naylor/Photo Researchers; **10.2.4** M. I. Walker/Photo Researchers; **10.2.5** Susan M. Barns; **10.2.6** National Library of Medicine; **10.3** Robert Clark Photography, robertclark.com; **10.4.2** Image courtesy Andreas Holzenburg and Stanislav Vitha, Dept. of Biology and Microscopy & Imaging Center, Texas A&M University; **10.4.3** E.H Newcomb & WP Wergin/Biological Photo Service; **10.12.2** Christine L. Case, Skyline College; **10.13.2** From "Architecture of the photosynthetic oxygen-evolving center."; . . . K. N. Ferreira,T. M. Iverson, K. Maghlaoui, J. Barber, S. Iwata. *Science*. 2004 Mar 19;303(5665):1831–8. Epub 2004 Feb 5; **10.21.2** David Muench/Corbis; **10.21.3** Dave Bartruff/Corbis.

Chapter 11 **11.1** Winfried Wisniewski/Corbis; **11.3.2** A. Dale Kaiser, Stanford University; **11.8.2** From "High-resolution crystal structure of an engineered human beta2-adrenergic G protein-coupled receptor." V. Cherezov, D. M. Rosenbaum, M. A. Hanson, S. G. F. Rasmussen, F. S. Thian, T. S. Kobilka, H-J. Choi, P. Kuhn, W. I. Weis, B. K. Kobilka, R. C. Stevens. *Science*. 2007 Nov 23;318(5854):1258–65. Epub 2007 Oct 25; **11.17.3** Matheos D, Metodiev M, Muller E, Stone D, Rose MD. Pheromone-induced polarization is dependent on the Fus3p MAPK acting through the formin Bni1p. *J Cell Biol*. 2004 April; 165(1);99–109; Fig.9. Reproduced by copyright permission of The Rockefeller University Press; **11.20.2** Gopal Murti/Photo Researchers; **11.22.2** *Development* 127, 5245–5252 (2000). Mesenchymal cells engulf and clear apoptotic footplate cells in macrophageless PU.1 null mouse embryos. William Wood, Mark Turmaine, Roberta Weber, Victoria Camp, Richard A. Maki, Scott R. McKercher and Paul Martin; **11.22.3** *Development* 127, 5245–5252 (2000). Mesenchymal cells engulf and clear apoptotic footplate cells in macrophageless PU.1 null mouse embryos. William Wood, Mark Turmaine, Roberta Weber, Victoria Camp, Richard A. Maki, Scott R. McKercher and Paul Martin; **11.22.4** *Development* 127, 5245–5252 (2000). Mesenchymal cells engulf and clear apoptotic footplate cells in macrophageless PU.1 null mouse embryos. William Wood, Mark Turmaine, Roberta Weber, Victoria Camp, Richard A. Maki, Scott R. McKercher and Paul Martin.

Chapter 12 **12.1** Jan-Michael Peters/Silke Hauf; **12.2.2** Biophoto Associates/Photo Researchers; **12.2.3** C.R. Wyttenbach/Biological Photo Service; **12.2.4** Biophoto/Science Source/Photo Researchers; **12.3.2** John Murray; **12.4.2** Biophoto/Photo Researchers; **12.5.2** Biophoto/Photo Researchers; **12.7.2** Conly L. Rieder; **12.7.3** Conly L. Rieder; **12.7.4** Conly L. Rieder; **12.7.6** Conly L. Rieder; **12.7.7** Conly L. Rieder; **12.7.8** Conly L. Rieder; **12.8.2** J. Richard McIntosh, University of Colorado at Boulder; **12.8.3** Reproduced by permission from Matthew Schibler, from *Protoplasma* 137. Copyright © 1987: 29–44 by Springer-Verlag GmbH & Co KG; **12.10.2** Don W. Fawcett/Photo Researchers; **12.10.3** B. A. Palevitz, Courtesy of E. H. Newcomb, University of Wisconsin; **12.11.2** Elisabeth Pierson, FNWI-Radboud University Nijmegen, Pearson Science; **12.11.3** Elisabeth Pierson, FNWI-Radboud University Nijmegen, Pearson Science; **12.11.4** Elisabeth Pierson, FNWI-Radboud University Nijmegen, Pearson Science; **12.11.5** Elisabeth Pierson, FNWI-Radboud University Nijmegen, Pearson Science; **12.11.6** Elisabeth Pierson, FNWI-Radboud University Nijmegen, Pearson Science; **12.18.2** Guenter Albrecht-Buehler; **12.19.2** Lan Bo Chen; **12.19.3** Lan Bo Chen; **12.21** Anne Weston, LRI, CRUK, Wellcome Images; **p. 245** USDA/ARS/Agricultural Research Service.

Chapter 13 **13.1** Steve Granitz/WireImage/Getty Images; **13.2.2** Roland Birke/OKAPIA/Photo Researchers; **13.2.3** SuperStock; **13.3.2** Veronique Burger/Phanie Agency/Photo Researchers; **13.3.3** CNRI/Photo Researchers; **13.11.2** Mark Petronczki and Maria Siomos; **13.12.2** John Walsh, Micrographia.com.

Chapter 14 **14.1** Mendel Museum, Augustinian Abbey, Brno; **14.14.2** Altrendo nature/Getty Images; **14.14.3** PictureNet Corporation/Corbis; **14.15.2** Photodisc/Getty Images; **14.15.3** Photodisc/Getty Images; **14.15.4** Anthony Loveday; **14.15.5** Anthony Loveday; **14.16.2** Rick Guidotti and Diane McLean/Positive Exposure; **14.17.2** Michael Ciesielski Photography; **14.18** Douglas C. Pizac/AP Images; **14.19.2** CNRI/Photo Researchers; **p. 285** Norma Jubinville.

Chapter 15 **15.1** David C. Ward; **15.3.2** From: "Learning to Fly: Phenotypic Markers in *Drosophila*." A poster of common phenotypic markers used in *Drosophila* genetics. Jennifer Childress, Richard Behringer, and Georg Halder. 2005. *Genesis* 43(1). Cover illustration; **15.3.3** From: "Learning to Fly: Phenotypic Markers in *Drosophila*." A poster of common phenotypic markers used in *Drosophila* genetics. Jennifer Childress, Richard Behringer, and Georg Halder. 2005. *Genesis* 43(1). Cover illustration; **15.5** Andrew Syred/Photo Researchers; **15.8.2** Dave King/Dorling Kindersley; **15.15.2** Lauren Shear/SPL/Photo Researchers; **15.15.3** CNRI/SPL/Photo Researchers; **15.18** Geoff Kidd/Photo Researchers; **p. 304** James K. Adams, Biology, Dalton State College, Dalton, Georgia.

Chapter 16 **16.1** National Institutes of Health; **16.3.2** Oliver Meckes/Photo Researchers; **16.6.2** Courtesy of the Library of Congress; **16.6.3** From "The Double Helix" by James D. Watson, Atheneum Press, N.Y., 1968, p. 215. © 1968. Courtesy CSHL Archive; **16.12.2** Jerome Vinograd; **16.12.3** From D. J. Burks and P. J. Stambrook, *The Journal of Cell Biology 77* (1978). 762, fig. 6 by copyright permission of The Rockefeller University Press. Photo provided by P. J. Stambrook; **16.21** Peter Lansdorp; **16.22.2** S. C. Holt/Biological Photo Service; **16.22.3** Victoria E. Foe; **16.22.4** Barbara Hamkalo; **16.22.5** From J. R. Paulsen and U. K. Laemmli, *Cell* 12 (1977):817–828; **16.22.6** Biophoto/Photo Researchers; **16.23.2** From: Multicolor Spectral Karyotyping of Human Chromosomes. E. Schröck, S. du Manoir, T. Veldman, B. Schoell, J. Wienberg, M. A. Ferguson-Smith, Y. Ning, D. H. Ledbetter, I. Bar-Am, D. Soenksen, Y. Garini, T. Ried. *Science*. 1996 Jul 26;273(5274):494–7; **16.23.3** From: The new cytogenetics: blurring the boundaries with molecular biology; . . . M. R. Speicher, N. P. Carter. *Nat Rev Genet*. 2005 Oct;6(10):782–92; **p. 324** Thomas A. Steitz, Yale University.

Chapter 17 **17.1** Deutscher Fotodienst GmbH; **17.6.2** Keith V. Wood, University of California, San Diego; **17.6.3** AP Images; **17.16.2** Thomas Steitz; **17.17.2** Joachim Frank; **17.21.2** B. Hamkalo and O. Miller, Jr; **17.25.2** Reprinted with permission from O. L. Miller , Jr., B. A. Hamkalo, and C. A Thomas, Jr., *Science* 169 (1970):392. Copyright © 1970 American Association for the Advancement of Science.

Chapter 18 **18.1** Reproduced by permission from Cook O, Biehs B, Bier E. Brinker and optomotor-blind act coordinately to initiate development of the L5 wing vein primordium in *Drosophila*. *Development*. 2004 May: 131(9):2113–24; **18.12.2** From: The new cytogenetics: blurring the boundaries with molecular biology; M. R. Speicher, N. P. Carter. *Nat Rev Genet*. 2005 Oct;6(10):782–92; **18.16.2** Mike Wu; **18.16.3** Hans Pfletschinger/Peter Arnold/PhotoLibrary; **18.20.2** F. R. Turner, Indiana University; **18.20.3** F. R. Turner, Indiana University; **18.21.2** Wolfgang Driever, University of Freiburg, Freiburg, Germany; **18.21.3** Wolfgang Driever, University of Freiburg, Freiburg, Germany; **18.22.2** Ruth Lehmann, The Whitehead Institution; **18.22.3** Ruth Lehmann, The Whitehead Institution; **18.26** Roy Kaltschmidt, Lawrence Berkeley National Laboratory.

Chapter 19 **19.1** Science Photo Library/Photo Researchers; **19.2.2** Peter von Sengbusch/Botanik; **19.2.3** Peter von Sengbusch/Botanik; **19.2.4** Peter von Sengbusch/Botanik; **19.3.2** Robley C. Williams/Biological Photo Service; **19.3.3** R.C. Valentine and H.G. Pereira, "Antigens and Structure of the Adenovirus," *Journal of Molecular Biology* 13: 13–20 (1965); **19.3.4** Hazel Appleton, Health Protection Agency Centre for Infections/Photo Researchers; **19.3.5** Robley C. Williams/ Biological Photo Service; **19.8.2** C. Dauguet/Institute Pasteur/Photo Researchers; **19.8.3** C. Dauguet/Institute Pasteur/Photo Researchers; **19.8.4** C. Dauguet/Institute Pasteur/Photo Researchers; **19.8.5** C. Dauguet/Institute Pasteur/Photo Researchers; **19.8.6** C. Dauguet/Institute Pasteur/Photo Researchers; **19.9.2** NIBSC/Photo Researchers; **19.9.3** Seo Myung-gon/AP Images; **19.9.4** National Museum of Health and Medicine/Armed Forces Institute of Pathology; **19.10.2** Dennis E. Mayhew; **19.10.3** Thomas A. Zitter; **19.10.4** A. Vogler/Shutterstock.

Chapter 20 **20.1** Reproduced with permission from R.F. Service, *Science* (1998) 282:396–3999. Copyright 1998 American Association for the Advancement of Science. Incyte Pharmaceuticals, Inc., Palo Alto, CA; **20.5.2** L. Brent Selinger, Pearson Science; **20.9.2** Repligen Corporation; **20.14** Ethan Bier; **20.15.2** Reproduced with permission from R.F. Service, *Science* (1998) 282:396–3999. Copyright 1998 American Association for the Advancement of Science. Incyte Pharmaceuticals, Inc., Palo Alto, CA; **20.20** Pat Sullivan/AP Images; **20.24.2** Brad DeCecco Photography; **20.24.3** Brad DeCecco Photography; **20.25.2** Steve Helber/AP Images.

Chapter 21 **21.1** Karen Huntt/Corbis; **21.5.2** From "The genetic landscape of a cell." M. Costanzo, et al. *Science*. 2010 Jan 22;327(5964):425–31; **21.6** GeneChip Human Genome U133 Plus 2.0 Array, image courtesy of Affymetrix; **21.8.2** AP Images; **21.8.3** Courtesy of Virginia Walbot, Stanford University; **21.11.2** Courtesy of O. L. Miller Jr., Dept. of Biology, University of Virginia; **21.17.2** Shu W, Cho JY, Jiang Y, Zhang M, Weisz D, Elder GA, Schmeidler J, De Gasperi R, Sosa MA, Rabidou D, Santucci AC, Perl D, Morrisey E, Buxbaum JD. Altered ultrasonic vocalization in mice with a disruption in the *Foxp2* gene. : *Proc Natl Acad Sci U S A*. 2005 Jul 5; 102(27):9643–8; Fig. 3. Image supplied by Joseph Buxbaum; **21.17.3** Shu W, Cho JY, Jiang Y, Zhang M, Weisz D, Elder GA, Schmeidler J, De Gasperi R, Sosa MA, Rabidou D, Santucci AC, Perl D, Morrisey E, Buxbaum JD. Altered ultrasonic vocalization in mice with a disruption in the *Foxp2* gene. : *Proc Natl Acad Sci U S A*. 2005 Jul 5;102(27):9643–8; Fig. 3. Image supplied by Joseph Buxbaum; **21.17.4** Shu W, Cho JY, Jiang Y, Zhang M, Weisz D, Elder GA, Schmeidler J, De Gasperi R, Sosa MA, Rabidou D, Santucci AC, Perl D, Morrisey E, Buxbaum JD. Altered ultrasonic vocalization in mice with a disruption in the *Foxp2* gene. : *Proc Natl Acad Sci U S A*. 2005 Jul 5; 102(27):9643–8; Fig. 3. Image supplied by Joseph Buxbaum; **21.17.5** Joe McDonald/Corbis.

Chapter 22 **22.1** Olivier Grunewald; **22.2.2** PD image: Skeleton of the "Rhinoceros unicorne de Java" in the Paris Museum of Natural History. From G. Cuvier, Recherches sur les ossemens fossiles, *Atlas*, pl. 17. 1836; **22.2.3** Eileen Tweedy/Picture Desk/Kobal Collection; **22.2.4** Wayne Lynch/AGE Fotostock; **22.2.5** Neg./Transparency no. 330300. Courtesy Dept. of Library Services, American Museum of Natural History; **22.4** Michael S. Yamashita/Corbis; **22.5.2** George Richmond/ARCHIV/Photo Researchers; **22.5.3** National Maritime Museum; **22.6.2** Michel Gunther/PhotoLibrary; **22.6.3** David Hosking/FLPA; **22.6.4** David Hosking/Alamy; **22.7** Darwin's Tree of life sketch, MS.DAR.121:p.36. Reproduced with the permission of the Cambridge University Library; **22.9.2** Gerard Schulz/Naturphoto; **22.9.3** Robert Sarno/iStockphoto; **22.9.4** Paul Rapson/Alamy; **22.9.5** Izaokas Sapiro/Shutterstock; **22.9.6** YinYang/iStockphoto; **22.9.7** floricica buzlea/iStockphoto; **22.10** Laura Jesse, Extension Entomologist, Iowa State University; **22.11.2** Richard Packwood/Oxford Scientific/Jupiter Images; **22.12.2** E. S. Ross, California Academy of Sciences; **22.12.3** Mark Taylor/Nature Picture Library; **22.13.1** Scott P. Carroll; **22.16.2** Dr. Keith Wheeler/Photo Researchers; **22.16.3** Lennart Nilsson/Scanpix; **22.18.2** Visible Earth (http://visibleearth.nasa.gov/), NASA; **22.19.2** Chris Linz, Thewissen lab, Northeastern Ohio Universities College of Medicine; **22.19.3** Chris Linz, Thewissen lab, Northeastern Ohio Universities College of Medicine; **22.19.4** Chris Linz, Thewissen lab, Northeastern Ohio Universities College of Medicine; **22.19.5** Chris Linz, Thewissen lab, Northeastern Ohio Universities College of Medicine.

Chapter 23 **23.1** Rosemary B. Grant; **23.3.2** Erick Greene, University of Montana; **23.3.3** Erick Greene, University of Montana; **23.4.2** Janice Britton-Davidian, ISEM, UMR 5554 CNRS, Universite Montpellier II. Reprinted by permission from *Nature,* Vol. 403, 13 January 2000, p. 158. © 2000 Macmillan Magazines Ltd.; **23.4.3** Janice Britton-Davidian, ISEM, UMR 5554 CNRS, Universite Montpellier II. Reprinted by permission from *Nature,* Vol. 403, 13 January 2000, p. 158. © 2000 Macmillan Magazines Ltd.; **23.4.4** Steve Gorton/Dorling Kindersley; **23.5.2** New York State Department of Environmental Conservation; **23.6.2** Gary Schultz/Photoshot; **23.6.3** James L. Davis/ProWildlife; **23.11.2** William Ervin/SPL/Photo Researchers; **23.12.2** Jan Visser; **23.14.2** John Visser/Photoshot; **23.15** Dave Blackey/PhotoLibrary; **23.16.2** Allison M. Welch; **23.19** Merlin D. Tuttle, Bat Conservation International, www.batcon.org.

Chapter 24 **24.1** Mark Jones/AGE Fotostock; **24.2.2** Malcolm Schuyl/Alamy; **24.2.3** Wave Rf/PhotoLibrary; **24.2.4** Robert Kneschke/iStockphoto; **24.2.5** Justin Horrocks/iStockphoto; **24.2.6** Photodisc/Getty Images; **24.2.7** Photodisc/Getty Images; **24.2.8** Photodisc/Getty Images; **24.2.9** Masterfile; **24.3.a2** Joe McDonald/Photoshot; **24.3.b2** Joe McDonald/Corbis; **24.3.c2** USDA/APHIS/Animal and Plant Health Inspection Service; **24.3.d2** Stephen Krasemann/Photo Researchers; **24.3.e1** Michael Dietrich/

imagebroker/Alamy; **24.3.f2** Takahito Asami; **24.3.g2** William E. Ferguson; **24.3.h2** Charles W. Brown; **24.3.i2** Photodisc/Getty Images; **24.3.j2** Corbis; **24.3.k2** DawnYL/Fotolia; **24.3.l2** Kazutoshi Okuno; **24.4.2** CLFProductions/Shutterstock; **24.4.3** Boris Karpinski/Alamy; **24.4.4** Troy Maben/AP Images; **24.6.2** John Shaw/Bruce Coleman/Photoshot; **24.6.3** Michael Fogden/Bruce Coleman/Photoshot; **24.6.4** Corbis; **24.7.2** From: Morphology, performance, fitness: functional insight into a post-Pleistocene radiation of mosquitofish. R. B. Langerhans. *Biology Letters* 2009;5(4):488–491; **24.7.3** From: Morphology, performance, fitness: functional insight into a post-Pleistocene radiation of mosquitofish. R. B. Langerhans. *Biology Letters* 2009;5(4):488–491; **24.8.2** Visible Earth (http://visibleearth.nasa.gov/), NASA; **24.8.3** Arthur Anker; **24.12** Ole Seehausen; **24.13.2** Jeroen Speybroeck, Research Institute for Nature and Forest, Belgium; **24.13.3** Jeroen Speybroeck, Research Institute for Nature and Forest, Belgium; **24.16.2** Ole Seehausen; **24.16.3** Ole Seehausen; **24.16.4** Ole Seehausen; **24.18** Jason Rick; **24.20.2** Reprinted by permission from *Nature* Bradshaw HD, Schemske DW. Allele substitution at a flower colour locus produces a pollinator shift in monkeyflowers. *Nature*. 2003 November 12; 426(6963):176–8 Copyright © 2003. Macmillan Magazines Limited; **24.20.3** Reprinted by permission from *Nature* Bradshaw HD, Schemske DW. Allele substitution at a flower colour locus produces a pollinator shift in monkeyflowers. *Nature*. 2003 November 12; 426(6963):176–8 Copyright © 2003. Macmillan Magazines Limited; **24.20.4** Reprinted by permission from *Nature* Bradshaw HD, Schemske DW. Allele substitution at a flower colour locus produces a pollinator shift in monkeyflowers. *Nature*. 2003 November 12; 426(6963):176–8 Copyright © 2003. Macmillan Magazines Limited; **24.20.5** Reprinted by permission from *Nature* Bradshaw HD, Schemske DW. Allele substitution at a flower colour locus produces a pollinator shift in monkeyflowers. *Nature*. 2003 November 12; 426(6963):176–8 Copyright © 2003. Macmillan Magazines Limited.

Chapter 25 25.1 Gerhard Boeggemann; **p. 507.UN1** Rebecca Hunt; **25.2.2** UPI Photo/Landov; **25.3.2** Courtesy of F. M. Menger and Kurt Gabrielson, Emory University; **25.3.3** M. Hanczyc; **25.4.10** Specimen No 12478, Markus Moser, Staatliches Museum fur Naturkunde Stuttgart; **25.4.2** S. M. Awramik/Biological Photo Service; **25.4.3** Sinclair Stammers/Photo Researchers; **25.4.4** Andrew H. Knoll, Harvard University; **25.4.5** Lisa-Ann Gershwin, University of California-Berkeley, Museum of Paleontology; **25.4.6** Chip Clark; **25.4.7** Ted Daeschler/Academy of Natural Sciences/VIREO; **25.4.8** Roger Jones; **25.4.9** Seelevel.com; **25.11.2** Shuhai Xiao, Tulane University; **25.11.3** Shuhai Xiao, Tulane University; **25.20.2** Gerald D. Carr; **25.20.3** Gerald D. Carr; **25.20.4** Gerald D. Carr; **25.20.5** Gerald D. Carr; **25.20.6** Gerald D. Carr; **25.20.7** Bruce G. Baldwin; **25.21.2** Jean Kern; **25.21.3** Jean Kern; **25.22** Juniors Bildarchiv/Alamy; **25.23** Reprinted from "The Origin of Form" by Sean B. Carroll, *Natural History*, November 2005. Burke, A.C. 2000, "*Hox*" Genes and the Global Patterning of the Somitic Mesoderm. In Somitogenesis. C. Ordahl (ed.) "Current Topics in Developmental Biology", Vo. 47. Academic Press; **25.25.1** Oxford Scientific/PhotoLibrary; **25.25.3** Shapiro MD, Marks ME, Peichel CL, Blackman BK, Nereng KS, Jonsson B, Schluter D, Kingsley DM. Genetic and developmental basis of evolutionary pelvic reduction in threespine sticklebacks. *Nature*. Erratum. 2006 Feb 23;439(7079):1014; Fig. 1; **25.25.4** Shapiro MD, Marks ME, Peichel CL, Blackman BK, Nereng KS, Jonsson B, Schluter D, Kingsley DM. Genetic and developmental basis of evolutionary pelvic reduction in threespine sticklebacks. *Nature*. Erratum. 2006 Feb 23;439(7079):1014; Fig. 1.

Chapter 26 26.1 Ken Griffiths/NHPA/Photoshot; **26.2.2** Ryan McVay/Photodisc/Getty Images; **26.2.3** Neil Fletcher/Dorling Kindersley; **26.2.4** Dorling Kindersley; **26.17.2** Ed Heck; **26.17.3** Courtesy Dept. of Library Services, American Museum of Natural History.

Chapter 27 27.1 Bonnie K. Baxter, Great Salt Lake Institute, Westminster College, Utah; **27.2.2** CDC; **27.2.3** Dr. Kari Lounatmaa/Photo Researchers; **27.2.4** Stem Jems/Photo Researchers; **27.3.2** L. Brent Selinger, Pearson Science; **27.4.2** Dr. Immo Rantala/Photo Researchers; **27.5.2** Kwangshin Kim/Photo Researchers; **27.6.2** Julius Adler; **27.7.2** S. W. Watson; **27.7.3** Norma J. Lang/Biological Photo Service; **27.8.2** Huntington Potter, Byrd Alzheimer's Institute and University of South Florida, David Dressler, Oxford University and Balliol College; **27.9.2** H.S. Pankratz, T.C. Beaman/Biological Photo Service; **27.12.2** Charles C. Brinton, Jr., University of Pittsburgh; **27.14.2** Susan M. Barns; **27.16** Jack Dykinga/Stone/Getty Images; **27.17.3** L. Evans Roth/Biological Photo Service; **27.17.5** Yuichi Suwa; **27.17.7** National Library of Medicine; **27.17.9** From "Scanning electron microscopy of fruiting body formation by myxobacteria." P. L. Grilione and J. Pangborn. *J. Bacteriol.* 1975 December; 124(3): 1558–1565; **27.17.13** Photo Researchers; **27.17.15** Moredon Animal Health/SPL/Photo Researchers; **27.17.17** CNRI/SPL/Photo Researchers; **27.17.19** Culture Collection CCALA, Institute of Botany, Academy of Sciences Dukelska, Czech Republic; **27.17.21** Paul Hoskisson, Strathclyde Institute of Pharmacy and Biomedical Sciences, Glasgow, Scotland; **27.17.22** David M. Phillips/Photo Researchers; **27.18.2** Pascale Frey-Klett; **27.19** Ken Lucas/Biological Photo Service; **27.20.2** Scott Camazine/Photo Researchers; **27.20.3** David M. Phillips/Photo Researchers; **27.20.4** James Marshall/The Image Works; **27.21** Metabolix; **27.21.3** Courtesy of Exxon Mobil Corporation; **27.21.4** Seelevel.com.

Chapter 28 28.1 Brian S. Leander; **28.3.3** Joel Mancuso, University of California, Berkeley; **28.3.5** M. I. Walker/NHPA/Photoshot; **28.3.7** Howard Spero, University of California-Davis; **28.3.8** NOAA; **28.3.10** Kim Taylor/Nature Picture Library; **28.3.11** David J. Patterson/micro*scope; **28.3.13** Tom Stack/PhotoLibrary; **28.4.2** David M. Phillips/The Population Council/Photo Researchers; **28.5.2** David J. Patterson; **28.6.2** Meckes/Ottawa/Photo Researchers; **28.7.2** D. J. Patterson, L. Amaral-Zettler, M. Peglar and T. Nerad, http://micro*scope.mbl.edu; **28.8.2** Guy Brugerolle, Universitad Clearmont, Ferrand; **28.9.2** Virginia Institute of Marine Science; **28.10.2** Masamichi Aikawa, Tokai University School of Medicine, Japan; **28.11.2** M. I. Walker/Photo Researchers; **28.12.2** Centers for Disease Control & Prevention; **28.13.2** Steve Gschmeissner/Photo Researchers; **28.14.2** Stephen Durr; **28.15.2** Colin Bates; **28.16.2** J. Robert Waaland/Biological Photo Service; **28.17.2** Fred Rhoades; **28.18.2** Robert Brons/Biological Photo Service; **28.19.2** Eva Nowack; **28.20.2** D. P. Wilson, Eric & David Hosking/Photo Researchers; **28.20.3** Michael D. Guiry; **28.20.4** Biophoto Associates/Photo Researchers; **28.20.6** Michael Yamashita/IPN/Aurora Photos; **28.20.7** David Murray/Dorling Kindersley; **28.21.2** Laurie Campbell/NHPA; **28.21.3** Marine Sciences, University of Puerto Rico; **28.22.2** William L. Dentler; **28.24.2** George Barron; **28.24.3** Mikel Tapia (www.argazkik.com); **28.25.2** Robert Kay; **28.25.3** Robert Kay; **28.26.2** Kevin Carpenter and Patrick Keeling.

Chapter 29 29.1 Martin Rugner/AGE Fotostock; **29.2.2** S. C. Mueller and R. M. Brown, Jr; **29.3.2** Natural Visions; **29.3.3** Linda Graham, University of Wisconsin-Madison; **29.5.11** Ed Reschke; **29.5.12** CDC; **29.5.3** Linda Graham, University of Wisconsin-Madison; **29.5.4** Karen S. Renzaglia; **29.5.6** Alan S. Heilman; **29.5.7** Michael Clayton, University of Wisconsin-Madison; **29.5.9** David John Jones (mybitoftheplanet.com); **29.6.2** Charles H. Wellman; **29.6.3** Charles H. Wellman; **29.8.2** Laurie Knight (www.laurieknight.net); **29.9.2** Linda Graham, University of Wisconsin-Madison; **29.9.3** Alvin E. Staffan/National Audubon Society/Photo Researchers; **29.9.4** Hidden Forest; **29.9.6** Hidden Forest; **29.9.8** Tony Wharton, Frank Lane Picture Agency/Corbis; **p. 609** From "Mosses and Other Bryophytes, an Illustrated Glossary" (2006) by Bill and Nancy Malcolm; **29.11.1** Brian Lightfoot/AGE Fotostock; **29.11.2** Chris Lisle/Corbis; **29.15.2** Jody Banks, Purdue University; **29.15.3** Murray Fagg, Australian National Botanic Gardens; **29.15.4** Helga & Kurt Rasbach; **29.15.6** Jon Meier/iStockphoto; **29.15.8** Milton Rand/Tom Stack & Associates; **29.15.9** Francisco Javier Yeste Garcia (www.flickr.com/photos/fryega/); **29.15.10** Francisco Javier Yeste Garcia (www.flickr.com/photos/fryega/); **29.16.2** The Open University; **p. 616, top** Ed Reschke; **p. 616, bottom** Michael Clayton.

Chapter 30 30.1 National Museum of Natural History, Smithsonian Institution; **30.5.1** Johannes Greyling/iStockphoto; **30.5.3** CC-BY-SA photo: Kurt Stueber (www.biolib.de); **30.5.4** Jeroen Peys/iStockphoto; **30.5.5** Michael Clayton; **30.5.6** Thomas Schoepke; **30.5.7** Bob Gibbons/FLPA; **30.5.8** Raymond Gehman/Corbis; **30.5.9** Adam Jones/Getty Images; **30.5.10** David Muench/Corbis; **30.5.11** Gunter Marx Photography/Corbis; **30.5.12** Royal Botanic Gardens Sydney; **30.5.13** Jaime Plaza/ AP Images; **30.5.14** Mario Verin/Photolibrary; **30.8.2** Dave King/Dorling Kindersley; **30.8.3** Andy Crawford/Dorling Kindersley; **30.8.4** Dave King/Dorling Kindersley; **30.8.5** Maria Dryfhout/iStockphoto; **30.8.6** Peter Rees/Getty Images; **30.9.2** PIXTAL/AGE Fotostock; **30.9.3** Hans Dieter Brandl, Frank Lane Picture Agency/Corbis; **30.9.4** Scott Camazine/Photo Researchers; **30.9.5** Derek Hall/Dorling Kindersley; **30.11.2** David L. Dilcher; **30.13.1** Howard Rice/Dorling Kindersley; **30.13.2** Jack Scheper, Floridata.com; **30.13.3** Stephen McCabe; **30.13.4** Andrew Butler/Dorling Kindersley; **30.13.6** Eric Crichton/Dorling Kindersley; **30.13.7** John Dransfield; **30.13.8** Dorling Kindersley; **30.13.10** Terry W. Eggers/Corbis; **30.13.11** CC-BY-SA photo: Artslave; **30.13.12** Matthew Ward/Dorling Kindersley; **30.13.13** Tony Wharton, Frank Lane Picture Agency/Corbis; **30.13.14** Howard Rice/Dorling Kindersley; **30.13.15** Gerald D. Carr; **30.14** kkaplin/Shutterstock; **30.15** D. Wilder; **30.16.2** NASA's Earth Observatory; **30.16.3** NASA's Earth Observatory.

Chapter 31 31.1 Georg Müller; **31.2.2** Fred Rhoades/Mycena Consulting; **31.2.3** Hans Reinhard/Taxi/Getty Images; **31.2.4** George Barron, University of Guelph, Canada; **31.4.2** © N. Allin & G.L. Barron, University of Guelph/Biological Photo Service; **31.6.2** Biophoto Associates/Photo Researchers; **31.6.3** Popovaphoto/dreamstime; **31.7.2** Stephen J. Kron; **31.9.2** Dirk Redecker, Robin Kodner, and Linda E. Graham. Glomalean Fungi from the Ordovician. *Science* 15 September 2000; 289: 1920–1921; **31.10.2** CDC; **31.11.2** John Taylor; **31.11.3** Ray Watson; **31.11.5** Reproduced by permission from Kiers ET, van der Heijden MG. Mutualistic stability in the arbuscular mycorrhizal symbiosis: exploring hypotheses of evolutionary cooperation. *Ecology*. 2006 July: 87(7):1627–36: Fig. 1a. Image by Marcel van der Heijden, Swiss Federal Research Station for Agroecology and Agriculture. Copyright © 2006, Ecological Society of America; **31.11.6** Frank Young/Papilio/Corbis; **31.11.7** Phil Dotson/Photo Researchers; **31.12.2** William E. Barstow; **31.13.2** Antonio D'Albore/iStockphoto; **31.13.3** Alena Kubátová (http://botany.natur.cuni.cz/cs/sbirka-kultur-hub-ccf); **31.13.4** Ed Reschke/Peter Arnold/PhotoLibrary; **31.13.5** George Barron; **31.14.2** G.L. Barron/Biological Photo Service; **31.15.2** M. F. Brown/Biological Photo Service; **31.16.2** Viard/Jacana/Photo Researchers; **31.16.3** Douglas Adams/iStockphoto; **31.17.2** Fred Spiegel; **31.18.2** Fletcher and Baylis/Photo Researchers; **31.18.3** Michael Fogden/Photolibrary; **31.18.4** Frank Paul/Alamy; **31.19.2** Biophoto Associates/Photo Researchers; **31.20** University of Tennessee Entomology and Plant Pathology; **31.22** Mark Bowler/Photo Researchers; **31.23.2** Ralph Lee Hopkins/Getty Images; **31.23.3** Geoff Simpson/naturepl.com; **31.23.4** Wild-Worlders of Europe/Benvie/naturepl.com; **31.24.2** Eye of Science/Photo Researchers; **31.25.2** Alamy; **31.25.3** Peter Chadwick/Dorling Kindersley; **31.25.4** Hecker-Sauer/AGE Fotostock; **31.26.2** Vance T. Vredenburg, San Francisco State University; **31.27.2** Christine Case.

Chapter 32 32.1 Jeff Hunter/Image Bank/Getty Images; **p. 656** Biological Photo Service; **32.4.2** © The Museum Board of South Australia 2004 Photographer: Dr. J. Gehling; **32.4.3** © The Museum Board of South Australia 2004 Photographer: Dr. J. Gehling; **32.5** J. Sibbick/The Natural History Museum, London; **32.6.2** Wikramanayake AH, Hong M, Lee PN, Pang K, Byrum CA, Bince JM, Xu R, Martindale MQ. An ancient role for nuclear beta-catenin in the evolution of axial polarity and germ layer segregation. *Nature*. 2003 Nov 27;426(6965):446–50; Fig. 2, 3 and 4; **32.6.3** Wikramanayake AH, Hong M, Lee PN, Pang K, Byrum CA, Bince JM, Xu R, Martindale MQ. An ancient role for nuclear beta-catenin in the evolution of axial polarity and germ layer segregation. *Nature*. 2003 Nov 27;426(6965):446–50; Fig. 2, 3 and 4; **32.6.4** Wikramanayake AH, Hong M, Lee PN, Pang K, Byrum CA, Bince JM, Xu R, Martindale MQ. An ancient role for nuclear beta-catenin in the evolution of axial polarity and germ layer segregation. *Nature*. 2003 Nov 27;426(6965):446–50; Fig. 2, 3 and 4; **32.6.5** Wikramanayake AH, Hong M, Lee PN, Pang K, Byrum CA, Bince JM, Xu R, Martindale MQ. An ancient role for nuclear beta-catenin in the evolution of axial polarity and germ layer segregation. *Nature*. 2003 Nov 27; 426(6965):446–50; Fig. 2, 3 and 4; **32.12** Kent Wood/Photo Researchers; **32.13.2** Hecker/Sauer/AGE Fotostock.

Chapter 33 33.1 C. Wolcott Henry III/National Geographic/Getty Images; **33.3.1** Andrew J. Martinez/Photo Researchers; **33.3.2** Robert Brons/Biological Photo Service; **33.3.3** Teresa (Zubi) Zuberbühler; **33.3.4** Stephen Dellaporta; **33.3.5** Gregory G. Dimijian/Photo Researchers; **33.3.6** Ed Robinson/Pacific Stock/Photolibrary; **33.3.7** Hecker/Sauer/AGE Fotostock; **33.3.8** W. I. Walker/Photo Researchers; **33.3.9** Kåre Telnes/Image Quest Marine; **33.3.10** PD image: Anilocra/Taken by Neil Campbell, University of Aberdeen, Scotland, UK; **33.3.11** Erling Svensen/UWPhoto ANS; **33.3.12** Peter Funch; **33.3.13** photonimo/iStockphoto; **33.3.14** Peter Batson/Image Quest Marine; **33.3.15** Reinhart Mobjerg Kristensen; **33.3.16** Erling Svensen/

UWPhoto ANS; **33.3.17** Andrew Syred/Photo Researchers; **33.3.18** Reproduced with permission from A. Eizinger and R. Sommer, Max Planck Institut fur entwicklungsbiologie, Tubingen. Copyright 2000 American Association for the Advancement of Science. Cover 278(5337) 17 Oct 97; **33.3.19** Thomas Stromberg; **33.3.20** Tim Flach/Stone/Getty Images; **33.3.21** Heather Angel/Natural Visions; **33.3.22** Robert Harding World Imagery/Alamy; **33.3.23** Robert Brons/Biological Photo Service; **33.4.2** Andrew J. Martinez/Photo Researchers; **33.7.2** Andrew J. Martinez/Photo Researchers; **33.7.3** Robert Brons/Biological Photo Service; **33.7.4** Commonwealth of Australia (GBRMPA); **33.7.5** Neil G. McDaniel/Photo Researchers; **33.8.2** Robert Brons/Biological Photo Services; **33.9.2** Ed Robinson/Pacific Stock/PhotoLibrary; **33.11.2** CDC; **33.12.2** Eye of Science/Photo Researchers; **33.13.2** W. I. Walker/Photo Researchers; **33.14.2** Hecker/Sauer/AGE Fotostock; **33.14.3** Kåre Telnes/Image Quest Marine; **33.16** Jeff Foott/Tom Stack and Associates; **33.17.2** Amruta Bhelke/Dreamstime.com; **33.17.3** Corbis; **33.19** H. W. Pratt/Biological Photo Service; **33.21.2** photonimo/iStockphoto; **33.21.3** Mark Conlin/Image Quest Marine; **33.21.4** Jonathan Blair/Corbis; **33.22.2** Photograph courtesy of the U.S. Bureau of Fisheries (1919), and Illinois State Museum; **33.22.3** © Zoological Society of London (ZSL); **33.23.2** Peter Batson/Image Quest Marine; **33.24.2** A.N.T./NHPA/Photoshot; **33.25** Astrid & Hanns-Frieder Michler/Photo Researchers; **33.26.2** Reproduced with permission from A. Eizinger and R. Sommer, Max Planck Institut fur entwicklungsbiologie, Tubingen. Copyright 2000 American Association for the Advancement of Science. Cover 278(5337) 17 Oct 97; **33.27.2** SPL/Photo Researchers; **33.28** Collection of Dan Cooper; **33.29.2** Grenier JK, Garber TL, Warren R, Whitington PM, Carroll S. Evolution of the entire arthropod *Hox* gene set predated the origin and radiation of the onychophoran/arthropod clade. *Curr Biol.* 1997 Aug 1;7(8):547–53; Fig. 3c; **33.31** Mark Newman/FLPA; **33.32.2** Tim Flach/Stone/Getty Images; **33.32.3** Andrew Syred/Photo Researchers; **33.32.4** Eric Lawton/iStockphoto; **33.34.2** Premaphotos/Nature Picture Library; **33.34.3** Tom McHugh/Photo Researchers; **33.36** Meul/ARCO/Nature Picture Library; **33.37.2** John Shaw/Tom Stack and Associates; **33.37.3** John Shaw/Tom Stack and Associates; **33.37.4** John Shaw/Tom Stack and Associates; **33.37.5** John Shaw/Tom Stack and Associates; **33.37.6** John Shaw/Tom Stack and Associates; **33.38.2** Dr. John Brackenbury/Photo Researchers; **33.38.3** Perry Babin; **33.38.4** PREMAPHOTOS/Nature Picture Library; **33.38.5** Dante Fenolio/Photo Researchers; **33.38.6** John Cancalosi/Nature Picture Library; **33.38.7** Hans Christoph Kappel/Nature Picture Library; **33.38.8** Michael & Patricia Fogden/CORBIS; **33.38.9** CC-BY-SA photo: Bruce Marlin (www.cirrusimage.com/fly_whale-tail.htm); **33.39.2** Maximilian Weinzierl/Alamy; **33.39.3** Peter Herring/Image Quest Marine; **33.39.4** Peter Parks/Image Quest Marine; **33.40.2** Andrey Nekrasov/Image Quest Marine; **33.41** Daniel Janies; **33.42** Jeff Rotman/Photo Researchers; **33.43** Robert Harding World Imagery/Alamy; **33.44** Jurgen Freund/Nature Picture Library; **33.45** Hal Beral/Corbis.

Chapter 34 **34.1** Xian-guang H, Aldridge RJ, Siveter DJ, Siveter DJ, Xiang-hong F. New evidence on the anatomy and phylogeny of the earliest vertebrates. *Proceedings of the Royal Society of London-B-Biological Sciences,* Sept. 22, 2002; 269 (1503) 1865–1869; Fig. 1c; **34.4.2** Oxford Scientific/PhotoLibrary; **34.5.2** Robert Brons/Biological Photo Service; **34.8.2** Nanjing Institute of Geology and Palaeontology; **34.9.2** Tom McHugh/Photo Researchers; **34.10.2** A Hartl/AGE Fotostock; **34.10.3** Marevision/AGE Fotostock; **34.14.2** The Field Museum, #GEO82014; **34.15.2** Carlos Villoch/Image Quest Marine; **34.15.3** Masa Ushioda/Image Quest Marine; **34.15.4** Andy Murch/V&W/Image Quest Marine; **34.17.2** James D. Watt/Image Quest Marine; **34.17.3** Jez Tryner/Image Quest Marine; **34.17.4** George Grall/Getty Images; **34.17.5** Fred McConnaughey/Photo Researchers; **34.18.2** From "The oldest articulated osteichthyan reveals mosaic gnathostome characters." M. Zhu et al. *Nature.* 2009 Mar 26;458(7237):469–74; **34.19** Arnaz Mehta; **34.20.2** © Ted Daeschler/Academy of Natural Sciences/VIREO; **34.20.3** © Ted Daeschler/Academy of Natural Sciences/VIREO; **34.20.4** © Kalliopi Monoyios; **34.22.2** Alberto Fernandez/AGE Fotostock America, Inc. **34.22.3** Michael Fogden/Bruce Coleman/Photoshot; **34.22.4** Michael Fogden/Bruce Coleman/Photoshot; **34.23.2** John Cancalosi/Peter Arnold/PhotoLibrary; **34.23.3** Stephen Dalton/Photo Researchers; **34.23.4** Hans Pfletschinger/Peter Arnold/PhotoLibrary; **34.24** Michael Fogden/OSF/PhotoLibrary; **34.27** Nobumichi Tamura; **34.28** Michael Fogden/OSF/PhotoLibrary; **34.29.2** Natural Visions/Alamy; **34.29.3** Matt T. Lee; **34.29.5** Medford Taylor/National Geographic Image Collection; **34.29.6** Carl & Ann Purcell/Corbis; **34.30.2** Visceralimage/dreamstime; **34.30.3** Janice Sheldon; **34.32** Russell Mountford/Alamy; **34.33** DLILLC/Corbis; **34.34** Yufeng Zhou/iStockphoto; **34.35.1** McPHOTO/AGE Fotostock; **34.35.2** paolo barbanera/AGE Fotostock; **34.36** Gianpiero Ferrari/FLPA; **34.36** Gianpiero Ferrari/FLPA; **34.38.2** D. Parer and E. Parer Cook/Auscape International Proprietary Ltd. **34.38.3** Mervyn Griffiths/Commonwealth Scientific and Industrial Research Organization; **34.39.2** John Cancalosi/Alamy; **34.39.3** Wells Bert & Babs/OSF/PhotoLibrary; **34.42** Frans Lanting/Corbis; **34.44.2** Kevin Schafer/AGE Fotostock America, Inc. **34.44.3** J & C Sohns/Photolibrary; **34.45.2** Morales/AGE Fotostock America, Inc. **34.45.3** Anup Shah/ImageState/Alamy Images; **34.45.4** T. J. Rick/Nature Picture Library; **34.45.5** E. A. Janes/AGE Fotostock America, Inc. **34.45.6** Frans Lanting/Corbis; **34.47** Fossilized bone, partial skeleton, of Ardipithecus ramidus, articulated, with bones laid in their approximate positions. Housed in National Museum of Ethiopia, Addis Ababa. Photo © T. White 2009, From "Ardipithecus ramidus and the Paleobiology of Early Hominids." T. White et al. *Science.* 2009 Oct 2;326(5949):75–86; **34.48.2** John Reader/SPL/Photo Researchers; **34.48.3** John Gurche Studios; **34.49.2** Alan Walker @ National Museums of Kenya. Printed with permission; **34.51** David L Brill/Brill Atlanta; **34.52** C. Henshilwood & F. d'Errico/Professor Chris Henshilwood.

Chapter 35 **35.1** PD photo: John Walker (www.fourmilab.ch); **35.3** Robert & Linda Mitchell; **35.4.2** CC-BY-SA photo: Forest & Kim Starr; **35.4.3** Rob Walls/ Alamy; **35.4.4** YinYang/iStockphoto; **35.4.5** Robert Holmes/Corbis; **35.4.6** Geoff Tompkinson/Science Photo Library/Photo Researchers; **35.5.2** Donald Gregory Clever; **35.5.3** Gusto Productions/SPL/Photo Researchers; **35.5.4** Dorling Kindersley; **35.5.5** Aflo Foto Agency/Alamy; **35.7.2** Neil Cooper/Alamy; **35.7.3** Martin Ruegner/ jupiterimages; **35.7.4** Mike Zens/Corbis; **35.7.5** Jerome Wexler/Photo Researchers; **35.7.6** Kathy Piper/iStockphoto; **35.9.2** Purdue Extension Entomology; **35.10.2** Brian Capon; **35.10.4** © Clouds Hill Imaging/www.lastrefuge.co.uk; **35.10.6** Graham Kent, Pearson Science; **35.10.7.** Graham Kent, Pearson Science; **35.10.9** N.C. Brown Center for Ultrastructure Studies, SUNY-Environmental Science & Forestry, Syracuse, NY; **35.10.11** Graham Kent; **35.10.12** Reproduced with permission from "Plant Cell Biology on CD," by B E S Gunning, www.plantcellbiologyonCD.com; **35.10.13** Professor Ray F. Evert; **35.13.2** From "Arabidopsis TCP20 links regulation of growth and cell division control pathways." C. Li et al. *Proc Natl Acad Sci U S A.* 2005 Sep 6;102(36):12978–83. Epub 2005 Aug 25. Photo: Peter Doerner; **35.14.2** Natalie B. Bronstein; **35.14.3** Ed Reschke; **35.14.4** Ed Reschke; **35.15.2** Michael Clayton; **35.16.2** Michael Clayton; **35.17.2** Ed Reschke; **35.17.3** Ed Reschke; **35.18.2** Ed Reschke; **35.18.3** Ed Reschke; **35.19.2** Michael Clayton; **35.19.3** Alison W. Roberts; **35.21.2** Dr. Edward R. Cook; **35.23** California Historical Society Collection (CHS-1177), University of Southern California on behalf of the USC Specialized Libraries and Archival Collections; **p. 756** Reproduced by permission from Janet Braam, *Cell* 60 (9 February 1990): Cover. Copyright ©1990 Cell Press. Image courtesy of Elsevier Sciences Ltd. **35.25.2** From "Microtubule plus-ends reveal essential links between intracellular polarization and localized modulation of endocytosis during division-plane establishment in plant cells." P. Dhonukshe et al. *BMC Biol.* 2005 Apr 14;3:11; **35.26.2** From "The tangled-1 mutation alters cell division orientations throughout maize leaf development without altering leaf shape." L. G. Smith et al. *Development.* 1996 Feb;122(2):481–9; **35.26.3** From "The tangled-1 mutation alters cell division orientations throughout maize leaf development without altering leaf shape." L. G. Smith et al. *Development.* 1996 Feb;122(2):481–9; **35.28** From figure 1a in U. Mayer et al., *Development* 117 (1): 149–162. © 1993 The Company of Biologists Ltd. **35.29.2** From "Microtubule plus-ends reveal essential links between intracellular polarization and localized modulation of endocytosis during division-plane establishment in plant cells." P. Dhonukshe et al. *BMC Biol.* 2005 Apr 14;3:11; **35.29.3** B. Wells and K. Roberts; **35.30.2** Reproduced by permission from figure 1 in D. Hareven et al, *Cell* 84 (5): 735–744. Copyright © 1996. by Elsevier Science Ltd. **35.30.3** Reproduced by permission from figure 1 in D. Hareven et al, *Cell* 84 (5): 735–744. Copyright © 1996. by Elsevier Science Ltd. **35.31.2** Reproduced by permission from Figure 2g in Hung et al, *Plant Physiology* 117:73–84. Copyright © 1998 by the American Society of Plant Biologists. Image courtesy of John Schiefelbein/Univesity of Michigan; **35.32.2** Dr. Gerald D. Carr, PhD; **35.33.2** Dr. E. M. Meyerowitz and John Bowman, *Development* 112 1991:1–231.2. Division of Biology, California Institute of Technolgy; **35.33.3** Dr. E. M. Meyerowitz and John Bowman, *Development* 112 1991:1–231.2. Division of Biology, California Institute of Technolgy.

Chapter 36 **36.1** Peggy Heard/FLPA/Alamy; **36.3.2** Rolf Rutishauser; **36.3.3** Rolf Rutishauser; **36.5.2** Dr. Jeremy Burgess/SPL/Photo Researchers; **p. 770, top** Nigel Cattlin/Holt Studios International/Photo Researchers; **p. 770, bottom** Nigel Cattlin/Holt Studios International/Photo Researchers; **36.11** Scott Camazine/Photo Researchers; **36.14.2** Graham Kent; **36.14.3** Graham Kent; **36.16.2** Mlane/Dreamstime.com; **36.16.3** Kate Shane, Southwest School of Botanical Medicine; **36.16.4** Frans Lanting/Corbis; **36.16.5** Natalie Bronstein; **36.16.6** Andrew de Lory/Dorling Kindersley; **36.16.7** Danita Delimont/Alamy; **36.19.2** M. H. Zimmermann, courtesy of Professor P. B. Tomlinson, Harvard University; **36.19.3** M. H. Zimmermann, courtesy of Professor P. B. Tomlinson, Harvard University; **36.19.4** M. H. Zimmermann, courtesy of Professor P. B. Tomlinson, Harvard University; **36.20.2** From "A coiled-coil interaction mediates cauliflower mosaic virus cell-to-cell movement." L. Stavolone et al. *Proc Natl Acad Sci U S A.* 2005 Apr 26;102(17):6219–24. Epub 2005 Apr 18.

Chapter 37 **37.1** Chris Mattison/Alamy; **37.2.2** USDA/ARS/Agricultural Research Service; **37.4** National Oceanic and Atmospheric Administration NOAA; **37.5** U.S. Geological Survey, Denver; **37.6** Kevin Horan/Stone/Getty Images; **37.9.2** White et al., Plant Physiology, June 2003; **37.11.2** Scimat/Photo Researchers; **37.11.3** E. H. Newcomb and S. R. Tandon/Biological Photo Service; **37.13.2** Hugues B. Massicotte, University of Northern British Columbia Ecosystem Science and Management Program, Prince George, B.C., Canada; **37.13.3** Mark Brundrett (http://mycorrhizas.info); **37.13.4** Mark Brundrett (http://mycorrhizas.info); **37.14.2** Elizabeth J. Czarapata/The Park People; **37.15.1** Wolfgang Kaehler/Corbis; **37.15.2** Ruud de Man/iStockphoto; **37.15.3** Kevin Schafer/Corbis; **37.15.4** Gary W. Carter/Corbis; **37.15.5** Kim Taylor and Jane Burton/Dorling Kindersley; **37.15.6** Biophoto Associates/Photo Researchers; **37.15.7** Philip Blenkinsop/Dorling Kindersley; **37.15.8** Paul A. Zahl/Photo Researchers; **37.15.9** Fritz Polking, Frank Lane Picture Agency/Corbis.

Chapter 38 **38.1** Pierre-Michel Blais; **38.3.2** Ed Reschke/Peter Arnold/PhotoLibrary; **38.3.3** Ed Reschke/Peter Arnold/PhotoLibrary; **38.3.4** David Scharf/Peter Arnold/PhotoLibrary; **38.4.2** Marianne Wiora; **38.4.3** Stephen Dalton/NHPA Limited/Photoshot; **38.4.4** Bjorn Rorslett Photographe; **38.4.5** Bjorn Rorslett Photographe; **38.4.7** Doug Backlund; **38.4.8** Martin Pieter Heigan; **38.4.9** Rolf Nussbaumer/Nature Picture Library; **38.4.10** Merlin D. Tuttle, Bat Conservation International, www.batcon.org; **38.5** Photo: © W. Barthlott/W. Rauh; **38.11.2** Kevin Schafer/Alamy; **38.11.3** Nature Production; **38.11.4** Brian Gordon Green/National Geographic Image Collection; **38.11.5** Steve Bloom Images/Alamy; **38.11.6** Aaron McCoy/Botanica/Photolibrary; **38.11.7** California Department of Food and Agriculture's Plant Health and Pest Prevention Services; **38.11.8** Kim A. Cabrera Photographer; **38.11.9** Steve Shattuck/CSIRO Entomology; **38.11.10** Alan Williams/ Alamy; **38.12.11** Dennis Frates/Alamy; **38.13.2** Marcel E. Dorken; **38.13.3** Marcel E. Dorken; **38.13.4** Nobumitsu Kawakubo, Gifu University, Japan; **38.14.2** Bruce Iverson, Photomicrography; **38.14.3** Bruce Iverson, Photomicrography; **38.14.4** Meriel G. Jones, University of Liverpool, UK; **38.15.2** Sinclair Stammers/Photo Researchers; **38.16.2** Andrew McRobb/Dorling Kindersley; **38.16.3** Andrew McRobb/Dorling Kindersley; **38.17** John Van Hasselt/Corbis.

Chapter 39 **39.1** *Plant Physiol.* 1999 Jul;120(3): Cover. By permission of the American Society of Plant Physiologists. Illustration by Niemeyer MI and Fernandez MC; **39.2** Natalie B. Bronstein; **39.7.2** "Regulation of Polar Auxin transport ATPIN1 in Arabidopsis Vascular Tissue," by Leo Galweiler, et al. *Science* 18 DEC 1998, vol. 282; **39.7.3** "Regulation of Polar Auxin transport ATPIN1 in Arabidopsis Vascular Tissue," by Leo Galweiler, et al. *Science* 18 DEC 1998, vol. 282; **39.9.2** Malcolm B. Wilkins, University of Glasgow, Glasgow, Scotland, U.K. **39.9.3** Malcolm B. Wilkins, University of Glasgow, Glasgow, Scotland, U.K. **39.9.4** Malcolm B. Wilkins, University of Glasgow, Glasgow, Scotland, U.K. **39.10.2** Dr. Richard Amasino; **39.10.3** Fred Jensen; **39.12.2** Mia Molvray; **39.12.3** Karen E. Koch; **39.14.2** Kurt Stepnitz, DOE

Plant Research Laboratory, Michigan State University; **39.14.3** Joe Kieber, University of North Carolina; **39.15.2** Ed Reschke; **39.16.3** Malcolm B. Wilkins, University of Glasgow, Glasgow, Scotland, U.K. **39.16.4** Malcolm B. Wilkins, University of Glasgow, Glasgow, Scotland, U.K. **39.17.2** Malcolm B. Wilkins, University of Glasgow, Glasgow, Scotland, U.K. **39.17.3** Malcolm B. Wilkins, University of Glasgow, Glasgow, Scotland, U.K. **39.17.4** Malcolm B. Wilkins, University of Glasgow, Glasgow, Scotland, U.K. **39.17.5** Malcolm B. Wilkins, University of Glasgow, Glasgow, Scotland, U.K. **39.17.6** Malcolm B. Wilkins, University of Glasgow, Glasgow, Scotland, U.K. **39.20.2** Malcolm B. Wilkins, University of Glasgow, Glasgow, Scotland, U.K. **39.20.3** Malcolm B. Wilkins, University of Glasgow, Glasgow, Scotland, U.K. **39.24.2** Michael Evans, Ohio State University; **39.24.3** Michael Evans, Ohio State University; **39.24.4** Michael Evans, Ohio State University; **39.24.5** Michael Evans, Ohio State University; **39.25** Reproduced by permission from Janet Braam, *Cell* 60 (9 February 1990): Cover. Copyright © 1990 Cell Press. Image courtesy of Elsevier Sciences Ltd. **39.26.2** Martin Shields/Photo Researchers; **39.26.3** Martin Shields/Photo Researchers; **39.26.4** From K. Esau. "Anatomy of Seed Plants," 2nd ed. (New York: John Wiley and Sons, 1977), fig. 19.4, p.358; **39.26.5** From K. Esau. "Anatomy of Seed Plants," 2nd ed. (New York: John Wiley and sons, 1977), fig. 19.4, p.358; **39.27.2** J. L. Basq and M. C. Drew; **39.27.3** J. L. Basq and M. C. Drew; **39.29.2** New York State Agricultural Experiment Station (NYSAES)/Cornell.

Chapter 40 **40.1** JOEL SARTORE/National Geographic Stock; **40.2.2** Brandon Cole/PhotoLibrary; **40.2.3** Duncan Usher/Alamy; **40.2.4** Frank Greenaway/Dorling Kindersley/Getty Images; **40.4.2** Susumu Nishinaga/Photo Researchers; **40.4.3** Eye of Science/Photo Researchers; **40.4.4** Susumu Nishinaga/Photo Researchers; **40.5.7** CNRI/SPL/Photo Researchers; **40.5.11** Dr. Gopal Murti/SPL/Photo Researchers; **40.5.13** Chuck Brown/Photo Researchers; **40.5.15** Nina Zanetti; **40.5.17** Nina Zanetti; **40.5.19** Nina Zanetti; **40.5.21** Alamy; **40.5.24** Nina Zanetti; **40.5.26** Ed Reschke/Peter Arnold Images/PhotoLibrary; **40.5.28** Manfred Kage/Peter Arnold/Photolibrary; **40.5.31** Ulrich Gartner; **40.5.33** Thomas Deerinck/National Center for Microscopy and Imaging Research, University of California, San Diego; **40.10.2** Patricio Robles Gil/naturepl.com; **40.10.3** Matt T. Lee; **40.13** Robert Ganz; **40.18** Jeff Rotman/Alamy.

Chapter 41 **41.1** Michael deYoung/Corbis; **41.2** Roland Seitre/Peter Arnold/PhotoLibrary; **41.3** Stefan Huwiler/Rolf Nussbaumer Photography/Alamy; **41.5.2** cameilia/Shutterstock; **41.6.2** Hervey Bay Whale Watch (www.herveybaywhalewatch.com.au); **41.6.4** Thomas Eisner; **41.6.5** Photo Lennart Nilsson/Scanpix; **41.6.6** Gunter Ziesler/Peter Arnold/PhotoLibrary; **41.11.2** Visuals Unlimited/Corbis; **41.17.2** Fritz Polking/Peter Arnold, Inc./Alamy; **41.17.3** EyeWire Collection/Photodisc/Getty Images; **41.22** Photo courtesy of The Jackson Laboratory, Bar Harbor, Maine; **41.23** Wolfgang Kaehler/Corbis.

Chapter 42 **42.1** Stephen Dalton/Photo Researchers; **42.2.2** Reinhard dirscheri/PhotoLibrary; **42.2.3** Eric Grave/Photo Researchers; **42.10.2** From "Human Histology Photo CD." Image courtesy Indigo Instruments (www.indigo.com); **42.10.3** Photo Lennart Nilsson/Scanpix; **42.16** Biophoto Associates/Photo Researchers; **42.18.2** Eye of Science/Photo Researchers; **42.22.2** Peter Batson/Image Quest Marine; **42.22.3** Olgysha/Shutterstock; **42.22.4** Jez Tryner/Image Quest Marine; **42.24.2** Prepared by Dr. Hong Y. Yan, University of Kentcky and Dr. Peng Chai, University of Texas; **42.25.2** Motta & Macchiarelli/Anatomy Dept., Univ. La Sapienza, Rome/Photo Researchers; **42.27.2** Hans-Rainer Duncker, University of Giessen, Germany.

Chapter 43 **43.1** Biology Media/Science Source/Photo Researchers; **43.4** Dominique Ferrandon; **43.21.2** Steve Gschmeissner/Photo Researchers; **43.23** CNRI/Photo Researchers; **43.26** The Laboratory of Structural Cell Biology, headed by Stephen C. Harrison, Harvard Medical School/HHMI.

Chapter 44 **44.1** David Wall/Alamy; **44.4** Mark Conlin/Image Quest Marine; **44.5.2** Dr. John Crowe, University of California, Davis; **44.5.3** Dr. John Crowe, University of California, Davis; **44.9** AFP/Getty Images; **44.14.5** Steve Gschmeissner/Photo Researchers; **44.17** John Cancalosi/Peter Arnold Images/PhotoLibrary; **44.18** Michael Fodgen/OSF/PhotoLibrary.

Chapter 45 **45.1** Ralph A. Clevenger/Corbis; **p. 974** Stuart Wilson/Photo Researchers; **45.3** Volker Witte; **45.18.2** Astier/BSIP/Photo Researchers; **45.19.2** Photoshot Holdings Ltd/Alamy; **45.19.3** Jurgen & Christine Sohns/FLPA.

Chapter 46 **46.1** David Doubilet/Getty Images; **46.2** David Wrobel; **46.4** Chris Wallace Photography, photographersdirect.com; **46.5.2** P. de Vries, courtesy of David Crews; **46.6** Andy Sands/naturepl.com; **46.7** John Cancalosi/Peter Arnold/PhotoLibrary; **46.17.2** Photo Lennart Nilsson/Scanpix; **46.17.3** Photo Lennart Nilsson/Scanpix; **46.17.4** Photo Lennart Nilsson/Scanpix.

Chapter 47 **47.1** Photo Lennart Nilsson/Scanpix; **47.4.2** Vacquier VD, Payne JE. Methods for quantitating sea urchin sperm-egg binding. *Exp Cell Res.* 1973 Nov;82(1):227–35; **47.4.3** Vacquier VD, Payne JE. Methods for quantitating sea urchin sperm-egg binding. *Exp Cell Res.* 1973 Nov;82(1):227–35; **47.4.4** Vacquier VD, Payne JE. Methods for quantitating sea urchin sperm-egg binding. *Exp Cell Res.* 1973 Nov;82(1):227–35; **47.4.5** Vacquier VD, Payne JE. Methods for quantitating sea urchin sperm-egg binding. *Exp Cell Res.* 1973 Nov;82(1):227–35; **47.4.7** Hafner, M., Petzelt, C., Nobiling, R., Pawley, J., Kramp, D. and G. Schatten. Wave of Free Calcium at Fertilization in the Sea Urchin Egg Visualized with Fura-2. *Cell Motil. Cytoskel.*, 9:271–277 (1988); **47.4.8** Hafner, M., Petzelt, C., Nobiling, R., Pawley, J., Kramp, D. and G. Schatten. Wave of Free Calcium at Fertilization in the Sea Urchin Egg Visualized with Fura-2. *Cell Motil. Cytoskel.*, 9:271–277 (1988); **47.4.9** Hafner, M., Petzelt, C., Nobiling, R., Pawley, J., Kramp, D. and G. Schatten. Wave of Free Calcium at Fertilization in the Sea Urchin Egg Visualized with Fura-2. *Cell Motil. Cytoskel.*, 9:271–277 (1988); **47.4.10** Hafner, M., Petzelt, C., Nobiling, R., Pawley, J., Kramp, D. and G. Schatten. Wave of Free Calcium at Fertilization in the Sea Urchin Egg Visualized with Fura-2. *Cell Motil. Cytoskel.*, 9:271–277 (1988); **47.6.2** George von Dassow; **47.6.3** George von Dassow; **47.6.4** George von Dassow; **47.6.5** George von Dassow; **47.7.2** Jürgen Berger/Max Planck Institute for Developmental Biology, Tübingen, Germany; **47.7.3** Andrew J. Ewald, Johns Hopkins Medical School; **47.9.2** Charles A. Ettensohn; **47.13.2** Huw Williams; **47.13.3** Thomas Poole, SUNY Health Science Center; **47.14.2** Dr. Keith Wheeler/Photo Researchers; **47.17.2** Hiroki Nishida, *Developmental Biology* 121 (1987): 526. Reprinted by permission of Academic Press; **47.17.3** Hiroki Nishida, *Developmental Biology* 121 (1987): 526. Reprinted by permission of Academic Press; **47.18.2** J. E. Sulston and H. R. Horvitz, *Dev. Biol.* 56 (1977):110–156; **47.19.2** Adapted from Strome (International Review of Cytology 114: 81–123, 1989); **47.20.2** Adapted from Strome (International Review of Cytology 114: 81–123, 1989); **47.20.3** Adapted from Strome (International Review of Cytology 114: 81–123, 1989); **47.20.4** Adapted from Strome (International Review of Cytology 114: 81–123, 1989); **47.20.5** Adapted from Strome (International Review of Cytology 114: 81–123, 1989); **47.24.2** Kathryn W. Tosney, University of Michigan; **47.25.2** Dennis Summerbell.

Chapter 48 **48.1** Marinethemes.com; **48.2.2** David Fleetham/Alamy; **48.6.2** Thomas Deerinck; **48.13.2** Bear, Connors, and Paradiso, "Neuroscience: Exploring the Brain" © 1996, p. 43; **48.16.2** Edwin R. Lewis, University of California at Berkeley.

Chapter 49 **49.1** Brainbow mouse cerebellum. Image by Tamily Weissman, Harvard University. The Brainbow mouse was produced by Livet J, Weissman TA, Kang H, Draft RW, Lu J, Bennis RA, Sanes JR, Lichtman JW. *Nature* (2007) 450:56–62; **49.6.2** N. Kedersha/Photo Researchers; **49.9** Larry Mulvehill/Corbis; **49.14.2** From "A functional MRI study of happy and sad affective states induced by classical music."; . . . M. T. Mitterschiffthaler et al. *Hum Brain Mapp.* 2007 Nov;28(11):1150–62; **49.16.2** Marcus E. Raichle, M.D., Washington University Medical Center; **p. 1075** From "Dr. Harlow's Case of Recovery from the passage of an Iron Bar through the Head." H. Bigelow. *Am. Journal of the Med. Sci.* July 1850;XXXIX. Images from the History of Medicine (NLM); **49.21** Image by Sebastian Jessberger. Fred H. Gage, Laboratory of Genetics LOG-G, The Salk Institute for Biological Studies; **49.24.2** Martin M. Rotker/Photo Researchers.

Chapter 50 **50.1** Kenneth Catania; **50.6.1** CSIRO; **50.6.2** R. A. Steinbrecht; **50.7.2** James Gerholdt/Photolibrary; **50.7.3** Splashdown Direct/ OSF/PhotoLibrary; **50.9.2** From Richard Elzinga, "Fundamentals of Entomology" 3rd. © 1987, p. 185. Reprinted by permission of Prentice-Hall, Upper Saddle River, NJ; **50.10.2** SPL/Photo Researchers; **50.16.2** USDA/APHIS Animal and Plant Health Inspection Service; **50.17.3** STEVE GSCHMEISSNER/SPL/Photo Researchers; **50.21** From "Gene therapy for red-green colour blindness in adult primates."; . . . K. Mancuso et al. *Nature.* 2009 Oct 8; 461(7265):784–7. Photo: Neitz Laboratory; **50.26.2** Clara Franzini-Armstrong, University of Pennsylvania; **50.27.2** Courtesy of Dr. H. E. Huxley; **50.27.3** Courtesy of Dr. H. E. Huxley; **50.27.4** Courtesy of Dr. H. E. Huxley; **50.33** George Cathcart Photography, photographersdirect.com; **50.38** Dave Watts/NHPA/Photo Researchers; **50.39** Vance A. Tucker.

Chapter 51 **51.1** Michael Nichols/National Geographic/Getty Images; **51.3** Susan Lee Powell; **51.5.2** Kenneth Lorenzen, UC Davis; **51.7.2** Thomas McAvoy/Life Magazine/Getty Images; **51.7.3** Operation Migration Inc.; **51.9.2** Lincoln Brower, Sweet Briar College; **51.9.3** Lincoln Brower, Sweet Briar College; **51.9.4** Lincoln Brower, Sweet Briar College; **51.11** Clive Bromhall/OSF/PhotoLibrary; **51.12.2** Alissa Crandall/Corbis; **51.12.3** Richard Wrangham; **51.15.2** Matt T. Lee; **51.15.3** David Osborn/Alamy; **51.15.4** Bill Schmoker; **51.16.2** James D Watt/Image Quest Marine; **51.17** Courtesy of Gerald S. Wilkinson; from G.S. Wilkinson and G. N. Dodson, in J. Choe and B. Crespi (eds)., "The Evolution of Mating Systems in Insects and Arachnids," Cambridge University Press, Cambridge (1997), pp. 310–328; **51.18** Cyril Laubscher/Dorling Kindersley; **51.21** Martin Harvey/Peter Arnold Images/PhotoLibrary; **51.22** Erik Svensson, Lund University, Sweden; **51.23.2** Robert Pickett/Corbis; **51.24** Lowell L. Getz and Lisa Davis; **51.25** Rory Doolin; **51.27** Jennifer Jarvis; **51.29.2.** Stephen J. Krasemann/Peter Arnold/PhotoLibrary.

Chapter 52 **52.1** Dr. Paul A. Zahl/Photo Researchers; **52.2.2** James D. Watt/Stephen Frink Collection/Alamy; **52.2.3** Gianni Tortoli/Photo Researchers; **52.2.4** Tom Bean/Corbis; **52.2.5** B. Tharp/Photo Researchers; **52.2.6** Yann Arthus-Bertrand/Corbis; **52.2.7** NASA/Goddard Space Flight Center; **52.11.1** JTB Photo Communications, Inc./Alamy; **52.11.2** imagebroker/Alamy; **52.12.2** Frans Lanting/Corbis; **52.12.4** Gordon Whitten/Corbis; **52.12.6** Wolfgang Kaehler/Corbis; **52.12.8** The California Chaparral Institute (http://californiachaparral.net). Photo supplied by Richard Halsey; **52.12.10** Tom Bean/Image Bank/ Getty Images; **52.12.12** Shutterstock; **52.12.14** Kennan Ward/Corbis; **52.12.16** Darrell Gulin/Corbis; **52.16.1** Allen Russell/PhotoLibrary; **52.16.2** AfriPics.com/Alamy; **52.16.3** David Tipling/Nature Picture Library; **52.16.4** Ron Watts/Corbis; **52.16.5** Photononstop/SuperStock; **52.16.6** James Randklev/Image Bank/Getty Images; **52.16.7** Stuart Westmorland/Corbis; **52.16.8** Stuart Westmorland/Corbis; **52.16.9** Digital Vision/Getty Images; **52.16.10** William Lange/Woods Hole Oceanographic Institution; **52.17.2** Geoff Dann/Dorling Kindersley; **52.19.2** Peter Llewellyn/Alamy; **52.21** Daniel Mosquin.

Chapter 53 **53.1** Arpat Ozgul; **53.2** Todd Pusser/Naturepl.com; **53.4.2** Bernard Castelein/Nature Picture Library/Alamy; **53.4.3** Frans Lanting/Corbis; **53.4.4** Niall Benvie/Corbis; **53.8.2** Hansjoerg Richter/iStockphoto; **53.11** Photodisc/White/PhotoLibrary; **53.12** Tom Bean/Corbis; **53.13.2** H. Willcox/Wildlife Picture/Peter Arnold/Photolibrary; **53.14.2** Jean Louis Batt/Taxi/Getty Images; **53.14.3** Christine Osborne/ Corbis; **53.14.4** Edward Parker/Alamy; **53.17.2** fotoVoyager/iStockphoto; **53.17.3** Adrian Bailey/Aurora Photos; **53.17.4** Joe Raedle/Getty Images; **53.17.5** Patrick Clayton, www.fisheyeguyphotography.com; **53.17.6** JOZSEF SZENTPETERI/NGS Image Collection; **53.17.7** Andrew Syred/Photo Researchers; **53.17.8** Nicholas Bergkessel, Jr./ Photo Researchers; **53.19.2** Joe McDonald/Corbis; **53.20.2** Robert Kay; **53.21.2** Niclas Fritzén.

Chapter 54 **54.1** Hal Beral VWPics/SuperStock; **54.2.2** Joseph T. Collins/Photo Researchers; **54.2.3** National Museum of Natural History/Smithsonian Institution; **p. 1196** Frank W Lane/FLPA; **54.5.2** Barry Mansell/Nature Picture Library; **54.5.3** Fogden/Corbis; **54.5.4** Stephen J. Krasemann/Photo Researchers; **54.5.5** Robert Pickett/Papilio/Alamy; **54.5.6** Edward S. Ross, California Academy of Sciences; **54.5.7** © James K. Lindsey; **54.6** Douglas Faulkner/Photo Researchers; **54.7.2** Fogden/Corbis; **54.7.3** Dan Janzen, Department of Biology, University of Pennsylvania; **54.8** Peter Johnson/Corbis; **54.9.2** Sally D. Hacker; **54.12** Cedar Creek Ecosystem Science Reserve, University of Minnesota; **54.17.2** Genny Anderson; **54.19** SuperStock; **54.21.2** Ron Landis Photography, www.ronlandisphotography.co; **54.21.3** Scott T. Smith/Corbis; **54.22.2** Charles Mauzy/Corbis; **54.22.3** Keith Boggs; **54.22.4** Glacier Bay National Park Photo/Glacier Bay National Park and Preserve; **54.22.5** Terry Donnelly, Mary Liz Austin; **54.24.2** R. Grant Gilmore, Dynamac Corporation; **54.24.3** Lance Horn, National Undersea Research Center, University of North Carolina-Wilmington; **54.29** Nelish Pradhan, Bates College, Lewiston, ME; **54.30** Josh Spice.

Chapter 55 **55.1** Hassan Basagic; **55.2** Stone Nature Photography/Alamy; **55.3** Justus de Cuveland/AGE Fotostock; **55.16.2** Hubbard Brook Research Foundation/USDA Forest Service; **55.16.3** USDA Forest Service; **55.17.2** Mark Gallagher;

55.17.3 Mark Gallagher; **55.18.2** U.S. Department of Energy; **55.19.2** Stewart Rood, University of Lethbridge; **55.19.3** Photo provided by Kissimmee Division staff, South Florida Water Management District (WPB); **55.19.4** Tim Day, Xcluder Pest Proof Fencing Company; **55.19.5** Daniel H. Janzen, University of Pennsylvania; **55.19.6** Bert Boekhoven; **55.19.7** Jean Hall/Holt Studios/Photo Researchers; **55.19.8** Kenji Morita/Environment Division, Tokyo Kyuei Co., Ltd.

Chapter 56 **56.1** Stephen J Richards; **56.2** Wayne Lawler/Ecoscene/Corbis; **56.4.2** Neil Lucas/Nature Picture Library; **56.4.3** Mark Carwardine/Still Pictures/Peter Arnold/ PhotoLibrary; **56.4.4** Nazir Foead; **56.5** Merlin D. Tuttle, Bat Conservation International, www.batcon.org; **56.6** Scott Camazine/Photo Researchers; **56.7** Michael Edwards/Getty Images; **56.8.2** Bruce Cowell, www.brucecowellphotographer .com; **56.8.3** Robert Ginn/PhotoEdit Inc. **56.9** Benezeth Mutayoba, photo provided by the University of Washington; **56.10** Richard Vogel/Liaison/Getty Images; **56.13.2** William Ervin/SPL/Photo Researchers; **56.14** Craighead Environmental Research Institute; **56.15.2** Tim Thompson/Corbis; **56.15.3** Chuck Bargeron, University of Georgia; **56.15.4** William D. Boyer, USDA Forest Service; **56.16.2** Yann Arthus-Bertrand/Corbis; **56.16.3** James P. Blair/National Geographic Image Collection; **56.17** R. O. Bierregaard, Jr., Biology Dept., University of North Carolina, Charlotte; **56.18** SPL/Photo Researchers; **56.21.2** Edwin Giesbers/naturepl.com; **56.22.2** Mark Chiappone and Steven Miller, Center for Marine Science, University of North Carolina-Wilmington, Key Largo, FL; **56.23** Nigel Cattlin/Photo Researchers; **56.24.2** NASA; **56.24.3** NASA; **56.26** Erich Hartmann/Magnum Photos; **56.28** Prof. William H. Schlesinger; **56.31.2** NASA; **56.31.3** NASA; **56.33.2** Serge de Sazo/ Photo Researchers; **56.33.3** Hilde Jensen, University of Tubingen/Nature Magazine/ AP Photo; **56.33.4** Gabriel Rojo/Nature Picture Library; **56.33.5** Titus Lacoste/Getty Images. Appendix A **p. A-5** OMIKRON/Science Source/Photo Researchers; **p. A-8** John N. A. Lott/Biological Photo Service; **p. A-10, top** Biophoto/Photo Researchers; **p. A-10, 2nd from top** Conly L. Rieder; **p. A-11** USDA/ARS/Agricultural Research Service; **p. A-35** Peter Kitin; Appendix E **p. E-1, top left** Dr. Kari Lounatmaa/ Photo Researchers; **p. E-1, bottom left** Eye of Science/Photo Researchers; **p. E-1, middle** M. I. Walker/NHPA/Photoshot; **p. E-1 right** Kathy Piper/iStockphoto; **p. E-2, left** Douglas Adams/iStockphoto; **p. E-2, right** McPHOTO/AGE Fotostock.

Illustration and Text Credits

4.6b, 9.9, 17.17b and **c** are adapted from C. K. Matthews and K. E. van Holde, *Biochemistry,* 2nd ed. Copyright © 1996 Pearson Education, Inc., publishing as Pearson Benjamin Cummings. **4.7, 6.6b, 11.7, 11.12, 17.11, 18.25, 20.8, 21.9, and 21.10** are adapted from W. M. Becker, J. B. Reece, and M. F. Poenie, *The World of the Cell,* 3rd ed. Copyright © 1996 Pearson Education, Inc., publishing as Pearson Benjamin Cummings. **Table 6.1a** Adapted from W. M. Becker, L. J. Kleinsmith, and J. Hardin, *The World of the Cell,* 4th ed. p. 753. Copyright © 2000 Pearson Education, Inc., publishing as Pearson Benjamin Cummings. **6.8** and **6.23a** and cell organelle drawings in **6.11** and **6.12** are adapted from illustrations by Tomo Narashima in E. N. Marieb, *Human Anatomy and Physiology,* 5th ed. Copyright © 2001 Pearson Education, Inc., publishing as Pearson Benjamin Cummings. **6.9a, 50.12,** and **50.13** are also from *Human Anatomy and Physiology,* 5th ed. Copyright © 2001 Pearson Education, Inc., publishing as Pearson Benjamin Cummings. **30.4, 30.13i,** and **39.13** are adapted from M. W. Nabors, *Introduction to Botany,* Copyright © 2004 Pearson Education, Inc., publishing as Pearson Benjamin Cummings. **42.30a, 46.16, 49.8, 49.10, 50.26, and 50.30** are adapted from E. N. Marieb, *Human Anatomy and Physiology,* 4th ed. Copyright © 1998 Pearson Education, Inc., publishing as Pearson Benjamin Cummings. **42.30a** From Campbell et al., *Biology: Concepts and Connections,* 6th ed., fig. 22.10, p. 462. Copyright © 2009 Pearson Education, Inc., publishing as Pearson Benjamin Cummings. **43.8** Adapted from Gerard J. Tortora, Berdell R. Funke, and Christine L. Case. 1998. *Microbiology: An Introduction,* 6th ed. Copyright © 1998 Pearson Education, Inc., publishing as Pearson Benjamin Cummings. **44.8** and **51.8** are adapted from L. G. Mitchell, J. A. Mutchmor, and W. D. Dolphin. *Zoology.* Copyright © 1988 Pearson Education, Inc., publishing as Pearson Benjamin Cummings.

Chapter 1 **1.25** Map provided courtesy of David W. Pfennig, University of North Carolina at Chapel Hill; **1.27** Data in bar graph based on D. W. Pfennig et al. 2001. Frequency-dependent Batesian mimicry. *Nature 410:*323.

Chapter 2 **2.2 (bottom)** Reprinted by permission of Macmillan Publishers Ltd. *Nature.* M.E. Frederickson et al. 'Devil's gardens' bedeviled by ants, *437:*495, 9/22/05. Copyright © 2005.

Chapter 3 **3.8a** Adapted from *Scientific American,* Nov. 1998, p.102.

Chapter 5 **5.12** Adapted from *Biology: The Science of Life,* 3rd ed. by Robert Wallace et al. Copyright © 1991. Reprinted by permission of Pearson Education, Inc; **5.15 and 5.20F** PDB ID 1CGD: J. Bella, B. Brodsky, and H. M. Berman. 1995. Hydration structure of a collagen peptide. *Structure 3*: 893–906; **5.18** Adapted from D. W. Heinz et al. 1993. How amino-acid insertions are allowed in an alpha-helix of T4 lysozyme. *Nature 361:*561; **5.20D** PDB ID 3GS0: Palaninathan, S.K., Mohamedmohaideen, N.N., Orlandini, E., Ortore, G., Nencetti, S., Lapucci, A., Rossello, A., Freundlich, J.S., Sacchettini, J.C. 2009. Novel transthyretin amyloid fibril formation inhibitors: synthesis, biological evaluation, and X-ray structural analysis. *Public Library of Science One 4*: e6290–e6290; **5.20G, 21.10b, Un42.1** PDB ID 2HHB: G. Fermi, M. F. Perutz, B. Shaanan, R. Fourme. 1984. The crystal structure of human deoxyhaemoglobin at 1.74 A resolution. *J.Mol.Biol. 175*: 159–174.

Chapter 7 **7.9** PDB ID 3HAO: N. H. Joh, A. Oberai, D. Yang, J. P. Whitelegge, J. U. Bowie. 2009. Similar energetic contributions of packing in the core of membrane and water-soluble proteins. *J. Am.Chem.Soc. 131*: 10846–10847.

Chapter 8 **8.18** PDB ID 3elf: D H. Juers, B. Rob, M. L. Dugdale, N. Rahimzadeh, C. Giang, M. Lee, B. W. Matthews, R.E. Huber. 2009. Direct and indirect roles of His-418 in metal binding and in the activity of beta-galactosidase (E. coli). *Protein Sci. 18:* 1281–1292; **8.19** PDB ID 1MDY: P. C. Ma, M.A. Rould, H. Weintraub, C.O. Pabo. 1994. Crystal structure of MyoD bHLH domain-DNA complex: perspectives on DNA recognition and implications for transcriptional activation. *Cell* (Cambridge, Mass.) *77*: 451–459; **8.20** Figures 4a and 4e from "A common allosteric site and mechanism in caspases" by J.M. Scheer et al., in *Proceedings of the National Academy of Sciences, 103,* no. 20: 7595–7600, May 16, 2006. Copyright © 2006 National Academy of Sciences, U.S.A. Used by permission.

Chapter 9 **9.5** From *Molecular Biology of the Cell,* 4th edition, by Bruce Alberts et al., fig. 2.69, p. 92. Copyright © 2002 by Bruce Alberts, Alexander Johnson, Julian Lewis, Martin Raff, Keith Roberts, and Peter Walter. Used by permission

Chapter 10 **10.13b** Figure 1a from "Architecture of the photosynthetic Oxygen-evolving center" by K.N. Ferreira et al., in *Science, 303,* No. 5665: 1831–1838, March 19, 2004. Copyright © 2004, The American Association for the Advancement of Science. Reprinted with permission from AAAS; **10.15** Adapted from Richard and David Walker. Energy, Plants, and Man, fig. 4.1, p. 69. Sheffield: University of Sheffield. Oxygraphics http:// www.oxygraphics.co.uk. © Richard Walker. Used with permission.

Chapter 12 **12.13** From *Molecular Biology of the Cell,* 4th edition, by Bruce Alberts et al., fig. 18.41, p. 1059. Copyright © 2002 by Bruce Alberts, Alexander Johnson, Julian Lewis, Martin Raff, Keith Roberts, and Peter Walter. Used by permission.

Chapter 17 **17.13** Figure 10-45 from *Principles of Cell and Molecular Biology,* 2nd edition, by Valerie M. Kish and Lewis J. Kleinsmith. Copyright © 1995 HarperCollins College Publishers. Reprinted by permission of Pearson Education.

Chapter 18 **18.15a** Figure 1d from "An Abundant Class of Tiny RNAs with Probable Regulatory Roles in *Caenorhabditis elegans"* by N.C. Lau et al., in *Science, 294,* No. 5543: 858–862, Oct. 1, 2001. Copyright © 2001, The American Association for the Advancement of Science. Reprinted with permission from AAAS.

Chapter 20 **20.10** Figure 15.24, p. 481, from *Genetics,* 5th ed., by Peter Russell. Copyright © 1998 Pearson Education, Inc., publishing as Pearson Benjamin Cummings. Used by permission of the publisher.

Chapter 21 **21.2** Adapted from a figure by Chris A. Kaiser and Erica Beade; **21.5b** Figure 2B from "The Genetic Landscape of a Cell" by M. Costanzo et al., in *Science, 327,* No. 5964: 425–431, Jan. 22, 2010. Copyright © 2010, The American Association for the Advancement of Science. Reprinted with permission from AAAS; **21.17** Adapted from an illustration by William McGinnis in Peter Radetsky, "The homeobox: Something very precious that we share with flies, from egg to adult." Bethesda, MD: Howard Hughes Medical Institute, 1992, p. 92. Reprinted by permission of William McGinnis; **21.18** Adapted from M. Akam, *"Hox* genes and the evolution of diverse body plans," *Philosophical Transactions of the Royal Society B,* 1995, *349:* 313–319, fig. 3. © Royal Society of London. Reprinted by permission.

Chapter 22 **22.8** © Utako Kikutani 2007. Used with permission; **22.13** Figure 4a from "Host Race Radiation in the Soapberry Bug: Natural History with the History" by S.P. Carroll and C. Boyd, in Evolution, Vol. 46 No. 4, p. 1060, Aug. 1992. Reproduced with permission of Blackwell Publishing Ltd.; **22.14** Reprinted from *Trends in Microbiology, 16,* Issue 8: 361–369. B.A. Diep and M. Otto, "The role of virulence determinants in community-associated MRSA pathogenesis," Copyright © 2008, with permission from Elsevier.

Chapter 23 **23.5** Figure 3 from "Genetic mechanisms for adapting to a changing environment" by D.A. Powers et al., from *Annual Review of Genetics, 25,* Dec. 1991. Copyright © 1991 by Annual Reviews. Reprinted by permission; **23.11** Figure 20.6 (maps only) from *Discover Biology,* Second Edition by Michael L. Cain, Hans Damman, Robert A. Lue & Carol Kaesuk Loon, Editors. Copyright © 2002 by Sinauer Associates, Inc. Used by permission of W.W. Norton & Company, Inc.; **23.12** Reprinted by permission from Macmillan Publishers Ltd.: *Nature.* E. Postma and A. J. van Noordijk, Gene flow maintains a large genetic difference in clutch size at a small spatial scale, *433,* 1/6/05. Copyright © 2005; **23.14** Adapted from many sources including D. J. Futuyma. 2005. *Evolution,* fig. 11.3. Sunderland, MA: Sinauer Associates and from R. L. Carroll, 1988. *Vertebrate Paleontology and Evolution.* W.H. Freeman & Co.; **23.16** Adapted from A. M. Welch et al. 1998. Call duration as an indicator of genetic quality in male gray tree frogs. *Science* 280:1928–1930; **23.17** Adapted from A. C. Allison. 1961. Abnormal hemoglobin and erythrocyte enzyme-deficiency traits. In *Genetic Variation in Human Populations,* ed. G.A. Harrison. Oxford: Elsevier Science and from S. I. Hay et al., A world malaria map: *Plasmodium falciparum* endemicity in 2007. *PLoS Medicine 6:* fig. 3, p. 291; **23.18** Figure 2a from "Frequency-Dependent Natural Selection in the Handedness of Scale-Eating Cichlid Fish" by Michio Hori in *Science, 260,* No. 5105: 216–219, April 9, 1993. Copyright © 1993, The American Association for the Advancement of Science. Reprinted with permission from AAAS; **Un 23.2** Data from R. K. Koehn and T. J. Hilbish. 1987. The adaptive importance of genetic variation. *American Scientist* 75: 134–141.

Chapter 24 **24.7** Figure 3 from "Ecological Speciation in *Gambusia* Fishes" by R.B. Langerhans et al., from *Evolution, 61,* No. 9, July 2007, published by The Society for the Study of Evolution. Copyright © 2007 R.B. Langerhans, M.E. Gifford, E.O. Joseph. Reprinted by permission; **24.9** From figure 2 in "Correspondence between sexual isolation and allozyme differentiation" *Proceedings of the National Academy of Science, 87:* 2715–2719, 1990, p. 2718. Copyright © 1990 Stephen G. Tilley, Paul A. Verrell, Steven J. Arnold. Used with permission; **24.10a** Adapted from D. M. B. Dodd, 1989. Reproductive isolation as a consequence of adaptive divergence in *Drosophila pseudoobscura. Evolution* 43: 1308–1311; **24.13** *Hybrid Zone and the Evolutionary Process* edited by R.G. Harrison (1993): Map of Bombina hybrid zone (p. 263) and figure 10.1 (p. 278) from chapter "Analysis of hybrid zones with bombina" by J. M. Szymura. By permission of Oxford University Press; **24.15** Reprinted by permission of Macmillan Publishers Ltd.: *Nature.* G. P. Saetre et al. A sexually selected character displacement in flycatchers reinforces premating isolation, *387:*589–591, fig. 2, 6/5/97. Copyright © 1997; **24.19b** From fig. 2 in L. H. Rieseberg et al. 1996. Role of Gene Interactions in Hybrid Speciation: Evidence from Ancient and Experimental Hybrids. *Science* 272: 741–745. Copyright © 1996. Reprinted with permission from AAAS.

Chapter 25 **25.3a** Graph from "Experimental Models of Primitive Cellular Compartments Encapsulation Growth and Division" by M.M. Hanczyc et al., in *Science, 302,* No. 5645: 618–622. Oct. 24, 2003. Copyright © 2003, The American Association for the Advancement of Science. Reprinted with permission from AAAS; **25.5** From Don L. Eicher, *Geologic Time,* 1st edition, © 1968. Printed and Electronically reproduced by permission of Pearson Education, Inc., Upper Saddle River, New Jersey; **25.6a–d** Adapted from many sources including D. J. Futuyma. 2005. *Evolution,* fig. 4.10. Sunderland, MA: Sinauer Associates and from R. L. Carroll, 1988. *Vertebrate*

Paleontology and Evolution. W.H. Freeman & Co.; **25.6e** Adapted from Luo et al. 2001. A new mammalia form from the Early Jurassic and evolution of mammalian characteristics. *Science* 292: 1535; **25.7** Adapted from D. J. Des Marais. September 8, 2000. When did photosynthesis emerge on Earth? *Science* 289: 1703–1705; **25.8** Reprinted by permission from Macmillan Publishers Ltd.: *Nature*. L. R. Kump. The rise of atmospheric oxygen, *451*:277–278, 1/17/08. Copyright © 2008; **25.13** Map adapted from http://gelology.er.usgs.gov/eastern/plates.html; **25.15** Graph created from D. M. Raup and J. J. Sepkoski, Jr. 1982. Mass extinctions in the marine fossil record. *Science*. 215: 1501–1503 and J. J. Sepkoski, Jr. 1984. A kinetic model of Phanerozoic taxonomic diversity. III. Post-Paleozoic families and mass extinctions. *Paleobiology*. Vol 10, No. 2, pp. 246–267 in D. J. Futuyma, fig. 7.3a, p. 143 and fig. 7.6, p. 145, Sunderland, MA: Sinauer Associates; **25.17** From Mayhew, P. J. et al. 2008. A long-term association between global temperature and biodiversity, origination and extinction in the fossil record. *Proceedings of the Royal Society B, 275:* 47–53, fig. 3b. Reprinted by permission; **25.18** Figure 3 from "Anatomical and ecological constraints on Phanerozoic animal diversity in the marine realm" by R.K. Bambach et al., in *Proceedings of the National Academy of Sciences, 99,* no. 10: 6854–6859, May 14, 2002. Copyright © 2002 National Academy of Sciences, U.S.A. Used by permission; **25.19** Adapted from Hickman, Roberts, and Larson.1997, *Zoology*, 10th ed, Wm. C. Brown, fig. 31.1; **25.24** Reprinted by permission from Macmillan Publishers Ltd: *Nature*. M. Ronshaugen et al. Hox protein mutation and macroevolution of the insect body plan, *415*:914–917, fig. 1a. Copyright © 2002; **25.26** Adapted from M. Strickberger, 1990. *Evolution,* Boston: Jones & Bartlett.

Chapter 26 **26.6** Figure 1 from "Which Whales Are Hunted? A Molecular Genetic Approach to Monitoring Whaling" by C.S. Baker and S.R. Palumbi, in *Science, 265,* No. 5178:1538–1539, Sep. 9, 1994. Copyright © 1994, The American Association for the Advancement of Science. Reprinted with permission from AAAS; **26.12** With kind permission from Springer Science+Business Media: *Development Genes and Evolution,* "The evolution of the hedgehog gene family in chordates: insights from amphioxus hedgehog," vol. 209, 1999, pp. 40–47, Jan. 1999, S.M. Shimeld, fig. 3; **26.19** Figure 4.3c, p. 124 from *Molecular Markers, Natural History, and Evolution,* 2nd edition, by John Avise. Copyright © 2004 Sinauer Associates. Used by permission; **26.20** Figure from "Timing the Ancestor of the HIV-1 Pandemic Strains" by B. Korber et al., in *Science, 288,* No. 5472: 1789–1796, Jun 9, 2000. Copyright © 2000, The American Association for the Advancement of Science. Reprinted with permission from AAAS; **26.21** Figure 4.1, p. 45, "The three domains of life" by S.L. Baldauf et al., from *Assembling the Tree of Life,* edited by Joel Cracraft and Michael Donoghue. By permission of Oxford University Press, Inc. **26.22** Adapted from S. Blair Hedges. The origin and evolution of model organisms. *Nature Reviews Genetics 3*:838–848, fig. 1, p. 840.

Chapter 27 **27.10** Reprinted by permission from Macmillan Publishers Ltd.: *Nature*. V.S. Cooper and R.E. Lenski. The population genetics of ecological specialization in evolving *E. coli* populations, *407*:736–739, fig. 1. Copyright © 2000; **27.18** Graph created from data in C. Calvaruso et al. 2006. Root-associated bacteria contribute to mineral weathering and to mineral nutrition in trees: A budgeting analysis. *Applied and Environmental Microbiology* 72: 1258–1266.

Chapter 28 **28.2** Reprinted from *Trends in Genetics,18,* No. 11, J.M. Archibald and P.J. Keeling, "Recycled plastids: a 'green movement' in eukaryotic evolution," Copyright © 2002, with permission from Elsevier; **28.11** Figure 12.7, p. 350, from *Microbiology*, by R. W. Bauman. Copyright © 2004 Pearson Education, Inc., publishing as Benjamin Cummings; **28.23** Data from A. Stechman and T. Cavalier-Smith. 2002. Rooting the eukaryote tree by using a derived gene fusion, *Science* 297: 89–91; **28.28** Reprinted by permission from Macmillan Publishers Ltd.: *Nature*. M.J. Behrenfeld et al. Climate-driven trends in contemporary ocean productivity, *44:* 752–755, fig. 3. Copyright © 2006.

Chapter 29 **29.10** Source: R. D. Bowden. 1991. Inputs, outputs and accumulation of nitrogen in an early successional moss *(Polytrichum)* ecosystem. *Ecological Monographs* 61: 207–223; **29.14** Adapted from Raven et al. *Biology of Plants*, 6th ed., fig. 19.7, W. H. Freeman & Co.

Chapter 30 **30.12a** From P. R. Crane. 1985. Phylogenetic analysis of seed plants and the origin of angiosperms. *Annals of the Missouri Botanical Garden, 72:*716–793, fig. 11a, p. 738. Used by permission of Missouri Botanical Garden Press; **30.12b** Figure 2.3, p. 28, from *Phylogeny and Evolution of Angiosperms* by Douglas E. Soltis et al. Copyright © 2005 Sinauer Associates. Used by permission. **Table 30.1** Adapted from Randy Moore et al., *Botany*, 2nd ed. Dubuque, IA: Brown, 1998, Table 2.2, p. 37.

Chapter 31 **31.21** Figures 4 and 5 from "Fungal endophytes limit pathogen damage in a tropical tree" by A.E. Arnold et al., in *Proceedings of the National Academy of Sciences, 100,* no. 26: 15652–15653, December 23, 2003. Copyright © 2003 National Academy of Sciences, U.S.A. Used by permission; **31.26** Figure 1 from "Reversing introduced species effects: Experimental removal of introduced fish leads to rapid recovery of a declining frog" by Vance T. Vredenburg from *Proceedings of the National Academy of Sciences 101:* 7646–7650. Copyright © 2004 National Academy of Sciences, U.S.A. Used by permission.

Chapter 33 **33.22** C. Lydeard et al., The Global Decline of Nonmarine Mollusks, *BioScience,* Vol. 54, No. 4: 321–330. © 2004, American Institute of Biological Sciences. Used by permission. All rights reserved. (Updated data are from International Union for Conservation of Nature, 2008.); **33.29a** Reprinted from *Current Biology, 7,* Issue 8, J.K. Grenier, S. Carroll et al., "Evolution of the entire arthropod *Hox* gene set predated the origin and radiation of the onychophoran/arthropod clade," p. 551, fig. 2a, Copyright © 1987, with permission from Elsevier.

Chapter 34 **34.8b** From "Fossil sister group of craniates: Predicted and found" by J. Mallatt and J. Chen, from *Journal of Morphology, 258,* Issue 1, May 15, 2003. Copyright © 2003 Wiley-Liss, Inc. Reprinted with permission; **34.12** From *Vertebrates: Comparative Anatomy, Function, Evolution,* 3/e by Kenneth Kardong. © 2002 McGraw-Hill Science/Engineering/Mathematics. Reprinted by permission of The McGraw-Hill Companies, Inc. **34.18 (bottom)** Reprinted by permission of Macmillan Publishers Ltd.: *Nature*. M. Zhu et al. The oldest articulated osteichthyan reveals mosaic gnathostome characters, *458*:469–474. Copyright © 2009; **34.21 (left)** Reproduced by permission of the Royal Society of Edinburgh from Transactions of the Roya Society of Edinburgh: Earth Sciences, volume 87 (1996), pp. 363–421; **34.21 (right)** Reprinted by permission of Macmillan Publishers Ltd.: *Nature*. N.H. Shubin et al. The pectoral fin of *Tiktaalik roseae* and the origin of the tetrapod limb, *440*:768, fig. 4. Copyright © 2006; **34.37a** Adapted from many sources including D. J. Futuyma. 2005. *Evolution,* 1/e, fig. 4.10. Sunderland, MA: Sinauer Associates and from R. L. Carroll, 1988. *Vertebrate Paleontology and Evolution*. W.H. Freeman & Co. **34.47** Drawn from many photos of fossils. Some sources are *O. tugenensis* photo in Michael Balter, Early hominid sows division, *ScienceNow*, Feb. 22, 2001, © 2001 American Association for the Advancement of Science. *A. garhi*, and *H. neanderthalensis* adapted from *The Human Evolution Coloring Book. K platyops* drawn from photo in Meave Leakey et al., New hominid genus from eastern Africa shows diverse middle Pliocene lineages, *Nature,* March 22, 2001, 410: 433. *P. boisei* drawn from a photo by David Bill. *H. ergaster* drawn from a photo at www.inhandmuseum.com. *S. tchadensis* drawn from a photo in Michel Brunet et al., A new hominid from the Upper Miocene of Chad, Central Africa, *Nature,* July 11, 2002, 418: 147, fig. 1b; **34.50 (a/b)** Reprinted by permission of Macmillan Publishers Ltd.: *Nature*. I.V. Ovchinnikov et al. Molecular analysis of Neanderthal DNA from the northern Caucasus, *404:* 492, fig.3a and b. Copyright © 2000.

Chapter 35 **35.21** Figure 2b from "Mongolian Tree Rings and 20th-Century Warming" by G.C. Jacoby et al., in *Science, 273,* No. 5276: 771–773, Aug. 9, 1996. Copyright © 1996, The American Association for the Advancement of Science. Reprinted with permission from AAAS.

Chapter 39 **39.16 (top)** Adapted from M. Wilkins. 1988. *Plant Watching,* Facts of File Publ. **39.28** Figure "No Free Lunch" from "Plant Biology: New fatty acid-based signals: A lesson from the plant world" by Edward Farmer in *Science, 276,* No. 5314: 912–913, May 9, 1997. Copyright © 1997, The American Association for the Advancement of Science. Reprinted with permission from AAAS.

Chapter 40 **40.14** Figure 2 from "Thermoregulation in a brooding female Indian python, Python molurus bivittatus" by V.H. Hutchison et al., in *Science, 151,* No. 3711: 694–695, Feb. 11, 1966. Copyright © 1966, The American Association for the Advancement of Science. Reprinted with permission from AAAS; **40.15** Figure 7 from "Thermoregulation in Endothermic Insects" by Bernd Heinrich, in *Science, 185,* No. 4153: 747–756, August 30, 1974. Copyright © 1974, The American Association for the Advancement of Science. Reprinted with permission from AAAS; **40.21** Adapted from figures 2b and 2c from "The circadian clock stops ticking during deep hibernation in the European hamster" by F.G. Revel et al., in *Proceedings of the National Academy of Sciences, 104,* no. 34: 13816–13820, Aug. 21, 2007. Copyright © 2007 National Academy of Sciences, U.S.A. Used by permission.

Chapter 41 **41.9a** Figure 23.1 from *Human Anatomy and Physiology,* 8e, by Elaine Marieb and Katja Hoehn. Copyright © 2010 Pearson Education, Inc., publishing as Pearson Benjamin Cummings. Used by permission of the publisher; **41.9b** Figure 22-1 from Rhoades, *Human Physiology,* 3e. © 1996 Brooks/Cole, a part of Cengage Learning, Inc. Reproduced by permission. www.cengage.com/permissions; **41.21** Figure "Appetite Controllers" from "Cellular Warriors at the Battle of the Bulge" by Jean Marx, from *Science, 299:* p. 86, Feb. 7, 2003. Illustration by Kathleen Sutliff. Copyright © 2003, The American Association for the Advancement of Science. Reprinted with permission from AAAS.

Chapter 42 **42.20** Reprinted by permission from Macmillan Publishers Ltd.: *Nature*. D.J. Rader and A. Daugherty. Translating molecular discoveries into new therapies for atherosclerosis, *451*:904–913, fig. 1, 2/21/08. Copyright © 2008; **42.21** From J.C. Cohen et al., "Sequence variations in PCSK9, low LDL, and protection against coronary heart disease," *New England Journal of Medicine*. 2006 Mar 23; *354:*1264–72, fig. 1A. Copyright © 2006 Massachusetts Medical Society. Used by permission. All rights reserved; **42.26** Adapted from "Surface properties in relation to atelectasis and hyaline membrane disease" by M.E. Avery and J. Mead, from A.M.A. *American Journal of Diseases of Children 97*:517–523 (June 1959). Copyright © 1959 American Medical Association. Used by permission. All rights reserved.

Chapter 43 **43.5** From figures 2a and 4a in Phoebe Tzou et al., "Constitutive expression of a single antimicrobial peptide can restore wild-type resistance to infection in immuno-deficient *Drosophila* mutants," *PNAS, 99:* 2152–2157. Copyright © 2002 National Academy of Sciences, U.S.A. Used with permission; **43.7** Figures 20.4 and 20.5 from *Human Anatomy and Physiology,* 8e, by Elaine Marieb and Katja Hoehn. Copyright © 2010 Pearson Education, Inc., publishing as Pearson Benjamin Cummings. Used by permission of the publisher.

Chapter 44 **44.6** Kangaroo rat data adapted from K. B. Schmidt-Nielson. 1990. *Animal Physiology: Adaptation and Environment,* 4th ed., p. 339. Cambridge: Cambridge University Press; **44.7a** Adapted from K. B. Schmidt-Nielsen et al. 1958. Extrarenal salt excretion in birds. *American Journal of Physiology* 193: 101–107; **44.14B and 44.15** Figure 25.3b from *Human Anatomy and Physiology,* 8e, by Elaine Marieb and Katja Hoehn. Copyright © 2010 Pearson Education, Inc., publishing as Pearson Benjamin Cummings. Used by permission of the publisher; **44.21** Table 1 from "Requirement of human renal water channel aquaporin-2 for vasopressin-dependent concentration in urine" by P.M. Deen et al., in *Science, 264,* No. 5155: 92–95, April 1, 1994. Copyright © 1994, The American Association for the Advancement of Science. Reprinted with permission from AAAS; **44EOC** From W.S. Beck et al., *Life: An Introduction to Biology,* 3rd edition, copyright © 1991. Reprinted and electronically reproduced by permission of Pearson Education, Inc., Upper Saddle River, New Jersey.

Chapter 46 **46.9** Reprinted by permission of Macmillan Publishers Ltd.: *Nature*. R.R. Snook and D.J. Hosken, "Sperm death and dumping in *Drosophila*," *428:*939–941, fig. 2. Copyright © 2004.

Chapter 47 **47.16** Figures 1.10 and 8.25 from *Principles of Development* by Lewis Wolpert (1998). By permission of Oxford University Press, Inc. **47.17a** Figure 21-70 from *Molecular Biology of the Cell,* 4th ed, by Bruce Alberts, used with permission from Garland Science - Books, permission conveyed through Copyright Clearance Center, Inc. Adapted for use from "Cell commitment and gene expression in the axolotl embryo" by T.J. Mohun from *Cell 22:* 9–15 (1980), by permission of the author; **47.17b** Reprinted from *Developmental Biology, 121,* Issue 2, Hiroki Nishida, "Cell lineage analysis in ascidian embryos by intracellular injection of a tracer enzyme: III. Up to the tissue restricted stage," p. 526, Copyright © 1987, with permission from Elsevier; **47.18** From *Molecular Biology of the Cell,* 4th edition, by Bruce Alberts et al., fig. 21.17, p. 1172. Copyright © 2002 by Bruce Alberts, Alexander Johnson, Julian Lewis, Martin Raff, Keith Roberts, and Peter Walter. Used by permission; **47.23 (experiment, results left)** Figures 1.10 and 8.25 from *Principles of Development* by Lewis Wolpert (1998). By permission of Oxford University Press, Inc. **47.23 (results, right)** Figure 15.12, p. 604, from *Developmental Biology,* 5th edition, by Gilbert et al. Copyright © 1997 Sinauer Associates. Used by permission; **47.26** Figure 23.1 from *Human Anatomy and Physiology,* 8e, by Elaine Marieb and Katja Hoehn. Copyright © 2010 Pearson Education, Inc., publishing as Pearson Benjamin Cummings. Used by permission of the publisher.

Chapter 48 **48.11** Adapted from G. Matthews, 2003. *Cellular Physiology of Nerve and Muscle,* 4th edition, fig. 6-2d, p. 61. Cambridge, MA: Blackwell Scientific Publications. Reprinted by permission of Wiley Blackwell; **48.18** Table 1 from "Opiate Receptor: Demonstration in nervous tissue" by C.B. Pert and S.H. Snyder in *Science, 179,* No. 4077: 1011–1014, March 9, 1973. Copyright © 1973, The American Association for the Advancement of Science. Reprinted with permission from AAAS.

Chapter 49 **49.11** Adapted from L. M. Mukhametov. 1984. Sleep in marine mammals. In *Sleep Mechanisms,* by A. A. Borbély and J. L. Valatx (eds.). Munich: Springer-Verlag, pp 227–238; **49.12** Figure 2a from "Transplanted suprachiasmatic nucleus determines circadian period" by M.R. Ralph et al., in *Science, 247,* No. 4945: 975–978, Feb. 23, 1990. Copyright © 1990, The American Association for the Advancement of Science. Reprinted with permission from AAAS; **49.18** Adapted from E. D. Jarvis et al. 2005. Avian brains and a new understanding of vertebrate brain evolution. *Nature Reviews Neuroscience* 6: 151–159, fig. 1c; **49.22** Adapted from Gottesman, I. I. and Wolfgram, D. (1991). *Schizophrenia genesis: The origins of madness,* fig. 10, p. 96. New York: Freeman.

Chapter 50 **50.17A** Figure 15.4(a) from *Human Anatomy and Physiology,* 8e, by Elaine Marieb and Katja Hoehn. Copyright © 2010 Pearson Education, Inc., publishing as Pearson Benjamin Cummings. Used by permission of the publisher; **50.17B** Figure 15.15 from *Human Anatomy and Physiology,* 8e, by Elaine Marieb and Katja Hoehn. Copyright © 2010 Pearson Education, Inc., publishing as Pearson Benjamin Cummings. Used by permission of the publisher; **50.23** Reprinted by permission from Macmillan Publishers Ltd.: *Nature.* K.L. Mueller et al. The receptors and coding logic for bitter taste, *434:*225–229, fig. 4b. Copyright © 2005; **50.24a** Figure 15.23(a) and (b) from *Human Anatomy and Physiology,* 8e, by Elaine Marieb and Katja Hoehn. Copyright © 2010 Pearson Education, Inc., publishing as Pearson Benjamin Cummings. Used by permission of the publisher; **50.34** Grasshopper adapted from Hickman et al. 1993. *Integrated Principles of Zoology,* 9th ed., Fig. 22.6, p. 518. New York: McGraw-Hill Higher Education. © 1993 The McGraw-Hill Companies; **50.40** Figure 4 from "Locomotion: Energy Cost of Swimming, Flying, and Running" by K. Schmidt-Nielsen, in *Science, 177,* No. 4045: 222–228, July 21, 1972. Copyright © 1972, The American Association for the Advancement of Science. Reprinted with permission from AAAS.

Chapter 51 **51.2b** Figure 20, p. 28, from *Study of Instinct* by N. Tinbergen (1989). By permission of Oxford University Press, Inc.; **51.4** Reprinted by permission from Macmillan Publishing Ltd.: *Nature Reviews: Genetics:* M.B. Sokolowski. Drosophila: Genetics meets behavior, *2:*881, fig. 1. Copyright © 2001; **51.10** Figure 3a from "Prospective and retrospective learning in honeybees" by M. Biurfa and J. Benard, in *International Journal of Comparative Psychology, 19,* Issue 3: 358–367, 2006; Reprinted with permission; **51.13** Figure 2a from "Evolution of foraging behavior in Drosophila by density-dependent selection" by M.B. Sokolowski et al., in *Proceedings of the National Academy of Sciences, 94,* no. 14: 7373–7377, July 8, 1997. Copyright © 2007 National Academy of Sciences, U.S.A. Used by permission; **51.19** K. Witte and N. Sawka. 2003. Sexual imprinting on a novel trait in the dimorphic zebra finch: sexes differ. *Animal Behaviour* 65: 195–203. Art adapted from http://www.uni-bielefeld.de/biologue/vhf/KW/Forschungsprojekte2.html; **51.23 (left, bottom)** With kind permission from Springer Science+Business Media: *Genetica,* "The inheritance of mating songs in two cryptic, sibling lacewing species," vol. 116, 2002, pp. 269–289, C. S. Henry, fig. 2; **51.26 (top)** Adapted from a photograph by Jonathan Blair in Alcock, 2002. *Animal Behavior,* 7th ed. Sinauer Associates, Inc., Publiishers; **51.26 (bottom)** Reprinted by permission from Macmillan Publishers Ltd.: *Nature.* P. Berthold et al. Rapid microevolution of migratory behaviour in a wild bird species, *360:*668, fig. 1, 12/17/92. Copyright © 1992.

Chapter 52 **52.7** Adapted from L. Roberts. 1989. How fast can trees migrate? *Science* 243: 736, fig. 2. © 1989 by the American Association for the Advancement of Science; **52.8** Reprinted by permission from Macmillan Publishers Ltd.: *Nature.* C. Parmesan et al. Poleward shift of butterfly species' ranges associated with regional warming, *399:*579–583, fig. 3. Copyright © 1999; **52.9** Adapted from Heinrich Walter and Siegmar-Walter Breckle. 2003. *Walter's Vegetation of the Earth,* fig. 16, p. 36. Springer-Verlag, © 2003; **52.17** Figure 1.7, p. 9 from *Kangaroos, Their Ecology and Management in the Sheep Rangelands of Australia,* edited by Graeme Caughley, Neil Shepherd, Jeff Short. © Cambridge University Press 1987. Reprinted with the permission of Cambridge University Press; **52.19** Map adapted from R. L. Smith. 1974. *Ecology and Field Biology,* fig. 11.19, p. 353. Harper and Row Publishers. Map updated from D. A. Sibley. 2000. National Audubon Society *The Sibley Guide to Birds,* Alfred A. Knopf: New York; **52.20** Data from W. J. Fletcher. 1987. Interactions among subtidal Australian sea urchins, gastropods and algae: effects of experimental removals. *Ecological Monographs* 57: 89–109; **Un. 52.2** Data from J. Clausen, D. D. Keck, and W. M. Hiesey. 1948. Experimental studies on the nature of species. III. Environmental responses of climatic races of *Achillea.* Carnegie Institution of Washington Publication 581.

Chapter 53 **53.5** Adapted from P. W. Sherman and M. L. Morton. 1984. "Demography of Belding's ground squirrels," *Ecology,* Vol. 65, No. 5, p. 1622, fig. 1a, 1984. Copyright © 1984 Ecological Society of America. Used by permission; **53.15** Adapted from J. T. Enright. 1976. Climate and population regulation: The biogeographer's dilemma. *Oecologia* 24: 295–310; **53.16** Figure 3.5b, page 59, from *Soay Sheep: Dynamics and Selection in an Island Population* edited by T.H. Clutton-Brock and J.M. Pemberton. Copyright © Cambridge University Press 2004. Reprinted with the permission of Cambridge University Press; **53.18** Data courtesy of Rolf O. Peterson, Michigan Technological University; **53.23** Data from U. S. Census Bureau International Data Base; **53.24** Data from U. S. Census Bureau International Data Base; **53.25** Data from U. S. Census Bureau International Data Base 2008; **53.26** Source: Used courtesy of UNEP/GRID-Arendel at http://maps.grida.no/go/graphic/energy_consumption_per_capita_2004; **Tables 53.1** and **53.2** Data from P. W. Sherman and M. L. Morton, "Demography of Belding's ground squirrels," *Ecology,* Vol. 65, No. 5, p. 1622, fig. 1a, 1984. Copyright © 1984 Ecological Society of America.

Chapter 54 **54.2** From "The anoles of La Palma: aspects of their ecological relationships" by A.S. Rand and E. E. Williams in *Breviora 327:* 1–19, 1969. Copyright © 1969 by the President and Fellows of Harvard College. Used with permission from the Museum of Comparative Zoology, Harvard University; **54.9** Data for graph from S. D. Hacker and M. D. Bertness. 1999. Experimental evidence for factors maintaining plant species diversity in a New England salt marsh. *Ecology* 80: 2064–2073; **54.11** Adapted from N. Fierer and R. B. Jackson. 2006. The diversity and biogeography of soil bacterial communities. *Proceedings of the National Academy of Sciences USA 103:* 626–631 fig. 1a; **54.14** Adapted from E. A. Knox. 1970. Antarctic marine ecosystems. In *Antarctic Ecology,* ed. M. W. Holdgate, 69–96. London: Academic Press; **54.15** Adapted from D. L. Breitburg et al. 1997. Varying effects of low dissolved oxygen on trophic interactions in an estuarine food web. *Ecological Monographs* 67: 490. Copyright © 1997 Ecological Society of America; **54.16** Adapted from B. Jenkins. 1992. Productivity, disturbance and food web structure at a local spatial scale in experimental container habitats. *Oikos* 65: 252. Copyright © 1992 Oikos, Sweden; **54.17** Adapted from R. T. Paine. 1966. Food web complexity and species diversity. *American Naturalist* 100: 65–75; **54.18** From J. A. Estes et al. 1998. Killer whale predation on sea otters linking oceanic and nearshore ecosystems. *Science 282:* 474, fig. 1. Copyright © 1998 by the American Association for the Advancement of Science; **54.20** Graph adapted from A. R. Townsend et al. 1997. The intermediate disturbance hypothesis, refugia, and diversity in streams. *Limnology and Oceanography* 42:938–949. **54.22** Adapted from R. L. Crocker and J. Major. 1955. Soil Development in relation to vegetation and surface age at Glacier Bay, Alaska. *Journal of Ecology* 43: 427–448; **54.23** Data from F. S. Chapin, III, et al. 1994. Mechanisms of primary succession following deglaciation at Glacier Bay, Alaska. *Ecological Monographs* 64: 149–175; **54.25** Adapted from D. J. Currie. 1991. Energy and large-scale patterns of animal- and plant-species richness. *American Naturalist* 137: 27–49; **54.26** Adapted from F. W. Preston. 1960. Time and space and the variation of species. *Ecology 41:* 611–627; **54.28** Adapted from F. W. Preston. 1962. The canonical distribution of commonness and rarity. *Ecology 43:*185–215, 410–432.

Chapter 55 **55.4** and **Un 55.1** Adapted from D. L. DeAngelis. 1992. *Dynamics of Nutrient Cycling and Food Webs.* New York: Chapman & Hall; **55.7** Adapted from National Oceanic and Atmospheric Administration's National Data Buoy Center Voluntary Observing Ship Project, www.vos.noaa.gov/MWL/dec_06/Images/OCP_fig4.jpg; **55.8** Figure 2 from "Nitrogen, phosphorus, and eutrophication in the coastal marine environment" by J.H. Ryther, in *Science, 171,* No. 3975: 1008–1013, Mar. 12, 1971. Copyright © 1971, The American Association for the Advancement of Science. Reprinted with permission from AAAS; **55.9** Fig. 4.1 p. 82 from *Communities and Ecosystems,* 1e by Robert H. Whittaker. Copyright © 1970 Robert H. Whittaker. Reprinted by permission of Pearson Education; **55.14** From *The Economy of Nature,* 4th edition. © 1997 by W. H. Freeman and Company. Used with permission; **55.15a** Adapted from J.A. Trofymow et al., *The Canadian Intersite Decomposition Experiment: Project and Site Establishment Report,* Information Report BC-X-378, page 2, Natural Resources Canada, Canadian Forest Service (1998). Reproduced with permission from the Minister of Public Works and Government Services, Canada, 2010; **55.15b** Adapted from: T.R. Moore et al., Litter decomposition rates in Canadian forests, *Global Change Biology* 5: 75–82 (1999), copyright © 2001, 1998 Blackwell Science Ltd. Reproduced with permission from the Minister of Public Works and Government Services, Canada, 2010 and Wiley Blackwell; **55.18b** Data adapted from Wu, W-M et al. 2006. Pilot-scale in situ bioremediation of uranium in a highly contaminated aquifer. 2. Reduction of U(VI) and geochemical control of a U(VI) bioavailability. *Environ Sci. Technol. 40:* 3986–3995, fig. 1D; **Table 55.1** Data from Menzel and Ryther. 1961. *Deep Sea Ranch* 7: 276–281.

Chapter 56 **56.12** Figure 19.1 from *Ecology,* 5e by C.J. Krebs. Copyright © 2001 Pearson Education, Inc. Reprinted by permission; **56.13** Figure 2 from "Tracking the Long-Term Decline and Recovery of an Isolated Population" by R.L. Westemeier et al., in *Science, 282,* No. 5394: 1695–1698, Nov. 27, 1998. Copyright © 1998, The American Association for the Advancement of Science. Reprinted with permission from AAAS; **56.19** Reprinted by permission from Macmillan Publishers, Ltd.: *Nature.* N. Myers et al. Biodiversity hotspots for conservation priorities, *403:* 853–858, fig. 1, 2/24/00. Copyright © 2000; **56.20** Reprinted from W. D. Newmark, 1985. Legal and biotic boundaries of western North American national parks: a problem of congruence. *Biological Conservation* 33: 197–208, fig. 1, p. 199. © 1985, with permission from Elsevier; **56.21a** Map adapted from W. Purves and G. Orians, *Life, The Science of Biology,* 5th ed., fig. 55.23, p. 1239. © 1998 by Sinauer Associates, Inc. Reprinted with permission; **56.27** CO_2 data from www.esrl.noaa.gov/gmd/ccgg/trends. Temperature data from www.giss.nasa.gov/gistemps/graphs/Fig.A.lrg.gif; **56.29** Data from ozonewatch.gsfc.nasa.gov/facts/history/html; **56.32** Data from Instituto Nacional de Estadistica y Censos de Costa Rica and Centro Centroamericano de Poblacion, Universidad de Costa Rica.

Glossary

Pronunciation Key

Pronounce

ā	as in	ace
a/ah		ash
ch		chose
ē		meet
e/eh		bet
g		game
ī		ice
i		hit
ks		box
kw		quick
ng		song
ō		robe
o		ox
oy		boy
s		say
sh		shell
th		thin
ū		boot
u/uh		up
z		zoo

′ = primary accent

′ = secondary accent

5′ cap A modified form of guanine nucleotide added onto the 5′ end of a pre-mRNA molecule.

A site One of a ribosome's three binding sites for tRNA during translation. The A site holds the tRNA carrying the next amino acid to be added to the polypeptide chain. (A stands for aminoacyl tRNA.)

ABC hypothesis A model of flower formation identifying three classes of organ identity genes that direct formation of the four types of floral organs.

abiotic (ā′-bī-ot′-ik) Nonliving; referring to the physical and chemical properties of an environment.

abortion The termination of a pregnancy in progress.

abscisic acid (ABA) (ab-sis′-ik) A plant hormone that slows growth, often antagonizing the actions of growth hormones. Two of its many effects are to promote seed dormancy and facilitate drought tolerance.

absorption The third stage of food processing in animals: the uptake of small nutrient molecules by an organism's body.

absorption spectrum The range of a pigment's ability to absorb various wavelengths of light; also a graph of such a range.

abyssal zone (uh-bis′-ul) The part of the ocean's benthic zone between 2,000 and 6,000 m deep.

acanthodian (ak′-an-thō′-dē-un) Any of a group of ancient jawed aquatic vertebrates from the Silurian and Devonian periods.

accessory fruit A fruit, or assemblage of fruits, in which the fleshy parts are derived largely or entirely from tissues other than the ovary.

acclimatization (uh-klī′-muh-tī-zā′-shun) Physiological adjustment to a change in an environmental factor.

acetyl CoA Acetyl coenzyme A; the entry compound for the citric acid cycle in cellular respiration, formed from a fragment of pyruvate attached to a coenzyme.

acetylcholine (as′-uh-til-kō′-lēn) One of the most common neurotransmitters; functions by binding to receptors and altering the permeability of the postsynaptic membrane to specific ions, either depolarizing or hyperpolarizing the membrane.

acid A substance that increases the hydrogen ion concentration of a solution.

acid precipitation Rain, snow, or fog that is more acidic than pH 5.2.

acoelomate (uh-sē′-lō-māt) A solid-bodied animal lacking a cavity between the gut and outer body wall.

acrosomal reaction (ak′-ruh-sōm′-ul) The discharge of hydrolytic enzymes from the acrosome, a vesicle in the tip of a sperm, when the sperm approaches or contacts an egg.

acrosome (ak′-ruh-sōm) A vesicle in the tip of a sperm containing hydrolytic enzymes and other proteins that help the sperm reach the egg.

actin (ak′-tin) A globular protein that links into chains, two of which twist helically about each other, forming microfilaments (actin filaments) in muscle and other kinds of cells.

action potential An electrical signal that propagates (travels) along the membrane of a neuron or other excitable cell as a nongraded (all-or-none) depolarization.

action spectrum A graph that profiles the relative effectiveness of different wavelengths of radiation in driving a particular process.

activation energy The amount of energy that reactants must absorb before a chemical reaction will start; also called free energy of activation.

activator A protein that binds to DNA and stimulates gene transcription. In prokaryotes, activators bind in or near the promoter; in eukaryotes, activators generally bind to control elements in enhancers.

active immunity Long-lasting immunity conferred by the action of B cells and T cells and the resulting B and T memory cells specific for a pathogen. Active immunity can develop as a result of natural infection or immunization.

active site The specific region of an enzyme that binds the substrate and that forms the pocket in which catalysis occurs.

active transport The movement of a substance across a cell membrane against its concentration or electrochemical gradient, mediated by specific transport proteins and requiring an expenditure of energy.

adaptation Inherited characteristic of an organism that enhances its survival and reproduction in a specific environment.

adaptive immunity A vertebrate-specific defense that is mediated by B lymphocytes (B cells) and T lymphocytes (T cells). It exhibits specificity, memory, and self-nonself recognition. Also called acquired immunity.

adaptive radiation Period of evolutionary change in which groups of organisms form many new species whose adaptations allow them to fill different ecological roles in their communities.

addition rule A rule of probability stating that the probability of any one of two or more mutually exclusive events occurring can be determined by adding their individual probabilities.

adenosine triphosphate *See* ATP (adenosine triphosphate).

adenylyl cyclase (uh-den′-uh-lil) An enzyme that converts ATP to cyclic AMP in response to an extracellular signal.

adhesion The clinging of one substance to another, such as water to plant cell walls by means of hydrogen bonds.

adipose tissue A connective tissue that insulates the body and serves as a fuel reserve; contains fat-storing cells called adipose cells.

adrenal gland (uh-drē′-nul) One of two endocrine glands located adjacent to the kidneys in mammals. Endocrine cells in the outer portion (cortex) respond to adrenocorticotropic hormone (ACTH) by secreting steroid hormones that help maintain homeostasis during long-term stress. Neurosecretory cells in the central portion (medulla) secrete epinephrine and norepinephrine in response to nerve signals triggered by short-term stress.

adrenocorticotropic hormone (ACTH) A tropic hormone that is produced and secreted by the anterior pituitary and that stimulates the production and secretion of steroid hormones by the adrenal cortex.

aerobic respiration A catabolic pathway for organic molecules, using oxygen (O_2) as the final electron acceptor in an electron transport chain and ultimately producing ATP. This is the most efficient catabolic pathway and is carried out in most eukaryotic cells and many prokaryotic organisms.

age structure The relative number of individuals of each age in a population.

aggregate fruit A fruit derived from a single flower that has more than one carpel.

AIDS (acquired immunodeficiency syndrome) The symptoms and signs present during the late stages of HIV infection, defined by a specified reduction in the number of T cells and the appearance of characteristic secondary infections.

alcohol fermentation Glycolysis followed by the reduction of pyruvate to ethyl alcohol, regenerating NAD^+ and releasing carbon dioxide.

aldosterone (al-dos′-tuh-rōn) A steroid hormone that acts on tubules of the kidney to regulate the transport of sodium ions (Na^+) and potassium ions (K^+).

algae A diverse grade of photosynthetic protists, including unicellular and multicellular forms. Algal species are included in three of the five eukaryote supergroups (Chromalveolata, Rhizaria, and Archaeplastida).

alimentary canal (al′-uh-men′-tuh-rē) A complete digestive tract, consisting of a tube running between a mouth and an anus.

allele (uh-lē′-ul) Any of the alternative versions of a gene that may produce distinguishable phenotypic effects.

allergen An antigen that triggers an exaggerated immune response.

allopatric speciation (al′-uh-pat′-rik) The formation of new species in populations that are geographically isolated from one another.

allopolyploid (al′-ō-pol′-ē-ployd) A fertile individual that has more than two chromosome sets as a result of two different species interbreeding and combining their chromosomes.

allosteric regulation The binding of a regulatory molecule to a protein at one site that affects the function of the protein at a different site.

alpha (α) helix (al′-fuh hē′-liks) A coiled region constituting one form of the secondary structure of proteins, arising from a specific pattern of hydrogen bonding between atoms of the polypeptide backbone (not the side chains).

alternation of generations A life cycle in which there is both a multicellular diploid form, the sporophyte, and a multicellular haploid form, the gametophyte; characteristic of plants and some algae.

alternative RNA splicing A type of eukaryotic gene regulation at the RNA-processing level in which different mRNA molecules are produced from the same primary transcript, depending on which RNA segments are treated as exons and which as introns.

altruism (al′-trū-iz-um) Selflessness; behavior that reduces an individual's fitness while increasing the fitness of another individual.

alveolate (al-vē′-uh-let) A protist with membrane-bounded sacs (alveoli) located just under the plasma membrane.

alveolus (al-vē′-uh-lus) (plural, **alveoli**) One of the dead-end air sacs where gas exchange occurs in a mammalian lung.

Alzheimer's disease (alts′-hī-merz) An age-related dementia (mental deterioration) characterized by confusion and memory loss.

amacrine cell (am′-uh-krin) A neuron of the retina that helps integrate information before it is sent to the brain.

amino acid (uh-mēn′-ō) An organic molecule possessing both a carboxyl and an amino group. Amino acids serve as the monomers of polypeptides.

amino group A chemical group consisting of a nitrogen atom bonded to two hydrogen atoms; can act as a base in solution, accepting a hydrogen ion and acquiring a charge of 1+.

aminoacyl-tRNA synthetase An enzyme that joins each amino acid to the appropriate tRNA.

ammonia A small, toxic molecule (NH_3) produced by nitrogen fixation or as a metabolic waste product of protein and nucleic acid metabolism.

ammonite A member of a group of shelled cephalopods that were important marine predators for hundreds of millions of years until their extinction at the end of the Cretaceous period (65.5 million years ago).

amniocentesis (am′-nē-ō-sen-tē′-sis) A technique associated with prenatal diagnosis in which amniotic fluid is obtained by aspiration from a needle inserted into the uterus. The fluid and the fetal cells it contains are analyzed to detect certain genetic and congenital defects in the fetus.

amniote (am′-nē-ōt) Member of a clade of tetrapods named for a key derived character, the amniotic egg, which contains specialized membranes, including the fluid-filled amnion, that protect the embryo. Amniotes include mammals as well as birds and other reptiles.

amniotic egg An egg that contains specialized membranes that function in protection, nourishment, and gas exchange. The amniotic egg was a major evolutionary innovation, allowing embryos to develop on land in a fluid-filled sac, thus reducing the dependence of tetrapods on water for reproduction.

amoeba (uh-mē′-buh) A protist grade characterized by the presence of pseudopodia.

amoebocyte (uh-mē′-buh-sīt′) An amoeba-like cell that moves by pseudopodia and is found in most animals. Depending on the species, it may digest and distribute food, dispose of wastes, form skeletal fibers, fight infections, or change into other cell types.

amoebozoan (uh-mē′-buh-zō′-an) A protist in a clade that includes many species with lobe- or tube-shaped pseudopodia.

amphibian Member of the tetrapod class Amphibia, including salamanders, frogs, and caecilians.

amphipathic (am′-fē-path′-ik) Having both a hydrophilic region and a hydrophobic region.

amplification The strengthening of stimulus energy during transduction.

amygdala (uh-mig′-duh-luh) A structure in the temporal lobe of the vertebrate brain that has a major role in the processing of emotions.

amylase (am′-uh-lās′) An enzyme that hydrolyzes starch (a glucose polymer from plants) and glycogen (a glucose polymer from animals) into smaller polysaccharides and the disaccharide maltose.

anabolic pathway (an′-uh-bol′-ik) A metabolic pathway that consumes energy to synthesize a complex molecule from simpler molecules.

anaerobic respiration (an-er-ō′-bik) A catabolic pathway in which inorganic molecules other than oxygen accept electrons at the "downhill" end of electron transport chains.

analogous Having characteristics that are similar because of convergent evolution, not homology.

analogy (an-al′-uh-jē) Similarity between two species that is due to convergent evolution rather than to descent from a common ancestor with the same trait.

anaphase The fourth stage of mitosis, in which the chromatids of each chromosome have separated and the daughter chromosomes are moving to the poles of the cell.

anatomy The structure of an organism.

anchorage dependence The requirement that a cell must be attached to a substratum in order to initiate cell division.

androgen (an′-drō-jen) Any steroid hormone, such as testosterone, that stimulates the development and maintenance of the male reproductive system and secondary sex characteristics.

aneuploidy (an′-yū-ploy′-dē) A chromosomal aberration in which one or more chromosomes are present in extra copies or are deficient in number.

angiosperm (an′-jē-ō-sperm) A flowering plant, which forms seeds inside a protective chamber called an ovary.

angiotensin II A peptide hormone that stimulates constriction of precapillary arterioles and increases reabsorption of NaCl and water by the proximal tubules of the kidney, increasing blood pressure and volume.

anhydrobiosis (an-hī′-drō-bī-ō′-sis) A dormant state involving loss of almost all body water.

animal pole The point at the end of an egg in the hemisphere where the least yolk is concentrated; opposite of vegetal pole.

anion (an′-ī-on) A negatively charged ion.

anterior Pertaining to the front, or head, of a bilaterally symmetrical animal.

anterior pituitary A portion of the pituitary that develops from nonneural tissue; consists of endocrine cells that synthesize and secrete several tropic and nontropic hormones.

anther In an angiosperm, the terminal pollen sac of a stamen, where pollen grains containing sperm-producing male gametophytes form.

antheridium (an-thuh-rid′-ē-um) (plural, **antheridia**) In plants, the male gametangium, a moist chamber in which gametes develop.

anthropoid (an′-thruh-poyd) Member of a primate group made up of the monkeys and the apes (gibbons, orangutans, gorillas, chimpanzees, bonobos, and humans).

antibody A protein secreted by plasma cells (differentiated B cells) that binds to a particular antigen; also called immunoglobulin. All antibodies have the same Y-shaped structure and in their monomer form consist of two identical heavy chains and two identical light chains.

anticodon (an′-tī-kō′-don) A nucleotide triplet at one end of a tRNA molecule that base-pairs with a particular complementary codon on an mRNA molecule.

antidiuretic hormone (ADH) (an′-tī-dī-yū-ret′-ik) A peptide hormone, also known as vasopressin, that promotes water retention by the kidneys. Produced in the hypothalamus and released from the posterior pituitary, ADH also functions in the brain.

antigen (an′-ti-jen) A substance that elicits an immune response by binding to receptors of B cells, antibodies, or of T cells.

antigen presentation The process by which an MHC molecule binds to a fragment of an intracellular protein antigen and carries it to the cell surface, where it is displayed and can be recognized by a T cell.

antigen receptor The general term for a surface protein, located on B cells and T cells, that binds to antigens, initiating adaptive immune responses. The antigen receptors on B cells are called B cell receptors, and the antigen receptors on T cells are called T cell receptors.

antigen-presenting cell A cell that upon ingesting pathogens or internalizing pathogen proteins generates peptide fragments that are bound by class II MHC molecules and subsequently displayed on the cell surface to T cells. Macrophages, dendritic cells, and B cells are the primary antigen-presenting cells.

antiparallel Referring to the arrangement of the sugar-phosphate backbones in a DNA double helix (they run in opposite $5' \rightarrow 3'$ directions).

aphotic zone (ā′-fō′-tik) The part of an ocean or lake beneath the photic zone, where light does not penetrate sufficiently for photosynthesis to occur.

apical bud (ā′-pik-ul) A bud at the tip of a plant stem; also called a terminal bud.

apical dominance (ā′-pik-ul) Tendency for growth to be concentrated at the tip of a plant shoot, because the apical bud partially inhibits axillary bud growth.

apical ectodermal ridge (AER) A thickened area of ectoderm at the tip of a limb bud that promotes outgrowth of the limb bud.

apical meristem (ā′-pik-ul mār′-uh-stem) Embryonic plant tissue in the tips of roots and buds of shoots. The dividing cells of an apical meristem enable the plant to grow in length.

apicomplexan (ap′-ē-kom-pleks′-un) A protist in a clade that includes many species that parasitize animals. Some apicomplexans cause human disease.

apomixis (ap′-uh-mik′-sis) The ability of some plant species to reproduce asexually through seeds without fertilization by a male gamete.

apoplast (ap′-ō-plast) Everything external to the plasma membrane of a plant cell, including cell walls, intercellular spaces, and the space within dead structures such as xylem vessels and tracheids.

apoptosis (ā-puh-tō′-sus) A type of programmed cell death, which is brought about by activation of enzymes that break down many chemical components in the cell.

aposematic coloration (ap′-ō-si-mat′-ik) The bright warning coloration of many animals with effective physical or chemical defenses.

appendix A small, finger-like extension of the vertebrate cecum; contains a mass of white blood cells that contribute to immunity.

aquaporin A channel protein in the plasma membrane of a plant, animal, or microorganism cell that specifically facilitates osmosis, the diffusion of free water across the membrane.

aqueous solution (ā′-kwē-us) A solution in which water is the solvent.

arachnid A member of a major arthropod group, the chelicerates. Arachnids include spiders, scorpions, ticks, and mites.

arbuscular mycorrhiza (ar-bus′-kyū-lur mī′-kō-rī′-zuh) Association of a fungus with a plant root system in which the fungus causes the invagination of the host (plant) cells' plasma membranes.

arbuscular mycorrhizal fungus A symbiotic fungus whose hyphae grow through the cell wall of plant roots and extend into the root cell (enclosed in tubes formed by invagination of the root cell plasma membrane).

Archaea (ar′-kē′-uh) One of two prokaryotic domains, the other being Bacteria.

Archaeplastida (ar′-kē-plas′-tid-uh) One of five supergroups of eukaryotes proposed in a current hypothesis of the evolutionary history of eukaryotes. This monophyletic group, which includes red algae, green algae, and land plants, descended from an ancient protist ancestor that engulfed a cyanobacterium. *See also* Excavata, Chromalveolata, Rhizaria, and Unikonta.

archegonium (ar-ki-gō′-nē-um) (plural, **archegonia**) In plants, the female gametangium, a moist chamber in which gametes develop.

archenteron (ar-ken′-tuh-ron) The endoderm-lined cavity, formed during gastrulation, that develops into the digestive tract of an animal.

archosaur (ar′-kō-sōr) Member of the reptilian group that includes crocodiles, alligators and dinosaurs, including birds.

arteriole (ar-ter′-ē-ōl) A vessel that conveys blood between an artery and a capillary bed.

artery A vessel that carries blood away from the heart to organs throughout the body.

arthropod A segmented ecdysozoan with a hard exoskeleton and jointed appendages. Familiar examples include insects, spiders, millipedes, and crabs.

artificial selection The selective breeding of domesticated plants and animals to encourage the occurrence of desirable traits.

ascocarp The fruiting body of a sac fungus (ascomycete).

ascomycete (as′-kuh-mī′-sēt) Member of the fungal phylum Ascomycota, commonly called sac fungus. The name comes from the saclike structure in which the spores develop.

ascus (plural, **asci**) A saclike spore capsule located at the tip of a dikaryotic hypha of a sac fungus.

asexual reproduction The generation of offspring from a single parent that occurs without the fusion of gametes (by budding, division of a single cell, or division of the entire organism into two or more parts). In most cases, the offspring are genetically identical to the parent.

assisted migration The translocation of a species to a favorable habitat beyond its native range for the purpose of protecting the species from human-caused threats.

assisted reproductive technology A fertilization procedure that generally involves surgically removing eggs (secondary oocytes) from a woman's ovaries after hormonal stimulation, fertilizing the eggs, and returning them to the woman's body.

associative learning The acquired ability to associate one environmental feature (such as a color) with another (such as danger).

aster A radial array of short microtubules that extends from each centrosome toward the plasma membrane in an animal cell undergoing mitosis.

astrocyte A glial cell with diverse functions, including providing structural support for neurons, regulating the interstitial environment, facilitating synaptic transmission, and assisting in regulating the blood supply to the brain.

atherosclerosis A cardiovascular disease in which fatty deposits called plaques develop in the inner walls of the arteries, obstructing the arteries and causing them to harden.

atom The smallest unit of matter that retains the properties of an element.

atomic mass The total mass of an atom, which is the mass in grams of 1 mole of the atom.

atomic nucleus An atom's dense central core, containing protons and neutrons.

atomic number The number of protons in the nucleus of an atom, unique for each element and designated by a subscript to the left of the elemental symbol.

ATP (adenosine triphosphate) (a-den′-ō-sēn trī-fos′-fāt) An adenine-containing nucleoside triphosphate that releases free energy when its phosphate bonds are hydrolyzed. This energy is used to drive endergonic reactions in cells.

ATP synthase A complex of several membrane proteins that functions in chemiosmosis with adjacent electron transport chains, using the energy of a hydrogen ion (proton) concentration gradient to make ATP. ATP synthases are found in the inner mitochondrial membranes of eukaryotic cells and in the plasma membranes of prokaryotes.

atrial natriuretic peptide (ANP) (ā′-trē-ul na′-trē-yū-ret′-ik) A peptide hormone secreted by cells of the atria of the heart in response to high blood pressure. ANP's effects on the kidney alter ion and water movement and reduce blood pressure.

atrioventricular (AV) node A region of specialized heart muscle tissue between the left and right atria where electrical impulses are delayed for about 0.1 second before spreading to both ventricles and causing them to contract.

atrioventricular (AV) valve A heart valve located between each atrium and ventricle that prevents a backflow of blood when the ventricle contracts.

atrium (ā′-trē-um) (plural, **atria**) A chamber of the vertebrate heart that receives blood from the veins and transfers blood to a ventricle.

autocrine Referring to a secreted molecule that acts on the cell that secreted it.

autoimmune disease An immunological disorder in which the immune system turns against self.

autonomic nervous system (ot′-ō-nom′-ik) An efferent branch of the vertebrate peripheral nervous system that regulates the internal environment; consists of the sympathetic, parasympathetic, and enteric divisions.

autopolyploid (ot′-ō-pol′-ē-ployd) An individual that has more than two chromosome sets that are all derived from a single species.

autosome (ot′-ō-sōm) A chromosome that is not directly involved in determining sex; not a sex chromosome.

autotroph (ot′-ō-trōf) An organism that obtains organic food molecules without eating other organisms or substances derived from other organisms. Autotrophs use energy from the sun or from oxidation of inorganic substances to make organic molecules from inorganic ones.

auxin (ôk′-sin) A term that primarily refers to indoleacetic acid (IAA), a natural plant hormone that has a variety of effects, including cell elongation, root formation, secondary growth, and fruit growth.

average heterozygosity (het′-er-ō-zī-gō′-si-tē) The percentage, on average, of a population's loci that are heterozygous in members of the population.

avirulent Describing a pathogen that can mildly harm, but not kill, the host.

axillary bud (ak′-sil-ār-ē) A structure that has the potential to form a lateral shoot, or branch. The bud appears in the angle formed between a leaf and a stem.

axon (ak′-son) A typically long extension, or process, of a neuron that carries nerve impulses away from the cell body toward target cells.

B cells The lymphocytes that complete their development in the bone marrow and become effector cells for the humoral immune response.

Bacteria One of two prokaryotic domains, the other being Archaea.

bacterial artificial chromosome (BAC) A large plasmid that acts as a bacterial chromosome and can carry inserts of 100,000 to 300,000 base pairs (100–300 kb).

bacteriophage (bak-tēr′-ē-ō-fāj) A virus that infects bacteria; also called a phage.

bacteroid A form of the bacterium *Rhizobium* contained within the vesicles formed by the root cells of a root nodule.

balancing selection Natural selection that maintains two or more phenotypic forms in a population.

bark All tissues external to the vascular cambium, consisting mainly of the secondary phloem and layers of periderm.

Barr body A dense object lying along the inside of the nuclear envelope in cells of female mammals, representing a highly condensed, inactivated X chromosome.

basal angiosperm A member of one of three clades of early-diverging lineages of flowering plants. Examples are *Amborella*, water lilies, and star anise and its relatives.

basal body (bā′-sul) A eukaryotic cell structure consisting of a "9 + 0" arrangement of microtubule triplets. The basal body may organize the microtubule assembly of a cilium or flagellum and is structurally very similar to a centriole.

basal metabolic rate (BMR) The metabolic rate of a resting, fasting, and nonstressed endotherm at a comfortable temperature.

basal taxon In a specified group of organisms, a taxon whose evolutionary lineage diverged early in the history of the group.

base A substance that reduces the hydrogen ion concentration of a solution.

basidiocarp Elaborate fruiting body of a dikaryotic mycelium of a club fungus.

basidiomycete (buh-sid′-ē-ō-mī′-sēt) Member of the fungal phylum Basidiomycota, commonly called club fungus. The name comes from the club-like shape of the basidium.

basidium (plural, **basidia**) (buh-sid′-ē-um, buh-sid′-ē-ah) A reproductive appendage that produces sexual spores on the gills of mushrooms (club fungi).

Batesian mimicry (bāt′-zē-un mim′-uh-krē) A type of mimicry in which a harmless species looks like a species that is poisonous or otherwise harmful to predators.

behavior Individually, an action carried out by muscles or glands under control of the nervous system in response to a stimulus; collectively, the sum of an animal's responses to external and internal stimuli.

behavioral ecology The study of the evolution of and ecological basis for animal behavior.

benign tumor A mass of abnormal cells with specific genetic and cellular changes such that the cells are not capable of surviving at a new site and generally remain at the site of the tumor's origin.

benthic zone The bottom surface of an aquatic environment.

benthos (ben′-thōz) The communities of organisms living in the benthic zone of an aquatic biome.

beta (β) pleated sheet One form of the secondary structure of proteins in which the polypeptide chain folds back and forth. Two regions of the chain lie parallel to each other and are held together by hydrogen bonds between atoms of the polypeptide backbone (not the side chains).

beta oxidation A metabolic sequence that breaks fatty acids down to two-carbon fragments that enter the citric acid cycle as acetyl CoA.

bicoid A maternal effect gene that codes for a protein responsible for specifying the anterior end in *Drosophila melanogaster*.

bilateral symmetry Body symmetry in which a central longitudinal plane divides the body into two equal but opposite halves.

bilaterian (bī′-luh-ter′-ē-uhn) Member of a clade of animals with bilateral symmetry and three germ layers.

bile A mixture of substances that is produced in the liver and stored in the gallbladder; enables formation of fat droplets in water as an aid in the digestion and absorption of fats.

binary fission A method of asexual reproduction by "division in half." In prokaryotes, binary fission does not involve mitosis, but in single-celled eukaryotes that undergo binary fission, mitosis is part of the process.

binomial The two-part, latinized format for naming a species, consisting of the genus and specific epithet; a binomen.

biodiversity hot spot A relatively small area with numerous endemic species and a large number of endangered and threatened species.

bioenergetics (1) The overall flow and transformation of energy in an organism. (2) The study of how energy flows through organisms.

biofilm A surface-coating colony of one or more species of prokaryotes that engage in metabolic cooperation.

biofuel A fuel produced from dry organic matter or combustible oils produced by plants.

biogenic amine A neurotransmitter derived from an amino acid.

biogeochemical cycle Any of the various chemical cycles, which involve both biotic and abiotic components of ecosystems.

biogeography The study of the past and present geographic distribution of species.

bioinformatics The use of computers, software, and mathematical models to process and integrate biological information from large data sets.

biological augmentation An approach to restoration ecology that uses organisms to add essential materials to a degraded ecosystem.

biological clock An internal timekeeper that controls an organism's biological rhythms. The biological clock marks time with or without environmental cues but often requires signals from the environment to remain tuned to an appropriate period. *See also* circadian rhythm.

biological magnification A process in which retained substances become more concentrated at each higher trophic level in a food chain.

biological species concept Definition of a species as a group of populations whose members have the potential to interbreed in nature and produce viable, fertile offspring, but do not produce viable, fertile offspring with members of other such groups.

biology The scientific study of life.

biomanipulation An approach that applies the top-down model of community organization to alter ecosystem characteristics. For example, ecologists can prevent algal blooms and eutrophication by altering the density of higher-level consumers in lakes instead of by using chemical treatments.

biomass The total mass of organic matter comprising a group of organisms in a particular habitat.

biome (bī′-ōm) Any of the world's major ecosystem types, often classified according to the predominant vegetation for terrestrial biomes and the physical environment for aquatic biomes and characterized by adaptations of organisms to that particular environment.

bioremediation The use of organisms to detoxify and restore polluted and degraded ecosystems.

biosphere The entire portion of Earth inhabited by life; the sum of all the planet's ecosystems.

biotechnology The manipulation of organisms or their components to produce useful products.

biotic (bī-ot′-ik) Pertaining to the living factors—the organisms—in an environment.

bipolar cell A neuron that relays information between photoreceptors and ganglion cells in the retina.

bipolar disorder A depressive mental illness characterized by swings of mood from high to low; also called manic-depressive disorder.

birth control pill A chemical contraceptive that inhibits ovulation, retards follicular development, or alters a woman's cervical mucus to prevent sperm from entering the uterus.

blade (1) A leaflike structure of a seaweed that provides most of the surface area for photosynthesis. (2) The flattened portion of a typical leaf.

blastocoel (blas′-tuh-sēl) The fluid-filled cavity that forms in the center of a blastula.

blastocyst (blas′-tuh-sist) The blastula stage of mammalian embryonic development, consisting of an inner cell mass, a cavity, and an outer layer, the trophoblast. In humans, the blastocyst forms 1 week after fertilization.

blastomere An early embryonic cell arising during the cleavage stage of an early embryo.

blastopore (blas′-tō-pōr) In a gastrula, the opening of the archenteron that typically develops into the anus in deuterostomes and the mouth in protostomes.

blastula (blas′-tyū-luh) A hollow ball of cells that marks the end of the cleavage stage during early embryonic development in animals.

blood A connective tissue with a fluid matrix called plasma in which red blood cells, white blood cells, and cell fragments called platelets are suspended.

blue-light photoreceptor A type of light receptor in plants that initiates a variety of responses, such as phototropism and slowing of hypocotyl elongation.

body cavity A fluid- or air-filled space between the digestive tract and the body wall.

body plan In animals, a set of morphological and developmental traits that are integrated into a functional whole—the living animal.

Bohr shift A lowering of the affinity of hemoglobin for oxygen, caused by a drop in pH. It facilitates the release of oxygen from hemoglobin in the vicinity of active tissues.

bolus A lubricated ball of chewed food.

bone A connective tissue consisting of living cells held in a rigid matrix of collagen fibers embedded in calcium salts.

book lung An organ of gas exchange in spiders, consisting of stacked plates contained in an internal chamber.

bottleneck effect Genetic drift that occurs when the size of a population is reduced, as by a natural disaster or human actions. Typically, the surviving population is no longer genetically representative of the original population.

bottom-up model A model of community organization in which mineral nutrients influence community organization by controlling plant or phytoplankton numbers, which in turn control herbivore numbers, which in turn control predator numbers.

Bowman's capsule (bō′-munz) A cup-shaped receptacle in the vertebrate kidney that is the initial, expanded segment of the nephron where filtrate enters from the blood.

brachiopod (bra′-kē- uh-pod′) A marine lophophorate with a shell divided into dorsal and ventral halves. Brachiopods are also called lamp shells.

brain Organ of the central nervous system where information is processed and integrated.

brainstem A collection of structures in the vertebrate brain, including the midbrain, the pons, and the medulla oblongata; functions in homeostasis, coordination of movement, and conduction of information to higher brain centers.

branch point The representation on a phylogenetic tree of the divergence of two or more taxa from a common ancestor. A branch point is usually shown as a dichotomy in which a branch representing the ancestral lineage splits (at the branch point) into two branches, one for each of the two descendant lineages.

brassinosteroid A steroid hormone in plants that has a variety of effects, including inducing cell elongation, retarding leaf abscission, and promoting xylem differentiation.

breathing Ventilation of the lungs through alternating inhalation and exhalation.

bronchiole (brong′-kē-ōl′) A fine branch of the bronchi that transports air to alveoli.

bronchus (brong′-kus) (plural, **bronchi**) One of a pair of breathing tubes that branch from the trachea into the lungs.

brown alga A multicellular, photosynthetic protist with a characteristic brown or olive color that results from carotenoids in its plastids. Most brown algae are marine, and some have a plantlike body (thallus).

bryophyte (brī′-uh-fīt) An informal name for a moss, liverwort, or hornwort; a nonvascular plant that lives on land but lacks some of the terrestrial adaptations of vascular plants.

budding Asexual reproduction in which outgrowths from the parent form and pinch off to live independently or else remain attached to eventually form extensive colonies.

buffer A solution that contains a weak acid and its corresponding base. A buffer minimizes changes in pH when acids or bases are added to the solution.

bulk feeder An animal that eats relatively large pieces of food.

bulk flow The movement of a fluid due to a difference in pressure between two locations.

bundle-sheath cell In C_4 plants, a type of photosynthetic cell arranged into tightly packed sheaths around the veins of a leaf.

C_3 plant A plant that uses the Calvin cycle for the initial steps that incorporate CO_2 into organic material, forming a three-carbon compound as the first stable intermediate.

C_4 plant A plant in which the Calvin cycle is preceded by reactions that incorporate CO_2 into a four-carbon compound, the end product of which supplies CO_2 for the Calvin cycle.

calcitonin (kal′-si-tō′-nin) A hormone secreted by the thyroid gland that lowers blood calcium levels by promoting calcium deposition in bone and calcium excretion from the kidneys; nonessential in adult humans.

callus A mass of dividing, undifferentiated cells growing in culture.

calorie (cal) The amount of heat energy required to raise the temperature of 1 g of water by 1°C; also the amount of heat energy that 1 g of water releases when it cools by 1°C. The Calorie (with a capital C), usually used to indicate the energy content of food, is a kilocalorie.

Calvin cycle The second of two major stages in photosynthesis (following the light reactions), involving fixation of atmospheric CO_2 and reduction of the fixed carbon into carbohydrate.

CAM plant A plant that uses crassulacean acid metabolism, an adaptation for photosynthesis in arid conditions. In this process, carbon dioxide entering open stomata during the night is converted to organic acids, which release CO_2 for the Calvin cycle during the day, when stomata are closed.

Cambrian explosion A relatively brief time in geologic history when many present-day phyla of animals first appeared in the fossil record. This burst of evolutionary change occurred about 535–525 million years ago and saw the emergence of the first large, hard-bodied animals.

cAMP *See* cyclic AMP (cAMP).

canopy The uppermost layer of vegetation in a terrestrial biome.

capillary (kap′-il-ār′-ē) A microscopic blood vessel that penetrates the tissues and consists of a single layer of endothelial cells that allows exchange between the blood and interstitial fluid.

capillary bed A network of capillaries in a tissue or organ.

capsid The protein shell that encloses a viral genome. It may be rod-shaped, polyhedral, or more complex in shape.

capsule (1) In many prokaryotes, a dense and well-defined layer of polysaccharide or protein that surrounds the cell wall and is sticky, protecting the cell and enabling it to adhere to substrates or other cells. (2) The sporangium of a bryophyte (moss, liverwort, or hornwort).

carbohydrate (kar′-bō-hī′-drāt) A sugar (monosaccharide) or one of its dimers (disaccharides) or polymers (polysaccharides).

carbon fixation The initial incorporation of carbon from CO_2 into an organic compound by an autotrophic organism (a plant, another photosynthetic organism, or a chemoautotrophic prokaryote).

carbonyl group (kar-buh-nēl′) A chemical group present in aldehydes and ketones and consisting of a carbon atom double-bonded to an oxygen atom.

carboxyl group (kar-bok′-sil) A chemical group present in organic acids and consisting of a single carbon atom double-bonded to an oxygen atom and also bonded to a hydroxyl group.

cardiac cycle (kar′-dē-ak) The alternating contractions and relaxations of the heart.

cardiac muscle A type of striated muscle that forms the contractile wall of the heart. Its cells are joined by intercalated disks that relay the electrical signals underlying each heartbeat.

cardiac output The volume of blood pumped per minute by each ventricle of the heart.

cardiovascular system A closed circulatory system with a heart and branching network of arteries, capillaries, and veins. The system is characteristic of vertebrates.

carnivore An animal that mainly eats other animals.

carotenoid (kuh-rot′-uh-noyd′ An accessory pigment, either yellow or orange, in the chloroplasts of plants and in some prokaryotes. By absorbing wavelengths of light that chlorophyll cannot, carotenoids broaden the spectrum of colors that can drive photosynthesis.

carpel (kar′-pul) The ovule-producing reproductive organ of a flower, consisting of the stigma, style, and ovary.

carrier In genetics, an individual who is heterozygous at a given genetic locus for a recessively inherited disorder. The heterozygote is generally phenotypically normal for the disorder but can pass on the recessive allele to offspring.

carrying capacity The maximum population size that can be supported by the available resources, symbolized as *K*.

cartilage (kar′-til-ij) A flexible connective tissue with an abundance of collagenous fibers embedded in chondroitin sulfate.

Casparian strip (ka-spār′-ē-un) A water-impermeable ring of wax in the endodermal cells of plants that blocks the passive flow of water and solutes into the stele by way of cell walls.

catabolic pathway (kat′-uh-bol′-ik) A metabolic pathway that releases energy by breaking down complex molecules to simpler molecules.

catalyst (kat′-uh-list) A chemical agent that selectively increases the rate of a reaction without being consumed by the reaction.

catastrophism (kuh-tas′-truh-fiz′-um) The principle that events in the past occurred suddenly and were caused by different mechanisms than those operating today. *See* uniformitarianism.

catecholamine (kat′-uh-kōl′-uh-mēn) Any of a class of neurotransmitters and hormones, including the hormones epinephrine and norepinephrine, that are synthesized from the amino acid tyrosine.

cation (cat′-ī-on) A positively charged ion.

cation exchange A process in which positively charged minerals are made available to a plant when hydrogen ions in the soil displace mineral ions from the clay particles.

cDNA library A gene library containing clones that carry complementary DNA (cDNA) inserts. The library includes only the genes that were transcribed in the cells whose mRNA was isolated to make the cDNA.

cecum (sē′-kum) (plural, **ceca**) The blind pouch forming one branch of the large intestine.

cell body The part of a neuron that houses the nucleus and most other organelles.

cell cycle An ordered sequence of events in the life of a cell, from its origin in the division of a parent cell until its own division into two. The eukaryotic cell cycle is composed of interphase (including G_1, S, and G_2 subphases) and M phase (including mitosis and cytokinesis).

cell cycle control system A cyclically operating set of molecules in the eukaryotic cell that both triggers and coordinates key events in the cell cycle.

cell division The reproduction of cells.

cell fractionation The disruption of a cell and separation of its parts by centrifugation at successively higher speeds.

cell plate A membrane-bounded, flattened sac located at the midline of a dividing plant cell, inside which the new cell wall forms during cytokinesis.

cell wall A protective layer external to the plasma membrane in the cells of plants, prokaryotes, fungi, and some protists. Polysaccharides such as cellulose (in plants and some protists), chitin (in fungi), and peptidoglycan (in bacteria) are important structural components of cell walls.

cell-mediated immune response The branch of adaptive immunity that involves the activation of cytotoxic T cells, which defend against infected cells.

cellular respiration The catabolic pathways of aerobic and anaerobic respiration, which break down organic molecules and use an electron transport chain for the production of ATP.

cellular slime mold A type of protist characterized by unicellular amoeboid cells and aggregated reproductive bodies in its life cycle.

cellulose (sel′-yū-lōs) A structural polysaccharide of plant cell walls, consisting of glucose monomers joined by β glycosidic linkages.

Celsius scale (sel′-sē-us) A temperature scale (°C) equal to $^5/_9$(°F – 32) that measures the freezing point of water at 0°C and the boiling point of water at 100°C.

central canal The narrow cavity in the center of the spinal cord that is continuous with the fluid-filled ventricles of the brain.

central nervous system (CNS) The portion of the nervous system where signal integration occurs; in vertebrate animals, the brain and spinal cord.

central vacuole In a mature plant cell, a large membranous sac with diverse roles in growth, storage, and sequestration of toxic substances.

centriole (sen′-trē-ōl) A structure in the centrosome of an animal cell composed of a cylinder of microtubule triplets arranged in a 9 + 0 pattern. A centrosome has a pair of centrioles.

centromere (sen′-trō-mēr) In a duplicated chromosome, the region on each sister chromatid where they are most closely attached to each other by proteins that bind to specific DNA sequences; this close attachment causes a constriction in the condensed chromosome. (An uncondensed, unduplicated chromosome has a single centromere, identified by its DNA sequence.)

centrosome (sen′-trō-sōm) A structure present in the cytoplasm of animal cells that functions as a microtubule-organizing center and is important during cell division. A centrosome has two centrioles.

cephalization (sef′-uh-luh-zā′-shun) An evolutionary trend toward the concentration of sensory equipment at the anterior end of the body.

cercozoan An amoeboid or flagellated protist that feeds with threadlike pseudopodia.

cerebellum (sār′-ruh-bel′-um) Part of the vertebrate hindbrain located dorsally; functions in unconscious coordination of movement and balance.

cerebral cortex (suh-rē′-brul) The surface of the cerebrum; the largest and most complex part of the mammalian brain, containing nerve cell bodies of the cerebrum; the part of the vertebrate brain most changed through evolution.

cerebral hemisphere The right or left side of the cerebrum.

cerebrospinal fluid (suh-rē′-brō-spī′-nul) Blood-derived fluid that surrounds, protects against infection, nourishes, and cushions the brain and spinal cord.

cerebrum (suh-rē′-brum) The dorsal portion of the vertebrate forebrain, composed of right and left hemispheres; the integrating center for memory, learning, emotions, and other highly complex functions of the central nervous system.

cervix (ser′-viks) The neck of the uterus, which opens into the vagina.

chaparral A scrubland biome of dense, spiny evergreen shrubs found at midlatitudes along coasts where cold ocean currents circulate offshore; characterized by mild, rainy winters and long, hot, dry summers.

chaperonin (shap′-er-ō′-nin) A protein complex that assists in the proper folding of other proteins.

character An observable heritable feature that may vary among individuals.

character displacement The tendency for characteristics to be more divergent in sympatric populations of two species than in allopatric populations of the same two species.

checkpoint A control point in the cell cycle where stop and go-ahead signals can regulate the cycle.

chelicera (kē-lih′-suh-ruh) (plural, **chelicerae**) One of a pair of clawlike feeding appendages characteristic of chelicerates.

chelicerate (kē-lih-suh′-rāte) An arthropod that has chelicerae and a body divided into a cephalothorax and an abdomen. Living chelicerates include sea spiders, horseshoe crabs, scorpions, ticks, and spiders.

chemical bond An attraction between two atoms, resulting from a sharing of outer-shell electrons or the presence of opposite charges on the atoms. The bonded atoms gain complete outer electron shells.

chemical energy Energy available in molecules for release in a chemical reaction; a form of potential energy.

chemical equilibrium In a chemical reaction, the state in which the rate of the forward reaction equals the rate of the reverse reaction, so that the relative concentrations of the reactants and products do not change with time.

chemical reaction The making and breaking of chemical bonds, leading to changes in the composition of matter.

chemiosmosis (kem′-ē-oz-mō′-sis) An energy-coupling mechanism that uses energy stored in the form of a hydrogen ion gradient across a membrane to drive cellular work, such as the synthesis of ATP. Under aerobic conditions, most ATP synthesis in cells occurs by chemiosmosis.

chemoautotroph (kē′-mō-ot′-ō-trōf) An organism that obtains energy by oxidizing inorganic substances and needs only carbon dioxide as a carbon source.

chemoheterotroph (kē′-mō-het′-er-ō-trōf) An organism that requires organic molecules for both energy and carbon.

chemoreceptor A sensory receptor that responds to a chemical stimulus, such as a solute or an odorant.

chiasma (plural, **chiasmata**) (kī-az′-muh, kī-az′-muh-tuh) The X-shaped, microscopically visible region where crossing over has occurred earlier in prophase I between homologous nonsister chromatids. Chiasmata become visible after synapsis ends, with the two homologs remaining associated due to sister chromatid cohesion.

chitin (kī′-tin) A structural polysaccharide, consisting of amino sugar monomers, found in many fungal cell walls and in the exoskeletons of all arthropods.

chlorophyll (klōr′-ō-fil) A green pigment located in membranes within the chloroplasts of plants and algae and in the membranes of certain prokaryotes. Chlorophyll *a* participates directly in the light reactions, which convert solar energy to chemical energy.

chlorophyll *a* A photosynthetic pigment that participates directly in the light reactions, which convert solar energy to chemical energy.

chlorophyll *b* An accessory photosynthetic pigment that transfers energy to chlorophyll *a*.

chloroplast (klōr′-ō-plast) An organelle found in plants and photosynthetic protists that absorbs sunlight and uses it to drive the synthesis of organic compounds from carbon dioxide and water.

choanocyte (kō-an′-uh-sīt) A flagellated feeding cell found in sponges. Also called a collar cell, it has a collar-like ring that traps food particles around the base of its flagellum.

cholesterol (kō-les′-tuh-rol) A steroid that forms an essential component of animal cell membranes and acts as a precursor molecule for the synthesis of other biologically important steroids, such as many hormones.

chondrichthyan (kon-drik′-thē-an) Member of the class Chondrichthyes, vertebrates with skeletons made mostly of cartilage, such as sharks and rays.

chordate Member of the phylum Chordata, animals that at some point during their development have a notochord; a dorsal, hollow nerve cord; pharyngeal slits or clefts; and a muscular, post-anal tail.

chorionic villus sampling (CVS) (kōr′-ē-on′-ik vil′-us) A technique associated with prenatal diagnosis in which a small sample of the fetal portion of the placenta is removed for analysis to detect certain genetic and congenital defects in the fetus.

Chromalveolata One of five supergroups of eukaryotes proposed in a current hypothesis of the evolutionary history of eukaryotes. Chromalveolates may have originated by secondary endosymbiosis and include two large protist clades, the alveolates and the stramenopiles. *See also* Excavata, Rhizaria, Archaeplastida, and Unikonta.

chromatin (krō′-muh-tin) The complex of DNA and proteins that makes up eukaryotic chromosomes. When the cell is not dividing, chromatin exists in its dispersed form, as a mass of very long, thin fibers that are not visible with a light microscope.

chromosome (krō′-muh-sōm) A cellular structure carrying genetic material, found in the nucleus of eukaryotic cells. Each chromosome consists of one very long DNA molecule and associated proteins. (A bacterial chromosome usually consists of a single circular DNA molecule and associated proteins. It is found in the nucleoid region, which is not membrane bounded.) *See also* chromatin.

chromosome theory of inheritance A basic principle in biology stating that genes are located at specific positions (loci) on chromosomes and that the behavior of chromosomes during meiosis accounts for inheritance patterns.

chylomicron (kī′-lō-mī′-kron) A lipid transport globule composed of fats mixed with cholesterol and coated with proteins.

chyme (kīm) The mixture of partially digested food and digestive juices formed in the stomach.

chytrid (kī′-trid) Member of the fungal phylum Chytridiomycota, mostly aquatic fungi with flagellated zoospores that represent an early-diverging fungal lineage.

ciliate (sil′-ē-it) A type of protist that moves by means of cilia.

cilium (sil′-ē-um) (plural, **cilia**) A short appendage containing microtubules in eukaryotic cells. A motile cilium is specialized for locomotion or moving fluid past the cell; it is formed from a core of nine outer doublet microtubules and two inner single microtubules (the "9 + 2" arrangement) ensheathed in an extension of the plasma membrane. A primary cilium is usually nonmotile and plays a sensory and signaling role; it lacks the two inner microtubules (the "9 + 0" arrangement).

circadian rhythm (ser-kā′-dē-un) A physiological cycle of about 24 hours that persists even in the absence of external cues.

***cis-trans* isomer** One of several compounds that have the same molecular formula and covalent bonds between atoms but differ in the spatial arrangements of their atoms owing to the inflexibility of double bonds; formerly called a geometric isomer.

citric acid cycle A chemical cycle involving eight steps that completes the metabolic breakdown of glucose molecules begun in glycolysis by oxidizing acetyl CoA (derived from pyruvate) to carbon dioxide; occurs within the mitochondrion in eukaryotic cells and in the cytosol of prokaryotes; together with pyruvate oxidation, the second major stage in cellular respiration.

clade (klayd) A group of species that includes an ancestral species and all of its descendants.

cladistics (kluh-dis′-tiks) An approach to systematics in which organisms are placed into groups called clades based primarily on common descent.

class In Linnaean classification, the taxonomic category above the level of order.

cleavage (1) The process of cytokinesis in animal cells, characterized by pinching of the plasma membrane. (2) The succession of rapid cell divisions without significant growth during early embryonic development that converts the zygote to a ball of cells.

cleavage furrow The first sign of cleavage in an animal cell; a shallow groove around the cell in the cell surface near the old metaphase plate.

climate The long-term prevailing weather conditions at a given place.

climograph A plot of the temperature and precipitation in a particular region.

cline A graded change in a character along a geographic axis.

clitoris (klit′-uh-ris) An organ at the upper intersection of the labia minora that engorges with blood and becomes erect during sexual arousal.

cloaca (klō-ā′-kuh) A common opening for the digestive, urinary, and reproductive tracts found in many nonmammalian vertebrates but in few mammals.

clonal selection The process by which an antigen selectively binds to and activates only those lymphocytes bearing receptors specific for the antigen. The selected lymphocytes proliferate and differentiate into a clone of effector cells and a clone of memory cells specific for the stimulating antigen.

clone (1) A lineage of genetically identical individuals or cells. (2) In popular usage, an individual that is genetically identical to another individual. (3) As a verb, to make one or more genetic replicas of an individual or cell. *See also* gene cloning.

cloning vector In genetic engineering, a DNA molecule that can carry foreign DNA into a host cell and replicate there. Cloning vectors include plasmids and bacterial artificial chromosomes (BACs), which move recombinant DNA from a test tube back into a cell, and viruses that transfer recombinant DNA by infection.

closed circulatory system A circulatory system in which blood is confined to vessels and is kept separate from the interstitial fluid.

cnidocyte (nī′-duh-sīt) A specialized cell unique to the phylum Cnidaria; contains a capsule-like organelle housing a coiled thread that, when discharged, explodes outward and functions in prey capture or defense.

cochlea (kok′-lē-uh) The complex, coiled organ of hearing that contains the organ of Corti.

codominance The situation in which the phenotypes of both alleles are exhibited in the heterozygote because both alleles affect the phenotype in separate, distinguishable ways.

codon (kō′-don) A three-nucleotide sequence of DNA or mRNA that specifies a particular amino acid or termination signal; the basic unit of the genetic code.

coefficient of relatedness The fraction of genes that, on average, are shared by two individuals.

coelom (sē′-lōm) A body cavity lined by tissue derived only from mesoderm.

coelomate (sē′-lō-māt) An animal that possesses a true coelom (a body cavity lined by tissue completely derived from mesoderm).

coenocytic fungus (sē′-no-si′-tic) A fungus that lacks septa and hence whose body is made up of a continuous cytoplasmic mass that may contain hundreds or thousands of nuclei.

coenzyme (kō-en′-zīm) An organic molecule serving as a cofactor. Most vitamins function as coenzymes in metabolic reactions.

coevolution The joint evolution of two interacting species, each in response to selection imposed by the other.

cofactor Any nonprotein molecule or ion that is required for the proper functioning of an enzyme. Cofactors can be permanently bound to the active site or may bind loosely and reversibly, along with the substrate, during catalysis.

cognition The process of knowing that may include awareness, reasoning, recollection, and judgment.

cognitive map A neural representation of the abstract spatial relationships between objects in an animal's surroundings.

cohesion The linking together of like molecules, often by hydrogen bonds.

cohesion-tension hypothesis The leading explanation of the ascent of xylem sap. It states that transpiration exerts pull on xylem sap, putting the sap under negative pressure or tension, and that the cohesion of water molecules transmits this pull along the entire length of the xylem from shoots to roots.

cohort A group of individuals of the same age in a population.

coitus (kō′-uh-tus) The insertion of a penis into a vagina; also called sexual intercourse.

coleoptile (kō′-lē-op′-tul) The covering of the young shoot of the embryo of a grass seed.

coleorhiza (kō′-lē-uh-rī′-zuh) The covering of the young root of the embryo of a grass seed.

collagen A glycoprotein in the extracellular matrix of animal cells that forms strong fibers, found extensively in connective tissue and bone; the most abundant protein in the animal kingdom.

collecting duct The location in the kidney where processed filtrate, called urine, is collected from the renal tubules.

collenchyma cell (kō-len′-kim-uh) A flexible plant cell type that occurs in strands or cylinders that support young parts of the plant without restraining growth.

colloid A mixture made up of a liquid and particles that (because of their large size) remain suspended rather than dissolved in that liquid.

colon (kō′-len) The largest section of the vertebrate large intestine; functions in water absorption and formation of feces.

commensalism (kuh-men′-suh-lizm) A symbiotic relationship in which one organism benefits but the other is neither helped nor harmed.

communication In animal behavior, a process involving transmission of, reception of, and response to signals. The term is also used in connection with other organisms, as well as individual cells of multicellular organisms.

community All the organisms that inhabit a particular area; an assemblage of populations of different species living close enough together for potential interaction.

community ecology The study of how interactions between species affect community structure and organization.

companion cell A type of plant cell that is connected to a sieve-tube element by many plasmodesmata and whose nucleus and ribosomes may serve one or more adjacent sieve-tube elements.

competitive exclusion The concept that when populations of two similar species compete for the same limited resources, one population will use the resources more efficiently and have a reproductive advantage that will eventually lead to the elimination of the other population.

competitive inhibitor A substance that reduces the activity of an enzyme by entering the active site in place of the substrate, whose structure it mimics.

complement system A group of about 30 blood proteins that may amplify the inflammatory response, enhance phagocytosis, or directly lyse extracellular pathogens.

complementary DNA (cDNA) A double-stranded DNA molecule made *in vitro* using mRNA as a template and the enzymes reverse transcriptase and DNA polymerase. A cDNA molecule corresponds to the exons of a gene.

complete digestive tract A digestive tube that runs between a mouth and an anus; also called an alimentary canal.

complete dominance The situation in which the phenotypes of the heterozygote and dominant homozygote are indistinguishable.

complete flower A flower that has all four basic floral organs: sepals, petals, stamens, and carpels.

complete metamorphosis The transformation of a larva into an adult that looks very different, and often functions very differently in its environment, than the larva.

compound A substance consisting of two or more different elements combined in a fixed ratio.

compound eye A type of multifaceted eye in insects and crustaceans consisting of up to several thousand light-detecting, focusing ommatidia.

concentration gradient A region along which the density of a chemical substance increases or decreases.

conception The fertilization of an egg by a sperm in humans.

condom A thin, latex rubber or natural membrane sheath that fits over the penis to collect semen.

conduction The direct transfer of thermal motion (heat) between molecules of objects in direct contact with each other.

cone A cone-shaped cell in the retina of the vertebrate eye, sensitive to color.

conformer An animal for which an internal condition conforms to (changes in accordance with) changes in an environmental variable.

conidium (plural, **conidia**) A haploid spore produced at the tip of a specialized hypha in ascomycetes during asexual reproduction.

conifer Member of the largest gymnosperm phylum. Most conifers are cone-bearing trees, such as pines and firs.

conjugation (kon′-jū-gā′-shun) (1) In prokaryotes, the direct transfer of DNA between two cells that are temporarily joined. When the two cells are members of different species, conjugation results in horizontal gene transfer. (2) In ciliates, a sexual process in which two cells exchange haploid micronuclei but do not reproduce.

connective tissue Animal tissue that functions mainly to bind and support other tissues, having a sparse population of cells scattered through an extracellular matrix.

conodont An early, soft-bodied vertebrate with prominent eyes and dental elements.

conservation biology The integrated study of ecology, evolutionary biology, physiology, molecular biology, and genetics to sustain biological diversity at all levels.

contraception The deliberate prevention of pregnancy.

contractile vacuole A membranous sac that helps move excess water out of certain freshwater protists.

control element A segment of noncoding DNA that helps regulate transcription of a gene by serving as a binding site for a transcription factor. Multiple control elements are present in a eukaryotic gene's enhancer.

controlled experiment An experiment in which an experimental group is compared with a control group that varies only in the factor being tested.

convection The mass movement of warmed air or liquid to or from the surface of a body or object.

convergent evolution The evolution of similar features in independent evolutionary lineages.

convergent extension A process in which the cells of a tissue layer rearrange themselves in such a way that the sheet of cells becomes narrower (converges) and longer (extends).

cooperativity A kind of allosteric regulation whereby a shape change in one subunit of a protein caused by substrate binding is transmitted to all the other subunits, facilitating binding of additional substrate molecules to those subunits.

copepod (cō′-puh-pod) Any of a group of small crustaceans that are important members of marine and freshwater plankton communities.

coral reef Typically a warm-water, tropical ecosystem dominated by the hard skeletal structures secreted primarily by corals. Some coral reefs also exist in cold, deep waters.

corepressor A small molecule that binds to a bacterial repressor protein and changes the protein's shape, allowing it to bind to the operator and switch an operon off.

cork cambium (kam′-bē-um) A cylinder of meristematic tissue in woody plants that replaces the epidermis with thicker, tougher cork cells.

corpus callosum (kor′-pus kuh-lō′-sum) The thick band of nerve fibers that connects the right and left cerebral hemispheres in mammals, enabling the hemispheres to process information together.

corpus luteum (kor′-pus lū′-tē-um) A secreting tissue in the ovary that forms from the collapsed follicle after ovulation and produces progesterone.

cortex (1) The outer region of cytoplasm in a eukaryotic cell, lying just under the plasma membrane, that has a more gel-like consistency than the inner regions due to the presence of multiple microfilaments. (2) In plants, ground tissue that is between the vascular tissue and dermal tissue in a root or eudicot stem.

cortical nephron In mammals and birds, a nephron with a loop of Henle located almost entirely in the renal cortex.

corticosteroid Any steroid hormone produced and secreted by the adrenal cortex.

cotransport The coupling of the "downhill" diffusion of one substance to the "uphill" transport of another against its own concentration gradient.

cotyledon (kot′-uh-lē′-dun) A seed leaf of an angiosperm embryo. Some species have one cotyledon, others two.

countercurrent exchange The exchange of a substance or heat between two fluids flowing in opposite directions. For example, blood in a fish gill flows in the opposite direction of water passing over the gill, maximizing diffusion of oxygen into and carbon dioxide out of the blood.

countercurrent multiplier system A countercurrent system in which energy is expended in active transport to facilitate exchange of materials and generate concentration gradients.

covalent bond (kō-vā′-lent) A type of strong chemical bond in which two atoms share one or more pairs of valence electrons.

craniate A chordate with a head.

crassulacean acid metabolism (CAM) An adaptation for photosynthesis in arid conditions, first discovered in the family Crassulaceae. In this process, a plant takes up CO_2 and incorporates it into a variety of organic acids at night; during the day, CO_2 is released from organic acids for use in the Calvin cycle.

crista (plural, **cristae**) (kris′-tuh, kris′-tē) An infolding of the inner membrane of a mitochondrion. The inner membrane houses electron transport chains and molecules of the enzyme catalyzing the synthesis of ATP (ATP synthase).

critical load The amount of added nutrient, usually nitrogen or phosphorus, that can be absorbed by plants without damaging ecosystem integrity.

crop rotation The practice of planting nonlegumes one year and legumes in alternating years to restore concentrations of fixed nitrogen in the soil.

cross-fostering study A behavioral study in which the young of one species are placed in the care of adults from another species.

crossing over The reciprocal exchange of genetic material between nonsister chromatids during prophase I of meiosis.

cross-pollination In angiosperms, the transfer of pollen from an anther of a flower on one plant to the stigma of a flower on another plant of the same species.

crustacean (kruh-stā′-shun) A member of a subphylum of mostly aquatic arthropods that includes lobsters, crayfishes, crabs, shrimps, and barnacles.

cryptic coloration Camouflage that makes a potential prey difficult to spot against its background.

culture A system of information transfer through social learning or teaching that influences the behavior of individuals in a population.

cuticle (kyū′-tuh-kul) (1) A waxy covering on the surface of stems and leaves that prevents desiccation in terrestrial plants. (2) The exoskeleton of an arthropod, consisting of layers of protein and chitin that are variously modified for different functions. (3) A tough coat that covers the body of a nematode.

cyclic AMP (cAMP) Cyclic adenosine monophosphate, a ring-shaped molecule made from ATP that is a common intracellular signaling molecule (second messenger) in eukaryotic cells. It is also a regulator of some bacterial operons.

cyclic electron flow A route of electron flow during the light reactions of photosynthesis that involves only photosystem I and that produces ATP but not NADPH or O_2.

cyclin (sī′-klin) A cellular protein that occurs in a cyclically fluctuating concentration and that plays an important role in regulating the cell cycle.

cyclin-dependent kinase (Cdk) A protein kinase that is active only when attached to a particular cyclin.

cystic fibrosis (sis′-tik fī-brō′-sis) A human genetic disorder caused by a recessive allele for a chloride channel protein; characterized by an excessive secretion of mucus and consequent vulnerability to infection; fatal if untreated.

cytochrome (sī′-tō-krōm) An iron-containing protein that is a component of electron transport chains in the mitochondria and chloroplasts of eukaryotic cells and the plasma membranes of prokaryotic cells.

cytogenetic map A map of a chromosome that locates genes with respect to chromosomal features distinguishable in a microscope.

cytokine (sī′-tō-kīn′) Any of a group of small proteins secreted by a number of cell types, including macrophages and helper T cells, that regulate the function of other cells.

cytokinesis (sī′-tō-kuh-nē′-sis) The division of the cytoplasm to form two separate daughter cells immediately after mitosis, meiosis I, or meiosis II.

cytokinin (sī′-tō-kī′-nin) Any of a class of related plant hormones that retard aging and act in concert with auxin to stimulate cell division, influence the pathway of differentiation, and control apical dominance.

cytoplasm (sī′-tō-plaz′-um) The contents of the cell bounded by the plasma membrane; in eukaryotes, the portion exclusive of the nucleus.

cytoplasmic determinant A maternal substance, such as a protein or RNA, that when placed into an egg influences the course of early development by regulating the expression of genes that affect the developmental fate of cells.

cytoplasmic streaming A circular flow of cytoplasm, involving interactions of myosin and actin filaments, that speeds the distribution of materials within cells.

cytoskeleton A network of microtubules, microfilaments, and intermediate filaments that extend throughout the cytoplasm and serve a variety of mechanical, transport, and signaling functions.

cytosol (sī′-tō-sol) The semifluid portion of the cytoplasm.

cytotoxic T cell A type of lymphocyte that, when activated, kills infected cells as well as certain cancer cells and transplanted cells.

dalton A measure of mass for atoms and subatomic particles; the same as the atomic mass unit, or amu.

data Recorded observations.

day-neutral plant A plant in which flower formation is not controlled by photoperiod or day length.

decapod A member of the group of crustaceans that includes lobsters, crayfishes, crabs, and shrimps.

decomposer An organism that absorbs nutrients from nonliving organic material such as corpses, fallen plant material, and the wastes of living organisms and converts them to inorganic forms; a detritivore.

deductive reasoning A type of logic in which specific results are predicted from a general premise.

deep-sea hydrothermal vent A dark, hot, oxygen-deficient environment associated with volcanic activity on or near the seafloor. The producers in a vent community are chemoautotrophic prokaryotes.

de-etiolation The changes a plant shoot undergoes in response to sunlight; also known informally as greening.

dehydration reaction A chemical reaction in which two molecules become covalently bonded to each other with the removal of a water molecule.

deletion (1) A deficiency in a chromosome resulting from the loss of a fragment through breakage. (2) A mutational loss of one or more nucleotide pairs from a gene.

demographic transition In a stable population, a shift from high birth and death rates to low birth and death rates.

demography The study of changes over time in the vital statistics of populations, especially birth rates and death rates.

denaturation (dē-nā′-chur-ā′-shun) In proteins, a process in which a protein loses its native shape due to the disruption of weak chemical bonds and interactions, thereby becoming biologically inactive; in DNA, the separation of the two strands of the double helix. Denaturation occurs under extreme (noncellular) conditions of pH, salt concentration, or temperature.

dendrite (den′-drīt) One of usually numerous, short, highly branched extensions of a neuron that receive signals from other neurons.

dendritic cell An antigen-presenting cell, located mainly in lymphatic tissues and skin, that is particularly efficient in presenting antigens to helper T cells, thereby initiating a primary immune response.

density The number of individuals per unit area or volume.

density dependent Referring to any characteristic that varies with population density.

density independent Referring to any characteristic that is not affected by population density.

density-dependent inhibition The phenomenon observed in normal animal cells that causes them to stop dividing when they come into contact with one another.

deoxyribonucleic acid (DNA) (dē-ok′-sē-rī′-bō-nū-klā′-ik) A nucleic acid molecule, usually a double-stranded helix, in which each polynucleotide strand consists of nucleotide monomers with a deoxyribose sugar and the nitrogenous bases adenine (A), cytosine (C), guanine (G), and thymine (T); capable of being replicated and determining the inherited structure of a cell's proteins.

deoxyribose (dē-ok′-si-rī′-bōs) The sugar component of DNA nucleotides, having one fewer hydroxyl group than ribose, the sugar component of RNA nucleotides.

depolarization A change in a cell's membrane potential such that the inside of the membrane is made less negative relative to the outside. For example, a neuron membrane is depolarized if a stimulus decreases its voltage from the resting potential of −70 mV in the direction of zero voltage.

dermal tissue system The outer protective covering of plants.

desert A terrestrial biome characterized by very low precipitation.

desmosome A type of intercellular junction in animal cells that functions as a rivet, fastening cells together.

determinate cleavage A type of embryonic development in protostomes that rigidly casts the developmental fate of each embryonic cell very early.

determinate growth A type of growth characteristic of most animals and some plant organs, in which growth stops after a certain size is reached.

determination The progressive restriction of developmental potential in which the possible fate of each cell becomes more limited as an embryo develops. At the end of determination, a cell is committed to its fate.

detritivore (deh-trī′-tuh-vōr) A consumer that derives its energy and nutrients from nonliving organic material such as corpses, fallen plant material, and the wastes of living organisms; a decomposer.

detritus (di-trī′-tus) Dead organic matter.

deuteromycete (dū′-tuh-rō-mī′-sēt) Traditional classification for a fungus with no known sexual stage.

deuterostome development (dū′-tuh-rō-stōm′) In animals, a developmental mode distinguished by the development of the anus from the blastopore; often also characterized by radial cleavage and by the body cavity forming as outpockets of mesodermal tissue.

development The events involved in an organism's changing gradually from a simple to a more complex or specialized form.

diabetes mellitus (dī′-uh-bē′-tis mel′-uh-tus) An endocrine disorder marked by an inability to maintain glucose homeostasis. The type 1 form results from autoimmune destruction of insulin-secreting cells; treatment usually requires daily insulin injections. The type 2 form most commonly results from reduced responsiveness of target cells to insulin; obesity and lack of exercise are risk factors.

diacylglycerol (DAG) (dī-a′-sil-glis′-er-ol) A second messenger produced by the cleavage of the phospholipid PIP_2 in the plasma membrane.

diaphragm (dī′-uh-fram′) (1) A sheet of muscle that forms the bottom wall of the thoracic cavity in mammals. Contraction of the diaphragm pulls air into the lungs. (2) A dome-shaped rubber cup fitted into the upper portion of the vagina before sexual intercourse. It serves as a physical barrier to the passage of sperm into the uterus.

diapsid (dī-ap′-sid) Member of an amniote clade distinguished by a pair of holes on each side of the skull. Diapsids include the lepidosaurs and archosaurs.

diastole (dī-as′-tō-lē) The stage of the cardiac cycle in which a heart chamber is relaxed and fills with blood.

diastolic pressure Blood pressure in the arteries when the ventricles are relaxed.

dicot A term traditionally used to refer to flowering plants that have two embryonic seed leaves, or cotyledons. Recent molecular evidence indicates that dicots do not form a clade; species once classified as dicots are now grouped into eudicots, magnoliids, and several lineages of basal angiosperms.

differential gene expression The expression of different sets of genes by cells with the same genome.

differentiation The process by which a cell or group of cells become specialized in structure and function.

diffusion The spontaneous movement of a substance down its concentration or electrochemical gradient, from a region where it is more concentrated to a region where it is less concentrated.

digestion The second stage of food processing in animals: the breaking down of food into molecules small enough for the body to absorb.

dihybrid (dī′-hī′-brid) An organism that is heterozygous with respect to two genes of interest. All the offspring from a cross between parents doubly homozygous for different alleles are dihybrids. For example, parents of genotypes *AABB* and *aabb* produce a dihybrid of genotype *AaBb*.

dihybrid cross A cross between two organisms that are each heterozygous for both of the characters being followed (or the self-pollination of a plant that is heterozygous for both characters).

dikaryotic (dī′-kār-ē-ot′-ik) Referring to a fungal mycelium with two haploid nuclei per cell, one from each parent.

dinoflagellate (dī′-nō-flaj′-uh-let) Member of a group of mostly unicellular photosynthetic algae with two flagella situated in perpendicular grooves in cellulose plates covering the cell.

dinosaur Member of an extremely diverse clade of reptiles varying in body shape, size, and habitat. Birds are the only extant dinosaurs.

dioecious (dī-ē′-shus) In plant biology, having the male and female reproductive parts on different individuals of the same species.

diploblastic Having two germ layers.

diploid cell (dip′-loyd) A cell containing two sets of chromosomes ($2n$), one set inherited from each parent.

diplomonad A protist that has modified mitochondria, two equal-sized nuclei, and multiple flagella.

directional selection Natural selection in which individuals at one end of the phenotypic range survive or reproduce more successfully than do other individuals.

disaccharide (dī-sak′-uh-rīd) A double sugar, consisting of two monosaccharides joined by a glycosidic linkage formed by a dehydration reaction.

dispersal The movement of individuals or gametes away from their parent location. This movement sometimes expands the geographic range of a population or species.

dispersion The pattern of spacing among individuals within the boundaries of a population.

disruptive selection Natural selection in which individuals on both extremes of a phenotypic range survive or reproduce more successfully than do individuals with intermediate phenotypes.

distal tubule In the vertebrate kidney, the portion of a nephron that helps refine filtrate and empties it into a collecting duct.

disturbance A natural or human-caused event that changes a biological community and usually removes organisms from it. Disturbances, such as fires and storms, play a pivotal role in structuring many communities.

disulfide bridge A strong covalent bond formed when the sulfur of one cysteine monomer bonds to the sulfur of another cysteine monomer.

DNA (deoxyribonucleic acid) (dē-ok′-sē-rī′-bō-nū-klā′-ik) A double-stranded, helical nucleic acid molecule, consisting of nucleotide monomers with a deoxyribose sugar and the nitrogenous bases adenine (A), cytosine (C), guanine (G), and thymine (T); capable of being replicated and determining the inherited structure of a cell's proteins.

DNA ligase (lī′-gās) A linking enzyme essential for DNA replication; catalyzes the covalent bonding of the 3′ end of one DNA fragment (such as an Okazaki fragment) to the 5′ end of another DNA fragment (such as a growing DNA chain).

DNA methylation The presence of methyl groups on the DNA bases (usually cytosine) of plants, animals, and fungi. (The term also refers to the process of adding methyl groups to DNA bases.)

DNA microarray assay A method to detect and measure the expression of thousands of genes at one time. Tiny amounts of a large number of single-stranded DNA fragments representing different genes are fixed to a glass slide and tested for hybridization with samples of labeled cDNA.

DNA polymerase (puh-lim′-er-ās) An enzyme that catalyzes the elongation of new DNA (for example, at a replication fork) by the addition of nucleotides to the 3′ end of an existing chain. There are several different DNA polymerases; DNA polymerase III and DNA polymerase I play major roles in DNA replication in *E. coli*.

DNA replication The process by which a DNA molecule is copied; also called DNA synthesis.

domain (1) A taxonomic category above the kingdom level. The three domains are Archaea, Bacteria, and Eukarya. (2) A discrete structural and functional region of a protein.

dominant allele An allele that is fully expressed in the phenotype of a heterozygote.

dominant species A species with substantially higher abundance or biomass than other species in a community. Dominant species exert a powerful control over the occurrence and distribution of other species.

dopamine A neurotransmitter that is a catecholamine, like epinephrine and norepinephrine.

dormancy A condition typified by extremely low metabolic rate and a suspension of growth and development.

dorsal Pertaining to the top of an animal with radial or bilateral symmetry.

dorsal lip The region above the blastopore on the dorsal side of the amphibian embryo.

double bond A double covalent bond; the sharing of two pairs of valence electrons by two atoms.

double circulation A circulatory system consisting of separate pulmonary and systemic circuits, in which blood passes through the heart after completing each circuit.

double fertilization A mechanism of fertilization in angiosperms in which two sperm cells unite with two cells in the female gametophyte (embryo sac) to form the zygote and endosperm.

double helix The form of native DNA, referring to its two adjacent antiparallel polynucleotide strands wound around an imaginary axis into a spiral shape.

Down syndrome A human genetic disease usually caused by the presence of an extra chromosome 21; characterized by developmental delays and heart and other defects that are generally treatable or non-life-threatening.

Duchenne muscular dystrophy (duh-shen′) A human genetic disease caused by a sex-linked recessive allele; characterized by progressive weakening and a loss of muscle tissue.

duodenum (dū′-uh-dēn′-um) The first section of the small intestine, where chyme from the stomach mixes with digestive juices from the pancreas, liver, and gallbladder as well as from gland cells of the intestinal wall.

duplication An aberration in chromosome structure due to fusion with a fragment from a homologous chromosome, such that a portion of a chromosome is duplicated.

dynamic stability hypothesis The concept that long food chains are less stable than short chains.

dynein (dī′-nē-un) In cilia and flagella, a large motor protein extending from one microtubule doublet to the adjacent doublet. ATP hydrolysis drives changes in dynein shape that lead to bending of cilia and flagella.

E site One of a ribosome's three binding sites for tRNA during translation. The E site is the place where discharged tRNAs leave the ribosome. (E stands for exit.)

ecdysozoan Member of a group of animal phyla identified as a clade by molecular evidence. Many ecdysozoans are molting animals.

ecdysteroid A steroid hormone, secreted by the prothoracic glands, that triggers molting in arthropods.

echinoderm (i-kī′-nō-derm) A slow-moving or sessile marine deuterostome with a water vascular system and, in larvae, bilateral symmetry. Echinoderms include sea stars, brittle stars, sea urchins, feather stars, and sea cucumbers.

ecological footprint The aggregate land and water area required by a person, city, or nation to produce all of the resources it consumes and to absorb all of the wastes it generates.

ecological niche (nich) The sum of a species' use of the biotic and abiotic resources in its environment.

ecological species concept A definition of species in terms of ecological niche, the sum of how members of the species interact with the nonliving and living parts of their environment.

ecological succession Transition in the species composition of a community following a disturbance; establishment of a community in an area virtually barren of life.

ecology The study of how organisms interact with each other and their environment.

ecosystem All the organisms in a given area as well as the abiotic factors with which they interact; one or more communities and the physical environment around them.

ecosystem ecology The study of energy flow and the cycling of chemicals among the various biotic and abiotic components in an ecosystem.

ecosystem engineer An organism that influences community structure by causing physical changes in the environment.

ecosystem service A function performed by an ecosystem that directly or indirectly benefits humans.

ecotone The transition from one type of habitat or ecosystem to another, such as the transition from a forest to a grassland.

ectoderm (ek′-tō-durm) The outermost of the three primary germ layers in animal embryos; gives rise to the outer covering and, in some phyla, the nervous system, inner ear, and lens of the eye.

ectomycorrhiza (ek′-tō-mī′-kō-rī′-zuh) Association of a fungus with a plant root system in which the fungus surrounds the roots but does not cause invagination of the host (plant) cells' plasma membranes.

ectomycorrhizal fungus A symbiotic fungus that forms sheaths of hyphae over the surface of plant roots and also grows into extracellular spaces of the root cortex.

ectoparasite A parasite that feeds on the external surface of a host.

ectopic Occurring in an abnormal location.

ectoproct A sessile, colonial lophophorate; also called a bryozoan.

ectothermic Referring to organisms for which external sources provide most of the heat for temperature regulation.

Ediacaran biota (ē′-dē-uh-keh′-run bī-ō′-tuh) An early group of soft-bodied, multicellular eukaryotes known from fossils that range in age from 565 million to 550 million years old.

effective population size An estimate of the size of a population based on the numbers of females and males that successfully breed; generally smaller than the total population.

effector cell (1) A muscle cell or gland cell that performs the body's response to stimuli as directed by signals from the brain or other processing center of the nervous system. (2) A lymphocyte that has undergone clonal selection and is capable of mediating an adaptive immune response.

egg The female gamete.

egg-polarity gene A gene that helps control the orientation (polarity) of the egg; also called a maternal effect gene.

ejaculation The propulsion of sperm from the epididymis through the muscular vas deferens, ejaculatory duct, and urethra.

ejaculatory duct In mammals, the short section of the ejaculatory route formed by the convergence of the vas deferens and a duct from the seminal vesicle. The ejaculatory duct transports sperm from the vas deferens to the urethra.

electrocardiogram (ECG or EKG) A record of the electrical impulses that travel through heart muscle during the cardiac cycle.

electrochemical gradient The diffusion gradient of an ion, which is affected by both the concentration difference of an ion across a membrane (a chemical force) and the ion's tendency to move relative to the membrane potential (an electrical force).

electrogenic pump An active transport protein that generates voltage across a membrane while pumping ions.

electromagnetic receptor A receptor of electromagnetic energy, such as visible light, electricity, or magnetism.

electromagnetic spectrum The entire spectrum of electromagnetic radiation, ranging in wavelength from less than a nanometer to more than a kilometer.

electron A subatomic particle with a single negative electrical charge and a mass about 1/2,000 that of a neutron or proton. One or more electrons move around the nucleus of an atom.

electron microscope (EM) A microscope that uses magnets to focus an electron beam on or through a specimen, resulting in a practical resolution of a hundredfold greater than that of a light microscope using standard techniques. A transmission electron microscope (TEM) is used to study the internal structure of thin sections of cells. A scanning electron microscope (SEM) is used to study the fine details of cell surfaces.

electron shell An energy level of electrons at a characteristic average distance from the nucleus of an atom.

electron transport chain A sequence of electron carrier molecules (membrane proteins) that shuttle electrons down a series of redox reactions that release energy used to make ATP.

electronegativity The attraction of a given atom for the electrons of a covalent bond.

electroporation A technique to introduce recombinant DNA into cells by applying a brief electrical pulse to a solution containing the cells. The pulse creates temporary holes in the cells' plasma membranes, through which DNA can enter.

element Any substance that cannot be broken down to any other substance by chemical reactions.

elimination The fourth and final stage of food processing in animals: the passing of undigested material out of the body.

embryo sac (em′-brē-ō) The female gametophyte of angiosperms, formed from the growth and division of the megaspore into a multicellular structure that typically has eight haploid nuclei.

embryonic lethal A mutation with a phenotype leading to death of an embryo or larva.

embryophyte Alternate name for land plants that refers to their shared derived trait of multicellular, dependent embryos.

emergent properties New properties that arise with each step upward in the hierarchy of life, owing to the arrangement and interactions of parts as complexity increases.

emigration The movement of individuals out of a population.

enantiomer (en-an′-tē-ō-mer) One of two compounds that are mirror images of each other and that differ in shape due to the presence of an asymmetric carbon.

endangered species A species that is in danger of extinction throughout all or a significant portion of its range.

endemic (en-dem′-ik) Referring to a species that is confined to a specific geographic area.

endergonic reaction (en′-der-gon′-ik) A nonspontaneous chemical reaction, in which free energy is absorbed from the surroundings.

endocrine gland (en′-dō-krin) A ductless gland that secretes hormones directly into the interstitial fluid, from which they diffuse into the bloodstream.

endocrine system The internal system of communication involving hormones, the ductless glands that secrete hormones, and the molecular receptors on or in target cells that respond to hormones; functions in concert with the nervous system to effect internal regulation and maintain homeostasis.

endocytosis (en′-dō-sī-tō′-sis) Cellular uptake of biological molecules and particulate matter via formation of vesicles from the plasma membrane.

endoderm (en′-dō-durm) The innermost of the three primary germ layers in animal embryos; lines the archenteron and gives rise to the liver, pancreas, lungs, and the lining of the digestive tract in species that have these structures.

endodermis In plant roots, the innermost layer of the cortex that surrounds the vascular cylinder.

endomembrane system The collection of membranes inside and surrounding a eukaryotic cell, related either through direct physical contact or by the transfer of membranous vesicles; includes the plasma membrane, the nuclear envelope, the smooth and rough endoplasmic reticulum, the Golgi apparatus, lysosomes, vesicles, and vacuoles.

endometriosis (en′-dō-mē-trē-ō′-sis) The condition resulting from the presence of endometrial tissue outside of the uterus.

endometrium (en′-dō-mē′-trē-um) The inner lining of the uterus, which is richly supplied with blood vessels.

endoparasite A parasite that lives within a host.

endophyte A fungus that lives inside a leaf or other plant part without causing harm to the plant.

endoplasmic reticulum (ER) (en′-dō-plaz′-mik ruh-tik′-yū-lum) An extensive membranous network in eukaryotic cells, continuous with the outer nuclear membrane and composed of ribosome-studded (rough) and ribosome-free (smooth) regions.

endorphin (en-dōr′-fin) Any of several hormones produced in the brain and anterior pituitary that inhibit pain perception.

endoskeleton A hard skeleton buried within the soft tissues of an animal.

endosperm In angiosperms, a nutrient-rich tissue formed by the union of a sperm with two polar nuclei during double fertilization. The endosperm provides nourishment to the developing embryo in angiosperm seeds.

endospore A thick-coated, resistant cell produced by some bacterial cells when they are exposed to harsh conditions.

endosymbiont theory The theory that mitochondria and plastids, including chloroplasts, originated as prokaryotic cells engulfed by an ancestral eukaryotic cell. The engulfed cell and its host cell then evolved into a single organism.

endosymbiosis A process in which a unicellular organism (the "host") engulfs another cell, which lives within the host cell and ultimately becomes an organelle in the host cell. *See also* endosymbiont theory.

endothelium (en′-dō-thē′-lē-um) The simple squamous layer of cells lining the lumen of blood vessels.

endothermic Referring to organisms that are warmed by heat generated by their own metabolism. This heat usually maintains a relatively stable body temperature higher than that of the external environment.

endotoxin A toxic component of the outer membrane of certain gram-negative bacteria that is released only when the bacteria die.

energetic hypothesis The concept that the length of a food chain is limited by the inefficiency of energy transfer along the chain.

energy The capacity to cause change, especially to do work (to move matter against an opposing force).

energy coupling In cellular metabolism, the use of energy released from an exergonic reaction to drive an endergonic reaction.

enhancer A segment of eukaryotic DNA containing multiple control elements, usually located far from the gene whose transcription it regulates.

enteric division One of three divisions of the autonomic nervous system; consists of networks of neurons in the digestive tract, pancreas, and gallbladder; normally regulated by the sympathetic and parasympathetic divisions of the autonomic nervous system.

entropy A measure of disorder, or randomness.

enzymatic hydrolysis The process in digestion that splits macromolecules from food by the enzymatic addition of water.

enzyme (en′-zīm) A macromolecule serving as a catalyst, a chemical agent that increases the rate of a reaction without being consumed by the reaction. Most enzymes are proteins.

enzyme-substrate complex A temporary complex formed when an enzyme binds to its substrate molecule(s).

epicotyl (ep′-uh-kot′-ul) In an angiosperm embryo, the embryonic axis above the point of attachment of the cotyledon(s) and below the first pair of miniature leaves.

epidemic A general outbreak of a disease.

epidermis (1) The dermal tissue system of nonwoody plants, usually consisting of a single layer of tightly packed cells. (2) The outermost layer of cells in an animal.

epididymis (ep′-uh-did′-uh-mus) A coiled tubule located adjacent to the mammalian testis where sperm are stored.

epigenetic inheritance Inheritance of traits transmitted by mechanisms not directly involving the nucleotide sequence of a genome.

epinephrine (ep′-i-nef′-rin) A catecholamine that, when secreted as a hormone by the adrenal medulla, mediates "fight-or-flight" responses to short-term stresses; also released by some neurons as a neurotransmitter; also known as adrenaline.

epiphyte (ep′-uh-fīt) A plant that nourishes itself but grows on the surface of another plant for support, usually on the branches or trunks of trees.

epistasis (ep′-i-stā′-sis) A type of gene interaction in which the phenotypic expression of one gene alters that of another independently inherited gene.

epithelial tissue (ep′-uh-thē′-lē-ul) Sheets of tightly packed cells that line organs and body cavities as well as external surfaces.

epithelium An epithelial tissue.

epitope A small, accessible region of an antigen to which an antigen receptor or antibody binds; also called an antigenic determinant.

EPSP *See* excitatory postsynaptic potential.

equilibrium potential (E_{ion}) The magnitude of a cell's membrane voltage at equilibrium; calculated using the Nernst equation.

erythrocyte (eh-rith′-ruh-sīt) A blood cell that contains hemoglobin, which transports oxygen; also called a red blood cell.

erythropoietin (EPO) (eh-rith′-rō-poy′-uh-tin) A hormone that stimulates the production of erythrocytes. It is secreted by the kidney when body tissues do not receive enough oxygen.

esophagus (eh-sof′-uh-gus) A muscular tube that conducts food, by peristalsis, from the pharynx to the stomach.

essential amino acid An amino acid that an animal cannot synthesize itself and must be obtained from food in prefabricated form.

essential element A chemical element required for an organism to survive, grow, and reproduce.

essential fatty acid An unsaturated fatty acid that an animal needs but cannot make.

essential nutrient A substance that an organism cannot synthesize from any other material and therefore must absorb in preassembled form.

estradiol (es′-truh-dī′-ol) A steroid hormone that stimulates the development and maintenance of the female reproductive system and secondary sex characteristics; the major estrogen in mammals.

estrogen (es′-trō-jen) Any steroid hormone, such as estradiol, that stimulates the development and maintenance of the female reproductive system and secondary sex characteristics.

estrous cycle (es′-trus) A reproductive cycle characteristic of female mammals except humans and certain other primates, in which the nonpregnant endometrium is reabsorbed rather than shed, and sexual response occurs only during mid-cycle at estrus.

estuary The area where a freshwater stream or river merges with the ocean.

ethylene (eth′-uh-lēn) A gaseous plant hormone involved in responses to mechanical stress, programmed cell death, leaf abscission, and fruit ripening.

etiolation Plant morphological adaptations for growing in darkness.

euchromatin (yū-krō′-muh-tin) The less condensed form of eukaryotic chromatin that is available for transcription.

eudicot (yū-dī′-kot) Member of a clade that contains the vast majority of flowering plants that have two embryonic seed leaves, or cotyledons.

euglenid (yū′-glen-id) A protist, such as *Euglena* or its relatives, characterized by an anterior pocket from which one or two flagella emerge.

euglenozoan Member of a diverse clade of flagellated protists that includes predatory heterotrophs, photosynthetic autotrophs, and pathogenic parasites.

Eukarya (yū-kar′-ē-uh) The domain that includes all eukaryotic organisms.

eukaryotic cell (yū′-ker-ē-ot′-ik) A type of cell with a membrane-enclosed nucleus and membrane-enclosed organelles. Organisms with eukaryotic cells (protists, plants, fungi, and animals) are called eukaryotes.

eumetazoan (yū′-met-uh-zō′-un) Member of a clade of animals with true tissues. All animals except sponges and a few other groups are eumetazoans.

eurypterid (yur-ip′-tuh-rid) An extinct carnivorous chelicerate; also called a water scorpion.

Eustachian tube (yū-stā′-shun) The tube that connects the middle ear to the pharynx.

eutherian (yū-thēr′-ē-un) Placental mammal; mammal whose young complete their embryonic development within the uterus, joined to the mother by the placenta.

eutrophic lake (yū-trōf′-ik) A lake that has a high rate of biological productivity supported by a high rate of nutrient cycling.

eutrophication A process by which nutrients, particularly phosphorus and nitrogen, become highly concentrated in a body of water, leading to increased growth of organisms such as algae or cyanobacteria.

evaporation The process by which a liquid changes to a gas.

evaporative cooling The process in which the surface of an object becomes cooler during evaporation, a result of the molecules with the greatest kinetic energy changing from the liquid to the gaseous state.

evapotranspiration The total evaporation of water from an ecosystem, including water transpired by plants and evaporated from a landscape, usually measured in millimeters and estimated for a year.

evo-devo Evolutionary developmental biology; a field of biology that compares developmental processes of different multicellular organisms to understand how these processes have evolved and how changes can modify existing organismal features or lead to new ones.

evolution Descent with modification; the idea that living species are descendants of ancestral species that were different from the present-day ones; also defined more narrowly as the change in the genetic composition of a population from generation to generation.

evolutionary tree A branching diagram that reflects a hypothesis about evolutionary relationships among groups of organisms.

Excavata One of five supergroups of eukaryotes proposed in a current hypothesis of the evolutionary history of eukaryotes. Excavates have unique cytoskeletal features, and some species have an "excavated" feeding groove on one side of the cell body. *See also* Chromalveolata, Rhizaria, Archaeplastida, and Unikonta.

excitatory postsynaptic potential (EPSP) An electrical change (depolarization) in the membrane of a postsynaptic cell caused by the binding of an excitatory neurotransmitter from a presynaptic cell to a postsynaptic receptor; makes it more likely for a postsynaptic cell to generate an action potential.

excretion The disposal of nitrogen-containing metabolites and other waste products.

exergonic reaction (ek′-ser-gon′-ik) A spontaneous chemical reaction, in which there is a net release of free energy.

exocytosis (ek′-sō-sī-tō′-sis) The cellular secretion of biological molecules by the fusion of vesicles containing them with the plasma membrane.

exon A sequence within a primary transcript that remains in the RNA after RNA processing; also refers to the region of DNA from which this sequence was transcribed.

exoskeleton A hard encasement on the surface of an animal, such as the shell of a mollusc or the cuticle of an arthropod, that provides protection and points of attachment for muscles.

exotoxin (ek′-sō-tok′-sin) A toxic protein that is secreted by a prokaryote or other pathogen and that produces specific symptoms, even if the pathogen is no longer present.

expansin Plant enzyme that breaks the cross-links (hydrogen bonds) between cellulose microfibrils and other cell wall constituents, loosening the wall's fabric.

exponential population growth Growth of a population in an ideal, unlimited environment, represented by a J-shaped curve when population size is plotted over time.

expression vector A cloning vector that contains a highly active bacterial promoter just upstream of a restriction site where a eukaryotic gene can be inserted, allowing the gene to be expressed in a bacterial cell. Expression vectors are also available that have been genetically engineered for use in specific types of eukaryotic cells.

external fertilization The fusion of gametes that parents have discharged into the environment.

extinction vortex A downward population spiral in which inbreeding and genetic drift combine to cause a small population to shrink and, unless the spiral is reversed, become extinct.

extracellular digestion The breakdown of food in compartments that are continuous with the outside of an animal's body.

extracellular matrix (ECM) The meshwork surrounding animal cells, consisting of glycoproteins, polysaccharides, and proteoglycans synthesized and secreted by the cells.

extraembryonic membrane One of four membranes (yolk sac, amnion, chorion, and allantois) located outside the embryo that support the developing embryo in reptiles and mammals.

extreme halophile An organism that lives in a highly saline environment, such as the Great Salt Lake or the Dead Sea.

extreme thermophile An organism that thrives in hot environments (often 60–80°C or hotter).

extremophile An organism that lives in environmental conditions so extreme that few other species can survive there. Extremophiles include extreme halophiles ("salt lovers") and extreme thermophiles ("heat lovers").

F factor In bacteria, the DNA segment that confers the ability to form pili for conjugation and associated functions required for the transfer of DNA from donor to recipient. The F factor may exist as a plasmid or be integrated into the bacterial chromosome.

F plasmid The plasmid form of the F factor.

F_1 generation The first filial, hybrid (heterozygous) offspring arising from a parental (P generation) cross.

F_2 generation The offspring resulting from interbreeding (or self-pollination) of the hybrid F_1 generation.

facilitated diffusion The passage of molecules or ions down their electrochemical gradient across a biological membrane with the assistance of specific transmembrane transport proteins, requiring no energy expenditure.

facilitation An interaction in which one species has a positive effect on the survival and reproduction of another species without the intimate association of a symbiosis.

facultative anaerobe (fak′-ul-tā′-tiv an′-uh-rōb) An organism that makes ATP by aerobic respiration if oxygen is present but that switches to anaerobic respiration or fermentation if oxygen is not present.

family In Linnaean classification, the taxonomic category above genus.

fast block to polyspermy The depolarization of the egg plasma membrane that begins within 1–3 seconds after a sperm binds to an egg membrane protein. The depolarization lasts about 1 minute and prevents additional sperm from fusing with the egg during that time.

fast-twitch fiber A muscle fiber used for rapid, powerful contractions.

fat A lipid consisting of three fatty acids linked to one glycerol molecule; also called a triacylglycerol or triglyceride.

fate map A territorial diagram of embryonic development that displays the future derivatives of individual cells and tissues.

fatty acid A carboxylic acid with a long carbon chain. Fatty acids vary in length and in the number and location of double bonds; three fatty acids linked to a glycerol molecule form a fat molecule, also known as a triacylglycerol or triglyceride.

feces (fē′-sēz) The wastes of the digestive tract.

feedback inhibition A method of metabolic control in which the end product of a metabolic pathway acts as an inhibitor of an enzyme within that pathway.

fermentation A catabolic process that makes a limited amount of ATP from glucose (or other organic molecules) without an electron transport chain and that produces a characteristic end product, such as ethyl alcohol or lactic acid.

fertilization (1) The union of haploid gametes to produce a diploid zygote. (2) The addition of mineral nutrients to the soil.

fetus (fē′-tus) A developing mammal that has all the major structures of an adult. In humans, the fetal stage lasts from the 9th week of gestation until birth.

fiber A lignified cell type that reinforces the xylem of angiosperms and functions in mechanical support; a slender, tapered sclerenchyma cell that usually occurs in bundles.

fibroblast (fī′-brō-blast) A type of cell in loose connective tissue that secretes the protein ingredients of the extracellular fibers.

fibronectin An extracellular glycoprotein secreted by animal cells that helps them attach to the extracellular matrix.

filament In an angiosperm, the stalk portion of the stamen, the pollen-producing reproductive organ of a flower.

filtrate Cell-free fluid extracted from the body fluid by the excretory system.

filtration In excretory systems, the extraction of water and small solutes, including metabolic wastes, from the body fluid.

fimbria (plural, **fimbriae**) A short, hairlike appendage of a prokaryotic cell that helps it adhere to the substrate or to other cells.

first law of thermodynamics The principle of conservation of energy: Energy can be transferred and transformed, but it cannot be created or destroyed.

fission The separation of an organism into two or more individuals of approximately equal size.

fixed action pattern In animal behavior, a sequence of unlearned acts that is essentially unchangeable and, once initiated, usually carried to completion.

flaccid (flas′-id) Limp. Lacking turgor (stiffness or firmness), as in a plant cell in surroundings where there is a tendency for water to leave the cell. (A walled cell becomes flaccid if it has a higher water potential than its surroundings, resulting in the loss of water.)

flagellum (fluh-jel′-um) (plural, **flagella**) A long cellular appendage specialized for locomotion. Like motile cilia, eukaryotic flagella have a core with nine outer doublet microtubules and two inner single microtubules (the "9 + 2" arrangement) ensheathed in an extension of the plasma membrane. Prokaryotic flagella have a different structure.

florigen A flowering signal, probably a protein, that is made in leaves under certain conditions and that travels to the shoot apical meristems, inducing them to switch from vegetative to reproductive growth.

flower In an angiosperm, a specialized shoot with up to four sets of modified leaves, bearing structures that function in sexual reproduction.

fluid feeder An animal that lives by sucking nutrient-rich fluids from another living organism.

fluid mosaic model The currently accepted model of cell membrane structure, which envisions the membrane as a mosaic of protein molecules drifting laterally in a fluid bilayer of phospholipids.

follicle (fol′-uh-kul) A microscopic structure in the ovary that contains the developing oocyte and secretes estrogens.

follicle-stimulating hormone (FSH) A tropic hormone that is produced and secreted by the anterior pituitary and that stimulates the production of eggs by the ovaries and sperm by the testes.

follicular phase That part of the ovarian cycle during which follicles are growing and oocytes maturing.

food chain The pathway along which food energy is transferred from trophic level to trophic level, beginning with producers.

food vacuole A membranous sac formed by phagocytosis of microorganisms or particles to be used as food by the cell.

food web The interconnected feeding relationships in an ecosystem.

foot (1) The portion of a bryophyte sporophyte that gathers sugars, amino acids, water, and minerals from the parent gametophyte via transfer cells. (2) One of the three main parts of a mollusc; a muscular structure usually used for movement. *See also* mantle, visceral mass.

foraging The seeking and obtaining of food.

foram (foraminiferan) An aquatic protist that secretes a hardened shell containing calcium carbonate and extends pseudopodia through pores in the shell.

forebrain One of three ancestral and embryonic regions of the vertebrate brain; develops into the thalamus, hypothalamus, and cerebrum.

fossil A preserved remnant or impression of an organism that lived in the past.

founder effect Genetic drift that occurs when a few individuals become isolated from a larger population and form a new population whose gene pool composition is not reflective of that of the original population.

fovea (fō′-vē-uh) The place on the retina at the eye's center of focus, where cones are highly concentrated.

fragmentation A means of asexual reproduction whereby a single parent breaks into parts that regenerate into whole new individuals.

frameshift mutation A mutation occurring when nucleotides are inserted in or deleted from a gene and the number inserted or deleted is not a multiple of three, resulting in the improper grouping of the subsequent nucleotides into codons.

free energy The portion of a biological system's energy that can perform work when temperature and pressure are uniform throughout the system. The change in free energy of a system (ΔG) is calculated by the equation $\Delta G = \Delta H - T\Delta S$, where ΔH is the change in enthalpy (in biological systems, equivalent to total energy), T is the absolute temperature, and ΔS is the change in entropy.

frequency-dependent selection Selection in which the fitness of a phenotype depends on how common the phenotype is in a population.

fruit A mature ovary of a flower. The fruit protects dormant seeds and often aids in their dispersal.

functional group A specific configuration of atoms commonly attached to the carbon skeletons of organic molecules and involved in chemical reactions.

G protein A GTP-binding protein that relays signals from a plasma membrane signal receptor, known as a G protein-coupled receptor, to other signal transduction proteins inside the cell.

G protein-coupled receptor (GPCR) A signal receptor protein in the plasma membrane that responds to the binding of a signaling molecule by activating a G protein. Also called a G protein-linked receptor.

G_0 phase A nondividing state occupied by cells that have left the cell cycle, sometimes reversibly.

G_1 phase The first gap, or growth phase, of the cell cycle, consisting of the portion of interphase before DNA synthesis begins.

G_2 phase The second gap, or growth phase, of the cell cycle, consisting of the portion of interphase after DNA synthesis occurs.

gallbladder An organ that stores bile and releases it as needed into the small intestine.

game theory An approach to evaluating alternative strategies in situations where the outcome of a particular strategy depends on the strategies used by other individuals.

gametangium (gam′-uh-tan′-jē-um) (plural, **gametangia**) Multicellular plant structure in which gametes are formed. Female gametangia are called archegonia, and male gametangia are called antheridia.

gamete (gam′-ēt) A haploid reproductive cell, such as an egg or sperm. Gametes unite during sexual reproduction to produce a diploid zygote.

gametogenesis The process by which gametes are produced.

gametophore (guh-mē′-tō-fōr) The mature gamete-producing structure of a moss gametophyte.

gametophyte (guh-mē′-tō-fīt) In organisms (plants and some algae) that have alternation of generations, the multicellular haploid form that produces haploid gametes by mitosis. The haploid gametes unite and develop into sporophytes.

gamma-aminobutyric acid (GABA) An amino acid that functions as a CNS neurotransmitter in the central nervous system of vertebrates.

ganglia (gang′-glē-uh) (singular, **ganglion**) Clusters (functional groups) of nerve cell bodies in a centralized nervous system.

ganglion cell A type of neuron in the retina that synapses with bipolar cells and transmits action potentials to the brain via axons in the optic nerve.

gap junction A type of intercellular junction in animal cells, consisting of proteins surrounding a pore that allows the passage of materials between cells.

gas exchange The uptake of molecular oxygen from the environment and the discharge of carbon dioxide to the environment.

gastric juice A digestive fluid secreted by the stomach.

gastrovascular cavity A central cavity with a single opening in the body of certain animals, including cnidarians and flatworms, that functions in both the digestion and distribution of nutrients.

gastrula (gas′-trū-luh) An embryonic stage in animal development encompassing the formation of three layers: ectoderm, mesoderm, and endoderm.

gastrulation (gas′-trū-lā′-shun) In animal development, a series of cell and tissue movements in which the blastula-stage embryo folds inward, producing a three-layered embryo, the gastrula.

gated channel A transmembrane protein channel that opens or closes in response to a particular stimulus.

gated ion channel A gated channel for a specific ion. The opening or closing of such channels may alter a cell's membrane potential.

gel electrophoresis (ē-lek′-trō-fōr-ē′-sis) A technique for separating nucleic acids or proteins on the basis of their size and electrical charge, both of which affect their rate of movement through an electric field in a gel made of agarose or another polymer.

gene A discrete unit of hereditary information consisting of a specific nucleotide sequence in DNA (or RNA, in some viruses).

gene annotation Analysis of genomic sequences to identify protein-coding genes and determine the function of their products.

gene cloning The production of multiple copies of a gene.

gene expression The process by which information encoded in DNA directs the synthesis of proteins or, in some cases, RNAs that are not translated into proteins and instead function as RNAs.

gene flow The transfer of alleles from one population to another, resulting from the movement of fertile individuals or their gametes.

gene pool The aggregate of all copies of every type of allele at all loci in every individual in a population. The term is also used in a more restricted sense as the aggregate of alleles for just one or a few loci in a population.

gene therapy The introduction of genes into an afflicted individual for therapeutic purposes.

gene-for-gene recognition A widespread form of plant disease resistance involving recognition of pathogen-derived molecules by the protein products of specific plant disease resistance genes.

genetic drift A process in which chance events cause unpredictable fluctuations in allele frequencies from one generation to the next. Effects of genetic drift are most pronounced in small populations.

genetic engineering The direct manipulation of genes for practical purposes.

genetic map An ordered list of genetic loci (genes or other genetic markers) along a chromosome.

genetic profile An individual's unique set of genetic markers, detected most often today by PCR or, previously, by electrophoresis and nucleic acid probes.

genetic recombination General term for the production of offspring with combinations of traits that differ from those found in either parent.

genetic variation Differences among individuals in the composition of their genes or other DNA segments.

genetically modified (GM) organism An organism that has acquired one or more genes by artificial means; also known as a transgenic organism.

genetics The scientific study of heredity and hereditary variation.

genome (jē′-nōm) The genetic material of an organism or virus; the complete complement of an organism's or virus's genes along with its noncoding nucleic acid sequences.

genome-wide association study A large-scale analysis of the genomes of many people having a certain phenotype or disease, with the aim of finding genetic markers that correlate with that phenotype or disease.

genomic imprinting A phenomenon in which expression of an allele in offspring depends on whether the allele is inherited from the male or female parent.

genomic library A set of cell clones containing all the DNA segments from a genome, each within a plasmid, BAC, or other cloning vector.

genomics (juh-nō′-miks) The study of whole sets of genes and their interactions within a species, as well as genome comparisons between species.

genotype (jē′-nō-tīp) The genetic makeup, or set of alleles, of an organism.

genus (jē′-nus) (plural, **genera**) A taxonomic category above the species level, designated by the first word of a species' two-part scientific name.

geographic variation Differences between the gene pools of geographically separate populations or population subgroups.

geologic record The division of Earth's history into time periods, grouped into three eons—Archaean, Proterozoic, and Phanerozoic—and further subdivided into eras, periods, and epochs.

germ layer One of the three main layers in a gastrula that will form the various tissues and organs of an animal body.

gestation (jes-tā′-shun) Pregnancy; the state of carrying developing young within the female reproductive tract.

gibberellin (jib′-uh-rel′-in) Any of a class of related plant hormones that stimulate growth in the stem and leaves, trigger the germination of seeds and breaking of bud dormancy, and (with auxin) stimulate fruit development.

glans The rounded structure at the tip of the clitoris or penis that is involved in sexual arousal.

glia (glial cells) Cells of the nervous system that support, regulate, and augment the functions of neurons.

global climate change Increase in temperature and change in weather patterns all around the planet, due mostly to increasing atmospheric CO_2 levels from the burning of fossil fuels. The increase in temperature, called global warming, is a major aspect of global climate change.

global ecology The study of the functioning and distribution of organisms across the biosphere and how the regional exchange of energy and materials affects them.

glomeromycete (glō′-mer-ō-mī′-sēt) Member of the fungal phylum Glomeromycota, characterized by a distinct branching form of mycorrhizae called arbuscular mycorrhizae.

glomerulus (glō-mār′-yū-lus) A ball of capillaries surrounded by Bowman's capsule in the nephron and serving as the site of filtration in the vertebrate kidney.

glucagon (glū′-kuh-gon) A hormone secreted by pancreatic alpha cells that raises blood glucose levels. It promotes glycogen breakdown and release of glucose by the liver.

glucocorticoid A steroid hormone that is secreted by the adrenal cortex and that influences glucose metabolism and immune function.

glutamate An amino acid that functions as a neurotransmitter in the central nervous system.

glyceraldehyde 3-phosphate (G3P) (glis′-er-al′-de-hīd) A three-carbon carbohydrate that is the direct product of the Calvin cycle; it is also an intermediate in glycolysis.

glycogen (glī′-kō-jen) An extensively branched glucose storage polysaccharide found in the liver and muscle of animals; the animal equivalent of starch.

glycolipid A lipid with one or more covalently attached carbohydrates.

glycolysis (glī-kol′-uh-sis) A series of reactions that ultimately splits glucose into pyruvate. Glycolysis occurs in almost all living cells, serving as the starting point for fermentation or cellular respiration.

glycoprotein A protein with one or more covalently attached carbohydrates.

glycosidic linkage A covalent bond formed between two monosaccharides by a dehydration reaction.

gnathostome (na′-thu-stōm) Member of the vertebrate subgroup possessing jaws.

golden alga A biflagellated, photosynthetic protist named for its color, which results from its yellow and brown carotenoids.

Golgi apparatus (gol′-jē) An organelle in eukaryotic cells consisting of stacks of flat membranous sacs that modify, store, and route products of the endoplasmic reticulum and synthesize some products, notably noncellulose carbohydrates.

gonads (gō′-nadz) The male and female sex organs; the gamete-producing organs in most animals.

grade A group of organisms that share the same level of organizational complexity or share a key adaptation.

graded potential In a neuron, a shift in the membrane potential that has an amplitude proportional to signal strength and that decays as it spreads.

Gram stain A staining method that distinguishes between two different kinds of bacterial cell walls; may be used to help determine medical response to an infection.

gram-negative Describing the group of bacteria that have a cell wall that is structurally more complex and contains less peptidoglycan than the cell wall of gram-positive bacteria. Gram-negative bacteria are often more toxic than gram-positive bacteria.

gram-positive Describing the group of bacteria that have a cell wall that is structurally less complex and contains more peptidoglycan than the cell wall of gram-negative bacteria. Gram-positive bacteria are usually less toxic than gram-negative bacteria.

granum (gran′-um) (plural, **grana**) A stack of membrane-bounded thylakoids in the chloroplast. Grana function in the light reactions of photosynthesis.

gravitropism (grav′-uh-trō′-pizm) A response of a plant or animal to gravity.

gray matter Regions of dendrites and clustered neuron cell bodies within the CNS.

green alga A photosynthetic protist, named for green chloroplasts that are similar in structure and pigment composition to those of land plants. Green algae are a paraphyletic group, some of whose members are more closely related to land plants than they are to other green algae.

greenhouse effect The warming of Earth due to the atmospheric accumulation of carbon dioxide and certain other gases, which absorb reflected infrared radiation and reradiate some of it back toward Earth.

gross primary production (GPP) The total primary production of an ecosystem.

ground tissue system Plant tissues that are neither vascular nor dermal, fulfilling a variety of functions, such as storage, photosynthesis, and support.

growth An irreversible increase in size or biomass.

growth factor (1) A protein that must be present in the extracellular environment (culture medium or animal body) for the growth and normal development of certain types of cells. (2) A local regulator that acts on nearby cells to stimulate cell proliferation and differentiation.

growth hormone (GH) A hormone that is produced and secreted by the anterior pituitary and that has both direct (nontropic) and tropic effects on a wide variety of tissues.

guard cells The two cells that flank the stomatal pore and regulate the opening and closing of the pore.

gustation The sense of taste.

guttation The exudation of water droplets from leaves, caused by root pressure in certain plants.

gymnosperm (jim′-nō-sperm) A vascular plant that bears naked seeds—seeds not enclosed in protective chambers.

hair cell A mechanosensory cell that alters output to the nervous system when hairlike projections on the cell surface are displaced.

half-life The amount of time it takes for 50% of a sample of a radioactive isotope to decay.

Hamilton's rule The principle that for natural selection to favor an altruistic act, the benefit to the recipient, devalued by the coefficient of relatedness, must exceed the cost to the altruist.

haploid cell (hap′-loyd) A cell containing only one set of chromosomes (n).

Hardy-Weinberg principle The principle that frequencies of alleles and genotypes in a population remain constant from generation to generation, provided that only Mendelian segregation and recombination of alleles are at work.

haustorium (plural, **haustoria**) (ho-stōr′-ē-um, ho-stōr′-ē-uh) In certain symbiotic fungi, a specialized hypha that can penetrate the tissues of host organisms.

heart A muscular pump that uses metabolic energy to elevate the hydrostatic pressure of the circulatory fluid (blood or hemolymph). The fluid then flows down a pressure gradient through the body and eventually returns to the heart.

heart attack The damage or death of cardiac muscle tissue resulting from prolonged blockage of one or more coronary arteries.

heart murmur A hissing sound that most often results from blood squirting backward through a leaky valve in the heart.

heart rate The frequency of heart contraction (in beats per minute).

heat The total amount of kinetic energy due to the random motion of atoms or molecules in a body of matter; also called thermal energy. Heat is energy in its most random form.

heat of vaporization The quantity of heat a liquid must absorb for 1 g of it to be converted from the liquid to the gaseous state.

heat-shock protein A protein that helps protect other proteins during heat stress. Heat-shock proteins are found in plants, animals, and microorganisms.

heavy chain One of the two types of polypeptide chains that make up an antibody molecule and B cell receptor; consists of a variable region, which contributes to the antigen-binding site, and a constant region.

helicase An enzyme that untwists the double helix of DNA at replication forks, separating the two strands and making them available as template strands.

helper T cell A type of T cell that, when activated, secretes cytokines that promote the response of B cells (humoral response) and cytotoxic T cells (cell-mediated response) to antigens.

hemoglobin (hē′-mō-glō′-bin) An iron-containing protein in red blood cells that reversibly binds oxygen.

hemolymph (hē′-mō-limf′) In invertebrates with an open circulatory system, the body fluid that bathes tissues.

hemophilia (hē′-muh-fil′-ē-uh) A human genetic disease caused by a sex-linked recessive allele resulting in the absence of one or more blood-clotting proteins; characterized by excessive bleeding following injury.

hepatic portal vein A large vessel that conveys nutrient-laden blood from the small intestine to the liver, which regulates the blood's nutrient content.

herbivore (hur′-bi-vōr′) An animal that mainly eats plants or algae.

herbivory An interaction in which an organism eats parts of a plant or alga.

heredity The transmission of traits from one generation to the next.

hermaphrodite (hur-maf′-ruh-dīt′) An individual that functions as both male and female in sexual reproduction by producing both sperm and eggs.

hermaphroditism (hur-maf′-rō-dī-tizm) A condition in which an individual has both female and male gonads and functions as both a male and female in sexual reproduction by producing both sperm and eggs.

heterochromatin (het′-er-ō-krō′-muh-tin) Eukaryotic chromatin that remains highly compacted during interphase and is generally not transcribed.

heterochrony (het′-uh-rok′-ruh-nē) Evolutionary change in the timing or rate of an organism's development.

heterocyst (het′-er-ō-sist) A specialized cell that engages in nitrogen fixation in some filamentous cyanobacteria; also called a *heterocyte*.

heterokaryon (het′-er-ō-kār′-ē-un) A fungal mycelium that contains two or more haploid nuclei per cell.

heteromorphic (het′-er-ō-mōr′-fik) Referring to a condition in the life cycle of plants and certain algae in which the sporophyte and gametophyte generations differ in morphology.

heterosporous (het-er-os′-pōr-us) Referring to a plant species that has two kinds of spores: microspores, which develop into male gametophytes, and megaspores, which develop into female gametophytes.

heterotroph (het′-er-ō-trōf) An organism that obtains organic food molecules by eating other organisms or substances derived from them.

heterozygote advantage Greater reproductive success of heterozygous individuals compared with homozygotes; tends to preserve variation in a gene pool.

heterozygous (het′-er-ō-zī′-gus) Having two different alleles for a given gene.

hexapod An insect or closely related wingless, six-legged arthropod.

hibernation A long-term physiological state in which metabolism decreases, the heart and respiratory system slow down, and body temperature is maintained at a lower level than normal.

high-density lipoprotein (HDL) A particle in the blood made up of thousands of cholesterol molecules and other lipids bound to a protein. HDL scavenges excess cholesterol.

hindbrain One of three ancestral and embryonic regions of the vertebrate brain; develops into the medulla oblongata, pons, and cerebellum.

histamine (his′-tuh-mēn) A substance released by mast cells that causes blood vessels to dilate and become more permeable in inflammatory and allergic responses.

histone (his′-tōn) A small protein with a high proportion of positively charged amino acids that binds to the negatively charged DNA and plays a key role in chromatin structure.

histone acetylation The attachment of acetyl groups to certain amino acids of histone proteins.

HIV (human immunodeficiency virus) The infectious agent that causes AIDS. HIV is a retrovirus.

holdfast A rootlike structure that anchors a seaweed.

holoblastic (hō′-lō-blas′-tik) Referring to a type of cleavage in which there is complete division of the egg; occurs in eggs that have little yolk (such as those of the sea urchin) or a moderate amount of yolk (such as those of the frog).

homeobox (hō′-mē-ō-boks′) A 180-nucleotide sequence within homeotic genes and some other developmental genes that is widely conserved in animals. Related sequences occur in plants and yeasts.

homeostasis (hō′-mē-ō-stā′-sis) The steady-state physiological condition of the body.

homeotic gene (hō-mē-o′-tik) Any of the master regulatory genes that control placement and spatial organization of body parts in animals, plants, and fungi by controlling the developmental fate of groups of cells.

hominin (hō′-mi-nin) A member of the human branch of the evolutionary tree. Hominins include *Homo sapiens* and our ancestors, a group of extinct species that are more closely related to us than to chimpanzees.

homologous chromosomes (hō-mol′-uh-gus) A pair of chromosomes of the same length, centromere position, and staining pattern that possess genes for the same characters at corresponding loci. One homologous chromosome is inherited from the organism's father, the other from the mother. Also called homologs, or a homologous pair.

homologous structures Structures in different species that are similar because of common ancestry.

homology (hō-mol′-ō-jē) Similarity in characteristics resulting from a shared ancestry.

homoplasy (hō′-muh-play′-zē) A similar (analogous) structure or molecular sequence that has evolved independently in two species.

homosporous (hō-mos′-puh-rus) Referring to a plant species that has a single kind of spore, which typically develops into a bisexual gametophyte.

homozygous (hō′-mō-zī′-gus) Having two identical alleles for a given gene.

horizontal cell A neuron of the retina that helps integrate the information that is sent to the brain.

horizontal gene transfer The transfer of genes from one genome to another through mechanisms such as transposable elements, plasmid exchange, viral activity, and perhaps fusions of different organisms.

hormone In multicellular organisms, one of many types of secreted chemicals that are formed in specialized cells, travel in body fluids, and act on specific target cells in other parts of the body, changing the target cells' functioning. Hormones are thus important in long-distance signaling.

hornwort A small, herbaceous, nonvascular plant that is a member of the phylum Anthocerophyta.

host The larger participant in a symbiotic relationship, often providing a home and food source for the smaller symbiont.

host range The limited number of species whose cells can be infected by a particular virus.

human chorionic gonadotropin (hCG) (kōr′-ē-on′-ik gō-na′-dō-trō′-pin) A hormone secreted by the chorion that maintains the corpus luteum of the ovary during the first three months of pregnancy.

Human Genome Project An international collaborative effort to map and sequence the DNA of the entire human genome.

humoral immune response (hyū′-mer-ul) The branch of adaptive immunity that involves the activation of B cells and that leads to the production of antibodies, which defend against bacteria and viruses in body fluids.

humus (hyū′-mus) Decomposing organic material that is a component of topsoil.

Huntington's disease A human genetic disease caused by a dominant allele; characterized by uncontrollable body movements and degeneration of the nervous system; usually fatal 10 to 20 years after the onset of symptoms.

hybrid Offspring that results from the mating of individuals from two different species or from two true-breeding varieties of the same species.

hybrid zone A geographic region in which members of different species meet and mate, producing at least some offspring of mixed ancestry.

hybridization In genetics, the mating, or crossing, of two true-breeding varieties.

hydration shell The sphere of water molecules around a dissolved ion.

hydrocarbon An organic molecule consisting only of carbon and hydrogen.

hydrogen bond A type of weak chemical bond that is formed when the slightly positive hydrogen atom of a polar covalent bond in one molecule is attracted to the slightly negative atom of a polar covalent bond in another molecule or in another region of the same molecule.

hydrogen ion A single proton with a charge of 1+. The dissociation of a water molecule (H_2O) leads to the generation of a hydroxide ion (OH^-) and a hydrogen ion (H^+); in water, H^+ is not found alone but associates with a water molecule to form a hydronium ion.

hydrolysis (hī-drol′-uh-sis) A chemical reaction that breaks bonds between two molecules by the addition of water; functions in disassembly of polymers to monomers.

hydronium ion A water molecule that has an extra proton bound to it; H_3O^+, commonly represented as H^+.

hydrophilic (hī′-drō-fil′-ik) Having an affinity for water.

hydrophobic (hī′-drō-fō′-bik) Having no affinity for water; tending to coalesce and form droplets in water.

hydrophobic interaction A type of weak chemical interaction caused when molecules that do not mix with water coalesce to exclude water.

hydroponic culture A method in which plants are grown in mineral solutions rather than in soil.

hydrostatic skeleton A skeletal system composed of fluid held under pressure in a closed body compartment; the main skeleton of most cnidarians, flatworms, nematodes, and annelids.

hydroxide ion A water molecule that has lost a proton; OH^-.

hydroxyl group (hī-drok′-sil) A chemical group consisting of an oxygen atom joined to a hydrogen atom. Molecules possessing this group are soluble in water and are called alcohols.

hymen A thin membrane that partly covers the vaginal opening in the human female. The hymen is ruptured by sexual intercourse or other vigorous activity.

hyperpolarization A change in a cell's membrane potential such that the inside of the membrane becomes more negative relative to the outside. Hyperpolarization reduces the chance that a neuron will transmit a nerve impulse.

hypersensitive response A plant's localized defense response to a pathogen, involving the death of cells around the site of infection.

hypertension A disorder in which blood pressure remains abnormally high.

hypertonic Referring to a solution that, when surrounding a cell, will cause the cell to lose water.

hypha (plural, **hyphae**) (hī′-fuh, hī′-fē) One of many connected filaments that collectively make up the mycelium of a fungus.

hypocotyl (hī′-puh-cot′-ul) In an angiosperm embryo, the embryonic axis below the point of attachment of the cotyledon(s) and above the radicle.

hypothalamus (hī′-pō-thal′-uh-mus) The ventral part of the vertebrate forebrain; functions in maintaining homeostasis, especially in coordinating the endocrine and nervous systems; secretes hormones of the posterior pituitary and releasing factors that regulate the anterior pituitary.

hypothesis (hī-poth′-uh-sis) A testable explanation for a set of observations based on the available data and guided by inductive reasoning. A hypothesis is narrower in scope than a theory.

hypotonic Referring to a solution that, when surrounding a cell, will cause the cell to take up water.

imbibition The physical adsorption of water onto the internal surfaces of structures.

immigration The influx of new individuals into a population from other areas.

immune system An animal body's system of defenses against agents that cause disease.

immunization The process of generating a state of immunity by artificial means. In active immunization, also called vaccination, an inactive or weakened form of a pathogen is administered, inducing B and T cell responses and immunological memory. In passive immunization, antibodies specific for a particular microbe are administered, conferring immediate but temporary protection.

immunodeficiency A disorder in which the ability of an immune system to protect against pathogens is defective or absent.

immunoglobulin (Ig) (im′-yū-nō-glob′-yū-lin) Any of the class of proteins that function as antibodies. Immunoglobulins are divided into five major classes that differ in their distribution in the body and antigen disposal activities.

imprinting In animal behavior, the formation at a specific stage in life of a long-lasting behavioral response to a specific individual or object. *See also* genomic imprinting.

***in situ* hybridization** A technique using nucleic acid hybridization with a labeled probe to detect the location of a specific mRNA in an intact organism.

***in vitro* fertilization (IVF)** (vē′-trō) Fertilization of oocytes in laboratory containers followed by artificial implantation of the early embryo in the mother's uterus.

***in vitro* mutagenesis** A technique used to discover the function of a gene by cloning it, introducing specific changes into the cloned gene's sequence, reinserting the mutated gene into a cell, and studying the phenotype of the mutant.

inclusive fitness The total effect an individual has on proliferating its genes by producing its own offspring and by providing aid that enables other close relatives to increase production of their offspring.

incomplete dominance The situation in which the phenotype of heterozygotes is intermediate between the phenotypes of individuals homozygous for either allele.

incomplete flower A flower in which one or more of the four basic floral organs (sepals, petals, stamens, or carpels) are either absent or nonfunctional.

incomplete metamorphosis A type of development in certain insects, such as grasshoppers, in which the young (called nymphs) resemble adults but are smaller and have different body proportions. The nymph goes through a series of molts, each time looking more like an adult, until it reaches full size.

indeterminate cleavage A type of embryonic development in deuterostomes in which each cell produced by early cleavage divisions retains the capacity to develop into a complete embryo.

indeterminate growth A type of growth characteristic of plants, in which the organism continues to grow as long as it lives.

induced fit Caused by entry of the substrate, the change in shape of the active site of an enzyme so that it binds more snugly to the substrate.

inducer A specific small molecule that binds to a bacterial repressor protein and changes the repressor's shape so that it cannot bind to an operator, thus switching an operon on.

induction The process in which one group of embryonic cells influences the development of another, usually by causing changes in gene expression.

inductive reasoning A type of logic in which generalizations are based on a large number of specific observations.

inflammatory response An innate immune defense triggered by physical injury or infection of tissue involving the release of substances that promote swelling, enhance the infiltration of white blood cells, and aid in tissue repair and destruction of invading pathogens.

inflorescence A group of flowers tightly clustered together.

ingestion The first stage of food processing in animals: the act of eating.

ingroup A species or group of species whose evolutionary relationships we seek to determine.

inhibin A hormone produced in the male and female gonads that functions in part by regulating the anterior pituitary by negative feedback.

inhibitory postsynaptic potential (IPSP) An electrical change (usually hyperpolarization) in the membrane of a postsynaptic neuron caused by the binding of an inhibitory neurotransmitter from a presynaptic cell to a postsynaptic receptor; makes it more difficult for a postsynaptic neuron to generate an action potential.

innate behavior Animal behavior that is developmentally fixed and under strong genetic control. Innate behavior is exhibited in virtually the same form by all individuals in a population despite internal and external environmental differences during development and throughout their lifetimes.

innate immunity A form of defense common to all animals that is active immediately upon exposure to pathogens and that is the same whether or not the pathogen has been encountered previously.

inner cell mass An inner cluster of cells at one end of a mammalian blastocyst that subsequently develops into the embryo proper and some of the extraembryonic membranes.

inner ear One of three main regions of the vertebrate ear; includes the cochlea (which in turn contains the organ of Corti) and the semicircular canals.

inositol trisphosphate (IP_3) (in-ō′-suh-tol) A second messenger that functions as an intermediate between certain signaling molecules and a subsequent second messenger, Ca^{2+}, by causing a rise in cytoplasmic Ca^{2+} concentration.

inquiry The search for information and explanation, often focusing on specific questions.

insertion A mutation involving the addition of one or more nucleotide pairs to a gene.

insulin (in′-suh-lin) A hormone secreted by pancreatic beta cells that lowers blood glucose levels. It promotes the uptake of glucose by most body cells and the synthesis and storage of glycogen in the liver and also stimulates protein and fat synthesis.

integral protein A transmembrane protein with hydrophobic regions that extend into and often completely span the hydrophobic interior of the membrane and with hydrophilic regions in contact with the aqueous solution on one or both sides of the membrane (or lining the channel in the case of a channel protein).

integrin In animal cells, a transmembrane receptor protein with two subunits that interconnects the extracellular matrix and the cytoskeleton.

integument (in-teg′-yū-ment) Layer of sporophyte tissue that contributes to the structure of an ovule of a seed plant.

integumentary system The outer covering of a mammal's body, including skin, hair, and nails, claws, or hooves.

intercalated disk (in-ter′-kuh-lā′-ted) A specialized junction between cardiac muscle cells that provides direct electrical coupling between the cells.

interferon (in′-ter-fēr′-on) A protein that has antiviral or immune regulatory functions. Interferon-α and interferon-β, secreted by virus-infected cells, help nearby cells resist viral infection; interferon-γ, secreted by T cells, helps activate macrophages.

intermediate disturbance hypothesis The concept that moderate levels of disturbance can foster greater species diversity than low or high levels of disturbance.

intermediate filament A component of the cytoskeleton that includes filaments intermediate in size between microtubules and microfilaments.

internal fertilization The fusion of eggs and sperm within the female reproductive tract. The sperm are typically deposited in or near the tract.

interneuron An association neuron; a nerve cell within the central nervous system that forms synapses with sensory and/or motor neurons and integrates sensory input and motor output.

internode A segment of a plant stem between the points where leaves are attached.

interphase The period in the cell cycle when the cell is not dividing. During interphase, cellular metabolic activity is high, chromosomes and organelles are duplicated, and cell size may increase. Interphase often accounts for about 90% of the cell cycle.

intersexual selection Selection whereby individuals of one sex (usually females) are choosy in selecting their mates from individuals of the other sex; also called mate choice.

interspecific competition Competition for resources between individuals of two or more species when resources are in short supply.

interspecific interaction A relationship between individuals of two or more species in a community.

interstitial fluid The fluid filling the spaces between cells in most animals.

intertidal zone The shallow zone of the ocean adjacent to land and between the high- and low-tide lines.

intracytoplasmic sperm injection (ICSI) The fertilization of an egg in the laboratory by the direct injection of a single sperm.

intrasexual selection Selection in which there is direct competition among individuals of one sex for mates of the opposite sex.

introduced species A species moved by humans, either intentionally or accidentally, from its native location to a new geographic region; also called non-native or exotic species.

intron (in′-tron) A noncoding, intervening sequence within a primary transcript that is removed from the transcript during RNA processing; also refers to the region of DNA from which this sequence was transcribed.

invasive species A species, often introduced by humans, that takes hold outside its native range.

inversion An aberration in chromosome structure resulting from reattachment of a chromosomal fragment in a reverse orientation to the chromosome from which it originated.

invertebrate An animal without a backbone. Invertebrates make up 95% of animal species.

ion (ī′-on) An atom or group of atoms that has gained or lost one or more electrons, thus acquiring a charge.

ion channel A transmembrane protein channel that allows a specific ion to diffuse across the membrane down its concentration or electrochemical gradient.

ionic bond (ī-on′-ik) A chemical bond resulting from the attraction between oppositely charged ions.

ionic compound A compound resulting from the formation of an ionic bond; also called a salt.

IPSP *See* inhibitory postsynaptic potential.

iris The colored part of the vertebrate eye, formed by the anterior portion of the choroid.

isomer (ī′-sō-mer) One of several compounds with the same molecular formula but different structures and therefore different properties. The three types of isomers are structural isomers, *cis-trans* isomers, and enantiomers.

isomorphic Referring to alternating generations in plants and certain algae in which the sporophytes and gametophytes look alike, although they differ in chromosome number.

isopod A member of one of the largest groups of crustaceans, which includes terrestrial, freshwater, and marine species. Among the terrestrial isopods are the pill bugs, or wood lice.

isotonic (ī′-sō-ton′-ik) Referring to a solution that, when surrounding a cell, causes no net movement of water into or out of the cell.

isotope (ī′-sō-tōp′) One of several atomic forms of an element, each with the same number of protons but a different number of neutrons, thus differing in atomic mass.

iteroparity Reproduction in which adults produce offspring over many years; also known as repeated reproduction.

joule (J) A unit of energy: 1 J = 0.239 cal; 1 cal = 4.184 J.

juxtaglomerular apparatus (JGA) (juks′-tuh-gluh-mār′-yū-ler) A specialized tissue in nephrons that releases the enzyme renin in response to a drop in blood pressure or volume.

juxtamedullary nephron In mammals and birds, a nephron with a loop of Henle that extends far into the renal medulla.

karyogamy (kār′-ē-og′-uh-mē) In fungi, the fusion of haploid nuclei contributed by the two parents; occurs as one stage of sexual reproduction, preceded by plasmogamy.

karyotype (kār′-ē-ō-tīp) A display of the chromosome pairs of a cell arranged by size and shape.

keystone species A species that is not necessarily abundant in a community yet exerts strong control on community structure by the nature of its ecological role or niche.

kidney In vertebrates, one of a pair of excretory organs where blood filtrate is formed and processed into urine.

kilocalorie (kcal) A thousand calories; the amount of heat energy required to raise the temperature of 1 kg of water by 1°C.

kin selection Natural selection that favors altruistic behavior by enhancing the reproductive success of relatives.

kinetic energy (kuh-net′-ik) The energy associated with the relative motion of objects. Moving matter can perform work by imparting motion to other matter.

kinetochore (kuh-net′-uh-kōr) A structure of proteins attached to the centromere that links each sister chromatid to the mitotic spindle.

kinetoplastid A protist, such as a trypanosome, that has a single large mitochondrion that houses an organized mass of DNA.

kingdom A taxonomic category, the second broadest after domain.

***K*-selection** Selection for life history traits that are sensitive to population density; also called density-dependent selection.

labia majora A pair of thick, fatty ridges that encloses and protects the rest of the vulva.

labia minora A pair of slender skin folds that surrounds the openings of the vagina and urethra.

labor A series of strong, rhythmic contractions of the uterus that expels a baby out of the uterus and vagina during childbirth.

lactation The continued production of milk from the mammary glands.

lacteal (lak′-tē-ul) A tiny lymph vessel extending into the core of an intestinal villus and serving as the destination for absorbed chylomicrons.

lactic acid fermentation Glycolysis followed by the reduction of pyruvate to lactate, regenerating NAD^+ with no release of carbon dioxide.

lagging strand A discontinuously synthesized DNA strand that elongates by means of Okazaki fragments, each synthesized in a $5' \rightarrow 3'$ direction away from the replication fork.

lancelet Member of the clade Cephalochordata, small blade-shaped marine chordates that lack a backbone.

landscape An area containing several different ecosystems linked by exchanges of energy, materials, and organisms.

landscape ecology The study of how the spatial arrangement of habitat types affects the distribution and abundance of organisms and ecosystem processes.

large intestine The portion of the vertebrate alimentary canal between the small intestine and the anus; functions mainly in water absorption and the formation of feces.

larva (lar′-vuh) (plural, **larvae**) A free-living, sexually immature form in some animal life cycles that may differ from the adult animal in morphology, nutrition, and habitat.

larynx (lār′-inks) The portion of the respiratory tract containing the vocal cords; also called the voice box.

lateral geniculate nucleus One of a pair of structures in the brain that are the destination for most of the ganglion cell axons that form the optic nerves.

lateral inhibition A process that sharpens the edges and enhances the contrast of a perceived image by inhibiting receptors lateral to those that have responded to light.

lateral line system A mechanoreceptor system consisting of a series of pores and receptor units along the sides of the body in fishes and aquatic amphibians; detects water movements made by the animal itself and by other moving objects.

lateral meristem (mār′-uh-stem) A meristem that thickens the roots and shoots of woody plants. The vascular cambium and cork cambium are lateral meristems.

lateral root A root that arises from the pericycle of an established root.

lateralization Segregation of functions in the cortex of the left and right cerebral hemispheres.

law of conservation of mass A physical law stating that matter can change form but cannot be created or destroyed. In a closed system, the mass of the system is constant.

law of independent assortment Mendel's second law, stating that each pair of alleles segregates, or assorts, independently of each other pair during gamete formation; applies when genes for two characters are located on different pairs of homologous chromosomes or when they are far enough apart on the same chromosome to behave as though they are on different chromosomes.

law of segregation Mendel's first law, stating that the two alleles in a pair segregate (separate from each other) into different gametes during gamete formation.

leading strand The new complementary DNA strand synthesized continuously along the template strand toward the replication fork in the mandatory $5' \rightarrow 3'$ direction.

leaf The main photosynthetic organ of vascular plants.

leaf primordium A finger-like projection along the flank of a shoot apical meristem, from which a leaf arises.

learning The modification of behavior based on specific experiences.

lens The structure in an eye that focuses light rays onto the photoreceptors.

lenticel (len′-ti-sel) A small raised area in the bark of stems and roots that enables gas exchange between living cells and the outside air.

lepidosaur (leh-pid′-uh-sōr) Member of the reptilian group that includes lizards, snakes, and two species of New Zealand animals called tuataras.

leptin A hormone produced by adipose (fat) cells that acts as a satiety factor in regulating appetite.

leukocyte (lū′-kō-sīt′) A blood cell that functions in fighting infections; also called a white blood cell.

Leydig cell (lī′-dig) A cell that produces testosterone and other androgens and is located between the seminiferous tubules of the testes.

lichen The mutualistic association between a fungus and a photosynthetic alga or cyanobacterium.

life cycle The generation-to-generation sequence of stages in the reproductive history of an organism.

life history The traits that affect an organism's schedule of reproduction and survival.

life table An age-specific summary of the survival pattern of a population.

ligament A fibrous connective tissue that joins bones together at joints.

ligand (lig′-und) A molecule that binds specifically to another molecule, usually a larger one.

ligand-gated ion channel A transmembrane protein containing a pore that opens or closes as it changes shape in response to a signaling molecule (ligand), allowing or blocking the flow of specific ions; also called an ionotropic receptor.

light chain One of the two types of polypeptide chains that make up an antibody molecule and B cell receptor; consists of a variable region, which contributes to the antigen-binding site, and a constant region.

light microscope (LM) An optical instrument with lenses that refract (bend) visible light to magnify images of specimens.

light reactions The first of two major stages in photosynthesis (preceding the Calvin cycle). These reactions, which occur on the thylakoid membranes of the chloroplast or on membranes of certain prokaryotes, convert solar energy to the chemical energy of ATP and NADPH, releasing oxygen in the process.

light-harvesting complex A complex of proteins associated with pigment molecules (including chlorophyll *a*, chlorophyll *b*, and carotenoids) that captures light energy and transfers it to reaction-center pigments in a photosystem.

lignin (lig′-nin) A hard material embedded in the cellulose matrix of vascular plant cell walls that provides structural support in terrestrial species.

limiting nutrient An element that must be added for production to increase in a particular area.

limnetic zone In a lake, the well-lit, open surface waters far from shore.

linear electron flow A route of electron flow during the light reactions of photosynthesis that involves both photosystems (I and II) and produces ATP, NADPH, and O_2. The net electron flow is from H_2O to $NADP^+$.

linkage map A genetic map based on the frequencies of recombination between markers during crossing over of homologous chromosomes.

linked genes Genes located close enough together on a chromosome that they tend to be inherited together.

lipid (lip′-id) Any of a group of large biological molecules, including fats, phospholipids, and steroids, that mix poorly, if at all, with water.

littoral zone In a lake, the shallow, well-lit waters close to shore.

liver A large internal organ in vertebrates that performs diverse functions, such as producing bile, maintaining blood glucose level, and detoxifying poisonous chemicals in the blood.

liverwort A small, herbaceous, nonvascular plant that is a member of the phylum Hepatophyta.

loam The most fertile soil type, made up of roughly equal amounts of sand, silt, and clay.

lobe-fin Member of the vertebrate clade Sarcopterygii, osteichthyans with rod-shaped muscular fins, including coelacanths, lungfishes, and tetrapods.

local regulator A secreted molecule that influences cells near where it is secreted.

locomotion Active motion from place to place.

locus (lō′-kus) (plural, **loci**) A specific place along the length of a chromosome where a given gene is located.

logistic population growth Population growth that levels off as population size approaches carrying capacity.

long-day plant A plant that flowers (usually in late spring or early summer) only when the light period is longer than a critical length.

long-term memory The ability to hold, associate, and recall information over one's lifetime.

long-term potentiation (LTP) An enhanced responsiveness to an action potential (nerve signal) by a receiving neuron.

loop of Henle The hairpin turn, with a descending and ascending limb, between the proximal and distal tubules of the vertebrate kidney; functions in water and salt reabsorption.

lophophore (lof′-uh-fōr) In some lophotrochozoan animals, including brachiopods, a crown of ciliated tentacles that surround the mouth and function in feeding.

lophotrochozoan Member of a group of animal phyla identified as a clade by molecular evidence. Lophotrochozoans include organisms that have lophophores or trochophore larvae.

low-density lipoprotein (LDL) A particle in the blood made up of thousands of cholesterol molecules and other lipids bound to a protein. LDL transports cholesterol from the liver for incorporation into cell membranes.

lung An infolded respiratory surface of a terrestrial vertebrate, land snail, or spider that connects to the atmosphere by narrow tubes.

luteal phase That portion of the ovarian cycle during which endocrine cells of the corpus luteum secrete female hormones.

luteinizing hormone (LH) (lū′-tē-uh-nī′-zing) A tropic hormone that is produced and secreted by the anterior pituitary and that stimulates ovulation in females and androgen production in males.

lycophyte (lī′-kuh-fīt) An informal name for a member of the phylum Lycophyta, which includes club mosses, spike mosses, and quillworts.

lymph The colorless fluid, derived from interstitial fluid, in the lymphatic system of vertebrates.

lymph node An organ located along a lymph vessel. Lymph nodes filter lymph and contain cells that attack viruses and bacteria.

lymphatic system A system of vessels and nodes, separate from the circulatory system, that returns fluid, proteins, and cells to the blood.

lymphocyte A type of white blood cell that mediates immune responses. The two main classes are B cells and T cells.

lysogenic cycle (lī′-sō-jen′-ik) A type of phage replicative cycle in which the viral genome becomes incorporated into the bacterial host chromosome as a prophage, is replicated along with the chromosome, and does not kill the host.

lysosome (lī′-suh-sōm) A membrane-enclosed sac of hydrolytic enzymes found in the cytoplasm of animal cells and some protists.

lysozyme (lī′-sō-zīm) An enzyme that destroys bacterial cell walls; in mammals, found in sweat, tears, and saliva.

lytic cycle (lit′-ik) A type of phage replicative cycle resulting in the release of new phages by lysis (and death) of the host cell.

macroclimate Large-scale patterns in climate; the climate of an entire region.

macroevolution Evolutionary change above the species level. Examples of macroevolutionary change include the origin of a new group of organisms through a series of speciation events and the impact of mass extinctions on the diversity of life and its subsequent recovery.

macromolecule A giant molecule formed by the joining of smaller molecules, usually by a dehydration reaction. Polysaccharides, proteins, and nucleic acids are macromolecules.

macronutrient An essential element that an organism must obtain in relatively large amounts. *See also* micronutrient.

macrophage (mak′-rō-fāj) A phagocytic cell present in many tissues that functions in innate immunity by destroying microbes and in acquired immunity as an antigen-presenting cell.

magnoliid Member of the angiosperm clade that is most closely related to the combined eudicot and monocot clades. Extant examples are magnolias, laurels, and black pepper plants.

major depressive disorder A mood disorder characterized by feelings of sadness, lack of self-worth, emptiness, or loss of interest in nearly all things.

major histocompatibility complex (MHC) molecule A host protein that functions in antigen presentation. Foreign MHC molecules on transplanted tissue can trigger T cell responses that may lead to rejection of the transplant.

malignant tumor A cancerous tumor containing cells that have significant genetic and cellular changes and are capable of invading and surviving in new sites. Malignant tumors can impair the functions of one or more organs.

Malpighian tubule (mal-pig′-ē-un) A unique excretory organ of insects that empties into the digestive tract, removes nitrogenous wastes from the hemolymph, and functions in osmoregulation.

mammal Member of the class Mammalia, amniotes that have hair and mammary glands (glands that produce milk).

mammary gland An exocrine gland that secretes milk to nourish the young. Mammary glands are characteristic of mammals.

mandible One of a pair of jaw-like feeding appendages found in myriapods, hexapods, and crustaceans.

mantle One of the three main parts of a mollusc; a fold of tissue that drapes over the mollusc's visceral mass and may secrete a shell. *See also* foot, visceral mass.

mantle cavity A water-filled chamber that houses the gills, anus, and excretory pores of a mollusc.

map unit A unit of measurement of the distance between genes. One map unit is equivalent to a 1% recombination frequency.

marine benthic zone The ocean floor.

mark-recapture method A sampling technique used to estimate the size of animal populations.

marsupial (mar-sū′-pē-ul) A mammal, such as a koala, kangaroo, or opossum, whose young complete their embryonic development inside a maternal pouch called the marsupium.

mass extinction The elimination of a large number of species throughout Earth, the result of global environmental changes.

mass number The sum of the number of protons and neutrons in an atom's nucleus.

mast cell A vertebrate body cell that produces histamine and other molecules that trigger inflammation in response to infection and in allergic reactions.

mate-choice copying Behavior in which individuals in a population copy the mate choice of others, apparently as a result of social learning.

maternal effect gene A gene that, when mutant in the mother, results in a mutant phenotype in the offspring, regardless of the offspring's genotype. Maternal effect genes, also called egg-polarity genes, were first identified in *Drosophila melanogaster*.

matter Anything that takes up space and has mass.

maximum likelihood As applied to molecular systematics, a principle that states that when considering multiple phylogenetic hypotheses, one should take into account the hypothesis that reflects the most likely sequence of evolutionary events, given certain rules about how DNA changes over time.

maximum parsimony A principle that states that when considering multiple explanations for an observation, one should first investigate the simplest explanation that is consistent with the facts.

mechanoreceptor A sensory receptor that detects physical deformation in the body's environment associated with pressure, touch, stretch, motion, or sound.

medulla oblongata (meh-dul′-uh ōb′-long-go′-tuh) The lowest part of the vertebrate brain, commonly called the medulla; a swelling of the hindbrain anterior to the spinal cord that controls autonomic, homeostatic functions, including breathing, heart and blood vessel activity, swallowing, digestion, and vomiting.

Medusa (plural, **medusae**) (muh-dū′-suh) The floating, flattened, mouth-down version of the cnidarian body plan. The alternate form is the polyp.

megapascal (MPa) (meg′-uh-pas-kal′) A unit of pressure equivalent to about 10 atmospheres of pressure.

megaphyll (meh′-guh-fil) A leaf with a highly branched vascular system, characteristic of the vast majority of vascular plants. *See* microphyll.

megaspore A spore from a heterosporous plant species that develops into a female gametophyte.

meiosis (mī-ō′-sis) A modified type of cell division in sexually reproducing organisms consisting of two rounds of cell division but only one round of DNA replication. It results in cells with half the number of chromosome sets as the original cell.

meiosis I The first division of a two-stage process of cell division in sexually reproducing organisms that results in cells with half the number of chromosome sets as the original cell.

meiosis II The second division of a two-stage process of cell division in sexually reproducing organisms that results in cells with half the number of chromosome sets as the original cell.

melanocyte-stimulating hormone (MSH) A hormone produced and secreted by the anterior pituitary with multiple activities, including regulating the behavior of pigment-containing cells in the skin of some vertebrates.

melatonin A hormone that is secreted by the pineal gland and that is involved in the regulation of biological rhythms and sleep.

membrane potential The difference in electrical charge (voltage) across a cell's plasma membrane due to the differential distribution of ions. Membrane potential affects the activity of excitable cells and the transmembrane movement of all charged substances.

memory cell One of a clone of long-lived lymphocytes, formed during the primary immune response, that remains in a lymphoid organ until activated by exposure to the same antigen that triggered its formation. Activated memory cells mount the secondary immune response.

menopause The cessation of ovulation and menstruation marking the end of a human female's reproductive years.

menstrual cycle (men′-strū-ul) In humans and certain other primates, a type of reproductive cycle in which the nonpregnant endometrium is shed through the cervix into the vagina; also called the uterine cycle.

menstrual flow phase That portion of the uterine (menstrual) cycle when menstrual bleeding occurs.

menstruation The shedding of portions of the endometrium during a uterine (menstrual) cycle.

meristem (mār′-uh-stem) Plant tissue that remains embryonic as long as the plant lives, allowing for indeterminate growth.

meristem identity gene A plant gene that promotes the switch from vegetative growth to flowering.

meroblastic (mār′-ō-blas′-tik) Referring to a type of cleavage in which there is incomplete division of a yolk-rich egg, characteristic of avian development.

mesoderm (mez′-ō-derm) The middle primary germ layer in a triploblastic animal embryo; develops into the notochord, the lining of the coelom, muscles, skeleton, gonads, kidneys, and most of the circulatory system in species that have these structures.

mesohyl (mez′-ō-hīl) A gelatinous region between the two layers of cells of a sponge.

mesophyll (mez′-ō-fil) Leaf cells specialized for photosynthesis. In C_3 and CAM plants, mesophyll cells are located between the upper and lower epidermis; in C_4 plants, they are located between the bundle-sheath cells and the epidermis.

messenger RNA (mRNA) A type of RNA, synthesized using a DNA template, that attaches to ribosomes in the cytoplasm and specifies the primary structure of a protein. (In eukaryotes, the primary RNA transcript must undergo RNA processing to become mRNA.)

metabolic pathway A series of chemical reactions that either builds a complex molecule (anabolic pathway) or breaks down a complex molecule to simpler molecules (catabolic pathway).

metabolic rate The total amount of energy an animal uses in a unit of time.

metabolism (muh-tab′-uh-lizm) The totality of an organism's chemical reactions, consisting of catabolic and anabolic pathways, which manage the material and energy resources of the organism.

metagenomics The collection and sequencing of DNA from a group of species, usually an environmental sample of microorganisms. Computer software sorts partial sequences and assembles them into genome sequences of individual species making up the sample.

metamorphosis (met′-uh-mōr′-fuh-sis) A developmental transformation that turns an animal larva into either an adult or an adult-like stage that is not yet sexually mature.

metanephridium (met′-uh-nuh-frid′-ē-um) (plural, **metanephridia**) An excretory organ found in many invertebrates that typically consists of tubules connecting ciliated internal openings to external openings.

metaphase The third stage of mitosis, in which the spindle is complete and the chromosomes, attached to microtubules at their kinetochores, are all aligned at the metaphase plate.

metaphase plate An imaginary structure located at a plane midway between the two poles of a cell in metaphase on which the centromeres of all the duplicated chromosomes are located.

metapopulation A group of spatially separated populations of one species that interact through immigration and emigration.

metastasis (muh-tas′-tuh-sis) The spread of cancer cells to locations distant from their original site.

methanogen (meth-an′-ō-jen) An organism that produces methane as a waste product of the way it obtains energy. All known methanogens are in domain Archaea.

methyl group A chemical group consisting of a carbon bonded to three hydrogen atoms. The methyl group may be attached to a carbon or to a different atom.

microclimate Climate patterns on a very fine scale, such as the specific climatic conditions underneath a log.

microevolution Evolutionary change below the species level; change in the allele frequencies in a population over generations.

microfilament A cable composed of actin proteins in the cytoplasm of almost every eukaryotic cell, making up part of the cytoskeleton and acting alone or with myosin to cause cell contraction; also known as an actin filament.

micronutrient An essential element that an organism needs in very small amounts. *See also* macronutrient.

microphyll (mī′-krō-fil) In lycophytes, a small leaf with a single unbranched vein. *See* megaphyll.

micropyle A pore in the integuments of an ovule.

microRNA (miRNA) A small, single-stranded RNA molecule, generated from a hairpin structure on a precursor RNA transcribed from a particular gene. The miRNA associates with one or more proteins in a complex that can degrade or prevent translation of an mRNA with a complementary sequence.

microspore A spore from a heterosporous plant species that develops into a male gametophyte.

microtubule A hollow rod composed of tubulin proteins that makes up part of the cytoskeleton in all eukaryotic cells and is found in cilia and flagella.

microvillus (plural, **microvilli**) One of many fine, finger-like projections of the epithelial cells in the lumen of the small intestine that increase its surface area.

midbrain One of three ancestral and embryonic regions of the vertebrate brain; develops into sensory integrating and relay centers that send sensory information to the cerebrum.

middle ear One of three main regions of the vertebrate ear; in mammals, a chamber containing three small bones (the malleus, incus, and stapes) that convey vibrations from the eardrum to the oval window.

middle lamella (luh-mel′-uh) In plants, a thin layer of adhesive extracellular material, primarily pectins, found between the primary walls of adjacent young cells.

migration A regular, long-distance change in location.

mineral In nutrition, a simple nutrient that is inorganic and therefore cannot be synthesized in the body.

mineralocorticoid A steroid hormone secreted by the adrenal cortex that regulates salt and water homeostasis.

minimum viable population (MVP) The smallest population size at which a species is able to sustain its numbers and survive.

mismatch repair The cellular process that uses specific enzymes to remove and replace incorrectly paired nucleotides.

missense mutation A nucleotide-pair substitution that results in a codon that codes for a different amino acid.

mitochondrial matrix The compartment of the mitochondrion enclosed by the inner membrane and containing enzymes and substrates for the citric acid cycle, as well as ribosomes and DNA.

mitochondrion (mī′-tō-kon′-drē-un) (plural, **mitochondria**) An organelle in eukaryotic cells that serves as the site of cellular respiration; uses oxygen to break down organic molecules and synthesize ATP.

mitosis (mī-tō′-sis) A process of nuclear division in eukaryotic cells conventionally divided into five stages: prophase, prometaphase, metaphase, anaphase, and telophase. Mitosis conserves chromosome number by allocating replicated chromosomes equally to each of the daughter nuclei.

mitotic (M) phase The phase of the cell cycle that includes mitosis and cytokinesis.

mitotic spindle An assemblage of microtubules and associated proteins that is involved in the movement of chromosomes during mitosis.

mixotroph An organism that is capable of both photosynthesis and heterotrophy.

model organism A particular species chosen for research into broad biological principles because it is representative of a larger group and usually easy to grow in a lab.

molarity A common measure of solute concentration, referring to the number of moles of solute per liter of solution.

mold Informal term for a fungus that grows as a filamentous fungus, producing haploid spores by mitosis and forming a visible mycelium.

mole (mol) The number of grams of a substance that equals its molecular weight in daltons and contains Avogadro's number of molecules.

molecular clock A method for estimating the time required for a given amount of evolutionary change, based on the observation that some regions of genomes evolve at constant rates.

molecular mass The sum of the masses of all the atoms in a molecule; sometimes called molecular weight.

molecular systematics A scientific discipline that uses nucleic acids or other molecules to infer evolutionary relationships between different species.

molecule Two or more atoms held together by covalent bonds.

molting A process in ecdysozoans in which the exoskeleton is shed at intervals, allowing growth by the production of a larger exoskeleton.

monoclonal antibody (mon′-ō-klōn′-ul) Any of a preparation of antibodies that have been produced by a single clone of cultured cells and thus are all specific for the same epitope.

monocot Member of a clade consisting of flowering plants that have one embryonic seed leaf, or cotyledon.

monogamous (muh-nog′-uh-mus) Referring to a type of relationship in which one male mates with just one female.

monohybrid An organism that is heterozygous with respect to a single gene of interest. All the offspring from a cross between parents homozygous for different alleles are monohybrids. For example, parents of genotypes *AA* and *aa* produce a monohybrid of genotype *Aa*.

monohybrid cross A cross between two organisms that are heterozygous for the character being followed (or the self-pollination of a heterozygous plant).

monomer (mon′-uh-mer) The subunit that serves as the building block of a polymer.

monophyletic (mon′-ō-fī-let′-ik) Pertaining to a group of taxa that consists of a common ancestor and all of its descendants. A monophyletic taxon is equivalent to a clade.

Glossary

monosaccharide (mon′-ō-sak′-uh-rīd) The simplest carbohydrate, active alone or serving as a monomer for disaccharides and polysaccharides. Also known as simple sugars, monosaccharides have molecular formulas that are generally some multiple of CH_2O.

monosomic Referring to a diploid cell that has only one copy of a particular chromosome instead of the normal two.

monotreme An egg-laying mammal, such as a platypus or echidna. Like all mammals, monotremes have hair and produce milk, but they lack nipples.

morphogen A substance, such as Bicoid protein in *Drosophila*, that provides positional information in the form of a concentration gradient along an embryonic axis.

morphogenesis (mōr′-fō-jen′-uh-sis) The cellular and tissue-based processes by which an animal body takes shape.

morphological species concept A definition of species in terms of measurable anatomical criteria.

moss A small, herbaceous, nonvascular plant that is a member of the phylum Bryophyta.

motor neuron A nerve cell that transmits signals from the brain or spinal cord to muscles or glands.

motor protein A protein that interacts with cytoskeletal elements and other cell components, producing movement of the whole cell or parts of the cell.

motor system An efferent branch of the vertebrate peripheral nervous system composed of motor neurons that carry signals to skeletal muscles in response to external stimuli.

motor unit A single motor neuron and all the muscle fibers it controls.

movement corridor A series of small clumps or a narrow strip of quality habitat (usable by organisms) that connects otherwise isolated patches of quality habitat.

MPF Maturation-promoting factor (or M-phase-promoting factor); a protein complex required for a cell to progress from late interphase to mitosis. The active form consists of cyclin and a protein kinase.

mucus A viscous and slippery mixture of glycoproteins, cells, salts, and water that moistens and protects the membranes lining body cavities that open to the exterior.

Müllerian mimicry (myū-lār′-ē-un) Reciprocal mimicry by two unpalatable species.

multifactorial Referring to a phenotypic character that is influenced by multiple genes and environmental factors.

multigene family A collection of genes with similar or identical sequences, presumably of common origin.

multiple fruit A fruit derived from an entire inflorescence.

multiplication rule A rule of probability stating that the probability of two or more independent events occurring together can be determined by multiplying their individual probabilities.

muscle tissue Tissue consisting of long muscle cells that can contract, either on its own or when stimulated by nerve impulses.

mutagen (myū′-tuh-jen) A chemical or physical agent that interacts with DNA and can cause a mutation.

mutation (myū-tā′-shun) A change in the nucleotide sequence of an organism's DNA or in the DNA or RNA of a virus.

mutualism (myū′-chū-ul-izm) A symbiotic relationship in which both participants benefit.

mycelium (mī-sē′-lē-um) The densely branched network of hyphae in a fungus.

mycorrhiza (mī′-kō-rī′-zuh) (plural, **mycorrhizae**) A mutualistic association of plant roots and fungus.

mycosis (mī-kō′-sis) General term for a fungal infection.

myelin sheath (mī′-uh-lin) Wrapped around the axon of a neuron, an insulating coat of cell membranes from Schwann cells or oligodendrocytes. It is interrupted by nodes of Ranvier, where action potentials are generated.

myofibril (mī′-ō-fī′-bril) A longitudinal bundle in a muscle cell (fiber) that contains thin filaments of actin and regulatory proteins and thick filaments of myosin.

myoglobin (mī′-uh-glō′-bin) An oxygen-storing, pigmented protein in muscle cells.

myosin (mī′-uh-sin) A type of motor protein that associates into filaments that interact with actin filaments to cause cell contraction.

myotonia (mī′-uh-tō′-nī-uh) Increased muscle tension, characteristic of sexual arousal in certain human tissues.

myriapod (mir′-ē-uh-pod′) A terrestrial arthropod with many body segments and one or two pairs of legs per segment. Millipedes and centipedes are the two major groups of living myriapods.

NAD^+ Nicotinamide adenine dinucleotide, a coenzyme that cycles easily between oxidized (NAD^+) and reduced (NADH) states, thus acting as an electron carrier.

$NADP^+$ Nicotinamide adenine dinucleotide phosphate, an electron acceptor that, as NADPH, temporarily stores energized electrons produced during the light reactions.

natural family planning A form of contraception that relies on refraining from sexual intercourse when conception is most likely to occur; also called the rhythm method.

natural killer cell A type of white blood cell that can kill tumor cells and virus-infected cells as part of innate immunity.

natural selection A process in which individuals that have certain inherited traits tend to survive and reproduce at higher rates than other individuals *because of* those traits.

negative feedback A form of regulation in which accumulation of an end product of a process slows the process; in physiology, a primary mechanism of homeostasis, whereby a change in a variable triggers a response that counteracts the initial change.

negative pressure breathing A breathing system in which air is pulled into the lungs.

nematocyst (nem′-uh-tuh-sist′) In a cnidocyte of a cnidarian, a capsule-like organelle containing a coiled thread that when discharged can penetrate the body wall of the prey.

nephron (nef′-ron) The tubular excretory unit of the vertebrate kidney.

neritic zone The shallow region of the ocean overlying the continental shelf.

nerve A fiber composed primarily of the bundled axons of PNS neurons.

nerve net A weblike system of neurons, characteristic of radially symmetrical animals, such as hydras.

nervous system The fast-acting internal system of communication involving sensory receptors, networks of nerve cells, and connections to muscles and glands that respond to nerve signals; functions in concert with the endocrine system to effect internal regulation and maintain homeostasis.

nervous tissue Tissue made up of neurons and supportive cells.

net ecosystem production (NEP) The gross primary production of an ecosystem minus the energy used by all autotrophs and heterotrophs for respiration.

net primary production (NPP) The gross primary production of an ecosystem minus the energy used by the producers for respiration.

neural crest In vertebrates, a region located along the sides of the neural tube where it pinches off from the ectoderm. Neural crest cells migrate to various parts of the embryo and form pigment cells in the skin and parts of the skull, teeth, adrenal glands, and peripheral nervous system.

neural plasticity The capacity of a nervous system to change with experience.

neural tube A tube of infolded ectodermal cells that runs along the anterior-posterior axis of a vertebrate, just dorsal to the notochord. It will give rise to the central nervous system.

neurohormone A molecule that is secreted by a neuron, travels in body fluids, and acts on specific target cells, changing their functioning.

neuron (nyūr′-on) A nerve cell; the fundamental unit of the nervous system, having structure and properties that allow it to conduct signals by taking advantage of the electrical charge across its plasma membrane.

neuropeptide A relatively short chain of amino acids that serves as a neurotransmitter.

neurotransmitter A molecule that is released from the synaptic terminal of a neuron at a chemical synapse, diffuses across the synaptic cleft, and binds to the postsynaptic cell, triggering a response.

neutral theory The hypothesis that much evolutionary change in genes and proteins has no effect on fitness and therefore is not influenced by natural selection.

neutral variation Genetic variation that does not provide a selective advantage or disadvantage.

neutron A subatomic particle having no electrical charge (electrically neutral), with a mass of about 1.7×10^{-24} g, found in the nucleus of an atom.

neutrophil The most abundant type of white blood cell. Neutrophils are phagocytic and tend to self-destruct as they destroy foreign invaders, limiting their life span to a few days.

nitric oxide (NO) A gas produced by many types of cells that functions as a local regulator and as a neurotransmitter.

nitrogen cycle The natural process by which nitrogen, either from the atmosphere or from decomposed organic material, is converted by soil bacteria to compounds assimilated by plants. This incorporated nitrogen is then taken in by other organisms and subsequently released, acted on by bacteria, and made available again to the nonliving environment.

nitrogen fixation The conversion of atmospheric nitrogen (N_2) to ammonia (NH_3). Biological nitrogen fixation is carried out by certain prokaryotes, some of which have mutualistic relationships with plants.

nociceptor (nō′-si-sep′-tur) A sensory receptor that responds to noxious or painful stimuli; also called a pain receptor.

node A point along the stem of a plant at which leaves are attached.

node of Ranvier (ron′-vē-ā′) Gap in the myelin sheath of certain axons where an action potential may be generated. In saltatory conduction, an action potential is regenerated at each node, appearing to "jump" along the axon from node to node.

nodule A swelling on the root of a legume. Nodules are composed of plant cells that contain nitrogen-fixing bacteria of the genus *Rhizobium*.

noncompetitive inhibitor A substance that reduces the activity of an enzyme by binding to a location remote from the active site, changing the enzyme's shape so that the active site no longer effectively catalyzes the conversion of substrate to product.

nondisjunction An error in meiosis or mitosis in which members of a pair of homologous chromosomes or a pair of sister chromatids fail to separate properly from each other.

nonequilibrium model A model that maintains that communities change constantly after being buffeted by disturbances.

nonpolar covalent bond A type of covalent bond in which electrons are shared equally between two atoms of similar electronegativity.

nonsense mutation A mutation that changes an amino acid codon to one of the three stop codons, resulting in a shorter and usually nonfunctional protein.

norepinephrine A catecholamine that is chemically and functionally similar to epinephrine and acts as a hormone or neurotransmitter; also known as noradrenaline.

norm of reaction The range of phenotypes produced by a single genotype, due to environmental influences.

Northern blotting A technique that enables specific nucleotide sequences to be detected in samples of mRNA. It involves gel electrophoresis of RNA molecules and their transfer to a membrane (blotting), followed by nucleic acid hybridization with a labeled probe.

northern coniferous forest A terrestrial biome characterized by long, cold winters and dominated by cone-bearing trees.

no-till agriculture A plowing technique that minimally disturbs the soil, thereby reducing soil loss.

notochord (nō′-tuh-kord′) A longitudinal, flexible rod made of tightly packed mesodermal cells that runs along the anterior-posterior axis of a chordate in the dorsal part of the body.

nuclear envelope In a eukaryotic cell, the double membrane that surrounds the nucleus, perforated with pores that regulate traffic with the cytoplasm. The outer membrane is continuous with the endoplasmic reticulum.

nuclear lamina A netlike array of protein filaments that lines the inner surface of the nuclear envelope and helps maintain the shape of the nucleus.

nucleariid Member of a group of unicellular, amoeboid protists that are more closely related to fungi than they are to other protists.

nuclease An enzyme that cuts DNA or RNA, either removing one or a few bases or hydrolyzing the DNA or RNA completely into its component nucleotides.

nucleic acid (nū-klā′-ik) A polymer (polynucleotide) consisting of many nucleotide monomers; serves as a blueprint for proteins and, through the actions of proteins, for all cellular activities. The two types are DNA and RNA.

nucleic acid hybridization The process of base pairing between a gene and a complementary sequence on another nucleic acid molecule.

nucleic acid probe In DNA technology, a labeled single-stranded nucleic acid molecule used to locate a specific nucleotide sequence in a nucleic acid sample. Molecules of the probe hydrogen-bond to the complementary sequence wherever it occurs; radioactive, fluorescent, or other labeling of the probe allows its location to be detected.

nucleoid (nū′-klē-oyd) A non-membrane-bounded region in a prokaryotic cell where the DNA is concentrated.

nucleolus (nū-klē′-ō-lus) (plural, **nucleoli**) A specialized structure in the nucleus, consisting of chromosomal regions containing ribosomal RNA (rRNA) genes along with ribosomal proteins imported from the cytoplasm; site of rRNA synthesis and ribosomal subunit assembly. *See also* ribosome.

nucleosome (nū′-klē-ō-sōm′) The basic, bead-like unit of DNA packing in eukaryotes, consisting of a segment of DNA wound around a protein core composed of two copies of each of four types of histone.

nucleotide (nū′-klē-ō-tīd′) The building block of a nucleic acid, consisting of a five-carbon sugar covalently bonded to a nitrogenous base and one or more phosphate groups.

nucleotide excision repair A repair system that removes and then correctly replaces a damaged segment of DNA using the undamaged strand as a guide.

nucleotide-pair substitution A type of point mutation in which one nucleotide in a DNA strand and its partner in the complementary strand are replaced by another pair of nucleotides.

nucleus (1) An atom's central core, containing protons and neutrons. (2) The organelle of a eukaryotic cell that contains the genetic material in the form of chromosomes, made up of chromatin. (3) A cluster of neurons.

nutrition The process by which an organism takes in and makes use of food substances.

obligate aerobe (ob′-lig-et ār′-ōb) An organism that requires oxygen for cellular respiration and cannot live without it.

obligate anaerobe (ob′-lig-et an′-uh-rōb) An organism that only carries out fermentation or anaerobic respiration. Such organisms cannot use oxygen and in fact may be poisoned by it.

ocean acidification Decreasing pH of ocean waters due to absorption of excess atmospheric CO_2 from the burning of fossil fuels.

oceanic pelagic zone Most of the ocean's waters far from shore, constantly mixed by ocean currents.

odorant A molecule that can be detected by sensory receptors of the olfactory system.

Okazaki fragment (ō′-kah-zah′-kē) A short segment of DNA synthesized away from the replication fork on a template strand during DNA replication. Many such segments are joined together to make up the lagging strand of newly synthesized DNA.

olfaction The sense of smell.

oligodendrocyte A type of glial cell that forms insulating myelin sheaths around the axons of neurons in the central nervous system.

oligotrophic lake A nutrient-poor, clear lake with few phytoplankton.

ommatidium (ōm′-uh-tid′-ē-um) (plural, **ommatidia**) One of the facets of the compound eye of arthropods and some polychaete worms.

omnivore An animal that regularly eats animals as well as plants or algae.

oncogene (on′-kō-jēn) A gene found in viral or cellular genomes that is involved in triggering molecular events that can lead to cancer.

oocyte A cell in the female reproductive system that differentiates to form an egg.

oogenesis (ō′-uh-jen′-uh-sis) The process in the ovary that results in the production of female gametes.

oogonium (ō′-uh- gō′-nē-em) (plural, **oogonia**) A cell that divides mitotically to form oocytes.

oomycete (ō′-uh-mī′-sēt) A protist with flagellated cells, such as a water mold, white rust, or downy mildew, that acquires nutrition mainly as a decomposer or plant parasite.

open circulatory system A circulatory system in which fluid called hemolymph bathes the tissues and organs directly and there is no distinction between the circulating fluid and the interstitial fluid.

operator In bacterial and phage DNA, a sequence of nucleotides near the start of an operon to which an active repressor can attach. The binding of the repressor prevents RNA polymerase from attaching to the promoter and transcribing the genes of the operon.

operculum (ō-per′-kyuh-lum) In aquatic osteichthyans, a protective bony flap that covers and protects the gills.

operon (op′-er-on) A unit of genetic function found in bacteria and phages, consisting of a promoter, an operator, and a coordinately regulated cluster of genes whose products function in a common pathway.

opisthokont (uh-pis′-thuh-kont′) Member of the diverse clade Opisthokonta, organisms that descended from an ancestor with a posterior flagellum, including fungi, animals, and certain protists.

opposable thumb A thumb that can touch the ventral surface of the fingertips of all four fingers.

opsin A membrane protein bound to a light-absorbing pigment molecule.

optic chiasm The place where the two optic nerves meet and axons representing distinct sides of the visual field are segregated from one another before reaching the brain.

optimal foraging model The basis for analyzing behavior as a compromise between feeding costs and feeding benefits.

oral cavity The mouth of an animal.

orbital The three-dimensional space where an electron is found 90% of the time.

order In Linnaean classification, the taxonomic category above the level of family.

organ A specialized center of body function composed of several different types of tissues.

organ identity gene A plant homeotic gene that uses positional information to determine which emerging leaves develop into which types of floral organs.

organ of Corti The actual hearing organ of the vertebrate ear, located in the floor of the cochlear duct in the inner ear; contains the receptor cells (hair cells) of the ear.

organ system A group of organs that work together in performing vital body functions.

organelle (ōr-guh-nel′) Any of several membrane-enclosed structures with specialized functions, suspended in the cytosol of eukaryotic cells.

organic chemistry The study of carbon compounds (organic compounds).

organismal ecology The branch of ecology concerned with the morphological, physiological, and behavioral ways in which individual organisms meet the challenges posed by their biotic and abiotic environments.

organogenesis (ōr-gan′-ō-jen′-uh-sis) The process in which organ rudiments develop from the three germ layers after gastrulation.

orgasm Rhythmic, involuntary contractions of certain reproductive structures in both sexes during the human sexual response cycle.

origin of replication Site where the replication of a DNA molecule begins, consisting of a specific sequence of nucleotides.

orthologous genes Homologous genes that are found in different species because of speciation.

osculum (os′-kyuh-lum) A large opening in a sponge that connects the spongocoel to the environment.

osmoconformer An animal that is isoosmotic with its environment.

osmolarity (oz′-mō-lār′-uh-tē) Solute concentration expressed as molarity.

osmoregulation Regulation of solute concentrations and water balance by a cell or organism.

osmoregulator An animal that controls its internal osmolarity independent of the external environment.

osmosis (oz-mō′-sis) The diffusion of free water across a selectively permeable membrane.

osteichthyan (os′-tē-ik′-thē-an) Member of a vertebrate clade with jaws and mostly bony skeletons.

outer ear One of three main regions of the ear in reptiles (including birds) and mammals; made up of the auditory canal and, in many birds and mammals, the pinna.

outgroup A species or group of species from an evolutionary lineage that is known to have diverged before the lineage that contains the group of species being studied. An outgroup is selected so that its members are closely related to the group of species being studied, but not as closely related as any study-group members are to each other.

oval window In the vertebrate ear, a membrane-covered gap in the skull bone, through which sound waves pass from the middle ear to the inner ear.

ovarian cycle (ō-vār′-ē-un) The cyclic recurrence of the follicular phase, ovulation, and the luteal phase in the mammalian ovary, regulated by hormones.

ovary (ō′-vuh-rē) (1) In flowers, the portion of a carpel in which the egg-containing ovules develop. (2) In animals, the structure that produces female gametes and reproductive hormones.

oviduct (ō′-vuh-duct) A tube passing from the ovary to the vagina in invertebrates or to the uterus in vertebrates, where it is also known as a fallopian tube.

oviparous (ō-vip′-uh-rus) Referring to a type of development in which young hatch from eggs laid outside the mother's body.

ovoviviparous (ō′-vō-vī-vip′-uh-rus) Referring to a type of development in which young hatch from eggs that are retained in the mother's uterus.

ovulation The release of an egg from an ovary. In humans, an ovarian follicle releases an egg during each uterine (menstrual) cycle.

ovule (o′-vyūl) A structure that develops within the ovary of a seed plant and contains the female gametophyte.

oxidation The complete or partial loss of electrons from a substance involved in a redox reaction.

oxidative phosphorylation (fos′-fōr-uh-lā′-shun) The production of ATP using energy derived from the redox reactions of an electron transport chain; the third major stage of cellular respiration.

oxidizing agent The electron acceptor in a redox reaction.

oxytocin (ok′-si-tō′-sen) A hormone produced by the hypothalamus and released from the posterior pituitary. It induces contractions of the uterine muscles during labor and causes the mammary glands to eject milk during nursing.

P generation The true-breeding (homozygous) parent individuals from which F_1 hybrid offspring are derived in studies of inheritance; P stands for "parental."

P site One of a ribosome's three binding sites for tRNA during translation. The P site holds the tRNA carrying the growing polypeptide chain. (P stands for peptidyl tRNA.)

***p53* gene** A tumor-suppressor gene that codes for a specific transcription factor that promotes the synthesis of proteins that inhibit the cell cycle.

paedomorphosis (pē′-duh-mōr′-fuh-sis) The retention in an adult organism of the juvenile features of its evolutionary ancestors.

pain receptor A sensory receptor that responds to noxious or painful stimuli; also called a nociceptor.

paleoanthropology The study of human origins and evolution.

paleontology (pā′-lē-un-tol′-ō-jē) The scientific study of fossils.

pancreas (pan′-krē-us) A gland with exocrine and endocrine tissues. The exocrine portion functions in digestion, secreting enzymes and an alkaline solution into the small intestine via a duct; the ductless endocrine portion functions in homeostasis, secreting the hormones insulin and glucagon into the blood.

pandemic A global epidemic.

Pangaea (pan-jē′-uh) The supercontinent that formed near the end of the Paleozoic era, when plate movements brought all the landmasses of Earth together.

parabasalid A protist, such as a trichomonad, with modified mitochondria.

paracrine Referring to a secreted molecule that acts on a neighboring cell.

paralogous genes Homologous genes that are found in the same genome as a result of gene duplication.

paraphyletic (pār′-uh-fī-let′-ik) Pertaining to a group of taxa that consists of a common ancestor and some, but not all, of its descendants.

parareptile A basal group of reptiles, consisting mostly of large, stocky quadrupedal herbivores. Parareptiles died out in the late Triassic period.

parasite (pār′-uh-sīt) An organism that feeds on the cell contents, tissues, or body fluids of another species (the host) while in or on the host organism. Parasites harm but usually do not kill their host.

parasitism (pār′-uh-sit-izm) A symbiotic relationship in which one organism, the parasite, benefits at the expense of another, the host, by living either within or on the host.

parasympathetic division One of three divisions of the autonomic nervous system; generally enhances body activities that gain and conserve energy, such as digestion and reduced heart rate.

parathyroid gland One of four small endocrine glands, embedded in the surface of the thyroid gland, that secrete parathyroid hormone.

parathyroid hormone (PTH) A hormone secreted by the parathyroid glands that raises blood calcium level by promoting calcium release from bone and calcium retention by the kidneys.

parenchyma cell (puh-ren′-ki-muh) A relatively unspecialized plant cell type that carries out most of the metabolism, synthesizes and stores organic products, and develops into a more differentiated cell type.

parental type An offspring with a phenotype that matches one of the true-breeding parental (P generation) phenotypes; also refers to the phenotype itself.

Parkinson's disease A progressive brain disease characterized by difficulty in initiating movements, slowness of movement, and rigidity.

parthenogenesis (par'-thuh-nō'-jen'-uh-sis) A form of asexual reproduction in which females produce offspring from unfertilized eggs.

partial pressure The pressure exerted by a particular gas in a mixture of gases (for instance, the pressure exerted by oxygen in air).

passive immunity Short-term immunity conferred by the transfer of antibodies, as occurs in the transfer of maternal antibodies to a fetus or nursing infant.

passive transport The diffusion of a substance across a biological membrane with no expenditure of energy.

pathogen An organism, virus, viroid, or prion that causes disease.

pattern formation The development of a multicellular organism's spatial organization, the arrangement of organs and tissues in their characteristic places in three-dimensional space.

peat Extensive deposits of partially decayed organic material often formed primarily from the wetland moss *Sphagnum*.

pedigree A diagram of a family tree with conventional symbols, showing the occurrence of heritable characters in parents and offspring over multiple generations.

pelagic zone The open-water component of aquatic biomes.

penis The copulatory structure of male mammals.

PEP carboxylase An enzyme that adds CO_2 to phosphoenolpyruvate (PEP) to form oxaloacetate in mesophyll cells of C_4 plants. It acts prior to photosynthesis.

pepsin An enzyme present in gastric juice that begins the hydrolysis of proteins.

pepsinogen The inactive form of pepsin secreted by chief cells located in gastric pits of the stomach.

peptide bond The covalent bond between the carboxyl group on one amino acid and the amino group on another, formed by a dehydration reaction.

peptidoglycan (pep'-tid-ō-glī'-kan) A type of polymer in bacterial cell walls consisting of modified sugars cross-linked by short polypeptides.

perception The interpretation of sensory system input by the brain.

pericycle The outermost layer in the vascular cylinder, from which lateral roots arise.

periderm (pār'-uh-derm') The protective coat that replaces the epidermis in woody plants during secondary growth, formed of the cork and cork cambium.

peripheral nervous system (PNS) The sensory and motor neurons that connect to the central nervous system.

peripheral protein A protein loosely bound to the surface of a membrane or to part of an integral protein and not embedded in the lipid bilayer.

peristalsis (pār'-uh-stal'-sis) (1) Alternating waves of contraction and relaxation in the smooth muscles lining the alimentary canal that push food along the canal. (2) A type of movement on land produced by rhythmic waves of muscle contractions passing from front to back, as in many annelids.

peristome A ring of interlocking, tooth-like structures on the upper part of a moss capsule (sporangium), often specialized for gradual spore discharge.

peritubular capillary One of the tiny blood vessels that form a network surrounding the proximal and distal tubules in the kidney.

permafrost A permanently frozen soil layer.

peroxisome (puh-rok'-suh-sōm') An organelle containing enzymes that transfer hydrogen atoms from various substrates to oxygen (O_2), producing and then degrading hydrogen peroxide (H_2O_2).

petal A modified leaf of a flowering plant. Petals are the often colorful parts of a flower that advertise it to insects and other pollinators.

petiole (pet'-ē-ōl) The stalk of a leaf, which joins the leaf to a node of the stem.

pH A measure of hydrogen ion concentration equal to $-\log [H^+]$ and ranging in value from 0 to 14.

phage (fāj) A virus that infects bacteria; also called a bacteriophage.

phagocytosis (fag'-ō-sī-tō'-sis) A type of endocytosis in which large particulate substances or small organisms are taken up by a cell. It is carried out by some protists and by certain immune cells of animals (in mammals, mainly macrophages, neutrophils, and dendritic cells).

pharyngeal cleft (fuh-rin'-jē-ul) In chordate embryos, one of the grooves that separate a series of pouches along the sides of the pharynx and may develop into a pharyngeal slit.

pharyngeal slit (fuh-rin'-jē-ul) In chordate embryos, one of the slits that form from the pharyngeal clefts and communicate to the outside, later developing into gill slits in many vertebrates.

pharynx (fār'-inks) (1) An area in the vertebrate throat where air and food passages cross. (2) In flatworms, the muscular tube that protrudes from the ventral side of the worm and ends in the mouth.

phase change A shift from one developmental phase to another.

phenotype (fē'-nō-tīp) The observable physical and physiological traits of an organism, which are determined by its genetic makeup.

pheromone (fār'-uh-mōn) In animals and fungi, a small molecule released into the environment that functions in communication between members of the same species. In animals, it acts much like a hormone in influencing physiology and behavior.

phloem (flō'-em) Vascular plant tissue consisting of living cells arranged into elongated tubes that transport sugar and other organic nutrients throughout the plant.

phloem sap The sugar-rich solution carried through a plant's sieve tubes.

phosphate group A chemical group consisting of a phosphorus atom bonded to four oxygen atoms; important in energy transfer.

phospholipid (fos'-fō-lip'-id) A lipid made up of glycerol joined to two fatty acids and a phosphate group. The hydrocarbon chains of the fatty acids act as nonpolar, hydrophobic tails, while the rest of the molecule acts as a polar, hydrophilic head. Phospholipids form bilayers that function as biological membranes.

phosphorylated intermediate A molecule (often a reactant) with a phosphate group covalently bound to it, making it more reactive (less stable) than the unphosphorylated molecule.

photic zone (fō'-tic) The narrow top layer of an ocean or lake, where light penetrates sufficiently for photosynthesis to occur.

photoautotroph (fō'-tō-ot'-ō-trōf) An organism that harnesses light energy to drive the synthesis of organic compounds from carbon dioxide.

photoheterotroph (fō'-tō-het'-er-ō-trōf) An organism that uses light to generate ATP but must obtain carbon in organic form.

photomorphogenesis Effects of light on plant morphology.

photon (fō'-ton) A quantum, or discrete quantity, of light energy that behaves as if it were a particle.

photoperiodism (fō'-tō-pēr'-ē-ō-dizm) A physiological response to photoperiod, the relative lengths of night and day. An example of photoperiodism is flowering.

photophosphorylation (fō'-tō-fos'-fōr-uh-lā'-shun) The process of generating ATP from ADP and phosphate by means of chemiosmosis, using a proton-motive force generated across the thylakoid membrane of the chloroplast or the membrane of certain prokaryotes during the light reactions of photosynthesis.

photoreceptor An electromagnetic receptor that detects the radiation known as visible light.

photorespiration A metabolic pathway that consumes oxygen and ATP, releases carbon dioxide, and decreases photosynthetic output. Photorespiration generally occurs on hot, dry, bright days, when stomata close and the O_2/CO_2 ratio in the leaf increases, favoring the binding of O_2 rather than CO_2 by rubisco.

photosynthesis (fō'-tō-sin'-thi-sis) The conversion of light energy to chemical energy that is stored in sugars or other organic compounds; occurs in plants, algae, and certain prokaryotes.

photosystem A light-capturing unit located in the thylakoid membrane of the chloroplast or in the membrane of some prokaryotes, consisting of a reaction-center complex surrounded by numerous light-harvesting complexes. There are two types of photosystems, I and II; they absorb light best at different wavelengths.

photosystem I (PS I) A light-capturing unit in a chloroplast's thylakoid membrane or in the membrane of some prokaryotes; it has two molecules of P700 chlorophyll *a* at its reaction center.

photosystem II (PS II) One of two light-capturing units in a chloroplast's thylakoid membrane or in the membrane of some prokaryotes; it has two molecules of P680 chlorophyll *a* at its reaction center.

phototropism (fō'-tō-trō'-pizm) Growth of a plant shoot toward or away from light.

phragmoplast (frag'-mō-plast') An alignment of cytoskeletal elements and Golgi-derived vesicles that forms across the midline of a dividing plant cell.

phyllotaxy (fil'-uh-tak'-sē) The pattern of leaf attachment to the stem of a plant.

PhyloCode Proposed system of classification of organisms based on evolutionary relationships: Only groups that include a common ancestor and all of its descendants are named.

phylogenetic species concept A definition of species as the smallest group of individuals that share a common ancestor, forming one branch on the tree of life.

phylogenetic tree A branching diagram that represents a hypothesis about the evolutionary history of a group of organisms.

phylogeny (fī-loj'-uh-nē) The evolutionary history of a species or group of related species.

phylum (fī'-lum) (plural, **phyla**) In Linnaean classification, the taxonomic category above class.

physical map A genetic map in which the actual physical distances between genes or other genetic markers are expressed, usually as the number of base pairs along the DNA.

physiology The processes and functions of an organism.

phytochrome (fī'-tuh-krōm) A type of light receptor in plants that mostly absorbs red light and regulates many plant responses, such as seed germination and shade avoidance.

phytoremediation An emerging technology that seeks to reclaim contaminated areas by taking advantage of some plant species' ability to extract heavy metals and other pollutants from the soil and to concentrate them in easily harvested portions of the plant.

pilus (plural, **pili**) (pī'-lus, pī'-lī) In bacteria, a structure that links one cell to another at the start of conjugation; also known as a sex pilus or conjugation pilus.

pineal gland (pī'-nē-ul) A small gland on the dorsal surface of the vertebrate forebrain that secretes the hormone melatonin.

pinocytosis (pī'-nō-sī-tō'-sis) A type of endocytosis in which the cell ingests extracellular fluid and its dissolved solutes.

pistil A single carpel or a group of fused carpels.

pith Ground tissue that is internal to the vascular tissue in a stem; in many monocot roots, parenchyma cells that form the central core of the vascular cylinder.

pituitary gland (puh-tū'-uh-tār'-ē) An endocrine gland at the base of the hypothalamus; consists of a posterior lobe, which stores and releases two hormones produced by the hypothalamus, and an anterior lobe, which produces and secretes many hormones that regulate diverse body functions.

placenta (pluh-sen'-tuh) A structure in the pregnant uterus for nourishing a viviparous fetus with the mother's blood supply; formed from the uterine lining and embryonic membranes.

placental transfer cell A plant cell that enhances the transfer of nutrients from parent to embryo.

placoderm A member of an extinct group of fishlike vertebrates that had jaws and were enclosed in a tough outer armor.

planarian A free-living flatworm found in ponds and streams.

plasma (plaz'-muh) The liquid matrix of blood in which the blood cells are suspended.

plasma cell The antibody-secreting effector cell of humoral immunity. Plasma cells arise from antigen-stimulated B cells.

plasma membrane The membrane at the boundary of every cell that acts as a selective barrier, regulating the cell's chemical composition.

plasmid (plaz'-mid) A small, circular, double-stranded DNA molecule that carries accessory genes separate from those of a bacterial chromosome; in DNA cloning, used as vectors carrying up to about 10,000 base pairs (10 kb) of DNA. Plasmids are also found in some eukaryotes, such as yeasts.

plasmodesma (plaz'-mō-dez'-muh) (plural, **plasmodesmata**) An open channel through the cell wall that connects the cytoplasm of adjacent plant cells, allowing water, small solutes, and some larger molecules to pass between the cells.

plasmodial slime mold (plaz-mō'-dē-ul) A type of protist that has amoeboid cells, flagellated cells, and a plasmodial feeding stage in its life cycle.

plasmodium A single mass of cytoplasm containing many diploid nuclei that forms during the life cycle of some slime molds.

plasmogamy (plaz-moh'-guh-mē) In fungi, the fusion of the cytoplasm of cells from two individuals; occurs as one stage of sexual reproduction, followed later by karyogamy.

plasmolysis (plaz-mol'-uh-sis) A phenomenon in walled cells in which the cytoplasm shrivels and the plasma membrane pulls away from the cell wall; occurs when the cell loses water to a hypertonic environment.

plastid One of a family of closely related organelles that includes chloroplasts, chromoplasts, and amyloplasts. Plastids are found in cells of photosynthetic eukaryotes.

plate tectonics The theory that the continents are part of great plates of Earth's crust that float on the hot, underlying portion of the mantle. Movements in the mantle cause the continents to move slowly over time.

platelet A pinched-off cytoplasmic fragment of a specialized bone marrow cell. Platelets circulate in the blood and are important in blood clotting.

pleiotropy (plī'-o-truh-pē) The ability of a single gene to have multiple effects.

pluripotent Describing a cell that can give rise to many, but not all, parts of an organism.

point mutation A change in a single nucleotide pair of a gene.

polar covalent bond A covalent bond between atoms that differ in electronegativity. The shared electrons are pulled closer to the more electronegative atom, making it slightly negative and the other atom slightly positive.

polar molecule A molecule (such as water) with an uneven distribution of charges in different regions of the molecule.

polarity A lack of symmetry; structural differences in opposite ends of an organism or structure, such as the root end and shoot end of a plant.

pollen grain In seed plants, a structure consisting of the male gametophyte enclosed within a pollen wall.

pollen tube A tube that forms after germination of the pollen grain and that functions in the delivery of sperm to the ovule.

pollination (pol'-uh-nā'-shun) The transfer of pollen to the part of a seed plant containing the ovules, a process required for fertilization.

poly-A tail A sequence of 50–250 adenine nucleotides added onto the 3' end of a pre-mRNA molecule.

polygamous Referring to a type of relationship in which an individual of one sex mates with several of the other.

polygenic inheritance (pol'-ē-jen'-ik) An additive effect of two or more genes on a single phenotypic character.

polymer (pol'-uh-mer) A long molecule consisting of many similar or identical monomers linked together by covalent bonds.

polymerase chain reaction (PCR) (puh-lim'-uh-rās) A technique for amplifying DNA *in vitro* by incubating it with specific primers, a heat-resistant DNA polymerase, and nucleotides.

polynucleotide (pol'-ē-nū'-klē-ō-tīd) A polymer consisting of many nucleotide monomers in a chain. The nucleotides can be those of DNA or RNA.

polyp The sessile variant of the cnidarian body plan. The alternate form is the medusa.

polypeptide (pol'-ē-pep'-tīd) A polymer of many amino acids linked together by peptide bonds.

polyphyletic (pol'-ē-fī-let'-ik) Pertaining to a group of taxa derived from two or more different ancestors.

polyploidy (pol'-ē-ploy'-dē) A chromosomal alteration in which the organism possesses more than two complete chromosome sets. It is the result of an accident of cell division.

polyribosome (polysome) (pol'-ē-rī'-buh-sōm') A group of several ribosomes attached to, and translating, the same messenger RNA molecule.

polysaccharide (pol'-ē-sak'-uh-rīd) A polymer of many monosaccharides, formed by dehydration reactions.

polytomy (puh-lit'-uh-mē) In a phylogenetic tree, a branch point from which more than two descendant taxa emerge. A polytomy indicates that the evolutionary relationships between the descendant taxa are not yet clear.

pons A portion of the brain that participates in certain automatic, homeostatic functions, such as regulating the breathing centers in the medulla.

population A group of individuals of the same species that live in the same area and interbreed, producing fertile offspring.

population dynamics The study of how complex interactions between biotic and abiotic factors influence variations in population size.

population ecology The study of populations in relation to their environment, including environmental influences on population density and distribution, age structure, and variations in population size.

positional information Molecular cues that control pattern formation in an animal or plant embryonic structure by indicating a cell's location relative to the organism's body axes. These cues elicit a response by genes that regulate development.

positive feedback A form of regulation in which an end product of a process speeds up that process; in physiology, a control mechanism in which a change in a variable triggers a response that reinforces or amplifies the change.

positive pressure breathing A breathing system in which air is forced into the lungs.

posterior Pertaining to the rear, or tail end, of a bilaterally symmetrical animal.

posterior pituitary An extension of the hypothalamus composed of nervous tissue that secretes oxytocin and antidiuretic hormone made in the hypothalamus; a temporary storage site for these hormones.

postzygotic barrier (pōst′-zī-got′-ik) A reproductive barrier that prevents hybrid zygotes produced by two different species from developing into viable, fertile adults.

potential energy The energy that matter possesses as a result of its location or spatial arrangement (structure).

predation An interaction between species in which one species, the predator, eats the other, the prey.

pregnancy The condition of carrying one or more embryos in the uterus.

prepuce (prē′-pyūs) A fold of skin covering the head of the clitoris or penis.

pressure potential (Ψ_P) A component of water potential that consists of the physical pressure on a solution, which can be positive, zero, or negative.

prezygotic barrier (prē′-zī-got′-ik) A reproductive barrier that impedes mating between species or hinders fertilization if interspecific mating is attempted.

primary cell wall In plants, a relatively thin and flexible layer that surrounds the plasma membrane of a young cell.

primary consumer An herbivore; an organism that eats plants or other autotrophs.

primary electron acceptor In the thylakoid membrane of a chloroplast or in the membrane of some prokaryotes, a specialized molecule that shares the reaction-center complex with a pair of chlorophyll *a* molecules and that accepts an electron from them.

primary growth Growth produced by apical meristems, lengthening stems and roots.

primary immune response The initial adaptive immune response to an antigen, which appears after a lag of about 10 to 17 days.

primary oocyte (ō′-uh-sīt) An oocyte prior to completion of meiosis I.

primary producer An autotroph, usually a photosynthetic organism. Collectively, autotrophs make up the trophic level of an ecosystem that ultimately supports all other levels.

primary production The amount of light energy converted to chemical energy (organic compounds) by the autotrophs in an ecosystem during a given time period.

primary structure The level of protein structure referring to the specific linear sequence of amino acids.

primary succession A type of ecological succession that occurs in an area where there were originally no organisms present and where soil has not yet formed.

primary transcript An initial RNA transcript from any gene; also called pre-mRNA when transcribed from a protein-coding gene.

primary visual cortex The destination in the occipital lobe of the cerebrum for most of the axons from the lateral geniculate nuclei.

primase An enzyme that joins RNA nucleotides to make a primer during DNA replication, using the parental DNA strand as a template.

primer A short stretch of RNA with a free 3′ end, bound by complementary base pairing to the template strand and elongated with DNA nucleotides during DNA replication.

primitive streak A thickening along the future anterior-posterior axis on the surface of an early avian or mammalian embryo, caused by a piling up of cells as they congregate at the midline before moving into the embryo.

prion An infectious agent that is a misfolded version of a normal cellular protein. Prions appear to increase in number by converting correctly folded versions of the protein to more prions.

problem solving The cognitive activity of devising a method to proceed from one state to another in the face of real or apparent obstacles.

producer An organism that produces organic compounds from CO_2 by harnessing light energy (in photosynthesis) or by oxidizing inorganic chemicals (in chemosynthetic reactions carried out by some prokaryotes).

product A material resulting from a chemical reaction.

production efficiency The percentage of energy stored in assimilated food that is not used for respiration or eliminated as waste.

progesterone A steroid hormone that prepares the uterus for pregnancy; the major progestin in mammals.

progestin Any steroid hormone with progesterone-like activity.

progymnosperm (prō′-jim′-nō-sperm) An extinct seedless vascular plant that may be ancestral to seed plants.

prokaryotic cell (prō′-kār′-ē-ot′-ik) A type of cell lacking a membrane-enclosed nucleus and membrane-enclosed organelles. Organisms with prokaryotic cells (bacteria and archaea) are called prokaryotes.

prolactin A hormone produced and secreted by the anterior pituitary with a great diversity of effects in different vertebrate species. In mammals, it stimulates growth of and milk production by the mammary glands.

proliferative phase That portion of the uterine (menstrual) cycle when the endometrium regenerates and thickens.

prometaphase The second stage of mitosis, in which the nuclear envelope fragments and the spindle microtubules attach to the kinetochores of the chromosomes.

promiscuous Referring to a type of relationship in which mating occurs with no strong pair-bonds or lasting relationships.

promoter A specific nucleotide sequence in the DNA of a gene that binds RNA polymerase, positioning it to start transcribing RNA at the appropriate place.

prophage (prō′-fāj) A phage genome that has been inserted into a specific site on a bacterial chromosome.

prophase The first stage of mitosis, in which the chromatin condenses into discrete chromosomes visible with a light microscope, the mitotic spindle begins to form, and the nucleolus disappears but the nucleus remains intact.

prostaglandin (pros′-tuh-glan′-din) One of a group of modified fatty acids secreted by virtually all tissues and performing a wide variety of functions as local regulators.

prostate gland (pros′-tāt) A gland in human males that secretes an acid-neutralizing component of semen.

protease An enzyme that digests proteins by hydrolysis.

proteasome A giant protein complex that recognizes and destroys proteins tagged for elimination by the small protein ubiquitin.

protein (prō′-tēn) A biologically functional molecule consisting of one or more polypeptides folded and coiled into a specific three-dimensional structure.

protein kinase An enzyme that transfers phosphate groups from ATP to a protein, thus phosphorylating the protein.

protein phosphatase An enzyme that removes phosphate groups from (dephosphorylates) proteins, often functioning to reverse the effect of a protein kinase.

proteoglycan (prō′-tē-ō-glī′-kan) A large molecule consisting of a small core protein with many carbohydrate chains attached, found in the extracellular matrix of animal cells. A proteoglycan may consist of up to 95% carbohydrate.

proteomics (prō′-tē-ō′-miks) The systematic study of the full protein sets (proteomes) encoded by genomes.

protist An informal term applied to any eukaryote that is not a plant, animal, or fungus. Most protists are unicellular, though some are colonial or multicellular.

protocell An abiotic precursor of a living cell that had a membrane-like structure and that maintained an internal chemistry different from that of its surroundings.

proton (prō′-ton) A subatomic particle with a single positive electrical charge, with a mass of about 1.7×10^{-24} g, found in the nucleus of an atom.

proton pump An active transport protein in a cell membrane that uses ATP to transport hydrogen ions out of a cell against their concentration gradient, generating a membrane potential in the process.

protonema (plural, **protonemata**) A mass of green, branched, one-cell-thick filaments produced by germinating moss spores.

protonephridium (prō′-tō-nuh-frid′-ē-uhm) (plural, **protonephridia**) An excretory system, such as the flame bulb system of flatworms, consisting of a network of tubules lacking internal openings.

proton-motive force The potential energy stored in the form of a proton electrochemical gradient, generated by the pumping of hydrogen ions (H^+) across a biological membrane during chemiosmosis.

proto-oncogene (prō′-tō-on′-kō-jēn) A normal cellular gene that has the potential to become an oncogene.

protoplast The living part of a plant cell, which also includes the plasma membrane.

protoplast fusion The fusing of two protoplasts from different plant species that would otherwise be reproductively incompatible.

protostome development In animals, a developmental mode distinguished by the development of the mouth from the blastopore; often also characterized by spiral cleavage and by the body cavity forming when solid masses of mesoderm split.

provirus A viral genome that is permanently inserted into a host genome.

proximal tubule In the vertebrate kidney, the portion of a nephron immediately downstream from Bowman's capsule that conveys and helps refine filtrate.

pseudocoelomate (sū′-dō-sē′-lō-māt) An animal whose body cavity is lined by tissue derived from mesoderm and endoderm.

pseudogene (sū′-dō-jēn) A DNA segment very similar to a real gene but which does not yield a functional product; a DNA segment that formerly functioned as a gene but has become inactivated in a particular species because of mutation.

pseudopodium (sū′-dō-pō′-dē-um) (plural, **pseudopodia**) A cellular extension of amoeboid cells used in moving and feeding.

pterophyte (ter′-uh-fīt) An informal name for a member of the phylum Pterophyta, which includes ferns, horsetails, and whisk ferns and their relatives.

pterosaur Winged reptile that lived during the Mesozoic era.

pulmocutaneous circuit A branch of the circulatory system in many amphibians that supplies the lungs and skin.

pulmonary circuit The branch of the circulatory system that supplies the lungs.

pulse The rhythmic bulging of the artery walls with each heartbeat.

punctuated equilibria In the fossil record, long periods of apparent stasis, in which a species undergoes little or no morphological change, interrupted by relatively brief periods of sudden change.

Punnett square A diagram used in the study of inheritance to show the predicted genotypic results of random fertilization in genetic crosses between individuals of known genotype.

pupil The opening in the iris, which admits light into the interior of the vertebrate eye. Muscles in the iris regulate its size.

purine (pyū′-rēn) One of two types of nitrogenous bases found in nucleotides, characterized by a six-membered ring fused to a five-membered ring. Adenine (A) and guanine (G) are purines.

pyrimidine (puh-rim′-uh-dēn) One of two types of nitrogenous bases found in nucleotides, characterized by a six-membered ring. Cytosine (C), thymine (T), and uracil (U) are pyrimidines.

quantitative character A heritable feature that varies continuously over a range rather than in an either-or fashion.

quaternary structure (kwot-er-nār-ē) The particular shape of a complex, aggregate protein, defined by the characteristic three-dimensional arrangement of its constituent subunits, each a polypeptide.

R plasmid A bacterial plasmid carrying genes that confer resistance to certain antibiotics.

radial cleavage A type of embryonic development in deuterostomes in which the planes of cell division that transform the zygote into a ball of cells are either parallel or perpendicular to the vertical axis of the embryo, thereby aligning tiers of cells one above the other.

radial symmetry Symmetry in which the body is shaped like a pie or barrel (lacking a left side and a right side) and can be divided into mirror-imaged halves by any plane through its central axis.

radiation The emission of electromagnetic waves by all objects warmer than absolute zero.

radicle An embryonic root of a plant.

radioactive isotope An isotope (an atomic form of a chemical element) that is unstable; the nucleus decays spontaneously, giving off detectable particles and energy.

radiolarian A protist, usually marine, with a shell generally made of silica and pseudopodia that radiate from the central body.

radiometric dating A method for determining the absolute age of rocks and fossils, based on the half-life of radioactive isotopes.

radula A straplike scraping organ used by many molluscs during feeding.

***ras* gene** A gene that codes for Ras, a G protein that relays a growth signal from a growth factor receptor on the plasma membrane to a cascade of protein kinases, ultimately resulting in stimulation of the cell cycle.

ratite (rat′-īt) Member of the group of flightless birds.

ray-finned fish Member of the class Actinopterygii, aquatic osteichthyans with fins supported by long, flexible rays, including tuna, bass, and herring.

reabsorption In excretory systems, the recovery of solutes and water from filtrate.

reactant A starting material in a chemical reaction.

reaction-center complex A complex of proteins associated with a special pair of chlorophyll *a* molecules and a primary electron acceptor. Located centrally in a photosystem, this complex triggers the light reactions of photosynthesis. Excited by light energy, the pair of chlorophylls donates an electron to the primary electron acceptor, which passes an electron to an electron transport chain.

reading frame On an mRNA, the triplet grouping of ribonucleotides used by the translation machinery during polypeptide synthesis.

receptacle The base of a flower; the part of the stem that is the site of attachment of the floral organs.

receptor potential An initial response of a receptor cell to a stimulus, consisting of a change in voltage across the receptor membrane proportional to the stimulus strength.

receptor tyrosine kinase (RTK) A receptor protein spanning the plasma membrane, the cytoplasmic (intracellular) part of which can catalyze the transfer of a phosphate group from ATP to a tyrosine on another protein. Receptor tyrosine kinases often respond to the binding of a signaling molecule by dimerizing and then phosphorylating a tyrosine on the cytoplasmic portion of the other receptor in the dimer. The phosphorylated tyrosines on the receptors then activate other signal transduction proteins within the cell.

receptor-mediated endocytosis (en′-dō-sī-tō′-sis) The movement of specific molecules into a cell by the inward budding of vesicles containing proteins with receptor sites specific to the molecules being taken in; enables a cell to acquire bulk quantities of specific substances.

recessive allele An allele whose phenotypic effect is not observed in a heterozygote.

reciprocal altruism Altruistic behavior between unrelated individuals, whereby the altruistic individual benefits in the future when the beneficiary reciprocates.

recombinant chromosome A chromosome created when crossing over combines DNA from two parents into a single chromosome.

recombinant DNA A DNA molecule made *in vitro* with segments from different sources.

recombinant type (recombinant) An offspring whose phenotype differs from that of the true-breeding P generation parents; also refers to the phenotype itself.

rectum The terminal portion of the large intestine, where the feces are stored prior to elimination.

red alga A photosynthetic protist, named for its color, which results from a red pigment that masks the green of chlorophyll. Most red algae are multicellular and marine.

redox reaction (rē′-doks) A chemical reaction involving the complete or partial transfer of one or more electrons from one reactant to another; short for **red**uction-**ox**idation reaction.

reducing agent The electron donor in a redox reaction.

reduction The complete or partial addition of electrons to a substance involved in a redox reaction.

reflex An automatic reaction to a stimulus, mediated by the spinal cord or lower brain.

refractory period (rē-frakt′-ōr-ē) The short time immediately after an action potential in which the neuron cannot respond to another stimulus, owing to the inactivation of voltage-gated sodium channels.

regulator An animal for which mechanisms of homeostasis moderate internal changes in a particular variable in the face of external fluctuation of that variable.

regulatory gene A gene that codes for a protein, such as a repressor, that controls the transcription of another gene or group of genes.

reinforcement In evolutionary biology, a process in which a process in which natural selection strengthens prezygotic barriers to reproduction, thus reducing the chances of hybrid formation. Such a process is likely to occur only if hybrid offspring are less fit than members of the parent species.

relative abundance The proportional abundance of different species in a community.

relative fitness The contribution an individual makes to the gene pool of the next generation, relative to the contributions of other individuals in the population.

renal cortex The outer portion of the vertebrate kidney.

renal medulla The inner portion of the vertebrate kidney, beneath the renal cortex.

renal pelvis The funnel-shaped chamber that receives processed filtrate from the vertebrate kidney's collecting ducts and is drained by the ureter.

renin-angiotensin-aldosterone system (RAAS) A hormone cascade pathway that helps regulate blood pressure and blood volume.

repetitive DNA Nucleotide sequences, usually noncoding, that are present in many copies in a eukaryotic genome. The repeated units may be short and arranged tandemly (in series) or long and dispersed in the genome.

replication fork A Y-shaped region on a replicating DNA molecule where the parental strands are being unwound and new strands are being synthesized.

repressor A protein that inhibits gene transcription. In prokaryotes, repressors bind to the DNA in or near the promoter. In eukaryotes, repressors may bind to control elements within enhancers, to activators, or to other proteins in a way that blocks activators from binding to DNA.

reproductive isolation The existence of biological factors (barriers) that impede members of two species from producing viable, fertile offspring.

reproductive table An age-specific summary of the reproductive rates in a population.

reptile Member of the clade of amniotes that includes tuataras, lizards, snakes, turtles, crocodilians, and birds.

residual volume The amount of air that remains in the lungs after forceful exhalation.

resource partitioning The division of environmental resources by coexisting species such that the niche of each species differs by one or more significant factors from the niches of all coexisting species.

respiratory pigment A protein that transports oxygen in blood or hemolymph.

response (1) In cellular communication, the change in a specific cellular activity brought about by a transduced signal from outside the cell. (2) In feedback regulation, a physiological activity triggered by a change in a variable.

resting potential The membrane potential characteristic of a nonconducting excitable cell, with the inside of the cell more negative than the outside.

restriction enzyme An endonuclease (type of enzyme) that recognizes and cuts DNA molecules foreign to a bacterium (such as phage genomes). The enzyme cuts at specific nucleotide sequences (restriction sites).

restriction fragment A DNA segment that results from the cutting of DNA by a restriction enzyme.

restriction fragment length polymorphism (RFLP) A single nucleotide polymorphism (SNP) that exists in the restriction site for a particular enzyme, thus making the site unrecognizable by that enzyme and changing the lengths of the restriction fragments formed by digestion with that enzyme. A RFLP can be in coding or noncoding DNA.

restriction site A specific sequence on a DNA strand that is recognized and cut by a restriction enzyme.

reticular formation (re-tik′-yū-ler) A diffuse network of neurons in the core of the brainstem that filters information traveling to the cerebral cortex.

retina (ret′-i-nuh) The innermost layer of the vertebrate eye, containing photoreceptor cells (rods and cones) and neurons; transmits images formed by the lens to the brain via the optic nerve.

retinal The light-absorbing pigment in rods and cones of the vertebrate eye.

retrotransposon (re′-trō-trans-pō′-zon) A transposable element that moves within a genome by means of an RNA intermediate, a transcript of the retrotransposon DNA.

retrovirus (re′-trō-vī′-rus) An RNA virus that replicates by transcribing its RNA into DNA and then inserting the DNA into a cellular chromosome; an important class of cancer-causing viruses.

reverse transcriptase (tran-skrip′-tās) An enzyme encoded by certain viruses (retroviruses) that uses RNA as a template for DNA synthesis.

reverse transcriptase–polymerase chain reaction (RT-PCR) A technique for determining expression of a particular gene. It uses reverse transcriptase and DNA polymerase to synthesize cDNA from all the mRNA in a sample and then subjects the cDNA to PCR amplification using primers specific for the gene of interest.

Rhizaria (rī-za′-rē-uh) One of five supergroups of eukaryotes proposed in a current hypothesis of the evolutionary history of eukaryotes; a morphologically diverse protist clade that is defined by DNA similarities. *See also* Excavata, Chromalveolata, Archaeplastida, and Unikonta.

rhizobacterium A soil bacterium whose population size is much enhanced in the rhizosphere, the soil region close to a plant's roots.

rhizoid (rī′-zoyd) A long, tubular single cell or filament of cells that anchors bryophytes to the ground. Unlike roots, rhizoids are not composed of tissues, lack specialized conducting cells, and do not play a primary role in water and mineral absorption.

rhizosphere The soil region close to plant roots and characterized by a high level of microbiological activity.

rhodopsin (rō-dop′-sin) A visual pigment consisting of retinal and opsin. Upon absorbing light, the retinal changes shape and dissociates from the opsin.

rhythm method A form of contraception that relies on refraining from sexual intercourse when conception is most likely to occur; also called natural family planning.

ribonucleic acid (RNA) (rī′-bō-nū-klā′-ik) A type of nucleic acid consisting of a polynucleotide made up of nucleotide monomers with a ribose sugar and the nitrogenous bases adenine (A), cytosine (C), guanine (G), and uracil (U); usually single-stranded; functions in protein synthesis, gene regulation, and as the genome of some viruses.

ribose The sugar component of RNA nucleotides.

ribosomal RNA (rRNA) (rī′-buh-sō′-mul) RNA molecules that, together with proteins, make up ribosomes; the most abundant type of RNA.

ribosome (rī′-buh-sōm′) A complex of rRNA and protein molecules that functions as a site of protein synthesis in the cytoplasm; consists of a large and a small subunit. In eukaryotic cells, each subunit is assembled in the nucleolus. *See also* nucleolus.

ribozyme (rī′-buh-zīm) An RNA molecule that functions as an enzyme, such as an intron that catalyzes its own removal during RNA splicing.

RNA interference (RNAi) A technique used to silence the expression of selected genes. RNAi uses synthetic double-stranded RNA molecules that match the sequence of a particular gene to trigger the breakdown of the gene's messenger RNA.

RNA polymerase An enzyme that links ribonucleotides into a growing RNA chain during transcription, based on complementary binding to nucleotides on a DNA template strand.

RNA processing Modification of RNA primary transcripts, including splicing out of introns, joining together of exons, and alteration of the 5′ and 3′ ends.

RNA splicing After synthesis of a eukaryotic primary RNA transcript, the removal of portions of the transcript (introns) that will not be included in the mRNA and the joining together of the remaining portions (exons).

rod A rodlike cell in the retina of the vertebrate eye, sensitive to low light intensity.

root An organ in vascular plants that anchors the plant and enables it to absorb water and minerals from the soil.

root cap A cone of cells at the tip of a plant root that protects the apical meristem.

root hair A tiny extension of a root epidermal cell, growing just behind the root tip and increasing surface area for absorption of water and minerals.

root pressure Pressure exerted in the roots of plants as the result of osmosis, causing exudation from cut stems and guttation of water from leaves.

root system All of a plant's roots, which anchor it in the soil, absorb and transport minerals and water, and store food.

rooted Describing a phylogenetic tree that contains a branch point (often, the one farthest to the left) representing the most recent common ancestor of all taxa in the tree.

rough ER That portion of the endoplasmic reticulum with ribosomes attached.

round window In the mammalian ear, the point of contact where vibrations of the stapes create a traveling series of pressure waves in the fluid of the cochlea.

***r*-selection** Selection for life history traits that maximize reproductive success in uncrowded environments; also called density-independent selection.

rubisco (rū-bis′-kō) Ribulose bisphosphate (RuBP) carboxylase, the enzyme that catalyzes the first step of the Calvin cycle (the addition of CO_2 to RuBP).

ruminant (rū′-muh-nent) An animal, such as a cow or a sheep, with multiple stomach compartments specialized for an herbivorous diet.

S phase The synthesis phase of the cell cycle; the portion of interphase during which DNA is replicated.

saccule In the vertebrate ear, a chamber in the vestibule behind the oval window that participates in the sense of balance.

salicylic acid (sal′-i-sil′-ik) A signaling molecule in plants that may be partially responsible for activating systemic acquired resistance to pathogens.

salivary gland A gland associated with the oral cavity that secretes substances that lubricate food and begin the process of chemical digestion.

salt A compound resulting from the formation of an ionic bond; also called an ionic compound.

saltatory conduction (sol′-tuh-tōr′-ē) Rapid transmission of a nerve impulse along an axon, resulting from the action potential jumping from one node of Ranvier to another, skipping the myelin-sheathed regions of membrane.

sarcomere (sar′-kō-mēr) The fundamental, repeating unit of striated muscle, delimited by the Z lines.

sarcoplasmic reticulum (SR) (sar′-kō-plaz′-mik ruh-tik′-yū-lum) A specialized endoplasmic reticulum that regulates the calcium concentration in the cytosol of muscle cells.

saturated fatty acid A fatty acid in which all carbons in the hydrocarbon tail are connected by single bonds, thus maximizing the number of hydrogen atoms that are attached to the carbon skeleton.

savanna A tropical grassland biome with scattered individual trees and large herbivores and maintained by occasional fires and drought.

scaffolding protein A type of large relay protein to which several other relay proteins are simultaneously attached, increasing the efficiency of signal transduction.

scanning electron microscope (SEM) A microscope that uses an electron beam to scan the surface of a sample, coated with metal atoms, to study details of its topography.

schizophrenia (skit′-suh-frē′-nē-uh) A severe mental disturbance characterized by psychotic episodes in which patients have a distorted perception of reality.

Schwann cell A type of glial cell that forms insulating myelin sheaths around the axons of neurons in the peripheral nervous system.

science An approach to understanding the natural world.

scion (sī′-un) The twig grafted onto the stock when making a graft.

sclereid (sklār′-ē-id) A short, irregular sclerenchyma cell in nutshells and seed coats. Sclereids are scattered throughout the parenchyma of some plants.

sclerenchyma cell (skluh-ren′-kim-uh) A rigid, supportive plant cell type usually lacking a protoplast and possessing thick secondary walls strengthened by lignin at maturity.

scrotum A pouch of skin outside the abdomen that houses the testes; functions in maintaining the testes at the lower temperature required for spermatogenesis.

second law of thermodynamics The principle stating that every energy transfer or transformation increases the entropy of the universe. Usable forms of energy are at least partly converted to heat.

second messenger A small, nonprotein, water-soluble molecule or ion, such as a calcium ion (Ca^{2+}) or cyclic AMP, that relays a signal to a cell's interior in response to a signaling molecule bound by a signal receptor protein.

secondary cell wall In plant cells, a strong and durable matrix that is often deposited in several laminated layers around the plasma membrane and provides protection and support.

secondary consumer A carnivore that eats herbivores.

secondary endosymbiosis A process in eukaryotic evolution in which a heterotrophic eukaryotic cell engulfed a photosynthetic eukaryotic cell, which survived in a symbiotic relationship inside the heterotrophic cell.

secondary growth Growth produced by lateral meristems, thickening the roots and shoots of woody plants.

secondary immune response The adaptive immune response elicited on second or subsequent exposures to a particular antigen. The secondary immune response is more rapid, of greater magnitude, and of longer duration than the primary immune response.

secondary oocyte (ō′-uh-sīt) An oocyte that has completed the first of the two meiotic divisions.

secondary production The amount of chemical energy in consumers' food that is converted to their own new biomass during a given time period.

secondary structure Regions of repetitive coiling or folding of the polypeptide backbone of a protein due to hydrogen bonding between constituents of the backbone (not the side chains).

secondary succession A type of succession that occurs where an existing community has been cleared by some disturbance that leaves the soil or substrate intact.

secretion (1) The discharge of molecules synthesized by a cell. (2) The discharge of wastes from the body fluid into the filtrate.

secretory phase That portion of the uterine (menstrual) cycle when the endometrium continues to thicken, becomes more vascularized, and develops glands that secrete a fluid rich in glycogen.

seed An adaptation of some terrestrial plants consisting of an embryo packaged along with a store of food within a protective coat.

seed coat A tough outer covering of a seed, formed from the outer coat of an ovule. In a flowering plant, the seed coat encloses and protects the embryo and endosperm.

seedless vascular plant An informal name for a plant that has vascular tissue but lacks seeds. Seedless vascular plants form a paraphyletic group that includes the phyla Lycophyta (club mosses and their relatives) and Pterophyta (ferns and their relatives).

selective permeability A property of biological membranes that allows them to regulate the passage of substances across them.

self-incompatibility The ability of a seed plant to reject its own pollen and sometimes the pollen of closely related individuals.

semelparity Reproduction in which an organism produces all of its offspring in a single event; also known as big-bang reproduction.

semen (sē′-mun) The fluid that is ejaculated by the male during orgasm; contains sperm and secretions from several glands of the male reproductive tract.

semicircular canals A three-part chamber of the inner ear that functions in maintaining equilibrium.

semiconservative model Type of DNA replication in which the replicated double helix consists of one old strand, derived from the parental molecule, and one newly made strand.

semilunar valve A valve located at each exit of the heart, where the aorta leaves the left ventricle and the pulmonary artery leaves the right ventricle.

seminal vesicle (sem′-i-nul ves′-i-kul) A gland in males that secretes a fluid component of semen that lubricates and nourishes sperm.

seminiferous tubule (sem′-i-nif′-er-us) A highly coiled tube in the testis in which sperm are produced.

senescence (se-nes′-ens) The growth phase in a plant or plant part (as a leaf) from full maturity to death.

sensitive period A limited phase in an animal's development when learning of particular behaviors can take place; also called a critical period.

sensor In homeostasis, a receptor that detects a stimulus.

sensory adaptation The tendency of sensory neurons to become less sensitive when they are stimulated repeatedly.

sensory neuron A nerve cell that receives information from the internal or external environment and transmits signals to the central nervous system.

sensory reception The detection of a stimulus by sensory cells.

sensory receptor An organ, cell, or structure within a cell that responds to specific stimuli from an organism's external or internal environment.

sensory transduction The conversion of stimulus energy to a change in the membrane potential of a sensory receptor cell.

sepal (sē'-pul) A modified leaf in angiosperms that helps enclose and protect a flower bud before it opens.

septum (plural, **septa**) One of the cross-walls that divide a fungal hypha into cells. Septa generally have pores large enough to allow ribosomes, mitochondria, and even nuclei to flow from cell to cell.

serial endosymbiosis A hypothesis for the origin of eukaryotes consisting of a sequence of endosymbiotic events in which mitochondria, chloroplasts, and perhaps other cellular structures were derived from small prokaryotes that had been engulfed by larger cells.

serotonin (ser'-uh-tō'-nin) A neurotransmitter, synthesized from the amino acid tryptophan, that functions in the central nervous system.

set point In homeostasis in animals, a value maintained for a particular variable, such as body temperature or solute concentration.

seta (sē'-tuh) (plural, **setae**) The elongated stalk of a bryophyte sporophyte.

sex chromosome A chromosome responsible for determining the sex of an individual.

sex-linked gene A gene located on either sex chromosome. Most sex-linked genes are on the X chromosome and show distinctive patterns of inheritance; there are very few genes on the Y chromosome.

sexual dimorphism (dī-mōr'-fizm) Differences between the secondary sex characteristics of males and females.

sexual reproduction A type of reproduction in which two parents give rise to offspring that have unique combinations of genes inherited from both parents via the gametes.

sexual selection A form of selection in which individuals with certain inherited characteristics are more likely than other individuals to obtain mates.

Shannon diversity An index of community diversity symbolized by H and represented by the equation $H = -(p_A \ln p_A + p_B \ln p_B + p_C \ln p_C + \ldots)$, where A, B, C . . . are species, p is the relative abundance of each species, and ln is the natural logarithm.

shared ancestral character A character, shared by members of a particular clade, that originated in an ancestor that is not a member of that clade.

shared derived character An evolutionary novelty that is unique to a particular clade.

shoot system The aerial portion of a plant body, consisting of stems, leaves, and (in angiosperms) flowers.

short tandem repeat (STR) Simple sequence DNA containing multiple tandemly repeated units of two to five nucleotides. Variations in STRs act as genetic markers in STR analysis, used to prepare genetic profiles.

short-day plant A plant that flowers (usually in late summer, fall, or winter) only when the light period is shorter than a critical length.

short-term memory The ability to hold information, anticipations, or goals for a time and then release them if they become irrelevant.

sickle-cell disease A recessively inherited human blood disorder in which a single nucleotide change in the β-globin gene causes hemoglobin to aggregate, changing red blood cell shape and causing multiple symptoms in afflicted individuals.

sieve plate An end wall in a sieve-tube element, which facilitates the flow of phloem sap in angiosperm sieve tubes.

sieve-tube element A living cell that conducts sugars and other organic nutrients in the phloem of angiosperms; also called a sieve-tube member. Connected end to end, they form sieve tubes.

sign stimulus An external sensory cue that triggers a fixed action pattern by an animal.

signal In animal behavior, transmission of a stimulus from one animal to another. The term is also used in the context of communication in other kinds of organisms and in cell-to-cell communication in all multicellular organisms.

signal peptide A sequence of about 20 amino acids at or near the leading (amino) end of a polypeptide that targets it to the endoplasmic reticulum or other organelles in a eukaryotic cell.

signal transduction The linkage of a mechanical, chemical, or electromagnetic stimulus to a specific cellular response.

signal transduction pathway A series of steps linking a mechanical, chemical, or electrical stimulus to a specific cellular response.

signal-recognition particle (SRP) A protein-RNA complex that recognizes a signal peptide as it emerges from a ribosome and helps direct the ribosome to the endoplasmic reticulum (ER) by binding to a receptor protein on the ER.

silent mutation A nucleotide-pair substitution that has no observable effect on the phenotype; for example, within a gene, a mutation that results in a codon that codes for the same amino acid.

simple fruit A fruit derived from a single carpel or several fused carpels.

simple sequence DNA A DNA sequence that contains many copies of tandemly repeated short sequences.

single bond A single covalent bond; the sharing of a pair of valence electrons by two atoms.

single circulation A circulatory system consisting of a single pump and circuit, in which blood passes from the sites of gas exchange to the rest of the body before returning to the heart.

single nucleotide polymorphism (SNP) A single base-pair site in a genome where nucleotide variation is found in at least 1% of the population.

single-lens eye The camera-like eye found in some jellies, polychaete worms, spiders, and many molluscs.

single-strand binding protein A protein that binds to the unpaired DNA strands during DNA replication, stabilizing them and holding them apart while they serve as templates for the synthesis of complementary strands of DNA.

sinoatrial (SA) node A region in the right atrium of the heart that sets the rate and timing at which all cardiac muscle cells contract; the pacemaker.

sister chromatids Two copies of a duplicated chromosome attached to each other by proteins at the centromere and, sometimes, along the arms. While joined, two sister chromatids make up one chromosome. Chromatids are eventually separated during mitosis or meiosis II.

sister taxa Groups of organisms that share an immediate common ancestor and hence are each other's closest relatives.

skeletal muscle A type of striated muscle that is generally responsible for the voluntary movements of the body.

sliding-filament model The idea that muscle contraction is based on the movement of thin (actin) filaments along thick (myosin) filaments, shortening the sarcomere, the basic unit of muscle organization.

slow block to polyspermy The formation of the fertilization envelope and other changes in an egg's surface that prevent fusion of the egg with more than one sperm. The slow block begins about 1 minute after fertilization.

slow-twitch fiber A muscle fiber that can sustain long contractions.

small interfering RNA (siRNA) One of multiple small, single-stranded RNA molecules generated by cellular machinery from a long, linear, double-stranded RNA molecule. The siRNA associates with one or more proteins in a complex that can degrade or prevent translation of an mRNA with a complementary sequence. In some cases, siRNA can also block transcription by promoting chromatin modification.

small intestine The longest section of the alimentary canal, so named because of its small diameter compared with that of the large intestine; the principal site of the enzymatic hydrolysis of food macromolecules and the absorption of nutrients.

smooth ER That portion of the endoplasmic reticulum that is free of ribosomes.

smooth muscle A type of muscle lacking the striations of skeletal and cardiac muscle because of the uniform distribution of myosin filaments in the cells; responsible for involuntary body activities.

social learning Modification of behavior through the observation of other individuals.

sociobiology The study of social behavior based on evolutionary theory.

sodium-potassium pump A transport protein in the plasma membrane of animal cells that actively transports sodium out of the cell and potassium into the cell.

soil horizon A soil layer with physical characteristics that differ from those of the layers above or beneath.
solute (sol'-yūt) A substance that is dissolved in a solution.
solute potential (Ψ_S) A component of water potential that is proportional to the molarity of a solution and that measures the effect of solutes on the direction of water movement; also called osmotic potential, it can be either zero or negative.
solution A liquid that is a homogeneous mixture of two or more substances.
solvent The dissolving agent of a solution. Water is the most versatile solvent known.
somatic cell (sō-mat'-ik) Any cell in a multicellular organism except a sperm or egg or their precursors.
somite One of a series of blocks of mesoderm that exist in pairs just lateral to the notochord in a vertebrate embryo.
soredium (plural, **soredia**) In lichens, a small cluster of fungal hyphae with embedded algae.
sorus (plural, **sori**) A cluster of sporangia on a fern sporophyll. Sori may be arranged in various patterns, such as parallel lines or dots, which are useful in fern identification.
Southern blotting A technique that enables specific nucleotide sequences to be detected in samples of DNA. It involves gel electrophoresis of DNA molecules and their transfer to a membrane (blotting), followed by nucleic acid hybridization with a labeled probe.
spatial learning The establishment of a memory that reflects the environment's spatial structure.
spatial summation A phenomenon of neural integration in which the membrane potential of the postsynaptic cell is determined by the combined effect of EPSPs or IPSPs produced nearly simultaneously by different synapses.
speciation (spē'-sē-ā'-shun) An evolutionary process in which one species splits into two or more species.
species (spē'-sēz) A population or group of populations whose members have the potential to interbreed in nature and produce viable, fertile offspring, but do not produce viable, fertile offspring with members of other such groups.
species diversity The number and relative abundance of species in a biological community.
species richness The number of species in a biological community.
species-area curve The biodiversity pattern that shows that the larger the geographic area of a community is, the more species it has.
specific heat The amount of heat that must be absorbed or lost for 1 g of a substance to change its temperature by 1°C.
spectrophotometer An instrument that measures the proportions of light of different wavelengths absorbed and transmitted by a pigment solution.
sperm The male gamete.
spermatheca (sper'-muh-thē'-kuh) In many insects, a sac in the female reproductive system where sperm are stored.
spermatogenesis The continuous and prolific production of mature sperm cells in the testis.
spermatogonium (plural, **spermatogonia**) A cell that divides mitotically to form spermatocytes.
sphincter (sfink'-ter) A ringlike band of muscle fibers that controls the size of an opening in the body, such as the passage between the esophagus and the stomach.
spiral cleavage A type of embryonic development in protostomes in which the planes of cell division that transform the zygote into a ball of cells are diagonal to the vertical axis of the embryo. As a result, the cells of each tier sit in the grooves between cells of adjacent tiers.
spliceosome (splī'-sō-sōm) A large complex made up of proteins and RNA molecules that splices RNA by interacting with the ends of an RNA intron, releasing the intron and joining the two adjacent exons.
spongocoel (spon'-jō-sēl) The central cavity of a sponge.
spontaneous process A process that occurs without an overall input of energy; a process that is energetically favorable.
sporangium (spōr-an'-jē-um) (plural, **sporangia**) A multicellular organ in fungi and plants in which meiosis occurs and haploid cells develop.
spore (1) In the life cycle of a plant or alga undergoing alternation of generations, a haploid cell produced in the sporophyte by meiosis. A spore can divide by mitosis to develop into a multicellular haploid individual, the gametophyte, without fusing with another cell. (2) In fungi, a haploid cell, produced either sexually or asexually, that produces a mycelium after germination.
sporocyte A diploid cell, also known as a spore mother cell, that undergoes meiosis and generates haploid spores.
sporophyll (spō'-ruh-fil) A modified leaf that bears sporangia and hence is specialized for reproduction.
sporophyte (spō-ruh-fīt') In organisms (plants and some algae) that have alternation of generations, the multicellular diploid form that results from the union of gametes. The sporophyte produces haploid spores by meiosis that develop into gametophytes.
sporopollenin (spōr-uh-pol'-eh-nin) A durable polymer that covers exposed zygotes of charophyte algae and forms the walls of plant spores, preventing them from drying out.
stabilizing selection Natural selection in which intermediate phenotypes survive or reproduce more successfully than do extreme phenotypes.
stamen (stā'-men) The pollen-producing reproductive organ of a flower, consisting of an anther and a filament.
standard metabolic rate (SMR) Metabolic rate of a resting, fasting, and nonstressed ectotherm at a particular temperature.
starch A storage polysaccharide in plants, consisting entirely of glucose monomers joined by α glycosidic linkages.
start point In transcription, the nucleotide position on the promoter where RNA polymerase begins synthesis of RNA.
statocyst (stat'-uh-sist') A type of mechanoreceptor that functions in equilibrium in invertebrates by use of statoliths, which stimulate hair cells in relation to gravity.
statolith (stat'-uh-lith') (1) In plants, a specialized plastid that contains dense starch grains and may play a role in detecting gravity. (2) In invertebrates, a dense particle that settles in response to gravity and is found in sensory organs that function in equilibrium.
stele (stēl) The vascular tissue of a stem or root.
stem A vascular plant organ consisting of an alternating system of nodes and internodes that support the leaves and reproductive structures.
stem cell Any relatively unspecialized cell that can produce, during a single division, one identical daughter cell and one more specialized daughter cell that can undergo further differentiation.
steroid A type of lipid characterized by a carbon skeleton consisting of four fused rings with various chemical groups attached.
sticky end A single-stranded end of a double-stranded restriction fragment.
stigma (plural, **stigmata**) The sticky part of a flower's carpel, which receives pollen grains.
stimulus In feedback regulation, a fluctuation in a variable that triggers a response.
stipe A stemlike structure of a seaweed.
stock The plant that provides the root system when making a graft.
stoma (stō'-muh) (plural, **stomata**) A microscopic pore surrounded by guard cells in the epidermis of leaves and stems that allows gas exchange between the environment and the interior of the plant.
stomach An organ of the digestive system that stores food and performs preliminary steps of digestion.
stramenopile A protist in which a "hairy" flagellum (one covered with fine, hairlike projections) is paired with a shorter, smooth flagellum.
stratum (strah'-tum) (plural, **strata**) A rock layer formed when new layers of sediment cover older ones and compress them.
striated muscle Muscle in which the regular arrangement of filaments creates a pattern of light and dark bands.
strigolactones A class of plant hormone that inhibits shoot branching, triggers the germination of parasitic plant seeds, and stimulates the association of plant roots with mycorrhizal fungi.
strobilus (strō-bī'-lus) (plural, **strobili**) The technical term for a cluster of sporophylls known commonly as a cone, found in most gymnosperms and some seedless vascular plants.
stroke The death of nervous tissue in the brain, usually resulting from rupture or blockage of arteries in the head.
stroke volume The volume of blood pumped by a heart ventricle in a single contraction.
stroma (strō'-muh) The dense fluid within the chloroplast surrounding the thylakoid membrane and containing ribosomes and DNA; involved in the synthesis of organic molecules from carbon dioxide and water.

stromatolite Layered rock that results from the activities of prokaryotes that bind thin films of sediment together.

structural isomer One of several compounds that have the same molecular formula but differ in the covalent arrangements of their atoms.

style The stalk of a flower's carpel, with the ovary at the base and the stigma at the top.

substrate The reactant on which an enzyme works.

substrate feeder An animal that lives in or on its food source, eating its way through the food.

substrate-level phosphorylation The enzyme-catalyzed formation of ATP by direct transfer of a phosphate group to ADP from an intermediate substrate in catabolism.

sugar sink A plant organ that is a net consumer or storer of sugar. Growing roots, shoot tips, stems, and fruits are examples of sugar sinks supplied by phloem.

sugar source A plant organ in which sugar is being produced by either photosynthesis or the breakdown of starch. Mature leaves are the primary sugar sources of plants.

sulfhydryl group A chemical group consisting of a sulfur atom bonded to a hydrogen atom.

suprachiasmatic nucleus (SCN) A group of neurons in the hypothalamus of mammals that functions as a biological clock.

surface tension A measure of how difficult it is to stretch or break the surface of a liquid. Water has a high surface tension because of the hydrogen bonding of surface molecules.

surfactant A substance secreted by alveoli that decreases surface tension in the fluid that coats the alveoli.

survivorship curve A plot of the number of members of a cohort that are still alive at each age; one way to represent age-specific mortality.

suspension feeder An aquatic animal, such as a sponge, clam, or baleen whale, that feeds by sifting small organisms or food particles from the water.

sustainable agriculture Long-term productive farming methods that are environmentally safe.

sustainable development Development that meets the needs of people today without limiting the ability of future generations to meet their needs.

swim bladder In aquatic osteichthyans, an air sac that enables the animal to control its buoyancy in the water.

symbiont (sim'-bē-ont) The smaller participant in a symbiotic relationship, living in or on the host.

symbiosis An ecological relationship between organisms of two different species that live together in direct and intimate contact.

sympathetic division One of three divisions of the autonomic nervous system; generally increases energy expenditure and prepares the body for action.

sympatric speciation (sim-pat'-rik) The formation of new species in populations that live in the same geographic area.

symplast In plants, the continuum of cytoplasm connected by plasmodesmata between cells.

synapse (sin'-aps) The junction where a neuron communicates with another cell across a narrow gap via a neurotransmitter or an electrical coupling.

synapsid Member of an amniote clade distinguished by a single hole on each side of the skull. Synapsids include the mammals.

synapsis (si-nap'-sis) The pairing and physical connection of duplicated homologous chromosomes during prophase I of meiosis.

systematics A scientific discipline focused on classifying organisms and determining their evolutionary relationships.

systemic acquired resistance A defensive response in infected plants that helps protect healthy tissue from pathogenic invasion.

systemic circuit The branch of the circulatory system that supplies oxygenated blood to and carries deoxygenated blood away from organs and tissues throughout the body.

systems biology An approach to studying biology that aims to model the dynamic behavior of whole biological systems based on a study of the interactions among the system's parts.

systole (sis'-tō-lē) The stage of the cardiac cycle in which a heart chamber contracts and pumps blood.

systolic pressure Blood pressure in the arteries during contraction of the ventricles.

T cells The class of lymphocytes that mature in the thymus; they include both effector cells for the cell-mediated immune response and helper cells required for both branches of adaptive immunity.

taproot A main vertical root that develops from an embryonic root and gives rise to lateral (branch) roots.

tastant Any chemical that stimulates the sensory receptors in a taste bud.

taste bud A collection of modified epithelial cells on the tongue or in the mouth that are receptors for taste in mammals.

TATA box A DNA sequence in eukaryotic promoters crucial in forming the transcription initiation complex.

taxis (tak'-sis) An oriented movement toward or away from a stimulus.

taxon (plural, **taxa**) A named taxonomic unit at any given level of classification.

taxonomy (tak-son'-uh-mē) A scientific discipline concerned with naming and classifying the diverse forms of life.

Tay-Sachs disease A human genetic disease caused by a recessive allele for a dysfunctional enzyme, leading to accumulation of certain lipids in the brain. Seizures, blindness, and degeneration of motor and mental performance usually become manifest a few months after birth, followed by death within a few years.

technology The application of scientific knowledge for a specific purpose, often involving industry or commerce but also including uses in basic research.

telomerase An enzyme that catalyzes the lengthening of telomeres in eukaryotic germ cells.

telomere (tel'-uh-mēr) The tandemly repetitive DNA at the end of a eukaryotic chromosome's DNA molecule. Telomeres protect the organism's genes from being eroded during successive rounds of replication. *See also* repetitive DNA.

telophase The fifth and final stage of mitosis, in which daughter nuclei are forming and cytokinesis has typically begun.

temperate broadleaf forest A biome located throughout midlatitude regions where there is sufficient moisture to support the growth of large, broadleaf deciduous trees.

temperate grassland A terrestrial biome that exists at midlatitude regions and is dominated by grasses and forbs.

temperate phage A phage that is capable of replicating by either a lytic or lysogenic cycle.

temperature A measure of the intensity of heat in degrees, reflecting the average kinetic energy of the molecules.

template strand The DNA strand that provides the pattern, or template, for ordering, by complementary base pairing, the sequence of nucleotides in an RNA transcript.

temporal summation A phenomenon of neural integration in which the membrane potential of the postsynaptic cell in a chemical synapse is determined by the combined effect of EPSPs or IPSPs produced in rapid succession.

tendon A fibrous connective tissue that attaches muscle to bone.

terminator In bacteria, a sequence of nucleotides in DNA that marks the end of a gene and signals RNA polymerase to release the newly made RNA molecule and detach from the DNA.

territoriality A behavior in which an animal defends a bounded physical space against encroachment by other individuals, usually of its own species.

tertiary consumer (ter-shē-ār'-ē) A carnivore that eats other carnivores.

tertiary structure The overall shape of a protein molecule due to interactions of amino acid side chains, including hydrophobic interactions, ionic bonds, hydrogen bonds, and disulfide bridges.

testcross Breeding an organism of unknown genotype with a homozygous recessive individual to determine the unknown genotype. The ratio of phenotypes in the offspring reveals the unknown genotype.

testis (plural, **testes**) The male reproductive organ, or gonad, in which sperm and reproductive hormones are produced.

testosterone A steroid hormone required for development of the male reproductive system, spermatogenesis, and male secondary sex characteristics; the major androgen in mammals.

tetanus (tet'-uh-nus) The maximal, sustained contraction of a skeletal muscle, caused by a very high frequency of action potentials elicited by continual stimulation.

tetrapod A vertebrate clade whose members have limbs with digits. Tetrapods include mammals, amphibians, and birds and other reptiles.

thalamus (thal'-uh-mus) An integrating center of the vertebrate forebrain. Neurons with cell bodies in the thalamus relay neural input to specific areas in the cerebral cortex and regulate what information goes to the cerebral cortex.

thallus (plural, **thalli**) A seaweed body that is plantlike, consisting of a holdfast, stipe, and blades, yet lacks true roots, stems, and leaves.

theory An explanation that is broader in scope than a hypothesis, generates new hypotheses, and is supported by a large body of evidence.

thermal energy *See* heat.

thermocline A narrow stratum of abrupt temperature change in the ocean and in many temperate-zone lakes.

thermodynamics (ther′-mō-dī-nam′-iks) The study of energy transformations that occur in a collection of matter. *See* first law of thermodynamics; second law of thermodynamics.

thermoreceptor A receptor stimulated by either heat or cold.

thermoregulation The maintenance of internal body temperature within a tolerable range.

theropod Member of a group of dinosaurs that were bipedal carnivores.

thick filament A filament composed of staggered arrays of myosin molecules; a component of myofibrils in muscle fibers.

thigmomorphogenesis A response in plants to chronic mechanical stimulation, resulting from increased ethylene production. An example is thickening stems in response to strong winds.

thigmotropism (thig-mo′-truh-pizm) A directional growth of a plant in response to touch.

thin filament A filament consisting of two strands of actin and two strands of regulatory protein coiled around one another; a component of myofibrils in muscle fibers.

threatened species A species that is considered likely to become endangered in the foreseeable future.

threshold The potential that an excitable cell membrane must reach for an action potential to be initiated.

thrombus A fibrin-containing clot that forms in a blood vessel and blocks the flow of blood.

thylakoid (thī′-luh-koyd) A flattened, membranous sac inside a chloroplast. Thylakoids often exist in stacks called grana that are interconnected; their membranes contain molecular "machinery" used to convert light energy to chemical energy.

thymus (thī′-mus) A small organ in the thoracic cavity of vertebrates where maturation of T cells is completed.

thyroid gland An endocrine gland, located on the ventral surface of the trachea, that secretes two iodine-containing hormones, triiodothyronine (T_3) and thyroxine (T_4), as well as calcitonin.

thyroxine (T_4) One of two iodine-containing hormones that are secreted by the thyroid gland and that help regulate metabolism, development, and maturation in vertebrates.

Ti plasmid A plasmid of a tumor-inducing bacterium (the plant pathogen *Agrobacterium*) that integrates a segment of its DNA (T DNA) into a chromosome of a host plant. The Ti plasmid is frequently used as a vector for genetic engineering in plants.

tidal volume The volume of air a mammal inhales and exhales with each breath.

tight junction A type of intercellular junction between animal cells that prevents the leakage of material through the space between cells.

tissue An integrated group of cells with a common structure, function, or both.

tissue system One or more tissues organized into a functional unit connecting the organs of a plant.

Toll-like receptor (TLR) A membrane receptor on a phagocytic white blood cell that recognizes fragments of molecules common to a set of pathogens.

tonicity The ability of a solution surrounding a cell to cause that cell to gain or lose water.

top-down model A model of community organization in which predation influences community organization by controlling herbivore numbers, which in turn control plant or phytoplankton numbers, which in turn control nutrient levels; also called the trophic cascade model.

topoisomerase A protein that breaks, swivels, and rejoins DNA strands. During DNA replication, topoisomerase helps to relieve strain in the double helix ahead of the replication fork.

topsoil A mixture of particles derived from rock, living organisms, and decaying organic material (humus).

torpor A physiological state in which activity is low and metabolism decreases.

torsion In gastropods, a developmental process in which the visceral mass rotates up to 180°, causing the animal's anus and mantle cavity to be positioned above its head.

totipotent (tō′-tuh-pōt′-ent) Describing a cell that can give rise to all parts of the embryo and adult, as well as extraembryonic membranes in species that have them.

trace element An element indispensable for life but required in extremely minute amounts.

trachea (trā′-kē-uh) The portion of the respiratory tract that passes from the larynx to the bronchi; also called the windpipe.

tracheal system In insects, a system of branched, air-filled tubes that extends throughout the body and carries oxygen directly to cells.

tracheid (trā′-kē-id) A long, tapered water-conducting cell found in the xylem of nearly all vascular plants. Functioning tracheids are no longer living.

trait One of two or more detectable variants in a genetic character.

trans fat An unsaturated fat, formed artificially during hydrogenation of oils, containing one or more *trans* double bonds.

transcription The synthesis of RNA using a DNA template.

transcription factor A regulatory protein that binds to DNA and affects transcription of specific genes.

transcription initiation complex The completed assembly of transcription factors and RNA polymerase bound to a promoter.

transcription unit A region of DNA that is transcribed into an RNA molecule.

transduction (1) A process in which phages (viruses) carry bacterial DNA from one bacterial cell to another. When these two cells are members of different species, transduction results in horizontal gene transfer. (2) In cellular communication, the conversion of a signal from outside the cell to a form that can bring about a specific cellular response; also called *signal transduction*.

transfer RNA (tRNA) An RNA molecule that functions as a translator between nucleic acid and protein languages by carrying specific amino acids to the ribosome, where they recognize the appropriate codons in the mRNA.

transformation (1) The conversion of a normal animal cell to a cancerous cell. (2) A change in genotype and phenotype due to the assimilation of external DNA by a cell. When the external DNA is from a member of a different species, transformation results in horizontal gene transfer.

transgenic Pertaining to an organism whose genome contains a gene introduced from another organism of the same or a different species.

translation The synthesis of a polypeptide using the genetic information encoded in an mRNA molecule. There is a change of "language" from nucleotides to amino acids.

translocation (1) An aberration in chromosome structure resulting from attachment of a chromosomal fragment to a nonhomologous chromosome. (2) During protein synthesis, the third stage in the elongation cycle, when the RNA carrying the growing polypeptide moves from the A site to the P site on the ribosome. (3) The transport of organic nutrients in the phloem of vascular plants.

transmission The passage of a nerve impulse along axons.

transmission electron microscope (TEM) A microscope that passes an electron beam through very thin sections stained with metal atoms and is primarily used to study the internal ultrastructure of cells.

transpiration The evaporative loss of water from a plant.

transport epithelium One or more layers of specialized epithelial cells that carry out and regulate solute movement.

transport protein A transmembrane protein that helps a certain substance or class of closely related substances to cross the membrane.

transport vesicle A small membranous sac in a eukaryotic cell's cytoplasm carrying molecules produced by the cell.

transposable element A segment of DNA that can move within the genome of a cell by means of a DNA or RNA intermediate; also called a transposable genetic element.

transposon A transposable element that moves within a genome by means of a DNA intermediate.

transverse (T) tubule An infolding of the plasma membrane of skeletal muscle cells.

triacylglycerol (trī-as′-ul-glis′-uh-rol) A lipid consisting of three fatty acids linked to one glycerol molecule; also called a fat or triglyceride.

triiodothyronine (T_3) (trī′-ī-ō′-dō-thī′-rō-nēn) One of two iodine-containing hormones that are secreted by the thyroid gland and that help regulate metabolism, development, and maturation in vertebrates.

trimester In human development, one of three 3-month-long periods of pregnancy.

triple response A plant growth maneuver in response to mechanical stress, involving slowing of stem elongation, thickening of the stem, and a curvature that causes the stem to start growing horizontally.

triplet code A genetic information system in which a set of three-nucleotide-long words specify the amino acids for polypeptide chains.

triploblastic Possessing three germ layers: the endoderm, mesoderm, and ectoderm. Most eumetazoans are triploblastic.

trisomic Referring to a diploid cell that has three copies of a particular chromosome instead of the normal two.

trochophore larva (trō′-kuh-fōr) Distinctive larval stage observed in some lophotrochozoan animals, including some annelids and molluscs.

trophic efficiency The percentage of production transferred from one trophic level to the next.

trophic structure The different feeding relationships in an ecosystem, which determine the route of energy flow and the pattern of chemical cycling.

trophoblast The outer epithelium of a mammalian blastocyst. It forms the fetal part of the placenta, supporting embryonic development but not forming part of the embryo proper.

tropic hormone A hormone that has an endocrine gland or cells as a target.

tropical dry forest A terrestrial biome characterized by relatively high temperatures and precipitation overall but with a pronounced dry season.

tropical rain forest A terrestrial biome characterized by relatively high precipitation and temperatures year-round.

tropics Latitudes between 23.5° north and south.

tropism A growth response that results in the curvature of whole plant organs toward or away from stimuli due to differential rates of cell elongation.

tropomyosin The regulatory protein that blocks the myosin-binding sites on actin molecules.

troponin complex The regulatory proteins that control the position of tropomyosin on the thin filament.

true-breeding Referring to organisms that produce offspring of the same variety over many generations of self-pollination.

tubal ligation A means of sterilization in which a woman's two oviducts (fallopian tubes) are tied closed to prevent eggs from reaching the uterus. A segment of each oviduct is removed.

tube foot One of numerous extensions of an echinoderm's water vascular system. Tube feet function in locomotion and feeding.

tumor-suppressor gene A gene whose protein product inhibits cell division, thereby preventing the uncontrolled cell growth that contributes to cancer.

tundra A terrestrial biome at the extreme limits of plant growth. At the northernmost limits, it is called arctic tundra, and at high altitudes, where plant forms are limited to low shrubby or matlike vegetation, it is called alpine tundra.

tunicate Member of the clade Urochordata, sessile marine chordates that lack a backbone.

turgid (ter′-jid) Swollen or distended, as in plant cells. (A walled cell becomes turgid if it has a lower water potential than its surroundings, resulting in entry of water.)

turgor pressure The force directed against a plant cell wall after the influx of water and swelling of the cell due to osmosis.

turnover The mixing of waters as a result of changing water-temperature profiles in a lake.

turnover time The time required to replace the standing crop of a population or group of populations (for example, of phytoplankton), calculated as the ratio of standing crop to production.

twin study A behavioral study in which researchers compare the behavior of identical twins raised apart with that of identical twins raised in the same household.

tympanic membrane Another name for the eardrum, the membrane between the outer and middle ear.

uniformitarianism The principle that mechanisms of change are constant over time. *See* catastrophism.

Unikonta (yū′-ni-kon′-tuh) One of five supergroups of eukaryotes proposed in a current hypothesis of the evolutionary history of eukaryotes. This clade, which is supported by studies of myosin proteins and DNA, consists of amoebozoans and opisthokonts. *See also* Excavata, Chromalveolata, Rhizaria, and Archaeplastida.

unsaturated fatty acid A fatty acid that has one or more double bonds between carbons in the hydrocarbon tail. Such bonding reduces the number of hydrogen atoms attached to the carbon skeleton.

urea A soluble nitrogenous waste produced in the liver by a metabolic cycle that combines ammonia with carbon dioxide.

ureter (yū-rē′-ter) A duct leading from the kidney to the urinary bladder.

urethra (yū-rē′-thruh) A tube that releases urine from the mammalian body near the vagina in females and through the penis in males; also serves in males as the exit tube for the reproductive system.

uric acid A product of protein and purine metabolism and the major nitrogenous waste product of insects, land snails, and many reptiles. Uric acid is relatively nontoxic and largely insoluble.

urinary bladder The pouch where urine is stored prior to elimination.

uterine cycle The changes that occur in the uterus during the reproductive cycle of the human female; also called the menstrual cycle.

uterus A female organ where eggs are fertilized and/or development of the young occurs.

utricle In the vertebrate ear, a chamber in the vestibule behind the oval window that opens into the three semicircular canals.

vaccination *See* immunization.

vaccine A harmless variant or derivative of a pathogen that stimulates a host's immune system to mount defenses against the pathogen.

vacuole (vak′-yū-ōl′) A membrane-bounded vesicle whose specialized function varies in different kinds of cells.

vagina Part of the female reproductive system between the uterus and the outside opening; the birth canal in mammals. During copulation, the vagina accommodates the male's penis and receives sperm.

valence The bonding capacity of a given atom; usually equals the number of unpaired electrons required to complete the atom's outermost (valence) shell.

valence electron An electron in the outermost electron shell.

valence shell The outermost energy shell of an atom, containing the valence electrons involved in the chemical reactions of that atom.

van der Waals interactions Weak attractions between molecules or parts of molecules that result from transient local partial charges.

variation Differences between members of the same species.

vas deferens In mammals, the tube in the male reproductive system in which sperm travel from the epididymis to the urethra.

vasa recta The capillary system in the kidney that serves the loop of Henle.

vascular cambium A cylinder of meristematic tissue in woody plants that adds layers of secondary vascular tissue called secondary xylem (wood) and secondary phloem.

vascular plant A plant with vascular tissue. Vascular plants include all living plant species except liverworts, mosses, and hornworts.

vascular tissue Plant tissue consisting of cells joined into tubes that transport water and nutrients throughout the plant body.

vascular tissue system A transport system formed by xylem and phloem throughout a vascular plant. Xylem transports water and minerals; phloem transports sugars, the products of photosynthesis.

vasectomy The cutting and sealing of each vas deferens to prevent sperm from entering the urethra.

vasocongestion The filling of a tissue with blood, caused by increased blood flow through the arteries of that tissue.

vasoconstriction A decrease in the diameter of blood vessels caused by contraction of smooth muscles in the vessel walls.

vasodilation An increase in the diameter of blood vessels caused by relaxation of smooth muscles in the vessel walls.

vector An organism that transmits pathogens from one host to another.

vegetal pole The point at the end of an egg in the hemisphere where most yolk is concentrated; opposite of animal pole.

vegetative reproduction Cloning of plants by asexual means.

vein (1) In animals, a vessel that carries blood toward the heart. (2) In plants, a vascular bundle in a leaf.

ventilation The flow of air or water over a respiratory surface.

ventral Pertaining to the underside, or bottom, of an animal with radial or bilateral symmetry.

ventricle (ven′-tri-kul) (1) A heart chamber that pumps blood out of the heart. (2) A space in the vertebrate brain, filled with cerebrospinal fluid.

venule (ven′-yūl) A vessel that conveys blood between a capillary bed and a vein.

vernalization The use of cold treatment to induce a plant to flower.

vertebrate A chordate animal with a backbone, including sharks and rays, ray-finned fishes, coelacanths, lungfishes, amphibians, reptiles, and mammals.

vesicle (ves′-i-kul) A membranous sac in the cytoplasm of a eukaryotic cell.

vessel A continuous water-conducting micropipe found in most angiosperms and a few nonflowering vascular plants.

vessel element A short, wide water-conducting cell found in the xylem of most angiosperms and a few nonflowering vascular plants. Dead at maturity, vessel elements are aligned end to end to form micropipes called vessels.

vestigial structure A feature of an organism that is a historical remnant of a structure that served a function in the organism's ancestors.

villus (plural, **villi**) (1) A finger-like projection of the inner surface of the small intestine. (2) A finger-like projection of the chorion of the mammalian placenta. Large numbers of villi increase the surface areas of these organs.

viral envelope A membrane, derived from membranes of the host cell, that cloaks the capsid, which in turn encloses a viral genome.

viroid (vī′-royd) A plant pathogen consisting of a molecule of naked, circular RNA a few hundred nucleotides long.

virulent Describing a pathogen against which an organism has little specific defense.

virulent phage A phage that replicates only by a lytic cycle.

virus An infectious particle incapable of replicating outside of a cell, consisting of an RNA or DNA genome surrounded by a protein coat (capsid) and, for some viruses, a membranous envelope.

visceral mass One of the three main parts of a mollusc; the part containing most of the internal organs. *See also* foot, mantle.

visible light That portion of the electromagnetic spectrum that can be detected as various colors by the human eye, ranging in wavelength from about 380 nm to about 750 nm.

vital capacity The maximum volume of air that a mammal can inhale and exhale with each breath.

vitamin An organic molecule required in the diet in very small amounts. Many vitamins serve as coenzymes or parts of coenzymes.

viviparous (vī-vip′-uh-rus) Referring to a type of development in which the young are born alive after having been nourished in the uterus by blood from the placenta.

voltage-gated ion channel A specialized ion channel that opens or closes in response to changes in membrane potential.

vulva Collective term for the female external genitalia.

water potential (Ψ) The physical property predicting the direction in which water will flow, governed by solute concentration and applied pressure.

water vascular system A network of hydraulic canals unique to echinoderms that branches into extensions called tube feet, which function in locomotion and feeding.

wavelength The distance between crests of waves, such as those of the electromagnetic spectrum.

wetland A habitat that is inundated by water at least some of the time and that supports plants adapted to water-saturated soil.

white matter Tracts of axons within the CNS.

wild type The phenotype most commonly observed in natural populations; also refers to the individual with that phenotype.

wilting The drooping of leaves and stems as a result of plant cells becoming flaccid.

wobble Flexibility in the base-pairing rules in which the nucleotide at the 5′ end of a tRNA anticodon can form hydrogen bonds with more than one kind of base in the third position (3′ end) of a codon.

xerophyte A plant adapted to an arid climate.

X-linked gene A gene located on the X chromosome; such genes show a distinctive pattern of inheritance.

X-ray crystallography A technique used to study the three-dimensional structure of molecules. It depends on the diffraction of an X-ray beam by the individual atoms of a crystallized molecule.

xylem (zī′-lum) Vascular plant tissue consisting mainly of tubular dead cells that conduct most of the water and minerals upward from the roots to the rest of the plant.

xylem sap The dilute solution of water and dissolved minerals carried through vessels and tracheids.

yeast Single-celled fungus that reproduces asexually by binary fission or by the pinching of small buds off a parent cell. Some species exhibit cell fusion between different mating types.

yolk Nutrients stored in an egg.

zero population growth (ZPG) A period of stability in population size, when additions to the population through births and immigration are balanced by subtractions through deaths and emigration.

zona pellucida The extracellular matrix surrounding a mammalian egg.

zone of polarizing activity (ZPA) A block of mesoderm located just under the ectoderm where the posterior side of a limb bud is attached to the body; required for proper pattern formation along the anterior-posterior axis of the limb.

zoned reserve An extensive region that includes areas relatively undisturbed by humans surrounded by areas that have been changed by human activity and are used for economic gain.

zoonotic pathogen A disease-causing agent that is transmitted to humans from other animals.

zoospore Flagellated spore found in chytrid fungi and some protists.

zygomycete (zī′-guh-mī′-sēt) Member of the fungal phylum Zygomycota, characterized by the formation of a sturdy structure called a zygosporangium during sexual reproduction.

zygosporangium (zī′-guh-spōr-an′-jē-um) In zygomycete fungi, a sturdy multinucleate structure in which karyogamy and meiosis occur.

zygote (zī′-gōt) The diploid cell produced by the union of haploid gametes during fertilization; a fertilized egg.

Index

NOTE: A page number in regular type indicates where a topic is discussed in text (topic may also be in a figure on that page); a **bold** page number indicates where a term is bold and defined; an *f* following a page number indicates a figure (topic may also be discussed in text on that page); a *t* following a page number indicates a table (topic may also be discussed in text on that page).

Index

D

E

F

G

H

M

U

V

W